Evolutionary Analysis

FIFTH EDITION

Jon C. Herron
University of Washington

Scott Freeman
University of Washington

With contributions by

Jason Hodin
Hopkins Marine Station of Stanford University and University of Washington

Brooks Miner
Cornell University

Christian Sidor
University of Washington

PEARSON

Boston Columbus Indianapolis New York San Francisco Upper Saddle River
Amsterdam Cape Town Dubai London Madrid Milan Munich Paris Montreal Toronto
Delhi Mexico City Sao Paulo Sydney Hong Kong Seoul Singapore Taipei Tokyo

Editor-in-Chief: Beth Wilbur
Senior Acquisitions Editor: Michael Gillespie
Executive Director of Development: Deborah Gale
Project Editor: Laura Murray
Assistant Editor: Eddie Lee
Manager, Text Permissions: Tim Nicholls
Text Permissions Specialist: Kim Schmidt, S4Carlisle Publishing Services
Director of Production: Erin Gregg
Managing Editor: Michael Early
Production Project Manager: Lori Newman
Production Management Services: Progressive Publishing Alternatives
Copyeditor: Chris Thillen
Design Manager: Mark Ong
Cover and Interior Designer: Mark Ong
Art Developer, Illustrator: Robin Green
Senior Photo Editor: Travis Amos
Photo Research and Permissions Management: Bill Smith Group
Content Producer: Daniel Ross
Media Project Manager: Shannon Kong
Director of Marketing: Christy Lesko
Executive Marketing Manager: Lauren Harp
Manufacturing Buyer: Christy Hall
Cover and Text Printer: LSC Communications

Cover Photo Credit: Auke Holwerda / Getty Images

Credits and acknowledgments for materials borrowed from other sources and reproduced, with permission, in this textbook appear on the appropriate page within the text **[or on p. 822]**.

Copyright ©2014, 2007, 2004, 2001, 1998 by Jon C. Herron and Scott Freeman. Published by Pearson Education, Inc. All rights reserved. Manufactured in the United States of America. This publication is protected by Copyright, and permission should be obtained from the publisher prior to any prohibited reproduction, storage in a retrieval system, or transmission in any form or by any means, electronic, mechanical, photocopying, recording, or likewise. To obtain permission(s) to use material from this work, please submit a written request to Pearson Education, Inc., Permissions Department, 1900 E. Lake Ave., Glenview, IL 60025. For information regarding permissions, call (847) 486-2635.

Readers may view, browse, and/or download material for temporary copying purposes only, provided these uses are for noncommercial personal purposes. Except as provided by law, this material may not be further reproduced, distributed, transmitted, modified, adapted, performed, displayed, published, or sold in whole or in part, without prior written permission from the publisher.

Many of the designations used by manufacturers and sellers to distinguish their products are claimed as trademarks. Where those designations appear in this book, and the publisher was aware of a trademark claim, the designations have been printed in initial caps or all caps.

Library of Congress Cataloging-in-Publication Data is available upon request.

ISBN 10: 0-321-61667-7; ISBN 13: 978-0-321-61667-8 (Student Edition)
ISBN 10: 0-321-92799-0; ISBN 13: 0-321-92799-6 (Instructor's Review Copy)
ISBN 10: 0-321-92816-4; ISBN 13: 978-0-321-92816-0 (Books a la Carte Edition)

Library of Congress Control Number: 2013943730

www.pearsonhighered.com

23 2022

Brief Contents

PART 1 INTRODUCTION 1

CHAPTER 1 A Case for Evolutionary Thinking: Understanding HIV 1

CHAPTER 2 The Pattern of Evolution 37

CHAPTER 3 Evolution by Natural Selection 73

CHAPTER 4 Estimating Evolutionary Trees 109

PART 2 MECHANISMS OF EVOLUTIONARY CHANGE 147

CHAPTER 5 Variation Among Individuals 147

CHAPTER 6 Mendelian Genetics in Populations I: Selection and Mutation 179

CHAPTER 7 Mendelian Genetics in Populations II: Migration, Drift, and Nonrandom Mating 233

CHAPTER 8 Evolution at Multiple Loci: Linkage and Sex 291

CHAPTER 9 Evolution at Multiple Loci: Quantitative Genetics 329

PART 3 ADAPTATION 369

CHAPTER 10 Studying Adaptation: Evolutionary Analysis of Form and Function 369

CHAPTER 11 Sexual Selection 407

CHAPTER 12 The Evolution of Social Behavior 455

CHAPTER 13 Aging and Other Life-History Characters 491

CHAPTER 14 Evolution and Human Health 535

CHAPTER 15 Genome Evolution and the Molecular Basis of Adaptation 581

PART 4 THE HISTORY OF LIFE 609

CHAPTER 16 Mechanisms of Speciation 609

CHAPTER 17 The Origins of Life and Precambrian Evolution 645

CHAPTER 18 Evolution and the Fossil Record 691

CHAPTER 19 Development and Evolution 735

CHAPTER 20 Human Evolution 769

Contents

Preface ix

PART 1
INTRODUCTION 1

CHAPTER **1**
A Case for Evolutionary Thinking: Understanding HIV 1

1.1 The Natural History of the HIV Epidemic 2
1.2 Why Does HIV Therapy Using Just One Drug Ultimately Fail? 9
1.3 Are Human Populations Evolving as a Result of the HIV Pandemic? 15
1.4 Where Did HIV Come From? 18
1.5 Why Is HIV Lethal? 23
Computing Consequences 1.1 *When did HIV move from chimpanzees to humans?* 24
Summary 31 • Questions 31
Exploring the Literature 32 • Citations 33

CHAPTER **2**
The Pattern of Evolution 37

2.1 Evidence of Microevolution: Change through Time 39
2.2 Evidence of Speciation: New Lineages from Old 44
2.3 Evidence of Macroevolution: New Forms from Old 49
2.4 Evidence of Common Ancestry: All Life-Forms Are Related 55
2.5 The Age of Earth 62
Computing Consequences 2.1 *A closer look at radiometric dating* 65
Summary 66 • Questions 67
Exploring the Literature 68 • Citations 69

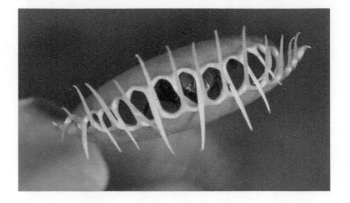

CHAPTER **3**
Evolution by Natural Selection 73

3.1 Artificial Selection: Domestic Animals and Plants 74
3.2 Evolution by Natural Selection 77
3.3 The Evolution of Flower Color in an Experimental Snapdragon Population 79
3.4 The Evolution of Beak Shape in Galápagos Finches 81
Computing Consequences 3.1 *Estimating heritabilities despite complications* 84
3.5 The Nature of Natural Selection 90
3.6 The Evolution of Evolutionary Biology 94
3.7 Intelligent Design Creationism 97
Summary 104 • Questions 105
Exploring the Literature 106 • Citations 106

CHAPTER **4**
Estimating Evolutionary Trees 109

4.1 How to Read an Evolutionary Tree 110
4.2 The Logic of Inferring Evolutionary Trees 114
4.3 Molecular Phylogeny Inference and the Origin of Whales 123
Computing Consequences 4.1 *Calculating the likelihood of an evolutionary tree* 129
Computing Consequences 4.2 *Neighbor joining: A distance matrix method* 130

4.4 Using Phylogenies to Answer Questions 137
Summary 141 • Questions 141
Exploring the Literature 143 • Citations 143

PART 2
MECHANISMS OF EVOLUTIONARY CHANGE 147

CHAPTER 5
Variation Among Individuals 147

5.1 Three Kinds of Variation 148
Computing Consequences 5.1 Epigenetic inheritance and evolution 154
5.2 Where New Alleles Come From 157
5.3 Where New Genes Come From 161
Computing Consequences 5.2 Measuring genetic variation in natural populations 162
5.4 Chromosome Mutations 166
5.5 Rates and Fitness Effects of Mutations 169
Summary 174 • Questions 175
Exploring the Literature 176 • Citations 176

CHAPTER 6
Mendelian Genetics in Populations I: Selection and Mutation 179

6.1 Mendelian Genetics in Populations: Hardy–Weinberg Equilibrium 180
Computing Consequences 6.1 Combining probabilities 185
Computing Consequences 6.2 The Hardy–Weinberg equilibrium principle with more than two alleles 189
6.2 Selection 191
Computing Consequences 6.3 A general treatment of selection 194 
Computing Consequences 6.4 Statistical analysis of allele and genotype frequencies using the χ^2 (chi-square) test 198
Computing Consequences 6.5 Predicting the frequency of the CCR5-Δ32 allele in future generations 201
6.3 Patterns of Selection: Testing Predictions of Population Genetics Theory 201
Computing Consequences 6.6 An algebraic treatment of selection on recessive and dominant alleles 204
Computing Consequences 6.7 Stable equilibria with heterozygote superiority and unstable equilibria with heterozygote inferiority 208
6.4 Mutation 216
Computing Consequences 6.8 A mathematical treatment of mutation as an evolutionary mechanism 218
Computing Consequences 6.9 Allele frequencies under mutation–selection balance 220
Computing Consequences 6.10 Estimating mutation rates for recessive alleles 222
6.5 An Engineering Test of Population Genetics Theory 224
Computing Consequences 6.11 Predicting the frequency of Medea *across generations* 226
Summary 227 • Questions 227
Exploring the Literature 229 • Citations 231

CHAPTER 7
Mendelian Genetics in Populations II: Migration, Drift, and Nonrandom Mating 233

7.1 Migration 234
Computing Consequences 7.1 An algebraic treatment of migration as an evolutionary process 236
Computing Consequences 7.2 Selection and migration in Lake Erie water snakes 238
7.2 Genetic Drift 240
Computing Consequences 7.3 The probability that a given allele will be the one that drifts to fixation 248
Computing Consequences 7.4 Effective population size 251
Computing Consequences 7.5 The rate of evolutionary substitution under genetic drift 256

7.3 Genetic Drift and Molecular Evolution 260
7.4 Nonrandom Mating 275
Computing Consequences 7.6 Genotype frequencies in an inbred population 279
7.5 Conservation Genetics of the Florida Panther 283
Summary 285 • Questions 285
Exploring the Literature 287 • Citations 288

CHAPTER 8
Evolution at Multiple Loci: Linkage and Sex 291

8.1 Evolution at Two Loci: Linkage Equilibrium and Linkage Disequilibrium 292
Computing Consequences 8.1 The coefficient of linkage disequilibrium 295
Computing Consequences 8.2 Hardy–Weinberg analysis for two loci 296
Computing Consequences 8.3 Sexual reproduction reduces linkage disequilibrium 301
8.2 Practical Reasons to Study Linkage Disequilibrium 307
Computing Consequences 8.4 Estimating the age of the GBA–84GG *mutation* 309
8.3 The Adaptive Significance of Sex 314
Computing Consequences 8.5 A demographic model of the maintenance of males in the nematode Caenorhabditis elegans 317
Summary 324 • Questions 325
Exploring the Literature 326 • Citations 327

CHAPTER 9
Evolution at Multiple Loci: Quantitative Genetics 329

9.1 The Nature of Quantitative Traits 330
9.2 Identifying Loci That Contribute to Quantitative Traits 334
Computing Consequences 9.1 Genetic mapping and LOD scores 338
9.3 Measuring Heritable Variation 343
Computing Consequences 9.2 Additive genetic variation versus dominance genetic variation 345
9.4 Measuring Differences in Survival and Reproductive Success 348 
Computing Consequences 9.3 The selection gradient and the selection differential 349
9.5 Predicting the Evolutionary Response to Selection 350
9.6 Modes of Selection and the Maintenance of Genetic Variation 356
9.7 The Bell-Curve Fallacy and Other Misinterpretations of Heritability 360
Summary 365 • Questions 365
Exploring the Literature 367 • Citations 367

PART 3
ADAPTATION 369

CHAPTER 10
Studying Adaptation: Evolutionary Analysis of Form and Function 369

10.1 All Hypotheses Must Be Tested: Oxpeckers Reconsidered 370
10.2 Experiments 373
Computing Consequences 10.1 A primer on statistical testing 377
10.3 Observational Studies 378
10.4 The Comparative Method 382
Computing Consequences 10.2 Calculating phylogenetically independent contrasts 384
10.5 Phenotypic Plasticity 387
10.6 Trade-Offs and Constraints 389
10.7 Selection Operates on Different Levels 397
10.8 Strategies for Asking Interesting Questions 401
Summary 402 • Questions 402
Exploring the Literature 404 • Citations 405

CHAPTER 11
Sexual Selection 407

11.1 Sexual Dimorphism and Sex 408
11.2 Sexual Selection on Males: Competition 417
11.3 Sexual Selection on Males: Female Choice 423
Computing Consequences 11.1 Runaway sexual selection 430
11.4 Sexual Selection on Females 438
11.5 Sexual Selection in Plants 441
11.6 Sexual Dimorphism in Humans 444
Summary 448 • Questions 448
Exploring the Literature 450 • Citations 451

CHAPTER 12
The Evolution of Social Behavior 455

12.1 Four Kinds of Social Behavior 456
12.2 Kin Selection and Costly Behavior 459
Computing Consequences 12.1 Calculating relatedness as the probability of identity by descent 461
12.3 Multilevel Selection and Cooperation 471
Computing Consequences 12.2 Different perspectives on the same evolutionary process 473
12.4 Cooperation and Conflict 477
12.5 The Evolution of Eusociality 483
Summary 486 • Questions 487
Exploring the Literature 488 • Citations 489

CHAPTER 13
Aging and Other Life-History Characters 491

13.1 Basic Issues in Life-History Analysis 493
13.2 Why Do Organisms Age and Die? 495
Computing Consequences 13.1 Late-acting deleterious mutations are weakly selected 501
Computing Consequences 13.2 Alleles conferring early benefits and late costs can be adaptive 504
13.3 How Many Offspring Should an Individual Produce in a Given Year? 513
13.4 How Big Should Each Offspring Be? 517
13.5 Conflicts of Interest between Life Histories 522
13.6 Life Histories in a Broader Evolutionary Context 525
Summary 530 • Questions 530
Exploring the Literature 532 • Citations 532

CHAPTER 14
Evolution and Human Health 535

14.1 Evolving Pathogens: Evasion of the Host's Immune Response 537
14.2 Evolving Pathogens: Antibiotic Resistance 545
14.3 Evolving Pathogens: Virulence 548
14.4 Tissues as Evolving Populations of Cells 553
14.5 Selection Thinking Applied to Humans 556
14.6 Adaptation and Medical Physiology: Fever 564
14.7 Adaptation and Human Behavior: Parenting 567
Computing Consequences 14.1 Is cultural evolution Darwinian? 569
Summary 575 • Questions 575
Exploring the Literature 577 • Citations 577

CHAPTER 15
Genome Evolution and the Molecular Basis of Adaptation 581

15.1 Diversity among Genomes 582
15.2 Mobile Genetic Elements 586
15.3 The Evolution of Mutation Rates 591
15.4 Gene Duplication and Gene Families 594
15.5 The Locus of Adaptation in Natural Populations 601
Summary 606 • Questions 606
Exploring the Literature 607 • Citations 608

PART 4

THE HISTORY OF LIFE 609

CHAPTER 16

Mechanisms of Speciation 609

16.1 Species Concepts 610
16.2 Mechanisms of Isolation 616
16.3 Mechanisms of Divergence 623
16.4 Hybridization and Gene Flow between Species 629
16.5 What Drives Diversification? 637
Summary 640 • Questions 641
Exploring the Literature 642 • Citations 643

CHAPTER 17

The Origins of Life and Precambrian Evolution 645

17.1 What Was the First Living Thing? 647
17.2 Where Did the First Living Thing Come From? 655
17.3 What Was the Last Common Ancestor of All Extant Organisms and What Is the Shape of the Tree of Life? 663
17.4 How Did LUCA's Descendants Evolve into Today's Organisms? 678
Summary 683 • Questions 684
Exploring the Literature 686 • Citations 686

CHAPTER 18

Evolution and the Fossil Record 691

18.1 The Nature of the Fossil Record 692
18.2 Evolution in the Fossil Record 696
Computing Consequences 18.1 *Evolutionary trends* 706
18.3 Taxonomic and Morphological Diversity over Time 707
18.4 Mass and Background Extinctions 709
18.5 Macroevolution 719
18.6 Fossil and Molecular Divergence Timing 727
Summary 730 • Questions 731
Exploring the Literature 732 • Citations 732

CHAPTER 19

Development and Evolution 735

19.1 The Divorce and Reconciliation of Development and Evolution 736
19.2 Hox Genes and the Birth of Evo-Devo 738
19.3 Post Hox: Evo-Devo 2.0 744
19.4 Hox Redux: Homology or Homoplasy? 763
19.5 The Future of Evo-Devo 764
Summary 765 • Questions 766
Exploring the Literature 766 • Citations 767

CHAPTER 20

Human Evolution 769

20.1 Relationships among Humans and Extant Apes 770
20.2 The Recent Ancestry of Humans 780
20.3 Origin of the Species *Homo sapiens* 790
Computing Consequences 20.1 *Using allele frequencies and linkage disequilibrium to date the modern human expansion from Africa* 797
20.4 The Evolution of Distinctive Human Traits 802
Summary 807 • Questions 807
Exploring the Literature 809 • Citations 810

Glossary 815

Credits 822

Index 830

Preface

Evolutionary biology has changed dramatically during the 15 years we have worked on *Evolutionary Analysis*. As one measure of this change, consider that when the first edition went to press, the genomes of just five cellular organisms had been sequenced: three bacteria, one archaean, and one eukaryote. As the fifth edition goes to press, Erica Bree Rosenblum and colleagues reported in the *Proceedings of the National Academy of Sciences* (110: 9385–9390) that they had sequenced the genomes of 29 strains of a single species, the chytrid fungus *Batrachochytrium dendrobatidis*. This work was part of an effort to unravel the evolutionary history of an emerging pathogen that has decimated amphibian populations around the world and driven some species to extinction. The avalanche of sequence data has allowed evolutionary biologists to answer long-standing questions with greatly increased depth and clarity. In Chapter 20, *Human Evolution,* for example, we discuss a recent analysis of differences among genomic regions in the evolutionary relationships among humans, chimpanzees, and gorillas. For some questions, the answers have changed completely. In the fourth edition we noted that available sequence data provided no support for the hypothesis that modern humans and Neandertals interbred. But in the fifth edition we describe genomic analyses suggesting that the two lineages interbred after all.

Evolutionary Analysis provides an entry to this dynamic field of study for undergraduates majoring in the life sciences. We assume that readers have completed much or all of their introductory coursework and are beginning to explore in more detail areas of biology relevant to their personal and professional lives. Therefore, throughout the book we attempt to show the relevance of evolution to all of biology and to real-world problems.

Since the first edition, our primary goal has been to encourage readers to think like scientists. We present evolutionary biology not as a collection of facts but as an ongoing research effort. When exploring an issue, we begin with questions. Why are untreated HIV infections typically fatal? Why do purebred Florida panthers show such poor health, and what can be done to save their dwindling population? Why do mutation rates decrease with genome size among some kinds of organisms, but increase with genome size among others? We use such questions to engage students' curiosity and to motivate discussions of background information and theory. These discussions enable us to frame alternative hypotheses, consider how they can be tested, and make predictions. We then present and analyze data, consider its implications, and highlight new questions for future research. The analytical and technical skills readers learn from this approach are broadly applicable, and will stay with them long after the details of particular examples have faded.

New to This Edition

Many of the research areas we cover are advancing at a rate we would not have dreamed possible just a few years ago. We have looked closely at every chapter to both improve how we are teaching today's students and to thoroughly update our coverage.

- We have enhanced our traditional emphasis on scientific reasoning by including a data graph, evolutionary tree, or other piece of evidence to accompany the photo on the first page of every chapter. These one-page case studies engage students as active readers and help them become skilled at working with and interpreting data.
- We have enhanced our strong coverage of tree thinking by thoroughly revising Chapter 4. Consistent with the ever-growing use of phylogenetic analysis by scientists, we incorporate more phylogenies throughout the book. Among the new examples are a tree-based discussion of evolution of vertebrate eyes (Chapter 3); a new case study reconstructing the history of a patient's cancer (Chapter 14); and phylogeny-based reconstructions of the fish-tetrapod transition, the dinosaur-bird transition, and the origin of mammals (Chapter 18). Frequent practice at tree thinking helps students develop this essential skill.

Every chapter contains something new. Most of the new material is from the recent literature.

- Chapter 1 includes updated statistics on the status of the HIV pandemic, newer thinking on how HIV causes AIDS, new data on the origin of HIV, and new ideas and evidence on why HIV is lethal.
- Chapter 2 has a new organization featuring sections on evidence for microevolution, speciation, macroevolution, and common ancestry; discussions of why evolution at each level is relevant to humans outside of textbooks and classrooms; evidence of macroevolution presented using evolutionary trees showing the order in which derived traits are inferred to have evolved; and several new examples, including a terrestrial fish that does not like to swim.
- Chapter 3 brings new evidence on the evolution of development in the beaks of Darwin's finches, a new example of exaptation featuring carnivorous plants, and new coverage of the evolution of complex organs featuring a phylogeny-based discussion of the evolution of vertebrate eyes.
- Chapter 4 has been completely rewritten to offer an improved introduction to tree thinking; more detailed explanations of parsimony, maximum likelihood, and Bayesian phylogeny inference; and new examples of phylogenies used to answer interesting questions—such as identifying the surprising infectious agent responsible for a sexually transmitted tumor in dogs.
- Chapter 5 includes a new section on kinds of variation, featuring new and detailed examples of genetic variation and environmental variation; genotype-by-environment interaction; improved discussion of the mechanisms and consequences of mutation; new examples of gene duplication; and covers rates and fitness effects of mutation in a dedicated section with new examples and data.
- Chapter 6 is bookended with a powerful new example in which genetic engineers made a new gene, accurately predicted its effects on individuals carrying it, introduced it into a population, and used population genetics theory to accurately predict how its frequency would change over a span of 20 generations.
- Chapter 7 is bookended with a new example of conservation genetics involving the Florida panther. The chapter also includes a new example illustrating the founder effect in Polynesian field crickets; improved coverage of the inter-

action of drift and selection, the neutral theory, and the nearly neutral theory; and a new introduction to coalescence.

- Chapter 8 carries a new example—on Crohn's disease in humans—showing how linkage disequilibrium due to genetic hitchhiking can lead to spurious associations between genotype and phenotype and a revised and updated section on the adaptive significance of sex, featuring recent experiments using *C. elegans* as a model organism.
- Chapter 9 has improved narrative coherence due to the inclusion throughout the chapter of examples on the quantitative genetics of performance and prize winnings in thoroughbred racehorses.
- Chapter 10 includes an improved primer on statistical hypothesis testing, using research on the evolution of wild barley populations in response to a warming climate, and a new example of comparative research involving color in feather lice.
- Chapter 11 improves our coverage of the evolution of female choice by presenting the Fisher-Kirkpatrick-Lande model as the null hypothesis. New examples and data consider Bateman's principle in a hermaphrodite; female preferences in genetically modified zebrafish; correlated displays and preferences in Hawaiian crickets; and sexual selection in humans.
- Chapter 12 features enhanced coverage, with examples, of the four basic kinds of social behavior; improved coverage of kin selection and spite; a new section on multilevel selection and the evolution of cooperation; and several data sets on human social behavior.
- Chapter 13 has new examples on telomeres and aging; the evolution of menopause; life history traits and biological invasion; and life history traits and vulnerability to extinction.
- Chapter 14 discusses new evidence, from genome architecture, on the origin of influenza A; a new example using phylogenetic analysis to reconstruct the history of a cancer; and updated coverage of diseases of civilization, including a dramatic example from Iceland and new material on obesity.
- Chapter 15 has been completely rewritten, bringing new sections on the evolution of genome architecture; the evolution of mutation rates and gene families; and updated treatment of mobile genetic elements and the molecular basis of adaptation.
- Chapter 16 features new sections on mechanisms of divergence; hybridization and gene flow; and drivers of diversification. The chapter includes new examples illustrating the application of species concepts; updated coverage of vicariance in snapping shrimp; and new examples on mechanisms of isolation—including temporal isolation in a moth and single-gene speciation in snails.
- Chapter 17 incorporates updated coverage of the effort to create self-replicating RNAs and of the prebiotic synthesis of activated nucleotides.
- Chapter 18 has greatly expanded coverage of evolutionary transitions, featuring phylogeny-based reconstructions of the fish-tetrapod transition, the dinosaur-bird transition, and the origin of mammals; a new section on taxonomic and morphological diversity over time; updated treatment of mass extinctions, including the Permian-Triassic extinction; and a new section on fossil and molecular divergence timing.

- Chapter 19 has been completely rewritten. It includes revised coverage of Hox genes and detailed discussions of deep homology, developmental constraints and trade-offs, and the evolution of novel traits.
- Chapter 20 discusses new evidence from complete genomes on incomplete lineage sorting among humans, chimps, and gorillas, and on genetic differences between these species; new evidence, also from complete genomes, on hybridization between modern humans and Neandertals and between modern humans and Denisovans; and updated coverage of the human fossil record and the evolution of spoken language.

Hallmark Features

While fully updating this edition, we also maintained core strengths for which this book is recognized.

- We continue to strive for clarity of presentation, ensuring each chapter contains a coherent, accessible narrative that students can follow.
- We remain committed to strong information design and a tight integration between the text and illustration program. Nearly all phylogenies are presented horizontally, with time running from left to right, because research has shown this makes it easier for students to interpret them correctly.
- Boxes contain detailed explorations of quantitative issues discussed in the main text. These are called *Computing Consequences,* after physicist Richard Feynman's concise description of the scientific method: "First, we guess . . . No! Don't laugh—it's really true. Then we compute the consequences of the guess to see if this law that we guessed is right—what it would imply. Then we compare those computation results to nature—or, we say, to experiment, or experience—we compare it directly with observation to see if it works. If it disagrees with experiment, it's wrong."

All chapters end with a set of questions that encourage readers to review the material, apply concepts to new issues, and explore the primary literature.

Additional Resources for Instructors and Students

At the Pearson Instructor Resource Center, you can download JPEG and PowerPoint files containing all of the line art, tables, and photos from the book. You can access versions with and without labels to best suit your needs.

A thorough test bank and TestGen software is available to help you generate tests. Each chapter has dozens of multiple choice, short answer, and essay questions.

The updated Companion Website has been revised and updated to reflect the new edition. The website can be found at: www.pearsonhighered.com/herron

Activities such as case studies and simulations challenge students to pose questions, formulate hypotheses, design experiments, analyze data, and draw conclusions. Many of these activities accompany downloadable software programs that allow students to conduct their own virtual investigations. Students will also find chapter study quizzes that allow them to check their understanding of key ideas in each chapter.

Acknowledgments

Evolutionary Analysis is a team effort. The book owes its existence and quality to the generosity and talents of a large community of colleagues, students, and friends. They have reviewed chapters; made suggestions; answered our questions; shared their photos, data, and insights; and lent us their expertise in countless other ways. Getting to spend time with and learn from such smart and interesting people is the best part of our job.

For the fifth edition we have had the great fortune to work with three extraordinary contributors.

- Brooks Miner, *Cornell University,* wrote the entirely new Chapter 15 and extensively revised and updated Chapter 16.
- Christian Sidor, *University of Washington,* thoroughly revised and expanded Chapter 18.
- Jason Hodin, *Hopkins Marine Station of Stanford University* and *University of Washington,* wrote the entirely new Chapter 19.

Mark Ong provided the beautiful and rational design of the new edition. Robin Green designed and produced art that is both engaging and effective, and is responsible for the coherent integration of the illustrations with the text.

The editorial, production, and marketing team at Pearson Education has offered steadfast guidance and support: Michael Gillespie, Senior Acquisitions Editor—Biology; Beth Wilbur, VP and Editor-in-Chief of Biology and Environmental Science; Paul Corey, President—Science, Business, and Technology; Lauren Harp, Executive Marketing Manager; Lori Newman, Production Project Manager; Deborah Gale, Executive Director of Development—Biology; Laura Murray, Project Editor; and Eddie Lee, Assistant Editor.

Our preparation of the fifth edition has been guided by thoughtful, detailed, and constructive critiques by

Mirjana Brockett, *Georgia Institute of Technology*
Jeremy Brown, *LSU*
Michael Emerman, *University of Washington*
Charles Fenster, *University of Maryland*
Matthew Hahn, *University of Indiana*
Michael E. Hellberg, *LSU*
Christopher Hess, *Butler University*
Gene Hunt, *Smithsonian Institution*
Ben Kerr, *University of Washington*
Craig Lending, *SUNY Brockport*
Carlos MacHado, *University of Maryland*
Kurt McKean, *University at Albany*
James Mullins, *University of Washington*
Christopher Parkinson, *University of Central Florida*
Yale Passamaneck, *University of Hawaii*
Bruno Pernet, *California State University, Long Beach*
Thomas Ray, *University of Oklahoma*
Doug Schemske, *Michigan State University*
Billie J. Swalla, *University of Washington*
Sara Via, *University of Maryland*
Rebecca Zufall, *University of Houston*

Finally, we extend a special thank you to Christopher Parkinson and his students at the University of Central Florida and to Carol E. Lee and her students at the University of Wisconsin. Both groups class-tested preliminary versions of the chapters and provided insightful feedback that improved the final drafts.

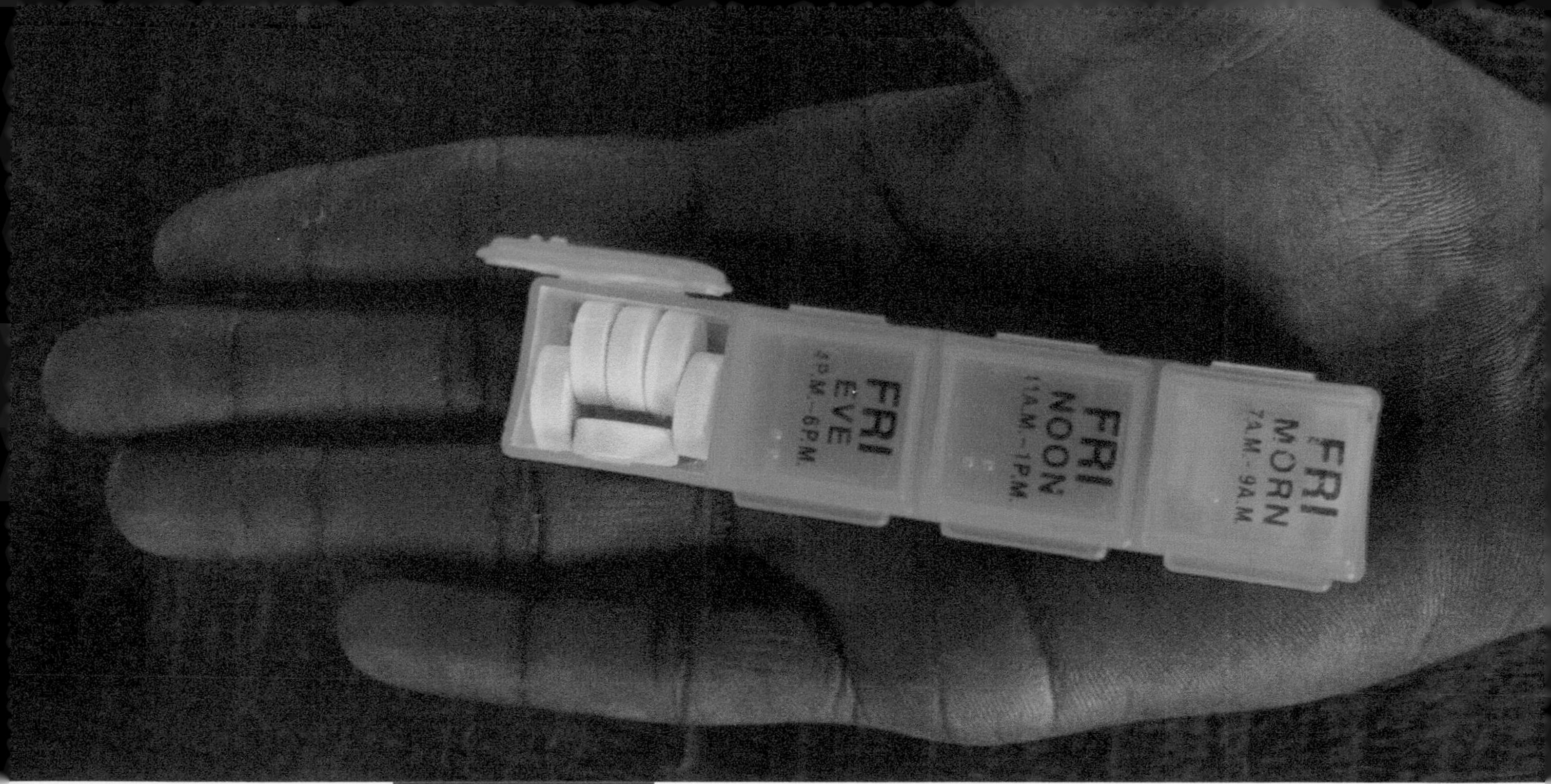

1

A Case for Evolutionary Thinking: Understanding HIV

Why study evolution? An incentive for Charles Darwin (1859) was that understanding evolution can help us know ourselves. "Light will be thrown," he wrote, "on the origin of man and his history." The allure for Theodosius Dobzhansky (1973), an architect of our modern view of evolution, was that evolutionary biology is the conceptual foundation for all of life science. "Nothing in biology makes sense," he said, "except in the light of evolution." The motive for some readers may simply be that evolution is a required course. This, too, is a valid inducement.

Here we suggest an additional reason to study evolution: The tools and techniques of evolutionary biology offer crucial insights into matters of life and death. To back this claim, we explore the evolution of HIV (human immunodeficiency virus). Infection with HIV causes AIDS (acquired immune deficiency syndrome)—sometimes, as shown at right, despite triple-drug therapy.

Our main objective in Chapter 1 is to show that evolution matters outside of labs and classrooms. However, a deep look at HIV will serve other goals as well. It will illustrate the kinds of questions evolutionary biologists ask, show how an evolutionary perspective can inform research throughout biology, and introduce concepts that we will explore in detail elsewhere in the book.

Multidrug therapies have, for some patients, transformed HIV from fatal to treatable. Such therapies work best for conscientious patients, but still may fail. The data below are from 2,800 patients on triple-drug therapy (Nachega et al. 2007).

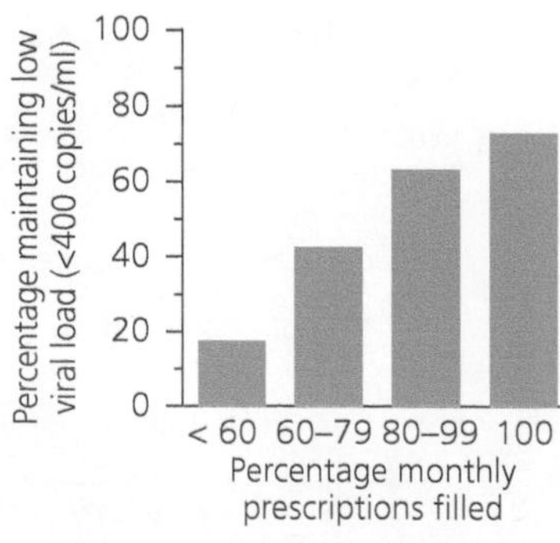

HIV makes a compelling case study because it illustrates public health issues likely to influence the life of every reader. It is an emerging pathogen. It rapidly evolves drug resistance. And, of course, it is deadly. AIDS is among the 10 leading causes of death worldwide (Lopez et al. 2006; WHO 2008).

Here are the questions we address:

- What is HIV, how does it spread, and how does it cause AIDS?
- Why do therapies using just one drug, and sometimes therapies using multiple drugs, work well at first but ultimately fail?
- Are human populations evolving as a result of the HIV pandemic?
- Where did HIV come from?
- Why are untreated HIV infections usually fatal?

While one of these questions contains the word *evolution*, some of the others may appear unrelated to the subject. But evolutionary biology is devoted to understanding how populations change over time and how new forms of life arise. These are the issues targeted by our queries about HIV and AIDS. In preparation to address them, the first section covers some requisite background.

As a case study, HIV will demonstrate how evolutionary biologists study adaptation and diversity.

1.1 The Natural History of the HIV Epidemic

AIDS was recognized in 1981, when doctors in the United States reported rare forms of pneumonia and cancer among men who have sex with men (Fauci 2008). The virus responsible, HIV, was identified shortly thereafter (Barré-Sinoussi et al. 1983; Gallo et al. 1984; Popovic et al. 1984). Nearly always fatal, HIV/AIDS was devastating for those infected. But few physicians or researchers foresaw the magnitude of the epidemic to come **(Figure 1.1)**.

Indeed, many were optimistic about the prospects for containing HIV (Walker and Burton 2008). Smallpox had been declared eradicated in 1980 (Moore et

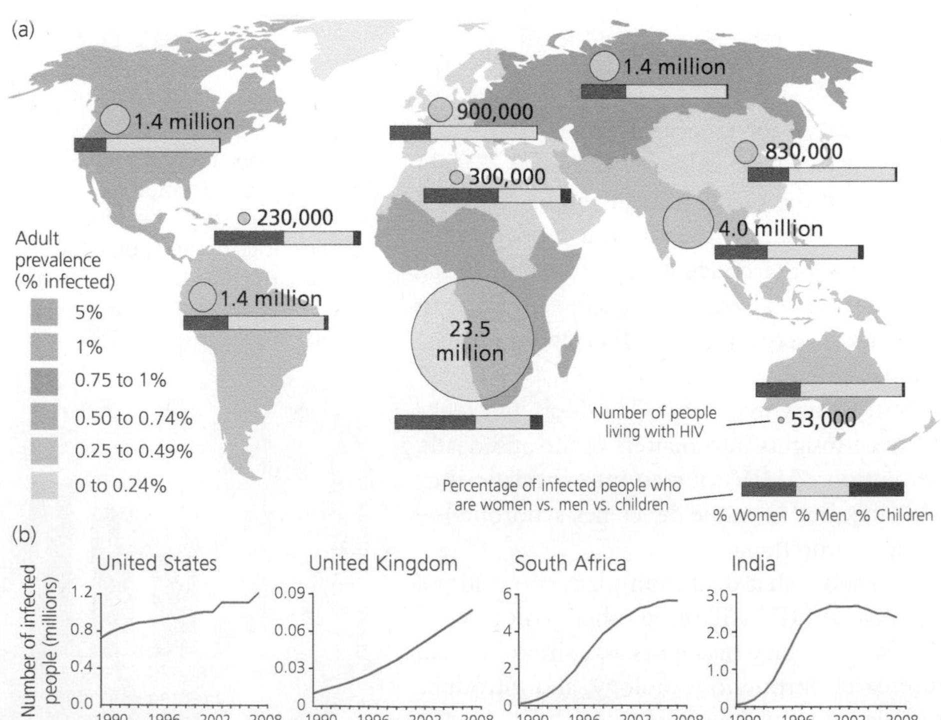

Figure 1.1 The HIV/AIDS pandemic The map (a) shows the geographic distribution of HIV infections. The color of each region indicates the fraction of adults infected with HIV (UNAIDS 2012b). The areas of the circles are proportional to the number of individuals living with HIV (UNAIDS 2012b). The bars divide individuals living with HIV by sex and age (UNAIDS 2008). The graphs (b) show the growth of the epidemic from 1990 to 2008 in four countries. Prepared with data from UNAIDS (2008).

al. 2006), and vaccines and antibiotics had brought many other infectious diseases under control. In 1984 the U.S. Secretary of Health and Human Services, Margaret Heckler, predicted that an AIDS vaccine would be ready for testing in two years. Actual events have, of course, played out rather differently.

HIV has infected over 65 million people (UNAIDS 2010, 2012a). Roughly 30 million have died of the opportunistic infections that characterize AIDS. The disease is the cause of about 3.1% of all deaths worldwide (WHO 2008/2011). AIDS is responsible for fewer deaths than heart disease (12.8%), strokes (10.8%), and lower respiratory tract infections (6.1%)—common agents of death among the elderly. But it causes more deaths than tuberculosis (2.4%), lung and other respiratory cancers (2.4%), and traffic accidents (2.1%).

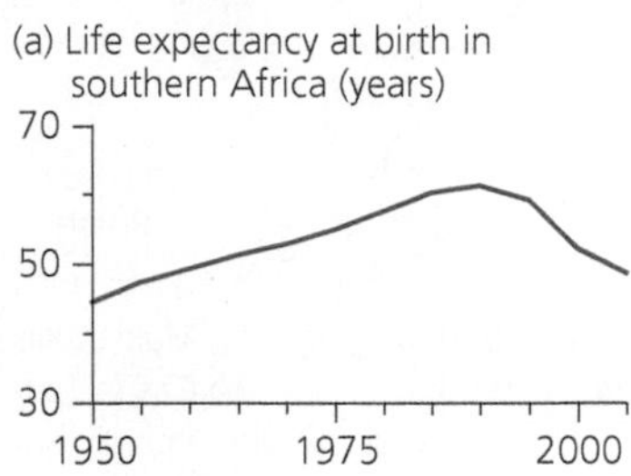

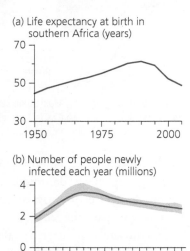

Figure 1.2 Long-term trends in HIV/AIDS (a) In southern Africa, the epidemic has caused a sharp reduction in life expectancy at birth. From UNAIDS (2008). (b) Worldwide, the annual number of new infections has been falling since the late 1990s. Red line shows best estimate; gray band shows range of estimates. From UNAIDS (2012).

Figure 1.1 summarizes the global AIDS epidemic. The map reveals substantial variation among regions in the number of people living with HIV, the percentage of the population infected, and the proportion of infected individuals who are women versus men versus children. The graphs show that the number of people infected has peaked in some countries but continues to climb in others.

The epidemic has been most devastating in sub-Saharan Africa, where 1 in 20 adults is living with HIV (UNAIDS 2008). Worst hit is Swaziland, with 26% of adults infected, followed by Botswana at 24%; Lesotho, 23%; and South Africa, 18%. Across southern Africa, life expectancy at birth has dropped below 50, a level last seen in the early 1960s **(Figure 1.2a)**. The good news is that the annual rate of new infections in sub-Saharan Africa has been falling for over a decade (UNAIDS 2012). This has meant that the global rate of new infections has been falling as well (Figure 1.2b).

In developed countries, overall infection rates are much lower than in sub-Saharan Africa (UNAIDS 2008). In western and central Europe, 0.3% of adults are infected. In Canada the rate is 0.4%, and in the United States it is 0.6%. For certain risk groups, however, infection rates rival those in southern Africa. Among men who have sex with men, the infection rate is 12% in London, 18% in New York City, and 24% in San Francisco (CDC 2005; Dodds et al. 2007; Scheer et al. 2008). Among injection drug users, the infection rate is 12% in France, 13% in Canada, and 16% in the United States (Mathers et al. 2008).

How Does HIV Spread, and How Can It Be Slowed?

A new HIV infection starts when a bodily fluid carries the virus from an infected person directly onto a mucous membrane or into the bloodstream of an uninfected person. HIV travels via semen, vaginal and rectal secretions, blood, and breast milk (Hladik and McElrath 2008). It can move during heterosexual or homosexual sex, oral sex, needle sharing, transfusion with contaminated blood products, other unsafe medical procedures, childbirth, and breastfeeding.

An HIV infection can be contracted only from someone else who already has it.

HIV has spread by different routes in different regions **(Figure 1.3, next page)**. In sub-Saharan Africa and parts of south and southeast Asia, heterosexual sex has been the most common mode of transmission. In other regions, including Europe and North America, male–male sex and needle sharing among injection drug users have predominated. Certain activities are particularly risky. For example, data on men who have sex with men in Victoria, Australia, show that having receptive anal intercourse with casual partners without the protection of a condom is a dangerous behavior. Individuals who report practicing it are nearly 60 times as likely to be infected with HIV as individuals who do not report practicing it (Read et al. 2007).

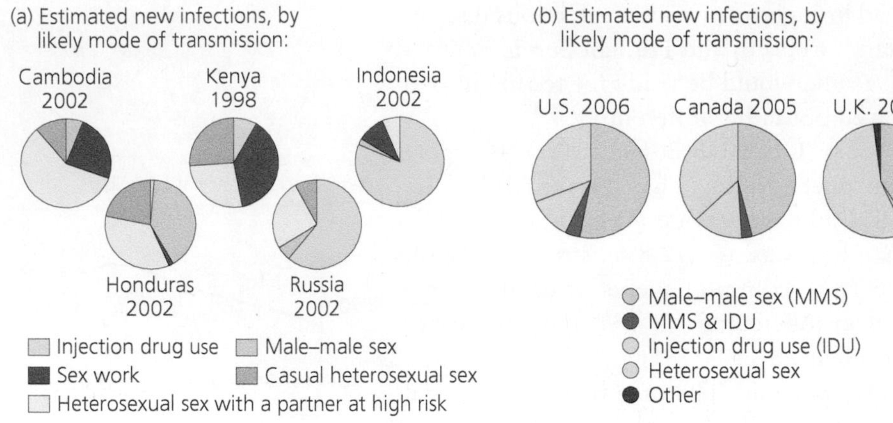

Figure 1.3 HIV's main routes of transmission in various regions (a) From Pisani et al. (2003). (b) From Hall et al. (2008), Public Health Agency of Canada (2006), Health Protection Agency (2008). The authors of the reports on Canada and the United Kingdom note that many of the individuals who contracted HIV through heterosexual sex likely did so in sub-Saharan Africa. See also UNAIDS (2008).

Clinical studies in which volunteers are randomly assigned to treatment versus control groups have identified medical interventions that reduce the rate of HIV transmission. Use of antiviral drugs, for example, lowers the risk that infected mothers will pass the virus to their infants by about 40% (Suksomboon et al. 2007). Antivirals are similarly effective in reducing transmission among men who have sex with men (Grant et al. 2010). Circumcision reduces the risk that men will contract HIV by about half (Bailey et al. 2007; Gray et al. 2007). Antiviral vaginal gels are comparably beneficial for women (Abdool Karim et al. 2010).

The value of encouraging people to change their behavior is less clear. Behavioral change undoubtedly has the potential to curtail transmission. Consistent use of condoms, for example, may reduce the risk of contracting HIV by 80% or more (Pinkerton and Abramson 1997; Weller and Davis 2002). And there are apparent success stories. In Uganda, for instance, a campaign discouraging casual sex and promoting condom use and voluntary HIV testing is thought to have substantially reduced the local AIDS epidemic (Slutkin et al. 2006; but see Oster 2009). On the other hand, the results of randomized controlled trials have been somewhat disappointing. A study of over 4,000 HIV-negative men who have sex with men in the United States offered extensive one-on-one counseling to members of the experimental group and conventional counseling to the control group (Koblin et al. 2004). As hoped, the experimental subjects engaged in fewer risky sexual behaviors than the controls. However, the rates at which the experimentals versus the controls contracted HIV were not statistically distinguishable.

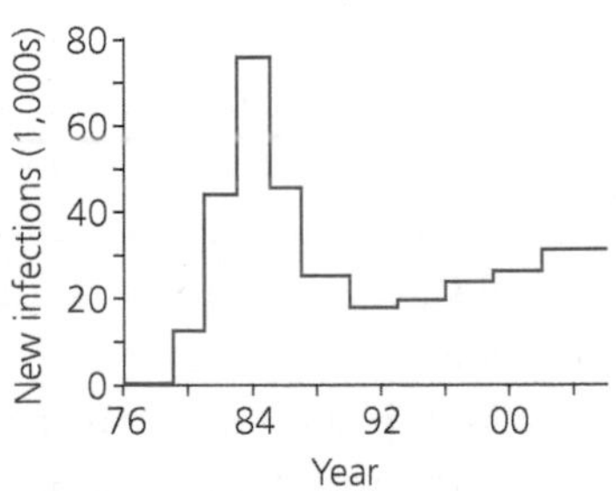

Figure 1.4 New HIV infections among men who have sex with men in the United States From Hall et al. (2008).

There is clearly no room for complacency. The graph in **Figure 1.4** tracks the number of new infections each year among men who have sex with men in the United States. After falling from the mid 1980s to the early 1990s, the annual number of new infections has since been rising steadily. The same thing seems to be happening elsewhere (Hamers and Downs 2004; Giuliani et al. 2005). Results of surveys suggest that the introduction of effective long-term drug therapies, which for some individuals has at least temporarily transformed HIV into a manageable chronic illness, has also prompted an increase in risky sexual behavior (Crepaz, Hart, and Marks 2004; Kalichman et al. 2007).

What Is HIV?

Like all viruses, HIV is an intracellular parasite incapable of reproducing on its own. HIV invades specific types of cells in the human immune system. The virus hijacks the enzymatic machinery, chemical materials, and energy of the host cells to make copies of itself, killing the host cells in the process.

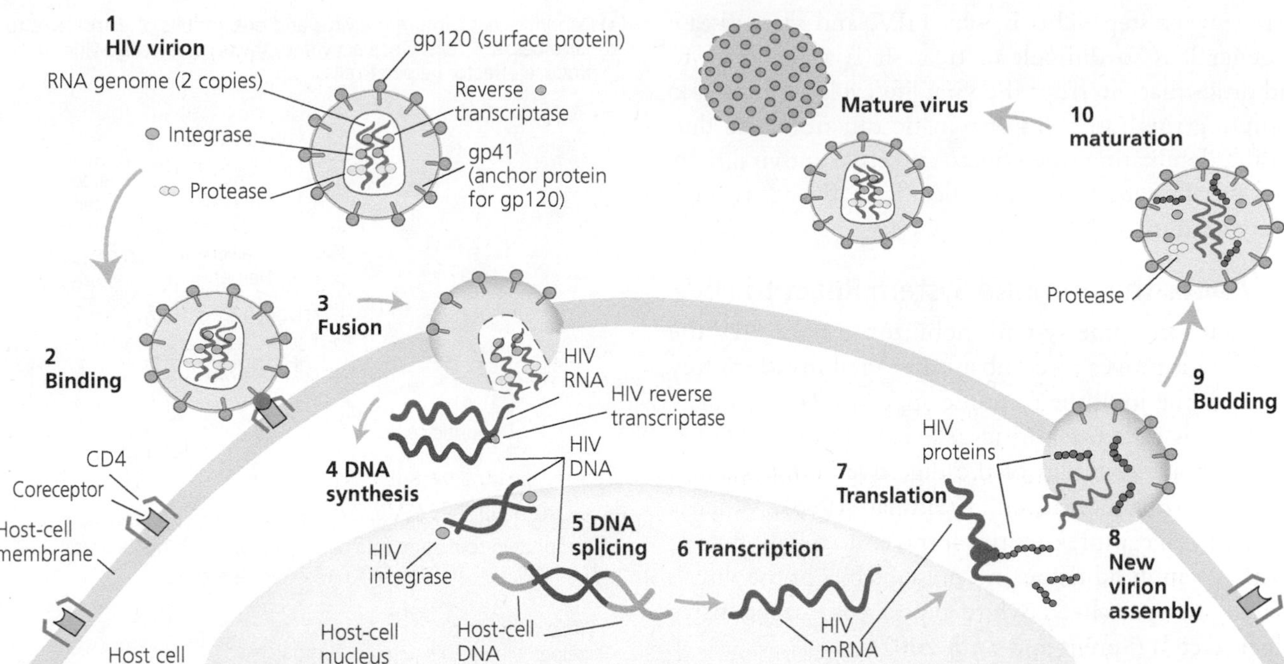

Figure 1.5 **The life cycle of HIV** (**1**, upper left) HIV's extracellular form, known as a virion, encounters a host cell (usually a helper T cell). **(2)** HIV's gp120 surface protein binds first to CD4, then to a coreceptor (usually CCR5; sometimes CXCR4) on the surface of the host cell. **(3)** The HIV virion fuses with the host cell; HIV's RNA genome and enzymes enter the host cell's cytoplasm. **(4)** HIV's reverse transcriptase enzyme synthesizes HIV DNA from HIV's RNA template. **(5)** HIV's integrase enzyme splices HIV's DNA genome into the host cell's genome. **(6)** HIV's DNA genome is transcribed into HIV mRNA by the host cell's RNA polymerase. **(7)** HIV's mRNA is translated into HIV precursor proteins by host cell's ribosomes. **(8)** A new generation of virions assembles at the membrane of the host cell. **(9)** New virions bud from the host cell's membrane. **(10)** HIV's protease enzyme cleaves precursors into mature viral proteins, allowing the new virions to mature.

Figure 1.5 outlines HIV's life cycle in more detail (Nielsen et al. 2005; Ganser-Pornillos et al. 2008). The life cycle includes an extracellular phase and an intracellular phase. During the extracellular, or infectious phase, the virus moves from one host cell to another and can be transmitted from host to host. The extracellular form of a virus is called a virion or virus particle. During the intracellular, or replication phase, the virus replicates.

HIV initiates its replication phase by latching onto two proteins on the surface of a host cell. After adhering first to CD4, HIV attaches to a second protein, called a coreceptor. This leads to fusion of the virion's envelope with the host's cell membrane and spills the contents of the virion into the cell. The contents include the virus's genome (two copies of a single-stranded RNA molecule) and two viral enzymes: reverse transcriptase, which transcribes the virus's RNA genome into DNA; and integrase, which splices this DNA genome into the host cell's genome.

HIV is a parasite that afflicts cells of the human immune system. HIV virions enter host cells by binding to proteins on their surface, then use the host cells' own machinery to make new virions.

Once HIV's genome has infiltrated the host cell's DNA, the host cell's RNA polymerase transcribes the viral genome into viral mRNA. The host cell's ribosomes synthesize viral proteins. New virions assemble at the host cell's membrane, then bud off into the bloodstream or other bodily fluid. Inside the new virions, HIV's protease enzyme cleaves precursors of various viral proteins into functional forms, allowing the virions to mature. The new virions are now ready to invade new cells in the same host or to move to a new host.

A notable feature of HIV's life cycle is that the virus uses the host cell's own enzymatic machinery—its polymerases, ribosomes, and tRNAs, and so on—in almost every step. This is why HIV, and viral disease in general, is so difficult to treat. It is a challenge to find drugs that interrupt the viral life cycle without also disrupting the host cell's enzymatic functions and thus causing debilitating side effects. Effective antiviral therapies usually target enzymes specific to the virus, such as reverse transcriptase and integrase.

How Does the Immune System React to HIV?

A patient's immune system mobilizes to fight HIV the same way it moves to combat other viral invaders. Key aspects of the immune response appear in **Figure 1.6**.

Sentinels called dendritic cells patrol vulnerable tissues, such as the lining of the digestive and reproductive tracts (Banchereau and Steinman 1998). When a dendritic cell captures a virus, it travels to a lymph node or other lymphoid tissue and presents bits of the virus's proteins to specialized white blood cells called naive helper T cells (Sprent and Surh 2002).

Naive helper T cells carry highly variable proteins called T-cell receptors. When a dendritic cell presents a helper T cell with a bit of viral protein that binds to the T cell's receptor, the helper T cell activates. It grows and divides, producing daughter cells called effector helper T cells. Effector helper T cells help coordinate the immune response.

Effector helper T cells issue commands, in the form of molecules called cytokines, that help mobilize a variety of immune cells to join the fight. They induce B cells to mature into plasma cells, which produce antibodies that bind invading virions and mark them for elimination (McHeyzer-Williams et al. 2000). They activate killer T cells, which destroy infected host cells (Williams and Bevan 2007). And they recruit macrophages (not shown), which destroy virus particles or kill infected cells (Seid et al. 1986; Abbas et al. 1996).

Most effector helper T cells die within a few weeks. However, a few survive and become memory helper T cells (Harrington et al. 2008). If the same pathogen invades again, the memory cells produce a new population of effector helper T cells.

How Does HIV Cause AIDS?

As we noted earlier, HIV invades host cells by first latching onto two proteins on the host cell's surface. The first of these is CD4; the second is a called a coreceptor. Different strains of HIV exploit different coreceptors, but most strains responsible for new infections use a protein called CCR5. Cells that carry both CD4 and CCR5 on their membranes, and are thus vulner-

(a) Dendritic cells capture the virus and present bits of its proteins to naive helper T cells. Once activated, these naive cells divide to produce effector helper T cells.

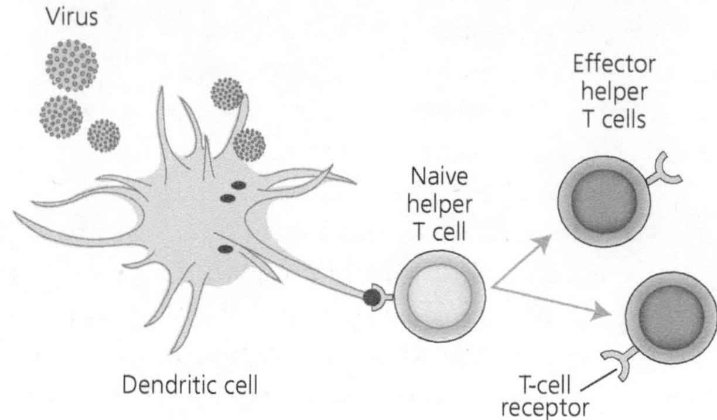

(b) Effector helper T cells stimulate B cells displaying the same bits of viral protein to mature into plasma cells, which make antibodies that bind and in some cases inactivate the virus.

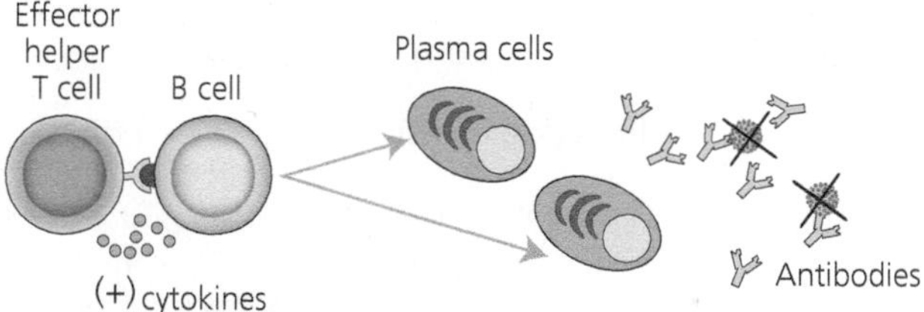

Effector helper T cells also help activate killer T cells, which destroy host cells infected with the virus.

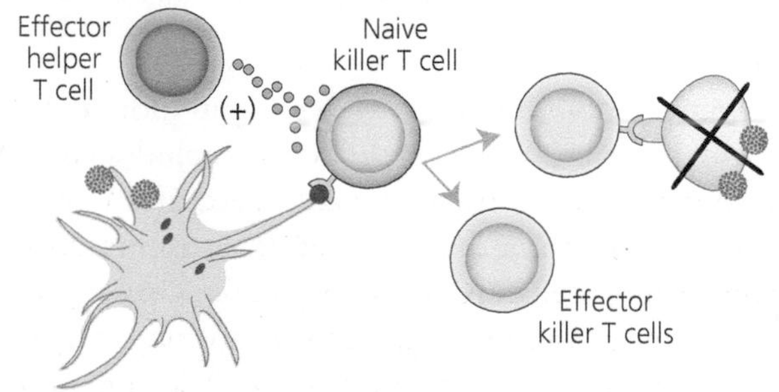

(c) Most effector T cells are short lived, but a few become long-lived memory helper T cells.

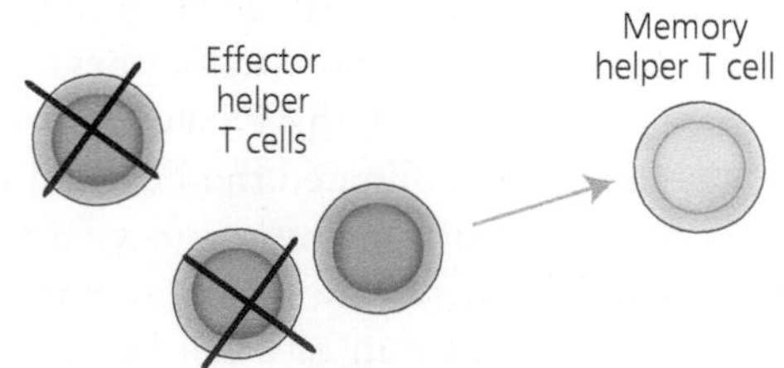

Figure 1.6 How the immune system fights a viral infection After NIAID (2003) and Watkins (2008).

able to HIV, include macrophages, effector helper T cells, and memory helper T cells **(Figure 1.7)**.

The progress of an HIV infection can be monitored by periodically measuring the concentration of HIV virions in the patient's bloodstream and the concentration of CD4 T cells in the patient's bloodstream and in the lymphoid (immune system) tissues associated with the mucous membranes of the gut. A typical untreated infection progresses through three phases.

In the acute phase, HIV virions enter the host's body and replicate explosively. The concentration of virions in the blood climbs steeply **(Figure 1.8)**. The concentrations of CD4 T cells plummet—especially in the lymphoid tissues of the gut. During this time, the host may show general symptoms of a viral infection. The acute phase ends when viral replication slows and the concentration of virions in the bloodstream drops. The host's CD4 T-cell counts recover somewhat.

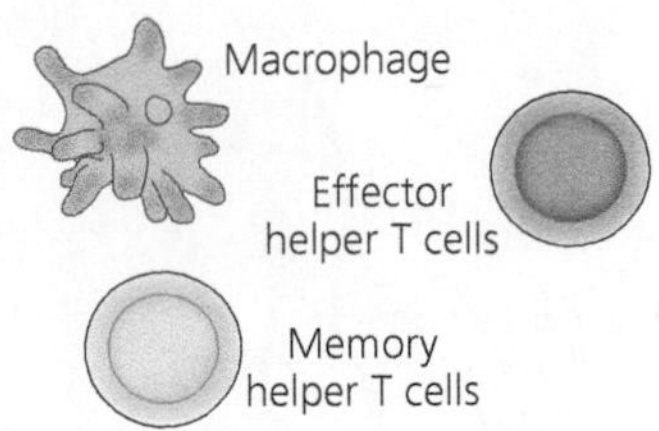

Figure 1.7 Immune system cells that carry both CD4 and CCR5 on their membranes, and are thus vulnerable to HIV. Data from UNAIDS (2008).

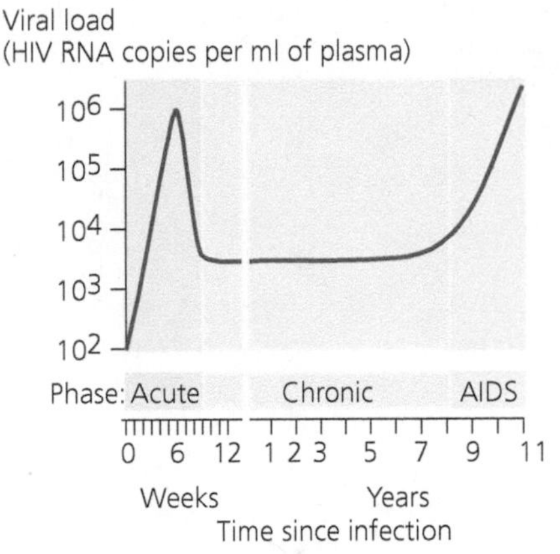

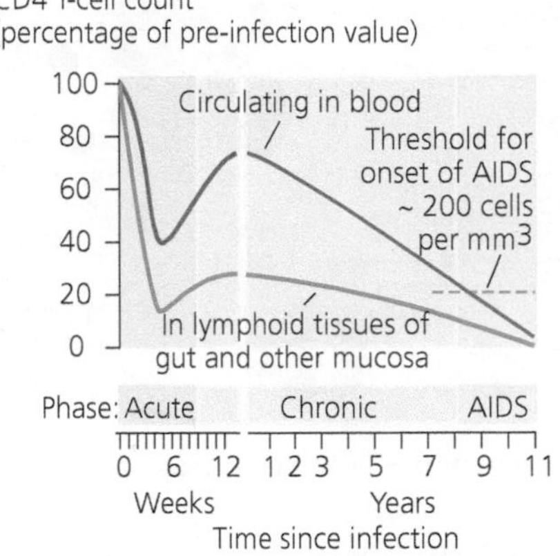

Figure 1.8 Typical clinical course of an untreated HIV infection By the time the concentration of CD4 T cells in the blood stream falls below about 200 cells per cubic millimeter, the patient's immune system begins to collapse. After Bartlett and Moore (1998), Brenchley et al. (2006), Pandrea et al. (2008).

During the chronic phase, the patient usually has few symptoms. HIV continues to replicate, however. The concentration of virions in the blood may stabilize for a while, but eventually rises again. Concentrations of CD4 T cells fall.

The AIDS phase begins when the concentration of CD4 T cells in the blood drops below 200 cells per cubic millimeter. By now the patient's immune system has begun to collapse and can no longer fend off a variety of opportunistic viruses, bacteria, and fungi that rarely cause problems for people with robust immune systems. Without effective anti-HIV drug therapy, a patient diagnosed with AIDS can expect to live less than three years (Schneider et al. 2005).

The mechanisms by which an HIV infection depletes the patient's CD4 T cells and undermines the patient's immune system are complex. Despite a quarter century of research, they remain incompletely understood (Pandrea et al. 2008; Douek et al. 2009; Silvestri 2009). The simple infection and destruction of host CD4 T cells may explain their precipitous loss during the acute phase of infection. But the immune system has an impressive capacity to regenerate these cells. Furthermore, during the chronic phase no more than one CD4 T cell in a hundred is directly infected. There must be more to the story.

Figure 1.9 (next page) outlines key events thought to lead from HIV infection to AIDS (Appay and Sauce 2008; Pandrea et al. 2008; Douek et al. 2009; Silvestri 2009). HIV's attack on the CD4 T cells in the gut (top) initiates a vicious cycle. This attack not only destroys a large fraction of the patient's helper T cells, it also damages other tissues in the gut that help provide a barrier between gut bacteria

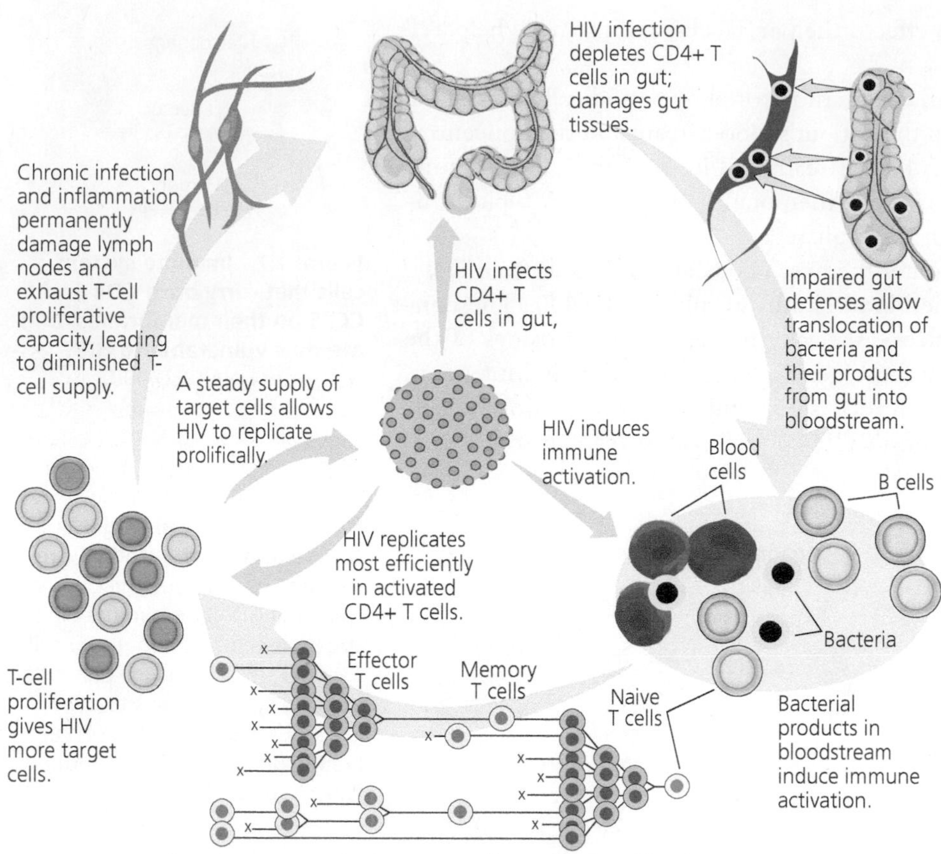
HIV infection depletes CD4+ T cells in gut; damages gut tissues.
Impaired gut defenses allow translocation of bacteria and their products from gut into bloodstream.
Blood cells
B cells
Bacteria
Bacterial products in bloodstream induce immune activation.
HIV induces immune activation.
HIV infects CD4+ T cells in gut,
A steady supply of target cells allows HIV to replicate prolifically.
HIV replicates most efficiently in activated CD4+ T cells.
Chronic infection and inflammation permanently damage lymph nodes and exhaust T-cell proliferative capacity, leading to diminished T-cell supply.
T-cell proliferation gives HIV more target cells.
Effector T cells
Memory T cells
Naive T cells
Immune system activation causes effector T-cell proliferation.

Figure 1.9 A model for how HIV causes AIDS To read the figure, start at the top, with HIV depleting CD4+ T cells in the gut. Then proceed clockwise. Direct effects of the virus are indicated by smaller pink arrows in the center; indirect effects by larger tan arrows around the outside. After Appay and Sauce (2008); Pandrea et al. (2008); Douek et al. (2009); Silvestri (2009).

and the bloodstream. The weakening of this barrier lets bacteria and their products move (translocate) from the gut into the blood (Figure 1.9, upper right).

The translocation of bacterial products into the blood triggers a high level of immune activation, to which the HIV infection itself also contributes (Biancotto et al. 2008). As we saw in Figure 1.6, activation of the immune system induces B cells and T cells to proliferate. This aggressive immune response has benefits, at least temporarily. For example, the anti-HIV killer T cells it yields help restrain HIV's replication. This and the production of new helper T cells allow the patient's concentrations of CD4 T cells to recover somewhat (Figure 1.8). But in the case of HIV, a strong immune response comes with heavy costs. The reason is that HIV replicates most efficiently in activated CD4 T cells. In other words, the immune system's best efforts to douse the HIV infection just add fuel to the fire.

HIV directly and indirectly induces immune activation, then replicates in activated immune system cells. When the ongoing battle damages the immune system to the point that it can no longer produce enough T cells to function properly, AIDS begins.

A major battleground in the ongoing fight between HIV and the immune system is the patient's lymph nodes (Lederman and Margolis 2008). The lymph nodes are, among other things, the places where naive T cells are activated. Chronic infection and inflammation eventually damages the lymph nodes irreversibly and exhausts the immune system's capacity to generate new T cells. As the patient's T-cell concentrations inexorably fall, the immune system loses its ability to fight other pathogens. The ultimate result is AIDS.

How might HIV be stopped before it leads to AIDS? The obvious answer is to prevent it from replicating. The first drug to do so, azidothymidine, or AZT, was approved for therapeutic use in 1987 (De Clercq 2009). Clinical experience with AZT, and every antiviral developed since, brings us to the first of our organizing questions. Why do single drugs offer only temporary benefits?

1.2 Why Does HIV Therapy Using Just One Drug Ultimately Fail?

To fight a virus, researchers often look for drugs that inhibit enzymes that are special to the virus and crucial to its life cycle. Such drugs should, in principle, hobble the virus and have limited side effects. For HIV, potential targets include the virus's protease, integrase, and reverse transcriptase (see Figure 1.5). AZT, the first drug approved to fight HIV, interferes with reverse transcriptase.

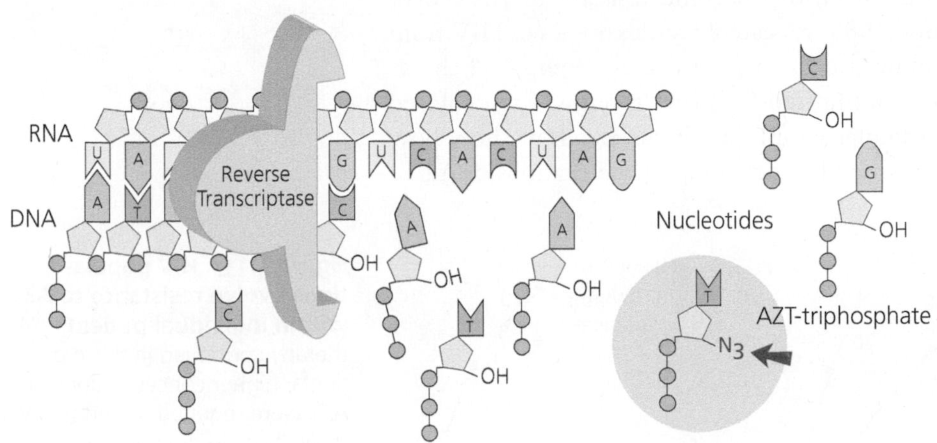

Figure 1.10 How AZT blocks reverse transcription HIV's reverse transcriptase enzyme uses nucleotides from the host cell to build a DNA strand complementary to the virus's RNA strand. AZT mimics a normal nucleotide well enough to fool reverse transcriptase, but lacks the attachment site for the next nucleotide in the chain.

Figure 1.10 shows what reverse transcriptase does. The enzyme uses the virus's RNA as a template to construct a complementary strand of viral DNA. Reverse transcriptase makes the DNA with building blocks—nucleotides—stolen from the host cell.

The figure also shows how AZT stops reverse transcription. Azidothymidine is similar in its chemical structure to the normal nucleotide thymidine—so similar that AZT fools reverse transcriptase into picking it up and incorporating it into the growing DNA strand. There is, however, a crucial difference between normal thymidine and AZT. Where thymidine has a hydroxyl group (–OH), AZT has an azide group ($-N_3$). The hydroxyl group that AZT lacks is precisely where reverse transcriptase would attach the next nucleotide to the growing DNA molecule. Reverse transcriptase is now stuck. Unable to add more nucleotides, it cannot finish its job. AZT thus interrupts the pathway to new viral proteins and new virions.

AZT is incorporated by HIV's reverse transcriptase into the viral DNA strand, where the drug prevents the enzyme from adding more nucleotides. However, alterations in the structure of reverse transcriptase can make viral replication less vulnerable to disruption.

In early tests AZT worked, halting the loss of T cells in AIDS patients. The drug caused serious side effects, because it sometimes fools the patient's own DNA polymerase and thereby interrupts normal DNA synthesis. But it appeared to promise substantially slower immune deterioration. By 1989, however, after only a few years of use, patients stopped responding to treatment. Their T-cell counts again began to fall. What went wrong?

Does AZT Alter the Patient's Physiology?

In principle, AZT could lose its effectiveness in either or both of two ways. One is that the patient's own cellular physiology could change. After it enters a cell, AZT has to be phosphorylated by the cell's own thymidine kinase enzyme to become biologically active. Perhaps long-term exposure to AZT causes a cell to make less thymidine kinase. If so, AZT would become less effective over time.

Patrick Hoggard and colleagues (2001) tested this hypothesis by periodically checking the intracellular concentrations of phosphorylated AZT in a group of patients taking the same dose of AZT for a year. The data refute the hypothesis. The concentrations of phosphorylated AZT did not change over time.

Does AZT Alter the Population of Virions Living in the Patient?

The other way AZT could lose its effectiveness is that the population of virions living inside the patient could change so that the virions themselves would be resistant to disruption by AZT.

To find out whether populations of virions become resistant to AZT over time, Brendan Larder and colleagues (1989) repeatedly took samples of HIV from patients and grew the virus on cultured cells in petri dishes. **Figure 1.11** shows data for two patients the researchers monitored for many months. Each colored curve in the graphs represents a particular sample. Each curve falls to show how rapidly HIV's ability to replicate is curbed by increasing concentrations of AZT.

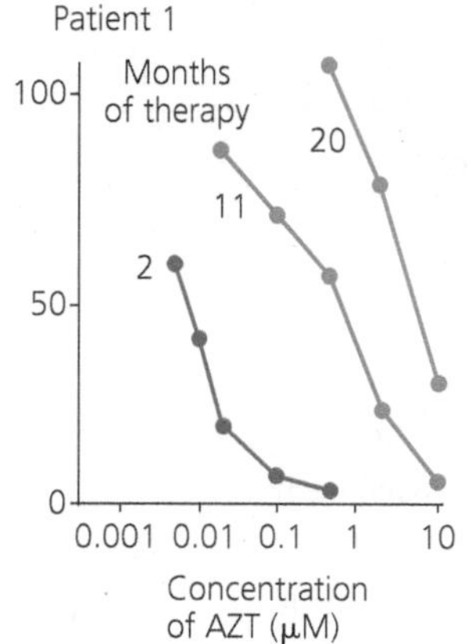

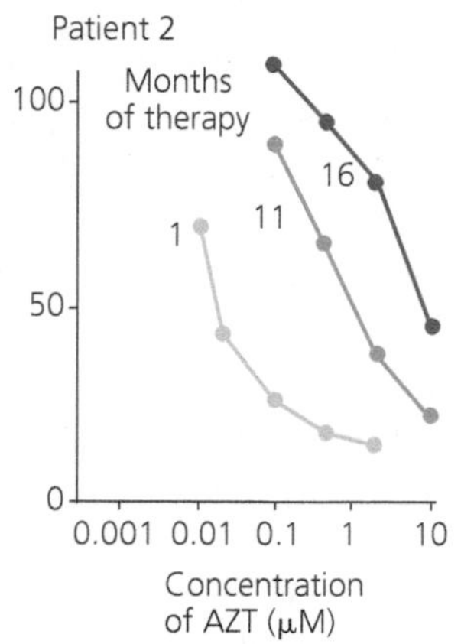

Figure 1.11 HIV populations evolve resistance to AZT within individual patients As therapy continued in these patients, higher concentrations of AZT were required to curtail HIV's replication. Redrawn from Larder et al. (1989).

Examine the three curves for Patient 1. Virions sampled from this patient after he had been taking AZT for two months were still susceptible to the drug. At moderate concentrations of AZT, the virions lost their ability to replicate almost entirely. Virions sampled from the patient after 11 months on AZT were partially resistant. They could be stopped, but it took about 10 times as much AZT to do it. Virions taken after 20 months on AZT were highly resistant. They were completely unaffected by AZT concentrations that stopped the first sample and could still replicate fairly well at concentrations that stopped the second sample.

The data for Patient 2 tell the same story. Populations of virions within individual patients change to become resistant to AZT. In other words, the populations evolve.

In many patients taking AZT, drug-resistant populations of HIV evolve within just six months **(Figure 1.12)**.

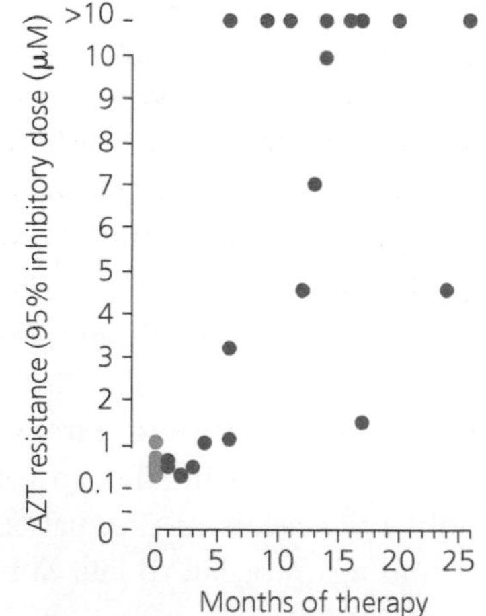

Figure 1.12 In many patients, AZT resistance evolves within six months This graph plots resistance in 39 patients checked at different times. Redrawn from Larder et al. (1989).

What Makes HIV Resistant to AZT?

What is the difference between a resistant virion versus a susceptible one? To answer this question, consider a thought experiment. If we wanted to engineer an HIV virion capable of replicating in the presence of AZT, what would we do? We would have to modify the virus's reverse transcriptase enzyme so that it either avoids inserting AZT molecules into the growing DNA strand in the first place or, having inserted an AZT molecule, is more likely to take it back out so that the DNA strand can continue to grow **(Figure 1.13)**.

In practice, we could expose large numbers of HIV virions to a mutagenic chemical or ionizing radiation. This would generate strains of HIV with altered nucleotide sequences in their genomes—and thus altered amino acid sequences in their proteins. If we generated enough mutants, at least a few would carry changes in the active site of the reverse transcriptase molecule—the part that recognizes nucleotides, adds them to the growing DNA strand, and corrects mistakes. If one of the reverse transcriptases with an altered binding site were less likely to mistake AZT for the normal nucleotide, or more likely to remove AZT after insertion, then the mutant variant of HIV would be able to continue replicating in the presence of the drug. If we treated our population of mutant virions with AZT, HIV strains unable to replicate in the presence of AZT would decline in numbers, and the resistant strain would become common.

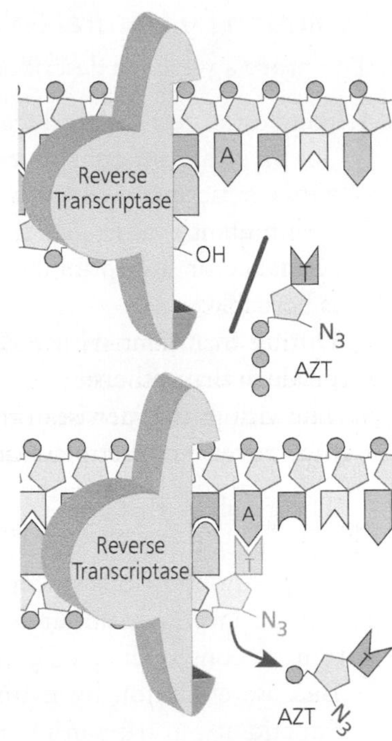

Figure 1.13 Two ways reverse transcriptase could resist AZT A resistant enzyme might avoid AZT (top) or, having inserted AZT into the growing DNA strand, take it back out (bottom).

The steps involved in this thought experiment are just what happens inside the bodies of HIV patients like the ones followed by Larder and colleagues. How do we know? In studies similar to Larder's, researchers took repeated samples of HIV virions from patients receiving AZT. The researchers found that viral strains present late in treatment were genetically different from viral strains that had been present before treatment in the same hosts. The mutations associated with AZT resistance were often the same from patient to patient (St. Clair et al. 1991; Mohri et al. 1993; Shirasaka et al. 1993) and caused amino acid changes in reverse transcriptase's active site **(Figure 1.14)**.

The altered reverse transcriptase enzymes still pick up AZT and insert it into the growing DNA strand, but they are more likely to subsequently remove the AZT and, therefore, be able to continue building the DNA copy (Boyer et al. 2001). Possession of such a modified reverse transcriptase enables HIV virions to replicate in the presence of AZT.

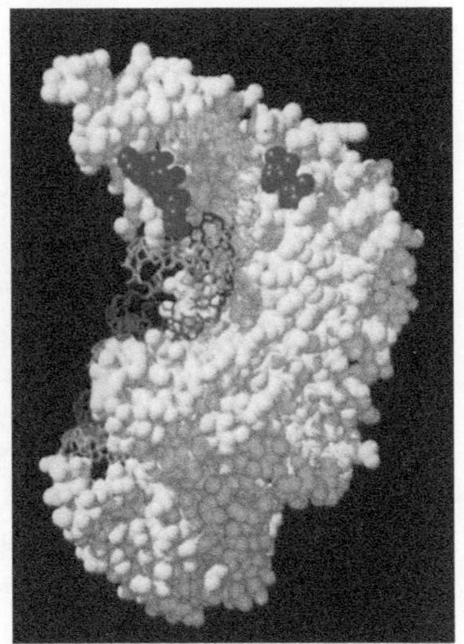

Figure 1.14 AZT-resistant reverse transcriptase The green and blue strands are the template RNA and growing DNA. The solid spheres show the structure of reverse transcriptase. Amino acid changes correlated with resistance to AZT are in red. They are in the enzyme's active site. By Lori Kohlstaedt; from Cohen (1993).

Note that, unlike the situation in our thought experiment, no conscious manipulation took place. How, then, did the change in the viral strains occur?

The answer is that, despite having some ability to correct transcription errors, reverse transcriptase is prone to mistakes. Over half the DNA transcripts it makes contain at least one error—one mutation—in their nucleotide sequence (Hübner et al. 1992; Wain-Hobson 1993). Because thousands of generations of HIV replication take place within each patient during an infection, a single strain of HIV produces enormous numbers of reverse transcriptase variants in every host.

Simply because of their numbers, it is a virtual certainty that one or more of these variants contains an amino acid substitution that improves reverse transcriptase's ability to recognize and remove AZT. If the patient takes AZT, the replication of unaltered HIV variants is suppressed, but the resistant mutants will still be able to synthesize some DNA and produce new virions. As the resistant virions propagate and the nonresistant virions fail, the fraction of the virions in the patient's body that are resistant to AZT increases over time. Furthermore, each new generation in the viral population contains virions with additional mutations, some of which may further enhance the ability of reverse transcriptase to function in the presence of AZT. Because they reproduce faster, the virions that carry these new mutations will also increase in frequency.

This process of change over time in the composition of the viral population is called evolution by natural selection. It has occurred so consistently in patients taking AZT that use of AZT alone has long been abandoned as an AIDS therapy.

Evolution by Natural Selection

The process we have described involves four steps **(Figure 1.15)**:

1. Replication errors produce mutations in the reverse transcriptase gene. Virions carrying different reverse transcriptase genes produce versions of the reverse transcriptase enzyme that vary in their resistance to AZT.
2. The mutant virions pass their reverse transcriptase genes, and thus their AZT resistance or susceptibility, to their offspring. In other words, AZT resistance is heritable.
3. During treatment with AZT, some virions are better able to survive and reproduce than others.
4. The virions that persist in the presence of AZT are the ones with mutations in their reverse transcriptase genes that confer resistance.

Heritable traits that lead to survival and abundant reproduction spread in populations; heritable traits that lead to reproductive failure disappear. This is evolution by natural selection.

The result is that the composition of the viral population within the host changes over time. Virions resistant to AZT comprise an ever larger fraction of the population; virions susceptible to AZT become rare. There is nothing mysterious or purposeful about evolution by natural selection; it just happens. It is an automatic consequence of heritable differences in replication.

Because evolution by natural selection is an automatic consequence of cold arithmetic, it can happen in any population in which the four steps occur. That is, it can happen in any population in which there is heritable variation in reproductive success. We will see many examples in the chapters to come.

One measure of whether we understand a process is whether we can control it. If we truly understand the mechanism of evolution by natural selection as it operates inside the bodies of HIV patients, we should be able to find a way to

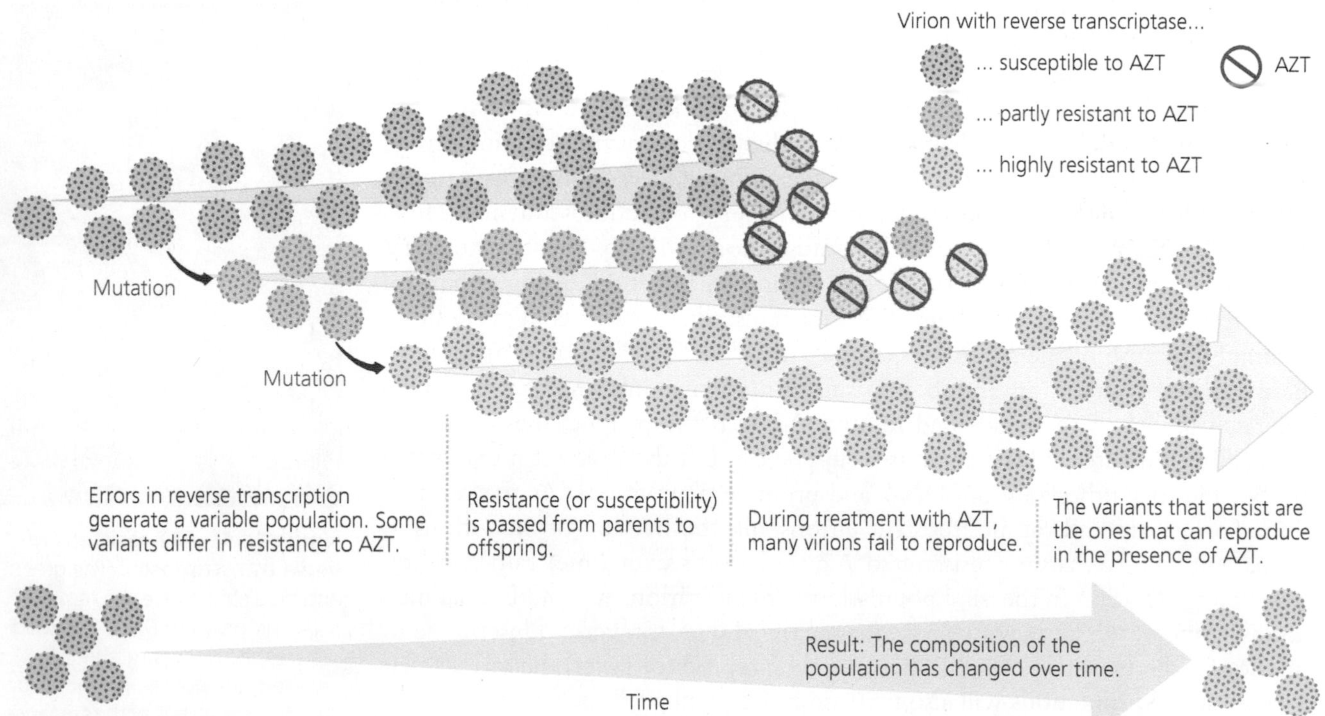

Figure 1.15 Evolution by natural selection, as illustrated by the evolution of resistance to AZT in an HIV population After Richman (1998).

stop it—or at least slow it down. We next consider how understanding the evolution of resistance allowed researchers to devise more effective therapies.

Understanding Evolution Helps Researchers Design Better Therapies

Since AZT was introduced, the number of drugs approved for treatment of HIV has grown to over two dozen (De Clercq 2009). The categories of drugs in use, in order of the stage of HIV's life cycle they are intended to disrupt, include

- **Coreceptor inhibitors.** These bar HIV from entering host cells in the first place by preventing them from latching onto the host cell's CCR5 molecules.
- **Fusion inhibitors.** These bar HIV from entering host cells by interfering with HIV's gp120 or gp41 proteins.
- **Reverse transcriptase inhibitors.** Some, like AZT, inhibit reverse transcriptase by mimicking the normal building blocks of DNA. Others inhibit reverse transcriptase by interfering with the enzyme's active site.
- **Integrase inhibitors.** These block HIV's integrase from inserting HIV's DNA into the host genome, preventing the transcription of new viral RNAs.
- **Protease inhibitors.** These prevent HIV's protease enzyme from cleaving viral precursor proteins to produce mature components for new virions.

Experience so far indicates that when any antiretroviral drug is used alone, the outcome will be the same as we have seen with AZT. The virus population in the host quickly evolves resistance (see, for example, St. Clair et al. 1991; Condra et al. 1996; Ala et al. 1997; Deeks et al. 1997; Doukhan and Delwart 2001).

Why do HIV populations evolve resistance so easily? With any single drug, just one or a few mutations in the gene for the targeted viral protein can render the virus resistant. With its high mutation rate, short generation time, and large population size, HIV generates so many mutant genomes that variants with the crucial combination of mutations are likely to be present much of the time. When the HIV population in a patient harbors genetic variation for replication in the presence of a drug, and the patient takes the drug, the population evolves.

This analysis suggests that the way to improve anti-HIV therapy is to increase the number of mutations that must be present in a virion's genome to render the virion resistant. The more mutations needed for resistance, the lower the probability that they will occur together in a single virion. In other words, we need a strategy to reduce the genetic variation for resistance. Without genetic variation—without differences in survival and reproduction that are passed from one generation to the next—the viral population cannot evolve.

By reducing the genetic variation for resistance in populations of virions, cocktails of drugs that target different points in HIV's life cycle limit the evolution of resistant strains. This, in turn, has dramatically improved patient survival.

The simplest way to raise the number of mutations required for resistance is to use two or more drugs at once. For this to work, a mutation that renders a virion resistant to one drug must not render it resistant to the others. Indeed, in the best scenario, a mutation that makes HIV resistant to one of the drugs will simultaneously make the virus more susceptible to the others (see St. Clair et al. 1991).

Treatment cocktails using combinations of drugs have, in fact, proven much more effective than single drugs used alone. Robert Murphy and colleagues (2008) tracked the viral loads of 100 patients who, as their first treatment for HIV, used a cocktail including two reverse transcriptase inhibitors and a protease inhibitor. Seven years later, 61 of the patients were still participating in the study, and 58 had viral loads under 50 copies per ml of blood—low enough to be

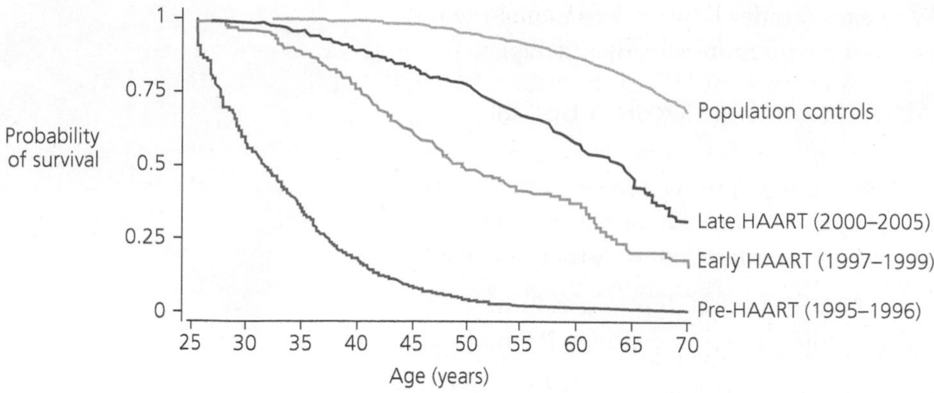

Figure 1.16 Treatment with multiple drugs prolongs the lives of patients with AIDS Redrawn from Lohse et al. (2007).

undetectable in standard tests. Results like these have earned regimens including three or more drugs that block HIV in two or more different ways the nickname highly active antiretroviral therapy, or HAART. (For more information on drug combinations used in HAART, see Hammer et al. 2008.)

Nicolai Lohse and colleagues (2007) followed 3,990 HIV-infected patients in Denmark from 1995 to 2005. **Figure 1.16** tracks the survival of the patients during three treatment eras and compares them to the survival of 379,872 controls from the general population. HAART dramatically improved patient survival, and it got better over time—as more drugs became available, and as researchers, doctors, and patients learned how best to deploy them. Understanding how resistance evolves has helped prolong lives.

Unfortunately, even the best drug cocktails do not cure HIV infection. A reservoir of viable HIV genomes remains in the body, hidden in resting white blood cells and other tissues (Maldarelli et al. 2007; Brennan et al. 2009). As a result, when patients go off HAART, their viral loads climb rapidly (Chun et al. 1999; Davey et al. 1999; Kaufmann et al. 2004).

The Evolution of HIV Strains Resistant to Multiple Drugs

Because HAART cannot eradicate HIV, the evolution of strains resistant to multiple drugs is a constant threat for patients. Richard Harrigan and colleagues (2005) followed 1,191 patients on HAART. By the end of three years, the HIV populations in 25% of the patients had evolved resistance to at least one antiretroviral drug. The HIV populations in some patients were resistant to both reverse transcriptase inhibitors and protease inhibitors. Not surprisingly, patients with drug-resistant strains of HIV face a higher risk of death (Hogg et al. 2006).

Some patients have the bad luck to become infected with HIV strains that are already drug resistant (Johnson et al. 2008). Other patients inadvertently allow their HIV populations to evolve by failing to follow their treatment regimens strictly enough. Any time the concentration of a drug in the patient's body falls to levels that allow partially resistant virions to replicate, there is an opportunity for fully resistant mutants to appear. When the concentration of the drug rises again, such mutants will enjoy a strong selective advantage.

Harrigan and colleagues (2005) gauged patient adherence to treatment by calculating the percentage of prescription refills each patient picked up. **Figure 1.17** plots the hazard ratio for the evolution of multidrug-resistant HIV as a function of refill percentage. The hazard ratio is the fraction of patients in a given adherence category who evolved resistant HIV divided by the fraction of patients in

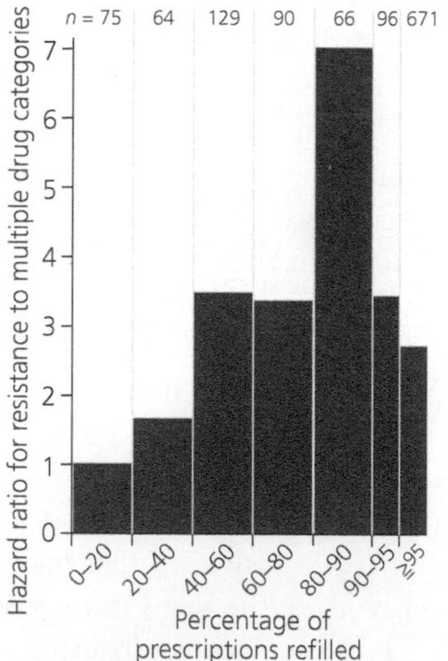

Figure 1.17 HIV evolves resistance to multiple drugs most readily in patients who take most, but not all of their prescribed doses Redrawn from Harrigan et al. (2005).

the 0 to 20% refill category who evolved resistant HIV. For an example of how to read the graph, patients who picked up 40% to 60% of their refills were between three and four times as likely to evolve resistant HIV. Patients who refilled most, but not all, of their prescriptions evolved resistant HIV at the highest rate. Given what we know about how evolution by natural selection works, the explanation is straightforward. Patients who took few of their doses subjected their HIV populations to weak selection. Patients who took all their doses shut down virtually all viral replication. Patients who took most, but not all, of their doses subjected their HIV populations to strong selection, but allowed some viral replication—thus creating permissive conditions for evolution.

Sometimes HIV populations evolve resistance even in patients taking multidrug cocktails. The risk is highest in patients who fill most, but not all, of their prescriptions.

One reason patients fail to take all of their prescribed doses is that antiretrovirals cause serious side effects. Among the reasons patients dropped out of Murphy's (2008) study were changes in body fat distribution, liver damage, elevated cholesterol, diarrhea, and joint pain. Anti-HIV therapies that are easily tolerated and that permanently suppress viral replication remain a goal of ongoing research.

We noted earlier that evolution by natural selection will happen in any population in which there are differences among individuals that are passed from parents to offspring and that influence survival and procreation. Variants associated with reproductive success automatically become common while variants associated with failure disappear. This broad applicability brings us to our next question.

1.3 Are Human Populations Evolving as a Result of the HIV Pandemic?

In Section 1.2, we saw how the HIV population inside a patient evolves in response to AZT. The drug influences which genetic variants of HIV survive and reproduce. Strains that do well despite the drug become common; strains that do poorly because of the drug become rare. In Section 1.1, we saw that the HIV pandemic is influencing which members of the human population survive and reproduce—particularly in southern Africa, where infection rates are high. This raises the question: Will human populations change over time in response to the pandemic? That is, will we evolve?

The answer depends on whether the humans who survive the pandemic owe their good fortune, at least in part, to genetic characteristics they can transmit to their offspring. If there are heritable differences among those who succumb versus those who live on, then traits conducive to surviving HIV will rise in frequency. Whether such differences exist is of more than academic interest. If we can identify genetic variants that confer resistance to HIV, then understanding how they work might suggest strategies for fighting the virus.

For a population to evolve, it must harbor genetic differences among individuals.

How might we discover genetic variants that make their carriers resistant to HIV? One way is to look for people who have not contracted the virus despite repeated exposure, or who remain healthy despite being infected. In the early 1990s, several laboratories demonstrated that both kinds of individuals exist (see Cao et al. 1995). By studying them, researchers have uncovered genetic variants that offer at least some protection against HIV (see An and Winkler 2010).

A Missing Coreceptor

In 1996, several groups of researchers identified the cell surface protein CCR5 as an important coreceptor for HIV (Deng et al., 1996, and Dragic et al., 1996, were first into print). Rong Liu and coworkers (1996) and Michel Samson and associates (1996), among others, immediately guessed that resistant individuals might have unusual forms of CCR5 that thwart HIV's entry into host cells.

To test this hypothesis, Liu and colleagues examined the gene that encodes CCR5 from two individuals who had been repeatedly exposed to HIV but remained uninfected. Samson and colleagues looked at the gene from three HIV-infected individuals who were long-term survivors. As predicted, both of Liu's subjects were homozygotes for a mutant form of the gene and one of Samson's subjects was a heterozygote. Because the mutant form is distinguished by a 32-base-pair deletion in the normal sequence of DNA, it has come to be known as the *Δ32* allele (*Δ* is the Greek letter delta).

Investigating further, Liu showed that the version of CCR5 encoded by the mutant allele fails to appear on the surface of the cell. Samson showed that cells making only the *Δ32* form are nearly impervious to invasion by the strains of HIV responsible for most new infections. Samson also found that individuals carrying one or two copies of the *Δ32* allele were substantially less common among Europeans infected with HIV than among the European population at large. Together these results indicated that the *Δ32* allele confers strong (though not perfect) protection against HIV, a conclusion later confirmed by research that followed initially uninfected high-risk subjects over time (Marmor et al. 2001).

In human populations, some individuals carry alleles that make them resistant to infection with HIV.

To find out how common the *Δ32* allele is in various human populations, Samson and colleagues took DNA samples from a large number of individuals of northern European, Japanese, and African heritage, examined the gene for CCR5 in each individual, and calculated the frequency of the normal and *Δ32* alleles in each population. The mutant allele turned out to be present at a relatively high frequency of 9% in the Europeans, but was completely absent among the individuals of Asian or African descent. Subsequent research has confirmed this result. The *CCR5-Δ32* allele is common in northern Europe and declines dramatically in frequency toward both the south and the east **(Figure 1.18)**.

Curiously, the frequency of the best-known protective allele is highest in regions with low rates of HIV infection.

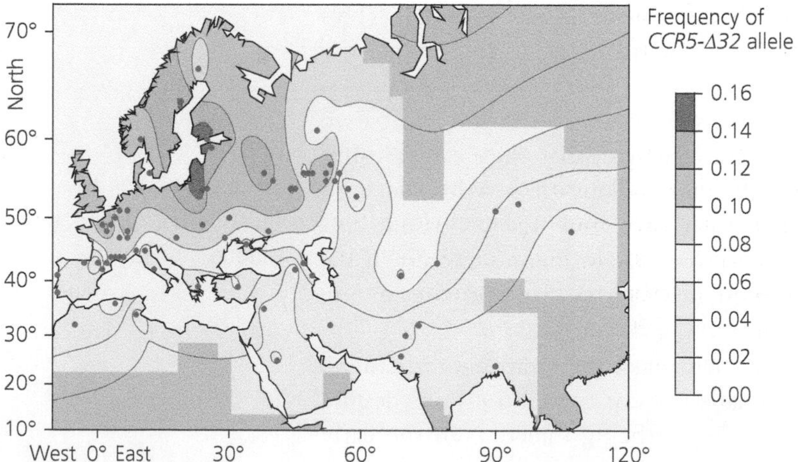

Figure 1.18 The frequency of the *CCR5-Δ32* allele in the Old World Blue dots indicate populations analyzed; colors between contour lines indicate inferred allele frequencies. Areas masked with dark gray are too far from the sources of data for reliable inferrence. From Novembre et al. (2005).

The data on the *CCR5-Δ32* allele show that human populations harbor heritable variation for resistance to HIV, but this variation will influence who lives and who dies only if HIV is present. Comparing the map of *Δ32* frequency in Figure 1.18 with the map of HIV prevalence in Figure 1.1 reveals a striking disconnect. The *Δ32* allele is common in a part of the world where HIV is rare, and HIV is rampant in parts of the world where the *Δ32* allele is rare. Do humans in sub-Saharan Africa vary in resistance to HIV?

Genetic Variation for HIV Resistance in African Populations

If different versions of the gene for HIV's coreceptor, CCR5, influence the risk of contracting the virus, perhaps the same is true of different versions of the gene for the virus's main receptor. As was shown in Figure 1.5, this is the cell-surface protein CD4. The most common allele of the gene for CD4 contains the nucleotide C at position 868. An alternative version, called *C868T*, has the nucleotide T instead, resulting in the substitution of the amino acid tryptophan for arginine. With a frequency of over 15%, this allele is fairly common among Kenyans.

Julius Oyugi and colleagues (2009) followed a group of Kenyan female sex workers who were uninfected when they volunteered to participate in the study. Among these women, 29 had genotype *CC* and 16 had genotype *CT*. (Individuals with genotype *TT* also exist, but there were too few of them in the group to allow for meaningful analysis.) **Figure 1.19** traces the percentage of women with each genotype who remained uninfected as a function of time. The women with genotype *CT* contracted HIV significantly more quickly.

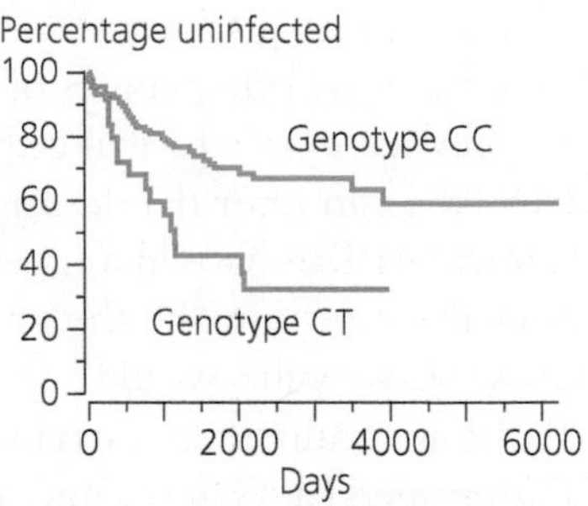

Figure 1.19 Among Kenyan female sex workers, genotype for the cell-surface protein CD4 influences risk of contracting HIV Redrawn from Oyugi et al. (2009).

Oyugi's study, as well as others (see Gonzalez et al. 1999; Winkler et al. 2004; Pelak et al. 2010), shows that African populations harbor heritable variation for resistance to HIV. Indeed, some scientists have suggested that one reason the rates at which individuals are contracting HIV have begun to fall in some of the worst-hit regions (see Figure 1.2b) is that many of the most susceptible individuals are already infected (Nagelkerke et al. 2009). If the African epidemic continues, we can expect genetic variants that confer resistance to become more common while variants that confer susceptibility dwindle. In other words, human populations in Africa are likely evolving in response to mortality imposed by HIV.

It appears that HIV is too new a human disease to have triggered substantial evolutionary change in human populations.

Evolutionary changes take time. It is probably too soon to expect the HIV pandemic to have produced measurable changes in the frequencies of particular genetic variants (Schliekelman et al. 2001; Ramaley et al. 2002; Cromer et al. 2010). Later in the book we will develop a model of evolution that allows us to predict the rate at which populations will change (Chapter 6).

A Missing Protective Allele

In addition to the protective genetic variants known to exist in human populations, biologists have discovered loss-of-function mutations that increase our susceptibility to HIV infection. This implies that back-mutations restoring the lost functions might make us more resistant. As an example, consider retrocyclin.

Retrocyclin is among a number of proteins vertebrate cells make that block stages of the retroviral life cycle (see Wolf and Goff 2008; Ortiz et al. 2009). The evolution of these intrinsic defenses is not surprising given the long history of retroviral infections suffered by our forebears. Evidence of this history litters our DNA. Roughly 8% of the human genome consists of remnants of retroviruses that inserted themselves into our ancestors' chromosomes (IHGSC 2001).

Retrocyclin is the human version of a protein, called theta defensin, that was originally discovered in rhesus macaques and subsequently found in Old World monkeys, lesser apes, and orangutans (Tang et al. 1999; Nguyen et al. 2003). Theta defensin is a small, circular protein made by joining two copies of a smaller linear precursor. Although human cells have the gene for retrocyclin's precursor, they ordinarily cannot make the protein. This is because our version of the gene—along with those carried by chimpanzees and gorillas—is disabled by a loss-of-function mutation that creates a premature stop codon. Our cells transcribe mRNA from the retrocylin gene but cannot translate it into protein.

Alexander Cole and colleagues (2002) synthesized retrocyclin and showed that it protects cultured human CD4+ T cells from HIV infection. This experiment suggests that if our cells could make retrocyclin, it would help us fight the virus.

Nitya Venkataraman and colleagues (2009) genetically engineered human cells to give them working copies of the gene for the retrocyclin precursor. The modified cells were able to make the precursor and process it into functional retrocyclin. We can infer that a simple back-mutation, involving just two nucleotide substitutions that would restore our retrocyclin gene to its original sequence, would allow us to make the protein. It is a reasonable hypothesis that the ability to make retrocyclin would confer some resistance to HIV.

Humans carry a loss-of-function mutation in the retrocyclin gene that appears to increase our susceptiblity to HIV. An allele with the back-mutation might be protective, but no such allele is known.

Although individuals carrying it might enjoy higher rates of survival in regions being ravaged by HIV, a functional version of the retrocyclin gene has yet to be found in humans. The missing functional retrocyclin allele is a useful reminder that without genetic variation, populations cannot evolve.

Practical Applications

The hunt for genetic variants resistant to HIV has yielded practical benefits. The discovery of *CCR5-Δ32* homozygotes—who lack functional CCR5, suffer few ill effects, and are resistant to sexually transmitted strains of HIV—suggested a new strategy for antiretroviral therapy: use of drugs that bind to CCR5 and thus stop HIV from latching onto its coreceptor. The first CCR5 blocker, maraviroc, was approved for patients in the United States in 2007 (Hammer et al. 2008).

Other treatments may be coming. For example, experiments in tissue culture by Phalguni Gupta and colleagues (2012) suggest that retrocyclin RC-101, used as a vaginal microbicide, could help block sexual transmission of HIV.

An Unresolved Mystery

Astute readers will have noticed that we have left a glaring question unanswered. Why is the *CCR5-Δ32* allele common in Europe and rare everywhere else? The current HIV pandemic is both too young and too geographically incongruous to explain the pattern. Samson and coworkers (1996) offered two general explanations: (1) The *CCR5-Δ32* allele may have been recently favored by natural selection in European populations; or (2) the allele could have risen to high frequency by chance, in a process called genetic drift. Researchers have proposed a variety of more specific hypotheses under each scenario. Later, we will revisit the puzzle of *CCR5-Δ32's* history and geographic distribution (Chapters 6 and 8). For now, we turn to a puzzle concerning the history of HIV itself.

1.4 Where Did HIV Come From?

Viruses, like other organisms, come from reproduction of their kind. So where did the first HIV virions come from? In preparation for discussing the analyses that have revealed the answer, we tell the story of a doctor accused of murder.

Louisiana v. Richard J. Schmidt

Nurse Janet Trahan had a multifarious relationship with Dr. Richard Schmidt (*State of Louisiana v. Richard J. Schmidt* 1997). She was his colleague at a Louisiana hospital, she was his romantic partner during a 10-year affair, and she was his patient. She eventually ended the affair, but continued as Schmidt's patient and colleague. Shortly after she broke off the affair, Trahan accepted Schmidt's offer to visit her apartment to administer her regular vitamin injection. Within a few weeks, Trahan experienced symptoms of a generalized viral infection, and five months after the injection she tested positive for HIV. On the theory that Schmidt had instructed his nurse to draw extra blood from an HIV-positive patient he saw on the day he visited Trahan's apartment, then intentionally injected the infected blood into Trahan, the local district attorney's office prosecuted Schmidt for second degree attempted murder.

To help them make their case, the prosecutors contacted evolutionary biologist David Hillis. Hillis and colleagues analyzed HIV from Trahan and Schmidt's patient and estimated their evolutionary relationships (Metzker et al. 2002).

Reconstructing Evolutionary History

Hillis's team took advantage of the fact that HIV diversifies rapidly, even within individual hosts. As we saw in Section 1.2, HIV's high mutation rate generates considerable genetic variation. Over time, the abundance of different variants changes as a result of selection imposed by antiviral therapy. The HIV population also changes as a result of selection by the host's immune system. And it even changes when mutations that have no effect on HIV's survival become more common—or dwindle—simply due to chance.

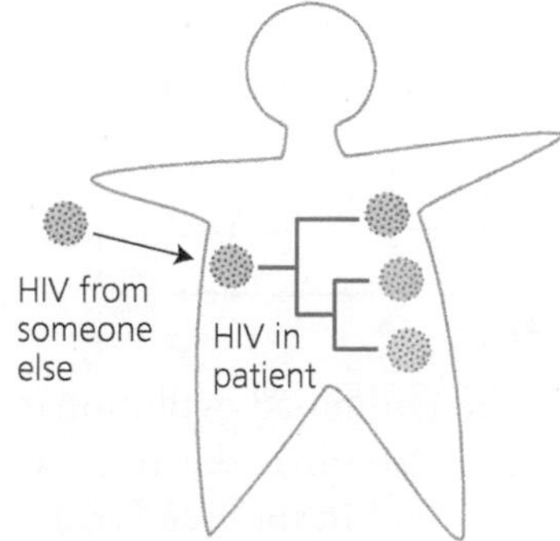

Figure 1.20 HIV evolves rapidly and diversifies within individual hosts We can summarize the pattern of diversification with an evolutionary tree.

We can summarize HIV's history of diversification within a patient with an evolutionary tree. The tree in **Figure 1.20** summarizes the evolutionary history that was depicted in detail in Figure 1.15. The tree grew from the root at the left toward the branch tips at the right. Each split in the tree represents the generation of a new variant; the original variant continues on one branch and the novel form continues on the other. The branch tips represent different living or extinct variants. Reading backwards from the tips to the root, we can see lineages that have a common ancestor merge with one another. Note that the middle and bottom variants in our tree share a more recent common ancestor with each other than either does with the top variant. By definition, we say that variants sharing a more recent common ancestor are more closely related to each other.

Figure 1.21 shows a small portion of the estimated evolutionary tree for HIV variants sampled from a real patient. The branch tips on the right represent individual virions collected from the patient at different times.

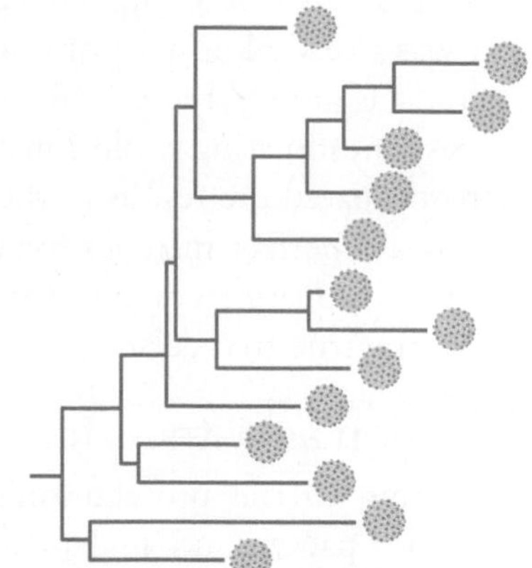

Figure 1.21 A small portion from the evolutionary tree for the HIV population within a real host Redrawn from Shankarappa et al. (1999).

To prepare the tree in Figure 1.21, Raj Shankarappa and colleagues (1999) estimated the relationships among the different virions by comparing the nucleotide sequences in their genes. The methods for estimating evolutionary trees from genetic sequences are complex in their details. We will cover them in some depth in later chapters. The basic principle, however, is straightforward. The researchers arrange the sequences and the order of branching so that less-divergent sequences are on neighboring branches. In the particular tree shown in the figure, the genetic difference between any two virions is represented by the total horizontal distance traveled when tracing down the tree from one virion's branch tip, through one or more shared branch points, and back up the tree to the other virion's branch tip.

Each time an HIV virion moves from one host to another to establish a new infection, it becomes the root of a new evolutionary tree. **Figure 1.22a (next page)** depicts this process for three individuals in a chain of transmission.

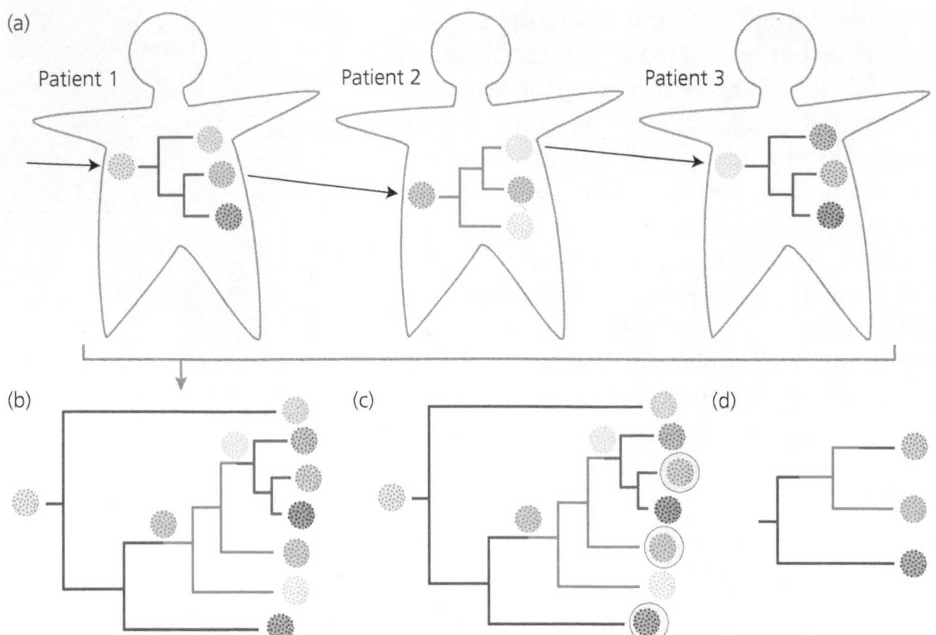

Figure 1.22 **Evolutionary trees for a chain of HIV transmission** (a) As HIV moves from host to host, it diversifies inside each patient. (b) In the evolutionary tree for the entire chain, the tree for the last link arises as a branch within the tree for the middle link, which in turn arises as a branch within the first link. (c) For most transmission chains, we are likely to have access to only one or a few virions from each host. (d) Nonetheless, in a tree estimated from the available samples, consecutive links in the chain will appear as closest relatives on the tree.

The combined evolutionary tree spanning all three patients appears in Figure 1.22b. Note that the tree for the last link in the chain—Patient 3—arises as a branch within the tree for the middle link. And that the tree for the middle link arises, in turn, as a branch within the tree for the first link. As shown in Figure 1.22c, researchers analyzing a chain of transmission will typically have access to just one or a few variants from each patient. Even so, in an evolutionary tree estimated from these samples, the virions taken from patients close together in the chain will appear as closest relatives (Figure 1.22d).

We know this method of reconstructing chains of transmission works for HIV. Thomas Leitner and colleagues (1996) tested it against a transmission chain that was already known from detailed information about the pattern and timing of contacts among the patients involved. The researchers used a variety of techniques for estimating evolutionary trees from the patients' viral genetic sequences, then compared the results to the known chain. The reconstructed histories were not always perfect matches for the truth. This result serves as a useful reminder that evolutionary trees are estimates, not revelations. But the discrepancies between the true tree versus the estimated trees, when they appeared, were minor.

Evolution as Witness for the Prosecution

According to the prosecution's theory of the crime in *Louisiana v. Schmidt,* Schmidt's patient and Trahan were consecutive links in a chain of transmission. To check this prediction, Hillis and colleagues analyzed the genetic sequences of viruses collected from Schmidt's patient, from Trahan, and from a number of other HIV-infected individuals from the city where they both lived **(Figure 1.23a)**. If Schmidt is innocent, then Trahan likely contracted HIV from someone other than Schmidt's patient. The HIVs from Schmidt's patient and from Trahan would probably not be closely related (Figure 1.23b). If, on the other hand, Schmidt is guilty, then the evolutionary tree reconstructed from the sequences will show his patient's HIV and Trahan's HIV as closest relatives (Figure 1.23c).

A portion of the evolutionary tree that Hillis and colleagues estimated from the HIV sequence data appears in **Figure 1.24**.

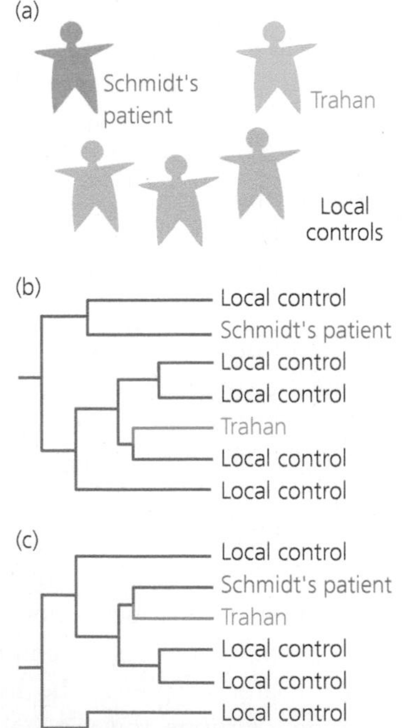

Figure 1.23 **Reconstructing evolutionary history tests the prosecution's theory of a crime** (a) The individuals from whom HIV samples were collected. (b) The viral evolutionary tree predicted if the suspect is innocent. (c) The evolutionary tree predicted if the suspect is guilty.

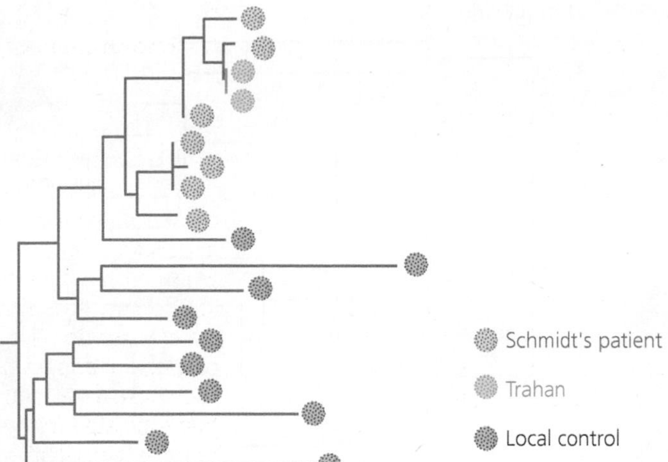

Figure 1.24 Incriminating evidence This reconstructed evolutionary history is consistent with the prosecution's charges. Redrawn from Metzker et al. (2002).

The tree is consistent with the prosecution's hypothesis. Not only does it show that the viruses from Schmidt's patient and from Trahan are closest relatives, it shows that Trahan's viruses occupy a branch arising within the tree for the viruses from Schmidt's patient. This is what we would expect if Schmidt's patient was the donor of the HIV that infected Trahan (Scaduto et al. 2010).

An evolutionary tree, also known as a phylogeny, shows the relationships among a group of viruses or organisms.

It is important to note that this tree, by itself, does not prove Schmidt's guilt (Pillay et al. 2007). Perhaps Trahan contracted HIV from Schmidt's patient without Schmidt's involvement. She might, for example, have participated in the patient's care herself and accidently pricked her own finger with a needle after drawing his blood. Or perhaps there were intervening links between Schmidt's patient and Trahan whose viral strains were not included in the analysis (see Learn and Mullins 2003 for an example). Such additional links might be revealed by the inclusion of more local controls. But the totality of evidence, which included a great deal more than just the evolutionary tree, was sufficient to convince the jury. Schmidt is now serving 50 years at hard labor (*State of Louisiana v. Richard J. Schmidt* 2000).

The Origin of HIV

The procedure Hillis and colleagues used to help convict Richard Schmidt can be applied on a larger scale to estimate the relationships among more distantly related organisms. We are now ready use it to trace the origin of HIV.

The first clue to where HIV came from is that the virus's genome and life cycle are similar to those of the simian immunodeficiency viruses (SIVs), a family of viruses that infect a variety of primates. A logical hypothesis is that HIV is derived from one of the SIVs and that the global AIDS epidemic started when this SIV moved from its primate host into humans. To test this hypothesis, Beatrice Hahn and colleagues sequenced genes from several SIVs and compared them to genes found in a variety of HIV strains (Gao et al. 1999; Hahn et al. 2000).

Hahn's reconstruction appears in **Figure 1.25a (next page)**. The black branches represent viral lineages that parasitize a variety of Old World monkeys. The colored branches trace lineages that infect chimpanzees and humans. Note that the HIV lineages in orange arise from within the tree of chimp SIVs. The chimp SIV branches (blue) arise, in turn, from within the tree of monkey SIVs. We got HIV from chimps, probably as a result of butchering them for food. Chimps got SIV from monkeys.

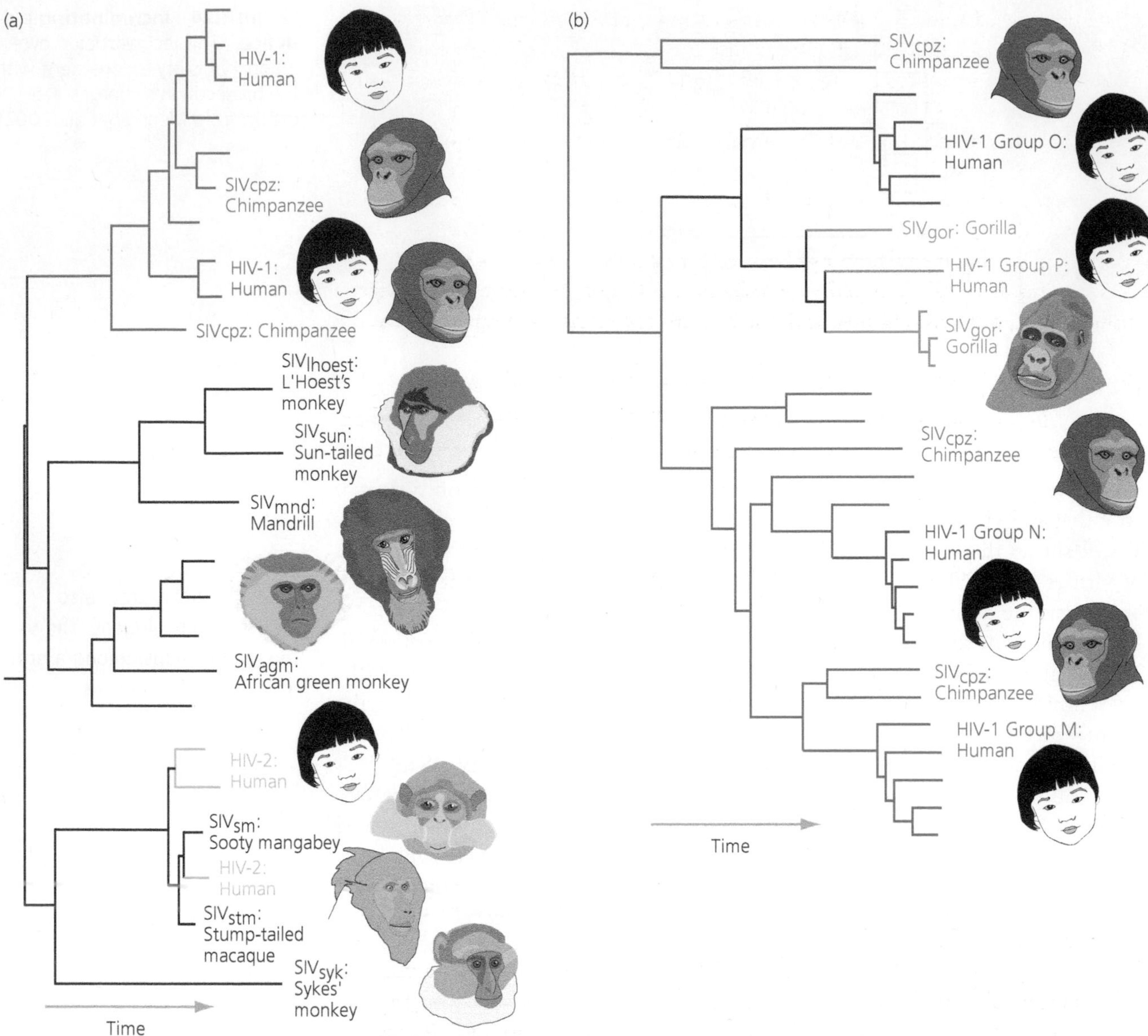

Figure 1.25 The origin of HIV (a) Estimated evolutionary tree for immunodeficiency viruses from humans, chimpanzees, and African monkeys. Redrawn from Hahn et al. (2000). (b) Estimated evolutionary tree for immunodeficiency viruses from humans, chimpanzees, and gorillas. Redrawn from Plantier et al. (2009).

The orange branches represent variants of the human virus that has been the topic of this chapter. In the tree it is called by its more proper name, HIV-1. This is to distinguish it from a different kind of HIV that appears at the bottom of the figure. HIV-2 circulates primarily in West Africa and is less virulent than HIV-1. Humans contracted HIV-2 from monkeys, most likely sooty mangabeys hunted for food or kept as pets. (The stump-tailed macaque virus included in the tree was obtained from a captive animal that had contracted SIV from a sooty mangabey.)

The two main types of HIV, HIV-1 and HIV-2, were transmitted to humans from different sources. HIV-2 originated in sooty mangabeys, and HIV-1 was transmitted to humans from chimpanzees.

The tree in Figure 1.25b, estimated by Jean-Christophe Plantier and colleagues (2009) gives more details on the history of HIV-1. Note that strains of HIV-1 appear on distinct branches arising from within the chimp SIVs. This is evidence that SIV has jumped from great apes to humans at least three times. HIV-1 Group M, in orange, is responsible for 95% of HIV infections (Sharp and Hahn 2008).

The branches leading to HIV-1 Groups O and P and to gorilla SIV are shown in black to indicate uncertainty about just how the ancestors of these viruses traveled from chimpanzees to their current hosts. A number of scenarios are consistent with the tree (Takehisa et al. 2009). Related strains of chimp SIV could have separately infected humans (becoming HIV-1 Group O) and gorillas, after which the gorilla strain jumped to humans (becoming HIV-1 Group P). Or a single strain of chimp SIV could have jumped to gorillas, after which the gorilla strain jumped to humans twice. It is even possible that a single strain of chimp SIV jumped to humans (becoming the ancestor of HIV-1 Groups O and P), after which the human strain jumped to gorillas twice. Given the nature of the interactions among humans, chimps, and gorillas, this last scenario seems unlikely.

Each major subgroup of HIV-1 originated in an independent transmission event from chimpanzees to humans.

When Did SIV Move from Chimpanzees to Humans?

Work on this question has focused on the group M branch of HIV, at the bottom of the tree in Figure 1.25b. This is because group M is responsible for the bulk of the global AIDS pandemic. Several groups of researchers have used sequence data from various group M strains to estimate the age of their last common ancestor (see **Computing Consequences 1.1, next page**). Despite considerable uncertainty, the best estimate is that the last common ancestor of the group M HIV-1 viruses lived in the 1930s or earlier. Corroboration comes from an analysis of the two earliest-known samples of HIV-1 (Worobey et al. 2008). Both of these are from patients who lived in Kinshasa, Democratic Republic of Congo; one sample was collected in 1959, the other in 1960. Both are group M viruses, but they are quite different from each other. One is most closely related to subtype A, the other to subtype D. In other words, by 1960 strains of pandemic HIV-1 had already had time to diverge considerably from their common ancestor. This common ancestor could, in principle, have lived in either a chimpanzee or a human. However, the available evidence is most consistent with a human host (Hillis 2000; Rambaut et al. 2001; Sharp et al. 2001). The implication is that the group M strains of HIV-1 originated in a transfer of SIV from chimps to humans that happened more than 70 years ago.

The common ancestor of HIV-1 group M, the strain primarily responsible for the AIDS pandemic, probably lived between 1915 and 1941.

One medical lesson of the fact that HIV-1 is derived from SIV is that the study of primate immunodeficiency virus infections is likely to yield insights into AIDS. A key difference between SIV infections in monkeys versus SIV infections in chimps and HIV-1 infections in humans is that when monkey SIVs infect their natural hosts, they generally cause little or no overt disease (Keele et al. 2009; Paiardini et al. 2009). What makes SIV_{cpz}, and especially HIV, unusual?

1.5 Why Is HIV Lethal?

This final section concerns the most difficult of the questions we consider in the chapter. There is no definitive explanation for the nearly universal lethality of HIV-1. Scientists have made progress, however. Here, we briefly review three approaches that apply an evolutionary perspective to the problem.

A Correlation between Lethality and Transmission?

Dying of AIDS is clearly bad for the host. If there is heritable variation among humans in their resistance to HIV and AIDS, then as discussed in Section 1.3, we can expect that resistance will spread throughout the human population as

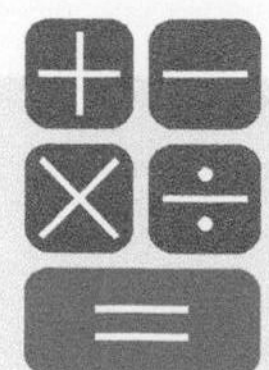

COMPUTING CONSEQUENCES 1.1

When did HIV move from chimpanzees to humans?

Here we outline the method Bette Korber and colleagues (2000) used to estimate the age of the common ancestor of the group M strains of HIV-1. The essence of the method can be expressed in an analogy. If we know two cars started at the same place and time, that they are now 240 miles apart, and that one has been driving east and the other west at 60 miles per hour, we can infer that they have been on the road for 2 hours.

From gene sequences, Korber's team estimated the genetic distances among 159 viral samples. The distances are summarized by the unrooted tree diagram in Figure 1.26a. Each twig represents a gene from a single virion. The distance along the tree from the tip of one twig to the tip of another indicates the genetic difference between two virions. The dot in the center represents the common ancestor of the virions at the tips. The tree shows how far apart the virions are now.

Next, Korber and colleagues drew a scatterplot showing the genetic distance from the common ancestor to each virion as a function of the year the virion was collected (Figure 1.26b). The more recently a sample was collected, the greater its genetic divergence from the common ancestor. The plot includes the statistical best-fit line through the data. The slope of this line shows us how fast the virions are traveling.

Finally, the researchers extrapolated the best-fit line back in time to estimate the year in which a virus would need to have been collected to show a genetic difference of zero versus the common ancestor (Figure 1.26c). In other words, they extrapolated back to the date of the common ancestor itself. The best-fit line hits zero at 1931. This tells us how long the virions have been on the move.

Extrapolating beyond the data is dangerous, and there may be biases in the data due to sampling error. The true relationship between sequence divergence and time could fall anywhere between the gray lines in the figure. Korber and associates estimate with 95% confidence that the common ancestor of their group M virions lived sometime between 1915 and 1941, indicated by the red bar on the horizontal axis in Figure 1.26c. Additional analyses have produced estimates that were similar (Salemi et al. 2001; Sharp et al. 2001; Yusim et al. 2001) or slightly earlier (Worobey et al. 2008).

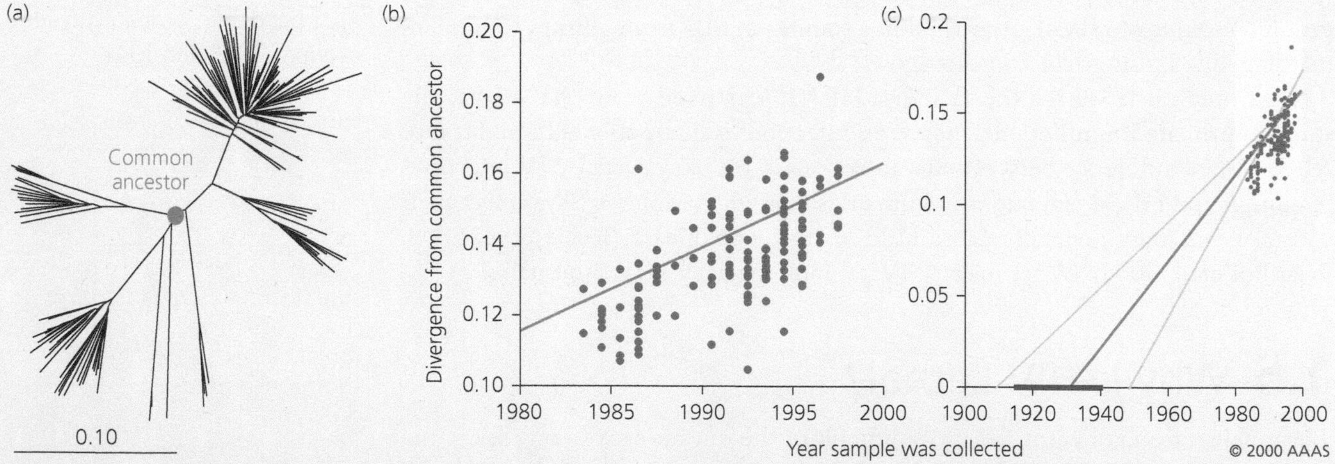

Figure 1.26 Dating the common ancestor of HIV-1 strains in group M (a) An unrooted evolutionary tree for 159 group M HIVs. The tip of each twig represents a virion; the distance traveled from one tip to another represents the genetic difference between the two virions. The orange dot marks the common ancestor of all group M strains. (b) This scatterplot shows the genetic difference between each HIV sample in (a) versus the common ancestor as a function of the date the sample was collected. The statistical best-fit line is in orange. (c) Extrapolating the best-fit line in (b) back to zero genetic difference gives an estimate of the date when the common ancestor lived. From Korber et al. (2000).

From "Timing the ancestor of the HIV-1 pandemic strains." *Science* 288: 1789–1796. Reprinted with permission from AAAS.

generations pass. However, the organism we want to focus on here is not the host, but the virus. Is killing the host not also bad for the virus? After all, when the host dies, the virions living inside the host die too.

Evolutionary logic suggests an answer if we recognize that even for a benign pathogen, all hosts eventually die. To persist beyond the life span of the host, a viral population must colonize new hosts. Thus a second level of natural selection is acting on HIV. The first level is the one we have already explored: There are differences among virions in their ability to survive and reproduce within a given host. The second level occurs when viral strains differ in their ability to move from one host to another. Strains that are good at getting transmitted will become more common over time; strains that are bad will disappear (Anderson and May 1982; Ewald 1983).

Rare strains of HIV-1 exist that kill their hosts more slowly than common strains (Deacon et al. 1995; Geffin et al. 2000; Rhodes et al. 2000; Tobiume et al. 2002; Churchill et al. 2006). The fact that these milder strains are rare suggests that they are seldom transmitted from one host to another. HIV-2 is also less damaging to its hosts than most strains of HIV-1 (Marlink et al. 1994). It, too, is transmitted at lower rates (Kanki et al. 1994). These observations hint that some damage to the host is inevitable if a strain of HIV is to be transmitted readily.

Christophe Fraser and colleagues (2007) evaluated this hypothesis in more detail by analyzing two data sets. The first data set, collected in Amsterdam before the development of highly effective antiretroviral therapy, involved 123 men who have sex with men. These data allowed Fraser and colleagues to estimate the effect of viral load during the asymptomatic phase of infection on the time until progression to AIDS (**Figure 1.27a**). Note that duration of AIDS-free infection is the dependent variable: The higher the viral load, the more quickly patients develop symptoms. The second data set, collected in Zambia, involved over 1,000 heterosexual couples in which one partner had HIV and the other did not. These

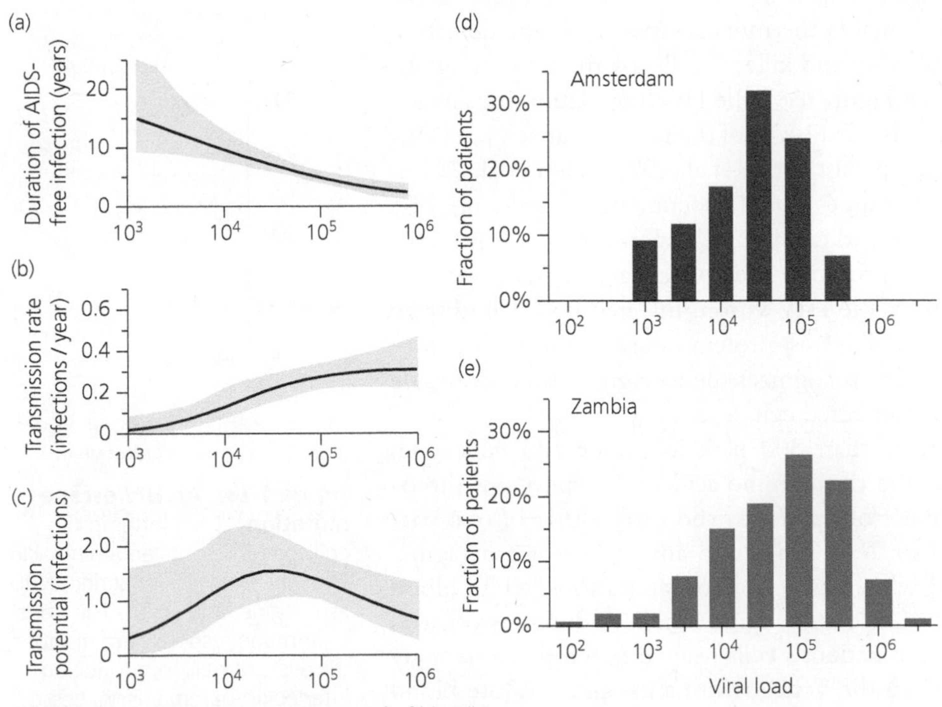

Figure 1.27 Viral load, progression to AIDS, and transmission The graphs on the left show three key properties of an HIV infection as a function of the concentration of virions in the patient's bloodstream during the asymptomatic phase: (a) the average duration of the asymptomatic phase; (b) the rate at which the virus is transmitted to new hosts; and (c) the transmission potential. Lines show mean value; gray areas show 95% confidence interval. The histograms on the right show the variation in viral load among the patients in (d) Amsterdam and (e) Zambia. Redrawn from Fraser et al. (2007).

data allowed the researchers to estimate the effect of viral load on the rate of transmission (Figure 1.27b). The higher the viral load, the greater the rate at which the infected partner passed the virus on. Finally, the researchers multiplied the duration of AIDS-free infection (in years) by the rate of transmission (in new infections per year) to estimate the transmission potential (in new infections) as a function of viral load (Figure 1.27c). For a hypothetical population in which most individuals are uninfected and spread of the virus is not limited by the rate at which people change partners, the HIV strains that establish the most new infections are the ones that maintain intermediate viral loads in their hosts.

One possible reason HIV infections are fatal: Traits that enhance the virus's ability to infect new hosts, such as the maintenance of a moderately high viral load, may also predispose HIV to kill.

Fraser's calculations suggest that there may be an optimum viral load for HIV transmission. For the patients in Amsterdam and Zambia on which the calculations are based, the actual average viral loads are on the same order of magnitude as the predicted optimum (Figure 1.27d and e).

Left unexplained is why viral loads of this magnitude are damaging to the host. This is a puzzle, given that natural hosts of monkey SIVs maintain similarly high viral loads without getting sick (Pandrea et al. 2008). For additional insights, we take a closer look at the evolution of viral populations inside individual hosts.

Shortsighted Evolution?

In Section 1.2, we covered evidence that HIV populations within individual hosts evolve resistance to AZT and other antiretroviral drugs. HIV populations also evolve the ability to evade the host's immune response. The immune system attacks HIV with antibodies and killer T cells (see Figure 1.6). These eliminate many of the virions in the HIV population, but not all. Due to errors during replication, the HIV population is genetically variable. Some of the variants are less susceptible than others to the immune system's assault.

Antibodies and killer T cells recognize HIV and HIV-infected cells by binding to epitopes—short pieces of viral protein displayed on the surface of the virion or the infected cell. These epitopes are encoded in HIV's genes. Mutations in the genes change the epitopes and may enable the mutant virion to evade detection by the host's current arsenal of antibodies and killer T cells. By the time a new infection enters the chronic phase (see Figure 1.8), the HIV population has already changed. Variants easily targeted by the first wave of the immune attack have disappeared; variants less easily detected persist (Price et al. 1997; Allen et al. 2000).

Research by A. J. Leslie and colleagues (2004) documents an example of a mutation that helps HIV virions evade the immune response in some patients. The mutation affects an epitope in a protein, called p24, that is a component of the capsule that surrounds the core of the HIV virion. Infected host cells display this epitope on their surface, along with a host protein called a human leucocyte antigen, or HLA. When a killer T-cell recognizes the foreign epitope alongside the self HLA protein, it destroys the infected cell.

In a survey of virions from more than 300 patients, Leslie and colleagues found that in most strains of HIV, the third amino acid in the epitope is threonine. However, in most HIV strains from patients who carry either of two particular alleles of HLA-B—*B5801* or *B57*—the third amino acid is asparagine. Experiments in test tubes showed why. Leslie and colleagues took white blood cells from a patient with the *B5801* allele and exposed them to different versions of the p24 epitope **(Figure 1.28)**. The patient's cells reacted much more strongly to the version with threonine than to the version with asparagine. White blood cells from patients with the *B57* allele showed a similar pattern.

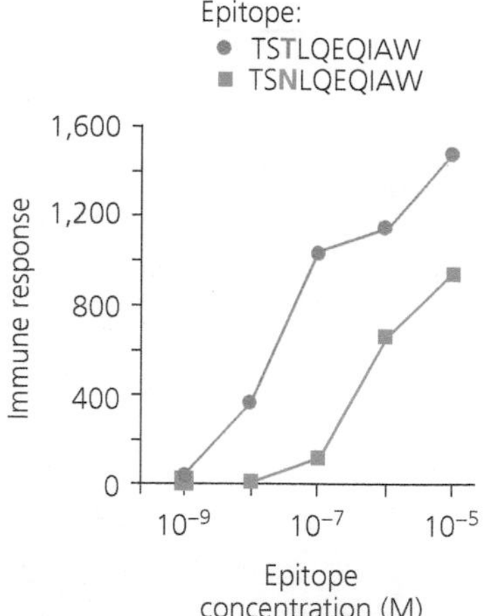

Figure 1.28 An HIV escape mutation Each letter in the epitope represents an amino acid: T for threonine; S for serine; N for asparagine; and so on. The units of immune response are number of cells, per million, producing interleukin-gamma. From Leslie and colleagues (2004).

The evolution of the HIV population appears to contribute to the death of the host in at least three ways. First, the continuous evolution toward novel epitopes enables the viral population to stay far enough ahead of the immune response to avoid elimination.

Second, the viral population within many hosts evolves toward ever more aggressive replication. Ryan Troyer and colleagues (2005) took sequential HIV samples from several untreated patients. The researchers grew the virions from each sample on white blood cells from an uninfected donor. To each culture dish, the researchers added one of four control strains of HIV against which the virions from the patient would have to compete. In each dish, the viral strain that could replicate most efficiently became numerically predominant. Troyer and colleagues assessed the competitive fitness of the virions from the patients' samples based on their overall performance against the four control strains. The results appear in **Figure 1.29**. Each color represents the sequential samples from a particular patient. In seven of eight cases, the competitive fitness of the patient's virions steadily increased over time. The longer a patient harbors an HIV population, the more damaging the virions in the population become.

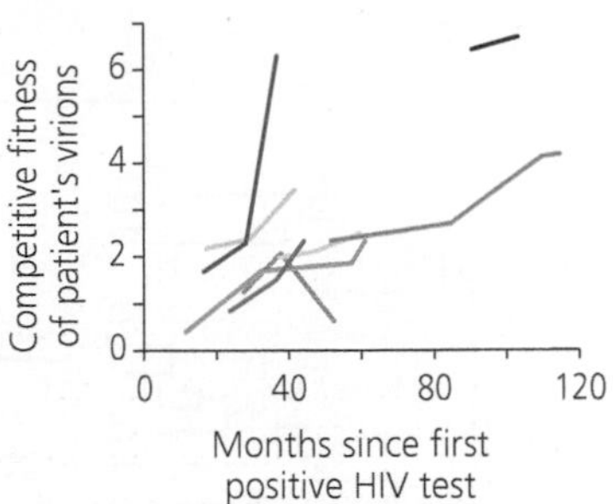

Figure 1.29 HIV populations in many hosts evolve toward more aggressive replication Redrawn from Troyer et al. (2005).

Third, in at least half of all hosts, strains of HIV evolve that can infect naive T cells (Shankarappa et al. 1999; Moore et al. 2004). An HIV virion's ability to infect a given cell type is determined by the coreceptor the virion uses. Early in most HIV infections, most virions use CCR5 as their coreceptor. CCR5 is found on macrophages and on regulatory, resting, and effector T cells (see Figure 1.6). As the infection progresses and the HIV population evolves, virions often emerge that exploit a different coreceptor. These strains, called X4 viruses, use a protein called CXCR4. CXCR4 is found on naive T cells.

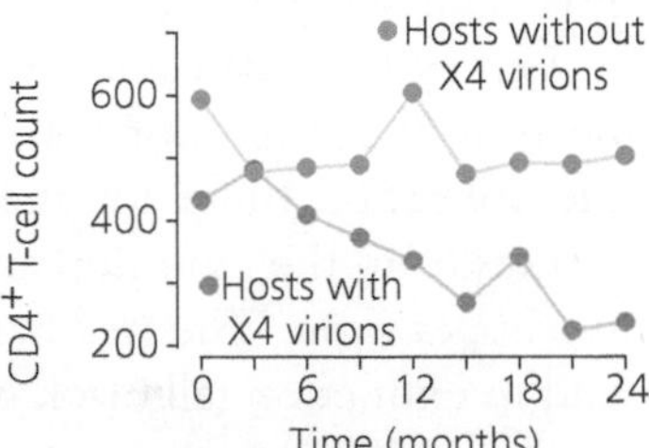

Figure 1.30 HIV strains that use CXCR4 as a coreceptor hasten the collapse of their hosts' immune systems Redrawn from Blaak et al. (2000).

Because naive T cells are the progenitors of memory and effector T cells, the emergence of virions that can infect and kill naive T cells is typically bad news for the host. Hetty Blaak and colleagues (2000) sampled the viral populations of 16 HIV patients to determine whether they contained X4 virions. Then, for a time span running from a year before to a year after the date of this sample, the researchers calculated the average helper T-cell counts in the blood of patients with X4 viruses versus the patients without. The results appear in **Figure 1.30**. The average T-cell counts in the patients without X4 viral strains held fairly steady over time; the average counts in the patients with X4 strains fell. When virions arise that undermine the immune system's ability to replenish its stock of T cells, they apparently accelerate the immune system's demise.

The evolution of the HIV population within a host is shortsighted (Levin and Bull 1994; Levin 1996). The virions do not look to the future and anticipate that as their population evolves, it will ultimately kill its host. Virions are just tiny, unthinking, molecular machines. Evolution by natural selection does not look to the future either. It just happens automatically.

A second possible reason HIV infection is fatal: During an infection, the viral population evolves—to evade the immune system, replicate more rapidly, and use a different coreceptor.

A Surprising Clue

Besides abetting the lethality of HIV infection, evolution within hosts has given researchers an unexpected clue to another contributory factor. This clue was uncovered by Philippe Lemey and colleagues (2007). Lemey reanalyzed an extraordinary data set collected by Raj Shankarappa and colleagues in the laboratory of James Mullins (1999). Shankarappa periodically collected HIV samples from nine patients from the time they first tested positive until, 6 to 12 years later, seven of them had developed AIDS. For each HIV sample, Shankarappa read the

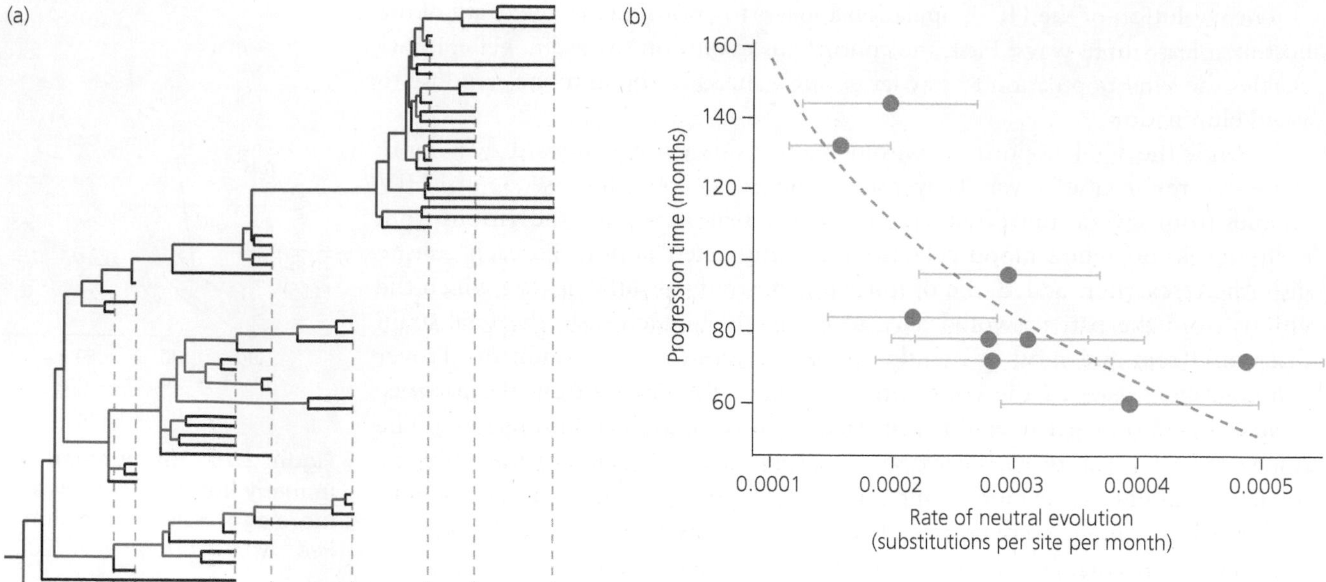

Figure 1.31 Neutral evolution and progression to AIDS (a) A partial evolutionary tree for the HIV virions sampled from a single patient. The red line traces the sole surviving lineage. (b) The relationship across patients between progress to AIDS and the rate at which synonymous mutations accumulate in the surviving lineage. Redrawn from Lemey et al. (2007).

sequence of nucleotides for the gene that encodes two of the proteins found on HIV's envelope.

Lemey first estimated the evolutionary tree for the viruses in each patient. A portion of the tree for Patient 1 appears in **Figure 1.31a**. The dashed lines mark the collection times for the samples. The tree traces the diversification of the viruses from the common ancestor on the lower left to the surviving lineages on the upper right. The tree has many side branches (shown in blue) that ultimately end in extinction (all black lines except the ones at upper right). This is probably because HIV populations accumulate many lineages burdened with damaging mutations. The tree also has one lineage (shown in red) that persists throughout the infection. The trees for the other eight patients showed similar patterns.

Lemey then looked at the mutations that accumulated along the red branch, separately tallying the rate of evolution for non-synonymous versus synonymous changes. Non-synonymous mutations alter the sequence of amino acids in the encoded protein and thus may be subject to natural selection. Synonymous mutations do not alter the sequence of amino acids and are thus more likely to be neutral (but see Ngandu et al. 2008).

We might have predicted that the speed at which the patients progressed to AIDS would be connected to the rate at which non-synonymous mutations accumulate. Immune systems doing a better job of blocking HIV's replication might, for example, impose stronger selection for altered protein structure and progress more slowly. In fact, Lemey found no such association.

Instead, as shown in Figure 1.31b, Lemey found a relationship between progression to AIDS and the rate at which synonymous mutations accumulate. Synonymous evolution is generally the result of chance. New variants arise by mutation and, usually, disappear again shortly thereafter. A few variants, however, have a different fate. Simply by luck, rather than by providing a survival advantage, they rise to high frequency.

Patients in which the HIV population shows a higher rate of nonadaptive evolution progress more rapidly to AIDS....

Why would patients' whose HIV lineages experience more rapid neutral evolution advance more quickly to AIDS? This odd result begins to make sense when considered in light of our current understanding, outlined in Figure 1.9, of how HIV causes AIDS. The infection triggers, directly and indirectly, a heightened state of immune activation characterized by extensive proliferation of helper T cells. But helper Ts are the cells in which HIV replicates most efficiently. This sets up a vicious cycle that causes extensive collateral damage to the immune system. Patients whose immune systems activate most aggressively against HIV develop AIDS more quickly (Sousa et al. 2002). They also likely experience higher rates of viral replication. More replication means more mutations, and more mutations mean a higher rate of neutral evolution.

...This is consistent with a model of AIDS in which heightened immune activation offers HIV more opportunities to replicate and is thus among the factors that leads to disease....

Lemey's finding is thus consistent with a model of AIDS in which the problem is not so much the virus as the immune system's response to it. This model is also supported by studies of monkeys that are natural hosts of SIVs (see, for example, Silvestri et al. 2005). Despite high rates of viral replication, natural hosts of SIV avoid chronic immune activation and also avoid AIDS (Pandrea et al. 2008).

These results raise another question. Does the failure of humans to either tolerate or fight off HIV result from something unusual about HIV as a retrovirus, something unusual about humans as a host, or both?

Unusual Features of HIV or Humans?

Comparison of HIV versus a variety of SIVs suggests that HIV has unusual properties. Among these is the possession of an extra gene, called *vpu* (see Schindler et al. 2006; Kirchhoff 2009). This extra gene appears to have arisen in the common ancestor of a group of closely related SIVs that infect monkeys in the genus *Cercopithecus*. Subsequently, the *vpu* gene was picked up by an ancestor of chimpanzee SIV when two different strains of SIV infected the same animal (Bailes et al. 2003; Kirchhoff 2009).

...And this, in turn, suggests a third reason HIV infection is fatal: Both the virus and its host have unusual properties.

One thing the Vpu protein does in many of the retroviruses that make it, including HIV-1 group M, is block the action of a host protein called tetherin (Sauter et al. 2009). Tetherin, as its name suggests, ties maturing virions to the membrane of the host cell, thereby preventing their release (Neil et al. 2008; Perez-Caballero et al. 2009).

Among the unusual features of HIV as a virus is the ability to make a protein, Vpu, that blocks a host defensive protein, tetherin.

In SIVs in which Vpu does not interfere with tetherin, another viral protein, called Nef, often plays this role. Human tetherin is resistant to Nef. This resistance is due to the loss, resulting from a deletion in the tetherin gene, of five amino acids from the protein (Zhang et al. 2009).

That HIV-1 group M has evolved a version of Vpu that blocks tetherin is likely one of the reasons group M has spread so rapidly. And perhaps the possession of a Vpu that blocks tetherin, which presumably allows greater production of virions from infected cells, conferred an advantage on viral strains that induce higher levels of immune activation in their hosts (Kirchhoff 2009).

Just as comparative studies have revealed unusual features of HIV-1 as a retrovirus, so too have they revealed unusual features of humans as a host. Research on a host-cell protein called TRIM5α provides an example.

TRIM5α, like tetherin, blocks retroviral replication. Precisely how it does so is unclear, but TRIM5α interacts with the contents of the invading virion after they have entered the cell and before reverse transcription of the viral genome begins (Johnson and Sawyer 2009). TRIM5α's ability to restrict retroviral infection was discovered by Matt Stremlau and colleagues (2004), who demonstrated that the version of the protein made by rhesus macaques can disrupt the life cycle of HIV. Somewhat perversely, human TRIM5α cannot block HIV's replication nearly as well.

Among the unusual features of humans as a host is that our TRIM5α protein is relatively ineffective in blocking HIV's replication, apparently because the protein is adapted to fight an extinct virus that afflicted our ancestors.

Sara Sawyer and colleagues (2005) suspected that retroviruses and their mammalian hosts are engaged in a long-running arms race. The viruses are constantly evolving to evade their hosts' TRIM5α, and the hosts are constantly evolving to counteract the evasion. The researchers compared the sequences of the TRIM5α genes from 20 different primates, including humans. They found that the evolutionary history of primate TRIM5α has featured an extraordinarily high rate of diversification among species in the amino acid sequence of the protein. The changes have been concentrated in a small patch in a region of the gene called the SPRY domain, which the researchers suspected is the point of contact between TRIM5α and the viral components it acts on. In experiments with genetically engineered cells, Sawyer and colleagues (2005) confirmed that swapping the rhesus monkey version of the SPRY patch into the human TRIM5α gene gave the encoded protein the ability to block HIV replication. Conversely, swapping the human version of the patch into the rhesus monkey TRIM5α diminished its potency against the virus. These results are consistent with Sawyer's arms-race hypothesis.

Sawyer's experiments suggest that modest changes in the human gene for TRIM5α might substantially improve our resistance to HIV-1. Indeed, in a similar study, Melvyn Yap and colleagues (2005) found that the mere substitution of glutamine for arginine at position 332 substantially improved human TRIM5α's ability to stop HIV. Why, then, does our TRIM5α not have glutamine at position 332? Shari Kaiser and colleagues (2007) hypothesized that human TRIM5α is adapted to fight not the recently arrived HIV, but an extinct retrovirus that once plagued our ancestors.

Kaiser and colleagues suspected that the extinct virus in question was PtERV1. This virus is known from the many copies of its genome, disabled by mutations, that persist in the chromosomes of chimpanzees and gorillas. By examining copies with different mutations, Kaiser was able to infer the sequence of the PtERV1's *gag* gene. This is the gene that encodes the viral proteins disrupted by TRIM5α. Kaiser modified a mouse retrovirus by replacing its *gag* gene with the one from PtERV1, then tested the virus's ability to infect target cells that make no TRIM5α, normal human TRIM5α, and human TRIM5α with glutamine at position 332. The results appear in **Figure 1.32**. Human TRIM5α does, in fact, confer strong resistance to PtERV1. And it confers much stronger resistance to PtERV1 than does human TRIM5α with glutamine. Note that the pattern of resistance to HIV-1 is just the opposite: TRIM5α with glutamine confers stronger resistance than the normal human protein. This trade-off is consistent with the notion that humans are susceptible to HIV in part because of past selection for resistance to PtERV1.

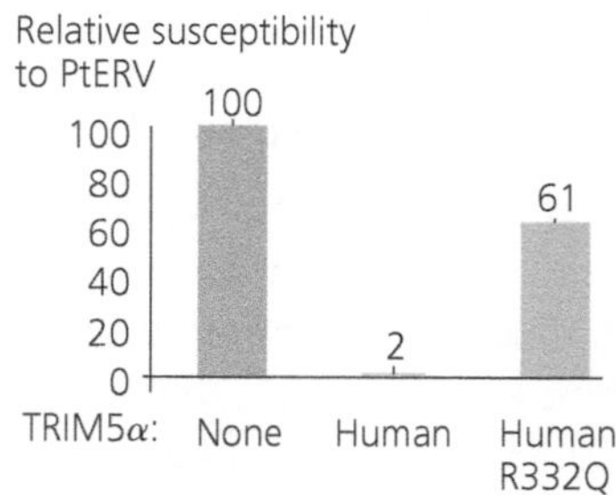

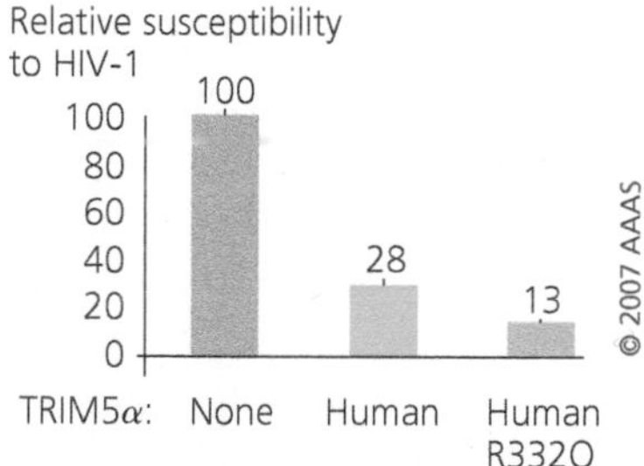

Figure 1.32 Human susceptibility to an extinct retrovirus R332Q is the version of TRIM5α with glutamine instead of arginine at position 332. From Kaiser et al. (2007).

From "Restriction of an extinct retrovirus by the human TRIM5a antiviral protein." *Science* 316: 1756–1758. Reprinted with permission from AAAS.

The explanations for HIV's lethality that we have considered in this section are mutually compatible. The grim course of infection with the virus may best be viewed as resulting from the combined effects of selection for high transmissibility, shortsighted evolution within patients, and unusual features of both the virus and its host.

In this section, and throughout the chapter, an evolutionary analysis has—as promised—given us crucial insights into matters of life and death. We will see many examples in the rest of the book.

SUMMARY

The story of HIV demonstrates that evolutionary analysis has practical applications outside of textbooks and classrooms. HIV/AIDS has killed some 30 million people, most of them Africans.

Each time an HIV virion invades a host cell, the virion reverse-transcribes its RNA genome into a DNA copy that serves as the template for the next generation of virus particles. Because reverse transcription is error prone, an HIV population quickly develops substantial genetic diversity. Some genetic variants replicate rapidly while others die. As a result, the composition of the population changes over time. That is, the population evolves.

HIV populations within patients quickly evolve resistance to any single antiretroviral drug and can even evolve resistance to multidrug cocktails. Without effective antiretroviral therapy, HIV populations also continuously evolve to evade the host's immune response, a process that ultimately contributes to the collapse of the immune system and the onset of AIDS.

Just as HIV populations evolve in response to selection imposed by their hosts, so too host populations may evolve in response to selection imposed by the virus. Human populations harbor genetic variation for susceptibility to HIV infection. If, during the AIDS epidemic, susceptible individuals die at higher rates than resistant individuals, then genetic composition of these populations will change over time. The search for genetic variation for resistance to HIV has led to the development of new antiretroviral drugs.

HIV belongs to a family of viruses that infect a variety of primates. Evolutionary trees based on genetic comparisons reveal that HIV-1 jumped to humans from chimpanzees and has done so more than once. HIV-1 also may have jumped to humans from gorillas. The strains responsible for the bulk of the HIV/AIDS pandemic, HIV-1 group M, have a common ancestor that lived several decades ago.

Comparison of HIV to evolutionarily related viruses, and of humans to related hosts, has provided insight into why HIV infection is lethal. HIV-1 possesses a gene not present in most SIVs. This gene may have made it advantageous for another of HIV's genes to lose its ability to suppress immune activation in the host. Immune activation, in turn, plays a crucial role in progression to AIDS. Humans may be particularly susceptible to HIV due to a genetic change that evolved in our ancestors because it conferred resistance to another retrovirus that is now extinct.

By focusing in this chapter on adaptation and diversification in HIV, we introduced topics that will resonate throughout the text: mutation and variation, competition, natural selection, evolutionary tree reconstruction, lineage diversification, and applications of evolutionary theory to scientific and human problems.

QUESTIONS

1. In an editoral published on March 28, 2009, *The Lancet* quoted Benedict XVI on Africa's battle with HIV/AIDS. The problem, the Pope said, "cannot be overcome by the distribution of condoms. On the contrary, they increase it." Not surprisingly, this statement generated some controversy. Consider the relevant scientific evidence. Is the Pope's first statement correct? How about his second statement? How do we know?
2. Review the process by which the HIV population inside a human host evolves resistance to the drug AZT. What traits of HIV contribute to its rapid evolution? How might a similar scenario explain the evolution of antibiotic resistance in a population of bacteria?
3. In the early 1990s, researchers began to find AZT-resistant strains of HIV-1 in recently infected patients who had never received AZT. How can this be?
4. Given the risk of evolution of resistance, why do you think the two patients shown in Figure 1.11 were not given high doses of AZT immediately, rather than starting them with low doses?
5. The idea behind multidrug therapy for HIV is to increase the number of mutations required for resistance and thus reduce genetic variation in the viral population for survival in the presence of drugs. Could we achieve the same effect by using antiretroviral drugs in sequence instead of simultaneously? Why or why not?
6. Some physicians have advocated "drug holidays" as a way of helping HIV patients cope with the side effects of multidrug therapy. Under this plan, every so often the patient would stop taking drugs for a while. From an evolutionary perspective, does this seem like a good idea or a bad idea? Justify your answer.

7. In a monograph published in 1883, Alexander Graham Bell wrote that "natural selection no longer influences mankind to any great extent." Do you agree? What is your evidence?
8. Design a study to test our prediction that human populations will evolve in response to selection imposed by HIV. Where would you conduct it? What data would you collect? How would you present your results?
9. When did HIV enter the human population, and from what source? How do we know?
10. Suppose that HIV were the ancestor of the SIVs, instead of the other way around. If immunodeficiency viruses were originally transmitted from humans to monkeys and chimpanzees, make a sketch of what Figure 1.25a would look like.
11. Recall that we discussed two different types of selection in this chapter: selection of different virus strains within one host and selection of those virus strains that are able to transmit themselves from host to host. Now consider the hypothesis, traditionally championed by biomedical researchers, that disease-causing agents naturally evolve into more benign forms as the immune systems of their hosts evolve more efficient responses to them. Is the evidence we have reviewed on the evolution of HIV within and among hosts consistent with this hypothesis? Why or why not?
12. Not all viruses are dangerous. (The common cold is an example.) HIV, however, is nearly 100% lethal. Describe three major hypotheses for why HIV is so highly lethal.
13. What is the evidence that the evolution of our ancestors to resist infection with an extinct retrovirus necessarily left us vulnerable to HIV-1? Can you think of a genetic change that might simultaneously protect an individual against both HIV-1 and PtERV1?
14. Authors in various fields often make interesting statements about evolution:
    a. A traditional view, particularly in parasitology and medicine, was that relationships between parasites and their hosts inevitably evolve toward peaceful coexistence (see Ewald 1983). Among the arguments for this view was that a parasite population is likely to survive longer if its host remains unharmed. Are the traditional view and the argument for it consistent with what you know about HIV—and about other diseases and parasites? What experiments do they suggest?
    b. HIV is a tiny, robotic, molecular machine. Many science fiction books describe robots that evolve to become intelligent and conscious (and, usually, seek freedom, develop emotions, and start wars with humans). Under what conditions could robots actually evolve? Is it necessary that the robots reproduce, for example?

EXPLORING THE LITERATURE

15. We discussed where HIV came from and when the pandemic started. But we did not much discuss how. For an investigation into this question, see:

    de Sousa, J. D., V. Muller, et al. 2010. High GUD incidence in the early 20th century created a particularly permissive time window for the origin and initial spread of epidemic HIV strains. *PLoS One* 5: e9936.

16. Drug resistance has evolved in a wide variety of viruses, bacteria, and other parasites. This paper describes the evolution of drug resistance in the bacterium that causes tuberculosis:

    Blower, S. M., and T. Chou. 2004. Modeling the emergence of "hot zones": Tuberculosis and the amplification dynamics of drug resistance. *Nature Medicine* 10: 1111–1116.

    This paper concerns hepatitis B virus (HBV):

    Shaw, T. A., A. Bartholomeusz, and S. Locarnini. 2006. HBV drug resistance: Mechanisms, detection and interpretation. *Journal of Hepatology* 44: 593–606.

17. Question 5 concerned treating HIV by using multiple drugs sequentially instead of simultaneously. For a clinical trial comparing these strategies, see:

    Gulick, R. M., J. W. Mellors, et al. 1998. Simultaneous vs sequential initiation of therapy with Indinavir, Zidovudine, and Lamivudine for HIV-1 infection: 100-week follow-up. *JAMA* 280: 35.

18. Our discussions of retrocyclin and TRIM5α revealed cases in which small genetic changes would render humans more resistant to HIV. For a third example, see:

    Gupta, R. K., S. Hue, et al. 2009. Mutation of a single residue renders human tetherin resistant to HIV-1 Vpu-mediated depletion. *PLoS Pathogens* 5: e1000443.

19. Antiretroviral drugs have, for many patients, transformed HIV into a treatable chronic condition. But a treatment is not a cure; patients have to remain on drug therapy for life. To date, exactly one patient with HIV has been cured. To read about how, see:

    Hutter, G., D. Nowak, et al. 2009. Long-term control of HIV by CCR5 Delta32/Delta32 stem-cell transplantation. *New England Journal of Medicine* 360: 692–698.

20. The discovery of genetic variation for resistance to HIV has suggested ways that HIV infection might be treated with gene therapy. For an example, see:

    Perez, E. E., J. Wang, et al. 2008. Establishment of HIV-1 resistance in $CD4^+$ T cells by genome editing using zinc-finger nucleases. *Nature Biotechnology* 26: 808–816.

    Hutter, G., D. Nowak, et al. 2009. Long-term control of HIV by CCR5 Delta32/Delta32 stem-cell transplantation. *New England Journal of Medicine* 360: 692–698.

21. For a paper that uses estimated evolutionary trees to examine the mechanism of HIV transmission among men who have sex with men, see:

Butler, D. M., W. Delport, et al. 2010. The origins of sexually transmitted HIV among men who have sex with men. *Science Translational Medicine* 2: 18re1.

22. Stanley Trask and colleagues hypothesize that most HIV-1 transmissions in sub-Saharan Africa occur between married couples. That is, the husband acquires HIV and then passes it to his wife, or the wife acquires HIV and then passes it to her husband. The researchers then use a reconstructed evolutionary tree to test their hypothesis. Think about how this test might work. What would the tree look like if the hypothesis is true? If it is false? Then look up Trask et al.'s paper:

Trask, S. A., C. A. Derdeyn, U. Fideli, et al. 2002. Molecular epidemiology of human immunodeficiency virus type 1 transmission in a heterosexual cohort of discordant couples in Zambia. *Journal of Virology* 76: 397–405.

Also see a follow-up of this topic, which found that risk of acquiring HIV from an infected partner is higher if the two partners share certain immune system alleles:

Dorak, M., J. Tang, J. Penman-Aguilar, et al. 2004. Transmission of HIV-1 and HLA-B allele-sharing within serodiscordant heterosexual Zambian couples. *Lancet* 363: 2137–2139.

23. AIDS has generated a number of controversial fringe theories. One of these contends that HIV originated not from wild chimpanzees but from an experimental oral polio vaccine that was derived from chimpanzee cell cultures and was administered to many Africans during the late 1950s. Researchers have managed to obtain samples of the polio vaccine that was used in Africa in the 1950s. By sequencing ribosomal RNAs present in the vaccines, they were able to test whether the species used to prepare the vaccine was really chimpanzee, as proposed. In addition, information is available on whether and where there are wild chimpanzee populations that harbor the SIV type that is most closely related to HIV. See:

Berry, N., A. Jenkins, J. Martin, et al. 2005. Mitochondrial DNA and retroviral RNA analyses of archival oral polio vaccine (OPV CHAT) materials: Evidence of macaque nuclear sequences confirms substrate identity. *Vaccine* 23: 1639–1648.

Keele, B. F., F. Van Heuverswyn, et al. 2006. Chimpanzee reservoirs of pandemic and nonpandemic HIV-1. *Science* 313: 523–526.

CITATIONS

Abbas, A. K., K. M. Murphy, and A. Sher. 1996. Functional diversity of helper T lymphocytes. *Nature* 383: 787–793.

Abdool Karim, Q., S. S. Abdool Karim, et al. 2010. Effectiveness and safety of tenofovir gel, an antiretroviral microbicide, for the prevention of HIV infection in women. *Science* 329: 1168–1174.

Ala, P. J., E. E. Huston, et al. 1997. Molecular basis of HIV-1 protease drug resistance: Structural analysis of mutant proteases complexed with cyclic urea inhibitors. *Biochemistry* 36: 1573–1580.

Allen, T. M., D. H. O'Connor, et al. 2000. Tat-specific cytotoxic T lymphocytes select for SIV escape variants during resolution of primary viraemia. *Nature* 407: 386–390.

An, P., and C. A. Winkler. 2010. Host genes associated with HIV/AIDS: Advances in gene discovery. *Trends in Genetics* 26: 119–131.

Anderson, R. M., and R. M. May. 1982. Coevolution of hosts and parasites. *Parasitology* 85: 411–426.

Appay, V., and D. Sauce. 2008. Immune activation and inflammation in HIV-1 infection: Causes and consequences. *Journal of Pathology* 214: 231–241.

Bailes, E., F. Gao, et al. 2003. Hybrid origin of SIV in chimpanzees. *Science* 300: 1713.

Bailey, R. C., S. Moses, et al. 2007. Male circumcision for HIV prevention in young men in Kisumu, Kenya: A randomised controlled trial. *Lancet* 369: 643–656.

Banchereau, J., and R. M. Steinman. 1998. Dendritic cells and the control of immunity. *Nature* 392: 245–252.

Barré-Sinoussi, F., J. Chermann, et al. 1983. Isolation of a T-lymphotrophic retrovirus from a patient at risk for acquired immune-deficiency syndrome (AIDS). *Science* 220: 868–871.

Bartlett, J. G., and R. D. Moore. 1998. Improving HIV therapy. *Scientific American* 279 (July): 84–89.

Bell, A. G. 1883. Upon the formation of a deaf variety of the human race. *Memoirs of the National Academy of Sciences*: 2 (4th memoir).

Biancotto, A., S. J. Iglehart, et al. 2008. HIV-1 induced activation of $CD4^+$ T cells creates new targets for HIV-1 infection in human lymphoid tissue ex vivo. *Blood* 111: 699–704.

Blaak, H., A. B. van't Wout, et al. 2000. In vivo HIV-1 infection of $CD45RA^+CD4^+$ T cells is established primarily by syncytium-inducing variants and correlates with the rate of $CD4^+$ T cell decline. *Proceedings of the National Academy of Sciences USA* 97: 1269–1274.

Boyer, P. L., S. G. Sarafianos, et al. 2001. Selective excision of AZTMP by drug-resistant human immunodeficiency virus reverse transcriptase. *Journal of Virology* 75: 4832–4842.

Brenchley, J. M., D. A. Price, and D. C. Douek. 2006. HIV disease: Fallout from a mucosal catastrophe? *Nature Immunology* 7: 235–239.

Brennan, T. P., J. O. Woods, et al. 2009. Analysis of human immunodeficiency virus type 1 viremia and provirus in resting $CD4^+$ T cells reveals a novel source of residual viremia in patients on antiretroviral therapy. *Journal of Virology* 83: 8470–8481.

Cao, Y., L. Qin, et al. 1995. Virologic and immunologic characterization of long-term survivors of human immunodeficiency virus type 1 infection. *New England Journal of Medicine* 332: 201–208.

Centers for Disease Control (CDC). 2005. HIV prevalence, unrecognized infection, and HIV testing among men who have sex with men—five U.S. cities, June 2004–April 2005. *Morbidity and Mortality Weekly Report* 54 (24): 597–601.

Chun, T.-W., R. T. Davey, Jr., et al. 1999. Re-emergence of HIV after stopping therapy. *Nature* 401: 874–875.

Churchill, M. J., D. I. Rhodes, et al. 2006. Longitudinal analysis of human immunodeficiency virus type 1 *nef*/long terminal repeat sequences in a cohort of long-term survivors infected from a single source. *Journal of Virology* 80: 1047–1052.

Cohen, J. 1993. AIDS research: The mood is uncertain. *Science* 260: 1254–1261.

Cole, A. M., T. Hong, et al. 2002. Retrocyclin: A primate peptide that protects cells from infection by T- and M-tropic strains of HIV-1. *Proceedings of the National Academy of Sciences USA* 99: 1813–1818.

Condra, J. H., D. J. Holder, et al. 1996. Genetic correlates of *in vivo* viral resistance to indinavir, a human immunodeficiency virus type 1 protease inhibitor. *Journal of Virology* 70: 8270–8276.

Crepaz, N., T. A. Hart, and G. Marks. 2004. Highly active antiretroviral therapy and sexual risk behavior: A meta-analytic review. *JAMA* 292: 294–236.

Cromer, D., S. M. Wolinsky, and A. R. McLean. 2010. How fast could HIV change gene frequencies in the human population? *Proceedings of the Royal Society B* 277: 1981–1989.

Darwin, C. R. 1859. *On the Origin of Species by Means of Natural Selection, Or the Preservation of the Favoured Races in the Struggle for Life.* London: John Murray.

Davey, R. T. Jr., N. Bhat, et al. 1999. HIV-1 and T cell dynamics after interruption of highly active antiretroviral therapy (HAART) in patients with a history of sustained viral suppression. *Proceedings of the National Academy of Sciences USA* 96: 15109–15114.

Deacon, N. J., A. Tsykin, et al. 1995. Genomic structure of an attenuated quasi species of HIV-1 from a blood transfusion donor and recipients. *Science* 270: 988–991.

De Clercq, E. 2009. Anti-HIV drugs: 25 compounds approved within 25 years after the discovery of HIV. *International Journal of Antimicrobial Agents* 33: 307–320.

Deeks, S. G., M. Smith, et al. 1997. HIV-1 protease inhibitors. *Journal of the American Medical Association* 277: 145–153.

Deng, H., R. Liu, et al. 1996. Identification of a major co-receptor for primary isolates of HIV-1. *Nature* 381: 661–666.

Dobzhansky, T. 1973. Nothing in biology makes sense except in the light of evolution. *American Biology Teacher* 35: 125–129.

Dodds, J. P., A. M. Johnson, et al. 2007. A tale of three cities: Persisting high HIV prevalence, risk behaviour and undiagnosed infection in community samples of men who have sex with men. *Sexually Transmitted Infections* 83: 392–396.

Douek, D. C., M. Roederer, and R. A. Koup. 2009. Emerging concepts in the immunopathogenesis of AIDS. *Annual Review of Medicine* 60: 471–484.

Doukhan, L., and E. Delwart. 2001. Population genetic analysis of the protease locus of human immunodeficiency virus type 1 quasispecies undergoing drug selection, using a denaturing gradient-heteroduplex tracking assay. *Journal of Virology* 75: 6729–6736.

Dragic, T., V. Litwin, et al. 1996. HIV-1 entry into $CD4^+$ cells is mediated by the chemokine receptor CC-CKR-5. *Nature* 381: 667–673.

Ewald, P. W. 1983. Host-parasite relations, vectors, and the evolution of disease severity. *Annual Review of Ecology and Systematics* 14: 465–485.

Fauci, A. 2008. 25 years of HIV. *Nature* 453: 289–290.

Fraser, C., T. D. Hollingsworth, et al. 2007. Variation in HIV-1 set-point viral load: Epidemiological analysis and an evolutionary hypothesis. *Proceedings of the National Academy of Sciences USA* 104: 17441–17446.

Gallo, R., S. Salahuddin, et al. 1984. Frequent detection and isolation of cytopathic retroviruses (HTIL-III) from patients with AIDS and at risk for AIDS. *Science* 224: 500–503.

Ganser-Pornillos, B. K., M. Yeager, and W. I. Sundquist. 2008. The structural biology of HIV assembly. *Current Opinion in Structural Biology* 18: 203–217.

Gao, F., E. Bailes, et al. 1999. Origin of HIV-1 in the chimpanzee *Pan troglodytes troglodytes*. *Nature* 397: 436–441.

Geffin, R., D. Wolf, et al. 2000. Functional and structural defects in HIV type 1 *nef* genes derived from pediatric long-term survivors. *AIDS Research and Human Retroviruses* 16: 1855–1868.

Giuliani, M., A. D. Carlo, et al. 2005. Increased HIV incidence among men who have sex with men in Rome. *AIDS* 19: 1429–1431.

Gonzalez, E., M. Bamshad, et al. 1999. Race-specific HIV-1 disease-modifying effects associated with CCR5 haplotypes. *Proceedings of the National Academy of Sciences USA* 96: 12004–12009.

Grant, R. M., J. R. Lama, et al. 2010. Preexposure chemoprophylaxis for HIV prevention in men who have sex with men. *New England Journal of Medicine* 363: 2587–2599.

Gray, R. H., G. Kigozi, et al. 2007. Male circumcision for HIV prevention in men in Rakai, Uganda: A randomised trial. *Lancet* 369: 657–666.

Gupta, P., D. Ratner, et al. 2012. Retrocyclin RC-101 blocks HIV-1 transmission across cervical mucosa in an organ culture. *Journal of Acquired Immune Deficiency Syndromes* 60: 455–461.

Hahn, B. H., G. M. Shaw, et al. 2000. AIDS as a zoonosis: Scientific and public health implications. *Science* 287: 607–614.

Hall, H. I., R. Song, et al. 2008. Estimation of HIV incidence in the United States. *JAMA* 300: 520–529.

Hamers, F. F., and A. M. Downs. 2004. The changing face of the HIV epidemic in Western Europe: What are the implications for public health policies? *Lancet* 364: 83–94.

Hammer, S. M., J. J. Eron, Jr., et al. 2008. Antiretroviral treatment of adult HIV infection: 2008 recommendations of the International AIDS Society-USA panel. *JAMA* 300: 555–570.

Harrigan, P. R., R. S. Hogg, et al. 2005. Predictors of HIV drug-resistance mutations in a large antiretroviral-naive cohort initiating triple antiretroviral therapy. *Journal of Infectious Diseases* 191: 339–347.

Harrington, L. E., K. M. Janowski, et al. 2008. Memory CD4 T cells emerge from effector T-cell progenitors. *Nature* 452: 356–360.

Health Protection Agency. 2008. HIV in the United Kingdom: 2008 report. Available at http://www.hpa.org.uk/.

Hillis, D. M. 2000. AIDS. Origins of HIV. *Science* 288: 1757–1759.

Hladik, F., and M. J. McElrath. 2008. Setting the stage: Host invasion by HIV. *Nature Reviews Immunology* 8: 447–457.

Hogg, R. S., D. R. Bangsberg, et al. 2006. Emergence of drug resistance is associated with an increased risk of death among patients first starting HAART. *PLoS Medicine* 3: e356.

Hoggard, P. G., J. Lloyd, et al. 2001. Zidovudine phosphorylation determined sequentially over 12 months in human immunodeficiency virus-infected patients with or without previous exposure to antiretroviral agents. *Antimicrobial Agents and Chemotherapy* 45: 976–980.

Hübner, A., M. Kruhoffer, et al. 1992. Fidelity of human immunodeficiency virus type 1 reverse transcriptase in copying natural RNA. *Journal of Molecular Biology* 223: 595–600.

International Human Genome Sequencing Consortium (IHGSC) 2001. Initial sequencing and analysis of the human genome. *Nature* 409: 860–921.

Johnson, J. A., J. F. Li, et al. 2008. Minority HIV-1 drug resistance mutations are present in antiretroviral treatment-naive populations and associate with reduced treatment efficacy. *PLoS Medicine* 5: e158.

Johnson, W. E., and S. L. Sawyer. 2009. Molecular evolution of the antiretroviral TRIM5 gene. *Immunogenetics* 61: 163–176.

Kaiser, S. M., H. S. Malik, and M. Emerman. 2007. Restriction of an extinct retrovirus by the human TRIM5α antiviral protein. *Science* 316: 1756–1758.

Kalichman, S. C., L. Eaton, et al. 2007. Beliefs about treatments for HIV/AIDS and sexual risk behaviors among men who have sex with men, 1997–2006. *Journal of Behavioral Medicine* 30: 497–503.

Kanki P. J., K. U. Travers, et al. 1994. Slower heterosexual spread of HIV-2 than HIV-1. *Lancet* 343: 943–946.

Kaufmann, D. E., M. Lichterfeld, et al. 2004. Limited durability of viral control following treated acute HIV infection. *PLoS Medicine* 2: e36.

Keele, B. F., J. H. Jones, et al. 2009. Increased mortality and AIDS-like immunopathology in wild chimpanzees infected with SIVcpz. *Nature* 460: 515–519.

Kirchhoff, F. 2009. Is the high virulence of HIV-1 an unfortunate coincidence of primate lentiviral evolution? *Nature Reviews Microbiology* 7: 467–476.

Koblin, B., M. Chesney, and T. Coates. 2004. Effects of a behavioural intervention to reduce acquisition of HIV infection among men who have sex with men: The EXPLORE randomised controlled study. *Lancet* 364: 41–50.

Korber, B., M. Muldoon, et al. 2000. Timing the ancestor of the HIV-1 pandemic strains. *Science* 288: 1789–1796.

Lancet editorial. 2009. Redemption for the pope? *Lancet* 373: 1054.

Larder, B. A., G. Darby, and D. D. Richman. 1989. HIV with reduced sensitivity to Zidovudine (AZT) isolated during prolonged therapy. *Science* 243: 1731–1734.

Learn, G. H., and J. I. Mullins. 2003. The microbial forensic use of HIV sequences. In *HIV Sequence Compendium 2003*, ed. T. Leitner, B. Foley, et al. Los Alamos: Theoretical Biology and Biophysics Group, Los Alamos National Laboratory, LA-UR number 04-7420. Available at http://www.hiv.lanl.gov/content/sequence/HIV/REVIEWS/reviews.html.

Lederman, M. M., and L. Margolis. 2008. The lymph node in HIV pathogenesis. *Seminars in Immunology* 20: 187–195.

Leitner, T., D. Escanilla, et al. 1996. Accurate reconstruction of a known HIV-1 transmission history by phylogenetic tree analysis. *Proceedings of the National Academy of Sciences USA* 93: 10864–10869.

Lemey, P., S. L. Kosakovsky Pond, et al. 2007. Synonymous substitution rates predict HIV disease progression as a result of underlying replication dynamics. *PLoS Computational Biology* 3: e29.

Leslie, A. J., K. J. Pfafferott, et al. 2004. HIV evolution: CTL escape mutation and reversion after transmission. *Nature Medicine* 10: 282–289.

Levin, B. R. 1996. The evolution and maintenance of virulence in microparasites. *Emerging Infectious Diseases* 2: 93–102.

Levin, B. R., and J. J. Bull. 1994. Short-sighted evolution and the virulence of pathogenic microorganisms. *Trends in Microbiology* 2: 76–81.

Liu, R., W. A. Paxton, et al. 1996. Homozygous defect in HIV-1 coreceptor accounts for resistance of some multiply-exposed individuals to HIV-1 infection. *Cell* 86: 367–377.

Lohse, N., A. B. Hansen, et al. 2007. Survival of persons with and without HIV infection in Denmark, 1995–2005. *Annals of Internal Medicine* 146: 87–95.

Lopez, A. D., C. Mathers, et al. 2006. Global and regional burden of disease and risk factors, 2001: Systematic analysis of population health data. *Lancet* 367: 1747–1757.

Maldarelli, F., S. Palmer, et al. 2007. ART suppresses plasma HIV-1 RNA to a stable set point predicted by pretherapy viremia. *PLoS Pathogens* 3: e46.

Marlink, R., P. Kanki, et al. 1994. Reduced rate of disease development after HIV-2 infection as compared to HIV-1. *Science* 265: 1587–1590.

Marmor, M., H. W. Sheppard, et al. 2001. Homozygous and heterozygous *CCR5-Δ32* genotypes are associated with resistance to HIV infection. *Journal of Acquired Immune Deficiency Syndromes* 27: 472–481.

Mathers, B. M., L. Degenhardt, et al. 2008. Global epidemiology of injecting drug use and HIV among people who inject drugs: A systematic review. *Lancet* 372: 1733–1745.

McHeyzer-Williams, M. G., L. J. McHeyzer-Williams, et al. 2000. Antigen-specific immunity. Th cell-dependent B cell responses. *Immunologic Research* 22: 223–236.

Metzker, M. L., D. P. Mindell, et al. 2002. Molecular evidence of HIV-1 transmission in a criminal case. *Proceedings of the National Academy of Sciences USA* 99: 14292–14297.

Mohri, H., M. K. Singh, et al. 1993. Quantitation of Zidovudine-resistant human immunodeficiency virus type 1 in the blood of treated and untreated patients. *Proceedings of the National Academy of Sciences USA* 90: 25–29.

Moore, J. P., S. G. Kitchen, et al. 2004. The CCR5 and CXCR4 coreceptors—central to understanding the transmission and pathogenesis of human immunodeficiency virus type 1 infection. *AIDS Research and Human Retroviruses* 20: 111–126.

Moore, Z. S., J. Seward, and J. M. Lane. 2006. Smallpox. *Lancet* 367: 425–435.

Murphy, R. L., B. A. da Silva, et al. 2008. Seven-year efficacy of a lopinavir/ritonavir-based regimen in antiretroviral-naive HIV-1-infected patients. *HIV Clinical Trials* 9: 1–10.

Nachega, J., M. Hislop, et al. 2007. Adherence to nonnucleoside reverse transcriptase inhibitor-based HIV therapy and virologic outcomes. *Annals of Internal Medicine* 146: 564–573.

Nagelkerke, N. J. D., S. J. de Vlas, et al. 2009. Heterogeneity in host HIV susceptibility as a potential contributor to recent HIV prevalence declines in Africa. *AIDS* 23: 125–130.

National Institute of Allergy and Infectious Diseases (NIAID). 2003. *Understanding the Immune System: How It Works.* NIH Publication No. 03-5423. Available at http://www.niaid.nih.gov/publications/immune/ the_immune_system.pdf.

Neil, S. J., T. Zang, and P. D. Bieniasz. 2008. Tetherin inhibits retrovirus release and is antagonized by HIV-1 Vpu. *Nature* 451: 425–430.

Ngandu, N. K., K. Scheffler, et al. 2008. Extensive purifying selection acting on synonymous sites in HIV-1 Group M sequences. *Virology Journal* 5: 160.

Nguyen, T. X., A. M. Cole, and R. I. Lehrer. 2003. Evolution of primate θ-defensins: A serpentine path to a sweet tooth. *Peptides* 24: 1647–1654.

Nielsen, M. H., F. S. Pedersen, and J. Kjems. 2005. Molecular strategies to inhibit HIV-1 replication. *Retrovirology* 2: 10.

Novembre, J., A. P. Galvani, and M. Slatkin. 2005. The geographic spread of the CCR5 Delta32 HIV-resistance allele. *PLoS Biology* 3: e339.

Ortiz, M., N. Guex, et al. 2009. Evolutionary trajectories of primate genes involved in HIV pathogenesis. *Molecular Biology and Evolution* 26: 2865–2875.

Oster, E. 2009. Routes of infection: Exports and HIV incidence in sub-Saharan Africa. *National Bureau of Economic Research Working Paper* Number 13610. Available at http://faculty.chicagobooth.edu/emily.oster/papers/ index.html.

Oyugi, J. O., F. C. Vouriot, et al. 2009. A common CD4 gene variant is associated with an increased risk of HIV-1 infection in Kenyan female commercial sex workers. *Journal of Infectious Diseases* 199: 1327–1334.

Paiardini, M., I. Pandrea, et al. 2009. Lessons learned from the natural hosts of HIV-related viruses. *Annual Review of Medicine* 60: 485–495.

Pandrea, I., D. L. Sodora, et al. 2008. Into the wild: Simian immunodeficiency virus (SIV) infection in natural hosts. *Trends in Immunology* 29: 419–428.

Pelak, K., D. B. Goldstein, et al. 2010. Host determinants of HIV-1 control in African Americans. *Journal of Infectious Diseases* 201: 1141–1149.

Perez-Caballero, D., T. Zang, et al. 2009. Tetherin inhibits HIV-1 release by directly tethering virions to cells. *Cell* 139: 499–511.

Pillay, D., A. Rambaut, et al. 2007. HIV phylogenetics. *British Medical Journal* 335: 460–461.

Pinkerton, S. D., and P. R. Abramson. 1997. Effectiveness of condoms in preventing HIV transmission. *Social Science & Medicine* 44: 1303–1312.

Pisani, E., G. P. Garnett, et al. 2003. Back to basics in HIV prevention: Focus on exposure. *British Medical Journal* 326: 1384–1387.

Plantier, J. C., M. Leoz, et al. 2009. A new human immunodeficiency virus derived from gorillas. *Nature Medicine* 15: 871–872.

Popovic, M., M. Sarngadharan, et al. 1984. Detection, isolation, and continuous production of cytopathic retroviruses (HTLV-III) from patients with AIDS and pre-AIDS. *Science* 224: 497–500.

Price, D. A., P. J. R. Goulder, et al. 1997. Positive selection of HIV-1 cytotoxic T lymphocyte escape variants during primary infection. *Proceedings of the National Academy of Sciences USA* 94: 1890–1895.

Public Health Agency of Canada. 2006. Estimates of the number of people living with HIV in Canada, 2005. Available at http://www.phacaspc.gc.ca/media/nr-rp/2006/20060731-hiv-vih-eng.php.

Ramaley, P. A., N. French, et al. 2002. Chemokine-receptor genes and AIDS risk. *Nature* 417: 140.

Rambaut, A., D. L. Robertson, et al. 2001. Human immunodeficiency virus. Phylogeny and the origin of HIV-1. *Nature* 410: 1047–1048.

Read, T. R. H., J. Hocking, et al. 2007. Risk factors for incident HIV infection in men having sex with men: A case-control study. *Sexual Health* 4: 35–39.

Rhodes, D. I., L. Ashton, et al. 2000. Characterization of three *nef*-defective human immunodeficiency virus type 1 strains associated with long-term nonprogression. *Journal of Virology* 74: 10581–10588.

Richman, D. D. 1998. How drug resistance arises. *Scientific American* 279: 88.

St. Clair, M. H., J. L. Martin, et al. 1991. Resistance to ddI and sensitivity to AZT induced by a mutation in HIV-1 reverse transcriptase. *Science* 253: 1557–1559.

Salemi, M., K. Strimmer, et al. 2001. Dating the common ancestor of SIV_{cpz} and HIV-1 group M and the origin of HIV-1 subtypes using a new method to uncover clock-like molecular evolution. *FASEB Journal* 15: 276–278.

Samson, M., F. Libert, et al. 1996. Resistance to HIV-1 infection in Caucasian individuals bearing mutant alleles of the CCR-5 chemokine receptor gene. *Nature* 382: 722–725.

Sauter, D., M. Schindler, et al. 2009. Tetherin-driven adaptation of Vpu and Nef function and the evolution of pandemic and nonpandemic HIV-1 strains. *Cell Host Microbe* 6: 409–421.

Sawyer, S. L., L. I. Wu, et al. 2005. Positive selection of primate TRIM5α identifies a critical species-specific retroviral restriction domain. *Proceedings of the National Academy of Sciences USA* 102: 2832–2837.

Scaduto, D. I., J. M. Brown, et al. 2010. Source identification in two criminal cases using phylogenetic analysis of HIV-1 DNA sequences. *Proceedings of the National Academy of Sciences, USA* 107: 21242–21247.

Scheer, S., T. Kellogg, et al. 2008. HIV is hyperendemic among men who have sex with men in San Francisco: 10-year trends in HIV incidence, HIV prevalence, sexually transmitted infections and sexual risk behaviour. *Sexually Transmitted Infections* 84: 493–498.

Schindler, M., J. Münch, et al. 2006. Nef-mediated suppression of T cell activation was lost in a lentiviral lineage that gave rise to HIV-1. *Cell* 125: 1055–1067.

Schliekelman, P., C. Garner, and M. Slatkin. 2001. Natural selection and resistance to HIV. *Nature* 411: 545–546.

Schneider, M. F., S. J. Gange, et al. 2005. Patterns of the hazard of death after AIDS through the evolution of antiretroviral therapy: 1984–2004. *AIDS* 19: 2009–2018.

Seid, J. M., M. Liberto, et al. 1986. T cell-macrophage interactions in the immune response to herpes simplex virus: The significance of interferon-gamma. *Journal of General Virology* 67: 2799–2802.

Shankarappa, R., J. B. Margolick, et al. 1999. Consistent viral evolutionary changes associated with the progression of human immunodeficiency virus type 1 infection. *Journal of Virology* 73: 10489–10502.

Sharp, P. M., and B. H. Hahn. 2008. AIDS: Prehistory of HIV-1. *Nature* 455: 605–606.

Sharp, P. M., E. Bailes, et al. 2001. The origins of acquired immune deficiency syndrome viruses: Where and when? *Philosophical Transactions of the Royal Society of London* B 356: 867–876.

Shirasaka, T., R. Yarchoan, et al. 1993. Changes in drug sensitivity of human immunodeficiency virus type 1 during therapy with azidothymidine, didieoxycytidine, and dideoxyinosine: An in vitro comparative study. *Proceedings of the National Academy of Sciences USA* 90: 562–566.

Silvestri, G. 2009. Immunity in natural SIV infections. *Journal of Internal Medicine* 265: 97–109.

Silvestri, G., A. Fedanov, et al. 2005. Divergent host responses during primary simian immunodeficiency virus SIVsm infection of natural sooty mangabey and nonnatural rhesus macaque hosts. *Journal of Virology* 79: 4043–4054.

Slutkin, G., S. Okware, et al. 2006. How Uganda reversed its HIV epidemic. *AIDS and Behavior* 10: 351–360.

Sousa, A. E., J. Carneiro, et al. 2002. CD4 T cell depletion is linked directly to immune activation in the pathogenesis of HIV-1 and HIV-2 but only indirectly to the viral load. *Journal of Immunology* 169: 3400–3406.

Sprent, J., and C. D. Surh. 2002. T cell memory. *Annual Review of Immunology* 20: 551–579.

State of Louisiana v. Richard J. Schmidt. 1997. K97–249. LA Court of Appeal, 3rd Circuit. 699 So.2d 488.

State of Louisiana v. Richard J. Schmidt. 2000. 99-1412. LA Court of Appeal, 3rd Circuit. 771 So.2d 131.

Stremlau, M., C. M. Owens, et al. 2004. The cytoplasmic body component TRIM5α restricts HIV-1 infection in Old World monkeys. *Nature* 427: 848–853.

Suksomboon, N., N. Poolsup, and S. Ket-Aim. 2007. Systematic review of the efficacy of antiretroviral therapies for reducing the risk of mother-to-child transmission of HIV infection. *Journal of Clinical Pharmacy and Therapeutics* 32: 293–311.

Takehisa, J., M. H. Kraus, et al. 2009. Origin and biology of simian immunodeficiency virus in wild-living western gorillas. *Journal of Virology* 83: 1635–1648.

Tang, Y.-Q., J. Yuan, et al. 1999. A cyclic antimicrobial peptide produced in primate leukocytes by the ligation of two truncated α-defensins. *Science* 286: 498–502.

Tobiume, M., M. Takahoko, et al. 2002. Inefficient enhancement of viral infectivity and CD4 downregulation by human immunodeficiency virus type 1 nef from Japanese long-term nonprogressors. *Journal of Virology* 76: 5959–5965.

Troyer, R. M., K. R. Collins, et al. 2005. Changes in human immunodeficiency virus type 1 fitness and genetic diversity during disease progression. *Journal of Virology* 79: 9006–9018.

UNAIDS (Joint United Nations Programme onf HIV/AIDS). 2008. Report on the global HIV/AIDS epidemic 2008. Available at http://www.unaids.org/en/resources/documents/2008/.

UNAIDS. 2010. Global report fact sheet. Available at http://www.unaids.org/globalreport/Press_kit.htm.

UNAIDS 2012a. Global fact sheet: World AIDS Day 2012. Available at http://www.unaids.org/en/resources/presscentre/factsheets/.

UNAIDS. 2012b. Global report: UNAIDS report on the global AIDS epidemic. Available at http://www.unaids.org/en/resources/documents/2012/.

Venkataraman, N., A. L. Cole, et al. 2009. Reawakening retrocyclins: Ancestral human defensins active against HIV-1. *PLoS Biology* 7: e95.

Wain-Hobson, S. 1993. The fastest genome evolution ever described: HIV variation in situ. *Current Opinion in Genetics and Development* 3: 878–883.

Walker, B. D., and D. Burton. 2008. Toward an AIDS vaccine. *Science* 320: 760–764.

Watkins, D. I. 2008. The vaccine search goes on. *Scientific American* 299: 69–76.

Weller, S., and K. Davis. 2002. Condom effectiveness in reducing heterosexual HIV transmission. *Cochrane Database of Systematic Reviews* Issue 1: Article No. CD003255.

Williams, M. A., and M. J. Bevan. 2007. Effector and memory CTL differentiation. *Annual Review of Immunology* 25: 171–192.

Winkler, C., P. An, and S. J. O'Brien. 2004. Patterns of ethnic diversity among the genes that influence AIDS. *Human Molecular Genetics* 13 (Spec. No 1): R9–19.

Wolf, D., and S. P. Goff. 2008. Host restriction factors blocking retroviral replication. *Annual Review of Genetics* 42: 143–163.

World Health Organization. (WHO). 2008/2011. The top ten causes of death. Fact Sheet No 310. Available at http://www.who.int/mediacentre/factsheets/fs310/en/ index.html.

Worobey, M., M. Gemmel, et al. 2008. Direct evidence of extensive diversity of HIV-1 in Kinshasa by 1960. *Nature* 455: 661–664.

Yap, M. W., S. Nisole, and J. P. Stoye. 2005. A single amino acid change in the SPRY domain of human TRIMα leads to HIV-1 restriction. *Current Biology* 15: 73–78.

Yusim, K., M. Peeters, et al. 2001. Using human immunodeficiency virus type 1 sequences to infer historical features of the acquired immune deficiency syndrome epidemic and human immunodeficiency virus evolution. *Philosophical Transactions of the Royal Society of London* B 356: 855–866.

Zhang, F., S. J. Wilson, et al. 2009. Nef proteins from simian immunodeficiency viruses are tetherin antagonists. *Cell Host & Microbe* 6: 54–67.

© 2008 Nature Publishing Group

2

The Pattern of Evolution

Where do we come from, we humans and the myriad other organisms, turtles included, that share our planet? Biologists have established the answer, but members of the general public remain divided.

For an international view of public sentiment, Jon D. Miller and colleagues (2006) assembled data from recent surveys conducted in 32 European countries, the United States, and Japan. All the polls had included this question:

> True or False? Human beings, as we know them, developed from earlier species of animals.

Iceland, Sweden, Denmark, and France ranked highest. In these countries, 80% or more of adults affirmed descent with modification **(Figure 2.1, next page)**. Japan was next, at 78%. The United States ranked second to last: 40% accepted evolution, 39% rejected it, and 21% were unsure. (Turkey, where scarcely one in four adults accepted evolution, was last.)

All readers, even in countries with high levels of acceptance, are bound to find themselves in conversation with individuals who remain unconvinced or reject evolution outright. Some of the doubters will be school board members, legislators, or teachers (see Berkman et al. 2008; Forrest 2008; Branch and Scott 2009). This alone is sufficient reason for covering the evidence for evolution.

This 220-million-year-old fossil turtle, *Odontochelys semitestacea*, has a full shell covering its belly (above) and expanded ribs protecting its back (below). Its anatomy is consistent with a hypothesis of turtle origins based on how the shell develops in living embryonic turtles. From Li et al. (2008); see also Burke (2009), Nagashima et al. (2009).

Reprinted by permission from Macmillan Publishers Ltd: *Nature* 456: 497–501, copyright 2008.

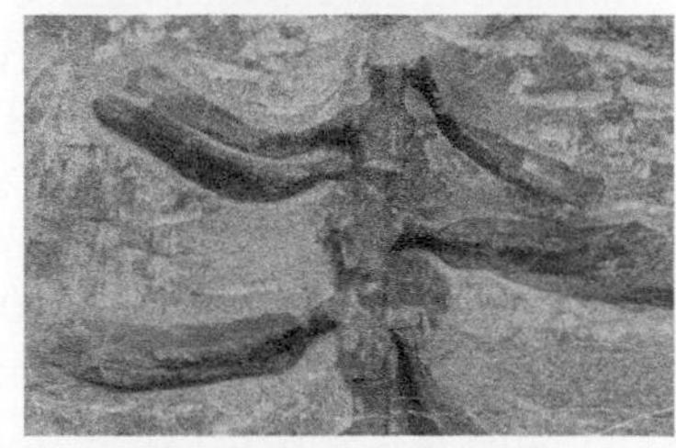

© 2008 Nature Publishing Group

But the chapter serves other purposes as well. It summarizes the pattern of life's history that the theory of evolution was developed to explain. It introduces concepts, such as homology and deep time, that will be important throughout the book. And it complements our discussion of HIV (Chapter 1) in providing additional examples of why evolution matters outside classrooms and textbooks.

We will organize our presentation around the points of difference between two competing models of the history of life on Earth. The first model is the theory of special creation **(Figure 2.2a, facing page)**. In this view, species are immutable—unchanged since their origin—and variation among individuals is limited. All species were created separately and are thus genealogically unrelated to each other. In the traditional version of special creation, the Earth and its living creatures are young—with a beginning as recent as 6,000 years ago (Ussher 1658; Brice 1982). As expressed by John Ray (1686), the first scientist to give a biological definition of species, "One species never springs from the seed of another."

When the English naturalist Charles Darwin began to study biology seriously, as a college student in the 1820s, the theory of special creation was still the leading explanation in Europe for the origin of species. Scholars had, however, begun to challenge it. The notion of biological change had been proposed by several authors in the late 1700s and early 1800s, including Georges-Louis Leclerc (often referred to by his title, Comte de Buffon), Erasmus Darwin (Charles's grandfather), and Jean-Baptiste Lamarck (Eiseley 1958; Desmond and Moore 1991).

By the time Darwin began working on the problem himself, in the 1830s, dissatisfaction with special creation had begun to grow in earnest. Research in the biological and geological sciences was advancing rapidly, and the data clashed with special creation's claims.

The idea of evolution had thus been under discussion for decades. But it was Darwin's most famous book, *On the Origin of Species by Means of Natural Selection*, first published in 1859, that convinced the scientific community that it was true—that Earth's species are the products of descent with modification from a common ancestor (Mayr 1964). Darwin had worked on the material for more than 20 years before publishing. Drawing on his own work and that of others, he had gathered an overwhelming collection of detailed evidence from a variety of fields of biology. His masterful presentation was persuasive. Within a decade, the fact of evolution had achieved general acceptance among scientists.

The theory of descent with modification includes five elements missing from special creation (Figure 2.2b). First, species are not immutable, but change through time. In a population of birds, for example, the average beak size may change from one generation to the next. This is known as **microevolution**. Second, lineages split and diverge, thereby increasing the number of species. An ancestral species of birds, for instance, may give rise to two distinct descendant species. This is called **speciation**. Third, over long periods of time, novel forms of life can derive from earlier forms. Tetrapods, for example, arose from a lineage of fish. This kind of dramatic change over time is called **macroevolution**. Fourth, species are derived not independently, but from common—that is, shared—ancestors. All species are thus genealogically related. "I should infer," Darwin said in *On the Origin of Species* (1859, p. 484), "that probably all the organic beings which have ever lived on this earth have descended from some one primordial form." Finally, the Earth and life are considerably more than 6,000 years old.

The five sections of the chapter review evidence that supports each of these elements in Darwin's view of life's history. The evidence—some of it presented

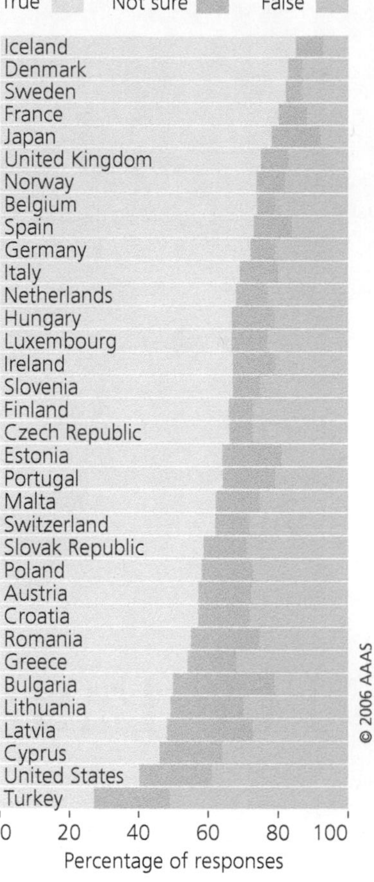

Figure 2.1 Acceptance of evolution in 34 countries Sample sizes range from 500 to 2,146; most are near 1,000. Redrawn from Miller et al. (2006).

From "Science communication. Public acceptance of evolution." *Science* 313: 765–766. Reprinted with permission from AAAS.

The theory of special creation and the theory of descent with modification make different assertions about species, where they came from, and how they are related, as well as different assertions about the age of Earth. These assertions can be checked against evidence.

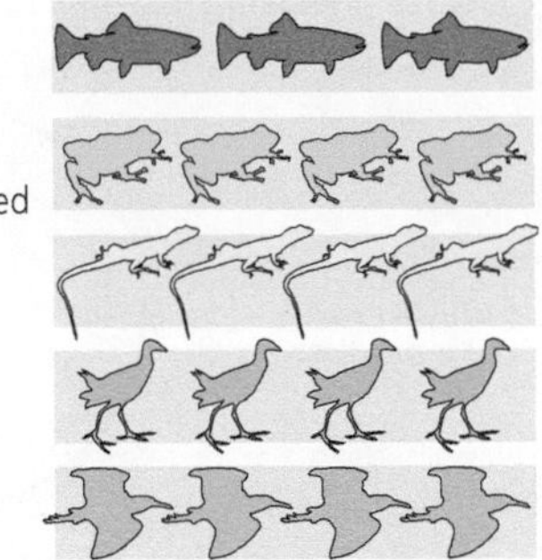

• Species do not change
• Lineages do not split
• Each species is separately created

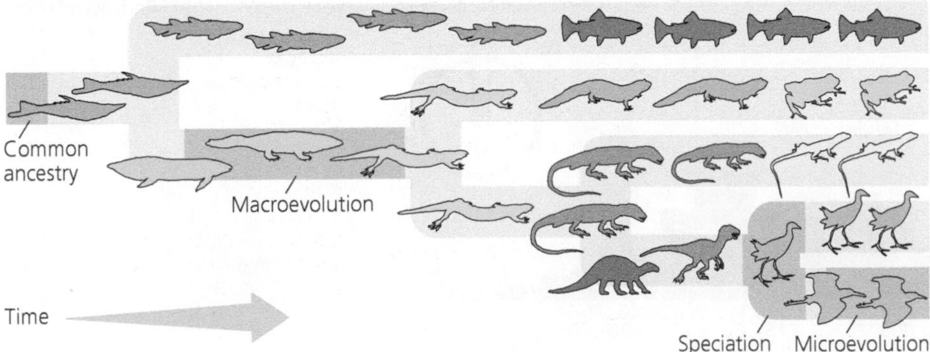

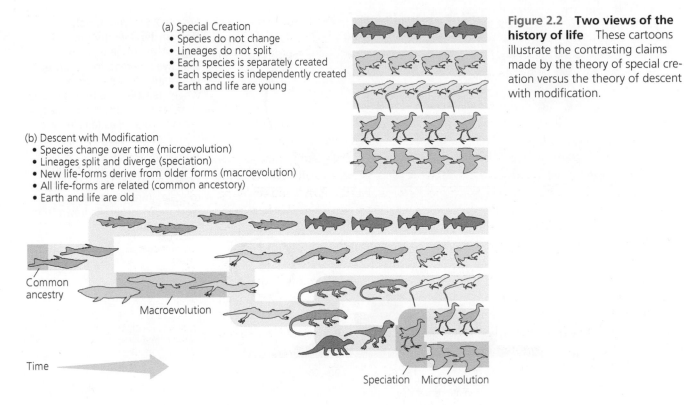

Figure 2.2 Two views of the history of life These cartoons illustrate the contrasting claims made by the theory of special creation versus the theory of descent with modification.

by Darwin, much of it accumulated since—has convinced virtually all scientists who study life that Darwin was right. Darwin called the pattern he saw "descent with modification." It has since come to be known as evolution.

2.1 Evidence of Microevolution: Change through Time

Numerous lines of evidence demonstrate that populations of organisms change across generations. Here we review data from selective breeding, direct observation of natural populations, and the anatomy of living species.

Evidence from Selective Breeding

That a population of organisms can change over time can be demonstrated by anyone with sufficient patience. The trick is selective breeding, also known as artificial selection. Each generation, the experimenter examines the population and chooses as breeders only those individuals with the most desirable characteristics.

Ted Garland and colleagues bred strains of mice that voluntarily run extraordinary distances on exercise wheels (Swallow et al. 1998). From a large population of lab mice, the researchers established four high-runner lines and four control lines, each consisting of 10 mated pairs. The scientists let the pairs breed, gave the offspring exercise wheels, and recorded the distance each mouse ran per day.

For the high-runner lines, the researchers selected from each family the male and female that ran the greatest distance. These they paired at random—except that siblings were not allowed to breed with each other—to produce the next generation. For the control lines, the researchers used as breeders a male and female from each family chosen at random.

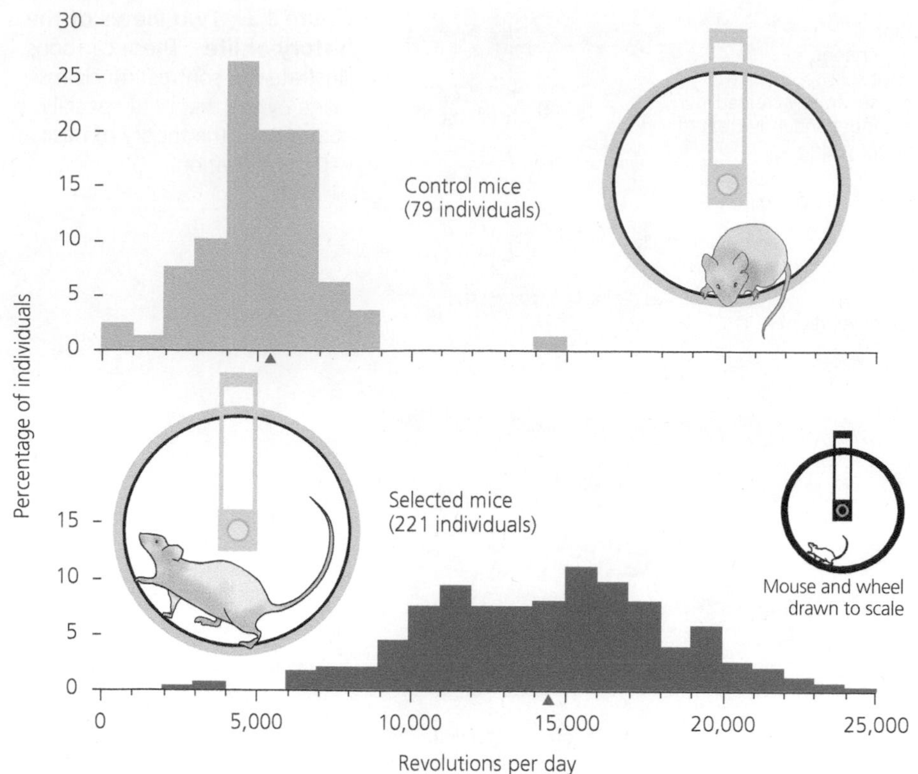

Figure 2.3 Microevolution under selective breeding After 24 generations of selective breeding for distance voluntarily run on an exercise wheel, female mice from selected lineages ran nearly three times as far, on average, as female mice from control lineages (14,458 versus 5,205 revolutions per day, as indicated by orange triangles). Mice and wheels are not shown to scale; in reality, the wheels were much larger. Redrawn from Garland (2003).

After 24 generations of selective breeding, the mice in the selected versus control lines clearly differed in their propensity for voluntary exercise (**Figure 2.3**). The females in the high-runner lineages traveled, on average, 2.78 times as far each day as the females in the control lineages (Garland 2003). There are, of course, two ways that the high runners might accomplish this. They could spend more time running, or they could run faster. It turns out that mostly what they do is run faster (Garland et al. 2011).

Garland's high-runner mice also differ from the controls in genotype (Garland et al. 2002; Kelly et al. 2010), physiology (Malisch et al. 2008; Meek et al. 2009), and morphology (Yan et al. 2008).

Perhaps most dramatic, however, are neurological differences (Rhodes et al. 2005). The high-runner mice resemble humans with attention deficit/hyperactivity disorder (ADHD). For example, Ritalin—a drug used to treat ADHD in humans—reduces the intensity of exercise in high-runner mice but has little effect on control mice.

Observations on living organisms provide direct evidence of microevolution by showing that populations and species change over time.

The extent to which lineages of organisms can be sculpted using artificial selection is dramatically illustrated by the fact that most domesticated plants and animals come in an abundance of distinct pure-breeding varieties. In each case, the distinctive varieties derive from common stock.

All breeds of dogs, for example, are descended from wolves (Wayne and Ostrander 2007; vonHoldt et al. 2010). The enthusiasm with which they hybridize demonstrates that, despite their differences, all dogs still belong to a single species. Compared to their wolf ancestors, dogs exhibit an astonishing diversity of sizes, shapes, colors, and behaviors (**Figure 2.4**). The differences persist when dogs of different breeds are reared together, showing that breed characteristics are the result of genetic divergence (see also Akey et al. 2010; Boyko et al. 2010; Shearin and Ostrander 2010).

Figure 2.4 **Dogs illustrate the capacity of lineages to change under artificial selection** The ancestors of dogs looked like today's wolves (left). A papillon and a great dane (right) show the diversity among dog breeds.

Evidence from Direct Observation of Natural Populations

Laboratories, farms, and kennels are not the only places where populations of organisms evolve rapidly enough that we can watch it happen. In recent decades, biologists have documented numerous examples of microevolution in natural populations. Looking just at studies of animals, for example, Andrew Hendry and colleagues (2008) reviewed thousands of measurements of change across generations in dozens of different species. For 17 species, there was definitive evidence that the modification across generations was at least partly due to changes in genes. A similar volume of research reports observations of microevolution in plants (Bone and Farres 2001).

Research on field mustard by Steven Franks and colleagues (2007) provides an elegant example. Field mustard, *Brassica rapa*, is a small herb closely related to turnip and broccoli rabe **(Figure 2.5)**. Franks and colleagues studied a natural population in coastal Southern California.

Figure 2.5 **Field mustard, *Brassica rapa***

The plants in this population depend on a rainy season that runs from winter into spring. Individuals germinate, grow, flower, set seed, and die within a single year. For several years in the mid 1990s the seasonal rains lasted into late spring, giving the plants unusually long growing seasons. In long growing seasons individuals that delay flowering, and thereby grow larger before reproducing, make more seeds. For several years in the early 2000s, in contrast, the seasonal rains ended early. In short growing seasons, individuals that flower early, and thereby avoid dying of dehydration before setting seed, enjoy higher reproductive success. Franks and colleagues (2007) hypothesized that during the drought, their field mustard population had evolved an earlier average flowering time.

Fortunately, the researchers had a collection of seeds they had gathered from the population in 1997, before the drought struck. This collection was effectively a fossil record, but one from which seeds would germinate and grow. Franks and colleagues gathered more seeds in 2004 for an experimental comparison.

Under ideal conditions in a greenhouse, Franks and colleagues first grew a refresher generation of mustard plants from each collection. This step was needed to eliminate any differences between the seeds from 1997 versus 2004 caused by the longer dormancy endured by the 1997 seeds or by the different environments experienced by the parents that produced the two sets of seeds. Using the offspring of the refresher generation as breeding stock, the researchers then produced purebred 1997 experimental plants, purebred 2004 experimental plants, and experimental hybrids. They watered all plants to mimic a long growing season and counted the days each plant took from germination to first open flower.

Figure 2.6 summarizes the variation in time to first flower among individuals in the three strains. As Franks and colleagues had predicted, the 2004 purebreds had, on average, the shortest flowering time, the hybrids were next, and the 1997 purebreds had the longest flowering time.

Note that the experimental plants were the third generation in which all individuals had been grown together in the same environment. This means that any differences among the 1997 purebreds versus the hybrids versus the 2004 purebreds must be due to differences in the plants' genes. Franks and colleagues concluded that their mustard population had, indeed, evolved during the drought.

Later in this section, and throughout the book, we will consider additional examples in which natural populations evolved while biologists were watching.

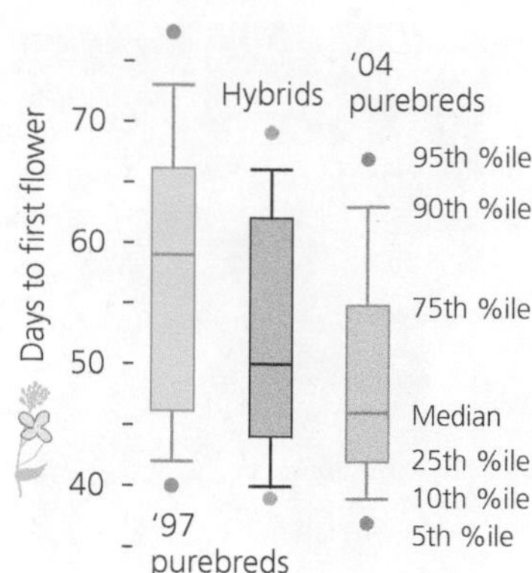

Figure 2.6 Microevolution documented in a natural population of field mustard The effect of strain on flowering time is statistically significant at $p < 0.001$. Redrawn from Franks et al. (2007).

Evidence from Living Anatomy: Vestigial Structures

By the time Darwin began working on "the species question," comparative anatomists had described a number of curious traits called **vestigial structures**. A vestigial structure is a useless or rudimentary version of a body part that has an important function in other, closely allied, species. Darwin argued that vestigial traits are difficult to explain under the theory of special creation, but readily interpretable under the theory of descent with modification.

Figure 2.7 Vestigial structures (a) The brown kiwi, a flightless bird, has tiny, useless wings. (b) The royal python (*Python regius*) has a tiny remnant hindlimb, called a spur, on either side of its vent.

Figure 2.7 shows examples of vestigial structures. The North Island brown kiwi, *Apteryx mantelli*, a flightless bird, has tiny, stubby wings. The royal python, *Python regius*, has remnant hindlimbs, represented internally by rudimentary hips and leg bones and externally by minute spurs. The evolutionary interpretation is that kiwis and pythons are descended, with modification, from ancestors whose wings or hind legs were fully formed and functional.

Humans have vestigial structures too. We have, for example, a tiny tailbone called the coccyx **(Figure 2.8a)**. We also have muscles attached to our hair follicles that contract to make our body hair stand on end when we are cold or frightened (Figure 2.8b). If we were hairy, like chimpanzees, this would increase the loft of our fur to keep us warm or make us look bigger and more intimidating to enemies (Figure 2.8c). But we are not hairy, so we just get goose bumps. Our goose bumps imply that we are descended from ancestors that were hairier than we are. Likewise, our tiny tailbones imply descent from ancestors with tails.

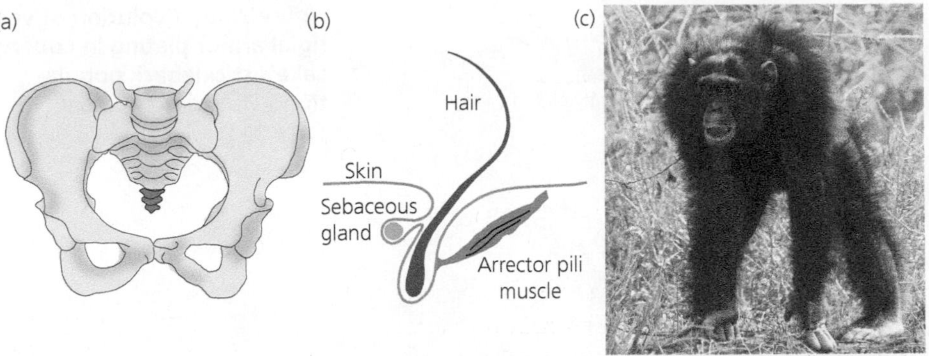

Figure 2.8 Human vestigial structures (a) The coccyx, a rudimentary tailbone. (b) The *arrector pili*, a muscle at the base of each hair follicle. When it contracts, the hair stands up. (c) In a chimpanzee, contraction of the *arrectores pilorum* increases the loft of the fur.

Vestigial traits also occur at the molecular level. Humans have one such trait on chromosome 6. It is a DNA sequence that looks like a gene for the enzyme CMAH (CMP-N-acetylneuraminic acid hydroxylase), except that it is disabled by a 92-base-pair deletion (Chou et al. 1998). Most mammals, including chimpanzees, bonobos, gorillas, and orangutans, make CMAH in abundance, but humans cannot (Chou et al. 2002). CMAH converts an acidic sugar, destined for display on the surface of cells, from one form to another. Because of our inability to make CMAH, we humans have a different biochemical signature on our cell membranes. This appears to explain why humans and chimpanzees are largely immune to each other's malaria parasites (Martin et al. 2005).

Both the anatomy and genomes of living organisms show evidence of descent with modification in the form of reduced or useless parts.

In a survey of the human genome, Zhengdong Zhang and colleagues (2010a) found 75 more examples of human genes that are disrupted by devastating mutations and whose functions have been lost from our species' biochemical tool kit. These nonfunctional genes are difficult to reconcile with the notion that humans were created in their present form. But they are readily explicable if humans are descended from ancestors that made CMAH and the other 75 proteins.

In some cases, the interpretation of vestigial traits as evidence of evolution can be tested. The threespine stickleback, *Gasterosteus aculeatus*, is a small fish that lives in coastal ocean waters throughout the Northern Hemisphere and readily invades freshwater (Bell and Foster 1994). Marine sticklebacks carry heavy body armor, featuring bony plates protecting their sides and pelvic fins modified into spines **(Figure 2.9a)**. Freshwater sticklebacks, however, often carry light armor. They have fewer bony plates, and their pelvic structures may be reduced to vestiges or absent altogether (Figure 2.9b). Freshwater and marine sticklebacks typically interbreed without hesitation, confirming their close relationship.

(a) Fully armored

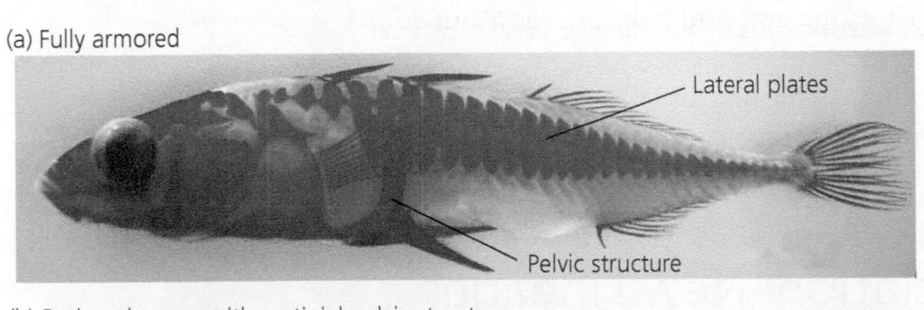

(b) Reduced armor with vestigial pelvic structure

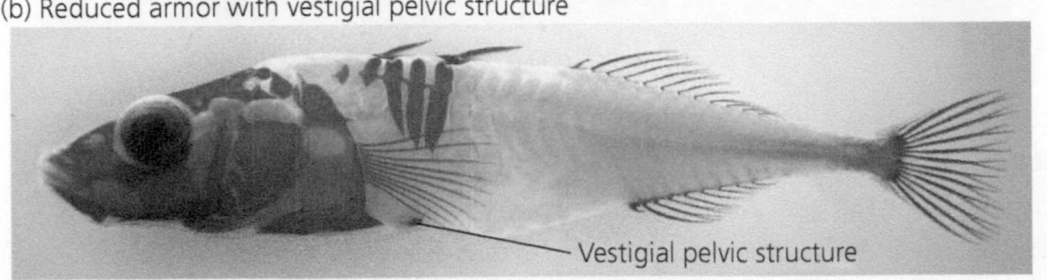

Figure 2.9 Reduced armor in freshwater sticklebacks (a) Marine sticklebacks are heavily armored with bony plates and pelvic spines. (b) Freshwater sticklebacks often have light armor plating and reduced or absent pelvic structures. [These particular individuals are offspring produced in a laboratory breeding experiment (Cresko et al. 2007) and are used here for illustration.] Photos courtesy of William A. Cresko.

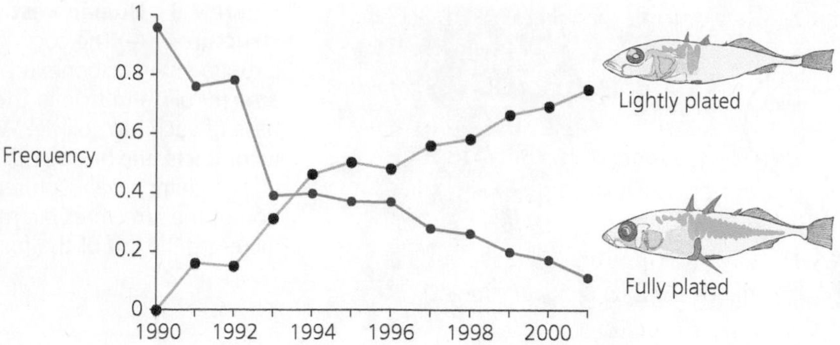

Figure 2.10 Evolution of vestigial armor plating in Loberg Lake's stickleback population Plotted from data in Table 1 in Bell et al. (2004).

The differences in body armor between marine versus freshwater sticklebacks are largely controlled by just two genes. The gene responsible for variation in armor plating encodes a protein called ectodysplasin (Colosimo et al. 2005). The gene responsible for variation in pelvic spines encodes pitx1 (Shapiro et al. 2004). Each gene has two alleles: one associated with heavy armor and one with light.

If the alleles for light plating and a reduced pelvis exist in marine stickleback populations, then marine populations invading freshwater might evolve toward the lightly armored form rapidly enough to watch it happen. Michael Bell and colleagues (2004) documented just such a swift transition **(Figure 2.10)**. Loberg Lake, Alaska, was poisoned in 1982 so that it could be restocked with trout and salmon for recreational fishing. By 1988, the lake had been invaded by marine sticklebacks from nearby Cook Inlet. Bell monitored the new Loberg population from 1990 through 2001. In less than a dozen years, the composition of the population changed from over 95% fully plated fish to over 75% lightly plated. In Loberg Lake, and probably elsewhere, freshwater sticklebacks with vestigial armor are, indeed, the modified descendants of heavily armored marine fish.

Why Microevolution Matters

We have reviewed three kinds of evidence for microevolution. Species are not fixed entities. Their characteristics can and do change across generations.

Microevolution is important in human affairs because it alters the nature of the many organisms we interact with. These include domestic plants and animals (for examples, see Driscoll et al. 2009; Tian et al. 2009), wild organisms we eat (Allendorf and Hard 2009), microbes that cause disease (Taubes 2008; Knapp et al. 2010), parasites (Sibley and Hunt 2003), pests (Mallet 1989), and commensal organisms such as the bacteria that inhabit our guts and help us digest our food (Hehemann et al. 2010).

In addition, our interactions with other organisms sometimes cause our own populations to evolve, leading to genetic differences among individuals who live in different places and have divergent lifestyles (Hancock et al. 2010).

2.2 Evidence of Speciation: New Lineages from Old

Having established that species change over time, we now consider whether they also spawn new species. Before we discuss the evidence, is helpful to define what **species** are and to establish a way of telling them apart. Achieving consensus on a definition has been difficult for biologists, in part because the utility of various criteria for distinguishing species depends on the type of organism under study and the problem to be solved (de Quieroz 2007).

In this chapter, we use a definition codified by Ernst Mayr (1942). Species are populations, or groups of populations, within and among which individuals actually or potentially interbreed and outside of which they do not interbreed. The virtue of this definition, known as the biological species concept, is that we can let the organisms themselves tell us whether they belong to the same species. If individuals from different populations have the opportunity to mate but are disinclined to do so, or if such individuals mate but fail to produce healthy, fertile offspring, then the individuals belong to different species.

The evidence we will examine comes from both laboratory experiments and observations of natural populations.

Evidence from Laboratory Experiments

Siobain Duffy and colleagues (2007) set up an experiment to encourage speciation in populations of a virus called $\phi 6$ (**Figure 2.11**). $\phi 6$ is an RNA virus that infects the bacterium *Pseudomonas syringae*. Different strains of the virus interbreed when they infect the same host cell and exchange bits of genetic material.

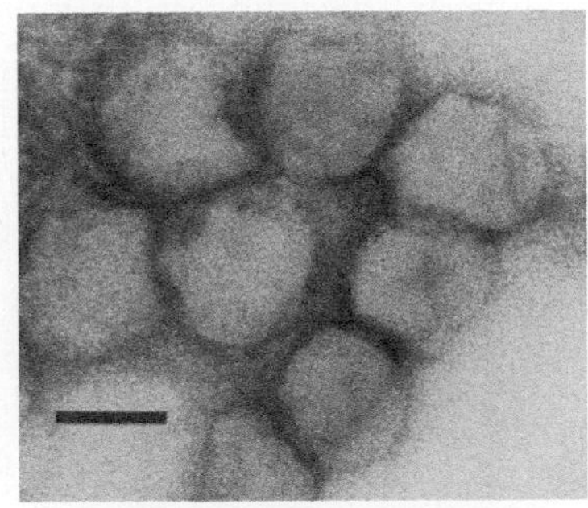

Figure 2.11 Bacteriophage ϕ6 Scale bar = 50 nm. Electron micrograph by Hans Ackermann.

Wild-type $\phi 6$ virions, $\phi 6_{WT}$, can infect four different varieties of *P. syringae*. While studying the virus, Duffy and colleagues had found a spontaneously occurring mutant strain with an extended host range. This strain, $\phi 6_{broad}$, carries an altered version of the gene for a protein called P3. P3 enables the virus to attach to host cells and initiate an infection. The modified P3 allows $\phi 6_{broad}$ to infect an additional variety of *P. syringae* plus the bacterium *Pseudomonas pseudoalcaligenes*.

Duffy and her team set up four experimental populations of $\phi 6_{broad}$ and grew them on *P. pseudoalcaligenes*. Every day for a month, the researchers transferred a random sample of a few hundred virions from each experimental population to a new petri dish of uninfected hosts. This procedure allowed the viral populations to cycle through a total of about 150 generations.

Duffy expected the experimental populations to evolve by natural selection. Because RNA viruses have high mutation rates, new mutant strains of $\phi 6$ would arise. Most would fare poorly in competition with the other individuals in the population, and would remain rare or disappear. However, due to the large number of virions that proliferate in a dish of bacteria, and thus the large variety of mutants getting a tryout, the researchers anticipated that a few mutant strains would arise that could replicate especially quickly in *P. pseudoalcaligenes*. Over the month, these strains would become common. This is, indeed, what happened.

Of particular interest is a mutant strain that appeared in experimental population 1 about midway through the month. The researchers detected this strain while periodically checking whether virions from each of the experimental populations could still infect the five different strains of *P. syringae*. The new strain, which the researchers called $\phi 6_{E1narrow}$, could not. It carried another alteration in the gene for P3. This alteration had two consequences. It endowed virions with an enhanced ability to infect *P. pseudoalcaligenes* and a resulting competitive advantage that allowed them to rise to high frequency (**Figure 2.12**). And it rendered them incapable of infecting any strain of *P. syringae*.

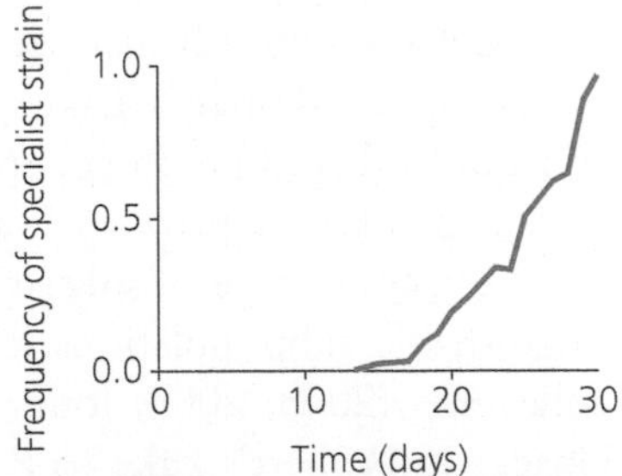

Figure 2.12 Evolution in an experimental population of bacteriophage ϕ6 The specialist strain that appeared midway through the experiment is capable of infecting only *P. pseudoalcaligenes*. Redrawn from Duffy et al. (2007).

By the biological species concept, $\phi 6_{WT}$ and $\phi 6_{E1narrow}$ are different species. They cannot infect the same host, and so they are incapable of interbreeding. Duffy's experiment demonstrates one way viruses can switch hosts. And it shows that, contrary to John Ray's claim, a species can spring from the seed of another.

Similar experiments with more complex organisms, most frequently insects, have been conducted over the past several decades [for reviews, see Rice and Hostert (1993); Florin and Ödeen (2002)]. A study by Dianne Dodd (1989) on the fruit fly *Drosophila pseudoobscura* provides an example.

Dodd worked with lab populations descended from a common ancestral wild stock. Four populations had been evolving for a year on a starch-based diet and four on a maltose-based diet. Both diets were stressful for the flies, so it had taken months, and several generations of evolution, for the populations to thrive.

Dodd ran mating trials to gauge whether flies adapted to different diets were inclined to interbreed. In each trial, she placed virgin males and females from a starch population together with virgins from a maltose population, then watched the flies mate. Most of the matings, 602 of 904, were between flies from the same food-based population **(Figure 2.13)**. Dodd's flies had not yet separated into different species, but their aversion to mating with flies from populations adapted to a different diet suggests they were moving in that direction.

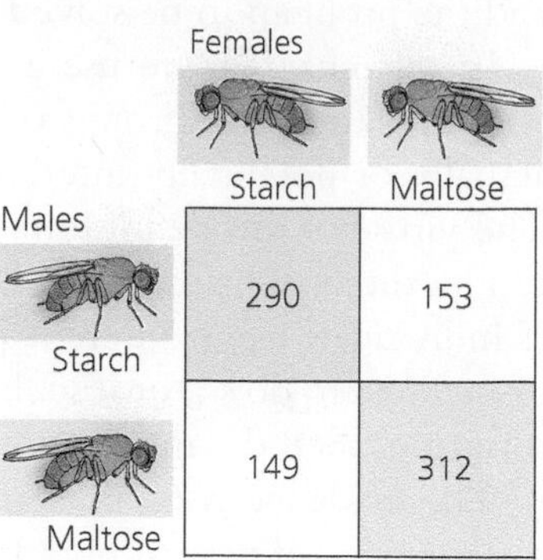

Figure 2.13 Mating preferences of fruit flies from populations adapted to different diets Boxes the number of matings between each type of couple. Data from Dodd (1989).

Dodd's results, which are typical for experiments of this type, may at first seem somewhat disappointing (Florin and Ödeen 2002). Unlike Duffy's experiment with viruses, laboratory natural selection on insects has generally produced only partial reproductive isolation. On the other hand, barriers to interbreeding may in most organisms accumulate slowly. The evolution of complete reproductive isolation may require hundreds of thousands, if not millions, of years (Coyne and Orr 1997; Gourbière and Mallet 2010). Instead of being disappointed, we should perhaps be impressed that partial isolation evolves fast enough for us to watch.

Evidence from Natural Populations

A lesson from experiments like Dodd's is that speciation is typically not a sudden event, but a gradual process (Nosil et al. 2009). **Figure 2.14a** illustrates a version of the speciation process thought to be common. If it is, then we should find examples in nature of populations in all stages. This we do, as Andrew Hendry and colleagues (2009) illustrate by reviewing research on threespine sticklebacks.

Speciation starts with a single population in which there is variation among individuals. Because variation is ubiquitous, virtually any population could serve as an example. Hendry and colleagues use the stickleback population in Robert's Lake on Vancouver Island, Canada. Among other traits, the fish vary in the relative length of their gill rakers (Figure 2.14b). The rakers form a sieve inside a stickleback's mouth that the fish can use to catch tiny planktonic prey such as copepods. In Robert's Lake, sticklebacks with long rakers for their body size consume a diet richer in copepods. Fish with short rakers eat a higher proportion of non-planktonic prey, such as insect larvae gleaned from the lake bottom.

Observations on living organisms provide direct evidence of speciation by showing that populations can diverge to the point that their individuals can no longer interbreed.

The second stage of speciation involves a population divided into readily distinguishable subpopulations that nonetheless still interbreed. David Berner and colleagues (2008, 2009) found such a situation when they compared the sticklebacks in Robert's Lake to the fish living in Robert's Creek, which drains the lake. The creek fish are larger than the lake fish (Figure 2.14c). The lake fish have relatively longer gill rakers and draw a larger portion of their diet from plankton. However, the transition in body types between lake fish versus creek fish is smooth rather than abrupt. Genetic analysis confirms that lake fish and creek fish sometimes mate with each other.

The third stage of speciation features distinct populations with limited interbreeding. Paxton Lake on Texada Island, Canada, is home to two kinds of

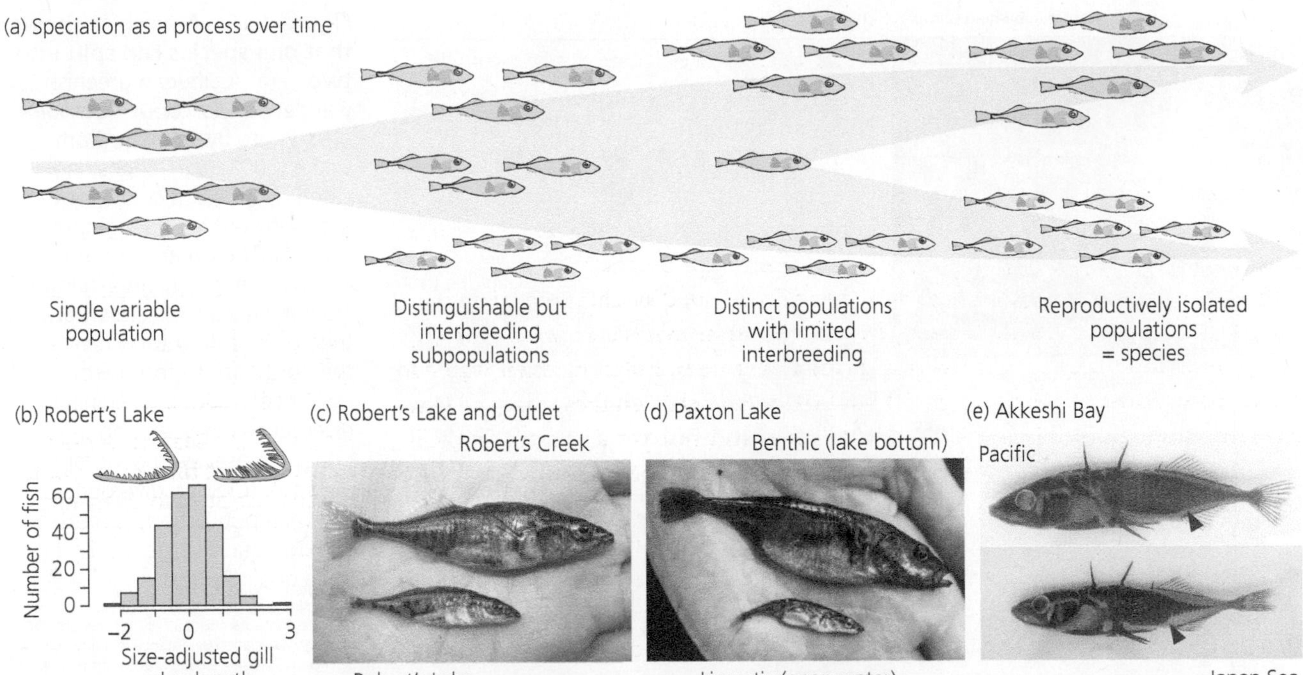

Figure 2.14 Threespine sticklebacks illustrate stages of speciation (a) Speciation is typically a gradual process. After Hendry et al. (2009) and Nosil et al. (2009). (b) Variation in gill raker length among sticklebacks in Robert's Lake. Diagram from Berner et al. (2008); graph redrawn from Hendry et al. (2009). (c) Individual sticklebacks representing subpopulations from Robert's Lake and Robert's Creek. Photo by Daniel Berner; see Berner et al. (2009), Hendry et al. (2009). (d) Individuals representing the two populations that share Paxton Lake. Photo by Todd Hatfield; from Schluter (2010). (e) Individuals representing the two species that share Akkeshi Bay. From Kitano et al. (2007).

sticklebacks (see Schluter 2010). One kind, called the benthic form, is large with short gill rakers (Figure 2.14d). It specializes on lake-bottom invertebrates. The other kind, the limnetic form, is small with long gill rakers. It specializes on plankton. Genetic studies show that limnetics and benthics mate with each other on occasion. Their hybrid offspring, though viable and fertile, compete poorly with purebred forms. Some experts consider the two forms to be different species.

Speciation ends with distinct populations whose reproductive isolation is irreversible. This state is rare in threespine sticklebacks, but it has been achieved by the two types that inhabit Akkeshi Bay, Japan (Kitano et al. 2007). One is the same type that occurs elsewhere in the Pacific Ocean; the other is found only in the Japan Sea (Figure 2.14e). The two types show similar lifestyles and diets, but clear differences in appearance—including different patterns of armor plating on their tails—and in mating behavior. They also exhibit clear genetic differences, including different sex chromosomes (Kitano et al. 2009). The two forms hybridize only rarely, and when they do, the male offspring are sterile. Akkeshi Bay's two sticklebacks meet Mayr's definition of species.

The fact that biologists have found threespine stickleback populations in all four stages is consistent with the hypothesis that speciation is constantly taking place in nature. Many other organisms show similar patterns. In a few cases, populations show all four stages in a geographically continuous loop. These, known as ring species, serve as our final piece of evidence for speciation.

Ring species offer particularly compelling evidence that one species can split into two. Our example is the Siberian greenish warbler (*Phylloscopus trochiloides*,

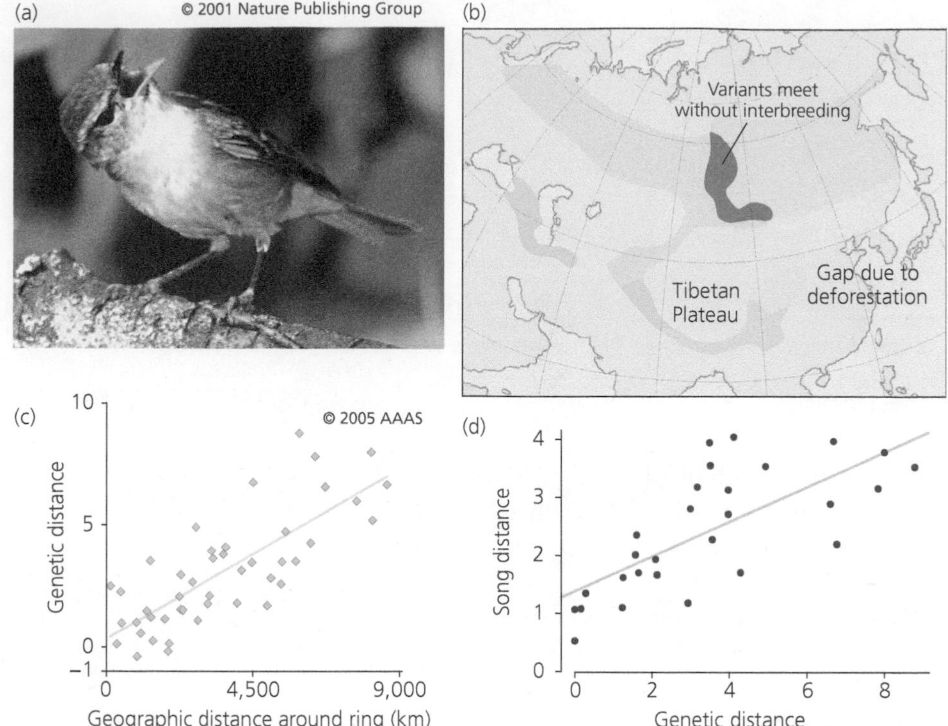

Figure 2.15 A warbler shows that one species can split into two (a) A Siberian greenish warbler (*Phylloscopus trochiloides*). Photo by D. Irwin, from Wake (2001). (b) Ranges, indicated by different colors, of the greenish warbler's geographic variants. The variants interbreed everywhere they meet around the Tibetan Plateau, except where the northwestern form meets the northeastern form. From Irwin et al. (2005). (c) Genetic divergence between populations increases with geographic distance. (d) Song divergence between populations increases with genetic divergence. From Irwin et al. (2008).

(a) Reprinted by permission from Macmillan Publishers Ltd: *Nature* 409: 299–300, copyright 2001. (c) From "Speciation by distance in a ring species." *Science* 307: 414–416. Reprinted with permission from AAAS.

Figure 2.15a), whose range forms a ring around the Tibetan Plateau (Figure 2.15b). Although their songs increase in complexity from south to north around both sides of this loop, the birds view each other as conspecifics—and interbreed—everywhere they meet (Irwin et al. 2001; Wake 2001). The exception is in central Siberia, where the northeastern and northwestern forms decline to mate.

Darren Irwin and colleagues (2005) present genetic evidence that there are no other biological boundaries, aside from the one in central Siberia, between one form of greenish warbler and another. All belong to a single population that circles around on itself. However, genetic differences among the birds increase with geographic distance (Figure 2.15c). And their songs, crucial in mating rituals, diverge with genetic distance (Figure 2.15d; Irwin et al. 2008). Irwin argues that today's greenish warblers are descended from a southern population that expanded northward in two directions. By the time the two fronts met generations later, the birds were modified enough to be mutually disinterested in romance.

Greenish warblers show that over space and time, a species can gradually divide into two. For another well-documented ring species, the salamander *Ensatina eschscholtzii*, see Kuchta et al. (2009) and Pereira and Wake (2009).

Why Speciation Matters

Siobain Duffy's study of host switching in phage $\phi 6$ shows the relevance of speciation to humans. Nathan Wolfe and colleagues (2007) review the origins of 25 major human diseases. Several are caused by pathogens that originated in other animals, switched hosts, and evolved into new species that infect only humans. Diphtheria is likely derived from a disease of domesticated herbivores (Martin et al. 2003), measles from a disease of cattle (Westover and Hughes 2001), and smallpox from a virus of rodents or camels (Hughes et al. 2010). For early warning of emerging epidemics, Wolfe and colleagues advocate better surveillance of pathogens infecting people who are in regular contact with wild animals.

2.3 Evidence of Macroevolution: New Forms from Old

Figure 2.16 An extinct deer

The evidence we presented for microevolution and speciation came from experiments and observations on living organisms. Now, as we consider whether new life-forms descend from old, we add a new source of evidence: fossils. A **fossil** is any trace of an organism that lived in the past. The worldwide collection of fossils is called the **fossil record**.

The simple fact that fossils exist, and that most of them preserve forms unlike species alive today, indicates that life has changed over time. Here we focus on this and more specific ways the fossil record offers evidence of macroevolution.

Extinction and Succession

In 1801, comparative anatomist Georges Cuvier published a list of 23 species known only from fossils. His point was to challenge the hypothesis that unusual forms in the fossil record would eventually be found alive, once European scientists had visited all parts of the globe. It was unlikely, Cuvier argued, that mastodons and other enormous creatures still lived but had escaped detection.

Controversy over the fact of extinction ended after 1812, when Cuvier published his analysis of the Irish elk (**Figure 2.16**). Fossils of this huge ice-age deer had been found across northern Europe and the British Isles. Other scientists had suggested that the elk belonged to a living species, such as the American moose or the European reindeer. These claims were more reasonable than they might seem today. Specimens of exotic animals, or even reliable descriptions, were scarce. Cuvier's anatomical analysis showed that the Irish elk was neither a moose, nor a reindeer, nor a member of any other extant species (Gould 1977). Subsequent centuries of paleontological research have revealed the extinction of the vast majority of the species that have lived on Earth (Erwin 2008).

An early 18th-century paleontologist named William Clift was the first to publish a related observation that was later confirmed and elaborated by Darwin (Darwin 1859; Dugan 1980; Eiseley 1958). Fossil and living organisms in the same geographic region resemble each other, but are distinct from organisms found in other areas (**Figure 2.17**). Clift studied the extinct mammals of Australia and noted that they were marsupials, similar to creatures alive in Australia today. Darwin analyzed the fossil glyptodonts he excavated in Argentina and noted their correspondence to the living armadillos found there now. The mammalian faunas of the two continents are markedly different, yet each continent's extant fauna is strikingly similar to the continent's recent fossil forms. The general pattern of correspondence between fossil and living forms from the same locale came to be known as the **law of succession**.

Extinction and succession are the patterns we would predict if present-day species are descended with modification from ancestors that lived before them in the same region.

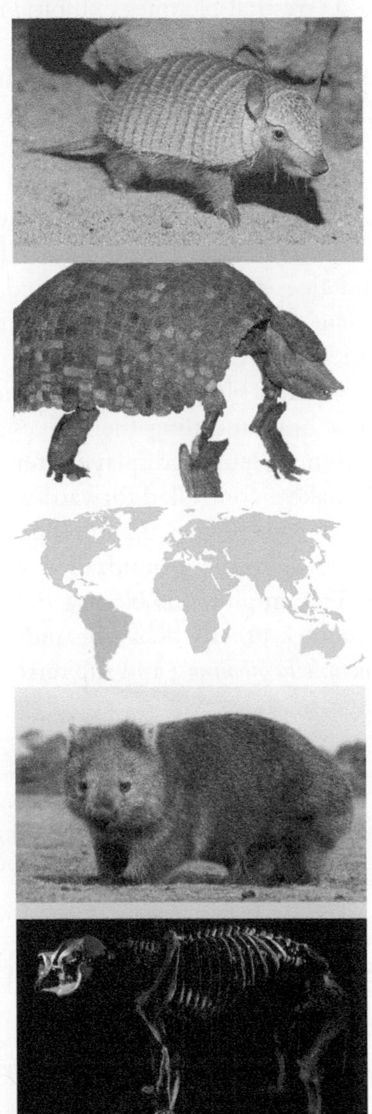

Figure 2.17 Succession The pygmy armadillo (top) is similar to fossil glyptodonts. Wombats in Australia (bottom) resemble the extinct marsupial *Diprotodon* (Dugan 1980).

Transitional Forms

If novel life-forms are, indeed, descended with modification from earlier forms, then the fossil record should capture evidence of transmutations in progress. We should find transitional species showing a mix of features, including traits typical of ancestral populations and novel traits seen later in descendants. Before discussing fossil **transitional forms**, it will be helpful to consider a living example.

A Living Transitional Form

Before meeting this creature, we must visit another. The Pacific leaping blenny (*Alticus arnoldorum*) is a terrestrial fish **(Figure 2.18a)**. Leaping blennies spend their larval phase in the plankton, but then leave the ocean and move into the supratidal zone—the wave-splashed area just above the high tide line. There they feed on algae scraped from rocks, claim territories around moisture-trapping tunnels, court mates, lay eggs, and defend their nests (Bhikajee and Green 2002; Bhikajee et al. 2006). S. Tonia Hsieh (2010) reports that when knocked into the water, the fish immediately jump back out. They never voluntarily go for a swim.

Terrestrial blennies exhibit a number of traits that help them survive on land. They breathe air through both their gills and skin (Martin and Lighton 1989; Brown et al. 1992). They can climb up a vertical sheet of plexiglass (Hsieh 2010). They can move across the rocks at speeds of more than 1.5 meters per second.

To learn just how they manage this last feat, Hsieh (2010) shot high-speed video of *Alticus arnoldorum* and two other blennies. These others, commonly known as rockskippers, can likewise breathe air but spend much less time out of water. *Praealticus labrovittatus* feeds on land at low tide, but stays close to the water and flees there when disturbed. Hsieh considers it amphibious. *Blenniella gibbifrons* typically appears on land only between waves. Hsieh considers it aquatic. Hsieh filmed all three blennies moving in air along a damp wooden track.

Aquatic blennies that find themselves out of water curl their tail around to their head and fling themselves into the air; but once airborne, they flail. Amphibious blennies display better coordination. They curl their tail, then push off to make a controlled forward hop. Terrestrial blennies, however, are in a class by themselves (Figure 2.18b). They curl their tail, twist it so as to press their fanned tail fin onto the ground, then vault forward up to three body lengths at a time.

The amphibious blenny *Praealticus labrovittatus* is our living transitional form **(Figure 2.19)**. Its behavior and physiology place it between the aquatic blenny *Blenniella gibbifrons* and the terrestrial blenny *Alticus arnoldorum*.

Figure 2.18 A terrestrial fish (a) A Pacific leaping blenny (*Alticus arnoldorum*) on a damp wooden racetrack in the lab. (b) Time series showing a single leap. The white spot is a fixed point. All photos by S. Tonia Hsieh.

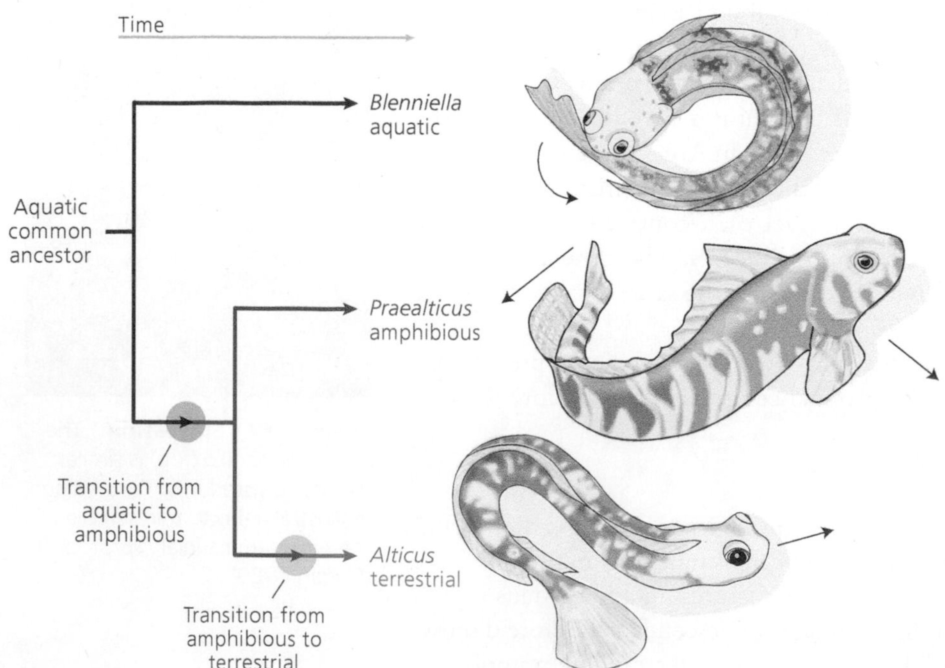

Figure 2.19 Amphibious blennies represent a transitional form in the inferred history of speciation and macroevolution leading to terrestrial blennies Horizontal lines represent lineages. Splits in these lines represent speciation events. Circles mark the evolution of novel traits. After Hsieh (2010).

By calling the amphibious blenny a transitional form, we are not claiming that it is a descendant of the aquatic blenny or the ancestor of the terrestrial blenny. These claims would be problematic, given that all three species are alive today.

Instead, we are claiming that the distribution of traits among the blennies is consistent with the hypothesis that all three are derived from a common ancestor. This ancestor was probably, like most blennies, aquatic. At some point the ancestral lineage split—that is, it underwent speciation. One of its daughter lineages remained aquatic and is represented today by *Blenniella*. The other evolved novel traits, including coordinated hopping, that made it amphibious. Later, the amphibious lineage also split. One of its daughter lineages remained amphibious and is represented today by *Praealticus*. The other evolved additional novel traits, including tail twisting, that made it terrestrial. It is represented today by *Alticus*.

What makes the amphibious blenny *Praealticus* a transitional form? It is derived from, and thus represents, a lineage that had evolved some, but not all, of the novel traits that transform an aquatic blenny into a terrestrial one. It shows that an intermediate species, with only some of these traits, is viable. And it indicates that coordinated hopping evolved before tail twisting.

A Fossil Transitional Bird

Few transitional fossils were known in Darwin's day, and he argued that they should be rare. The fossil record has grown, however, and many have been found.

The most famous transitional form is *Archaeopteryx* **(Figure 2.20a)**, discovered shortly after Darwin published *On the Origin of Species* (see Christiansen

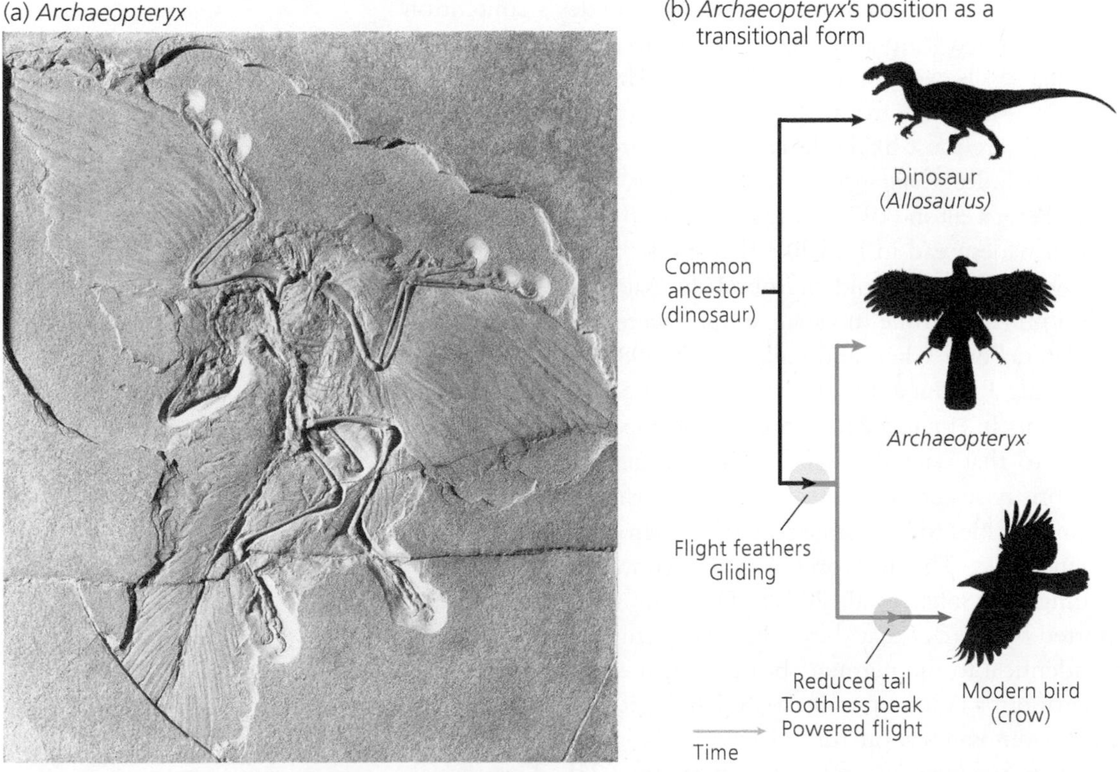

Figure 2.20 A bird with a dinosaur's skeleton (a) *Archaeopteryx* had flight feathers like a modern bird's and a dinosaur-like skeleton with teeth and a long tail. Museum für Naturkunde. (b) The distribution of traits in dinosaurs, birds, and *Archaeopteryx* is consistent with the idea that they share an ancestor. (Phylogeny simplified from Lloyd et al. 2008; Hu et al. 2009. *Archaeopteryx* reconstruction after Longrich 2006.)

and Bonde 2004). *Archaeopteryx* was a crow-sized animal that lived 145 to 150 million years ago in what is now Germany. It sported essentially modern flight feathers. At least one specimen preserves some of the feathers' original chemistry (Bergmann et al. 2010). *Archaeopteryx* appears to have been adept at gliding, if not at rudimentary powered flight (Longrich 2006; Nudds and Dyke 2010). *Archaeopteryx*'s feathers and aeronautical abilities identify it as a bird (Padian and Chiappe 1998; Alonso et al. 2004). The creature's skeleton, however, is so reptilian—with teeth, three-clawed hands, and a long, bony tail—that specimens have been mistaken for remains of the dinosaur *Compsognathus* (see Wellnhofer 1988).

Among the first to note the skeletal similarities of dinosaurs and birds was Darwin's friend and champion Thomas Henry Huxley (1868). Huxley suggested that *Archaeopteryx*, with its mixture of traits, documents an evolutionary transition.

Archaeopteryx was not in the direct line of descent from dinosaurs to modern birds (Hu et al. 2009). Instead, *Archaeopteryx* shows a combination of traits consistent with the hypothesis that it shared a common ancestor with both kinds of animals (Figure 2.20b). *Archaeopteryx* indicates that birds are derived from dinosaurs. And it reveals that birds evolved their birdness piecewise. Feathers came first, followed by the skeletal and muscular modifications associated with modern powered flight (Garner et al. 1999).

Transitional Forms Allow Predictive Tests of Evolutionary Hypotheses

Transitional forms give us a way to test specific hypotheses about macroevolution by making predictions that we can confirm or refute by digging for fossils. In this way, transitional forms offer their most powerful evidence for evolution.

Archaeopteryx serves as the starting point for an example. Huxley's contention that *Archaeopteryx* was descended from a lineage of dinosaurs led to a prediction that additional transitional fossils eventually would be unearthed.

John Ostrom (1973, 1974) argued, from detailed anatomical analyses, that the dinosaurs from which birds are most likely derived were theropods—a group of bipedal carnivores that includes *Compsognathus*, *Velociraptor*, and *Tyrannosaurus rex*. Robert Bakker and Peter Galton (1974) noted that if Ostrom is correct, then "feathers may have been widespread in bird-like theropods." The undiscovered fossil record, in other words, should hold theropod dinosaurs with feathers in various stages of evolution. At the time, no such animals were known.

Transitional fossils document the past existence of species displaying mixtures of traits typical of distinct groups of organisms. Sometimes transitional forms are predicted before they are found, allowing biologists to test hypotheses about macroevolution.

Some two decades later, Pei-ji Chen and colleagues (1998) reported the discovery, in China's Liaoning Province, of a theropod called *Sinosauropteryx*. The fossils, one of which appears in **Figure 2.21a**, were the most exquisitely preserved dinosaur remains found to that date (Unwin 1998). *Sinosauropteryx*, about the size of a chicken, bore bristly structures on its neck, back, flanks, and tail. Many paleontologists took these bristles to be simple feathers (Chen et al. 1998; Unwin 1998; Currie and Chen 2001). This interpretation was controversial (see Geist and Feduccia 2000; Lingham-Soliar et al. 2007). That the bristles are, indeed, feathers is now supported by the fact that they contain structures, revealed under electron microscopy, identical to the pigment-bearing organelles (melanosomes) in the feathers of modern birds (Vinther et al. 2008; Zhang et al. 2010b).

Many more feathered dinosaurs soon turned up (see Norell and Xu 2005). Among the most striking is *Similicaudipteryx*, shown in Figure 2.21b. This animal, which grew to about the size of a goose, wore vaned feathers on both its tail and forelimbs (He et al. 2008; Xu et al. 2010). Unlike the flight feathers of modern birds, *Similicaudipteryx*'s feathers are symmetrical.

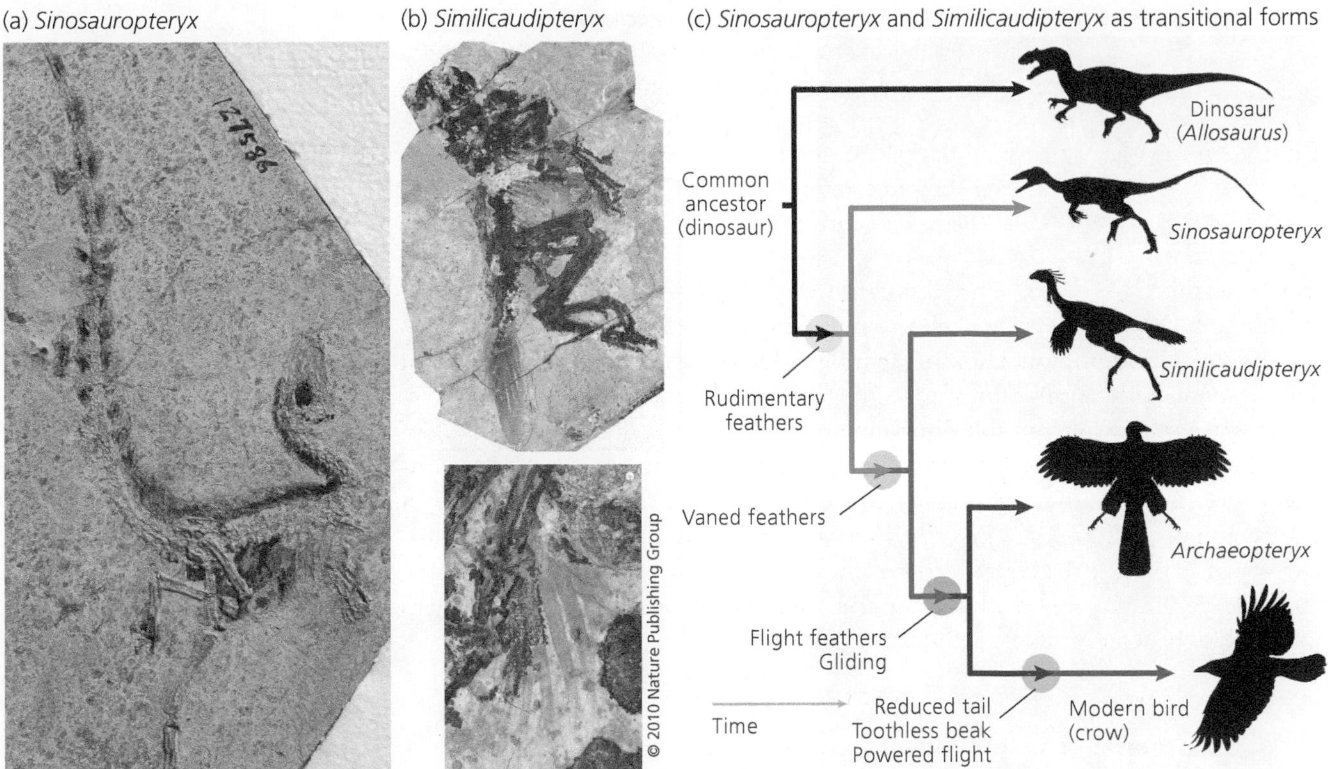

Figure 2.21 Dinosaurs with feathers (a) *Sinosauropteryx* had rudimentary feathers on its neck, back, flanks, and tail. From Chen et al. (1998). (b) *Similicaudipteryx* had vaned feathers on its tail (top) and forelimbs (bottom). The specimen shown here is a young juvenile. From Xu et al. (2010). (c) The distribution of traits among the species shown is consistent with the hypothesis that birds evolved from dinosaurs. (Evolutionary tree simplified from Lloyd et al. 2008; Hu et al. 2009.)

(b) Reprinted by permission from Macmillan Publishers Ltd: *Nature* 464: 1338–1341, © 2010.

Sinosauropteryx and *Similicaudipteryx* confirm the prediction that paleontologists would find theropod dinosaurs with transitional feathers (Figure 2.21c). That feathers occur in well over a dozen species ranging from dromaeosaurs to tyrannosaurs confirms the prediction that feathers would turn out to be widespread in theropods (Norell and Xu 2005; Xu and Zhang 2005; Zhang et al. 2008; Hu et al. 2009; Xu et al. 2010). The various theropod feathers now known match intermediate stages predicted by a model of feather evolution based on how feathers develop in extant birds (Prum and Brush 2002; Xu et al. 2009; McKellar et al. 2011). On this and other evidence, there is little doubt that birds are descended with modification from dinosaurs (Prum 2002; Padian and de Ricqlès 2009).

Indeed, it has become difficult to say just what a bird is and what distinguishes it from an ordinary theropod. It used to be easy: If it had feathers, it was a bird. Under that definition, however, *Sinosauropteryx*, *Similicaudipteryx*, and even some tyrannosaurs were birds (Xu et al. 2004). A more restrictive, but reasonable, definition is that if it has feathers and can fly, or if it is descended from an animal that had feathers and could fly, then it is a bird. Even by this criterion, it may turn out that dromaeosaurs like *Velociraptor*, a predator much loved by movie makers, was a bird (see Turner et al. 2007; Lloyd et al. 2008; Hu et al. 2009).

A Transitional Turtle

Another transitional fossil appears on the first page of the chapter. To understand its significance, it helps to know a bit about how turtles are put together. One of

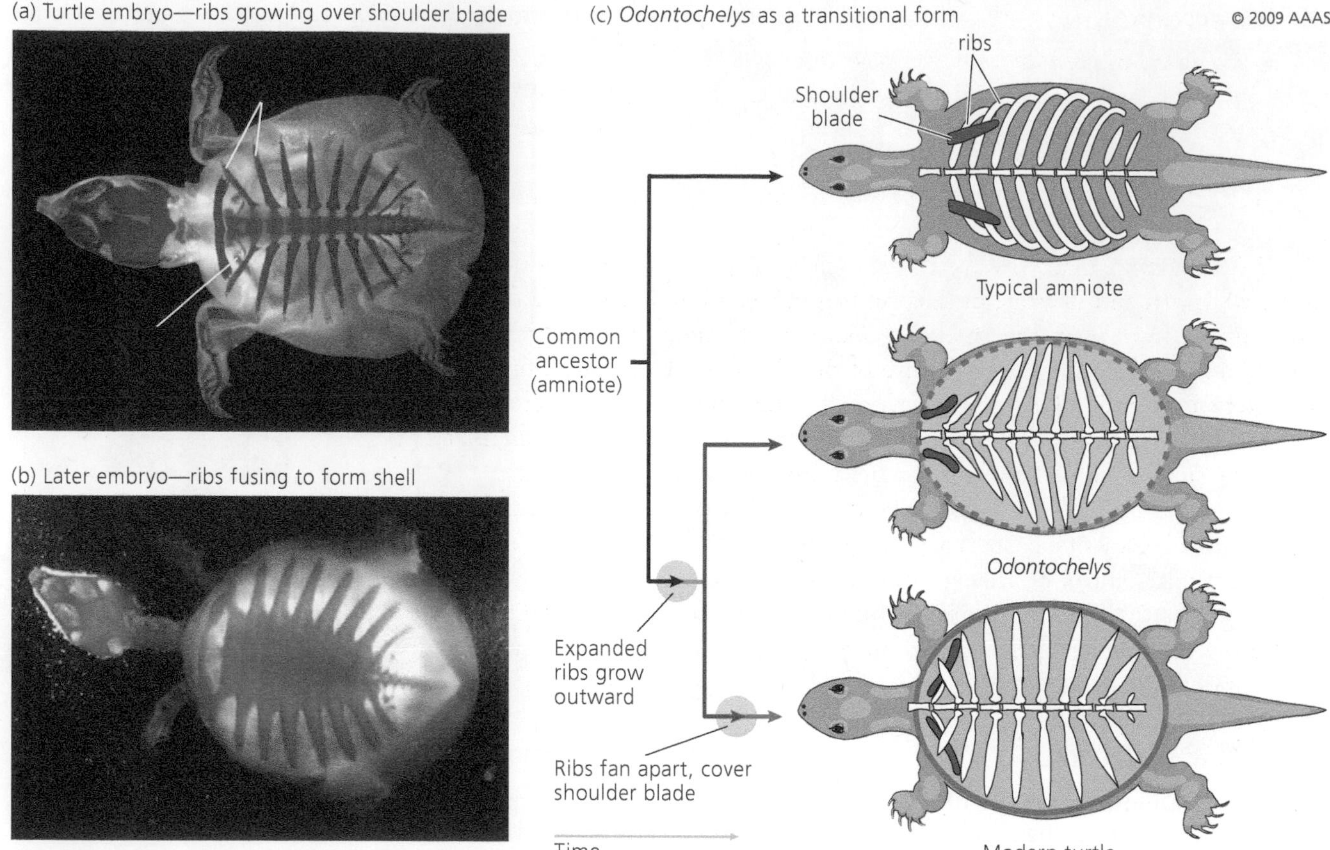

Figure 2.22 How the turtle got its shell (a) Embryo showing ribs growing over shoulder blade. From Sanchez-Villagra (2009). (b) Later embryo showing ribs fusing to form shell. Courtesy of the RIKEN Center for Developmental Biology. (c) *Odontochelys* (shown on first page of chapter) represents an intermediate stage in shell evolution. Redrawn from Nagashima et al. (2009). For more on the relationship between turtles and other amniotes, see Lyson et al. (2010).

(c) From "Evolution of the turtle body plan by the folding and creation of new muscle connections." *Science* 325: 193–196. Reprinted with permission from AAAS.

the key events during development that makes a turtle a turtle has been detailed Hiroshi Nagashima and colleagues (2009).

Turtles belong to the amniotes—a group that includes mammals, reptiles, and birds. In a typical amniote, the ribs follow the curve of the body wall as they travel outward from the spine. As a result, the ribs pass under the shoulder blade. In turtles, the ribs ignore the curving body wall **(Figure 2.22a)**. As the ribs travel outward they fan apart, pushing the body wall in front of them and forcing it to fold back on itself. This allows the ribs to pass over the shoulder blade. Later, the ribs expand and fuse to form the shell on the turtle's back (Figure 2.22b).

Nagashima and colleagues argue that *Odontochelys*, the fossil on the first page of the chapter, represents a predictable intermediate step in the transmutation of a typical amniote into a turtle (Figure 2.22c). *Odontochelys* has expanded ribs that grow outward, but do not fan apart and thus do not travel over the shoulder blade. *Odontochelys* thus helps document the descent with modification of turtles from a typical amniote ancestor (see also Burke 2009; Lyson et al. 2010).

We have presented just a few transitional fossils, but paleontologists have found a great many more (see Luo 2007; Prothero 2008). We will encounter some of them—including intermediates between fish and tetrapods and between great apes and humans—in later chapters.

Why Macroevolution Matters

Macroevolution matters in our everyday lives because our own bodies are its products. As we will see in the next section, our deep ancestry traces back to fish and beyond. Otherwise puzzling aspects of our anatomy and physiology begin to make sense when viewed in an evolutionary context (Shubin 2009).

Consider hiccups. Most of us think of hiccups as an annoying neurological quirk. A hiccup, however, is a highly coordinated event (Straus et al. 2003). It begins with a sharp intake of breath caused by a strong activation of the muscles in the neck ribs and diaphragm that control inhalation and a simultaneous inhibition of the muscles in the chest and abdomen that control exhalation. This is quickly followed by the active closure of the glottis—which causes the hiccup's characteristic sound. These activities are all under the control of neural circuits in the brain stem. Proposed explanations for why we exhibit this peculiar ability, which we share with other mammals, are numerous and controversial.

Christian Straus and colleagues (2003) argue that the best explanation is that hiccups are a legacy of our macroevolutionary past. We inherited the neurological circuits that control breathing—and hiccups—from our distant ancestors (see Vasilakos et al. 2005). These include amphibians that breathed with gills as juveniles and with lungs as adults. When modern tadpoles breathe water, they pump it across their gills while closing the glottis to keep it out of their lungs.

We may have retained the capacity to hiccup because the neural circuits involved have been repurposed for the control of vital functions such as suckling or normal breathing.

2.4 Evidence of Common Ancestry: All Life-Forms Are Related

Our evidence for macroevolution necessarily included some evidence for common ancestry. The amphibious blenny *Praealticus*, for example, connects extant aquatic and terrestrial blennies not in an ancestor–descendant relationship but as related descendants of a shared ancestor.

The theory of descent with modification ultimately connects all organisms to a single common ancestor. Humans, butterflies, lettuce, and bacteria all trace their lineages back to the same primordial stock. The crucial evidence for universal common ancestry is homology.

Homology

As the fields of comparative anatomy and comparative embryology developed in the early 1800s, one of the most striking results to emerge was that fundamental similarities underlie the obvious physical differences among species. Early researchers called the phenomenon **homology**—literally, the study of likeness. Richard Owen, Britain's leading anatomist, defined homology as "the same organ in different animals under every variety of form and function."

Structural Homology

A famous example of homology comes from work by Owen and by Georges Cuvier, the founder of comparative anatomy. They described extensive similarities among vertebrate skeletons and organs. Referring to Owen and Cuvier's work, Darwin (1859, p. 434) wrote:

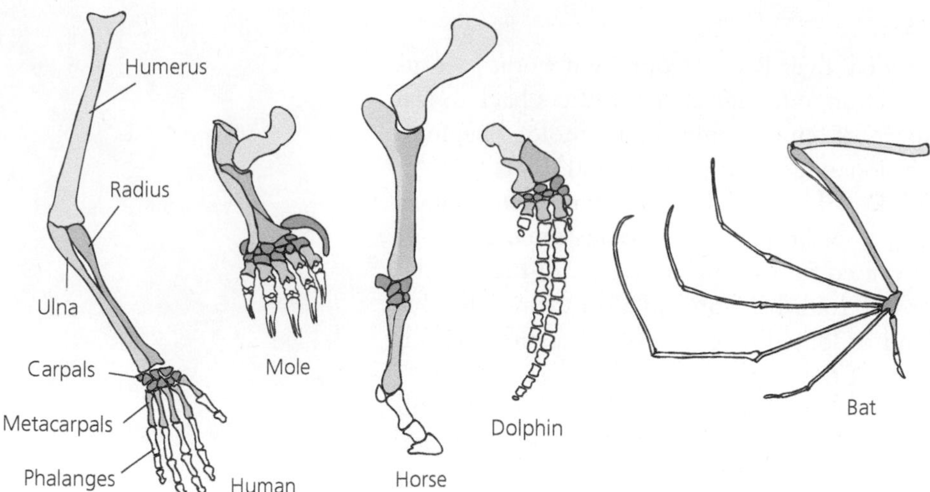

Figure 2.23 Structural homologies These vertebrate forelimbs are used for different functions, but have the same sequence and arrangement of bones. In this illustration, homologous bones are colored in the same way and are labeled on the human arm.

> What could be more curious than that the hand of a man, formed for grasping, that of a mole for digging, the leg of the horse, the paddle of the porpoise, and the wing of the bat, should all be constructed on the same pattern, and should include the same bones, in the same relative positions?

His point was that the underlying design of these vertebrate forelimbs is similar, even though their function and appearance are different **(Figure 2.23)**.

This makes the similarity in design among vertebrate forelimbs different from, say, that between a shark and a whale **(Figure 2.24)**. Both shark and whale have a streamlined shape, short fins or flippers for steering, and a strong tail for propulsion. These similarities in form make sense in view of their function: fast movement in water. Human engineers use the same features in watercraft. In contrast, the internal similarity between forelimbs with radically different functions seems arbitrary. Would an engineer design tools for grasping, digging, running, swimming, and flying using the same structural elements in the same arrangement?

Darwin himself (1862) analyzed the anatomy of orchid flowers and showed that, despite their diversity in shape and in the pollinators they attract, they are constructed from the same set of component pieces. Like vertebrate forelimbs, the flowers have the same parts in the same relative positions **(Figure 2.25)**.

Figure 2.24 Nonhomologous similarities This shark and Orca both have streamlined shapes, powerful tails, and short fins or flippers, even though one is a fish and the other a mammal. These similarities all make sense in the context of their function and are not homologous.

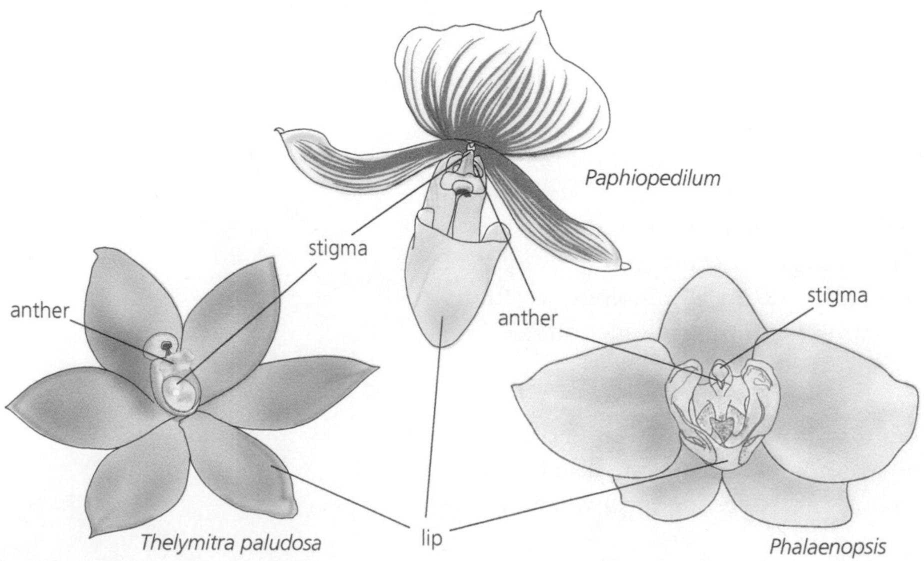

Figure 2.25 More structural homologies Orchid flowers are diverse in size and shape, but are comprised of elements that are similar in structure and orientation. After Darwin (1862).

What causes these similarities in construction despite differences in form and function? Darwin argued that descent from a common ancestor is the most logical explanation. He argued that the orchids in Figure 2.25 are similar because they share a common ancestor. Likewise, the tetrapods in Figure 2.23 have similar forelimbs because they are descended from a single lineage, from which they inherited the fundamental design of their appendages.

Using Homology to Test the Hypothesis of Common Ancestry

We can use homologous traits shared among species to test Darwin's hypothesis of common ancestry. We will show the logic using evolutionary novelties shared among imaginary snail species derived with modification from a single lineage.

Figure 2.26a shows the evolutionary history. The common ancestor is the lineage of squat-shelled blue snails at far left. This lineage underwent speciation (1). One of the daughter lineages persisted to the present with no further changes in its shell (2). The other lineage evolved elongated shells (3). The lineage with elongated shells underwent speciation (4). One daughter lineage evolved bands on its shell (5), then persisted to the present with no further changes (6). The other daughter evolved pink shells (7), then split (8). One daughter lineage evolved high-spired shells (9). The other persisted with no further changes (10). The high-spired lineage split (11). One daughter lineage persisted with no further changes (12). The other evolved spikes on its shell (13), then persisted with no further changes (14). These events yielded the five extant species at far right.

Figure 2.26b shows the novel shell traits shared by the four species that exhibit them. Note that these traits are shared in a nested pattern. The species with spikes is nested within the set of species with high spires. The set of species with high spires is nested within the set of species with pink shells. And the set of species with pink shells is nested within the set of species with elongated shells.

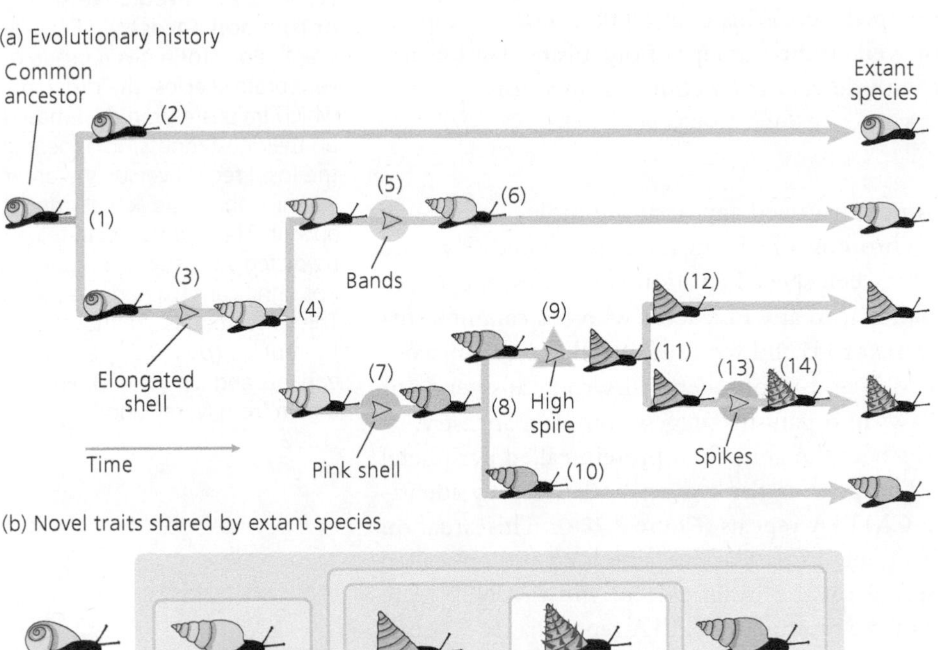

Figure 2.26 Descent with modification produces nested sets of shared traits (a) The evolutionary history of a suite of hypothetical snail species. See text for explanation of numbers. (b) The novel traits shared by the extant species.

Our hypothetical snails demonstrate that the theory of descent with modification from common ancestors makes a prediction. Extant organisms should share nested sets of novel traits.

And, indeed, they do. For example, humans are nested within the apes—a group of species that have large brains and no tails. The apes, in turn, are nested within the primates—which have grasping hands, and feet, with flat nails instead of claws. The primates are nested within the mammals—defined by hair and feeding milk to their young. The mammals are nested within the tetrapods, the tetrapods within the vertebrates, and so on. The nested pattern of traits shared among extant species thus confirms a prediction of Darwin's theory.

But we can go further. Look again at Figure 2.26 and compare part (b) to part (a). Notice that the most deeply nested sets are defined by traits, such as spikes, that evolved relatively late. If we start with one of these sets and work our way out across the progressively larger sets that enclose it, we encounter additional traits that evolved ever earlier in time. Spikes were preceded by high spires. High spires, in turn, were preceded by pink shells. And pink shells were preceded by elongated shells. Even if we had only the five extant species and did not know their evolutionary history, we could still use the nesting of the traits they share to predict the order in which the traits should appear in the fossil record. We could then check the fossil record to see if our prediction is correct.

Mark Norell and Michael Novacek (1992) performed such tests on two dozen groups of vertebrates. Representative results appear in **Figure 2.27**. In six cases, such as the duck-billed dinosaurs, there was no significant correlation between the predicted order in which traits arose versus the actual order (Figure 2.27a). However, in the other 18 cases, including the reptiles and the elephants and kin, the correlation was significant or strongly so (Figure 2.27b and c).

More sophisticated methods of assessing the correspondence between trait-based reconstructions of evolutionary history versus the order traits appear in the fossil record have since been developed (see Wills et al. 2008). The correspondence is generally high, at least for well-studied groups of organisms that fossilize readily. This pattern is consistent with descent from common ancestors.

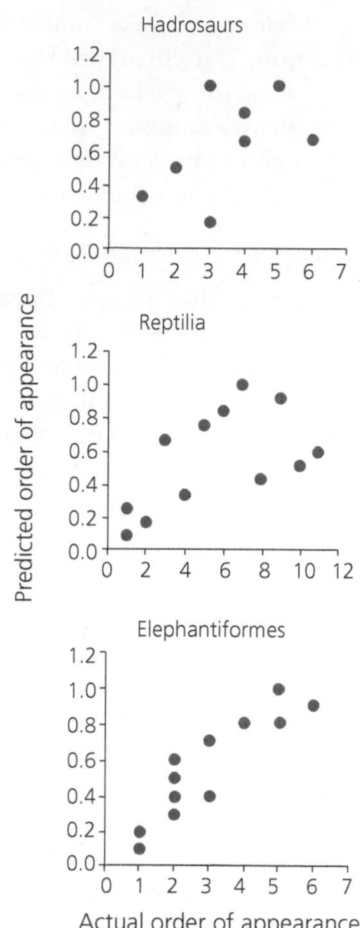

Figure 2.27 Predictive tests of common ancestry Each graph plots, for a given group of vertebrate species, the order in which traits are predicted, based on their nestedness, to appear in the fossil record versus the order in which the traits actually do appear. The correlation between predicted versus actual order is not significant for hadrosaurs. The correlation is significant for reptiles ($p < 0.05$) and elephants and kin ($p < 0.05$). Redrawn from Norell and Novacek (1992).

Molecular Homology

Curious similarities unrelated to functional need appear at the molecular level as well. Consider a genetic flaw on chromosome 17 in humans. Shared flaws are especially useful in distinguishing between special creation versus descent from a common ancestor. The reason is familiar to any instructor who has caught a student cheating on an exam. If A sat next to B and wrote identical correct answers, it tells us little. But if A sat next to B and wrote identical wrong answers, our suspicions rise. Likewise, shared flaws in organisms suggest common ancestry.

The flaw on chromosome 17 sits near the gene for a protein called peripheral myelin protein-22, or PMP-22. The gene is flanked on both sides by identical sequences of DNA, called the CMT1A repeats **(Figure 2.28a)**. This situation arose when the distal repeat, which contains part of the gene for a protein called COX10, was duplicated and inserted on the other side of the PMP-22 gene (Reiter et al. 1997). The presence of the proximal CMT1A repeat has to be considered a genetic flaw because it occasionally lines up with the distal repeat during meiosis, resulting in unequal crossing over (Figure 2.28b; Lopes et al. 1998). Among the products are a chromosome with two copies of PMP-22 and a chromosome that is missing the PMP-22 gene altogether. If either of these abnormal

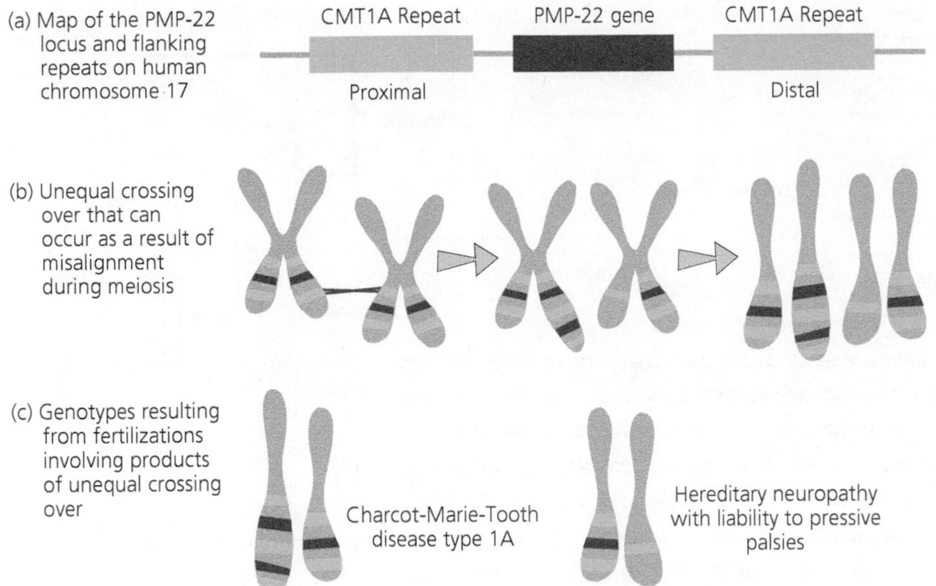

Figure 2.28 A genetic flaw that humans share with chimpanzees (a) The proximal CMT1A repeat, near the gene for PMP-22, is a duplication of the distal repeat on the other side of the gene. (b) The proximal repeat can align with the distal repeat during meiosis, resulting in unequal crossing over. (c) The genotypes that result from the unequal crossing over are associated with neurological disorders.

chromosomes participates in a fertilization, the resulting zygote is predisposed to neurological disease (Figure 2.28c). Individuals with three copies of PMP-22 suffer from Charcot-Marie-Tooth disease type 1A. Individuals with only one copy of PMP-22 suffer from hereditary neuropathy with liability to pressive palsies.

Motivated by the hypothesis that humans share a more recent common ancestor with the chimpanzees than either humans or chimps do with any other species, Marcel Keller and colleagues (1999) examined the chromosomes of common chimpanzees, bonobos (also known as pygmy chimpanzees), gorillas, orangutans, and several other primates. Both common chimps and bonobos share with us the paired CMT1A repeats that can induce unequal crossing over. The proximal repeat is absent, however, in gorillas, orangutans, and all other species the researchers examined. This result is difficult to explain under the view that humans and chimpanzees were separately created. But it makes sense under the hypothesis that humans are a sister species to the two chimpanzees. All three species inherited the proximal repeat from a recent common ancestor, just as three of the snail species in Figure 2.26 inherited pink shells.

A Predictive Test of Common Ancestry Using Molecular Homologies

Our second example of molecular homology concerns another kind of genetic quirk that might be considered a flaw: **processed pseudogenes**. Before we explain what processed pseudogenes are, note that most genes in the human genome consist of small coding bits, or **exons**, separated by noncoding intervening sequences, or **introns**. After a gene is transcribed into messenger RNA, the introns have to be spliced out before the message can be translated into protein. Note also that the human genome is littered with **retrotransposons**, retrovirus-like genetic elements that jump from place to place in the genome via transcription to RNA, reverse transcription to DNA, and insertion at a new site (see Luning Prak and Kazazian 2000). Some of the retrotransposons in our genome are active and encode functional reverse transcriptase.

Now we can explain that processed pseudogenes are nonfunctional copies of normal genes that originate when processed mRNAs are accidentally reverse transcribed to DNA by reverse transcriptase, then inserted back into the genome

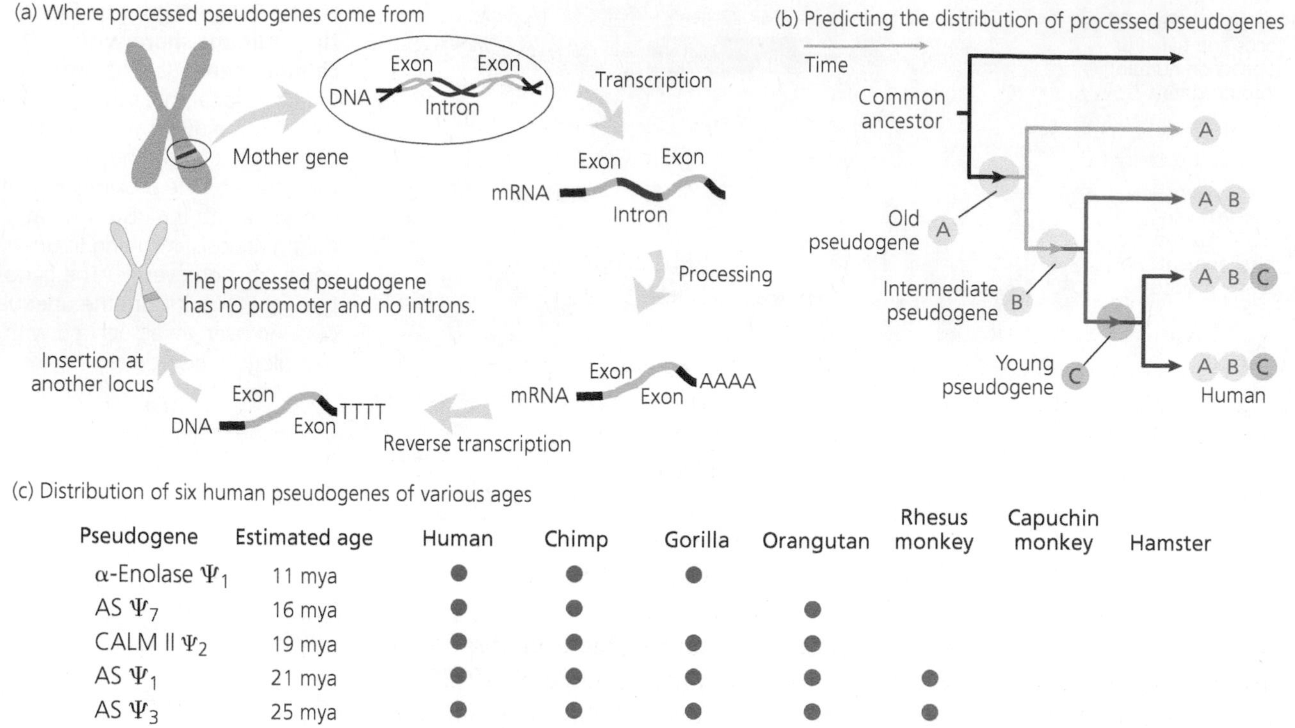

(c) Distribution of six human pseudogenes of various ages

Pseudogene	Estimated age	Human	Chimp	Gorilla	Orangutan	Rhesus monkey	Capuchin monkey	Hamster
α-Enolase Ψ_1	11 mya	●	●	●				
AS Ψ_7	16 mya	●	●		●			
CALM II Ψ_2	19 mya	●	●	●	●			
AS Ψ_1	21 mya	●	●	●	●	●		
AS Ψ_3	25 mya	●	●	●	●	●		
CALM II Ψ_3	36 mya	●	●	●	●	●	●	

Figure 2.29 Processed pseudogenes test the hypothesis of common ancestry (a) Processed pseudogenes arise when processed messenger RNAs are reverse transcribed and inserted into the genome; the mutations they have accumulated indicate their age. (b) If Darwin's hypothesis of common ancestry is correct, then older processed pseudogenes will occur in a broader range of species. (c) The distributions of processed pseudogenes are consistent with this prediction.

at a new location **(Figure 2.29a)**. Processed pseudogenes are readily distinguished from their mother genes because they lack both introns and promoters.

A useful feature of processed pseudogenes is that we can estimate their ages. Because processed pseudogenes have no function, they tend to accumulate mutations. The older a processed pseudogene, the more mutations it will have accumulated. By comparing the sequence of a processed pseudogene with that of its mother gene, we can estimate the number of mutations the pseudogene has accumulated. And from the mutations, we can estimate the pseudogene's age.

We can use the ages of pseudogenes to devise a test of Darwin's view of life's history. If species are related by descent from a common ancestor, then older processed pseudogenes should be shared by a greater variety of species. The logic behind this claim is illustrated in Figure 2.29b. The earlier the ancestor in which a processed pseudogene arose, the more descendant species will have inherited it. Some descendants may have lost the pseudogene by deletion of the entire sequence, but if we examine enough species the overall pattern should be clear.

The theory of descent with modification from common ancestors makes a testable prediction about the distribution of evolutionary novelties among species: They should form nested sets. That they do is evidence of common ancestry.

Felix Friedberg and Allen Rhoads (2000) estimated the ages of six processed pseudogenes in the human genome. The ages ranged from 11 million years to 36 million years. The researchers then looked for the same six processed pseudogenes in the genomes of the chimpanzee, gorilla, orangutan, rhesus monkey, black-capped capuchin monkey, and hamster. The results, shown in Figure 2.29c, are consistent with our prediction. Humans share the youngest of the six pseudogenes only with the African great apes (chimpanzee and gorilla). We share the four pseudogenes of intermediate age with an increasing diversity of primates (although the 16-million-year-old pseudogene appears to have been lost in gorillas). Finally, we share the oldest pseudogene with the African great apes, the Asian great ape (orangutan), the Old World monkey (rhesus), and the New World monkey (capuchin). These processed pseudogenes are molecular homologies, whose distribution among primates is evidence for common ancestry.

Universal Molecular Homologies

The molecular homologies we have discussed so far have been confined to small numbers of species. Advances in molecular genetics have revealed, however, many fundamental similarities among organisms.

Prominent among them is the genetic code. With minor exceptions (Knight et al. 2001), all organisms studied to date use the same nucleotide triplets, or codons, to specify the same amino acids to be incorporated into proteins **(Figure 2.30)**. This is why genetic engineers can, for example, take the gene for green fluorescent protein from a jellyfish, transfer it into the fertilized eggs of a monkey, and get green fluorescent baby monkeys (Yang et al. 2008).

Like the forelimbs in Figure 2.23, the genetic code appears highly evolved (Judson and Haydon 1999). The pattern of codon assignments to amino acids reduces ill effects of point mutations and translation errors (Freeland et al. 2003) and facilitates rapid evolution of proteins by selection (Zhu and Freeland 2006).

Also like the forelimbs, however, many details of the code have clearly arisen as a result of something other than functional necessity. An enormous number of alternative codes are theoretically possible, some of which would work better than the real one (Koonin and Novozhilov 2009; Kurnaz et al. 2010).

Furthermore, having a unique genetic code might offer distinct advantages. For example, if humans used a different genetic code from chimpanzees, we would not have been susceptible to the chimpanzee virus that jumped to humans and became HIV (see Chapter 1). When the virus attempted to replicate inside human cells, its proteins would have been garbled during translation.

If alternative genetic codes are possible, and if using them would be advantageous, then why do virtually all organisms use the same one? Darwin provided a logical answer a century before the genetic code was deciphered: All organisms inherited their fundamental internal machinery from a common ancestor.

First base	RNA Codon (Second base G)	Amino acid	Abbreviation	Third base
U	UGU	Cysteine	C	U
	UGC	Cysteine	C	C
	UGA	Stop		A
	UGG	Tryptophan	W	G
C	CGU	Arginine	R	U
	CGC	Arginine	R	C
	CGA	Arginine	R	A
	CGG	Arginine	R	G
A	AGU	Serine	S	U
	AGC	Serine	S	C
	AGA	Arginine	R	A
	AGG	Arginine	R	G
G	GGU	Glycine	G	U
	GGC	Glycine	G	C
	GGA	Glycine	G	A
	GGG	Glycine	G	G

Figure 2.30 A universal molecular homology: the genetic code In almost every organism, the same nucleotide triplets, or codons, specify the same amino acids to be incorporated into proteins. This chart shows a portion of the code. (The entire code appears in Chapter 5, Figure 5.21.)

The Modern Concept of Homology

Darwin's interpretation of homology has become deeply embedded in biological thinking. So deeply, in fact, that the interpretation has become the definition. Under Owen's definition, homology referred to curious similarity in structure despite differences in function. Now many biologists define homology as similarity due to the inheritance of traits from a common ancestor (Abouheif 1997; Mindell and Meyer 2001).

Why Common Ancestry Matters

Common ancestry is the conceptual foundation upon which all of modern biology, including biomedical science, is built. Because we are descended from the same ancestral lineage as monkeys, mice, baker's yeast, and even bacteria, we share with these organisms numerous homologies in the internal machinery of our cells. This is why studies of other organisms can teach us about ourselves.

Consider work on mice and yeast by Kriston McGary and colleagues (2010) in the lab of Edward Marcotte. The researchers knew that because mice and yeast are derived from a common ancestor, we find not only many of the same genes in both creatures, but many of the same groups of genes working together to carry out biological functions—what we might call gene teams. The scientists thus guessed that a good place to look for genes associated with mammalian diseases would be on mouse gene teams whose members are also teammates in yeast.

Using a database of genes known to occur in both mice and yeast, McGary and colleagues first identified gene teams as sets of genes associated with a particular phenotype. In mice the phenotype might be a disease. In yeast it might be sensitivity to a particular drug. The researchers then looked for mouse and yeast gene teams with overlapping membership.

Among the pairs of overlapping teams they found was a mouse team of eight genes known to be involved in the development of blood vessels (angiogenesis) and a yeast team of 67 genes known to influence sensitivity to the drug lovastatin. These teams formed a pair because of the five genes that belonged to both.

The connection between the two teams suggested that both might be larger than previously suspected, and that more than just five genes might play for both. In particular, the 62 genes from the yeast lovastatin team not already known to belong to the mouse angiogenesis team might, in fact, be members. Starting with this list of 62 candidates, the researchers conducted experiments in frogs revealing a role in angiogenesis for at least five of the genes. Three more genes on the list turned out to have been identified already as angiogenesis genes, but had not been flagged as such in the researchers' database. Eight hits in 62 tries is a much higher success rate than would have been expected had the researchers simply chosen genes at random and tested their influence on angiogenesis.

In other words, McGary and colleagues used genetic data from yeast, an organism with neither blood nor blood vessels, to identify genes in mammals that influence blood vessel growth. Researchers in Marcotte's lab have since exploited the overlap between the yeast lovastatin team and the mouse angiogenesis team to identify an antifungal drug as an angiogenesis inhibitor that may be useful in treating cancer (Cha et al. 2012). That the theory of descent with modification is such a powerful research tool indicates that it has a thing or two going for it.

2.5 The Age of Earth

By the time Darwin began writing *On the Origin of Species*, data from geology had challenged a key tenet of the theory of special creation: that Earth has existed for less than 10,000 years. Much of this evidence was grounded in **uniformitarianism**. First articulated by James Hutton in the late 1700s, uniformitarianism is the claim that geological processes taking place now worked similarly in the past. It was a direct challenge to catastrophism, the hypothesis that today's geological formations resulted from catastrophic events in the past on a scale never observed now. Research inspired by uniformitarianism led Hutton, and later Charles Lyell, to infer that Earth was unimaginably old **(Figure 2.31)**. These early geologists measured the rate of ongoing rock-forming processes such as the deposition of mud, sand, and gravel at beaches and river deltas and the accumulation of marine shells (the precursors of limestone). Based on these observations, it was clear that vast stretches of time were required to produce the immense rock formations being mapped in the British Isles and Europe.

For a more recently documented example, consider the age of Earth's Atlantic Ocean (Hazen 2010). The Atlantic was formed when the supercontinent Pangaea

Figure 2.31 (opposite) The geologic time scale (a) The last 541 million years. The sequence of eons, eras, periods, and epochs shown on the left of this diagram was established through the techniques of relative dating. Each named interval of time is associated with a distinctive fossil flora and fauna. The absolute ages included here were added much later, when radiometric dating became available. The abbreviation Ma stands for millions of years ago. Redrawn from Gradstein and Ogg (2009); dates from ICS (2012). The evolutionary tree shows the minimum possible ages, estimated from fossils, of the divergences of the lineages leading to some extant organisms of interest (Benton et al. 2009; see also Hedges et al. 2006). (b) The entire history of Earth. Again, the eons on the left were defined based on relative dates, and the absolute ages are based on radiometric dating. (Earliest fossil cells: Knoll and Barghoorn 1977, Javaux et al. 2010; eukaryotes: Han and Runnegar 1992, Lamb et al. 2009; multicellular eukaryotes: Bengtson et al. 2009; animals: Yin et al. 2007, Maloof et al. 2010; vertebrates: Zhang and Hou 2004; land plants: Rubinstein et al. 2010; flowering plants: Sun et al. 1998, Royer et al. 2010.)

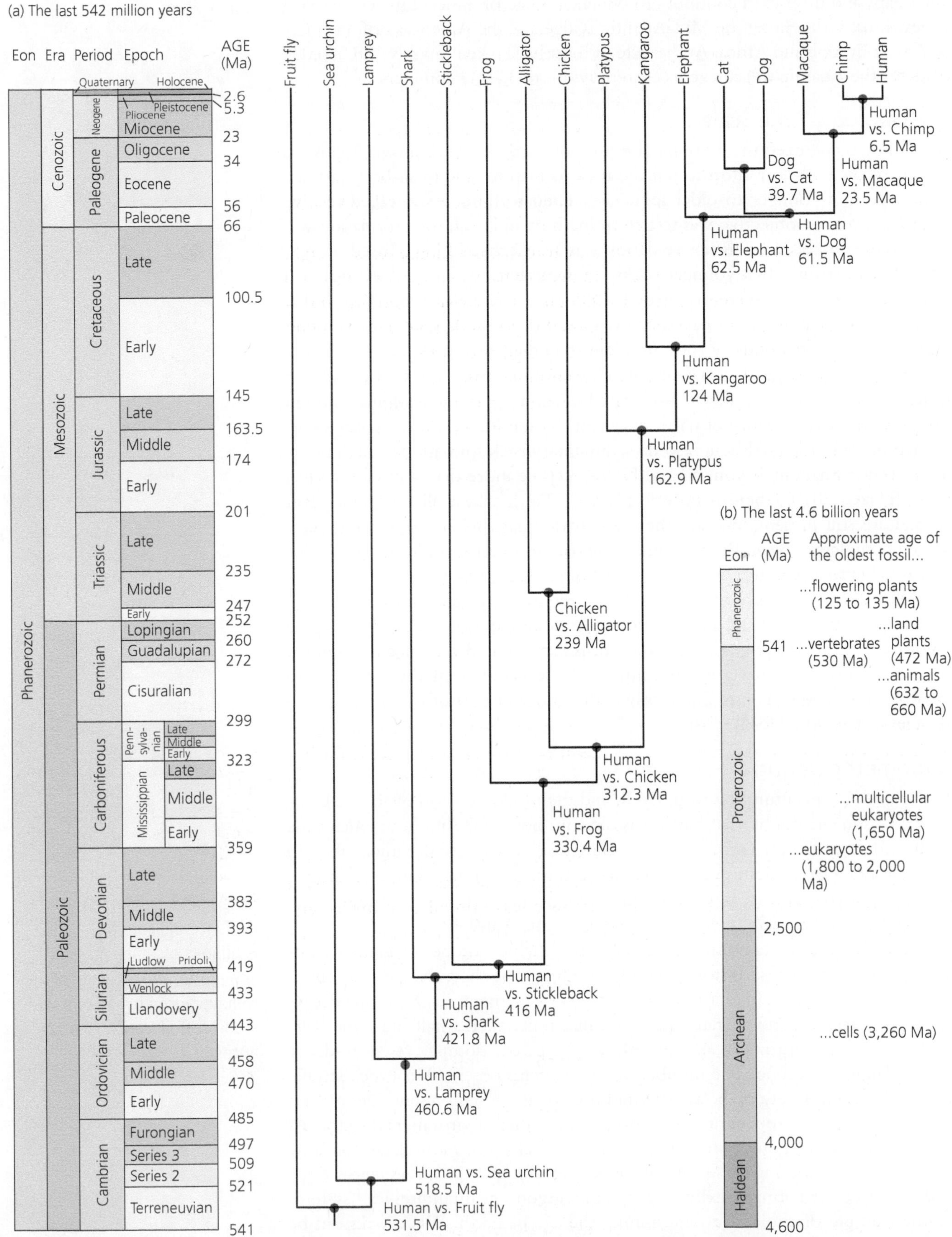
(a) The last 542 million years
Eon Era Period Epoch
AGE (Ma)
Phanerozoic
Cenozoic
Mesozoic
Paleozoic
Quaternary
Holocene
Pleistocene
Pliocene
Neogene
Miocene
Paleogene
Oligocene
Eocene
Paleocene
Cretaceous
Late
Early
Jurassic
Late
Middle
Early
Triassic
Late
Middle
Early
Permian
Lopingian
Guadalupian
Cisuralian
Carboniferous
Penn-sylva-nian
Late
Middle
Early
Mississippian
Late
Middle
Early
Devonian
Late
Middle
Early
Silurian
Ludlow
Pridoli
Wenlock
Llandovery
Ordovician
Late
Middle
Early
Cambrian
Furongian
Series 3
Series 2
Terreneuvian
2.6
5.3
23
34
56
66
100.5
145
163.5
174
201
235
247
252
260
272
299
323
359
383
393
419
433
443
458
470
485
497
509
521
541
Fruit fly
Sea urchin
Lamprey
Shark
Stickleback
Frog
Alligator
Chicken
Platypus
Kangaroo
Elephant
Cat
Dog
Macaque
Chimp
Human
Human vs. Chimp 6.5 Ma
Human vs. Macaque 23.5 Ma
Dog vs. Cat 39.7 Ma
Human vs. Dog 61.5 Ma
Human vs. Elephant 62.5 Ma
Human vs. Kangaroo 124 Ma
Human vs. Platypus 162.9 Ma
Chicken vs. Alligator 239 Ma
Human vs. Chicken 312.3 Ma
Human vs. Frog 330.4 Ma
Human vs. Stickleback 416 Ma
Human vs. Shark 421.8 Ma
Human vs. Lamprey 460.6 Ma
Human vs. Sea urchin 518.5 Ma
Human vs. Fruit fly 531.5 Ma
(b) The last 4.6 billion years
Eon
AGE (Ma)
Approximate age of the oldest fossil...
Phanerozoic
Proterozoic
Archean
Haldean
541
2,500
4,000
4,600
...flowering plants (125 to 135 Ma)
...land plants (472 Ma)
...vertebrates (530 Ma)
...animals (632 to 660 Ma)
...multicellular eukaryotes (1,650 Ma)
...eukaryotes (1,800 to 2,000 Ma)
...cells (3,260 Ma)

split apart, and this ocean continues to widen at a rate of 2.5 centimeters per year as new crust forms along the Mid-Atlantic Ridge and the Americas drift ever farther from Europe and Africa. At that rate, it must have taken roughly 148,000,000 years for the Atlantic to achieve its current width of 3,700 kilometers.

The Geologic Time Scale

Once they recognized the extreme antiquity of Earth, Hutton and his followers embarked on a 50-year effort to put the rock formations and fossil-bearing strata of Europe in a younger-to-older sequence. Their technique was called relative dating, because its objective was to determine how old each rock formation was compared to other strata. Relative dating was an exercise in logic based on reasonable assumptions: that younger rocks are deposited on top of older; that lava and sedimentary rocks were originally laid down in a horizontal position so that any tipping or bending must have occurred later; that boulders, cobbles, or other fragments found in a body of rock are older than their host rock; and so on.

Using these rules, geologists established the chronology of relative dates known as the geologic time scale. They also created the concept of the geologic column, which is a geologic history of Earth based on a composite, older-to-younger sequence of rock strata. Although some sedimentary rock formations—such as the Green River Shale in Wyoming and Utah—expose more than a million annual layers (Hazen 2010), there is no one place on Earth where all rock strata ever formed are still present. Instead, there are always gaps where some strata have eroded completely away. But by combining data from different locations, geologists are able to assemble a complete record of geologic history.

Included with the geologic time scale in Figure 2.31a are ages now known from radiometic dating and an evolutionary tree showing the currently accepted relationships among a few familiar extant organisms. The divergence times noted on the evolutionary tree are minimum ages estimated from the fossil record (Benton et al. 2009). Figure 2.31b shows the entire history of Earth along with the ages of a few key fossil "firsts."

Radiometric Dating

By the mid-19th century, Hutton, Lyell, and their followers had established, beyond a reasonable doubt, that Earth was old. But how old? And how much time has passed since life on Earth began? Marie Curie's discovery of radioactivity in the early 1900s gave scientists a way to answer these questions. Using a technique called radiometric dating, physicists and geologists began to assign absolute ages to the relative dates established by the geologic time scale.

Several independent lines of evidence indicate that Earth is billions of years old—old enough for the diversity of life to have arisen by descent with modification from a common ancestor.

The technique for radiometric dating uses unstable isotopes of naturally occurring elements. These isotopes decay, meaning that they change into either different elements or different isotopes of the same element. Each isotope decays at a particular and constant rate, measured in a unit called a half-life. One half-life is the amount of time it takes for 50% of the parent isotope present to decay into its daughter isotope. The number of decay events observed in a rock sample over time depends only on how many radioactive atoms are present. Decay rates are not affected by temperature, moisture, or any other environmental factor. As a result, radioactive isotopes function as natural clocks. For more details on how rocks can be dated using these natural clocks, see **Computing Consequences 2.1**.

Because of their long half-lives, potassium–argon and uranium–lead systems are the isotopes of choice for determining the age of Earth. What rocks can be

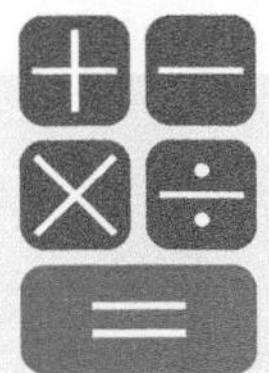

COMPUTING CONSEQUENCES 2.1

A closer look at radiometric dating

Radiometric dating lets geologists assign absolute ages to rocks. First the half-life of a radioactive isotope is determined by putting a sample in an instrument that records decay events over time. (For long-lived isotopes, of course, researchers must extrapolate from data collected over a short time.) Then the ratio of parent-to-daughter isotopes in a sample of rock is measured, often with a mass spectrometer. Once the half-life of the parent isotope and the current ratio of parent-to-daughter isotopes are known, the time elapsed since the rock formed can be calculated (Figure 2.32).

A key assumption is that the ratio of parent-to-daughter isotopes present when the rock was formed is known. This assumption can be tested. Potassium–argon dating, for example, is used for volcanic rocks. We can predict that, initially, no daughter isotope, argon-40, will be present. That is because argon-40 is a gas that bubbles out of liquid rock. It begins to accumulate only after solidification. Observations of recent lava flows confirm that this is true. Expressed as a ratio of percentages, the ratio of potassium-40 to argon-40 in newly hardened basalts, lavas, and ashes is, as predicted, 100:0 (see Damon 1968; Faure 1986).

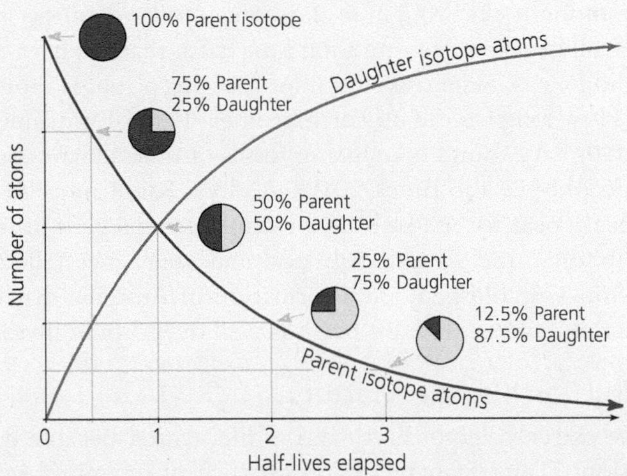

Figure 2.32 Radioactive decay Many radioactive isotopes decay through a series of intermediates until a stable daughter isotope is produced. Researchers measure the ratio of parent isotope to daughter isotope in a rock sample, then use a graph like this to convert the measured ratio to the half-lives elapsed. Multiplying the number of half-lives by the length of a half-life yields an estimate for the absolute age of the rock.

Of the many radioactive elements present in Earth's crust, the isotopes listed in Table 2.1 are the most useful. They are common enough to be present in measurable quantities and do not readily migrate into or out of rocks after their initial formation. In many cases more than one isotope system can be used on the same rocks or fossils, providing an independent check on the date.

Table 2.1 Parent and daughter isotopes used in radiometric dating

Method	Parent isotope	Daughter isotope	Half-life of parent (years)	Effective dating range (years)	Materials commonly dated
Rubidium–strontium	Rb-87	Sr-87	47 billion	10 million–4.6 billion	Potassium-rich minerals such as feldspar and hornblende; volcanic and metamorphic rock
Uranium–lead	U-238	Pb-206	4.5 billion	10 million–4.6 billion	Zircons, uraninite, and uranium ores such as pitchblende; igneous and metamorphic rock
Uranium–lead	U-235	Pb-207	71.3 million	10 million–4.6 billion	Same as above
Thorium–lead	Th-232	Pb-208	14.1 billion	10 million–4.6 billion	Zircons, uraninite
Potassium–argon	K-40	Ar-40	1.3 billion	100,000–4.6 billion	Potassium-rich minerals such as biotite and muscovite, volcanic rock
Carbon-14	C-14	N-14	5,730	100–100,000	Any carbon-bearing material, such as bones, wood, shells, cloth, and animal droppings

tested to determine when Earth first formed? Current models of Earth's formation predict that the planet was molten for much of its early history, which makes answering this question difficult. If we assume that all of the components of our solar system were formed at the same time, however, two classes of candidate rocks become available to date the origin of Earth: moon rocks and meteorites. Both uranium–lead and potassium–argon dating systems place the age of the moon rocks brought back by the *Apollo* astronauts at 4.53 billion years. Also, virtually every meteorite found on Earth that has been dated yields an age of 4.6 billion years. Scientists thus infer that our planet is about 4.6 billion years old.

How long has life on Earth been evolving? Emmanuelle Javaux and colleagues (2010) have found convincing fossils of unicellular organisms that are 3.2 billion years old (see also Buick 2010). Andrew Knoll and Elso Barghoorn (1977) found what appear to be fossils of dividing bacterial cells in rocks that are 3.26 billion years old. And Abigail Allwood and colleagues (2009) have analyzed a 3.43-billion-year-old geological formation in Australia that was likely a reef built by microorganisms. Life on Earth is well over 3 billion years old.

Why Earth's Age Matters

The extreme age of Earth and of life matter, because descent with modification is slow. The instantaneous appearance of organisms postulated by adherents of special creation was compatible with the 6,000-year-young Earth many of them claimed. Darwin rejected this age in favor of contemporary estimates based on geological processes. These estimates counted Earth's age in hundreds of millions of years. Only over such a vast span could gradual changes generate the present diversity of life from a common ancestral stock.

Darwin was gravely distressed when physicist William Thomson (Lord Kelvin) argued that the Sun, and by implication the Earth, was not more than 20 million years old. Unable to offer a rebuttal, Darwin, in a letter to Alfred Russell Wallace, compared Thomson's calculation to "an odious spectre" (Darwin 1887, page 146).

We now know that Darwin need not have lost sleep on Thomson's account. Thomson assumed the Sun was fueled by combustion—nuclear fusion had yet to be discovered. As Thomson himself recognized might happen, the discovery of a new source of heat rendered his calculations moot (see Turnbull 1935).

Radiometery has now revealed that life has existed at least 175 times longer than Thomson's calculation allowed, and more than 580,000 times longer than supposed by advocates of special creation. Three and a half billion years is ample time for descent with modifcation to do its work.

SUMMARY

Before Charles Darwin published *On the Origin of Species* in 1859, special creation was the leading explanation for where Earth's organisms came from. Arguing from a trove of carefully documented evidence, Darwin advocated a different view of the pattern of life's history. All organisms are descended, with modification, from a common ancestor. Darwin's evidence was sufficiently persuasive that within a decade the fact of evolution had achieved general acceptance among biologists.

The evidence showing that Darwin's view is correct has only grown since then. We reviewed evidence on each point of difference between descent with modification versus special creation.

The results of laboratory selection experiments, the selective breeding of domestic plants and animals, direct observations of change in natural populations, and vestigial structures all demonstrate that populations change across generations. Change in a population from one generation to the next is called microevolution.

The results of laboratory experiments involving selection for divergence, observations of natural populations with different degrees of reproductive isolation, and studies of ring species establish that lineages split and diverge. The separation of one species into two is called speciation.

Extinction, succession, and transitional forms—including transitional fossils predicted before they were found—show that over long time spans, new life-forms can arise from old lineages. This is macroevolution.

Structural and genetic homologies indicate, and predictive tests using the fossil record and shared genetic flaws confirm, that all of Earth's organisms are related by common ancestry. There is but one tree of life.

Finally, relative and absolute dating show that the Earth and life are billions of years old.

In addition to establishing the fact of evolution, Darwin had set himself a second goal in *On the Origin of Species*. This was to elucidate the mechanism responsible for evolution. The mechanism Darwin identified was, of course, natural selection. In contrast to the rapid acceptance of the fact of descent with modification, natural selection was not widely accepted as the mechanism of adaptive evolution until the 1930s.

Natural selection is the subject of the next chapter.

QUESTIONS

1. Review the kinds of evidence for evolution analyzed in Sections 2.1–2.5. List the sources of evidence that were available to Darwin and those that appeared later. For example, of the evidence for microevolution discussed in Section 2.1, Darwin knew, and wrote, about divergent strains of domestic plants and animals and about vestigial structures. However, in Darwin's day no one had ever directly observed change across generations in natural populations. For each section, indicate which evidence you consider strongest and which you consider weakest. Explain why.
2. Consider the experiment described in Section 2.1 in which Ted Garland and colleagues bred mice to run long distances on exercise wheels. We presented the results as evidence that two dozen generations of selective breeding had altered the experimental population. How does the control strain support this interpretation? If Garland had simply compared the behavior of the 24th experimental generation to the behavior of the first experimental generation, would the evidence for evolution be as strong? Explain.
3. In addition to dogs, list at least two or three other species of domestic plants or animals you are familiar with in which selective breeding has resulted in distinctive pure-breeding varieties. How could you verify, in each case on your list, that the varieties are, in fact, all descended from a common ancestor?
4. Look back at Figure 2.14d, which shows the two kinds of threespine sticklebacks that live in Paxon Lake. There used to be a similar limnetic/benthic pair in Enos Lake (see Hendry et al. 2009). However, recent studies have revealed that the two forms in Enos Lake have recently merged into a single highly variable population. How does this bear on the claim that the two forms in Paxton Lake are different species? How does it bear on the claim that varying degrees of divergence among stickleback populations provide evidence for speciation?
5. Figures 2.20 through 2.22 show examples of transitional fossils. If Darwin's theory of evolution is correct, and all organisms are descended with modification from a common ancestor, predict some other examples of transitional forms that should have existed and that might have produced fossils. If such fossils are someday found, will that strengthen the hypothesis that such transitional species once existed? Conversely, if such fossils have not been found, does this weaken the hypothesis that the transitional species once existed?
6. The transitional fossils in Figure 2.21 demonstrate that dinosaurs evolved feathers long before they evolved flight. Clearly, feathers did not evolve for their aerodynamic advantages. What else, besides aerodynamics, do feathers do for birds today? What advantages might feathers have offered for dinosaurs? Can you think of a way to test your hypothesis?
7. Section 2.4 presented two definitions of homology: the classical definition articulated by Richard Owen and the modern definition favored by many contemporary biologists. Look at the vestigial organs shown in Figure 2.7. Is the tiny wing of a brown kiwi homologous to the wing of an eagle? Are the spurs of a rubber boa homologous to the hindlimbs of a kangaroo? By which definition of homology?
8. Analogy and homology are important concepts used in comparing species. Traits are homologous if they are derived, evolutionarily and developmentally, from the same source structure. Traits are analogous if they have

similar functions but are derived, evolutionarily and developmentally, from different source structures. A classic example of analogous structures is insect wings and bat wings. Which of the following pairs of structures are analogous and which are homologous?

a. The dorsal fins of a porpoise and a salmon
b. The flippers of a porpoise and the pectoral fins (front fins) of a salmon
c. The jointed leg of a ladybird beetle and a robin
d. A rhesus monkey's tail and a human's coccyx
e. The bright red bracts (modified leaves) of a poinsettia and the green leaves of a rose
f. Red bracts on a poinsettia and red petals on a rose

9. Figure 2.26 is an evolutionary tree showing the relationships among five living species of snails. Draw a genealogical pedigree of your family or a friend's family, starting with the oldest and continuing to the youngest generation. Compare the pedigree to the evolutionary tree. How are evolutionary trees and pedigrees similar? How are they different?
10. According to the evolutionary tree shown in Figure 2.26, is the snail with spikes on its shell more closely related to the snail with a plain high-spired pink shell or to the snail with a simple elongated pink shell? Why?
11. According to the evolutionary tree in Figure 2.31, are cats more closely related to elephants or humans? Why?
12. In the early 20th century, radiometric dating allowed geologists to assign absolute ages to most fossil-bearing strata. The absolute dates turned out to be entirely consistent with the relative dating done in the early 19th century. What does this result say about the assumptions behind relative dating?

EXPLORING THE LITERATURE

13. Darwin's *On the Origin of Species* still stands as one of the most influential books written in the last two centuries. The first edition is the best version to read. The full text of *On the Origin,* along with virtually everything else Darwin wrote, is available free online at:

    http://darwin-online.org.uk/

14. For an artificial selection experiment similar to Garland's on mice—that also yielded a dramatic response to selection—see:

    Weber, K. E. 1996. Large genetic change at small fitness cost in large populations of *Drosophila melanogaster* selected for wind tunnel flight: Rethinking fitness surfaces. *Genetics* 144: 205–213.

15. Additional examples of microevolution documented in natural populations come from studies of soapberry bugs:

    Carroll, S. P., and C. Boyd. 1992. Host race radiation in the soapberry bug: Natural history with the history. *Evolution* 46: 1052–1069.

    Carroll, S. P., H. Dingle, and S. P. Klassen. 1997. Genetic differentiation of fitness-associated traits among rapidly evolving populations of the soapberry bug. *Evolution* 51: 1182–1188.

    Carroll, S. P., J. E. Loye, et al. 2005. And the beak shall inherit—evolution in response to invasion. *Ecology Letters* 8: 944–951.

    studies of snakes:

    Phillips, B. L., and R. Shine. 2006. An invasive species induces rapid adaptive change in a native predator: Cane toads and black snakes in Australia. *Proceedings of the Royal Society of London B* 273: 1545–1550.

    and studies of sticklebacks:

    Kitano, J., D. I. Bolnick, et al. 2008. Reverse evolution of armor plates in the threespine stickleback. *Current Biology* 18: 769–774.

16. To better understand the evolution of the dog under domestication, Russian biologist Dmitry K. Belyaev initiated a long-term project to domesticate the silver fox by selectively breeding for tameability. For a review of this work, see:

    Trut, L., I. Oskina, and A. Kharlamova. 2009. Animal evolution during domestication: The domesticated fox as a model. *BioEssays* 31: 349–360.

17. Our examples of speciation all involved a single ancestral lineage splitting into two. New species also sometimes arise by hybridization between two ancestral lineages:

    Mavárez, J., C. A. Salazar, et al. 2006. Speciation by hybridization in *Heliconius* butterflies. *Nature* 441: 868–871.

18. We interpreted the amphibious blenny as a transitional form between the aquatic and terrestrial blennies. Whether the terrestrial blenny (*Alticus*) will also prove to be a transitional form is unknowable. It is tempting to see it as the future ancestor of a lineage of bizarre, fully terrestrial fish. Such a lineage has evolved before. We are among its progeny. However, the supratidal zone may be as far out of the water as any marine fish will get. See:

    Graham, J. B., and H. J. Lee. 2004. Breathing air in air: In what ways might extant amphibious fish biology relate to prevailing concepts about early tetrapods, the evolution of vertebrate air breathing, and the vertebrate land transition? *Physiological and Biochemical Zoology* 77: 720–731.

19. Pterosaurs arise in the fossil record about 210 million years ago (Unwin 2003). For the next 55 million years, most were built like the *Rhamphorhyncus* in **Figure 2.33a** (Lü et al. 2010). They had long tails (red arrow) and relatively small skulls with two separate openings, called the antorbital fenestra and the nasal fenestra (orange arrows). They had neck ribs and elongated fifth toes (not visible in this fossil).

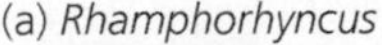
(a) *Rhamphorhyncus*

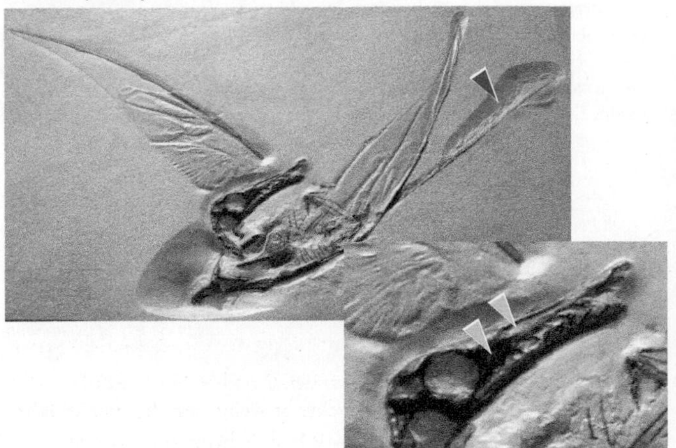

(b) *Pterodactylus*

Figure 2.33 Death on the wing (a) The pterosaur *Rhamphorhynchus* had a long tail (red arrow) and a skull with separate antorbital and nasal openings (orange). Photo by Ryan Somma. (b) The pterodactyl *Pterodactylus* had a short tail (blue) and a skull with a single nasoantorbital opening (black). Photo by Daderot.

Pterodactyls first appear about 155 million years ago (Unwin 2003). Within 30 million years, they replaced all earlier pterosaurs and then persisted until 65 million years ago—when they went extinct along with the non-avian dinosaurs. Most pterodactyls were built like the *Pterodactylus* in Figure 2.33b (Lü et al. 2010). They had short tails (blue arrow) and large skulls with a single nasoantorbital fenestra (black arrow). Their neck ribs and fifth toes were reduced or absent.

Detailed anatomical analyses have yielded results consistent with the macroevolutionary hypothesis that pterodactyls are derived from an earlier lineage of pterosaurs (Unwin 2003). From this it can be predicted that the fossil record should contain flying reptiles intermediate in form between the early pterosaurs and the later pterodactyls.

Write down a prediction of what these transitional forms might look like, and the age of the rocks in which they are most likely to be found. Then take a look at the photos of *Darwinopterus modularis*:

Lü, J., D. M. Unwin, et al. 2010. Evidence for modular evolution in… *Proceedings of the Royal Society of London B* 277: 383–389. (We have truncated the title to avoid revealing too much.)

20. For a feathered dinosaur that has influenced expert opinion on why dinosaurs evolved feathers in the first place, see:

Hu, D., L. Hou, et al. 2009. A *pre-Archaeopteryx* troodontid theropod from China with long feathers on the metatarsus. *Nature* 461: 640–643.

Witmer, L. M. 2009. Palaeontology: Feathered dinosaurs in a tangle. *Nature* 461: 601–602.

For a reconstruction of the color of this dinosaur's feathers, see:

Li, Q., K. Q. Gao, et al. 2010. Plumage color patterns of an extinct dinosaur. *Science* 327: 1369–1372.

21. For an example of a molecular vestigial trait predicted before it was found, see:

Meredith, R. W., J. Gatesy, et al. 2009. Molecular decay of the tooth gene *Enamelin* (*ENAM*) mirrors the loss of enamel in the fossil record of placental mammals. *PLoS Genetics* 5: e1000634.

CITATIONS

Abouheif, E. 1997. Developmental genetics and homology: A hierarchical approach. *Trends in Ecology and Evolution* 12: 405–408.

Akey, J. M., A. L. Ruhe, et al. 2010. Tracking footprints of artificial selection in the dog genome. *Proceedings of the National Academy of Sciences, USA* 107: 1160–1165.

Allendorf, F. W., and J. J. Hard. 2009. Human-induced evolution caused by unnatural selection through harvest of wild animals. *Proceedings of the National Academy of Sciences, USA* 106: 9987–9994.

Allwood, A. C., J. P. Grotzinger, et al. 2009. Controls on development and diversity of Early Archean stromatolites. *Proceedings of the National Academy of Sciences, USA* 106: 9548–9555.

Alonso, P. D., A. C. Milner, et al. 2004. The avian nature of the brain and inner ear of *Archaeopteryx*. *Nature* 430: 666–669.

Bakker, R. T., and P. M. Galton. 1974. Dinosaur monophyly and a new class of vertebrates. *Nature* 248: 168–172.

Bell, M. A., W. E. Aguirre, and N. J. Buck. 2004. Twelve years of contemporary armor evolution in a threespine stickleback population. *Evolution* 58: 814–824.

Bell, M. A., and S. A. Foster, eds. 1994. *The Evolutionary Biology of the Threespine Stickleback*. Oxford, England: Oxford University Press.

Bengtson, S., V. Belivanova, et al. 2009. The controversial "Cambrian" fossils of the Vindhyan are real but more than a billion years older. *Proceedings of the National Academy of Sciences, USA* 106: 7729–7734.

Benton, M. J., P. C. J. Donoghue, and R. J. Asher. 2009. Calibrating and constraining molecular clocks. In *The TimeTree of Life*, ed. S. B. Hedges and S. Kumar. New York: Oxford University Press, 35–85.

Bergmann, U., R. W. Morton, et al. 2010. *Archaeopteryx* feathers and bone chemistry fully revealed via synchrotron imaging. *Proceedings of the National Academy of Sciences, USA* 107: 9060–9065.

Berkman, M. B., J. S. Pacheco, and E. Plutzer. 2008. Evolution and creationism in America's classrooms: A national portrait. *PLoS Biology* 6: e124.

Berner, D., D. C. Adams, et al. 2008. Natural selection drives patterns of lake-stream divergence in stickleback foraging morphology. *Journal of Evolutionary Biology* 21: 1653–1665.

Berner, D., A. C. Grandchamp, and A. P. Hendry. 2009. Variable progress toward ecological speciation in parapatry: Stickleback across eight lake-stream transitions. *Evolution* 63: 1740–1753.

Bhikajee, M., and J. M. Green. 2002. Behaviour and habitat of the Indian Ocean amphibious blenny, *Alticus monochrus*. *African Zoology* 37: 221–230.

Bhikajee, M., J. M. Green, and R. Dunbrack. 2006. Life history characteristics of *Alticus monochrus*, a supratidal blenny of the southern Indian Ocean. *African Zoology* 41: 1–7.

Bone, E., and A. Farres. 2001. Trends and rates of microevolution in plants. *Genetica* 112: 165–182.

Boyko, A. R., P. Quignon, et al. 2010. A simple genetic architecture underlies morphological variation in dogs. *PLoS Biology* 8: e1000451.

Branch, G., and E. Scott. 2009. The latest face of creationism. *Scientific American* 300: 92–99.

Brice, W. R. 1982. Bishop Ussher, John Lightfoot and the age of creation. *Journal of Geological Education* 30: 18–26.

Brown, C. R., M. S. Gordon, and K. L. M. Martin. 1992. Aerial and aquatic oxygen uptake in the amphibious Red Sea rockskipper fish, *Alticus kirki* (Family Blenniidae). *Copeia* 1992: 1007–1013.

Buick, R. 2010. Early life: Ancient acritarchs. *Nature* 463: 885–886.

Burke, A. C. 2009. Turtles… …again. *Evolution and Development* 11: 622–624.

Cha, H. J., M. Byrom, et al. 2012. Evolutionarily repurposed networks reveal the well-known antifungal drug thiabendazole to be a novel vascular disrupting agent. *PLoS Biology* 10: e1001379.

Chen, P.-J., Z. M. Dong, and S. N. Zhen. 1998. An exceptionally well-preserved theropod dinosaur from the Yixian Formation of China. *Nature* 391: 147–152.

Chou, H.-H., T. Hayakawa, S. Diaz, et al. 2002. Inactivation of CMP-N-acetylneuraminic acid hydroxylase occurred prior to brain expansion during human evolution. *Proceedings of the National Academy of Sciences, USA* 99: 11736-11741.

Chou, H.-H., H. Takematsu, et al. 1998. A mutation in human CMP-sialic acid hydroxylase occurred after the *Homo–Pan* divergence. *Proceedings of the National Academy of Sciences, USA* 95: 11751–11756.

Christiansen, P., and N. Bonde. 2004. Body plumage in *Archaeopteryx*: A review and new evidence from the Berlin specimen. *Comptes Rendus Palevol* 3: 99–118.

Colosimo, P. F., K. E. Hosemann, et al. 2005. Widespread parallel evolution in sticklebacks by repeated fixation of Ectodysplasin alleles. *Science* 307: 1928–1933.

Coyne, J. A., and H. A. Orr. 1997. "Patterns of speciation in *Drosophila*" revisited. *Evolution* 51: 295–303.

Cresko, W. A., K. L. McGuigan, et al. 2007. Studies of threespine stickleback developmental evolution: Progress and promise. *Genetica* 129: 105–126.

Currie, P. J., and P. J. Chen. 2001. Anatomy of *Sinosauropteryx prima* from Liaoning, northeastern China. *Canadian Journal of Earth Sciences* 38: 1705–1727.

Damon, P. E. 1968. Potassium–argon dating of igneous and metamorphic rocks with applications to the basin ranges of Arizona and Sonora. In *Radiometric Dating for Geologists*, ed. E. I. Hamilton and R. M. Farquhar. London: Interscience Publishers.

Darwin, C. R. 1859. *On the Origin of Species by Means of Natural Selection, First Edition*. London: John Murray.

Darwin, C. R. 1862. *The Various Contrivances by Which Orchids are Fertilized by Insects*. London: John Murray.

Darwin, Francis (ed.). 1887. *The life and letters of Charles Darwin, including an autobiographical chapter*. London: John Murray. Volume 3.

De Queiroz, K. 2007. Species concepts and species delimitation. *Systematic Biology* 56: 879–886.

Desmond, A., and J. Moore. 1991. *Darwin*. New York: Norton.

Dodd, D. M. B. 1989. Reproductive isolation as a consequence of adaptive divergence in *Drosophila pseudoobscura*. *Evolution* 43: 1308–1311.

Driscoll, C. A., D. W. Macdonald, and S. J. O'Brien. 2009. From wild animals to domestic pets, an evolutionary view of domestication. *Proceedings of the National Academy of Sciences, USA* 106: 9971–9978.

Duffy, S., C. L. Burch, and P. E. Turner. 2007. Evolution of host specificity drives reproductive isolation among RNA viruses. *Evolution* 61: 2614–2622.

Dugan, K. G. 1980. Darwin and *Diprotodon*: The Wellington Caves fossils and the law of succession. *Proceedings of the Linnean Society of New South Wales* 104: 265–272.

Eiseley, L. 1958. *Darwin's Century*. Garden City, NY: Anchor Books.

Erwin, D. H. 2008. Extinction as the loss of evolutionary history. *Proceedings of the National Academy of Sciences, USA* 105 (Suppl 1): 11520–11527.

Faure, G. 1986. *Principles of Isotope Geology*. New York: John Wiley & Sons.

Florin, A. B., and A. Ödeen. 2002. Laboratory environments are not conducive for allopatric speciation. *Journal of Evolutionary Biology* 15: 10–19.

Forrest, B. 2008. Still creationism after all these years: Understanding and counteracting intelligent design. *Integrative and Comparative Biology* 48: 189.

Franks, S. J., J. C. Avise, et al. 2008. The resurrection initiative: Storing ancestral genotypes to capture evolution in action. *BioScience* 58: 870–873.

Franks, S. J., S. Sim, and A. E. Weis. 2007. Rapid evolution of flowering time by an annual plant in response to a climate fluctuation. *Proceedings of the National Academy of Sciences, USA* 104: 1278–1282.

Freeland, S. J., T. Wu, and N. Keulmann. 2003. The case for an error minimizing standard genetic code. *Origins of Life and Evolution of Biospheres* 33: 457–477.

Friedberg, F., and A. R. Rhoads. 2000. Calculation and verification of the ages of retroprocessed pseudogenes. *Molecular Phylogenetics and Evolution* 16: 127–130.

Garland, T., Jr. 2003. Selection experiments: An under-utilized tool in biomechanics and organismal biology. In *Vertebrate Biomechanics and Evolution*, ed. V. L. Bels, J.-P. Gasc, and A. Casinos. Oxford, England: BIOS Scientific Publishers, 23-56.

Garland, T., Jr., S. A. Kelly, et al. 2011. How to run far: Multiple solutions and sex-specific responses to selective breeding for high voluntary activity levels. *Proceedings of the Royal Society of London B* 278: 574–581.

Garland, T., M. T. Morgan, et al. 2002. Evolution of a small-muscle polymorphism in lines of house mice selected for high activity levels. *Evolution* 56: 1267–1275.

Garner, J. P., G. K. Taylor, and A. L. R. Thomas. 1999. On the origins of birds: The sequence of character acquisition in the evolution of avian flight. *Proceedings of the Royal Society of London B* 266: 1259–1266.

Geist, N. R., and A. Feduccia. 2000. Gravity-defying behaviors: Identifying models for protaves. *American Zoologist* 40: 664–675.

Gould, S. J. 1977. *Ever Since Darwin: Reflections in Natural History*. New York: Norton.

Gourbière, S., and J. Mallet. 2010. Are species real? The shape of the species boundary with exponential failure, reinforcement, and the "missing snowball." *Evolution* 64: 1–24.

Gradstein, F. M., and J. G. Ogg. 2009. The geologic time scale. In *The TimeTree of Life*, ed. S. B. Hedges and S. Kumar. New York: Oxford University Press, 26–34.

Han, T. M., and B. Runnegar. 1992. Megascopic eukaryotic algae from the 2.1-billion-year-old Negaunee Iron-Formation, Michigan. *Science* 257: 232–235.

Hancock, A. M., D. B. Witonsky, et al. 2010. Human adaptations to diet, subsistence, and ecoregion are due to subtle shifts in allele frequency. *Proceedings of the National Academy of Sciences, USA* 107 (Suppl 2): 8924–8930.

Hazen, R. M. 2010. How old is Earth, and how do we know? *Evolution: Education and Outreach* 3: 198–205.

He, T., X. L. Wang, and Z. H. Zhou. 2008. A new genus and species of caudipterid dinosaur from the Lower Cretaceous Jiufotang Formation of Western Liaoning, China. *Vertebrata PalAsiatica* 46: 178–189.

Hedges, S. B., J. Dudley, and S. Kumar. 2006. TimeTree: A public knowledge-base of divergence times among organisms. *Bioinformatics* 22: 2971–2972.

Hehemann, J. H., G. Correc, et al. 2010. Transfer of carbohydrate-active enzymes from marine bacteria to Japanese gut microbiota. *Nature* 464: 908–912.

Hendry, A. P., D. I. Bolnick, et al. 2009. Along the speciation continuum in sticklebacks. *Journal of Fish Biology* 75: 2000–2036.

Hendry, A. P., T. J. Farrugia, and M. T. Kinnison. 2008. Human influences on rates of phenotypic change in wild animal populations. *Molecular Ecology* 17: 20–29.

Hsieh, S. T. 2010. A locomotor innovation enables water–land transition in a marine fish. *PLoS One* 5: e11197.

Hu, D., L. Hou et al. 2009. A pre-*Archaeopteryx* troodontid theropod from China with long feathers on the metatarsus. *Nature* 461: 640–643.

Hughes, A. L., S. Irausquin, and R. Friedman. 2010. The evolutionary biology of poxviruses. *Infection, Genetics and Evolution* 10: 50–59.

Huxley, T. H. 1868. On the animals which are most nearly intermediate between birds and reptiles. *Geological Magazine* 5: 357–365.

ICS (International Commission on Stratigraphy). 2012. International Chronostratigraphic Chart. Available at http://www.stratigraphy.org/column.php?id=Chart/Time%20Scale.

Irwin, D. E., S. Bensch, and T. D. Price. 2001. Speciation in a ring. *Nature* 409: 333-337.

Irwin, D. E., S. Bensch, et al. 2005. Speciation by distance in a ring species. *Science* 307: 414-416.

Irwin, D. E., M. P. Thimgan, and J. H. Irwin. 2008. Call divergence is correlated with geographic and genetic distance in greenish warblers (*Phylloscopus trochiloides*): A strong role for stochasticity in signal evolution? *Journal of Evolutionary Biology* 21: 435–448.

Javaux, E. J., C. P. Marshall, and A. Bekker. 2010. Organic-walled microfossils in 3.2-billion-year-old shallow-marine siliciclastic deposits. *Nature* 463: 934–938.

Judson, O. P., and D. Haydon. 1999. The genetic code: What is it good for? An analysis of the effects of selection pressures on genetic codes. *Journal of Molecular Evolution* 49: 539–550.

Keller, M. P., B. A. Seifried, and P. F. Chance. 1999. Molecular evolution of the CMT1A-REP region: A human- and chimpanzee-specific repeat. *Molecular Biology and Evolution* 16: 1019–1026.

Kelly, S. A., D. L. Nehrenberg, et al. 2010. Genetic architecture of voluntary exercise in an advanced intercross line of mice. *Physiological Genomics* 42: 190–200.

Kitano, J., S. Mori, and C. L. Peichel. 2007. Phenotypic divergence and reproductive isolation between sympatric forms of Japanese threespine sticklebacks. *Biological Journal of the Linnean Society* 91: 671–685.

Kitano, J., J. A. Ross, et al. 2009. A role for a neo-sex chromosome in stickleback speciation. *Nature* 461: 1079–1083.

Knapp, C. W., J. Dolfing, et al. 2010. Evidence of increasing antibiotic resistance gene abundances in archived soils since 1940. *Environmental Science & Technology* 44: 580–587.

Knight, R. D., S. J. Freeland, and L. F. Landweber. 2001. Rewiring the keyboard: Evolvability of the genetic code. *Nature Reviews Genetics* 2: 49–58.

Knoll, A. H., and E. S. Barghoorn. 1977. Archean microfossils showing cell division from the Swaziland system of South Africa. *Science* 198: 396–398.

Koonin, E. V., and A. S. Novozhilov. 2009. Origin and evolution of the genetic code: The universal enigma. *IUBMB Life* 61: 99–111.

Kuchta, S. R., D. S. Parks, et al. 2009. Closing the ring: Historical biogeography of the salamander ring species *Ensatina eschscholtzii*. *Journal of Biogeography* 36: 982–995.

Kurnaz, M. L., T. Bilgin, and I. A. Kurnaz. 2010. Certain non-standard coding tables appear to be more robust to error than the standard genetic code. *Journal of Molecular Evolution* 70: 13–28.

Lamb, D. M., S. M. Awramik, et al. 2009. Evidence for eukaryotic diversification in the ~1800-million-year-old Changzhougou Formation, North China. *Precambrian Research* 173: 93–104.

Li, C., X.-C. Wu, et al. 2008. An ancestral turtle from the Late Triassic of southwestern China. *Nature* 456: 497–501.

Lingham-Soliar, T., A. Feduccia, and X. Wang. 2007. A new Chinese specimen indicates that 'protofeathers' in the Early Cretaceous theropod dinosaur *Sinosauropteryx* are degraded collagen fibres. *Proceedings of the Royal Society B* 274: 1823–1829.

Lloyd, G. T., K. E. Davis, et al. 2008. Dinosaurs and the cretaceous terrestrial revolution. *Proceedings of the Royal Society B* 275: 2483–2490.

Longrich, N. 2006. Structure and function of hindlimb feathers in *Archaeopteryx lithographica*. *Paleobiology* 32: 417.

Lopes, J., N. Ravisé, et al. 1998. Fine mapping of de novo CMT1A and HNPP rearrangements within CMT1A-REPs evidences two distinct sex-dependent mechanisms and candidate sequences involved in recombination. *Human Molecular Genetics* 7: 141–148.

Lü, J., D. M. Unwin, et al. 2010. Evidence for modular evolution in a long-tailed pterosaur with a pterodactyloid skull. *Proceedings of the Royal Society of London B* 277: 383–389.

Luning Prak, E. T., and H. H. Kazazian, Jr. 2000. Mobile elements and the human genome. *Nature Reviews Genetics* 1: 135–144.

Luo, Z. X. 2007. Transformation and diversification in early mammal evolution. *Nature* 450: 1011–1019.

Lyson, T. R., G. S. Bever, et al. 2010. Transitional fossils and the origin of turtles. *Biology Letters* 6: 830–833.

Malisch, J. L., C. W. Breuner, et al. 2008. Circadian pattern of total and free corticosterone concentrations, corticosteroid-binding globulin, and physical activity in mice selectively bred for high voluntary wheel-running behavior. *General and Comparative Endocrinology* 156: 210–217.

Mallet, J. 1989. The evolution of insecticide resistance: Have the insects won? *Trends in Ecology & Evolution* 4: 336–340.

Maloof, A. C., C. V. Rose, et al. 2010. Possible animal-body fossils in pre-Marinoan limestones from South Australia. *Nature Geoscience* 3: 653–659.

Martin, J. F., C. Barreiro, et al. 2003. Ribosomal RNA and ribosomal proteins in corynebacteria. *Journal of Biotechnology* 104: 41–53.

Martin, K. L. M., and J. R. B. Lighton. 1989. Aerial CO_2 and O_2 exchange during terrestrial activity in an amphibious fish, *Alticus kirki* (Blenniidae). *Copeia* 1989: 723–727.

Martin, M. J., J. C. Rayner, et al. 2005. Evolution of human-chimpanzee differences in malaria susceptibility: Relationship to human genetic loss of N-glycolylneuraminic acid. *Proceedings of the National Academy of Sciences, USA* 102: 12819–12824.

Mayr, E. 1942. *Systematics and the Origin of Species*. New York: Columbia University Press.

Mayr, E. 1964. *Introduction to: Darwin, C., 1859. On the Origin of Species: A Facsimile of the First Edition*. Cambridge, MA: Harvard University Press.

McGary, K. L., T. J. Park, et al. 2010. Systematic discovery of nonobvious human disease models through orthologous phenotypes. *Proceedings of the National Academy of Sciences, USA* 107: 6544–6549.

McKellar, R. C., B. D. Chatterton, et al. 2011. A diverse assemblage of Late Cretaceous dinosaur and bird feathers from Canadian amber. *Science* 333: 1619–1622.

Meek, T. H., B. P. Lonquich, et al. 2009. Endurance capacity of mice selectively bred for high voluntary wheel running. *Journal of Experimental Biology* 212: 2908–2917.

Miller, J. D., E. C. Scott, and S. Okamoto. 2006. Science communication. Public acceptance of evolution. *Science* 313: 765–766.

Mindell, D. P., and A. Meyer. 2001. Homology evolving. *Trends in Ecology and Evolution* 16: 434–440.

Nagashima, H., F. Sugahara, et al. 2009. Evolution of the turtle body plan by the folding and creation of new muscle connections. *Science* 325: 193–196.

Norell, M. A., and M. J. Novacek. 1992. The fossil record and evolution: Comparing cladistic and paleontologic evidence for vertebrate history. *Science* 255: 1690–1693.

Norell, M. A., and Xing Xu. 2005. Feathered dinosaurs. *Annual Review of Earth and Planetary Sciences* 33: 277–299.

Nosil, P., L. J. Harmon, and O. Seehausen. 2009. Ecological explanations for (incomplete) speciation. *Trends in Ecology and Evolution* 24: 145–156.

Nudds, R. L., and G. J. Dyke. 2010. Narrow primary feather rachises in *Confuciusornis* and *Archaeopteryx* suggest poor flight ability. *Science* 328: 887–889.

Ostrom, J. H. 1973. The ancestry of birds. *Nature* 242: 136.

Ostrom, J. H. 1974. *Archaeopteryx* and the origin of flight. *Quarterly Review of Biology* 49: 27–47.

Padian, K., and L. M. Chiappe. 1998. The origin and early evolution of birds. *Biological Reviews* 73: 1–42.

Padian, K., and A. de Ricqlès. 2009. L'origine et l'évolution des oiseaux: 35 années de progrès. *Comptes Rendus Palevol* 8: 257–280.

Pereira, R. J., and D. B. Wake. 2009. Genetic leakage after adaptive and nonadaptive divergence in the *Ensatina eschscholtzii* ring species. *Evolution* 63: 2288–2301.

Prothero, D. R. 2008. What missing link? *New Scientist*, March 1, 2008, 35–41.

Prum, R. O. 2002. Why ornithologists should care about the theropod origin of birds. *The Auk* 119: 1–17.

Prum, R. O., and A. H. Brush. 2002. The evolutionary origin and diversification of feathers. *Quarterly Review of Biology* 77: 261–295.

Ray, J. 1686. *Historia Plantarum*. London. (Vol. 1, p. 40; translated by E. T. Silk, as reproduced in Beddall 1957.)

Reiter, L. T., T. Murakami, et al. 1997. The human COX10 gene is disrupted during homologous recombination between the 24 kb proximal and distal CMT1A-REPs. *Human Molecular Genetics* 6: 1595–1603.

Rhodes, J. S., T. J. Garland, and S. C. Gammie. 2003. Patterns of brain activity associated with variation in voluntary wheel-running behavior. *Behavioral Neuroscience* 117: 1243–1256.

Rice, W. R., and E. E. Hostert. 1993. Laboratory experiments on speciation: What have we learned in 40 years? *Evolution* 47: 1637–1653.

Royer, D. L., I. M. Miller, et al. 2010. Leaf economic traits from fossils support a weedy habit for early angiosperms. *American Journal of Botany* 97: 438–445.

Rubinstein, C. V., P. Gerrienne, et al. 2010. Early Middle Ordovician evidence for land plants in Argentina (eastern Gondwana). *New Phytologist* 188: 365–369.

Sanchez-Villagra, M. R., H. Muller, et al. 2009. Skeletal development in the Chinese soft-shelled turtle *Pelodiscus sinensis* (Testudines: Trionychidae). *Journal of Morphology* 270: 1381–1399.

Schluter, D. 2010. Resource competition and coevolution in sticklebacks. *Evolution: Education and Outreach* 3: 54–61.

Shapiro, M. D., M. E. Marks, et al. 2004. Genetic and developmental basis of evolutionary pelvic reduction in threespine sticklebacks. *Nature* 428: 717–723.

Shearin, A. L., and E. A. Ostrander. 2010. Canine morphology: Hunting for genes and tracking mutations. *PLoS Biology* 8: e1000310.

Shubin, N. 2009. *Your Inner Fish*. New York: Vintage Books.

Sibley, C. H., and S. Y. Hunt. 2003. Drug resistance in parasites: Can we stay ahead of the evolutionary curve? *Trends in Parasitology* 19: 532–537.

Straus, C., K. Vasilakos, et al. 2003. A phylogenetic hypothesis for the origin of hiccough. *BioEssays* 25: 182–188.

Sun, G., D. L. Dilcher, et al. 1998. In search of the first flower: A Jurassic angiosperm, *Archaefructus*, from northeast China. *Science* 282: 1692.

Swallow, J. G., P. A. Carter, and T. Garland, Jr. 1998. Artificial selection for increased wheel-running behavior in house mice. *Behavior Genetics* 28: 227–237.

Taubes, G. 2008. The bacteria fight back. *Science* 321: 356–361.

Tian, F., N. M. Stevens, and E. S. Buckler IV. 2009. Tracking footprints of maize domestication and evidence for a massive selective sweep on chromosome 10. *Proceedings of the National Academy of Sciences, USA* 106 (Suppl 1): 9979–9986.

Turnbull, C. 1935. Kelvin and the age of the Sun. *Nature* 136: 761.

Turner, A. H., P. J. Makovicky, and M. A. Norell. 2007. Feather quill knobs in the dinosaur *Velociraptor*. *Science* 317: 1721.

Unwin, D. M. 1998. Feathers, filaments, and theropod dinosaurs. *Nature* 391: 119–120.

Unwin, D. M. 2003. On the phylogeny and evolutionary history of pterosaurs. In *Evolution and Palaeobiology of Pterosaurs*, ed. E. Buffetaut and J.-M. Mazin. London, UK: Geological Society, 139–190, Special Publications 217.

Ussher, James. 1658. *The Annals of the World*. London: Printed by E. Tyler for J. Crook and G. Bedell.

Vasilakos, K., R. J. Wilson, et al. 2005. Ancient gill and lung oscillators may generate the respiratory rhythm of frogs and rats. *Journal of Neurobiology* 62: 369–385.

Vinther, J., D. E. Briggs, et al. 2008. The colour of fossil feathers. *Biology Letters* 4: 522–525.

vonHoldt, B. M., J. P. Pollinger, et al. 2010. Genome-wide SNP and haplotype analyses reveal a rich history underlying dog domestication. *Nature* 464: 898–902.

Wake, D. B. 2001. Speciation in the round. *Nature* 409: 299–300.

Wayne, R. K., and E. A. Ostrander. 2007. Lessons learned from the dog genome. *Trends in Genetics* 23: 557–567.

Wellnhofer, P. 1988. A new specimen of *Archaeopteryx*. *Science* 240: 1790–1792.

Westover, K. M., and A. L. Hughes. 2001. Molecular evolution of viral fusion and matrix protein genes and phylogenetic relationships among the Paramyxoviridae. *Molecular Phylogenetics and Evolution* 21: 128–134.

Wills, M. A., P. M. Barrett, and J. F. Heathcote. 2008. The modified gap excess ratio (GER*) and the stratigraphic congruence of dinosaur phylogenies. *Systematic Biology* 57: 891–904.

Wolfe, N. D., C. P. Dunavan, and J. Diamond. 2007. Origins of major human infectious diseases. *Nature* 447: 279–283.

Xu, X., M. A. Norell, et al. 2004. Basal tyrannosauroids from China and evidence for protofeathers in tyrannosauroids. *Nature* 431: 680–684.

Xu, X., and F. Zhang. 2005. A new maniraptoran dinosaur from China with long feathers on the metatarsus. *Naturwissenschaften* 92: 173–177.

Xu, X., X. Zheng, and H. You. 2009. A new feather type in a nonavian theropod and the early evolution of feathers. *Proceedings of the National Academy of Sciences, USA* 106: 832–834.

Xu, X., X. Zheng, and H. You. 2010. Exceptional dinosaur fossils show ontogenetic development of early feathers. *Nature* 464: 1338–1341.

Yan, R. H., J. L. Malisch, et al. 2008. Selective breeding for a behavioral trait changes digit ratio. *PLoS ONE* 3: e3216.

Yang, S. H., P. H. Cheng, et al. 2008. Towards a transgenic model of Huntington's disease in a non-human primate. *Nature* 453: 921–924.

Yin, L., M. Zhu, et al. 2007. Doushantuo embryos preserved inside diapause egg cysts. *Nature* 446: 661–663.

Zhang, X. G., and X. G. Hou. 2004. Evidence for a single median fin-fold and tail in the Lower Cambrian vertebrate, *Haikouichthys ercaicunensis*. *Journal of Evolutionary Biology* 17: 1162–1166.

Zhang, F., S. L. Kearns, et al. 2010b. Fossilized melanosomes and the colour of Cretaceous dinosaurs and birds. *Nature* 463: 1075–1078.

Zhang, F., Z. Zhou, et al. 2008. A bizarre Jurassic maniraptoran from China with elongate ribbon-like feathers. *Nature* 455: 1105–1108.

Zhang, Z. D., A. Frankish, et al. 2010a. Identification and analysis of unitary pseudogenes: Historic and contemporary gene losses in humans and other primates. *Genome Biology* 11: R26.

Zhu, W., and S. Freeland. 2006. The standard genetic code enhances adaptive evolution of proteins. *Journal of Theoretical Biology* 239: 63–70.

3

Evolution by Natural Selection

In his masterwork, *On the Origin of Species by Means of Natural Selection,* Charles Darwin (1859) scrutinized evidence on the pattern of life's history and came to a conclusion that startled many of his contemporaries. Earth's organisms were not independently created, but are instead descended with modification from a single common ancestor or a few.

Darwin knew as well as anyone, however, that the mere recognition of a pattern does not constitute a scientific theory. "Such a conclusion," he wrote (p. 3), "even if well founded, would be unsatisfactory, until it could be shown how the innumerable species inhabiting this world have been modified."

In other words, if we are to claim any understanding of life's history, we must explain not only what happened but also the mechanism responsible. What is the process that yields the pattern we call evolution? Darwin's answer, natural selection, is the subject of this chapter.

Natural selection is sufficiently straightforward that at least two authors discovered it well before Darwin (Darwin 1872). In 1813, W. C. Wells used it to explain how human populations on different continents came to differ in their physical appearance and resistance to disease. In 1831, Patrick Matthew discussed it in a treatise on farming trees for lumber with which to build ships. Neither work was widely read, and neither came to Darwin's attention until after he had published the first edition of *The Origin.*

Using plasticine models painted to mimic natural variation in coat color, Sacha Vignieri and colleagues (2010) found that conspicuous mice suffer more predator attacks.

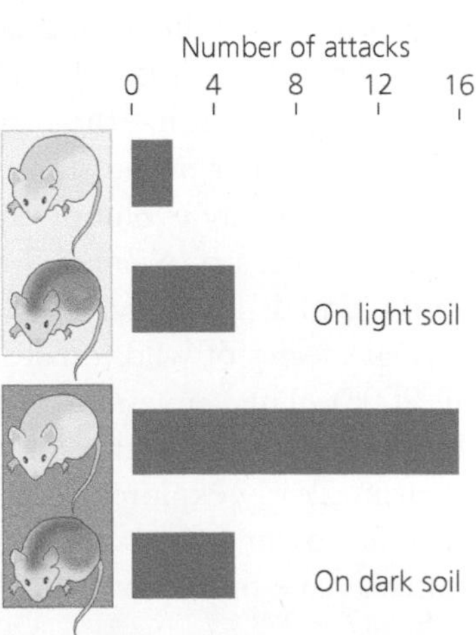

A third author, Alfred Russel Wallace, discovered natural selection independently while Darwin was incubating his ideas. Indeed, it was his receipt of a manuscript sent to him by Wallace that finally prompted Darwin to go public. Papers by Darwin and Wallace were read before the Linnean Society of London in 1858, and Darwin published his book the following year.

Straightforward though it may be, natural selection features subtleties that warrant careful attention. Natural selection is a process in which events that befall individuals alter the collective properties of populations, requiring us to think statistically. And natural selection depends on aspects of genetics that were not understood in Darwin's time—except by Gregor Mendel, whose work on garden peas and the mechanism of inheritance was ignored by virtually everyone.

For these and other reasons, biologists greeted natural selection with considerably greater skepticism than the fact of evolution itself (see Mayr and Provine 1980; Gould 1982; Bowler 2002). Lamarckism and a variety of other hypothetical processes remained popular until the 1930s. It took the rediscovery of Mendel's ideas in 1900 and over three decades of work on population genetics before natural selection was widely accepted as the mechanism of adaptive evolution.

The first section of the chapter sets the stage for our discussion of natural selection by considering artificial selection—the selective breeding of domestic plants and animals. The second section develops the theory of natural selection as a set of claims about populations and a consequence that automatically follows if the claims are true. The third and fourth sections cover in detail examples of research in which the claims and the consequence have been verified. The remaining sections consider natural selection's subtleties, progress since Darwin's time in our understanding of natural selection, and objections to natural selection that continue to be raised by present-day adherents to the theory of special creation.

3.1 Artificial Selection: Domestic Animals and Plants

To understand the mechanism of evolution in nature, Darwin studied the mechanism of evolution under domestication. That is, he studied the method breeders use to modify their crops and livestock. Darwin's favorite domestic organism was the pigeon. He bred pigeons himself to learn the experts' techniques. To refine a breed of pigeon so that, for example, the birds' tail feathers fan more spectacularly, breeders employ artificial selection. They scrutinize their flocks and select the individuals with the most desirable traits. These the breeders cross to produce the next generation. If the desirable traits are passed from parents to offspring, then the next generation—the progeny of the selected birds—will show the desirable traits in a higher proportion than last year's flock.

To increase the frequency of desirable traits in their stocks, plant and animal breeders employ artificial selection.

We will study evolution under domestication by considering the tomato. The domestic tomato, *Solanum lycopersicum*, occurs around the world, both in cultivation and as a weedy escapee. It is closely related to, and can interbreed with, several species of wild tomatoes, all found in western South America (Spooner et al. 2005). The domestic tomato was first cultivated by Native Americans before Europeans arrived in the New World (Tanksley 2004). It traveled to Europe with returning early explorers and spread around the globe from there (Albala 2002).

The power of artificial selection is evident in **Figure 3.1**. All species of wild tomato have fruits that, like the currant tomato on the left, are typically less than

Figure 3.1 Wild and domestic tomatoes Wild tomatoes have tiny fruit, like that of the currant tomato on the left. Domestic tomatoes are descended from tiny-fruited ancestors, but as a result of artificial selection have large fruit, like that of the Red Giant on the right. From Frary et al. (2000).

a centimeter across and weigh just a few grams (Frary et al. 2000). The ancestor of the domestic tomato probably had similarly tiny fruit. Modern varieties of domestic tomato, like the Red Giant on the right, have fruit 15 cm or more across that can weigh more than a kilogram. Descent with modification, indeed.

The disparity in fruit size between wild versus domestic tomatoes is largely due to genetic differences. Tomatoes carry a gene called *fw2.2* (Tanksley 2004). The gene encodes a protein made during early fruit development (Frary et al. 2000). The protein's job is to repress cell division. The more of the protein a plant makes, the smaller its fruit (Liu et al. 2003). Changes in the nucleotide sequence in the *fw2.2* promoter—the gene's on-off switch—alter the timing of production and the amount of repressor made (Cong et al. 2002; Nesbitt and Tanksley 2002). Every wild tomato tested has carried alleles of *fw2.2* associated with high production of the repressor and small fruit (Tanksley 2004). Every cultivated tomato has carried alleles associated with low production of the repressor and large fruit.

Anne Frary and colleagues (2000), working in the lab of Steven Tanksley, demonstrated the influence of *fw2.2* on fruit size with an elegant experiment. The researchers used genetic engineering to insert copies of a small-fruited allele into domestic tomatoes. The fruit on the left in **Figure 3.2** is from an unmanipulated plant; the fruit on the right is from a sibling that has been genetically modified to carry the wild, small-fruited allele. The fruits differ in size by about 30%.

Figure 3.2 A genetically determined difference in fruit size The tomato on the left carries only domestic alleles of the *fw2.2* gene. Its sibling on the right carries copies of the wild allele. From Frary et al. (2000).

Tanksley envisions a scenario in which early tomato farmers noticed variation in fruit size among their plants (Nesbitt and Tanksley 2002; Tanksley 2004). Some of this variation was due to the plants' possession of different alleles of the *fw2.2* gene. Large-fruited alleles might have been present as rare variants before domestication, or they might have arisen as new mutations in cultivated populations. Because the farmers preferred larger tomatoes, year after year they planted their fields with seeds from the largest fruit of the previous crop. By this discipline, the farmers eventually eliminated small-fruited alleles from their stocks.

Bin Cong and colleagues (2008) identified a second gene, called *fas*, that influences fruit size by controlling the number of compartments in the mature fruit. Alleles associated with fewer compartments are common in wild tomatoes and medium-sized domestic varieties. Alleles associated with more compartments are

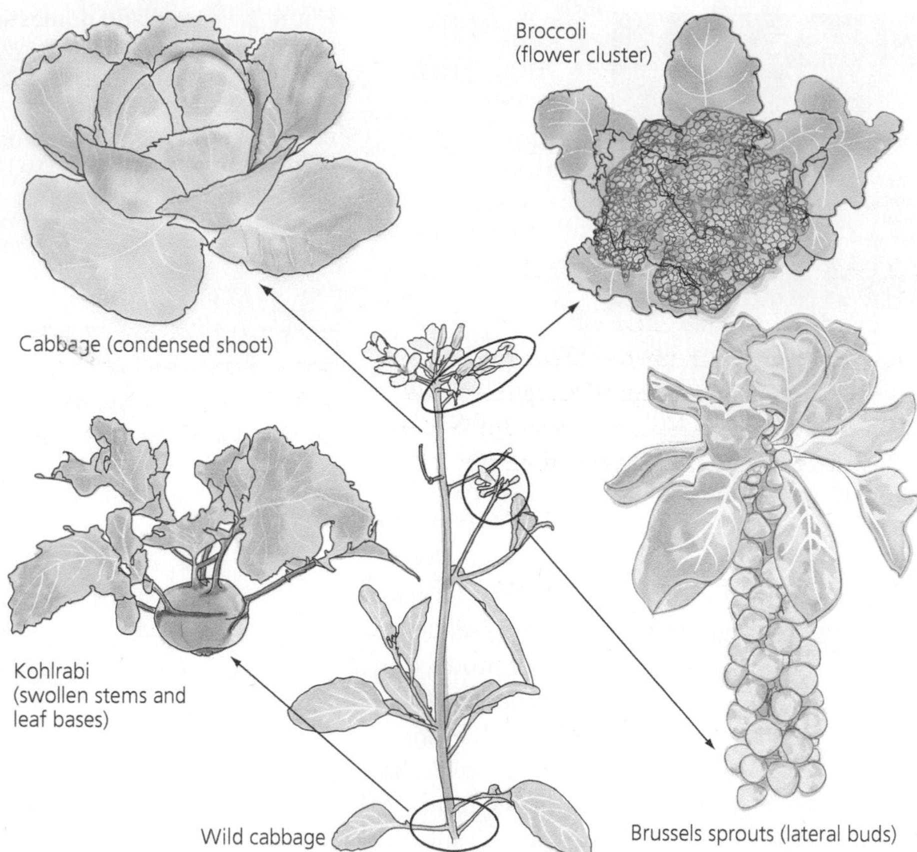

Figure 3.3 Wild and domestic varieties of *Brassica oleracea* Cauliflower (*Brassica oleracea botrytis*), broccoli (*Brassica oleracea italica*), brussels sprouts (*Brassica oleracea gemmifera*), kale (*Brassica oleracea acephala*), and kohlrabi (*Brassica oleracea gongylodes*) are all derived from wild cabbage (*Brassica oleracea oleracea*). After Niklaus (1997).

common in large domestic varieties. Cong and colleagues infer that the *fas* alleles associated with more compartments and large fruit derive from a mutation that arose in a domesticated tomato population. This novel source of variation gave farmers new opportunities to selectively cultivate large-fruited plants.

Farmers practicing artificial selection can, of course, change more than the size of a plant's fruit. The vegetables shown in **Figure 3.3**—broccoli, brussels sprouts, cauliflower, kale, and kohlrabi—are strikingly different in architecture. Yet all are varieties of the wild cabbage, *Brassica oleracea*, from which they are derived.

The dramatic differences between wild versus domestic varieties raises a question. If traits like large fruit in tomatoes evolve so readily under domestication, why have they not evolved in nature? The likely answer is that organisms with traits favored by humans would fare badly in the wild. Imagine a chihuahua living among wolves. Regarding tomatoes, Tanksley (2004) argues that in the wild, small fruits are better because they are more easily carried by the small animals that disperse the seeds. As far as we know, this hypothesis has not been tested.

The general hypothesis that traits favored under domestication are deleterious in the wild has, however, been tested in other organisms. In rare cases, traits evolved under domestication are advantageous in nature. By interbreeding with domestic dogs, for example, North American gray wolves have acquired a genetic variant conferring a black coat that benefits individuals living in forests (Anderson et al. 2009). In the vast majority of cases, however, traits selected under domestication are disastrous in nature (Frankham 2008). Fifth-generation farm salmon released into the wild to compete with stream fish, for instance, had an average lifetime reproductive success totaling just 16% that of their native cousins (Fleming et al. 2000). The fate organisms in nature brings us to natural selection.

3.2 Evolution by Natural Selection

Darwin realized that a process much like artificial selection happens automatically in nature. His theory of evolution by **natural selection**, which he laid out in his introduction to *On the Origin of Species*, can be stated as a short set of postulates and a consequence that follows if the postulates are true. Darwin considered the rest of the book "one long argument" in the theory's support (1859, p. 459). Darwin's postulates—claims about the nature of populations—are as follows:

1. The individuals within a population differ from one another.
2. The differences are, at least in part, passed from parents to offspring.
3. Some individuals are more successful at surviving and reproducing than others.
4. The successful individuals are not merely lucky; instead, they succeed because of the variant traits they have inherited and will pass to their offspring.

Darwin and Wallace realized that a process similar to artificial selection happens automatically in nature.

If all four postulates hold, then the composition of the population inevitably changes from one generation to the next.

Figure 3.4 shows how Darwin's theory, with the postulates phrased in slighty different language, might play out in a population of field mice that has recently invaded a white sand beach.

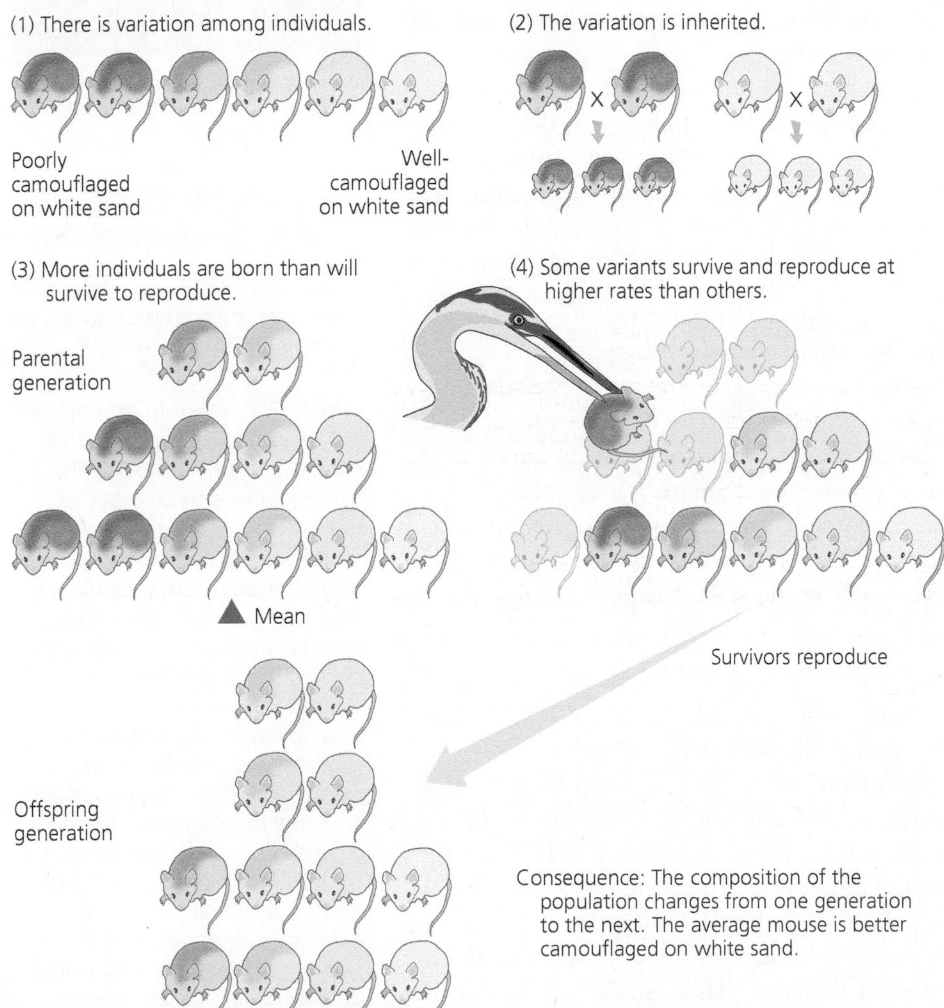
(1) There is variation among individuals.
Poorly camouflaged on white sand
Well-camouflaged on white sand
(2) The variation is inherited.
x
x
(3) More individuals are born than will survive to reproduce.
Parental generation
Mean
(4) Some variants survive and reproduce at higher rates than others.
Survivors reproduce
Offspring generation
Mean
Consequence: The composition of the population changes from one generation to the next. The average mouse is better camouflaged on white sand.

Figure 3.4 Darwin's theory of evolution by natural selection Darwin's theory consists of a short set of claims about populations of organisms and a logical outcome that follows, as a matter of simple mathematics, if the four postulates are true. These cartoons show how the theory might work in a population of field mice that have recently invaded a white sand beach and are exposed to predation by herons. If the mice vary in coat color, and if herons eat the most conspicuous mice, and if the survivors pass their coat color to their offspring, then the population will show a higher proportion of inconspicuous mice each generation. Inspired by Hoekstra et al. (2006), Mullen and Hoekstra (2008), Mullen et al. (2009), and Vignieri et al. (2010).

The logic is straightforward. Inherited traits conducive to survival and reproduction are passed to offspring frequently and thus become ever more common; inherited traits conducive to death without issue are passed to offspring rarely and thus become ever more scarce. As a result, the characteristics of the population change slightly with each succeeding generation. This gradual change in the population is Darwinian evolution.

Natural selection is a process that results in descent with modification, or evolution.

While straightforward, the logic contains a subtlety that can cause confusion. To understand how natural selection works, we have to think statistically. The selection itself—the surviving and reproducing—happens to individuals, but what changes is populations. Individual beach mice, for example, either get eaten or escape detection because of their coat color. But individuals do not change the color of their coats. And they produce offspring whose coats, on average, look just like mom and dad's. The population does not change because the individual mice want to change or need to change. Instead, the population changes as a result of simple, cold arithmetic. Some mice make babies, and their traits persist into the next generation. Other mice fail to make babies, and their traits vanish.

Darwin thought of individuals whose variant traits improve their chances of living and procreating as having been naturally selected, just as a tomato with large fruits is artificially selected as seed stock by a farmer. It is crucial to recognize, however, that during natural selection there is no conscious intent and no ultimate goal in mind. During natural selection, it just happens to be the case that individuals with particular variant traits do better.

Darwin described individuals who are better at surviving and reproducing, and whose offspring make up a greater percentage of the population in the next generation, as more fit. He thus gave the everyday English words *fit* and *fitness* new meanings. **Darwinian fitness** is an individual's ability to survive and reproduce.

The individuals that are fittest in the Darwinian sense are not always the ones we would ordinarily think of as being the most physically fit. This distinction is illustrated by data that Meritxell Genovart and colleagues (2010) collected during a government-run culling program designed to control an exploding population of yellow-legged gulls **(Figure 3.5a)**. The blue bars in Figure 3.5b show the relative numbers of birds with low, normal, or high muscle condition among 506 gulls shot by human hunters—presumably a random sample of the population. The green bars show the proportions among 122 gulls killed by trained falcons and hawks. Compared to the random sample, gulls with low muscle condition are, as we might expect, overrepresented among the birds taken by raptors. However, birds with high muscle condition are also overrepresented. During predation, the gulls with the best physical fitness did not have the highest Darwinian fitness.

(a) A yellow-legged gull

(b) Selective predation by trained raptors

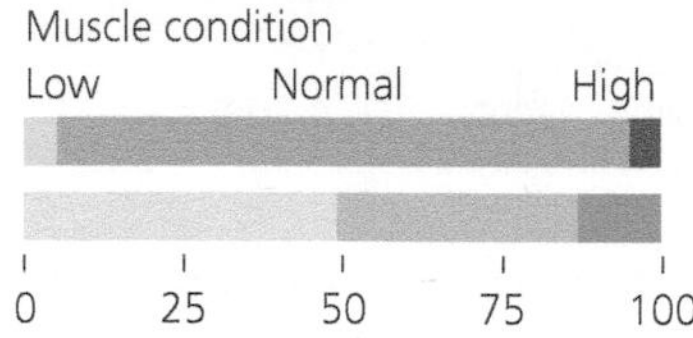

Figure 3.5 Darwinian versus physical fitness (a) A yellow-legged gull (*Larus michahellis*). (b) The relative frequencies of gulls with low, normal, and high muscle condition among two samples from a large population. Blue bars represent a random sample, among which gulls with both low and high physical fitness are rare. Green bars represent a sample taken by raptors. The predators killed both a disproportionately high number of gulls with low muscle condition and a disproportionately high number of gulls with high muscle condition. Exceptional physical fitness failed to confer exceptional Darwinian fitness. Plotted from data in Table 3 of Genovart et al. (2010).

Note that fitness is relative. It refers to how well an individual survives and reproduces compared to other individuals of its species. A trait that increases an organism's fitness relative to individuals lacking it, such as a well-camouflaged pelt, is called an **adaptation**. Such a trait is also said to be **adaptive**.

One of the most attractive aspects of the theory of evolution by natural selection is that each of the four postulates, and the consequence, can be verified independently. That is, the theory is testable. There are neither hidden assumptions nor facets that have to be accepted uncritically. In the next two sections, we examine each of the four postulates, and the predicted consequence, by reviewing two studies: an experiment on snapdragons and an ongoing study of finches in the Galápagos Islands off the coast of Ecuador. These studies show that the theory of evolution by natural selection has been verified by direct observation.

3.3 The Evolution of Flower Color in an Experimental Snapdragon Population

Kristina Niovi Jones and Jennifer Reithel (2001) wanted to know whether natural selection by choosy bumblebees could drive the evolution of flower color in snapdragons. To find out, they established an experimental population of 48 individuals in which they made sure Darwin's postulates 1 and 2 were true. Then they monitored the plants and their offspring to see whether postulates 3 and 4, and the predicted consequence, were true as well.

Postulate 1: Individuals Differ from One Another

The snapdragons in Jones and Reithel's population varied in flower color. Three-quarters of the plants had flowers that were almost pure white, with just two spots of yellow on the lower lip. The rest had flowers that were yellow all over.

Postulate 2: The Variation Is Inherited

The variation in color among Jones and Reithel's plants was due to differences in the plants' genotypes for a single gene. The gene has two alleles, which we will call *S* and *s*. Individuals with either genotype *SS* or *Ss* have white flowers with just two spots of yellow. Individuals with genotype *ss* are yellow all over. Among the 48 plants in the experimental population, 12 were *SS*, 24 were *Ss*, and 12 were *ss*. **Figure 3.6a** shows the variation in phenotype among Jones and Reithel's snapdragons, and the variation in genotype responsible for it.

Testing Postulate 3: Do Individuals Vary in Their Success at Surviving or Reproducing?

Although Jones and Reithel ran their experiment in a meadow in Colorado, they kept their snapdragons in pots and made sure all of the plants survived.

The researchers did not intervene, however, to help the snapdragons reproduce. Instead, they let free-living bumblebees pollinate the plants. To gauge the plants' success at reproducing by exporting pollen, Jones and Reithel tracked the number of times bees visited each flower. To gauge the plants' success at reproducing by making seeds, the researchers counted the seeds produced from each fruit. Consistent with Darwin's third postulate, the plants showed considerable variation in reproductive success, both as pollen donors and as seed mothers.

Testing Postulate 4: Is Reproduction Nonrandom?

Jones and Reithel expected one color to attract more bees than the other, but they did not know which color it would be. The yellow spots on otherwise white snapdragons are thought to serve as nectar guides, helping bumblebees find the reward the flower offers. Entirely yellow flowers lack nectar guides and so might be less attractive to bees. Or they might be more visible against the background vegetation and thus more attractive. Jones and Reithel found that white flowers attracted twice as many bee visits as yellow flowers (Figure 3.6b, left).

Reproductive success through seed production was less strongly associated with color than was success through pollen donation. Nonetheless, the white plants were somewhat more robust than the yellow plants and so produced, on average, slightly more seeds per fruit (Figure 3.6b, right).

Consistent with Darwin's fourth postulate, reproductive success was not random. At both pollen export and seed production, white plants beat yellow.

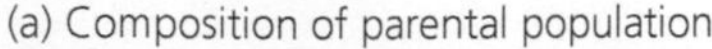

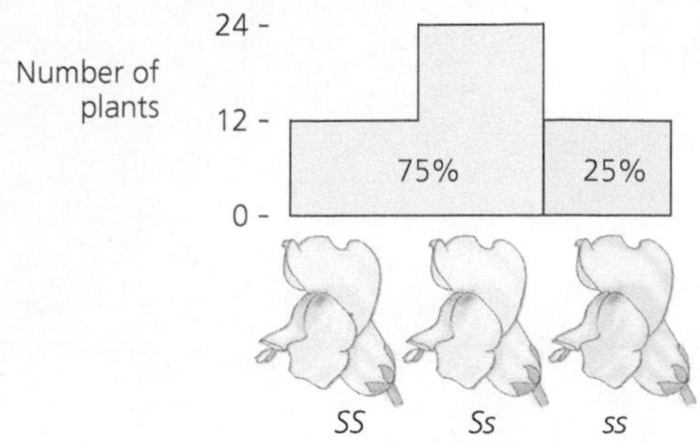

(b) Differences in reproductive success through male function (left) and female function (right)

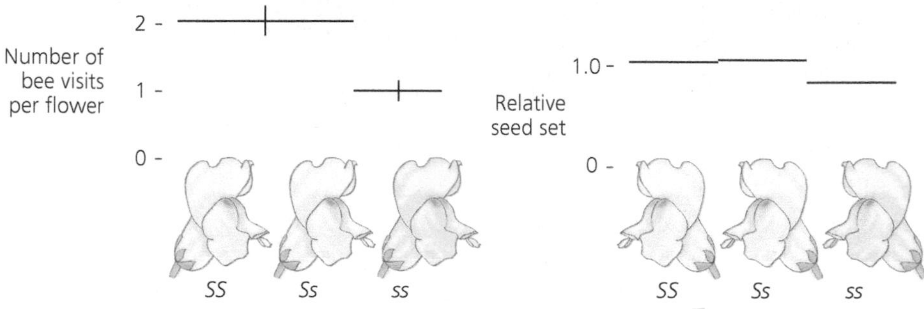

(c) Composition of offspring population

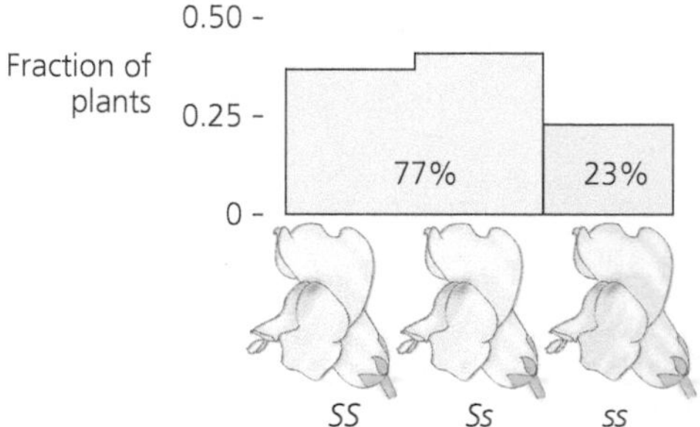

Figure 3.6 Evolution by natural selection demonstrated in an experimental population of snapdragons (a) The plants in the parental population vary in flower color. This variation in phenotype is due to variation in genotype. The graph shows the number of plants in the population with each of the three possible genotypes. (b) The white plants are more successful at reproducing. They are visited by bumblebees twice as often (left), and make more seeds (right). (c) Because plants with white flowers are more successful at passing on their genes, they occupy a larger fraction of the population in the next generation. Prepared from data in Jones and Reithel (2001). [In (b) left, the vertical bars show the size of the standard error; they indicate the accuracy of the researchers' estimate of the mean number of bee visits. In (b) right, the values for relative seed set were calculated as the fraction of seeds actually produced by plants with a particular genotype divided by the fraction of seeds expected based on the frequencies of the genotypes.]

Testing Darwin's Prediction: Did the Population Evolve?

The bumblebees that volunteered to participate in Jones and Reithel's experiment played the same role that Darwin himself played in breeding pigeons: They selected particular individuals in the target population and granted them high reproductive success. Since white plants had higher reproductive success than yellow, and since flower color is determined by genes, the next generation of snapdragons should have had a higher proportion of white flowers.

Indeed, the next generation did have a higher proportion of white flowers (Figure 3.6c). Among the plants in the starting population, 75% had white flowers; among their offspring, 77% had white flowers. The snapdragon population evolved as predicted. An increase of two percentage points in the proportion of white flowers might not seem like much. But modest changes can accumulate over many generations. With Jones and Reithel's population evolving at this rate, it would not take many years for white flowers to all but take over.

The theory of evolution by natural selection is testable. When researchers set up a plant population in which postulates 1 and 2 were true, they found that postulates 3 and 4 also were true—as was Darwin's prediction that the population would evolve as a result.

Jones and Reithel's experiment shows that Darwin's theory works, at least in experimental populations when researchers have made certain that Darwin's first two postulates hold. But does the theory work in completely natural populations, in which researchers have manipulated nothing? To find out, we turn to research on Darwin's finches in the Galápagos Islands.

3.4 The Evolution of Beak Shape in Galápagos Finches

Peter Grant, Rosemary Grant, and their colleagues have been studying finches in the Galápagos Archipelago since 1973 (see P. R. Grant 1999; B. R. Grant and P. R. Grant 1989, 2003; P. R. Grant and B. R. Grant 2002a, 2002b, 2005, 2006; B. R. Grant 2003). Collectively called Darwin's finches, the birds are derived from a small flock of dome-nested finches that invaded the archipelago, most likely from the Caribbean, 2 to 3 million years ago (Sato et al. 2001; Burns et al. 2002). The descendants of this flock comprise 13 species that inhabit the Galápagos, plus a 14th on Cocos Island. Close examination of the evolutionary tree in **Figure 3.7** reveals that all of these species are closely related. The deepest split on the tree separates two lineages of warbler finches that still recognize each other as potential mates and are thus classified (despite each having its own name) as belonging to a single species. The third-deepest split separates two lineages of sharp-beaked ground finches that are likewise considered a single species. Consistent with their close kinship, all species of Darwin's finches are similar in size and coloration. They range from 4 to 6 inches in length and from brown to black in color. They do, however, show remarkable variation in beak size and shape.

The beak is the primary tool used by birds in feeding, and the enormous range of beak morphologies among the Galápagos finches reflects the diversity of foods they eat. The warbler finches (*Certhidea olivacea* and *Certhidea fusca*) feed on insects, spiders, and nectar; woodpecker and mangrove finches (*C. pallida* and *C. heliobates*) use twigs or cactus spines as tools to pry insect larvae or termites from dead wood; several ground finches in the genus *Geospiza* pluck ticks from iguanas and tortoises in addition to eating seeds; the vegetarian finch (*Platyspiza crassirostris*) eats leaves and fruit.

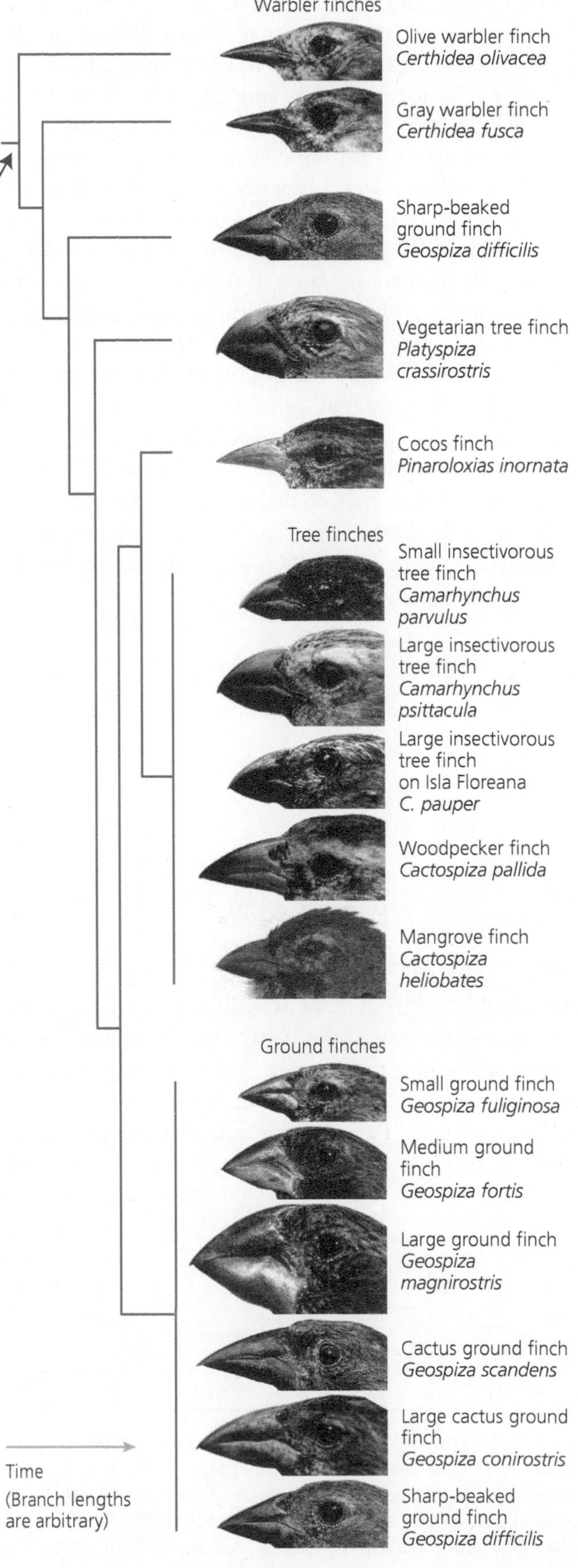

Figure 3.7 Diversity in Darwin's finches These finches are all descended from a common ancestral population (red arrow) that traveled from the Caribbean to the Galápagos Archipelago. The evolutionary tree, estimated from similarities and differences in DNA sequences by Kenneth Petren and colleagues (2005), shows the sometimes complex relationships among the major groups. The photos, from Petren et al. (1999) and Greg Lasley, show the extensive variation among species in beak size and shape.

For a test of the theory of evolution by natural selection, we focus on data Grant and Grant and colleagues have gathered on the medium ground finch, *Geospiza fortis*, on Isla Daphne Major **(Figure 3.8)**.

Daphne Major's location and tiny size make it a superb natural laboratory. Like all islands in the Galápagos, it is the top of a volcano **(Figure 3.9)**. It rises from the sea to a maximum elevation of just 120 meters. It has a main crater and a small secondary crater. Only one spot on the island is both flat enough and large enough to pitch a camp. It takes a mere 20 minutes to walk from the campsite all the way around the main crater's rim and back to camp. Despite the equatorial location, the climate is seasonal. A warmer, wetter season from January through May alternates with a cooler, drier season from June through December. The vegetation consists of dry forest and scrub along with several species of cactus.

The medium ground finches on Daphne Major make an ideal study population. Few finches migrate onto or off of the island, and the population is small enough to be studied exhaustively. In an average year, the population consists of about 1,200 individuals. By 1977, Grant and Grant's team had captured and marked more than half of them; since 1980, virtually 100% of the population has been marked. Medium ground finches live up to 16 years (Grant and Grant 2000). Their generation time is 4.5 years (Grant and Grant 2002a).

Figure 3.8 The medium ground finch, *Geospiza fortis* (top) An adult male; (bottom) an adult female. Peter R. Grant, Princeton University.

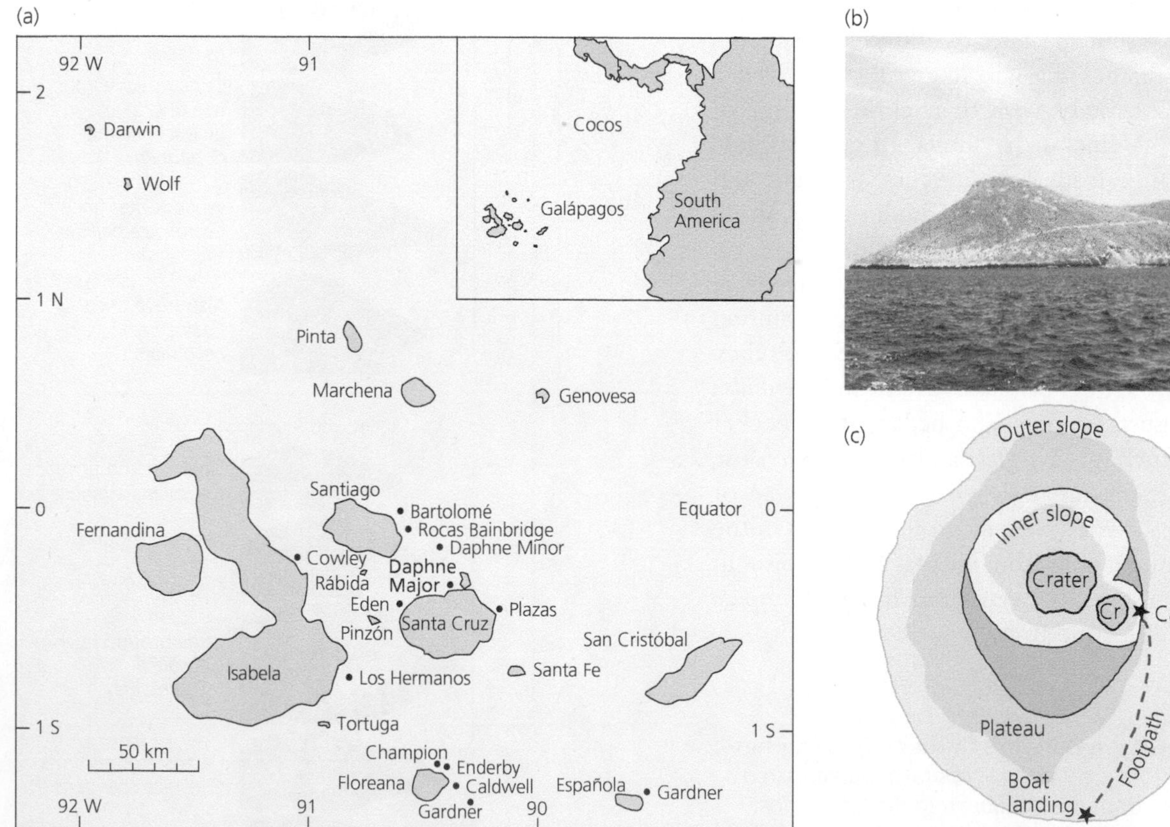

Figure 3.9 The Galápagos Archipelago and Daphne Major (a) Cocos Island and the Galapágos Archipelago, home of Darwin's finches. Isla Daphne Major is a tiny speck between Santa Cruz and Santiago. (b) Daphne Major, seen from a boat approaching the island. Visible as a faint white line running upward from left to right is the footpath that runs from the boat landing (at the waterline) to the campsite (on the rim of the crater). Courtesy of Robert D. Podolsky. (c) A map of Daphne Major. Note the island's tiny size. Redrawn from Boag and Grant (1984a).

Medium ground finches are primarily seed eaters. The birds crack seeds by grasping them at the base of the bill and applying force. Grant and Grant and their colleagues have shown that both within and across finch species, beak size is correlated with the size of seeds harvested. In general, birds with bigger beaks eat larger seeds, and birds with smaller beaks eat smaller seeds. This is because birds with different beak sizes are able to handle different sizes of seeds more efficiently (Bowman 1961; Grant et al. 1976; Abbott et al. 1977; Grant 1981).

Testing Postulate 1: Is the Finch Population Variable?

The researchers mark every finch they catch by placing on its legs a numbered aluminum band and three colored plastic bands. This allows the scientists to identify individual birds in the field. The scientists also weigh each bird and measure its wing length, tail length, beak width, beak depth, and beak length. All of these traits show diversity. For example, when Grant and Grant plotted beak depth in the Daphne Major population, the data indicated that the trait varies considerably **(Figure 3.10)**. All of the finch characteristics Grant and Grant have measured conform to Darwin's first postulate. Variation among the individuals within populations is virtually universal (see Chapter 5).

Some *Geospiza fortis* have beaks that are only half as deep as those of other individuals.

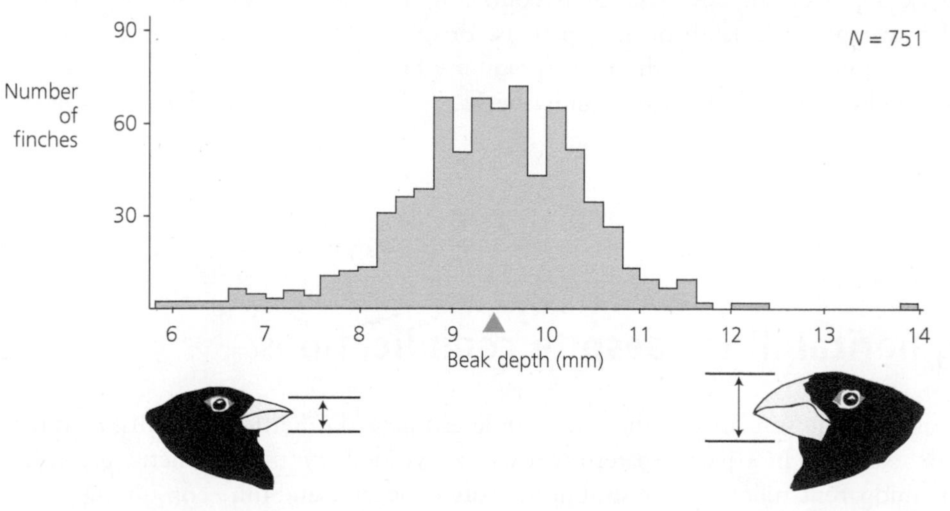

Figure 3.10 Variation in beak depth in medium ground finches This histogram shows the distribution of beak depth in medium ground finches on Daphne Major in 1976. A few birds have shallow beaks; a few birds have deep beaks; most birds have medium beaks. *N* stands for sample size; the blue triangle along the horizontal axis indicates the mean, or average. Redrawn from Boag and Grant (1984b).

Testing Postulate 2: Is Some of the Variation among Individuals Heritable?

Within the Daphne Major population, individual finches could vary in beak depth because the environments they have experienced are different, or because their genotypes are different, or both. There are several ways that environmental variation could cause the variation in beak depth documented in Figure 3.10. Variation in the amount of food that individual birds happened to have received as chicks could lead to variation in beak depth among adults. Injuries or abrasion against hard seeds or rocks could also affect beak size and shape.

To determine whether at least part of the variation among finch beaks is genetically based, and thus passed from parents to offspring, Peter Boag, a colleague of Peter Grant and Rosemary Grant, estimated the **heritability** of beak depth.

The heritability of a trait is defined as the fraction of the variation in a population that is due to differences in genes. It can take any value between 0 and 1. We will develop the theory of heritability estimation more fully later (in Chapter

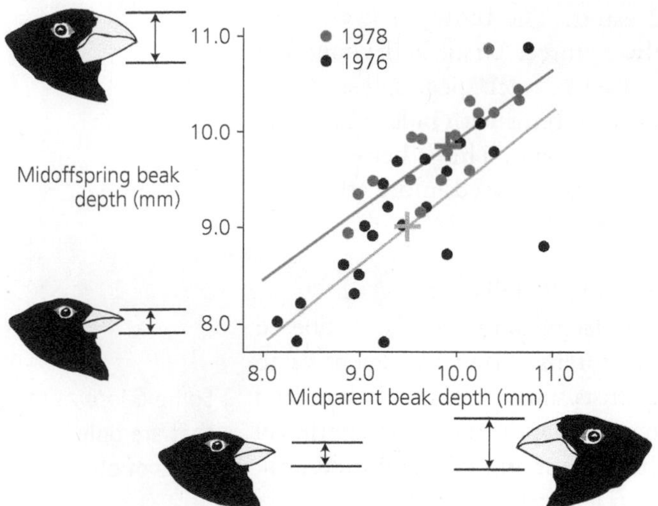

Figure 3.11 Heritability of beak depth in *Geospiza fortis* This graph shows the relationship between the beak depth of parents and their offspring. Midparent value is the average of the mother and father; midoffspring value is the average of the offspring. The lines in the graph are statistical best-fit lines. Data for both 1978 and 1976 show a strong relationship between the beak depth of parents and their offspring. Redrawn from Boag (1983).

9). For now, we note that if the differences among individuals are due to differences in the alleles they have inherited, then offspring will resemble their parents.

Boag compared the average beak depth of families of *G. fortis* young, after they had attained adult size, to the average beak depth of their parents. Boag's data reveal a strong correspondence among relatives. As the plot in **Figure 3.11** shows, parents with shallow beaks tend to have chicks with shallow beaks, and

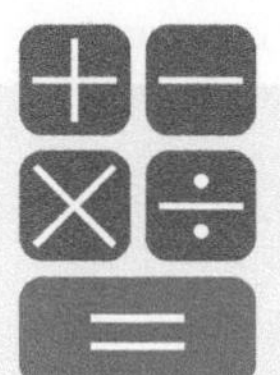

COMPUTING CONSEQUENCES 3.1

Estimating heritabilities despite complications

Heritabilities are estimated by measuring the similarity of traits among close relatives. For example, in a plot of midoffspring trait values versus midparent trait values such as Figure 3.11, the slope of the best-fit line—which ranges from 0 if there is no resemblance to 1 if there is perfect resemblance—gives an estimate of the heritability. The idea is that genes run in families. If the variation in phenotype among individuals is due in part to variation in genotype, then relatives will tend to resemble one another. But a number of confounding issues can complicate this approach. We consider four such issues here: misidentified paternity, conspecific nest parasitism, shared environments, and maternal effects.

Misidentified paternity In many species of birds, even socially monogamous birds like medium ground finches, females sometimes have extrapair sex. This means that a chick's social father is not always its biological father. If researchers simply assume that the social father at a nest is the biological father of all the chicks, they may underestimate the heritability. Misidentified paternity can be avoided by using genetic paternity tests, but they are expensive and time consuming.

Conspecific nest parasitism In some species of birds, females sneak into each other's nests and lay extra eggs. This means that even the social mother at a nest might not be the biological parent of all the chicks. Again, researchers may underestimate the heritability. As with misidentified paternity, this problem can be avoided by using genetic tests.

Shared environments Relatives share their environment as well as their genes, and any correlation that is due to their shared environment inflates the estimate of heritability. For example, birds tend to grow larger when well fed as chicks. But the bountiful breeding territories are often claimed and defended by the largest adults in the population. Young from these territories will tend to become the largest adults in the next generation. As a result, a researcher might measure a strong relationship between parent and offspring beak and parents with deep beaks tend to have chicks with deep beaks. This is evidence that a large proportion of the observed variation in beak depth is genetically based and can be transmitted to offspring (Boag and Grant 1978; Boag 1983).

In finches, the beak depths of parents and offspring are similar. This suggests that some alleles tend to produce shallow beaks, while other alleles tend to produce deeper beaks.

Boag himself would be the first to note that caution is warranted in interpreting these data. Environments shared by family members, maternal effects, conspecific nest parasitism, and misidentified paternity can cause graphs like the one in Figure 3.11 to exaggerate, or to underplay, the heritability of traits. However, Lukas Keller and colleagues (2001) have used modern genetic analyses of *G. fortis* to eliminate most of these confounding factors (**Computing Consequences 3.1**). It is clear that Darwin's second postulate is true for the medium ground finches on Daphne Major. A substantial fraction of the variation in beak size is due to variation in genotype and is transferred across generations.

We do not know the identity of the genes responsible for variation in beak size among medium ground finches. However, Otger Campàs and colleagues (2010), extending earlier work by Arhat Abzhanov and coworkers (2004), offered a clue.

Campàs and colleagues focused on a growth factor, bone morphogenic protein 4 (BMP4), known to be active during embryonic development. BMP4 is a signaling molecule that helps sculpt the shape of bird beaks (Wu et al. 2004). For each of the six species of ground finch, the researchers treated three embryos of a particular developmental age with a fluorescent probe that binds to messenger RNA made by *Bmp4*, the gene encoding BMP4. For each embryo, the scientists

body size and then claim a high heritability for these traits, when in reality there is none. In this case, the real relationship is between the environments that parents and their young each experienced as chicks.

In many species, this problem can be circumvented with what are called cross-fostering, common garden, or reciprocal-transplant experiments. In birds, such studies involve stealing eggs and placing them in the nests of randomly assigned foster parents. Measurements of young, taken once they are grown, are compared with the data from their biological parents. This experimental treatment removes any bias in the analysis created because parents and offspring share environments.

Maternal effects Even cross-fostering experiments cannot remove environmental effects due to differences in the nutrient stores or hormonal contents of eggs. These are called maternal effects. They can be largely avoided by estimating heritabilities from the resemblance between offspring and their fathers only.

Lukas Keller and colleagues (2001) have made painstaking estimates of the heritability of morphological traits in Daphne Major's medium ground finches. The researchers used genetic analyses to confirm the parentage of all the chicks in their sample. They found no evidence of conspecific nest parasitism, but did find that 20% of the chicks had been fathered by extrapair males. Excluding these chicks from their data, Keller and colleagues estimated that the heritability of beak depth is 0.65 (with a standard error of 0.15). That is, about 65% of the variation among finches in beak depth is due to differences in genes. This estimate is uncontaminated by extrapair paternity, conspecific nest parasitism, and maternal effects. It might, however, contain some error due to shared environments.

The Galápagos researchers have been unable to perform a cross-fostering experiment on Darwin's finches. Because the Galápagos are a national park, experiments that manipulate individuals beyond catching and marking are forbidden. But the finches themselves have run a sort of cross-fostering experiment: As mentioned earlier, about 20% of the chicks have been raised by males who are not their biological fathers. If some of the resemblance between parents and offspring is due to shared environments, then these chicks should resemble their social fathers. Using data on the social fathers and their foster offspring, Keller and colleagues calculated the "heritability" of beak depth. It was less than 0.2 and was not statistically distinguishable from zero. The data suggest that shared environments have little influence on the resemblance among relatives' beaks.

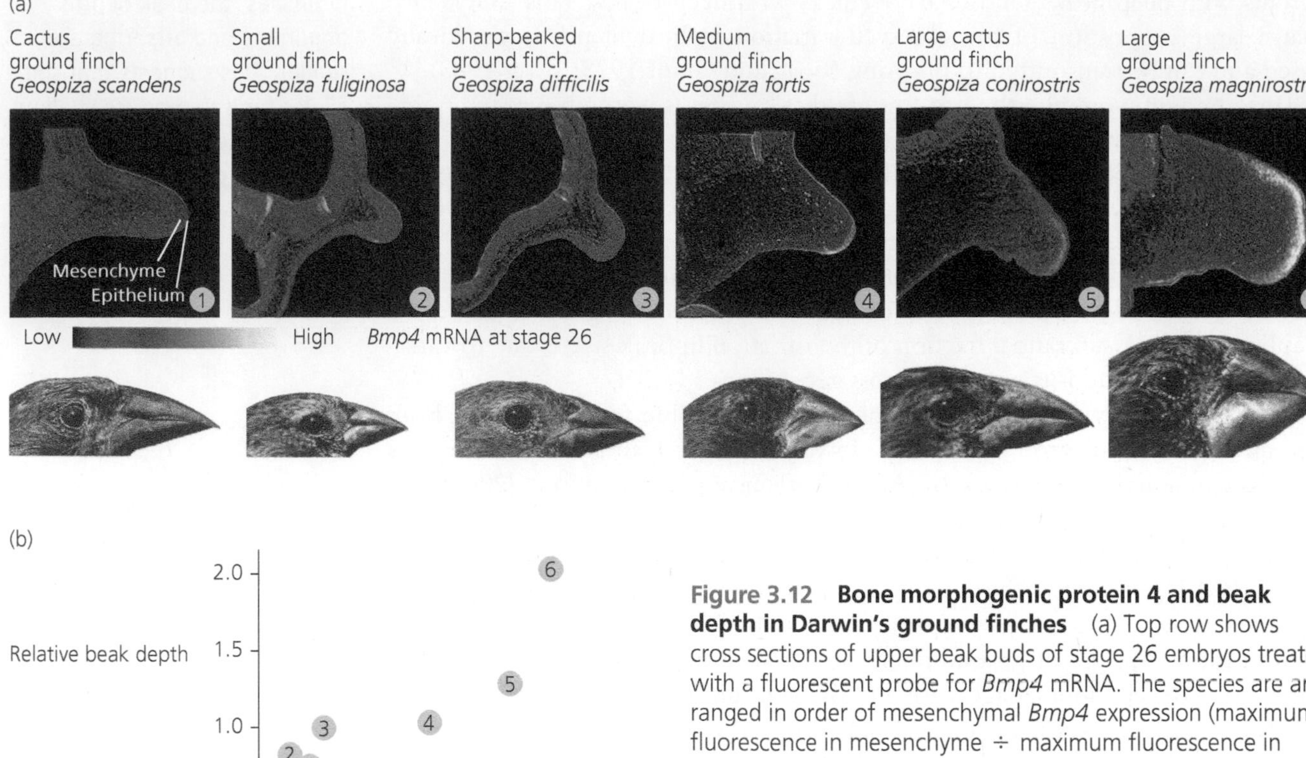

Figure 3.12 Bone morphogenic protein 4 and beak depth in Darwin's ground finches (a) Top row shows cross sections of upper beak buds of stage 26 embryos treated with a fluorescent probe for *Bmp4* mRNA. The species are arranged in order of mesenchymal *Bmp4* expression (maximum fluorescence in mesenchyme ÷ maximum fluorescence in epithelium). Bottom row shows beaks of adult birds. Embryos from Campàs et al. (2010); adult finch photos from Petren et al. (1999). (b) Beak depth versus mesenchymal *Bmp4* expression. All values on both axes scaled relative to *Geospiza difficilis*. Redrawn from Campàs et al. (2010).

quantified *Bmp4* expression in the upper beak bud by measuring the maximum fluorescence in the epithelium (the outer layer of the bud) and the mesenchyme (the tissue under the epithelium), then dividing the latter by the former. This procedure gave a standardized value for mesenchymal *Bmp4* expression.

The top row of photos in **Figure 3.12a** shows a stained beak bud for each species, arranged by increasing mesenchymal *Bmp4* expression. The bottom row shows the beaks of adult birds. The graph in Figure 3.12b demonstrates that species with higher *Bmp4* expression have deeper beaks. Abzhanov and colleagues (2004) suggested that the different species of ground finches harbor alternate versions of one or more of the genes that determine when, where, and how strongly the gene encoding BMP4 is activated.

The genetic mechanisms responsible for variation among individuals within species may or may not be the same as those responsible for differences between species (see McGregor et al. 2007). But a reasonable hypothesis would be that genetically encoded differences in *Bmp4* expression are responsible for some of the variation in beak size among Daphne Major's medium ground finches.

Testing Postulate 3: Do Individuals Vary in Their Success at Surviving and Reproducing?

Because Grant and Grant and their colleagues have been monitoring the finches on Daphne Major every year since 1973, two members of the team, Peter Boag and Laurene Ratcliffe, were on the island in 1977 to witness a terrible drought

(Boag and Grant 1981; Grant 1999). Instead of the usual 130 mm of rain during the wet season, the island got only 24 mm. The plants made few flowers and few seeds. The medium ground finches did not even try to breed. Over a span of 20 months, 84% of the *Geospiza fortis* on Daphne Major disappeared **(Figure 3.13a)**. The researchers inferred that most died of starvation. The decline in numbers was simultaneous with a decline in the availability of the seeds the birds eat (Figure 3.13b); 38 emaciated birds were actually found dead; none of the missing birds reappeared the next year. Clearly, only a fraction of the population survived to reproduce in 1978.

Mortality on this scale is not unusual. Rosemary Grant has shown that 89% of *Geospiza conirostris* individuals die before they breed (Grant 1985). Trevor Price and coworkers (1984) determined that 19% and 25% of the *G. fortis* on Daphne Major died during droughts in 1980 and 1982.

In fact, in every natural population studied, more offspring are produced each generation than survive to breed. In a population of constant size, each parent, over its lifetime, leaves an average of one offspring that survives to breed. But the reproductive capacity of organisms is vastly higher than this. Darwin (1859) picked the elephant to illustrate this point, because it was the slowest breeder then known among animals. He calculated that if all the descendants of a single pair survived and reproduced, then after 750 years there would be 19 million of them. The numbers are even more startling for rapid breeders. Dodson (1960) calculated that if all the descendants of a pair of starfish survived and reproduced, then after just 16 years they would exceed 10^{79}, the estimated number of electrons in the visible universe. The only thing that saves us from being buried in starfish and elephants is massive mortality.

Similarly, data show that in most populations, some individuals that survive to breed do better at mating and making offspring than others. Darwin's third postulate is universally true.

Testing Postulate 4: Are Survival and Reproduction Nonrandom?

Darwin's fourth claim was that the individuals who survive and go on to reproduce, or who reproduce the most, are those with certain variations. Did a nonrandom, or naturally selected, subset of the medium ground finch population survive the 1977 drought? The answer is yes.

As the drought wore on, the number as well as the types of seeds available changed (Figure 3.13c). Finches on Daphne Major eat seeds from a variety of plants. The seeds range from small and soft to large and hard. The small, soft seeds, easy to crack, are the birds' favorites. During the drought, as at other times, the finches ate the small, soft seeds first. Once most of those were gone, the large, hard fruits of a plant called *Tribulus cistoides* became a key food item. Only large birds with deep,

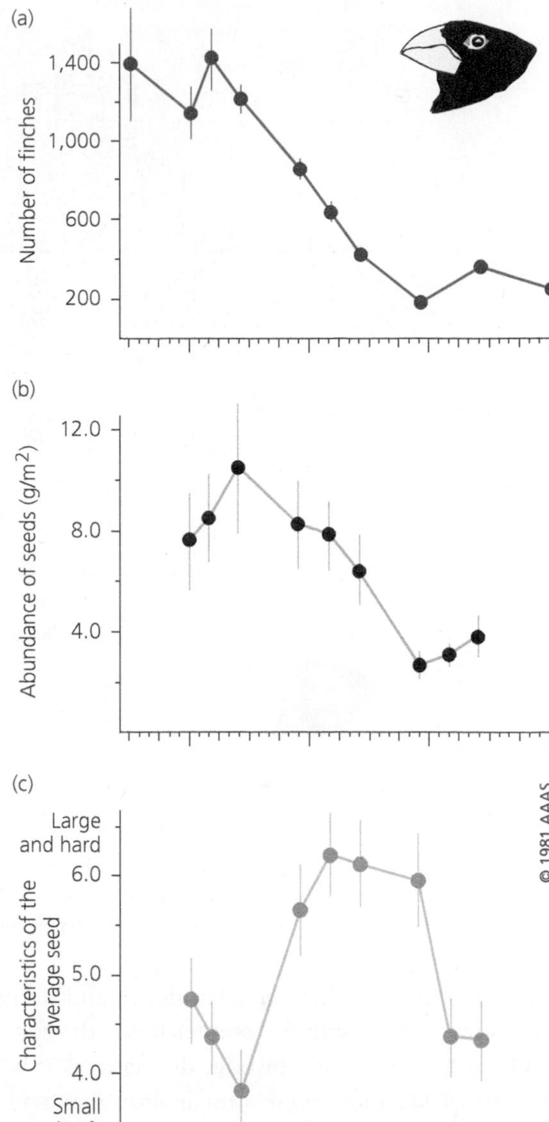

Figure 3.13 Death by starvation during the 1977 drought (a) The number of medium ground finches on Daphne Major before, during, and after the drought. The vertical lines through each data point represent the standard error, an indication of the variation in census estimates. (b) The abundance of seeds on Daphne Major before, during, and after the drought. (c) The characteristics of the average seed available as food to medium ground finches before, during, and after the drought. The hardness index plotted on the vertical axis is a special measure created by Boag and Grant (1981). Redrawn from Boag and Grant (1981).

From "Intense natural selection in a population of Darwin's finches (Geospizinae) in the Galápagos." *Science* 214: 82–85. Reprinted with permission from AAAS.

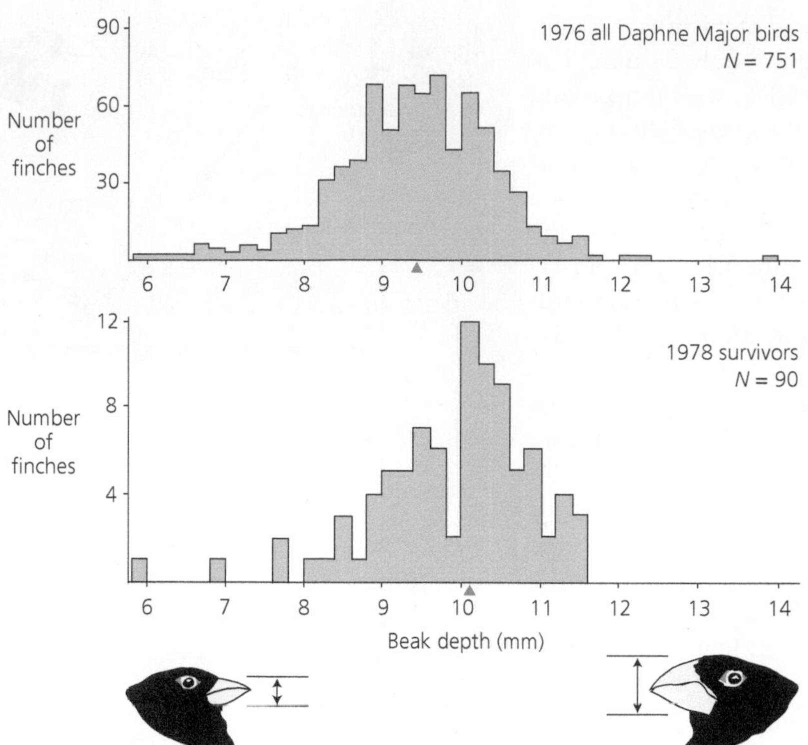

Figure 3.14 Beak depth before and after the drought These histograms show the distribution of beak depth in medium ground finches on Daphne Major before and after the drought of 1977. The blue triangles indicate the population means. Redrawn from Boag and Grant (1984b).

narrow beaks can successfully crack and eat *Tribulus* fruits. The rest were left to turn over rocks and scratch the soil in search of the few remaining smaller seeds.

The top graph in **Figure 3.14** shows the beak sizes of a large random sample of the birds on Daphne Major the year before the drought. The bottom graph shows the beak sizes of a random sample of 90 birds who survived. The survivors had deeper beaks, on average, than the pre-drought birds (and, we can infer, than the birds that starved). Because beak depth and body size are correlated, and because large birds tend to win fights over food, the survivors had larger body sizes too.

During a terrible drought, finches with larger, deeper beaks had an advantage in feeding, and thus in surviving.

The 1977 selection episode, dramatic as it was, was not an isolated event. In 1980 and 1982 there were similar droughts, and individuals with deep beaks and large body size were again naturally selected (Price et al. 1984). Then, in 1983, an influx of warm surface water off the South American coast, called an El Niño, created a wet season with 1,359 mm of rain on Daphne Major—almost 57 times as much as in 1977. This environmental shift led to a superabundance of small, soft seeds and, subsequently, to strong selection for smaller body size (Gibbs and Grant 1987). In wet years, small birds with shallow beaks survive better and reproduce more because they harvest small seeds more efficiently. Larger birds are favored in drought years, but smaller birds are favored in wet years. Natural selection is dynamic.

Testing Darwin's Prediction: Did the Population Evolve?

All four of Darwin's postulates are true for the medium ground finch population on Daphne Major. Darwin's theory therefore predicts a change in the composition of the population from one generation to the next. When the deep-beaked birds who survived the drought of 1977 bred to produce a new generation, they should have passed their genes for deep beaks to their offspring. **Figure 3.15** confirms that they did. The chicks hatched in 1978 had deeper beaks, on average, than the 1976 chicks. The population evolved.

Because of the drought, the finch population evolved. Selection occurs within generations; evolution occurs between generations.

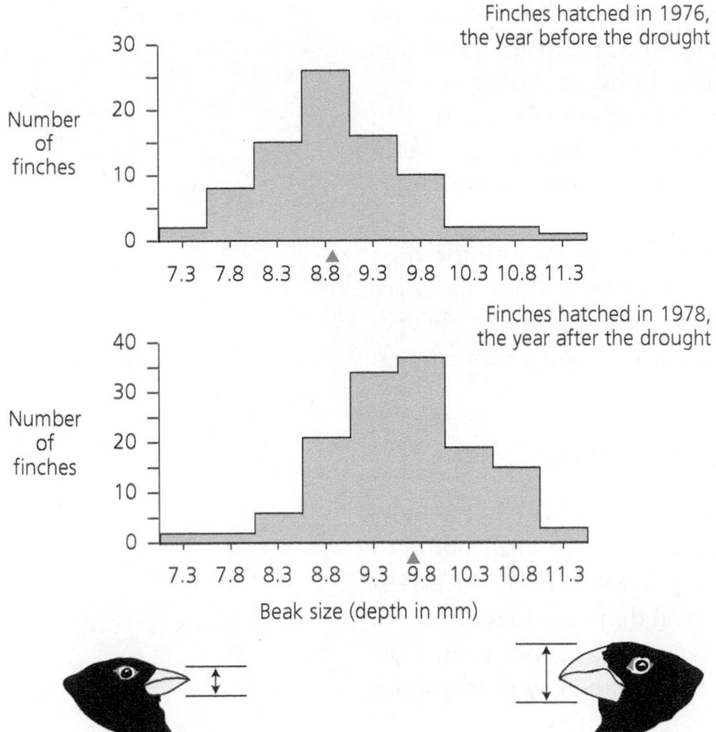

Figure 3.15 Beak depth in the finches hatched the year before the drought versus the year after the drought The red triangles represent population means. Redrawn from Grant and Grant (2003).

Peter Grant and Rosemary Grant and colleagues have continued to monitor the Daphne Major finch population since the 1970s. Thanks to unpredictable changes in the climate and bird community, and consequent changes in the island's plant community, the researchers have witnessed selection events when deep-beaked individuals survived at higher rates and selection events when shallow-beaked individuals survived at higher rates.

Figure 3.16 tracks changes in the population averages for three traits across three decades. Each trait is a statistical composite of measurable characteristics, like beak depth. For example, "PC1 beak size" (Figure 3.16a) combines beak depth, beak length, and beak width. If there had been no evolution in beak size, the 95% confidence intervals for all data points would have overlapped the gray band—the 95% confidence interval for 1973, the first year with complete data. That many do not overlap means the population showed statistically detectable evolution.

Figure 3.16a shows, first, that during the drought of 1977 the finch population evolved a significantly larger average beak size. This change is indicated in blue. The figure further shows that the population remained at this large mean beak size until the mid-1980s, then evolved back to the average size it started with. There the population stayed until another drought struck.

The drought of 2003 and 2004 was as bad as the one of 1977 (Grant and Grant 2006). Again many medium ground finches starved. This time, however, the medium ground finches faced competition from a substantial population of large ground finches (*Geospiza magnirostris*) that had recently appeared on the island. The large ground finches dominated access to, and ate,

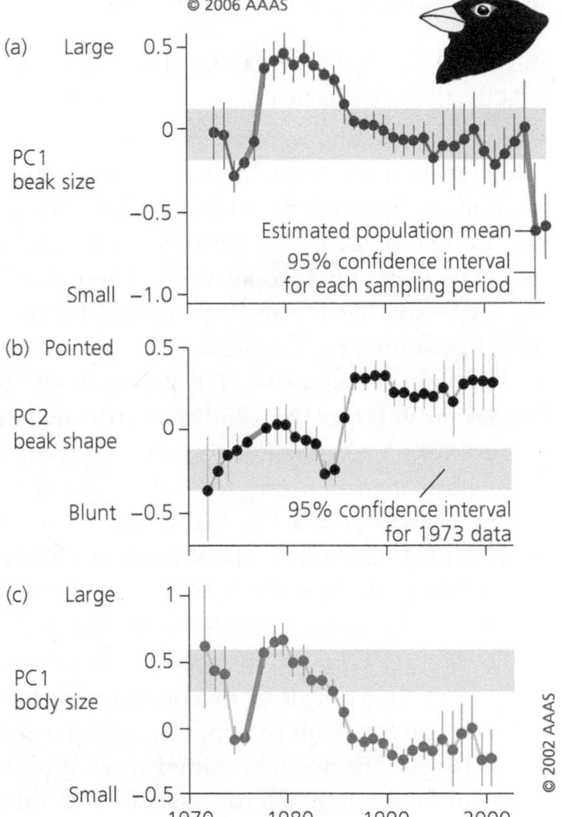

Figure 3.16 A finch population evolves Mean adult beak and body dimensions. Drought years in blue. Data from Grant and Grant (2002, 2006).

(a) From "Evolution of character displacement in Darwin's finches." *Science* 313: 224–226. (b, c) From "Unpredictable evolution in a 30-year study of Darwin's finches." *Science* 296: 707–711. Reprinted with permission from AAAS.

the *Tribulus* fruits that the large-beaked medium ground finches had survived on in 1977. As a result, medium ground finches with large beaks died at higher rates than those with small beaks, and the population evolved toward smaller beaks.

The medium ground finch population also showed substantial evolution in mean beak shape and mean body size (Figure 3.16b and c). The average bird in 2001 had a significantly sharper beak, and was significantly smaller, than the average bird in the mid 1970s (Grant and Grant 2002).

Grant and Grant's study shows that Darwin's mechanism of evolution can be documented in natural populations. When all four of Darwin's postulates are true, populations evolve. The study also shows that small evolutionary changes over short time spans can accumulate into larger changes over longer time spans.

3.5 The Nature of Natural Selection

Although the theory of evolution by natural selection can be stated concisely, tested rigorously in natural populations, and validated, it can be challenging to understand thoroughly. One reason is that evolution is a statistical process—a change in the trait distributions of populations. Statistical thinking does not come naturally to most people, and a number of widely shared ideas about natural selection are incorrect. Our goal in this section is to cover some key points about how selection does, and does not, operate.

Natural Selection Acts on Individuals, but Its Consequences Occur in Populations

When snapdragons were selected by bumblebees, or finches were selected as a result of food shortages, none of the selected individuals changed. Some snapdragons simply reproduced more than competing plants. Some finches survived the drought while others perished. What changed were the characteristics of the snapdragon and finch populations. There was a higher frequency of white-flower genotypes among the seeds produced by the snapdragons and a larger average beak size among the finches.

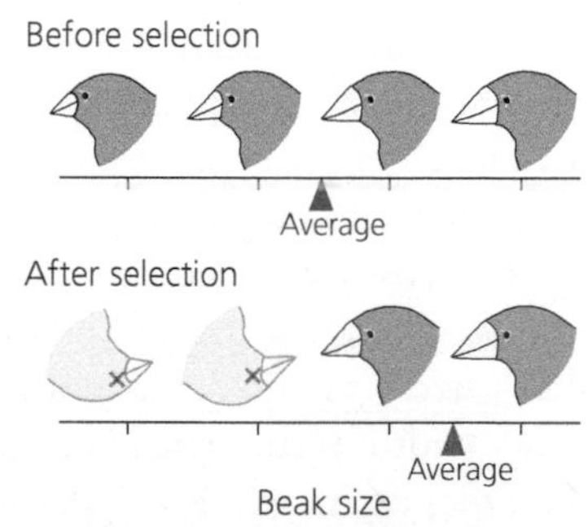

Figure 3.17 Natural selection happens to individuals, but what changes is populations During the drought on Daphne Major, individuals did not change their beak sizes; they simply lived or died. What changed was the average beak size, a characteristic of the population.

The effort of cracking *Tribulus* seeds did not make the beaks of individual finches grow larger. Nor did the birds' need for more food, or their desire for bigger beaks, make their beaks grow. Instead, the average beak depth in the finch population increased due to the simple, if cruel, fact that more small-beaked finches died **(Figure 3.17)**. Likewise, the effort of attracting pollinators had no effect on the pigments in the flowers of individual snapdragons. Instead, the proportion of white versus yellow flowers changed simply because white plants exported more pollen and made more seeds.

To be sure, exposure to particular environmental circumstances alters the phenotypes of individuals in myriad ways. Spending time in the sun, for example, induces many humans to deposit more melanin in their skin—that is, to tan. But such changes are not transmitted to offspring. A woman who sunbathes while pregnant does not give birth to a baby with darker skin. What she and the father transmit to the baby is not a tan, but a heritable tanning capacity.

Natural Selection Acts on Phenotypes, but Evolution Consists of Changes in Allele Frequencies

Finches with large bodies and deep beaks would have survived at higher rates during the drought even if all of the variation in the population had been environmental in origin (that is, if heritabilities had been zero). But no evolution would have occurred. Selection would have altered the frequencies of the phenotypes in the population, but in the next generation the phenotype distribution might have gone back to what it was before selection occurred **(Figure 3.18a)**.

Only when the survivors of selection pass their successful phenotypes to their offspring, via genotypes that help determine phenotypes, does natural selection cause populations to change from one generation to the next (Figure 3.18b). On Daphne Major, the variation in finch phenotypes that selection acted on had a genetic basis. As a result, the new phenotypic distribution seen among the survivors persisted into the next generation.

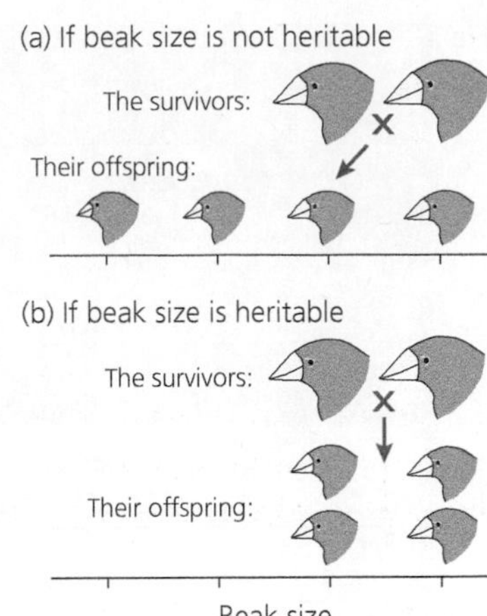

Figure 3.18 Populations evolve only if traits are heritable If variation is due to differences in genotype, then the survivors of selection pass their successful phenotypes to their offspring.

Natural Selection Is Not Forward Looking

Offspring are descended from the survivors of selection imposed by conditions that prevailed before the offspring were born. The finches that hatched on Daphne Major in 1978 were better adapted to drought, on average, than the finches that hatched in 1976. But if the environment had changed again during the lifetime of the 1978 birds, they might not have been any better adapted to the new conditions than the 1976 birds were to a shortage of small, soft seeds.

Students new to evolution may harbor the misconception that organisms can be adapted in advance to future conditions, or that selection can anticipate environmental changes that will occur during future generations. This is impossible. Evolution by natural selection involves no conscious entity with foresight. It is an unthinking, unfeeling mathematical process. And as a result, evolving populations always lag at least a generation behind changes in the environment.

Although Selection Acts on Existing Traits, New Traits Can Evolve

Selection itself generates no new genetic variation, adaptive or otherwise. Differences in survival or reproduction occur only among variants that already exist. The starvation of small-beaked individuals, for example, does not instantaneously create more variation in beak size among finches. In particular, it does not create finches with big beaks optimal for cracking *Tribulus* fruits. Starvation merely winnows the breeding population down to the largest-beaked birds already alive.

This might seem to imply that under natural selection, new traits cannot evolve. But the evolution of new traits is, in fact, possible. There are two reasons: The first applies to all species, the second to species that reproduce sexually. During reproduction in all species, random mutations produce new alleles. During reproduction in sexual species, meiosis and fertilization recombine existing alleles into new genotypes. Mutation and recombination yield new suites of traits that selection may subsequently sort among.

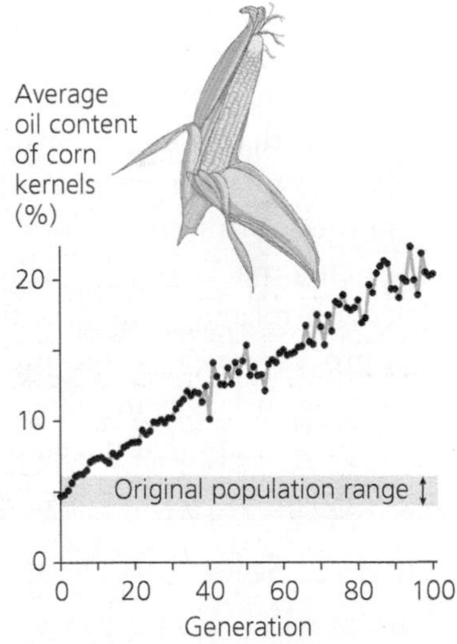

Figure 3.19 Persistent long-term selection can result in dramatic changes in traits Data from the Illinois Long-Term Selection Experiment document the increase in oil content in corn kernels during 100 generations of artificial selection. The average for the 100th generation lies far outside the range of the founding generation. Modified from Moose et al. (2004).

Consider, for example, an artificial selection study run at the University of Illinois (Moose et al. 2004). Since the study began in 1896, with 163 ears of corn, researchers have been sowing for next year's crop only seeds from the plants with the highest oil content in their kernels. In the starting population, oil content ranged from 4% to 6% by weight. After 100 generations of selection, the average oil content in the population was about 20% **(Figure 3.19)**. That is, a typical plant in the present population has over three times the oil content of the most oil-rich plant in the founding population. Mutation, recombination, and selection together produced a new phenotype.

Persistent natural selection can lead to the evolution of new functions for existing behaviors, structures, or genes. Carnivorous plants provide examples. The

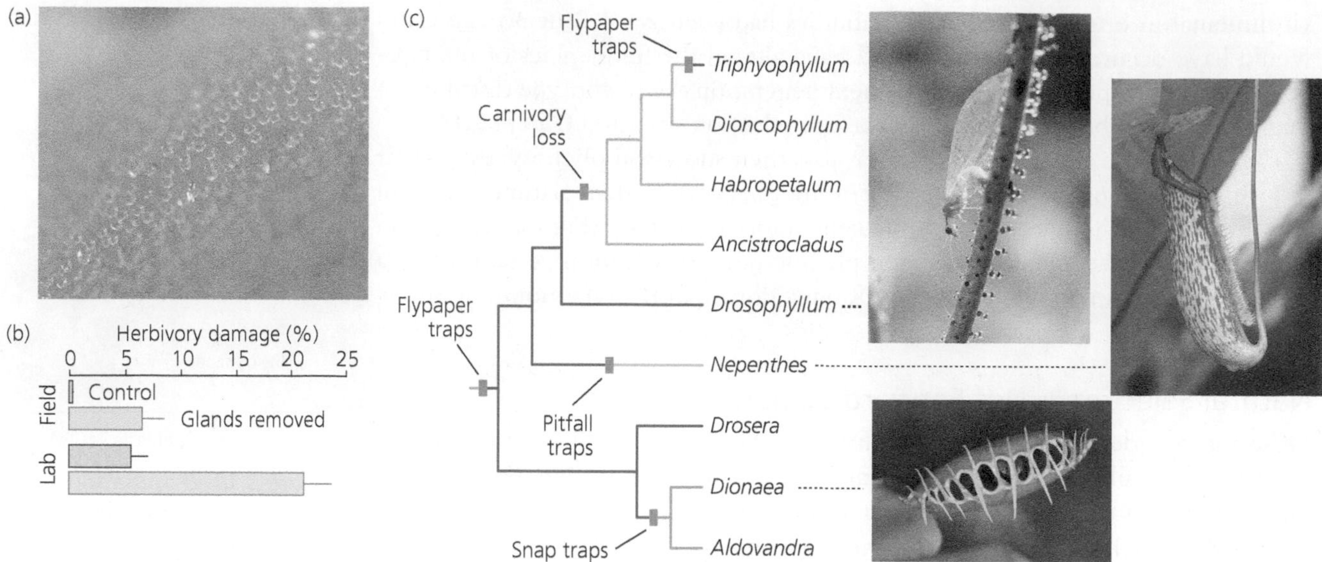

Figure 3.20 Evolutionary novelties in carnivorous plants (a) Leaves of the butterwort (*P. moranensis*) are covered with glandular trichomes that exude a sticky liquid. (b) *P. moranensis*'s trichomes serve in defense as well as trapping prey. Redrawn from Alcalá et al. (2010). (c) In the order Caryophyllales, lineages with flypaper traps have evolved pitfall traps and snap traps. Phylogeny redrawn from Heubl et al. (2006).

butterwort *Pinguicula moranensis* captures small insects in a sticky liquid exuded in droplets from glandular hairs, or trichomes, covering its leaves **(Figure 3.20a)**. Glandular trichomes are common herbivore deterrents in plants, suggesting that butterworts first became carnivorous when an existing defensive trait began serving a novel role in prey capture. Consistent with this hypothesis, Raúl Alcalá and colleagues (2010) found that butterworts denuded of trichomes suffered increased damage from herbivores, in both the field and the lab (Figure 3.20b).

A trait that is used in a novel way is known as an **exaptation** (see Gregory 2008). Exaptations represent happenstance. An exaptation enhances an individual's fitness fortuitously, not because natural selection is conscious or foresighted.

Such a trait may eventually be elaborated into a completely new structure by selection related to its new function. Additional modifications that arise during this process are called **secondary adaptations**. The Venus fly trap (*Dionaea muscipula*), with leaves modified into snap traps, and the monkey cups (*Nepenthes*), with pitfall traps that develop from tendrils on leaf tips, evolved from an ancestor with flypaper traps like the butterwort's (Figure 3.20c; Heubl et al. 2006).

Natural Selection Does Not Lead to Perfection

The previous paragraphs argue that populations evolving by natural selection become better adapted over time. It is equally important to realize, however, that evolution does not result in organisms that are perfect.

One reason is that populations may face contradictory patterns of selection. Consider the male mosquitofish (*Gambusia affinis*), whose anal fin is modified to serve as a copulatory organ, or gonopodium. Brian Langerhans and colleagues (2005) found that females prefer males with larger gonopodia. But when predators attack, a big gonopodium is literally a drag, slowing a male's escape. A perfect male would be irresistible to females and fleet enough to evade any predator. Alas, no male can be both. Instead, each population evolves a phenotype that strikes a compromise between opposing agents of selection **(Figure 3.21)**.

Figure 3.21 No guy is perfect These males sport gonopodia that attract mates but hinder escape. The lower male is from a high-predation population. From Langerhans et al. (2005).

Another reason organisms are not perfect is that evolution by natural selection shapes their bodies by culling from available variants. The result is body plans that can look cobbled together from spare parts rather than rationally designed. Compare, for example, a flatfish to a skate (Coyne 2009). Both have compressed bodies, cryptic coloration, and the habit of lying on the bottom, all apparently adaptations that enhance survival by reducing predation. The bodies of skates are compressed dorsoventrally and retain their symmetry. Flatfish, however, are tipped over on their sides. They begin life as upright, symmetrical fry. As they mature, one eye migrates to the other side of the head, the body becomes asymmetrical, and the fish lie down. Next to skates, flatfish look decidedly jury-rigged.

Natural selection occurs among extant variants and cannot simultaneously optimize all traits. It leads to adaptation, not perfection.

Natural Selection Is Nonrandom, but It Is Not Progressive

Evolution by natural selection is sometimes characterized as a random or chance process, but this is wrong. Mutation and recombination, the processes that generate genetic variation, are random with respect to the changes they produce in phenotypes. But natural selection, the automatic sorting among variant phenotypes and genotypes, is the opposite of random. It is, by definition, the nonrandom superiority at survival and reproduction of some variants over others. This is why evolution by natural selection increases adaptation to the environment.

Nonrandom selection as it occurs in nature is, however, completely free of conscious intent. Darwin came to regret using the phrase "naturally selected," because to some readers it implied decision by a sentient entity. But natural selection is not the work of an invisible hand. It just happens. The undirected quality of natural selection is evident in the data in Figure 3.16. As conditions on the island changed, the finch population first evolved one way, then another.

A related point is that while the complexity, degree of organization, and specialization of organisms have tended to increase over geologic time, evolution is not progressive in the sense of moving toward a predetermined goal. Evolving populations improve only in that their average adaptation to the environment increases. There is no inexorable trend toward more advanced forms of life. Complex traits are often lost, and many organisms are simpler than their ancestors. Contemporary tapeworms, for example, have no digestive system. Snakes evolved from ancestors with legs. Early fossil birds had teeth, but living birds do not.

Unfortunately, a progressivist view of evolution dies hard. Even Darwin had to remind himself to "never use the words higher or lower" when discussing evolutionary relationships. All extant and fossil organisms trace their ancestry back to the same primordial lineage. And all, in their time, were adapted to their environments, able to survive and reproduce. None is "higher" or "lower."

There is no such thing as a higher or lower plant or animal.

Fitness Is Not Circular

The theory of evolution by natural selection is sometimes criticized by nonbiologists as tautological, or circular in its reasoning. The supposed circularity is captured in the phrase "the survival of the fittest." The fittest are, by definition, those who survive and reproduce. The phrase—coined by Herbert Spencer (1864, p. 444) and later adopted by Darwin (e.g., 1868, p. 6)—can thus be rendered as "the survival of the survivors." Which must be true, but explains nothing.

The key to resolving the issue is to recognize that "survival of the fittest" is an oversimplified, and thus misleading, characterization of Darwin's theory. The essential feature of natural selection is that certain heritable variants do better than others. As long as a nonrandom subset of the population survives at higher rates and leaves more offspring, evolution will result. In the snapdragon and finch examples, researchers not only determined that survival and reproduction were nonrandom, they also uncovered why some individuals did better than others.

It should also be clear that Darwinian fitness is not an abstract quantity. It can be measured in nature. This is done by counting the offspring that individuals produce, or by observing their ability to survive a selection event, and comparing each individual's performance to that of others in the population. These are independent and objective criteria for assessing fitness. When heritable variants are found to be associated with differences in fitness, populations evolve as predicted.

Selection Acts on Individuals, Not for the Good of the Species

One of the most pervasive misconceptions about natural selection, especially selection on animal behavior, is that individual organisms perform actions for the good of the species. Self-sacrificing, or altruistic, acts do occur in nature. When mammalian predators approach, Belding's ground squirrels draw attention to themselves by giving alarm calls. Lion mothers sometimes nurse cubs that are not their own. But traits cannot evolve by natural selection unless they increase the fitness of the genes responsible for them relative to the fitness of other genes (see Chapter 12). This happens when, for example, the beneficiaries of generocity are kin or can be counted on to repay the favor.

Individuals do not do things for the good of their species. They behave in a way that maximizes their genetic contribution to future generations.

The idea that animals will do things for the good of the species is so ingrained that we will make the same point a second way. Lions live in social groups called prides. Coalitions of males fight to take over prides. If a new coalition defeats and expels a pride's males, the newcomers quickly kill the pride's unweaned cubs. These cubs are unrelated to the killers. Killing the cubs increases the new males' fitness because pride females become fertile again sooner and will conceive offspring by the new males (Packer and Pusey 1983, 1984). Infanticide is widespread in animals. Clearly, behavior like this does not exist for the good of the species. Rather, infanticide exists because, under certain conditions, it enhances the fitness of individuals who perform the behavior relative to those who do not.

3.6 The Evolution of Evolutionary Biology

Because evolution by natural selection is a general organizing feature of living systems, Darwin's theory ranks as one of the great ideas in intellectual history. Its impact on biology is like that of Newton's laws on physics, Copernicus's Sun-centered theory of the universe on astronomy, and the theory of plate tectonics on geology. In the words of evolutionary geneticist Theodosius Dobzhansky (1973), "Nothing in biology makes sense except in the light of evolution."

For all its scope and power, however, the theory of evolution by natural selection was not universally accepted by biologists until some 70 years after it was initially proposed. As originally formulated by Darwin, the theory had three serious problems.

Variation

Because Darwin knew nothing about mutation, he had no idea how variability was generated in populations. Thus he could not answer critics who maintained that the amount of variability in populations was strictly limited and that natural selection would grind to a halt when variability ran out. It was not until the early 1900s, when geneticists such as Thomas Hunt Morgan began experimenting with fruit flies, that biologists began to appreciate the continuous and universal nature of mutation. Morgan and colleagues showed that mutations occur in every generation and in every trait.

Inheritance

Because Darwin knew nothing about genetics, he had no idea how variations are passed to offspring (see Charlesworth and Charlesworth 2009). Biologists did not understand inheritance until Mendel's work with peas was rediscovered and verified, 35 years after its original publication. Mendel's laws of segregation and independent assortment confirmed the mechanism behind Darwin's postulate 2, which states that some of the variation seen in populations is heritable.

Before Mendel's laws became known, many biologists thought inheritance worked like pigments in paint. Advocates of this hypothesis, called **blending inheritance**, argued that favorable variants would merge into existing traits and be lost. In 1867, Scottish engineer Fleeming Jenkin published a mathematical treatment of blending inheritance along with a thought experiment. If a dark-skinned sailor were stranded on an equatorial island inhabited by light-skinned people, Jenkin's model predicted that no matter how advantageous dark skin might be (in, say, reducing skin cancer), the population would never become dark-skinned. If the dark-skinned sailor had children with a light-skinned woman, the kids would be brown-skinned. If they, in turn, had children with light-skinned people, their children would be light-brown-skinned, and so on. Conversely, if a light-skinned sailor were stranded on a northern island inhabited by dark-skinned people, blending inheritance argued that, no matter how advantageous light skin might be (in, say, facilitating the synthesis of vitamin D with energy from UV light), the population would never become light. Under blending inheritance, new variants are diluted away. For populations to evolve by natural selection, favorable new variations have to be passed to offspring intact and remain discrete.

We now understand that phenotypes blend in some traits, including skin color, but genotypes do not. **Figure 3.22** shows why for skin color. Color is determined mainly by pigments produced in cells called melanocytes (Figure 3.22a). Melanocytes make eumelanin, a brownish-black pigment, when alpha melanocyte-stimulating hormone (α-MSH) binds to their melanocortin 1 receptors (MC1Rs; Figure 3.22b). Melanocytes make pheomelanin, a reddish-yellow pigment, when their MC1Rs are dysfunctional or when they are blocked by agouti signaling protein (ASP; Figure 3.22c). Variation in human coloration has been tied to allelic variation in both the gene for MC1R and the gene for ASP (Harding et al. 2000; Schaffer and Bolognia 2001; Kanetsky et al. 2002). For example, homozygotes for the *Arg151Cys* allele of the gene for MC1R almost always have red hair and fair skin (Smith et al. 1998). The effects of alleles in determining phenotype may blend. An individual with just one copy of the *Arg151Cys* allele, for instance, may have intermediate coloration. But the alleles themselves are passed on intact to offspring, and two *Arg151Cys* heterozygotes can have a homozygous red-haired offspring. Inheritance is thus particulate, not blending.

Jenkin's hypothetical population would, in fact, become increasingly darker or lighter skinned if selection were strong and mutation continually added darker- or lighter-skinned variants to the population via changes in the genes that regulate the production of melanins.

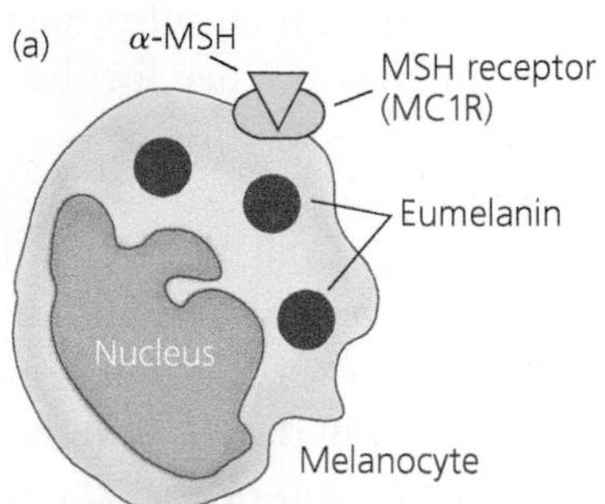

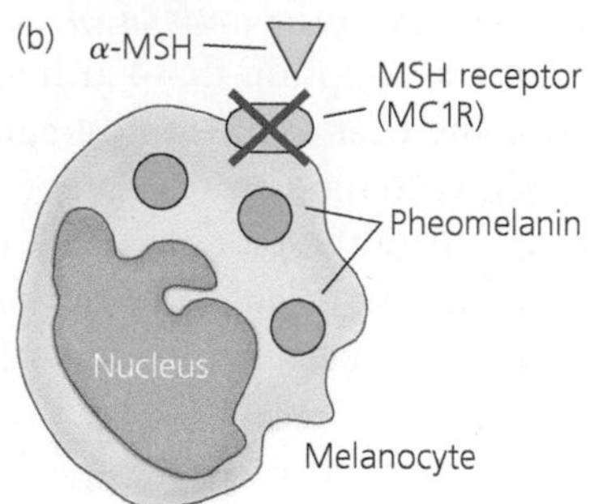

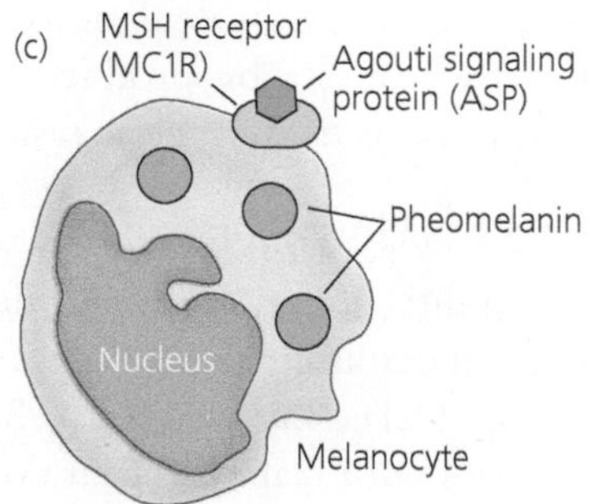

Figure 3.22 Regulation of skin color in humans Color is determined largely by pigments made by cells called melanocytes. The organelles that produce and store the pigments are called melanosomes. (a) Melanocytes make eumelanin, a dark brown pigment, when alpha melanocyte-stimulating hormone (α-MSH) binds to its receptor (MC1R). (b and c) Melanocytes make pheomelanin, a reddish yellow pigment, if MC1R is dysfunctional or blocked by agouti signaling protein (ASP). After Schaffer and Bolognia (2001).

Darwin himself struggled with the problem of inheritance, eventually adopting an incorrect view based on the work of French biologist Jean-Baptiste Lamarck. In the early 19th century, Lamarck proposed that species evolve through the inheritance of changes wrought in individuals. Lamarck's idea was a breakthrough: It recognized that species have changed through time and proposed a mechanism to explain how. His theory was wrong, however, because offspring do not inherit phenotypic changes acquired by their parents. If people bulk up by lifting weights, their offspring are not more powerful. If giraffes stretch for leaves, it has no consequence for the reach of their offspring.

Time

Physicist William Thomson (Lord Kelvin) published papers in the early 1860s estimating the age of Earth at 15–20 million years. His analyses were based on measurements of the Sun's heat and the temperature of Earth. Because fire was the only known heat source, Thomson assumed that the Sun was combusting like a giant lump of coal and slowly burning out. Likewise, geologists and physicists believed the surface of Earth was gradually cooling. This notion was based on the assumption that Earth was changing from a molten state to a solid one by radiating heat to the atmosphere—a view apparently supported by measurements of higher temperatures deeper down in mineshafts. These data allowed Thomson to calculate the rate of radiant cooling.

Thomson calculated that the transition from a hot to cold Sun and hot to cold Earth left a narrow window of time when life on Earth was possible. The window was too small to allow the gradual changes of Darwinism to accumulate, and thus supported a role for special creation in explaining adaptation and diversity.

The discovery of radioactive isotopes early in the 20th century changed all that. Thomson's calculations were correct, but his assumptions were wrong. Earth's heat is a by-product of radioactive decay, not radiant cooling, and the Sun's heat is from nuclear fusion, not combustion.

The Modern Synthesis

Variability, inheritance, and time posed such difficult problems that the first 70 years of evolutionary biology were fraught with turmoil (see Provine 1971; Mayr 1980, 1991). But between 1932 and 1953, a series of landmark books integrated genetics with Darwin's four postulates and led to a reformulation of the theory of evolution. This reformulation, known as the modern synthesis or the evolutionary synthesis, was a consensus grounded in two propositions. The first was that gradual evolution results from small genetic changes that rise and fall in frequency under natural selection. The second was that the origin of species and higher taxa, or macroevolution, can be explained in terms of natural selection acting on individuals, or microevolution.

The modern synthesis resolved decades of controversy over the validity of evolution by natural selection.

With the synthesis, Darwin's original four postulates and their outcome could be restated along the following lines:

1. Individuals vary as a result of mutation creating new alleles, and segregation and independent assortment shuffling alleles into new combinations.
2. Individuals pass their alleles on to their offspring intact.
3. In every generation, some individuals are more successful at surviving and reproducing than others.
4. The individuals most successful at surviving and reproducing are those with the alleles and allelic combinations that best adapt them to their environment.

The outcome is that alleles associated with higher fitness increase in frequency from one generation to the next.

Darwin ended the introduction to the first edition of *On the Origin of Species* with a statement that still represents the consensus view of evolutionary biologists (Darwin 1859, p. 6): "Natural Selection has been the main but not exclusive means of modification." We now think of evolution in terms of changes in the frequencies of the alleles responsible for traits like beak depth and flower color. We are more keenly aware of other processes that cause evolutionary change in addition to natural selection. (Chapters 6 and 7 explore these processes in detail.) But the Darwinian view of life, as a competition between individuals with varying abilities to survive and reproduce, has proven correct in almost every detail.

3.7 Intelligent Design Creationism

Scientific controversy over the fact of evolution ended in the late 1800s, when the evidence simply overwhelmed the critics (see Chapter 2). Whether natural selection was the primary process responsible for both adaptation and diversity remained an open question until the 1930s, when the works of the modern synthesis provided a mechanistic basis for Darwin's four postulates and unified micro- and macroevolution. Evolution by natural selection is now considered the great unifying idea in biology.

The theory of special creation was rejected over a century ago, but creationists now want it taught in public schools.

Although scientific dispute over the validity of evolution by natural selection ended long ago, a political and legal controversy continues (Scott and Matzke 2007; Forrest 2008; Padian and Matzke 2009).

History of the Controversy in the United States

The Scopes Trial of 1925 is perhaps the controversy's most celebrated event (see Gould 1983, essay 20; Larson 1997). John T. Scopes was a biology teacher who gave his students a reading about Darwinian evolution. This assignment violated the State of Tennessee's Butler Act, which prohibited teaching evolution in public schools. William Jennings Bryan, a famous politician and a fundamentalist orator, served as prosecutor. Clarence Darrow, a renowned defense attorney, led Scopes's defense in cooperation with the American Civil Liberties Union.

Although Scopes was convicted and fined $100, the trial was widely perceived as a triumph for evolution. Bryan suggested, while on the stand as a witness, that the six days of creation described in Genesis 1:1–2:4 may each have lasted far longer than 24 hours. This idea was considered a grave inconsistency, and therefore a blow to the integrity of the creationist viewpoint.

Figure 3.23 Guilty as charged John T. Scopes (right) and John R. Kia, one of his lawyers.

The Tennessee Supreme Court overturned the conviction on the basis of a technicality. This decision was a disappointment to Scopes and his defense team **(Figure 3.23)**, who had hoped to appeal the case to the U.S. Supreme Court and have the Butler Act declared unconstitutional.

The Butler Act stayed on the books until 1967. It was not until 1968, in *Epperson v. Arkansas*, that the U.S. Supreme Court struck down laws that prohibit the teaching of evolution. The court based its ruling on the U.S. Constitution's separation of church and state. In response, fundamentalist religious groups in the United States reformulated their arguments as "creation science" and demanded equal time for what they insisted was an alternative theory for the origin of species. By the late 1970s, 26 state legislatures were debating equal-time legislation (Scott 1994). Arkansas and Louisiana passed such laws, only to have them struck down in state courts. The Louisiana law was then appealed to the U.S. Supreme Court, which decided in 1987 (*Edwards v. Aquillard*) that because creationism is essentially a religious idea, teaching it in the public schools violates the First Amendment. Two justices, however, formally wrote that it would still be acceptable for teachers to present alternative theories to evolution (Scott 1994).

Opponents of evolution responded by dropping the words *creation* and *creator* from their literature. But they called for equal time either for teaching that no evolution has occurred or for teaching a proposal called intelligent design theory, which infers a designer from the perfection of adaptation in extant organisms (Scott 1994; Schmidt 1996). In 2005, *Kitzmiller et al. v. Dover Area School District* was tried in Dover, Pennsylvania. The district had enacted a policy requiring that biology students "be made aware of gaps/problems in Darwin's theory and of other theories of evolution including, but not limited to, intelligent design." A group of parents sued the district on the grounds, again, that the policy violates the First Amendment. The court agreed (Goodstein 2005; Jones 2005).

The complexity and perfection of organisms is a timeworn objection to evolution by natural selection. Darwin was aware of it. In his *Origin* he devoted a section of the chapter "Difficulties on Theory" to "Organs of extreme perfection." How can natural selection, by sorting random changes in the genome, produce elaborate and integrated traits such as, for example, the vertebrate eye?

Perfection and Complexity in Nature

English cleric William Paley (1802) promoted special creation with a line of reasoning now known as the argument from design (Dawkins 1986). Paley imagined finding a watch—a complex, precise machine—and inferring that it must be the handiwork of a skilled craftsman. As with the watch, Paley said, so too with the vertebrate eye. He held that organisms are so well engineered they must be the work of a conscious designer.

The argument from design contends that adaptations—traits that increase the fitness of individuals that possess them—must result from the actions of a conscious entity.

Because we see perfection and complexity in the natural world, evolution by natural selection seems to defy credulity. There are two concerns. The first is whether random changes can lead to order. They can. Mutations are chance events, so the generation of variation in a population is random. But selection on this variation is nonrandom. It distills the variants that increase fitness from those that do not. And adaptations—structures or behaviors that increase fitness—are what we perceive as ordered, complex, or even perfect in the natural world. Natural selection produces the appearance of design without a designer.

The second concern is that the theory of evolution predicts that traits evolve in small increments, and that each new step increases the fitness of the individuals that show it. This scenario is plausible for relatively simple traits, like beaks. It is easy to imagine that a modestly enlarged beak as an advantageous intermediate stage on the way to a greatly enlarged beak. But what about complex organs with many intricately interdependent parts? What good is half an eye?

As it turns out, half an eye is sometimes quite useful. Evidence for this comes from the tremendous diversity of light-sensing organs borne by extant creatures, many of which are considerably simpler than the standard vertebrate eye.

Figure 3.24 shows the light-sensing organs on the heads of five chordates. All contain a type of light-detecting cell called a ciliary photoreceptor (Lamb et al. 2007). In the lancelet these are found behind a cup of pigmented cells in a structure called the frontal eye (Lacalli et al. 1994). In larval sea squirts, a small cluster of them are surrounded by a single pigmented cell in a structure called an ocellus.

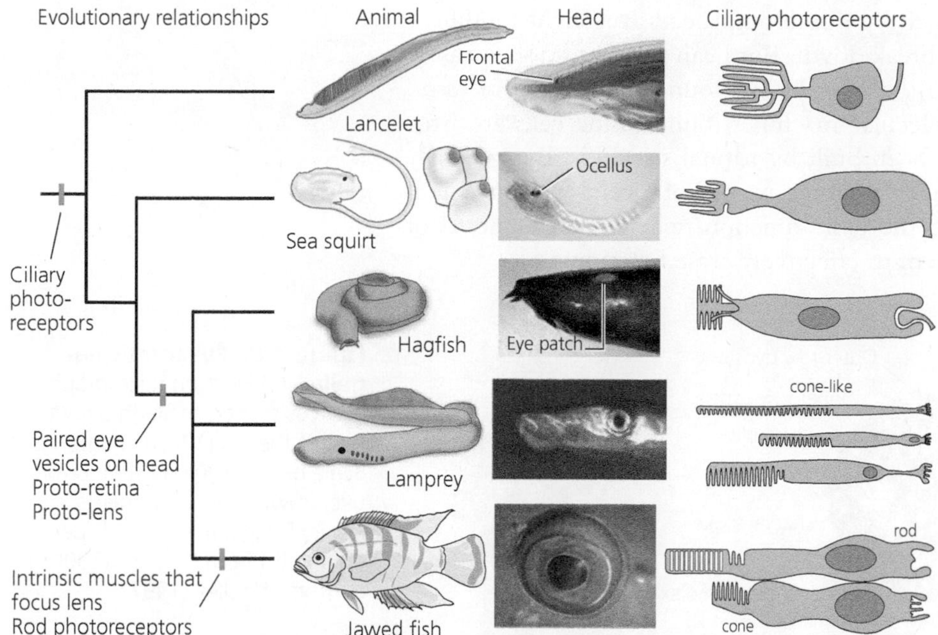

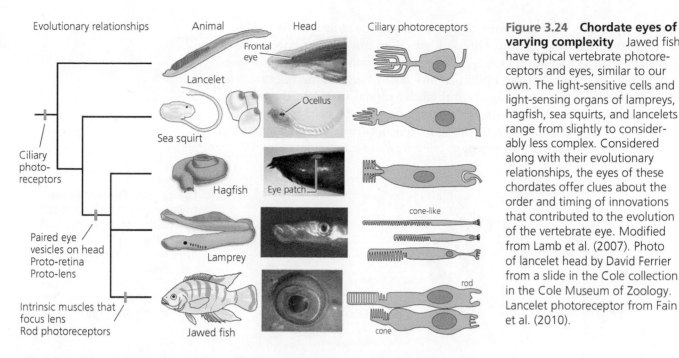

Figure 3.24 Chordate eyes of varying complexity Jawed fish have typical vertebrate photoreceptors and eyes, similar to our own. The light-sensitive cells and light-sensing organs of lampreys, hagfish, sea squirts, and lancelets range from slightly to considerably less complex. Considered along with their evolutionary relationships, the eyes of these chordates offer clues about the order and timing of innovations that contributed to the evolution of the vertebrate eye. Modified from Lamb et al. (2007). Photo of lancelet head by David Ferrier from a slide in the Cole collection in the Cole Museum of Zoology. Lancelet photoreceptor from Fain et al. (2010).

In hagfish the photoreceptors sit in rudimentary retinas, under layers of translucent skin, in paired eye patches. Because hagfish behave as if nearly blind, the eye patches are thought to function in regulating circadian rhythms. In lampreys a diversity of cone-like ciliary photoreceptors appear in the retinas of camara eyes with a lens, an iris, and exterior muscles that swivel the eye and focus it by changing the shape of the cornea. In jawed fish, two distinct types of ciliary photoreceptors, rods and cones, appear in the retinas of eyes that, in addition to a lens and iris, contain interior muscles that focus the eye by altering the shape of the lens.

Based on the evolutionary relationships of these and other organisms, the structure of their photoreceptors and light-sensing organs, the way the organs develop and the genes involved, Trevor Lamb and colleagues (2007) have developed a detailed hypothesis for the evolution of the vertebrate eye. A few of the key innovations are indicated on the evolutionary tree in Figure 3.24.

Much remains to be learned. For example, because the evolutionary relationships among hagfish, lampreys, and jawed fish are unresolved, it is unclear how we should interpret the hagfish's eyes. Anatomical analyses suggest that lampreys and jawed fish share a more recent common ancestor with each other than either does with hagfish. This would imply that hagfish eyes represent a transitional form between the ocelli of sea squirts and the camera eyes of lampreys. However, genetic analyses suggest that hagfish and lampreys are closest relatives (Smith et al. 2010). This would imply that hagfish eyes are reduced from camera-like eyes.

Lampreys, hagfish, sea squirts, and lancelets nonetheless demonstrate that eyes simpler than our own serve as contemporary adaptations to the problem of sensing light. They make it plausible, as Darwin argued in his section on extreme perfection, that the vertebrate eye evolved by incremental improvement. (For more about eye evolution, see Salvini-Plawen and Mayr 1977; Nilsson and Pelger 1994; Gehring 2004; Fernald 2004; Oakley and Pankey 2008; Nilsson 2009.)

The argument from design is wrong.

The Argument from Biochemical Design

Summarizing his views on perfection and complexity, Darwin wrote (1859, p. 189): "If it could be demonstrated that any complex organ existed, which could not possibly have been formed by numerous, successive, slight modifications, my theory would absolutely break down. But I can find out no such case."

Creationist Michael Behe (1996) believes he has found a profusion of cases. Behe claims that many of the molecular machines found inside cells are irreducibly complex and cannot have been built by natural selection. By *irreducibly complex*, Behe (p. 39) means "a single system composed of several well-matched, interacting parts that contribute to the basic function, wherein the removal of any one of the parts causes the system to effectively cease functioning."

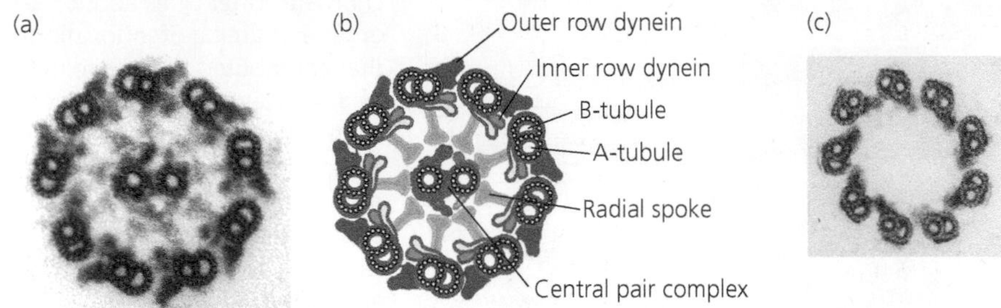

Figure 3.25 Eukaryotic flagella (a) Electron micrograph of cross section through a flagellum of the alga *Chlamydomonas*. Scale bar = 100 nm. (b) Interpretive drawing of components in (a). (c) Flagellum of an eel sperm. (a) and (b) from Mitchell (2000); (c) from Woolley (1997).

Among the examples Behe offers is the eukaryotic cilium (also known, when it is long, as a flagellum). **Figures 3.25a and 3.25b** show a cross section of the stalk, or axoneme, of one of these cellular appendages. Its main components are microtubules, made of the proteins α-tubulin and β-tubulin. At its core are two singlet microtubules bound by a protein bridge. Surrounding these are nine doublet microtubules connected to the central pair by protein spokes. Neighboring doublets are connected to each other by an elastic protein called nexin. The cilium is powered by dynein motors on the doublet microtubules. As the motors on each doublet crawl up their neighbor, they cause the entire axoneme to bend.

According to Behe (1998), the components of the cilium "combine to perform one task, ciliary motion, and all of these proteins must be present for the cilium to function. If the tubulins are absent, then there are no filaments to slide; if the dynein is missing, then the cilium remains rigid and motionless; if nexin or the other connecting proteins are missing, then the axoneme falls apart when the filaments slide." Because he thinks the cilium is irreducibly complex, Behe finds it implausible that this structure arose by a stepwise process in which each step is an incremental improvement over what came before. Having concluded that the cilium cannot have arisen by natural selection, he infers that it was designed.

Intelligent design creationism is a modern version of the argument from design.

At least two of Behe's three assertions about the cilium are amenable to scientific investigation (Felsenstein 2007). We can test his claims that (1) the cilium is irreducibly complex; and (2) irreducibly complex biological systems cannot evolve by natural selection. Both claims are wrong.

The Eukaryotic Cilium Is Not Irreducibly Complex

The eukaryotic cilium is certainly not irreducibly complex in an evolutionary sense. This is demonstrated by organisms with cilia simpler in structure than the one pictured in Figures 3.24a and b (see Miller 1999). Figure 3.24c, for example, shows an eel sperm's flagellum. It is fully functional, despite lacking the central pair of singlet microtubules, the spokes, and the outer row of dynein motors. The cilium is not even irreducibly complex in a mechanical sense. Its mechanical reducibility is shown by a mutation, known as *pf14*, in the single-celled alga *Chlamydomonas*. The flagella of cells with this mutation lack spokes. Although the *pf* in *pf14* stands for paralyzed flagella, the flagella of mutant cells still function under the right chemical conditions and genetic backgrounds (Frey et al. 1997).

Irreducibly Complex Systems Can Evolve by Natural Selection

Even if the cilium were irreducibly complex, Behe would still be wrong to conclude that it cannot have been built by natural selection. Behe's argument assumes that evolution by natural selection builds molecular machines and their components from scratch, and that the individual component proteins are useless until the entire structure has been assembled in its final form. In fact, evolution by natural selection cobbles together molecular machines from preexisting and functional component proteins that it co-opts for new roles (True and Carrol 2002). If the components of complex molecular machines are recruited from other jobs, then we no longer have to explain how the individual components were maintained by selection while the machine evolved from scratch.

By studying populations of digital organisms, Richard Lenski and colleagues (2003) showed that evolution by natural selection can, in fact, build complex machines in just this way. A digital organism is a self-replicating computer program living in a virtual world. Each of the organisms in the virtual world has a genome composed of a series of simple instructions—low-level scraps of computer code. There are some two dozen possible instructions in all, and they can be strung together in any order and repeated any number of times. Most possible sequences of instructions do nothing. Some allow an organism to copy itself. Still others allow an organism to take numbers as inputs, perform logical functions on them, and produce meaningful outputs. The researchers started with a large population of identical organisms whose modest-sized genomes allowed them to replicate themselves but not to perform logical functions. Replication was imperfect, meaning that occasionally one or more of the instructions in the genome was replaced with another chosen at random, or an instruction was inserted or deleted at random. The organisms had to compete for the chance to run their instructions and reproduce. If an organism appeared that could correctly perform one or more logical functions, it was rewarded with additional running time.

Like the classical argument from design, the claims of intelligent design creationism are demonstrably wrong.

The capacity to perform simple logical functions evolved first. Complex functions evolved later, building on the simple ones and co-opting them for new purposes. In genomes capable of performing the most complex function, many of the individual instructions were crucial; deleting them destroyed the organism's ability to perform the function. Intriguingly, some of the mutations on the path to the most complex function were initially harmful. That is, they disrupted the machinery for one or more simple functions. But they set the stage for later mutations that helped assemble new and more complex functions from old.

A striking demonstration of protein co-option in real organisms comes from the crystallins of animal eye lenses (True and Carrol 2002). Crystallins are water-soluble proteins that form densely packed, transparent, light-refracting arrays constituting about a third of the mass of the lens. Animal eyes contain an astonishing diversity of crystallins. Some, such as the α and $\beta\gamma$ crystallins, are widely distributed across the vertebrates and must have evolved early. These ancient crystallins evolved from duplicate copies of genes for proteins with other functions. Other crystallins are unique to particular taxa and must have evolved recently. Most of these recently evolved crystallins are similar or identical to

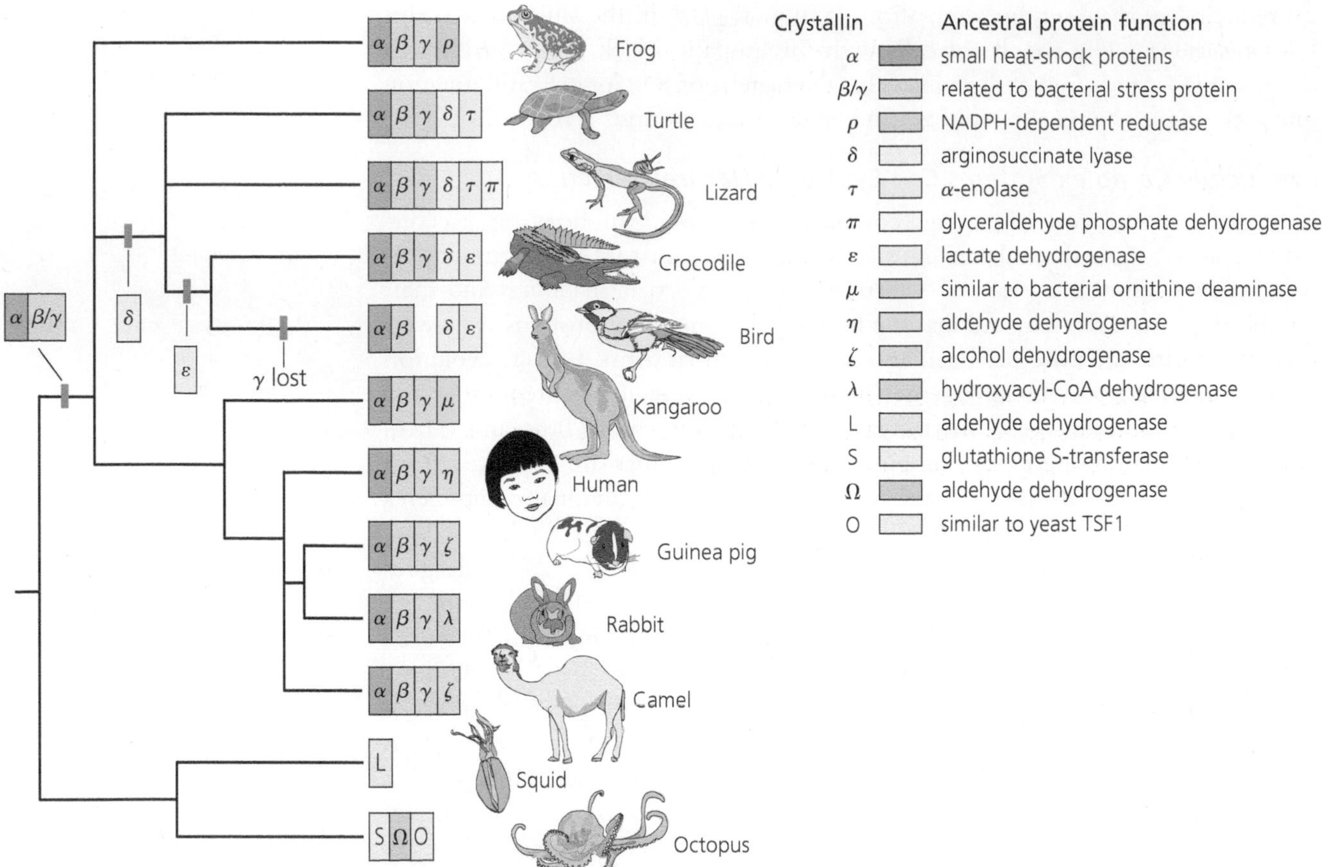

Figure 3.26 Gene co-option in the crystallins of animal eye lenses Crystallin proteins are major components of the lenses in animal eyes. All are derived from proteins with other functions. In some cases crystallins are encoded by duplicates of the genes for the proteins they are derived from; in others crystallins are encoded by the same genes. The evolutionary tree shows the relationships among various animals. Color-coded Greek letters indicate crystallins found in the lenses of each animal. The table lists the proteins the crystallins are derived from. Redrawn from True and Carroll (2002).

enzymes that function outside the eye **(Figure 3.26)**. Some, in fact, *are* enzymes that function outside the eye. That is, in some cases a single gene encodes a single protein that functions as an enzyme in some tissues and as a crystallin in the lens. The ε crystallin in chickens, for instance, is a metabolic enzyme called lactate dehydrogenase B. Additional examples of proteins co-opted for new functions come from the antifreeze proteins in the blood of Arctic and Antarctic marine fishes (Baardsnes and Davies 2001; Fletcher et al. 2001).

Crystallins and antifreeze proteins have switched roles during their evolution, but have not been incorporated into complex molecular machines. However, most components of the molecular machines Behe cites are homologous to proteins with other functions. The microtubules and dyneins of the eukaryotic cilium, for example, are similar to components of the spindle apparatus used in cell division. And work on simple examples such as crystallins has paved the way for progress on more challenging problems. Researchers have begun reconstructing the evolutionary origins of complex molecular machines and metabolic pathways. Examples include the Krebs citric acid cycle (Meléndez-Hevia et al. 1996; Huynen et al. 1999), the cytochrome *c* oxidase proton pump (Musser and Chan 1998), the blood-clotting cascade (Krem and Di Cera 2002), and various bacterial flagella (Pallen and Matzke 2006; Liu and Ochman 2007).

Behe is right that we have not yet worked out in detail the evolutionary histories of the molecular machines he takes as examples of irreducible complexity. He would have us give up and attribute them all to miracles. But that is no way to make progress. Ironically, Behe began claiming that the origins of cellular biochemistry would never be deciphered just as the techniques and data required to do so were becoming available. Among these are automated DNA sequencers and the whole-genome sequences they are providing. We predict that in the coming decades, all of Behe's examples of irreducible complexity will yield to evolutionary analysis.

Other Objections

Here are three additional arguments that creationists use regularly, with responses from an evolutionary perspective (see Gish 1978; Kitcher 1982; Futuyma 1983; Gould 1983 essays 19, 20, 21; Dawkins 1986; Swinney 1994):

1. Evolution by natural selection is unscientific because it is not falsifiable and because it makes no testable predictions. Each of Darwin's postulates is independently testable, so the theory meets the criterion that ideas must be falsifiable to be considered scientific. Also, the claim that evolutionary biologists do not make predictions is false. Paleontologists routinely (and correctly) predict which strata will bear fossils of certain types (a spectacular example was that fossil marsupial mammals would be found in Antarctica); Peter Grant and Rosemary Grant have used statistical techniques based on evolutionary theory to correctly predict the amount and direction of change in finch characteristics during selection events in the late 1980s and early 1990s (Grant and Grant 1993, 1995). Scientific creationism, on the other hand, is an oxymoron. In the words of one of its leading advocates, Dr. Duane Gish (1978, p. 42): "We cannot discover by scientific investigations anything about the creative processes used by God."

2. Because organisms progress from simpler to more complex forms, evolution violates the second law of thermodynamics. Although the second law has been stated in various ways since its formulation in the 19th century, the most general version is "Natural processes tend to move toward a state of greater disorder" (Giancoli 1995). The second law is focused on the concept of entropy. This is a quantity that measures the state of disorder in a system. The second law, restated in terms of entropy, is "The entropy of an isolated system never decreases. It can only stay the same or increase" (Giancoli 1995).

The key to understanding the second law's relevance to evolution is the word *isolated*. The second law is true only for closed systems. Organisms, however, live in an open system: Earth, where photosynthetic life-forms capture the radiant energy of the Sun and convert it to chemical energy that they and other organisms can use. Because energy is constantly being added to living systems, the second law does not apply to their evolution.

A similar objection is William Dembski's (2002) assertion that natural selection cannot lead the evolution of complex and meaningful genetic information because it is no better than a random search. He stakes this claim on a set of results in theoretical computer science called the no free lunch theorems. These show that averaged over all possible problems, no set of rules for finding a solution is better than any other, including random trial and error.

Joe Felsenstein (2007) has pointed out, however, that the no free lunch theorems, while mathematically correct, are irrelevant to populations evolving by natural selection. In an evolving population, natural selection amounts to a set of rules for finding genetic sequences that improve fitness: First, make small random changes to generate sequences similar to the ones you already have; then, test the sequences for fitness and keep the best performers. The no free lunch theorems establish that, averaged over all possible ways that fitness might be related to DNA sequence, the natural selection method works no better than random search. But among all the possible ways that fitness might be related to DNA sequence, most are the equivalent of assigning a random fitness value to every unique sequence. Changing a single nucleotide in a sequence would give it a completely unrelated fitness value. The real biological world does not work like that. In the real world, similar sequences often have similar fitnesses. And natural selection demonstrably results in increased adaptation.

If Dembski were correct in applying the no free lunch theorems to biology, then no trait in any population would ever systematically evolve toward higher average fitness. Not only would natural selection fail to produce adaptation, but artificial selection would fail to increase the frequency of desirable traits in populations of crops, livestock, or pets. In fact, of course, selection predictably and demonstrably increases the mean fitness of populations.

3. No one has ever seen a new species formed, so evolution is unproven. And because evolutionists say that speciation is too slow to be directly observed, evolution is unprovable and thus based on faith. First, it is simply wrong that the only way to establish that something happened is to observe it directly. Imagine that you and two friends are stranded on an otherwise deserted island. You find one friend face down with a knife in his back, and you know that you had nothing to do with the murder. Although you did not directly observe the killing, you can, beyond a reasonable doubt, infer the identity of the guilty party. We make inferences of this sort all the time in everyday life. They are common in science as well. We cannot observe atoms directly, for example, but we have considerable evidence by which to infer that they exist.

Second, although speciation is a slow process, it is ongoing and can be studied. And its consequences can be predicted and verified. Elsewhere in the book we discuss examples that include speciation directly observed in a laboratory population of viruses and data consistent with speciation in progress in natural populations of birds and fish (Chapter 2). We also discuss research of scientists who used the theory of descent with modification to make and confirm predictions about macroevolution (Chapter 2), and additional experimental and observational studies of speciation in action (Chapter 16).

SUMMARY

Before Darwin began to work on the origin of species, many scientists had become convinced that species change through time. The unique contribution made by Darwin and Wallace was to realize that the process of natural selection provided a mechanism for this pattern, which Darwin termed descent with modification.

Evolution by natural selection is the logical outcome of four facts: (1) Individuals vary in most or all traits; (2) some of this variation is genetically based and can be passed on to offspring; (3) more offspring are born than can survive to breed, and of those that do breed, some are more successful than others; and (4) the individuals that reproduce the most are a nonrandom, or more fit, subset of the general population. This selection process causes changes in the genetic makeup of populations over time, or evolution.

QUESTIONS

1. In everyday English, the word *adaptation* means an adjustment to environmental conditions. How is the evolutionary definition of adaptation different from the everyday English sense?
2. **a.** Describe Darwin's four postulates in your own words. What would have happened in the snapdragon experiment if any of the four had *not* been true?
   b. If Darwin's four postulates are true for a given population, is there any way that evolution cannot happen? What does this imply about whether evolution is or is not occurring in most populations today?
3. Think about how the finch bill data demonstrate Darwin's postulates.
   a. What would Figure 3.10 have looked like if bill depth was not variable?
   b. What would Figure 3.15 look like if bill depth was variable but the variation was not heritable?
   c. In Figure 3.11, why is the line drawn from 1978 data, after the drought, higher on the vertical axis than the line drawn from 1976 data, before the drought?
4. According to the text, it is correct to claim that most finches died from starvation during the 1977 drought because "there was a strong correspondence between population size and seed availability." Do you accept this hypothesis? If so, why don't the data in Figure 3.13 show a perfect correspondence between when seed supply started falling and when population size started to drop?
5. A common creationist criticism of the finch study is, "But it's just a little change in beak shape. Nothing really new has evolved." Or put a different way, "It's just microevolution and not macroevolution." The finch team continues to spend a great deal of effort on their project—traveling thousands of miles to the remote Galápagos every year, just to try to band an entire population of birds and all their nestlings and measure their bills. How would you respond to the creationists' criticisms? Do you think the ongoing 30-year-effort of the finch bill project has been worthwhile? Is it useful to try to document microevolution, and does it tell us anything about how macroevolution might work?
6. Suppose that you are starting a long-term study of a population of annual, flowering plants isolated on a small island. Reading some recent papers has convinced you that global warming will probably cause long-term changes in the amount of rain the island receives. Outline the observations and experiments you would need to do to document whether natural selection occurs in your study population over the course of your research. What traits would you measure, and why?
7. At the end of an article on how mutations in variable number tandem repeat (VNTR) sequences of DNA are associated with disease, Krontiris (1995, p. 1683) writes: "The VNTR mutational process may actually be positively selected; by culling those of us in middle age and beyond, evolution brings our species into fighting trim."
   (T.G. Krontiris. 1995. "Minisatellites and Human Disease." Science 269. Reprinted with permission of the AAAS.)
   This researcher proposes that natural selection on humans favors individuals who die relatively early in life. His logic is that the trait of dying from VNTR mutations is beneficial and should spread because the population as a whole becomes younger and healthier as a result. Can this hypothesis be true, given that selection acts on individuals? Explain.
8. Describe three major objections to Darwin's theory in the 19th century that were eventually resolved by discoveries by other scientists in the 20th century. What does this tell us about the utility of a theory that cannot yet answer all questions but that appears to be better than all alternative theories?
9. Many working scientists are relatively uninterested in the history of their fields. Did the historical development of evolutionary biology, reviewed in Section 3.6, help you understand the theory better? Why or why not? Do you think it is important for practicing scientists to spend time studying history?
10. **a.** Describe Behe's argument of "irreducible complexity." Is it a logical argument? How does it apply to the bacterial flagellum or the vertebrate eye?
    b. Opponents of intelligent design refer to irreducible complexity as an "argument from personal incredulity" (i.e., "I personally can't imagine how this could have evolved, so it must not have evolved."). What is the logical flaw of an argument from personal incredulity? Do you think it is fair to characterize irreducible complexity in this way?
11. In 1995, the Alabama School Board, after reviewing high school biology texts, voted to require that this disclaimer be posted on the inside front cover of the approved book (National Public Radio 1995):

    > This textbook discussed evolution, a controversial theory some scientists present as a scientific explanation for the origin of living things, such as plants, animals, and humans. No one was present when life first appeared on Earth; therefore, any statement about life's origins should be considered as theory, not fact.

    Do you accept the last sentence in this statement? Does the insert's point of view pertain to other scientific theories, such as the cell theory, the atomic theory, the theory of plate tectonics, and the germ theory of disease?

12. In his final opinion on the Dover intelligent design trial of 2005, Judge John E. Jones wrote (in part): "To be sure, Darwin's theory of evolution is imperfect. However, the fact that a scientific theory cannot yet render an explanation on every point should not be used as a pretext to thrust an untestable alternative hypothesis grounded in religion into the science classroom or to misrepresent well-established scientific propositions."

    Do you agree with Judge Jones? Why or why not? (For more information on this trial, see *Kitzmiller v. Dover Area School District*, item 21.)

13. A 2005 poll of U.S. adults found that 42% of the respondents believe that life on Earth "has existed in its present form since the beginning of time." Given the evidence for evolution by natural selection, comment on why so few people in the United States accept it.

EXPLORING THE LITERATURE

14. For more on the genetic control of beak shape in Darwin's finches, see:

    Mallarino, R., P. R. Grant, et al. 2011. Two developmental modules establish 3D beak-shape variation in Darwin's finches. *Proceedings of the National Academy of Sciences USA* 108: 4057–4062.

15. Hundreds of viruses, bacteria, fungi, insects, and weeds have evolved resistance to drugs, herbicides, fungicides, or pesticides, providing examples of evolution in action. In many cases, we know the molecular mechanisms of the evolutionary changes involved. Think about how the evidence from these studies compares with the evidence for evolution in Darwin's finches and HIV:

    Davies, J. 1994. Inactivation of antibiotics and the dissemination of resistance genes. *Science* 264: 375–382.

    Gaines, T. A., W. Zhang, et al. 2010. Gene amplification confers glyphosate resistance in *Amaranthus palmeri*. *Proceedings of the National Academy of Sciences USA* 107: 1029–1034.

    Van Rie, J., W. H. McGaughey, et al. 1990. Mechanism of insect resistance to the microbial insecticide *Bacillus thuringiensis*. *Science* 247: 72–74.

16. We mentioned flatfish as an example of apparently jury-rigged design. For a transitional fossil that sheds light on how the strange body plan of flatfish evolved, see:

    Friedman, M. 2008. The evolutionary origin of flatfish asymmetry. *Nature* 454: 209–212.

17. For more on the evolution of animal eyes, see the special issue of *Evolution: Education & Outreach*, published in October 2008 (vol. 1, no. 4).

18. For more on the evolution of carnivorous plants, see:

    Gibson, T. C., and D. M. Waller. 2009. Evolving Darwin's "most wonderful" plant: Ecological steps to a snap-trap. *New Phytologist* 183: 575–587.

19. For detailed critical discussions of intelligent design creationism, see:

    Pennock, R.T. (ed.). 2001. *Intelligent Design Creationism and Its Critics*. Cambridge, MA: MIT Press. (See especially Chapter 10 by M. J. Behe, Chapter 11 by Philip Kitcher, and Chapter 12 by M. J. Brauer and D. R. Brumbaugh.)

    Young, M., and T. Edis (eds.). 2004. *Why Intelligent Design Fails: A Scientific Critique of the New Creationism*. New Brunswick, NJ: Rutgers University Press.

20. For a detailed account of the 2005 Dover School Board intelligent design case (*Kitzmiller v. Dover Area School District*), see:

    Padian, K., and N. Matzke. 2009. Darwin, Dover, "Intelligent Design" and textbooks. *Biochemical Journal* 417: 29–42.

    For the decision in the 2005 Dover School Board intelligent design case (*Kitzmiller vs. Dover School District*), see: http://www.pamd.uscourts.gov/kitzmiller/kitzmiller_342.pdf

CITATIONS

Abbott, I., L. K. Abbott, and P. R. Grant. 1977. Comparative ecology of Galápagos ground finches (*Geospiza* Gould): Evaluation of the importance of floristic diversity and interspecific competition. *Ecological Monographs* 47: 151–184.

Abzhanov, A., M. Protas, et al. 2004. *Bmp4* and morphological variation of beaks in Darwin's finches. *Science* 305: 1462–1465.

Albala, K. 2002. *Eating Right in the Renaissance*. Berkeley: University of California Press. Page 237.

Alcalá, R. E., N. A. Mariano, et al. 2010. An experimental test of the defensive role of sticky traps in the carnivorous plant *Pinguicula moranensis* (Lentibulariaceae). *Oikos* 119: 891–895.

Anderson, T. M., B. M. vonHoldt, et al. 2009. Molecular and evolutionary history of melanism in North American gray wolves. *Science* 323: 1339–1343.

Baardsnes, J., and P. L. Davies. 2001. Sialic acid synthase: The origin of fish type III antifreeze protein? *Trends in Biochemical Sciences* 26: 468–469.

Behe, M. 1996. *Darwin's Black Box: The Biochemical Challenge to Evolution*. New York: Free Press/Simon and Schuster.

Behe, M. 1998. Molecular machines: Experimental support for the design inference. *Cosmic Pursuit* (Spring): 27–35. http://www.discovery.org/a/54.

Boag, P. T. 1983. The heritability of external morphology in Darwin's ground finches (*Geospiza*) on Isla Daphne Major, Galápagos. *Evolution* 37: 877–894.

Boag, P. T., and P. R. Grant. 1978. Heritability of external morphology in Darwin's finches. *Nature* 274: 793–794.

Boag, P. T., and P. R. Grant. 1981. Intense natural selection in a population of Darwin's finches (Geospizinae) in the Galápagos. *Science* 214: 82–85.

Boag, P. T., and P. R. Grant. 1984a. Darwin's finches (*Geospiza*) on Isla Daphne Major, Galápagos: Breeding and feeding ecology in a climatically variable environment. *Ecological Monographs* 54: 463–489.

Boag, P. T., and P. R. Grant. 1984b. The classical case of character release: Darwin's finches (*Geospiza*) on Isla Daphne Major, Galápagos. *Biological Journal of the Linnean Society* 22: 243–287.

Bowler, P. J. 2002. Evolution: History. In *Encyclopedia of Life Sciences*. Macmillan Publishers Ltd. Nature Publishing Group. Available at http://www.els.net.

Bowman, R. I. 1961. Morphological differentiation and adaptation in the Galápagos finches. *University of California Publications in Zoology* 58: 1–302.

Burns, K. J., S. J. Hackett, et al. 2002. Phylogenetic relationships and morphological diversity in Darwin's finches and their relatives. *Evolution* 56: 1240–1252.

Campàs, O., R. Mallarino, et al. 2010. Scaling and shear transformations capture beak shape variation in Darwin's finches. *Proceedings of the National Academy of Sciences USA* 107: 3356–3360.

Charlesworth, B., and D. Charlesworth. 2009. Darwin and genetics. *Genetics* 183: 757–766.

Cong, B., L. S. Barrero, and S. D. Tanksley. 2008. Regulatory change in YABBY-like transcription factor led to evolution of extreme fruit size during tomato domestication. *Nature Genetics* 40: 800–804.

Cong, B., J. Liu, and S. D. Tanksley. 2002. Natural alleles at a tomato fruit size quantitative trait locus differ by heterochronic regulatory mutations. *Proceedings of the National Academy of Sciences, USA* 99: 13606–13611.

Coyne, J. A. 2009. *Why Evolution Is True*. Oxford: Oxford University Press.

Darwin, C. R. 1859. *On the Origin of Species by Means of Natural Selection*, 1st ed. London: John Murray.

Darwin, C. R. 1868. *The Variation of Animals and Plants Under Domestication*. London: John Murray.

Darwin, C. R. 1872. *The Origin of Species by Means of Natural Selection*, 6th ed. London: John Murray.

Dawkins, R. 1986. *The Blind Watchmaker*. Essex: Longman Scientific.

Dembski, W. A. 2002. *No Free Lunch: Why Specified Complexity Cannot Be Purchased Without Intelligence*. Lanham, MA: Rowman & Littlefield.

Dobzhansky, T. 1973. Nothing in biology makes sense except in the light of evolution. *American Biology Teacher* 35: 125–129.

Dodson, E. O. 1960. *Evolution: Process and Product*. New York: Reinhold.

Fain, G. L., R. Hardie, and S. B. Laughlin. 2010. Phototransduction and the evolution of photoreceptors. *Current Biology* 20: R114–R124.

Felsenstein, J. 2007. Has natural selection been refuted? The arguments of William Dembski. *Reports of the National Center for Science Education* 27: 20–26.

Fernald, R. D. 2004. Evolving eyes. *International Journal of Developmental Biology* 48: 701–705.

Fleming, I. A., K. Hindar, et al. 2000. Lifetime success and interactions of farm salmon invading a native population. *Proceedings of the Royal Society B* 267: 1517–1523.

Fletcher, G. L., C. L. Hew, and P. L. Davies. 2001. Antifreeze proteins of teleost fishes. *Annual Review of Physiology* 63: 359–390.

Forrest, B. 2008. Still creationism after all these years: Understanding and counteracting intelligent design. *Integrative and Comparative Biology* 48: 189–201.

Frankham, R. 2008. Genetic adaptation to captivity in species conservation programs. *Molecular Ecology* 17: 325–333.

Frary, A., T. C. Nesbitt, et al. 2000. *fw2.2*: A quantitative trait locus key to the evolution of tomato fruit size. *Science* 289: 85–88.

Frey, Erica, C. J. Brokaw, and C. K. Omoto. 1997. Reactivation at low ATP distinguishes among classes of paralyzed flagella mutants. *Cell Motility and the Cytoskeleton* 38: 91–99.

Futuyma, D. J. 1983. *Science on Trial: The Case for Evolution*. New York: Pantheon.

Gehring, W. J. 2004. Historical perspective on the development and evolution of eyes and photoreceptors. *International Journal of Developmental Biology* 48: 707–717.

Genovart, M., N. Negre, et al. 2010. The young, the weak and the sick: Evidence of natural selection by predation. *PLoS ONE* 5: e9774.

Giancoli, D. C. 1995. *Physics: Principles with Applications*. Englewood Cliffs, NJ: Prentice Hall.

Gibbs, H. L., and P. R. Grant. 1987. Oscillating selection on Darwin's finches. *Nature* 327: 511–513.

Gish, D. T. 1978. *Evolution: The Fossils Say No!* San Diego: Creation-Life Publishers.

Goodstein, L. 2005. Judge bars "Intelligent Design" from Pa. classes. *New York Times*, December 20. http://www.nytimes.com/2005/12/21/education/21evolution.html.

Gould, S. J. 1982. Introduction to reprinted edition of Dobzhansky, T. G. 1937. *Genetics and the Origin of Species*. New York: Columbia University Press.

Gould, S. J. 1983. *Hen's Teeth and Horse's Toes*. New York: Norton.

Grant, B. R. 1985. Selection on bill characters in a population of Darwin's finches: *Geospiza conirostris* on Isla Genovesa, Galápagos. *Evolution* 39: 523–532.

Grant, B. R. 2003. Evolution in Darwin's finches: A review of a study on Isla Daphne Major in the Galápagos Archipelago. *Zoology* 106: 255–259.

Grant, B. R., and P. R. Grant. 1989. *Evolutionary Dynamics of a Natural Population*. Chicago: University of Chicago Press.

Grant, B. R., and P. R. Grant. 1993. Evolution of Darwin's finches caused by a rare climatic event. *Proceedings of the Royal Society of London B* 251: 111–117.

Grant, B. R., and P. R. Grant. 2003. What Darwin's finches can teach us about the evolutionary origin and regulation of biodiversity. *BioScience* 53 (10): 965–975.

Grant, P. R. 1981. The feeding of Darwin's finches on *Tribulus cistoides* (L.) seeds. *Animal Behavior* 29: 785–793.

Grant, P. R. 1999. *Ecology and Evolution of Darwin's Finches*, 2nd ed. Princeton: Princeton University Press.

Grant, P. R., and B. R. Grant. 1995. Predicting microevolutionary responses to directional selection on heritable variation. *Evolution* 49: 241–251.

Grant, P. R., and B. R. Grant. 2000. Non-random fitness variation in two populations of Darwin's finches. *Proceedings of the Royal Society of London B* 267: 131–138.

Grant, P. R., and B. R. Grant. 2002a. Unpredictable evolution in a 30-year study of Darwin's finches. *Science* 296: 707–711.

Grant, P. R., and B. R. Grant. 2002b. Adaptive radiation of Darwin's finches. *American Scientist* 90 (2): 130–139.

Grant, P. R., and B. R. Grant. 2005. Darwin's finches. *Current Biology* 15: R614–R615.

Grant, P. R., and B. R. Grant. 2006. Evolution of character displacement in Darwin's finches. *Science* 313: 224–226.

Grant, P. R., B. R. Grant, et al. 1976. Darwin's finches: Population variation and natural selection. *Proceedings of the National Academy of Sciences USA* 73: 257–261.

Gregory, T. R. 2008. The evolution of complex organs. *Evolution: Education and Outreach* 1: 358–389.

Harding, R. M., E. Healy, et al. 2000. Evidence for variable selective pressures at MC1R. *American Journal of Human Genetics* 66: 1351–1361.

Heubl, G., G. Bringmann, and H. Meimberg. 2006. Molecular phylogeny and character evolution of carnivorous plant families in Caryophyllales—revisited. *Plant Biology* 8: 821–830.

Hoekstra, H. E., R. J. Hirschmann, et al. 2006. A single amino acid mutation contributes to adaptive beach mouse color pattern. *Science* 313: 101–104.

Huynen, M. A., T. Dandekar, and P. Bork. 1999. Variation and evolution of the citric-acid cycle: A genomic perspective. *Trends in Microbiology* 7: 281–291.

Jones, J. E., III. 2005. *Tammy Kitzmiller v. Dover Area School District, Memorandum Opinion*. U.S. District Court for the Middle District of Pennsylvania, Case No. 04cv2688.

Jones, K. N., and J. S. Reithel. 2001. Pollinator-mediated selection on a flower color polymorphism in experimental populations of *Antirrhinum* (Scrophulariaceae). *American Journal of Botany* 88: 447–454.

Kanetsky, P. A., J. Swoyer, et al. 2002. A polymorphism in the agouti signaling protein gene is associated with human pigmentation. *American Journal of Human Genetics* 70: 770–775.

Keller, L. F., P. R. Grant, et al. 2001. Heritability of morphological traits in Darwin's finches: Misidentified paternity and maternal effects. *Heredity* 87: 325–336.

Kitcher, P. 1982. *Abusing Science: The Case Against Creationism.* Cambridge, MA: MIT Press.

Krem, M. M., and E. Di Cera. 2002. Evolution of enzyme cascades from embryonic development to blood coagulation. *Trends in Biochemical Sciences* 27: 67–74.

Krontiris, T. G. 1995. Minisatellites and human disease. *Science* 269: 1682–1683.

Lacalli, T. C., N. D. Holland, and J. E. West. 1994. Landmarks in the anterior central nervous system of amphioxus larvae. *Philosophical Transactions of the Royal Society B* 344: 165–185.

Lamb, T. D., S. P. Collin, and E. N. J. Pugh. 2007. Evolution of the vertebrate eye: Opsins, photoreceptors, retina and eye cup. *Nature Reviews Neuroscience* 8: 960–976.

Langerhans, R. B., C. A. Layman, and T. J. DeWitt. 2005. Male genital size reflects a tradeoff between attracting mates and avoiding predators in two live-bearing fish species. *Proceedings of the National Academy of Sciences USA* 102: 7618–7623.

Larson, E. J. 1997. *Summer for the Gods: The Scopes Trial and America's Continuing Debate Over Science and Religion.* Cambridge, MA: Harvard University Press.

Lenski, R. E., C. Ofria, et al. 2003. The evolutionary origin of complex features. *Nature* 423: 139–144.

Liu, J., B. Cong, and S. D. Tanksley. 2003. Generation and analysis of an artificial gene dosage series in tomato to study the mechanisms by which the cloned quantitative trait locus *fw2.2* controls fruit size. *Plant Physiology* 132: 292–299.

Liu, R., and H. Ochman. 2007. Stepwise formation of the bacterial flagellar system. *Proceedings of the National Academy of Sciences USA* 104: 7116–7121.

Mayr, E. 1980. Prologue: Some thoughts on the history of the evolutionary synthesis. In *The Evolutionary Synthesis: Perspectives on the Unification of Biology,* ed. E. Mayr and W. B. Provine. Cambridge, MA: Harvard University Press, 1–48.

Mayr, E. 1991. *One Long Argument: Charles Darwin and the Genesis of Modern Evolutionary Thought.* Cambridge, MA: Harvard University Press.

Mayr, E., and W. B. Provine (eds.). 1980. *The Evolutionary Synthesis: Perspectives on the Unification of Biology.* Cambridge, MA: Harvard University Press.

McGregor, A. P., V. Orgogozo, et al. 2007. Morphological evolution through multiple *cis*-regulatory mutations at a single gene. *Nature* 448: 587–590.

Meléndez-Hevia, E., T. G. Waddell, and M. Cascante. 1996. The puzzle of the Krebs citric acid cycle: Assembling the pieces of chemically feasible reactions, and opportunism in the design of metabolic pathways during evolution. *Journal of Molecular Evolution* 43: 293–303.

Miller, K. R. 1999. *Finding Darwin's God: A Scientist's Search for Common Ground Between God and Evolution.* New York: Cliff Street Books.

Mitchell, D. R. 2000. *Chlamydomonas* flagella. *Journal of Phycology* 36: 261–273.

Moose, S. P., J. W. Dudley, and T. R. Rocheford. 2004. Maize selection passes the century mark: A unique resource for 21st century genomics. *Trends in Plant Science* 9: 358–364.

Mullen, L. M., and H. E. Hoekstra. 2008. Natural selection along an environmental gradient: A classic cline in mouse pigmentation. *Evolution* 62: 1555–1570.

Mullen, L. M., S. N. Vignieri, et al. 2009. Adaptive basis of geographic variation: Genetic, phenotypic and environmental differences among beach mouse populations. *Proceedings of the Royal Society B* 276: 3809–3818.

Musser, S. M., and S. I. Chan. 1998. Evolution of the cytochrome *c* oxidase proton pump. *Journal of Molecular Evolution* 46: 508–520.

National Public Radio. 1995. Evolution disclaimer to be placed in Alabama textbooks. *Morning Edition,* Transcript 1747, Segment 13.

Nesbitt, T. C., and S. D. Tanksley. 2002. Comparative sequencing in the genus *Lycopersicon*: Implications for the evolution of fruit size in the domestication of cultivated tomatoes. *Genetics* 162: 365–379.

Niklaus, K. J. 1997. *The Evolutionary Biology of Plants.* Chicago: University of Chicago Press.

Nilsson, D.-E. 2009. The evolution of eyes and visually guided behaviour. *Philosophical Transactions of the Royal Society of London B* 364: 2833–2847.

Nilsson, D.-E., and S. Pelger. 1994. A pessimistic estimate of the time required for an eye to evolve. *Proceedings of the Royal Academy of London B* 256: 53–58.

Oakley, T. H., and M. S. Pankey. 2008. Opening the "Black Box": The genetic and biochemical basis of eye evolution. *Evolution: Education and Outreach* 1: 390–402.

Packer, C., and A. E. Pusey. 1983. Adaptations of female lions to infanticide by incoming males. *American Naturalist* 121: 716–728.

Packer, C., and A. E. Pusey. 1984. Infanticide in carnivores. In *Infanticide,* ed. G. Hausfater and S. B. Hrdy. New York: Aldine, 31–42.

Padian, K., and N. Matzke. 2009. Darwin, Dover, "Intelligent Design" and textbooks. *Biochemical Journal* 417: 29–42.

Paley, W. 1802. *Natural Theology: Or, Evidence of the Existence and Attributes of the Deity, Collected from the Appearances of Nature.* London: R. Faulder.

Pallen, M. J., and N. J. Matzke. 2006. From *The Origin of Species* to the origin of bacterial flagella. *Nature Reviews Microbiology* 4: 784–790.

Petren, K., B. R. Grant, and P. R. Grant. 1999. A phylogeny of Darwin's finches based on microsatellite DNA length variation. *Proceedings of the Royal Society of London B* 266: 321–329.

Petren, K., P. R. Grant, et al. 2005. Comparative landscape genetics and the adaptive radiation of Darwin's finches: The role of peripheral isolation. *Molecular Ecology* 14: 2943–2957.

Price, T. D., P. R. Grant, et al. 1984. Recurrent patterns of natural selection in a population of Darwin's finches. *Nature* 309: 787–789.

Provine, W. B. 1971. *The Origins of Theoretical Population Genetics.* Chicago: University of Chicago Press.

Salvini-Plawen, L. V., and E. Mayr. 1977. On the evolution of photoreceptors and eyes. *Evolutionary Biology* 10: 207–263.

Sato, A., H. Tichy, et al. 2001. On the origin of Darwin's finches. *Molecular Biology and Evolution* 18: 299–311.

Schaffer, J. V., and J. L. Bolognia. 2001. The melanocortin-1 receptor: Red hair and beyond. *Archives of Dermatology* 137: 1477–1485.

Schmidt, K. 1996. Creationists evolve new strategy. *Science* 273: 420–422.

Scott, E. C. 1994. The struggle for the schools. *Natural History* 7: 10–13.

Scott, E. C., and N. J. Matzke. 2007. Biological design in science classrooms. *Proceedings of the National Academy of Sciences, USA* 104 (Suppl 1): 8669–8676.

Smith, J. J., N. R. Saha, and C. T. Amemiya. 2010. Genome biology of the cyclostomes and insights into the evolutionary biology of vertebrate genomes. *Integrative and Comparative Biology* 50: 130–137.

Smith, R., E. Healy, et al. 1998. Melanocortin 1 receptor variants in an Irish population. *Journal of Investigative Dermatology* 111: 119–122.

Spencer, H. 1864. *The Principles of Biology, Vol. 1.* London: Williams and Norgate.

Spooner, D. M., I. E. Peralta, and S. Knapp. 2005. Comparison of AFLPs with other markers for phylogenetic inference in wild tomatoes [*Solanum* L. section *Lycopersicon* (Mill.) Wettst.] *Taxon* 54: 43–61.

Swinney, S. 1994. *Evolution: Fact or Fiction.* Kansas City, MO: 1994 Staley Lecture Series, KLJC Audio Services.

Tanksley, S. D. 2004. The genetic, developmental, and molecular bases of fruit size and shape variation in tomato. *Plant Cell* 16: S181–S189.

True, J. R., and S. B. Carroll. 2002. Gene co-option in physiological and morphological evolution. *Annual Review of Cell and Developmental Biology* 18: 53–80.

Vignieri, S. N., J. G. Larson, and H. E. Hoekstra. 2010. The selective advantage of crypsis in mice. *Evolution* 64: 2153–2158.

Woolley, D. M., 1997. Studies on the eel sperm flagellum. I. The structure of the inner dynein arm complex. *Journal of Cell Science* 110: 85–94.

Wu, P., T.-X. Jiang, et al. 2005. Molecular shaping of the beak. *Science* 305: 1465–1466.

4

Estimating Evolutionary Trees

Scott Baker and Stephen Palumbi (1994) traveled to Japan with a portable genetics lab. They visited retail food markets and bought samples of whale meat. Back in their hotel room, the scientists extracted and copied mitochondrial DNA from the samples. They later sequenced the DNA and compared it to genetic material from known specimens. The result of their analysis was an inferred evolutionary history, or phylogenetic tree. Part of this evolutionary tree is shown at right. Sample 19b appeared to have come from a humpback whale. Additional evidence suggested the humpback was from the North Pacific. Unless the meat from this whale had been in storage for decades, it likely came from an individual that had been harvested in violation of international whaling treaties.

Humpback whales are endangered. It is illegal to hunt them under international whaling treaties. Photo © Chris Huss.

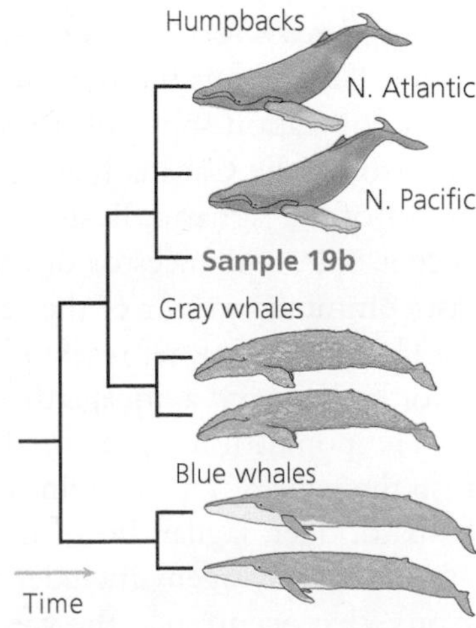

Forensic identification of unknown specimens is but one of many kinds of problems that can be addressed by reconstructing evolutionary relationships among species, populations, or individual organisms. The reconstruction of evolutionary relationships, and the kinds of questions it can answer, are our topics for this chapter.

When biologists reconstruct evolutionary relationships, they typically summarize their results the way Baker and Plumbi did: with tree diagrams. Evolutionary trees have become so ubiquitous that it is crucial for all students to understand them in some detail. We therefore open the chapter with a section on how to read evolutionary trees. This is followed by sections that cover methods biologists use to reconstruct evolutionary history from the morphology of organisms and from genetic data. For the latter, we briefly consider studies providing empirical evidence on the accuracy of evolutionary reconstructions. We close with a section on using evolutionary trees to answer interesting questions.

4.1 How to Read an Evolutionary Tree

An **evolutionary tree**, also known as a **phylogenetic tree** or a **phylogeny**, is a diagram showing the history of divergence and evolutionary change leading from a single ancestral lineage to a suite of descendants. In other words, it depicts a group of organisms' genealogical relationships as they are understood according to Darwin's theory of descent with modification from common ancestors.

How to Read Time on an Evolutionary Tree

In 1837 or 1838, in a notebook he used to record his thoughts on the transmutation of species, Darwin himself sketched the earliest phylogenies. On page 36 of the notebook, for example, he drew the evolutionary tree shown in **Figure 4.1a**. The line Darwin labeled "1" represents the **root** of the tree, the species that is the common ancestor of species A, B, C, and D (and of the lineages represented by unlabeled branches). The sole illustration in Darwin's 490-page *On the Origin of Species* was an evolutionary tree, part of which we have redrawn in Figure 4.1b. Here, the root is at the bottom of the tree. Moving upward from the root, the lines trace the divergences leading from the common ancestor to its descendants, living and extinct. Darwin's evolutionary trees, like all other phylogenies, should be thought of as having grown over time, like the visible portion of a real tree. At first just a single shoot reaches up from the root. Soon, however, the shoot begins to bifurcate, producing branches. These, too, bifurcate, producing smaller branches and finally twigs. The shape of the mature tree thus records a history of the tree's development. In this history, time flows from the root along the trunk, through the branches, toward the twigs.

Most evolutionary trees in this book are drawn like the one in **Figure 4.2**. Here, the root is on the left and time proceeds from left to right. The splitting branches trace the evolutionary history leading to eight living species of wild cats, all derived from a single ancestral lineage represented by the root. Starting at the root and reading across the tree, the first branch point we encounter is **node** 1. Node 1 represents the most recent common ancestor of the eight living species. This population split into two. One daughter population became the common ancestor of the Canada lynx and the bobcat. The other became the common ancestor of the jaguarundi, snow leopard, tiger, jaguar, lion, and leopard. The most recent common ancestor of the lynx and bobcat is represented by node 6. The last common ancestor of the remaining extant species is represented by node 2.

The population represented by node 6 split, producing a lineage that evolved into the lynx and a lineage that evolved into the bobcat.

The population represented by node 2 split, producing a lineage that evolved into the jaguarundi and a lineage that became the common ancestor of the snow leopard, tiger, jaguar, lion, and leopard.

The cat phylogeny includes some of the evolutionary modifications, or **transitions**, that occurred as the various cat lineages diverged. For example, sometime

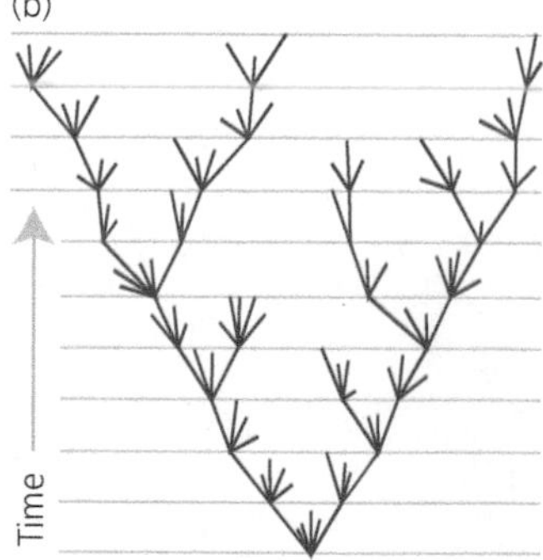

Figure 4.1 Darwin's evolutionary trees (a) An evolutionary tree Darwin sketched in his *Notebook B: Transmutation of Species* (1987–1838). (b) A portion of the evolutionary tree that was the only illustration in *On the Origin of Species* (1859).

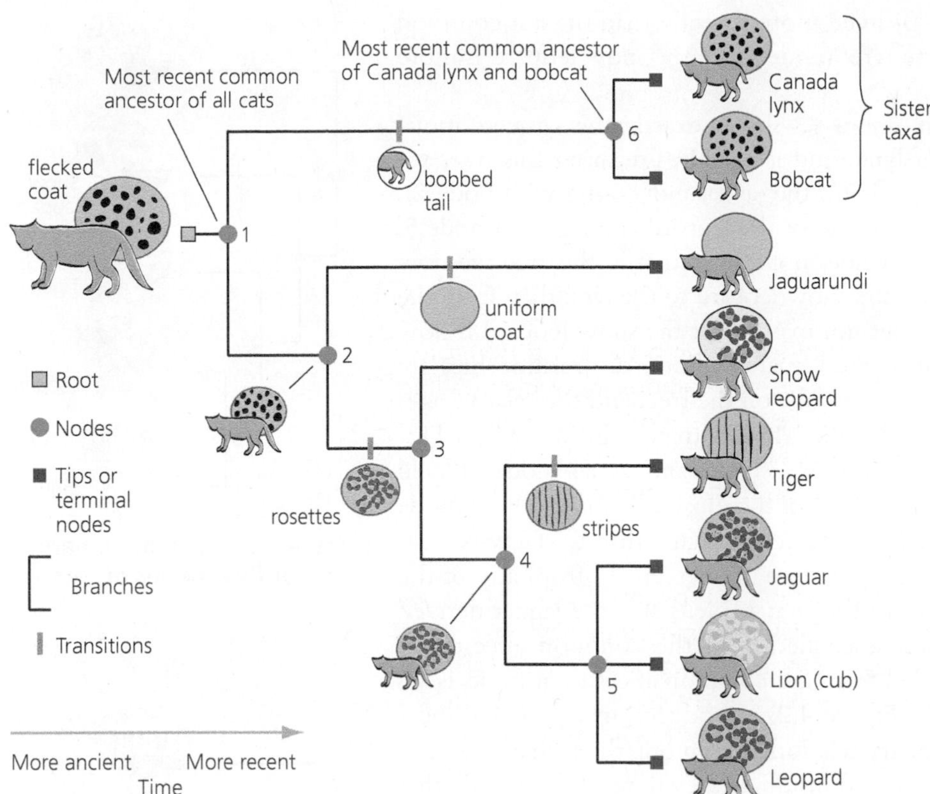

Figure 4.2 An evolutionary tree for eight species of cats After Werdelin and Olsson (1997).

between its divergence from the other lineages at node 1 and its own diversification at node 6, the common ancestor of the lynx and bobcat evolved a bobbed tail. Sometime after its divergence at node 2 from the lineage that would become the snow leopard, tiger, jaguar, lion, and leopard, the ancestor of the jaguarundi evolved a spotlessly uniform coat. And sometime after its divergence at node 2, but before its diversification at node 3, the common ancestor of the snow leopard, tiger, jaguar, lion, and leopard evolved rosettes—clusters of spots featuring a central fleck surrounded by smaller flecks.

Phylogenies, also known as evolutionary trees, are hypotheses about the history of descent with modification from a common ancestor that produced a set of species or other taxa. Time flows along a phylogeny from the root toward the branch tips.

We excerpted this cat phylogeny from a larger tree reconstructed by Lars Werdelin and Lennart Olsson (1997) for a paper they called "How the Leopard Got Its Spots." According to Werdelin and Olsson's hypothesis, the leopard got its spots by descent with modification, most recently from ancestors with a coat like the leopard's own and more distantly from ancestors with flecked coats.

How to Read Relationships on an Evolutionary Tree

The **evolutionary relationships** among species on a phylogeny are defined by the relative time elapsed since they last shared common ancestors. For example, in Figure 4.2 the last ancestor the lynx shares with the bobcat is represented by node 6. The last ancestor either the lynx or bobcat shares with any other species is represented by node 1. The node 6 population lived more recently than the node 1 population. Because they share a more recent last common ancestor with each other than either shares with any other species, the lynx and bobcat are considered **sister species**. They are each other's closest living relatives.

Lineages that share more recent common ancestors are considered more closely related.

With this concept of relatedness, we can use the evolutionary tree in Figure 4.2 to answer other questions. For example, is the snow leopard more closely related to the Canada lynx or to the jaguar? The last common ancestor of the snow leopard and the jaguar (node 3) lived more recently than the last common ancestor of the snow leopard and the lynx (node 1), so the snow leopard is more closely related to the jaguar.

The simplified cat phylogeny in **Figure 4.3** shows exactly the same relationships for the snow leopard, Canada lynx, and jaguar. In preparing this tree, we pruned out the bobcat (along with node 6), the jaguarundi (along with node 2), and the lion. We also swiveled the branches for the leopard and jaguar at node 5, to put the jaguar below the leopard. It does not matter that in this tree we now have to cross fewer nodes to get from the snow leopard to the lynx than from the snow leopard to the jaguar. And it does not matter that the snow leopard is now closer on the page to the lynx than to the jaguar. Evolutionary relationships are defined solely by the order of the branch points on the tree, and the relative ages of the common ancestors this order identifies. In this tree, as in the one in Figure 4.2, the last common ancestor of the snow leopard and jaguar (node 3) lived more recently than the last common ancestor of the snow leopard and lynx (node 1), so the snow leopard is more closely related to the jaguar than to the lynx.

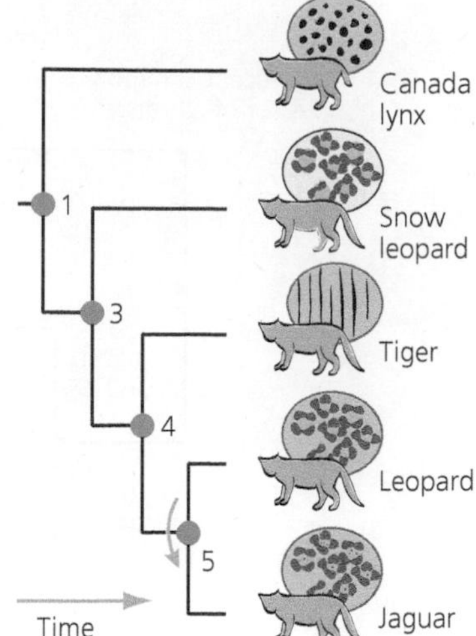

Figure 4.3 An evolutionary tree for five species of cats

None of the living cats at the tips of the tree are descended from any of the others. The leopard is not derived from the jaguar. Neither is the jaguar derived from the leopard. Instead, both are descended from the common ancestor at node 5. And all the cats are descended from the common ancestor at node 1.

The three-way split at node 5 in Figure 4.2, reproduced in **Figure 4.4**, indicates uncertainty about the evolutionary relationships among the jaguar, the lion, and the leopard. Werdelin and Olsson had insufficient evidence to conclude that any two of these cats were more closely related to each other than to the third species, so the researchers showed all three lineages diverging simultaneously.

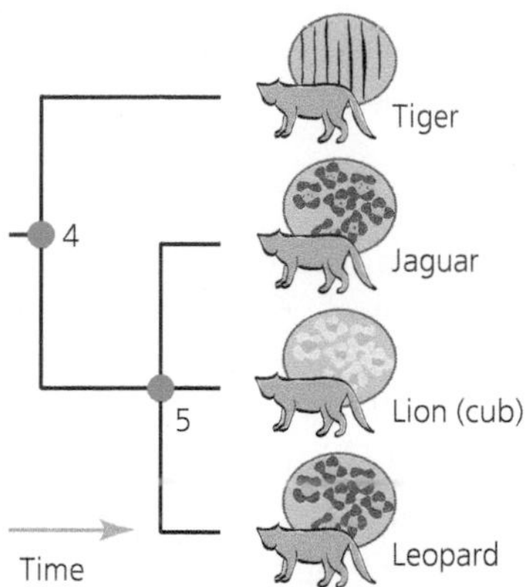

Figure 4.4 An evolutionary tree for four species of cats

Evolutionary Trees Do Not Show Everything

Any particular evolutionary tree contains only what the author who prepared it deemed relevant to the analysis at hand. It is important to keep in mind that a great many details of evolutionary history are missing from any given phylogeny.

For example, the tree in Figure 4.2 shows some of the changes in coat color and tail length that occurred during the evolution of eight species of cats. It does not show changes in body size, or the evolution of the mane in lions. Nor does it show any number of genetic, behavioral, or physiological changes that occurred. The absence of transitions marked on a branch thus does not imply that no evolution occurred. And the timing of the transitions that are marked on a phylogeny is generally known less precisely than the diagram might seem to imply. The placement of a particular transition on a particular branch indicates only that the transition occurred somewhere between the nodes at the branch's ends.

Furthermore, the tree in Figure 4.2 shows the relationships among only eight species of cats. It does not include the Eurasian lynx, more closely related to the Canada lynx than to the bobcat (Johnson et al. 2006). It does not include the clouded leopard, more closely related to the snow leopard, tiger, jaguar, lion, and leopard than to the jaguarundi. Nor does it include dozens of other cat species, living and extinct. A phylogeny speaks only about relationships among taxa (named groups of orgnanisms) it includes; it says nothing about taxa not included.

Evolutionary Trees Can Be Drawn in Various Styles

Because the order of branching carries all the information a phylogeny contains about relative relatedness, evolutionary trees can be drawn in many different

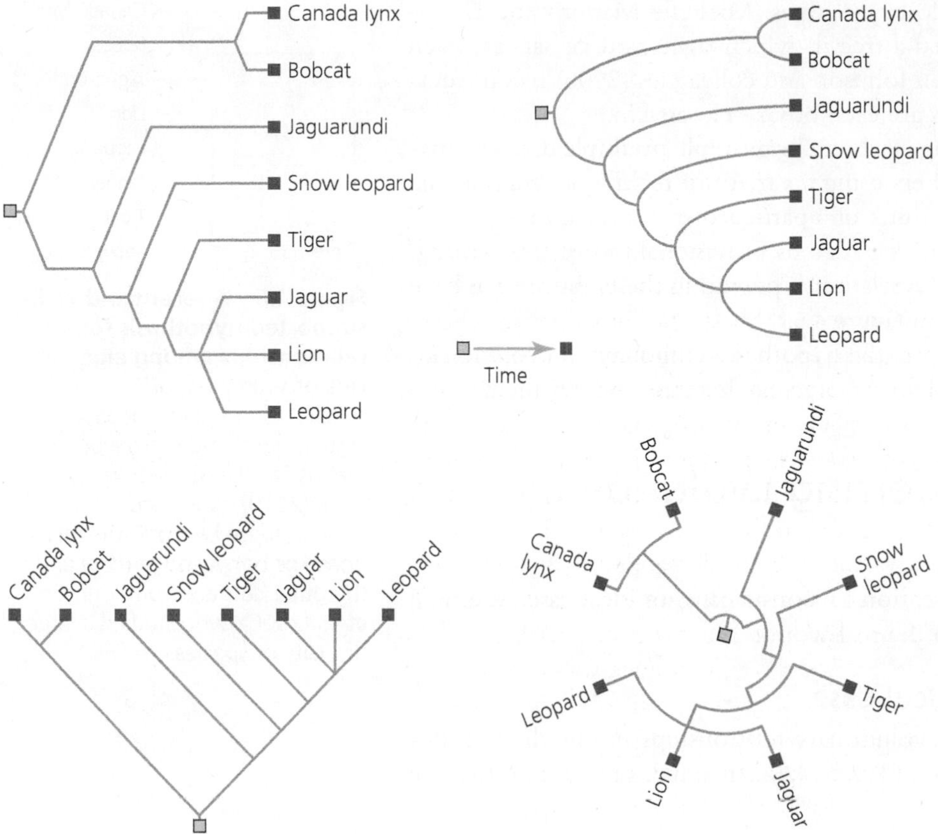

Figure 4.5 Four different versions of the same evolutionary tree All show the same evolutionary relationships among the eight extant cat species as is depicted in Figure 4.2. In all branches on all trees, time flows from the root toward the tips. After Gregory (2008). Trees prepared using PHYLIP-drawgram (Felsenstein 2009).

styles. **Figure 4.5** presents the cat phylogeny from Figure 4.2 in four alternative styles. All four show the same evolutionary history and the same relationships among the cats as the tree in Figure 4.2. In all branches on all trees, time flows from the root toward the tips.

Evolutionary relationships are depicted solely by the order of branching in a phylogeny.

Evolutionary Trees Are Hypotheses

Only under extremely rare circumstances do we know the evolutionary history of populations or species by direct observation. These usually involve lineages maintained and managed in laboratories (see, for example, Hillis et al. 1992). The rest of the time, history is unrecorded and we are left to infer it from what clues we can gather and piece together. We consider data and methods used to infer phylogenies shortly. But it bears emphasizing that an evolutionary tree is virtually never a revealed truth. Instead, it is a hypothesis based on a particular data set that has been analyzed with a particular technique.

Consider the cat phylogeny, assembled by Werdelin and Olsson, that we have looked at in this section. Neither Werdelin and Olsson, nor we, nor anyone else, knows the true evolutionary history of the leopard and its kin. Not only do we not know the true history of coat colors, we do not know the true pattern of branching on the tree from the common ancestor to today's cats. The best Werdelin and Olsson could do was use the available evidence to identify the most plausible scenarios.

This distinction between our hypotheses and the (almost always) unknowable truth helps explain why researchers using different data sets and different methods have inferred somewhat different phylogenies for the cats. Warren Johnson and Stephen O'Brien (1997), for example, reconstructed a phylogeny in which lions and tigers are each other's closest relatives. Michelle Mattern and Deborah McLennan (2000) reconstructed a tree in which tigers and jaguars are each other's closest relatives. And Warren Johnson and colleagues (2006) reconstructed a tree in which jaguars and lions are each other's closest kin.

Only when the evidence is strong—when, for example, multiple data sets analyzed by different teams of researchers using a variety of techniques support the same hypothesis—can we begin to think of a particular evolutionary tree as well established. Even then, it should be regarded as provisional. As of this writing, the best-supported hypothesis for the relationships among the eight cats we have been considering is the phylogeny in **Figure 4.6**.

Having considered how to interpret the hypotheses embodied in evolutionary trees, we now turn to the logic and methods researchers use to infer them.

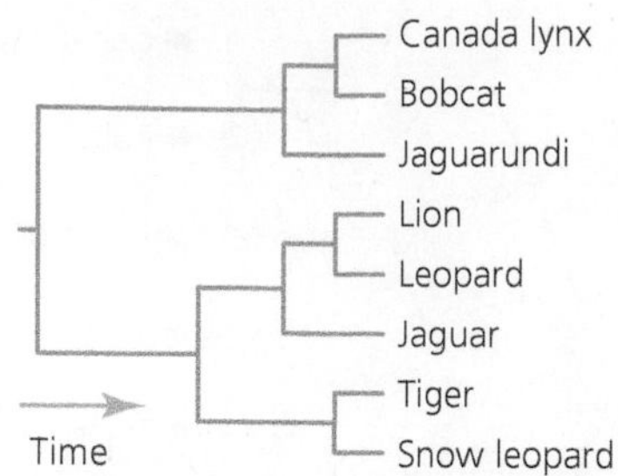

Figure 4.6 A recent and well-supported hypothesis for the relationships among eight species of wild cats The relationships among the lion, leopard, jaguar, tiger, and snow leopard (genus *Panthera*) are from Davis et al. (2010), who focused specifically on these big cats. The relationships between *Panthera* and the other cats are from Johnson et al. (2006), who studied a larger diversity of species.

4.2 The Logic of Inferring Evolutionary Trees

We begin our exploration in this section by considering an ideal case. We then consider complications that arise in the real world.

Phylogeny Inference in an Ideal Case

Imagine that we want to infer the evolutionary relationships among the four fictitious bird species on the right side of **Figure 4.7a**. Imagine, in addition, that we know the following:

- The four species are descended with modification from the common ancestor on the left side of Figure 4.7a. Note that this ancestor is an undecorated bird with a short beak. Knowledge of the common ancestor's characteristics allows us to identify the long beak and various decorations adorning the four species on the right as evolutionary novelties particular to this group of birds.
- Each of the evolutionary novelties evolved exactly once. This knowedge allows us to interpret the novelties as evidence of common ancestry.
- Once each of the novelties evolved in a lineage, it was never lost.

Under these circumstances, inferring the evolutionary history of the four species is straightforward (see Felsenstein 1982).

First we note which of the evolutionary novelties, also known as **derived characters**, are unique to one species and which are shared (Figure 4.7b). The characters unique to one species—long bills and dark tails—must have evolved after the lineages leading to these species diverged from the lineages leading to the other species. Otherwise, other species would have long bills and dark tails too. This conclusion allows us to begin drawing the evolutionary tree for the birds—starting from the twigs and moving backward. As shown in Figure 4.7c, we can add a twig leading to the long-billed bird and a twig leading to the dark-tailed bird, and we can mark each twig with a transition indicating the appearance of the unique novel character.

Now we look at the shared derived characters. Note that orange wing tips are shared by exactly two species. This evidence identifies the orange-tipped birds as sister species, and tells us that orange tips must have evolved after the lineage leading to the orange-tipped species diverged from the lineages leading to the other species. Otherwise, other species would have orange tips too. Continuing

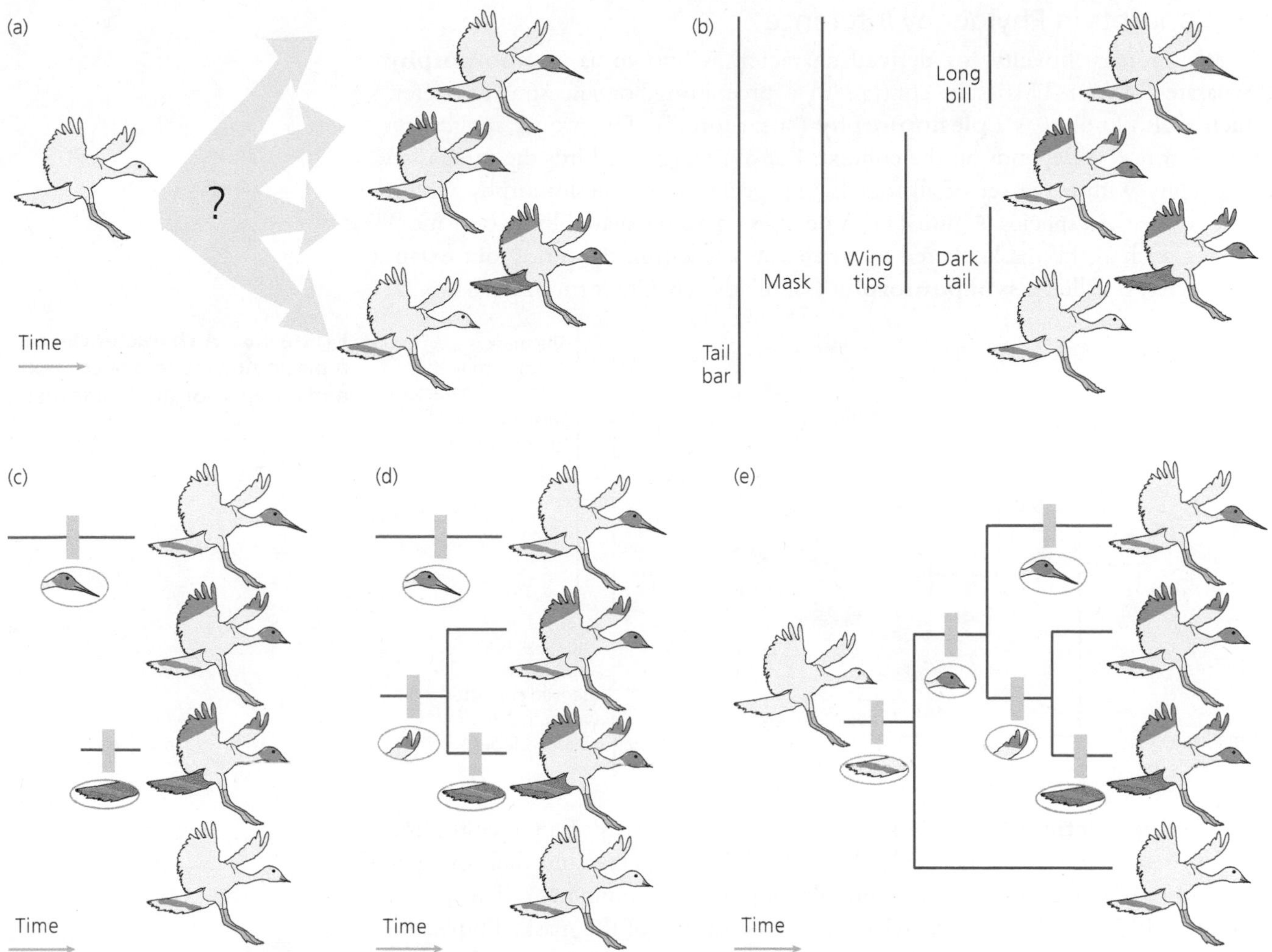

Figure 4.7 Inferring evolutionary relationships in an ideal case

to draw our tree backward in time, as shown in Figure 4.7d, we can now add a twig leading to the light-tailed orange-tipped species, connect it to the twig leading to the dark-tailed orange-tipped species, and place these connected twigs on a branch marked with the transition to orange tips.

That masks are shared by the two orange-tipped birds and the long-billed bird tells us that these three species are more closely related to each other than any of them is to the remaining species. And the fact that all four species have tail bars tells us that this trait must have evolved before any of the lineages leading to the four birds diverged from each other. We can now complete our tree. The finished phylogeny appears in Figure 4.7e.

In an ideal case, we can infer the evolutionary history of a set of species from their nested sets of shared evolutionary innovations.

We noted elsewhere (in Section 2.4) that descent with modification from common ancestors automatically produces species displaying nested sets of shared evolutionary novelties, that the order of nesting allows us to predict the order in which the novelties evolved, and that we can test Darwin's theory of descent with modification by checking such predictions against the fossil record. Here we have taken this logic just a step further. We have used nested sets of shared derived characters to reconstruct the history of diversification among lineages. That is, we have used shared derived characters to infer a phylogeny.

Key Concepts in Phylogeny Inference

An evolutionary novelty, or derived character, is known as an **apomorphy** ("separate form"). This is in contrast to a preexisting, or ancestral character, which is also known as a **plesiomorphy** ("near form"). The proper application of these concepts depends on the context. For our imaginary birds the mask is an apomorphy within the set of all four living species, but a plesiomorphy within the set of masked species **(Figure 4.8)**. A derived character shared by two or more lineages, such as the masks shared by three lineages within the set of four extant bird species, is called a **synapomorphy** ("similarly separate form").

The mask is a plesiomorphy (ancestral character) within the set of masked species.
The mask is an apomorphy (derived character) within the set of all four species.
Time

Figure 4.8 A character can be a plesiomorphy in one context and an apomorphy in another

A **monophyletic** group, also known as a **clade**, consists of an ancestor and all of its descendants. For our birds, the orange-tipped species and their orange-tipped common ancestor form a monophyletic group **(Figure 4.9a)**. This group is nested within another monophyletic group consisting of the masked birds and their masked common ancestor (Figure 4.9b). And this group, in turn, is nested within another monophyletic group consisting of the bar-tailed birds and their bar-tailed common ancestor (Figure 4.9c). A group consisting of an ancestor and some, but not all, of its descendants, such as the light-tailed birds and their light-tailed common ancestor in Figure 4.9c, is described as **paraphyletic**. A group that contains some, but not all, of an ancestor's descendants, and that also excludes the ancestor, such as the living masked light tails, is called **polyphyletic**.

Shared derived traits identify monophyletic groups—sets of taxa that include an ancestor and all of its descendants.

We can now concisely state a fundamental principle of phylogeny inference, advocated by the German entomologist Willi Hennig (1966): Synapomorphies identify monophyletic groups.

(a)
Time
(b)
Time
(c)
Time

Figure 4.9 Monophyletic groups Each consists of an ancestor and all its descendants.

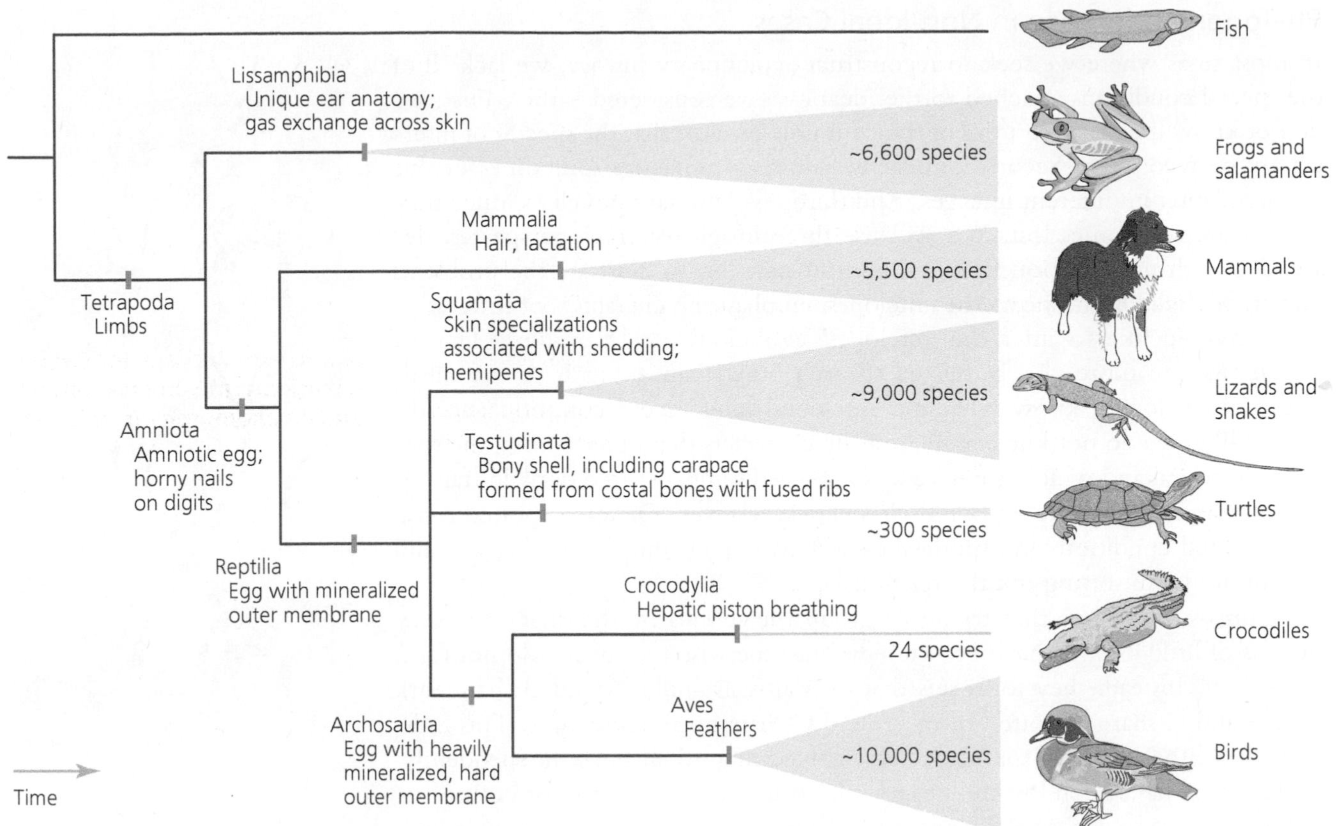

Figure 4.10 Major monophyletic groups of tetrapods Gray triangles at branch tips represent diversifications within monophyletic groups that could, space permitting, be represented by multifarious evolutionary trees. Based on Gans and Clark (1976), Meylan (2001), Alibardi and Maderson (2003), Kearney (2003), Mindell and Brown (2005), Carroll (2007), Laurin and Reisz (2007), Claessens (2009), Hoffmann et al. (2010), Laurin (2011), and Laurin and Gauthier (2011).

Figure 4.10 shows the major monophyletic groups of living terrestrial vertebrates along with some of the synapomorphies that distinguish them. Dogs, for example, belong to the mammals, with which they share hair and lactation. They also belong to the amniotes, a larger group with which they share an egg with an amniotic membrane. And, finally, dogs belong to the tetrapods, an even larger group with which they share limbs.

Note that the deepest branch point within the Reptilia is represented as a three-way split, or **polytomy**, rather than one bifurcation followed by another. We have drawn the tree this way due to lingering uncertainty about the evolutionary relationships among the Squamata, Testudinata, and Archosauria. Some data sets and analyses suggest that the lizard and snake lineage and the bird and crocodile lineage are more closely related to each other than either is to the turtle lineage (Werneburg and Sanchez-Villagra 2009; Lyson et al. 2010). Others suggest that the bird and crocodile lineage and the turtle lineage are more closely related to each other than either is to the lizard and snake lineage (Iwabe et al. 2005; Hugall et al. 2007). Still others suggest that the lizard and snake lineage and the turtle lineage are more closely related to each other than either is to the bird and crocodile lineage (Becker et al. 2011). This confusion demonstrates, if any demonstration were needed, that real biological history rarely conforms to our ideal case. We now consider ways to infer phylogenies despite real-world complications.

Uncertainty about the order of branching in a phylogeny is indicated by polytomies—nodes where a lineage splits into more than two descendant lineages simultaneously.

Phylogeny Inference in Non-ideal Cases

In most cases where we seek to reconstruct evolutionary history, we lack all of the special conditions attached to the ideal case we considered earlier. First, we do not know the characteristics of the common ancestor that the species of interest are derived from. Second, similar evolutionary novelties sometimes evolve independently in different lineages. And third, evolutionary novelties, once they evolve, are sometimes lost. We will use three imaginary antelope species, descended with modification from a common ancestor, to illustrate the problems that arise. **Figure 4.11a** shows the antelopes, emphasizing characters of interest.

We would like to infer the antelopes' evolutionary relationships. That is, which two are more closely related to each other than either is to the third? Because we do not know what the antelopes' most recent common ancestor looked like, we do not know which of the characters that differ among the species is ancestral and which is derived. Is a spotted rump a shared derived trait—a synapomorphy—identifying B and C as closest relatives? Or was a spotted rump the ancestral condition, and spotlessness a derived trait unique to species A and thus of no use in sorting out the relationships?

Figure 4.11b shows that trying to group the species by the characters they share is of little use. All have brown legs, but otherwise the species do not form nested sets. Instead, they form sets that overlap willy-nilly. A and B share dark tails; B and C share a spotted rump; A and C share horns and masks. This indicates that either some of the characters evolved more than once independently, or that some have been lost in lineages whose ancestors had them, or both.

How are we to proceed? The next two subsections introduce outgroup analysis and parsimony analysis, two strategies that together provide one of several methods that can help us untangle the antelopes' evolutionary past.

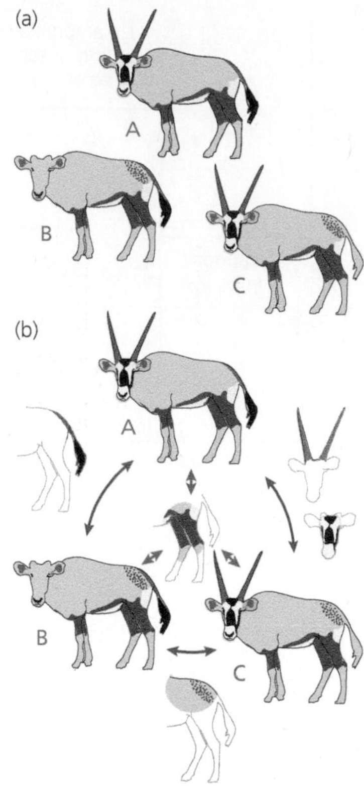

Figure 4.11 Three imaginary antelope species illustrate a non-ideal case for phylogeny inference

Outgroup Analysis

Outgroup analysis involves including in our historical reconstruction one or more additional species (Maddison et al. 1984). These should be relatives of the ingroup—the species whose relationships we wish to infer—but less closely related to them than the members of the ingroup are to each other. This ensures from the outset that in our finished reconstruction, the ingroup will be monophyletic.

In the simplest possible outgroup analysis, illustrated here, we add an outgroup of just one species **(Figure 4.12)**. We assume there has been no evolutionary change in the outgroup's lineage since it diverged from the lineage that gave rise to the ingroup. As we will soon see, this assumption allows us to make inferences about the characteristics of the ingroup's most recent common ancestor.

There are three possible resolutions of the relationships among the three species in our ingroup. A and B could be sister species, A and C could be sister species, or B and C could be sister species. We take these as hypotheses and compare them using parsimony analysis, the first of several criteria we will dicusss.

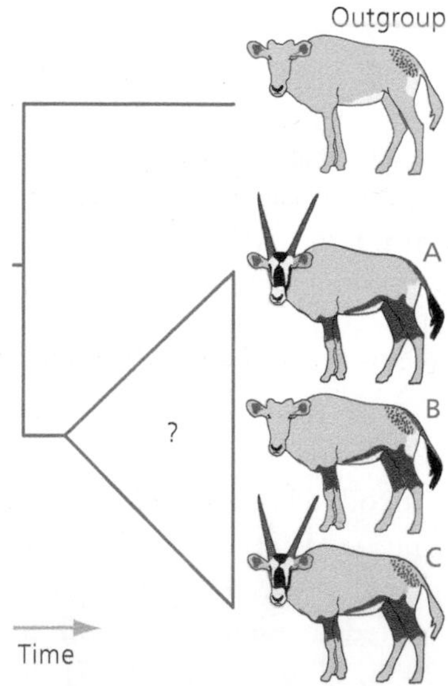

Figure 4.12 Outgroup analysis Adding another species to our reconstruction allows us to make inferences about the most recent common ancestor of the ingroup.

Parsimony Analysis

Under **parsimony** analysis, we prefer the hypothesis that requires the fewest evolutionary changes in the characters of interest (see Felsenstein 2004). We evaluate each character on each possible tree, looking for the simplest evolutionary scenario that can explain the distribution of the character states among the species at the tips. We add up the total number of evolutionary changes required by each hypothesis, and identify the hypothesis for which the total is lowest.

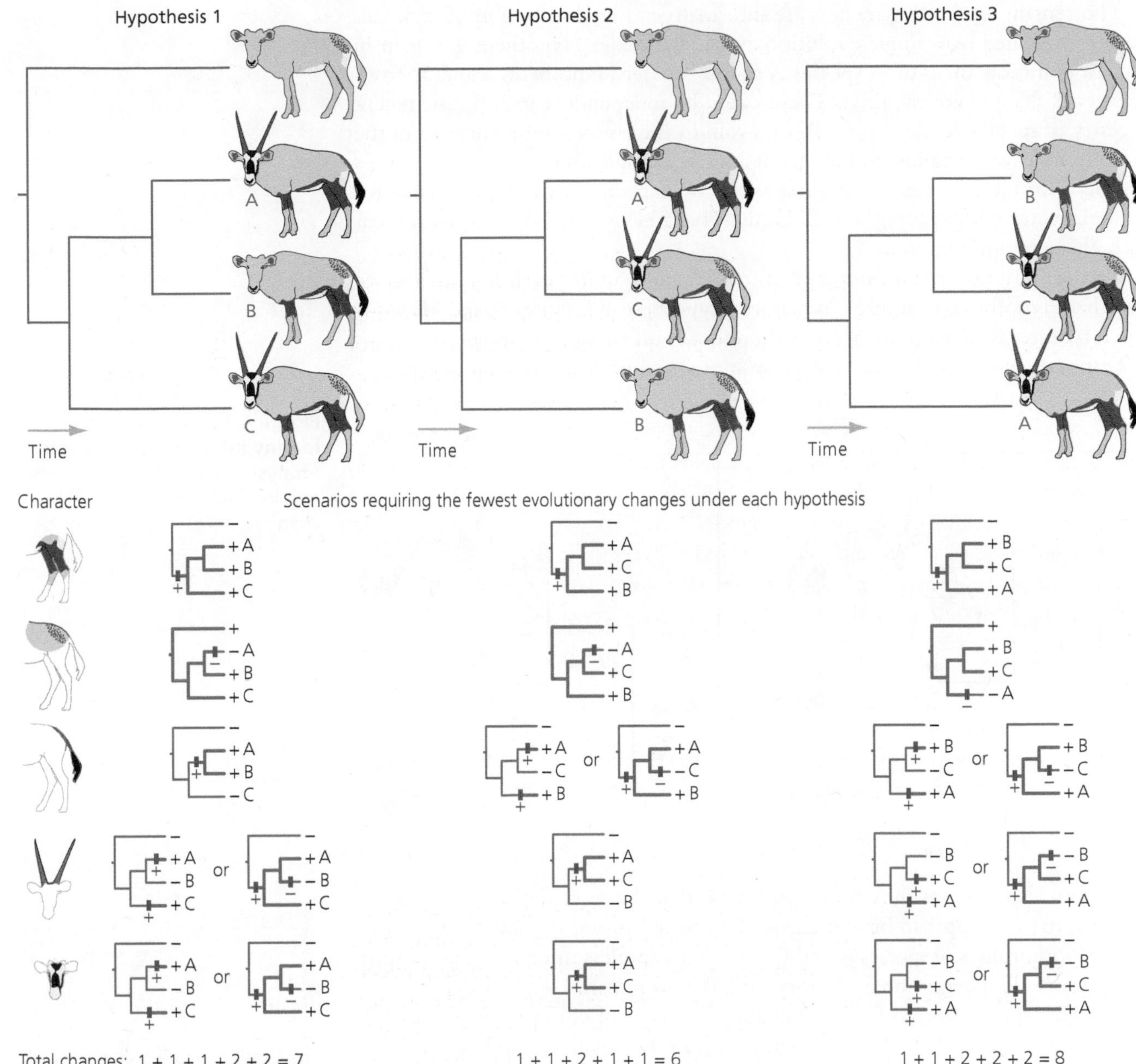

Figure 4.13 Parsimony analysis Each column evaluates one of our three hypotheses. Plus signs show the presence or gain of a character; minus signs indicate absence or loss.

The three possible evolutionary trees for our antelope species are shown at the top of **Figure 4.13**. The simplest scenarios for each character on each tree are shown in the lower portion of the figure.

Brown legs occur in all three of our ingroup antelopes, so each of our three hypotheses requires just one evolutionary change: the appearance of brown legs in the last common ancestor of the ingroup. Spotted rumps occur in the outgroup and all but one of the ingroup species, so all three hypotheses again require just one evolutionary change: the loss of rump spots in the recent ancestry of species A. Because the simplest scenarios require the same number of changes under all three hypotheses, striped legs and spotted rumps are **uninformative characters** for our analysis.

The remaining three characters are informative. The distribution of dark tails can be explained by a single evolutionary change under Hypothesis 1: a gain in the last common ancestor of species A and B. Under Hypotheses 2 and 3, however, two changes are required. These could be independent gains in the recent ancestry of species A and species B, or a gain in the last common ancestor of the ingroup followed by a loss in the recent ancestry of species C.

The distributions of horns and masks can be explained by a single change for each character under Hypothesis 2. Under Hypotheses 1 and 3, two changes for each character are required.

Counting the minimum number of total changes under each hypothesis, we find that Hypothesis 1 requires 7 changes, Hypothesis 2 requires 6, and Hypothesis 3 requires 8. Parsimony analysis therefore leads us to favor Hypothesis 2 as our estimate of the evolutionary relationships among our ingroup antelopes.

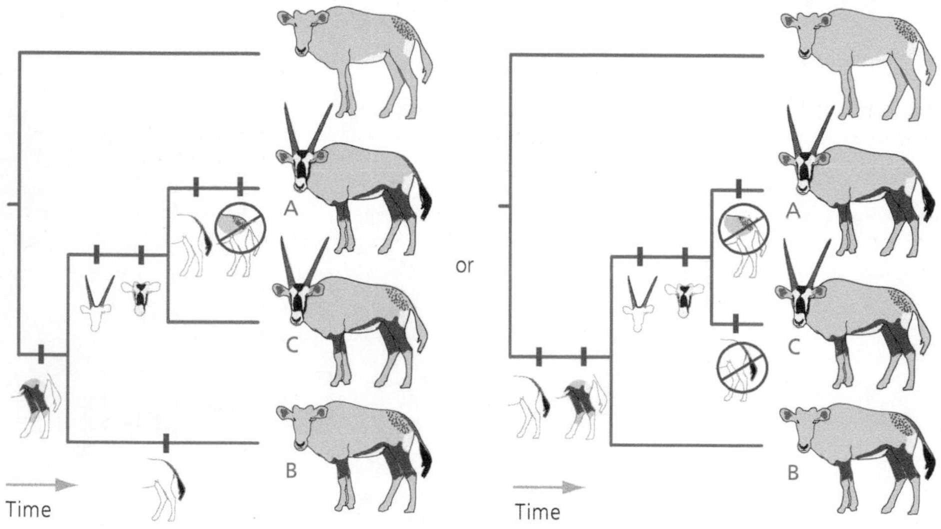

Figure 4.14 Antelope phylogeny inferred by parsimony analysis These two hypotheses require the fewest evolutionary changes.

We can also infer, as shown in **Figure 4.14**, that the most recent common ancestor of the ingroup had brown legs and a spotted rump, that horns and masks are synapomorphies of species A and C, and that a spotless rump is an apomorphy of species A. The evolutionary history of dark tails remains ambiguous.

It might be tempting to further conclude that the last common ancestor of all four species looked just like the outgroup. Recall, however, that we assumed this resemblance from the start. In fact, an equally parsimonious scenario is that the last common ancestor of all four species had brown legs and that this character was subsequently lost on the lineage leading to the outgroup. The last common ancestor of all four species may also have had a dark tail.

Most real examples do not work like our ideal case. Different characters often suggest different evolutionary relationships. In such circumstances, we consider all possible evolutionary trees as viable hypotheses and compare them using any of several criteria. One criterion is parsimony—the minimum amount of evolutionary change implied by a tree.

Parsimony analysis allows us to make inferences about the answers to questions that might otherwise be intractable. In reality, of course, evolutionary history may not always proceed according to the simplest possible scenario. Whether parsimony analysis is justified, and on what grounds, is largely beyond the scope of this chapter. Some biologists view it as justified by Ockham's razor, the principle that all else being equal, simpler explanations are better. Others view it as justified because of its properties as a statistical estimator of the unknown phylogeny (but see Felsenstein 1978). Still others view it as justified because, when used on known evolutionary trees, it yields reasonably accurate reconstructions. For further exploration of this issue, and for other methods of parsimony analysis, discussions by Joseph Felsenstein (1982, 1983, 2004) are good places to start.

The Number of Possible Trees

In our antelope example, we inferred the relationships among the brown-legged antelopes by considering all possible trees and choosing the one that required the fewest evolutionary changes (Figure 4.13). Because our ingroup had only three species, there were only three possible trees to consider. However, the number of possible trees increases rapidly with the number of species (see Felsenstein 2004). With four species in the ingroup, there are 15 possible bifurcating trees. With five species, there are 105. With 10 species, there are 34,459,425—more than the number of seconds in a year. It should be obvious why, in practice, most parsimony analyses are carried out with the assistance of computers.

Convergence and Reversal

The independent appearance in different lineages of similar derived characters is called **convergent evolution**. The loss of derived traits in a lineage, resulting in a return to the ancestral condition, is called **reversal**. As we saw in the antelope example, both phenomena result in conflicting patterns of shared derived characters that can mislead us in our efforts to reconstruct evolutionary history. Similarity in character states due to convergence and/or reversal is called **homoplasy**.

Many of the complications in inferring phylogenies result from convergent evolution and reversal. The independent appearance of a character's state in more than one place on a phylogeny is called homoplasy.

Morphological similarities like those shown in **Figure 4.15** can arise independently, by convergent evolution, when lineages experience similar patterns of natural selection due to similar environmental challenges. Both peacocks (Figure 4.15a) and male peacock spiders (Figure 4.15b) carry colorful fans on their backs and raise them to attract mates (Hill 2009; Otto and Hill 2010; Dakin and Montgomerie 2011). Both caimans (Figure 4.15c) and hippos (Figure 4.15d) have their eyes, nostrils, and ears at the tops of their skulls. This character is thought to be adaptive because it lets them see, smell, and hear above the water while

Figure 4.15 Convergent evolution These pairs of species have adapted to similar challenges. (b) Photo by Jurgen Otto.

remaining mostly submerged (Osburn 1903; Caldicott et al. 2005). Other examples of convergent evolution include the wings of bats and birds, the streamlined shapes of sharks and whales, and the elongated bodies of snakes and legless lizards.

Reversal may occur when formerly adaptive derived characters are lost because environmental changes have rendered them more costly than beneficial (Fong et al. 1995; Hall and Colegrave 2008). Snakes and legless lizards, in their (convergent) loss of limbs, illustrate reversal (see Skinner et al. 2008; Skinner and Lee 2009). Other reversals include the loss of eyes in cave-dwelling animals, of teeth in birds, and of body armor in freshwater threespine sticklebacks. In a rare double reversal, the lineage leading to Guenther's marsupial frog (*Gastrotheca guentheri*) re-evolved true teeth set in the lower jaw—a trait lost in the common ancestor of all modern frogs at least 230 million years ago (Wiens 2011).

We can save ourselves trouble in estimating phylogenies when we can identify examples of convergence and reversal ahead of time, and thereby avoid apparent conflict among the patterns of shared derived characters in our data sets. In other words, it is useful to try to restrict our analyses to evolutionary innovations that are shared because they are homologous—inherited from a common ancestor—and that are present in all of the common ancestor's descendants. Characters that meet these criteria are synapomorphies that identify monophyletic groups.

Recall that homology is similarity due to common ancestry. Synapomorphies are a particular category of homologous traits. They are similarities derived from common ancestors shared only by a subset of the species under consideration. Simply put, in the context of the lineages being studied, they are evolutionary novelties.

In some cases, convergence and reversal are easy to identify. No one would mistake the mating displays of the peacock and the peacock spider, for example, as evidence of common ancestry. One uses feathers, and the other a pair of abdominal flaps. What makes their similarity striking is not that it holds up to detailed scrutiny, but that one display belongs to a bird and the other to a spider. Likewise, the loss of legs in snakes is readily discovered by noting the vestigial limbs that remain in some species and the existence of fossil snakes with legs (Tchernov et al. 2000; Rieppel et al. 2003).

In other cases, homoplasy is exposed only by careful investigation. The magnificent tree frog (*Litoria splendida*) and African clawed frog (*Xenopus laevis*) secrete identical versions of the skin toxin caerulein, a short protein consisting of 10 amino acids. Production of caerulein might easily be taken for a synapomorphy. However, the protein is encoded by a different gene in each species (Roelants et al. 2010). In the tree frog, caerulein is encoded by a locus that arose as a duplication of the gastrin gene. In the clawed frog, caerulein is encoded by a locus that arose as a duplication of the cholecystokinin gene. Production of the same toxin in these frogs is not a synapomorphy, but a stunning example of convergence.

In still other cases, unfortunately, homoplasy can be discovered only by reconstructing a phylogeny and finding that the distribution of some characters cannot be explained without convergence, reversal, or both (Wake et al. 2011). In other words, homoplasy is a fact of life in phylogeny inference. It represents "noise" in the data sets used to reconstruct evolutionary history. Homoplastic traits are analogous to the faulty or misleading measurements that are present in almost every data set used in science. The best way to avoid being misled by the noise of homoplasy is to analyze many independent characters in reconstructing evolutionary relationships instead of just one or a few. Richard Mooi and colleagues (2000) used a parsimony analysis of 24 structural characters to estimate the evolutionary tree of fossil sand dollars shown in **Figure 4.16**. The most parsimonious tree required just one transition each for 21 of the characters. These transitions are marked in blue. The remaining three characters, marked by other colors, were homoplastic. Each of them required two changes in state.

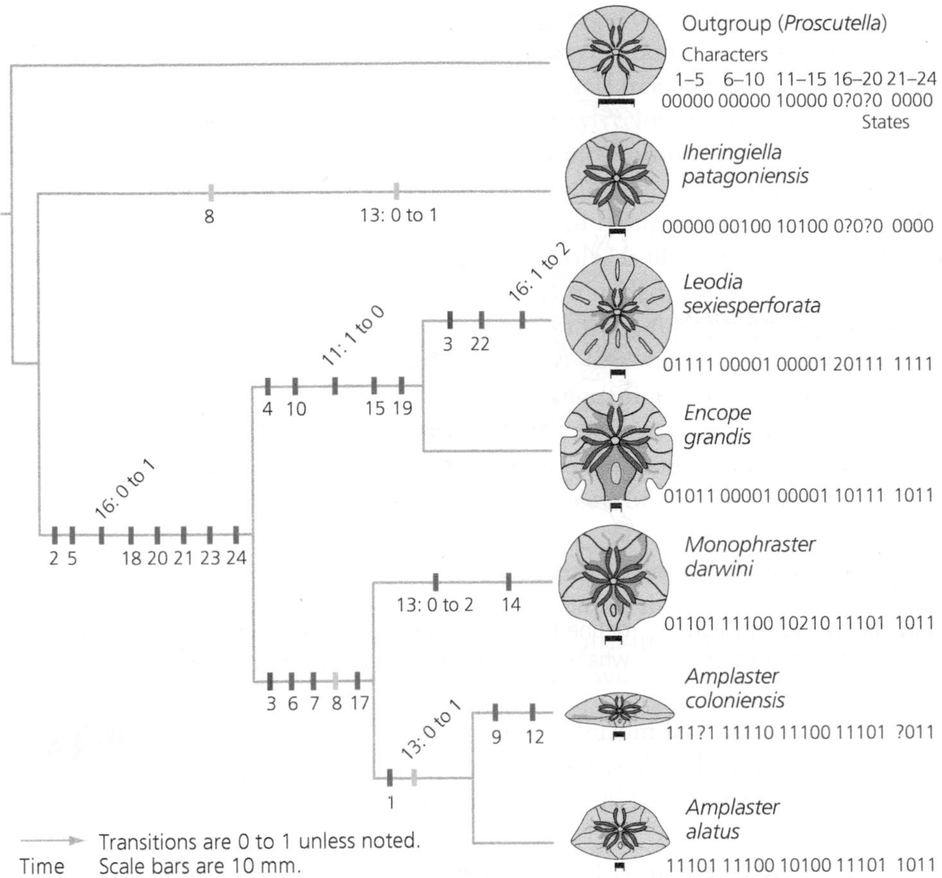

Figure 4.16 Phylogeny of fossil sand dollars estimated by parsimony analysis Most of the anatomical characters used to infer this evolutionary tree have two possible states; a few have three. The string of numbers next to each sand dollar lists its state for each character. Redrawn from Mooi et al. (2000).

Convergence and reversal are particularly common when the characters analyzed are nucleotides at particular sites in a DNA sequence. Here, there are only four possible states for a given character—A, T, G, or C—and a switch from one state to another, as a result of mutation, can happen easily. We discuss the use of DNA sequences in phylogeny estimation in the next section. We do so by exploring a case study in some detail: the evolutionary origin of whales.

4.3 Molecular Phylogeny Inference and the Origin of Whales

Whales, dolphins, and porpoises, along with a number of extinct species known only from fossils, form a monophyletic group. Some researchers call this clade Cetacea (Uhen 2010). Others call it Cetaceamorpha, reserving the name Cetacea exclusively for the smaller clade containing only modern forms (Spaulding et al. 2009). For simplicity, we use the term Cetacea in its more inclusive sense.

What Is a Cetacean?

Cetacea is identified as a monophyletic group by a number of synapomorphies in the skull (Uhen 2007, 2010). Cetaceans have an enlarged, thickened, and dense auditory bulla, a bony shell at the base of the skull that surrounds the structures of the ear. This is thought to be an adaptation for hearing underwater (Nummela et al. 2007). Cetaceans have skulls with a narrow postorbital/temporal region. And they have an elongated snout that places the front teeth (the incisors and

Next, we briefly review methods for inferring phylogenies from DNA sequences using examples from artiodactyls and cetaceans. Many of the methods we consider are also used for protein sequences. Our discussion is intended as an overview. For practical tutorials, see Baldauf (2003), Harrison and Langdale (2006), and Hall (2011). For more theoretically detailed treatments, see Graur and Li (2000) and Felsenstein (2004).

Aligning Sequences

The methods we discuss take as input a set of homologous sequences. In other words, the data are examples of the same gene from different lineages, descended from a common ancestral copy. Most of the methods also require that the sequences are aligned. This means that any insertions or deletions that have occurred in some lineages but not in others have been identified, and the sequences have been shifted to bring them into register. The result is that not only are the sequences homologous, but every site within the sequences is homologous.

Our examples will use sequences from exon 7 of the gene for a milk protein called β-casein (Gatesy et al. 1999). Among other insertions and deletions, the cow sequence, in comparison to the whale sequence, has a 3-nucleotide-long deletion starting at site 61. **Figure 4.21** shows the two animals' sequences in the neighborhood of the deletion before and after alignment. The effect of the alignment on how well the sequences match may be less easy to see in the nucleotide sequences than in the amino acid sequences they encode.

Nucleotide sequence before alignment

```
              50                60                70
whale: ... GGG CCA ATC CCT TAC CCT ATT CTT ACA CAA AAC ...
  cow: ... GGG CCC ATC CCT AAC AGC CTC CCA CAA AAC ...
```

After alignment

```
              50                60                70
whale: ... GGG CCA ATC CCT TAC CCT ATT CTT ACA CAA AAC ...
  cow: ... GGG CCC ATC CCT AAC --- AGC CTC CCA CAA AAC ...
```

Encoded amino acid sequence before alignment

whale: ...	Gly	Pro	Ile	Pro	Tyr	Pro	Ile	Leu	Thr	Gln	Asn	...
cow: ...	Gly	Pro	Ile	Pro	Asn	Ser	Leu	Pro	Gln	Asn	...	

After alignment

whale: ...	Gly	Pro	Ile	Pro	Tyr	Pro	Ile	Leu	Thr	Gln	Asn	...
cow: ...	Gly	Pro	Ile	Pro	Asn	—	Ser	Leu	Pro	Gln	Asn	...

Figure 4.21 Sequences before and after alignment Compare these short stretches of the gene for β-casein, and the amino acids it encodes, before and after the right portion of the cow sequence is shifted three sites to compensate for a deletion.

Researchers typically align sequences using a mix of computer software and human judgment (see Liu et al. 2009). Proper alignment is crucial. If sequences are poorly aligned, information they contain regarding evolutionary history may be lost. Moreover, faulty alignment can be misleading. Karen Wong and colleagues (2008) showed that different alignments of the same sequences, produced by different software, can lead to the reconstruction of different phylogenies. Conversely, more accurate alignments lead to more accurate reconstructions of evolutionary relationships (Ogden and Rosenberg 2006; Wang et al. 2011).

The first step in reconstructing evolutionary history using sequence data is to align the sequences. We then compare alternative phylogenetic hypotheses using any of several criteria, including parsimony.

To illustrate different methods of phylogeny inference, we will use the eight aligned sequences shown in **Figure 4.22**. A small piece of the β-casein gene, these represent a fraction of a much larger data set analyzed by John Gatesy and

Figure 4.22 Sequence data for phylogeny inference This table shows 60 nucleotides of aligned sequence (sites 141 through 200) from exon 7 of the β-casein gene. The data are from six artiodactyls, a whale (the dolphin *Lagenorhynchus obscurus*), and a rhinoceros as the outgroup (Gatesy et al. 1999). An X at a site indicates an ambiguously identified nucleotide.

colleagues (1999). We will show trees estimated from the sequences in the figure, all of which are consistent with the tree Gatesy estimated from β-casein sequences 1,100 nucleotides long. Some readers may wish to reproduce our analyses.

Evaluating Alternative Phylogenies with Parsimony

One way to estimate evolutionary trees from sequence data is to treat each site in sequence as an independent character and look for synapomorphies that identify monophyletic groups. That is, we can analyze nucleotides at sequence sites the same way we analyzed morphological characters in Section 4.2.

Examination of the sequences in Figure 4.22 reveals that some sites are uninformative. At site 142, for example, all eight taxa have the same character state, G. At site 192 only one taxon, camels, differs from the other seven. Other sites are informative. Site 166 features what appears to be a shared derived character state, C, identifying whales and hippos as sister lineages. And this synapomorphy is nested inside another one, a T at site 162, that appears to identify cows, deer, whales, and hippos as a monophyletic group. However, the characters at site 162 are in conflict with those at site 177. There, T appears to be a shared derived character identifying whales, hippos, pigs, and peccaries as a monophyletic group. We are clearly dealing with a non-ideal case.

When we encountered conflict among morphological characters, we turned to parsimony analysis. We can do the same with sequence data (Felsenstein 1988). We consider all possible trees as hypotheses, determine the minimum amount of evolution required to explain the distribution of nucleotides at each site on each tree, and seek the tree that requires the least change overall. **Figure 4.23** evaluates three of our 60 molecular characters, the nucleotides at sites 162, 166, and 177, on two of the 10,395 possible trees. Tree (a), in which whales are artiodactyls and form a monophyletic group with hippos, requires six nucleotide substitutions. Tree (b), in which whales are merely kin to artiodactyls, requires nine. We have made some arbitrary choices between convergence and reversal. The reader may find other equally parsimonious scenarios for some sites.

Using a computer program (Felsenstein 2009) to find the most parsimonious among all 10,395 trees for all 60 characters, we learn that the tree in Figure 4.23a is the winner, requiring 41 substitutions. The tree in Figure 4.23b requires 47.

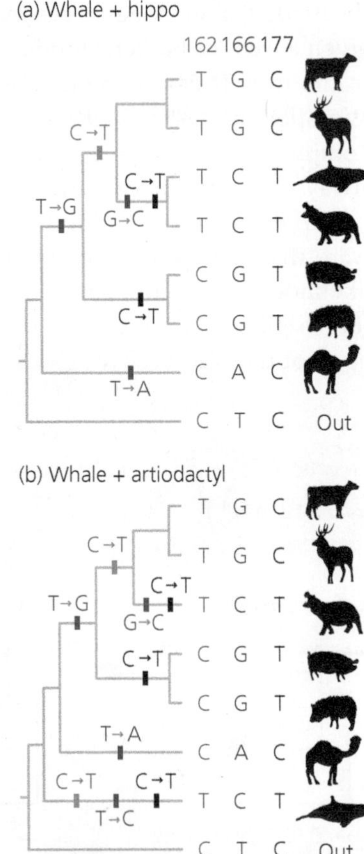

Figure 4.23 Parsimony analysis of three molecular characters on two trees Tree (a) requires six evolutionary changes, whereas tree (b) requires nine. For these characters, tree (a) is more parsimonious.

Evaluating Alternative Phylogenies with Likelihood

Parsimony is not the only criterion we can use to evaluate possible trees and identify the ones that offer the best estimate of evolutionary relationships. Another commonly used metric is likelihood (Felsenstein 1981). The calculations involved in a likelihood analysis are cumbersome, but the fundamental idea is straightforward. Our explanation of the idea will make more sense if we first look at the kind of tree it will allow us to produce.

Figure 4.24 displays a tree estimated using likelihood and the data in Figure 4.22. The tree has a feature not yet seen in the phylogenies we have shown. It displays information not only in the order of branching but also in the branch lengths. These are proportional to the number of nucleotide substitutions per site estimated to have occurred on each branch. The peccary, for example, is estimated to have accumulated more substitutions than has the pig since their lineages diverged. Estimating the branch lengths is integral to a likelihood analysis.

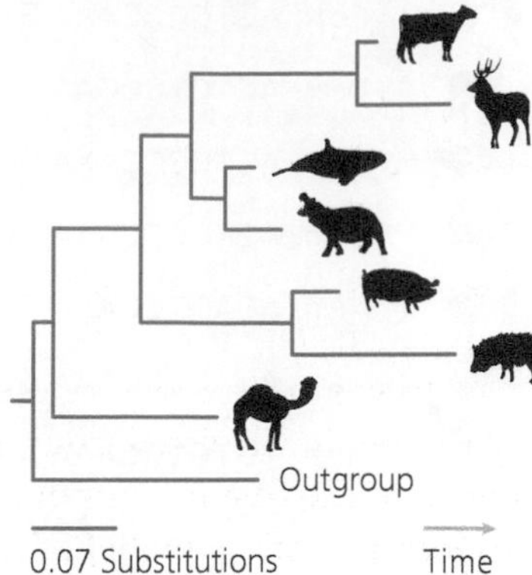

Figure 4.24 A maximum likelihood phylogeny Estimated using PhyML, with all options at their default settings (Chevenet et al. 2006; Dereeper et al. 2008; Guindon et al. 2010).

The logic of the analysis is as follows. If we have a proposed evolutionary tree, with branch lengths measured in expected number of substitutions per site, and a model of sequence evolution—a set of numbers describing the rates at which the various possible substitutions occur—we can calculate the probability of evolving the particular set of sequences we have found in our data. The probability of the data given a tree, its branch lengths, and a model of evolution is called the **likelihood** of the tree. It can be written as $L(\text{tree}) = P(\text{data}|\text{tree, branch lengths, model})$. We offer an overview of the steps required to calculate the likelihood of a tree in **Computing Consequences 4.1**. It will come as no surprise that biologists virtually always perform the calculations with computers.

To reconstruct evolutionary history using likelihood, biologists run software that adjusts the branch lengths of each possible tree to maximize the tree's likelihood, then compares the trees to find the one whose likelihood is highest (Huelsenbeck and Crandall 1997). The winner is the maximum likelihood estimate of the phylogeny. It is the tree with the best chance of producing the data.

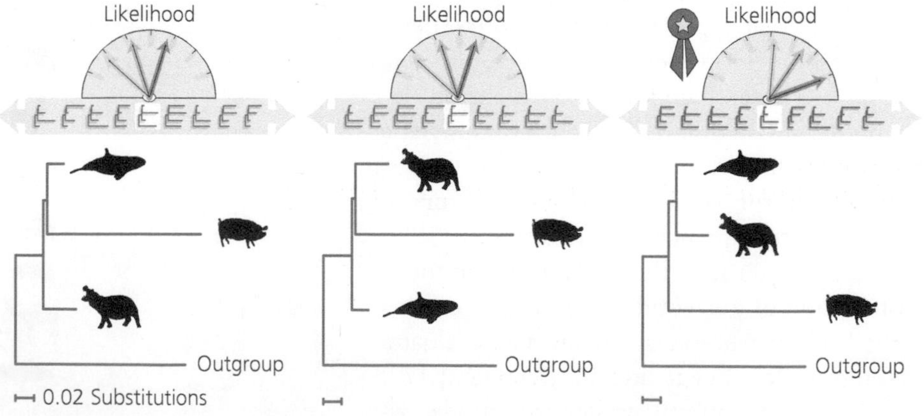

Figure 4.25 Finding the tree with the highest likelihood Before the trees are compared, the branch lengths of each are adjusted to maximize the likelihood.

Figure 4.25 shows a simple example. It uses the data for the whale, hippo, pig, and outgroup from Figure 4.22. With just three species in the ingroup, there are three possible evolutionary trees. After the branch lengths have been optimized for each tree, we find that the last tree has the highest likelihood. It indicates that whales and hippos are more closely related to each other than either is to pigs, and is consistent with the phylogenies for all eight taxa in Figures 4.24 and 4.23a.

Using maximum likelihood as a criterion for comparing phylogenies, we prefer the tree with the highest probability of producing our data.

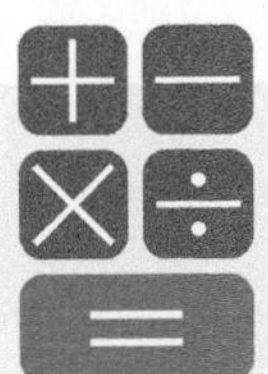

COMPUTING CONSEQUENCES 4.1

Calculating the likelihood of an evolutionary tree

Here we give an overview of how the likelihood of a tree is calculated. For the details, see Felsenstein (2004).

As shown in Figure 4.26a, the likelihood of an evolutionary tree is calculated from the tree itself, the branch lengths on the tree in units of expected number of substitutions per site, a model of sequence evolution (a set of numbers giving the probability of each possible substitution), and the aligned sequences that are the data.

As shown in Figure 4.26b, the likelihood of the tree is the probability of the data given the tree, the branch lengths, and the model.

This, in turn, is the product of the probability of the nucleotides at site 1 in our set of sequences, the probability of the nucleotides at site 2, and so on.

The probability of the nucleotides at a given site is the sum of the probabilities of each possible combination of nucleotides that can be assigned to the common ancestors in the tree. With three species in the ingroup, the tree includes two ingroup common ancestors. This means there are 16 possible assignments of nucleotides. The probability of any given assignment is a function of the branch lengths and the substitution probabilities.

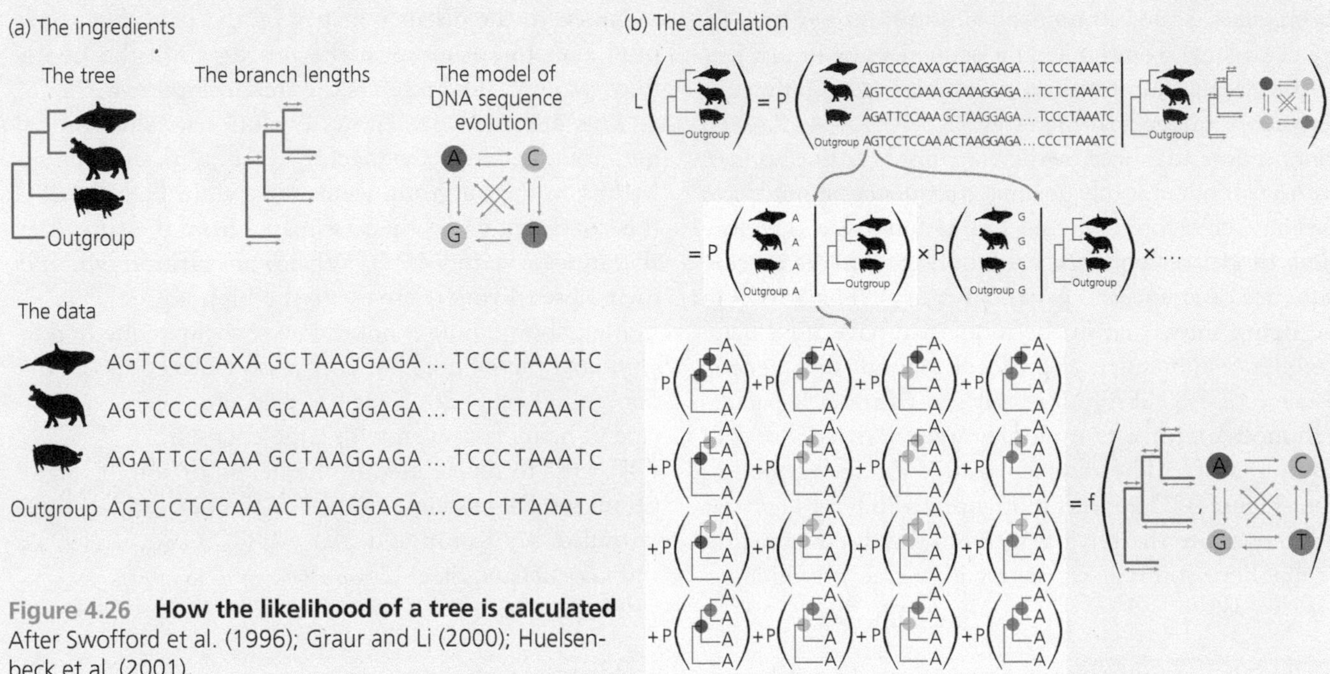

Figure 4.26 How the likelihood of a tree is calculated After Swofford et al. (1996); Graur and Li (2000); Huelsenbeck et al. (2001).

It is important to keep in mind that the likelihood of a tree is always calculated using a specific model of sequence evolution. If we switch to a different model, we may find that a different tree becomes the maximum likelihood estimate of the phylogeny.

Searching for the Best of All Possible Trees

The procedure for estimating a phylogeny by maximizing either parsimony or likelihood begins, as we have said, with the assumption that all possible trees are legitimate hypotheses. The only way to be certain of finding the best tree is to check them all. If we have more than a modest number of species, however,

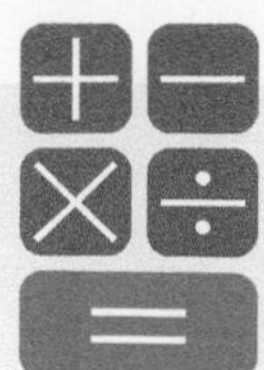

COMPUTING CONSEQUENCES 4.2

Neighbor joining: A distance matrix method

Distance matrix methods of phylogeny inference take their name from a first step they share that distinguishes them from the methods, including parsimony and maximum likelihood, that we discuss in the main text. This first step is to calculate, from the data, a table of pairwise evolutionary divergences (Swofford et al. 1996). We then estimate the evolutionary tree from this distance matrix. One distance matrix method that is commonly used to analyze sequence data is **neighbor joining** (Saitou and Nei 1987).

Like many other methods, neighbor joining takes as input **genetic distances**, or evolutionary sequence divergences, scaled in units of substitutions per site. To calculate these from nucleotide sequences, we must recognize that as substitutions accumulate, the difference between sequences saturates at 75% (Figure 4.27a). To see this, note that because there are just four nucleotides, even a pair of randomly generated sequences should have the same nucleotide at one site in four. We therefore use a model of sequence evolution to convert percent sequence differences to genetic distances. The arrows in the figure show that in the model we have used here, a sequence difference of 65% corresponds to a genetic distance of 1.8 substitutions per site (Kimura 2-parameter model with a transition/transversion ratio of 2.0; Kimura 1980). The distance matrix for the sequences from Figure 4.22 appears in Figure 4.27b.

To estimate the phylogeny from the distance matrix by neighbor joining, we start with a tree in which no pair of taxa is more closely related to each other than either member is to any other species (Figure 4.27c). This tree, in our case an eight-way polytomy, is unrooted. **Unrooted trees** do not encode information about which direction time flows along their branches. We will root the tree later using the outgroup.

We now try grouping a pair of taxa, such as the deer and cow, as sisters (Figure 4.27d). We estimate the total length of the resulting partially resolved tree from the distance matrix. After trying all possible pairs, we pick the grouping that results in the shortest total tree length. We consider this winning pair resolved, and replace its members in the distance matrix by the pair's last common ancestor. We repeat the process until the tree is fully resolved. The result is the tree in Figure 4.27e.

This unrooted tree shows evolutionary divergence, but not direction. Connecting a root to the lineage leading to the outgroup yields the tree in Figure 4.27f. It is consistent with others estimated from the same data (Figures 4.23a and 4.24). Whales are artiodactyls, and their closest living relatives are the hippos.

Neighbor joining is not the most accurate method of phylogeny inference, but it is reasonably good (Guindon and Gascuel 2003). It has the considerable advantage of being fast, even with large data sets.

For the formulae used in the neighbor-joining algorithm, see Felsenstein (2004). For the derivation of the formulae, see Saitou and Nei (1987), Gascuel (1994), and Gascuel and Steel (2006).

checking all possible trees is impractical. This is because the number of possible trees increases rapidly with the number of taxa in the ingroup. With seven species in the ingroup, as we have in the data in Figure 4.22, there are 10,395 possible trees. With 20 species, there are 8,200,794,532,637,891,559,375. Biologists commonly want to estimate phylogenies for much larger numbers of taxa than this. Even with the fastest computers, we cannot check all possible trees.

The problem is something like trying to find the highest point in a national park while blindfolded. We could, in principle, walk a tight grid across the entire park. After visiting every square meter, we would know for certain which point is highest. But if the park is large, an exhaustive search would take a long time.

One way we could speed our quest is by attending to clues that entire regions of the park can be ruled out. We need not search underwater, for example. Another way is by always walking upward. This plan's flaw is that it might leave

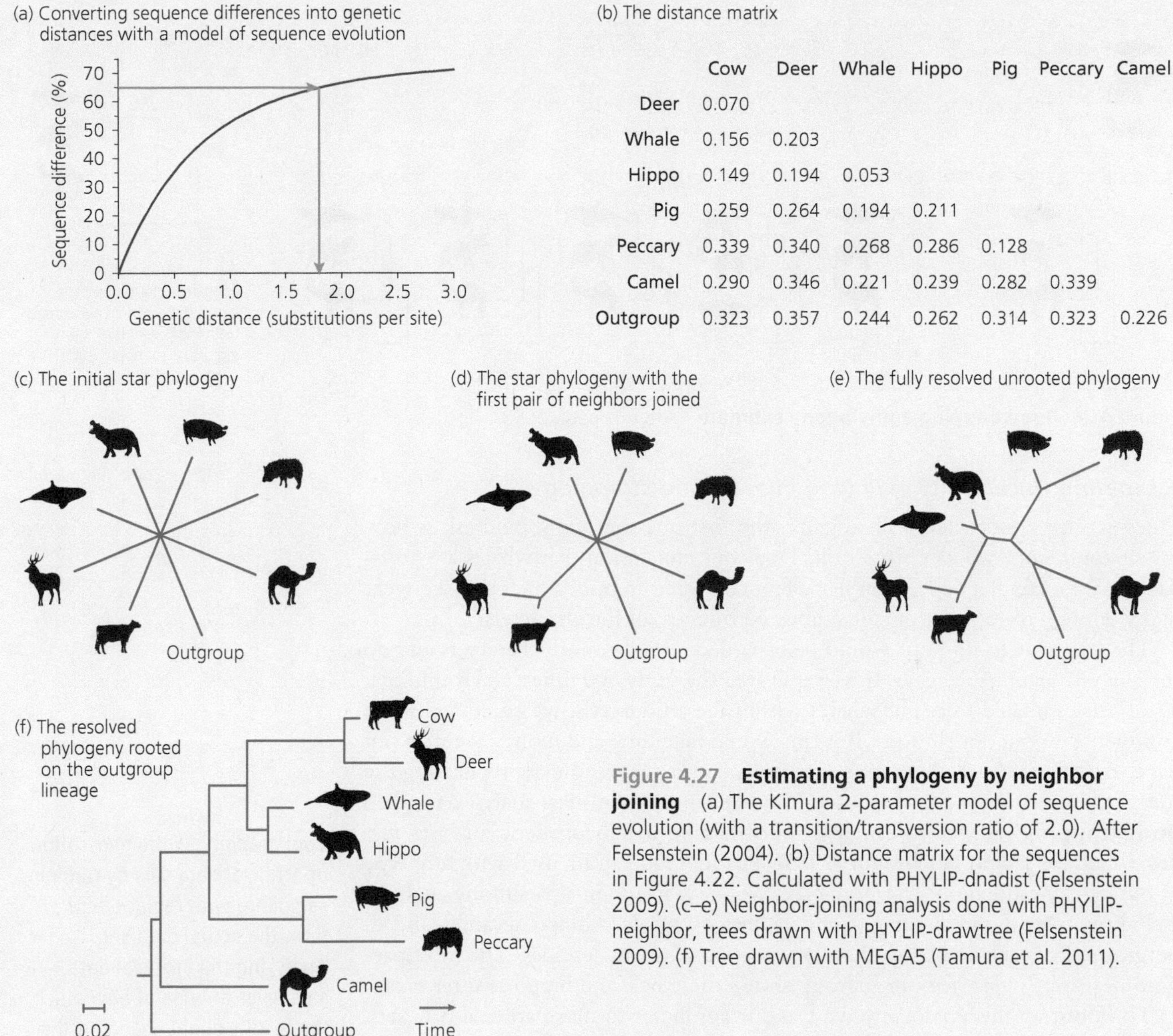

	Cow	Deer	Whale	Hippo	Pig	Peccary	Camel
Deer	0.070						
Whale	0.156	0.203					
Hippo	0.149	0.194	0.053				
Pig	0.259	0.264	0.194	0.211			
Peccary	0.339	0.340	0.268	0.286	0.128		
Camel	0.290	0.346	0.221	0.239	0.282	0.339	
Outgroup	0.323	0.357	0.244	0.262	0.314	0.323	0.226

Figure 4.27 **Estimating a phylogeny by neighbor joining** (a) The Kimura 2-parameter model of sequence evolution (with a transition/transversion ratio of 2.0). After Felsenstein (2004). (b) Distance matrix for the sequences in Figure 4.22. Calculated with PHYLIP-dnadist (Felsenstein 2009). (c–e) Neighbor-joining analysis done with PHYLIP-neighbor, trees drawn with PHYLIP-drawtree (Felsenstein 2009). (f) Tree drawn with MEGA5 (Tamura et al. 2011).

us trapped on a hill, unaware of higher peaks elsewhere in the park. But we could overcome this by doing several searches, each starting from a different place.

Authors of phylogeny inference software use analogous strategies to speed the search for the best of all possible trees (Felsenstein 2004). One strategy, called **branch and bound**, eliminates groups of trees from consideration upon discovering that all their members are worse than the best tree found so far. Other strategies, collectively called **heuristic searches**, look for trees superior to the current leader by rearranging the leader in various ways and evaluating the results. To avoid being trapped on a local peak, the search may be repeated from different starting points. Or the search may be started with a tree we have reason to believe may already be close the best possible tree, because it was constructed with an algorithm that usually performs fairly well (Guindon and Gascuel 2003). An example of such an algorithm is discussed in **Computing Consequences 4.2**.

The number of possible trees is usually so vast that we cannot come close to evaluating them all. Instead, we must use computational shortcuts to find the best hypothesis.

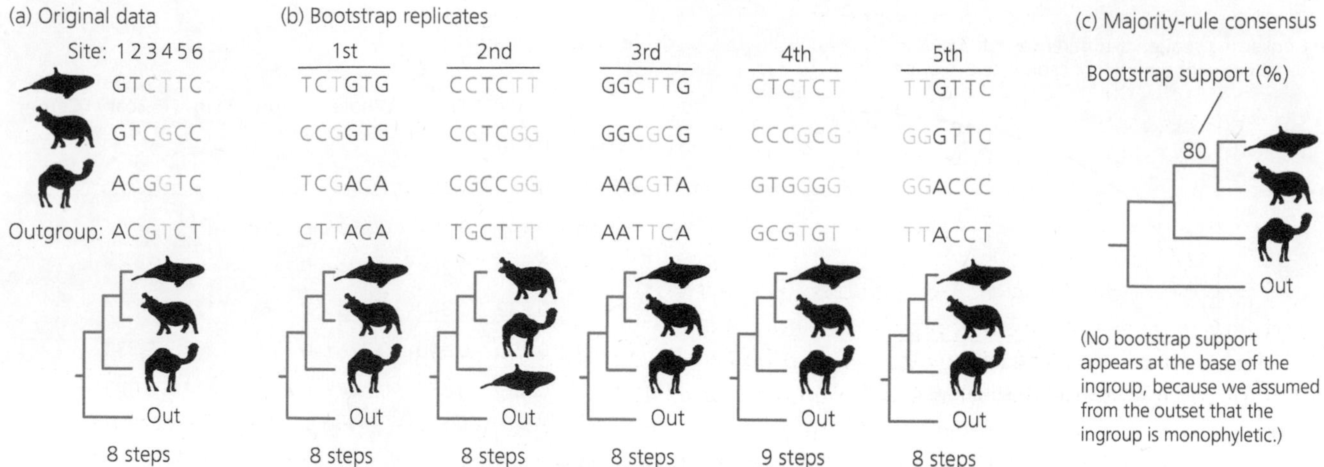

Figure 4.28 Bootstrapping a phylogeny estimate After Baldauf (2003).

Estimating Uncertainty in Phylogenies by Bootstrapping

Once we have estimated a phylogeny, the first question we should ask is how much confidence we can place in it. Does our conclusion that whales are artiodactyls, for example, depend on just a few characters in our data set that we were lucky enough to capture, or is it supported by most of the characters?

The best way to find out would be to collect data for more characters and do the analysis again. And again. If we replicated the study 100 times, and found that 97 of our estimated trees put whales within the artiodactyls, we could feel some confidence in our conclusions. But this would take time and money we may not have. A fast and cheap alternative is to use a computer to simulate replicating the study. One such method, often used in other kinds of statistical analyses, is called **bootstrapping**. A way to employ bootstrapping in phylogenetic analyses was developed by Joseph Felsenstein (1985). His method can be used with any type of phylogeny estimation. In **Figure 4.28**, we show its use in a parsimony analysis.

Bootstrapping is the generation of artificial data sets by random sampling, with replacement, from the actual data set. Analyzing the artificial data sets gives an idea of how much the results might change if the study were replicated many times.

Figure 4.28a shows an imaginary data set for a whale, a hippo, a camel, and an outgroup. For each animal we have a sequence of six nucleotides. The most parsimonious phylogeny for our ingroup shows the whale and hippo as sister taxa.

To bootstrap this phylogeny we use our computer to make artifical data sets, called bootstrap replicates, from our original data set (Figure 4.28b). We do this by sampling at random, with replacement, from our original characters—the six sites, each of which is shown in a different color. Notice that in our first replicate, site 1 (black) happened to get chosen twice while site 3 (orange) got left out altogether. In our second replicate, sites 4 (olive) and 6 (dark blue) got chosen twice while sites 1 (black) and 5 (light blue) got left out.

We then estimate the phylogeny from each of the bootstrap replicates using the same method we used on the original data. For our 1st, 3rd, 4th, and 5th replicates, the most parsimonious tree has the whale and hippo as closest relatives. For replicate two, the most parsimonious tree has the camel and hippo as sisters.

Finally, we draw a tree, called the majority-rule consensus phylogeny, containing all the monophyletic groups that appear in at least half of our bootstrap replicates (Figure 4.28c). The only clade within our ingroup that meets this criterion is the one with the whale and hippo as sister taxa. We label the node at the base of this clade with the percentage of replicates in which it appeared. This number, the bootstrap support for the clade, estimates the confidence we can have that the presence of the clade in our reconstruction would hold up to modest changes in the characters we sampled. In other words, high bootstrap support indicates that the clade is a strong winner across our entire data set (Baldauf 2003).

When researchers bootstrap evolutionary trees, they typically generate 100 or more replicates. **Figure 4.29** shows bootstrap support, based on 1,000 replicates, for several monophyletic groups in a maximum likelihood phylogeny estimated from β-casein sequences 1,100 nucleotides long. There is a strong signal in this data set (99% bootstrap support) indicating that whales belong to a monophyletic group with cows, deer, hippos, pigs, and peccaries—embedded within the artiodactyls. There is also strong support (100%) indicating that whales are members of a more exclusive clade with cows, deer, and hippos. There is weaker support (59%) for the hypothesis that whales and hippos are sister taxa.

How closely this estimate resembles the true evolutionary history of the eight species depends on how well the β-casein gene represents the rest of the genome. Analysis of different genes may yield different trees (see White et al. 2009).

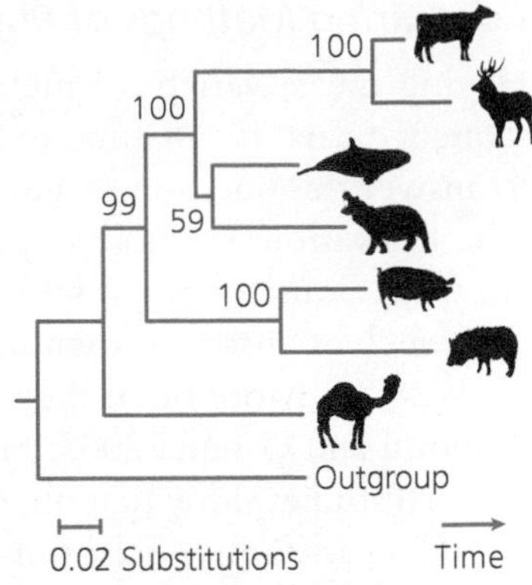

Figure 4.29 Percentage of bootstrap support on a maximum likelihood phylogeny Tree estimated from 1,100-nucleotide-long sequences from the β-casein gene (Gatesy et al. 1999). Bootstrap analysis by PhyML, all options at default (Guindon and Gascuel 2003).

Bayesian Phylogeny Inference

When we used maximum likelihood as our criterion for evaluating trees, we calculated for each possible tree the probability of the data given the tree, $P(\text{data}|\text{tree})$. This quantity is not the same as the probability of the tree given the data, $P(\text{tree}|\text{data})$. And the probability of the tree given the data, also known as the **posterior probability** of the tree, is what we really want to know. Bayes' theorem provides a way to calculate it, by means of this formula (Bayes 1763):

$$P(\text{tree}|\text{data}) = \frac{P(\text{data}|\text{tree})P(\text{tree})}{P(\text{data})}$$

The first term in the numerator on the right is the likelihood of the tree. The second term, $P(\text{tree})$, is the **prior probability** of the tree. It is the probability we assigned to the tree before taking into account the data. The denominator, P(data), is the prior probability of the data. It is the sum of the values we obtain by multiplying each possible tree's likelihood by its prior probability (Huelsenbeck et al. 2001). Thus we can calculate what we want—the probability of the tree given the data—only if we are willing to supply prior probabilities for all possible trees. Some biologists are comfortable doing this, and typically specify equal prior probabilities for all possible trees. Others are not (Felsenstein 2004).

Because the number of possible trees is usually enormous, we typically cannot calculate their posterior probabilities analytically. However, we can find the trees that have non-negligible posterior probabilities, and estimate what those probabilities are, by using computer software that simulates sampling trees from a population in which each possible tree is represented at a frequency equal to its posterior probability (Huelsenbeck et al. 2001). The algorithm employed moves from tree to tree, spending more time with trees that have higher likelihoods and higher prior probabilities, and periodically takes a snapshot of the tree it is with. After gathering a large number of snapshots, the computer can estimate posterior probabilities by how often each possible tree appears in snapshot collection.

The appeal of **Bayesian phylogeny inference** is that its results are easy to interpret. **Figure 4.30** shows results from the 1,100-nucleotide β-casein sequences used earlier. Figure 4.30a gives the posterior probabilities of the three trees that appeared in the snapshots. These sum to 1. Figure 4.30b shows the consensus tree based on majority rule among the snapshots. Each clade is marked with the sum of the posterior probabilities of the trees it occurs in—its **clade credibility**.

(a) Trees with non-neglibible posterior probabilities (pp)

pp = 0.86 pp = 0.135 pp = 0.005

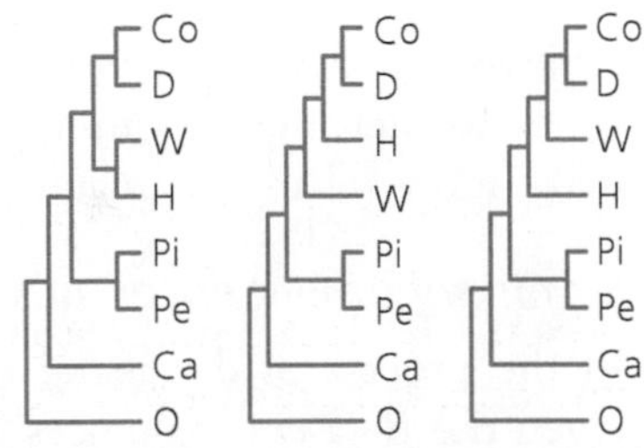

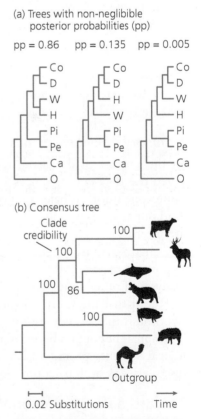

Figure 4.30 Bayesian phylogeny inference Tree estimated from same data as in Figure 4.29, with MrBayes (Ronquist and Huelsenbeck 2003).

Comparing Methods of Phylogeny Inference

Having met a variety of methods for reconstructing evolutionary history from sequence data, it is natural to ask how well they work. Researchers have sought to answer this question by generating known evolutionary histories and checking whether various methods of phylogeny inference can recover the true evolutionary trees. Sometimes the known evolutionary histories are for lineages of organisms, such as viruses, grown in the lab (Hillis et al. 1992; Hillis et al. 1994; Sousa et al. 2008). More often they are for sequences evolved by computer simulation (Guidon and Gascuel 2003; Hall 2005; Kolaczkowski and Thornton 2005).

Such studies show that under optimal conditions, all the methods we have discussed—parsimony, maximum likelihood, neighbor joining, and Bayesian inference—recover the branching pattern in the true tree with accuracies approaching 100%. All the methods do better with the sequences that have diverged from each other substantially (0.35 or more substitutions per site) but not so much that their divergence is approaching saturation (less than about 65% sequence difference). Within this range, all the methods do better with longer sequences.

Which method is best? Neighbor joining is not quite as good as the others, although its speed makes it a useful adjunct to the other methods. Parsimony, maximum likelihood, and Bayesian inference all have strengths and weaknesses (Felsenstein 1978; Kolaczkowski and Thornton 2005). For this reason researchers often use a variety of methods and check whether the trees they produce are consistent with each other (Huelsenbeck and Hillis 1993; Hillis et al. 1994).

Phylogeny inference is an active area of research, and new methods are constantly being devised. For example, Kevin Liu and colleagues (Liu et al. 2009; Liu et al. 2011) have developed a likelihood-based method that simultaneously estimates both sequence alignments and evolutionary trees. Scott Edwards (2009) reviews the development of methods for combining evidence from different genes to increase the number of independent characters analyzed and thereby make more reliable estimates of the relationships among species.

Summary of Evidence Thus Far on Cetacean Evolution

In the process of discussing techniques for analyzing sequence data, we have given the β-casein sequences a thorough workout. All of our trees point to the same conclusion. Whales are not merely kin to artiodactyls, they are artiodactyls. And our analyses provide some support for the hypothesis that whales and hippos are sister taxa. A close relationship between hippos and whales has a certain intuitive charm because hippos are semiaquatic and whales fully so. The whale–hippo hypothesis received additional strong support from Gatesy and colleagues' (1999) analyses of other genes besides β-casein and other taxa.

Recall that the morphological evidence we reviewed tended to support the hypothesis that whales are kin to artiodactyls rather than members of the clade. The apparent conflict between the implications of the molecular versus morphological evidence prompted a continuing hunt of other sources of data. This led to discoveries in both paleontology and molecular biology.

Toward a Resolution on Whales

The new molecular data clarifying the phylogeny of whales emerged before the fossil finds. The molecular data in question are the presence or absence of DNA sequences that occasionally insert themselves into new locations in a genome. The genetic elements involved are called **SINEs** and **LINEs**, for Short or Long

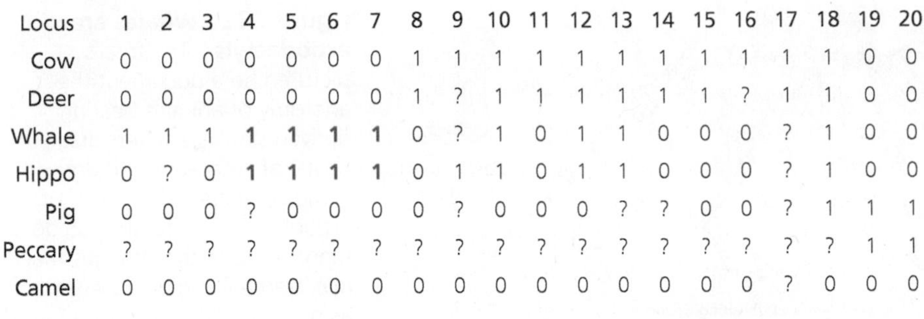

Locus	1	2	3	4	5	6	7	8	9	10	11	12	13	14	15	16	17	18	19	20
Cow	0	0	0	0	0	0	0	1	1	1	1	1	1	1	1	1	1	1	0	0
Deer	0	0	0	0	0	0	0	1	?	1	1	1	1	1	1	?	1	1	0	0
Whale	1	1	1	**1**	**1**	**1**	**1**	0	?	1	0	1	1	0	0	0	?	1	0	0
Hippo	0	?	0	**1**	**1**	**1**	**1**	0	1	1	0	1	1	0	0	0	?	1	0	0
Pig	0	0	0	?	0	0	0	0	?	0	0	0	?	?	0	0	?	1	1	1
Peccary	?	?	?	?	?	?	?	?	?	?	?	?	?	?	?	?	?	?	1	1
Camel	0	0	0	0	0	0	0	0	0	0	0	0	0	0	0	0	?	0	0	0

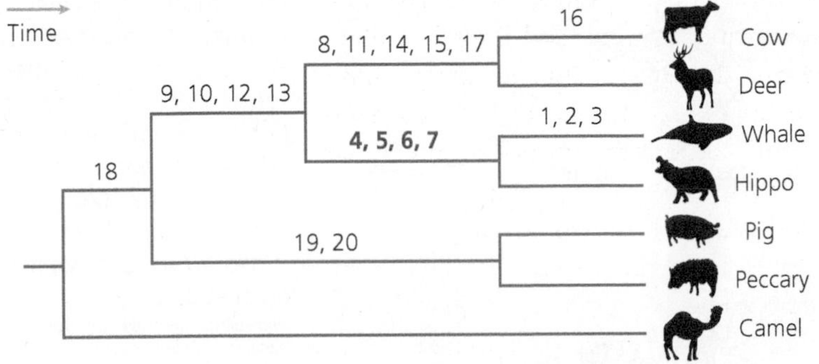

Figure 4.31 Nearly perfect phylogenetic characters? This table shows the presence (1) or absence (0) of a SINE or LINE at 20 loci in the genomes of six artiodactyls and a whale (Baird's beaked whale, *Berardius bairdii*). Question marks (?) indicate loci that are questionable in some taxa. Data are from Nikaido et al. (1999). The phylogenetic tree was produced by a parsimony analysis of these 20 characters. The presence of a SINE or LINE at loci 4–7 defines a clade of whales and hippos.

INterspersed Elements. The presence or absence of a particular SINE or LINE at a homologous location in the genomes of two different species can be used as a trait in phylogeny inference.

David Hillis (1999) outlined the potential advantages of using SINEs and LINEs in phylogeny inference. It is well established that transposition events, in which a parasitic genetic element inserts itself in a new location in the host genome, are relatively rare. As a result, it is unlikely that two homologous SINEs would insert themselves into two independent host lineages at exactly the same location. This kind of convergence is possible but improbable. Reversal to the ancestral condition is also unlikely, because the loss of a SINE or LINE can usually be detected. When SINEs and LINEs are lost, it is common also to observe the associated loss of part of the host genome. As a result, researchers can usually tell if a particular parasitic gene is absent or has been lost. If convergence and reversal are rare or can be identified, then homoplasy is unlikely. SINEs and LINEs should be extraordinarily reliable characters to use in phylogeny inference.

Some molecular characters come close to presenting ideal cases for phylogeny inference.

What do SINEs and LINEs have to say about whale evolution? Masato Nikaido and colleagues (1999) answered this question by analyzing 20 different SINEs and LINEs found in the genomes of artiodactyls. The data for the taxa we have been considering are given in **Figure 4.31**, along with a tree that shows how these data map onto the whale + hippo tree. Look at each of the 20 genes in turn, and note that the presence or absence of each SINE or LINE acts as a synapomorphy that identifies exactly one clade in the phylogeny. Stated another way, there is no homoplasy at all in this data set and thus no conflict among the characters when they are mapped onto the tree. The analysis is remarkably clean and strongly corroborates the conclusion from the DNA sequence studies.

Not long after Nikaido and coworkers published their conclusions, two research teams simultaneously announced fossil finds that were characterized as "one of the most important events in the past century of vertebrate palaeontology" (de Muizon 2001). The oldest of the fossils came from 48-million-year-old

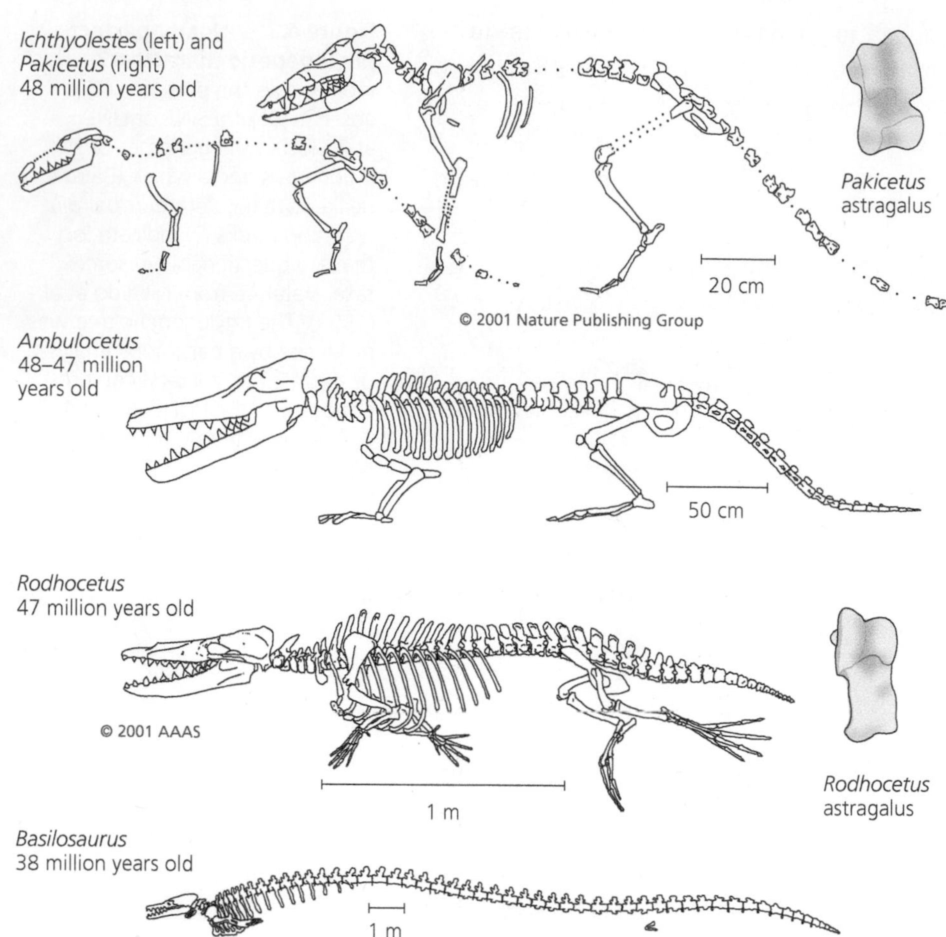

Figure 4.32 Whales are artiodactyls The fossils pictured here document the anatomy of animals bearing derived skull characters diagnostic of cetaceans and derived ankle characters diagnostic of artiodactyls. The fossils also document some of the changes that occurred early in whale evolution as members of this lineage made the transition from land to water. *Ichthyolestes* and *Pakicetus*, plus *Pakicetus* astragalus after Thewissen et al. (2001). *Ambulocetus* redrawn from Thewissen et al. (1994). *Rodhocetus* redrawn from Gingerich et al. (2001). *Rodhocetus* astragalus after Uhen (2010). *Basilosaurus* redrawn from Gingerich et al. (1990).

Ichthyolestes and *Pakicetus*, plus *Pakicetus* astragalus reprinted by permission from Macmillan Publishers Ltd: *Nature*, J. G. M. Thewissen et al., 2001, "Skeletons of terrestrial cetaceans and the relationship of whales to artiodactyles," *Nature* 413: 277–281, Figures 1 and 2. © 2001 Macmillan Magazines Limited. *Rodhocetus* from "Origins of whales from early artiodactyls: Hands and feet of Eocene Protocetidae from Pakistan," *Science* 293:2239–2242, Figure 3, illustration Douglas Boyer. © 2001 AAAS. Reprinted with permission from AAAS.

rocks and represented two species: the fox-sized *Ichthyolestes pinfoldi* and the wolf-sized *Pakicetus attocki* (Thewissen et al. 2001). Both were long-legged, long-tailed creatures that were clearly terrestrial. Both species show the synapomorphies in the skulls and ear bones that identify the cetaceans, as well as the pulley-like astragalus that diagnoses the artiodactyls **(Figure 4.32)**. The same characteristics are also present in two slightly more recent species, *Artiocetus clavis* (not shown here) and *Rodhocetus kasrani*, dated to 47 million years ago (Gingerich et al. 2001). Taken together, the suite of new fossils confirm what the molecular evidence has been telling us. Whales are artiodactyls (Uhen 2010). As shown in the figure, the fossil record now documents that the transition to an aquatic lifestyle took place in a lineage of artiodactyls that became today's whales, dolphins, and porpoises. Recent analyses of the fossil record have also identified an extinct group of semiaquatic artiodactyls as the sister group to hippos (Boisserrie et al. 2005). This report creates a link between the ancestors of today's hippopotamuses and the ancestors of today's whales, and suggests that both may descend from the same semiaquatic ancestor (Orliac et al. 2010).

Different kinds of evidence on evolutionary relationships can be used to cross-check each other.

Remaining Issues

It would give us pleasure to be able to report that the morphological and molecular evidence on the evolutionary affinities of whales are in now in complete accord. Alas, we cannot do so.

Michelle Spaulding and colleagues (2009) assembled the most comprehensive data set yet to evaluate the relationships among whales and their kin. The data included several hundred physical characters and tens of thousands of molecular characters. The taxa included 33 living species and 48 species known from fossils. Using the entire data set, the researchers reconstructed a tree in which whales are both artiodactyls and the closest living relatives of hippos. However, when the researchers analyzed only the roughly 600 characters that fossilize—and gave no more weight to astragalus shape than to any other character—their reconstruction yielded a tree in which the cetaceans branch outside the artiodactyls.

The consensus among specialists is that the molecular evidence—and the fossil astragali—are probably giving us the correct answer (see Thewissen et al. 2009). This implies that in the case of whales and their kin, morphological characters show considerable homoplasy.

4.4 Using Phylogenies to Answer Questions

The first three sections of this chapter were focused on methods. Having discussed how researchers estimate evolutionary trees from data, we now turn to examples showing how they use evolutionary trees to answer interesting questions. Our first question concerns whether tumor cells can escape the individual in which they arise.

Can Tumor Cells Move from Patient to Patient?

Canine transmissible venereal tumor (CTVT) is a cancer that grows on the genitalia of dogs. As its name suggests, CTVT is contagious. Dogs contract it by copulating with other dogs that already have it. Other cancers exist that are known to be induced by contagious agents. Cervical cancer in humans, for example, is caused by infection with HPV, the human papilloma virus (Schiffman et al. 2007); a vaccine is now available that protects against the strains responsible for about 70% of all cases (Muñoz et al. 2004). But many researchers who study CTVT long suspected that this tumor was different. They believed the transmissible agent was the tumor cells themselves.

Figure 4.33a shows how this hypothesis can be tested by reconstructing an evolutionary tree of tumors and the dogs they are growing on. If CTVT moves from dog to dog when tumor cells rub off one dog and stick to another, then the tumors will be more closely related to each other than to the dogs they are living on. If, on the other hand, each tumor is an abnormal growth of the patient's own cells, induced by an as yet undiscovered virus, then each tumor and its dog will be closest relatives.

Claudio Murgia and colleagues (2006) performed just this test. The researchers collected genetic samples from 11 CTVT patients and their tumors. They examined each sample's DNA at 21 highly variable regions called microsatellites. They used the genetic variation they observed to calculate pairwise genetic distances among the samples. Then the researchers estimated the phylogeny of the samples by neighbor joining. The resulting tree appears in Figure 4.33b. The tumors are all more closely related to each other than any is to the dog it is growing on. This result is consistent with the hypothesis that the tumor cells move from dog to dog. In other words, in an evolutionary sense, CTVT is a lineage of dogs (or wolves—see Rebbeck et al. 2009) that have ceased to live independently. Instead, they survive as parasites on other individuals and reproduce by cloning.

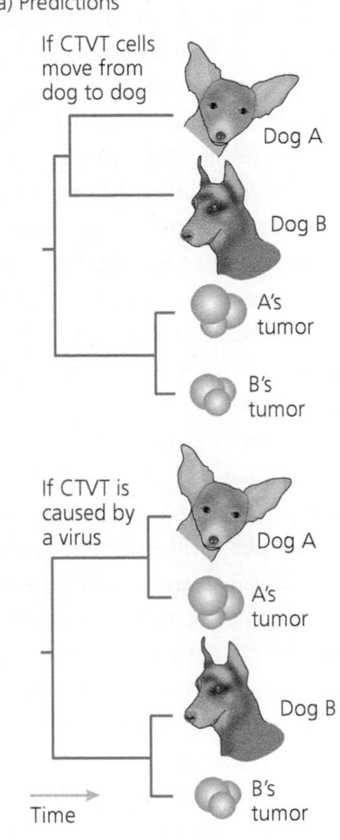

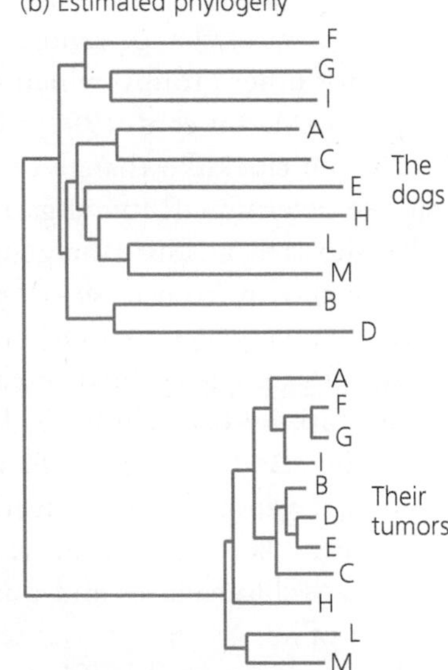

Figure 4.33 Relationships between tumors and the dogs they live on (a) The trees predicted under different hypotheses. (b) The tree estimated from genetic data. Redrawn from Murgia et al. (2006).

When Did Humans Start Wearing Clothes?

Some of our most basic questions about the history of life concern when major events occurred. Some evolutionary events can be dated from the fossil record. But what options are available when fossil data are missing? In at least some instances, it should be possible to address questions about timing by analyzing molecular traits that change at a steady rate. This hypothesis, called the **molecular clock**, originated with Emile Zuckerkandl and Linus Pauling (1962).

There are reasons to expect that some types of DNA sequences change in a clocklike fashion. Many mutations change an individual's DNA but not its phenotype. In most cases, mutations like these are not exposed to natural selection. Instead, these **neutral** changes evolve by a random process called genetic drift. As we will discuss elsewhere, the neutral theory of molecular evolution predicts that neutral changes in DNA should accumulate in populations at a rate equal to the mutation rate. If the mutation rate stays reasonably constant, and if generation times remain similar, then the number of neutral molecular differences between two taxa should be proportional to the age of their most recent common ancestor. By counting distinct neutral mutations observed in two species and multiplying by a calibration rate representing how frequently neutral changes occur per million years, researchers can estimate when the species diverged.

Although the possibility of dating events from estimates of genetic divergence is tantalizing, there are several important caveats. For example, it is critical to realize that the mutation rate to neutral alleles will vary from gene to gene and lineage to lineage, and even from base to base. For reasons explained in another chapter, silent site changes in the third positions of codons are more likely to be neutral with respect to fitness, and thus to accumulate at a clocklike rate, than replacement changes that occur at the first and second positions in codons. And if allele frequencies change rapidly due to strong selection at a particular gene, it is unlikely that the mutations involved are accumulating in a clocklike fashion. Finally, rates of change calibrated for a particular gene and lineage are unlikely to work for other groups, which may have different generation times and selection histories (Martin et al. 1992; Martin and Palumbi 1993; Hillis et al. 1996).

Even if clocklike change occurs in a particular gene and lineage, how can the rate be determined? Investigators have to rely on the fossil or geological records. The idea is to measure the genetic distance between two taxa whose divergence date is known from fossil or geological data and then to use this calibration to date the divergence times of groups that have no fossil record.

As an example of how researchers use molecular clocks to date events, consider work by Ralf Kittler and colleagues (2003, 2004) on the origin of human body lice. Body lice (*Pediculus humanus corporis*) are similar to head lice (*Pediculus humanus capitis*). Body lice feed on the body but live in clothing **(Figure 4.34a)**, while head lice feed on the scalp and live in hair. Both species are restricted to humans—chimpanzees and our other close relatives have their own specialized species of lice.

Kittler reasoned that if human body lice are adapted to live in clothing, then they must have diverged from human head lice about the time humans began wearing clothes. Based on sequence data from head and body lice collected from humans in 12 different populations from around the world, they estimated the average percentage of bases that differed between head lice and body lice.

To convert this estimate of genetic divergence into chronological divergence, the biologists analyzed homologous sequences in lice that parasitize chimps.

(a) Body lice making babies

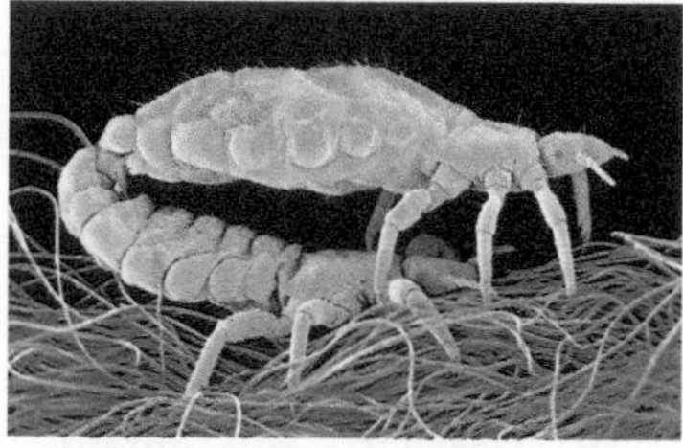

(b) Phylogeny of head and body lice

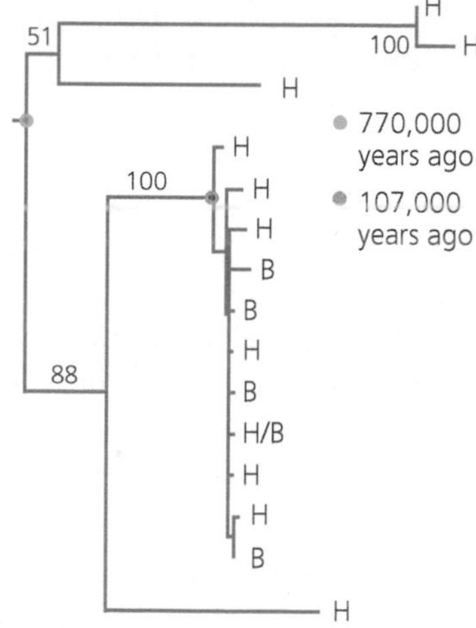

Figure 4.34 The evolution of body lice (a) Body lice (*Pediculus humanus corporis*). (b) An evolutionary tree of head and body lice. The last common ancestor of all extant body lice lived approximately 107,000 years ago. Redrawn from Kittler et al. (2003, 2004).

They took from other sources, including the fossil record, an estimate that the common ancestor of humans and chimps lived about 5.5 million years ago. By assuming that the head lice of humans and chimps diverged at the same time their host species did, the group estimated the percentage of bases that change per million years. When they multiplied this rate by the amount of divergence observed among body lice, they got an estimate of approximately 107,000 years ago for the origin of body lice, and thus of clothing (Figure 4.34b).

How Did the Seychelles Chameleon Cross the Indian Ocean?

The effort to understand where organisms live and how they came to be there is called **biogeography**. Biogeographers ask questions about why certain species are found in certain parts of the world and how geographic distributions have changed through time. When researchers turn to phylogenies for help in answering these types of questions, the research program is called **phylogeography**. We will use phylogeographic approaches to study the origin and radiation of human populations elsewhere (Chapter 20). Here we introduce the strategy by following the trail of a curious lizard.

Figure 4.35 The Seychellean tiger chameleon Photo by Hans Stieglitz.

The Seychellean tiger chameleon (*Calumma tigris*), shown in **Figure 4.35**, is a native of the Seychelles islands. The Seychelles are located in the Indian Ocean, about 1,100 km northeast of Madagascar and some 1,600 km east of Africa. How did a small lizard travel to such a remote outpost?

Because chameleons are poor swimmers, and even worse fliers, there would seem to be two possibilities. One is that the chameleon simply stayed put and rode the islands to their current position. The Seychelles were once part of the supercontinent of Gondwana. When Gondwana broke up, the chameleon could have traveled as a passenger as the Seychelles drifted out to sea. The second possibility is that the chameleon reached the Seychelles by floating on a raft of vegetation, either from Madagascar or Africa. The first scenario is called the vicariance hypothesis; the second is called the dispersal hypothesis.

Ted Townsend and colleagues (2011) sought to distinguish between these hypotheses by reconstructing a phylogeny and using a molecular clock. The logic is as follows. The breakup of Gondwana happened in stages. Some 160 million years ago, the Indigascar landmass, which would later become India, Madagascar,

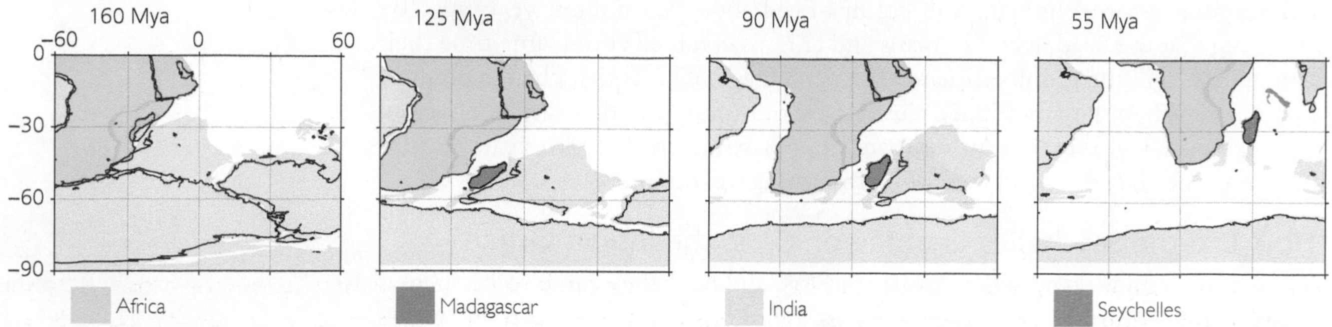

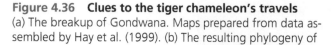

(b) The phylogeny of continents and islands

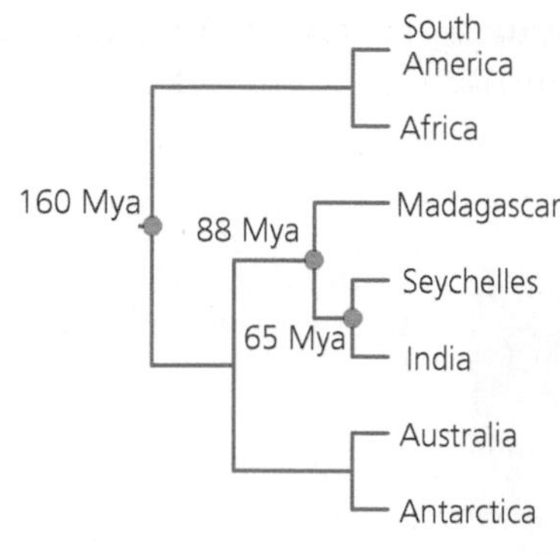

(c) The phylogeny of the Seychellean tiger chameleon and its kin

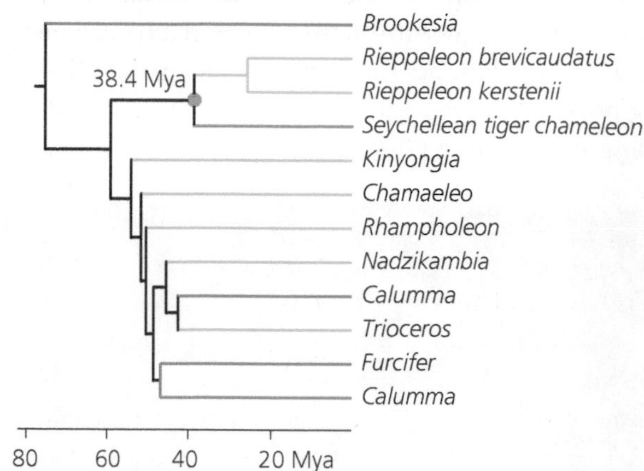

Figure 4.36 Clues to the tiger chameleon's travels (a) The breakup of Gondwana. Maps prepared from data assembled by Hay et al. (1999). (b) The resulting phylogeny of landmasses. (c) A phylogeny of the Seychellean chameleon and kin, with molecular clock estimates of divergence time. Redrawn from Townsend et al. (2011).

and the Seychelles, separated from Africa. About 88 million years ago, Madagascar separated from the India–Seychelles landmass. Finally, about 65 million years ago, India and the Seychelles separated from each other **(Figure 4.36a)**. If the tiger chameleon went along for the ride, we might expect that it would be most closely related to other chameleons from India or Madagascar, and that its last common ancestor with them would date to at least 65 to 88 million years ago (Figure 4.36b). This is what Peng Zhang and Marvalee Wake (2009) found for Seychellean caecilians. If, on the other hand, the tiger chameleon reached the Seychelles by raft, it might share a much more recent common ancestor with chameleons from Madagascar or Africa, depending on where it came from.

Townsend and colleagues used sequence data to reconstruct a phylogeny of 42 species of chameleons and used a molecular clock to estimate their divergence times. A simplified phylogeny appears in Figure 4.36c. The Seychellean chameleon's closest living relatives are African. Its last common ancestor with these kin lived roughly 40 million years ago. This rules out the vicariance hypothesis, instead suggesting that the tiger chameleon traveled by sea.

Confirmation that chameleons can disperse by raft over open ocean comes from the chameleons of the Comoros archipelago and Reunion island (Raxworthy et al. 2002). The Comoros and Reunion are volcanic, and have never been connected to any mainland. These are impressive journeys for such small lizards.

SUMMARY

Evolutionary trees, also known as phylogenies, summarize estimated histories of descent and diversification. Time flows along an evolutionary tree from the root, the common ancestor of the lineages depicted, toward the tips, the most recent descendants. Lineages on a phylogeny that share more recent last common ancestors are considered to be more closely related.

Estimating phylogenies is central to much of contemporary evolutionary biology. In the absence of convergence and reversal, we can reconstruct the history of a group of organisms by identifying shared derived characters—also known as synapomorphies. These identify groups of taxa that are all more closely related to each other than any is to taxa outside the group. Such a set, consisting of all descendants of a common ancestor, is called a monophyletic group, or clade.

Convergence and reversal create conflict among shared derived characters, thereby complicating our efforts to reconstruct history. To estimate phylogenies under these circumstances, we consider all possible evolutionary trees for the taxa of interest. We evaluate each tree for the minimum number of evolutionary changes it requires—also known as its parsimony—and choose the tree or trees that require the least evolution.

Molecular characters, such as nucleotide or amino acid sequences, are a common kind of evidence used in estimating phylogenies. After aligning a set of sequences, we can treat each site as a character. We search the forest of all possible trees, seeking the most parsimonious. Or we can use other criteria to identify the tree that offers the best estimate of evolutionary history. We can use likelihood, the probability of the data given the tree and a model of sequence evolution. If we are willing to specify prior probabilities for all possible trees, we can also use Bayesian posterior probability, the probability of the tree given the data.

The number of possible trees is typically so vast that even with fast computers, it is impossible to search for it exhaustively. Researchers instead use a variety of shortcuts, including starting our search at a tree estimated from genetic distance data, to increase our chances of finding the best tree in a reasonable amount of time.

Biologists use estimated phylogenies to answer questions in fields ranging from medicine to conservation.

QUESTIONS

1. According to the evolutionary tree in **Figure 4.37**, which is more closely related to rodents: shrews and moles, or primates? Explain how the tree shows this.
2. According to the evolutionary tree in Figure 4.37, which lived earlier: the last common ancestor of whales and pigs, or the last common ancestor of whales and bats? Explain how the tree shows this.
3. Sketch a version of the tree in Figure 4.37 in which whales, hippos, pigs, perissodactyls, carnivores, and pangolins appear in a different vertical order on the page, yet are depicted as having the same relationships.
4. In the tree in Figure 4.37, identify a monophyletic group to which the aardvark belongs. Who else is a member? Identify a larger clade that it also belongs to.
5. What is a synapomorphy?
6. High-crowned teeth that are well suited for grazing are found in some rodents, rabbits and hares, most even-toed hoofed animals, horses (which are perissodactyls), and elephants. According to the evolutionary tree in Figure 4.37, are high-crowned teeth a synapomorphy or a product of convergent evolution?

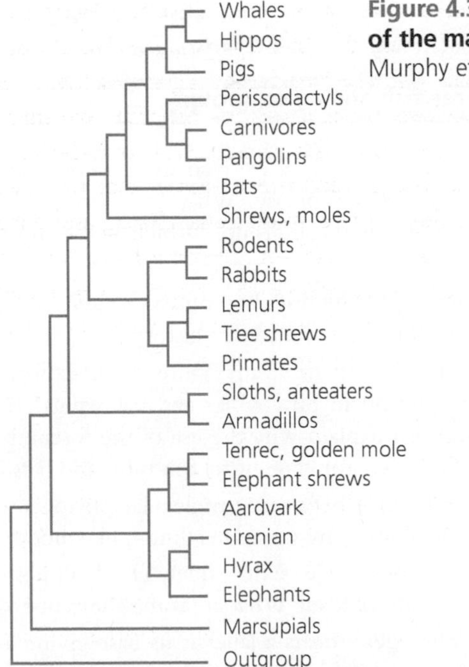

Figure 4.37 Phylogeny of the mammals From Murphy et al. (2001).

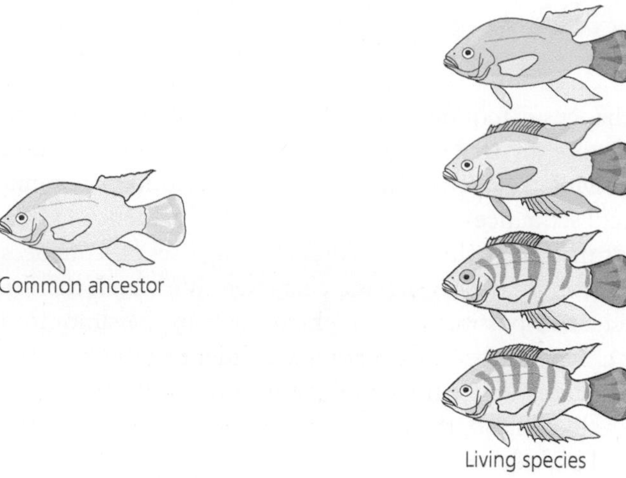

Figure 4.38 Four fish and their common ancestor

7. Assuming the four living species in **Figure 4.38** evolved from their common ancestor without convergence or reversal, reconstruct their evolutionary relationships and label the transitions. Which evolved first: stripes, or spiky back fins? How do you know?

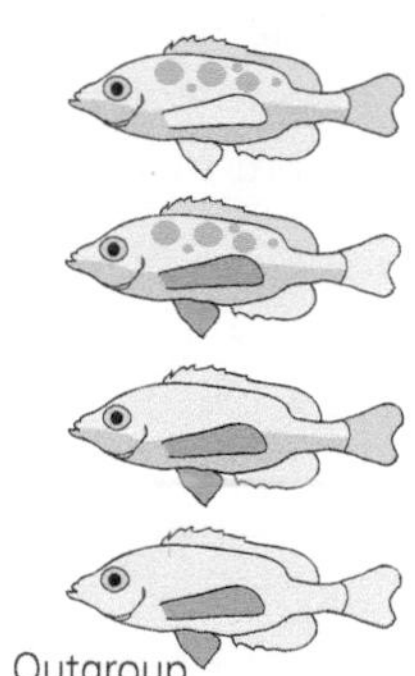

Figure 4.39 Three fish and an outgroup

8. The four fish in **Figure 4.39** evolved from a common ancestor with some homoplasy. What are the possible evolutionary trees for the ingroup? Which is the most parsimonious?
9. What is homoplasy? Why does homoplasy make it more challenging to estimate evolutionary history?
10. Referring to the information in Figure 4.10, explain why the bones found in bird wings and bat wings are homologous. Then explain why the use of the forelimb for powered flight is a convergent trait in birds and bats.
11. What is the difference between a molecular phylogeny reconstructed by parsimony versus maximum likelihood?
12. Why is it seldom possible to exhaustively check all possible trees for a suite of taxa? What are some shortcuts?
13. A clade in a phylogeny bears a label at its base giving a bootstrap support of 97%. What does this mean?

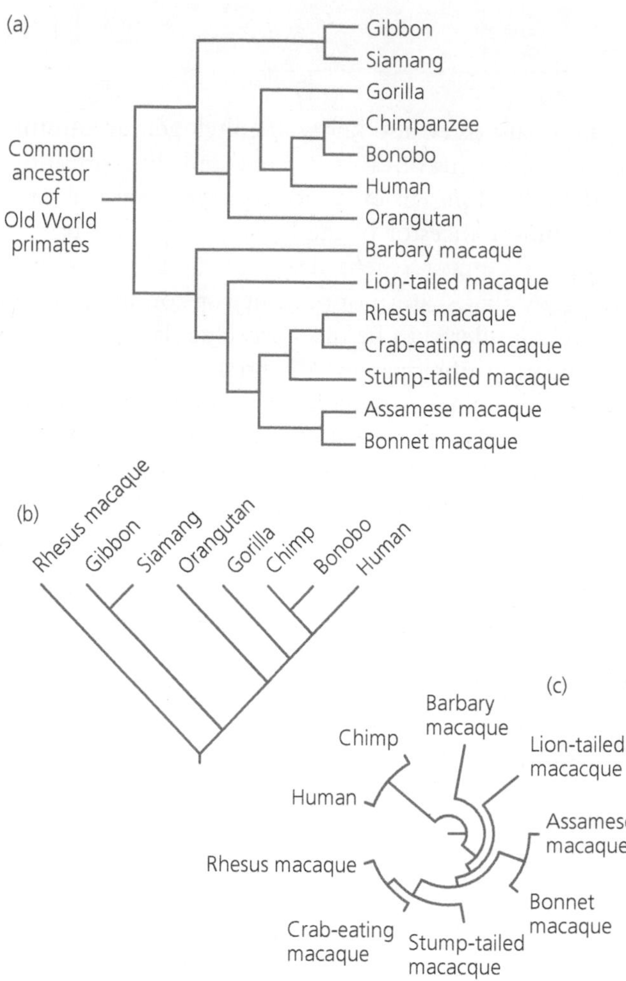

Figure 4.40 Phylogenies showing the relationships of some Old World primates (Branch lengths are not scaled.)

14. Examine the three primate phylogenies shown in **Figure 4.40**. Do the three phylogenies show the same relationships and the same order of branching? Do the phylogenies appear to lend different levels of support to the misconception that humans are the "highest" species of primate? Explain.
15. Historically, some scientists hypothesized that dogs are derived from wolves. Other scientists thought that some breeds of dogs were derived from wolves, while others were derived from other species of wild canids, such as jackals. Sketch the evolutionary trees for wolves, dogs, jackals, and coyotes under each hypothesis. Explain why the trees look different.
16. Sketch the tree you would expect for dogs, wolves, jackals, and coyotes if dogs are derived from wolves, but different breeds are derived from different wolf populations. (Include several lineages of wolves and dogs.)
17. Darwin maintained that among living species, there is no such thing as a higher (more evolved) or lower (less evolved) animal or plant. Explain what he meant.

EXPLORING THE LITERATURE

18. For a striking example of convergent evolution, see:

    Rosenblum, E. B., H. Rompler, et al. 2010. Molecular and functional basis of phenotypic convergence in white lizards at White Sands. *Proceedings of the National Academy of Sciences, USA* 107: 2113–2117.

19. For an example of convergent character loss, see:

    Meredith, R. W., J. Gatesy, et al. 2009. Molecular decay of the tooth gene *Enamelin* (ENAM) mirrors the loss of enamel in the fossil record of placental mammals. *PLoS Genetics* 5: e1000634.

20. Could we use molecular phylogenies to confirm that birds evolved from within the dinosaurs? For an attempt to recover molecular sequences from a dinosaur and place them on a phylogeny, see:

    Organ, C. L., M. H. Schweitzer, et al. 2008. Molecular phylogenetics of mastodon and *Tyrannosaurus rex*. *Science* 320: 499.

    This study has proven controversial. For an overview, see:

    Dalton, R. 2008. Fresh doubts over *T-rex* chicken link. *Nature* 454: 1035–1035.

    For a reanalysis of the raw data, see:

    Bern, M., B. S. Phinney, and D. Goldberg. 2009. Reanalysis of *Tyrannosaurus rex* mass spectra. *Journal of Proteome Research* 8: 4328–4332.

    Finally, for sequences recovered from another dinosaur and placed on a phylogeny, see:

    Schweitzer, M. H., W. Zheng, et al. 2009. Biomolecular characterization and protein sequences of the Campanian hadrosaur *B. canadensis*. *Science* 324: 626–631.

21. In Chapter 1, we discussed the use of a molecular phylogeny as evidence in a criminal case involving a doctor who intentionally infected a patient with HIV. For two more cases in which molecular phylogenies helped convict men accused of knowingly transmitting HIV, see:

    Scaduto, D. I., J. M. Brown, et al. 2010. Source identification in two criminal cases using phylogenetic analysis of HIV-1 DNA sequences. *Proceedings of the National Academy of Sciences, USA* 107: 21242–21247.

22. For an estimated phylogeny used to trace the origin of the human malaria parasite, see:

    Liu, W., Y. Li, et al. 2010. Origin of the human malaria parasite *Plasmodium falciparum* in gorillas. *Nature* 467: 420–425.

    Compare to an earlier result. Which is more credible?

    Rich, S. M., F. H. Leendertz, et al. 2009. The origin of malignant malaria. *Proceedings of the National Academy of Sciences, USA* 106: 14902–14907.

23. The phylogeny in Figure 4.10 features an unresolved polytomy for lizards and snakes versus turtles versus crocodilians and birds. For a late-breaking study that attempts to resolve this polytomy, see:

    Crawford, N. G., B. C. Faircloth, et al. 2012. More than 1000 ultraconserved elements provide evidence that turtles are the sister group of archosaurs. *Biology Letters* 8: 783–786.

24. Figure 4.33 offers evidence that CTVT spreads from host to host. For phylogenetic evidence on one mechanism by which the tumor lineage has evolved, despite reproducing asexually, see:

    Rebbeck, C. A., A. M. Leroi, and A. Burt. 2011. Mitochondrial capture by a transmissible cancer. *Science* 331: 303

CITATIONS

Alibardi, L., and P. F. Maderson. 2003. Observations on the histochemistry and ultrastructure of the epidermis of the tuatara, *Sphenodon punctatus* (Sphenodontida, Lepidosauria, Reptilia): A contribution to an understanding of the lepidosaurian epidermal generation and the evolutionary origin of the squamate shedding complex. *Journal of Morphology* 256: 111–133.

Bajpai, S., and P. D. Gingerich. 1998. A new Eocene archaeocete (Mammalia, Cetacea) from India and the time of origin of whales. *Proceedings of the National Academy of Science, USA* 95: 15464–15468.

Baker, C. S., and S. R. Palumbi. 1994. Which whales are hunted? A molecular-genetic approach to monitoring whaling. *Science* 265: 1538–1539.

Baldauf, S. L. 2003. Phylogeny for the faint of heart: A tutorial. *Trends in Genetics* 19: 345–351.

Bayes, T. 1763. An essay towards solving a problem in the doctrine of chances. *Philosophical Transactions of the Royal Society* 53: 370–418.

Becker, R. E., R. A. Valverde, and B. I. Crother. 2011. Proopiomelanocortin (POMC) and testing the phylogenetic position of turtles (Testudines). *Journal of Zoological Systematics and Evolutionary Research* 49: 148–159.

Boisserie, J.-R., F. Lihoreau, and M. Brunet. 2005. The position of Hippopotamidae within Cetartiodactyla. *Proceedings of the National Academy of Sciences, USA* 102: 1537-1541.

Caldicott, D. G., D. Croser, et al. 2005. Crocodile attack in Australia: An analysis of its incidence and review of the pathology and management of crocodilian attacks in general. *Wilderness & Environmental Medicine* 16: 143–159.

Campbell, K. L., J. E. Roberts, et al. 2010. Substitutions in woolly mammoth hemoglobin confer biochemical properties adaptive for cold tolerance. *Nature Genetics* 42: 536–540.

Carroll, R. L. 2007. The Palaeozoic ancestry of salamanders, frogs and caecilians. *Zoological Journal of the Linnean Society* 150 (Suppl 1): 1–140.

Chevenet, F., C. Brun, et al. 2006. TreeDyn: Towards dynamic graphics and annotations for analyses of trees. *BMC Bioinformatics* 7: 439.

Claessens, L. P. 2009. A cineradiographic study of lung ventilation in *Alligator mississippiensis*. *Journal of Experimental Zoology. Part A, Ecological Genetics and Physiology* 311: 563–585.

Dakin, R., and R. Montgomerie. 2011. Peahens prefer peacocks displaying more eyespots, but rarely. *Animal Behaviour* 82: 21–28.

Darwin, C. R. 1837–1838. *Notebook B: [Transmutation of species (1837–1838)]*. CUL-DAR121. Transcribed by K. Rookmaaker. (Darwin Online, http://darwin-online.org.uk/).

Darwin, C. R. 1859. *On the Origin of Species*. London: John Murray.

Davis, B. W., G. Li, and W. J. Murphy. 2010. Supermatrix and species tree methods resolve phylogenetic relationships within the big cats, *Panthera* (Carnivora: Felidae). *Molecular Phylogenetics and Evolution* 56: 64–76.

de Muizon, D. 2001.Walking with whales. *Nature* 413: 259–260.

Dereeper, A., V. Guignon, et al. 2008. Phylogeny.fr: Robust phylogenetic analysis for the non-specialist. *Nucleic Acids Research* 36: W465–W469.

Edwards, S. V. 2009. Is a new and general theory of molecular systematics emerging? *Evolution* 63: 1–19.

Felsenstein, J. 1978. Cases in which parsimony or compatibility methods will be positively misleading. *Systematic Zoology* 27: 401–410.

Felsenstein, J. 1981. Evolutionary trees from DNA sequences: A maximum likelihood approach. *Journal of Molecular Evolution* 17: 368–376.

Felsenstein, J. 1982. Numerical methods for inferring evolutionary trees. *Quarterly Review of Biology* 57: 379–404.

Felsenstein, J. 1983. Parsimony in systematics: Biological and statistical issues. *Annual Review of Ecology and Systematics* 14: 313–333.

Felsenstein, J. 1985. Confidence-limits on phylogenies—An approach using the bootstrap. *Evolution* 39: 783–791.

Felsenstein, J. 1988. Phylogenies from molecular sequences: Inference and reliability. *Annual Review of Genetics* 22: 521–565.

Felsenstein, J. 2004. *Inferring Phylogenies*. Sunderland, MA: Sinauer.

Felsenstein, J. 2009. *PHYLIP (Phylogeny Inference Package)*, Version 3.69. Distributed by the author. Department of Genome Sciences, University of Washington, Seattle. http://evolution.genetics.washington.edu/phylip/

Flower, W. H. 1883. On whales, past and present, and their probable origin. *Nature* 28: 199–202, 226–230.

Fong, D. W., T. C. Kane, and D. C. Culver. 1995. Vestigialization and loss of nonfunctional characters. *Annual Review of Ecology and Systematics* 26: 249–268.

Gans, C., and B. Clark. 1976. Studies on ventilation of *Caiman crocodilus* (Crocodilia: Reptilia). *Respiration Physiology* 26: 285–301.

Gascuel, O. 1994. A Note on Sattath and Tversky's, Saitou and Nei's, and Studier and Keppler's algorithms for inferring phylogenies from evolutionary distances. *Molecular Biology and Evolution* 11: 961–963.

Gascuel, O., and M. Steel. 2006. Neighbor-joining revealed. *Molecular Biology and Evolution* 23: 1997–2000.

Gatesy, J., M. Milinkovitch, et al. 1999. Stability of cladistic relationships between Cetacea and higher-level Artiodactyl taxa. *Systematic Biology* 48: 6–20.

Gingerich, P. D., B. H. Smith, and E. L. Simons. 1990. Hind limbs of Eocene *Basilosaurus*: Evidence of feet in whales. *Science* 249: 154–157.

Gingerich, P. D. 2001. *Research on the origin and early evolution of whales (Cetacea)*. http://www-personal.umich.edu/~gingeric/PDGwhales/ Whales.htm.

Gingerich, P. D., M. ul Haq, et al. 2001. Origin of whales from early artiodactyls: Hands and feet of Eocene Protocetidae from Pakistan. *Science* 293: 2239–2242.

Guindon, S., J. F. Dufayard, et al. 2010. New algorithms and methods to estimate maximum-likelihood phylogenies: Assessing the performance of PhyML 3.0. *Systematic Biology* 59: 307–321.

Guindon, S., and O. Gascuel. 2003. A simple, fast, and accurate algorithm to estimate large phylogenies by maximum likelihood. *Systematic Biology* 52: 696–704.

Graur, D., and W.-H. Li. 2000. *Fundamentals of Molecular Evolution*. 2nd ed. Sunderland, MA: Sinauer.

Green, R. E., J. Krause, et al. 2010. A draft sequence of the Neandertal genome. *Science* 328: 710–722.

Gregory, T. R. 2008. Understanding evolutionary trees. *Evolution: Education and Outreach* 1: 121–137.

Hall, A. R., and N. Colegrave. 2008. Decay of unused characters by selection and drift. *Journal of Evolutionary Biology* 21: 610–617.

Hall, B. G. 2005. Comparison of the accuracies of several phylogenetic methods using protein and DNA sequences. *Molecular Biology and Evolution* 22: 792–802.

Hall, B. G. 2011. *Phylogenetic Trees Made Easy: A How-To Manual*. 4th ed. Sunderland, MA: Sinauer.

Harrison, C. J., and J. A. Langdale. 2006. A step by step guide to phylogeny reconstruction. *Plant Journal* 45: 561–572.

Hay, W. W., R. DeConto, et al. 1999. Alternative global cretaceous paleogeography. In *The Evolution of Cretaceous Ocean/Climate Systems*, ed. E. Barrera and C. Johnson. Geological Society of America Special Paper 332, 1–47. Mapping facility available at http://www.odsn.de/odsn/services/paleomap/paleomap.html.

Hennig, W. 1966. *Phylogenetic Systematics*. Translated by D. D. Davis and R. Zangerl. Urbana: University of Illinois Press.

Hill, D. E. 2009. Euophryine jumping spiders that extend their third legs during courtship (Araneae: Salticidae: Euophryinae: *Maratus*, *Saitis*). *Peckhamia* 74.1: 1–27.

Hillis, D. M. 1999. SINEs of the perfect character. *Proceedings of the National Academy of Sciences, USA* 96: 9979–9981.

Hillis, D. M., J. J. Bull, et al. 1992. Experimental phylogenetics: Generation of a known phylogeny. *Science* 255: 589–592.

Hillis, D. M., J. P. Huelsenbeck, and C. W. Cunningham. 1994. Application and accuracy of molecular phylogenies. *Science* 264: 671–677.

Hillis, D. M., B. K. Mable, and C. Moritz. 1996. Applications of molecular systematics: The state of the field and a look to the future. In *Molecular Systematics*, ed. D. M. Hillis, C. Moritz, and B. K. Mable. Sunderland, MA: Sinauer, 515–543.

Hoffmann, M., C. Hilton-Taylor, et al. 2010. The impact of conservation on the status of the world's vertebrates. *Science* 330: 1503–1509.

Huelsenbeck, J. P., and K. A. Crandall. 1997. Phylogeny estimation and hypothesis testing using maximum likelihood. *Annual Review of Ecology and Systematics* 28: 437–466.

Huelsenbeck, J. P., and D. M. Hillis. 1993. Success of phylogenetic methods in the four-taxon case. *Systematic Biology* 42: 247–264.

Huelsenbeck, J. P., and F. Ronquist. 2001. MRBAYES: Bayesian inference of phylogeny. *Bioinformatics* 17: 754–755.

Huelsenbeck, J. P., F. Ronquist, et al. 2001. Bayesian inference of phylogeny and its impact on evolutionary biology, Supporting Online Material. *Science* 294: 2310–2314.

Hugall, A. F., R. Foster, and M. S. Lee. 2007. Calibration choice, rate smoothing, and the pattern of tetrapod diversification according to the long nuclear gene RAG-1. *Systematic Biology* 56: 543–563.

Iwabe, N., Y. Hara, et al. 2005. Sister group relationship of turtles to the bird-crocodilian clade revealed by nuclear DNA-coded proteins. *Molecular Biology and Evolution* 22: 810–813.

Johnson, W. E., E. Eizirik, et al. 2006. The late Miocene radiation of modern Felidae: A genetic assessment. *Science* 311: 73–77.

Johnson, W. E., and S. J. O'Brien. 1977. Phylogenetic reconstruction of the Felidae using 16S rRNA and NADH-5 mitochondrial genes. *Journal of Molecular Evolution* 44: S98–S116.

Kearney, M. 2003. Systematics of the Amphisbaenia (Lepidosauria: Squamata) based on morphological evidence from recent and fossil forms. *Herpetological Monographs* 17: 1–74.

Kimura, M. 1980. A simple method for estimating evolutionary rate of base substitutions through comparative studies of nucleotide sequences. *Journal of Molecular Evolution* 16:111–120.

Kittler, R., M. Kayser, and M. Stoneking. 2003. Molecular evolution of *Pediculus humanus* and the origin of clothing. *Current Biology* 13: 1414–1417.

Kittler, R., M. Kayser, and M. Stoneking. 2004. Erratum: Molecular evolution of *Pediculus humanus* and the origin of clothing. *Current Biology* 14: 2309.

Kolaczkowski, B., and J. W. Thornton. 2004. Performance of maximum parsimony and likelihood phylogenetics when evolution is heterogeneous. *Nature* 431: 980–984.

Lari, M., E. Rizzi, et al. 2011. The complete mitochondrial genome of an 11,450-year-old Aurochsen (*Bos primigenius*) from Central Italy. *BMC Evolutionary Biology* 11: 32.

Laurin, M. 2011. Terrestrial vertebrates. Stegocephalians: Tetrapods and other digit-bearing vertebrates. Version April 21, 2011. Available at http://tolweb.org/Terrestrial_Vertebrates/14952/2011.04.21 in The Tree of Life Web Project, http://tolweb.org/

Laurin, M., and J. A. Gauthier. 2011. Amniota. Mammals, reptiles (turtles, lizards, *Sphenodon*, crocodiles, birds) and their extinct relatives. Version April 22, 2011. Available at http://tolweb.org/Amniota/14990/2011.04.22 in The Tree of Life Web Project, http://tolweb.org/

Laurin, M., and R. R. Reisz. 2007. Synapsida. Mammals and their extinct relatives. Version April 6, 2007. Available at http://tolweb.org/Synapsida/14845/2007.04.06 in The Tree of Life Web Project, http://tolweb.org/

Liu, K., S. Raghavan, et al. 2009. Rapid and accurate large-scale coestimation of sequence alignments and phylogenetic trees. *Science* 324: 1561–1564.

Liu, K., T. J. Warnow, et al. 2011. SATé-II: Very fast and accurate simultaneous estimation of multiple sequence alignments and phylogenetic trees. *Systematic Biology* 61: 90–106.

Luckett, W. P., and N. Hong. 1998. Phylogenetic relationships between the orders Artiodactyla and Cetacea: A combined assessment of morphological and molecular evidence. *Journal of Mammalian Evolution* 5: 127–182.

Lyson, T. R., G. S. Bever, et al. 2010. Transitional fossils and the origin of turtles. *Biology Letters* 6: 830–833.

Maddison, W. P., M. J. Donoghue, and D. R. Maddison. 1984. Outgroup analysis and parsimony. *Systematic Zoology* 33: 83–103.

Martin, A. P., G. J. P. Naylor, and S. R. Palumbi. 1992. Rates of mitochondrial DNA evolution in sharks are slow compared with mammals. *Nature* 357: 153–155.

Martin, A. P., and S. R. Palumbi. 1993. Body size, metabolic rate, generation time, and the molecular clock. *Proceedings of the National Academy of Sciences, USA* 90: 4087–4091.

Mattern, M. Y., and D. A. McLennan. 2000. Phylogeny and speciation of felids. *Cladistics* 16: 232–253.

Meylan, P. A. 2001. Testudines. Turtles, tortoises and terrapins. Version May 31, 2001. Available at http://tolweb.org/Testudines/14861/2001.05.31 in The Tree of Life Web Project, http://tolweb.org/

Mindell, D. P., and J. W. Brown. 2005. Neornithes. Modern birds. Version December 14, 2005 (under construction). Available at http://tolweb.org/Neornithes/15834/2005.12.14 in The Tree of Life Web Project, http://tolweb.org/

Mooi, R., S. Martinez, and S. G. Parma. 2000. Phylogenetic systematics of Tertiary monophorasterid sand dollars (Clypeasteroida: Echinoidea) from South America. *Journal of Paleontology* 74: 263–281.

Muñoz, N., F. X. Bosch, et al. 2004. Against which human papillomavirus types shall we vaccinate and screen? The international perspective. *International Journal of Cancer* 111: 278–285.

Murgia, C., J. K. Pritchard, et al. 2006. Clonal origin and evolution of a transmissible cancer. *Cell* 126: 477–487.

Murphy, W. J., E. Eizirik, et al. 2001. Resolution of the early placental mammal radiation using Bayesian phylogenetics. *Science* 294: 2348–2351.

Nikaido, M., A. P. Rooney, and N. Okada. 1999. Phylogenetic relationships among cetartiodactyls based on insertions of short and long interspersed elements: Hippopotamuses are the closest extant relatives of whales. *Proceedings of the National Academy of Sciences, USA* 96: 10261–10266.

Nummela, S., J. G. Thewissen, et al. 2007. Sound transmission in archaic and modern whales: Anatomical adaptations for underwater hearing. *Anatomical Record: Advances in Integrative Anatomy and Evolutionary Biology* 290: 716–733.

Ogden, T. H., and M. S. Rosenberg. 2006. Multiple sequence alignment accuracy and phylogenetic inference. *Systematic Biology* 55: 314–328.

O'Leary, M. A., and J. H. Geisler. 1999. The position of Cetacea within Mammalia: Phylogenetic analysis of morphological data from extinct and extant taxa. *Systematic Biology* 48: 455–490.

Orliac, M., J. R. Boisserie, et al. 2010. Early Miocene hippopotamids (Cetartiodactyla) constrain the phylogenetic and spatiotemporal settings of hippopotamid origin. *Proceedings of the National Academy of Sciences, USA* 107: 11871–11876.

Osburn, R. C. 1903. Adaptation to aquatic, arboreal, fossorial and cursorial habits in mammals. I. Aquatic adaptations. *American Naturalist* 37: 651–665.

Otto, J. C., and D. E. Hill. 2010. Observations of courtship display by a male *Maratus amabilis* Karsch 1878 (Araneae: Salticidae). *Peckhamia* 79.1: 1–16.

Raxworthy, C. J., M. R. Forstner, and R. A. Nussbaum. 2002. Chameleon radiation by oceanic dispersal. *Nature* 415: 784–787.

Rebbeck, C. A., R. Thomas, et al. 2009. Origins and evolution of a transmissible cancer. *Evolution* 63: 2340–2349.

Rieppel, O., H. Zaher, et al. 2003. The anatomy and relationships of *Haasiophis terrasanctus*, a fossil snake with well-developed hind limbs from the mid-cretaceous of the Middle East. *Journal of Paleontology* 77: 536–558.

Roelants, K., B. G. Fry, et al. 2010. Identical skin toxins by convergent molecular adaptation in frogs. *Current Biology* 20: 125–130.

Ronquist, F., and J. P. Huelsenbeck. 2003. MRBAYES 3: Bayesian phylogenetic inference under mixed models. *Bioinformatics* 19: 1572–1574.

Saitou, N., and M. Nei. 1987. The neighbor-joining method—A new method for reconstructing phylogenetic trees. *Molecular Biology and Evolution* 4: 406–425.

Schaeffer, B. 1948. The origin of a mammalian ordinal character. *Evolution* 2: 164–175.

Schiffman, M., P. E. Castle, et al. 2007. Human papillomavirus and cervical cancer. *Lancet* 370: 890–907.

Skinner, A., and M. S. Y. Lee. 2009. Body-form evolution in the scincid lizard clade *Lerista* and the mode of macroevolutionary transitions. *Evolutionary Biology* 36: 292–300.

Skinner, A., M. S. Y. Lee, and M. N. Hutchinson. 2008. Rapid and repeated limb loss in a clade of scincid lizards. *BMC Evolutionary Biology* 8: 310.

Sousa, A., L. Ze-Ze, et al. 2008. Exploring tree-building methods and distinct molecular data to recover a known asymmetric phage phylogeny. *Molecular Phylogenetics and Evolution* 48: 563–573.

Spaulding, M., M. A. O'Leary, and J. Gatesy. 2009. Relationships of Cetacea (Artiodactyla) among mammals: Increased taxon sampling alters interpretations of key fossils and character evolution. *PLoS ONE* 4: e7062.

Swofford, D. L., G. J. Olsen, et al. 1996. Phylogenetic inference. In *Molecular Systematics*, ed. D. M. Hillis, C. Moritz, et al. Sunderland, MA: Sinauer, 407–514.

Tamura, K., D. Peterson, et al. 2011. MEGA5: Molecular evolutionary genetics analysis using maximum likelihood, evolutionary distance, and maximum parsimony methods. *Molecular Biology and Evolution* 28: 2731-2739.

Tchernov, E., O. Rieppel, et al. 2000. A fossil snake with limbs. *Science* 287: 2010–2012.

Thewissen, J. G. M., L. N. Cooper, et al. 2009. From land to water: The origin of whales, dolphins, and porpoises. *Evolution: Education and Outreach* 2: 272–288.

Thewissen, J. G. M., and S. T. Hussain. 1993. Origin of underwater hearing in whales. *Nature* 361: 444–445.

Thewissen, J. G. M., S. T. Hussain, and M. Arif. 1994. Fossil evidence for the origin of aquatic locomotion in archaeocete whales. *Science* 263: 210–212.

Thewissen, J. G. M., and S. I. Madar. 1999. Ankle morphology of the earliest cetaceans and its implications for the phylogenetic relations among ungulates. *Systematic Biology* 48: 21–30.

Thewissen, J. G. M., S. I. Madar, and S. T. Hussain. 1998. Whale ankles and evolutionary relationships. *Nature* 395: 452.

Thewissen, J. G. M., E. M. Williams, et al. 2001. Skeletons of terrestrial cetaceans and the relationship of whales to artiodactyls. *Nature* 413: 277–281.

Townsend, T. M., K. A. Tolley, et al. 2011. Eastward from Africa: Palaeocurrent-mediated chameleon dispersal to the Seychelles islands. *Biology Letters* 7: 225–228.

Uhen, M. D. 2007. Evolution of marine mammals: Back to the sea after 300 million years. *Anatomical Record: Advances in Integrative Anatomy and Evolutionary Biology* 290: 514–522.

Uhen, M. D. 2010. The origin(s) of whales. *Annual Review of Earth and Planetary Sciences* 38: 189–219.

Van Valen, L. M. 1966. Deltatheridia, a new order of mammals. *Bulletin of the American Museum of Natural History* 132: 1–126.

Wake, D. B., M. H. Wake, and C. D. Specht. 2011. Homoplasy: From detecting pattern to determining process and mechanism of evolution. *Science* 331: 1032–1035.

Wang, L. S., J. Leebens-Mack, et al. 2011. The impact of multiple protein sequence alignment on phylogenetic estimation. *IEEE–ACM Transactions on Computational Biology and Bioinformatics* 8: 1108–1119.

Werdelin, L., and L. Olsson. 1997. How the leopard got its spots: A phylogenetic view of the evolution of felid coat patterns. *Biological Journal of the Linnean Society* 62: 383–400.

Werneburg, I., and M. R. Sanchez-Villagra. 2009. Timing of organogenesis support basal position of turtles in the amniote tree of life. *BMC Evolutionary Biology* 9: 82.

White, M. A., C. Ané, et al. 2009. Fine-scale phylogenetic discordance across the house mouse genome. *PLoS Genetics* 5: e1000729.

Wiens, J. J. 2011. Re-evolution of lost mandibular teeth in frogs after more than 200 million years, and re-evaluating Dollo's law. *Evolution* 65: 1283–1296.

Wong, K. M., M. A. Suchard, and J. P. Huelsenbeck. 2008. Alignment uncertainty and genomic analysis. *Science* 319: 473–476.

Zhang, P., and M. H. Wake. 2009. A mitogenomic perspective on the phylogeny and biogeography of living caecilians (Amphibia: Gymnophiona). *Molecular Phylogenetics and Evolution* 53: 479–491.

Zuckerkandl, E., and L. Pauling. 1962. Molecular disease, evolution and genic heterogeneity. In *Horizons in Biochemistry*, ed. M. Kasha and B. Pullman. New York: Academic Press, 189–225.

5

Variation Among Individuals

We humans show tremendous variation in body size and a great variety of other traits. Indeed, the ways in which we differ from one another are so numerous and so obvious that we have little trouble distinguishing among the thousands of people we meet over a lifetime.

Ann McKellar and Andrew Hendry (2009) wondered why humans seem to be more variable than other species. One possibility is that humans *are* more variable. Another possibility, however, is that we just pay more attention to the variation among people. We tend to perceive humans as individuals, but other organisms as examples of their type (see Nettle 2010). McKellar and Hendry scoured the literature for data that would let them calculate coefficients of variation for body length in animal species and for height in human populations. The coefficient of variation (CV), defined as the standard deviation of a sample divided by the mean, allows us to compare the variablity present in sets of things as different as apples and oranges. The graph at right compares the distribution of CVs for male body length in 210 animal species and male height in 99 human populations. The patterns for female length and height look the same. The data show that, compared to other animals, the variation in height among humans is actually rather modest. Whether we notice it or not, variation among individuals is ubiquitous.

Tracy McGrady (6 feet, 8 inches) and Muggsy Bogues (5 feet, 3 inches) illustrate variation in height. Data compiled by Ann McKellar and Andrew Hendry (2009) show that human height variation is modest compared to body length variation in other animals.

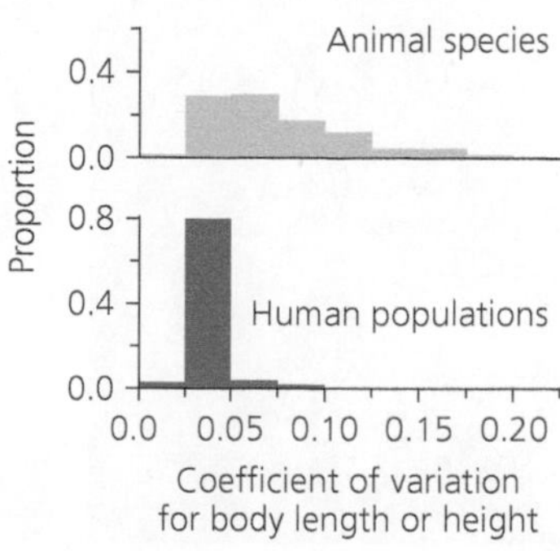

Variation among individuals is the raw material for evolution. **Figure 5.1** shows more examples. But to serve as raw material for evolution, the variation must exhibit a particular property. It must be transmitted genetically from parents to offspring.

In this chapter we explore variation in detail. In the first section we consider three different kinds of variation, the mechanisms behind them, and the role they play in evolution. In Sections 5.2, 5.3, and 5.4 we consider the ultimate source of genetic variation: the mutations that generate new alleles, new genes, and new chromosomes. Finally, in Section 5.5, we discuss the rates at which mutations occur, how they alter the fitness of the individuals that carry them, and how rates and fitness effects influence the long-term evolution of populations.

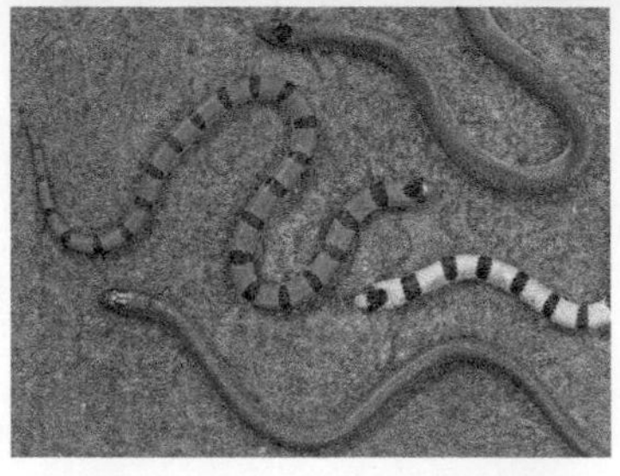

Figure 5.1 Variation among individuals Top, whiskerbrush (*Linanthus ciliatus*). Photo by Eric Knapp. Center, bat stars (*Asterina miniata*). Bottom, variable ground snakes (*Sonora semiannulata*). Photo by Alison Davis Rabosky and Christian Cox.

5.1 Three Kinds of Variation

Throughout this chapter, and the rest of the book, it will be useful to distinguish three different kinds of variation among individuals. These are **genetic variation**, **environmental variation**, and **genotype-by-environment interaction**. Some of these terms may be unfamiliar, but the reader is likely already acquainted with the kinds of variation they describe.

Consider the variation in skin color among humans. Newborns show differences in skin color as a result of differences in the genes they have inherited from their parents (Parra 2007). This is genetic variation.

In addition, individuals change their color upon exposure to sunlight—that is, they tan (Miyamura et al. 2011). Identical twins may have matching skin tones until one spends winter break sunbathing in the Caribbean while the other stays home to study in the library. This is environmental variation.

Finally, people differ in their ability to tan as a result of differences in the genes they have inherited from their parents (Han et al. 2008; Nan et al. 2009). When two friends with similar skin colors spend a day at the beach, one may turn brown while the other turns red. This is genotype-by-environment interaction.

Before exploring these three kinds of variation in more detail, it will be helpful to review the fundamental machinery of life.

The Machinery of Life

Much of the structure of living things is provided by proteins. Proteins also carry out much of the work of being alive. It is because they contain different proteins that the cells shown in **Figure 5.2** are able to perform different functions. Red blood cells contain large amounts of a protein called hemoglobin, which picks

(a) Red blood cells

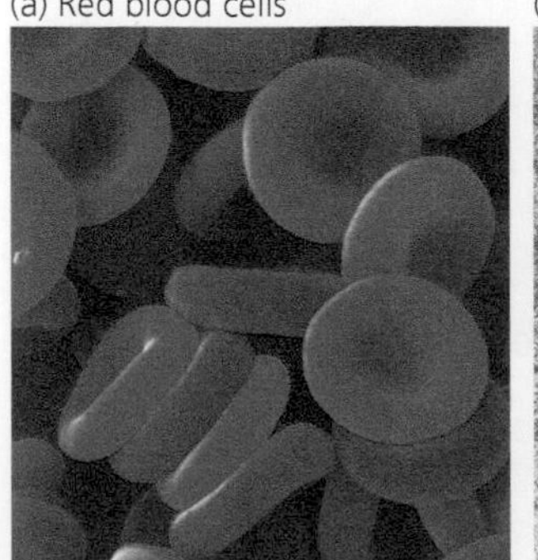

(b) Goblet cell

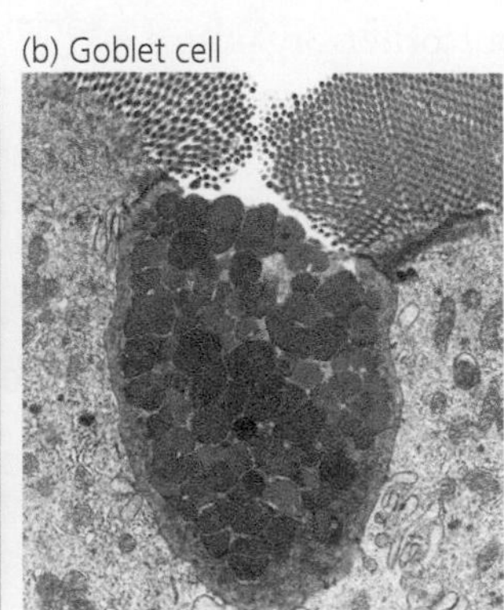

(c) Rod cells

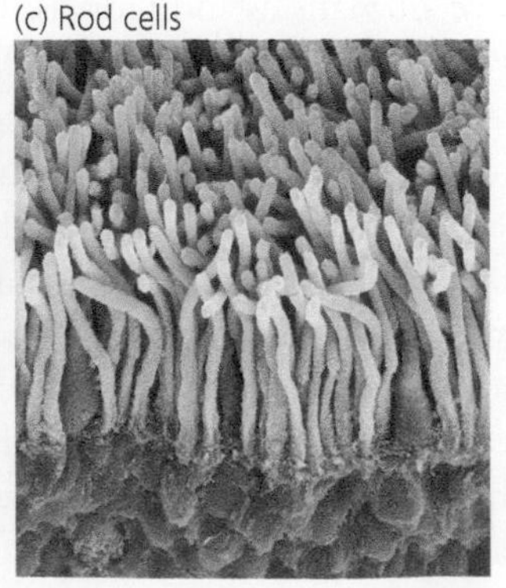

Figure 5.2 Different kinds of cells are distinguished by the proteins they make (a) Human red blood cells are filled with hemoglobin. (b) Human goblet cell filled with mucigen (red). (c) Human rod cells (blue) contain rhodopsin.

up oxygen when the cells pass through the lungs and drops it off when they pass through other tissues. Goblet cells, found in the lining of the small intestine, manufacture granules, of a protein called mucigen. When a goblet cell releases these granules, they combine with water to make mucin, a component of the mucus that lubricates the intestine. Rod cells, which serve as light detectors in the eye, contain a light-absorbing protein called rhodopsin.

Proteins are chains of amino acids. The 20 different amino acids that serve as components of proteins have diverse chemical properties. The properties of the amino acids in a given protein, and the order in which they are strung together, cause the protein to coil and fold into a characteristic three-dimensional shape. A model of the shape of hemoglobin appears in **Figure 5.3**. A protein's three-dimensional shape is crucial to its ability to perform its life-sustaining functions.

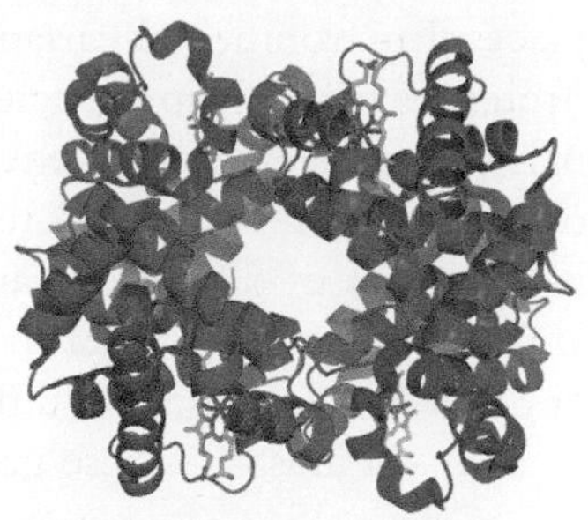

Figure 5.3 A three-dimensional model of hemoglobin From Richard Weeler (Zephyris) 2007.

Organisms carry instructions for how to build the proteins they use, as well as for when, where, and in what quantities to make them, in their genetic material. The genetic material is RNA in some viruses. In other viruses, and all cellular forms of life, the genetic material is DNA. For the remainder of this section, we will focus on DNA-based organisms.

A DNA molecule is shaped like a twisted ladder, or double helix **(Figure 5.4)**. Each half of the ladder is a chain of nucleotides. Each nucleotide consists of a phosphate group, a sugar called deoxyribose, and a base. The four bases found in DNA—adenine, thymine, guanine, and cytosine—give the four kinds of DNA nucleotides the abbreviated names A, T, G, and C. The bases in opposing strands pair to form the rungs of the ladder. A always pairs with T. C always pairs with G. Because of this, the two opposing strands in a DNA molecule are complementary. If we know the sequence of A's, T's, G's, and C's on one strand, we can infer the sequence on the opposing strand.

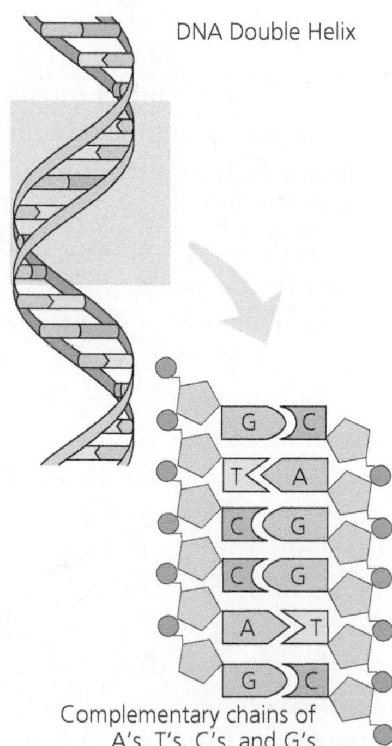

Figure 5.4 The DNA double-helix

The protein-specifying information carried by a DNA molecule is encoded in the order of the A's, T's, G's, and C's along its nucleotide strands. In portions of the molecule called coding regions, cells read the sequence of bases (along one strand or the other) as a series of three-letter words. Each word, or codon, specifies that a particular amino acid goes in a particular position in the encoded protein. A nucleotide sequence encoding plans for making a protein is called a gene.

Genes in organisms are embedded in long DNA molecules called chromosomes. A typical chromosome carries numerous genes. The physical location of a gene on a chromosome is called the gene's locus. **Figure 5.5** shows the loci of a few of the hundreds of protein-encoding genes on human chromosome 7.

The number of chromosomes, their sizes, the genes they contain, and the loci where those genes appear are similar across individuals in a species. Humans, for example, have 23 pairs of chromosomes containing roughly 22,500 protein-encoding genes (Pertea and Salzberg 2010). In contrast, chimpanzees have 24 pairs of chromosomes, dogs have 39 pairs, and wine grapes have 19.

The corpus of genetic instructions carried by an individual is called its **genome**. When organisms reproduce, they copy their genomes and bestow copies on their offspring.

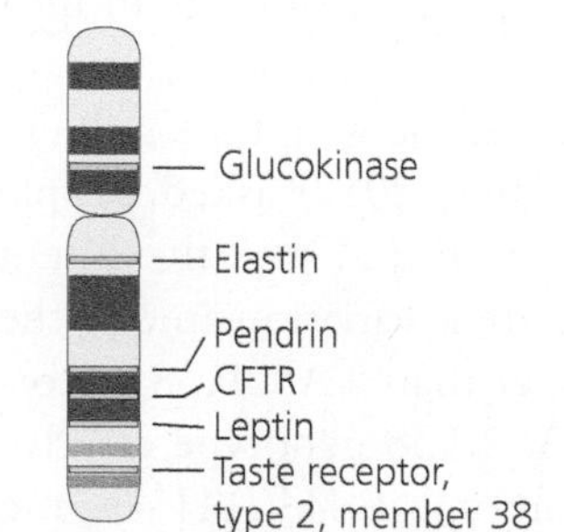

Figure 5.5 Some of the genes on human chromosome 7 From the National Center for Biotechnology Information.

With this background in mind, we can consider the mechanisms responsible for genetic variation, environmental variation, and genotype-by-environment interaction. Genetic variation is the result of differences among individuals that are encoded in their DNA. The instructions written in different individuals' genes may specify different versions of particular proteins, or they may specify the manufacture of certain proteins in different quantities or different times or places. Environmental variation arises when external factors influence how much protein is made from particular genes, or how the proteins work. When individuals experience different environments, they make different amounts of proteins or show differences in protein function. Genotype-by-environment interaction is the result of differences among individuals that are encoded into their DNA and that make them differ in their sensitivity to environmental influences. Different individuals thus react differently to a changing environment.

We will illustrate these generalizations with examples.

Genetic Variation

Humans show considerable variation in their perception of taste. One way to demonstrate this variation is to offer people small quantities of the chemical phenylthiocarbamide (PTC). Some individuals find it intensely bitter and unpleasant; others can scarcely taste it at all (Wooding 2006).

An experience of taste begins at a taste receptor protein in the surface membrane of a taste receptor cell in a taste bud on the tongue (Yarmolinsky et al. 2009). Taste receptor proteins have diverse shapes and chemical properties. Each kind of taste receptor protein binds with a subset of the chemicals in food, corresponding to sweet, sour, salty, umami (savory), and bitter flavors **(Figure 5.6a)**. When the right chemical binds to the receptors on its membrane, a taste receptor cell sends nerve impulses to the brain. The brain integrates messages from taste receptors across the tongue and generates a conscious sensation of flavor.

The receptor proteins responsible for bitter flavors are the type 2 taste receptors (TAS2Rs). The one that binds PTC is TAS2R38, encoded by a gene on chromosome 7 (see Figure 5.5). The coding region of the gene specifies a sequence of 333 amino acids. Un-kyung Kim and colleagues (2003) examined the TAS2R38 genes of a number of individuals. They found three places where different versions of the gene encode different amino acids. The 49th codon in the gene may specify either proline or alanine, the 262nd codon may specify either alanine or valine, and the 296th codon either valine or isoleucine. Different versions of a gene are called **alleles**. The most common TAS2R38 alleles, named for the amino acids they specify at the variable sites, are *AVI* and *PAV*.

Everyone has two chromosome 7s: one inherited from their mother, the other from their father. The two chromosomes may carry the same allele of the TAS2R38 gene, or they may carry different alleles. The combination of alleles an individual carries is called his or her **genotype**. Considering just alleles *AVI* and *PAV*, the three possible genotypes are *AVI/AVI*, *AVI/PAV*, and *PAV/PAV*.

The suite of traits an individual exhibits is called his or her **phenotype**. The aspect of phenotype we are interested in here is sense of taste. To show that TAS2R38 genotype influences sensory phenotype, Richard Newcomb and colleagues (2012) asked people with different genotypes to taste a standard PTC solution and rate the intensity of the flavor. Figure 5.6b presents the results. There is variation among the subjects plotted on each graph, showing that factors other than TAS2R38 genotype influence an individual's sensitivity to PTC. But TAS2R38 genotype clearly matters (see also Bufe et al. 2005). Individuals with genotype *PAV/PAV* are most sensitive, those with *AVI/AVI* are least sensitive, and those with *AVI/PAV* fall in between. Switching just 3 of the 333 amino acids in TAS2R38 changes the protein's shape and/or chemical properties enough to alter either the protein's ability to bind PTC (Figure 5.6c), its ability to trigger a nerve impulse in response to binding, or both (see Biarnés et al. 2010).

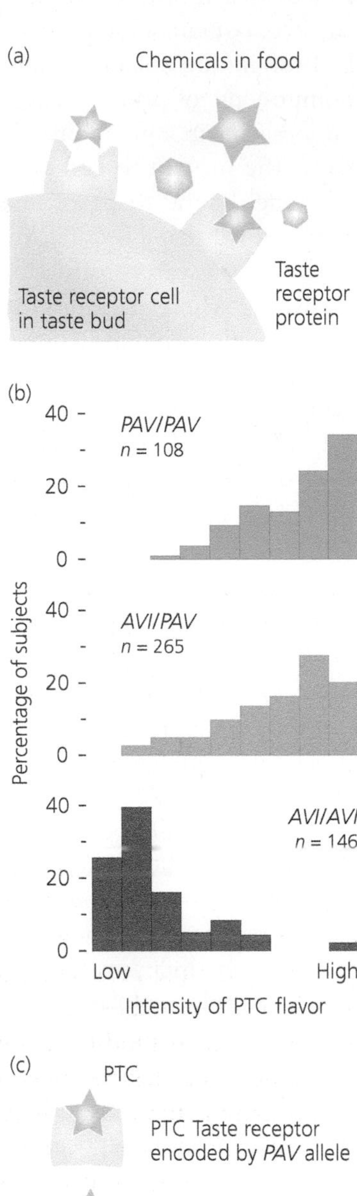

Figure 5.6 Genetic variation for bitter taste perception (a) Taste receptors bind chemicals in food. (b) PTC tastes different to people with different versions of TAS2R38, perhaps because (c) the version encoded by allele *PAV* binds PTC more strongly. Graphs from Newcomb et al. (2012).

PTC does not naturally occur in food. The ability to taste it might seem unimportant. However, different versions of TAS2R38 also respond differently to other bitter flavors (Sandell and Breslin 2006). Among these flavors are chemicals found in broccoli and its relatives, including mustard greens, turnips, and horseradish. People with genotype *PAV/PAV* rate these vegetables as more bitter than do people with genotype *AVI/AVI*. There is some evidence that *AVI/AVI* individuals eat more vegetables than individuals with other genotypes (Tepper 2008; Duffy et al. 2010; but see Gorovic et al. 2011).

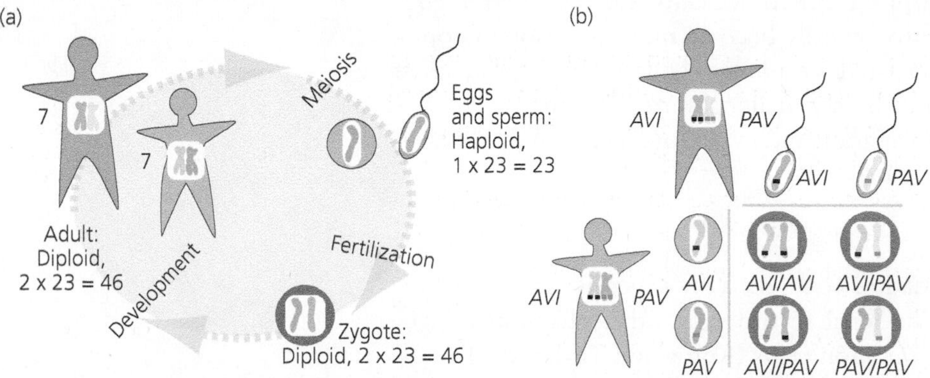

Figure 5.7 The human life cycle (a) We spend most of our lives in a diploid phase, carrying a complete set of chromosomes received via the egg and a complete set received via the sperm. Each gamete we make receives one member of each chromosome pair. (b) We can thus use a Punnett square to calculate the odds that a zygote will have a particular genotype.

To the extent that differences among individuals in the ability to taste bitter flavors are due to differences in genotype, they are transmitted from parents to offspring. **Figure 5.7a** shows the human life cycle. The figure does the bookkeeping for chromosome counts, highlighting chromosome 7 as an example. For most of our life cycle, our cells carry two chromosome 7s. When we make gametes, each egg or sperm receives a copy, selected at random, of one chromosome 7 or the other. When egg and sperm unite, they yield a zygote that once again has two chromosome 7s. If the gametes were produced by parents who both carried allele *AVI* on one chromosome 7 and allele *PAV* on the other, then all three genotypes are possible among the offspring (Figure 5.7b). In a large sample of offspring, we expect the genotypes to occur in a 1:2:1 ratio.

Genetic variation consists of differences among individuals that are encoded in the genome and transmitted from parents to offspring.

Given that the *PAV* allele tends to make people who carry it dislike broccoli and related vegetables, and that eating vegetables is good for one's health, we might expect that individuals with allele *PAV* would be less likely to survive and reproduce, and that the allele would disappear. Consider, however, that vegetables contain natural toxins—an adaptive trait that discourages animals from eating them. Consuming a healthy diet thus requires balancing one's intake of nutritious plants against one's ingestion of the toxins they contain, some of which taste bitter. That alleles *PAV* and *AVI* are both common in human populations all over the world suggests that historically the best genotype for survival and reproduction has been *AVI/PAV* (Wooding et al. 2004).

Stephen Wooding (2005, 2006) speculates that although the version of the PTC receptor encoded by allele *AVI* is less sensitive to the toxins in broccoli and its kin, it is perhaps more sensitive to toxins found in other plants. If this is so, then individuals with genotype *AVI/PAV* would be able to detect a wider variety of toxins in their food than individuals with either genotype *AVI/AVI* or *PAV/PAV*. Such individuals might have an advantage in seeking a nutritious diet that avoids an overdose of any particular plant toxin. Note that this is a hypothesis, not an established fact. It will have to be tested with careful experiments.

Genetic Variation and Evolution

We have already discussed other examples of genetic variation. We have looked at genetic variation among HIV virions in their susceptibility to the antiretroviral drug AZT as well as genetic variation among humans in the susceptibility to HIV infection. We have considered genetic variation among sticklebacks in the extent of body armor and genetic variation in fruit size in tomatoes. We will see many more examples throughout the book.

We have also discussed the role of genetic variation in evolution. Because genes are passed from parents to offspring, genetic variants associated with higher survival and reproductive success automatically become more common in populations over time, while variants associated with untimely death and reproductive failure disappear. Genetic variation is the raw material for evolution.

But there is more to the story of variation and evolution. That is why we now turn to environmental variation.

Environmental Variation

Our example of environmental variation concerns a prey species, the water flea, and its predator, a larval insect. The fossil record shows that phantom midges have been eating water fleas for 145 million years (Kotov and Taylor 2011).

Water fleas are tiny freshwater crustaceans that inhabit lakes and ponds all over the world (Lampert 2011). Among the traits that make water fleas useful for the study of environmental variation is that when conditions are auspicious they reproduce by cloning, switching to sexual reproduction only when conditions deteriorate. Also useful is that certain environmental cues trigger changes in their morphology, physiology, and behavior. Together these characteristics make it possible for researchers to expose genetically identical water fleas to different cues and get a pure look at how changes in the environment influence phenotype.

The water flea *Daphnia pulex* suffers substantial predation by phantom midge larvae, but only at certain times and places. *Daphnia pulex* is capable of developing a morphology that is well defended against phantom midges **(Figure 5.8a)**. It can nearly double the strength and thickness of its carapace and grow ridges, called neckteeth, on the back of its head (Laforsch et al. 2004). These defenses are costly, however (Hammill et al. 2008). Apparently as a result, *D. pulex* has evolved the capacity to grow anti-midge armor only when it smells midges. The water fleas in Figure 5.8 look different not because they carry different genes, but because they have been exposed to different chemical environments.

The chemical the water flea can detect emanating from phantom midges remains to be identified. Biologists refer to it by the generic term *kairomone*. *Daphnia pulex*'s growth of armor in response to phantom midge kairomone is an example of an **inducible defense**.

Hitoshi Miyakawa and colleagues (2010) suspected that to grow anti-midge armor, a water flea has to boost its production of a variety of proteins involved in development. The researchers exposed *Daphnia pulex* to kairomone from the phantom midge *Chaoborus flavicans* and compared them to genetically identical unexposed individuals. The researchers looked at the production, or **expression**, of dozens of candidate proteins they had reason to think, from earlier research, might play a role in *Daphnia*'s inducible defenses. They measured protein production indirectly, by quantifying the production of messenger RNA, the molecular intermediary that carries genetic information from the DNA in the nucleus to the ribosomes in the cytoplasm where proteins are built.

Environmental variation consists of differences among individuals due to exposure to different environments. One way environments can influence phenotype is by altering gene expression.

(a)

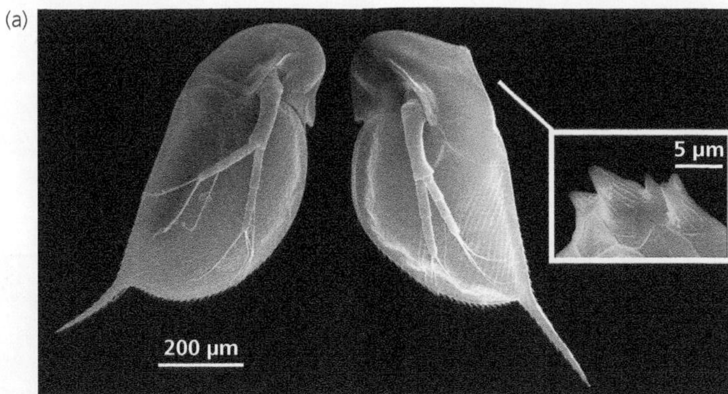

(b)

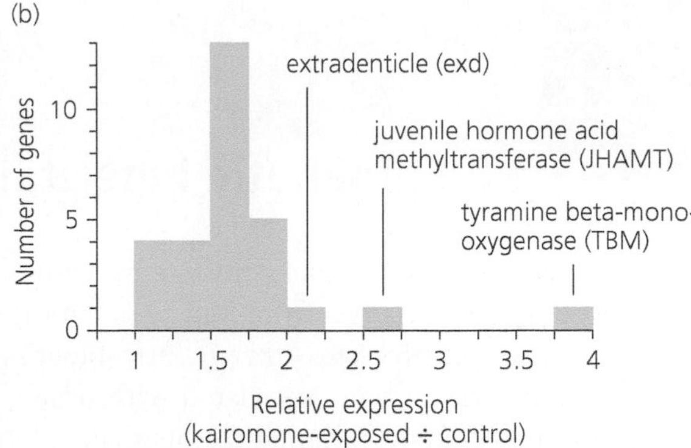

Figure 5.8 Inducible defenses in *Daphnia* (a) Juveniles that smell phantom midge larvae grow neck teeth and other defenses. Photo by Christian Laforsch. (b) This involves boosting production of many proteins. From data in Miyakawa et al. (2010).

The graph in Figure 5.8b shows the expression of 29 candidate genes in kairomone-exposed *D. pulex* relative to their expression in unexposed individuals. In every case, the exposed water fleas made more messenger RNA and thus presumably more protein. The proteins with the largest increase in expression upon exposure to kairomone were extradenticle (exd), juvenile hormone acid methyltransferase (JHAMT), and tyramine beta-monooxygenase (TBM). Exd acts during development to influence the identity of appendages in arthropods. JHAMT is an enzyme required for the synthesis of juvenile hormone, a major regulator of arthropod development. TBM is an enzyme that catalyzes the synthesis of neurotransmitters, chemicals used by nerve cells to send messages to each other.

Many details remain to be discovered, but Miyakawa's results show that the mechanism by which *D. pulex* changes its phenotype when it smells phantom midges involves changes in the production of a variety of proteins.

Environmental Variation and Evolution

Many other organisms alter the identity or quantity of the proteins they make in response to changes in the environment, thereby altering their phenotype. Human athletes living at low altitude, but training at simulated high altitude, produce more vascular endothelial growth factor (VEGF) than athletes living and training at low altitude (Hoppeler and Vogt 2001). The extra VEGF stimulates the growth of capillaries in the muscles. Environmental variation is ubiquitous.

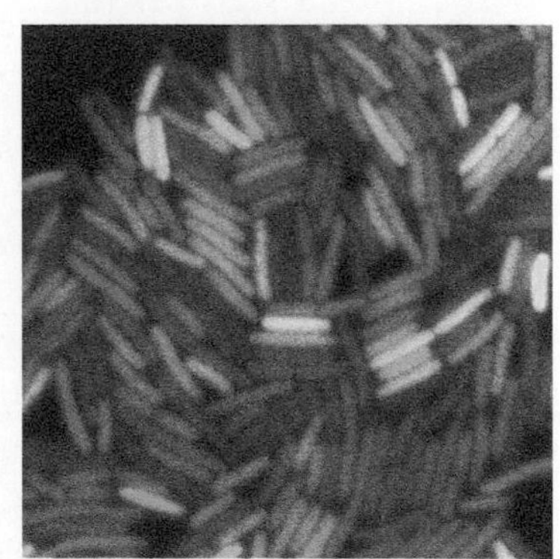

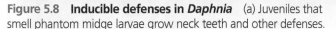

Figure 5.9 Random variation in protein production in genetically identical bacteria These cells are different colors because they are making different amounts of two fluorescent proteins. From Elowitz et al. (2002).

The non-genetic influences on protein expression, and thus phenotype, even include chance. The *Escherichia coli* bacteria in **Figure 5.9** are genetically identical. Michael Elowitz and colleagues (2002) inserted into the DNA of their common ancestor two copies of the gene for green fluorescent protein (GFP). The two copies encode distinct variants of GFP that emit different colors of light when they fluoresce. They are controlled by identical promoters—the switches that turn genes on or off. A bacterium making equal amounts of both versions of GFP would be yellow, a cell making more of one version would be green, and a cell making more of the other version would be orange. The explanation for the diversity of colors in the photo is random variation in the interactions between the promoters and the regulatory proteins that activate and deactivate them.

Despite its ubiquity, environmental variation supplies no raw material for evolution. This is because environmentally induced changes in phenotype are not transmitted to future generations. Whether a water flea born by clonal reproduction has neckteeth is determined not by the genes she inherits, but by the

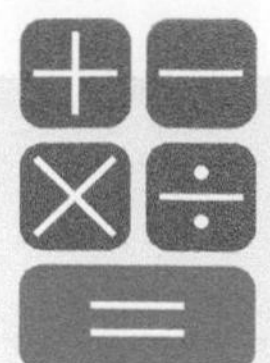

COMPUTING CONSEQUENCES 5.1

Epigenetic inheritance and evolution

Under Darwin's theory of evolution by natural selection, populations change from one generation to the next if there is particulate—that is, non-blending—inheritance of variable traits associated with fitness. There are examples of environmental factors that induce phenotypic changes subsequently transmitted to offspring (Cropley et al 2006; Li et al. 2011). Could such environmental variation provide raw material for evolution?

One mechanism of non-genetic inheritance involves the chemical modification—attachment of a methyl group—of cytosine nucleotides in DNA (Richards 2006). These and other modifications, sometimes referred to as **epigenetic marks**, are managed by enzymes encoded in the genome and can influence phenotype by altering gene expression. Epigenetic marks may be transmitted from parent to offspring because they are copied during DNA replication or even because they trigger behavior by parents that provokes their fresh establishment in offspring (Danchin et al. 2011). In some cases, including that of the toadflax variant shown in Figure 5.10, epigenetic marks can be propagated for many generations (Cubas et al. 1999; Johannes et al. 2009). Researchers working with bacteria and plants have found evidence that a few generations of selection on populations in which most of the variation is epigenetic, rather than genetic, can produce lineages with measurably distinct phenotypes (Adam et al. 2008; Hauben et al. 2009). Together these facts suggest that environmentally induced epigenetic differences could, in principle, serve as raw material for evolution.

This suggestion is tempered by consideration of the crucial functions epigenetic marks serve in organisms. There are at least three (Feng et al. 2010; Shea et al. 2011). Epigenetic marks silence transposons, integrated viruses, and other genomic parasites. They allow individuals to communicate to their offspring, and sometimes their grand-offspring, useful information about the state of the environment they are likely to encounter (Whittle et al. 2009; Scoville et al. 2011). And in multicellular organisms, they facilitate and maintain cell differentiation. Nicholas Shea and colleagues (2011) point out that the latter two functions require that epigenetic marks be periodically reset. This may explain why many epigenetic marks induced by environmental factors appear to be reprogrammed at some point in the life cycle or to decay over several generations (Feng et al. 2010; Paszkowski and Grossniklaus 2011; Shea et al. 2011). The impermanence of most epigenetic marks precludes a substantial contribution by epigenetic variation to long-term evolution (Slatkin 2009).

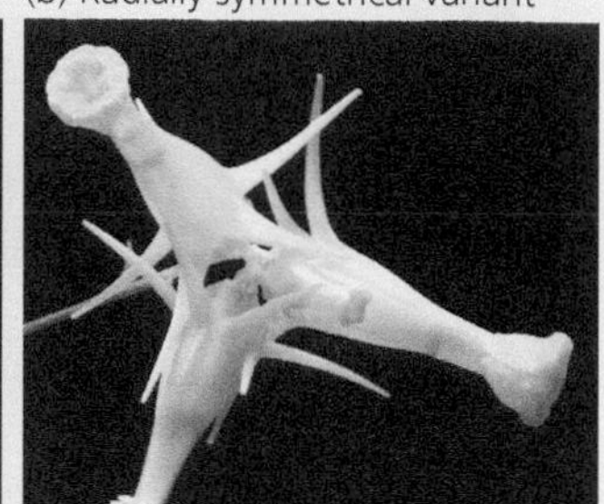

Figure 5.10 **A heritable epigenetic variant** (a) Typical flowers of common toadflax (*Linaria vulgaris*). (b) Radially symmetrical flowers from a plant of the same species. This variant was described by Linnaeus over 250 years ago and has been inherited since (Paun et al. 2010). It is caused by epigenetic marks that prevent expression of a gene called *Lcyc* (Cubas et al. 1999). From Grant-Downton and Dickinson (2006).

presence or absence of kairomones. (For exceptions to this general rule, and consideration of their implications, see **Computing Consequences 5.1**).

This is not to say that how the relationship between genotype and phenotype is altered by the environment is irrelevant to descent with modification. Indeed, as Theodosius Dobzhansky pointed out in 1937, "Selection deals not with the genotype as such, but with its dynamic properties, its reaction norm, which is the sole criterion of fitness in the struggle for existence." To understand this claim, we turn to genotype-by-environment interaction.

Genotype-by-Environment Interaction

We will start with another example of environmental variation, this time in the leopard gecko **(Figure 5.11a)**. In leopard geckos, as in many other reptiles (see Bull 1980), an individual's sex is determined largely by the temperature at which it incubates while developing in the egg. Leopard geckos that develop at cool or hot temperatures become female, whereas those that grow at intermediate temperatures tend to be male (Figure 5.11b). Development along the female versus the male pathway involves changes in the production of a variety of proteins (Shoemaker and Crews 2009). For example, expression of the gene *Sox9* in the gonad ceases earlier in leopard geckos developing as females than as males (Valleley et al. 2001). *Sox9* encodes a transcription factor that directs the expression of other genes and thus commits the gonad to being a testis versus an ovary.

Turk Rhen and colleagues (2011) wanted to know whether leopard geckos harbor genetic variation in the threshold temperatures for developing as female versus male. Analyzing data from 13 generations of geckos maintained in a breeding colony, the researchers compared the sex ratios among offspring that shared a father and hatched from eggs that had incubated at different temperatures.

Each of the dozen lines in Figure 5.11c represents the offspring of a particular father who had anywhere from 7 to 27 offspring reared at each temperature. Some offspring were full siblings and others were half sibs (with different mothers). The ends of each line show the sex ratio among the offspring reared at each temperature. Analyzing offspring with the same father but a variety of mothers allowed the researchers to statistically factor out variation due to non-genetic influences (called maternal effects) mothers might have had on their offspring via hormones, proteins, or messenger RNAs they might have placed in the eggs.

Note the variation among paternal families in the effect of temperature on sex ratio. For some families, such as the one highlighted in green, incubation at 30°C versus 32.5°C had little effect on sex ratio. For other families, such as the one in orange, incubation at different temperatures had a dramatic effect. The pattern of phenotypes an individual may develop upon exposure to different environments is called its **reaction norm**. The reaction norms in Figure 5.11c, and data on the

Genotype-by-environment interactions consists of differences among individuals, encoded in the genome, in the way the environment influences phenotype.

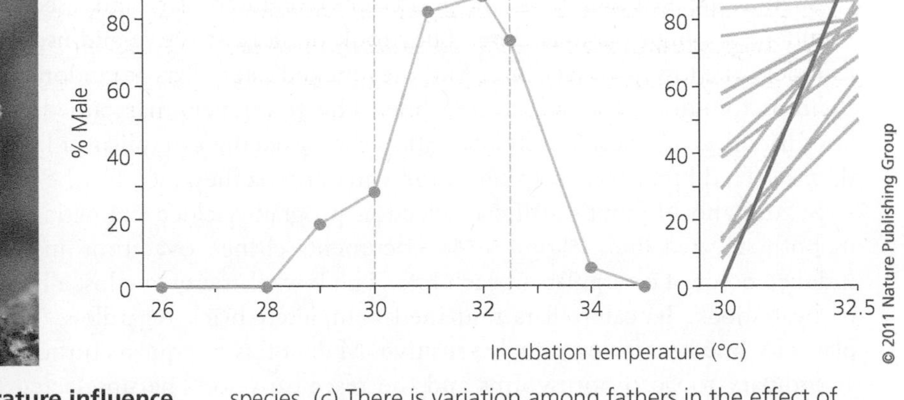

Figure 5.11 Both genotype and temperature influence sex in leopard geckos (a) A leopard gecko (*Eublepharis macularius*). (b) Individuals incubated at intermediate temperature are more likely to be male. This graph, drawn from data in Viets et al. (1993), shows the average relationship for the species. (c) There is variation among fathers in the effect of temperature on the sex ratio of their offspring. Redrawn from Rhen et al. (2011).

(c) T. Rhen, A. Schroeder, J.T. Sakata, V. Huang, D. Crews. "Segregating variation for temperature-dependent sex determination in a lizard." *Heredity* 106: 649–660. Copyright © 2011 Nature Publishing Group. Reprinted with permission.

offspring of many more sires in the breeding colony, reveal significant variation in temperature sensitivity due to the genes bequeathed by different fathers. This is a striking example of genotype-by-environment interaction.

The identities of the gene or genes responsible for variation in temperature sensitivity, and the proteins they encode, remain to be discovered.

Genotype-by-Environment Interaction and Evolution

Like genetic variation and environmental variation, genotype-by-environment interaction is common. Many instances have been documented in humans. **Figure 5.12** shows an example. Among people with genotype *ll* for the serotonin transporter gene, maltreatment in childhood has little effect on the probability of major depression in early adulthood, whereas among people with genotype *ss* maltreatment increases the probability of depression substantially (Caspi et al. 2003; Brown and Harris 2008; Caspi et al. 2010). The serotonin transporter is a cell-surface protein that mops up the neurotransmitter serotonin after nerve cells in the brain have used it to send messages to each other. The two alleles of the gene encode identical versions of the protein, but the *l* allele specifies production of the transporter in higher quantities. Individuals with genotype *ss* make less serotonin transporter, and are more sensitive to trauma in childhood.

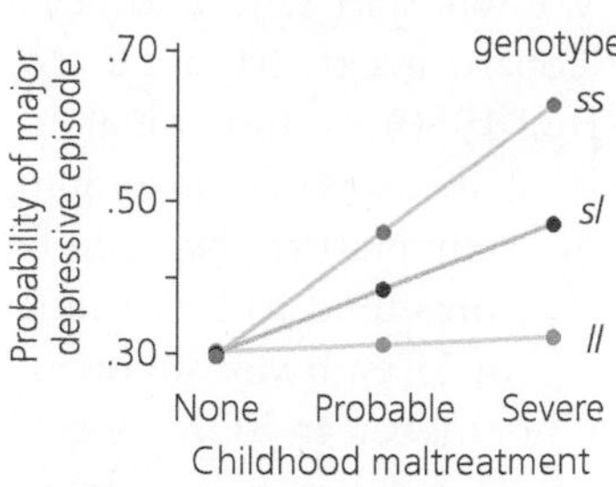

Figure 5.12 Human genotype-by-environment interaction People with different genotypes differ in sensitivity to maltreatment during childhood. Redrawn from Caspi et al. (2003).

An organism that develops different phenotypes in different environments is said to exhibit **phenotypic plasticity**. When populations harbor genetic variation for environmental sensitivity, populations can evolve greater or lesser plasticity. Documentation of this claim comes from an elegant study by Yuichiro Suzuki and Frederik Nijhout on tobacco hornworms (*Manduca sexta*).

Ordinary tobacco hornworm caterpillars are green **(Figure 5.13a)**. Suzuki and Nijhout (2006) worked with a laboratory strain in which the caterpillars are black (Figure 5.13b). They are black, that is, unless they are exposed to high temperature shortly before molting. After a heat shock, they may emerge from molting with green coloration. Figure 5.13c shows the variation in color among individuals in Suzuki and Nijhout's laboratory population following exposure to 42°C. Note that some individuals are highly sensitive to heat shock. These are the ones with a color score of 3. They turned nearly as green as ordinary caterpillars, which have a color score of 4. Other individuals are insensitive to heat shock. They remained black, earning a color score of 0. This variation in sensitivity is an example of genotype-by-environment interaction.

(a) Typical caterpillar

(b) Black strain caterpillar

(c) Response of black caterpillars to heat shock

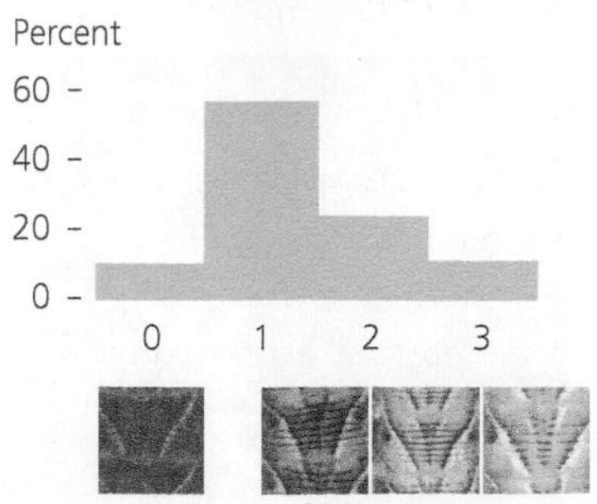

Figure 5.13 Insect genotype-by-environment interaction (a) A normal tobacco hornworm. (b) A black mutant. From Pennisi (2006). (c) Some black mutants turn green after heat shock. Redrawn from Suzuki and Nijhout (2006).

Suzuki and Nijhout took the most sensitive caterpillars and used them as founders of a high-plasticity line. They took the least sensitive and used them as founders of a low-plasticity line. And they picked caterpillars at random and used them as founders of an unselected line. The researchers maintained the three lines for 13 generations. Each generation they gave the caterpillars a heat shock, then selected breeders according to the same criteria they used for the founders.

Suzuki and Nijhout's artificial selection program yielded dramatic evolution in both selected lines. **Figure 5.14a** documents change over time in the heat-induced color of caterpillars in each line. The low-plasticity line lost all sensitivity to heat shock. Its caterpillars remained completely black regardless. The high-plasticity line became extremely sensitive. Many of its caterpillars turned as green as ordinary tobacco hornworms and the rest nearly so. The unselected line, as expected, retained roughly the same sensitivity over time.

Figure 5.14b compares the reaction norms of the three lines in the 13th generation. To prepare this graph, Suzuki and Nijhout reared caterpillars at a variety

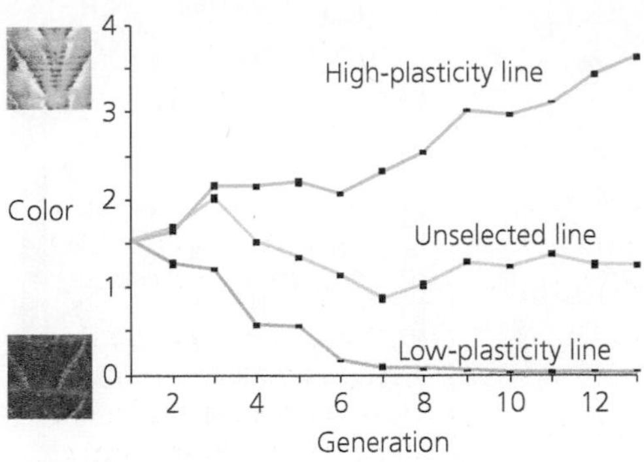

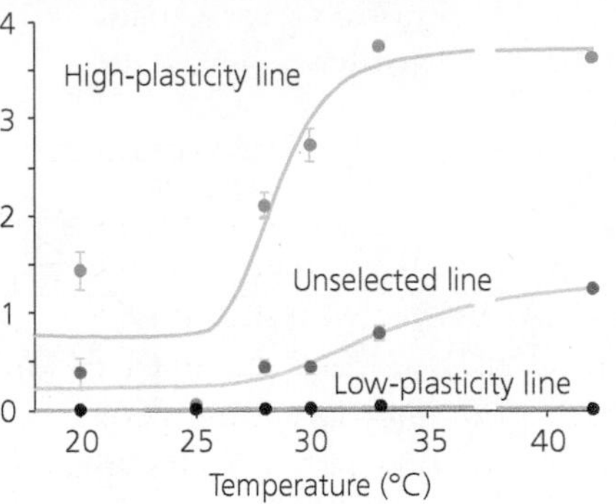

© 2006 AAAS

Figure 5.14 Genotype-by-environment interaction and the evolution of reaction norms (a) Black mutant strains of tobacco hornworms evolve in response to artificial selection for sensitivity to heat shock. (b) Lines selected for different sensitivities have different color versus temperature reaction norms. From Suzuki and Nijhout (2006).

From "Evolution of a polyphenism by genetic accommodation." *Science* 311: 650–652. Reprinted with permission from AAAS.

of temperatures without heat shock and recorded their colors. They also reared caterpillars at 28°C and heat shocked them at 42°C. The reaction norms are markedly different. Even without heat shock, the high-plasticity line developed different colors at different temperatures. Its caterpillars were mostly black if reared at 25°C or less, and green if reared at 33°C. The low-plasticity line was invariably black. The unselected line is intermediate, showing modest change in color at a somewhat higher threshold than the high-plasticity line.

With subsequent breeding experiments, Suzuki and Nijhout (2008) demonstrated that the distinctive reaction norms of the high- and low-plasticity lines are due to genetic differences at one locus of large effect and many loci of smaller effect. The identity of the loci, and the proteins they encode, remain unknown.

Suzuki and Nijhout's work demonstrates that genotype-by-environment interaction can serve as raw material for the evolution of different reaction norms. A population living in a variable environment in which different phenotypes are adaptive at different times and places can evolve a plastic response that allows individuals to develop phenotypes suitable for the conditions in which they find themselves. And a population living in an environment where the same phenotype is always adaptive can evolve low sensitivity. We can now see why Theodosius Dobzhansky viewed reaction norms as key traits influencing fitness.

Because of their heritable component, genetic variation and genotype-by-environment interaction are raw material for evolution by natural selection.

Looking Ahead

In this section we have seen that genetic variation and genotype-by-environment interaction serve as raw material for evolution. Genotype-by-environment interaction furthermore allows evolution of the pattern of environmental variation. Both genetic variation and genotype-by-environment interaction are ultimately due to differences in the genome. In the coming sections, we consider how differences in the genome arise. Changes in the genome, known as **mutations**, range in size from substitutions of one base for another to gains and losses of chromosomes. We will start small and work our way up.

5.2 Where New Alleles Come From

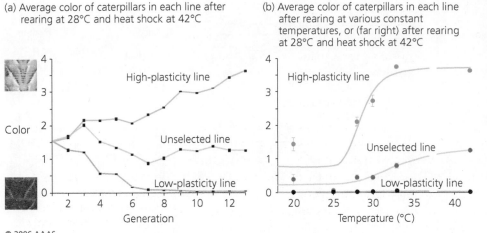

New alleles arise from alterations to existing alleles. Complete coverage of the mechanisms that can make such alterations happen is beyond the scope of this chapter. However, we can get a general understanding of mutation and its

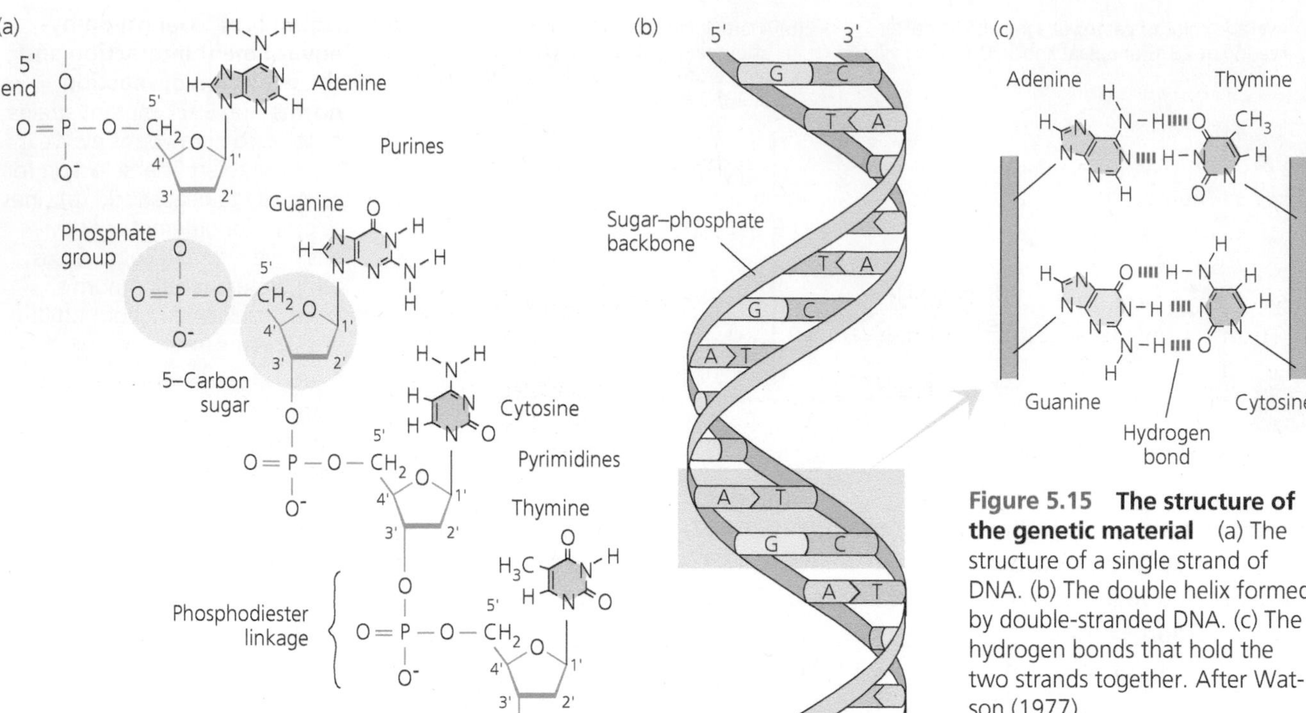

Figure 5.15 The structure of the genetic material (a) The structure of a single strand of DNA. (b) The double helix formed by double-stranded DNA. (c) The hydrogen bonds that hold the two strands together. After Watson (1977).

consequences by considering the structure and function of DNA in somewhat more detail than we have already.

As shown in **Figure 5.15a**, a single-stranded DNA molecule is built on a sugar–phosphate backbone held together by covalent bonds called phosphodiester linkages. Double-stranded DNA consists of two such strands twisted around each other (Figure 5.15b). The two strands are held together by hydrogen bonds between the purines (A and G) and pyrimidines (T and C) on opposite strands. A and T form two hydrogen bonds, whereas G and C form three (Figure 5.15c).

When James Watson and Francis Crick deduced that this was the structure of DNA, they realized that complementary base pairing provides a mechanism for copying the hereditary material (Watson and Crick 1953a, 1953b). As **Figure 5.16** shows, the two strands unzip via the release of their hydrogen bonds. Each strand then serves as a template for duplication of the other. The result is two identical double-stranded DNAs. The enzymes that copy DNA in cells are called DNA polymerases. The first to be discovered was isolated by Arthur Kornberg in 1960.

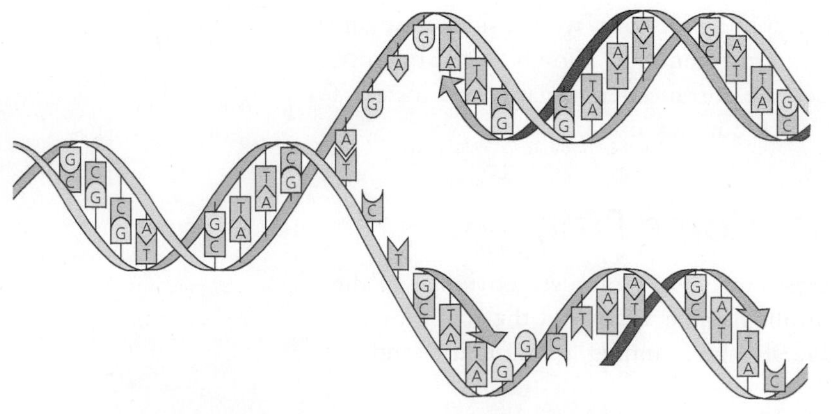

Figure 5.16 DNA replicates by complementary base pairing Modified from an illustration by Madeleine Price Ball.

Mutations creating new alleles can arise as a result of alterations to DNA that escape repair before or during replication, or because of errors that occur during replication itself and escape repair afterward.

For an example mutation due to DNA alteration, a cytosine that has already been chemically modified by the addition of a methyl group (CH_3) will sometimes spontaneously react with water, lose an amine group (NH_2), and thereby transform into thymine (**Figure 5.17**; Lindahl 1993; Arnheim and Calabrese 2009). If the resulting mismatched T-G base pair is not recognized and corrected before replication, one of the resulting DNA molecules will possess a T-A pair in substitution for the ancestral C-G pair **(Figure 5.18)**.

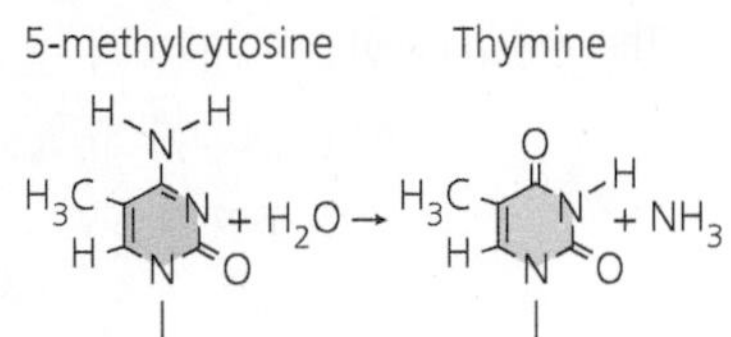

Figure 5.17 Spontaneous deamination Methylated cytosines can react with water to become thymines.

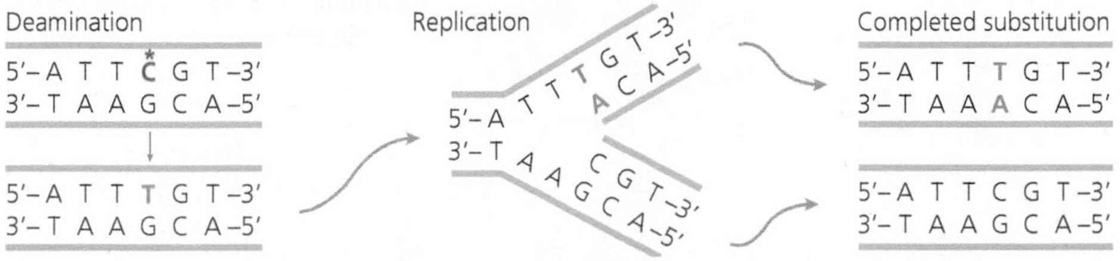

Deamination

5′–A T T Č G T–3′
3′–T A A G C A–5′

5′–A T T T G T–3′
3′–T A A G C A–5′

Replication

5′–A T T T G T–3′
A C A–5′
3′–T A A G C A–5′
C G T–3′

Completed substitution

5′–A T T T G T–3′
3′–T A A A C A–5′

5′–A T T C G T–3′
3′–T A A G C A–5′

Figure 5.18 Mutation by deamination If a cytosine that has turned into a thymine goes undetected until the next replication, the mutation becomes permanent.

For an example of mutation due to copying error, during replication the template and nascent DNA strands can become misaligned, particularly where the same base is repeated many times, resulting in the insertion or deletion of nucleotides (Pearson et al. 2005; Garcia-Diaz and Kunkel 2006). If the insertion or deletion evades repair before the next replication, the insertion or deletion becomes permanent in one of the daughter DNA molecules **(Figure 5.19)**.

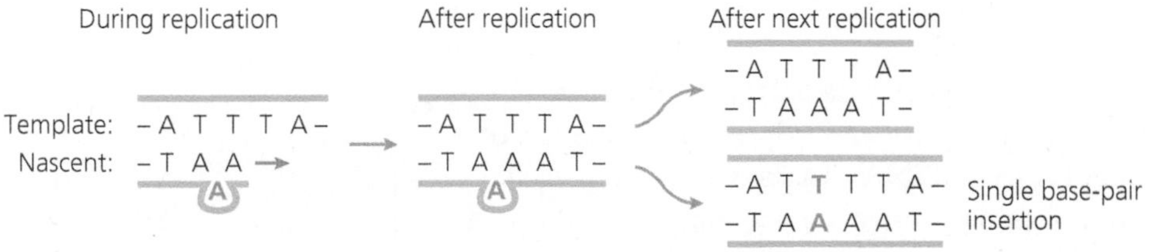

During replication

Template: –A T T T A–
Nascent: –T A A →
A

After replication

–A T T T A–
–T A A A T–
A

After next replication

–A T T T A–
–T A A A T–

–A T T T T A–
–T A A A A T–
Single base-pair insertion

Figure 5.19 Mutation by misalignment The template and nascent DNA strand can slip out of register at repeat sites, resulting in duplications and deletions.

We have noted that alterations to DNA due to chemical degradation and replication errors must evade correction to become persistent mutations. DNA alterations still susceptible to repair, like those shown at the left of Figures 5.18 and 5.19, are known as **premutations**. Premutations are appallingly common. A typical mammalian cell suffers roughly 20,000 cases of spontaneous chemical decay in its genome each day, and 100,000 replication errors per division (Preston et al. 2010). Fortunately, most premutations are identified and fixed by a battery of enzymes encoded in the genome. These include mismatch repair and proofreading enzymes. The value of these enzymes is revealed by their loss. Tina Albertson and colleagues (2009) studied genetically engineered mice whose DNA polymerases lacked the ability to proofread and correct newly synthesized DNA. The mice suffered high rates of cancer and had dramatically reduced life spans **(Figure 5.20)**. Deficient DNA maintenance and repair also appears to be an underlying defect in a variety of human cancers (Loeb 2011).

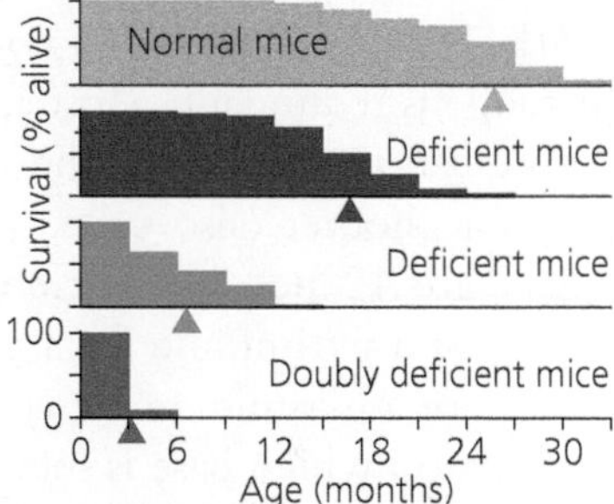

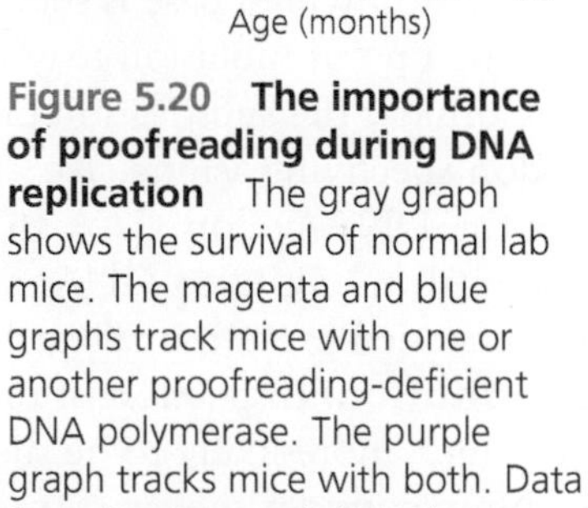

Figure 5.20 The importance of proofreading during DNA replication The gray graph shows the survival of normal lab mice. The magenta and blue graphs track mice with one or another proofreading-deficient DNA polymerase. The purple graph tracks mice with both. Data from Preston et al. (2010).

How Mutations Alter Protein Function

Mutations to new alleles can influence phenotypes if they alter the expression and/or function of proteins. When a cell makes a protein from the instructions

(a) The flow of genetic information

	Example				
DNA	C A A	T T C	C C G	A C A	G G A
Transcription ↓	↓	↓	↓	↓	↓
mRNA	G U U	A A G	G G C	U G U	C C U
Translation ↓	↓	↓	↓	↓	↓
Protein	Valine	Lysine	Glycine	Cysteine	Proline

(b) The RNA genetic code

First base	Second base: U	Second base: C	Second base: A	Second base: G	Third base
U	UUU Phenylalanine	UCU Serine	UAU Tyrosine	UGU Cysteine	U
	UUC Phenylalanine	UCC Serine	UAC Tyrosine	UGC Cysteine	C
	UUA Leucine	UCA Serine	UAA Stop	UGA Stop	A
	UUG Leucine	UCG Serine	UAG Stop	UGG Tryptophan	G
C	CUU Leucine	CCU Proline	CAU Histidine	CGU Arginine	U
	CUC Leucine	CCC Proline	CAC Histidine	CGC Arginine	C
	CUA Leucine	CCA Proline	CAA Glutamine	CGA Arginine	A
	CUG Leucine	CCG Proline	CAG Glutamine	CGG Arginine	G
A	AUU Isoleucine	ACU Threonine	AAU Asparagine	AGU Serine	U
	AUC Isoleucine	ACC Threonine	AAC Asparagine	AGC Serine	C
	AUA Isoleucine	ACA Threonine	AAA Lysine	AGA Arginine	A
	AUG Start (Methionine)	ACG Threonine	AAG Lysine	AGG Arginine	G
G	GUU Valine	GCU Alanine	GAU Aspartic acid	GGU Glycine	U
	GUC Valine	GCC Alanine	GAC Aspartic acid	GGC Glycine	C
	GUA Valine	GCA Alanine	GAA Glutamic acid	GGA Glycine	A
	GUG Valine	GCG Alanine	GAG Glutamic acid	GGG Glycine	G

Codon Amino acid — nonpolar | polar | basic | acidic

Figure 5.21 The genetic code (a) Genetic information flows from DNA to mRNA to protein. Note that RNA contains a nitrogenous base called uracil instead of thymine. An adenine in DNA specifies a uracil in RNA. (b) In the genetic code, each of the 64 mRNA codons specifies an amino acid or the start or end of a transcription unit. Note that in many instances, changing the third base in a codon does not change the message.

encoded in a gene, it follows a two-step process **(Figure 5.21a)**. In the first step, called transcription, it uses the DNA as a template and copies the sequence of bases into a complementary messenger RNA (mRNA). In the second step, called translation, it uses the mRNA as a template to construct a protein.

The cell reads mRNA as a series of three-letter codons, each specifying a particular amino acid. The genetic code used by most organisms appears in Figure 5.21b. Because there are 64 codons to specify 20 amino acids (plus start and stop), the code is redundant. Most amino acids are specified by more than one codon.

We can now see why the smallest possible mutation, the substitution of one base for another (also called a **point mutation**) can have a variety of effects. There are 12 possible nucleotide substitutions. Substitution of a purine for a purine or a pyrimidine for a pyrimidine is called a **transition**. Substitution of a purine for a pyrimidine, or vice versa, is called a **transversion (Figure 5.22)**. Depending on which base is substituted for which, in which position, and in which codon, a point mutation may have no effect, a subtle effect, or a drastic effect.

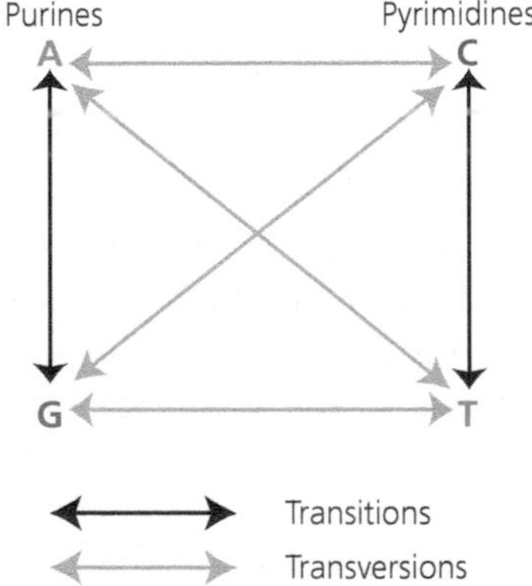

Figure 5.22 Transitions and transversions

Consider the substitution of any base for any other in the third position of any codon specifying valine. An example would be an A-to-T transversion changing the DNA codon CAA to CAT. This changes the complementary mRNA codon from GUU to GUA, which still specifies valine. Such a mutation leaves the encoded protein unaltered, and is thus known as a **synonymous** (or **silent**) **substitution**. Inspection of the code will reveal that third-position substitutions are much more likely to be silent than first- and second-position substitutions.

Now consider the substitution of an alternative nucleotide into the first position of a codon specifying alanine. An example would be a C-to-G transversion changing the DNA codon from CGA to GGA. This changes the complementary mRNA codon from GCU to CCU, which specifies proline. A mutation that changes the amino specified by a codon is known as a **nonsynonymous** (or **replacement**) **substitution**. Switching an amino acid may alter the function of a protein. For example, having a proline versus an alanine as the 49th amino acid in taste receptor TAS2R38 influences a person's ability to taste bitter flavors.

A mutation is any change in sequence in the genome of an organism. Some mutations alter the phenotype; others do not.

Finally, consider a substitution that can occur in the third position of the codon specifying tryptophan. A C-to-T transition changes the DNA codon from ACC to ACT. This changes the mRNA codon from UGG to UGA. UGA is a stop codon. It signals that the protein is complete and no more amino acids should be added. A mutation that introduces a premature stop codon is called a **nonsense mutation**. Nonsense mutations often render the encoded protein nonfunctional (Yamaguchi-Kabata et al. 2008). Many humans carry loss-of-function nonsense mutations in both of their copies of the gene for the muscle protein alpha-actinin-3 (North et al. 1999). The nonsense allele is overrepresented in elite endurance athletes and underrepresented in elite sprint and strength athletes (Niemi and Majamaa 2005; Roth et al. 2008).

Changing the meaning of a codon is not the only way a point mutation can alter protein function or expression. Many genes in eukaryotes contain intervening sequences, or **introns**, embedded among the coding sequences, or **exons**. The introns are transcribed into the mRNA and must be spliced out before translation. Mutations in splice sites can prevent introns from being excised, resulting in production of abnormal proteins. Janna Nousbeck et al. (2011) discovered a splice-site mutation in humans that causes adermatoglyphia—the absence of fingerprints (**Figure 5.23**). Mutations in the promoter regions of genes, non-coding sequences that play a role in gene regulation, can alter gene expression.

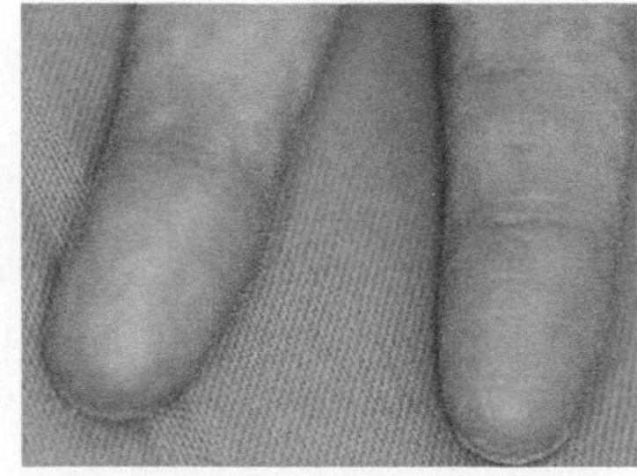

Figure 5.23 Some people lack fingerprints The condition is called adermatoglyphia, and is informally known as immigration delay disease. From Nousbeck et al. (2011).

Like point mutations, insertions and deletions (collectively called **indels**) vary in their effects. And as with point mutations, the genetic code shows why. Insertion or deletion in a coding region of one, two, or any other number of nucleotides not a multiple of three results in a shift of the codon reading frame. This changes the meaning of every codon downstream from the mutation.

The mutational mechanisms we have considered in this section stock populations with a diversity of alleles. In the next few chapters, we will be concerned with the relative frequencies of different alleles in populations. **Computing Consequences 5.2 (next page)** shows how allele frequencies can be quantified.

5.3 Where New Genes Come From

As with mutations that create new alleles, many mechanisms generate new genes (Long et al. 2003; Kaessmann 2010). We cannot cover them all, but we can get a sense of where new genes come from by considering a few examples.

Gene Duplication

Two mechanisms of **gene duplication** are thought to be among the most common sources of new genes. The first is **unequal crossing over**, an error in the genetic recombination that happens during meiosis. In normal crossing over, homologous chromosomes (the maternal and paternal members of a pair) align side by side during prophase of meiosis I and exchange stretches of DNA containing the same loci. In unequal crossing over, the homologous chromosomes align incorrectly. This can happen if the same nucleotide sequence occurs in more than

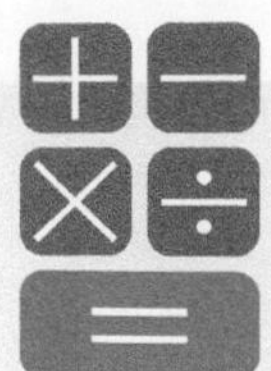

COMPUTING CONSEQUENCES 5.2

Measuring genetic variation in natural populations

Classical views of genetic variation held that one allele of each gene should confer higher fitness than all others. Natural selection should preserve the best allele, called the wild type, and eliminate the rest—which were considered mutants and expected to be rare.

Pioneering work by Harris (1966) and Lewontin and Hubby (1966) revealed that most populations harbor many alleles. The deeper biologists have looked for allelic variation, the more they have found. Today, evolutionary biologists recognize that the vast majority of natural populations harbor substantial genetic variation.

Determining Genotypes

The first step in measuring the diversity of alleles present at a particular gene is to determine the genotypes of a large sample of individuals in a population. To do so, biologists usually look directly at the DNA.

As an example, consider the human *CCR5* gene, which encodes a cell-surface protein used as a coreceptor by sexually transmitted strains of HIV-1. One *CCR5* allele has a 32-base-pair deletion in the coding sequence. As a result, the encoded protein is shortened and nonfunctional. We will call the functional allele *CCR5-+*, or just +, and the allele with the 32-base-pair deletion *CCR5-Δ32*, or just *Δ32*. Individuals with genotype +/+ are susceptible to infection with HIV-1, individuals with genotype +/*Δ32* are susceptible but may progress to AIDS more slowly, and individuals with genotype *Δ32*/*Δ32* are resistant.

Upon learning of the *CCR5-Δ32* allele, AIDS researchers immediately wanted to know how common it is. Michel Samson and colleagues (1996) developed a genotype test. Researchers extract DNA from a sample of the subject's cells, then use a polymerase chain reaction (PCR) to make copies of a region of the gene, several hundred base pairs long, that contains the site of the 32-base-pair deletion. (PCR duplicates a targeted sequence many times over by using a test-tube DNA replication system in which specifically tailored primer sequences direct DNA polymerase to copy just the locus of interest.) Finally, the researchers cut the duplicated sequences with a restriction enzyme and run the resulting fragments on an electrophoresis gel.

Results appear in Figure 5.24. Both alleles yield two DNA fragments. The fragments from a *CCR5-+* allele are 332 and 403 base pairs long. Those from a *CCR5-Δ32* allele are 332 and 371 base pairs long. Homozygotes have just two bands in their lane on the gel, whereas heterozygotes have three bands.

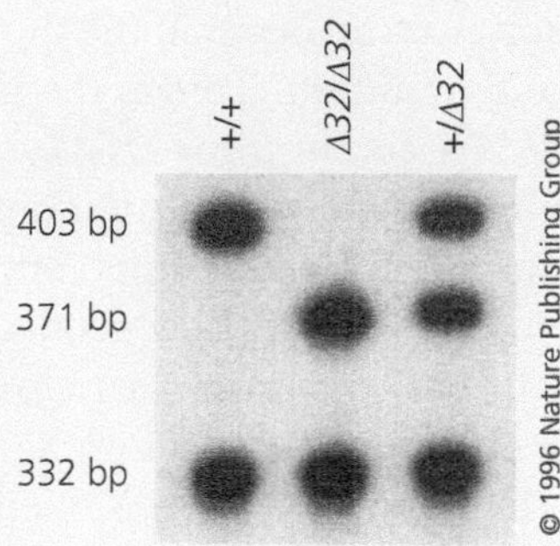

Figure 5.24 Determining CCR5 genotypes Each lane holds fragments made from DNA of one person. The locations of the spots, or bands, show the fragments' sizes. Samson et al. (1996). Reprinted by permission from Macmillan Publishers Ltd: *Nature* 382: 722–725, copyright 1996.

Calculating Allele Frequencies

Now that we can determine who carries the *Δ32* allele, we want to know how common it is. To find out, we use data on the genotypes of individuals to calculate the **frequency** of the *Δ32* allele in populations. The frequency of an allele is its fractional representation among all the gene copies present in a population.

We will calculate the frequency of the *Δ32* allele in an Ashkenazi population studied by Jeremy Martinson and colleagues (1997). Martinson's sample contained 43 individuals: 26 with genotype +/+, 16 with genotype +/*Δ32*, and 1 with genotype *Δ32*/*Δ32*. The simplest way to calculate allele frequencies is to count allele copies. Martinson and colleagues tested 43 individuals. Each individual carries 2 gene copies, so the researchers tested a total of $2 \times 43 = 86$ copies. Of these 86 gene copies, 18 were copies of the *Δ32* allele: 1 from each of the 16 heterozygotes, and 2 from the single homozygote. The frequency of the *Δ32* allele in the Ashkenazi sample is the number of *Δ32* copies divided by the total number of gene copies:

$$\frac{18}{86} = 0.209$$

or 20.9%. We can check our work by calculating the frequency of the + allele. It is

$$\frac{(52 + 16)}{86} = 0.791$$

or 79.1%. If our calculations are correct, the frequencies of the two alleles should sum to one, which they do.

An alternative method of figuring the allele frequencies is to calculate them from the genotype frequencies. Martinson and colleagues tested 43 individuals, so the genotype frequencies are as follows:

$+/+$	$+/\Delta 32$	$\Delta 32/\Delta 32$
$\frac{26}{43} = 0.605$	$\frac{16}{43} = 0.372$	$\frac{1}{43} = 0.023$

The frequency of the $\Delta 32$ allele is the frequency of $\Delta 32/\Delta 32$ plus half the frequency of $+/\Delta 32$:

$$0.023 + \tfrac{1}{2}(0.372) = 0.209$$

This is the same value we got by the first method. It is an unusually high frequency for the $\Delta 32$ allele.

How Much Genetic Diversity Exists in a Typical Population?

Studies on allelic diversity, similar to the work on the frequency of the $\Delta 32$ allele in humans, have been done in a wide variety of populations. Biologists use two statistics to summarize these types of data: the mean heterozygosity, and the percentage of polymorphic genes. The **mean heterozygosity** can be interpreted in two equivalent ways: as the average frequency of heterozygotes across loci, or as the fraction of genes that are heterozygous in the genotype of the average individual. The **percentage of polymorphic loci** is the fraction of genes in a population that have at least two alleles.

Early efforts to study allelic diversity used a technique called allozyme electrophoresis. Researchers isolated proteins from a large sample of individuals, separated the proteins in an electrophoresis gel, and then stained the gel to visualize the proteins produced by a particular gene. If the allelic proteins in a population were different enough in amino acid sequence that they had different sizes or charges, then they would migrate differently in the gel and appear as distinct bands.

Allozyme electrophoresis studies showed that most natural populations harbor substantial genetic variation. Figure 5.25 summarizes data on mean heterozygosities from invertebrates, vertebrates, and plants. As a broad generalization, in a typical natural population, between 33% and 50% of the genes that code for enzymes are polymorphic, and the average individual is heterozygous at 4%–15% of its genes (Mitton 1997).

Because not every change in the DNA sequence of a gene produces an electrophoretically distinguishable protein, methods that directly examine the DNA of alleles are much more powerful for revealing diversity. Among the most intensively studied genes is the one associated with cystic fibrosis in humans. This gene encodes a cell-surface protein, called the cystic fibrosis transmembrane conductance regulator (CFTR), that is expressed in the mucous membranes lining the intestines and lungs. Individuals homozygous for loss-of-function mutations have cystic fibrosis and suffer chronic infections. Geneticists have found more than 1,400 different disease-causing mutations at this one locus (Kreindler 2010). These alleles include amino acid substitutions, frameshifts, splice-site mutations, nonsense mutations, indels, and promoter variants (CFMD 2011).

This work and similar studies show that the amount of genetic variation in most populations is high (Ju et al. 2011; Nelson et al. 2012). Among the full genomes of 1,092 humans, Goncalo Abecasis and colleagues (2012) found 38 million single-nucleotide polymorphisms and over 1.4 million insertions and deletions. The old view of genetic diversity, under which little variation was expected in most populations, is wrong.

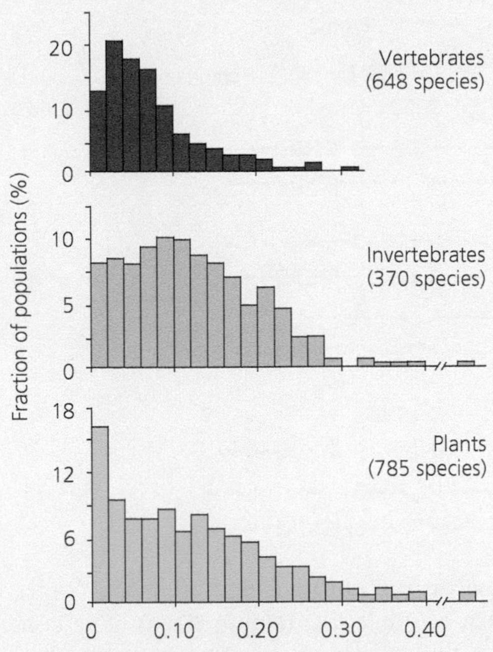

Figure 5.25 Analysis of proteins reveals that most populations harbor considerable genetic diversity These histograms show the distribution of enzyme heterozygosities among species of animals and plants. For example, about 7% of all plant species have a heterozygosity between 0.10 and 0.12. Data from John C. Avise (1994).

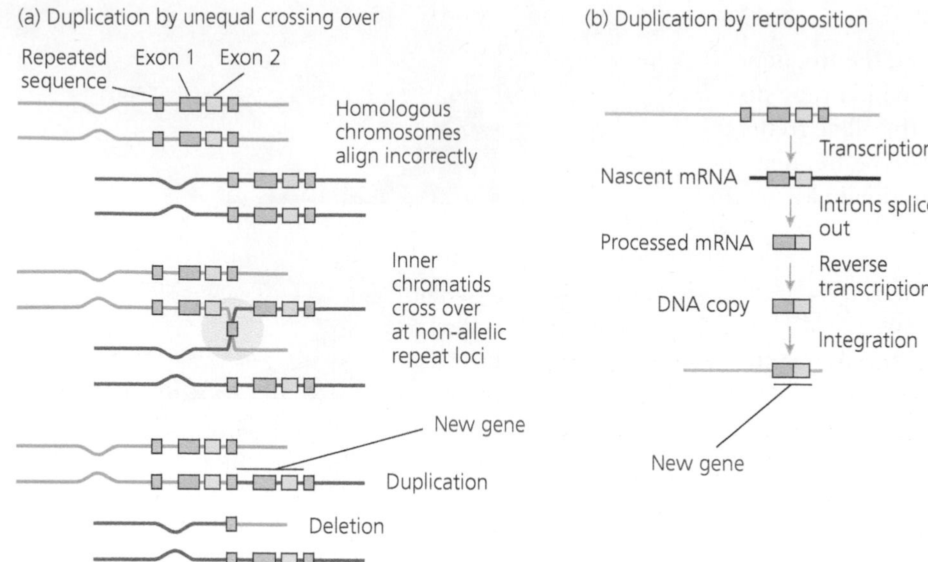

Figure 5.26 Two mechanisms of gene duplication (a) Unequal crossing over during meiosis. (b) Retroposition.

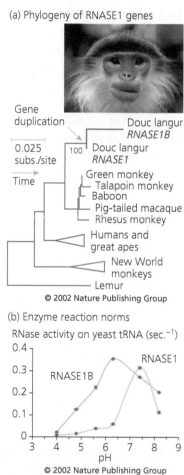

one place on the chromosome, as when copies of a transposable element have inserted at multiple loci (**Figure 5.26a**). The consequence of misalignment is that the DNA segments exchanged are out of register. One of the participating chromatids ends up with a duplication, while the other sustains a deletion.

The second mechanism, depicted in Figure 5.26b, is called **retroposition** or **retroduplication**. Retroposition begins when a processed messenger RNA, from which the introns have been spliced out, is reverse-transcribed by the enzyme reverse transcriptase to form a double-stranded segment of DNA. If this DNA segment integrates into one of the main chromosomes, the genome acquires a duplicated copy of the original gene. In many cases the new copy is a nonfunctional **pseudogene**, because it lacks regulatory sequences that cause it to be transcribed. If, however, the duplicate inserts near existing regulatory sequences, subsequently acquires them via a transposable element insertion, or evolves them from scratch, it may become a functional gene.

Retroposition and unequal crossing over leave distinctive footprints in the genome. Among other clues to their origin, retroduplicated genes lack introns and are usually found far from the original gene. Genes that were duplicated by unequal crossing over, in contrast, contain the same introns as their parental genes and are found in tandem with them on the same chromosome.

A New Gene Generated by Unequal Crossing Over

An example of a gene created by unequal crossing over comes from work by Jianzhi Zhang and colleagues (2002) on the douc langur (*Pygathrix nemaeaus*) of Southeast Asia. The douc has an unusual diet for a monkey: It eats leaves. The leaves are fermented by bacteria living in the monkey's foregut. Further along the digestive tract, the monkey digests the bacteria and absorbs the nutrients they contain. Like ruminants, which have a similar diet and digestive strategy, the douc maintains a relatively high concentration of RNASE1 in its foregut. RNASE1 is an enzyme, made by the pancreas, that breaks down RNA. This liberates the nitrogen in the RNA for recycling by the monkey's own metabolism.

Zhang and colleagues examined the genes for RNASE1 in douc langurs and other primates. Most primates have just one locus encoding RNASE1, but doucs have two. Zhang named the second enzyme RNASE1B. **Figure 5.27a** displays an

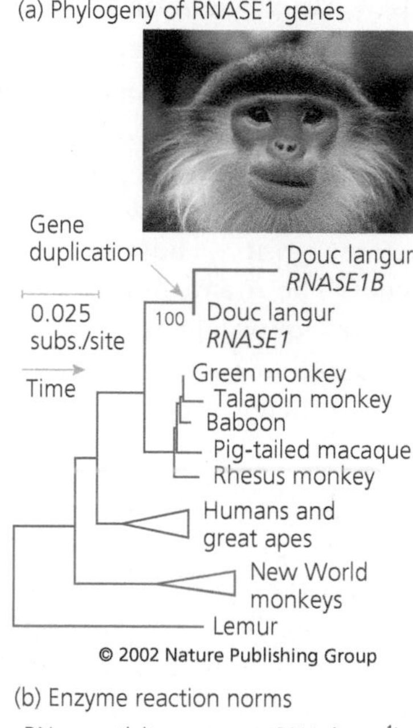

Figure 5.27 A new gene from unequal crossing over (a) Phylogeny of RNASE1 genes. (b) RNASE1 reaction norms.

(a, b) Reprinted by permission from Macmillan Publishers Ltd: Zhang, J., Y. P. Zhang, and H. F. Rosenberg. 2002. "Adaptive evolution of a duplicated pancreatic ribonuclease gene in a leaf-eating monkey." *Nature Genetics* 30: 411–415.

evolutionary tree, based on nucleotide sequence data, for the singleton RNASE1 genes of 15 primates, plus the two RNASE1 genes of doucs. The langur genes are each other's closest relatives. The simplest explanation is that the RNASE1B gene in doucs arose as a recent duplication of the monkey's RNASE1 gene. The two langur genes each have an intron, and their introns are nearly identical. This suggests that the duplication arose by unequal crossing over.

Zhang and colleagues found nine amino acid substitutions distinguishing the proteins encoded by the douc langur's two RNASE1 genes. To see whether the enzymes had diverged in function, the researchers tested their RNase activity at a variety of pHs. The reaction norms appear in Figure 5.27b. RNASE1B is more active at the relatively low pHs characteristic of the douc langur small intestine. RNASE1, on the other hand, retains an optimal pH similar to that of other primate RNASE1s. In addition, the researchers found that douc RNASE1B has lost the ability to break down double-stranded RNA, a capacity seen in other RNASE1s and thought to play a role in defense against viruses.

These results are consistent with the hypothesis that following the duplication creating RNASE1B, the new gene evolved to encode an enzyme specialized for the digestive demands of the douc langur's unusual diet, while the parental gene retained its ancestral, generalist function. Similar duplications and specializations of RNASE1 genes appear to have occurred independently in ruminants (Zhang 2003) and African leaf-eating monkeys (Zhang 2006; Yu et al. 2010).

Genes that are duplicated within a genome and later diverge in function, like RNASE1 and RNASE1B in douc langurs, are described as **paralogous**. Paralogous genes can be contrasted with **orthologous** genes. These are genes that are derived from a common ancestral sequence and separated by a speciation event, like RNASE1 in douc langurs and RNASE1 in humans.

(a) Who's a good girl?

(b) *Fgf4* retrogene genotypes in dogs of various breeds

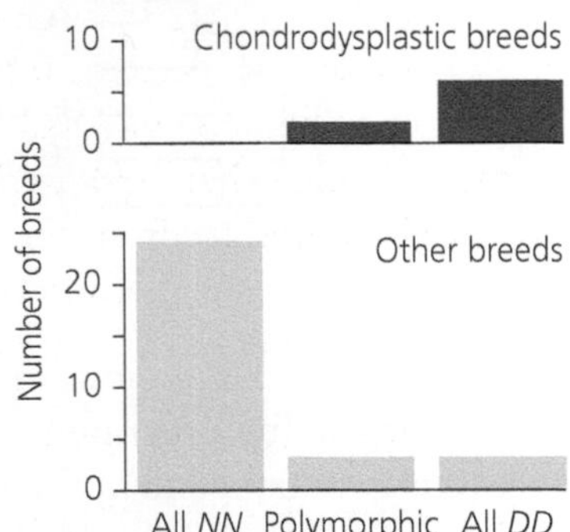

Figure 5.28 A new copy of the gene *fgf4* from retroposition (a) Like other Welsh corgis, Juno has chondrodysplasia—a condition characterized by short legs. (b) Across dog breeds, chondrodysplasia is strongly associated with the presence of a retroduplicated copy of the gene for fibroblast growth factor 4. Prepared using data from Parker and colleagues (2009).

A New Gene Generated by Retroposition

An example of a new gene created by retroposition comes from the work of Heidi Parker and colleagues (2009) on dogs. The researchers sought to find the gene or genes responsible for chondrodysplasia, the short-legged condition characteristic of corgis, dachshunds, bassets, and a variety of other breeds **(Figure 5.28a)**. Scanning the dog genome for differences between breeds with and without chondrodysplasia, Parker and colleagues found, on chromosome 18, a duplicate copy of the gene for fibroblast growth factor 4 (*fgf4*). Possession of the duplicate is strongly associated with chondrodysplasia. Figure 5.28b sorts breeds with and without the condition into three categories. In most breeds all individuals tested either carried the duplicate on both copies of chromosome 18 (*DD*), or on neither copy (*NN*). A few breeds were polymorphic, meaning that different individuals had different genotypes. (The breeds classified as Other in which all dogs carried two copies of the duplicate—cairn terrier, Norwich terrier, and shih tzu—have short legs but do not meet a stringent definition of chondrodysplasia.)

The duplicate copy of *fgf4* lacks introns and is located some distance from the parent copy. These and other clues indicate that the duplicate arose by retroposition. Parker and colleagues found that in dogs that carry it, the duplicate gene is expressed in the joint-forming cartilage in the long bones of puppies, as is the parent copy. It appears that the duplicate acquired a promoter by serendipitously inserting into the middle of a transposable element. Fine control over when and where the encoded protein is actually made likely involves regulatory sequences in the untranslated portion of the mRNA.

All the chondroplastic breeds that Parker and colleagues tested carried the same duplicate gene at the same locus. It thus appears that the duplicate arose only once, before the development of today's breeds. It has subsequently been driven to high frequency in modern short-legged breeds by artificial selection.

Duplicated Genes and Evolution

Gene duplication accounts for a substantial fraction of the genomic variation among individuals, and thus the raw material for evolution. Recent estimates suggest that more of the genome is affected by copy number variation than by differences derived from point mutations (Mills et al. 2011). Variation among individuals in gene copy number can, by itself, serve as a substrate for adaptive evolution (Perry et al. 2007). And serial duplication followed by gene divergence has generated families of functionally related genes that, as in the case of olfactory receptor genes, can include hundreds of members (Young et al. 2008).

Gene duplication is an important category of mutation for evolution. An increase in copy number may itself be adaptive. Duplication followed by divergence in function generates gene families.

New Genes from Scratch

Although most new genes are born as duplicates of existing genes, some genes appear to be born from noncoding DNA. David Knowles and Aoife McLysaght (2009) found evidence for three examples in humans. We will briefly discuss one.

C22orf45 is a gene of unknown function unique to humans. It is transcribed in a variety of tissues, and its mRNA is known to be translated into protein.

Even though the gene is unique to humans, similar nucleotide sequences occur at the homologous locus in chimpanzees, gorillas, orangutans, gibbons, and macaques. The sequences in all of these nonhuman primates contain elements, including premature stop codons, that would substantially alter the encoded protein were the sequence transcribed and translated. One of the premature stop codons is shared by all five nonhuman species **(Figure 5.29)**.

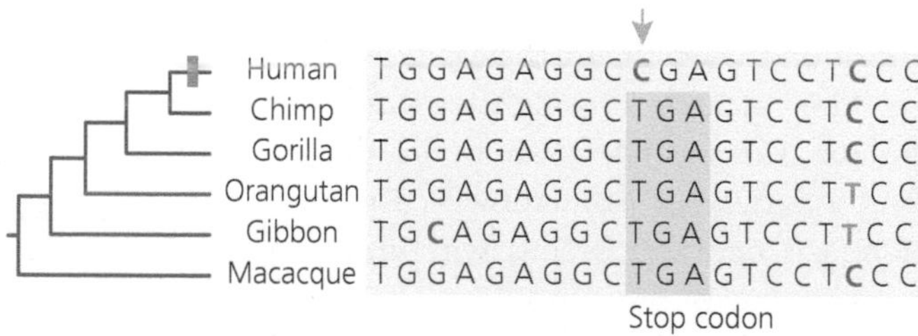

Figure 5.29 A new human gene from noncoding DNA In other apes and in macaques, the sequence homologous to the human gene *C22orf45* contains a premature stop codon. From Knowles and McLysaght (2009).

The most parsimonious explanation for this pattern is that the sequence was noncoding in the last common ancestor of humans and the other five species, and became a coding gene in the lineage leading to modern humans.

Having discussed mutations that create new alleles and new genes, we now turn to the most drastic mutations: those that alter large portions of chromosomes or even the entire genome.

5.4 Chromosome Mutations

The mutations discussed thus far occur on the scale of a single base pair in DNA to segments containing tens of thousands of base pairs. These alterations pale in comparison to mutations that alter the gross morphology of chromosomes. Some of these mutations affect only gene order and organization; others produce duplications or deletions that affect the total amount of genetic material. They can also involve the entire DNA molecule or segments of varying sizes. Here we focus on two types of chromosome alterations that are particularly important in evolution.

Inversions

Chromosome **inversions** often result from a multistep process that starts when radiation causes two double-strand breaks in a chromosome. After breakage, a chromosome segment can detach, flip, and reanneal in its original location. As Figure 5.30 shows, gene order along the chromosome is now inverted.

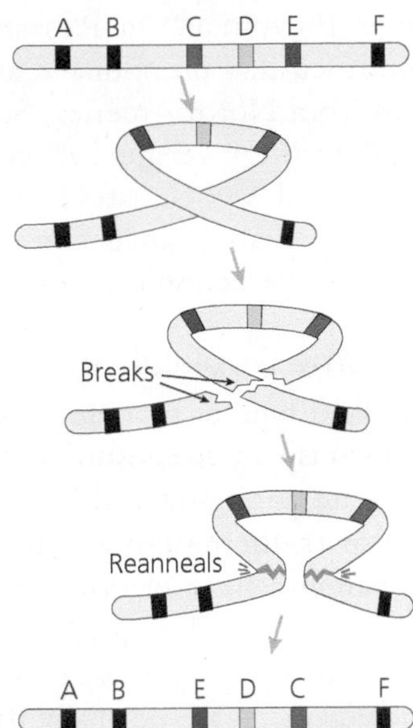

Figure 5.30 Chromosome inversion Inversions result when a chromosome segment breaks in two places and reanneals with the internal segment reversed. Note the order, before and after, of the genes labeled C, D, and E.

In addition to involving much larger stretches of DNA than point mutations and gene duplications, inversions have different consequences. Inversions affect a phenomenon known as genetic **linkage**. Linkage is the tendency for alleles of different genes to assort together at meiosis. Genes on the same chromosome tend to be more tightly linked (that is, more likely to be inherited together) than genes on nonhomologous chromosomes. Similarly, the closer together genes are on a chromosome, the tighter the linkage. Crossing over at meiosis, on the other hand, breaks up allele combinations and reduces linkage (see Chapter 8).

When inversions are heterozygous, meaning that one chromosome copy contains an inversion and the other does not, the inverted sequences cannot align properly when homologs synapse during prophase of meiosis I. Successful crossing-over events are rare. The result is that alleles inside the inversion are locked so tightly together that they are inherited as a single "supergene."

Inversions are common in *Drosophila.* Are they important in evolution? To answer this question, consider a series of inversions found in populations of *Drosophila subobscura.* This fruit fly is native to western Europe, North Africa, and the Middle East, and has six chromosomes. Five of these chromosomes are **polymorphic** for at least one inversion (Prevosti et al. 1988), meaning that chromosomes with and without the inversions exist. Biologists have known since the 1960s that the frequencies of these inversions vary regularly with latitude and climate. This type of regular change in the frequency of an allele or an inversion over a geographic area is called a **cline**. Several authors have argued that different inversions must contain specific combinations of alleles that function well together in cold, wet weather or hot, dry conditions. But is the cline really the result of natural selection on the supergenes? Or could it be a historical accident, caused by differences in the founding populations long ago?

A natural experiment has settled the issue. In 1978 *D. subobscura* showed up in the New World for the first time, initially in Puerto Montt, Chile, and then four years later in Port Townsend, Washington, USA. Several lines of evidence argue that the North American population is derived from the South American one. For example, of the 80 inversions present in Old World populations, precisely the same subset of 19 is found in both Chile and Washington State. Within a few years of their arrival on each continent, the *D. subobscura* populations had expanded extensively along each coast and developed the same clines in inversion frequencies found in the Old World (Figure 5.31). The clines are even correlated with the same general changes in climate type: from wet marine environments to mediterranean climates to desert and dry steppe habitats (Prevosti et al. 1988; Ayala et al. 1989). This is strong evidence that the clines result from natural selection and are not due to historical accident.

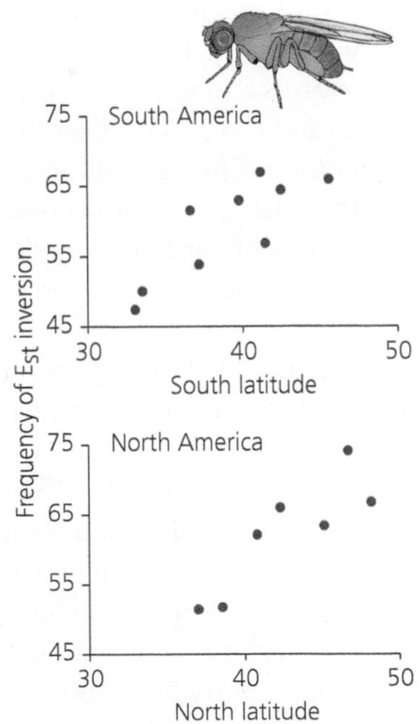

Figure 5.31 Inversion frequencies form clines in *Drosophila subobscura* From data in Prevosti et al. (1988); see also Balanyà et al. (2003).

Which genes are locked in the inversions, and how do they affect adaptation to changes in climate? In the lab, *D. subobscura* lines bred for small body size tend to become homozygous for the inversions found in the dryer, hotter part of the range (Prevosti 1967). Research by George Gilchrist and colleagues (2004) has confirmed that pronounced and parallel clines in body size exist in fly populations from North America, South America, and Europe. These results hint that alleles in the inversions affect body size, and that natural selection favors large flies in cold, wet climates and small flies in hot, dry areas. The fly study illustrates a key point about inversions: They are an important class of mutations because they affect selection on groups of alleles.

Chromosomal inversions suppress recombination and thus help maintain combinations of alleles across nearby loci.

Genome Duplication

The final type of mutation we will consider occurs at the largest scale possible: entire sets of chromosomes. For example, if homologous chromosomes fail to segregate during meiosis I or if sister chromatids do not separate properly during meiosis II, the resulting cells may have double the number of chromosomes of the parent cell. In plants, because the germ line is not segregated, similar mutations can occur during the mitotic cell divisions that lead up to gamete formation. Mutations like these can lead to the formation of a diploid gamete in species where gametes are normally haploid.

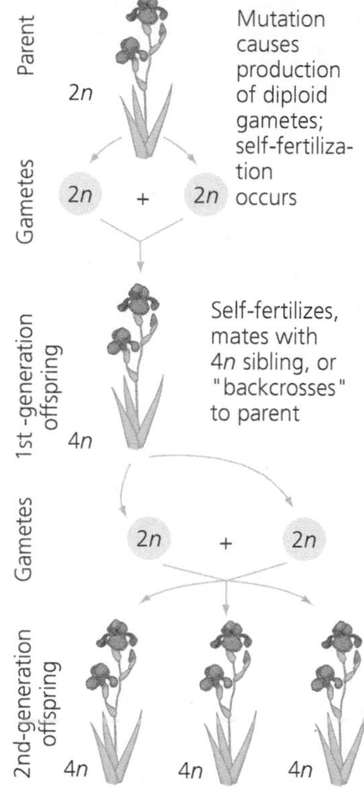

Figure 5.32 How tetraploid plants are produced

Figure 5.32 shows one possible outcome of a chromosome-doubling mutation. In the diagram, the individual that produces diploid gametes contains both male and female reproductive structures and can self-fertilize. When it does so, a tetraploid (4*n*) offspring results. If this offspring self-fertilizes when it matures, or if it mates with its parent or a tetraploid sibling that also produces diploid gametes, then a population of tetraploids can become established.

Organisms with more than two chromosome sets are said to be **polyploid**. Polyploid organisms can be tetraploid (4*n*), hexaploid (6*n*), octoploid (8*n*), or more. Polyploidy is common in plants and rare in animals—probably because self-fertilization is more common in plants than animals. Nearly half of all flowering plant species and the vast majority of the ferns are descended from ancestors where polyploidization occurred. In animals, polyploidy occurs in taxa like earthworms and some flatworms where individuals contain both male and female gonads and can self-fertilize. It is also present in animal groups that are capable of producing offspring without fertilization, through a process called parthenogenesis. In some species of beetles, sow bugs, moths, shrimp, goldfish, and salamanders, a type of parthenogenesis occurs that can lead to chromosomal doubling.

There are at least two reasons that polyploidy is an important type of mutation in evolution. First, it can lead to new species being formed. Second, it alters cell size, cell geometry, and gene dosage, and thus may endow individuals with new phenotypes that allow them to colonize and adapt to new environments.

Polyploidy and Speciation

To see why genome duplication can lead to speciation, imagine the outcome of matings between individuals in a tetraploid population and the most closely related diploid population. If individuals from the two populations mate, they produce triploid offspring. When these individuals mature and meiosis occurs, the homologous chromosomes cannot synapse correctly, because they are present in an odd number. As a result, the vast majority of the gametes produced by triploids end up with the wrong number of chromosomes and fail to survive. Triploid individuals have extremely low fertility.

Duplications of the entire genome are an important mechanism of speciation—particularly in plants.

In contrast, when tetraploid individuals continue to self-fertilize or mate among themselves, then fully fertile tetraploid offspring will result. In this way, natural selection should favor polyploids that are reproductively isolated from their parent population. Diploid and tetraploid populations that are genetically isolated are on their way to becoming separate species.

Genome Duplication and Adaptation

Justin Ramsey (2011) tested the hypothesis that polyploidy facilitates the colonization of, and adaptation to, new environments by performing a common garden experiment with wild yarrow **(Figure 5.33a)**. Along the coast of northern California, where Ramsey worked, yarrow populations with different ploidy occupy distinct habitats. Tetraploid populations live in coastal grasslands, conifer forests, and mountain meadows. Hexaploid plants live in sand dunes and oak woodlands. Because tetraploid plants occasionally produce hexaploid offspring, Ramsey suspected that the hexaploid populations were derived from tetraploid ancestors, and that the increase in ploidy aided their shift to drier habitats.

Ramsey grew tetraploid plants, first-generation hexaploid plants from tetraploid parents (neo-hexaploids), and hexaploid plants—all from wild populations—next to each other in sand dunes. He monitored them for three years.

The data on survival, displayed in 5.33b, show that hexaploids, which ordinarily live in dunes, are better adapted to dunes than tetraploids are. This is no surprise. The key result is that the neo-hexaploids did substantially better than the tetraploids, though not as well as the ordinary hexaploids. The data plotted in Figure 5.33c show that neo-hexaploids were also intermediate in flowering time.

Ramsey's results are consistent with the idea that changes in ploidy, by themselves, can alter phenotypes in a way that makes individuals better adapted to new environments. And they are also consistent with the idea that once a population with a novel ploidy level has colonized a new habitat, it can evolve by natural selection to become even better adapted.

This and the other research we have covered has illuminated the myriad ways that mutation supplies raw material for evolution. There are two more things we will need to know about mutation in the next several chapters: how often mutations happen, and how they affect the fitness of the individuals that carry them. These are these the issues we turn to in the final section of this chapter.

(a) Yarrow

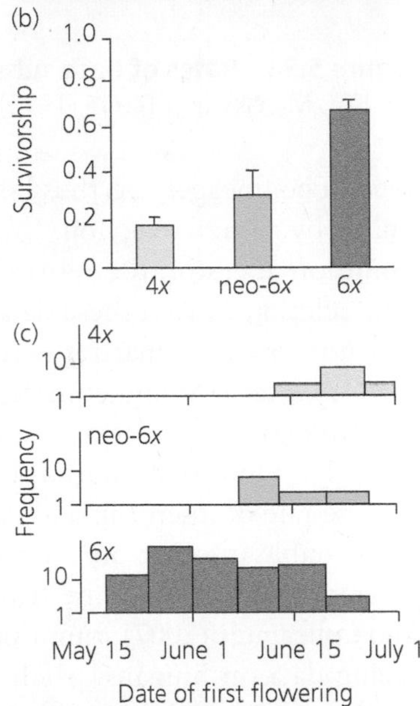

Figure 5.33 Genome duplication as a mechanism of adaptation (a) Wild yarrow (*Achillea borealis*). (b) Survival of plants with different ploidy in a common garden in dunes. (c) Flowering time distributions in the same plants. From Ramsey (2011).

5.5 Rates and Fitness Effects of Mutations

The rates and fitness effects of mutations have been hard to study, because mutations are rare and their consequences are often—though by no means always—subtle. Recently, however, advances in DNA sequencing and genetic engineering have allowed researchers to begin investigating these issues with new precision.

Mutation Rates

Traditionally, geneticists have estimated mutation rates by studying genes that, when disrupted, yield easily observable phenotypic changes (see Nachman 2004). Now that genes, and even whole genomes, can be sequenced quickly and cheaply, researchers can measure mutation rates more directly and on a larger scale.

For example, in a type of study called a **mutation accumulation** experiment, Stephan Ossowski and colleagues (2010) sequenced the genomes of five lineages of thale cress (*Arabidopsis thaliana*) derived from an already-sequenced common ancestor. Each of the five lineages had been grown under optimal conditions for 30 generations, and propagated each generation from a single randomly chosen

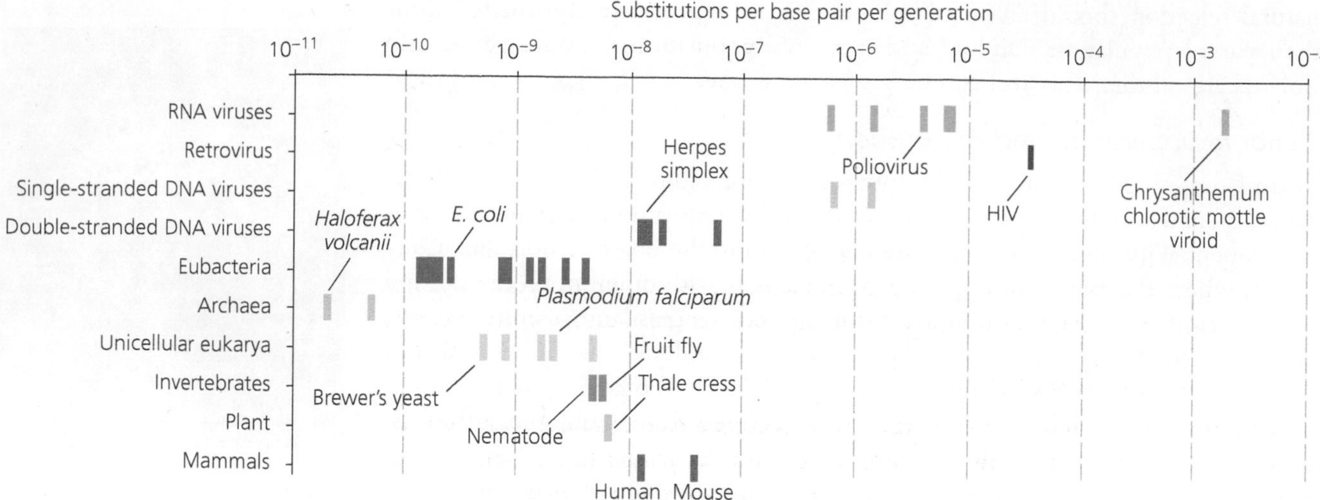

Figure 5.34 Rates of base substitution in various organisms Prepared using estimates compiled by Lynch (2010a) and, for HIV, Mansky and Temin (1995).

seed. The lineages had thus been allowed to accumulate mutations that were not culled by natural selection. In comparing the genomes of the descendants to their common ancestor, the researchers found 99 base substitutions and 17 insertions and deletions. From these data, the researchers estimated mutation rates.

Figure 5.34 summarizes estimates of the rate of base substitutions in thale cress and other taxa. Some estimates are based on whole genomes and others on representative genes. Mutation rates are diverse, spanning several orders of magnitude. Different kinds of organisms have different rates. Mutation rate clearly evolves.

The rate of insertions and deletions relative to base substitutions seems roughly comparable in other organisms to that in thale cress, although it too shows considerable variation. In the fruit fly *Drosophila melanogaster*, Cathy Haag-Liautard and colleagues (2007) found one indel for every three base substitutions. In analyzing data on humans, Michael Lynch (2010b) estimated that one indel occurs for every 17 base substitutions. In the nematode *Caenorhabditis elegans*, Dee Denver and colleagues (2004) found slightly more indels than base substitutions.

There are fewer directly measured rates of gene duplication, but given the variation in gene copy number among individuals in most populations, they must be high (Schrider and Hahn 2010). Kendra Lipinski and colleagues (2011) compared 10 lineages of the nematode *C. elegans* from a mutation-accumulation experiment like Ossowski's on thale cress. They estimated that duplications occur at a rate of 1.2×10^{-7} per gene per generation. A given gene in the *C. elegans* genome is more likely to be duplicated from one generation to the next than a given nucleotide within the gene is to experience a substitution.

All of these data indicate that mutation rates are low in cellular organisms. Genomes, however, are large. Multiplying the human base substitution rate in Figure 5.34, roughly a dozen mutations per billion base pairs per generation, by the 3.2 billion base pairs in a haploid human genome (Lander et al. 2001) suggests that every human inherits about three dozen point mutations via each of the gametes that united to form the zygote he or she grew from.

Research by Donald Conrad et al. (2011) corroborates this calculation. The researchers sequenced the genomes of both parents and a daughter in each of two families **(Figure 5.35)**. In the first family, they found that the daughter had

Family 1
Western/Northern European
from Utah, USA

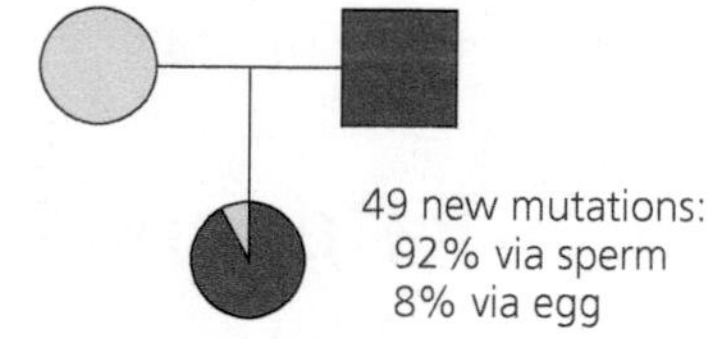

Family 2
Yoruban
from Ibadan, Nigeria

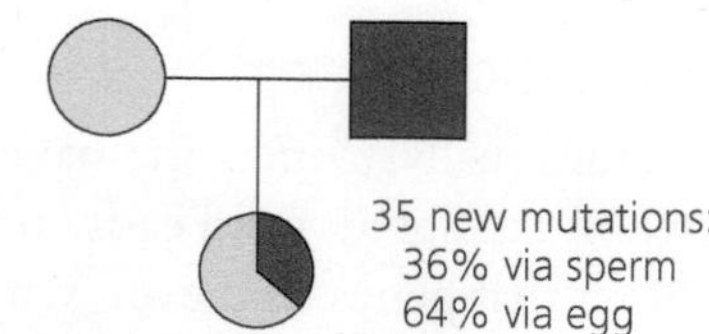

Figure 5.35 Everyone is a mutant New germ-line mutations found by sequencing the genomes of parents and an offspring in two families. From Conrad et al. (2011).

inherited 49 new germ-line mutations, most of them in the haploid genome she received from her father. In the second family, they found the daughter had inherited 35 new germ-line mutations, most of them in the haploid genome she received from her mother. The team's analysis covered about 80% of the genome, and they used stringent criteria that, they estimate, allowed them to identify only 70 to 75% of new germ-line mutations. Given these limitations, the number of mutations they found is close to what we would expect based on the mutation rate in Figure 5.34. The results also hint at differences in the mutation rate for the male versus female germ line, and even more strongly at differences in rate among individuals. More data will be required before clear patterns emerge.

Our knowledge of mutation rates is increasing dramatically thanks to genome sequencing. Mutation rates vary among loci in the genome, among individuals, and among kinds of organisms.

Fitness Effects of Mutations

Ronald Fisher predicted, in 1930, that most mutations altering fitness are deleterious (see Orr 2005). The most complete data on how mutations affect fitness come from viruses. Biologists have used a technique called site-directed mutagenesis to introduce random point mutations into the genomes of viruses, inserted the altered genomes into the cells of the viruses' hosts, and compared the fitness of the mutant viral strains—typically their population growth rate under optimal conditions—to that of the strain they were prepared from (Sanjuán 2010).

Figure 5.36a shows a typical result, from Joan Peris and colleagues (2010). The graph plots the distribution of relative fitness effects of 100 random point mutations in the genome of bacteriophage f1, a single-stranded DNA virus of *E. coli*. The relative fitness of lethal mutations and the ancestral strain are, by definition, 0 and 1. Nearly a quarter of the mutations are lethal. Over 40% are neutral. Of the rest, most are deleterious. Only two mutations are demonstrably beneficial.

Figure 5.36b displays data from a similar experiment with a cellular organism, brewer's yeast (*Saccharomyces cerevisiae*). Crucial to the design of the study is that

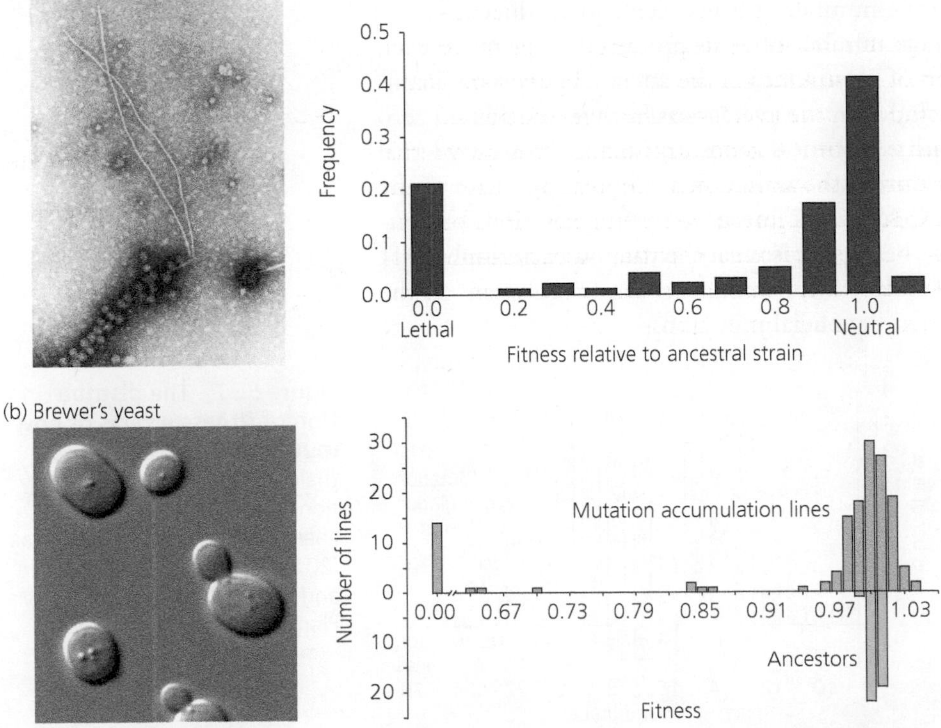

Figure 5.36 The distribution of fitness effects of new mutations (a) Fitness effects of 100 random mutations engineered into the genome of a virus. Redrawn from Peris et al. (2010). (b) Fitnesses of 144 mutation accumulation lineages in brewer's yeast compared to ancestral lineages. Redrawn from Hall and Joseph (2010).

yeast has a life cycle with diploid and haploid phases. It switches from the diploid phase to the haploid phase by undergoing meiosis, and from the haploid phase to the diploid phase by fertilization. Within each phase, it propagates by mitosis.

David Hall and Sarah Joseph (2010) started with a diploid strain homozygous at virtually all loci in the genome and used it as the ancestral stock for 144 mutation accumulation lines. They grew the lines under optimal conditions in the lab. Every other day, Hall and Joseph transferred a single cell from each line to fresh medium. Any mutant that happened to arise and get transferred to a new dish was thus protected from competition with nonmutant strains. Any mutations lethal to diploid cells would be lost, as would some mutations that dramatically reduce mitotic growth rate before their first chance of being transferred. All other kinds of mutations have the potential of being captured and preserved.

By the 50th transfer, the mutation accumulation lines had been propagated for 1,012 cell generations. Hall and Joseph induced the yeast to undergo meiosis and make haploid spores. The researchers separated the spores, placed them on fresh medium, and measured the growth rates of the resulting colonies. Figure 5.36b compares the relative growth rates of the 144 mutation accumulation strains to 42 colonies started from the ancestral stock. Fourteen of the mutation accumulation lines carried defects that rendered them unable to survive as haploids. Seventy-five lines had growth rates statistically indistinguishable from that of the ancestral stock. Thirty-five lines carried deleterious but nonlethal mutations. Twenty lines, about 14%, carried beneficial mutations.

Figure 5.37 shows data on a multicellular organism grown under natural conditions. Mathew Rutter, Frank Shaw, and Charles Fenster (2010) studied 100 thale cress mutation accumulation lines produced by the same experiment that generated the lines sequenced by Stephan Ossowski and colleagues discussed at the beginning of this section. They planted 70 seedlings from each line, plus 504 individuals representing the premutation genotype, in a field. The plants had to compete with each other and with the local plant community. After the plants set seed, the researchers counted the number of fruits each had produced.

The figure compares the average number of fruits produced by plants in each lineage to the average production of six sublines of the ancestral genotype. Plants that died before fruiting were included in the averages as having contributed zero fruit. The absence of lineages with zero fitness is unsurprising, because any lethal mutations would have been lost during the mutation accumulation phase of the experiment. That many of the experimental lineages carried deleterious or neutral mutations is also unsurprising, because it is what the data on bacteriophage f1 and brewer's yeast led us to expect. It is surprising, however, that so many of the mutation accumulation lines carried beneficial mutations.

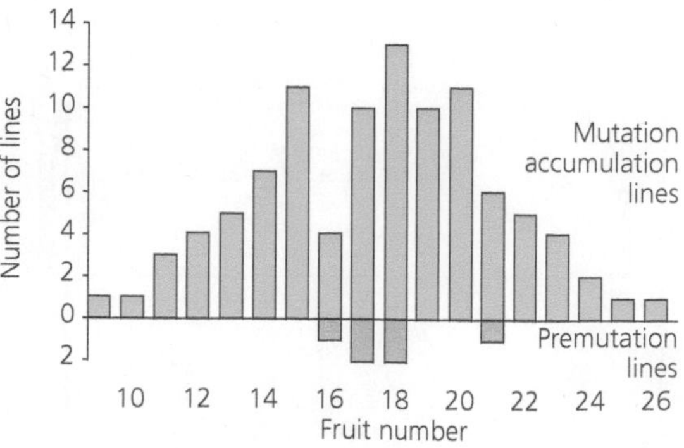

Figure 5.37 The distribution of fitness effects of new mutations Fitnesses of 100 thale cress mutation accumulation lines compared to control lines. Redrawn from Rutter et al. (2010). Photo by Janne Lempe and Suresh Balasubramanian/Max Planck Institute.

Most mutations are neutral, deleterious, or lethal. However, a detectable fraction of mutations are beneficial.

Rutter and colleagues offer two intriguing explanations for the high proportion of beneficial mutations among their thale cress lines. The first is that plants do not have an isolated germ line. The seeds produced by consecutive plant generations are thus separated by many generations of somatic cell division. Cell lineages carrying deleterious mutations produced during mitosis will be outcompeted by other cell lineages, and thus less likely to participate in the production of seeds. Furthermore, many plant loci are expressed in pollen, providing another chance for natural selection to weed out deleterious mutations. In other words, using a mutation accumulation protocol with plants might offer less protection against natural selection than we might have thought.

The scientists' second suggestion is that the ancestral lineages used in the experiment were from a stock that had been grown in the lab for 50 years. After evolving to adapt to laboratory conditions, this stock may have been poorly adapted for life in the natural field where the experiment was done. Research on viruses has shown that in lineages that are poorly adapted to begin with, a higher proportion of new mutations prove to be beneficial (Silander et al. 2007).

The studies we have reviewed demonstrate two patterns that appear to be general (Eyre-Walker and Keightley 2007). First, mutations come in four kinds: lethal, deleterious, neutral, and beneficial. Second, lethal and deleterious mutations outnumber beneficial mutations, usually by a considerable margin. These two patterns have important evolutionary consequences.

Mutations and Evolution

That lethal and deleterious mutations outnumber beneficial mutations means that a population not experiencing natural selection will show declining average fitness over time. That beneficial mutations occasionally appear, however, means that a population under selection can show increasing average fitness over time.

Demonstration of these claims comes from work by Larissa Vassilieva and colleagues (2000) and Suzanne Estes and Michael Lynch (2003) on the nematode *Caenorhabditis elegans* **(Figure 5.38)**. *C. elegans* have both male and female gonads and can self-fertilize. This characteristic allowed Dee Denver and colleagues (2004) to set up mutation accumulation lines from a single common ancestor and propagate them from a single worm each generation. Each line was maintained in the most benign environment possible, with optimal temperature and humidity, minimal crowding of individuals, abundant food, and no predators or parasites. This treatment insulated the worms as much as possible from natural selection.

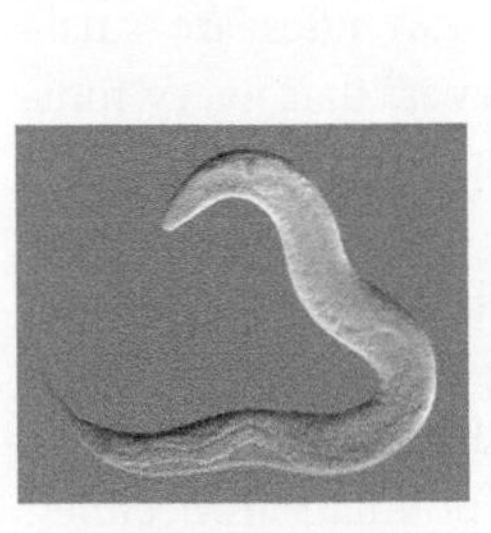
Caenorhabditis elegans

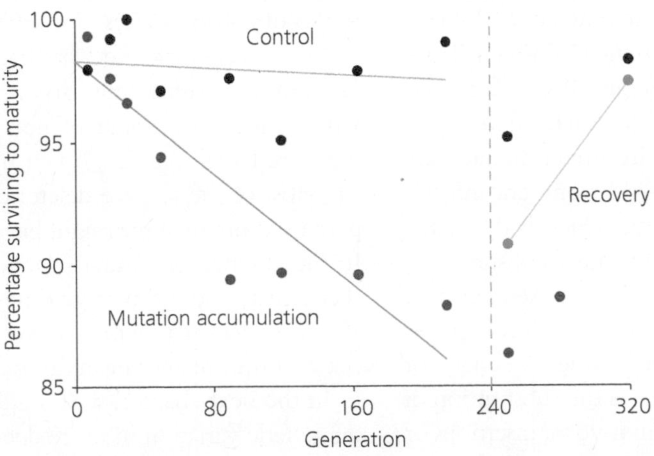

Figure 5.38 The balance between mutation and natural selection Nematode lineages insulated from natural selection declined in fitness relative to control lineages as they accumulated deleterious mutations. The lineages recovered upon exposure to natural selection due to the elimination of deleterious mutations and the preservation of favorable ones. Redrawn from Vassilieva et al. (2000), Estes and Lynch (2003). Photo by Bob Goldstein.

Over a span of 214 *C. elegans* generations, Vassilieva and colleagues periodically assessed the rate at which individuals in the mutation accumulation lines survived to adulthood. The red dots and best-fit line in Figure 5.38 show the data. As mutations piled up, the genetic quality of the worm population declined. Some lineages died out altogether.

The researchers simultaneously maintained control lines, founded from the same common ancestor but propagated from large numbers of individuals each generation. Any new mutants that appeared in these lines were in competition with nonmutant worms. Mutants with poor survival were less likely to be represented in future generations. As shown by the black dots and best-fit line, this continuous natural selection maintained the genetic quality of the control lines.

Mutation, by itself, tends to erode the genetic quality of populations. However, mutation supplies the genetic variation necessary for evolution, including adaptive evolution. Thus mutation in combination with natural selection can allow a population to maintain or even increase its mean fitness across generations.

After the 240th generation, Estes and Lynch set up duplicates of the mutation accumulation lines, which they call recovery lines. They started each line with a single individual, but thereafter propagated the lines from large numbers of individuals each generation. In this way, they restored the conditions required for natural selection.

The only source of genetic variation in the recovery lines was new mutations that appeared as the experiment progressed. But any new mutation that conferred a higher rate of survival had a better-than-average chance of being represented in the next generation. As shown by the blue dots and line in the figure, the recovery populations evolved quickly. Within 80 generations, their average survival rate had risen to match that of the control populations.

All populations experience a small but steady input of mutations, most of them deleterious. It is the action of natural selection, culling damaging variants and preserving the advantageous ones, that saves populations from inexorable decline. Evolutionary geneticists from Herman Muller (1950) to Michael Lynch (2010b) have noted the many ways modern humans are insulated from natural selection (see, for example, Ulizzi et al. 1998). The implications for the future are ominous, and the obvious solutions unappealing. The balance between mutation and selection is a topic we will return to in the chapters to come.

SUMMARY

In this chapter we explored two of the four claims made by the theory of evolution by natural selection. The first claim is that there is variation among individuals. Individuals differ because they carry different genes (genetic variation), because they have experienced different environments (environmental variation), and because the different genes they carry cause them to react differently to the environment (genotype-by-environment interaction). Genetic variation and genotype-by-environment interaction satisfy the theory's second claim, which is that some of the variation is transmitted from parents to offspring.

Genetic information is encoded by the sequence of bases in DNA (and RNA). The ultimate source of genetic variation is thus changes in nucleotide sequences, or mutations. Mutations can be as small as the substitution of one nucleotide for another. They can involve insertions or deletions of short or long runs of bases. They can entail the duplication of entire genes, making possible the evolution of new functions. Or they can be as large as rearrangements of chromosomes or doubling of the genome.

On a per-nucleotide basis, mutation rates are small. Genomes are sufficiently large, however, that every individual carries a number of new mutations. Some mutations are lethal; typically, even more are neutral or nearly so. Most of the rest are deleterious. But some mutations, perhaps more than we might have expected, are beneficial. In the absence of natural selection, deleterious mutations accumulate and the average fitness of populations declines. When selection is acting, however, beneficial mutations can accumulate and mean fitness can rise.

In the next chapter, we consider the action of selection on genetic variation in more detail.

QUESTIONS

1. What is the difference between genetic variation, environmental variation, and genotype-by-environment interaction? Give examples of each. Try to think of potential examples not covered in this chapter.
2. We noted on the first page of the chapter that humans vary considerably in height. State a hypothesis about whether this reflects genetic variation, environmental variation, or genotype-by-environment interaction (any hypothesis is okay). What kinds of evidence might settle the question? Are there experiments that, at least in principle, would decide the matter? Would it be easier to do them with another species, such as mice?
3. Because you are studying different subjects, the diversity of knowledge among you and your classmates is larger now than it was at the beginning of the school year. What kind of variation is this? Could the diversity in knowledge serve as raw material for evolution of the campus population? Why or why not?
4. What are reaction norms, and why do they matter? Draw your own reaction norm for mood as a function of the temperature outside. What kind of variation allows reaction norms to evolve?
5. Consider the nucleotide sequence TGACTAACGGCT. Transcribe this sequence into mRNA. Use the genetic code to translate it into a string of amino acids. Give an example of a point mutation, an insertion, a deletion, a frameshift mutation, a synonymous substitution, a nonsynonymous substitution, and a nonsense mutation. Which of your examples seem likely to dramatically influence protein function? Which seem likely to have little effect? Why?
6. Consider a population containing the following genotypes: *Aa*, *Aa*, *AA*, *aA*, *aa*, *Aa*, *aa*, *aA*, *aa*, *Aa*. What is the frequency of genotype *aa*? Allele *A*? Allele *a*? Can you tell which genotype is most advantageous? Can you tell whether *Aa* resembles *AA* or *aa*? Why or why not?
7. How many redheads live in a village of 250 people, where the frequency of red hair is 0.18?
8. Diagram two processes through which genes can be duplicated. How can you tell whether a duplicate copy of a gene arose by unequal crossing over or retroposition?
9. If a gene gets retroduplicated, how can you distinguish the original gene from the copy?
10. How do chromosome inversions happen? What consequences do they have for the evolution of populations?
11. Diagram the sequence of events that leads to the formation of second-generation polyploid individuals in plants that can self-fertilize.
12. Discuss factors that might cause mutation rates to vary among individuals in populations, and among species.
13. Which kind of mutation is most common: lethal, nonlethal but deleterious, neutral, or beneficial? Draw a graph to illustrate your answer. According to the graph, do most mutations have large or small effects on fitness?
14. Compare and contrast the evolutionary roles of point mutations, chromosome inversions, gene duplications, and polyploidization.
15. Suppose a silent mutation occurs in an exon that is part of the gene for TAS2R38 in a human. Has a new allele been created? Defend your answer.
16. The amino acid sequences encoded by the red and green visual pigment genes found in humans are 96% identical (Nathans et al. 1986). These two genes are found close together on the X chromosome, while the gene for the blue pigment is located on chromosome 7. Among primates, only Old World monkeys, the great apes, and humans have a third pigment gene—New World monkeys have only one X-linked pigment gene. Comment on the following three hypotheses:
    - One of the two visual pigment loci on the X chromosome originated in a gene duplication event.
    - The gene duplication event occurred after New World and Old World monkeys had diverged from a common ancestor, which had two visual pigment genes.
    - Human males with a mutated form of the red or green pigment gene experience the same color vision of our male primate ancestors.
17. Chromosome number can evolve by smaller-scale changes than duplication of entire chromosome sets. For example, domestic horses have 64 chromosomes per diploid set while Przewalski's horse, an Asian subspecies, has 66. Przewalski's horse is thought to have evolved from an ancestor with $2n = 64$ chromosomes. The question is: Where did its extra chromosome pair originate? It seems unlikely that an entirely new chromosome pair was created from scratch in Przewalski's horse. To generate a hypothesis explaining the origin of the new chromosome in Przewalski's horse, examine the adjacent figure. The drawing at right shows how certain chromosomes synapse in the hybrid offspring of a domestic horse–Przewalski's horse mating (Short et al. 1974). The remaining chromosomes show a normal 1:1 pairing. Do you think this sort of gradual change in chromosome number involves a change in the actual number of genes present, or just rearrangement of the same number of genes?

EXPLORING THE LITERATURE

18. Some evolutionary geneticists have suggested that the genetic code has been shaped by natural selection to minimize the deleterious consequences of mutations. For an entry into the literature on this issue, see:

Caporaso, J. G., M. Yarus, and R. Knight. 2005. Error minimization and coding triplet/binding site associations are independent features of the canonical genetic code. *Journal of Molecular Evolution* 61: 597–607.

Freeland, S. J., and L. D. Hurst. 1998. Load minimization of the genetic code: History does not explain the pattern. *Proceedings of the Royal Society London B* 265: 2111–2119.

Freeland, S. J., and L. D. Hurst. 1998. The genetic code is one in a million. *Journal of Molecular Evolution* 47: 238–248.

Knight, R. D., S. J. Freeland, and L. F. Landweber. 1999. Selection, history and chemistry: The three faces of the genetic code. *Trends in Biochemical Sciences* 24: 241–247.

19. We discussed temperature-dependent sex determination in geckos. Many other reptiles have environmental sex determination as well. For a paper exploring why temperature-dependent sex determination might be adaptive, see:

Warner, D. A., and R. Shine. 2008. The adaptive significance of temperature-dependent sex determination in a reptile. *Nature* 451: 566–568.

20. Figure 5.12 presented evidence that people with certain genotypes for the serotonin transporter gene are more sensitive to maltreatment during childhood. For an exploration of possible benefits associated with the sensitive genotype, see:

Homberg, J. R., and K. P. Lesch. 2011. Looking on the bright side of serotonin transporter gene variation. *Biological Psychiatry* 69: 513–519.

21. For an example of chromosomal rearrangements maintaining a supergene with multiple alleles, see:

Joron, M., L. Frezal, et al. 2011. Chromosomal rearrangements maintain a polymorphic supergene controlling butterfly mimicry. *Nature* 477: 203–206.

22. For an estimate of the number of loss-of-function mutations present in the genome of a typical human, see:

MacArthur, D. G., and C. Tyler-Smith. 2010. Loss-of-function variants in the genomes of healthy humans. *Human Molecular Genetics* 19: R125–R130.

23. Are there circumstances under which it is good to have a high mutation rate? See:

Gentile, C. F., S. C. Yu, et al. 2011. Competition between high- and higher-mutating strains of *Escherichia coli*. *Biology Letters* 7: 422–424.

24. For evidence that new genes may evolve from scratch more often than previously thought, see:

Carvunis, A. R., T. Rolland, et al. 2012. Proto-genes and *de novo* gene birth. *Nature* 487: 370–374.

CITATIONS

Abecasis, G. R., A. Auton, et al. 2012. An integrated map of genetic variation from 1,092 human genomes. *Nature* 491: 56–65.

Adam, M., B. Murali, et al. 2008. Epigenetic inheritance based evolution of antibiotic resistance in bacteria. *BMC Evolutionary Biology* 8: 52.

Albertson, T. M., M. Ogawa, et al. 2009. DNA polymerase ε and δ proofreading suppress discrete mutator and cancer phenotypes in mice. *Proceedings of the National Academy of Sciences USA* 106: 17101–17104.

Arnheim, N., and P. Calabrese. 2009. Understanding what determines the frequency and pattern of human germline mutations. *Nature Reviews Genetics* 10: 478–488.

Avise, J. C. 1994. *Molecular Markers, Natural History and Evolution.* New York: Chapman & Hall.

Ayala, F. J., L. Serra, and A. Prevosti. 1989. A grand experiment in evolution: The *Drosophila subobscura* colonization of the Americas. *Genome* 31: 246–255.

Balaynà, J., L. Serra, et al. 2003. Evolutionary pace of chromosomal polymorphism in colonizing populations of *Drosophila subobscura*: An evolutionary time series. *Evolution* 57: 1837–1845.

Biarnés, X., A. Marchiori, et al. 2010. Insights into the binding of phenyltiocarbamide (PTC) agonist to its target human TAS2R38 bitter receptor. *PLoS ONE* 5: e12394.

Brown, G. W., and T. O. Harris. 2008. Depression and the serotonin transporter 5-HTTLPR polymorphism: A review and a hypothesis concerning gene-environment interaction. *Journal of Affective Disorders* 111: 1–12.

Bufe, B., P. A. S. Breslin, et al. 2005. The molecular basis of individual differences in phenylthiocarbamide and propylthiouracil bitterness perception. *Current Biology* 15: 322–327.

Bull, J. J. 1980. Sex determination in reptiles. *Quarterly Review of Biology* 55: 3–21.

Caspi, A., A. R. Hariri, et al. 2010. Genetic sensitivity to the environment: The case of the serotonin transporter gene and its implications for studying complex diseases and traits. *American Journal of Psychiatry* 167: 509–527.

Caspi, A., K. Sugden, et al. 2003. Influence of life stress on depression: Moderation by a polymorphism in the 5-HTT gene. *Science* 301: 386–389.

CFMD (Cystic Fibrosis Mutation Database). 2011. CFMDB Statistics. Available at http://www.genet.sickkids.on.ca/cftr/StatisticsPage.html.

Conrad, D. F., J. E. Keebler, et al. 2011. Variation in genome-wide mutation rates within and between human families. *Nature Genetics* 43: 712–714.

Cropley, J. E., C. M. Suter, et al. 2006. Germ-line epigenetic modification of the murine Avy allele by nutritional supplementation. *Proceedings of the National Academy of Sciences USA* 103: 17308–17312.

Cubas, P., C. Vincent, and E. Coen. 1999. An epigenetic mutation responsible for natural variation in floral symmetry. *Nature* 401: 157–161.

Danchin, E., A. Charmantier, et al. 2011. Beyond DNA: Integrating inclusive inheritance into an extended theory of evolution. *Nature Reviews Genetics* 12: 475–486.

Denver, D. R., K. Morris, et al. 2004. High mutation rate and predominance of insertions in the *Caenorhabditis elegans* nuclear genome. *Nature* 430: 679–682.

Dobzhansky, T. 1937. *Genetics and the Origin of Species.* New York: Columbia University Press.

Duffy, V. B., J. E. Hayes, et al. 2010. Vegetable intake in college-aged adults is explained by oral sensory phenotypes and TAS2R38 genotype. *Chemosensory Perception* 3: 137–148.

Elowitz, M. B., A. J. Levine, et al. 2002. Stochastic gene expression in a single cell. *Science* 297: 1183–1186.

Estes, S., and M. Lynch. 2003. Rapid fitness recovery in mutationally degraded lines of *Caenorhabditis elegans*. *Evolution* 57: 1022–1030.

Eyre-Walker, A., and P. D. Keightley. 2007. The distribution of fitness effects of new mutations. *Nature Reviews Genetics* 8: 610–618.

Feng, S., S. E. Jacobsen, and W. Reik. 2010. Epigenetic reprogramming in plant and animal development. *Science* 330: 622–627.

Fisher, R. A. 1930. *The Genetical Theory of Natural Selection*. Oxford: Oxford University Press.

Garcia-Diaz, M., and T. A. Kunkel. 2006. Mechanism of a genetic glissando: Structural biology of indel mutations. *Trends in Biochemical Sciences* 31: 206–214.

Gilchrist, G. W., R. B. Huey, et al. 2004. A time series of evolution in action: A latitudinal cline in wing size in South American *Drosophila subobscura*. *Evolution* 58: 768–780.

Gorovic, N., S. Afzal, et al. 2011. Genetic variation in the hTAS2R38 taste receptor and brassica vegetable intake. *Scandinavian Journal of Clinical & Laboratory Investigation* 71: 274–279.

Grant-Downton, R. T., and H. G. Dickinson. 2006. Epigenetics and its implications for plant biology 2. The "epigenetic epiphany": Epigenetics, evolution and beyond. *Annals of Botany* 97: 11–27.

Haag-Liautard, C., M. Dorris, et al. 2007. Direct estimation of per nucleotide and genomic deleterious mutation rates in *Drosophila*. *Nature* 445: 82–85.

Hall, D. W., and S. B. Joseph. 2010. A high frequency of beneficial mutations across multiple fitness components in *Saccharomyces cerevisiae*. *Genetics* 185: 1397–1409.

Hammill, E., A. Rogers, and A. P. Beckerman. 2008. Costs, benefits and the evolution of inducible defenses: A case study with *Daphnia pulex*. *Journal of Evolutionary Biology* 21: 705–715.

Han, J., P. Kraft, et al. 2008. A genome-wide association study identifies novel alleles associated with hair color and skin pigmentation. *PLoS Genetics* 4: e1000074.

Harris, H. 1966. Enzyme polymorphisms in man. *Proceedings of the Royal Society of London B* 164: 298–310.

Hauben, M., B. Haesendonckx, et al. 2009. Energy use efficiency is characterized by an epigenetic component that can be directed through artificial selection to increase yield. *Proceedings of the National Academy of Sciences, USA* 106: 20109–20114.

Hoppeler, H., and M. Vogt. 2001. Muscle tissue adaptations to hypoxia. *Journal of Experimental Biology* 204: 3133–3139.

Johannes, F., E. Porcher, et al. 2009. Assessing the impact of transgenerational epigenetic variation on complex traits. *PLoS Genetics* 5: e1000530.

Ju, Y. S., J. I. Kim, et al. 2011. Extensive genomic and transcriptional diversity identified through massively parallel DNA and RNA sequencing of eighteen Korean individuals. *Nature Genetics* 43: 745–752.

Kaessmann, H. 2010. Origins, evolution, and phenotypic impact of new genes. *Genome Research* 20: 1313–1326.

Kim, U.-K., E. Jorgenson, et al. 2003. Positional cloning of the human quantitative trait locus underlying taste sensitivity to phenylthiocarbamide. *Science* 299: 1221–1225.

Knowles, D. G., and A. McLysaght. 2009. Recent *de novo* origin of human protein-coding genes. *Genome Research* 19: 1752–1759.

Kornberg, A. 1960. Biologic synthesis of deoxyribonucleic acid. *Science* 131: 1503–1508.

Kotov, A. A., and D. J. Taylor. 2011. Mesozoic fossils (>145 Mya) suggest the antiquity of the subgenera of *Daphnia* and their coevolution with chaoborid predators. *BMC Evolutionary Biology* 11: 129.

Kreindler, J. L. 2010. Cystic fibrosis: Exploiting its genetic basis in the hunt for new therapies. *Pharmacology & Therapeutics* 125: 219–229.

Laforsch, C., W. Ngwa, et al. 2004. An acoustic microscopy technique reveals hidden morphological defenses in *Daphnia*. *Proceedings of the National Academy of Sciences USA* 101: 15911–15914.

Lampert, W. 2011. Daphnia: Development of a model organism in ecology and evolution. In *Excellence in Ecology, Book 21*, ed. O. Kinne. Oldendorf/Luhe, Germany: International Ecology Institute.

Lander, E. S., A. Heaford, et al. 2001. Initial sequencing and analysis of the human genome. *Nature* 409: 860–921.

Lewontin, R. C., and J. L. Hubby. 1966. A molecular approach to the study of genetic heterozygosity in natural populations. II. Amount of variation and degree of heterozygosity in natural populations of *Drosophila pseudoobscura*. *Genetics* 54: 595–609.

Li, C. C., J. E. Cropley, et al. 2011. A sustained dietary change increases epigenetic variation in isogenic mice. *PLoS Genetics* 7: e1001380.

Lindahl, T. 1993. Instability and decay of the primary structure of DNA. *Nature* 362: 709–715.

Lipinski, K. J., J. C. Farslow, et al. 2011. High spontaneous rate of gene duplication in *Caenorhabditis elegans*. *Current Biology* 21: 306–310.

Loeb, L. A. 2011. Human cancers express mutator phenotypes: Origin, consequences and targeting. *Nature Reviews Cancer* 11: 450–457.

Long, M., E. Betrán, et al. 2003. The origin of new genes: Glimpses from the young and old. *Nature Reviews Genetics* 4: 865–875.

Lynch, M. 2010a. Evolution of the mutation rate. *Trends in Genetics* 26: 345–352.

Lynch, M. 2010b. Rate, molecular spectrum, and consequences of human mutation. *Proceedings of the National Academy of Sciences, USA* 107: 961–968.

Mansky, L. M., and H. M. Temin. 1995. Lower in vivo mutation rate of human immunodeficiency virus type 1 than that predicted from the fidelity of purified reverse transcriptase. *Journal of Virology* 69: 5087–5094.

Martinson, J. J., N. H. Chapman, et al. 1997. Global distribution of the CCR5 gene 32-base-pair deletion. *Nature Genetics* 16: 100–1103.

McKellar, A. E., and A. P. Hendry. 2009. How humans differ from other animals in their levels of morphological variation. *PLoS ONE* 4: e6876.

Mills, R. E., K. Walter, et al. 2011. Mapping copy number variation by population-scale genome sequencing. *Nature* 470: 59–65.

Mitton, J. B. 1997. *Selection in Natural Populations*. Oxford: Oxford University Press.

Miyakawa, H., M. Imai, et al. 2010. Gene up-regulation in response to predator kairomones in the water flea, *Daphnia pulex*. *BMC Developmental Biology* 10: 45.

Miyamura, Y., S. G. Coelho, et al. 2011. The deceptive nature of UVA tanning versus the modest protective effects of UVB tanning on human skin. *Pigment Cell & Melanoma Research* 24: 136–147.

Muller, H. J. 1950. Our load of mutations. *American Journal of Human Genetics* 2: 111–176.

Nachman, M. W. 2004. Haldane and the first estimates of the human mutation rate. *Journal of Genetics* 83: 231–233.

Nan, H., P. Kraft, et al. 2009. Genome-wide association study of tanning phenotype in a population of European ancestry. *Journal of Investigative Dermatology* 129: 2250–2257.

Nathans, J., D. Thomas, and D. S. Hogness. 1986. Molecular genetics of human color vision: The genes encoding blue, green, and red pigments. *Science* 232: 193–202.

Nelson, M. R., D. Wegmann, et al. 2012. An abundance of rare functional variants in 202 drug target genes sequenced in 14,002 people. *Science* 337: 100–104.

Nettle, D. 2010. Understanding of evolution may be improved by thinking about people. *Evolutionary Psychology* 8: 205–228.

Newcomb, R. D., M. B. Xia, and D. R. Reed. 2012. Heritable differences in chemosensory ability among humans. *Flavour* 1: 9.

Niemi, A.-K., and K. Majamaa. 2005. Mitochondrial DNA and ACTN3 genotypes in Finnish elite endurance and sprint athletes. *European Journal of Human Genetics* 13: 965–969.

North, K. N., N. Yang, et al. 1999. A common nonsense mutation results in alpha-actinin-3 deficiency in the general population. *Nature Genetics* 21: 353–354.

Nousbeck, J., B. Burger, et al. 2011. A mutation in a skin-specific isoform of SMARCAD1 causes autosomal-dominant adermatoglyphia. *American Journal of Human Genetics* 89: 302–307.

Orr, H. A. 2005. The genetic theory of adaptation: A brief history. *Nature Reviews Genetics* 6: 119–127.

Ossowski, S., K. Schneeberger, et al. 2010. The rate and molecular spectrum of spontaneous mutations in *Arabidopsis thaliana*. *Science* 327: 92–94.

Parker, H. G., B. M. VonHoldt, et al. 2009. An expressed fgf4 retrogene is associated with breed-defining chondrodysplasia in domestic dogs. *Science* 325: 995–998.

Parra, E. J. 2007. Human pigmentation variation: Evolution, genetic basis, and implications for public health. *American Journal of Physical Anthropology* (Suppl) 45: 85–105.

Paszkowski, J., and U. Grossniklaus. 2011. Selected aspects of transgenerational epigenetic inheritance and resetting in plants. *Current Opinion in Plant Biology* 14: 195–203.

Paun, O., R. M. Bateman, et al. 2010. Stable epigenetic effects impact adaptation in allopolyploid orchids (Dactylorhiza: Orchidaceae). *Molecular Biology and Evolution* 27: 2465–2473.

Pearson, C. E., K. Nichol Edamura, and J. D. Cleary. 2005. Repeat instability: Mechanisms of dynamic mutations. *Nature Reviews Genetics* 6: 729–742.

Peris, J. B., P. Davis, et al. 2010. Distribution of fitness effects caused by single-nucleotide substitutions in bacteriophage f1. *Genetics* 185: 603–609.

Perry, G. H., N. J. Dominy, et al. 2007. Diet and the evolution of human amylase gene copy number variation. *Nature Genetics* 39: 1256–1260.

Pertea, M., and S. L. Salzberg. 2010. Between a chicken and a grape: Estimating the number of human genes. *Genome Biology* 11: 206.

Preston, B. D., T. M. Albertson, and A. J. Herr. 2010. DNA replication fidelity and cancer. *Seminars in Cancer Biology* 20: 281–293.

Prevosti, A. 1967. Inversion heterozygosity and selection for wing length in *Drosophila subobscura*. *Genetical Research Cambridge* 10: 81–93.

Prevosti, A., G. Ribo, et al. 1988. Colonization of America by *Drosophila subobscura*: Experiment in natural populations that supports the adaptive role of chromosomal-inversion polymorphism. *Proceedings of the National Academy of Science, USA* 85: 5597–5600.

Ramsey, J. 2011. Polyploidy and ecological adaptation in wild yarrow. *Proceedings of the National Academy of Sciences, USA* 108: 7096–7101.

Rhen, T., A. Schroeder, et al. 2011. Segregating variation for temperature-dependent sex determination in a lizard. *Heredity* 106: 649–660.

Richards, E. J. 2006. Inherited epigenetic variation—Revisiting soft inheritance. *Nature Reviews Genetics* 7: 395–401.

Roth, S. M., S. Walsh, et al. 2008. The ACTN3 R577X nonsense allele is underrepresented in elite-level strength athletes. *European Journal of Human Genetics* 16: 391–394.

Rutter, M. T., F. H. Shaw, and C. B. Fenster. 2010. Spontaneous mutation parameters for *Arabidopsis thaliana* measured in the wild. *Evolution* 64: 1825–1835.

Samson, M., F. Libert, et al. 1996. Resistance to HIV-1 infection in Caucasian individuals bearing mutant alleles of the CCR5 chemokine receptor gene. *Nature* 382: 722–725.

Sandell, M. A., and P. A. Breslin. 2006. Variability in a taste-receptor gene determines whether we taste toxins in food. *Current Biology* 16: R792–R794.

Sanjuán, R. 2010. Mutational fitness effects in RNA and single-stranded DNA viruses: Common patterns revealed by site-directed mutagenesis studies. *Philosophical Transactions of the Royal Society B* 365: 1975–1982.

Schrider, D. R., and M. W. Hahn. 2010. Gene copy-number polymorphism in nature. *Proceedings of the Royal Society B* 277: 3213–3221.

Scoville, A. G., L. L. Barnett, et al. 2011. Differential regulation of a MYB transcription factor is correlated with transgenerational epigenetic inheritance of trichome density in *Mimulus guttatus*. *New Phytologist* 191: 251–263.

Shea, N., I. Pen, and T. Uller. 2011. Three epigenetic information channels and their different roles in evolution. *Journal of Evolutionary Biology* 24: 1178–1187.

Shoemaker, C. M., and D. Crews. 2009. Analyzing the coordinated gene network underlying temperature-dependent sex determination in reptiles. *Seminars in Cell & Developmental Biology* 20: 293–303.

Short, R. V., A. C. Chandley, et al. 1974. Meiosis in interspecific equine hybrids. II. The Przewalski horse/domestic horse hybrid. *Cytogenetics and Cell Genetics* 13: 465–478.

Silander, O. K., O. Tenaillon, and L. Chao. 2007. Understanding the evolutionary fate of finite populations: The dynamics of mutational effects. *PLoS Biology* 5: e94.

Slatkin, M. 2009. Epigenetic inheritance and the missing heritability problem. *Genetics* 182: 845–850.

Suzuki, Y., and H. F. Nijhout. 2006. Evolution of a polyphenism by genetic accommodation. *Science* 311: 650–652.

Suzuki, Y., and H. F. Nijhout. 2008. Genetic basis of adaptive evolution of a polyphenism by genetic accommodation. *Journal of Evolutionary Biology* 21: 57–66.

Tepper, B. J. 2008. Nutritional implications of genetic taste variation: The role of PROP sensitivity and other taste phenotypes. *Annual Review of Nutrition* 28: 367–388.

Ulizzi, L., P. Astolfi, and L. A. Zonta. 1998. Natural selection in industrialized countries: A study of three generations of Italian newborns. *Annals of Human Genetics* 62: 47–53.

Valleley, E. M., E. J. Cartwright, et al. 2001. Characterisation and expression of *Sox9* in the Leopard gecko, *Eublepharis macularius*. *Journal of Experimental Zoology Part B: Molecular and Developmental Evolution* 291: 85–91.

Vassilieva, L. L., A. M. Hook, and M. Lynch. 2000. The fitness effects of spontaneous mutations in *Caenorhabditis elegans*. *Evolution* 54: 1234–1246.

Viets, B. E., A. Tousignant, et al. 1993. Temperature-dependent sex determination in the leopard gecko, *Eublepharis macularius*. *Journal of Experimental Zoology* 265: 679–683.

Watson, J. D. 1977. *Molecular Biology of the Gene*. 3rd ed. Menlo Park, CA: W. A. Benjamin.

Watson, J. D., and F. H. C. Crick. 1953a. A structure for deoxyribose nucleic acid. *Nature* 171: 737–738.

Watson, J. D., and F. H. C. Crick. 1953b. Genetical implications of the structure of deoxyribonucleic acid. *Nature* 171: 964–967.

Whittle, C. A., S. P. Otto, et al. 2009. Adaptive epigenetic memory of ancestral temperature regime in *Arabidopsis thaliana*. *Botany* 87: 650–657.

Wooding, S. 2005. Evolution: A study in bad taste? *Current Biology* 15: R805–R807.

Wooding, S. 2006. Phenylthiocarbamide: A 75-year adventure in genetics and natural selection. *Genetics* 172: 2015–2023.

Wooding, S., U. K. Kim, et al. 2004. Natural selection and molecular evolution in PTC, a bitter-taste receptor gene. *American Journal of Human Genetics* 74: 637–646.

Yamaguchi-Kabata, Y., M. K. Shimada, et al. 2008. Distribution and effects of nonsense polymorphisms in human genes. *PLoS ONE* 3: e3393.

Yarmolinsky, D. A., C. S. Zuker, and N. J. Ryba. 2009. Common sense about taste: From mammals to insects. *Cell* 139: 234–244.

Young, J. M., R. M. Endicott, et al. 2008. Extensive copy-number variation of the human olfactory receptor gene family. *American Journal of Human Genetics* 83: 228–242.

Yu, L., X. Y. Wang, et al. 2010. Adaptive evolution of digestive RNASE1 genes in leaf-eating monkeys revisited: New insights from ten additional colobines. *Molecular Biology and Evolution* 27: 121–131.

Zhang, J. 2003. Parallel functional changes in the digestive RNases of ruminants and colobines by divergent amino acid substitutions. *Molecular Biology and Evolution* 20: 1310–1317.

Zhang, J. 2006. Parallel adaptive origins of digestive RNases in Asian and African leaf monkeys. *Nature Genetics* 38: 819–823.

Zhang, J., Y. P. Zhang, and H. F. Rosenberg. 2002. Adaptive evolution of a duplicated pancreatic ribonuclease gene in a leaf-eating monkey. *Nature Genetics* 30: 411–415.

6

Mendelian Genetics in Populations I: Selection and Mutation

Darwin's theory of evolution by natural selection provides a mechanistic explanation of descent with modification that is supported by considerable evidence. However, as Darwin himself recognized, the theory is incomplete without an accurate understanding of the mechanism of inheritance (see Darwin 1868). That understanding has been provided by Mendelian and molecular genetics. With Darwin's insights and modern genetics, we have the tools we need to develop a more complete model of the mechanism of evolution.

Population genetics, the subject of this chapter (as well as Chapters 7 and 8), integrates evolution by natural selection with Mendelian genetics (for a history, see Provine 1971). The crucial insight of population genetics is that changes in the relative abundance of traits in a population can be tied to changes in the relative abundance of the genetic variants that influence them. A decline over several decades in the frequency of dark-colored Soay sheep in St. Kilda, Scotland, for example, is tied to a decline in the frequency of the dominant allele responsible for dark coloration. From a population geneticist's perspective, evolution can be defined as change across generations in the frequencies of alleles. Population genetics provides the theoretical foundation for much of our modern understanding of evolution.

Some Soay sheep are light, others dark, due to alleles of the gene for tyrosinase-related protein 1 (TRYP1). Recently, the frequency of the *G* allele, which confers dark color, has declined—and with it the frequency of dark sheep (Graph from Gratten et al. 2008; photo by Arpat Ozgul).

Graph from "A localized negative genetic correlation constrains microevolution of coat color in wild sheep." *Science* 319: 318–320. Reprinted with permission from AAAS.

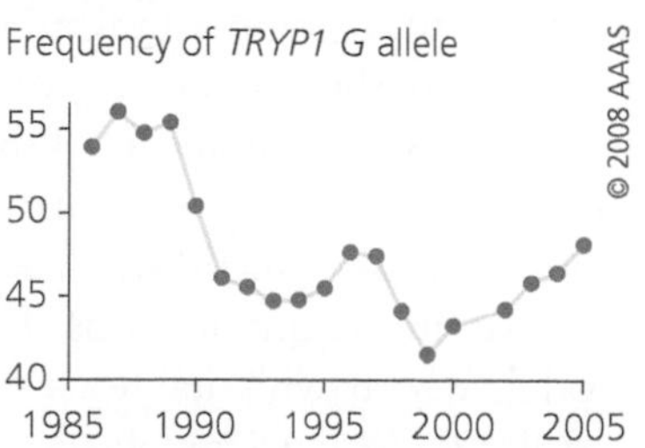

Scientists evaluate theories by using them to make predictions, then checking whether the predictions come true. Some of the clearest tests of theory are found in engineering. When Neil Armstrong and Buzz Aldrin traveled to the surface of the moon and back in July of 1969, they demonstrated that NASA's engineers understand a thing or two about thrust, inertia, and gravity **(Figure 6.1)**. In this chapter, we present data from predictive tests of population genetics theory. At the end, we tell the story of a team of genetic engineers who designed and built a new gene, introduced it into a population of fruit flies, and used population genetics theory to predict the trajectory of its changing frequency 20 generations into the future.

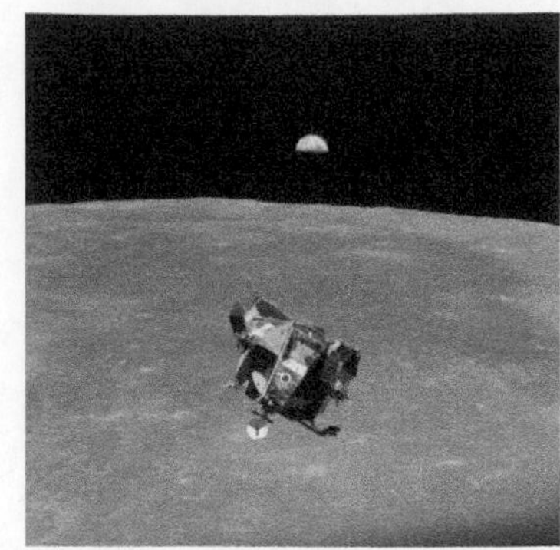

Figure 6.1 Engineering success demonstrates the value of theory *Apollo 11*'s lunar module "Eagle," carrying Neil Armstrong and Buzz Aldrin, returns from the surface of the moon to dock with the command service module. The photo was taken on 21 July 1969 by command module pilot Michael Collins. Note Earth rising in the background.

Our first task, however, is to introduce the fundamentals of population genetics. In Section 6.1, we introduce an algebraic model that allows us to track Mendelian alleles across generations in an idealized population under simplifying assumptions. This model will show us circumstances under which evolution does not occur. In Section 6.2, we relax one of the simplifying assumptions and learn to predict how populations evolve under natural selection. In Section 6.3, we look at data that puts a variety of predictions of evolution by natural selection to the test. In Section 6.4, we relax another assumption to look at mutation as a mechanism of evolution. In Section 6.5, we close with the genetic engineering story. Throughout the chapter, we use population genetics theory to address practical issues arising from human diseases and human evolution. The first of these issues involves human evolution in response to the HIV epidemic.

6.1 Mendelian Genetics in Populations: Hardy–Weinberg Equilibrium

Most people are susceptible to HIV. Their best hope of avoiding infection is to avoid contact with the virus. There are, however, a few individuals who remain uninfected despite repeated exposure. In 1996, AIDS researchers discovered that at least some of this variation in susceptibility has a genetic basis (see Chapters 1 and 5). The gene responsible encodes a cell-surface protein called CCR5. CCR5 is the handle exploited by most sexually transmitted strains of HIV-1 as a means of binding to white blood cells. A mutant allele of the CCR5 gene, called *CCR5-Δ32,* has a 32-base-pair deletion that destroys the encoded protein's ability to function. Individuals who inherit two copies of this allele have no CCR5 on the surface of their cells and are therefore resistant to HIV-1. Given that individuals homozygous for *CCR5-Δ32* are much less likely to contract HIV, we might ask whether the global AIDS epidemic will cause an increase in the frequency of the *Δ32* allele in human populations. If so, how fast will it happen?

Before we can hope to answer such questions, we need to understand how the *CCR5-Δ32* allele would behave without the AIDS epidemic. In other words, we need to develop a null model for the behavior of genes in populations. This null model should specify, under the simplest possible assumptions, what will happen across generations to the frequencies of alleles and genotypes. The model should apply not just to humans, but to any population of organisms that are both diploid and sexual. In this first section of the chapter, we develop such a model and explore its implications. In the next section we add natural selection to the model, which will enable us to address our questions about the AIDS epidemic and the *CCR5-Δ32* allele.

Population genetics begins with a model of what happens to allele and genotype frequencies in an idealized population. Once we know how Mendelian genes behave in the idealized population, we will be able to explore how they behave in real populations.

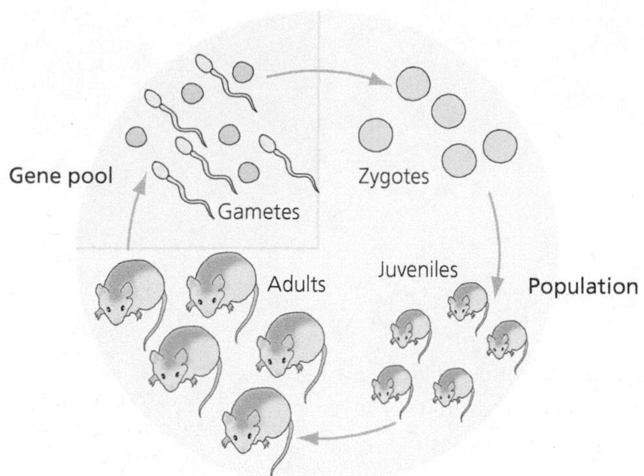

Figure 6.2 The life cycle of an idealized population The labels highlight the stages that will be important in our development of population genetics.

We develop our model by scaling Mendelian genetics up from the level of families, where the reader has used it until now, to the level of populations. We illustrate the model with an idealized population of mice (**Figure 6.2**). A **population** is a group of interbreeding individuals and their offspring. The crucial events in the life cycle of a population are these: Adults produce gametes, gametes combine to make zygotes, zygotes develop into juveniles, and juveniles grow up to become the next generation of adults. We want to track the fate of Mendelian genes in a population. We want to know whether particular alleles or genotypes become more common or less common across generations, and why.

Imagine that the mice in Figure 6.2 have in their genome a Mendelian locus, the A locus, with two alleles: *A* and *a*. We can begin tracking these alleles at any point in the life cycle. We then follow them through one complete turn of the cycle, from one generation to the next, to see if their frequencies change.

A Simulation

Our task of following alleles around the life cycle will be simplest if we start with the gametes produced by the adults when they mate. We will assume that the adults choose their mates at random. A useful mental trick is to picture random mating happening like this: We take all the eggs and sperm produced by all the adults in the population, dump them together in a barrel, and stir. This barrel is known as the **gene pool**. Each sperm in the gene pool swims about at random until it collides with an egg, whereupon the egg and sperm fuse to make a zygote. Something rather like this actually happens in sea urchins and other marine creatures that simply release their gametes onto the tide. For other organisms, like mice and humans, this picture is obviously a simplification.

The adults in our mouse population are diploid, so each carries two copies of the A locus. But the adults made their eggs and sperm by meiosis. Following Mendel's law of segregation, each gamete received just one copy of the A locus. Imagine that 60% of the eggs and sperm received a copy of allele *A*, and 40% received a copy of allele *a*. That is, the frequency of allele *A* in the gene pool is 0.6, and the frequency of allele *a* is 0.4.

Starting with the eggs and sperm that constitute the gene pool, our model tracks alleles through zygotes and adults and into the next generation's gene pool.

What happens when eggs meet sperm? For example, what fraction of the zygotes they produce have genotype *AA*? And once these zygotes develop into juveniles, grow up, and spawn, what are the frequencies of alleles *A* and *a* in the next generation's gene pool?

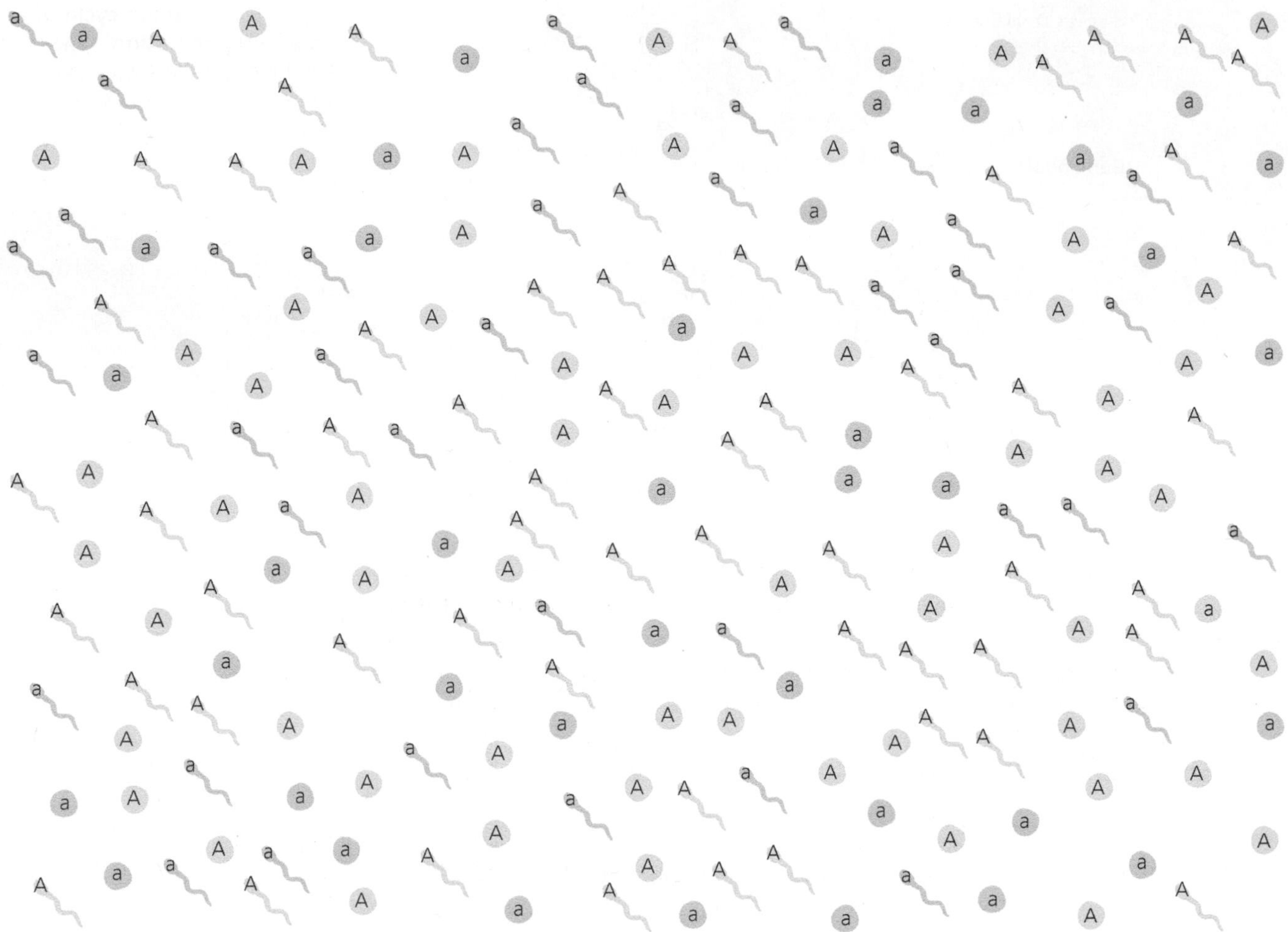

Figure 6.3 A gene pool with frequencies of 0.6 for allele *A* and 0.4 for allele *a*

One way to find out is by simulation. We can close our eyes and put a finger down on **Figure 6.3** to choose an egg. Perhaps it carries a copy of allele *A*. Now we close our eyes and put down a finger to choose a sperm. Perhaps it carries a copy of allele *a*. If we combine these gametes, we get a zygote with genotype *Aa*. We encourage the reader to carry out this process to make a large sample of zygotes—50, say, or even 100. We have paused to do so as we write. Among the 100 zygotes we made, 34 had genotype *AA*, 57 had *Aa*, and 9 had *aa*.

Now let us imagine that all these zygotes develop into juveniles, and that all the juveniles survive to adulthood. Imagine, furthermore, that when the adults reproduce, they all donate the same number of gametes to the gene pool. We can choose any number of gametes we like for the standard donation, so we will choose 10 to make the arithmetic easy. We are not worried about whether a particular adult makes eggs or sperm; instead, we are simply counting gametes:

Our 34 *AA* adults together make 340 gametes: 340 carry allele *A*; none carry allele *a*.
Our 57 *Aa* adults together make 570 gametes: 285 carry allele *A*; 285 carry allele *a*.
Our 9 *aa* adults together make 90 gametes: none carry allele *A*; 90 carry allele *a*.

Summing the gametes carrying copies of each allele, we get 625 carrying *A* and 375 carrying *a*, for a total of 1,000. The frequency of allele *A* in the new gene pool is 0.625; the frequency of allele *a* is 0.375.

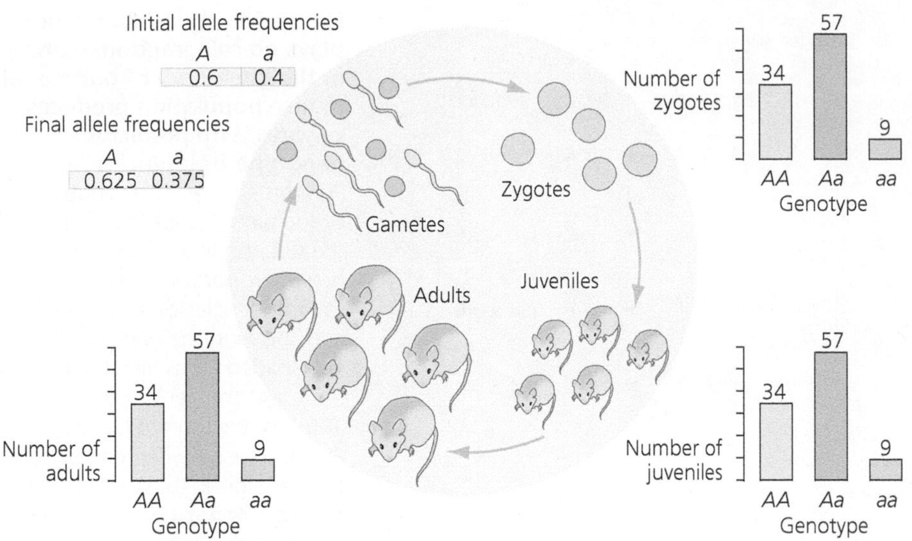

Figure 6.4 Allele and genotype frequencies throughout the life cycle in a numerical simulation We made the zygotes by picking gametes at random from the gene pool in Figure 6.3 and assumed that all the zygotes survived. The reader's results, on repeating this exercise, will likely be somewhat different.

We have followed the alleles around one complete turn of the population's life cycle and found that their ending frequencies are somewhat different from their starting frequencies **(Figure 6.4)**. In other words, our population has evolved.

The genotype frequencies among the zygotes in the reader's sample, and the frequencies of the alleles in the reader's next generation, will almost certainly be somewhat different from ours. Indeed, we carried out the simulation two more times ourselves and got different results each time. In our second simulation, we got zygotes in proportions of 41% *AA*, 44% *Aa*, and 15% *aa*. The allele frequencies in the next generation's gene pool were 0.63 for *A* and 0.37 for *a*. In our third simulation, we got zygotes in proportions of 34% *AA*, 49% *Aa*, and 17% *aa*. The allele frequencies in the next generation were 0.585 for *A* and 0.415 for *a*.

In simulated populations, allele frequencies change somewhat across generations. This is evolution resulting from blind luck.

Our three results are not wildly divergent, but neither are they identical. In two cases the frequency of allele *A* rose, whereas in one it fell. We got different results because in each simulation, blind luck in picking gametes from the gene pool gave us a different number of zygotes with each genotype. The fact that blind luck can cause a population to evolve unpredictably is an important result of population genetics. This mechanism of evolution is called **genetic drift**. (We will return to drift in Chapter 7.) For now, however, we are interested not in whether evolution is sometimes unpredictable, but whether it is ever predictable. We want to know what would have happened in our simulations if chance had played no role.

A Numerical Calculation

We can discover the luck-free result of combining eggs and sperm to make zygotes by using a Punnett square. Punnett squares, invented by Reginald Crundall Punnett, are more typically used in Mendelian genetics to predict the genotypes among the offspring of a particular male and female. **Figure 6.5**, for example, shows the Punnett square for a mating between an *Aa* female and an *Aa* male. We write the genotypes of the eggs made by the female, in the proportions we expect her to make them, along the side of the square. We write the genotypes of the sperm made by the male, in the proportions we expect him to make them, along the top. Then we fill in the boxes in the square to get the genotypes of the zygotes. This Punnett square predicts that among the offspring of an *Aa* female and an *Aa* male, one-quarter will be *AA*, one-half *Aa*, and one-quarter *aa*.

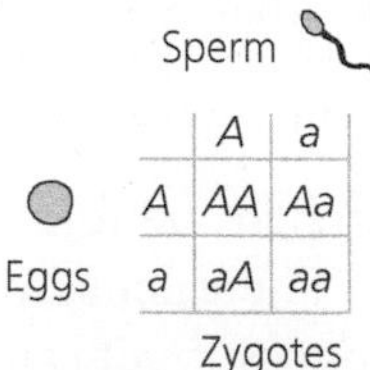

Figure 6.5 Punnett square for a cross between two heterozygotes This device makes accurate predictions about the genotype frequencies among the zygotes because the genotypes of the eggs and sperm are represented in the proportions in which the parents produce them.

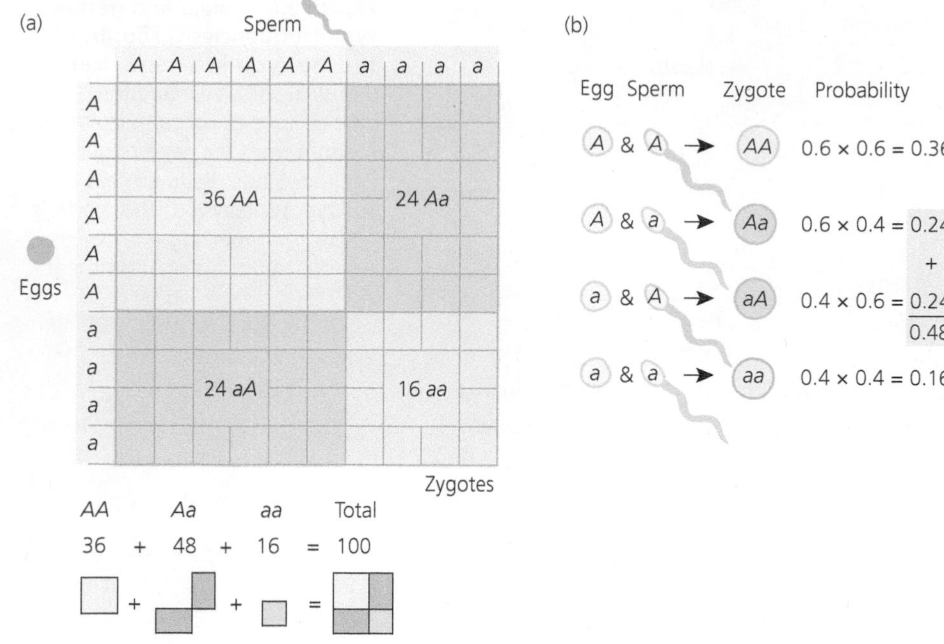

Figure 6.6 When blind luck plays no role, random mating in the gene pool of our model mouse population produces zygotes with predictable genotype frequencies (a) A Punnett square. The genotypes of the gametes are listed along the left and top edges of the box in proportions that reflect the frequencies of *A* and *a* eggs and sperm in the gene pool. The shaded areas inside the box represent the genotypes among 100 zygotes formed by random encounters between gametes in the gene pool. (b) We can also calculate genotype frequencies among the zygotes by multiplying allele frequencies. (See Computing Consequences 6.1.)

We can use the same device to predict the genotypes among the offspring of an entire population **(Figure 6.6a)**. The trick is to write the egg and sperm genotypes along the side and top of the Punnett square in proportions that reflect their frequencies in the gene pool. Sixty percent of the eggs carry copies of allele *A* and 40% carry copies of allele *a*, so we have written six *A*'s and four *a*'s along the side of the square. Likewise, for the sperm, we have written six *A*'s and four *a*'s along the top. Filling in the boxes in the square, we find that among 100 zygotes in our population, we can expect 36 *AA*'s, 48 *Aa*'s, and 16 *aa*'s. Note that our population Punnett square has predicted genotype proportions different from the 1:0, 1:1, or 1:2:1 ratios that appear in single-family Punnett squares.

The Punnett square in Figure 6.6a suggests that we could also predict the genotype frequencies among the zygotes by multiplying probabilities. Figure 6.6b shows the four possible combinations of egg and sperm, the resulting zygotes, and a calculation specifying the probability of each (see also **Computing Consequences 6.1**). For example, if we look into the gene pool and pick an egg to watch, there is a 60% chance that it will have genotype *A*. When a sperm comes along to fertilize the egg, there is a 60% chance that the sperm will have genotype *A*. The probability that we will witness the production of an *AA* zygote is therefore

$$0.6 \times 0.6 = 0.36$$

If we watched the formation of all the zygotes, 36% of them would have genotype *AA*. The calculations in Figure 6.6b show that random mating in the gene pool produces zygotes in the following proportions:

AA	*Aa*	*aa*
0.36	0.48	0.16

(The *Aa* category includes heterozygotes produced by combining either an *A* egg with an *a* sperm or an *a* egg with an *A* sperm.) Notice that

$$0.36 + 0.48 + 0.16 = 1$$

This confirms that we have accounted for all of the zygotes.

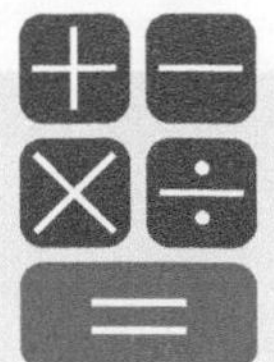

COMPUTING CONSEQUENCES 6.1

Combining probabilities

The combined probability that two independent events will occur together is the product of their individual probabilities. For example, the probability that a tossed penny will come up heads is $\frac{1}{2}$. The probability that a tossed dime will come up heads is also $\frac{1}{2}$. If we toss both together, the outcome for the penny is independent of that for the dime. Thus the probability of getting heads on the penny and heads on the dime is $\frac{1}{2} \times \frac{1}{2} = \frac{1}{4}$.

The combined probability that one or the other of two mutually exclusive events will occur is the sum of their individual probabilities. When rolling a die we can get a one or we can get a two (among other possibilities), but we cannot get both a one and a two at once. The individual probability of each outcome is $\frac{1}{6}$. The combined probability of rolling either a one or a two is therefore $\frac{1}{6} + \frac{1}{6} = \frac{1}{3}$.

We now let the zygotes grow to adulthood, and we let the adults produce gametes to make the next generation's gene pool. When chance plays no role, will the frequencies of alleles A and a in the new gene pool change from one generation to the next?

If we assume, as we did before, that 100 adults each make 10 gametes, then

The 36 AA adults together make 360 gametes: 360 carry allele A; none carry allele a.

The 48 Aa adults together make 480 gametes: 240 carry allele A; 240 carry allele a.

The 16 aa adults together make 160 gametes: none carry allele A; 160 carry allele a.

Summing the gametes carrying each allele, we get 600 carrying copies of A and 400 carrying copies of a, for a total of 1,000. The frequency of allele A in the new gene pool is 0.6; the frequency of allele a is 0.4.

As **Figure 6.7a** shows graphically, we can also calculate the composition of the new gene pool using frequencies. Because adults of genotype AA constitute

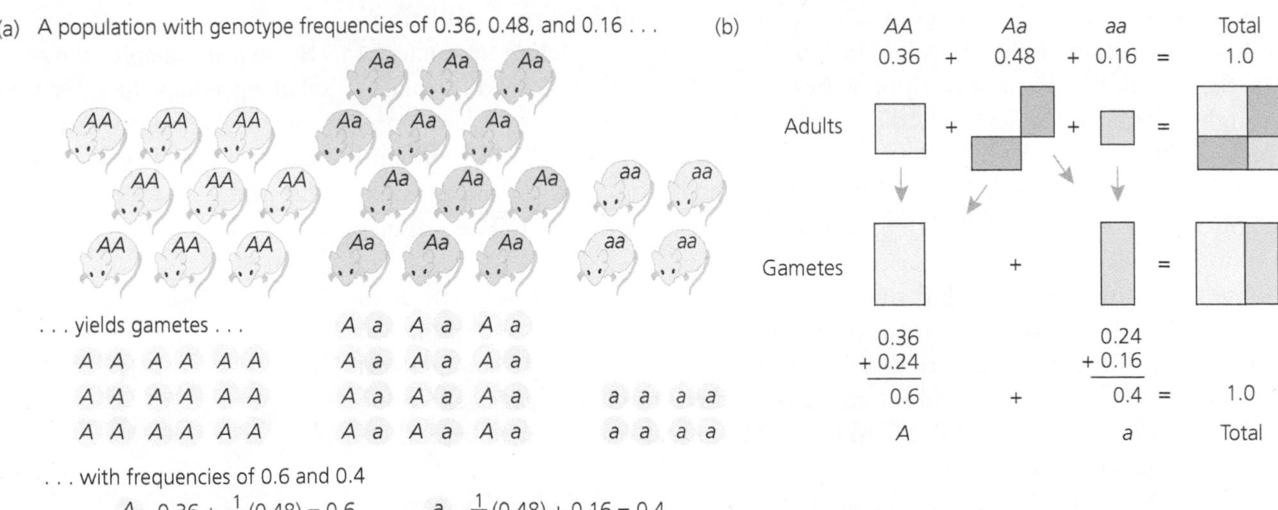

Figure 6.7 When the adults in our model mouse population make gametes, they produce a gene pool in which the allele frequencies are identical to the ones we started with a generation ago (a) Calculations using frequencies. (b) A geometrical representation. The area of each box represents the frequency of an adult or gamete genotype. Note that half the gametes produced by *Aa* adults carry allele *A*, and half carry allele *a*.

36% of the population, they will make 36% of the gametes. All of these gametes carry copies of allele A. Likewise, adults of genotype Aa constitute 48% of the population and will make 48% of the gametes. Half of these gametes carry copies of allele A. So the total fraction of the gametes in the gene pool that carry copies of A is

$$0.36 + (\tfrac{1}{2})0.48 = 0.6$$

The figure also shows a calculation establishing that the fraction of the gametes in the gene pool that carry copies of allele a is 0.4. Notice that

$$0.6 + 0.4 = 1$$

This confirms that we have accounted for all of the gametes. Figure 6.7b shows a geometrical representation of the same calculations.

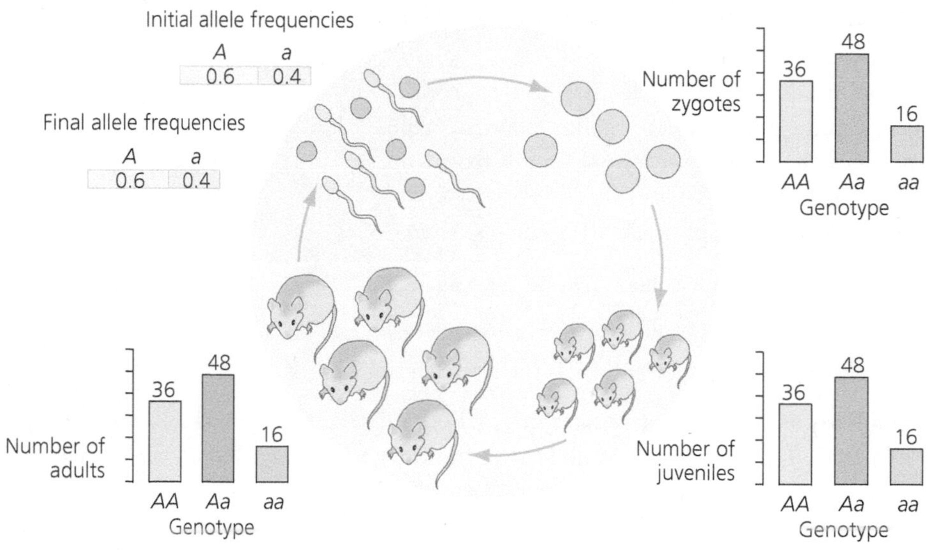

Figure 6.8 When blind luck plays no role in our model population, the allele frequencies do not change from one generation to the next We made the zygotes with the Punnett square in Figure 6.6 and assumed that all the zygotes survived.

We have come full circle **(Figure 6.8)**. And this time, unlike in our simulations, we have arrived precisely where we began. We started with allele frequencies of 60% for A and 40% for a in our population's gene pool. We followed the alleles through zygotes, juveniles, and adults and into the next generation's gene pool. The allele frequencies in the new gene pool are still 60% and 40%. When blind luck plays no role, the allele frequencies for A and a in our population are in equilibrium: They do not change from one generation to the next. The population does not evolve.

Numerical examples show that when blind luck plays no role, allele frequencies remain constant from one generation to the next.

The first biologist to work a numerical example, tracing the frequencies of Mendelian alleles from one generation to the next in an ideal population, was G. Udny Yule in 1902. He started with a gene pool in which the frequencies of two alleles were 0.5 and 0.5 and showed that in the next generation's gene pool, the allele frequencies were still 0.5 and 0.5. The reader may want to reproduce Yule's calculations as an exercise.

Like us, Yule concluded that the allele frequencies in his imaginary population were in equilibrium. Yule's conclusion was both groundbreaking and correct, but he took it a bit too literally. He had worked only one example, and he believed that allele frequencies of 0.5 and 0.5 represented the only possible equilibrium state for a two-allele system. For example, Yule believed that if a single copy of allele A appeared as a mutation in a population whose gene pool otherwise contained only copies of a, then the A allele would automatically increase in frequency until copies of it constituted one-half of the gene pool. Yule argued this claim during the discussion that followed a talk given in 1908 by none other than Reginald Punnett. Punnett thought that Yule was wrong, but he did not know how to prove it.

We have already demonstrated, of course, that Punnett was correct in rejecting Yule's claim. Our calculations showed that a population with allele frequencies of 0.6 and 0.4 is in equilibrium too. What Punnett wanted, however, is a general proof. This proof should show that any allele frequencies, so long as they sum to 1, will remain unchanged from one generation to the next.

Punnett took the problem to his mathematician friend Godfrey H. Hardy, who produced the proof in short order (Hardy 1908). Hardy simply repeated the calculations that Yule had performed, using variables in place of the specific allele frequencies that Yule had assumed. Hardy's calculation of the general case indeed showed that any allele frequencies can be in equilibrium.

The General Case

For our version of Hardy's general case, we again work with our imaginary mouse population. We are concerned with a single locus with two alleles: A_1 and A_2. We use capital letters with subscripts because we want our calculation to cover cases in which the alleles are codominant as well as cases in which they are dominant and recessive. The three possible diploid genotypes are A_1A_1, A_1A_2, and A_2A_2.

As in our simulations and numerical example, we will start with the gene pool and follow the alleles through one complete turn of the life cycle. The gene pool will contain some frequency of A_1 gametes and some frequency of A_2 gametes. We will call the frequency of A_1 in the gene pool p and the frequency of A_2 in the gene pool q. There are only two alleles in the population, so

$$p + q = 1$$

The first step is to let the gametes in the gene pool combine to make zygotes. **Figure 6.9a** shows the four possible combinations of egg and sperm, the zygotes they produce, and a calculation specifying the probability of each. For example, if we pick an egg to watch at random, the chance is p that it will have genotype A_1. When a sperm comes along to fertilize the egg, the chance is p that the sperm will have genotype A_1. The probability that we will witness the production of an A_1A_1 zygote is therefore

$$p \times p = p^2$$

If we watched the formation of all the zygotes, p^2 of them would have genotype A_1A_1. The calculations in Figure 6.9a show that random mating in our gene pool produces zygotes in the following proportions:

$$\begin{array}{ccc} A_1A_1 & A_1A_2 & A_2A_2 \\ p^2 & 2pq & q^2 \end{array}$$

Figure 6.9b shows a Punnett square that yields the same genotype frequencies. The Punnett square also shows geometrically that

$$p^2 + 2pq + q^2 = 1$$

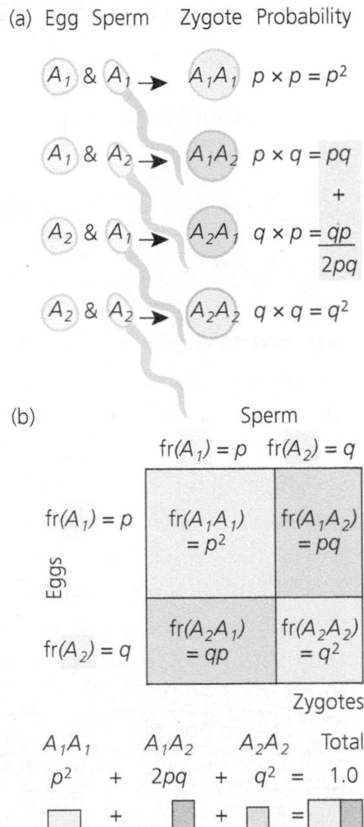

Figure 6.9 The general case for random mating in our model population (a) We can predict the genotype frequencies among the zygotes by multiplying the allele frequencies. (b) A Punnett square. The variables along the left and top edges of the box represent the frequencies of A and a eggs and sperm in the gene pool. The expressions inside the box represent the genotype frequencies among zygotes formed by random encounters between gametes in the gene pool.

This confirms that we have accounted for all the zygotes. The same result can be demonstrated algebraically by substituting $(1 - p)$ for q in the expression $p^2 + 2pq + q^2$, then simplifying.

The challenge now is to prove algebraically that there was nothing special about our numerical examples. Any allele frequencies will remain constant from generation to generation.

We have gone from the allele frequencies in the gene pool to the genotype frequencies among the zygotes. We now let the zygotes develop into juveniles, let the juveniles grow up to become adults, and let the adults produce gametes to make the next generation's gene pool.

We can calculate the frequency of allele A_1 in the new gene pool as follows. Because adults of genotype A_1A_1 constitute a proportion p^2 of the population, they will make p^2 of the gametes. All of these gametes carry copies of allele A_1. Likewise, adults of genotype A_1A_2 constitute a proportion $2pq$ of the population, and will make $2pq$ of the gametes. Half of these gametes carry copies of allele A_1. So the total fraction of the gametes in the gene pool that carry copies of A_1 is

$$p^2 + (\tfrac{1}{2})2pq = p^2 + pq$$

We can simplify the expression on the right by substituting $(1 - p)$ for q. This gives

$$\begin{aligned} p^2 + pq &= p^2 + p(1 - p) \\ &= p^2 + p - p^2 \\ &= p \end{aligned}$$

Figure 6.10 shows this calculation graphically. The figure also shows a calculation establishing that the fraction of the gametes in the gene pool that carry copies of allele A_2 is q. We assumed at the outset that p and q sum to 1, so we know that we have accounted for all the gametes.

Once again, we have come full circle and are back where we started. We started with allele frequencies of p and q in our population's gene pool. We followed the alleles through zygotes and adults and into the next generation's gene pool. The allele frequencies in the new gene pool are still p and q. The allele frequencies p and q can be stable at any values at all between 0 and 1, as long as they sum to 1. In other words, any allele frequencies can be in equilibrium, not just $p = q = 0.5$ as Yule thought.

Our model has shown that our idealized population does not evolve. This conclusion is known as the Hardy–Weinberg equilibrium principle.

This is a profound result. At the beginning of the chapter we defined evolution as change in allele frequencies in populations. The calculations we just performed show, given simple assumptions, that in populations following the rules of Mendelian genetics, allele frequencies do not change.

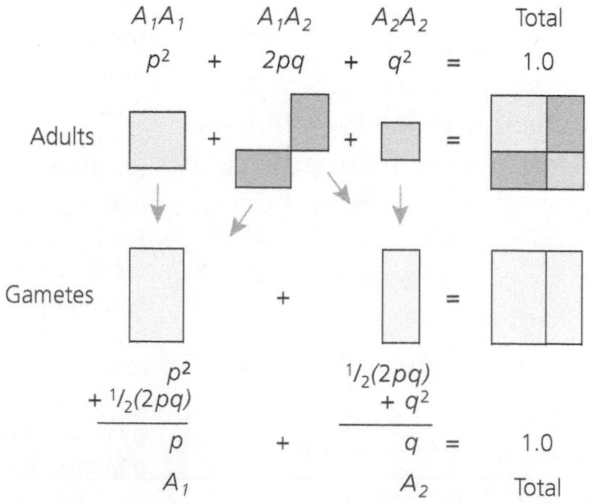

A_1A_1 A_1A_2 A_2A_2 Total

$p^2 + 2pq + q^2 = 1.0$

Adults

Gametes

$p^2 + \frac{1}{2}(2pq) = p$ + $\frac{1}{2}(2pq) + q^2 = q$ = 1.0

A_1 A_2 Total

Figure 6.10 A geometrical representation of the general case for the allele frequencies produced when the adults in our model population make gametes The area of each box represents the frequency of an adult or gamete genotype.

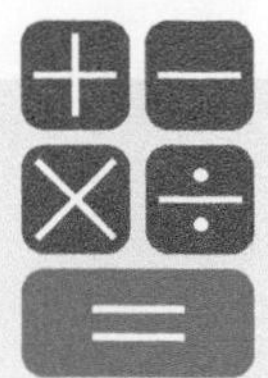

COMPUTING CONSEQUENCES 6.2

The Hardy–Weinberg equilibrium principle with more than two alleles

Imagine a single locus with several alleles. We can call the alleles A_i, A_j, A_k, and so on, and we can represent the frequencies of the alleles in the gene pool with the variables p_i, p_j, p_k, and so on. The formation of a zygote with genotype A_iA_i requires the union of an A_i egg with an A_i sperm. Thus, the frequency of the homozygous genotype A_iA_i is p_i^2. The formation of a zygote with genotype A_iA_j requires either the union of an A_i egg with an A_j sperm, or an A_j egg with an A_i sperm. Thus, the frequency of the heterozygous genotype A_iA_j is $2p_ip_j$.

For example, if there are three alleles with frequencies p_1, p_2, and p_3 such that

$$p_1 + p_2 + p_3 = 1$$

then the genotype frequencies are given by

$$(p_1 + p_2 + p_3)^2 = p_1^2 + p_2^2 + p_3^2 + 2p_1p_2 + 2p_1p_3 + 2p_2p_3$$

and the allele frequencies do not change from generation to generation.

We have presented this result as the work of Hardy (1908). It was derived independently by Wilhelm Weinberg (1908) and has become known as the Hardy–Weinberg equilibrium principle. Some evolutionary biologists refer to it as the Hardy–Weinberg–Castle equilibrium principle, because William Castle (1903) worked a numerical example and stated the general equilibrium principle nonmathematically five years before Hardy and Weinberg explicitly proved the general case (see Provine 1971). The Hardy–Weinberg equilibrium principle yields two fundamental conclusions:

- **Conclusion 1:** The allele frequencies in a population will not change, generation after generation.
- **Conclusion 2:** If the allele frequencies in a population are given by p and q, the genotype frequencies will be given by p^2, $2pq$, and q^2.

We get an analogous result if we generalize the analysis from the two-allele case to the usual case of a population containing many alleles at a locus (see **Computing Consequences 6.2**).

What Use Is the Hardy–Weinberg Equilibrium Principle?

It may seem puzzling that in a book about evolution we have devoted so much space to a proof apparently showing that evolution does not happen. Evolution does, of course, happen—we saw it happen in this chapter in our own simulations. What makes the Hardy–Weinberg equilibrium principle useful is that it rests on a specific set of simple assumptions. When one or more of these assumptions is violated, the Hardy–Weinberg conclusions no longer hold.

We left some of the assumptions unstated when we developed our null model of Mendelian alleles in populations. We can now make them explicit. The crucial assumptions are as follows:

1. There is no selection. All members of our model population survived at equal rates and contributed equal numbers of gametes to the gene pool. When this assumption is violated—when individuals with some genotypes survive and reproduce at higher rates than others—the frequencies of alleles may change from one generation to the next.

2. There is no mutation. In the model population, no copies of existing alleles were converted by mutation into copies of other existing alleles, and no new alleles were created. When this assumption is violated, and, for example, some alleles have higher mutation rates than others, allele frequencies may change from one generation to the next.

3. There is no migration. No individuals moved into or out of the model population. When this assumption is violated, and individuals carrying some alleles move into or out of the population at higher rates than individuals carrying other alleles, allele frequencies may change from one generation to the next.

4. There are no chance events that cause individuals with some genotypes to pass more of their alleles to the next generation than others. That is, blind luck plays no role. We saw the influence of blind luck in our simulations. We avoided it in our analysis of the general case by assuming that the eggs and sperm in the gene pool collided with each other at their actual frequencies of p and q and that no deviations were caused by chance. Another way to state this assumption is that the model population was infinitely large. When this assumption is violated, and by chance some individuals contribute more alleles to the next generation than others, allele frequencies may change from one generation to the next. This mechanism of allele frequency change is called, as we said earlier, genetic drift.

5. Individuals choose their mates at random. We explicitly set up the gene pool to let gametes find each other at random. In contrast to assumptions 1 through 4, when this assumption is violated—when, for example, individuals prefer to mate with other individuals of the same genotype—allele frequencies do not change from one generation to the next. Genotype frequencies may change, however. Such shifts in genotype frequency, in combination with a violation of one of the other four assumptions, can influence the evolution of populations.

The Hardy–Weinberg equilibrium principle becomes useful when we list the assumptions we made about our idealized population. By providing a set of explicit conditions under which evolution does not happen, the Hardy–Weinberg analysis identifies the mechanisms that can cause evolution in real populations.

By furnishing a list of ideal conditions under which populations will not evolve, the Hardy–Weinberg equilibrium principle identifies the set of events that can cause evolution in the real world **(Figure 6.11)**. This is how the Hardy–Weinberg principle serves as a null model. Biologists can measure allele and genotype frequencies in nature, and determine whether the Hardy–Weinberg conclusions

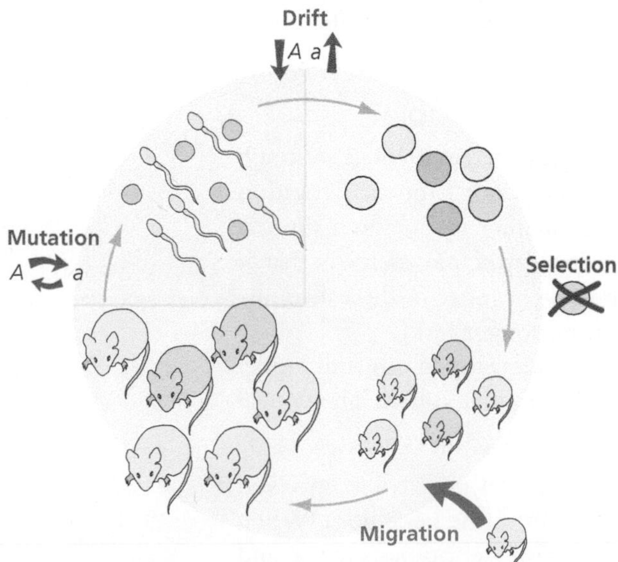

Figure 6.11 Summary of the mechanisms of evolution Four processes can cause allele frequencies to change from one generation to the next. Selection occurs when individuals with different genotypes survive or make gametes at different rates. Migration occurs when individuals move into or out of the population. Mutation occurs when mistakes during meiosis turn copies of one allele into copies of another. Genetic drift occurs when blind chance allows gametes with some genotypes to participate in more fertilizations than gametes with other genotypes.

hold. A population in which they hold is said to be in **Hardy–Weinberg equilibrium**. If a population is not in Hardy–Weinberg equilibrium—if the allele frequencies change from generation to generation or if the genotype frequencies cannot, in fact, be predicted by multiplying the allele frequencies—then one or more of the Hardy–Weinberg model's assumptions are being violated. Such a discovery does not, by itself, tell us which assumptions are being violated, but it indicates that further research may be rewarded with interesting discoveries.

In the remaining sections of Chapter 6, we consider how violations of assumptions 1 and 2 affect the two Hardy–Weinberg conclusions, and we explore empirical research on selection and mutation as mechanisms of evolution. (In Chapter 7, we consider violations of assumptions 3, 4, and 5.)

Changes in the Frequency of the *CCR5-Δ32* Allele

We began this section by asking whether we can expect the frequency of the *CCR5-Δ32* allele to change in human populations. Now that we have developed a null model for how Mendelian alleles behave in populations, we can give a partial answer. As long as individuals of all CCR5 genotypes survive and reproduce at equal rates, as long as no mutations convert some CCR5 alleles into others, as long as no one moves from one population to another, as long as populations are infinitely large, and as long as people choose their mates at random, then no, the frequency of the *CCR5-Δ32* allele will not change.

This answer is, of course, thoroughly unsatisfying. It is unsatisfying because none of the assumptions will be true in any real population. We asked the question in the first place precisely because we expect *Δ32/Δ32* individuals to survive the AIDS epidemic at higher rates than individuals with either of the other two genotypes. In the next two sections, we will see that our null model, the Hardy–Weinberg equilibrium principle, provides a framework that allows us to assess with precision the importance of differences in survival.

6.2 Selection

Our analysis in Section 6.1 was motivated by a desire to predict whether the frequency of the *CCR5-Δ32* allele will change as a result of the AIDS epidemic. We started by scaling Mendelian genetics up from single crosses to whole populations. This is the first step in integrating Mendelism with Darwin's theory of evolution by natural selection. The next step is to add differences in survival and reproductive success. Doing so makes the algebra a bit more complicated. But it also lets us glimpse the predictive strength of population genetics.

In the model we used to derive the Hardy–Weinberg equilibrium principle, first on our list of assumptions was that all individuals survive at equal rates and contribute equal numbers of gametes to the gene pool. Systematic violations of this assumption are examples of **selection**. Selection happens when individuals with particular phenotypes survive to sexual maturity at higher rates than those with other phenotypes, or when individuals with particular phenotypes produce more offspring during reproduction than those with other phenotypes. The bottom line in either kind of selection is differential reproductive success. Some individuals have more offspring than others. Selection can lead to evolution when the phenotypes that exhibit differences in reproductive success are heritable—that is, when certain phenotypes are associated with certain genotypes.

First on the list of assumptions about our idealized population was that individuals survive at equal rates and have equal reproductive success. We now explore what happens to allele frequencies when this assumption is violated.

Population geneticists often assume that phenotypes are determined strictly by genotypes. They might, for example, think of pea plants as being either tall or short, such that individuals with the genotypes *TT* and *Tt* are tall and individuals with the genotype *tt* are short. Such a view is at least roughly accurate for some traits, including the examples we use in this chapter.

When phenotypes fall into discrete classes that appear to be determined strictly by genotypes, we can think of selection as if it acts directly on the genotypes. We can then assign a particular level of lifetime reproductive success to each genotype. In reality, most phenotypic traits are not, in fact, strictly determined by genotype. Pea plants with the genotype *TT*, for example, vary in height. This variation is due to genetic differences at other loci and to differences in the environments where the pea plants grew. We will consider such complications elsewhere (see Chapter 9). For the present, however, we adopt the simple view.

When we think of selection as if it acts directly on genotypes, its defining feature is that some genotypes contribute more alleles to future generations than others. In other words, there are differences among genotypes in fitness.

Our task in this section is to incorporate selection into the Hardy–Weinberg analysis. We begin by asking whether selection can change the frequencies of alleles in the gene pool from one generation to the next. In other words, can violation of the no-selection assumption lead to a violation of conclusion 1 of the Hardy–Weinberg equilibrium principle?

Adding Selection to the Hardy–Weinberg Analysis: Changes in Allele Frequencies

We start with a numerical example showing that selection can, indeed, change the frequencies of alleles. Imagine that in our population of mice there is a locus, the B locus, that affects the probability of survival. Assume that the frequency of allele B_1 in the gene pool is 0.6 and the frequency of allele B_2 is 0.4 **(Figure 6.12)**. After random mating, we get genotype frequencies for B_1B_1, B_1B_2, and B_2B_2 of 0.36, 0.48, and 0.16. The rest of our calculations will be simpler if we give the population of zygotes a finite size, so imagine that there are 100 zygotes:

B_1B_1	B_1B_2	B_2B_2
36	48	16

These zygotes are represented by a bar graph on the upper right in the figure. We will follow the individuals that develop from these zygotes. Those that survive to adulthood will breed to produce the next generation's gene pool.

We incorporate selection by stipulating that the genotypes differ in survival. All of the B_1B_1 individuals survive, 75% of the B_1B_2 individuals survive, and 50% of the B_2B_2 individuals survive. As shown in Figure 6.12, we now have 80 adults:

B_1B_1	B_1B_2	B_2B_2
36	36	8

If we assume that each survivor donates 10 gametes to the new gene pool, then

The 36 B_1B_1 adults together make 360 gametes: 360 carry B_1; none carry B_2.

The 36 B_1B_2 adults together make 360 gametes: 180 carry B_1; 180 carry B_2.

The 8 B_2B_2 adults together make 80 gametes: none carry B_1; 80 carry B_2.

A numerical example shows that when individuals with some genotypes survive at higher rates than individuals with other genotypes, allele frequencies can change from one generation to the next. In other words, our model shows that natural selection causes evolution.

Summing the gametes carrying copies of each allele, we get 540 carrying copies of B_1 and 260 carrying copies of B_2, for a total of 800. The frequency of allele B_1

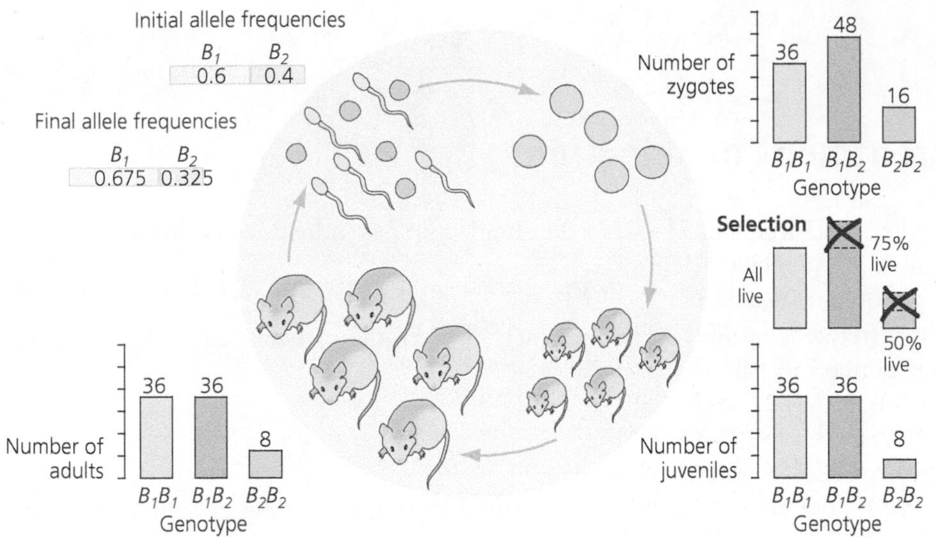

Figure 6.12 Selection can cause allele frequencies to change across generations This figure follows our model mouse population from one generation's gene pool to the next generation's gene pool. The bar graphs show the number of individuals of each genotype in the population at any given time. Selection, in the form of differences in survival among juveniles, causes the frequency of allele B_1 to increase.

in the new gene pool is $\frac{540}{800} = 0.675$; the frequency of allele B_2 is $\frac{260}{800} = 0.325$. The frequency of allele B_1 has risen by an increment of 7.5 percentage points. The frequency of allele B_2 has dropped by the same amount.

Violation of the no-selection assumption has resulted in violation of conclusion 1 of the Hardy–Weinberg analysis. The population has evolved.

We used strong selection to make a point in our numerical example. Rarely in nature are differences in survival rates large enough to cause such dramatic change in allele frequencies in a single generation. If selection continues for many generations, however, even small changes in allele frequency in each generation can add up to substantial changes over the long run.

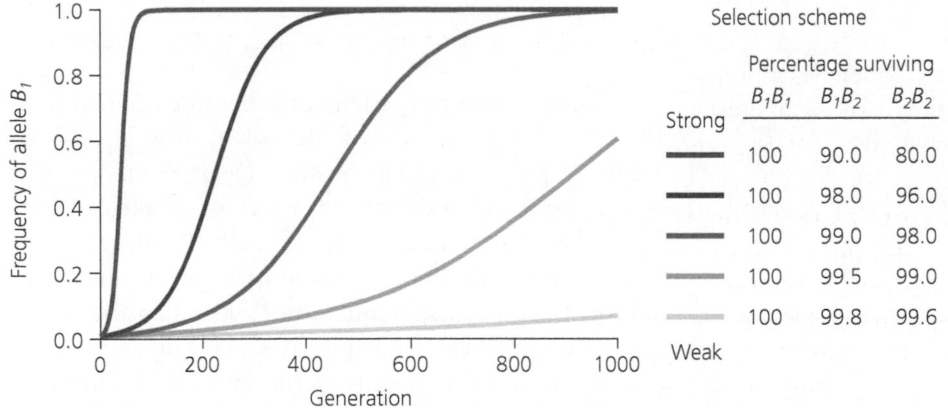

Figure 6.13 Persistent selection can produce substantial changes in allele frequencies over time Each curve shows the change in allele frequency over time under a particular selection intensity.

Figure 6.13 illustrates the cumulative change in allele frequencies that can be wrought by selection. The figure is based on a model population similar to the one we used in the preceding numerical example, except that the initial allele frequencies are 0.01 for B_1 and 0.99 for B_2. The red line shows the change in allele frequencies when the survival rates are 100% for B_1B_1, 90% for B_1B_2, and 80% for B_2B_2. The frequency of allele B_1 rises from 0.01 to 0.99 in less than 100 generations. Under weaker selection schemes, the frequency of B_1 rises more slowly, but still inexorably. (See **Computing Consequences 6.3** for a general algebraic treatment incorporating selection into the Hardy–Weinberg analysis.)

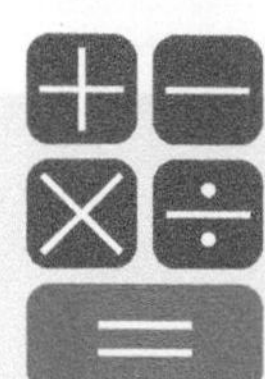

COMPUTING CONSEQUENCES 6.3

A general treatment of selection

Here we develop equations that predict allele frequencies in the next generation, given allele frequencies in this generation and fitnesses for the genotypes. We start with a gene pool in which allele A_1 is at frequency p and allele A_2 is at frequency q. We allow gametes to pair at random to make zygotes of genotypes A_1A_1, A_1A_2, and A_2A_2 at frequencies p^2, $2pq$, and q^2, respectively. We incorporate selection by imagining that A_1A_1 zygotes survive to adulthood at rate w_{11}, A_1A_2 zygotes survive at rate w_{12}, and A_2A_2 zygotes survive at rate w_{22}. All survivors produce the same number of offspring. Therefore, a genotype's survival rate is proportional to the genotype's lifetime reproductive success, or fitness. We thus refer to the survival rates as fitnesses. The average fitness for the whole population, $\overline{w}$, is given by

$$\overline{w} = p^2w_{11} + 2pqw_{12} + q^2w_{22}$$

[To see this, note that we can calculate the average of the numbers 1, 2, 2, and 3 as $\frac{(1 + 2 + 2 + 3)}{4}$ or as $(\frac{1}{4} \times 1) + (\frac{1}{2} \times 2) + (\frac{1}{4} \times 3)$. Our expression for the average fitness is of the second form: We multiply the fitness of each genotype by its frequency in the population and sum the results.]

We now calculate the genotype frequencies among the surviving adults (right before they make gametes). The new frequencies of the genotypes are

A_1A_1	A_1A_2	A_2A_2
$\frac{p^2w_{11}}{\overline{w}}$	$\frac{2pqw_{12}}{\overline{w}}$	$\frac{q^2w_{22}}{\overline{w}}$

(We have to divide by the average fitness in each case to ensure that the new frequencies still sum to 1.)

Finally, we let the adults breed, and calculate the allele frequencies in the new gene pool:

- For the A_1 allele: A_1A_1 individuals contribute $\frac{p^2w_{11}}{\overline{w}}$ of the gametes, all of them A_1, and A_1A_2 individuals contribute $\frac{2pqw_{12}}{\overline{w}}$ of the gametes, half of them A_1. The new frequency of A_1 is thus
$$\frac{p^2w_{11} + pqw_{12}}{\overline{w}}$$
- For the A_2 allele: A_1A_2 individuals contribute $\frac{2pqw_{12}}{\overline{w}}$ of the gametes, half of them A_2; A_2A_2 individuals contribute $\frac{q^2w_{22}}{\overline{w}}$ of the gametes, all of them A_2. So the new frequency of A_2 is
$$\frac{pqw_{12} + q^2w_{22}}{\overline{w}}$$

The reader should confirm that the new frequencies of A_1 and A_2 sum to 1.

It is instructive to calculate the change in the frequency of allele A_1 from one generation to the next. This value, Δp, is the new frequency of A_1 minus the old frequency of A_1:

$$\begin{aligned}\Delta p &= \frac{p^2w_{11} + pqw_{12}}{\overline{w}} - p \\ &= \frac{p^2w_{11} + pqw_{12}}{\overline{w}} - \frac{p\overline{w}}{\overline{w}} \\ &= \frac{p^2w_{11} + pqw_{12} - p\overline{w}}{\overline{w}} \\ &= \frac{p}{\overline{w}}(pw_{11} + qw_{12} - \overline{w})\end{aligned}$$

The final expression is a useful one, because it shows that the change in frequency of allele A_1 is proportional to $(pw_{11} + qw_{12} - \overline{w})$. The quantity $(pw_{11} + qw_{12} - \overline{w})$ is sometimes called the **average excess** of allele A_1. It is equal to the average fitness of allele A_1 when paired at random with other alleles $(pw_{11} + qw_{12})$ minus the average fitness of the population $(\overline{w})$. When the average excess of allele A_1 is positive, A_1 will increase in frequency. In other words, if the average A_1-carrying individual has higher-than-average fitness, then the frequency of allele A_1 will rise.

The change in the frequency of allele A_2 from one generation to the next is

$$\begin{aligned}\Delta q &= \frac{pqw_{12} + q^2w_{22}}{\overline{w}} - q \\ &= \frac{q}{\overline{w}}(pw_{12} + qw_{22} - \overline{w})\end{aligned}$$

Empirical Research on Allele Frequency Change by Selection

Douglas Cavener and Michael Clegg (1981) documented a cumulative change in allele frequencies over many generations in a laboratory-based natural selection experiment on the fruit fly (*Drosophila melanogaster*). Fruit flies, like most other animals, make an enzyme that breaks down ethanol, the poisonous active ingredient in beer, wine, and rotting fruit. This enzyme is called alcohol dehydrogenase, or ADH. Cavener and Clegg worked with populations of flies that had two alleles at the ADH locus: Adh^F and Adh^S. (The F and S refer to whether the protein encoded by the allele moves quickly or slowly through an electrophoresis gel.)

The scientists kept two experimental populations on food spiked with ethanol and two control populations of flies on normal, nonspiked food. The researchers picked the breeders for each generation at random. This is why we are calling the project a natural selection experiment: Cavener and Clegg set up different environments for their different populations, but the researchers did not themselves directly manipulate the survival or reproductive success of individual flies.

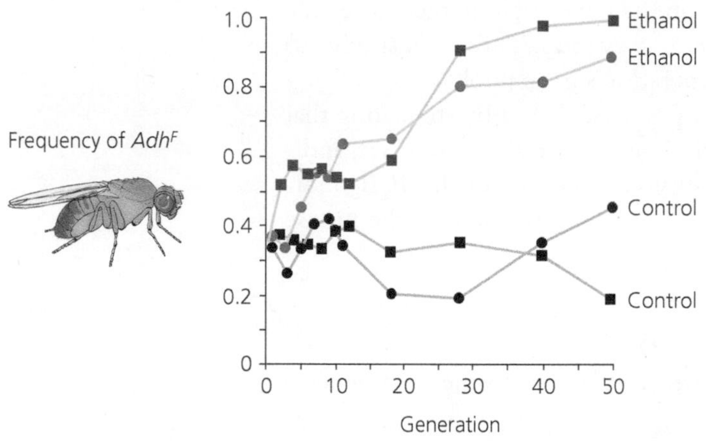

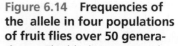

Figure 6.14 Frequencies of the allele in four populations of fruit flies over 50 generations The black squares and circles represent control populations living on normal food; the magenta squares and orange circles represent experimental populations living on food spiked with ethanol. From Cavener and Clegg (1981).

Empirical research on fruit flies is consistent with our conclusion that natural selection can cause allele frequencies to change.

Every several generations, Cavener and Clegg took a random sample of flies from each population, determined their ADH genotypes, and calculated the allele frequencies. The results appear in **Figure 6.14**. The control populations showed no large or consistent long-term change in the frequency of the Adh^F allele. The experimental populations, in contrast, showed a rapid and largely consistent increase in the frequency of Adh^F (and, of course, a corresponding decrease in the frequency of Adh^S). Hardy–Weinberg conclusion 1 appears to hold true in the control populations, but is clearly not valid in the experimental populations.

Can we identify for certain which of the assumptions of the Hardy–Weinberg analysis is being violated? The only difference between the two kinds of populations is that the experimentals have ethanol in their food. This suggests that it is the no-selection assumption that is being violated in the experimental populations. Flies carrying the Adh^F allele appear to have higher lifetime reproductive success (higher fitness) than flies carrying the Adh^S allele when ethanol is present in the food. Cavener and Clegg note that this outcome is consistent with the fact that alcohol dehydrogenase extracted from Adh^F homozygotes breaks down ethanol at twice the rate of alcohol dehydrogenase extracted from Adh^S homozygotes. Whether flies with the Adh^F allele have higher fitness because they have higher rates of survival or because they produce more offspring is unclear.

Adding Selection to the Hardy–Weinberg Analysis: The Calculation of Genotype Frequencies

The calculations and example we have just discussed show that selection can cause allele frequencies to change across generations. Selection invalidates conclusion 1 of the Hardy–Weinberg analysis. We now consider how selection affects conclusion 2 of the Hardy–Weinberg analysis. In a population under selection, can we still calculate the genotype frequencies by multiplying the allele frequencies?

Often, we cannot. As before, we use a population with two alleles at a locus affecting survival: B_1 and B_2. We assume that the initial frequency of each allele in the gene pool is 0.5 **(Figure 6.15)**. After random mating, we get genotype frequencies for B_1B_1, B_1B_2, and B_2B_2 of 0.25, 0.5, and 0.25. The rest of our calculations will be simpler if we give the population of zygotes a finite size, so imagine there are 100 zygotes:

B_1B_1	B_1B_2	B_2B_2
25	50	25

These zygotes are represented by a bar graph on the upper right in the figure. We will follow the individuals that develop from these zygotes. Those that survive to adulthood will breed to produce the next generation's gene pool.

As in our first selection example, we incorporate selection by stipulating that the genotypes differ in their rates of survival. This time, 60% of the B_1B_1 individuals survive, all of the B_1B_2 individuals survive, and 60% of the B_2B_2 individuals survive. As shown in Figure 6.15, there are now 80 adults in the mouse population:

B_1B_1	B_1B_2	B_2B_2
15	50	15

If we assume that each surviving adult donates 10 gametes to the next generation's gene pool, then

The 15 B_1B_1 adults together make 150 gametes: 150 carry B_1; none carry B_2.

The 50 B_1B_2 adults together make 500 gametes: 250 carry B_1; 250 carry B_2.

The 15 B_2B_2 adults together make 150 gametes: none carry B_1; 150 carry B_2.

Summing the gametes carrying each allele, we get 400 carrying B_1 and 400 carrying B_2, for a total of 800. Both alleles are still at a frequency of 0.5. Despite strong selection against homozygotes, the frequencies of the alleles have not changed; the population has not evolved.

But let us calculate frequencies of the three genotypes among the surviving adults. These frequencies are as follows:

B_1B_1	B_1B_2	B_2B_2
$\frac{15}{80} = 0.1875$	$\frac{50}{80} = 0.625$	$\frac{15}{80} = 0.1875$

These genotype frequencies reveal that violation of the no-selection assumption has resulted in violation of conclusion 2 of the Hardy–Weinberg analysis. We can no longer calculate the genotype frequencies among the adult survivors by multiplying the frequencies of the alleles. For example:

Natural selection can also drive genotype frequencies away from the values predicted under the Hardy–Weinberg equilibrium principle.

$$\begin{array}{ccc} \textit{Frequency of } B_1B_1 & & (\textit{Frequency of } B_1)^2 \\ 0.1875 & \neq & (0.5)^2 = 0.25 \end{array}$$

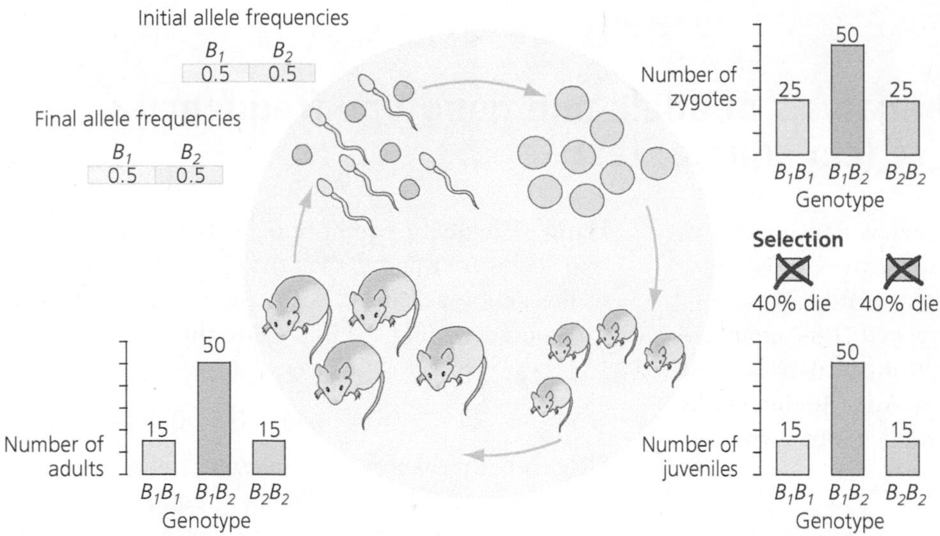

Figure 6.15 Selection can change genotype frequencies so that they cannot be calculated by multiplying the allele frequencies When 40% of the homozygotes in this population die, the allele frequencies do not change. But among the survivors, there are more heterozygotes than predicted under Hardy–Weinberg equilibrium.

We used strong selection in our numerical example to make a point. In fact, selection is rarely strong enough to produce, in a single generation, such a large violation of Hardy–Weinberg conclusion 2. Even if it does, a single bout of random mating will immediately put the genotypes back into Hardy–Weinberg equilibrium. Nonetheless, researchers sometimes find violations of Hardy–Weinberg conclusion 2 that seem to be the result of selection.

Empirical Research on Selection and Genotype Frequencies

Our example comes from research by Atis Muehlenbachs and colleagues (2008), working in the laboratory of Patrick Duffy, on genetic variation for the outcome of falciparum malaria during pregnancy. Falciparum malaria is caused by infection with the single-celled parasite *Plasmodium falciparum*. When a pregnant woman contracts the disease, the parasites invade the placenta via the mother's circulatory system (Karumanchi and Haig 2008). This triggers placental inflammation and may also interfere with placental development (Umbers et al. 2011). The potential complications include spontaneous abortion, premature delivery, low birth weight, and higher risk of infant death.

Pregnancy itself brings an increased risk of malaria infection, particularly a woman's first pregnancy (Karumanchi and Haig 2008). During a first bout of placental malaria, women develop antibodies that confer partial resistance during later pregnancies. Some 125 million women who live in areas affected by malaria become pregnant each year, and malaria infection during pregnancy is estimated to be responsible for an annual toll of 100,000 infant deaths (Umbers et al. 2011).

Muehlenbachs and colleagues (2008) suspected that the outcome of placental malaria hinges on the fetus's genotype at the locus encoding vascular endothelial growth factor receptor 1 (VEGFR1), also known as *fms*-like tyrosine kinase 1 (Flt1). Fetal cells in the placenta release a soluble form of this protein, sVEGFR1, into the mother's circulation. By interacting with vascular endothelial growth factor, VEGFR1 influences both placental development and inflammation.

Copies of the gene for VEGFR1 vary in the length of a two-nucleotide repeat in a region that is transcribed to mRNA but not translated. Alleles cluster into a short group (*S* alleles) and a long group (*L* alleles). Cultured cord blood cells with genotypes *SS* and *SL* produce more VEGFR1 than do *LL* cells.

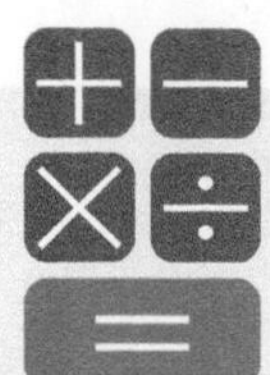

COMPUTING CONSEQUENCES 6.4

Statistical analysis of allele and genotype frequencies using the χ^2 (chi-square) test

Here we use data from Muehlenbachs and colleagues (2008) to illustrate a method for determining whether genotype frequencies deviate from Hardy–Weinberg equilibrium. The researchers surveyed Tanzanian infants born to first-time mothers during malaria season. The genotype counts (provided by Atis Muehlenbachs and Patrick Duffy, personal communication) were

SS	*SL*	*LL*
16	50	10

The analysis proceeds in five steps:

1. Calculate the allele frequencies. The sample of 76 infants is also a sample of 152 gene copies. All 32 copies carried by the *SS* infants are *S*, as are 50 of the copies carried by the *SL* infants. Thus, the frequency of *S* is

$$\frac{32 + 50}{152} = 0.54$$

The frequency of *L* is

$$\frac{50 + 20}{152} = 0.46$$

2. Calculate the genotype frequencies expected under Hardy–Weinberg equilibrium. If the frequencies of two alleles are p and q, then the expected frequencies of the genotypes are p^2, $2pq$, and q^2. The expected frequencies among the infants are thus

SS	*SL*	*LL*
$0.54^2 = 0.29$	$2 \cdot 0.54 \cdot 0.46 = 0.5$	$0.46^2 = 0.21$

3. Calculate the expected number of infants of each genotype under Hardy–Weinberg equilibrium. This is simply the expected frequency of each genotype multiplied by the total number of infants, 76:

SS	*SL*	*LL*
$0.29 \cdot 76 = 22$	$0.5 \cdot 76 = 38$	$0.21 \cdot 76 = 16$

These expectations are different from the numbers observed (16, 50, and 10). The actual sample contains more heterozygotes and fewer homozygotes. Is it plausible that a difference this large between expectation and reality could arise by chance? Or is the difference statistically significant? Our null hypothesis is that the difference is simply due to chance.

Working with newborn babies of first-time mothers in Muheza, Tanzania, where malaria is a perennial scourge, Muehlenbachs and colleagues (2008) tested their hypothesis in part by using the Hardy–Weinberg equilibrium principle.

The researchers first determined the allele frequencies among 163 infants born from October through April, when the rate of placental malaria was at its annual low. The frequencies were

S	*L*
0.555	0.445

If the population of infants was in Hardy–Weinberg equilibrium, then multiplying these allele frequencies will allow us to predict the genotype frequencies:

SS	*SL*	*LL*
$0.555^2 = 0.308$	$2 \cdot 0.555 \cdot 0.445 = 0.494$	$0.445^2 = 0.198$

These predicted frequencies are, in fact, close to the actual genotype frequencies among the off-season infants:

SS	*SL*	*LL*
$\frac{49}{163} = 0.301$	$\frac{83}{163} = 0.509$	$\frac{31}{163} = 0.190$

The true frequency of heterozygotes is slightly higher than predicted, and the frequencies of homozygotes are slightly lower, but the discrepancies are modest. The infants thus conform to conclusion 2 of the Hardy–Weinberg analysis.

4. Calculate a test statistic. We will use one devised in 1900 by Karl Pearson. It is called chi-square (χ^2).

$$\chi^2 = \sum \frac{(\text{observed} - \text{expected})^2}{\text{expected}}$$

where the symbol $\sum$ indicates a sum taken across all the classes considered. In our data there are three classes: the three genotypes. For our data set

$$\chi^2 = \frac{(16 - 22)^2}{22} + \frac{(50 - 38)^2}{38} + \frac{(10 - 16)^2}{16} = 7.68$$

5. Determine whether the test statistic is significant. χ^2 is defined such that it gets larger as the difference between the observed and expected values gets larger. How likely is it that we could get a χ^2 as large as 7.68 by chance? Most statistical textbooks have a table giving the answer. In Zar's (1996) book. it is called "Critical values of the chi-square distribution."

To use this table, we need to calculate a number called the degrees of freedom for the test statistic. This value for χ^2 is the number of classes minus the number of independent values calculated from the data for use in determining the expected values. For our χ^2 there are three classes: the genotypes. We calculated two values from the data for use in determining the expected values: the total number of individuals, and the frequency of allele *S*. (We also calculated the frequency of *L*, but it is not independent of the frequency of *S*, because the two must sum to 1.) Thus the number of degrees of freedom is 1. (Another formula for calculating the degrees of freedom in χ^2 tests for Hardy–Weinberg equilibrium is

$$df = k - 1 - m$$

where *k* is the number of classes and *m* is the number of independent allele frequencies estimated.)

According to the table, the critical value of χ^2 for one degree of freedom and $P = 0.05$ is 3.841. This means there is a 5% chance under the null hypothesis of getting $\chi^2 \geq 3.841$. The probability under the null hypothesis of getting $\chi^2 \geq 7.68$ is therefore (considerably) less than 5%. We reject the null hypothesis and assert that our χ^2 is statistically significant at $P < 0.05$. (In fact, $P < 0.006$.)

The χ^2 test tells us that among infants born during malaria season, the alleles of the gene for VEGFR1 are not in Hardy–Weinberg equilibrium. This indicates that one or more assumptions of the Hardy–Weinberg analysis has been violated. By itself, however, it does not tell us which are being violated, or how.

Muehlenbachs and colleagues then determined the allele frequencies among 76 infants born from May through September, when the rate of placental malaria was at its annual high. The frequencies were nearly the same as among the off-season newborns:

S	*L*
0.539	0.461

If this segment of the population was, like their off-season counterparts, in Hardy–Weinberg equilibrium, then multiplying the allele frequencies will again allow us to predict the genotype frequencies:

SS	*SL*	*LL*
$0.539^2 = 0.291$	$2 \cdot 0.539 \cdot 0.461 = 0.497$	$0.461^2 = 0.213$

This time the predicted values are a poor fit to the actual frequencies:

SS	*SL*	*LL*
$\frac{16}{76} = 0.211$	$\frac{50}{76} = 0.658$	$\frac{10}{76} = 0.132$

There are substantially more heterozygotes than expected, and substantially fewer homozygotes. This discrepancy between prediction and data is statistically significant (see **Computing Consequences 6.4**). The genotypes of the infants born during peak malaria season are in violation of Hardy–Weinberg conclusion 2.

The discovery that genotype frequencies in a population are not in Hardy–Weinberg equilibrium may be a clue that natural selection is at work.

On this and other evidence, Muehlenbachs and colleagues (2008) believe the best explanation for the missing homozygotes is that they did not survive fetal development. A fetus's chance of surviving depends on both its own genotype and whether its mother contracts malaria (**Figure 6.16**). If the mother does not contract malaria, *SS* infants do somewhat better than others. If, however, the mother does contract malaria, *SL* infants do substantially better than others. Overall, when malaria is common, heterozygotes survive at the highest rate. Consistent with this explanation, where malaria is absent, *S* alleles occur at high frequency.

Figure 6.16 Probability of fetal survival as a function of genotype and placental malaria Inferred from the patterns in maternal and newborn genotype frequencies in Muehlenbachs et al. (2008).

Changes in the Frequency of the *CCR5-Δ32* Allele Revisited

We are now in a position to give a more satisfying answer to the question we raised at the beginning of Section 6.1: Will the AIDS epidemic cause the frequency of the *CCR5-Δ32* allele to increase in human populations? The AIDS epidemic could, in principle, cause the frequency of the allele to increase rapidly, but at present it appears that it will probably not do so in any real population. This conclusion is based on the three model populations depicted in **Figure 6.17** (see **Computing Consequences 6.5** for the algebra). Each model is based on different assumptions about the initial frequency of the *CCR5-Δ32* allele and the prevalence of HIV infection. Each graph shows the predicted change in the frequency of the *Δ32* allele over 40 generations, or approximately 1,000 years.

The model population depicted in Figure 6.17a offers a scenario in which the frequency of the *Δ32* allele could increase rapidly. In this scenario, the initial frequency of the *CCR5-Δ32* allele is 20%. One-quarter of the individuals with genotype +/+ or +/*Δ32* contract AIDS and die without reproducing, whereas all of the *Δ32*/*Δ32* individuals survive. The 20% initial frequency of *Δ32* is approximately equal to the highest frequency reported for any population, a sample of Ashkenazi Jews studied by Martinson et al. (1997). The mortality rates approximate the situation in Botswana, Namibia, Swaziland, and Zimbabwe, where up to 25% of individuals between the ages of 15 and 49 are infected with HIV (UNAIDS 1998). In this model population, the frequency of the *Δ32* allele increases by as much as a few percentage points each generation. By the end of 40 generations, the allele is at a frequency of virtually 100%. Thus, in a human population that combined the highest reported frequency of the *Δ32* allele with the highest reported rates of infection, the AIDS epidemic could cause the frequency of the allele to increase rapidly.

At present, however, no known population combines a high frequency of the *Δ32* allele with a high rate of HIV infection. In northern Europe, many populations have *Δ32* frequencies between 0.1 and 0.2 (Martinson et al. 1997; Stephens et al. 1998), but HIV infection rates are under 1% (UNAIDS 1998). A model population reflecting these conditions is depicted in Figure 6.17b. The initial frequency of the *Δ32* allele is 0.2, and 0.5% of the +/+ and +/*Δ32* individuals contract AIDS and die without reproducing. The frequency of the *Δ32* allele hardly changes at all. Selection is too weak to cause appreciable evolution in such a short time.

In parts of sub-Saharan Africa, as many as a quarter of all individuals of reproductive age are infected with HIV. However, the *Δ32* allele is virtually absent (Martinson et al. 1997). A model population reflecting this situation is depicted in Figure 6.17c. The initial frequency of the *Δ32* allele is 0.01, and 25% of the +/+ and +/*Δ32* individuals contract AIDS and die without reproducing. Again, the frequency of the *Δ32* allele hardly changes at all. When the *Δ32* allele is at low

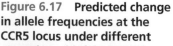

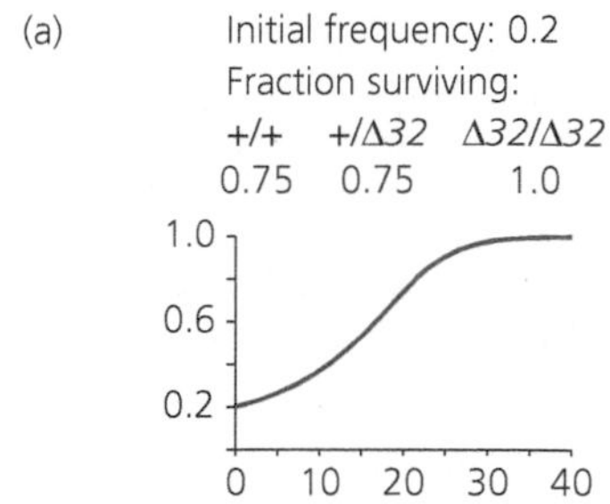

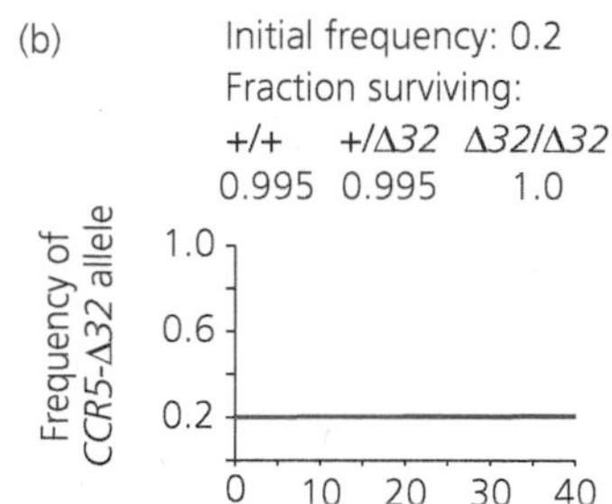

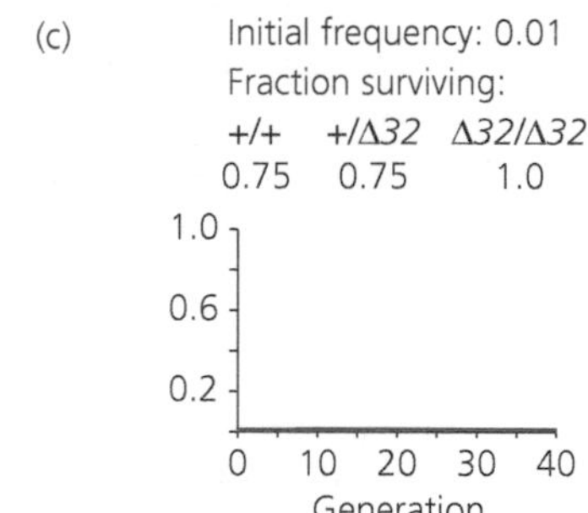

Figure 6.17 Predicted change in allele frequencies at the CCR5 locus under different scenarios (a) If the initial frequency of *CCR5-Δ32* is high and many people become infected with HIV, allele frequencies can change rapidly. (b) In Europe allele frequencies are high, but infection rates are low. (c) In parts of Africa the infection rates are high, but allele frequencies are low.

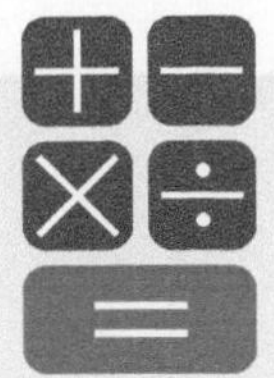

COMPUTING CONSEQUENCES 6.5

Predicting the frequency of the *CCR5-Δ32* allele in future generations

Let q_g be the frequency of the *CCR5-Δ32* allele in the present generation. Based on Computing Consequences 6.3, we can write an equation predicting the frequency of the allele in the next generation, given estimates of the survival rates (fitnesses) of individuals with each genotype. The equation is

$$q_{g+1} = \frac{(1 - q_g)q_g w_{+\Delta} + q_g^2 w_{\Delta\Delta}}{(1 - q_g)^2 w_{++} + 2(1 - q_g)q_g w_{+\Delta} + q_g^2 w_{\Delta\Delta}}$$

where q_{g+1} is the frequency of the $\Delta 32$ allele in the next generation, w_{++} is the fitness of individuals homozygous for the normal allele, $w_{+\Delta}$ is the fitness of heterozygotes, and $w_{\Delta\Delta}$ is the fitness of individuals homozygous for the *CCR5-Δ32* allele.

After choosing a starting value for the frequency of the $\Delta 32$ allele, we plug it and the estimated fitnesses into the equation to generate the frequency of the $\Delta 32$ allele after one generation. We then plug this resulting value into the equation to get the frequency of the allele after two generations, and so on.

frequency, most copies are in heterozygotes. Because heterozygotes are susceptible to infection, these copies are hidden from selection.

Our exploration of natural selection has given us tools we can use to predict the future of human populations.

The analysis we have just described is based on a number of simplifying assumptions. We have assumed, for example, that all HIV-infected individuals die without reproducing. In fact, however, many HIV-infected individuals have children. We have also assumed that the death rate is the same in heterozygotes as in +/+ homozygotes. In reality, although heterozygotes are susceptible to HIV infection, they appear to progress more slowly to AIDS (Dean et al. 1996). As a result, the fitness of heterozygotes may actually be higher than that of +/+ homozygotes. We challenge the reader to explore the evolution of human populations under a variety of selection schemes, to see how strongly our simplifying assumptions affect the predicted course of evolution. For analyses of more complex models of human evolution in response to selection imposed by AIDS, see models by Schliekelman et al. (2001; but also Ramaley et al. 2002), by Sullivan et al. (2001), and by Cromer et al. (2010).

6.3 Patterns of Selection: Testing Predictions of Population Genetics Theory

In the 1927 case of *Buck v. Bell*, the United States Supreme Court upheld the state of Virginia's sterilization statute by a vote of eight to one. Drafted on the advice of eugenicists, the law was intended to improve the genetic quality of future generations by allowing the forced sterilization of individuals afflicted with hereditary forms of insanity, feeblemindedness, and other mental defects. The court's decision in *Buck v. Bell* reinvigorated a compulsory sterilization movement dating from 1907 (Kevles 1995). By 1940, 30 states had enacted sterilization laws, and by 1960 over 60,000 people had been sterilized without their consent (Reilly 1991; Lane 1992). In hindsight, the evidence that these individuals suffered from hereditary diseases was weak. But what about the evolutionary logic behind compulsory sterilization? If the genetic assumptions had been correct, would sterilization have been an effective means of reducing the incidence of undesirable traits?

Before we try to answer this question, it will be helpful to address a more general one. How well does the theory of population genetics actually work? We developed this theory in Sections 6.1 and 6.2. The final product is a model of how allele frequencies change in response to natural selection (Figures 6.12 and 6.13, Computing Consequences 6.3 and 6.5). If our model is a good one, it should accurately predict the direction and rate of allele frequency change under a variety of selection schemes. It should work, for example, whether the allele favored by selection is dominant or recessive, common or rare. It should work whether selection favors heterozygotes or homozygotes. It should even predict what will happen when a particular allele is favored by selection under some circumstances and disfavored in others.

In this section, we will find out how well our model works. Using the theory we have developed to predict the course of evolution under different patterns of selection, we compare our predictions to empirical data from experimental populations. We then return to our question about the effectiveness of eugenic sterilization in changing the composition of populations.

Selection on Recessive and Dominant Alleles

For our first test, we focus on whether our theory accurately predicts changes in the frequencies of recessive and dominant alleles. Our example comes from the work of Peter Dawson (1970). Dawson had been studying a laboratory colony of flour beetles **(Figure 6.18)** and had identified a gene we will call the l locus. This locus has two alleles: + and *l*. Individuals with genotype +/+ or +/*l* are phenotypically normal, whereas individuals with genotype *l*/*l* do not survive. In other words, *l* is a recessive lethal allele.

Figure 6.18 Flour beetles, *Tribolium castaneum* Courtesy of Susan J. Brown, Professor/Kansas State University, Kansas.

Dawson collected heterozygotes from his beetle colony and used them to establish two new experimental populations. Because all the founders were heterozygotes, the initial frequencies of the two alleles were 0.5 in both populations. Because *l*/*l* individuals have zero fitness, Dawson expected his populations to evolve toward ever lower frequencies of the *l* allele and ever higher frequencies of the + allele. He let his two populations evolve for a dozen generations, each generation measuring the frequencies of the two alleles.

Dawson used the equations derived in Computing Consequences 6.3 and the method described in Computing Consequences 6.5 to make a quantitative prediction of the course of evolution in his populations. We can reproduce this prediction with a straightforward numerical calculation like the ones we performed in Figures 6.12 and 6.13. Imagine a gene pool in which alleles + and *l* are both at a frequency of 0.5. If we combine gametes at random to make 100 zygotes, we get the three genotypes in the following numbers:

+/+	+/*l*	*l*/*l*
25	50	25

Now we imagine that all the *l*/*l* individuals die and that everyone else survives to breed. Finally, imagine that each of the survivors donates 10 gametes to the new gene pool:

The 25 +/+ survivors together make 250 gametes: 250 carry +; none carry *l*.

The 50 +/*l* survivors together make 500 gametes: 250 carry +; 250 carry *l*.

This gives us 500 copies of the + allele and 250 copies of the *l* allele for a total of 750. In this new gene pool, the frequency of the + allele is 0.67, and the frequency of the *l* allele is 0.33. We have gone from the gene pool in generation zero to the gene pool in generation one. The frequency of the + allele has risen, and the frequency of the *l* allele has fallen.

To get from generation one's gene pool to generation two's gene pool, we just repeat the exercise. We combine the gametes in generation one's gene pool at random to make 100 zygotes—45 +/+, 44 +/*l*, and 11 *l*/*l*—and so on. The only problem with using pencil-and-paper numerical calculations to predict evolution is that chasing the alleles around and around the life cycle all the way to generation 12 is a tedious job.

With a computer, however, predicting how Dawson's population will evolve is quick and easy. We can use a spreadsheet application to set up the required calculations ourselves (see Computing Consequences 6.3 and 6.5), or we can use any of a variety of population genetics programs that are already set up to do the calculations for us. Such programs take starting allele frequencies and genotype fitnesses as input and use the model we have developed in this chapter to produce predicted allele frequencies in future generations as output. We encourage the reader to get one of these programs and experiment with it.

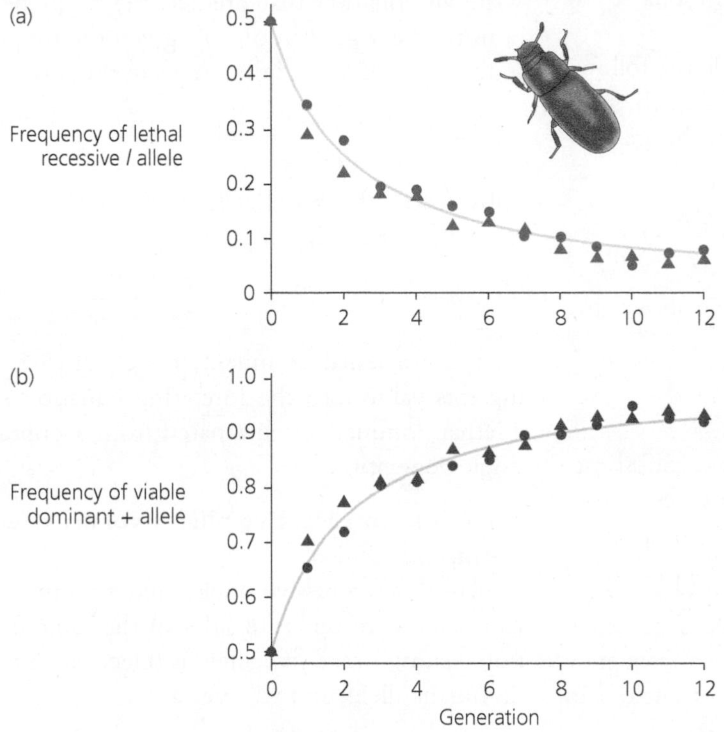

Figure 6.19 Evolution in laboratory populations of flour beetles (a) The decline in frequency of a lethal recessive allele (blue symbols) matches the theoretical prediction (blue curve) almost exactly. As the allele becomes rare, the rate of evolution slows dramatically. (b) This graph plots the increase in frequency of the corresponding dominant allele. Redrawn from Dawson (1970).

The prediction for Dawson's experiment appears as a curve in each of the graphs in **Figure 6.19**. The curve in the top graph predicts the falling frequency of the *l* allele; equivalently, the curve in the bottom graph predicts the rising frequency of the + allele. Our theory predicts that evolution will be rapid at first but will slow as the experiment proceeds.

Empirical research on flour beetles shows that predictions made with population genetics models are accurate, at least under laboratory conditions.

Dawson's data appear in the graphs as colored circles and triangles. They match our theoretical predictions closely. This tight fit between prediction and data may seem unsurprising, even mundane. It should not. It should be astonishing. We

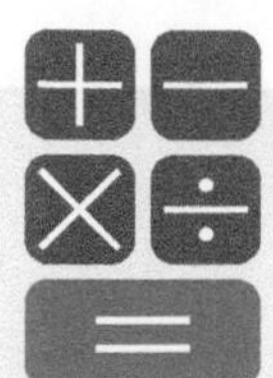

COMPUTING CONSEQUENCES 6.6

An algebraic treatment of selection on recessive and dominant alleles

Here we develop equations that illuminate the differences between selection on recessive versus dominant alleles. Imagine a single locus with two alleles. Let p be the frequency of the dominant allele A, and let q be the frequency of the recessive allele a.

Selection on the recessive allele

Let the fitnesses of the genotypes be given by

w_{AA}	w_{Aa}	w_{aa}
1	1	$1 - s$

where s, called the **selection coefficient**, represents the strength of selection against homozygous recessives relative to the other genotypes. (Selection in favor of homozygous recessives can be accommodated by choosing a negative value for s.)

Based on Computing Consequences 6.3, the following equation gives the frequency of allele a in the next generation, q', given the frequency of a in this generation and the fitnesses of the three genotypes:

$$q' = \frac{pqw_{Aa} + q^2w_{aa}}{\overline{w}} = \frac{pqw_{Aa} + q^2w_{aa}}{p^2w_{AA} + 2pqw_{Aa} + q^2w_{aa}}$$

Substituting the fitness values from the table above, and $(1 - q)$ for p, then simplifying, gives

$$q' = \frac{q(1 - sq)}{1 - sq^2}$$

If a is a lethal recessive, then s is equal to 1. Substituting this value into the preceding equation gives

$$q' = \frac{q(1 - q)}{1 - q^2} = \frac{q(1 - q)}{(1 - q)(1 + q)} = \frac{q}{(1 + q)}$$

A little experimentation shows that once a recessive lethal allele becomes rare, further declines in frequency are slow. For example, if the frequency of allele a in this generation is 0.01, then in the next generation its frequency will be approximately 0.0099.

Selection on the dominant allele

Let the fitnesses of the genotypes be given by

w_{AA}	w_{Aa}	w_{aa}
$1 - s$	$1 - s$	1

where s, the selection coefficient, represents the strength of selection against genotypes containing the dominant allele relative to homozygous recessives. (Selection in favor of genotypes containing the dominant allele can be accommodated by choosing a negative value of s.)

Based on Computing Consequences 6.3, we can write an equation that predicts the frequency of allele A in the next generation, p', given the frequency of A in this generation and the fitness of the three genotypes:

$$p' = \frac{p^2w_{AA} + pqw_{Aa}}{\overline{w}} = \frac{p^2w_{AA} + pqw_{Aa}}{p^2w_{AA} + 2pqw_{Aa} + q^2w_{aa}}$$

Substituting the fitnesses from the table, and $(1 - p)$ for q, then simplifying, gives

$$p' = \frac{p(1 - s)}{1 - 2sp + sp^2}$$

If A is a lethal dominant, s is equal to 1. Substituting this value into the foregoing equation shows that a lethal dominant is eliminated from a population in a single generation.

Selection on recessive alleles versus selection on dominant alleles

Selection on recessive alleles and selection on dominant alleles are opposite sides of the same coin. Selection against a recessive allele is selection in favor of the dominant allele, and vice versa.

used a simple model of the mechanism of evolution combining the fundamental insights of Gregor Mendel with those of Charles Darwin to predict how a population would change over 12 generations. If the creatures in question had been humans instead of flour beetles, it would have meant forecasting events that will happen in 300 years. And Dawson's data show that our prediction was not just reasonably accurate, but spot on. If we had a theory that worked like that for picking stocks or racehorses—well, we could have retired years ago. Our model has passed its first test.

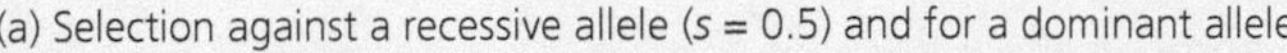

(a) Selection against a recessive allele (s = 0.5) and for a dominant allele

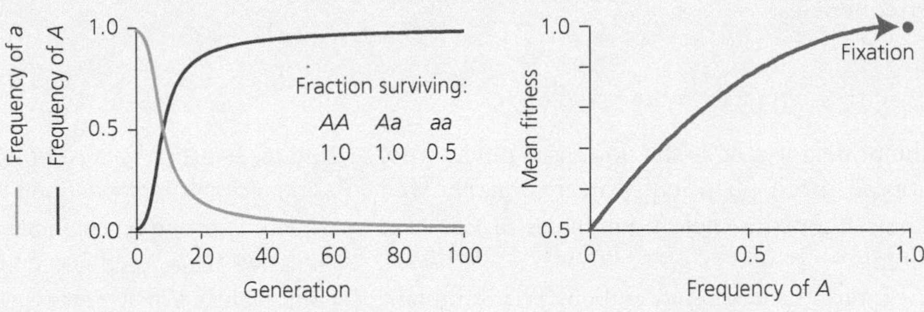

(b) Selection for a recessive allele and against a dominant allele (s = 0.6)

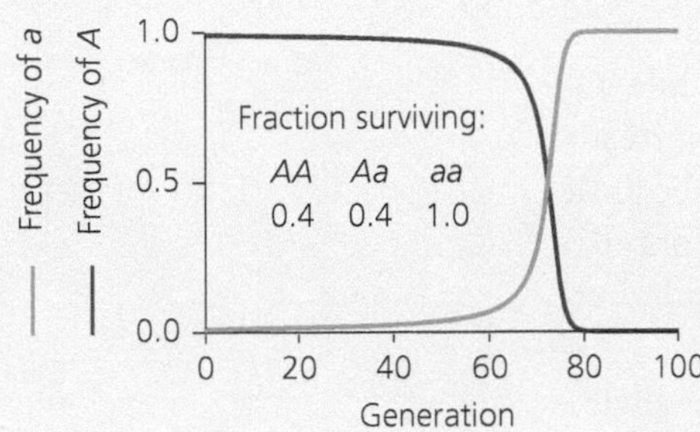

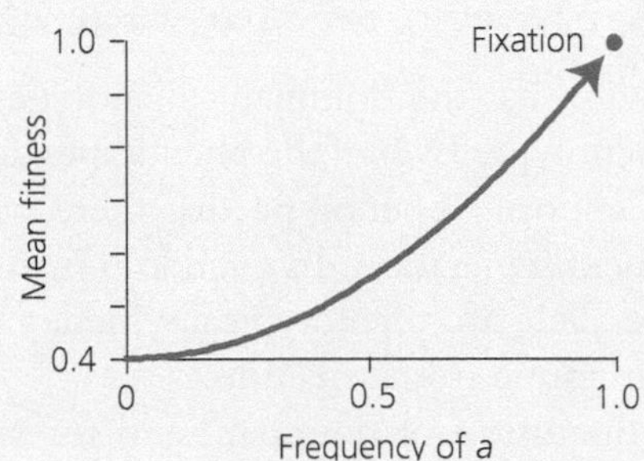

Figure 6.20 Evolution in model populations under selection on recessive and dominant alleles Graphs on the left show changes in allele frequencies over time. Graphs on the right show adaptive landscapes: Changes in population mean fitness as a function of allele frequencies.

Figure 6.20a (left) shows 100 generations of evolution in a model population with selection against a recessive allele and in favor of the dominant allele. At first, the allele frequencies change rapidly. As the recessive allele becomes rare, however, the rate of evolution slows dramatically. When the recessive allele is rare, most copies in the population are in heterozygous individuals, where they are effectively hidden from selection.

The figure also shows (right) the mean fitness of the population (see Computing Consequences 6.3) as a function of the frequency of the dominant allele. As the dominant allele goes from rare to common, the mean fitness of the population rises. Mean fitness is maximized when the favored allele reaches a frequency of 100%. Graphs of mean fitness as a function of allele frequency are often referred to as adaptive landscapes.

Figure 6.20b (left) shows 100 generations of evolution in a model population with selection in favor of a recessive allele and against the dominant allele. At first, the allele frequencies change slowly. The recessive allele is rare, most copies present are in heterozygotes, and selection cannot see it. However, as the recessive allele becomes common enough that a substantial fraction of homozygotes appear, the rate of evolution increases dramatically. Once the pace of evolution accelerates, the favorable recessive allele quickly achieves a frequency of 100%. That is, the recessive allele becomes fixed in the population.

The figure also shows (right) the mean fitness of the population (see Computing Consequences 6.3) as a function of the frequency of the recessive allele. As the recessive allele goes from rare to common, the mean fitness of the population rises. Mean fitness is maximized when the favored allele reaches a frequency of 100%.

An algebraic treatment of selection on recessive and dominant alleles appears in **Computing Consequences 6.6.** Even without the algebra, we can draw some important conclusions by reflecting further on Dawson's experiment.

Dawson's experiment shows that dominance and allele frequency interact to determine the rate of evolution. When a recessive allele is common (and a dominant allele is rare), evolution by natural selection is rapid. In contrast, when a recessive allele is rare, and a dominant allele is common, evolution by natural selection is slow. The Hardy–Weinberg equilibrium principle explains why.

First imagine a recessive allele that is common: Its frequency is, say, 0.95. The dominant allele thus has a frequency of 0.05. By multiplying the allele frequencies, we can calculate the genotype frequencies:

AA	Aa	aa
$0.05^2 = 0.0025$	$2 \cdot 0.05 \cdot 0.95 = 0.095$	$0.95^2 = 0.9025$

Roughly 10% of the individuals in the population have the dominant phenotype, while 90% have the recessive phenotype. Both phenotypes are reasonably well represented, and if they differ in fitness, then the allele frequencies in the next generation may be substantially different.

Natural selection is most potent as a mechanism of evolution when it is acting on common recessive alleles (and rare dominant alleles). When a recessive allele is rare, most copies are hidden in heterozygotes and protected from selection.

Now imagine a recessive allele that is rare: Its frequency is 0.05. The dominant allele thus has a frequency of 0.95. The genotype frequencies are

AA	Aa	aa
$0.95^2 = 0.9025$	$2 \cdot 0.95 \cdot 0.05 = 0.095$	$0.05^2 = 0.0025$

Approximately 100% of the population has the dominant phenotype, while approximately 0% has the recessive phenotype. Even if the phenotypes differ greatly in fitness, there are so few of the minority phenotype that there will be little change in allele frequencies in the next generation. In a random mating population, most copies of a rare recessive allele are phenotypically hidden inside heterozygous individuals and thereby immune from selection.

As a final consideration in our discussion of dominant and recessive alleles, note that selection may favor or disfavor both kinds of variants. We emphasize this point because many people new to population genetics expect that dominant alleles are necessarily beneficial and thus tend to rise in frequency. While it is certainly true that some dominant alleles are beneficial, many others are deleterious. For example, Eileen Shore and colleagues (2006) identified a dominant mutation, located in a gene encoding a receptor for bone morphogenic protein, as the cause of fibrodysplasia ossificans progressiva, a rare and severely disabling condition in which skeletal muscle and connective tissue transform inexorably into bone. In all, some 30% of the alleles known to cause human diseases are autosomal dominants (López-Bigas et al. 2006). The terms *dominant* and *recessive* describe the relationship between genotype and phenotype, not the relationship between genotype and fitness.

Selection on Heterozygotes and Homozygotes

In our next two tests, we focus on whether our model can accurately predict what happens when selection favors heterozygotes or homozygotes. Both tests will use data on laboratory populations of fruit flies (*Drosophila melanogaster*).

Selection Favoring Heterozygotes

Our first example comes from research by Terumi Mukai and Allan Burdick (1959). Like Dawson, Mukai and Burdick studied evolution at a single locus with two alleles. We will call the alleles V, for viable, and L for lethal. This is because flies with genotype VV or VL are alive, whereas flies with genotype LL are dead. The researchers used heterozygotes as founders to establish two experimental populations with initial allele frequencies of 0.5. They let the populations evolve for 15 generations, each generation measuring the frequency of allele V.

So far, Mukai and Burdick's experiment sounds just like Dawson's. If it is, then our theory predicts that V will rise in frequency—rapidly at first, then more

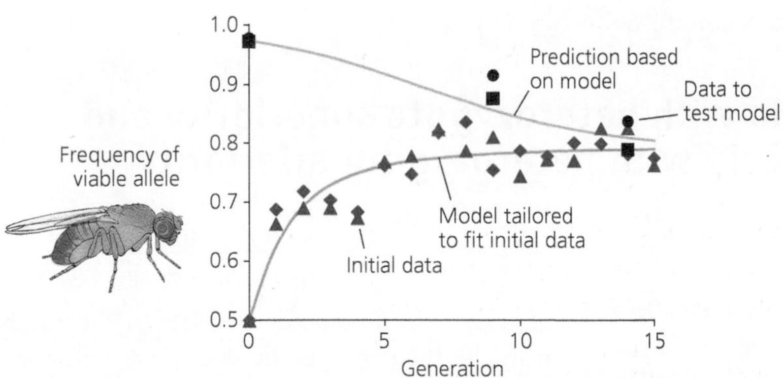

Figure 6.21 Evolution in four laboratory populations of fruit flies When homozygous, one allele is viable and the other lethal. Nonetheless, populations with a frequency of 0.5 for both alleles (red) evolved toward an intermediate equilibrium. The black populations represent a test of the hypothesis that heterozygotes enjoy the highest fitness. From data in Mukai and Burdick (1959).

slowly. By generation 15 it should reach a frequency of over 94%. But that is not what happened.

Mukai and Burdick's data appear in **Figure 6.21**, represented by the red symbols. As expected, the frequency of V increased rapidly over the first few generations. However, in both populations the rate of evolution slowed long before the viable allele approached a frequency of 0.94. Instead, V seemed to reach an equilibrium, or unchanging state, at a frequency of about 0.79.

How could this happen? An equilibrium frequency of 0.79 for the viable allele means that the lethal allele has an equilibrium frequency of 0.21. How could natural selection maintain a lethal allele at such a high frequency in this population? Mukai and Burdick argue that the most plausible explanation is **heterozygote superiority**, also known as **overdominance**. Under this hypothesis, heterozygotes have higher fitness than either homozygote. At equilibrium, the selective advantage enjoyed by the lethal allele when it is in heterozygotes exactly balances the obvious disadvantage it suffers when it is in homozygotes.

A little experimentation with a computer should allow the reader to confirm that Mukai and Burdick's hypothesis explains their data nicely. The red curve in Figure 6.21 represents evolution in a model population in which the fitnesses of the three genotypes are as follows:

VV	VL	LL
0.735	1.0	0

This theoretical curve matches the data closely.

Note that in this case the fit between theory and data does not represent a rigorous test of our model. That is because we examined the data first, then tweaked the fitnesses in the model to make its prediction fit. That is a bit like shooting at a barn and then painting a target around the bullet hole. Mukai and Burdick's flies did, however, provide an opportunity for a test of our model that is rigorous. And Mukai and Burdick performed it.

The researchers established two more experimental populations, this time with the initial frequency of the viable allele at 0.975. If the genotype fitnesses are, indeed, those required to make our model fit the red data points in Figure 6.21, then this time our model predicts that the frequency of the V allele should fall. As before, it should ultimately reach an equilibrium near 0.79. The predicted fall toward equilibrium is shown by the blue curve in Figure 6.21. Mukai and Burdick's data appear in the figure as blue symbols. The data match the prediction closely. Our model has passed its second test.

Research on fruit flies shows that natural selection can act to maintain two alleles at a stable equilibrium. One way this can happen is when heterozygotes have superior fitness.

Mukai and Burdick's flies have shown us something new. In all our previous examples, selection has favored one allele or the other. Under such circumstances

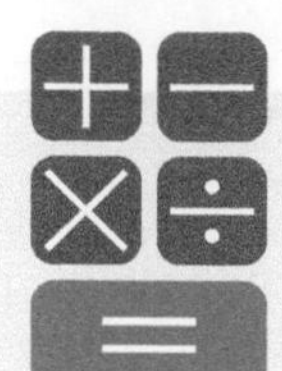

COMPUTING CONSEQUENCES 6.7

Stable equilibria with heterozygote superiority and unstable equilibria with heterozygote inferiority

Here we develop algebraic and graphical methods for analyzing evolution at loci with overdominance and underdominance. Imagine a population in which allele A_1 is at frequency p and allele A_2 is at frequency q. In Computing Consequences 6.3, we developed an equation describing the change in p from one generation to the next under selection:

$$\Delta p = \frac{p}{\overline{w}}(pw_{11} + qw_{12} - \overline{w})$$

$$= \frac{p}{\overline{w}}(pw_{11} + qw_{12} - p^2w_{11} - 2pqw_{12} - q^2w_{22})$$

Substituting $(1 - q)$ for p in the first and third terms in the expression in parentheses gives

$$\Delta p = \frac{p}{\overline{w}}[(1 - q)w_{11} + qw_{12} - (1 - q)^2w_{11} - 2pqw_{12} - q^2w_{22}]$$

which, after simplifying and factoring out q, becomes

$$\Delta p = \frac{pq}{\overline{w}}(w_{12} + w_{11} - qw_{11} - 2pw_{12} - qw_{22})$$

Now, by definition, the frequency of allele A_1 is at equilibrium when $\Delta p = 0$. The equation above shows that $\Delta p = 0$ when $p = 0$ or $q = 0$. These two equilibria are unsurprising. They occur when one allele or the other is absent from the population. The equation also gives a third condition for equilibrium, which is

$$w_{12} + w_{11} - qw_{11} - 2pw_{12} - qw_{22} = 0$$

Substituting $(1 - p)$ for q and solving for p gives

$$\hat{p} = \frac{w_{22} - w_{12}}{w_{11} - 2w_{12} + w_{22}}$$

where $\hat{p}$ is the frequency of allele A_1 at equilibrium. Finally, let the genotype fitnesses be as follows:

A_1A_1	A_1A_2	A_2A_2
$1 - s$	1	$1 - t$

Positive values of the selection coefficients s and t represent overdominance; negative values represent underdominance. Substituting the fitnesses into the previous equation and simplifying gives

$$\hat{p} = \frac{t}{s + t}$$

For example, when $s = 0.4$ and $t = 0.6$, heterozygotes have superior fitness, and the equilibrium frequency for allele A_1 is 0.6. When $s = -0.4$ and $t = -0.6$, heterozygotes have inferior fitness, and the equilibrium frequency for allele A_1 is also 0.6.

Another useful method for analyzing equilibria is to plot Δp as a function of p. Figure 6.20a shows such a plot for the two numerical examples we just calculated. Both curves show that $\Delta p = 0$ when $p = 0$, $p = 1$, or $p = 0.6$.

The curves in **Figure 6.22a** also allow us to determine whether an equilibrium is stable or unstable. Look at the red curve; it describes a locus with heterozygote superiority. Notice that when p is greater than 0.6, Δp is negative. This means that when the frequency of allele A_1 exceeds its equilibrium value, the population will move back toward equilibrium in the next generation. Likewise, when p is less than 0.6, Δp is positive. When

our model predicts that sooner or later the favored allele will reach a frequency of 100%, and the disfavored allele will disappear. By keeping a population at an equilibrium in which both alleles are present, however, heterozygote superiority can maintain genetic diversity indefinitely. For an algebraic treatment of heterozygote superiority, see **Computing Consequences 6.7**.

Selection Favoring Homozygotes

Our second example comes from work by G. G. Foster and colleagues (1972). These researchers set up experiments to demonstrate how populations evolve when heterozygotes have lower fitness than either homozygote. Foster and colleagues used fruit flies with compound chromosomes.

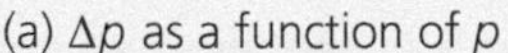

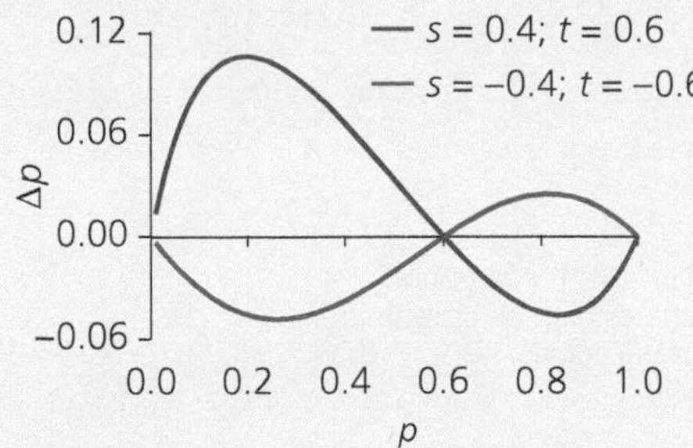

(b) Mean fitness as a function of p for overdominance

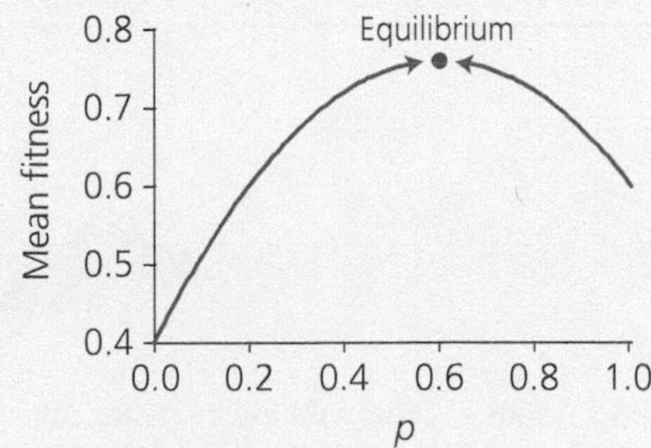

(c) Mean fitness as a function of p for underdominance

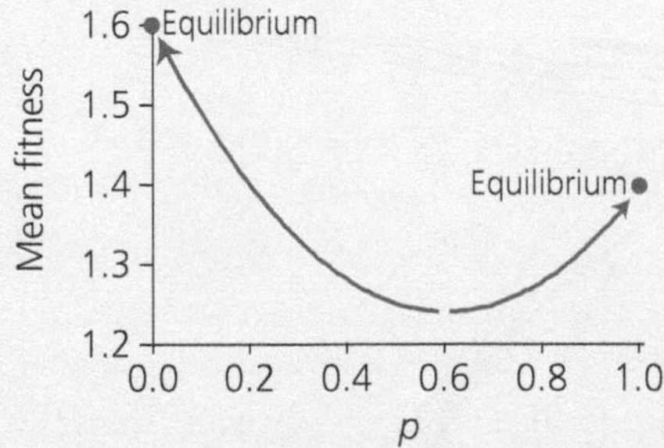

Figure 6.22 **A graphical analysis of stable and unstable equilibria at loci with overdominance and underdominance** (a) A plot of Δp as a function of p. (b) and (c) Adaptive landscapes.

the frequency of allele A_1 is below its equilibrium value, the population will move back toward equilibrium in the next generation. The "internal" equilibrium for a locus with heterozygote superiority is stable.

Figure 6.22b shows an adaptive landscape for a locus with heterozygote superiority. The graph plots population mean fitness as a function of the frequency of allele A_1. Mean fitness is low when A_1 is absent, and relatively low when A_1 is fixed. As the allele frequency moves from either direction toward its stable equilibrium, the population mean fitness rises to a maximum.

Now, look at the blue curve in Figure 6.22a. It describes a locus with heterozygote inferiority. If p rises even slightly above 0.6, p will continue to rise toward 1.0 in subsequent generations; if p falls even slightly below 0.6, p will continue to fall toward 0 in subsequent generations. The internal equilibrium for a locus with heterozygote inferiority is unstable.

Figure 6.22c shows an adaptive landscape for a locus with heterozygote inferiority. Population mean fitness is lowest when the frequency of allele A_1 is at its unstable internal equilibrium. As the allele frequency moves away from this equilibrium in either direction, mean fitness rises.

A comparison of the adaptive landscape in Figure 6.22c with those in Figure 6.22b and Figure 6.20 offers a valuable insight. As a population evolves in response to selection, the mean fitness of the individuals in the population tends to rise. Selection does not, however, always maximize mean fitness in a global sense. Depending on the initial allele frequencies, the population depicted in Figure 6.22c may evolve toward either fixation or loss of A_1. If the allele becomes fixed, the population will be at a stable equilibrium, but the population's mean fitness will be substantially lower than it would be if the allele were lost.

Compound chromosomes are homologous chromosomes that have swapped entire arms, so that one homolog has two copies of one arm, and the other homolog has two copies of the other arm **(Figure 6.23a and b, next page)**. During meiosis, compound chromosomes may or may not segregate. As a result, four kinds of gametes are produced in equal numbers: gametes with both homologous chromosomes, gametes with just one member of the pair, gametes with the other member of the pair, and gametes with neither member of the pair (Figure 6.23c). When two flies with compound chromosomes mate with each other, one-quarter of their zygotes have every chromosome arm in the correct dose and are thus viable (Figure 6.23d). The other three-quarters have too many or too few of copies of one or both chromosome arms and are thus inviable. When a fly

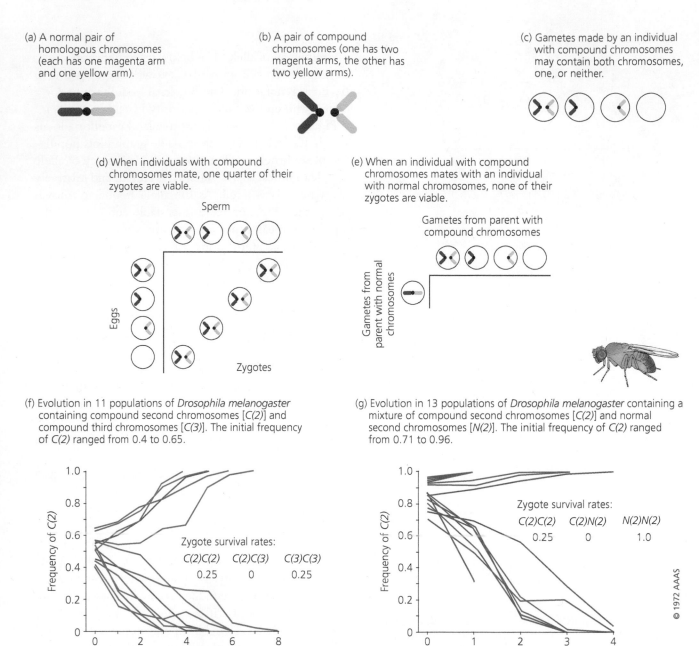

(a) A normal pair of homologous chromosomes (each has one magenta arm and one yellow arm).

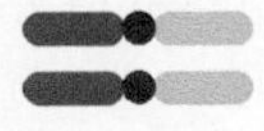

(b) A pair of compound chromosomes (one has two magenta arms, the other has two yellow arms).

(c) Gametes made by an individual with compound chromosomes may contain both chromosomes, one, or neither.

(d) When individuals with compound chromosomes mate, one quarter of their zygotes are viable.

Sperm

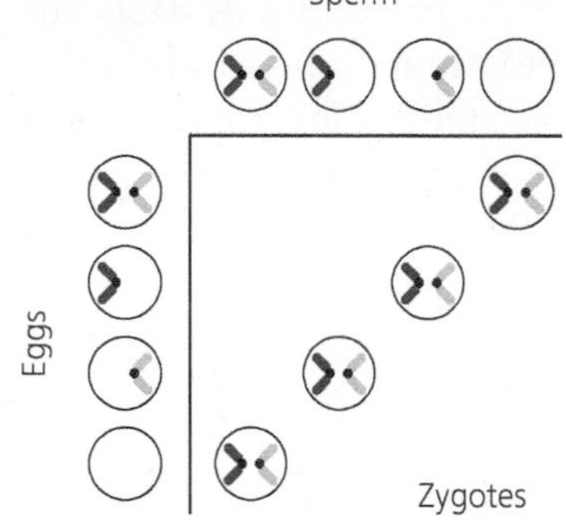

(e) When an individual with compound chromosomes mates with an individual with normal chromosomes, none of their zygotes are viable.

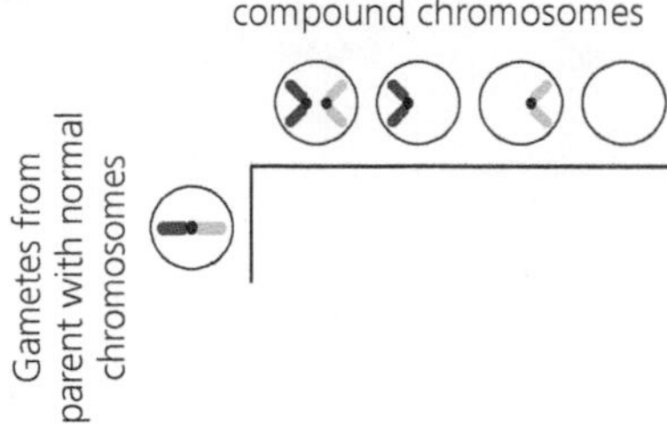

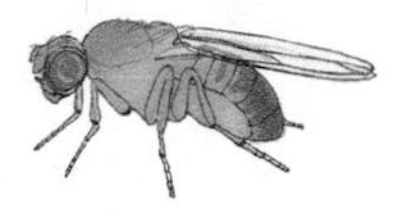

(f) Evolution in 11 populations of *Drosophila melanogaster* containing compound second chromosomes [*C(2)*] and compound third chromosomes [*C(3)*]. The initial frequency of *C(2)* ranged from 0.4 to 0.65.

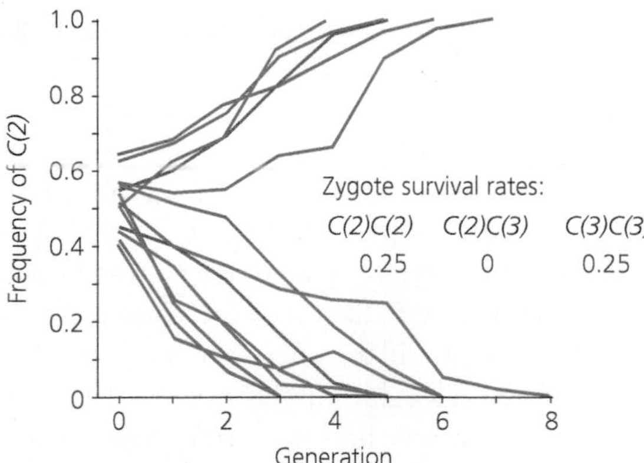

(g) Evolution in 13 populations of *Drosophila melanogaster* containing a mixture of compound second chromosomes [*C(2)*] and normal second chromosomes [*N(2)*]. The initial frequency of *C(2)* ranged from 0.71 to 0.96.

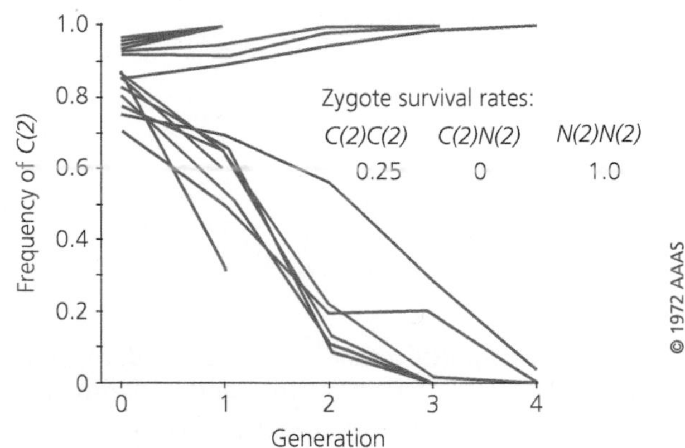

Figure 6.23 An experiment designed to show how populations evolve when heterozygotes have lower fitness than either homozygote (a–e) The experimental design makes clever use of compound chromosomes. (f and g) The data (orange and purple) match the theoretical predictions (gray). Redrawn with permission from Foster et al. (1972).

From "Chromosome rearrangements for the control of insect pests." *Science* 176:875–880. Reprinted with permission from AAAS.

with compound chromosomes mates with a fly with normal chromosomes, none of the zygotes they make are viable (Figure 6.23e).

Foster and colleagues established two sets of laboratory populations. In the first set of populations, some of the founders had compound second chromosomes [*C(2)*] and others had compound third chromosomes [*C(3)*]. Note that if two flies with compound second chromosomes mate, one-quarter of their offspring survive. Likewise, if two flies with compound third chromosomes mate, one-quarter of their offspring survive. But if a fly with compound second chromosomes (and

It is also possible for heterozygotes to have inferior fitness.

normal third chromosomes) mates with a fly with compound third chromosomes (and normal second chromosomes), none of their offspring survive. For purposes of analysis, then, we can treat the second and third chromosome as though they are alleles of a single locus. Thus the founders consisted of *C(2)C(2)* homozygotes and *C(3)C(3)* homozygotes. Based on the zygote viabilities we just described, the fitnesses of the possible offspring genotypes in the mixed population are

$C(2)C(2)$	$C(2)C(3)$	$C(3)C(3)$
0.25	0	0.25

In other words, the genotypes exhibit strong **underdominance**.

The algebraic analysis described in Computing Consequences 6.7 predicts that such a mixed population will be in genetic equilibrium, with both alleles present, when the frequency of *C(2)* is exactly 0.5. This equilibrium is unstable, however. If the frequency of *C(2)* ever gets above 0.5, then it should quickly rise to 1.0. Likewise, if the frequency of *C(2)* ever dips below 0.5, it should quickly fall to zero. Experimentation with a computer should allow the reader to reproduce this behavior.

Intuitively, the reason for the behavior is as follows. Heterozygotes are inviable, so the adults in the population are all homozygotes. Imagine first a situation in which *C(2)C(2)* individuals are common and *C(3)C(3)* individuals are rare. If the flies mate at random, then most matings will involve *C(2)C(2)* flies mating with each other, or *C(2)C(2)* flies mating with *C(3)C(3)* flies. Only rarely will *C(3)C(3)* flies mate with their own kind. Consequently, most *C(3)C(3)* flies will have zero reproductive success, and the frequency of *C(2)* will climb toward 1.0. Now imagine that *C(3)C(3)* individuals are common and *C(2)C(2)* individuals are rare. Under random mating, most matings involve *C(3)C(3)* flies mating with each other, or *C(3)C(3)* flies mating with *C(2)C(2)* flies. As a result, most of the *C(2)C(2)* flies will have zero reproductive success, and the frequency of *C(2)* will fall toward zero.

Foster and colleagues set up 11 mixed populations, with *C(2)* frequencies ranging from about 0.4 to about 0.65, then monitored their evolution for up to eight generations. Predictions for the evolution of populations with initial C(2) frequencies of 0.45 and 0.55 appear as gray lines in the graph in Figure 6.23f. The data from Foster et al.'s flies appear as orange lines. There is some deviation between prediction and result, probably due to genetic drift. That is, in a few of the experimental populations the frequency of *C(2)* started above 0.5 but ultimately fell to zero. In all 11 populations, however, once the frequency of *C(2)* had moved substantially away from 0.5, it continued moving in the same direction until it hit zero or one.

In the researchers' second set of populations, some founders had compound second chromosomes [*C(2)*] and others had normal second chromosomes [*N(2)*]. If two flies with compound second chromosomes mate, one-quarter of their offspring are viable. If a fly with compound second chromosomes mates with a fly with normal second chromosomes, none of their offspring is viable. If two flies with normal second chromosomes mate, all of their offspring are viable. Again, for purposes of analysis, we can treat each chromosome as though it were a single allele. Thus the founders consisted of *C(2)C(2)* homozygotes and *N(2)N(2)* homozygotes. The fitnesses of the genotypes in the mixed population are

$C(2)C(2)$	$C(2)N(2)$	$N(2)N(2)$
0.25	0	1.0

As in the first set of populations, the genotypes exhibit strong underdominance. This time, however, one kind of homozygote has much higher fitness than the other.

The algebraic analysis described in Computing Consequences 6.7 predicts an unstable equilibrium when the frequency of *C(2)* is exactly 0.8. If the frequency of *C(2)* ever gets above 0.8, then it should quickly rise to 1.0. Likewise, if the frequency of *C(2)* ever dips below 0.8, it should quickly fall to zero. Experimentation with a computer should allow the reader to reproduce this prediction.

The intuitive explanation is as follows. Heterozygotes are inviable, so the adults in the population are all homozygotes. Imagine first that *C(2)C(2)* individuals are common and *N(2)N(2)* individuals are rare. If the flies mate at random, then almost all matings will involve *C(2)C(2)* flies mating with each other, or *C(2)C(2)* flies mating with *N(2)N(2)* flies. Only very rarely will *N(2)N(2)* flies mate with their own kind. Consequently, most *N(2)N(2)* flies will have zero reproductive success, and the frequency of *C(2)* will climb to 1.0. Now imagine that there are enough *N(2)N(2)* flies present that appreciable numbers of them do mate with each other. These matings will produce four times as many offspring as matings between *C(2)C(2)* flies. Consequently, the frequency of *N(2)* will climb to 1.0 and the frequency of *C(2)* will fall to zero.

Foster and colleagues set up 13 mixed populations, with *C(2)* frequencies ranging from 0.71 to 0.96, then monitored their evolution for up to four generations. Predictions for the evolution of populations with initial *C(2)* frequencies of 0.75 and 0.85 appear as gray lines in the graph in Figure 6.23g. The data appear as purple lines. Qualitatively, the outcome matches the theoretical prediction nicely. In populations with higher initial *C(2)* frequencies, *C(2)* quickly rose to fixation, while in populations with lower initial *C(2)* frequencies, *C(2)* was quickly lost. The exact location of the unstable equilibrium turned out to be approximately 0.9 instead of 0.8. Foster and colleagues note that their *C(2)C(2)* flies carried recessive genetic markers, bred into them to allow for easy identification. They suggest that these markers reduced the relative fitness of the *C(2)C(2)* flies below the value of 0.25 inferred solely on the basis of their compound chromosomes.

Our model's predictions were not as accurate for Foster et al.'s experiments as they were for Dawson's and Mukai and Burdick's. Nonetheless, the model performed well. It predicted something we have not seen before: an unstable equilibrium above which the frequency of an allele would rise and below which it would fall. It predicted that the unstable equilibrium would be higher in Foster et al.'s second set of populations than in their first. And its predictions about the rate of evolution were roughly correct. Our model has passed its third test.

When heterozygotes have inferior fitness, one allele tends to go to fixation while the other allele is lost. However, different populations may lose different alleles.

Foster et al.'s experiments demonstrate that heterozygote inferiority leads to a loss of genetic diversity within populations. By driving different alleles to fixation in different populations, however, heterozygote inferiority may help maintain genetic diversity among populations.

Frequency-Dependent Selection

For our fourth and final test of population genetics theory, we will see whether our model can predict the evolutionary outcome when the fitness of individuals with a particular phenotype depends on their frequency in the population. Our example, from the work of Luc Gigord, Mark Macnair, and Ann Smithson (2001), concerns a puzzling color polymorphism in the Elderflower orchid (*Dactylorhiza sambucina*).

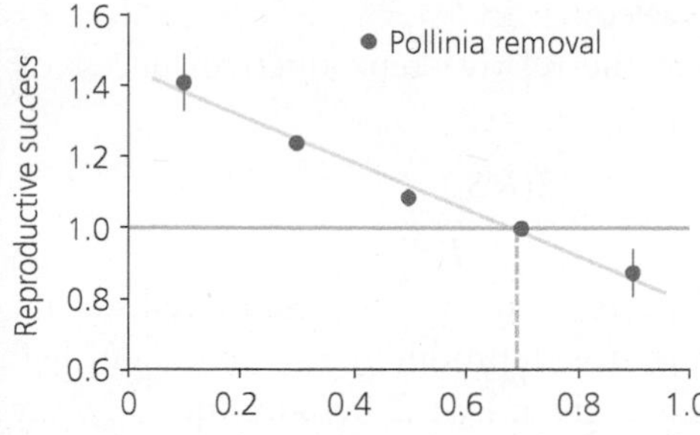

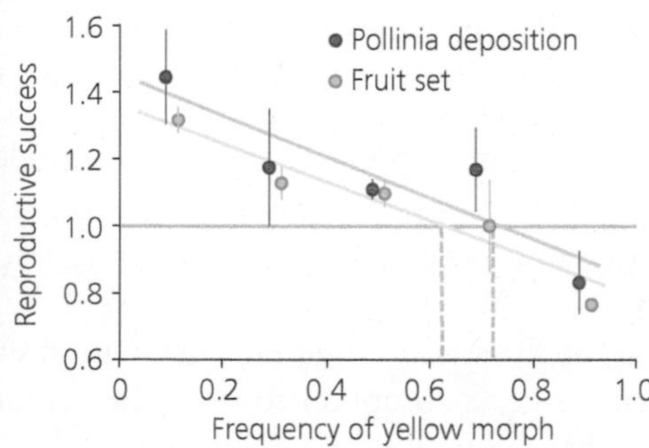

Figure 6.24 Frequency-dependent selection in Elderflower orchids (a) A mixed population. Some plants have yellow flowers, others have purple flowers. (b) Through male function, yellow flowers have higher fitness than purple flowers when yellow is rare, but lower fitness than purple flowers when yellow is common. (c) Through female function, yellow flowers have higher fitness than purple flowers when yellow is rare, but lower fitness than purple flowers when yellow is common. The dashed vertical lines show the predicted frequency of yellow flowers, which matches the frequency in natural populations. From Gigord et al. (2001).

Elderflower orchids come in yellow and purple **(Figure 6.24a)**. Populations typically include both colors, though yellow is usually more common. The flowers attract bumblebees, which are the orchid's main pollinators. But the bees that visit Elderflower orchids are always disappointed. To the bees the orchid's colorful flowers appear to advertise a reward, but in fact they offer nothing. The puzzle Gigord and colleagues wanted to solve was this: How can two distinct deceptive advertisements persist together in Elderflower orchid populations?

The researchers' hypothesis grew from earlier observations by Smithson and Macnair (1997). When naive bumblebees visit a stand of orchids to sample the flowers, they tend to alternate between colors. If a bee visits a purple flower first and finds no reward, it looks next in a yellow flower. Finding nothing there either, it tries another purple one. Disappointment sends it back to a yellow, and so on, until the bee gives up and leaves. Because bumblebees tend to visit equal numbers of yellow and purple flowers, orchids with the less common of the two colors receive more visits per plant. If more pollinator visits translate into higher reproductive success, then the rare-color advantage could explain why both colors persist. Selection by bumblebees favors yellow until it becomes too common, then it favors purple. This is an example of **frequency-dependent selection**.

Selection can also maintain two alleles in a population if each allele is advantageous when it is rare.

To test their hypothesis, Gigord and colleagues collected and potted wild orchids, then placed them in the orchids' natural habitat in 10 experimental arrays of 50 plants each. The frequency of yellow flowers varied among arrays, with two arrays at each of five frequencies: 0.1, 0.3, 0.5, 0.7, and 0.9. The researchers monitored the orchids for removal of their own pollinia (pollen-bearing structures), for deposition of pollinia from other individuals, and for fruit set. From their data, Gigord and colleagues estimated the reproductive advantage of yellow flowers, relative to purple, via both male and female function.

The resulting estimates of relative reproductive success, plotted as a function of the frequency of yellow flowers, appear in Figure 6.24b and c. Consistent with the researchers' hypothesis, yellow-flowered orchids enjoyed higher reproductive success than purple-flowered plants when yellow was rare and suffered lower reproductive success when yellow was common.

Gigord and colleagues calculated the relative reproductive success of yellow orchids as

$$RRS_y = \frac{2(RS_y)}{RS_y + RS_p}$$

where RS_y and RS_p are the absolute reproductive success of yellow and purple orchids. The relationship between relative reproductive success via male function and the frequency of yellow flowers is given by the best-fit line in Figure 6.21b. It is

$$RRS_y = -0.66F_y + 1.452$$

where F_y is the frequency of yellow flowers.

We can incorporate this relationship into a population genetics model. We might imagine, for example, that flower color is determined by two alleles at a single locus and that yellow is recessive to purple. We set the starting frequency of the yellow allele to an arbitrary value. We assign fitnesses to the three genotypes as we have before, except that the fitnesses change each generation with the frequency of yellow flowers. When we use a computer to track the evolution of our model population, we discover that the frequency of the yellow allele moves rapidly to equilibrium at an intermediate value. This value is precisely the allele frequency at which yellow flowers have a relative fitness of 1. We get the same result if we imagine that yellow flowers are dominant. Again the equilibrium value for the yellow allele is the frequency at which yellow and purple flowers have equal fitness.

The dashed vertical lines in Figure 6.24b and c indicate the predicted equilibrium frequencies Gigord and colleagues calculated for each of their fitness measures. The predictions are 61%, 69%, and 72% yellow flowers. The researchers surveyed 20 natural populations in the region where they had placed their experimental arrays. The actual frequency of yellow flowers, 69 ± 3%, is in good agreement with the predicted frequency. Our model has passed its fourth test.

Gigord et al.'s study of Elderflower orchids demonstrates that frequency-dependent selection can have an effect similar to heterozygote superiority. Both patterns of selection can maintain genetic diversity in populations.

Compulsory Sterilization

The theory of population genetics, despite its simplifying assumptions, allows us to predict the course of evolution. Our four tests show that the model we have developed works remarkably well. So long as we know the starting allele frequencies and genotype fitnesses, the model can predict how allele frequencies will change, under a variety of selection schemes, many generations into the future. The requisite knowledge is easiest to get, of course, for experimental populations living under controlled conditions in the lab. But Gigord et al.'s study of Elderflower orchids shows that the model can even make fairly accurate predictions in natural populations. Given its success in the four tests, it is reasonable to use our model to consider the evolutionary consequences of a eugenic sterilization program. The proponents of eugenic sterilization sought to reduce the fitness of particular genotypes to zero and thereby to reduce the frequency of alleles responsible for undesirable phenotypes. Would their plan have worked?

We can use population genetics models to evaluate whether eugenic sterilization could have accomplished the aims of its proponents, had their assumptions about the heritability of traits been correct. The answer depends on the frequency of the alleles in question, and on the criteria for success.

The phenotype that caught the eugenicists' attention perhaps more than any other was feeblemindedness. The Royal College of Physicians in England defined a feebleminded individual as "One who is capable of earning his living under favorable circumstances, but is incapable from mental defect existing from birth or from an early age (a) of competing on equal terms with his normal fellows or (b) of managing himself and his affairs with ordinary prudence" (see Goddard 1914). Evidence presented in 1914 by Henry H. Goddard, who was the director of research at the Training School for Feebleminded Girls and Boys in Vineland, New Jersey, convinced many eugenicists that strength of mind behaved like a simple Mendelian trait (see Paul and Spencer 1995). Normal-mindedness was believed to be dominant and feeblemindedness recessive.

A recessive genetic disease is not a promising target for a program that would eliminate it by sterilizing affected individuals. As Figures 6.19a and 6.20a show, rare recessive alleles decline in frequency slowly, even under strong selection. On the other hand, eugenicists did not believe that feeblemindedness was especially rare (Paul and Spencer 1995). Indeed, they believed that feeblemindedness was alarmingly common and increasing in frequency. Edward M. East (1917) estimated the frequency of feeblemindedness at three per thousand. Henry H. Goddard reported a frequency of 2% among New York schoolchildren. Tests of American soldiers during World War I suggested a frequency of nearly 50% among white draftees.

We will assume a frequency for feeblemindedness of 1% and reproduce a calculation reported by R. C. Punnett (1917) and revisited by R. A. Fisher (1924). Let *f* be the purported allele for feeblemindedness, with frequency q. If 1% of the population has genotype *ff*, then, by the Hardy–Weinberg equilibrium principle, the initial frequency of *f* is

$$q = \sqrt{0.01} = 0.1$$

If all affected individuals are sterilized, then the fitness of genotype *ff* is zero (or, equivalently, the selection coefficient for genotype *ff* is 1). Using the equation developed in Computing Consequences 6.6, we can calculate the value of q in successive generations, and from q we can calculate the frequency of genotype *ff*.

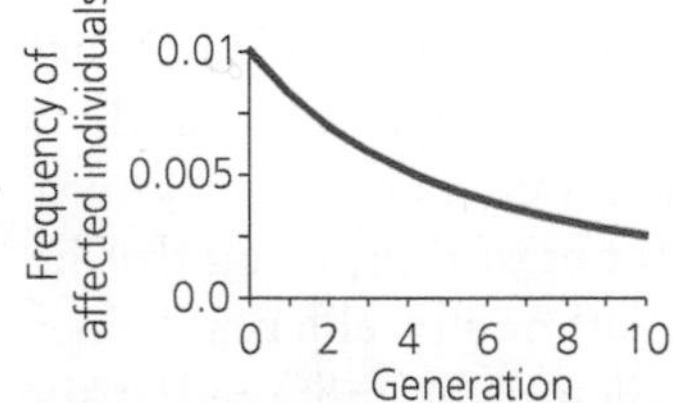

Figure 6.25 Predicted evolution due to sterilization The graph shows the change in the frequency of homozygotes for a putative allele for feeblemindedness under a eugenic sterilization program that prevents homozygous recessive individuals from reproducing.

The result appears in **Figure 6.25**. Over 10 generations, about 250 years, the frequency of affected individuals declines from 0.01 to 0.0025.

Whether geneticists saw this calculation as encouraging or discouraging depended on whether they saw the glass as partially empty or partially full. Some looked at the numbers, saw that it would take a very long time to completely eliminate feeblemindedness, and argued that compulsory sterilization was such a hopelessly slow solution that it was not worth the effort. Others, such as Fisher, dismissed this argument as "anti-eugenic propaganda." Fisher noted that after just one generation, the frequency of affected individuals would drop from 100 per 10,000 to 82.6 per 10,000. "In a single generation," he wrote, "the load of public expenditure and personal misery caused by feeblemindedness . . . would be reduced by over 17 percent." Fisher also noted that most copies of the allele for feeblemindedness are present in heterozygous carriers rather than affected individuals. Along with East, Punnett, and others, Fisher called for research into methods for identifying carriers.

While their evolutionary logic was sound, the eugenicists' models were built on dubious genetic hypotheses. It is not entirely fair to use modern standards to criticize Goddard's research on the genetics of feeblemindedness. Mendelian genetics was in its infancy. Still, looking back after nearly a century, we can see that Goddard's evidence was deeply flawed. We will consider three problems.

First, the individuals whose case studies he reports are a highly diverse group. Some have Down syndrome; some have other developmental challenges. At least one is deaf and appears to be the victim of a woefully inadequate education. Some appear to have been deposited at Goddard's training school by widowed fathers who felt that children from a prior marriage were a liability in finding a new wife. Some may just have behaved differently than the directors of the school thought they should. Concluding the first case report in his book, Goddard writes of a 16-year-old who has been at the school for seven years:

> Gertrude is a good example of that type of girl who, loose in the world, makes so much trouble. Her beauty and attractiveness and relatively high [intelligence] would enable her to pass almost anywhere as a normal child and yet she is entirely incapable of controlling herself and would be led astray most easily. It is fortunate for society that she is cared for as she is.

Second, Goddard's methods for collecting data were prone to distortion. He sent caseworkers to collect pedigrees from the families of the students at the training school. The caseworkers relied on hearsay and subjective judgments to assess the strength of mind of family members—many of whom were long since deceased.

Third, Goddard's method of analysis stacked the cards in favor of his conclusion. He first separated his 327 cases into various categories: definitely hereditary cases; probably hereditary cases; cases caused by accidents; and cases with no assignable cause. He apparently placed cases in his "definitely hereditary" group only when they had siblings, recent ancestors, or other close kin also classified as feebleminded. When he later analyzed the data to determine whether feeblemindedness was a Mendelian trait, Goddard analyzed only the data from his "definitely hereditary" group. Given how he had filtered the data ahead of time, it is not too surprising that he concluded that feeblemindedness is Mendelian.

Although feeblemindedness is not among them, many genetic diseases are now known to be inherited as simple Mendelian traits. Yet eugenic sterilization has few advocates. One reason is that most serious genetic diseases are recessive and very rare; sterilization of affected individuals would have little impact on the frequency at which new affected individuals are born. A second reason is that mainstream attitudes about reproductive rights have changed to favor individual autonomy over societal mandates (Paul and Spencer 1995). A third reason is that, as we discuss in the next section, a growing list of disease alleles are suspected or known to be maintained in populations by heterozygote superiority. It would be futile and possibly ill advised to try to reduce the frequency of such alleles by preventing affected individuals from reproducing.

6.4 Mutation

Cystic fibrosis is among the most common serious genetic diseases among people of European ancestry, affecting approximately 1 newborn in 2,500. Cystic fibrosis is inherited as an autosomal recessive trait. Affected individuals suffer chronic infections with the bacterium *Pseudomonas aeruginosa* and ultimately sustain severe lung damage (Pier et al. 1997). At present, most individuals with cystic fibrosis live into their thirties or forties (Elias et al. 1992), but until recently few survived to reproductive age. Although cystic fibrosis was lethal for most of human history, in some populations as many as 4% of individuals are carriers. How can alleles that cause a lethal genetic disease remain this common?

Our consideration of heterozygote superiority in the previous section hinted at one possible answer. Another potential answer is that new disease alleles are constantly introduced into populations by mutation. Before we can evaluate the relative merits of these two hypotheses for explaining the persistence of any particular disease allele, we need to discuss mutation in more detail.

Elsewhere in the book (see Chapter 5), we presented mutation as the source of all new alleles and genes. In its capacity as the ultimate source of all genetic variation, mutation provides the raw material for evolution. Here, we consider the importance of mutation as a mechanism of evolution. How rapidly does mutation cause allele frequencies to change over time? How strongly does mutation affect the conclusions of the Hardy–Weinberg analysis?

Adding Mutation to the Hardy–Weinberg Analysis: Mutation as an Evolutionary Mechanism

Mutation by itself is generally not a rapid mechanism of evolution. To see why, return to our model population of mice. Imagine a locus with two alleles, *A* and *a*, with initial frequencies of 0.9 and 0.1. *A* is the wild-type allele, and *a* is a recessive loss-of-function mutation. Furthermore, imagine that copies of *A* are converted by mutation into new copies of *a* at the rate of 1 copy per 10,000 per generation. This is a very high mutation rate, but it is within the range of mutation rates known. Back-mutations that restore function are much less common than loss-of-function mutations, so we will ignore mutations that convert copies of *a* into new copies of *A*. Finally, imagine that all mutations happen while the adults are making gametes to contribute to the gene pool.

Second on the list of assumptions for the Hardy–Weinberg equilibrium principle was that there are no mutations. We now explore what happens to allele frequencies when this assumption is violated.

Figure 6.26 follows the allele and genotype frequencies through one turn of the life cycle. Among the zygotes, juveniles, and adults, the genotypes are in Hardy–Weinberg proportions:

AA	*Aa*	*aa*
0.81	0.18	0.01

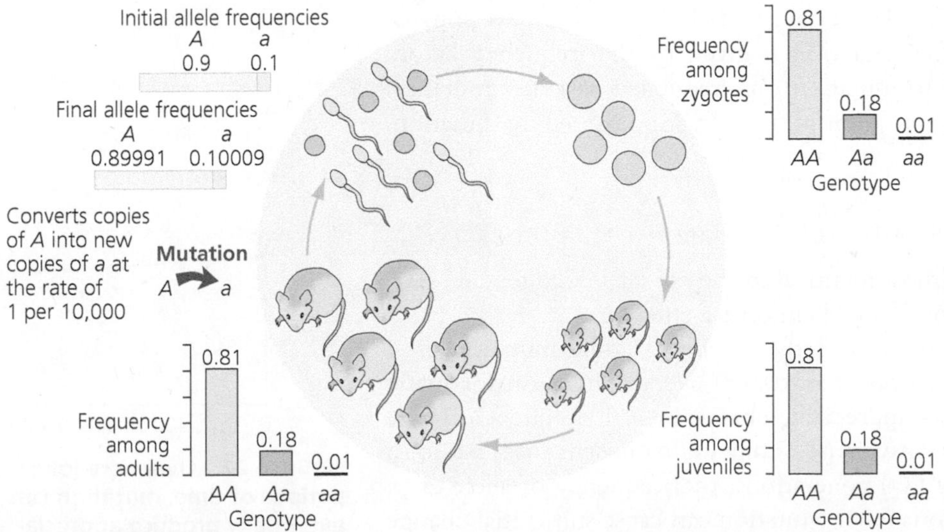

Figure 6.26 Mutation is a weak mechanism of evolution In a single generation in our model population, mutation produces virtually no change in allele and genotype frequencies.

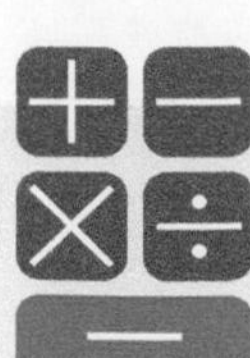

COMPUTING CONSEQUENCES 6.8

A mathematical treatment of mutation as an evolutionary mechanism

Imagine a single locus with two alleles: a wild-type allele, A, and a recessive loss-of-function mutation, a. Let μ be the rate of mutation from A to a. Assume that the rate of back-mutation from a to A is negligible. If the frequency of A in this generation is p, then its frequency in the next generation is given by

$$p' = p - \mu p$$

If the frequency of a in this generation is q, then its frequency in the next generation is given by

$$q' = q + \mu p$$

The change in p from one generation to the next is

$$\Delta p = p' - p$$

which simplifies to

$$\Delta p = -\mu p$$

After n generations, the frequency of A is approximately

$$p_n = p_0 e^{-\mu n}$$

where p_n is the frequency of A in generation n, p_0 is the frequency of A in generation 0, and e is the base of the natural logarithms.

Readers familiar with calculus can derive the last equation as follows. First, assume that a single generation is an infinitesimal amount of time, so that we can rewrite the equation $\Delta p = -\mu p$ as

$$\frac{dp}{dg} = -\mu p$$

Now divide both sides by p, and multiply both sides by dg to get

$$\left(\frac{1}{p}\right)dp = -\mu dg$$

Finally, integrate the left side from frequency p_0 to p_n and the right side from generation 0 to n:

$$\int_{p_0}^{p_n}\left(\frac{1}{p}\right)dp = \int_0^n -\mu dg$$

and solve for p_n.

Now the adults make gametes. Were it not for mutation, the allele frequencies in the new gene pool would be

A	a
0.9	0.1

But mutation converts 1 of every 10,000 copies of allele A into a new copy of allele a. The frequency of A after mutation is given by the frequency before mutation minus the fraction lost to mutation; the frequency of a after mutation is given by the frequency before mutation plus the fraction gained by mutation. That is,

A	a
$0.9 - (0.0001 \cdot 0.9) = 0.89991$	$0.1 + (0.0001 \cdot 0.9) = 0.10009$

The new allele frequencies are almost identical to the old allele frequencies. As a mechanism of evolution, mutation has had almost no effect.

Almost no effect is not the same as exactly no effect. Could mutation of A into a, occurring at the rate of 1 copy per 10,000 every generation for many generations, eventually result in an appreciable change in allele frequencies? The graph in **Figure 6.27** provides the answer (see **Computing Consequences 6.8** for a mathematical treatment). After 1,000 generations, the frequency of allele A in our model population will be about 0.81. Mutation can cause substantial change in allele frequencies, but it does so slowly.

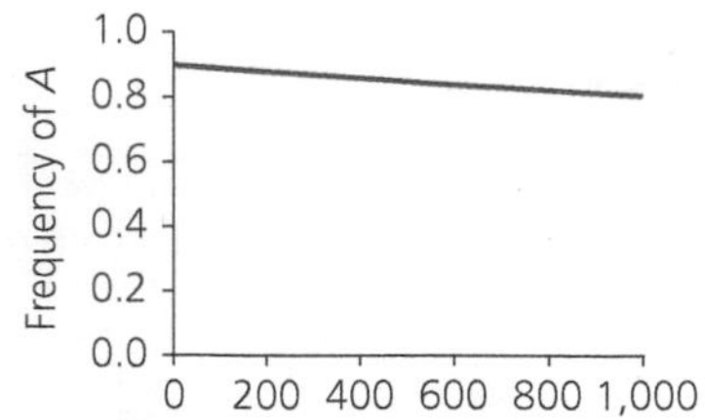

Figure 6.27 Over very long periods of time, mutation can eventually produce appreciable changes in allele frequency

As mutation rates go, the value we used in our model, 1 per 10,000 per generation, is very high. For most genes, mutation is an even less efficient mechanism of allele frequency change.

Hardy–Weinberg analysis shows that mutation is a weak mechanism of evolution.

Mutation and Selection

Although mutation alone usually does not cause appreciable changes in allele frequencies, this does not mean that mutation is unimportant in evolution. In combination with selection, mutation is a crucial piece of the evolutionary process. This point is demonstrated by an experiment conducted by Mingcai Zhang, Priti Azad, and Ron Woodruff (2011), who showed that fruit fly populations with virtually no initial genetic variation accumulate novel alleles quickly enough to allow rapid adaptive evolution.

Zhang and colleagues began with a stock of *Drosophila melanogaster* that had been propagated by single-pair sibling matings for over 150 generations. As we will see later (Chapter 7), this kind of intense inbreeding results in rapid loss of genetic variation. Screening at loci that are highly variable in most populations confirmed that all flies in the inbred stock were essentially genetically identical.

The researchers next reared larvae from their inbred stock on food spiked with table salt (NaCl) at concentrations ranging from 1% to 6%. At least a few larvae survived to adulthood at concentrations up to 4%, but all the larvae died at 5%.

Then Zhang and colleagues used flies from the inbred stock as founders for six experimental populations, which they maintained for 30 generations. They kept the population sizes large, establishing each generation with 200 pairs of randomly chosen adults from the previous generation. The researchers kept two of the populations under benign conditions, and four on food spiked with salt. They distributed the salty food in patches with different concentrations, but all of it was stressful for the flies. The conditions of the experiment allowed the populations to evolve by natural selection to adapt to their new environments, but the populations would do so only if they accrued genetic variation via mutation.

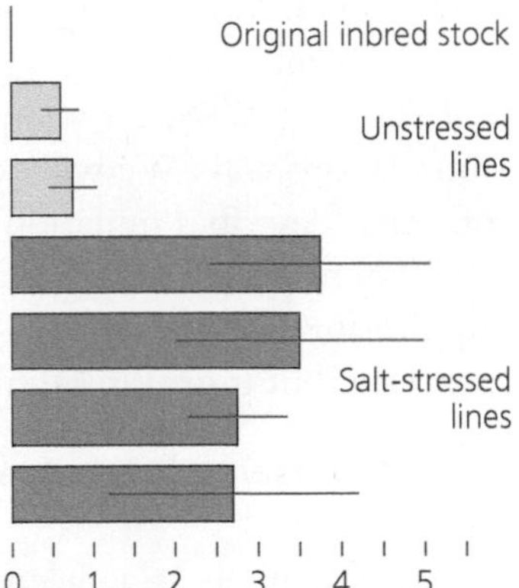

Mean number of surviving offspring from vials with five pairs of parents and medium containing 5% NaCl (±standard error)

Figure 6.28 Adaptive evolution resulting from natural selection on novel mutations Bars show the salt tolerance of flies in the thirtieth generation of fruit fly lineages founded from a single genetically homogenous stock. Prepared from data in Zhang et al. (2011).

Finally, the researchers assessed the salt tolerance of the thirtieth generation in each of the six populations. Survival data for larvae reared on food spiked with 5% salt appear in **Figure 6.28**. The original inbred stock appears first, as a reminder that for the founding flies, 5% salt was 100% lethal. The unstressed lines appear next. Even though theses lines had evolved under benign conditions, both included a few individuals that could survive in 5% salt. This result demonstrates the accumulation of genetic variation by mutation in the absence of selection. The salt-stressed lines appear last. They contained higher proportions of individuals that could survive 5% salt. This result demonstrates adaptive evolution as a result of mutation and natural selection in combination. Further evidence that the stressed lines harbored alleles for salt tolerance at higher frequency than the unstressed lines came when Zhang and colleagues attempted to rear larvae on 6% salt. A few individuals from each of the salt-stressed lines survived, but all the flies from the unstressed lines died.

The experiment by Zhang and colleagues reinforces one of the messages we discussed earler (Chapter 5). Without mutation, evolution would eventually grind to a halt. Mutation is the ultimate source of genetic variation.

Research with fruit flies illustrates that while mutation itself is only a weak mechanism of evolution, it nonetheless supplies the raw material on which natural selection acts.

Mutation–Selection Balance

Unlike the mutations that allowed the evolution of increased salt tolerance in Zhang et al.'s fruit fly populations, many mutations are at least mildly deleterious.

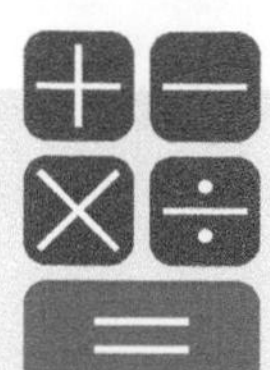

COMPUTING CONSEQUENCES 6.9

Allele frequencies under mutation–selection balance

Here we derive equations for predicting the equilibrium frequencies of deleterious alleles under mutation–selection balance. Imagine a single locus with two alleles, A_1 and A_2, with frequencies p and q. A_1 is the wild type; A_2 is deleterious. Let μ be the rate at which copies of A_1 are converted into copies of A_2 by mutation. Assume that the rate of back-mutation is negligible.

Selection continuously removes copies of A_2 from the population, while mutation continuously creates new copies. We want to calculate the frequency of A_2 at which these processes cancel each other. Following Felsenstein (1997), we will perform our calculation in a roundabout way. We will develop an equation in terms of p that describes mutation–selection balance for allele A_1. Then we will solve the equation for q to get the equilibrium frequency of A_2. This approach may seem perverse, but it greatly simplifies the algebra.

Mutation–selection balance for a deleterious recessive

Imagine that A_2 is a deleterious recessive allele, such that the genotype fitnesses are given by

w_{11}	w_{12}	w_{22}
1	1	$1 - s$

where the selection coefficient s gives the strength of selection against A_2.

First, we write an equation for p^*, the frequency of allele A_1 after selection has acted, but before mutations occur. From Computing Consequences 6.3, this is

$$p^* = \frac{p^2 w_{11} + pqw_{12}}{p^2 w_{11} + 2pqw_{12} + q^2 w_{22}}$$

Substituting the fitnesses from the table above, and $(1 - p)$ for q, then simplifying gives

$$p^* = \frac{p}{1 - s(1 - p)^2}$$

Next we write an expression for p', the frequency of allele A_1 after mutations occur. Mutations convert a fraction μ of the copies of A_1 into copies of A_2, leaving behind a fraction $(1 - \mu)$. Thus

$$p' = (1 - \mu)p^* = \frac{(1 - \mu)p}{1 - s(1 - p)^2}$$

Finally, when mutation and selection are in balance, p' is equal to p, the frequency of allele A_1 that we started with:

$$\frac{(1 - \mu)p}{1 - s(1 - p)^2} = p$$

This simplifies to

$$(1 - p)^2 = \frac{\mu}{s}$$

Substituting q for $(1 - p)$ and solving for q yields an equation for $\hat{q}$, the equilibrium frequency of allele A_2 under mutation–selection balance:

$$\hat{q} = \sqrt{\frac{\mu}{s}}$$

If A_2 is a lethal recessive, then $s = 1$, and the equilibrium frequency of A_2 is equal to the square root of the mutation rate.

Mutation–selection balance for a lethal dominant allele

Imagine that A_2 is a lethal dominant allele, such that the genotype fitnesses are given by

w_{11}	w_{12}	w_{22}
1	0	0

Now the expression for p^* simplifies to

$$p^* = 1$$

which makes sense because, by definition, selection removes all copies of the lethal dominant A_2 from the population. Now the expression for p' is

$$p' = 1 - \mu$$

and the equilibrium condition is

$$1 - \mu = p$$

Substituting $(1 - q)$ for p and simplifying gives

$$\hat{q} = \mu$$

In other words, the equilibrium frequency of A_2 is equal to the mutation rate.

Selection tends to eliminate such mutations from populations. Deleterious alleles persist, however, because they are continually created anew. When the rate at which copies of a deleterious allele are being eliminated by selection is exactly equal to the rate at which new copies are being created by mutation, the frequency of the allele is at equilibrium. This situation is called **mutation–selection balance**.

What is the frequency of the deleterious allele at equilibrium? If the allele is recessive, its equilibrium frequency, $\hat{q}$, is given by

$$\hat{q} = \sqrt{\frac{\mu}{s}}$$

where μ is the mutation rate, and s, the selection coefficient, is a number between 0 and 1 expressing the strength of selection against the allele (see **Computing Consequences 6.9** for a derivation). This equation captures with economy what intuition tells us about mutation–selection balance. If the selection coefficient is small (the allele is only mildly deleterious) and the mutation rate is high, then the equilibrium frequency of the allele will be relatively high. If the selection coefficient is large (the allele is highly deleterious) and the mutation rate is low, then the equilibrium frequency of the allele will be low.

Research by Brunhilde Wirth and colleagues (1997) on patients with spinal muscular atrophy provides an example. Spinal muscular atrophy is a neurodegenerative disease characterized by weakness and wasting of the muscles that control voluntary movement. It is caused by deletions in a locus on chromosome 5 called the telomeric survival motor neuron gene (*telSMN*). In some cases, the disease may be exacerbated by additional mutations in a nearby gene. Spinal muscular atrophy is, after cystic fibrosis, the second most common lethal autosomal recessive disease in Caucasians (McKusick et al. 1999).

Collectively, the loss-of-function alleles of *telSMN* have a frequency of about 0.01 in the Caucasian population. Wirth and colleagues estimate that the selection coefficient is about 0.9. With such strong selection against them, we would expect that disease-causing alleles would slowly but inexorably disappear from the population. How, then, do they persist at a frequency of 1 in 100?

At the same time selection removes deleterious alleles from a population, mutation constantly supplies new copies. In some cases, this balance between mutation and selection may explain the persistence of deleterious alleles in populations.

One possibility is that the disease alleles are being kept in the population by a balance between mutation and selection. If we substitute the allele frequency and selection coefficient for $\hat{q}$ and s in the equation above, and then solve for μ, we find that this scenario requires a mutation rate of about 9.0×10^{-5} ($= 0.9 \times 10^{-4}$) mutations per *telSMN* allele per generation.

Wirth and colleagues analyzed the chromosomes of 340 individuals with spinal muscular atrophy, and the chromosomes of their parents and other family members. They found that 7 of the 340 affected individuals carried a new mutation not present in either parent. These numbers allowed the scientists to estimate directly the mutation rate at the *telSMN* locus (see **Computing Consequences 6.10**). Their estimate is 1.1×10^{-4}. This directly measured mutation rate is in good agreement with the rate predicted under the hypothesis of a mutation–selection balance. Wirth and colleagues conclude that mutation–selection balance provides a sufficient explanation for the persistence of spinal muscular atrophy alleles.

It is possible that the parents of spinal muscular atrophy patients are unusually susceptible to mutations in *telSMN*. Ideally, we would determine the mutation rate by comparing the genotypes of parents and offspring in the general population. However, such a study would require an extremely large sample size.

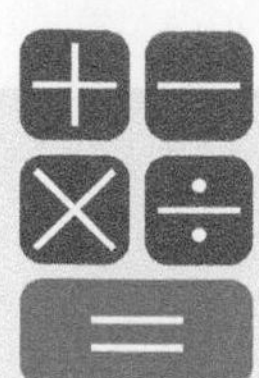

COMPUTING CONSEQUENCES 6.10

Estimating mutation rates for recessive alleles

Here, we present the method used by Brunhilde Wirth and colleagues (1997) to estimate mutation rates for recessive alleles. The key information required is the fraction of affected individuals that carry a brand-new mutant allele. With modern molecular techniques, this fraction can be obtained by direct examination of the chromosomes of affected individuals and their relatives.

Let q be the frequency of recessive loss-of-function allele a. Ignoring the extremely rare individuals with two new mutant copies, there are two ways to be born with genotype *aa*:

1. An individual can be the offspring of two carriers. The probability of this outcome for a given birth is the product of (a) the probability that an offspring of two carriers will be affected; (b) the probability that the mother is a carrier; and (c) the probability that the father is a carrier. This probability is given by

$$\left[\frac{1}{4}\right] \times [2q(1-q)] \times [2q(1-q)]$$

2. An individual can be the offspring of one carrier and one homozygous dominant parent and can receive allele a from the affected parent and a new mutant copy of a from the unaffected parent. The probability of this outcome for a given birth is the product of (a) the probability that an offspring of one carrier will receive that carrier's mutant allele; (b) the probability that the mother is a carrier; (c) the probability that the father is the homozygous dominant; and (d) the mutation rate plus the same probability for the scenario in which the father is the carrier and the mother is the homozygous dominant:

$$\left\{\left[\frac{1}{2}\right] \times [2q(1-q)] \times [(1-q)^2] \times [\mu]\right\} + \left\{\left[\frac{1}{2}\right] \times [2q(1-q)] \times [(1-q)^2] \times [\mu]\right\} = [2q(1-q)] \times [(1-q)^2] \times [\mu]$$

With these probabilities, we can write an expression for r, the fraction of affected individuals that carry one new mutant allele. This is the second probability divided by the sum of the second probability and the first.

Simplified just a bit, we have

$$r = \frac{2q(1-q)(1-q)^2\mu}{2q(1-q)(1-q)^2\mu + q(1-q)q(1-q)}$$

Simplifying further yields

$$r = \frac{2(1-q)\mu}{2(1-q)\mu + q}$$

Finally, assume that q is small, so that $(1-q)$ is approximately equal to one. This assumption gives

$$r = \frac{2\mu}{2\mu + q}$$

which can be solved for μ:

$$\mu = \frac{rq}{2 - 2r}$$

The mutation rate for spinal muscular atrophy

In Caucasian populations, spinal muscular atrophy affects about 1 infant in 10,000, implying that the frequency of the mutant allele is

$$q = \sqrt{0.0001} = 0.01$$

Wirth and colleagues examined the chromosomes of 340 affected patients and their family members. The researchers discovered that seven of their patients had a new mutant allele not present in either parent. Thus

$$r = \frac{7}{340} = 0.021$$

Substituting these values for q and r into the equation for μ gives the estimate

$$\mu = \frac{(0.021)(0.01)}{2 - 2(0.021)} = 0.00011$$

The mutation rate for cystic fibrosis

In Caucasian populations, cystic fibrosis affects about 1 infant in 2,500. Wirth and colleagues cite data from other authors to establish that only 2 of about 30,000 cystic fibrosis patients studied proved to have a new mutant allele not present in either parent. These figures give an estimated mutation rate of

$$\mu = 6.7 \times 10^{-7}$$

Are the Alleles That Cause Cystic Fibrosis Maintained by a Balance between Mutation and Selection?

Cystic fibrosis is caused by recessive loss-of-function mutations in a locus on chromosome 7 that encodes a protein called the cystic fibrosis transmembrane conductance regulator (CFTR). CFTR is a cell-surface protein expressed in the mucous membrane lining the intestines and lungs. Gerald Pier and colleagues (1997) demonstrated that one of CFTR's key functions is to enable cells of the lung lining to ingest and destroy *Pseudomonas aeruginosa* bacteria (see also Campodónico et al. 2008). **Figure 6.29** shows a snapshot of this process (Bajmoczi et al. (2009). In individuals with cystic fibrosis, *P. aeruginosa* cause chronic lung infections that eventually lead to severe lung damage **(Figure 6.30)**.

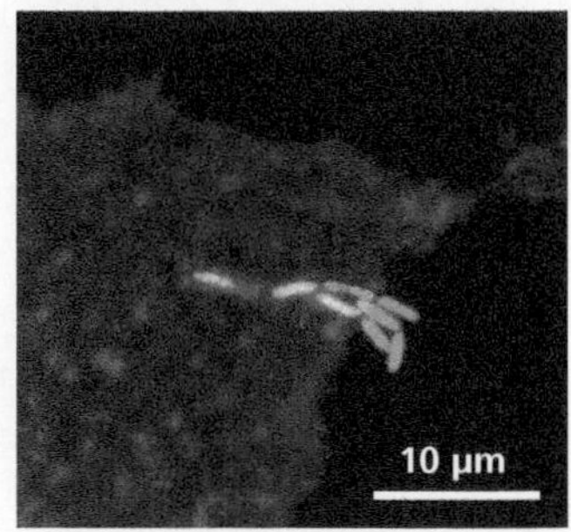

Figure 6.29 A lung epithelial cell ingesting *Pseudomonas aeruginosa* The lung cell is red; the bacteria are green. The bacteria on the left are already inside the cell. They are surrounded by halos of fluorescently labeled CFTR. From Bajmoczi et al. (2009).

Selection against the alleles that cause cystic fibrosis appears to be strong. Until recently, few affected individuals survived to reproductive age; those that do survive are often infertile. And yet the alleles that cause cystic fibrosis have a collective frequency of approximately 0.02 among people of European ancestry.

Could cystic fibrosis alleles be maintained at a frequency of 0.02 by mutation–selection balance? If we assume a selection coefficient of 1 and use the equation derived in Computing Consequences 6.9, the mutation rate creating new disease alleles would have to be 4×10^{-4}. The actual mutation rate for cystic fibrosis alleles appears to be considerably lower than that: approximately 6.7×10^{-7} (see Computing Consequences 6.10). We can conclude that a steady supply of new mutations cannot, by itself, explain the maintenance of cystic fibrosis alleles at a frequency of 0.02.

In other cases, the frequency of a deleterious allele may be too high to explain by mutation–selection balance. This may be a clue that heterozygotes have superior fitness.

Our discussion of heterozygote superiority suggests an alternative explanation (Figure 6.21 and Computing Consequences 6.7). Perhaps the fitness cost suffered by cystic fibrosis alleles when they are in homozygotes is balanced by a fitness advantage they enjoy when they are in heterozygotes.

Gerald Pier and colleagues (1998) hypothesized that cystic fibrosis heterozygotes might be resistant to typhoid fever and therefore have superior fitness. Typhoid fever is caused by *Salmonella typhi* bacteria (also known as *Salmonella enterica* serovar *typhi*). The bacteria initiate an infection by crossing the layer of epithelial cells that line the gut. Pier and colleagues suggested that *S. typhi* bacteria infiltrate the gut by exploiting the CFTR protein as a point of entry. If so, then heterozygotes, which have fewer copies of CFTR on the surface of their cells, should be less vulnerable to infiltration.

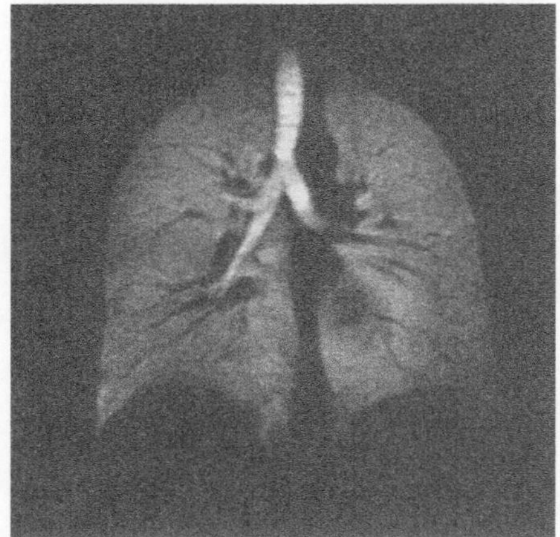

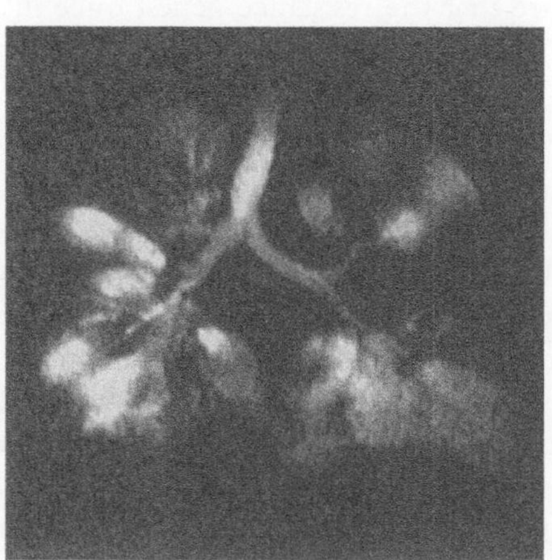

Figure 6.30 Lung damage in cystic fibrosis Left, normal lungs. Right, lungs ravaged by the bacterial infections that accompany cystic fibrosis. Photos by James R. MacFall, Duke University Medical Center.

Pier and colleagues tested their hypothesis by constructing mouse cells with three different CFTR genotypes: homozygous wild-type cells; heterozygotes with one functional CFTR allele and one allele containing the most common human cystic fibrosis mutation, a single-codon deletion called *ΔF508;* and homozygous *ΔF508* cells. The researchers exposed these cells to *S. typhi*, then measured the number of bacteria that got inside cells of each genotype. The results were dramatic **(Figure 6.31a)**. As the researchers predicted, homozygous *ΔF508* cells were almost totally resistant to infiltration by *S. typhi*, while homozygous wild-type cells were highly vulnerable. Heterozygous cells were partially resistant; they accumulated 86% fewer bacteria than did the wild-type cells. These results are consistent with the hypothesis that cystic fibrosis disease alleles are maintained in human populations because heterozygotes have superior fitness during typhoid fever epidemics.

Also consistent with the hypothesis are two more recent discoveries by Pier and coworkers. First, Jeffrey Lyczak and Pier (2002) found that *S. typhi* bacteria manipulate the gut cells of their hosts, causing the cells to display more CFTR protein on their membranes and easing the bacteria's entry. This helps explain why cells that cannot make CFTR are resistant to invasion. Second, Lyczak, Carolyn Cannon, and Pier (2002), using data compiled from the literature, found an apparent association across 11 European countries between the severity of tyhoid fever outbreaks and the frequency a generation later, among CFTR mutations, of the *ΔF508* allele (Figure 6.31b).

Pier et al.'s research serves as another example in which an evolutionary analysis has proved valuable in biomedical research.

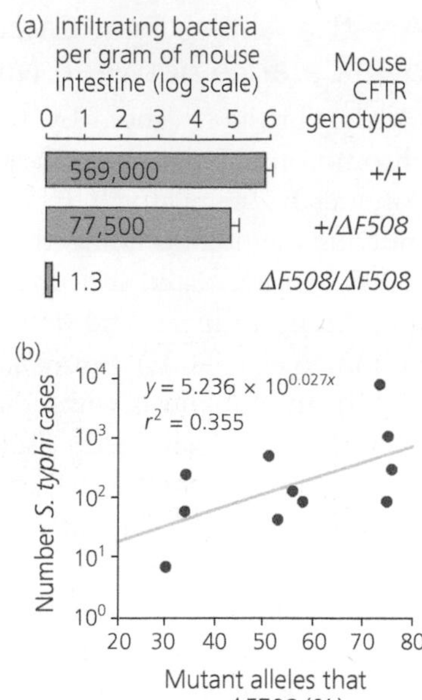

Figure 6.31 Heterozygotes for the *ΔF508* allele are resistant to typhoid fever (a) The rate at which *Salmonella typhi* infiltrate cultured mouse cells with different CFTR genotypes. From Pier et al. (1998). (b) The severity of typhoid fever outbreaks in 11 European countries versus the frequency of *ΔF508*, among cystic fibrosis mutations, in the generation born following the outbreak. From Lyczak et al. (2002).

6.5 An Engineering Test of Population Genetics Theory

Chun-Hong Chen and colleagues (2007), working in the laboratory of Bruce Hay, sought methods to confer malaria resistance on free-living mosquitoes. Their concern was not for the mosquitoes, but for people. If the mosquitoes are resistant to malaria, they cannot transmit the disease to humans.

The task the researchers had set for themselves was one of evolutionary engineering. Genetic variants that make mosquitoes resistant to malaria were already known. The challenge was to ensure that the resistance genes, once introduced into a wild population, would rise to high frequency. Chen and colleagues had an idea for how to do this, which they put to the test in laboratory populations of fruit flies.

The researchers designed a new gene that they expected would carry a strong selective advantage. The gene was a synthetic example of a kind of genetic element, called a *Medea*, that also occurs naturally. *Medea* is an acronym for Maternal-effect dominant embryonic arrest. It is also the name of the title character in a play by Euripides about a mother who murders her own children.

Chen's synthetic *Medea* includes two sets of instructions **(Figure 6.32a)**. One causes mothers that carry the element to infuse their eggs with a poison. The other allows embryos that carry the element to make an antidote. If mother and baby both carry the gene, the baby lives (Figure 6.32b). If mother carries the gene but the baby does not, the baby dies (Figure 6.32c).

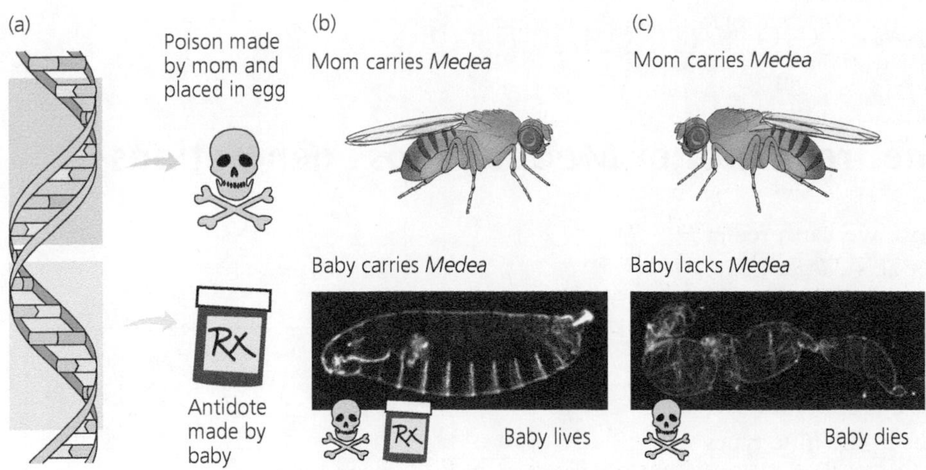

Figure 6.32 An artificial *Medea* gene (a) The gene encodes both a poison and its antidote. (b) Mothers carrying *Medea* make the poison and put it into their eggs. If the baby carries *Medea*, it makes the antidote and lives. (c) If the baby does not carry *Medea*, it dies. Photos from Kambris et al. (2003).

An embryonic fly's fate is determined by the genotypes of both its mother and its father. We will call the two alleles a fly can carry M (for *Medea*) and + (for wild type, or lack of *Medea*). Punnett squares predict that if mom is a heterozygote and dad is a ++ homozygote, half the babies will die. If mom and dad are both heterozygotes, a quarter of the babies will die **(Figure 6.33)**.

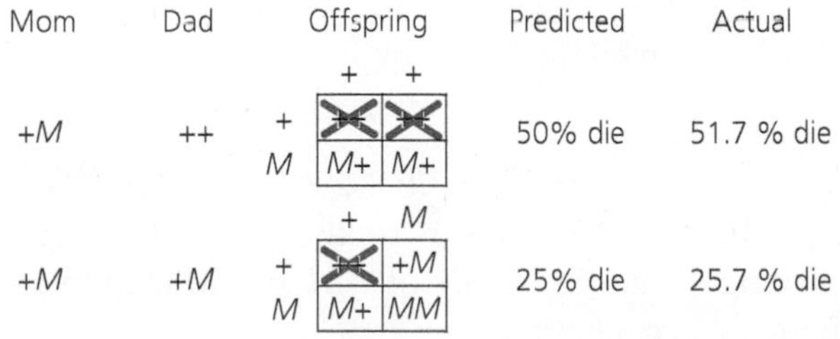

Mom	Dad	Offspring	Predicted	Actual
+*M*	++	(dad: + +) + row: ~~++~~ ~~++~~; *M* row: *M*+ *M*+	50% die	51.7 % die
+*M*	+*M*	(dad: + *M*) + row: ~~++~~ +*M*; *M* row: *M*+ *MM*	25% die	25.7 % die

Figure 6.33 Punnett squares for crosses in which the mother carries *Medea* and some of the offspring do not. Data from Chen et al. (2007).

For all other matings, all the babies live. In experimental matings, Chen and colleagues found these predictions to be accurate.

We claimed earlier that Chen and colleagues expected their synthetic Medea gene to carry a strong selective advantage. This claim may be counterintuitive for a gene that causes mothers to kill their offspring. But note that the offspring that die all lack *Medea*. The selective advantage accrues not to the individuals that carry the gene, but to the gene itself. *Medea* is a selfish allele that, given the chance, kills individuals that do not carry it. If introduced into a population, *Medea* should inexorably rise in frequency.

Chen and colleagues introduced their synthetic *Medea* into laboratory fruit fly populations at a frequency of 0.25. They used the basic population genetics theory we have discussed in this chapter to predict the trajectory of the gene's rise in frequency. They assumed an infinitely large population with random mating and no mutation or migration. They assumed no fitness costs of *Medea* other than the mortality inflicted by the maternal poison on embryos lacking the gene. The only complication in Chen's model, compared to the models we have used throughout the chapter, is that an individual's fitness depends not just on its own genotype but also on the genotypes of its parents. That meant Chen had to track genotype frequencies and account for the progeny from random matings, rather

An attempt to design a gene that would rise to high frequency in a predictable way in an insect population provides a strong test of our understanding of the mechanisms of evolution.

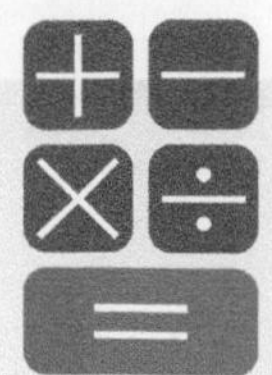

COMPUTING CONSEQUENCES 6.11

Predicting the frequency of *Medea* across generations

Here we provide an overview of how we can predict changes in the frequency of Chen et al.'s *Medea* gene. Because an individual's fitness depends on both its own genotype and the genotypes of its parents, we have to keep track of genotype frequencies and matings.

Let P, Q, and R be the frequencies of genotypes ++, +M, and MM. Random mating results in the nine types of matings shown in **Figure 6.34**. Each type of mating occurs at the frequency shown in the upper left of its square. For example, matings between ++ females and ++ males, shown in square (1), occur at frequency P^2.

We want to know the frequency of each genotype in the next generation. Consider genotype ++. Zygotes with this genotype are conceived in matings (1), (2), (4), and (5). All the zygotes conceived in mating type (1) are ++, as are half of those conceived in mating types (2) and (4) and a quarter of those conceived in mating type (5). The frequency of genotype ++ at conception is thus

$$P^2 + \tfrac{1}{2}PQ + \tfrac{1}{2}QP + \tfrac{1}{4}Q^2$$

Before the zygotes develop into larvae and hatch, however, all the ++ offspring conceived in mating types (4) and (5) die. The frequency of ++ individuals in the next generation is the number of surviving ++ individuals divided by the total number of survivors,

Mother \ Father	P ++	Q +M	R MM
P ++	(1) P^2; father + +; mother +: ++ ++; mother +: ++ ++	(2) PQ; father + M; mother +: ++ +M; mother +: ++ +M	(3) PR; father M M; mother +: +M +M; mother +: +M +M
Q +M	(4) QP; father + +; mother +: ~~++~~ ~~++~~; mother M: M+ M+	(5) Q^2; father + M; mother +: ~~++~~ +M; mother M: M+ MM	(6) QR; father M M; mother +: +M +M; mother M: MM MM
R MM	(7) RP; father + +; mother M: M+ M+; mother M: M+ M+	(8) RQ; father + M; mother M: M+ MM; mother M: M+ MM	(9) R^2; father M M; mother M: MM MM; mother M: MM MM

Figure 6.34 A Punnett square of Punnett squares

which is given by

$$\frac{P^2 + \tfrac{1}{2}PQ}{P^2 + PQ + PR + \tfrac{1}{2}QP + \tfrac{3}{4}Q^2 + QR + RP + RQ + R^2}$$

We leave it to the reader to write expressions for the frequency of +M and MM individuals in the next generation.

Once we have a generation's genotype frequencies, we can readily calculate the allele frequencies.

than simply track allele frequencies and account for the progeny from random unions of gametes (**Computing Consequences 6.11**).

Chen and colleagues' prediction, and the data for 7 populations, some maintained for 15 generations and some for 20, appear in **Figure 6.35**. Their predic-

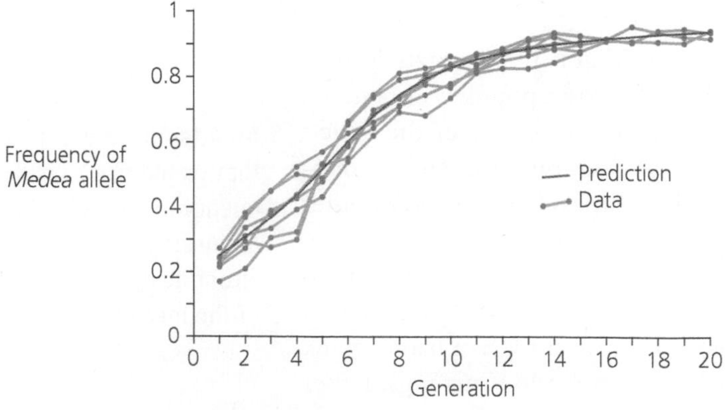

Figure 6.35 Predicted and actual evolution of laboratory populations harboring an engineered gene The prediction is in gray; the data are in orange. From Chen et al. (2007).

tion, as we have come to expect, was spot on. Take a moment to reflect on what Chen and colleagues accomplished. They designed and built a new gene, accurately predicted the effect it would have on the individuals that carry it, introduced the gene into populations, and accurately predicted how the populations would change over the course of 20 generations. The success of this evolutionary engineering project demonstrates that, like NASA's engineers know something about physics, we know a thing or two about inheritance and descent with modification.

Population genetics is a theory that works.

In the next chapter, we will continue to use the Hardy–Weinberg equilibrium principle to explore additional mechanisms of evolution.

SUMMARY

Population genetics represents a synthesis of Mendelian genetics and Darwinian evolution and is concerned with the mechanisms that cause allele frequencies to change from one generation to the next. The Hardy–Weinberg equilibrium principle is a null model that provides the conceptual framework for population genetics. It shows that under simple assumptions—no selection, no mutation, no migration, no genetic drift, and random mating—allele frequencies do not change. Furthermore, genotype frequencies can be calculated from allele frequencies.

When any one of the first four assumptions is violated, allele frequencies may change across generations. Selection, mutation, migration, and genetic drift are thus the four mechanisms of evolution. Nonrandom mating does not cause allele frequencies to change and is thus not a mechanism of evolution. It can, however, alter genotype frequencies and thereby affect the course of evolution.

Population geneticists can measure allele and genotype frequencies in real populations. Thus, biologists can test whether allele frequencies are stable across generations and whether the genotype frequencies conform to Hardy–Weinberg expectations. If either of the conclusions of the Hardy–Weinberg analysis is violated, then one or more of the assumptions does not hold. The nature of the deviation from Hardy–Weinberg expectations does not, by itself, identify the faulty assumption. We can, however, often infer which mechanisms of evolution are at work based on other characteristics of the populations under study.

Selection occurs when individuals with some genotypes are more successful at getting copies of their genes into future generations than are individuals with other genotypes. Selection is a powerful mechanism of evolution. It can cause allele frequencies to change from one generation to the next and can take genotype frequencies away from Hardy–Weinberg equilibrium. Some patterns of selection tend to drive alleles to fixation or to loss; other patterns of selection serve to maintain allelic diversity in populations. Population genetics theory allows us to make accurate predictions about both the direction and the rate of evolution under a variety of patterns of selection.

Alone, mutation is a weak evolutionary mechanism. Mutation does, however, provide the genetic variation that is the raw material for evolution. In some cases, a steady supply of new mutant alleles can counterbalance selection against those same alleles and thereby serve to hold allele frequencies at equilibrium.

QUESTIONS

1. List the five conditions that must be true for a population to be in Hardy–Weinberg equilibrium. Why is it useful to know the conditions that prevent evolution? For each condition, specify whether violation of that assumption results in changes in genotype frequencies, allele frequencies, or both.

2. Why was it important that G. H. Hardy used variables in his mathematical treatment of changes in population allele frequencies across generations? Would it have been equally useful to simply work several more examples with different specific allele frequencies?

3. Name the phenomenon being described in each of these (hypothetical) examples, and describe how it is likely to affect allele frequencies in succeeding generations.
   a. A beetle species is introduced to an island covered with dark basaltic rock. On this dark background, dark beetles, *TT* or *Tt*, are much more resistant to predation than are light-colored beetles, *tt*. The dark beetles have a large selective advantage. Both alleles are relatively common in the group of beetles released on the new island.
   b. Another beetle population, this time consisting of mostly light beetles and just a few dark beetles, is introduced onto a different island with a mixed substrate of light sand, vegetation, and black basalt. On this island, dark beetles have only a small selective advantage.
   c. A coral-reef fish has two genetically determined types of male. One kind of male is much smaller than the other, and sneaks into larger males' nests to fertilize their females' eggs. When small males are rare, they have a selective advantage over large males. However, if there are too many small males, large males switch to a more aggressive strategy of nest defense, and small males lose their advantage.
   d. In a tropical plant, *CC* and *Cc* plants have red flowers and cc plants have yellow flowers. However, *Cc* plants have defective flower development and produce very few flowers.
   e. In a species of bird, individuals with genotype *MM* are susceptible to avian malaria, *Mm* birds are resistant to avian malaria, and *mm* birds are resistant to avian malaria, but the *mm* birds are also vulnerable to avian pox.
4. In Muehlenbachs et al.'s study of placental malaria, why was it important that they studied infants born during both high and low malaria season? Can you think of any other possible explanations for their data?
5. Black color in horses is governed primarily by a recessive allele at the A locus. *AA* and *Aa* horses are nonblack colors such as bay, while *aa* horses are black all over. (Other loci can override the effect of the A locus, but we will ignore that complication.) In an online conversation, one person asked why there are relatively few black horses of the Arabian breed. One response was, "Black is a rare color because it is recessive. More Arabians are bay or gray because those colors are dominant." Discuss the merits and/or problems with this argument. (Assume that the *A* and *a* alleles are in Hardy–Weinberg equilibrium, which was probably true at the time of this discussion.) Generally, what does the Hardy–Weinberg model show us about the impact that an allele's dominance or recessiveness has on its frequency?
6. In humans, the COL1A1 locus codes for a certain collagen protein found in bone. The normal allele at this locus is denoted with *S*. A recessive allele *s* is associated with reduced bone mineral density and increased risk of fractures in both *Ss* and *ss* women. A study of 1,778 women showed that 1,194 were *SS*, 526 were *Ss*, and 58 were *ss* (Uitterlinden et al. 1998). Are these two alleles in Hardy–Weinberg equilibrium in this population? How do you know? What information would you need to determine whether the alleles will be in Hardy–Weinberg equilibrium in the next generation?
7. We used Figure 6.14 as an example of how the frequency of an allele (in fruit flies) does not change in unselected (control) populations but does change in response to selection. However, look again at the unselected control lines in Figure 6.14. The frequency of the allele in the two control populations did change a little, moving up and down over time. Which assumption of the Hardy–Weinberg model is most probably being violated? If this experiment were repeated, what change in experimental design would reduce this deviation from Hardy–Weinberg equilibrium?
8. Most animal populations have a 50:50 ratio of males to females. This does not have to be so; it is theoretically possible for parents to produce predominantly male offspring or predominantly female offspring. Imagine a population with a male-biased sex ratio, say, 70% males and 30% females. Which sex will have an easier time finding a mate? As a result, which sex will probably have higher average fitness? Which parents will have higher fitness—those that produce mostly males or those that produce mostly females? Now imagine the same population with a female-biased sex ratio, and answer the same questions. What sort of selection is probably maintaining the 50:50 sex ratio seen in most populations?
9. Discuss how each of the following recent developments—resulting from improvements in medicine,

technology, pubic health, and from evolution—may affect the frequency of alleles that cause cystic fibrosis (CF).

a. Many women with CF now survive long enough to have children. (CF causes problems with reproductive ducts, but many CF women can bear children nonetheless. CF men are usually sterile.)
b. Typhoid fever in developed nations has declined to very low levels since 1900.
c. In some populations, couples planning to have children are now routinely screened for the most common CF alleles.
d. Drug-resistant typhoid fever has recently appeared in several developing nations.

10. Kerstin Johannesson and colleagues (1995) studied two populations of a marine snail living in the intertidal zone on the shore of Ursholmen Island. Each year, the researchers determined the allele frequencies for the enzyme aspartate aminotransferase (don't worry about what this enzyme does). Their data are shown in the graphs in **Figure 6.36**. The first year of the study was 1987. In 1988, a bloom of toxic algae (tan bars) killed all of the snails in the intertidal zone across the entire island. That is why there are no data for 1988 and 1989. Although the snails living in the intertidal zone were exterminated by the bloom, snails of the same species living in the splash zone just above the intertidal survived unscathed. By 1990, the intertidal zone had been recolonized by splash-zone snails. Your challenge in this question is to develop a coherent explanation for the data in the graphs. In each part, be sure to name the evolutionary mechanism involved (selection, mutation, migration, or drift).

a. Why was the frequency of the Aat^{120} allele higher in both populations in 1990 than it was in 1987? Name the evolutionary mechanism, and explain.
b. Why did the allele frequency decline in both populations from 1990 through 1993? Name the evolutionary mechanism, and explain.
c. Why are the curves traced by the 1990–1993 data for the two populations generally similar but not exactly identical? Name the evolutionary mechanism, and explain.
d. Predict what would happen to the allele frequencies if we followed these two populations for another 100 years (assuming there are no more toxic algal blooms). Explain your reasoning.

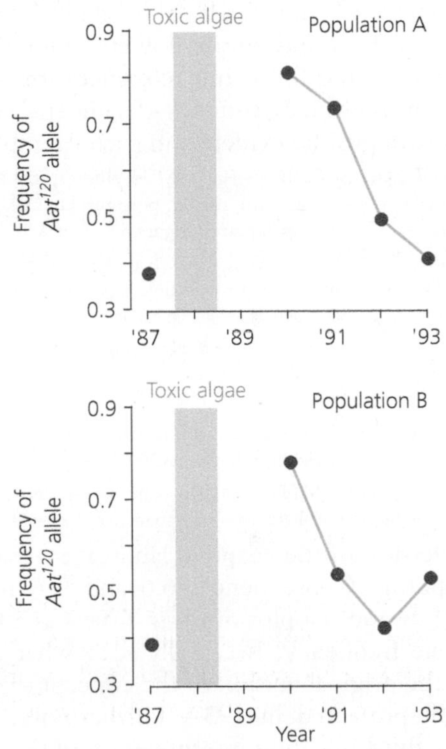

Figure 6.36 Changes over time in the frequency of an allele in two intertidal populations of a marine snail From Johannesson et al. (1995).

EXPLORING THE LITERATURE

11. The photo and graph on the first page of this chapter document the evolution of coat color in a population of Soay sheep. The cause of this evolutionary change been the subject of some controversy. See the following papers:

Ozgul, A., S. Tuljapurkar, et al. 2009. The dynamics of phenotypic change and the shrinking sheep of St. Kilda. *Science* 325: 464–467.

Maloney, S. K., A. Fuller, and D. Mitchell. 2009. Climate change: Is the dark Soay sheep endangered? *Biology Letters* 5: 826–829.

Gratten, J., A. J. Wilson, et al. 2010. No evidence for warming climate theory of coat colour change in Soay sheep: A comment on Maloney et al. *Biology Letters* 6: 678–679.

Maloney, S. K., A. Fuller, and D. Mitchell. 2010. A warming climate remains a plausible hypothesis for the decrease in dark Soay sheep. *Biology Letters* 6: 680–681.

For an analysis of the evolution of coat pattern in the same sheep population, see:

Gratten, J., J. G. Pilkington, et al. 2012. Selection and microevolution of coat pattern are cryptic in a wild population of sheep. *Molecular Ecology* 21: 2977–2990

12. Often, the first step in a study of genetic variation is to evaluate deviations from Hardy–Weinberg equilibrium. Read the following paper to explore how careful examination of Hardy–Weinberg equilibrium is necessary for assessing gene–disease associations in humans.

Trikalinos, T. A., G. Salanti, et al. 2006. Impact of violations and deviations in Hardy–Weinberg equilibrium on postulated gene-disease associations. *American Journal of Epidemiology* 163: 300–309.

13. In the Elderflower orchid, we saw that frequency-dependent selection tends to maintain the presence of both yellow and purple flowers in mixed populations. See the following references for additonal cases of possible frequency-dependent selection. How strong is the evidence in each example?

Cox, C. L., and A. R. Davis Rabosky. In press. Spatial and temporal drivers of phonotypic diversity in polymorphic snakes. *American Naturalist.* (These snakes appear in Figure 5.1, page 148.)

Eizaguirre, C., T. L. Lenz, et al. 2012. Rapid and adaptive evolution of MHC genes under parasite selection in experimental vertebrate populations. *Nature Communications* 3: 621.

Faurie, C., and M. Raymond. 2005. Handedness, homicide and negative frequency-dependent selection. *Proceedings of the Royal Society of London B* 272: 25–28.

Hori, M. 1993. Frequency-dependent natural selection in the handedness of scale-eating cichlid fish. *Science* 260: 216–219.

Sinervo, B., and C. M. Lively. 1996. The rock-paper-scissors game and the evolution of alternative male strategies. *Nature* 380: 240–243.

14. The version of the adaptive landscape presented in Computing Consequences 6.6 and 6.7, in which the landscape is a plot of mean fitness as a function of allele frequency, is actually somewhat different from the original version of the concept that Sewall Wright presented in 1932. Furthermore, there is even a third common interpretation of the adaptive landscape idea. For a discussion of the differences among the three versions, see Chapter 9 in

Provine, W. B. 1986. *Sewall Wright and Evolutionary Biology*. Chicago: University of Chicago Press.

For Sewall Wright's response to Provine's history, see:

Wright, S. 1988. Surfaces of selective value revisited. *American Naturalist* 131: 115–123.

Wright's original 1932 paper is reprinted in Chapter 11 of:

Wright, S. 1986. *Evolution: Selected Papers,* ed. W. B. Provine. Chicago: University of Chicago Press.

15. If you have access to the earliest volumes of the *Journal of Heredity*, read:

Bell, Alexander Graham. 1914. How to improve the race. *Journal of Heredity* 5: 1–7.

Keep in mind that population genetics was in its infancy; Mendelism had yet to be integrated with natural selection. What was accurate and inaccurate in Bell's understanding of the mechanisms of evolution? Would the policy Bell advocated actually have accomplished his aims? Why or why not? If so, would it have done so for the reasons Bell thought it would?

16. For an example in which strong natural selection caused rapid change in allele frequencies in wild populations, see:

Rank, N. E., and E. P. Dahlhoff. 2002. Allele frequency shifts in response to climate change and physiological consequences of allozyme variation in a montane insect. *Evolution* 56: 2278–2289.

17. For an example in which strong selection for insecticid resistance caused rapid change in allele frquencies, see:

Mathias, D. K., E. Ochomo, et al. 2011. Spatial and temporal variation in the kdr allele L1014S in Anopheles gambiae s.s. and phenotypic variability in susceptibility to insecticides in Western Kenya. *Malaria Journal* 10: 10.

18. For another example of a human population taken out of Hardy–Weinberg equlibrium, apparently by strong selection, see:

Mead, S., M. P. H. Stumpf, et al. 2003. Balancing selection at the prion protein gene consistent with prehistoric kurulike epidemics. *Science* 300: 640–643.

Hedrick, P. W. 2003. A heterozygote advantage. *Science* 302: 57.

Mead, S., J. Whitfield, et al. 2008. Genetic susceptibility, evolution and the kuru epidemic. *Philosophical Transactions of the Royal Society B* 363: 3741–3746.

19. Patients with cystic fibrosis (CF) are chronically infected with *Pseudomonas aeruginosa* bacteria. Their immune systems are engaged in a constant battle with the bacteria. In addition, they take powerful antibiotics to help keep the bacterial populations under control. Consider the consequences for the bacteria. How would you expect a *P. aeruginosa* population to evolve in the environment found inside a CF patient's lungs? What novel traits would you expect to appear? Make some predictions, then see the following paper (we are withholding the full title to avoid giving too much away):

Oliver, A., R. Cantón, et al. 2000. High frequency of . . . in cystic fibrosis lung infection. *Science* 288: 1251–1253.

20. As discussed in this chapter, the chemokine receptor CCR5 is the major means by which HIV gains entry to human white blood cells. CCR5 is also important in susceptibility to other important diseases. One example is described in the following article. Consider how CCR5's multiple role in different emerging diseases may affect its evolution, and the implications for medical treatments.

Glass, W. G., D. H. McDermott, et al. 2006. CCR5 deficiency increases risk of symptomatic West Nile virus infection. *Journal of Experimental Medicine* 203: 35–40.

CITATIONS

Much of the population genetics material in this chapter is modeled after presentations in the following:

Crow, J. F. 1983. *Genetics Notes.* Minneapolis, MN: Burgess Publishing.

Felsenstein, J. 1997. *Theoretical Evolutionary Genetics.* Seattle, WA: ASUW Publishing, University of Washington.

Felsenstein, J. 2011. *Theoretical Evolutionary Genetics.* Seattle, WA: http://evolution.genetics.washington.edu/pgbook/pgbook.html.

Griffiths, A. J. F., J. H. Miller, et al. 1993. *An Introduction to Genetic Analysis.* New York: W. H. Freeman.

Templeton, A. R. 1982. Adaptation and the integration of evolutionary forces. In *Perspectives on Evolution,* ed. R. Milkman. Sunderland, MA: Sinauer, 15–31.

Here is the list of all other citations in this chapter:

Bajmoczi, M., M. Gadjeva, et al. 2009. Cystic fibrosis transmembrane conductance regulator and caveolin-1 regulate epithelial cell internalization of *Pseudomonas aeruginosa. American Journal of Physiology—Cell Physiology* 297: C263–C277.

Campodónico, V. L., M. Gadjeva, et al. 2008. Airway epithelial control of *Pseudomonas aeruginosa* infection in cystic fibrosis. *Trends in Molecular Medicine* 14: 120–133.

Castle, W. E. 1903. The laws of heredity of Galton and Mendel, and some laws governing race improvement by selection. *Proceedings of the American Academy of Arts and Sciences* 39: 223–242.

Cavener, D. R., and M. T. Clegg. 1981. Multigenic response to ethanol in *Drosophila melanogaster. Evolution* 35: 1–10.

Chen, C.-H., H. Huang, et al. 2007. A synthetic maternal-effect selfish genetic element drives population replacement in *Drosophila. Science* 316: 597–600.

Cromer, D., S. M. Wolinsky, and A. R. McLean. 2010. How fast could HIV change gene frequencies in the human population? *Proceedings of the Royal Society B* 277: 1981–1989.

Darwin, C. R. 1868. *The Variation of Animals and Plants Under Domestication.* London: John Murray.

Dawson, P. S. 1970. Linkage and the elimination of deleterious mutant genes from experimental populations. *Genetica* 41: 147–169.

Dean, M., M. Carrington, et al. 1996. Genetic restriction of HIV-1 infection and progression to AIDS by a deletion allele of the CKR5 structural gene. *Science* 273: 1856–1862.

East, E. M. 1917. Hidden feeblemindedness. *Journal of Heredity* 8: 215–217.

Elias, S., M. M. Kaback, et al. 1992. Statement of the American Society of Human Genetics on cystic fibrosis carrier screening. *American Journal of Human Genetics* 51: 1443–1444.

Fisher, R. A. 1924. The elimination of mental defect. *Eugenics Review* 16: 114–116.

Foster, G. G., M. J. Whitten, et al. 1972. Chromosome rearrangements for the control of insect pests. *Science* 176: 875–880.

Gigord, L. D. B., M. R. Macnair, and A. Smithson. 2001. Negative frequency-dependent selection maintains a dramatic flower color polymorphism in the rewardless orchid *Dactylorhiza sambucina* (L.) Soò. *Proceedings of the National Academy of Sciences, USA* 98: 6253–6255.

Goddard, H. H. 1914. *Feeblemindedness: Its Causes and Consequences.* New York: Macmillan.

Gratten, J., A. J. Wilson, et al. 2008. A localized negative genetic correlation constrains microevolution of coat color in wild sheep. *Science* 319: 318–320.

Hardy, G. H. 1908. Mendelian proportions in a mixed population. *Science* 28: 49–50.

Johannesson, K., B. Johannesson, and U. Lundgren. 1995. Strong natural selection causes microscale allozyme variation in a marine snail. *Proceedings of the National Academy of Sciences, USA* 92: 2602–2606.

Kambris, Z., H. Bilak, et al. 2003. *DmMyD88* controls dorsoventral patterning of the *Drosophila* embryo. *EMBO Reports* 4: 64–69.

Karumanchi, S. A., and D. Haig. 2008. Flt1, pregnancy, and malaria: Evolution of a complex interaction. *Proceedings of the National Academy of Sciences, USA* 105: 14243–14244.

Kevles, D. J. 1995. *In the Name of Eugenics: Genetics and the Uses of Human Heredity.* Cambridge, MA: Harvard University Press.

Lane, H. 1992. *The Mask of Benevolence: Disabling the Deaf Community.* New York: Vintage Books.

López-Bigas, N., B. J. Blencowe, and C. A. Ouzounis. 2006. Highly consistent patterns for inherited human diseases at the molecular level. *Bioinformatics* 22: 269–277.

Lyczak, J. B., C. L. Cannon, and G. B. Pier. 2002. Lung infections associated with cystic fibrosis. *Clinical Microbiology Reviews* 15: 194–222.

Lyczak, J. B., and G. B. Pier. 2002. *Salmonella enterica* serovar *typhi* modulates cell surface expression of its receptor, the cystic fibrosis transmembrane conductance regulator, on the intestinal epithelium. *Infection and Immunity* 70: 6416–6423.

Martinson, J. J., N. H. Chapman, et al. 1997. Global distribution of the CCR5 gene 32-base-pair deletion. *Nature Genetics* 16: 100–103.

McKusick, Victor A., et al. 1999. Spinal muscular atrophy I. Record 253300 in *Online Mendelian Inheritance in Man.* Center for Medical Genetics, Johns Hopkins University (Baltimore, MD) and National Center for Biotechnology Information, National Library of Medicine (Bethesda, MD). Available at http://www.ncbi.nlm.nih.gov/omim/.

Muehlenbachs, A., M. Fried, et al. 2008. Natural selection of *FLT1* alleles and their association with malaria resistance *in utero. Proceedings of the National Academy of Sciences, USA* 105: 14488–14491.

Mukai, T., and A. B. Burdick. 1959. Single gene heterosis associated with a second chromosome recessive lethal in *Drosophila melanogaster. Genetics* 44: 211–232.

Paul, D. B., and H. G. Spencer. 1995. The hidden science of eugenics. *Nature* 374: 302–304.

Pier, G. B., M. Grout, and T. S. Zaidi. 1997. Cystic fibrosis transmembrane conductance regulator is an epithelial cell receptor for clearance of *Pseudomonas aeruginosa* from the lung. *Proceedings of the National Academy of Sciences, USA* 94: 12088–12093.

Pier, G. B., M. Grout, et al. 1998. *Salmonella typhi* uses CFTR to enter intestinal epithelial cells. *Nature* 393: 79–82.

Provine, W. B. 1971. *The Origins of Theoretical Population Genetics.* Chicago: University of Chicago Press.

Punnett, R. C. 1917. Eliminating feeblemindedness. *Journal of Heredity* 8: 464–465.

Ramaley, P. A., N. French, et al. 2002. Chemokine-receptor genes and AIDS risk. *Nature* 471: 140.

Reilly, P. 1991. *The Surgical Solution: A History of Involuntary Sterilization in the United States.* Baltimore, MD: Johns Hopkins University Press.

Schliekelman, P., C. Garner, and M. Slatkin. 2001. Natural selection and resistance to HIV. *Nature* 411: 545–546.

Shore, E. M., M. Xu, et al. 2006. A recurrent mutation in the BMP type I receptor ACVR1 causes inherited and sporadic fibrodysplasia ossificans progressiva. *Nature Genetics* 38: 525–527.

Smithson, A., and M. R. Macnair. 1997. Negative frequency-dependent selection by pollinators on artificial flowers without rewards. *Evolution* 51: 715–723.

Stephens, J. C., D. E. Reich, et al. 1998. Dating the origin of the AIDS-resistance allele by the coalescence of haplotypes. *American Journal of Human Genetics* 62: 1507–1515.

Sullivan, A. D., J. Wigginton, and D. Kirschner. 2001. The coreceptor mutation influences the dynamics of HIV epidemics and is selected for by HIV. *Proceedings of the National Academy of Sciences, USA* 98: 10214–10219.

Uitterlinden A. G., H. Burger, et al. 1998. Relation of alleles of the collagen type IA1 gene to bone density and the risk of osteoporotic fractures in postmenopausal women. *New England Journal of Medicine* 338: 1016–1021.

Umbers, A. J., E. H. Aitken, and S. J. Rogerson. 2011. Malaria in pregnancy: Small babies, big problem. *Trends in Parasitology* 27: 168–175.

UNAIDS. 1998. AIDS epidemic update: December 1998. (Geneva, Switzerland). Available at http://www.unaids.org.

United States Supreme Court. 1927. *Buck v. Bell*, 274 U.S. 200.

Weinberg, W. 1908. Ueber den nachweis der vererbung beim menschen. *Jahreshefte des Vereins für Vaterländische Naturkunde in Württemburg* 64: 368–382. English translation in Boyer, S. H. 1963. *Papers on Human Genetics*. Englewood Cliffs, NJ: Prentice Hall.

Wirth, B., T. Schmidt, et al. 1997. De novo rearrangements found in 2% of index patients with spinal muscular atrophy: Mutational mechanisms, parental origin, mutation rate, and implications for genetic counseling. *American Journal of Human Genetics* 61: 1102–1111.

Yule, G. U. 1902. Mendel's laws and their probable relations to intraracial heredity. *New Phytologist* 1: 193–207; 222–238.

Zar, J. H. 1996. *Biostatistical Analysis*. 3rd ed. Upper Saddle River, NJ: Prentice Hall.

Zhang, M. C., P. Azad, and R. C. Woodruff. 2011. Adaptation of *Drosophila melanogaster* to increased NaCl concentration due to dominant beneficial mutations. *Genetica* 139: 177–186.

7

Mendelian Genetics in Populations II: Migration, Drift, & Nonrandom Mating

Florida's state animal was on the verge of extinction. The Florida panther (*Puma concolor coryi*) once ranged across southeastern North America from Louisiana to South Carolina and as far north as Tennessee. By 1995, however, the big cat had seen its confines shrink, due to habitat loss and hunting, to a pair of tiny and dwindling patches of swampland in Florida's southern tip (Johnson et al. 2010). Scarcely two dozen individuals remained (McBride et al. 2008).

To make matters worse, the surviving panthers were suffering poor health (Roelke et al. 1993). They showed high rates of heart defects, undescended testes, skeletal malformations, infectious and parasitic diseases, and, as documented at right, sperm abnormalities associated with infertility. Exposure to environmental toxins may have contributed to some of these problems, but the primary causes appeared to be genetic.

Wildlife managers decided that intervention offered the best hope of saving the Florida panther (see Johnson et al. 2010). Their diagnosis of the panther's underlying problem, and the treatment they prescribed, involves three population genetics phenomena introduced earlier (in Chapter 6) but not discussed in depth: migration, genetic drift, and nonrandom mating.

Among other health problems, the Florida panther had high rates of defective sperm compared to other *Puma* populations and other cats. Bars show mean ± se. Photo by Rodney Cammauf, NPS; graph from Roelke et al. (1993).

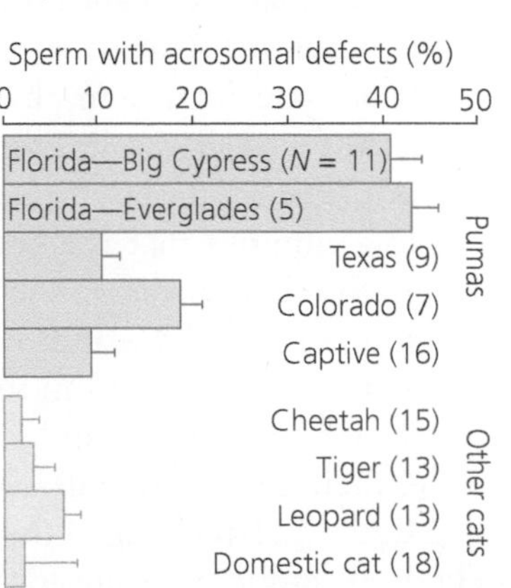

We identified migration, drift, and nonrandom mating as factors in the evolution of populations when we developed the Hardy–Weinberg equilibrium principle. When a population has no selection, no mutation, no migration, an infinite number of individuals, and random choice of mates by all, then (1) the allele frequencies do not change from one generation to the next, and (2) the genotype frequencies can be calculated by multiplying the allele frequencies. We looked (in Chapter 6) at what happens when we relax the assumptions of no selection and no mutation. In this chapter, we explore what happens when we relax the assumptions of no migration, infinite population size, and random mating.

We look at migration in Section 7.1. We devote Sections 7.2 and 7.3 to genetic drift and its role in molecular evolution. We cover nonrandom mating in Section 7.4. With these sections as background, we return in Section 7.5 to the Florida panther and consider the intervention that may prevent its demise.

7.1 Migration

Migration, in an evolutionary sense, is the movement of alleles between populations. This use of the term *migration* is distinct from its more familiar meaning, which refers to the seasonal movement of individuals. To evolutionary biologists, migration means gene flow: the transfer of alleles from the gene pool of one population to the gene pool of another population. Migration can be caused by anything that moves alleles far enough to go from one population to another. Mechanisms of gene flow range from the long-distance dispersal of juvenile animals to the transport of pollen, seeds, or spores by wind, water, or animals. The amount of migration between populations varies enormously among species, depending on the mobility of individuals or propagules across the life cycle.

Adding Migration to the Hardy–Weinberg Analysis: Gene Flow as a Mechanism of Evolution

To investigate the role of gene flow in evolution, consider a simple model of migration. Imagine two populations: one on a continent, the other on an island (**Figure 7.1**). The island population is tiny relative to the continental population, so any migration from the island to the continent will be inconsequential for the continent's allele and genotype frequencies. Migration, and the accompanying gene flow, thus effectively go one way, from the continent to the island. Consider a single locus with two alleles, A_1 and A_2. Can migration take the allele and genotype frequencies on the island away from Hardy–Weinberg equilibrium?

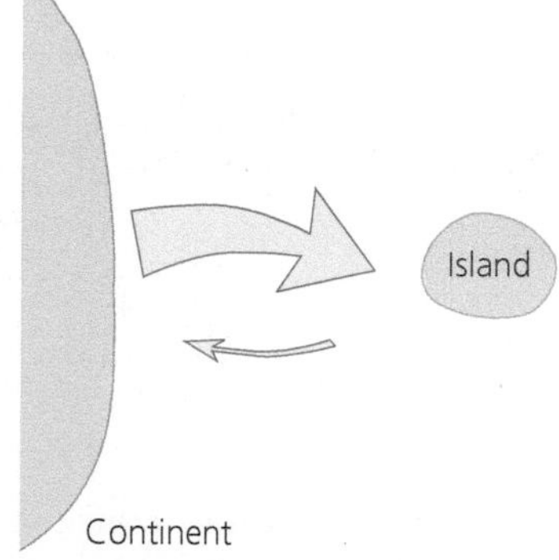

Figure 7.1 The one-island model of migration Arrows show relative gene flow between island and continental populations. Alleles arriving on the island from the continent represent a relatively large fraction of the island gene pool, whereas alleles arriving on the continent from the island represent a small fraction of the continental gene pool.

To see that they can, imagine that before migration, the frequency of A_1 on the island is 1.0 (that is, A_1 is fixed—see **Figure 7.2**). When gametes in the gene pool combine at random, the genotype frequencies among the zygotes are 1.0 for A_1A_1, 0 for A_1A_2, and 0 for A_2A_2. Imagine that there are 800 zygotes, which we will let develop into juveniles and grow to adulthood.

Now suppose that the continental population is fixed for allele A_2 and that before the individuals on the island reach maturity, 200 individuals migrate from the continent to the island. After migration, 80% of the island population is from the island, and 20% is from the continent. The new genotype frequencies are 0.8 for A_1A_1, 0 for A_1A_2, and 0.2 for A_2A_2. When individuals on the island reproduce, their gene pool will have allele frequencies of 0.8 for A_1 and 0.2 for A_2.

Migration has changed the allele frequencies in the island population, violating Hardy–Weinberg conclusion 1. Before migration, the island frequency of A_1 was

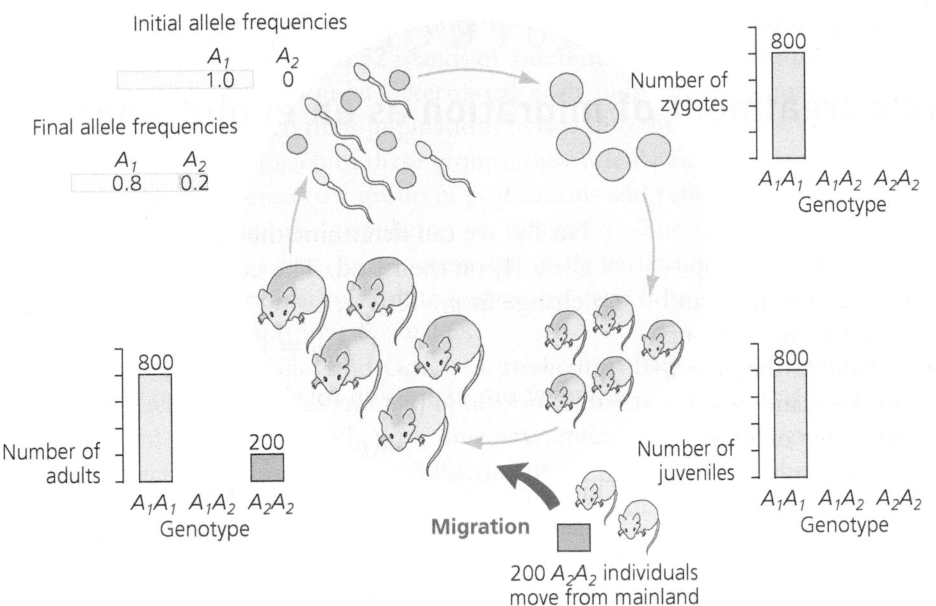

Figure 7.2 Migration can alter allele and genotype frequencies This diagram follows an imaginary island population of mice from one generation's gene pool (initial allele frequencies) to the next generation's gene pool (final allele frequencies). The bar graphs show the number of individuals of each genotype in the population at any given time. Migration, in the form of individuals arriving from a continental population fixed for allele A_2, increases the frequency of allele A_2 in the island population.

1.0; after migration, the frequency of A_1 is 0.8. The island population has evolved as a result of migration. For an algebraic treatment of migration as a mechanism of allele frequency change, see **Computing Consequences 7.1 (next page)**.

Migration is a potent mechanism of evolution. In practice, migration is most important in preventing populations from diverging.

Migration has also produced genotype frequencies among the adults on the island that violate Hardy–Weinberg conclusion 2. Under the Hardy–Weinberg equilibrium principle, a population with allele frequencies of 0.8 and 0.2 should have genotype frequencies of 0.64, 0.32, and 0.04. Compared to these expected values, the post-migration island population has an excess of homozygotes and a deficit of heterozygotes. A single bout of random mating will, of course, put the population back into Hardy–Weinberg equilibrium for genotype frequencies.

Migration as a Homogenizing Evolutionary Process

Migration in our model made the island population more similar to the mainland population than it otherwise would have been. This is the general effect of migration: It tends to homogenize allele frequencies across populations.

How far would the homogenization ultimately proceed? The algebraic models developed in Computing Consequences 7.1 show that gene flow among populations will eventually equalize their allele frequencies. In other words, if not opposed by another mechanism of evolution, migration will homogenize allele frequencies across populations completely.

Barbara Giles and Jérôme Goudet (1997) documented the homogenizing effect of gene flow on populations of red bladder campion, *Silene dioica*. Red bladder campion is an insect-pollinated perennial wildflower **(Figure 7.3)**. The populations that Giles and Goudet studied occupy islands in the Skeppsvik Archipelago, Sweden. These islands are mounds of material deposited by glaciers during the last ice age and left underwater when the ice melted. The area on which the islands sit is rising at a rate of 0.9 centimeters per year. As a result of this geological uplift, new islands are constantly rising out of the water. The Skeppsvik Archipelago thus contains dozens of islands of different ages.

Figure 7.3 Red bladder campion *Silene dioica* is a perennial wildflower.

Red bladder campion seeds are transported by wind and water, and the plant is among the first to colonize new islands. Campion populations grow to several thousand individuals. There is gene flow among islands because of seed dispersal

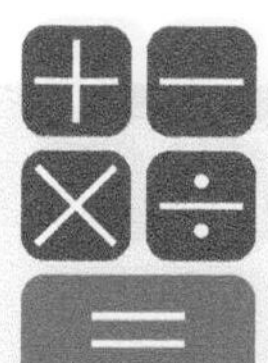

COMPUTING CONSEQUENCES 7.2

Selection and migration in Lake Erie water snakes

As described in the main text, the genetics of color pattern in Lake Erie water snakes can be roughly approximated by a single locus with a dominant allele for the banded pattern and a recessive allele for the unbanded one (King 1993a). Selection by predators on the islands favors unbanded snakes. If the fitness of unbanded individuals is defined as 1, then the relative fitness of banded snakes is between 0.78 and 0.90 (King and Lawson 1995). Why has selection not eliminated banded snakes from the islands? Here we calculate the effect migration has when it introduces new copies of the banded allele into the island population every generation.

King and Lawson (1995) lumped all the island snakes into a single population, because snakes appear to move among islands much more often than from the mainland to the islands. King and Lawson used genetic techniques to estimate that 12.8 snakes move from the mainland to the islands every generation. The scientists estimated that the total island population is between 523 and 4,064 snakes, with a best estimate of 1,262. Migrants thus represent a fraction of 0.003 to 0.024 of the population each generation, with a best estimate of 0.01.

With King and Lawson's estimates of selection and migration, we can calculate the equilibrium allele frequencies in the island population at which the effects of selection and migration exactly balance each other. Let A_1 represent the dominant allele for the banded pattern, and A_2 the recessive allele for the unbanded pattern. Let p represent the frequency of A_1, and q the frequency of A_2. Following Computing Consequences 6.3, we create individuals by random mating, then let selection act. After selection (but before migration), the new frequency of allele A_2 is

$$q^\star = \frac{pqw_{12} + q^2w_{22}}{\overline{w}}$$

where w_{12} is the fitness of A_1A_2 heterozygotes, w_{22} is the fitness of A_2A_2 homozygotes, and $\overline{w}$ is the mean fitness of all the individuals in the population, given by $(p^2w_{11} + 2pqw_{12} + q^2w_{22})$.

For our first calculation, we will use $w_{11} = w_{12} = 0.84$, and $w_{22} = 1$. A relative fitness of 0.84 for banded snakes is the midpoint of the range within which King and Lawson (1995) estimated the true value to fall. This gives

$$q^\star = \frac{pq(0.84) + q^2}{[p^2(0.84) + 2pq(0.84) + q^2]}$$

Substituting $(1 - q)$ for p gives

$$q^\star = \frac{(1-q)q(0.84) + q^2}{[(1-q)^2(0.84) + 2(1-q)q(0.84) + q^2]}$$

$$= \frac{0.84q + 0.16q^2}{0.84 + 0.16q^2}$$

If selection favors unbanded snakes on the islands, then we would expect that the island populations would consist entirely of unbanded snakes. Why is this not the case? The answer, at least in part, is that in every generation several banded

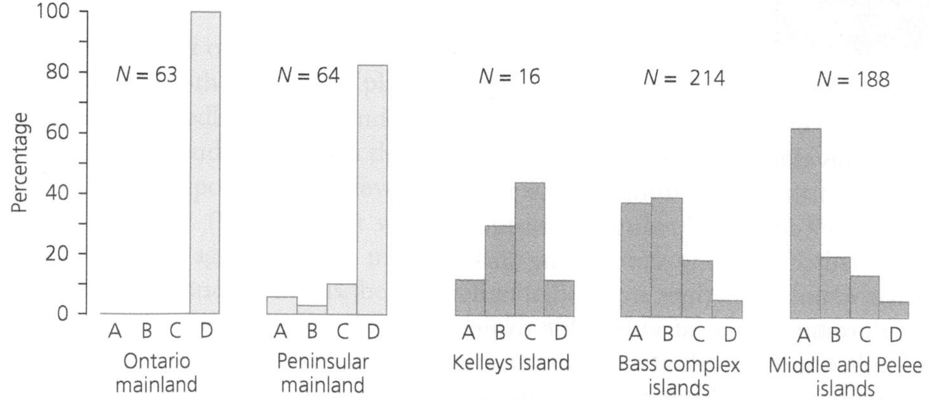

Figure 7.6 Variation in color pattern within and between populations These histograms show frequency of different color patterns in various populations. Category A snakes are unbanded; category B and C snakes are intermediate; category D snakes are strongly banded. Snakes on the mainland tend to be banded; snakes on the islands tend to be unbanded or intermediate. From Camin and Ehrlich (1958).

Now we allow migration, with the new migrants representing, in this first calculation, a fraction 0.01 of the island's population (King and Lawson's best estimate). None of the new migrants carry allele A_2, so the new frequency of A_2 is

$$q' = (0.99)\frac{0.84q + 0.16q^2}{0.84 + 0.16q^2}$$

The change in q from one generation to the next is

$$\Delta q = q' - q = (0.99)\frac{0.84q + 0.16q^2}{0.84 + 0.16q^2} - q$$

Plots of Δq as a function of q appear in **Figure 7.7**. The red curve (b) is for the function we just calculated. It shows that if q is greater than 0.05 and less than 0.93 in this generation, then q will be larger in the next generation (Δq is positive). If q is less than 0.05 or greater than 0.93 in this generation, then q will be smaller in the next generation (Δq is negative). The points where the curve crosses the horizontal axis, where $\Delta q = 0$, are the equilibrium points. The upper equilibrium point is stable: If q is less than 0.93, then q will rise in the next generation; if q is greater than 0.93, then it will fall in the next generation. Thus a middle-of-the-road prediction, given King and Lawson's estimates of selection and gene flow, is that the equilibrium frequency of the unbanded allele in the island population will be 0.93.

Curve (a) is a high-end estimate; it uses fitnesses of 0.78 for A_1A_1, 0.78 for A_1A_2, and 1 for A_2A_2, and a migration rate of 0.003 (0.3% of every generation's population are migrants). It predicts an equilibrium at $q = 0.99$. Curve (c) is a low-end estimate; it uses fitnesses of 0.90 for A_1A_1, 0.90 for A_1A_2, and 1 for A_2A_2, and a migration rate of 0.024 (2.4% of every generation's population are migrants). It predicts an equilibrium at $q = 0.64$.

King and Lawson's estimate of the frequency of A_2 is 0.73. This value is toward the low end of our range of predictions. Our calculation is a relatively simple one. It leaves out many factors, including recent changes in the population sizes of both the water snakes and their predators, and recent changes in the frequencies of banded versus unbanded snakes. For more details, see King and Lawson (1995) and Hendry et al. (2001).

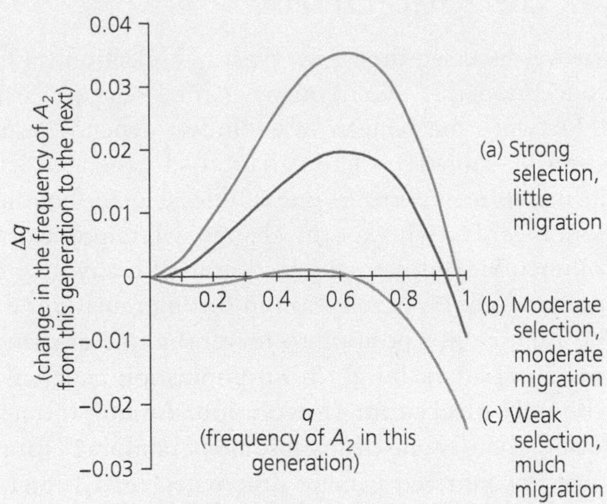

Figure 7.7 The combined effects of selection and migration on allele frequencies in island water snakes The curves show Δq as a function of q for different combinations of migration and selection. See text for details.

snakes move from the mainland to the islands. The migrants bring with them copies of the allele for banded coloration. When the migrant snakes interbreed with the island snakes, they contribute these copies to the island gene pool. In this example, natural selection is acting as an evolutionary mechanism in opposition to migration, preventing the island population from being driven to the same allele frequency seen in the mainland population.

Migration of individuals from the mainland to islands appears to be preventing the divergence—due to selection—of island versus mainland populations of Lake Erie water snakes.

For an algebraic treatment of the opposing influences of migration and selection on the Lake Erie water snakes, see **Computing Consequences 7.2**.

In summary, migration is the movement of alleles from population to population. Within a participating population, migration can cause allele frequencies to change from one generation to the next. For small populations receiving immigrants from large source populations, migration can be a potent mechanism of evolution. Across groups of populations, gene flow tends to homogenize allele frequencies, thus preventing the evolutionary divergence of populations, unless it is balanced by an opposing mechanism of evolution.

7.2 Genetic Drift

When we discussed the logic of natural selection (in Chapter 3), we refuted the misconception that evolution by natural selection is a random process. To be sure, Darwin's mechanism of evolution depends on the generation of random variation by mutation. The variation so generated is random in the sense that when mutation substitutes one amino acid for another in a protein, it does so without regard to whether the change will improve or damage the protein's ability to function. But natural selection itself is anything but random. It is precisely the nonrandomness of selection in sorting mutations that leads to adaptation.

We are now in a position to revisit the role of chance in evolution. Arguably, the most important insight from population genetics is that natural selection is not the only mechanism of evolution. Among the nonselective mechanisms of evolution, there is one that is absolutely random. That mechanism is genetic drift. We first encountered genetic drift when (in Chapter 6) we simulated drawing gametes from a gene pool to make zygotes. We found that blind luck produced different outcomes in different trials. Genetic drift does not lead to adaptation, but it does lead to changes in allele frequencies. In the Hardy–Weinberg model, genetic drift results from violation of the assumption of infinite population size.

A Model of Genetic Drift

To begin our exploration of how genetic drift works, we return to a simulation like the one we used earlier. Imagine an ideal population that is finite—in fact, small—in size. As usual, we are focusing on a single locus with two alleles, A_1 and A_2. Imagine that in the present generation's gene pool, allele A_1 is at frequency 0.6, and allele A_2 is at frequency 0.4 **(Figure 7.8a, upper left)**. We let the gametes in this gene pool combine at random to make exactly 10 zygotes. These 10 zygotes will constitute the entire population for the next generation.

We can simulate the production of 10 zygotes from our gene pool with a physical model. The gene pool appears in Figure 7.8b. It includes 100 gametes. Sixty of these eggs and sperm carry allele A_1; 40 carry A_2. We make each zygote by closing our eyes and putting a finger down to choose a random egg, then closing our eyes and putting a finger down to choose a random sperm. (The chosen gametes remain in the gene pool and can be chosen again. We are imagining that our gene pool is much bigger than what we can see in the illustration, and that removing a few gametes has no effect on the allele frequencies.) We are pausing to choose gametes as we write. The genotypes of the 10 zygotes are

A_2A_1	A_1A_1	A_1A_1	A_1A_1	A_2A_2
A_1A_1	A_2A_2	A_1A_2	A_1A_1	A_1A_1

Counting the genotypes, we have A_1A_1 at a frequency of 0.6, A_1A_2 at a frequency of 0.2, and A_2A_2 at a frequency of 0.2 (Figure 7.8a). Counting the allele copies, we see that when these zygotes develop into juveniles, which then grow up and reproduce, the frequency of allele A_1 in the new gene pool will be 0.7, and the frequency of allele A_2 will be 0.3 (Figure 7.8a).

We have completed one turn of the life cycle of our model population. Nothing much seems to have happened, but note that both conclusions of the Hardy–Weinberg equilibrium principle have been violated. The allele frequencies have changed from one generation to the next, and we cannot calculate the genotype frequencies by multiplying the allele frequencies.

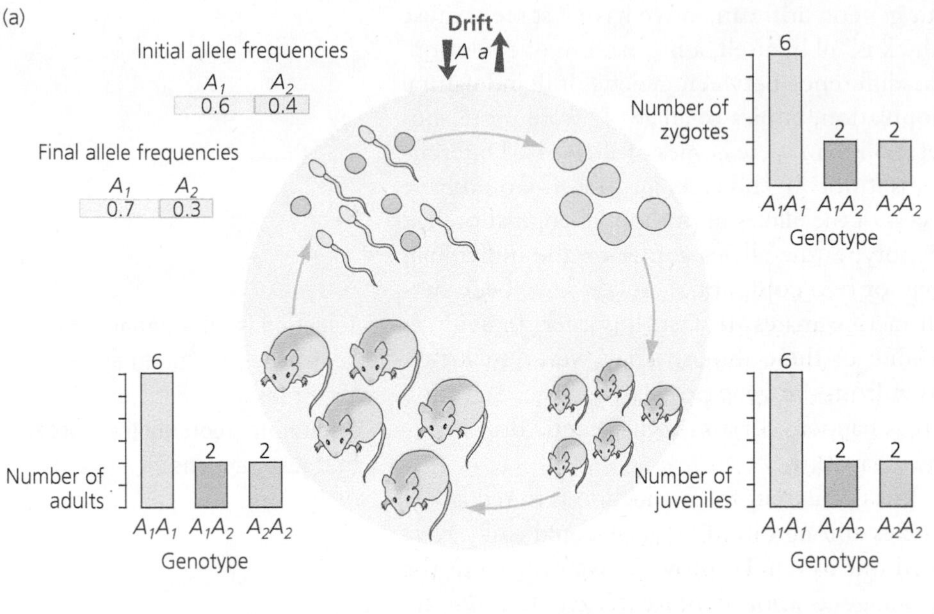

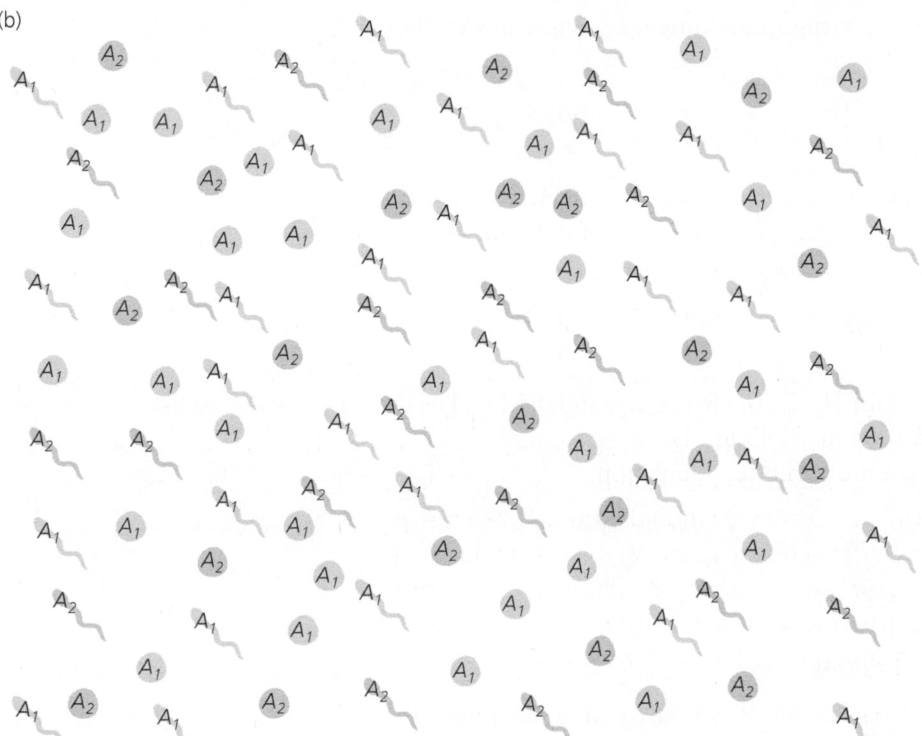

Figure 7.8 Chance events can alter allele and genotype frequencies (a) This diagram follows an imaginary population of 10 mice from one generation's gene pool (initial allele frequencies) to the next generation's gene pool (final allele frequencies). The bar graphs show the number of individuals of each genotype in the population at any given time. Genetic drift, in the form of sampling error in drawing gametes from the initial gene pool (b) to make zygotes, increases the frequency of allele A_1 Note that many other outcomes are also possible.

Our population has failed to conform to the Hardy–Weinberg principle simply because the population is small. In a small population, chance events produce outcomes that differ from theoretical expectations. The chance events in our simulated population were the blind choices of gametes to make zygotes. We picked gametes carrying copies of A_1 and A_2 not in their exact predicted ratio of 0.6 and 0.4, but in a ratio that just happened to be a bit richer in A_1 and a bit poorer in A_2. This kind of random discrepancy between theoretical expectations and actual results is called **sampling error.** Sampling error in the production of zygotes from a gene pool is called **random genetic drift,** or just genetic drift. Because it is nothing more than a cumulative effect of random events, genetic

In populations of finite size, chance events—in the form of sampling error in drawing gametes from the gene pool—can cause evolution.

drift cannot produce adaptation. But genetic drift can, as we have just seen, cause allele frequencies to change. Blind luck is, all by itself, a mechanism of evolution.

Sometimes it is difficult to see the difference between genetic drift and natural selection. In our simulated small population, copies of allele A_1 were more successful at getting into the next generation than were copies of allele A_2. Differential reproductive success is selection, is it not? In this case, it is not. It would have been selection if the differential success of the alleles in our model population had been explicable in terms of the phenotypes the alleles confer on the individuals that carry them. Individuals with one or two copies of A_1 might have been better at surviving, finding food, or attracting mates. In fact, however, individuals carrying copies of allele A_1 were none of these things. They were just lucky. Their gametes happened to get drawn from the gene pool more often. Selection is differential reproductive success that happens for a reason. Genetic drift is differential reproductive success that just happens.

Selection is differential reproductive success that happens for a reason; genetic drift is differential reproductive success that just happens.

Another way to see that genetic drift is different from selection is to recognize that the genotype and allele frequencies among our 10 zygotes could easily have been different from what they turned out to be. To prove it, we can repeat the exercise of drawing gametes from our gene pool to make 10 zygotes. We are again pausing to choose gametes as we write. This time, the genotypes of the zygotes are

A_1A_1	A_1A_1	A_1A_1	A_2A_1	A_1A_2
A_2A_2	A_1A_2	A_1A_1	A_2A_1	A_2A_2

Among this set of zygotes, the genotype frequencies are 0.4 for A_1A_1, 0.4 for A_1A_2, and 0.2 for A_2A_2. The allele frequencies are 0.6 for A_1 and 0.4 for A_2.

Repeating the exercise a third time produces these zygotes:

A_1A_1	A_1A_1	A_1A_1	A_1A_2	A_1A_1
A_1A_2	A_2A_1	A_2A_2	A_2A_2	A_2A_2

Now the genotype frequencies are 0.4 for A_1A_1, 0.3 for A_1A_2, and 0.3 for A_2A_2, and the allele frequencies are 0.55 for A_1 and 0.45 forA_2.

Here is a summary of the results from our model population:

	Frequency of A_1
In the gene pool	0.6
In the first set of 10 zygotes	0.7
In the second set of 10 zygotes	0.6
In the third set of 10 zygotes	0.55

The three sets of zygotes have shown us that if we start with a gene pool in which allele A_1 is at a frequency of 0.6 and make a population of just 10 zygotes, the frequency of A_1 may rise, stay the same, or fall. In fact, the new frequency of A_1 among a set of 10 zygotes drawn from our gene pool could turn out to be anywhere from 0 to 1.0, although outcomes at the extremes of this range are not likely. The graph in **Figure 7.9** shows the theoretical probability of each possible outcome. Overall, there is about an 18% chance that the frequency of allele A_1 will stay at 0.6, about a 40.5% chance that it will drop to a lower value, and about a 41.5% chance that it will rise to a higher value. But do not just take our word for it; use the gene pool in Figure 7.8b to make a few batches of zygotes. The results are likely to be different each time. Again, the point is that genetic drift is evolution that simply happens by chance.

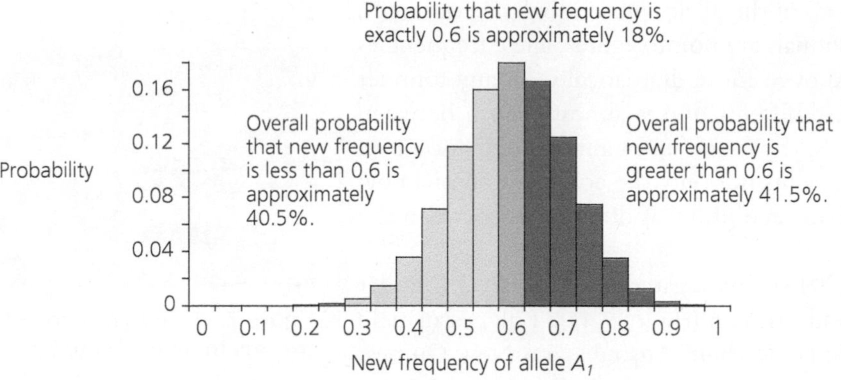

Figure 7.9 The possible outcomes in our model population of 10 mice When we make 10 zygotes by drawing from a gene pool in which alleles A_1 and A_2 have frequencies of 0.6 and 0.4, the single most probable outcome is that the allele frequencies will remain unchanged. However, the chance of this happening is only about 18%.

Genetic Drift and Population Size

Genetic drift is fundamentally the result of finite population size. If we draw gametes from our gene pool to make a population of more than 10 zygotes, the allele frequencies among the zygotes will get closer to the values predicted by the Hardy–Weinberg equilibrium principle. Closing our eyes and pointing at a book quickly becomes tedious, so we used a computer to simulate drawing gametes to make not just 10, but 250 zygotes **(Figure 7.10a)**. As the computer drew each gamete, it gave a running report of the frequency of A_1 among the zygotes it had made so far. At first this running allele frequency fluctuated wildly. For example, the first zygote turned out to have genotype A_2A_2, so the running frequency of allele A_1 started at zero. The next several zygotes were mostly A_1A_1 and A_1A_2, which sent the running frequency of allele A_1 skyrocketing to 0.75. As the cumulative number of zygotes made increased, the frequency of allele A_1 in the new generation bounced around less and less, gradually settling toward the expected value of 0.6. The deviations from expectation that we see along the way to a large number of zygotes are random, as illustrated by the graphs in Figure 7.10b and c. These graphs show two more sets of draws to make 250 zygotes. In each, the allele frequency in the new generation fluctuates wildly at first, but in a unique pattern. As in the first graph, however, the allele frequency in the new generation always eventually settles toward the theoretically predicted value of 0.6.

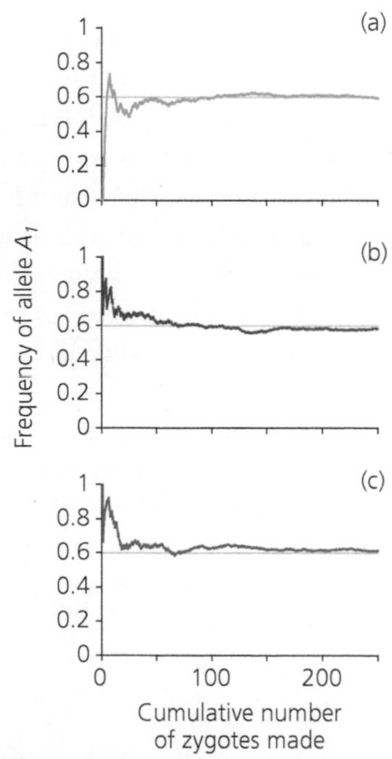

Figure 7.10 A simulation of drawing alleles from a gene pool, run three times At first the new frequency of allele A_1 fluctuates considerably, in a unique trajectory for each run. As the number of zygotes made increases, however, the new frequency of A_1 settles toward the expected value of 0.6.

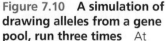

Our simulations demonstrate that sampling error diminishes as sample size increases. If we kept drawing gametes forever, to make an infinitely large population of zygotes, the frequency of allele A_1 among the zygotes would be exactly 0.6. Genetic drift is a powerful evolutionary mechanism in small populations, but its power declines in larger populations. We will return to this point later.

Empirical Research on Sampling Error as a Mechanism of Evolution: The Founder Effect

To observe genetic drift in nature, the best place to look is in small populations. Populations are often small when they have just been founded by a group of individuals that have moved, or been moved, to a new location. The allele frequencies in the new population are likely, simply by chance, to be different from what they were in the source population. This is called the **founder effect.**

Genetic drift is most important in small populations.

The founder effect is a direct result of sampling error. For example, if 35 different alleles are present at a single locus in a continental population of lizards, but just 15 individuals are riding on a mat of vegetation that floats to a remote island (see Censky et al. 1998 for a documented example), the probability is zero that the new island population will contain all of the alleles present on the continent. If, by chance, any of the founding individuals are homozygotes, allele frequencies in the new population will have shifted even more dramatically. In any founder event, some degree of random genetic differentiation is almost certain between old and new populations. In other words, the founding of a new population by a small group of individuals typically represents not only the colonization of a new patch of habitat but also the instantaneous evolution of differences between the new population and the old one.

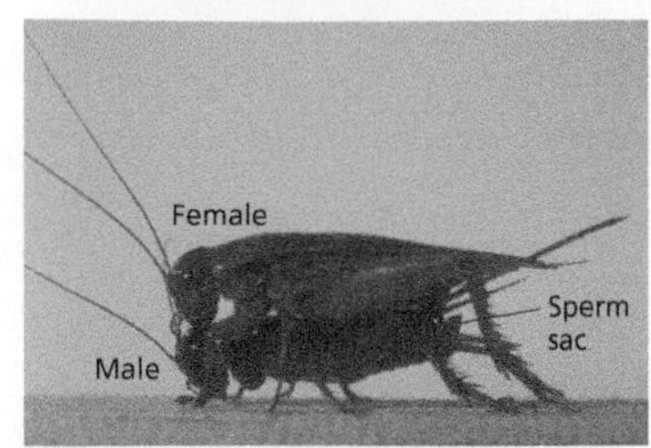

Figure 7.11 Polynesian field crickets in love Photo by Gerald McCormack, Cook Islands Natural Heritage Project.

Robin Tinghitella and colleagues (2011) investigated evolution via founder effects in populations of Polynesian field crickets **(Figure 7.11)**. Polynesian field crickets (*Teleogryllus oceanicus*) are native to northern Australia and New Guinea. The crickets are also found on islands across the Pacific, including Hawaii. How do crickets cross the ocean? Over short distances, they might fly or raft on floating vegetation. To reach destinations as remote as Hawaii, however, they would almost certainly need boats. Their first opportunity to travel by boat to Vanuatu, Fiji, and points east would have been some 3,000 years ago, as stowaways—or invited guests—of the Polynesians.

If Polynesian crickets dispersed across the Pacific by hopping from island to island aboard boats, then each island's population would likely have been founded by a small number of individuals. And these founders would have carried with them just a subset of the genetic variants that were present on the island they embarked from. Unless there is ongoing migration after the invasion of new islands, cricket populations should harbor ever fewer alleles with greater distance from the Australian continent.

Tinghitella and colleagues determined the genotypes of 394 crickets from 19 populations at seven microsatellite loci. Microsatellites are regions of noncoding DNA with many easily identifiable alleles. The alleles are distinguished by the number of times a short sequence of nucleotides is repeated. The results for a representative locus, called Totri 9a, appear in **Figure 7.12**. All the bar graphs are

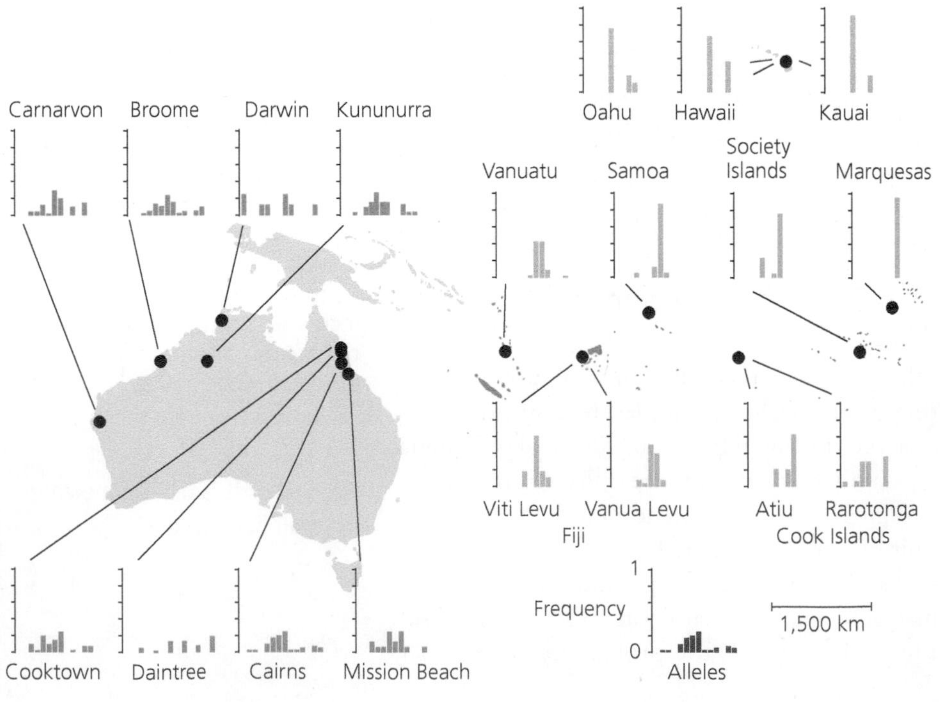

Figure 7.12 Allelic diversity at a representative microsatellite locus in Pacific field cricket populations from Australia, Oceania, and Hawaii The bar graphs show the frequencies of alleles of the Totri 9a locus in eight populations from Australia (orange), eight populations from Oceania (purple), and three populations from Hawaii (green). Sample sizes range from 5 to 25 individuals per population. Redrawn from Tinghitella et al. (2011).

plotted on the same scale. The height of each bar represents the frequency of a particular allele.

The Australian populations appear in orange. Each harbors numerous alleles, all of them at fairly low frequency. The populations from Vanuatu, Fiji, Samoa, the Cook Islands, the Society Islands, and the Marquesas appear in purple. They harbor fewer alleles, some of them at higher frequency. The populations from Hawaii, the remotest of the islands the researchers sampled, harbor just two or three alleles each, one of them at high frequency.

The overall pattern across all seven loci was the same. The cricket populations from Hawaii carried significantly less allelic diversity than those from Oceania. The populations from Oceania, in turn, harbored significantly less allelic diversity than those from Australia. This pattern is consistent with dispersal aboard Polynesian boats, and with genetic drift in the form of the founder effect.

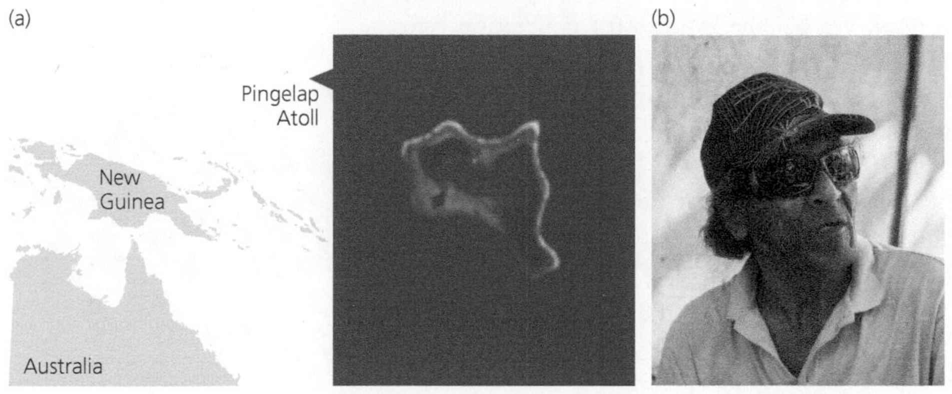

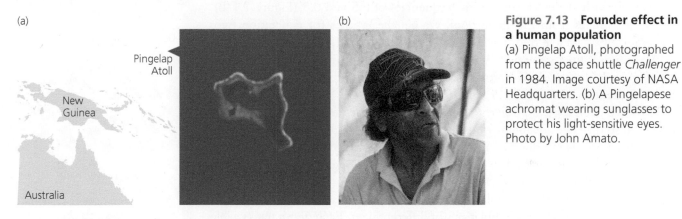

Figure 7.13 Founder effect in a human population
(a) Pingelap Atoll, photographed from the space shuttle *Challenger* in 1984. Image courtesy of NASA Headquarters. (b) A Pingelapese achromat wearing sunglasses to protect his light-sensitive eyes. Photo by John Amato.

Founder effects are often seen in genetically isolated human populations. For example, the Pingelapese people of the Eastern Caroline Islands, located about 2,700 miles southwest of Hawaii, are descended from 20 survivors of a typhoon and subsequent famine that devastated Pingelap Atoll, shown in **Figure 7.13a**, in about 1775 (Sheffield 2000). Among the survivors was a heterozygous carrier of a recessive loss-of-function allele of the *CNGB3* gene (Sundin et al. 2000). This gene, located on chromosome eight, encodes one component of a protein crucial to the function of cone cells, the photoreceptors in the retina that give us color vision. We know this survivor was a carrier because four generations after the typhoon, homozygotes for the mutant allele began to appear among his descendants. These individuals have achromatopsia, a condition characterized by complete color blindness, extreme sensitivity to light, and poor visual acuity (Figure 7.13b). Achromatopsia is rare in most populations, affecting less than 1 person in 20,000 (Winick et al. 1999). Among today's 3,000 Pingelapese, however, about 1 in 20 are achromats.

When a new population is founded by a small number of individuals, it is likely that chance alone will cause the allele frequencies in the new population to be different from those in the source population. This is the founder effect.

The high frequency of the achromatopsia allele among the Pingelapese is probably not due to any selective advantage it confers on either heterozygotes or homozygotes. Instead, the high frequency of the allele is simply due to chance. Sampling error by the typhoon, a founder effect, left the allele at a frequency of at least 2.5%. Further genetic drift in subsequent generations carried it still higher, to its current frequency of more than 20%.

Our examples from Polynesian crickets and the Pingelapese illustrate not only the founder effect, but the cumulative nature of genetic drift. In the next section we consider the cumulative consequences of genetic drift in more detail.

Random Fixation of Alleles and Loss of Heterozygosity

We have seen that genetic drift can change allele frequencies in a single generation, and that drift is even more powerful as a mechanism of evolution when its effects are compounded over many generations. We can further investigate the cumulative effects of genetic drift with the same physical model we have used before: closing our eyes and picking gametes from a paper gene pool. Our starting point will be the gene pool in **Figure 7.14a**, with alleles A_1 and A_2 at frequencies of 0.6 and 0.4. We will call the parents who produced this gene pool generation zero. As we did before, we now blindly select gametes to simulate the production of 10 zygotes by random mating. This time, the allele frequencies among the newly formed zygotes turn out to be 0.5 for A_1 and 0.5 for A_2. We will call these zygotes generation one. The reader's own results will likely vary.

To continue the simulation for another generation, we need to set up a new gene pool, with alleles A_1 and A_2 at frequencies of 0.5 and 0.5 (Figure 7.14b). Drawing gametes from this gene pool, we get the zygotes for generation two. Generation two's allele frequencies happen to be 0.4 for A_1 and 0.6 forA_2.

We now set up a gene pool with alleles A_1 and A_2 at frequencies of 0.4 and 0.6 (Figure 7.14c) and draw zygotes to make generation three. Generation three's allele frequencies are 0.45 for A_1 and 0.55 forA_2.

Now we need a gene pool with alleles A_1 and A_2 at frequencies of 0.45 and 0.55, and so on. The advantage of using a computer to simulate drawing gametes from gene pools is rapidly becoming apparent. We can have the computer run the simulation for us generation after generation for as long as we like, then plot graphs tracing the frequency of allele A_1 over time.

Graphs in **Figure 7.15a**, b, and c show the results of 100 successive generations of genetic drift in simulated populations of different sizes. Each graph tracks allele frequencies in eight populations. Every population starts with allele frequencies of 0.5 for A_1 and 0.5 for A_2. The populations tracked in graph (a) have just 4 individuals each, the populations tracked in graph (b) have 40 individuals each, and the populations tracked in graph (c) have 400 individuals each. Three patterns are evident:

1. Because the fluctuations in allele frequency from one generation to the next are caused by random sampling error, every population follows a unique evolutionary path.
2. Genetic drift has a more rapid and dramatic effect on allele frequencies in small populations than in large populations.
3. Given sufficient time, genetic drift can produce substantial changes in allele frequencies even in populations that are fairly large.

Note that if genetic drift is the only evolutionary mechanism at work in a population—if there is no selection, no mutation, and no migration—then sampling error causes allele frequencies to wander between 0 and 1. This wandering is particularly apparent in the population whose evolution is highlighted in the graph in Figure 7.15b. During the first 25 generations, allele A_1's frequency rose from 0.5 to over 0.9. Between generations 25 and 40 it dropped back to 0.5. Between generations 40 and 80 the frequency bounced between 0.5 and 0.8. Then the frequency of A_1 dropped precipitously, so that by generation 85 it hit 0 and A_1 disappeared from the population altogether. The wandering of allele frequencies produces two important and related effects: (1) Eventually alleles drift to fixation or loss, and (2) the frequency of heterozygotes declines.

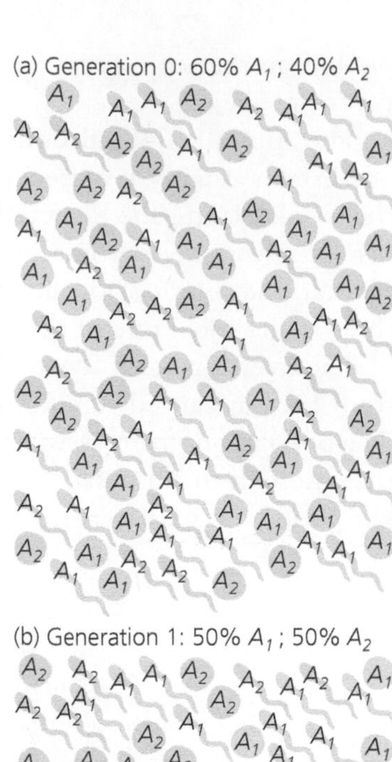

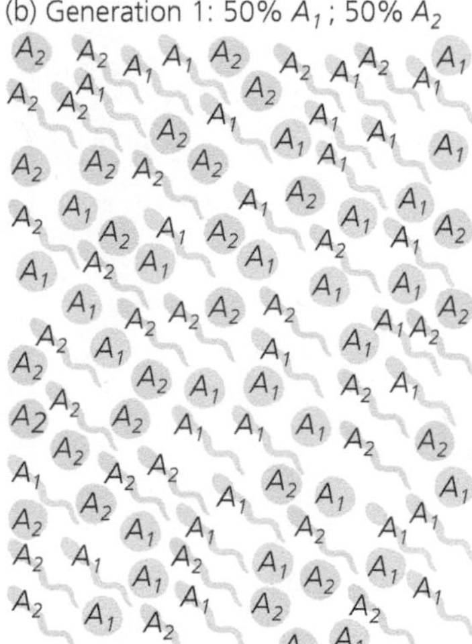

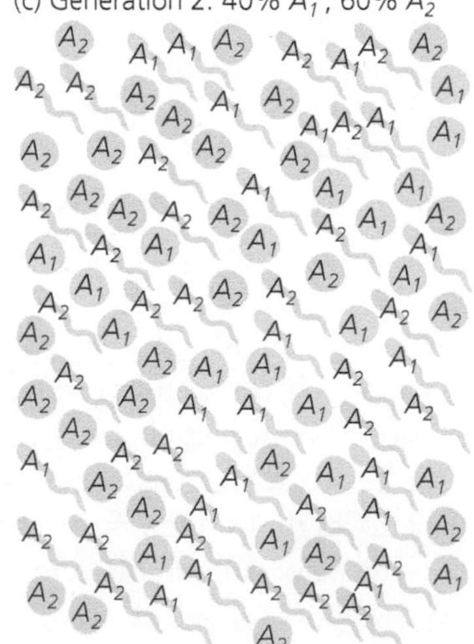

Figure 7.14 Modeling the cumulative effects of drift The gametes that make each generation's zygotes are drawn, with sampling error, from the previous generation's gene pool.

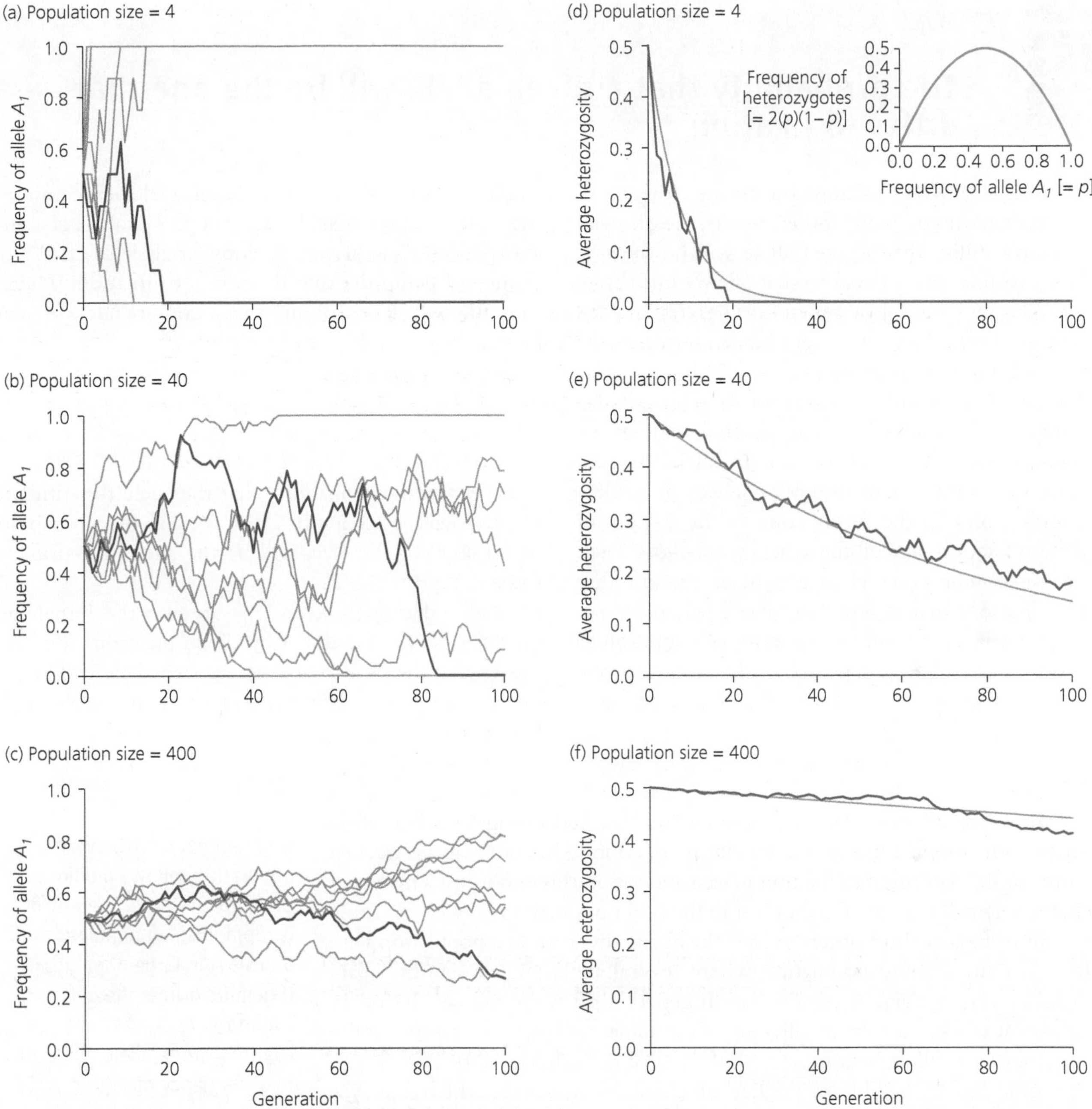

Figure 7.15 Simulations of genetic drift in populations of different sizes Plots (a), (b), and (c) show the frequency of allele A_1 across 100 generations. Eight populations are tracked in each plot, one of them highlighted in red. Plots (d), (e), and (f) show the average frequency of heterozygotes over 100 generations in the same sets of simulated populations. The gray curves represent the rate of decline predicted by theory. The inset in plot (d) shows the frequency of heterozygotes in a population in Hardy–Weinberg equilibrium, calculated as $2(p)(1-p)$, where p is the frequency of allele A_1. Collectively, the graphs in this figure show that (1) genetic drift leads to random fixation of alleles and loss of heterozygosity; and (2) drift is a more potent mechanism of evolution in small populations.

Random Fixation of Alleles

As any allele drifts between frequencies of 0 and 1.0, sooner or later it will meet an inevitable fate: Its frequency will hit one boundary or the other. If the allele's frequency hits 0, then the allele is lost forever (unless it is reintroduced by mutation or migration). If the allele's frequency hits 1, then the allele is said to be

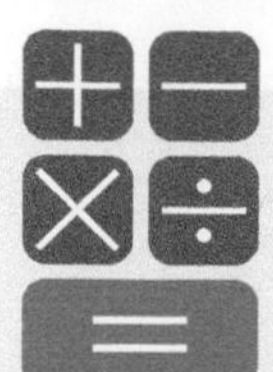

COMPUTING CONSEQUENCES 7.3

The probability that a given allele will be the one that drifts to fixation

Sewall Wright (1931) developed a detailed theory of genetic drift. Among many other results, he showed that the probability that a given allele will be the one that drifts to fixation is equal to that allele's initial frequency. Wright's model of genetic drift is beyond the scope of this book, but we can provide an intuitive explanation of fixation probabilities.

Imagine a population of N individuals. This population contains a total of $2N$ gene copies. Imagine that every one of these gene copies is a unique allele. Assume that drift is the only mechanism of evolution at work.

At some point in the future, one of the $2N$ alleles will drift to fixation, and all the others will be lost. Each allele must have an equal chance of being the one that drifts to fixation; that is what we meant when we assumed that drift is the only mechanism of evolution at work. So we have $2N$ alleles, each with an equal probability of becoming fixed. Each allele's chance must therefore be $\frac{1}{2N}$.

Now imagine that instead of each allele being unique, there are x copies of allele A_1, y copies of allele A_2, and z copies of allele A_3. Each copy of allele A_1 has a $\frac{1}{2N}$ chance of being the one that drifts to fixation. Therefore, the overall probability that a copy of allele A_1 will be the allele that drifts to fixation is

$$x \times \frac{1}{2N} = \frac{x}{2N}$$

Likewise, the probability that the allele that drifts to fixation will be a copy of A_2 is $\frac{y}{2N}$, and the probability that a copy of allele A_3 will be the allele that drifts to fixation is $\frac{z}{2N}$.

Notice that $\frac{x}{2N}$, $\frac{y}{2N}$, and $\frac{z}{2N}$ are also the initial frequencies of A_1, A_2, and A_3 in the population. We have shown that the probability that a given allele will be the one that drifts to fixation is equal to that allele's initial frequency.

fixed, also forever. Among the eight populations tracked in Figure 7.15a, allele A_1 drifted to fixation in five and to loss in three. Among the populations tracked in Figure 7.15b, A_1 drifted to fixation in one and loss in three. It is just a matter of time before A_1 will become fixed or lost in the other populations as well. As some alleles drift to fixation and others to loss, the allelic diversity in a population falls.

Now imagine a finite population where several alleles are present at a particular locus: A_1, A_2, A_3, A_4, and so on. If genetic drift is the only evolutionary mechanism at work, then eventually one of the alleles will drift to fixation. At the same moment one allele becomes fixed, the last of the others will be lost.

We would like to be able to predict which alleles will meet which fate. We cannot do so with certainty, but we can give odds. Sewall Wright (1931) proved that the probability that any given allele in a population will be the one that drifts to fixation is equal to that allele's initial frequency (see **Computing Consequences 7.3**). If, for example, we start with a finite population in which A_1 is at a frequency of 0.73, and A_2 is at a frequency of 0.27, there is a 73% chance that the allele that drifts to fixation will be A_1 and a 27% chance that it will be A_2.

Under genetic drift, every population follows a unique evolutionary path. Genetic drift is rapid in small populations and slow in large populations. If genetic drift is the only evolutionary process at work, eventually one allele will drift to a frequency of 1 (that is, to fixation) and all other alleles will be lost.

Loss of Heterozygosity

As allele frequencies in a finite population drift toward fixation or loss, the frequency of heterozygotes decreases. Graphs (d), (e), and (f) in Figure 7.15 show the decline in the frequency of heterozygotes in our simulated populations.

To see why the frequency of heterozygotes declines, look at the inset in graph (d). The inset plots the frequency of heterozygotes in a random mating population as a function of p, the frequency of allele A_1. Random mating allows us to calculate the frequency of heterozygotes as $2(p)(1 - p)$. The frequency of heterozygotes has its highest value, 0.5, when A_1 is at frequency 0.5. As the frequency of A_1 drops toward 0 or rises toward 1, the frequency of heterozygotes falls. If the frequency of A_1 reaches 0 or 1, the frequency of heterozygotes falls to 0.

Now look at graphs (a), (b), and (c). In any given generation, the frequency of A_1 may move toward or away from 0.5 in any particular population (so long as A_1 has not already been fixed or lost). Thus the frequency of heterozygotes in any particular population may rise or fall. But the overall trend across all populations is for allele frequencies to drift away from intermediate values and toward 0 or 1. So the average frequency of heterozygotes, across populations, should tend to fall.

Finally, look at graphs (d), (e), and (f). In each graph, the blue line tracks the frequency of heterozygotes averaged across the eight populations. The frequency of heterozygotes indeed tends to fall, rapidly in small populations and slowly in large populations. Eventually one allele or the other will become fixed in every population, and the average frequency of heterozygotes will fall to 0.

The frequency of heterozygotes in a population is sometimes called its **heterozygosity.** We would like to be able to predict just how fast the heterozygosity of finite populations can be expected to decline. Sewall Wright (1931) showed that, averaged across many populations, heterozygosity obeys the relationship

As alleles drift to fixation or loss, the frequency of heterozygotes in the population declines.

$$H_{g+1} = H_g[1 - \tfrac{1}{2N}]$$

where H_{g+1} is the heterozygosity in the next generation, H_g is the heterozygosity in this generation, and N is the number of individuals in the population. The value of $[1 - \frac{1}{2N}]$ is always between $\frac{1}{2}$ and 1, so the expected frequency of heterozygotes in the next generation is always less than the frequency of heterozygotes in this generation. In Figure 7.15, the gray curves in graphs (d), (e), and (f) show the declines in heterozygosity predicted by Wright's equation.

We can assess the differentiation among a set of populations due to genetic drift by calculating F_{ST}, a statistic we mentioned earlier. It is defined as follows:

$$F_{ST} = \frac{H_T - H_S}{H_T}$$

where H_T is the expected heterozygosity under Hardy–Weinberg equilibirum in a total population created by combining all of our separate populations, and H_S is the average across separate populations (also known as subpopulations) in their expected heterozygosities. At the start of the simulation depicted in graph (a), F_{ST} is zero, because both H_T and H_S are 0.5. By the end, F_{ST} is 1, because—with all subpopulations fixed—H_S is 0. F_{ST} is sometimes called the **fixation index.**

To appreciate just one implication of the inevitable loss of heterozygosity in finite populations, imagine you are managing a captive population of an endangered species. Suppose there are just 50 breeding adults in zoos around the world. Even if you could transport adults or semen to accomplish random mating, you would still see a loss in heterozygosity of 1% every generation due to genetic drift.

An Experiment on Random Fixation and Loss of Heterozygosity

Our discussion of random fixation and heterozygosity loss has so far been based on simulated populations and mathematical equations. Peter Buri (1956) studied these phenomena empirically, in laboratory populations of the fruit fly *Drosophila melanogaster.* Adopting an approach used earlier by Kerr and Wright (1954), Buri established 107 populations of flies, each with eight females and eight males. All the founders were heterozygotes for an eye-color gene called *brown*. They all had the genotype bw^{75}/bw. Thus, in all 107 populations, the initial frequency of the bw^{75} allele was 0.5. Buri maintained these populations for 19 generations. For every population in every generation, Buri kept the population size at 16 by picking eight females and eight males at random to be the breeders for the next generation.

What results would we predict? If neither allele bw^{75} nor allele *bw* confers a selective advantage, we expect the frequency of allele bw^{75} to wander at random by genetic drift in every population. Nineteen generations should be enough, in populations of 16 individuals, for many populations to become fixed for one allele or the other. Because allele bw^{75} has an initial frequency of 0.5, we expect it to be lost about as often as it becomes fixed. As bw^{75} is drifting toward fixation or loss in each population, we expect the average heterozygosity across all populations to decline. The rate of decline should follow Wright's equation, given in the previous section.

Buri's results confirm these predictions. Each small graph in **Figure 7.16** is a histogram summarizing the allele frequencies in all 107 populations in a particular generation. The horizontal axis represents the frequency of the bw^{75} allele, and the vertical axis represents the number of populations showing each frequency. The frequency of bw^{75} was 0.5 in all populations in generation zero, which is not shown in the figure. After one generation of genetic drift, most populations still had an allele frequency near 0.5, although one had an allele frequency as low as 0.22 and another had an allele frequency as high as 0.69. As the frequency of bw^{75} rose in some populations and fell in others, the distribution of allele frequencies rapidly spread out. In generation four, the frequency of bw^{75} hit 1 in a population for the first time. In generation six, the frequency of bw^{75} hit 0 in a population for the first time. As the allele frequency reached 0 or 1 in ever more populations, the distribution of frequencies became U-shaped. By the end of the experiment, bw^{75} had been lost in 30 populations and had become fixed in 28. The 30:28 ratio of losses to fixations is very close to the 1:1 ratio we would predict under genetic drift. During Buri's experiment there was dramatic evolution in nearly all 107 of the fruit fly populations, but natural selection had nothing to do with it.

The genetic properties of *brown* were such that Buri could identify all three genotypes from their phenotypes. Thus Buri was able to directly assess the frequency of heterozygotes in each population. All the founding flies

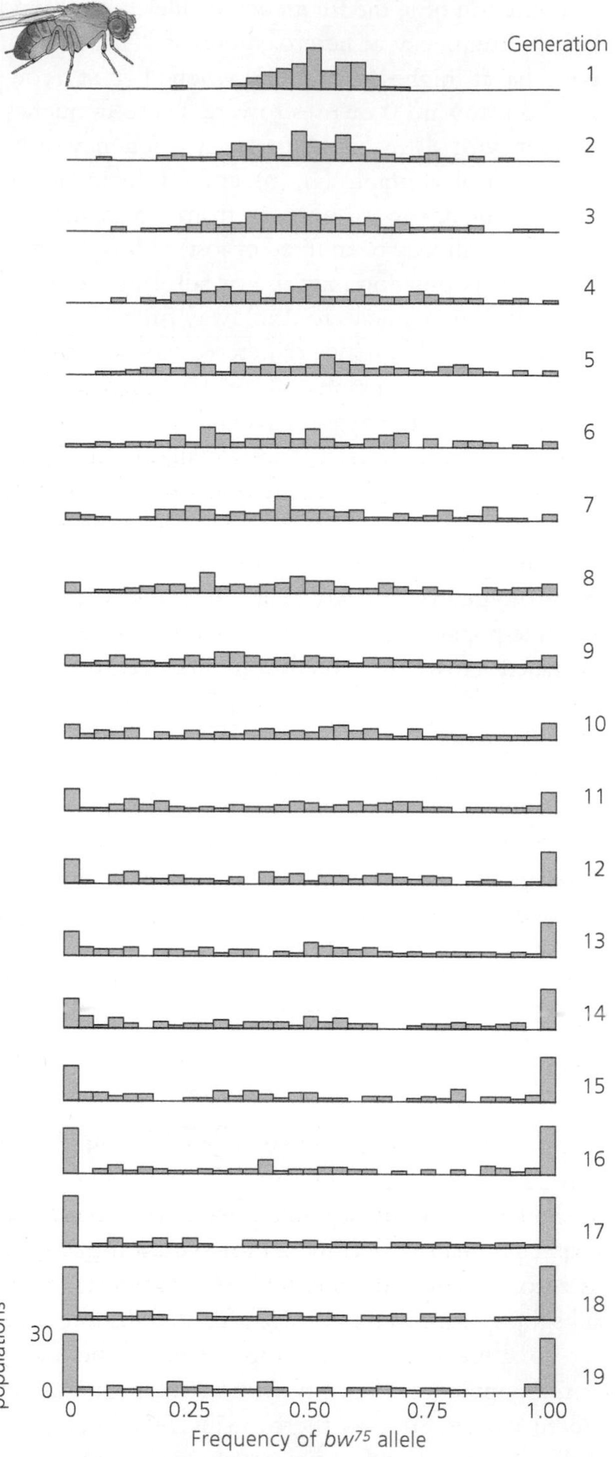

Figure 7.16 Drift in 107 populations of 16 flies Each histogram summarizes allele frequencies in all 107 populations in a particular generation. The horizontal axis represents the frequency of the bw^{75} allele; the vertical axis represents the number of populations showing each frequency. The frequency of bw^{75} was 0.5 in all populations in generation zero (not shown). By generation 19, bw^{75} had been lost from 30 populations, and fixed at a frequency of 100% in 28 populations. From data in Buri (1956), after Ayala and Kiger (1984).

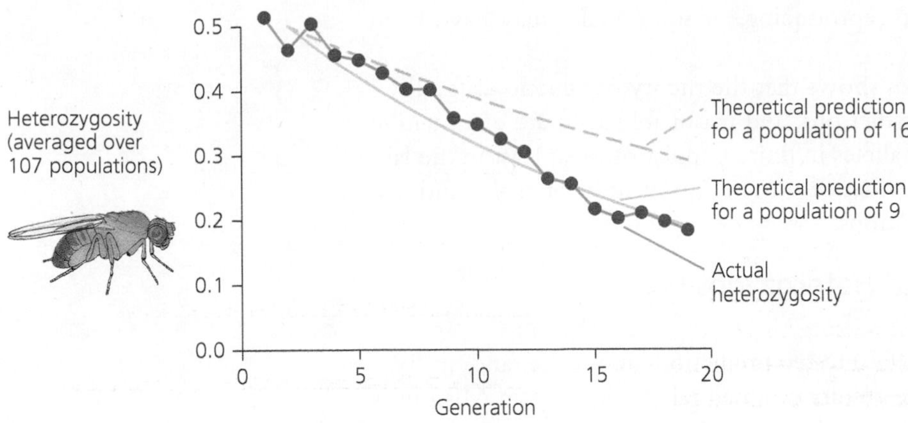

Figure 7.17 Buri's drift experiment summarized The frequency of heterozygotes declined with time in Buri's experimental populations. This graph demonstrates that (1) heterozygosity decreases across generations in small populations; and (2) although all the populations had an actual size of 16 flies, their effective size was roughly 9. Replotted from data in Buri (1956), after Hartl (1981).

were heterozygotes, so the heterozygosity in generation one was 0.5. Every generation thereafter, Buri noted the frequency of heterozygotes in each population, then took the average across all 107 populations. **Figure 7.17** tracks these values for average heterozygosity over the 19 generations of the study. Look first at the red dots, which show the actual data. Consistent with our theoretical prediction, the average frequency of heterozygotes steadily declined.

Empirical studies confirm that under genetic drift alleles become fixed or lost, and the frequency of heterozygotes declines. Indeed, these processes often happen faster than predicted.

The fit between theory and results is not perfect, however. The dashed gray curve in the figure shows the predicted decline in heterozygosity, using Wright's equation and a population size of 16. The actual decline in heterozygosity was more rapid than expected. The solid gray curve shows the predicted decline for a population size of 9; it fits the data well. Buri's populations lost heterozygosity as though they contained only 9 individuals instead of 16. In other words, the **effective population size** in Buri's experiment was 9 (see **Computing Consequences 7.4**). Among the explanations are that some of the flies in each population may

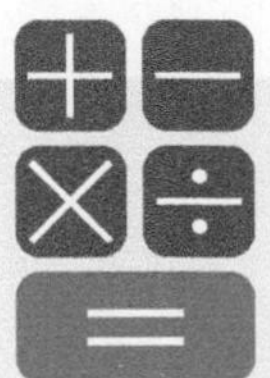

COMPUTING CONSEQUENCES 7.4

Effective population size

The effective population size is the size of an ideal theoretical population that would lose heterozygosity at the same rate as an actual population of interest. The effective population size is virtually always smaller than the actual population size. In Buri's experiment, two possible reasons for the difference in effective versus actual population size are that (1) some of the flies in each bottle died (by accident) before reproducing, and (2) fruit flies exhibit sexual selection by both male–male combat and female choice (see Chapter 11)—either of which could have prevented some males from reproducing.

The effective population size is particularly sensitive to differences in the number of reproductively active females versus males. When there are different numbers of each sex in a population, the effective population size N_e can be estimated as

$$N_e = \frac{4N_m N_f}{(N_m + N_f)}$$

where N_m is the number of males and N_f is the number of females.

To see how strongly an imbalanced sex ratio can reduce the effective population size, use the formula to show that when there are 5 males and 5 females, $N_e = 10$; when there is 1 male and 9 females, $N_e = 3.6$; and when there is 1 male and 1,000 females, $N_e = 4$. Consider the logistical problems involved in maintaining a captive breeding program for a species in which the males are extremely aggressive and will not tolerate each other's presence.

have died due to accidents before reproducing, or some males may have been rejected as mates by the females.

Buri's experiment with fruit flies shows that the theory of genetic drift allows us to make accurate qualitative predictions, and reasonably accurate quantitative predictions, about the behavior of alleles in finite populations—at least in the lab. In the next section, we consider evidence on random fixation of alleles and loss of heterozygosity in natural populations.

Random Fixation and Loss of Heterozygosity in Natural Populations

Alan Templeton and colleagues (1990) tested predictions about the random fixation of alleles by documenting the results of a natural experiment in Missouri's Ozark Mountains. Although now largely covered in oak–hickory forest, the Ozarks were part of a desert during an extended period of hot, dry climate that lasted from 8,000 to 4,000 years ago. The desert that engulfed the Ozarks was contiguous with the desert of the American Southwest. Many southwestern desert species expanded their ranges eastward into the Ozarks. Among them was the collared lizard **(Figure 7.18)**. When the warm period ended, the collared lizard's range retracted westward and the Ozarks were largely overgrown with savannas. Within these mixed woodlands and grasslands, however, on exposed rocky outcrops, were small remnants of desert habitat called glades. Living in these glades were relict populations of collared lizards.

Figure 7.18 A collared lizard (*Crotaphytus collaris*). Photo by Alan R. Templeton.

Every five years or so, wildfires swept the Ozark savannas (Templeton et al. 2001). This periodic burning was essential to the maintenance of the savanna plant community. We know this because of what happened after European settlers arrived. First the Europeans clear-cut the Ozark woodlands. Then, starting in about 1950, they suppressed all fires. These interventions allowed the oak–hickory forest that covers the area today to invade the savannas. And they allowed eastern red cedars to invade the glades.

The invasion of the glades by eastern red cedars, and the other woody plants that followed, was bad news for collared lizards (Templeton et al. 2001). The cedars partially overgrew many of the glades, drastically reducing their size. The oak–hickory forest between the glades was even worse. Its dense understory prevented the lizards from migrating from one glade to another. Most of the lizard populations, even some separated by just 50 meters of oak–hickory forest, were sufficiently isolated from each other that there was little or no gene flow among them. And the relict populations in the few remaining glades were tiny; most harbored no more than a few dozen lizards.

Because of the small size and genetic isolation of the glade populations, Templeton and colleagues (1990) predicted that Ozark collared lizards would bear a strong imprint of genetic drift. Within each population, most loci should be fixed for a single allele, and genetic variation should be very low. Which allele became fixed in any particular population should be a matter of chance, however, so there should be considerable genetic diversity among populations.

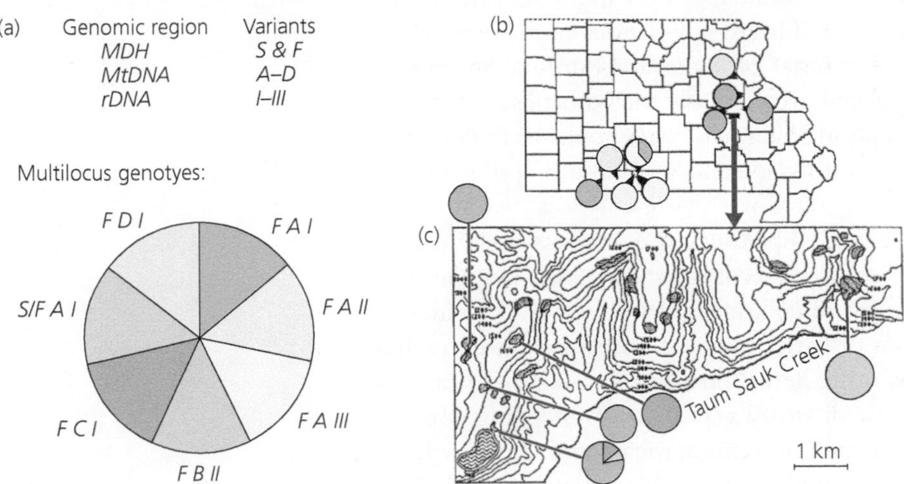

Figure 7.19 Genetic variation in Ozark glade populations of collared lizard (a) Key to seven distinct multilocus genotypes, each characterized by a malate dehydrogenase (MDH) genotype [with "slow" (*S*) and "fast" (*F*) alleles], a mitochondrial DNA haplotype (A–D), and a ribosomal DNA genotype (I–III). (b) Genetic compositions of nine glade populations in southern Missouri. (c) Expanded map of a small piece of (b), with compositions of five more populations. From Templeton et al. (1990).

Templeton and colleagues (1990) assayed glade populations for genetic variation. The researchers screened lizards for their genotypes at a variety of enzyme loci, for their ribosomal DNA genotypes, and for their mitochondrial DNA genotypes. They identified among the lizards seven distinct multilocus genotypes **(Figure 7.19a)**. Confirming the predicted consequences of isolation and small population size, most glade populations were fixed for a single multilocus genotype, and different genotypes were fixed in different glades (Figure 7.19b and c).

Empirical data from a natural experiment confirm that due to drift, small isolated populations lose their genetic diversity.

Templeton and colleagues (2001) believed that the nearly complete loss of genetic diversity in the glade populations had doomed the Ozark collared lizards to extinction. This extinction would happen one glade at a time and have any of a number of proximate causes. If a pathogen appeared that could infect and kill one of the lizards in a glade, it could infect and kill all lizards in the glade—because they were virtually identical. As the biological and physical environment changed, the lizard populations would be unable to evolve in response—because genetic variation is the raw material for adaptive evolution. And if an adaptation did evolve in one of the populations, it would be unable to spread to other glades because the lizards were unable to cross the oak–hickory forests that divided them. Templeton and colleagues surveyed 130 Ozark glades. Consistent with their expectations, two-thirds of them were already devoid of collared lizards.

If Templeton and colleagues were right, simple measures could save the Ozark collared lizards. One is the relocation of lizards to repopulate the empty glades. In the 1980s, with the cooperation of the Missouri Department of Conservation, Templeton and colleagues established three new populations in the Stegall Mountain Natural Area, a former ranch with many glades but no lizards. The lizards in the new populations thrived but did not migrate, neither from population to population nor to any of the empty glades. As long as the oak–hickory forest was in the way, the populations would remain isolated and suffer the long-term consequences of genetic drift. Starting in 1994, the Missouri Department of Conservation and the United States Forest Service began using controlled burns to clear the oak–hickory forest at Stegall Mountain. The lizards responded almost immediately, moving among populations and colonizing many of the empty glades. This behavior should restore the genetic diversity of the glade populations and dramatically improve the collared lizard's prospects for long-term survival.

Jennifer Brisson, Jared Strasburg, and Templeton (2003) monitored the results of a controlled experiment at Taum Sauk Mountain State Park, 80 km from Stegall Mountain. They compared the collared lizard populations occupying glades in an area that had been treated with a series of controlled burns versus populations occupying glades in an unburned area. Consistent with observations at Stegall Mountain, the burned area supported a much larger population of lizards, and the lizards there moved from glade to glade and colonized empty glades at much higher rates.

In their research on collared lizards, Templeton and colleagues documented the random fixation of alleles and loss of heterozygosity in small populations. Working with plants, Andrew Young and colleagues (1996) reviewed evidence of these processes among populations of various sizes. The researchers compiled data from the literature on three flowering herbs and a tree. From these data they plotted two measures of overall genetic diversity against census breeding population size. The first measure was genetic polymorphism, the fraction of loci within the genome that have at least two alleles with frequencies higher than 0.01. The

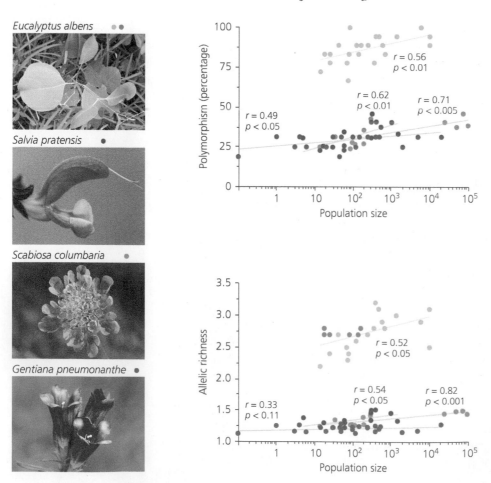

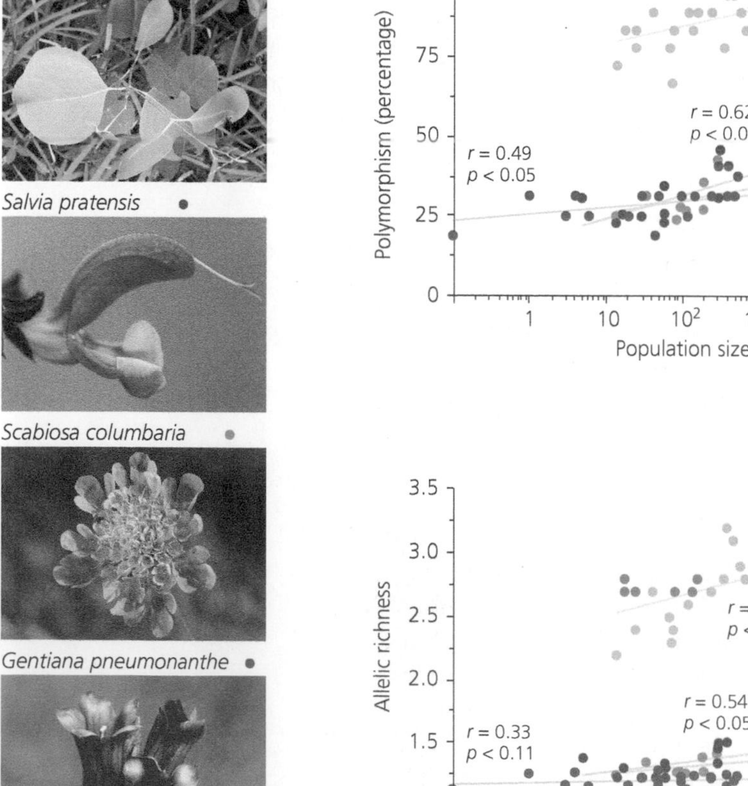

Figure 7.20 Population size and genetic diversity Each data point on these scatterplots represents a population of flowering plants. Polymorphism, plotted on the vertical axis of the top graph, is the proportion of allozyme loci at which the frequency of the most common allele in the population is less than 0.99. In other words, polymorphism is the fraction of alleles that are substantially polymorphic. Allelic richness, plotted on the vertical axis of the bottom graph, is the average number of alleles per locus. The statistic r, the Pearson correlation coefficient, varies from 0 (no association between variables) to 1 (perfect correlation). P specifies the probability that the correlation coefficient is significantly different from zero. *Eucalyptus albens* is a tree; dark green dots represent small populations isolated by less than 250 m. *Salvia pratensis*, *Scabiosa columbaria*, and *Gentiana pneumonanthe* are flowering herbs. After Young et al. (1996) and sources therein.

second was allelic richness, the average number of alleles per locus. Both measures are related to heterozygosity. Imagine a single locus in a randomly mating population. As the number of alleles at the locus increases and as the fraction of those alleles that have substantial frequencies increases, the frequency of heterozygotes at the locus increases as well. If, on the other hand, the locus is fixed for a single allele, then no individual in the population will be a heterozygote. Genetic polymorphism, allelic richness, and heterozygosity rise and fall together. Because genetic drift is more pronounced in small populations than in large ones, and because drift results in the loss of heterozygosity, Young and colleagues predicted that small populations would have lower levels of polymorphism and allelic richness. Plots of the data appear in **Figure 7.20**. Consistent with the prediction, in almost every case smaller populations did indeed harbor less genetic diversity.

The studies by Templeton et al. and Young et al. show that in at least some natural populations, genetic drift leads—as predicted—to random fixation and reduced heterozygosity. The loss of genetic diversity in small populations is of particular concern to conservation biologists, for two reasons. First, genetic diversity is the raw material for adaptive evolution. Imagine a species reduced to a few remnant populations by habitat destruction or some other environmental change. Genetic drift may rob the remnant populations of their potential to evolve in response to a changing environment at precisely the moment the environment is changing most drastically. Second, a loss of heterozygosity also entails an increase in homozygosity. Increased homozygosity often leads to reduced fitness in experimental populations (see, for example, Polans and Allard 1989; Barrett and Charlesworth 1991). Presumably this involves the same mechanism as inbreeding depression: It exposes deleterious alleles to selection. We consider inbreeding depression in Section 7.4.

The Rate of Evolution by Genetic Drift

The theory and experiments we have discussed in this section establish that sampling error can be an important mechanism of evolution. Next we consider the rate of evolution when genetic drift is the only process at work.

First, we need to define what we mean by the rate of evolution at a single locus. We will take the rate of evolution to be the rate at which new alleles created by mutation are substituted for other alleles already present. **Figure 7.21** illustrates the process of **substitution** and distinguishes substitution from mutation. The figure follows a gene pool of 10 alleles for 20 generations. Initially, all of the alleles are identical (light green dots). In the fourth generation, a new allele appears

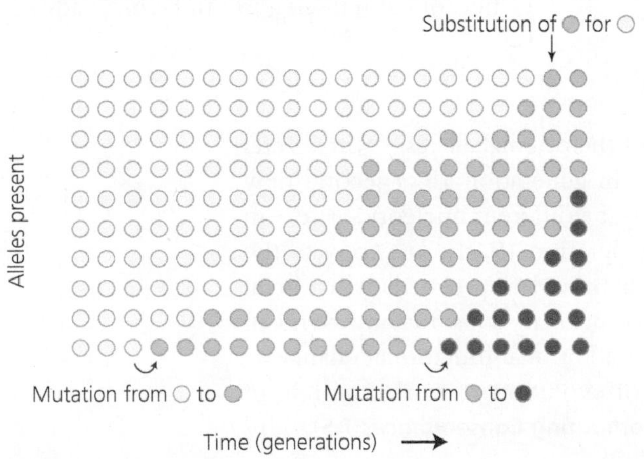

Figure 7.21 Mutation versus substitution Mutation is the creation of a new allele; substitution is the fixation of the new allele, with or without additional mutational change. This graph shows the 10 alleles present in each of 20 successive generations in a hypothetical population of five individuals. During the time covered, the dark green allele was substituted for the light green allele. The blue allele may ultimately be substituted for the dark green allele, or it may be lost.

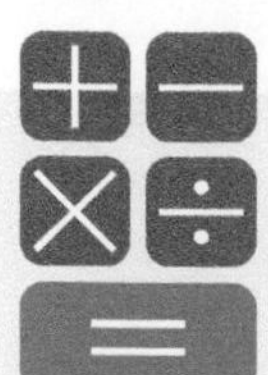

COMPUTING CONSEQUENCES 7.5

The rate of evolutionary substitution under genetic drift

Here we show a calculation establishing that when genetic drift is the only mechanism of evolution at work, the rate of evolutionary substitution is equal to the mutation rate (Kimura 1968).

Imagine a diploid population of size N. Within this population are $2N$ alleles of the locus of interest, where by *alleles* we mean copies of the gene, regardless of whether they are identical. Let v be the rate of selectively neutral mutations per allele per generation, and assume that each mutation creates an allele that has not previously existed in the population. Then every generation, there will be

$$2Nv$$

new alleles created by mutation. Because by assumption all new alleles are selectively neutral, genetic drift is the only process at work. Each new allele has the same chance of drifting to fixation as any other allele in the population. That chance, equal to the frequency of the new allele, is

$$\frac{1}{2N}$$

Therefore, each generation the number of new alleles that are created by mutation and are destined to drift to fixation is

$$2Nv \times \frac{1}{2N} = v$$

The same argument applies to every generation. Therefore, the rate of evolution at the locus of interest is v substitutions per generation.

Mutation, selection, and drift in molecular evolution

It will be useful for our discussion of molecular evolution to explore in more detail what we mean by v, the rate of neutral mutations. Imagine that the locus of interest is a gene encoding a protein that is L amino acids long. Let u be the rate of mutations per codon per generation. The overall rate of mutation at our locus is given by

$$\mu = uL(d + a + f) = uLd + uLa + uLf$$

where d is the fraction of codon changes that are deleterious, a is the fraction that are selectively advantageous, f is the fraction that are selectively neutral, and $d + a + f = 1$. Note that the rightmost term, uLf, is equal to our earlier v.

In showing that the rate of substitution is equal to v, we assumed that d and a are both equal to zero. In any real population, of course, many mutations are deleterious and d is not zero. This does not change our calculation of the substitution rate. Deleterious alleles are eliminated by natural selection and do not contribute to the rate of evolutionary substitution.

Proponents of the neutral theory hold that a is approximately equal to zero and that f is much larger than a. Therefore, they predict that evolutionary substitution will be dominated by neutral mutations and drift and will occur at the rate $v = uLf$, as we have calculated.

Proponents of the selectionist theory hold that a is too large to ignore and that the rate of evolutionary substitution will be significantly influenced by the action of natural selection in favor of advantageous alleles.

(dark green dot), created by a mutation in one of the original alleles. Over several generations, this allele drifts to high frequency. In generation 15, a second new allele appears (blue dot), created by a mutation—at a different nucleotide site—in a descendant of the first dark green allele. In generation 19, the last copy of the original allele is lost. At this point, we can say that the dark green allele has been substituted for the light green allele. Thus, by evolutionary substitution, we mean the fixation of a new mutation, with or without additional mutational change.

When genetic drift is the only mechanism of evolution at work, the rate of substitution is equal to the mutation rate (see **Computing Consequences 7.5**). This is true regardless of the population size, because two effects associated with population size cancel each other out: More mutations occur in a larger population, but in a large population each new mutation has a smaller chance of drifting to fixation. Under genetic drift, large populations generate and maintain more genetic variation than small populations, but populations of all sizes accumulate substitutions at the same rate.

Of course, mechanisms of evolution other than drift are often at work. We can allow some natural selection into our model and still get a similar result. Imagine that some mutations are deleterious while others are selectively neutral. The deleterious mutations will be eliminated by natural selection and will never become fixed. The rate of substitution will then be equal to the rate at which neutral mutations occur.

Evolutionary biologists are divided on the relevance of this calculation to real populations. All agree that one kind of mutation, and thus one kind of selection, has been left out (see Computing Consequences 7.5). Some mutations are selectively advantageous and are swept to fixation by natural selection more surely and much faster than drift would ever carry them. Evolutionists are of two minds, however, over how often this happens.

Proponents of the neutral theory, long championed by Motoo Kimura (1983), hold that advantageous mutations are exceedingly rare and that most alleles of most genes are selectively neutral. Neutralists predict that for most genes in most populations, the rate of evolution will be equal to the neutral mutation rate.

Proponents of the selectionist theory, most strongly championed by John Gillespie (1991), hold that advantageous mutations are common enough that they cannot be ignored. Selectionists predict that for many genes in most populations, the rate of substitution will reflect the action of natural selection on advantageous mutations. The neutralist–selectionist debate has largely been fought in the arena of molecular evolution, so that is where we go, in the next section, to explore it. Before doing so, however, it will be worthwhile to consider how genetic drift and natural selection interact.

When mutation, genetic drift, and selection interact, three processes occur: (1) Deleterious alleles appear and are eliminated by selection; (2) neutral mutations appear and are fixed or lost by chance; and (3) advantageous alleles appear and are swept to fixation by selection. The relative importance of (2) and (3) in determining the overall substitution rate is a matter of debate.

Genetic Drift versus Natural Selection

In an ideal population of infinite size, natural selection favoring one allele over others will inexorably carry the favored allele to fixation. If the same beneficial allele occurs in a finite population, however, sampling error will cause the allele's frequency to fluctuate at random around the trajectory it would have taken in a population of infinite size. Sometimes the allele's frequency will rise, and sometimes it will fall. The allele may even go extinct. Likewise, in an infinitely large population, selection favoring heterozygotes will maintain multiple alleles at equilibrium frequencies indefinitely. In a finite population, genetic drift may cause one allele to become fixed and the other to be lost. Whether drift or selection plays the predominant role in determining the evolutionary outcome will depend on both the size of the population and the strength of selection.

Stephen S. Rich and colleagues (1979) studied the interplay between natural selection and genetic drift in laboratory populations of the red flour beetle (*Tribolium castaneum*). As the name suggests, red flour beetles are typically red. But not always. Rich and colleagues took advantage of genetic variation for color at the *b* locus. The wild-type allele is called b^+. A mutant allele that can be maintained in lab populations is called *b*. Beetles with genotype b^+b^+ are red, beetles with genotype b^+b are brown, and beetles with genotype *bb* are black.

As part of a larger experiment, Rich and colleagues set up 24 populations of flour beetles in which the initial frequencies of allele b^+ and allele b were both 0.5. They started 12 populations with 50 males and 50 females and 12 with 5 males and 5 females. They maintained the populations at these sizes for 20 generations, each generation choosing adults at random to serve as breeders for the next generation. Every generation, they examined 240 randomly chosen individuals from each population to assess the frequencies of the two alleles.

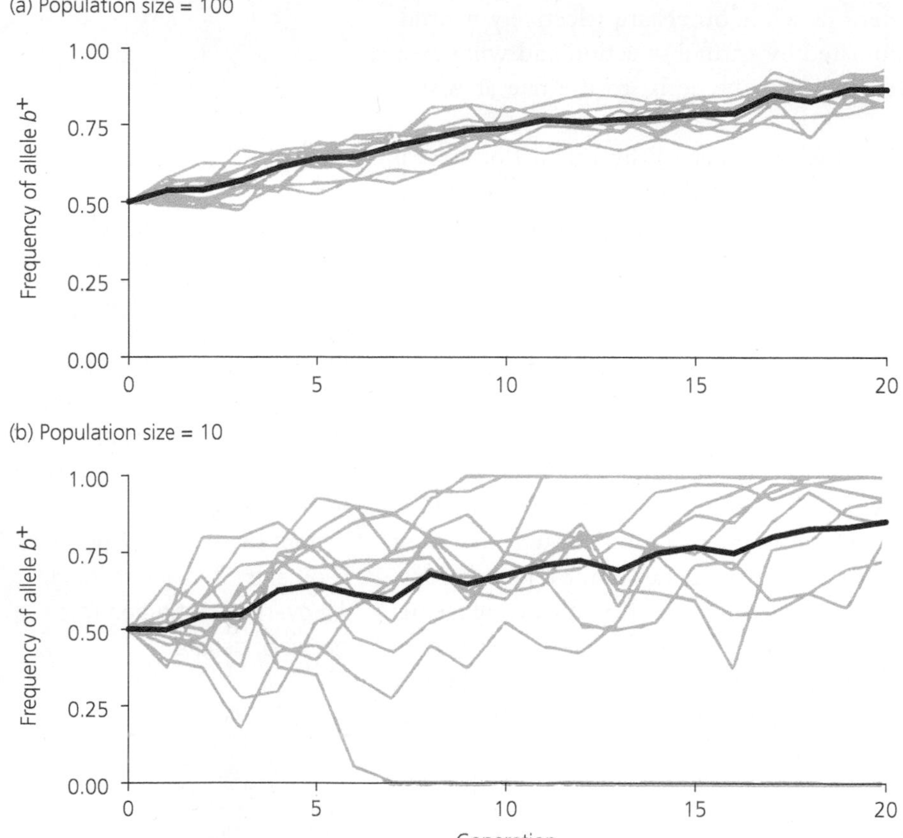

Figure 7.22 Natural selection and random genetic drift in flour beetle populations of different size The black lines, which trace the average allele frequency for each set of 12 populations, show that the strength of selection was similar in large (a) versus small (b) populations. The tan lines, which trace individual populations, show that genetic drift played a larger role in the evolution of the small populations. Modified from Rich et al. (1979).

Figure 7.22 displays the data. The tan lines trace the frequency of allele b^+ in the individual populations. The black lines trace the average frequency of b^+ across all 12 populations in each set. Three patterns are notable.

First, the black lines show an overall trend toward higher frequencies of allele b^+. In analyzing data across the entire experiment, Rich and colleagues found the trend statistically significant. This pattern is consistent with natural selection. The researchers estimated the relative fitnesses of genotypes b^+b^+, b^+b, and bb to be 1, 0.95, and 0.9. The data are also consistent with a model in which heterozygotes have the highest fitness, but the equilibrium frequency of b^+ is fairly high.

Second, the tan lines document considerable variation in allele frequencies among populations. This pattern is consistent with genetic drift.

Third, comparison of graphs (a) and (b) reveals that the small populations traveled considerably more diverse evolutionary paths than did the large ones. In these populations, drift was predominant. Indeed, in one of them the allele associated with higher fitness went extinct while the allele associated with lower fitness became fixed. In the large populations, however, selection was predominant.

When populations are subject to both selection and genetic drift, smaller populations follow more diverse evolutionary paths.

Just how large an advantage or liability must an allele carry, in a population of a given size, for selection to overcome drift and play a role in determining the allele's fate? One way to approach this question is to consider the likely fate of a novel allele created by mutation. In a diploid population of N individuals, a new allele has a frequency of $\frac{1}{2N}$. This means that unless the population is tiny, the new allele will be rare. If genetic drift is the only mechanism of evolution at work, the allele's chance of eventually reaching fixation is, as we showed in Computing Consequences 7.3, equal to its frequency: $\frac{1}{2N}$. In populations evolving by drift alone, most new mutations disappear shortly after they arise.

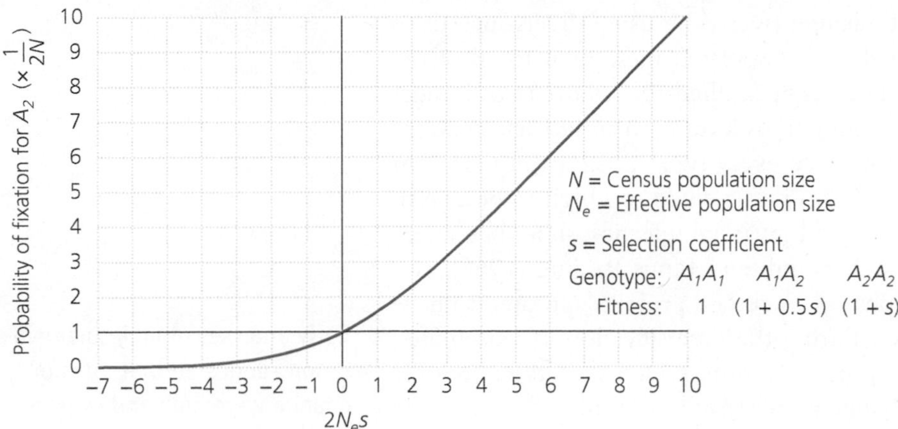

Figure 7.23 Genetic drift and natural selection The red curve shows the probability that a new allele created by mutation will rise to fixation, relative to its chances under genetic drift, as a function of effective population size and the strength of selection. After Charlesworth (2009).

It is possible to calculate the extent to which selection improves or impairs a novel mutation's probability of rising to fixation. The math required is beyond the scope of this text (Kimura 1964), but the result can be summarized graphically. **Figure 7.23** presents an example (Charlesworth 2009). The vertical axis shows the probability that a new allele created by mutation will ultimately rise to fixation. This probability is scaled in multiples of what it would be in a population evolving by drift alone $\left(\frac{1}{2N}\right)$. The horizontal axis shows a composite variable combining the effective population size (N_e) and the strength of selection (s). This variable is scaled in multiples of $2N_es$.

If twice the effective population size multiplied by the selection coefficient is less than −1 or greater than 1, then selection substantially alters an allele's chances of loss or fixation.

If $2N_es$ is equal to zero, as it would be in the absence of selection, then a new allele's chance of becoming fixed is $\left(\frac{1}{2N}\right)$. If $2N_es$ is equal to 5, on the other hand, then a new allele's chance of becoming fixed is five times higher. There are two ways for $2N_es$ to be high: large effective population size or strong selection. In a population with an effective size of 10,000 a selection coefficient of just 0.0005 would increase a new mutant's chances of becoming fixed by factor of 10. On the other hand, 10 times a tiny chance is still a tiny chance.

If $2N_es$ is negative, as in the case of selection against a deleterious mutation, the chance of fixation is less than it would be in a population evolving by drift alone. If $2N_es$ is less than −5, a new allele has virtually no chance of becoming fixed. If the population is small, however, or if selection is weak, a deleterious mutation may have nearly the same chance of fixation as a neutral or beneficial one.

In summary, genetic drift is a nonadaptive mechanism of evolution. As a result of chance sampling error, allele frequencies can change from one generation to the next. Drift can lead to the fixation of some alleles, the loss of others, and an overall decline in genetic diversity. Drift is most influential in small populations, when selection is weak, and when its effects are compounded across generations.

7.3 Genetic Drift and Molecular Evolution

The study of molecular evolution began in the mid-1960s, when biochemists succeeded in determining the amino acid sequences of hemoglobin, cytochrome *c*, and other abundant and well-studied proteins found in humans and other vertebrates. These sequences provided the first opportunity for evolutionary biologists to compare the amount and rate of molecular change among species.

Early workers in the field made several striking observations. Foremost among them were calculations by Motoo Kimura (1968). Kimura took the number of sequence differences in the well-studied proteins of humans versus horses and converted them to rates of sequence change over time using divergence dates estimated from the fossil record. He then extrapolated these rates to all of the protein-coding loci in the genome. The result implied that as the two lineages diverged from their common ancestor, mutations leading to amino acid replacements had, on average, risen to fixation once every two years. Given that most mutations are thought to be deleterious, this rate seemed too high to be due to natural selection. Beneficial mutations fixed by natural selection should be rare.

A second observation, by Emil Zuckerkandl and Linus Pauling (1965), was that the rate of amino acid sequence change in certain proteins appeared to have been constant over time, or clocklike, during the diversification of vertebrates. This too seemed inconsistent with natural selection, which should be episodic and correlated with environmental change rather than with time.

Early analyses of molecular evolution suggested that rates of change were high and constant through time. These conclusions appeared to be in conflict with what might be expected under natural selection.

In short, early data on molecular evolution did not match expectations derived from the notion that most evolutionary change was due to natural selection. But if natural selection does not explain evolution at the molecular level, then what process is responsible for rapid, clocklike sequence change? Many researchers believe the answer is genetic drift.

The Neutral Theory of Molecular Evolution

Kimura (1968, 1983) formulated the **neutral theory** of molecular evolution to explain the observed patterns of amino acid sequence divergence. To understand the neutral theory's central claim, note that with respect to effect on fitness, there are three kinds of mutations. Some mutations are deleterious, some are neutral, and some are beneficial. Mutations that are deleterious tend to be eliminated by natural selection and thus contribute little to molecular evolution. Mutations that are neutral (or nearly so—more on that later) rise and fall in frequency as a result of genetic drift. Many are lost, but some become fixed. Mutations that are beneficial are often lost to drift while still at low frequency, but otherwise tend to rise to fixation as a result of natural selection. Kimura's neutral theory holds that effectively neutral mutations that rise to fixation by drift vastly outnumber beneficial mutations that rise to fixation by natural selection. Genetic drift, not natural selection, is thus the mechanism responsible for most molecular evolution.

The neutral theory models the fate of new alleles that were created by mutation and whose frequencies change by genetic drift. It claims to explain most evolutionary change at the level of nucleotide sequences.

Based on his view that drift dominates sequence evolution, and on the calculation detailed in Computing Consequences 7.5, Kimura postulated that the rate of molecular evolution is, to a good approximation, equal to the mutation rate.

Kimura's theory was startling to many evolutionary biologists. Given that drift has a larger influence on allele frequencies in small populations than in large ones, the absence of an effect of population size on the rate of evolution was counterintuitive. So was the assertion that sequence evolution by natural selection was so rare, compared to evolution by drift, as to be insignificant.

Although Kimura's theory appeared to explain why the amino acid sequences of hemoglobin, cytochrome *c*, and other proteins change steadily over time, the theory was inspired by limited amounts of data. How did the neutral theory hold up, once large volumes of DNA sequence data became available?

Patterns in DNA Sequence Divergence

During the late 1970s and 1980s, biologists mined growing databases of DNA sequences to analyze the amounts and rates of change in different loci. They began to see patterns that varied by the type of sequence examined. The most basic distinction was between coding versus noncoding sequences. Coding sequences contain instructions for tRNAs, rRNAs, or proteins; noncoding sequences include introns, regions that flank coding regions, regulatory sites, and pseudogenes. What predictions does the neutral theory make about the rate and pattern of change in different types of sequences, and have they been verified or rejected?

Pseudogenes Establish a Canonical Rate of Neutral Evolution

Pseudogenes are functionless stretches of DNA that result from gene duplications (see Chapter 5). Because they do not encode proteins, mutations in pseudogenes should be neutral with respect to fitness. When such mutations achieve fixation in populations, it should happen solely as a result of drift. Pseudogenes are thus considered a paradigm of neutral evolution (Li et al. 1981). As predicted by the neutral theory, the divergence rates recorded in pseudogenes—which should be equal to the neutral mutation rate v—are among the highest seen for loci in nuclear genomes (Li et al. 1981; Li and Graur 2000). This finding is consistent with the neutral theory's explanation for evolutionary change at the molecular level. It also quantifies the rate of evolution due to drift. For humans versus chimps, this rate is about 2.5×10^{-8} mutations per nucleotide site per generation (Nachman and Crowell 2000). How do rates of change in other types of sequences compare to the standard, or canonical, rate?

The evolution of pseudogenes conforms to the assumptions and predictions of the neutral theory.

Silent Sites Change Faster than Replacement Sites in Most Coding Loci

Recall (from Chapter 5) that bases in DNA are read in three-letter codons, and that the genetic code contains considerable redundancy. In the portion of the code shown in **Figure 7.24a**, two codons specify phenylalanine, two specify leucine, and four code for serine. As shown in Figure 7.24b, base-pair changes

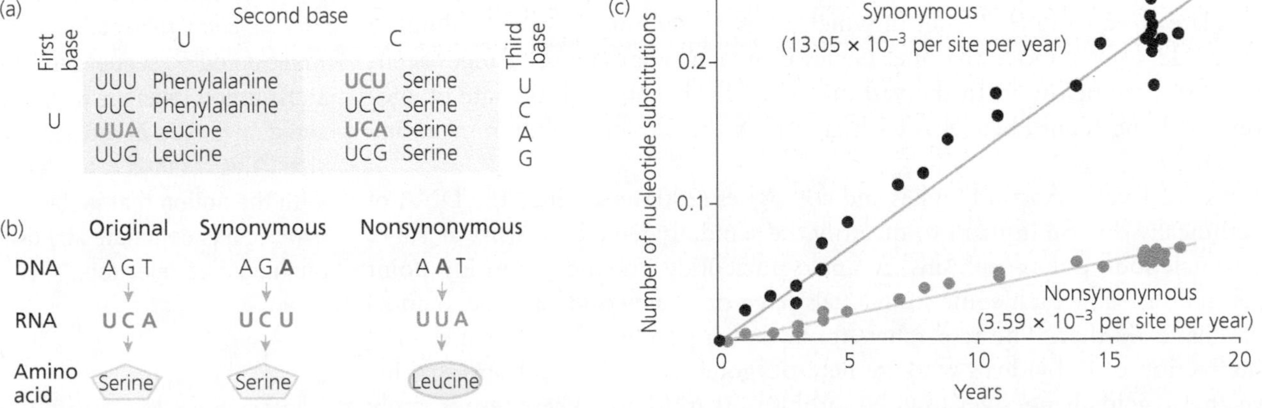

Figure 7.24 Molecular evolution in influenza viruses is consistent with the neutral theory Because the genetic code is redundant (a), there are two kinds of point mutations (b). The neutral theory predicts that both will accumulate by drift, but synonymous substitutions will accumulate faster. (c) Data from the flu virus. From Gojobori et al. (1990).

may or may not lead to amino acid sequence changes. DNA sequence changes that do not result in amino acid changes are called **silent-site** (or **synonymous**) mutations; sequence changes that do result in an amino acid change are called **replacement** (or **nonsynonymous**) mutations.

Figure 7.24c presents data on the rate of silent versus replacement substitution in a gene belonging to the influenza virus, based on comparisons of flu viruses collected over a span of 20 years with a reference sample collected in 1968 (Gojobori et al. 1990). Both kinds of substitution accumulated in a linear, clocklike fashion, but the rate of evolution for silent changes is much higher than the rate of evolution for replacement changes.

This pattern accords with the neutral theory. Silent changes are not exposed to natural selection on protein function, because they do not alter the amino acid sequence. New alleles created by silent mutations should thus increase or decrease in frequency largely as a result of drift. Replacement mutations, in contrast, change the amino acid sequences of proteins. If most of these alterations are deleterious, then most of them should be eliminated by natural selection without ever becoming common enough to be detected. This type of natural selection is called **negative** or **purifying selection,** as opposed to **positive selection** on beneficial mutations. Less frequently, replacement mutations occur that have no effect on protein function and may be fixed by drift.

Natural selection against deleterious mutations is called negative selection.

Natural selection favoring beneficial mutations is called positive selection.

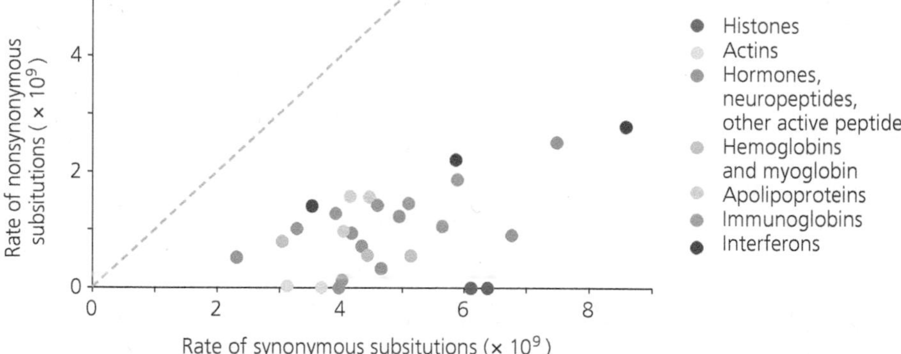

Figure 7.25 Rates of nucleotide substitution vary among genes and among sites within genes Data points report rates of replacement and silent substitutions in protein-coding genes compared between humans and either mice or rats. Units are substitutions per site per billion years. The number of codons compared per gene ranges from 28 to 435. Data from Li and Graur (1991).

Molecular biologists have compared the rate of replacement versus silent substitutions in a great variety of coding loci. In **Figure 7.25**, the dashed line marks where the data would fall if the nonsynonymous and synonymous substitutions accumulate at equal rates. Genes in which nonsynonymous changes accumulate faster would appear above the line. Genes in which synonymous changes accumulate faster fall below it. In the vast majority of genes studied, the rate of evolution involving silent changes is far higher than the rate of evolution involving replacements.

In most coding sequences, substitution rates are higher at silent sites than at replacement sites. This result is consistent with the notion that molecular evolution is dominated by drift and negative selection.

In a similar vein, Austin Hughes and colleagues (2003) examined the DNA of 102 ethnically diverse humans to quantify the standing genetic diversity at 1,442 single-nucleotide polymorphisms. A single-nucleotide polymorphism is a point in the genome at which some individuals have one nucleotide and other individuals have another. The researchers found lower standing diversity, measured as the fraction of individuals who are heterozygotes, for polymorphisms that involve amino acid changes versus polymorphisms that do not. These results imply that most single-nucleotide mutations that swap one amino acid for another are deleterious and held at low frequency by negative selection.

These observations are consistent with the patterns predicted if most mutations are either deleterious or neutral and drift dominates molecular evolution. They support the central tenet of the neutral theory.

Variation among Loci: Evidence for Functional Constraints

The data in Figure 7.25 contain another important pattern. When homologous coding sequences from humans and rodents are compared, some loci are found to be nearly identical, while others have undergone rapid divergence. This result turns out to be typical. Rates of molecular evolution vary widely among loci.

The key to explaining this pattern is that genes responsible for the most vital cellular functions appear to have the lowest rates of replacement substitutions. Histone proteins, for example, interact with DNA to form structures called nucleosomes. These protein–DNA complexes are a major feature of the chromatin fibers in eukaryotic cells. Changes in the amino acid sequences of histones disrupt the structural integrity of the nucleosome and have ill consequences for DNA transcription and synthesis. In contrast, genes less vital to the cell, and thus under less stringent functional constraints, show more rapid rates of replacement substitutions. When functional constraints are lower, a larger fraction of replacement mutations are neutral with respect to fitness and may fix by drift.

Nearly Neutral Mutations

Although the neutral theory appeared to account for several important patterns in molecular evolution, data indicating clocklike change in proteins compared across species presented a problem. The issue was that the neutral mutation rate v should vary among species as a function of generation time, not clock time. Over a given interval of clock time, more neutral substitutions should accumulate in species with short generation times than in species with long generation times. Contrary to expectation, at least some protein sequence comparisons reveal clocklike change in absolute time—independent of differences in generation time among the species compared. The data points in **Figure 7.26** fall along lines, despite comparing humans to species with drastically different generation times.

To account for this observation, Tomoko Ohta and Motoo Kimura (1971; Ohta 1972, 1977) considered how the probability of fixation for a novel mutation depends on the effective population size and the strength of selection. We looked at an example of this relationship in Figure 7.23. If the product of twice the effective population size and the selection coefficient is sufficiently close to zero—because the population is tiny, selection is weak, or both—the probability

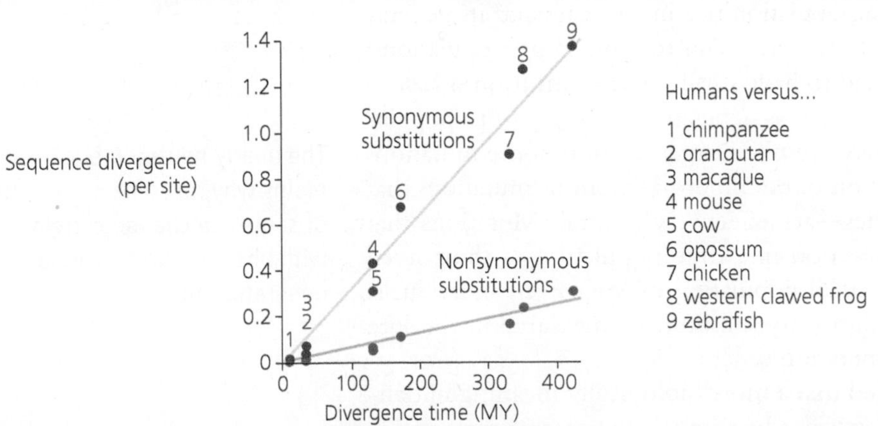

Figure 7.26 The vertebrate molecular clock ticks in calendar time, not generation time The data points, showing sequence divergence versus clock time, fall on lines—regardless of whether they compare humans with other species with long generation times (chimpanzees, orangutans) or short generation times (mouse, zebrafish). Each point represents an average for over 4,000 genes. From Nei et al. (2010).

of fixation is roughly the same as it would be if the mutation had no effect on fitness at all. The allele's frequency will evolve primarily as a result of genetic drift. In population genetics models of evolution in finite populations, neutral alleles and nearly neutral alleles behave the same way.

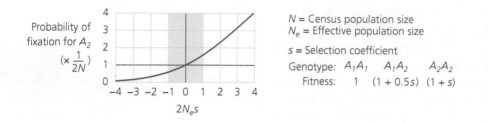

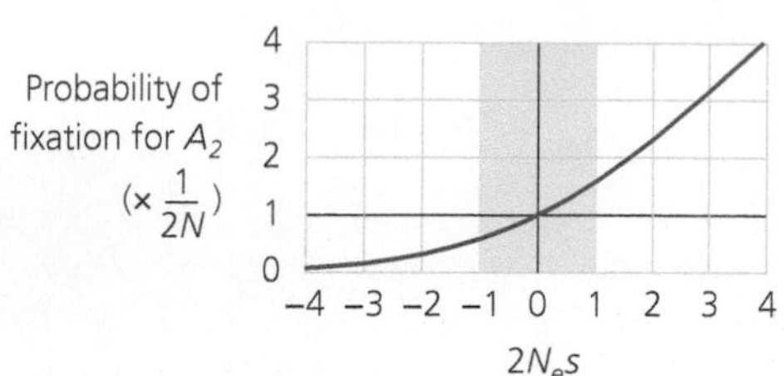

N = Census population size
N_e = Effective population size
s = Selection coefficient

Genotype:	A_1A_1	A_1A_2	A_2A_2
Fitness:	1	$(1 + 0.5s)$	$(1 + s)$

Figure 7.27 A definition of nearly neutral evolution The green band shows a range of values for $2N_es$ over which the frequency of a new mutation changes mostly by genetic drift. After Charlesworth (2009).

We have reproduced part of Figure 7.23 in **Figure 7.27**. Examination of the figure will reveal that the threshold value of $|2N_es|$ below which we will call a mutation nearly neutral is somewhat arbitrary. It also depends on how the selection coefficient is defined. Ohta and Kimura's (1971) analysis suggests that, with the selection coefficient defined as in the figure, a reasonable criterion is $|2N_es| \leq 1$, or $|s| \leq \frac{1}{2N_e}$. This range is covered by the green band in the graph.

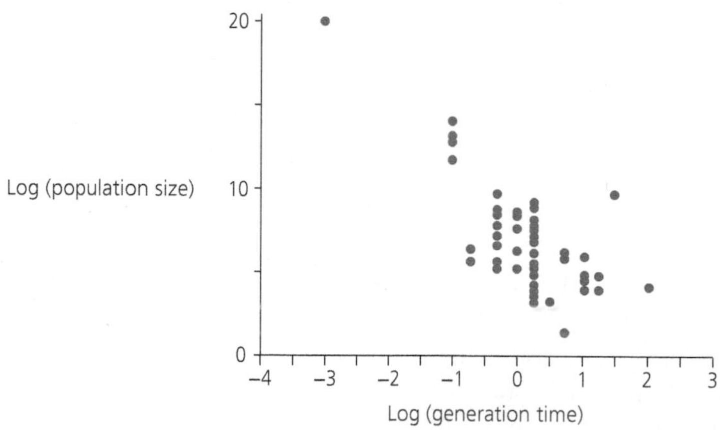

Figure 7.28 Population size versus generation time Across organisms, as generation time goes up, population size goes down. Statistical tests confirm the strong inverse correlation displayed in this log-log plot. From Chao and Carr (1993).

How does the consideration of nearly neutral mutations explain the observation that molecular clocks tick in absolute time rather than in the number of generations? As Lin Chao and David Carr (1993) have shown, there is a strong negative correlation between average population size in a species and its generation time. Species with short generation times tend to have large populations; species with long generation times tend to have small populations **(Figure 7.28)**.

This is important because, Ohta argued, as generation time goes up, population size, and thus $|2N_es|$, go down. As a result, a larger fraction of the mutations that arise—in particular, a larger fraction of the mildly deleterious mutations that are typically abundant in most species—are effectively neutral. Mutations that would be eliminated by purifying selection in a large population of short-lived individuals instead evolve by drift a small population of long-lived individuals. This tends to equalize the rate of evolutionary substitution, measured in absolute time, across species with different generation times.

The nearly neutral model explains why, in some cases, rates of sequence change correlate with absolute time instead of generation time.

Matsatoshi Nei (2005) has suggested that a more biologically meaningful definition of a neutral mutation would consider how much the mean fitness of the population would change were the mutation to become fixed. If s is defined as in Figure 7.27, Nei would call a mutation effectively neutral if $|s| \leq 0.002$. In population genetics models of extremely large populations, selection this weak can drive an allele to fixation. But the time required for it to do so may be unrealistically long (Nei et al. 2010). Furthermore, for an allele so weakly associated with fitness, the strength and even direction of selection are likely to change over time, across different environments, and on different genetic backgrounds.

The Neutral Theory as a Null Hypothesis: Detecting Natural Selection on DNA Sequences

Since their inception, the neutral and nearly neutral theories have been controversial (see Berry 1996; Ohta and Kreitman 1996). Discussion has focused on the claims by Kimura (1983) and King and Jukes (1969) that the number of beneficial mutations fixed by positive natural selection is inconsequential compared to the number of mutations that change in frequency under the influence of drift. Is this claim accurate? How can we determine that natural selection has been responsible for changes observed at the molecular level?

The neutralist–selectionist controversy is a debate about the relative importance of drift and positive selection in explaining molecular evolution.

When researchers compare homologous DNA sequences among individuals and want to explain the differences they observe, they routinely use the neutral theory as a null hypothesis. The neutral theory specifies the rates and patterns of sequence change that occur in the absence of natural selection. If the changes that are actually observed are significantly different from the predictions made by the neutral theory, and if a researcher can defend the proposition that the sequences in question have functional significance for the organism, then there is convincing evidence that natural selection has caused molecular evolution.

Here we examine a few of the strategies being used to detect molecular evolution due to natural selection. We begin with studies of replacement changes, then explore evidence that many silent-site mutations are also under selection.

Selection on Replacement Mutations

We noted earlier that according to the neutral theory, silent mutations are expected to evolve largely by genetic drift. Replacement mutations are expected either to be deleterious, in which case they are eliminated by negative selection and we will not see them, or to be neutral, in which case they, too, evolve by drift. If the neutral theory is wrong for a particular gene, however, and replacement mutations are advantageous, then they will be rapidly swept to fixation by positive selection. Thus, to find out whether replacements within a particular gene are deleterious, neutral, or advantageous, we can compare two sequences and calculate the rate of nonsynonymous substitutions per site (d_N) and the rate of synonymous substitutions per site (d_S). If we take their ratio, we will get

$$\frac{d_N}{d_S} < 1 \text{ when replacements are deleterious,}$$

$$\frac{d_N}{d_S} = 1 \text{ when replacements are neutral, and}$$

$$\frac{d_N}{d_S} > 1 \text{ when replacements are advantageous}$$

When sequences evolve by drift and negative selection, synonymous substitutions outnumber replacement substitutions. When sequences evolve by drift and positive selection, replacement substitutions outnumber synonymous substitutions.

Austin Hughes and Masatoshi Nei (1988) tested the neutral theory by estimating the ratio of replacement to silent substitutions in genes vital to immune function. When mammalian cells are infected by a bacterium or a virus, they respond by displaying pieces of bacterial or viral protein on their surfaces. Immune system cells then kill the infected cell, which prevents the bacterium or virus inside the cell from replicating. The membrane proteins that display pathogen proteins are encoded by a cluster of genes called the major histocompatibility complex, or MHC. The part of an MHC protein that binds to the foreign peptide is called the antigen recognition site (ARS). Hughes and Nei (1988) studied sequence changes in the ARS of MHC loci in humans and mice.

When Hughes and Nei compared alleles from the MHC complexes of 12 different humans and counted the number of differences observed in silent versus replacement sites, they found significantly more replacement-site than silent-site changes. The same pattern occurred in the ARS of mouse MHC genes, although the differences were not as great. This pattern could result only if the replacement changes were selectively advantageous. The logic here is that positive selection causes replacement changes to spread through the population much more quickly than neutral alleles can spread by chance.

It is important to note, however, that Hughes and Nei found this pattern only in the ARS. Other exons within the MHC showed more silent than replacement changes, or no difference. At sites other than the ARS, then, they could not rule out the null hypothesis that sequence change is dominated by drift.

In many examples, replacement substitutions outnumber synonymous substitutions—a signature of positive selection.

Research by Gavin Huttley and colleagues (2000) on *BRCA1*, a gene associated with breast cancer, provides another example. *BRCA1* encodes a protein that participates in the repair of damaged DNA (see O'Connell 2010) and in the regulation of programmed cell death during neural development (Pulvers and Huttner 2009). Huttley and colleagues sequenced exon 11 from the *BRCA1* genes of a variety of mammals, then inferred the rates of nonsynonymous and synonymous substitution along the branches of the evolutionary tree that connects the extant species to their common ancestors **(Figure 7.29)**. Along most branches of the phylogeny the value of $\frac{d_N}{d_S}$ was less than one, consistent with the neutral theory. On the branches connecting humans and chimpanzees to their common ancestor, however, $\frac{d_N}{d_S}$ was significantly greater than one. This suggests that the sequence of exon 11 has been under positive selection in the ancestors of today's humans and

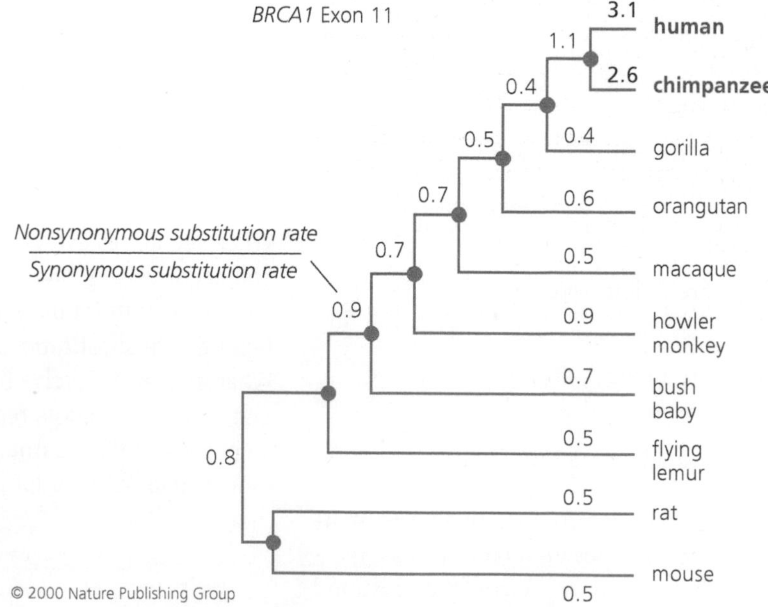

Figure 7.29 Positive selection on the *BRCA1* gene in humans and chimpanzees On most branches of this phylogeny, the ratio of replacement to silent substitution rates is less than one, consistent with neutral evolution. On the branches leading to humans and chimps, however, the ratio is significantly greater than one—consistent with positive selection.

Reprinted by permission from Macmillan Publishers Ltd: Huttley, G. A., E. Easteal, M. C. Southey, et al. 2000. *Nature Genetics* 25: 410–413.

chimps. The selective agent responsible remains unknown, although Pulvers and Huttner (2009) speculate that it may involve brain size.

Comparing Silent and Replacement Changes within and between Species. The research by Hughes and Nei and by Huttley and colleagues provides clear examples of gene segments where neutral substitutions do not predominate. Thanks to the efforts of numerous researchers, many other loci have been found where replacement substitutions outnumber silent substitutions.

Even though the $\frac{d_N}{d_S}$ criterion for detecting positive selection has been useful, Paul Sharp (1997) notes that it is extremely conservative. Replacement substitutions will outnumber silent substitutions only when positive selection has been strong. In a comparison of 363 homologous loci in mice and rats, for example, only one showed an excess of replacement over silent changes. But as Sharp notes (1997, p. 111), "It would be most surprising if this were the only one of these genes that had undergone adaptive changes during the divergence of the two species." Are more sensitive methods for detecting natural selection available?

Researchers have developed statistical tests for detecting positive selection that are more sensitive than the simple ratio of nonsynonymous to synonymous substitution.

John McDonald and Martin Kreitman (1991) invented a test for natural selection that is widely used. The McDonald–Kreitman, or MK, test is based on the neutral theory's assertion that all standing variation at both silent sites and replacement sites consists of neutral alleles evolving by drift (see Fay 2011). If this assertion is true, then the ratio of nonsynonymous to synonymous substitutions between closely related species, $\frac{d_N}{d_S}$, should be the same as the ratio of synonymous to nonsynonymous polymorphisms within species, $\frac{p_N}{p_S}$. A **polymorphism** is a locus at which different individuals in a population carry different alleles. Positive selection on nonsynonymous substitutions within species can elevate $\frac{d_N}{d_S}$ above $\frac{p_N}{p_S}$ because beneficial mutations rise quickly to fixation within populations. They thus contribute only briefly to polymorphism, but permanently and cumulatively to interspecific divergence.

McDonald and Kreitman's initial use of this test compared sequence data from the alcohol dehydrogenase (*Adh*) gene of 12 *Drosophila melanogaster*, 6 *D. simulans*, and 12 *D. yakuba* individuals. *Adh* was an interesting locus to study because fruit flies feed on rotting fruit that may contain toxic concentrations of ethanol, and the alcohol dehydrogenase enzyme catalyzes the conversion of ethanol to a nontoxic product. Because of the enzyme's importance to these species, and because ethanol concentrations vary among food sources, it is reasonable to suspect that the locus is under selection when populations begin exploiting different fruits.

In an attempt to sample as much within-species variation as possible, the individuals chosen for the study were from geographically widespread locations. McDonald and Kreitman aligned the *Adh* sequences from each individual in the study and identified sites where a base differed from the most commonly observed nucleotide, or what is called the consensus sequence. The researchers counted differences as fixed if they were present in all individuals from a particular species, and as polymorphisms if they were present in only some individuals from a particular species. Differences that were fixed in one species and polymorphic in another were counted as polymorphic.

McDonald and Kreitman found that 29% of the differences that were fixed between species were replacement substitutions. Within species, however, only 5% of the polymorphisms in the study represented replacements. Rather than being the same, these ratios show an almost sixfold, and statistically significant, difference ($p = 0.006$). This is strong evidence against the neutral model's prediction.

McDonald and Kreitman's interpretation is that the differences in replacement mutations fixed in different species are selectively advantageous. They suggest that these mutations occurred after *D. melanogaster*, *D. simulans*, and *D. yakuba* had diverged and spread rapidly to fixation due to positive selection in the differing environments occupied by these species.

Using the MK test, natural selection has now been detected in loci from plants, protists, and a variety of animals (Escalante et al. 1998; Purugganan and Suddith 1998). With an extension of the MK test applied to 35 genes in *D. simulans* and *D. yakuba*, Nick Smith and Adam Eyre-Walker (2002) estimated that 45% of all amino acid substitutions between the genomes of the two species were fixed by positive selection. With an extension applied to the genomes of humans and chimpanzees, Carlos Bustamante and colleagues (2005) identified 304 human genes that have evolved under positive selection.

Which Loci Are under Strong Positive Selection? Thanks to studies employing the Hughes and Nei analysis, the MK test, and other strategies, generalizations are beginning to emerge concerning the types of loci where positive natural selection has been particularly strong (Yang and Bielawski 2000; Vallender and Lahn 2004; Nielsen 2005; Nielsen et al. 2005). Replacement substitutions appear to be particularly abundant in loci involved in arms races between pathogens and their hosts (for example, Hughes and Nei 1989), in loci with a role in reproductive conflicts such as sperm competition and egg–sperm interactions (Swanson and Vacquier 1998; Dorus et al. 2004), and in recently duplicated genes that have attained new functions (Zhang et al. 1998). Positive selection has also been detected in genes involved in sex determination, gametogenesis, sensory perception, interactions between symbionts, tumor suppression, and programmed cell death as well as in genes that code for certain enzymes or regulatory proteins.

Positive selection seems to be particularly common in genes involved in biological conflict.

As data accumulate from genome-sequencing projects in closely related species, such as humans and chimpanzees, the number and quality of comparative studies are exploding. Even before the era of genome sequencing began, however, it became clear that silent substitutions, as well as replacement changes, are subject to natural selection.

Selection on "Silent" Mutations

The term *silent mutation* was coined to reflect two aspects of base changes at certain positions of codons: They do not result in a change in the amino acid sequence of the protein product, and they are not exposed to natural selection. The second proposition had to be discarded, however, in the face of data on phenomena known as codon bias, hitchhiking, and background selection. How can mutations that do not alter an amino acid sequence be affected by natural selection?

Direct Selection on Synonymous Mutations: Codon Bias and Other Factors. Most of the 20 amino acids are encoded by more than one codon. We have emphasized that changes among redundant codons do not cause changes in the amino acid sequences of proteins, and we have implied that these silent changes are neutral with respect to fitness. If this were strictly true, we would expect codon usage to be random, and in a given species each codon in a suite of synonymous codons to be present in proportions that reflect the G+C content of the species' genome. But early sequencing studies confirmed that codon usage is highly nonrandom **(Figure 7.30)**. This phenomenon is known as **codon bias.**

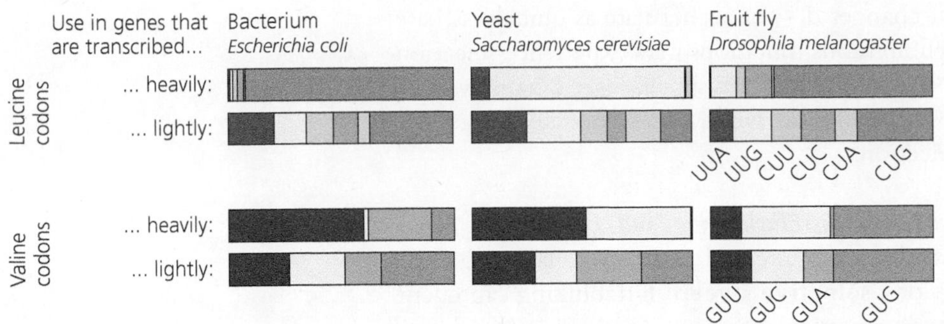

Figure 7.30 Codon bias Bars show relative use of possible codons for two amino acids in genes with different transcription levels. Lightly transcribed genes use all available codons in roughly equal amounts. Heavily transcribed genes tend to use one or two codons to the exclusion of others. Drawn from data in Sharp et al. (1998).

Several important patterns have emerged from studies of codon bias. Codon bias is strongest in highly expressed genes—such as those for the proteins found in ribosomes—and weak to nonexistent in rarely expressed genes. In addition, the suite of codons that are used most frequently correlates strongly with the most abundant species of tRNA in the cell **(Figure 7.31)**.

The leading hypothesis to explain these observations is natural selection for translational efficiency (Sharp and Li 1986; Sharp et al. 1988; Akashi 1994). The logic here is that if a "silent" mutation in a highly expressed gene creates a codon that is rare in the pool of tRNAs, the mutation will be selected against. The selective agent is the speed and accuracy of translation. Speed and accuracy are especially important when the proteins encoded by particular genes are turning over rapidly and the corresponding genes must be transcribed continuously. It is reasonable, then, to observe the strongest codon bias in highly expressed genes.

Codon bias suggests that some synonymous mutations are not selectively neutral.

Selection against certain synonymous mutations represents a form of negative selection; it slows the rate of molecular evolution. As a result, codon bias may

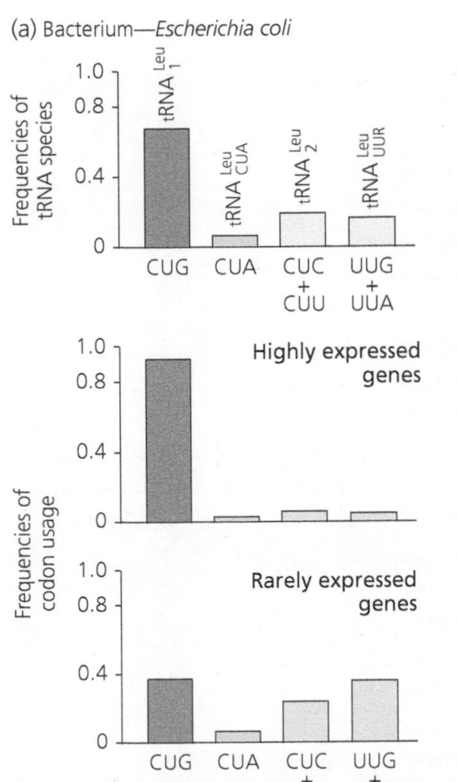

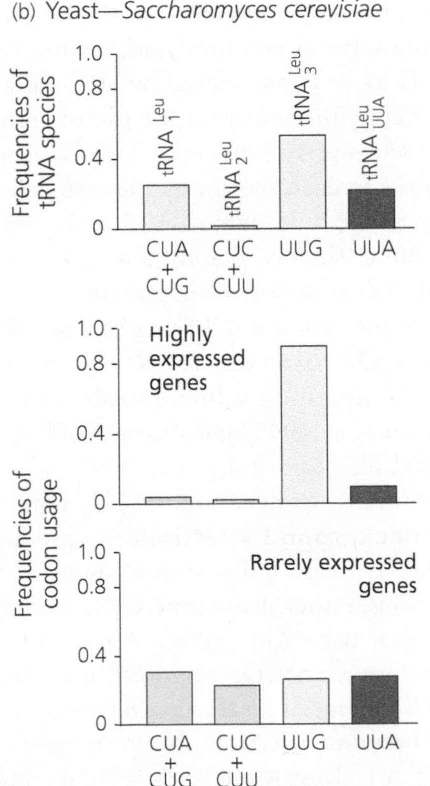

Figure 7.31 Codon bias correlates with the relative frequencies of tRNA species The bar chart in the top row of both (a) and (b) shows the frequencies of four different tRNA species that carry leucine in *E. coli* (a) and the yeast *Saccharomyces cerevisiae* (b). The bar charts in the middle and bottom rows report the frequency of the mRNA codons corresponding to each of these tRNA species in the same organisms. The mRNA codons were measured in two different classes of genes: those that are highly transcribed (middle) and those that are rarely transcribed (bottom). The data show that codon usage correlates strongly with tRNA availability in highly expressed genes, but not at all in rarely expressed genes. Redrawn from Li and Graur (1991).

explain the observation that silent changes do not accumulate as quickly as base changes in pseudogenes. Other synonymous mutations may experience selection as a result of their effects on mRNA stability or exon splicing (see Chamary et al. 2006). The general message here is that not all redundant sequence changes are "silent" with respect to natural selection.

Indirect Effects on Synonymous Mutations: Hitchhiking and Background Selection. Another phenomenon that affects the rate and pattern of change at silent sites is referred to as **hitchhiking,** or a **selective sweep.** Hitchhiking can occur when strong positive selection acts on a particular amino acid change. As a favorable mutation increases in frequency, neutral or even slightly deleterious mutations closely linked to the favored site will increase in frequency along with the beneficial locus. These linked mutations are swept along by selection and may even ride to fixation. Note that this process occurs when only recombination fails to break up the linkage between the hitchhiking sites and the site under selection.

A striking example of hitchhiking happened on the fourth chromosome of fruit flies. The *Drosophila* fourth chromosome is unusual because it shows no recombination. The entire chromosome is inherited like a single gene.

Andrew Berry and colleagues (1991) sequenced a 1.1-kb region of the fourth chromosome in 10 *Drosophila melanogaster* and 9 *D. simulans*. The region includes the introns and exons of a gene that is expressed in fly embryos and called *cubitus interruptus Dominant* (*ciD*). Within it Berry et al. found no differences whatsoever among the *D. melanogaster* individuals surveyed. The entire 1.1 kb of sequence was identical in all 10 individuals. Among the *D. simulans* they found only one base difference. In other words, there was almost no polymorphism in this region. In contrast, when the researchers compared the sequences between the two species, they found 54 substitutions.

Other chromosomes surveyed in the same individuals showed normal amounts of polymorphism. These latter data serve as a control and confirm that the lack of variation in and around the *ciD* locus is not caused by an unusual sample of individuals. Rather, there is something unusual about the fourth chromosome.

Berry et al. suggest that recent selective sweeps cleaned out all or most of the variation on the fourth chromosome in each species. An advantageous mutation anywhere on the fourth chromosome would eliminate all within-species polymorphism as it rose to fixation. New variants, like the one polymorphism observed in the *D. simulans* sampled, will arise only through mutation. In this way, selective sweeps leave a footprint in the genome: a striking lack of polymorphism within linkage groups. Similar footprints have been found in other chromosomal regions where the frequency of recombination is low, including the ZFY locus of the human Y chromosome (Dorit et al. 1995) and a variety of loci in *D. melanogaster* and other flies (for example, see Nurminsky et al. 1998).

Selection at nearby sites can influence the evolutionary fate of synonymous mutations.

Has hitchhiking produced all of these regions of reduced polymorphism? Probably not. Another process, called **background selection,** can produce a similar pattern (Charlesworth et al. 1993). Background selection results from negative selection against deleterious mutations, rather than positive selection for advantageous mutations. Like hitchhiking, it occurs in regions of reduced recombination. The idea here is that selection against deleterious mutations removes closely linked neutral mutations and yields a reduced level of polymorphism.

Although hitchhiking and background selection are not mutually exclusive, their effects can be distinguished in at least some cases. Hitchhiking results in dramatic reductions in polymorphism as an occasional advantageous mutation quickly sweeps through a population. Background selection causes a slow, steady decrease in polymorphism as frequent deleterious mutations remove individuals from the population.

Status of the Neutral Theory

The neutral theory of molecular evolution explains the clocklike evolution of nucleotide sequences we saw in Figures 7.24 and 7.26. It also explains why silent substitutions outnumber replacement substitutions in most genes, as we saw in Figures 7.24 and 7.25. And the neutral theory serves as a null hypothesis that allows researchers to identify examples of positive selection on nucleotide sequences, as illustrated in Figure 7.29. By all these criteria, the neutral theory of molecular evolution is extraordinarily useful.

As a null hypothesis for detecting positive selection in molecular evolution, the neutral theory has been highly successful.

What about the theory's fundamental claim that the vast majority of nucleotide changes that become fixed in populations are selectively neutral and that molecular evolution is largely due to genetic drift? To assess this claim, we need (1) data for as many substitutions as possible in as many species as possible, and (2) a breakdown of the proportion of substitutions that are neutral versus deleterious versus beneficial. The data we need are accumulating. To assemble the information summarized in **Figure 7.32**, Justin Fay combed the literature to compile estimates of α, the fraction of amino acid substitutions driven by positive selection. He included data on 38 species for which multiple genes have been studied.

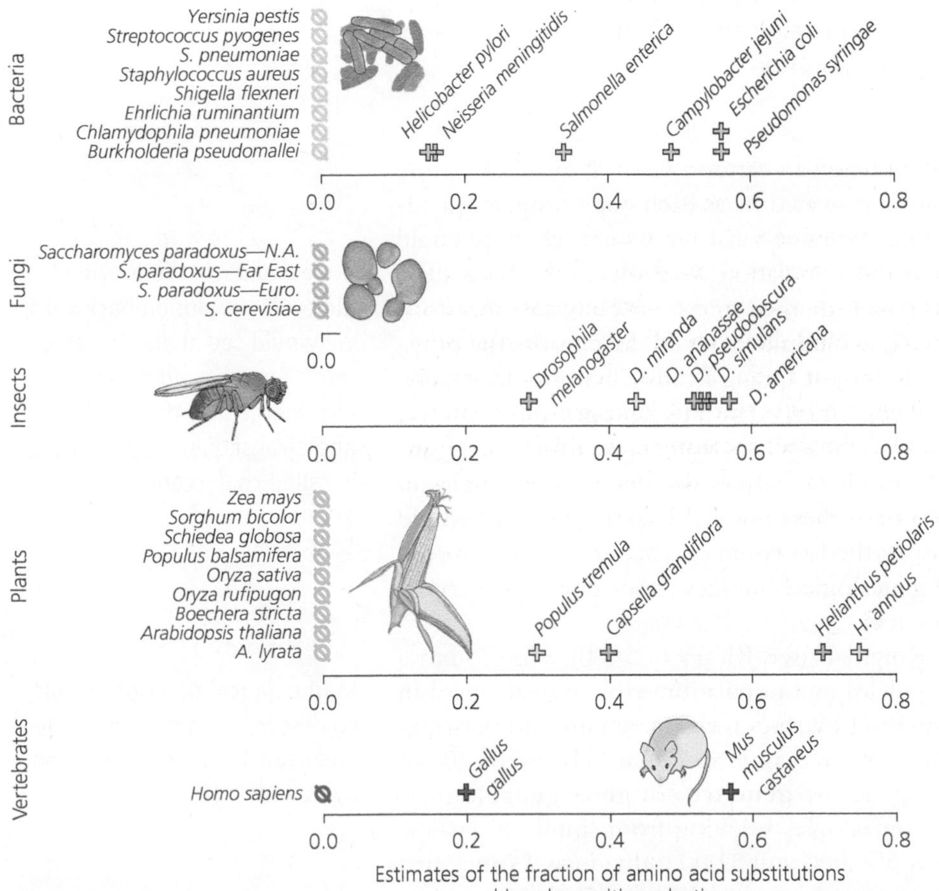

Figure 7.32 Estimates for different species of the fraction of amino acid substitutions driven by positive selection Null signs indicate a lack of statistically significant evidence, based on the McDonald–Kreitman test, for positive selection. Light plus signs indicate that there is conflicting evidence from different studies. The agents of selection are generally unknown. Data from Fay (2011).

Alpha can be estimated from a McDonald–Kreitman test as the elevation of $\frac{d_N}{d_S}$ above $\frac{p_N}{p_S}$:

$$\alpha = 1 - \frac{\left(\frac{p_N}{p_S}\right)}{\left(\frac{d_N}{d_S}\right)}$$

Plus signs appear in the graph only for those species for which the McDonald–Kreitman test gave statistically significant evidence of positive selection.

At first glance, the data appear to refute—at least for some species—the neutral theory's claim that selectively neutral mutations that rise to fixation by drift vastly outnumber beneficial mutations that rise to fixation by natural selection. Fay argues, however, that it is too early to draw such a conclusion. The McDonald–Kreitman test does not distinguish between positive selection and other mechanisms that can lead to elevated levels of nonsynonymous divergence between species (see also Hughes 2007; Nei et al. 2010). One alternative is reduced population size, which can lead to fixation by drift of mildly deleterious mutations. Another is hitchhiking. If an unknown number of linked deleterious substitutions ride to fixation with a single positively selected one, the true proportion of substitutions driven by positive selection is obscured.

Data are now accumulating that will allow researchers to evaluate the neutral theory's claim that most molecular evolution is dominated by negative selection and drift. For now, the issue is undecided.

The jury is still out on the neutral theory's fundamental claim.

Coalescence

Before closing our discussion of genetic drift and molecular evolution, we want to mention another area of research in which sequence data and the null model of genetic drift are being put to productive use. This is the study of coalescence. Here we consider coalescence as a tool for estimating effective population size, although it has a great variety of other applications.

Coalescence Defined

Figure 7.21 showed an evolving population in cartoon form. New alleles arose by mutation and became more common over time as each copy propagated additional copies into future generations. Imagine what we would see if we could reverse the flow of time and watch the population de-evolve. The blue allele would become rarer as descendent copies merged into their common ancestors. So, too, would the dark green allele. The blue allele would disappear as the original copy merged into the dark green copy it sprang from. Then the dark green allele would disappear as the original copy merged into its light green progenitor.

If we could run the movie of molecular evolution backward, we would see alleles become less divergent and eventually merge into their common ancestral sequence. This process is called coalescence.

Now imagine that we have a sample of real alleles from a population of organisms. Each represents an unbroken lineage of copies descended from copies in ever earlier generations. If we could trace these lineages back in time, we would see them merge until only one lineage, the last common ancestor of our sampled alleles, remained. The merging of genealogical lineages as we trace allele copies backward in time is called **coalescence.**

The term was coined by John Kingman (see Kingman 2000), who found a way to simulate the coalescence of alleles in a population evolving backward in time by genetic drift. Among his method's virtues is that it requires no information about the rest of the population other than its size (see Felsenstein 2004). The result is an evolutionary tree of genes—a **gene tree** or **gene genealogy.**

Mathematical descriptions of coalescence provide an efficient means of simulating evolution by genetic drift.

Figure 7.33a shows several gene genealogies resulting from simulated coalescence of seven alleles in populations of 1,000 and 5,000 individuals. Notice first that every one of the simulated gene trees is unique. We are modeling genetic

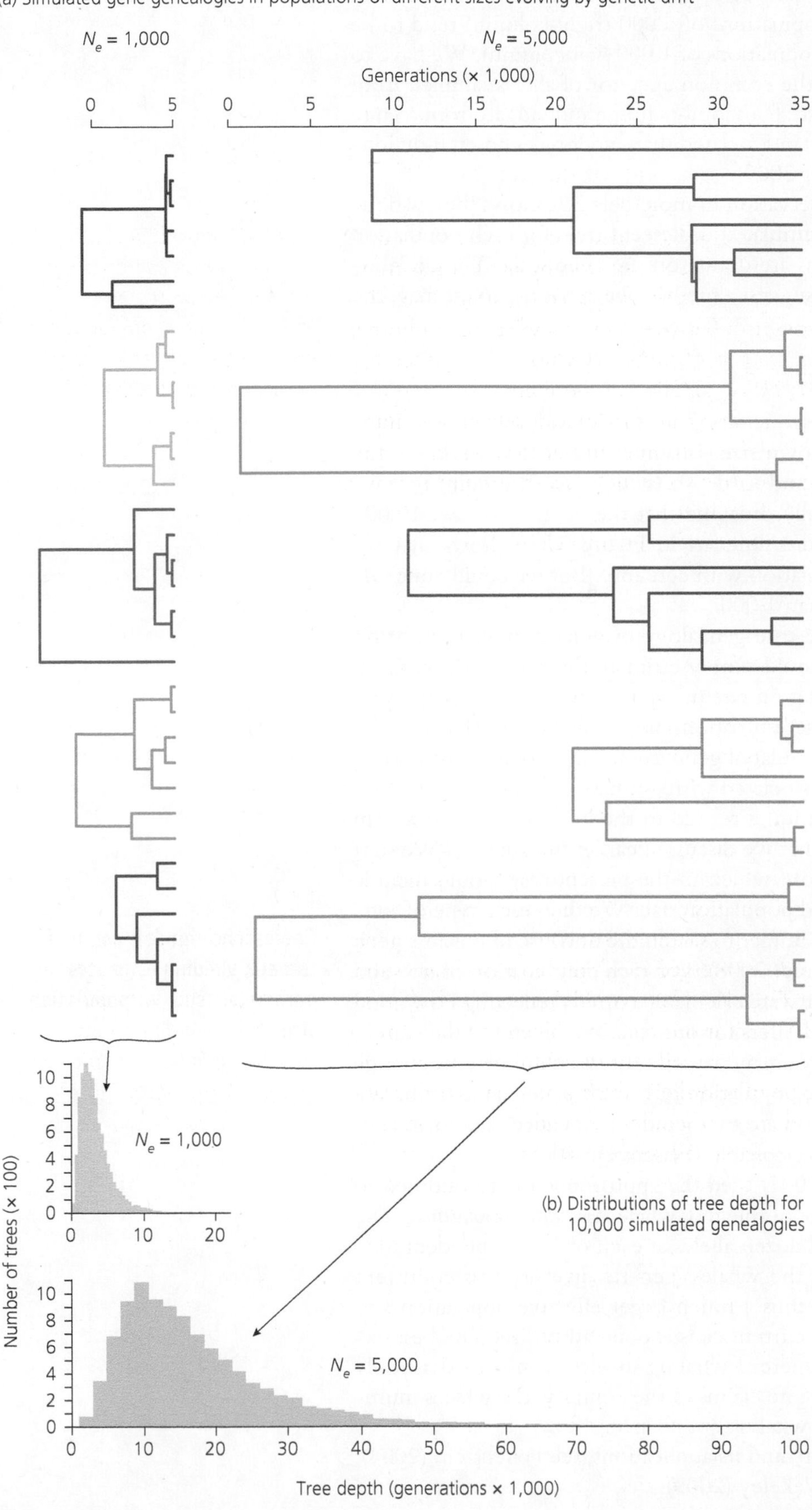

Figure 7.33 Gene genealogies produced by simulation of coalescence (a) Examples of genealogies produced by simulating genetic drift running backward in time. The five trees in the column on the left are examples of the results from simulating the coalescence of seven alleles in a population of 1,000 individuals. The five trees in the column on the right are examples of the results from simulating the coalescence of seven alleles from a population of 5,000 individuals. All trees are drawn to the same scale. (b) Distributions of tree depth, or time (in generations) back to the most recent common ancestor, for 10,000 simulated trees from a population of 1,000 and a population of 5,000. Simulations performed and distributions generated by *Mesquite* (Maddison 2011; Maddison and Maddison 2011). Trees drawn by PHYLIP (Felsenstein 2009a).

drift, so the differences among trees are due to chance events. Second, notice that the coalescent trees for alleles in populations of 5,000 (right column) tend to be deeper than the trees for alleles in populations of 1,000 (left column). We have to travel further back in time to find the common ancestor of alleles sampled from a large population. This makes sense. Randomly chosen individuals from a large population are likely to be more distantly related than randomly chosen individuals from a small population (Kuhner 2009).

Figure 7.33b documents this observation in more detail. It shows the distribution of tree depths among 10,000 simulated coalescent trees for each population size. Like the trees, the distributions are drawn on the same scale. The distributions overlap, but they nonetheless suggest a method we could use to estimate the effective size of a real population.

Coalescence Applied

Imagine we had a sample of allelic sequences from seven randomly chosen individuals from a population of unknown size. Imagine further that we knew the true genealogy of the seven alleles and its depth (which would require that we also knew the mutation rate). Finally, imagine that the depth was, say, 10,000 generations. Comparing the two distributions in Figure 7.33b shows that we could not infer the size of our population with certainty. But we could conclude that 5,000 is a much better guess than 1,000.

Of course, we do not know the true genealogy of our seven alleles, nor do we know the mutation rate. We could simply estimate the tree (with methods discussed in Chapter 4) and the mutation rate (using data discussed in Chapter 5). It would be tempting to make these estimations and treat the results as true for purposes of comparison with our simulated gene trees. The problem with doing so is that it ignores the uncertainty associated with estimation (Felsenstein 2009b).

A better approach is to use techniques related to the likelihood and Bayesian methods for inferring phylogenies that we discussed earlier (Chapter 4). We start with a model of molecular evolution, which in the present case would include parameters for the mutation rate and population size. We then use a type of software called a coalescent genealogy sampler to search the universe of possible gene trees and parameter values (see Kuhner 2008). For each combination of tree and parameter values it considers, the software calculates a metric reflecting how good an explanation that particular model offers for our data. At the end of the search, which is long and computationally demanding, the software can give us a range of plausible values for the size of the population our set of alleles came from. We can increase the accuracy of our estimate by including sequence data for sets of alleles at as many independent loci as possible (Felsenstein 2006).

Coalescence models can be fit to data, yielding estimates for parameters such as population size.

Elizabeth Alter and colleagues (2007) used the approach we have outlined to estimate the effective population size of the gray whale (*Eschrichtius robustus*). The researchers analyzed data for several dozen alleles at each of 10 independent loci. Coalescence analysis indicated that the whales' genetic diversity was consistent with deeper gene genealogies, and thus a much larger effective population size than would be expected from their current census population size. The best explanation is the obvious one. Commercial whaling in the recent past drastically reduced the population size and, despite claims to the contrary, the whales' numbers have yet to return to their pre-whaling abundance.

For more on the coalescent theory and its applications, see Felsenstein (2004). For a book-length treatment, see Wakeley (2009).

7.4 Nonrandom Mating

We have so far considered what happens in populations when we relax the assumptions of no migration and no genetic drift. The final assumption of the Hardy–Weinberg analysis is that individuals in the population mate at random. In this section, we relax that assumption and allow individuals to mate nonrandomly. Nonrandom mating does not, by itself, cause evolution. Nonrandom mating can nonetheless have profound indirect effects on evolution.

The most common type of nonrandom mating, and the kind we focus on here, is **inbreeding**. Inbreeding is mating among genetic relatives. The effect of inbreeding on the genetics of a population is to increase the frequency of homozygotes compared to what is expected under Hardy–Weinberg assumptions.

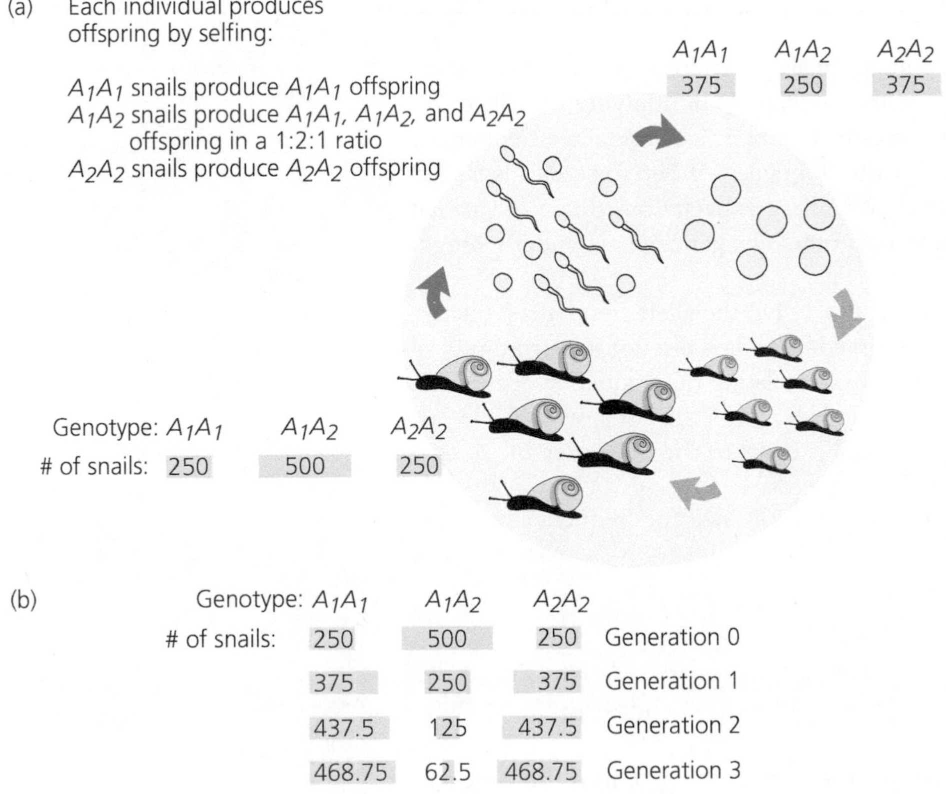

Figure 7.34 Inbreeding alters genotype frequencies (a) This figure follows the genotype frequencies in an imaginary population of 1,000 snails from one generation's adults (lower left) to the next generation's zygotes (upper right). The frequencies of both allele A_1 and A_2 are 0.5. The colored bars show the number of individuals with each genotype. Every individual reproduces by selfing. Homozygotes produce homozygous offspring and heterozygotes produce both heterozygous and homozygous offspring, so the frequency of homozygotes goes up and the frequency of heterozygotes goes down. (b) These bar charts show what will happen to the genotype frequencies if this population continues to self for two more generations. The tan portions of the bars show the decrease in heterozygosity and the increase in homozygosity due to inbreeding.

To see how this happens, consider the most extreme example of inbreeding: self-fertilization, or selfing. Imagine a population in Hardy–Weinberg equilibrium with alleles A_1 and A_2 at frequencies of 0.5 each. The frequency of A_1A_1 individuals is 0.25, that of A_1A_2 individuals is 0.5, and that of A_2A_2 individuals is 0.25 **(Figure 7.34a)**. Imagine there are 1,000 individuals in the population: 250 A_1A_1, 500 A_1A_2, and 250 A_2A_2. If all the individuals reproduce by selfing, homozygous parents will produce all homozygous offspring while heterozygous parents will produce half homozygous and half heterozygous offspring. Among 1,000 offspring in our population, there will be 375 A_1A_1, 250 A_1A_2, and 375 A_2A_2. If selfing continues for two more generations, then, among every 1,000 individuals in the final generation, there will be 468.75 homozygotes of each type and 62.5 heterozygotes (Figure 7.34b). The frequency of heterozygotes has been halved every generation, and the frequency of homozgyotes has increased.

Inbreeding decreases the frequency of heterozygotes and increases the frequency of homozygotes compared to expectations under Hardy–Weinberg assumptions.

Table 7.1 Changes in genotype frequency with selfing

The frequency of allele A_1 is p and the frequency of allele A_2 is q. Note that allele frequencies do not change from generation to generation—only the genotype frequencies. After Crow (1983).

Generation	Frequency of A_1A_1	A_1A_2	A_2A_2
0	p^2	$2pq$	q^2
1	$p^2 + (pq/2)$	pq	$q^2 + (pq/2)$
2	$p^2 + (3pq/4)$	$pq/2$	$q^2 + (3pq/4)$
3	$p^2 + (7pq/8)$	$pq/4$	$q^2 + (7pq/8)$
4	$p^2 + (15pq/16)$	$pq/8$	$q^2 + (15pq/16)$

Conclusion 2 of the Hardy–Weinberg analysis is violated when individuals self: We cannot predict the genotype frequencies by multiplying the allele frequencies. Note that in generation three, in Figure 7.34b, the allele frequencies are still 0.5 for A_1 and 0.5 forA_2. Yet the frequency of heterozygotes is far less than 2(0.5)(0.5). Compared to Hardy–Weinberg expectations, there is a deficit of heterozygotes and an excess of homozygotes. The general case under selfing is shown algebraically in **Table 7.1.**

What about Hardy–Weinberg conclusion 1? Do the allele frequencies change from generation to generation under inbreeding? They did not in our numerical example. We can check the general case by calculating the frequency of allele A_1 in the gene pool produced by the population shown in the last row of Table 7.1. The frequency of allele A_1 in the gene pool is equal to the frequency of

A_1A_1 adults in the population $\left(= p^2 + \dfrac{15pq}{16}\right)$ plus half the frequency of

$A_1A_2\left(= \dfrac{1}{2}\left[\dfrac{pq}{8}\right]\right)$. That gives

$$p^2 + \frac{15pq}{16} + \frac{1}{2}\left[\frac{pq}{8}\right] = p^2 + \frac{15pq}{16} + \frac{pq}{16} = p^2 + pq$$

Now substitute $(1 - p)$ for q to give $p^2 + p(1 - p) = p$. This is the same frequency for allele A_1 that we started out with at the top of Table 7.1. Although inbreeding does cause genotype frequencies to change from generation to generation, it does not cause allele frequencies to change. Inbreeding by itself, therefore, is not a mechanism of evolution. As we will see, however, inbreeding can have important evolutionary consequences.

Empirical Research on Inbreeding

Because inbreeding can produce a large excess of homozygotes, Hardy–Weinberg analysis can be used to detect inbreeding in nature. As an example, we consider research on California sea otters (*Enhydra lutris*). Sea otters **(Figure 7.35)** were once abundant on the West Coast of North America from Alaska to Baja California. They were nearly wiped out, however, by the fur trade in the 18th and 19th centuries. At its lowest, the California otter population numbered fewer than 50 individuals (Lidicker and McCollum 1997). The good news is that since

Figure 7.35 A sea otter feeding in a kelp bed near Monterey, California

they were placed under protection in 1911, the California otters have been making a comeback. By the end of the 20th century, there were some 1,500 of them.

Because of the bottleneck, the California otter population harbors less genetic diversity than before the fur hunters arrived (Larson et al. 2002). William Lidicker and F. C. McCollum (1997) investigated whether the reduced size and density of the otter population also led to inbreeding.

Lidicker and McCollum determined the genotypes of a number of California otters for each of 31 allozyme loci. One of them was the PAP locus (1-phenylalanyl-1-proline peptidase), for which the otter population harbored two alleles: *S* (for slow) and *F* (for fast). Among a sample of 33 otters, the number of individuals with each genotype were

SS	*SF*	*FF*
16	7	10

The sample of 33 otters includes 66 alleles. The frequencies of *S* and *F* are

$$S: \quad \frac{2(16) + 7}{66} \approx 0.6 \qquad F: \quad \frac{7 + 2(10)}{66} \approx 0.4$$

If the otter population were in Hardy–Weinberg equilibrium, the genotype frequencies would be

$$SS: \ (0.6)^2 = 0.36 \qquad SF: \ 2(0.6)(0.4) = 0.48 \qquad FF: \ (0.4)^2 = 0.16$$

Data revealing a deficit of heterozygotes and an excess of homozygotes may be evidence of inbreeding.

The actual frequencies, however, were

$$SS: \ \frac{16}{33} = 0.485 \qquad SF: \ \frac{7}{33} = 0.212 \qquad FF: \ \frac{10}{33} = 0.303$$

There are more homozygotes and fewer heterozygotes than expected in a population where individuals are mating at random. Lidicker and McCollum also

Table 7.2 Observed and expected number of heterozygotes for California and Alaska sea otters

The numbers given here are means across 31 loci for 74 otters from California and 9 from Alaska. For each population, the observed number of individuals with a particular kind of genotype is compared to the number expected under Hardy–Weinberg conditions of random mating and no mutation, selection, migration, or genetic drift.

	California	Alaska
Heterozygotes observed	**4.6%**	**6.8%**
Heterozygotes expected	**7.2%**	**7.7%**

Source: Lidicker and McCollum (1997).

determined the PAP genotypes of six sea otters from Alaska, where the population experienced a less severe bottleneck. Their sample size was small, but the Alaskan otters showed no evidence of missing heterozygotes (1 had genotype *SS*; 3 had *SF*; 2 had *FF*).

Table 7.2 gives the mean frequencies of heterozygotes across all 31 loci for all the otters Lidicker and McCollum examined. The overall results are consistent with the results for the PAP locus. The California otter population shows a substantial deficit of heterozygotes. This is consistent with inbreeding.

Strictly speaking, the excess of homozygotes shows only that one or more of the Hardy–Weinberg assumptions are being violated in the otter population. In principle, a deficit of heterozygotes could result from selection against them and in favor of homozygotes. The appearance of a heterozygote deficit could also arise if the California otters, which Lidicker and McCollum treated as a single population, actually comprise two separate populations with different allele frequencies. Lidicker and McCollum consider these alternative explanations, however, and conclude that inbreeding is more plausible. They recommend that recovering otter populations be monitored for evidence of inbreeding depression, a phenomenon we discuss later in this section.

General Analysis of Inbreeding

So far our treatment of inbreeding has been limited to self-fertilization and sibling mating. But inbreeding can also occur as matings among more distant relatives, such as cousins. Inbreeding that is less extreme than selfing produces the same effect as selfing—it increases the proportion of homozygotes—but at a slower rate. For a general mathematical treatment of inbreeding, population geneticists use a conceptual tool called the **coefficient of inbreeding.** This quantity is symbolized by F, and is defined as the probability that the two alleles in an individual are identical by descent (meaning that both alleles came from the same ancestor allele in some previous generation). **Computing Consequences 7.6** shows that in an inbred population that otherwise obeys Hardy–Weinberg assumptions, the genotype frequencies are

A_1A_1	A_1A_2	A_2A_2
$p^2(1 - F) + pF$	$2pq(1 - F)$	$q^2(1 - F) + qF$

The reader can verify these expressions by substituting the values $F = 0$, which gives the original Hardy–Weinberg genotype ratios, and $F = 0.5$, which represents selfing and gives the ratios shown for generation 1 in Table 7.1.

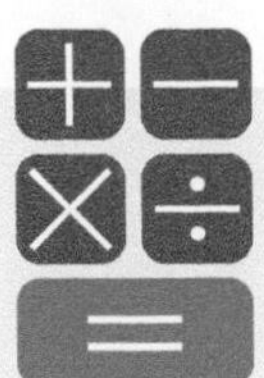

COMPUTING CONSEQUENCES 7.6

Genotype frequencies in an inbred population

Here we add inbreeding to the Hardy–Weinberg analysis. Imagine a population with two alleles at a single locus: A_1 and A_2, with frequencies p and q. We can calculate the genotype frequencies in the next generation by letting gametes find each other in the gene pool, as we would for a random mating population. The twist added by inbreeding is that the gene pool is not thoroughly mixed. Once we have picked an egg to watch, for example, we can think of the sperm in the gene pool as consisting of two fractions: a fraction $(1 - F)$ carrying alleles that are not identical by descent to the one in the egg, and the fraction F carrying alleles that are identical by descent to the one in the egg (because they were produced by relatives of the female that produced the egg). The calculations of genotype frequencies are as follows:

- A_1A_1 homozygotes: There are two ways we might witness the creation of an A_1A_1 homozygote. The first way is that we pick an egg that is A_1 (an event with probability p) and watch it get fertilized by a sperm that is A_1 by chance, rather than by common ancestry. The frequency of unrelated A_1 sperm in the gene pool is $p(1 - F)$, so the probability of getting a homozygote by chance is

  $$p \times p(1 - F) = p^2(1 - F)$$

  The second way to get a homozygote is to pick an egg that is A_1 (an event with probability p) and watch it get fertilized by a sperm that is A_1 because of common ancestry (an event with probability F). The probability of getting a homozygote this way is pF. The probability of getting an A_1A_1 homozygote by either the first way or the second way is the sum of their individual probabilities:

  $$p^2(1 - F) + pF$$

- A_1A_2 heterozygotes: There are two ways we might get an A_1A_2 heterozygote. The first way is to pick an egg that is A_1 (an event with probability p) and watch it get fertilized by an unrelated sperm that is A_2. The frequency of A_2 unrelated sperm is $q(1 - F)$, so the probability of getting a heterozygote this first way is $pq(1 - F)$. The second way is to pick an egg that is A_2 (probability: q) and watch it get fertilized by an unrelated sperm that is A_1 [probability: $p(1 - F)$]. The probability of getting a heterozygote the second way is $qp(1 - F)$. The probability of getting a heterozygote by either the first way or the second way is the sum of their individual probabilities:

  $$pq(1 - F) + qp(1 - F) = 2pq(1 - F)$$

- A_2A_2 homozygotes: We can get an A_2A_2 homozygote either by picking an A_2 egg (probability: q) and watching it get fertilized by an unrelated A_2 sperm [probability: $q(1 - F)$], or by picking an A_2 egg (probability: q) and watching it get fertilized by a sperm that is A_2 because of common ancestry (probability: F). The overall probability of getting an A_2A_2 homozygote is

  $$q^2(1 - F) + qF$$

The reader may wish to verify that the genotype frequencies sum to 1.

The same logic applies when many alleles are present in the gene pool. Then, the frequency of any homozygote A_iA_i is given by

$$p_i^2(1 - F) + p_iF$$

and the frequency of any heterozygote A_iA_j is given by

$$2p_ip_j(1 - F)$$

where p_i is the frequency of allele A_i and p_j is the frequency of allele A_j.

The last expression states that the fraction of individuals in a population that are heterozygotes (that is, the population's heterozygosity) is proportional to $(1 - F)$. If we compare the heterozygosity of an inbred population, H_F, with that of a random mating population, H_0, then the relationship will be

$$H_F = H_0(1 - F)$$

Anytime F is greater than 0, the frequency of heterozygotes is lower in an inbred population than it is in a random mating population.

Computing *F*

To measure the degree of inbreeding in actual populations, we need a way to calculate F. Doing this directly requires a pedigree—a diagram showing the genealogical relationships of individuals. **Figure 7.36a** shows a pedigree leading to a focal female who is the daughter of half-siblings. She is inbred because her parents share a common ancestor in her grandmother. For the focal female to have gene copies that are identical by descent, the following would have to have happened (reading clockwise from the focal female): The female's mother passed to her, via the egg, a copy of the same gene copy the mother received from the grandmother (an event with probability $\frac{1}{2}$); the father received from the grandmother a copy of the same gene copy the mother received from the grandmother (probability $\frac{1}{2}$); the focal female received from her father, via the sperm, a copy of the same gene copy the father received from the grandmother (probability $\frac{1}{2}$). F is the probability of all three events happening together, or $(\frac{1}{2})^3 = \frac{1}{8}$. Figure 7.36b shows that for an offspring of full sibs, there are two loops passing through a common ancestor, each with three internal links. F in this case is thus $\frac{1}{8} + \frac{1}{8} = \frac{1}{4}$.

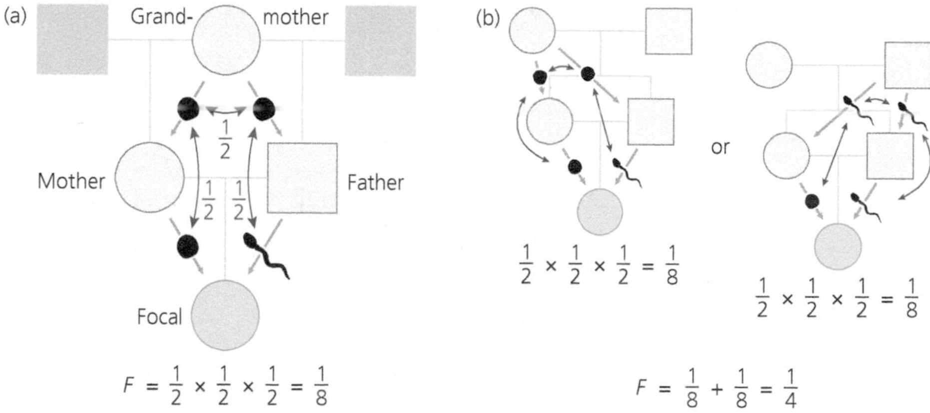

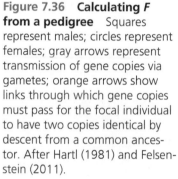

Figure 7.36 Calculating *F* from a pedigree Squares represent males; circles represent females; gray arrows represent transmission of gene copies via gametes; orange arrows show links through which gene copies must pass for the focal individual to have two copies identical by descent from a common ancestor. After Hartl (1981) and Felsenstein (2011).

Inbreeding Depression

Although inbreeding does not directly change allele frequencies, it can still affect evolution. Among the most important consequences of inbreeding for evolution is **inbreeding depression** (see Charlesworth and Willis 2009).

Inbreeding may lead to reduced mean fitness if it generates offspring homozygous for deleterious alleles.

Inbreeding depression usually results from the exposure of deleterious recessive alleles to selection. To see how this works, consider the extreme case illustrated by loss-of-function mutations. These alleles are often recessive, because a single wild-type allele can still generate enough functional protein, in most instances, to produce a normal phenotype. Even though they may have no fitness consequences at all in heterozygotes, loss-of-function mutations can be lethal in homozygotes. By increasing the proportion of individuals in a population that

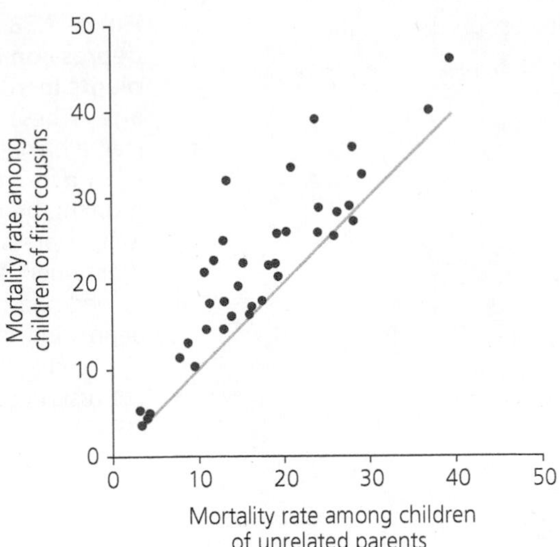

Figure 7.37 Inbreeding depression in humans Each dot represents childhood mortality rates for a human population. The horizontal axis represents mortality rates for children of unrelated parents; the vertical axis represents mortality rates for children of first cousins. The gray line shows where the points would fall if mortality rates for the two kinds of children were equal. Plotted from data in Bittles and Neel (1994).

are homozygotes, inbreeding increases the frequency with which deleterious recessives affect phenotypes. Inbreeding depression refers to the effect these alleles have on the average fitness of offspring in the population.

Studies on humans have shown that inbreeding does, in fact, expose deleterious recessive alleles, and data from numerous studies consistently show that children of first cousins have higher mortality rates than children of unrelated parents **(Figure 7.37)**. Strong inbreeding depression has also been frequently observed in captive populations of animals (for example, Hill 1974; Ralls et al. 1979).

Perhaps the most powerful studies of inbreeding depression in natural populations concern flowering plants, in which the inbreeding can be studied experimentally. In many angiosperms, selfed and outcrossed offspring can be produced from the same parent through hand pollination. In experiments like these, inbreeding depression can be defined as

$$\delta = 1 - \frac{w_s}{w_o}$$

where w_s and w_o are the fitnesses of selfed and outcrossed progeny, respectively. This definition makes levels of inbreeding depression comparable across species. Three patterns are starting to emerge from experimental studies.

First, inbreeding effects are often easiest to detect when plants undergo some sort of environmental stress. For example, when Michele Dudash (1990) compared the growth and reproduction of selfed and outcrossed rose pinks (*Sabatia angularis*), the plants showed some inbreeding depression when grown in the greenhouse or garden, but their performance diverged more strongly when they were planted in the field. Lorne Wolfe (1993) got a similar result with a waterleaf (*Hydrophyllum appendiculatum*): Selfed and outcrossed individuals had equal fitness when grown alone, but differed significantly when grown under competition. And in the common annual called jewelweed (*Impatiens capensis*), McCall et al. (1994) observed the strongest inbreeding effects on survival when an unplanned insect outbreak occurred during the course of their experiment.

Second, inbreeding effects are more likely to show up later in the life cycle **(Figure 7.38, next page)**—not, for example, during the germination or seedling stage. Why? Wolfe (1993) suggests that maternal effects—specifically, the seed mother's influence on offspring phenotype through provisioning of seeds—can mask the influence of deleterious recessives until later in the life cycle.

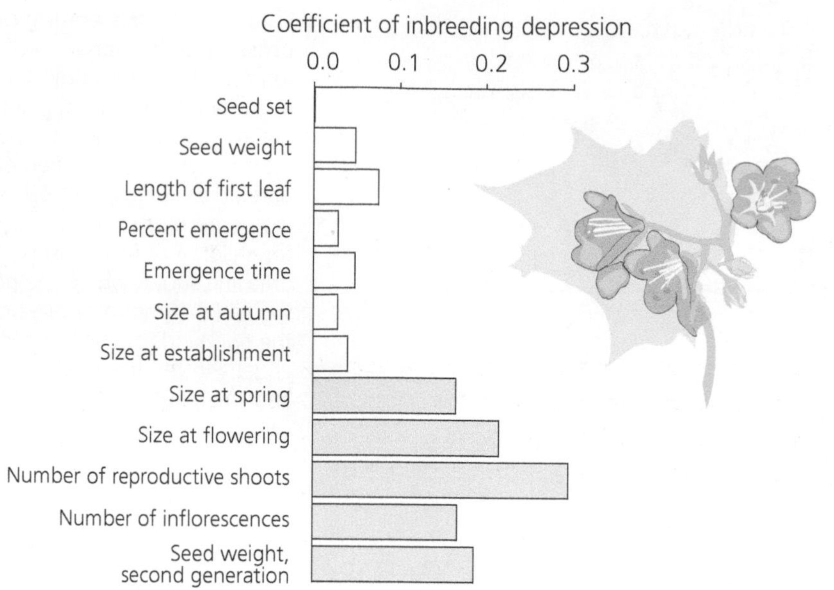

Figure 7.38 Inbreeding depression in flowering plants increases as individuals age These data are for waterleaf, a biennial. The open bars show data from the first year of growth; the filled bars indicate traits expressed in the second year (when the plants mature, flower, and die). Inbreeding depression is much more pronounced in the second year than the first. Redrawn from Wolfe (1993).

Third, inbreeding depression varies among family lineages. Michele Dudash and colleagues (1997) compared the growth and reproductive performance of inbred versus outcrossed individuals from each of several families in two annual populations of the herb *Mimulus guttatus*. Some families showed inbreeding depression; others showed no discernible effect of type of mating; still others showed improved performance under inbreeding.

Inbreeding depression has been documented in natural populations of animals as well. Long-term studies in two separate populations of a bird called the great tit (*Parus major*) have shown that inbreeding depression can have strong effects on reproductive success. When Paul Greenwood and coworkers (1978) defined inbred matings as those between first cousins or more closely related individuals, they found that the survival of inbred nestlings was much lower than that of outbred individuals. Similarly, A. J. van Noordwijk and W. Scharloo (1981) showed that in an island population of tits, there is a strong relationship between the level of inbreeding in a pair and the number of eggs in a clutch that fail to hatch **(Figure 7.39)**. More recently, Keller et al. (1994) found that outbred individuals in a population of song sparrows in British Columbia, Canada, were much more likely than inbred individuals to survive a severe winter.

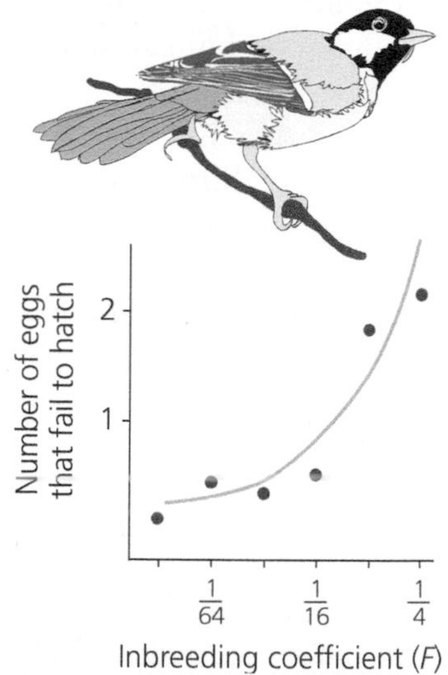

Figure 7.39 Inbreeding increases egg failure in great tits From van Noordwijk and Scharloo (1981).

Given the theory and data we have reviewed on inbreeding depression, it is not surprising that animals and plants have evolved mechanisms to avoid it. Mechanisms of inbreeding avoidance include mate choice, genetically controlled self-incompatibility, and dispersal. But under some circumstances, inbreeding may be unavoidable. In small populations, for example, the number of potential mates for any particular individual is limited. If a population is small and remains small for many generations, and if the population receives no migrants from other populations, then eventually all the individuals in the population will be related to each other even if mating is random. Thus small populations eventually become inbred, and the individuals in them may suffer inbreeding depression. This can be a problem for rare and endangered species, and it creates a challenge for the managers of captive breeding programs, as we describe in Section 7.5.

In summary, nonrandom mating does not, by itself, alter allele frequencies. It is not, therefore, a mechanism of evolution. Nonrandom mating does, however, alter the frequencies of genotypes. It can thereby change the distribution of phenotypes in a population and alter the pattern of natural selection and the evolution of the population. For example, inbreeding increases the frequency of homozygotes and decreases the frequency of heterozygotes. This can expose deleterious recessive alleles to selection, leading to inbreeding depression.

7.5 Conservation Genetics of the Florida Panther

We opened this chapter with the case of the Florida panther, a once abundant big cat that in the mid-1990s appeared to be destined for extinction. Like a great many other vulnerable and endangered species, the panther's worst enemies are humans with plows, bulldozers, and guns. Yet habitat loss and hunting are not the panther's only problem.

The cat was placed under the protection of the State of Florida in 1958 and listed as endangered by the federal government in 1967 (Pimm et al. 2006). In the 1980s, after a time during which the panther was thought to be extinct, government and citizen groups sought to aid the panther's recovery by protecting additional habitat, changing the way its prey are managed, and building highway underpasses to reduce road deaths (Culver et al. 2008; Johnson et al. 2010). Nonetheless, the panther's population size hovered at less than three dozen from the mid-1989s to the mid-1990s. Something else was now threatening the survival of the Florida panther, but what?

Our discussion of migration, genetic drift, and nonrandom mating has given us the tools to understand the likely answer. Human activity did two things to the panther. First, it directly reduced the size of the cat's population. Second, it isolated the cat from other puma populations that it is closely related to—and that it once interbred with.

A small population with little or no gene flow is precisely the place where genetic drift is most influential. And genetic drift results in random fixation and declining heterozygosity. If some of the alleles that become fixed are deleterious recessives, then the average fitness of individuals will be reduced. A reduction in fitness due to genetic drift is reminiscent of inbreeding depression. In fact, it is inbreeding depression. Reduced heterozygosity due to drift and increased homozygosity due to inbreeding are two sides of the same coin. In a small population all individuals are related, and there is no choice but to mate with kin.

Michael Lynch and Wilfried Gabriel (1990) have proposed that an accumulation of deleterious recessives (a phenomenon known as genetic load) can lead to the extinction of small populations. They noted that when exposure of deleterious mutations produces a reduction in population size, the effectiveness of drift is increased. The speed and proportion of deleterious mutations going to fixation subsequently increases, which further decreases population size. Lynch and Gabriel termed this synergistic interaction between mutation, population size, and drift a "mutational meltdown."

The Florida panther appeared to be trapped in just such a scenario. As the population dwindled, the cats began to display conditions we mentioned in the introduction—heart defects, low sperm counts, and susceptibility to infection—that looked like symptoms of inbreeding depression. This inbreeding depression reduced individual reproductive success and caused the remnant population to

A loss of allelic diversity under genetic drift appears to have caused inbreeding depression in Florida panthers.

continue its decline. The continued decline in population size led to even more drift, which led to worse inbreeding depression, and so on. The cats had fallen into an "extinction vortex" (see Soulé and Mills 1998).

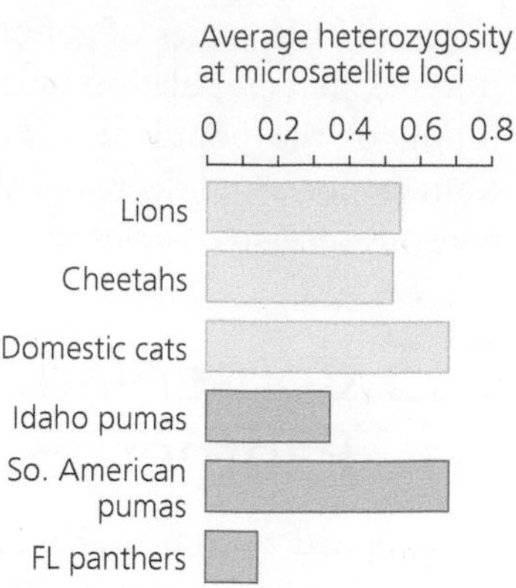

Figure 7.40 Genetic variation in Florida panthers relative to other puma populations and other cats Drawn from data in Driscoll et al. (2002).

To test this hypothesis, Carlos Driscoll and colleagues (2002) assessed the genetic diversity of Florida panthers relative to other puma populations and other species of cats at several different kinds of loci. **Figure 7.40** shows a typical result. Florida panthers have substantially lower heterozygosity than other populations or species of cat.

Philip Hedrick and colleagues (Culver et al. 2008) compared the genetic variation of present-day Florida panthers to that of museum specimens collected in the 1890s. Although their sample sizes were small, their results were consistent with Driscoll's. Present-day panthers have microsatellite heterozygosities roughly a third of those shown by museum specimens.

Hedrick and colleagues solved Sewall Wright's equation for the decline in heterozygosity across generations for N_e, plugged in the heterozygosities from 1890 and today along with generation times ranging from 4 to 6 years, and calculated how small the effective population size must be to reduce heterozygosity by two-thirds over the course of a century. The answer was fewer than 10. If the bottleneck in the Florida panther's population size was shorter and more recent, the breeding population may at one point have consisted of just two individuals.

In sum, consistent with the extinction vortex hypothesis, the Florida panther is genetically depauperate compared to both its own ancestral population and other present-day populations.

The final test of the extinction vortex hypothesis was to use it to develop a conservation strategy. If the problem for the Florida panther is reduced genetic diversity, then the solution is gene flow. Migrants from other populations should bring with them the alleles that have been lost in Florida. Reintroduction of these lost alleles should reverse the effects of drift and eliminate inbreeding depression. Natural migration of panthers into Florida ceased long ago. But in 1995, managers trapped eight Texas pumas and released them in southern Florida.

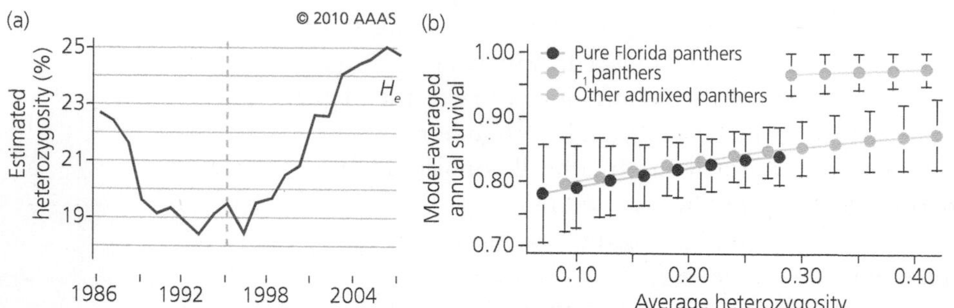

Figure 7.41 Genetic restoration of the Florida panther (a) Heterozygosity has increased since the introduction of Texas pumas. (b) So, as a result, has survival. From Johnson et al. (2010) and Benson et al. (2011).

(a) From Johnson, W. E., Onorato, D. P., et al. 2010. "Genetic restoration of the Florida panther." *Science* 329: 1641–1645. Reprinted with permission from AAAS.

The plan seems to be working. Warren Johnson and colleagues (2010) report that the Texas and Florida panthers are interbreeding, and that heterozygosity is rising **(Figure 7.41a)**. John Benson and colleagues (2011) report that higher heterozygosity has led to improved survival (Figure 7.41b). And the population has risen to over 100 cats. The Florida panther is not completely back in the woods yet, but its chances of avoiding extinction have improved.

Arranged migration of panthers from Texas to Florida appears to be replenishing the allelic diversity of the Florida population and alleviating inbreeding depression.

In this chapter and the previous one, we have traveled a great distance by analyzing evolution at one locus at a time. In the next chapter, we will begin to consider two or more loci at once.

SUMMARY

Among the important implications of the Hardy–Weinberg equilibrium principle is that natural selection is not the only mechanism of evolution. In this chapter, we examined violations of three assumptions of the Hardy–Weinberg analysis first introduced in Chapter 6 and considered their effects on allele and genotype frequencies.

Migration, in its evolutionary meaning, is the movement of alleles from one population to another. When allele frequencies are different in the source population than in the recipient population, migration causes the recipient population to evolve. As a mechanism of evolution, migration tends to homogenize allele frequencies across populations. In doing so, it may tend to eliminate adaptive differences between populations that have been produced by natural selection.

Genetic drift is evolution that occurs due to sampling error in the production of a finite number of zygotes from a gene pool. Just by chance, allele frequencies change from one generation to the next. Genetic drift is more dramatic in smaller populations than in large ones. Over many generations, drift results in an inexorable loss of genetic diversity. If some of the alleles that become fixed are deleterious recessives, genetic drift can result in a reduction in the fitness of individuals in the population.

The neutral theory of molecular evolution suggests that genetic drift is the most important mechanism of evolution at the level of DNA sequences. The neutral theory explains the clocklike evolution observed in some genes and serves as a null hypothesis for detecting the action of positive natural selection.

Nonrandom mating does not directly change allele frequencies and is thus not, strictly speaking, a mechanism of evolution. However, nonrandom mating does influence genotype frequencies. For example, inbred populations have more homozygotes and fewer heterozygotes than otherwise comparable populations in which mating is random. An increase in homozygosity often exposes deleterious recessive alleles and results in a reduction in fitness known as inbreeding depression.

As illustrated by the Florida panther, the phenomena discussed in this chapter find practical application in conservation efforts. Drift can rob small remnant populations of genetic diversity, resulting in inbreeding depression and greater risk of extinction. Migration can sometimes restore lost genetic diversity, improving a population's chances for long-term survival.

QUESTIONS

1. Conservation managers often try to purchase corridors of undeveloped habitat so that larger preserves are linked into networks. Why? What genetic goals do you think the conservation managers are aiming to accomplish?
2. The graph in **Figure 7.42** shows F_{ST}, a measure of genetic differentiation between populations as a function of geographic distance. The data are from human populations in Europe. Genetic differentiation has been calculated based on loci on the autosomes (inherited from both parents), the mitochondrial chromosome (inherited only from the mother), and the Y chromosome (inherited only from the father). Note that the patterns are different for the three different kinds of loci. Keep in mind that migration tends to homogenize allele frequencies across populations. Develop a hypothesis to explain why allele frequencies are more homogenized

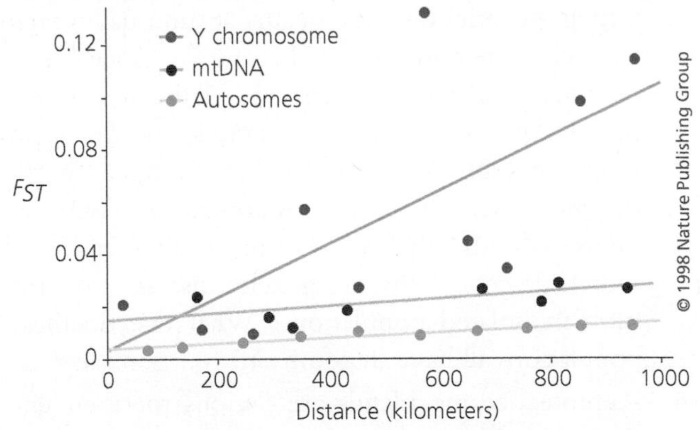

Figure 7.42 Genetic distance between human populations as a function of geographic distance Colors indicate that genetic distance (F_{ST}) has been calculated based on three different kinds of loci. From Seielstad et al. (1998).

Reprinted by permission from Macmillan Publishers Ltd: Seielstad, M. T., E. Minch, and L. L. Cavalli-Sforza. 1998. Genetic evidence for a higher female migration rate in humans. *Nature Genetics* 20: 278–280.

across populations for autosomal and mitochondrial loci than for Y-chromosome loci. Then go to the library and look up the following paper, to see if your hypothesis is similar to the one favored by the biologists who prepared the graph:

Seielstad, M. T., E. Minch, and L. L. Cavalli-Sforza. 1998. Genetic evidence for . . . in humans. *Nature Genetics* 20: 278–280. [Part of title deleted to encourage readers to develop their own hypotheses.]

3. Consider three facts: (i) Loss of heterozygosity may be especially detrimental at MHC loci, because allelic variability at these loci enhances disease resistance; (ii) Microsatellite loci show that the gray wolves on Isle Royale, Michigan, are highly inbred (Wayne et al. 1991); (iii) This wolf population crashed during an outbreak of canine parvovirus during the 1980s. How might these facts be linked? What other hypotheses could explain the data? How could you test your ideas?
4. If you were a manager charged with conserving the collared lizards of the Ozarks, one of your tasks might be to reintroduce the lizards into glades in which they have gone extinct. When reintroducing lizards to a glade, you will have a choice between using only individuals from a single extant glade population or from several extant glade populations. What would be the evolutionary consequences of each choice, for both the donor and recipient populations? Which strategy will you follow, and why?
5. Bodmer and McKie (1995) review several cases, similar to achromatopsia in the Pingelapese, in which genetic diseases occur at unusually high frequency in populations that are, or once were, relatively isolated. An enzyme deficiency called hereditary tyrosinemia, for example, occurs at an unusually high rate in the Chicoutimi region north of Quebec City in Canada. A condition called porphyria is unusually common in South Africans of Dutch descent. Why are genetic diseases so common in isolated populations? What else do these populations all have in common?
6. Remote oceanic islands are famous for their endemic species—unique forms that occur nowhere else (see Quammen 1996 for a gripping and highly readable account). Consider the roles of migration and genetic drift in the establishment of new species on remote islands.
   a. How do plant and animal species become established on remote islands? Do you think island endemics are more likely to evolve in some groups of plants and animals than others?
   b. Consider a new population that has just arrived at a remote island. Is the population likely to be large or small? Will founder effects, genetic drift, and additional waves of migration from the mainland play a relatively large or a small role in the evolution of the new island population (compared to a similar population on an island closer to the mainland)? Do your answers help explain why unusual endemic species are more common on remote islands than on islands close to the mainland?
7. By using the start codon AUG as a guidepost, researchers can determine whether substitutions in pseudogenes correspond to silent changes or replacement changes. In contrast to most other loci, the rate of silent and replacement changes is identical in pseudogenes. Explain this observation in light of the neutral theory of evolution.
8. When researchers compare a gene in closely related species, why is it logical to infer that positive natural selection has taken place if replacement substitutions outnumber silent substitutions?
9. What is codon bias? Why is the observation of nonrandom codon use evidence that certain codons might be favored by natural selection? If you were given a series of gene sequences from the human genome, how would you determine whether codon usage is random or nonrandom?
10. Sequences are now available for both the human and the chimpanzee genomes. Outline how you would analyze homologous genes in the two species to determine which of the observed sequence differences result from drift and which result from selection.
11. Recall that the fourth chromosome of *Drosophila melanogaster* does not recombine during meiosis. The lack of genetic polymorphism on this chromosome has been interpreted as the product of a selective sweep. If the fourth chromosome had normal rates of recombination, would you expect the level of polymorphism to be different? Why?
12. As we have seen, inbreeding can reduce offspring fitness by exposing deleterious recessive alleles. However, some animal breeders practice generations of careful inbreeding within a family, or "line breeding," and surprisingly many of the line-bred

animals, from champion dogs to prize cows, have normal health and fertility. How can it be possible to continue inbreeding for many generations without experiencing inbreeding depression due to recessive alleles? (Hint: Consider some of the differences between animal breeders and natural selection in the wild.) Generally, if a small population continues to inbreed for many generations, what will happen to the frequency of the deleterious recessive alleles over time?

13. In the mid-1980s, conservation biologists reluctantly recommended that zoos should not try to preserve captive populations of all the endangered species of large cats. For example, some biologists recommended ceasing efforts to breed the extremely rare Asian lion, the beautiful species seen in Chinese artwork. In place of the Asian lion, the biologists recommended increasing the captive populations of other endangered cats, such as the Siberian tiger and Amur leopard. By reducing the number of species kept in captivity, the biologists hoped to increase the captive population size of each species to several hundred, preferably at least 500. Why did the conservation biologists think that this was so important as to be worth the risk of losing the Asian lion forever?

14. In this chapter we saw that in many cases, gene frequencies in small populations change at different rates than in large populations. As a review, state how the following processes tend to vary in speed and effects in small versus large populations. (Assume the typical relationship of population size and generation time.)

Selection
Migration
Genetic drift
Inbreeding
New mutations per individual
New mutations per generation in the whole population
Substitution of a new mutation for an old allele
Fixation of a new mutation

EXPLORING THE LITERATURE

15. For a paper that explores migration as a homogenizer of allele frequencies among human populations, see:

Parra, E. J., A. Marcini, et al. 1998. Estimating African-American admixture proportions by use of population-specific alleles. *American Journal of Human Genetics* 63: 1839–1851.

For genetic analysis of a long-isolated human population with low genetic diversity, see:

Gómez-Pérez, L., M. A. Alfonso-Sánchez, et al. 2011. Alu polymorphisms in the Waorani tribe from the Ecuadorian Amazon reflect the effects of isolation and genetic drift. *American Journal of Human Biology* 23: 790–795.

16. Human genome sequences are being examined with a variety of new techniques to assess the role of positive natural selection in recent human evolution. For a start on this literature, see:

Voight, B. F., S. Kudaravalli, et al. 2006. A map of recent positive selection in the human genome. *PLoS Biology* 4 (3): e72.

17. For another example like the research on collared lizards by Templeton and colleagues (1990), in which biologists took advantage of a natural experiment to test predictions about the effect of genetic drift on genetic diversity, see:

Eldridge, M. D. B., J. M. King, et al. 1999. Unprecedented low levels of genetic variation and inbreeding depression in an island population of the black-footed rock-wallaby. *Conservation Biology* 13: 531–541.

18. We mentioned in Section 7.4 that inbreeding depression is a concern for biologists trying to conserve endangered organisms with small population sizes. Inbreeding depression turns out to vary among environments and among families. In addition, new genetic techniques are enabling more precise measures of inbreeding in wild populations that have unknown genealogies. For more information, see:

Spielman, D., B. W. Brook, and R. Frankham. 2004. Most species are not driven to extinction before genetic factors impact them. *Proceedings of the National Academy of Sciences, USA* 101 (42): 15261–15264.

Hedrick, P. W., and S. T. Kalinowski. 2000. Inbreeding depression in conservation biology. *Annual Review of Ecology, Evolution, and Systematics* 31: 139–162.

Liberg, O., H. Andren, et al. 2005. Severe inbreeding depression in a wild wolf (*Canis lupus*) population. *Biology Letters* 1: 17–20.

19. An essential step of any conservation program is to determine the minimum population size necessary to make the extinction of a species unlikely over the long term. The following papers explore this question:

Lande, R. 1995. Mutation and conservation. *Conservation Biology* 9: 782–791.

Lynch, M. 1996. A quantitative genetic perspective on conservation issues. In *Conservation Genetics: Case Histories from Nature*, ed. J. C. Avise and J. Hamrick. New York: Chapman and Hall, 471–501.

20. For another story of genetic rescue, and a realistic view of long-term prospects, see:

Bouzat, J. L., J. A. Johnson, et al. 2009. Beyond the beneficial effects of translocations as an effective tool for the genetic restoration of isolated populations. *Conservation Genetics* 10: 191–201.

21. Cheetahs have long been cited as a classic example of a species whose low genetic diversity put it at increased risk of extinction. Other researchers have debated the validity of this view. For a start on the literature, see:

Menotti-Raymond, M., and S. J. O'Brien. 1993. Dating the genetic bottleneck of the African cheetah. *Proceedings of the National Academy of Sciences, USA* 90: 3172–3176.

Merola, M. 1994. A reassessment of homozygosity and the case for inbreeding depression in the cheetah, *Acinonyx jubatus:* Implications for conservation. *Conservation Biology* 8: 961–971.

22. For an intriguing hypothesis about how genetic drift might lead to the evolution of fundamental differences in the molecular machinery of different kinds of organisms, see:

Fernández, A., and M. Lynch. 2011. Non-adaptive origins of interactome complexity. *Nature* 474: 502–505.

23. In animals, the rate of sequence change appears to vary as a function of metabolic rate as well as generation time. Gillooly and colleagues have recently attempted to unify these data with the original classic neutral model of evolution. According to their model, the molecular clock ticks at one substitution "per unit of mass-specific metabolic energy" rather than per unit time. Here is Gillooly's paper, along with two of the original papers that raised the issue of metabolic rate:

Gillooly, J. F., A. P. Allen, et al. 2005. The rate of DNA evolution: Effects of body size and temperature on the molecular clock. *Proceedings of the National Academy of Sciences, USA* 102: 140–145.

Martin, A. P., G. J. P. Naylor, and S. R. Palumbi. 1992. Rates of mitochondrial DNA evolution in sharks are slow compared with mammals. *Nature* 357: 153–155.

Martin, A. P., and S. R. Palumbi. 1993. Body size, metabolic rate, generation time, and the molecular clock. *Proceedings of the National Academy of Sciences, USA* 90: 4087–4091.

24. For evidence that languages experience founder effects analogous to those seen in gene pools, read:

Atkinson, Q. D. 2011. Phonemic diversity supports a serial founder effect model of language expansion from Africa. *Science* 332: 346–349.

25. For a test of the neutral theory's prediction that the rate of evolution is equal to the mutation rate, see:

Sanjuám. R. 2012. From molecular genetics to phylodynamics: Evolutionary relevance of mutation rates across viruses. *PLoS Pathogens* 8: e1002685.

CITATIONS

Much of the population genetics material in this chapter is modeled after presentations in the following:

Crow, J. F. 1983. *Genetics Notes.* Minneapolis, MN: Burgess Publishing.

Felsenstein, J. 2011. *Theoretical Evolutionary Genetics.* Seattle, WA. Published by the author. Available at http://evolution.genetics.washington.edu/pgbook/pgbook.html

Griffiths, A. J. F., J. H. Miller, et al. 1993. *An Introduction to Genetic Analysis.* New York: W. H. Freeman.

Halliburton, R. 2004. *Introduction to Population Genetics.* Upper Saddle River, NJ: Pearson.

Maynard Smith, J. 1998. *Evolutionary Genetics.* Oxford: Oxford University Press.

Roughgarden, J. 1979. *Theory of Population Genetics and Evolutionary Ecology: An Introduction.* New York: Macmillan.

Templeton, A. R. 1982. Adaptation and the integration of evolutionary forces. In *Perspectives on Evolution,* ed. R. Milkman. Sunderland, MA: Sinauer, 15–31.

Here is the list of all other citations in this chapter:

Akashi, H. 1994. Synonymous codon usage in *Drosophila melanogaster*: Natural selection and translational accuracy. *Genetics* 144: 927–935.

Alter, S. E., E. Rynes, and S. R. Palumbi. 2007. DNA evidence for historic population size and past ecosystem impacts of gray whales. *Proceedings of the National Academy of Sciences, USA* 104: 15162–15167.

Ayala, F. J., and J. A. Kiger, Jr. 1984. *Modern Genetics.* Menlo Park, CA: Benjamin Cummings.

Barrett, S. C. H., and D. Charlesworth. 1991. Effects of a change in the level of inbreeding on the genetic load. *Nature* 352: 522–524.

Benson, J. F., J. A. Hostetler, et al. 2011. Intentional genetic introgression influences survival of adults and subadults in a small, inbred felid population. *Journal of Animal Ecology* 80: 958–967.

Berry, A. 1996. Non-non-Darwinian evolution. *Evolution* 50: 462–466.

Berry, A., J. W. Ajioka, and M. Kreitman. 1991. Lack of polymorphism on the *Drosophila* fourth chromosome resulting from selection. *Genetics* 129: 1111–1117.

Bittles, A. H., and J. V. Neel. 1994. The costs of human inbreeding and their implications for variations at the DNA level. *Nature Genetics* 8: 117–121.

Brisson, J. A., J. L. Strassburg, and A. R. Templeton. 2003. Impact of fire management on the ecology of collared lizard (*Crotaphytus collaris*) populations living on the Ozark Plateau. *Animal Conservation* 6: 247–254.

Bodmer, W., and R. McKie. 1995. *The Book of Man.* New York: Scribner.

Buri, P. 1956. Gene frequency in small populations of mutant *Drosophila. Evolution* 10: 367–402.

Bustamante, C. D., A. Fledel-Alon, et al. 2005. Natural selection on protein-coding genes in the human genome. *Nature* 437: 1153–1157.

Camin, J. H., and P. R. Ehrlich. 1958. Natural selection in water snakes (*Natrix sipedon* L.) on islands in Lake Erie. *Evolution* 12: 504–511.

Censky, E. J., K. Hodge, and J. Dudley. 1998. Over-water dispersal of lizards due to hurricanes. *Nature* 395: 556.

Chamary, J. V., J. L. Parmley, and L. D. Hurst. 2006. Hearing silence: Non-neutral evolution at synonymous sites in mammals. *Nature Reviews Genetics* 7: 98–108.

Chao, L., and D. E. Carr. 1993. The molecular clock and the relationship between population size and generation time. *Evolution* 47: 688–690.

Charlesworth, B. 2009. Effective population size and patterns of molecular evolution and variation. *Nature Reviews Genetics* 10: 195–205.

Charlesworth, B., M. T. Morgan, and D. Charlesworth. 1993. The effects of deleterious mutations on neutral molecular variation. *Genetics* 134: 1289–1303.

Charlesworth, D., and J. H. Willis. 2009. The genetics of inbreeding depression. *Nature Reviews Genetics* 10: 783–796.

Culver, M., P. W. Hedrick, et al. 2008. Estimation of the bottleneck size in Florida panthers. *Animal Conservation* 11: 104–110.

Dorit, R. L., H. Akashi, and W. Gilbert. 1995. Absence of polymorphism at the ZFY locus on the human Y chromosome. *Science* 268: 1183–1185.

Dorus, S., P. D. Evans, et al. 2004. Rate of molecular evolution of the seminal protein gene *SEMG2* correlates with levels of female promiscuity. *Nature Genetics* 12: 1326–1329.

Driscoll, C. A., M. Menotti-Raymond, et al. 2002. Genomic microsatellites as evolutionary chronometers: A test in wild cats. *Genome Research* 12: 414–423.

Dudash, M. R. 1990. Relative fitness of selfed and outcrossed progeny in a self-compatible, protandrous species, *Sabatia angularis* L. (Gentianaceae): A comparison in three environments. *Evolution* 44: 1129–1139.

Dudash, M. R., D. E. Carr, and C. B. Fenster. 1997. Five generations of enforced selfing and outcrossing in *Mimulus guttatus*: Inbreeding depression variation at the population and family level. *Evolution* 51: 54–65.

Escalante, A. A., A. A. Lal, and F. J. Ayala. 1998. Genetic polymorphism and natural selection in the malaria parasite *Plasmodium falciparum*. *Genetics* 149: 189–202.

Fay, J. C. 2011. Weighing the evidence for adaptation at the molecular level. *Trends in Genetics* 27: 343–349.

Felsenstein, J. 2004. *Inferring Phylogenies*. Sunderland, MA: Sinauer.

Felsenstein, J. 2006. Accuracy of coalescent likelihood estimates: Do we need more sites, more sequences, or more loci? *Molecular Biology and Evolution* 23: 691–700.

Felsenstein, J. 2009a. *PHYLIP* (Phylogeny Inference Package), Version 3.69. Distributed by the author. Department of Genome Sciences, University of Washington, Seattle. Available at http://evolution.genetics.washington.edu/phylip/.

Felsenstein, J. 2009b. Coalescents in full bloom. *Evolution* 63: 3275–3276.

Giles, B. E., and J. Goudet. 1997. Genetic differentiation in *Silene dioica* metapopulations: Estimation of spatiotemporal effects in a successional plant species. *American Naturalist* 149: 507–526.

Gillespie, J. H. 1991. *The Causes of Molecular Evolution*. New York: Oxford University Press.

Gojobori, T., E. N. Etsuko, and M. Kimura. 1990. Molecular clock of viral evolution, and the neutral theory. *Proceedings of the National Academy of Sciences USA* 87: 10015–10018.

Greenwood, P. J., P. H. Harvey, and C. M. Perrins. 1978. Inbreeding and dispersal in the great tit. *Nature* 271: 52–54.

Hartl, D. L. 1981. *A Primer of Population Genetics*. Sunderland, MA: Sinauer.

Hendry, A. P., T. Day, and E. B. Taylor. 2001. Population mixing and the adaptive divergence of quantitative traits in discrete populations: A theoretical framework for empirical tests. *Evolution* 55: 459–466.

Hill, J. L. 1974. *Peromyscus*: Effect of early pairing on reproduction. *Science* 186: 1042–1044.

Hughes, A. L. 2007. Looking for Darwin in all the wrong places: The misguided quest for positive selection at the nucleotide sequence level. *Heredity* 99: 364–373.

Hughes, A. L., and M. Nei. 1988. Pattern of nucleotide substitution at major histocompatibility complex class I loci reveals overdominant selection. *Nature* 335: 167–170.

Hughes, A. L., and M. Nei. 1989. Nucleotide substitution at major histocompatibility complex class-II loci—Evidence for overdominant selection. *Proceedings of the National Academy of Sciences, USA* 86: 958–962.

Hughes, A. L., B. Packer, et al. 2003. Widespread purifying selection at polymorphic sites in human protein-coding loci. *Proceedings of the National Academy of Sciences, USA* 100: 15754–15757.

Huttley, G. A., E. Easteal, et al. 2000. Adaptive evolution of the tumor suppressor BRCA1 in humans and chimpanzees. *Nature Genetics* 25: 410–413.

Johnson, W. E., D. P. Onorato, et al. 2010. Genetic restoration of the Florida panther. *Science* 329: 1641–1645.

Keller, L., P. Arcese, et al. 1994. Selection against inbred song sparrows during a natural population bottleneck. *Nature* 372: 356–357.

Kerr, W. E., and S. Wright. 1954. Experimental studies of the distribution of gene frequencies in very small populations of *Drosophila melanogaster*. I. Forked. *Evolution* 8: 172–177.

Kimura, M. 1964. Diffusion models in population genetics. *Journal of Applied Probability* 1: 177–232.

Kimura, M. 1968. Evolutionary rate at the molecular level. *Nature* 217: 624–626.

Kimura, M. 1983. *The Neutral Theory of Molecular Evolution*. New York: Cambridge University Press.

King, J. L., and T. H. Jukes. 1969. Non-Darwinian evolution. *Science* 164: 788–798.

King, R. B. 1993a. Color pattern variation in Lake Erie water snakes: Inheritance. *Canadian Journal of Zoology* 71: 1985–1990.

King, R. B. 1993b. Color-pattern variation in Lake Erie water snakes: Prediction and measurement of natural selection. *Evolution* 47: 1819–1833.

King, R. B., and R. Lawson. 1995. Color-pattern variation in Lake Erie water snakes: The role of gene flow. *Evolution* 49: 885–896.

Kingman, J. F. 2000. Origins of the coalescent. 1974–1982. *Genetics* 156: 1461–1463.

Kuhner, M. K. 2009. Coalescent genealogy samplers: Windows into population history. *Trends in Ecology and Evolution* 24: 86–93.

Larson, S., R. Jameson, et al. 2002. Loss of genetic diversity in sea otters (*Enhydra lutris*) associated with the fur trade of the 18th and 19th centuries. *Molecular Ecology* 11: 1899–1903.

Li, W.-H., T. Gojobori, and M. Nei. 1981. Pseudogenes as a paradigm of neutral evolution. *Nature* 292: 237–239.

Li, W.-H., and D. Graur. 1991. *Fundamentals of Molecular Evolution*. Sunderland, MA: Sinauer.

Li, W.-H., and D. Graur. 2000. *Fundamentals of Molecular Evolution*, 2nd ed. Sunderland, MA: Sinauer.

Lidicker, W. Z., and F. C. McCollum. 1997. Allozymic variation in California sea otters. *Journal of Mammology* 78: 417–425.

Lynch, M., and W. Gabriel. 1990. Mutation load and the survival of small populations. *Evolution* 44: 1725–1737.

Maddison, W.P. 2011. *Coalescence Package for Mesquite*. Version 2.75. Available at http://mesquiteproject.org.

Maddison, W. P., and D. R. Maddison. 2011. *Mesquite: A Modular System for Evolutionary Analysis*. Version 2.75. Available at http://mesquiteproject.org.

McBride, R. T., R. T. McBride, et al. 2008. Counting pumas by categorizing physical evidence. *Southeastern Naturalist* 7: 381–400.

McCall, C., D. M. Waller, and T. Mitchell-Olds. 1994. Effects of serial inbreeding on fitness components in *Impatiens capensis*. *Evolution* 48: 818–827.

McDonald, J. H., and M. Kreitman. 1991. Adaptive protein evolution at the Adh locus in *Drosophila*. *Nature* 351: 652–654.

Nachman, M. W., and S. L. Crowell. 2000. Estimate of the mutation rate per nucleotide in humans. *Genetics* 156: 297–304.

Nei, M. 2005. Selectionism and neutralism in molecular evolution. *Molecular Biology and Evolution* 22: 2318–2342.

Nei, M., Y. Suzuki, and M. Nozawa. 2010. The neutral theory of molecular evolution in the genomic era. *Annual Review of Genomics and Human Genetics* 11: 265–289.

Nielsen, R. 2005. Molecular signatures of natural selection. *Annual Review of Genetics* 39: 197–218.

Nielsen, R., C. Bustamante, et al. 2005. A scan for positively selected genes in the genomes of humans and chimpanzees. *PLoS Biology* 3: e170.

Nurminsky, D. I., M. V. Nurminskaya, et al. 1998. Selective sweep of a newly evolved sperm-specific gene in *Drosophila*. *Nature* 396: 572–575.

O'Connell, M. J. 2010. Selection and the cell cycle: Positive Darwinian selection in a well-known DNA damage response pathway. *Journal of Molecular Evolution* 71: 444–457.

Ohta, T. 1972. Evolutionary rate of cistrons and DNA divergence. *Journal of Molecular Evolution* 1: 150–157.

Ohta, T. 1977. Extension to the neutral mutation random drift hypothesis. In *Molecular Evolution and Polymorphism*, ed. M. Kimura. Mishima, Japan: National Institute of Genetics, 148–176.

Ohta, T., and M. Kimura. 1971. On the constancy of the evolutionary rate of cistrons. *Journal of Molecular Evolution* 1:18–25.

Ohta, T., and M. Kreitman. 1996. The neutralist–selectionist debate. *BioEssays* 18: 673–683.

Pimm, S. L., L. Dollar, and O. L. Bass Jr. 2006. The genetic rescue of the Florida panther. *Animal Conservation* 9: 115–122.

Polans, N. O., and R. W. Allard. 1989. An experimental evaluation of the recovery potential of ryegrass populations from genetic stress resulting from restriction of population size. *Evolution* 43: 1320–1324.

Pulvers, J. N., and W. B. Huttner. 2009. *Brca1* is required for embryonic development of the mouse cerebral cortex to normal size by preventing apoptosis of early neural progenitors. *Development* 136: 1859–1868.

Purugganan, M. D., and J. I. Suddith. 1998. Molecular population genetics of the *Arabidopsis CAULIFLOWER* regulatory gene: Nonneutral evolution and naturally occurring variation in floral homeotic evolution. *Proceedings of the National Academy of Sciences, USA* 95: 8130–8134.

Quammen, D. 1996. *The Song of the Dodo*. New York: Touchstone.

Ralls, K., K. Brugger, and J. Ballou. 1979. Inbreeding and juvenile mortality in small populations of ungulates. *Science* 206: 1101–1103.

Rich, S. S., A. E. Bell, and S. P. Wilson. 1979. Genetic drift in small populations of *Tribolium*. *Evolution* 33: 579–584.

Roelke, M. E., J. S. Martenson, and S. J. O'Brien. 1993. The consequences of demographic reduction and genetic depletion in the endangered Florida panther. *Current Biology* 3: 340–350.

Seielstad, M. T., E. Minch, and L. L. Cavalli-Sforza. 1998. Genetic evidence for a higher female migration rate in humans. *Nature Genetics* 20: 278–280.

Sharp, P. M. 1997. In search of molecular Darwinism. *Nature* 385: 111–112.

Sharp, P. M., E. Cowe, et al. 1988. Codon usage patterns in *Escherichia coli, Bacillus subtilis, Saccharomyces cerevisiae, Schizosaccharomyces pombe, Drosophila melanogaster*, and *Homo sapiens*: A review of the considerable within-species diversity. *Nucleic Acids Research* 16: 8207–8211.

Sharp, P. M., and W.-H. Li. 1986. An evolutionary perspective on synonymous codon usage in unicellular organisms. *Journal of Molecular Evolution* 24: 28–38.

Sheffield, V. C. 2000. The vision of Typhoon Lengkieki. *Nature Medicine* 6: 746–747.

Smith, N. G. C., and A. Eyre-Walker. 2002. Adaptive protein evolution in *Drosophila*. *Nature* 415: 1022–1024.

Soulé, M. E., and L. S. Mills. 1998. No need to isolate genetics. *Science* 282: 1658–1659.

Sundin, O. H., J.-M. Yang, et al. 2000. Genetic basis of total colourblindness among the Pingelapese islanders. *Nature Genetics* 25: 289–293.

Swanson, W. J., and V. D. Vacquier. 1998. Concerted evolution in an egg receptor for a rapidly evolving abalone sperm protein. *Science* 281: 710–712.

Templeton, A. R., R. J. Robertson, et al. 2001. Disrupting evolutionary processes: The effect of habitat fragmentation on collared lizards in the Missouri Ozarks. *Proceedings of the National Academy of Sciences, USA* 98: 5426–5432.

Templeton, A. R., K. Shaw, et al. 1990. The genetic consequences of habitat fragmentation. *Annals of the Missouri Botanical Garden* 77: 13–27.

Tinghitella, R. M., M. Zuk, et al. 2011. Island hopping introduces Polynesian field crickets to novel environments, genetic bottlenecks and rapid evolution. *Journal of Evolutionary Biology* 24: 1199–1211.

Vallender, E. J., and B. T. Lahn. 2004. Positive selection on the human genome. *Human Molecular Genetics* 13: R245–R254.

van Noordwijk, A. J., and W. Scharloo. 1981. Inbreeding in an island population of the great tit. *Evolution* 35: 674–688.

Wakeley, J. 2009. *Coalescent Theory: An Introduction*. Greenwood Village, CO: Roberts and Company.

Wayne, R. K., N. Lehman, et al. 1991. Conservation genetics of the endangered Isle Royale gray wolf. *Conservation Biology* 5: 41–51.

Winick, J. D., M. L. Blundell, et al. 1999. Homozygosity mapping of the achromatopsia locus in the Pingelapese. *American Journal of Human Genetics* 64: 1679–1685.

Wolfe, L. M. 1993. Inbreeding depression in *Hydrophyllum appendiculatum*: Role of maternal effects, crowding, and parental mating history. *Evolution* 47: 374–386.

Wright, S. 1931. Evolution in Mendelian populations. *Genetics* 16: 97–159.

Yang, Z., and J. P. Bielawski. 2000. Statistical methods for detecting molecular adaptation. *Trends in Ecology and Evolution* 15: 496–503.

Young, A., T. Boyle, and T. Brown. 1996. The population genetic consequences of habitat fragmentation for plants. *Trends in Ecology and Evolution* 11: 413–418.

Zhang, J. Z., H. F. Rosenberg, and M. Nei. 1998. Positive Darwinian selection after gene duplication in primate ribonuclease genes. *Proceedings of the National Academy of Sciences, USA* 95: 3708–3713.

Zuckerkandl, E., and L. Pauling. 1965. Evolutionary divergence and convergence in proteins. In *Evolving Genes and Proteins*, ed. V. Bryson and H. J. Vogel. New York: Academic Press, 97–165.

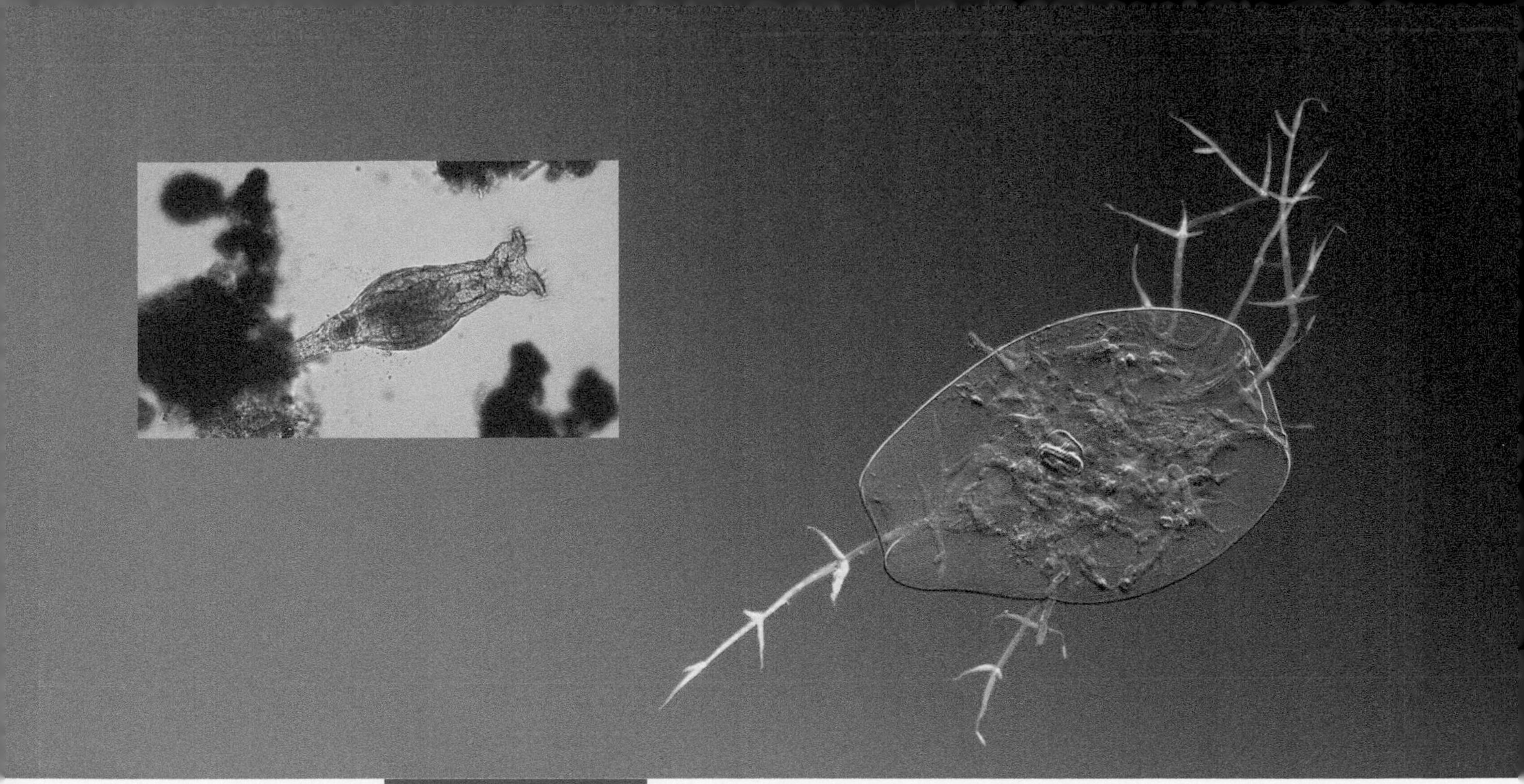

8

Evolution at Multiple Loci: Linkage and Sex

Bdelloid rotifers reproduce without sex. Females make daughters that are, except for mutations, clones of their mother. A variety of evidence suggests that a common ancestor of the bdelloids gave up sex tens of millions of years ago (Mark Welch et al. 2009; Schurko et al. 2009; Birky 2010). The bdelloids have since diversified into more than 450 female-only species.

Reproducing asexually ought to leave bdelloids vulnerable to pathogens (see Lively 2010). If a pathogen evolves the ability to infect one rotifer, it can also infect that individual's kin. And, indeed, bdelloids suffer lethal infections by specialist fungi (Wilson and Sherman 2010). The photo above shows a rotifer (*Habrotrocha elusa*) that, after accidentally swallowing spores of one such fungus (*Rotiferophthora angustispora*), is being digested from the inside. Fungal filaments are now breaking through the rotifer's skin to release a new generation of spores.

The bdelloids, however, have a means of escape. As shown in the graph at right, their resistance to dessication far exceeds that of their enemies. Christopher Wilson and Paul Sherman (2010) inoculated rotifer cultures with fungus, waited three days, then dried them. When dried, bdelloids enter a state of suspended animation called anhydrobiosis. The longer Wilson and Sherman waited to revive the rotifers with water, the fewer viable fungi remained.

Killer fungi (above) threaten bdelloid rotifers (inset). Superior dessication resistance lets the rotifers escape (below)—and, perhaps, forgo sex. Photos by K. Loeffler, K. T. Hodge, and C. Wilson; D. H. Zanette. Graph redrawn from Wilson and Sherman (2010).

From "Anciently asexual bdelloid rotifers escape lethal fungal parasites by drying up and blowing away." *Science* 327: 574–576. Reprinted with permission from AAAS.

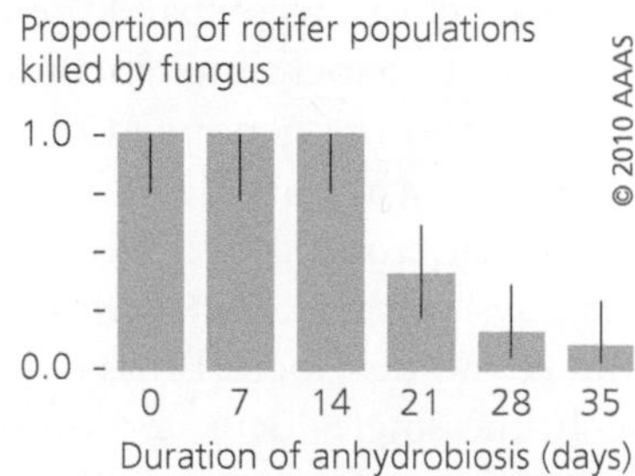

Desiccated rotifers are tiny and light, and readily disperse by wind. Wilson and Sherman suggest that by drying up and blowing away, bdelloids in nature can temporarily escape their fungal killers. By achieving a defense their pathogens have not been able to match, the bdelloids appear to have won an arms race.

Many other hosts are not so lucky. Instead, they and their pathogens experience perpetual cycles of evolving defense and offense. This, in turn, may help explain why so few other organisms have joined the bdelloids in totally eschewing sex. The connection between host–pathogen arms races and sex is, to say the least, not immediately apparent. It becomes clear from consideration of the consequences of sex at the level of genes in populations.

Earlier we introduced basic population genetics, built on the Hardy–Weinberg equilibrium principle (Chapters 6 and 7). The models we discussed are elegant and powerful. As with many theories, however, basic population genetics buys its elegance at the price of simplification. The models we used track allele frequencies at just one locus at a time. We were thus able to consider only the evolution of traits that are (or appear to be) controlled by a single gene. The genomes of real organisms, of course, contain hundreds or thousands of loci. And many traits are determined by the combined influence of numerous genes. Among such traits are the ability to infect a host, or to defend oneself against infection.

In Chapter 8, we take our models of the mechanics of evolution closer to real organisms by considering two or more loci simultaneously. Our first step in that direction, the subject of Section 8.1, is an extension of the Hardy–Weinberg analysis that follows two loci at a time. The two-locus model will tell us when we can use the single-locus models developed in earlier chapters to make predictions and when we must take into account the confounding influence of selection at other loci.

Our discussion of the two-locus version of Hardy–Weinberg analysis, which introduces terms like *linkage disequilibrium,* may at first seem dauntingly abstract. But effort invested in understanding it will produce two surprising payoffs. These are the subjects of Sections 8.2 and 8.3. First, the two-locus model provides tools we can use to reconstruct the history of genes and populations. We use these tools to address, among other issues, an unresolved question from our earlier discussions (in Chapters 1, 5, and 6) of *CCR5-Δ32*, the allele that protects against HIV: Where did the *Δ32* allele come from, and why does it occur only in Europe? Second, the two-locus model provides insight into the adaptive significance of sexual reproduction.

8.1 Evolution at Two Loci: Linkage Equilibrium and Linkage Disequilibrium

In this section, we expand the one-locus version of Hardy–Weinberg analysis to consider two loci simultaneously. In principle, we could focus on any pair of loci in an organism's genome. Our discussion will be easier to understand, however, if we focus on a pair of loci located sufficiently close together on the same chromosome that crossing over between them is rare. That is, we consider two loci that are physically linked **(Figure 8.1)**. We will imagine that locus A has two alleles, *A* and *a*, and that locus B has two alleles, *B* and *b*.

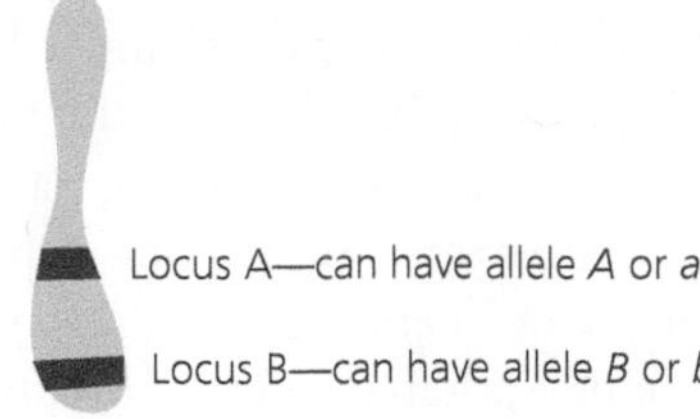

Figure 8.1 Linked loci

In the single-locus version of Hardy–Weinberg analysis, we were concerned primarily with tracking allele frequencies. In the two-locus version, we are concerned with tracking both allele frequencies and chromosome frequencies. Note that the assumptions we made in the previous paragraph allow four different chromosome genotypes: *AB*, *Ab*, *aB*, and *ab*. The multilocus genotype of a chromosome or gamete is sometimes referred to as its **haplotype** (a term that comes from the contraction of *haploid genotype*).

Our main goal is to determine whether selection at the A locus will interfere with our ability to use the models of earlier chapters to make predictions about evolution at the B locus. The answer is: sometimes—depending on whether the loci are in linkage equilibrium or linkage disequilibrium. We will define linkage equilibrium and linkage disequilibrium shortly.

When we use population genetics models to analyze evolution at a particular locus, do we need to worry about the effects of selection at other loci? Only if the locus of interest and the other loci are in linkage disequilibrium.

A Numerical Example

A numerical example illustrates key concepts and helps us define terms. **Figure 8.2** shows two hypothetical populations, each with a gene pool containing 25 chromosomes. In studying the genetic structure of these populations, the first thing we might do is calculate allele frequencies. In population 1, for example, 15 of the 25 chromosomes carry allele *A* at locus A. Thus the frequency of *A* is $\frac{15}{25} = 0.6$. The same is true for population 2. In fact, the allele frequencies at both loci are identical in the two populations. If we were studying locus A only, or locus B only, we would conclude that the two populations are identical.

But the populations are not identical. This we discover when we calculate the chromosome frequencies. In population 1, for example, 12 of the 25 chromosomes carry haplotype *AB*, giving this haplotype a frequency of 0.48. In population 2, on the other hand, the frequency of *AB* chromosomes is 11 of 25, or 0.44. This is the first lesson of two-locus Hardy–Weinberg analysis: Populations can have identical allele frequencies but different chromosome frequencies.

(a) Population 1 is in linkage equilibrium

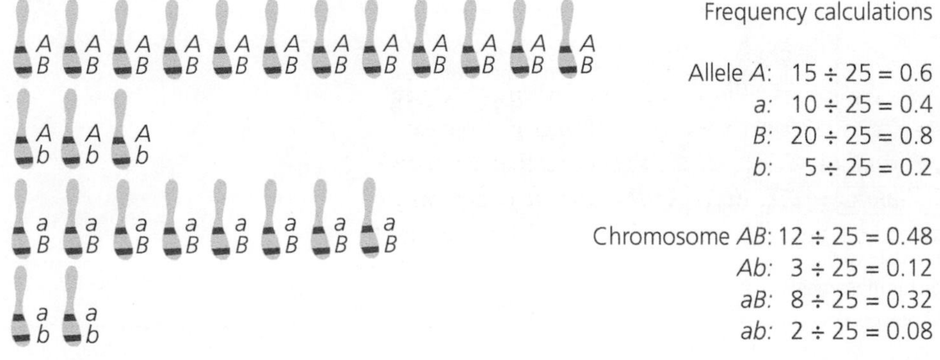

(b) Population 2 is in linkage disequilibrium

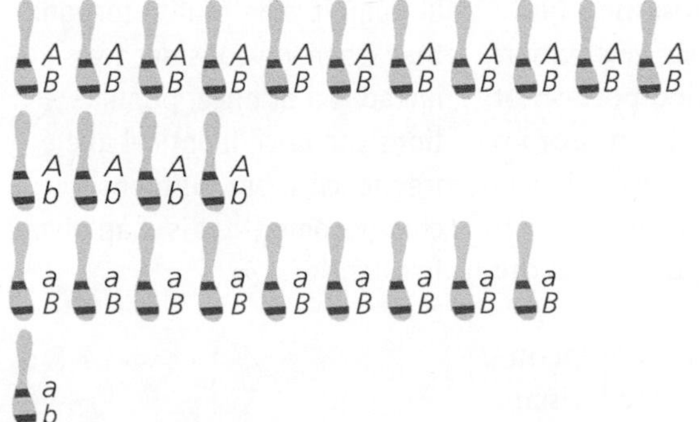

Frequency calculations

Allele *A*: 15 ÷ 25 = 0.6
a: 10 ÷ 25 = 0.4
B: 20 ÷ 25 = 0.8
b: 5 ÷ 25 = 0.2

Chromosome *AB*: 11 ÷ 25 = 0.44
Ab: 4 ÷ 25 = 0.16
aB: 9 ÷ 25 = 0.36
ab: 1 ÷ 25 = 0.04

Figure 8.2 Populations with identical allele frequencies, but different chromosome frequencies (a) In population 1 the frequency of allele *B* among *A*-bearing chromosomes (12 of 15, or 0.8) is the same as it is among *a*-bearing chromosomes (8 of 10, or 0.8). (b) In population 2 the frequencies of *B* among *A*-bearing versus *a*-bearing chromosomes are different (11 of 15, or 0.73, versus 9 or 10, or 0.9). Population 2 is said to be in linkage disequilibrium.

Another way to see the difference between the populations in Figure 8.2 is to calculate the frequency of allele *B* on chromosomes carrying allele *A* versus chromosomes carrying allele *a*. In population 1 there are 15 chromosomes carrying *A*, 12 of which carry *B*. The frequency of *B* on *A* chromosomes is thus $\frac{12}{15} = 0.8$. In the same population there are 10 chromosomes carrying *a*, 8 of which carry *B*. The frequency of *B* on *a* chromosomes is thus $\frac{8}{10} = 0.8$. In population 1, then, the frequency of *B* is the same on chromosomes carrying *A* as it is on chromosomes carrying *a*. The same is not true for the population 2. There, the frequency of *B* is 0.73 on *A* chromosomes, but 0.9 on *a* chromosomes.

(a) Population 1 is in linkage equilibrium

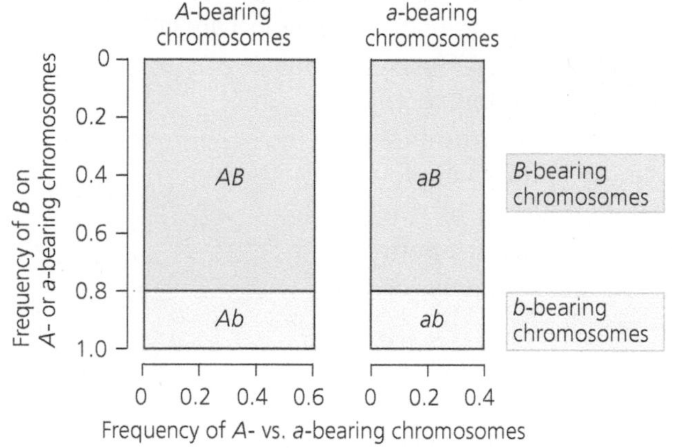

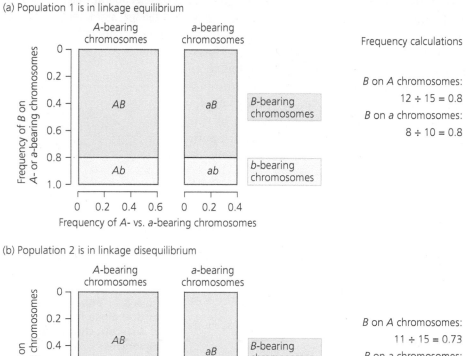

(b) Population 2 is in linkage disequilibrium

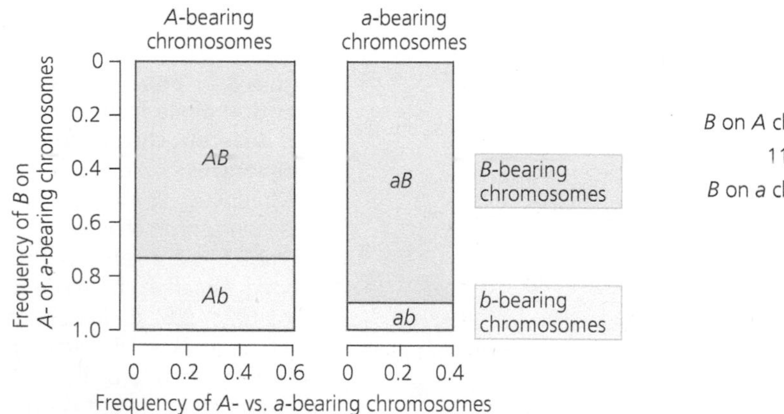

Figure 8.3 A graphical representation of populations with identical allele frequencies but different chromosome frequencies These are the same populations shown in Figure 8.2. The width of each bar represents the frequencies of *A*- versus *a*-bearing chromosomes. The shading of each bar represents the frequencies of *B*- versus *b*-bearing chromosomes.

The bar graphs in **Figure 8.3** provide a visual representation of the difference between the populations. The widths of the two bars in each graph represent the frequencies of *A*-bearing chromosomes versus *a*-bearing chromosomes. Note that the combined widths of the two bars must equal 1, so if one bar gets wider, the other must get narrower. The darkly shaded versus lightly shaded portion of each bar represents the frequency of allele *B* versus allele *b* on the chromosomes in question. The graphs show at a glance what we discovered earlier by calculation. In population 1 the frequency of *B* is the same on *A* chromosomes as on *a* chromosomes—the same fraction is shaded in both bars. In population 2 the frequency of *B* is lower on *A* chromosomes than on *a* chromosomes.

To understand linkage disequilibrium, it is helpful to recognize that when we consider two linked loci at once, populations can have identical allele frequencies, but different chromosome (that is, haplotype) frequencies.

The differences we have identified between our populations may seem inconsequential. Imagine, however, that individuals with genotype *AABB* are resistant

COMPUTING CONSEQUENCES 8.1

The coefficient of linkage disequilibrium

The coefficient of linkage disequilibrium, symbolized by D, is defined as

$$g_{AB}g_{ab} - g_{Ab}g_{aB}$$

where g_{AB}, g_{ab}, g_{Ab}, and g_{aB} are the frequencies of AB, ab, Ab, and aB chromosomes.

To see why D is called the coefficient of linkage disequilibrium, recall that when two loci are in linkage equilibrium, the allele frequencies at one locus are independent of allele frequencies at the other locus. Let p and q be the frequencies of A and a, and let s and t be the frequencies of B and b. If a population is in linkage equilibrium, then $g_{AB} = ps$, $g_{Ab} = pt$, $g_{aB} = qs$, and $g_{ab} = qt$. And furthermore,

$$D = psqt - ptqs = 0$$

If, on the other hand, the population is in linkage disequilibrium, then $g_{AB} \neq ps$, $g_{Ab} \neq pt$, $g_{aB} \neq qs$, and $g_{ab} \neq qt$. And $D \neq 0$.

The maximum value that D can assume is 0.25, when AB and ab are the only chromosomes present and each has a frequency of 0.5. The minimum value that D can assume is -0.25, when Ab and aB are the only chromosomes present and each is at a frequency of 0.5. Thus calculating D is a useful way to quantify the degree of linkage disequilibrium in a population.

to a lethal pathogen and all others are susceptible. Random sampling of gametes from population 1's gene pool will produce a higher frequency of resistant individuals than will random sampling from population 2's gene pool. For this and other reasons, the differences between the populations could matter after all.

Linkage Disequilibrium Defined

In population 1 in Figures 8.2 and 8.3, locus A and locus B are in linkage equilibrium. In population 2, the loci are in linkage *dis*equilibrium. Two loci in a population are in **linkage equilibrium** when the genotype of a chromosome at one locus is independent of its genotype at the other locus. This means that knowing the genotype of the chromosome at one locus is of no use at all in predicting the genotype at the other. Two loci are in **linkage disequilibrium** when there is a nonrandom association between a chromosome's genotype at one locus and its genotype at the other locus. If we know the genotype of a chromosome at one locus, it provides a clue about the genotype at the other.

When genotypes at one locus are independent of genotypes at another locus, the two loci are in linkage equilibrium. Otherwise, the loci are in linkage disequilibrium.

These definitions are rather abstract. More concretely, the following conditions are true for a pair of loci if, and only if, they are in linkage equilibrium:

1. The frequency of B on chromosomes carrying allele A is equal to the frequency of B on chromosomes carrying allele a.
2. The frequency of any chromosome haplotype can be calculated by multiplying the frequencies of the constituent alleles. For example, the frequency of AB chromosomes can be calculated by multiplying the frequency of allele A and the frequency of allele B.
3. The quantity D, known as the coefficient of linkage disequilibrium, is equal to zero. D is calculated as

$$g_{AB}g_{ab} - g_{Ab}g_{aB}$$

where g_{AB}, g_{ab}, g_{Ab}, and g_{aB} are the frequencies of AB, ab, Ab, and aB chromosomes (see **Computing Consequences 8.1**).

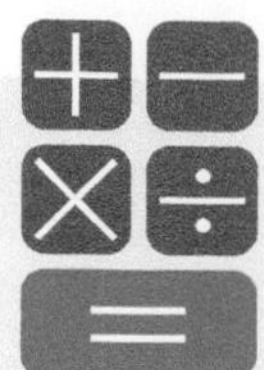

COMPUTING CONSEQUENCES 8.2

Hardy–Weinberg analysis for two loci

Here we develop the two-locus version of the Hardy–Weinberg equilibrium principle. We show that when an ideal population is in linkage equilibrium, the chromosome frequencies do not change across generations.

In the single-locus Hardy–Weinberg analysis introduced in Chapter 6, we followed allele frequencies for a complete turn of the life cycle, from the gene pool into zygotes, then juveniles, then adults, and from adults into the next generation's gene pool. We use a similar strategy here, except that we are tracking not allele frequencies but chromosome frequencies. The chromosomes in our organisms contain two loci: the A locus, with alleles *A* and *a*; and the B locus, with alleles *B* and *b*. (We do not intend these symbols to necessarily imply a dominant–recessive relationship between alleles. We use them only because they make the equations easier to read than alternative notations.) There are four kinds of chromosomes: *AB*, *Ab*, *aB*, and *ab*.

Imagine an ideal population in whose gene pool chromosomes *AB*, *Ab*, *aB*, and *ab* are present at frequencies g_{AB}, g_{Ab}, g_{aB}, and g_{ab}, respectively. If the gametes in the gene pool combine at random to make zygotes, among the possible zygote genotypes is *AB*/*AB*. Its frequency will equal the probability that a randomly chosen egg contains an *AB* chromosome multiplied by the probability that a randomly chosen sperm contains an *AB* chromosome, or $g_{AB} \times g_{AB}$. Another possible zygote genotype is *AB*/*Ab*. Its frequency will be $2 \times g_{AB} \times g_{Ab}$. This expression contains a 2 because there are two ways to make an *AB*/*Ab* zygote: An *AB* egg can be fertilized by an *Ab* sperm, or an *Ab* egg can be fertilized by an *AB* sperm. Overall, there are 10 possible zygote genotypes. Their frequencies are

AB*/*AB	***Ab*/*Ab***	***aB*/*aB***	***ab*/*ab***	***AB*/*Ab***
$g_{AB}g_{AB}$	$g_{Ab}g_{Ab}$	$g_{aB}g_{aB}$	$g_{ab}g_{ab}$	$2g_{AB}g_{Ab}$
AB*/*aB	***AB*/*ab***	***Ab*/*aB***	***Ab*/*ab***	***aB*/*ab***
$2g_{AB}g_{aB}$	$2g_{AB}g_{ab}$	$2g_{Ab}g_{aB}$	$2g_{Ab}g_{ab}$	$2g_{aB}g_{ab}$

If we allow these zygotes to grow to adulthood without selection, then the genotype frequencies among the adults will be the same as they are among the zygotes.

We have followed the chromosome frequencies from gene pool to zygotes to juveniles to adults. We can now calculate the chromosome frequencies in the next generation's gene pool. Consider chromosome *AB*. Gametes containing *AB* chromosomes can be produced by 5 of the 10 adult genotypes. The adults that can make *AB* gametes, together with the allotment of *AB* gametes they contribute to the new gene pool, are

Adult	***AB* gametes contributed**	**Notes**
AB/*AB*	$g_{AB}g_{AB}$	
AB/*Ab*	$(\frac{1}{2})(2g_{AB}g_{Ab})$	
AB/*aB*	$(\frac{1}{2})(2g_{AB}g_{aB})$	
AB/*ab*	$(1 - r)(\frac{1}{2})(2g_{AB}g_{ab})$	r = recombination rate
Ab/*aB*	$(r)(\frac{1}{2})(2g_{Ab}g_{aB})$	r = recombination rate

We have already established, by calculation and with bar graphs, that the first condition is true for the top population in Figure 8.2 but false for the bottom population. The reader should verify that the second and third conditions are likewise true for the top population but false for the bottom one.

The Two-Locus Version of Hardy–Weinberg Analysis

We can perform a two-locus version of the Hardy–Weinberg analysis that is analogous to the single-locus version we performed earlier (in Chapter 6). We assume no selection, no mutation, no migration, infinite population size, and random mating, and we follow chromosome frequencies through one complete turn of our population's life cycle, from gametes in the gene pool to zygotes to juveniles to adults and back to gametes in the gene pool. This calculation is given in **Computing Consequences 8.2**. It provides our first piece of evidence

Under Hardy–Weinberg assumptions, chromosome frequencies remain unchanged from one generation to the next, but only if the loci in question are in linkage equilibrium. If the loci are in linkage disequilibrium, the chromosome frequencies move closer to linkage equilibrium each generation.

The first row in this table is straightforward. AB/AB adults constitute a fraction $g_{AB}g_{AB}$ of the population. They therefore contribute a fraction $g_{AB}g_{AB}$ of the gametes in the population's gene pool. All of these gametes are AB. The second row is also straightforward: AB/Ab adults constitute a fraction $2g_{AB}g_{Ab}$ of the population and therefore contribute $2g_{AB}g_{Ab}$ of the gametes in the gene pool, half of them AB. The third row is straightforward as well. Only the last two rows of the table require explanation.

Adults of genotype AB/ab will produce gametes containing AB chromosomes only when meiosis occurs without crossing over between the A locus and the B locus. When no crossing over occurs, half of the gametes produced by AB/ab adults carry AB chromosomes. If r is the rate of crossing over, or recombination, between the A locus and the B locus, then the allotment of AB gametes contributed to the gene pool by AB/ab individuals is $(1 - r)(\frac{1}{2})(2g_{AB}g_{ab})$.

Adults of genotype Ab/aB produce gametes containing AB chromosomes only when meiosis occurs with crossing over between the A locus and the B locus. When crossing over occurs, half of the gametes produced by Ab/aB adults carry AB chromosomes. If r is the rate of crossing over, then the allotment of AB gametes contributed to the gene pool by Ab/aB individuals is $(r)(\frac{1}{2})(2g_{Ab}g_{aB})$.

We can now write an expression for g_{AB}', the frequency of AB chromosomes in the new gene pool:

$$
\begin{aligned}
g_{AB}' &= g_{AB}g_{AB} + (\tfrac{1}{2})(2g_{AB}g_{Ab}) + (\tfrac{1}{2})(2g_{AB}g_{aB}) \\
&\quad + (1 - r)(\tfrac{1}{2})(2g_{AB}g_{ab}) + (r)(\tfrac{1}{2})(2g_{Ab}g_{aB}) \\
&= g_{AB}g_{AB} + g_{AB}g_{Ab} + g_{AB}g_{aB} \\
&\quad + g_{AB}g_{ab} - rg_{AB}g_{ab} + rg_{Ab}g_{aB} \\
&= g_{AB}(g_{AB} + g_{Ab} + g_{aB} + g_{ab}) \\
&\quad - r(g_{AB}g_{ab} - g_{Ab}g_{aB})
\end{aligned}
$$

We can simplify this expression further by noting that $(g_{AB} + g_{Ab} + g_{aB} + g_{ab}) = 1$, and that $g_{AB}g_{ab} - g_{Ab}g_{aB}$ is D, defined in the text and Computing Consequences 8.1. This gives us

$$g_{AB}' = g_{AB} - rD$$

We leave it to the reader to derive the expressions for the other three chromosome frequencies, which are

$$g_{Ab}' = g_{Ab} + rD$$

$$g_{aB}' = g_{aB} + rD$$

$$g_{ab}' = g_{ab} - rD$$

The expressions for g_{AB}', g_{Ab}', g_{aB}', and g_{ab}' show that when a population is in linkage equilibrium—when $D = 0$—the chromosome frequencies do not change from one generation to the next. When, on the other hand, the population is in linkage disequilibrium—when $D \neq 0$—the chromosome frequencies do change from one generation to the next. Whether a given chromosome's frequency rises or falls depends on whether D is positive or negative. The first population geneticist to report this result was H. S. Jennings (1917).

We should note that allele frequencies at a pair of loci can be in linkage disequilibrium even when the loci are on different chromosomes. For loci on different chromosomes, it is appropriate to speak of gamete frequencies rather than chromosome frequencies. The Hardy–Weinberg analysis for such a situation is identical to the one we have developed here, except that r is always equal to exactly $\frac{1}{2}$.

that linkage equilibrium is important in evolution. If the two loci in our ideal population are in linkage equilibrium, then under Hardy–Weinberg conditions chromosome frequencies will not change from one generation to the next. If, instead, the loci are in linkage disequilibrium, then the chromosome frequencies will change across generations.

What Creates Linkage Disequilibrium in a Population?

Three mechanisms can create linkage disequilibrium in a random-mating population: selection on multilocus genotypes, genetic drift, and population admixture. We consider each of these mechanisms in turn. As we mentioned earlier, the mechanisms that create linkage disequilibrium may be easier to visualize if the reader imagines how they would apply to a pair of loci that are not assorting independently because they are physically linked.

In random-mating populations, three mechanisms create linkage disequilibrium: selection on multilocus genotypes, genetic drift, and population admixture.

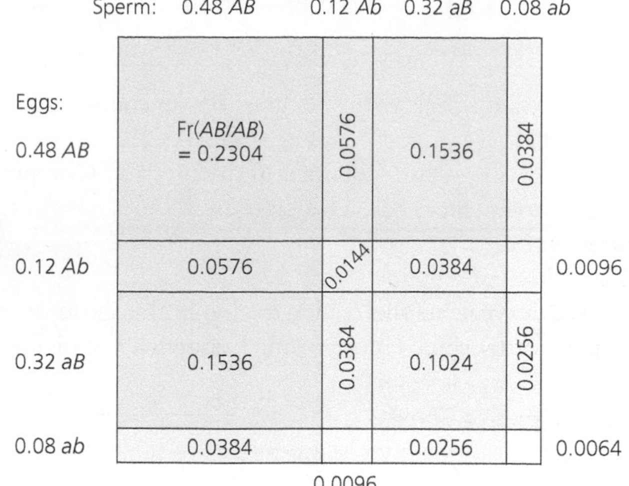

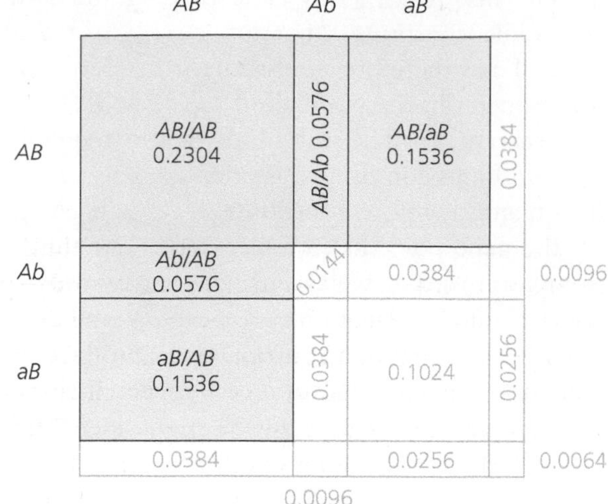

Figure 8.4 Selection on multilocus genotypes can create linkage disequilibrium (a) The expected frequencies of zygotes produced by random mating among the individuals in the population in linkage equilibrium from Figure 8.2a. (b) The genotypes that survive after predators kill all individuals with fewer than three capital-letter alleles in their genotype. The population of survivors is in linkage disequilibrium because some possible genotypes are missing.

Selection on Multilocus Genotypes Can Create Linkage Disequilibrium

To see how selection on multilocus genotypes can create linkage disequilibrium, start with the population whose gene pool is shown in Figure 8.2a. Locus A and locus B are in linkage equilibrium. Imagine that the gametes in this gene pool combine at random to make zygotes. The 10 kinds of zygotes produced, and their expected frequencies, appear in the grid in **Figure 8.4a**. Because 32% of the eggs are *aB*, for example, and 32% of the sperm are *aB*, we predict that the frequency of *aB*/*aB* zygotes will be $0.32 \times 0.32 = 0.1024$.

Now let the zygotes develop into adults, and assign phenotypes as follows: Individuals with genotype *ab*/*ab* have a size of 10. For other genotypes, every copy of *A* or *B* adds 1 unit to the individual's size. For example, *aB*/*aB* individuals have a size of 12, and *AB*/*Ab* individuals have a size of 13. Finally, imagine that predators catch and eat every individual whose size is less than 13. The survivors, which represent 65.28% of the original population, appear in color in the grid in Figure 8.4b.

In the population of survivors, locus A and locus B are in linkage disequilibrium. Perhaps the easiest way to see this is to calculate the frequency of allele *a* and allele *b*. Here is one way to calculate the frequency of *a*: Of the survivors, the fraction carrying copies of allele *a* is $(0.1536 + 0.1536)/0.6528 \approx 0.47$. All of these carriers of allele *a* are heterozygotes. Therefore, the frequency of allele *a* in the population of survivors is $0.5 \times 0.47 \approx 0.24$. The frequency of *b* is approximately 0.09. If our two loci were in linkage equilibrium, then, by criterion 2 on our list, the frequency of *ab* chromosomes among the survivors would be $0.24 \times 0.09 \approx 0.02$. In fact, the frequency of *ab* chromosomes is 0. Because a nonrandom subset of multilocus genotypes survived, our two loci are in linkage disequilibrium.

As an exercise, the reader should demonstrate that the loci are in linkage disequilibrium by criteria 1 and 3 as well.

Genetic Drift Can Create Linkage Disequilibrium

To see how genetic drift can create linkage disequilibrium, look at the scenario diagrammed in **Figure 8.5**. This scenario starts with a gene pool in which the only chromosomes present are *AB* and *Ab* (Figure 8.5a). In other words, copies of allele *a* do not exist in this population. Locus A and locus B are in linkage equilibrium.

Now imagine that in a single *Ab* chromosome, a mutation converts allele *A* into allele *a*. This creates a single *ab* chromosome (Figure 8.5b).

The mutation also puts the population in linkage disequilibrium because there is now a possible chromosome haplotype—*aB*—that is missing. The missing haplotype could be created by another mutation or by recombination during meiosis in an *AB*/*ab* diploid, but it may be many generations before either happens.

Finally, imagine that selection favors allele *a* over allele *A*, so that *a* increases in frequency and *A* decreases (Figure 8.4c). This increases the degree of linkage disequilibrium between locus A and locus B.

The reader may wonder why we are ascribing the linkage disequilibrium created in this scenario to genetic drift, when the key events seem to be mutation and selection. The reason is that the scenario, as we described it, could happen only in a finite population. In an infinite population, the mutation converting allele *A* into allele *a* would happen not once, but many times each generation, on both *AB* and *Ab* chromosomes. At no point would *aB* chromosomes be missing. Selection favoring *a* over *A* would simultaneously increase the frequency of both *ab* and *aB* chromosomes. Locus A and locus B would never be in linkage disequilibrium. Because our scenario can create linkage disequilibrium only in a finite population, the crucial evolutionary mechanism at work is genetic drift. It was sampling error that caused the mutation creating allele *a* to happen only once, and in an *Ab* chromosome.

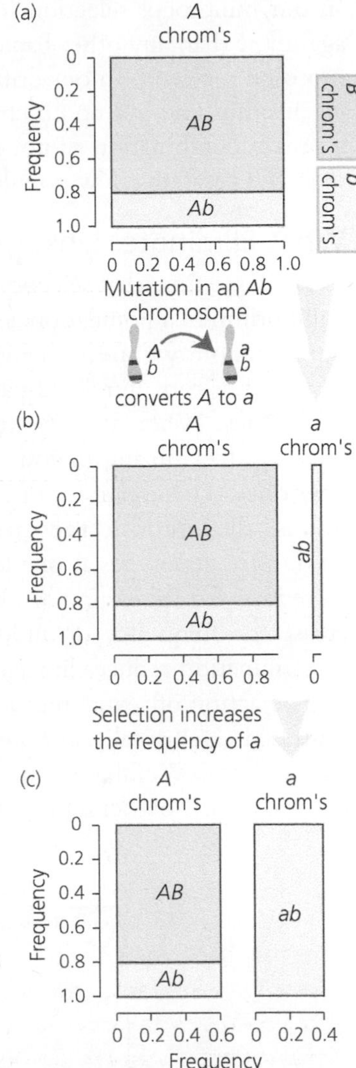

Figure 8.5 Genetic drift can create linkage disequilibrium (a) Chromosome frequencies in a finite population in which only allele *A* is present; (b) after a mutation creates a copy of allele *a*; and (c) after selection in favor of *a* increases the frequency of *ab* chromosomes.

Population Admixture Can Create Linkage Disequilibrium

Finally, to see how population admixture can create linkage disequilibrium, imagine two gene pools **(Figure 8.6)**. In one, there are 60 *AB* chromosomes, 20 *Ab* chromosomes, 15 *aB* chromosomes, and 5 *ab* chromosomes. In the other, there are 10 *AB*, 40 *Ab*, 10 *aB*, and 40 *ab* chromosomes. Locus A and locus B are in linkage equilibrium in both gene pools, as the first two bar graphs in the figure show. Now combine the two gene pools. This produces a new gene pool in which there are 70 *AB*, 60 *Ab*, 25 *aB*, and 45 *ab* chromosomes. In this new gene pool, locus A and locus B are in linkage disequilibrium.

Population admixture, genetic drift, and selection on multilocus genotypes can all create linkage disequilibrium because they can all produce populations in which some chromosome haplotypes are underrepresented and others overrepresented, compared to what their frequencies would be under linkage equilibrium.

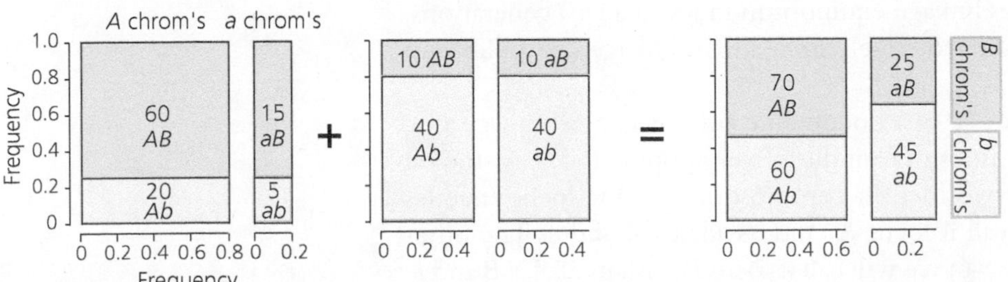

Figure 8.6 Population admixture can create linkage disequilibrium The first two bar charts represent chromosome frequencies in two populations, each in linkage equilibrium. Mixed together, these yield the third population—which is in linkage disequilibrium.

In our multilocus selection scheme, for example, selection acted more strongly against *ab* than any other haplotype because no individual containing an *ab* chromosome survived. In our drift scenario, a chance event led to the creation of an *ab* chromosome but no *aB* chromosome. In our population admixture example, a simple combination of populations with different allele and chromosome frequencies created a new population with an excess of *AB* and *ab* chromosomes.

What Eliminates Linkage Disequilibrium from a Population?

At the same time that selection, drift, and admixture may be creating linkage disequilibrium in a population, sexual reproduction inexorably reduces it. By sexual reproduction, we mean meiosis with crossing over and outbreeding. The union of gametes from unrelated parents brings together chromosomes with different haplotypes. When the zygotes grow to adulthood and themselves reproduce, crossing over during meiosis breaks up old combinations of alleles and creates new ones. The creation of new combinations of alleles during sexual reproduction is called **genetic recombination.** Because genetic recombination tends to randomize genotypes at one locus with respect to genotypes at another, it tends to reduce the frequency of overrepresented chromosome haplotypes and to increase the frequency of underrepresented haplotypes. In other words, genetic recombination reduces linkage disequilibrium.

One mechanism reduces linkage disequilbrium: genetic recombination resulting from meiosis and outbreeding (that is, sex).

The action of sexual reproduction in reducing linkage disequilibrium is demonstrated algebraically in **Computing Consequences 8.3**. The analysis shows that under Hardy–Weinberg assumptions, the rate of decline in linkage disequilibrium between a pair of loci is proportional to the rate of recombination between them.

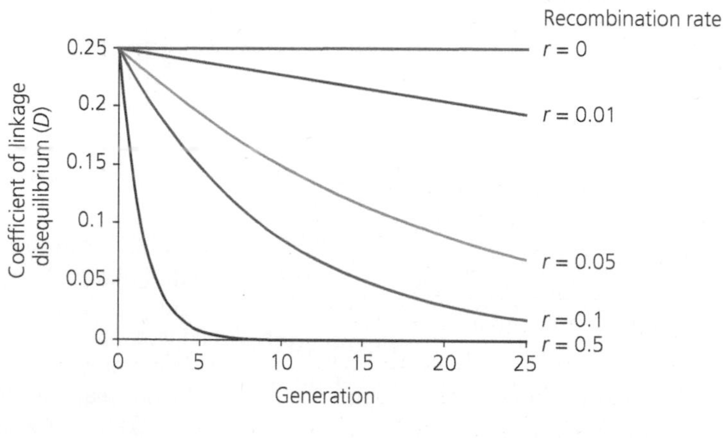

Recombination rate
$r = 0$
$r = 0.01$
$r = 0.05$
$r = 0.1$
$r = 0.5$
Coefficient of linkage disequilibrium (D)
0.25
0.2
0.15
0.1
0.05
0
Generation
0 5 10 15 20 25

Figure 8.7 With sexual reproduction and random mating, linkage disequilibrium falls over time This graph shows the level of linkage disequilibrium between two loci over 25 generations in random-mating populations with different rates of recombination, r. The populations start with the coefficient of linkage disequilibrium, D, at its maximum possible value, 0.25. After Hedrick (1983).

Predictions of the rate of decline for different rates of recombination, r, appear in **Figure 8.7**. Each curve in the figure shows the decline in linkage disequilibrium, according to the equation $D' = D(1 - r)$, for a different value of r. With $r = 0.5$, which corresponds to the free recombination of loci on different chromosomes, the population reaches linkage equilibrium in less than 10 generations. With $r = 0.01$, which corresponds to closely linked loci, linkage disequilibrium persists for many generations.

Michael Clegg and colleagues (1980) documented the decay of linkage disequilibrium in laboratory populations of fruit flies. Every population they studied harbored two alleles at each of two loci on chromosome 3. One locus encodes the enzyme esterase-c; we will call it locus A, and its alleles A and a. The other locus encodes the enzyme esterase-6; we will call it locus B, and its alleles B and b.

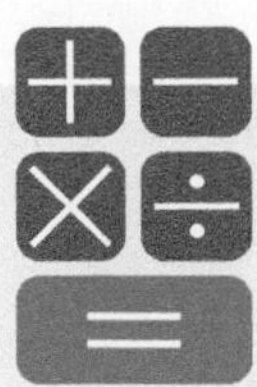

COMPUTING CONSEQUENCES 8.3

Sexual reproduction reduces linkage disequilibrium

Here we show that the level of linkage disequilibrium inexorably declines in a random-mating sexual population. We do so by starting with the definition of D, given in the text and Computing Consequences 8.1, and deriving an expression for D', the coefficient of linkage disequilibrium in the next generation.

By the definition of D,

$$D' = g_{AB}'g_{ab}' - g_{Ab}'g_{aB}'$$

Substituting the expressions for $g_{AB}', g_{ab}', g_{Ab}'$, and g_{aB}' that were derived in Computing Consequences 8.2 gives

$$\begin{aligned} D' &= [(g_{AB} - rD)(g_{ab} - rD)] \\ &\quad - [(g_{Ab} + rD)(g_{aB} + rD)] \\ &= [g_{AB}g_{ab} - g_{AB}rD - g_{ab}rD + (rD)^2] \\ &\quad - [g_{Ab}g_{aB} + g_{Ab}rD + g_{aB}rD + (rD)^2] \\ &= g_{AB}g_{ab} - g_{AB}rD - g_{ab}rD + (rD)^2 \\ &\quad - g_{Ab}g_{aB} - g_{Ab}rD - g_{aB}rD - (rD)^2 \end{aligned}$$

Canceling and rearranging terms gives

$$\begin{aligned} D' &= g_{AB}g_{ab} - g_{Ab}g_{aB} - g_{AB}rD \\ &\quad - g_{ab}rD - g_{Ab}rD - g_{aB}rD \\ &= (g_{AB}g_{ab} - g_{Ab}g_{aB}) - rD(g_{AB} \\ &\quad + g_{ab} + g_{Ab} + g_{aB}) \end{aligned}$$

Finally, the expression $(g_{AB}g_{ab} - g_{Ab}g_{aB})$ is equal to D, and the expression $(g_{AB} + g_{ab} + g_{Ab} + g_{aB})$ is equal to 1, so we have

$$D' = D - rD = D(1 - r)$$

Recall that r is the rate of recombination during meiosis, which is always between 0 and $\frac{1}{2}$. This means that $(1 - r)$ is always between $\frac{1}{2}$ and 1. Thus, unless there is no recombination at all between a pair of loci, the linkage disequilibrium between them will move closer to 0 every generation. The higher the rate of recombination between the loci, the faster the population reaches linkage equilibrium.

Clegg and colleagues set up fly populations with only AB and ab chromosomes, each at a frequency of 0.5. They also set up populations with only Ab and aB chromosomes, again at frequencies of 0.5. Thus every population was initially in complete linkage disequilibrium, with either $D = 0.25$ or $D = -0.25$.

The researchers maintained their fly populations for 48 to 50 generations, at sizes of approximately 1,000 individuals, and let the flies mate as they pleased. Every generation or two, the researchers sampled each population to determine the frequencies of the four chromosome haplotypes and calculated the level of linkage disequilibrium between the two loci. For reasons beyond the scope of our discussion, the researchers measured linkage disequilibrium not with D, but with a related statistic called the correlation of allelic state. The correlation of allelic state, r, is defined as follows:

$$r = \frac{D}{\sqrt{pqst}}$$

where p and q are the frequencies of A and a, and s and t are the frequencies of B and b. There is no one-to-one relationship between values of D and the correlation of allelic state, but as a general rule we can say that as linkage disequilibrium in a population declines, and as D moves from 0.25 or -0.25 toward 0, the correlation of allelic state declines as well, moving toward 0 from 1.0 or -1.0. Clegg and colleagues predicted that this is just what they would see in their freely mating fruit fly populations.

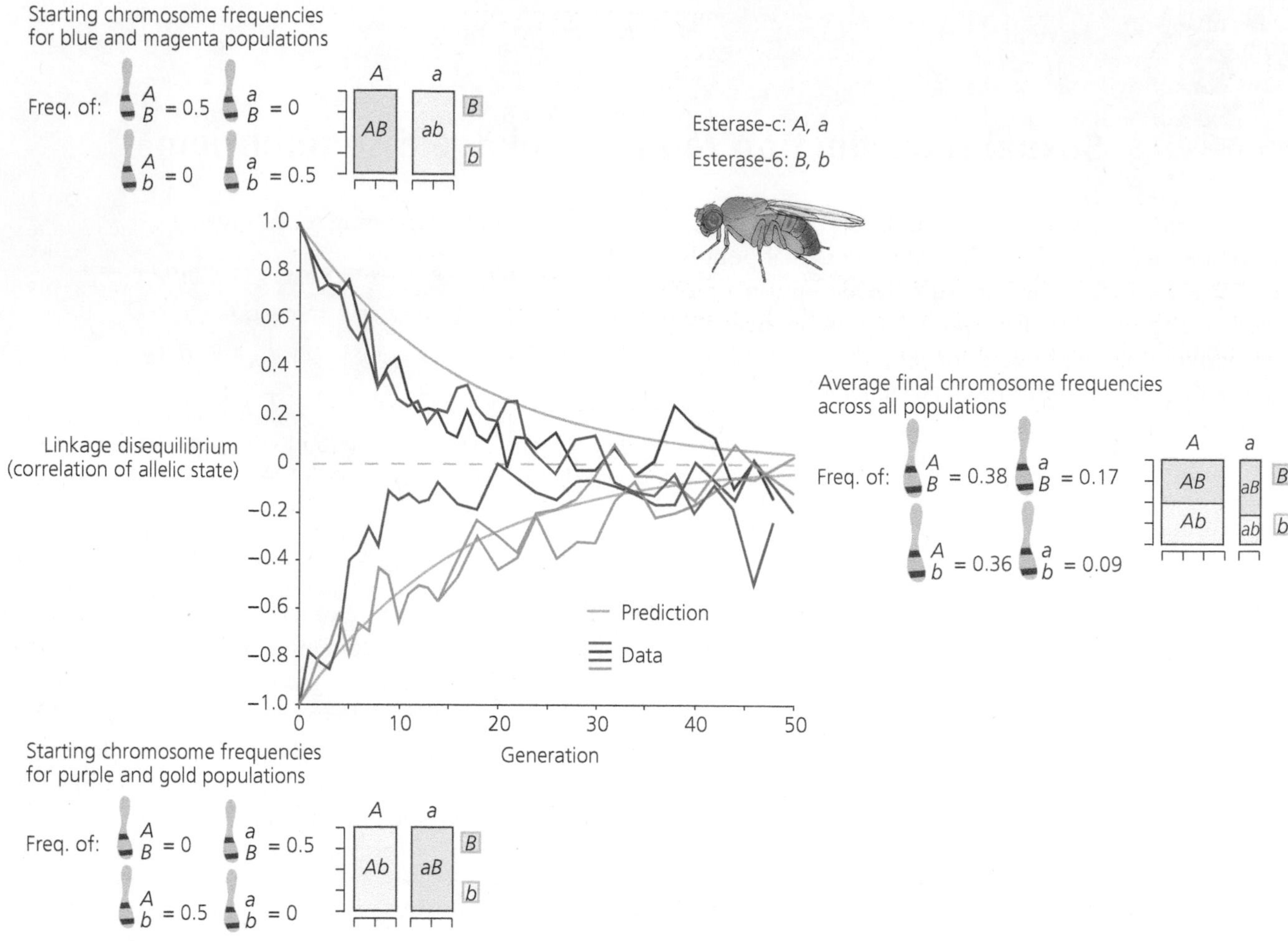

Figure 8.8 An empirical demonstration that sexual reproduction reduces linkage disequilibrium Each of several populations of fruit flies began in complete linkage disequilibrium (bar graphs at upper and lower left). Over 50 generations, all populations approached linkage equilibrium (bar graph at right). Redrawn from Clegg et al. (1980); frequencies in the bar graph at right are inferred from data therein.

The results appear in **Figure 8.8**. The smooth gray curves show the predicted pattern of decline; the jagged colored lines show the data. As predicted, crossing over during meiosis created the missing chromosome haplotypes, and the linkage disequilibrium between the loci declined. Indeed, linkage disequilibrium declined somewhat faster than predicted. Clegg and colleagues believe that the faster-than-expected decline was the result of heterozygote superiority at the enzyme loci they were studying. Heterozygote superiority would increase the frequency of individuals heterozygous for both loci and thus provide more opportunities for crossing over to break down nonrandom associations between alleles at one locus and alleles at the other.

Why Does Linkage Disequilibrium Matter?

We have defined linkage disequilibrium as a nonrandom association between genotypes at different loci. We have identified multilocus selection, genetic drift, and population admixture as evolutionary mechanisms that can create it. We have seen that sexual reproduction reduces it, restoring populations to a state of linkage equilibrium. And we have demonstrated that in an ideal Hardy–Weinberg population that is in linkage equilibrium, chromosome frequencies do not change from one generation to the next. We have not, however, addressed what we said was to be this section's primary goal: determining whether selection at a locus can interfere with our ability to use single-locus models to predict the course of evolution at other loci. We are ready to do so now.

The Bad News about Linkage Disequilibrium

The bad news is that if locus A and locus B are in linkage disequilibrium, then selection at locus A changes the frequencies of the alleles at locus B. This means that a single-locus population genetics model looking only at locus B will make inaccurate predictions about evolution.

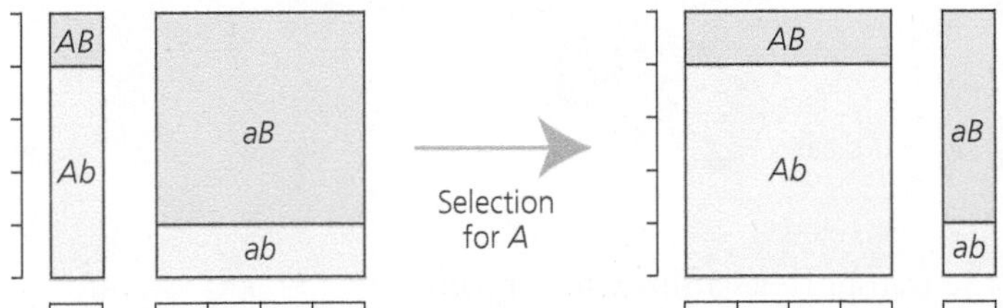

Figure 8.9 Linkage disequilibrium, selection, and allele frequencies at a linked locus In a population in linkage disequilibrium, selection in favor of allele *A* at locus A changes the frequency of allele *B* at locus B.

Figure 8.9 illustrates how selection on locus A can change allele frequencies at locus B. Before selection, allele *b* is at low frequency overall. However, among *A*-bearing chromosomes, which are relatively rare, the frequency of *b* is high. Selection in favor of allele *A* increases the frequency of both *AB* and *Ab* chromosomes. But because there are more *Ab* chromosomes than *AB* chromosomes, the result at locus B is that the frequency of allele *b* rises.

When a pair of loci are in linkage disequilibrium, selection at one locus can change allele frequencies at the other locus. This means that single-locus models may make inaccurate predictions.

Note that in this scenario, selection has acted only at locus A. Genotypes at locus B had no effect on fitness. Instead, the frequency of allele *b* rose simply because copies of the allele got carried along for the ride. If we had been monitoring only locus B, and watching the frequency of allele *b* rise over time, we might erroneously have concluded that the target of selection was locus B itself. Allele *b* could even have been mildly deleterious, instead of merely neutral, if the selective advantage conferred by allele *A* were sufficiently high.

We introduced this phenomenon earlier (in Chapter 7), under the name **genetic hitchhiking.** Hitchhiking leads to the most depressing lesson of the two-locus version of Hardy–Weinberg analysis: Because of linkage disequilibrium, single-locus studies can yield misleading conclusions.

An example comes from the locus on human chromosome 5 that encodes the ergothioneine transporter. This gene caught the attention of human geneticists because one of its alleles, called *L503F*, is statistically associated with an individual's risk of developing Crohn's disease, a serious autoimmune disorder of the digestive tract (Peltekova et al. 2004). Allele *L503F* has in its coding region a single nucleotide substitution—a C-to-T transition—that results in the substitution of a phenylalanine for a leucine at amino acid position 503. As shown in **Figure 8.10**, individuals with genotype *TT* are about 1.5 times more likely to exhibit Crohn's disease than individuals with genotype *CC* (Wang et al. 2011).

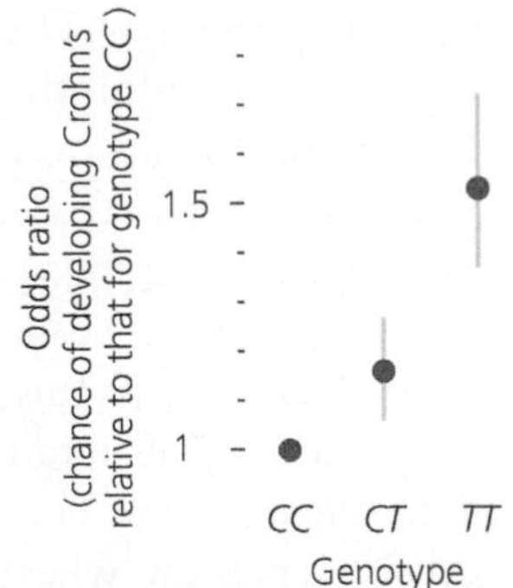

Figure 8.10 Ergothioneine transporter genotype is statistically associated with the risk of Crohn's disease Whiskers show 95% confidence intervals. Redrawn from Wang et al. (2011).

Ergothioneine is made by fungi but found in most plants and animals (Gründemann et al. 2005). Plants absorb it via their roots; animals obtain it in their diet. The ergothioneine transporter enables cells to take up the chemical and sequester it. This capacity appears to be adaptive because ergothioneine is an antioxidant. It protects DNA, proteins, and lipids from oxidative stress (Paul and Snyder 2010).

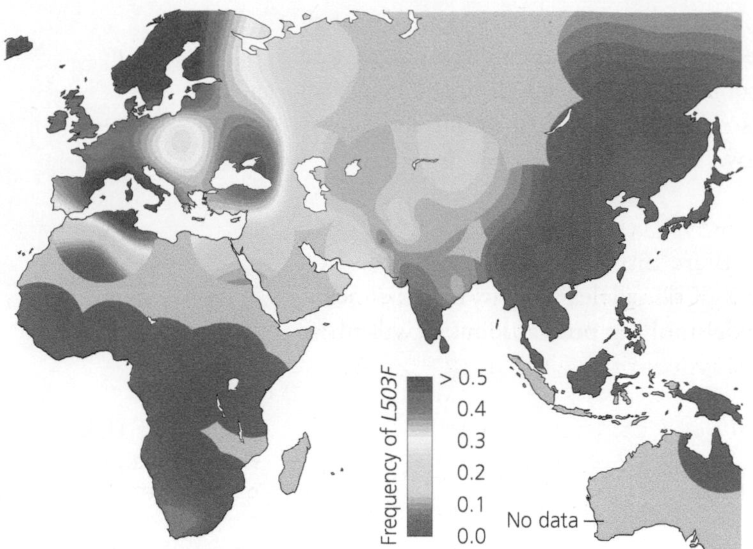

Figure 8.11 The frequency of ergothioneine transporter allele *L503F* in Old World human populations Based on 1,726 individuals from 85 populations. From Huff et al. (2012).

The substitution in allele *L503F* increases the transporter's affinity for ergothioneine and boosts the efficiency of uptake (Taubert et al. 2005). Individuals with genotype *TT* accumulate higher concentrations of ergothioneine in their tissues than individuals with genotype *CC* (Taubert et al. 2009). The obvious implication is that elevated concentrations of ergothioneine may contribute to Crohn's disease. Biologists have expended some effort trying to ascertain whether this is the case and, if so, what the physiological mechanism might be (Ey et al. 2007; Petermann et al. 2009). There is, however, an alternative explanation.

The ergothioneine transporter gene is in linkage disequilibrium with several other nearby genes, any of which could be the locus at which allelic variation influences the risk of Crohn's disease (Silverberg et al. 2007). Chad Huff and colleagues (2012) examine this explanation in detail.

Figure 8.11 shows the frequency of *L503F* across the Old World. In parts of Europe, the frequency is 50% or more. Beyond the Middle East, it is low. Huff and colleagues hypothesize that *L503F* appeared as a unique mutation in a single copy of chromosome 5. The allele then rose to high frequency due to natural selection, carrying with it the multilocus genotype of which it happened to be a part. By a scenario similar to that in Figure 8.5, this event created linkage disequilibrium that has since been decaying under the influence of recombination. Based on the current level of linkage disequilibrium, and on estimates of the recombination rate, Huff and colleagues calculate the age of the *L503F* allele to be about 12,000 years. We discuss this method of estimating allele age in Section 8.2.

Huff and colleagues note that the estimated age of *L503F* makes the appearance of the allele roughly contemporaneous with the earliest evidence of farming. The first farmers grew wheat, barley, peas, and lentils, all of which are low in ergothioneine. Huff and colleagues conjecture that this shift in diet made *L503F*, which might otherwise have been lost to drift or negative selection, adaptive.

Huff and colleagues suggest that two genes located near the ergothioneine transporter gene are better candidates as causal factors influencing the risk of Crohn's disease **(Figure 8.12a)**. *IRF1* encodes interferon regulatory factor 1. *IL5* encodes interleukin 5. Both gene products play a role in the immune response.

Huff and colleagues determined the frequency of chromosomes carrying *L503F* among 1,868 patients with Crohn's disease and 5,540 controls. The researchers

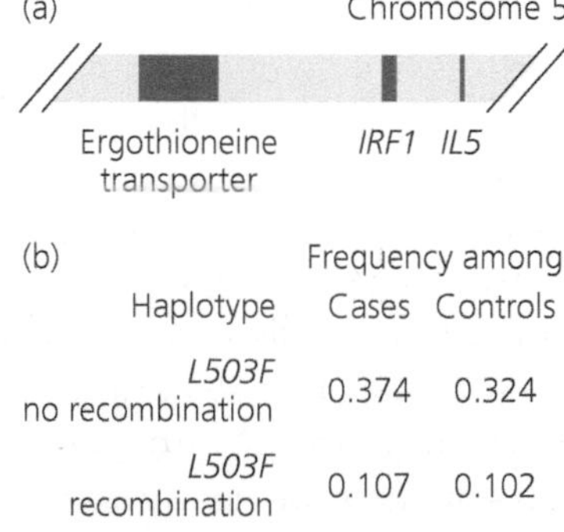

(b)

Haplotype	Frequency among Cases	Controls
L503F no recombination	0.374	0.324
L503F recombination	0.107	0.102

Figure 8.12 Recombination breaks the association between *L503F* and Crohn's disease (a) Map of chromosome 5 near the ergothioneine transporter gene. (b) The association between Crohn's and *L503F* chromosomes in which there has not been recombination between the ergothioneine transporter gene and *IL5* is statistically significant ($p = 2.6 \times 10^{-8}$). The association between Crohn's and *L503F* chromosomes in which there has been recombination is not significant ($p = 2.1$). From Huff et al (2012).

split the *L503F* chromosomes into two groups: (1) those in which the original multilocus genotype where *L503F* arose has been disrupted by recombination between the ergothioneine transporter gene and *IL5* and (2) those in which that haplotype has not been disrupted. As shown in Figure 8.12b, nonrecombinant *L503F* chromosomes occur at significantly higher frequency in Crohn's cases than in controls. Recombinant *L503F* chromosomes occur at nearly the same rate in both groups. That is, recombination between the ergothioneine transporter gene and *IL5* breaks the statistical association between the *L503F* allele and Crohn's disease. This is consistent with the hypothesis that the apparent connection between Crohn's and ergothioneine is a spurious result of linkage disequilibrium, and that the true culprit is an as-yet-unidentified allele of *IRF1* or *IL5*.

The Good News about Linkage Disequilibrium

The good news is that if linkage disequilibrium is absent—if locus A and locus B are in linkage equilibrium—then selection on locus A has no effect whatsoever on allele frequencies at locus B.

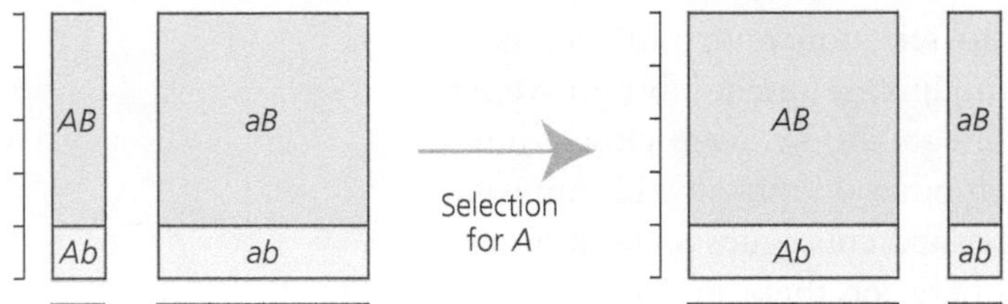

Figure 8.13 Linkage equilibrium, selection, and allele frequencies at a linked locus In a population in linkage equilibrium, selection in favor of allele *A* at locus A has no effect on the frequency of allele *B* at locus B.

Look at **Figure 8.13**. Selection in favor of allele A again eliminates many *aB* chromosomes. But because the frequency of B is the same among *A*-bearing chromosomes as among *a*-bearing chromosomes, every copy of allele *B* that is lost is replaced by another copy of *B*. If selection at locus A has no effect on allele frequencies at locus B, then it will not interfere with our use of single-locus models to analyze locus B's evolution.

Still better news is that in random-mating populations, sex is so good at eliminating linkage disequilibrium that most pairs of loci are in linkage equilibrium most of the time. Research by Elisabeth Dawson and colleagues (2002) illustrates this claim. Dawson and colleagues surveyed linkage disequilibrium among 1,504 marker loci on human chromosome 22 in a population of European families. The marker loci were sites showing allelic variation in single nucleotides or small insertions or deletions. **Figure 8.14a** shows the average level of linkage

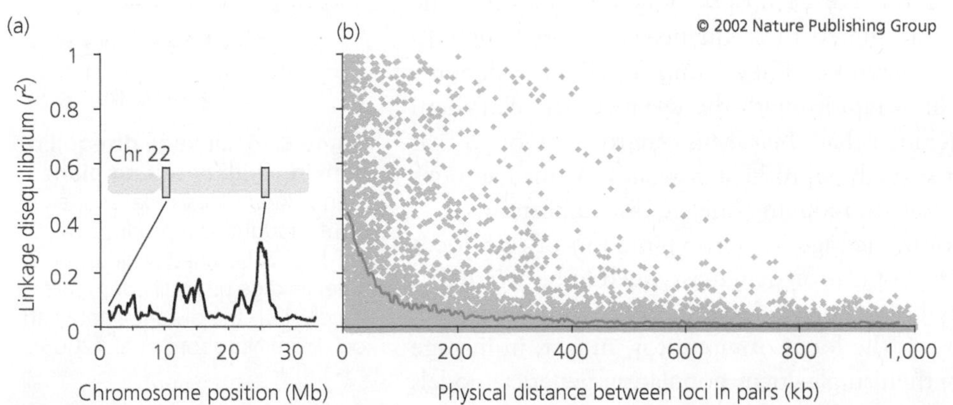

Figure 8.14 On human chromosome 22, most pairs of loci are in linkage equilibrium (a) Calculations of the average linkage disequilibrium (squared correlation of allelic state) among nearby loci reveal localized peaks. (b) However, the disequilibrium between loci falls with distance. From Dawson et al. (2002).

Reprinted by permission from Macmillan Publishers Ltd: Dawson, E., G. R. Abecasis, S. Bumpstead, et al. 2002. "A first-generation linkage disequilibrium map of human chromosome 22. *Nature* 418: 544–548.

disequilibrium among marker loci within a 1.7-megabase-wide window as it slides along the long arm of the chromosome. The graph shows that nearby loci tend to be in linkage disequilibrium with each other in some regions. However, Figure 8.14b shows that the amount of linkage disequilibrium among 18,736 pairs of loci falls rapidly with the physical distance between the loci in the pairs. Pairs of loci separated by more than 1,000 kilobases—roughly 3% of the length of the long arm of the chromosome—show virtually no linkage disequilibrium.

When a pair of loci are in linkage equilibrium, selection at one locus has no effect on allele frequencies at the other, and we can use single-locus models with confidence. Fortunately, sex is so good at reducing linkage disequilibrium that most pairs of loci are in linkage equilibrium most of the time.

Gavin Huttley and colleagues (1999) surveyed the entire human genome for linkage disequilibrium among short tandem repeat loci. This type of locus is a spot where a short nucleotide sequence is repeated several times. Such loci typically have several alleles. Huttley and colleagues conducted 200,000 pairwise tests of linkage disequilibrium involving 5,000 loci distributed across all 22 autosomes. Like Dawson, Huttley and colleagues found several places where neighboring loci exhibit substantial linkage disequilibrium. One such region, known from earlier studies, is an area on chromosome 6 containing the human leukocyte antigen (HLA) loci. The HLA loci encode proteins the immune system uses to recognize foreign invaders. The HLA loci are under strong selection, and the disequilibrium among them is probably due to selection on multilocus genotypes. Overall, however, pairs of loci exhibiting linkage disequilibrium were in the minority. The pairs most likely to show disequilibrium were closely linked physically—that is, situated near enough to each other on the same chromosome that crossing over between them is rare. Huttley and colleagues focused on pairs of loci close enough that crossing over occurs between them in 4% or fewer of meiotic cell divisions. Of these pairs, just 4% exhibited linkage disequilibrium.

The International HapMap Consortium (2005) assembled a database of over 1 million loci in the human genome with allelic variation at single nucleotides, and determined the complete genotypes at these sites for 269 individuals in four populations. Researchers have used these data to analyze patterns of linkage disequilibrium across the entire genome (see McVean et al. 2005 for an overview). They have confirmed the general findings of earlier studies. There are blocks of sequence throughout the genome in which nearby loci are in linkage disequilibrium, but on a larger scale disequilibrium falls rapidly with distance between loci (see De La Vega et al. 2005; International HapMap Consortium 2005).

In a similar study, Magnus Nordborg and colleagues (2005) surveyed the genome of the plant *Arabidopsis thaliana,* a member of the mustard family. One might expect the *Arabidopsis* genome to harbor considerable linkage disequilibrium, even among loci far apart or on different chromosomes. This is because *Arabidopsis* typically self-fertilizes, which leads to increased homozygosity and reduced opportunity for recombination. For each of 96 plants from around the world, Nordborg and colleagues determined the sequences of short fragments of 876 loci scattered throughout the genome. They found that linkage disequilibrium between pairs of loci declines rapidly with the distance between them **(Figure 8.15)**. The researchers concluded that *Arabidopsis* outcrosses enough that its genome resembles that of other sexually reproducing species. A small amount of recombination goes a long way toward reducing linkage disequilibrium.

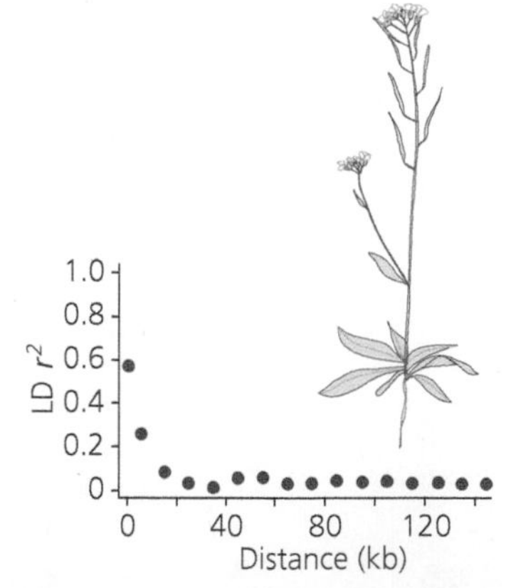

Figure 8.15 Linkage disequilibrium in a highly selfing plant Even in *Arabidopsis thaliana*, a plant thought to reproduce by self-fertilization nearly all of the time, linkage disequilibrium falls rapidly with the distance between loci. From Nordborg et al. (2005).

We can summarize the take-home message in our exploration of two-locus Hardy–Weinberg analysis as follows. Population geneticists need to be aware that any particular locus of interest may be in linkage disequilibrium with other loci, especially other loci located nearby. If the locus of interest is, in fact, in linkage disequilibrium with another locus, then single-locus population genetics models may yield inaccurate predictions. Nonetheless, in freely mating populations, most pairs of loci can be expected to be in linkage equilibrium. In general, we can expect that single-locus models will work well most of the time.

8.2 Practical Reasons to Study Linkage Disequilibrium

In the introduction to this chapter, we promised rewards awaiting readers who mastered the abstractions of Section 8.1. Two such rewards are these: Measurements of linkage disequilibrium provide clues that are useful in reconstructing the history of genes and populations; and linkage disequilibrium can be used to identify alleles that recently have been favored by positive selection.

Reconstructing the History of Genes and Populations

Ashkenazi Jews suffer a high incidence of some 20 rare genetic diseases (see Bray et al. 2010). Among them is type 1 Gaucher disease, caused by mutations in a gene on chromosome 1. The gene encodes the enzyme glucocerebrosidase, also known as acid β-glucosidase, or GBA (Beutler 1993). GBA is found inside lysosomes within cells, where it breaks down the lipid glucocerebroside for recycling. When GBA activity is low or absent, glucocerebroside accumulates in cells. For this reason, Gaucher disease is categorized as a lysosomal storage disorder. Symptoms include enlargement of the liver and spleen, anemia, and fragile bones.

While Gaucher disease occurs worldwide, it is most common among Ashkenazim, of whom 1 in 19 is a carrier (Strom et al. 2004). Most Ashkenazi carriers harbor a nonsynonymous substitution altering amino acid 370 in GBA (Beutler et al. 1991). This allele is also found in other populations. A substantial minority of Ashkenazi carriers, however, harbor an allele exclusive to the Ashkenazim. This exclusive allele, called *84GG*, has an extra guanine inserted at nucleotide position 84. The insertion results in a frameshift and complete loss of GBA function.

George Diaz and colleagues (2000) wanted to know how long the *GBA-84GG* mutation has been circulating among the Ashkenazim. The researchers started by identifying a marker locus on chromosome 1 that is in linkage disequilibrium with the GBA locus (**Figure 8.16a**). This locus, called D1S305, is a short tandem repeat polymorphism. The linkage disequilibrium between D1S305 and the GBA locus is diagrammed in Figure 8.16b. The frequency of the *8-repeat* allele of D1S305 is at a frequency of just 24% among chromosomes carrying the + allele at the GBA locus, but 59% among chromosomes carrying the *84GG* allele.

How did this linkage disequilibrium arise? As discussed in Section 8.1, there are three possibilities: selection on multilocus genotypes, genetic drift, and

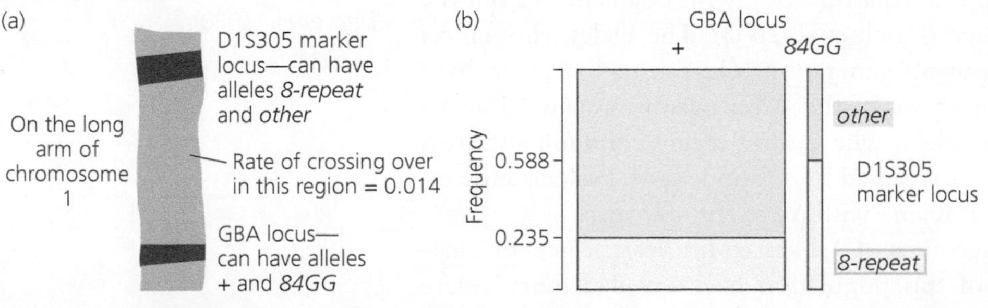

Figure 8.16 A marker for dating the most recent common ancestor of extant copies of *GBA-84GG* (a) D1S305 is a short tandem repeat locus near the GBA locus on chromosome 1. (b) D1S305 and the GBA locus are in linkage disequilibrium. After Diaz et al. (2000).

population admixture. Selection on multilocus genotypes is an unlikely candidate, because D1S305 is a noncoding locus and its alleles appear to be selectively neutral. Population admixture is also unlikely, because it would require a source population in which the frequency of the *84GG* allele is much higher than it is in the Ashkenazim, and no such population exists. That leaves genetic drift.

Because the list of mechanisms that create linkage disequilibrium is short, the presence of linkage disequilibrium in a population provides clues to the population's past.

Diaz and colleagues believe that, similar to the scenario in Figure 8.5, all copies of *84GG* now circulating among the Ashkenazim are derived from a single common ancestor that carried the *8-repeat* allele at D1S305. The ancestral copy of *84GG* might have been a new mutation that arose in an Ashkenazi individual. It might have been a mutation carried into the Ashkenazi population by a single heterozygous migrant from a population in which *84GG* was subsequently lost. Or it might have been the only copy to survive the genetic drift that accompanied a bottleneck, or reduction in the size of the Ashkenazi population.

When there was only one copy of *84GG* in the population, the linkage disequilibrium between the GBA locus and D1S305 was complete. The frequency of the *8-repeat* allele on chromosomes carrying *84GG* was 100%. Soon, however, *84GG* had produced multiple descendants and the disequilibrium began to break down. Crossing over swapped other alleles at the short tandem repeat locus onto chromosomes carrying *84GG*. Eventually the frequency of the *8-repeat* allele will be the same on *84GG* chromosomes as it is on normal chromosomes.

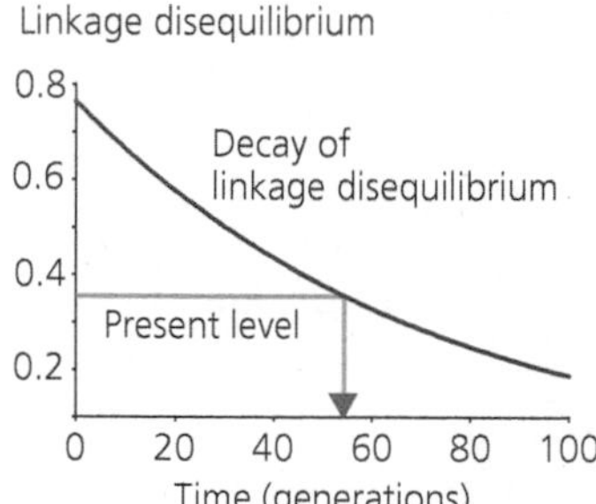

Figure 8.17 Dating the most recent common ancestor of the *GBA-84GG* allele The purple curve shows the decay in linkage disequilibrium between the GBA locus and D1S305 since the time of the most recent common ancestor. Linkage disequilibrium is quantified as the difference between the frequency of the *8-repeat* allele on normal chromosomes versus *84GG* chromosomes. The orange lines show the present level of linkage disequilibrium and the inferred number of generations elapsed since the most recent common ancestor. If the generation time is 25 years, then the most recent common ancestor of extant *84GG* alleles existed about 1,375 years ago. Based on data and calculations in Diaz et al. (2000).

Diaz and colleagues measured linkage disequilibrium as the difference between the frequency of the *8-repeat* allele on *84GG* chromosomes versus normal chromosomes. They used the rate of crossing over between the GBA locus and D1S305 to estimate the rate at which the disequilibrium is decaying (**Figure 8.17**, purple curve). Then they used the rate of decay and the present level of disequilibrium to calculate that the most recent common ancestor of all extant *84GG* alleles existed between 750 and 2,325 years ago, with a best estimate of 1,375 years ago (Figure 8.17, orange lines; see **Computing Consequences 8.4** for details).

Analysis of linkage disequilibrium can thus allow us to reconstruct the history of an allele. But it can do more, because the history of alleles illuminates the history of populations. As we have mentioned, the Ashkenazim carry at unusually high frequency alleles causing a number of other genetic disorders, including Tay-Sachs disease, Fanconi anemia type C, and elevated risk of breast cancer. Neil Risch and colleagues (2003) reviewed efforts, including that of Diaz and colleagues, to estimate the ages of the most recent common ancestors of 11 Ashkenazi disease alleles. The ages fall roughly into three categories: about 12 generations old, about 50 generations old, and more than 100 generations old.

These dates are broadly consistent with the history of the Ashkenazi population. The Ashkenazim trace their recent ancestry to eastern and central Europe and their more ancient ancestry to the Middle East. The disease alleles whose most recent common ancestors are over 100 generations old can be explained by a founder effect accompanying the departure of Jewish populations from the Middle East 2,000 to 3,000 years ago (Risch et al. 2003). The alleles whose most recent common ancestors are about 50 generations old can be explained by a founder effect associated with the arrival of the Ashkenazim in central Europe 1,000 to 1,500 years ago. And the alleles whose most recent common ancestors are about 12 generations old can be explained by a founder effect associated with the arrival of the Ashkenazim in Lithuania within the last 400 years.

While they generally confirm accounts of Ashkenazi history, the genetic analyses suggest that the migrations of this population have entailed more severe

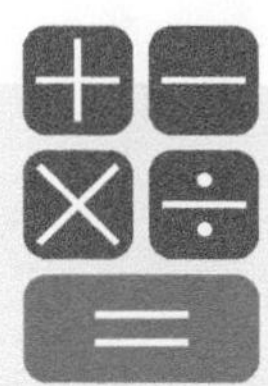

COMPUTING CONSEQUENCES 8.4

Estimating the age of the *GBA-84GG* mutation

Here we outline the calculation George Diaz and colleagues (2000) used to estimate the age of the most recent common ancestor of all copies of the *GBA-84GG* mutation in the Ashkenazim (see also Slatkin and Rannala 2000). The GBA locus is on chromosome 1 (see Figure 8.16). For our purposes it has two alleles: the normal allele, +, and *84GG*. Nearby is the short tandem repeat locus D1S305, with alleles *8* (for the number of repeats) and *other*.

The GBA locus and D1S305 are in linkage disequilibrium: The frequency of allele *8* is higher on chromosomes carrying *84GG* than on chromosomes carrying +. Assuming this disequilibrium is not being maintained by selection on multilocus genotypes, it will be in the process of decaying due to recombination. Diaz and colleagues developed an equation to describe this decay.

They started with an equation that predicts the frequency of *8* on *84GG* chromosomes in any given generation from its frequency in the generation before. The researchers assumed that *84GG* chromosomes are rare enough that they virtually always pair with + chromosomes, that the frequency of *8* on + chromosomes is constant over time, and that there is no mutation. The equation is

$$X_t = (1 - c)X_{t-1} + cY$$

where X_t is the frequency of *8* on *84GG* chromosomes in generation t, Y is the frequency of *8* on + chromosomes, and c is the rate of crossing over between the GBA locus and the short tandem repeat locus. The first term on the right accounts for the *84GG–8* chromosomes that do not experience recombination. The second accounts for the *84GG* chromosomes that receive a copy of the *8* as a result of recombination.

Subtracting Y from both sides gives

$$X_t - Y = (1 - c)X_{t-1} - Y + cY$$

Factoring $-Y$ out of the two rightmost terms gives

$$X_t - Y = (1 - c)X_{t-1} - Y(1 - c)$$

And now factoring $(1 - c)$ out on the right gives

$$X_t - Y = (1 - c)(X_{t-1} - Y)$$

We can think of $(X_t - Y)$ as a measure of linkage disequilibrium, because if GBA and D1S305 were in equilibrium it would be zero. Our equation thus describes the decay of disequilibrium. Compare it to the equation we derived in Computing Consequences 8.3.

By our equation, each generation the difference between the frequency of *8* on *84GG* versus + chromosomes declines by a factor of $(1 - c)$. This implies that

$$X_t - Y = (1 - c)^t(X_0 - Y)$$

where X_0 is the frequency of *8* on *84GG* chromosomes in the generation in which the *84GG* mutation last appeared in the population as just a single gene copy. Note that when the population contained only one copy of *84GG* (which was on an *84GG–8* chromosome), the frequency of *8* on *84GG* chromosomes was 100%. In other words, $X_0 = 1$. Thus

$$X_t - Y = (1 - c)^t(1 - Y)$$

Dividing both sides by $(1 - Y)$ and taking the natural logarithm of both sides allows us to solve for t:

$$\ln\left[\frac{(X_t - Y)}{(1 - Y)}\right] = \ln[(1 - c)^t]$$

$$\ln\left[\frac{(X_t - Y)}{(1 - Y)}\right] = t\ln(1 - c)$$

$$t = \frac{1}{\ln(1 - c)} \times \ln\left[\frac{(X_t - Y)}{(1 - Y)}\right]$$

All we have to do to estimate t is plug estimates of X_t, Y, and c into this equation.

Based on a sample of 85 + chromosomes and 58 *84GG* chromosomes, Diaz and colleagues estimated that $X_t = 0.588$ and $Y = 0.235$. Their estimate of the recombination rate, c, was 0.014. These values put the time of the last common ancestor of all copies of *84GG* at 55 generations ago. If the generation time is 25 years, the last common ancestor existed about 1,375 years ago. This estimate is sensitive to the recombination rate, which is small and hard to measure accurately. Allowing for error in the recombination rate, Diaz and colleagues concluded that the *84GG* carrier of the ancestral allele probably lived between 750 and 2,325 years ago.

bottlenecks than might otherwise have been appreciated. There is debate over whether genetic drift is sufficient to explain the high frequencies of all Ashkenazi disease alleles. Some researchers have concluded that it is (Behar et al. 2004; Slatkin 2004; Bray et al. 2010); others invoke selection favoring heterozygotes for at least some categories of alleles (Cochran et al. 2006).

History of the CCR5 Gene

Although not associated with a genetic disease, another allele that is surprisingly common among the Ashkenazim, as well as other European populations, is *CCR5-Δ32*. This allele (which we discussed in Chapters 1, 5, and 6) is a loss-of-function mutation at the CCR5 locus. It protects homozygotes against sexually transmitted strains of HIV-1. Unresolved issues from our earlier discussions are where the *Δ32* allele came from and why is it common only in Europe.

J. Claiborne Stephens and colleagues (1998) addressed these issues with an analysis similar to that by Diaz and colleagues on *GBA-84GG*. Stephens and colleagues found that the CCR5 locus is in strong linkage disequilibrium with two nearby marker loci, both noncoding and apparently neutral. Most chromosomes carrying the *Δ32* allele at the CCR5 locus also carry a specific haplotype at the marker loci. This suggests that the *Δ32* allele arose just once, as a unique mutation. As the *Δ32* allele subsequently rose to high frequency, the marker alleles that happened to be linked with it came along for the ride.

The linkage disequilibrium between the CCR5 locus and the marker loci is no longer perfect. Since *Δ32* appeared, recombination and/or additional mutations have put the allele into new haplotypes. Stephens and colleagues used estimates of the rates of crossing over and mutation to calculate how fast the disequilibrium is breaking down, then used the result to estimate the age of the last common ancestor of all extant *Δ32* copies. They concluded that the common ancestor lived 275 to 1,875 years ago, with a best estimate of about 700 years ago.

This tantalizing result implied that the frequency of *Δ32* in Europe had risen from virtually zero to 15% or more in roughly 30 generations. Such a rapid climb can be explained most readily by strong natural selection. What might have been the selective agent that gave the *Δ32* allele such an advantage? The obvious suspects were epidemic diseases. One was the Black Death (Stephens et al. 1998), which swept Europe during the 14th century and killed a third of the population. Another was smallpox (Lalani et al. 1999; Galvani and Slatkin 2003).

John Novembre, Alison Galvani, and Montgomery Slatkin (2005) developed a population genetics simulation of how an allele might increase in frequency as it spread across Europe. Their model included parameters describing the distance individuals move in a lifetime and the strength of selection. Consistent with the epidemic hypothesis, the model suggested that *Δ32* could not have achieved its present distribution as fast as it apparently did without the aid of natural selection.

The story did not end there, however. Pardis Sabeti and colleagues (2005) pointed out that Stephens and colleagues had based their calculations on a genetic map that turned out to be flawed. The marker loci Stephens used are closer to the CCR5 locus, and the recombination rates between the markers and CCR5 are therefore lower than first thought. Using a larger set of genetic markers, Sabeti and colleagues calculated that the common ancestor of all extant copies of the *Δ32* allele lived between 3,150 and 7,800 years ago, with a best estimate of 5,075 years ago. On this and other evidence, Sabeti and colleagues argued that the current frequency and distribution of *Δ32* can be explained by genetic drift.

Corroboration came from Susanne Hummel and colleagues (2005; see also Hedrick and Verrelli 2006), who recovered DNA sequences from the CCR5 locus in the skeletons of 17 Bronze Age Europeans. The skeletons' owners lived 2,900 years ago in what is now northwestern Germany and were buried in Lichtenstein Cave. Four were heterozygous carriers of the $\Delta 32$ allele. This confirms that the allele is at least a few thousand years old. It also puts the Bronze Age frequency of the allele at about 12%—well within the modern range. Given this older age for $\Delta 32$, Novembre's model no longer rules out Sabeti's genetic drift hypothesis (Novembre and Han 2012).

In summary, the $\Delta 32$ allele appears to have been created by a unique mutation that occurred in Europe within the past several thousand years. The allele does not occur outside Europe, either because the mutation creating it has never occurred in a non-European population or because when the mutation has occurred outside Europe, it has not been favored by selection. Whether the allele was ever favored by selection in Europe—and if so, why—remains uncertain.

Detecting Positive Selection

We have seen that a unique mutation, by the mere fact of its birth, puts its locus in linkage disequilibrium with nearby markers. This linkage disequilibrium immediately begins to break down. This means that when we find a locus in linkage disequilibrium with nearby markers, we suspect that the locus may harbor a young allele. If a young allele is at high frequency, we suspect that during its short life the allele has been favored by positive natural selection. Using this logic, Pardis Sabeti, David Reich, and colleagues (2002) developed a general method for identifying alleles recently favored by selection. Sabeti and colleagues demonstrated their method by applying it to the G6PD locus in humans.

The G6PD locus, located on the X chromosome, encodes an essential housekeeping enzyme called glucose-6-phosphate dehydrogenase (Ruwende and Hill 1998). The locus is highly variable. Hundreds of alleles are known, distinguishable by the encoded protein's biochemical properties. Dozens of these variants reach frequencies of 1% or more. And many common alleles have reduced enzymatic activity **(Figure 8.18)**. Indeed, with 400 million people affected worldwide,

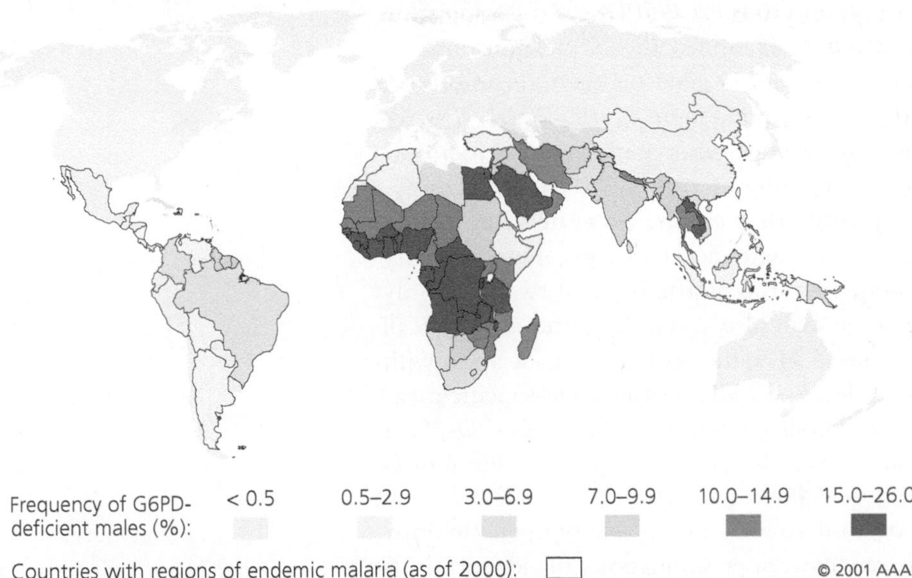

Figure 8.18 Glucose-6-phosphate dehydrogenase deficiency is common, especially in regions with malaria This map shows the frequency of G6PD deficiency in various parts of the world. Regions where malaria is common are outlined in black. Redrawn from Luzzatto and Notaro (2001) and Centers for Disease Control, Division of Parasitic Diseases (2001).

From "Protecting against bad air." *Science* 293: 442–443. Reprinted with permission from AAAS.

G6PD deficiency is the most common enzyme deficiency known. Individuals with mild G6PD deficiency often have no symptoms, but individuals with more severe deficiencies can suffer episodes in which their red blood cells rupture, a condition known as acute hemolytic anemia.

Why is this potentially serious genetic condition so common? The geographic distribution of G6PD deficiency suggests that it confers some resistance to malaria. This inference is supported by epidemiological evidence. For example, individuals carrying the allele *G6PD-202A*, a reduced-activity variant common in Africa, have a substantially lower risk of suffering severe malaria. Sabeti, Reich, and colleagues reasoned that if the *G6PD-202A* allele confers resistance to malaria, then it should bear the signature of recent positive selection.

To see what this signature might look like, it helps to think about alleles that have not experienced recent selection. Imagine a new mutant allele appearing in a finite population. The allele is unique and consequently exists only in a single haplotype, physically linked to the particular alleles found at other loci on the chromosome in which it arose. If it is neutral, meaning that its frequency evolves by genetic drift, our new allele will experience one of three fates. It may disappear. It may persist but remain rare. Or it may persist and gradually drift to high frequency. If our allele persists, its association with a particular haplotype will break down under the influence of recombination. The farther away the other loci are, the more rapidly the association will break down.

In a population evolving by mutation and genetic drift, we can therefore expect to find three kinds of alleles. Some alleles will be rare and, because they are young, strongly associated with a particular haplotype. Some alleles will be rare and, because they are old, weakly associated with a particular haplotype. And some alleles will be common and, because they are old, weakly associated with a particular haplotype. What we do not expect to find is alleles that are common and strongly associated with a particular haplotype.

When an allele at a coding locus is in linkage disequilibrium with alleles at nearby neutral marker loci, we can infer that the coding allele is relatively young. When a young allele is at high frequency, we can infer that it has recently been favored by positive selection.

The signature of recent positive selection is thus a high frequency combined with strong association with a particular haplotype. The higher the frequency, and the farther the association extends from the locus of interest, the stronger the recent selection must have been.

The *G6PD-202A* allele has a frequency of about 18% in the three African populations Sabeti and colleagues studied. To assess *G6PD-202A*'s association with a particular haplotype, the researchers examined the X chromosomes of 230 men. First the scientists looked at the *G6PD* gene on each chromosome. They found nine distinct alleles of the gene, among them *G6PD-202A*. The researchers next determined each X chromosome's genotype for marker loci outside the *G6PD* gene, located at distances ranging up to 413,000 base pairs away.

Sabeti and colleagues calculated a quantity they call the **extended haplotype homozygosity,** or **EHH.** A given allele's extended haplotype homozygosity to a particular outside distance x is defined as the probability that two randomly chosen chromosomes carrying the allele will also carry the same alleles at all marker loci out to x. The higher an allele's EHH, the stronger its association with a particular haplotype. Another way to describe EHH is as an allele-specific measure of linkage disequilibrium. Sabeti and colleagues found that the *G6PD-202A* allele had higher linkage disequilibrium, extending farther away from the *G6PD* gene, than any of the other alleles **(Figure 8.19a)**.

Is this a strong enough signature of positive selection to rule out genetic drift? To find out, Sabeti and colleagues ran computer simulations of genetic drift,

© 2002 Nature Publishing Group

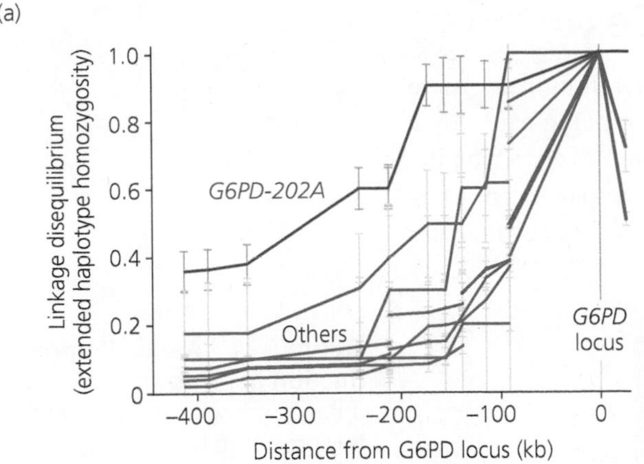

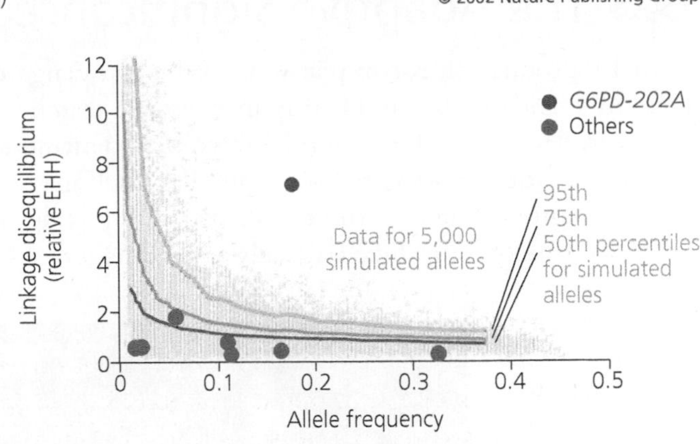

Figure 8.19 The signature of recent positive selection (a) *G6PD-202A* (red) has higher linkage disequilibrium, extending farther, than other alleles (blue) of *G6PD*. (b) *G6PD-202A*'s disequilibrium, and its high frequency, distinguish it from other alleles of *G6PD* and from neutral alleles in simulated populations evolving by drift. (An allele's relative EHH is its EHH at the most distant marker divided by the average EHH at that marker of all other alleles.) From Sabeti et al. (2002). Reprinted by permission from Macmillan Publishers Ltd: Sabeti, P. C., D. E. Reich, J. M. Higgins, et al. 2002. "Detecting recent positive selection in the human genome from haplotype structure." *Nature* 419: 832–837.

producing several thousand replicates of their actual data set. The frequencies of the alleles in these simulated data sets, and their EHH values at 413 kb, are plotted in the graph in Figure 8.19b. They form a gray cloud that clings to the horizontal and vertical axes. Consistent with the verbal argument we made earlier, neutral alleles evolving by drift can have a high frequency, or high linkage disequilibrium, but not both. *G6PD-202A* and the other alleles from the actual data set are also plotted in the graph. The other alleles, shown in blue, fall well within the gray cloud. Their numbers are easily explainable by drift. But *G6PD-202A*, shown in red, is clearly an outlier. Sabeti and colleagues concluded that *G6PD-202A* has recently been favored by natural selection. Matthew Saunders and colleagues (2005), analyzing additional data, reached the same conclusion.

The method developed by Sabeti and colleagues is applicable to other loci. Dong-Dong Wu and colleagues (2010) used it to show that an allele of a gene involved in skeletal development has recently been favored by positive selection in humans, primarily in European populations **(Figure 8.20)**. A different allele of the same gene appears to have been favored in Asian populations (see also Wu et al. 2012). The selective agent responsible is unclear. Failure of the method to reveal positive selection on *CCR5-Δ32* was among the results that suggested to Sabeti and colleagues (2005) that *Δ32* achieved its current frequency by drift.

Benjamin Voight, Jonathan Pritchard, and colleagues (2006) used an extension of Sabeti et al.'s method to scan the entire human genome for loci showing evidence of recent positive selection in East Asians, Europeans, and West Africans (Yoruba). Among the loci bearing, in one or more of these groups, the signature of positive selection—in the form of high frequency and high linkage disequilibrium—are genes involved in sperm motility and fertilization, olfaction, skin color, skeletal development, and carbohydrate metabolism.

We have shown that an understanding of linkage disequilibrium yields powerful tools for reconstructing the history of alleles and for detecting positive selection. An additional reward we promised readers was that understanding linkage disequilibrium would help us understand the adaptive significance of sexual reproduction. The mystery of sex is the subject of the next section.

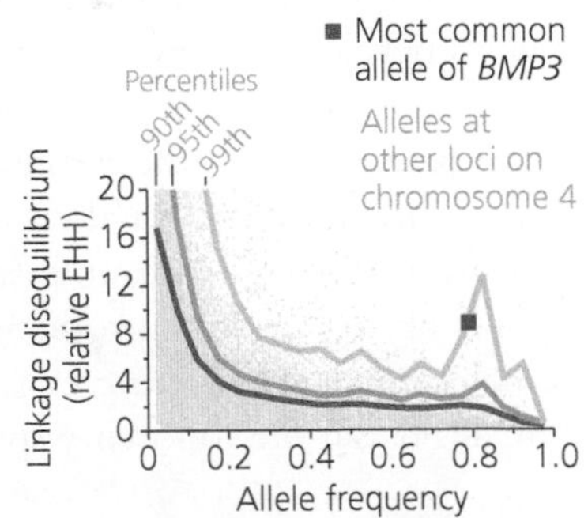

Figure 8.20 Evidence of recent positive selection on the most common allele of the gene *BMP3* *BMP3* encodes bone morphogenic protein 3, also known as osteogenin. The most common *BMP3* allele has a combination of high linkage disequilibrium with nearby markers and high frequency that is unusual among other loci on chromosome 4, and is unlikely to be due to drift. From Wu et al. (2010).

8.3 The Adaptive Significance of Sex

Sexual reproduction is complicated, costly, and dangerous. Searching for a mate takes time and energy, and it may increase the searcher's risk of being eaten by a predator. Once found, a potential mate may demand additional exertion or investment before agreeing to cooperate. Sex itself may expose the parties to sexually transmitted diseases. And for all that, the mating may prove to be infertile. Why not simply reproduce asexually instead **(Figure 8.21)**?

Figure 8.21 Asexual reproduction in an aphid The large aphid is giving birth to a daughter, produced by parthenogenesis, that is genetically identical to its mother. In the fall, the aphids will switch to sexual reproduction. Photo by MedievalRich.

This question sounds odd to our ears, because for us reproducing asexually is not an option. But for many organisms it is an option, at least in a physiological sense. They are capable of both sexual and asexual reproduction and regularly switch between the two. Many aphid species, for example, have spring and summer populations composed entirely of asexual females. These females feed on plant juices and, without the participation of males, produce live-born young genetically identical to their mothers. This mode of reproduction, in which offspring develop from unfertilized eggs, is called **parthenogenesis.** In the fall aphids change modes, producing males and sexual females. These mate, and the females lay overwintering eggs from which a new generation of parthenogenetic females hatch the following spring.

Many species are capable of both sexual and asexual reproduction.

(a) *Volvox* (b) *Hydra*

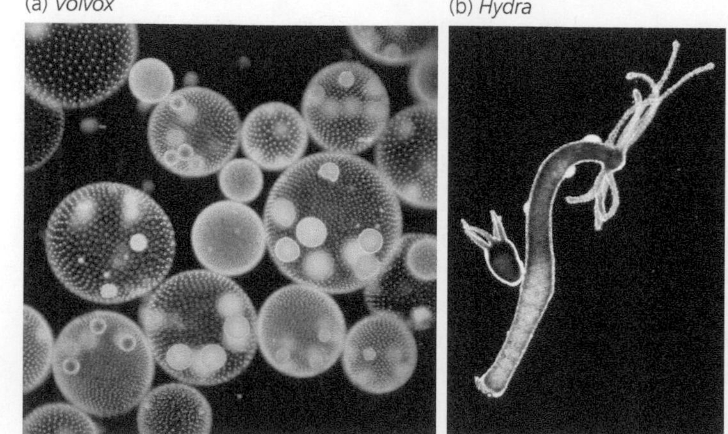

(c) Strawberry

Figure 8.22 Organisms with two modes of reproduction (a) *Volvox*, a freshwater alga. Each large sphere is an adult. Most are cloning themselves by growing daughters inside that are genetically identical to the parent. The individuals containing red spheres—encysted zygotes—are reproducing sexually as females. They have been fertilized by sperm from sexual males. (b) *Hydra*, a freshwater invertebrate. The crown of tentacles at the upper right surround the hydra's mouth. Along the body below the mouth are three gonads. Below the gonads is an asexual bud. (c) Strawberry. This plant is reproducing sexually via fruits with seeds, and asexually via runners.

Many other organisms are capable of both sexual and asexual reproduction **(Figure 8.22)**. *Volvox,* for example, like aphids, alternate between sexual and asexual phases (Figure 8.22a). *Hydra* can reproduce sexually and asexually at the same time (Figure 8.22b). So can the many species of plants that reproduce both by developing flowers that exchange pollen with other individuals and by sending out runners (Figure 8.22c).

Which Reproductive Mode Is Better?

The existence of two different modes of reproduction in the same population raises the question of whether one mode will replace the other over time. John Maynard Smith (1978) approached this question by developing a null model. The null model explores, under the simplest possible assumptions, the evolutionary fate of a population in which some females reproduce sexually and others reproduce asexually. Maynard Smith made just two assumptions:

1. A female's reproductive mode does not affect how many offspring she makes.
2. A female's reproductive mode does not affect the probability that her offspring will survive.

Maynard Smith also noted that all the offspring of a parthenogenetic female are themselves female, whereas the offspring of a sexual female are a mixture, typically with equal numbers of daughters and sons.

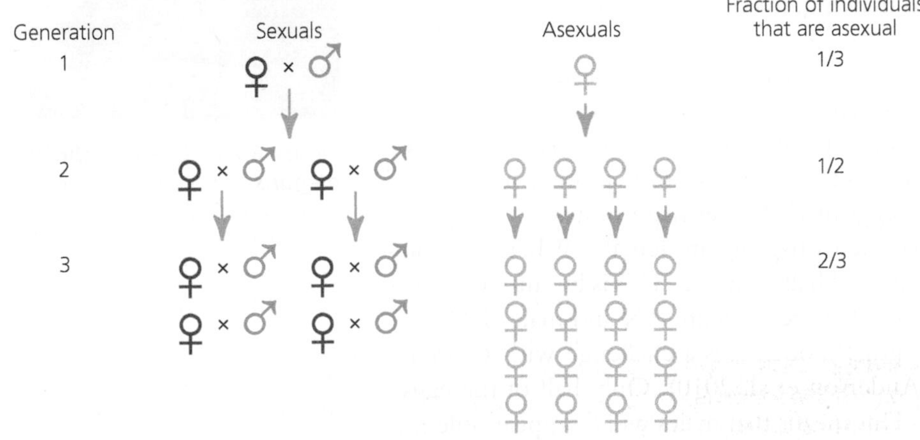

Figure 8.23 The reproductive advantage of asexual females The population imagined here is founded by a sexual female, a sexual male, and an asexual female. Each generation, each sexual female makes two daughters and two sons. Each asexual female makes four daughters. Under these simple assumptions, the fraction of individuals in the population that are asexual females increases every generation. After John Maynard Smith (1978).

In a population conforming to Maynard Smith's assumptions, asexual females produce twice as many grandchildren as sexual females **(Figure 8.23)**. This means that asexual females will constitute a larger fraction of the population each generation. Ultimately, asexual females should completely take over. In principle, all that would be required is for a mutation to produce a single asexual female in an otherwise exclusively sexual population. From the moment the mutation occurred, the population would be destined to be overwhelmed by asexuals.

And yet such asexual takeovers do not seem to have happened very often. The vast majority of multicellular species are sexual, and there are many species, like aphids, *Volvox*, and hydra, in which sexual and asexual reproduction stably coexist. Maynard Smith's model demonstrates, as he intended it to, that these facts represent a paradox for evolutionary theory.

The persistence of sex is a paradox, because a simple model shows that asexual females should rapidly take over any population.

Obviously, sex must confer benefits that allow it to persist despite the strong reproductive advantage offered by parthenogenesis. But what are these benefits?

Jennifer Anderson and colleagues (2010) maintained 10 distinct strains of *C. elegans* in the lab for a dozen generations, then assessed the frequency of males. Over half the strains either lost males altogether or maintained them only at low frequencies. Other strains, however, maintained males at higher frequencies. One maintained them at a frequency of 35%. This variation in the equilibrium frequency of males indicates that the product of male fertilization success and the relative fitness of outcrossed progeny is often below, but sometimes above, the threshold for male maintenance. It also suggests that the genetic composition of a population matters. Other researchers have documented genetic variation among strains for male fertilization success (Teotónio et al. 2006; Murray et al. 2011). The variation is attributable to differences in both male mating ability and hermaphrodite receptivity (Wegewitz et al. 2008).

Males can persist in a population of facultatively sexual females if they have sufficiently high fertilization success, if they produce offspring that survive at a sufficiently elevated rate, or both.

Of particular use to researchers, mutations are known that can peg the frequency of males in a *C. elegans* population at 0 or 0.5 (see Anderson et al. 2010). For example, a mutation that renders hermaphrodites unable to make sperm, effectively converting them into females, maintains males at a frequency of 0.5. And a mutation that is lethal to males eliminates them from a population.

The experiments Morran and colleagues (2009, 2011) have conducted with *C. elegans* identify conditions under which populations maintain high frequencies of males. We use these experiments to introduce hypotheses about the benefits of sex. We then look at evidence from other organisms as well. Before we do all of this, we need to consider what sex does to genes in populations.

The Population Genetics Consequences of Sex

When population geneticists talk about sex, what they usually mean, and what we mean here, is reproduction involving (1) meiosis with crossing over and (2) matings between unrelated individuals, such as occur during random mating. Organisms that reproduce by cloning themselves, such as summer aphids, lack both meiosis and outcrossing. Organisms that reproduce by self-fertilization, such as selfing hermaphroditic *C. elegans,* have meiosis but lack outcrossing.

In concert, meiosis and outcrossing result in allelic segregation and genetic recombination. If we follow a lineage of allele copies through several generations, in every generation the allele will be part of a different multilocus genotype. At its own locus, our allele may be part of a homozygote in one generation and a heterozygote in the next. At other loci, our allele will be linked to different variants each generation. For example, a particular allele for blue eyes may be part of a genotype that includes alleles for blond hair in one generation and part of a genotype that includes alleles for brown hair in the next generation.

Sex, to a population geneticist, means allelic segregation and genetic recombination.

If we follow the whole population across generations, segregation will tend to restore the population to Hardy–Weinberg equilibrium. (This was a central conclusion of Chapter 6.) Genetic recombination will tend to restore the population to linkage equilibrium. This was a central conclusion of Section 8.1.

A population genetics theory that would explain the maintenance of sex via segregation must include a mechanism that takes populations away from Hardy–Weinberg equilibrium and a reason that the restoration of equilibrium is beneficial (see Agrawal 2009). A population genetics theory that would explain the maintenance of sex via genetic recombination must likewise include a mechanism that takes populations away from linkage equilibrium and a reason that the restoration of equilibrium is beneficial (Felsenstein 1988). Evolutionary biologists have devoted considerably more attention to theories that explain the

In a population genetics analysis, sex does only two things: It restores Hardy–Weinberg equilibrium, and it restores linkage equilibrium.

maintenance of sex via its effect on linkage disequilibrium (see Otto 2003). These theories are our primary focus here. We will, however, keep the effect of sex on Hardy–Weinberg equilibrium in mind as well.

Mutation, Drift, Inbreeding, and Sex

Morran and colleagues (2009) worked with a strain of *C. elegans,* called CB4856, that maintains males at a frequency of 10% to 20%. The researchers set up replicate populations and watched them evolve by natural selection for 50 generations. The scientists kept their research subjects in an environment that required the worms to cross rugged terrain to find food. This requirement imposed selection against deleterious mutations. At the end of 50 generations, the researchers measured the mean fitness of each population, relative to that of the ancestor, by assessing the population's growth rate in competition with a reference strain.

The replicate populations the researchers used were of three kinds. The first kind were fixed for a mutation lethal to males and thus consisted of obligately selfing hermaphrodites. The second kind were wild type and consisted of hermaphrodites and males. The third kind were fixed for a mutation that renders hermaphrodites unable to make sperm and thus consisted of equal numbers of obligately outcrossing females and males.

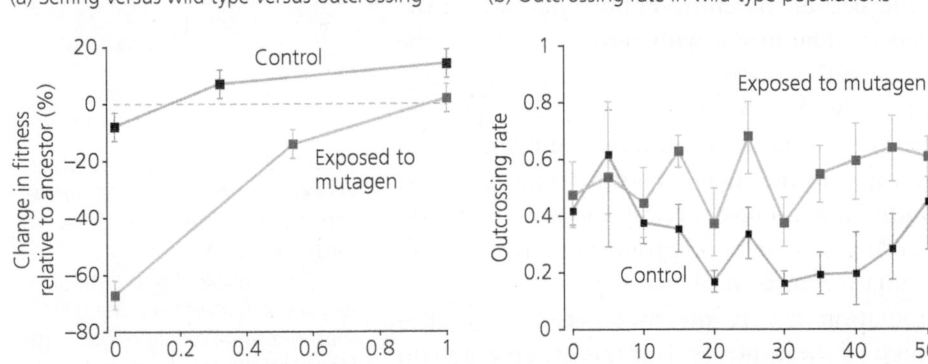

Figure 8.26 Elevated mutation rate selects for outcrossing in *C. elegans* (a) Mean relative fitness ±2 s.e. of populations after 50 generations of laboratory natural selection. Five replicates per symbol. (b) Evolution of outcrossing rate (the fraction of individuals produced by outcrossing) in wild-type populations. From Morran et al. (2009).

Reprinted by permission from Macmillan Publishers Ltd: Morran, L. T., M. D. Parmenter, and P. C. Phillips. 2009. "Mutation load and rapid adaptation favour outcrossing over self-fertilization." *Nature* 462: 350–352.

The black symbols in **Figure 8.26a** show what happened to the mean fitness of the three kinds of populations. The obligately selfing populations (with an outcrossing rate of zero) suffered a decline in fitness relative to their ancestor. The likely explanation for this loss is the accumulation of deleterious mutations. In contrast, the wild-type and obligately outcrossing populations (with outcrossing rates of 0.3 and 1) showed modest gains in relative fitness.

The orange symbols in Figure 8.26a show what happened to the same three kinds of populations when Morran and colleagues dosed the worms with a chemical mutagen every other generation. This treatment increased the mutation rate by a factor of four. Now the obligately selfing populations suffered a dramatic loss of fitness and the wild-type populations a substantial loss. Only the obligately outcrossing populations maintained a fitness equivalent to that of their ancestor.

In the wild-type populations, composed of hermaphrodites and males, the outcrossing rate—the fraction of offspring fathered by males—evolved. This appears in the higher outcrossing rates for the mutated versus control wild-type populations in Figure 8.26a and in the time-series plots in Figure 8.26b. A higher mutation rate selected for more frequent outcrossing, and thus more males.

Morran and colleagues' results suggest that outcrossing is, at least in part, an adaptation for maintaining fitness in the face of deleterious mutations. The problem created by deleterious mutations, and the solution offered by sex, are perhaps best appreciated by imagining an asexual female that reproduces by cloning.

If a parthenogenetic female sustains a deleterious mutation in her germ line, she will pass it to all her offspring. They, in turn, will pass it to all their offspring. The female's lineage will be hobbled by the mutation forever. The only hope of escape is if one of her descendants is lucky enough to experience either a back-mutation or an additional mutation that compensates for the first. In the meantime, the descendants are at risk of sustaining additional deleterious mutations.

Now, following H. J. Muller (1964), imagine a population of asexual individuals that occasionally sustain deleterious mutations. Because the mutations envisioned by Muller are deleterious, they will be selected against. The frequency of each mutant allele in the population will reflect the mutation rate, the strength of selection, and, if the population is finite, genetic drift (see Chapters 6 and 7).

At any given time, the population may include individuals that carry no mutations, individuals that carry one mutation, individuals that carry two mutations, and so on. Because the population is asexual, we can think of these groups as distinct subpopulations and plot the relative number of individuals in each subpopulation in a histogram (**Figure 8.27**). The number of individuals in each group may be quite small, depending on the size of the entire population and on the balance between mutation and selection. The group with zero mutations is the one whose members, on average, enjoy the highest fitness. But if this group is small, then in any given generation chance events may conspire to prevent the reproduction of all individuals in the group. If this happens just once, then the zero-mutation subpopulation is lost, and the members of the one-mutation group are now the highest-fitness individuals. The only way the zero-mutation group will reappear is if a member of the one-mutation group sustains the back-mutation that converts into a zero-mutation individual.

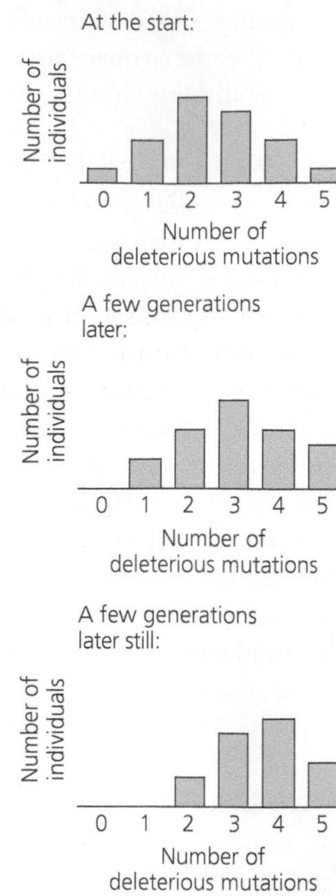

Figure 8.27 Muller's ratchet: Asexual populations accumulate deleterious mutations Each histogram shows a snapshot of a finite asexual population. In any given generation, the class with the fewest deleterious mutations may be lost by drift. Because forward-mutation to deleterious alleles is more likely than back-mutation to wild-type alleles, the distribution slides inexorably to the right. After Maynard Smith (1988).

With the demise of the zero-mutation group, the members of the one-mutation subpopulation enjoy the highest mean fitness. But this group may also be quite small and may be lost by chance in any given generation. Again, the loss of the group by drift is much easier than its re-creation by back-mutation. Each time the most fit group is lost, it is as if a ratchet has turned one click. As the ratchet clicks away, and highest-fitness group after highest-fitness group is lost from the population, the average fitness of the population declines over time. The burden imposed by the accumulating mutations is known as the **genetic load**. Eventually, the genetic load carried by the asexual population becomes so high that the population goes extinct.

Sex breaks Muller's ratchet. If the no-mutation group is lost by chance in any given generation, it can be reconstituted by outcrossing and recombination. If two individuals mate, each carrying a single deleterious mutation, one-quarter of their offspring will be mutation free. Even individuals who do not possess a single mutation-free chromosome between them can generate mutation-free chromosomes by mating. Outcrossing followed by meiosis with recombination in the offspring will do the job, so long as the mutations occur at different loci.

Sex may be advantageous because it re-creates favorable multilocus genotypes that have been lost to drift. The genes for sex then ride to high frequency in the high-fitness genotypes they help to create.

The crux of Muller's ratchet is that linkage disequilibrium is created by drift. Particular multilocus genotypes are at lower-than-linkage-equilibrium frequencies because chance events have eliminated them. These missing multilocus genotypes are the zero-mutation genotype, then the one-mutation genotype, and so on. Sex reduces linkage disequilibrium by re-creating the missing genotypes. The genes responsible for sex are maintained in populations because they help to create zero-mutation genotypes. As these zero-mutation genotypes increase in frequency, the genes for sex hitchhike to high frequency with them.

Haigh (1978; reviewed in Maynard Smith 1988) developed and explored an explicit mathematical model of Muller's ratchet. Not surprisingly, the most critical parameter in the model is population size. In populations of 10 or fewer individuals, drift is a potent mechanism of evolution and the ratchet turns rapidly. In populations of more than 1,000, drift is a weak mechanism of evolution and the ratchet does not turn at all. Also important are the mutation rate and the impact of deleterious mutations. The ratchet operates fastest with mildly deleterious mutations. This is because severely deleterious mutations are eliminated by selection before drift can carry them to fixation.

The situation is somewhat more complicated when we compare selfing to outcrossing hermaphrodites, because both have meiosis. The appearance of a new deleterious mutation renders the locus where it occurs heterozygous. Selfing then leads to a rapid increase in homozygosity. The original allele and the deleterious mutation thus quickly segregate into homozygotes. In effect, this reduces the mutation rate by half but doubles the effect of each mutation (Lynch et al. 1995; Schultz and Lynch 1997). Otherwise, the conditions under which Muller's ratchet turns are essentially the same in selfing populations as in asexual populations (Heller and Maynard Smith 1978). The ratchet still operates fastest with mildly deleterious mutations, and sex still breaks it by recreating missing genotypes.

To assess whether Muller's ratchet operates in nature, Lee Henry and colleagues (2012) took advantage of a natural experiment. They studied six asexual species of *Timema* walking stick and the six sexual species from which the asexuals are independently derived **(Figure 8.28a)**. The estimated ages of the asexual species range from 400,000 to 1,850,000 years. The researchers predicted that by now the asexuals should have accumulated a considerable load of deleterious mutations.

Henry and colleagues examined the sequences in all 12 species of three genes thought to be subject to strong purifying selection. For each gene, they calculated the ratio of nonsynonymous to synonymous substitutions among the asexual versus sexual species. The results appear in Figure 8.28b. Consistent with the researcher's expectation that the genes are subject to purifying selection, the overall ratio of nonsynonymous to synonymous substitutions was low. Consistent with the operation of Muller's ratchet, the asexual species accumulated far more nonsynonymous mutations than the sexual species. Also consistent with Muller's ratchet, the nonsynonymous mutations accumulated by the asexual species were more likely, on average, to encode a replacement amino acid with chemical properties substantially different from those of the original. This suggests that the mutations were, indeed, deleterious.

The examples we have discussed demonstrate that Muller's ratchet works in theory, in laboratory experiments, and in nature. By counteracting the ratchet, sex can confer benefits. Working with a more general model of disequilibrium arising from deleterious mutations, Peter Keightley and Sarah Otto (2006) found sex beneficial even in—indeed, more so in—large populations.

Drift is not the only source of disequilibrium, however. We next consider whether disequilibrium generated by selection can also lead to situations in which the reduction of disequilibrium confers benefits.

(a) *Timema* walking stick

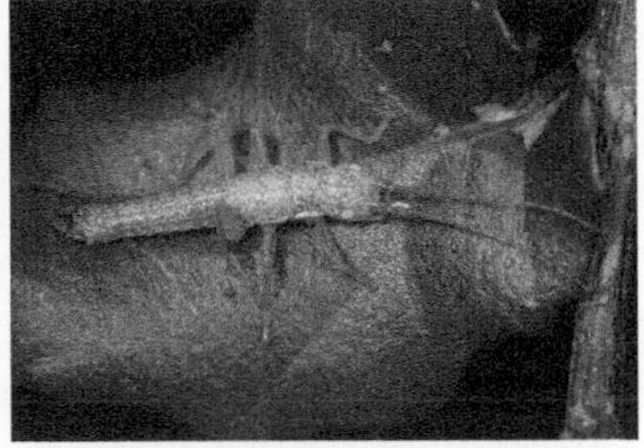

(b) Ratio of nonsynonymous to synonymous substitutions

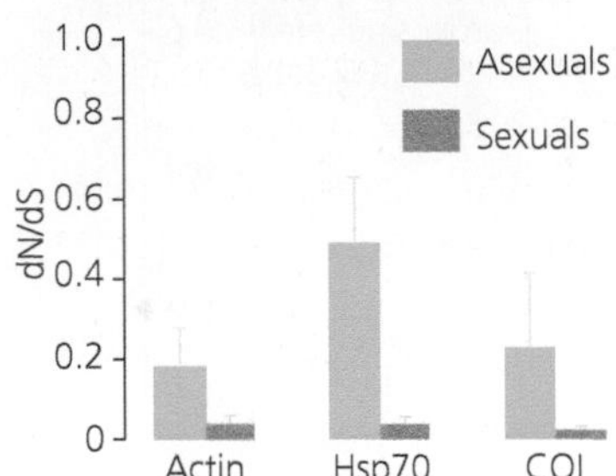

Figure 8.28 Muller's ratchet in asexual walking sticks (a) A *Timema* walking stick. Photo by Joyce Gross. (b) The ratio of nonsynonymous to synonymous substitutions in three highly conserved genes among six asexual species and the six sexual species they are derived from. From Henry et al. (2012).

Selection, Parasites, Environmental Change, and Sex

Building on the earlier work with *C. elegans* strain CB4856, Levi Morran and colleagues (2011) set up new replicate populations of wild-type worms and watched them evolve in the lab. The scientists first treated all the worms with a chemical mutagen for three generations to introduce genetic variation. They subsequently cultivated the populations in an environment that required the worms to cross a lawn of pathogenic bacteria and a stripe of antibiotic to find food. The pathogenic bacteria were *Serratia marcescens,* which cause an often-lethal infection in *C. elegans* that eat them. Every few generations, the researchers assessed the frequency of males and inferred the total fraction of the population that had been produced by outcrossing versus selfing parents. They tracked changes in the outcrossing rate for 30 generations. The results appear in **Figure 8.29**.

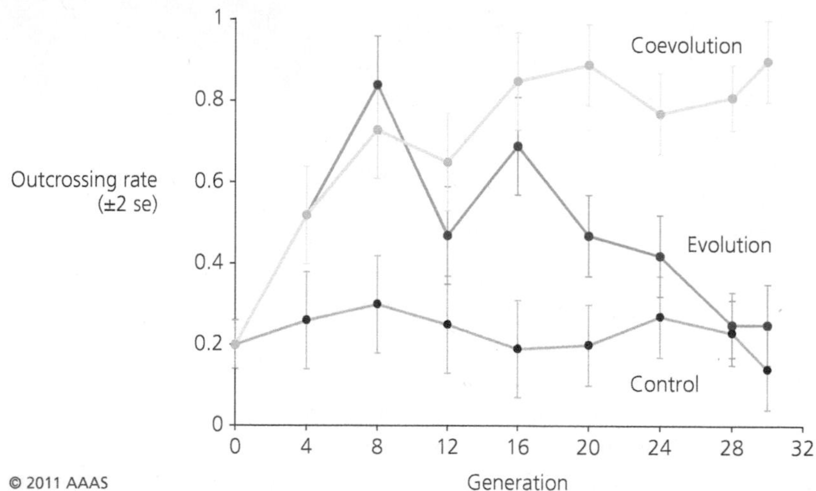

© 2011 AAAS

Figure 8.29 Pathogens select for outcrossing in *C. elegans* Frequency of outcrossing in *C. elegans* populations evolving by natural selection in the lab. *Control* populations were exposed to heat-killed pathogenic bacteria, *evolution* populations to pathogenic bacteria from a stock population, and *coevolution* populations to pathogens under selection for their ability to infect and kill the worms. From Morran et al. (2011).

From "Running with the Red Queen: host-parasite coevolution selects for biparental sex." *Science* 333: 216–218. Reprinted with permission from AAAS.

First look at the *control* populations (black). The controls were exposed to heat-killed pathogenic bacteria, which are not dangerous to *C. elegans*. Across the entire experiment, the controls maintained males at a frequency of roughly 20%.

Now look at the *evolution* populations (red). These populations were exposed to pathogenic bacteria drawn from a stock population. The pathogens killed many worms, imposing selection for resistance to infection. The worms' outcrossing rate evolved as well. It rose over eight generations to a peak of over 80%. Then it slowly fell back to the level of outcrossing displayed by the controls.

Finally, look at the *coevolution* populations (green). These were exposed to pathogenic bacteria harvested from the dead carcasses of worms that had been infected and killed in the previous generation. This imposed selection on the worms for resistance to infection and on the bacteria for the ability to infect and kill. This time the outcrossing rate rose across the first 20 generations to a high of about 90%, where it remained for the duration of the experiment.

Morran's evolution treatment suggests that sex is beneficial in populations subject to directional selection. The mechanism is similar to Muller's ratchet. In a finite population, some advantageous genotypes will be missing due to sampling error. Sex is adaptive because it recreates these missing genotypes through segregation and recombination (Otto and Lenormand 2002).

Selection—on any sort of trait—seems to favor recombination.

Once optimal genotypes arise, however, sex becomes disadvantageous. This is because segregation and recombination break up the advantageous genotypes they recently helped to create (see Otto 2009). This may explain why, by the end of Morran's experiment, the rate of outcrossing in the *evolution* populations had fallen back to where it started.

Morran's *coevolution* treatment suggests that sex remains beneficial indefinitely in populations subject to an ever-changing selection regime. In the *coevolution* treatment, the worms and their pathogens were locked in an ongoing arms race. As the worm population was evolving ever-better defenses, the pathogen population was simultaneously evolving ever-better means of infecting the worms.

Figure 8.30 illustrates how this situation can make sex perpetually beneficial. Both resistance and infectivity are likely influenced by genotype at a variety of loci. Imagine a pathogen population with two genotypes, one common and one rare (Figure 8.30a). Imagine a host population with genetic variation for resistance, such that hosts resistant to pathogen genotype I are susceptible to pathogen genotype II and vice versa (Figure 8.30b). The host population will evolve resistance to whichever pathogen genotype is more common (Figure 8.30c). In response, the pathogen population will evolve toward a higher frequency of whichever genotype is currently rare. Sex remains adaptive because, as during simple directional selection, segregation and recombination help recreate genotypes that are, at any given time, both rare or missing and favored by selection.

The notion that sex is adaptive during perpetual arms races between biological antagonists is called the Red Queen hypothesis (for reviews, see Seger and Hamilton 1988; Lively 2010). The hypothesis is named for a character in Lewis Carroll's *Through the Looking Glass* who runs as fast as she can just to stay in one place.

The crux of the Red Queen hypothesis is disequilibrium created by selection. Particular genotypes are at lower-than-equilibrium frequencies because individuals who carry them have recently fared poorly. Soon, however, these genotypes become advantageous. Sex reduces linkage disequilibrium by re-creating the missing genotypes. The genes responsible for sex are maintained in populations because they hitchhike to high frequency with genotypes they recreate.

Similar scenarios favoring sex can also arise if the environment perpetually changes (Sturtevant and Mather 1938, reviewed in Felsenstein 1988), or if individuals perpetually migrate among different environments (Agrawal 2009).

Curtis Lively (1992) investigated whether parasites select in favor of sex in nature. Lively studied the freshwater snail *Potamopyrgus antipodarum,* which lives in lakes and streams throughout New Zealand and is the host of over a dozen species of parasitic trematode worms. The trematodes typically castrate their host by eating its gonads. Trematodes thus exert strong selection on snail populations for resistance to infection. Most populations of the snail contain two kinds of females: obligately sexual females that produce a mixture of male and female offspring, and obligately parthenogenetic females whose daughters are clones of their mother. Both kinds of female must have an ovary to reproduce; the difference is that the eggs of the parthenogenetic females do not have to be fertilized. The proportion of sexuals versus asexuals varies from population to population. So does the frequency of trematode infection. If an evolutionary arms race between the snails and the trematodes selects in favor of sex in the snails, then sexual snails should be more common in populations with higher trematode infection rates.

Lively sampled snails from 66 lakes and determined the sex and infection status of each individual. He used the frequency of males in each population as an index of the frequency of sexual females, on the logic that sexual females are the only source of males. Lively found that, as predicted, a higher proportion of the females are sexual in more heavily parasitized populations. This result, shown in

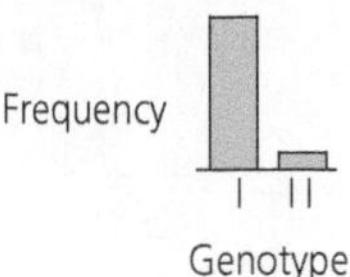

(b) Host population

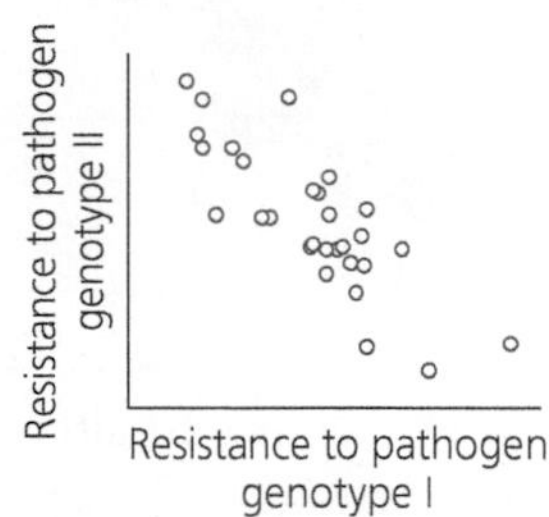

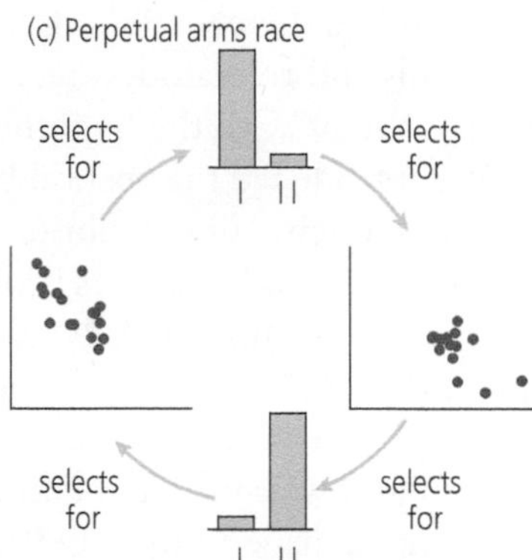

Figure 8.30 A host parasite arms race can make sex beneficial Hosts resistant to parasite genotype I are necessarily susceptible to parasite genotype II, and vice versa. As the parasite population evolves in response to the hosts, it first selects for hosts resistant to parasite genotype I, then for hosts resistant to parasite genotype II. Genes for sex ride to high frequency in the currently more fit genotypes they help to create.

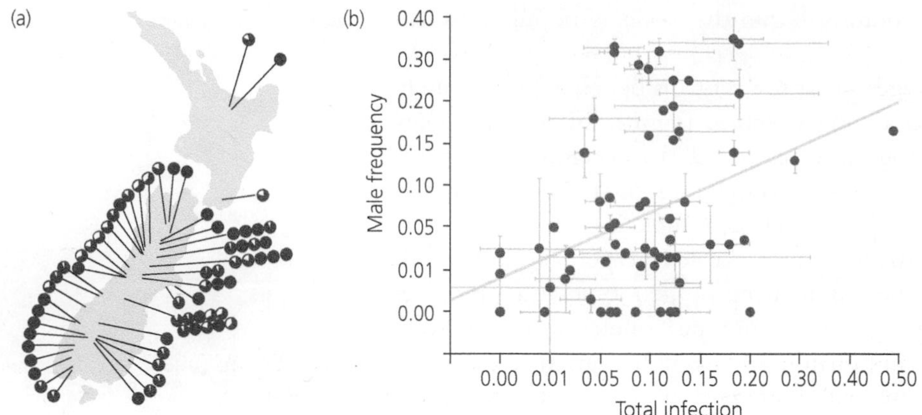

Figure 8.31 The frequency of sexual individuals in populations of a host snail is positively correlated with the frequency of its trematode parasites (a) Pie diagrams show the frequency of males (white) in 66 lakes. (b) The frequency of males versus the proportion of snails infected with trematodes, with best-fit line. Males are more frequent in populations where more snails are infected. From Lively (1992).

Figure 8.31, is open to alternative causal interpretations. One is the Red Queen hypothesis. Another is that males are more susceptible to infection. Experiments rule out the latter, because in the lab males are no more vulnerable to the parasite than females (Lively 1989). Overall, the result in Figure 8.31, as well as others on the snails and trematodes (see, for example, Jokela et al. 2009; King et al. 2009) are consistent with the hypothesis that parasites select in favor of sex.

We introduced this section by asking why sex is so common. We have learned that by reducing disequilibrium, sex both helps maintain fitness despite deleterious mutation and facilitates evolution in response to selection. We might now ask why sex is not universal. Recall the bdelloid rotifers we discussed at the start of the chapter. They have evolved a foolproof method of evading their pathogens. When their environment turns disagreeable they enter a state of suspended animation and wait until conditions improve. And they are extraordinarily good at repairing damage to their DNA (Gladyshev and Meselson 2008). Together these unusual traits apparently make sex superfluous—or at least worth less than its cost.

In environments that are changing—especially because they include enemies that are evolving—sex may be advantageous because it re-creates favorable multilocus genotypes that were recently eliminated by selection. Again, the genes for sex then ride to high frequency in the high-fitness genotypes they help to create.

SUMMARY

Single-locus models are powerful, but potentially oversimplified. Extension of the Hardy–Weinberg analysis to two loci reveals complications. When genotypes at one locus are nonrandomly associated with genotypes at the other, the loci are in linkage disequilibrium. Even under Hardy–Weinberg assumptions, chromosome frequencies change across generations. Furthermore, selection on one locus can alter allele frequencies at the other, and single-locus models may make inaccurate predictions. When genotypes at one locus are independent of genotypes at the other locus, however, the loci are in linkage equilibrium. Chromosome frequencies do not change across generations. Selection on one locus has no effect on allele or genotype frequencies at the other, and we can use single-locus models to make predictions about evolution.

In a random-mating population, linkage disequilibrium can be created by selection on multilocus genotypes, genetic drift, and population admixture. These mechanisms create an excess of some chromosome haplotypes and a deficit of others. Linkage disequilibrium is reduced by sexual reproduction. Sex brings together chromosomes with different haplotypes, and crossing over during meiosis allows the chromosomes to exchange genes. This genetic recombination tends to break up overrepresented haplotypes and create underrepresented haplotypes.

Measurements of linkage disequilibrium are useful in inferring the history of alleles. If an allele is in linkage disequilibrium with nearby neutral marker loci, we can infer that the allele is relatively young. If we have an estimate of the rate at which disequilibrium between the allele and the nearby neutral marker loci breaks down, then we can use the strength of the persisting disequilibrium to estimate the allele's age. If an allele is both young, as indicated by linkage disequilibrium, and present at high frequency, then we can infer that the allele has recently been favored by positive natural selection.

Knowing that sexual reproduction reduces linkage disequilibrium provides a key to understanding why sexual reproduction persists. Simple theoretical arguments suggest that asexual reproduction should sweep to fixation in any population in which it appears. Empirical observations and experiments indicate, however, that sex confers substantial benefits. These benefits can be found in the population genetics consequences of sex. When drift or selection has reduced the frequency of particular multilocus genotypes below their expected levels under linkage equilibrium, sex can be favored because it re-creates the missing genotypes.

QUESTIONS

1. Describe the three mathematical consequences of linkage equilibrium. That is, what three equations about genotype and chromosome frequencies will be true if a population is in linkage equilibrium? What is *D*, and how is it calculated?
2. Figure 8.4 presented an example of selection favoring certain multilocus genotypes. The chapter text demonstrated that, after selection, the population failed criterion 2 for linkage equilibrium. Now test the same population in some different ways:
   a. What is the frequency of *B* on chromosomes that are carrying allele *A*? What is the frequency of *B* on chromosomes carrying allele *a*? Does the population meet criterion 1?
   b. What is *D*, the coefficient of linkage disequilibrium? Does the population meet criterion 3?
   c. From the postselection population in Figure 8.4b, develop a bar graph like the ones in Figure 8.3. Does this bar graph confirm that the postselection population is in linkage disequilibrium?
3. In horses, the basic color of the coat is governed by the E locus. *EE* and *Ee* horses can make black pigment, while *ee* horses are a reddish chestnut. A different locus, the R locus, can cause roan, a scattering of white hairs throughout the basic coat color. However, the roan allele has a serious drawback: *RR* embryos always die during fetal development. *Rr* embryos survive and are roan, while *rr* horses survive and are not roan. The E locus and the R locus are tightly linked.

   Suppose that several centuries ago, a Spanish galleon with a load of conquistadors' horses was shipwrecked by a large grassy island. Just by chance, the horses that survived the shipwreck and swam to shore were 20 chestnut roans (*eeRr*) and 20 nonroan homozygous blacks (*EErr*). On the island, they interbred with each other and established a wild population. The island environment exerts no direct selection on either locus.
   a. What was *D*, the coefficient of linkage disequilibrium, in the initial population of 20 horses? Was the initial population in linkage equilibrium or not? If not, what chromosomal genotypes were underrepresented?
   b. Do you expect the frequency of the chestnut allele, *e*, to increase or decrease in the first crop of foals? Would your answer be different if the founding population had been just 10 horses (5 of each color)? Explain your reasoning.
   c. If you could travel to this island today, can you predict what *D* would be now? Do you have predictions about whether more horses will be roan versus nonroan, or chestnut versus black? If not, explain what further information you would need.
4. Imagine a population of pea plants that is in linkage equilibrium for two linked loci, flower color (P = purple, p = pink) and pollen shape (L = long, l = round).
   a. What sort of selection event would create linkage disequilibrium? For example, will selection at just one locus (e.g., all red-flowered plants die) create linkage disequilibrium? How about selection at two loci (e.g., red-flowered plants die, and long-pollen plants die)? How about selection on a certain combination of genotypes at two loci (e.g., only plants that are both red-flowered and have long pollen grains die)?
   b. Now imagine a population that is already in linkage disequilibrium for these two loci. Will selection for purple flowers affect evolution of pollen shape? How is your answer different from that to part a, and why?
5. Would it be possible for male *C. elegans* to persist if the proportion of eggs they collectively fertilize is less than their own frequency in the population? How?
6. In 1992, Spolsky, Phillips, and Uzzell reported genetic evidence that asexually reproducing lineages of a salamander species have persisted for about 5 million years. Is this surprising? Why or why not? Speculate about what sort of environment these asexual salamanders live in, and whether their population sizes are typically small (say, less than 100) or large (say, more than 1,000).
7. How can you identify an allele that has experienced recent strong positive selection?
8. Populations of rats exposed to the poison warfarin rapidly evolve resistance. The gene for warfarin resistance is located on rat chromosome 1. Michael Kohn and colleagues (2000) surveyed rats in five German rat populations known to vary in their recent exposure to warfarin and in their resistance. The researchers determined the genotype of each rat at a number of marker loci near the warfarin resistance gene. For each population, the

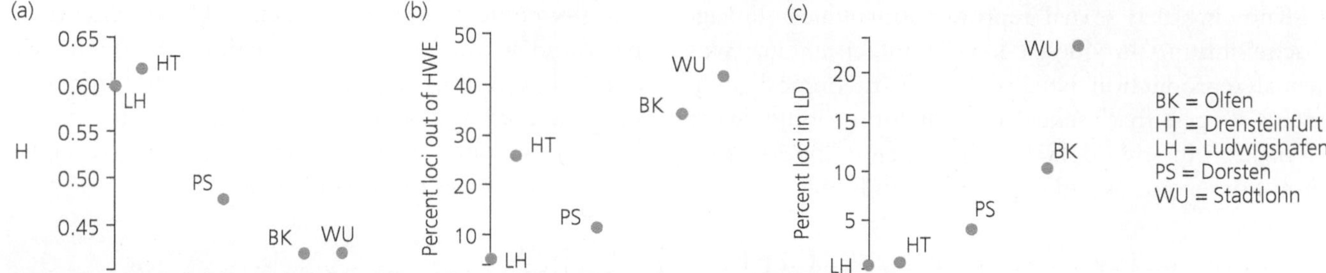

The missing horizontal axes plot warfarin resistance. Does resistance increase or decrease from left to right across these graphs?

Figure 8.32 Population genetics data on five rat populations Redrawn from Kohn et al. (2000).

researchers calculated the average heterozygosity (H) among the marker loci, the fraction of loci that were out of Hardy–Weinberg equilibrium (HWE), and the fraction of marker–locus pairs that were in linkage disequilibrium (LD). Their results appear in **Figure 8.32**. Based on these graphs, rank the five populations in order, from lowest to highest, for exposure to warfarin and resistance. Explain your reasoning.

9. Describe the major hypotheses for the cause of high frequency of the *CCR5-Δ32* allele among European populations. Why is the age of the allele relevant for distinguishing among the hypotheses? Do we know how old this allele is, and if so, what is the evidence?

10. *Volvox* (Figure 8.22a) are abundant and active in lakes during the spring and summer. During winter they are inactive, existing in a resting state. During most of the spring and summer, *Volvox* reproduce asexually; but at times they switch and reproduce sexually instead. When would you predict that *Volvox* would be sexual: spring, early summer, or late summer? Explain your reasoning.

11. In mammals, sex is determined by the X and Y chromosomes. Females are XX; males are XY. The Y chromosome contains a gene that causes development of testes, which then causes the embryo to become male. The Y chromosome does not undergo crossing over with the X during spermatogenesis in males, but the two X's cross over with each other during oogenesis in females.

a. The Y chromosome is thought to have once been the same size as the large, fully functional X chromosome. But during the evolution of the mammals, the Y chromosome seems to have accumulated an enormous number of deleterious mutations. It has also lost almost all of its genes and has shrunk to a rudimentary chromosome containing just the testis-determining gene, a few other genes, and some nonfunctional remnants of other genes. Why has this occurred?

b. Birds use a reverse system, in which females have two different chromosomes (called WZ in birds) and males have two of the same kind of chromosome (ZZ). In birds, sex is determined by a gene on the W chromosome that causes ovary formation, which then causes the bird embryo to become female. Would you predict one of these chromosomes might have accumulated mutations in the same way that the Y chromosome has? If so, which one?

c. Some plants also have genetically determined sex but are polyploid. Should their sex chromosomes show accumulation of mutations?

EXPLORING THE LITERATURE

12. To read more about the accumulation of mutations in sex chromosomes, see:

Berlin, S., and H. Ellegren. 2006. Fast accumulation of nonsynonymous mutations on the female-specific W chromosome in birds. *Journal of Molecular Evolution* 62: 66–72.

Gerrard, D. T., and D. A. Filatov. 2005. Positive and negative selection on mammalian Y chromosomes. *Molecular Biology and Evolution* 22: 1423–1432.

13. Many human pathogens, including bacteria and eukaryotes, are capable of both asexual reproduction and genetic recombination (that is, sex in the population genetics sense). The frequency of recombination in a pathogen can have medical implications. (Think about how fast resistance to multiple antibiotics will evolve in a population of bacteria that has recombination versus one that does not.) How can we tell whether a pathogen population is engaging in genetic recombination or is predominantly clonal? Recombination is such a powerful mechanism for reducing linkage disequilibrium that the amount of disequilibrium in a population gives a clue. See:

Xu, J. 2004. The prevalence and evolution of sex in microorganisms. 2004. *Genome* 47: 775–780.

Maynard Smith, J., N. H. Smith, et al. 1993. How clonal are bacteria? *Proceedings of the National Academy of Sciences, USA* 90: 4384–4388.

Burt, A., D. A. Carter, et al. 1996. Molecular markers reveal cryptic sex in the human pathogen *Coccidioides immitis*. *Proceedings of the National Academy of Sciences, USA* 93: 770–773.

Gräser, Y., et al. 1996. Molecular markers reveal that population structure of the human pathogen *Candida albicans* exhibits both clonality and recombination. *Proceedings of the National Academy of Sciences USA* 93: 12473–12477.

14. Most biologists assume that genetic recombination in bacteria, via processes such as conjugation and transduction, is equivalent to eukaryotic sex—and that it is favored by selection for similar reasons. For an argument that recombination in bacteria is nothing like eukaryotic sex and evolved for entirely different reasons, see:

Redfield, R. J. 2001. Do bacteria have sex? *Nature Reviews Genetics* 2: 634–639.

15. For late-breaking news of a possible selective agent contributing to selection on *CCR5-Δ32*, see:

Alonzo, F., L. Kozhaya, et al. 2013. CCR5 is a receptor for *Staphylococcus aureus* leukotoxin ED. *Nature* 493: 51–55.

16. For evidence, based on linkage disequilibrium, that a genetic variant allowing some humans to digest milk sugar in adulthood has recently been favored by selection, see:

Bersaglieri, T., P. C. Sabeti, et al. 2004. Genetic signatures of recent positive selection at the lactase gene. *American Journal of Human Genetics* 74: 1111-1120.

For evidence that different alleles confer lactase persistence in different populations, see:

Gallego Romero, I., C. Basu Mallick, et al. 2012. Herders of Indian and European cattle share their predominant allele for lactase persistence. *Molecular Biology and Evolution* 29: 249–260.

Tishkoff, S. A., F. A. Reed, et al. 2007. Convergent adaptation of human lactase persistence in Africa and Europe. *Nature Genetics* 39: 31–40.

17. For evidence that sex is beneficial for organisms that move among different habitats, using an experimental system similar to Morran's *C. elegans*, see:

Becks, L., and A. F. Agrawal. 2010. Higher rates of sex evolve in spatially heterogeneous environments. *Nature* 468: 89–92.

CITATIONS

Please note that much of the population genetics in this chapter is modeled after presentations in the following:

Cavalli-Sforza, L. L., and W. F. Bodmer. 1971. *The Genetics of Human Populations*. San Francisco: W. H. Freeman.

Felsenstein, J. 1997. *Theoretical Evolutionary Genetics*. Seattle, WA: ASUW Publishing, University of Washington.

Felsenstein, J. 1988. Sex and the evolution of recombination. In *The Evolution of Sex*, ed. R. E. Michod and B. R. Levin. Sunderland, MA: Sinauer, 74–86.

Hartl, D. L. 1981. *A Primer of Population Genetics*. Sunderland, MA: Sinauer.

Maynard Smith, J. 1998. *Evolutionary Genetics*. 2nd ed. Oxford: Oxford University Press.

Here is the listing of all citations in this chapter:

Agrawal, A. F. 2009. Spatial heterogeneity and the evolution of sex in diploids. *American Naturalist* 174 (Suppl 1): S54–S70.

Anderson, J. L., L. T. Morran, and P. C. Phillips. 2010. Outcrossing and the maintenance of males within *C. elegans* populations. *Journal of Heredity* 101 (Suppl 1): S62–S74.

Behar, D., M. F. Hammer, et al. 2004. MtDNA evidence for a genetic bottleneck in the early history of the Ashkenazi Jewish population. *European Journal of Human Genetics* 12: 355–364.

Beutler, E. 1993. Gaucher disease as a paradigm of current issues regarding single gene mutations of humans. *Proceedings of the National Academy of Sciences, USA* 90: 5384-5390.

Beutler, E., T. Gelbart, et al. 1991. Identification of the second common Jewish Gaucher disease mutation makes possible population-based screening for the heterozygous state. *Proceedings of the National Academy of Sciences, USA* 88: 10544–10547.

Birky, C. W. 2010. Positively negative evidence for asexuality. *Journal of Heredity* 101: S42–S45.

Bray, S. M., J. G. Mulle, et al. 2010. Signatures of founder effects, admixture, and selection in the Ashkenazi Jewish population. *Proceedings of the National Academy of Sciences, USA* 107: 16222–16227.

Carroll, L. 1872. *Through the Looking-Glass*. London: Macmillan.

Centers for Disease Control, Division of Parasitic Diseases. 2001. Malaria. Available at http://www.dpd.cdc.gov/dpdx/HTML/Malaria.asp?body=Frames/M-R/Malaria/body_Malaria_page2.htm.

Clegg, M. T., J. F. Kidwell, and C. R. Horch. 1980. Dynamics of correlated genetic systems. V. Rates of decay of linkage disequilibria in experimental populations of *Drosophila melanogaster*. *Genetics* 94: 217–234.

Cochran, G., J. Hardy, and H. Harpending. 2006. Natural history of Ashkenazi intelligence. *Journal of Biosocial Science* 38: 659–693.

Dawson, E., G. R. Abecasis, et al. 2002. A first-generation linkage disequilibrium map of human chromosome 22. *Nature* 418: 544–548.

De La Vega, F. M., H. Isaac, et al. 2005. The linkage disequilibrium maps of three human chromosomes across four populations reflect their demographic history and a common underlying recombination pattern. *Genome Research* 15: 454–462.

Diaz, G. A., B. D. Gelb, et al. 2000. Gaucher disease: The origins of the Ashkenazi Jewish N370S and 84GG acid β-glucosidase mutations. *American Journal of Human Genetics* 66: 1821–1832.

Ey, J., E. Schömig, and D. Taubert. 2007. Dietary sources and antioxidant effects of ergothioneine. *Journal of Agricultural and Food Chemistry* 55: 6466–6474.

Felsenstein, J. 1988. Sex and the evolution of recombination. In *The Evolution of Sex*, ed. R. E. Michod and B. R. Levin. Sunderland, MA: Sinauer, 74–86.

Galvani, A. P., and M. Slatkin. 2003. Evaluating plague and smallpox as historical selective pressures for the *CCR5-Δ32* HIV-resistance allele. *Proceedings of the National Academy of Sciences, USA* 100: 15276–15279.

Gladyshev, E., and M. Meselson. 2008. Extreme resistance of bdelloid rotifers to ionizing radiation. *Proceedings of the National Academy of Sciences, USA* 105: 5139–5144.

Gründemann, D., S. Harlfinger, et al. 2005. Discovery of the ergothioneine transporter. *Proceedings of the National Academy of Sciences, USA* 102: 5256–5261.

Haigh, J. 1978. The accumulation of deleterious mutations in a population: Muller's ratchet. *Theoretical Population Biology* 14: 251–267.

Hedrick, P. W. 1983. *Genetics of Populations*. Boston: Science Books International.

Hedrick, P. W., and B. C. Verrelli. 2006. "Ground truth" for selection on *CCR5-Δ32*. *Trends in Genetics* 22: 293–296.

Heller, J., and J. Maynard Smith. 1979. Does Muller's Ratchet work with selfing? *Genetical Research* 8: 269–294.

Henry, L., T. Schwander, and B. J. Crespi. 2012. Deleterious mutation accumulation in asexual *Timema* stick insects. *Molecular Biology and Evolution* 29: 401–408.

Huff, C. D., D. Witherspoon, et al. 2012. Crohn's disease and genetic hitchhiking at *IBD5*. *Molecular Biology and Evolution* 29: 101–111.

Hummel, S., D. Schmidt, et al. 2005. Detection of the *CCR5-Δ32* HIV resistance gene in Bronze Age skeletons. *Genes and Immunity* 6: 371–374.

Huttley, G. A., M. W. Smith, et al. 1999. A scan for linkage disequilibrium across the human genome. *Genetics* 152: 1711–1722.

The International HapMap Consortium. 2005. A haplotype map of the human genome. *Nature* 437: 1299–1320.

Jennings, H. S. 1917. The numerical results of diverse systems of breeding, with respect to two pairs of characters, linked or independent, with special relation to the effects of linkage. *Genetics* 2: 97–154.

Jokela, J., M. F. Dybdahl, and C. M. Lively. 2009. The maintenance of sex, clonal dynamics, and host-parasite coevolution in a mixed population of sexual and asexual snails. *American Naturalist* 174 (Suppl 1): S43–S53.

Keightley, P. D., and S. P. Otto. 2006. Interference among deleterious mutations favors sex and recombination in finite populations. *Nature* 443: 89–92.

King, K. C., L. F. Delph, et al. 2009. The geographic mosaic of sex and the Red Queen. *Current Biology* 19: 1438–1441.

Kohn, M. H., H.-J. Pelz, and R. K. Wayne. 2000. Natural selection mapping of the warfarin-resistance gene. *Proceedings of the National Academy of Science, USA* 97: 7911–7915.

Lalani, A. S., J. Masters, et al. 1999. Use of chemokine receptors by poxviruses. *Science* 286: 1968–1971.

Lively, C. M. 1989. Adaptation by a parasitic trematode to local populations of its snail host. *Evolution* 43: 1663–1671.

Lively, C. M. 1992. Parthenogenesis in a freshwater snail: Reproductive assurance versus parasitic release. *Evolution* 46: 907–913.

Lively, C. M. 2010. A review of Red Queen models for the persistence of obligate sexual reproduction. *Journal of Heredity* 101: S13–S20.

Luzzatto, L., and R. Notaro. 2001. Protecting against bad air. *Science* 293: 442–443.

Lynch, M., J. Conery, and R. Burger. 1995. Mutational melt-downs in sexual populations. *Evolution* 49: 1067–1080.

Maynard Smith, J. 1978. *The Evolution of Sex*. Cambridge, UK: Cambridge University Press.

Maynard Smith, J. 1988. The evolution of recombination. In *The Evolution of Sex*, ed. R. E. Michod and B. R. Levin. Sunderland, MA: Sinauer, 106–125.

McVean, G., C. C. A. Spencer, and R. Chaix. 2005. Perspectives on Human Genetic Variation from the HapMap Project. *PLoS Genetics* 1: e54.

Mark Welch, D. B., C. Ricci, and M. Meselson. 2009. Bdelloid rotifers: Progress in understanding the success of an evolutionary scandal. In *Lost Sex: The Evolutionary Biology of Parthenogenesis*, ed. I. Schön, K. Martens, and P. van Dijk. Heidelberg: Springer, 259–279.

Morran, L. T., M. D. Parmenter, and P. C. Phillips. 2009. Mutation load and rapid adaptation favour outcrossing over self-fertilization. *Nature* 462: 350–352.

Morran, L. T., O. G. Schmidt, et al. 2011. Running with the Red Queen: Host–parasite coevolution selects for biparental sex. *Science* 333: 216–218.

Muller, H. J. 1964. The relation of recombination to mutational advance. *Mutation Research* 1: 2–9.

Murray, R. L., J. L. Kozlowska, and A. D. Cutter. 2011. Heritable determinants of male fertilization success in the nematode *Caenorhabditis elegans*. *BMC Evolutionary Biology* 11: 99.

Nordborg, M.,T. T. Hu, et al. 2005. The pattern of polymorphism in *Arabidopsis thaliana*. *PLoS Biology* 3: e196.

Novembre, J., A. P. Galvani, and M. Slatkin. 2005. The geographic spread of the *CCR5-Δ32* HIV-resistance allele. *PLoS Biology* 3: e339.

Novembre, J., and E. Han. 2012. Human population structure and the adaptive response to pathogen-induced selection pressures. *Philosophical Transactions of the Royal Society B* 367: 878–886.

Otto, S. P. 2003. The advantages of segregation and the evolution of sex. *Genetics* 164: 1099–1118.

Otto, S. P. 2009. The evolutionary enigma of sex. *American Naturalist* 174 (Suppl 1): S1–S14.

Otto, S. P., and T. Lenormand. 2002. Resolving the paradox of sex and recombination. *Nature Reviews Genetics* 3: 252–261.

Otto, S. P., C. Sassaman, and M. W. Feldman. 1993. Evolution of sex determination in the conchostracan shrimp *Eulimnadia texana*. *American Naturalist* 141: 329–337.

Paul, B. D., and S. H. Snyder. 2010. The unusual amino acid L-ergothioneine is a physiologic cytoprotectant. *Cell Death and Differentiation* 17: 1134–1140.

Peltekova, V. D., R. F. Wintle, et al. 2004. Functional variants of OCTN cation transporter genes are associated with Crohn disease. *Nature Genetics* 36: 471–475.

Petermann, I., C. M. Triggs, et al. 2009. Mushroom intolerance: A novel diet–gene interaction in Crohn's disease. *British Journal of Nutrition* 102: 506–508.

Risch, N., H. Tang, et al. 2003. Geographic distribution of disease mutations in the Ashkenazi Jewish population supports genetic drift over selection. *American Journal of Human Genetics* 72: 812–822.

Ruwende, C., and A. Hill. 1998. Glucose-6-phosphate dehydrogenase deficiency and malaria. *Journal of Molecular Medicine* 76: 581–588.

Sabeti, P. C., D. E. Reich, et al. 2002. Detecting recent positive selection in the human genome from haplotype structure. *Nature* 419: 832–837.

Sabeti, P. C., E. Walsh, et al. 2005. The case for selection at *CCR5-Δ32*. *PLoS Biology* 3: e378.

Saunders, M. A., M. Slatkin, et al. 2005. The extent of linkage disequilibrium caused by selection on G6PD in humans. *Genetics* 171: 1219–1229.

Schultz, S. T., and M. Lynch. 1997. Mutation and extinction: The role of variable mutational effects, synergistic epistasis, beneficial mutations, and degree of outcrossing. *Evolution* 51: 1363–1371.

Schurko, A. M., M. Neiman, and J. M. Logsdon. 2009. Signs of sex: What we know and how we know it. *Trends in Ecology and Evolution* 24: 208–217.

Seger, J., and W. D. Hamilton. 1988. Parasites and sex. In *The Evolution of Sex*, ed. R. E. Michod and B. R. Levin. Sunderland, MA: Sinauer, 176–193.

Silverberg, M. S., R. H. Duerr, et al. 2007. Refined genomic localization and ethnic differences observed for the IBD5 association with Crohn's disease. *European Journal of Human Genetics* 15: 328–335.

Slatkin, M. 2004. A population-genetic test of founder effects and implications for Ashkenazi Jewish diseases. *American Journal of Human Genetics* 75: 282–293.

Slatkin, M., and B. Rannala. 2000. Estimating allele age. *Annual Review of Genomics and Human Genetics* 1: 225–249.

Spolsky, C. M., C. A. Phillips, and T. Uzzell. 1992. Antiquity of clonal salamander lineages revealed by mitochondrial DNA. *Nature* 356: 706–710.

Stephens, J. C., D. E. Reich, et al. 1998. Dating the origin of the *CCR5-Δ32* AIDS-resistance allele by the coalescence of haplotypes. *American Journal of Human Genetics* 62: 1507–1515.

Stewart, A. D., and P. C. Phillips. 2002. Selection and maintenance of androdioecy in *Caenorhabditis elegans*. *Genetics* 160: 975–982.

Strom, C. M., B. Crossley, et al. 2004. Molecular screening for diseases frequent in Ashkenazi Jews: Lessons learned from more than 100,000 tests performed in a commercial laboratory. *Genetics in Medicine* 6: 145–152.

Sturtevant, A. H., and K. Mather. 1938. The interrelations of inversions, heterosis, and recombination. *American Naturalist* 72: 447–452.

Taubert, D., G. Grimberg, et al. 2005. Functional role of the *503F* variant of the organic cation transporter OCTN1 in Crohn's disease. *Gut* 54: 1505–1506.

Taubert, D., N. Jung, et al. 2009. Increased ergothioneine tissue concentrations in carriers of the Crohn's disease risk-associated *503F* variant of the organic cation transporter OCTN1. *Gut* 58: 312–314.

Teotónio, H., D. Manoel, and P. C. Phillips. 2006. Genetic variation for outcrossing among *Caenorhabditis elegans* isolates. *Evolution* 60: 1300–1305.

Voight, B. F., Sridhar Kudaravalli, et al. 2006. A map of recent positive selection in the human genome. *PLoS Biology* 4: e72.

Wang, J., X. Wang, et al. 2011. Contribution of the IBD5 locus to inflammatory bowel disease: A meta-analysis. *Human Genetics* 129: 597–609.

Wegewitz, V., H. Schulenburg, and A. Streit. 2008. Experimental insight into the proximate causes of male persistence variation among two strains of the androdioecious *Caenorhabditis elegans* (Nematoda). *BMC Ecology* 8: 12.

Wilson, C. G., and P. W. Sherman. 2010. Anciently asexual bdelloid rotifers escape lethal fungal parasites by drying up and blowing away. *Science* 327: 574–576.

Wu, D. D., W. Jin, et al. 2010. Evidence for positive selection on the *Osteogenin* (*BMP3*) gene in human populations. *PLoS ONE* 5: e10959.

Wu, D. D., G. M. Li, et al. 2012. Positive selection on the osteoarthritis-risk and decreased-height associated variants at the GDF5 gene in East Asians. *PLoS ONE* 7: e42553.

9

Evolution at Multiple Loci: Quantitative Genetics

Moving toward the rail as they passed the grandstand for the first time, the thoroughbreds running the 137th Kentucky Derby sorted themselves into a living histogram. We can see at a glance how they varied in speed over the start of the race. The graph at right is a histogram showing how the horses varied in distance yet to run when the eventual winner, the slow-starting Animal Kingdom under jockey John Velazquez, crossed the finish line. At the moment captured in the photo, Animal Kingdom and his rider, wearing a red helmet and green jersey, were in 14th place. With a late charge they nabbed for the horse's owners a purse of $1,411,800. Provided he lives to retirement, Animal Kingdom will likely earn the owners another fortune in stud fees.

Thoroughbred race horses, including the entrants in the 2011 Kentucky Derby, vary in speed. Graph plotted from official results published in the *Daily Racing Form* (drf.com).

Whether money spent on thoroughbred breeding fees is a good investment surely depends on whether variation in racing performance is at least partly due to differences in the genes the horses inherit from their parents. Only if performance is heritable will selective breeding lead to the evolution of better runners.

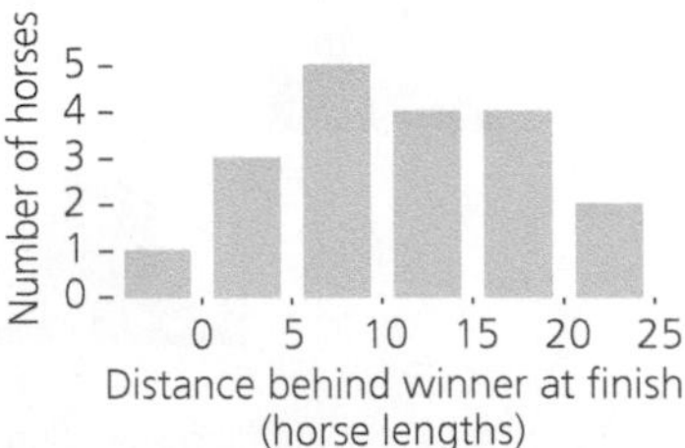

We can predict whether and how populations will evolve, sometimes with considerable precision, thanks to the population genetics models discussed earlier (Chapters 6, 7, and 8). However, we can use these models only when we are content to analyze the evolution of just one or two loci at a time. That is often, as in the case of thoroughbreds, not good enough. Many interesting traits, including race performance, are determined by the combined influence of alleles at many loci. Worse still, when studying such traits, we often do not know the identities of the particular loci involved.

This chapter introduces **quantitative genetics,** the branch of evolutionary biology that provides tools for analyzing the evolution of multilocus traits. The material we present features some abstract ideas, but there are payoffs. Our discussion of quantitative genetics gives us insight into the evolution of most ecologically important traits. It also allows us to assess the wisdom of paying thoroughbred stud fees, and to debunk erroneous claims about the cause of differences among human ethnic groups in mean IQ score.

In Section 9.1 we explore the nature of multilocus, continuously variable traits, and in Section 9.2 we see how researchers can identify some of the underlying genes. In Sections 9.3, 9.4, and 9.5 we cover methods for estimating how much of the variation in a trait is heritable, how to quantify the strength of selection, and how to predict the evolutionary response. In Section 9.6 we discuss the evolutionary mechanisms that maintain genetic variation in populations. Finally, in Section 9.7, we debunk erroneous claims about differences between populations.

9.1 The Nature of Quantitative Traits

Throughout our coverage of evolutionary genetics, we have been discussing traits whose phenotypes come in discrete categories. An Elderflower orchid is either purple or yellow; a person either has cystic fibrosis or does not. We might call characters such as these **qualitative traits,** because we can assign individuals to categories just by looking at them, or perhaps by using a simple genetic test.

Traits with discrete phenotypes are special examples; most traits in most organisms show continuous variation. Examples of human traits with continuous variation are height, athletic ability, and intelligence **(Figure 9.1)**. For traits with

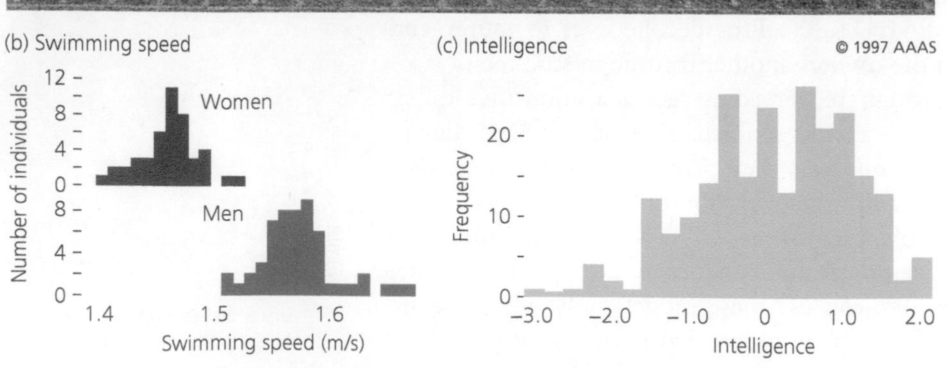

Figure 9.1 Quantitative traits in humans (a) Students and faculty at the University of Connecticut line up by height to form a living histogram. Women are wearing white; men are in blue. (b) Variation in average speed over a distance of 1,500 meters for swimmers at the U.S. national championship. Plotted from USA Swimming (2002). (c) Variation in intelligence, or general cognitive ability, assessed as a statistical composite—called a principal component score—of performance in a variety of tests. Participants were Swedish twins. For more details, see discussion of twins in Section 9.3. Redrawn from McClearn et al. (1997).

(c) From "Substantial genetic influence on cognitive abilities in twins 80 or more years old," *Science* 276: 1560–1563, Figure 1A. Reprinted with permission from AAAS.

continuous variation, we cannot assign individuals to discrete phenotypic categories by simple inspection. Instead, we have to take measurements. For this reason, characters with continuously distributed phenotypes are called **quantitative traits.** Quantitative traits are determined by the combined influence of (1) the genotype at many different loci, and (2) the environment.

Quantitative traits are traits for which the distribution of phenotypes is continuous rather than discrete.

Early in the 20th century, there was considerable debate among biologists over whether Gregor Mendel's model of genetics can be applied to quantitative traits (see Provine 1971). Among the first researchers to provide convincing affirmative evidence was Edward East (1916). East worked with populations of longflower tobacco (*Nicotiana longiflora*). The trait he studied was the length of the corolla, the part of the flower formed by the petals. In longflower tobacco the corolla is shaped like a tube.

East started with two pure-breeding strains of *Nicotiana,* one with short corollas and the other with long corollas. He crossed individuals from these parental strains to produce F_1 hybrids, then let the F_1 hybrids self-fertilize to produce an F_2 generation. Before looking at East's data on corolla length in the F_1s and F_2s, let us first make predictions under alternative models based on Mendelian genetics.

The simplest Mendelian model is one in which corolla length is determined by a single locus with two alleles. We will imagine the alleles are codominant, so there are three phenotypes. Genotype *aa* will produce short flowers, *aA* medium flowers, and *AA* long flowers **(Figure 9.2a)**. By this model, East's first cross is between *aa* and *AA* parents (first row). All the F_1s will have genotype *aA*, and all will have medium flowers (second row). When the F_1s self-fertilize, the F_2s they produce will have genotypes *aa*, *aA*, and *AA* in proportions $\frac{1}{4} : \frac{1}{2} : \frac{1}{4}$ (third row). The reader can reproduce this prediction with a 2×2 Punnett square.

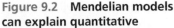

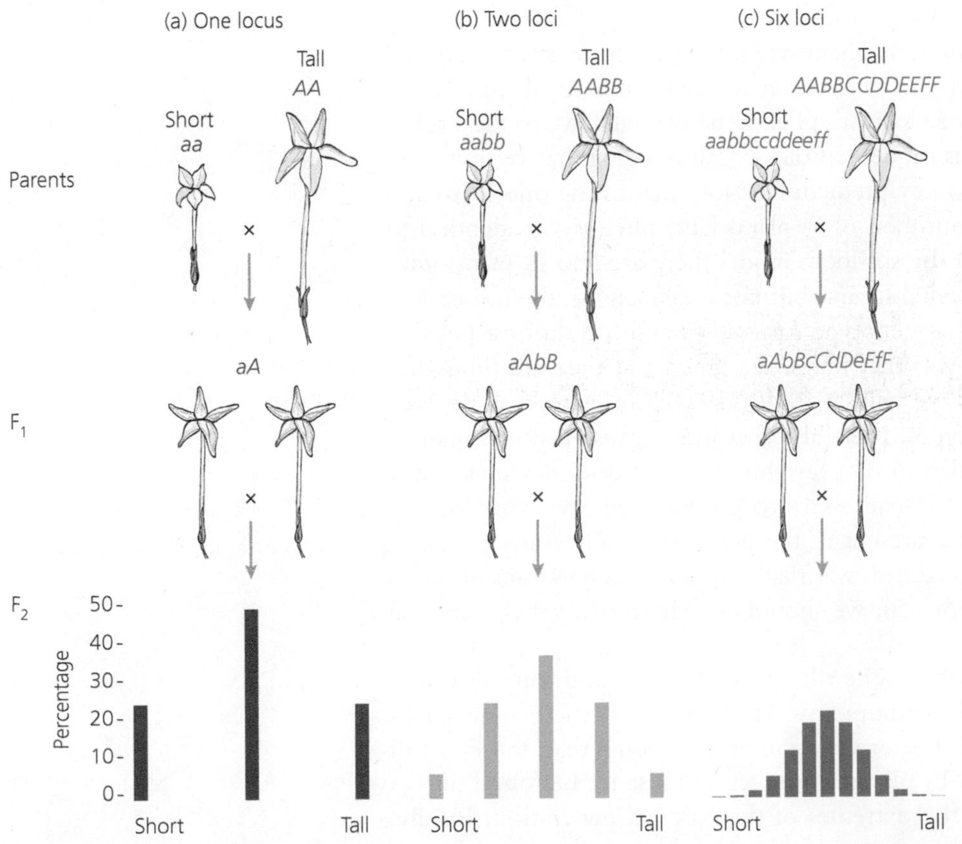

Figure 9.2 Mendelian models can explain quantitative traits (a), (b), and (c) show the predicted genotypes and phenotypes for parental, F_1, and F_2 longflower tobacco plants under Mendelian models in which corolla length is determined by the alleles at one, two, and six loci. See text for more details. After East (1916); Ayala (1982).

East knew from his prior experience with longflower tobacco that this model is much too simple. In most populations corolla length is highly variable, and the variations form a continuum, not three distinct phenotypes. However, a straightforward modification of our simple Mendelian model will improve its prospects for accurately predicting the results of East's crosses.

Instead of imagining that corolla length is determined by the alleles at a single locus, we will imagine it is determined by the alleles at two loci (Figure 9.2b). We still have our first locus, with alleles *a* and *A*, but now we add a second, with alleles *b* and *B*. As with the first locus, each copy of an allele designated by a capital letter independently contributes to a longer corolla. In our new model there are five phenotypes, associated with genotypes that have 0, 1, 2, 3, or 4 capital-letter alleles. East's parental cross, *aabb* × *AABB*, yields F_1s with genotype *aAbB* and medium corollas (first and second rows). When the F_1s self-fertilize, the F_2s they make will have phenotypes ranging from short to long in proportions $\frac{1}{16} : \frac{4}{16} : \frac{6}{16} : \frac{4}{16} : \frac{1}{16}$ (third row). The reader can reproduce this prediction with a 4 × 4 Punnett square (the gametes from an *aAbB* plant are *ab*, *aB*, *Ab*, and *AB*).

The two-locus model is a step in the right direction, but still produces F_2s with discrete phenotypes. We can fix this by skipping ahead to a model in which corolla length is determined by several loci. Figure 9.2c shows the predictions for a model with six. This model yields 13 phenotypes, associated with genotypes that have 0 to 12 capital-letter alleles. East's parental cross in this model is *aabbccddeeff* × *AABBCCDDEEFF* (first row). The F_1s will have genotype *aAbBcCdDeEfF* and medium corollas (second row). The F_2s they produce will have corolla length phenotypes running from short to long in proportions $\frac{1}{4096} : \frac{12}{4096} : \frac{66}{4096} : \frac{220}{4096} : \frac{495}{4096} : \frac{792}{4096} : \frac{924}{4096} : \frac{792}{4096} : \frac{495}{4096} : \frac{220}{4096} : \frac{66}{4096} : \frac{12}{4096} : \frac{1}{4096}$ (last row). This prediction can be reproduced with a 64 × 64 Punnett square, but we suspect the reader will just take our word for it.

Quantitative traits are consistent with Mendelian genetics. They are influenced by the combined effects of the genotype at many loci.

By the time there are 13 different phenotypes, they begin to grade into one another, and with real plants it will be difficult to assign individuals to discrete categories. Instead, to characterize a plant's phenotype we will have to use a ruler. In other words, in our six-locus model, corolla length is a quantitative trait.

Now we come to East's two key predictions. Note that in the one-locus and two-locus models, substantial numbers of F_2 plants have phenotypes identical to those of the parental strains. In the six-locus model there are also F_2 plants with phenotypes identical to the parental strains, but not very many of them. Just 1 in 4,096 F_2 plants, for example, has genotype *aabbccddeeff* and the shortest possible corollas. East's first prediction was that unless we breed and measure thousands of plants, the range of variation we will see in the F_2s will not extend all the way to the original parental phenotypes. Note also that just because parental phenotypes do not appear in a population of a few hundred F_2s does not mean they have been lost forever. The alleles necessary to produce genotypes *aabbccddeeff* and *AABBCCDDEEFF* are still present in the population. They are just all in heterozygotes. East's second prediction was that with a few generations of selective breeding for short or tall corollas, we should be able to recover the original parental phenotypes.

East's data appear in **Figure 9.3**. When he crossed short- and long-flowered parents, they produced F_1s with medium flowers. When he let the F_1s self-fertilize, the F_2s they made showed greater variation in phenotype than the F_1s. But because he examined only 454 F_2 plants, not several thousand, he found no F_2s with phenotypes approaching the extremes of the parental generation. Finally,

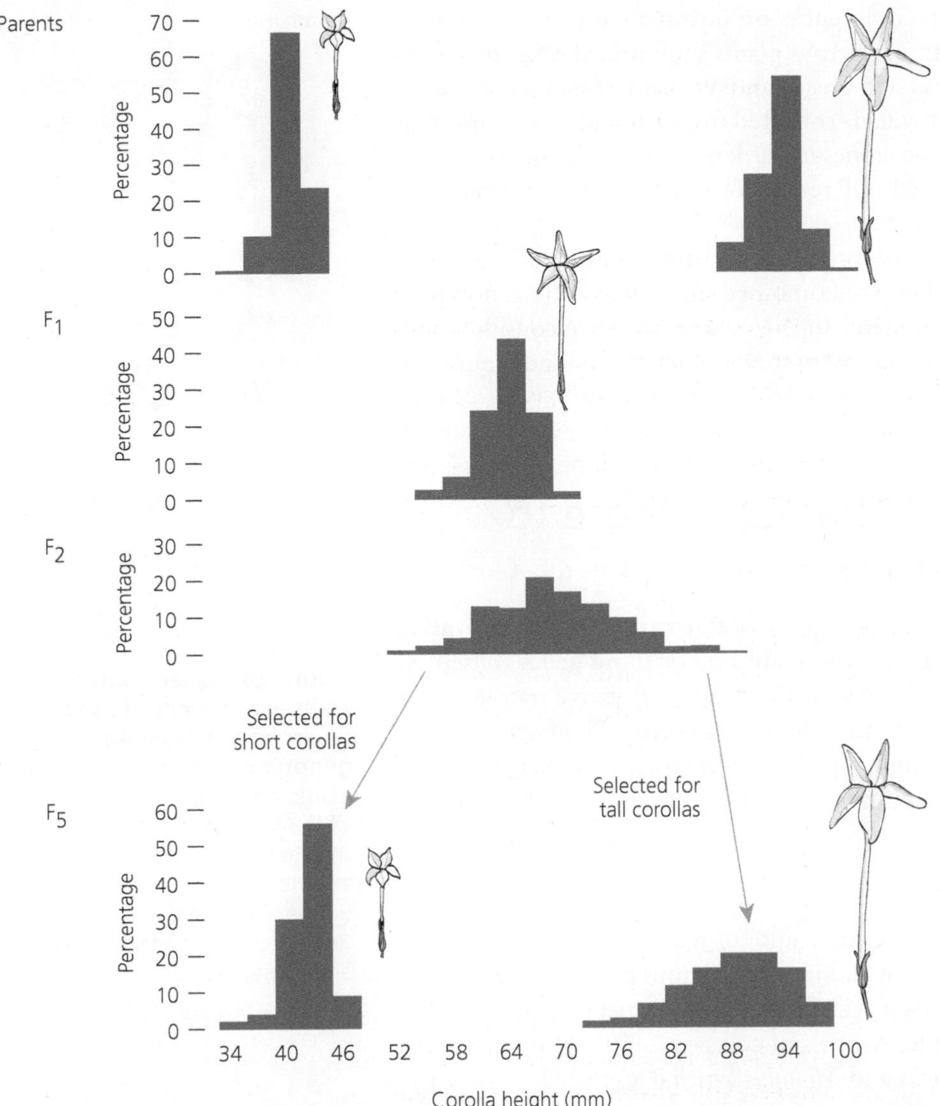

Figure 9.3 Edward East's data confirm the predictions of the Mendelian model in Figure 9.2c East crossed short and long-flowered parental plants to produce medium-flowered F_1s. He then selfed the F_1s, which produced an F_2 generation that was more variable than the F_1 generation, but did not approach the extremes of the parental strains. Finally, East recovered the parental phenotypes by selectively breeding from the F_2 plants. Drawn from data in Table 1 of East (1916); after Ayala (1982).

starting with the F_2 plants, East selectively bred for short corollas and long corollas. By the time he reached the F_5 generation, most plants in his selected lines had corolla lengths within the ranges of the original parents. East's data confirm his predictions. His experiment, and others like it, established that quantitative traits are determined by the combined influence of Mendelian alleles at many loci.

The fit between East's data and his Mendelian model is impressive, but there is one respect in which it is not perfect. Under a strict interpretation of even the six-locus model, the short parental plants should all have been exactly the same. Likewise, the long parental plants and the F_1s should all have been exactly the same. If, as we have assumed, all the plants in each of these groups have the same genotype, then they should also all have the same phenotype. But, of course, they do not. One possibility is that despite coming from true-breeding stocks, East's parental plants were not all homozygous at all loci. More important, there is some phenotypic variation even among genetically identical plants. The reason is that each plant, even in East's experimental garden, was exposed to a unique environment. Some got a little more water, others got a little more sun. These small differences in environment produced small differences in phenotype.

Quantitative traits are also influenced by the environment.

The influence of environmental differences on quantitative phenotypes is especially clear in **Figure 9.4**. The three yarrow plants shown in the figure are genetically identical. Jens Clausen, David Keck, and William Hiesey (1948) grew them from cuttings of a single individual, collected from a population living at an elevation of 50 feet. They reared the clones in gardens at elevations of 100, 4,600, and 10,000 feet. The environmental differences associated with elevation had a dramatic impact on the plants' heights.

The data presented in this section substantiate a claim we made at the outset. Quantitative traits are determined by the combined influence of the genotype at many different loci and the environment. In the next section, we consider modern techniques for identifying some of the particular loci that underlie quantitative variation.

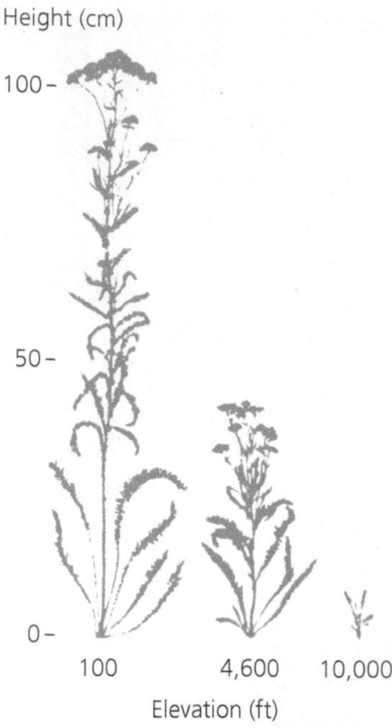

Figure 9.4 Quantitative traits are influenced by the environment as well as genotype These three yarrow plants were grown from cuttings of the same individual and are thus genetically identical. Reared at different altitudes, they show dramatic differences in height. Reprinted from Clausen, Keck, and Hiesey (1948).

9.2 Identifying Loci That Contribute to Quantitative Traits

Portions of the genome that influence quantitative traits are called **quantitative trait loci,** or **QTLs.** A given QTL may contain one or more genes. Often we would like to identify the QTLs behind an interesting quantitative trait. Modern genetic and statistical methods make it possible to do so. We will review two such methods: QTL mapping, and investigation of candidate loci. Our examples come from a study of the genetics of adaptation in monkeyflowers, a study of human personality, and a study of genetic factors contributing to a human disease.

QTL Mapping

QTL mapping is the collective name for a suite of related techniques that employ marker loci to scan chromosomes and identify regions containing genes that contribute to a quantitative trait. We will illustrate QTL mapping with an example from research by H. D. Bradshaw, Jr. and colleagues (1998) on two species of monkeyflowers, *Mimulus cardinalis* and *Mimulus lewisii* **(Figure 9.5)**.

Mimulus cardinalis and *M. lewisii* are sister species. They have overlapping ranges in the Sierra Nevada of California. They hybridize readily in the lab and produce

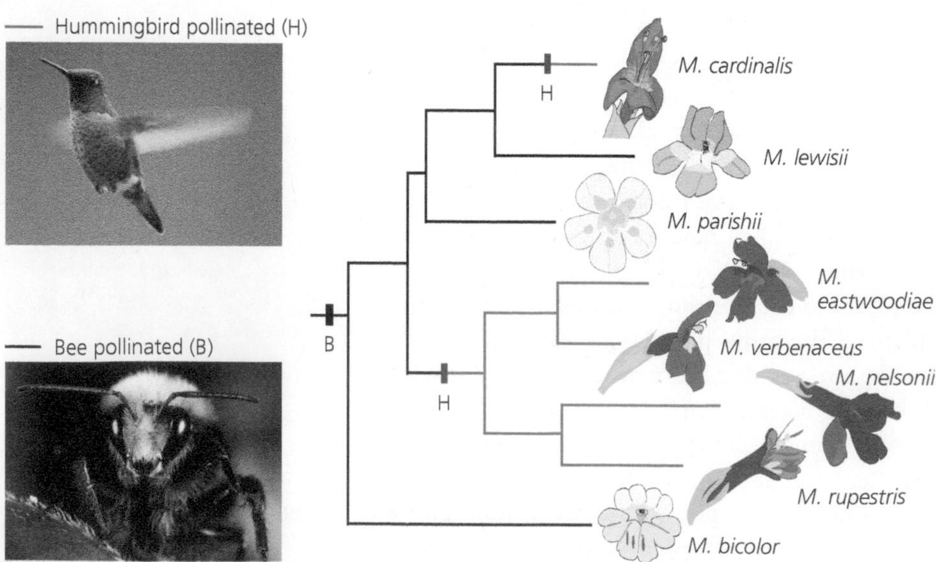

Figure 9.5 A phylogeny of *Mimulus cardinalis, Mimulus lewisii,* and kin The common ancestor of these species was pollinated by bees. Pollination by hummingbirds evolved twice: once in the common ancestor of *M. eastwoodiae* and kin, and once in *M. cardinalis*. We would like to know the genes involved and their effects. After Beardsley et al. (2003).

Figure 9.6 ***Mimulus lewisii, Mimulus cardinalis*, and their F_1 and F_2 descendants** Photo (a) shows *M. lewisii*, photo (b) shows an F_1 hybrid, and photo (c) shows *M. cardinalis*. The remaining photos (d–l) show F_2 hybrids produced by crosses between F_1s. The F_2s show wide variation in their floral characters. Reprinted from Schemske and Bradshaw (1999).

fertile offspring **(Figure 9.6)**. Yet hybrids have never been found in the field. The reason is that the two monkeyflowers attract different pollinators. *Mimulus cardinalis* is pollinated by hummingbirds; *M. lewisii* is pollinated by bees.

The difference in pollinators between *M. lewisii* versus *M. cardinalis* is reflected in their flowers. Bees do not see well in the red part of the visible spectrum and need a platform to land on before they can crawl into a flower and forage. Hummingbirds, in contrast, see red well, have long, narrow beaks, and hover while sipping nectar. *M. lewisii* has a prominent landing pad, while *M. cardinalis* has an elongated tube with a nectar reward at the end. Indeed, the flowers of *M. lewisii* and *M. cardinalis* conform to classical bee- and bird-pollinated colors and shapes.

As the phylogeny in Figure 9.5 shows, the most recent common ancestor of *M. lewisii* and *M. cardinalis* was likely pollinated by bees (Beardsley et al. 2003). This implies that many characteristics of *M. lewisii*'s bee-pollinated flower are ancestral and that *M. cardinalis*'s more tubular, reddish, hummingbird-pollinated flower is derived. The questions are: What genes are responsible for the radical makeover of *M. cardinalis*'s flower? How many of them are there? How strong are their effects? QTL mapping offers a way to find out.

Bradshaw and colleagues' interest in these questions was motivated in part by fascination with the plants and in part by a debate in theoretical evolutionary genetics. Starting with Ronald Fisher (1930), most evolutionary geneticists have held that the alleles driven to fixation by natural selection, and thus responsible for the adaptive differences between species, virtually all exert small effects on the phenotype. A minority of theoreticians, most prominently H. Allen Orr, have instead contended that some alleles fixed during adaptive evolution exhibit large phenotypic effects (Orr and Coyne 1992; Orr 1998, 1999). Bradshaw and coworkers saw monkeyflowers as an ideal test case. If Orr's view is correct, the researchers would find QTLs with obvious effects on floral phenotype; if Fisher's view is correct, they would find only QTLs with subtle effects.

Bradshaw and colleagues crossed *M. lewisii* and *M. cardinalis* to make F_1 hybrids (Figure 9.6a–c), then crossed the F_1s to make 465 F_2 individuals. The F_2s show a diversity of floral phenotypes (Figure 9.6d–l). This result is similar to Edward East's (shown in Figure 9.3) and has a similar genetic explanation. The parental forms, *M. lewisii* and *M. cardinalis*, were essentially homozygous at all loci influencing floral appearance. As a result, the F_1s were all heterozygous. The F_2s are the product of genetic recombination among the F_1 heterozygotes. At any given locus, a given F_2 may be a homozygote for the *M. lewisii* allele, a heterozygote, or a homozygote for the allele. Bradshaw and colleagues scored all 465 F_2 plants for each of a dozen floral characters that differ between the two species (**Table 9.1**).

Table 9.1 Flowers of *Mimulus cardinalis* versus *Mimulus lewisii*

Traits scored by Bradshaw et al. (1998) are listed in the first column, grouped by function in pollination. The direction of the difference between species is listed in the second and third columns. After Bradshaw et al. (1998).

Characteristic	*M. cardinalis*	*M. lewisii*	Notes
Pollinator attraction			
Purple pigment (anthocyanins) in petals	high	low	
Yellow pigment (carotenoids) in petals	high	low	The yellow pigment in *M. lewisii* petals is arranged in stripes called nectar guides, which are interpreted as a runway for bees as they land on wide petals.
Lateral petal width	high	low	
Corolla width	low	high	
Corolla projected area	low	high	
Upper petal reflexing	high	low	
Lateral petal reflexing	high	low	
Pollinator reward			
Nectar volume	high	low	Higher nectar volume in *M. cardinalis* is probably adaptive simply because birds drink more than bees.
Pollinator efficiency			
Stamen (male structure) length	high	low	The difference in stamen and pistil length is important. In *M. cardinalis* these structures extend beyond the flower and touch the hummingbird's forehead as the bird feeds.
Pistil (female structure) length	high	low	
Corolla aperture width	low	high	
Corolla aperture height	high	low	

Bradshaw and colleagues also determined the genotype of each F_2 plant at 66 marker loci randomly distributed across the monkeyflower genome. A marker locus is a known site in the genome where the nucleotide sequence varies among chromosomes and where a simple genetic test will identify different alleles. Bradshaw and colleagues chose marker loci at which *M. cardinalis* are all homozygous for one allele and *M. lewisii* are all homozygous for another. This meant that the F_1 plants were all heterozygous, and that the F_2s could be homozygous for the *M. lewisii* allele, heterozygous, or homozygous for the *M. cardinalis* allele.

To map QTLs in the *Mimulus* genome, the researchers examined the F_2 population for statistical associations between genotype at marker loci and phenotype. If phenotype was associated with genotype at a particular marker locus, they could interpret the association as evidence that a QTL influencing the trait of interest is located near the marker.

To see the logic of QTL mapping, imagine a marker locus at which the *lewisii* allele is M_L and the *cardinalis* allele is M_C. Imagine also a quantitative trait locus that influences one of the monkeyflower floral traits. We will call the *lewisii* allele Q_L and the *cardinalis* allele Q_C. Consider first a case in which the QTL and the marker locus are physically linked—that is, close together on the same chromosome. The *M. lewisii* parent had genotype M_LQ_L/M_LQ_L and the *M. cardinalis* parental plant genotype M_CQ_C/M_CQ_C, where M_CQ_C indicates a two-locus genotype on a single chromosome **(Figure 9.7)**. The F_1 plants all had genotype M_LQ_L/M_CQ_C. In rare cases crossing over will occur between the QTL and the marker locus, but except for these, the F_2 population will consist of plants with three genotypes: M_LQ_L/M_LQ_L, M_LQ_L/M_CQ_C, and M_CQ_C/M_CQ_C. Plants homozygous for the *lewisii* marker allele will tend toward the *lewisii* phenotype, heterozygotes will have intermediate phenotypes, and plants homozgyous for the *cardinalis* marker allele will tend toward the *cardinalis* phenotype (top right). In other words, the marker locus and the QTL are in linkage disequilibrium (see Chapter 8). Of the four possible chromosome genotypes, only two are present. This linkage disequilibrium reveals itself in a nonrandom association between the genotype at the marker locus and the phenotype influenced by the QTL.

We can detect the presence and location of loci influencing a quantitative trait by crossing parents from populations with fixed differences. Among the grandoffspring, we look for associations between phenotype and genotype at marker loci.

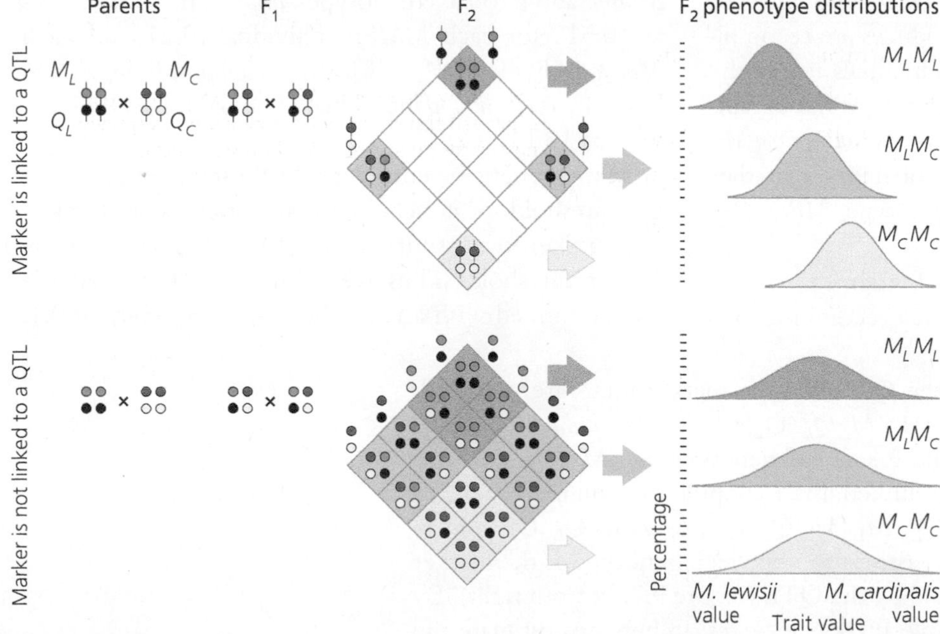

Figure 9.7 The logic of QTL mapping In QTL mapping, researchers cross parents from different species to produce F_1 hybrids, then self or intercross the F_1 hybrids to produce a large population of F_2s (see Figure 9.6). For each F_2 individual the researchers measure the phenotype for the quantitative trait of interest and the genotype at marker loci distributed across the genome. Examining the entire F_2 population, researchers compare individuals with different genotypes at each marker locus. If phenotypes differ among individuals with different genotypes at a particular marker (top right), then we can infer that the marker locus sits near a locus that contributes to the quantitative trait.

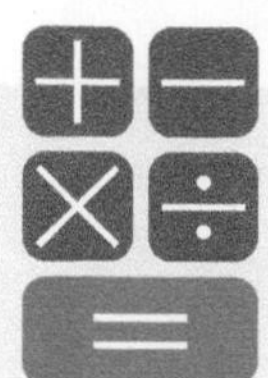

COMPUTING CONSEQUENCES 9.1

Genetic mapping and LOD scores

Here we use a simple example to illustrate the statistical reasoning behind a method of genetic mapping that employs likelihoods. We first consider a *qualitative* phenotypic trait controlled by a single locus with two codominant alleles. This means that we can infer genotype from phenotype. We want to know whether the locus for this trait is linked to a particular marker locus.

Let the alleles at the locus affecting phenotype be *P* and *p*, and the alleles at the marker locus be *M* and *m*. Imagine that we have crossed a parent homozygous for allele *P* at the trait locus and allele *M* at the marker locus with a parent homozygous for allele *p* and allele *m*. The F_1 offspring of this cross are heterozygous for both loci. We then crossed two F_1 individuals to produce an F_2 individual, which turned out to be homozygous for both allele *P* and allele *M*. We want to know whether this outcome constitutes evidence that the trait locus and the marker locus are linked. To do so, we will calculate the **likelihood** of producing a double homozygote from our F_1 cross. The likelihood of a particular outcome is its probability given a model of the process that produced it. We will consider two models: one in which the loci are linked with a recombination frequency of 0.1, and one in which they are not linked.

Linkage model with $r = 0.1$: Under this model the loci are linked with the distance between them such that 10% of the gametes meiosis produces are recombinants. The genotype for our F_1 individuals under this model was *MP*/*mp*. These individuals produce gametes in the following proportions: 45% *MP*, 5% *Mp*, 5% *mP*, and 45% *mp*. Thus the probability that they would produce an offspring of genotype *MP*/*MP* is $0.45 \times 0.45 = 0.2025$.

Free-recombination model: Under this model the loci are unlinked. As a result, they recombine during meiosis 50% of the time. The genotype for our F_1 individuals under this model was *M*/*m P*/*p*. These individuals produce gametes in the following proportions: 25% *MP*, 25% *Mp*, 25% *mP*, and 25% *mp*. Thus the probability that they would produce an offspring of genotype *M*/*M P*/*P* is $0.25 \times 0.25 = 0.0625$.

If we divide the likelihood under the linkage model by the likelihood under the free-recombination model we get the **likelihood ratio**, which is 3.24. This calculation shows that our double-homozygote F_2 offspring is over three times more likely under the linkage model than under the free-recombination model. In other words, our F_2 offspring provides evidence that the trait locus and the marker locus are linked.

But it does not provide much evidence; it is only a single individual. We need to look at many F_2 offspring and assess the strength of the evidence they provide collectively. To do this we take the logarithm of the likelihood ratio, which gives us a value called the **LOD** score, short for *logarithm of the odds*. The LOD score for our first F_2 individual is 0.511. Now we can calculate the LOD scores for other individuals and sum them to get an overall LOD score. Imagine that we examine nine more F_2s and find that two have genotype *MMPP*, four have genotype *MmPp*, one has genotype *Mmpp*, and two have genotype *mmpp*. The LOD scores are 0.511 for each *MMPP* individual, 0.215 for each *MmPp* individual, −0.444 for the *Mmpp* individual, and 0.511 for each *mmpp* individual. The overall LOD score for our 10 F_2s is 2.97.

Many geneticists consider a LOD score of 3.0 to be the threshold for concluding that a trait locus is linked to a marker locus. Our overall LOD score falls just shy of this threshold. Thus we cannot conclude that the loci are linked with a recombination frequency of 0.1.

Consider now a case in which the QTL and the marker locus are unlinked. The *M. lewisii* parent had genotype $M_L/M_L Q_L/Q_L$ and the *M. cardinalis* parental plant genotype $M_C/M_C Q_C/Q_C$. The F_1s all had genotype $M_L/M_C\ Q_L/Q_C$. Because the QTL and the marker are unlinked, the F_2 population will include plants with nine genotypes: $M_L/M_L\ Q_L/Q_L$, $M_L/M_L\ Q_L/Q_C$, $M_L/M_C\ Q_L/Q_L$, and so on. Among the F_2s, there will be no association between genotype at the marker and phenotype for the trait influenced by the QTL (Figure 9.7, bottom right). By looking for associations with multiple marker loci, researchers can estimate the

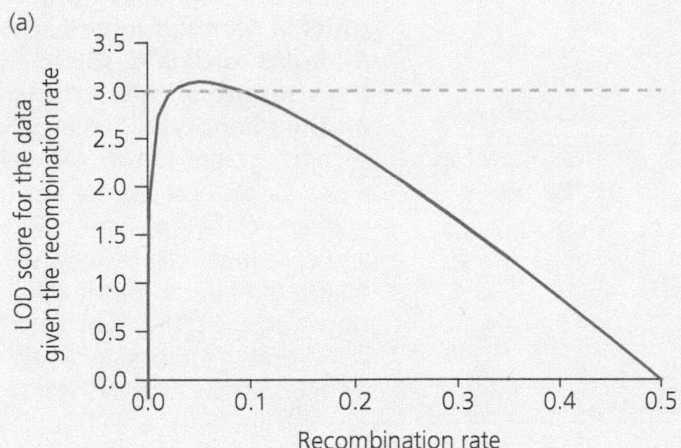

© 1988 Nature Publishing Group

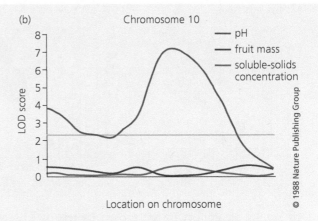

Figure 9.8 **Results from QTL mapping studies are often summarized with plots of LOD score** LOD score measures the degree to which we can better explain the data with a model in which a locus influencing phenotype is linked to the marker loci examined versus a model without such linkage. Where LOD score crosses a threshold chosen by the researchers, they conclude there is evidence of linkage. (a) LOD score versus recombination rate for our hypothetical example. (b) LOD score as a function of chromosomal location for three quantitative traits in tomatoes. From Paterson et al. (1988).

(b) Reprinted by permission from Macmillan Publishers Ltd: A. H. Paterson, E. S. Lander, et al., 1988, "Resolution of quantitative traits into Mendelian factors by using a complete linkage map of restriction fragment length polymorphisms," *Nature* 335: 721–726, Figure 3.

But what about other recombination frequencies? We could just as well have chosen a recombination frequency of 0.05 for our linkage model, or 0.2, or any value between 0 and 0.5. Figure 9.8a plots the overall LOD score for our data under all possible linkage models. The model under which the LOD score reaches its maximum value, 3.10, is one in which the recombination rate is 0.05. We can conclude that the trait locus and the marker locus are linked, and that the best estimate for the recombination rate between them is 0.05.

If the genetic basis of our trait is more complex, and we are trying to assess whether a QTL influencing the trait is linked to our marker, the data will consist of phenotypes for the trait of interest and genotypes for the marker. Our model will include parameters for the mean phenotypes of individuals with different QTL genotypes, the variance among individuals with the same QTL genotype, and the recombination rate. We seek the suite of parameter values that maximizes the the LOD score for linkage. The value of the recombination parameter estimates the position of the QTL; the values of the other parameters estimate its effects.

By using data for multiple marker loci, researchers can scan chromosomes for evidence of multiple QTLs. Often they summarize their analyses with plots of LOD score as a function of the chromosomal location of the QTL under the linkage model. Figure 9.8b, for example, shows LOD score as a function of location on chromosome 10 for three quantitative traits in tomatoes—fruit mass, concentration of soluble solids, and pH—in a study by Andrew Paterson and colleagues (1988). Taking 2.4 as their threshold LOD score for detecting a single QTL on a chromosome, the researchers found strong evidence for a locus in the middle of chromosome 10 that influences pH. There may also be a QTL influencing pH near the left end of the chromosome. There is no evidence of QTLs on chromosome 10 that influence fruit mass or soluble solids.

The difference between the analysis for our simplified example and the analysis for a quantitative trait using multiple markers is largely in the details of the probability calculations. For an overview of QTL mapping, see Tanksley (1993). For a detailed treatment, see Lynch and Walsh (1998).

number of QTLs, their locations, and the strength of their influence on phenotype. For more on QTL mapping, see **Computing Consequences 9.1**.

In practice, at most of the marker loci that Bradshaw and colleagues used, one allele was dominant and the other recessive. As a result, it was possible to distinguish only two genotypes: homozygous recessive versus other. Nonetheless, it was still possible to look for, and find, associations between marker locus genotypes and floral phenotypes. For each of the dozen floral traits the researchers scored, they found between one and six QTLs that influence flower phenotype.

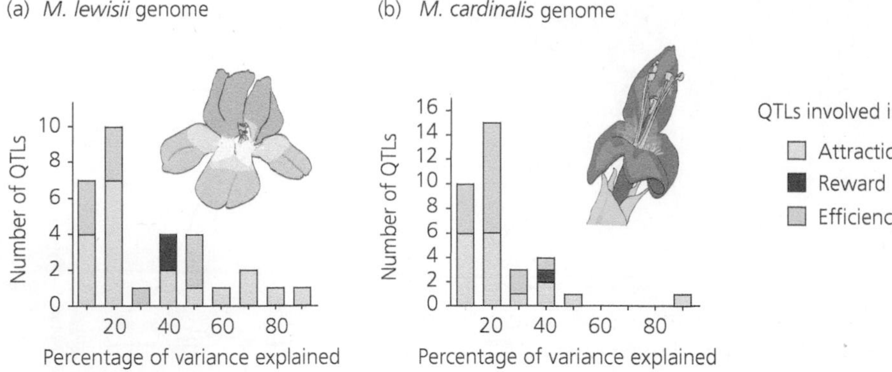

Figure 9.9 QTLs for floral traits in *Mimulus lewisii* and *Mimulus cardinalis*, sorted by the strength of their effects on the phenotype There are separate graphs for *M. lewisii* and *M. cardinalis* because most of the marker loci used by Bradshaw and colleagues can be detected only in one species or the other. However, the QTLs probably occur in both genomes. Redrawn from Bradshaw et al. (1998).

Differences in genotype for most of these QTLs explained only a modest amount of the variation in flower phenotype, as indicated by the fact that most of the QTLs fall in bars on the left side of the graphs in **Figure 9.9**. However, some of the QTLs Bradshaw and colleagues found had pronounced effects on the appearance of the flowers. For 9 of the 12 floral traits, there was at least one QTL at which differences in genotype explained more than 25% of the variation in phenotype. These QTLs with large phenotypic effects fall in the bars on the center and right portions of the graphs in the figure.

To confirm that the QTLs that Bradshaw and colleagues identified were, indeed, the loci subject to selection during the diversification of the two species, Douglas Schemske and Bradshaw (1999) reared a large series of F_2 individuals in the greenhouse and recorded the amount of purple pigment, yellow pigment, and nectar in their flowers, along with overall flower size. Then they planted the individuals in a natural habitat where both species of monkeyflowers naturally coexist, and recorded which pollinators visited which flowers. Their data revealed a strong trend. Bees prefer large flowers and avoid flowers with a high concentration of yellow pigments. Hummingbirds, in contrast, tended to visit the most nectar-rich flowers and those with the highest amounts of purple pigment.

By collecting tissues from each F_2 individual planted in the field and determining which QTL markers they contained, the researchers were able to calculate that an allele associated with increased concentration of yellow pigments reduced bee visits by 80%, while an allele responsible for increasing nectar production doubled hummingbird visits. It is reasonable to surmise that changes in the frequencies of these alleles, driven by differential success in attracting hummingbirds as pollinators, were the mechanism behind the diversification of the two monkeyflowers.

Bradshaw and Schemske (2003) reproduced an event that might have been the first step in the evolution of *M. cardinalis* from a bee-pollinated ancestor. They bred plants that had the *M. lewisii* genotype across virtually the entire genome, except that they carried the *M. cardinalis* genotype at a QTL, mentioned in the previous paragraph, that strongly influences the amount of yellow pigment in the petals. This single genetic change turns *M. lewisii*'s petals from purplish-pink to pale yellow-orange **(Figure 9.10)**. When they monitored these plants in a natural habitat, the researchers found that the *M. lewisii* plants with the novel genotype were considerably less attractive to bees than the wild type, but dramatically more attractive to hummingbirds. The yellow-orange *M. lewisii* were still much less attractive to hummingbirds than are wild-type *M. cardinalis*, but they had been given a good start down the path toward switching pollinators. The results of

(a) Wild-type *M. lewisii*

(b) *M. lewisii* with *M. cardinalis* genotype at the *YUP* locus

© 2007 Nature Publishing Group

Figure 9.10 A novel allele at a single locus can dramatically alter a flower's attractiveness to different pollinators The monkeyflowers shown here are full siblings. Genetically they are virtually identical, except that they carry different alleles at the *YUP* locus. The flower in (a) is about 6 times more attractive to bees than the flower in (b). The flower in (b) is roughly 70 times more attractive to hummingbirds than the flower in (a). From Bradshaw and Schemske (2003).

Reprinted by permission from Macmillan Publishers Ltd: *Nature* 426: 176–178, copyright 2007.

Bradshaw and Schemske's monkeyflower research are consistent with Orr's view that adaptive evolution often involves the selective fixation of alleles with large effects. The results are inconsistent with Fisher's view.

QTL mapping can reveal the number of genomic regions that influence a quantitative trait, the magnitude of their effects on phenotype, and their location in the genome. It cannot, however, tell us the identity of the loci and the proteins they encode. To determine these, researchers must evaluate candidate loci.

Candidate Loci

To detect a QTL and find its location, we look for an association between a marker locus genotype and phenotype. To learn the identity of a QTL and the protein it encodes, we have to look for associations between a coding-locus genotype, the structure and function of the locus's gene product, and phenotype. Sometimes we know to evaluate a particular coding locus because we already know something about the function of its gene product and suspect it may play a role in phenotype. At other times we know to evaluate a particular coding locus because its location matches that of a QTL we have mapped with markers.

Our first example comes from a study of human personality, most aspects of which are quantitative traits. Jonathan Benjamin and colleagues (1996) were interested in a personality trait called novelty seeking. Novelty seeking, assessed with questionnaires, is highly variable among individuals and has the familiar bell-shaped distribution. People with high novelty-seeking scores tend to be more impulsive, excitable, and exploratory, whereas people with low scores tend to be reflective, stoic, and rigid (Ebstein et al. 1996). Benjamin and colleagues had reason to suspect that some of the variation in novelty seeking might be associated with allelic variation in the gene for the D4 dopamine receptor, or D4DR.

We can confirm that a particular locus influences a quantitative trait by looking for associations between genotype and phenotype.

D4DR is a neurotransmitter receptor. It sits on the surface of neurons in the brain, waiting for messages from other cells in the form of dopamine, a neurotransmitter. Benjamin and colleagues knew that neurons using D4DR as a receptor participate in thought and emotion in humans and exploratory behavior in animals. And they knew that one of the coding regions in the gene for D4DR harbors a 48-base-pair tandem repeat. The number of repeats varies from two to eight. Alleles with different numbers have distinguishable physiological effects.

To see if the *D4DR* gene is a QTL influencing novelty seeking, Benjamin and colleagues determined the novelty-seeking scores and *D4DR* genotypes of 315 people. The researchers divided the *D4DR* alleles they found into two categories: short (*S*) for alleles with two to five repeats versus long (*L*) for alleles with six to eight repeats. They divided their subjects into two categories by their *D4DR* genotypes: a group with two short alleles (*SS*) versus a group with at least one long allele (*SL* or *LL*). When the researchers compared the novelty-seeking scores for the two groups, they found that people with at least one long allele scored three points higher, on average, than people with two short alleles (**Figure 9.11**). This difference is slight; it explains only 3% to 4% of the variation in novelty seeking. Some *SS* individuals scored much higher than some *SL* or *LL* individuals. But genotype at the locus encoding D4DR does appear to exert some influence on this personality trait (see also Schinka et al. 2002).

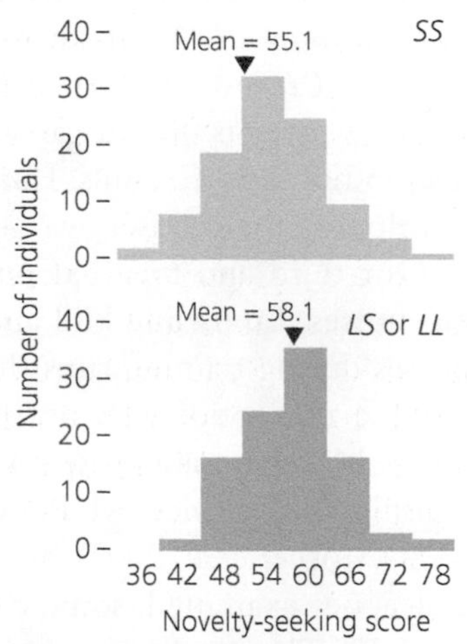

Figure 9.11 Identification of a quantitative trait locus influencing a personality trait Sequence variation at the D4 dopamine receptor locus can be reduced to two categories of alleles: short (*S*) and long (*L*). Individuals with genotype *LS* or *LL* tend to score slightly but significantly higher on psychological tests of novelty seeking. Redrawn from Benjamin et al. (1996).

Our second example comes from a study of Crohn's disease, a chronic digestive disorder characterized by inflammation of the intestine. Crohn's disease tends to run in families, suggesting that it may be caused, in part, by genetic factors. Furthermore, its frequency has increased in recent decades, suggesting that it may also be caused, in part, by environmental factors, such as bacterial pathogens (Hugot et al. 2001) or even reduced exposure to intestinal worms (Moreels and Pelckmans 2005; Summers et al. 2005a, 2005b). The severity of Crohn's disease ranges from mild to extreme. We can think of it as a quantitative trait.

Yasunori Ogura and colleagues (2001) knew, from mapping studies using marker loci, that a QTL influencing susceptibility to Crohn's disease is located in a particular region of chromosome 16. This region contains several coding loci, some of which had been evaluated previously and shown to play no role in the disease. Ogura and colleagues investigated a gene called NOD2. NOD2 encodes a protein that helps regulate the immune response to bacteria in the intestine (Kobayashi et al. 2005). It makes sense that genotypes at the NOD2 locus might be associated with the risk of developing Crohn's disease.

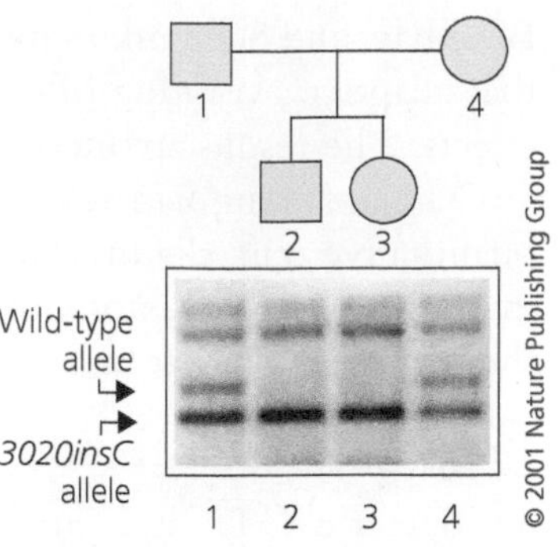

Figure 9.12 Transmission of a suspect allele in a family with Crohn's disease The parents in this family are unaffected; both children have Crohn's disease. The electrophoresis gel shows DNA fragments from PCR amplification of a portion of the NOD2 gene. Both parents are carriers of the *3020insC* allele; both children are homozygotes. These data are consistent with the hypothesis that *3020insC* homozygotes are at higher risk for Crohn's disease. From Ogura et al. (2001).

Reprinted by permission from Macmillan Publishers Ltd: *Nature* 411: 603–606, copyright 2001.

Ogura and colleagues sequenced the NOD2 genes from a dozen patients with Crohn's disease. In three patients they found a frameshift mutation, a single-nucleotide insertion that creates, downstream, a premature stop codon. This allele, called *3020insC*, encodes a truncated protein with compromised function.

To see whether the *3020insC* allele increases the risk of Crohn's disease, the researchers performed what is called a transmission disequilibrium test (see Lynch and Walsh 1998). They screened a large sample of Crohn's disease patients and found 56 independent cases in which a patient had a parent who carried one copy of *3020insC* and thus had the opportunity to pass the allele to the patient (see **Figure 9.12** for a family with two such parents). If the allele plays no role in the disease, then we would expect the heterozygous parent to have transmitted the allele to the patient in half the cases. If, on the other hand, the allele does play a role, we would expect the parents to have transmitted the allele in more than half the cases. In fact, the parent transmitted the allele in 39 cases and failed to transmit it in only 17. A chi-square test shows that this result is statistically significant.

The *3020insC* allele is clearly not the sole cause of Crohn's disease. Many Crohn's patients do not carry the allele, and many individuals who carry the allele do not have Crohn's. But the allele does appear to increase a person's risk of developing the disease.

Our third and final example comes from a study of speed in thoroughbred racehorses. Emmeline Hill and colleagues (2012) measured the top speeds of 85 horses during training runs that simulated races. They genotyped the horses for a single-nucleotide polymorphism in a noncoding region of the gene for myostatin. Myostatin is a protein known to regulate muscle mass in a variety of mammals, including mice (McPherron et al. 1997), cattle (McPherron and Lee 1997), dogs (Mosher et al. 2007), and humans (Schuelke et al. 2004). At the site Hill and colleagues examined, some copies of the gene carry nucleotide C, while others carry T. The results appear in **Figure 9.13**. By a statistically significant margin, horses with genotype *CC* had faster top speeds than horses with genotype *TT*. Heterozygotes were intermediate.

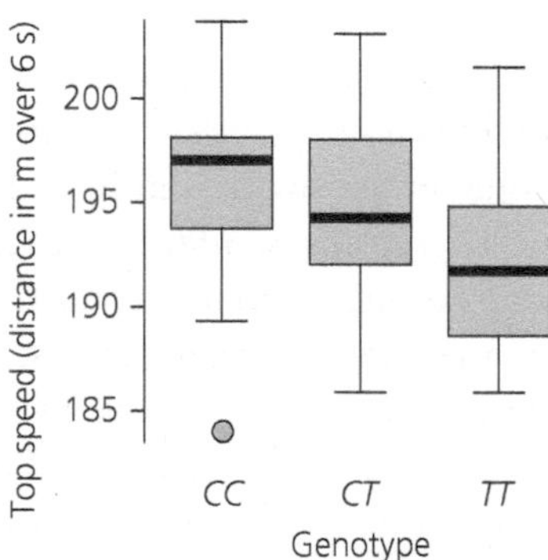

Figure 9.13 A QTL for top speed in thoroughbreds Box plots showing median, interquartile range, range, and an outlier (pink) are for top speed of 85 horses genotyped for a single-nucleotide polymorphism in intron 1 of the myostatin gene. Top speed is measured as distance traveled over the 6 seconds before reaching maximum velocity. From Hill et al. (2012).

Note that while Hill et al.'s study documents genetic variation for a component of racing performance, it does not show that money invested in breeding fees is well spent. There is more to winning than sheer speed. Endurance matters too. Earlier work by Hill and colleagues (2010) had already revealed that among race-winning horses *CC* individuals did best at short races, *CT* individuals at middle distances, and *TT* individuals at long races. However, in comparing horses that had won a race to horses that had not, Hill and colleagues found no difference in allele frequency—suggesting that earnings may be unrelated to genotype.

The examples we have discussed show that it is possible to trace a quantitative trait to the genomic regions and, sometimes, to the Mendelian loci that influence it. This effort, however, is time consuming and expensive. We need tools that allow us to analyze and understand the genetics and evolution of quantitative traits even when we do not know the identities of the many specific genes involved. These tools are the subject of the next three sections of the chapter.

9.3 Measuring Heritable Variation

Recall the basic tenets of Darwin's theory of evolution by natural selection: If there is heritable variation among the individuals in a population, and if there are differences in survival and/or reproductive success among the variants, then the population will evolve. Quantitative genetics includes tools for measuring heritable variation, tools for measuring differences in survival and/or reproductive success, and tools for predicting the evolutionary response to selection. In this section, we focus on the first challenge—measuring heritable variation.

Quantitative genetics allows us to analyze evolution by natural selection in traits controlled by many loci.

Imagine a population whose individuals show continuous variation in a trait. For example, imagine a human population in which there is continuous variation in height. Continuously variable traits are typically normally distributed, so that a histogram has a bell-shaped curve. Assuming our human population follows this pattern, a few people are short, many people are more or less average in height, and a few people are tall. We want to know: Is height heritable?

It is worth thinking carefully about what this question means. Questions about heritability are often expressed in terms of nature versus nurture. But such questions are meaningful only if they concern comparisons among individuals. It makes no sense to focus only on a woman on the far left of Figure 9.1a (page 330) and ask, without reference to the other individuals, whether this woman is 5 feet tall because of her genes (nature) or because of her environment (nurture). She had to have both genes and an environment to be alive and of any height at all. She did not get 3 feet of her height from her genes, and 2 feet from her environment, so that $3 + 2 = 5$. Instead, she got all of her 5 feet from the activity of her genes operating within her environment. For this single student we cannot, even in principle, disentangle the influence of nature and nurture.

The only kind of question it makes sense to ask is comparative: Is the shortest woman shorter than the tallest woman because they have different genes, grew up in different environments, or both? This we can answer. In principle, for example, we could take an identical twin of the short woman and raise her in the environment experienced by the tall woman. If the twin still grew up to be 5 feet tall, we would know that the difference between the shortest and tallest women is due entirely to differences in their genes. If the twin grew up to be 6 feet 2 inches tall, we would know that the difference between the shortest and tallest women is due entirely to differences in their environments. If the twin grew up to be somewhere between 5 feet and 6 feet 2 inches, it would indicate that the difference between the women is partly due to differences in their genes and partly due to differences in their environments. Considering the whole population, we can ask: What fraction of the variation in height among the students is due to variation in their genes, and what fraction is due to variation in their environments?

The first step in a quantitative genetics analysis is to determine the extent to which the trait in question is heritable. That is, we must partition the total phenotypic variation (V_P) into a component attributable to genetic variation (V_G) and a component attributable to environmental variation (V_E).

The fraction of the total variation in a trait that is due to variation in genes is called the **heritability** of the trait. The total variation in a trait is referred to as the **phenotypic variation** and is symbolized by V_P. Variation among individuals that is due to variation in their genes is called **genetic variation** and is symbolized by V_G. Variation among individuals due to variation in their environments is called **environmental variation** and is symbolized by V_E. Thus, we have

$$\text{heritability} = \frac{V_G}{V_P} = \frac{V_G}{V_G + V_E}$$

More precisely, this fraction is known as the **broad-sense heritability,** or degree of genetic determination. We will define the narrow-sense heritability shortly. Heritability is always a number between 0 and 1.

Estimating Heritability from Parents and Offspring

Before we wade any deeper into symbolic abstractions, note that if the variation among individuals is due to variation in their genes, then offspring will resemble their parents. It is easy, in principle, to check whether they do. We make a scatterplot with offspring's trait values represented on the vertical axis, and their parents' trait values on the horizontal axis **(Figure 9.14)**. We have two parents for every offspring, so we use the **midparent** value—the average of the parents. If we have more than one offspring per family, we use a **midoffspring** value. We draw the best-fit line through the points. If offspring do not resemble their parents, then the slope of the best-fit line will be near 0 (Figure 9.14a). This is evidence that the variation among individuals is due to variation in their environments, not variation in their genes. If offspring strongly resemble their parents, the slope of the best-fit line will be near 1 (Figure 9.14c); this is evidence that variation among individuals is due to variation in their genes, not variation in their environments. Most traits in most populations fall somewhere in the middle—offspring show moderate resemblance to their parents (Figure 9.14b); this is evidence that the variation among individuals is partly due to variation in their environments and partly due to variation in their genes. Figure 9.14d shows data for an actual population of students.

The examples in Figure 9.14 illustrate that the slope of the best-fit line for a plot of midoffspring versus midparents is a number between 0 and 1 that reflects the degree to which variation in a population is due to variation in genes. In other words, we can take the slope of the best-fit line as an estimate of the heritability. If we determine the best-fit line using the method of least-squares linear regression, which minimizes the sum of the squared vertical distances between the points and the line, then the slope represents a version of the heritability symbolized by h^2 and known as the **narrow-sense heritability.** Least-squares linear regression is the standard method taught in introductory statistics texts and used by statistical software packages to determine best-fit lines. (For readers familiar with statistics, it may prevent some confusion if we note that h^2 is not the fraction of the variation among the offspring that is explained by variation in the parents. That quantity would be r^2. Instead, h^2 is an estimate of the fraction of the variation among the *parents* that is due to variation in their genes.)

To explain the difference between narrow-sense heritability and broad-sense heritability, we need to distinguish between two components of genetic variation: additive genetic variation versus dominance genetic variation. Additive genetic variation (V_A) is variation among individuals due to the additive effects of genes, whereas dominance genetic variation (V_D) is variation among individuals due to gene interactions such as dominance (see **Computing Consequences 9.2**).

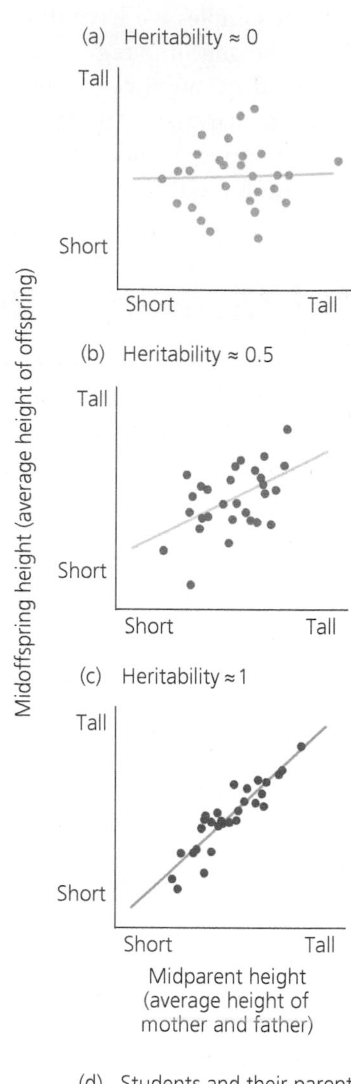

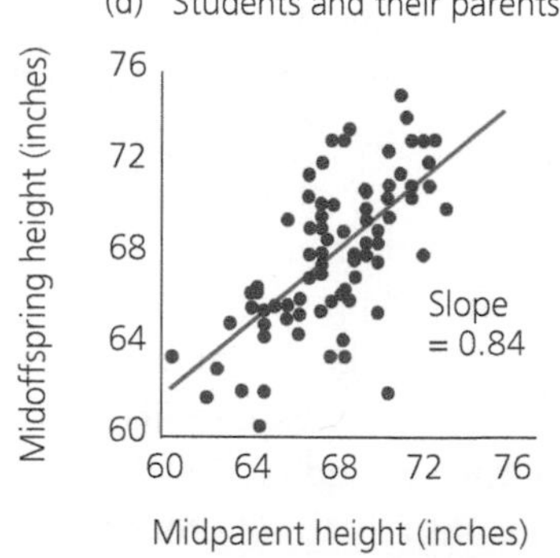

Figure 9.14 Scatterplots of offspring height as a function of parent height These graphs show data for hypothetical populations in which offspring resemble their parents (a) not at all, (b) moderately, and (c) strongly; and for an actual population (d) of students in an evolution course at a U.S. university.

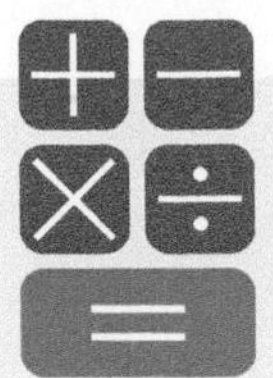

COMPUTING CONSEQUENCES 9.2

Additive genetic variation versus dominance genetic variation

Here we use a numerical example to distinguish additive versus dominance genetic variation. For simplicity, we analyze a single locus with two alleles as if it were a quantitative trait. We assume there is no environmental variation: An individual's phenotype is determined solely and exactly by its genotype. The alleles at the locus are A_1 and A_2, each has a frequency of 0.5, and the population is in Hardy–Weinberg equilibrium. We consider two situations: (1) the alleles are codominant; (2) allele A_2 is dominant over allele A_1.

Situation 1: Codominant alleles

A_1A_1 individuals have a phenotype of 1. In A_1A_2 and A_2A_2 individuals, each copy of A_2 adds 0.5 to the phenotype. At the left in Figure 9.15a is a histogram showing the distribution of phenotypes in the population. At the center and right are scatterplots that allow us to analyze the genetic variation. The x-axis represents genotype, calculated as the number of copies of A_2. The y-axis represents phenotype. The horizontal gray line shows the mean phenotype for the population (= 1.5). The plot at center shows that the total genetic variation V_G is a function of the deviations of the data points from the population mean (green arrows). We can quantify V_G by calculating the sum of the squared deviations. The plot at right shows the best-fit line through the data points (blue). The additive genetic variation V_A is defined as that fraction of the total genetic variation explained by the best-fit line (orange arrows). In this case, the best-fit line explains all the genetic variation, so $V_G = V_A$. There is no dominance genetic variation.

Situation 2: Recessive and dominant alleles

This time, A_1A_1 individuals again have a phenotype of 1. The effects of substituting copies of A_2 for copies of A_1 are not strictly additive, however: The first copy of A_2 (which makes the genotype A_1A_2) changes the phenotype from 1 to 2. The second copy of A_2 (which makes the genotype A_2A_2) does not alter the phenotype any further. At left in Figure 9.15b is a histogram showing the distribution of phenotypes in the population. At center and right are scatterplots that allow us to analyze the genetic variation. The plot at center shows that the total genetic variation V_G is a function of the deviations of the data points (green arrows) from the population mean (gray line) of 1.75. The plot at right shows the best-fit line through the data points (blue). The additive genetic variation V_A is that fraction of the total genetic variation that is explained by the best-fit line (orange arrows). The dominance genetic variation V_D is that fraction of the total genetic variation left unexplained by the best-fit line (red arrows). In this case, the best-fit line explains only part of the genetic variation, so $V_G = V_A + V_D$.

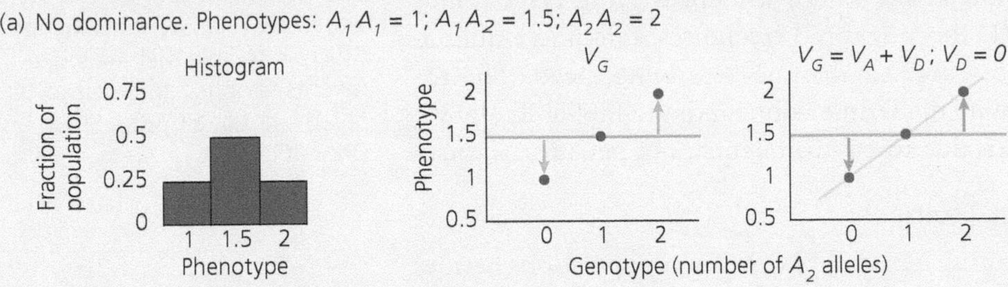

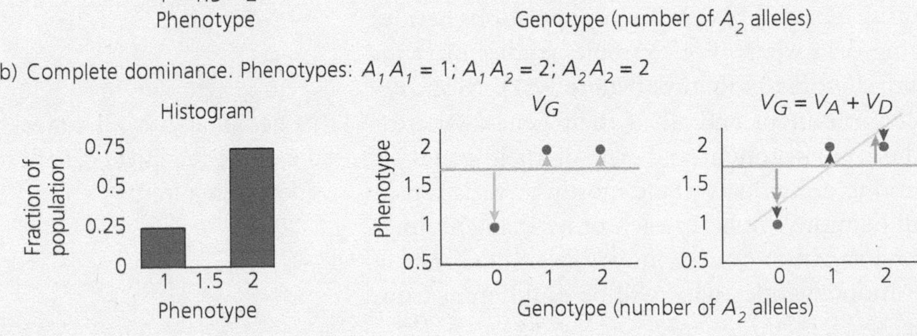

Figure 9.15 Additive genetic variation versus dominance genetic variation in a trait controlled by two alleles at a single locus (a) No dominance. (b) Complete dominance. After Felsenstein (2011).

The total genetic variation is the sum of the additive and dominance genetic variation: $V_G = V_A + V_D$. The broad-sense heritability, defined earlier, is V_G/V_P. The narrow-sense heritability, h^2, is defined as follows:

$$h^2 = \frac{V_A}{V_P} = \frac{V_A}{V_A + V_D + V_E}$$

When evolutionary biologists mention heritability without noting whether they are using the term in the broad or narrow sense, they almost always mean the narrow-sense heritability. We use the narrow-sense heritability in the rest of this discussion. It is the narrow-sense heritability, h^2, that allows us to predict how a population will respond to selection—because it describes the extent to which offspring resemble their parents.

When estimating the heritability of a trait in a population, it is important to keep in mind that offspring can resemble their parents for reasons other than the genes the offspring inherit. Environments run in families too. Among humans, for example, some families exercise more than others, and different families eat different diets. Our estimate of heritability will be accurate only if we can make sure there is no correlation between the environments experienced by parents and those experienced by their offspring. We obviously cannot do so in a study of humans. In an animal study, however, we could collect all the offspring at birth, then distribute them at random with foster parents. In a plant study, we could place seeds at random locations in a field.

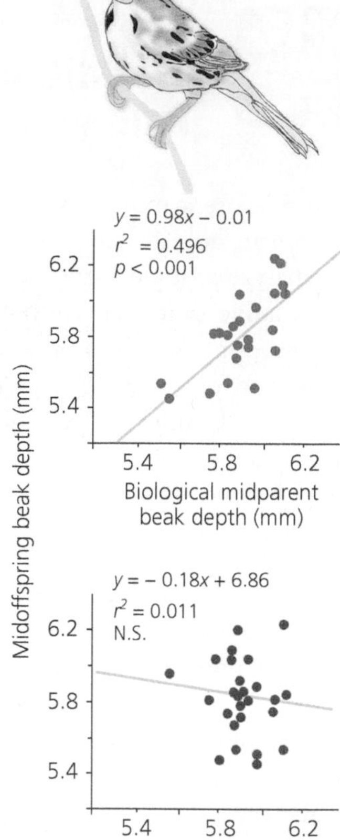

Figure 9.16 A field experiment on the heritability of beak size in song sparrows The top scatterplot shows the relationship between midoffspring beak depth and biological midparent beak depth. The bottom scatterplot shows the relationship between midoffspring beak depth and foster midparent beak depth. Chicks resemble their biological parents strongly, and their foster parents not at all. From Smith and Dhondt (1980).

James Smith and André Dhondt (1980), for example, studied song sparrows (*Melospiza melodia*) to determine the heritability of beak size. They collected young from natural nests, sometimes as eggs and sometimes as hatchlings, and moved them to the nests of randomly chosen foster parents. When the chicks grew up, Smith and Dhondt calculated midoffspring values for the chicks and midparent values for both the biological and foster parents. Graphs of offspring beak depth versus parental beak depth appear in **Figure 9.16**. The chicks resembled their biological parents strongly. They resembled their foster parents not at all—that is, the regression slope was not distinguishable from zero. These results show that virtually all the variation in beak depth in this population is due to variation in genes. Smith and Dhondt estimated that the heritability of beak depth is 0.98.

Alastair J. Wilson and Andrew Rambaut (2008) assembled pedigrees and lifetime prize winnings for some 2,500 thoroughbred racehorses. Lifetime winnings are, of course, of more interest to a breeder than peak running speed. The researchers estimated that the heritability of lifetime winnings is just 0.095 ± 0.034. Most of the variation in winnings is due to environmental, not genetic, factors.

Estimating Heritability from Twins

There are other methods for estimating heritability of traits in populations besides calculating the regression of offspring on parents. For example, studies of twins can be used. The logic of twin studies works as follows **(Figure 9.17)**. Monozygotic (identical) twins share their environment and all of their genes, whereas dizygotic (fraternal) twins share their environment and half of their genes. If heritability is high, and variation among individuals is due mostly to variation in genes, then monozygotic twins will be more similar to each other than are dizygotic twins. If heritability is low, and variation among individuals is due mostly to variation in environments, then monozygotic twins will be as different from each other as dizygotic twins.

The heritability, h^2, is a measure of the (additive) genetic variation in a trait.

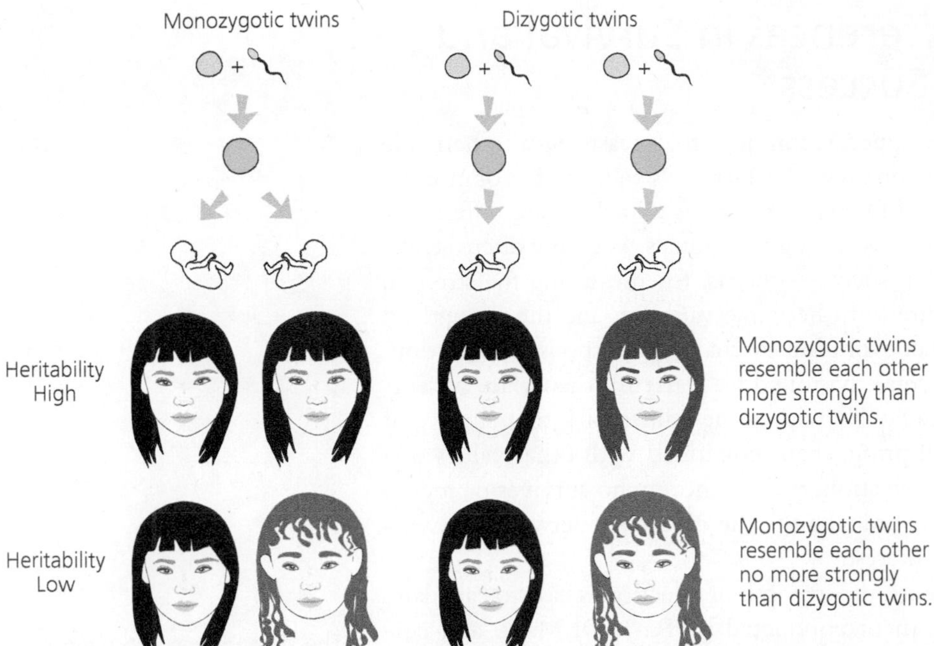
Monozygotic twins
Dizygotic twins
Heritability High
Heritability Low
Monozygotic twins resemble each other more strongly than dizygotic twins.
Monozygotic twins resemble each other no more strongly than dizygotic twins.

Figure 9.17 Estimating heritability from twin studies Monozygotic twins develop from a single zygote and thus share all their genes. Dizygotic twins develop from separate zygotes and share half their genes. If the heritability of a trait is high, monozygotic twins will resemble each other more strongly than dizygotic twins.

Gerald McClearn and colleagues (1997) used a twin study to estimate the heritability of general cognitive ability, a measure of intelligence, in a Swedish population. Figure 9.1c (page 330) shows the distribution of this trait among 110 pairs of monozygotic twins and 130 pairs of same-sex dizygotic twins. All of the study subjects were at least 80 years old. The monozygotic twins tended to resemble each other much more strongly in general cognitive ability than did the dizygotic twins. McClearn and colleagues estimated that the heritability of general cognitive ability in their study population is between 0.29 and 0.75, with a best estimate of 0.62.

One might expect that the heritability of a trait like cognitive ability would decline with the age of the population studied, because environmental differences have had a longer time to exert their effects. Rather surprisingly, however, the heritability of a variety of psychological traits—including intelligence—tends to increase with age (Bergen et al. 2007). One possible explanation is that individuals with different genotypes may actively choose different life experiences (Kendler and Baker 2007).

Comparisons between different kinds of twins assume that the environments shared by members of monozygotic twin pairs and the environments shared by members of dizygotic twin pairs are equally similar. This assumption may not be entirely justified. Comparisons of different kinds of twins are also complicated by the fact that monozygotic twins, unlike dizygotic twins, often share a placenta. The best kind of twin study is to look at the singleton children of pairs of monozygotic twins (see Lynch and Walsh 1998). Such children grow up as if they were merely first cousins, but genetically they are half sibs. If the trait of interest is heritable, there will be more variation among such pairs of "half sibs" than there is between the half sibs within pairs.

We can estimate the heritability from comparisons among relatives, including twins. If a trait is heritable, monozygotic twins will resemble each other more strongly than dizygotic twins.

For detailed treatments of methods for measuring heritability, see Falconer (1989) and Lynch and Walsh (1998). Data on the heritability of traits are frequently misinterpreted, particularly when the species under study is humans. We will return to this issue in Section 9.7.

9.4 Measuring Differences in Survival and Reproductive Success

In the preceding section we developed techniques for measuring the heritable variation in quantitative traits, the first tenet of Darwin's theory of evolution by natural selection. The next tenet of Darwin's theory is that there are differences in survival and/or reproductive success among individuals. We now discuss techniques for measuring differences in success—that is, for measuring the strength of selection. Once we can measure both heritable variation and the strength of selection, we will be able to predict evolutionary change in response to selection.

The kind of differences in success envisioned by Darwin's theory are systematic differences. On average, individuals with some values of a trait survive at higher rates, or produce more offspring, than individuals with other values of a trait. To measure the strength of selection, we first note who survives or reproduces and who fails to do so. Then we quantify the difference between the winners and the losers in the trait of interest.

In selective-breeding experiments, the strength of selection is easy to calculate. Consider, for example, an experiment conducted by R. J. Di Masso and colleagues (1991). These researchers set out to breed mice with longer tails. They wanted to know how the developmental program that constructs a mouse's tail would change under selection. Would a mouse embryo make a longer tail by elongating the individual vertebrae or by adding extra ones? Every generation, the researchers measured the tails of all the mice in their population. Then they picked the mice with the longest tails and let them breed among themselves to produce the next generation.

The second step in a quantitative genetic analysis is to measure the strength of selection on the trait in question. One measure is the selection differential, S, equal to the difference between the mean of the selected individuals and the mean of the entire population.

To see how to quantify the strength of selection, suppose the researchers picked as breeders the one-third of the mice whose tails are the longest. The simplest measure of the strength of selection is the difference between the mean tail length of the breeders and the mean tail length of the entire population **(Figure 9.18a)**. This measure of selection is called the selection differential and is symbolized by S.

A second measure of the strength of selection is useful because of its broad applicability. This measure is called the selection gradient (Lande and Arnold 1983). As we describe how the selection gradient is calculated, it may not appear at first

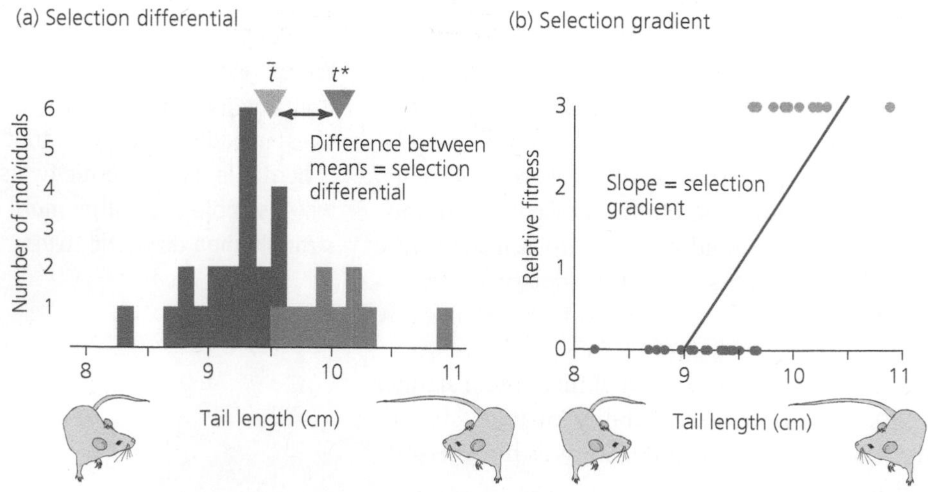

Figure 9.18 Measuring the strength of selection (a) The variation in tail length in a fictional population of mice. Orange bars represent mice chosen as breeders. Gray triangle indicates the average tail length for the entire population; orange triangle indicates the average tail length for breeders. The difference between these averages is the selection differential. (b) Scatterplot for the same population showing relative fitness versus tail length. Orange dots represent breeders. The slope of the best-fit line (red) is the selection gradient.

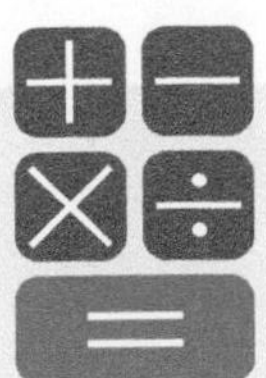

COMPUTING CONSEQUENCES 9.3

The selection gradient and the selection differential

The selection differential is an intuitively straightforward measure of the strength of selection: It is the difference between the mean of a trait among the survivors and the mean of the trait among the entire population. The selection gradient, while more abstract, has several advantages. Among them is that the selection gradient can be calculated for a wider variety of fitness measures. Because it is closely related to the selection differential, the selection gradient has some of the same intuitive appeal.

Here we show that in our example on tail length in mice (Figure 9.18), the selection gradient for tail length t is equal to the selection differential for tail length divided by the variance of tail length. Imagine that we have 30 mice in our population. First, note that the selection differential is

$$S = t^{\star} - \bar{t}$$

where $t^{\star}$ is the mean tail length of the 10 mice we kept as breeders, and $\bar{t}$ is the mean tail length of the entire population of 30 mice.

The selection gradient is the slope of the best-fit line for relative fitness w as a function of tail length. The slope of the best-fit line in linear regression is given by the covariance of y and x divided by the variance of x:

$$\text{slope} = \frac{\text{cov}(y, x)}{\text{var}(x)}$$

The covariance of y and x is defined as

$$\text{cov}(y, x) = \frac{1}{n}\sum_{i=1}^{n}(y_i - \bar{y})(x_i - \bar{x})$$

and the variance of x is defined as

$$\text{var}(x) = \frac{1}{n}\sum_{i=1}^{n}(x_i - \bar{x})^2$$

where n is the number observations, $\bar{y}$ is the mean value of y, and $\bar{x}$ is the mean value of x. The selection gradient for t is therefore:

$$\text{selection gradient} = \frac{\text{cov}(w, t)}{\text{var}(t)}$$

Thus, we need to show that cov $(w, t) = t^{\star} - \bar{t}$.

Because (by definition) the mean relative fitness is 1, we can write

$$\begin{aligned}
\text{cov}(w, t) &= \frac{1}{30}\sum_{i=1}^{30}(w_i - 1)(t_i - \bar{t}) \\
&= \frac{1}{30}\sum_{i=1}^{30}(w_i t_i) - \frac{1}{30}\sum_{i=1}^{30}(w_i \bar{t}) \\
&\quad - \frac{1}{30}\sum_{i=1}^{30}(t_i) + \frac{1}{30}\sum_{i=1}^{30}(\bar{t}) \\
&= \frac{1}{30}\sum_{i=1}^{30}(w_i t_i) - \bar{t} - \bar{t} + \bar{t} \\
&= \frac{1}{30}\sum_{i=1}^{30}(w_i t_i) - \bar{t} \\
&= t^{\star} - \bar{t}
\end{aligned}$$

The last step may not be transparent. To see that

$$\frac{1}{30}\sum_{i=1}^{30}(w_i t_i) = t^{\star}$$

note that for the first 20 mice $w_i = 0$, and for the last 10 mice $w_i = 3$. This means that

$$\begin{aligned}
\frac{1}{30}\sum_{i=1}^{30}(w_i t_i) &= \frac{1}{30}\sum_{i=21}^{30}(3t_i) \\
&= \frac{3}{30}\sum_{i=21}^{30}(t_i) \\
&= \frac{1}{10}\sum_{i=21}^{30}(t_i) = t^{\star}
\end{aligned}$$

glance that the selection gradient and the selection differential have much to do with each other. In fact, they are closely related, and each can be converted into the other. If we are analyzing selection on a single trait like tail length, then the selection gradient is equal to the selection differential divided by the variance in tail length (see **Computing Consequences 9.3**).

A second (and related) measure of the strength of selection is the selection gradient.

The selection gradient can be calculated as follows:

1. Assign absolute fitnesses to the mice in the population. We will think of fitness as survival to reproductive age. In our population, one-third of the mice survived long enough to reproduce. (This does not necessarily mean that the short-tailed mice were actually killed, just that they were removed from the breeding population; as far as our breeding population is concerned, the short-tails did not breed so they did not survive long enough to reproduce.) The short-tailed two-thirds of the mice have a fitness of 0, and the long-tailed one-third have a fitness of 1.
2. Convert the absolute fitnesses to relative fitnesses. The mean fitness of the population is 0.33 (if, for example, there are 30 mice in the population, the mean is $[\{20 \times 0\} + \{10 \times 1\}]/30 = 0.33$). We calculate each mouse's relative fitness by dividing its absolute fitness (0 or 1) by the mean fitness (0.33). The short-tailed mice have a relative fitness of 0; the long-tailed mice have a relative fitness of 3.
3. Make a scatterplot of relative fitness as a function of tail length, and calculate the slope of the regression line (Figure 9.18b). The slope of this best-fit line is the selection gradient.

The selection gradient is the slope of the regression of relative fitness on phenotype.

An advantage of the selection gradient is that we can calculate it for any measure of fitness, not just survival. We might, for example, measure fitness in a natural population of mice as the number of offspring weaned. If we first calculate each mouse's relative fitness (by dividing its number of offspring by the mean number of offspring), then plot relative fitness as a function of tail length and calculate the slope of the regression line, then that slope is the selection gradient.

In their mice, Di Masso et al. selected for long tails in 18 successive generations. The mice in the 18th generation had tails more than 10% longer than mice in a control population. The long-tailed mice had 28 vertebrae in their tails, compared to 26 or 27 vertebrae for the controls. The developmental program had been altered to make more vertebrae, not to elongate individual vertebrae.

9.5 Predicting the Evolutionary Response to Selection

Once we know the heritability and the selection differential, we can predict the evolutionary response to selection. Here is the equation for doing so:

$$R = h^2S$$

where R is the predicted response to selection, h^2 is the heritability, and S is the selection differential.

Once we know h^2 and S, we can use them to predict the response to selection as $R = h^2S$.

The logic of this equation is shown graphically in **Figure 9.19**. This figure shows a scatterplot of midoffspring values as a function of midparent values, just like the scatterplots in Figure 9.14. The scatterplot in Figure 9.19 represents tail lengths in a population of 30 families of mice. The plot includes the regression line of midoffspring on midparent, whose slope estimates the heritability h^2.

Look first at the x-axis. $\overline{P}$ is the average midparent value for the entire population. $P^\star$ is the average of the 10 largest midparent values. The difference between $P^\star$ and $\overline{P}$ is the selection differential (S) that we would have applied to

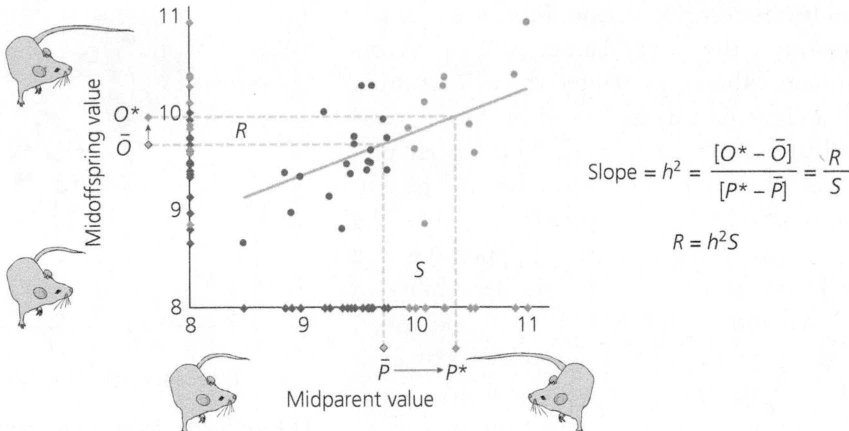

Figure 9.19 Predicting the response to selection, R Midoffspring and midparent values appear as dots on the scatterplot and diamonds on the axes. Orange symbols show the 10 families with the largest midparent values. $\bar{P}$ is the average midparent for the entire population; P^* is the average midparent of the orange families. $\bar{O}$ is the average midoffspring for the entire population; O^* is the average midoffspring for the orange families. After Falconer (1989).

this population had we picked as our breeders only the 10 pairs of parents with the largest midparent values.

Now look at the y-axis. $\bar{O}$ is the average midoffspring value for the entire population. O^* is the average midoffspring value for the 10 pairs of parents with the largest midparent values. The difference between O^* and $\bar{O}$ is the evolutionary response (R) we would have gotten as a result of selecting as breeders the 10 pairs of parents with the largest midparent values.

The slope of a line can be calculated as the rise over the run. If we compare the population averages with selection versus the averages without selection, we have a rise of $(O^* - \bar{O})$ over a run of $(P^* - \bar{P})$, so

$$h^2 = \frac{(O^* - \bar{O})}{(P^* - \bar{P})} = \frac{R}{S}$$

In other words, $R = h^2 S$.

We now have a set of tools for studying the evolution of quantitative traits under natural selection. We can estimate how much of the variation in a trait is due to variation in genes, quantify the strength of selection that results from differences in survival or reproduction, and put these two together to predict how much the population will change from one generation to the next.

Alpine Skypilots and Bumblebees

As an example of the questions biologists answer with quantitative genetics, we review Candace Galen's (1996) research on flower size in the alpine skypilot (*Polemonium viscosum*), a perennial Rocky Mountain wildflower **(Figure 9.20a)**. Galen studied populations on Pennsylvania Mountain, Colorado, including populations growing at the timberline and populations in the higher-elevation tundra. At the timberline, skypilots are pollinated by a diversity of insects, including flies, small solitary bees, and some bumblebees. In the tundra, they are pollinated almost exclusively by bumblebees (Figure 9.20b). The flowers of tundra skypilots are, on average, 12% larger in diameter than those of timberline skypilots. Previously, Galen (1989) had documented that larger flowers attract more visits from bumblebees and that skypilots that attract more bumblebees produce more seeds.

Galen wanted to know whether the selection on flower size imposed by bumblebees is responsible for the larger flowers of tundra skypilots. If it is, she also wanted to know how long it would take for selection by bumblebees to increase the average flower size in a skypilot population by 12%—the difference between tundra and timberline flowers.

(a)

(b)

Figure 9.20 An alpine skypilot and a bumblebee (a) Alpine skypilot (*Polemonium viscosum*). (b) Bumblebee (*Bombus* sp.).

Galen worked with a small-flowered timberline population. First, she estimated the heritability of flower size. She measured the flower diameters of 144 skypilots and collected their seeds. She germinated these and planted the 617 resulting seedlings at random in the same habitat their parents had lived in. Seven years later, 58 had matured and Galen could measure their flowers. This let her plot offspring flower diameter (corolla flare) as a function of maternal, or seed-parent, diameter **(Figure 9.21)**. The slope of the regression line is approximately 0.5. The slope of the regression line for offspring versus a single parent (as opposed to the midparent) is an estimate of $\frac{1}{2}h^2$ (see Falconer 1989). Thus the heritability of flower size in the timberline skypilots is roughly $2 \times 0.5 = 1$. Note, however, that the scatter in Figure 9.21 makes the true slope uncertain. Galen's analysis indicated that she could safely conclude only that the heritability of flower size is between 0.2 and 1. That is, at least 20% of the phenotypic variation in skypilot flower size is due to additive genetic variation.

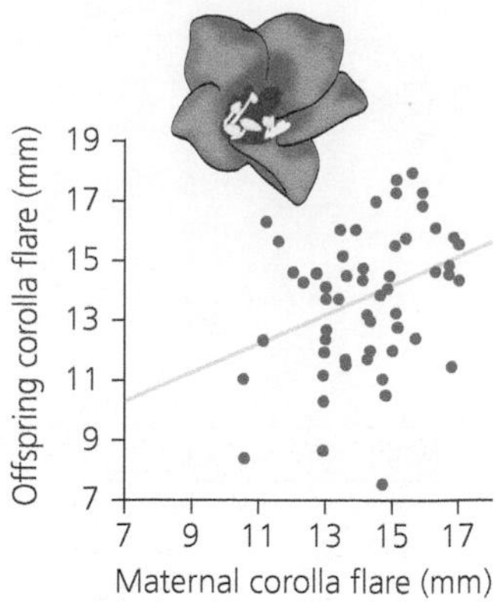

Figure 9.21 Estimating the heritability of flower size (corolla flare) in alpine skypilots This scatterplot shows offspring corolla flare as a function of maternal plant corolla flare for 58 skypilots. The slope of the best-fit line is 0.5. Redrawn from Galen (1996).

Next Galen estimated the strength of selection imposed by bumblebees. Recall that bumblebees prefer to visit larger flowers and that more bumblebee pollinators means more seeds. Galen built a screen-enclosed cage at her study site, moved 98 soon-to-flower skypilots into it, and added bumblebees. The cage kept the bees in and other pollinators out. When the caged plants flowered, Galen measured their flowers. Later she collected their seeds, germinated them in the lab, and planted the seedlings at random back in the parental habitat. Six years later, Galen counted the surviving offspring of the original caged plants. Using the number of surviving 6-year-old offspring as her measure of fitness, Galen plotted relative fitness as a function of flower size and calculated the slope of the best-fit line **(Figure 9.22)**. This slope, 0.13, is the selection gradient resulting from bumblebee pollination. Multiplying the selection gradient by the variance in flower size, 5.66, gives the selection differential: $S = 0.74$ mm. The average flower size was 14.2 mm. Thus the selection differential can also be expressed as $\frac{0.74}{14.2} = 0.05$, or 5%. Roughly, this means that when skypilots attempt to reproduce by attracting bumblebees, the plants that succeed have flowers 5% larger than average.

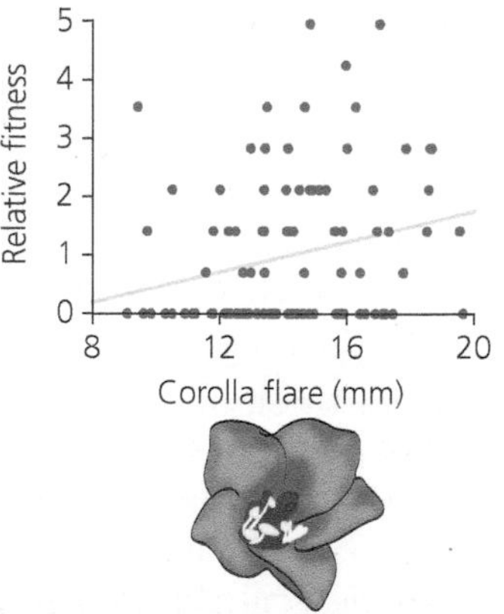

Figure 9.22 Estimating the selection gradient for alpine skypilots pollinated by bumblebees This scatterplot shows relative fitness (number of surviving 6-year-old offspring divided by average number of surviving 6-year-old offspring) as a function of maternal flower size (corolla flare). The slope of the best-fit line is 0.13. Prepared with data provided by Candace Galen.

Galen performed two control experiments to confirm that bumblebees select for larger flowers. In one, she pollinated skypilots by hand (without regard to flower size); in the other, she allowed skypilots to be pollinated by all other natural pollinators except bumblebees. In neither control was there any relationship between flower size and fitness. Only bumblebees select for larger flowers.

Galen's data allowed her to predict how the population of timberline skypilots should respond to selection by bumblebees. The scenario she imagined was that a population of timberline skypilots that had been pollinated by a variety of insects moved (by seed dispersal) to the tundra, where the plants are now pollinated exclusively by bumblebees. Using the low-end estimate that $h^2 = 0.2$, and the estimate that $S = 0.05$, Galen predicted that the response to selection would be $R = h^2S = 0.2 \times 0.05 = 0.01$. Using the high-end estimate that $h^2 = 1$, and the estimate that $S = 0.05$, Galen predicted that the response to selection would be $R = h^2S = 1 \times 0.05 = 0.05$. In other words, a single generation of selection by bumblebees should produce an increase of 1% to 5% in the average flower size of a population of timberline skypilots moved to the tundra.

Galen's prediction was, therefore, that flower size would evolve rapidly under selection by bumblebees. Is this prediction correct? Recall the experiment described earlier, in which Galen reared offspring of timberline skypilots that had been pollinated by hand and offspring of timberline skypilots that been pollinated

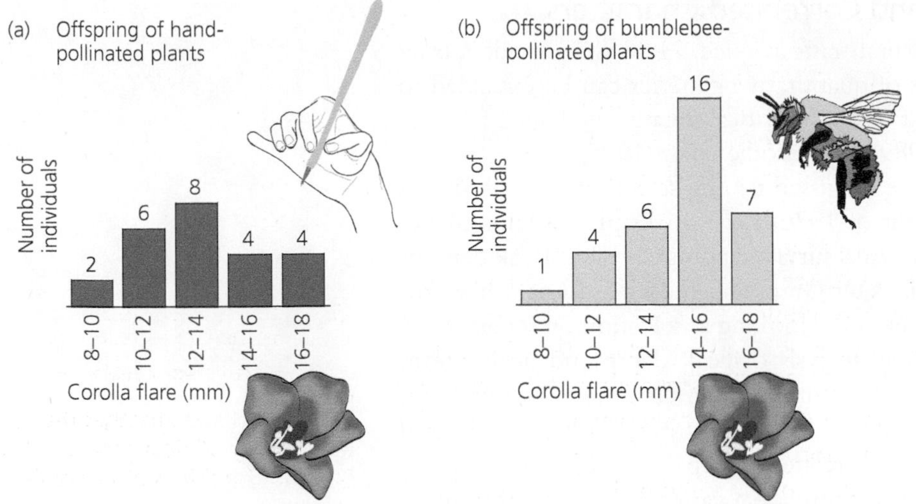

Figure 9.23 Measuring the evolutionary response to selection in alpine skypilots These histograms show the distribution of flower size (corolla flare) in the offspring of hand-pollinated skypilots (a; mean = 13.1 mm) and bumblebee-pollinated skypilots (b; mean = 14.4 mm). Redrawn from Galen (1996).

exclusively by bumblebees. Galen calculated the mean flower size of each group and found that the offspring of bumblebee-pollinated skypilots had flowers that were, on average, 9% larger than those of hand-pollinated skypilots **(Figure 9.23)**. Her prediction was correct: Skypilots show a strong and rapid response to selection. In fact, the response is even larger than Galen predicted.

Galen concluded that the 12% difference in flower size between timberline and tundra skypilots can be plausibly explained by the fact that timberline skypilots are pollinated by a diversity of insects, whereas tundra skypilots are pollinated almost exclusively by bumblebees. Timberline skypilots can set seed even if bumblebees avoid them, but tundra skypilots cannot. Furthermore, it would take only a few generations of bumblebee-only pollination for a population of timberline skypilots to evolve flowers that are as large as those of tundra skypilots.

Thoroughbred Racehorses and Breeding Fees

We have already noted that the heritability of lifetime prize winnings in thoroughbreds is about 0.095. While most of the variation is due to differences in training, jockeys, races entered, and other environmental factors—including luck—there is nonetheless some genetic variation. This suggests that there are horses with good genes that breeders could buy.

Horse breeders do not employ mass selection, like biologists breeding for long tails in mice or bumblebees pollinating skypilots. We cannot calculate a simple selection differential. However, Alastair Wilson and Andrew Rambaut (2008) used data on 554 stallions serving as studs and the heritability of winnings to predict the lifetime winnings of the studs' hypothetical offspring. Predicted winnings ranged from \$57,500 to \$140,000. The stud fees commanded by the stallions ranged from nominal to \$500,000. **Figure 9.24** shows the relationship between predicted winnings and stud fee. The slope of the best-fit line is just 0.02, and not significantly different from zero. This suggests that every additional dollar spent on stud fees returns about 2 cents in additional prize winnings. Winning prizes is not the only way a horse can earn money for its owners, but breeders do not appear to be getting good value for their investments.

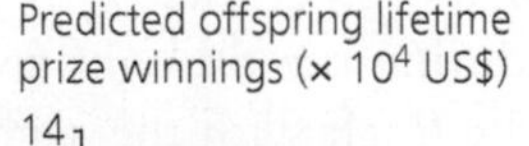

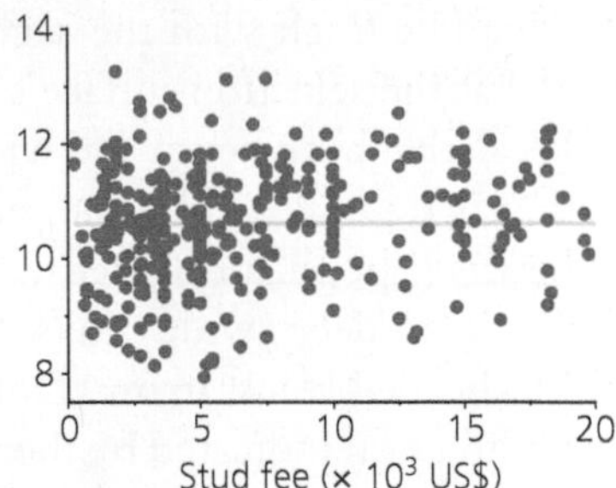

Figure 9.24 Do more expensive studs father bigger winners? No. A small number of very high stud fees have been omitted for clarity. From Wilson and Rambaut (2008).

In Sections 9.3, 9.4, and 9.5, we have used the tools of quantitative genetics to analyze the evolution of just one trait at a time. The tools we have developed can be generalized, however, to analyze the simultaneous evolution of multiple traits.

Selection on Multiple Traits and Correlated Characters

Selection in nature often acts on several traits at once. Here we provide a brief introduction to how the techniques of quantitative genetics can be extended to analyze selection on multiple traits. For mathematical details, see Lande and Arnold (1983), Phillips and Arnold (1989), and Brodie et al. (1995).

Earlier in the book (Chapter 3), we discussed natural selection on beak size in Darwin's finches. During the drought of 1976–1977 on Daphne Major Island, medium ground finches with deeper beaks survived at higher rates. Beak depth is heritable, so the population evolved. Peter Grant and Rosemary Grant have reanalyzed the data from this selection episode, looking at selection on several traits simultaneously. Here, we discuss their analysis of beak depth and beak width. We present a qualitative overview; for the numbers, see Grant and Grant (1995).

The medium ground finches of Daphne Major vary both in beak depth and beak width. These traits are strongly correlated. Deep beaks tend to be wide; shallow beaks tend to be narrow. While there are a variety of reasons this might be so, the one that most interests us here is that the traits are genetically correlated. That is, finches that inherit genes for deep beaks tend also to inherit genes for wide beaks, and vice versa. Depth and width may be genetically correlated because the same genes influence both traits, or because selection in the past has favored particular combinations of alleles of depth genes and of width genes.

During the drought of 1976–1977, when food was scarce and many finches starved, selection acted on both beak depth and width. If we were looking at just one of these characteristics, we could measure the strength of selection as the slope of the regression line relating fitness to beak size. This is the selection gradient. To look at both characteristics at once, we can measure the strength of selection as the slope of the regression plane relating fitness to both beak depth and beak width. This slope is the two-dimensional selection gradient.

Look at the three-dimensional diagram in **Figure 9.25a**. Beak depth is represented on one horizontal axis and width on the other. Fitness is represented on the vertical axis. The surface given by the blue grid is the regression plane. The fitness at each corner of this selection surface is marked with a blue triangle. Selection during the drought favored birds with beaks that were both deep and narrow. The finch with the highest chance of surviving would have been a bird located at the selection surface's right rear corner.

Recall that beak depth and width are genetically correlated, however. This correlation is represented by the gray oval, which encompasses the variation found in the finch population. Because this variation tends to run from shallow, narrow beaks to deep, wide beaks, the population can evolve most readily along the double-headed black arrow. During the drought, selection pushed most strongly in the direction indicated by the wide dark blue arrow. But because of the genetic correlation between beak depth and beak width, the poplation actually evolved along a route between the dark blue arrow and the path of least resistance. Selection favored deep beaks more strongly than it favored narrow beaks. As a result, the population average moved toward a beak that was deeper and wider than it was before the drought. This change is represented by the wide orange arrow.

When characters are genetically correlated, selection on one can drag the other along for the ride.

Three-dimensional graphs can be difficult to interpret, so we have included Figures 9.25b–d, which illustrate the same analysis with two-dimensional graphs. Figure 9.25b shows the selection gradient on beak depth with width held constant. Selection favored deeper beaks. Figure 9.25c shows the selection gradient on beak width with depth held constant. Selection favored narrower beaks.

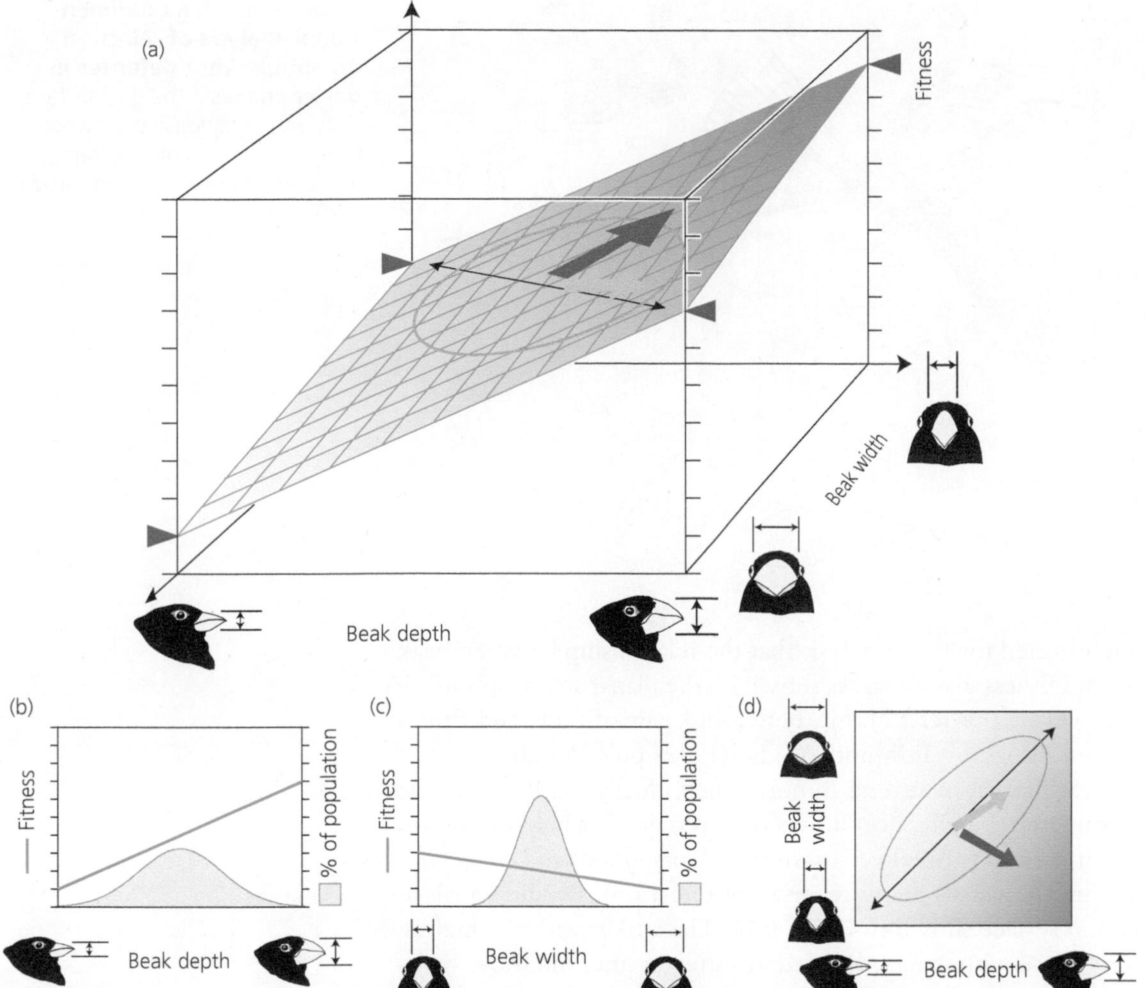

Figure 9.25 A multidimensional analysis of selection on beak size in medium ground finches (a) The grid plane shows the relationship between fitness and both beak depth and beak width. Birds with deep and narrow beaks had highest fitness. (b–d) show the same scenario in two-dimensional graphs.

Figure 9.25d shows the correlation between beak width and beak depth (gray oval), where the intensity of blue color across the graph indicates fitness. Selection pushed the population average toward a bird with a deep, narrow beak (dark blue arrow), but because of the genetic correlation between depth and width, the population tended toward the path of least resistance (double-headed black arrow). Selection favored increased depth more strongly than decreased width, so the population moved toward a deeper, wider beak (orange arrow).

Grant and Grant's analysis of selection on finch beaks illustrates the advantages of looking at several traits at once, of using selection gradients to measure the strength of selection, and of recognizing that traits may be genetically correlated with each other. Imagine that we were to look only at beak width, and calculate the selection differential. The average survivor had a wider beak than did the average bird alive before the drought. The selection differential, the difference between the population mean before and after selection, would suggest that selection favored wider beaks. But the multidimensional analysis reveals that this was not the case. Beak width was selected against, but was dragged along for the ride as a result of stronger selection on beak depth.

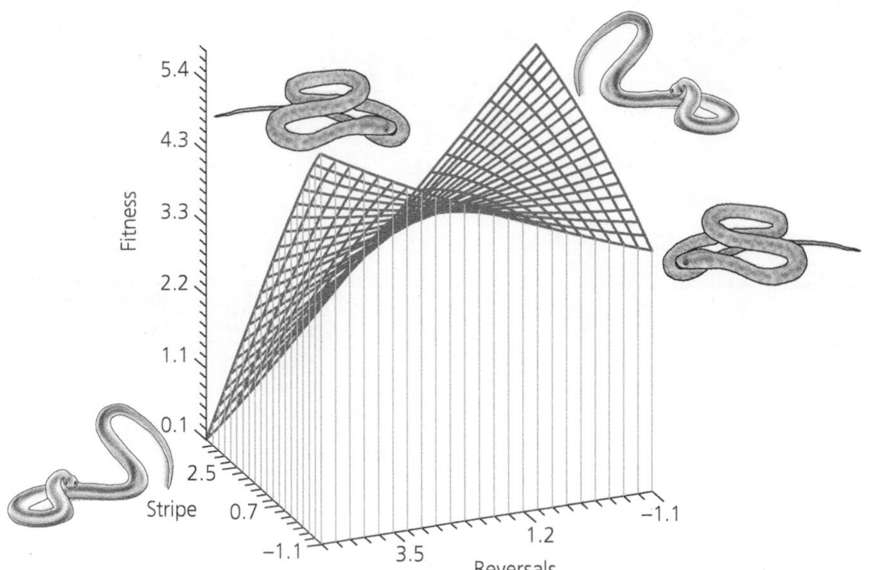

Figure 9.26 A multidimensional analysis of selection on antipredator defenses in garter snakes The grid surface shows the relationship between fitness and both color pattern and evasive behavior. From Brodie (1992).

Grant and Grant assumed for their analysis that the relationship between beak depth, beak width, and fitness was linear, as shown by the planar selection surface in Figure 9.25a. However, the relationship between a pair of traits and fitness can have other shapes. Work by Edmund Brodie (1992) provides an example. Brodie monitored the survival of several hundred individually marked juvenile garter snakes. He estimated the effect on fitness of two traits that help the snakes evade predators: color pattern (striped versus unstriped or spotted) and escape behavior (straight-line escape versus many reversals of direction). Brodie's analysis produced the selection surface shown in **Figure 9.26**. The snakes with the highest rates of survival were those with stripes that fled in a straight line, and those without stripes that performed many reversals of direction. Snakes with other combinations of traits were apparently more easily spotted by predators and eaten.

Given a selection surface, we can follow the evolution of a population by tracking the position of the average individual. In general, a population is expected to evolve so as to move up the steepest slope from its present location. As Grant and Grant's study of finch beaks demonstrated, however, correlations among traits may prevent a population from following this route. Selection surfaces like those shown in Figures 9.25a and 9.26 are sometimes referred to as adaptive landscapes. This term has a complex history, however, and several different meanings (see Chapter 9 in Provine 1986; Chapter 11 of Wright 1986; Wright 1988).

9.6 Modes of Selection and the Maintenance of Genetic Variation

In our discussions of selection on quantitative traits, we have assumed that the relationship between phenotype and fitness is simple. In our mice, long tails were better than short tails; in skypilots, bigger flowers were better than smaller flowers. Before leaving the topic of selection on quantitative traits, we note that the relationship between phenotype and fitness may be complex. A variety of patterns, or modes of selection, are possible.

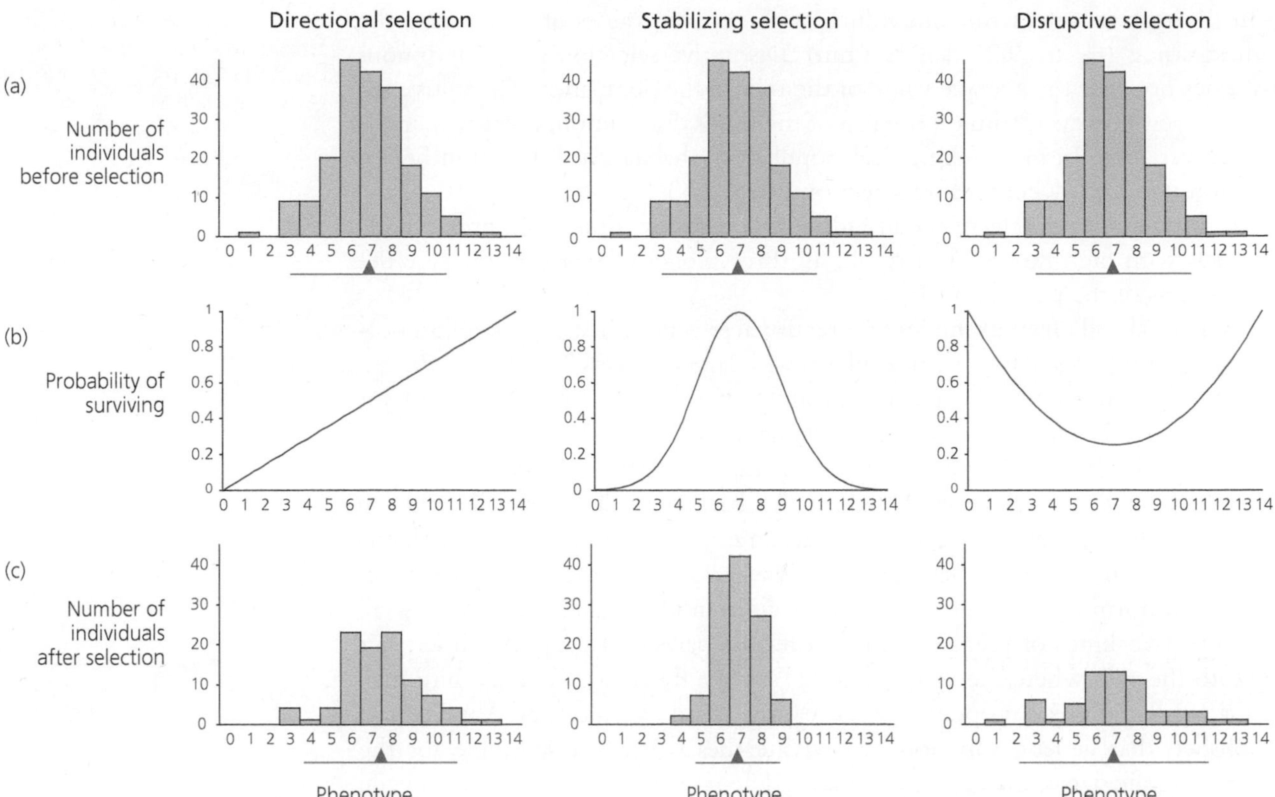

Figure 9.27 Three modes of selection Each column represents a mode of selection. Row (a) shows the distribution of a phenotypic trait in a hypothetical population before selection. Row (b) shows different patterns of selection. Row (c) shows the distribution of the trait in the survivors. Under each histogram, the blue triangle shows the mean of the population and the blue bar shows the variation (±2 standard deviations from the mean). After Cavalli-Sforza and Bodmer (1971).

Figure 9.27 shows three distinct modes of selection acting on a hypothetical population. Each column represents a different mode. The histogram in the top row shows the distribution of values for a phenotypic trait before selection. The graph in the middle row shows the relationship between phenotype and fitness, plotted as the probability that a given phenotype will survive. The histogram in the bottom row shows the distribution of phenotypes among the survivors. The triangle and bar below each histogram show the mean and variation in the population. (The bar representing variation encompasses ±2 standard deviations around the mean, or approximately 95% of the individuals in the population.)

Selection on a population may take any of several forms. Directional selection and stabilizing selection tend to reduce the amount of variation in a population; disruptive selection tends to increase the amount of variation.

In **directional selection,** fitness consistently increases (or decreases) with the value of a trait (Figure 9.27, first column). Directional selection on a continuous trait changes the average value of the trait in the population. In the hypothetical population shown, the mean phenotype before selection was 6.9, whereas the mean phenotype after selection was 7.4. Directional selection also reduces the variation in a population, although often not dramatically. In our population, the standard deviation before selection was 1.92, whereas after selection it was 1.89.

In **stabilizing selection,** individuals with intermediate values of a trait have highest fitness (Figure 9.27, middle column). Stabilizing selection on a continuous trait does not alter the average value of the trait in the population. Stabilizing selection does, however, reduce the number of individuals in the tails of the trait's distribution, thereby reducing the variation. In our hypothetical population, the standard deviation before selection was 1.92, whereas after selection it was 1.04.

In **disruptive selection,** individuals with extreme values of a trait have the highest fitness (Figure 9.27, last column). Disruptive selection on a continuous trait does not alter the average value of the trait in the population. Disruptive selection does, however, trim off the top of the trait's distribution, thereby increasing the variance. In our hypothetical population, the standard deviation before selection was 1.92, whereas after selection it was 2.33.

All three modes of selection cull individuals with low fitness and preserve individuals with high fitness. As a result, all three modes of selection increase the mean fitness of the population.

We have already seen examples of directional selection. In alpine skypilots pollinated by bumblebees, for instance, plants with larger flowers have higher fitness. And in medium ground finches, the drought of 1976–1977 on Daphne Major selected for birds with deeper beaks (see Chapter 3).

Research by Arthur Weis and Warren Abrahamson (1986) provides an elegant example of stabilizing selection. Weis and Abrahamson studied a fly called *Eurosta solidaginis*. The female in this species injects an egg into a bud of the tall goldenrod, *Solidago altissima*. After hatching, the fly larva digs into the stem and induces the plant to form a protective gall. As it develops inside its gall, the larva may fall victim to two kinds of predators. First, a female parasitoid wasp may insert her egg into the gall, where the wasp larva will eat the fly larva. Second, a bird may spot the gall and break it open, again to eat the larval fly. Weis and Abrahamson established that genetic variation among the flies is partly responsible for the variation in the size of the galls they induce. The researchers also collected several hundred galls and, by dissecting them, learned the fate of the larva inside each.

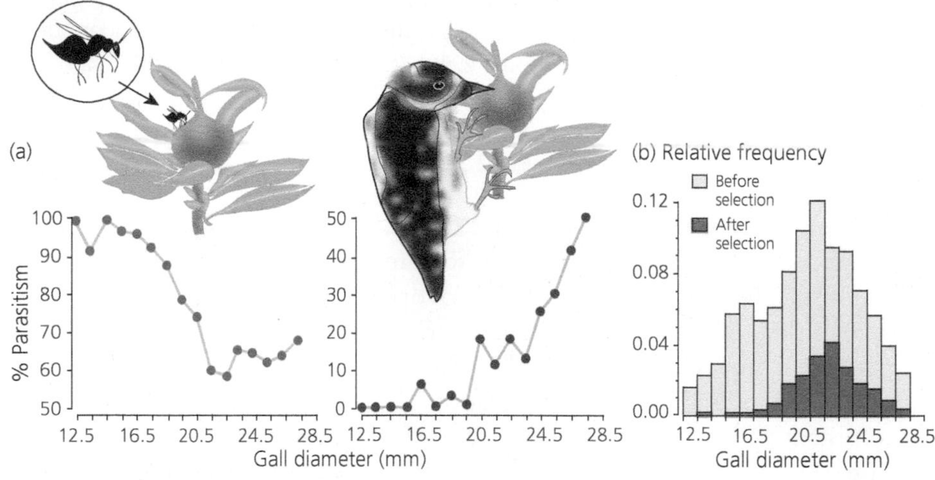

Figure 9.28 Stabilizing selection on a gall-making fly
(a) Parasitoid wasps kill fly larvae in small galls at higher rates than larvae in large galls (left). Birds kill larvae in large galls at higher rates than they kill larvae in small galls (right). (b) The distribution of gall sizes before (tan + red portion of bars) and after (red portion) selection by parasitoids and birds. Overall, fly larvae in medium galls survived at the highest rates. From Weis and Abrahamson (1986).

Weis and Abrahamson discovered that parasitoid wasps impose on the gall-making flies strong directional selection favoring larger galls **(Figure 9.28a, left)**. Nearly all larvae in galls under about 16 mm in diameter were killed by wasps, whereas larvae in larger galls had at least some chance of surviving. However, the researchers also found that birds impose on the gall makers strong directional selection favoring smaller galls (Figure 9.28a, right). Selection by birds almost, though not entirely, counterbalances selection by wasps on gall size. Figure 9.28b shows the distribution of sizes among the galls before and after selection.

Research by Thomas Bates Smith (1993) provides an example of disruptive selection. Bates Smith studied an African finch called the black-bellied seedcracker.

Birds in this species exhibit two distinct beak sizes: large and small. The birds in the two groups specialize on different kinds of seeds. Bates Smith followed the fate of over 200 juvenile birds. The graphs in **Figure 9.29** show the distribution of beak sizes among all juveniles and among juveniles that survived to adulthood. The survivors were the birds whose bills were either relatively large or relatively small. Birds with medium beaks did not survive. (Note that an element of stabilizing selection appears here too: Except in the case of birds with extremely long bills, the birds with the most extreme phenotypes did not survive.)

Evolutionary biologists generally assume that directional selection and stabilizing selection are common, whereas disruptive selection is rare. If the preponderance of directional and stabilizing selection is real, however, it presents a puzzle. Recall from Figure 9.27 that both directional and stabilizing selection reduce the phenotypic variation in a population. If the trait in question is heritable, these modes of selection will reduce the genetic variation in the population too. Eventually, the genetic variation in any trait related to fitness should be eliminated, and the population should reach an equilibrium at which the mean value of the trait, the variation in the trait, and the mean fitness of the population all cease to change. The puzzle is that populations typically exhibit substantial genetic variation, even in traits closely related to fitness. How is this variation maintained?

Here are three possible solutions to the puzzle of how genetic variation for fitness is maintained:

1. Most populations are not in evolutionary equilibrium with respect to directional and/or stabilizing selection. In any population there is a steady, if slow, supply of new favorable mutations creating genetic variation for fitness-related traits. While favorable mutations are rising in frequency, but have not yet become fixed, the population will exhibit genetic variation for fitness. We will call this the "Fisher's fundamental theorem hypothesis." It was Ronald Fisher who first showed mathematically that the rate at which the mean fitness of a population increases is proportional to the additive genetic variation for fitness, a result he called the fundamental theorem of natural selection.
2. In most populations, there is a balance between deleterious mutations and selection. In any population, there is a steady supply of new deleterious mutations. We showed earlier that unless the mutation rate is high or selection is weak, selection will keep any given deleterious allele at low frequency (Chapter 6). But quantitative traits are determined by the combined influence of many loci of small effect. Thus selection on the alleles at any single locus affecting a quantitative trait may be very weak, allowing substantial genetic variation to persist at the equilibrium between mutation and selection.
3. Disruptive selection, or patterns of selection with similar effects, may be more common than is generally recognized. Other patterns of selection that can maintain genetic variation in populations include frequency-dependent selection, in which rare phenotypes (and genotypes) have higher fitness than common phenotypes, and selection imposed by a fluctuating environment.

All three hypotheses are controversial and have been the subject of considerable theoretical and empirical research (see, for example, Barton and Turelli 1989). A detailed discussion is beyond the scope of this text. We can, however, briefly review an intriguing experiment by Santiago Elena and Richard Lenski.

Elena and Lenski (1997) studied six populations of the bacterium *Escherichia coli* established from a common ancestral culture. Each population was founded by a

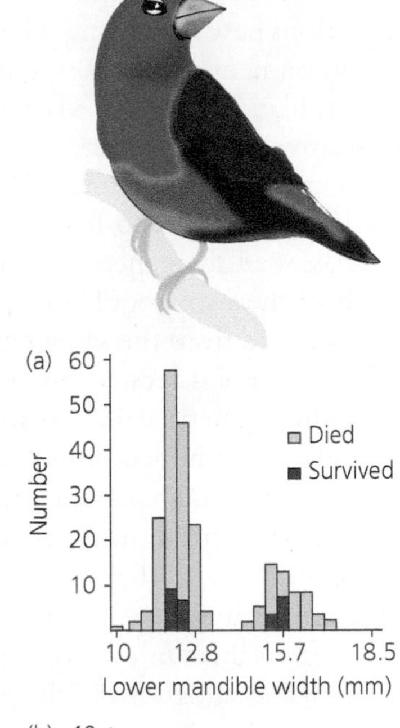

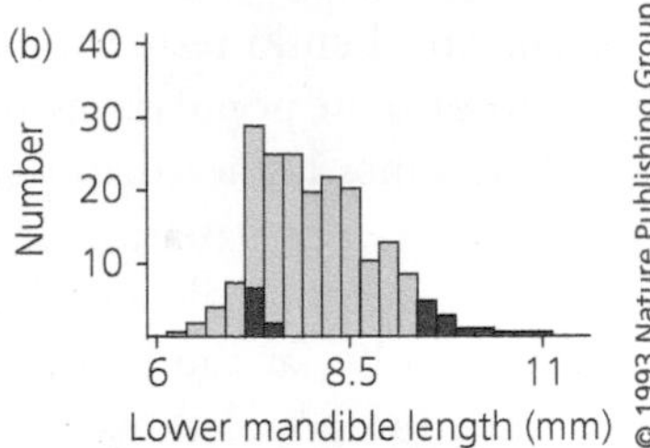

© 1993 Nature Publishing Group

Figure 9.29 Disruptive selection on bill size in the black-bellied seedcracker (*Pyrenestes o. ostrinus*) Each graph shows the distribution of lower bill widths (a) or lengths (b) in a population of black-bellied seedcrackers, an African finch. The light-colored portion of each bar represents juveniles that did not survive to adulthood; the dark-colored portion represents juveniles that did survive. The survivors were those individuals whose bills were either relatively large or relatively small. Redrawn from Bates Smith (1993).

Reprinted by permission from Macmillan Publishers Ltd: Bates Smith, T. 1993. Disruptive selection and the genetic basis of bill size polymorphism in the African finch Pyrenestes. *Nature* 363: 618–620.

single bacterium, so within any given population all genetic variation had arisen as a result of new mutations. The six populations had been evolving in a constant lab environment for 10,000 generations. During this time, the mean fitness of each population, assessed via competition experiments, had increased by over 50% relative to their common ancestor. However, much of the increase in fitness had occurred in the first few thousand generations. After 10,000 generations, the populations appeared to have arrived at an evolutionary equilibrium. Elena and Lenski assessed the genetic variation in fitness among the various strains present in each of their six populations and found it to be significant. On average, two strains selected from the same population differed in fitness by about 4%.

Elena and Lenski tested the Fisher's fundamental theorem hypothesis by using the standing genetic variation in fitness within each population to predict how much additional improvement in fitness should occur in a further 500 generations of evolution. Depending on the assumptions used, the researchers predicted an additional increase in fitness of between 4% and 50%. In fact, between generations 10,000 and 10,500, none of the six populations showed any significant increase in mean fitness. Elena and Lenski concluded that the genetic variation in fitness within their *E. coli* populations is not the result of a continuous supply of new favorable mutations in the process of rising to fixation.

Elena and Lenski tested the mutation–selection balance hypothesis by noting that two of their populations had evolved extraordinarily high mutation rates—some 100 times higher than the other four populations and the common ancestor. If the genetic variation in fitness within each population is maintained by mutation–selection balance, then the two populations with high mutation rates should show by far the highest standing genetic variation in fitness. One of the high-mutation populations, indeed, exhibited much higher genetic variation for fitness than the low-mutation populations. But the other one did not. Elena and Lenski concluded that the genetic variation in fitness within their populations is probably not the result of a balance between deleterious mutations and selection.

Finally, Elena and Lenski tested the hypothesis that frequency-dependent selection is maintaining the genetic variation for fitness in their populations. The researchers used competition experiments to determine whether the various strains present in each population enjoyed a fitness advantage when rare. They found that a typical *E. coli* strain, when rare, did indeed have a fitness edge of about 2% relative to its source population. Furthermore, across the six bacterial populations, the intensity of frequency-dependent selection was significantly correlated with the amount of standing genetic variation for fitness. Elena and Lenksi noted that the three hypotheses they tested are mutually compatible. Nonetheless, the researchers concluded that the best explanation for the variation in fitness in their populations is frequency-dependent selection. Whether this conclusion applies to other populations and other organisms remains to be seen.

9.7 The Bell-Curve Fallacy and Other Misinterpretations of Heritability

We promised, in the introduction to this chapter, that our discussion of quantitative genetics would enable us to debunk erroneous claims about the causes of differences in IQ scores among various ethnic groups. We are now ready to make good on that promise.

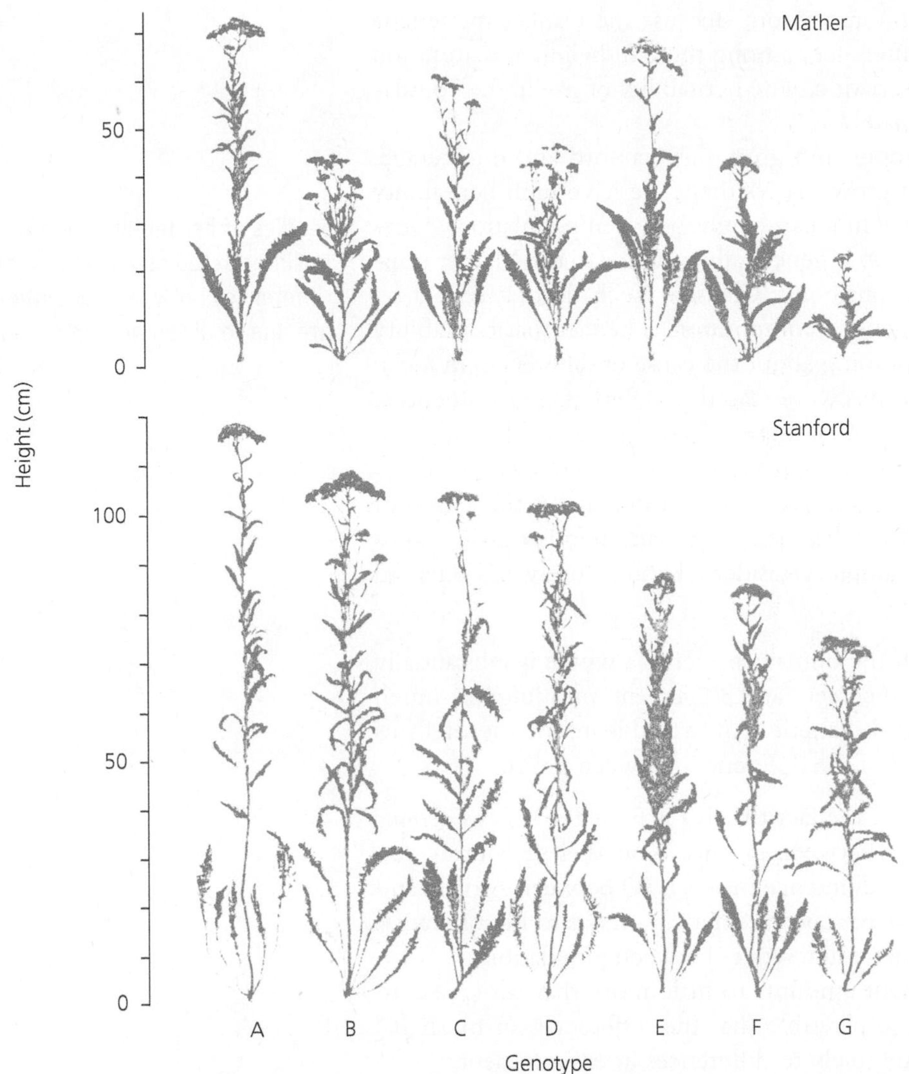

Figure 9.30 High heritability within populations tells us nothing about the cause of differences between population means We know the variation in height among the plants within each of these populations is entirely due to differences in their genes, because the plants grew in experimental common gardens where all experienced the same environment. The plants in the Stanford population are taller, on average, than the plants in the Mather population. Does this mean that the Stanford population is genetically superior to the Mather population? No: We know these two populations are genetically identical because they were grown from cuttings of the same seven plants. Reprinted from Clausen, Keck, and Hiesey (1948).

A key point is that the formula for heritability includes both genetic variation, V_G, and environmental variation, V_E. Any estimate of heritability is, therefore, specific to a particular population living in a particular environment. As a result, heritability tells us nothing about the causes of differences between populations living in different environments. We can illustrate this point with data from Jens Clausen, David Keck, and William Hiesey (1948). Clausen, Keck, and Hiesey studied *Achillea*, a perennial wildflower. *Achillea* will grow from cuttings, making it possible to create duplicates—clones—of a single individual. The researchers collected seven plants from a wild population and took two cuttings from each.

They grew one cutting from each plant in an experimental garden at Mather, California **(Figure 9.30, top row)**. As the cuttings grew up side by side, they lived in the same soil, got the same amount of water, the same amount of sunlight, and so on. Because the plants experienced virtually the same environment, differences among them in height at maturation are almost entirely due to genetic variation. The heritability of size in the population grown at Mather is approximately 1.

Clausen, Keck, and Hiesey grew the second cutting from each plant in an experimental garden at Stanford, California (Figure 9.30, bottom row). As the cuttings grew up side by side, they lived in the same soil, got the same amount of water, the same amount of sunlight, and so on. Because the plants experienced virtually the same environment, differences among them in height at maturation are almost entirely due to genetic variation. The heritability of size in the population grown at Stanford is approximately 1.

Notice that the plants in the population grown at Stanford are, on average, taller than those in the population grown at Mather. We have high heritability in both populations, and a difference in mean height between populations. Does this mean that the Stanford population is genetically superior to the Mather population with respect to height? Of course not; Clausen, Keck, and Hiesey set up the populations to be identical in genetic composition. The fact that heritability is high in each population tells us nothing about the cause of differences in mean between them, because the populations were reared in different environments.

Studies of heritability are often misinterpreted as implying that differences between populations are due to differences in genes.

Unsupported Claims about IQ

The mistaken notion that heritability can tell us something about the causes of differences between population means has been especially pernicious in claims about human intelligence. As an example, consider Charles Murray and Richard J. Herrnstein (1994). First they note:

> Most scholars accept that I.Q. in the human species as a whole is substantially heritable, somewhere between 40 percent and 80 percent, meaning that much of the observed variation in I.Q. is genetic. And yet this information tells us nothing for sure about the origin of the differences between groups.

This is incorrect. As we have shown, the fact that IQ is heritable tells us *nothing at all* about the origin of differences between groups. Nonetheless, Murray and Herrnstein proceed to argue that the difference in mean IQ between African and European Americans—the mean for European Americans tends to be somewhat higher—is at least partly due to genetic differences between populations.

Murray and Herrnstein's argument amounts to little more than an appeal to personal incredulity. They find it implausible that the differences in mean IQ between ethnic groups could be due solely to differences in environment.

Such an appeal is no substitute for science. A scientific approach to Murray and Herrnstein's hypothesis would be to conduct a common garden experiment: Rear European Americans and African Americans together in an environment typically experienced by European Americans, and then compare their IQ scores. This design, and the reciprocal experiment, in which everyone is reared in an environment typically experienced by African Americans, is shown in **Figure 9.31**.

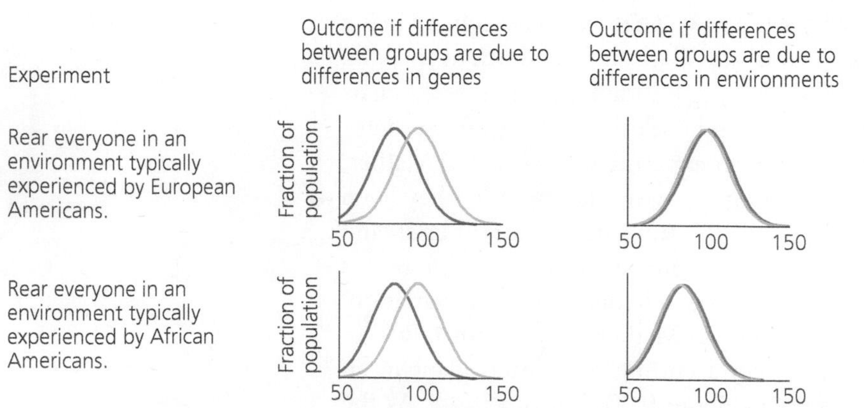

Figure 9.31 An experiment that would test Murray and Herrnstein's claim The left column describes two experimental treatments. The middle and right columns show the predicted outcomes under the hypothesis that differences between groups are due to differences in genes versus the predicted outcomes under the hypothesis that differences between groups are due to differences in environments.

We cannot do this experiment with humans. It might be suggested that we could approximate it by studying European American and African American children that have been adopted into similar families. But the children would still differ in appearance and might be treated differently by their parents, teachers, peers, and so on. In other words, even though they lived in similar families, the children might experience very different environments. Because we cannot do the definitive experiment, we simply have no way to assess whether genetics has anything to do with the difference in IQ score between ethnic groups.

But experiments like the one in Figure 9.31 have been done with plants and animals. It is instructive to look at the results. For example, Sara Via conducted common garden experiments with the pea aphid *Acyrthosiphon pisum*. Via measured the fecundity, or reproductive output, of aphids from a population living on clover and a population living on alfalfa **(Figure 9.32a)**. The clover aphids had somewhat higher mean fecundity. Is this difference in the means due to genetic differences between the populations? When Via reared the same genotypes of clover and alfalfa aphids in common gardens on both crops, she found that while clover aphids have higher fecundity on clover, alfalfa aphids have higher fecundity on alfalfa (Figure 9.32b).

The only way to determine the cause of differences between populations is to rear individuals from each of the populations in identical environments.

This result was unanticipated in the experimental design we outlined in Figure 9.31. It reveals genetic differences between clover and alfalfa aphids in how each group responds to the environment. And it reveals that each population has higher fecundity on the crop it came from. This unanticipated outcome demonstrates that hypothetical claims about the causes of differences between populations are no substitute for experimental results. What would happen if we could do this kind of experiment with African and European Americans? No one has any idea.

Finally, it is worth noting that heritability also tells us nothing about the role of genes in shaping phenotypic traits that are shared by all members of a population. There is virtually no variation among humans in number of noses. The

(a) Clover aphids have slightly higher mean fecundity

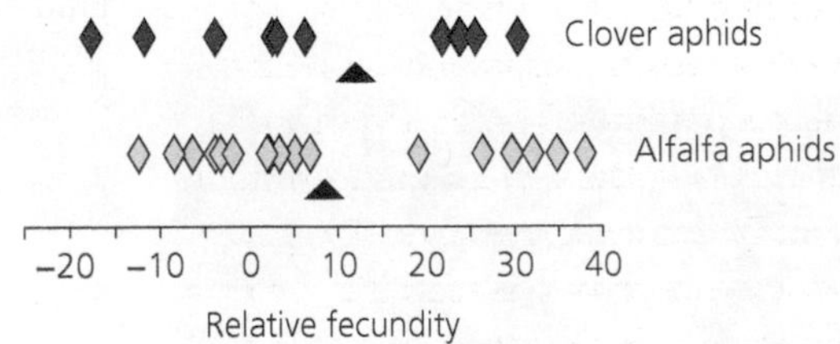

(b) A common garden experiment reveals local adaptation

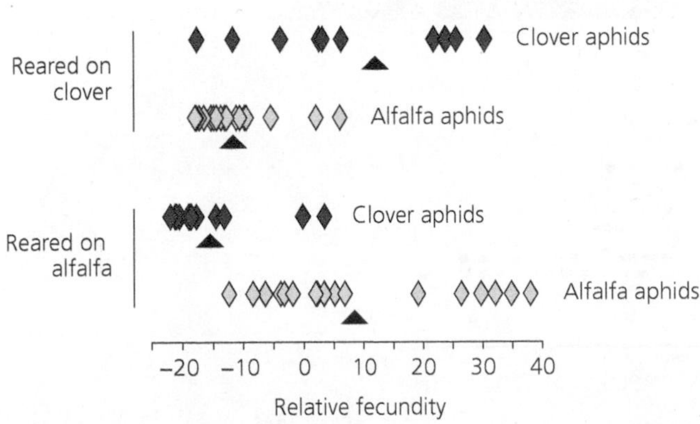

Figure 9.32 A common garden experiment with pea aphids Diamonds show fecundities of various aphid clonal genotypes. Black triangles show population means. Prepared with data presented in Via (1991).

heritability of nose number is undefined, because $V_A/V_P = 0/0$. This obviously does not mean that our genes are not important in determining how many noses we have.

Why Measure Heritability?

So what good does it do us to measure the heritability of a trait? If the heritability is greater than zero, it tells us that selecting on the trait will cause a population to evolve. Conversely, if the heritability is less than 1, it tells us that altering the environment can shift a population's trait distribution. The latter is clearly more pertinent to anyone interested in improving human intelligence. Knowing that the heritability of IQ in human populations is between 0.4 and 0.8 tells us nothing about the causes of differences among groups, but it does tell us that all groups have the capacity to respond to changes in their environment.

A striking illustration of how a change in the environment can alter a trait with substantial environmental variation comes from an analysis of thoroughbred racing performance by Thilo Pfau and colleagues (2009). The graph in **Figure 9.33** plots the speed of the winner, and the all-time record, for the Derby Stakes, an annual horse race in Great Britain. Between 1897 and 1910, the record time for the race dropped dramatically. The improvement had nothing to do with the horses or their genes. Instead, it was due to a change in the riding style used by the jockeys. The modern style, in which the jockey rides high on the horse's back, squats in the stirrups, and moves relative to the horse as the animal runs, substantially reduces the biomechanical cost to the horse of carrying the rider.

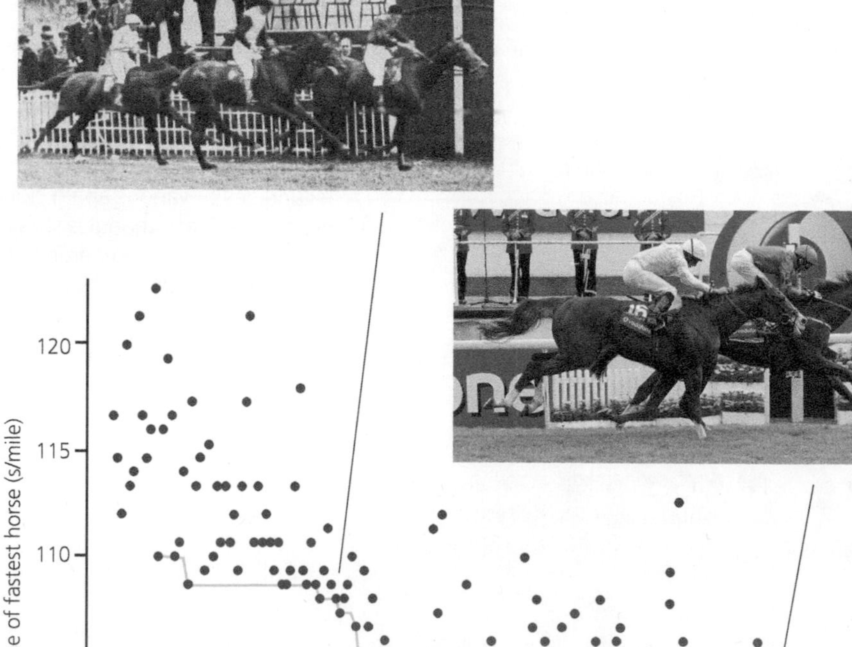

© 2009 AAAS

Figure 9.33 Winning and all-time record speeds for the Derby Stakes The striking improvement in the race record between 1897 and 1910 was due to a change in the riding style employed by the jockeys. From Pfau et al. (2009).

From "Modern riding style improves horse racing times." *Science* 325: 289. Reprinted with permission from AAAS.

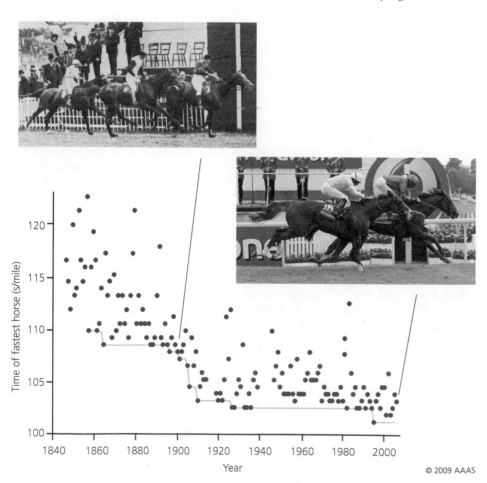

SUMMARY

Quantitative traits show continuous variation among individuals. They are influenced by the genotype at many loci as well as the environment.

Sometimes we can identify loci that contribute to a quantitative trait. We start with phenotypically distinct parental strains or species in which we have identified marker loci where different alleles are fixed in each parental population. We then generate a large population of F_2 individuals and look for associations between the genotype at the marker loci versus phenotype. Such associations indicate that the marker is linked to a locus that influences the trait of interest. If known protein-encoding genes are nearby, they may warrant further investigation.

Often we do not know the identity of the loci that influence a quantitative trait. Quantitative genetics gives us tools for analyzing the evolution of such traits, anyway. Heritability, the fraction of overall variation due to genetic causes, can be estimated by examining similarities among relatives. The strength of selection can be measured by analyzing the relationship between phenotypes and fitness. When we know both the heritability of a trait and the strength of selection on the trait, we can predict how the population will evolve in response to selection.

Selection on quantitative traits can follow a variety of patterns, including directional selection, stabilizing selection, and disruptive selection. Directional selection and stabilizing selection reduce genetic variation in populations. Nonetheless, genetic variation persists in most populations, even for traits closely related to fitness.

Genetic variation for fitness-related traits may persist because most populations are not in equilibrium, because there is a balance between mutation and selection, or because disruptive selection (and related patterns, like frequency-dependent selection) are more common than generally recognized.

Estimates of heritability are often misinterpreted. Heritabilities tell us nothing about the cause of differences between population means. What they do tell us is whether a population will respond to selection and/or to changes in the environment.

QUESTIONS

1. Degree of antisocial behavior is a quantitative trait in human males. Avshalon Caspi and colleagues (2002) used data on several hundred men to investigate the relationship between antisocial behavior and two factors. The first factor was genotype at the locus that encodes the enzyme monoamine oxidase A (MAOA). MAOA acts in the brain, where it breaks down a variety of the neurotransmitters nerve cells use to communicate with each other. The gene for MAOA is located on the X chromosome. Due to genetic differences in the gene's promoter, some men have low MAOA activity and others have high MAOA activity. The second factor was the experience of maltreatment during childhood. Based on a variety of evidence, the researchers determined whether each man had experienced no maltreatment, probable maltreatment, or severe maltreatment. The data are summarized in **Figure 9.34**.
   - **a.** Is the variation among men in antisocial behavior at least partly due to differences in genotype? Explain.
   - **b.** Is the variation among men in antisocial behavior at least partly due to differences in environment? Explain.
   - **c.** Do men with different genotypes respond the same way to changes in the environment? Explain.
   - **d.** Is antisocial behavior heritable? Explain.

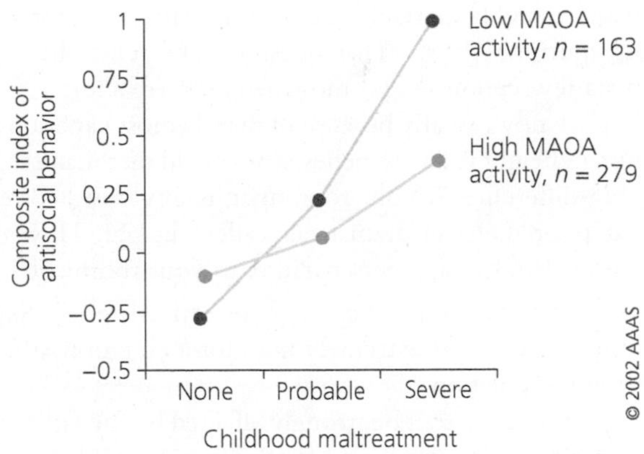

Figure 9.34 Degree of antisocial behavior among men with different levels of MAOA activity as a function of childhood maltreatment From Caspi, et al. (2002).

From "Role of genotype in the cycle of violence in maltreated children." *Science* 297: 851–854, Figure 1. Reprinted with permission from AAAS.

   - **e.** Do these data influence your opinion about how men who exhibit antisocial behavior should be treated and/or punished?

2. The serotonin transporter is a cell-surface protein that recycles the neurotransmitter serotonin after it has been used to carry a message between nerve cells in the brain. There are two alleles of the serotonin transporter gene:

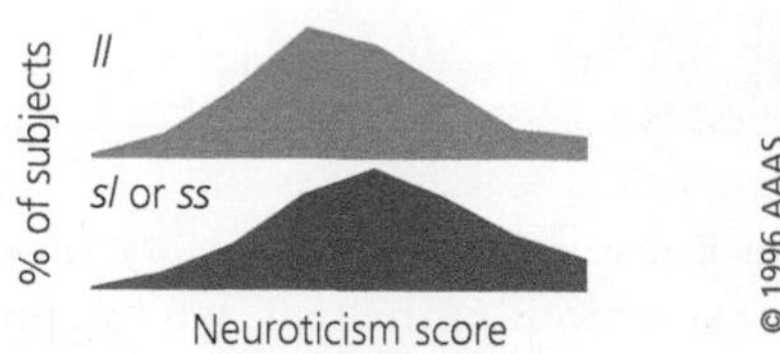

Figure 9.35 Distribution of neuroticism score among people with different genotypes of the serotonin transporter gene Redrawn from Lesch et al. (1996).

From "Association of anxiety-related traits with a polymorphism in the serotonin transporter gene regulatory regions," *Science* 274: 1527–1531, Figure 3. Reprinted with permission from AAAS.

s and *l*. Klaus-Peter Lesch and colleagues (1996) found that people with genotypes *ls* and *ss*, score slightly, but significantly, higher than people with genotype *ll* on psychological tests of neuroticism (see **Figure 9.35**).

a. Are these data consistent with the hypothesis that the serotonin transporter gene is a QTL that influences neuroticism? Explain.

b. Is the serotonin transporter gene the gene for neuroticism? Explain.

c. Can you think of another plausible explanation, in which the serotonin transporter gene plays no role at all in neuroticism? Explain.

3. An owner of racing greyhounds asks you how she can identify some of the loci and alleles that distinguish winners from losers. Describe, in as much detail as possible, a research program that might reveal this information.

4. Suppose you are telling your roommate that you learned in biology class that within any given human population, height is highly heritable. Your roommate, who is studying nutrition, says, "That doesn't make sense, because just a few centuries ago most people were shorter than they are now, clearly because of diet. If most variation in human height is due to genes, how could diet make such a big difference?" Your roommate is obviously correct that poor diet can dramatically affect height. How do you explain this apparent paradox to your roommate?

5. Now consider heritability in more general terms. Suppose heritability is extremely high for a certain trait in a certain population.

   a. First, can the trait be strongly affected by the environment despite the high heritability value? To answer this question, suppose that all the individuals within a certain population have been exposed all their lives to the same level of a critical environmental factor. Will the heritability value reflect the fact that the environment is very important?

   b. Second, can the heritability value change if the environment changes? To answer this question, imagine that the critical environmental factor changes such that different individuals are now exposed to different levels of this environmental factor. What happens to variation in the trait in the whole population? What happens to the heritability value?

6. A dog breeder has asked you for advice. The breeder keeps Alaskan huskies, which she races in sledding events. She would like to breed huskies that run faster. The table below gives data on the running speeds (m/s) of 15 families of dogs in the breeder's kennel.

Family	Midparent	Midoffspring
1	12.7	10.8
2	7.6	8.0
3	14.4	8.0
4	4.3	9.7
5	11.3	6.6
6	12.5	6.2
7	8.9	12.5
8	8.2	7.4
9	6.3	3.4
10	12.7	6.7
11	13.9	7.9
12	7.3	13.6
13	5.9	7.4
14	12.8	12.1
15	12.5	11.3

   a. Use a piece of graph paper to prepare a scatterplot of midoffspring values versus midparent values. Approximately what is the heritability of running speed in the breeder's dog population?

   b. If the breeder selectively breeds her dogs, will the next generation run substantially faster than the dogs she has now?

   c. What else would you suggest the breeder should try if she wants to win more races?

7. Imagine that the dog breeder in Question 6 were to pick just the five pairs of parents with the highest midparent values and use them as breeders for the next generation of dogs.

   a. Calculate the selection differential and the selection gradient the breeder has imposed on her population.

   b. Use your estimate of the heritability from Question 6 and the selection differential you just calculated to predict the response to selection. What is the predicted average running speed of the dogs in the next generation?

   c. How does your predicted speed of the next generation compare to the actual average running speed of offspring of the fastest five families? Discuss.

8. In our discussion of Weis and Abrahamson's work on goldenrod galls (data plotted in Figure 9.28), we mentioned that the researchers established that there is heritable variation among flies in the size of the galls they induce. How do you think Weis and Abrahamson did this? Describe the necessary experiment in as much detail as possible.

9. Given the strength of selection that bumblebees exert on alpine skypilots, why haven't flower corollas in the tundra population evolved to be even larger than they are now? Develop at least two hypotheses, and describe how you could you test your ideas.

10. a. Describe, in your own words, the three major modes of selection and their general effects on population means and on population variation.

b. Which mode of selection is at work on gall size of the gall-making flies?

c. If parasitoid wasps became extinct, which mode of selection would affect the next generation of gall-making flies? Predict what would happen to average gall size in subsequent generations.

EXPLORING THE LITERATURE

11. For an analysis identifying QTLs involved in the local adaptation of pea aphids to different crops, and consideration of the role genetic correlations may play in speciation, see:

Hawthorne, D. J., and S. Via. 2001. Genetic linkage of ecological specialization and reproductive isolation in pea aphids. *Nature* 412: 904–907.

12. For a study in which researchers used quantitative genetics to predict how behavior in a bird might evolve in response to global warming, see:

Pulido, F., P. Berthold, et al. 2001. Heritability of the timing of autumn migration in a natural bird population. *Proceedings of the Royal Society of London B* 268: 953–959.

13. How far and how fast can directional selection on a quantitative trait shift the distribution of the trait in a population? For one answer, see:

Weber, K. E. 1996. Large genetic change at small fitness cost in large populations of *Drosophila melanogaster* selected for wind tunnel flight: Rethinking fitness surfaces. *Genetics* 144: 205–213.

14. As genome-wide mapping techniques become increasingly sophisticated and cost effective, QTLs are rapidly being identified for many human diseases. For some examples, see:

Kissebah, A. H., G. E. Sonnenberg, et al. 2000. Quantitative trait loci on chromosomes 3 and 17 influence phenotypes of the metabolic syndrome. *Proceedings of the National Academy of Sciences, USA* 97: 14478–14483.

Arya, R., R. Duggirala, et al. 2004. Evidence of a novel quantitative-trait locus for obesity on chromosome 4p in Mexican Americans. *American Journal of Human Genetics* 74: 272–282.

Nyholt, D. R., K. I. Morley, et al. 2005. Genomewide significant linkage to migrainous headache on chromosome 5q21. *American Journal of Human Genetics* 77: 500–512.

15. For further information on genetic and environmental effects on human intelligence, see:

Greenwood, P. M., and R. Parasuraman. 2003. Normal genetic variation, cognition, and aging. *Behavioral and Cognitive Neuroscience Reviews* 2: 278–306.

Posthuma, D., M. Luciano, et al. 2005. A genomewide scan for intelligence identifies quantitative trait loci on 2q and 6p. *American Journal of Human Genetics* 77: 318–326.

16. For study mapping quantitiative trait loci that influence burrowing behavior in a pair of closely related mice, see:

Weber, J. N., B. K. Peterson, and H. E. Hoekstra. 2013. Discrete genetic modules are responsible for complex burrow evolution in *Peromyscus* mice. *Nature* 493: 402–405.

CITATIONS

Please note that much of the quantitative genetics in this chapter is modeled after presentations in the following:

Ayala, F. J. 1982. *Population and Evolutionary Genetics: A Primer.* Menlo Park, CA: Benjamin Cummings.

Falconer, D. S. 1989. *Introduction to Quantitative Genetics.* New York: John Wiley & Sons.

Felsenstein, J. 2011. *Theoretical Evolutionary Genetics.* Seattle, WA: http://evolution.genetics.washington.edu/pgbook/pgbook.html

Jorde, L. B., J. C. Carey, et al. 1999. *Medical Genetics.* 2nd ed. St. Louis: Mosby.

Lynch, M., and B. Walsh. 1998. *Genetics and Analysis of Quantitative Traits.* Sunderland, MA: Sinauer.

Maynard Smith, J. 1998. *Evolutionary Genetics.* 2nd ed. Oxford: Oxford University Press.

Here is the listing of all citations in this chapter:

Ayala, F. J. 1982. *Population and Evolutionary Genetics: A Primer.* Menlo Park, CA: Benjamin Cummings.

Barton, N. H., and M. Turelli. 1989. Evolutionary quantitative genetics: How little do we know? *Annual Review of Genetics* 23: 337–370.

Bates Smith, T. 1993. Disruptive selection and the genetic basis of bill size polymorphism in the African finch *Pyrenestes. Nature* 363: 618–620.

Beardsley, P. M., A. Yen, and R. G. Olmstead. 2003. AFLP phylogeny of *Mimulus*, section Erythranthe and the evolution of hummingbird pollination. *Evolution* 57: 1397–1410.

Benjamin, J., L. Li, et al. 1996. Population and familial association between the D4 dopamine receptor gene and measures of novelty seeking. *Nature Genetics* 12: 81–84.

Bergen, S. E., C. O. Gardner, and K. S. Kendler. 2007. Age-related changes in heritability of behavioral phenotypes over adolescence and young adulthood: A meta-analysis. *Twin Research & Human Genetics* 10: 423–433.

Bradshaw, H. D., Jr., K. G. Otto, et al. 1998. Quantitative trait loci affecting differences in floral morphology between two species of monkeyflower (*Mimulus*). *Genetics* 149: 367–382.

Bradshaw, H. D., Jr., and D. W. Schemske. 2003. Allele substitution at a flower colour locus produces a pollinator shift in monkeyflowers. *Nature* 426: 176–178.

Brodie, E. D., III. 1992. Correlational selection for color pattern and anti-predator behavior in the garter snake *Thamnophis ordinoides*. *Evolution* 46: 1284–1298.

Brodie, E. D., III, A. J. Moore, and F. J. Janzen. 1995. Visualizing and quantifying natural selection. *Trends in Ecology and Evolution* 10: 313–318.

Caspi, A., J. McClay, et al. 2002. Role of genotype in the cycle of violence in maltreated children. *Science* 297: 851–854.

Cavalli-Sforza, L. L., and W. F. Bodmer. 1971. *The Genetics of Human Populations*. San Francisco: W. H. Freeman.

Clausen, J., D. D. Keck, and W. M. Hiesey. 1948. *Experimental Studies on the Nature of Species. III. Environmental Responses of Climatic Races of* Achillea. Washington, D.C.: Carnegie Institution of Washington Publication No. 581, 45–86.

Di Masso, R. J., G. C. Celoria, and M. T. Font. 1991. Morphometric traits and femoral histomorphometry in mice selected for body conformation. *Bone and Mineral* 15: 209–218.

East, E. M. 1916. Studies on size and inheritance in *Nicotiania*. *Genetics* 1: 164–176.

Ebstein, R. P., O. Novick, et al. 1996. Dopamine D4 receptor (D4DR) exon III polymorphism associated with the human personality trait of novelty seeking. *Nature Genetics* 12: 78–80.

Elena, S. F., and R. E. Lenski. 1997. Long-term experimental evolution in *Escherichia coli*. VII. Mechanisms maintaining genetic variability within populations. *Evolution* 51: 1058–1067.

Falconer, D. S. 1989. *Introduction to Quantitative Genetics*. New York: John Wiley & Sons.

Felsenstein, J. 2011. *Theoretical Evolutionary Genetics*. Seattle, WA: http://evolution.genetics.washington.edu/pgbook/pgbook.html

Fisher, R. A. 1930. *The Genetical Theory of Natural Selection*. Oxford, UK: Clarendon Press.

Galen, C. 1989. Measuring pollinator-mediated selection on morphometric floral traits: Bumblebees and the alpine sky pilot, *Polemonium viscosum*. *Evolution* 43: 882–890.

Galen, C. 1996. Rates of floral evolution: Adaptation to bumblebee pollination in an alpine wildflower, *Polemonium viscosum*. *Evolution* 50: 120–125.

Grant, P. R., and B. R. Grant. 1995. Predicting microevolutionary responses to directional selection on heritable variation. *Evolution* 49: 241–251.

Hill, E. W., J. Gu, et al. 2010. A sequence polymorphism in MSTN predicts sprinting ability and racing stamina in thoroughbred horses. *PLoS ONE* 5: e8645.

Hill, E. W., R. G. Fonseca, et al. 2012. MSTN genotype (g.66493737C/T) association with speed indices in thoroughbred racehorses. *Journal of Applied Physiology* 112: 86–90.

Hugot, J.-P., M. Chamaillard, et al. 2001. Association of NOD2 leucine-rich repeat variants with susceptibility to Crohn's disease. *Nature* 411: 599–603.

Kendler, K. S., and J. H. Baker. 2007. Genetic influences on measures of the environment: A systematic review. *Psychological Medicine* 37: 615–626.

Kobayashi, K. S., M. Chamaillard, et al. 2005. Nod2-dependent regulation of innate and adaptive immunity in the intestinal tract. *Science* 307: 731–734.

Lande, R., and S. J. Arnold. 1983. The measurement of selection on correlated characters. *Evolution* 37: 1210–1226.

Lesch, K.-P., D. Bengel, et al. 1996. Association of anxiety-related traits with a polymorphism in the serotonin transporter gene regulatory region. *Science* 274: 1527–1531.

Lynch, M., and B. Walsh. 1998. *Genetics and Analysis of Quantitative Traits*. Sunderland, MA: Sinauer.

McClearn, G. E., B. Johansson, et al. 1997. Substantial genetic influence on cognitive abilities in twins 80 or more years old. *Science* 276: 1560–1563.

McPherron A. C., A. M. Lawler, and S.-J. Lee. 1997. Regulation of skeletal muscle mass in mice by a new TGF-beta superfamily member. *Nature* 387: 83–90.

McPherron, A. C., and S.-J. Lee. 1997. Double muscling in cattle due to mutations in the myostatin gene. *Proceedings of the National Academy of Sciences, USA* 94: 12457–12461.

Moreels, T. G., and P. A. Pelckmans. 2005. Gasterointestinal parasites: Potential therapy for refractory inflammatory bowel diseases. *Inflammatory Bowel Disease* 11: 178–184.

Mosher, D. S., P. Quignon, et al. 2007. A mutation in the myostatin gene increases muscle mass and enhances racing performance in heterozygote dogs. *PLoS Genetics* 3: e79.

Murray, C., and R. J. Herrnstein. 1994. Race, genes and I.Q.—An apologia. *The New Republic* 211 (October 31): 27–37.

Ogura, Y., D. K. Bonen, et al. 2001. A frameshift mutation in NOD2 associated with susceptibility to Crohn's disease. *Nature* 411: 603–606.

Orr, H. A. 1998. The population genetics of adaptation: The distribution of factors fixed during adaptive evolution. *Evolution* 52: 935–949.

Orr, H. A. 1999. The evolutionary genetics of adaptation: A simulation study. *Genetical Research, Cambridge* 74: 207–214.

Orr, H. A., and J. A. Coyne. 1992. The genetics of adaptation: A reassessment. *American Naturalist* 140: 725–742.

Paterson, A. H., E. S. Lander, et al. 1988. Resolution of quantitative traits into Mendelian factors by using a complete linkage map of restriction fragment length polymorphisms. *Nature* 335: 721–726.

Pfau, T., A. Spence, et al. 2009. Modern riding style improves horse racing times. *Science* 325: 289.

Phillips, P. C., and S. J. Arnold. 1989. Visualizing multivariate selection. *Evolution* 43: 1209–1222.

Provine, W. B. 1971. *The Origins of Theoretical Population Genetics*. Chicago: University of Chicago Press.

Provine, W. B. 1986. *Sewall Wright and Evolutionary Biology*. Chicago: University of Chicago Press.

Schemske, D. W., and H. D. Bradshaw, Jr. 1999. Pollinator preference and the evolution of floral traits in monkeyflowers (*Mimulus*). *Proceedings of the National Academy of Science USA* 96: 11910–11915.

Schinka, J. A., E. A. Letsch, and F. C. Crawford. 2002. DRD4 and novelty seeking: Results of meta-analyses. *American Journal of Medical Genetics* (Neuropsychiatric Genetics) 114: 643–648.

Schuelke, M., K. R. Wagner, et al. 2004. Myostatin mutation associated with gross muscle hypertrophy in a child. *New England Journal of Medicine* 350: 2682–2688.

Smith, J. M. N., and A. A. Dhondt. 1980. Experimental confirmation of heritable morphological variation in a natural population of song sparrows. *Evolution* 34: 1155–1160.

Summers, R. W., D. E. Elliot, et al. 2005a. *Trichuris suis* therapy in Crohn's disease. *Gut* 54: 87–90.

Summers, R. W., D. E. Elliot, et al. 2005b. *Tichuris suis* therapy for active ulcerative colitis: A randomized controlled trial. *Gastroenterology* 128: 825–832.

Tanksley, S. D. 1993. Mapping polygenes. *Annual Review of Genetics* 27: 205–233.

USA Swimming. 2002. http://www.usaswimming.org/fast_times/template.pl?opt=results&eventid=471.

Via, S. 1991. The genetic structure of host plant adaptation in a spatial patchwork: Demographic variability among reciprocally transplanted pea aphid clones. *Evolution* 45: 827–852.

Weis, A. E., and W. G. Abrahamson. 1986. Evolution of host-plant manipulation by gall makers: Ecological and genetic factors in the *Solidago-Eurosta* system. *American Naturalist* 127: 681–695.

Wilson, A. J., and A. Rambaut. 2008. Breeding racehorses: What price good genes? *Biology Letters* 4: 173–175.

Wright, S. 1986. *Evolution: Selected papers*. William B. Provine, editor. Chicago: University of Chicago Press.

Wright, S. 1988. Surfaces of selective value revisited. *American Naturalist*. 131: 115–121.

10

Studying Adaptation: Evolutionary Analysis of Form and Function

Why do the flowers of the South African iris *Lapeirousia oreogena* have white arrowheads on their petals? Dennis Hansen and colleagues (2012) suspected they knew the answer. Similar high-contrast markings appear on the petals of many other kinds of flowers. Called nectar guides, these markings are generally thought to help pollinators locate rewards hidden in the flowers. If pollinators find well-marked flowers more attractive, the plants that display them should enjoy higher reproductive success.

Hansen and colleagues tested this hypothesis with a straightforward experiment. They used black ink to fill in the white arrowheads on experimental flowers. They applied similar amounts of ink to the dark areas of control flowers, but left the white arrowheads alone. The researchers predicted that if the nectar guide hypothesis is correct, control flowers would be more attractive than experimental flowers to the long-proboscid flies (*Prosoeca* sp.) that pollinate the iris.

The biologists monitored pairs of experimental and control flowers in a natural iris population. Whenever a fly came to visit, they noted which flower the fly visited first. As documented in the figure at right, the flies showed a clear preference for flowers with a full complement of white markings. The preference increased with the difference between the control versus experimental flower.

The long-proboscid flies that pollinate the iris *Lapeirousia oreogena* prefer flowers with six nectar guides over flowers with three, one, or none. Redrawn from Hansen et al. (2012).

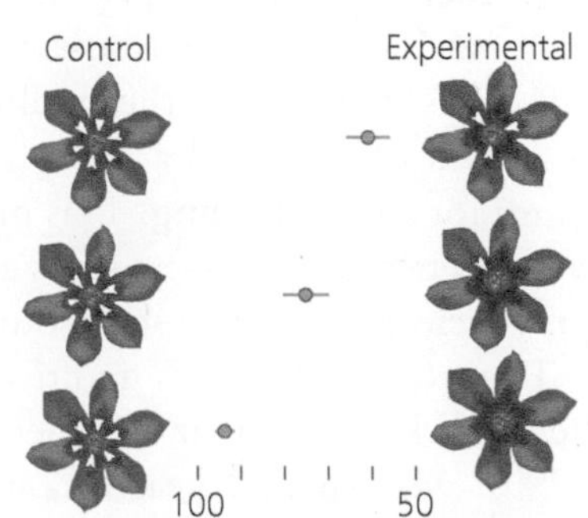

Additional experiments showed that control flowers had higher reproductive success than arrowhead-free flowers, via both male (pollen) and female (seeds) function. Collectively, the data indicate that Hansen and colleagues were right.

The explanation of organismal design is among the triumphs of the theory of evolution by natural selection. Individuals in previous generations varied in heritable aspects of their design, and the ones with the best designs passed on their genes in greater numbers. A trait, or integrated suite of traits, that increases the fitness of its possessor is called an **adaptation** and is said to be **adaptive.**

Demonstrating that the traits of organisms are indeed adaptations has been one of the major activities of evolutionary biology since the time of Darwin (Mayr 1983). Roughly speaking, to demonstrate that a trait is an adaptation, we need first to determine what a trait is for and then show that individuals possessing the trait contribute more genes to future generations than individuals lacking it.

The adaptive significance of some traits may seem obvious. Eyes are manifestly devices for detecting objects at a distance by gathering and analyzing light; in many animal species, individuals with good eyesight will be better able to find food and avoid predators than individuals with poor eyesight. Other traits offer more subtle advantages. Understanding their adaptive significance requires considerable ingenuity and effort. Still other traits, or trait values, may not be adaptive at all.

This chapter explores the analysis of adaptation. In the first section, we consider a cautionary tale on the dangers of being seduced by obvious explanations. No hypothesis for the adaptive value of a trait should be accepted simply because it is plausible and charming (Gould and Lewontin 1979). Instead, all hypotheses should be tested. This can be done by using them to make predictions, then checking whether the predictions are correct. In the next three sections we consider a variety of methods evolutionary biologists use to test hypotheses about adaptations, including experiments, observational studies, and the comparative method. These are followed by three sections in which we also explore complexities of biological form and function that continue to make the study of adaptation a challenging and active area of research. We close with a set of strategies for asking interesting questions.

A plausible hypothesis about the adaptive value of a trait is the beginning of a careful study, not the end.

10.1 All Hypotheses Must Be Tested: Oxpeckers Reconsidered

Oxpeckers call for an explanation **(Figure 10.1)**. Why do these birds flock to large mammals? Why do their hosts tolerate them? Most readers either have heard, or will quickly conceive, the traditional answer. The oxpeckers are looking for an easy meal of ticks and a safe place to eat it. Their hosts are happy to oblige in return for a free cleaning. This mutually advantageous association sometimes runs even deeper, as when the birds apparently minister to their mammalian benefactors by cleaning their open wounds.

The trouble with this traditional answer is that on careful observation, neither the oxpeckers nor their hosts seem to believe it. This we know from Paul Weeks (1999), who spent a year in Zimbabwe watching red-billed oxpeckers feed on domestic cattle. Weeks was able to establish that oxpeckers indeed sometimes eat ticks, because he found tick parts in the pellets of indigestible material that the birds occasionally regurgitate. But he seldom witnessed oxpeckers eating ticks.

Figure 10.1 Oxpecker on an impala The association between oxpeckers and large mammals has traditionally been considered mutually beneficial. The oxpeckers get an easy meal of ticks and a safe place to eat it; their hosts receive a free cleaning. But does the traditional view stand up to careful scrutiny?

More often, Weeks watched the birds ignore ticks that the biologist himself could plainly see. Instead, the oxpeckers devoted more than 85% of their feeding time to three activities: licking blood from open wounds; probing their hosts' ears, apparently for wax; and scissoring their beaks through their hosts' hair, apparently gleaning and eating dead skin. The hosts, in turn, seemed anything but pleased to have the oxpeckers around. When oxpeckers were bothering their wounds or poking in their ears, the cattle tried, once or twice per minute and with limited success, to shoo the birds away.

To better understand what oxpeckers do, Weeks (2000) set up an oxpecker exclusion experiment. He divided a small herd of cattle, at random, into two groups. One group he allowed oxpeckers to visit as usual. The other group he protected from oxpeckers by paying an assistant to chase the birds away. Weeks ran the experiment for a month, then switched the treatments and ran the experiment for another month. Finally, he shuffled the cattle to form two new groups and ran the experiment for a third month. At the beginning and end of each month, Weeks counted the ticks on every ox.

The graph in **Figure 10.2** shows, for each month, the change in tick load for cattle with oxpeckers versus cattle without. If oxpeckers serve their hosts by eating ticks, the change in tick load should be worse—more positive or less negative—for cattle without oxpeckers than for cattle with oxpeckers. This is what happened in trials one and three. But the opposite happened in trial two, and there was no significant difference between cattle with versus without oxpeckers in any trial. Oxpeckers have no discernible effect on their hosts' tick loads.

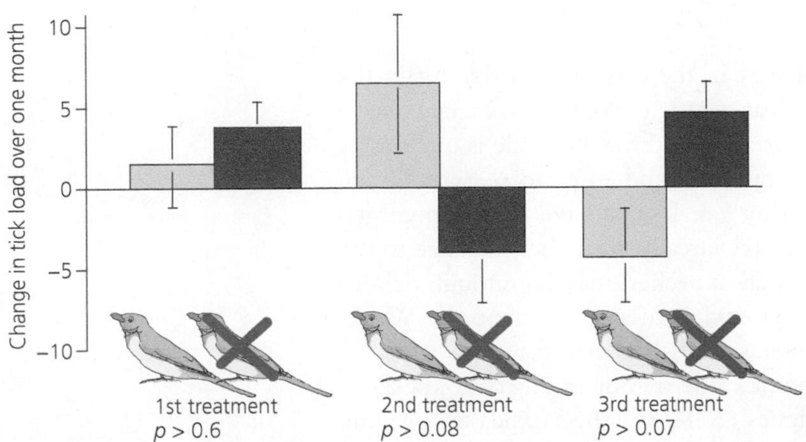

Figure 10.2 Red-billed oxpeckers have no effect on the tick loads of cattle This graph shows, for each of three month-long trials, the change in tick load from the beginning of the month to the end for domestic cattle exposed versus unexposed to oxpeckers. There is no clear pattern in the data, and none of the comparisons are statistically significant. Each bar includes data for 8 to 11 cattle. Redrawn from Weeks (2000).

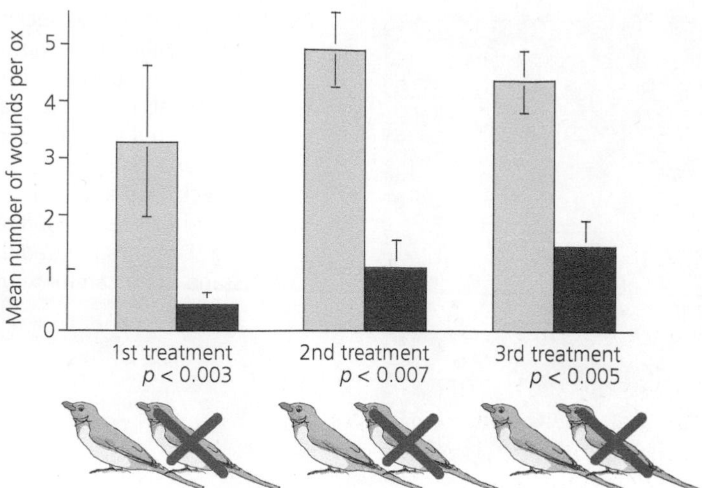

Figure 10.3 Red-billed oxpeckers maintain open wounds on their hosts This graph shows, for each of three month-long trials, the mean number of open wounds per individual on cattle exposed versus unexposed to oxpeckers. Each bar includes data for 8 to 11 cattle. Redrawn from Weeks (2000).

Before and after each month-long trial, Weeks also counted the number of open wounds on each ox. These results appear in **Figure 10.3**, and this time they are clear. Cattle exposed to oxpeckers have, on average, many more open wounds than cattle protected from the birds. We have already mentioned Weeks's observation that oxpeckers spend a considerable fraction of their feeding time drinking blood from open wounds. He also saw the oxpeckers enlarge existing wounds, and found that wounds took longer to heal when oxpeckers were present.

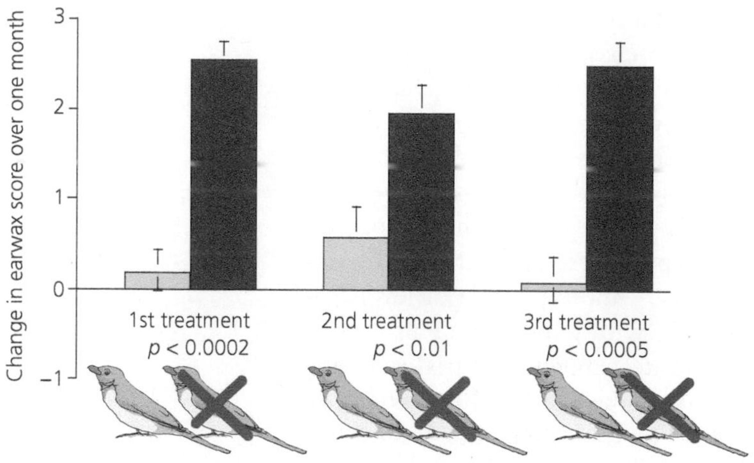

Figure 10.4 Red-billed oxpeckers remove their hosts' earwax This graph shows, for each of three month-long trials, the change in the amount of wax in the ears of cattle exposed versus unexposed to oxpeckers. Whether having their earwax removed is good or bad for the hosts is unclear. Each bar includes data for 8 to 11 cattle. Redrawn from Weeks (2000).

Finally, Weeks scored the amount of wax in the ears of each ox. Again the results, which appear in **Figure 10.4**, are clear. Cattle exposed to oxpeckers have considerably less earwax. Whether this is good or bad for the cattle is unclear.

Weeks concluded that oxpeckers are vampires and eaters of earwax. Even when they do eat ticks, oxpeckers prefer adult females that have already engorged themselves with blood—that is, ticks that have already done their damage to the host. Weeks acknowledges the possibility that oxpeckers may eat enough ticks to provide a benefit for other hosts or in other environments. For the cattle Weeks studied, however, oxpeckers appear to be less mutualists than parasites.

One flaw in Weeks's study is that domestic cattle are not among the oxpeckers' native hosts. Alan McElligott and colleagues (2004) watched oxpeckers feeding on a pair of black rhinoceroses. Consistent with their behavior in Weeks's study, the birds spent most of their foraging time at open wounds. Moreover, most of the wounds on the two rhinos were injuries the researchers watched the oxpeckers create. On the available evidence, it appears that the conventional wisdom about oxpeckers and their hosts is wrong.

The oxpecker example demonstrates that we cannot uncritically accept a hypothesis about the adaptive significance of a behavior, or any other trait, simply because it is plausible or because everyone knows that it must be true. Instead, we must subject all hypotheses to rigorous testing.

All hypotheses must be tested.

Other considerations to keep in mind when studying adaptations are these:

- Differences among populations or species are not always adaptive. There are two species of oxpecker; one has red bills, the other yellow. It is possible that each color is adaptive for the species that displays it. It is also possible, however, that the difference is not adaptive at all. Mutations causing different colors may have become fixed in the two oxpeckers by genetic drift. At the molecular level, much of the variation among individuals, populations, and species may be selectively neutral (see Chapter 7).
- Not every trait of an organism, or every use of a trait by an organism, is an adaptation. While feeding on large mammals, oxpeckers may sometimes meet a potential mate. This does not necessarily mean that feeding on large mammals evolved because it creates mating opportunities.
- Not every adaptation is perfect. Feeding on the blood and earwax of large mammals may provide oxpeckers with high-quality meals. But because many large mammals migrate long distances, it may also expose oxpeckers to the risk of an unpredictable food supply.

Alternative explanations must also be considered.

In the next three sections, we review three methods evolutionary biologists use to test hypotheses about the adaptive significance of traits. The first of these sections concerns experiments, the second looks at observational studies, and the third explores the comparative method.

10.2 Experiments

Experiments are among the most powerful tools in science. A well-designed experiment allows us to isolate the effect that a single, well-defined factor has on the phenomenon in question. We have reviewed a variety of experiments in earlier chapters. An experiment on mice, for example, compared two lineages of mice derived from a common ancestor to isolate the effect of selective breeding on voluntary wheel running. Another experiment, this time on fruit flies, isolated the effect of population size on genetic drift. Here our focus is on the process of planning and interpreting experiments. We have chosen our example because it illustrates several aspects of good experimental design.

Experiments are the most powerful method for testing hypotheses. A good experiment restricts the difference between study groups to a single variable.

What Is the Function of the Wing Markings and Wing-Waving Display of the Tephritid Fly *Zonosemata*?

The tephritid fly *Zonosemata vittigera* has dark bands on its wings. When disturbed, the fly holds its wings perpendicular to its body and waves them up and down. Entomologists had noticed that this display seems to mimic the leg-waving, territorial threat display of jumping spiders (species in the family Salticidae). These

Figure 10.5 A sheep in wolf's clothing? This photograph shows the tephritid fly *Zonosemata vittigera* (right) facing one of its predators, the jumping spider *Phidippus apacheanus* (left). Photo by Erick Greene, University of Montana.

entomologists suggested that, because jumping spiders are fast and have a nasty bite, a fly mimicking a jumping spider might be avoided by other predators. Erick Greene and colleagues (1987) had a different idea. Because jumping spiders are *Zonosemata*'s major predators **(Figure 10.5)**, Greene and colleagues proposed that the fly uses its wing-waving display to intimidate the jumping spiders themselves. The fly, in other words, is a sheep in wolf's clothing. Mimicry of a predator's behavior by its own prey had never before been recorded.

Both mimicry explanations are plausible hypotheses about the adaptive value of the fly's wing-waving display, but unless we test them they are just good stories. Can these hypotheses be tested rigorously? Greene and his coworkers (1987) sought to do so with an experiment.

The first step in any evolutionary analysis is to phrase the question as precisely as possible. In this case: Do the wing markings and the wing waving of *Zonosemata vittigera* mimic the threat displays that jumping spiders use on each other, and thereby allow the flies to escape predation? Stating a question precisely makes it easier to design an experiment that will provide a clear answer.

The researchers' next step was to list alternative explanations for the behavior. Good experiments test as many competing hypotheses as possible (Platt 1964). Note that each of the following is a biologically realistic explanation, not an implausible straw man proposed just to give the impression of rigor.

Good experimental designs test the predictions made by several alternative hypotheses.

Hypothesis 1: The flies do not mimic jumping spiders. This is a distinct possibility, because other fly species have dark wing bands and wing-flicking displays that do not deter predators. In many species, the flies use their markings and displays during courtship.

Hypothesis 2: The flies mimic jumping spiders, but the flies behave like spiders to deter other, nonspider predators. Other fly predators that might be intimidated by a jumping spider, or a jumping spider mimic, include other kinds of spiders, assassin bugs, praying mantises, and lizards.

Hypothesis 3: The flies mimic jumping spiders, and this mimicry functions specifically to deter predation by the jumping spiders themselves.

To test these alternatives, Greene and colleagues needed flies with some features, but not all, of the *Zonosemata* display. The biologists found they could cut the wings off a *Zonosemata* fly and glue them back on with household glue. And they could cut the wings off a *Zonosemata* fly and replace them with the wings of a housefly (*Musca domestica*), which are clear and unmarked. Remarkably, the surgically altered *Zonosemata* still waved their wings normally and could even fly.

In an ideal experiment, the control and experimental groups are treated identically except for exactly one factor.

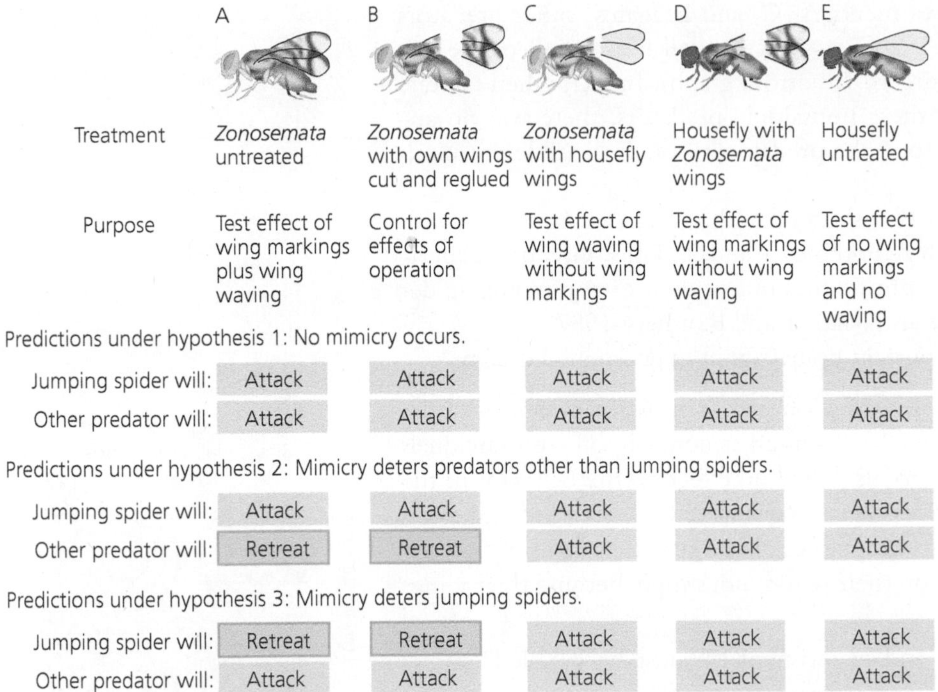

A
B
C
D
E
Treatment
Zonosemata untreated
Zonosemata with own wings cut and reglued
Zonosemata with housefly wings
Housefly with Zonosemata wings
Housefly untreated
Purpose
Test effect of wing markings plus wing waving
Control for effects of operation
Test effect of wing waving without wing markings
Test effect of wing markings without wing waving
Test effect of no wing markings and no waving
Predictions under hypothesis 1: No mimicry occurs.
Jumping spider will: Attack Attack Attack Attack Attack
Other predator will: Attack Attack Attack Attack Attack
Predictions under hypothesis 2: Mimicry deters predators other than jumping spiders.
Jumping spider will: Attack Attack Attack Attack Attack
Other predator will: Retreat Retreat Attack Attack Attack
Predictions under hypothesis 3: Mimicry deters jumping spiders.
Jumping spider will: Retreat Retreat Attack Attack Attack
Other predator will: Attack Attack Attack Attack Attack

Figure 10.6 Surgical treatments used in experiments testing the function of wing-waving display The predicted outcomes when different predators encounter flies with different treatments. Note that each hypothesis makes a unique suite of predictions. (The predictions listed for hypotheses 2 and 3 assume that both *Zonosemata*'s wing markings and wing waving are necessary for effective mimicry.)

Greene and colleagues created a total of five experimental groups of flies **(Figure 10.6)**. The five treatments distinguish among the three hypotheses, because each hypothesis makes a different suite of predictions about what will happen in encounters between predators and flies. The treatments also allow the researchers to determine whether both the wing markings and the wing-waving display are important in mimicry. This is a powerful experimental design.

To run the experiment, Greene and coworkers had to measure the responses of jumping spiders and other predators to the five types of experimental flies. When confronted with a test fly, would the predators retreat, stalk and attack, or kill? The researchers starved 20 jumping spiders from 11 different species for two days. Then they presented one of each of the five experimental fly types to each spider, in random order. The researchers made these presentations in a test arena and recorded each jumping spider's most aggressive response during a 5-minute interval. There was a clear difference: Jumping spiders tended to retreat from flies that gave the wing-waving display with marked wings, but attacked flies that lacked either wing markings, wing waving, or both **(Figure 10.7, top graph)**.

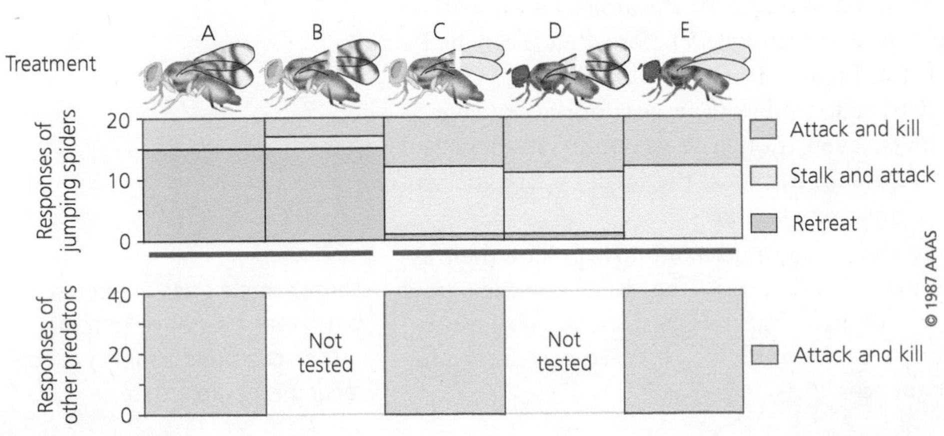

Figure 10.7 Tephritid flies mimic jumping spiders to avoid spider predation Bar heights represent the number of predators showing each response to the fly types listed in Figure 10.6. Red bars indicate that statistical analysis showed the jumping spider responses to treatments A and B to be significantly different from the responses to C, D, and E. From Greene et al. (1987).

From "A tephritid fly mimics the territorial displays of its jumping spider predators," *Science* 236: 310–312. Reprinted with permission from AAAS.

When the researchers tested treatments A, C, and E against other predators (nonsalticid spiders, assassin bugs, mantises, and whiptail lizards), all of the test flies were captured and eaten (Figure 10.7, bottom graph). In fact, when Greene and colleagues placed flies before these nonsalticid predators, there was no appreciable difference in the time it took the predators to catch the different kinds of flies.

Comparison of Figures 10.6 and 10.7 shows that the results are consistent with hypothesis 3, but inconsistent with hypotheses 1 and 2. Thus, Greene's experiment supports the hypothesis that tephritid flies mimic their own jumping-spider predators to avoid being eaten (see also Mather and Roitberg 1987).

The Greene study illustrates important points about experimental design:

- Defining and testing effective control groups is critical. In Greene's study, groups A and B (Figures 10.6 and 10.7) served as controls. These individuals demonstrated that the wing surgery itself had no effect on the behavior of the flies or the spiders. When the *Zonosemata* in group C were attacked and eaten by jumping spiders, Green and colleagues could be sure that this was because the flies no longer had markings on their wings, not simply because their wings had been cut and glued.
- All of the treatments (controls and experimentals) must be handled exactly alike. It was critical that Greene and colleagues used the same test arena, the same time interval, and the same definitions of predator response in each test. Using standard conditions allows a researcher to avoid bias and increase the precision of the data **(Figure 10.8)**. Think about the problems that could arise if a different test arena were used for each of the five treatment groups.
- Randomization is a key technique for equalizing other, miscellaneous effects among control and experimental groups. In essence, it is another way to avoid bias. Greene and colleagues presented the different kinds of test flies to the spiders and other predators in random order. What problems could arise if they had presented the five types of flies in the same sequence to each spider?
- Repeating the test on many individuals is essential. It is almost universally true in experimental (and observational) work that larger sample sizes are better. This is because any single test result is contaminated by at least a small amount of random environmental variation. With enough replicates, these random differences cancel each other out.

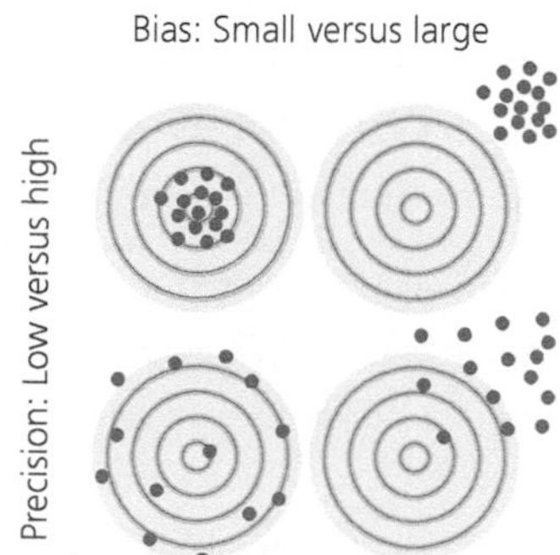

Figure 10.8 Bias and precision When designing an experiment or set of observations with the goal of estimating some quantity, it is important to minimize bias and maximize precision. In this illustration the quantity being estimated in a study is represented by the bull's-eye, and the data points collected are represented by red dots. Techniques like standardizing the experimental conditions and randomizing other factors help to minimize bias and increase precision. Note that our ability to measure precision depends on having a large number of data points.

Replicated experiments or observations do two things:

- They reduce the amount of distortion in the estimate caused by unusual individuals or circumstances. For example, 4 of the 10 *Zonosemata* with marked wings that were attacked were captured and killed before they even had a chance to display (groups A and B in Figures 10.6 and 10.7). Because Greene and colleagues were using standardized conditions, it was not acceptable to simply throw out these data points, even though they might represent bad luck. If events like this really do represent bad luck, they will be rare and will not bias the result as long as the sample size is large.
- Replicated experiments allow researchers to understand how precise their estimate is by measuring the amount of variation in the data. Knowing how precise the data are allows the use of statistical tests. Statistical tests, in turn, allow us to quantify the probability that the result we observed was simply due to chance (see **Computing Consequences 10.1**).

Large sample sizes are better, but researchers have to trade off the costs and benefits of collecting ever more data.

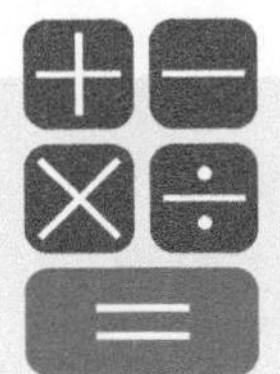

COMPUTING CONSEQUENCES 10.1

A primer on statistical hypothesis testing

The goal of many experimental and observational studies is to determine whether there is a meaningful difference between two groups, such as a treatment group versus a control group. Statistical analysis allows researchers to make this assessment in a formal and quantitative manner. The statistical approach we discuss here is known as hypothesis testing. For detailed treatments, we recommend Freedman et al. (1978), Zar (1999), and Whitlock and Schluter (2009).

As our example, we will analyze the results of an experiment by Eviatar Nevo and colleagues (2012). The scientists wanted to know whether wild barley populations in Israel have evolved earlier flowering times in recent decades as a result of a warming climate. We have chosen this example because it is amenable to the use of a simple statistical proceedure called a sign test. In practice, Nevo and colleagues performed additional analyses that made more compete use of their data.

A proper statistical analysis begins as soon as a question has been posed. The first step is to formulate a **null hypothesis**. The null hypothesis usually amounts to an assertion that nothing interesting is going on. For Nevo and colleagues, the null hypothesis was that wild barley populations have not evolved earlier flowering times. The alternative hypothesis, that the populations have evolved earlier flowering, is the one that drove the team's research. But we approach the alternative hypothesis by attempting to rule out the null hypothesis.

To answer their question, Nevo and colleagues ran a straightforward experiment. In a greenhouse, they grew seeds collected from 10 geographically distinct wild barley populations in 2008 alongside seeds derived from samples of the same 10 populations taken in 1980. For each individual plant, the researchers counted the days from germination to flowering. For each of the 10 populations, the scientists compared the mean flowering times of plants from the 2008 versus 1980 samples.

If the null hypothesis is correct, each population's 2008 mean will be the same as its 1980 mean. If the null hypothesis is wrong and the alternative is correct, then the 2008 means will be smaller than the 1980 means.

The mean flowering times appear in Figure 10.9. In all 10 populations, the 2008 plants flowered several days earlier, on average, than the 1980 plants. This result appears to refute the null hypothesis. But note that the mean flowering times are estimates based on samples, not true means for entire populations. Chance events, in the form of sampling error, have undoubtedly made the estimates somewhat different from the true means. It is possible that the apparent difference between the 2008 versus the 1980 populations is due to sampling error, not evolution. Possible, but not probable. The sign test will allow us to put a number on the probability.

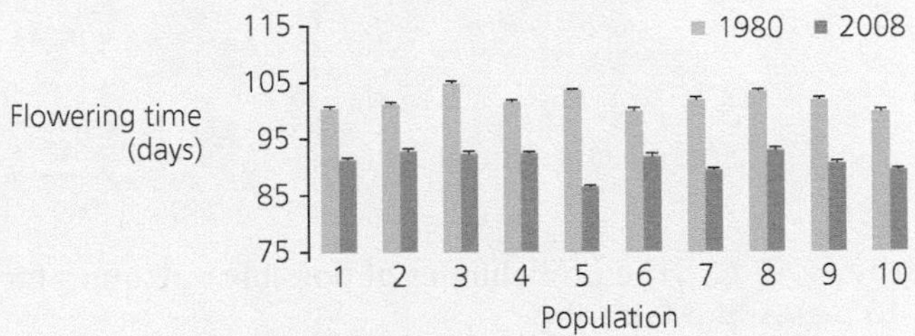

Figure 10.9 Flowering time in wild barley populations Heights of bars represent the mean flowering time, in days, of representatives of 10 wild barley populations sampled in 1980 and again in 2008. From Nevo et al. (2012).

The second step of our statistical analysis is to calculate the value of a test statistic from the data. A test statistic is a quantity that reflects the difference between the pattern we see in the data versus the pattern we would expect to see if the null hypothesis were true. For a sign test, the test statistic is called X. It is simply the number of populations in which the mean flowering time of the 2008 plants was less than the mean time of the 1980 plants. For Nevo's data, $X = 10$.

The third step in our analysis is to use a model of the experiment to calculate the probabilities of all possible values of the test statistic under the assumption that the null hypothesis is true. We can model a single run of the experiment by flipping a coin 10 times, once for each population. If the null hypothesis is true, the only cause of differences between the mean for the 2008 versus 1980 plants is sampling error. Sampling error in any given population is as likely to make the 2008 mean higher than the 1980 mean as to make it lower. Counting the heads in 10 tosses is thus equivalent to running the experiment once and calculating our test statistic.

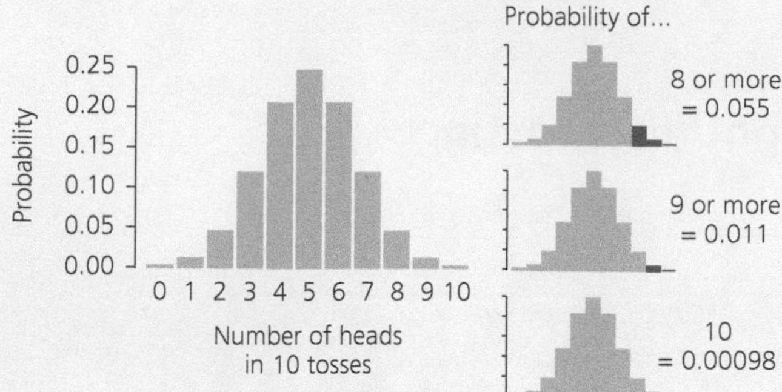

Figure 10.10 The probabilities of possible outcomes for 10 tosses of a fair coin

We have just flipped a coin 10 times and gotten five heads. That, by itself, does not tell us much. But if we simulate many replicates of the experiment, we can begin to see how much more often we get 5 heads than 6 heads, or 7 heads, or 2. The large graph in Figure 10.10 shows the calculated probability for all possible outcomes of 10 coin flips of a fair coin. The probability of getting 5 heads is just under 0.25. For 6 heads, it is just over 0.2. For 2 heads, it is a bit over 0.04.

The fourth step in our analysis is to use the distribution of possible outcomes to calculate the probability, under the null hypothesis, of values of the test statistic at least as extreme as the one we got. The small graphs in Figure 10.10 show that the probability of 8 or more heads is 0.055, the probability of 9 or more heads is 0.011, and the probability of 10 heads is 0.00098. These probabilities are known as *p*-values. After analyzing Nevo's data with the sign test, we can report that $p = 0.00098$, or that $p < 0.001$. Nevo's results would be possible, but highly improbable, if the null hypothesis were true. This does not prove the alternative hypothesis correct, but is certainly consistent with it.

The fifth and final step of our analysis is to decide whether to call our results "statistically significant." There is no hard-and-fast rule. By convention, however, if the *p*-value for a statistical test is less than 0.05, biologists consider it significant. This convention represents a comprise between the risk of erroneously rejecting the null hypothesis even though it is true (a type I error) versus erroneously failing to reject the null hypothesis even though it is false (a type II error). Keep in mind, however, that "$p < 0.05$" is not synonymous with "the null hypothesis is wrong," or with "the alternative hypothesis is correct." Indeed, Ronald Fisher, who first used the term *significant* in association with *p*-values, interpreted a value of less than 0.05 as merely indicating that the experiment that produced it was worthy of being repeated (see Goodman 2008).

For the data depicted in the top graph in Figure 10.7, Greene and colleagues' null hypothesis was that the true rate of attack by jumping spiders was the same for all types of flies. The researchers chose a test statistic that reflects differences among the observed rates. They calculated that under the null hypothesis, the probability of seeing rates at least as different as those between flies waving marked wings (groups A and B) versus other kinds of flies (groups C, D, and E) was less than 0.01. For all other comparisons, the *p*-values were greater than 0.1.

Although the hypothesis-testing approach to statistical analysis is widely used by biologists and other scientists, it is controversial among statisticians. For criticism and an introduction to alternative approaches, see Berger and Berry (1988) and Dienes (2011).

In sum, Greene et al.'s experimental design was successful because it allowed independent tests of the effect that predator type, wing type, and wing display have on the ability of *Zonosemata* flies to escape predation. Experiments are the most powerful means of testing hypotheses about adaptation. In the next section, we consider how careful observational studies can sometimes be nearly as good.

10.3 Observational Studies

Some hypotheses about adaptations are difficult or impossible to test with experiments. It is hard to imagine, for example, how we could do a controlled experiment to test alternative hypotheses about why giraffes have long necks. To do so, we would have to be able to make giraffes that are identical in all respects except the lengths of their necks. Experiments may also be inappropriate when a hypothesis makes predictions about how organisms will behave in nature. When experiments are impractical or inappropriate, careful observations can sometimes yield sufficient information to evaluate a hypothesis.

When an experiment is impractical, a careful observational study may be the next best method for evaluating a hypothesis.

Behavioral Thermoregulation

The vast majority of organisms are ectothermic. An ectotherm's body temperature is determined by the temperature of its environment. As **Figure 10.11** demonstrates for nematode worms (*Caenorhabditis elegans*), body temperature has a profound effect on an ectotherm's physiological performance and thus fitness.

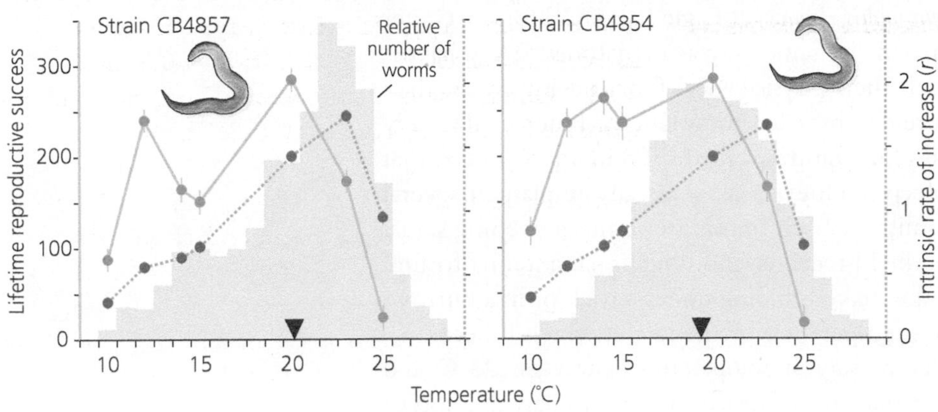

Figure 10.11 Fitness of two strains of nematode worm (*Caenorhabditis elegans*) as a function of body temperature Dots show mean ±se of two fitness measures for 15 worms at each temperature. Arrows mark means of gray distributions—which display temperatures chosen by approximately 1,100 worms placed on a thermal gradient. Redrawn from Anderson et al. (2011).

In the lab, Jennifer Anderson and colleagues (2011) estimated two measures of fitness for strains of *C. elegans* at temperatures ranging from 10°C to 25°C. The first measure was the lifetime reproductive success of individuals (orange symbols and solid lines); the second was the intrinsic rate of increase of lineages (blue symbols and dotted lines). As temperature climbs, fitness first rises and then plateaus. Above the mid 20s, the worms overheat and collapse. Intrinsic rate of increase shows a higher optimal temperature than lifetime reproductive success. This is likely because worms develop faster and thus have shorter generation times at higher temperatures. As generation time falls, intrinsic rate of increase rises but lifetime reproductive success is unaffected (see Huey and Berrigan 2001).

The relationship between physiological performance and temperature is called a thermal performance curve. The shapes of the *C. elegans* thermal performance curves for fitness are typical of the curves for physiological processes—ranging from hearing efficiency to sprint speed to endurance—in many organisms (Huey and Kingsolver 1989). Given the sensitivity of physiology and fitness to temperature, we can predict that ectotherms will exhibit behavioral thermoregulation. That is, we predict that ectotherms will move around in their environments and maintain themselves at or near the temperature at which they perform the best.

When, for example, Anderson and colleagues (2011) placed large numbers of *C. elegans* on a thermal gradient at an uncomfortably warm temperature, the worms did not just passively accept the consequences. Instead, nearly all moved to cooler locations (Figure 10.11, gray bars). As shown by the black triangles, on average the worms prefer temperatures near where their fitness is highest. Intriguingly, some strains—such as CB4857—prefer temperatures where their intrinsic rate of increase is highest, whereas other strains—such as CB4854—prefer temperatures where their lifetime reproductive success is highest.

Do nematodes show behavioral thermoregulation in nature like they do in the lab? The question is hard to answer for an organism as small as *C. elegans*, but more tractable for larger creatures like lizards and snakes. To demonstrate behavioral thermoregulation in nature, we must show (1) that the animal in question is choosing particular temperatures more often than it would encounter those temperatures if it simply moved at random through its environment, and (2) that its choice of temperatures is adaptive.

Do Garter Snakes Make Adaptive Choices When Looking for a Nighttime Retreat?

Ray Huey and colleagues (1989b) made a detailed study of the thermoregulatory behavior of the garter snake (*Thamnophis elegans*) at Eagle Lake, California. Garter snakes are affected by temperature in the same way as nematodes. They have a range of temeratures within which their physiological function improves, then plateus, then crashes as temperature climbs. Also important for Huey et al's study are the lowest and highest critical temperatures, called CTmin and CTmax, that snakes can endure briefly and survive. Huey et al. surgically implanted several snakes with miniature radio transmitters. Each transmitter emits a beeping signal that allows a biologist with a handheld receiver and directional antenna to find the implanted snake, even when the snake is hiding under a rock or in a burrow. The transmitter reports the snake's temperature by the rate at which it beeps.

Garter snakes in the lab prefer to stay at temperatures between 28°C and 32°C. Huey and colleagues found that snakes in nature do a remarkable job of thermoregulating in the same range. **Figure 10.12** shows the body temperatures of two implanted snakes, each over the course of a 24-hour day. Both snakes kept their temperature within or near the preferred range. How do snakes manage to thermoregulate so well? The two shown in the figure spent the day under or near rocks. Other options include moving up and down a burrow or staying on the surface while shuttling between sunshine and shade.

Huey and colleagues compared the relative merits of each of these thermoregulatory strategies by monitoring the environmental temperature under rocks of various sizes, and at various depths in a burrow, and by monitoring the temperature of a model snake left on the surface in the sun or shade **(Figure 10.13)**. For a snake under a rock, the thickness of the rock proves critical. A snake under a thin rock (Figure 10.13a) would not only get dangerously cold at night, but would

A good observational study seeks to find circumstances in nature that resemble an experiment.

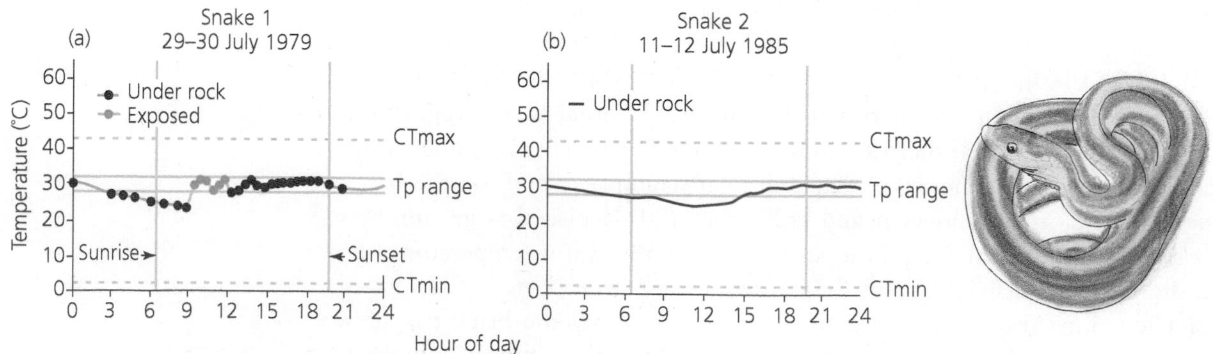

Figure 10.12 Body temperatures of garter snakes in nature (a) Snake 1 spent part of the day under a rock and part of the day in the sun. (b) Snake 2, monitored at short enough intervals that the data points, if shown, would run together, spent the entire day under a rock. Tp is the preferred temperature range, measured in the lab. CTmax and CTmin are the extremes snakes can survive. Both snakes kept their temperature near 30°C for the entire day. From Huey et al. (1989b).

(a) Under a thin rock

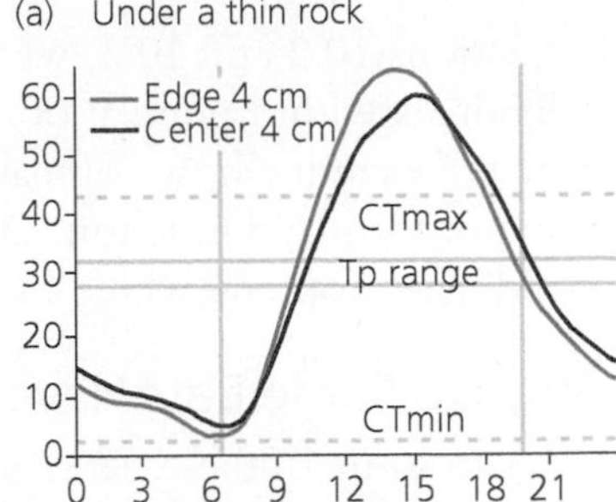

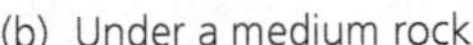

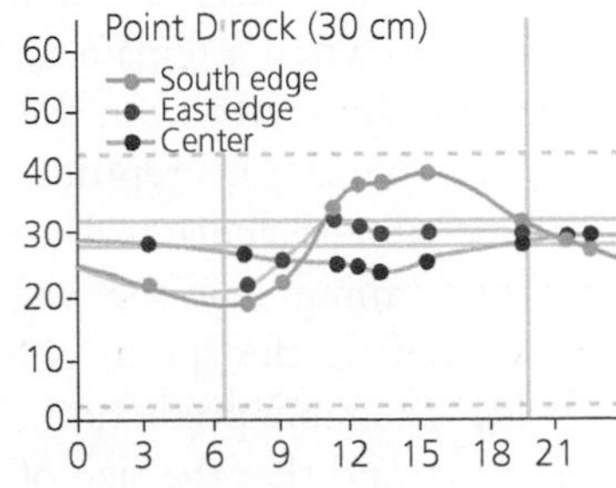

(c) Under a thick rock

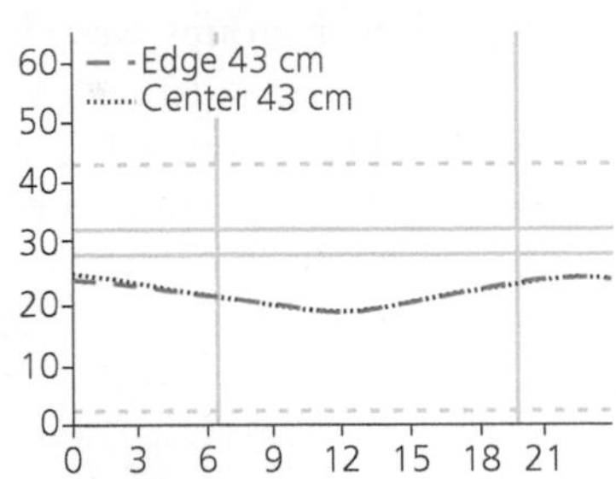

(d) In a burrow

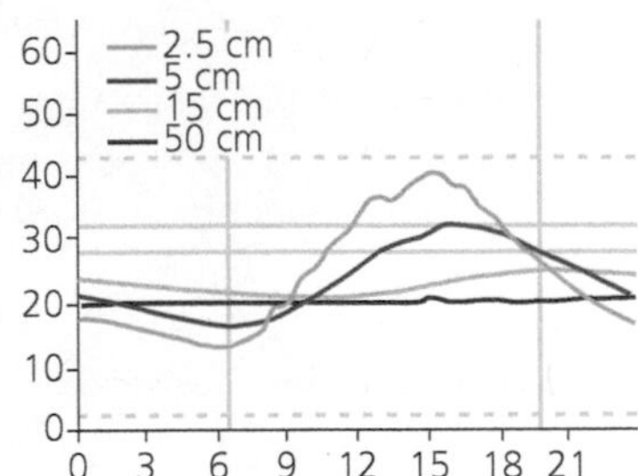

(e) For a model on the surface

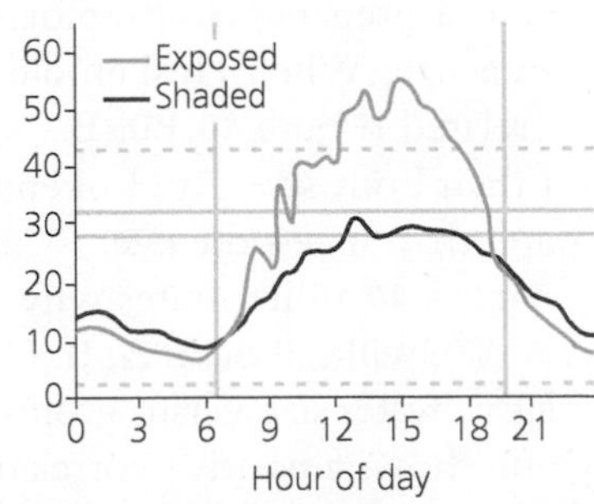

Figure 10.13 Environmental temperatures available to garter snakes at Eagle Lake Graphs show the daily cycle of temperatures in various places a snake might go. Under a thin (4 cm) rock (a), it is cold at night and hot during the day. Under a medium (30 cm) rock (b), there is virtually always a spot within the range of temperatures preferred by snakes. Under a thick (43 cm) rock (c), it is cool all the time. In a burrow (d), it is cool at night and cool to warm in the daytime—depending on depth within the burrow. On the surface (e), it is cold at night and just right to hot in the daytime—depending on whether a snake is in shade or direct sunlight. From Huey et al. (1989b).

fatally overheat in the daytime. A snake under a thick rock (Figure 10.13c) would remain safe all day, but would never reach its preferred temperature. Rocks of medium thickness are just right (Figure 10.13b). By moving around under the rock, a snake under a rock of medium thickness can stay close to or within its preferred temperature range for the entire day. A snake moving up and down a burrow could do reasonably well (Figure 10.13d), but would get colder at night than a snake under a medium-sized rock. Finally, a snake on the surface could thermoregulate effectively in the daytime by moving between sun and shade, but would get dangerously cold at night (Figure 10.13e). Putting these observations together, it appears that snakes have many options for thermoregulation during the daytime, as long as they avoid thin rocks or direct sun in the afternoon. At night, however, the best place to be is under a rock of medium thickness.

Most garter snakes do, in fact, retreat under rocks at night. Under the hypothesis of behavioral thermoregulation, Huey and colleagues predicted that snakes would choose their nighttime retreats adaptively. That is, they predicted that snakes would select rocks of medium thickness. Huey et al. tested their prediction by comparing the availability of rocks of different sizes at Eagle Lake to the sizes of the rocks actually chosen as nighttime retreats by radio-implanted snakes (**Table 10.1**). Thin, medium, and thick rocks are equally available, so if the snakes chose their nocturnal retreats at random, they should be found equally often under rocks of each size. In fact, however, the garter snakes are almost always found under medium rocks or thick rocks. The fact that snakes avoid thin rocks is good evidence that the snakes are active behavioral thermoregulators.

What made the observational study by Huey and colleagues effective in testing the hypothesis that garter snakes thermoregulate is the care with which the researchers monitored the snakes' environment. By determining the options available to snakes, and measuring the frequency of each option in the environment, the researchers were able to show that the snakes they observed were not simply picking their retreats at random, but were instead making an adaptive choice. In the next section, we consider a kind of observational study that looks at adaptations on a broader scale. Biologists using the comparative method evaluate hypotheses by looking at patterns of evolution among species.

Table 10.1 Rocks available versus rocks chosen by snakes

Thin, medium, and thick rocks are equally abundant at Eagle Lake, but garter snakes retreating under rocks at night show a strong preference for medium ($p < 0.05$; χ^2 test with thin and thick rocks combined because of small expected values).

	Thin (< 20 cm)	Medium (20–40 cm)	Thick (> 40 cm)
Rocks available	**32.4%**	**34.6%**	**33%**
Rocks chosen	**7.7%**	**61.5%**	**30.8%**

Source: From Table 1 in Huey et al. (1989b).

10.4 The Comparative Method

In Sections 10.2 and 10.3, we considered how experiments and observations on individuals within populations can be used to test hypotheses about adaptation. Here we examine how comparisons among species can be used to study the evolution of form and function. Our first example comes from a group of bats called the Megachiroptera, which includes the fruit bats and flying foxes **(Figure 10.14)**.

Figure 10.14 A grey-headed flying fox (*Pteropus poliocephalus*)

Why Do Some Bats Have Bigger Testes than Others?

Males in some of these bat species have larger testes for their body size than others. Based on work on a variety of other animals, David Hosken (1998) hypothesized that large testes are an adaptation for sperm competition. Sperm competition occurs when a female mates with two or more males during a single estrus cycle, and the sperm from the different males are in a race to the egg. One way a male can increase his reproductive success in the face of sperm competition is to produce larger ejaculates. By entering more sperm into the race, he increases his odds of winning. And the way to produce larger ejaculates is to have larger testes.

To evaluate the sperm competition hypothesis, Hosken needed to use it to develop a testable prediction. Hosken knew that fruit bats and flying foxes roost in groups, and that the size of a typical group varies dramatically among species, from two or three individuals to tens of thousands. Hosken reasoned that females living in larger groups would have more opportunities for multiple matings and that males living in larger groups would thus experience greater sperm competition. Hosken predicted that whenever a bat species evolves a preference for roosting in larger groups, its males will also evolve larger testes for their body size.

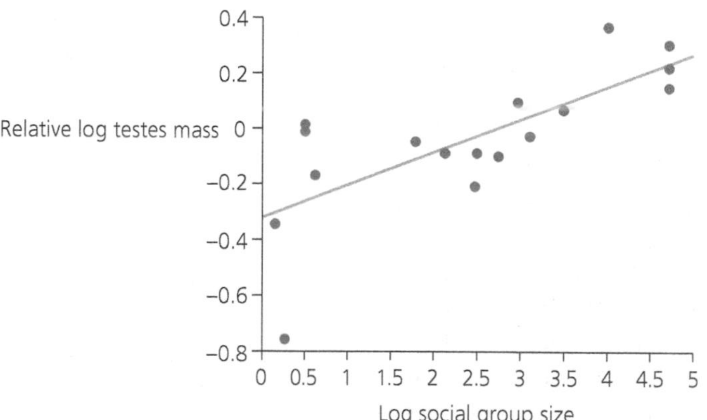

Figure 10.15 Variation in testis size among fruit bats and flying foxes This graph plots relative testes size (that is, testes size adjusted for body size) as a function of roost group size for 17 species of fruit bats and flying foxes. The best-fit line (the regression line) is in gray. From Hosken (1998).

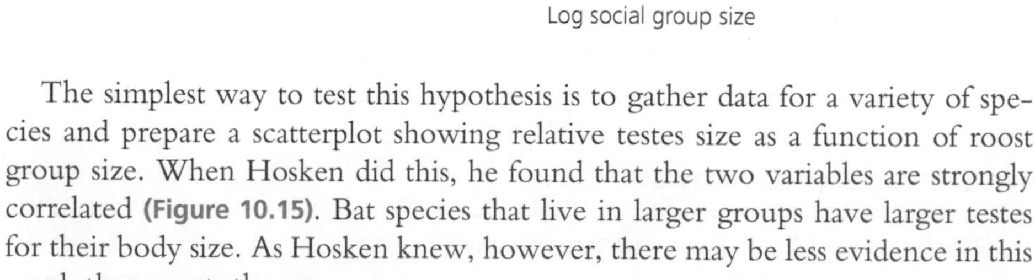

The simplest way to test this hypothesis is to gather data for a variety of species and prepare a scatterplot showing relative testes size as a function of roost group size. When Hosken did this, he found that the two variables are strongly correlated **(Figure 10.15)**. Bat species that live in larger groups have larger testes for their body size. As Hosken knew, however, there may be less evidence in this graph than meets the eye.

The comparative method seeks to evaluate hypotheses by testing for patterns across species, such as correlations among traits, or correlations between traits and features of the environment.

Figure 10.16 illustrates why. Imagine, for simplicity, a graph for only six species. We will call them A, B, C, D, E, and F. Figure 10.16a shows a scatterplot for relative testes size versus group size. Like the real scatterplot in Figure 10.15, this graph shows a positive correlation between the two traits. Now imagine that the evolutionary relationships among our six species are as shown in the phylogeny in

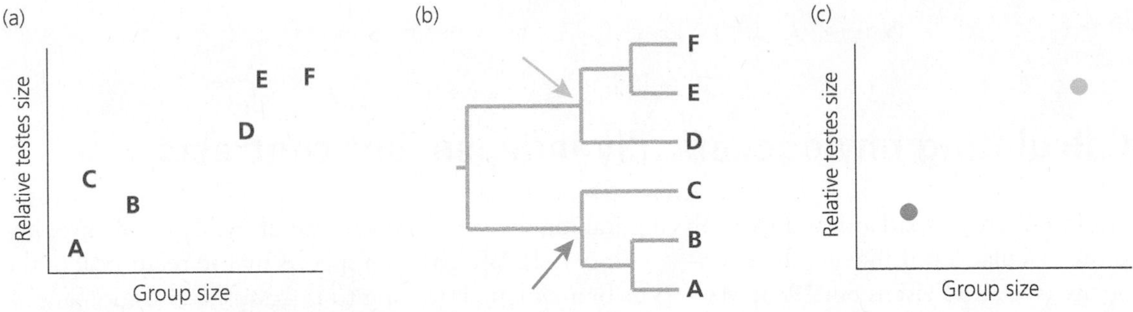

Figure 10.16 A simple scatterplot may provide only weak evidence that two traits evolve in tandem See text for explanation. After Lauder et al. (1995).

Figure 10.16b. A, B, and C are all closely related to each other, as are D, E, and F. It may be that A, B, and C all inherited their small group sizes and their small testes from their common ancestor (blue arrow). Likewise, it may be that D, E, and F all inherited their large group sizes and large testes from their common ancestor (green arrow). The possibility that our six species inherited their traits from just two common ancestors deflates the strength of our evidence considerably.

When we use a scatterplot as the basis for claims about nature, we want all the data points to be independent of each other. If they are independent, then each makes a separate statement for or against our claim. Furthermore, independence of the data points is a requirement for traditional statistical tests. To make sure our scatterplot accurately reflects the nature of the evidence, we should thus replace the points for species A, B, and C with a single point representing their common ancestor, and we should do the same with the points for species D, E, and F.

The graph in Figure 10.16c shows the result. It may be true that group size and testes size evolve together, and that sperm competition is the reason. But a scatterplot with only two data points is weak evidence on which to base such a claim.

Joe Felsenstein (1985) developed a better way to evaluate cross-species correlations among traits. What we look at in Felsenstein's method are patterns of divergence as sister species evolve independently away from their common ancestors. **Figure 10.17** shows a graphical interpretation of the method's basic approach.

The first thing we need is a phylogeny for the species we are studying. Figure 10.17a shows a hypothetical phylogeny for five extant species. We will call these species M through Q. The phylogeny also includes the common ancestors that lived at all the nodes on the tree. These are species R, S, T, and U. Note that there are four places on this phylogeny where sister species diverged from a common ancestor; each is indicated by a different color. For example, M and N are sister species that diverged from common ancestor S. Likewise, S and O are sister species that diverged from common ancestor T. What we want to know is this: When species diverge from a common ancestor, does the species that evolves larger group sizes also evolve larger testes?

We can answer this question by first plotting all the pairs of sister species on a scatterplot with lines connecting their data points (Figure 10.17b). We then slide each pair (without stretching or tilting their connector) until the left point rests on the origin (Figure 10.17c). Finally, we can erase the points at the origin and the connecting lines. We are left with a scatterplot with four data points (Figure 10.17d). Each data point represents the divergence, or contrast, that arose between a pair of sister species as they evolved away from their common ancestor. Because each contrast represents a separate divergence, they are collectively

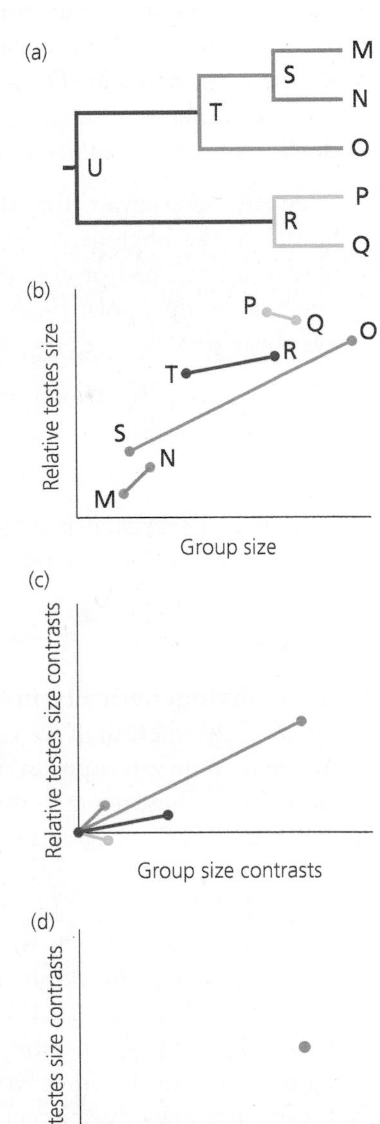

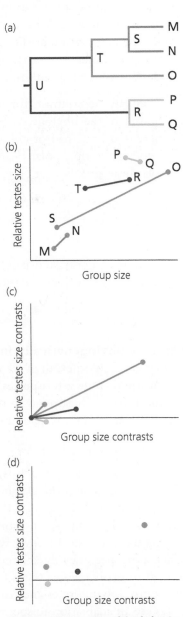

Figure 10.17 Graphical depiction of Felsenstein's phylogenetically independent contrasts See text for explanation.

COMPUTING CONSEQUENCES 10.2

Calculating phylogenetically independent contrasts

Here we use an example from Garland and Adolph (1994) to illustrate the calculation of independent contrasts from a phylogeny (see also Felsenstein 1985; Martins and Garland 1991; Garland et al. 1999; Garland et al. 2005). The phylogeny we will use appears in Figure 10.18. It shows the relationships among polar bears, grizzly bears, and black bears, and gives the body mass and home range of each. (For a more recent bear phylogeny, see Hailer et al. 2012.) We will calculate independent contrasts for both traits as follows:

1. Calculate the contrasts for pairs of sibling species at the tips of the phylogeny. In our three-species tree, there is just one pair of siblings in which both species reside at the tips: polar bears and grizzly bears. The polar bear–grizzly bear contrast for body mass is

$$265 - 251 = 14$$

The polar bear–grizzly bear contrast for range is

$$116 - 83 = 33$$

2. Prune each contrasted pair from the tree, and estimate the trait values for their common ancestor by taking the weighted average of the descendants' phenotypes. Weight each species by the reciprocal of the branch length leading to it from the common ancestor. We are pruning polar bears and grizzlies from the tree and estimating the body mass and home range of their common ancestor A. The branch lengths from A to its descendants are both two units long. Thus, the weighted average for body mass is

$$\text{Mass of A} = \frac{(\frac{1}{2})265 + (\frac{1}{2})251}{(\frac{1}{2}) + (\frac{1}{2})} = 258$$

The weighted average for home range is

$$\text{Range of A} = \frac{(\frac{1}{2})116 + (\frac{1}{2})83}{(\frac{1}{2}) + (\frac{1}{2})} = 99.5$$

3. Lengthen the branch leading to the common ancestor of each pruned pair by adding to it the product of the branch lengths from the common ancestor to its descendants, divided by their sum. We are lengthening the branch leading to A. The new length is

$$3 + \frac{2 \times 2}{2 + 2} = 4$$

known as **phylogenetically independent contrasts**. If the contrasts are correlated with each other, then we can conclude that when a species evolved a larger group size than its sister species, it also tended to evolve larger testes. In practice, we must make adjustments to the data before we can do statistical tests to evaluate the strength of any patterns. These adjustments are described in Computing Consequences 10.2.

Hosken (1998) repeated his analysis of testes size and group size in bats using Felsenstein's method, which is called the method of phylogenetically independent contrasts. Figure 10.19a shows a phylogeny of the 17 bat species whose data Hosken analyzed. Figure 10.19b shows a plot of the contrasts in relative testes size versus the contrasts in group size. There is a significant positive correlation among the contrasts. In other words, when analyzed correctly the data show that when a bat species evolved larger roosting group sizes than its sister species, it also tended to evolve larger testes for its body size. Hosken concluded that the evidence from flying foxes and fruit bats is consistent with the hypothesis that large testes are an adaptation to sperm competition.

Proper application of the comparative method requires knowledge of the evolutionary relationships among the species under study.

With data on 57 species representing all kinds of bats, Scott Pitnick and colleagues (2006) used phylogenetically independent contrasts to show that across species there is a negative association between testis size and brain size. That

4. Continue down the tree, calculating contrasts, estimating phenotypes of common ancestors, and lengthening branches. Our only remaining contrast is between species A and black bears. We need not estimate the phenotype of species B or lengthen the branch leading to it, because B is at the root of our tree. The species A–black bear contrast for mass is

$$258 - 93 = 165$$

The species A–black bear contrast for home range is

$$99.5 - 57 = 42.5$$

5. Divide each contrast by its standard deviation to yield the standardized contrasts. The standard deviation for a contrast is the square root of the sum of its (adjusted) branch lengths. The standard deviation for the polar bear–grizzly bear contrast is

$$\sqrt{2 + 2} = 2$$

The standard deviation for the species A–black bear contrast is

$$\sqrt{4 + 5} = 3$$

The standardized contrasts for our example are given in Figure 10.18.

Once we have calculated the standardized contrasts, we can use them to prepare a scatterplot and to perform traditional statistical tests.

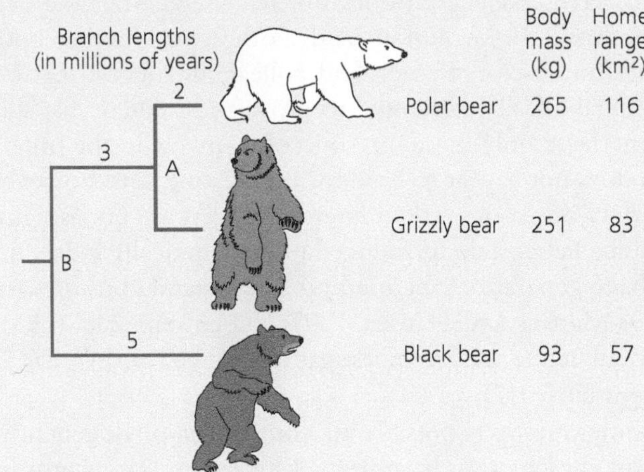

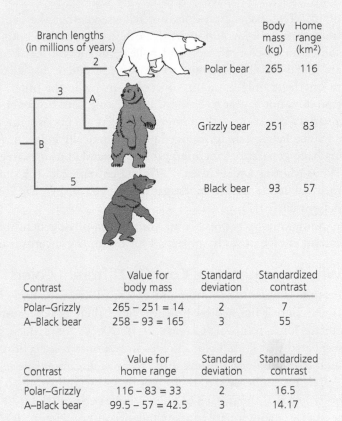

Contrast	Value for body mass	Standard deviation	Standardized contrast
Polar–Grizzly	265 – 251 = 14	2	7
A–Black bear	258 – 93 = 165	3	55

Contrast	Value for home range	Standard deviation	Standardized contrast
Polar–Grizzly	116 – 83 = 33	2	16.5
A–Black bear	99.5 – 57 = 42.5	3	14.17

Figure 10.18 An example showing how the data are adjusted when calculating phylogenetically independent contrasts From Garland and Adolph (1994).

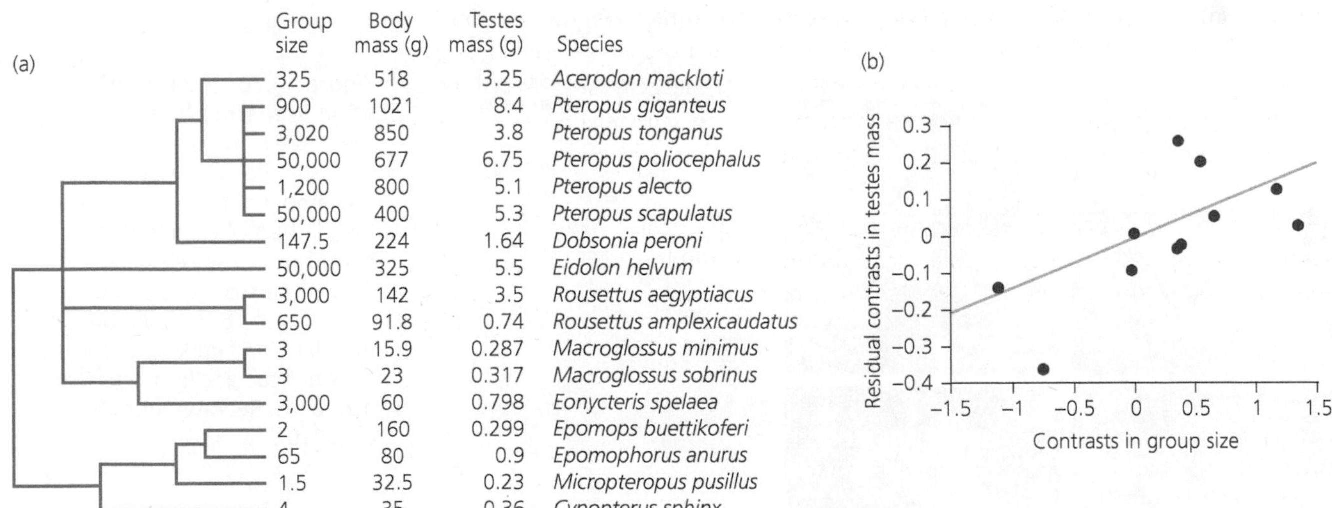

Group size	Body mass (g)	Testes mass (g)	Species
325	518	3.25	*Acerodon mackloti*
900	1021	8.4	*Pteropus giganteus*
3,020	850	3.8	*Pteropus tonganus*
50,000	677	6.75	*Pteropus poliocephalus*
1,200	800	5.1	*Pteropus alecto*
50,000	400	5.3	*Pteropus scapulatus*
147.5	224	1.64	*Dobsonia peroni*
50,000	325	5.5	*Eidolon helvum*
3,000	142	3.5	*Rousettus aegyptiacus*
650	91.8	0.74	*Rousettus amplexicaudatus*
3	15.9	0.287	*Macroglossus minimus*
3	23	0.317	*Macroglossus sobrinus*
3,000	60	0.798	*Eonycteris spelaea*
2	160	0.299	*Epomops buettikoferi*
65	80	0.9	*Epomophorus anurus*
1.5	32.5	0.23	*Micropteropus pusillus*
4	35	0.36	*Cynopterus sphinx*

Figure 10.19 Correlated evolution of group size and testes size in fruit bats and flying foxes (a) A phylogeny for 17 species of bats, showing roost group size, body mass, and testes mass for each species. (b) Independent contrasts for relative testes size versus group size. The points on this graph show that when a bat species evolved larger (or smaller) group sizes than its sister species, it also tended to evolve larger (or smaller) testes ($p = 0.027$). From Hosken (1998).

is, bat species that have evolved larger testes have also tended to evolve smaller brains. This association does not tell us whether the evolution of one of these traits drove the evolution of the other, or whether both traits were driven by some third factor. Pitnick and colleagues suggest that because brains and testes are metabolically expensive to grow and operate, bat species that invest heavily in one have only limited resources to invest in the other. The negative correlation does not appear to hold for fruit-eating bats, however, perhaps because—for a variety of reasons—their energy budgets are not as tight as those of other bats.

Since Felsenstein introduced phylogenetically independent contrasts, researchers have generalized the method and expanded its uses (see, for example, Grafen 1989; Martins and Hansen 1997; Felsenstein 2008). Comparative methods remain an active area of investigation (Revell and Collar 2009; Stone et al. 2011; Felsenstein 2012).

Sometimes it is possible to control for phylogenetic relationships simply by careful choice of study subjects. Our example concerns feather lice.

Why Do Feather Lice Come in Different Colors?

Feather lice are a common and costly affliction of birds (Clayton et al. 1999). Birds control the lice by preening, and their beaks feature adaptations that facilitate louse removal (Clayton et al. 2005). Sarah Bush and colleagues (2010) suspected that louse color evolves in response to selection imposed by preening such that louse populations come to be well camouflaged.

The researchers tested their hypothesis by comparing 16 pairs of closely related body lice that infest pairs of related light versus dark birds. Each pair of lice represents an evolutionarily independent contrast. **Figure 10.20a** shows an example. The louse *Neopsittaconirums albus,* which infests a white cockatoo, is white. The closely related louse *Neopsittaconirums borgiolii,* which infests a black cockatoo, is black. Each louse is well camouflaged against the feathers of its own host, but poorly camouflaged against the other's host.

Bush and colleagues photographed the lice and used software to measure the light-versus-dark color, or luminosity, of each. Consistent with their hypothesis, the lice from light hosts were, on average, significantly lighter (Figure 10.20b).

(a) Lice from light and dark birds on feathers of light and dark birds

(b) Mean luminosity of body and head lice from light and dark birds

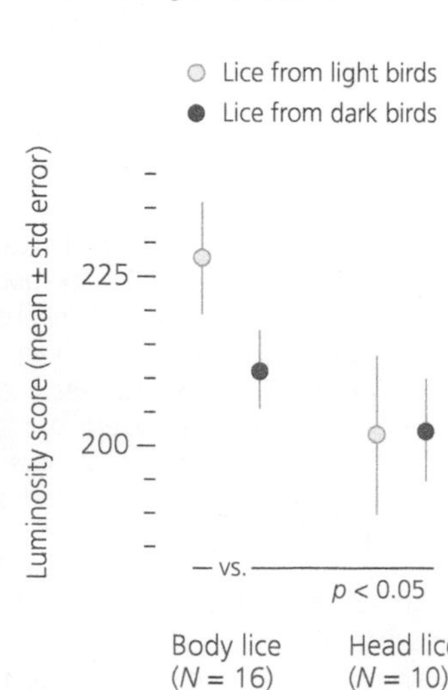

Figure 10.20 Adaptive coloration in feather lice (a) Body lice of the sulfur-crested cockatoo (left) and the yellow-tailed black cockatoo (right) photographed on the feathers of each bird. Louse photos from Bush et al. (2010). (b) Dark versus light color of members of 16 pairs of body lice from light and dark birds and 10 pairs of head lice from light and dark birds. Redrawn from Bush et al. (2010).

Bush and colleagues also compared 10 pairs of closely related head lice that infest pairs of related light versus dark birds. Head lice are much less vulnerable than body lice to removal by a preening host. This time—again, consistent with the researchers' hypothesis—there was no difference in color between the lice from light versus dark birds (Figure 10.20b).

We have now considered three methods biologists use to evaluate hypotheses about adaptation. In the next two sections of the chapter, we turn to complexities in organismal form and function that are active areas of current research. In the examples we discuss, researchers use experiments, observational studies, and the comparative method to investigate hypotheses about phenotypic plasticity (Section 10.5) and trade-offs and constraints on adaptation (Section 10.6).

When formulating and testing hypotheses about adaptation, biologists must keep in mind that organisms, and the lives they live, are complex.

10.5 Phenotypic Plasticity

Throughout much of this book, we treat phenotypes as though they were determined solely and immutably by genotypes. We know, however, that phenotypes are often strongly influenced by the environment as well. Elsewhere we have discussed methods for estimating how much of the phenotypic variation among individuals is due to variation in genotypes and how much is due to variation in environments (Chapter 9). Here, we focus on the interplay between genotype, environment, and phenotype.

Another way to say that an individual's phenotype is influenced by its environment is to say that its phenotype is plastic. When phenotypes are plastic, individuals with the same genotype may have different phenotypes if they live in different environments. Phenotypic plasticity is itself a trait that can evolve, and it may or may not be adaptive. As with the other traits we have discussed, to demonstrate that an example of phenotypic plasticity is adaptive, we must first determine its function, then show that individuals who have it achieve higher fitness than individuals who lack it.

Phenotypic Plasticity in the Behavior of Water Fleas

To illustrate phenotypic plasticity, we present the water flea, *Daphnia magna*. *Daphnia magna* is a tiny filter-feeding crustacean that lives in freshwater lakes (**Figure 10.21**). Conveniently for evolutionary biologists, *Daphnia* reproduce asexually most of the time. In other words, *Daphnia* clone themselves. This makes them ideal for studies of phenotypic plasticity, because researchers can grow genetically identical individuals in different environments and compare their phenotypes.

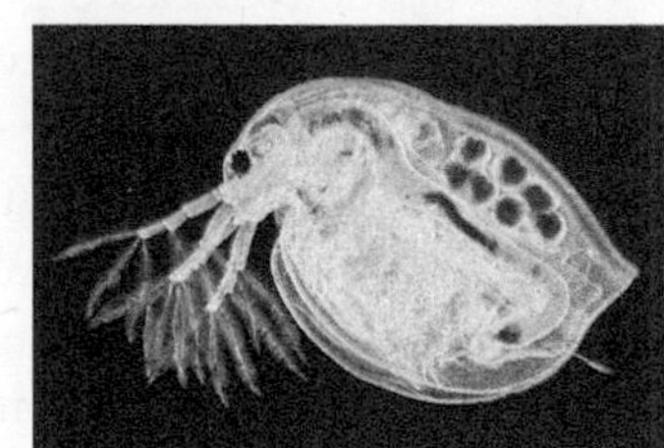

Figure 10.21 A water flea, *Daphnia magna* The branched appendages are antennae; the water flea uses them like oars for swimming. The dark object nearby is an eyespot. Also visible through the transparent carapace are the intestine and other internal organs, plus several darkly colored eggs. Enlarged about 10×.

Luc De Meester (1996) studied phenotypic plasticity in *D. magna*'s phototactic behavior. An individual is positively phototactic if it swims toward light and negatively phototactic if it swims away from light. De Meester measured the phototactic behavior typical of different genotypes of *D. magna*. In each single test, De Meester placed 10 genetically identical individuals in a graduated cylinder, illuminated them from above, gave them time to adjust to the change in environment, and then watched to see where in the column they swam. De Meester summarized the results by calculating an index of phototactic behavior. The index can range in value from -1 to $+1$. A value of -1 means that all the *Daphnia* in the test swam to the bottom of the column, away from the light. A value of $+1$ means that all the *Daphnia* in the test swam to the top of the column, toward the light. An intermediate value indicates a mixed result.

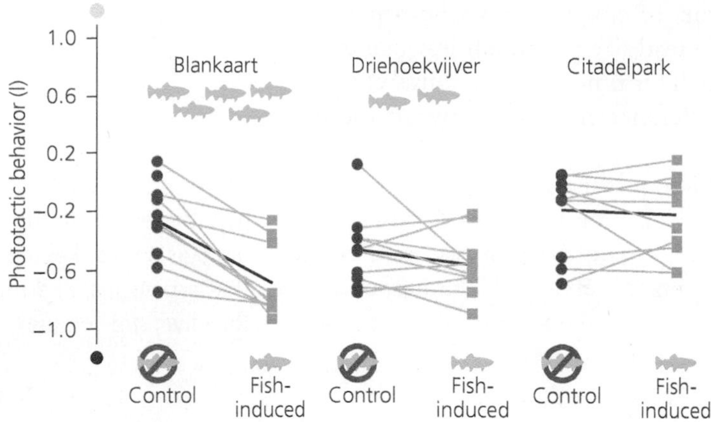

Figure 10.22 Variation in phototactic behavior in *Daphnia magna* The data are for *Daphnia* from three lakes in Belgium with different densities of fish. Each symbol represents the mean result from three to five tests of the phototactic behavior of a single genotype. Gray lines connect tests of the same genotype. Black lines connect means across genotypes. Redrawn from De Meester (1996).

De Meester measured the phototactic behavior of 10 *Daphnia* genotypes (also called clones) from each of three lakes. The data, indicated by maroon dots in **Figure 10.22**, show that most *Daphnia* tend somewhat to avoid light. They also show that each lake harbors considerable genetic variation in phototactic behavior.

De Meester also measured the phototactic behavior of the same 30 *Daphnia* genotypes in water that previously had been occupied by fish. The results are indicated by the orange squares in Figure 10.22. The dot and square for each genotype are connected by a line. These lines are called **reaction norms;** they show a genotype's change in phenotype across a range of environments. *Daphnia magna*'s phototactic behavior is phenotypically plastic. In Lake Blankaart, in particular, most *Daphnia* genotypes score considerably lower on the phototactic index when tested in the presence of chemicals released by fish.

Genetically identical individuals reared in different environments may be different in form, physiology, or behavior. Such individuals demonstrate phenotypic plasticity.

Finally, and most importantly, De Meester's results demonstrate that phenotypic plasticity is a trait that can evolve. Recall that a trait can evolve in a population only if the population harbors genetic variation for the trait. Each of the *Daphnia* populations De Meester studied contains genetic variation for phenotypic plasticity. That is, some genotypes alter their behavior more than others in the presence versus absence of fish (see Figure 10.22). Genetic variation for phenotypic plasticity is called **genotype-by-environment interaction.**

Has phenotypic plasticity evolved in the *Daphnia* populations De Meester studied? It apparently has. The average genotype in Lake Blankaart shows considerably more phenotypic plasticity than the average genotype in either of the other lakes. Blankaart is the only one of the lakes with a sizable population of fish. Fish are visual predators, and they eat *Daphnia.* A reasonable interpretation is that predation by fish selects in favor of *Daphnia* that avoid well-lit areas when fish are present.

Christophe Cousyn, De Meester, and colleagues (2001) tested this hypothesis by taking advantage of the fact that *Daphnia* produce resting eggs that remain viable even after being buried in sediment for decades. The researchers took sediment cores from Oud Heverlee Pond, a small human-made lake constructed in 1970. From sediments of three different depths, representing distinct episodes in the history of the pond, the researchers hatched *Daphnia* clones. Each set of clones is a sample from the population's past. The researchers measured the phototactic behavior of the reawakened genotypes in the presence and absence of chemicals released by fish.

When there is genetic variation for the degree or pattern of phenotypic plasticity, plasticity itself can evolve. Plasticity is adaptive when it allows individuals to adjust their phenotype so as to increase their fitness in the particular environment in which they find themselves.

The people who built Oud Heverlee Pond began stocking it with planktivorous fish in 1973. They stocked it heavily until the mid-1980s, then less heavily through the late 1980s. Cousyn, De Meester, and colleagues predicted that the *Daphnia* population in the pond would have evolved in response to fish predation, and that genotypes preserved in resting eggs from the period of heavy stocking would show greater phenotypic plasticity in phototactic behavior than earlier or later genotypes.

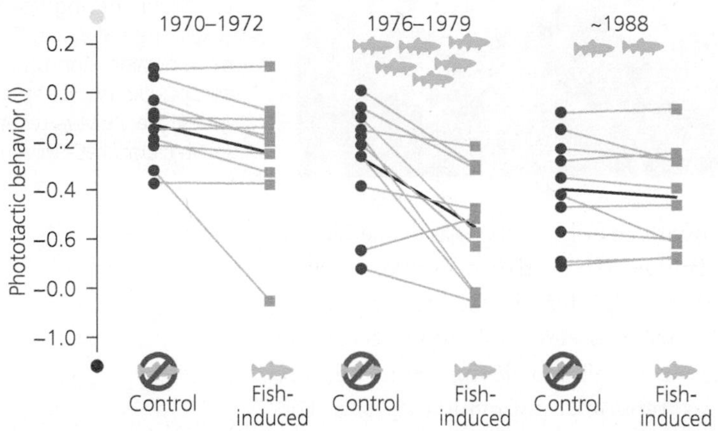

Figure 10.23 Evolution of phototactic behavior in *Daphnia magna* Connected symbols show behavior of genotypes in the absence versus presence of chemicals released by fish; black lines show means across clones. The first sample is from before Oud Heverlee Pond was stocked with plankivorous fish, the middle sample from the era of heavy stocking, and the last from a time of reduced stocking. Redrawn from Cousyn et al. (2001).

The results appear in **Figure 10.23**. As predicted, the water flea population in Oud Heverlee changed over time. Clones from the period of heaviest fish stocking show the greatest shift in behavior across environments. They stay out of the light when they smell predators.

Phenotypic plasticity is widespread, and perhaps underappreciated as an adaptation. As Theodosius Dobzhansky pointed out in 1937 (page 170), "Selection deals not with the genotype as such, but with its dynamic properties, its reaction norm, which is the sole criterion of fitness in the struggle for existence."

10.6 Trade-Offs and Constraints

It is impossible for any population of organisms to evolve optimal solutions to all selective challenges at once. We have mentioned examples of trade-offs in passing. In Section 10.4, for example, we noted that large testes help bats win at sperm competition but appear to impose metabolic costs that lead to the evolution of smaller and less energetically demanding brains. In an earlier passage (Chapter 3), we lamented the male mosquitofish whose large gonopodium entices mates but slows his escape from predators. In this section, we explore additional factors that limit adaptive evolution. These include trade-offs, functional constraints, and lack of genetic variation.

It is impossible to build a perfect organism. Organismal design reflects a compromise among competing demands.

Female Flower Size in a Begonia: A Trade-Off

The tropical plant *Begonia involucrata* is **monoecious**—that is, there are separate male and female flowers on the same plant. The flowers are pollinated by bees. As the bees travel among male flowers gathering pollen, they sometimes also transfer pollen from male flowers to female flowers. The male flowers offer the bees a reward, in the form of the pollen itself. The female flowers offer nothing; instead they get pollinated by deceit (Ågren and Schemske 1991). Not surprisingly, bees make more and longer visits to male flowers than to female flowers.

(a)

(b)

Figure 10.24 Begonia involucrata (a) Both male (left) and female flowers feature white or pinkish petaloid sepals framing yellow anthers or stigmas. Stigmas of females resemble anthers. (b) This inflorescence, or stalk, is unusual in having flowers of both sexes open at once. Typically, the male flowers open first. Courtesy of Douglas W. Schemske, Michigan State University, and Jon Ågren, Uppsala University.

The female flowers resemble the male flowers in color, shape, and size **(Figure 10.24a)**. This resemblance is presumably adaptive. Given that bees avoid female flowers in favor of male flowers, the rate at which female flowers are visited should depend on how closely they mimic male flowers. The ability to attract pollinators should, in turn, influence fitness through female function, because seed set is limited by pollen availability. Presumption is not evidence, however. There are other possibilities.

(a) Stabilizing selection hypothesis

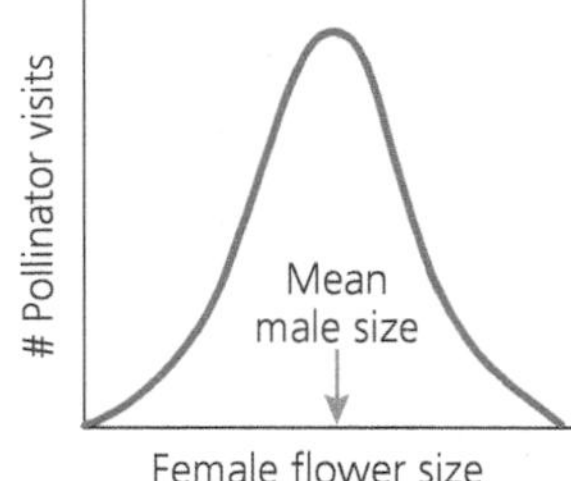

(b) Directional selection hypothesis

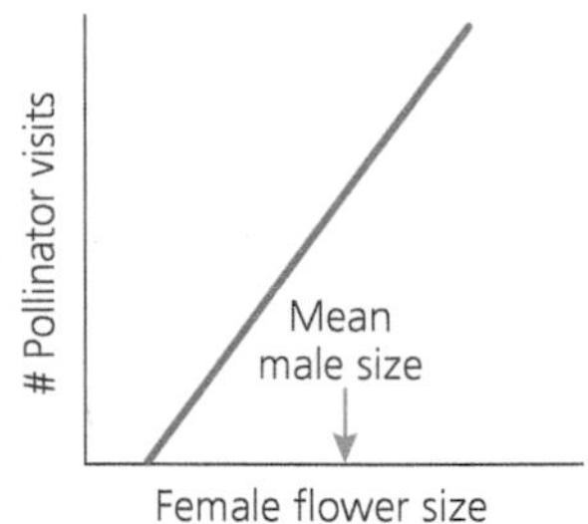

Figure 10.25 Two hypotheses about the pattern of selection on female flower size in *Begonia involucrata*

Doug Schemske and Jon Ågren (1995) sought to distinguish between two hypotheses about how bees might select on female flower size:

Hypothesis 1: The more closely female flowers mimic typical male flowers, the more often they will trick bees into visiting. Selection on female flowers is stabilizing, and the best phenotype for females is identical to the mean phenotype of males **(Figure 10.25a)**.

Hypothesis 2: The more closely female flowers mimic the most rewarding male flowers, the more often they will succeed in duping bees. If larger male flowers offer bigger rewards, then selection on female flowers is directional, and bigger flowers are always favored over smaller flowers (Figure 10.25b).

Schemske and Ågren made artificial flowers of three different sizes **(Figure 10.26a)**, arrayed equal numbers of each in the forest, and watched to see how often bees approached and visited them. The results were clear: The larger the flower, the more bee approaches and visits it attracted (Figure 10.26b). Selection by bees on female flowers is strongly directional.

Taken at face value, this result suggests that female flower size in *Begonia involucrata* is maladaptive. Selection by bees favors larger flowers, yet female flowers are no bigger than those of males. Why are female flowers not huge? One solution to this paradox is that *B. involucrata* simply lacks genetic variation for female

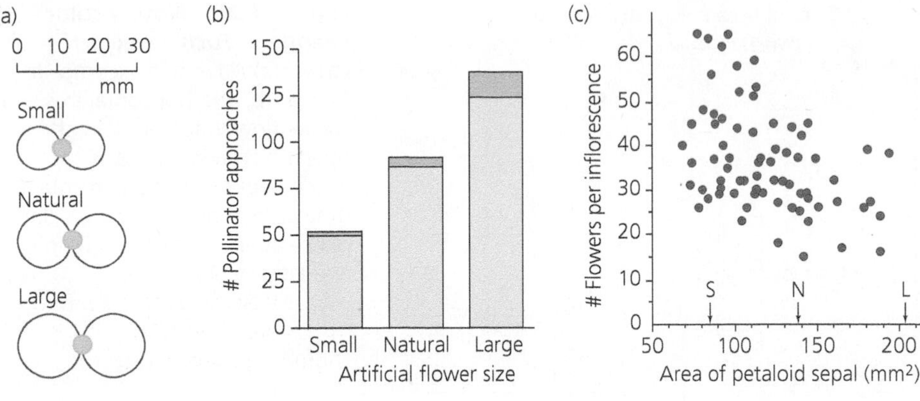

Figure 10.26 Selection on female flower size in *Begonia involucrata* (a) Three sizes of artificial flowers. (b) Pollinator preference as a function of flower size. Tan represents bees that approached artificial flowers; orange represents pollinators that actually visited. (c) A statistically significant trade-off between flower size and number of female flowers per inflorescence. From Schemske and Ågren (1995).

flowers that are substantially larger than male flowers. Schemske and Ågren have no direct evidence on this suggestion; *B. involucrata* is a perennial that takes a long time to reach sexual maturity, so quantitative genetic experiments are difficult.

Another solution is that focusing on individual female flowers gives too narrow a view of selection. Schemske and Ågren expanded their focus from individual flowers to inflorescences (see Figure 10.24b). When the researchers measured the size and number of the female flowers on 74 inflorescences, they discovered a trade-off. The larger the female flowers, the fewer flowers there are (Figure 10.26c). Such a trade-off makes intuitive sense. If an individual plant has a finite supply of energy and nutrients to invest in flowers, it can slice this pie into a few large pieces or many small pieces—but not into many large pieces. Inflorescences with more flowers may be favored by selection for two reasons. First, bees may be more attracted to inflorescences with more flowers. Second, more female flowers means greater potential seed production. Schemske and Ågren hypothesize that female flower size in *B. involucrata* has been determined by a compromise between selection for larger individual flowers and more flowers per inflorescence.

Resources devoted to one body part or function may be resources stolen from another part or function.

Flower Color Change in a Fuchsia: A Constraint

Fuchsia excorticata, also known as the Kotukutuku, is a bird-pollinated tree endemic to New Zealand (Delph and Lively 1989). Its flowers hang downward like bells **(Figure 10.27).**

Figure 10.27 *Fuchsia excorticata* This bird-pollinated tree is native to New Zealand. Why do its flowers change color? By Lynda Delph, Indiana University.

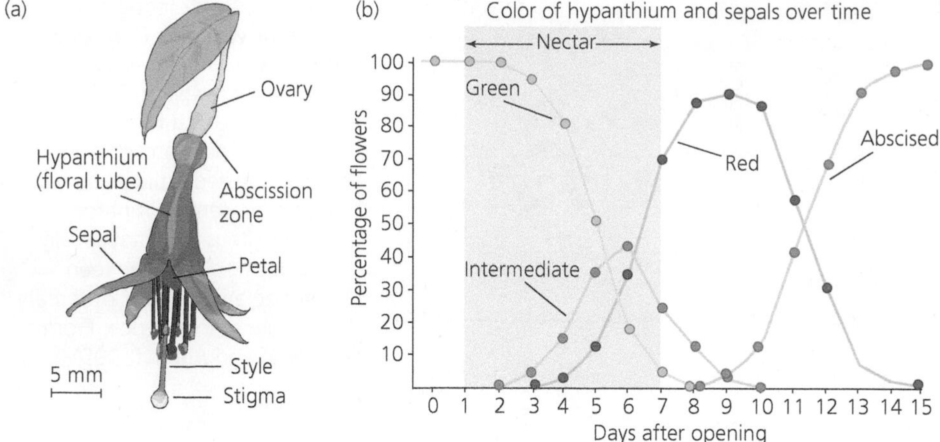

Figure 10.28 Flower color change in *Fuchsia excorticata* (a) A *Fuchsia excorticata* flower. (b) The horizontal axis shows flower age, in days after opening. The vertical axis and graph lines show the percentage of flowers that are in each color phase at each age. From Delph and Lively (1989).

Because a Kotukutuku flower hangs downward, its ovary is at the top **(Figure 10.28a)**. The body of the bell consists of the hypanthium, or floral tube, and the sepals. The style resembles an elongated clapper. It is surrounded by shorter stamens and a set of reduced petals.

The hypanthium and sepals are the most conspicuously showy parts of the flower. They remain green for about 5.5 days after the flower opens, then begin to turn red (Figure 10.28b). The transition from green to red lasts about 1.5 days, at the end of which the hypanthium and sepals are fully red. The red flowers remain on the tree for about five days. The red flowers then separate from the ovary at the abscission zone and drop from the tree.

Pollination occurs during the green phase and into the intermediate phase, but it is complete by the time the flowers are fully red. The flowers produce nectar on days 1 through 7 (Figure 10.28b). Most flowers have exported more than 90% of their pollen by the end of that time. The stigmas are receptive to pollen at least until the second day of the fully red phase, but rarely does pollen arriving after the first day of the red phase actually fertilize eggs. Not surprisingly, bellbirds and other avian pollinators strongly prefer green flowers and virtually ignore nectarless red flowers (Delph and Lively 1985).

Why do the flowers of this tree change color? A general answer, supported by research in a variety of plants, is that color change serves as a cue to pollinators, alerting them that the flowers are no longer offering a reward (for a review, see Delph and Lively 1989). By paying attention to this cue, pollinators can increase their foraging efficiency; they do not waste time looking for nonexistent rewards. Individual plants benefit in return, because when pollinators forage efficiently they also transfer pollen efficiently. They do not deposit viable pollen on unreceptive stigmas, and they do not deposit nonviable pollen on receptive stigmas.

This answer is only partially satisfying, however. Why does *F. excorticata* not just drop its flowers immediately after pollination is complete? Dropping the flowers would give an unambiguous signal to pollinators that a reward is no longer being offered, and it would be metabolically cheaper than maintaining the red flowers for several days. Retention of the flowers beyond the time of pollination seems maladaptive.

Lynda Delph and Curtis Lively (1989) consider two hypotheses for why *F. excorticata* keeps its flowers (and changes them to red) instead of just letting them drop. The first is that red flowers may still attract pollinators to the tree displaying them, if not to the red flowers themselves. Once drawn to the tree, pollinators could then forage on the green flowers still present. Thus, retention of the red flowers could increase the overall pollination efficiency of the individual tree retaining them.

If this hypothesis is correct, then green flowers surrounded by red flowers should receive more pollen than green flowers not surrounded. Delph and Lively tested this prediction by removing red flowers from some trees but not from others, and from some branches within trees but not from others. The researchers then compared the amount of pollen deposited on green flowers in red-free trees and branches versus red-retaining trees and branches. They found no significant differences. The pollinator-attraction hypothesis does not explain the retention of the red flowers in *F. excorticata.*

The second hypothesis Delph and Lively consider is that a physiological constraint prevents *F. excorticata* from dropping its flowers any sooner than it does. This physiological constraint is the growth of pollen tubes. After a pollen grain lands on a stigma, the pollen germinates. The germinated pollen grain grows a tube down through the style to the ovary. The pollen grain's two sperm travel through this tube to the ovary, where one of the sperm fertilizes an egg. The growth of pollen tubes takes time, especially in a plant like *F. excorticata*, which has long styles. If the plant were to drop its flowers before the pollen tubes had time to reach the ovaries, the result would be the same as if the flowers had never been pollinated at all.

Delph and Lively pollinated 40 flowers by hand. After 24 hours, they plucked 10 of the flowers, dissected them, and examined them under a microscope to see whether the pollen tubes had reached the ovary. After 48 hours, they plucked and dissected 10 more flowers, and so on. The results appear in **Table 10.2**. It takes about three days for the pollen tubes to reach the ovary.

Table 10.2 Pollen tube growth in *Fuchsia excorticata*

Days since pollination	1	2	3	4
Percentage of 10 flowers with pollen tubes in ovary	0	20%	100%	100%

Source: After Delph and Lively (1989).

This result is consistent with the physiological constraint hypothesis. *F. excorticata* cannot start the process of dropping a flower until about three days after the flower is finished receiving pollen. Dropping a flower involves forming a structure called an abscission zone between the ovary and the flower (Figure 10.28a). The abscission zone consists of several layers of cells that form a division between the ovary and the flower. In *F. excorticata*, the growth of the abscission layer takes at least 1.5 days. The plant is therefore constrained to retain its flowers for at least 4.5 days after pollination ends. In fact, the plant retains its flowers for approximately five days.

Traits or behaviors that would appear to be adaptive may, in fact, be physiologically or mechanically problematic.

Delph and Lively suggest that flower color change in *F. excorticata* is an adaptation that evolved to compensate for the physiological constraints that necessitate flower retention. Given that the plant had to retain its flowers, selection favored individuals offering cues that allow their pollinators to distinguish the receptive versus unreceptive flowers on their branches. The pollinators deposit the incoming pollen onto receptive stigmas only, and they carry away only outgoing pollen that is viable.

Host Shifts in an Herbivorous Beetle: Constrained by Lack of Genetic Variation?

In several previous chapters, we have noted that genetic variation is the raw material for evolution by natural selection. Because natural selection is the process that produces adaptations, genetic variation is also the raw material from which adaptations are molded. Conversely, populations of organisms may be prevented from evolving particular adaptations simply because they lack the necessary genetic variation to do so.

Here is an extreme example: Pigs have not evolved the ability to fly. We can imagine that flying might well be adaptive for pigs. It would enable them to escape from predators and to travel farther in search of their favorite foods. Pigs do not fly, however, because the vertebrate developmental program lacks genetic variation for the growth of both a trotter and a wing from the same shoulder. Other vertebrates have evolved the ability to fly, of course. But in bats and in birds, the developmental program has been modified to convert the entire forelimb from a leg to a wing; in neither group does an entirely new limb sprout from the body. Too bad for pigs.

Pig flight makes a vivid example, but in the end it is a trivial one. The wished-for adaptation is unrealistic. Douglas Futuyma and colleagues sought to determine whether lack of genetic variation has constrained adaptation in a realistic and meaningful example (Funk et al. 1995; Futuyma et al. 1995; references therein).

Futuyma and colleagues studied host plant use by herbivorous leaf beetles in the genus *Ophraella*. Among these small beetles, each species feeds, as larvae and adults, on the leaves of one or a few closely related species of composites (plants in the sunflower family, the Asteraceae). Each species of host plant makes a unique mixture of toxic chemicals that serve as defenses against herbivores. For the beetles, the ability to live on a particular species of host plant is a complex adaptation that includes the ability to recognize the plant as an appropriate place to feed and lay eggs, as well as the ability to detoxify the plant's chemical defenses.

An estimate of the phylogeny for 12 species of leaf beetle appears in **Figure 10.29**. The figure also lists the host plant for each beetle species. The evolutionary history of the beetle genus has included several shifts from one host plant to another. Four of the host shifts were among relatively distantly related plant species.

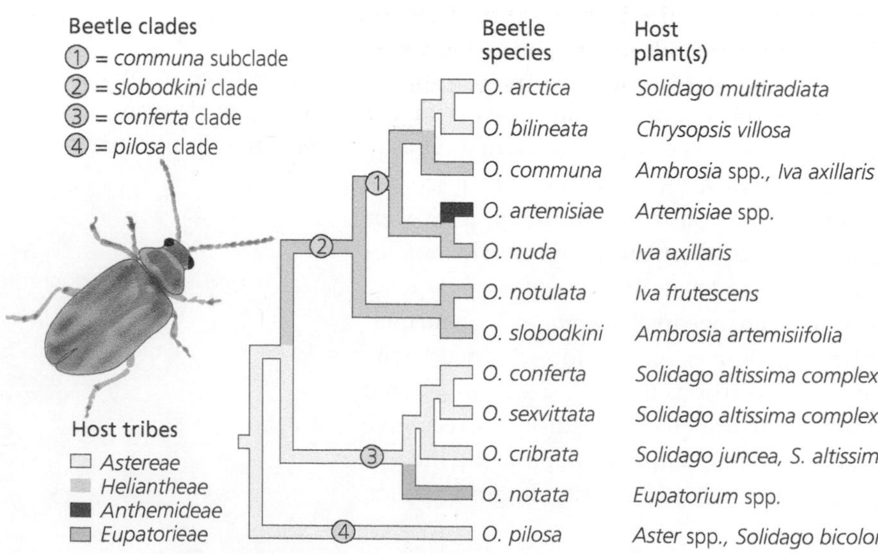

Figure 10.29 Phylogeny of the leaf beetles, genus *Ophraella* The numbers on the tree define the major branches (clades) of beetles. The shading of branches indicates the tribes of host species. The evolutionary history of the beetle genus has included four host shifts across tribes. From Funk et al. (1995); see also Futuyma et al. (1995).

These involved switches from a plant in one tribe of the Asteraceae to a plant in another tribe and are indicated in the figure by changes in the shading of the phylogeny. Other shifts involved movement to a new host in the same genus as the ancestral host, or in a genus closely related to that of the ancestral host.

Each combination of a beetle species and the host plant used by one of its relatives represents a plausible evolutionary scenario for a host shift that might have happened, but did not. For example, the beetle *Ophraella arctica* might have switched to the host *Iva axillaris*. Futuyma and colleagues have attempted to elucidate why some host shifts have actually happened while others have remained hypothetical.

Here are two hypotheses:

Hypothesis 1: All host shifts are genetically possible. That is, every beetle species harbors sufficient genetic variation in its feeding and detoxifying mechanisms to allow at least some individuals to feed and survive on every potential host species. If a few individuals can feed and survive, they can be the founders for a new population of beetles that will evolve to become well adapted to the new host. Because all host shifts are genetically possible, the pattern of actual host shifts has been determined by ecological factors and by chance. Ecological factors might include the abundance of the various host species within the geographic ranges of the beetle species and the predators and competitors associated with each host species.

Hypothesis 2: Most host shifts are genetically impossible. That is, most beetle species lack sufficient genetic variation in their feeding and detoxifying mechanisms to allow any individuals to feed and survive on any but a few of the potential host species. The pattern of actual host shifts has been largely determined by what was genetically possible. Genetically possible host shifts have happened; genetically impossible host shifts have not.

We have presented these hypotheses as mutually exclusive. In fact, the truth is almost certainly that the actual pattern of host shifts has resulted from a mixture of genetic constraints, ecological factors, and chance. What Futuyma and colleagues were looking for was concrete evidence that genetic constraints have been at least part of the picture.

Futuyma and colleagues used a quantitative genetic approach (see Chapter 9) to determine how much genetic variation the beetles harbor for feeding and surviving on other potential hosts. The researchers examined various combinations of four of the beetle species listed in Figure 10.29 with six of the host plants. Their tests revealed little genetic variation in most beetle species for feeding and surviving on most potential host species. In 18 of 39 tests of whether larvae or adults of a beetle species would recognize and feed on a potential host plant, the researchers found no evidence of genetic variation for feeding. In 14 of 16 tests of whether larvae could survive on a potential host plant, the researchers found no evidence of genetic variation for survival. These results suggest that hypothesis 2 is at least partially correct. Many otherwise plausible host shifts appear to be genetically impossible.

Futuyma and colleagues performed an additional test of hypothesis 2 by looking for patterns in their data on genetic variation for larval and adult feeding. If hypothesis 2 is correct, then a beetle species is more likely to show genetic variation for feeding on a potential new host if the new host is a close relative of the beetle's present host.

Table 10.3 Summary of tests for genetic variation in larval or adult feeding on potential host plants

(a) Tests for genetic variation in larval or adult feeding, by relationship among host plants

	Genetic Variation?	
Beetle tested for feeding on a plant that is...	**Yes**	**No**
...in the same tribe as the beetle's actual host	7	1
...in a different tribe than the beetle's actual host	14	17

Conclusion: Genetic variation for feeding is more likely to be found when a beetle is tested on a potential host that is closely related to its actual host.

(b) Tests for genetic variation in larval or adult feeding, by relationship among beetles

	Genetic Variation?	
Beetle tested for feeding on a plant that is...	**Yes**	**No**
...the host of a beetle in the same major clade	12	4
...the host of a beetle in a different major clade	9	14

Conclusion: Genetic variation for feeding is more likely to be found when a beetle is tested on a potential host that is the actual host of a closely related beetle.

Source: From Table 7 in Futuyma et al. (1995).

Futuyma et al.'s data confirm this prediction **(Table 10.3a)**.

Likewise, if hypothesis 2 is correct, then a beetle species is more likely to show genetic variation for feeding on a potential new host if the new host is the actual host of one of the beetles' close relatives.

Futuyma et al.'s data also confirm this prediction (Table 10.3b).

Futuyma and colleagues conclude that hypothesis 2 is at least partially correct. The history of host shifts in the beetle genus *Ophraella* has been constrained by the availability of genetic variation for evolutionary change.

Populations sometimes lack the genetic variation that would provide the raw material to evolve particular adaptations.

Host Shifts in Feather Lice: Constrained by Dispersal Ability?

In the study we have just discussed, Futuyma and colleagues sought to show that host shifts are sometimes constrained by lack of genetic variation. The alternative explanations for why some host shifts have happened and others have not are ecological factors and chance. Dale Clayton and Kevin Johnson (2003) have identified a case in which shifts appear constrained by an ecological factor.

Clayton and Johnson analyzed the history of host shifts in the feather lice that infest doves. These ectoparasites include lice that live on wing feathers (genus *Columbicola*) and lice that live on body feathers (genus *Physconelloides*). **Figure 10.30a** compares the evolutionary trees for several dove species versus their wing-feather lice. The phylogenies are not congruent, indicating that wing-feather lice have switched host species frequently. Figure 10.30b compares the evolutionary trees for the same dove species versus their body-feather lice. This time the phylogenies are highly congruent, indicating that body-feather lice have not switched host species. Instead, they have simply gone along for the ride, speciating only when their hosts have speciated.

Why have wing-feather lice switched host species often while body-feather lice have not? Experiments in which Clayton and colleagues (2003) transferred

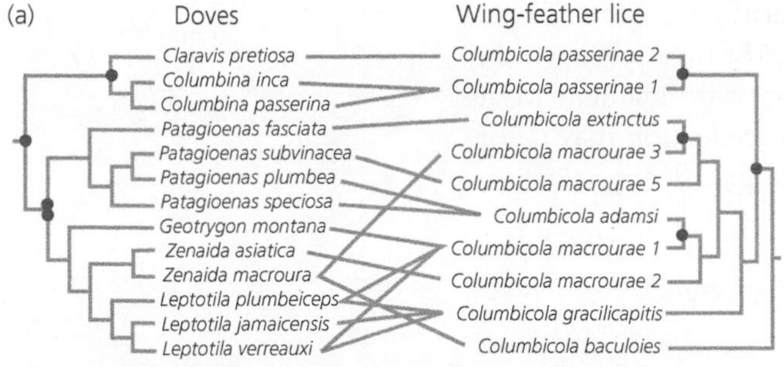

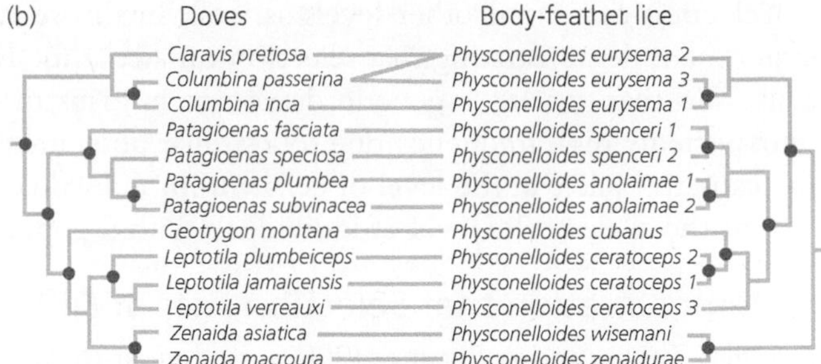

Figure 10.30 Phylogenetic congruence and discord for doves and their feather lice (a) The tree on the left is for doves; the tree on the right is for their wing-feather lice (genus *Columbicola*). Lines connect the parasite species to the bird species they infect. The many crossing lines indicate frequent host shifts in the evolutionary history of the lice. (b) The tree on the left is for doves; the tree on the right is for their body-feather lice (genus *Physconelloides*). Lines connect the parasite species to the bird species they infect. The absence of crossing lines indicates that the lice have not changed hosts. Instead, they have gone along for the ride, diverging when their hosts have. Redrawn from Clayton and Johnson (2003).

feather lice to novel hosts suggest that many host switches are genetically possible. Transplanted lice attach and feed on novel hosts. They can also evade the host's preening as long as their new host is similar in body size to their native host. Instead of being constrained by lack of variation for the ability to survive on novel hosts, Clayton and Johnson think that body-feather lice simply have fewer chances to switch host species. This is because body-feather lice disperse among individual hosts less readily than wing-feather lice do. Field observations by Noah Whiteman and colleagues (2004) support this contention. These researchers looked for wing and body lice from Galápagos doves on Galápagos hawks. The two parasite species are equally common on doves, their native host, but on hawks dove-wing lice are much more common than dove-body lice.

One way feather lice move from one host to another is via direct bodily contact between the two birds. Another way is by hitching a ride on the legs of a parasitic hippoboscid fly, as shown in **Figure 10.31**. The flies are less host-specific than lice, so a stowaway louse may find itself deposited on a novel host. Published records suggest that wing-feather lice hitch rides on flies much more often than body-feather lice. Apparently the reason body-feather lice have so rarely switched host species is that they could not get a lift.

Figure 10.31 Dispersal via a lousy fly Wing-feather lice hitching a ride on a parasitic fly. After Clayton et al. (2004).

In this section and the previous one, we have examined complications of organismal form and function that must be taken into account when studying adaptation. In the next section, we consider another kind of complication that must sometimes be taken into account—a complication in the action of natural selection itself.

10.7 Selection Operates on Different Levels

In the examples we have discussed, both in this chapter and in the book, we have been concerned with natural selection operating at the level of individuals within populations. At this level it is the birth, reproduction, and death of individual organisms that determines which alleles become common and which disappear. If an allele influences phenotype such that the average individual carrying it has greater than average reproductive success, then the allele's frequency will rise; otherwise, its frequency will fall.

Selection can act at other levels as well. In this section, we first describe an experiment demonstrating that selection can act at the level of organelles within cells. We then discuss a genetic disease in humans that may be maintained at unexpectedly high frequency due to a similar phenomenon. Selection may favor the causative allele at the level of cells within individuals despite strong selection against the allele at the level of individuals within populations.

A Demonstration that Selection Acts at Different Levels

Douglas Taylor and colleagues (2002) used yeast (*Saccaromyces cerevisiae*) and their mitochondria to demonstrate that selection can act simultaneously at different levels. The researchers chose yeast because normal yeast cells can harvest energy in two ways: by fermentation and by respiration. This means that for yeast, unlike for most eukaryotes, the ability to respire is not essential for life.

Respiration in yeast cells, as in other eukaryotes, is carried out by mitochondria. Mitochondria occasionally sustain large deletions in their genomes that render them unable to respire. These non-respiratory mitochondria are intracellular parasites. They extract energy and material from their host cells and provide nothing in return. For most eukaryotic cells, having exclusively parasitic mitochondria would be fatal. For yeast cells, however, it is merely disadvantageous. It limits the yeast cells to harvesting energy by fermentation, but it does not kill them.

Taylor and colleagues established yeast populations in which the founding individuals contained both normal and parasitic mitochondria. That is, each cell in the yeast population was itself home to a genetically variable population of mitochondrial genomes. These mitochondrial genomes replicate independently of the host cell's nuclear genome. Some mitochondrial genomes may replicate more rapidly than others. This means that the population of mitochondrial genomes within a cell can evolve by natural selection, just as any other population can. Mitochondrial genomes that replicate rapidly will become common; mitochondrial genomes that replicate slowly will become rare.

Organisms harbor populations of cells, organelles, and nucleotide sequences. Selection can operate within these populations.

At the level of mitochondrial genomes within a yeast cell, selection favors parasites over normal mitochondria. This is because parasitic mitochondrial genomes can replicate faster. If we stayed inside a yeast cell and tracked the evolution of its mitochondrial population, we would expect parasitic mitochondria to inexorably increase in frequency.

At the level of yeast cells within a petri dish, however, selection favors the ability to respire. Yeast cells that can respire can harvest energy more quickly and thus replicate faster. If we stayed inside a petri dish and tracked the evolution of its yeast population, we would expect respiration-competent yeast cells to inexorably increase in frequency.

Selection at the level of mitochondria within yeast cells is thus in opposition to selection at the level of yeast cells within petri dishes. What is the ultimate outcome? That depends on the relative intensities of selection at the two different levels.

Taylor and colleagues maintained their yeast cultures for 150 generations at three population sizes: small (about 10 yeast cells), medium (about 250 yeast cells), and large (about 18,000 cells). In small populations, selection among yeast cells is largely insignificant. Sampling error during propagation is the primary determinant of which yeast lineages persist and which disappear. In small populations, therefore, Taylor and colleagues predicted that selection at the level of mitochondria within cells would dominate. They expected that at the end of their experiment, most yeast cells in their small cultures would contain exclusively parasitic mitochondria. In large populations, however, selection among yeast cells is important. Sampling error during propagation is negligible, and speed of growth and replication is the primary determinant of which yeast lineages persist and which disappear. In large populations, therefore, the researchers expected that yeast cells containing exclusively parasitic mitochondria would be rare.

As a control, the researchers established yeast cultures in which the founding individuals contained both chloramphenicol-susceptible and chloramphenicol-resistant mitochondria. Chloramphenicol susceptibility was selectively neutral under the conditions of the experiment. Averaged across populations, Taylor and colleagues expected the frequency of yeast cells containing exclusively susceptible mitochondria to remain at intermediate frequency.

The results, shown in **Figure 10.32**, confirm the researchers' predictions. In small populations most yeast cells contained exclusively parasitic mitochondria, whereas in large populations few yeast cells contained exclusively parasitic mitochondria. In control cultures the frequency of yeast cells containing exclusively chloramphenicol-susceptible mitochondria, averaged across populations, remained at intermediate levels.

This experiment demonstrates that when selection among yeast cells is relatively weak, selection among mitochondria within yeast cells can lead to the fixation of traits that decrease the mean fitness of the yeast population. Had we not recognized that selection acts on different levels, we might have mistakenly

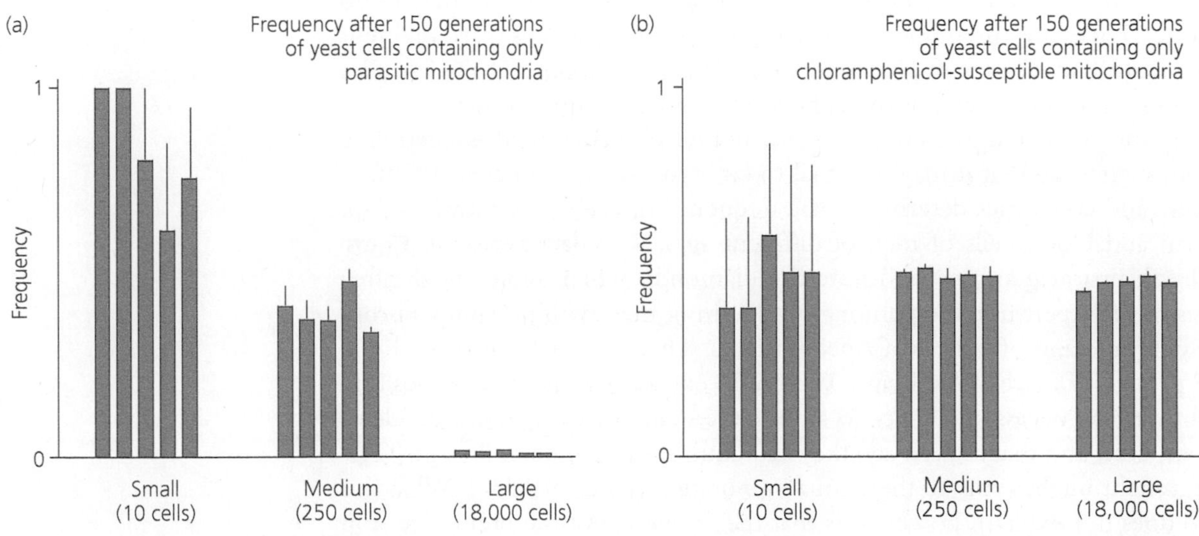

Figure 10.32 Selection at the level of cells in populations versus selection at the level of mitochondria inside cells (a) Each bar represents the average of five experimental populations started with yeast cells containing a mixture of normal versus parasitic mitochondria. Among mitochondria within yeast cells, selection favors parasites, because they replicate faster. This selective advantage was constant across experiments. Among yeast cells within populations, selection favors yeast containing normal mitochondria, because they can harvest energy by respiration as well as fermentation. This selective advantage varies among yeast cultures maintained at different population sizes; it is weakest in small populations and strongest in large populations. Parasitic mitochondria thrive in small yeast populations but fall to low frequency in large yeast populations. (b) Each bar represents the average of four or five control populations started with yeast cells containing a mixture of chloramphenicol-resistant versus chloramphenicol-susceptible mitochondria. Chloramphenicol resistance is selectively neutral at both levels of selection. From Taylor et al. (2002).

concluded that loss of the ability to respire is somehow adaptive for yeast cells living in small populations. Instead, we can see that a trait that is maladaptive for the organism that carries it may be driven to fixation by selection at lower levels. Taylor and colleagues believe their work will also provide insight into the progression of degenerative genetic diseases associated with the accumulation, within tissues, of mutant mitochondria.

Just as a yeast cell is home to an evolving population of self-replicating mitochondrial genomes, an animal is home to an evolving population of self-replicating cellular genomes. Recent research has suggested that evolution of cell populations can explain puzzling features of a human genetic disease.

Multilevel Selection in Apert Syndrome?

Apert syndrome is a genetic disease caused by a mutation in the gene for fibroblast growth factor receptor 2 (FGFR2). The manifestations are severe, including premature fusion of the suture joints in the skull, facial malformation, and fusion of the fingers and toes. Given that the condition is dominant and affected individuals have low fitness, it is not surprising that most cases are caused by new mutations. Less expected is that the mutant allele virtually always comes from the father. The risk of Apert syndrome increases with the father's age, and the incidence among newborns suggests a rather high mutation rate.

To better understand these facts, Anne Goriely and colleagues (2003) used molecular techniques to assess the frequency of mutant alleles among the sperm produced by men of different ages. The most common Apert mutation is a single-nucleotide substitution at nucleotide 755, the middle position in codon 252, in the gene for FGFR2. The normal codon reads TCG, specifying the amino acid serine. There are three possible substitutions for C, yielding three different codons. TGG encodes tryptophan and causes Apert syndrome. TTG encodes leucine and results in either a normal phenotype or a condition known as Crouzon syndrome. TAG is a stop codon and has not been documented as a germline mutation, suggesting that it may be lethal to embryos (see Arman et al. 1999).

Goriely and colleagues determined the frequency of each substitution among the sperm and blood cells of men of different ages. The data appear in **Figure 10.33**. Look first at graph (a) and note the diamonds, which represent the frequencies of the Apert mutation among the sperm of men with no family history of the syndrome and of fathers of Apert patients. There is much variation, but a general pattern of increase with age. We might explain the increase by positing that the mutation occassionally occurs among the stem cell lineages that divide to produce sperm, and that as men age they accumulate more mutant lineages. The same argument might apply to the Crouzon mutation (Figure 10.33b). What this scenario does not explain, however, is that the frequency of the premature stop mutation does not increase with age (Figure 10.33c). It also fails to explain why none of the frequencies change with age in blood cells, as shown by the squares.

Selection at the level of cells, organelles, or sequences may be in opposition to selection at the level of whole organisms.

Goriely and colleagues think there is a better explanation, one that accounts for all of these puzzling features (see also Crow 2003). They suggest that all three substitutions occur only rarely in the stem cell lineages that yield both sperm and blood. When either the Apert mutation or the Crouzon mutation occurs in spermatogonia, however, it enables the cells that carry it to divide more rapidly, essentially giving rise to small cancers that continue to produce sperm. Because the mutant cells enjoy higher reproductive success, the population of spermatogonia—which now harbors genetic variation—evolves. The frequency of mutant

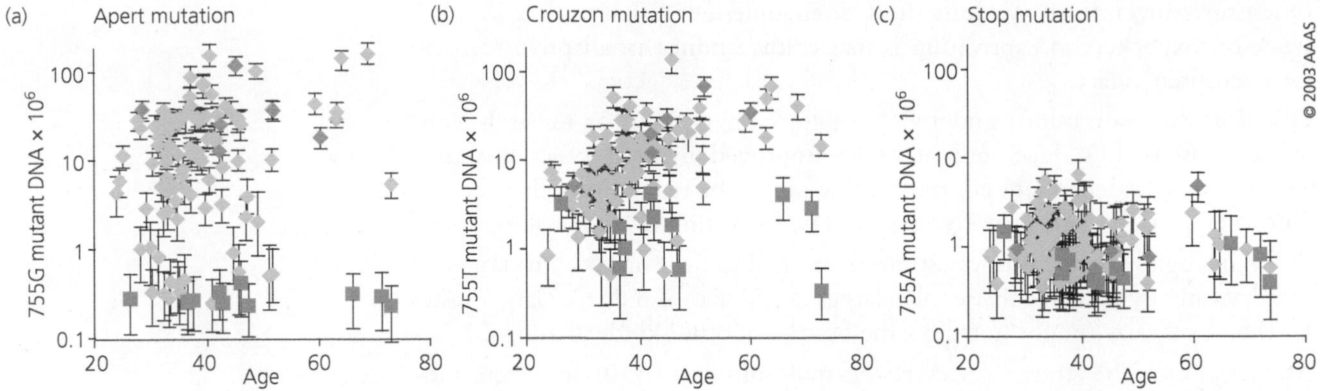

Figure 10.33 Change with men's age in the frequencies of mutations among cells in two tissues (a) The frequency of the Apert syndrome mutation rises among sperm but not blood cells. (b) The frequency of the Crouzon syndrome mutation likewise rises among sperm but not blood cells. (c) The frequency of a nonsense mutation does not change with age. All three mutations are single-nucleotide substitutions at position 755 in the gene for FGFR2. These patterns make sense if the Apert and Crouzon mutations increase proliferation in the stem cells that make sperm. From Goriely et al. (2003).

From "Evidence for selective advantage of pathogenic FGFR2 mutations in the male germ line," *Science* 301: 643–646, Figure 3. Reprinted with permission from AAAS.

stem cells, and thus the frequency of mutant sperm, increases over time. Consistent with this explanation is that the Apert and Crouzon mutations are both gain-of-function mutations. They enable the FGFR2 protein to bind its ligands more strongly and to bind a greater variety of molecules. Also consistent is that *FGFR2* is expressed in spermatogonial stem cells (Goriely et al. 2005).

If Goriely and colleagues are right, then the frequency of Apert syndrome in the human population is determined by a balance between opposing patterns of selection acting at different levels. When the Apert mutation occurs within the testes of an individual man, selection acting at the level of cells within tissues causes the frequency of the mutation to increase. When babies are born with Apert syndrome, however, selection at the level of individuals within populations causes the frequency of the mutation to fall.

10.8 Strategies for Asking Interesting Questions

Learning how to ask good questions is as important as learning how to answer them.

We began this chapter with a review of approaches evolutionary biologists use when testing hypotheses about organismal form and function. However, testing a hypothesis is the second half of a good research project. The first half is formulating a hypothesis in the first place. Formulating hypotheses worthy of testing means asking interesting questions, then making educated guesses about the answers. We close the chapter with a brief list of strategies for asking good questions about evolution:

- Study natural history. Descriptive studies can lead to the discovery of new patterns that need explanation. Some things in nature just leap out and demand explanation, such as the oxpecker's odd habits, or the wing-waving display of *Zonosemata*. Some of the most compelling science happens when a researcher simply picks an organism and sets out to learn about it.

- Question conventional wisdom. It is often untested. What makes Weeks's work on oxpeckers so captivating is that it undermines an adaptive scenario long accepted as fact.
- Question the assumptions underlying a popular hypothesis or research technique. Felsenstein's development of an improved method of comparative analysis, using independent contrasts, grew out of the recognition that the traditional approach to comparative research was violating its own assumptions.
- Draw analogies that transfer questions from field to field or taxon to taxon. If fruit bats and flying foxes that evolve larger group sizes also evolve larger testes for their body size, might not the same be true of other kinds of animals?
- Ask why not. The studies reviewed on trade-offs and constraints were motivated by researchers who thought their study organisms were failing to do something that might be adaptive.

SUMMARY

Among the major activities of evolutionary biology is analyzing the form and function of organisms to determine whether and why particular traits are adaptive. To establish that a trait is adaptive, researchers must formulate hypotheses about how the trait is used and why individuals possessing the trait have higher fitness than those lacking it. Then, because no hypothesis should be accepted simply because it is plausible, researchers must put their hypotheses to test. Researchers test hypotheses by using them to make predictions, then collecting data to determine whether the predictions are correct.

Researchers use a variety of approaches in collecting data to test hypotheses. The most powerful method is the controlled experiment. Controlled experiments involve groups of organisms that are identical but for a single variable of interest. The experimental variable can then be confidently identified as the cause of any differences in survival and reproductive success among the groups. When controlled experiments are impractical, careful observational studies can yield data valuable for testing hypotheses. Finally, comparisons among species can be used to confirm or refute predictions, so long as researchers take into account the shared evolutionary history of the species under study.

When analyzing adaptations, we do well to keep in mind that organisms are complicated. Individuals may be phenotypically plastic, so that genetically identical individuals reared in different environments have different phenotypes. The function of a particular trait may change over evolutionary time, and it may reflect a compromise among competing environmental or physiological demands. Finally, populations may simply lack the genetic variation required to become perfectly adapted to their environments.

We also do well to keep in mind that selection can act at multiple levels. When selection at the level of organisms within populations is relatively weak, selection at the level of organelles within cells or cells within tissues can drive the evolution of traits that are maladaptive for the organisms that carry them. These and other complications are the subjects of current research by evolutionary biologists.

QUESTIONS

1. Describe in your own words the difference between an experimental study, an observational study, and a comparative study. What sorts of questions are they each suited for (i.e., why don't researchers always use the experimental method)? Give an example of each type of study from this chapter.
2. What were Futuyma and colleagues' two hypotheses to explain why leaf beetles have not colonized all possible species of host plants? What did the researchers do to test the hypotheses? How do their results illuminate the general question of whether all traits in all organisms are adaptive?

3. For which of the following studies would you recommend the use of Felsenstein's method of phylogenetically independent contrasts? Why?
   a. A comparison of feather parasite burden and beak shape in different species of birds.
   b. An experiment that tests whether birds whose beak shapes are experimentally altered will end up with greater parasite loads (similar to Clayton et al.'s study).
   c. An observational study that measures the correlations among beak shape of individual birds with their preening behavior, and with their parasite loads.
4. What is an evolutionary trade-off? Why do they occur? Give two examples. How does the occurrence of trade-offs illuminate the general question of whether all traits are adaptive?
5. What is an evolutionary constraint? Why do they occur? Give two examples. How does their occurrence illuminate the general question of whether all traits are adaptive?
6. How does Apert syndrome explain why some traits occur that may be maladaptive?
7. As a review, list all the reasons you can think of that may cause a given trait not to be adaptive, despite the action of positive natural selection on the trait.
8. a. Why was it important that Greene and colleagues tested tephritid flies whose wings had been cut off and then glued back on?
   b. Why did they do the wing-cutting experiments at all? For example, why didn't they just compare intact tephritid flies to houseflies?
   c. Why was it important that the five types of flies were presented to each spider in random order?
9. In Huey et al.'s experiment, snakes often chose thick rocks despite the associated risk of being too cool. Outline two hypotheses for why snakes sometimes choose thick rocks. Are your hypotheses testable? Do both hypotheses assume that the behavioral trait of choosing thick rocks is adaptive?
10. Geckos are unusual lizards in that they are active at night. Describe the difficulties a gecko would face in trying to use behavior to regulate its temperature at night. Would you predict that geckos have an optimal temperature for sprinting that is the same, higher, or lower than that of a typical diurnal lizard? Huey et al. (1989a) found that the geckos they studied had optimal temperatures that are the same as those of typical diurnal lizards (a finding in conflict with the researchers' own hypothesis). Can you think of an explanation?
11. Suppose that fish were introduced into Lake Citadelpark, one of the lakes De Meester studied. What do you predict will happen to phenotypic plasticity in the *Daphnia* of Lake Citadelpark? Outline the observations you would need to make to test your prediction.
12. Throughout this chapter, we have stressed this fundamental question: How can we test whether a given trait is adaptive or not? As a further exercise, think about the costs and benefits of being a certain body size. For example, a mouse can easily survive a 30-foot fall. A human falling 30 feet would probably be injured, and an elephant falling 30 feet would probably be killed. Finally, a recent study of the bone strength of *Tyrannosaurus rex* revealed that if a fast-running *T. rex* ever tripped, it would probably die (Farlow, Smith, and Robinson 1995). Given these costs, why has large body size ever evolved? Can you think of some costs of small body size? How would you test your ideas?
13. Imagine you are an explorer who has just discovered two previously unknown large islands. Each island has a population of a species of shrub unknown elsewhere. On Island A, the shrubs have high concentrations of certain poisonous chemicals in their leaves. On Island B, the shrubs have nonpoisonous, edible leaves. The islands differ in many ways—for instance, Island A has less rainfall and a colder winter than Island B, and it has some plant-eating insects that are not found on Island B. Island A also has a large population of muntjacs, a small tropical deer that loves to eat shrubs. You suspect the muntjacs have been the selective force that has caused evolution of leaf toxins. How could you test this hypothesis? What alternative hypotheses can you think of? What data would disprove your hypothesis, and what data would disprove the other hypotheses?
14. Consider skin color in humans. Does this trait show genetic variation? Phenotypic plasticity? Genotype-by-environment interaction? Give examples documenting each phenomenon. Could phenotypic plasticity for skin color evolve in human populations? How?
15. The example on begonias (Section 10.6) illustrated that organisms are frequently caught between opposing agents of selection. Each of the following examples also illustrates a tug-of-war between several agents of natural selection. For each example, hypothesize about what selective patterns may maintain the trait described and what selective patterns may oppose it.
   a. A male moose grows new antlers, made of bone, each year.
   b. Douglas fir trees often grow to more than 60 feet tall.
   c. A termite's gut is full of cellulose-digesting microorganisms.
   d. Maple trees lose all of their leaves in the autumn.
   e. A male moth has huge antennae, which can detect female pheromones.
   f. A barnacle attaches itself permanently to a rock when it matures.
16. Schemske and Ågren (1995) used artificial flowers instead of real flowers in their experiment (Section 10.6).

What were the advantages of using artificial flowers? (There are at least two important ones.) What were the disadvantages?

17. P1 is a virus that infects bacteria. It often resides for long periods as a plasmid inside a bacterial cell, replicating and being transmitted to descendants of the original cell. Among the genes in P1's genome are two loci that form what is called an addiction module. One of the genes (called doc, which stands for "death on cure") encodes a small protein that is both highly poisonous to bacterial cells and chemically stable. The other gene (called phd, which stands for "prevent host death") encodes a protein that serves as an antidote to the poison. This antidote is chemically unstable because it is degraded by a bacterial protease. If a bacterial cell containing the P1 plasmid divides and produces a daughter cell that does not contain the plasmid, that daughter cell's protoplasm still contains the poison and the antidote. The antidote breaks down quickly, and the poison persists. The daughter cell then dies. Explain why it might be selectively advantageous for the P1 virus to carry an addiction module. How does selection at the level of the virus and its genes affect selection at the level of the bacteria that harbor them? (For more information on this example, see Lehnherr et al. 1993; Lehnherr and Yarmolinsky 1995.)

18. An exercise used in some graduate programs is to have students list 20 questions they would like to answer. Groups of students discuss the questions and help each other sort out which would be most interesting to pursue. There are many criteria for deciding that a question is interesting. Is it new? Does it address a large or otherwise important issue? Would pursuing it lead to other questions? Is it feasible, or would development of a new technique make it feasible? Try this exercise yourself.

EXPLORING THE LITERATURE

19. An important aspect of evaluating scientific papers is to consider other explanations for the data that the authors might have overlooked. See if you can think of alternate explanations for the data presented in the following papers:

Benkman, C. W., and A. K. Lindholm. 1991. The advantages and evolution of a morphological novelty. *Nature* 349: 519–520.

Soler, M., and A. P. Møller. 1990. Duration of sympatry and coevolution between the great spotted cuckoo and its magpie host. *Nature* 343: 748–750.

Finally, see the following review of Soler and Møller's work for an example of how scientific criticism can result in better science by all involved:

Lotem, A., and S. I. Rothstein. 1995. Cuckoo-host coevolution: From snapshots of an arms race to the documentation of microevolution. *Trends in Ecology and Evolution* 10: 436–437.

20. Male sticklebacks sometimes steal eggs from other males' nests to rear as their own. Sievert Rohwer suggested that egg stealing is a courtship strategy. Males often eat eggs out of their own nests, in effect robbing the reproductive investment made by their mates and using the proceeds to fund their own reproductive activities. Females, in consequence, should prefer to lay eggs in nests already containing the eggs of other females, thereby reducing the risk to their own. A female preference for nests already containing eggs would mean that males without eggs in their nests could increase their attractiveness by stealing eggs from other males. See:

Rohwer, S. 1978. Parent cannibalism of offspring and egg raiding as a courtship strategy. *American Naturalist* 112: 429–440.

For a related phenomenon in birds, see:

Gori, D. F., S. Rohwer, and J. Caselle. 1996. Accepting unrelated broods helps replacement male yellow-headed blackbirds attract females. *Behavioral Ecology* 7: 49–54.

21. For a dramatic example of phenotypic plasticity in which an herbivorous insect uses the chemical defenses of its host as a cue for the development of defenses against its predators, see:

Greene, Erick. 1989. A diet-induced developmental polymorphism in a caterpillar. *Science* 243: 643–646.

See this paper for a discussion of an organism's "choice" about when to change phenotype:

Nishimura, Kinya. 2006. Inducible plasticity: Optimal waiting time for the development of an inducible phenotype. *Evolutionary Ecology Research* 8 (3): 553–559.

22. Among the challenges faced by parasites is moving from one host to another. This challenge is a particularly potent agent of selection for parasites in which every individual must spend different parts of its life cycle in different hosts. What adaptations might you expect to find in parasites to facilitate dispersal from host to host? For dramatic examples in which parasites manipulate their hosts' behavior or appearance, see:

Tierney, J. F., F. A. Huntingford, and D. W. T. Crompton. 1993. The relationship between infectivity of *Schistocephalus solidus* (Cestoda) and antipredator behavior of its intermediate host, the three-spined stickleback, *Gasterosteus aculeatus*. *Animal Behavior* 46: 603–605.

Lafferty, K. D., and A. K. Morris. 1996. Altered behavior of parasitized killifish increases susceptibility to predation by bird final hosts. *Ecology* 77: 1390–1397.

Bakker, T. C. M., D. Mazzi, and S. Zala. 1997. Parasite-induced changes in behavior and color make *Gammarus pulex* more prone to fish predation. *Ecology* 78: 1098–1104.

House, P. K., A. Vyas, and R. Sapolsky. 2011. Predator cat odors activate sexual arousal pathways in brains of *Toxoplasma gondii* infected rats. *PLoS One* 6: e23277.

23. Brown-headed cowbirds (*Molothrus ater*) lay their eggs in other birds' nests, a behavior called nest parasitism. When this strategy succeeds, the host birds accept the cowbird egg as one of their own and rear the cowbird chick. When it fails, the host birds recognize the cowbird egg as an imposter and eject it from the nest. Why do any host species accept cowbird eggs in their nests? Given the obvious cost of rearing a chick of another species, acceptance seems maladaptive. Evolutionary biologists have proposed two competing hypotheses to explain why some host species accept cowbird eggs. The evolutionary lag hypothesis posits that species that accept cowbird eggs simply have not yet evolved ejection behavior. Either the host species lack genetic variation that would allow them to evolve ejection behavior, or the host species have been exposed to cowbird nest parasitism only recently and therefore have not had sufficient time for such behavior to evolve. The evolutionary equilibrium hypothesis posits that host species that accept cowbird eggs do so because they face a fundamental mechanical constraint: Their bills are too small to allow them to grasp a cowbird egg, and if they tried to puncture the cowbird egg they would destroy too many of their own eggs in the process. Given this constraint, host species have evolved a strategy that makes the best of a bad situation. Think about how you would test each of the competing hypotheses. Then see:

Rohwer, S., and C. D. Spaw. 1988. Evolutionary lag versus bill-size constraints: A comparative study of the acceptance of cowbird eggs by old hosts. *Evolutionary Ecology* 1988: 27–36.

Rohwer, S., C. D. Spaw, and E. Røskaft. 1989. Costs to northern orioles of puncture-ejecting parasitic cowbird eggs from their nests. *Auk* 106: 734–738.

Røskaft, E., S. Rohwer, and C. D. Spaw. 1993. Cost of puncture ejection compared with costs of rearing cowbird chicks for northern orioles. *Ornis Scandinavica* 24: 28–32.

Sealy, S. G. 1996. Evolution of host defenses against brood parasitism: Implications of puncture-ejection by a small passerine. *Auk* 113: 346–355.

Given that some host species eject cowbird eggs by first puncturing the egg, then lifting it out of the nest, what adaptations would you expect to find in cowbird eggs? Would these adaptations carry any costs? See:

Spaw, C. D., and S. Rohwer. 1987. A comparative study of eggshell thickness in cowbirds and other passerines. *Condor* 89: 307–318.

Picman, J. 1997. Are cowbird eggs unusually strong from the inside? *Auk* 114: 66–73.

24. For additional examples in which evolutionary biologists used a comparative approach employing independent contrasts to address interesting questions, see:

Downes, S. J., and M. Adams. 2001. Geographic variation in antisnake tactics: The evolution of scent-mediated behavior in a lizard. *Evolution* 55: 605–615.

Pearce, E., and R. Dunbar. 2011. Latitudinal variation in light levels drives human visual system size. *Biology Letters* 8: 90–93.

Iwaniuk, A. N., S. M. Pellis, and I. Q. Whishaw. 1999. Brain size is not correlated with forelimb dexterity in fissiped carnivores (Carnivora): A comparative test of the principle of proper mass. *Brain, Behavior, and Evolution* 54: 167–180.

25. For an experiment addressing why humans, who have much-reduced body hair compared to our closest relatives, have not lost our body hair entirely, see:

Dean, I., and M. T. Siva-Jothy. 2011. Human fine body hair enhances ectoparasite detection. *Biology Letters* 8: 358–361.

CITATIONS

Ågren, J., and D. W. Schemske. 1991. Pollination by deceit in a Neotropical monoecious herb, *Begonia involucrata*. *Biotropica* 23: 235–241.

Anderson, J. L., L. Albergotti, et al. 2011. Does thermoregulatory behavior maximize reproductive fitness of natural isolates of *Caenorhabditis elegans*? *BMC Evolutionary Biology* 11: 157.

Arman, E., R. Haffner-Krausz, et al. 1999. Fgfr2 is required for limb outgrowth and lung-branching morphogenesis. *Proceedings of the National Academy of Sciences, USA* 96: 11895–11899.

Berger, J. O., and D. A. Berry. 1988. Statistical analysis and the illusion of objectivity. *American Scientist* 76: 159–165.

Bush, S. E., D. Kim, et al. 2010. Evolution of cryptic coloration in ectoparasites. *American Naturalist* 176: 529–535.

Clayton, D. H., S. E. Bush, et al. 2003. Host defense reinforces host–parasite cospeciation. *Proceedings of the National Academy of Sciences, USA* 100: 15694–15699.

Clayton, D. H., S. E. Bush, and K. P. Johnson. 2004. Ecology of congruence: Past meets present. *Systematic Biology* 53: 165–173.

Clayton, D. H., and K. P. Johnson. 2003. Linking coevolutionary history to ecological process: Doves and lice. *Evolution* 57: 2335–2341.

Clayton, D. H., P. L. M. Lee, et al. 1999. Reciprocal natural selection on host-parasite phenotypes. *American Naturalist* 154: 261–270.

Clayton, D. H., B. R. Moyer, et al. 2005. Adaptive significance of avian beak morphology for ectoparasite control. *Proceedings of the Royal Society of London B* 272: 811–817.

Cousyn, C., L. De Meester, et al. 2001. Rapid, local adaptation of zooplankton behavior to changes in predation pressure in the absence of neutral genetic changes. *Proceedings of the National Academy of Sciences, USA* 98: 6256–6260.

Crow, J. F. 2003. There's something curious about paternal-age effects. *Science* 301: 606–607.

Delph, L. F., and C. M. Lively. 1985. Pollinator visits to floral colour phases of *Fuchsia excorticata*. *New Zealand Journal of Zoology* 12: 599–603.

Delph, L. F., and C. M. Lively. 1989. The evolution of floral color change: Pollinator attraction versus physiological constraints in *Fuchsia excorticata*. *Evolution* 43: 1252–1262.

De Meester, L. 1996. Evolutionary potential and local genetic differentiation in a phenotypically plastic trait of a cyclical parthenogen, *Daphnia magna*. *Evolution* 50: 1293–1298.

Dienes, Z. 2011. Bayesian versus Orthodox statistics: Which side are you on? *Perspectives on Psychological Science* 6: 274–290.

Dobzhansky, T. 1937. *Genetics and the Origin of Species*. New York: Columbia University Press.

Farlow, J. D., M. B. Smith, and J. M. Robinson. 1995. Body mass, bone "strength indicator," and cursorial potential of *Tyrannosaurus rex*. *Journal of Vertebrate Paleontology* 15: 713–725.

Felsenstein, J. 1985. Phylogenies and the comparative method. *American Naturalist* 125: 1–15.

Felsenstein, J. 2008. Comparative methods with sampling error and within species variation: Contrasts revisited and revised. *American Naturalist* 171: 713–725.

Felsenstein, J. 2012. A comparative method for both discrete and continuous characters using the threshold model. *American Naturalist* 179: 145–156.

Freedman, D., R. Pisani, and R. Purves. 1978. *Statistics*. New York: Norton.

Funk, D. J., D. J. Futuyma, et al. 1995. A history of host associations and evolutionary diversification for *Ophraella* (Coleoptera: Chrysomelidae): New evidence from mitochondrial DNA. *Evolution* 49: 1008–1017.

Futuyma, D. J., M. C. Keese, and D. J. Funk. 1995. Genetic constraints on macroevolution: The evolution of host affiliation in the leaf beetle genus *Ophraella*. *Evolution* 49: 797–809.

Garland, T., Jr., and S. C. Adolph. 1994. Why not do two-species comparative studies: Limitations on inferring adaptation. *Physiological Zoology* 67: 797–828.

Garland, T., Jr., A. F. Bennett, and E. L. Rezende. 2005. Phylogenetic approaches in comparative physiology. *Journal of Experimental Biology* 208: 3015–3035.

Garland, T., Jr., P. E. Midford, and A. R. Ives. 1999. An introduction to phylogenetically based statistical methods, with a new method for confidence intervals on ancestral values. *American Zoologist* 39: 374–388.

Goodman, Steven. 2008. A dirty dozen: Twelve *P*-value misconceptions. *Seminars in Hematology* 45: 135–140.

Goriely, A., G. A. T. McVean, et al. 2003. Evidence for selective advantage of pathogenic FGFR2 mutations in the male germ line. *Science* 301: 643–646.

Goriely, A., G. A. T. McVean, et al. 2005. Gain-of-function amino acid substitutions drive positive selection of FGFR2 mutations in human spermatogonia. *Proceedings of the National Academy of Sciences, USA* 102: 6051–6056.

Gould, S. J., and R. C. Lewontin. 1979. The spandrels of San Marco and the Panglossian paradigm: A critique of the adaptationist programme. *Proceedings of the Royal Society of London B* 205: 581–598.

Grafen, A. 1989. The phylogenetic regression. *Philosophical Transactions of the Royal Society of London B* 326: 119–157.

Greene, E., L. J. Orsak, and D. W. Whitman. 1987. A tephritid fly mimics the territorial displays of its jumping spider predators. *Science* 236: 310–312.

Hailer, F., V. E. Kutschera, et al. 2012. Nuclear genomic sequences reveal that polar bears are an old and distinct bear lineage. *Science* 336: 344–347.

Hansen, D. M., T. Van der Niet, and S. D. Johnson. 2012. Floral signposts: Testing the significance of visual "nectar guides" for pollinator behaviour and plant fitness. *Proceedings of the Royal Society B* 279: 634–639.

Hosken, D. J. 1998. Testes mass in megachiropteran bats varies in accordance with sperm competition theory. *Behavioral Ecology and Sociobiology* 44: 169–177.

Huey, R. B., and D. Berrigan. 2001. Temperature, demography, and ectotherm fitness. *American Naturalist* 158: 204–210.

Huey, R. B., and J. G. Kingsolver. 1989. Evolution of thermal sensitivity of ectotherm performance. *Trends in Ecology and Evolution* 4: 131–135.

Huey, R. B., P. H. Niewiarowski, et al. 1989a. Thermal biology of nocturnal ectotherms: Is sprint performance of geckos maximal at low body temperatures? *Physiological Zoology* 62: 488–504.

Huey, R. B., C. R. Peterson, et al. 1989b. Hot rocks and not-so-hot rocks: Retreat-site selection by garter snakes and its thermal consequences. *Ecology* 70: 931–944.

Lauder, G. V., R. B. Huey, et al. 1995. Systematics and the study of organismal form and function. *BioScience* 45: 696–704.

Lehnherr, H., E. Maguin, et al. 1993. Plasmid addiction genes of bacteriophage P1: *doc*, which causes death on curing of prophage, and *phd*, which prevents host death when prophage is retained. *Journal of Molecular Biology* 233: 414–428.

Lehnherr, H., and Y. B. Yarmolinsky. 1995. Addiction protein Phd of plasmid prophage P1 is a substrate of the ClpXP serine protease of *Escherichia coli*. *Proceedings of the National Academy of Sciences, USA* 92: 3274–3277.

Martins, E. P., and T. Garland, Jr. 1991. Phylogenetic analyses of the correlated evolution of continuous characters: A simulation study. *Evolution* 45: 534–557.

Martins, E. P., and T. F. Hansen. 1997. Phylogenies and the comparative method: A general approach to incorporating phylogenetic information into the analysis of interspecific data. *American Naturalist* 149: 646–667. Erratum: Martins, E. P., and T. F. Hansen. 1999. Erratum. *American Naturalist* 153: 448.

Mather, M. H., and B. D. Roitberg. 1987. A sheep in wolf's clothing: Tephritid flies mimic spider predators. *Science* 236: 308–310.

Mayr, E. 1983. How to carry out the adaptationist program? *American Naturalist* 121: 324–334.

McElligott, A. G., I. Maggini, et al. 2004. Interactions between red-billed oxpeckers and black rhinos in captivity. *Zoo Biology* 23: 347–354.

Nevo, E., Y. B. Fu, et al. 2012. Evolution of wild cereals during 28 years of global warming in Israel. *Proceedings of the National Academy of Sciences, USA* 109: 3412–3415.

Pitnick, S., K. E. Jones, and G. S. Wilkinson. 2006. Mating system and brain size in bats. *Proceedings of the Royal Society B* 273: 719–724.

Platt, J. R. 1964. Strong inference. *Science* 146: 347–353.

Revell, L. J., and D. C. Collar. 2009. Phylogenetic analysis of the evolutionary correlation using likelihood. *Evolution* 63: 1090–1100.

Schemske, D. W., and J. Ågren. 1995. Deceit pollination and selection on female flower size in *Begonia involucrata*: An experimental approach. *Evolution* 49: 207–214.

Stone, G. N., S. Nee, and J. Felsenstein. 2011. Controlling for non-independence in comparative analysis of patterns across populations within species. *Philosophical Transactions of the Royal Society of London B* 366: 1410–1424.

Taylor, D. R., C. Zeyl, and E. Cooke. 2002. Conflicting levels of selection in the accumulation of mitochondrial defects in *Saccharomyces cerevisiae*. *Proceedings of the National Academy of Sciences, USA* 99: 3690–3694.

Weeks, P. 1999. Interactions between red-billed oxpeckers, *Buphagus erythrorhynchus*, and domestic cattle, *Bos taurus*, in Zimbabwe. *Animal Behaviour* 58: 1253–1259.

Weeks, P. 2000. Red-billed oxpeckers: Vampires or tickbirds? *Behavioral Ecology* 11: 154–160.

Whiteman, N. K., D. Santiago-Alarcon, et al. 2004. Differences in straggling rates between two genera of dove lice (Insecta: Phthiraptera) reinforce population genetic and cophylogenetic patterns. *International Journal for Parasitology* 34: 1113–1119.

Whitlock, M. C., and D. Schluter. 2009. *The Analysis of Biological Data*. Greenwood Village, CO: Roberts and Company.

Zar, J. H. 1999. *Biostatistical Analysis*. 4th ed. Upper Saddle River, NJ: Prentice Hall.

11

Sexual Selection

Male and female túngara frogs (*Physalaemus pustulosus*) play different roles in courtship. Males serenade females in choruses, while females shop for a mate with an especially beguiling song (Ryan 1985).

The túngara love song consists of a sweeping whine ("túnnnnng!") that may be followed by as many as seven short chucks ("gara!"). To sing it, the males employ specialized anatomical features that include the enormous vocal sac on view in the photo above and modifications of the larynx (Gridi-Papp et al. 2006).

Female túngaras placed between loudspeakers playing different calls tend to approach the speaker emitting the more complex song (Akre et al. 2011). The data appear at right. "Túng-gara-gara" is much more seductive than "túng." Unfortunately for the crooners, precisely the same preference is shown by frog-eating bats (*Trachops cirrhosus*). Even among frogs, romance is fraught with danger.

As in túngara frogs, male and female animals often differ strikingly in size, appearance, and behavior. In marine iguanas, for example, males weigh twice as much as females. Males become intensely territorial during the breeding season, while females remain gregarious throughout the year. In red-collared widowbirds, the adults of the two sexes sport plumage so distinct it would be easy to mistake them for different species. Males are jet black, carry tail feathers twice

Female túngaras prefer more complex male calls. So do bats. Photo by Alexander T. Baugh. Graph from Akre et al. (2011).

From "Signal perception in frogs and bats and the evolution of mating signals." *Science* 333: 751–752. Reprinted with permission from AAAS.

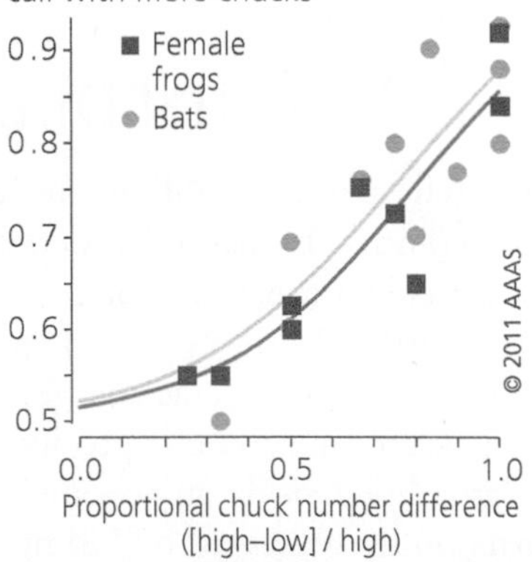

(a) Great frigate birds

(b) Greater kudu

(c) Giant scale insects

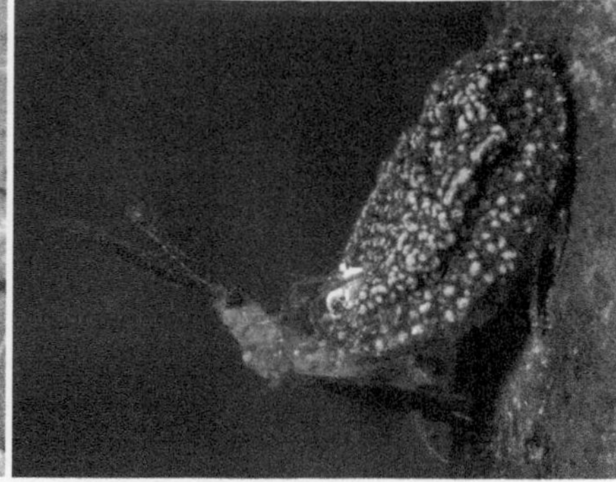

Figure 11.1 Male versus female differences (a) In great frigate birds, the male (right) has an inflatable red throat pouch. The female (left) does not. (b) In greater kudu, the male (right), has horns, a mane, and a beard. (c) In these giant scale insects, the male (lower left) and female (upper right) are so different it would be difficult to guess that they belong together had the photographer not caught them copulating. Photo by Dan L. Perlman.

the length of their own bodies, and wear a bright red collar. Females dress in cryptic brown colors and have short tails and no collars. In stalk-eyed flies, both sexes protrude their eyes on the ends of long, thin stalks, but males have longer eyestalks than females. In some species of pipefish, females display blue stripes and skin folds on their bellies. Males lack these ornaments. The photos in **Figure 11.1** provide additional examples.

Figure 11.2 Male versus female differences in humans

In humans, too, females and males differ conspicuously. Our differences exceed the obvious and essential ones in genitalia and reproductive organs. They are found in the appearance of our faces, the sound of our voices, the distribution of our body fat and body hair, our size, and our behavior **(Figure 11.2)**.

A difference between the males and females of a species is called a **sexual dimorphism.** In this chapter, we ask why sexual dimorphism occurs in such a great variety of organisms. It is a question Charles Darwin (1871) wrote half a book about, and it has captivated evolutionary biologists ever since.

A difference between the sexes is called a sexual dimorphism.

The first section of the chapter discusses potential explanations for why sexual dimorphism evolves and introduces the concept of sexual selection. The second and third sections look at how sexual selection on males leads to the evolution of exclusively male traits used in combat and mating displays. The fourth section looks at how sexual selection on females leads to the evolution of exclusively female traits. The fifth section considers sexual selection in plants. The chapter closes with a section on the role of sexual selection in the evolution of sexually dimorphic traits in humans.

11.1 Sexual Dimorphism and Sex

Elsewhere in the book, we have explained the traits of organisms with the theory of evolution by natural selection. We have seen, for example, how natural selection sculpts the beaks of medium ground finches on Daphne Major (Chapter 3). During droughts, when small, soft seeds are scarce, birds with bigger beaks can more readily crack large, hard seeds and thus get more to eat. When the big-beaked survivors reproduce, they pass genes for big beaks to their offspring. Natural selection can account for a great variety of other traits as well, from the bars on fly wings to the hiding places chosen by garter snakes (Chapter 10).

(a) Purple-throated carib (*Eulampis jugularis*)

(b) Hollyhock weevil (*Rhopalapion longirostre*)

Figure 11.3 Examples of sexual dimorphism attributable to natural selection Photo (b) by Gilles San Martin.

The theory of evolution by natural selection can readily explain some cases of sexual dimorphism (Hedrick and Temeles 1989). Divergent traits may be adaptive for the two sexes for ecological reasons, as when males and females eat different foods. This appears to be the case in the purple-throated carib **(Figure 11.3a)**. Given a choice, both sexes of this hummingbird prefer to sip nectar from the flowers of *Heliconia caribaea* (Temeles et al. 2009). In nature, however, the males—which are larger and socially dominant—tend to monopolize this favored food (Temeles et al. 2005). Females are, instead, more strongly associated with the flowers of *Heliconia bihai* (Temeles et al. 2000; Temeles and Kress 2003). *Heliconia bihai* has longer and more tightly curved flowers than *H. caribaea*. The longer and more tightly curved beaks of female purple-throated caribs appear to be adaptive because they enable the females to exploit *H. bihai* flowers more easily.

Divergent traits may also be adaptive for intrinsic reasons, as when males and females play different roles in reproduction. This appears to be the case in the hollyhock weevil (Figure 11.3b). The females of this beetle use their elongated snouts to bore holes deep into the buds of their host plant, into which they deposit their eggs (Wilhelm et al. 2011).

However, a great many other cases of sexual dimorphism present considerably more challenging puzzles. To see why, try to imagine how we might use natural selection to account for the peacock's train **(Figure 11.4)**. Two problems

Figure 11.4 Peafowl The hen is on the left.

arise. First, if a fan of iridescent eyespots improves the survival or fecundity of a peafowl, why do the hens not have trains too? Second, how could a fan of iridescent eyepots improve survival or fecundity in the first place? As with complex songs of túngara frogs, trains probably make peacocks easier for predators to find and catch. Furthermore, elaborate plumage requires considerable energy to grow, maintain, and drag around (see Walther and Clayton 2005). Energy spent on feathers is energy that cannot be spent on making offspring. It is therefore less than immediately apparent how the theory of evolution by natural selection might explain why peacocks and peahens are different. And it is less than obvious why the birds' most striking trait, the elaborate train, exists at all.

"The sight of a feather in a peacock's tail," Darwin (1860) complained to his friend Asa Gray, "whenever I gaze at it, makes me sick!"

Sexual dimorphism presents a challenging evolutionary puzzle.

Sex Provides Another Explanation

As Darwin himself was later the first to recognize, sex provides another solution to the puzzle of sexual dimorphism. Consider, for a moment, life without sex. For organisms that reproduce without sex (see Chapter 8), getting genes into the next generation is straightforward, if not always easy. The two big challenges are surviving to adulthood, then reproducing. Sex adds a third major challenge: finding a member of the opposite sex and persuading him or her to cooperate.

Darwin realized that individuals vary not only in their success at surviving and reproducing but also in their success at persuading members of the opposite sex to mate. About birds, for example, Darwin wrote,

> Inasmuch as the act of courtship appears to be with many birds a prolonged and tedious affair, so it occasionally happens that certain males and females do not succeed during the proper season, in exciting each other's love, and consequently do not pair. (1871, page 107)

Failing to obtain a mate, like dying young, can severely curtail an individual's genetic contribution to future generations.

Darwin had already applied the label natural selection to differential reproductive success due to variation among individuals in survival and fecundity. Differential reproductive success due to variation among individuals in success at getting mates, he called **sexual selection.** We can develop a theory of evolution by sexual selection that is logically equivalent to the theory of evolution by natural selection: If there is heritable variation in a trait that affects the ability to obtain mates, then variants conducive to success will become more common over time.

Asymmetries in Sexual Reproduction

If sexual selection is to explain differences between the sexes, it will have to act on the sexes differently. Angus John Bateman (1948) argued that it often does. The logic he developed to support his claim was later refined by Robert Trivers (1972). It hinges on a simple observation: In many animals, eggs (or pregnancies) are more expensive than ejaculates.

More generally, mothers typically make a larger parental investment in each offspring than fathers do. By **parental investment,** we mean energy and time expended constructing and caring for the offspring. Ultimately, this investment is measured in fitness. Parental investment increases the reproductive success of the offspring receiving it. At the same time, it decreases the remaining reproductive success that the parent may achieve in the future by way of additional offspring.

Consider the parental investments made by orangutans (**Figure 11.5**). Adult orangutans of opposite sex tolerate each other's company for one purpose only (Nowak 1991). After a brief tryst, including a copulation that lasts about 15 minutes, the lovers go their separate ways. If a pregnancy results, then the mother, who weighs about 40 kg, will carry the fetus for 8 months, give birth to a 1-kg baby, nurse it for about 3 years, and continue to protect it until it reaches the age of 7 or 8. For the father, who weighs about 70 kg, the beginning and end of parental investment is a few grams of semen—which he can replace in a matter of hours or days. In their pattern of parental investment, orangutans are typical mammals. In more than 90% of mammal species, females provide substantial parental care and males provide little or none (Woodroffe and Vincent 1994).

Figure 11.5 Parental investment by mothers versus fathers Orangutan mothers invest considerably more time and energy in each offspring than orangutan fathers do.

Because mammalian mothers provide such intensive parental care, mammals may offer a somewhat extreme example of disparity in parental investment. In most animal species, neither parent cares for the young. Mated pairs of parents just make eggs, fertilize them, and leave them. But in these species, too, females usually make a larger investment in each offspring than males. Eggs are typically large and yolky, with a big supply of stored energy and nutrients. Think of a sea turtle's eggs, some of which are as large as a hen's eggs. Most sperm, on the other hand, are little more than DNA with a propeller. Even when a single ejaculate delivers hundreds of millions of sperm, the ejaculate seldom represents more than a tiny fraction of the investment contained in a clutch of eggs.

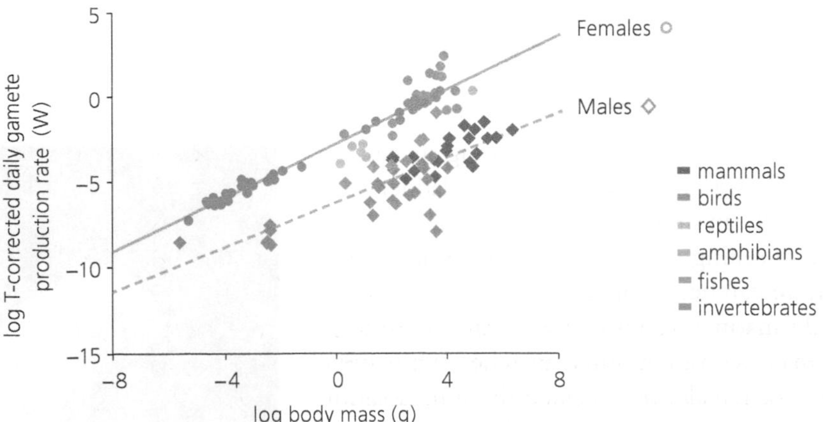

Figure 11.6 Investment in gametes by mothers versus fathers The vertical axis plots daily energy expenditure in eggs or sperm, calculated in Watts and corrected for differences in temperature, for adults in a variety of animals. The horizontal axis represents body mass in grams. Note that both axes use logarithmic scales. From Hayward and Gillooly (2011).

April Hayward and James Gillooly (2011) combed the literature for data that would allow them to estimate the daily energetic investment in gametes by females versus males in a variety of animals. The results appear in the log-log plot in **Figure 11.6**. Hayward and Gillooly calculate that during the breeding season a typical female devotes to the production of eggs about three times the energy required to maintain her basal metabolism. A typical male, on the other hand, devotes to the production of sperm about one thousandth the energy required for basal metabolism. Including the additional components of seminal fluid raises his total expenditure on ejaculates to four thousandths the energy spent on basal metabolism. Making babies generally costs mom considerably more than it costs dad.

The key to explaining sexual dimorphism is in recognizing that sexual reproduction imposes different patterns of selection on females versus males.

Asymmetric Limits on Reproductive Success

When eggs are more expensive than ejaculates—when mothers make a larger parental investment than fathers—the factors limiting lifetime reproductive success

In the second, each barrel contained two males and six females. The second experiment probably mimics natural conditions more closely than the first. It takes a female less time to produce a clutch of eggs than it takes a male to rear them to hatching (Berglund et al. 1989). As a result, at any given time there are more females with eggs to lay than males with space to accept them.

Jones and colleagues used genetic tests to determine each offspring's mother. As in the newt study, this enabled the biologists to determine the number of mates and the number of offspring for each adult.

The results for pipefish are similar to those for newts, except that the roles of the sexes are reversed. In pipefish, more females than males failed to mate (Figure 11.12a and d). Consequently, the reproductive success of females was more skewed than the reproductive success of males and had a sharper distinction between the winners and the losers (Figure 11.12b and e). Most important, it was in females that reproductive success depended most strongly on mating success (Figure 11.12c and f). In broad-nosed pipefish, sexual selection is a more potent force in the evolution of females than it is in the evolution of males. Heritable traits that are associated, in females, with failure to mate will tend to disappear, while heritable traits associated with mating success will become more common. We need to keep this result in mind as we consider the behavioral consequences of asymmetrical limits on fitness.

Behavioral Consequences of Asymmetrical Limits on Fitness

An asymmetry in the factors that limit reproductive success for females versus males allows us to predict differences in mating behavior. Consider the pattern in rough-skinned newts. For males, reproductive success is limited by access to mates, and at any given time there are more males than females in the pond looking for love. Under such circumstances, we can predict that males will compete among themselves for opportunities to fertilize eggs. For females, in contrast, reproductive success is limited by capacity to make eggs, mating involves the commitment of a large investment, and there is an excess of willing partners. We can expect that females will be selective about which partners they accept.

More generally, when sexual selection is strong for one sex and weak for the other, we can predict that:

- Members of the sex subject to strong sexual selection will be competitive.
- Members of the sex subject to weak sexual selection will be choosy.

These predictions have been confirmed in a great variety of animal species. We will look at some examples shortly.

In making these general predictions, we have used inclusive language for a reason. It is easy to get carried away with generalities, as Bateman and many who followed appear, in hindsight, to have done (see Knight 2002). Bateman thought that greater sexual selection on males than on females is inherent in maleness and femaleness as such. He and others therefore assumed that the optimal strategy for males, in virtually any species, would be to mate with as many females as possible, and that the optimal strategy for females would be to choose one male and mate with him only. These assumptions have often turned out to be wrong. We will see, later in the chapter, that males often have good reasons to stick with one mate and that females often have good reasons to be promiscuous.

Theory predicts that when one sex is subject to sexual selection and the other is not, the members of the sex experiencing selection will compete over mates and the members of the other sex will be choosy.

Furthermore, as Jones et al.'s pipefish study shows, greater sexual selection on males than on females is not inherent in the identity of the sexes themselves.

When access to mates is limiting for females instead of males, we predict that females will compete over access to males and that males will be choosy.

Competition for mates in one sex and choosiness in the other can play out in two ways. First, members of the competitive sex may fight among themselves, head-to-head, claw-to-claw, or antler-to-antler. Sometimes they fight over direct control of mates, sometimes they fight over control of a resource vital to mates, and sometimes they just fight. The members of the other sex then mate with the winners. This form of sexual selection is called **intrasexual selection,** because the key event that determines reproductive success (the fighting) involves interactions among the members of a single sex. Second, instead of fighting the members of the competitive sex may advertise for mates by singing, dancing, or showing off bright colors. The members of the other sex then choose the individual with the best display. This form of sexual selection is called **intersexual selection,** because the key event that determines reproductive success (the choosing) involves an interaction between members of the two sexes.

In the next two sections, we look at examples of intrasexual and intersexual selection on males. We then look at sexual selection on females.

11.2 Sexual Selection on Males: Competition

Sexual selection by male–male competition often occurs when individual males can monopolize access to females. Males may monopolize females through direct control of the females themselves or through control of some resource important to females, such as feeding territory or nest sites. Male–male competition can also occur for no apparent reason beyond simply impressing females. In this section, we consider examples of research into three forms of male–male competition: outright combat, sperm competition, and infanticide.

Combat

Outright combat is the most obvious form of male–male competition for mates. Intrasexual selection involving male–male combat over access to mates can favor morphological traits including large body size, weaponry, and armor. Male–male combat also selects for tactical cleverness.

Male–male competition can take the form of combat over access to females.

Our example of male–male combat comes from the marine iguanas (*Amblyrhynchus cristatus*) of the Galápagos Islands **(Figure 11.13)**. Marine iguanas have a

Figure 11.13 A Galápagos marine iguana These unusual lizards make their living foraging on algae in the intertidal zone.

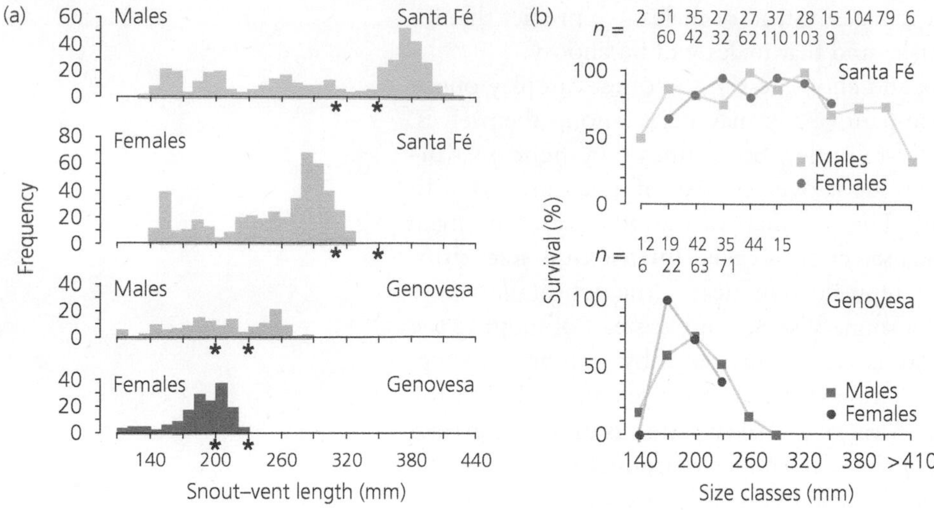

Figure 11.14 Natural selection on body size in marine iguanas (a) Size distributions of iguanas on Genovesa and Santa Fé. Asterisks mark maximum sizes at which iguanas maintained their weight in two different years (1991–1992 and 1992–1993). From Wikelski et al. (1997). (b) Survival rates of marked individuals of different sizes (snout–vent length, mm) from March 1991 to March 1992 on Genovesa and from February 1990 to February 1992 on Santa Fé. *n* = sample sizes. From Wikelski and Trillmich (1997).

lifestyle unique among the lizards. They make their living grazing on algae in the intertidal zone. Between bouts of grazing, they bask on rocks at the water's edge. Basking aids digestion and warms them for their next foray into the cold water. Marine iguanas grow to different sizes on different islands (see Wikelski 2005), but on any given island the males get larger than the females **(Figure 11.14a)**.

The size dimorphism in marine iguanas is an excellent example for the study of sexual selection, because we know a great deal about how marine iguana size is affected by natural selection (Wikelski et al. 1997; Wikelski and Trillmich 1997). Martin Wikelski and Fritz Trillmich documented natural selection on iguana body size by monitoring the survival of marked individuals on two islands over one to two years. Natural selection was harsher on Genovesa than on Santa Fé, but it was happening on both. Moreover, selection was stabilizing. Medium-sized iguanas survived at higher rates than either small or large iguanas (Figure 11.14b).

Potential agents of natural selection on size are few. The iguanas do not compete with other species and have virtually no predators. Other than reproduction, about all they have to contend with is competition for food among themselves.

Larger iguanas can harvest more algae and thus gather more energy, but they also expend more energy on metabolism. Wikelski and colleagues (1997) found during two different years that small iguanas ran a net energy surplus, but large iguanas ran a net energy deficit. Consistent with the hypothesis that the availability of food limits body size, the largest iguanas on Santa Fé and Genovesa lost weight during both 1991–1992, a bad year for algae, and 1992–1993, a fairly good year (see also Wikelski and Thom 2000). The largest sizes at which iguanas were able to maintain their weight are indicated by the asterisks in Figure 11.14a.

Now compare Figure 11.14a with Figure 11.14b. The maximum sizes at which iguanas could sustain their weight are close to the optimal sizes for survival. The largest females on each island are near the optimal size for survival, but the largest males are much larger than the optimal size. The large body size of male marine iguanas is thus an evolutionary puzzle. It is a challenge to explain it with natural selection, because Wikelski and Trillmich have documented natural selection acting against it. It is exactly the kind of puzzle for which Darwin invoked sexual selection.

Male marine iguanas fight over territories where females congregate. Large iguanas win more fights, claim better territories, and thus get to copulate with more females.

As we discussed earlier, a crucial issue in sexual selection is the relative parental investment per offspring made by females versus males. In marine iguanas, the

Figure 11.15 Male marine iguanas in combat Note the number painted on the individual on the right; he is participating in a study. By Martin C. Wikelski.

parental investment by females is much larger. Each female digs a nest on a beach away from the basking and feeding areas, buries her eggs, guards the nest for a few days, and then abandons it (Rauch 1988). Males provide no parental care at all. So parental investment by females consists mostly of producing eggs, and parental investment by males consists entirely of producing ejaculates. Females lay a single clutch of one to six eggs each year, into which they put about 20% of their body mass (Rauch 1985; Rauch 1988; Wikelski and Trillmich 1997). Compared to the female investment, the cost of the single ejaculate needed to fertilize all the eggs in a clutch is trivial. This difference in investment suggests that the maximum potential reproductive success of males is much higher than that of females. Number of mates will limit the lifetime reproductive success of males, but not females.

The iguanas' mating behavior is consistent with these inferences. Females copulate only once each reproductive season. Martin Wikelski, Silke Bäurle, and their field assistants followed several dozen marked females on Genovesa. The researchers watched the females from dawn to dusk every day during the entire month-long mating season in 1992–1993 and 1993–1994 (Wikelski and Bäurle 1996). They also watched the marked females from dawn to dusk every day during the subsequent nesting seasons. Every marked female that dug a nest and laid eggs had been seen copulating, but no marked female had been seen copulating more than once. Male iguanas, in contrast, attempt to copulate many times with many different females. But the opportunity to copulate with females is a privilege that a male iguana has to fight for **(Figure 11.15)**.

Before the mating season each year, male iguanas stake out territories on the rocks where females bask between feeding bouts. In these small, densely packed territories **(Figure 11.16a)**, males attempt to claim and hold ground by ousting male interlopers. Confrontations begin with head-bobbing threats and escalate to chases and head pushing. If neither backs down, fights can end with bites leaving serious injuries on the head, neck, flanks, and legs (Trillmich 1983). Males that hold territories are more attractive to females than males that do not (Trillmich 1983; Rauch 1985; Partecke et al. 2002). Because only some males manage to claim territories, because some owners maintain their claims longer than others, and because females prefer some territories and owners over others, there is extreme variation among males in copulatory mating success (Figure 11.16b).

Because claiming and holding a territory requires combat, bigger males tend to win. In the iguana colony Krisztina Trillmich (1983) studied on Camaaño Islet, the male that got 45 copulations, far more than any other male, was iguana 59. His neighbor, iguana 65, was the second most successful with 10 copulations. Both of their territories were females' favorite early-morning and late-afternoon basking

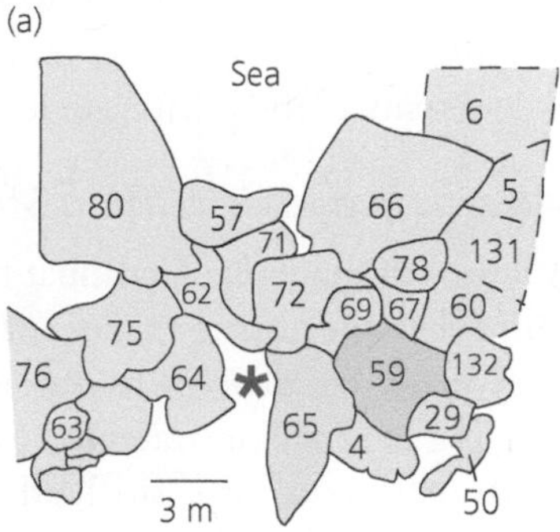

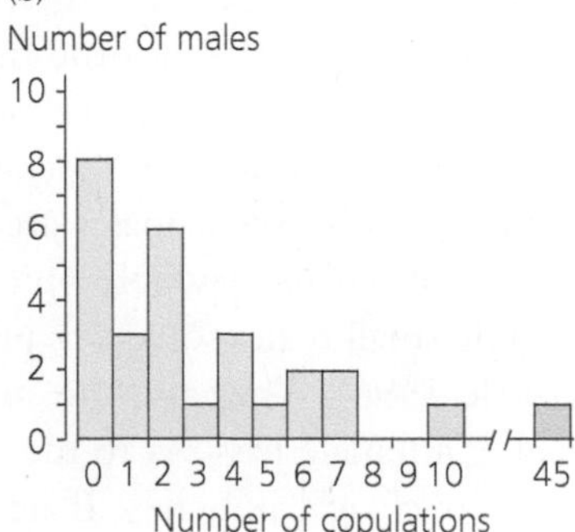

Figure 11.16 Mating success in male marine iguanas (a) Iguana mating territories on Camaaño Islet, Galápagos. Numbers identify territory owners. Asterisk shows where Krisztina Trillmich sat to watch. From Trillmich (1983). (b) Variation in copulations obtained by males on territories shown in (a). Note the break in the horizontal scale; the most successful male, iguana 59, got more than four times as many copulations as any rival. Includes only males that held a territory at least briefly during the mating season. From Trillmich (1983).

places. Trillmich reported that iguana 59 was the largest male in the colony; that to claim his territory, he had to eject four other males who tried to take it; and that during his tenure, he lost parts of it to four neighboring males who were pushing their territories in from the sides. Wikelski and coworkers studied iguana colonies on Genovesa and Santa Fé (Wikelski et al. 1996; Wikelski and Trillmich 1997). Consistent with Krisztina Trillmich's observations, these researchers found that the mean size of males that actually got to copulate was significantly larger than the mean size of all males that tried to copulate **(Figure 11.17)**.

If we assume that body size is heritable in marine iguanas, then we have variation, heritability, and differential mating success. These are the elements of evolution by sexual selection. We thus have an explanation for why male marine iguanas get so much bigger than the optimal size for survival. Male iguanas get big because bigger males get more mates and pass on more of their big-male genes.

Male–male combat, analogous to that in marine iguanas, happens in a great variety of species. In addition to large size, this kind of sexual selection leads to the evolution of other traits that are assets in combat, such as weaponry and armor. Male–male combat can also lead to the evolution of alternative mating strategies.

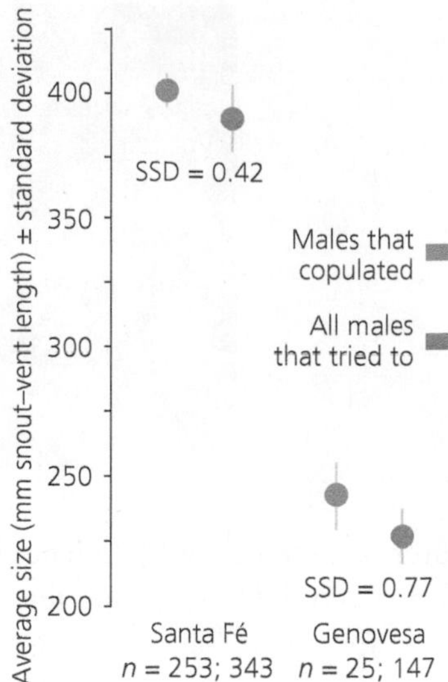

Figure 11.17 Sexual selection differentials for body size in marine iguanas SSD = standardized selection differential. SSD is the difference between the males that copulated versus all the males that tried to, expressed in standard deviations of the distribution of all males that tried. Both SSDs are positive ($p < 0.05$), indicating that males who got to copulate were, on average, larger. n = sample size. After Wikelski and Trillmich (1997).

Alternative Male Mating Strategies

Victory in male–male combat typically goes to the large, strong, and well armed. But what about the smaller males? Is their only chance at fitness to survive until they grow big enough to win brawls? Often small males attempt to mate by employing alternative strategies. Sometimes they succeed.

In marine iguanas, small adult males are ousted from the mating territories on the basking grounds. But many do not give up; they continue trying to convince females to copulate with them. The small males are only rarely successful, but they do get about 5% of the matings in the colony (Wikelski et al. 1996). Small males attempting to mate are often harassed by other males. This happens to large territorial males too, but it happens more often to small males. Furthermore, copulations by small males are more likely to be disrupted before the male has time to ejaculate—which typically happens after about 3 minutes **(Figure 11.18)**.

The small males solve this problem by ejaculating ahead of time (Wikelski and Bäurle 1996). They use the stimulation of an attempted copulation, or even of seeing a female pass by, to induce ejaculation. The males then store the ejaculate in their cloacal pouches. If he gets a chance to mate, a small male transfers his stored ejaculate to the female at the beginning of copulation. Martin Wikelski and Silke Bäurle examined the cloacae of a dozen females caught immediately after copulations that had lasted less than 3 minutes. None of these females had copulated earlier that mating season, but 10 of the 12 females had old ejaculates in their cloacae that must have been transferred during the short copulation.

The sperm in these old ejaculates were viable. From dawn to dusk every day for about a month, Wikelski and Bäurle watched five of the females until they laid their eggs. None of the five copulated again, but all laid fertilized eggs.

Prior ejaculation appears to be a strategy practiced more often by small nonterritorial males than by large territorial males. Wikelski and Bäurle caught 13 nonterritorial and 13 territorial males at random; 85% of the nonterritorial males had stored ejaculates in their cloacal pouches, versus only 38% of the territorial males ($p < 0.05$). This difference is unlikely to result from more frequent copulation by territorial males, because even territorial males copulate only about once every 6 days (Wikelski et al. 1996).

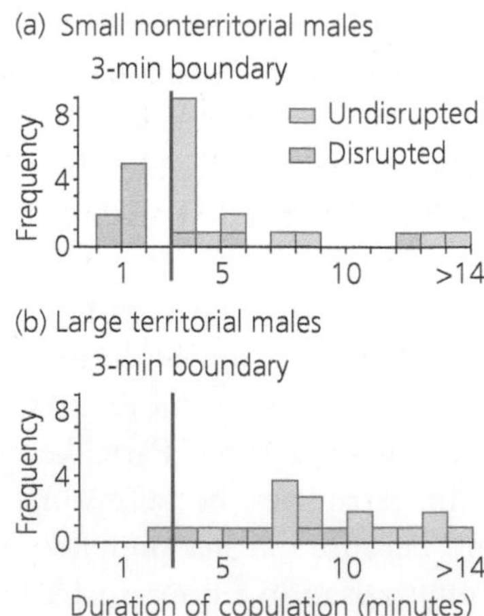

Figure 11.18 Length of copulations by male iguanas Small males (a) are more often disrupted before ejaculating. From Wikelski and Bäurle (1996).

Alternative, or sneaky, male mating strategies also have evolved in a variety of other species. In coho salmon, *Oncorhynchus kisutch,* for example, males return from the sea to spawn and die at one or the other of two distinct ages (Gross 1984, 1985, 1991). One group, called hooknoses, returns at 18 months. They are large, armed with enlarged hooked jaws, and armored with cartilaginous deposits along their backs. The other group, called jacks, returns at 6 months. They are small, poorly armed, and poorly armored.

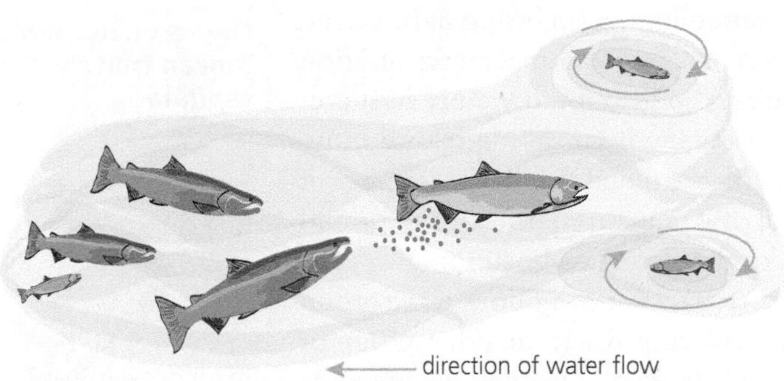

Figure 11.19 Alternative mating strategies in coho The big fish at right (upstream) is a female that has built a nest and is ready to lay eggs. Downstream are three hooknose males and a jack that have opted to fight and have sorted themselves by size. Two sneaky jacks lurk nearby. After Gross (1991).

When a female coho is ready to mate, she digs a nest and lays her eggs in it. As she prepares the nest, males congregate. The males use one of two strategies in trying to fertilize the female's eggs **(Figure 11.19)**. Some males fight for a position close to the female. These fighters quickly sort themselves out by size, making a line downstream of the nest. When the female lays her eggs, the males spawn over them in order. The first male to spawn fertilizes the most eggs. Males using the other strategy do not fight for position but instead look for a hiding place near the female, perhaps in a shallow or behind a rock. When the female lays her eggs, these sneakers attempt to dart out and spawn over the eggs.

When there is intense competition among males over access to mates, alternative sneaky mating strategies sometimes evolve.

Among hooknoses, those that adopt the fighting strategy are more successful. Among jacks, those that adopt the sneaky strategy are more successful. The relative fitness of hooknoses versus jacks depends, in part, on the frequency of each type of male in the breeding population.

There is an important distinction between the iguana example and the coho example. In marine iguanas, the small nonterritorial males appear to be making the best of a bad situation while they grow big enough to fight for a territory. In coho, a male irreversibly becomes either a hooknose or a jack. Which strategy a male coho pursues depends on a mixture of environmental and genetic factors.

Sperm Competition

Male–male competition does not necessarily stop when copulation is over. The real determinant of a male's mating success is not whether he copulates, but whether his sperm fertilize eggs. If an animal has internal fertilization, and if a female mates with two or more different males within a short period, then the sperm from the males will be in a race to the eggs. Indeed, females may produce litters or clutches in which different offspring are fathered by different males. Batches of offspring with multiple fathers have been documented in a variety of animals, including squirrels (Boellstorff et al. 1994), bears (Schenk and Kovacs 1995), birds (Gibbs et al. 1990), lizards (Olsson et al. 1994), and spiders (Watson 1991). It happens in humans too; Smith (1984) reviews several reports of twins with different fathers.

Male–male competition can take the form of sperm competition. If a female mates with two or more males, the male whose sperm win the race to the eggs has higher reproductive success.

Given sperm competition, what traits contribute to victory? One useful trait might simply be the production of large ejaculates containing many sperm. If sperm competition is something of a lottery, then the more tickets a male buys, the better his chances of winning. This hypothesis has been tested by Matthew Gage (1991) with the Mediterranean fruit fly, *Ceratitis capitata* (**Figure 11.20**). Gage's experiment was based on the observation that, although ejaculates are cheap, they are not free (see, for example, Nakatsuru and Kramer 1982). Gage reasoned that if male Mediterranean fruit flies are subject to any constraints on sperm production, they might benefit from conserving sperm, using during each copulation only the minimum number necessary to ensure complete fertilization of the female's eggs. However, if larger ejaculates contribute to victory in sperm competition, males whose sperm are at risk of competition should release more sperm during copulation than males whose sperm are not at risk. If the number of sperm released is unimportant to the outcome of competition, then males should release the same number of sperm regardless of the risk of competition.

Figure 11.20 A Mediterranean fruit fly, *Ceratitis capitata*

Gage raised and mated male medflies under two sets of conditions. One group of 20 he raised by themselves and allowed to mate in private; the other group of 20 he raised in the company of another male and allowed to mate in the presence of that second male. Immediately after each mating, Gage dissected the females and counted the number of sperm the males had released. Males raised and mated in the presence of a potential rival ejaculated more than $2\frac{1}{2}$ times as many sperm (average $\pm$ standard error $= 3{,}520 \pm 417$) as males raised and mated in isolation ($1{,}379 \pm 241$), a highly significant difference ($p < 0.0001$). Gage's interpretation was that large ejaculates do contribute to victory in sperm competition and that male medflies dispense their sperm to balance the twin priorities of ensuring successful fertilization and conserving sperm.

In addition to large ejaculates, sperm competition has apparently led to various other adaptations. Males may guard their mates, prolong copulation, deposit a copulatory plug, or apply pheromones that reduce the female's attractiveness (Gilbert 1976; Beecher and Beecher 1979; Sillén-Tullberg 1981; Thornhill and Alcock 1983; Schöfl and Taborsky 2002). During copulation in many species of damselflies, the male uses special structures on his penis to scoop out sperm left by the female's previous mates (**Figure 11.21**; Waage 1984, 1986). R. E. Hooper and M. T. Siva-Jothy (1996) used genetic paternity tests to show that this strategy is highly effective. In the damselfly species they studied, the second male to mate with a female fertilized nearly all of the eggs produced during her first postcopulatory bout of oviposition.

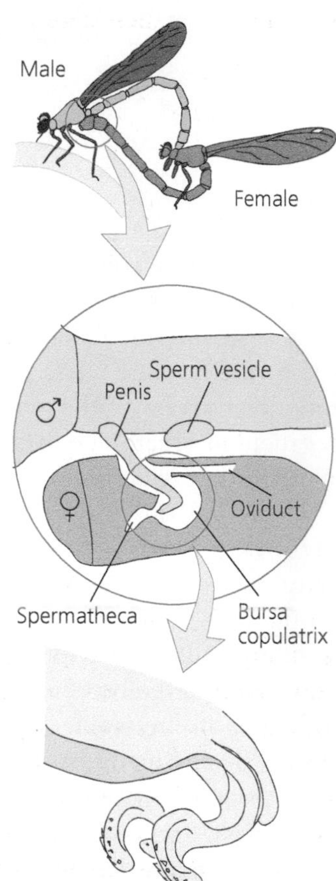

Figure 11.21 Sperm competition in damselflies During copulation (top), the male uses the barbed horns on his penis (bottom) to remove sperm left by the female's previous mates. Redrawn from Waage (1984).

Infanticide

In some species of mammals, competition between males continues even beyond conception. One example, discovered by B. C. R. Bertram (1975) and also studied by Craig Packer and Anne Pusey (reviewed in Packer et al. 1988), happens in lions. The basic social unit for lions is the pride. The core of a pride is a group of closely related females—mothers, daughters, sisters, nieces, aunts—and their cubs. Also in the pride is a small group of adult males; two or three is a typical number. The males are usually related to each other but not to the adult females. This system is maintained because females reaching sexual maturity stay in the pride they were born into, whereas newly mature males move to another pride.

The move for young adult males from one pride to another is no stroll in the park. The adult males already resident in the new pride resist the invaders. That is why males stay with their other male kin: Each group, the residents and the newcomers, forms a coalition. The residents fight the newcomers, sometimes violently, over the right to live in the pride. If the residents win, they stay in the pride and the newcomers search for a different pride to take over. If the residents lose, they are evicted, and the newcomers have exclusive access to the pride's females—exclusive, that is, until another coalition of younger, stronger, or more numerous males comes along and kicks them out. Pusey and Packer found that the average time a coalition of males holds a pride is a little over two years. Because residence in a pride is the key to reproductive success in lions, males in a victorious coalition quickly begin trying to father cubs. One impediment to quick fatherhood, however, is the presence of still-nursing cubs fathered by males of the previous coalition. That is because females do not return to breeding condition until after their cubs are weaned.

Figure 11.22 Lion infanticide A male lion killing another male's cub.

How can the males overcome this problem? They frequently employ the obvious, if grisly, solution: They kill any cubs in the pride that have not been weaned **(Figure 11.22)**. Packer and Pusey have shown that this strategy causes the cubs' mothers to return to breeding condition an average of eight months earlier than they otherwise would. Infanticide by males is the cause of about 25% of all cub deaths in the first year of life and over 10% of all lion mortality.

Male–male competition does not always end at conception. It can take the form of infanticide. By killing other males' cubs, male lions gain more opportunities to mate.

Infanticide improves the males' reproductive prospects but is obviously detrimental to the reproductive success of the females. The females try to protect their own interests in this bad situation (Packer and Pusey 1983). They defend their cubs from infanticidal males, occasionally at the cost of their own lives. Nonetheless, Packer and Pusey report that young cubs rarely survive more than two months in the presence of a new coalition of males. With this shift in focus to female reproductive strategy, we leave the subject of male–male conflict and move to the other side of sexual selection on males: female choice.

11.3 Sexual Selection on Males: Female Choice

In a great variety of species male reproductive success is limited by opportunities to mate, but males are unable to monopolize either females themselves or any resource vital to females. In many such species, the males advertise for mates. Females typically inspect advertisements of several males before they choose a mate. Sexual selection by mate choice leads to the evolution of elaborate courtship displays.

Charles Darwin first asserted that female choice is an important mechanism of selection in 1871, in *The Descent of Man, and Selection in Relation to Sex*. Although widely accepted today, the notion that females actively discriminate among individual males was controversial for decades. Most evolutionary biologists thought that female discrimination was limited to choosing a male of the right species (see Trivers 1985). Beyond this function, male courtship displays were thought to function primarily in overcoming a general female reluctance to mate. Once ready to mate, a female would accept any male at hand.

When males cannot monopolize access to females, they often compete by advertising for mates. Although biologists were long skeptical that females discriminate among the advertising males, female choice is now well established.

We begin this section with two sets of experiments demonstrating that females are in fact highly selective, actively choosing particular males from among the many available. We then consider whether female choosiness is adaptive. Potential benefits to a choosy female include acquiring genes for her offspring that will make her sons attractive, acquiring genes for her offspring that confer general fitness benefits, and acquiring resources offered by males. Alternatively, females may prefer male displays that exploit preexisting sensory biases built into the females' nervous systems.

Female Choice in Red-Collared Widowbirds

Our first experiment demonstrating active female choice comes from the work of Sarah Pryke and Steffan Andersson (2005) on the red-collared widowbird (*Euplectes ardens*). During the breeding season, red-collared widowbirds are highly dimorphic. Adult males are jet black with long tail feathers and a crimson collar **(Figure 11.23)**. Adult females are streaked with yellow and brown, and they have normal-length tail feathers.

Figure 11.23 A male red-collared widowbird During the breeding season, adult males wear jet-black plumage, long tail feathers, and a red collar.

Inspired by a classic study by Malte Andersson (1982) on long-tailed widowbirds, Pryke and Andersson captured 120 males before they had claimed nesting territories and assigned every other one to the control group or the experimental group. The researchers trimmed the tail feathers of the control birds to a maximum length of 20 cm, just 2 cm below the population average. They trimmed the tail feathers of the experimental birds to a maximum length of 12.5 cm. The researchers weighed the birds and measured their legs. The researchers then put a unique set of colored ankle bands on each of the birds and released them.

Pryke and Andersson recaptured about half of the males at various times during the breeding season and reweighed them. For each one they calculated an

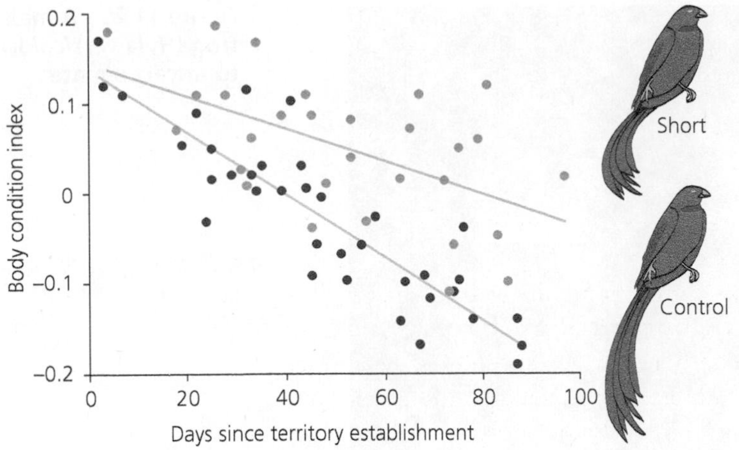

Figure 11.24 Long tail feathers are a ball and chain for male red-collared widowbirds This graph shows the decline in body condition (weight relative to linear size) throughout the breeding season for males with experimentally shortened tail feathers versus controls. The control males lost weight much more quickly. From Pryke and Andersson (2005).

index of body condition that reflects the bird's mass relative to its leg length. The data appear in **Figure 11.24**. The control males, who retained their long tails, lost weight much faster than the experimental males. This despite the fact that the control males spent less time flying around and showing off their tails to females. It appears that long tail feathers are simply expensive to drag around and maintain. In other words, natural selection probably acts against them. As with large body size in male marine iguanas, long tail feathers in widowbirds are just the sort of dimorphic trait Darwin invoked sexual selection to explain.

Could long tail feathers be a signal that males use to intimidate their rivals in the competition to claim territories? Apparently not. Forty-eight of the males with shortened feathers were able to acquire territories versus 43 of the control males, roughly the same success rate for each group. And the territories the two groups claimed were of similar size and quality. Pryke and Andersson concluded that long tail feathers did not evolve as a result of intrasexual selection.

Why, then, do male red-collared widowbirds have long tail feathers? Pryke and Andersson suspected from the beginning that the reason was that female red-collared widowbirds think long tails are sexy. The researchers monitored each male's territory throughout the breeding season, noting the number of females he enticed to nest there. As shown in **Figure 11.25**, on average the control males attracted nearly three times as many mates as the males with experimentally shortened feathers. We would need genetic tests to be certain, but presumably the control males enjoyed higher reproductive success as well.

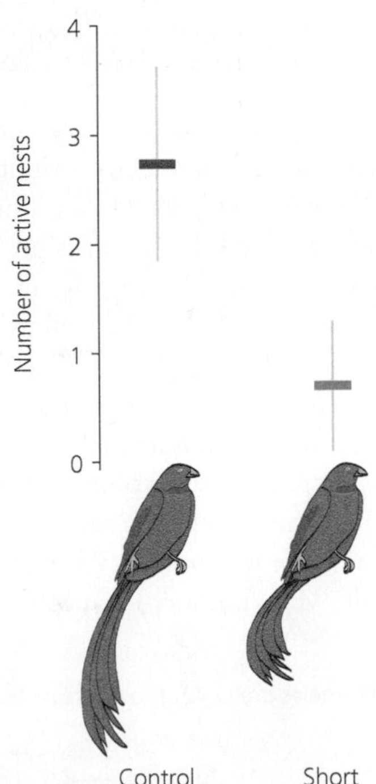

Figure 11.25 Female red-collared widowbirds prefer long-tailed males Symbols show mean ±standard deviation number of females that nested in territories of control versus short males. From Pryke and Andersson (2005).

Prkye and Andersson's experiment corroborates Darwin's contention that females are choosy. They actively discriminate among males of their own species and select particular kinds for their mates.

Female Choice in Gray Tree Frogs

Our second set of experiments demonstrating active female choice comes from the work of H. Carl Gerhardt and colleagues on gray tree frogs. Gray tree frogs, *Hyla versicolor,* live in woodlands in the eastern United States. During breeding season, the males serenade the females. The love song of the gray tree frog is a series of calls, each call consisting of a number of pulses, or trills. Some males give long calls with many trills; other males give short calls with few trills. Gerhardt and colleagues suspected, for at least two reasons, that female gray tree frogs discriminate among potential mates on the basis of their songs. First, they knew that when a male hears others joining him to make a chorus, he sometimes

Figure 11.26 A male gray tree frog (*Hyla versicolor*) singing to attract a mate

increases the length of his calls. Second, several times in the field the researchers had watched females go right past one singer to mate with a more distant one. The male in **Figure 11.26** is doing his best to sound that attractive. The researchers hypothesized that females prefer to mate with longer-calling males.

Gerhardt, Miranda Dyson, and Steven Tanner (1996) caught female gray tree frogs and tested their preferences in the lab. In one experiment, the researchers released females between a pair of loudspeakers **(Figure 11.27a)**. Each speaker played a computer-synthesized mating call. To make the experiment conservative, the researchers made the call they expected to be less attractive louder, either by increasing the volume on that speaker or by releasing the female closer to it. Then they waited to see which speaker the female would approach. They found that 30 of 40 females (75%) preferred long calls to short calls, even when the short calls were louder. In another experiment, the researchers released female frogs

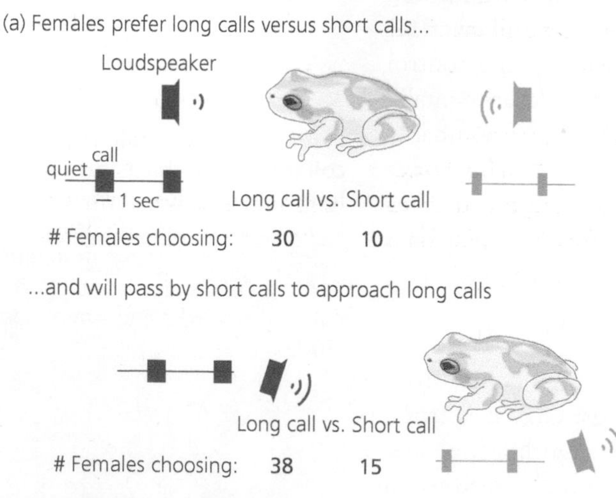

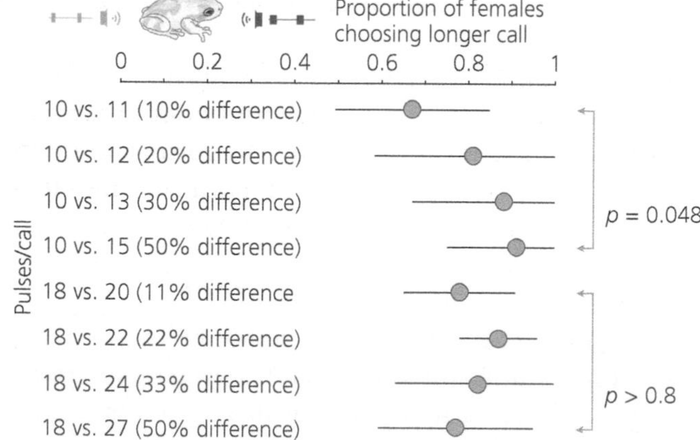

Figure 11.27 Preferences of female gray tree frogs (a) Most females prefer long to short calls, even when the short calls are initially louder ($p < 0.001$). Most females will pass a loudspeaker playing short calls to approach one playing long calls ($p < 0.001$). After Gerhardt et al. (1996). (b) Females discriminate most strongly among short calls. Each dot represents the proportion of females, in a group of 16 to 61 individuals, that preferred a long call over a short call in a paired choice test. The whiskers show the 95% confidence limits. When the shorter call had just 10 pulses, significantly more females preferred a call that was 50% longer.

facing two loudspeakers. The closer speaker played short calls, while the more distant speaker played long calls. The scientists found that 38 of 53 females (72%) went past the short-calling speaker to approach the long-calling speaker.

Gerhardt, Steven Tanner, and colleagues (2000) assessed female tastes more precisely. They gave females a choice between short calls and calls that were 10%, 20%, 30% and 50% longer. The longer the call, the more females liked it (Figure 11.27b). However, when offered a medium call and calls that were 11%, 22%, 33%, and 50% longer, females were less picky. Majorities preferred longer calls, but calls 50% longer were no more attractive than calls 11% longer (see also Bush et al. 2002). These results are consistent with observations made under more natural conditions by Joshua Schwartz, Bryant Buchanan, and Gerhardt (2001). In the midst of a chorus of singing males and other background noise, females are less able to make fine distinctions among males. But they do continue to discriminate, especially against short calls, and call length continues to explain a modest but statistically significant fraction of the variation in mating success among males.

The experiments of Gerhardt and colleagues show that female gray tree frogs are choosy and prefer, as predicted, males giving longer calls. And they show that longer-calling males are more likely to find mates.

Female choice, as illustrated by red-collared widowbirds and gray tree frogs, is thought to be the selective agent responsible for the evolution of a great variety of male advertisement displays—from the gaudy tail feathers of the peacock to the chirping of crickets to the leg tufts of wolf spiders. Some male displays, like those of peacocks, are loud and clear; others are more subtle. In barn swallows, for example, a mere 2 cm added to—or subtracted from—a male's tail feathers can alter his attractiveness enough to dramatically influence his reproductive success (Møller 1988). Why should the females care about such a small difference? And for that matter, why should females care about any of the advertisements, even loud ones, that males use to attract mates?

Choosy females raise a question. Why should females prefer one male display over another?

Models of Female Preference

Evolutionary biologists have devised numerous hypotheses about how and why female preferences evolve (see Mead and Arnold 2004; Jones and Ratterman 2009). We consider four explanations, starting with the simplest.

The Traits by Which Females Choose Their Mates May Be Arbitrary

The first scenario we consider derives from an idea advanced by Ronald Fisher in 1915 and is perhaps traceable to a remark made by T. H. Morgan in 1903 (reviewed in Andersson 1994). It makes the fewest assumptions possible for a mechanism with any potential to describe the evolution of a male display and a female preference for it. It can thus be considered a null model for sexual selection by female choice (Kirkpatrick and Ryan 1991; Prum 2010).

The simplest hypothesis for female choice is that female preferences are arbitrary.

Imagine a population with genetic variation among individuals for a display expressed in males and a preference expressed in females. Both the display and the preference can be arbitrary in the sense that they are neutral with regard to natural selection. The display can be any trait detectable by females. The preference need not require sophisticated perceptual or mental faculties on the part of females. It can involve any aspect of biology that makes a female more likely to mate with males showing particular values of the display versus other males. The ultimate source of the genetic variation in both the display and the preference is random genetic mutation.

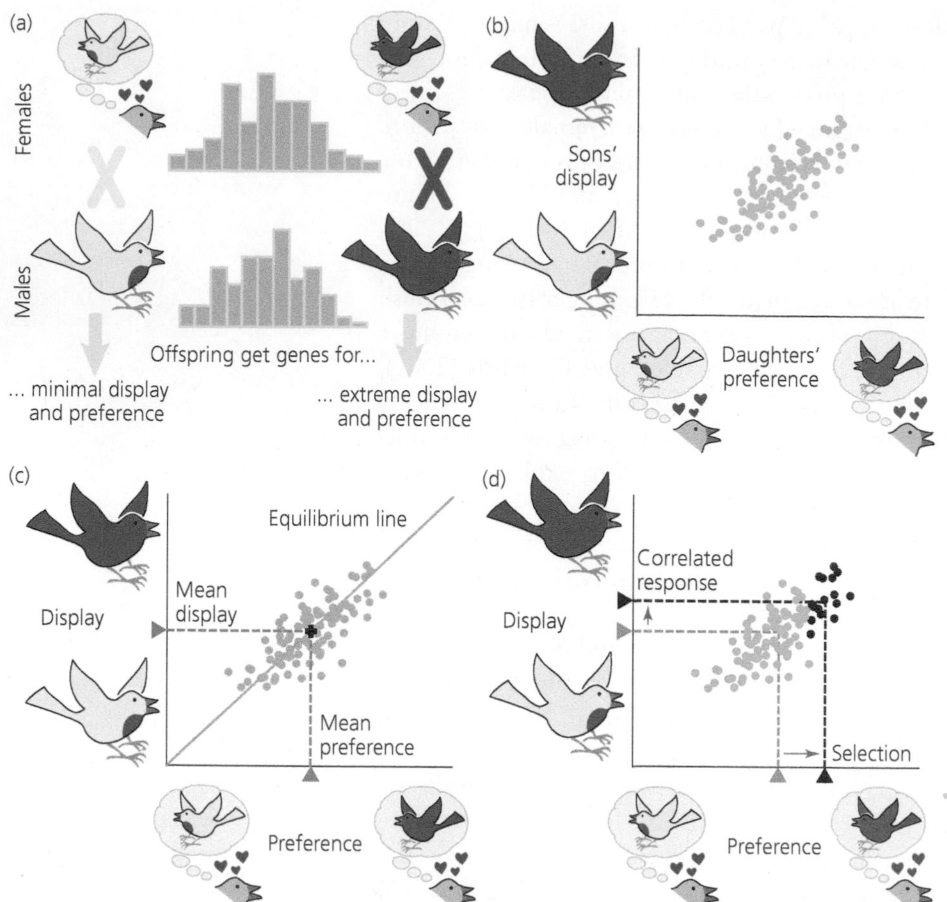

Figure 11.28 A simple model of evolution by female choice (a) Genetic variation in a female preference and a male display leads to assortative mating. (b) Assortative mating leads to a genetic correlation. Each point represents the midoffspring (that is, mean) values for a family. After Arnold (1983). (c) Populations in which the population mean display matches the population mean preference are at evolutionary equilibrium. Triangles represent means for the particular population shown. After Kirkpatrick (1987). (d) Selection on either the display or the preference leads to a correlated response in the other. Here, the black dots represent families chosen as breedstock because the daughters show extreme preferences. Triangles show the population means before and after selection. After Falconer (1989).

With these features alone, our model population already exhibits several interesting properties. To see the first of these, let the females choose mates **(Figure 11.28a)**. Females that favor minimal displays choose minimally decorated males, females that like medium displays pick moderately decorated males, and females that prefer extreme displays take extravagantly decorated males. In other words, the individuals in our simple population show assortative mating for preference and display.

To see the second property, consider the genetic variants our mated pairs will transmit to their offspring. Offspring that receive from their fathers variants conferring minimal displays also receive from their mothers variants conferring a preference for minimal displays. And offspring that receive from their father genes for extreme displays also receive from their mother variants conferring a preference for extreme displays. Across generations, preference and display will become genetically correlated (Figure 11.28b). If we examine the offspring in a number of families, we will find that the sons' displays are similar to the daughters' preferences.

To see the third property, reflect on the pattern of selection in the population. If the average male display matches the average female preference, then selection is stabilizing. Males with deviant displays are less likely to find mates. And females with deviant preferences have sons that are less likely to find mates. This means that there is a line of equilibrium, shown in Figure 11.28c. As long as a population's average display and average preference place it on the line, the population will not evolve.

Finally, to see the fourth property, consider what will happen if we impose selection on the population. We might, for example, choose as breeding stock only families whose females prefer extreme displays (Figure 11.28d). Because display and preference are genetically correlated, selecting for female preference yields a correlated response in display. Likewise, selecting for male display yields a correlated response in preference.

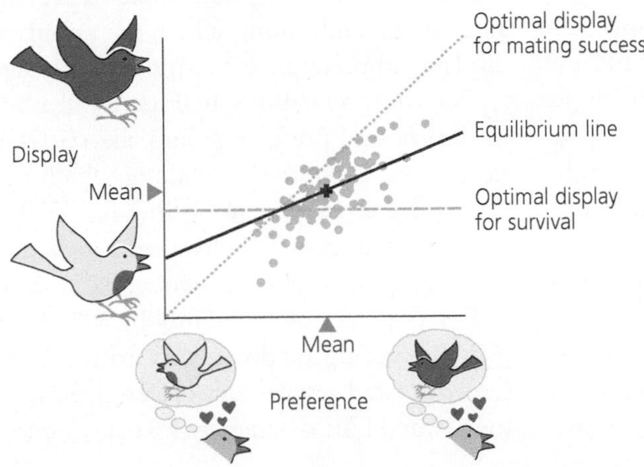

Figure 11.29 A simple model of evolution by female choice with natural selection on the display Populations are at evolutionary equilibrium when the mean display is a compromise between the average female preference versus the optimum for survival. After Kirkpatrick and Ryan (1991); Prum (2010).

The most obvious deficiency in our model is that the display is neutral with respect to natural selection. After all, Darwin devised the theory of evolution by sexual selection to explain the existence of displays that appear deleterious to survival. We can incorporate natural selection on the display by imagining that there is a value of the display trait that is optimal for survival **(Figure 11.29)**. This rotates the line of equilibrium. Now a population will not evolve as long as the average display is a compromise between the average preference and the optimal value for survival.

Following a verbal description by Ronald Fisher (1958), Russell Lande (1981) and Mark Kirkpatrick (1982) devised equations that describe a population evolving according to this model. In exploring how these equations behave, both researchers confirmed an intriguing phenomenon (see Prum 2010 for a review). Imagine that the population is displaced such that its mean display and preference are no longer on the equilibrium line. This might happen by genetic drift or any other mechanism of evolution. When it does, two outcomes are possible.

The population may return to the equilibrium line at some distance from the point it occupied before displacement. The population moves to a new equilibrium point because, as we showed in Figure 11.28d, selection on either the display or the character produces a correlated response in the other.

Alternatively, the population may evolve away from the line indefinitely. This second outcome is called runaway selection. It happens when correlated responses to selection carry the population further from equilibrium than it was immediately after the displacement. See **Computing Consequences 11.1** for a more detailed graphical explanation.

Female preferences for elaborate displays can evolve simply as a by-product of genetic variation for an arbitrary male trait and a corresponding female preference.

The rather startling implication is this: All that may be required for any genetically variable male trait to evolve a considerable distance away from the optimal value for survival is a genetically variable female preference and a bit of genetic drift. In other words, females may prefer long tail feathers, or longer, more complex love songs, simply because they prefer them.

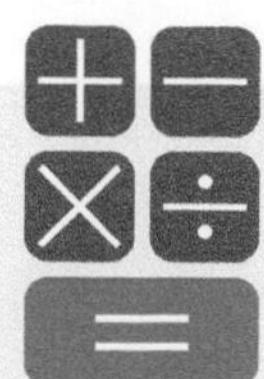

COMPUTING CONSEQUENCES 11.1

Runaway sexual selection

Here we present a brief graphical explanation of runaway sexual selection. For more detailed accounts, see Arnold (1983), Kirkpatrick and Ryan (1991), and Prum (2010). For full mathematical treatment, see Lande (1981) and Kirkpatrick (1982).

Recall (from Chapter 9) that populations evolve only when they harbor genetic variation. In particular, recall (from Figure 9.25) that when characters are correlated, populations are constrained to evolve more readily toward some combinations of trait values than others.

Imagine a model population, as described in the main text, displaced by genetic drift from equilibrium. Figure 11.30 shows four possibilities. Each graph shows the equilibrium line, data points representing families, and a dashed line along which the population means for display and preference can most readily evolve.

For the populations in Figure 11.30a the variation in display and preference and the correlation between them is such that the lines along which the populations can evolve cross the equilibrium line. Whether the current mean display is below or above the equilibrium line, the populations can evolve back to equilibrium.

For populations in Figure 11.30b, as the mean display evolves up or down toward the equilibrium line, the correlated response in preference carries it ever further away. This is runaway sexual selection.

(a)

Display

Equilibrium line

Preference

Display

Equilibrium line

Preference

(b)

Display

Equilibrium line

Preference

Display

Equilibrium line

Preference

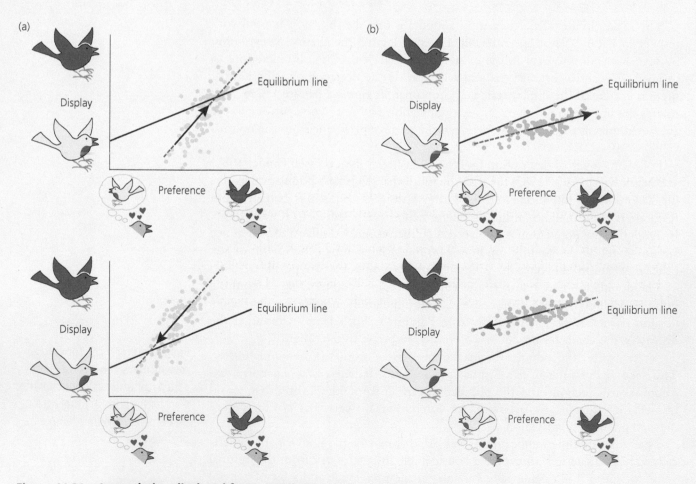

Figure 11.30 A population displaced from equilibrium may or may not return

Does the null model describe a process that happens in real populations? We can get some insight into whether it might by checking its key features.

Females sometimes show preferences for male traits that appear, at least at first glance, to be arbitrary. Aaron Owen and colleagues (2012), for example, tested whether female zebrafish (*Danio rerio*) prefer red GloFish males versus normal males **(Figure 11.31)**. Red GloFish are zebrafish genetically engineered to produce a red fluorescent protein. Across four experiments involving females reared in a variety of environments, most but not all females spent more time near a GloFish male than near a normal male.

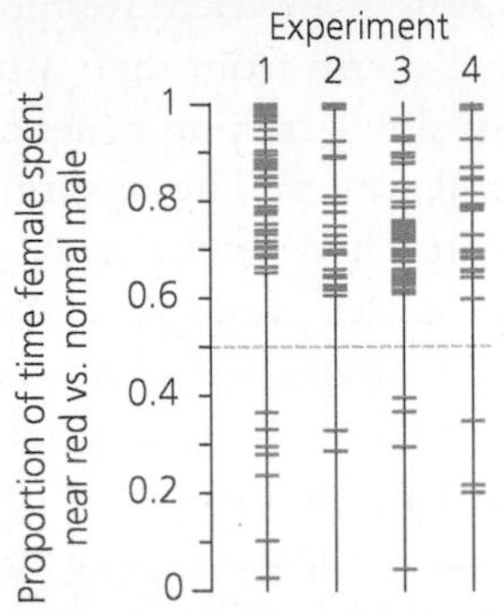

Figure 11.31 Female zebrafish prefer red fluorescent males Each tick represents a female. In all experiments, females given a choice spent, on average, significantly more time near a red fluorescent male than a normal male. Graph redrawn from Owen et al. (2012).

Species with choosy females have been documented to show genetic correlations between preference and display, such that selecting on one yields a response in the other. Research by Gerald Wilkinson and Paul Reillo (1994) on stalk-eyed flies provides an example. Stalk-eyed flies carry their eyes on the ends of long, thin stalks. In both sexes bigger flies have longer eyestalks, but males have longer stalks for their size than females. By day the flies are solitary and forage for rotting plants. In the evening, the flies congregate beneath overhanging stream banks, where they cling in small groups to exposed root hairs and spend the night **(Figure 11.32a)**. At dawn and dusk, flies roosting together often mate with each other.

Wilkinson and Reillo collected stalk-eyed flies (*Cyrtodiopsis dalmanni*) in Malaysia and established three laboratory populations. In one, the unselected line, the researchers chose their breeding stock at random. In another, the long-selected line, the researchers chose females at random and males for their extra-long stalks. In the third, the short-selected line, the researchers chose females at random and males for their extra-short stalks. After 13 generations, the populations had diverged substantially in eyestalk length.

Wilkinson and Reillo then performed paired-choice tests to assay female preferences. In each test, five females shared a cage with two males. One male was from the long-selected line, the other from the short-selected line. The males were separated by a clear plastic barrier, and each had his own artificial root hair on which to roost. In the center of the barrier was a hole, just large enough to allow the females to pass back and forth but too small for the males, with their longer eyestalks, to fit. Wilkinson and Reillo watched to see which male attracted more females. In both the control and the long-selected lines, more females chose to roost with the long-stalked male. In the short-selected line, however, more females chose to roost with the short-stalked male (Figure 11.32b). Artificial selection for short eyestalks in males had changed the mating preferences of females.

Female preferences and male displays can be self-reinforcing. The spotted cucumber beetle (*Diabrotica undecimpunctata howardi*) illustrates. The female

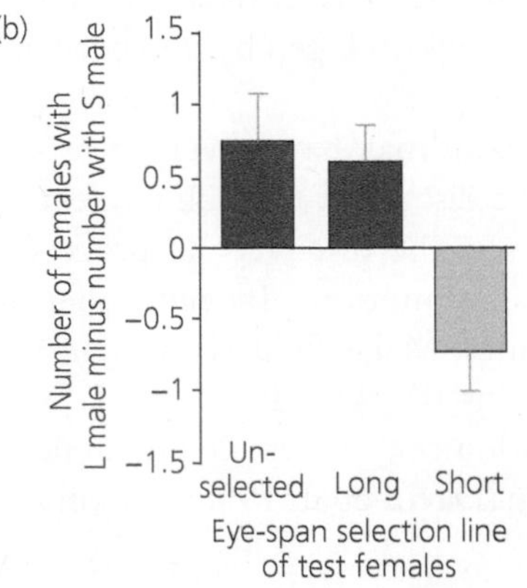

Figure 11.32 Correlated evolution of display and preference (a) Stalk-eyed flies (*Cyrtodiopsis whitei*) on a root hair. The largest is male; the rest female. Gerald S.Wilkinson, University of Maryland. (b) Females from the unselected and long lines preferred long-stalked males ($p = 0.0033$; $p = 0.005$). Those from the short line preferred short-stalked males ($p = 0.023$). From Wilkinson and Reillo (1994).

discriminates among males to some extent by how they smell (Brodt et al. 2006). For the most part, however, she decides whether to allow a male to fertilize her eggs based on how he behaves during copulation (**Figure 11.33**). While they copulate, the male uses his antennae to stroke the female (Tallamy et al. 2002). The faster his movements, the more likely she is to relax the muscle that allows him to deposit his sperm in her reproductive tract. Douglas Tallamy and colleagues (2003) assigned females at random to be paired with either a fast male or a slow male. Because they had no choice, most of the females eventually accepted sperm from their assigned mate. Females mated with fast-stroking males versus slow-stroking males laid similar numbers of eggs and saw their eggs hatch at similar rates. The offspring of the two groups of females developed at similar rates, reached similar sizes, and enjoyed similar fecundities. The offspring sired by fast males versus slow males survived to adulthood at a somewhat higher rate, but the difference was not statistically significant. The only significant differences Tallamy and colleagues could find between the offspring of fast versus slow males were that the sons of fast-stroking males grew up to be fast strokers themselves, and that—apparently as a result—the sons of fast-stroking males had their sperm accepted by females at over twice the rate as sons of slow-stroking males.

Figure 11.33 Spotted cucumber beetles in love The male is on the left.

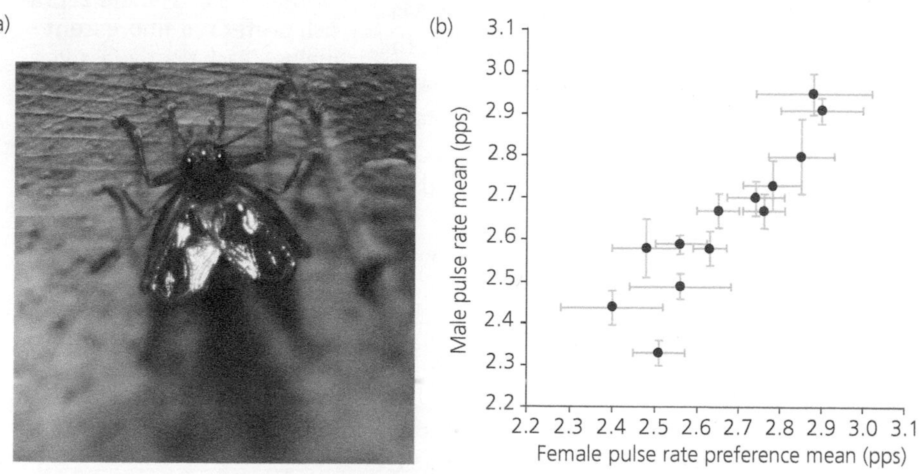

Figure 11.34 A cross-population correlation between display and preference in the Hawaiian cricket Each point represents the average pulse rate of male songs versus the average preference for female preferences. Error bars show 95% confidence intervals. From Grace and Shaw (2011). Photo by Jaime Grace.

Finally, it appears that populations can evolve along a line of equilibrium on which display and preference are matched. Jaime Grace and Kerry Shaw (2011) measured the mean pulse rate of male mating songs and the mean preferences of females in 13 populations of the cricket *Laupala cerasina* on the Big Island of Hawaii (**Figure 11.34a**). Figure 11.34b plots mean display against mean preference. The males' songs and the females' preferences are strongly correlated. This pattern is consistent with the notion that the populations diverged by the mechanism described in the Lande-Kirkpatrick null model.

Of course, Grace and Shaw's cricket populations may have diverged by some other mechanism. In all of the examples we have discussed, the null model could be ruled out—at least as the sole explanation—by the discovery of processes at work that are not included among the model's assumptions. Indeed, other processes are known to be at work in stalk-eyed flies. Males fight with each other, and the one with longer eyestalks typically wins (Burkhardt and de la Motte 1983, 1987; Panhuis and Wilkinson 1999). In addition, the female preference for long-stalked males may not be entirely arbitrary (David et al. 1998; Wilkinson et al. 1998; Wilkinson and Taper 1999).

Richard Prum (2010) has argued, however, that when we lack such evidence that other factors are at work we should provisionally assume that female preferences are arbitrary and that preferences and displays have evolved by Lande-Kirkpatrick sexual selection. Based on a comparative analysis of the elaborate courtship rituals of manakins, Prum (1997) concluded that the null model offers a plausible explanation for the evolution of many elements of male display.

The additional explanations for female choice that we consider next start with the null model and add features.

Choosy Females May Have Preexisting Sensory Biases

Perhaps the most obvious feature to add to the model is natural selection on the female preference. Under this new feature, there is an optimal preference that confers on females maximal survival and fecundity. This addition collapses the equilibrium line to a single point **(Figure 11.35)**. The population will remain unchanged only if the mean preference is the optimal preference and the mean display is a compromise between the optimal display for mating success and the optimal display for survival.

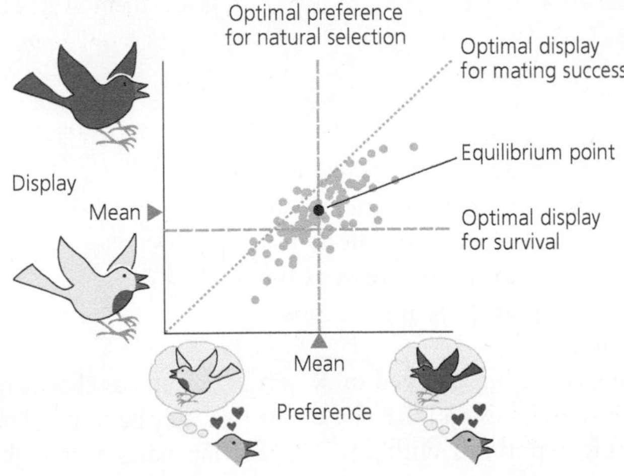

Figure 11.35 Natural selection on female preference reduces the equilibrium condition to a single point After Kirkpatrick and Ryan (1991).

One reason natural selection might act on preference is that females use their sensory organs and nervous systems for many other purposes than just discriminating among potential mates. Selection for such abilities as avoiding predators, finding food, and identifying members of the same species may result in sensory biases that make females particularly responsive to certain cues (see Enquist and Arak 1993). This pattern of responsiveness can select on males to display those cues. In other words, the preexisting bias, or sensory exploitation, hypothesis holds that female preferences evolve first and that male mating displays follow.

Research by Heather Proctor on the water mite *Neumania papillator* illustrates possible sensory exploitation (1991, 1992). Members of this species are small freshwater animals that live amid aquatic plants and make their living by ambushing copepods. Water mites have simple eyes that can detect light but cannot form images. Instead of vision, water mites rely heavily on smell and touch. Both males and females hunt copepods by adopting a posture that Proctor calls net-stance. The hunting mite stands on its four hind legs on an aquatic plant, rears up, and spreads its four front legs to form a sort of net. The mite waits until it detects vibrations in the water that might be produced by a swimming copepod, then turns toward the source of the vibrations and clutches at it.

Mating in *Neumania papillator* does not involve copulation. Instead, the male attaches sperm-bearing structures called spermatophores to an aquatic plant, then attempts to induce the female to accept them. He does this by fanning water across the spermatophores toward the female. The moving water carries to the female pheromones released by the spermatophores. When the female smells the pheromones, she may pick up the spermatophores.

(a)

(b)

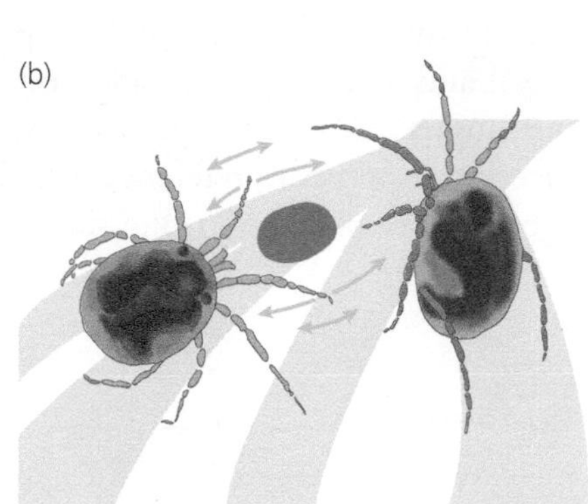

Figure 11.36 Courtship in the water mite, *Neumania papillator* (a) The female (on the right) is in net-stance, waiting to ambush a copepod; the male has found her and is now trembling his legs. (b) The female has turned toward the male in response to the trembling. The male has deposited spermatophores and is now fanning water across them. After Proctor (1991).

Male water mites search for females by moving about on aquatic vegetation. When a male smells a female, he walks in a circle while lifting and trembling his front legs **(Figure 11.36a)**. If the male has detected a female that is still there, not just the scent of one that has recently left, the female typically turns toward the trembling male. Often she also clutches at him. At this point, the male deposits his spermatophores and begins to fan (Figure 11.36b).

Sometimes choosy females may simply be responding to courting males as though the males were prey.

Proctor suspected that male leg-trembling during courtship evolved in *N. papillator* because it mimics the vibrations produced by copepods and thereby elicits predatory behavior from the female. She tested this hypothesis with a series of experiments in which she watched water mites under a microscope. First, Proctor measured the frequency of vibrations produced by trembling males and compared it to the frequency of vibrations produced by copepods. Water mites tremble their legs at frequencies of 10 to 23 cycles per second, well within the copepod range of 8 to 45 cycles per second. Second, Proctor observed the behavior during net-stance of female water mites when they were alone, when they were with copepods, and when they were with males. Females in net-stance rarely turned and never clutched unless copepods or males were present, and the behavior of females toward males was similar to their behavior toward copepods. Third, Proctor observed the responses to male mites of hungry females versus well-fed females. Hungry females turned toward males, and clutched them, significantly more often than well-fed females. All of these results are consistent with the hypothesis that male courtship trembling evolved to exploit the predatory behavior of females.

Males employing leg trembling during courtship probably benefit in several ways. First, males appear to use the female response to trembling to determine whether a female is actually present. Proctor observed that a male that has initiated courtship by trembling is much more likely to deposit spermatophores if the female clutches him than if she does not. Second, trembling appears to allow males to distinguish between receptive females versus unreceptive ones. Proctor observed that a male has a strong tendency to deposit spermatophores for the first female he encounters that remains in place after he initiates courtship, but that virgin females are more likely to remain in place than are nonvirgins. Third, males appear to use the female response to trembling to determine which direction the female is facing. Proctor observed that males deposit their spermatophores in front of the female more often than would be expected by chance. These benefits mean that a male that trembles should get more of his spermatophores picked up by females than would a hypothetical male that does not tremble. In other words, a male that trembles would enjoy higher mating success. This conclusion is consistent with the hypothesis that trembling evolved by sexual selection.

A key prediction of the sensory exploitation hypothesis is that net-stance evolved before male trembling. Proctor tested this hypothesis by using a suite of morphological characters to estimate the phylogeny of *Neumania papillator* and several related water mites. (Methods for estimating phylogenies were discussed in Chapter 4.) She noted which species have net-stance and which species have male courtship trembling. She then inferred the places on the phylogeny at which net-stance and courtship trembling are most likely to have evolved, based on the assumption that simpler evolutionary scenarios are more probable.

Proctor concluded that one of two evolutionary scenarios is most likely to be correct **(Figure 11.37)**. In the first scenario, net-stance and courtship trembling both evolved at the base of the branch that includes all species with either or both of these traits, and trembling was subsequently lost once (Figure 11.37a). In the second scenario, net-stance evolved at the base of the branch, and courtship trembling subsequently evolved twice (Figure 11.37b). The first scenario supplies insufficient evidence to test the prediction that net-stance evolved before trembling. We simply cannot tell, under this scenario, whether one trait evolved before the other, or the two traits evolved simultaneously. The second scenario is consistent with the prediction that net-stance evolved first. Which scenario is closest to the truth remains unknown. However, given the phylogenetic evidence in combination with the data from her observations of water mite behavior, Proctor concludes that sensory exploitation is the best explanation for the evolution of courtship trembling.

For another example in which male courtship displays appear to have evolved to exploit preexisting female sensory biases, see the work of John Christy and colleagues on fiddler crabs (Christy et al. 2003; Kim et al. 2007). Males in some species build sand hoods over the burrows from which they court females **(Figure 11.38)**. The team has evidence that these hoods are attractive to females because they look like good places to seek shelter from predators.

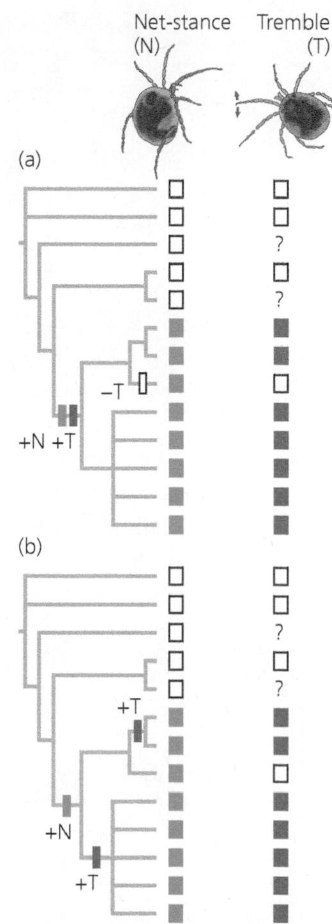

Figure 11.37 A phylogeny of the water mite *Neumania papillator* and several related species The boxes at the tips indicate which species have net-stance and which have male courtship trembling. Color indicates the trait is present; white indicates the trait is absent. The two versions of the tree show the most likely scenarios for the evolution of these traits. Redrawn from Proctor (1992).

Figure 11.38 A fiddler crab ready to attract a mate The sand hood over his burrow helps entice females to visit. Photo by John H. Christy, STRI.

Choosy Females May Benefit Directly through the Acquisition of Resources

Another scenario in which natural selection can act on female preference is when the exercise of the preference during mating confers advantage on females. In many species the males provide food, parental care, or some other resource that is beneficial to the female and her young. If it is possible to distinguish good providers from poor ones, then choosy females reap a direct benefit in the form of the resource provided. Such is the case in the hangingfly (*Bittacus apicalis*), studied by Randy Thornhill (1976). Hangingflies live in the woods of eastern North

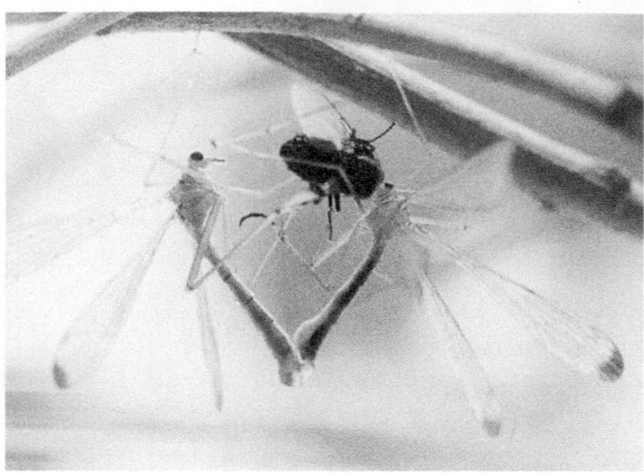

Figure 11.39 Courtship and mating in hangingflies A female (right) copulates with a male while eating a blowfly he has captured and presented to her. Photo by Randy Thornhill, University of New Mexico.

America, where they hunt for other insects. After a male catches an insect, he hangs from a twig and releases a pheromone to attract females. When a female approaches, the male presents his prey. If she accepts it, the pair copulates while she eats **(Figure 11.39)**.

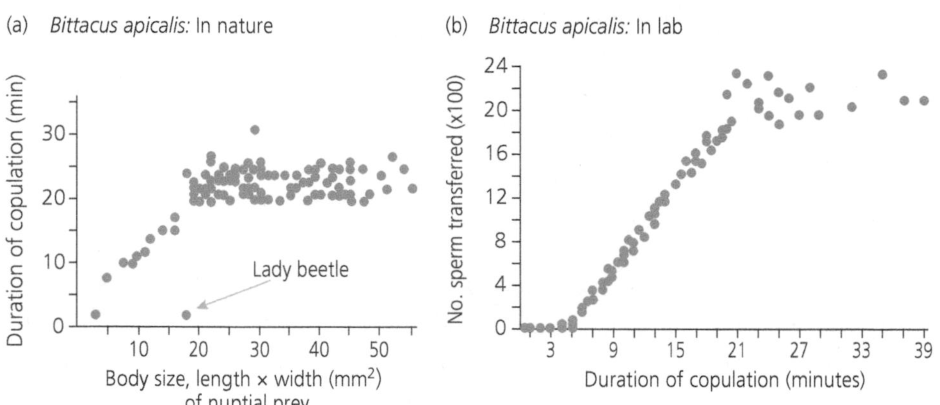

Figure 11.40 Female choice in hangingflies (a) The larger the gift the male presents, the longer the pair copulates. Copulation ends after about 20 minutes, even if the female is still eating. (b) The longer a pair copulates, the more sperm the female allows the male to transfer. The male must present a gift that takes at least 5 minutes to eat or the female breaks off the copulation without accepting any sperm. From Thornhill (1976).

The larger the prey, the longer it takes her to eat it, and the longer the pair copulates **(Figure 11.40a)**. The longer the pair copulates, the more sperm the female accepts from the male (Figure 11.40b). If she finishes her meal in less than 20 minutes, the female breaks off the copulation and flies away looking for another male and another meal. The female's preference for males bearing large gifts benefits her in two ways: (1) It provides her with more nutrients, allowing her to lay more eggs; and (2) it saves her from the need to hunt for herself. Hunting is dangerous, and males die in spider webs at more than twice the rate of females. The males behave in accord with the same kind of economic analysis: If the female is still eating after accepting all the sperm she can, the male grabs his gift back and flies off to look for a second female to share it with.

Choosy females sometimes get food or other resources from their mates.

Choosy Females May Get Better Genes for Their Offspring

The final scenario we consider is one in which natural selection acts on female preference not directly, as in our previous two scenarios, but indirectly via the production of high-fitness offspring. Perhaps the displays presented by males are indicators of genetic quality. If males with attractive displays are genetically superior to males giving less attractive displays, then choosy females will secure better genes for their offspring (Fisher 1915; Williams 1966; Zahavi 1975). In mathematical models of the good genes hypothesis, the required genetic correlation

Choosy females sometimes get better genes for their offspring.

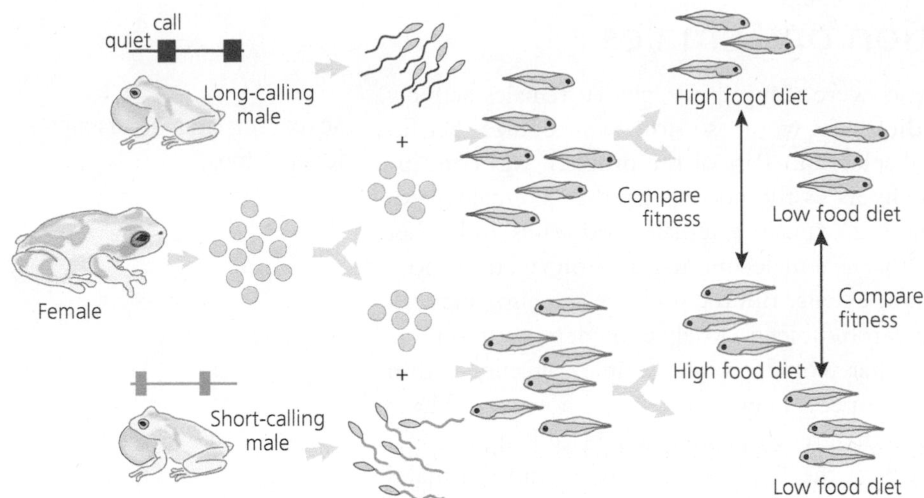

Figure 11.41 An experiment to determine whether male gray tree frogs that give long calls are genetically superior to males that give short calls Overall, the experiment included batches of eggs from 20 different females fertilized with sperm from 25 different pairs of males.

between display and general viability tends to break down over time (Kirkpatrick 1987). If the genetic correlation is between display and resistance to disease, however, it can be maintained by the perpetual arms race between host and pathogen (Hamilton and Zuk 1982).

Allison Welch and colleagues (1998) used an elegant experiment to investigate whether male gray tree frogs giving long calls are genetically superior to males giving short calls **(Figure 11.41)**. During two breeding seasons, the researchers collected unfertilized eggs from wild females. They divided each female's clutch into separate batches of eggs, then fertilized one batch of eggs with sperm from a long-calling male and the other batch of eggs with sperm from a short-calling male. They reared some of the tadpoles from each batch of eggs on a generous diet and the others on a restricted diet.

This experimental design allowed Welch and colleagues to compare the fitness of tadpoles that were maternal half-siblings—that is, tadpoles with the same mother but different fathers. When comparing tadpoles fathered by long-calling males versus short-calling males, the researchers did not have to worry about uncontrolled differences in the genetic contribution of the mothers, because the mothers were the same.

Welch and colleagues measured five aspects of offspring performance related to fitness: larval growth rate (faster is better); time as a tadpole (shorter is better); mass at metamorphosis (bigger is better); larval survival; and post-metamorphic growth (faster is better). The results of their comparisons appear in **Figure 11.42**. In 18 comparisons between the offspring of long-calling males versus short-calling males, there was either no significant difference or better performance by the offspring of long callers. The offspring of short callers never did better. Overall, the data indicate that the offspring of long-calling males have significantly higher fitness. This result is consistent with the good genes hypothesis. At least one genetic difference between long-calling males versus short-calling males is that they feed more voraciously as tadpoles, even when a predator is present (Doty and Welch 2001). The nature of other genetic differences between long-calling frogs versus short-calling frogs is a subject of ongoing research.

The overarching lesson of our exploration of female choice is that a variety of mechanisms can cause the evolution of male mating displays, even displays that are costly for survival, and that different mechanisms appear to be at work in different species.

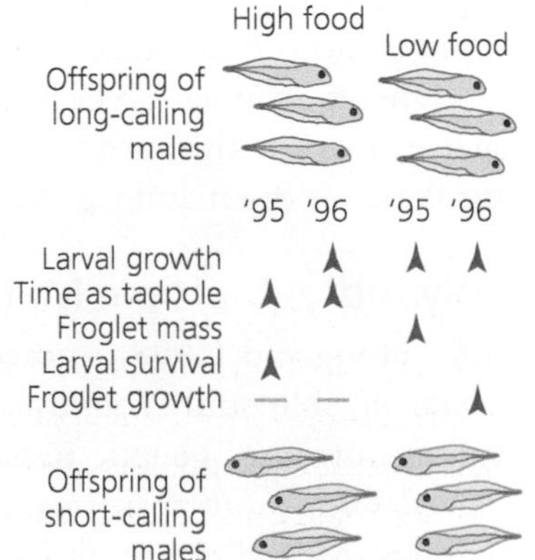

Figure 11.42 Fitness of offspring of long-calling versus short-calling males Arrowheads indicate that offspring of long callers did better. Blanks indicate there was no significant difference. Dashes indicate no data were taken. Overall, offspring of long callers had significantly higher fitness than their paternal half-sibs fathered by short callers ($p < 0.0008$). After Welch and colleagues (1998).

11.4 Sexual Selection on Females

Sexual selection occurs among females too.

We mentioned earlier that biologists were slow to accept that females actively discriminate among their suitors. Biologists were also slow to recognize that females themselves experience sexual selection. Part of the problem was conceptual. When a single insemination delivers vastly more sperm than a female needs to fertilize her eggs, and when mating exposes a female to diseases and other dangers, biologists could see little that a female might gain from mating more than once, let alone with more than one male. But the problem was also practical. Few biologists could watch their research subjects closely enough to witness multiple matings. Modern genetic tests that would reveal multiple paternity within clutches or litters had not been developed. There were hints, however. There were species of birds and fish, such as the greater painted snipe (see Knight 2002) and the broad-nosed pipefish, in which the females are showier and the males provide the parental care. And there were biologists, such as Sarah Hrdy (1979), who watched as females in the species they studied pursued multiple mates.

When biologists began using genetic tests, they discovered that multiple mating is rampant. Lisle Gibbs and colleagues (1990), for example, tested two families of red-winged blackbirds. In both nests some of the chicks had been fathered not by the mother's social mate, but by the male who owned the territory next door. Elizabeth Gray (1997) used DNA fingerprints to assess the frequency of extrapair copulation in an entire population of red-winged backbirds in eastern Washington state. She estimated that in a given breeding season, between 50% and 64% of all nests contained at least one chick sired by a male other than its social parent. With similar methods, biologists have discovered that many socially monogamous birds engage in frequent extrapair copulations.

Genetic analyses have, in fact, revealed that females in a great many animal species mate with more than one male (see Knight 2002). True monogamy, it turns out, is the mating system that is rare. The obvious question is: Why?

Polyandry: Multiple Mating by Females

John Hoogland (1998) studied polyandry in Gunnison's prairie dogs (*Cynomys gunnisoni*). His study is unusual, and valuable, because it relied on careful observation instead of genetic tests. Genetic tests can identify females that have conceived with more than one male. Direct observation can, in addition, identify matings that did not result in conception or birth. Prairie dogs mate in burrows, so Hoogland did not witness copulations. But he could infer from their aboveground behavior what the prairie dogs were doing. Among other things, before or after copulating, male prairie dogs often bark a unique call.

During 15,000 person-hours of observation over a 7-year study, Hoogland and his assistants gathered data on the mating behavior and reproductive success of more than 200 female prairie dogs. Hoogland found that 65% of these females mated with more than one male, and a few mated with as many as four or five. The females have good reason to seek multiple mates. The probability of getting pregnant and giving birth was 92% for females that had only one or two mates, but 100% for females that had three or more. Hoogland thinks the reason is that some males are either permanently sterile or temporarily depleted of sperm. Female prairie dogs are fertile only once a year, and for a very short time. By mating with more than one male, a female increases her chances of receiving enough viable sperm to fertilize all her eggs.

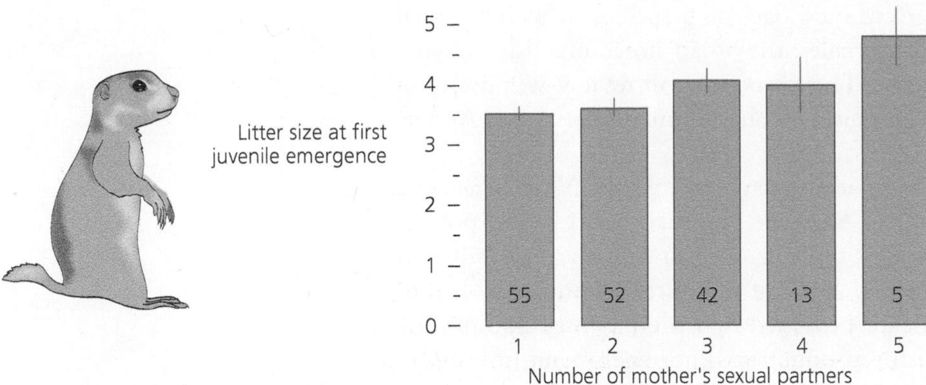

Figure 11.43 Female Gunnison's prairie dogs increase their reproductive success by mating with multiple males Bars show average litter sizes, ±1 standard error, on the day the pups came aboveground for the first time. Numbers on bars show the number of females in each category. Litter size was correlated with number of partners ($r = 0.226$; $p = 0.003$). Redrawn from Hoogland (1998).

Among females that gave birth, litter size on the day the pups first came aboveground increased with number of mates **(Figure 11.43)**. This pattern may in part be due to associations with a third factor, female body size. Larger females can support larger pregnancies. They may also attract more attention from amorous males, or be better able to resist a mate's attempts to prevent them from seeking other mates. But associations with body size are not the whole story, because when Hoogland used statistical methods to remove them, females with more mates still had larger litters. This might be due to increased genetic diversity among pups with different fathers. The more genetically diverse a female's pups, the less likely all will be killed by the same disease or parasite. Or perhaps sperm competition increases the chance that the sperm that fertilize eggs are of high genetic quality. Although the precise reasons remain to be discovered, Hoogland's data show that female Gunnison's prairie dogs experience sexual selection.

Females often benefit from mating with more than one male.

In some species, benefits garnered by mating with multiple males have led to the evolution of special mechanisms that help females avoid mating twice with the same partner. For example, when given a choice, female crickets (*Gryllodes sigillatus*) show a clear preference for mating with novel males versus previous mates. Do the females avoid mating twice with the same male by remembering what he smells like? Or do the females recognize their own lingering scent? Tracie Ivy and colleagues (2005) let female crickets choose between a novel male and the inbred brother of a male they had already mated with. The females did not discriminate between the males. But when the researchers gave females a choice between a novel male and a male who had mated with the female's own inbred sister, most chose the novel male. Ivy and colleagues concluded that female crickets scent-mark their mates so that they can avoid them in the future.

When Sexual Selection Is Stronger for Females than for Males

In Section 11.1 we argued that knowing the relative strength of sexual selection on males versus females allows us to predict how the sexes will approach an opportunity to mate. The sex subject to stronger selection should be competitive; the other should be choosy. We saw these predictions confirmed when sexual selection is stronger for males. What happens when it is stronger for females?

Species in which males invest more in each offspring, and are thus a limiting resource for females, provide a valuable opportunity to test the rules of sexual selection.

Sexual selection is most likely to be stronger for females than males when males provide parental care. When fathers care for young, male parental investment per offspring may be comparable to, or even greater than, female parental investment. Species with male parental care include humans, many fish, about 5% of frogs, and over 90% of birds. When males actually do invest more per offspring than females, access to mates will be a limiting resource for females.

In Figure 11.12 (page 415), we examined data on a species in which sexual selection is stronger for females than for males: the broad-nosed pipefish, *Syngnathus typhle*. We return to pipefish now. The work we will review was done on *S. typhle* and another pipefish, *Nerophis ophidion*, by Gunilla Rosenqvist, Anders Berglund, and colleagues.

Recall that in pipefish males provide all the parental care. In *N. ophidion*, the male has a brood patch on his belly; in *S. typhle*, the male has a brood pouch. In both species, the female lays her eggs directly onto or into the male's brood structure. The male supplies the eggs with oxygen and nutrients until they hatch.

Although the extensive parental care provided by male pipefish requires energy, the pivotal currency for pipefish reproduction is not energy but time (Berglund et al. 1989). Females of both *N. ophidion* and *S. typhle* can make eggs faster than males can rear them to hatching. As a result, access to male brood space limits female reproductive success. If the theory of sexual selection we have developed is correct, then in these pipefish the females should compete over access to mates, and the males should be choosy.

In *N. ophidion*, females are larger than males and have two traits males lack: dark blue stripes and skin folds on their bellies. These traits appear to function primarily as advertisements for attracting mates. Females develop skin folds during the breeding season and lose them after, and in captivity females develop skin folds only when males are present (Rosenqvist 1990). In paired-choice tests **(Figure 11.44a)**, *N. ophidion* males are choosy, preferring larger females (Figure 11.44b) and females with larger skin folds (Figure 11.44c). Females, in contrast,

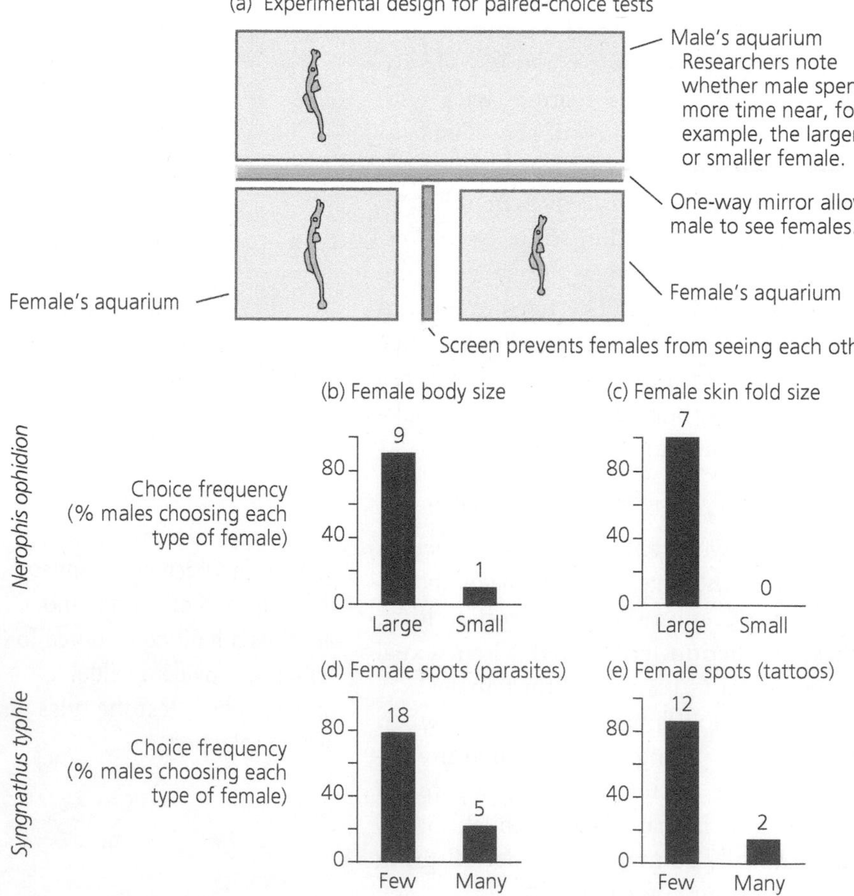

Figure 11.44 Male choice in pipefish (a) In paired-choice tests, researchers place a male pipefish in an aquarium from which he can see two females. The researchers infer which female the male would prefer as a mate from where he spends more of his time. In (b–d) the numbers above the bars indicate the number of males tested. After Rosenqvist and Johansson (1995). (b) Male pipefish prefer large females ($p = 0.022$). Replotted from Rosenqvist (1990). (c) Male pipefish prefer females with large skin folds ($p = 0.016$). Replotted from Rosenqvist (1990). (d) Male pipefish prefer females with fewer black spots caused by parasites ($p < 0.05$). Replotted from Rosenqvist and Johansson (1995). (e) Males still prefer females with fewer spots, even when the spots are tattooed onto parasite-free females ($p < 0.01$). Replotted from Rosenqvist and Johansson (1995).

appear to be less choosy. In paired-choice tests, females showed no tendency to discriminate between males of different sizes (Berglund and Rosenqvist 1993).

In *S. typhle,* the males and females are similar in size and appearance. Females, however, can change their color to intensify the zigzag pattern on their sides (Berglund et al. 1997; Bernet et al. 1998). The females compete with each other over access to males (Berglund 1991) and while doing so display their dark colors. Females initiate courtship and mate more readily than males (Berglund and Rosenqvist 1993). Males are choosy (Rosenqvist and Johansson 1995). In paired-choice tests (Figure 11.44a), male *S. typhle* prefer females showing fewer of the black spots that indicate infection with a parasitic worm, whether the black spots were actually caused by parasites (Figure 11.44d) or were tattooed onto the females (Figure 11.44e). This choosiness benefits the males directly, because females with fewer parasites lay more eggs for the males to fertilize and rear.

The mating behavior of pipefish males and females is consistent with the theory of sexual selection. Other examples of "sex-role-reversed" species whose behavior appears to support the theory include moorhens (Petrie 1983), spotted sandpipers (Oring et al. 1991a,b; 1994), giant waterbugs (see Andersson 1994), and some species of katydids (Gwynne 1981; Gwynne and Simmons 1990).

11.5 Sexual Selection in Plants

Sexual selection theory can be applied to plants as well as animals.

Plants are sometimes sexually dimorphic (Renner and Ricklefs 1995). Orchids in the genus *Catasetum* provide a dramatic example. Individual plants produce male and female flowers at different times. The flowers of the opposite sex are so different that early orchid systematists placed individuals with male flowers **(Figure 11.45a)** in one genus and individuals with female flowers (Figure 11.45b)

Figure 11.45 Sexual dimorphism in plants In the orchid *Catasetum barbatum*, the flowers of males (a) and females (b) are strikingly different. Photo (b) from Sharon Dahl. In the herb *Wurmbea dioica* (c), males make larger flowers. Redrawn from Vaughton and Ramsey (1998).

in another. The herb *Wurmbea dioica,* from Australia, provides a more typical example. The males make larger flowers than the females (Figure 11.45c). We have seen that sexual selection can explain sexual dimorphism in animals. Can it also explain sexual dimorphism in plants?

Many of the ideas we have developed about sexual selection in the context of animal mating can, in fact, be applied to plants (Bateman 1948; Willson 1979; but see Grant 1995). In plants, mating involves the movement of pollen from one individual to another. The recipient, the seed parent, must produce a fruit. As a result, the seed parent may make a larger reproductive investment per seed than the pollen donor, which must make only pollen. When pollen is transported from individual to individual by animals, a plant's access to mates is a function of its access to pollinators. Based on the principles of sexual selection in animals, we can hypothesize that access to pollinators limits the reproductive success of pollen donors to a greater extent than it limits the reproductive success of seed parents.

Maureen Stanton and colleagues (1986) tested this hypothesis in wild radish (*Raphanus raphanistrum*). Wild radish is a self-incompatible annual herb that is pollinated by a variety of insects, including honeybees, bumblebees, and butterflies. Many natural populations of wild radish contain a mixture of white-flowered and yellow-flowered individuals. Flower color is determined by a single locus: White (*W*) is dominant to yellow (*w*). Stanton and colleagues set up a study population with eight homozygous white plants (*WW*) and eight yellow plants (*ww*). The scientists monitored the number of pollinator visits to plants of each color, then measured reproductive success through female and male function.

Measuring reproductive success through female function was easy: The researchers just counted the number of fruits produced by each plant of each color. Measuring reproductive success through male function was harder; in fact, it was not possible at the level of individual plants. Note, however, that a yellow seed parent (*ww*) will produce yellow offspring (*ww*) if it mated with a yellow pollen donor (*ww*), but white offspring (*Ww*) if it mated with a white pollen donor (*WW*). Thus, by rearing the seeds produced by the yellow seed parents and noting the color of their flowers, Stanton and colleagues could compare the population-level reproductive success of white versus yellow pollen donors. The relative reproductive success of pollen donors through yellow seed parents should be a reasonable estimate of the pollen donors' relative reproductive success through seed parents of both colors. The scientists repeated their experiment three times.

As Stanton and colleagues expected from previous research, the yellow-flowered plants got about three-quarters of the pollinator visits **(Figure 11.46a)**. If reproductive success is limited by pollinator visits, the yellow-flowered plants should also have gotten about three-quarters of the reproductive success. This was true for reproductive success through pollen donation (Figure 11.46c), but not for reproductive success through seed production (Figure 11.46b). Reproductive success through seed production was simply proportional to the number of plants of each type. These results are consistent with the typical pattern in animals: The reproductive success of males is more limited by access to mates than is the reproductive success of females. The results also suggest that the evolution of showy flowers that attract pollinators has been driven more by their effect on male versus female reproductive success (Stanton et al. 1986).

If it is true, in general, that the number of pollinator visits is more important to male than to female reproductive success, then in animal-pollinated plant species with separate male and female flowers, the flowers should be dimorphic and the

(a) Pollinator discrimination

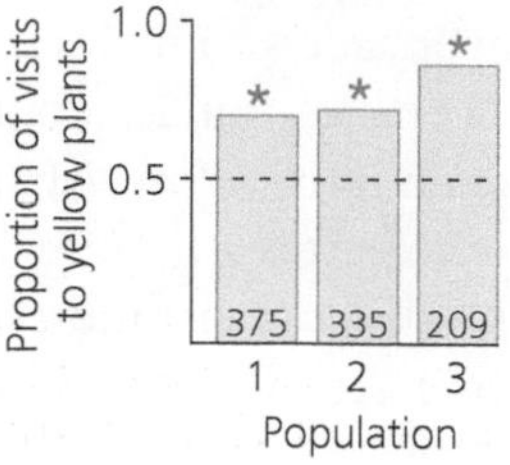

(b) Maternal function

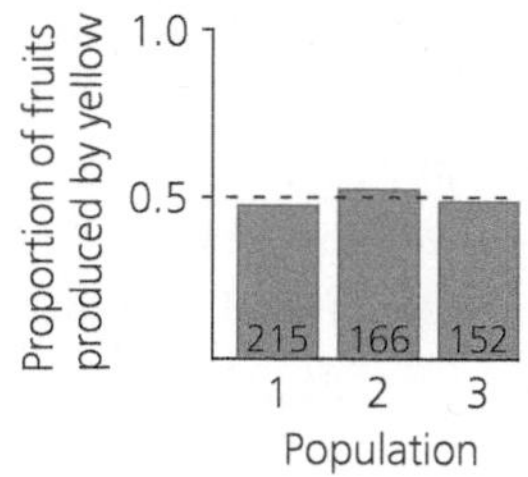

(c) Paternal function

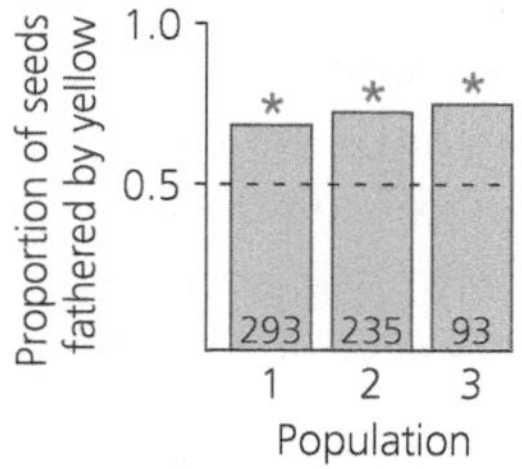

© 1986 AAAS

Figure 11.46 Reproductive success through pollen donation is more strongly affected by the number of pollinator visits than is reproductive success through the production of seeds The numbers inside each bar represent the number of pollinator visits (a), the number of fruits (b), and the number of seeds (c) examined. Bars marked with an asterisk have heights significantly different from 0.5 ($p < 0.0001$). (a) In populations with equal numbers of white and yellow flowers, yellow got most of the visits. (b) Despite the inequality in visits, white- and yellow-flowered plants produced equal numbers of fruits. (c) Most of the seeds, however, were fathered by yellow plants. From Stanton et al. (1986).

From "Floral evolution: Attractiveness to pollinators increases male fitness," *Science* 232: 1625–1727, Figure 1. Reprinted with permission from AAAS.

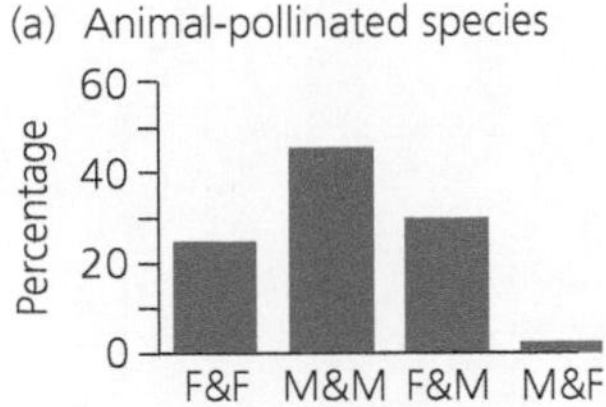

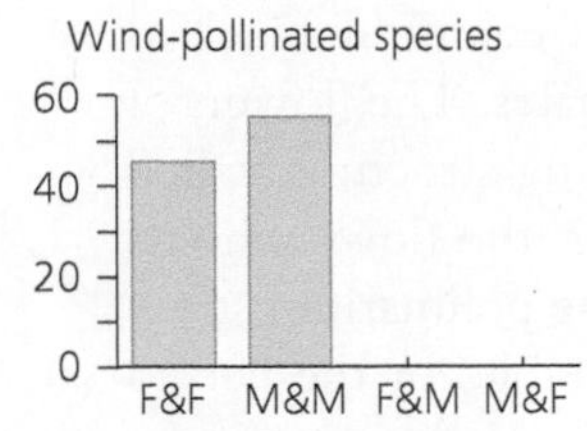

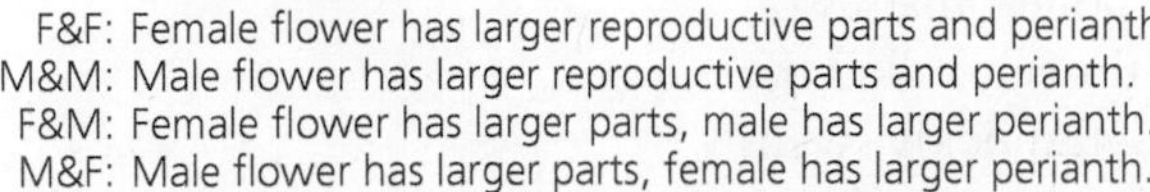

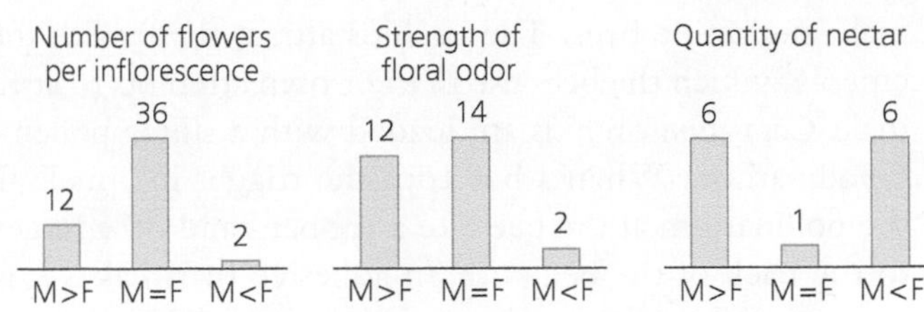

M>F: Male investment is greater than female investment.
M=F: Male investment and female investment are equal.
M<F: Male investment is less than female investment.

Figure 11.47 Patterns of sexual dimorphism in plants with separate male and female flowers (a) In 29% of 42 animal-pollinated species (left), the female has larger reproductive parts but the male has the larger perianth, a pattern significantly different from that for wind-pollinated species ($p < 0.001$). From Delph et al. (1996). (b) When animal-pollinated plants have flowers sexually dimorphic for investment in pollinator attraction, investment by males tends to be larger for two of three traits studied ($p < 0.01$; $p = 0.01$; N.S.). Drawn from data in Delph et al. (1996).

male flowers should be more attractive. Lynda Delph and colleagues (1996) tested this hypothesis with a survey of animal- and wind-pollinated plants, including both dioecious species (separate male and female individuals) and monoecious species (separate male and female flowers on the same individual).

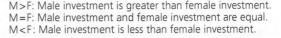

Delph and her coauthors first noted that the showiest parts of a flower, the petals and sepals that together form the perianth, serve not only to attract pollinators but also to protect the reproductive structures when the flower is developing in the bud. If protection were the only function of the perianth, then the sex that has the bigger reproductive parts should always have the bigger perianth. This was the case in all 11 wind-pollinated species Delph and colleagues measured **(Figure 11.47a, right)**. If, however, pollinator attraction is also important, and more important to males than to females, then there should be species in which the female flowers have bigger reproductive parts, but the male flowers have bigger perianths. This was the case in 29% of the 42 animal-pollinated plants Delph and colleagues measured (Figure 11.37a, left). Furthermore, in species that are dimorphic, male function tends to draw a greater investment in number of flowers per inflorescence and strength of floral odor, although not in quantity of nectar (Figure 11.37b). These results are consistent with the hypothesis that sexual selection, via pollinator attraction, is often stronger for male than for female flowers.

When male reproductive success is limited by access to pollinators, but female reproductive success is not, male flowers may evolve showier displays than female flowers.

Can sexual selection explain the particular examples of sexual dimorphism we introduced at the beginning of this section? Recall that in the herb *Wurmbea dioica,* males make larger flowers than females. This plant is pollinated by bees, butterflies, and flies. Glenda Vaughton and Mike Ramsey (1998) found that bees and butterflies visit larger flowers at higher rates than smaller flowers. As a result, pollen is removed from large flowers more quickly than from small flowers. Males with large flowers may benefit from exporting their pollen more quickly if a head start allows their pollen to beat the pollen of other males in the race to females' ovules. In addition, larger male flowers make more pollen, giving the pollen donor more chances to win. For females, larger flowers probably do not confer any benefit. Female flowers typically receive more than four times the pollen needed to fertilize all their ovules, and seed production is therefore not limited by pollen. These patterns are consistent with the hypothesis that sexual selection on males is responsible for the sexual dimorphism in flower size.

Orchids in the genus *Catasetum,* the plants with the dramatically dimorphic flowers, have an unusual pollination system. They are pollinated exclusively by male euglossine bees. The orchids attract them with fragrant chemicals, such as cineole, which the bees use in their own attempts to attract mates. The flowers of male *Catasetum* orchids are loaded with a single pollen-bearing structure, called a pollinarium. When a bee trips the trigger in a male flower, the flower shoots the pollinarium at the bee like a rubber band off a finger. The pollinarium sticks to the back of the bee with an adhesive that makes it impossible for the bee to remove. When the bee later visits a female flower, one of the pollen masses on the pollinarium lodges in the receptive structure on the female flower, called the stigmatic cleft, and is torn from the pollinarium. The stigmatic cleft quickly swells shut. This means that a female flower typically receives pollen from just one male.

Gustavo Romero and Craig Nelson (1986) observed bees pollinating *Catasetum ochraceum*. The researchers found that after being shot with a pollinarium by one male flower, the bees avoided visiting other male flowers but continued to forage in female flowers. Romero and Nelson offer the following scenario to account for the sexual dimorphism in the flowers of *C. ochraceum*. At any given time, there are many more male flowers blooming than female flowers. This, combined with the fact that females accept pollen from only one male, means there is competition among male flowers over opportunities to mate. The competition is further intensified because a second pollinarium attached to a bee would probably interfere with the first. A male flower that has attracted a bee and loaded it with a pollinarium would be at a selective advantage if it could prevent the bee from visiting another male flower. It is therefore adaptive for male flowers to train bees to avoid other male flowers, so long as they do not also train the bees to avoid female flowers. If this scenario is correct, forcible attachment of the pollinarium to the bee and sexually dimorphic flowers make sense together, and both are due to competition for mates—that is, to sexual selection.

11.6 Sexual Dimorphism in Humans

Like the other organisms discussed in this chapter, humans are sexually dimorphic. **Figure 11.48** documents just one of our differences. Humans vary considerably in height both within and among populations, but almost everywhere the average man is about 10% taller than the average woman. We now ask whether the human dimorphism in size is the result of sexual selection. It is a difficult question to answer because sexual selection concerns mating behavior. The evolutionary significance of human behavior is hard to study for at least two reasons:

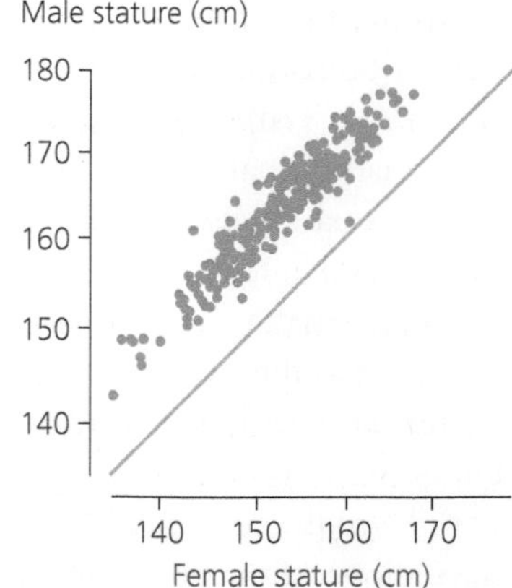

Figure 11.48 Women and men differ in height For each of more than 200 human societies, the average height of men versus women. The diagonal is the line of equality. From Rogers and Mukherjee (1992).

- Human behavior is driven by a complex combination of culture and biology. Studies based on the behavior of people in any one culture provide no means of disentangling these two influences. Cross-cultural studies can identify universal traits or broad patterns of behavior, either of which may warrant biological explanations. Cultural diversity is rapidly declining, however, and some biologists feel that it is no longer possible to do a genuine cross-cultural study.
- Ethical and practical considerations prohibit many of the kinds of experiments we might conduct on individuals of other species. This means that most studies of human behavior are observational. Observational studies can identify correlations between variables, but they offer little evidence of cause and effect.

Human behavior is fascinating, however, and we therefore proceed, with caution, to briefly consider the question of sexual selection and body size in humans.

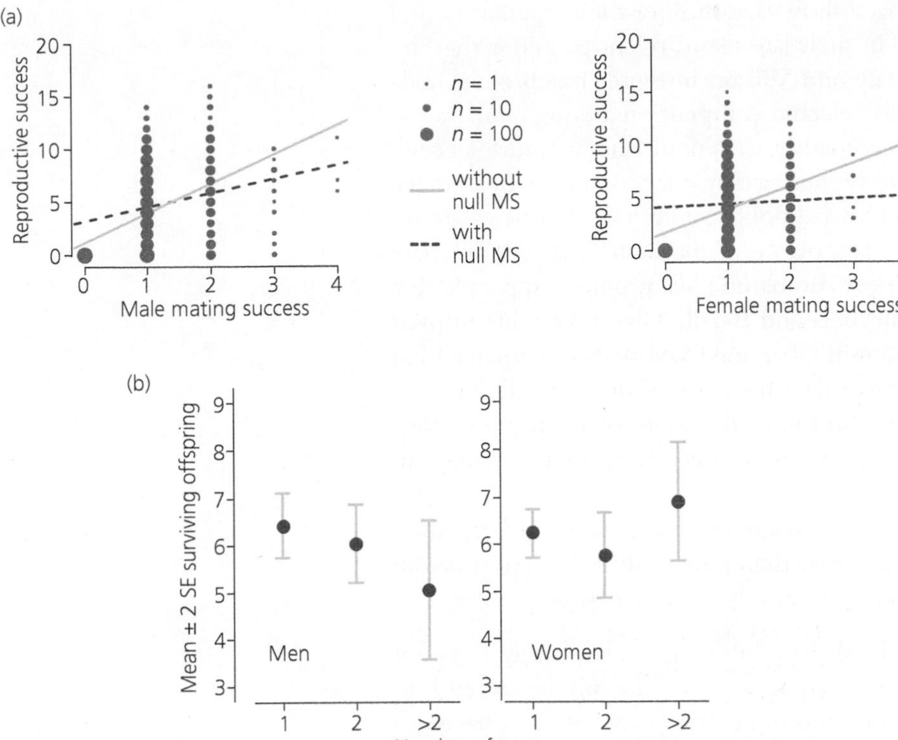

Figure 11.49 Sexual selection on men and women (a) Pre-industrial Finnish. From Courtiol et al. (2012). (b) The Pimbwe, a horticultural society in Tanzania. MS is mating success. From Borgerhoff Mulder (2009).

The most basic knowledge of human reproductive biology suggests that the reproductive success of men is more likely to be associated with mating success than is the reproductive success of women. The data in **Figure 11.49a**, from a pre-industrial Finnish society, are consistent with this prediction. Among individuals with at least one marriage, additional marriages increase the reproductive success of men but add little to the reproductive success of women. The data in Figure 11.49b, however, are not consistent with the prediction. These data come from a marginalized horticultural society in Tanzania. Clearly, the association between mating success and reproductive success varies from population to population. We should not be surprised to see humans pursuing different reproductive strategies in different circumstances (Brown et al. 2009). Indeed, even within our own culture, most of us know individuals who seem to be pursing reproductive success through monogamy and parental investment and others who seem to be pursuing reproductive success through high mating success.

If we assume, for the sake of argument, that sexual selection is typically stronger for men than for women, the most obvious kind of sexual selection to look at is male–male competition. This is because male–male competition drives the evolution of large male size in a great variety of other species. Men do, on occasion, compete among themselves over access to mates. But so do women. Do men compete more intensely?

On the reasoning that homicide is an unambiguous indication of conflict, and that virtually all homicides are reported to the police, Martin Daly and Margo Wilson (1988) assembled data on rates of same-sex homicide from a variety of modern and traditional cultures. In all of these cultures, men kill men at much higher rates than women kill women. In the culture with the most balanced rates of male–male versus female–female killings, men committed 85% of the same-sex homicides. In several cultures, men committed all of the same-sex homicides.

Data from the United States and Canada show that most of the perpetrators, and victims, of male–male homicides are in their late teens, twenties, and early thirties. On these and other grounds, Daly and Wilson interpret much male–male homicide as a manifestation of sexually selected competition among men.

If Daly and Wilson's interpretation is correct, then men who are more successful in male–male combat should have higher mating success and higher fitness, at least in premodern cultures without formal police and criminal justice systems. Napoleon Chagnon (1988) reported data on the Yąnomamö that confirm this prediction, at least for one culture. The Yąnomamö are a premodern people that live in the Amazon rain forest in Venezuela and Brazil. They take pride in their ferocity. Roughly 40% of the adult men in Chagnon's sample had participated in a homicide, and roughly 25% of the mortality among adult men was due to homicide. The Yąnomamö refer to men who have killed as *unokais*. Chagnon's data show that *unokais* have significantly more wives, and significantly more children, than non-*unokais* **(Figure 11.50a)**.

The Yąnomamö fight with clubs, arrows, spears, machetes, and axes. It would be reasonable to predict that *unokais* are larger than non-*unokais*. Chagnon (1988) reports, however, that

> Personal, long-term familiarity with all the adult males in this study does not encourage me to conclude at this point that they could easily be sorted into two distinct groups on the basis of obvious biometric characters, nor have detailed anthropometric studies of large numbers of Yąnomamö males suggested this as a very likely possibility.

Stephen Beckerman and colleagues (2009) conducted a similar study of the Waorani, a traditional society that lives in the rain forest of Ecuador. The Waorani are even more violent than the Yąnomamö. Over 40% of all deaths are due to homicide. Among the Waorani, in contrast to the Yąnomamö, zealous warriors had fewer wives and surviving children, although the differences were not statistically significant.

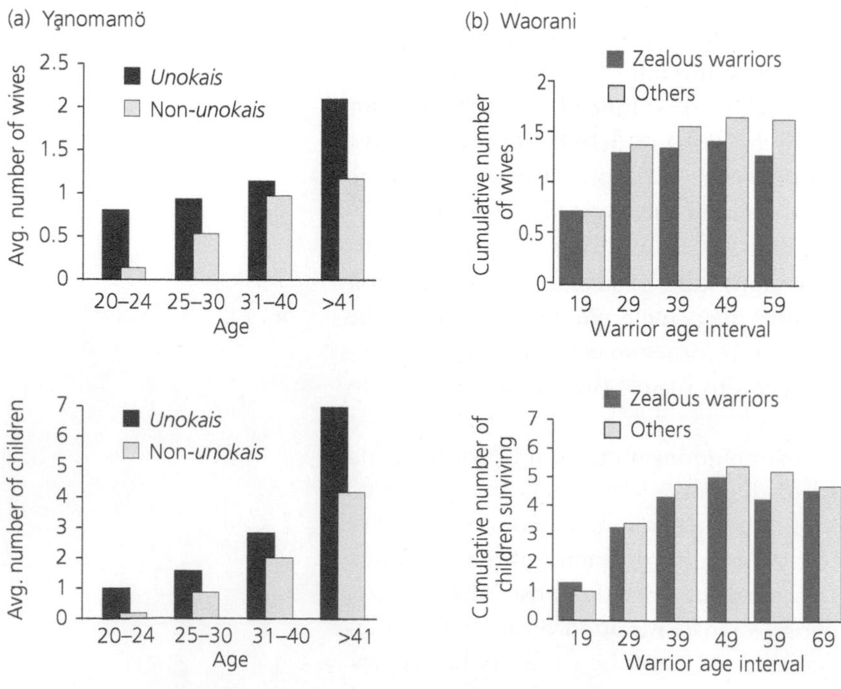

Figure 11.50 Male–male aggression, mating success, and reproductive success in two traditional societies
(a) Among the Yąnomamö, *unokais* (killers) are more successful than non-*unokais* ($p < 0.00001$). Plotted from data in Chagnon (1988).
(b) Among the Waorani, zealous warriors are less successful than others. The differences shown here were not statistically significant. From Beckerman et al. (2009).

In summary, what little information we have on male–male combat in traditional societies is contradictory on whether aggressiveness, let alone large size, are associated with mating and reproductive success.

B. Pawlowski and colleagues (2000) investigated the hypothesis that the sexual dimorphism in human body size is a result of female choice. The researchers gathered data from medical records of 3,201 Polish men. They used statistical techniques to remove the effects of various confounding variables. Pawlowski and colleagues then compared bachelors to married men. The married men were taller by a slight but statistically significant margin. In addition, men with one or more children were significantly taller than childless men **(Figure 11.51)**. The exception to this pattern was the group of men in their fifties, within which there was no difference in height between fathers versus childless men. Pawlowski and colleagues note that the men in their fifties reached marrying age shortly after World War II, when the ratio of women to men in Poland was unusually high. The researchers speculate that the men in their fifties had experienced less intense sexual selection through female choice than is the norm.

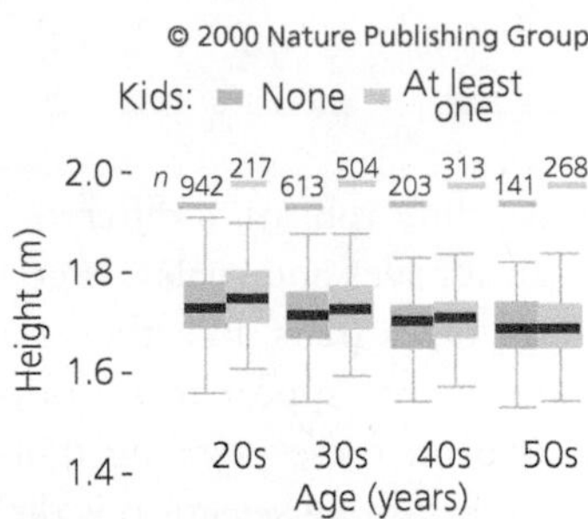

Figure 11.51 Men with children are taller, on average, than childless men The heavy horizontal bars show, for a sample of Polish men, the average height of individuals in each age class, by whether they have children. The colored boxes show ±1 standard deviation around the mean; the whiskers show the range about the mean that includes 95% of the men. *n* is the number of men in each category. Though small, the differences between men with versus without children are statistically significant for men in their twenties ($p = 0.005$), thirties ($p = 0.001$), and forties ($p = 0.002$). The difference is not significant for men in their fifties ($p = 0.863$). From Pawlowski et al. (2000).

Reprinted by permission from Macmillan Publishers Ltd: B. Pawlowski, R. I. M. Dunbar, and A. Lipowicz, 2000, "Tall men have more reproductive success," *Nature* 403: 156, Figure 1b. Copyright © 2000 Macmillan Magazines Limited.

Additional evidence suggesting that female choice favors tall men comes from a study by Ulrich Mueller and Allan Mazur (2001). Mueller and Mazur surveyed members of the class of 1950 from the United States Military Academy at West Point. Among these career officers, unlike in many other more diverse populations, height was not associated with social status or socioeconomic success. Height was, however, associated with reproductive success. The tallest men had, over their lifetimes, more wives, and younger second wives, than other men. As a result, the tallest men had more children.

Daniel Nettle (2002) examined the relationship between height and reproductive success in women. Analyzing data from a large national health survey in Britain, Nettle found a weak but significant effect. Unlike in men, selection on women is stabilizing. Women of slightly less than average height had more children than either shorter or taller women. The cause appears to be that women of moderate height were healthier, on average, than extremely short or extremely tall women. That is, the higher fitness of slightly shorter-than-average women is due to natural selection. (See also Guégan and colleagues 2000).

Rebecca Sear and Frank Marlowe (2009) examined the effect of height on mating success among the Hadza, a hunter-gatherer society in Tanzania. Unlike in modern societies, among the Hadza they found no evidence that couples mate assortatively for height, or that height is associated with number of marriages.

Based on the data we have reviewed, the evolutionary significance of sexual size dimorphism in humans remains unresolved. The studies we have discussed are observational, and the associations they documented either small or inconsistent. Thus the evidence they provide about causation is suggestive at best. It is also possible that we humans simply inherited our sexual size dimorphism from our ancestors, who were more sexually dimorphic in size than we are (McHenry 1992). To settle the issue, data are needed from a larger number of cultures on the relationship between body size, number of mates, survival, and reproductive success for both women and men. Preferably, the data would come from hunter-gatherer cultures, whose members live lifestyles ancestral for our species. The most challenging factor, given the high rate of extrapair paternity in some cultures (Cerda-Flores et al. 1999; Scelza 2011), is to accurately measure the reproductive success of men. Modern techniques have made it feasible, in principle, to collect such data. However, much research remains to be done.

It is unclear whether sexual selection helps maintain the sexual dimorphism in body size in humans. Males compete for mates, but larger males do not necessarily win. Females are choosy, and limited data suggest a slight preference for taller men in some cultures but not others.

SUMMARY

Sexual dimorphism, a difference in form or behavior between females and males, is common. The difference often involves traits, like the enormous tail feathers of the peacock, that appear to be opposed by natural selection. To explain these puzzling traits, Darwin invoked sexual selection. Sexual selection is differential reproductive success resulting from variation in mating success.

Mating success is often a more important determinant of fitness for one sex than for the other. Often, but by no means always, it is males whose reproductive success is limited by mating opportunities, and females whose reproductive success is limited by other resources.

The members of the sex experiencing strong sexual selection typically compete among themselves over access to mates. This competition may involve direct combat, gamete competition, infanticide, or advertisement.

The members of the sex whose reproductive success is limited by resources rather than matings are typically choosy. This choosiness may be arbitrary with respect to natural selection, it may be the result of a preexisting sensory bias, or it may provide the chooser with direct benefits such as food or indirect benefits such as better genes for its offspring.

Females, like males, often benefit from having multiple mates. When access to mates is limiting to females, they may compete over choosy males.

The theory of sexual selection was developed to explain sexual dimorphism in animals, but it applies to plants as well. In plants, access to pollinators is sometimes more limiting to reproductive success via pollen donation than to reproductive success via seed production. This can lead to the evolution of sexual dimorphism in which male flowers are showier than female flowers.

Humans, like many other animals, are sexually dimorphic. Patterns of sexual selection in humans vary from population to population. So far, the evidence is equivocal on whether sexual selection is the explanation for our dimorphism in body size.

QUESTIONS

1. Under what conditions will sexual selection produce different traits in the two sexes (i.e., sexual dimorphism)? Why is one sex often "choosy" while the other is "showy"?
2. What is the difference between intersexual selection and intrasexual selection? What kinds of traits do they each tend to produce? Give three examples of each.
3. In marine iguanas and red-collared widowbirds, what evidence is there that sexual selection acts contrary to natural selection? That is, what is the evidence that the sexually selected trait may reduce survival? What does this imply about survival rates of "attractive" males in many species, as compared to less attractive or less competitive males?
4. What are four reasons that females may choose males with particular traits and reject other males? Give one example for each. Does she always benefit from her choice?
5. Which parts of a flower are under more intensive selection for "showiness" (size, scent, color), the seed-producing parts or the pollinator-producing parts? Why? In plants that have separate genders, are the showier flowers found on male plants or female plants? In the sex that has less showy flowers, why does it have flowers at all?
6. In our discussion of rough-skinned newts, we inferred that tail crests in males evolved by sexual selection. Why is this a reasonable inference? Do you think the mechanism of sexual selection was male–male competition or female choice? Why? Design an experiment to find out.
7. Figure 11.12 shows the results of the experiment by Jones et al. (2000), in which broad-nosed pipefish mated in barrels in the lab. Each barrel contained either 4 males and 4 females, or 2 males and 6 females. Jones and colleagues also did experiments in which each barrel contained 6 males and 2 females. What do you think the analogous graphs from these experiments looked like? Why?
8. Males in many species often attempt to mate with strikingly inappropriate partners. Ryan (1985), for example, describes male túngara frogs clasping other males. Some orchids mimic female wasps and are pollinated by amorous male wasps—who have to be fooled twice for the strategy to work. Would a female túngara or a female wasp make the same mistake? Why or why not? (Think of general explanations that are applicable to a wide range of species.)
9. Male butterflies and moths commonly drink from puddles, a behavior known as puddling. Scott Smedley and Thomas Eisner (1996) report a detailed physiological analysis of puddling in the moth *Gluphisia septentrionis*. A male *G. septentrionis* may puddle for hours at a time. He rapidly processes huge amounts of water, extracting the sodium and

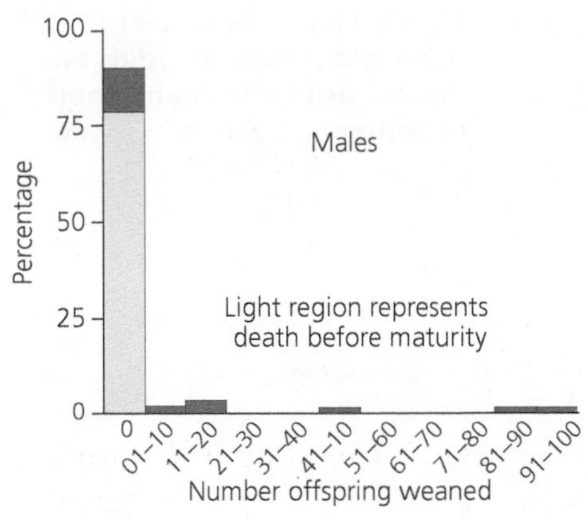

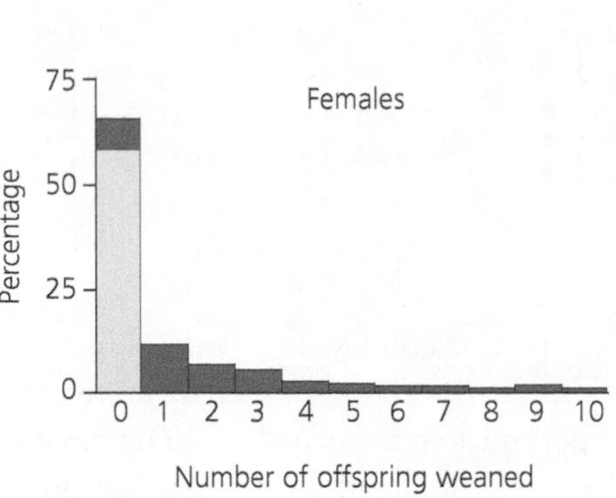

Figure 11.52 Distributions of lifetime reproductive success in male and female elephant seals From Le Boeuf and Reiter (1988).

expelling the excess liquid in anal jets (see Smedley and Eisner's paper for a dramatic photo). The male moth will later give his harvest of sodium to a female during mating. The female will then put much of the sodium into her eggs. Speculate on the role this gift plays in the moth's mating ritual and in the courtship roles taken by the male and the female. How would you test your ideas?

10. The graphs in **Figure 11.52** show the variation in lifetime reproductive success of male versus female elephant seals (Le Boeuf and Reiter 1988). Note that the scales on the horizontal axes are different. Why is the variation in reproductive success so much more extreme in males than females? Draw a graph showing your hypothesis for the relationship between number of mates and reproductive success for male and female elephant seals. Why do you think male elephant seals are four times larger than females? Why aren't males even bigger?

11. What sex would you guess is the sage grouse pictured in **Figure 11.53**? What is it doing and why? Do you think this individual provides parental care? What else can you guess about the social system of this species?

Figure 11.53 Sage grouse

12. In some species of deep-sea anglerfish, the male lives as a symbiont permanently attached to the female (see Gould 1983, essay 1). The male is tiny compared to the female. Many of the male's organs, including the eyes, are reduced, though the testes remain large. Other structures, such as the jaws and teeth, are modified for attachment to the female. The circulatory systems of the two sexes are fused, and the male receives all of his nutrition from the female via the shared bloodstream. Often, two or more males are attached to a single female. What are the costs and benefits of the male's symbiotic habit for the male? For the female? What limits the lifetime reproductive success of each sex—the ability to gather resources, or the ability to find mates? Do you think that the male's symbiotic habit evolved as a result of sexual selection or natural selection? (It may be helpful to break the male symbiotic syndrome into separate features, such as staying with a single female for life, physical attachment to the female, reduction in body size, and nutritional dependence on the female.)

13. The scatterplot in **Figure 11.54** shows the relationship between the importance of attractiveness in mate choice (as reported by subjects responding to a questionnaire) and the prevalence of six species of parasites (including leprosy, malaria, and filaria) in 29 cultures (Gangestad 1993; Gangestad and Buss 1993). (Statistical techniques have been used to remove the effects of latitude, geographic region, and mean income.) What is the pattern in the graph? Does this pattern make sense from an evolutionary perspective? One of the parasitic diseases is schistosomiasis. There is evidence that resistance to schistosomiasis is heritable (Abel et al. 1991). What do women gain (evolutionarily) by choosing an attractive mate? What do men gain (evolutionarily) by choosing an attractive mate? Can you offer a cultural explanation that could also account for this pattern?

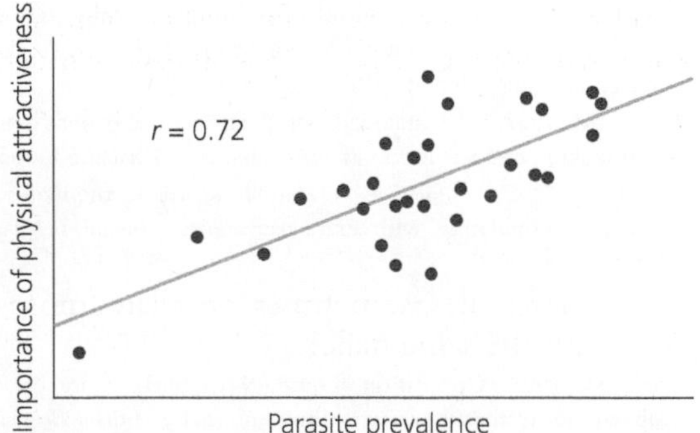

Figure 11.54 Importance of physical attractiveness in mate choice versus parasite prevalence in 29 human cultures Redrawn from Gangestad (1993).

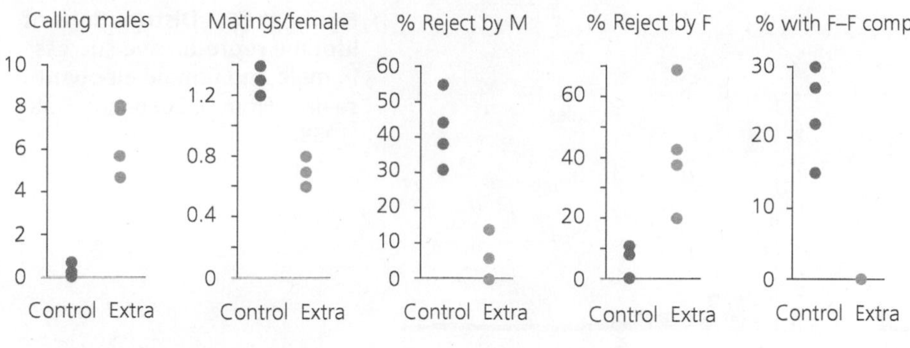

Figure 11.55 Behavior of male and female katydids under control versus extra-food conditions Data from Gwynne and Simmons (1990).

14. In many katydids, the male delivers his sperm to the female in a large spermatophore that contains nutrients the female eats (for a photo, see Gwynne 1981). The female uses these nutrients in the production of eggs. Darryl Gwynne and L. W. Simmons (1990) studied the behavior of caged populations of an Australian katydid under low-food (control) and high-food (extra) conditions. Some of their results are graphed in **Figure 11.55**. (The graph shows the results from four sets of replicate cages; calling males = number of males calling at any given time; matings/female = number of times each female mated; % reject by M = fraction of the time a female approached a male for mating and was rejected; % reject by F = fraction of the time a female approached a male but then rejected him before copulating; % with F–F comp = fraction of matings in which one or more females were seen fighting over the male.) Based on the graphs, when were the females choosy and the males competitive? When were the males choosy and the females competitive? Why?

EXPLORING THE LITERATURE

15. If a single insemination provides all the sperm necessary to fertilize an entire clutch of eggs, then what do females gain by engaging in extrapair copulations? Earlier we presented data showing that female Gunnison's prairie dogs that mate with multiple males are more likely to get pregnant and have larger litters. For more hypotheses and tests, see:

Kempenaers, B., G. R. Verheyn, et al. 1992. Extra-pair paternity results from female preference for high-quality males in the blue tit. *Nature* 357: 494–496.

Madsen, T., R. Shine, et al. 1992. Why do female adders copulate so frequently? *Nature* 355: 440–441.

Gray, E.M. 1996. Female control of offspring paternity in a western population of red-winged blackbirds (*Agelaius phoeniceus*). *Behavioral Ecology and Sociobiology* 38: 267–278.

Gray, E. M. 1997. Do female red-winged blackbirds benefit genetically from seeking extra-pair copulations? *Animal Behaviour* 53: 605–623.

Gray, E. M. 1997. Female red-winged blackbirds accrue material benefits from copulating with extra-pair males. *Animal Behaviour* 53: 625–629.

16. Why do females sometimes copulate more than once with the same male?

Petrie, M. 1992. Copulation frequency in birds: Why do females copulate more than once with the same male? *Animal Behaviour* 44: 790–792.

17. In many species, the male guards the female after copulation, following her closely, apparently in an effort to prevent sperm competition from other males that she may mate with later. In many insects, the male may even remain attached to the female in a copulatory position. See this paper for an interesting case of a grasshopper in which the male spends as long as 17 days mounted on the female:

Cueva del Castillo, R. 2003. Body size and multiple copulations in a neotropical grasshopper with an extraordinary mate-guarding duration. *Journal of Insect Behavior* 16: 503–522.

18. For a bizarre story about sexual selection in a hermaphroditic nudibranch, see:

Sekizawa, A., S. Seki, et al. 2013. Disposable penis and its replenishment in a simultaneous hermaphrodite. *Biology Letters*. doi:10.1098/rsbl.2012.1150.

19. Peacocks are among the most famous animals with an elaborate male mating display. For research on sexual selection in peacocks, see:

Loyau, A., M. S. Jalme, et al. 2005. Multiple sexual advertisements honestly reflect health status in peacocks (*Pavo cristatus*). *Behavioral Ecology and Sociobiology* 58: 552–557.

Loyau, A., M. S. Jalme, and G. Sorci. 2005. Intra- and intersexual selection for multiple traits in the peacock (*Pavo cristatus*). *Ethology* 111: 810–820.

Petrie, M. 1992. Peacocks with low mating success are more likely to suffer predation. *Animal Behaviour* 44: 585–586.

Petrie, M., T. Halliday, and C. Sanders. 1991. Peahens prefer peacocks with elaborate trains. *Animal Behaviour* 41: 323–331.

20. Figure 11.14 (page 418) shows that iguanas of intermediate size survive bad years at higher rates. For evidence of a surprising adaptation that may help big iguanas survive bad years, see:

Wikelski, M., and C. Thom. 2000. Marine iguanas shrink to survive El Niño. *Nature* 403: 37–38.

21. We presented a null model for the evolution of female choice under which female preferences are arbitrary. This model hinges on a genetic correlation between male display and female preference. For evidence of such a correlation in threespine sticklebacks, see:

Rick, I. P., M. Mehlis, and T. C. Bakker. 2011. Male red ornamentation is associated with female red sensitivity in sticklebacks. *PLoS One* 6: e25554.

22. What is the evidence that male displays and female preferences can be self-reinforcing? See these two papers for two very different stories:

Gustafsson, L., and A. Qvarnstrom. 2006. A test of the "sexy son" hypothesis: Sons of polygynous collared flycatchers do not inherit their fathers' mating status. *American Naturalist* 167: 297–302.

Gwinner, H., and H. Schwabl. 2005. Evidence for sexy sons in European starlings (*Sturnus vulgaris*). *Behavioral Ecology and Sociobiology* 58: 375–382.

23. Our discussion of sexual selection in plants focused mostly on the plants' perspective. See this paper for the perspective of the pollinators, who have their own preferences about which flowers they want to visit:

Abraham, J. N. 2005. Insect choice and floral size dimorphism: Sexual selection or natural selection? *Journal of Insect Behavior* 18: 743–756.

24. We have seen that in most cases, the "showier" sex will be the one that provides less parental care. See this paper for an interesting exception to the rule:

Heinsohn, R., S. Legge, and J. A. Endler. 2005. Extreme reversed sexual dichromatism in a bird without sex role reversal. *Science* 309: 617–619.

25. We discussed the hypothesis that male ornaments may evolve in response to a preexisting sensory bias of females. But why would females have a preexisting "sensory bias" in the first place? See these three papers for evidence that male ornaments and courtship behavior can evolve in response to female sensory biases that evolved originally for foraging for orange and red fruits and for grapes, respectively:

Grether, G. F., G. R. Kolluru, et al. 2005. Carotenoid availability affects the development of a colour-based mate preference and the sensory bias to which it is genetically linked. *Proceedings of the Royal Society of London B* 272: 2181–2188.

Smith, C., I. Barber, et al. 2004. A receiver bias in the origin of three-spined stickleback mate choice. *Proceedings of the Royal Society of London B 271*: 949–955.

Madden, J. R., and K. Tanner. 2003. Preferences for coloured bower decorations can be explained in a nonsexual context. *Animal Behaviour* 65: 1077–1083.

26. For a parasite that manipulates mate choice in its host, see:

Dass, S. A., A. Vasudevan, et al. 2011. Protozoan parasite *Toxoplasma gondii* manipulates mate choice in rats by enhancing attractiveness of males. *PLoS One* 6: e27229.

27. For evidence that human height is at an equilibrium set by contrasting patterns of selection in women versus men, see:

Stulp, G., B. Kuijper, et al. 2012. Intralocus sexual conflict over human height. *Biology Letters* 8: 976–978.

CITATIONS

Abel, L., F. Demenais, et al. 1991. Evidence for the segregation of a major gene in human susceptibility/resistance to infection by *Schistosoma mansoni*. *American Journal of Human Genetics* 48: 959–970.

Akre, K. L., H. E. Farris, et al. 2011. Signal perception in frogs and bats and the evolution of mating signals. *Science* 333: 751–752.

Andersson, M. 1982. Female choice selects for extreme tail length in a widowbird. *Nature* 299: 818–820.

Andersson, M. 1994. *Sexual Selection*. Princeton, NJ: Princeton University Press.

Arnold, S. J. 1983. Sexual selection: The interface of theory and empiricism. In P. Bateson, ed. *Mate Choice*. Cambridge: Cambridge University Press, 67–107.

Arnold, S. J. 1994. Bateman's principles and the measurement of sexual selection in plants and animals. *American Naturalist* 144: S126–S149.

Arnold, S. J., and D. Duvall. 1994. Animal mating systems: A synthesis based on selection theory. *American Naturalist* 143: 317–348.

Bateman, A. J. 1948. Intra-sexual selection in *Drosophila*. *Heredity* 2: 349–368.

Beckerman, S., P. I. Erickson, et al. 2009. Life histories, blood revenge, and reproductive success among the Waorani of Ecuador. *Proceedings of the National Academy of Sciences, USA* 106: 8134–8139.

Beecher, M. D., and I. M. Beecher. 1979. Sociobiology of bank swallows: Reproductive strategy of the male. *Science* 205: 1282–1285.

Berglund, A. 1991. Egg competition in a sex-role reversed pipefish: Subdominant females trade reproduction for growth. *Evolution* 45: 770–774.

Berglund, A., and G. Rosenqvist. 1993. Selective males and ardent females in pipefish. *Behavioral Ecology and Sociobiology* 32: 331–336.

Berglund, A., G. Rosenqvist, and P. Bernet. 1997. Ornamentation predicts reproductive success in female pipefish. *Behavioral Ecology and Sociobiology* 40: 145–150.

Berglund, A., G. Rosenqvist, and I. Svensson. 1989. Reproductive success of females limited by males in two pipefish species. *American Naturalist* 133: 506–516.

Bernet, P., G. Rosenqvist, and A. Berglund. 1998. Female-female competition affects female ornamentation in the sex-role reversed pipefish *Syngnathus typhle*. *Behaviour* 135: 535–550.

Bertram, C. R. 1975. Social factors influencing reproduction in wild lions. *Journal of Zoology* 177: 463–482.

Boellstorff, D. E., D. H. Owings, et al. 1994. Reproductive behavior and multiple paternity of California ground squirrels. *Animal Behaviour* 47: 1057–1064.

Borgerhoff Mulder, M. 2009. Serial monogamy as polygyny or polyandry? *Human Nature* 20: 130–150.

Brodt, J. F., D. W. Tallamy, and J. Ali. 2006. Female choice by scent recognition in the spotted cucumber beetle. *Ethology* 112: 300–306.

Brown, G. R., K. N. Laland, and M. B. Mulder. 2009. Bateman's principles and human sex roles. *Trends in Ecology and Evolution* 24: 297–304.

Burkhardt, D., and I. de la Motte. 1983. How stalk-eyed flies eye stalk-eyed flies: Observations and measurements of the eyes of *Cyrtodiopsis whitei* (Dopsidae, Diptera). *Journal of Comparative Physiology* 151: 407–421.

Burkhardt, D., and I. de la Motte. 1987. Physiological, behavioural, and morphometric data elucidate the evolutive significance of stalked eyes in Diopsidae (Diptera). *Entomologia Generalis* 12: 221–233.

Bush, S. L., H. C. Gerhardt, and J. Schul. 2002. Pattern recognition and call preferences in treefrogs (Anura: Hylidae): A quantitative analysis using a no-choice paradigm. *Animal Behaviour* 63: 7–14.

Cerda-Flores, R. M., S. A. Barton, et al. 1999. Estimation of nonpaternity in the Mexican population of Nuevo Leon: A validation study with blood group markers. *American Journal of Physical Anthropology* 109: 281–293.

Chagnon, N. A. 1988. Life histories, blood revenge, and warfare in a tribal population. *Science* 239: 985–992.

Christy, J. H., P. R. Y. Backwell, and U. Schober. 2003. Interspecific attractiveness of structures built by courting male fiddler crabs: Experimental evidence of a sensory trap. *Behavioral Ecology and Sociobiology* 53: 84–91.

Courtiol, A., J. E. Pettay, et al. 2012. Natural and sexual selection in a monogamous historical human population. *Proceedings of the National Academy of Sciences, USA* 109: 8044–8049.

Daly, M., and M. Wilson. 1988. *Homicide*. New York: Aldine de Gruyter.

Darwin, C. 1860. Letter to Asa Gray. In Darwin, F. (ed.). 1887. *The Life and Letters of Charles Darwin, Including an Autobiographical Chapter*. London: John Murray, Vol. 2, p. 296.

Darwin, C. 1871. *The Descent of Man, and Selection in Relation to Sex*. London: John Murray.

David, P., A. Hingle, et al. 1998. Male sexual ornament size but not asymmetry reflects condition in stalk-eyed flies. *Proceedings of the Royal Society of London B* 265: 2211–2216.

Delph, L. F., L. F. Galloway, and M. L. Stanton. 1996. Sexual dimorphism in flower size. *American Naturalist* 148: 299–320.

Doty, G. V., and A. M. Welch. 2001. Advertisement call duration indicates good genes for offspring feeding rate in gray tree frogs (*Hyla versicolor*). *Behavioral Ecology and Sociobiology* 49: 150–156.

Enquist, M., and A. Arak. 1993. Selection of exaggerated male traits by female aesthetic senses. *Nature* 361: 446–448.

Falconer, D. S. 1989. *Introduction to Quantitative Genetics*. New York: John Wiley & Sons.

Fisher, R. A. 1915. The evolution of sexual preference. *Eugenics Review* 7: 184–192.

Fisher, R. A. 1958. *The Genetical Theory of Natural Selection*. 2nd ed. New York: Dover.

Gage, M. J. G. 1991. Risk of sperm competition directly affects ejaculate size in the Mediterranean fruit fly. *Animal Behaviour* 42: 1036–1037.

Gangestad, S. W. 1993. Sexual selection and physical attractiveness: Implications for mating dynamics. *Human Nature* 4: 205–235.

Gangestad, S. W., and D. M. Buss. 1993. Pathogen prevalence and human mate preferences. *Ethology and Sociobiology* 14: 89–96.

Gerhardt, H. C., M. L. Dyson, and S. D. Tanner. 1996. Dynamic properties of the advertisement calls of gray tree frogs: Patterns of variability and female choice. *Behavioral Ecology* 7: 7–18.

Gerhardt, H. C., S. D. Tanner, et al. 2000. Female preference functions based on call duration in the gray tree frog (*Hyla versicolor*). *Behavioral Ecology* 11: 663–669.

Gibbs, H. L., P. J. Weatherhead, et al. 1990. Realized reproductive success of polygynous red-winged blackbirds revealed by DNA markers. *Science* 250: 1394–1397.

Gilbert, L. E. 1976. Postmating female odor in *Heliconius* butterflies: A male-contributed anti-aphrodisiac? *Science* 193: 419–420.

Gould, S. J. 1983. *Hen's Teeth and Horse's Toes*. New York: Norton.

Grace, J. L., and K. L. Shaw. 2011. Coevolution of male mating signal and female preference during early lineage divergence of the Hawaiian cricket, *Laupala cerasina*. *Evolution* 65: 2184–2196.

Grant, V. 1995. Sexual selection in plants: Pros and cons. *Proceedings of the National Academy of Sciences, USA* 92: 1247–1250.

Gray, E. M. 1997. Do female red-winged blackbirds benefit genetically from seeking extra-pair copulations? *Animal Behaviour* 53: 605–623.

Gridi-Papp, M., A. S. Rand, and M. J. Ryan. 2006. Animal communication: Complex call production in the túngara frog. *Nature* 441: 38.

Gross, M. R. 1984. Sunfish, salmon, and the evolution of alternative reproductive strategies and tactics in fishes. In *Fish Reproduction: Strategies and Tactics*, ed. G. W. Potts and R. J. Wootton. London: Academic Press, 55–75.

Gross, M. R. 1985. Disruptive selection for alternative live histories in salmon. *Nature* 313: 47–48.

Gross, M. R. 1991. Salmon breeding behavior and life history evolution in changing environments. *Ecology* 72: 1180–1186.

Guegan, J. F., A. T. Teriokhin, and F. Thomas. 2000. Human fertility variation, size-related obstetrical performance and the evolution of sexual stature dimorphism. *Proceedings of the Royal Society of London B* 267: 2529–2535.

Gwynne, D. T. 1981. Sexual difference theory: Mormon crickets show role reversal in mate choice. *Science* 213: 779.

Gwynne, D. T., and L. W. Simmons. 1990. Experimental reversal of courtship roles in an insect. *Nature* 346: 172–174.

Hamilton, W. D., and M. Zuk. 1982. Heritable true fitness and bright birds: A role for parasites? *Science* 218: 384–387.

Hayward, A., and J. F. Gillooly. 2011. The cost of sex: Quantifying energetic investment in gamete production by males and females. *PLoS One* 6: e16557.

Hedrick, A. V., and E. J. Temeles. 1989. The evolution of sexual dimorphism in animals: Hypotheses and tests. *Trends in Ecology and Evolution* 4: 136–138.

Hoogland, J. L. 1998. Why do female Gunnison's prairie dogs copulate with more than one male? *Animal Behaviour* 55: 351–359.

Hooper, R. E., and M. T. Siva-Jothy. 1996. Last male sperm precedence in a damselfly demonstrated by RAPD profiling. *Molecular Ecology* 5: 449–452.

Hrdy, S. B. 1979. Infanticide among animals: A review, classification, and examination of the implications for the reproductive strategies of females. *Ethology and Sociobiology* 1: 13–40.

Ivy, T. M., C. B. Weddle, and S. K. Sakaluk. 2005. Females use self-referent cues to avoid mating with previous mates. *Proceedings of the Royal Society of London B* 272: 2475–2478.

Jones, A. G., J. R. Arguello, and S. J. Arnold. 2002. Validation of Bateman's principles: A genetic study of sexual selection and mating patterns in the rough-skinned newt. *Proceedings of the Royal Society of London B* 269: 2533–2539.

Jones, A. G., and N. L. Ratterman. 2009. Mate choice and sexual selection: What have we learned since Darwin? *Proceedings of the National Academy of Sciences, USA* 106 (Suppl 1): 10001–10008.

Jones, A. G., G. Rosenqvist, et al. 2000. The Bateman gradient and the cause of sexual selection in a sex-role-reversed pipefish. *Proceedings of the Royal Society of London B* 267: 677–680.

Jones, A. G., G. Rosenqvist, et al. 2005. The measurement of sexual selection using Bateman's principles: An experimental test in the sex-role-reversed pipefish *Syngnathus typhle*. *Integrative and Comparative Biology* 45: 874–884.

Kim, T. W., J. H. Christy, and J. C. Choe. 2007. A preference for a sexual signal keeps females safe. *PLoS One* 2: e422.

Kirkpatrick, M. 1982. Sexual selection and the evolution of female choice. *Evolution* 82:1–12.

Kirkpatrick, M. 1987. Sexual selection by female choice in polygynous animals. *Annual Review of Ecology and Systematics* 18: 43–70.

Kirkpatrick, M., and M. J. Ryan. 1991. The evolution of mating preferences and the paradox of the lek. *Nature* 350: 33–38.

Knight, J. 2002. Sexual stereotypes. *Nature* 415: 254–256.

Lande, R. 1981. Models of speciation by sexual selection on polygenic traits. *Proceedings of the National Academy of Sciences, USA* 78: 3721–3725.

Le Boeuf, B. J., and J. Reiter. 1988. Lifetime reproductive success in northern elephant seals. In *Reproductive Success: Studies of Individual Variation in Contrasting Breeding Systems,* ed. T. H. Cutton-Brock. Chicago: University of Chicago Press, 344–362.

McHenry, H. M. 1992. Body size and proportions in early hominids. *American Journal of Physical Anthropology* 87: 407–431.

Mead, L. S., and S. J. Arnold. 2004. Quantitative genetic models of sexual selection. *Trends in Ecology and Evolution* 19: 264–271.

Møller, A. P. 1988. Female choice selects for male sexual tail ornaments in the monogamous swallow. *Nature* 332: 640–642.

Mueller, U., and A. Mazur. 2001. Evidence of unconstrained directional selection for male tallness. *Behavioral Ecology and Sociobiology* 50: 302–311.

Nakatsuru, K., and D. L. Kramer. 1982. Is sperm cheap? Limited male fertility and female choice in the lemon tetra (Pisces, Characidae). *Science* 216: 753–755.

Nettle, D. 2002. Women's height, reproductive success and the evolution of sexual dimorphism in modern humans. *Proceedings of the Royal Society of London B* 269: 1919–1923.

Nowak, R. M. 1991. *Walker's Mammals of the World.* Baltimore, MD: Johns Hopkins University Press.

Olsson, M., A. Gullberg, and H. Tegelstrom. 1994. Sperm competition in the sand lizard, *Lacerta agilis*. *Animal Behaviour* 48: 193–200.

Oring, L. W., M. A. Colwell, and J. M. Reed. 1991a. Lifetime reproductive success in the spotted sandpiper (*Actitis macularia*)—sex-differences and variance-components. *Behavioral Ecology and Sociobiology* 28: 425–432.

Oring, L. W., J. M. Reed, et al. 1991b. Factors regulating annual mating success and reproductive success in spotted sandpipers (*Actitis macularia*). *Behavioral Ecology and Sociobiology* 28: 433–442.

Oring, L. W., J. M. Reed, and S. J. Maxson. 1994. Copulation patterns and mate guarding in the sex-role reversed, polyandrous spotted sandpiper, *Actitis macularia*. *Animal Behaviour* 47: (5) 1065–1072.

Owen, M. A., K. Rohrer, and R. D. Howard. 2012. Mate choice for a novel male phenotype in zebrafish, *Danio rerio*. *Animal Behaviour* 83: 811–820.

Packer, C., L. Herbst, et al. 1988. Reproductive success of lions. In *Reproductive Success: Studies of Individual Variation in Contrasting Breeding Systems,* ed. T. H. Clutton-Brock. Chicago: University of Chicago Press, 263–283.

Packer, C., and A. E. Pusey. 1983. Adaptations of female lions to infanticide by incoming males. *American Naturalist* 121: 716–728.

Panhuis, T. M., and G. S. Wilkinson. 1999. Exaggerated male eye span influences contest outcome in stalk-eyed flies (Diopsidae). *Behavioral Ecology and Sociobiology* 46: 221–227.

Partecke, J., A. von Haeseler, and M. Wikelski. 2002. Territory establishment in lekking marine iguanas, *Amblyrhynchus cristatus*: Support for the hotshot mechanism. *Behavioral Ecology and Sociobiology* 51: 579–587.

Pawlowski, B., R. I. M. Dunbar, and A. Lipowicz. 2000. Tall men have more reproductive success. *Nature* 403: 156.

Pélissié, B., P. Jarne, and P. David. 2012. Sexual selection without sexual dimorphism: Bateman gradients in a simultaneous hermaphrodite. *Evolution* 66: 66–81.

Petrie, M. 1983. Female moorhens compete for small fat males. *Science* 220: 413–415.

Proctor, H. C. 1991. Courtship in the water mite, *Neumania papillator*: Males capitalize on female adaptations for predation. *Animal Behaviour* 42: 589–598.

Proctor, H. C. 1992. Sensory exploitation and the evolution of male mating behaviour: A cladistic test using water mites (Acari: Parasitengona). *Animal Behaviour* 44: 745–752.

Prum, R. O. 1997. Phylogenetic tests of alternative intersexual selection mechanisms: Trait macroevolution in a polygynous clade (Aves: Pipridae). *American Naturalist* 149: 668–692.

Prum, R. O. 2010. The Lande-Kirkpatrick mechanism is the null model of evolution by intersexual selection: Implications for meaning, honesty, and design in intersexual signals. *Evolution* 64: 3085–3100.

Pryke, S. R., and S. Andersson. 2005. Experimental evidence for female choice and energetic costs of male tail elongation in red-collared widowbirds. *Biological Journal of the Linnean Society* 86: 35–43.

Rauch, N. 1985. Female habitat choice as a determinant of the reproductive success of the territorial male marine iguana (*Amblyrhynchus cristatus*). *Behavioral Ecology and Sociobiology* 16: 125–134.

Rauch, N. 1988. Competition of marine iguana females *Amblyrhynchus cristatus* for egg-laying sites. *Behaviour* 107: 91–106.

Renner, S. S., and R. E. Ricklefs. 1995. Dioecy and its correlates in the flowering plants. *American Journal of Botany* 82: 596–606.

Rogers, A. R., and A. Mukherjee. 1992. Quantitative genetics of sexual dimorphism in human body size. *Evolution* 46: 226–234.

Romero, G. A., and C. E. Nelson. 1986. Sexual dimorphism in *Catasetum* orchids: Forcible pollen emplacement and male flower competition. *Science* 232: 1538–1540.

Rosenqvist, G. 1990. Male mate choice and female-female competition for mates in the pipefish *Nerophis ophidion*. *Animal Behaviour* 39: 1110–1115.

Rosenqvist, G., and K. Johansson. 1995. Male avoidance of parasitized females explained by direct benefits in a pipefish. *Animal Behaviour* 49: 1039–1045.

Ryan, M. J. 1985. *The Túngara Frog: A Study in Sexual Selection and Communication.* Chicago: University of Chicago Press.

Scelza, B. A. 2011. Female choice and extra-pair paternity in a traditional human population. *Biology Letters* 7: 889–891.

Schenk, A., and K. M. Kovacs. 1995. Multiple mating between black bears revealed by DNA fingerprinting. *Animal Behaviour* 50: 1483–1490.

Schöfl, G., and M. Taborsky. 2002. Prolonged tandem formation in firebugs (*Pyrrhocoris apterus*) serves mate-guarding. *Behavioral Ecology and Sociobiology* 52: 426–433.

Schwartz, J. J., B. W. Buchanan, and H. C. Gerhardt. 2001. Female mate choice in the gray treefrog (*Hyla versicolor*) in three experimental environments. *Behavioral Ecology and Sociobiology* 49: 443–455.

Sear, R., and F. W. Marlowe. 2009. How universal are human mate choices? Size does not matter when Hadza foragers are choosing a mate. *Biology Letters* 5: 606–609.

Sillén-Tullberg, B. 1981. Prolonged copulation: A male "postcopulatory" strategy in a promiscuous species, *Lygaeus equestris* (Heteroptera: Lygaeidae). *Behavioral Ecology and Sociobiology* 9: 283–289.

Smedley, S. R., and T. Eisner. 1996. Sodium: A male moth's gift to its offspring. *Proceedings of the National Academy of Sciences, USA* 93: 809–813.

Smith, R. L. 1984. Human sperm competition. In *Sperm Competition and the Evolution of Animal Mating Systems,* ed. R. L. Smith. Orlando, FL: Academic Press, 601–659.

Stanton, M. L., A. A. Snow, and S. N. Handel. 1986. Floral evolution: Attractiveness to pollinators increases male fitness. *Science* 232: 1625–1627.

Tallamy, D. W., M. B. Darlington, et al. 2003. Copulatory courtship signals male genetic quality in cucumber beetles. *Proceedings of the Royal Society of London B* 270: 77–82.

Tallamy, D. W., B. E. Powell, and J. A. McClafferty. 2002. Male traits under cryptic female choice in the spotted cucumber beetle (Coleoptera: Chrysomelidae). *Behavioral Ecology* 13: 511–518.

Tang-Martinez, Z., and T. B. Ryder. 2005. The problem with paradigms: Bateman's worldview as a case study. *Integrative and Comparative Biology* 45: 821–830.

Temeles, E. J., R. S. Goldman, et al. 2005. Foraging and territory economics of sexually dimorphic purple-throated caribs (*Eulampis jugularis*) on three *Heliconia* morphs. *Auk* 122: 187–204.

Temeles, E. J., C. R. Koulouris, et al. 2009. Effect of flower shape and size on foraging performance and trade-offs in a tropical hummingbird. *Ecology* 90: 1147–1161.

Temeles, E. J., and W. J. Kress. 2003. Adaptation in a plant-hummingbird association. *Science* 300: 630–633.

Temeles, E. J., I. L. Pan, et al. 2000. Evidence for ecological causation of sexual dimorphism in a hummingbird. *Science* 289: 441–443.

Thornhill, R. 1976. Sexual selection and nuptial feeding behavior in *Bittacus apicalis* (Insecta: Mecoptera). *American Naturalist* 110: 529–548.

Thornhill, R., and J. Alcock. 1983. *The Evolution of Insect Mating Systems.* Cambridge, MA: Harvard University Press.

Trillmich, K. G. K. 1983. The mating system of the marine iguana (*Amblyrhynchus cristatus*). *Zeitschrift für Tierpsychologie* 63: 141–172.

Trivers, R. L. 1972. Parental investment and sexual selection. In *Sexual Selection and the Descent of Man 1871–1971,* ed. B. Campbell. Chicago: Aldine, 136–179.

Trivers, R. 1985. *Social Evolution.* Menlo Park, CA: Benjamin Cummings.

Vaughton, G., and M. Ramsey. 1998. Floral display, pollinator visitation, and reproductive success in the dioecious perennial herb *Wurmbea dioica* (Liliaceae). *Oecologia* 115: 93–101.

Waage, J. K. 1984. Sperm competition and the evolution of Odonate mating systems. In *Sperm Competition and the Evolution of Animal Mating Systems,* ed. R. L. Smith. Orlando, FL: Academic Press, 251–290.

Waage, J. K. 1986. Evidence for widespread sperm displacement ability among Zygoptera (Odonata) and the means for predicting its presence. *Biological Journal of the Linnean Society* 28: 285–300.

Walther, B. A., and D. H. Clayton. 2005. Elaborate ornaments are costly to maintain: Evidence for high maintenance handicaps. *Behavioural Ecology* 16: 89–95.

Watson, P. J. 1991. Multiple paternity as genetic bet-hedging in female sierra dome spiders, *Linyphia litigiosa* (Linyphiidae). *Animal Behaviour* 41: 343–360.

Welch, A. M., R. D. Semlitsch, and H. Carl Gerhardt. 1998. Call duration as an indicator of genetic quality in male gray tree frogs. *Science* 280: 1928–1930.

Wilhelm, G., S. Handschuh, et al. 2011. Sexual dimorphism in head structures of the weevil *Rhopalapion longirostre* (Olivier 1807)(Coleoptera: Curculionoidea): A response to ecological demands of egg deposition. *Biological Journal of the Linnean Society* 104: 642–660.

Wikelski, M. 2005. Evolution of body size in Galapagos marine Iguanas. *Proceedings of the Royal Society of London B* 272: 1985–1993.

Wikelski, M., and S. Bäurle. 1996. Precopulatory ejaculation solves time constraints during copulations in marine iguanas. *Proceedings of the Royal Society of London B* 263: 439–444.

Wikelski, M., C. Carbone, and F. Trillmich. 1996. Lekking in marine iguanas: Female grouping and male reproductive strategies. *Animal Behaviour* 52: 581–596.

Wikelski, M., V. Carrillo, and F. Trillmich. 1997. Energy limits to body size in a grazing reptile, the Galápagos marine iguana. *Ecology* 78: 2204–2217.

Wikelski, M., and C. Thom. 2000. Marine iguanas shrink to survive El Niño. *Nature* 403: 37.

Wikelski, M., and F. Trillmich. 1997. Body size and sexual size dimorphism in marine iguanas fluctuate as result of opposing natural and sexual selection: An island comparison. *Evolution* 51: 922–936.

Wilkinson, G. S., D. C. Presgraves, and L. Crymes. 1998. Male eye span in stalk-eyed flies indicates genetic quality by meiotic drive suppression. *Nature* 391: 276–279.

Wilkinson, G. S., and P. R. Reillo. 1994. Female choice response to artificial selection on an exaggerated male trait in a stalk-eyed fly. *Proceedings of the Royal Society of London B* 255: 1–6.

Wilkinson, G. S., and M. Taper. 1999. Evolution of genetic variation for condition-dependent traits in stalk-eyed flies. *Proceedings of the Royal Society of London B* 266: 1685–1690.

Williams, G. C. 1966. *Adaptation and Natural Selection: A Critique of Some Current Evolutionary Thought.* Princeton University Press, Princeton, NJ.

Willson, M. F. 1979. Sexual selection in plants. *American Naturalist* 113: 777–790.

Woodroffe, R., and A. Vincent. 1994. Mother's little helpers: Patterns of male care in mammals. *Trends in Ecology and Evolution* 9: 294–297.

Zahavi, A. 1975. Mate selection—A selection for a handicap. *Journal of Theoretical Biology* 53: 205–214.

12

The Evolution of Social Behavior

Female house mice (*Mus domesticus*) aggressively defend their pups against same-species intruders. When Stephen Gammie and colleagues (2006) counted the seconds lactating mothers spent attacking intruding males during 3-minute trials, they found that some females defend their pups more aggresively than others. Based on the resemblance between mothers and daughters, the researchers estimated that about 60% of the variation among females in their outbred lab population was due to differences in genes.

Gammie and colleagues then randomly assigned mice to a control line and a selected line. Each generation, the researchers chose females at random from the control line and kept their offspring as breeders. For the selected line, they kept the offspring of the most aggressive female in each family. By the eighth generation, the females in the selected line were defending their pups significantly more vigorously than the females in the control line. On average they spent 15 seconds attacking intruding males, whereas the control females spent just 3.77. The brains of the selected mice showed altered expression of a number of genes, including the neuropeptide neurotensin (Gammie et al. 2007; Gammie et al. 2009).

The results of this artificial selection experiment, summarized in the graph at right, show that social behavior evolves by the same mechanism as other traits.

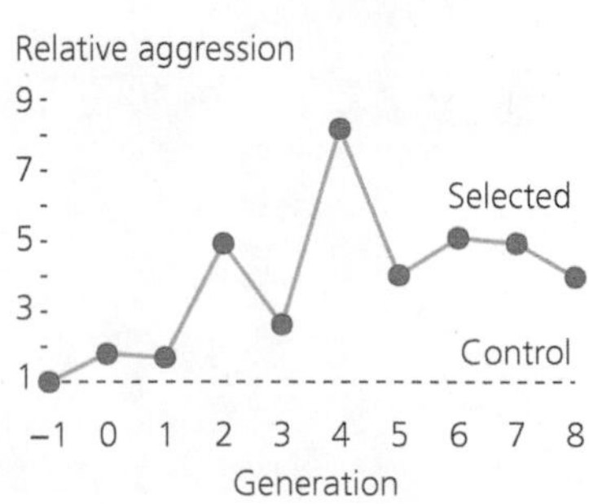

Maternal defense of pups evolves in response to artificial selection in house mice. In the graph, relative aggression of the selected line is calculated as the average duration of attacks on intruders divided by the average duration for the control line. Plotted with data from Gammie et al. (2006).

The evolution of social behavior—of interactions among members of the same species—warrants a chapter of its own for at least two reasons. First, because we humans are highly social animals, we tend to find social behavior inherently interesting. Second, because social interactions create opportunities for both cooperation and conflict, they raise intriguing challenges for evolutionary theory. The most prominent of these challenges, explaining the evolution of altruistic behavior, will occupy much of our attention.

We begin the chapter by considering four categories of simple social interaction, two of which are difficult to explain. Sections 12.2 and 12.3 explore two frameworks, kin selection and multilevel selection, that biologists have developed for resolving these difficulties. Section 12.4 looks at the tension between cooperation and conflict in a variety of social interactions. Finally, Section 12.5 examines the pinnacle of social evolution, eusociality.

12.1 Four Kinds of Social Behavior

The simplest possible social interaction involves two individuals: an actor and a recipient. When the actor does something with, to, or for the recipient, the outcome may be good or bad for either or both players. A good outcome increases a player's direct fitness, or genetic contribution to future generations via personal reproduction. A bad outcome decreases an individual's direct fitness.

Actor		Recipient	
+	→	+	Mutual benefit
+	→	−	Selfishness
−	→	+	Altruism
−	→	−	Spite

Figure 12.1 The four kinds of simple social interaction The plus and minus signs indicate whether an interaction increases or decreases an individual's fitness.

As shown in **Figure 12.1**, these considerations imply four kinds of simple social interaction (Hamilton 1964a). We follow the recommendations of Stuart West and colleagues (2007a) on the most appropriate terms to describe the interactions. In a **mutually beneficial** interaction, both the actor and recipient enjoy increased reproductive success. In a **selfish** interaction, the actor benefits at the expense of the recipient. In an **altruistic** interaction, the actor makes a sacrifice on behalf of the recipient. And in a **spiteful** interaction, the actor suffers a loss in order to impose a penalty on the recipient. We consider examples of each.

Mutual Benefit

There are four basic kinds of social interaction, defined by the fitness outcome for the actor and recipient.

The greater ani (*Crotophaga major,* **Figure 12.2a**) is a member of the cuckoo family that lives and nests near water in tropical forests ranging from Panama to Argentina (Riehl and Jara 2009). Adult males and females form mated pairs, but the pairs seldom nest alone. Instead, genetically unrelated couples nest together, typically in groups of two or three pairs (Riehl 2011). All the females lay eggs (Figure 12.2b), and all the adults work together to incubate the eggs, feed the

(a) A greater ani

(b) A communal nest

Figure 12.2 Greater anis nest communally (a) An adult. (b) A nest containing 11 eggs laid by three females. Photo by Christina Riehl.

chicks, and defend the nest. The greatest threat to eggs and nestlings comes from predators—mostly snakes, but occasionally monkeys. When a predator appears, the adult anis attempt to drive it away by mobbing.

To learn why mated pairs join forces instead of building their own single-family nests, Christina Riehl (2011, 2012) spent four years tracking several dozen nesting groups along the shores of Lake Gatún, Panama. She used genetic tests to match biological parents and offspring.

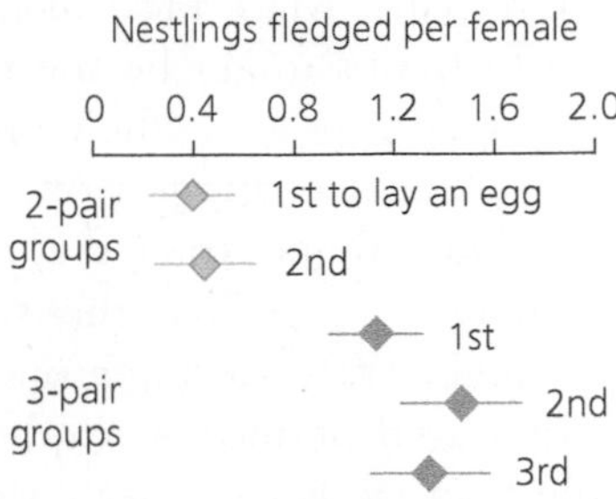

Figure 12.3 Reproductive success of individual female anis in nesting coalitions of two or three pairs Symbols show mean ± standard error (se). Redrawn from Riehl (2011).

Riehl saw only two pairs that tried to nest alone. In both cases, all the eggs were taken by predators and the parents consequently produced no offspring. In contrast, two-pair coalitions were more successful at defending their nests and fledging chicks, and three-pair coalitions did better still. Regardless of the order in which they laid their first egg, all the females in a coalition gained roughly equal reproductive success (**Figure 12.3**). The same was true for the males, even though some appeared to work considerably harder than others.

Life in a communal nest is less than perfectly harmonious. Before each female lays her first egg, she tosses out any eggs already laid by her coalition partners. The first female to lay in a group nest therefore always loses at least one egg. And it remains unknown why some males work harder than others, despite the lack of a fitness payoff.

Nonetheless, in its basic features **communal breeding** in greater anis meets our definition of mutual benefit. And the explanation for why mated pairs join coalitions is straightforward. Each pair achieves higher reproductive success by working together than any of the pairs could on its own.

The fitness benefit to the actor offers a general explanation for the evolution of mutual benefit. However, the frequent presence of conflict within mutually beneficial interactions, the risk of exploitation by cheaters, and other considerations make mutual benefit a rich field of investigation. We will revisit mutual benefit, conflict, and cheating later in the chapter.

Selfishness

Cane toad tadpoles (*Rhinella marina*, formerly known as *Bufo marinus*) routinely eat cane toad eggs (**Figure 12.4**). Michael Crossland and Richard Shine (2011) wanted to know whether the tadpoles are merely generalist consumers that eat conspecific eggs when they happen to come across them, or targeted cannibals that actively seek conspecific eggs to eat. The researchers put pairs of funnel traps in a pond in Australia, where cane toads are a troublesome invasive species. The pond was home to tadpoles of both cane toads and native frogs. Crossland and Shine chose one trap from each pair at random and baited it with cane toad eggs in a mesh container. They left the other trap empty as a control.

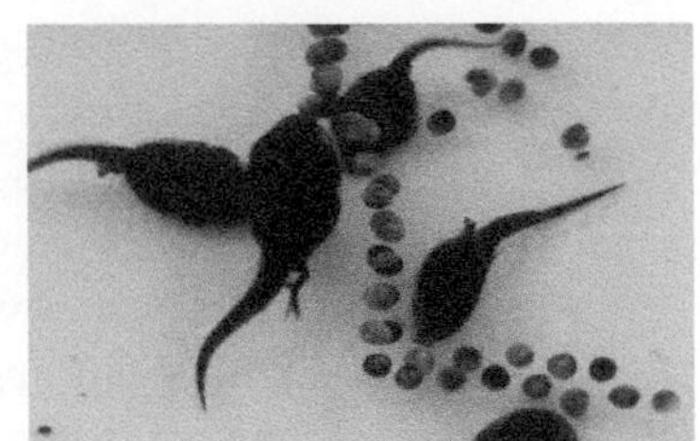

Figure 12.4 Cane toad tadpoles cannibalizing eggs Photo by Mattias Hagman.

When Crossland and Shine checked the traps the next day, they found 14 native frog tadpoles in the control traps versus 12 in the baited traps. Native frog tadpoles are not attracted to cane toad eggs. In contrast, the researchers found 70 cane toad tadpoles in the control traps versus 6,009 in the baited traps. Additional experiments showed that cane toad tadpoles are specifically attracted to chemicals released by cane toad eggs late in development. Crossland and Shine concluded that cane toad tadpoles are targeted cannibals.

Crossland and colleages (2012) identified the chemical cues released by eggs as toxins, called bufogenins, that cane toads make to deter heterospecific predators. While most predators avoid bufogenins, cane toad tadpoles find them irresistable. So irresistable, in fact, as to suggest a promising strategy for controlling the

Australian cane toad invasion. By baiting funnel traps with bufogenins, Crossland and colleagues found that they could clear a pond of cane toad tadpoles in a matter of days.

Crossland, Mark Hearnden, and colleagues (2011) suspected that the acquisition of toxins might be the reason cane toads evolved their cannibalistic bent. Cane toad eggs are sufficiently poisonous that tadpoles of the native frog *Litoria rothii* die after eating just one. Late-stage cane toad tadpoles, on the other hand, are sufficiently nontoxic that frog tadpoles can eat them with impunity. Contrary to the researchers' hypothesis, cane toad tadpoles that have cannibalized eggs are not significantly more poisonous to frogs than cane toad tadpoles that have not.

In search of another explanation for the cannibalism, the researchers reared cane toad tadpoles under a variety of conditions. They found that tadpoles can complete development and metamorphose into toads on a diet consisting solely of cannibalized eggs. And they found that the reduction in population density that results from cannibalism enables tadpoles to grow faster, metamorphose more quickly, and survive at higher rates. Cannibalism thus provides at least two benefits: nutrition and the reduction of future competition.

Cannibalism in cane toad tadpoles meets our definition of selfishness. The explanation for its evolution is as straightforward as that for mutual benefit. Selfish tadpoles achieve higher fitness than unselfish ones. The fitness benefit to the actor is the general explanation for the evolution of selfish behavior.

Altruism

Altruistic behavior can be tricky to identify. Research by Paul Sherman (1985) on Belding's ground squirrels shows why. Belding's ground squirrels (*Spermophilus beldingi*) live in groups in the mountains of the western United States. When a squirrel notices that the group is being stalked by a predator, it sometimes gives a loud, high-pitched call **(Figure 12.5)**. These warnings, known as alarm calls, alert nearby individuals and allow them to flee or dive for cover. They may also expose the caller to danger. But all is not always as it appears.

Figure 12.5 A Belding's ground squirrel gives an alarm call

Belding's ground squirrels give two kinds of alarms: They trill in response to mammals approaching on foot and whistle in response to hawks attacking on the wing. During 14 years of observation, Sherman and his assistants witnessed 30 natural predator attacks in which ground squirrels were captured and killed.

Sherman's data show that when squirrels spot an attacking hawk and whistle, the whistling squirrel is captured only 2% of the time while nonwhistling squirrels are captured 28% of the time. The squirrel raising the alarm reduces its own chances of dying, perhaps by informing the hawk that the caller has seen it. At the same time, the caller sows panic and confusion among the other squirrels. Whistles, in other words, are selfish.

When squirrels spot a stalking mammal and trill, however, the trilling squirrel is killed 8% of the time while non-trilling squirrels are killed just 4% of the time. The squirrel raising the alarm increases its own peril to the benefit of other squirrels nearby. Trills meet our definition of altruism.

The evolution of altruism is a challenge to understand because the actor suffers a fitness loss. Altruistic behavior appears to be common, however. Examples include slime molds that sacrifice themselves to help others reproduce, birds that help at their parents' nests, and humans who risk death to save others from fire or drowning. Why does altruism exist in nature? In upcoming sections of the chapter, we consider a variety of explanations.

Spite

Most bacteria make proteins, called bacteriocins, that are lethal to other members of the same species (Riley and Wertz 2002; Riley et al. 2003). The makers of a given bacteriocin are either immune to their own poison because they lack the poison's molecular target or resistant because they make an antidote. **Figure 12.6** shows a nutrient plate seeded with *E. coli* from two strains (Majeed et al. 2011). One strain makes a bacteriocin called colicin E2; the other makes colicin E7. Note that by day 5, many E2 colonies have merged, as have many E7 colonies. Blocks of E2 versus E7, however, are separated by bacteria-free bands. Any bacterial cell straying into these border zones is killed by the other strain's toxin.

Because bacteriocins require energy and materials to make, strains that produce them grow somewhat more slowly than otherwise identical strains that do not (Inglis et al. 2011). Furthermore, many strains that make bacteriocins release the toxins by rupturing their cell walls and spilling their contents to the outside (Riley and Wertz 2002; Morales-Soto and Forst 2011). By making and releasing bacteriocins, a bacterial cell therefore reduces its direct fitness. And because the weapons are lethal, a bacterial cell that constructs and deploys them reduces the direct fitness of susceptible recipient cells. Bacteriocin production thus meets our definition of spite (Gardner et al. 2004; West and Gardner 2010).

To investigate whether spiteful interactions among bacteria occur in nature, Hadas Hawlena and colleagues (2010) isolated 36 strains of the bacterium *Xenorhabdus bovienii* from two soil samples taken from an Illinois forest. *X. bovienii* is a lethal insect pathogen transmitted by nematode worms. The first 18 strains came from the first soil sample, while the remaining 18 strains came from the second sample. The locations of the two samples were separated by a distance of just four meters. Hawlena and colleagues tried all 1,260 pairwise combinations in which one bacterial strain served as actor (bacteriocin donor) and a different strain served as recipient (bacteriocin recipient). For each trial, the researchers induced cells of the actor strain to make bacteriocins, prepared a cell-free extract of their contents, and tested its ability to inhibit the growth of the recipient strain.

Figure 12.7 summarizes the remarkably consistent results. No strain inhibited any other strain from its own soil sample. And every strain from each sample inhibited all the strains from the other sample. Additional tests supported the researchers' interpretation that the inhibition was due to bacteriocins. That naturally occurring *X. bovienii* can make costly poisons that are deadly to members of the same species living just a few meters away suggests that wild bacteria indeed behave spitefully toward each other.

Like altruism, the evolution of spite is difficult to understand. Heritable behaviors causing fitness loss for the actor should disappear from populations. However, bacteriocin production and other examples show that spite occurs (Gardner and West 2006; Gardner et al. 2007). In the next section, we see that one of the explanations biologists have developed to explain altruism can also explain spite.

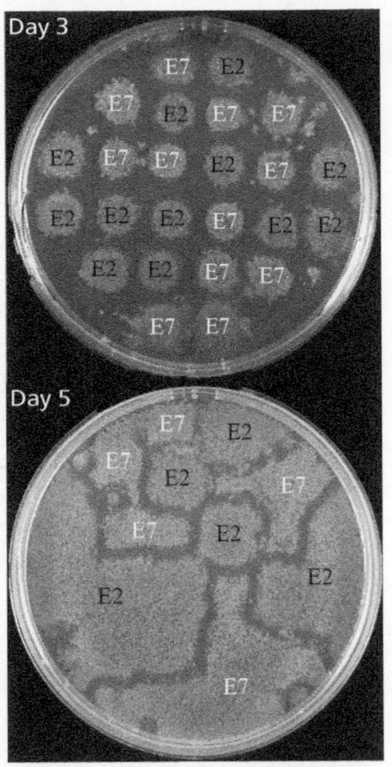

Figure 12.6 Two strains of *E. coli* that make mutually lethal bacteriocins From Majeed et al. (2011).

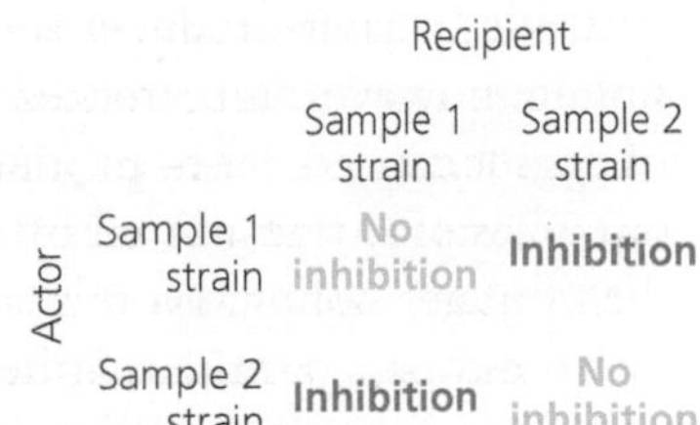

Recipient
Sample 1 strain
Sample 2 strain
Actor
Sample 1 strain
No inhibition
Inhibition
Sample 2 strain
Inhibition
No inhibition

Figure 12.7 Mutual spite among naturally occurring bacterial strains found in soil samples collected just 4 meters apart After Hawlena et al. (2010).

12.2 Kin Selection and Costly Behavior

Altruism is a central paradox of Darwinism. An allele that results in behavior benefiting other individuals at the expense of the allele's bearer would seem destined for elimination by natural selection. Charles Darwin (1859, p. 236) viewed the apparent existence of altruism as a "special difficulty, which at first appeared to me insuperable, and actually fatal to my whole theory." But he glimpsed a resolution. Selection could favor traits that result in decreased direct fitness if they increase the survival and reproductive success of close relatives. This crucial insight had to wait over a hundred years to be formalized and widely applied.

Inclusive Fitness

In 1964, William D. Hamilton devised a genetic model showing how an allele causing altruistic behavior can spread (Hamilton 1964a). His crucial insight was that an individual is seldom the sole repository of his or her genes. Copies occur in other individuals too, most predictably in kin. This means that the behavior of an individual toward others can influence the success of the actor's genes. A key parameter in Hamilton's model is the **relatedness,** r, of the actor and recipient. Biologists use somewhat different definitions of relatedness in different contexts, but by all definitions r is a measure of the genetic similarity between individuals.

Using his model, Hamilton derived a condition, called **Hamilton's rule,** under which altruism will increase in frequency:

$$Br - C > 0$$

where B is the benefit to the recipient and C is the cost to the actor. Both B and C are measured in units of surviving offspring. This simple inequality indicates that altruism is more likely to spread when the benefits to the recipient are great, the cost to the actor is low, and the participants are closely related.

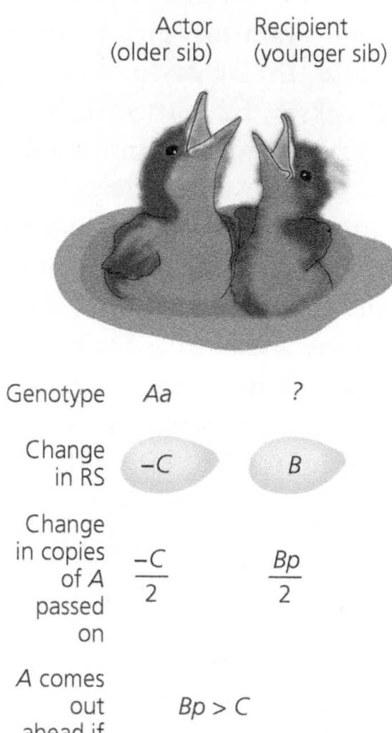

Figure 12.8 A simple derivation of Hamilton's rule See text for explanation. RS is reproductive success. After Maynard Smith (1998).

A simple derivation of Hamilton's rule, from John Maynard Smith (1998), appears in Figure 12.8. Imagine a rare dominant allele, A, in a large randomly mating population of birds that always lays two eggs. Because the allele is rare, nearly all copies are in heterozygotes. A's only influence on phenotype, relative to a, is that when it finds itself in an older sibling, A causes its carrier to beg less aggressively for food the parents bring to the nest. This diminishes the older sibling's prospects, ultimately reducing the number of eggs it can expect to produce over its lifetime by C. Because half the eggs lost would have carried a copy of A, the altruistic behavior reduces the number of copies transmitted by $\frac{C}{2}$. The younger sibling, however, is better fed. Its prospects rise, ultimately increasing the number of eggs it can expect to produce over its lifetime by B. This increases the number of copies of A transmitted by $\frac{Bp}{2}$, where p is the probability that allele A occurs in the younger sibling. Note that A has a net gain in copies transmitted if $Bp > C$.

To estimate p, note that because A is rare, it will occur in the younger sibling only if one or the other of the younger sibling's gene copies is **identical by descent** to the copy of A in the older sibling. The probability that a gene copy sampled from one individual is identical by descent to a gene copy present in another individual is a commonly used definition of relatedness. A method for calculating relatedness this way is described in Computing Consequences 12.1.

Personally costly behavior can evolve if the benefit to kin is sufficiently high.

The condition for A to increase in frequency is thus $Br > C$, which is Hamilton's rule. If the birds are monogamous, so that the chicks in a clutch are always full sibs, r is $\frac{1}{2}$. The older chick will value the survival and reproduction of its sibling at half the worth of its own. If they are half-sibs, r is $\frac{1}{4}$. The older chick will value the fitness of its sibling at a quarter the worth of its own.

More general derivations of Hamilton's rule—requiring fewer assumptions about, among other things, the number, frequency, and dominance of the genes involved—are possible (Grafen 1985; Gardner et al. 2011). The essence of all derivations is the inclusion of two components of an individual's genetic

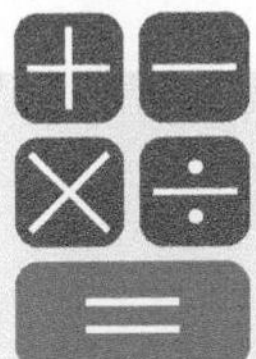

COMPUTING CONSEQUENCES 12.1

Calculating relatedness as the probability of identity by descent

Calculating the probability of identity by descent requires a pedigree including the actor and the recipient. Starting with the actor, all genealogical paths are traced through the pedigree to the recipient. For example, half-siblings share one parent and have one genealogical path with two steps, as shown in Figure 12.9a. Parents give half their genes to each offspring, so the probability that genes are identical by descent in each step is $\frac{1}{2}$. Put another way, the probability that a particular allele was passed from parent to actor is $\frac{1}{2}$. The probability that the same allele was transmitted from parent to recipient is $\frac{1}{2}$. The probability that this same allele was transmitted to both the actor and the recipient (making the alleles in actor and recipient identical by descent) is the product of these two independent probabilities, or $\frac{1}{4}$.

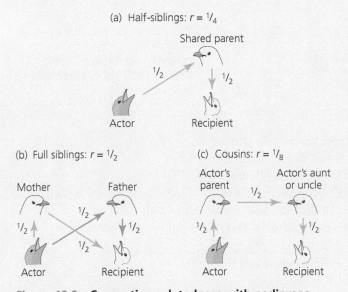

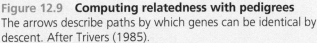

Figure 12.9 Computing relatedness with pedigrees The arrows describe paths by which genes can be identical by descent. After Trivers (1985).

Full siblings, on the other hand, share genes inherited from both parents. To calculate relatedness when actor and recipient are full siblings, we have to add the probabilities that genes are identical by descent through each genealogical path in the pedigree. In this case, we add the probability that genes are identical by descent through the mother to the probability that they are identical by descent through the father (see Figure 12.9b). This is $\frac{1}{4} + \frac{1}{4} = \frac{1}{2}$.

Using this method gives the following coefficients:

- First cousins, $\frac{1}{8}$ (Figure 12.9c)
- Parent to offspring, $\frac{1}{2}$
- Grandparent to grandchild, $\frac{1}{4}$
- Aunt or uncle to niece or nephew, $\frac{1}{4}$

The analyses we have just performed work for autosomal loci in sexual organisms and assume that no inbreeding has occurred. If the population is inbred, then the probabilities will be higher. But when studying populations in the field, investigators usually have no data on inbreeding and have to assume that individuals are completely outbred. On this basis, estimates of relatedness that are reported in the literature should be considered minimum estimates. Another uncertainty in calculating relatedness comes in assigning paternity in pedigrees. As we indicated elsewhere (Chapter 10), extrapair copulations are common in many species. If paternity is assigned on the basis of male–female pairing relationships and extrapair copulations go undetected, estimates of relatedness may be inflated.

When constructing genealogies is impractical, relatedness can be estimated from genetic data (Queller and Goodnight 1989). Microsatellites and other marker loci have proven useful for estimating relatedness in a wide variety of social insects (e.g., Peters et al. 1999).

contribution to future generations. **Direct fitness** results from reproduction an individual achieves on its own, without help from related individuals. **Indirect fitness** results from additional reproduction by relatives that is made possible by the individual's actions. The sum of its direct and indirect fitness is an individual's **inclusive fitness** (West et al. 2007b). Natural selection leading to the spread of alleles that increase the indirect component of fitness is called **kin selection.**

Fitness has both direct and indirect components. Inclusive fitness is the sum of both.

Alarm Calling in Black-Tailed Prairie Dogs

To see the power of kin selection in explaining the evolution of altruism, we go first to the Black Hills of South Dakota, where John Hoogland (1983, 1994, 1995) studied black-tailed prairie dogs (*Cynomys ludovicianus*). Prairie dogs are large squirrels that live in family groups called coteries. Each coterie holds a territory within a prairie dog town. Females typically remain in their birth coterie for life, whereas males disperse at maturity. Prairie dogs are prey to badgers, bobcats, coyotes, eagles, and falcons, and spend as much as half of their aboveground time standing watch. When it spies a predator, a prairie dog often sounds the alarm with the high-pitched bark that gave the species its common name **(Figure 12.10)**.

Figure 12.10 A black-tailed prairie dog barks an alarm

Despite logging over 50,000 person-hours watching individually marked prairie dogs in a town in South Dakota, Hoogland and his assistants were unable to document for certain whether prairie dog alarms are selfish or altruistic. Hoogland suspected that they are altruistic, however, and sought to determine whether the prairie dogs' calling behavior was consistent with the hypothesis that it evolved as a result of kin selection. Hoogland simulated predator attacks by having an assistant pull a stuffed badger through a prairie dog town on a sled while watching to see who gave alarm calls and who just dove for cover.

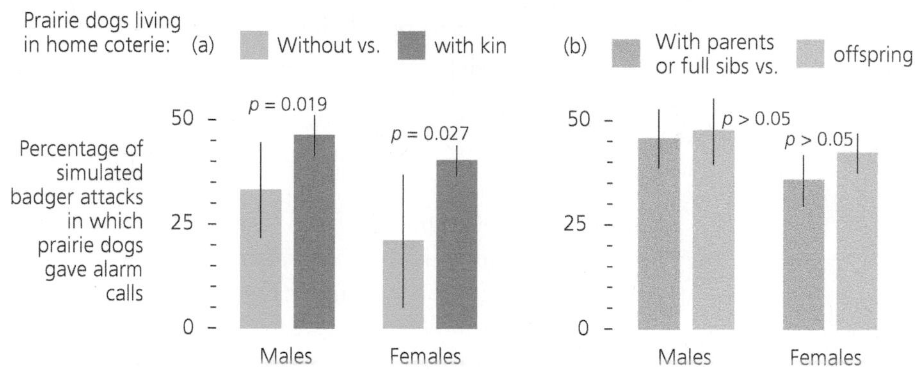

Figure 12.11 Black-tailed prairie dogs give alarms under circumstances predicted by kin selection theory (a) Rates of alarm calling by prairie dogs without versus with kin nearby. (b) Rates of alarm calling by males and females living with non-offspring kin versus offspring. Redrawn from Hoogland (1995).

Both male and female prairie dogs are more likely to give alarm calls if their coterie includes genetic kin **(Figure 12.11a)**. These calls are not simply a form of parental care: Individuals give calls nearly as often when the kin they live with are parents and siblings as when they are offspring (Figure 12.11b). Hoogland was even able to follow individual males across different stages of life, and saw them modify their rate of calling with changes in their proximity to kin **(Figure 12.12)**.

Prairie dogs are more likely to give alarm calls when close relatives are nearby.

Hoogland's data show that apparently altruistic alarm calls are not dispensed randomly. They are nepotistic. Self-sacrifice is directed at close relatives and thus

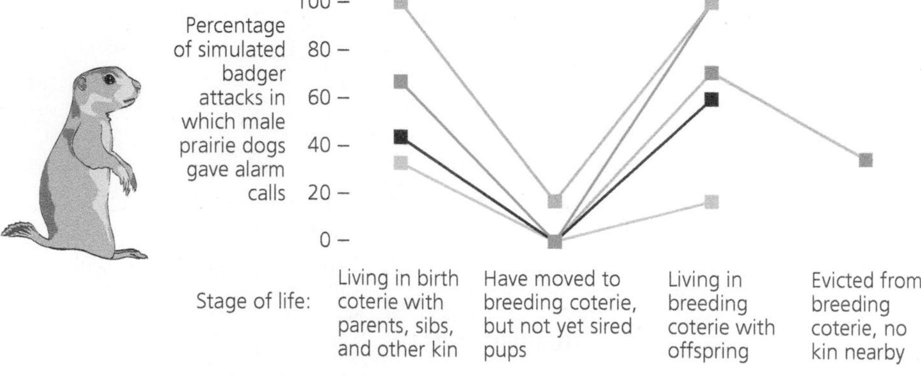

Figure 12.12 Male black-tailed prairie dogs change their alarm-calling behavior when their living situation changes This graph plots the rate of alarm calling by five individual males at different stages of life. The males increased or decreased their rate of calling according to whether kin were nearby. Redrawn from Hoogland (1995).

should result in indirect fitness gains. We turn next to a study in which researchers were able to quantify the costs, benefits, and indirect fitness gains more precisely.

Measuring Costs and Benefits for Adoptive Mother Squirrels

Stan Boutin and colleagues have followed a population of North American red squirrels (*Tamiasciurus hudsonicus*) near Kluane Lake, Canada, since 1987 (Boutin et al. 2006; McAdam et al. 2007). Each squirrel defends a feeding territory that, except when mothers are caring for kittens, it occupies alone. During spring and summer the researchers livetrap and tag the squirrels, monitor reproduction by the females, and track the growth and survival of their litters. From 1989 through 2008, the scientists gathered data on the survival and reproductive success of 6,793 juveniles from 2,230 litters birthed by 1,101 mothers (Gorrell et al. 2010).

This sample includes 34 litters orphaned during lactation, from which a kitten could have been adopted and nursed to weaning by one or more lactating females nearby **(Figure 12.13a)**. For seven of the litters, a genetic relative was among the available adoptive mothers. A kitten was adopted from five of these litters, always by the relative. For the other 27 litters, there were no relatives among the available adoptive mothers. No kittens were adopted from any of these. The association between kinship and adoption is consistent with Hamilton's rule. But the data allow a deeper look at why some related females adopted and others did not.

Because they knew how the size of a litter affects the probability that any given kitten will survive, Jamieson Gorrell and colleagues (2010) could estimate the direct fitness cost to a potential adoptive mother of adding a kitten to her litter. This cost is shown by the black line in Figure 12.13b. Adding a kitten reduces the chance that each of her existing offspring will survive (Humphries and Boutin 2000), so the mother's direct fitness cost increases with her original litter size.

And because Gorrell and colleagues had maternal genealogies for all the squirrels in their population—and some data from genetic tests—they were able to estimate the indirect fitness benefit to a potential adoptive mother of caring for a related kitten. This indirect benefit, shown by the gray lines in Figure 12.13b, is the probability that the adopted kitten will survive, which falls as a function of litter size, multiplied by the relatedness between the kitten and the mother. The figure includes three lines that cover the estimated values of r for each of the seven litters that might have been adopted by a relative. In five cases, r was at least 0.125, equivalent to the relationship between first cousins. In one case, r was at least 0.25, equivalent to half-siblings. In the remaining case, r was 0.368.

Female red squirrels adopt related kittens when the result is a net gain in the mother's inclusive fitness.

(a)

(b)

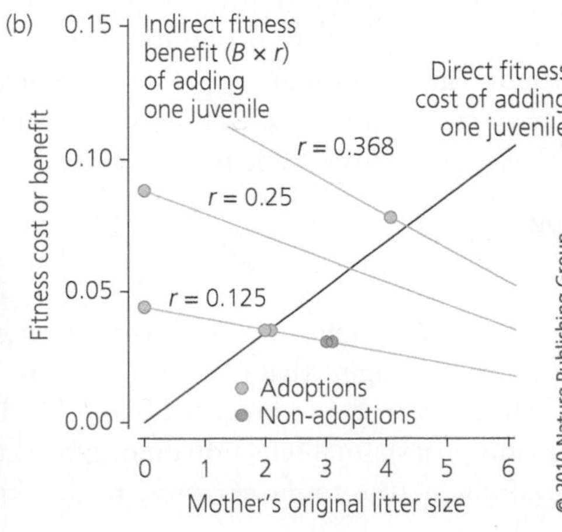

Figure 12.13 Inclusive fitness and adoption in red squirrels (a) A female red squirrel moving a juvenile from one nest to another. (b) Estimates of terms in Hamilton's rule for five adoptions and two non-adoptions by kin. From Gorrell et al. (2010).

(a) Reprinted by permission from Macmillan Publishers Ltd: *Nature Communications* 1: 22. Copyright 2010. (b) Reprinted by permission from the Macmillan Publishers Ltd. *Nature Communications*. Gorrell, J. C., A. G. McAdam, et al. "Adopting kin enhances inclusive fitness in asocial red squirrels," 1: 22. Copyright © 2010 The Nature Publishing Group.

The intersection between an indirect benefit line and the direct cost line is the break-even point for a given value of *r*. In the five cases where mothers adopted, the indirect benefit was above the break-even point, whereas in the two cases where they did not, the benefit was below the break-even point. Red squirrel mothers act as if they understand Hamilton's rule.

Kin Selection and Human Behavior

We humans often feel inclined toward helping our kin. Do we, too, act as though we understand Hamilton's rule?

Kin Selection and Inherited Wealth

Martin Smith and colleagues (1987) examined 1,000 wills written by 552 men and 448 women in Vancouver, British Columbia. The researchers assumed that humans care what happens to our wealth when we die because we want the resources we have accumulated to keep working on our genes' behalf. If this view is correct, Hamilton's rule should predict who people bequeath money to.

The researchers' first prediction was that people will leave a greater share to kin than to non-kin. Money left to relatives is more likely to help an individual's genes propagate than is money left to strangers. The only complication is how to classify spouses. A surviving spouse is not genetic kin, but might manage inherited wealth for the benefit of descendants shared with the deceased. Ignoring spouses, who received 37% of the wealth left behind, the average will writer bequeathed

55.3% of his or her wealth to kin, versus

7.7% to non-kin

This pattern is consistent with the prediction.

Smith and colleagues predicted that among genetic kin, larger slices of the pie would go to individuals more closely related to the deceased. Consisent with this prediction, the average citizen of Vancouver left

46.5% of their wealth to offspring and siblings ($r = 0.5$)

8.3% to grandkids, nieces, and nephews ($r = 0.25$)

0.6% to cousins ($r = 0.125$)

Finally, Smith and colleagues considered how much of the wealth divided among relatives with $r = 0.5$ should go to offspring versus siblings. Given that most people probably expect to live well past their prime reproductive years, and that most people are approximately the same age as their siblings, money left to offspring is more likely to encourage the production of another grandchild than money left to a sibling is to encourage the production of another niece or nephew. The researchers therefore predicted that more offspring would get more money than siblings. Consistent with this prediction, Vancouver will writers gave

38.6% of their wealth to offspring

7.9% to siblings

Smith and colleagues concluded that the data are in broad accord with kin selection theory. This does not require us to imagine that people composing wills routinely draw pedigrees, calculate direct costs and indirect benefits, and divide their wealth accordingly. It suggests only that natural selection among our ancestors left us with a tendency to feel more generous to more closely related kin.

A limitation of Smith and colleagues' study is that it concerns observational research done within a single culture. As a result, it leaves open the possibility that some uncontrolled variable, such as familiarity or local custom, offers a better explanation for the patterns in the data than kin selection. Next we consider an experimental study that partially avoids this problem.

Kin Selection and Trust

Lisa DeBruine (2002) asked 24 student volunteers to play a simple two-player bargaining game. **Figure 12.14a** summarizes each player's choices and the payoffs for each possible outcome. Player 1 first chooses whether to trust versus not trust Player 2. If Player 1 chooses to trust, then Player 2 chooses to be unselfish versus selfish. If Player 2 chooses to be unselfish, each player gets \$3. If Player 2 chooses to be selfish, Player 2 gets \$4 while Player 1 gets just \$1. If Player 1 chooses not to trust, Player 2 has no decision to make and each player gets \$2. The actual payouts varied somewhat from game to game, but the incentives and temptations always followed the same pattern. Trusting always gave Player 1 a chance to get more money at the risk of getting less. Being selfish always gave Player 2 the biggest reward and not being trusted the smallest.

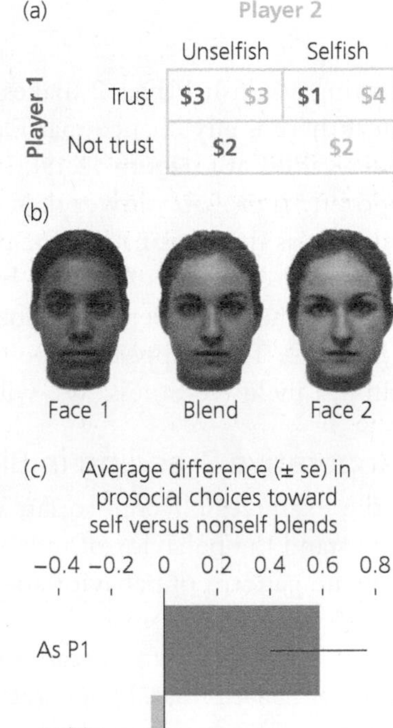

Figure 12.14 Facial resemblance and trust (a) Payoffs for each player in a simple game. Trusting and being unselfish are prosocial choices. (b) Volunteers played against virtual opponents whose faces were blends. (c) Volunteers were significantly more trusting toward self-blends ($p < 0.005$), but no more unselfish. Modified from DeBruine (2002).

DeBruine's volunteers played at a computer against a series of opponents whose pictures appeared on the screen before each game. Unbeknownst to the volunteers, their opponents were not students at remote computers, but avatars that DeBruine had programmed in advance. And each opponent's photo was a computer-generated blend of two individuals' faces (Figure 12.14b). Sometimes the volunteer's own face was a component in the blend, making the blend resemble a family member; at other times the volunteer's own face was not part of the blend. By the end, each volunteer had gotten three chances to trust a self-blended opponent and three chances to trust a nonself-blended opponent, plus three chances to be unselfish toward a self-blended opponent and three to be unselfish toward a nonself-blended opponent. On the logic that indirect fitness offers an extra incentive to trust and reduces the sting of betrayal, DeBruine predicted that her volunteers would be more prosocial toward self-blended versus nonself-blended opponents.

The results appear in Figure 12.14c. DeBruine took the number of prosocial choices each volunteer made as Player 1 (P1) against self-blended opponents and subtracted the number of prosocial choices they made against nonself-blended opponents. The average of these differences, represented by the top bar in the graph, was significantly greater than zero. The volunteers were more trusting toward opponents who looked like kin. However, when DeBruine did the same for the choices the volunteers made as Player 2 (P2), she found that the volunteers were no less selfish to opponents who looked like kin than to opponents who did not.

At first glance, this looks like a decidedly mixed result. But we have yet to carefully apply Hamilton's rule. Given the payouts shown in Figure 12.14c, we can calculate the expected inclusive fitness gain for each player making each choice. We start with Player 2 because the calculation is simpler. For an unselfish Player 2, the direct fitness gain is 3 and the indirect fitness gain through the opponent is $3r$, making the total inclusive fitness gain $3 + 3r$. For a selfish Player 2, the total inclusive fitness gain is $4 + r$. The two choices offer equivalent payoffs when

$$3 + 3r = 4 + r, \text{ or, after solving for } r,\ r = \tfrac{1}{2}$$

If the opponent is anything less than a full sibling, being selfish offers a higher inclusive payout. This is a fairly high threshold for nepotism.

To calculate Player 1's expected inclusive fitness gain, we need to know the probability that Player 2 will be unselfish. Let us assume for the sake of argument that this probability is zero. In this case, the expected inclusive fitness gain for a trusting Player 1 is $1 + 4r$. The expected inclusive fitness gain for an untrusting Player 1 is $2 + 2r$. The two strategies will be equivalent when

$$1 + 4r = 2 + 2r, \text{ or } r = \tfrac{1}{2}$$

Trusting a selfish Player 2 makes sense only if he or she is at least a full sibling. But if there is any chance that Player 2 will be unselfish, trusting makes sense at lower values of r **(Figure 12.15)**. In other words, the threshold for Player 1 to be nepotistic is probably lower than the threshold for Player 2 to be. DeBruine suggests this as the explanation for the fact that apparent relatedness influenced the volunteers' choices as Player 1, but not as Player 2. Indeed, it is somewhat puzzling that the volunteers were not both more trusting as Player 1 and more selfish as Player 2. There appears to be more at stake in these human social interactions than just inclusive fitness. We will return to this point later.

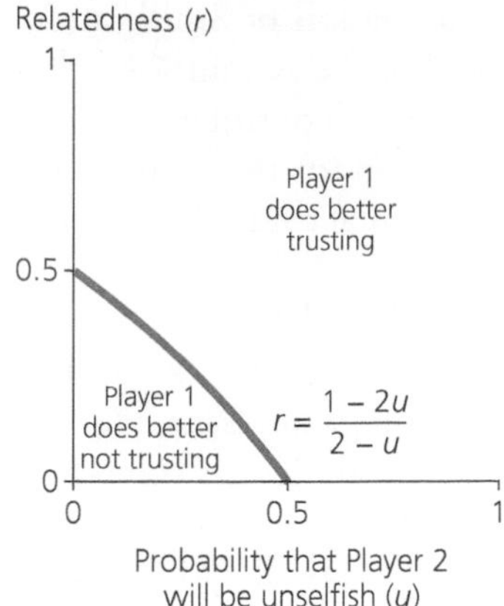

Figure 12.15 Player 1's threshold relatedness for trust as a function of the chance that Player 2 will be unselfish After DeBruine (2002).

Cooperative Breeding in Birds

In the examples covered so far, we have seen that Hamilton's rule can help us understand the behavior of individuals within species. To see how it can also illuminate patterns of behavior among species, we turn to **cooperative breeding** in birds. In species from a wide variety of bird families, young that are old enough to breed on their own instead remain and help their parents rear their brothers, sisters, or half-siblings. Helpers assist with nest building, nest defense, and/or food delivery to incubating parents and chicks.

Figure 12.16 White-winged choughs (*Corcorax melanorhamphos*) in a mud nest

In some species, including white-winged choughs **(Figure 12.16)**, breeders are incapable of fledging young without help (Heinsohn 1992). Although most white-winged chough groups consist of genetic relatives (Beck et al. 2008), adults occasionally kidnap recent fledglings from neighboring groups, care for them until they reach independence, and then recruit them as helpers (Heinsohn 1991).

As with prairie dogs, red squirrels, and humans, kin selection helps explain the behavior of individuals within cooperatively breeding species. For example, in the chestnut-crowned babbler (*Pomatostomus ruficeps*), helpers with a choice about which individuals to assist offer their care to kin (Browning et al. 2012).

Charlie Cornwallis and colleagues (2010) asked a broader question. They wanted to know whether kin selection helps explain evolutionary transitions from noncooperative breeding to cooperative breeding, and vice versa. The researchers focused on relatedness. The average relatedness among the maternal siblings in a family varies from 0.25 if each has a different father to 0.5 if they all have the same father. All else being equal, higher relatedness among the siblings in a family should facilitate the evolution of helping at the nest, and lower relatedness should facilitate the loss of helping.

All else is never equal, of course. Relatedness, r, is just one of three terms in Hamilton's rule. The benefit to the recipient, B, and the cost to the actor, C, matter as well. Costs and benefits are influenced by ecological factors and life history traits that vary dramatically across species (Hatchwell and Komdeur 2000). Cornwallis and colleagues knew of species, such as Australian magpies, with low relatedness among siblings that breed cooperatively (Hughes et al. 2003). And

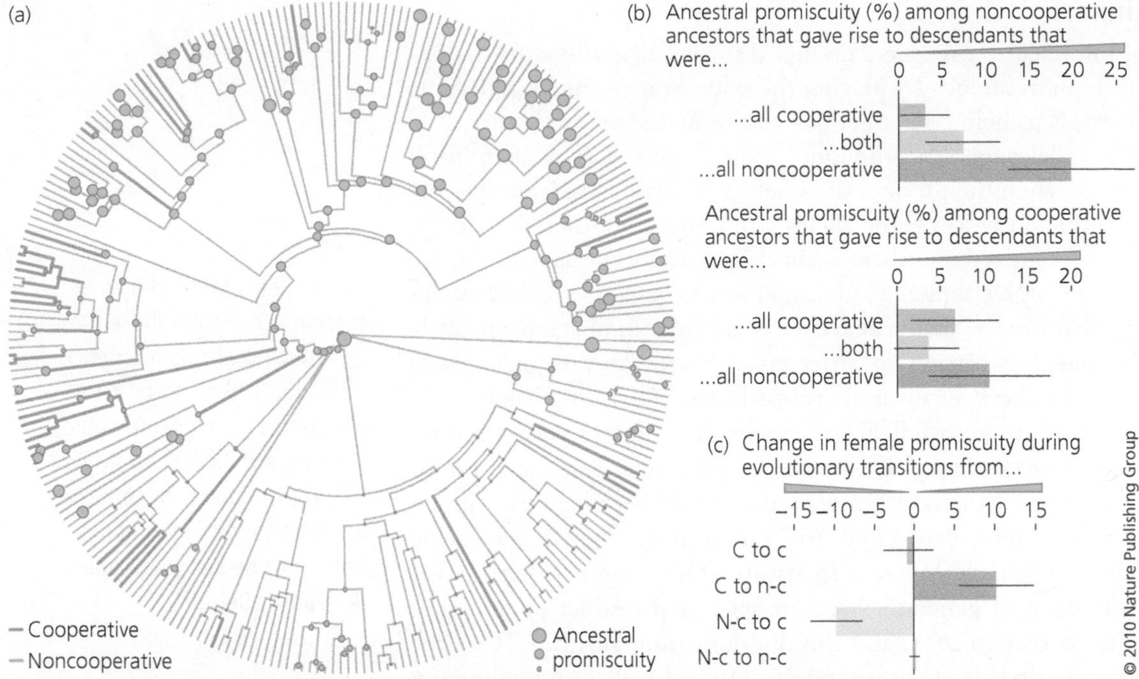

Figure 12.17 Female promiscuity and evolutionary transitions in breeding behavior in birds (a) Estimated phylogeny of 267 bird species. Known or inferred breeding system is shown by branch color. Inferred ancestral promiscuity is indicated by the size of the green dot at each node. (b) Mean ± se female promiscuity of ancestors that gave rise to descendants with different breeding systems. (c) Mean ± se change in female promiscuity during evolutionary transitions in breeding system. From Cornwallis et al. (2010).

Reprinted by permission from the Macmillan Publishers Ltd. *Nature*. Cornwallis, C. K., S. A. West, et al. "Promiscuity and the evolutionary transition to complex societies." *Nature* 466: 969–972. Copyright © 2010 The Nature Publishing Group.

they knew of species, such as puffins, with high relatedness among siblings that breed noncooperatively (Anker-Nilssen et al. 2008). The scientists nonetheless hypothesized that relatedness has a substantial influence on evolutionary trends.

Cooperative breeding evolves most readily in bird lineages in which potential helpers and beneficiaries tend to be more closely related to each other.

Cornwallis and colleagues estimated the evolutionary tree for 267 bird species for which the breeding system and degree of female promiscuity are known **(Figure 12.17a)**. Based on the evolutionary relationships, breeding systems, and levels of female promiscuity in the extant species, the researchers inferred the breeding systems and level of promiscuity for the ancestral lineages. In the reconstructed phylogeny, cooperative breeding evolved 33 times independently and was lost 20 times independently. When the researchers looked at noncooperative ancestors, they found that the ones that gave rise to only cooperative descendants were significantly less promiscuous than those that gave rise to only noncooperative descendants (Figure 12.17b, top; $p < 0.001$). The difference in promiscuity between cooperative ancestors that gave rise to only cooperative versus only noncooperative descendants was not statistically significant (Figure 12.17b, bottom; $p = 0.07$). Cornwallis and colleagues also found that transitions from cooperative to noncooperative breeding tend to be accompanied by increases in promiscuity, whereas transitions from noncooperative to cooperative breeding tend to be accompanied by decreases in promiscuity (Figure 12.17c).

These patterns are consistent with the hypothesis that low levels of female promiscuity facilitate the evolution of cooperative breeding. When maternal siblings in a nest are more closely related to each other, the investments older offspring make in helping their parents breed yield greater gains in indirect fitness.

Greenbeard Alleles

In the examples of kin selection we have discussed so far, alleles for altruistic behavior have risen to high frequency by playing the odds. Some of the kin that the alleles cause their carriers to help contain copies of the alleles; other recipients do not. But so long as the alleles for altruism induce their carriers to obey Hamilton's rule—helping kin only when the product of benefit and relatedness is sufficiently high and the cost is sufficiently low—the alleles win more often than they lose.

Hamilton (1964b) recognized that there is another mechanism that could drive an allele for altruism to high frequency, at least in principle. Richard Dawkins (1976) called this mechanism the **greenbeard effect.** Dawkins imagined an allele that simultaneously causes its carriers to grow green beards, to recognize green beards on others, and to behave altruistically toward them. Carriers of such an allele would not have to distinguish full siblings versus half-siblings versus cousins and adjust their behavior accordingly. Under the greenbeard effect, alleles for altruism would not have to play the odds. Instead, they could bet on a sure winner.

Most alleles favored by kin selection rise to high frequency by inducing altruism toward individuals likely to be carrying copies of the same allele. Greenbeard alleles would rise to high frequency by inducing altruism toward individuals certain to be carrying copies of the same allele.

We might expect the greenbeard effect to be little more than a theoretical curiosity, interesting in principle but rare in nature. The reason is that the effect requires a single allele to generate three complex and distinct phenotypes: the beard, the ability to recognize it, and the discriminating altruism. Rare the greenbeard effect may be, but it is not unknown. David Queller and colleagues (2003) reported an example that comes close to matching the scenario Hamilton and Dawkins had in mind.

Queller and colleagues studied the slime mold *Dictyostelium discoideum*. Individuals in this soil-dwelling species germinate from spores and spend most of their lives as free-living, single-celled amoebae **(Figure 12.18a, facing page)**. When food runs short, however, the amoebae send each other chemical signals and, under their influence, stream together to form a mass containing thousands of cells. This mass differentiates into a slug that travels about for a time, then transforms into a tall, thin stalk that supports a fruiting body. The cells in the fruiting body form spores, which disperse to new locations and begin the cycle anew. The cells in the stalk, some 20% of the amoebae that aggregated to form the collective, altruistically sacrifice themselves to support the reproduction and dispersal of the rest.

The particular allele Queller and colleagues studied is the wild-type allele of a gene called *csA*. Relative to loss-of-function mutations, the wild-type allele exhibits all the properties of a greenbeard allele. The protein encoded by *csA* sits on the surface of slime mold amoebae and sticks to other copies of itself on the surface of other cells. Thus the wild-type *csA* allele simultaneously specifies a trait (the protein) and the ability to recognize the trait in others (by adhesion). The remaining greenbeard trait is discriminating altruism.

Queller and colleagues mixed wild-type amoebae with amoebae carrying a knock-out allele of the *csA* gene, grew them on agar plates in the lab, then starved them to encourage them to stream together and make fruiting bodies. The researchers found that wild-type cells were disproportionately represented in the stalks (Figure 12.18b). The wild-type cells apparently ended up on the bottom, relegated to a supporting and nonreproductive role, because they stuck to each other more strongly. So far, the wild-type allele appears to be not a greenbeard but the opposite: an altruist that sacrifices itself for the beardless. In lab cultures growing on agar plates, it should quickly disappear.

The situation is reversed, however, when mixed cultures are grown in soil, their natural environment (Figure 12.18c). It is more difficult for amoebae to

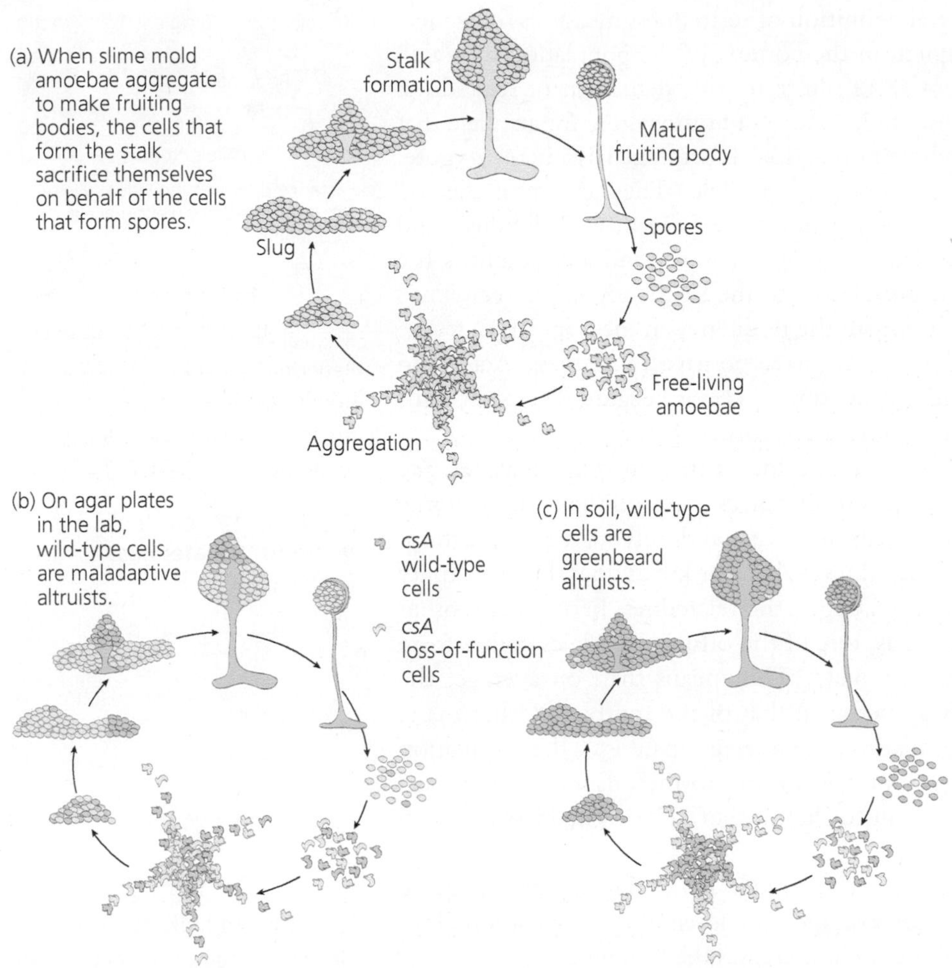
(a) When slime mold amoebae aggregate to make fruiting bodies, the cells that form the stalk sacrifice themselves on behalf of the cells that form spores.
Stalk formation
Mature fruiting body
Spores
Slug
Free-living amoebae
Aggregation
(b) On agar plates in the lab, wild-type cells are maladaptive altruists.
csA wild-type cells
csA loss-of-function cells
(c) In soil, wild-type cells are greenbeard altruists.

Figure 12.18 Life cycle of the slime mold *Dictyostelium discoideum* (a) When free-living amoebae aggregate to form a fruiting body, 20% of them sacrifice themselves on behalf of the others. The altruists form a stalk that supports the fruiting body proper. Only the amoebae in the fruiting body make spores. The spores disperse, germinate, and start a new life cycle. (b and c) The outcome in genetically mixed aggregates depends on the environment.

stream together in soil than it is on agar plates. Wild-type cells can stick to each other and pull each other along. Now wild-type cells are disproportionately represented among the spores in fruiting bodies as well as in the stalks. Knock-out cells are less adhesive and tend to get left out of aggregations altogether. Under natural conditions, then, the wild-type allele of *csA* renders its carriers preferentially altruistic toward other wild-type cells. Kin selection works not just at the level of individual organisms but also at the level of individual alleles.

Our final topic in this section is whether kin selection sheds light on the other kind of costly and thus puzzling behavior we introduced in Section 12.1: spite.

Kin Selection and Spite

We have seen how kin selection can help explain altruistic behavior. However, as Hamilton (1970) himself recognized, it can also help explain spite. Hamilton's rule states that an allele for a social behavior will be favored if $Br - C > 0$. In a spiteful interaction, the "benefit" to the recipient is actually a cost. This means that B is negative. However, Br can still be positive, and potentially larger than C, if r is also negative (Gardner and West 2004).

Negative Relatedness

How can relatedness be negative? If we are defining relatedness as the probability of identity by descent, then r cannot be negative. A probability cannot be less than zero. However, a more general definition of relatedness measures the genetic similarity of the actor and recipient in the context of the population to which they belong (Grafen 1985). **Figure 12.19** illustrates this definition of relatedness geometrically. In the example illustrated, Actor is a homozygote for an allele that is at fairly low frequency in the whole population. Recipient 1 is a heterozygote, and Recipient 2 is a homozygote for the other allele. The cartoon at the top shows the frequency of the allele in the genotypes of the three individuals and in the entire population. The relatedness between Actor and a recipient is the distance from the frequency in the population to the frequency in the recipient's genotype as a fraction of the distance from the frequency in the population to the frequency in Actor. Actor and Recipient 1 have positive relatedness. Actor and Recipient 2, however, have negative relatedness. This is because Recipient 2 sits on the opposite side of the population frequency from Actor.

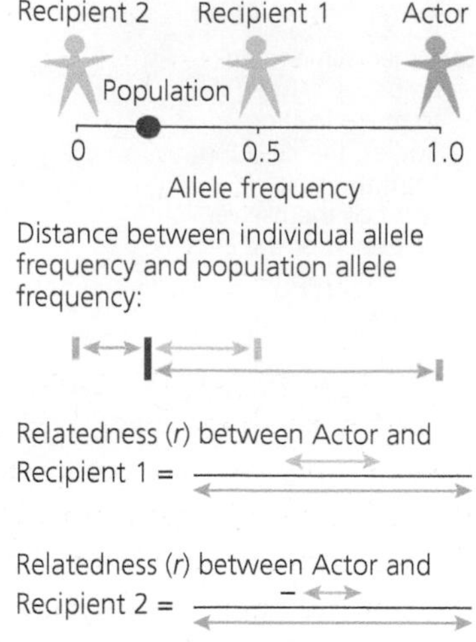

Figure 12.19 Grafen's geometric relatedness After Grafen (1985).

Figure 12.19 shows relatedness at a single locus. In principle, however, we could perform this calculation using genetic distances based on the entire genome (Grafen 1985). If we do so for individuals in a large randomly mating population, this definition of *r* yields the same relatedness values for kin as does the probability of identity by descent. Consider, for example, the relatedness between a mother and one of her offspring. At each locus, one of the offspring's alleles comes from the mother and one comes from her mate. This means that, on average, the offspring's genotype will be halfway between that of the mother and her mate. Under the assumption of random mating, the average mate is at the population mean. Thus the distance from the population mean to the offspring will be half the distance from the population mean to the mother. The relatedness between the mother and her offspring is $\frac{1}{2}$.

Now look back at the two recipients in Figure 12.19. Recipient 1 is more genetically similar to Actor than is the average individual in the population. This means that when Recipient 1 reproduces, it increases the frequency of Actor's allele in the population. If Actor helps Recipient 1 reproduce, it contributes to Actor's inclusive fitness (Gardner and West 2004). Recipient 2, on the other hand, is less genetically similar to Actor than is the average individual in the population. When Recipient 2 reproduces, it decreases the frequency of Actor's allele in the population. If Actor impedes Recipient 2 from reproducing, it contributes to Actor's inclusive fitness. This explains how spite can evolve.

Spite toward individuals less closely related to the actor than the average member of the population may increase the actor's inclusive fitness.

The Enemy of My Enemy Is My Friend

Spite can be thought of as a kind of altruism in which individuals with positive relatedness to the actor indirectly benefit when the actor pays a cost to reduce the fitness of competitors with negative relatedness to the actor. Hamilton's rule shows that this is possible in theory, but does it work in practice?

Recall the example of spite we discussed in Section 12.1. Bacteria, which compete among themselves over small distances, produce and release costly toxins that kill other members of the same species while leaving clonemates of the self-sacrificing producer unharmed. Farrah Bashey and colleagues (2012) sought to find out whether spiteful individuals of the bacterium *Xenorhabdus bovienii* improve the competitive success of their clonemates under seminatural conditions.

Xenorhabdus bovienii is a pathogen of insects. From soil, Bashey and colleagues isolated two strains. One, which we will call spiteful, makes bacteriocins that kill a third strain, the susceptible competitor. The other, which we will call benign, does not. Neither the spiteful nor the benign strain can kill a strain of

X. koppenhoeferi, the resistant competitor. Finally, neither the susceptible competitor nor the resistant competitor makes bacteriocins of its own.

Bashey and colleagues performed two controls. In the first, they inoculated the spiteful and benign strains by themselves into caterpillars and monitored the bacterial growth rates. When each is living alone, the two strains show no significant difference in growth **(Figure 12.20a)**. In the second, the researchers inoculated the spiteful strain into caterpillars along with the resistant competitor. They did the same with the benign strain. In competition with the resistant competitor, the benign strain grew significantly faster than the spiteful strain (Figure 12.20b).

To test the hypothesis that against a susceptible competitor bacteriocin production can improve the competitive prospects of clonemates of the spiteful actors, Bashey and colleagues inoculated the spiteful strain into caterpillars along with the susceptible competitor. They did the same with the benign strain. This time, the spiteful strain grew significantly faster than the benign strain (Figure 12.20c). This result is consistent with the notion that spiteful production of bacteriocins evolved by kin selection. So was the result of a second experiment involving a different pair of spiteful and benign strains.

Intriguingly, caterpillars inoculated with a mixture of the susceptible competitor and the spiteful strain survived longer than did caterpillars inoculated with either strain alone. In contrast, caterpillars inoculated with a mixture of the susceptible competitor and the benign strain died just as fast as caterpillars inoculated with either strain alone. It appears that spiteful interactions between pathogens can alter the course of infection for the host.

In sum, the examples in this section have demonstrated how kin selection helps explain the evolution of costly behaviors in a variety of organisms. In the next section, we explore an explanatory framework called multilevel selection.

(a) Control 1:
Spiteful and benign strains alone

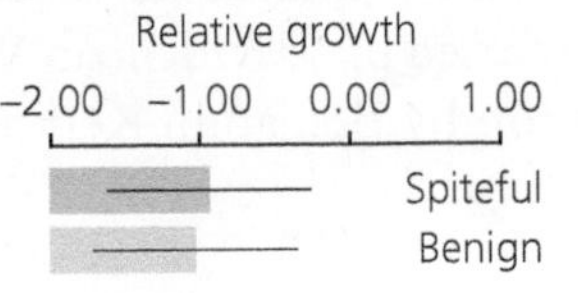

(b) Control 2:
Spiteful and benign strains grown with resistant competitor

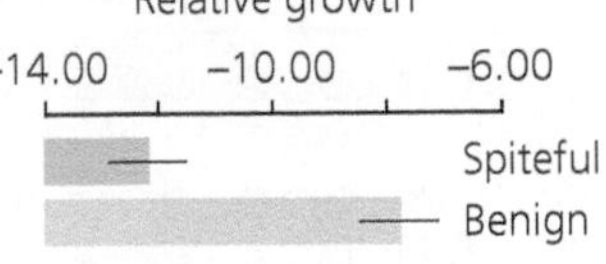

(c) Experiment:
Spiteful and benign strains grown with susceptible competitor

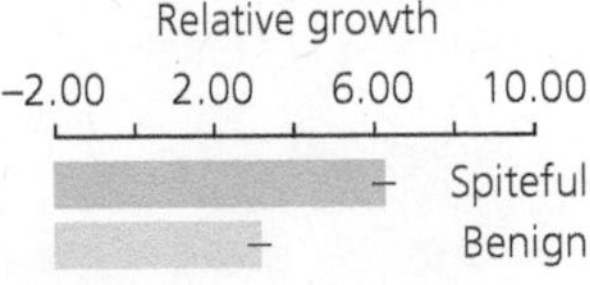

Figure 12.20 Spite alters the outcome of bacterial competition (a) $p = 0.85$. (b) $p < 0.0001$. (c) $p < 0.0001$. From Bashey et al. (2012).

12.3 Multilevel Selection and Cooperation

As with kin selection, the notion of **multilevel selection** can be traced back to Charles Darwin (see Marshall 2011). Among the problems Darwin tackled in *The Descent of Man* was the evolution of human morality. "[A]lthough a high standard of morality gives but a slight or no advantage to each individual man and his children over the other men of the same tribe," he wrote (1871, p. 166), a high frequency of moral individuals will "give an immense advantage to one tribe over another."

Following Stuart West and colleagues (2007a), we will use the term **cooperation** for behavioral traits that, like morality as Darwin described it, are beneficial or even costly to the actor, beneficial to the recipient, and selected for at least in part because the recipient benefits. Cooperation in this sense encompasses all cases of altruism and some cases of mutual benefit. The interesting cases of mutual benefit are those in which the recipient gains more than the actor. Like altruism, these cases of unbalanced mutual benefit raise questions about how a trait that does more for the fitness of others than it does for fitness of the individuals that carry it can spread in populations. In this section, we consider whether cooperation can evolve when it confers benefits on groups.

David Sloan Wilson (1975) developed a model showing that cooperators can increase in overall frequency when the higher productivity of groups with many cooperators outweighs the lower relative fitness of cooperators within groups.

A Numerical Example of Multilevel Selection

We introduce multilevel selection by means of a numerical example similar to one discussed by David Sloan Wilson (1975). Our treatment draws heavily on the exposition by Benjamin Kerr and Peter Godfrey-Smith (2002).

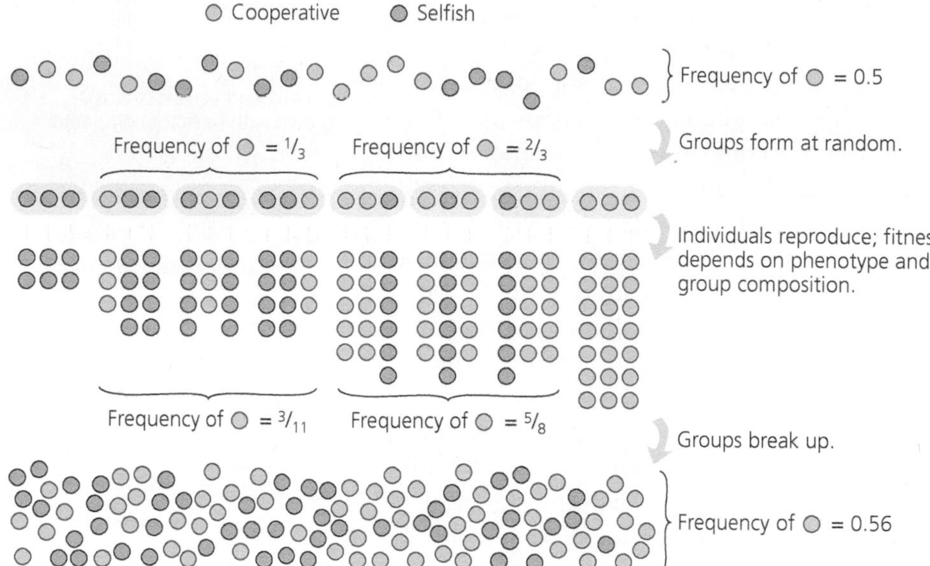

Figure 12.21 How cooperation can increase in frequency under multilevel selection Each generation the members of a population form groups of three at random. Individuals reproduce by cloning and die, after which the groups break up. An individual's reproductive success is a function of its own phenotype—cooperative or selfish—and the phenotypes of its group mates. Within mixed groups, the frequency of cooperation falls. However, group productivity increases dramatically with the number of cooperators in the group, and the overall population frequency of cooperation rises.

Imagine a population with two types of individuals, cooperative and selfish, at frequencies of 0.5 **(Figure 12.21)**. The individuals reproduce by cloning. Before reproducing, they form groups of three at random. If the population is very large, there will be groups with 0, 1, 2, or 3 cooperators in proportions of 1:3:3:1. Reproduction takes place within the groups. Afterward, the adults die and the offspring disperse. We want to know how frequency of cooperators changes across generations. This depends, of course, on their relative reproductive success.

An individual's success depends on its own behavior and that of its group mates. In groups with no cooperators, each individual has 2 offspring. In groups with one cooperator, the cooperator has 3 offspring and each selfish individual has 4. In groups with two cooperators, each cooperator has 5 offspring and the selfish individual has 6. In groups with three cooperators, each has 7 offspring.

Note that a cooperator always increases the reproductive success of everyone in its group, but can be exploited to additional advantage by selfish individuals. As a result, within mixed groups the frequency of cooperators falls from one generation to the next. In groups with one cooperator, we saw it fall from $\frac{1}{3}$ to $\frac{3}{11}$. In groups with two cooperators, it fell from $\frac{2}{3}$ to $\frac{5}{8}$.

Nonetheless, cooperators raise the productivity of their groups so much that when the offspring mingle, the frequency of cooperators in the offspring population as a whole is higher than it was in the parental population. Among the parents the overall frequency of cooperators was 0.5; among the offspring it is 0.56. This pattern of contrasting trends within groups versus a population is called **Simpson's paradox.**

If cooperators provide sufficient benefit to their social groups, cooperation may increase in frequency in the population at large, even if it decreases in frequency within groups.

Scenarios of this sort are described as models of multilevel selection. In comparisons of individuals within groups, selfish individuals have higher fitness than cooperators. In comparisons between groups within the population, groups with more cooperators produce more offspring. In our example, between-group selection outweighs within-group selection and cooperation spreads.

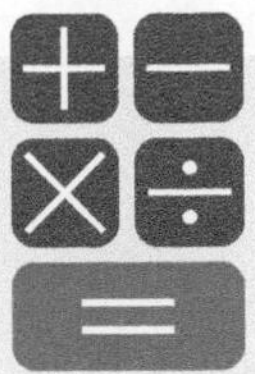

COMPUTING CONSEQUENCES 12.2

Different perspectives on the same evolutionary process

Here we follow Kerr and Godfrey-Smith (2002) in distinguishing two mathematically equivalent perspectives on the evolutionary model shown in Figure 12.21.

A fully useful model of an evolving population lets us predict change in the frequency of alleles or phenotypes from generation to generation indefinitely. To do so with models like the one in Figure 12.21, we need

- The initial frequencies of types of individuals.
- The rule by which social groups form. If this rule lets us determine the relative frequencies of all types of groups from the frequencies of types of individuals, then we need not know anything about the groups present in earlier generations.
- The set of rules by which reproductive success is determined. This is called the fitness structure.

Figure 12.22 shows two views of the fitness structure of the population in Figure 12.21. Figure 12.22a plots individual fitness as a function of group composition. Biologists who prefer this view tend to focus on fitness as an individual attribute with direct and indirect components, and to use the terms *individual selection* and *kin selection.* Figure 12.22b shows group productivity (left) and fraction of offspring that are cooperators (right) as functions of group composition. Biologists who prefer this view tend to emphasize that variation in group composition and group productivity are required for cooperation to spread, and to use the terms *multilevel selection* and *group selection.*

Note, however, that each view of the fitness structure contains all the information needed to draw the other, and both predict the same evolutionary change.

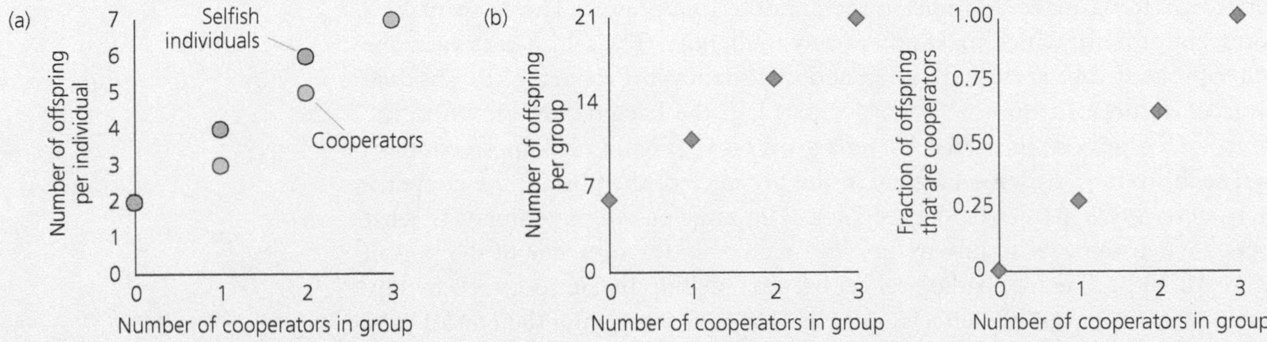

Figure 12.22 Two depictions of the same fitness structure Both parts show the fitness structure of the model in Figure 12.21. (a) Fitness of individuals as a function of group composition. (b) Productivity of groups (left) and proportion of offspring that are altruists (right) as a function of group composition. After Kerr and Godfrey-Smith (2002).

Many variants of multilevel selection are possible. The groups can be of any size. Instead of forming at random, they can be composed of kin or of individuals sharing genes or phenotypes. Behavior can be a polygenic trait. The organisms can be sexual. Fitness can depend on phenotype and group composition according to any pattern we choose (Kerr et al. 2004). Cooperation—sometimes even altruism—rises to fixation in some scenarios, disappears in some, and is maintained at intermediate frequency in still others (Kerr and Godfrey-Smith 2002).

Biologists vary in the words they use to describe the mechanisms of evolution that operate in populations structured into groups (West et al. 2007a). This can give the impression of more disagreement over process than actually exists (Lion 2011; Marshall 2011). How biologists would describe evolution in our example largely depends on how they do the accounting **(Computing Consequences 12.2)**.

Multilevel Selection and Cooperation in Bacteria

The example we discussed on the previous two pages shows that, in principle, cooperation can spread under multilevel selection. Does this happen in practice?

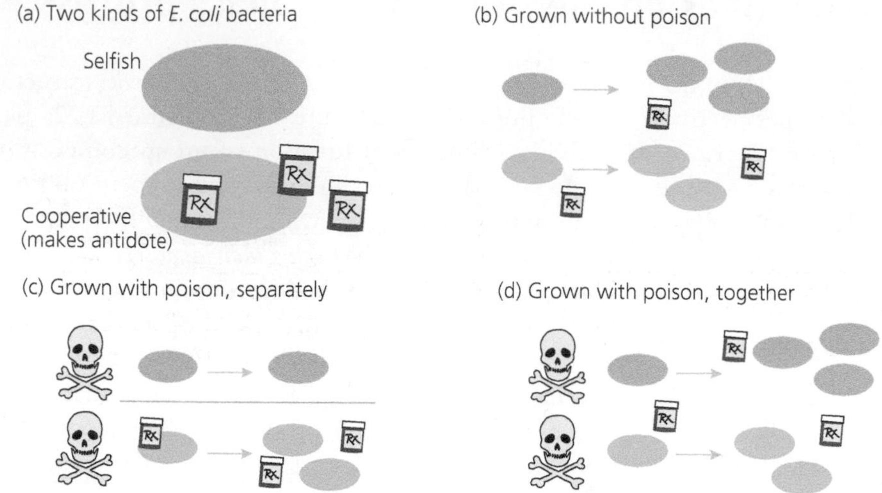

Figure 12.23 Bacteria genetically engineered to be selfish or cooperative (a) Otherwise identical to the selfish strain, cooperative bacteria make and release the antidote to a poison. The poison is the antibiotic chloramphenicol; the antidote is a molecule that triggers the expression of choramphenicol resistance. (b–d) Which strain grows faster depends on whether the other strain and the poison are present. After Chuang et al. (2009).

John Chuang and colleagues (2009) investigated experimentally. They genetically engineered strains of *E. coli* bacteria to be selfish or cooperative, like the creatures in our multilevel selection model. As shown in **Figure 12.23**, the cooperative bacteria make and release the antidote to a poison. They also make a fluorescent protein, which makes them easy to identify. The selfish bacteria make neither product, but are otherwise genetically identical. Producing the antidote and green fluorescent protein is costly, so when the bacteria are grown in the absence of the poison, the selfish bacteria grow faster (Figure 12.23b). In contrast, when each strain is grown separately in the presence of the poison, the cooperative bacteria grow faster (Figure 12.23c). The crux of the experiment is what happens when the two strains are grown together in the presence of the poison (Figure 12.23d). The freeloading selfish bacteria pick up the antidote released by the cooperators, escape the effects of the poison without paying the cost of production, and thereby grow faster than the cooperators.

Chuang and colleagues set up 21 replicates of their experiment. Each replicate consisted of 12 bacterial cultures with initial frequencies of cooperators ranging from 0 to 1. The cultures fixed for selfish individuals or cooperators served as controls. The 10 mixed cultures together represented a large bacterial population temporarily segregated into groups. The researchers added poison to the cultures, let them grow overnight, and then assessed each one's growth rate and ending frequency of cooperators. Finally, the researchers combined the 10 mixed cultures and assessed the change from start to finish in global cooperator frequency.

The results, which were the same across all 21 replicates, appear in **Figure 12.24** (next page). The higher the proportion of cooperators in any given culture, the faster the culture grew (Figure 12.24a). This result is consistent with the notion that the cooperators provided a common good to their social groups.

Within every mixed culture, the frequency of cooperators fell overnight (Figure 12.24b). This result is consistent with the notion that the selfish bacteria benefited from the common good without paying the cost of production. In other words, within any given culture, selection favored the selfish freeloaders.

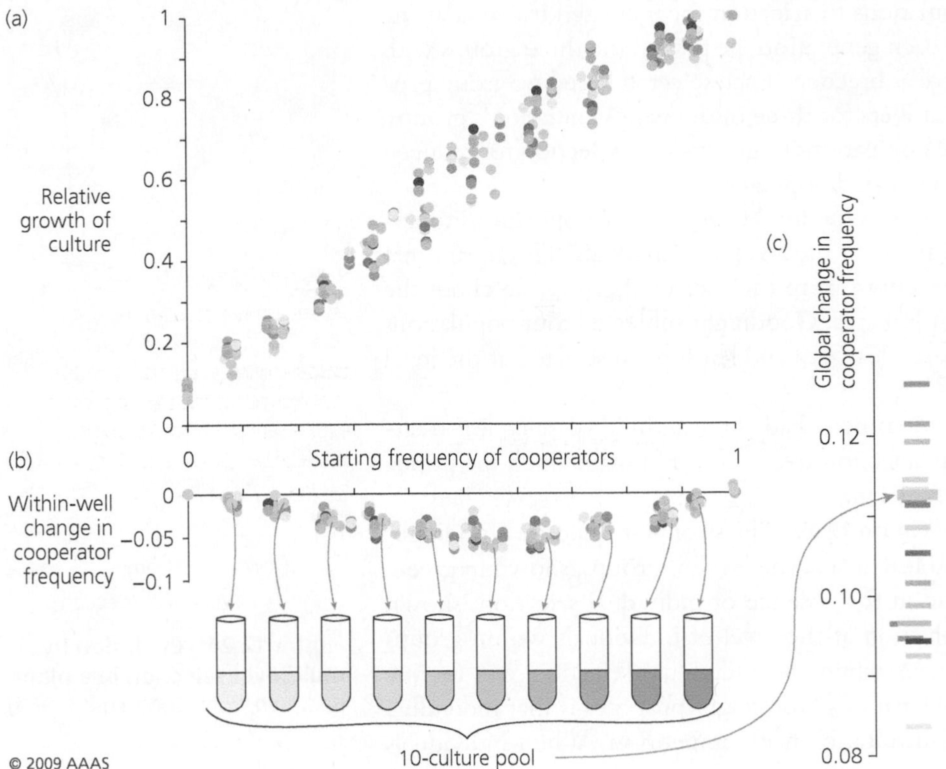

Figure 12.24 Multilevel selection favoring cooperative bacteria Colors represent replicates of the same experiment. (a) The higher the frequency of cooperators in a culture, the faster the culture grows. (b) Within individual cultures, the frequency of cooperators falls over time. (c) Nonetheless, due to the enhanced productivity of cultures with a high frequency of cooperators, the global frequency of cooperators rises over time. From Chuang et al. (2009).

From Chuang, J. S., O. Rivoire, and S. Leibler. 2009. "Simpson's paradox in a synthetic microbial system." *Science* 323: 272–275. Copyright © 2009 AAAS. Reprinted with permission.

When the scientists combined the mixed cultures in each replicate, however, they found that the global frequency of cooperators always rose (Figure 12.24c). This result can be described as a case where selection against cooperators within cultures was outweighed by selection among cultures favoring cooperators.

We can think of selection as happening among cultures only when there is sufficient variation from culture to culture in the frequency of cooperators. In the experiment we have described, there was variation among cultures because Chuang and colleagues deliberatiely set them up that way. However, the researchers hypothesized that adequate variation would arise by chance if each culture were established by a small number of randomly chosen founders. The researchers diluted a stock culture with a 10% frequency of cooperators so that the average aliquot drawn from it contained just two or three bacteria. They used aliquots to establish 288 subcultures. After letting the subcultures grow, the reseachers pooled them, diluted the resulting population, and drew samples again to start another set of 288 cultures. After five rounds of dilution and growth, the global frequency of cooperators had risen to over 95% **(Figure 12.25)**. Under the right conditions, cooperation can, indeed, spread in a context of multilevel selection.

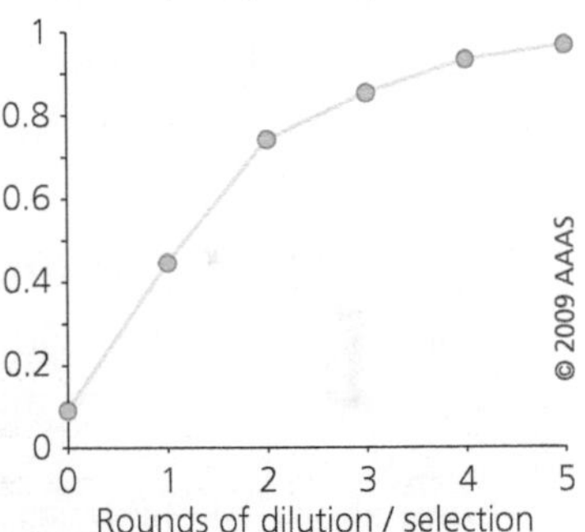

Figure 12.25 Evolution of altruism under multilevel selection in a bacterial population with randomly founded groups From Chuang et al. (2009).

From Chuang, J. S., O. Rivoire, and S. Leibler. 2009. "Simpson's paradox in a synthetic microbial system." *Science* 323: 272–275. Copyright © 2009 AAAS. Reprinted with permission.

Multilevel Selection in a Plant

The appearance of plants in a chapter on social behavior may at first glance seem surprising. However, plants compete both above and below the ground, and they show genetic variation in both tolerance for competition and the ability to suppress the growth of their neighbors (Willis et al. 2010; Wolf et al. 2011).

Inspired by Michael Wade's (1977) work on beetles, Charles Goodnight (1985) investigated multilevel selection in thale cress (*Arabidopsis thaliana*). He set up populations consisting of 144 individuals divided into 9 social groups of 16 plants. The plants in each group lived together in a tub of nutrient medium.

Goodnight subjected 12 populations to selection for increased leaf area at the level of groups. This meant that each generation, he picked the three groups with the highest mean leaf area to serve as breeders. Each selected breeding group produced the seeds that served as founders for three of the next generation's groups. Goodnight subjected another 12 populations to group-level selection for reduced leaf area and kept a further 12 unselected populations.

Within each set of 12 populations, Goodnight subjected 4 populations to additional selection for increased leaf area at the level of individuals. This meant that instead of choosing breeders at random from each selected group, he chose the eight individuals with the highest leaf area. Goodnight subjected four populations to individual selection for decreased leaf area and left four unselected at the level of individuals.

In total, then, Goodnight's experiment had three individual-selection treatments nested within three group-selection treatments. He maintained the populations through eight episodes of selection.

The results are summarized in **Figure 12.26**. The strongest response Goodnight saw was to selection for increased leaf area at the level of groups, shown in green. This response was most dramatic in the absence of individual selection, shown in the green point at center. Selection at the level of individuals within groups interfered with this response, even when the individual selection was for increased leaf area. Among the explanations Goodnight proposes is that individual selection favored traits involved in interference competition. When individuals that interefere with their neighbors are present in a group, the entire group may suffer. In contrast, selection at the level of groups may favor competitive restraint.

Benjamin Kerr and colleagues (2006) got a similar result in experiments with a bacteriophage. Limited migration among culture wells, which effectively divided the phage population into groups, led to the evolution of competitive restraint. Competitive restraint can be viewed as a kind of cooperation.

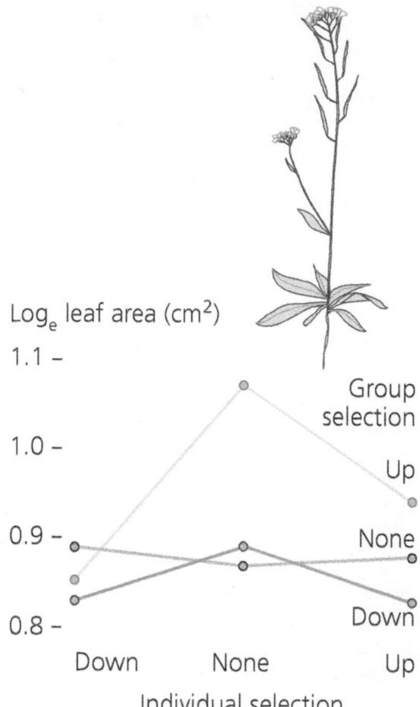

Figure 12.26 Evolution by multilevel selection in a plant Redrawn from Goodnight (1985).

Multilevel Selection and Human Morality

We began this section with Darwin's hypothesis that the human moral sense evolved in the context of multilevel selection. The hypothesis appears plausible, but it is difficult to test.

Until about 10,000 years ago, all humans lived as hunter-gatherers. Few traditional hunter-gatherer societies remain, and fewer still occupy habitats, harvest resources, and use tools unaffected by contact with outsiders. Nonetheless, studying present-day hunter-gatherers is the closest we can come to watching how our ancestors lived while fundamental human traits were evolving.

Kim Hill and colleagues (2011) compiled data on 32 such societies. Hunter-gatherers live in social groups, or bands, of six to several dozen people. The mean size of a band is 28.2 individuals. Both men and women move between bands. For a typical individual, fewer than 10% of the other adults in the band are parents, siblings, or first degree in-laws. The rest are friends.

To find out how bands form among the Hadza hunter-gatherers of Tanzania, Coren Apicella and colleagues (2012) measured cooperativity by asking individuals to play a public goods game. They gave each subject four honey sticks and offered them the option of anonymously donating as many as they wished to the public good. For every stick a subject donated, the researchers donated three. The researchers explained that after all the Hadza in a band had made their donatations, the researchers would divide the public pool of sticks evenly among the members of the band. Because individuals lost honey on their own donations but gained honey on the donations of others, we can think of individuals who made bigger donations as being more cooperative. When Apicella and colleagues compared the cooperativity scores of band members, they found that there was more variation among bands, and less variation within bands, than would be expected if bands formed randomly **(Figure 12.27)**. In other words, the Hadza are grouped according to cooperativity. When the researchers asked Hadza individuals who they wanted to live with in their next camp, cooperative individuals named each other more often than expected by chance. Uncooperative individuals tended to name each other as well. And the researchers found that genetically related individuals tended to resemble each other in cooperativity.

If sorting by cooperativity was common among ancestral humans, and if cooperativity is heritable, then cooperation may have spread in part because members of bands with a higher proportion of cooperators produced more children.

When college students in modern Western cultures play public goods games, they become more cooperative when groups of players are in competition with each other (Puurtinen and Mappes 2009). They also report experiencing more anger and guilt, emotions associated with moral behavior. Just comparing groups to each other, rather than having them compete, produces similar effects (Burton-Chellew and West 2012). Again these results are consistent with the hypothesis that human morality evolved in a context of multilevel selection. Unfortunately, college students in Western postindustrial cultures are known to be atypical representatives of human behavior (Henrich et al. 2010). More data are needed from a greater diversity of cultures.

The data we need most would show how human fitness is related to individual cooperativity and group composition. That is, we need the fitness structure.

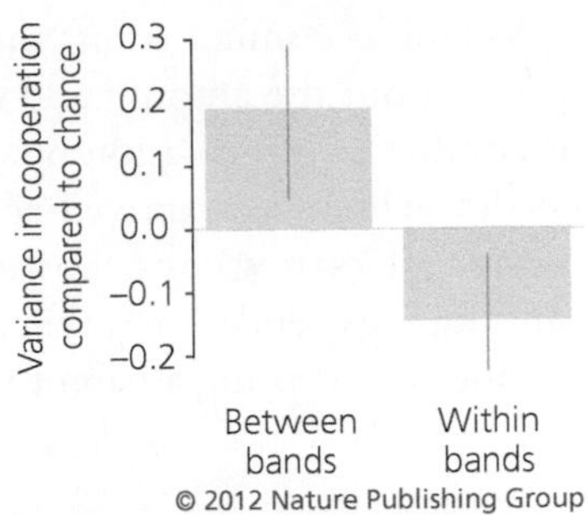

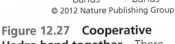

Figure 12.27 Cooperative Hadza band together There is more variation in cooperativity among bands, and less within bands, than would be expected if bands formed at random ($p < 0.05$). From Apicella et al. (2012).

Reprinted by permission from the Macmillan Publishers Ltd. *Nature*. From Apicella, C. L., F. W. Marlowe, et al. "Social networks and cooperation in hunter-gatherers." 481: 497–501. Copyright © 2012 The Nature Publishing Group.

12.4 Cooperation and Conflict

As we mentioned at the beginning of the chapter, social interactions generate both opportunities for cooperation and conflict among individuals. In this section we explore the interplay between cooperation and conflict in two social situations. The first is relationships between parents and their offspring.

Parent–Offspring Conflict

Parental care is a special case of providing fitness benefits for close relatives. Although kin selection can lead to close cooperation between related individuals such as parents and offspring, even close kin can be involved in conflicts when the costs and benefits of altruism change or when degrees of relatedness are not symmetrical. Robert Trivers (1974) was the first to point out that parents and offspring are expected to disagree about each other's fitness interests. Because parental care is so extensive in birds and mammals, conflicts over the amount of parental investment should be especially sharp in these taxa.

Conflict over Provisioning of Offspring

Weaning conflict is a well-documented example of parent–offspring strife. Aggressive and avoidance behaviors are common toward the end of nursing in a variety of mammals. Mothers ignore or push young away when they try to nurse, and offspring retaliate by screaming or by attacking their mothers **(Figure 12.28)**.

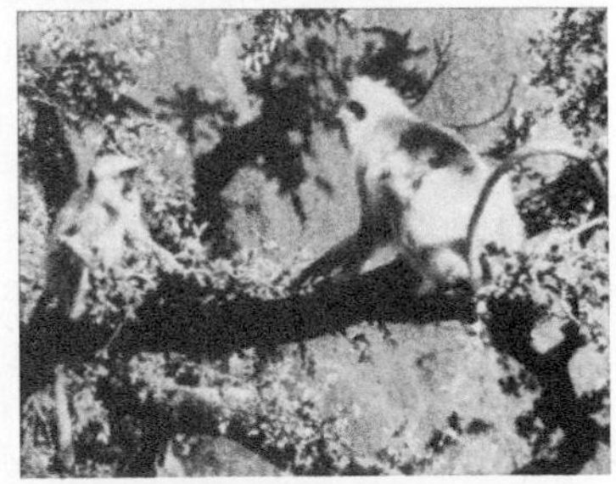

Figure 12.28 Weaning conflict This infant langur monkey (left) has just attempted to nurse from its mother (right). The mother refused to nurse. In response, the infant is screaming at her. In a moment, the infant will dash across the branch and slap its mom.

Weaning conflict is just one form of disagreement between parents and offspring about the amount and duration of parental provisioning. To see the logic of conflict over provisioning, imagine a mother bird feeding a clutch of chicks. Feeding chicks is adaptive because it increases the probability that the chicks will survive. It carries costs, however. Care expended on one offspring is care that cannot be expended on others. Caring for present offspring may even reduce the number of offspring a parent will be able to produce in the future.

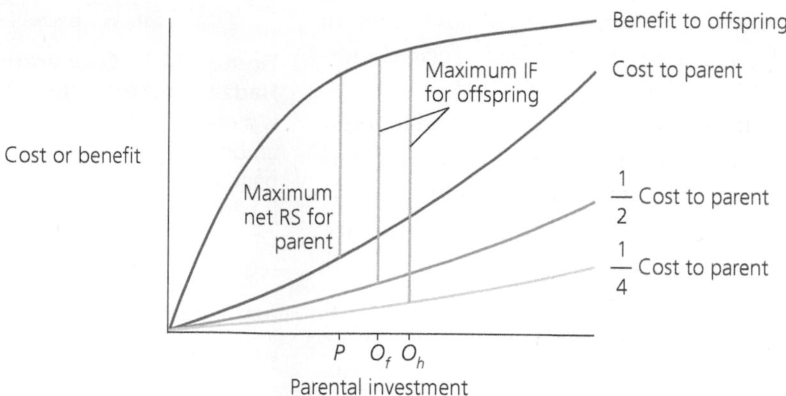

Figure 12.29 The logic of parent–offspring conflict over the amount of provisioning RS = reproductive success; IF = inclusive fitness. P, O_f, and O_h mark the optimal parental investment for the parent, an offspring whose lost siblings are full sibs, and an offspring whose lost siblings are half-sibs. After Trivers (1974).

Figure 12.29 presents a graphical analysis of this situation developed by Robert Trivers (1974). The blue line traces the benefit to an offspring of being fed as a function of the size of the parental investment. At first, the probability the offspring will survive increases steeply with the amount of provisioning. Eventually, however, the benefit levels off. Once the probability of survival is high, additional parental investment does little to raise it further. The upper orange line traces the cost to the parent in additional offspring that cannot be produced. The level of investment that maximizes the parent's reproductive success is the one with the largest difference between benefit and cost.

Now look at the situation from the perspective of one of the chicks. The benefit of being fed is the same for the chick as for the parent. The cost to the chick in lost siblings, however, is discounted for the chick relative to the parent. This is because siblings carry only a fraction of the chick's genes. Full siblings have a relatedness of $\frac{1}{2}$. The cost to the chick in lost full sibs is shown by the middle orange line in the figure. The parental investment that maximizes the offspring's inclusive fitness is larger than the investment that maximizes the parent's fitness. If the lost siblings are half-sibs, their relatedness is $\frac{1}{4}$, and the optimal parental investment for the chick is higher still. Any given chick in the nest, then, disagrees with the parent over how much food should go to itself versus its nestmates. A more detailed analysis appears in Lessells and Parker (1999).

Chicks attempt to influence the way their parents distribute food by begging **(Figure 12.30)**. This behavior, which is physiologically demanding and dangerous (Moreno-Rueda and Redondo 2011; Ibáñez-Álamo et al. 2012), provides an opportunity to test Trivers's analysis. If the analysis is right, chicks should beg more vigorously when their nestmates are less closely related. Giuseppe Boncoraglio and colleagues (2008, 2009) found this to be true in experimentally manipulated clutches of barn swallows (*Hirundo rustica*).

Many social interactions bring opportunities for both cooperation and conflict. Individuals may change their behavior as the costs and benefits to their inclusive fitness change.

For a broader comparative test, James Briskie and colleagues (1994) recorded the begging calls of chicks in a variety of bird species for which estimates of the

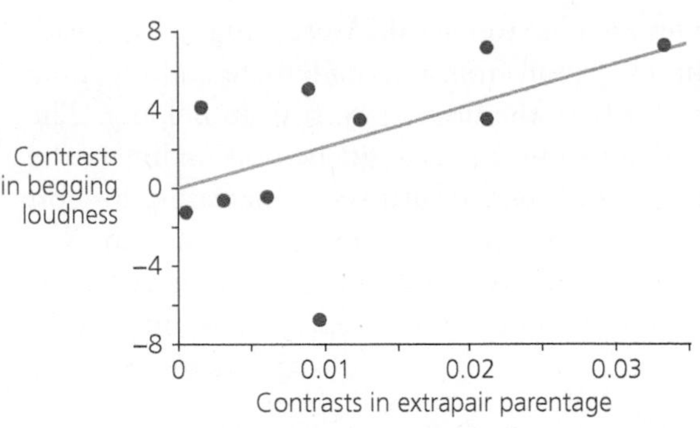

Figure 12.30 Loudness of begging versus extrapair parentage in birds Contrasts reflect evolutionary divergences between sister taxa. Species that evolve higher levels of extrapair parentage tend to evolve louder begging ($p = 0.022$). Redrawn from Briskie et al. (1994).

rate of extrapair paternity were known. Using a phylogney of the species in their data set, the researchers calculated, for pairs of taxa, the divergence in the loudness of begging and the divergence in the frequency of clutches sired by multiple fathers (see Section 10.4). As shown in Figure 12.30, these evolutionary changes are positively correlated. That is, when a bird species evolves a higher level of female promiscuity—and thus a lower average relatedness among the chicks in a nest—it also tends to evolve louder begging by chicks. This result is consistent with Trivers's analysis of parent–offspring conflict.

Harassment in White-Fronted Bee-Eaters

Another dramatic example of parent–offspring conflict occurs in the white-fronted bee-eaters. Steve Emlen and Peter Wrege (1992) have collected data suggesting that fathers occasionally coerce sons into helping to raise their siblings. They do this by harassing sons who are trying to raise their own young.

A variety of harassment behaviors are observed at bee-eater colonies. Individuals chase resident birds off their territory, physically prevent the transfer of food during courtship feeding, or repeatedly visit nests that are not their own before egg laying or hatching. During the course of their study, Emlen and Wrege observed 47 cases of harassment. Over 90% of the instigators were male, and over 70% were older than the targeted individual. In 58% of the episodes, the instigator and victim were kin. In fact, statistical tests show that harassment behavior is not targeted randomly, but is preferentially directed at close kin ($p < 0.01$; χ^2 test).

Emlen and Wrege interpret this behavior by proposing that instigators are actively trying to break up the nesting attempts of close kin. Furthermore, they suggest that instigators do this to recruit the targeted individuals as helpers at their own (the instigator's) nest.

What evidence do Emlen and Wrege present to support this hypothesis? In 16 of the 47 harassment episodes observed, the behavior actually resulted in recruitment: The harassed individuals abandoned their own nesting attempts and helped at the nest of the instigator. Of these successful events, 69% involved a parent and offspring and 62% involved a father and son. The risk of being recruited is clearly highest for younger males and for males with close genetic relatives breeding within their clan **(Figure 12.31)**.

These data raise the question of why sons do not resist harassment more effectively. Emlen and Wrege suggest that harassment can be successful because sons are equally related to their own offspring and to their siblings. Parents, in contrast, are motivated to harass because they are more closely related to their own offspring ($r = \frac{1}{2}$) than they are to their grandchildren ($r = \frac{1}{4}$). On average, each helper is responsible for an additional 0.47 offspring being raised. In comparison,

© 1992 Nature Publishing Group

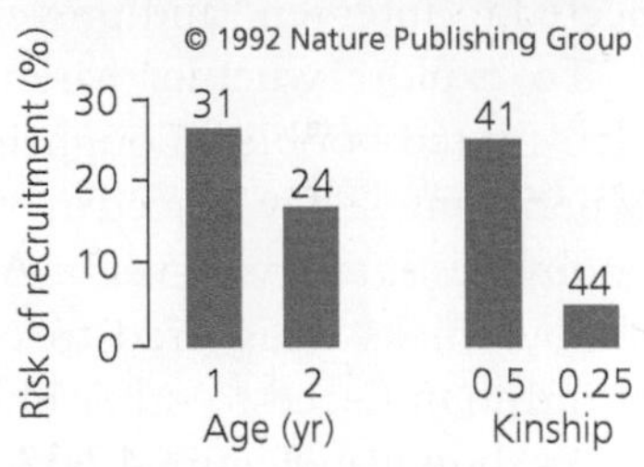

Figure 12.31 Bee-eaters recruit helpers who are younger and closely related Emlen and Wrege (1992) considered each paired male in a colony who had a same-age or older male present in its clan, and calculated the percentage of these individuals who were recruited as helpers after experiencing harassment. Kinship represents the relatedness of the targeted individual and the offspring in the instigator's nest.

(bottom) Reprinted by permission from Macmillan Publishers Ltd. *Nature*. Emlen, S. T., and P. H. Wrege. 1992. "Parent-offspring conflict and the recruitment of helpers among bee-eaters." *Nature* 356: 331–333. Copyright © 1992.

parents at a nest unaided by relatives are able to raise 0.51 offspring. This means that for a first-time breeder, the fitness payoff from finding a mate and breeding without helpers is only slightly greater than the fitness payoff from helping. The payoffs are close enough to suggest that parents can change the bottom line of the fitness accounting. Perhaps harassing a son tips the balance by increasing his cost of rearing young. Then helping becomes a more favorable strategy for the son than raising his own young. Emlen and Wrege's data imply that bee-eater fathers recognize sons and coerce them into serving the father's reproductive interests.

Siblicide

In certain species of birds and mammals, it is common for young siblings to kill each other while parents look on passively. How can this behavior be adaptive, given that the parents and siblings are related by a relatedness of one half?

Lynn Lougheed and David Anderson (1999) took an experimental approach to this question. They studied seabirds called the masked booby and the blue-footed booby **(Figure 12.32a and b)**. In both species, females normally lay a two-egg clutch. One chick hatches 2–10 days before the other. In the masked booby, the older chick pushes its younger sibling from the nest within a day or two of hatching. There the smaller chick quickly dies of exposure or is taken by a predator.

Siblicide is more complex in the blue-footed booby. The older chick does not always kill its sibling at hatching. During short-term food shortages, Anderson and Robert Ricklefs (1995) found that older chicks eat less, thereby helping the sib survive. But if famine continues, the older chick attacks and kills the younger one. Presumably, reduced competition raises the chick's odds of surviving.

Lougheed and Anderson (1999) wanted to understand whether parents play a role in these events. In both booby species, siblicide makes sense in light of the relatedness asymmetry between individuals (where $r = 1$) and their siblings (where $r = \frac{1}{2}$). But parents are equally related to each chick and would be expected to intervene and prevent attacks.

To explore whether parental behavior differs between masked boobies and blue-footed boobies, Lougheed and Anderson performed a reciprocal transplant experiment. They placed newly hatched masked booby chicks in blue-footed booby nests, and vice versa. As controls, they also monitored the fate of masked booby chicks transferred to other masked booby nests and blue-footed booby broods transferred to other blue-footed booby nests.

As shown in Figure 12.32c, the fate of the chicks varied dramatically across treatments. Chicks were more likely to die if they had a masked booby nestmate. This result is consistent with the observation that siblicide is virtually universal in this species. But chicks were also more likely to die if they had masked booby parents. Lougheed and Anderson suggest that masked booby parents tolerate siblicidal chicks, while blue-footed booby parents attempt to intervene and prevent the death of their younger offspring. This data set was the first to indicate that in some siblicidal species, parents act to defend their reproductive interests.

Why masked booby parents tolerate siblicide is unclear. The available evidence is consistent with the notion that the second egg has value as insurance against the failure of the first (Anderson 1990a). However, the egg would have more value if it yielded a surviving offspring. Masked booby parents appear to be as capable of feeding a second chick as blue-footed booby parents (Anderson 1990b; Anderson and Ricklefs 1992). Given the inclusive fitness masked booby chicks could often gain through a sibling, it is also unclear why they are invariably siblicidal.

(a) Masked boobies

(b) Blue-footed boobies

(c)

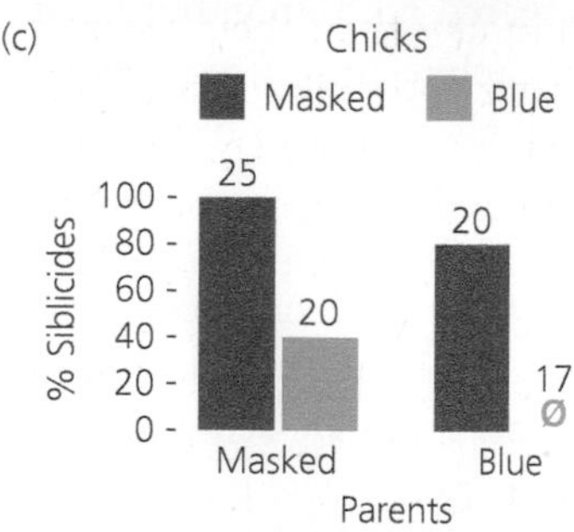

Figure 12.32 Two species of boobies (a and b) Masked and blue-footed boobies nest in adjacent colonies in the Galápagos Islands. (c) In experimental clutches, the rate of infanticide varies with the species of both parents and chicks. Data from Lougheed and Anderson (1999).

Cooperation and Conflict among Non-Kin

The social interactions we have so far considered in this section have involved kin. We now turn to cooperation and conflict among non-kin.

Reciprocity

Among the ways that non-kin cooperate is by trading favors (Trivers 1971). Doing a favor for another individual may be costly in the short term, but if the recipient eventually pays it back with a favor of greater value, the actor comes out ahead in the long run. The exchange of favors that ulitimately benefit both participants is called **reciprocity.**

Dorothy Cheney and colleagues (2010) found evidence of reciprocity among unrelated baboons (*Papio hamadryas ursinus*). Among the favors one baboon can offer another are grooming and providing support in aggressive encounters with a third baboon. Cheney and colleagues recorded the threat grunts of a number of individuals. A threat grunt is a call a baboon makes when attempting to intimidate a subordinate individual; it puts the subordinate on notice and helps the caller recruit allies. The researchers then played the recorded grunts back to other baboons. In some cases, the subject hearing the call had been groomed by the recorded caller at least 10 minutes before the playback. In other cases, the subject had been threatened by the caller at least 10 minutes before. The researchers noted whether the first movement the subject made after hearing the call was toward the caller. **Figure 12.33** shows the results. Listeners that had been groomed by the caller were significantly more likely to move closer than were listeners who had been threatened by the caller ($p = 0.004$).

Frequency of movement toward call
0 0.1 0.2 0.3 0.4 0.5
Caller groomed subject before trial
Caller threatened subject before trial

Figure 12.33 Reciprocity in baboons Baboons recently groomed by a recorded caller are more likely to move closer than are baboons recently threatened by the caller. Graph redrawn from Cheney et al. (2010).

A key feature of reciprocity is that while it is mutually beneficial in the long run, it leaves the individual who acts first vulnerable to a net loss in the short run. This can happen if the recipient cheats by failing to return the favor—a strategy that is adaptive any time the cheater can get away with it. One way actors can avoid being taken advantage of is to offer aid only to individuals they know well. Gloria Sabbatini and colleagues (2012) found that when captive capuchin monkeys (*Cebus apella*) are given a choice of who to share food with, the quality of their long-term social relationships with each of the potential partners is more important than their recent history of food sharing. Similarly, Ada Grabowska-Zhang and colleagues (2012) found that great tits (*Parus major*) are much more likely to help a neighbor defend its nest from predators if they are familiar with the neighbor from the previous year. Another way to avoid being taken advantage of is, as we will see shortly, to punish cheaters (Raihani et al. 2012).

Robert Trivers (1971) noted that reciprocity is most likely to evolve when the same individuals repeatedly interact with each other, have many opportunities to exchange favors, sometimes need and sometimes can offer favors, and have good memories. Accordingly, we expect reciprocity in long-lived, intelligent, social species with small group size, low rates of dispersal from the group, and a high degree of mutual dependence in group defense, foraging, or other activities. Reciprocity is generally thought to be fairly rare (Clutton-Brock 2009), which is a bit surprising given its intuitive appeal. Reciprocity seems straightforward to us humans because in our species it is common.

Reciprocity and Punishment in Humans

Robert Trivers (1971, 1985) has proposed that human emotions like moralistic aggression, gratitude, guilt, and trust are adaptations that evolved in response to selection for reciprocity. He suggests that these emotions function as scorekeeping mechanisms. Studies indicate that humans are particularly good at detecting cheaters in social exchanges (Stone et al. 2002; Sugiyama et al. 2002; Cosmides et al. 2010). Among the strategies we use to limit cheating is punishment.

Ernst Fehr and Simon Gächter (2002) asked students to play a public goods game, for cash prizes, in groups of four. Each player started with a stake of 20 points. Each round, players had the option to donate up to 20 points to a group pool. For each point donated, the group received 1.6 points that were divided evenly among the members (0.4 point to each). Thus each player lost points on their own donations, but gained points on everyone else's. If all four players donated 20 points, each received 32. Sometimes the game included an additional rule. At the end of every round, each player had the option of paying to punish any or all of their group mates. For each point a player paid, the punished individual lost three. Some groups played six rounds with punishment followed by six rounds without. Other groups played six rounds without punishment followed by six rounds with.

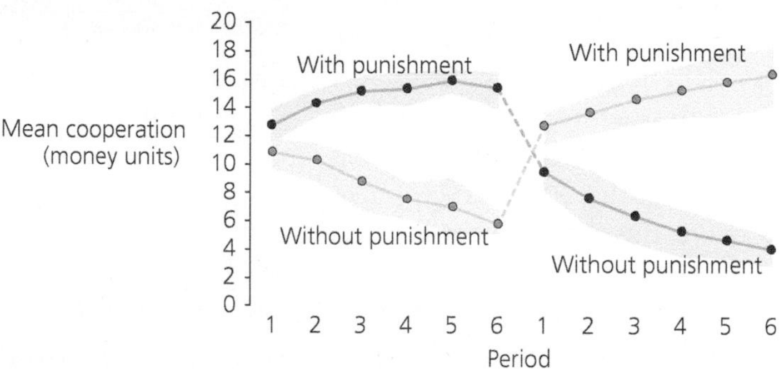

Figure 12.34 Punishment enhances cooperation in a public goods game When players were allowed to pay to punish each other, they achieved higher levels of cooperation. Redrawn from Fehr and Gächter (2002).

Figure 12.34 tracks the mean donation to the common good in both sets of groups throughout the entire tournament. When punishment was not an option, the mean donation started at an intermediate level and fell as players reacted to being taken advantage of. When punishment was allowed, the mean donation started at an intermediate level and rose as players paid to fine each other.

We noted earlier that college students in Western postindustrial societies are poor representatives of typical human behavior. It is important to recognize that the tendency to punish insufficient cooperation varies across cultures.

Nonrelatives may cooperate by exchanging favors. They may also punish each other for failure to cooperate.

Documentation of this variation comes from Joseph Henrich et al. (2006), who asked volunteers from 15 different societies to play an ultimatum game. The subjects ranged from traditional hunter-gatherers to Western college students. They played the game in pairs. The first player was asked to propose a division, in increments of 10%, of an amount of money equal to a day's wages. The second player had to decide, before hearing the offer, which offers would be acceptable and which would not. If the actual offer was acceptable, the players kept the amounts agreed to; otherwise neither got anything. If Player 2 sought only to maximize the return for participating, he would accept any offer greater than zero. If Player 1 could count on this and shared the same motive, she would always offer the smallest nonzero amount allowed: 10% of the stake.

Figure 12.35 summarizes the players' behavior. Note first that player behavior varied dramatically across cultures. The mean offer, shown by the black vertical lines, ranged from 25% to 50% of the stake. The rate of refusal of different offers,

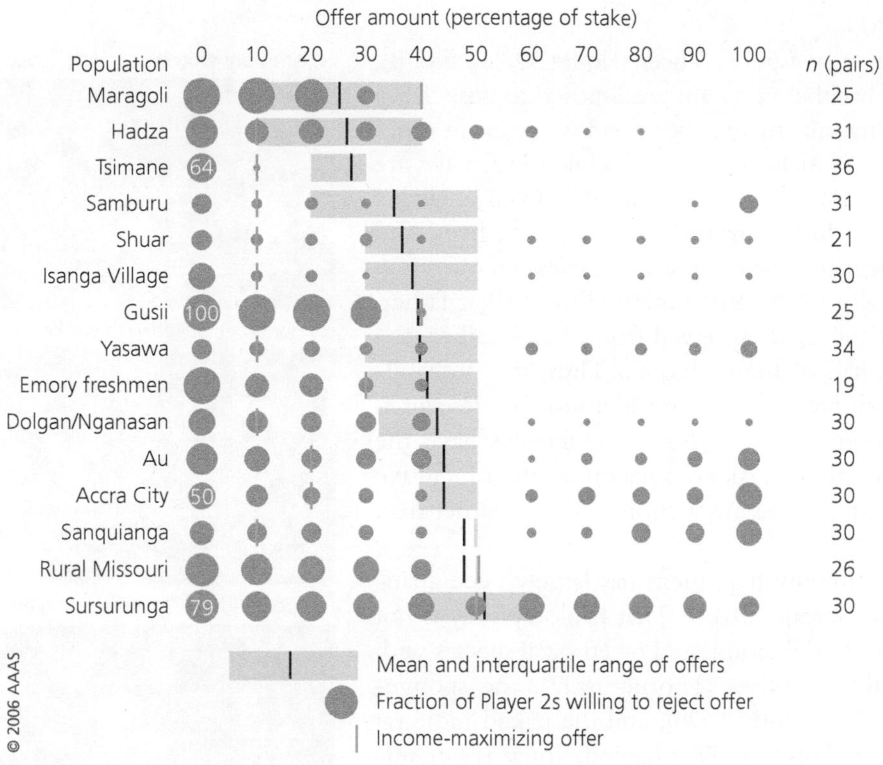

Figure 12.35 Cross-cultural variation, and cross-cultural universals, in player behavior in a simple economic game Volunteers from different cultures varied in how they played the ultimatum game. Members of all cultures, however, showed some willingness to punish unfair offers. And in all cultures the mean offer was higher than zero, the minimum offer allowed. From Henrich et al. (2006).

Henrich, J., R. McElreath, et al. 2006. "Costly punishment across human societies." *Science* 312: 1767–1770. Copyright © 2006 AAAS. Reprinted with permission.

shown by the areas of the green circles, varied even more. Tsimane players, for example, virtually always accepted any nonzero offer, whereas the members of several cultures frequently refused offers to give them the entire stake.

Note also, however, that nowhere did the volunteers play as if everyone sought only to maximize their short-term gain. In all but the Tsimane, nonzero offers were sometimes or always refused—at a cost to both participants. And in all cultures, including the Tsimane, the mean offer was considerably larger than the minimum amount allowed. A reasonable guess as to why is that the players feared, with some justification, that they would be punished for being unfair.

12.5 The Evolution of Eusociality

Darwin (1859) recognized that social insects represent an epitome of altruism, and thus present a special challenge for evolutionary theory. A colony of the leafcutter ant *Atta cephalotes,* for example, may contain millions of nonreproductive workers in a variety of castes (see Suen et al. 2011). Large soldiers defend the colony's nest and foraging territory. Medium workers gather leaves that serve as substrate for the fungus the ants cultivate for food. Small workers tend the fungus garden. They also care for the colony's young, all produced by a single queen.

Suzanne Batra coined the adjective **eusocial** to describe animals that, like leafcutter ants, have overlapping adult generations in which nonreproductive individuals participate in the cooperative care of young (Batra 1966; see Costa and Fitzgerald 2005). Animals exhibiting these characteristics vary in the degree to which the traits are elaborated. Likewise, biologists vary in their use of Batra's term (see Costa and Fitzgerald 2005; Crespi 2005; Lacey & Sherman 2005; Wcislo 2005). Here we consider efforts to explain the evolution of eusociality.

The Haplodiploidy Hypothesis

Eusociality was first recognized in ants, wasps, and bees **(Figure 12.36a and b).** William Hamilton (1972) proposed that these taxa are predisposed to eusociality by their unusual form of sex determination. In ants, bees, and wasps, as in other members of the order Hymenoptera, the males are haploid while the females are diploid. Males develop from unfertilized eggs; females develop from fertilized eggs. As a result of this system, called **haplodiploidy,** female ants, bees, and wasps are more closely related to their sisters than they are to their own offspring.

This follows because sisters share all of the genes inherited from their father, which is half their genome, and half the genes received from their mother (the colony's queen), which is the other half of their genome. Thus, the probability that homologous alleles in hymenopteran sisters are identical by descent is $(1 \times \frac{1}{2}) + (\frac{1}{2} \times \frac{1}{2}) = \frac{3}{4}$. To their own offspring, however, females have the usual relatedness of $\frac{1}{2}$. Hamilton argued that females can maximize their inclusive fitness by acting as workers and investing in the production of sisters, rather than by acting as reproductives.

Although once popular, the haplodiploidy hypothesis has largely been abandoned (Nowak et al. 2010; West and Gardner 2010). That haplodiploidy is not necessary for the evolution of eusociality is demonstrated by eusocial species with ordinary diploid inheritance, including termites (Thorne 1997), the sponge-dwelling shrimp shown in Figure 12.36c (Duffy 1996), and the naked mole rat shown in Figure 12.36d (Alexander et al. 1991). That haplodiploidy is not sufficient for the evolution of eusociality is demonstrated by the many haplodiploid taxa, including many hymenoptera, that are not eusocial. Finally, recent theoretical work has suggested that haplodiploidy may not even facilitate the evolution of eusociality (Gardner et al. 2012a, 2012b).

(a) Army ant workers carrying larvae

(b) Honeybee queen with workers

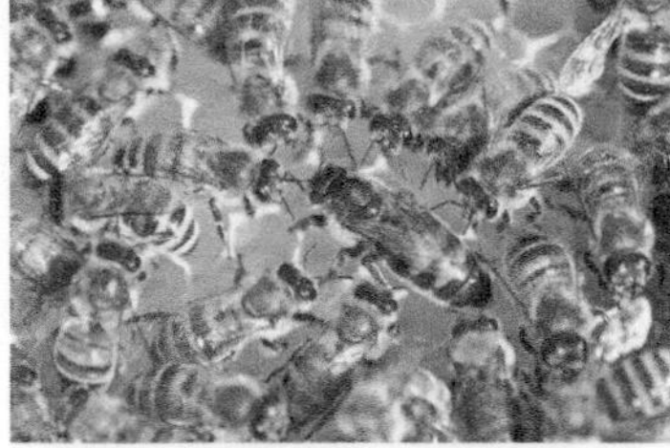

(c) Sponge-dwelling shrimp workers

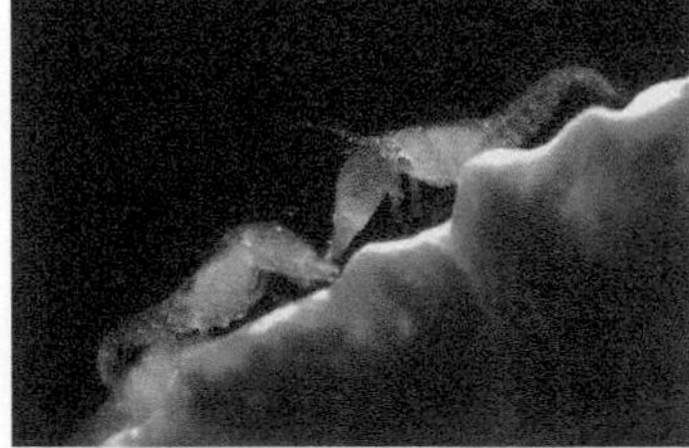

(d) Naked mole rats

Figure 12.36 Four eusocial animals (a) *Eciton* sp. (b) *Apis mellifera*. (c) *Synalpheus regalis*. Photo by J. Emmett Duffy. (d) *Heterocephalus glaber*.

The Monogamy Hypothesis

An alternative hypothesis currently under discussion is that lifelong monogamy facilitates the evolution of eusociality (Boomsma 2009; West and Gardner 2010). The idea behind the monogamy hypothesis is that when an individual can be certain that future siblings will be full siblings, a new brother or sister increases an individual's inclusive fitness just as much as an offspring does. The relatedness is $\frac{1}{2}$ in each case. Whether it is better for an individual to help its parents or to reproduce on its own depends only on the relative costs and benefits of each strategy. If a species' life history or the ecological conditions under which it lives make helping easier than reproducing, then eusociality is favored by selection.

Consistent with this notion, data compiled by William Hughes and colleagues (2008) indicate that in most or all of the nearly three dozen lineages in which eusociality has independently evolved, females mated with just one male. Monogamy clearly is not sufficient for the evolution of eusociality. Hughes and colleagues' analysis shows that monogamy is ancestral for all of the eusocial hymenoptera, but they belong to a clade with more solitary than social species (Hunt 2012). Monogamy may nonetheless be necessary for eusociality to evolve.

On the other hand, monogamy may not be required for eusociality at all. Instead, it may merely be correlated with other traits that are (Nonacs 2011).

The Ecology and Life-History Hypothesis

To understand which traits are most closely associated with the evolution of eusociality in hymenoptera, James Hunt (1999) analyzed the evolutionary tree

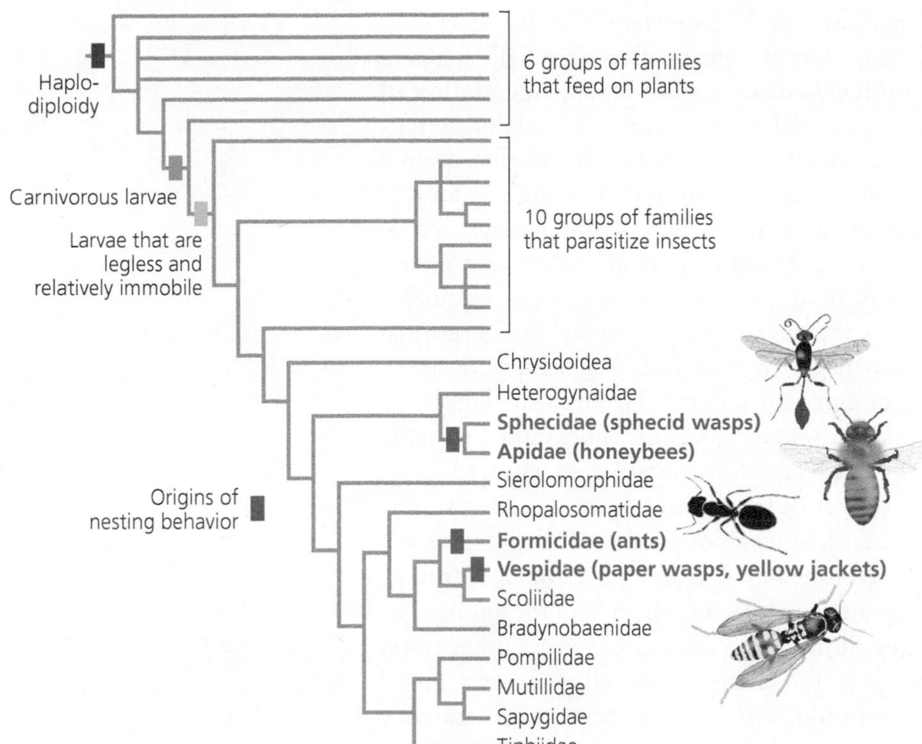

Figure 12.37 A phylogeny of the hymenoptera The taxa at the tips of this tree are either families or groups of families. Families that include eusocial species are indicated in bold blue type. (Not all of the species in these families are eusocial, however.) The colored boxes indicate points where certain key traits evolved. Modified from Hunt (1999).

shown in **Figure 12.37**. Because all hymenopterans are haplodiploid, Hunt could infer that this system of sex determination evolved at the root of the tree. Eusociality is found in just a few families of hymenopterans, however. Because these families are scattered around the tree, eusociality likely evolved not just once, but multiple times independently. Most important, Hunt noted that eusociality evolved only in groups that build complex nests and that care for their larvae for extended periods.

The association between nest building, care of larvae, and eusociality is important because it suggests that the primary factors favoring reproductive altruism in insects involve details of ecology and life history. Nest building and the need to supply larvae with a continuous supply of food make it difficult or impossible for a female to breed on her own (see Alexander et al. 1991). Also, when predation rates are high but young are dependent on parental care for a long period, then individuals who breed alone are unlikely to survive long enough to bring their young to adulthood (Queller 1989; Queller and Strassmann 1998). Finally, other eusocial lineages—including aphids, termites, thrips, shrimps, naked mole rats, and the sole eusocial beetle known—also occupy fortress nests (Wilson 2008; West and Gardner 2010).

Eusocial species present extreme examples of altruistic behavior. The evolution of eusociality hinges on ecological and life-history factors.

A Scenario for the Evolution of Eusociality in Vespid Wasps

Based on detailed study of the life histories of social wasps (see Hunt 2007), James Hunt (2012) has proposed a scenario for how some wasp lineages came to be eusocial. Wasps make good subjects for such an inquiry because extant species show a great diversity of social orgnizations, which Hunt takes to represent transitional forms. Some wasps are solitary. Each female reproduces on her own, and thus carries out all requisite maternal tasks, including building a nest, gathering food, laying eggs, and caring for young. Some wasps are facultatively eusocial. Each female has a choice whether to nest on her own or in cooperation with others, and females that begin their reproductive careers as helpers often eventually become breeders. Still other species are primitively eusocial. Most females spend most of their adult lives as either helpers or reproductives, but remain capable of switching roles. Finally, some wasps show advanced eusociality. Females exhibit distinct worker or queen phenotypes and in some cases are committed to one role for life.

Hunt suggests that facultative eusociality evolves because cooperative breeding is mutually beneficial. Reproductives get help defending the nest and raising young. Newly emerged helpers stay in their birth nest because food is available there. While in the nest, they respond to cues from larvae by exhibiting normal maternal care. The origin of worker behavior is thus readily explained.

Support for the notion that worker behavior is derived from normal maternal behavior comes from a study by Amy Toth, Hunt, and colleagues (2007) on gene expression in the brains of females in *Polistes metricus*, a primitively eusocial wasp. Workers showed expression similar to that of foundresses establishing new colonies and providing maternal care to their first batch of young. Worker expression was, however, distinct from that of established queens and that of females emerging in the fall and preparing to overwinter, neither of which care for young.

In contrast, Hunt asserts that the evolution of distinct queen and worker castes hinges on relatedness among nestmates. He finds multilevel selection a useful perspective. Alleles associated with reproductive altruism increase in frequency because colonies with a higher proportion of altruists produce more offspring. He expects this hypothesis to be testable once complete genome sequencing allows the identification of genetic variants associated with eusociality (see Woodard et al. 2011).

SUMMARY

Behavior evolves like any other aspect of phenotype. There are four kinds of simple social interaction, classified by their effect on the fitness of the actor and recipient: mutual benefit (+,+); selfishness (+,−); altruism (−,+); and spite (−,−). Because they involve the actor exhibiting a behavior that lowers its fitness, altruism and spite, in particular, require explanation.

One explanation for costly behavior arises from the recognition that the actor's behavior can influence the fate of copies of its genes that reside in other individuals. We expect natural selection to produce behavior that allows individuals to maximize their inclusive fitness. This is the sum of their genetic contribution to future generations via their own unassisted reproductive efforts (direct fitness) and via the assistance they provide to relatives (indirect fitness). Hamilton's rule states that a gene for altruism or spite will spread when $Br - C > 0$, where B is the benefit to the recipient, C is the cost to the actor, and r is the relatedness of the two individuals—a measure of genetic similarity. Hamilton's rule shows that the adaptive value of an altruistic act is highest when the cost to the actor is low, the benefit to the recipient is high, and the individuals are genetically similar, typically because they are kin. It also shows that the adaptive value of a spiteful act is highest when the cost to the actor is low, the "benefit" (damage inflicted) to the recipient is high, and the recipient is less genetically similar to the actor than to the average individual in the population.

Another way to look at the evolution of costly behavior is to recognize that selection acts on multiple levels. Altruistic behavior disfavored by selection within groups can still spread in the larger population if there is sufficient variation among groups in the frequency of altruists and if groups with more altruists enjoy sufficiently elevated productivity. For behaviors that take place in populations divided into social groups, kin selection models and multilevel selection models are often mathematically interchangeable.

Ongoing relationships among individuals often entail a mixture of cooperation and conflict. Parents and offspring may disagree over how much food offspring should receive and how many siblings they should share the parents' attention with. These conflicts make evolutionary sense when we realize that parents and offspring do not share the same relatedness with the interested parties. For example, a parent is equally related to two of its offspring, but an offspring is more related to itself than to its sibling. Given divergent interests, individuals in families may attempt to manipulate each other's behavior by altering the costs and benefits of different strategies. Unrelated individuals engaged in reciprocal exchanges that are mutually beneficial in the long run may attempt to discourage their partners from cheating by punishing them if they do.

Eusocial animals display extreme altruism; castes of nonreproductive workers spend their lives helping reproductives reproduce. Current efforts to explain the evolution of eusociality are focused as much on how the details of life history and ecology influence the costs and benefits of helping as on the role of relatedness among individuals.

Humans engage in a great variety of social interactions. Individual selection, kin selection, and multilevel selection all offer insights into human behavior. Compared to other species, humans seem particularly adept at maintaining reciprocity with unrelated individuals, at least in part by punishing free riders. Still, there is considerable variation in human behavior due to differences in culture.

QUESTIONS

1. Suppose adult bee-eaters could raise only 0.3 more offspring with a helper than without a helper. Would you still expect male bee-eaters to give in to the harassment of their fathers, or would male bee-eaters tend to fight off their fathers? Explain your reasoning.
2. When a Thomson's gazelle detects a stalking cheetah, the gazelle often begins bouncing up and down with a stiff-legged gait called stotting (see **Figure 12.38**). Stotting was originally assumed to be an altruistic behavior that distracts the cheetah from the gazelle's kin and also alerts the gazelle's kin to the presence of the predator, at considerable risk to the stotting gazelle. However, T. M. Caro reports that stotting does not seem to increase the gazelle's risk of being attacked. In fact, once a gazelle begins to stott, the cheetah often gives up the hunt.
   a. If Caro is right, how does C (the cost of stotting) for a gazelle compare to C (the cost of trilling) for a Belding's ground squirrel?
   b. Do you think stotting is altruistic, selfish, spiteful, or cooperative (mutualistic)? If you are not sure, what further studies could you do to answer this question?
   c. With this in mind, make a prediction about whether a gazelle will stott when no other gazelles are around, and then look up Caro's papers to see if your predictions is right.

Figure 12.38 A Thomson's gazelle stotting

Caro, T. M. 1986. The function of stotting in Thomson's gazelles: Some tests of the hypotheses. *Animal Behaviour* 34: 663–684.

Caro, T. M. 1994. Ungulate antipredator behaviour: Preliminary and comparative data from African bovids. *Behaviour* 128: 189–228.

3. The cubs of spotted hyenas often begin fighting within moments of birth, and often one hyena cub dies. The mother hyena does not interfere. How could such a behavior have evolved? For instance:
   a. From the winning sibling's point of view, what must B (benefit of siblicide) be, relative to C (cost of siblicide), to favor the evolution of siblicide?
   b. From the parent's point of view, what must B be, relative to C, for the parent to watch calmly rather than interfere?
   c. In general, when would you expect parents to evolve "tolerance of siblicide" (watching calmly while siblings kill each other without interfering).

   For more about the unusual social system of spotted hyenas and for studies of wild hyenas, see:

Frank, L G., S. E. Glickman, and P. Licht. 1991. Fatal sibling aggression, precocial development, and androgens in neonatal spotted hyenas. *Science* 252: 702–704.

Frank, L. G. 1997. Evolution of genital masculinization: Why do female hyaenas have such a large "penis"? *Trends in Ecology and Evolution* 12: 58–62.

Golla, W., H. Hofer, and M. L. East. 1999. Within-litter sibling aggression in spotted hyaenas: Effect of maternal nursing, sex and age. *Animal Behaviour* 58: 715–726.

4. Blue jays (*Cyanocitta cristata*) seem better than American robins (*Turdus migratorius*) at recognizing individuals. In one study (Schimmel and Wasserman 1994), blue jays raised with robins could distinguish strange from familiar robins better than the robins themselves. Do you think these species differ in occurrence of kin selection or reciprocity (or both)? Why?
5. When an interviewer asked evolutionary biologist J. B. S. Haldane if he would risk his life to save a drowning man, he reportedly answered, "No, but I would for two brothers or eight cousins." Explain his reasoning.
6. Look at Figure 12.21 on page 472.
   a. What would have happened to the frequency of cooperators in the global population if all groups had two cooperators and one selfish individual? Why?
   b. What would have happened to the frequency of cooperators in the global population if the cooperators and selfish individuals had sorted themselves into groups like the Hadza do? Why?
   c. Now look at Figure 12.25 on page 475. What would have happened to the global frequency of cooperators if, instead of diluting the bacterial culture and transferring a small number of individuals, the researchers had simply transferred a large aliquot? Why?
7. The text claims that eusociality has evolved several times independently within the hymenoptera. What is the evidence for this statement? If it is true, in what sense is eusociality in ants, bees, and wasps an example of convergent evolution? (See Chapter 4.)
8. Speculate about why some greater ani couples form 2-pair coalitions when 3-pair coalitions have higher reproductive success. How could you test your idea?
9. House sparrows often produce two successive broods of young. Males feed their first brood only briefly, but feed their second brood for much longer. Why do males feed first broods less than second broods? (Hint: Consider how *C*, the cost of feeding the current brood, changes.) How could you test your hypothesis? How is this situation analogous to weaning conflict in mammals?
10. Which is more common in human cultures—eusociality (look back at the three requirements of eusociality; can you think of any human cultures that fit?) or a helper-at-the-nest social system? Which do you think is generally more common in social animals? Why?
11. Human siblings often show intense sibling rivalry that typically declines during the teenage years. Suggest an evolutionary explanation for this pattern.
12. The writers of wills in Vancouver discounted distant kin rather more steeply than we might expect based on Hamilton's rule. Can you explain why this might be so? On the other hand, why invest in distant kin when closer kin are available?

EXPLORING THE LITERATURE

13. Given that an individual in a cooperatively breeding bird species is helping at its parents' nest instead of breeding on its own, is there anything else it can do to maximize its reproductive success? See:

    McDonald, P. G., and J. Wright. 2011. Bell miner provisioning calls are more similar among relatives and are used by helpers at the nest to bias their effort towards kin. *Proceedings of the Royal Society of London B* 278: 3403–3411.
14. We have noted that many social interactions are characterized by a mixure of cooperation and conflict. This is true even in eusocial insects. See:

    Hughes, W. O., and J. J. Boomsma. 2008. Genetic royal cheats in leaf-cutting ant societies. *Proceedings of the National Academy of Sciences, USA* 105: 5150–5153.
15. Özgür Gürek and colleagues asked student volunteers to play a public goods game in which different groups competed with each other. Individuals in some groups were allowed to punish each other; individuals in other groups were not. At first, most students chose to join non-punishing groups. As the tournament progressed, however, they changed their minds. See:

    Gurerk, O., B. Irlenbusch, and B. Rockenbach. 2006. The competitive advantage of sanctioning institutions. *Science* 312: 108–111.
16. For an example of humans using punishment to enforce cooperation in the real world, see:

    Mathew, S., and R. Boyd. 2011. Punishment sustains large-scale cooperation in prestate warfare. *Proceedings of the National Academy of Sciences, USA* 108: 11375–11380.
17. In addition to being alert to cheaters, humans are concerned with their own reputation. For documentation of how this affects human behavior, see:

    Francey, D., and R. Bergmüller. 2012. Images of eyes enhance investments in a real-life public good. *PLoS One* 7: e37397.
18. For another cross-cultural study documenting the influence of culture on human behavior, see:

    Henrich, J., J. Ensminger, et al. 2010. Markets, religion, community size, and the evolution of fairness and punishment. *Science* 327: 1480–1484.
19. For an account of a mollusk clever enough to cheat in social interactions and get away with it, see:

    Brown, C., M. P. Garwood, and J. E. Williamson. 2012. It pays to cheat: Tactical deception in a cephalopod social signalling system. *Biology Letters* 8: 729–732.
20. For a test of Hamilton's rule using robots, see:

    Waibel, M., D. Floreano, and L. Keller. 2011. A quantitative test of Hamilton's rule for the evolution of altruism. *PLoS Biology* 9: e1000615.

CITATIONS

Alexander, R. D., K. M. Noonan, and B. J. Crespi. 1991. The evolution of eusociality. In *The Biology of the Naked Mole Rat*, ed. P. W. Sherman, J. U. M. Jarvis, and R. D. Alexander. Princeton, NJ: Princeton University Press, 3–44.

Anderson, D. J. 1990a. Evolution of obligate siblicide in boobies. 1. A test of the insurance-egg hypothesis. *American Naturalist* 135: 334–350.

Anderson, D. J. 1990b. Evolution of obligate siblicide in boobies. 2: Food limitation and parent-offspring conflict. *Evolution* 44: 2069–2082.

Anderson, D. J., and R. E. Ricklefs. 1992. Brood size and food provisioning in masked and blue-footed boobies (*Sula* spp.). *Ecology* 73: 1363–1374.

Anderson, D. J., and R. E. Ricklefs. 1995. Evidence of kin-selected tolerance by nestlings in a siblicidal bird. *Behavioral Ecology and Sociobiology* 37: 163–168.

Anker-Nilssen, T., O. Kleven, et al. 2008. No evidence of extra-pair paternity in the Atlantic puffin *Fratercula arctica*. *Ibis* 150: 619–622.

Apicella, C. L., F. W. Marlowe, et al. 2012. Social networks and cooperation in hunter-gatherers. *Nature* 481: 497–501.

Bashey, F., S. K. Young, et al. 2012. Spiteful interactions between sympatric natural isolates of *Xenorhabdus bovienii* benefit kin and reduce virulence. *Journal of Evolutionary Biology* 25: 431–437.

Batra, S. W. T. 1966. Nests and social behavior of halictine bees of India (Hymenoptera: Halictidae). *Indian Journal of Entomology* 28: 375.

Beck, N. R., R. Peakall, and R. Heinsohn. 2008. Social constraint and an absence of sex-biased dispersal drive fine-scale genetic structure in white-winged choughs. *Molecular Ecology* 17: 4346–4358.

Boncoraglio, G., M. Caprioli, and N. Saino. 2009. Fine-tuned modulation of competitive behaviour according to kinship in barn swallow nestlings. *Proceedings of the Royal Society of London B* 276: 2117–2123.

Boncoraglio, G., and N. Saino. 2008. Barn swallow chicks beg more loudly when broodmates are unrelated. *Journal of Evolutionary Biology* 21: 256–262.

Boomsma, J. J. 2009. Lifetime monogamy and the evolution of eusociality. *Philosophical Transactions of the Royal Society B* 364: 3191–3207.

Boutin, S., L. A. Wauters, et al. 2006. Anticipatory reproduction and population growth in seed predators. *Science* 314: 1928–1930.

Briskie, J. V., C. T. Naugler, and S. M. Leech. 1994. Begging intensity of nestling birds varies with sibling relatedness. *Proceedings of the Royal Society of London B* 258: 73–78.

Browning, L. E., S. C. Patrick, et al. 2012. Kin selection, not group augmentation, predicts helping in an obligate cooperatively breeding bird. *Proceedings of the Royal Society of London B* 279: 3861–3869.

Burton-Chellew, M. N., and S. A. West. 2012. Pseudocompetition among groups increases human cooperation in a public-goods game. *Animal Behaviour* 84: 947–952.

Cheney, D. L., L. R. Moscovice, et al. 2010. Contingent cooperation between wild female baboons. *Proceedings of the National Academy of Sciences USA* 107: 9562–9566.

Chuang, J. S., O. Rivoire, and S. Leibler. 2009. Simpson's paradox in a synthetic microbial system. *Science* 323: 272–275.

Clutton-Brock, T. 2009. Cooperation between non-kin in animal societies. *Nature* 462: 51–57.

Cornwallis, C. K., S. A. West, et al. 2010. Promiscuity and the evolutionary transition to complex societies. *Nature* 466: 969–972.

Cosmides, L., H. C. Barrett, and J. Tooby. 2010. Colloquium paper: Adaptive specializations, social exchange, and the evolution of human intelligence. *Proceedings of the National Academy of Sciences, USA* 107 (Suppl 2): 9007–9014.

Costa, J. T., and T. D. Fitzgerald. 2005. Social terminology revisited: Where are we ten years later? *Annales Zoologici Fennici* 42: 559–564.

Crespi, B. J. 2005. Social sophistry: Logos and mythos in the forms of cooperation. *Annales Zoologici Fennici* 42: 569–571.

Crossland, M. R., T. Haramura, et al. 2012. Exploiting intraspecific competitive mechanisms to control invasive cane toads (*Rhinella marina*). *Proceedings of the Royal Society of London B* 279: 3436–3442.

Crossland, M. R., M. N. Hearnden, et al. 2011. Why be a cannibal? The benefits to cane toad, *Rhinella marina* [=*Bufo marinus*], tadpoles of consuming conspecific eggs. *Animal Behaviour* 82: 775–782.

Crossland, M. R., and R. Shine. 2011. Cues for cannibalism: Cane toad tadpoles use chemical signals to locate and consume conspecific eggs. *Oikos* 120: 327–332.

Darwin, C. R. 1859. *On the Origin of Species*. London: John Murray.

Darwin, C. R. 1871. *The Descent of Man and Selection in Relation to Sex*. London: John Murray.

Dawkins, R. 1976. *The Selfish Gene*. Oxford: Oxford University Press.

DeBruine, L. M. 2002. Facial resemblance enhances trust. *Proceedings of the Royal Society of London B* 269: 1307–1312.

Duffy, J. E. 1996. Eusociality in a coral-reef shrimp. *Nature* 381: 512–514.

Emlen, S. T., and P. H. Wrege. 1992. Parent–offspring conflict and the recruitment of helpers among bee-eaters. *Nature* 356: 331–333.

Fehr, E., and S. Gachter. 2002. Altruistic punishment in humans. *Nature* 415: 137–140.

Gammie, S. C., A. P. Auger, et al. 2007. Altered gene expression in mice selected for high maternal aggression. *Genes, Brain & Behavior* 6: 432–443.

Gammie, S. C., K. L. D'Anna, et al. 2009. Neurotensin inversely modulates maternal aggression. *Neuroscience* 158: 1215–1223.

Gammie, S. C., T. J. Garland, and S. A. Stevenson. 2006. Artificial selection for increased maternal defense behavior in mice. *Behavior Genetics* 36: 713–722.

Gardner, A., J. Alpedrinha, and S. A. West. 2012a. Haplodiploidy and the evolution of eusociality: Split sex ratios. *American Naturalist* 179: 240–256.

Gardner, A., J. Alpedrinha, and S. A. West. 2012b. Correction. *American Naturalist* 179: 554–555.

Gardner, A., I. C. Hardy, et al. 2007. Spiteful soldiers and sex ratio conflict in polyembryonic parasitoid wasps. *American Naturalist* 169: 519–533.

Gardner, A., and S. A. West. 2004. Spite and the scale of competition. *Journal of Evolutionary Biology* 17: 1195–1203.

Gardner, A., and S. A. West. 2006. Spite. *Current Biology* 16: R662–R664.

Gardner, A., S. A. West, and A. Buckling. 2004. Bacteriocins, spite and virulence. *Proceedings of the Royal Society of London B* 271: 1529–1535.

Gardner, A., S. A. West, and G. Wild. 2011. The genetical theory of kin selection. *Journal of Evolutionary Biology* 24: 1020–1043.

Goodnight, C. J. 1985. The influence of environmental variation on group and individual selection in a cress. *Evolution* 39: 545–558.

Gorrell, J. C., A. G. McAdam, et al. 2010. Adopting kin enhances inclusive fitness in asocial red squirrels. *Nature Communications* 1: 22.

Grabowska-Zhang, A. M., B. C. Sheldon, and C. A. Hinde. 2012. Long-term familiarity promotes joining in neighbour nest defence. *Biology Letters* 8: 544–546.

Grafen, A. 1985. A geometric view of relatedness. *Oxford Surveys in Evolutionary Biology* 2: 28–89.

Hamilton, W. D. 1964a. The genetical evolution of social behaviour. I. *Journal of Theoretical Biology* 7: 1–16.

Hamilton, W. D. 1964b. The genetical evolution of social behaviour. II. *Journal of Theoretical Biology* 7: 17–52.

Hamilton, W. D. 1970. Selfish and spiteful behaviour in an evolutionary model. *Nature* 228: 1218–1220.

Hamilton, W. D. 1972. Altruism and related phenomena, mainly in the social insects. *Annual Review of Ecology and Systematics* 3: 193–232.

Hatchwell, B. J., and J. Komdeur. 2000. Ecological constraints, life history traits and the evolution of cooperative breeding. *Animal Behaviour* 59: 1079–1086.

Hawlena, H., F. Bashey, et al. 2010. Spiteful interactions in a natural population of the bacterium *Xenorhabdus bovienii*. *American Naturalist* 175: 374–381.

Heinsohn, R. G. 1991. Kidnapping and reciprocity in cooperatively breeding white-winged choughs. *Animal Behaviour* 41: 1097–1100.

Heinsohn, R. G. 1992. Cooperative enhancement of reproductive success in white-winged choughs. *Evolutionary Ecology* 6: 97–114.

Henrich, J., S. J. Heine, and A. Norenzayan. 2010. The weirdest people in the world? *Behavioral and Brain Sciences* 33: 61-83; discussion 83–135.

Henrich, J., R. McElreath, et al. 2006. Costly punishment across human societies. *Science* 312: 1767–1770.

Hill, K. R., R. S. Walker, et al. 2011. Co-residence patterns in hunter-gatherer societies show unique human social structure. *Science* 331: 1286–1289.

Hoogland, J. L. 1983. Nepotism and alarm calling in the black-tailed prairie dog (*Cynomys ludovicianus*). *Animal Behaviour* 31: 472–479.

Hoogland, J. L. 1994. Nepotism and infanticide among prairie dogs. In *Infanticide and Parental Care*, ed. S. Parmigiani and F. S. vom Saal. Chur, Switzerland: Harwood Academic Publishers, 321-337.

Hoogland, J. L. 1995. *The Black-Tailed Prairie Dog: Social Life of a Burrowing Mammal*. Chicago: University of Chicago Press.

Hughes, J. M., P. B. Mather, et al. 2003. High levels of extra-group paternity in a population of Australian magpies *Gymnorhina tibicen*: Evidence from microsatellite analysis. *Molecular Ecology* 12: 3441–3450.

Hughes, W. O. H., B. P. Oldroyd, et al. 2008. Ancestral monogamy shows kin selection is key to the evolution of eusociality. *Science* 320: 1213–1216.

Humphries, M. M., and S. Boutin. 2000. The determinants of optimal litter size in free-ranging red squirrels. *Ecology* 81: 2867–2877.

Hunt, J. H. 1999. Trait mapping and salience in the evolution of eusocial vespid wasps. *Evolution* 53: 225–237.

Hunt, J. H. 2007. *The Evolution of Social Wasps*. New York: Oxford University Press.

Hunt, J. H. 2012. A conceptual model for the origin of worker behaviour and adaptation of eusociality. *Journal of Evolutionary Biology* 25: 1–19.

Ibáñez-Álamo, J. D., L. Arco, and M. Soler. 2012. Experimental evidence for a predation cost of begging using active nests and real chicks. *Journal of Ornithology* 153: 801–807.

Inglis, R. F., P. G. Roberts, et al. 2011. Spite and the scale of competition in *Pseudomonas aeruginosa*. *American Naturalist* 178: 276–285.

Kerr, B., and P. Godfrey-Smith. 2002. Individualist and multilevel perspectives on selection in structured populations. *Biology & Philosophy* 17: 477–517.

Kerr, B., P. Godfrey-Smith, and M. W. Feldman. 2004. What is altruism? *Trends in Ecology and Evolution* 19: 135–140.

Kerr, B., C. Neuhauser, et al. 2006. Local migration promotes competitive restraint in a host-pathogen "tragedy of the commons." *Nature* 442: 75–78.

Lacey, E. A., and P. W. Sherman. 2005. Redefining eusociality: Concepts, goals and levels of analysis. *Annales Zoologici Fennici* 42: 573–577.

Lessells, C. M., and G. A. Parker. 1999. Parent-offspring conflict: The full-sib–half-sib fallacy. *Proceedings of the Royal Society of London B* 266: 1637–1643.

Lion, S., V. A. Jansen, and T. Day. 2011. Evolution in structured populations: Beyond the kin versus group debate. *Trends in Ecology and Evolution* 26: 193–201.

Lougheed, L. W., and D. J. Anderson. 1999. Parent blue-footed boobies suppress siblicidal behavior of offspring. *Behavioral Ecology and Sociobiology* 45: 11–18.

Majeed, H., O. Gillor, et al. 2011. Competitive interactions in *Escherichia coli* populations: The role of bacteriocins. *ISME Journal* 5: 71–81.

Marshall, J. A. 2011. Group selection and kin selection: Formally equivalent approaches. *Trends in Ecology and Evolution* 26: 325–332.

Maynard Smith, J. 1998. *Evolutionary Genetics*. 2nd ed. Oxford: Oxford University Press.

McAdam, A. G., S. Boutin, et al. 2007. Life histories of female red squirrels and their contributions to population growth and lifetime fitness. *Ecoscience* 14: 362–369.

Morales-Soto, N., and S. A. Forst. 2011. The *xnp1* P2-like tail synthesis gene cluster encodes xenorhabdicin and is required for interspecies competition. *Journal of Bacteriology* 193: 3624–3632.

Moreno-Rueda, G., and T. Redondo. 2011. Begging at high level simultaneously impairs growth and immune response in southern shrike (*Lanius meridionalis*) nestlings. *Journal of Evolutionary Biology* 24: 1091–1098.

Nonacs, P. 2011. Monogamy and high relatedness do not preferentially favor the evolution of cooperation. *BMC Evolutionary Biology* 11: 58.

Nowak, M. A., C. E. Tarnita, and E. O. Wilson. 2010. The evolution of eusociality. *Nature* 466: 1057–1062.

Peters, J. M., D. C. Queller, et al. 1999. Mate number, kin selection and social conflicts in stingless bees and honeybees. *Proceedings of the Royal Society of London B* 266: 379–384.

Queller, D. C. 1989. The evolution of eusociality: Reproductive head starts of workers. *Proceedings of the National Academy of Sciences, USA* 86: 3224–3226.

Queller, D. C., and K. F. Goodnight. 1989. Estimating relatedness using genetic markers. *Evolution* 43: 258–275.

Queller, D. C., E. Ponte, et al. 2003. Single-gene greenbeard effects in the social amoeba *Dictyostelium discoideum*. *Science* 299: 105–106.

Queller, D. C., and J. E. Strassmann. 1998. Kin selection and social insects. *BioScience* 48: 165–175.

Puurtinen, M., and T. Mappes. 2009. Between-group competition and human cooperation. *Proceedings of the Royal Society of London B* 276: 355–360.

Raihani, N. J., A. Thornton, and R. Bshary. 2012. Punishment and cooperation in nature. *Trends in Ecology and Evolution* 27: 288–295.

Riehl, C. 2011. Living with strangers: Direct benefits favour non-kin cooperation in a communally nesting bird. *Proceedings of the Royal Society B* 278: 1728–1735.

Riehl, C. 2012. Mating system and reproductive skew in a communally breeding cuckoo: Hard-working males do not sire more young. *Animal Behaviour* 84: 707–714.

Riehl, C., and L. Jara. 2009. Natural history and reproductive biology of the communally breeding greater ani (*Crotophaga major*) at Gatún Lake, Panama. *Wilson Journal of Ornithology* 121: 679–687.

Riley, M. A., C. M. Goldstone, et al. 2003. A phylogenetic approach to assessing the targets of microbial warfare. *Journal of Evolutionary Biology* 16: 690–697.

Riley, M. A., and J. E. Wertz. 2002. Bacteriocins: Evolution, ecology, and application. *Annual Review of Microbiology* 56: 117–137.

Sabbatini, G., A. De Bortoli Vizioli, et al. 2012. Food transfers in capuchin monkeys: An experiment on partner choice. *Biology Letters* 8: 757–759.

Schimmel, K. L., and F. E. Wasserman. Individual and species preference in two passerine birds: Auditory and visual cues. 1994. *Auk* 111: 634–642.

Sherman, P. W. 1985. Alarm calls of Belding's ground squirrels to aerial predators: Nepotism or self-preservation? *Behavioral Ecology and Sociobiology* 17: 313–323.

Smith, M. S., B. J. Kish, and C. B. Crawford. 1987. Inheritance of wealth as human kin investment. *Ethology and Sociobiology* 8: 171–182.

Stone, V. E., L. Cosmides, et al. 2002. Selective impairment of reasoning about social exchange in a patient with bilateral limbic system damage. *Proceedings of the National Academy of Sciences, USA* 99: 11531–11536.

Suen, G., C. Teiling, et al. 2011. The genome sequence of the leaf-cutter ant *Atta cephalotes* reveals insights into its obligate symbiotic lifestyle. *PLoS Genetics* 7: e1002007.

Sugiyama, L. S., J. Tooby, and L. Cosmides. 2002. Cross-cultural evidence of cognitive adaptations for social exchange among the Shiwiar of Ecuadorian Amazonia. *Proceedings of the National Academy of Sciences, USA* 99: 11537–11542.

Thorne, B. L. 1997. Evolution of eusociality in termites. *Annual Review of Ecology and Systematics* 28: 27–54.

Trivers, R. L. 1971. The evolution of reciprocal altruism. *Quarterly Review of Biology* 46: 35–57.

Trivers, R. L. 1974. Parent-offspring conflict. *American Zoologist* 14: 249–264.

Trivers, R. L. 1985. *Social Evolution*. Menlo Park, CA: Benjamin Cummings.

Toth, A. L., K. Varala, et al. 2007. Wasp gene expression supports an evolutionary link between maternal behavior and eusociality. *Science* 318: 441–444.

Wade, M. J. 1977. An experimental study of group selection. *Evolution* 31: 134–153.

Wcislo, W. T. 2005. Social labels: We should emphasize biology over terminology and not vice versa. *Annales Zoologici Fennici* 42: 565–568.

West, S. A., and A. Gardner. 2010. Altruism, spite, and greenbeards. *Science* 327: 1341–1344.

West, S. A., A. S. Griffin, and A. Gardner. 2007a. Social semantics: Altruism, cooperation, mutualism, strong reciprocity and group selection. *Journal of Evolutionary Biology* 20: 415–432.

West, S. A., A. S. Griffin, and A. Gardner. 2007b. Evolutionary explanations for cooperation. *Current Biology* 17: R661–R672.

Willis, C. G., M. T. Brock, and C. Weinig. 2010. Genetic variation in tolerance of competition and neighbour suppression in *Arabidopsis thaliana*. *Journal of Evolutionary Biology* 23: 1412–1424.

Wilson, D. S. 1975. A theory of group selection. *Proceedings of the National Academy of Sciences, USA* 72: 143–146.

Wilson, E. O. 2008. One giant leap: How insects achieved altruism and colonial life. *BioScience* 58: 17–25.

Wolf, J. B., J. J. Mutic, and P. X. Kover. 2011. Functional genetics of intraspecific ecological interactions in *Arabidopsis thaliana*. *Philosophical Transactions of the Royal Society of London B* 366: 1358–1367.

Woodard, S. H., B. J. Fischman, et al. 2011. Genes involved in convergent evolution of eusociality in bees. *Proceedings of the National Academy of Sciences, USA* 108: 7472–7477.

13

Aging and Other Life-History Characters

Male blue-footed boobies seek to attract mates by showing off their feet (Velando et al. 2006). Foot color ranges from dull blue to bright green and is due in part to the concentration of the carotenoid pigment zeaxanthin in the skin. Bright green feet are more attractive to females.

The color of a male's feet can change rapidly depending on how well fed he is, and how healthy (Velando et al. 2006; Torres and Velando 2007). Maintaining enough pigment in his feet to stay in the mating game gets harder as he ages (Velando et al. 2010). As shown in the graph at right, however, males that take a year off from reproduction have brighter feet for their age. This result is open to alternative interpretations. Perhaps males that failed to reproduce last year increase the proportion of their energy budget they devote to advertising this year. Or perhaps males that are well rested, instead of exhausted from helping their mates rear chicks, simply have a larger budget overall. Either way, the pattern suggests that there is a conflict between reproductive investment this year versus future years. Males that strike the best compromise will enjoy the highest lifetime reproductive success.

Reproductive success is ultimately the trait on which natural selection always acts. Despite selection on the same trait, however, organisms go about the

Male blue-footed boobies that skipped breeding last year have brighter, more attractive feet for their age. Graph redrawn from Velando et al. (2010).

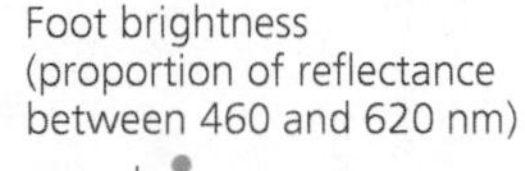

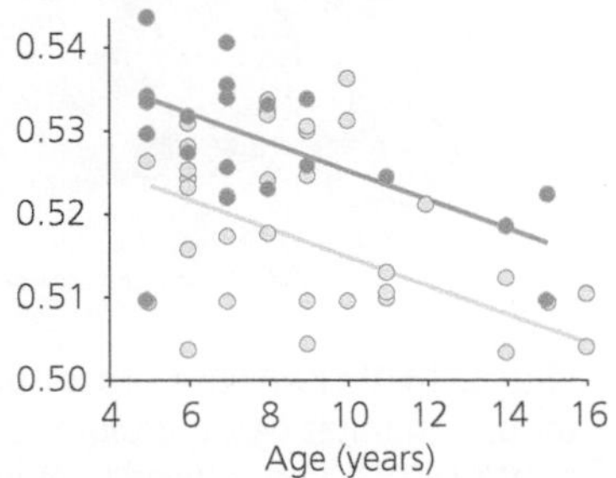

business of reproducing in a great variety of ways. A few examples will illustrate the tremendous diversity:

- Some mammals mature early and reproduce quickly, whereas others mature late and reproduce slowly. For example, female deer mice (*Peromyscus maniculatus*) mature at about seven weeks and have three or four litters of pups each year, whereas female black bears (*Ursus americanus*) mature at four or five years and produce cubs only once every two years (Nowak 1991).
- Plants have a wide range of reproductive life spans. Some, like the California poppy (*Eschscholzia californica*), live and flower for just a single season. Others, like the black cherry (*Prunus serotina*), flower yearly for decades.
- Some bivalves produce enormous numbers of tiny eggs, whereas others produce small numbers of large eggs (Strathmann 1987). The oyster *Crassostrea gigas* releases 10 to 50 million eggs in a single spawn, each 50–55 micrometers in diameter. The clam *Lasaea subviridis,* in contrast, broods fewer than 100 eggs at a time, each some 300 micrometers in diameter.

The branch of evolutionary biology that attempts to make sense of the diversity in reproductive strategies is called life-history analysis.

We might imagine that an organism perfected for reproduction would mature at birth, continuously produce high-quality offspring in large numbers, and live forever. Some actual organisms come close to this ideal in some respects, but all fall strikingly short in others.

For example, the female thrips egg mite (*Adactylidium* sp.) is mature at birth and already inseminated, having hatched inside her mother's body and mated with her brother (Elbadry and Tawfik 1966). But she produces just one clutch of offspring during a short life. She dies at the age of just four days, when her own offspring eat her alive from the inside **(Figure 13.1a)**.

Another example, the brown kiwi (*Apteryx australis mantelli*), produces high-quality offspring (Taborsky and Taborsky 1993). Female kiwis weigh about 6 pounds and lay eggs that weigh 1 pound (Figure 13.1b). The chicks that hatch from these huge eggs become largely self-reliant within a week. However, kiwi parents cannot produce these chicks continuously, and they cannot produce them in large numbers. It takes the female over a month to make each of the eggs

(a)

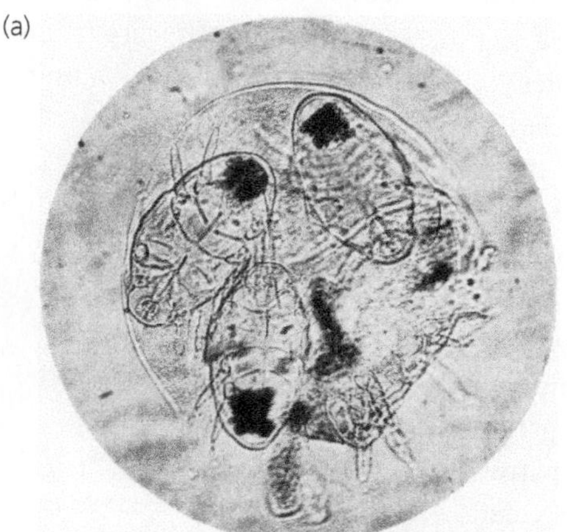

(b)

Figure 13.1 Extreme reproductive strategies (a) Having devoured their mother from the inside, three thrips egg mites (*Adactylidium* sp.) prepare to depart her empty cuticle. The mother's legs are visible at lower right (180×). Reproduced by permission from Elbadry and Tawfik (1966). (b) An X-ray of a female brown kiwi (*Apteryx australis mantelli*) ready to lay an egg.

in a typical two-egg clutch. The male has to incubate the eggs for about three months, during which time he loses some 20% of his body weight.

Organisms face fundamental trade-offs in their use of energy and time.

As the egg mite and the kiwi suggest, the laws of physics and biology impose fundamental trade-offs. The amount of energy an organism can harvest is finite, and biological processes take time. Energy and time devoted to one activity are energy and time that cannot be devoted to another. For example, an individual can allocate energy to growth for a long time, which may enable it to reach a larger size and ultimately produce more offspring. This benefit of size, however, is balanced by a cost. The time required to grow is time during which predators, diseases, or accidents may strike. An individual that takes the time to grow large thus incurs a greater risk of dying without reproducing at all. Elsewhere, we introduced the concept of trade-offs and discussed how they constrain the evolution of adaptations (Chapter 10). Whenever there is a trade-off between different components of fitness, we expect natural selection to favor individuals that allocate energy and time with an optimal balance between benefits and costs, thereby maximizing lifetime reproductive success. Because different balances are optimal in different environments, environmental variation is the source of much of the life-history variation seen among living organisms.

The first section of this chapter discusses the fundamental issues biologists consider when they analyze life histories. Sections 13.2 through 13.4 look at costs and benefits, and fitness trade-offs, as they apply to questions about why organisms age and die, about how many offspring individuals produce, and about offspring size. Section 13.5 explores what happens when the optimal life-history strategies for male versus female parents are in conflict. Finally, Section 13.6 places life-history analysis in a broader evolutionary context by considering the maintenance of genetic variation and evolutionary transitions in life history.

13.1 Basic Issues in Life-History Analysis

An example of a life history, one that we return to near the end of Section 13.2, appears in **Figure 13.2**. The figure follows the career of a hypothetical female Virginia opossum (*Didelphis virginiana*). As a baby, this female nursed for a little more than three months, was then weaned, and became independent. She continued to grow for another several months, reaching sexual maturity at an age of about 10 months. Shortly thereafter the female had her first litter, consisting of eight offspring. A few months later, she had a second litter, this time with seven offspring. At the age of 20 months, the female was killed by a predator.

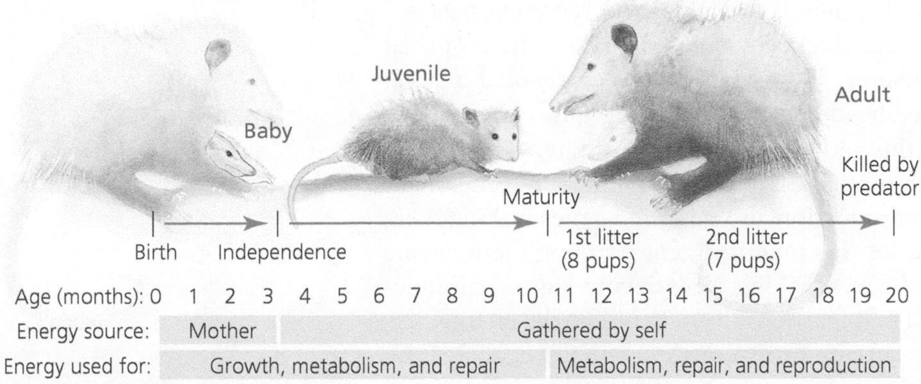

Figure 13.2 The life history of a hypothetical female Virginia opossum (*Didelphis virginiana*) This hypothetical female has a life history typical of female Virginia opossums living in the mainland United States (Austad 1988, 1993). Figure designed after Charnov and Berrigan (1993).

Figure 13.2 also indicates where the female opossum got her energy at different stages of her life, and the functions to which she allocated that finite energy supply. Before she became sexually mature, the female used her energy for growth, metabolic functions like thermoregulation, and the repair of damaged tissues. After she became sexually mature, the female stopped growing, thereafter using her energy for metabolism, repair, and reproduction.

Changes in life history are caused by changes in the allocation of energy.

Fundamentally, differences among life histories concern differences in the allocation of energy. For example, a different female opossum than the one shown in Figure 13.2 might stop allocating energy to growth at an earlier age, thereby reaching sexual maturity more quickly. This strategy involves a trade-off: The female also matures at a smaller size, which means that she will produce smaller litters. Still another female might, after reaching sexual maturity, allocate less energy to reproduction and more to repair, thereby keeping her tissues in better condition. Again there is a trade-off: Allocating less energy to reproduction means having smaller litters.

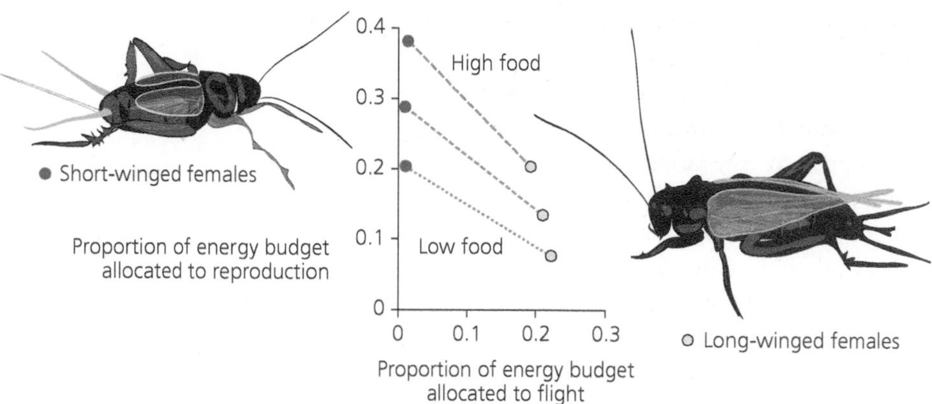

Figure 13.3 A trade-off between reproduction and flight capacity in female sand crickets (*Gryllus firmus*) Whether fed low, medium, or high quantities of food, long-winged females invest more in flight capacity during early adulthood, while short-winged females invest more in reproduction. Drawn from data presented in King et al. (2011a).

Elizabeth King and colleagues (2011a) documented a trade-off in the allocation of energy by female sand crickets. Natural populations of this insect contain two kinds of females: short-winged and long-winged **(Figure 13.3)**. Any given female develops one body or the other due to the influence of both genetic and environmental factors (Zera and Denno 1997). King and colleagues reared over 5,000 females from 63 paternal half-sib families at three different food levels ranging from near-starvation rations to all you can eat. A week after the females reached adulthood, the researchers dissected them and weighted their flight muscles, ovaries, and remaining tissues. The researchers used these and other data to estimate, for each female, the fraction of her total energy budget she allocated to reproduction versus flight capacity. The points in Figure 13.3 represent means.

Within each food level, short-winged females devoted a larger share of their energy budget to reproduction, whereas long-winged females devoted a larger share to flight capacity. The breeding design allowed King and colleagues to show that a significant fraction of this variation in allocation strategy was due to differences in genes. That is, at each food level, females that inherited genes making them devote more energy to reproduction also inherited genes making them devote less to flight. Likewise, females that inherited genes making them devote more energy to flight also inherited genes making them devote less to reproduction. Additional analyses confirmed that female sand crickets slice the energy they use for reproduction and for flight capacity from a single pie (King et al. 2011b).

Anthony Zera and colleagues have investigated biochemical and metabolic changes responsible for the differences in energy allocation between short versus long-winged females (see Zera 2005). They studied nearly pure-breeding stocks of short-winged and long-winged crickets produced by artificial selection. When long-winged females emerge as adults they have, in addition to well-developed flight muscles, elevated rates of fatty-acid synthesis. The bulk of the fatty acids they make are triglycerides, which they store to serve as fuel for their flight muscles (Zhao and Zera 2002; Zera and Zhao 2003). In contrast, short-winged females have not only poorly developed flight muscles, but lower rates of fatty-acid synthesis. The fatty acids they make are mostly phospholipids, rather than triglycerides. They put these phospholipids into their eggs.

These metabolic differences between short- versus long-winged crickets are due in part to differences in the activity of the enzyme $NADP^+$-isocitrate dehydrogenase ($NADP^+$-IDH). This enzyme produces NADPH, which is required for fatty-acid synthesis. Long-winged females have higher $NADP^+$-IDH activity than short-winged flies. Rudolf Schilder, Zera, and colleagues (2011) found that pure-breeding long-winged lines have higher $NADP^+$-IDH activity because they make more mRNA from the $NADP^+$-IDH gene, and thus more of the enzyme. The implication is that female sand crickets manage at least this aspect of the trade-off between reproduction and flight capability by means of gene regulation.

The differences in gene regulation between short- versus long-winged crickets are under hormonal control. Zera and Zhangwu Zhao (2004) demonstrated this by treating long-winged females with methoprene. Methoprene is a biochemical analog of juvenile hormone, a potent regulator of insect development. Treatment with methoprene induced the crickets to develop the short-winged phenotype.

For female sand crickets, the ability to fly is adaptive under some circumstances (see Zera 2005). If a female finds herself in a poor environment, she can seek a better one elsewhere. The ability to fly comes with a fitness cost, however. Because they receive less energy, the ovaries of long-winged females grow more slowly than those of short-winged females. As a result, long-winged females have lower early-life fecundity. Consistent with the view that this represents a fitness cost, by the end of their second week of adulthood, most long-winged females have broken down their flight muscles and shifted the remaining energy they contain to their ovaries (see King et al. 2011b).

In summary, sand crickets appear to face an inescapable trade-off in the allocation of energy to different activities crucial to fitness. They have genetic, developmental, and physiological mechanisms that allow different individuals to pursue divergent strategies, each of which is adaptive under certain circumstances.

Many other organisms face trade-offs in energy allocation. As in sand crickets, natural selection on life histories leads to adjustments in energy allocation that maximize the total lifetime production of offspring.

13.2 Why Do Organisms Age and Die?

Aging, or **senescence,** is a late-life decline in an individual's fertility and probability of survival (Partridge and Barton 1993). **Figure 13.4** (next page) documents aging in three animal species: a bird, a mammal, and an insect. All three show declines in both fertility and survival. All else being equal, aging reduces an individual's fitness. Aging should therefore be opposed by natural selection.

Aging should be opposed by natural selection.

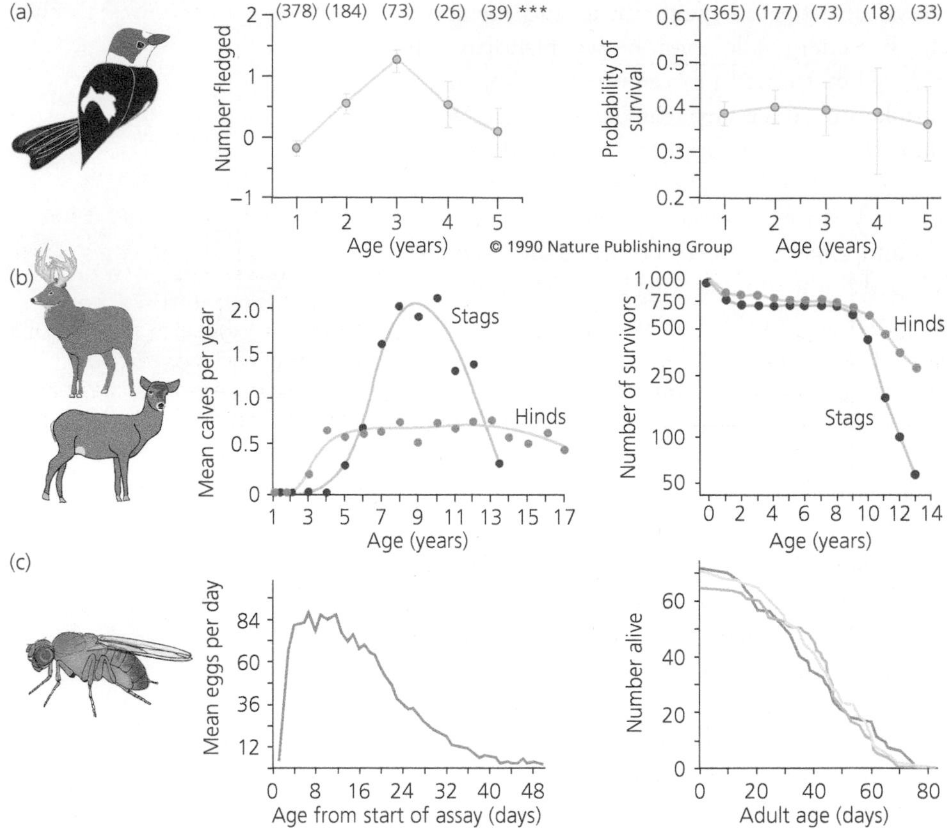

Figure 13.4 Aging in three animals (a) For female collared flycatchers, the number of young fledged each year declines after age 3 ($p < 0.001$). The probability of survival from year to year declines slightly, but not significantly, after age 2. Sample sizes in parentheses. From Gustafsson and Pärt (1990). (b) For male red deer, the number of calves fathered each year declines after age 9; for females, the number of calves produced declines after age 13. For both sexes, the probability of surviving from year to year plummets after age 9. From Clutton-Brock et al. (1988). (c) For female fruit flies (*Drosophila melanogaster*), the average number of eggs laid per day declines after 12 days. In three lab populations, the probability of surviving from day to day falls at about 20 days. Modified from Rose (1984).

(a) Reprinted by permission of Macmillan Publishers Ltd. Gustafsson, L., and T. Pärt. 1990. "Acceleration of senescence in the collared flycatcher (*Ficedula albicollis*) by reproductive costs." *Nature* 347: 279–281. Copyright © 1990.

We consider two theories on why aging persists. The first, called the rate-of-living theory, invokes an evolutionary constraint (see Chapter 10); it posits that populations lack the genetic variation to respond any further to selection against aging. The second, called the evolutionary theory, invokes, in part, a trade-off between the allocation of energy to reproduction versus repair.

The Rate-of-Living Theory of Aging

The rate-of-living theory of senescence holds that aging is caused by the accumulation of irreparable damage to cells and tissues (reviewed in Austad and Fischer 1991). Damage to cells and tissues is caused by errors during replication, transcription, and translation, and by the accumulation of poisonous metabolic by-products. Under the rate-of-living theory, all organisms have been selected to resist and repair cell and tissue damage to the maximum extent physiologically possible. They have reached the limit of biologically possible repair. In other words, populations lack the genetic variation that would enable them to evolve more effective repair mechanisms than they already have.

One theory holds that aging is a function of metabolic rate...

The rate-of-living theory makes two predictions: (1) Because cell and tissue damage is caused in part by the by-products of metabolism, aging rate should be correlated with metabolic rate; and (2) because organisms have been selected to resist and repair damage to the maximum extent possible, species should not be able to evolve longer life spans, whether subjected to natural or artificial selection.

Tests of the Rate-of-Living Theory

Steven Austad and Kathleen Fischer (1991) tested the first prediction, that aging will correlate with metabolic rate, with a comparative study of mammals. Using

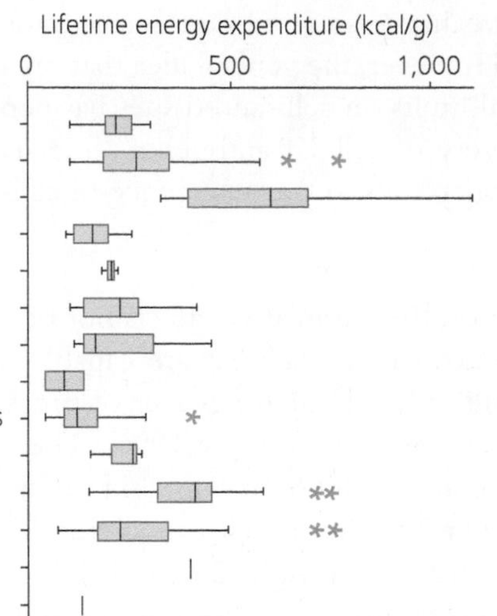

Figure 13.5 Variation among mammals in lifetime energy expenditure This box plot represents the range of lifetime energy expenditure within each of 14 orders of mammals. The vertical line dividing each box represents the median value for that order. The right and left ends of each box represent the 75th and 25th percentiles. The horizontal lines extending to the right and left of each box represent the range of values; the asterisks represent statistical outliers. From Austad and Fischer (1991).

data from the literature, Austad and Fischer calculated the energy expended per gram of tissue per lifetime for 164 species in 14 orders. According to the rate-of-living theory, all should expend about the same amount of energy per gram per lifetime, whether they burn it slowly over a long life or rapidly over a short one.

In fact, there is wide variation among mammals **(Figure 13.5)**. Energy expenditure ranges from 39 kilocalories per gram (kcal/g) per lifetime in an elephant shrew to 1,102 kcal/g per lifetime in a bat. Even within orders, energy expenditure varies greatly. Bat species burn 325 to 1,102 kcal/g per lifetime. As a group, bats have metabolic rates similar to those of other mammals of the same size, but life spans averaging nearly three times longer. Marsupial species burn 43 to 406 kcal/g per lifetime. As a group, marsupials have metabolic rates significantly lower than those of other mammals of the same size but life spans significantly shorter. These patterns contradict the rate-of-living theory.

...but data on variation in metabolic rate and aging among mammals refute this theory.

Leo Luckinbill and colleagues (1984) tested the second prediction, that species cannot evolve longer life spans, by artificially selecting for longevity in fruit flies (*Drosophila melanogaster*). Luckinbill et al. collected wild flies to establish four laboratory populations. In two, the researchers selected for early reproduction by gathering eggs from young adults (two to six days after eclosion) and using the individuals that hatched from these eggs as the next generation's breeders. Longevity in these populations did not change significantly across 13 generations **(Figure 13.6)**. In the other two, Luckinbill and colleagues selected for late reproduction by gathering eggs from old adults. "Old" meant 22 days after eclosion at the beginning of the experiment and 58 days after eclosion by the end. Longevity in these populations increased dramatically. At the beginning of the experiment, the average life span was about 35 days; by the end, it was about 60.

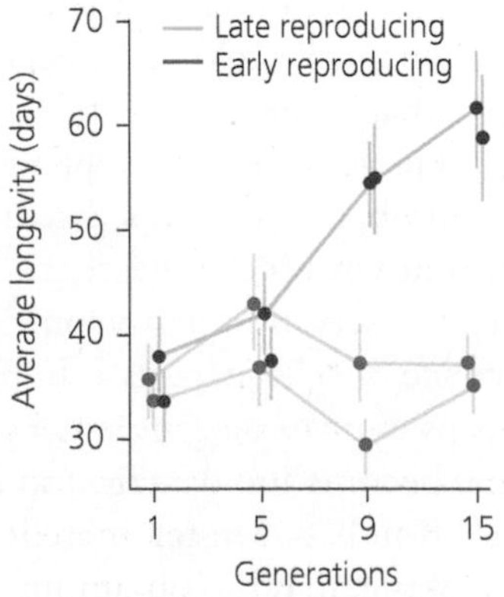

Figure 13.6 Artificial selection increases life span in fruit flies The graph shows average life span in each of four laboratory populations of *Drosophila melanogaster* over 13 generations of selection. The vertical lines show the 95% confidence intervals for the estimated population averages. From Luckinbill et al. (1984).

Other researchers running similar experiments have confirmed that life span increases in *Drosophila* populations in response to selection for late-life reproduction (Rose 1984; Partridge 1987; Partridge and Fowler 1992; Roper et al. 1993).

These results are consistent with the rate-of-living theory only if the long-lived populations have evolved lower metabolic rates. Phillip Service (1987) found that fruit flies selected for long life span indeed had lower metabolic rates than controls, but only in the first 15 days of life. It is not clear that an evolved difference in metabolic rate can explain an evolved difference in life span as large as that obtained by Luckinbill and colleagues.

The experiments and observations we have discussed largely contradict the predictions of the rate-of-living theory. However, the general idea that organisms age and die due to intrinsic physiological limits on cells and tissues has persisted. Its tenacity is due, in part, to the discovery of cellular and genetic mechanisms that appeared to link the senescence of organisms to the senescence of cells.

Senescence of Cells and Senescence of Organisms

One such mechanism in animals is based on the cumulative effects not of energy expenditure, but of cell division. Most normal animal cells are capable of duplicating their chromosomes and dividing only a limited number of times, after which the cells cease proliferating and eventually die (Campisi 1996). The exceptions are germ-line cells, some embryonic and blood stem cells, and cancer cells.

The mechanism that limits the number of times a cell can divide involves its telomeres (Donate and Blasco 2011; Shay and Wright 2011). Telomeres are tandem repeat sequences found at the ends of eukaryotic chromosomes. In vertebrates, including humans, the repeated sequence is TTAGGG. The telomeres and the proteins that bind to them prevent the ends of the chromosomes from being mistaken for strand breaks by the DNA repair machinery in the cell.

Every time a cell duplicates its DNA, the telomeres get shorter. The reason is that as DNA polymerase approaches the end of a linear chromosome, it cannot replicate the final stretch of the template strand that has a loose 3′ end. The lost bits can be restored by telomerase, an enzyme that uses an RNA strand as a template to append new repeats to DNA. In humans, telomerase is expressed in the germ line, early embryos, stem cells, and cancers, but not in most other cells. In cells lacking telomerase, the telomeres get shorter with every cell division.

When a cell's telomeres become too short, or if they sustain too much oxidative damage, sensor kinase enzymes activate a protein called p53 (Donehower 2009; Hewitt et al. 2012). p53 is a transcription factor that initiates the expression of a number of proteins that may put the cell into a permanently nondividing state, known as cell senescence, or induce it to undergo programmed cell death.

The observation that telomere shortening is associated with the replicative senescence of cells suggests a simple explanation for the aging and death of animals. Perhaps animals die, in part, because their telomeres are lost and their chromosomes become too damaged to function. If this is the case, then we should see an association between telomere length and longevity.

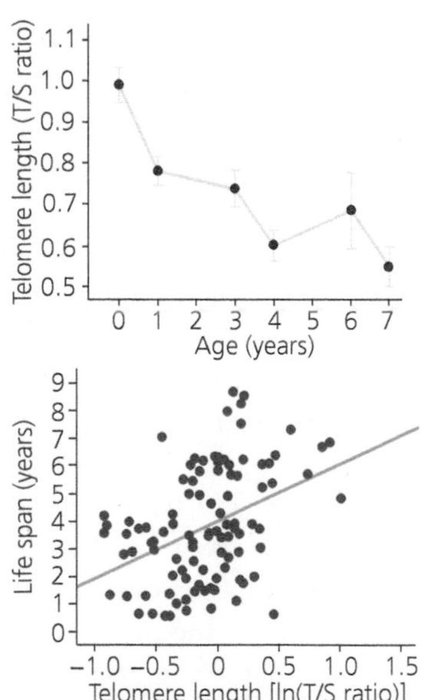

Figure 13.7 Telomeres and longevity in zebra finches (a) Mean telomere length ± standard error declines with age. (b) Finches with longer telomeres at 25 days of age tended to live longer ($p < 0.001$). The unit of measure for telomere length, T/S ratio, is the length of a telomere, T, divided by the length of a protein-coding gene used as a standard, S. From Heidinger et al. (2012).

Looking at variation among individuals within populations, researchers have found examples of such an association. Britt Heidinger and colleagues (2012) followed a cohort of captive zebra finches (*Taeniopygia guttata*) from 25 days after hatching until death from natural causes. The length of the telomeres in the finches' red blood cells—which retain their nuclei in birds—declined with age **(Figure 13.7a)**. And telomere length at the age of 25 days was significantly associated with longevity (Figure 13.7b). Gil Atzmon and colleagues (2010) compared a group of Ashkenazi Jewish subjects who were at least 100 years old, their children, and unrelated controls who matched the children's ages. Both the centenarians and their kids had longer telomeres than did the controls.

However, looking at variation among species in a broad diversity of mammals, Nuno Gomes and colleagues (2011) found precisely the opposite association. These researchers assessed telomere length in cultured connective tissue cells from 57 species, including whales, hoofed mammals, dogs, cats, bats, rodents, rabbits, primates, shrews, elephants, and an opossum. After removing the effects of evolutionary relationship, they found that longer-lived mammals tend to have shorter telomeres **(Figure 13.8)**. This pattern is not consistent with the notion that the longevity of organisms is limited by the longevity of their cells.

Indeed, Gomes and colleagues suggest that short telomeres facilitate long life span in mammals. They argue that the evolution of endothermy would have elevated the mutation rate in early mammals, increasing their risk of developing cancer. The repression of telomerase expression in most cells was adaptive because it enlisted the telomeres as accurate counters of cell divisions. Limiting divisions, in turn, reduced the probability that enough mutations would accumulate within cell lineages for cancer to arise. While short telomeres might thus have been adaptive, they entail a cost. Short telomeres have to be protected from oxidative damage to avoid triggering cell senescence or programmed death too soon.

Gomes and colleagues' analysis indicates that low telomerase expression and short telomeres were likely ancestral for the placental mammals. Elevated telomerase expression and long telomeres have evolved independently several times. Using telomerase to maintain long telomeres would reduce the need for protection from oxidation. Consistent with this idea, Gomes and colleagues found that short-lived mammals, and mammals with longer telomeres, are in fact less resistant to chemicals that induce oxidative damage. The downside may be an increased susceptibility to cancer. Mice are among the species with elevated telomerase and long telomeres. They are much smaller than humans and have much shorter life spans, yet suffer a similar incidence of cancer (Shay and Wright 2011).

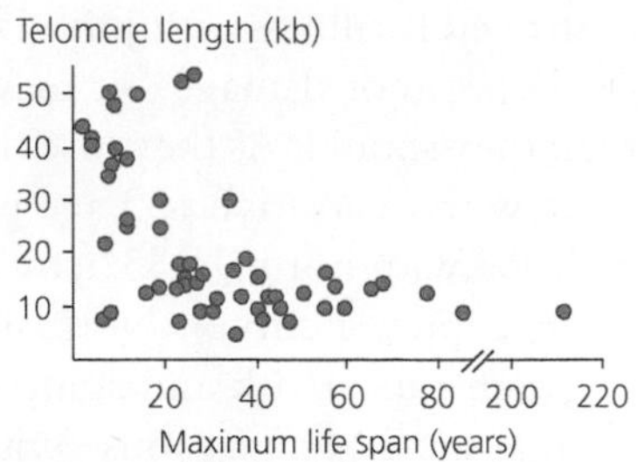

Figure 13.8 Longer-lived mammals have shorter telomeres Controlling for shared evolutionary history, mammals show a significant inverse association between telomere length and maximum life span ($p = 0.0032$). From Gomes et al. (2011).

A Trade-Off between Cancer Risk and Aging

Research on p53, which we mentioned earlier, suggests that whether a species is well or poorly protected against cancer, individuals face a trade-off between cancer risk and aging. Stuart Tyner and colleagues (2002) compared genetically manipulated mice with deficient, elevated, or normal p53 activity.

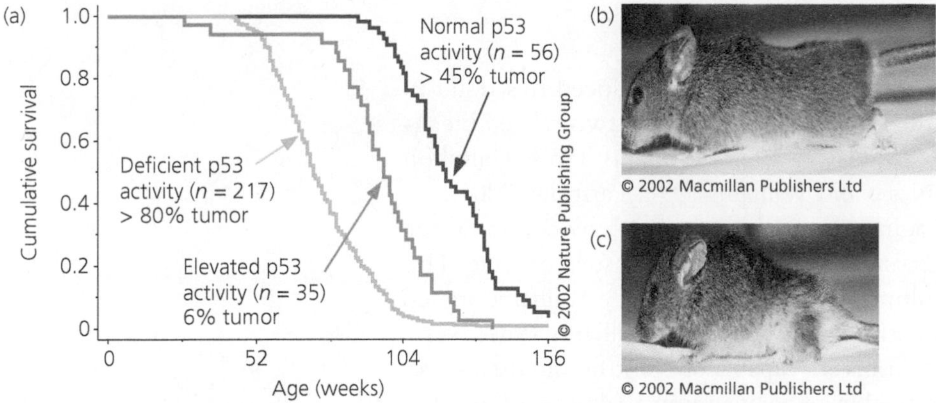

Figure 13.9 The tumor suppressor p53 trades off cancer risk with rate of aging (a) Survival of mice with different p53 activities. (b) A normal mouse looks healthy at age 104 weeks. (c) A mouse with elevated p53 looks old at the same age. From Tyner et al. (2002).

(a) Reprinted by permission from Macmillan Publishers Ltd: From Tyner, et al., 2002, p53 mutant mice that display early ageing-associated phenotypes. *Nature* 415: 45–53, Figure 2a, page 47. Copyright © 2002. (b and c) Reprinted by permission from Macmillan Publishers Ltd: *Nature* 415: 45–53. Copyright 2002.

Mice deficient in p53 are highly susceptible to cancer. As a result, such mice tend to die early, before showing other signs of old age **(Figure 13.9, blue curve)**. It appears p53 is failing to halt replication in cells whose DNA is badly damaged.

Mice with elevated p53 activity are highly resistant to cancer, but die of premature aging (orange curve). Tyner and colleagues think the explanation lies in p53's effects on stem cells. Stem cells play a key role in the maintenance and repair of organs and tissues. Their job is to divide and produce daughter cells that serve as replacements for mature cells that have worn out and died or been destroyed by illness or injury. If p53 is too active—if it is overly sensitive to signs of even minor damage—it may cause stem cells to stop dividing and die earlier than they should. As they lose their ability to maintain and repair their body parts, mice with overactive p53 age prematurely.

Mice with normal p53 activity suffer intermediate rates of cancer and live the longest (purple curve). Normal p53 activity appears to strike an optimal balance between cancer risk and aging rate. The photos in Figure 13.9b and (c) show a normal mouse and a mouse with elevated p53 activity at the same age.

p53 is an important cancer suppressor in humans as well (see Ferbeyre and Lowe 2002). Germ-line mutations in the *p53* gene cause increased susceptibility to a variety of cancers, and more cancers carry mutations in the *p53* gene, whether transmitted via the germ line or acquired somatically, than in any other gene.

Human populations harbor a nonsynonymous polymorphism at codon position 72 in the *p53* gene. One allele encodes arginine, the other proline. In combining data from numerous studies, Diana van Heemst and colleagues (2005) found that individuals homozygous for the proline allele have a somewhat higher risk of getting cancer than do arginine homozygotes. If they avoid dying of cancer or other causes until age 85, however, proline homozygotes tend to live longer.

To summarize, the rate-of-living theory and related hypotheses about aging have been scientific successes in the sense that they have stimulated considerable research and pointed the way to important discoveries. Among these discoveries is a paradox. Many populations, like Luckinbill's fruit flies, harbor genetic variation that would allow the evolution of longer life spans. And yet, longer life spans have not evolved. Furthermore, some species—such as mice—appear to have evolved short life spans in exchange for reduced maintenance costs. Another discovery is that organisms face trade-offs between the rate of aging and other traits as well. To see how the recognition of additional trade-offs helps solve the paradox of populations that age faster and have shorter life spans than their genetic variation might allow, we turn to the evolutionary theory of aging.

Many populations harbor genetic variation for longevity, yet longer life spans have not evolved.

The Evolutionary Theory of Aging

If selection can lead to longer life spans, why has it not produced this result in all populations? The evolutionary theory of senescence offers two related mechanisms to resolve this conundrum (Medawar 1952; Williams 1957; Hamilton 1966; Partridge and Barton 1993; Nesse and Williams 1995; Partridge 2001).

Under the evolutionary theory, aging is caused not so much by cell and tissue damage itself as by the failure of organisms to completely repair such damage. This failure leads to gradual decay and ultimate collapse. George C. Williams argued that complete repair ought to be physiologically possible (Williams 1957; Nesse and Williams 1995). Given that organisms are capable of constructing themselves from scratch, they should also be capable of maintaining their organs and tissues. Upkeep is, in principle, easier than manufacture. Indeed, organisms do have remarkable abilities to replace or repair damaged parts. Yet in many organisms, repair is incomplete. Under the evolutionary theory of senescence, the failure to completely repair damage is ultimately caused by either deleterious mutations or trade-offs between repair and reproduction.

Deleterious Mutations and Aging: The Mutation Accumulation Hypothesis

The mutation accumulation hypothesis notes that mutations causing premature senescence are selected against weakly. **Computing Consequences 13.1** shows why.

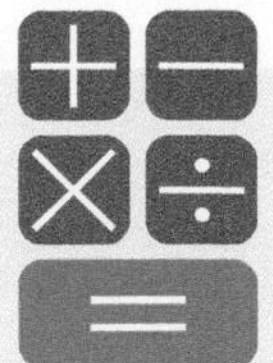

COMPUTING CONSEQUENCES 13.1

Late-acting deleterious mutations are weakly selected

Figure 13.10 uses a simple model of a hypothetical population of imaginary creatures to show how deleterious mutations can lead to the evolution of senescence. The figure follows the life histories of individuals from birth to death. Individuals are always at risk of death due to accidents, predators, and diseases. Except where noted, the probability that an individual will survive from one year to the next is 0.8. This leads to an exponential decline over time in the fraction of individuals still alive.

Figure 13.10a tracks individuals with the wild-type genotype. These mature at age 3 and die no later than 16. The columns of the table are as follows:

- The first column lists ages.
- The second column indicates the fraction of individuals still alive at each age. From age 1 onward, each number in the second column is simply the number immediately above it multiplied by 0.8. Few individuals survive to the age of 15. To keep the size of the table reasonable, we assume that all individuals that survive until their 16th birthday die before reproducing that year. This assumption affects only about 3% of the population.
- The third column shows yearly reproductive success (RS). Individuals mature at age 3. They each have one offspring every year for as long as they survive.
- The fourth column shows the expected reproductive success at each age. The expected reproductive success at age 5, for example, is simply the fraction of individuals that will survive to age 5 multiplied by the number of offspring each survivor will have at age 5. The sum of the numbers in this column gives the expected lifetime reproductive success.

The numbers in the table are plotted in the graph. The expected lifetime reproductive success of wild-type individuals is equal to the area of the shaded region. It is about 2.42.

Figure 13.10b depicts a mutation that causes death at age 14. In other words, the mutation causes premature senescence. All other aspects of life history are unchanged. The mutation is obviously deleterious, but how strongly will it be selected against? As shown in the table and graph, the expected lifetime reproductive success of individuals with the mutation is about 2.34. This is over 96% of the fitness of wild-type individuals. Because few individuals survive to age 14 anyway, individuals carrying the mutation causing death at 14 do not, on average, suffer much of a penalty. The mutation is selected against much less severely than it would be if it acted early in life.

(a) Wild type matures at age 3 and dies at age 16; before age 16, annual rate of survival = 0.8.

Age	Fraction of individuals surviving	RS of survivors	Expected RS for individuals
0	1.000	0	0.000
1	0.800	0	0.000
2	0.640	0	0.000
3	0.512	1	0.512
4	0.410	1	0.410
5	0.328	1	0.328
6	0.262	1	0.262
7	0.210	1	0.210
8	0.168	1	0.168
9	0.134	1	0.134
10	0.107	1	0.107
11	0.086	1	0.086
12	0.069	1	0.069
13	0.055	1	0.055
14	0.044	1	0.044
15	0.035	1	0.035

Expected lifetime RS: 2.419

1
0
0
5
10
15
Age
Fraction of individuals surviving
RS of survivors
Expected RS for individuals

(b) Mutation from wild type that causes death at age 14; before age 14, annual rate of survival = 0.8.

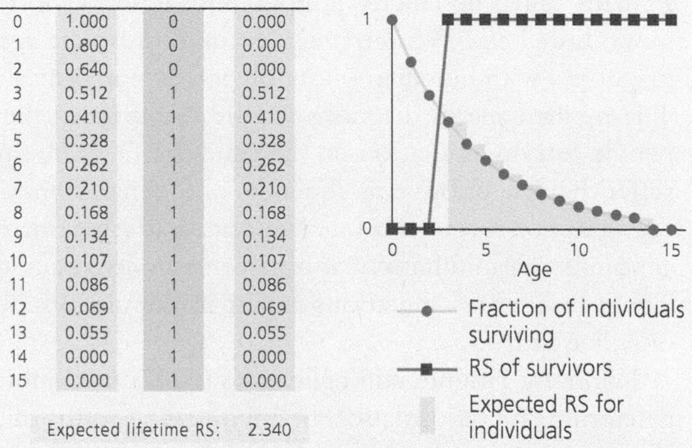

0	1.000	0	0.000
1	0.800	0	0.000
2	0.640	0	0.000
3	0.512	1	0.512
4	0.410	1	0.410
5	0.328	1	0.328
6	0.262	1	0.262
7	0.210	1	0.210
8	0.168	1	0.168
9	0.134	1	0.134
10	0.107	1	0.107
11	0.086	1	0.086
12	0.069	1	0.069
13	0.055	1	0.055
14	0.000	1	0.000
15	0.000	1	0.000

Expected lifetime RS: 2.340

Figure 13.10 **A simple model shows the modest fitness cost of a late-acting deleterious mutation**

Few individuals survive long enough to experience the ill effects of a late-acting deleterious mutation. This means that individuals who carry such a mutation—even a lethal one—suffer only a modest reduction in fitness.

To understand why populations have not evolved longer life spans, researchers explore how natural selection varies as a function of an individual's age.

At first glance, it may seem surprising that a mutation causing death could be only mildly deleterious. Many mutations that cause death are, in fact, highly deleterious. For the population of imaginary creatures considered in Computing Consequences 13.1, a mutation causing death at age 2, for example, would be selected against strongly. Individuals carrying such a mutation would have an expected lifetime reproductive success of zero. But mutations causing death after reproduction has begun are selected against less strongly. The later in life such mutations exert their deleterious effects, the more weakly they are selected against. Mutations selected against only weakly can persist in mutation–selection balance (see Chapter 6). The accumulation in populations of deleterious mutations whose effects occur only late in life is one evolutionary explanation for aging (Medawar 1952).

Natural selection is weak late in life, so alleles that cause aging are only mildly deleterious. They may persist in mutation–selection balance or rise to high frequency by drift.

What kind of mutation could cause death, but only at an advanced age? One possibility is a mutation that reduces an organism's ability to maintain itself in good repair. Humans provide an example. Among the kinds of cellular damage that humans (and other organisms) must repair are DNA mismatch errors. Mismatched nucleotide pairs can be created by mistakes during DNA replication, or they can be induced by chemical damage to DNA (Vani and Rao 1996). Repair of these errors is performed by a suite of special enzymes. Germ-line mutations in the genes that code for these enzymes can result in the accumulation of mismatch errors, which in turn can result in cancer.

Germ-line mutations in DNA mismatch repair genes cause a form of cancer in humans called hereditary nonpolyposis colon cancer (Eshleman and Markowitz 1996; Fishel and Wilson 1997). In one study, the age at which individuals were diagnosed with hereditary nonpolyposis colon cancer ranged from 17 to 92 years. The median age of diagnosis was 48 (Rodriguez Bigas et al. 1996). Thus, most people carrying mutations in the genes for DNA mismatch repair enzymes do not suffer the deleterious consequences of the mutations until well after the age when reproduction begins. In an evolutionary sense, hereditary nonpolyposis colon cancer is a manifestation of senescence that is caused by deleterious mutations. These deleterious mutations persist in populations because they reduce survival only late in life.

Kimberly Hughes and colleagues (2002) used inbreeding depression to detect deleterious mutations associated with aging in fruit flies (*Drosophila melanogaster*). If inbreeding depression is caused by deleterious recessive alleles, and if late-acting deleterious alleles are maintained at higher frequency under mutation–selection balance than early-acting deleterious alleles, then the severity of inbreeding depression should increase with age.

Hughes and colleagues tested this prediction by preparing 10 inbred stocks of fruit flies. Within each stock, all individuals were homozygous at most loci. The researchers then performed all 100 possible crosses among their inbred stocks. Ten of these crosses involved inbred lines crossed with themselves. The progeny of these crosses were inbred. The other 90 crosses involved crosses among inbred lines. The progeny of these crosses were outbred. That is, they were heterozygous at most loci. The researchers measured the reproductive success of the progeny at various ages. They calculated inbreeding depression as the difference in fitness between outbred versus inbred lines, divided by the fitness of outbred lines.

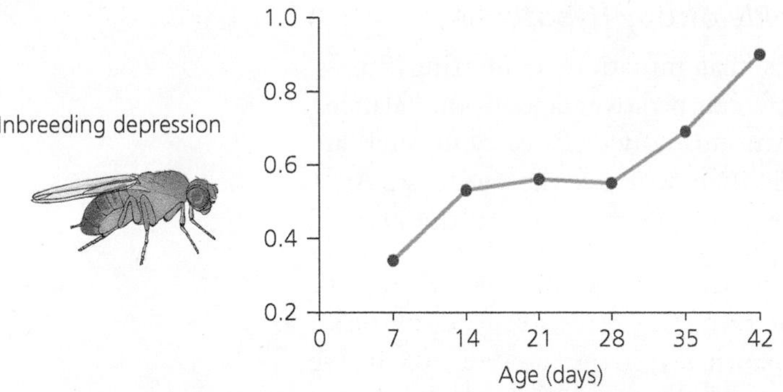

Figure 13.11 Inbreeding depression increases with age in fruit flies Drawn from data in Hughes et al. (2002).

The results, shown in **Figure 13.11**, reveal that, as predicted, inbreeding depression increases with age. The fruit fly stocks that Hughes and colleagues studied harbor deleterious mutations that contribute to senescence.

Juan Escobar and colleagues (2008) did a similar study with a self-compatible hermaphroditic snail. Inbreeding depression increased with age. The benefit of outcrossing among individuals from different populations increased with age as well, suggesting that populations harbor distinct mutations contributing to aging.

Deleterious mutations that contribute to aging can accumulate rapidly. David Reed and Edwin Bryant (2000) documented the accumulation of late-acting deleterious mutations in populations of houseflies (*Musca domestica*). Reed and Bryant established laboratory populations with wild flies. Each generation, the researchers allowed the adults to reproduce for only four or five days, then used the offspring produced during this window to establish the next generation. This procedure in essence limited the adult life span of every fly to less than a week.

Reed and Bryant reasoned that any late-acting deleterious mutations present in their populations would thus be rendered neutral. Some of these now-neutral alleles should drift to high frequency. Because neutral evolution proceeds as rapidly in large populations as in small ones (see Chapter 7), the researchers expected the effect to proceed at the same pace regardless of population size.

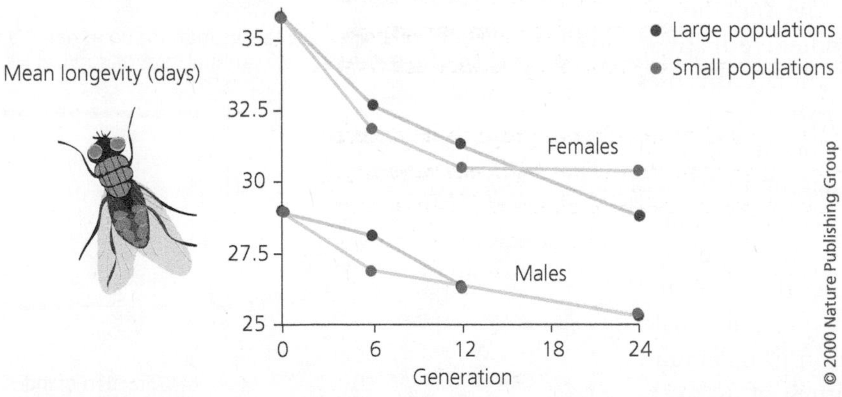

Figure 13.12 Life span declines rapidly in housefly populations allowed to breed only early in life This is likely due to the accumulation of late-acting deleterious mutations. Redrawn from Reed and Bryant (2000).

Reprinted by permission of Macmillan Publishers Ltd. D. Reed and E. Bryant. 2000. "The evolution of senescence under curtailed life span in laboratory populations of *Musca domestica* (housefly)." *Heredity* 85: 1115–1121. Copyright © 2000.

Reed and Bryant monitored the accumulation of late-acting deleterious mutations in their housefly populations by periodically allowing the flies to live out their natural life spans. As shown in **Figure 13.12**, in both large populations and small populations the natural longevity of the houseflies declined substantially over 24 generations. These results are consistent with the mutation accumulation hypothesis of senescence.

Trade-Offs and Aging: The Antagonistic Pleiotropy Hypothesis

The antagonistic pleiotropy hypothesis notes that mutations conferring fitness benefits early in life and fitness costs late in life can be advantageous on balance. **Computing Consequences 13.2** shows why. Among individuals carrying such an allele, more will experience the early benefit than will pay the late cost. As a result, the allele can increase the fitness of the average carrier even if the benefit seems modest and the cost seems high.

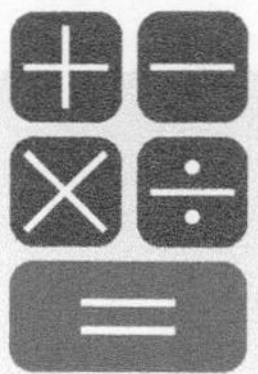

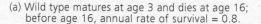

COMPUTING CONSEQUENCES 13.2

Alleles conferring early benefits and late costs can be adaptive

Figure 13.13 uses a simple model of a hypothetical population of imaginary creatures to show how a mutation with early benefits and late costs can lead to the evolution of senescence. The model is the same as the one we considered in Computing Consequences 13.1, except that the effects of the mutation we introduce are different—and slightly more complex.

Figure 13.13a tracks individuals with the wild-type genotype. These mature at age 3 and die no later than 16. The columns of the table show age, fraction surviving, yearly reproductive success of survivors, and expected reproductive success. The sum of the last column is the expected lifetime reproductive success for wild-type individuals. It is about 2.42. This sum is also represented by the shaded area in the graph.

Figure 13.13 depicts a mutation that affects two different life-history characters. That is, the mutation is pleiotropic. The mutation causes reproductive maturation at age 2 instead of age 3, and the mutation causes death at age 10. In other words, the mutation involves a trade-off between reproduction early in life and survival late in life. Its pleiotropic effects are antagonistic.

As shown in the table and graph, the expected lifetime reproductive success of individuals with the mutation is about 2.66. This is 1.1 times the expected lifetime reproductive success of wild-type individuals. Despite causing a rather drastic reduction in maximum longevity, the mutation increases the fitness of the average carrier. The average carrier gains 0.64 offspring by maturing early, but loses only 0.396 offspring by dying young. It thus comes out about a quarter of an offspring ahead of its wild-type peers. As our mutation rises in frequency, average life span in the population will fall.

(a) Wild type matures at age 3 and dies at age 16; before age 16, annual rate of survival = 0.8.

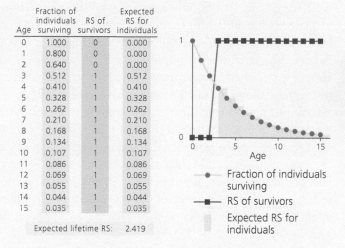

Age	Fraction of individuals surviving	RS of survivors	Expected RS for individuals
0	1.000	0	0.000
1	0.800	0	0.000
2	0.640	0	0.000
3	0.512	1	0.512
4	0.410	1	0.410
5	0.328	1	0.328
6	0.262	1	0.262
7	0.210	1	0.210
8	0.168	1	0.168
9	0.134	1	0.134
10	0.107	1	0.107
11	0.086	1	0.086
12	0.069	1	0.069
13	0.055	1	0.055
14	0.044	1	0.044
15	0.035	1	0.035

Expected lifetime RS: 2.419

(b) Mutation from wild type that causes maturation at age 2 and death at age 10; before age 10, annual rate of survival = 0.8.

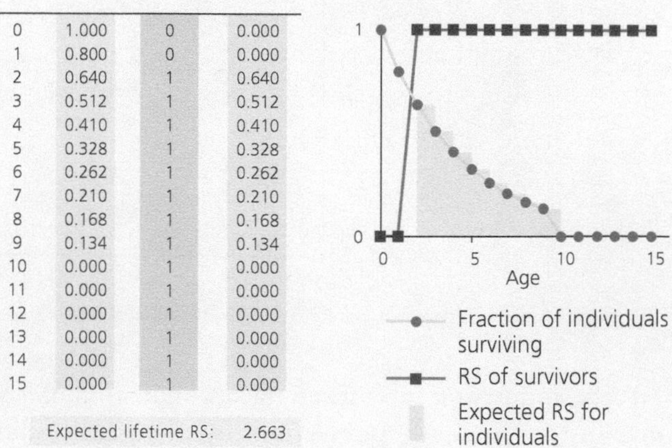

Age	Fraction of individuals surviving	RS of survivors	Expected RS for individuals
0	1.000	0	0.000
1	0.800	0	0.000
2	0.640	1	0.640
3	0.512	1	0.512
4	0.410	1	0.410
5	0.328	1	0.328
6	0.262	1	0.262
7	0.210	1	0.210
8	0.168	1	0.168
9	0.134	1	0.134
10	0.000	1	0.000
11	0.000	1	0.000
12	0.000	1	0.000
13	0.000	1	0.000
14	0.000	1	0.000
15	0.000	1	0.000

Expected lifetime RS: 2.663

Figure 13.13 A simple model shows the net fitness advantage of an allele with an early benefit and late cost

Alleles that influence more than one trait are said to be **pleiotropic**. An allele conferring both a benefit and a cost has pleiotropic effects that are antagonistic. Selection for alleles with pleiotropic effects that trade reproduction or survival late in life for increased reproductive success early in life is a second evolutionary explanation for aging (Williams 1957; Rose 1991). As such alleles increase in frequency in a population, the average life span declines.

Because natural selection is weaker late in life, alleles that enhance early-life reproduction may be favored even if they also hasten death.

What kind of mutation could increase reproduction early in life at the same time it reduced reproduction or survival late in life? Perhaps a mutation that causes less energy to be allocated to repair early in life and more energy to be allocated to reproduction (see Figure 13.2 on page 493). Researchers have found several genes that appear to have this kind of pleiotropic action (Leroi et al. 2005).

Research by David Walker and colleagues (2000) on the *age-1* gene in *Caenorhabditits elegans* provides an example. *C. elegans* is a tiny nematode worm, about 1 mm in length, that lives in soil and eats bacteria. In *C. elegans* the protein encoded by *age-1* plays a role in an intracellular signaling pathway involved in the control of development and the determination of stress resistance. The *age-1* gene product also plays a role in senescence. Mutations in the gene increase life span by as much as 80%. Carriers of such mutations appear to be otherwise normal. They develop at the same rate as wild-type worms, have similar activity levels, and achieve comparable total fertility.

Walker and colleagues sought more subtle effects on fitness of a mutant *age-1* allele called *hx546*. The researchers established laboratory populations of worms in which the individuals were genetically identical except that some were homozygous for the normal *age-1* allele whereas others were homozygous for *hx546*. All the worms were hermaphrodites and reproduced by self-fertilization. The researchers tracked the frequency of the *hx546* allele over 10–12 generations. If the allele were beneficial, its frequency would rise; if it were deleterious, its frequency would fall. Given that the only obvious difference between *hx546* worms and normal worms is that the *hx546* worms live considerably longer, one might expect that the allele would be advantageous.

Walker and colleagues first reared populations in which they gave the worms ample food. The researchers established two populations in which the starting frequency of *hx546* was 0.9, two in which it was 0.5, and two in which it was 0.1. Surprisingly, the frequency of the allele changed little over 10 generations, regardless of its starting frequency **(Figure 13.14a)**. The *hx546* allele was not advantageous, but neither was it deleterious. This result suggests that the benefit of longer life span was balanced by a roughly equivalent cost.

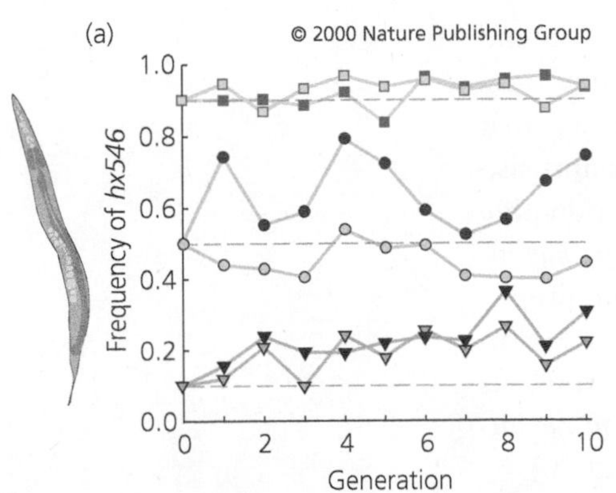

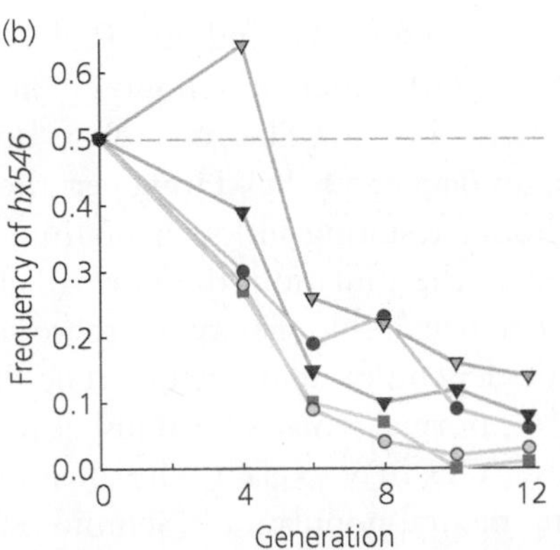

Figure 13.14 Under semi-natural conditions, the *age-1* gene in the nematode worm *C. elegans* exhibits antagonistic pleiotropy (a) Frequency of longevity allele *hx546* in populations reared in the lab with ample food. (b) Frequency of *hx546* in populations reared under semi-natural conditions characterized by periodic bouts of starvation. From Walker et al. (2000).

Reprinted with permissions of Macmillan Publishers, Ltd. D. W. Walker, et al., 2000. "Evolution of life span in *C. elegans*." *Nature* 405: 296–297. Copyright © 2000.

The true cost of carrying the *hx546* allele was revealed when Walker and colleagues reared populations under conditions that more closely resemble what *C. elegans* experiences in nature. Each time they established a new culture, they let the worms eat all of the bacteria in their petri dish. The researchers let the worms starve for four days, then gave them more bacteria to eat. Finally, the researchers used only eggs the worms produced in the first 24 hours after feeding resumed to establish the next culture. From start to finish, each starvation cycle lasted about 2 generations. Walker and colleagues tracked the frequency of *hx546* for six starvation cycles in five populations, all with a starting frequency of 0.5. The result was dramatic. The frequency of *hx546* plummeted to an average of 0.06 in 12 generations (Figure 13.14b). For the frequency of *hx546* to fall that far, the fitness of *hx546* worms must have been less than 80% that of normal worms.

Additional observations established that the only worms that reproduced during the 24 hours after feeding resumed in the experimental cultures were young adults. It must thus be during young adulthood that *hx546* is deleterious. The implication is that, compared to *hx546*, the normal allele acts in just the way the antagonistic pleiotropy hypothesis of senescence predicts: It increases the reproductive success of its carriers in young adulthood, at the cost of a shorter life span.

Researchers have documented trade-offs between reproduction early in life and longevity.

A gene in the fruit fly, *Drosophila melanogaster*, has similar effects. When Yi-Jyun Lin and colleagues (1998) discovered this gene, they named it *methuselah* because homozygotes for a mutant allele with reduced expression live 35% longer than normal flies. The *methuselah* mutation attracted notice because, in addition to increasing longevity, it enhances resistance to starvation, heat, and the herbicide paraquat (see Pennisi 1998). A cost-free mutation that extends life span and increases tolerance to a variety of stresses would challenge the evolutionary theory of senescence.

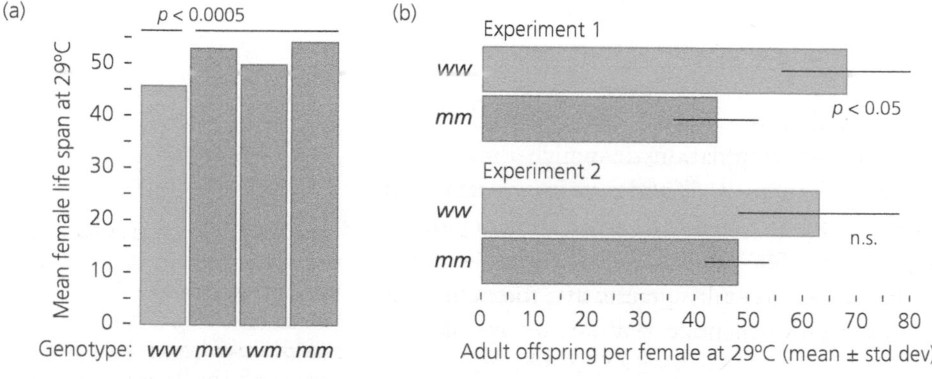

Figure 13.15 The *methuselah* gene controls a trade-off between reproductive success versus longevity and stress resistance (a) Normal female fruit flies (*ww*) die younger than *methuselah* mutants (*mw*, *wm*, *mm*). (b) Normal flies, however, have considerably higher lifetime reproductive success than homozygous mutants. Data from Mockett and Sohal (2006).

But is the *methuselah* mutation free of costs? To find out, Robin Mockett and Rajindar Sohal (2006) reared females with different genoptyes and counted the offspring each fly produced over her life span. The researchers confirmed that flies carrying the mutation live longer **(Figure 13.15a)**. However, the mutants also lay fewer eggs during early adulthood, resulting in lower lifetime reproductive success (Figure 13.15b). Compared to the mutation, the normal allele of *methuselah* thus appears to trade stress resistance and longevity for reproductive fitness. Intriguingly, the effects of genotype depend on temperature. The flies tracked in Figure 13.15 lived at 29°C. At 18°C, normal females lived just as long as mutants but had lower reproductive success. This may explain why alleles of the *methuselah* gene vary in frequency among natural populations (Schmidt et al. 2000).

In addition to identifying specific genes exhibiting antagonistic pleiotropy, biologists have found widespread evidence that there is indeed a trade-off between reproduction early in life and reproduction or survival late in life. Most of this evidence comes from the analysis of quantitative genetic or phenotypic trade-offs between traits. In the paragraphs that follow, we review two examples.

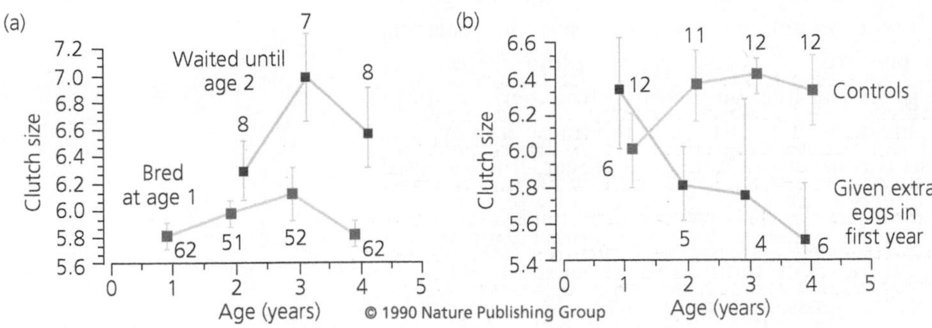

Figure 13.16 A trade-off between early versus late reproduction Mean clutch size ± std. error, with sample sizes, for (a) female collared flycatchers that first bred at different ages, and (b) females given extra eggs in first year versus controls. From Gustafsson and Pärt (1990).

Reprinted by permission of Macmillan Publishers Ltd. From Gustafsson, L., and T. Pärt. 1990. "Acceleration of senescence in the collared flycatcher (*Ficedula albicollis*) by reproductive costs." *Nature* 347: 279–281. Copyright © 1990.

Lars Gustafsson and Tomas Pärt (1990) studied trade-offs in a bird, the collared flycatcher (*Ficedula albicollis*), on the Swedish island of Gotland. Over a span of 10 years, Gustafsson and Pärt followed the life histories of individuals from hatching to death. Some female flycatchers begin breeding at age 1, whereas others wait until age 2. Females that breed at age 1 have smaller clutch sizes throughout life **(Figure 13.16a)**, indicating that there is a late-life cost to breeding early. To investigate further, Gustafsson and Pärt manipulated early reproductive effort by giving some first-year breeders extra eggs. The females given extra eggs had progressively smaller clutch sizes in subsequent years, whereas control females did not show reproductive senescence until age 4 (Figure 13.16b). Gustafsson and Pärt conclude that there is a trade-off in collared flycatchers between early-life and late-life reproduction. This suggests that if there is genetic variation for age at first breeding, there is likely antagonistic pleiotropy. Despite the trade-off, first-year breeders had higher lifetime reproductive success than second-year breeders (1.24 ± 0.08 versus 0.90 ± 0.14 offspring surviving to adulthood; $p < 0.05$).

Truman Young (1990) studied trade-offs in plants. Young reviewed data from the literature on the energy allocated to reproduction by closely related pairs of annuals versus perennials. Each bar in **Figure 13.17** shows the estimated allocation by annuals in multiples of the estimated allocation by perennials. The numbers range from 1.7 to 5.3. Annuals, which reproduce once and die, always allocate more energy to their sole bout of reproduction than perennials allocate to any given bout. This pattern indicates that there is a trade-off in plants between reproduction and survival. Annual plants enjoy enhanced reproduction in their first reproductive season at the expense of drastically accelerated senescence.

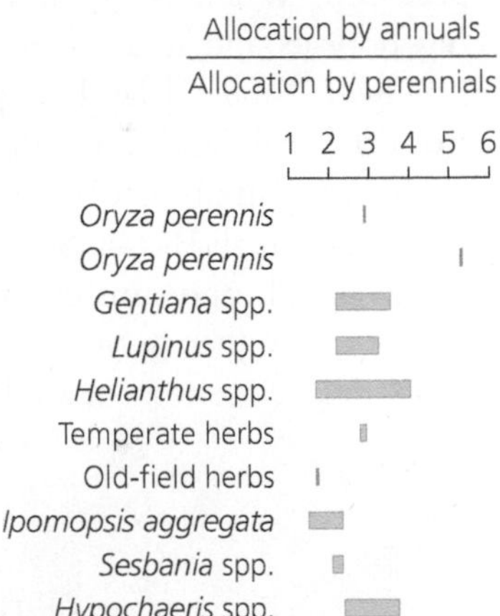

Figure 13.17 Reproductive allocation by annual versus perennial plants The comparisons for *Oryza perennis* (two studies) and *Ipomopsis aggregata* are within species; all other comparisons are between species. Prepared with data compiled and analyzed by Young (1990).

A Natural Experiment on Ecological Mortality and the Evolution of Aging

Steven Austad took advantage of a natural experiment to compare populations historically exposed to different rates of mortality caused by extrinsic factors such as predators, diseases, and accidents. We will call this kind of mortality ecological mortality, in contrast to mortality caused by processes intrinsic to the organism, like the wearing out of body parts (which we could call physiological mortality.)

The evolutionary theory of senescence predicts that populations with lower rates of ecological mortality will evolve delayed senescence (Austad 1993). What is the logic behind this prediction? Both of the evolutionary mechanisms that lead to senescence have reduced effectiveness in populations with lower ecological mortality rates.

In the case of late-acting deleterious mutations, lower ecological mortality means that a higher fraction of zygotes will live long enough to experience the deleterious effects. Late-acting deleterious mutations are thus more strongly selected against and will be held at lower frequency in mutation–selection balance.

In the case of mutations with pleiotropic effects, lower ecological mortality means that a higher fraction of zygotes will live long enough to experience both the early-life benefits and the late-life costs. The change in the fraction of zygotes experiencing the benefits and costs is more pronounced, however, for the costs. Thus mutations with pleiotropic effects are less strongly favored by selection.

All else being equal, if the evolutionary theory of aging is correct, individuals in populations with lower ecological mortality should show later senescence.

Austad (1993) studied the Virginia opossum **(Figure 13.18)**. He compared a population living in the mainland southeastern United States to a population living on Sapelo Island, located off the coast of Georgia. In the mainland population, opossums have high ecological mortality rates. In one study reviewed by Austad, more than half of all naturally occurring opossum deaths were caused by predators. When identifiable, two-thirds of the predators were mammals, including bobcats and feral dogs. Mammalian predators are absent on Sapelo Island, however. Sapelo Island supports an opossum population that has been isolated from the mainland population for 4,000–5,000 years. Other than the difference in mammalian predators, Sapelo Island differs little from Austad's mainland study site at Savannah River, South Carolina. The two habitats are similar in temperature, rainfall, opossum ectoparasite loads, and food available per opossum. The evolutionary theory of senescence predicts that the Sapelo Island opossums will show delayed senescence relative to the mainland opossums.

Figure 13.18 A Virginia opossum and her young Opossums, like other marsupials, have relatively short life spans for mammals. But opossums in some populations have shorter life spans than opossums in others. This finding suggests that rates of aging evolve in nature.

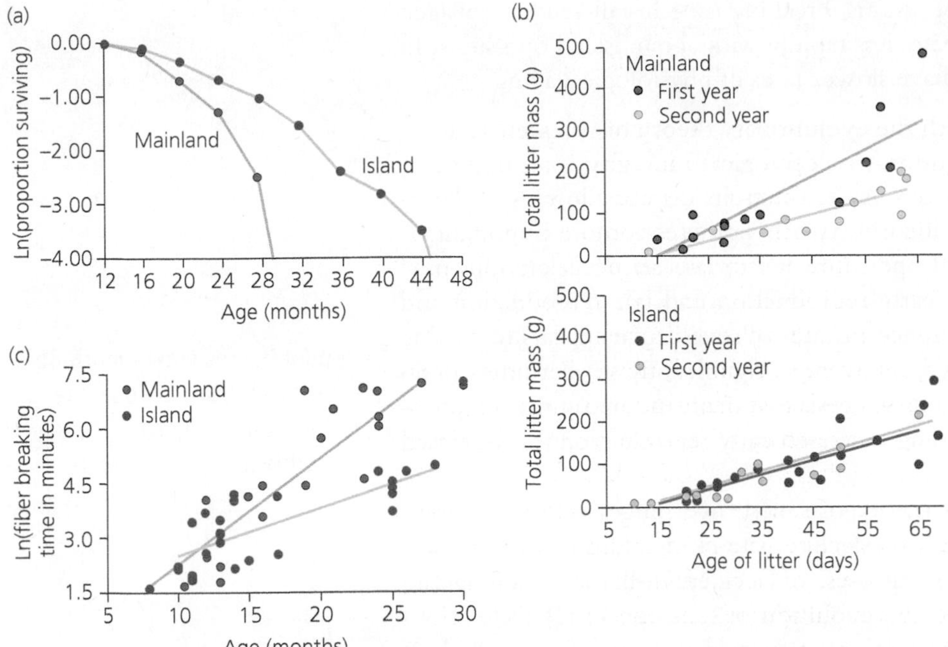

Figure 13.19 Female opossums on Sapelo Island age more slowly than female opossums on the mainland (a) The rate of survival versus age for mainland and island opossums. (b) Total litter mass as a function of litter age for females in their first and second years of reproduction. In mainland females, offspring in second-year litters grow more slowly than offspring in first-year litters ($p < 0.001$). In island females, second-year litters grow just as fast as first-year litters. (c) Tail collagen fiber-breaking time versus age. Increase with age is slower in island females than in mainland females ($p < 0.001$). From Austad (1993).

To test this prediction, Austad put radio collars on 34 island females and 37 mainland females and followed their life histories from birth until death. By three different measures, island females indeed show delayed senescence:

- Island females show delayed senescence in month-to-month probability of survival. This is indicated not by the fact that the island curve is higher than the mainland curve in **Figure 13.19a**, but by the shapes of the curves. For island opossums, ln(proportion surviving) versus age traces a relatively straight line. This means that their risk of dying in their 40th month of life is not much higher than their risk of dying in their 20th month. In other words, the island opossums do not seem to accumulate much wear and tear. For mainland opossums, however, ln(proportion surviving) versus age traces a precipitous downward curve. Their risk of dying in their 28th month (approximately 100%) is much higher than their risk of dying in their 20th month. The immediate cause of death for most mainland opossums is predation. But older opossums are much more vulnerable to predation, because they are getting stiff and slow. As a result, the average life span of island females is significantly longer than the average life span of mainland females (24.6 versus 20.0; $p < 0.02$).
- Island females show delayed senescence in reproductive performance (Figure 13.19b). Austad measured reproductive performance by monitoring the growth rates of litters of young. For mainland females, litters produced in the mother's second year of reproduction grew more slowly than litters produced in the mother's first year of reproduction. This difference indicates that second-year mothers are less efficient at nourishing their young. Island females show no such decline in performance with age.
- Island females show delayed senescence in connective tissue physiology (Figure 13.19c). As mammals age, the collagen fibers in their tendons develop cross-links between protein molecules. These cross-links reduce the flexibility of the tendons. The amount of cross-linking in a tendon can be determined by measuring how long it takes for collagen fibers from the tendon to break.

In both island and mainland opossums, breaking time in tail-tendon collagen increases with age, but it increases less rapidly with age in island opossums. In other words, island opossums have slower rates of physiological aging.

These results are all consistent with the evolutionary theory of senescence.

While they support the conclusion that ecological mortality is an important factor in the evolution of senescence, Austad's results do not allow us to determine which of the evolutionary theory's two hypotheses is more important. Is the more rapid aging of mainland opossums due to late-acting deleterious mutations, or to trade-offs between early reproduction and late reproduction and survival? At least part of the difference in rates of senescence appears to be due to trade-offs. Island opossums have, on average, significantly smaller litters (5.66 versus 7.61; $p < 0.001$). This finding suggests that mainland opossums are, physiologically and evolutionarily, trading increased early reproduction for decreased later reproduction and survival.

In populations where mortality rates are high, individuals tend to invest more heavily in early reproduction.

In summary, the evolutionary theory of senescence hinges on the observation that the power of natural selection declines late in life. This is because most individuals die—due to predators, diseases, or accidents—before reaching late life. Two mechanisms can lead to the evolution of senescence: (1) Deleterious mutations whose effects occur late in life can accumulate in populations; and (2) when there are trade-offs between reproduction and maintenance, selection may favor investing in early reproduction even at the expense of maintaining cells and tissues in good repair. The evolutionary theory of senescence has been successful in explaining variation in life history among populations and species.

Before leaving the topic of aging, we consider whether evolutionary theory can help explain an unusual aspect of the reproductive life histories of human females: menopause.

An Evolutionary Explanation for Menopause?

In humans, reproductive capacity declines earlier and more rapidly in women than in men **(Figure 13.20a)**. The early decline in the reproductive capacity of women is puzzling, especially given that other measures of women's physiological capacity decline much more slowly (Figure 13.20b). Why should women's reproductive systems shut down by age 50, while the rest of their organs and tissues are still in good repair?

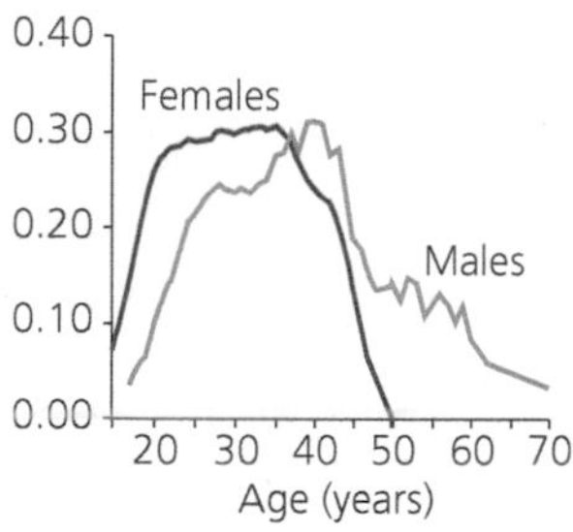

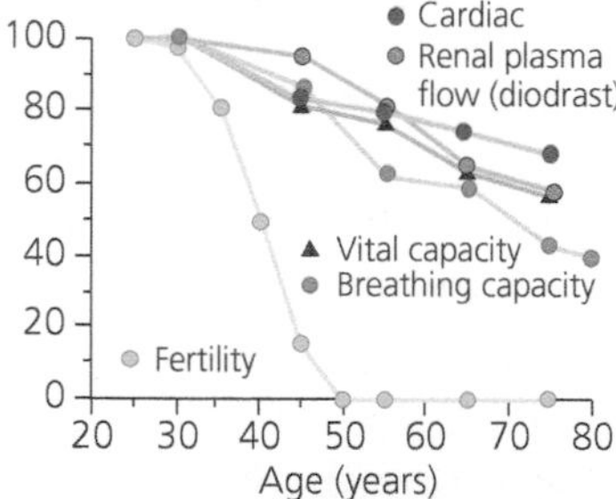

Figure 13.20 Menopause in humans (a) Probability that women and men will have a child born during the year that they are any given age. Data are from the Ache hunter-gatherers of Paraguay. From Hill and Hurtado (1996). (b) Functional capacity of various physiological systems in women versus age. From Hill and Hurtado (1991).

We consider two hypotheses. One hypothesis suggests that menopause is a nonadaptive artifact of our modern lifestyle (see Austad 1994). The other hypothesis suggests that menopause is a life-history adaptation associated with the contribution grandmothers make to feeding their grandchildren (Hawkes at al. 1989).

Advocates of the artifact hypothesis point out that archaeologists reconstructing the demography of ancient peoples have often concluded that in premodern cultures, virtually all adults died by age 50 or 55 (see Hill and Hurtado 1991). If death by age 50 or 55 was the rule for our hunter-gatherer ancestors, then the modern situation, in which individuals often live into their 80s and 90s, is unprecedented in our evolutionary history. Menopause cannot be an adaptation, because our hunter-gatherer ancestors never lived long enough to experience it.

When other mammals are kept in captivity and given modern medical care, they too live far longer than individuals of the same species do in nature. Furthermore, in captive mammals, females in at least some species show a decline in reproductive capacity well in advance of the decline in male reproductive capacity, and long before death. Data on reproductive capacity as a function of age in captive rats appear in **Figure 13.21**. They suggest that menopause in humans may need no other explanation than that our modern lifestyle has extended our life span beyond that experienced by our ancestors.

Critics of the artifact hypothesis point out that in contemporary hunter-gatherer societies, many individuals live into their 60s and 70s **(Figure 13.22)**. These data may be more reliable indicators of the demography of our hunter-gatherer ancestors than are archaeological reconstructions (Hill and Hurtado 1991; see also Austad 1994; Gurven and Kaplan 2007). If a substantial fraction of our female hunter-gatherer ancestors lived long enough to experience menopause, then menopause needs an evolutionary explanation.

Advocates of the grandmother hypothesis note that human children depend on their mothers for food for several years after weaning. This is true in contemporary hunter-gatherer cultures, particularly when mothers harvest foods that yield a high return for adults, but are difficult for children to process (Hawkes et al. 1989). Thus, a woman's ability to have additional children may be substantially limited by her need to provision her older, still-dependent children. Furthermore, as a woman gets older, several relevant trends are likely to occur: (1) The probability that she will live long enough to be able to nurture another baby from birth to independence declines, (2) the risks associated with pregnancy and childbirth rise, and (3) her own daughters will themselves start to have children. The grandmother hypothesis suggests that older women may reach a point at which they can get more additional copies of their genes into future generations by ceasing to reproduce themselves and instead helping to provision their weaned grandchildren so their daughters can have more babies. In other words, grandmothers face a trade-off between investment in children versus grandchildren.

Kristen Hawkes and colleagues (1989, 1997) studied postmenopausal women in the Hadza, a contemporary hunter-gatherer society in East Africa. If the grandmother hypothesis is correct, then women in their 50s, 60s, and 70s should continue to work hard at gathering food. If the grandmother hypothesis is wrong, then we might expect older women (who no longer have dependent children) to relax. In fact, older Hadza women work harder at foraging than any other group **(Figure 13.23a)**. Furthermore, for at least some crops at some times of year, older women are the most effective foragers (Figure 13.23b). Older women do with

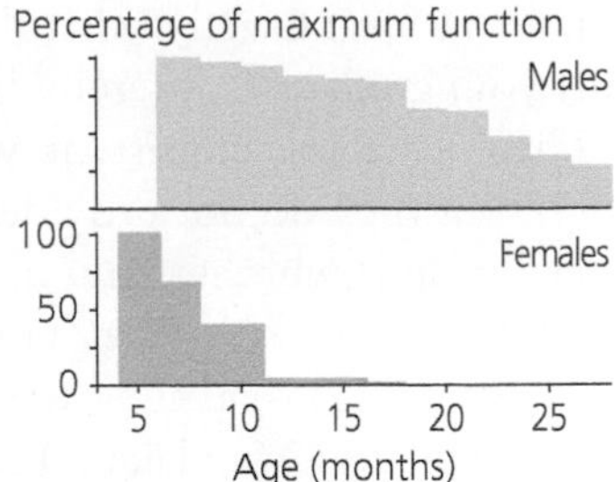

Figure 13.21 Reproductive capacity versus age in captive rats Data from Austad (1994).

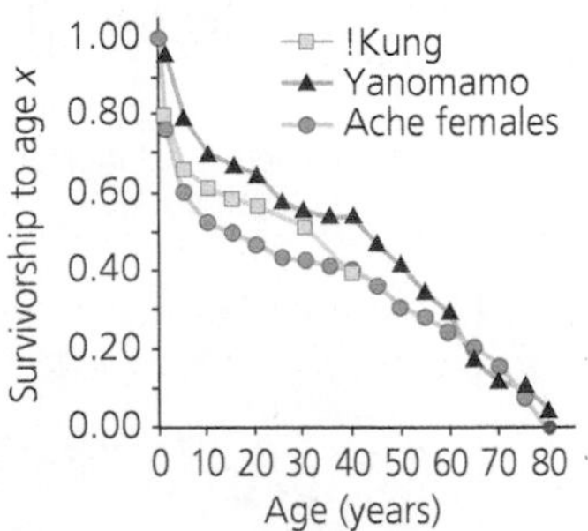

Figure 13.22 Fraction of individuals surviving versus age in three hunter-gatherer cultures From Hill and Hurtado (1991).

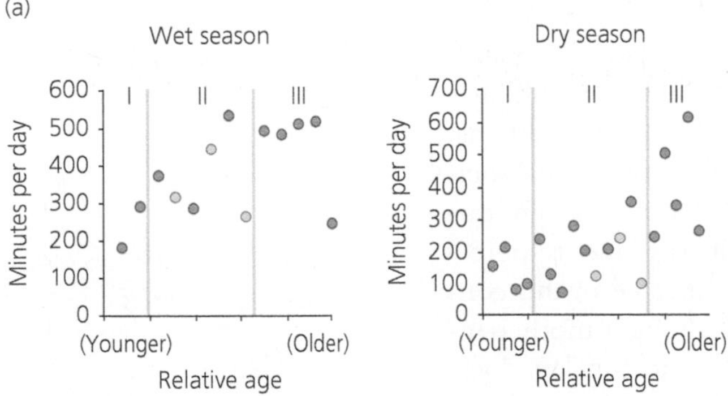

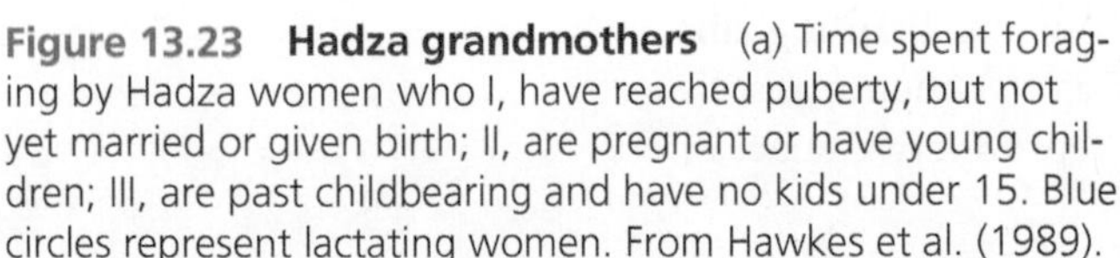

Figure 13.23 Hadza grandmothers (a) Time spent foraging by Hadza women who I, have reached puberty, but not yet married or given birth; II, are pregnant or have young children; III, are past childbearing and have no kids under 15. Blue circles represent lactating women. From Hawkes et al. (1989).

(b) This Hadza woman, about 65 years old, is using a digging stick and muscle power to dig tubers from under rocks. Digging tubers requires knowledge, skill, patience, strength, and experience, making Hadza grandmothers the most productive foragers. By James F. O'Connell, University of Utah.

their extra food exactly what the grandmother hypothesis predicts: They share it with young relatives, thereby improving the children's nutritional status.

These data are consistent with the grandmother hypothesis (Hawkes et al. 1998), but they do not provide a definitive test. As Austad (1994) points out, the crucial issue is whether daughters of helpful grandmothers are able to have more children, and whether the grandmothers thereby achieve higher inclusive fitness (see Chapter 12) than they would by trying to have more kids of their own.

Kim Hill and Magdalena Hurtado (1991, 1996) addressed this issue with data on the Ache hunter-gatherers of Paraguay. Hill and Hurtado's data show that the average 50-year-old woman has 1.7 surviving sons and 1.1 surviving daughters. The researchers calculate that by helping these children reproduce, the average Ache grandmother can gain the inclusive fitness equivalent of only 5% of an additional offspring of her own. This is scant support for the grandmother hypothesis.

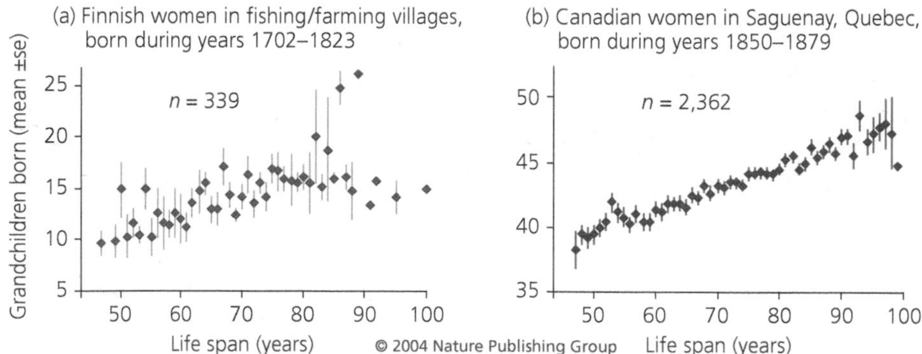

Figure 13.24 Number of grandchildren versus post-menopausal life span in pre-modern women In both populations, longer-lived women had more grandkids ($p = 0.0002$, both data sets). From Lahdenperä et al. (2004).

Reprinted by permission of Macmillan Publishers Ltd. Lahdenperä, M., V. Lummaa, et al. 2004. "Fitness benefits of prolonged post-reproductive lifespan in women." *Nature* 428: 178–181. Copyright © 2004.

Other researchers have sought to answer a slightly different question: Is it adaptive for women to remain alive after menopause? Mirkka Lahdenperä and colleagues (2004) analyzed large data sets, compiled from church and government records, on premodern Finnish and Canadian women. After controlling statistically for socioeconomic status, geography, and change over time in living conditions, they found that a typical woman acquired two additional grandchildren for every decade she survived past age 50 **(Figure 13.24)**. In contrast, Lorena Madrigal and Mauricio Meléndez-Obando (2008), analyzing data on Costa Rican women who lived during the 1500s through the 1900s, found that longer life span was associated with fewer grandchildren. It appears that survival past menopause is sometimes but not always advantageous. It is unclear whether investing in grandchildren is ever a better strategy than investing in additional children would be.

Michael Cant and Rufus Johnstone (2008) suggest an additional factor that might help explain the evolution of menopause. In a majority of traditional societies, women tend to live with their husband's family. This creates an opportunity for competition between women and their daughters-in-law over resources needed for reproduction. The daughters-in-law have more at stake in this conflict and are therefore more likely to win. This is because helping a mother-in-law reproduce offers no fitness benefit, whereas helping a daughter-in-law does.

The pattern of selection on menopause in modern industrial societies is likely different than it was for our ancestors. In analyzing data from the Framingham Heart Study, Sean Byars and colleagues (2010) found that age at menopause is heritable and that later menopause confers higher lifetime reproductive success. They predict later menopause in future generations.

13.3 How Many Offspring Should an Individual Produce in a Given Year?

Section 13.2 dealt, in part, with the allocation of energy to reproduction versus repair over an organism's entire life. In this section, we turn to the related issue of how much an organism should invest in any single episode of reproduction. Again we are concerned with trade-offs. First among them is this straightforward constraint: The more offspring a parent (or pair of parents) attempts to raise at once, the less time and energy the parent can devote to caring for each one.

Questions about the optimal number of offspring have been addressed most thoroughly by biologists studying clutch size in birds. It is easy to count the eggs in a nest and easy to manipulate clutch size by adding or removing eggs. Assuming that egg size is fixed, how many eggs should a bird lay in a single clutch?

Clutch Size in Birds

The simplest hypothesis for the evolution of clutch size, first articulated by David Lack (1947), is that selection will favor the clutch size that produces the most surviving offspring. Here we consider a simple mathematical formulation of this hypothesis (for a more detailed mathematical treatment, see Stearns 1992). The model assumes a fundamental trade-off in which the probability that any individual offspring will survive decreases with increasing clutch size. Many researchers have tested this assumed trade-off by adding eggs to nests; in most cases they have found that adding eggs indeed reduces the survival rate for individual chicks (see Stearns 1992). One explanation could be that the ability of the parents to feed any individual offspring declines as the number of offspring increases. In **Figure 13.25a**, we assume that the decline in offspring survival is a linear function of clutch size, but the model depends only on survival being a decreasing function. Given a function describing offspring survival, the number of surviving offspring from a clutch of a given size is just the product of the clutch size and the probability of survival (Figure 13.25b). The number of surviving offspring reaches a maximum at an intermediate clutch size. It is this most-productive clutch size that Lack's hypothesis predicts will evolve by natural selection.

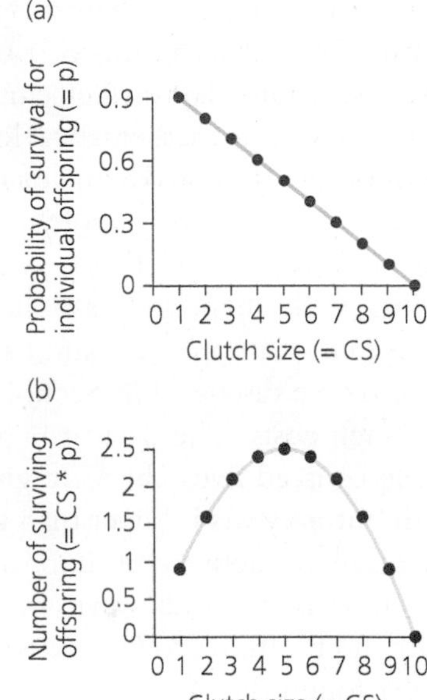

Figure 13.25 Lack's hypothesis on the evolution of clutch size (a) The probability that any given individual offspring will survive versus clutch size. (b) The number of surviving offspring per clutch, which is the number of eggs multiplied by the probability that any given individual offspring will survive.

Mark Boyce and C. M. Perrins (1987) tested Lack's hypothesis with data from a long-term study of great tits (*Parus major*) nesting in Wytham Wood, a research site near Oxford, England **(Figure 13.26)**. Combining data for 4,489 clutches monitored over the years 1960 through 1982, Boyce and Perrins plotted a histogram showing the distribution of clutch sizes in the Wytham Wood tit population. The mean clutch size was 8.53. Boyce and Perrins also determined the average number of surviving offspring from clutches of each size. This number was highest for clutches of 12 eggs. When researchers added three eggs to each

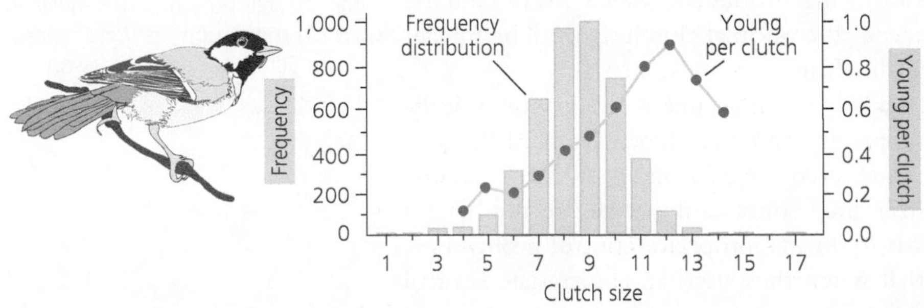

Figure 13.26 Lack's hypothesis tested with data on great tits (*Parus major*) The mean clutch size was 8.53. The number of surviving young per clutch was highest for clutches of 12 eggs. From Roff (1992); redrawn from Boyce and Perrins (1987).

of a large number of clutches, the most productive clutch size was still 12 (but see below). In other words, birds that produced smaller clutches apparently could have increased their reproductive success for the year by laying 12 eggs. Taken at face value, these data indicate that natural selection in Wytham Wood favors larger clutches than the birds in the population actually produce. Because the average clutch size was less than the most productive clutch size, the results are not consistent with Lack's hypothesis.

Lack's hypothesis predicts that parents will attempt to rear that number of young that maximizes the number of surviving offspring. Data indicate that parents often rear fewer offspring. Efforts to identify which of Lack's assumptions are violated have led to the discovery of additional trade-offs and improved estimates of lifetime fitness.

The literature on Lack's hypothesis is extensive, and many researchers have done studies similar to that of Boyce and Perrins (see reviews in Roff 1992 and Stearns 1992). The results of Boyce and Perrins are typical: Most studies have shown that birds lay smaller clutches than predicted. How can we explain this discrepancy? The mathematical logic of Lack's hypothesis is correct. The hypothesis must therefore make one or more implicit assumptions that often turn out to be wrong. Evolutionary biologists have identified and tested several assumptions implicit in Lack's hypothesis. We discuss three of them now.

First, Lack's hypothesis assumes there is no trade-off between a parent's reproductive effort in one year and its survival or reproductive performance in future years. As we discussed in Section 13.2, however, reproduction often entails exactly such costs. The data in Figure 13.16b, page 507, demonstrated that when female collared flycatchers are given an extra egg in their first year, their clutch size in future years is lower than that of control females. In a review of the literature on reproductive costs in birds, Mats Lindén and Anders Møller (1989) found that 26 of the 60 studies that looked for trade-offs between current reproductive effort and future reproductive performance found them. In addition, Lindén and Møller found that 4 of the 16 studies that looked for trade-offs between current reproductive effort and future survival found them. When reproduction is costly and selection favors withholding some reproductive effort for the future, the optimal clutch size may be less than the most productive clutch size.

Second, Lack's hypothesis assumes that the only effect of clutch size on offspring is in determining whether the offspring survive. Being part of a large clutch may, however, impose other costs on individual offspring than just reducing their probability of survival. Dolph Schluter and Lars Gustafsson (1993) added or removed eggs from the nests of collared flycatchers, put leg bands on the chicks that hatched from the nests, and then monitored the chicks' subsequent life histories. When the female chicks matured and built nests of their own, there was a strong relationship between the size of the clutches they produced and how much the clutch they were reared in had been manipulated **(Figure 13.27)**. Females reared in nests from which eggs had been removed produced larger clutches, whereas females reared in nests to which eggs had been added produced smaller clutches. This result indicates that clutch size affects not only offspring survival but also offspring reproductive performance. These data suggest that there is a trade-off between the quality and quantity of offspring produced. When larger clutches entail lower offspring reproductive success, the optimal clutch size will be smaller than the most numerically productive clutch size.

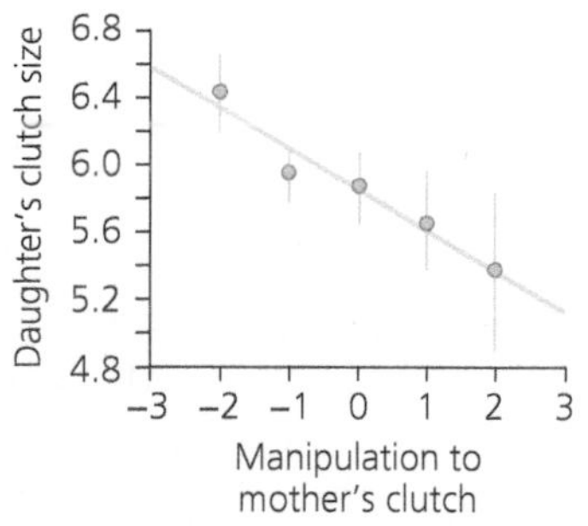

Figure 13.27 Does clutch size affect offspring reproductive performance? This graph shows the relationship between daughters' clutch sizes and the number of eggs added to or removed from the mothers' nests in which the daughters were reared. From Schluter and Gustafsson (1993).

Third, the discrepancy between Lack's hypothesis and the behavior of individual birds may sometimes be more apparent than real. When Richard Pettifor, Perrins, and R. H. McCleery (2001) reanalyzed the data on egg addition experiments used by Boyce and Perrins (1987), they concluded that Boyce and Perrins had compared their experimental birds to an inappropriate control group. Pettifor, Perrins, and McCleery found that when they used an appropriate control group, there was in fact no evidence that the birds that received extra eggs produced more surviving young than they would have had they been left alone. This result suggests that, in the observational data represented in Figure 13.26, the birds that laid fewer than 12 eggs did so because they had lower reproductive capacities—and that each bird was producing a clutch size that would optimize its own reproductive success.

Note that we have assumed that clutch size is fixed for any given genotype. In fact, clutch size is often phenotypically plastic (see Chapter 10). If clutch size is plastic, and if birds can predict whether they are going to have a good year or a bad year, we would predict that individuals will adjust their clutch size to the optimum value for each kind of year (for example, see Sanz and Moreno 1995).

Lack's Hypothesis Applied to Parasitoid Wasps

Although Lack's hypothesis often proves too simple to accurately predict clutch size, the examples we have reviewed demonstrate that is a useful null model. By explicitly specifying what we should expect to observe under minimal assumptions, Lack's hypothesis alerts us to interesting patterns we might not otherwise have noticed. This application of Lack's hypothesis is not limited to birds.

Lack's hypothesis is a useful null model for other organisms in addition to birds.

Eric Charnov and Samuel Skinner (1985) used Lack's hypothesis to explore the evolution of clutch size in parasitoid wasps. Parasitoid wasps use a stingerlike ovipositor to inject their eggs into the eggs or body cavity of a host insect. When the larval parasitoids hatch, they eat the host alive from the inside. The larvae then pupate inside the empty cuticle of the host and emerge as adults.

For a parasitoid, a host is like a nest. A female parasitoid can lay one or more eggs in a single host. The larvae compete among themselves for food, so there is a trade-off between clutch size and the survival of individual larvae. An added twist with insects is that adult size is plastic. In addition to reducing offspring survival, competition for food may result in larvae simply becoming smaller adults. The maternal fitness associated with a given clutch size must therefore be calculated as the product of the clutch size, the probability of survival of individual larvae, and the expected lifetime egg production by offspring of the size that will emerge.

Charnov and Skinner used this modified version of Lack's hypothesis to analyze the oviposition behavior of female parasitoid wasps in the species *Trichogramma embryophagum*. This wasp deposits its eggs in the eggs of a variety of host insects. Using data from the literature, Charnov and Skinner calculated maternal fitness as a function of clutch size for three different host species **(Figure 13.28a)**. Figure 13.28b plots the actual clutch sizes female wasps lay in each species of host egg against the most productive clutch sizes. The data indicate that female wasps shift their behavior in a manner appropriate to different hosts. Females lay fewer

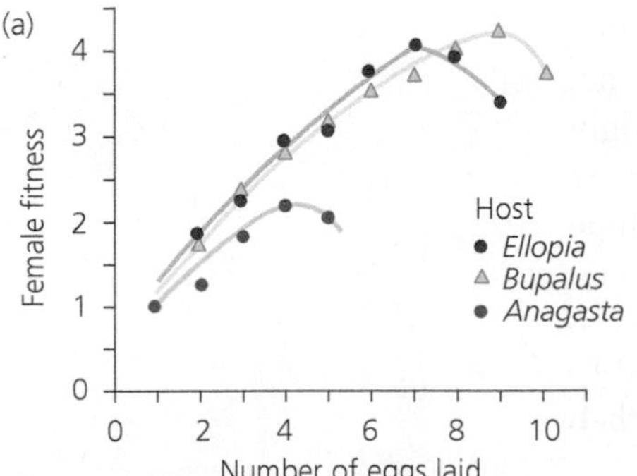

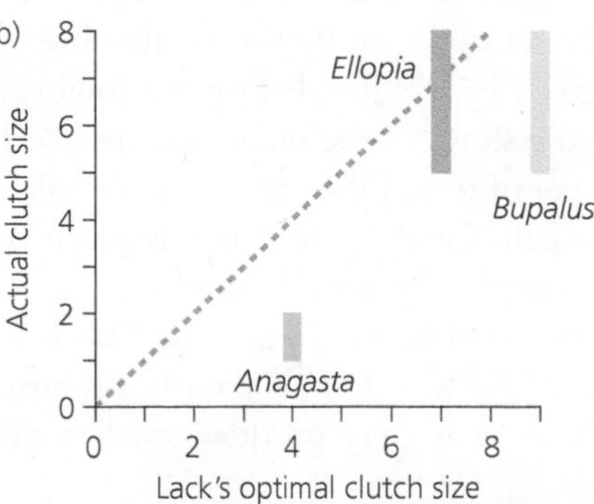

Figure 13.28 Lack's hypothesis applied to a parasitoid wasp (a) Fitness of female parasitoid wasps (*Trichogramma embryophagum*) versus clutch size in three different hosts. From Charnov and Skinner (1985). (b) Actual range of clutch sizes female parasitoids lay in each of the hosts. Dashed line marks actual = predicted. Drawn from data in Charnov and Skinner (1985).

eggs in the relatively poor hosts and more eggs in the relatively good hosts. As with many birds, however, female wasps tend to lay smaller clutches than those predicted by Lack's hypothesis.

Why do female wasps typically lay clutches smaller than the predicted sizes? Charnov and Skinner consider three reasons. Two of the three are similar to factors we discussed for birds. Larger clutch sizes may reduce offspring fitness in ways that Charnov and Skinner did not include in their calculations. And there may be trade-offs between a female's investment in a particular clutch and her own future survival or reproductive performance. Charnov and Skinner's third hypothesis is novel to parasitoid wasps.

Unlike birds, female parasitoid wasps may produce more than one clutch in rapid succession. Soon after she has laid one clutch, a female wasp may begin looking for another host to parasitize. The appropriate measure of a wasp's fitness with regard to clutch size may not be the discrete fitness she gains from a single clutch. Instead, it may be the rate at which her fitness rises as she searches for hosts and lays eggs in them. Readers familiar with behavioral ecology may recognize this as an optimal foraging problem.

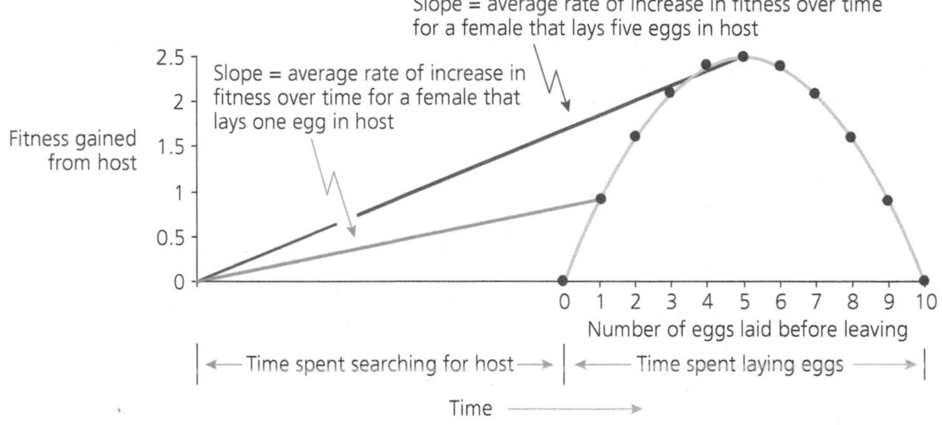

Figure 13.29 Rate of increase in parasitoid maternal fitness with time spent searching for hosts and laying eggs The horizontal axis represents the time spent by a female searching for a host egg and depositing a clutch. The vertical axis represents female fitness, in units of surviving offspring. The red dots show the relationship between number of surviving offspring and clutch size, as in Lack's hypothesis. After Charnov and Skinner (1985).

Figure 13.29 presents a graphical analysis of a female's rate of increase in fitness over time. The figure follows the female from the time she sets out to find a host egg until she leaves that host egg to look for another. While she is searching, the female gains no fitness. Once she finds a host and begins to lay eggs in it, however, her fitness begins to rise. The fitness she gets from a clutch of any given size is determined by a parabolic function, as in our original depiction of Lack's hypothesis (Figure 13.25). In this example, if a female leaves to look for a new host after laying just one egg, her total fitness gain from the first host is 0.9. Her average rate of fitness gain from the time she set out looking for the first host to the time she leaves to look for a second is given by 0.9 divided by the total elapsed time. This rate of fitness gain is equal to the slope of the diagonal line from the origin to the point representing a clutch size of one. Likewise, if the female stays to lay five eggs, her average rate of fitness gain for the whole trip is given by the slope of the upper diagonal line. In this example, the female would get the highest rate of fitness gain from this host if she left after laying four eggs. This is one egg less than the most productive clutch size. Thus, if female parasitoids are selected to maximize their rate of fitness increase, they may produce smaller clutches than those predicted by Lack's hypothesis.

To summarize, Lack's hypothesis is a useful starting point for the evolutionary analysis of clutch size. Assuming only that there is a trade-off between the number of offspring in a clutch and the survival of individual offspring, Lack's hypothesis predicts that parents will produce clutches of the size that maximizes the number of surviving offspring. This prediction is often violated, and actual clutches typically are smaller than expected. These violations indicate the possible presence of other trade-offs. Current parental reproductive effort may be negatively correlated with future parental survival or reproductive performance, or clutch size may be negatively correlated with offspring reproductive performance. Alternatively, a violation of predicted clutch size may indicate that we have chosen the wrong measure of parental fitness.

13.4 How Big Should Each Offspring Be?

In Section 13.3, we assumed that the size of individual offspring was fixed. We now relax that assumption. Given that an organism will invest a particular amount of energy in an episode of reproduction, we can ask whether that energy should be invested in many small offspring or a few large offspring.

A trade-off between the size and number of offspring should be fundamental. A pie can be sliced into many small pieces or a few large pieces, but it cannot be sliced into many large pieces. Biologists have found empirical evidence for a size-versus-number trade-off in a variety of taxa.

Mark Elgar (1990), for example, analyzed data from the literature on 26 families of fish **(Figure 13.30a)**. Larger fish produce bigger clutches, so Elgar had to use statistical techniques to remove the effect of variation among fish families in body size. The vertical axis shows relative clutch size, or the number of eggs per clutch adjusted for differences in body size among families. The horizontal axis shows relative egg volume, or egg size adjusted for differences in body size among families. Elgar found a clear negative correlation between clutch size and egg size. Fish that produce larger eggs produce fewer eggs per clutch.

David Berrigan (1991) performed a similar analysis of variation in egg size and number among species in fruit flies (Figure 10.30b). Larger fruit flies produce more and larger eggs, so Berrigan also had to use statistical techniques to remove the effect of variation in body size. The vertical axis shows relative egg number; the horizontal axis shows relative egg volume. Berrigan found a clear negative correlation between egg number and egg size. He found similar patterns in wasps and beetles.

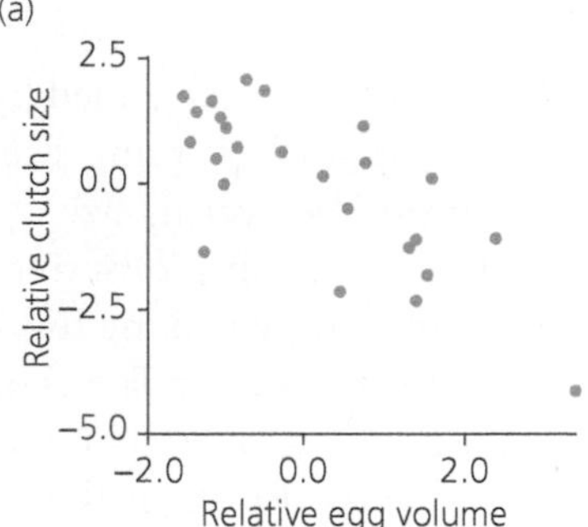

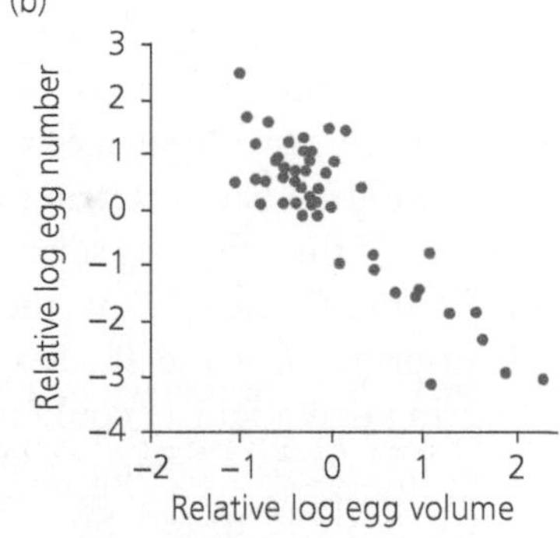

Figure 13.30 Trade-offs across taxa in size and number of offspring (a) A trade-off across 26 fish families. The negative correlation between size and number of eggs is statistically significant ($p < 0.001$). From Elgar (1990). (b) A trade-off across fruit fly species. The negative correlation is significant ($p < 0.001$). Provided by David Berrigan using data analyzed in Berrigan (1991).

Selection on Offspring Size

If selection on parents is forced by a fundamental constraint to strike a balance between the size and number of offspring, what is the optimal compromise? Christopher Smith and Stephen Fretwell (1974) offered a mathematical analysis of this question. Smith and Fretwell's analysis is based on two assumptions.

The first assumption is the trade-off between size and number of offspring. A graph depicting this assumption appears in **Figure 13.31a** (next page). The units we have used are arbitrary. The shape of the curve may vary from species to species. Here we have used the equation

$$\text{Number} = \frac{10}{\text{Size}}$$

The second assumption is that individual offspring will have a better chance of surviving if they are larger. There must be a minimum size below which offspring have no chance of survival. As offspring get larger, their probability of surviving rises. If survival probability approaches one, it must do so in a saturating fashion, because survival probability cannot exceed one. A graph depicting this assumption appears in Figure 13.31b. Again, we have used arbitrary units. The shape of the curve may vary from species to species. Here we have used the equation

$$\text{Survival} = 1 - \frac{1}{\text{size}}$$

Given the two assumptions, the analysis is simple: The expected fitness of a parent producing offspring of a particular size is the number of such offspring the parent can make multiplied by the probability that any individual offspring will survive. A plot of expected parental fitness versus offspring size, which appears in Figure 13.31c, reveals the size of offspring that gives the highest parental fitness. For example, given the equations and units used here, if a parent makes offspring of size 5, it can make two of them. Each has a probability of survival of 0.8. Thus, the expected fitness gained by the parent from this clutch is $2 \times 0.8 = 1.6$.

The optimal offspring size depends on the shapes of the relationships for offspring number versus size and offspring survival versus size. Often (though not always), the optimal offspring size is intermediate. In our example, parental fitness is maximized by making five offspring, each of size 2. The key point is that selection on parents often favors offspring smaller than the size favored by selection on offspring themselves. This identification of a potential conflict of interest between parents and offspring is the primary contribution of Smith and Fretwell's model.

The shape of the offspring survival curve is particularly important (Figure 13.31b). In Smith and Fretwell's model, survival probability increases with offspring size, but the rate of increase declines: That is, increasingly large offspring gain a progressively smaller survival benefit. This leads directly to the prediction of an intermediate offspring size that gives the highest parental fitness (Figure 13.31c). If the offspring survival curve were a linear relationship instead of a concave curve (Vance 1973), the model would predict selection favoring extremes of offspring size: the smallest offspring capable of development, or the largest offspring that a female could manufacture, rather than some optimal intermediate offspring size (see Levitan 1993, 1996; Podolsky and Strathmann 1996).

It is possible to test Smith and Fretwell's analysis empirically only if there is substantial variation in offspring size among parents within a population. Variation in offspring size is relatively small in most species (Stearns 1992). We review two studies that have confirmed both the assumptions and the conclusion of Smith and Fretwell's analysis. In one study, researchers took advantage of the large variation in egg size in a population of fish. In the other study, researchers took advantage of phenotypic plasticity in egg size in a beetle.

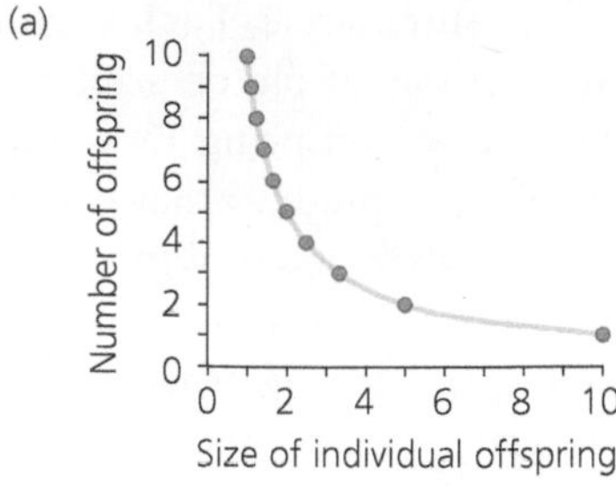

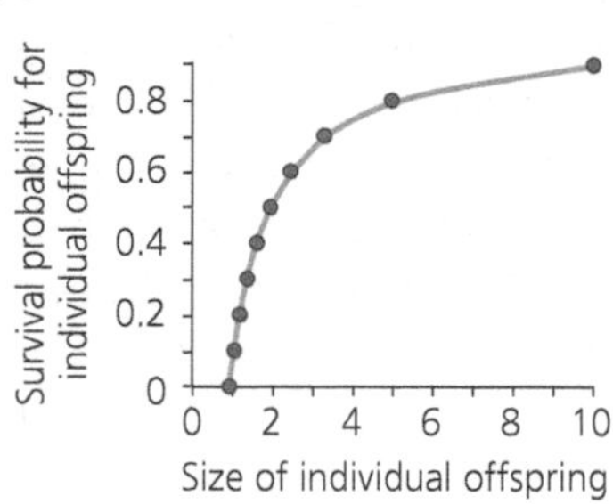

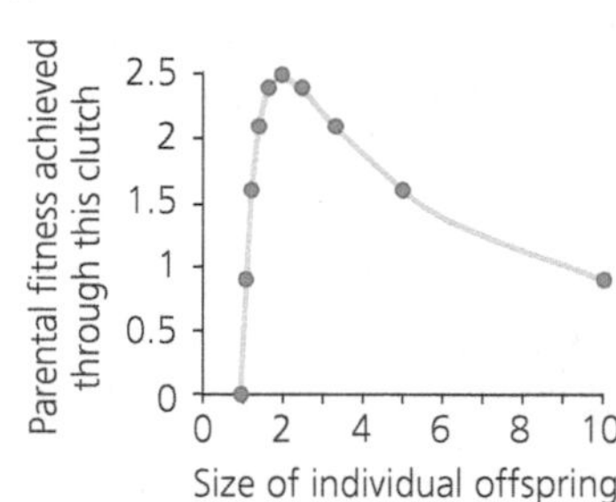

Figure 13.31 The optimal compromise between size and number of offspring (a) Assumption 1: There is a trade-off between size and number of offspring. (b) Assumption 2: Above a minimum size, the probability that any individual offspring will survive is an increasing function of its size. (c) Analysis: The parental fitness gained from a single clutch of offspring of a given size is the number of offspring in the clutch multiplied by the probability that any individual offspring will survive. After Smith and Fretwell (1974).

Selection on Offspring Size in a Population of Fish

Daniel Heath and colleagues (2003) studied chinook salmon (*Oncorhynchus tshawytscha*) at a commercial hatchery in British Columbia, Canada. When adult salmon return to the hatchery, workers harvest eggs from the females and fertilize them with sperm from the males. After the fry have hatched and grown for a time, the hatchery workers release them into natural rivers. When the fry mature, they return to the hatchery to continue the cycle. Eggs produced by female salmon at the hatchery range in mass from less than 0.15 g to more than 0.30 g.

Heath and colleagues tested the first assumption of Smith and Fretwell at the level of individual females. The plot in **Figure 13.32a** shows the relationship between female relative fecundity—that is, the number of eggs a female lays per kg of her body mass—and the average size of a female's eggs. For the year shown, and for the three other years the researchers analyzed, there was a trade-off between size and number of eggs.

Organisms face a trade-off between making many low-quality offspring or a few high-quality offspring. Selection on parents favors a compromise between the quality and quantity of offspring, but selection on individual offspring favors high quality.

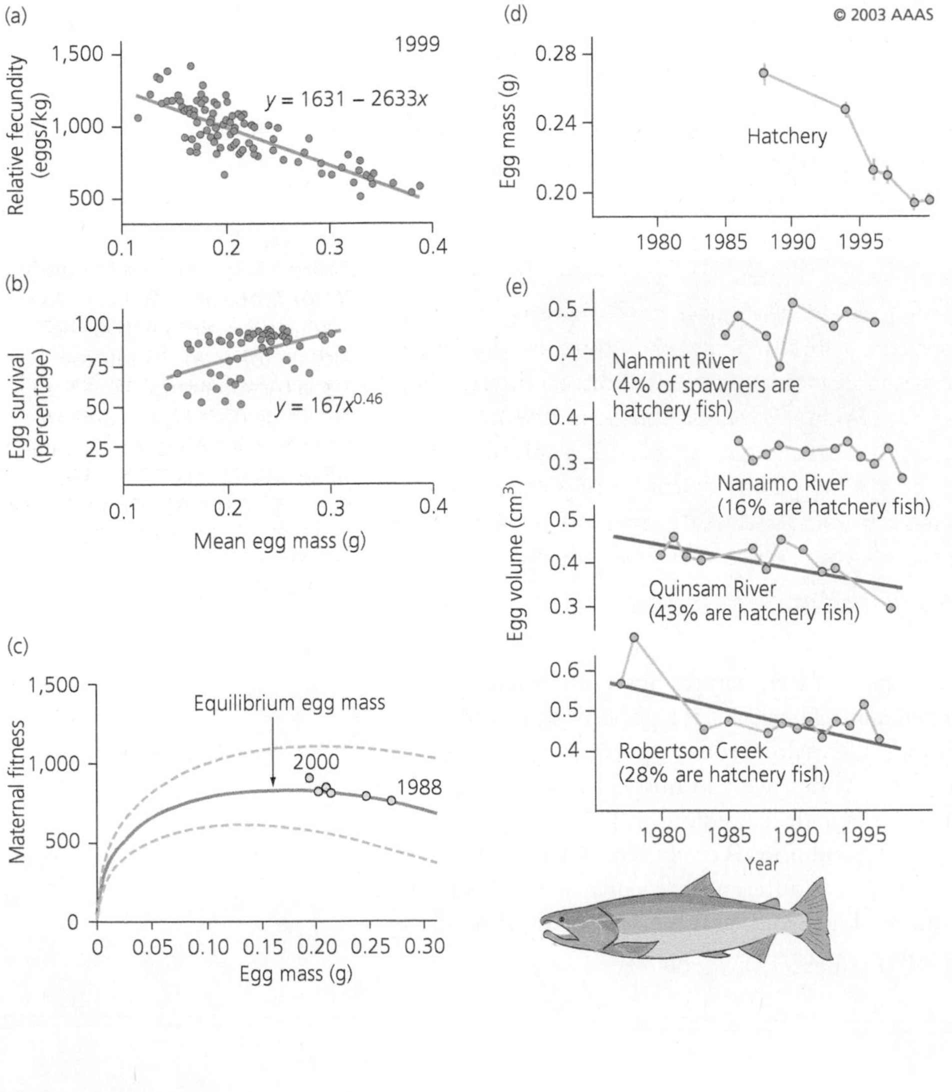

Heath and colleagues tested the second assumption of Smith and Fretwell by tracking the fates of individual fry. The plot in Figure 13.32b shows the probability of survival as a function of the mass of the egg a fry hatches from. For the year shown, and for the three other years the researchers analyzed, larger offspring survived at higher rates. Furthermore, the relationship between survival and size followed a concave curve, just as Smith and Fretwell assumed.

How do the trade-off between egg size versus number and the selection on fry combine to select on the egg size produced by the mothers? Heath and colleagues estimated the relationship between maternal fitness and egg size by multiplying the fitted curve in Figure 13.32a and the fitted curve in Figure 13.32b. The result appears in Figure 13.32c. Female fitness is maximal at intermediate egg sizes. The optimal egg size for females breeding at the hatchery was just over 0.15 g.

The optimal egg size for hatchery females turns out to be lower than the optimal egg size for females in the wild. This is because the hatchery provides a safe environment for young fry. Small fry, in particular, are more likely to survive in

© 2003 AAAS

Figure 13.32 Evolution of reduced egg size in hatchery salmon (a) Controlling for mass, female salmon face a trade-off between number vs. size of eggs. (b) Offspring survival increases with egg size. (Although the curve appears to be linear, it is the power function given by the equation.) (c) Maternal fitness, the product of relative fecundity and offspring survival, is maximized at intermediate egg size. The optimal egg size for hatchery salmon, however, is smaller than that for wild salmon. The points show mean values for salmon in a recently established hatchery population for each of several years. (d) The average egg size in the hatchery population evolved toward smaller sizes over a period of just a few years. (e) Egg size also declined in wild populations receiving substantial numbers of immigrants from the hatchery population (Quinsam River and Robertson Creek) but remained stable in wild populations receiving relatively few immigrants from the hatchery (Nahmint River and Nanaimo River). From Heath et al. (2003).

From D. D. Heath, et al. 2003. "Rapid evaluation of egg size in captive salmon." *Science* 299: 1738–1740. Reprinted with permission from AAAS.

the hatchery than they are in natural rivers. When small fry are more likely to survive, females that make more and smaller fry have higher reproductive success. This finding enabled Heath and colleagues to predict that the hatchery population, which was founded in the late 1980s from wild stock, should be evolving toward smaller average egg sizes. Data from the hatchery confirm this prediction, as can be seen from the data points in Figure 13.32c and the time series in Figure 13.32d. In other words, the commercial hatchery has been running an unintentional experiment that confirms the predictions of Smith and Fretwell's analysis.

The evolution of hatchery populations toward smaller egg sizes has implications for the conservation of wild salmon stocks. A widespread conservation strategy for salmon is to supplement wild populations with fish from hatcheries. This amounts to migration from hatchery populations into wild populations. Heath and colleagues analyzed data from four rivers on Vancouver Island in which chinook stocks are being supplemented with hatchery fish. The amount of supplementation varies among the four rivers: 4% of the fish spawning in the Nahmint River are migrants from the hatchery, as are 16% of the fish in the Nanaimo River, 28% of the fish in Roberston Creek, and 43% of the fish in Quinsam River. As the time series in Figure 13.32e show, the chinook populations in Quinsam River and Robertson Creek have been evolving toward smaller egg sizes since at least 1980. Gene flow from the hatchery appears to be driving the evolution of suboptimal egg size in heavily supplemented wild chinook populations.

Phenotypic Plasticity in Egg Size in a Beetle

Charles Fox and colleagues (1997) studied the seed beetle *Stator limbatus*. The females of this small beetle lay their eggs directly onto the surface of host seeds **(Figure 13.33)**. The larvae hatch and burrow into the seed. Inside, the larvae feed, grow, and pupate. They emerge from the seed as adults. *S. limbatus* is a generalist seed predator; it has been reared on the seeds of over 50 different host species.

Figure 13.33 **The seed beetle, *Stator limbatus*** This female is looking for a place to lay her eggs on the seeds of catclaw acacia (*Acacia greggii*) and blue palo verde (*Cercidium floridum*). Photo by Timothy A. Mousseau, University of South Carolina.

Fox and colleagues studied *S. limbatus* on two natural hosts: an acacia (*Acacia greggii*) and a palo verde (*Cercidium floridum*). The acacia is a good host; most larvae living in its seeds survive to adulthood. The palo verde is a poor host; fewer than half the larvae living in its seeds survive. When we add hosts of different quality to the Smith and Fretwell analysis, we get a clear prediction: Females should lay larger eggs on the poor host than on the good host. Recall (from Chapter 10) that when selection favors different phenotypes at different times or places, organisms sometimes evolve phenotypic plasticity. The Smith and Fretwell analysis predicts that *S. limbatus* should exhibit phenotypic plasticity in egg size.

Figure 13.34 explains why. As before, we assume a trade-off between size and number of offspring (Figure 13.34a). As before, we assume that there is a minimum size below which individual offspring do not survive. Above this size, the probability that any individual offspring will survive is an increasing function. Unlike before, we have two curves depicting this relationship: one for the good host, another for the bad host (Figure 13.34b). The minimum size for offspring survival is smaller on the good host. And survival is higher on the good host at all sizes above the minimum. When we calculate the parental fitness gained for a clutch of offspring of a given size—by multiplying the number of offspring in the clutch by the probability that they will survive—we get two curves as well (Figure 13.34c). The analysis shows that the optimal offspring size for the mother is bigger on the poor host than on the good host.

Fox and colleagues found that, as predicted, female *S. limbatus* adjust the size of the eggs they lay to the host on which they deposit them. When the researchers took newly emerged females from the same population and gave them only one kind of seed, females given palo verde seeds (the poor host) laid significantly larger eggs than females given acacia seeds, the good host **(Figure 13.35a)**. Confirming assumption 1 of Smith and Fretwell, these larger eggs came at the cost of fewer eggs produced over a lifetime (Figure 13.35b).

For females laying on poor seeds, the production of large eggs is adaptive. Fox et al. manipulated females into laying small eggs on poor seeds by keeping the females on good seeds until they laid their first egg, then moving them to poor seeds. Only 0.3% of the larvae hatching from small eggs on poor seeds survived to adulthood, whereas 24% of the larvae hatching from large eggs on poor seeds survived ($p < 0.0001$). Confirming assumption 2 of Smith and Fretwell, even among the large eggs on poor seeds, the probability of survival from egg to adult was positively correlated with egg size. For females laying on good seeds, the production of small eggs is adaptive. Given that nearly all larvae hatching on good seeds survive, females producing more and smaller eggs have higher lifetime reproductive success.

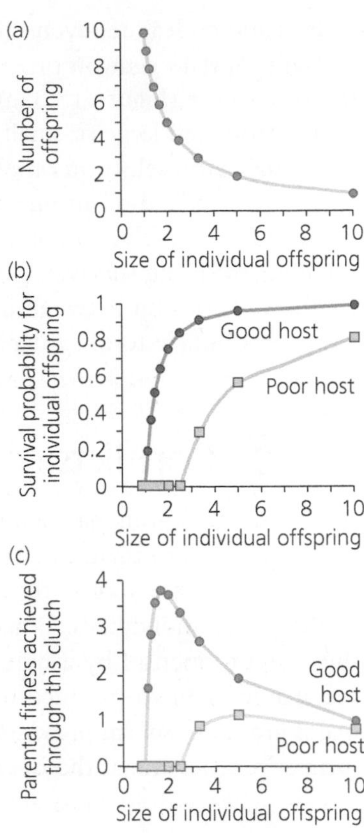

Figure 13.34 Offspring size and host quality (a) Assumed number vs. size trade-off. (b) Assumed pattern of survival versus size. (c) Parental fitness per clutch versus offspring size. After Smith and Fretwell (1974).

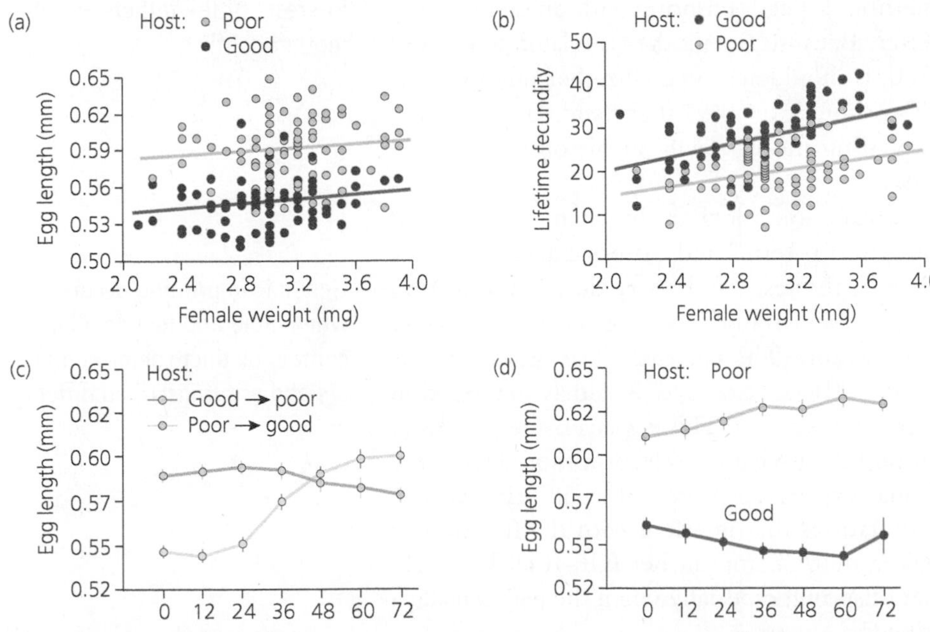

Figure 13.35 Phenotypic plasticity in egg size in *Stator limbatus* (a) Females laying on *Cercidium floridum* lay larger eggs for their size ($p < 0.001$). (b) Females laying on *A. greggii* have higher fecundity (number of eggs) for their size ($p < 0.001$). (c) Egg size versus time for the 72 hours after females were allowed to lay their first egg, then moved to a new host. (d) Control experiment for (c). Females stayed on one kind of seed for life. From Fox et al. (1997).

Fox and colleagues even showed that individual females that had started to lay size-appropriate eggs on one host could readjust their egg size when switched to the other host (Figure 13.35c). Control females left on one kind of seed consistently produced large or small eggs for life (Figure 13.35d).

In summary, selection on offspring size often involves a conflict of interest between parents and offspring. Because making larger offspring also means making fewer offspring, selection on parents can favor smaller offspring sizes than are optimal for offspring survival. The exact balance between size and number depends on the relationship between offspring size and survival. Poor environments pose a greater obstacle to offspring survival and thus favor larger offspring.

13.5 Conflicts of Interest between Life Histories

Analyzing trade-offs has helped explain much of the extraordinary life-history diversity among organisms. However, this view can sometimes obscure the fact that the life history of each organism unfolds in an ecological context that includes other individuals. For example, the opossum life history in Figure 13.2 shows reproduction by a hypothetical female but does not show the males that she mated with to produce offspring. This simplification might imply that the males are mere sperm providers and that their interests in the production of offspring are the same as the interests of the female. In fact, the reproductive interests of males and females will often be different. In this section, we discuss two such conflicts of interest and their evolutionary consequences.

Genetic Conflict between Mates: Genomic Imprinting

Opossums and other mammals that nourish their offspring through a placenta offer a surprising opportunity for conflict between the reproductive interests of females (which brood the offspring) and males (which do not). Consider the copies of a mammalian gene inherited from the father versus the mother. Why might these alleles be in conflict? In most mammals, females carry offspring from many different males in the course of a lifetime. Indeed, offspring with different fathers are frequently found in the same litter. Because the mother is related to each of these offspring equally, natural selection should act to equalize her physiological investment in each. Natural selection should, on the other hand, favor a father that can coerce the mother into investing more heavily in his offspring at the expense of offspring from other males.

When different males father offspring within the same litter or clutch, the reproductive interests of the fathers and the mother conflict.

Consistent with this prediction, at least some loci are biochemically marked (or imprinted) in mammals, to distinguish paternal and maternal alleles (Barlow 1995). This marking of alleles occurs in the testis and ovary during production of gametes. Imprinting affects the subsequent transcription of the marked genes within cells of the embryo after fertilization. The paternal allele of a hormone called insulin-like growth factor II (IGF-II), for example, is widely expressed in mice, while the maternal copy is hardly transcribed. This is a surprising pattern of gene expression for a diploid organism, because natural selection should favor the equal expression of both alleles. Equal expression protects the offspring against the effects of deleterious recessive mutations that interfere with the function of one allele (Hurst 1999). Why should a mother imprint her IGF-II alleles to reduce transcription of this gene in her offspring, especially when the paternal allele of the same gene is actively transcribed?

Genomic imprinting occurs when male and female alleles contain distinct chemical markers and are transcribed differently.

The answer hinges on the function of IGF-II and its interaction with other molecules. This hormone is a general stimulant to cell division and acts through a cell-surface protein called the type-1 IGF-II receptor. However, it happens that another abundant cell-surface protein in mice, called the cation-independent mannose-6-phosphate receptor (CI-MPR), also has a binding site for IGF-II (this alternative binding site is called the type-2 receptor). CI-MPR's function is completely unrelated to growth, and in mouse embryos it is transcribed only from the maternal allele.

David Haig and colleagues have proposed that this bizarre arrangement of hormones, receptors, and transcription patterns results from a tug-of-war between the interests of maternal and paternal alleles within the uterus. According to this interpretation, the paternally transcribed IGF-II is selected to maximize rates of cell division in the developing embryo. This increases growth rates and monopolizes the flow of maternal resources to the embryo through the placenta. The maternal IGF-II allele is turned off to conserve resources for future reproduction. In contrast, the maternally transcribed type-2 receptor is selected to bind excess paternal hormone, mitigate the effects of IGF-II overtranscription, and equalize the flow of resources to different embryos, while the paternal type-2 receptor allele is turned off to maximize the influence of the paternal IGF-II hormone on the mother (Haig and Graham 1991; Moore and Haig 1991).

Consistent with this interpretation, CI-MPR does not bind IGF-II in chickens and frogs; their embryos are provisioned before fertilization. Chicken and frog fathers have no opportunity to manipulate the distribution of maternal resources among offspring. This is a hint that the type-2 receptor of mammals evolved after the advent of the placenta, in response to selection that favored equalization of maternal resources among all offspring. Genomic imprinting has also been confirmed in flowering plants and may have been important in the evolution of the nutritive tissue called endosperm (see Haig and Westoby 1989, 1991).

The qualitative predictions of Haig's hypothesis for genomic imprinting have generally been confirmed, though there is some debate over whether quantitative variation in imprinting occurs and whether multiple paternity is necessary for imprinting to arise (Haig 1999; Hurst 1999; Spencer et al. 1999). For example, alleles could vary in the amount of transcription rather than being turned "on" or "off." Imprinting is known to be widespread in mammalian genomes (see references in Spencer et al. 1999), and the details of the imprinting mechanism and interactions among imprinted alleles could vary among genes and species.

Finally, it is important to note that mammals are not the only animals that have evolved placental development. For example, lizards (Guillette and Jones 1985), sharks (Wourms 1993), and numerous marine invertebrate groups (Strathmann 1987) have evolved structures like a placenta that transfer materials between the maternal body and the internally brooded offspring. Haig's hypothesis predicts that imprinted genes should be found in these groups, and that these will be genes that moderate the conflict among offspring within a brood as they compete for maternal resources. The hypothesis has not yet been tested in these groups, however (Spencer et al. 1999).

Physiological Conflict between Mates: Sexual Coevolution

Earlier we introduced the idea of adaptations arising in competing species, such as hosts and pathogens, that counteract each other's effects so that neither lineage shows a net gain in fitness (Chapter 8). In these circumstances, fitness evolves around a kind of dynamic equilibrium in which the environment, and thus the nature of selection acting on a population of organisms, is largely determined by interactions with other organisms and their adaptations (Van Valen 1973).

This idea can be extended to life-history adaptations arising within species as well. Experiments by William Rice and colleagues show that, where the reproductive interests of male and female fruit flies differ, sexual selection may favor adaptations that arise in one sex but are actually detrimental to the other sex. One of these adaptations involves the biochemistry of male seminal fluid, which has evolved to influence female behavior, such as egg-laying rate or the tendency to remate with another male (Fowler and Partridge 1989). These effects are beneficial to a male if his mate is likely to have multiple partners, because these adaptations will tend to increase the number of eggs that are fertilized by his sperm. Such seminal fluid is toxic and increases mortality of females, however (Fowler and Partridge 1989). Toxic effects favor the subsequent evolution of resistance among females, followed by more extreme adaptations among males to overcome female resistance. This iterated process has been called chase-away sexual selection (Rice 1987; Rice and Holland 1997; Holland and Rice 1998).

When mates are not monogamous, the life-history strategy that is optimal for one sex may be suboptimal for the other.

Direct evidence for this kind of antagonistic sexual adaptation comes from experiments conducted by Rice (1996). In these experiments, male flies competed with each other for matings with females. Females, in turn, were able to mate with multiple partners. The competition among males resulted in selection for traits such as high rate of remating with the same female and highly toxic seminal fluid. However, only male offspring were retained from each generation of experimental mating. After each round of selection, the selected males were mated to females from a control group in which competition for mates was not occurring. In this way, Rice kept the female response to male sexual adaptations static while the selected males competed with each other to overcome female defenses.

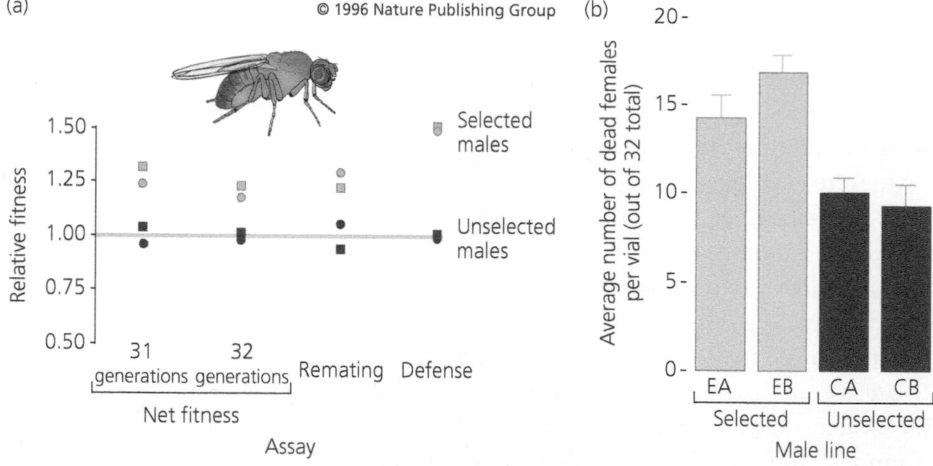

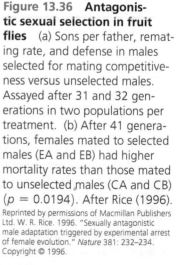

Figure 13.36 Antagonistic sexual selection in fruit flies (a) Sons per father, remating rate, and defense in males selected for mating competitiveness versus unselected males. Assayed after 31 and 32 generations in two populations per treatment. (b) After 41 generations, females mated to selected males (EA and EB) had higher mortality rates than those mated to unselected males (CA and CB) ($p = 0.0194$). After Rice (1996).

Reprinted by permissions of Macmillan Publishers Ltd. W. R. Rice. 1996. "Sexually antagonistic male adaptation triggered by experimental arrest of female evolution." *Nature* 381: 232–234. Copyright © 1996.

The results of 31–41 generations of such selection are shown in **Figure 13.36**. Compared to males in the control group, selected males had higher fitness (more sons born per male, shown as the net fitness assays in Figure 13.36a). The data on the right-hand side of Figure 13.36a suggest that two traits contributed to the higher fitness of selected males. They were more likely to remate with the same female, and they fertilized a much higher proportion of eggs when the female was remated to another male (defense assays, Fig. 13.36a). These benefits to male reproductive success came at a cost to females, however: After 41 generations of selection, the mortality rate for females mated to selected males was about 50% higher than the mortality rate for females mated to unselected males (Figure 13.36b). The experiment suggests that males and females are engaged in a reproductive arms race, and that males win if female countermeasures are prevented.

It is important to recognize, however, that this result is based on a particular mating system. By enforcing monogamous mating upon flies from the same source population for many generations, Holland and Rice (1999) showed that the effects of antagonistic sexual adaptation could actually be reversed: Monogamous lines of flies evolved lower sperm toxicity in males and lower resistance to sperm toxicity in females. These results make sense in light of the new relationship between male and female fitness: Males with only one lifetime mate depend on the fitness of that female alone to produce offspring, and they have no fear of being cuckolded by another male. These males should evolve less harmful life-history traits in order to increase their own fitness. The defenses evolved by females against toxic male seminal fluid (or other male life-history traits) become less beneficial as males evolve more benign traits of their own. If female resistance is expensive in time or energy, then resistance traits should be selected against (Holland and Rice 1999).

13.6 Life Histories in a Broader Evolutionary Context

In this final section of the chapter, we place life histories in a broader evolutionary context. We briefly consider examples of research addressing the maintenance of genetic variation in life-history traits and the importance of life-history variation in microevolutionary and macroevolutionary processes.

The Maintenance of Genetic Variation

Natural selection on a trait should reduce the genetic variation for the trait (Fisher 1930). A simple example illustrates why (Roff 1992). Imagine a series of loci, each with two alleles, that collectively affect a single trait correlated with fitness. At each locus, one allele ("0") contributes to the trait in such a way as to add zero units to the fitness of individuals, whereas the other allele ("1") contributes one unit to individual fitness. The genotype with the highest fitness is homozygous for allele "1" at all loci. Over time, selection should lead to the fixation of the "1" allele at each locus, and there will no longer be genetic variation in the trait.

Life-history traits, because of their intimate connection with reproduction, should be more closely correlated with fitness than other kinds of traits, including behavioral, physiological, or morphological traits (Mousseau and Roff 1987). Consequently, life-history traits should show less genetic variation—lower heritability—than other kinds of traits (for a discussion of heritability, see Chapter 9).

Life-history traits are closely correlated with fitness and have relatively low heritabilities.

To test this hypothesis, Mousseau and Roff (1987) assembled from the literature a sample of 1,120 estimates of the heritability of various traits. They summarized their data set by plotting cumulative frequency distributions for four kinds of traits. A cumulative frequency distribution is a running sum, moving across a histogram, of the heights of the bars. The more rapidly the curve in a cumulative frequency distribution rises to 1, the lower the mean of the histogram. Mousseau and colleagues compared estimated heritabilities of life-history traits, behavioral

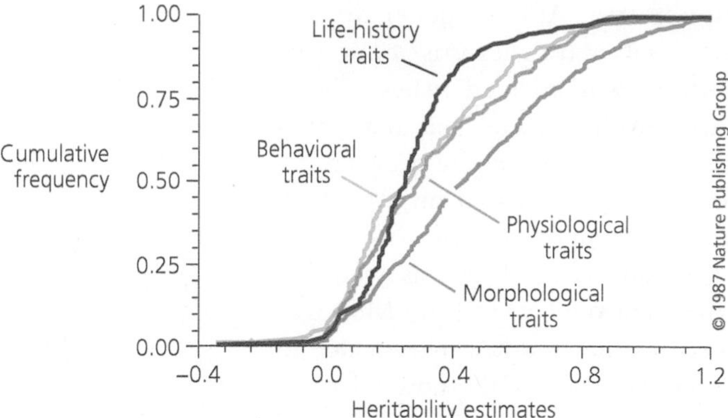

Figure 13.37 Heritabilities of four kinds of traits Life-history traits tend to have the lowest heritabilities. From Mousseau and Roff (1987).

Reprinted by permissions of Macmillan Publishers Ltd. From Mousseau, T. A., and D. A. Roff. 1987. "Natural selection and the heritability of fitness components." *Heredity* 1987: 181–197. Copyright © 1987.

traits, physiological traits, and morphological traits. They found that life-history traits indeed tend to have the lowest heritabilities **(Figure 13.37)**. This result is consistent with the expectation from our simple theoretical treatment (for an alternative interpretation, see Price and Schluter 1991).

Nonetheless, Mousseau and Roff's review documents that life-history traits typically have substantial genetic variation. What evolutionary forces maintain genetic variation in populations? The list of possibilities includes mutation, heterozygote advantage, frequency-dependent selection, and genotype-by-environment interaction in which different genotypes have higher fitness in different environments or at different times (see Chapters 6–10).

Figure 13.38 A colony of the sea squirt, *Botryllus schlosseri* Photo by Richard K. Grosberg, University of California-Davis.

Richard Grosberg (1988) studied the maintenance of genetic variation for life-history traits in a population of the sea squirt *Botryllus schlosseri* **(Figure 13.38)**. *B. schlosseri* is a colonial animal that lives attached to hard surfaces in shallow marine waters of the temperate zone. Colonies consist of a number of identical modules. The modules in a colony are physiologically connected, and their life histories are synchronous.

The population Grosberg studied contains two distinct life-history morphs. One morph is **semelparous:** Upon reaching sexual maturity, the modules in a colony reproduce once and die. The other morph is **iteroparous:** Colonies have at least three episodes of sexual reproduction before they die. In a series of experiments in which he grew sea squirts in a common environment and bred the morphs with each other, Grosberg demonstrated that the two morphs are genetically determined.

What maintains genetic variation for life-history morphs in this sea squirt population? Grosberg tracked the seasonal frequency of the two morphs over two years **(Figure 13.39)**. In both years, the semelparous morph dominated the population in the spring and early summer, whereas the iteroparous morph dominated in the late summer. This result indicates that the two morphs are maintained in the population by seasonal variation in selection. One important selective factor may be competitive interactions with another sea squirt (*Botryllus leachi*). This competitor, which becomes more abundant late in the summer, overgrows colonies of the semelparous *B. schlosseri* morph but not the iteroparous morph—a genotype-by-environment interaction.

Genotype-by-environment interaction for a variety of traits, including life-history traits, is a key factor in the phenomenon we consider next: biological invasions.

Biological Invasions

How does a species that is benign on one continent transmogrify into a pest when transported to another continent? Research by Lorne Wolfe and colleagues on the snowy campion, *Silene latifolia*, provides an example. Snowy campion is a small perennial herb native to Europe, where it is innocuous. Accidentally introduced into North America some 200 years ago, it has there become an agricultural weed. The traditional explanation is that when the plant moved to the New World, it left all of its natural enemies behind. These include, among many others, a fungus that attacks the plant's anthers and a seed-eating caterpillar. Freed from the burdens imposed by these mortal foes, snowy campion thrived in America.

Amy Blair and Wolfe (2004) suspected there was more to the story—that snowy campion had not just escaped its enemies but had also evolved. They tested their hypothesis by planting seeds from European and North American snowy campions together in a common garden in the United States. Since the plants would all experience the same environment, any differences in phenotype must be due to genotype. Consistent with the researchers' prediction, the European and American plants were not the same. The American plants germinated earlier, grew faster, made more flowers, and survived at higher rates than the European plants.

How could a plant that had been evolving in Europe for millions of years suddenly become so superior in North America? Blair and Wolfe had evidence to suggest that the answer involved a life-history tradeoff. With few enemies, a change in energy budget was

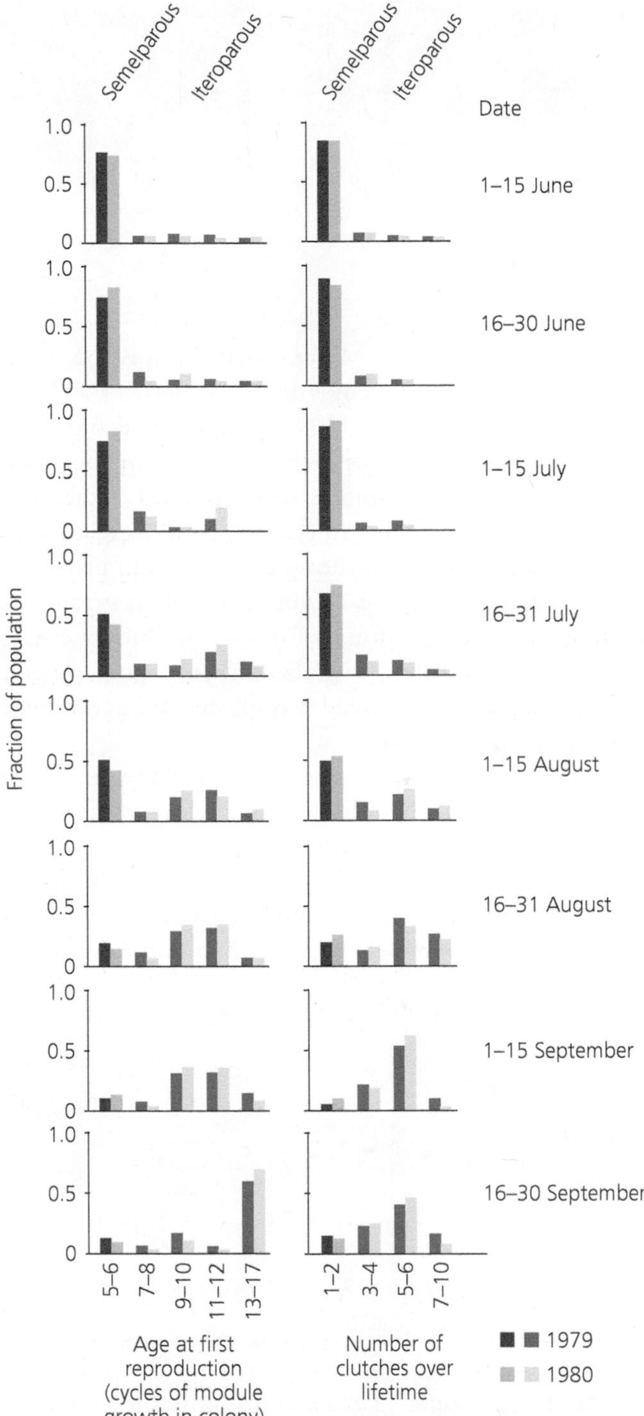

Figure 13.39 Annual cycles in the frequencies of two life-history morphs in a population of sea squirts Each of the bar graphs shows a frequency distribution for the population during a two-week period. Colonies of the semelparous morph (brown and tan bars) reproduce at an early age (age cycles of module growth in the colony) and produce only a single clutch of offspring. Colonies of the iteroparous morph (blue and green bars) reproduce at a late age and produce at least three clutches of offspring. From Grosberg (1988).

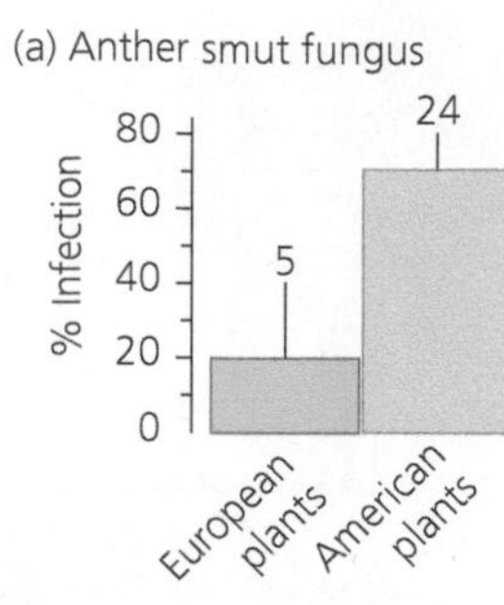

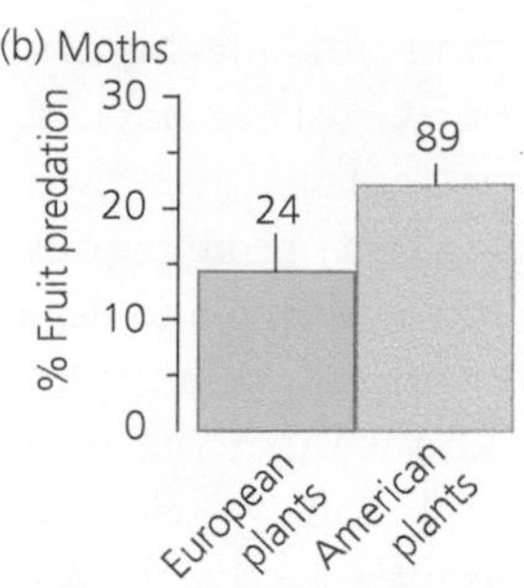

Figure 13.40 Life-history evolution in snowy campion Grown in Europe, American campions were more vulnerable to (a) diseases ($p = 0.008$) and (b) predators ($p = 0.042$) than European campions. Wolfe et al. divided plants by size for analysis; we have included only the largest. From Wolfe et al. (2004).

possible, even adaptive. Mutants that skimped on defense to invest more heavily in reproduction should enjoy higher fitness.

Wolfe and colleagues (2004) planted a second common garden experiment in Europe. Consistent with the results from the American garden, the campions from the two continents were strikingly different. This time, however, the difference was that the American plants were easy pickings for predators and pathogens. **Figure 13.40a** shows the American plants' greater susceptibility to anther smut fungus. Figure 13.40b shows their greater vulnerability to seed-eating moths. Upon repatriation to Europe, the snowy campion's evolved defenselessness was a grave liability. Wolfe and colleagues concluded that what had turned snowy campion into a weed was life-history evolution toward an optimal energy budget in a new habitat.

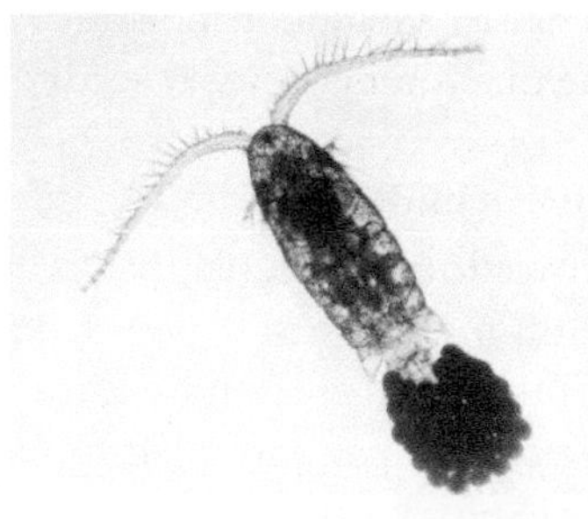

Figure 13.41 The copepod *Eurytemora affinis* Photo by Carol Eunmi Lee, University of Wisconsin.

Invading species sometimes encounter new life-history constraints, rather than escaping old ones. Carol E. Lee and colleagues have studied the invasion of freshwater habitats by the copepod *Eurytemora affinis* **(Figure 13.41)**, which ordinarily lives in saline estuaries and salt marshes. Lee and colleagues (2003) used a common garden experiment to compare freshwater *E. affinis* from Lake Michigan to individuals from a closely related saline population in the St. Lawrence marsh.

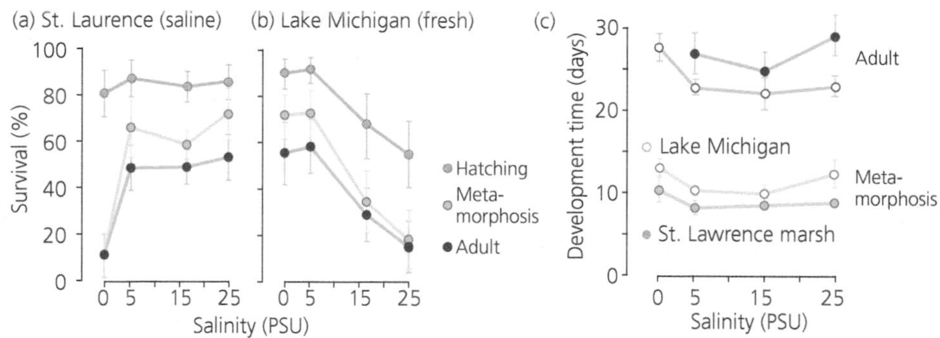

Figure 13.42 Evolved differences between closely related *E. affinis* populations from different habitats (a and b) Survival to different developmental stages as a function of salinity for individuals from the saline habitat and freshwater populations.
(c) Development time versus salinity. PSU = practical salinity units. From Lee et al. (2003).

The data in **Figure 13.42a** and (b) show the rates of survival to hatching, metamorphosis, and adulthood as a function of salinity. They reveal a trade-off between capacity to survive in salt water versus freshwater. When populations of *E. affinis* invade freshwater, they undergo dramatic evolution in osmoregulatory physiology (Lee et al. 2007, 2011, 2012).

The data in Figure 13.42c document differences in life history between the freshwater versus saline populations. Across all salinities, individuals from the freshwater population take longer to reach metamorphosis but less time to reach adulthood. Lee believes that slower development to metamorphosis is adaptive in freshwater because it allows juveniles to accumulate larger stores of key nutrients that are scarce in freshwater habitats. Like the snowy campion example, the copepod example suggests that genetic variation for life-history traits facilitates biological invasion.

Biological invasions are microevolutionary processes. Evidence suggests that differences in life-history traits among lineages may also play a role in macroevolutionary processes.

Life-History Traits and Vulnerability to Extinction

The largest terrestrial animals ever to walk the Earth were the sauropod dinosaurs (Sander and Clauss 2008). With body masses up to 800,000 kg, they were 10 times larger than the biggest terrestrial mammals. And they persisted for over 100 million years. What allowed the sauropods to evolve sizes so much bigger than mammals, and to survive for so long? Jan Werner and Eva Maria Griebeler (2011) suggest that a difference in life-history strategy between dinosaurs and mammals might be a crucial part of the answer.

Pursuing a hypothesis advanced by Christine Janis and Matthew Carrano (1992), Werner and Griebeler compiled data on body mass and clutch or litter size in dinosaurs, birds, and mammals. Because the sauropods were ground-nesting herbivores, the researchers focused on ground-breeding, herbivorous birds and herbivorous mammals.

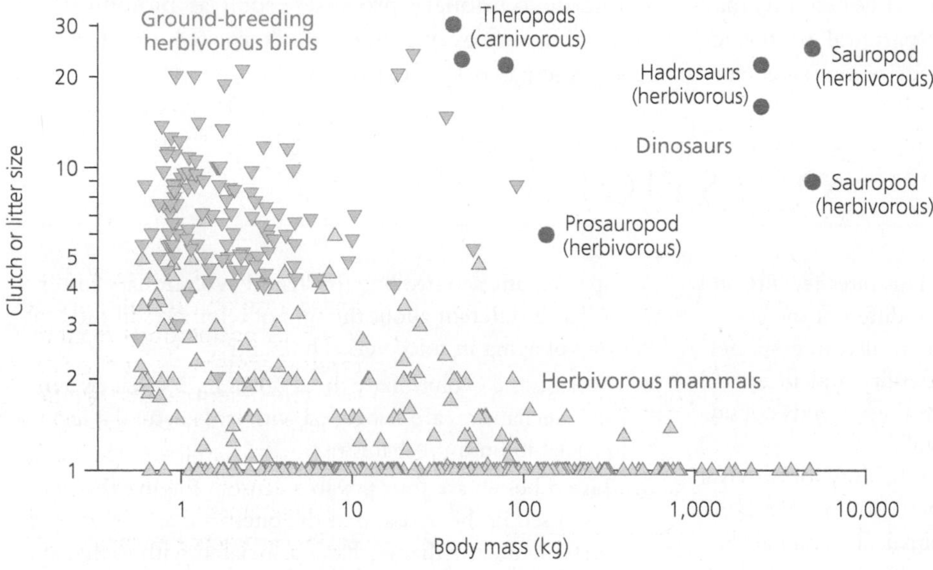

Figure 13.43 Clutch or litter size versus body mass in herbivorous birds, herbivorous mammals, and dinosaurs Plotted with data from Werner and Griebeler (2011).

Figure 13.43 reveals that as herbivorous mammals get larger, their litter sizes get smaller. The largest herbivorous mammals produce just one high-quality offspring at a time. In contrast, the clutch sizes of herbivorous birds do not change with body mass. Birds divide their reproductive investment into more offspring, but of lower quality. The limited data available suggests that dinosaurs, including the sauropods, resemble birds rather than mammals.

Werner and Griebeler argue that lineages that produce larger clutches are less vulnerable to extinction, because their higher reproductive capacity allows them to recover more quickly from population bottlenecks. And being less prone to extinction would have given sauropod lineages more opportunity to evolve enormous sizes. Life-history analysis thus illuminates a macroevolutionary trend.

Life-history differences among lineages may influence macroevolutionary patterns.

SUMMARY

Organisms face fundamental trade-offs. The amount of energy available is finite, and energy devoted to one function—such as growth or repair—cannot be devoted to others—such as reproduction. Furthermore, biological processes take time. An individual growing to a large size before maturing may gain higher fitness, but risks dying before ever reproducing. Fundamental trade-offs involving energy and time mean that every organism's life history is an evolutionary compromise.

Senescence evolves because natural selection is weaker late in life. Late-acting deleterious mutations can persist in populations under mutation–selection balance. And selection may favor increased investment in reproduction early in life at the expense of repair. Both mechanisms can result in a decline in reproductive performance and survival with age.

A trade-off between the number of offspring in a clutch and the survival of individual offspring constrains the evolution of clutch size. Additional constraints may involve trade-offs between present parental reproductive effort and future reproductive performance or survival, as well as trade-offs between clutch size and offspring reproductive performance.

A trade-off between the size and number of offspring constrains the evolution of offspring size. Selection on parents may favor smaller offspring than does selection on the offspring themselves.

Life-history traits may reflect conflicts of interest between individuals. These conflicts have led to the evolution of differential gene expression (imprinting) and sexually antagonistic traits in males and females.

Theory predicts that life-history traits should have low heritability because they are closely related to fitness. Life-history traits do tend to have lower heritability than other kinds of traits, but nonetheless they typically show substantial genetic variation. One mechanism demonstrated to maintain genetic variation in life-history traits is temporally varying selection.

Variation in life-history traits plays important roles in microevolutionary processes, such as biological invasions. It may even influence macroevolutionary processes, such as risk of extinction.

QUESTIONS

1. Look again at Figure 13.4, which illustrates fertility and survival as a function of age in three different species.
   a. What similarities are there across all three species? What is the general trend in fertility and in annual probability of survival? Why are these trends considered to be an evolutionary puzzle?
   b. Which species has the best probability of survival even in old age? This is a characteristic of this taxon. Do you remember another animal (a mammal, discussed later in this chapter) that has a similarly high probability of survival in old age? What trait do these long-lived animals have in common?
   c. In red deer, how do patterns of survival and reproduction vary with the two sexes? Why do you think these differences occur between males and females?
2. What are the two predictions of the rate-of-living theory of aging? What data exist to support or refute the two predictions?
3. What is a telomere? Describe how telomere shortening is thought to influence variation in life span among individuals. Is telomere length associated with life span in zebra finches? In elderly humans? In different mammal species, after controlling for shared evolutionary history? What is different about the role of telomeres in the biology of aging in mice versus humans?
4. What is the evolutionary theory of aging? What two major mechanisms are associated with it? Is natural selection crucial in both mechanisms?
5. Listed below are four possible causes of aging that were discussed in the text and in the questions above. As a review, name the theory that is associated with each cause, and describe whether selection for a longer life span is possible under each theory. What predictions does each theory make about the effect of ecological mortality (death due to external causes—predators, starvation, etc.) on aging rate?
   - "Wearing out" due to metabolic activity
   - Reduction in size of telomeres with each cell division
   - Mutations that have negative effects late in life
   - Mutations that have positive effects early and negative effects late in life
6. Most domestic female rabbits will get uterine cancer if they are not spayed. The cancer usually appears after the age of 2 years. Describe a hypothesis for why rabbits

have not evolved better defenses against uterine cancer. What do you think the average life span of a wild female rabbit might be? What do you think is a typical cause of death in wild rabbits? Why do you think that uterine cancer, and not (say) pancreatic cancer or throat cancer, is the most common cancer in female rabbits?

7. We have seen how aging can evolve due to two different phenomena: First, aging may evolve due to mutations that have deleterious effects only late in life. As a review, explain how such mutations could ever become common in a population. Second, aging may evolve due to mutations with pleiotropic effects that cause "trade-offs"—positive effects early and negative effects late. What would happen if a mutation arose with a reverse trade-off—that is, a mutation with negative effects early and positive effects late in life? Could such a mutation ever be selected for?

8. Does the *p53* gene in humans provide an example of antagonistic pleiotropy? Why or why not?

9. Look again at Figures 13.10 and 13.13, which show life-history trade-offs for a hypothetical species. Suppose you are studying these animals, and you discover a new mutation from the wild type that causes its carriers to have two offspring per year instead of one. The new mutation does not alter the age of maturation, which still occurs at 3 years. Your initial observations indicate that the new mutation may cause an early death, but you are not certain exactly how early. You do notice, however, that the new mutation is increasing in frequency and the wild-type allele is decreasing. Make a prediction about the minimum possible age of death of organisms that carry this mutation, and explain your reasoning.

10. Now suppose that during your research, you bring a large population of these animals into captivity. You notice that their annual survival rate immediately jumps from 0.80 to 0.95, primarily due to protection from predators. Make a prediction about whether the captive population will evolve changes in fertility or life span, simply due to this reduction in predation. Could this same process be occurring in zoo populations of captive animals today? Explain your reasoning.

11. Assuming that the grandmother hypothesis of menopause is correct, speculate on what aspects of a species' behavior and sociality may make menopause likely to evolve. For instance, is it important whether the species is highly social, or whether the species lives in kin groups? Might the age of independence of the young be important? Could menopause ever evolve in a species without parental care, such as aphids or willow trees? As fuel for thought, consider the likelihood of evolution of menopause in (1) orangutans, who live in small groups consisting simply of a female and her dependent young; (2) lions, in which females are very social and remain with their female kin for most of their lives; and (3) Arabian oryx, a species of antelope that lives in small family groups in arid deserts and must sometimes find distant waterholes known only to the older oryx.

12. As a review, describe why hatchery salmon may be evolving smaller egg size, and the implications for wild populations. What could hatchery managers do to reverse the effects on wild populations?

13. The examples of the chinook salmon and seed beetles indicate that females, in general, cannot produce many large eggs. Instead, they must choose between producing many small eggs or producing a few large eggs (and sometimes, in unfortunate cases, just a few small eggs). Explain, then, how it is possible for a queen honeybee to produce a very large number of relatively large eggs. (Hint: Consider what the other bees are doing.) Does this suggest a general way in which a female can escape from the size–number trade-off?

14. Two old science fiction movies, *Godzilla* (1998) and *Aliens* (1986), depict fictional large female carnivores. The *Godzilla* female lives off a large prey population of humans and fishes but has no assistance from others of her kind. In a few days, she produces hundreds of 7-ft-tall eggs, enough to fill Madison Square Garden. The *Aliens* female lives off a small prey population of a few dozen humans, is assisted by nonreproducing workers, and produces hundreds of large eggs in a few weeks. Comment on what is realistic and unrealistic about the life-history traits and egg production abilities of each of these fictional animals. If they were real, would they have long or short life spans? Why?

15. Dairy farmers are sometimes frustrated in their attempts to breed a better milk cow because heritability values for milk production and reproductive traits are low—generally below 0.10. In addition, those cows that produce the most milk tend to have longer intervals between birth of successive calves and require more breedings to a bull before the cow will conceive. Do these patterns make sense in light of evolutionary life-history theory? Explain.

16. What is Lack's hypothesis? Is it supported by most experimental data? If not, why not?

17. Many human generations ago, most women worldwide began childbearing in their mid-teens. Today, a large proportion of women worldwide delay childbearing until their 20s. Among college-educated women in developed nations, the trend in delaying reproduction has been taken even further; childbearing often is delayed until past age 30 due to education and career pressures. Suppose that most women worldwide were to delay childbearing until age 30, and that women were to continue to make this choice for many human generations. Make a prediction about how human life span and fertility might evolve in response.

EXPLORING THE LITERATURE

18. A near-starvation diet prolongs life span markedly in a variety of animals. See the following reviews for insight into the mechanism and implications:

Alic, N., and L. Partridge. 2011. Death and dessert: Nutrient signalling pathways and ageing. *Current Opinion in Cell Biology* 23: 738–743.

Nakagawa, S., M. Lagisz, et al. 2012. Comparative and meta-analytic insights into life extension via dietary restriction. *Aging Cell* 11: 401–409.

19. For a lizard with an extraordinary life history, see:

Karsten, K. B., L. N. Andriamandimbiarisoa, et al. 2008. A unique life history among tetrapods: An annual chameleon living mostly as an egg. *Proceedings of the National Academy of Sciences, USA* 105: 8980–8984.

20. For evidence that a life-history trait is evolving by natural selection in a human population, see:

Milot, E., F. M. Mayer, et al. 2011. Evidence for evolution in response to natural selection in a contemporary human population. *Proceedings of the National Academy of Sciences, USA* 108: 17040–17045.

21. For a trade-off between the growth of one body part versus another, see:

Maginnis, T. L. 2006. Leg regeneration stunts wing growth and hinders flight performance in a stick insect (*Sipyloidea sipylus*). *Proceedings of the Royal Society of London B* 273: 1811–1814.

22. Look back at the data on blue-footed boobies on the first page of the chapter. One interpretation is that taking a year off from breeding has rejuvenating effects for males. Imagine that you are a male blue-footed booby and that you contract an infection. If you are relatively young, with a long future ahead, what should you do? What if you are old, with a short future ahead? For hypotheses and an experimental test, see:

Velando, A., H. Drummond, and R. Torres. 2006. Senescent birds redouble reproductive effort when ill: Confirmation of the terminal investment hypothesis. *Proceedings of the Royal Society of London B* 273: 1443–1448.

23. Mating in spiders sometimes involves extreme interactions between males and females. And it can lead to extreme life-history strategies. Not for the faint of heart, nor the arachnophobic:

Lee, Q. Q., J. Oh, et al. 2012. Emasculation: Gloves-off strategy enhances eunuch spider endurance. *Biology Letters* 8: 733–735.

24. Graham Bell distinguished between the rate-of-living versus evolutionary theories of aging by comparing invertebrates that have a distinct soma and germ line with invertebrates that have no such division. According to the rate-of-living theory, both kinds of organisms will accumulate irreparable damage. According to the evolutionary theory, genes responsible for senescence will accumulate only in organisms with a disposable soma. See:

Bell, G. 1984. Evolutionary and nonevolutionary theories of senescence. *American Naturalist* 124: 600–603.

25. For tests of the evolutionary theory of aging employing comparisons between eusocial versus noneusocial insects and comparisons between castes of worker ants, see:

Keller, L., and M. Genoud. 1997. Extraordinary life spans in ants: A test of evolutionary theories of ageing. *Nature* 389: 958–960.

Chapuisat, M., and L. Keller. 2002. Division of labour influences the rate of ageing in weaver ant workers. *Proceedings of the Royal Society of London B* 269: 909–913.

26. Populations with high ecological mortality may not always evolve short life spans. See:

Reznick, D. N., M. J. Bryant, et al. 2004. Effect of extrinsic mortality on the evolution of senescence in guppies. *Nature* 431: 1095–1099.

Bronikowski, A. M., and D. E. L. Promislow. 2005. Testing evolutionary theories of aging in wild populations. *Trends in Ecology and Evolution* 20: 271–273.

Williams, P. D., T. Day, et al. 2006. The shaping of senescence in the wild. *Trends in Ecology and Evolution* 21: 458-463.

CITATIONS

Atzmon, G., M. Cho, et al. 2010. Genetic variation in human telomerase is associated with telomere length in Ashkenazi centenarians. *Proceedings of the National Academy of Sciences, USA* 107 (Suppl 1): 1710–1717.

Austad, S. N. 1988. The adaptable opossum. *Scientific American* (February): 98–104.

Austad, S. N. 1993. Retarded senescence in an insular population of Virginia opossums (*Didelphis virginiana*). *Journal of Zoology, London* 229: 695–708.

Austad, S. N. 1994. Menopause: An evolutionary perspective. *Experimental Gerontology* 29: 255–263.

Austad, S. N., and K. E. Fischer. 1991. Mammalian aging, metabolism, and ecology: Evidence from the bats and marsupials. *Journal of Gerontology* 46: B47–B53.

Barlow, D. P. 1995. Gametic imprinting in mammals. *Science* 270: 1610–1613.

Berrigan, D. 1991. The allometry of egg size and number in insects. *Oikos* 60: 313–321.

Blair, A. C., and L. M. Wolfe. 2004. The evolution of an invasive plant: An experimental study with *Silene latifolia*. *Ecology* 85: 3035–3042.

Boyce, M. S., and C. M. Perrins. 1987. Optimizing great tit clutch size in a fluctuating environment. *Ecology* 68: 142–153.

Byars, S. G., D. Ewbank, et al. 2010. Natural selection in a contemporary human population. *Proceedings of the National Academy of Sciences, USA* 107 (Suppl 1): 1787–1792.

Campisi, J. 1996. Replicative senescence: An old lives' tale? *Cell* 84: 497–500.

Cant, M. A., and R. A. Johnstone. 2008. Reproductive conflict and the separation of reproductive generations in humans. *Proceedings of the National Academy of Sciences, USA* 105: 5332–5336.

Charnov, E. L., and D. Berrigan. 1993. Why do female primates have such long life spans and so few babies? or Life in the slow lane. *Evolutionary Anthropology* 1: 191–194.

Charnov, E. L., and S. W. Skinner. 1985. Complementary approaches to the understanding of parasitoid oviposition decisions. *Environmental Entomology* 14: 383–391.

Clutton-Brock, T. H., S. D. Albon, and F. E. Guinness. 1988. Reproductive success in male and female red deer. In *Reproductive Success: Studies of Individual Variation in Contrasting Breeding Systems*, ed. T. H. Clutton-Brock. Chicago: University of Chicago Press, 325–343.

Donate, L. E., and M. A. Blasco. 2011. Telomeres in cancer and ageing. *Philosophical Transactions of the Royal Society of London B* 366: 76–84.

Donehower, L. A. 2009. Using mice to examine p53 functions in cancer, aging, and longevity. *Cold Spring Harbor Perspectives in Biology* 1: a001081.

Elbadry, E. A., and M. S. F. Tawfik. 1966. Life cycle of the mite *Adactylidium* sp. (Acarina: Pyemotidae), a predator of thrips eggs in the United Arab Republic. *Annals of the Entomological Society of America* 59: 458–461.

Elgar, M. A. 1990. Evolutionary compromise between a few large and many small eggs: Comparative evidence in teleost fish. *Oikos* 59: 283–287.

Escobar, J. S., P. Jarne, et al. 2008. Outbreeding alleviates senescence in hermaphroditic snails as expected from the mutation-accumulation theory. *Current Biology* 18: 906–910.

Eshleman, J. R., and S. D. Markowitz. 1996. Mismatch repair defects in human carcinogenesis. *Human Molecular Genetics* 5: 1489–1494.

Ferbeyre, G., and S. W. Lowe. 2002. The price of tumor suppression? *Nature* 415: 26–27.

Fishel, R., and T. Wilson. 1997. *MutS* homologs in mammalian cells. *Current Opinion in Genetics and Development* 7: 105–133.

Fisher, R. A. 1930. *The Genetical Theory of Natural Selection*. Oxford: Clarendon Press.

Fowler, K., and L. Partridge. 1989. A cost of mating in female fruit flies. *Nature* 338: 760–761.

Fox, C. W., M. S. Thakar, and T. A. Mousseau. 1997. Egg size plasticity in a seed beetle: An adaptive maternal effect. *American Naturalist* 149: 149–163.

Gomes, N. M., O. A. Ryder, et al. 2011. Comparative biology of mammalian telomeres: Hypotheses on ancestral states and the roles of telomeres in longevity determination. *Aging Cell* 10: 761–768.

Grosberg, R. K. 1988. Life-history variation within a population of the colonial ascidian *Botryllus schlosseri*. I. The genetic and environmental control of seasonal variation. *Evolution* 42: 900–920.

Guillette, L. J., Jr., and R. E. Jones. 1985. Ovarian, oviductal and placental morphology of the reproductively bimodal lizard, *Sceloporus aeneus*. *Journal of Morphology* 184: 85–98.

Gurven, M., and H. Kaplan. 2007. Longevity among hunter-gatherers: A cross-cultural examination. *Population and Development Review* 33: 321–365.

Gustafsson, L., and T. Pärt. 1990. Acceleration of senescence in the collared flycatcher (*Ficedula albicollis*) by reproductive costs. *Nature* 347: 279–281.

Haig, D. 1999. Multiple paternity and genomic imprinting. *Genetics* 151: 1229–1231.

Haig, D., and C. Graham. 1991. Genomic imprinting and the strange case of the insulin-like growth factor II receptor. *Cell* 64: 1045–1046.

Haig, D., and M. Westoby. 1989. Parent-specific gene expression and the triploid endosperm. *American Naturalist* 134: 147–155.

Haig, D., and M. Westoby. 1991. Genomic imprinting in endosperm: Its effect on seed development in crosses between species, and between different ploidies of the same species, and its implications for the evolution of apomixis. *Philosophical Transactions of the Royal Society of London B* 333: 1–13.

Hamilton, W. D. 1966. The moulding of senescence by natural selection. *Journal of Theoretical Biology* 12: 12–45.

Hawkes, K., J. F. O'Connell, and N. G. Blurton Jones. 1989. Hardworking Hadza grandmothers. In *Comparative Socioecology*, ed. V. Standen and R. A. Foley. Oxford: Blackwell Scientific Publications, 341–366.

Hawkes, K., J. F. O'Connell, and N. G. Blurton Jones. 1997. Hadza women's time allocation, offspring provisioning, and the evolution of long postmenopausal life spans. *Current Anthropology* 38: 551–577.

Hawkes, K., J. F. O'Connell, et al. 1998. Grandmothering, menopause, and the evolution of human life histories. *Proceedings of the National Academy of Sciences, USA* 95: 1336–1339.

Heath, D. D., J. W. Heath, C. A. Bryden, et al. 2003. Rapid evolution of egg size in captive salmon. *Science* 299: 1738–1740.

Heidinger, B. J., J. D. Blount, et al. 2012. Telomere length in early life predicts lifespan. *Proceedings of the National Academy of Sciences, USA* 109: 1743–1748.

Hewitt, G., D. Jurk, et al. 2012. Telomeres are favoured targets of a persistent DNA damage response in ageing and stress-induced senescence. *Nature Communications* 3: 708.

Hill, K., and A. M. Hurtado. 1991. The evolution of premature reproductive senescence and menopause in human females: An evaluation of the "grandmother hypothesis." *Human Nature* 2: 313–351.

Hill, K., and A. M. Hurtado. 1996. *Ache Life History: The Ecology and Demography of a Foraging People*. New York: Aldine de Gruyter.

Holland, B., and W. R. Rice. 1998. Perspective: Chase-away sexual selection: Antagonistic seduction versus resistance. *Evolution* 52: 1–7.

Holland, B., and W. R. Rice. 1999. Experimental removal of sexual selection reverses intersexual antagonistic coevolution and removes a reproductive load. *Proceedings of the National Academy of Sciences, USA* 96: 5083–5088.

Hughes, K. A., J. A. Alipaz, et al. 2002. A test of evolutionary theories of aging. *Proceedings of the National Academy of Sciences, USA* 99: 14286–14291.

Hurst, L. D. 1999. Is multiple paternity necessary for the evolution of genomic imprinting? *Genetics* 153: 509–512.

Janis, C. M., and M. Carrano. 1992. Scaling of reproductive turnover in archosaurs and mammals: Why are large terrestrial mammals so rare? *Annales Zoologici Fennici* 28: 201–216.

King, E. G., D. A. Roff, and D. J. Fairbairn. 2011a. The evolutionary genetics of acquisition and allocation in the wing dimorphic cricket, *Gryllus firmus*. *Evolution* 65: 2273–2285.

King, E. G., D. A. Roff, and D. J. Fairbairn. 2011b. Trade-off acquisition and allocation in *Gryllus firmus*: A test of the Y model. *Journal of Evolutionary Biology* 24: 256–264.

Lack, D. 1947. The significance of clutch size. *Ibis* 89: 302–352.

Lahdenperä, M., V. Lummaa, et al. 2004. Fitness benefits of prolonged post-reproductive lifespan in women. *Nature* 428: 178–181.

Lee, C. E., M. Kiergaard, et al. 2011. Pumping ions: Rapid parallel evolution of ionic regulation following habitat invasions. *Evolution* 65: 2229–2244.

Lee, C. E., M. Posavi, and G. Charmantier. 2012. Rapid evolution of body fluid regulation following independent invasions into freshwater habitats. *Journal of Evolutionary Biology* 25: 625–633.

Lee, C. E., J. L. Remfert, and Y. M. Chang. 2007. Response to selection and evolvability of invasive populations. *Genetica* 129: 179–192.

Lee, C. E., J. L. Remfert, and G. W. Gelembiuk. 2003. Evolution of physiological tolerance and performance during freshwater invasions. *Integrative and Comparative Biology* 43: 439–449.

Leroi, A. M., A. Bartke, et al. 2005. What evidence is there for the existence of individual genes with antagonistic pleiotropic effects? *Mechanisms of Ageing and Development* 126: 421–429.

Levitan, D. R. 1993. The importance of sperm limitation to the evolution of egg size in marine invertebrates. *American Naturalist* 141: 517–536.

Levitan, D. R. 1996. Predicting optimal and unique egg sizes in free-spawning marine invertebrates. *American Naturalist* 148: 174–188.

Lin, Y.-J., L. Seroude, and S. Benzer. 1998. Extended life-span and stress resistance in the *Drosophila* mutant *methuselah*. *Science* 282: 943–946.

Lindén, M., and A. P. Møller. 1989. Cost of reproduction and covariation of life history traits in birds. *Trends in Ecology and Evolution* 4: 367–371.

Luckinbill, L. S., R. Arking, et al. 1984. Selection for delayed senescence in *Drosophila melanogaster*. *Evolution* 38: 996–1003.

Madrigal, L., and M. Meléndez-Obando. 2008. Grandmothers' longevity negatively affects daughters' fertility. *American Journal of Physical Anthropology* 136: 223–229.

Medawar, P. B. 1952. *An Unsolved Problem in Biology*. London: H. K. Lewis.

Mockett, R. J., and R. S. Sohal. 2006. Temperature-dependent trade-offs between longevity and fertility in the *Drosophila* mutant, *methuselah*. *Experimental Gerontology* 41: 566–573.

Moore, T., and D. Haig. 1991. Genomic imprinting in mammalian development—a parental tug-of-war. *Trends in Genetics* 7: 45–49.

Mousseau, T. A., and D. A. Roff. 1987. Natural selection and the heritability of fitness components. *Heredity* 1987: 181–197.

Nesse, R. M., and G. C. Williams. 1995. *Why We Get Sick: The New Science of Darwinian Medicine*. New York: Vintage Books.

Nowak, R. M. 1991. *Walker's Mammals of the World*. 5th ed. Baltimore, MD: Johns Hopkins University Press.

Partridge, L. 1987. Is accelerated senescence a cost of reproduction? *Functional Ecology* 1: 317–320.

Partridge, L. 2001. Evolutionary theories of ageing applied to long-lived organisms. *Experimental Gerontology* 36: 641–650.

Partridge, L., and N. H. Barton. 1993. Optimality, mutation, and the evolution of ageing. *Nature* 362: 305–311.

Partridge, L., and K. Fowler. 1992. Direct and correlated responses to selection on age at reproduction in *Drosophila melanogaster*. *Evolution* 46: 76–91.

Pennisi, E. 1998. Singe gene controls fruit fly life-span. *Science* 282: 856.

Pettifor, R. A., C. M. Perrins, and R. H. McCleery. 2001. The individual optimization of fitness: Variation in reproductive output, including clutch size, mean nestling mass and offspring recruitment, in manipulated broods of great tits *Parus major*. *Journal of Animal Ecology* 70: 62–79.

Podolsky, R. D., and R. R. Strathmann. 1996. Evolution of egg size in free-spawners: Consequences of the fertilization-fecundity trade-off. *American Naturalist* 148: 160–173.

Price, T., and D. Schluter. 1991. On the low heritability of life-history traits. *Evolution* 45: 853–861.

Reed, D. H., and E. H. Bryant 2000. The evolution of senescence under curtailed life span in laboratory populations of *Musca domestica* (the housefly). *Heredity* 85: 115–121.

Rice, W. R. 1987. The accumulation of sexually antagonistic genes as a selective agent promoting the evolution of reduced recombination between primitive sex chromosomes. *Evolution* 41: 911–914.

Rice, W. R. 1996. Sexually antagonistic male adaptation triggered by experimental arrest of female evolution. *Nature* 381: 232–234.

Rice, W. R., and B. Holland. 1997. The enemies within: Intergenomic conflict, interlocus contest evolution (ICE), and the intraspecific Red Queen. *Behavioral Ecology and Sociobiology* 41: 1–10.

Rodriguez Bigas, M. A., P. H. U. Lee, et al. 1996. Establishment of a hereditary nonpolyposis colorectal cancer registry. *Diseases of the Colon & Rectum* 39: 649–653.

Roff, D. A. 1992. *The Evolution of Life Histories*. New York: Chapman & Hall.

Roper, C., P. Pignatelli, and L. Partridge. 1993. Evolutionary effects of selection on age at reproduction in larval and adult *Drosophila melanogaster*. *Evolution* 47: 445–455.

Rose, M. R. 1984. Laboratory evolution of postponed senescence in *Drosophila melanogaster*. *Evolution* 38: 1004–1010.

Rose, M. R. 1991. *Evolutionary Biology of Aging*. New York: Oxford University Press.

Sander, P. M., and M. Clauss. 2008. Paleontology. Sauropod gigantism. *Science* 322: 200–201.

Sanz, J. J., and J. Moreno. 1995. Experimentally induced clutch size enlargements affect reproductive success in the pied flycatcher. *Oecologia* 103: 358–364.

Schilder, R. J., A. J. Zera, et al. 2011. The biochemical basis of life history adaptation: Molecular and enzymological causes of NADP(+)-isocitrate dehydrogenase activity differences between morphs of *Gryllus firmus* that differ in lipid biosynthesis and life history. *Molecular Biology and Evolution* 28: 3381–3393.

Schluter, D., and L. Gustafsson. 1993. Maternal inheritance of condition and clutch size in the collared flycatcher. *Evolution* 47: 658–667.

Schmidt, P. S., D. D. Duvernell, and W. F. Eanes. 2000. Adaptive evolution of a candidate gene for aging in *Drosophila*. *Proceedings of the National Academy of Sciences, USA* 97: 10861–10865.

Service, P. 1987. Physiological mechanisms of increased stress resistance in *Drosophila melanogaster* selected for postponed senescence. *Physiological Zoology* 60: 321–326.

Shay, J. W., and W. E. Wright. 2011. Role of telomeres and telomerase in cancer. *Seminars in Cancer Biology* 21: 349–353.

Smith, C. C., and S. D. Fretwell. 1974. The optimal balance between size and number of offspring. *American Naturalist* 108: 499–506.

Spencer, H. G., A. G. Clark., and M. W. Feldman. 1999. Genetic conflicts and the evolutionary origin of genomic imprinting. *Trends in Ecology and Evolution* 14: 197–201.

Stearns, S. C. 1992. *The Evolution of Life Histories*. Oxford: Oxford University Press.

Strathmann, M. F. 1987. *Reproduction and Development of Marine Invertebrates of the Northern Pacific Coast*. Seattle: University of Washington Press.

Taborsky, M., and B. Taborsky. 1993. The kiwi's parental burden. *Natural History* 1993: 50–56.

Torres, R., and A. Velando. 2007. Male reproductive senescence: The price of immune-induced oxidative damage on sexual attractiveness in the blue-footed booby. *Journal of Animal Ecology* 76: 1161–1168.

Tyner, S. D., S. Venkatachalam, J. Choi, et al. 2002. p53 mutant mice that display early ageing-associated phenotypes. *Nature* 415: 45–53.

Vance, R. R. 1973. On reproductive strategies in marine benthic invertebrates. *American Naturalist* 107: 339–352.

Vani, R. G., and M. R. S. Rao. 1996. Mismatch repair genes of eukaryotes. *Journal of Genetics* 75: 181–192.

van Heemst, D., S. P. Mooijaart, et al. 2005. Variation in the human *TP53* gene affects old age survival and cancer mortality. *Experimental Gerontology* 40: 11–15.

Van Valen, L. 1973. A new evolutionary law. *Evolutionary Theory* 1: 1–30.

Velando, A., R. Beamonte-Barrientos, and R. Torres. 2006. Pigment-based skin colour in the blue-footed booby: An honest signal of current condition used by females to adjust reproductive investment. *Oecologia* 149: 535–542.

Velando, A., H. Drummond, and R. Torres. 2010. Senescing sexual ornaments recover after a sabbatical. *Biology Letters* 6: 194–196.

Walker, D. W., G. McColl, N. L. Jenkins, et al. 2000. Evolution of lifespan in *C. elegans*. *Nature* 405: 296–297.

Werner, J., and E. M. Griebeler. 2011. Reproductive biology and its impact on body size: Comparative analysis of mammalian, avian and dinosaurian reproduction. *PLoS One* 6: e28442.

Williams, G. C. 1957. Pleiotropy, natural selection, and the evolution of senescence. *Evolution* 11: 398–411.

Wolfe, L. M., J. A. Elzinga, and A. Biere. 2004. Increased susceptibility to enemies following introduction in the invasive plant *Silene latifolia*. *Ecology Letters* 7: 813–820.

Wourms, J. P. 1993. Maximization of evolutionary trends for placental viviparity in the spadenose shark, *Scoliodon laticaudus*. *Environmental Biology of Fish* 38: 269–294.

Young, T. P. 1990. Evolution of semelparity in Mount Kenya lobelias. *Evolutionary Ecology* 4: 157–171.

Zera, A. J. 2005. Intermediary metabolism and life history trade-offs: Lipid metabolism in lines of the wing-polymorphic cricket, *Gryllus firmus*, selected for flight capability vs. early age reproduction. *Integrative and Comparative Biology* 45: 511–524.

Zera, A. J., and R. F. Denno. 1997. Physiology and ecology of dispersal polymorphism in insects. *Annual Review of Entomology* 42: 207–230.

Zera, A. J., and Z. Zhao. 2003. Life-history evolution and the microevolution of intermediary metabolism: Activities of lipid-metabolizing enzymes in life-history morphs of a wing-dimorphic cricket. *Evolution* 57: 586–596.

Zera, A. J., and Z. Zhao. 2004. Effect of a juvenile hormone analogue on lipid metabolism in a wing-polymorphic cricket: Implications for the endocrine-biochemical bases of life-history trade-offs. *Physiological and Biochemical Zoology* 77: 255–266.

Zhao, Z., and A. J. Zera. 2002. Differential lipid biosynthesis underlies a tradeoff between reproduction and flight capability in a wing-polymorphic cricket. *Proceedings of the National Academy of Sciences, USA* 99: 16829–16834.

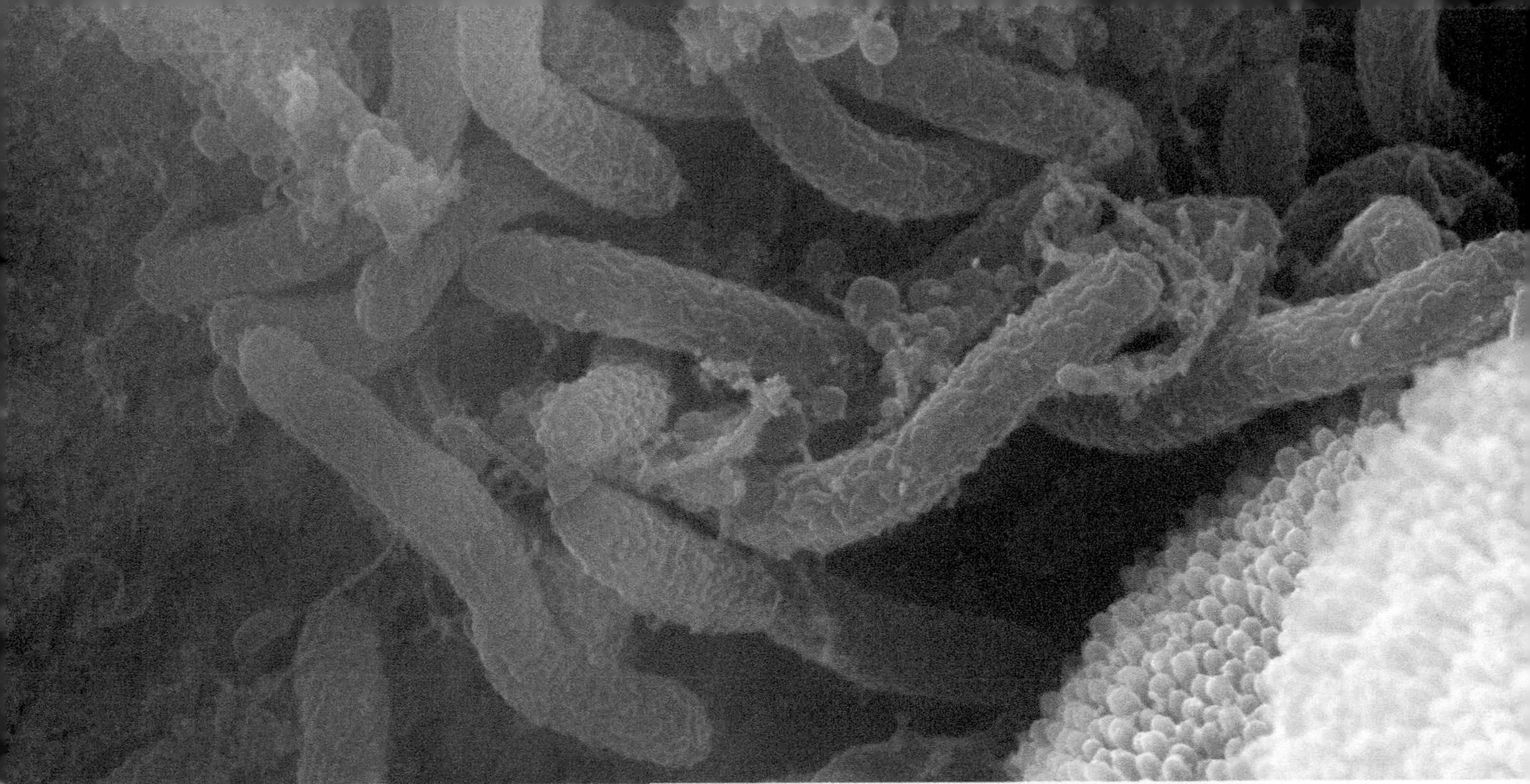

14

Evolution and Human Health

In 1854 a cholera epidemic struck central London. The disease, which causes severe diarrhea and dehydration, killed more than 500 people. In a famous act of medical detection, John Snow (1855) prepared a map of the affected neighborhood (see Tufte 1997). On it, he plotted the homes of the victims and the locations of the area's water pumps. The fatalities clustered around the Broad Street pump, at the center of Snow's map. Sealing the case were the deaths of two women in distant neighborhoods, who fell ill shortly after drinking water delivered by special arrangement from Broad Street. Although cholera's cause remained to be discovered, it was clearly associated with contaminated water.

In 1858, Louis Pasteur proposed that contagious diseases like cholera are caused by germs. Pasteur had been studying the fermentation of beer, wine, and milk, and he had been working to stop an epidemic of childbirth fever in a Paris maternity hospital. In a paper on lactic acid fermentation, Pasteur suggested that just as a particular microorganism is the cause of each kind of fermentation, so also might a particular microorganism be the cause of each infectious illness. Inspired by Pasteur, Robert Koch and others soon discovered the bacteria responsible for anthrax, wound infections, gonorrhea, typhoid fever, and tuberculosis. In 1883, Koch showed that cholera is caused by the bacterium *Vibrio cholerae*.

20,000× scanning electron micrograph of *Vibrio cholerae* in mouse intestine by Louisa Howard (see Krebs and Taylor 2011). John Snow's map redrawn from Gilbert (1958).

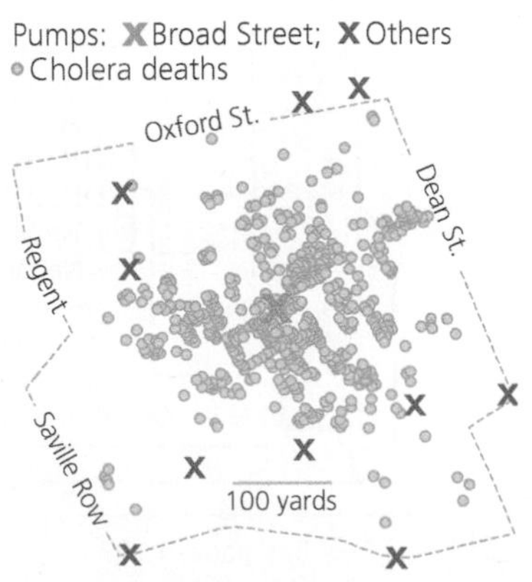

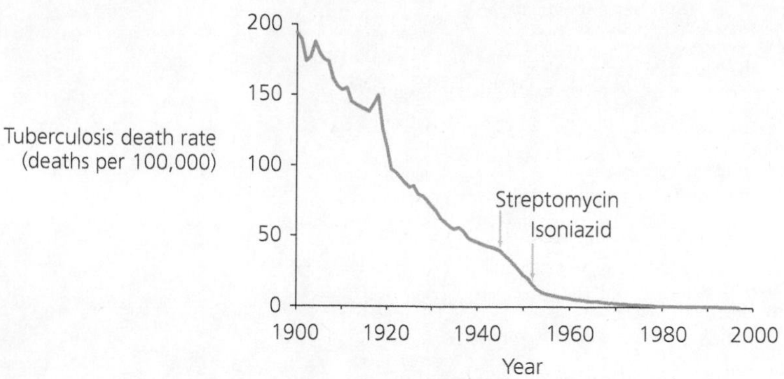

Figure 14.1 Tuberculosis death rate as a function of time in the United States Between 1900 and 1950, the TB death rate declined dramatically, largely as a result of improvements in sanitation and housing. The introduction of antibiotics at mid-century further hastened the decline.

The germ theory of disease was arguably the most important breakthrough in the development of modern medicine. It led to the identification of numerous pathogens, the development of antiseptic surgery by Joseph Lister, the discovery of antibiotics by Alexander Fleming and others, and dramatic improvements in sanitation. The impact of sanitation and antibiotics on public health can be seen in **Figure 14.1**, which plots the death rate due to tuberculosis in the United States from 1900 through 1997. From 1900 to 1945, the death rate dropped from nearly 200 per 100,000 to about 40. This decline was largely due to improvements in sanitation, housing, and nutrition. Then, in 1945, the death rate began falling more sharply still. The accelerated decline was due to the introduction of antibiotics, including streptomycin and isoniazid. By 1997, the tuberculosis death rate was fewer than 0.4 per 100,000, less than two-tenths of 1% what it was in 1900.

Charles Darwin published *On the Origin of Species* in 1859, the year after Pasteur proposed the germ theory. Evolutionary biology and modern medicine were born at the same time and have grown up in parallel. The relevance of evolutionary biology to human health is deep. Some kinds of evolutionary analysis, such as phylogenetics, are routinely used in health-related research. **Figure 14.2**, for example, shows a phylogeny of cholera samples estimated by Rene Hendriksen and colleagues (2011). It suggests Nepal as the geographic source of the bacteria responsible for a devastating cholera epidemic that ravaged Haiti in 2010, though other locations in South Asia and elsewhere cannot be definitively excluded (see Keim et al. 2011; Mutreja et al. 2011; Pun 2011; Hasan et al. 2012).

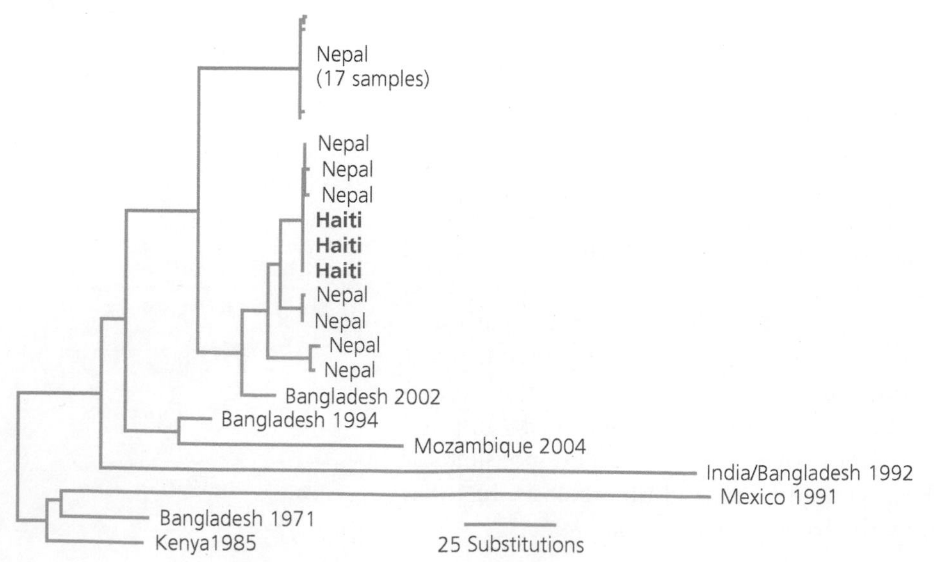

Figure 14.2 Evolutionary relationships among cholera strains This tree includes cholera bacteria collected in Nepal and Haiti in 2010, plus bacteria collected earlier in other regions. It is consistent with other evidence suggesting that cholera was inadvertently brought to Haiti from Nepal by United Nations peacekeeping troops (see Frerichs et al. 2012). Redrawn from Hendriksen et al. (2011).

Other kinds of evolutionary analysis, such as selection thinking, are gaining recognition as useful tools in health research. George C. Williams and Randolph Nesse have been leaders in the expanded application of evolutionary biology to human health. They are viewed as founders of a field they call Darwinian medicine (Williams and Nesse 1991; Nesse and Williams 1994, 1998).

Throughout this book we have highlighted medical applications of evolutionary analysis. We have discussed the evolution of HIV (Chapter 1). We have considered the impact of infectious diseases such as AIDS, malaria, and cystic fibrosis on the evolution of human populations (Chapters 5–8). We have explored senescence and menopause (Chapter 13). Here we devote an entire chapter to medical applications we have not elsewhere had the chance to address.

The chapter is divided into two parts. In Sections 14.1 through 14.4, we consider medical consequences of the fact that populations evolve. Examples include the evolution of pathogen populations and the evolution of cell populations within individual patients. In Sections 14.5 through 14.7, we turn our attention from pathogen and cell populations to the human animal as it has been shaped by natural selection. We consider applications of selection thinking (introduced in Chapter 10) in understanding puzzling aspects of human physiology and behavior. The message throughout is that evolutionary analysis is an invaluable tool for researchers and clinicians seeking to improve public health.

14.1 Evolving Pathogens: Evasion of the Host's Immune Response

The fundamental event in evolution is a change in the frequencies of genotypes within a population. There are two kinds of populations whose ongoing evolution is important in medicine: populations of pathogens, and populations of human cells within individual patients. We consider evolving pathogen populations first, then evolving populations of cells.

A population of pathogens and their host are, by definition, in conflict. The pathogens attempt to consume the host's tissues and convert them into more pathogens; the host attempts to limit the damage by slowing or killing the pathogens. When we become hosts, our bodies employ an impressive array of weapons against the invaders. Our immune systems can recognize billions of foreign proteins, mount an aggressive and multifaceted response, and remember the structure of the pathogen's proteins so as to mobilize more quickly next time. Pathogens are formidable enemies, however. Many have large population sizes, short generation times, and high mutation rates. These traits mean that pathogen populations evolve quickly. Any mutation that enables its possessors to evade or withstand the host's immune response should be strongly selected and should quickly increase in frequency. If we can understand how particular pathogens evolve in response to attack by the human immune system, we should be better able to intervene in the conflict and improve the odds of a favorable outcome.

Conflicts among organisms are inevitable. In the conflict between a parasite and its host, the host's immune system selects for parasites that can evade detection.

Walter Fitch and colleagues (1991) investigated whether selection imposed by the human immune system is responsible for detectable evolution in populations of influenza A viruses. If they could discern how and why flu populations have evolved in the past, perhaps they would be able to predict how currently circulating populations will evolve in the future. Such predictions would be of great help to the developers of flu vaccines, which have to be redesigned every year.

Flu Virus Evolution

Influenza A is responsible for annual flu epidemics and for occasional global pandemics, such as occurred in 1918, 1957, and 1968. Most of us think of flu as merely an annoyance—worse than a cold, certainly, but not as bad as chickenpox. In fact, flu can be deadly. In an ordinary flu season, the disease kills about 20,000 Americans. The 1918 pandemic flu was among the most devastating plagues in history. Within a period of months, it sickened some 20% of the world's population and killed between 50 and 100 million people (Kolata 1999; Johnson and Mueller 2002).

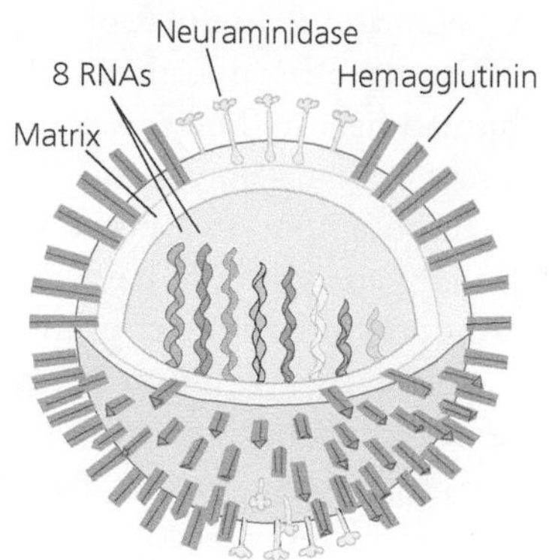

Figure 14.3 The influenza A virus The flu virus has two major surface proteins, hemagglutinin and neuraminidase. The viral genome is carried on eight separate pieces of RNA. Redrawn from Webster et al. (1992).

Influenza A, diagrammed in **Figure 14.3**, has a genome composed of eight RNA strands that encode a total of 13 proteins (Webster et al. 1992; Chen et al. 2001; Wise et al. 2009; Jagger et al. 2012). These proteins include polymerases, structural proteins, and coat proteins. The predominant coat protein is called hemagglutinin. Hemagglutinin initiates an infection by binding to sialic acid on the surface of a host cell (Laver et al. 1999).

Hemagglutinin is also the primary protein recognized, attacked, and remembered by the host's immune system. To stay alive, any given strain of influenza A must either find a steady supply of naive hosts who have never been exposed to its version of hemagglutinin or alter its hemagglutinin so that previously exposed hosts no longer recognize it. Walter Fitch and colleagues focused on mutations that alter the amino acids in hemagglutinin's antigenic sites **(Figure 14.4)**. **Antigenic sites** are the specific parts of a foreign protein that the immune system recognizes and remembers. The researchers hypothesized that flu strains with novel antigenic sites would enjoy a selective advantage.

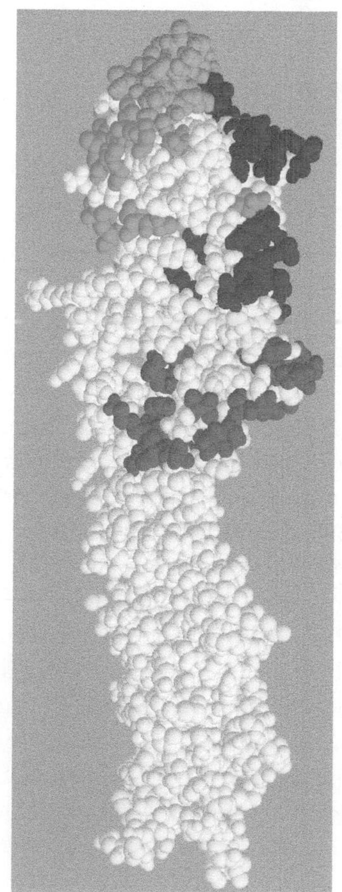

Figure 14.4 Hemagglutinin Five antigenic sites, regions recognized by the immune system, appear in color. From Plotkin and Dushoff (2003).

To test their hypothesis, the researchers examined the hemagglutinin genes of influenza A viruses that had been isolated from infected humans, and stored in freezers, between 1968 and 1987. Flu viruses evolve a million times faster than mammals, so the period of 20 years spanned by the frozen virus samples is equivalent to roughly four times the duration that separates humans from our common ancestor with the chimpanzees. In other words, the frozen flu samples constitute a fossil record—but one from which we can sequence genes.

From the sequences of the hemagglutinin genes, Fitch and colleagues estimated rate of evolution and the phylogeny of the frozen flu samples. The results appear in **Figure 14.5** (facing page). Two patterns are apparent. First, the flu strains accumulated nucleotide substitutions in their hemagglutinin genes at a steady rate, about 6.7×10^{-3} per nucleotide per year (Figure 14.5a). Second, most of the flu samples represent extinct side branches on the evolutionary tree (Figure 14.5b). The flu lineages that persisted into the 1980s were not a diverse assembly of strains descended from a variety of ancestors from the late 1960s and early 1970s. Instead the strains alive in the 1980s were close relatives, and all were descended from a single one of the late-1960s strains. The progeny of the other late-sixties and early-seventies strains had all died out.

What allowed the surviving lineage to endure while the other lineages perished? According to the researchers' hypothesis, it was nucleotide substitutions resulting in amino acid replacements in hemagglutinin's antigenic sites. From the nucleotide sequences, the researchers inferred all of the amino acid replacements that had occurred in the surviving lineage and in the extinct lineages. Then they noted whether each replacement had occurred in an antigenic site or a nonantigenic site. Fitch and colleagues predicted that, compared with the extinct lineages, the surviving lineage would have a higher fraction of its amino

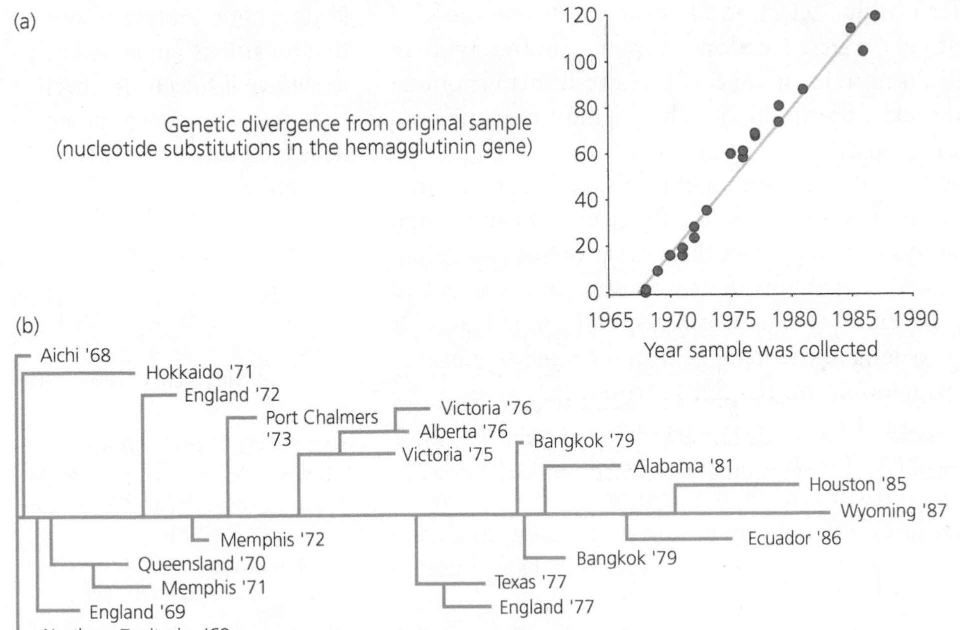

Figure 14.5 A phylogenetic analysis of frozen flu samples (a) The molecular evolution of the influenza A hemagglutinin gene as a function of time. The surviving lineage accumulated nucleotide substitutions at a constant rate. (b) A phylogeny of flu viruses isolated between 1968 and 1987. From Fitch et al. (1991).

acid replacements in its antigenic sites. The amino acid replacements, 109 in total, were distributed as follows:

	In antigenic sites	In nonantigenic sites
Surviving lineage	**33**	**10**
Extinct lineages	**31**	**35**

Consistent with the researchers' prediction, more than three-quarters of the surviving lineage's replacements had occurred in regions of hemagglutinin recognized by the immune system, compared with fewer than half of the extinct lineages' replacements. This association between a lineage's fate and the location of its replacements is statistically significant ($p = 0.002$).

Robin Bush, Walter Fitch, and colleagues (1999) followed up on this result by examining nucleotide substitutions in a phylogeny of hemagglutinin genes from 357 influenza A strains isolated between 1985 and 1996. The researchers took as their null hypothesis the neutral theory of molecular evolution (see Chapter 7). Recall that under the neutral theory, two processes dominate molecular evolution: (1) Mutations resulting in amino acid replacements are typically deleterious and are eliminated by selection, and (2) mutations to synonymous codons are neutral and may become fixed in the population by genetic drift. According to the neutral theory, when we look at the nucleotide substitutions that have occurred on an evolutionary tree, silent substitutions should outnumber replacements. Of the 331 nucleotide substitutions that Bush, Fitch, and colleagues analyzed, 191 (58%) were silent and 140 (42%) were replacement substitutions. This result is consistent with the neutral theory (see Figure 7.24, page 261).

However, the researchers also identified 18 codons in the hemagglutinin gene in which there had been significantly more replacement substitutions than silent substitutions. The ratios in these 18 codons ranged from 4 replacement substitutions and 0 silent substitutions to 20 replacement substitutions and 1 silent substitution. An excess of replacement substitutions over silent substitutions is not consistent with the neutral theory. Bush, Fitch, and colleagues concluded that these 18 codons had been under positive selection for changes in the encoded amino acid. All 18 of the positively selected codons were for amino acids in antigenic sites of the hemagglutinin protein. It appears that the human immune system does, indeed, exert strong selection on flu virus hemagglutinin genes and that virus populations evolve in response.

Phylogenetic analyses show that flu strains are more likely to survive if they have novel amino acid sequences in proteins recognized by the host's immune system.

This result is potentially useful to the makers of flu vaccines. Flu vaccines work by exposing the patient's immune system to killed flu viruses. Even though the viruses are dead, the immune system recognizes the viral proteins as foreign, mounts a response against them, and remembers their structure. In the event of a later infection by live viruses, the immune system can respond immediately. It can respond immediately, that is, as long as the hemagglutinin on the live invaders is similar enough to the hemagglutinin on the dead viruses that were in the vaccine. The problem is that flu populations evolve rapidly, and vaccines take months to prepare in large quantities. Vaccine makers must begin production well in advance of the flu season. That means their scientific advisors must try to predict which among recently circulating flu strains are most likely to be responsible for next season's epidemic, so that they know which strains to include in the vaccine.

Robin Bush, Catherine Bender, and colleagues (1999; see also Bush 2001) devised a way to predict which of the currently circulating flu strains is most likely to have surviving descendants in the future. The survivor, they reasoned, is most likely to be the currently circulating strain with the most mutations in the 18 codons known to be under positive selection (**Figure 14.6**). On this basis, the researchers were able to accurately "predict," for 9 of 11 recent flu seasons, which of each season's strains would be the one to survive while the rest became extinct.

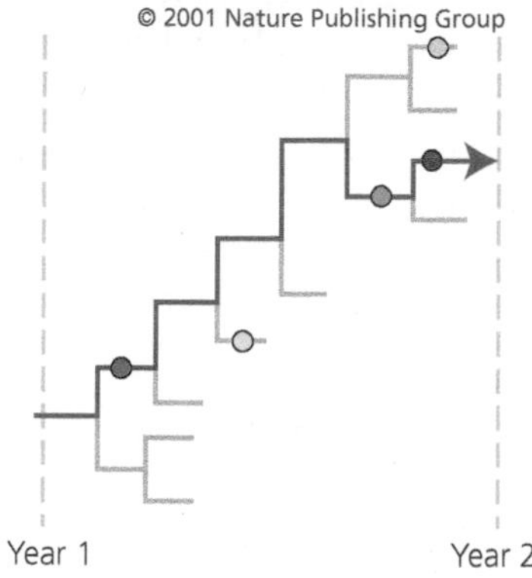

Figure 14.6 Predicting which lineages of flu will survive to cause future epidemics Mutations continually generate new lineages of flu, represented on an evolutionary tree as new branches. Among the lineages alive at any given time, one (red line) will ultimately survive; the rest (gray lines) will eventually go extinct. Which lineage will survive? Usually, it is the one with the most amino acid replacements in its hemagglutinin antigenic sites (indicated by colored dots). Redrawn from Bush (2001). Reprinted by permission of Macmillan Publishers Ltd. R. M. Bush. 2001. "Predicting adaptive evolution." *Nature Reviews Genetics* 2(5): 387–392. Copyright © 2001.

Bush, Bender, and colleagues are careful to note that predicting which of this season's flu strains will be the ancestor of future lineages is not the same as predicting which, if any, of this season's strains will be responsible for next season's epidemic. Nonetheless, the predictive technique devised by Bush, Bender, and colleagues adds an additional tool to the kit of forecasting methods already available. And it has spurred other researchers to look for ways to refine the technique (Ferguson and Anderson 2002; Plotkin et al. 2002; Ito et al. 2011).

The Origin of Pandemic Flu Strains

The fact that flu viruses with novel hemagglutinin genes appear to be at a selective advantage in evading their hosts' immune systems suggests a mechanism by which a strain could gain the ability to cause a pandemic. If a flu strain could somehow radically alter the structure of its hemagglutinin so that it was different from any hemagglutinin that had ever been seen by any human's immune system, then the strain could sweep the world and potentially infect everyone alive.

How could a flu strain radically alter the structure of its hemagglutinin? The organization of the influenza genome indicates a way (Figure 14.3). Recall that the flu genome has eight different RNA strands that encode a total of 13 different genes. If two flu strains simultaneously infect the same host cell, their genomes can recombine. That is, when new virions form, they may contain some RNA strands from strain 1 and other RNA strands from strain 2. Strain 1, for example, might produce offspring carrying strain 2's hemagglutinin gene.

The phylogeny in **Figure 14.7** provides evidence that flu strains do, in fact, swap genes. This phylogeny, by Owen Gorman and colleagues (1991), is based on nucleotide sequences of influenza nucleoprotein genes.

Phylogenetic analyses shed light on where, when, and how epidemics emerge.

Nucleoprotein is thought to be the viral protein most responsible for host specificity. The structure of a strain's nucleoprotein enables the strain to infect particular species of hosts. The structure of its nucleoprotein also tends to confine the strain to those species only. Phylogenies based on the nucleoprotein gene should therefore be reliable indicators of flu strain history.

The nucleoprotein phylogeny has several distinct **clades.** A clade, in this context, is a set of strains derived from a particular common ancestor. The distinct clades on the flu phylogeny include one that infects mainly horses, one that infects mainly humans, another that infects mainly pigs, and two that infect mainly birds. Unusual hosts within clades are marked with asterisks.

Look at the branch tips, their colors, and their labels. The colors indicate the species from which each strain was isolated. The labels give the year of isolation and the viral subtype. The subtype H3N2, for example, means hemagglutinin-3, neuraminidase-2. Neuraminidase, like hemagglutinin, is a coat protein. The numbers refer to groups of hemagglutinins or neuraminidases, which are defined by the ability of host antibodies to recognize them. The most important point for our purposes is that each hemagglutinin group constitutes a clade. That is, all H1s are more closely related to each other than to any H2 or H3 or H4. The same is true of the neuraminidases.

Find the human strains Human/Victoria/1968 (H2N2) and Human/Northern Territory/60/1968 (H3N2); they are set in boldface type. These strains have nucleoproteins that are closely related. They share a more recent common ancestor with each other than either does with all but one of the other strains on the tree. The two 1968 strains also have neuraminidases that are closely related. Both carry neuraminidase N2. But they have hemagglutinins that are distantly related. One carries H2, the other carries H3.

How is it possible that two flu strains can have some genes that are closely related and others that are distantly related? The simplest explanation is that flu strains can trade genes. An examination of the phylogeny will reveal numerous additional examples. Appearing at the bottom of the larger bird clade, for instance, are a mink strain and a bird strain with closely related nucleoproteins but distantly related hemagglutinins and neuraminidases.

Returning to the 1968 strains, note that before the global pandemic of that year, human flu viruses had never carried H3. This suggests that it was the acquisition of H3 from a nonhuman strain that allowed the

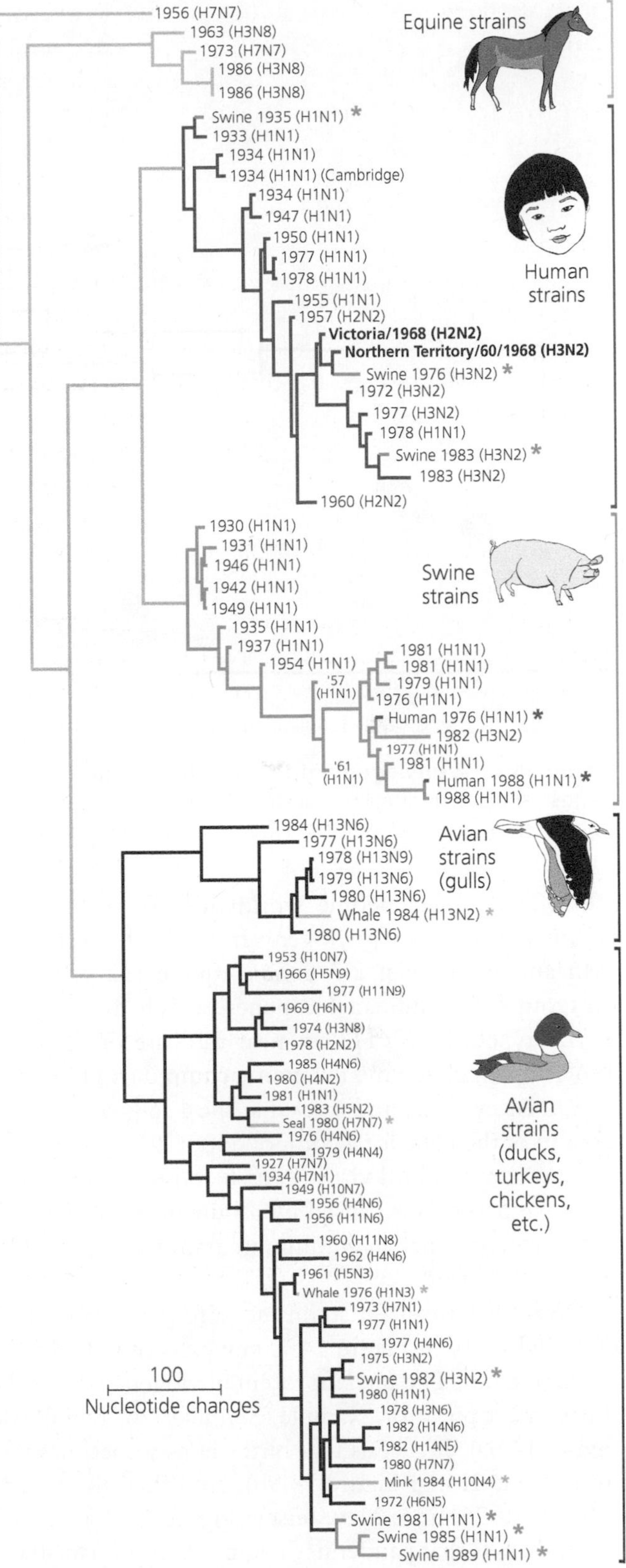

Figure 14.7 A phylogeny of flu virus nucleoprotein genes Indicated for each viral strain is the host species, the year of isolation, and the type of hemagglutinin and neuraminidase it carries. Redrawn from Gorman et al. (1991).

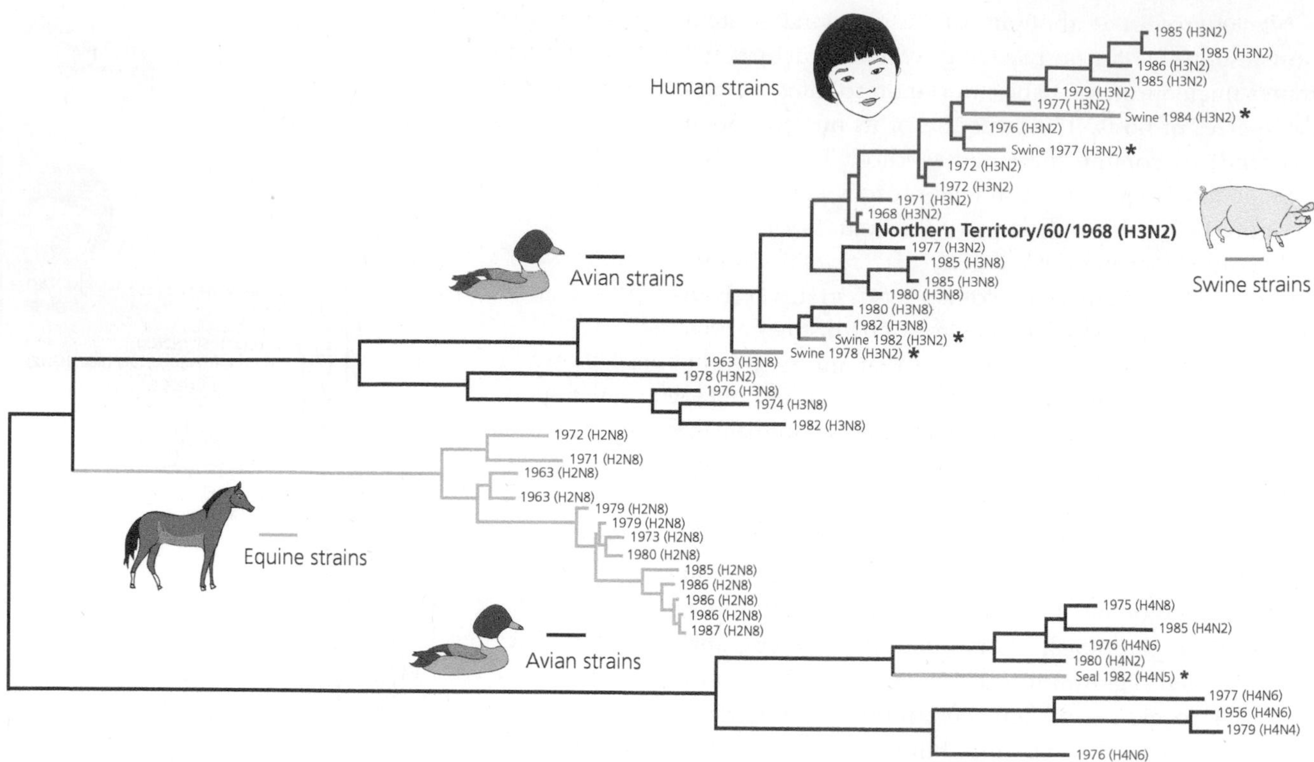

Figure 14.8 A phylogeny of flu virus hemagglutinin genes Indicated for each viral strain is the species it was isolated from, the year, and the type of hemagglutinin and neuraminidase it carries. The 1968 human flu appears to have acquired its hemagglutinin gene from a bird flu strain. Redrawn from Bean et al. (1992).

1968 flu to infect people worldwide. What was the source of the H3 gene? **Figure 14.8** shows a phylogeny from W. J. Bean and colleagues (1992) of human and nonhuman H3 genes; equine H2 genes and bird H4 genes are the outgroups. The human H3 genes branch from within the bird H3s. Northern Territory/60/1968's H3 sits near the base of the human clade. Apparently, the 1968 human pandemic flu strain acquired its H3 gene from a bird virus.

Similar evidence indicates that the 1968 pandemic strain also picked up a new version of the gene for a component of its polymerase enzyme called PB1. Again, the source was a bird virus (Parrish and Kawaoka 2005). Eleven years earlier, the 1957 pandemic was caused by a human strain that had replaced genes for three of its proteins—hemagglutinin, neuraminidase, and PB1—with copies it took up from a bird virus.

How do human flu strains acquire genes from bird strains? The nucleoprotein phylogeny in Figure 14.7, key parts of which are reproduced in **Figure 14.9**, contains clues. The nucleoprotein phylogeny reveals that human flu strains sometimes infect pigs (for example, Swine 1976, which was isolated in Hong Kong; Figure 14.9a). It reveals that bird strains sometimes infect pigs (Swine 1982, also from Hong Kong; Figure 14.9b). And that pig strains sometimes infect humans (Human 1976, from New Jersey; Figure 14.9c). A popular hypothesis among flu researchers is that pandemics begin when human strains and bird strains simultaneously infect a pig, swap genes with each other and perhaps with pig strains, and later move from pigs to people (Webster et al. 1992). Phylogenetic analysis by Gavin Smith and colleagues (2009) indicates that such a scenario was responsible for generating a new human strain of H1N1 influenza A that appeared in 2009.

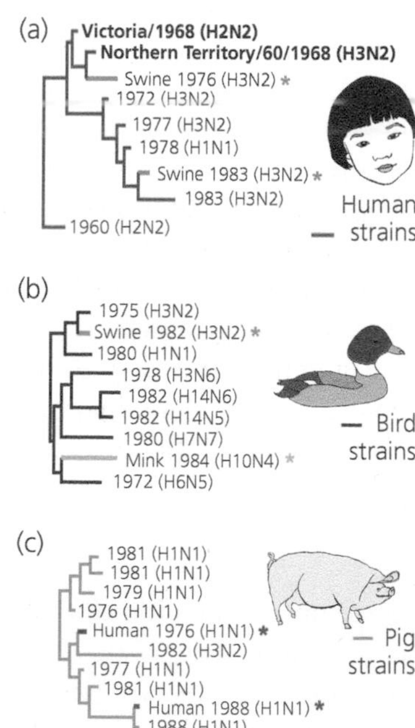

Figure 14.9 Branches from the nucleoprotein phylogeny revealing cross-species transmissions (a) Human flu can infect pigs. (b) Bird flu can infect pigs. (c) Pig flu can infect people.

Of course, the pandemic flu we most need to understand is the scourge of 1918. Toward this end Jeffery Taubenberger, Ann Reid, and colleagues (2005) have sequenced the genome of a flu strain recovered from the body of an Inuit woman who died in 1918 at Brevig Mission, Alaska, and was buried in permafrost. Terrence Tumpey and colleagues (2005) recreated live flu virus from the sequence, tested it in mice, and confirmed that it is extraordinarily deadly.

Reid, Taubenberger, and coworkers have analyzed the evolutionary history of each gene of the 1918 virus (Reid et al. 1999, 2000, 2002, 2004a; Basler et al. 2001; Taubenberger et al. 2005). The results for the nucleoprotein gene are typical (see Reid et al. 2004b). It is these we will focus on here.

An evolutionary tree of nucleoprotein genes, including the one from 1918, appears in **Figure 14.10a**. Like the nucleoprotein phylogeny in Figure 14.7, the genes on this tree form clades based on the animals they infect. The gene from the 1918 flu branches near the base of the clade that infects humans and pigs. This implies that all more recent human and pig strains are descended from the 1918 flu or one of its close relatives. The molecular clock plot in Figure 14.10b allows us to estimate the age of the common ancestor of the human and pig strains in the clade. This common ancestor lived shortly before the great pandemic.

Where did the virus that wrought havoc in 1918 and begat all subsequent human influenzas come from? Alas, the data yield no definitive answer. The problem is that the human–swine clade does not arise from within any other clade on the tree. The founder of the human–swine clade could have come from an earlier human or swine strain, a horse strain, a bird strain, or a strain that infects some other animal not represented in the tree at all. After completing a more extensive phylogenetic analysis, in which they reconstructed trees for each of eight genes from several thousand flu strains, Jiajie Zhang and colleagues (2011) reached a similar conclusion.

There are, however, additional clues in influenza genomes. **Figure 14.11**, from Mario dos Reis and colleagues (2009), compares the G+C content of various influenza genomes. G+C content is simply the fraction of nucleotides that are guanine or cytosine. The black and gray points represent influenza A strains that infect birds or have recently jumped from birds to humans. They show that strains adapted to birds have G+C contents close to 0.45. The green points represent influenza B strains that have been replicating in humans for a long time. They show that strains adapted to humans have G+C contents below 0.42. The red and blue points represent human and swine influenza A H1N1 strains. These show that since 1918, human and swine H1N1 lineages have been evolving from a birdlike G+C content toward a humanlike G+C content. Statistical analysis

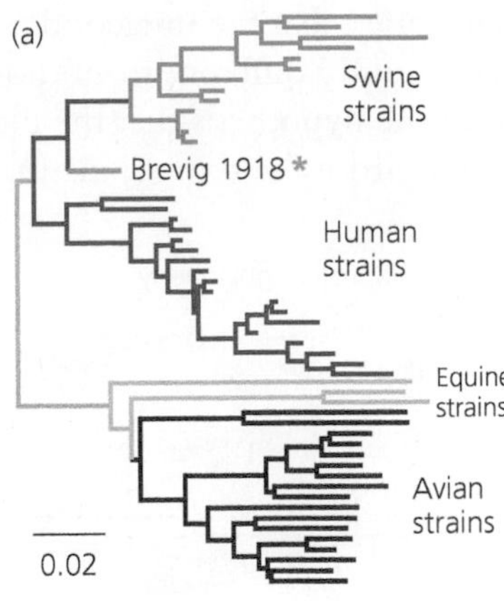

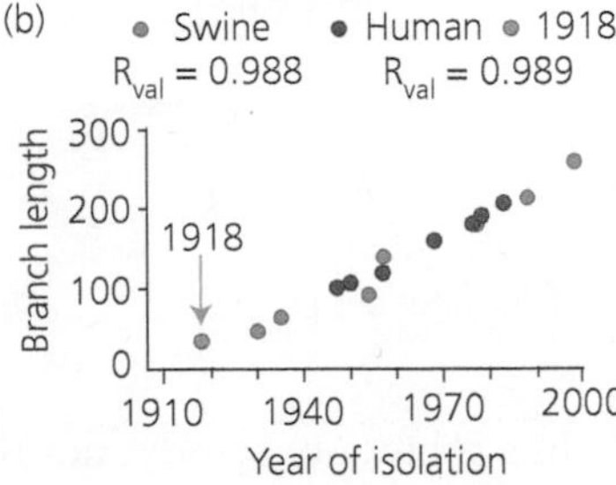

Figure 14.10 Evolutionary analysis of flu nucleoprotein genes (a) Phylogeny of various flu strains. (b) Molecular clock plot for human and swine strains. From Reid et al. (2004a).

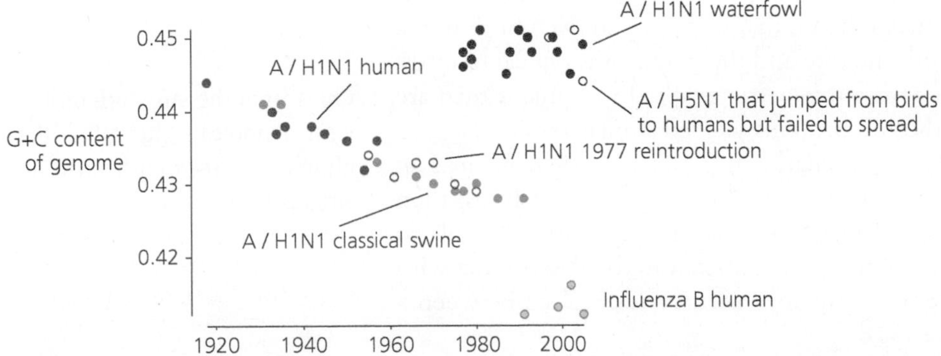

Figure 14.11 Guanine and cytosine content in influenza genomes The open red dots represent human H1N1 strains that reappeared in 1977 after a quarter-century of evolutionary stasis, most likely in a laboratory freezer (see Wertheim 2010). Their isolation times have been adjusted to account for time spent frozen. From dos Reis et al. (2009).

confirmed that the nucleotide substitution rates for human and swine lineages are significantly different from those rates for avian lineages. This result is consistent with the hypothesis that the ultimate reservoir from which the 1918 flu emerged was a bird influenza population.

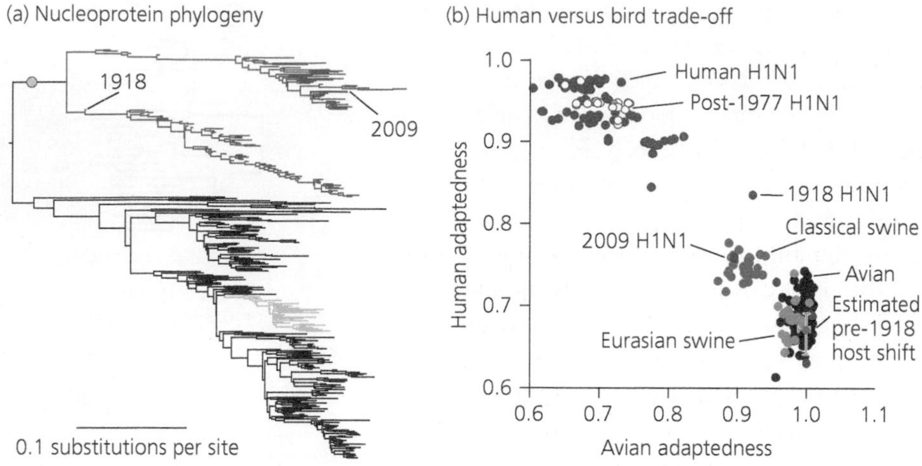

Figure 14.12 Adaptedness of flu nucleoproteins to humans versus birds (a) Phylogeny of the nucleoprotein genes analyzed. Colors represent hosts as in (b). (b) The trade-off in nucleoprotein adaptedness to humans versus birds. From dos Reis et al. (2011).

In a subsequent study, dos Reis and colleagues (2011) analyzed the nucleoprotein genes from 430 H1N1 influenza samples. **Figure 14.12a** shows an estimated phylogeny. Note that the 1918 sequence branches near the base of the human clade, that the human and classical swine clades are closest relatives, that the 2009 human sequence arises within the classical swine clade, and that the Eurasian swine clade arises from within the avian clade.

From the gene sequences, dos Reis and colleagues inferred the amino acid sequences. They scanned the amino acid sequences to identify sites where the distribution of amino acids found in human versus bird samples were significantly different. This analysis allowed them to calculate an index, for each nucleoprotein in the sample, of how well the protein was adapted to living in humans versus birds. The results appear in Figure 14.12b. The plot reveals a number of interesting patterns. First, the avian flu proteins are highly adapted to birds, whereas the human flu proteins range from poorly adapted to their host—including the 1918 and 2009 sequences—to well, but imperfectly adapted. Although time is not shown in the figure, more recent human flu proteins are—with some exceptions—better adapted to humans. Second, classical swine strains, which have been evolving in a mammal considerably longer than Eurasian swine strains, have proteins better adapted to humans. Third, the 2009 human strain, which jumped from pigs to humans, appears typical of classic swine strains. Fourth, the estimated sequence of the nucleoprotein in the flu that first jumped from birds into a mammal, indicated in orange in both the tree and the graph, was typical of birds. Finally, the 1918 human flu nucleoprotein is better adapted to humans than are swine proteins. Analyses of five other flu proteins yielded similar results.

Clues from the structure of flu virus genomes suggest that the ultimate source of influenza A virus is birds.

These patterns are consistent with a scenario in which the 1918 flu had an origin similar to that of the 2009 flu. The virus ultimately arose in birds, then jumped into pigs. After evolving better adaptedness to pigs, the virus was able to switch to humans. The exact timing of the host shifts cannot be determined with certainty. Dos Reis et al. (2009, 2011) estimate that they took place between a few years and a few decades before 1918.

As a result of the discoveries discussed in this section—the work of evolutionary biologists applying the tools of their discipline to a practical problem—an international team of researchers maintains constant surveillance of flu strains circulating in pigs, birds, and humans. Their goal is to spot new pandemic strains early enough to allow the production and distribution of large quantities of vaccine. The flu surveillance researchers keep an especially keen watch for recombinant strains and for strains that are moving from species to species.

14.2 Evolving Pathogens: Antibiotic Resistance

Antibiotics are chemicals that kill bacteria by disrupting biochemical processes. For patients, antibiotics are lifesaving drugs. For populations of bacteria, however, antibiotics are agents of selection. An antibiotic quickly sorts resistant bacteria (those that can tolerate the drugs) from susceptible ones (those that cannot). An evolutionary perspective suggests that antibiotics should be used judiciously; otherwise, these miracle drugs may undermine their own effectiveness.

There are numerous antibiotics and many molecular mechanisms whereby bacteria can become resistant (for reviews, see Nikaido 2009; MacLean et al. 2010). Some of these mechanisms of resistance involve losses of function. Resistance to isoniazid in the tuberculosis pathogen *Mycobacterium tuberculosis* provides an example. Isoniazid poisons bacteria by interfering with the production of components of the cell wall (Rattan et al. 1998). Before it can do so, however, isoniazid must be converted by a bacterial enzyme into its biologically active form. The conversion is performed by the enzyme catalase/peroxidase, encoded by a gene called *KatG*. Mutations in *KatG* that reduce or eliminate catalase/peroxidase activity render bacteria tolerant or immune to isoniazid's effects.

Other mechanisms of resistance involve gains of function. Numerous extrachromosomal elements of bacteria, such as plasmids and transposons, carry genes conferring resistance to one or more antibiotics. The plasmid *Tn3*, for example, found in *Escherichia coli*, contains a gene called *bla*. This gene encodes an enzyme, β-lactamase, that breaks down the antibiotic ampicillin.

Evidence That Antibiotics Select for Resistant Bacteria

Several lines of evidence show that antibiotics select in favor of resistant bacteria, and that bacterial populations evolve rapidly in response.

Evidence that antibiotics select in favor of resistant bacteria comes from a variety of studies. On the smallest scale are studies of bacterial evolution within individual patients. William Bishai and colleagues (1996) monitored an AIDS patient with tuberculosis. When they initially determined that the patient had tuberculosis, the researchers cultured bacteria from the patient's lungs and found them sensitive to a variety of antibiotics, including rifampin. They and other doctors treated the patient with rifampin in combination with several other drugs. The patient responded well to treatment. At one point, the patient was so nearly recovered that the researchers were unable to culture tuberculosis bacteria from his lungs. Soon, however, the patient relapsed and died. After his death, the researchers found that the tuberculosis bacteria in the patient's lungs had resurged. They screened these bacteria for resistance to antibiotics. The bacteria were still susceptible to most drugs, but they were resistant to rifampin. The researchers sequenced the *rpoB* gene from some of the resistant bacteria. In the gene, they found a point mutation known to confer rifampin resistance.

Did the rifampin-resistant strain of tuberculosis evolve in the patient's lungs, or had he been infected with a new strain that was already resistant when he got it? The researchers prepared genetic fingerprints of rifampin-sensitive bacteria from the patient's initial infection and rifampin-resistant bacteria from the patient's autopsy. Other than the *rpoB* point mutation, the genetic fingerprints of the two groups of bacteria were identical. The researchers examined over 100 other strains of bacteria from patients living in the same city at the same time. Only two had genetic fingerprints matching the strain that killed the patient, and neither was rifampin resistant. The simplest explanation for these results is that the *rpoB* point mutation occurred in bacteria living in the patient's lungs and ultimately rose to high frequency due to selection imposed by treatment with rifampin.

On a larger scale, researchers can compare the incidence of susceptible versus resistant bacterial strains among patients who are newly diagnosed, and thus have not been previously treated with antibiotics, versus patients who have relapsed after antibiotic treatment. If antibiotics select in favor of drug resistance, then a higher fraction of relapsed patients should harbor antibiotic-resistant bacteria. Alan Bloch and colleagues (1994) reported the results of a survey of tuberculosis patients conducted by the Centers for Disease Control. The results for bacterial susceptibility to isoniazid are as follows:

	New cases	Relapsed cases
Number with resistant bacteria	**243**	**41**
Number with susceptible bacteria	**2728**	**150**
Fraction resistant	**8.2%**	**21.5%**

These numbers are consistent with the notion that populations of bacteria within patients evolve in response to treatment.

Finally, on the largest scale, researchers can examine the relationship over time between the fraction of patients with resistant bacteria and the society-wide level of antibiotic use. If antibiotics select in favor of resistance, then the level of resistance should track antibiotic consumption. D. J. Austin and colleagues (1999) plotted data on penicillin resistance among *Pneumococcus* bacteria in children in Iceland **(Figure 14.13a)**. In the late 1980s and the early 1990s, the fraction of children whose bacteria were resistant to penicillin rose dramatically. In response, Icelandic public health authorities waged a campaign to reduce the use of penicillin. Between 1992 and 1995, the per capita consumption of penicillin by children dropped by about 13%. The level of penicillin resistance peaked at just under 20% in 1993, then fell below 15% by 1996. Lova Sun and colleagues (2012) combined data gathered in the United States from 1999 through 2007 to plot seasonal changes in aminopenicillin prescriptions and the frequency of ampicillin resistance in *E. coli*. Figure 14.13b shows that the frequency of antibiotic resistance tracks changes in the number of prescriptions. Once again, the data are consistent with the hypothesis that bacterial populations evolve in response to selection imposed by antibiotics.

(a)

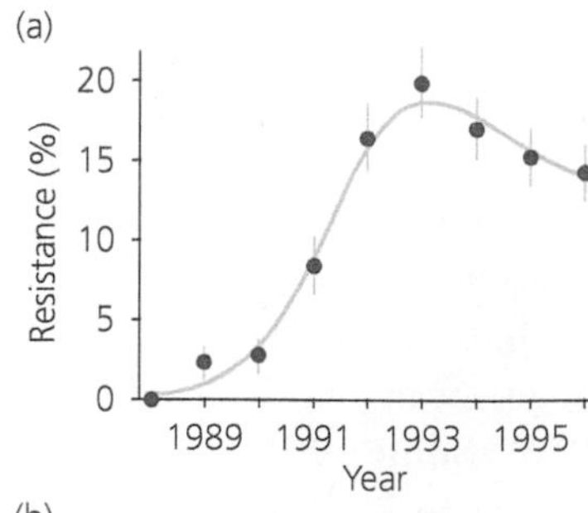

(b)

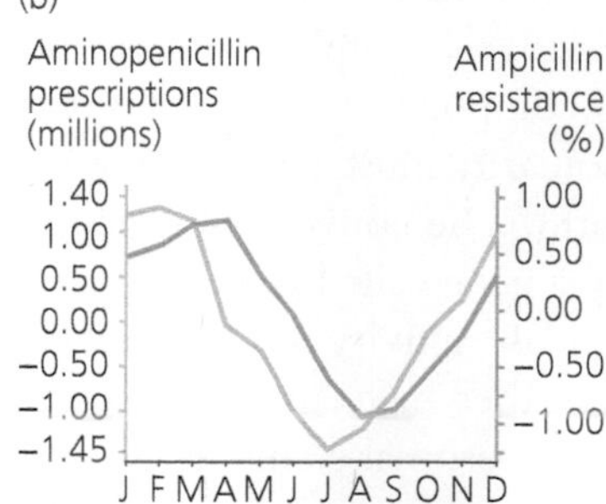

Figure 14.13 Antibiotic use and antibiotic resistance (a) The frequency of penicillin resistance among *Pneumococcus* bacteria in Icelandic children as a function of time. When public health officials waged a public campaign to reduce the use of penicillin, the incidence of resistant strains declined. Redrawn from Austin et al. (1999). (b) Seasonal pattern of change, relative to the mean, in aminopenicillin prescriptions and *E. coli* ampicillin resistance in the United States. Redrawn from Sun et al. (2012).

Evaluating the Costs of Resistance to Bacteria

Presumably, antibiotic resistance falls with declining antibiotic use because antibiotic resistance is costly to bacteria. If resistance comes at a cost, then when antibiotics are absent, sensitive bacteria will have higher fitness.

Costs of resistance are common (Andersson and Hughes 2010). When antibiotic resistance is conferred by loss-of-function mutations, costs can be levied by the loss of function itself. When resistance is conferred by gains of function, costs can be levied by the expense of maintaining new genes and proteins.

Costs to bacteria associated with antibiotic resistance should be good news for public health. If an antibiotic begins to lose its effectiveness because too many bacteria are resistant, doctors and patients could simply make a collective agreement to suspend the use of the antibiotic until bacterial populations have evolved back to the point that they are again dominated by susceptible strains.

However, while resistance typically comes at a cost, the cost does not always persist. Additional mutations elsewhere in the bacterial genome may compensate, making resistant bacteria equal in fitness to sensitive ones even when antibiotics are absent. Stephanie Schrag and colleagues (1997) investigated whether compensatory mutations could alleviate the cost of streptomycin resistance in *E. coli.*

Antibiotic resistance is generally assumed to be costly to bacteria; but over the long term, costs of resistance can be eliminated by natural selection.

Schrag and colleagues started with a population of streptomycin-sensitive *E. coli* and screened for new streptomycin-resistant mutants. Streptomycin interferes with protein synthesis by binding to a ribosomal protein encoded by the *rpsL* gene. Point mutations in the *rpsL* gene render bacteria resistant to streptomycin. In one set of experiments, the researchers competed newly streptomycin-resistant strains against identical strains restored to sensitivity by replacement of the mutant version of *rpsL* with the wild-type version. If resistance comes at a cost, at least in the short term, then in mixed cultures the streptomycin-sensitive bacteria should increase in frequency over time. This is exactly what happened **(Figure 14.14a)**.

In a second set of experiments, Schrag and colleagues let streptomycin-resistant strains evolve for many generations. The researchers again competed streptomycin-resistant strains against identical strains restored to sensitivity by genetic manipulation. If compensatory mutations had occurred and become fixed while the resistant strains were evolving, then in mixed cultures the streptomycin-sensitive strains should fail to increase in frequency over time. In fact, the result was even more dramatic: The sensitive strains decreased in frequency (Figure 14.14b).

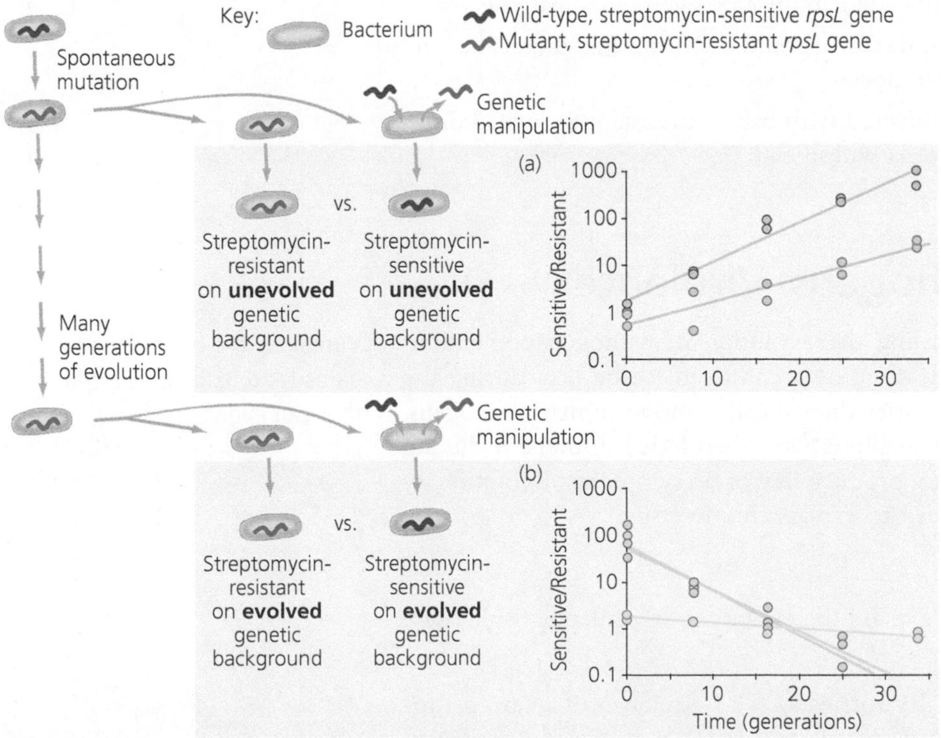

Figure 14.14 An assessment of the costs of antibiotic resistance to bacteria, over the short versus long term (a) New mutations conferring streptomycin resistance in *E. coli* come at a cost. (b) Given time for compensatory evolution, the cost of resistance disappears. Indeed, the resistance mutations become beneficial. See text for details. Scatterplots from Schrag et al. (1997).

Schrag and colleagues concluded that compensatory evolution had not only alleviated the cost of streptomycin resistance, it had created a multilocus genotype, or genetic background, on which the resistant allele of *rpsL* enjoyed a fitness advantage over the sensitive allele.

Judicious Use of Antibiotics

As Schrag et al.'s results show, there is no guarantee that an antibiotic can be restored to medical effectiveness simply by withdrawing it from use until bacterial populations have evolved back to sensitivity. If we want to maintain an arsenal of potent antibiotics for use when patients' lives are at stake, it would be wise to try to avoid letting bacterial populations evolve resistance in the first place.

Bacteriologist Stuart Levy (1998) recommends guidelines for the limitation of antibiotic resistance. The guidelines are intended to prevent people from contracting bacterial infections, restrict unnecessary uses of antibiotics that may select for resistance in potentially pathogenic bacterial bystanders, and ensure that, when antibiotics are used, they exterminate the targeted bacterial population before resistance evolves. Among Levy's guidelines are these:

- To avoid contracting foodborne bacteria, consumers should wash fruits and vegetables and avoid raw eggs and undercooked meat.
- Consumers should use antibacterial soaps and cleaners only when they are needed to prevent infection in patients with compromised immune systems.
- Patients should not request antibiotics for viral infections, such as colds or flu.
- When they take antibiotics, patients should complete the course of treatment. Patients should not save antibiotics prescribed for one infection and use them to treat another.
- To avoid spreading infections from patient to patient, doctors should wash their hands thoroughly between patients.
- Doctors should not prescribe unneeded antibiotics, even when patients ask.
- When they prescribe antibiotics, doctors should use drugs that target the narrowest possible range of bacterial species.
- Doctors should isolate patients infected with bacteria resistant to several drugs to reduce the risk that such bacteria will spread.

The best defense against antibiotic-resistant bacteria is to avoid letting bacterial populations evolve resistance in the first place.

14.3 Evolving Pathogens: Virulence

The final issue we consider regarding the evolution of pathogen populations is virulence. Virulence is the harm done by a pathogen to the host during the course of an infection. Virulence varies dramatically among human pathogens. Some pathogens, like cholera and smallpox, are often lethal. Others, including some herpes viruses and cold viruses, produce few or no symptoms. Evolutionary biologists investigating virulence seek to explain this diversity.

Virulence, the harm done by a parasite to its host, is a trait that can evolve.

How Virulence Evolves

There are three general models to explain the evolution of virulence (Bull 1994; Ewald 1994; Levin 1996):

1. **The coincidental evolution hypothesis.** The virulence of many pathogens in humans may not be a target of selection itself, but rather an accidental

by-product of selection on other traits. For example, tetanus is caused by a soil bacterium, *Claustridium tetani.* When tetanus bacteria find themselves inside a human wound, they can grow and divide. They also produce a potent neurotoxin, making tetanus infections highly lethal. However, tetanus bacteria ordinarily do not live in humans and are not transmitted by humans. The ability of these bacteria to produce tetanus toxin is probably the result of selection during their ordinary life in the soil, not selection inside human hosts.

2. **The shortsighted evolution hypothesis.** Pathogens may experience many generations of evolution by natural selection within an individual host before they have the opportunity to move to a new host. As a result, traits that enhance the within-host fitness of pathogen strains may rise to high frequency even if they are detrimental to transmission of the pathogen to new hosts. Poliovirus may provide an example. Ordinarily, polioviruses infect only cells that line the digestive tract, produce no symptoms, and are transmitted via feces. Occasionally, however, polio virions invade the cells of the nervous system. Acquiring the ability to invade the nervous system probably increases within-host fitness, because the virions that can do so have fewer intraspecific competitors. But the virions living in the nervous system are unlikely to ever be transmitted to a new host.
3. **The trade-off hypothesis.** Biologists traditionally believed that all pathogen populations would evolve toward ever-lower virulence. The reasoning was that damage to the host must ultimately be detrimental to the interests of the pathogens that live inside it. If the host dies, for example, the pathogens die with it. Thus, it was thought, more benign pathogens should enjoy higher lifetime reproductive success. This view was naive. Recall our earlier discussion of the evolution of aging (Chapter 13). There we concluded that genes that hasten the death of their carriers can nonetheless rise in frequency if they confer a sufficient enhancement of early-life reproductive success. As with genes, so too with pathogens. A strain can be virulent but nonetheless increase in frequency in the total pathogen population if, in the process of killing its hosts, it sufficiently increases its chances of being transmitted **(Figure 14.15)**. Natural selection should favor pathogens that strike an optimal balance between the costs and benefits of harming their hosts.

Figure 14.15 A parasite that kills its host before reproducing The *Cordyceps* fungus shown here invaded the body of a fly and grew inside. Finally, the fungus killed the fly and sprouted fruiting bodies that will release spores. Killing its host is not necessarily detrimental to the interests of a parasite.

We explored the shortsighted evolution hypothesis, as it applies to HIV, earlier in the book (Chapter 1). Here, we focus on the trade-off hypothesis. A key assumption of the trade-off hypothesis is that a pathogen cannot reproduce inside its host without doing the host some harm. Every offspring the pathogen makes is constructed with energy and nutrients stolen from the host. In addition, the pathogen produces metabolic wastes that the host must detoxify and eliminate. These are the reasons the host mounts an immune response against the pathogen—an expensive endeavor that compounds the host's costs but that may be better for the host than the alternative.

All else being equal, pathogens with higher within-host reproductive rates should be transmitted to new hosts at higher rates as well. But all else is equal only up to a point. Because reproducing faster within the host necessarily means harming the host more severely, it is possible for the pathogen to reproduce too fast. Reproducing too fast may mean debilitating the host so severely, or killing it so quickly, that the rate of transmission to new hosts is reduced.

According to the trade-off hypothesis, selection favors parasites that reproduce more quickly within their hosts—until the parasites begin to harm the hosts so severely that the probability of transmission begins to fall.

Sharon Messenger, Ian Molineux, and James Bull (1999) tested the trade-off hypothesis by using *E. coli* as the host and a virus, bacteriophage f1, as the pathogen. Phage f1 produces lasting, nonlethal infections in *E. coli.* The phage invades a bacterium and lives inside it as a plasmid. It induces the machinery of the host cell to produce new phage copies that are secreted from the cell as phage chromosomes encased in protein filaments. Production of new phages slows the growth rate of the host bacterium to about a third of normal. But when the host bacterium does divide, copies of phage f1 typically travel with both daughter cells. Thus phage f1 has two modes of transmission: It is transmitted vertically, from one host generation to the next, when the host cell divides, and it is transmitted horizontally, from one host to another, when secreted virions invade new hosts.

Messenger and colleagues maintained cultures of phage f1 in which they forced the viruses to alternate between the two modes of transmission. During the vertical transmission phase, the researchers prevented secreted virions from infecting new bacterial cells. The only way phages could spread was via the reproduction of their hosts. During the horizontal transmission phase, the researchers harvested secreted virions and introduced them to cultures of uninfected bacteria. Now the only way the phages could spread was via secretion.

The researchers maintained two sets of cultures. For one set they alternated 1-day-long vertical transmission phases with brief horizontal transmission phases. For the other set, they alternated 8-day-long vertical transmission phases with brief horizontal transmission phases. After 24 days, the researchers measured phage virulence and phage reproductive rate. They measured virulence as the growth rate of infected hosts, where lower growth rates indicated more virulent viruses. They measured phage reproductive rate as the rate of virion secretion from hosts, where more rapid secretion indicated faster phage reproduction.

Messenger and colleagues made two predictions. First they predicted that, across their cultures, they would find a correlation between phage virulence and phage reproduction rate. In other words, phages that induced their hosts to produce and secrete more phage copies would slow the growth of their hosts more severely. Second, they predicted that the cultures subjected to eight-day vertical transmission phases would evolve lower reproductive rates and lower virulence than the cultures subjected to one-day vertical transmission phases. Their reasoning here was that during the vertical transmission phase, natural selection should favor viral strains that allow their host bacteria to divide more quickly, whereas during the horizontal transmission phase, selection should favor viral stains that induce their host bacteria to secrete more viral copies. The eight-day cultures experienced more selection to allow their hosts to divide and less selection to induce secretion, so they should evolve lower virulence.

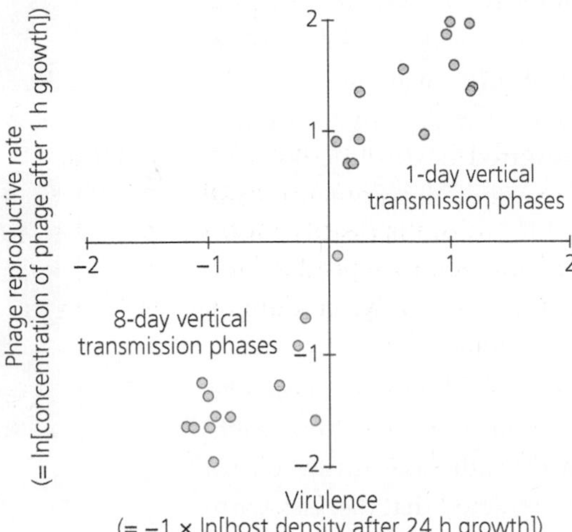

Figure 14.16 A trade-off between the virulence and within-host reproductive rate of a virus that infects *E. coli* When researchers gave the viruses more opportunities for horizontal transmission (red dots), the viruses evolved higher virulence and higher reproductive rates than viruses given fewer opportunities for horizontal transmission (blue dots). Redrawn from Messenger et al. (1999).

The results appear in **Figure 14.16**. The figure is a scatterplot showing the reproductive rate and virulence of each of 13 one-day cultures and 13 eight-day cultures. First, note the strong correlation across both experiments between viral reproductive rate and virulence. As Messenger and colleagues predicted, the phage strains that slow their hosts' growth rate most dramatically are the strains that reproduce more quickly within their hosts. Second, note that the eight-day cultures had lower reproductive rates and lower virulence than the one-day cultures. As the researchers predicted, different patterns of selection favor different levels of virulence. These results are consistent with the trade-off hypothesis.

Other researchers have also found experimental support for the trade-off hypothesis. Vaughn Cooper and colleagues (2002), for example, selected for rapid replication in populations of nuclear polyhedrosis virus by allowing transmission from host to host only during the early stages of infection. Compared to viral strains transmitted late, the early transmission strains not only had higher replication rates, they were also significantly more likely to kill their gypsy moth hosts.

Virulence in Human Pathogens

Paul Ewald (1993, 1994) considered how the trade-off hypothesis might apply to human pathogens. He used the hypothesis to guide his thinking about how the details of disease transmission should select for different levels of virulence. His key insight was that some pathogens thrive only so long as their hosts are reasonably healthy, whereas other pathogens thrive even when their hosts are severely ill. Here we discuss two of Ewald's specific predictions, along with the tests he devised using comparative data compiled from the literature.

Parasites transmitted by insect vectors or water can thrive even when their host is severely debilitated. As a result, they tend to be more virulent than parasites that are transmitted by direct contact.

Ewald's first prediction concerns the virulence of diseases like colds and flu, which are transmitted by direct contact between an infected person and an uninfected person, versus diseases like malaria, which are transmitted by insect vectors. Ewald reasoned that diseases transmitted by direct contact cannot afford to be virulent. If a host is so incapacitated by illness that she stays home and avoids contact with uninfected individuals, the pathogen has no chance to be transmitted. Vectorborne diseases, on the other hand, can afford to be highly virulent. An insect vector can carry pathogens away from even a severely debilitated host and might, in fact, be at less risk of being killed in the process. Ewald compiled data on the mortality rates of a variety of vectorborne and directly transmitted diseases. These data are consistent with Ewald's prediction (**Figure 14.17**). The vast majority of directly transmitted diseases have mortality rates under 0.1%, whereas more than 60% of vectorborne diseases have mortality rates of 0.1% or higher.

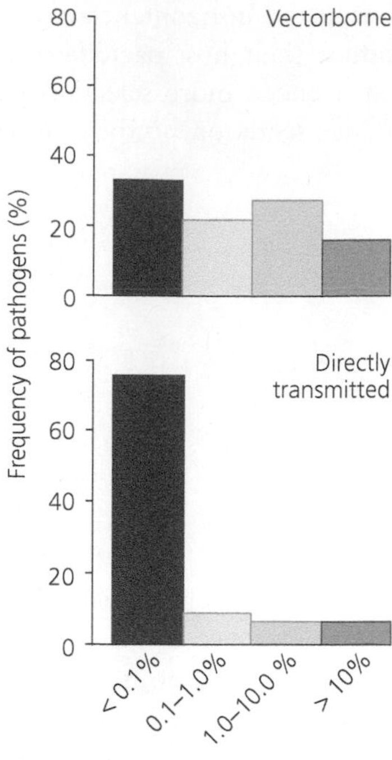

Figure 14.17 The virulence of vectorborne versus directly transmitted diseases Diseases carried from host to host by insects (top) are, on average, more virulent. From Ewald (1994).

Ewald's second prediction concerns bacteria that infect the digestive tract and cause diarrhea. These bacteria can typically be transmitted both directly from person to person and via contaminated water. Ewald reasoned that contaminated water can play the same role that insect vectors did in his first prediction. That is, when sewage enters the drinking water supply, even a severely incapacitated host can transmit bacteria to remote individuals over long distances. Ewald assembled data on roughly 1,000 outbreaks of illness caused by nine different kinds of bacteria. Some of the bacteria have a stronger tendency to be spread by direct contact; others have a stronger tendency to be spread by contaminated water. For each kind of bacteria, Ewald calculated both the fraction of outbreaks attributed to contaminated water and the victim mortality rate. He predicted that diseases with a higher frequency of waterborne transmission would be more virulent. The data, plotted in **Figure 14.18**, are consistent with Ewald's prediction. The most virulent of the nine bacteria in the study is classical *Vibrio cholerae*, the pathogen responsible for the waterborne, and lethal, cholera outbreak in London during 1854.

The trade-off hypothesis for the evolution of virulence implies that human behavior can affect the severity of human diseases. For example, when people dump untreated sewage directly into rivers, or when health-care workers fail to wash their hands thoroughly between patients, they create conditions that may select for increased virulence in human pathogens. Conversely, when people keep their drinking-water supplies pure, and when health-care workers avoid becoming inadvertent vectors, they create conditions that may select for reduced virulence in human pathogens.

We now turn our attention from evolving populations of pathogens to another group of evolving populations important to human health: populations of human cells inside individual humans. It may seem surprising to even propose that human cell populations can evolve, because we are used to thinking of all the cells in a human body as being genetically identical. In fact, however, there are mechanisms that produce genetic diversity among somatic cells, as well as conditions under which somatic cell populations can evolve.

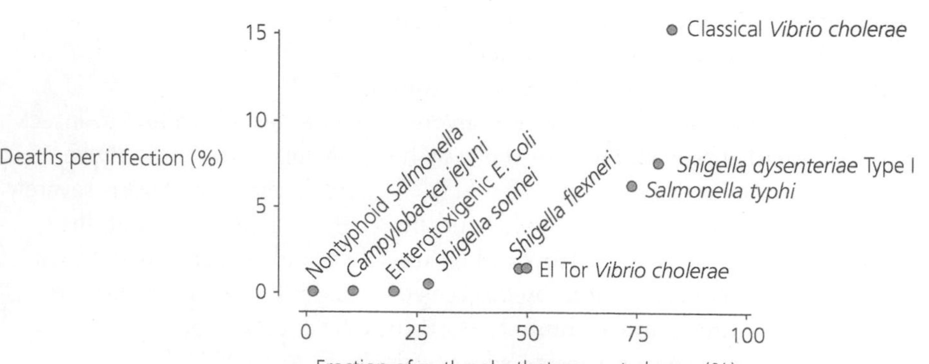

Figure 14.18 The virulence of intestinal bacteria, as a function of tendency toward waterborne transmission The higher the fraction of a disease's outbreaks that are waterborne, the more virulent the disease. Plotted with data from Ewald (1991), after Ewald et al. (1998).

14.4 Tissues as Evolving Populations of Cells

All the cells in an individual's body are descended from a common ancestor, the zygote. If, during the development of a tissue, a mutation occurs in a cell still capable of continued division, then we can think of the tissue as a population of reproducing cells with heritable genetic variation. If one of the genetic variants leads to increased cell survival or faster reproduction, then the tissue will evolve by natural selection, just like a population of free-living organisms.

A Patient's Spontaneous Recovery

Rochelle Hirschhorn and colleagues (1996) documented a case in which tissue evolution saved the life of a boy with a genetic disease. The disease is adenosine deaminase deficiency. Adenosine deaminase (ADA), encoded by a locus on chromosome 20, is a housekeeping enzyme normally made in all cells of the body. ADA's job is to recycle purines. Cells lacking ADA accumulate two poisonous metabolites—deoxyadenosine and deoxyadenosine triphosphate. The cells in the body most susceptible to these poisons are the lymphocytes, including the T cells and B cells vital to the immune system (Youssoufian 1996). Individuals who inherit loss-of-function mutations in both copies of the ADA gene have no T cells and have B cells that are nonfunctional or absent (Klug and Cummings 1997). In consequence, these individuals suffer from severe combined immunodeficiency. Without treatment, they usually die of opportunistic infections at an early age.

Both parents of the boy that Hirschhorn and colleagues studied carry, in heterozygous state, a recessive loss-of-function allele for ADA. One of the boy's older siblings inherited two loss-of-function alleles, made no ADA, and died of severe combined immunodeficiency at age 2. The boy himself also inherited both his parents' mutant alleles, and during his first five years he suffered the recurrent bacterial and fungal infections characteristic of severe combined immunodeficiency. Between the ages of 5 and 8, however, the boy spontaneously and mysteriously recovered. He was 12 years old when Hirschhorn and colleagues published their paper, and he had been clinically healthy for four years.

With a careful genetic analysis of the mother, father, and son, Hirschhorn and colleagues were able to reconstruct a plausible explanation for the boy's recovery. Although the boys' parents are both carriers for ADA deficiency, they are carriers for different loss-of-function alleles (**Figure 14.19**). Hirschhorn and colleagues showed that the son's blood cells are a genetic mosaic. The father's mutation is present in all of the boy's peripheral leukocytes (white blood cells) and lymphoid B cells. The mother's mutation is present in all of the boy's peripheral leukocytes, but absent in most of his B cells.

How could this happen? Hirschhorn and colleagues found evidence that the cell ancestral to most of the boy's existing lymphoid B cells had sustained a lucky back-mutation in the allele the boy inherited from his mother, thus spontaneously reverting to wild type, or normal. Over time, the descendants of this reverted cell apparently became more and more abundant in the boy's B-cell population. Eventually, reverted B cells became abundant enough, and made and released enough ADA, that the boy's clinical symptoms of ADA deficiency vanished.

Hirschhorn and colleagues believe that the increase in frequency of reverted B cells in the boy's B-cell population happened by natural selection. It is also possible that the increase happened by drift. It is likely, however, that reverted cells are at a distinct selective advantage. Because they make their own supply of a

© 1996 Nature Publishing Group

Wild type:

...G C G C C A C C A G C C C A G T...

Dad's mutation:

...G C G C C A C C A G C C C A **A** T...

Mom's mutation:

...G C **A** C C A C C A G C C C A G T...

Figure 14.19 A short piece of the gene for adenosine deaminase Point mutations causing loss of function appear in orange. Dad's mutation is in an intron/exon splice site. Mom's mutation results in an amino acid substitution. From Hirschhorn et al. (1996).

Reprinted by permission of Macmillan Publishers Ltd. Hirschorn, R., D. R. Yang, et al. 1996. "Spontaneous in vivo reversion to normal of an inherited mutation in a patient with adenosine deaminase deficiency." *Nature Genetics* 13: 290–295. Copyright © 1996.

crucial housekeeping enzyme, they should live longer than cells that have to pick up the enzyme after it has been released by the cells that make it.

Populations of cells inside an individual's body may exhibit genetic variation and differential fitness and thus may evolve.

The boy's story may have implications for the treatment of other individuals with ADA deficiency. The outlook for patients with ADA deficiency improved in recent decades. In 1987, researchers developed a form of injectable ADA that is an effective enzyme replacement treatment for many ADA patients (Hershfield et al. 1987). In the early 1990s, researchers began the first clinical experiments with somatic-cell gene therapy (Blaese et al. 1995; Bordignon et al. 1995). Somatic-cell gene therapy involves removing lymphocytes and/or bone marrow cells from the patient, inserting a functioning version of the ADA gene with its own promoter into their chromosomes, and returning the cells to the patient's body. In other words, gene therapy is an attempt to accomplish by design the reversion mutation that happened spontaneously in the boy studied by Hirschhorn and colleagues. The early trials have shown that the engineered cells can survive for years, and that they can grow and divide. In some early trials, gene therapy appears to have been responsible for dramatic improvements in the patients' clinical health.

As a precaution against the failure of gene therapy, researchers conducting gene-therapy trials have kept their patients on enzyme replacement therapy. Hirschhorn and colleagues suggest, however, that enzyme replacement may reduce the effectiveness of gene therapy. If enzyme replacement reduces the selective advantage the engineered cells have over ADA-deficient cells, it will slow the rate at which the patients' blood cell population evolves by natural selection. Consistent with this idea, Alessandro Aiuti and colleagues (2002b) report on an ADA gene therapy patient in which the genetically modified cells appeared to enjoy a stronger selective advantage when enzyme replacement therapy was stopped. For the short term, gene therapists will have to balance the benefits of encouraging rapid selective fixation of engineered cells against the risks of depriving patients of the insurance provided by continued enzyme replacement therapy. For the long term, researchers are working on new methods to give genetically modified cells a selective advantage inside patients (Aiuti et al. 2002a; Persons and Nienhuis 2002).

Reconstructing the History of a Cancer

Another medical context in which it is productive to view tissues as evolving populations of cells is cancer (Shibata et al. 1996). A cancer starts with a cell that has accumulated mutations that free it from the normal controls on cell division. The cell divides, and its offspring divide, and so on, to produce a large population of descendants—that is, a tumor. The cells in some kinds of cancer have extremely high mutation rates, allowing tumors to accumulate measurable genetic diversity. Some of these mutations may be adaptive, allowing cells that possess them to replicate more rapidly and increase in frequency in the tumor cell population.

Occasionally, a cell may leave the tumor it was born in and migrate elsewhere to initiate a new tumor. This process is called metastasis. The new tumor represents a new population of cells. Because it was founded by a single individual, this new population will have low genetic diversity. As it grows, however, the population will evolve. Like the population from which its founder came, the new tumor will accumulate genetic diversity as a result of mutation and genetic drift. As in the original tumor population, some of the new mutations may be adaptive.

Adopting this perspective, Yong Tao and colleagues (2011) reconstructed the history of cancer cell populations in the body of a patient. The patient, a woman

(a)

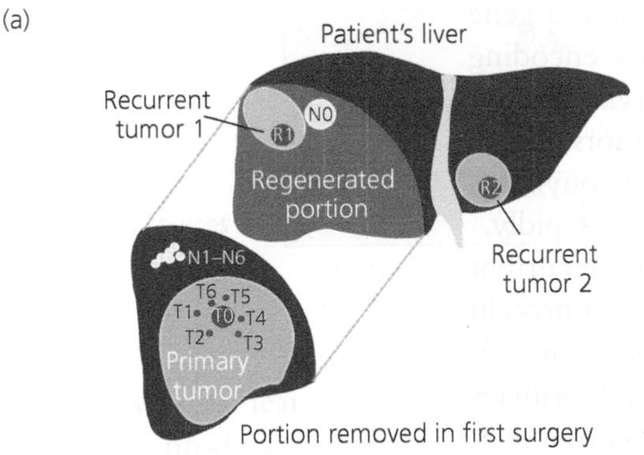

T0, T1–T6: Samples from the primary tumor (8 × 7 × 7 cm³)
R1: Sample from the recurrent tumor 1 (4.5 × 3 × 2.5 cm³)
R2: Sample from the recurrent tumor 2 (1.8 × 1.5 × 1 cm³)
N0, N1–N6: Samples from the adjacent non-tumor tissues

(b)

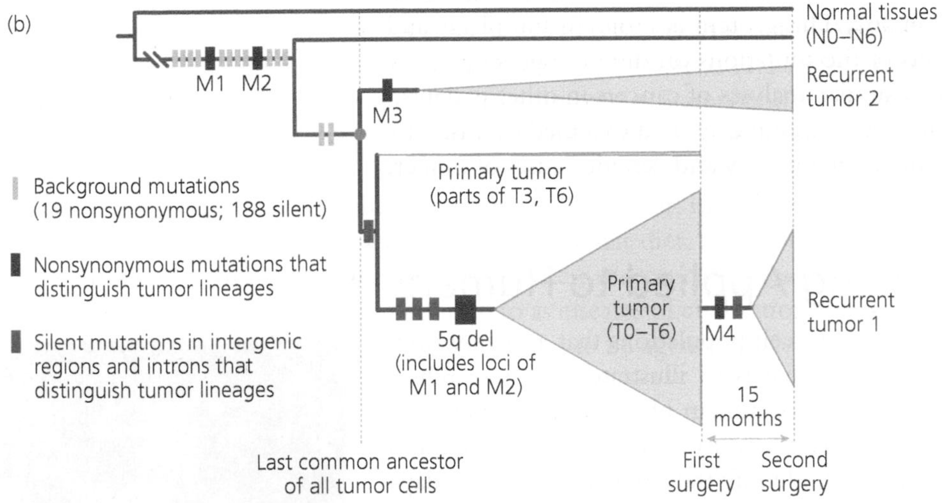

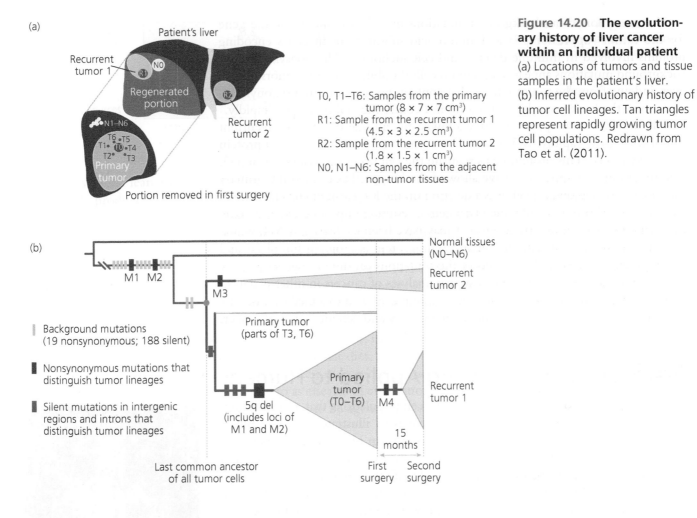

Figure 14.20 The evolutionary history of liver cancer within an individual patient (a) Locations of tumors and tissue samples in the patient's liver. (b) Inferred evolutionary history of tumor cell lineages. Tan triangles represent rapidly growing tumor cell populations. Redrawn from Tao et al. (2011).

in her 30s, was diagnosed with liver cancer associated with a hepatitis B infection. The cancer, a type known as hepatocellular carcinoma, consisted of liver cells dividing out of control. As shown in **Figure 14.20a**, the primary tumor—the first tumor discovered—encompassed much of the large lobe of the patient's liver. Surgeons removed the primary tumor along with a margin of apparently normal tissue. The patient's liver regenerated, as livers do. Fifteen months later, the surgeons removed two recurrent—that is, apparently new—tumors. One was in the regenerated portion of the large lobe; the other was in the small lobe.

Tao and colleagues used whole-genome sequencing, among other techniques, to identify mutations in tissue specimens collected from the primary tumor (T0–T6), the recurrent tumors (R1 and R2), and apparently normal nearby tissues (N0–N6). These mutations allowed the scientists to reconstruct the evolutionary tree of cell lineages shown in Figure 14.20b. The inferred tree reveals that the patient's tumors contained at least four distinct cell lineages that vary in their growth rates. It also indicates that "recurrent" tumor 2 was probably already present, though undetected, at the time of the first surgery.

The analytical tools evolutionary biologists use to reconstruct the history of populations can also be used to reconstruct the history of tissues and tumors within a patient's body.

During the patient's life before the divergence of the tumor lineages, the tumors' ancestors accumulated 188 silent mutations and 19 nonsynonymous mutations. Most of these alterations, which the researchers call background mutations, also appear in the seemingly normal tissues most closely related to the tumors.

of milk sugar in the diet of most mammals is mother's milk, so there would be no advantage, and probably some cost, in continuing to produce lactase after weaning. Indeed, individuals stop producing lactase around the age of weaning in most mammal species and in many populations of humans. However, for modern humans with lifelong access to cows' milk, continuing to produce lactase after weaning can be advantageous. In human populations with a long history of drinking fresh milk, and only in these populations, many individuals have a heritable ability to continue producing lactase as adults (Durham 1991).

Consider, however, that in a milk-drinking culture the ability to produce lactase is probably beneficial throughout life. Many diseases of civilization, on the other hand, strike only late in life. As a result, selection on genetic susceptibility to these diseases is probably weak (see Chapter 13). This pattern, combined with the fact that many novel aspects of our modern lifestyle date back just a few generations, implies that we should not expect that evolution by natural selection in human populations will have kept pace with our changing way of life. In other words, even the most sophisticated modern urbanites have bodies and brains that are at least partly designed for life in earlier eras—including the Stone Age.

The pace of cultural change has been so much faster than the pace of evolution by natural selection that modern humans are still to some extent adapted to life in the Stone Age.

The recognition that the environment we live in is different from the environment we are adapted to has at least two implications. It helps make sense of some otherwise puzzling features of our physiology, and it suggests ways to reduce some of the risks associated with modern life. Myopia provides an example of the former; breast cancer and obesity provide examples of the latter.

Myopia

In many populations the incidence of myopia, or nearsightedness, is 25% or more. Researchers have used twin studies to determine whether variation in vision has a genetic basis. For example, J. M. Teikari and colleagues (1991) assessed the similarities in vision between monozygotic versus dizygotic twins. If a trait is heritable, then monozygotic twins resemble each other more strongly than dizygotic twins (see Chapter 9). Teikari et al.'s data are as follows:

	Monozygotic twins	Dizygotic twins
Concordant pairs	**36**	**19**
Discordant pairs	**18**	**36**

Two-thirds of the monozygotic twin pairs were concordant—that is, both nearsighted or both normal—versus only one-third of the dizygotic twin pairs. These data suggest that nearsightedness is partially heritable.

But how could nearsightedness be even partially heritable? Many modern urbanites are legally blind without their corrective lenses. These people would be at a serious disadvantage if forced to live as hunter-gatherers. Surely natural selection among hunter-gatherers would quickly eliminate alleles associated with myopia. And if natural selection eliminated alleles that caused myopia among our hunter-gatherer ancestors, they cannot have passed such alleles to us.

The solution to the puzzle of myopia is to recognize that modern humans live a lifestyle that is as novel in its visual demands as in its diet and activity levels. Hunter-gatherers do not spend their childhoods indoors reading under artificial light. Perhaps the alleles that predispose some of us to myopia actually cause myopia only in a modern environment.

Human populations have not had time to adapt to the environment we live in; this can help explain some of our apparently maladaptive traits.

Evidence to evaluate this hypothesis comes from populations of people who have only recently adopted a modern lifestyle. Francis Young and colleagues

(Young et al. 1969; Sorsby and Young 1970) went to Barrow, Alaska, to measure the incidence of myopia among Inuit. The researchers chose this population because most of the families in it had moved to Barrow from isolated communities during and after World War II, drawn by the economic activity associated with a naval research laboratory, a radar station, and oil exploration. As the town of Barrow grew, a school system grew with it. Most of the children Young and colleagues examined attended formal, American-style schools and did a great deal of reading. Most of the adults over 35 had attended less formal, ungraded schools for a maximum of six years. Young et al.'s data on the incidence of myopia in younger versus older individuals are as follows:

Age	Number myopic	Number not myopic	Fraction myopic
6–35	**146**	**202**	**42%**
36–88	**8**	**152**	**5%**

The children, who had been much more strongly exposed to a modern visual lifestyle, had a substantially higher incidence of myopia.

Young et al.'s study was observational, not experimental, and the amount of schooling was just one of many differences between the environments of the children versus the adults. Nonetheless, the data are consistent with a variety of studies on humans and animals indicating that the shape of the growing eye is molded by visual experience (see Norton and Wildsoet 1999 for a review). This body of human and animal research suggests that myopia is caused by a combination of genetic susceptibility and close-in visual work. The older Inuit in Barrow had grown up in a visual environment more like that of our hunter-gatherer ancestors. The older adults had the same alleles they passed to their children and grandchildren, but they themselves were not myopic. In other words, myopia can be partially heritable because the alleles that predispose some modern humans to myopia do not cause myopia in a hunter-gatherer environment.

Breast Cancer

About one in eight North American women gets breast cancer during her lifetime. Some of these women die while still in their child-bearing years. Breast cancer is commonly thought to result from a combination of genetic susceptibility and environmental factors. But if we take a Darwinian view, breast cancer presents a puzzle (Cochran et al. 2000). If genes are responsible for a substantial fraction of breast cancers, natural selection should have eliminated breast cancer alleles from our ancestors' populations. And if breast cancer is caused by environmental factors to which our ancestors were long exposed, evolution by natural selection should have favored individuals immune to their effects.

So why is the rate of breast cancer so high? Here are two possible solutions to this puzzle:

- Breast cancer may be caused by a pathogen, such as a virus or bacterium. Viruses and bacteria are living organisms. As we have seen, pathogen populations evolve in response to selection imposed by the host's immune system. Thus we do not expect host populations to be able to evolve complete immunity to all diseases.
- Breast cancer may be a disease of civilization, like myopia. That is, it may be caused by the interaction between genes and novel environments that our ancestors were never exposed to.

Breast Cancer as a Viral Disease

Mice carry a virus, called mouse mammary tumor virus (MMTV), that causes the mouse equivalent of breast cancer. A small group of researchers have long suspected that MMTV, or something like it, can cause breast cancer in humans (Mason et al. 2011). We briefly consider two suggestive pieces of evidence.

The first piece of evidence comes from Yue Wang and colleagues (1995), working in the lab of Beatriz G.-T. Pogo. Wang and colleagues sought to extract from human tissue samples DNA sequences similar to a piece of one of MMTV's genes. The researchers analyzed 314 tissue samples from breast cancers and 107 tissue samples from normal breasts. Their results were as follows:

	Number of samples	MMTV positive	Fraction positive
Cancer tissue	**314**	**121**	**38.5%**
Normal tissue	**107**	**2**	**1.9%**

Sequences similar to MMTV's lurked in over a third of the samples from breast cancers but almost none of the samples from normal tissues. Pogo and colleagues later found viral particles in cultured breast cancer cells (see Pogo et al. 2010).

James Lawson and Benjamin Heng (2010) reviewed 23 studies in 11 countries attempting to replicate Wang et al.'s results. In 17 of these studies, researchers found evidence of MMTV or something like it in breast tumors. There was little evidence of MMTV in normal breast tissues. Chiara Mazzanti and colleagues (2011) used labeled probes to locate MMTV-like sequences in the nuclei of breast cancer cells. With other methods, they found that viral load was higher in more advanced tumors.

The second piece of evidence comes from work by T. H. M. Stewart and colleagues (2000). These researchers looked at rates of breast cancer among women in various countries across Europe. The researchers knew that the species of house mouse found in western Europe is *Mus domesticus*, whereas the species found in eastern Europe is *Mus musculus*. And they knew that *Mus domesticus* tends to be more heavily infected with MMTV. The researchers reasoned that if MMTV causes breast cancer in humans, than the rate of breast cancer should be higher in countries with *Mus domesticus* than in countries with *Mus musculus*. The data, shown in **Figure 14.24**, are consistent with this prediction.

All of the studies we have reviewed are observational. They demonstrate that MMTV is associated with a substantial fraction of breast cancers, but cannot reveal whether the virus is causing the cancers. One way researchers could investigate cause is with a randomized-controlled vaccine trial (Mason et al. 2011).

Although it may ultimately explain some cases of breast cancer, MMTV will certainly not explain them all. We should therefore also consider the hypothesis that breast cancer is, in part, a disease of civilization.

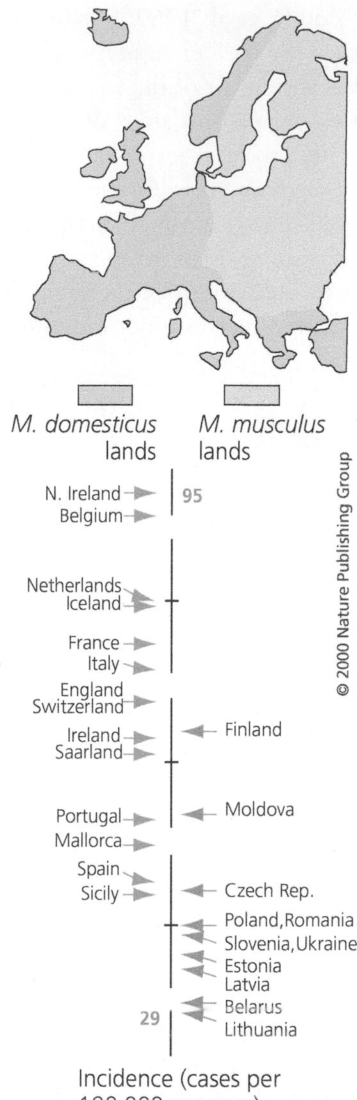

Figure 14.24 Incidence of breast cancer versus local house mouse species Rates of breast cancer, which range from 29 to 95 cases per 100,000 women per year, are higher where *Mus domesticus* lives. From Stewart et al. (2000).

Reprinted by permission of Macmillan Publishers Ltd. H. M. Stewart, R. D. Sage, A. F. R. Stewart, & D. W. Cameron. 2000. "Breast cancer incidence highest in the range of one species of house mouse, *Mus domesticus*." *British Journal of Cancer* 82: 446–451. Copyright © 2000.

Breast Cancer and Menstrual Cycling

The monthly menstrual cycling experienced by most modern women is usually considered normal. However, epidemiological evidence suggests that continuous menstrual cycling increases a woman's risk of breast cancer. A woman's risk of cancer is higher the earlier she begins to menstruate, the later she has her first child, and the less time she spends nursing (see, for example, Layde et al. 1989; Berkey et al. 1999). Menstrual cycling appears to elevate the risk of breast cancer because the combination of estrogen and progesterone present during the post-ovulatory phase of the cycle stimulates cell division in the lining of the milk ducts

Figure 14.25 Dogon women grinding millet with a mortar and pestle The Dogon are a traditional society located in Mali, northwestern Africa.

(Henderson et al. 1993). With more cell divisions come more opportunities for mutations that may create cancers. Given the high incidence of breast cancer among modern women—about one in eight in North America—it is worth knowing whether continuous menstrual cycling really is normal.

Beverly Strassmann (1999) spent two years observing menstrual cycling among the Dogon of Mali (**Figure 14.25**). The Dogon are a traditional society who use no contraceptives. Their culture is an easy one within which to study menstrual cycling because custom dictates that the women sleep in special menstrual huts while they are menstruating. Strassmann confirmed that menstruating women do, in fact, sleep in the huts, and that women sleeping in the huts are menstruating, by regularly collecting urine samples from 93 women for two and one-half months and checking for metabolites of estrogen and progesterone.

Strassmann then tracked visits to the huts by the Dogon women over a period of two years. She found that women between the ages of 20 and 35 spend little time cycling (**Figure 14.26a**). Instead, they are usually either pregnant or experiencing lactational amenorrhea—a suppression of cycling due to nursing. On any given day, about 25% of adult Dogon women are cycling, about 15% are pregnant, about 30% are in lactational amenorrhea, and about 30% are past menopause (Figure 14.26b). Strassmann estimates that the average Dogon woman has a total of about 100 menstrual cycles during her lifetime. This is less than one-third the number for a typical modern urban woman.

Strassmann's data suggest that women's bodies may not have been designed by natural selection to tolerate long periods of continuous menstrual cycling. If continuous cycling is not normal for women, then we can think of the high rates of breast cancer among modern women as another maladaptive consequence of life in a novel environment. Strassmann does not have data on the incidence of breast cancer among Dogon women, but she notes that among urban West African women, whose menstrual patterns parallel those of the Dogon, the breast cancer rate is about one-twelfth that among North American women.

Perhaps modern women should consider using hormonal treatments that maintain their bodies in a hormonal state more consistent with the state experienced

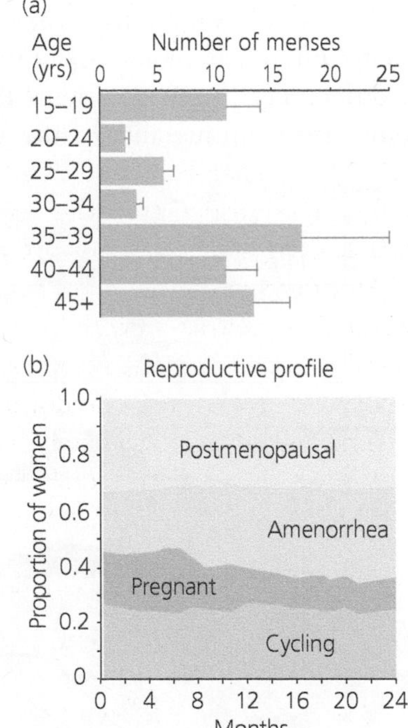

Figure 14.26 Menstrual cycling in Dogon women (a) Cycles during a two-year period. (b) Reproductive profile over time. From Strassmann (1999).

by our ancestors. Oral contraceptives reduce the risk of endometrial and ovarian cancer among modern women who use them, but they do not reduce the risk of breast cancer (Henderson et al. 1993). Darcy Spicer and colleagues have developed an oral contraceptive regimen that suppresses ovarian function, does not stimulate cell division in the breast, and contains sufficient concentrations of sex steroids to avoid adverse side effects such as accelerated osteoporosis (Spicer et al. 1991; Henderson et al. 1993; Pike et al. 2004). They predict that their strategy will be equally as effective as present oral contraceptives at reducing the risk of ovarian cancer, about half as effective at reducing the risk of endometrial cancer, and much better at reducing the risk of breast cancer. They have tested the regimen on small numbers of women, including carriers of mutations increasing the risk of breast cancer (Spicer et al. 1994; Weitzel et al. 2007). With minimal side effects, the treatment group showed significantly reduced mammographic density of breast tissue relative to controls. This is a positive sign, because mammographic density is associated with the risk of breast cancer (Boyd et al. 2011). More extensive trials are needed.

Human populations have not had time to adapt to the lifestyle we live; this may explain why we lack physiological defenses against many diseases of civilization.

Obesity

Waistlines in industrialized countries have been expanding for decades. **Figure 14.27a** shows that between the Civil War and World War II, the average military academy cadet in America went from borderline underweight to normal weight. Figure 14.27b shows that between the late 1970s and the middle 2000s, the average adult American went from borderline overweight to overweight. The graphs show variation in body mass index, calculated as weight (in kilograms) divided by height (in meters) squared. Individuals with higher body mass indexes are heavier for their height.

Some of the variation in body mass index among people within populations is due to differences in genes, but much of the variation between populations is due to differences in environments (Swinburn 2011). Evidence for the latter claim comes from immigrants to the United States (Goel et al. 2004). The average

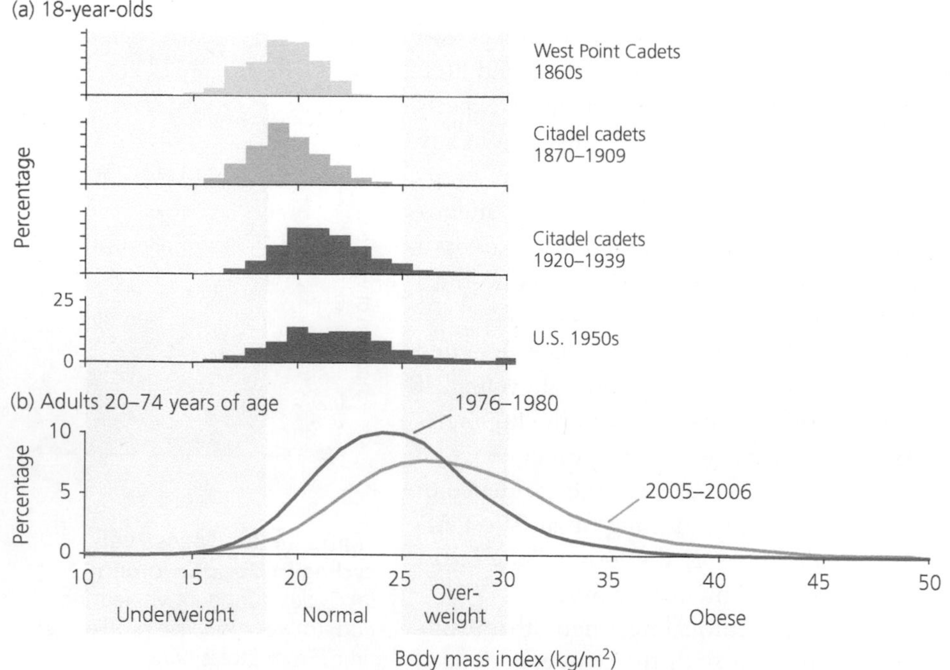

Figure 14.27 Change over time in the body mass index distributions of Americans (a) Military academy cadets between 1860 and 1939, plus the general population of 18-year-olds in the 1950s. Data from Komlos and Brabec (2011). (b) Adults in the late 1970s and mid-2000s. From Ogden et al. (2007).

immigrant is leaner than the average U.S.-born resident. However, the longer an immigrant lives in the United States, the more his or her body mass index tends to rise. Suspected environmental culprits are legion. They range from viral infections to endocrine disrupters to sleep debt (McAllister et al. 2009). Here we consider just one suspect: disruption of the microbial community in the human gut.

The microbes in our gut are a feature of our environment whose importance to our physiology and health has only recently begun to be fully appreciated (see Nicholson et al. 2012). The composition of the gut microbial community varies across populations, among individuals, and over time within individuals (Huttenhower et al. 2012; Yatsunenko et al. 2012). Among the aspects of a modern lifestyle that alter the composition of our gut community is the consumption of antibiotics (Dethlefsen et al. 2008; Dethlefsen and Relman 2010).

The consequences can be dire. For example, the use of antibiotics makes patients more susceptible to infection by *Clostridium difficile*, which can result in severe diarrhea and colitis (Loo et al. 2011). The traditional treatment for *Clostridium difficile* is more antibiotics, which can further damage the gut microbiota and leave the patient vulnerable to reinfection (Borody and Khoruts 2012).

An alternative approach involves trying to restore the gut community to its pre-antibiotic state (Borody and Khoruts 2012). A growing list of case histories suggests that fecal microbiota transplantation—that is, inoculation of the gut with microbes harvested from the stool of a healthy individual—is often effective.

A hint that our gut microbiota might play a role in obesity came from work by Peter Turnbaugh and colleagues (2006) on germ-free mice. Not surprisingly, given that they lack intestinal bacteria to help digest their food, germ-free mice are lean. Inoculating their gut with microbes from another mouse induces them to gain weight. Turnbaugh and colleagues found that germ-free mice inoculated with microbes from a genetically obese mouse gained more weight than those inoculated with microbes from a normal mouse (**Figure 14.28**).

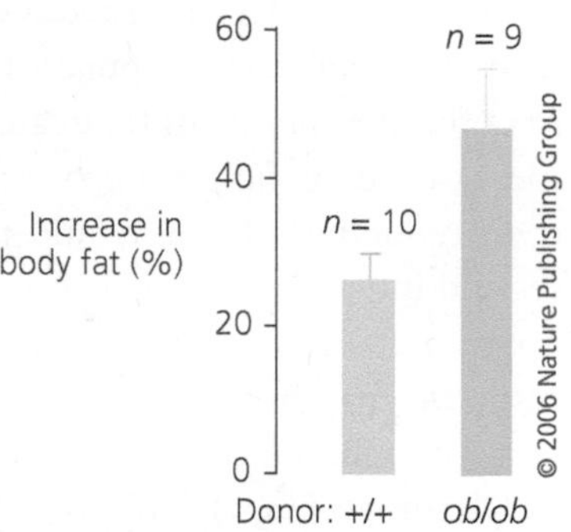

Figure 14.28 Weight gain over two weeks in germ-free mice inoculated with microbes from the guts of normal (+/+) versus genetically obese (*ob*/*ob*) mice The difference is significant at $p < 0.05$. From Turnbaugh et al. (2006).

Reprinted by permission of Macmillan Publishers Ltd. Turnbaugh, P. J., R. E. Ley, et al. 2006. "An obesity-associated gut microbiome with increased capacity for energy harvest." *Nature* 444: 1027–1131.

Obese versus lean people harbor different communities of gut microbes (Ley et al. 2006). Perhaps differences in genotype, diet, lifestyle, and environment that lead to obesity also lead to differences in the gut community. Or perhaps differences in the gut community are among the causes of obesity. To evaluate the latter hypothesis, Franck Thuny and colleagues (2010) followed 48 adults who required antibiotic treatment for bacterial endocarditis and an equal number of age-matched controls. A year after leaving the hospital, the patients who had been treated with a combination of vancomycin and gentamycin had gained enough weight to increase their body mass index by an average of 2.3 ($p = 0.03$), whereas the patients treated with other antibiotics and the controls showed no significant changes. This result suggests that disruption of the gut microbiota can contribute to obesity.

The best kind of study for establishing cause and effect is, as we have emphasized elsewhere, a randomized controlled trial. In the first such experiment to test fecal microbiota transplantation as a treatment for obesity, Anna Vrieze and colleagues (2012) enlisted 18 male volunteers with metabolic syndrome—obesity and insulin resistance—and assigned them at random to two groups of 9. The patients in the treatment group were inoculated with microbes from a lean donor. The controls were re-infused with their own microbes. Six weeks later, the treatment group showed a significant improvement in insulin sensitivity ($p < 0.05$). The controls showed no change. A follow-up study is under way (see Kootte et al. 2012). Stay tuned to the literature for more details.

Keeping in mind that the environment modern humans live in may not be the environment that we are adapted to, we devote the remaining two sections of this chapter to examples showing how researchers use an adaptationist framework to develop and test hypotheses about medical physiology and hypotheses about fundamental aspects of human behavior.

14.6 Adaptation and Medical Physiology: Fever

Many people consider the symptoms that accompany illness to be a nuisance. A common response to the fever associated with a cold or flu, for example, is to take aspirin, acetaminophen, or ibuprofen. These drugs reduce the fever, but they do not combat the virus that is causing the cold or flu. Here we ask whether taking drugs to reduce fever is a good idea. To answer the question, we need to know why people run a fever when they are sick.

An evolutionary perspective suggests two interpretations of fever. One is that fever may reflect manipulation of the host by the pathogen. Viruses or bacteria may release chemicals that cause the host to elevate its body temperature so as to increase the pathogen's growth or reproductive rate. If this hypothesis is correct, then reducing a fever would probably help the host combat the infection. The second interpretation is that fever may be an adaptive defense. The pathogen may grow and reproduce more slowly at higher temperatures, or the host's immune response may be more effective at higher temperatures. If this hypothesis is correct, then taking drugs that alleviate fever might be counterproductive.

An evolutionary perspective can help medical researchers develop hypotheses about physiological function.

Matthew Kluger has advocated the second hypothesis—that fever is an adaptive defense against disease. In 1974, Linda Vaughn, Harry Bernheim, and Kluger discovered that the desert iguana (*Dipsosaurus dorsalis*, **Figure 14.29a**) develops a

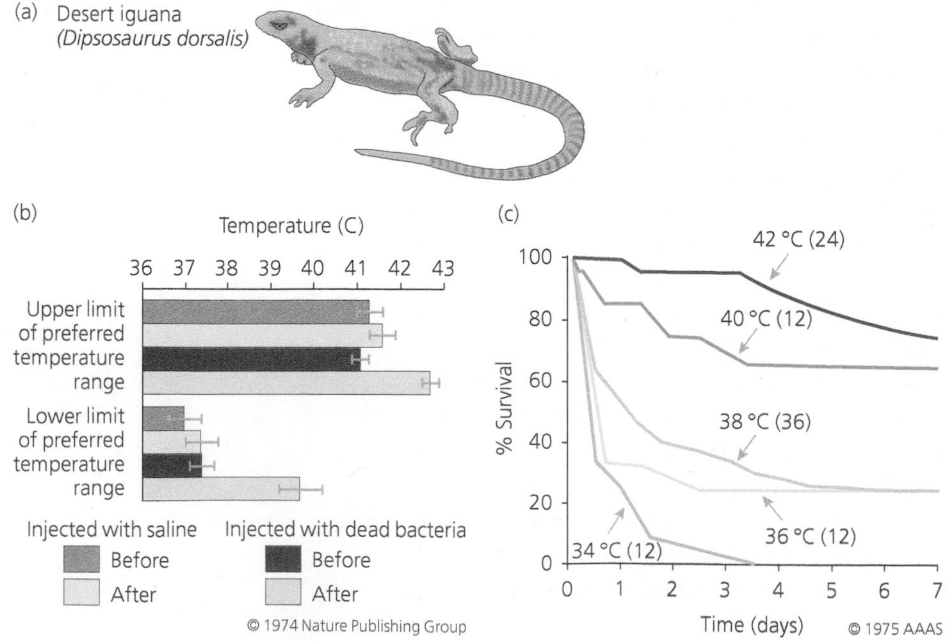

Figure 14.29 Behavioral fever in the desert iguana (*Dipsosaurus dorsalis*) (a) A desert iguana. (b) Average upper and lower preferred temperatures, or set points (± standard error), before and after injection with saline or dead bacteria. There was no significant change in either set point for the nine saline-injected lizards. The 10 bacteria-injected animals significantly increased both set points ($p < 0.001$). From Vaughn et al. (1974). (c) Survival of infected lizards (sample sizes in parens). From Kluger, Ringler, and Anver (1975).

(b) Reprinted by permission of Macmillan Publishers Ltd. L. K. Vaughn, H. A. Bernheim, and M. Kluger. 1974. "Fever in the lizard *Dipsosaurus dorsalis*. *Nature* 252. Copyright © 1974. (c) From M. J. Kluger, D. H. Ringler, & M. R. Anver. 1975. "Fever and survival." *Science* 188: 166–168. Reprinted with permission of the AAAS.

behavioral fever in response to infection with a bacterium called *Aeromonas hydrophila*. Recall that iguanas, being ectotherms, use behavior instead of physiology to regulate their body temperatures (Chapter 10). They move to hot spots to warm themselves and to cold spots to cool themselves. Vaughn et al. found that when they injected desert iguanas with dead bacteria, the lizards chose body temperatures about 2°C higher than they normally choose (Figure 14.29b).

Is behavioral fever an adaptive response to infection, or are the bacteria manipulating the iguanas? To distinguish between these hypotheses, Kluger, Daniel Ringler, and Miriam Anver (1975) infected desert iguanas with live bacteria, then prevented the lizards from thermoregulating by keeping them in fixed-temperature incubators. Most of the iguanas kept at temperatures mimicking behavioral fever survived, whereas most of the iguanas kept at lower temperatures died (Figure 14.29c). This result suggests that behavioral fever is, in fact, adaptive for desert iguanas infected with *A. hydrophila*.

If fever is an adaptive defense against *A. hydrophila*, then it would probably be a bad idea for infected lizards to take aspirin. This is not as silly a statement as it sounds, at least in one sense: The researchers found that the aspirin-like drug sodium salicylate reduces behavioral fever in iguanas just as it reduces physiological fever in mammals. Apparently thermoregulation is controlled by similar neurological mechanisms in both groups of animals. Bernheim and Kluger (1976) infected 24 desert iguanas with bacteria, then gave half of the infected lizards sodium salicylate. The researchers allowed all the iguanas to behaviorally thermoregulate. All of the control iguanas developed behavioral fever, and all but one of them survived the infection. Five of the medicated lizards developed behavioral fever despite the medication, and all of them survived. The other seven medicated iguanas failed to develop behavioral fever, and all of them died.

Since the mid 1970s, researchers have documented behavioral fever in a wide variety of reptiles, amphibians, fishes, and invertebrates. In several animal studies, researchers have shown that fever increases survival (see Kluger 1992 for a review). These results broadly support the hypothesis that fever is an adaptive response to infection.

Fever is much harder to study in endotherms than it is in ectotherms. Researchers cannot force endotherms to be at an arbitrary body temperature simply by putting them in an incubator. And as we will see shortly, drugs that reduce fever have effects on the immune system that are independent of fever.

In an attempt to disentangle the effects of the increased metabolic rate that accompanies a fever from the increase in body temperature per se, Manuel Banet used rats implanted with cooling devices and infected with *Salmonella enteritidis*. First, Banet (1979) implanted cooling devices in the brains of rats and used them to chill the hypothalamus. This technique induced the infected rats to develop extra-high fevers, without dramatically elevating their metabolic rates. The rats with extra-high fevers survived the infection at much lower rates than control rats. Second, Banet (1981a) cooled the spinal cords of infected rats. This procedure induced the rats to increase their metabolic rates while preventing them from elevating their body temperatures. The rats with high metabolic rates survived the infection at somewhat higher rates than did those with normal metabolic rates. Finally, Banet (1981b) carefully monitored the body temperatures and metabolic rates of a group of infected rats, some of which had implants but none of which were heated or cooled. Banet found that the rats that ran the highest fevers had the lowest rates of survival, but the rats that showed the highest metabolic rates had the highest rates of survival. Together, Banet's results suggest that moderate fever is beneficial to infected rats, that the benefits of fever may be associated not so much with elevated temperature itself as with increased metabolic rate or other effects on the immune system, and that high fever in rats is deleterious to survival.

It is unclear how the results with iguanas and rats might apply to humans. Fewer clinical studies have been done on this question than might be expected (Kluger 1992; Green and Vermeulen 1994; Carey 2010). We review just one.

Fever and the Common Cold

Neil Graham and colleagues (1990) intentionally infected 56 consenting adult volunteers with rhinovirus type 2, one of the viruses that can cause the common cold. Assigned to groups double-blind and at random, 14 subjects took a placebo. The rest took common over-the-counter antifever medications: 13 took ibuprofen, 15 took aspirin, and 14 took acetaminophen. The volunteers taking the placebo suffered less nose stuffiness **(Figure 14.30a)** and made more antibodies against the rhinovirus (Figure 14.30b) than did the volunteers taking antifever medicines. The reason for the reduced antibody response in volunteers taking medicine may be that the medicines prevented monocytes, a class of white blood cells, from moving from the blood to the infected tissues (Figure 14.30c). Once in the infected tissues, monocytes differentiate into macrophages, which help mount an immune response against the virus (Graham et al. 1990).

The simple interpretation is that the antifever medications interfered with the immune response to the common cold and therefore that fever is an adaptive defense against the disease. Kluger (1992) points out, however, that few of the subjects in the study ran a fever. Furthermore, the fraction of subjects taking the placebo who did run a fever (14%) was not significantly higher than the fraction of subjects taking medicine who ran a fever (7%). Kluger's interpretation is that few people infected with rhinovirus type 2 run a fever and thus that the study did not test the hypothesis that fever is adaptive.

The study by Graham and colleagues did show, however, that antifever medicines interfered with the immune response to the virus. This result demonstrates

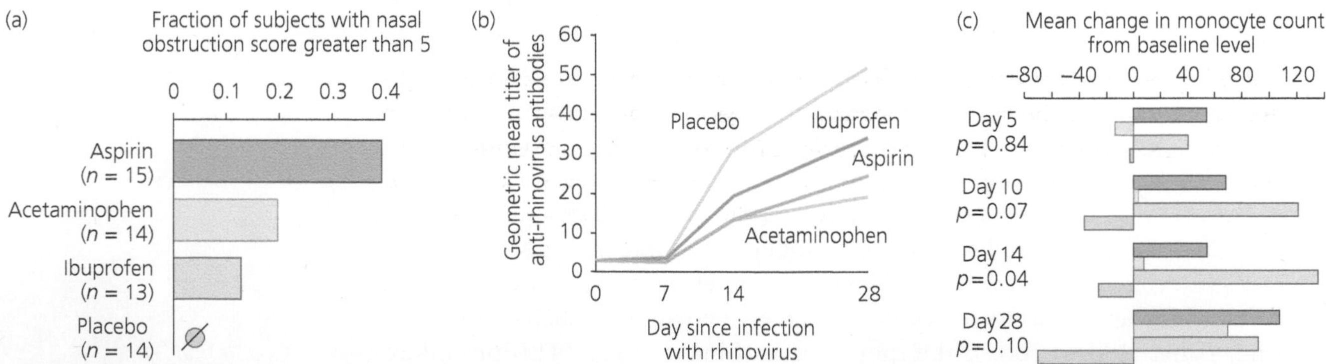

Figure 14.30 Do antifever medicines have any effect on the course of the common cold? (a) Volunteers taking a placebo had less-stuffy noses than volunteers taking one of three antifever medicines ($p = 0.02$). (b) Volunteers taking a placebo made more antibodies against the rhinovirus that caused their cold than did volunteers taking one of three antifever medicines. On day 28, the difference between the placebo group and the other three groups combined was significant at $p = 0.03$. (c) Monocytes are white blood cells that circulate to infected tissues, then leave the blood and differentiate into macrophages. Volunteers taking a placebo showed a drop in the concentration of monocytes in their blood over time (indicating that the cells had moved into the tissues), whereas volunteers taking one of three antifever medicines showed an increase in the concentration of monocytes. From Graham et al. (1990).

that antifever medicines have multiple physiological effects. As we noted earlier, this fact makes it extremely difficult to design studies on the adaptive significance of fever in mammals. Studies using traditional antifever medicines cannot separate fever from other aspects of the immune response.

Fever and Medical Practice

Jane Carey (2010) reviewed the literature on the value of treating fever in humans, concluding that it offers weak guidance. More research is needed on the adaptive significance of fever in humans and on the costs and benefits of using various antifever medications.

Even if researchers find clear evidence that fever in humans is, indeed, an adaptive response to some infections, no responsible doctor (or evolutionary biologist) would suggest that it is always a bad idea to suppress fever. First, fever may be an adaptive response against some pathogens, but not against others. Some bacteria or viruses may grow and reproduce faster at fever temperatures than at normal temperatures. In other words, the adaptive-response and pathogen-manipulation-of-host hypotheses may be mutually exclusive for any particular pathogen, but they are not mutually exclusive across all pathogens. Second, even when fever is beneficial, it carries costs as well (Nesse and Williams 1994). In the case of mild illness and low fever, sometimes the benefits of antifever medicines in alleviating symptoms and allowing people to continue their normal activities outweigh the costs of a somewhat diminished immune response. In the case of serious illness and high fever, the fever itself can deplete nutrient reserves and can even cause temporary or permanent tissue damage. Finally, in some circumstances fever may cause damage directly, unconnected with its role in infections. For example, experiments with animals and observational studies of humans suggest that fever following a stroke causes neurological damage and reduces the likelihood of survival (Azzimondi et al. 1995).

14.7 Adaptation and Human Behavior: Parenting

When using selection thinking to understand human behavior, evolutionary psychologists assume that the brain's properties as a regulator of behavior have been shaped by natural selection. The brain is a flexible machine, not a computer slavishly converting input to output according to some fixed program. It runs on a complex mix of conscious and unconscious perception, emotion, experience, and calculation, in pursuit of a variety of goals. But in the view of evolutionary psychologists,

> The ultimate objective of our conspicuously purposive physiology and psychology is not longevity or pleasure or self-actualization or health or wealth or peace of mind. It is fitness. Our appetites and ambitions and intellects and revulsions exist because of their historical contributions to this end. Our perceptions of self-interest have evolved as proximal tokens of expected gains and losses of fitness, "expected" being used here in its statistical sense of what would be anticipated on average from the cumulative evidence of the past (Daly and Wilson 1988b, page 10).

This adaptationist approach to human behavior requires caution. In its capacity as a regulator of behavior, the human brain is influenced by culture as well as by evolutionary history. Culture evolves by its own set of rules (see **Computing Consequences 14.1**). Furthermore, culture can manifestly induce individuals to behave in ways contrary to the interests of their genetic fitness. The mass suicide of 39 members of the Heaven's Gate cult in March of 1997, for example, defies adaptationist explanation.

The influence of culture on human behavior means that studies of behavior within a single society cannot disentangle the effects of culture from those of evolutionary history. To make a plausible claim that a psychological trait or pattern of behavior is a product of natural selection, evolutionary psychologists must show that the trait or pattern is broadly cross-cultural. Cross-cultural diversity has fallen dramatically during the last century. All but the most remote and isolated traditional societies have been contacted, and Western ideas and artifacts have spread virtually everywhere (see Diamond 1992). Some biologists feel it is no longer possible to conduct a genuine cross-cultural study. Others feel that such studies are still worth pursuing, particularly when new findings are combined with information extracted from databases of earlier anthropological research.

Another caveat for the study of human behavior is one we have already discussed. The environments most humans live in today are strikingly different from the environments all humans lived in for most of our evolutionary history. From the time of the earliest members of the genus *Homo*, over 2 million years ago (see Chapter 20), until the advent of agriculture, at about 8000 B.C., all humans lived in small groups and made their living by hunting and gathering. The pace of cultural change has been accelerating ever since, and many changes have likely been too fast for genetic evolution to keep up. As a result, it is of little use to ask why natural selection would have produced human behaviors, such as a willingness to ski down a mountainside at 75 miles per hour, that can occur only in a modern context. So long as we are careful to allow for our incomplete understanding of the hunter-gatherer lifestyle, however, it may make sense to ask why natural selection would have built into us a desire for the social rewards that we can attain, under the right circumstances, through dramatic demonstrations of superior athleticism and bravery. Scientists pursuing such an inquiry would formulate and test hypotheses about how a recipient of these social rewards, living in a hunter-gatherer society, might convert them into reproductive success.

Evolution and Parenthood

An evolutionary perspective can help researchers develop hypotheses about patterns of human behavior.

We now explore evolutionary psychology by considering aspects of parenthood. We begin with a prediction. On the assumption that the psychology of parenting has been shaped by natural selection, we can predict that human adults should direct more of their parental caregiving to their own genetic offspring than to the genetic offspring of others. We would make the same prediction about any organism that provides parental care. Care is expensive to the caregiver, and caregivers who reserve their efforts for their own genetic young should enjoy higher lifetime reproductive success than caregivers who are indiscriminate. The generality of this prediction gives us confidence that it is legitimate for human hunter-gatherers. And it has hidden subtleties, as an animal example will show.

Reed buntings (*Emberiza shoeniclus*) are small ground-nesting birds in which both males and females provide parental care. Most nesting pairs are socially monogamous: Each partner tends no other nest than the one they tend together.

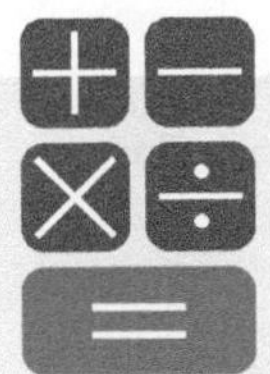

COMPUTING CONSEQUENCES 14.1

Is cultural evolution Darwinian?

The mechanisms of cultural evolution are beyond the scope of this chapter. In fact, they are probably beyond the scope of evolutionary biology altogether.

Richard Dawkins, in his 1976 book *The Selfish Gene,* suggested that we might develop a theory of cultural evolution by natural selection that works exactly like our theory of biological evolution. Central to this suggestion is the idea that natural selection is a generalizable process. Natural selection works on organisms because they have four key features: mutation, reproduction, inheritance, and differential reproductive success. In principle, natural selection should operate on any entities that have the same four properties.

Dawkins noted that elements of culture have these four properties, and thus should evolve by natural selection. A new word, song, idea, or style is analogous to a new allele created by mutation. The Shakers' austere and beautiful style of furniture design, for example, is an element of culture. A new piece of culture reproduces when other people adopt it and pass it on, as when a woodworker admires a Shaker table and imitates the design. Some bits of culture are more successful than others at getting themselves transmitted from person to person. Shaker-style furniture has achieved wider adoption than Shaker-style lifelong celibacy. Culture evolves as the relative frequencies of styles and ideas change.

Dawkins coined the term *meme* for the fundamental unit of cultural evolution. He saw the meme as analogous to the gene, the fundamental unit of biological evolution. Dawkins envisioned a detailed theory of population memetics that would be similar to the theory of population genetics (covered in Chapters 5 through 8). [For a more recent exposition on the potential explanatory power of this idea, see Dennett (1995).]

The trouble with Dawkins's suggestion, noted by Dawkins himself (see also the 1989 edition of his book), is that the effectiveness of natural selection as a mechanism of evolution depends not just on the property of inheritance but also on the details of how inheritance works. This fact was first recognized by Fleeming Jenkin, one of Darwin's critics, in 1867. In Darwin's time, the prevailing model of inheritance involved the blending in the offspring of infinitely divisible particles contributed by the parents. Jenkin pointed out that blending inheritance undermines evolution by natural selection because of the fate it implies for new variations. In a sexual population with blending inheritance, any new variation would quickly vanish, like a single drop of black paint dissolving into a bucket of white. Mendelian genetics rescues Darwin's theory, because Mendelian inheritance is particulate. Genes do not blend. A new recessive mutation can remain hidden in a population for generations. Eventually, the mutant allele may reach a high enough frequency that heterozygotes start to mate with each other, producing among their offspring a few homozygous recessives.

In correct form, then, the generalizable theory of evolution by natural selection applies to entities with mutation, reproduction, *particulate* inheritance, and differential reproductive success. The crucial question for the theory of cultural evolution by natural selection is whether memes are transmitted by particulate or blending inheritance. As Allen Orr (1996) puts it, "Do street fashion and high fashion segregate like good genes, or do they first mix before replicating in magazines or storefronts?" Nobody knows. If memes are transmitted by blending inheritance, then natural selection is, at best, a weak mechanism of cultural evolution. We need other mechanisms to explain cultural evolution.

Although biological evolution and cultural evolution may proceed by different mechanisms, this does not mean that either is irrelevant to the other. Cultural evolution can set the stage for biological evolution. Most humans, for example, stop producing the enzyme lactase in childhood, but the cultural practice of dairy farming led to the evolution of lifelong lactase production in many human populations (Durham 1991). Likewise, biological evolution can influence cultural evolution. For example, the division of the visible light spectrum into verbally distinguished colors follows cross-culturally universal patterns (Durham 1991). These patterns are determined by the way our eyes and brains encode visual information, indicating that the structure of our nervous systems has constrained cultural variation in color terminology. Cultural and biological evolution are distinct but interdependent.

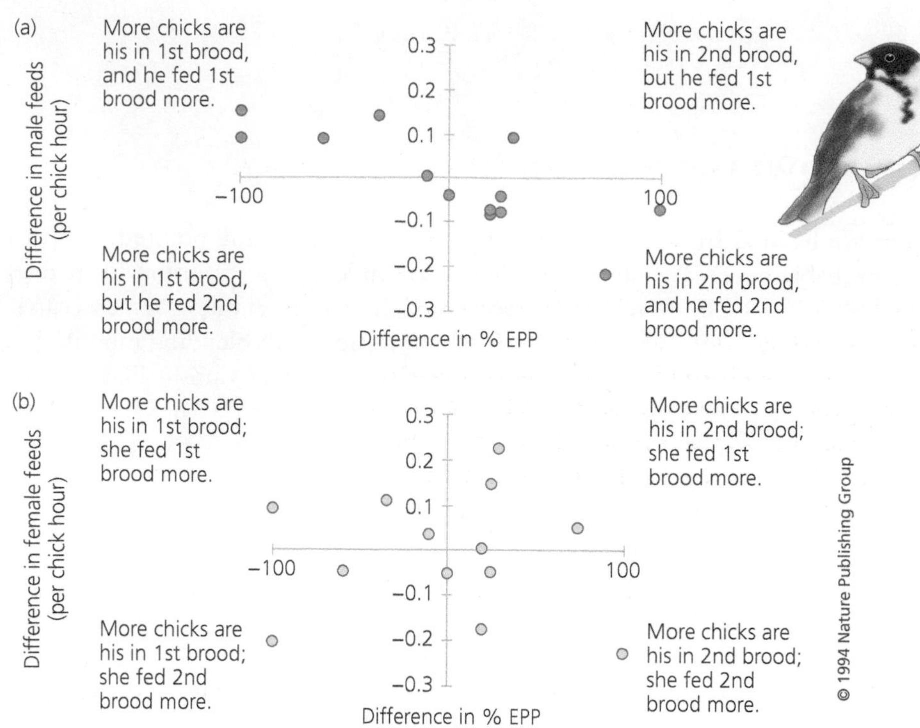
(a)
Difference in male feeds (per chick hour)
More chicks are his in 1st brood, and he fed 1st brood more.
More chicks are his in 2nd brood, but he fed 1st brood more.
More chicks are his in 1st brood, but he fed 2nd brood more.
More chicks are his in 2nd brood, and he fed 2nd brood more.
0.3
0.2
0.1
−0.1
−0.2
−0.3
−100
100
Difference in % EPP
(b)
Difference in female feeds (per chick hour)
More chicks are his in 1st brood; she fed 1st brood more.
More chicks are his in 2nd brood; she fed 1st brood more.
More chicks are his in 1st brood; she fed 2nd brood more.
More chicks are his in 2nd brood; she fed 2nd brood more.
0.3
0.2
0.1
−0.1
−0.2
−0.3
−100
100
Difference in % EPP
© 1994 Nature Publishing Group

Figure 14.31 Male reed buntings adjust parental effort (a) Each dot represents a male who raised two broods. The *x*-axis plots the difference between the two broods in the percentage of extrapair paternity (% EPP), or fraction of chicks sired by another male. The *y*-axis plots the difference in how frequently the male fed chicks in the first versus second brood. Most males fed chicks more often in the nest where they had sired a higher fraction ($p = 0.0064$). (b) Each dot represents a female who raised two broods. Females showed no relationship between parenting effort and the relative number of chicks sired by extrapair males. Redrawn from Dixon et al. (1994).

Reprinted by permission of Macmillan Publishers Ltd. Dixon, A., D. Ross, S. L. C. O'Malley, and T. Burke. 1994. "Paternal investment inversely related to degree of extra-pair paternity in the reed bunting." *Nature* 371: 698–700. Copyright © 1994.

Genetic testing by Andrew Dixon and colleagues (1994) revealed that there is more to the reed bunting mating system than meets the eye. They found that 55% of chicks were sired by males other than their mother's social mate and that 86% of all nests included at least one such chick. Dixon predicted that males, if they could tell what fraction of the chicks they had sired in any given nest, would adjust their parental effort accordingly. Dixon looked at the chick-feeding behavior of 13 pairs of buntings that raised two clutches of chicks in a single season. The males fed the chicks more often in the nest where they had sired a higher proportion of the chicks **(Figure 14.31a)**. The females, who were the genetic mothers of all the chicks in both nests, showed no such pattern (Figure 14.31b).

We have presented the reed bunting example because evolutionary biologists using Darwinism to understand human behavior are often accused of genetic determinism (see, for example, Lewontin 1980). Genetic determinism is the notion that fundamental characteristics of human societies are unchangeably programmed into our genes. Note, however, the sense in which genes do and do not determine the parental behavior of male reed buntings. A male's genotype does not specify a particular level of parental care that the male will provide no matter what. Instead, each male's genotype specifies a range of phenotypic plasticity in parental care (see Chapter 10). That is, the bird's brain has a mechanism that adjusts the effort the male expends in caring for a brood, based on cues that indicate his probable level of paternity in that brood. If a male's social or biological environment changes, he alters his level of parental care accordingly, as Figure 14.31a shows. The pattern of phenotypic plasticity in a trait is called the trait's reaction norm. Reaction norms for reed bunting parental care presumably vary from male to male—or at least did vary in ancestral populations. This genetic variation in reaction norms provides the raw material for the evolution of parental behavior. The average reaction norm of today's reed buntings appears to be adaptive. The average reaction norm might be described as "reed bunting nature."

Evolutionary psychologists studying human behavior are likewise interested in phenotypic plasticity—that is, reaction norms. They recognize that human reaction norms allow wide latitude for social and environmental circumstances to modify human behavior, and they recognize that reaction norms vary from person to person. What evolutionary psychologists do is formulate and test hypotheses about average human reaction norms.

Are humans as discriminating as male reed buntings in adjusting their provision of parental care? The question is difficult to study directly, at least in modern Western cultures, where most of the interactions between parents and children take place in private. Other cultures, however, are more amenable to study.

Mark Flinn (1988) conducted an extensive and detailed observational study of interactions between parents and offspring in a small village in rural Trinidad. Flinn interviewed all residents of the village to determine which people were genetically related and which people were living together. Then, once or twice a day for six months, Flinn walked a standard route through the village that took him within 20 meters of every house and public building. He started at a different, randomly determined point each day, so as not to regularly pass particular places at particular times. Each time he saw any of the village's 342 residents, Flinn recorded what the person was doing, who he or she was with, and the nature of the interaction they were having. The houses and buildings are all rather open, so Flinn was able to see much that went on inside as well as outside.

Fourteen of the village's 112 households included mothers who were the genetic mothers of all the resident children and fathers who were the genetic fathers of some of the resident children and the stepfathers of others. These 14 families included 28 genetic offspring and 26 step-offspring of the fathers. There can be no hidden differences between the genetic fathers and the stepfathers because the genetic fathers and the stepfathers are the same men.

Flinn calculated the amount of time the fathers spent with their children and the fraction of their interactions with their children that were agonistic. An agonistic interaction was one that "involved physical or verbal combat (e.g., spanking or arguing) or expressions of injury inflicted by another individual (e.g., screaming in pain or anguish or crying)" (Flinn 1988). Note that, overall, only 6% of the parent–offspring interactions Flinn saw were agonistic, and 94% of these involved only verbal exchanges. During his study, Flinn was not aware of any interactions between parents and children that would be considered physical child abuse. ("Screaming in pain or anguish" may sound like evidence of physical child abuse, but anyone who has spent time with a 2-year-old knows that this is not necessarily so.) In other words, Flinn's research concerns parent–offspring interactions that most anyone would consider normal.

Flinn found that the 14 fathers with both genetic offspring and step-offspring spent more of their time with their genetic offspring (**Figure 14.32a**). Furthermore, a smaller fraction of the father–genetic offspring interactions were agonistic (Figure 14.32b). These results are consistent with the prediction that parents discriminate among children on the basis of their genetic relationship with them.

This is an observational study, however, and there is a potentially confounding variable. The pattern in Flinn's data could be explained by the late arrival of the stepfathers in the lives of their stepchildren. Men might feel less affection and concern for their stepchildren simply because they joined the family when the stepchildren were older, whereas the men were already in the family when all of their own genetic children were born.

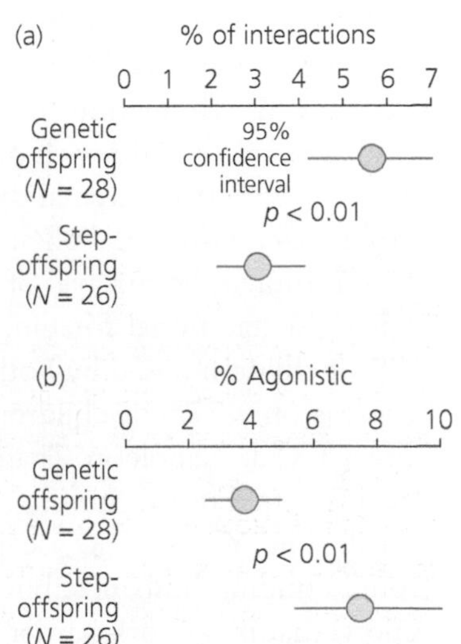

Figure 14.32 Fathers with both step- and biological children spend more time, and get along better, with their biological children (a) The fraction of their time (% of all observed interactions) that 14 fathers spent with their biological children versus their stepchildren. (b) The fraction of interactions between 14 fathers and their children that were agonistic (see text for definition). Data from Flinn (1988).

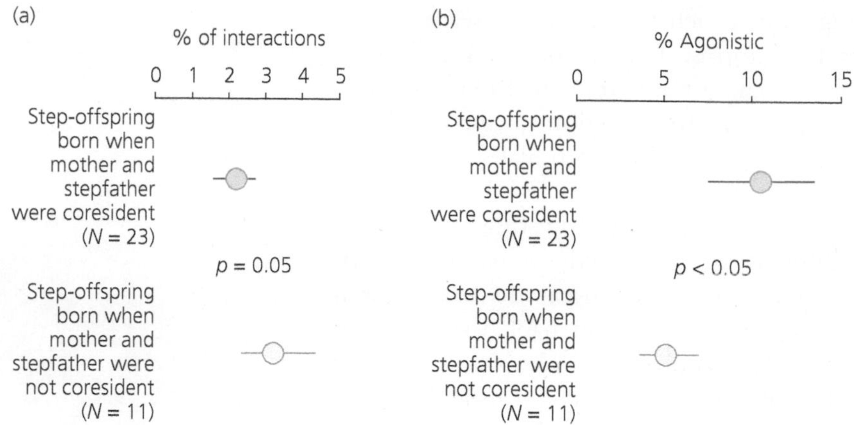

Figure 14.33 Stepfathers spend more time, and get along better, with their stepchildren if the stepchildren are born before the stepfather joins the family (a) Fraction of time (% of all observed interactions) that stepfathers spent with their stepchildren. (b) Fraction of interactions between stepfathers and their stepchildren that were agonistic (see text for definition). Data from Flinn (1988).

Flinn's data for this village, remarkably, include 23 stepchildren born when their mothers and stepfathers were already living together, plus 11 stepchildren born before their mothers and stepfathers moved in together. (This sample includes all the stepfathers in the village, not just those who also have genetic children living in the same house.) If a man's parental affection is simply a function of the fraction of the child's life during which the man has lived with the child, then the stepfathers who have lived with their stepchildren from the children's births should be more affectionate. In fact, the opposite appears to be true. The stepfathers spent more of their time, and had a lower fraction of agonistic interactions, with stepchildren born before the stepfathers joined the family **(Figure 14.33)**.

We noted earlier that studies within a single society offer no means of disentangling the influences of culture and evolutionary history. We could argue that the pattern of discrimination revealed in Flinn's study is simply a product of culture and has nothing to do with our species' adaptive history. Evidence is accumulating, however, that parental discrimination between own and others' genetic offspring is a cross-cultural phenomenon. For example, Kim Hill and Hillard Kaplan (1988) studied the survival of biological children versus stepchildren in the Ache Indians, a traditional foraging culture in Paraguay. Hill and Kaplan found that 81% of children raised by both biological parents survived to their 15th birthday, whereas only 57% of children raised by one biological parent and one stepparent survived. Napoleon Chagnon (1992; see also 1988) studied the Yąnomamö Indians, a traditional hunting, gathering, and gardening culture in Venezuela and Brazil. The Yąnomamö are polygynous, which means that women have little trouble finding husbands, but men often have difficulty finding wives. Chagnon reports that men work harder to find wives for their biological sons than for their stepsons. Frank Marlowe (1999) studied Hadza hunter-gatherers in Tanzania. He found that compared with stepfathers, genetic fathers spend more time near their children, and play with them, talk with them, and nurture them more. Kermyt Anderson and colleagues (1999) studied modern American men living in Albuquerque, New Mexico. They found that men invest more toward the college education of their genetic children than that of their stepchildren.

Parental Discrimination and Children's Health

Discrimination by parents against stepchildren becomes a public health issue when we consider its impact on the childrens' physiological state. Mark Flinn and Barry England (1995, 1997) went to another rural Caribbean village, this time in Dominica. They gave chewing gum to children, then asked the children

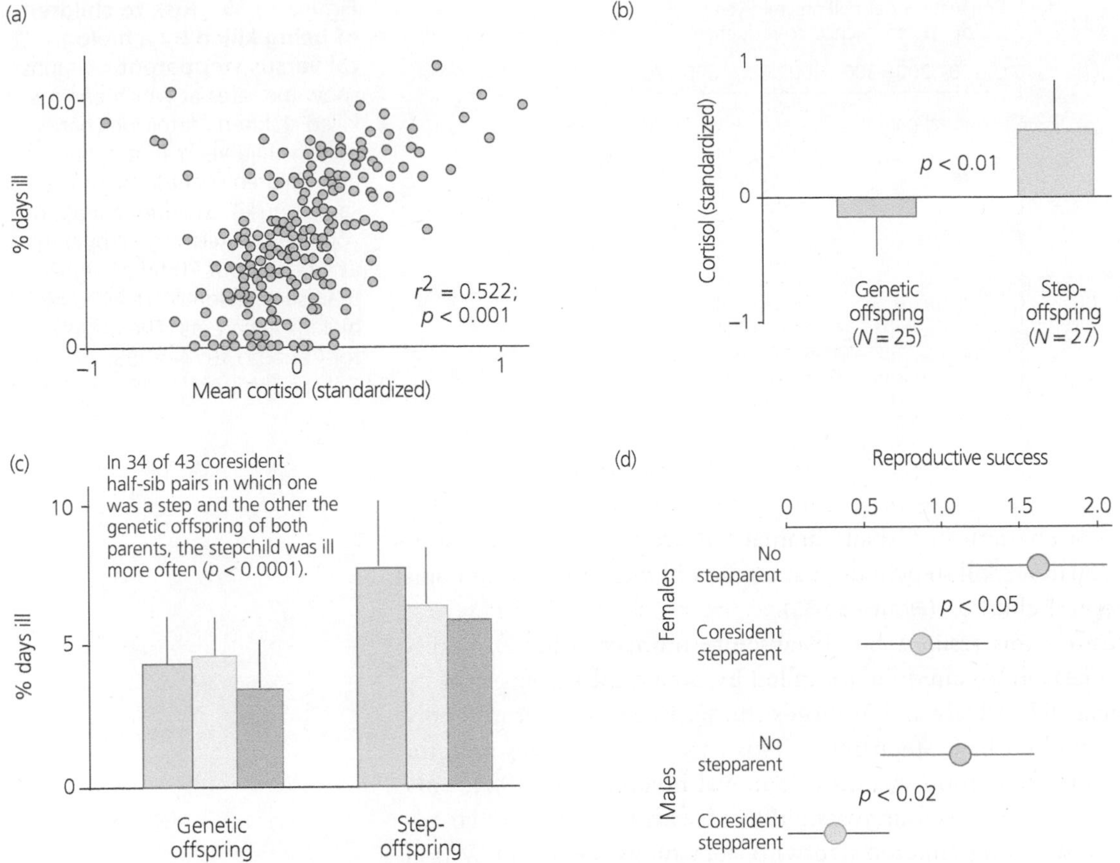

Figure 14.34 Stress, cortisol levels, illness, and reproductive success for stepchildren versus genetic children (a) Children with higher levels of the stress hormone cortisol in their blood get sick more often. Negative numbers indicate lower than average cortisol concentrations; positive numbers higher than average concentrations. (b) Step-offspring have higher concentrations of cortisol in their blood than do biological children. (a, b) From Flinn and England (1995). (c) The difference in health between biological children and stepchildren is larger than the differences attributable to socioeconomic status (different colors indicate different levels of status). From Flinn and England (1997). (d) Biological children have higher reproductive success during early adulthood (ages 18–28 for women, 20–30 for men) than do stepchildren. Data from Flinn (1988).

to provide saliva samples in which the researchers measured the concentration of cortisol. Cortisol is a hormone that animals produce when under stress. Over the short term, high levels of cortisol cause an animal to divert resources to immediate demands, for example by increasing metabolic rate and alertness and inhibiting growth and reproduction. Over the long term, chronically high cortisol can inhibit the immune system, deplete energy stores, and induce social withdrawal.

Flinn and England found that among children in the village they studied, individuals with relatively high concentrations of cortisol in their saliva were sick more often **(Figure 14.34a)**. Not surprisingly, it was the stepchildren who had the highest cortisol levels (Figure 14.34b) and higher frequencies of illness (Figure 14.34c). Ultimately, stepchildren had lower reproductive success during early adulthood (Figure 14.34d) and were more likely to leave town.

Researchers investigating Darwinian predictions have discovered patterns of human behavior with profound consequences for public health.

Martin Daly and Margo Wilson approached the public health consequences of parental discrimination by analyzing case files of homicides in which parents killed their children (Daly and Wilson 1988b; see also Daly and Wilson 1988a, 1994a, 1994b). Daly and Wilson predicted that children would be killed at a higher rate by stepparents than by biological parents.

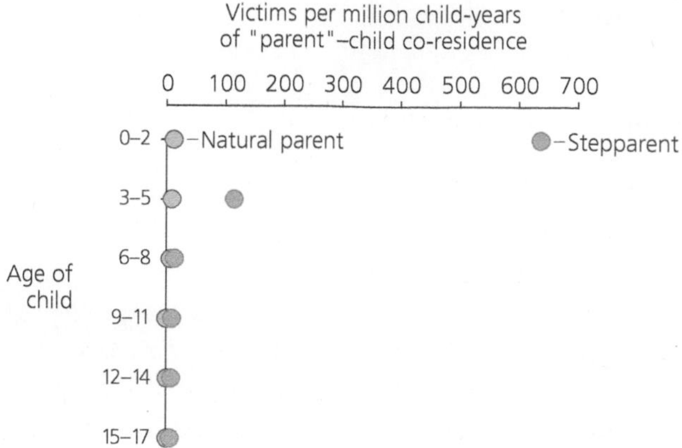

Figure 14.35 Risk to children of being killed by a biological versus stepparent Graphs show the rates at which parents killed children (homicides per million child-years that parents and children spent living in the same house). Children aged 2 or younger are killed by stepparents at a rate about 70 times higher than such children are killed by biological parents. The data are for Canada, 1974–1983. From Daly and Wilson (1988a, 1988b).

Data on murders of children in Canada dramatically confirm Daly and Wilson's prediction: Stepparents kill stepchildren at a much higher rate than biological parents kill biological children **(Figure 14.35)**.

It is worth discussing this result a bit. In absolute numbers (that is, simply counting up homicides), more children are killed by biological parents than by stepparents (341 versus 67 in Daly and Wilson's study). But this is because only a small minority of children have stepparents. This pattern is especially true for young children, the most common victims of parental homicide. In 1984, only 0.4% of Canadian children one to four years old lived with a stepparent. To adjust for the fact that few young children live with stepparents, Daly and Wilson reported the data in Figure 14.35 as rates: the number of homicides per million child-years that parents or stepparents and children spend living together. Epidemiologists often summarize the results of such a study by reporting a relative risk. Here the relative risk of homicide in stepchildren versus biological children is the rate at which stepparents kill stepchildren divided by the rate at which biological parents kill biological children. For children zero to two years old, the relative risk of parental homicide for stepchildren versus biological children is about 70. This is an extraordinarily high relative risk. For comparison, the relative risk of lung cancer in smokers versus nonsmokers is about 11.

Daly and Wilson do not suggest that killing stepchildren, in and of itself, is or ever was adaptive for humans. Anyone who kills someone else's child, even in a traditional hunter-gatherer society, is likely to suffer social penalties that outweigh any potential benefits of eliminating an unwelcome demand for stepparental investment. Instead, what Daly and Wilson suggest is adaptive is the combination of two traits: (1) an intellectual and psychological apparatus that perceives a personal interest in the distinction between one's own and others' genetic offspring, and (2) the emotional motivation to turn this perception into active discrimination between the two kinds of children. Whenever such an apparatus exists, individuals will, rarely, commit errors of excess. These errors of excess become Daly and Wilson's data.

Daly and Wilson's data come from an observational study in which it was impossible to control, as Flinn (1988) was able to, for differences between biological parents and stepparents. Nonetheless, they provide an argument that research conducted within a Darwinian framework can yield insights useful to public health workers and providers of social services.

SUMMARY

Evolutionary biology has numerous applications in medicine. This chapter has considered two general ways in which evolutionary analysis improves our understanding of issues relating to human health. First, we used our knowledge of the mechanisms of evolution to study pathogens and tumors. Second, we used selection thinking to address questions about human physiology and behavior.

Pathogens and their hosts are locked in a perpetual evolutionary arms race. Our immune systems and the drugs we take impose strong selection on the viruses and bacteria that infect our tissues. Because pathogens have short generation times, large population sizes, and often high mutation rates, populations of viruses and bacteria evolve quickly. Phylogenetic analysis helps us reconstruct the history of pathogen evolution, and in the case of flu, understand some of the mechanisms that create pathogen strains capable of causing epidemics. Selection thinking also helps us predict when pathogen populations will become resistant to drugs, whether drug resistance will persist in pathogen populations if drug use is suspended, and what makes some diseases virulent and others benign.

Humans, like other organisms, are a product of evolution by natural selection. As a result, selection thinking can help us understand aspects of our own form and function. Selection thinking suggests, for example, that symptoms of disease, such as fever, may be adaptive facets of our immune response. And aspects of our behavior with significant public health consequences, such as cross-culturally consistent patterns in the way we treat children, may be interpretable as psychological adaptations. It is also important to keep in mind that change in our environment in recent centuries has far outpaced the rate of adaptive evolution. Modern epidemics of myopia, breast cancer, and obesity may be the result of our exposure to novel environments.

Recognition that evolutionary biology is a medical science has, in some respects, been slow in arriving. We expect that interactions between evolutionary biologists and medical researchers will become more frequent and more productive in the years to come.

QUESTIONS

1. **a.** As a review, summarize the evidence discussed in this chapter that antibiotic resistance is due to evolution (i.e., due to new mutations that increase in frequency due to antibiotic exposure).

   b. What would health-care workers, patients, and healthy people do if they wanted antibiotic resistance to evolve as quickly as possible? Do you know of any cases where humans are (unintentionally) doing this?

2. Some biologists regard our bodies as small ecosystems that exert selective pressure for the evolution of invasive metastatic cancer. If this is true, why don't we all get cancer? (Hint: Consider the speed of evolution.) However, these same biologists believe that humans have certain genes that have evolved specifically to prevent cancer. How is it possible to have both strong selection for cancer and strong selection for anticancer genes? (Hint: Consider which population is under selection in each case.)

3. We have seen how the genetic diversity within a tumor can be used to estimate the tumor's history (see Figure 14.20).

   a. In what ways is this similar to the process of reconstructing the evolutionary history of organisms? In what ways is it different?

   b. Do the genetic traits we use to reconstruct a tumor's history need to be selectively neutral? Why or why not?

4. Pathogens require a minimum population size of potential hosts. If the host population is too small, in a short time the entire population has either been killed by the pathogen or has survived the initial infection and become immune. If this occurs, the pathogen dies out. What evolutionary changes in a pathogen might increase its ability to survive in a smaller population? For example, measles requires a host population of about 500,000 humans, while diphtheria can get by with only about 50,000 humans. Develop some hypotheses for why diphtheria can survive with just one-tenth the number of hosts. For example, how might these two diseases differ from each other in transmission rate, virulence, latency to infection, or mutation rate?

5. **a.** In the study of streptomycin resistance, why did Shrag and colleagues use genetically manipulated bacteria, instead of the original wild-type bacteria, to compare sensitive versus resistant strains?

   b. Summarize the key finding of Shrag et al.'s study. Why are these results worrying to the medical and veterinary professions?

6. Review the studies on fever that were presented in this chapter, and summarize each in a sentence or two. Do you agree with Kluger that the study on fever in human colds did not really test the adaptive fever hypothesis? If so, can you design an experiment that will truly test the hypothesis? Is your experiment ethical?
7. The male reed buntings in Dixon et al.'s study (Figure 14.31) seem to be consciously aware of genetic relationships and "trying" to increase their reproductive success. Can evolution cause reed buntings (and other animals) to behave as if they are aware of the evolutionary consequences of their actions, without actually being aware of them? Does your answer also apply to humans?
8. Daly and Wilson's data on infanticide risks might be explained by stepfathers having, on average, more violent personalities than biological fathers. Could this "violent personality" explanation also apply to Flinn's data from the Trinidad village? Why or why not? Daly and Wilson's study involved general data about a large number of families, whereas Flinn's study involved detailed data on a small number of families. What are the advantages and disadvantages of each kind of study?
9. An evolutionary biologist once hypothesized that if evolution has affected human social behavior, then a mother's brothers should take a particular interest in her children—more so than the father's brothers, and perhaps even more so than the father himself. Why did he hypothesize this? (As it turns out, there are many cultures in which men do, in fact, direct parental care primarily to their sisters' kids.)
10. In 1999, a mysterious outbreak of human encephalitis occurred in the northeastern United States. The cause was tentatively identified as St. Louis encephalitis virus. At the same time, an unusual number of dead birds were noticed along the northeastern Atlantic coast. **Figure 14.36** shows genetic relationships of three known encephalitis viruses (St. Louis, Japanese, and West Nile) and several viruses isolated from the birds, from two human patients that died, from one dead horse, and from mosquitoes. (Data compiled from Anderson et al. 1999 and Lanciotti et al. 1999.)
    a. Were the birds, the horse, and the humans all suffering from the same disease?
    b. Do you think the outbreak was caused by St. Louis encephalitis virus?
    c. Does this cladogram suggest how the disease might be spread?
11. It has now become clear that birds are the primary host of West Nile virus. If the virus reaches a human (or horse), it is not spread from human to human (nor from horse to horse) and is unlikely to be transferred back to birds. Is the virulence of the virus in humans and horses an example of coincidental evolution or shortsighted evolution? Explain your reasoning.
12. In a review on the effect of oral contraceptives (OCs) on various cancers, Pike and Spicer (2000) stated: "Direct observational studies of breast-cell proliferation in women taking OCs suggest that the total breast-cell proliferation is very similar over an OC cycle and a normal menstrual cycle. These results predict that breast cancer risk should not be substantially affected by OC use, as is observed."
    If it is correct that oral contraceptives have no more effect on breast cancer than does a normal menstrual cycle, does it follow that OCs do not affect the risk of breast cancer? [Hint: The risk as compared to what?]
13. An avian influenza virus of type H5N1 has recently evolved a "high pathogenicity" (hp) strain that causes severe illness in most wild birds (except ducks) as well as in domestic poultry. A few humans have been infected. The World Health Organization (WHO) currently inspects every human case with particular attention to how the patients contracted the virus. Why is this virus a cause for concern, and why are WHO officials so interested in each patient's source of infection?

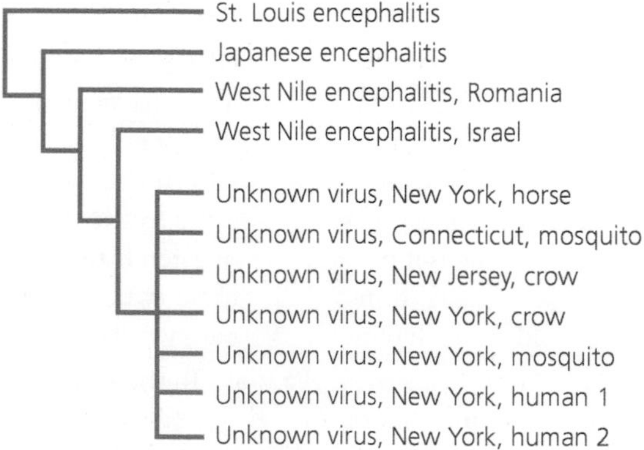

Figure 14.36 A phylogeny of encephalitis viruses isolated from various hosts in the northeastern United States during 1999 Based on data and analyses in Anderson et al. (1999) and Lanciotti et al. (1999).

EXPLORING THE LITERATURE

14. For insight into whether and how highly pathogenic strains of avian H5N1 influenza A virus might evolve a capacity for airborne transmission between humans, researchers in the laboratory of Ron A. M. Fouchier used experimental evolution to produce a laboratory strain capable of airborne transmission among ferrets. At the request of the U.S. National Science Advisory Board for Biosecurity, which feared that terrorists might use the results as a recipe for a biological weapon, the journal *Science* initially declined to publish the study. After months of controversy, and with the approval of a majority of the biosecurity board, *Science* published the study, and a companion paper, after all:

Herfst, S., E. J. Schrauwen, et al. 2012. Airborne transmission of influenza A/H5N1 virus between ferrets. *Science* 336: 1534–1541.

Russell, C. A., J. M. Fonville, et al. 2012. The potential for respiratory droplet-transmissible A/H5N1 influenza virus to evolve in a mammalian host. *Science* 336: 1541–1547.

These papers were part of a special section, including news and commentary, that *Science* has made freely available at

http://www.sciencemag.org/site/special/h5n1/index.xhtml

For more news, commentary, and a similar study in *Nature* that was part of the same debate, visit

http://www.nature.com/news/specials/mutantflu/index.html#research

15. For evidence that the community of microbes living in the human mouth has changed in association with historical shifts in diet, see:

Adler, C. J., K. Dobney, et al. 2013. Sequencing ancient calcified dental plaque shows changes in oral microbiota with dietary shifts of the Neolithic and Industrial revolutions. *Nature Genetics* 45: 450–455.

16. For a randomized clinical trial of fecal microbiota transplantation for the treatment of *Clostridium difficle* infection—that was stopped early because it produced such dramatic results—see:

van Nood, E., A. Vrieze, et al. 2013. Duodenal infusion of donor feces for recurrent *Clostridium difficile*. *New England Journal of Medicine* 368: 407–415.

17. For experimental evidence that gastric bypass surgery is effective in treating obesity because it alters the community of bacteria in the patient's gut, see:

Liou, A. P., M. Paziuk, et al. 2013. Conserved shifts in the gut microbiota due to gastric bypass reduce host weight and adiposity. *Science Translational Medicine* 5: 178ra41.

18. For a review of common notions about obesity that are unsupported or contradicted by evidence, see:

Casazza, K., K. R. Fontaine, et al. 2013. Myths, presumptions, and facts about obesity. *New England Journal of Medicine* 368: 446–454.

19. A crucial question in deciding whether modern women should use hormonal treatments that suppress menstruation is whether menstruation itself is adaptive. The issue is controversial. For an introduction to the controversy, see:

Profet, M. 1993. Menstruation as a defense against pathogens transported by sperm. *Quarterly Review of Biology* 68: 335–381.

Strassmann, B. I. 1996. The evolution of endometrial cycles and menstruation. *Quarterly Review of Biology* 71: 181–220.

20. We presented evidence in Section 14.5 that myopia is a disease of civilization and that the crucial change in lifestyle responsible for myopia is close visual work in childhood. For evidence that modern diets may also be involved in myopia, see:

Cordain, L., S. B. Eaton, et al. 2002. An evolutionary analysis of the aetiology and pathogenesis of juvenile-onset myopia. *Acta Ophthalmologica Scandinavica* 80: 125–135.

21. For evidence that acne is a disease of civilization, see:

Cordain, L., S. Lindeberg, et al. 2002. Acne vulgaris: A disease of Western civilization. *Archives of Dermatology* 138: 1584–1590.

Bek-Thomsen, M., H. B. Lomholt, and M. Kilian. 2008. Acne is not associated with yet-uncultured bacteria. *Journal of Clinical Microbiology* 46: 3355–3360.

Smith, R. N., N. J. Mann, et al. 2007. A low-glycemic-load diet improves symptoms in acne vulgaris patients: A randomized controlled trial. *American Journal of Clinical Nutrition* 86: 107–115.

22. For more on cancer as an evolutionary process, see:

Greaves, M. 2013. Cancer stem cells as "units of selection." *Evolutionary Applications* 6: 102–108.

CITATIONS

Aiuti, A., S. Slavin, et al. 2002a. Correction of ADA-SCID by stem cell gene therapy combined with nonmyeloablative conditioning. *Science* 296: 2410–2413.

Aiuti, A., S. Vai, et al. 2002b. Immune reconstitution in ADA-SCID after PBL gene therapy and discontinuation of enzyme replacement. *Nature Medicine* 8: 423–425.

Anderson, J. F., T. G. Andreadis, et al. 1999. Isolation of West Nile virus from mosquitoes, crows, and a Cooper's hawk in Connecticut. *Science* 286: 2331–2333.

Anderson, K., H. Kaplan, and J. Lancaster. 1999. Paternal care by genetic fathers and stepfathers I: Reports from Albuquerque men. *Evolution and Human Behavior* 20: 405–431.

Andersson, D. I., and D. Hughes. 2010. Antibiotic resistance and its cost: Is it possible to reverse resistance? *Nature Reviews Microbiology* 8: 260–271.

Austin, D. J., K. G. Kristinsson, and R. M. Anderson. 1999. The relationship between the volume of antimicrobial consumption in human communities and the frequency of resistance. *Proceedings of the National Academy of Sciences, USA* 96: 1152–1156.

Azzimondi, G., L. Bassein, et al. 1995. Fever in acute stroke worsens prognosis: A prospective study. *Stroke* 26: 2040–2043.

Banet, M. 1979. Fever and survival in the rat. *Pflügers Archive* 381: 35–38.

Banet, M. 1981a. Fever and survival in the rat. The effect of enhancing the cold defense response. *Experientia* 37: 985–986.

Banet, M. 1981b. Fever and survival in the rat. Metabolic rate versus temperature response. *Experientia* 37: 1302–1304.

Basler, C. F., A. H. Reid, et al. 2001. Sequence of the 1918 pandemic influenza virus nonstructural gene (NS) segment and characterization of recombinant viruses bearing the 1918 NS genes. *Proceedings of the National Academy of Sciences, USA* 98: 2746–2751.

Bean, W. J., Schell, M., et al. 1992. Evolution of the H3 influenza virus hemagglutinin from human and nonhuman hosts. *Journal of Virology* 66: 1129–1138.

Berkey, C. S., A. L. Frazier, et al. 1999. Adolescence and breast carcinoma risk. *Cancer* 85: 2400–2409.

Bernheim, H. A., and M. J. Kluger. 1976. Fever: Effect of drug-induced antipyresis on survival. *Science* 193: 237–239.

Bishai, W. R., N. M. H. Graham, et al. 1996. Rifampin-resistant tuberculosis in a patient receiving rifabutin prophylaxis. *New England Journal of Medicine* 334: 1573–1576.

Blaese, R. M., K. W. Culver, et al. 1995. T. lymphocyte-directed gene therapy for ADA-SCID: Initial trial results after 4 years. *Science* 270: 475–480.

Bloch, A. B., G. M. Cauthen, et al. 1994. Nationwide survey of drug-resistant tuberculosis in the United States. *Journal of the American Medical Association* 271: 665–671.

Bordignon, C., L. D. Notarangelo, et al. 1995. Gene therapy in peripheral blood lymphocytes and bone marrow for ADA-immunodeficient patients. *Science* 270: 470–475.

Borody, T. J., and A. Khoruts. 2012. Fecal microbiota transplantation and emerging applications. *Nature Reviews Gastroenterology & Hepatology* 9: 88–96.

Boyd, N. F., L. J. Martin, et al. 2011. Mammographic density and breast cancer risk: Current understanding and future prospects. *Breast Cancer Research* 13: 223.

Bull, J. J. 1994. Virulence. *Evolution* 48: 1423–1437.

Bush, R. M. 2001. Predicting adaptive evolution. *Nature Reviews Genetics* 2: 387–392.

Bush, R. M., C. A. Bender, et al. 1999. Predicting the evolution of human influenza A. *Science* 286: 1921–1925.

Bush, R. M., W. M. Fitch, et al. 1999. Positive selection on the H3 hemagglutinin gene of human influenza virus A. *Molecular Biology and Evolution* 16: 1457–1465.

Carey, J. V. 2010. Literature review: Should antipyretic therapies routinely be administered to patients with fever? *Journal of Clinical Nursing* 19: 2377–2393.

Chagnon, N. A. 1988. Male manipulations of kinship classifications of female kin for reproductive advantage. In *Human Reproductive Behavior: A Darwinian Perspective*, ed. L. Betzig, M. Borgerhoff Mulder, and P. Turke. Cambridge: Cambridge University Press, 23–48.

Chagnon, N. A. 1992. *Yąnomamö*. Fort Worth, TX: Harcourt Brace College Publishers.

Chen, W., P. A. Calvo, et al. 2001. A novel influenza A virus mitochondrial protein that induces cell death. *Nature Medicine* 7: 1306–1312.

Cochran, G. M., P. W. Ewald, and K. D. Cochran. 2000. Infectious causation of disease: An evolutionary perspective. *Perspectives in Biology and Medicine* 43: 406–448.

Cooper, V. S., M. H. Reiskind, et al. 2002. Timing of transmission and the evolution of virulence of an insect virus. *Proceedings of the Royal Society of London B* 269: 1161–1165.

Cordain, L., S. B. Eaton, et al. 2002. The paradoxical nature of hunter-gatherer diets: Meat-based, yet non-atherogenic. *European Journal of Clinical Nutrition* 56: S42–S52.

Cordain, L., R. W. Gotshall, and S. B. Eaton. 1997. Evolutionary aspects of exercise. *World Review of Nutrition and Dietetics* 81: 49–60.

Daly, M., and M. Wilson. 1988a. Evolutionary social psychology and family homicide. *Science* 242: 519–524.

Daly, M., and M. Wilson. 1988b. *Homicide*. New York: Aldine de Gruyter.

Daly, M., and M. I. Wilson. 1994a. Some differential attributes of lethal assaults on small children by stepfathers versus genetic fathers. *Ethology and Sociobiology* 15: 207–217.

Daly, M., and M. Wilson. 1994b. Stepparenthood and the evolved psychology of discriminative parental solicitude. In *Infanticide and Parental Care*, ed. S. Parmigiani and F. S. vom Saal. London: Harwood Academic Publishers, 121–133.

Dawkins, R. 1976. *The Selfish Gene*. 1st ed. Oxford: Oxford University Press.

Dawkins, R. 1989. *The Selfish Gene*. 2nd ed. Oxford: Oxford University Press.

Dennett, D. C. 1995. *Darwin's Dangerous Idea*. New York: Simon & Schuster.

Dethlefsen, L., S. Huse, et al. 2008. The pervasive effects of an antibiotic on the human gut microbiota, as revealed by deep 16S rRNA sequencing. *PLoS Biology* 6: e280.

Dethlefsen, L., and D. A. Relman. 2010. Incomplete recovery and individualized responses of the human distal gut microbiota to repeated antibiotic perturbation. *Proceedings of the National Academy of Sciences, USA* 108: 4554–4561.

Diamond, J. 1992. *The Third Chimpanzee*. New York: HarperCollins.

Dixon, A., D. Ross, et al. 1994. Paternal investment inversely related to degree of extra-pair paternity in the reed bunting. *Nature* 371: 698–700.

dos Reis, M., A. J. Hay, and R. A. Goldstein. 2009. Using non-homogeneous models of nucleotide substitution to identify host shift events: Application to the origin of the 1918 "Spanish" influenza pandemic virus. *Journal of Molecular Evolution* 69: 333–345.

dos Reis, M., A. U. Tamuri, et al. 2011. Charting the host adaptation of influenza viruses. *Molecular Biology and Evolution* 28: 1755–1767.

Durham, W. H. 1991. *Coevolution: Genes, Culture, and Human Diversity*. Stanford, CA: Stanford University Press.

Eaton, S. B., and L. Cordain. 1997. Evolutionary aspects of diet: Old genes, new fuels. *World Review of Nutrition and Dietetics* 81: 26–37.

Eaton, S. B., S. B. Eaton III, and M. J. Konner. 1997. Paleolithic nutrition revisited: A twelve-year retrospective on its nature and implications. *European Journal of Clinical Nutrition* 51: 207–216.

Ewald, P. W. 1991. Waterborne transmission and the evolution of virulence among gastrointestinal bacteria. *Epidemiology and Infection* 106: 83–119.

Ewald, P. W. 1993. The evolution of virulence. *Scientific American* (April): 86–93.

Ewald, P. W. 1994. *Evolution of Infectious Disease*. Oxford: Oxford University Press.

Ewald, P. W., J. B. Sussman, et al. 1998. Evolutionary control of infectious disease: Prospects for vectorborne and waterborne pathogens. *Mem Inst Oswaldo Cruz* 93: 567–576.

Ferguson, N. M., and R. M. Anderson. 2002. Predicting evolutionary change in the influenza A virus. *Nature Medicine* 8: 562–563.

Fitch, W. M., J. M. Leiter, et al. 1991. Positive Darwinian evolution in human influenza A viruses. *Proceedings of the National Academy of Sciences, USA* 88: 4270–4274.

Flinn, M. V. 1988. Step- and genetic-parent-offspring relationships in a Caribbean village. *Ethology and Sociobiology* 9: 335–369.

Flinn, M. V., and B. G. England. 1995. Childhood stress and family environment. *Current Anthropology* 36: 854–866.

Flinn, M. V., and B. G. England. 1997. Social economics of childhood glucocorticoid stress response. *American Journal of Physical Anthropology* 102: 33–53.

Frerichs, R. R., P. S. Keim, et al. 2012. Nepalese origin of cholera epidemic in Haiti. *Clinical Microbiology and Infection* 18: E158–E163.

Gerbault, P., A. Liebert, et al. 2011. Evolution of lactase persistence: An example of human niche construction. *Philosophical Transactions of the Royal Society B* 366: 863–877.

Gilbert, E. W. 1958. Pioneer maps of health and disease in England. *Geographical Journal* 124: 172–183.

Goel, M. S., E. P. McCarthy, et al. 2004. Obesity among U.S. immigrant subgroups by duration of residence. *JAMA* 292: 2860–2867.

Gorman, O. T., W. J. Bean, et al. 1991. Evolution of influenza A virus nucleoprotein genes: Implications for the origins of H1N1 human and classical swine viruses. *Journal of Virology* 65: 3704–3714.

Graham, N. M. H., C. J. Burrell, et al. 1990. Adverse effects of aspirin, acetaminophen, and ibuprofen on immune function, viral shedding, and clinical status in rhinovirus-infected volunteers. *Journal of Infectious Diseases* 162: 1277–1282.

Green, M. H., and C. W. Vermeulen. 1994. Fever and the control of Gram-negative bacteria. *Research in Microbiology* 145: 269–272.

Hasan, N. A., S. Y. Choi, et al. 2012. Genomic diversity of 2010 Haitian cholera outbreak strains. *Proceedings of the National Academy of Sciences, USA* 109: E2010–E2017.

Henderson, B. E., P. K. Ross, and M. C. Pike. 1993. Hormonal chemoprevention of cancer in women. *Science* 259: 633–638.

Hendriksen, R. S., L. B. Price, et al. 2011. Population genetics of *Vibrio cholerae* from Nepal in 2010: Evidence on the origin of the Haitian outbreak. *mBio* 2: e00157–11.

Hershfield, M. S., R. H. Buckley, et al. 1987. Treatment of adenosine deaminase deficiency with polyethylene glycol-modified adenosine deaminase. *New England Journal of Medicine* 316: 589–596.

Hill, K., and H. Kaplan. 1988. Trade-offs in male and female reproductive strategies among the Ache, part 2. In *Human Reproductive Behavior: A Darwinian Perspective*, ed. L. Betzig, M. Borgerhoff Mulder, and P. Turke. Cambridge: Cambridge University Press, 291–305.

Hirschhorn, R., D. R. Yang, et al. 1996. Spontaneous in vivo reversion to normal of an inherited mutation in a patient with adenosine deaminase deficiency. *Nature Genetics* 13: 290–295.

Huttenhower, C., D. Gevers, et al. 2012. Structure, function and diversity of the healthy human microbiome. *Nature* 486: 207–214.

Ito, K., M. Igarashi, et al. 2011. Gnarled-trunk evolutionary model of influenza A virus hemagglutinin. *PLoS One* 6: e25953.

Jagger, B. W., H. M. Wise, et al. 2012. An overlapping protein-coding region in influenza A virus segment 3 modulates the host response. *Science* 337: 199–204.

Johnson, N. P. A. S., and J. Mueller. 2002. Updating the accounts: Global mortality of the 1918–1920 "Spanish" influenza pandemic. *Bulletin of the History of Medicine* 76: 105–115.

Keim, P. S., F. M. Aarestrup, et al. 2011. Reply to "South Asia instead of Nepal may be the origin of the Haitian cholera outbreak strain." *mBio* 2: e00245–11.

Klug, W. S., and M. R. Cummings. 1997. *Concepts of Genetics*. 5th ed. Upper Saddle River, NJ: Prentice Hall.

Kluger, M. J. 1992. Fever revisited. *Pediatrics* 90: 846–850.

Kluger, M. J., D. H. Ringler, and M. R. Anver. 1975. Fever and survival. *Science* 188: 166–168.

Koch, R.1883. Der zweite bericht der deutschen cholerakommission. *Deutsche Medizinische Wochenschrift* 9: 743–744.

Kolata, G. 1999. *Flu*. New York: Farrar, Straus, and Giroux.

Komlos, J., and M. Brabec. 2011. The trend of BMI values of U.S. adults by deciles, birth cohorts 1882–1986 stratified by gender and ethnicity. *Economics and Human Biology* 9: 234–250.

Kootte, R. S., A. Vrieze, et al. 2012. The therapeutic potential of manipulating gut microbiota in obesity and type 2 diabetes mellitus. *Diabetes, Obesity, and Metabolism* 14: 112–120.

Krebs, S. J., and R. K. Taylor. 2011. Protection and attachment of *Vibrio cholerae* mediated by the toxin-coregulated pilus in the infant mouse model. *Journal of Bacteriology* 193: 5260–5270.

Laland, K. N., J. Odling-Smee, and S. Myles. 2010. How culture shaped the human genome: Bringing genetics and the human sciences together. *Nature Reviews Genetics* 11: 137–148.

Lanciotti, R. S., J. T. Roehrig, et al. 1999. Origin of West Nile virus responsible for an outbreak of encephalitis in the northeastern United States. *Science* 286: 2333–2337.

Laver, W. G., N. Bischofberger, and R. G. Webster. 1999. Disarming flu viruses. *Scientific American* 280 (January): 78–87.

Lawson, J. S., and B. Heng. 2010. Viruses and breast cancer. *Cancers* 2: 752–772.

Layde P. M., L. A. Webster, et al. 1989. The independent associations of parity, age at first full term pregnancy, and duration of breastfeeding with the risk of breast cancer. *Journal of Clinical Epidemiology* 42: 963–973.

Levin, B. R. 1996. The evolution and maintenance of virulence in microparasites. *Emerging Infectious Diseases* 2: 93–102.

Levy, S. B. 1998. The challenge of antibiotic resistance. *Scientific American* 278 (March): 46–53.

Lewontin, R. C. 1980. Sociobiology: Another biological determinism. *International Journal of Health Services* 10: 347–363.

Ley, R. E., P. J. Turnbaugh, et al. 2006. Microbial ecology—Human gut microbes associated with obesity. *Nature* 444: 1022–1023.

Loo, V. G., A. M. Bourgault, et al. 2011. Host and pathogen factors for *Clostridium difficile* infection and colonization. *New England Journal of Medicine* 365: 1693–1703.

MacLean, R. C., A. R. Hall, et al. 2010. The population genetics of antibiotic resistance: Integrating molecular mechanisms and treatment contexts. *Nature Reviews Genetics* 11: 405–414.

Marlowe, F. 1999. Showoffs or providers? The parenting effort of Hadza men. *Evolution and Human Behavior* 20: 391–404.

Mason, A. L., S. Y. Gilady, and J. R. Mackey. 2011. Mouse mammary tumor virus in human breast cancer red herring or smoking gun? *American Journal of Pathology* 179: 1588–1590.

Mazzanti, C. M., M. Al Hamad, et al. 2011. A mouse mammary tumor virus env-like exogenous sequence is strictly related to progression of human sporadic breast carcinoma. *American Journal of Pathology* 179: 2083–2090.

McAllister, E. J., N. V. Dhurandhar, et al. 2009. Ten putative contributors to the obesity epidemic. *Critical Reviews in Food Science and Nutrition* 49: 868–913.

Messenger, S. L., I. J. Molineux, and J. J. Bull. 1999. Virulence evolution in a virus obeys a trade-off. *Proceedings of the Royal Society of London B* 266: 397–404.

Mutreja, A., D. W. Kim, et al. 2011. Evidence for several waves of global transmission in the seventh cholera pandemic. *Nature* 477: 462–465.

Nesse, R. M., and G. C. Williams. 1994. *Why We Get Sick*. New York: Vintage Books.

Nesse, R. M., and G. C. Williams. 1998. Evolution and the origins of disease. *Scientific American* (November): 86–93.

Nicholson, J. K., E. Holmes, et al. 2012. Host-gut microbiota metabolic interactions. *Science* 336: 1262–1267.

Nikaido, H. 2009. Multidrug resistance in bacteria. *Annual Review of Biochemistry* 78: 119–146.

Norton, T. T., and C. Wildsoet. 1999. Toward controlling myopia progression? *Optometry and Vision Science* 76: 341–342.

Ogden C. L., M. D. Carroll, et al. 2007. *Obesity among adults in the United States—no change since 2003–2004*. NCHS data brief no 1. Hyattsville, MD: National Center for Health Statistics.

O'Keefe, J. H., R. Vogel, et al. 2010. Achieving hunter-gatherer fitness in the 21(st) century: Back to the future. *American Journal of Medicine* 123: 1082–1086.

Orr, H. A. 1996. Dennett's dangerous idea. *Evolution* 50: 467–472.

Palsdottir, A., A. Helgason, et al. 2008. A drastic reduction in the life span of cystatin *CL68Q* carriers due to life-style changes during the last two centuries. *PLoS Genetics* 4: e1000099.

Parrish, C. R., and Y. Kawaoka. 2005. The origins of new pandemic viruses: The acquisition of new host ranges by canine parvovirus and influenza A viruses. *Annual Review of Microbiology* 59: 553–586.

Pasteur, L. 1858. Mèmoire sur la fermentation appeleé lactique. *Annales de Chimie et de Physique 3e. sér.* 52: 404–418.

Persons, D. A., and A. W. Nienhuis. 2002. In vivo selection to improve gene therapy of hematopoietic disorders. *Current Opinion in Molecular Therapeutics* 4: 491–498.

Pike, M. C., C. L. Pearce, and A. H. Wu. 2004. Prevention of cancers of the breast, endometrium and ovary. *Oncogene* 23: 6379–6391.

Pike, M. C., and D. V. Spicer. 2000. Hormonal contraception and chemoprevention of female cancers. *Endocrine-Related Cancer* 7: 73–83.

Plotkin, J. B., and J. Dushoff. 2003. Codon bias and frequency-dependent selection on the hemagglutinin epitopes of influenza A virus. *Proceedings of the National Academy of Sciences, USA* 100: 7152–7157.

Plotkin, J. B., J. Dushoff, and S. A. Levin. 2002. Hemagglutinin sequence clusters and the antigenic evolution of influenza A virus. *Proceedings of the National Academy of Sciences, USA* 99: 6263–6268.

Pogo, B. G., J. F. Holland, and P. H. Levine. 2010. Human mammary tumor virus in inflammatory breast cancer. *Cancer* 116: 2741–2744.

Pontzer, H., F. F. Chehab, et al. 2012. Hunter-gatherer energetics and human obesity. *PLoS One* 7: e40503.

Pun, S. B. 2011. South Asia instead of Nepal may be the origin of the Haitian cholera outbreak strains. *mBio* 2: e00219–11.

Rattan, A., A. Kalia, and N. Ahmad. 1998. Multidrug-resistant *Mycobacterium tuberculosis*: Molecular perspectives. *Emerging Infectious Diseases* 4: 195–209.

Reid, A. H., T. G. Fanning, et al. 1999. Origin and evolution of the 1918 "Spanish" influenza virus hemagglutinin gene. *Proceedings of the National Academy of Sciences, USA* 96: 1651–1656.

Reid, A. H., T. G. Fanning, et al. 2000. Characterization of the 1918 "Spanish" influenza neuraminidase gene. *Proceedings of the National Academy of Sciences, USA* 97: 6785–6790.

Reid, A. H., T. G. Fanning, et al. 2002. Characterization of the 1918 "Spanish" influenza virus matrix gene segment. *Journal of Virology* 76: 10717–10723.

Reid, A. H., T. G. Fanning, et al. 2004a. Novel origin of the 1918 pandemic influenza virus nucleoprotein gene. *Journal of Virology* 78: 12462–12470.

Reid, A. H., J. K. Taubenberger, and T. G. Fanning. 2004b. Evidence of an absence: The genetic origins of the 1918 pandemic influenza virus. *Nature Reviews Microbiology* 2: 909–914.

Schrag, S. J., V. Perrot, and B. R. Levin. 1997. Adaptation to the fitness costs of antibiotic resistance in *Escherichia coli*. *Proceedings of the Royal Society of London B* S264: 1287–1291.

Shibata, D., W. Navidi, et al. 1996. Somatic microsatellite mutations as molecular tumor clocks. *Nature Medicine* 2: 676–681.

Smith, G. J., D. Vijaykrishna, et al. 2009. Origins and evolutionary genomics of the 2009 swine-origin H1N1 influenza A epidemic. *Nature* 459: 1122–1125.

Snow, J. 1855. *On the Mode of Communication of Cholera*. 2nd ed. London: John Churchill.

Sorsby, A., and F. A. Young. 1970. Transmission of refractive errors within Eskimo families. *American Journal of Optometry* 47: 244–249.

Spicer, D. V., D. Shoupe, and M. C. Pike. 1991. GnRH agonists as contraceptive agents: Predicted significantly reduced risk of breast cancer. *Contraception* 44: 289–310.

Spicer, D. V., G. Ursin, et al. 1994. Changes in mammographic densities induced by a hormonal contraceptive designed to reduce breast cancer risk. *Journal of the National Cancer Institute* 86: 431–436.

Stewart, T. H. M., R. D. Sage, et al. 2000. Breast cancer incidence highest in the range of one species of house mouse, *Mus domesticus*. *British Journal of Cancer* 82: 446–451.

Strassmann, B. I. 1999. Menstrual cycling and breast cancer: An evolutionary perspective. *Journal of Women's Health* 8: 193–202.

Sun, L., E. Y. Klein, and R. Laxminarayan. 2012. Seasonality and temporal correlation between community antibiotic use and resistance in the United States. *Clinical Infectious Diseases* 55: 687–694.

Swinburn, B. A., G. Sacks, et al. 2011. The global obesity pandemic: Shaped by global drivers and local environments. *Lancet* 378: 804–814.

Tao, Y., J. Ruan, et al. 2011. Rapid growth of a hepatocellular carcinoma and the driving mutations revealed by cell-population genetic analysis of whole-genome data. *Proceedings of the National Academy of Sciences, USA* 108: 12042–12047.

Taubenberger, J. K., A. H. Reid, et al. 2005. Characterization of the 1918 influenza virus polymerase genes. *Nature* 437: 889–893

Teikari, J. M., J. O'Donnell, et al. 1991. Impact of heredity in myopia. *Human Heredity* 41: 151–156.

Thuny, F., H. Richet, et al. 2010. Vancomycin treatment of infective endocarditis is linked with recently acquired obesity. *PLoS One* 5: e9074.

Tufte, E. R. 1997. *Visual Explanations*. Cheshire, CT: Graphics Press.

Tumpey, T. M., C. F. Basler, et al. 2005. Characterization of the reconstructed 1918 Spanish influenza pandemic virus. *Science* 310: 77–80.

Turnbaugh, P. J., R. E. Ley, et al. 2006. An obesity-associated gut microbiome with increased capacity for energy harvest. *Nature* 444: 1027–1131.

Vaughn, L. K., H. A. Bernheim, and M. Kluger. 1974. Fever in the lizard *Dipsosaurus dorsalis*. *Nature* 252: 473–474.

Vrieze, A., E. van Nood, et al. 2012. Transfer of intestinal microbiota from lean donors increases insulin sensitivity in subjects with metabolic syndrome. *Gastroenterology* 143: 913–916.e7.

Wang, Y., J. F. Holland, et al. 1995. Detection of mammary tumor virus ENV gene-like sequences in human breast cancer. *Cancer Research* 55: 5173–5179.

Webster, R. G., W. J. Bean, et al. 1992. Evolution and ecology of influenza A viruses. *Microbiological Reviews* 56: 152–179.

Weitzel, J. N., S. S. Buys, et al. 2007. Reduced mammographic density with use of a gonadotropin-releasing hormone agonist-based chemoprevention regimen in BRCA1 carriers. *Clinical Cancer Research* 13: 654–658.

Wertheim, J. O. 2010. The re-emergence of H1N1 influenza virus in 1977: A cautionary tale for estimating divergence times using biologically unrealistic sampling dates. *PLoS One* 5: e11184.

Williams, G. C., and R. M. Nesse. 1991. The dawn of Darwinian medicine. *Quarterly Review of Biology* 66: 1–22.

Wise, H. M., A. Foeglein, et al. 2009. A complicated message: Identification of a novel PB1-related protein translated from influenza A virus segment 2 mRNA. *Journal of Virology* 83: 8021–8031.

Yatsunenko, T., F. E. Rey, et al. 2012. Human gut microbiome viewed across age and geography. *Nature* 486: 222–227.

Young, F. A., W. R. Baldwin, et al. 1969. The transmission of refractive errors with Eskimo families. *American Journal of Optometry* 46: 676–685.

Youssoufian, H. 1996. Natural gene therapy and the Darwinian legacy. *Nature Genetics* 13: 255–256.

Zhang, J., A. M. Mamlouk, et al. 2011. PhyloMap: An algorithm for visualizing relationships of large sequence data sets and its application to the influenza A virus genome. *BMC Bioinformatics* 12: 248.

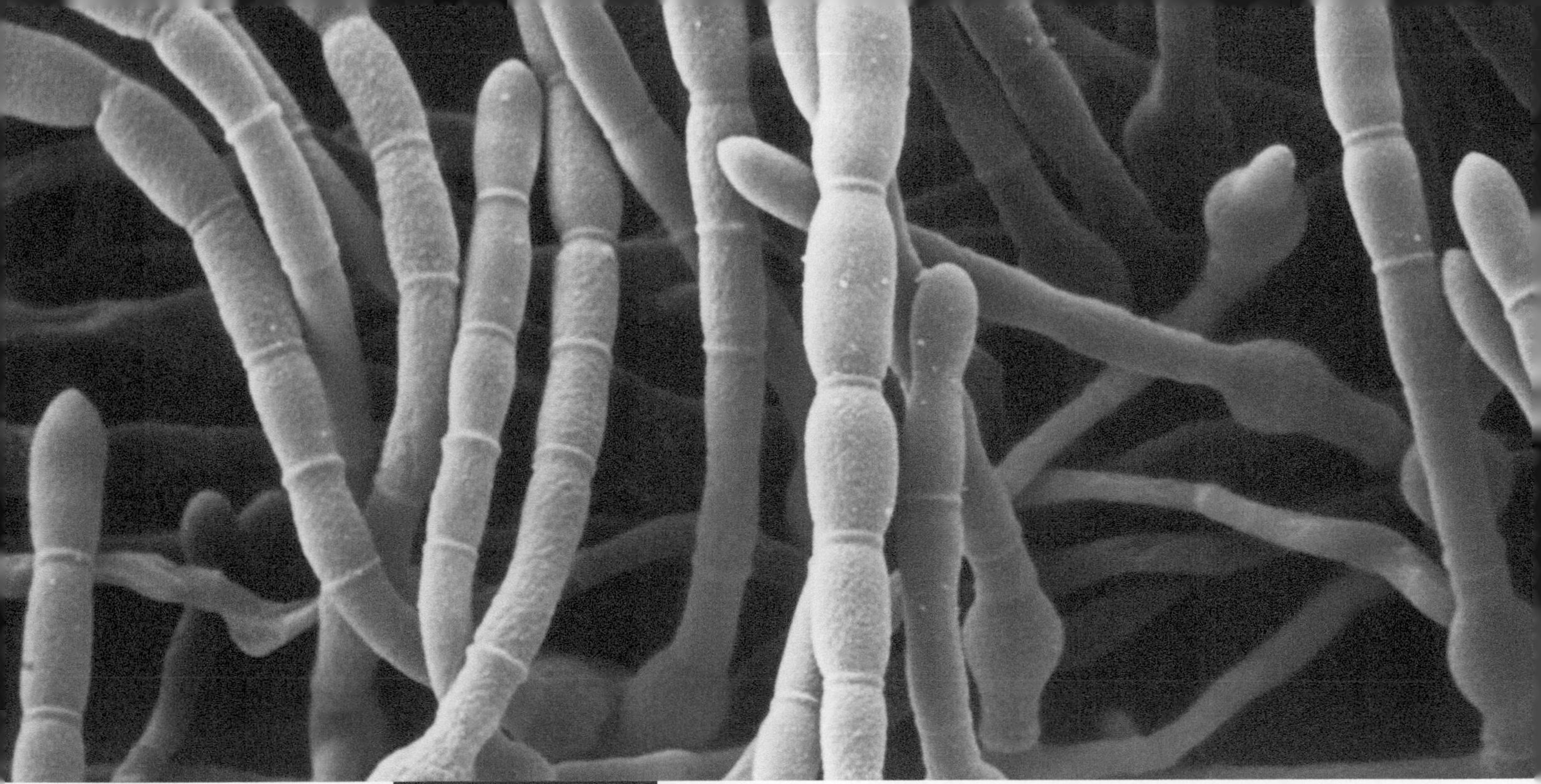

15

Genome Evolution and the Molecular Basis of Adaptation

Powdery mildews are fungal pathogens that grow and reproduce on living plants. They infect approximately 10,000 flowering plants worldwide, including important crops such as wheat, rice, and barley. When Pietro Spanu (2010) and colleagues sequenced the complete genomes of three powdery mildew species, they were surprised to discover that the genomes were at least three times larger than those of other closely related fungi that are not obligate pathogens. This despite the fact that the powdery mildew genomes have fewer functional genes than other fungi. Powdery mildew genomes are bloated because a large portion of their genome sequence (64% in the case of the barley mildew *Blumeria graminis*) comes from mobile genetic elements, which are parasitic self-replicating DNA entities we consider in detail later in this chapter.

Up to this point, we have focused on the evolution of traits in organisms by considering the evolutionary mechanisms (Chapters 5–9) and agents of natural selection (Chapters 10–14) that result in the adaptation of organisms to their environments. The outcomes of evolution that we have considered are the phenotypes of organisms: features of morphology, physiology, life history, and behavior that can be seen and measured. Here we broaden this approach by considering an additional feature of organisms that can now be studied: their genomes.

Blumeria graminis (*Bg*) and its kin (orange) have big genomes. Data from Spanu et al. (2010).

From "Genome expansion and gene loss in powdery mildew fungi reveal tradeoffs in extreme parasitism." *Science* 330: 1543–1546. Reprinted with permission from AAAS.

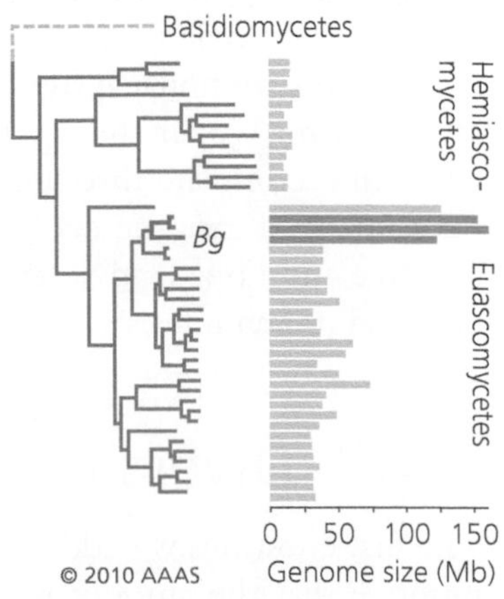

While not a phenotype in the classic sense, a genome can be observed and measured to illuminate important elements of its **genome architecture**—the structure and organization of the information encoded within. The development of high-throughput DNA sequencing technologies (see Mardis 2011; Niedringhaus et al. 2011) has enabled scientists to record the complete genome sequences of thousands of different species. Genome sequences—the ordered series of individual DNA bases encoded in the nucleus of each cell in an organism—have been obtained from organisms all across the web of life, from microbes living in hydrothermal vents on the ocean floor to the black cottonwood trees present in many forests of western North America.

Using these genome sequences, scientists have begun asking important questions about genome architecture, such as:

- How big or small are genomes, in terms of the number of DNA bases they contain? Is there variation among species in the size of genomes? Do more complex organisms have more genes?
- How is the information in a genome organized? How are genes separated from each other, and what exists between them?
- Much of the genetic information in multicellular organisms does not contain protein-coding genes. Where do noncoding sequences come from, and what do they do?
- Where do new genes come from? What evolutionary processes ensure the preservation of new genes?
- What genes underlie adaptations in nature, and where do they come from?

Questions like these are the subject of this chapter. As we will see, the answers are not always what we might have expected. The analysis of genomes has uncovered some of the most surprising findings of modern biology.

Section 15.1 presents an overview of genome diversity. Section 15.2 considers mobile genetic elements, the genomic parasites we mentioned earlier. Sections 15.3 and 15.4 look at the evolution of mutation rates and genome duplication. Finally, Section 15.5 examines recent studies of loci within genomes that are targets of selection.

Throughout our study of genome evolution, we should keep one concept in mind. Although this part of the book is entitled "Adaptation," many aspects of genomes may not have arisen as a result of natural selection. Instead, some genome features, particularly those of larger, more complex organisms, likely have resulted from other mechanisms of evolution, such as mutation and genetic drift (Lynch 2007; Lynch et al. 2011). Genome features that are not the result of natural selection are nonadaptive, but this does not necessarily mean they are bad for an organism. Under the right conditions, harmless but useless genetic information can accumulate through the neutral process of genetic drift. The challenge for scholars of genome evolution is to uncover the evolutionary processes, both adaptive and nonadaptive, that have led to the content, structure, and functions encoded in genomes.

15.1 Diversity among Genomes

The first questions we address concern the size and content of genomes. It seems intuitive that the amount of DNA in an organism's genome should bear some relationship to the organism's size and complexity, such as whether it is single celled or multicellular, or how many different cell types are created by the same genome. Yet genome sizes turn out to be more enigmatic than this simple hypothesis would suggest.

Variation in Genome Size

Long before genome sequencing was possible, biologists could measure the total amount of DNA found in a cell, called its C-value. They found that prokaryotes usually had smaller C-values than eukaryotes, but within eukaryotes there was little correlation between DNA quantity and an organism's perceived morphological complexity. The Japanese canopy plant (*Paris japonica*) has about 150 billion basepairs (bp) of DNA per cell (Pellicer et al. 2010). Humans, in contrast, have only 3.2 billion bp (Morton 1991), and the fruit fly *Drosophila melanogaster* has just 180 million bp (Adams et al. 2000). This puzzle, called the **C-value paradox,** was eventually explained by two phenomena: whole-genome duplications resulting in polyploidy (discussed in Chapter 5), and the existence of large portions of an organism's genome that are largely functionless from the cell's viewpoint.

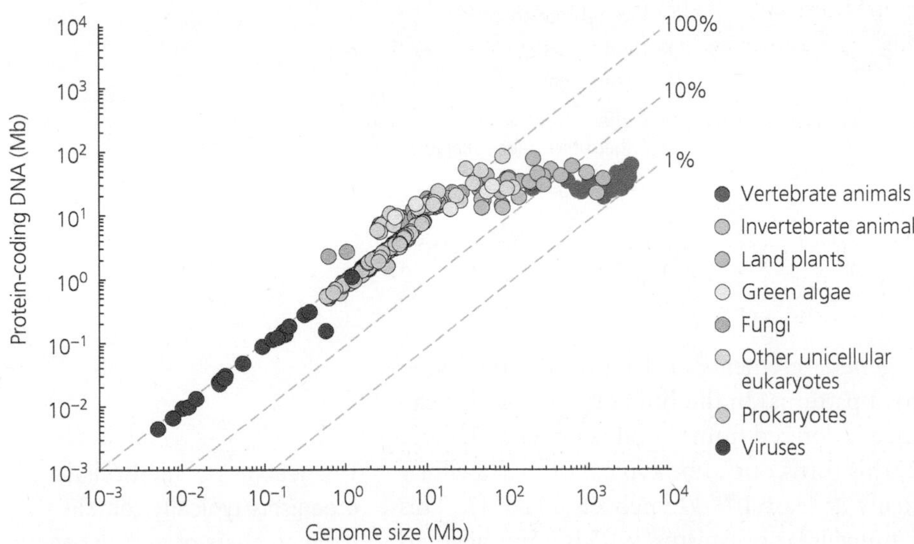

Figure 15.1 Contribution of protein-coding DNA to total genome size Both *x*- and *y*-axes are on a log scale, and the dashed lines illustrate proportional relationships that exist for points on each line. For example, points near the 10% line represent organisms for which approximately 10% of the DNA in the genome codes for proteins. Redrawn from Lynch (2006) and Lynch et al. (2011).

The relationship between an organism's genome size and the amount of its DNA that is **coding,** meaning it codes for functional proteins, is shown in **Figure 15.1**. At the smaller end of the spectrum, there is a clear relationship: Larger genomes have more DNA that codes for proteins (see the viruses, prokaryotes, and single-celled eukaryotes in the graph). There is no C-value paradox here. But for multicellular organisms, the paradox emerges. The size of a plant or animal's genome is unrelated to its body size or phylogenetic position. More puzzling, most plant and animal genomes are largely composed of **noncoding** DNA that does not directly code for proteins (for an exception see Ibarra-Laclette et al. 2013).

A multicellular organism's genome size is unrelated to its body size or phylogenetic position. Genomes of multicelluar organisms typically contain a great deal of noncoding DNA.

There are now reasons to believe that some noncoding DNA serves to regulate the expression of protein-coding genes. In complex organisms like mammals, RNA molecules transcribed from noncoding regions of the genome have roles in regulating gene expression, even though these RNAs do not code for proteins (Mercer et al. 2009).

Long before this realization, however, the discovery of **mobile genetic elements** was a critical milestone in our understanding of genome biology. Mobile genetic elements replicate and insert themselves in an organism's genome by hijacking the same cellular machinery that replicates and transcribes protein-coding DNA. In rare cases, mobile genetic elements disrupt the function of protein-coding genes because they happen to get inserted right in the middle of a coding sequence of DNA. These insertions result in obvious changes to an organism's phenotype. But in the vast majority of cases, mobile genetic elements have no effect on an organism's phenotype, even though taken together they often comprise most of an organism's genome.

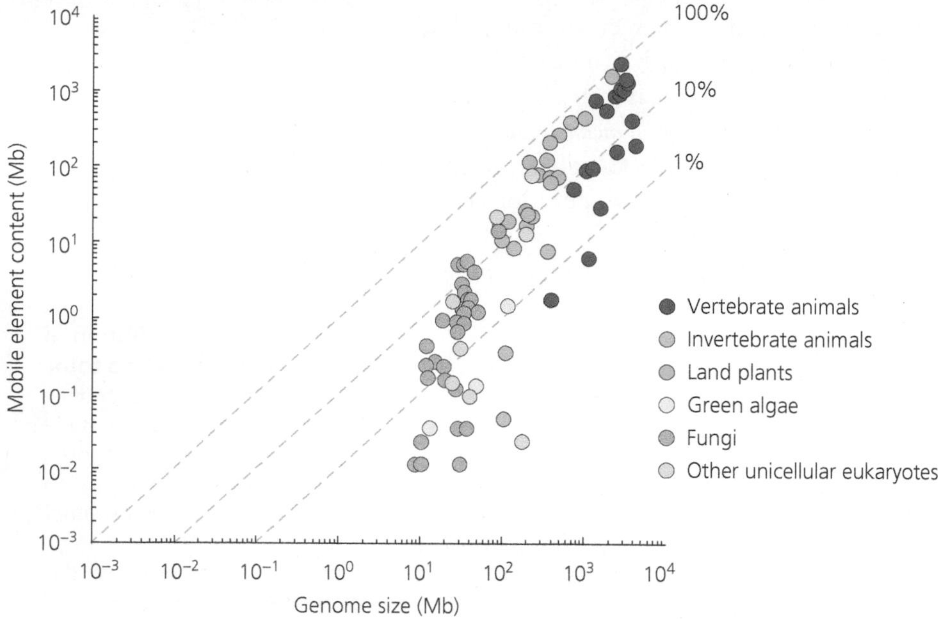

Figure 15.2 Contribution of mobile genetic element DNA to total genome size Both axes are on a log scale, and the diagonal lines illustrate specific proportional relationships, in this case between genome size and mobile element content. Prepared with data provided by Michael Lynch, after Lynch et al. (2011).

We discuss the biology of mobile genetic elements in detail in the following section, but for now consider the broad patterns. In the human genome, for example, some 45% of the DNA sequence is derived from mobile elements (Lander et al. 2001). As shown in **Figure 15.2**, this turns out to be typical of multicellular genomes, for which this value is usually at least 10% (Lynch et al. 2011). This pattern starkly contrasts with that for unicellular organisms, which often have no mobile elements in their genomes (Lynch 2006). Clearly, there are important differences among categories of organisms, and no account of genome biology and evolution is complete without considering the role of mobile elements.

The genomes of multicellular organisms typically contain large numbers of mobile genetic elements.

Variation in Genome Architecture

Mobile genetic elements and their remnants make up most of the **intergenic** regions of eukaryotic genomes—the space between protein-coding genes. But eukaryotic genomes also contain other important forms of noncoding DNA, the most predominant of which are **introns.**

Introns occur within the coding regions of genes and are transcribed into mRNA when genes are expressed, but are then removed and the remaining protein-coding **exons** spliced together before proteins are translated. In multicellular species, most genes contain introns, which are often much longer than the exons. Unicellular eukaryotes typically have 1–2 introns per genome, but the average among vertebrate animals, for example, is 5–8 introns per gene (Koonin 2009). A typical vertebrate genome thus contains thousands of introns. The net

Eukaryotic genomes typically contain introns, but the number of introns varies dramatically.

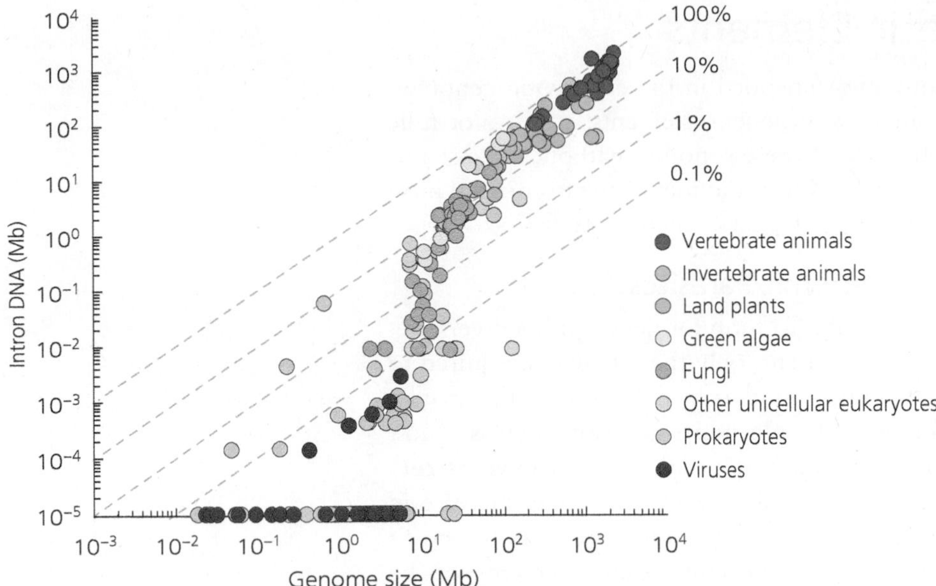

Figure 15.3 Contribution of introns to total genome size Both axes are on a log scale, and the diagonal lines illustrate specific proportional relationships. Many prokaryotes and viruses have no introns (points along *x*-axis), while almost all land plants and all animals have at least 10% intron content. Redrawn from Lynch (2006) and Lynch et al. (2011).

result is that, as documented in **Figure 15.3**, introns make up a striking proportion of noncoding DNA in multicellular genomes.

In comparing the genome sequences of hundreds of species, the clearest pattern to emerge is that there are essentially two kinds of genomes: the small, compact genomes of prokaryotes with short intergenic regions and no introns, and the large, expanded genomes of multicellular eukaryotes, which have vast intergenic regions and several introns per gene (**Figure 15.4**; Koonin 2009).

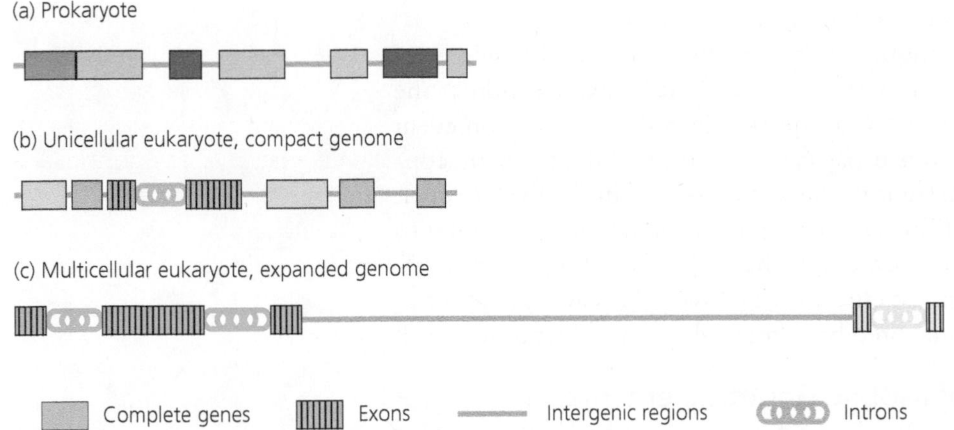

Figure 15.4 Generalized genome architectures for taxonomic groups (a) Prokaryotes have compact genomes with few mobile elements or introns. (b) Some eukaryotes have compact genomes that are similar to those of prokaryotes, but contain the occasional gene with introns. (c) Multicellular eukaryotes typically have genomes that are expanded, with long intergenic regions (often containing mobile elements) and frequent introns. Redrawn from Koonin (2009).

Where do introns come from? And are they adaptive? This remains one of the biggest mysteries in genome biology. It could be that introns are simply excess genetic material that serves no purpose in the organism, like mobile elements. But if introns do not do anything, their presence would likely be quite variable. Instead, the positions of introns are well conserved across distantly related organisms (Rodríguez-Trelles et al. 2006; Roy & Gilbert 2006). Because the evolutionary origin and biological mechanisms that create and maintain introns are not well understood, we can only speculate as to their function, if any.

15.2 Mobile Genetic Elements

Because much of the genetic information encoded in the eukaryotic genomes, including our own, is derived from mobile genetic elements, we cannot fully understand the structure and functioning of these genomes without understanding mobile elements. How do they work? Where did they come from? And most important, what effects do they have on the fitness of their hosts?

Mobile Genetic Elements Are Genomic Parasites

"Mobile genetic element" is really an umbrella term for genes with a diverse set of characteristics. Most mobile elements contain only the sequences required to **transpose,** or move, from one location to another in the genome; this is why they are often referred to as **transposable elements** or **transposons.** Most leave a copy of themselves behind when they move; these are known as **retrotransposons.** For these elements, transposition events lead to an increase in element abundance within the host genome. If transposition occurs in the host's germ line, then the increased abundance of the transposable element will be passed on to the host's offspring.

Transposition can increase the fitness of a mobile element among a population of genes within a genome, but how is the rest of the genome affected? This depends on the new location that the element transposes into. If the new location is an intergenic region, there may be no effect on the phenotype of the host. But if a mobile element lands within the coding sequence of a gene, deleterious knock-out mutations usually result. In humans, transposition events have resulted in many kinds of heritable diseases, including hemophilia, cystic fibrosis, and cancer (see Cordaux & Batzer 2009). Because they can disrupt coding sequences and may place an energetic burden on the cell, mobile genetic elements are most accurately characterized as genome parasites. Carrying mobile elements in the genome appears to be neutral at best and maladaptive at worst.

Mobile genetic elements are genomic parasites. They spread within genomes, sometimes with deleterious consequences for their host.

What is the key to their evolutionary success? The answer is that while selection at the level of hosts may select against mobile elements, selection at the level of the elements themselves favors their spread. Even if a transposition event reduces the survival and reproductive capacity of the host slightly, the extra copies of the mobile element now present in the gene pool can make up the deficit and result in the parasite's spread throughout the host population. According to models developed by Brian Charlesworth and Charles Langley (1989), mobile elements that replicate themselves most efficiently and with the least fitness cost to the host genome are, on balance, favored by natural selection and tend to spread.

The Evolutionary Impact of Mobile Genetic Elements

Once biologists recognized the nature of mobile elements and had characterized their diversity and distribution, the next challenge was to understand their dynamics. If mobile elements are parasitizing hosts, do host genomes have mechanisms to counter them? And how do mobile elements affect phenotypes?

Defending against the Spread of Mobile Elements

In most plants and many animals, the attachment of methyl ($-CH_3$) groups to DNA nucleotides, called **methylation,** is common and prevents the transcription of DNA into RNA. Interestingly, methylation is especially common in regions of the genome associated with mobile elements. This finding led to the hypothesis that organisms use methylation to restrict the proliferation of mobile elements by preventing them from transposing (Yoder et al. 1997). Several research groups supported this hypothesis by experimentally reducing the amount of methylation in an organism's genome and finding that this prodecure increases the activity of mobile genetic elements. This result holds for both plants (Miura et al. 2001) and mammals (Bourc'his and Bestor 2004), suggesting that DNA methylation of mobile elements is an ancient mechanism that predates the plant/animal divergence.

Host genomes have evolved a number of defenses against mobile genetic elements.

Following the finding that DNA methylation restricts the activity of mobile genetic elements, an additional mechanism was discovered, called RNA interference or RNAi. RNAi was first described in nematode worms as a general phenomenon in which the insertion into nematode cells of short sequences of double-stranded RNA that matched a particular gene would effectively silence expression of that gene (Fire et al. 1998). Biologists soon realized that small RNA molecules were an important component of gene regulation in many organisms.

In the ensuing flurry of research on small RNAs, Titia Sijen and Ronald Plasterk (2003) hypothesized that nematodes possess an innate RNA interference system that targets transposons in the germ line as an evolved defense against transposon proliferation in offspring. They checked whether wild-type nematodes had small interfering RNAs in their cells that target the nematode transposon *Tc1*. As shown by the bands in the wild-type lane of the electrophoresis gel in **Figure 15.5**, the scientists found many such RNAs. This result indicated that normal nematodes make small RNAs that likely silence *Tc1* transposons in the germ line.

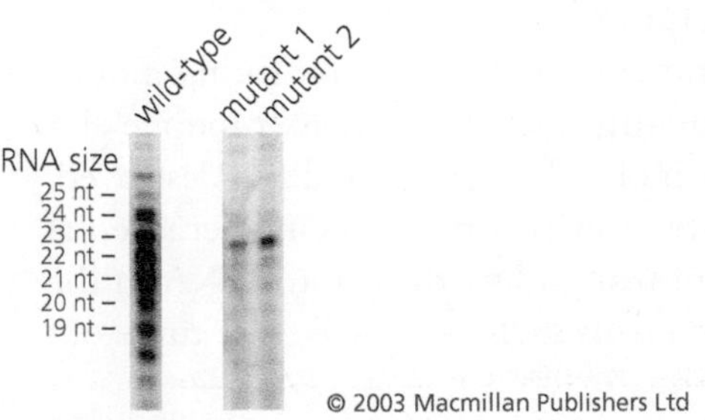

Figure 15.5 Nematode mutants are missing small RNAs targeted to transposons RNA molecules were sorted by size and assayed with a probe that bound to molecules targeting the transposon *Tc1* (black bands on gels). From Sijen and Plasterk (2003).
Reprinted by permission from Macmillan Publishers Ltd: *Nature* 426: 310–314. Copyright 2003.

Next Sijen and Plasterk guessed that some mutant nematode lineages had excessive transposon activity because their innate RNA interference mechanism was broken. Indeed, the scientists found that mutants with extra transposon activity had almost no small RNAs targeted to *Tc1* (mutant lanes in Figure 15.5). This was the first example of small RNAs suppressing mobile elements in a genome.

We now know that small RNA mechanisms for silencing mobile genetic elements exist in many kinds of organisms. For example, as shown in **Figure 15.6**, a large proportion of the small RNAs that appear during various stages of gamete development in mice and *Drosophila* have sequences that match to mobile genetic elements (Aravin et al. 2007; Brennecke et al. 2007; Kuramochi-Miyagawa et al. 2008; Tam et al. 2008; Watanabe et al. 2008).

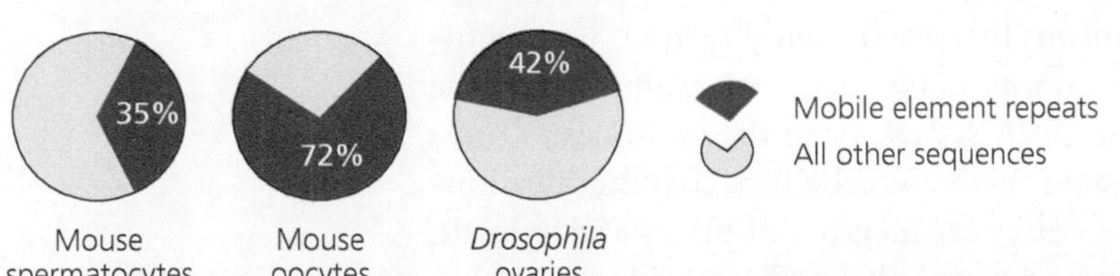

Figure 15.6 The fraction of small RNAs in germ-line cells that match mobile element sequences From Aravin et al. (2007), Watanabe et al. (2008), and Brennecke et al. (2007).

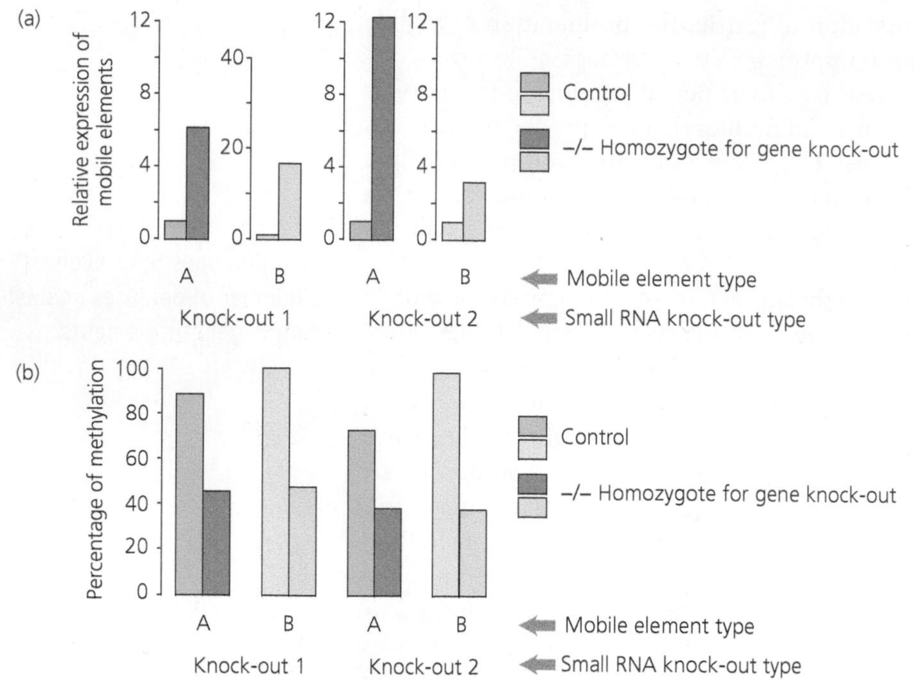

Figure 15.7 Importance of small RNA molecules in restricting the activity of mobile genetic elements
(a) Mobile element gene expression increases when key proteins of the mouse small RNA defense system are knocked out. Knock-out 1 redrawn from Aravin et al. (2007); knock-out 2 from Carmell et al. (2007). (b) Knocking out small RNA defense proteins also results in a large decrease in the amount of methylation of two kinds of mobile elements. Drawn from data presented in Kuramochi-Miyagawa et al. (2008).

As shown in **Figure 15.7a**, when critical proteins associated with these small RNA molecules are knocked out by researchers, the transcription of mobile element RNA often increases drastically (Aravin et al. 2007; Carmell et al. 2007). Following this finding came yet another surprise.

So far we have discussed an RNA interference method of combating mobile elements called **post-transcriptional silencing,** meaning that after mobile elements are transcribed into RNA, the RNA molecules are targeted and destroyed. Methylation, in contrast, is a form of **pre-transcriptional** silencing because the attached methyl groups prevent DNA from being transcribed into RNA in the first place. In mice and probably other mammals such as ourselves, it turns out that these two mechanisms are linked.

As documented in Figure 15.7b, when researchers knocked out critical components of retrotransposon RNA-silencing pathways during mouse gamete development, they found that methylation of retrotransposon DNA in the genome decreased dramatically (Aravin et al. 2007; Kuramochi-Miyagawa et al. 2008). This is a striking discovery: Small RNAs are responsible for targeting the destruction of mobile element RNA or DNA during transposition, but they are also responsible for targeting the methylation of mobile elements DNA in the genome, which prevents future transposition.

The Evolutionary Origin of Defenses against Mobile Elements

It is clear that both DNA methylation and RNA silencing are mechanisms that restrict the activity of mobile elements in most multicellular organisms. When traits are **evolutionarily conserved,** meaning the trait and its molecular machinery are broadly similar among organisms that are distantly related, we can infer that they were likely present in their common ancestor. For example, because elevated methylation of mobile element DNA is present in the genomes of most plants and animals **(Figure 15.8, facing page)** researchers believe that methylation-regulated gene expression existed in the common ancestor of plants and animals. Methylation is an ancient phenomenon (Feng et al. 2010; Zemach et al. 2010).

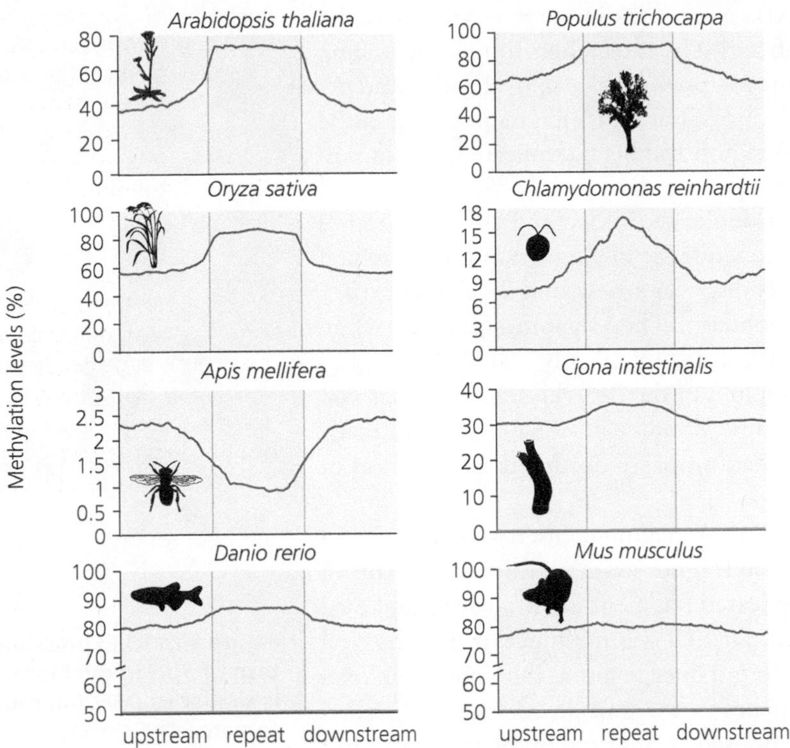

Figure 15.8 Methylation of repetitive mobile element DNA within genomes The center of each graph represents regions of repetitive DNA associated with mobile genetic elements. Note that the scale of the *y*-axis varies greatly among organisms. For all except the *Apis* bumblebee, methylation is higher in repetitive mobile element DNA than in nearby regions. Redrawn from Feng et al. (2010).

Given that methylation and small RNAs may date back to the origin of multicellularity, is it possible that these mechanisms originally evolved in response to the arrival of mobile genetic elements? Or did these mechanisms have a different origin, as innate regulators of gene expression, and were then co-opted via natural selection to also defend against the spread of parasitic DNAs? These are fascinating questions, but a definitive answer has been elusive.

Mobile Elements Can Affect Phenotypes: The Case of the Roma Tomato

An intriguing example of how a transposable element can affect phenotype comes from tomatoes. To explore the genetic differences between oval tomatoes and their round ancestors **(Figure 15.9)**, Han Xiao and collaborators collected DNA sequences from the region of the tomato genome known to influence fruit shape. The obvious difference between the oval and round lineages was the presence of an extra piece of DNA, 24,700 nucleotides in length, in the allele carried by plants with oval fruits. As shown in **Figure 15.10**, this long stretch of DNA was located inside another gene, called *DEFL1* (Xiao et al. 2008). The round-fruit

Figure 15.9 Round versus oval roma tomatoes

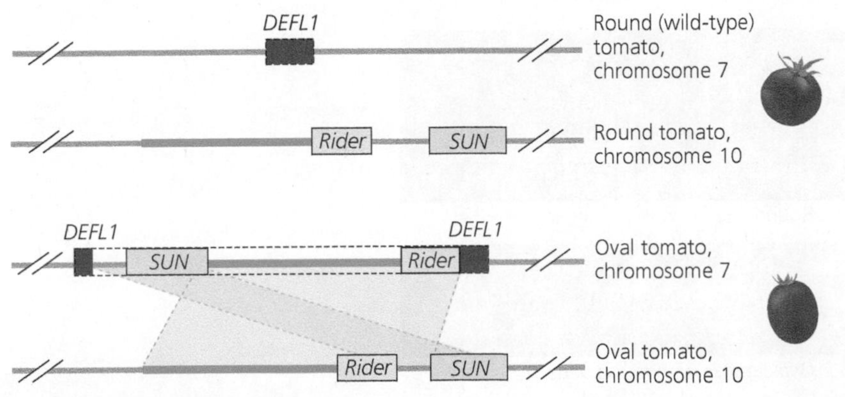

Figure 15.10 Duplication and transposition of the SUN locus in the oval tomato Wild-type plants, which produce round fruits, have a single occurrence of the *SUN* locus on chromosome 10. Lineages that produce oval fruits have a duplication of *SUN* that is inserted into *DEFL1* on chromosome 7. Modified from Xiao et al. (2008).

allele also had *DEFL1* but without the extra piece stuck in the middle. Where did the extra 24.7 kilobases of DNA come from? The researchers found a matching sequence on chromosome 10. This region was present in both oval and round tomatoes; oval tomatoes just have an extra copy of this DNA fragment sitting inside *DEFL1* on chromosome 7. The 24.7-kb region had been duplicated, a common mechanism of gene creation (see Chapter 5).

How was the 24.7-kb DNA fragment duplicated? Xiao and associates found that one end of the fragment contained the sequence of a retrotransposon, which they named *Rider* (often when researchers discover a new mobile element, they give it a name that implies motion or mobility). They hypothesized that when *Rider* had transposed from chromosome 10, it had mistakenly included a lengthy stretch of DNA that bordered it. This resulted in the 24.7-kb fragment that was inserted at the new location on chromosome 7, inside *DEFL1*. *Rider* had inadvertently caused gene duplication, and it had probably disrupted the function of *DEFL1* in the process.

To confirm their hypothesis that the DNA fragment inserted into *DEFL1* was responsible for oval fruits, Xiao and colleagues examined the expression of *DEFL1* and the genes present on the duplicated fragment in both round and oval tomatoes **(Figure 15.11)**. As they suspected, *DEFL1* was nonfunctional in the oval lineage, although it was expressed in the round lineage just at the time when oval fruits elongate. In addition, one of the genes in the transposed fragment, called *SUN* (van der Knaap et al. 2004), was expressed at this time in the oval fruits but not in the round fruits. The transposition event had caused *DEFL1* to be turned off, and another gene, *SUN*, to be turned on in its place.

To confirm their findings, Xiao and collaborators inserted just a piece of the transposed fragment from oval tomatoes, containing *SUN* and a portion of the *DEFL1* sequence that is just upstream, into round tomatoes. Their results showed the effect of *SUN* on fruit shape to be dosage dependent: The more *SUN* that was expressed, the more elongated the fruits got **(Figure 15.12)**.

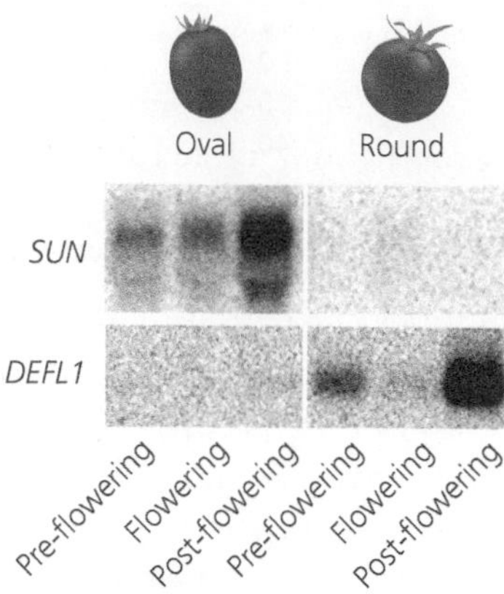

Figure 15.11 Expression pattern of *SUN* in oval tomatoes is similar to *DEFL1* in round tomatoes Gene expression was measured by Northern blots of RNA at three times during flower development. From Xiao et al. (2008).

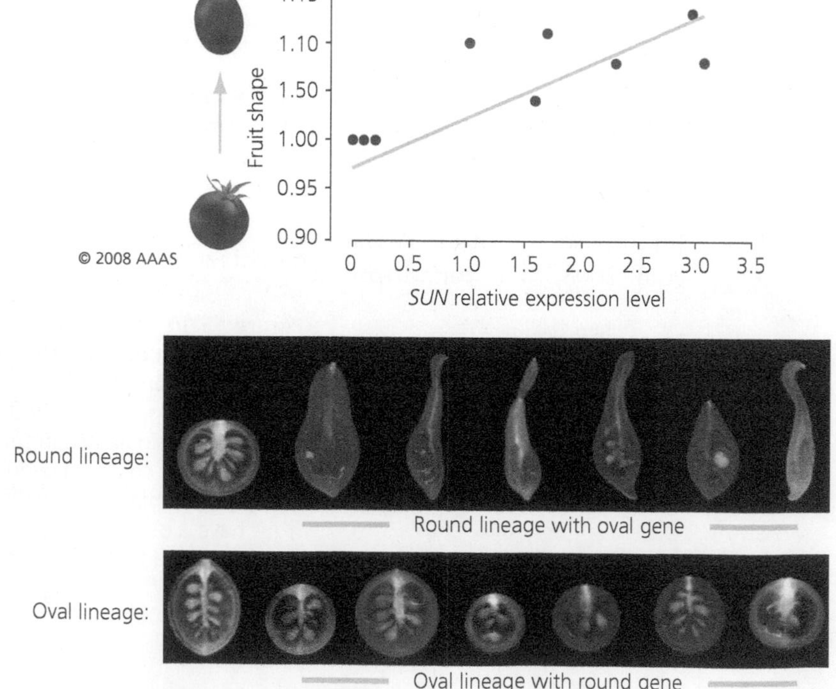

© 2008 AAAS

Figure 15.12 Effect of the *SUN* locus on fruit shape is dosage dependent Examples of fruits from oval and round lineages transformed with the *SUN* gene from the opposite type appear at bottom. From Xiao et al. (2008).

(Top) From "A retrotransposon-mediated gene duplication underlies morphological variation of tomato fruit." *Science* 319: 1527–1530. Reprinted with permission from AAAS.

The complete story becomes apparent when we put these lines of evidence together. When *SUN* was inserted into *DEFL1*, its expression pattern changed because *SUN*'s upstream regulatory elements changed. Gene regulation by nearby noncoding sequences is called cis-regulation. After transposition, *SUN* had the cis-regulatory elements of *DEFL1*, so it was expressed when *DEFL1* was normally expressed, during fruit development, and the result was oval fruits. This example illustrates three important features of gene and genome evolution:

Mobile genetic elements occasionally influence their host's phenotype.

- The *SUN* locus was modified by a mobile element. Mobile elements exist throughout multicellular genomes, but they rarely affect organisms' phenotypes.
- The specific modification of *SUN* was a duplication that created an extra copy of a locus present elsewhere in the genome. Gene duplication is a pervasive mode of genome evolution that we discuss in detail later in this chapter.
- The duplication of *SUN* resulted in a change in gene expression via the alteration of its cis-regulation. Changes in cis-regulation have been proposed as an important source of morphological novelty in evolution (see Wittkopp and Kalay 2012).

15.3 The Evolution of Mutation Rates

Mutation rates vary among organisms (see Chapter 5). This variation in mutation rates across the web of life shows some striking patterns. **Figure 15.13a**, for example, reveals that mutation rates decrease with increasing genome size across a variety of microorganisms. However, the four open circles—which were not used in calculating the best-fit line—hint at a different trend among many eukaryotes. This trend is confirmed by the data in Figure 15.13b. Why should mutation rates decrease with genome size in some organisms, but increase with genome size in others? Michael Lynch (2010) proposed a hypothesis based upon the mathematical logic of population genetics, suggesting that the lower limit on mutation rates is set by the interplay between natural selection and genetic drift.

Among most microorganisms, mutation rate decreases with genome size. However, among many eukaryotes—particularly multicellular ones—mutation rate increases with genome size.

The influence of genetic drift is stronger in smaller populations (see the theory and examples in Chapter 7). When natural selection and genetic drift jointly influence a phenotype, the ability of natural selection to favor an improved phenotype is opposed by the tendency of drift to cause a loss in genetic diversity for

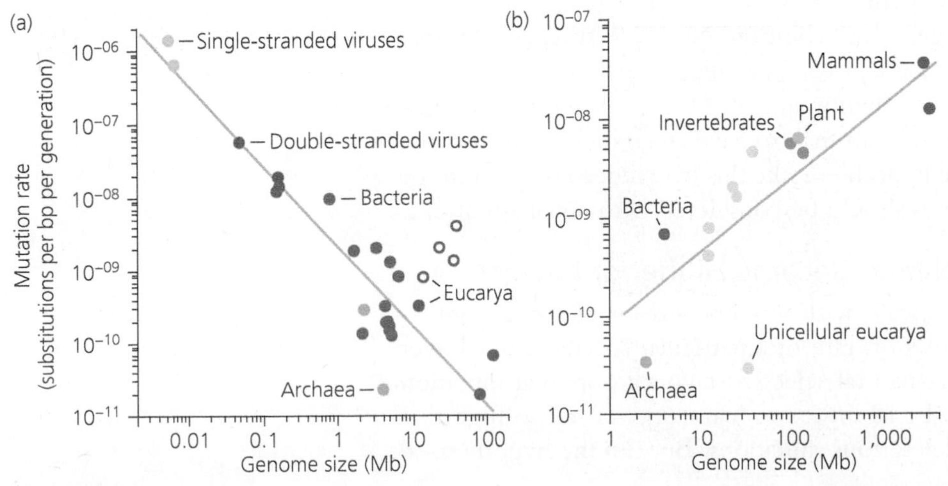

Figure 15.13 Scaling of mutation rates with genome size (a) For viruses, prokaryotes, and some unicellular eukaryotes, the relationship is negative; that is, organisms with larger genomes have lower per-basepair mutation rates. Data from Sung et al (2012). (b) For many eukaryotes, however, the relationship is positive. Redrawn from Saxer et al. (2012).

the phenotype. In the present case, the phenotype of interest is mutation rate. If we assume that selection for a low mutation rate is strong across all organisms, this selection will be opposed by an influence of genetic drift that increases for organisms with smaller effective population sizes. As effective population size goes down, the influence of drift goes up, limiting the ability of natural selection to produce the optimal phenotype—a low mutation rate.

For decades, intriguing hypotheses like this about broad patterns in molecular evolution have been untestable. For most of biology's history, scientists have not been able to measure baseline mutation rates, in part because they are so low. However, genome sequencing has permitted the collection of the requisite data.

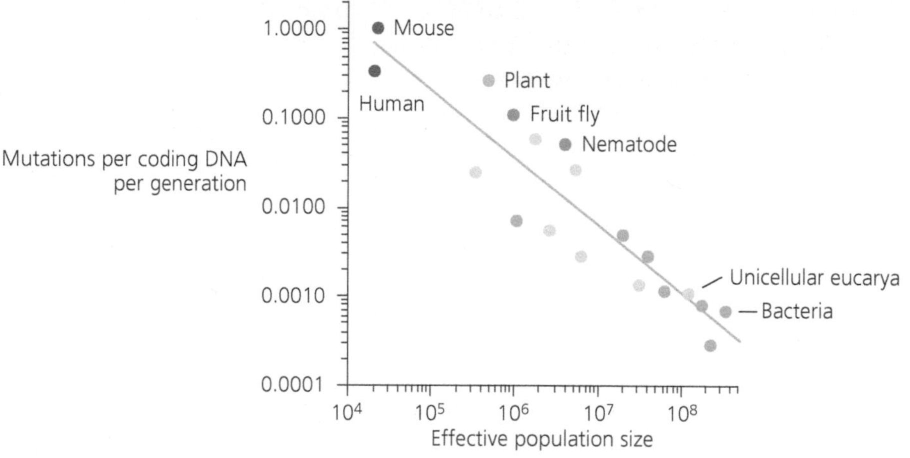

Figure 15.14 Scaling of mutation rates with effective population size Taxa with larger effective population sizes have lower mutation rates. Data from Sung et al. (2012).

Michael Lynch's hypothesis about the role of effective population size in determining the lower limit for mutation rates is supported by the observed relationship between these two values, shown in **Figure 15.14**. Organisms with larger body sizes, such as mammals and land plants, have smaller effective population sizes and greater mutation rates than smaller organisms like invertebrate animals and single-celled eukaryotes and prokaryotes.

The lower limit of mutation rate may be set by the interplay between natural selection and genetic drift.

Is Lynch's hypothesis that baseline mutation rates are determined by the balance between selection and drift, mediated by effective population size, correct? To fully evaluate the hypothesis would require manipulating the effective population sizes of different organisms and observing whether mutation rates evolved in response. Although such an experiment is impractical, Lynch's hypothesis highlights an important reality in genome evolution: Not all features of genomes and their evolution are necessarily adaptive. We might be tempted to devise adaptive explanations for all genomic phenomena we observe. We might, for example, speculate that it is adaptive for mammals to have an elevated mutation rate compared to land plants. Adaptive hypotheses like this may indeed turn out to be correct, but nonadaptive hypotheses should be considered with equal attention.

Can the Mutation Rate Evolve to Become Higher or Lower?

Regardless of how mutation rates scale with genome size or effective population size, we can ask why the mutation rate of a particular species is not lower or higher. It is easy to assume that natural selection favors an optimal mutation rate—one that is not so low as to be laborious to maintain, yet not so high that individuals suffer from frequent deleterious mutations. But can the hypothesized connection between mutation rate and fitness be tested experimentally? In fact, it can for organisms that have short enough generation times.

Lawrence Loeb and collaborators took an ingenious approach to studying the evolution of mutation rates by creating over 60 different strains of *E. coli* bacteria that each had a slightly different form of DNA polymerase I (PolI), an enzyme involved in DNA replication and repair (Loh et al. 2010). Because of the variation in PolI, the bacterial strains exhibited mutation rates that ranged from $\frac{1}{1{,}000}$ the normal *E. coli* mutation rate to 1,000 times greater than the normal rate. Although some researchers have proposed a trade-off between DNA replication fidelity and growth rate for single-celled organisms like *E. coli*, the strains that Loeb and colleagues created had similar growth rates. Therefore the researchers were able to test which mutation rate resulted in highest fitness in the absence of any growth-rate cost. To do so, they put all 66 *E. coli* strains together to compete for resources in several replicate populations for many generations.

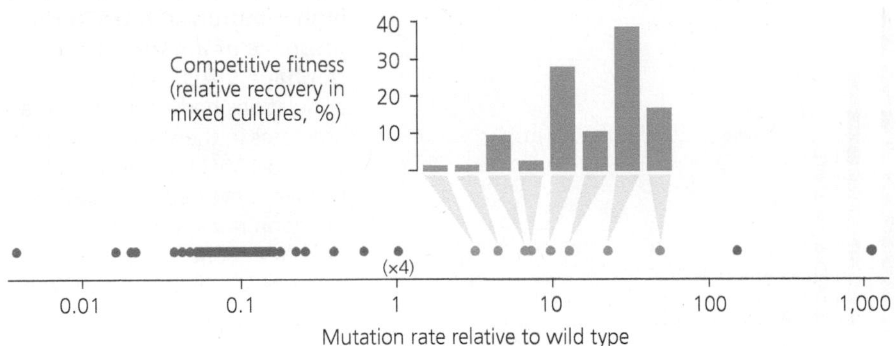

Figure 15.15 Evolution of mutation rate during competition among *E. coli* strains The range of mutation rates explored is shown along the horizontal scale at bottom. Each point represents a different *E. coli* strain. The graph shows the frequency of *E. coli* strains recovered after experimental competition among all strains. From Loh et al. (2010).

The results of the competition experiments are shown in **Figure 15.15**. In all replicates, the *E. coli* lineages that took over the population had mutation rates 10 to 47 times higher than wild-type *E. coli*. Bacterial strains with higher than normal, but not the absolute highest, mutation rates had optimal fitness.

Was this because strains with higher mutation rates were more likely during the 350 generations of the experiment to acquire beneficial mutations that raised their fitness? Loeb and colleagues compared the winning mutator strains from the end of the competition experiment against their direct ancestors, with matching mutation rates, from before the experiment **(Figure 15.16)**. The result was that pre-competition mutators had lower fitness than wild-type *E. coli*, but

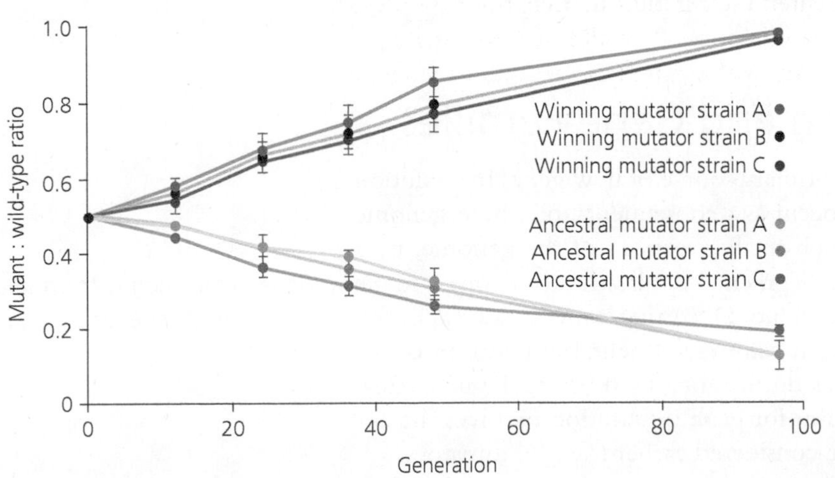

Figure 15.16 Relative fitness of mutator and ancestor strains Ancestor strains have the same mutator phenotype as their descendants that won the competition experiment, but have not experienced the competition. This shows that the mutator phenotype itself does not increase fitness. Instead, winning mutator strains must have acquired increased fitness during the course of the competition experiment. Redrawn from Loh et al. (2010).

post-competition winning mutators outcompeted the wild-type strain. This is convincing evidence that the winning strains were not superior at the beginning of the competition and must have acquired beneficial mutations during the 350 generations of competition.

The experiments by Loeb and colleagues demonstrate that under certain environmental conditions, a mutation rate higher than wild type is adaptive. However, this result cannot apply in all environments; otherwise, the mutation rate of wild-type *E. coli* would already have evolved to the higher rate. Some insight into this interplay can be gained from a study of bacteria evolving in the presence or absence of a parasitic virus. Csaba Pal and collaborators (2007) seeded 36 laboratory microcosms from the same starting population of the bacterium *Pseudomonas fluorescens* and grew them for 170 generations. They also established a parallel set of 36 microcosms in which the bacteria coexisted with a harmful virus.

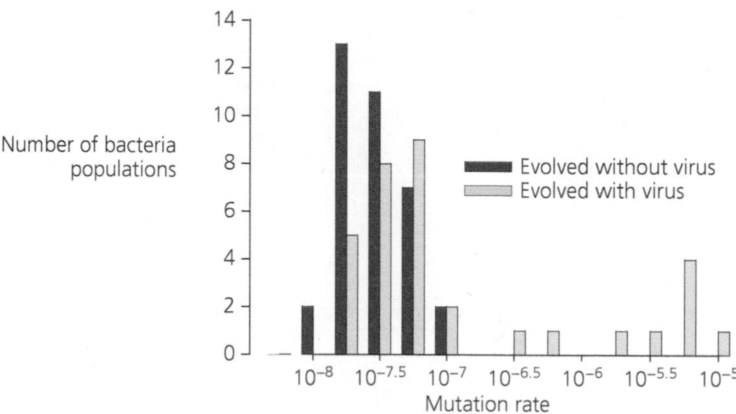

Figure 15.17 Bacteria evolve higher mutation rates in the presence of a harmful bacteriophage virus Although most populations had endpoint mutation rates in the starting range of 10^{-8} to 10^{-7}, the only populations that evolved increased mutation rates were those in the virus treatment. Drawn from data in Figure 2 of Pal et al. (2007).

The results appear in **Figure 15.17**. After 170 bacterial generations, the mutation rate of the bacteria had increased ten- to a hundredfold in many of the populations evolving with virus, but not in the populations evolving without virus (Pal et al. 2007). The fitness advantage of a higher mutation rate was environment specific: Bacteria with an elevated mutation rate had an advantage in the virus-containing environment, but not in the virus-free environment.

The preceding experiments demonstrate that bacterial mutation rates tend to evolve to elevated levels when in short-term competition or the presence of parasites. Clearly, genome-wide mutation rates can evolve by natural selection. It is not known whether these findings will extend to multicellular organisms.

Mutation rate is a trait that evolves by natural selection.

15.4 Gene Duplication and Gene Families

The duplication of existing genes is the primary source of new genes in evolution (see Chapter 5). Gene duplication can occur as a consequence of whole-genome duplication (polyploidization) or in much smaller sections of the genome, by a process called **segmental duplication.** Segmental duplication can copy a single complete gene, a piece of a gene, or a stretch of DNA that contains several genes. The most common molecular mechanism causing segmental duplication is unequal cross-over between chromosomes during meiosis (refer to Figure 5.26a, page 164). Another important mechanism for gene duplication involves the activity of mobile genetic elements, as we considered earlier for oval tomatoes.

Gene duplication is the primary source of new genes.

Although biologists have long appreciated the importance of gene duplication, understanding the evolutionary mechanisms that result in the preservation of duplicate genes is challenging. Initially, a duplication event is often selectively neutral because it does not affect phenotype. Neutral alleles will only rarely drift to fixation (see Chapter 7); this means most duplications are soon lost from a population due to drift. So why are some gene duplications preserved, and how do they go on to evolve changes in function?

Mechanisms for the Preservation of Duplicate Genes

We now consider several evolutionary hypotheses, or models, for how duplicated genes are preserved (Ohno 1970; Hahn 2009). All involve three phases, illustrated in **Figure 15.18** (Innan & Kondrashov 2010).

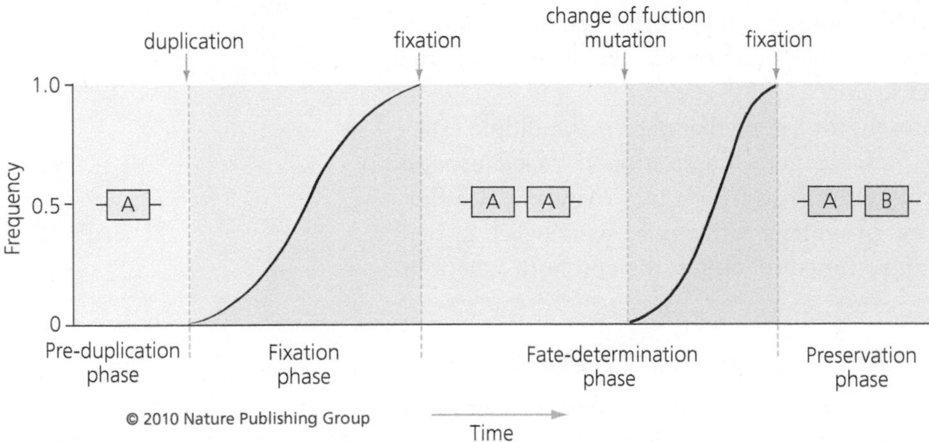

Figure 15.18 A model for the preservation of duplicate genes First, a duplicated allele must rise to fixation in a population. Next, one of the duplicates acquires a mutation that changes its function from A to B. If this new allele rises to fixation, then two genes with different functions have been preserved. Redrawn from Innan and Kondrashov (2010).

Reprinted by permission from Macmillan Publishers Ltd: Innan, H. & Kondrashov, F. (2010). "The evolution of gene duplications: classifying and distinguishing between models." *Nature Review Genetics* 11: 97–108.

After a gene is duplicated in a single individual, the new two-gene allele must achieve fixation within the population. Most duplications have no selective advantage and are lost due to drift, but in rare cases the duplicated allele will rise to fixation (yellow to tan transition). Following this, one of the gene copies acquires a mutation that changes its function. This new allele, comprised of copy A that performs the original function and copy B that performs a new function, then rises to fixation in the population (tan to blue). The reason for its preservation is typically attributed to selection: The new function of copy B increases fitness.

In the model known as **neofunctionalization,** the gene copy that mutates from the original acquires an entirely new function that by chance benefits the organism. Because two gene copies are present, one copy can continue to perform the ancestral function while the other acquires mutations that give it a new function **(Figure 15.19)**. This is a widely accepted mechanism for the preservation

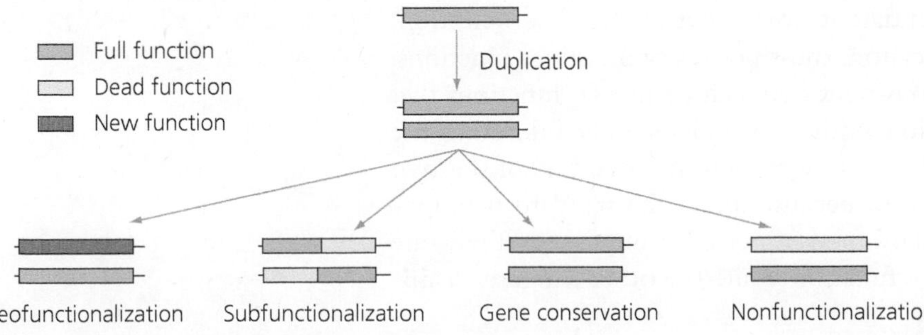

Figure 15.19 Possible outcomes of gene duplication Redrawn from Lynch (2007) and Hahn (2009).

of duplicate genes, but it is unknown how often neofunctionalization occurs relative to other mechanisms that have been proposed. In the related **subfunctionalization** model, the ancestral gene has two different functions. After the ancestral two-function gene is duplicated, one of the copies acquires a mutation that inhibits function A, while still performing function B effectively. Eventually the remaining two-function copy acquires a complementary mutation that inhibits function B, at which point each copy performs only a single function and both are required for survival of the organism.

Following a duplication event, copies of a gene can experience a variety of fates.

A third model involves a subfunctionalization event that allows for a new, or at least greatly improved, function to evolve. This model relies on the presence of an **adaptive constraint** in which multiple functions or phenotypes cannot both be optimized at the same time: A gene cannot get any better at performing function A without getting worse at performing function B, and vice versa. Because a single locus performs both functions, it can perform each function only moderately well, and it cannot improve either function via evolution by natural selection because of the adaptive constraint.

When an adaptive constraint is present for a gene that performs multiple functions, one possible mechanism to preserve gene duplication is called **escape from adaptive conflict** (Piatigorsky & Wistow 1991; Des Marais & Rausher 2008). In this model, gene duplication provides a pathway by which each of two genes can become specialized to a single function, and in the end both functions are performed better.

Escape from Adaptive Conflict in the Evolution of Antifreeze Proteins

Many fishes of the polar oceans have evolved specialized antifreeze proteins that inhibit the formation of ice crystals in body tissue, protecting the fish from death by freezing. Cheng Deng and colleagues were curious about the evolutionary origin of antifreeze proteins in the Antarctic eelpout *Lycodichthys dearborni* (see **Figure 15.20**), because its genes for antifreeze proteins (called *AFPIII*s) are similar to part of a different gene it possesses called sialic acid synthase B (*SAS-B*). Even though SAS-B is an enzyme that helps create sialic acids inside cells, and AFPIIIs are proteins secreted into blood plasma that bind to ice crystals, the researchers hypothesized that the *AFPIII*s were the result of a duplication of *SAS-B*. They also hypothesized that the mechanism by which *AFPIII*s evolved from duplicated *SAS-B* would fit the model of escape from adaptive conflict (Deng et al. 2010).

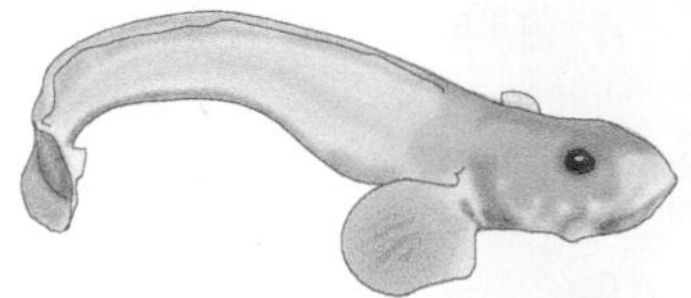

Figure 15.20 Antarctic eelpout *Lycodichthys dearborni* After a photo by Christina Cheng.

To test the first hypothesis, Deng's team first sequenced the region of the eelpout genome containing the *SAS-B* gene and the separate region on a different chromosome containing the *AFPIII* genes. Although *AFPIII* has sequence similarity with *SAS-B*, it is much shorter, suggesting that *AFPIII* resulted from a duplication of *SAS-B* followed by a large deletion **(Figure 15.21, facing page)**.

To support the escape-from-adaptive-conflict model, the researchers needed to satisfy three predictions that differentiate it from other models of duplication. First, the ancestral gene, before duplication, must possess two distinct functions. Second, there must be an adaptive constraint between these two functions that prevents improvement in one function without simultaneously inhibiting the other function. Third, the duplicated genes must each improve one of the two functions. This third prediction is critical because under the standard neofunctionalization model, the single ancestral function does not improve; rather, one daughter gene takes on a new adaptive function while the other daughter maintains the original function unchanged.

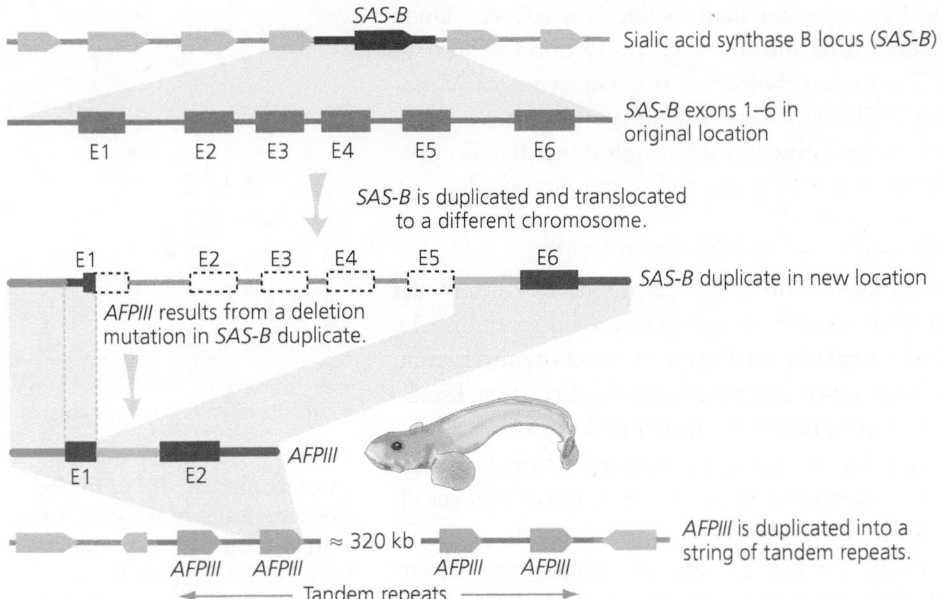

Figure 15.21 The origin of and duplication of AFPIII antifreeze proteins *SAS-B* was duplicated and translocated to a different locus, but a large portion of the locus was deleted during the process. The resulting *AFPIII* locus was then duplicated many times into a string of tandem repeats. Modified from Deng et al. (2010).

To assess the first prediction, the researchers inserted and expressed the *SAS-B* gene from eelpout in laboratory bacteria, generating the SAS-B protein. This protein is responsible for the creation of sialic acid, but it also turned out to have ice-binding affinity. An ice crystal always grows in a circular shape in pure water, but in the presence of SAS-B, ice crystal growth was inhibited and crystals were hexagonal **(Figure 15.22)**.

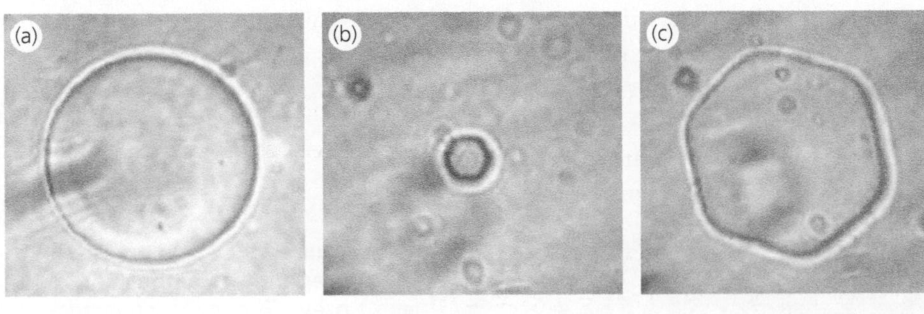

Figure 15.22 Ice-binding affinity of the ancestral SAS-B protein (a) Normal ice crystals grow in a circular shape in pure water. (b) In the presence of SAS-B, ice crystal growth was at first inhibited. (c) The crystals that eventually grew had a hexagonal shape. Images reprinted from Deng et al. (2010).

However, SAS-B also possesses its original function, catalyzing the formation of sialic acid, which it did even better than its ancestor protein, SAS-A **(Figure 15.23)**. Clearly the ancestral gene, before the duplication that led to the *AFPIII*s, had two functions: helping to create sialic acid and inhibiting ice crystals.

To evaluate the prediction that ice binding and catalyzing sialic acid synthesis are functions that constrain each other within *SAS-B*, the researchers changed the sequence of *SAS-B* slightly, altering four amino acids to match those in *AFPIII*. These changes were previously known to increase antifreeze properties. Sure enough, the slightly modified SAS-B was unable to catalyze sialic acid formation (see Figure 15.23). Making SAS-B a little bit more like an antifreeze protein destroyed its original function, helping to create sialic acid.

What about the third prediction, that after duplication both copies of the now subfunctionalized gene evolved adaptively to improved function? There is already evidence for this, as noted earlier: AFPIII inhibits ice formation better than

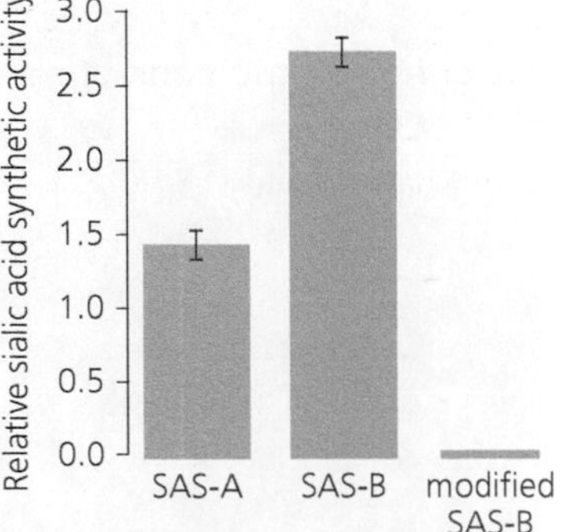

Figure 15.23 Catalytic activity of SAS-A and two versions of SAS-B Drawn from data in Figure S4B of Deng et al. (2010).

SAS-B, and SAS-B helps create sialic acid faster than its ancestor SAS-A. Deng and colleagues also tested this prediction in another way, by testing for positive selection (as discussed in Section 7.3 under "Selection on Replacement Mutations"). In the case of *SAS-B* and *AFPIII*, nonsynonomous substitutions were much more common than synonomous substitutions, suggesting that positive selection for altered function occurred for both genes following duplication.

Other Hypotheses for the Preservation of Duplicate Genes

The models for preserving gene duplicates discussed earlier each assume that initial duplication events are selectively neutral. Duplication events can also be initially adaptive if they increase gene expression. This is the case for the human salivary amylase gene, which encodes a protein expressed in saliva that breaks down starch in foods. As documented in **Figure 15.24**, individuals with more copies of the gene in their genome have higher concentrations of amylase in their saliva (Perry et al. 2007). People from populations with a history of high starch consumption have, on average, more copies of the amylase gene than people from populations with a history of low starch consumption. This study suggests that in populations that consumed high-starch foods, repeated duplication of the salivary amylase gene has been favored by natural selection as a mechanism to increase amylase concentrations in saliva.

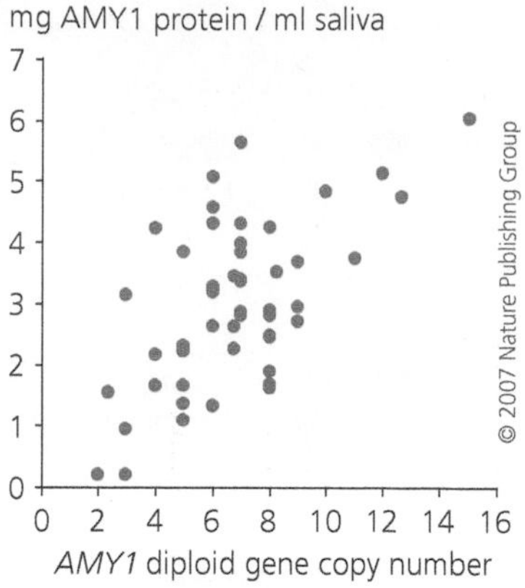

mg AMY1 protein / ml saliva

0 1 2 3 4 5 6 7

0 2 4 6 8 10 12 14 16

AMY1 diploid gene copy number

© 2007 Nature Publishing Group

Figure 15.24 People with more copies of *AMY1* in their genome have more AMY1 protein in their saliva Redrawn from Perry et al. (2007).

Reprinted by permission from Macmillan Publishers Ltd: Perry, G. H., N. J. Dominy, et al. 2007. "Diet and the evolution of human amylase gene copy number variation." *Nature Genetics* 39: 1256–1260.

Gene Families

Single gene duplications can be evolutionary endpoints, but in many cases repeated duplication events occur, as in the amylase example. This phenomenon results in **gene families**—groups of anywhere from two to hundreds of genes that are all descendants of a single ancestor gene. All genes in a gene family typically have similar structure and function. In most cases, though, they do not perform identical functions; rather, each produces a slightly different protein. Sometimes the existence of gene families is inferred to be adaptive; in other cases, the benefits of gene families are unknown.

Serial gene duplication can result in the evolution of gene families—groups of genes with similar functions derived from a common ancestor.

One striking example of gene family evolution comes from venomous cone snails **(Figure 15.25)**. All species of cone snails produce small molecules called

Figure 15.25 A cone snail preparing to ambush its prey The snail (*Conus striatus*) will harpoon the fish with a venomous barb. The poison will quickly immobilize the fish, which the snail will then swallow whole. The entire process will take about 15 seconds.

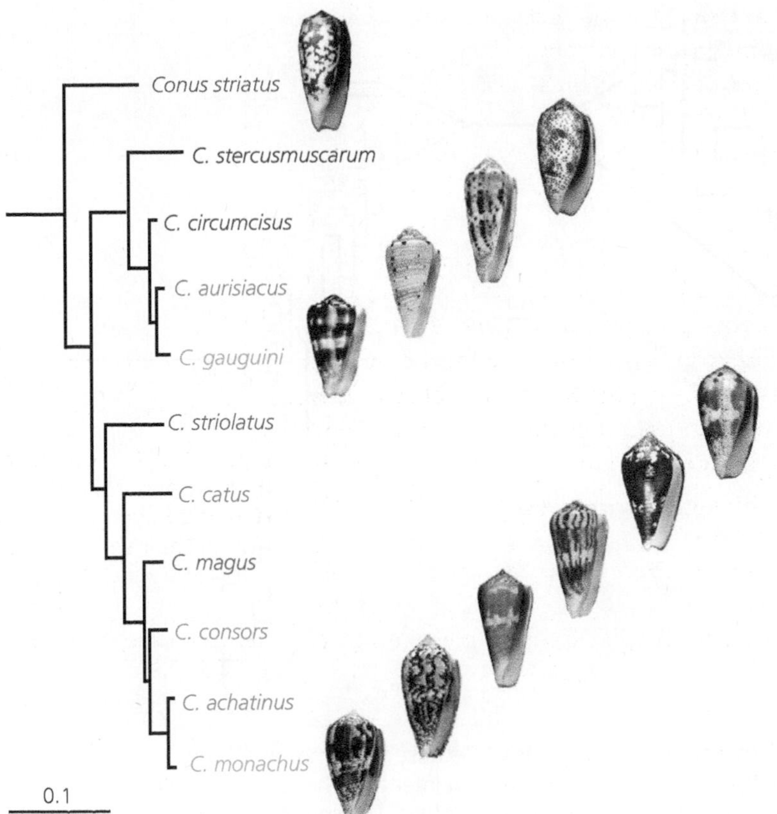

Figure 15.26 Relationships among *Conus* species Species tree created from mitochondrial and nuclear genes. Scale bar represents substitutions per site. Tree redrawn from Puillandre et al. (2010); photos from same source.

conotoxins that immobilize prey. There are literally hundreds of different conotoxin molecules, even within a single species, and researchers have hypothesized that the genes encoding conotoxins were generated by gene duplications. Nicolas Puillandre and colleagues explored the evolutionary history of one conotoxin gene family in a group of fish-hunting cone snails (Puillandre et al. 2010). To evaluate whether the conotoxin genes showed evidence of ancient gene duplications, they first constructed a phylogeny of 11 cone snail species based upon noncoding DNA sequences **(Figure 15.26)**.

They then collected DNA sequences for the genes encoding molecules known as A-conotoxins. From the 11 species in their study, they found over 90 different DNA sequences encoding A-conotoxins. The researchers used these sequences to create a phylogeny of A-conotoxin genes. This phylogeny of individual genes, known as a **gene tree,** could then be compared against the phylogeny of the cone snail species, known as a **species tree.**

Let us first consider the obvious features of the A-conotoxin gene tree **(Figure 15.27, next page)**. There are clearly four major phylogenetic groups ($\alpha3/5$, $\alpha4/4$, $\alpha4/7$, and κA), indicating more recent common ancestry within each group. Because all of the genes encoding $\alpha4/7$ conotoxins in each species, for example, are more closely related to each other than to genes encoding $\alpha3/5$ conotoxins, the genes encoding $\alpha4/7$ conotoxins must be descended from a common ancestor gene. This means that long before all of the $\alpha4/7$ and $\alpha3/5$ genes diversified, the common ancestor gene of both groups was duplicated. Because both groups contain genes from multiple cone snail species, the gene duplication that gave rise to $\alpha3/5$ and $\alpha4/7$ conotoxins must have occurred before the species themselves diverged. This is an important point: If the gene duplications of the $\alpha4/7$

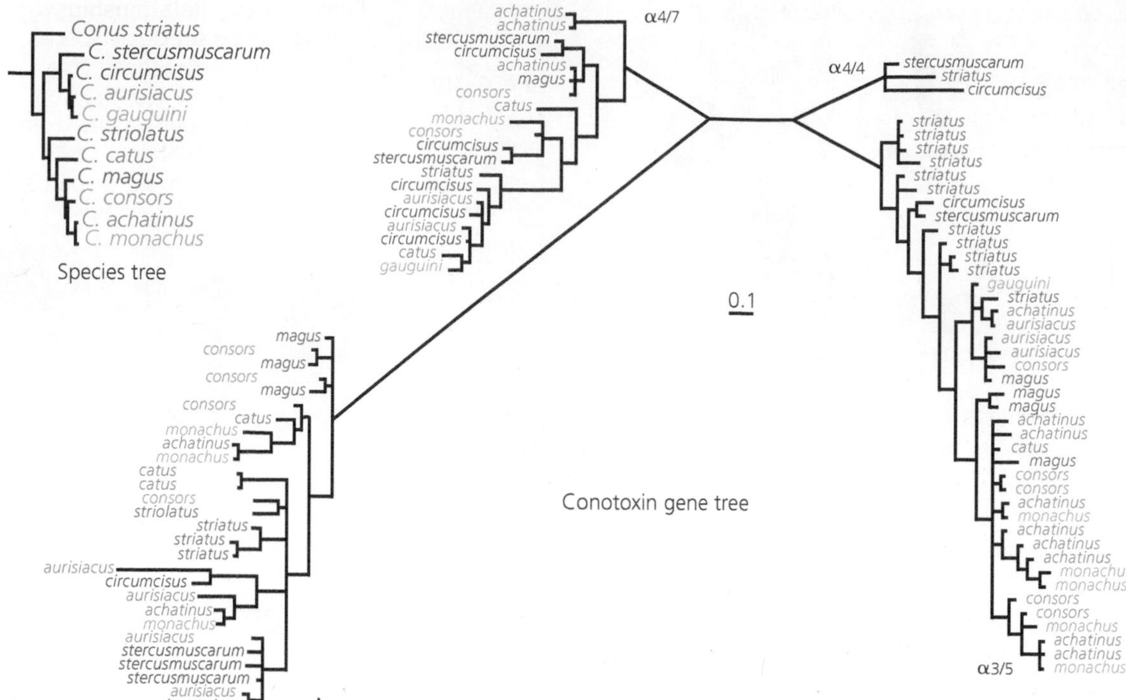

Figure 15.27 Relationships among conotoxin genes across species Because the conotoxin genes from within the same species do not group together on the phylogeny, we can infer that the conotoxin gene duplications must have occurred before these 11 species separated from each other. The species tree from Figure 15.26 is inset at upper left. Scale bar reresents substitutions per site. Redrawn from Puillandre et al. (2010).

conotoxins had occurred after the species separated, then the $\alpha 4/7$ genes from each species would form separate clades on the tree, instead of being all mixed together. This logic applies to all four of the major groups on the gene tree.

The gene tree led Puillandre and colleagues to conclude that three major gene duplication events, separating the four major clades, and several more recent duplications, had led to the diverse array of A-conotoxin genes in this group. But why? The answer is not definitive, but the researchers have two lines of evidence to suggest that the expansion of the A-conotoxin gene family was adaptive. First, there are functional differences among the four major groups. For example, α-conotoxins have a distinct structure and block nerve signals in the muscles of cone snail prey, while κA-conotoxins have a different structure and cause nerves to fire uncontrollably. If we assume that having multiple toxic peptides with different modes of action in the prey helps cone snails to more easily acquire food, then the functional expansion of the A-conotoxin gene family may have been adaptive. In addition, Puillandre and his collaborators looked at the rates of nonsynonymous versus synonomous nucleotide substitution in the conotoxin genes, just like in the antifreeze example discussed earlier (and as detailed in Chapter 7). All four of the major groups had $\frac{d_N}{d_S}$ ratios above 1.0, indicating positive selection favoring changes in the peptide structure.

This method of comparing gene trees and species trees, testing for positive selection, and exploring differences in function among gene products is common to many studies of the evolution of gene families. Because gene families are typically large and complex, researchers are often able to hint at adaptive hypotheses for the expansion of gene families, but it is difficult to establish definitively that all stages in gene family expansion were adaptive. It is quite possible that in many cases gene families expand solely due genetic drift: Because 2 copies of a gene, or 10 for that matter, are often no worse or better than 1, in principle it is reasonable for gene families to expand neutrally as well as in response to selection.

15.5 The Locus of Adaptation in Natural Populations

Evolutionary biologists have long been interested in pinpointing the exact DNA nucleotide changes responsible for adaptive phenotypic change in natural populations of organisms. In a sense this is the Holy Grail of modern evolutionary biology because it connects the molecular basis of inheritance, DNA, with evolution by natural selection in the wild. Although there are many examples of alleles that affect organismal phenotypes, and of phenotypes that affect fitness, the complete combination of allele, phenotype, and fitness effects in nature is a tall order. Fortunately, the molecular revolution in biology has enabled studies that connect all three of these elements to form a complete understanding of "the locus of adaptation."

Now that biologists can study variation in genes that affects fitness in nature, they have begun to address long-standing questions about the genetics of adaptation. Beyond simply asking which genes underlie adaptive phenotypes, we can develop and test hypotheses about the kinds of genetic changes that are most important for adaptive evolution. We can also formulate and assess hypotheses about the processes by which adaptive genetic variation arises and spreads in populations. Research along these lines is still in the early stages, but already scientists have begun to address important questions, such as:

- Where do the alleles for new adaptations come from? Do they already exist at low frequency in a population, or do they arise from new mutations?
- What are the phenotypic effects of adaptive mutations? Are single mutations with large phenotypic effects possible, or are multiple mutations with individually small phenotypic effects more typical?

The source and nature of genetic variation for novel adaptations is a current topic of research.

In this section, we will learn about some of the findings relevant to the first group of these questions. With regard to the second group, only a partial and somewhat circular answer is currently possible (Rockman 2012). We now know there are mutations with phenotypic effects large enough that we can detect them. The insertion of a copy of the *SUN* locus into the *DEFL1* locus in tomatoes (Figures 15.9 through 15.12) is just one of many examples we have discussed in this book. By their very nature, however, mutations with small effects on phenotypes are difficult to find and study. As a result, we currently know relatively little about them. Instead, biologists have focused on mutations with large phenotypic effects because they are the easiest to study. As we will see in the coming examples, single mutations with large phenotypic effects definitely do occur, and they may be widespread in nature. However, assessments of the relative contribution to adaptive evolution by mutations with large versus small effects is only just becoming possible (Rogers et al. 2012).

New Mutation versus Standing Genetic Variation

When a population of organisms is exposed to a novel selection pressure, natural selection favors the evolution of new adaptive phenotypes, but where does the genetic variation that encodes these phenotypes come from? If the phenotypes of individuals in a population are going to change over time, the alleles that encode the new phenotypes must come from somewhere. The short answer is that adaptive genetic variation, like all genetic variation, was originally generated by mutational processes (as we learned in Chapter 5). In the case of new adaptations, though, we might wonder whether the mutations that created alleles favored by a new selection pressure arose recently or in the distant past.

If the alleles that encode a novel adaptive phenotype appear after a new selective challenge arises, this is known as **adaptation from new mutation.** In this situation natural selection challenges a population, and by chance a new mutation arises that meets the challenge. The newly mutated allele, because it provides a fitness advantage, rises to fixation in the population. A different possibility is that alleles favored by the new selective challenge were already present at low frequency in the population before the challenge arose. These alleles, newly beneficial in the face of the selective challenge, could have been selectively neutral in the previous environment, or even mildly deleterious but present at low frequency due to chance or other factors. When the alleles that encode a novel adaptive phenotype were present in the population before the novel selection challenge arose, this is known as **adaptation from standing genetic variation.**

When natural populations evolve in response to changing selection pressure, do they rely on standing genetic variation or new mutations? The best answer we currently have is that both phenomena are possible.

Adaptation from New Mutation in Garter Snakes

In western North America there are several species of garter snake that make their living eating newts that possess a potent neurotoxin. This poison is called tetrodotoxin (TTX). TTX binds to voltage-gated sodium channels in nerves and muscles and blocks the movement of sodium ions across cell membranes. This inhibits nerve impulses and results in paralysis and even death. Yet multiple species of garter snakes are immune to the effects of TTX and can prey upon toxic newts with no negative effects. Because being able to eat toxic newts provides garter snakes with a prey resource that few other predators can eat, it is easy to see why TTX resistance would be favored by natural selection.

Shana Geffeney and collaborators determined that the physiological mechanism underlying TTX resistance in the common garter snake was sodium channels in skeletal muscle cell membranes that were unaffected by TTX (Geffeney et al. 2002). The researchers traced the genetic basis of TTX resistance to just a few key amino acid changes in the gene encoding the skeletal muscle sodium channel, $Na_V1.4$ (Geffeney et al. 2005). But where did the $Na_V1.4$ alleles responsible for TTX resistance come from? Were they present as standing genetic variation in the ancestor of all TTX-resistant snake species, or did they arise from new mutations in each snake lineage?

Chris Feldman and colleagues (2009) approached this question in a manner similar to that employed in earlier examples in this chapter: by comparing gene trees and species trees. They chose populations of three garter snake species in California that are resistant to TTX **(Figure 15.28)**. The common ancestor of these three species may or may not have been resistant to TTX, but newts with TTX

Thamnophis sirtalis, Willow Creek

T. atratus, Molino Creek

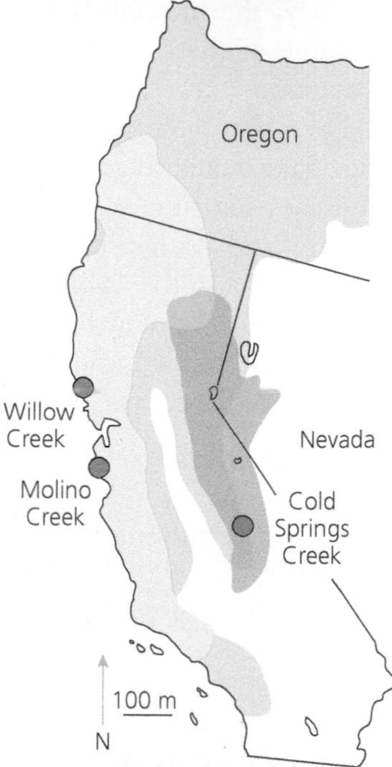

T. couchii, Cold Springs Creek

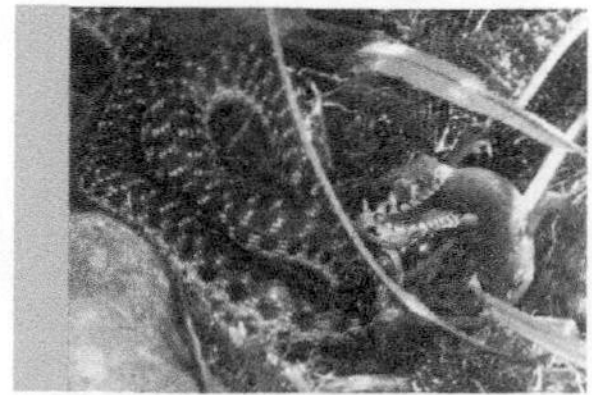

Figure 15.28 TTX-resistant *Thamnophis* garter snakes and their distributions Map redrawn from Feldman et al. (2009); photos from same source.

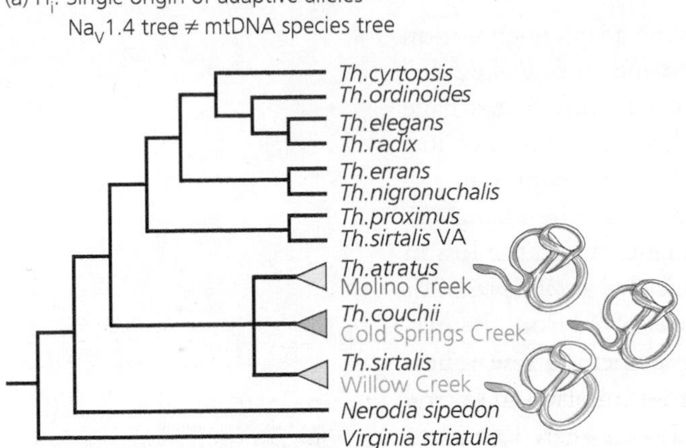

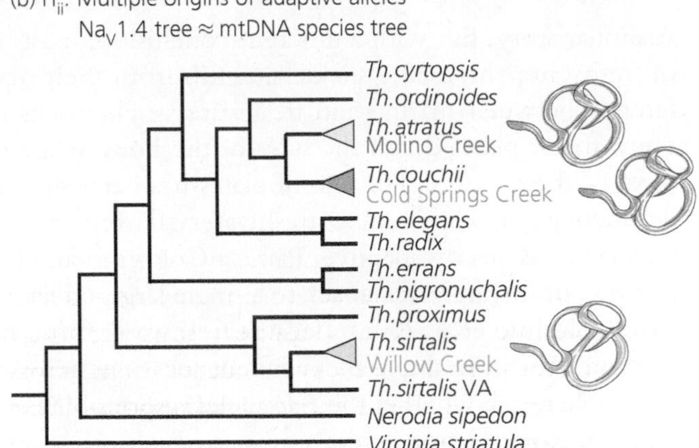

Figure 15.29 Alternative hypotheses for the origin of sodium channel alleles conferring resistance to TTX (a) A resistant allele arose once, in the common ancestor of all presently resistant snake species. (b) Resistance alleles arose independently in each resistant species. The hypothetical sodium channel gene tree in (b) matches a species tree reconstructed from mitochondrial DNA sequences. Redrawn from Feldman et al. (2009).

did exist at that time in history. It is possible that the alleles conferring TTX resistance first arose in the common ancestor and were passed down to each species. If this were the case, the nucleotide sequences of $Na_V1.4$ in each species would be similar, and the gene tree for $Na_V1.4$ would show the alleles for these species being each other's closest relatives **(Figure 15.29a)**. However, if the alleles conferring TTX resistance had arisen more recently from separate new mutations in each species, then the gene tree would match the species tree estimated from mtDNA, in which some TTX-resistant species are more closely related to non-TTX-resistant species than to the other TTX-resistant species (Figure 15.29b).

Feldman and his team sequenced $Na_V1.4$ alleles to generate the gene phylogeny for 14 species of garter snakes. They also measured TTX resistance in each species. Their results, shown in **Figure 15.30**, are consistent with the hypothesis that the alleles for TTX resistance arose separately in each species. The gene tree matches the species tree, and the amino acid sequences of the TTX-resistant sodium channels are different in each species. The amino acid changes must have arisen from independent mutations in each species.

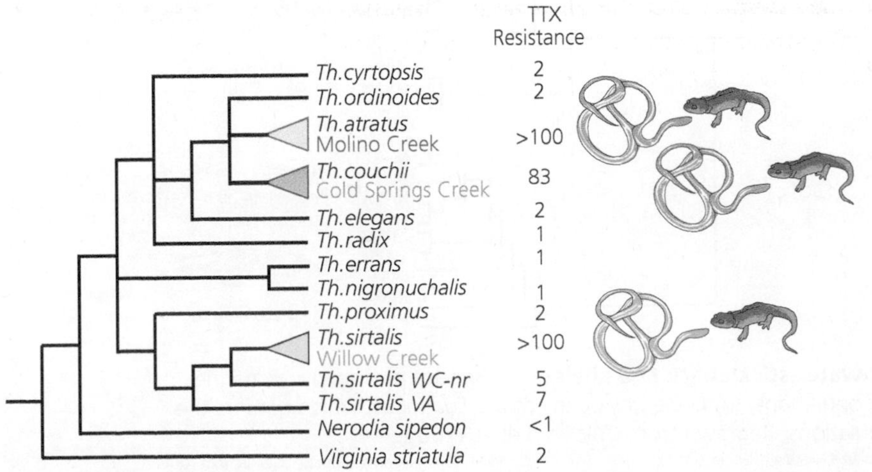

Figure 15.30 Phylogeny of *Thamnophis* sodium channel gene and resistance of each species to TTX Because this gene tree matches the species tree of Figure 15.29b, we can infer that the alleles for TTX resistance arose independently in each species. Redrawn from Feldman et al. (2009).

Adaptation from Standing Genetic Variation in Sticklebacks

A similar story, but with a different conclusion, exists for the adaptive evolution of freshwater threespine stickleback fish from their marine ancestors. A key difference between marine and freshwater sticklebacks is that marine forms have bony armor plates down the sides of the body while freshwater forms have lost most of these plates. The loss of plates has been shown to evolve rapidly in experimental introductions to freshwater (Barrett et al. 2008), so the change in phenotype is clearly adaptive. Pamela Colosimo and colleagues traced the loss in plates in freshwater stickleback to a single large-effect gene called *Ectodysplasin*, or *Eda* (Colosimo et al. 2005). Because freshwater forms have evolved from the marine ancestor in multiple independent locations across the globe, the researchers were able to ask whether the *Eda* alleles responsible for the loss in plates in several freshwater populations were similar to each other, indicating a shared ancestry supportive of adaptation from standing variation, or had arisen independently.

Colosimo and colleagues took an approach similar to that of the researchers who investigated TTX resistance in garter snakes. They collected DNA sequences from a number of populations, both freshwater with few bony plates and marine with many. They then constructed both a gene tree for the *Eda* locus that is responsible for the loss of plates and a population tree that shows the evolutionary history of relationships among the different groups. Because freshwater populations from different regions of the world represent separate invasions, most freshwater populations are more closely related to nearby marine populations than to geographically distant freshwater populations **(Figure 15.31a)**. But the phylogeny of *Eda* alleles is different. All but one of the low-plate alleles form a single clade that is separate from the marine alleles (Figure 15.31b). In every freshwater population except NAKA, the low-plate *Eda* alleles are descended from a single common ancestor allele that must have been present in the ancestral marine population before sticklebacks invaded freshwater. Rather than each population evolving a new mutation at the *Eda* locus that was favored in the transition to freshwater, adaptation at the *Eda* locus arose from standing genetic variation in the marine ancestor.

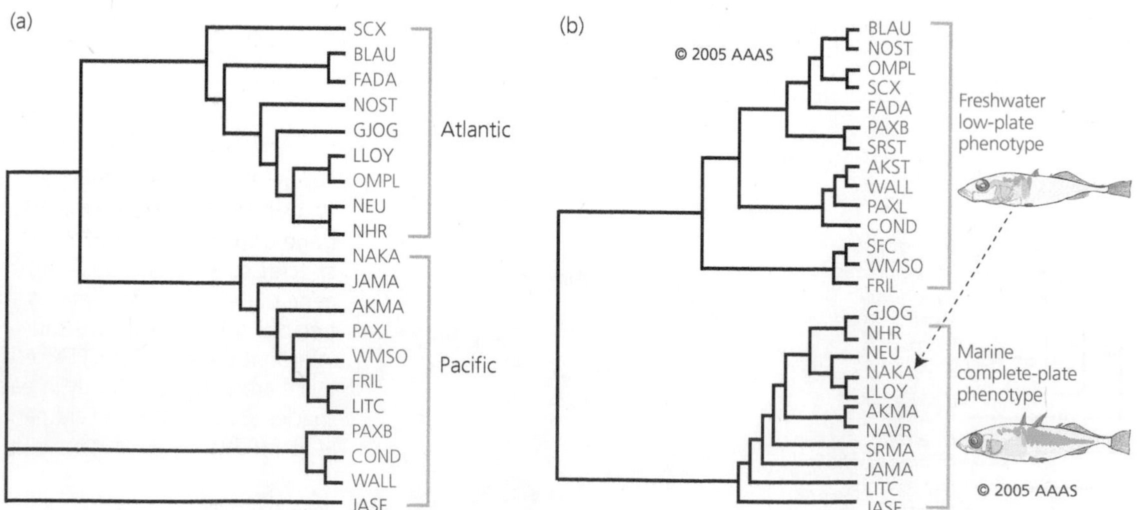

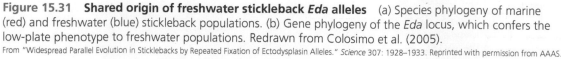

Figure 15.31 Shared origin of freshwater stickleback *Eda* alleles (a) Species phylogeny of marine (red) and freshwater (blue) stickleback populations. (b) Gene phylogeny of the *Eda* locus, which confers the low-plate phenotype to freshwater populations. Redrawn from Colosimo et al. (2005).

From "Widespread Parallel Evolution in Sticklebacks by Repeated Fixation of Ectodysplasin Alleles." *Science* 307: 1928–1933. Reprinted with permission from AAAS.

Which Is More Common?

The preceding studies are among the best examples for which the molecular basis of adaptive phenotypes has been traced to its origin. In one case adaptive alleles resulted from mutations that arose after the selective challenge favoring them had arrived. In the other case, the allele was present at low frequency before it became adaptive due to an altered environment. Biologists would like to know which scenario is more common, and what kinds of adaptations or species are more likely to have relied upon one or the other. While other examples exist, such as fur-color evolution from new mutations in beach mice (Linnen et al. 2009) and host shifts using standing variation in tephritid flies (Feder 2003), more studies on more organisms are needed before definitive judgments can be made.

In some cases the alleles involved in adaptive evolution arise from standing genetic variation. In other cases they arise from new mutations.

Adaptive Mutations of Small versus Large Effect

Sean Rogers and colleagues (2012) looked at threespine sticklebacks to investigate the relative importance, during morphological evolution, of adaptive mutations of small versus large effect. They took advantage of the fact that marine sticklebacks have invaded a great variety of freshwater habitats. Adaptation to freshwater involves more dramatic evolutionary change under some circumstances than others. For example, stickleback populations in lakes without prickly sculpin undergo more dramatic morphological evolution than stickleback populations in lakes with prickly sculpin. Among the reasons is that sculpin eat sticklebacks, thereby selecting for partial retention of the armor that sticklebacks bring from the ocean.

Based on an argument originally developed by Ronald Fisher (1930; see Orr 2005), Rogers and colleagues predicted that adaptation to lakes lacking sculpin would involve more mutations of large effect. Particularly if mutations affect more than one trait—that is, if they are pleiotropic—a novel allele of large effect is more likely to be maladaptive, and thus less likely to be adaptive, for an individual already close to the optimum phenotype than for an individual far away.

To test this prediction, the scientists mated marine sticklebacks with freshwater sticklebacks from two lakes with sculpin and two lakes without. They then performed a quantitative trait loci (QTL) analysis to identify genetic loci involved in morphological adaptation to freshwater (see Chapter 9). The distributions of QTL effect sizes appear in **Figure 15.32**. Consistent with the prediction, adaptation to

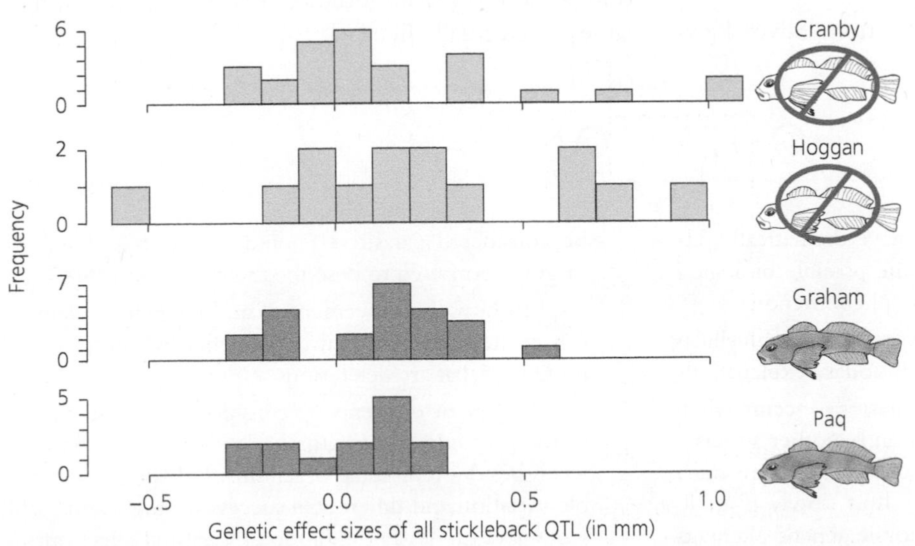

Figure 15.32 Distribution of effect sizes of QTL involved in morphological evolution of threespine sticklebacks in lakes with and without sculpin Redrawn from Rogers et al. (2012).

lakes lacking sculpin involved more alleles of large effect. The longer the adaptive trip, the bigger the steps taken along the way.

The effect sizes of alleles involved in adaptive evolution appear to be related to the size of the adaptive change.

As noted earlier, biologists are a long way from being able to make empirical generalizations about the distribution of effect sizes of the alleles involved in adaptive evolution. However, the study by Rogers and colleagues shows that it is possible to document circumstances under which the relative importance of mutations of large effect changes.

SUMMARY

Genome architecture refers to the structure, organization, and function of an organism's genetic information. Prokaryotes have small genomes with short intergenic regions and few introns, while multicellular eukaryotes have large expanded genomes with lengthy intergenic regions and many introns. Much of the sequence in multicellular genomes encodes for mobile genetic elements, which are selfish entities that transpose and reproduce themselves within a genome. Because mobile elements are genomic parasites that rarely benefit their host, organisms mount molecular defenses against them. In plants and animals, methylation prevents mobile elements from being transcribed, while small RNAs are responsible for post-transcriptional silencing. Mobile elements can affect phenotypes when their insertion alters or inhibits gene expression.

Mutation rates vary among multicellular species along with genome size; organisms with smaller effective population sizes have higher mutation rates. This may be due to the increased influence of genetic drift in smaller populations. Prokaryote mutation rates can evolve in response to environmental conditions such as competition or parasitism.

Most new genes arise via duplication. Several hypotheses account for how duplicate genes are preserved. These include neofunctionalization, in which one of the gene copies takes on a new function while the other maintains the ancestral function, and subfunctionalization, in which multiple functions of the ancestral gene are partitioned among the daughter genes. Pleiotropic genes sometimes have adaptive constraints among the functions they perform; these constraints can be removed via gene duplication by distributing functions across separate loci. Gene families are groups of genes descended from a single ancestral gene via multiple duplications. Genes within a family often have similar functions, but expanded families may be beneficial if each gene product varies slightly from the rest.

Researchers explore the evolutionary history of gene duplications by comparing gene trees with species trees. This method can also be used to infer the origin of alleles responsible for adaptive phenotypes in natural populations. Adaptive genetic variation can come from standing genetic variation or from new mutation. There are examples of both kinds in nature, but it is not yet known which is more common. Also under investigation is the relative importance of adaptive mutations of large versus small effect.

QUESTIONS

1. Two closely related plant species have dramatically different genome sizes. What are some possible biological mechanisms that could explain this phenomenon?
2. How do the genomes of prokaryotes and multicellular eukaryotes differ? What is the evolutionary explanation?
3. In interactions between species, parasitism occurs when one species gains a fitness benefit and another suffers a fitness loss. In contrast to predation, parasites are small relative to their hosts and kill the host slowly if at all. Based on this definition, should mobile genetic elements be considered parasites? Should biologists still use the original term used to describe them—"junk DNA"?
4. Explain how the movement of mobile genetic elements can create mutations that are beneficial to hosts as well mutations that are deleterious.
5. Mobile genetic elements are considered the best example of the prediction that natural selection can act on other levels besides individual organisms. Explain how heritable variation and differential success among transposable sequences can lead to evolution by natural selection.

6. Suppose a researcher discovers a new population of mice with higher expression of retrotransposon RNA than wild-type mice. Can you predict the behavior of other molecular mechanisms in this new mouse lineage that may be related to retrotransposon activity?
7. Consider Michael Lynch's hypothesis that mutation rates are determined by an interplay between natural selection and genetic drift. Now imagine an experiment in which you keep many different populations of *E. coli* bacteria, each at a different effective population size, for many thousands of generations. How would Lynch's hypothesis predict that mutation rates would evolve in populations of different sizes?
8. After a gene is duplicated, what eventual evolutionary outcomes are possible? What is the most likely outcome?
9. Suppose you discover a new gene family in your favorite study organism, which includes many similar but not identical genes that all arose from a single ancestral gene via duplication events. What data would you gather, and what calculations would you make, to explore the hypothesis that the expansion of the gene family was favored by natural selection (i.e., adaptive)?
10. All teleost fish species have two copies of the sodium channel gene *SCN4A*: *scn4aa* and *scn4ab*.
    a. From the evidence in **Figure 15.33a**, did the duplication of *SCN4A* happen just once in teleosts, or multiple times independently? Explain your reasoning.
    b. In two lineages of electric fish, the expression of *scn4aa* is restricted to the myogenic electric organ. From the evidence in Figure 15.33b, did this change in expression happen just once, or multiple times independently? Explain.
    c. From the evidence in Figure 15.33b, has *scn4aa* undergone adaptive evolution following the change in expression? Explain.
11. What kinds of information can be gained by comparing gene trees and species trees? What kinds of questions about evolutionary history can be answered?
12. What is the difference between adaptation from new mutation and adaptation from existing genetic variation? Why is this distinction important?

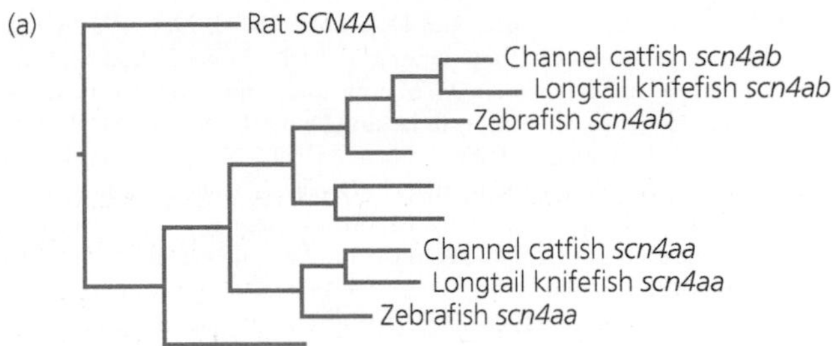

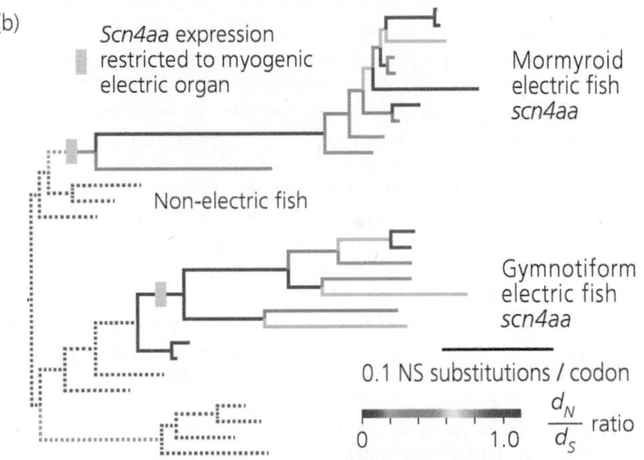

Figure 15.33 Sodium channel gene trees (a) *SCN4A* from rat and several fish. From Novak et al. (2006). (b) *Scn4aa* from two clades of electric fish and from several non-electric fish. From Arnegard et al. (2010).

EXPLORING THE LITERATURE

13. For clues about how and why introns form, see:

Kang, L., Z. Zhu, et al. 2012. Newly evolved introns in human retrogenes provide novel insights into their evolutionary roles. *BMC Evolutionary Biology* 12: 128.

14. In evolving populations, the fitness effects of beneficial mutations may change over time:

Barrick, J. E., D. S. Yu, et al. 2009. Genome evolution and adaptation in a long-term experiment with *Escherichia coli*. *Nature* 461: 1243–1247.

15. For evidence that the mutation rate of evolving populations influences evolutionary outcomes, see:

Racey, D., R. F. Inglis, et al. 2010. The effect of elevated mutation rates on the evolution of cooperation and virulence of *Pseudomonas aeruginosa*. *Evolution* 64: 515–521.

16. For evidence that transposable elements have played a role in the evolution of gene regulation, see:

Emera, D., C. Casola, et al. 2012. Convergent evolution of endometrial prolactin expression in primates, mice, and elephants through the independent recruitment of transposable elements. *Molecular Biology and Evolution* 29: 239–247.

17. For a clever use of parallel adaptation from standing genetic variation to identify loci associated with body size and weight in mice, see:

Akey, J. M. 2012. Parallel selection: Evolution's surprising predictability. *Current Biology* 22: R407–R409.

Chan, Y. F., F. C. Jones, et al. 2012. Parallel selection mapping using artificially selected mice reveals body weight control loci. *Current Biology* 22: 794–800.

CITATIONS

Adams, M. D., et al. 2000. The genome sequence of *Drosophila melanogaster*. *Science* 287: 2185–2195.

Aravin, A. A., R. Sachidanandam, et al. 2007. Developmentally regulated piRNA clusters implicate MILI in transposon control. *Science* 316: 744–747.

Arnegard, M. E., D. J. Zwickl, et al. 2010. Old gene duplication facilitates origin and diversification of an innovative communication system—twice. *Proceedings of the National Academy of Sciences, USA* 107: 22172–22177.

Barrett, R. D. H., S. M. Rogers, and D. Schluter 2008. Natural selection on a major armor gene in threespine stickleback. *Science* 322: 255–257.

Bourc'his, D., and T. H. Bestor. 2004. Meiotic catastrophe and retrotransposon reactivation in male germ cells lacking Dnmt3L. *Nature* 431: 96–99.

Brennecke, J., A. A. Aravin, et al. 2007. Discrete small RNA-generating loci as master regulators of transposon activity in *Drosophila*. Cell 128: 1089–1103.

Carmell, M. A., A. Girard, et al. 2007. MIWI2 is essential for spermatogenesis and repression of transposons in the mouse male germline. *Developmental Cell* 12: 503–514.

Charlesworth, B., and C. H. Langley. 1989. The population genetics of *Drosophila* transposable elements. *Annual Review of Genetics* 23: 251–287.

Colosimo, P. F., K. E. Hosemann, et al. 2005. Widespread parallel evolution in sticklebacks by repeated fixation of ectodysplasin alleles. *Science* 307: 1928–1933.

Cordaux, R., and M. A. Batzer. 2009. The impact of retrotransposons on human genome evolution. *Nature Reviews Genetics* 10: 691–703.

Deng, C., C.-H. C. Cheng, et al. 2010. Evolution of an antifreeze protein by neofunctionalization under escape from adaptive conflict. *Proceedings of the National Academy of Sciences, USA* 107: 21593–21598.

Des Marais, D. L., and M. D. Rausher. 2008. Escape from adaptive conflict after duplication in an anthocyanin pathway gene. *Nature* 454: 762–765.

Feder, J. L. 2003. Allopatric genetic origins for sympatric host-plant shifts and race formation in *Rhagoletis*. *Proceedings of the National Academy of Sciences, USA* 100: 10314–10319.

Feldman, C. R., E. D. Brodie, et al. 2009. The evolutionary origins of beneficial alleles during the repeated adaptation of garter snakes to deadly prey. *Proceedings of the National Academy of Sciences, USA* 106: 13415–13420.

Feng, S., S. J. Cokus, et al. 2010. Conservation and divergence of methylation patterning in plants and animals. *Proceedings of the National Academy of Sciences, USA* 107: 8689–8694.

Fire, A., S. Xu, et al. 1998. Potent and specific genetic interference by double-stranded RNA in *Caenorhabditis elegans*. *Nature* 391: 806–811.

Fisher, R. A. 1930. *The Genetical Theory of Natural Selection*. Oxford: Oxford University Press.

Geffeney, S., E. D. Brodie, et al. 2002. Mechanisms of adaptation in a predator-prey arms race: TTX-resistant sodium channels. *Science* 297: 1336–1339.

Geffeney, S. L., E. Fujimoto, et al. 2005. Evolutionary diversification of TTX-resistant sodium channels in a predator-prey interaction. *Nature* 434: 759–763.

Hahn, M. W. 2009. Distinguishing among evolutionary models for the maintenance of gene duplicates. *Journal of Heredity* 100: 605–617.

Ibarra-Laclette, E., E. Lyons, et al. 2013. Architecture and evolution of a minute plant genome. *Nature* doi:10.1038/nature12132.

Innan, H., and F. Kondrashov. 2010. The evolution of gene duplications: Classifying and distinguishing between models. *Nature Reviews Genetics* 11: 97–108.

Koonin, E. V. 2009. Evolution of genome architecture. *International Journal of Biochemistry & Cell Biology* 41: 298–306.

Kuramochi-Miyagawa, S., T. Watanabe, et al. 2008. DNA methylation of retrotransposon genes is regulated by Piwi family members MILI and MIWI2 in murine fetal testes. *Genes & Development* 22: 908 –917.

Lander, E. S., L. M. Linton, et al. 2001. Initial sequencing and analysis of the human genome. *Nature* 409: 860–921.

Linnen, C. R., E. P. Kingsley, et al. 2009. On the origin and spread of an adaptive allele in deer mice. *Science* 325: 1095–1098.

Loh, E., J. J. Salk, and L. A. Loeb. 2010. Optimization of DNA polymerase mutation rates during bacterial evolution. *Proceedings of the National Academy of Sciences, USA* 107: 1154–1159.

Lynch, M. 2006. Streamlining and simplification of microbial genome architecture. *Annual Review of Microbiology* 60: 327–349.

Lynch, M. 2007. *The Origins of Genome Architecture*. Sunderland, MA: Sinauer.

Lynch, M. 2010. Evolution of the mutation rate. *Trends in Genetics* 26: 345–352.

Lynch, M., L.-M. Bobay, et al. 2011. The repatterning of eukaryotic genomes by random genetic drift. *Annual Review of Genomics and Human Genetics* 12: 347–366.

Mardis, E. R. 2011. A decade's perspective on DNA sequencing technology. *Nature* 470: 198–203.

Mercer, T. R., M. E. Dinger, and J. S. Mattick. 2009. Long non-coding RNAs: Insights into functions. *Nature Reviews Genetics* 10: 155–159.

Miura, A., S. Yonebayashi, et al. 2001. Mobilization of transposons by a mutation abolishing full DNA methylation in *Arabidopsis*. *Nature* 411: 212–214.

Morton, N. E. 1991. Parameters of the human genome. *Proceedings of the National Academy of Sciences, USA* 88: 7474–7476.

Niedringhaus, T. P., D. Milanova, et al. 2011. Landscape of next-generation sequencing technologies. *Analytical Chemistry* 83: 4327–4341.

Novak, A. E., M. C. Jost, et al. 2006. Gene duplications and evolution of vertebrate voltage-gated sodium channels. *Journal of Molecular Evolution* 63: 208–221.

Ohno, S. 1970. *Evolution by Gene Duplication*. Berlin: Springer.

Orr, H. A. 2005. The genetic theory of adaptation: A brief history. *Nature Reviews Genetics* 6: 119–127.

Pal, C., M. D. Maciá, et al. 2007. Coevolution with viruses drives the evolution of bacterial mutation rates. *Nature* 450: 1079–1081.

Pellicer, J., M. F. Fay, and I. J. Leitch. 2010. The largest eukaryotic genome of them all? *Botanical Journal of the Linnean Society* 164: 10–15.

Perry, G. H., N. J. Dominy, et al. 2007. Diet and the evolution of human amylase gene copy number variation. *Nature Genetics* 39: 1256–1260.

Piatigorsky, J. and G. Wistow. 1991. The recruitment of crystallins: New functions precede gene duplication. *Science* 252: 1078–1079.

Puillandre, N., M. Watkins, and B. M. Olivera 2010. Evolution of *Conus* peptide genes: Duplication and positive selection in the A-superfamily. *Journal of Molecular Evolution* 70: 190–202.

Rockman, M. V. 2012. The QTN program and the alleles that matter for evolution: All that's gold does not glitter. *Evolution* 66: 1–17.

Rodríguez-Trelles, F., R. Tarrío, and F. J. Ayala. 2006. Origins and evolution of spliceosomal introns. *Annual Review of Genetics* 40: 47–76.

Rogers, S. M., P. Tamkee, et al. 2012. Genetic signature of adaptive peak shift in threespine stickleback. *Evolution* 66: 2439–2450.

Roy, S.W., and W. Gilbert. 2006. The evolution of spliceosomal introns: Patterns, puzzles and progress. *Nature Reviews Genetics* 7: 211–221.

Saxer, G., P. Havlak, et al. 2012. Whole genome sequencing of mutation accumulation lines reveals a low mutation rate in the social amoeba *Dictyostelium discoideum*. *PLoS One* 7: e46759.

Sijen, T., and R. H. A. Plasterk. 2003. Transposon silencing in the *Caenorhabditis elegans* germ line by natural RNAi. *Nature* 426: 310–314.

Spanu, P. D., J. C. Abbott, et al. 2010. Genome expansion and gene loss in powdery mildew fungi reveal tradeoffs in extreme parasitism. *Science* 330: 1543–1546.

Sung, W., M. S. Ackerman, et al. 2012. Drift-barrier hypothesis and mutation-rate evolution. *Proceedings of the National Academy of Sciences, USA* 109: 18488–18492.

Tam, O. H., A. A. Aravin, et al. 2008. Pseudogene-derived small interfering RNAs regulate gene expression in mouse oocytes. *Nature* 453: 534–538.

van der Knaap, E., A. Sanyal, et al. 2004. High-resolution fine mapping and fluorescence *in situ* hybridization analysis of *sun*, a locus controlling tomato fruit shape, reveals a region of the tomato genome prone to DNA rearrangements. *Genetics* 168: 2127–2140.

Watanabe, T., Y. Totoki, et al. 2008. Endogenous siRNAs from naturally formed dsRNAs regulate transcripts in mouse oocytes. *Nature* 453: 539–543.

Wittkopp, P. J., and G. Kalay. 2012. Cis-regulatory elements: Molecular mechanisms and evolutionary processes underlying divergence. *Nature Reviews Genetics* 13: 59–69.

Xiao, H., N. Jiang, et al. 2008. A retrotransposon-mediated gene duplication underlies morphological variation of tomato fruit. *Science* 319: 1527–1530.

Yoder, J. A., C. P. Walsh, and T. H. Bestor. 1997. Cytosine methylation and the ecology of intragenomic parasites. *Trends in Genetics* 13: 335–340.

Zemach, A., I. E. McDaniel, et al. 2010. Genome-wide evolutionary analysis of eukaryotic DNA methylation. *Science* 328: 916–919.

16

Mechanisms of Speciation

The tiger-striped leaf frog (*Phyllomedusa tomopterna*) is one of three dozen hylid tree frogs found at Santa Cecilia, Ecuador—the site representing the local species richness of the Amazon in the graph below. Graph from Wiens et al. (2011).

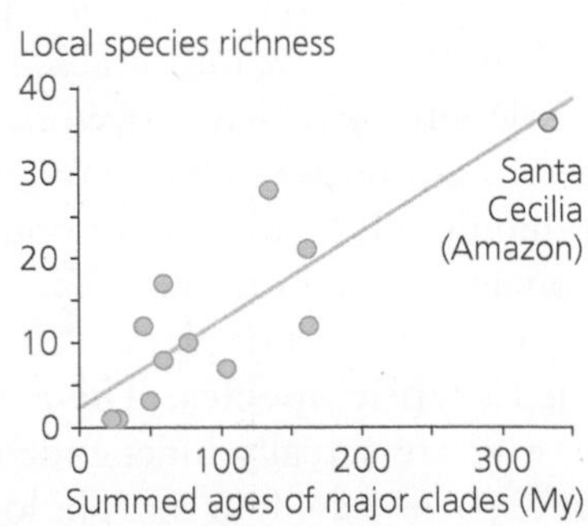

Where do new species come from, and why? This question and others related to speciation and biodiversity have intrigued and challenged biologists for decades. For example, species of tree frogs belonging to the family Hylidae are present throughout the tropics, but the Amazon region of South America is home to the greatest number of co-occurring species—why? An analysis by John Wiens and colleagues (2011) considered several hypotheses that might explain this pattern, such as climate variables like temperature and precipitation, or the overall rate at which new species form. Instead, they found that what mattered most was how long ago, and how many times, tree frogs had colonized an area. In places where tree frogs arrived earlier and repeatedly, such as Amazonia, there are more co-occurring species today.

No one knows how many different species are living on Earth. Slightly over 1.9 million species of animals, plants, fungi, and protists have been described thus far (Chapman 2009). Conservative estimates propose that the total number of described and undescribed species is about 9 million (Mora et al. 2011); some analyses suggest that it could be as high as 100 million. This chapter focuses on how these species came to be. More specifically, we explore how the mechanisms of evolutionary change can cause populations to diverge and form new species.

Speciation is among the most fundamental events in the history of life. It has occurred millions, if not billions, of times since life began over 3 billion years ago. In addition to its intrinsic importance, studying speciation has practical applications. Understanding what species are and how they form is central to efforts to preserve biodiversity. We begin with a critical question: What is a species?

16.1 Species Concepts

All human cultures recognize different types of organisms in nature and name them. These taxonomic or naming systems are based on the degree of similarity among organisms. People intuitively group like with like. The challenge to biologists has been to move beyond these informal judgments to a definition of species that is mechanistic and testable, and to a system for classifying the diversity of life that accurately reflects evolutionary history.

These goals have been difficult to achieve, even though most biologists agree on what a species is: the smallest evolutionarily independent unit. Evolutionary independence occurs when mutation, selection, gene flow, and drift operate on populations separately. Evolution consists of changes in allele frequencies, and species form a boundary to the spread of alleles. As a result, different species follow different evolutionary paths. By this definition, it is clear that the essence of speciation is lack of gene flow when populations are in contact with each other.

Species consist of interbreeding populations that evolve independently of other populations.

Although defining species by evolutionary independence sounds straightforward, it is often difficult in practice. The challenge is to establish practical criteria for identifying when populations are actually evolving independently. To illustrate this point, we consider the three most important species concepts currently in use. Each of the three agrees that species are evolutionarily independent units that are isolated by lack of gene flow, but employs a different criterion for determining that independence is actually in effect.

The Morphospecies Concept

In traditional cultures, people name species based on morphological similarities and differences. In biology, careful analyses of phenotypic differences are the basis of identifying **morphospecies.**

The great advantage of the morphospecies concept is that it is widely applicable. Morphospecies can be identified in individuals that are extinct or living, and in organisms that reproduce sexually or asexually. A disadvantage of the morphospecies concept is that when it is not applied carefully, species definitions can become arbitrary and idiosyncratic. In the worst-case scenario, species designations made by different researchers are not comparable. In addition, the concept can be difficult to apply in groups like bacteria, archaea, and many fungi that are small and have few measurable morphological characters.

Paleontologists have to work around other restrictions when identifying species. Fossil species that differed in color or the anatomy of soft tissues cannot be distinguished. Neither can populations that are similar in morphology but were strongly divergent in traits like songs, temperature or drought tolerance, habitat use, or courtship displays. Whether living or fossil, populations like these are called **cryptic species.** The adjective *cryptic* is appropriate because groups that were or are actually independent of one another appear to be members of the same species based on morphological similarity.

The Phylogenetic Species Concept

The phylogenetic species concept (PSC) focuses on a criterion for identifying species that is known as monophyly. Monophyletic groups are defined as lineages that contain all of the known descendants of a single common ancestor (for a more detailed discussion, see Chapter 4). Under the phylogenetic species concept, species are identified by estimating the phylogeny of closely related populations and finding the smallest monophyletic groups. On a tree like this, species form the tips. For example, if the taxa labeled A–G in **Figure 16.1** represent populations—as opposed to genera, families, orders, or other types of taxa—then they are the smallest monophyletic groups on the tree and represent distinct species. In contrast, if populations cannot be clearly distinguished in a phylogeny by unique, derived characters, then they will form clusters like the populations designated B_{1-3}, E_{1-2}, and G_{1-4}. The populations that make up these clusters would be considered part of the same species.

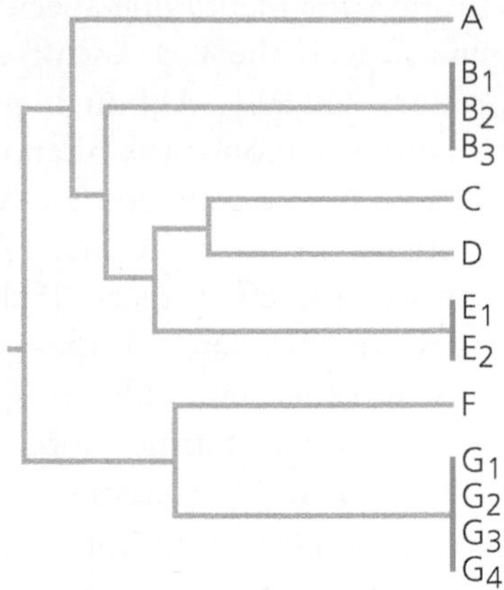

Figure 16.1 Phylogenetic species The taxa labeled A–G on the tips of this phylogeny represent distinct species. Groups labeled G_1, G_2, etc. represent populations of the same species.

The rationale behind the phylogenetic species concept is that traits can distinguish populations on a phylogeny only if the populations have been isolated in terms of gene flow and have diverged genetically, and possibly morphologically as well. Put another way, to be called separate phylogenetic species, populations must have been evolutionarily independent long enough for the diagnostic traits to have evolved. Populations within species have shared, derived traits that distinguish them from populations of other species (see Chapter 4).

The appeals of this approach are that it can be applied to any type of organism—including asexual and fossils—and that it is testable: Species are named on the basis of statistically significant differences in the traits used to estimate the phylogeny. The challenge comes with putting the phylogenetic species concept into practice, particularly when making the arbitrary decision of which specific traits to use in constructing the phylogeny. Different sources of information can lead to differing phylogenies, resulting in conflicting species designations under the PSC. In addition, it is widely recognized that instituting the phylogenetic species concept could easily double the number of named species and might create a great deal of confusion if traditional names and species identities are changed.

Proponents of the concept are not bothered by the prospect of recognizing many additional species. They claim that if a dramatic increase in the number of named species did occur, it would be necessary to reflect biological reality. As predicted, recent analyses have found that the PSC often distinguishes multiple cryptic species in populations that were formerly considered a single species (e.g., (Suatoni et al. 2006; Amato et al. 2007; Malenke et al. 2009; Griffiths et al. 2010). For organisms such as fungi, the PSC has been immensely useful in identifying possible species boundaries that can then be further explored experimentally (Giraud et al. 2008).

The Biological Species Concept

Under the biological species concept (BSC), the criterion for identifying evolutionary independence is **reproductive isolation.** Specifically, if populations of organisms do not hybridize regularly in nature, or if they fail to produce fertile offspring when they do, then they are reproductively isolated and considered separate species. The biological species concept has been widely accepted since Ernst Mayr championed it in 1942. It is used in practice by many biologists and is the legal definition employed in the U.S. Endangered Species Act of 1973, the flagship biodiversity legislation in the United States.

The great strength of the BSC is that reproductive isolation is a meaningful criterion for identifying species because it confirms lack of gene flow. Lack of gene flow is the key test of evolutionary independence in organisms that reproduce sexually. Although this criterion is compelling in concept and useful in some situations, it is often difficult to apply. For example, if nearby populations do not actually come in contact, it can be difficult for researchers to assess whether they are reproductively isolated. Instead, biologists have to make judgments to the effect that, "If these populations were to meet in the future, we believe that they are divergent enough already that they would not interbreed, so we will name them different species." In these cases, species designations cannot easily be tested. Furthermore, the biological species concept cannot be tested in fossil forms and is irrelevant to asexual populations. Despite these limitations, the biological species concept has proven to be the most useful because it provides a mechanistic framework that allows biologists to assess the evolutionary status of populations in a non-arbitrary manner (Schemske 2010).

What about Bacteria and Archaea?

Much of the research reviewed in this chapter focuses on events that lead to reproductive isolation and result in lack of gene flow. Indeed, the BSC treats reproductive isolation as the criterion of speciation. But in many eukaryotes and in all bacteria and archaea, reproduction takes place asexually. Thus, there is no exchange of genetic material when bacteria and archaea reproduce. When gene flow does occur between bacterial cells, it is limited to small segments of the genome, and the exchange does not have to go both ways. Gene flow in bacteria and archaea occurs in the absence of reproduction and can result in genetic recombination, in which new combinations of alleles are created.

Although biologists have confirmed that several processes can result in gene transfer between bacterial or archaeal species, researchers are just beginning to quantify the extent of gene flow in nature. One thing is certain, though: Alleles are routinely transferred between members of widely diverged bacterial and archaeal lineages. In some cases the species involved have genomes whose base sequences have diverged up to 16% (Cohan 1994, 1995). In contrast, genetic exchange between eukaryotes is generally limited to organisms whose genomes have diverged a total of 2% or less.

A key point here is that what most of us consider normal sex—meaning meiosis followed by the reciprocal exchange of homologous halves of genomes, among members of the same species—is unheard of in bacteria and archaea. As a result, gene flow plays a relatively minor role in homogenizing allele frequencies among bacteria populations. In these organisms the primary consequence of gene flow is that certain cells acquire alleles—via one-way flow from other cells—with high fitness advantages, such as sequences that confer antibiotic resistance or give the recipient cell the ability to use a new type of sugar or other energy source.

Based on these observations, Lawrence and Ochman (1998) have proposed that gaining novel alleles by lateral transfer is the primary mechanism of speciation in bacteria. Their hypothesis is that gene flow triggers divergence among bacterial populations, even though it prevents divergence among eukaryotes. If correct, it means that bacterial species may consist of cells that recently descended from a common ancestor and that have not experienced gene flow via lateral transfer.

More recent work comparing the complete genomes of strains of the same bacterial species has yielded a similar conclusion—specifically, that bacterial and archaeal species should be identified on the basis of gene content, or which genes are present (Konstantinidis and Tiedje 2005). The logic here is that bacterial and archaeal species are best defined in an ecological context, based on their ability to thrive in a particular environment. This ability, in turn, is dependent on which genes are present in the genomes of these organisms. Because the genes required for using particular sources of food are often acquired through gene flow, the ecological view coincides nicely with the idea that gene flow triggers speciation in bacteria and archaea, instead of impeding speciation as it does in eukaryotes.

Applying Species Concepts: Two Case Histories

Although it is probably unrealistic to insist on a single, all-purpose criterion for identifying species, the primary species concepts are productive when applied in appropriate situations. Consider, for example, how recent efforts to apply more than one species concept have improved our understanding of diversity in an abundant group of ocean-dwelling algae and informed efforts to preserve the largest ray species in the world.

Species can be identified by distinctive morphological traits, reproductive isolation, and/or phylogenetic independence. Each species concept has advantages and disadvantages.

Cryptic Species in Marine Phytoplankton

At the base of the marine food web are single-celled photosynthesizing plants and bacteria known as phytoplankton, or algae. In addition to being responsible for primary production in the world's oceans, phytoplankton sometimes explode in density in certain coastal areas, resulting in algal blooms. Whether a bloom is harmful to other living creatures depends on many factors, but the most dangerous are blooms of phytoplankton species that produce neurotoxins. These so-called harmful algal blooms occur in many regions of the world. Along the western coast of North America, a genus of diatoms called *Pseudo-nitzschia* is responsible for harmful blooms due to its production of domoic acid, a neurotoxin that can accumulate in shellfish and poison people who consume it.

Historically, the partitioning of *Pseudo-nitzschia* populations into species was based upon morphological appearance, typically under a light microscope. Alberto Amato and colleagues recently chose two *Pseudo-nitzschia* morphospecies for further investigation using modern microscopy and molecular tools (2007).

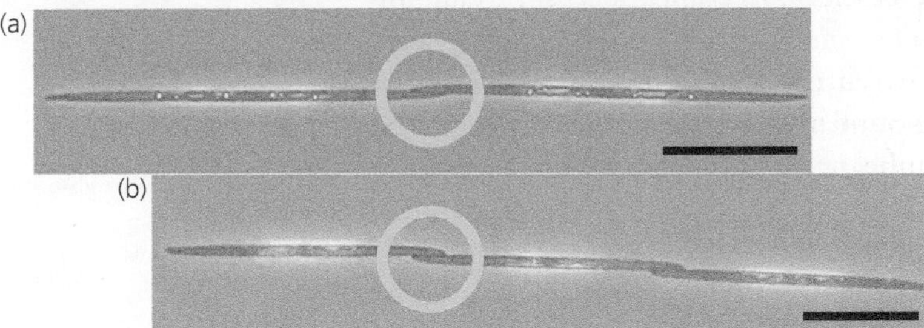

Figure 16.2 Morphospecies of *Pseudo-nitzschia* Historically, the two morphospecies shown in (a) and (b) were distinguished via light microscopy based upon the way cells overlap to form chains. Scale bars = 20 μm. From Amato et al. (2007).

These morphospecies are defined by their needle-shaped cells and the way that cells overlap to form chains **(Figure 16.2)**. But the differences are subtle. Are these really two separate species? Perhaps they are just two different cell types of the same species, or perhaps more than two distinct evolutionary lineages exist. To address this question, Amato and colleagues used multiple species concepts. To apply the morphospecies concept with increased rigor, they viewed algal cells under high-power transmission electron microscopy. To apply the phylogenetic species concept, they constructed phylogenies from DNA sequence data.

And, remarkably for phytoplankton, the researchers were able to induce sexual reproduction and conduct breeding experiments to apply the biological species concept in the same populations.

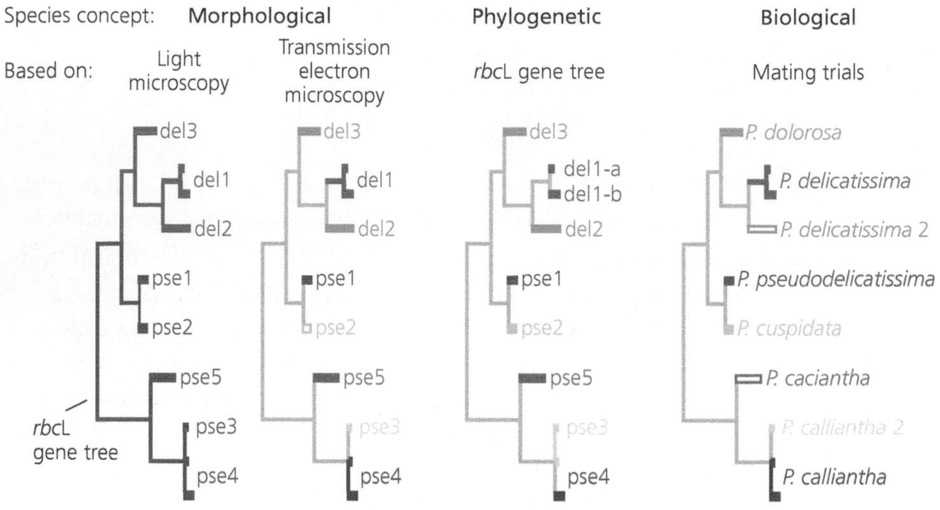

Figure 16.3 Agreement between species concepts Under three separate species concepts, the two historical morphospecies of *Pseudo-nitzschia* turn out to contain eight different species in total. Open boxes indicate absence of data. After Amato et al. (2007).

The outcome of the *Pseudo-nitzschia* study, summarized in **Figure 16.3**, highlights the power of applying multiple species concepts. Within one of the previously recognized morphospecies, three distinct lineages were apparent, and within the other, there were at least five. Most striking is the concurrence among results from different species concepts. Electron microscopy identified subtle morphological features that differentiated populations in the same way the gene trees did, and breeding experiments during the sexual phase corroborated these conclusions (Amato et al. 2007). With the new data, no matter which species concept was applied, the result was the same: There are eight different *Pseudo-nitzschia* species.

Diatoms are just one group of organisms where species diversity is turning out to be far greater than previously thought. This is an important realization, because it shows that employing multiple species concepts can help biologists recognize species diversity and organize research on its consequences. Production of the dangerous neurotoxin by *Pseudo-nitzschia* is highly variable during algal blooms, and part of this variability is due to differences among species (Thessen et al. 2009), so knowledge about species boundaries in this instance is not just an academic exercise. It carries practical significance directly relevant to ecosystem function and human health.

Conservation of a Critically Endangered Fish

The International Union for the Conservation of Nature lists almost 20,000 species as threatened with extinction worldwide. What is the best strategy for saving many of these species from being lost forever? Conservation biologists take many approaches, but a primary goal is the preservation of genetic diversity. If genetic diversity correlates with phenotypic diversity, then preserving multiple genetically distinct populations (whether or not they are identified as separate species) is more likely to lead to long-term survival in the face of environmental changes. This is the reasoning behind the concept of the **evolutionarily significant unit,** or **ESU,** defined in the U.S. Endangered Species Act. All else being equal, preserving more ESUs increases the likelihood of a species' persistence.

Of all marine species threatened with extinction as a result of overfishing, the European common skate (*Dipturus batis*) is perhaps closest to being lost. At up to 2.5 meters long, it is the largest species of skate, a group of cartilaginous fishes that group with rays and are closely related to sharks. Once widespread in the Northeast Atlantic, it was recognized over 30 years ago as being the first documented case of a fish species brought to the brink of extinction by commercial fishing (Brander 1981). Its last strongholds now exist along the western coasts of the British Isles and Norway.

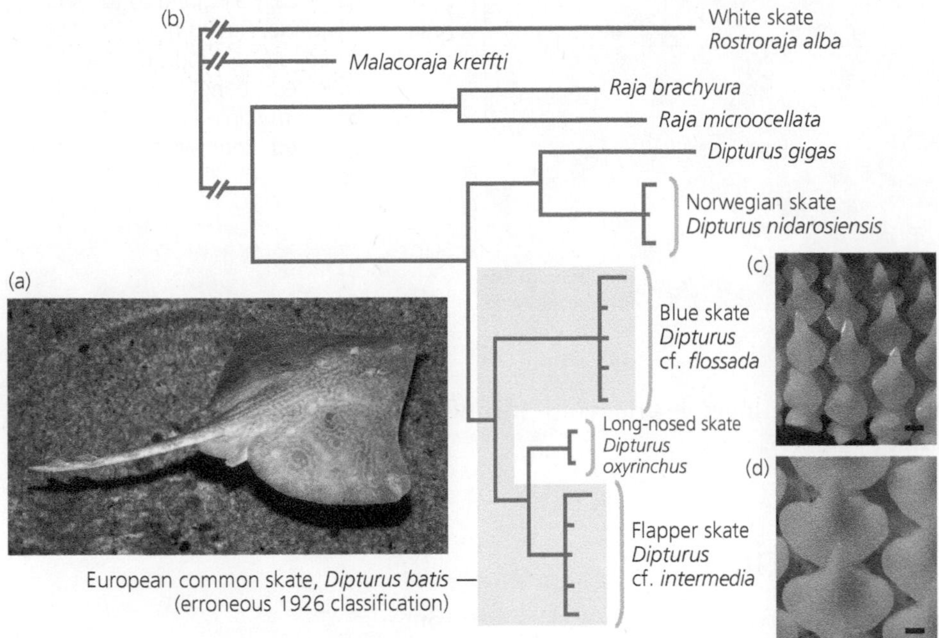

Figure 16.4 A molecular phylogeny reveals the existence of two common skate species (a) A European common skate, formerly classified as *Dipturus batis*. (b) A sequence-based phylogeny revealing distinct lineages of common skate that are not even each other's closest relatives. Redrawn from Iglésias et al. (2010). (c) and (d) Photos showing tooth shape, one of the morphological characters that distinguish the two lineages of common skate. The teeth shown are from the lower jaws of adult females. Scale bars = 1 mm. Photos by Samuel Iglésias.

The taxonomic history of the common skate has been marked with confusion because morphospecies are difficult to identify. Until the 1920s it was widely regarded as two distinct species, the blue skate and the flapper skate, but in 1926 these were combined into a single species and have been viewed that way ever since **(Figure 16.4a)**. Two recent studies used molecular data to challenge this single-species classification (Griffiths et al. 2010; Iglésias et al. 2010). Although the two studies were conducted independently and relied upon DNA sequence data from different loci, the findings matched. As shown in Figure 16.4b, there are two distinct species of common skate.

Informed by this new phylogenetic classification, one research group was then able to define morphological features that can be used to distinguish the two species (Iglésias et al. 2010). Among these features are the shape of the skates' teeth, shown in Figure 16.4c and d. The other group found a geographic pattern in the distribution of the two species—a northern clade and a southern clade, with almost no overlap **(Figure 16.5, next page)**.

These findings are important to the conservation of common skates. Not only are we now aware that two, not one, critically endangered species exist, but there is reason to believe that preserving the genetic variation of both species will increase the chances that at least one will persist. Because the range of temperatures experienced in the northern and southern habitats differs (Griffiths et al. 2010), each species may possess unique thermal adaptations that could prove essential

Reliable criteria for identifying species are essential for preserving biodiversity.

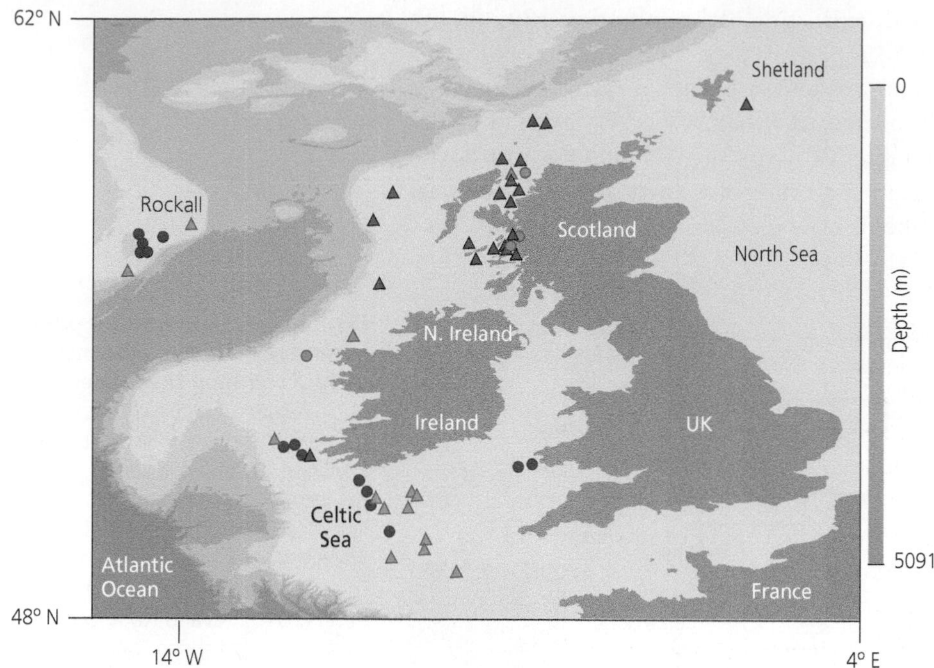

Figure 16.5 Northern and southern clades of common skate The two clades of the common skate, now thought to be separate species, display geographic separation with minimal overlap. Blue symbols represent skates genetically assigned to the northern clade; red and orange symbols represent individuals genetically assigned to the southern clade. Circles mark locations where a single individual was collected; triangles mark locations where multiple individuals were collected. Redrawn from Griffiths et al. (2010).

to survival in the future. In this case, employing multiple criteria for identifying species clarified both conservation and evolutionary issues.

16.2 Mechanisms of Isolation

Given that several tools are available for identifying species, we now turn our attention to the question of how species form. Classically, speciation has been hypothesized to be a three-stage process: an initial step that isolates populations, a second step that results in divergence in traits such as mating system or habitat use, and a final step that produces reproductive isolation. According to this model, the isolation and divergence steps were thought to take place over time and to occur while populations were located in different geographic areas. The final phase was hypothesized to occur when these diverged populations came back into physical contact—an event known as **secondary contact.**

Speciation via the process just described is widespread in nature. However, recent research has shown that in some instances speciation can follow a different path. For example, the isolation and divergence steps that initiate speciation have in some cases occurred at the same time and in the same place. In addition, it appears likely that in many speciation events, the third phase never occurs. Even so, the isolation/divergence/secondary contact hypothesis provides a useful framework for analyzing how speciation takes place.

The focus of this section is isolation, the first step in speciation. Isolation, put simply, is a reduction in gene flow between two populations. Isolation most commonly occurs as a result of geographic factors that cause populations to become physically separated. However, any factor that reduces the probability or effectiveness of interbreeding between two populations can cause isolation. Therefore, genetic events like changes in chromosome number can also cause isolation. Once gene flow is dramatically reduced or ceases, evolutionary independence begins and speciation may take place. Section 16.3 addresses how ecological forces

The speciation process begins when gene flow is disrupted and populations become genetically isolated.

cause genetically isolated populations to diverge. In Section 16.4, we consider what happens if secondary contact occurs.

Physical Isolation as a Barrier to Gene Flow

Gene flow tends to homogenize gene frequencies and reduce the differentiation of populations (for a detailed discussion, with models, see Chapter 7). You may recall the example of water snakes from mainland and island habitats in Lake Erie, and experiments that showed a selective advantage for unbanded snakes on island habitats. But because migration of banded forms from the mainland to the islands occurs regularly, and because banded and unbanded forms subsequently interbreed, the island populations did not completely diverge from mainland forms. Migration continually introduced alleles for bandedness, even though selection tended to eliminate them from the island populations.

Now consider a thought experiment: What would happen if lake currents changed in a way that effectively stopped the migration of banded forms from the mainland to the islands? Gene flow between the two populations would end, and the balance between migration and natural selection would change. The island population would be free to differentiate as a consequence of mutation, natural selection, and drift. These processes would occur among the island snakes independently of the processes acting among mainland forms.

This scenario illustrates a classical theory for how speciation begins, called the **allopatric model** (Mayr 1942, 1963). Translated literally, *allopatric* means "different country or homeland." The essence of allopatric speciation is that physical isolation creates an effective barrier to gene flow. Research has shown that in many cases, geographic isolation has been an important trigger for the second stage in the speciation process: genetic and ecological divergence.

Geographic isolation produces reproductive isolation, and thus genetic isolation.

Geographic isolation can come about via two mechanisms **(Figure 16.6)**. The first is dispersal across a physical barrier followed by colonization of a new habitat, such as a when a group of organisms rides a raft of vegetation to an island (Censky et al. 1998). The second is vicariance, in which an existing range is split by the appearance of a new physical barrier—which may be as small as a road or as large as an ocean (Clark et al. 2010; Chakrabarty et al. 2012).

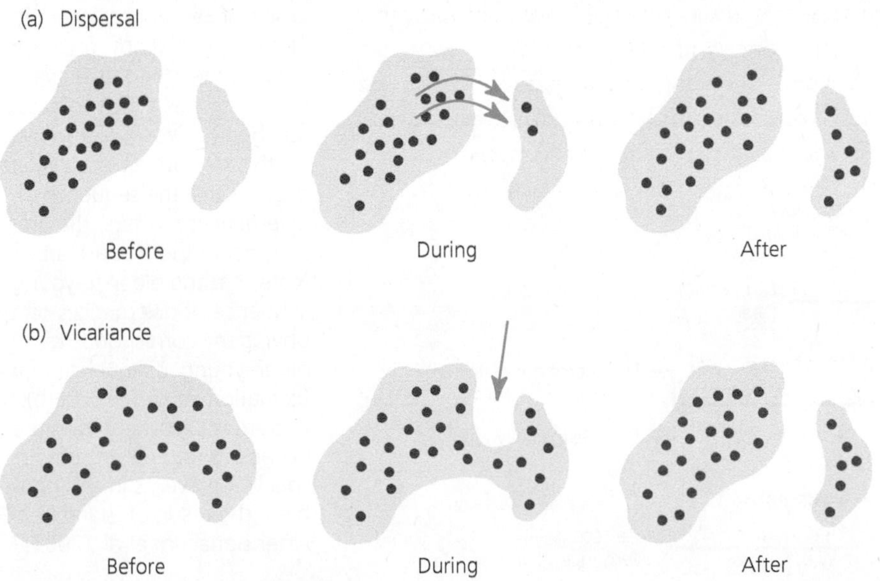

Figure 16.6 Isolation by dispersal and vicariance In the diagram of dispersal (a), the arrows indicate movement of individuals. In the diagram of vicariance (b), the arrows indicate an encroaching physical feature such as a river, glacier, lava flow, or new habitat.

Geographic Isolation through Dispersal and Colonization

One of the most spectacular adaptive radiations among insects is also a superb example of geographic isolation through dispersal. The Hawaiian drosophilids, close relatives of the fruit flies we have encountered before, include an estimated 1,000 species and are renowned for their exceptional ecological diversity. Hawaiian flies can be found from sea level to montane habitats and from dry scrub to rain forests. Food sources, especially the plant material used as the medium for egg laying and larval development, vary widely among species. One of the Hawaiian flies even lays its eggs in spiders, while another has aquatic larvae. In addition, many species have elaborate traits, such as patterns on their wings or modified head shapes, that are used in combat or in courtship displays **(Figure 16.7a)**.

The leading explanation for this diversity begins with dispersal and colonization. Many of the Hawaiian flies are island endemics, meaning that their range is restricted to a single island in the archipelago. If small populations of flies, or perhaps even single gravid females, disperse to new habitats or islands, then the colonists establish new populations that are physically cut off from the ancestral species. Divergence begins after the founding event, resulting from genetic drift and natural selection acting on the genes responsible for courtship displays and habitat use.

Populations can become geographically isolated when individuals colonize a new habitat.

(a)

(b)

(c)

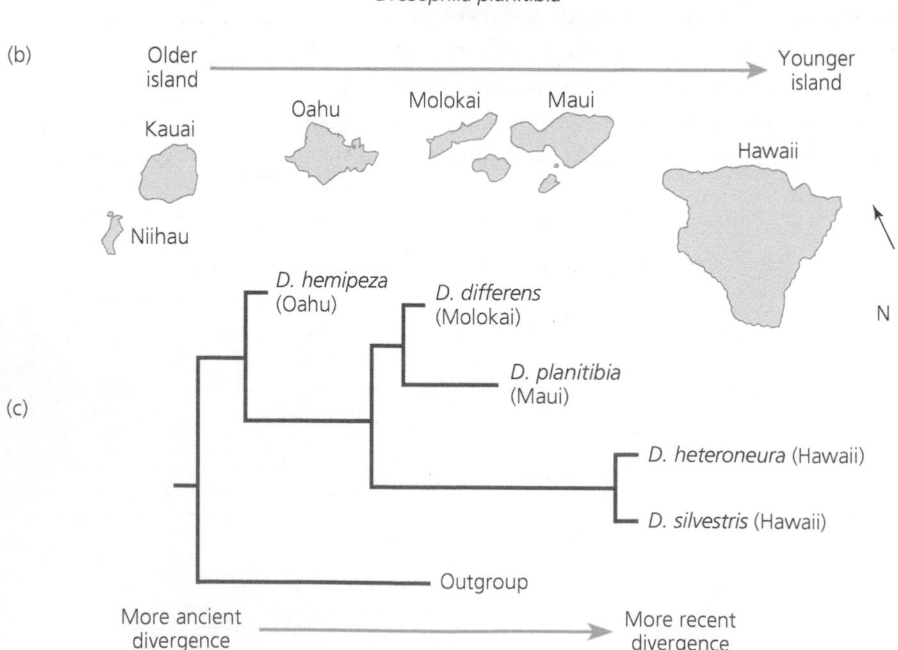

Figure 16.7 Evidence for speciation by dispersal and colonization events (a) *Drosophila hemipeza*, *D. planitibia*, and *D. heteroneura* illustrate the remarkable diversity—in body size, wing coloration, and other traits—among the *Drosophila* found in Hawaii. Photos by Karl Magnacca, DNA Barcoding Endemic Hawaiian Species Project, University of Hawaii at Hilo. (b) The Hawaiian islands are part of an archipelago that stretches from the island of Hawaii to the Emperor Seamounts near Siberia. The youngest landform in the chain is the island of Hawaii, which still has active volcanoes. (c) The five *Drosophila* species on this tree are a closely related group. Only the sequence of divergences matters; the branch lengths on the tree are arbitrary. Note that the older-to-younger sequence of divergences on the phylogeny corresponds to the older–younger sequence of island formation shown in part (b).This pattern is consistent with the hypothesis that at least some of the speciation events in this group were the result of island hopping. After Bonacum et al. (2005).

The logic of the dispersal-and-colonization hypothesis is compelling, but do we have evidence, other than endemism, that these events actually occurred? Because the geology of the Hawaiian islands is well known, the hypothesis makes a strong prediction about speciation patterns in flies. The Hawaiian islands are produced by a volcanic hotspot under the Pacific Ocean. The hotspot is stationary but periodically spews magma onto the Pacific Plate, forming islands. After islands form, continental drift carries them with the Pacific Plate to the north and west (Figure 16.7b). As time passes, the volcanic cones gradually erode down to atolls and submarine mountains.

The dispersal-and-colonization hypothesis makes two predictions based on these facts: (1) Closely related species should be found on adjacent islands; and (2) the sequence of branching events should correspond to the sequence of island formation. James Bonacum and coworkers (2005) used DNA sequence differences in a series of mitochondrial and nuclear genes to estimate the phylogeny of closely related Hawaiian flies and found exactly these patterns (Figure 16.7c). This is strong evidence that dispersal to new habitats triggered speciation. Similar patterns have been observed in the phylogenies of Hawaiian crickets (Mendelson and Shaw 2005) and Galápagos tortoises (Beheregaray et al. 2004).

As a mechanism for producing physical isolation and triggering speciation, the dispersal-and-colonization hypothesis is relevant to a wide variety of habitats in addition to oceanic islands. Hot springs, deep-sea vents, fens, bogs, caves, mountaintops, and lakes or ponds with restricted drainage also represent habitat islands (for example, Dawson and Hamner 2005). Dispersal to novel environments has proven to be a general mechanism for initiating speciation.

Geographic Isolation through Vicariance

Vicariance events split a species' distribution into two or more isolated ranges and discourage or prevent gene flow between them. There are many possible mechanisms of vicariance, ranging from slow processes, such as the rise of a mountain range, to rapid events such as a mile-wide lava flow that bisects a snail population.

Nancy Knowlton and colleagues studied a classic vicariance event: the recent separation of marine organisms on either side of Central America. Geological evidence has established that the Isthmus of Panama closed about 3 million years ago. As the isthmus rose and created a land bridge between North and South America, populations of marine organisms became separated on the Atlantic and Pacific sides. When the oceans were separated in this way, did the populations that ended up on either side speciate?

To address this question, Knowlton and coworkers (1993; Hurt et al. 2009) analyzed snapping shrimp (genus *Alpheus*) populations from either side of the isthmus **(Figure 16.8a)**. Based on the morphospecies concept, the populations they sampled appeared to represent seven closely related species pairs, or sister species, with one member of each pair found on each side of the land bridge. The phylogeny of these shrimp, estimated from DNA sequence data, confirms this hypothesis (Figure 16.8b). The species pairs from either side of the isthmus, reputed to be sisters on the basis of morphology, are indeed each others' closest relatives. This result is consistent with the vicariance hypothesis.

Furthermore, when Knowlton and coworkers put males and females of various species pairs together in aquaria and watched for aggressive or courtship interactions, the researchers found a strong correlation between the degree of genetic distance between species pairs and how interested the shrimp were in mating.

(a)

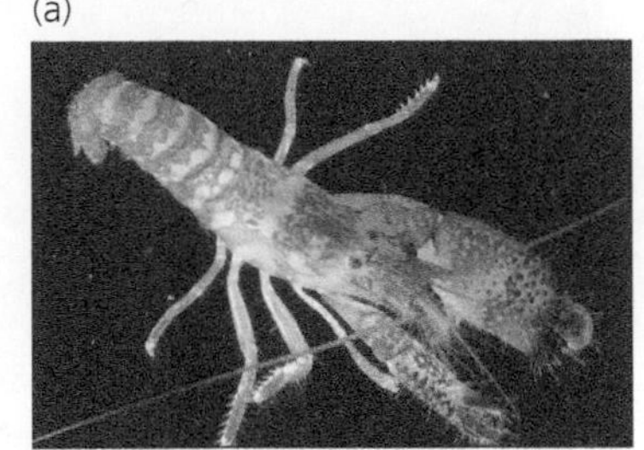

(b)

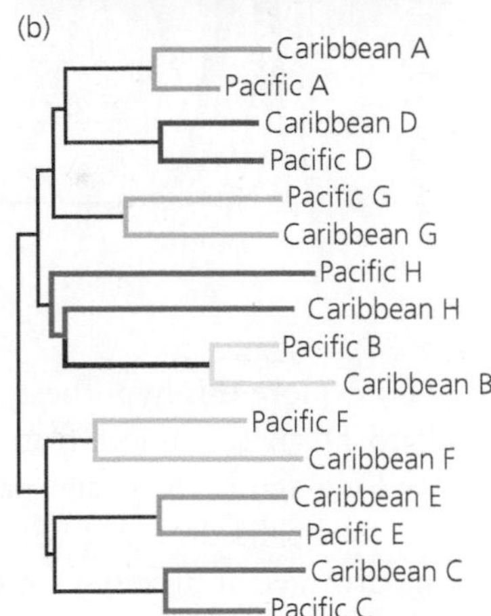

Figure 16.8 Snapping shrimp speciation due to vicariance (a) *Alpheus malleator*, found on the Pacific side of the Panamanian isthmus. Carl C. Hansen/Nancy Knowlton/Smithsonian Institution Photo Services. (b) Phylogeny estimated from sequence divergence in mitochondrial DNA. Morphological sister species from opposite sides of the isthmus are identified by letters and colors. Redrawn from Hurt et al. (2009).

Males and females from species with greater genetic divergence, indicative of longer isolation times, were less interested in one another. Finally, almost none of the pairs that formed during the courtship experiments produced fertile clutches. This last observation confirms that the Pacific and Caribbean populations are indeed separate species under all three of the species concepts we have reviewed.

Populations also can become geographically isolated when a species' former range is split into two or more distinct areas.

One of the most interesting aspects of the study, though, was that the data contradicted an original prediction made by the vicariance hypothesis. If the land bridge had formed rapidly, we would expect that genetic distances and degrees of reproductive isolation would be identical in all seven species pairs. This is not the case. For example, DNA sequence divergence between species pairs varied from about 6.5% to more than 19% (Knowlton and Weigt 1998).

What is going on? Upon reflection, a prediction of identical divergence was naive, because it is unlikely that the land bridge popped up suddenly. Instead, as the land rose and the ocean gradually split and retreated on either side, shrimp populations would become isolated in a staggered fashion, depending on the depth of water each occupies and how efficiently their larvae disperse. The ranges of deeper-water species, or those with less-motile larvae, would be split first.

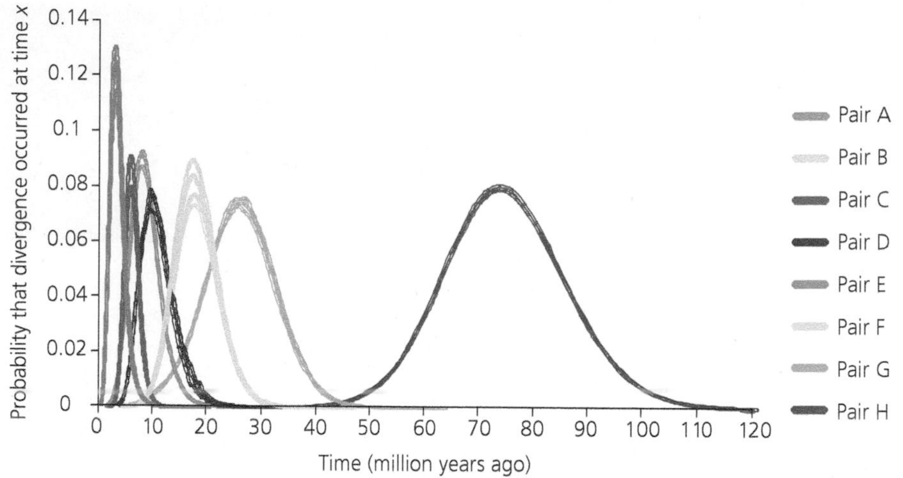

Figure 16.9 Estimated divergence times for Caribbean-Pacific snapping shrimp species pairs Pairs A, C, D, and E speciated more recently than the other four pairs. Consistent with the vicariance hypothesis due to the rising isthmus, the four spcies with older divergence times live in the deepest waters. Redrawn from Hurt et al. (2009).

To explore this hypothesis, Carla Hurt and colleagues applied statistical methods to DNA sequences from multiple loci in the shrimp genomes to estimate how long ago each species pair diverged (Hurt et al. 2009). As shown in **Figure 16.9**, four pairs diverged with similar timing and fairly recently. The other four pairs diverged at different times in the more distant past. Consistent with the hypothesis regarding water depth, the species with more distant divergence inhabit deep water, while the four species with the most recent divergence live in shallower water along the coast (see Knowlton and Weigt 1998). Note also that the timing of divergence in the four most recent species pairs is very similar. These are the lowest values observed, perhaps indicating "the final break" between the two oceans 3 million years ago.

An array of similar studies has convinced biologists that vicariance has been an important isolating mechanism and trigger for speciation in a wide variety of groups. Other well-studied examples include a seaway that separated the northern and southern portions of the Baja California peninsula about 1 million years ago (see Riginos 2005) and the fragmentation of habitats by glacial advances during the Pleistocene (Weir and Schluter 2004; Hoskin et al. 2005).

A Role for Mutation: Polyploidy and Other Chromosome Changes May Reduce Gene Flow

Theory predicts that populations may speciate after becoming physically isolated due to dispersal or vicariance, and data have confirmed that these events are common triggers for speciation. But it is also possible for gene flow between populations to be reduced even in the absence of physical isolation. For example, mutations resulting in polyploidization can produce instant reproductive isolation between parental and daughter populations (Chapter 5). Differences in ploidy, as when a tetraploid (four copies of each chromosome, or $4n$) lineage descends from a diploid ($2n$) ancestor, almost always cause reproductive isolation because of dysfunctional chromosome complements that result from crosses.

Changes in chromosome number isolate populations genetically.

How important is polyploidization as a mechanism of speciation? Biologists estimate that among the approximately 300,000 species of land plants, at least 2% to 4% are derived directly from polyploidization events. Although speciation by polyploidy is much less common in animals than in plants, researchers have documented a few instances in vertebrates such as freshwater fishes (Machordom and Doadrio 2001; Cunha et al. 2008), water frogs (Christiansen and Reyer 2009), and whiptail lizards (Lutes et al. 2011).

Speciation triggered by changes in chromosome number has been especially important in plants.

Changes in chromosome number less drastic than polyploidization may also be important in speciation, because crossing between lineages differing in chromosome number rarely results in fertile offspring due to problems during meiosis. For example, butterfly species of the genus *Agrodiaetus* have striking diversity in total chromosome number: from 10 to 134, depending on the species. Nikolai Kandul and colleagues hypothesized that differences in chromosome number are important in maintaining boundaries between *Agrodiaetus* species that share geographic ranges. This hypothesis is supported by the finding that recently diverged **sympatric** species—those that overlap geographically—are more likely to differ in chromosome number than distantly related allopatric species (Kandul et al. 2007). Sympatric species with matching chromosome number are rare, likely because coexisting populations are not able to diverge into separate species without a boundary to gene flow. In this case, differences in chromosome number provide the necessary boundary for isolation.

It is common to find small-scale chromosomal changes like these when the karyotypes of closely related species are compared. Although these mutations could be important in causing genetic divergence between populations, much of the extensive work on chromosomal differentiation done to date is only correlative. That is, many studies have measured chromosome differences in related species and claimed that chromosomal incompatibilities are responsible for isolation. But in many cases, the chromosome differences likely arose after speciation and occurred due to other causes, leading some researchers to conclude that speciation by chromosomal changes alone is very rare (Coyne and Orr 2004). Work continues on establishing causative links between small-scale karyotype differences and speciation (see Faria and Navarro 2010).

Other Mechanisms of Isolation

Physical isolation due to geography and reproductive incompatibility due to chromosomal differences are widespread in nature, but other forms of isolation exist. In plants, other common isolating mechanisms include differences in flowering time or visitation by different pollinators. Elsewhere we discussed the difference in flower morphology between two species of monkeyflowers (Chapter 9). One has flowers pollinated by bees, the other has flowers visited by hummingbirds. Because neither bees nor hummingbirds visit both species, there is no opportunity for pollen from one species to be deposited on a flower of the other species. The two species are genetically isolated by pollinator specialization.

Biologists increasingly appreciate that isolation takes many forms. The annual timing of reproduction, long known to be important for flowering plants, can cause isolation for animals too. The Japanese winter moth offers a compelling example. Satoshi Yamamoto and Teiji Sota (2009) found that in cold habitats, the moth *Inurois punctigera* consists of two genetically isolated populations that meet the criteria for distinct species under the BSC and PSC. The two species co-occur across Japan, so geography cannot account for their isolation. Instead, they are separated by the annual timing of reproduction. One species emerges as adults, breeds, and lays eggs in late autumn/early winter. The other does so in late winter/early spring (**Figure 16.10a**). Although the moths favor cool wintry conditions for emergence and reproduction, deep snow and extreme cold preclude them from being active during midwinter, so two temporally isolated species exist.

Although temporal isolation is a compelling hypothesis for Japanese winter moths, the early winter and late winter species could be isolated for other reasons as well. Perhaps each is differentially adapted to subtle environmental differences across the winter season. Or maybe they would not choose to interbreed if given the chance, due to pheromone differences unknown to us. If so, they still would be considered separate species, but reproductive timing would not be the only isolating mechanism.

Fortunately, the temporal isolation hypothesis can be tested by comparing populations across Japan because the intensity of midwinter varies with latitude. When the researchers studied southern moths that reproduced throughout the milder southern winter, they found that the isolation between the two species had disappeared (Figure 16.10b; Yamamoto and Sota 2009). The northerly sites

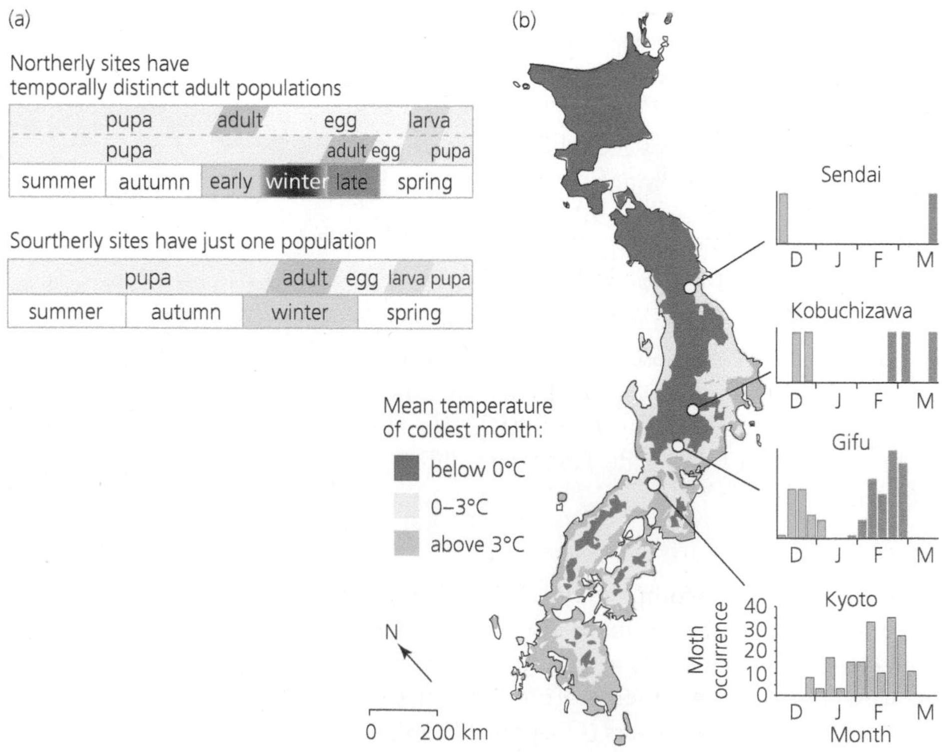

Figure 16.10 Temporal isolation among Japanese winter moth populations (a) Charts of the seasonal activities of *Inurois punctigera* show that at northerly sites, which have cold winters, the moths have two temporally isolated populations. At southerly sites, which have warmer winters, the moths show no temporal isolation. (b) Temporal frequency distributions of adult moths during the winter at four study sites (open circles) across Japan confirm that temporal isolation is correlated with latitude and winter cold. In the most southerly location with the highest minimum temperature, no temporal isolation exists, but at the other three sites there are temporally and genetically distinct populations. Redrawn from Yamamoto and Sota (2009).

harbor genetically distinct species (blue and lavender bars in Figure 16.10) that reproduce at different times, but the southernmost population showed no genetic differentiation (gray bars). The collapse of the two species into a single interbreeding population at the southern site supports the hypothesis that temporal isolation is the mechanism that separates the early versus late winter moth species.

One final example illustrates that isolation can take forms that no one would have imagined until their discovery. In Japanese land snails of the genus *Euhadra,* a single gene controls whether a snail's shell shows "left-handed" or "right-handed" coiling **(Figure 16.11)**. When a mutation arises that causes all of a left-handed snail's offspring to have shells coiled in reverse, the resulting right-handed snails are physically incapable of mating with left-handed snails because their genital openings do not line up correctly (Ueshima and Asami 2003). The phylogeny in Figure 16.11 suggests that the right-handed species *E. aomoriensis* has arisen by this mechanism more than once from within the species *E. quaesita.* In this unusual example, a single mutation causes instant reproductive isolation due to anatomical reproductive incompatibility.

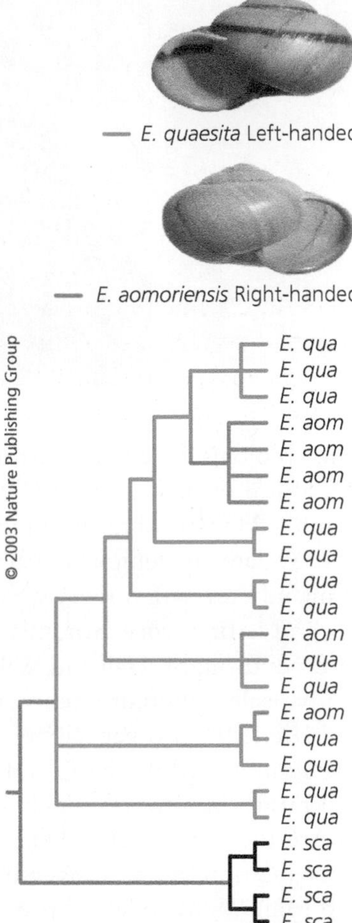

Figure 16.11 Speciation by reversal of shell-coiling direction On the branch tips *E. qua* is *Euhadra quaesita*; *E. aom* is *E. aomoriensis*; and *E. sca* is *E. scaevola* (an outgroup for the *E. quaesita/aomoriensis* clade). Photos from Davison et al. (2005). Tree redrawn from Ueshima and Asami (2003).

Reprinted by permission of Macmillan Publishers Ltd. From Ueshima, R., and T. Asami. 2003. "Evolution: Single-gene speciation by left-right reversal." *Nature* 425: 679. Copyright © 2003.

16.3 Mechanisms of Divergence

Mechanisms of isolation like geographic isolation, vicariance, polyploidy, and other chromosomal changes divide populations, which is necessary for speciation to occur. Yet we recognize species on a day-to-day basis not by the absence of gene flow, but by their distinct phenotypes. Even in closely related species like diatoms or the common skate, morphological differences between species become apparent when we look hard enough. What causes phenotypic divergence between closely related species? Isolation is a critical step, but it must be followed or paralleled by divergence for complete speciation to occur.

We now know that ecological factors, primarily those that result in differences in the direction and intensity of natural selection between closely related populations, are the most frequent divergence mechanisms in nature (Schluter 2009; Via 2009; Schemske 2010; Sobel et al. 2010). Biologists long speculated that genetic drift in the absence of natural selection might be an important pathway to speciation, and genetic drift once dominated discussions of speciation. However, theoretical support for speciation by drift is equivocal, and real-world examples of speciation by drift have proven difficult, if not impossible, to document. A majority of evolutionary biologists now agree that natural selection is the most important mechanism promoting divergence between populations.

Adaptation to Different Habitats

Natural selection can cause populations to diverge based on food preferences, habitats used, or other ecological differences.

The evidence that natural selection causes populations in differing habitats to diverge in phenotypes is indisputable. Across all taxonomic groups, from single-celled microbes to myriad plants and animals, biologists routinely find that ecological factors, both biotic (like competition and predation) and abiotic (like climate) are important agents of natural selection, and that populations adapt over time. When populations of a single species occupy multiple habitats with differing selection pressures, a frequent outcome is phenotypic and genetic divergence, which can lead to speciation if gene flow is limited.

Here we consider an example involving the yellow monkeyflower (*Mimulus guttatus*). In western North America, there are two distinct "ecotypes," which

Figure 16.12 Two ecotypes of *Mimulus guttatus* The coastal ecotype (left) is perennial and invests energy into vegetate growth while the inland ecotype (right) is an annual that produces flowers quickly. From Lowry and Willis (2010).

most biologists view as partway along the process toward speciation (recall our discussion of speciation in Chapter 2). Along the Pacific coast, *M. guttatus* has a perennial life history, surviving year-round in soil that is consistently moist—due to the relatively benign coastal climate—and flowering in late summer. However, at inland sites with hot dry summers, *M. guttatus* has an annual life history, flowering and setting seeds promptly in the early summer and then dying off during the summer drought (Hall and Willis 2006). These two life-history strategies result in dramatically different phenotypes: The coastal perennial ecotype invests considerable energy into vegetative growth early in the season and produces flowers in late summer, while the inland annual ecotype invests little in vegetative growth and produces flowers quickly in early summer **(Figure 16.12)**.

Are these divergent phenotypes the result of evolution by natural selection? This question can be answered with a common garden experiment. David Lowry and John Willis collected plants from coastal and inland sites and replanted them into both their native habitat and into the alternative habitat (2010). If natural selection favors particular life-history strategies in each habitat, the hypothesis predicts that survival and/or reproductive success will differ based upon life-history strategies. This is exactly what the researchers found. As shown by the dark symbols and lines in **Figure 16.13**, in each habitat type the plants with the highest survival and reproductive success were the plants native to that habitat.

Lowry and Willis also added another twist in this experiment by measuring the effects of a particular genome region on survival and reproductive success. Based on earlier studies, they knew that a particular region on one chromosome had a large effect on the life-history phenotype. Using controlled breeding, they were able to create *M. guttatus* lineages with genomes composed entirely of material from one of the parental populations, except for the known large-effect region, which came from the other parental population. The results for these genetic mosaics, represented by light symbols and lines in Figure 16.13, show that a mostly coastal perennial genome with the critical region from the annual population has a fitness phenotype that is intermediate (light blue line).

One reason this study is particularly compelling is that the critical genomic region is rearranged on the chromosome in perennial versus annual ecotypes. In the perennial genome, the DNA bases in this region read in one direction, but in the annual genome, the entire region is inverted and the bases read in the opposite direction. Chromosomal inversions like these are important because they suppress recombination between the two types: When the chromosomes line up during meiosis, the nucleotides in the inverted region do not match, and crossing

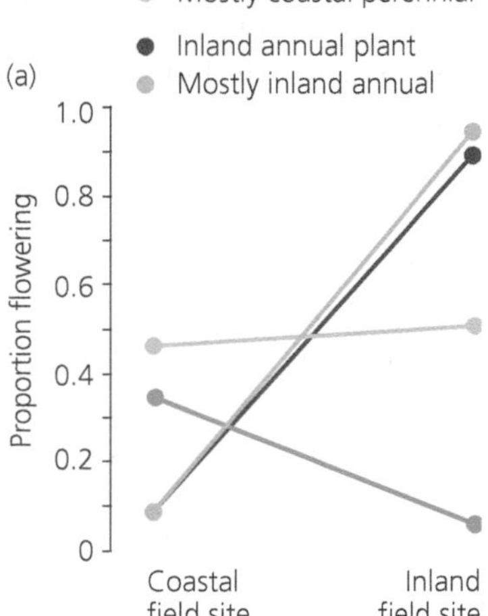

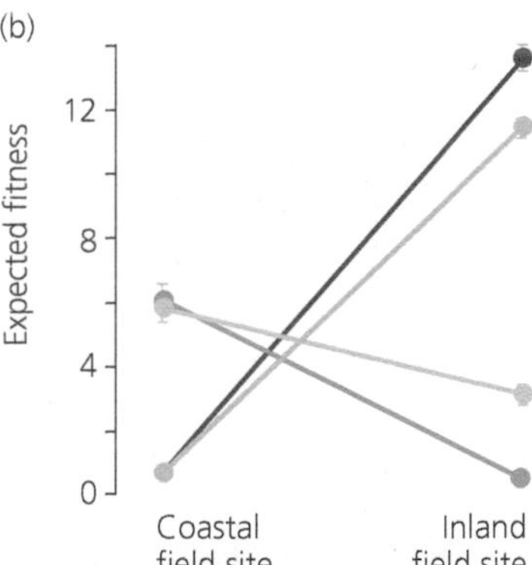

Figure 16.13 Fittness measures in a common garden experiment The dark symbols and lines represent the two *Mimulus* ecotypes found in nature. The light lines and symbols represent genetic mosaics. Redrawn from Lowry and Willis (2010).

over cannot take place in this region. This means that all of the genes in the inverted region (and there could be several) will always be inherited together and cannot be shuffled between the two ecotypes. Lowry and Willis's study is striking not simply because it shows that natural selection can cause populations to diverge in different habitats, but because an unusual chromosomal phenomenon is responsible for divergent phenotypes that may lead to reproductive isolation in nature. This result raises the possibility that chromosomal inversions may sometimes play an important role in adaptive divergence and perhaps speciation, a hypothesis that has been suggested by others.

Assortative Mating

Another way that populations can diverge in ways that promote speciation involves mating behavior. In species that reproduce only sexually, successful reproduction always involves finding and choosing a mate, leading to the opportunity for sexual selection to act on phenotypes involved in mating (see Chapter 11). Traits that influence the way species choose mates can lead to **assortative mating,** in which individuals with certain traits more often choose mates that share those same traits. When assortative mating is strong enough, it can contribute to reproductive isolation between populations that may lead to speciation.

Assortative mating can contribute to reproductive isolation.

In the Hawaiian cricket genus *Laupala,* females choose male mates based upon the songs they produce using specialized wing structures. The songs consist of a simple series of pulses, and the pulse rate is highly conserved within species but differs widely among species **(Figure 16.14a)**. Males of some species sing at a rapid

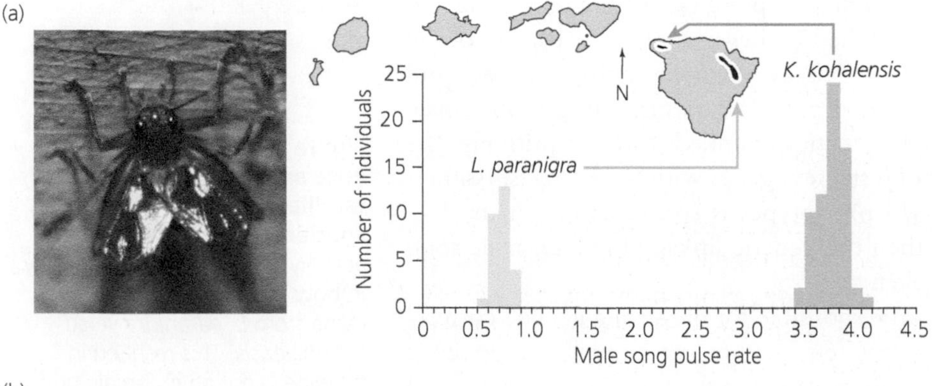

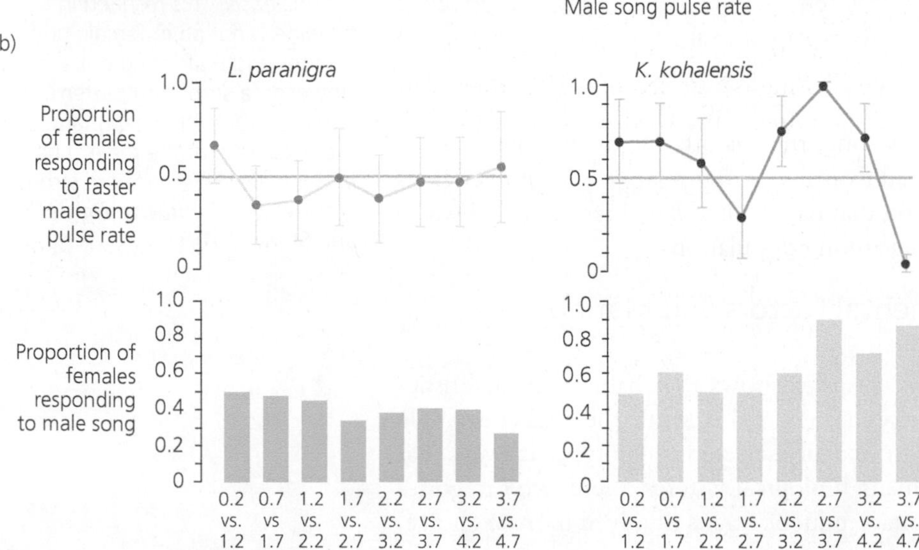

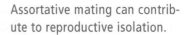

Figure 16.14 Male song and female preference in Hawaiian crickets (a) Males of the species *Laupala paranigra* and *L. kohalensis* have mating songs that pulse at different rates. Photo by Jaime Grace; graph redrawn from Shaw et al. (2007). (b) Female preference in the two species favors pulse rates closer to the true pulse rate in males of that species. Redrawn from Shaw (2000).

pulse rate, while males of other species have a much slower pulse rate. Females choose their mates largely based upon this pulse rate, so male song and female preference for song are traits important in maintaining reproductive isolation among the 38 species in this diverse genus (Mendelson and Shaw 2005). For example, males of *Laupala kohalensis* have a faster pulse rate than males of *Laupala paranigra* (Shaw et al. 2007). As shown in Figure 16.14b, females prefer songs that more closely match the pulse rate of males from their own species (Shaw 2000).

What is the basis for the divergence in male song and female preference between *L. kohalensis* and *L. paranigra*? Previous experiments had shown that both male song and female preference have a complex genetic basis controlled by many genes (Shaw et al. 2007). Why, then, is the genetic variation for male song and female preference so completely partitioned between the two species? Chris Wiley and Kerry Shaw (2010) explored the genetic connection between male song and female preference with breeding experiments. They created hybrids of the two species and backcrossed them repeatedly to *L. kohalensis* mates. This generated a set of hybrid families that were genetically more similar to *L. kohalensis* than *L. paranigra,* but that varied widely in male song and female preference genotypes and phenotypes **(Figure 16.15a)**. Some families happened to retain more *L. paranigra* genes for male song or female preference, while others lost most of these and had genes and phenotypes most similar to *L. kohalensis*. The researchers then measured male song and female preference within each unique family.

The results are astonishing. A strong and obvious relationship exists among families: Those with males that sang at faster pulse rates had females that more strongly preferred faster songs (Figure 16.15b). Although each family had a unique male song pulse rate, within a family the pulse rate of the male song and the females' preference were closely matched (Wiley and Shaw 2010). All of the genetic shuffling caused by repeated meiosis events over multiple crosses and backcrosses between the two species could not separate male song from female preference. The two traits are strongly genetically linked. Most surprisingly, the genetic linkage cannot be the result of just a few genes with large effects. Rather, the complex genetic basis of the two phenotypes is spread among many loci throughout the genome, and even then the genetic linkage between male song and female preference cannot be broken.

Although the molecular mechanism responsible for the strong genetic correlation between male song and female preference is not yet known, the presence of the correlation provides a compelling explanation for why Hawaiian *Laupala* crickets have diverged into so many different species. If every time a particular population evolved a slightly different male song it also evolved a slight difference in female preference favoring the new song, then assortative mating would keep this population reproductively isolated from its nearby relatives. *Laupala* crickets are a compelling example of speciation that has resulted largely from sexual selection rather than ecological differences among populations.

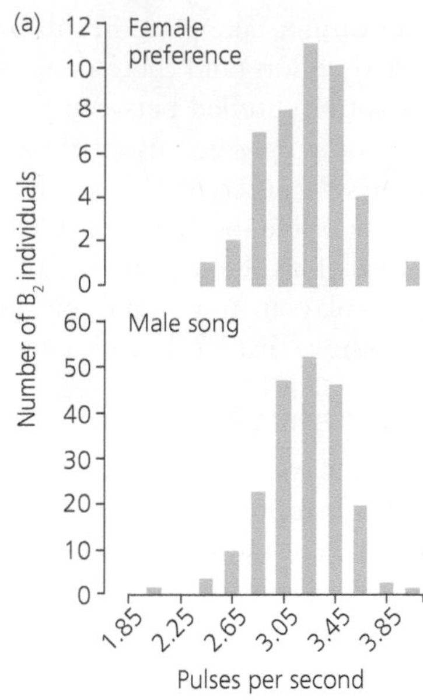

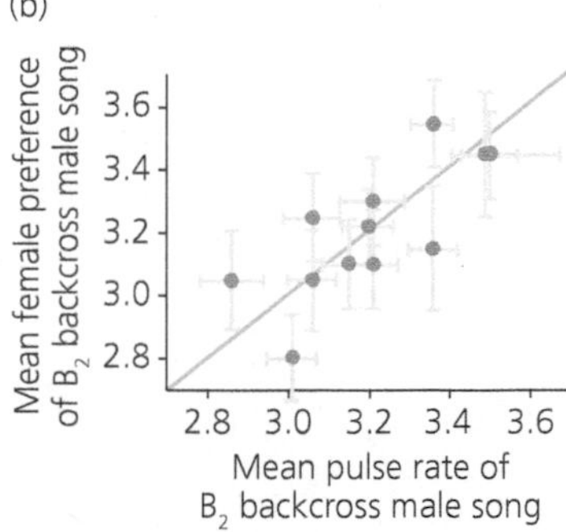

Figure 16.15 Female preference and male song in hybrid families of the two *Laupala* species Experimentally created hybrid families varied widely in how much of their genomes came from *L. paranigra* versus *L. kohalensis*. This resulted in (a) wide variation in female preferences and male song pulse rate. However, a surprising pattern emerges: (b) Within each family, the pulse rate of the male song and female preference are strongly correlated. Redrawn from Wiley and Shaw (2010).

Red Fish, Blue Fish: Environmental Factors and Mating Preferences Can Act Together

The monkeyflower example discussed earlier shows that habitat characteristics can drive phenotypic divergence between populations, while the cricket example shows that populations can diverge in sexually selected phenotypes in ways that promote reproductive isolation. In many biological systems, these two categories of divergence interact. Different habitats influence traits involved in mate choice, and populations therefore diverge in mating preferences as a result. Species of cichlid fish from Lake Victoria in Africa provide a system in which the interaction between habitat and mating phenotypes has been thoroughly explored.

In the water of Lake Victoria, blue wavelengths of light tend to get absorbed near the surface, while longer yellow and red wavelengths always penetrate much deeper, depending on overall water clarity. This means that the visual environment in deeper water is composed mostly of yellow and red light. Work by Yohey Terai and collaborators has shown that cichlid species inhabiting different depths have different alleles of a key opsin gene, called *LWS*, related to color vision. Species found in deeper water have *LWS* alleles that encode opsins that respond more strongly to red light than those in species found in shallower water (Terai et al. 2006). The researchers found evidence of strong natural selection on the *LWS* gene, suggesting that different underwater habitats have favored divergence of the *LWS* gene in ways that improve color vision for cichlids in each.

Cichlid species that live in shallow water near rocky shorelines exist in isolated locations around different islands within Lake Victoria. At many islands, two sympatric species are present: *Pundamilia pundamilia* and *Pundamilia nyererei*. At some islands the two species have more divergent phenotypes than at others **(Figure 16.16)**. In these locations, *P. pundamilia* males have blue-gray coloration that attracts females of their species (hereafter called the "blue species"), while *P. nyererei* males are yellow with a bright red dorsum (hereafter called the "red species"). However, a range of intermediate phenotypes are present at some islands. An important habitat characteristic that varies among islands is overall water clarity, because water clarity varies not only with depth but also among islands within the lake. At islands with lower water clarity, the two *Pundamilia* species look somewhat similar and coexist at the same shallow depths in the water, but at islands where the water is clearer, the difference in phenotypes is much more obvious. There is depth stratification also: The red species is found in deeper water and the blue species in shallow water.

At this point we have established that cichlid species from different depths typically have different *LWS* opsin alleles as a result of selection for improved color vision. We also know that male coloration differs between two particular *Pundamilia* species, and this difference is stronger at islands where the water is clearer and the two species separate into different depth environments. Might there be an evolutionary connection between these two observations? Ole Seehausen and colleagues hypothesized that there is, and that it occurred through a mechanism called speciation by sensory drive (2008). This hypothesis begins

Figure 16.16 Phenotypes of island *Pundamilia* cichlids in Lake Victoria *P. pundamilia* males have blue-gray coloration that attracts females of their species, while *P. nyererei* males have a bright red dorsum that attracts females of their species. Depending on the island and the water clarity associated with it, a range of intermediate phenotypes exist. Photos by Dr. Ole Seehausen. After Seehausen et al. (2008), van der Sluijs et al. (2010).

with the observation that adaptation to different habitats can result in divergence between populations in sensory phenotypes, such as visual perception of different colors of light. Under sensory drive, the differing sensory abilities of two populations then influence the way that individuals choose mates using the same sensory trait. In the case of *Pundamilia,* the hypothesis predicts the following:

1. A higher frequency of red-shifted alleles of the *LWS* opsin gene and predominantly red-colored males will be found in the species inhabiting deeper water.
2. Female preference for male color will also be associated with *LWS* alleles, such that females with red-shifted alleles choose red males as mates.
3. The strongest associations among water depth, *LWS* allele frequency, male color, and female preference will occur at sites with greater water clarity, where fish are spread out across a wider range of depths and there is greater difference between the light environments that the two species experience.

To test these predictions, Seehausen and his team collected male cichlids at a range of depths at five different islands that differed in overall water clarity. They measured male color and *LWS* opsin allele frequency at each depth at each island. They also measured female preference for male color in females from all depths for two of the islands.

The results, shown in **Figure 16.17**, support the sensory drive hypothesis (Seehausen et al. 2008). The divergence in both male color and *LWS* opsin allele

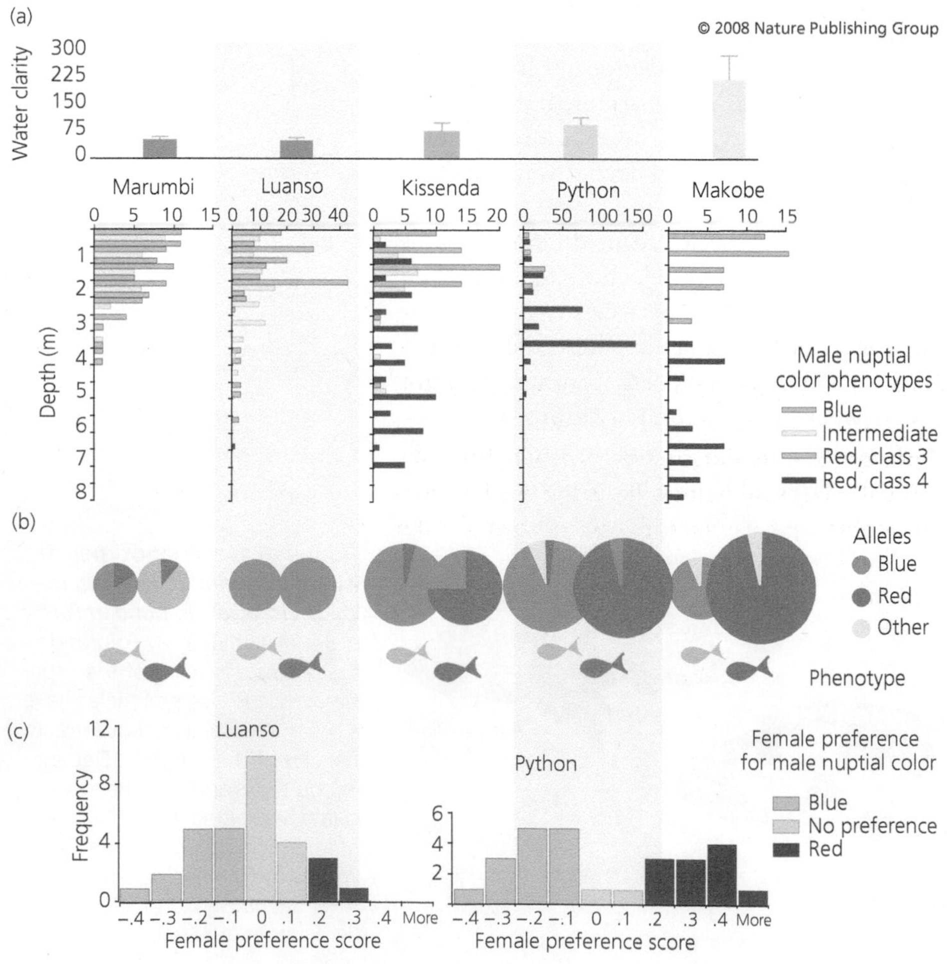

Figure 16.17 Speciation by sensory drive in cichlids Among five different islands, as water clarity increases (left to right), (a) male color and depth distribution become increasingly divergent between the two species, as does (b) *LWS* opsin allele frequency. (c) Finally, females from an island with clearer water showed much stronger color preferences in mates (right graph) than females from an island with murkier water (left). Redrawn from Seehausen et al. (2008).

Reprinted by permission of Macmillan Publishers Ltd. Seehausen, O., Y. Terai, et al. 2008. "Speciation through sensory drive in cichlid fish." *Nature* 455: 620–626. Copyright © 2008.

frequency between species was greater at sites with clearer water (Figure 16.17a,b). At islands with the murkiest water, the males of each species did not look as different, and a single *LWS* opsin allele class dominated in both species. Furthermore, females from the island with clearer water showed much stronger color preferences, indicating that assortative mating is stronger at clearer water sites (Figure 16.17c). This pattern represents a striking interaction between natural selection and sexual selection that has resulted in speciation. Natural selection favors divergence in color vision between cichlids in different underwater light environments, and this divergence influences females' choice of mates, which in turn results in dramatically different male coloration between the two species.

16.4 Hybridization and Gene Flow between Species

Even when two populations are considered separate species under one or more species concepts, there may be gene flow between them. When speciation occurs between geographically isolated populations, interbreeding between the two sister species can follow due to migration or geographic changes. This is referred to as **secondary contact.** For species that have always overlapped geographically, interbreeding and gene flow are always possible. In either case, the possible outcomes of hybridization between two species are diverse and consequential.

Hybridization occurs when recently diverged populations interbreed.

Hybridization between recently diverged species is common in plants. For example, over 700 of the plant species that have been introduced to the British Isles in the recent past have hybridized with native species at least occasionally, and about half of these native/nonnative matings produce fertile offspring (Abbott 1992). In at least some cases, the fate of these hybrid offspring determines the course of speciation. Will the hybrids thrive, interbreed with each of the parental populations, and eventually erase the divergence between them? Or will hybrids have new characteristics and create a distinct population of their own? What happens if hybrid offspring have reduced fitness relative to the parental populations?

Possible Outcomes of Contact between Closely Related Species

When two species overlap in geography, the potential for interbreeding and gene flow always exists. What keeps two sympatric species separate? Two general mechanisms are involved: (1) **prezygotic isolation,** in which hybrid offspring (zygotes) are never formed, usually because mating between species does not take place; and (2) **postzygotic isolation,** in which hybrid offspring suffer from inviability, sterility, or reduced fitness. When species do interbreed, possible evolutionary outcomes include reinforcement of prezygotic isolation, hybrid speciation, or stable hybrid zones, all of which are described shortly. A conceptual diagram of these outcomes is shown in **Figure 16.18** (next page).

Reinforcement

The geneticist Theodosius Dobzhansky (1937) reasoned that if populations have diverged during a period when the groups lived in different geographic areas, then any hybrid offspring that are produced should have markedly reduced fitness relative to individuals in each of the parental populations. The logic here is that if natural selection produced adaptations to distinct habitats, if sexual selection produced changes in the mating system, or if genetic drift led to the fixation of alleles

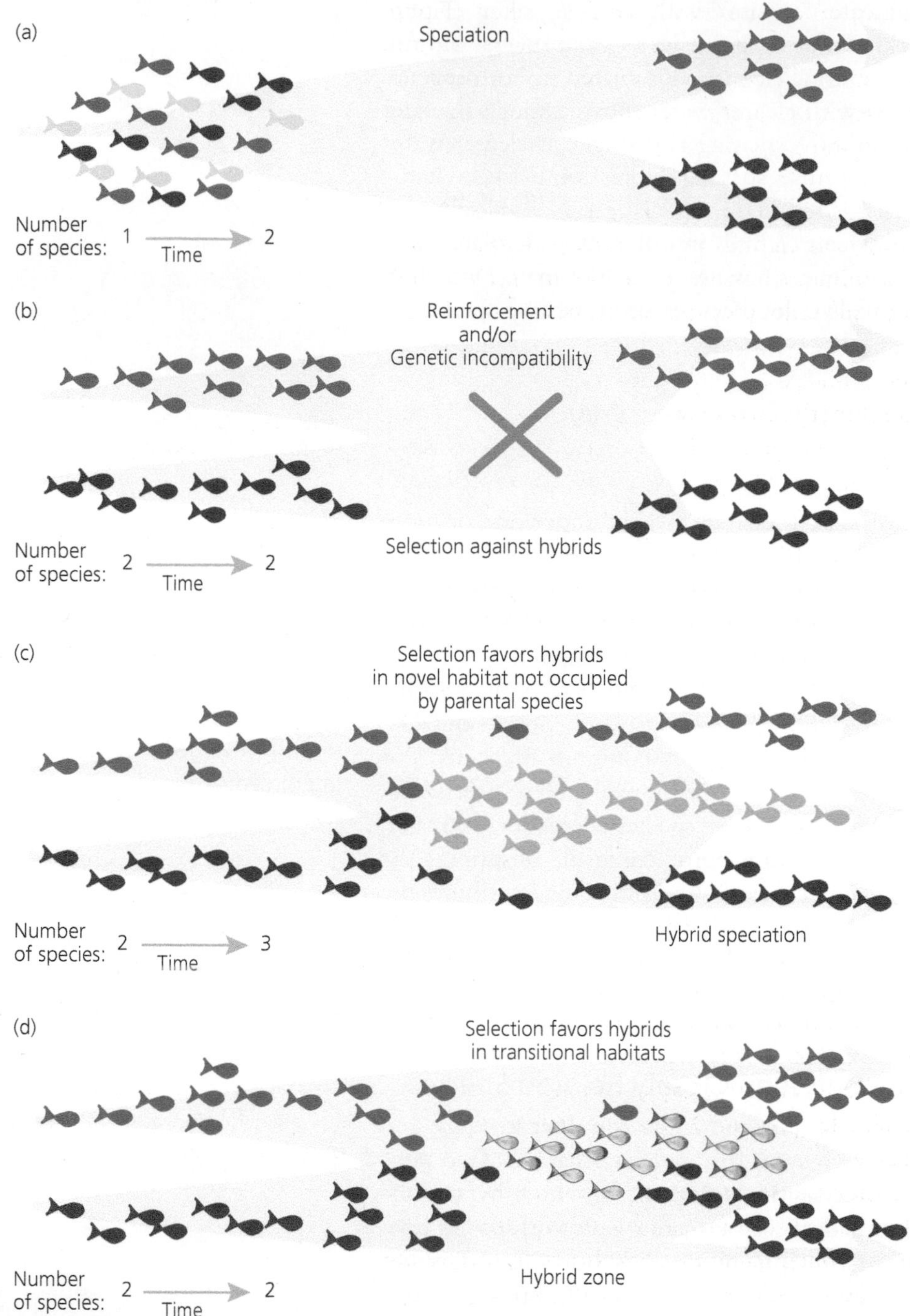

Figure 16.18 Possible outcomes of hybridization between species The process of speciation (top; compare to Figure 2.14 on page 47) can be followed by at least three different outcomes when hybridization occurs between species.

that do not work well together when heterozygous, then hybrid offspring would have low fitness. As a result, there should be strong natural selection in favor of assortative mating—meaning that selection should favor individuals that choose mates only from the same population. Selection that reduces the frequency of hybrids in this way is called **reinforcement** (Figure 16.18).

Reinforcement occurs when hybrid offspring have low fitness, and natural selection leads to assortative mating and the prezygotic isolation of populations.

The reinforcement hypothesis predicts that when closely related species come into contact and hybridize, a mechanism that reduces the occurrence of hybridization will evolve. For example, selection might favor mutations that alter aspects of mate choice or life history (such as the timing of breeding). Divergence in these traits prevents fertilization from occurring and results in prezygotic isolation of the two species. Hybridization can be prevented even when mating between two different species occurs. Here we must keep in mind the distinction between *premating* isolation and *prezygotic* isolation. Even when mating between species occurs, it does not necessarily result in fertilization and the formation of a zygote. Any trait that prevents fertilization can contribute to reinforcement.

Reinforcement of Postmating Gametic Isolation in Drosophila

An intriguing example of reinforcement comes from two *Drosophila* species inhabiting the islands of São Tomé and Príncipe, off the coast of Africa. *Drosophila yakuba* and *Drosophila santomea* occur in both sympatry and allopatry on the islands **(Figure 16.19)**. At sympatric locations, both species are present and come into regular contact with each other, while at allopatric sites only one of the two species is present. In the sympatric region, the two species hybridize regularly and form what is known as a hybrid zone. However, hybridization carries a reproductive cost because all hybrid males are sterile. Daniel Matute (2010) investigated the hypothesis that reinforcement has occurred between the two species in the hybrid zone where the two species coexist. Because reinforcement is the result of selection against hybrids in regions of sympatry, Matute's hypothesis predicts that populations of *D. yakuba* and *D. santomea* from sympatric areas will show greater prezygotic isolation than those that occur in allopatry.

To test his hypothesis, Matute conducted laboratory matings between the two species. He mated *D. santomea* females to *D. yakuba* males, and vice versa, and then counted the number of eggs that resulted from each mating. Most important, each female was known to come from a sympatric or an allopatric population. If postmating, prezygotic isolation has evolved via reinforcement, the result would be a mechanism in sympatric females that avoids the production of hybrid zygotes. For *D. yakuba* females, this is exactly what he found. Females of *D. yakuba* from the sympatric area laid fewer eggs after mating with a *D. santomea* male than did females from allopatric areas **(Figure 16.20a)**.

Matute also ran an experiment in which *D. yakuba* females mated with a male of each species before laying eggs: First a female mated with a male of the other species, then several days later with a male of her own species. The result was that females from allopatric populations produced about equal numbers of offspring sired by each male, but sympatric females produced offspring that were mostly fathered by the male of the same species (Figure 16.20b). *D. yakuba* females show evidence of reinforcement because they preferentially produce zygotes sired by males of the same species, even when mated first with a male of the other species (Matute 2010). This phenomenon, in which gametes from the same species are preferentially selected to produce zygotes, is known as **gametic isolation.**

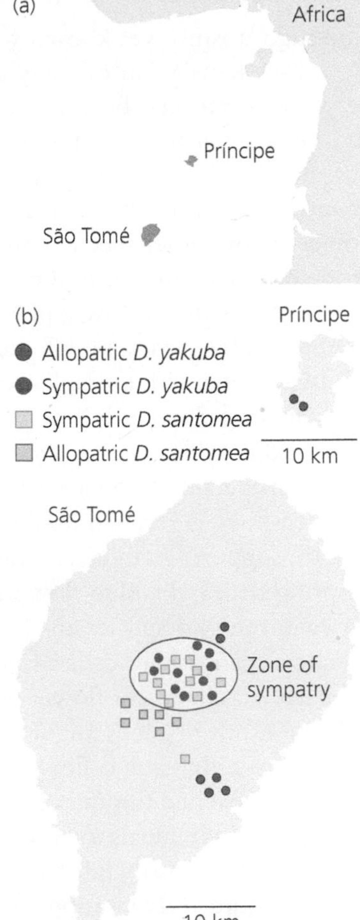

Figure 16.19 Both sympatric and allopatric populations of *Drosophila* occur on the island of São Tomé (a) Islands of São Tomé and Príncipe, near Africa. From www.planiglobe.com. (b) Sample locations. Redrawn from Matute (2010).

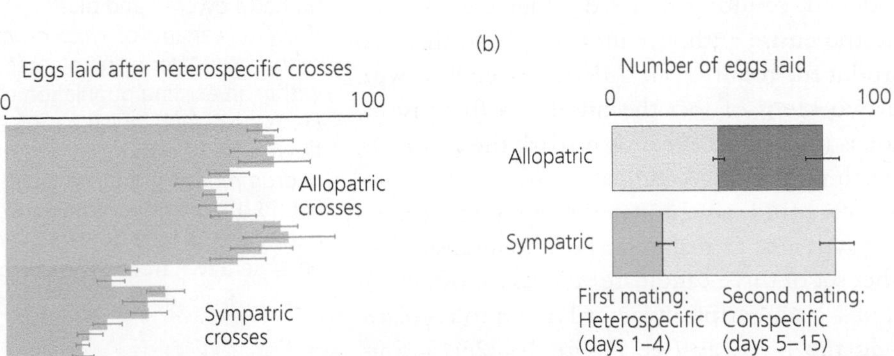

Figure 16.20 Reinforcement via gametic isolation (a) After mating with a heterospecific, females from sympatric regions laid fewer eggs. (b) After mating first with a heterospecific, followed by a conspecific, females from sympatric regions produced more eggs from the mating with the correct species. Redrawn from Matute (2010).

The Molecular Basis of Reinforcement of Premating Isolation in Phlox

Although it is not yet known what genes are responsible for gametic isolation in *D. yakuba* females, in other systems the reinforcement of reproductive isolation between species has been traced to specific loci. For example, *Phlox drummondii* and *Phlox cuspidata* are wildflower species with similar light blue flowers throughout most of their native ranges in Texas. Both are pollinated by the same array of moths and butterflies, meaning that hybridization due to pollen arriving at flowers of the other species could be common in regions of sympatry. However, in regions of sympatry, *P. drummondii* produces dark red flowers **(Figure 16.21a)**.

Could this novel flower phenotype have evolved to avoid hybridization with *P. cuspidata* by attracting different pollinator species? Donald Levin first explored this hypothesis with a simple experiment: He collected both light-blue-flowered and dark-red-flowered *P. drummondii* plants and placed them within a natural population of *P. cuspidata.* By measuring the species identity of the offspring that his experimental *P. drummondii* plants produced, he was able to assess whether dark red flowers resulted in fewer hybrid offspring. The results, shown in Figure 16.21b, support Levin's hypothesis: Red-flowered plants had many fewer hybrid offspring than light-blue-flowered plants (Levin 1985). Because hybrid offspring between the two species are known to be frequently sterile, it makes sense that selection against hybrids has favored the reinforcement of premating isolation in sympatry via divergent flower colors between the two species.

What kinds of genes might be responsible for the evolution of dark red flower color from a light-blue-flowered ancestor? Robin Hopkins and Mark Rausher (2011) approached this question by crossing the two species and then crossing the resulting F_1 individuals to create a population of F_2 hybrids. In the F_2 generation, they found four distinct flower colors, rather than a wide range of shades from red to blue **(Figure 16.22a, facing page)**. The occurrence of only four different colors, and the fact that the colors occurred in a 9:3:3:1 ratio, suggested that the flower-color phenotype was controlled by only two genes, each of which had dominant and recessive alleles. The researchers then measured the quantities of anthocyanin pigments, which affect flower color, and found that while light blue flowers contained the pigments peonidin, cyanidin, and maldivin, dark red flowers produced no maldivin and increased amounts of peonidin and cyanidin (Figure 16.22b).

What mutations underlie these differences in flower pigment content? Fortunately, the anthocyanin biosynthesis pathway is well understood (Figure 16.22c), and there is a clear split point between two branches of the pathway, one of which results in maldivin while the other results in peonidin and cyanidin. Hopkins and Rausher hypothesized that two different mutations were responsible for the switch from light blue to dark red flowers. One would adjust the biosynthesis pathway so that no maldivin is produced, resulting in a red rather than a blue flower. The other would upregulate the entire anthocyanin pathway so that increased amounts of pigment are produced, resulting in a darker overall flower color. The researchers called these two kinds of loci the hue locus (because it affects the color) and the intensity locus (because it affects how dark the color is).

Hopkins and Rausher considered three possible candidate genes for the hue locus (see Figure 16.22c), each of which could have mutated to a state that prevented the pathway from completing synthesis of maldivin. For the intensity locus, the researchers considered another set of three candidate genes, two of which were core enzymes of the pathway, plus a transcription factor known to regulate the expression of several enzymes in the pathway (see Figure 16.22c). Using

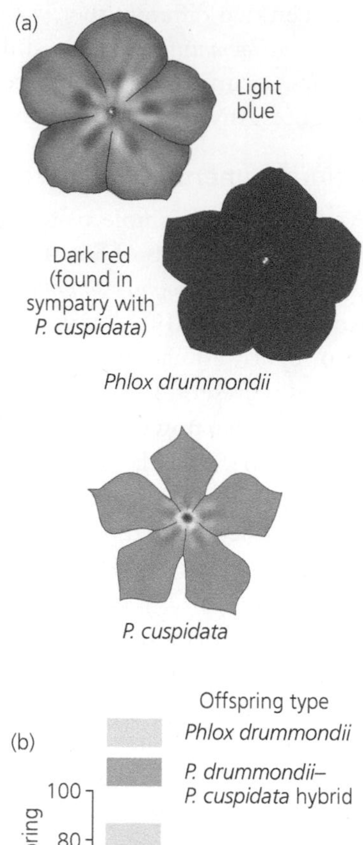

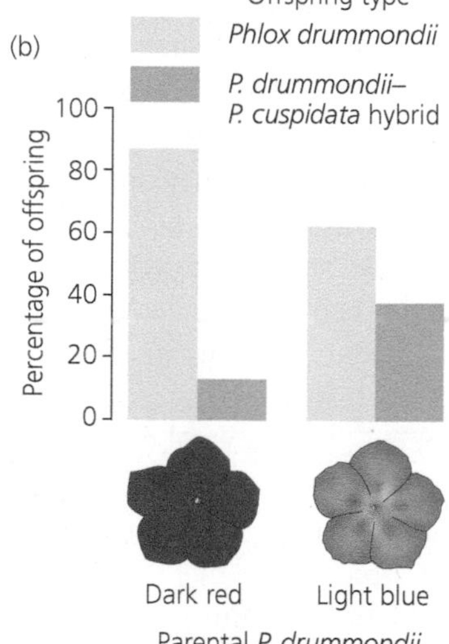

Figure 16.21 Flower color and hybridization frequency (a) Red-flowered and blue-flowered variants of *Phlox drummondii.* (b) After being planted within an existing population of *Phlox cuspidata*, which has blue flowers, red-flowered variants of *P. drummondii* produced fewer hybrid offspring sired by *P. cuspidata* than did blue-flowered variants. Redrawn from Levin (1985).

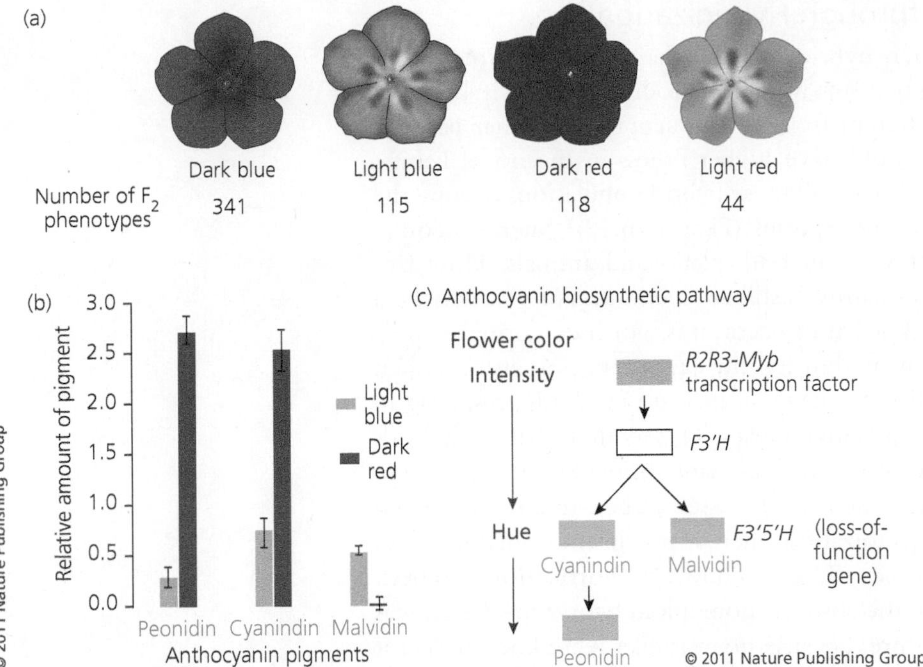

Figure 16.22 Flower color, pigment concentration, and candidate genes (a) Among F_2 hybrids, only four flower colors were present, suggesting control by two genes. (b) Dark red flowers differ from light blue flowers because they produce no maldivin and increased amounts of peonidin and cyanidin. (c) The authors investigated candidate genes in the anthocyanin pathway that could affect color intensity by upregulating the entire pathway, and candidate genes that could affect color hue by altering the balance between malvidin and cyanidin/peonidin. Redrawn from Hopkins and Rausher (2011).

(b and c) Reprinted by permission of Macmillan Publishers Ltd. Hopkins, R., and M. D. Rausher. 2011. "Identification of two genes causing reinforcement in the Texas wildflower *Phlox drummondi*." *Nature* 469: 411–414. Copyright © 2011.

molecular markers in each of the candidate genes, the researchers documented an association between floral hue and the *F3′5′H* locus, which they verified by measuring *F3′5′H* expression in blue and red flowers from both the F_2 generation and from natural populations **(Figure 16.23a)**. The transcription factor (called *Myb*) was associated with color intensity, which the researchers also verified by measuring expression in both F_2 and natural plants (Figure 16.23b).

The *Drosophila* and *Phlox* examples just described illustrate two important points about the process of reinforcement. First, it can occur at any point that prevents the formation of hybrid zygotes between two species. In the *Phlox* example, flower-color changes increase premating isolation, while in the *Drosophila* example, isolation occurs after mating, at the level of gametes. Second, the molecular basis of traits that contribute to reinforcement can be quite simple: Only two mutations are required to change from a light blue to a dark red flower, which effectively decreases the formation of hybrid offspring by *P. drummondii*.

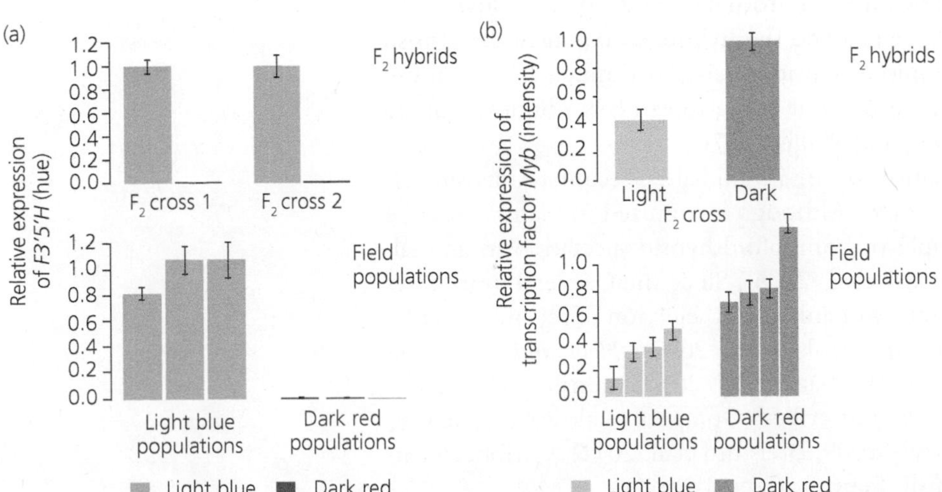

Figure 16.23 Gene expression correlates with flower hue and intensity (a) In both F_2 hybrids and their parental populations, expression of *F3′5′H* was observed in blue flowers only. (b) In addition, expression of the transcription factor *Myb* was higher in F_2 flowers with darker color (whether blue or red) and in the original dark-red-flowered populations. Redrawn from Hopkins and Rausher (2011).

Reprinted by permission of Macmillan Publishers Ltd. Hopkins, R., and M. D. Rausher. 2011. "Identification of two genes causing reinforcement in the Texas wildflower *Phlox drummondi*." *Nature* 469: 411–414. Copyright © 2011.

Formation of New Species through Hybridization

Reinforcement should occur when hybrid offspring have reduced fitness. But what happens to hybrid offspring that survive and reproduce well? If hybrid offspring occupy habitats that are different from those occupied by either parental population, the hybrid offspring could have higher fitness in the novel habitat than either of the parental species. If so, will these hybrid populations occupy the new environment and become distinct species (Figure 16.18)? Such outcomes have been documented in recent years in both plants and animals. How frequently hybrid speciation occurs in nature is still an open question—it may be a very rare event—but because it is known to occur, it is worth exploring.

First we must draw a distinction within hybrid speciation. Sometimes when species hybridize, the resulting offspring have an increase in ploidy, resulting in genetic incompatibilities with both parental species. This requires that the hybrids mate either with only themselves or with other similar hybrids. This is an example of **polyploid hybrid speciation,** which we discussed earlier. In contrast, when ploidy stays the same following hybridization, there is no chromosomal isolation between hybrids and parental species. This is known as **homoploid hybrid speciation.** Although definable in the abstract, homoploid hybrid speciation was once thought not to happen in nature, because no examples were known. Yet it is inherently difficult to detect, because a convincing case can be made only with genomic data. Today, genomic tools have let researchers begin investigating possible cases of homoploid hybrid speciation in both plants and animals.

One of the first compelling examples came from sunflowers. Loren Rieseberg and colleagues (1996) showed that experimentally produced hybrids of *Helianthus anuus* and *H. petiolaris* had a genetic composition that matched closely with a third natural species, *Helianthus anomalus*. This suggested that *H. anomalus* originated as a hybrid of the other two species. Further work by Rieseberg (2003) showed that two additional sunflowers, *Helianthus deserticola* and *Helianthus paradoxus,* are also hybrid species descended from the same parental species. Another example of homoploid hybrid speciation in plants comes from Oxford ragwort (*Senecio squalidus*) in the British Isles, which was introduced from Sicily in the 18th century. The parental species *S. aethnensis* and *S. chrysanthemifolius* have a natural hybrid zone in Sicily. Hybrids from this zone were introduced to the UK, resulting in a hybrid species that remained distinct from its parental species due to geographic isolation (James and Abbott 2005). Interestingly, the hybrid species *S. squalidus* shows dramatic differences in gene expression from its parental species, suggesting that changes in gene regulation have enabled the hybrids to colonize the British Isles (Hegarty et al. 2009). Homoploid hybrid speciation is much less common among plants than polyploid speciation, but some researchers count about 20 convincing cases to date (Rieseberg and Willis 2007).

Researchers have experimentally re-created a speciation event that occurred naturally via hybridization.

Does homoploid hybrid speciation occur in animals? Recent results suggest it is more common than once assumed. Although recognized in plants since the 1990s, the first compelling example of homoploid hybrid speciation in animals was not reported until 2005 (Schwarz et al. 2005). Since then, several more cases have been discovered in a diverse array of animals. A selection of recent examples includes tephritid fruit flies (Schwarz et al. 2005, 2007), *Heliconius* butterflies (Mavárez et al. 2006; Melo et al. 2009; Salazar et al. 2010; Merrill et al. 2011—but see also Brower 2010), *Lycaeides* butterflies (Gompert et al. 2006), yellow-rumped warblers (Brelsford and Irwin 2009; Brelsford et al. 2011), Caribbean bats (Larsen et al. 2010), and sculpin fish (Stemshorn et al. 2011).

Hybrid Origin of Audubon's Warbler

Audubon's warbler is one of the more intriguing cases of homoploid hybrid speciation because it is one of the few in tetrapods and the only known example in birds. There are four species within the yellow-rumped warbler complex, which have at various times been recognized either as separate species or as merely distinct subspecies. Audubon's warbler, *Dendroica auduboni,* occurs in temperate latitudes of western North America, while closely related species have other geographic ranges: *D. coronata* lives to the north in higher latitudes in Canada and Alaska, *D. nigrifrons* to the south in Mexico, and *D. goldmani* the farthest south, in Central America (see Figure 16.24). Based on previous research, Alan Brelsford and colleagues suspected that *D. auduboni* was a hybrid species. To test this hypothesis, they sampled warblers across the ranges of all four species, collecting both DNA and phenotypic measurements (Brelsford et al. 2011).

From the DNA samples, the researchers assessed variation in mitochondrial sequences and at markers in the nuclear genome. They then conducted a computational analysis that inferred the number of genetic clusters present in the data, analyzing the nuclear and mitochondrial data separately. If each species is distinct and none has arisen recently via hybridization, four genetic clusters would be expected. However, fewer genetic clusters, with overlap among species, could indicate a hybrid species origin.

The results, shown in **Figure 16.24**, support the hybrid origin hypothesis. *D. coronata, D. nigrifrons,* and *D. goldmani* form distinct genetic clusters for both nuclear and mitochondrial data, shown by the red, blue and purple bars. However, *D. auduboni* does not form a distinct genetic cluster. *D. audoboni* is, instead, composed of a combination of *D. coronata* and *D. nigrifrons* haplotypes. Furthermore,

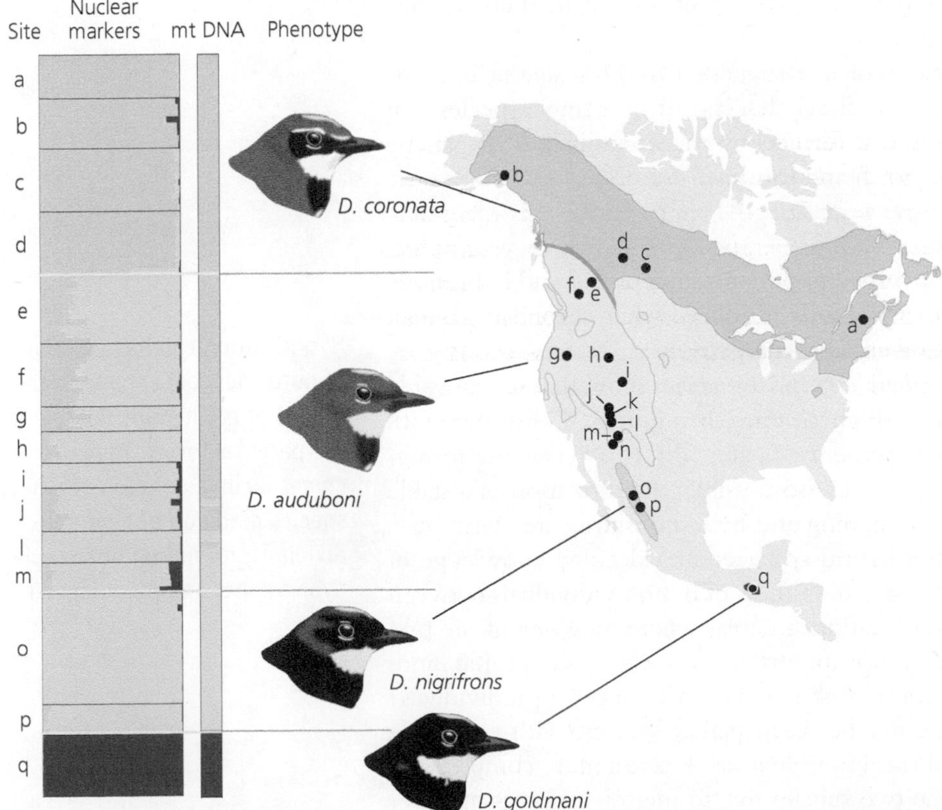

Figure 16.24 Geographic range and genetic clustering of *Dendroica* warblers Although *D. coronata, D. nigrifrons,* and *D. goldmani* each form entirely separate genetic clusters, *D. auduboni* does not form its own genetic cluster. It is instead identified as having nuclear DNA most closely related to *D. nigrifrons* and mitochondrial DNA most closely related to *D. coronata*. Redrawn from Brelsford et al. (2011).

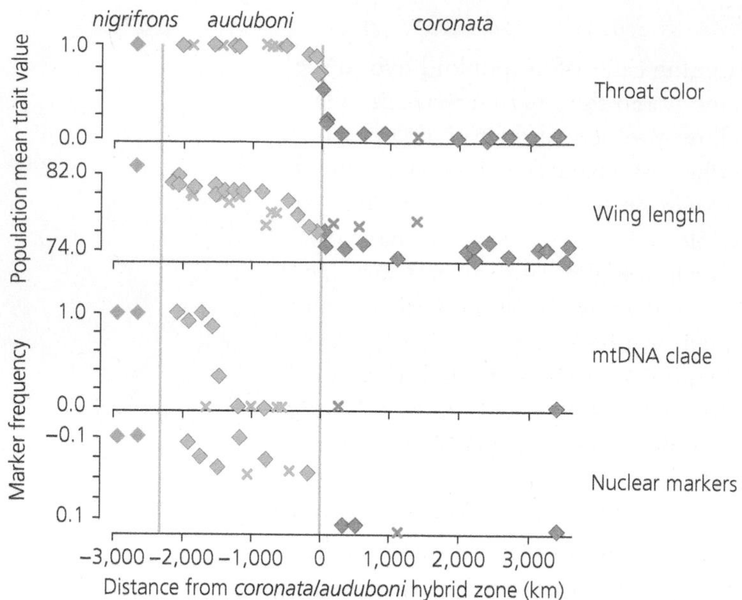

Figure 16.25 Geographic breakpoints within *Dendroica auduboni* differ among traits Some traits, such as wing length and nuclear DNA markers, change gradually across the range of *D. auduboni*, while others change abruptly but in different locations. X's represent coastal sampling sites; diamonds represent interior sites. Redrawn from Brelsford et al. (2011).

the geographic breakpoints within *D. auduboni* in the molecular and phenotypic data differ across traits and DNA regions. This is apparent in **Figure 16.25**, which shows the frequency, from south to north, of particular phenotypic and molecular characters. Some characters, such as throat color and mitochondrial DNA, have abrupt geographic breakpoints, although not at the same location. Throat color shows a geographic break at the junction of *D. auduboni* and *D. coronata,* while the mitochondrial DNA switches abruptly at a cryptic location within the *D. auduboni* range that is not associated with any differences in phenotype. Other characters, such as wing length and the frequencies of nuclear markers, change gradually across the range of *D. auduboni.*

The molecular and phenotypic evidence suggests that *D. auduboni* is a homoploid hybrid. Rather than being a direct descendant of a single species that became reproductively isolated, it was formed by the combination of genetic material from its neighbors to the north and south.

Hybrid Zones

A hybrid zone is a region where diverged populations interbreed and hybrid offspring are common. Hybrid zones are usually produced when secondary contact occurs between species that have diverged in allopatry.

Hybridization can have a variety of outcomes, depending on the fitness of the hybrids relative to the parental forms. The outcomes include creation of a new species made up of hybrid individuals, formation of a stable hybrid zone, and reinforcement.

Data from *Drosophila, Phlox, Helianthus,* and *Dendroica* show that it is possible for hybrid offspring to have lower or higher fitness than purebred offspring, with very different consequences: reinforcement of parental forms versus the formation of a new species. A third outcome is also possible: the formation of a stable hybrid zone where hybridization is ongoing and hybrid offspring are common.

To understand what maintains a hybrid zone requires detailed knowledge of hybrid fitness. Biologists often compare the fitness of hybrid individuals between locations within a hybrid zone and locations in areas where only one of the two parental species is present. However, it is important to consider the specific kinds of hybrids being tested. For example, first-generation hybrids (F_1 individuals) are heterozygous at all loci that differ between parent species, with one allele from each. Second-generation hybrids (F_2 individuals) have a more complex and variable mixture of alleles from the two species due to meiosis and crossing over when the F_1s produce gametes. Backcrosses (offspring of a cross between an F_1 and a parental species) have yet another mix of alleles. Because hybrid zones are inhabited by a variety of hybrid crosses, not just F_1 individuals, measuring fitness across hybrid types is the best way to study the dynamics of hybrid zones.

Diane Campbell, Carrie Wu, and colleagues took such an approach in studying a pair of mountain wildflowers in Colorado, *Ipomopsis aggregata* and *I. tenuituba*. These are sister species that show little genetic divergence (Wu and Campbell 2005). They coexist in many locations in western North America, but form stable hybrid zones only at some sites, particularly in Colorado (Aldridge 2005). Campbell and colleagues investigated hybrid fitness in nature by generating F_1, F_2, and backcross hybrids and planting them, along with the parental species, at two sites: one within a natural hybrid zone and another where only *I. aggregata* occur. They then measured lifetime fitness, which they summarized by calculating λ, the finite rate of population increase (Campbell et al. 2008).

The results of this experiment demonstrate the environmental dependence of hybrid fitness. At the site where only *I. aggregata* occurs naturally, all hybrid forms except for F_2s had lower fitness than pure *I. aggregata* individuals **(Figure 16.26, top)**. But at the site within the hybrid zone, no hybrid form had lower fitness, and many had higher fitness, than both parental species (Figure 16.26, bottom). Whether or not hybrid forms can persist depends entirely on their location. In appropriate transitional habitats, hybrid individuals come to dominate because of superior fitness, and the result is a stable hybrid zone. One surprising result of this study concerns the fitness of the F_2 individuals outside the hybrid zone. Many biologists had previously assumed that while F_1 hybrids often fared well in contact zones between two species, F_2 individuals would suffer reduced fitness due to genetic incompatibilities between the parental genomes that would be exposed following meiosis and gamete production in F_1 parents. However, the outcome for F_2 hybrids in this study runs directly counter to this assumption.

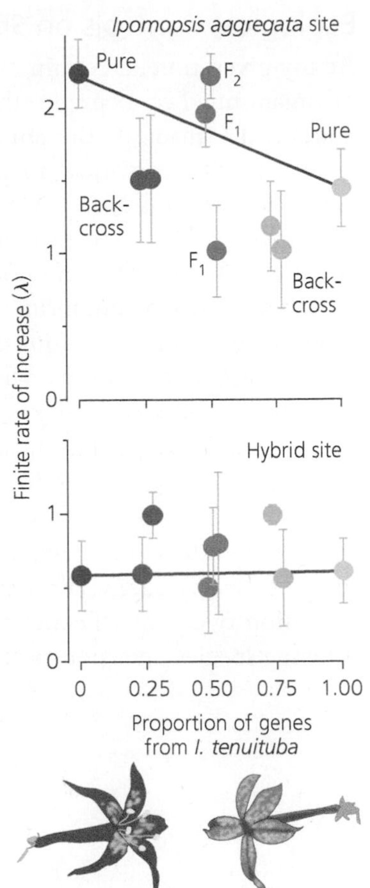

Figure 16.26 Population growth rates of *Ipomopsis* parental species and hybrids in two different locations At the site where only *I. aggregata* is found, pure *I. aggregata* individuals have the highest growth rates, and an increase in the proportion of genes from *I. tenuituba* results in a decrease in growth rate (top panel). However, at the site where hybrids are common, the two parental species have similar growth rates, and hybrids fare either the same or better. Redrawn from Campbell et al. (2008).

16.5 What Drives Diversification?

At this point it should be clear that the key ingredient required for the formation of new species is reproductive isolation between populations. Following or in concert with the requisite isolation, environmental factors commonly facilitate phenotypic divergence, at least in the cases where biologists understand the mechanism of divergence. These findings suggest that environmental heterogeneity could be a major factor in the diversification of life. However, it is unreasonable to draw such a conclusion from just a few examples, in part because diversification has not occurred at a constant rate. Rather, the rate of speciation has changed over time in particular groups of organisms, remaining relatively low for long time spans interspersed with periods of accelerated diversification (Hedges and Kumar 2009). These large-scale evolutionary patterns require an explanation.

What causes periods of rapid diversification? Do environmental factors influence the overall rate of speciation? These hard questions are at the forefront of current studies of diversification. Put simply, biologists want to know whether evolution over long spans of time is any different than evolution on the shorter timescales that we can observe directly. Researchers have begun to address these issues by combining genomic data with phylogenetic tools. While the field is far from consensus, a few recent examples highlight the power of such analyses.

Ecological Controls on Species Richness in Caribbean Lizards

At any given time, a certain number of species exist within any particular group of organisms. For example, there are currently 914 bird species in the United States and Canada. If the environment were to remain unchanged and we returned in 100 years, how many species would we find then? Would new ones have formed? Or is North America already full of bird species, with no remaining ecological niches to be exploited? These questions all stem from a single issue: We do not know whether enough time has passed for species richness in any given habitat to reach equilibrium. When a new habitat is first exploited by a certain group of organisms, an **adaptive radiation** often results, in which many new species form rapidly. Eventually the rate of diversification slows as the habitat fills with species, but how quickly does this occur? When is a radiation complete?

The primary challenge in answering these questions is that we have only a snapshot of current species diversity, without knowing when each species formed. Recently, researchers have begun to overcome this problem by using computational methods to infer when species were formed, based on the structure of the species' phylogeny. This knowledge can then be used to estimate the rate of speciation over time. Daniel Rabosky and Richard Glor took this approach to studying species diversification in *Anolis* lizards on islands of the Caribbean.

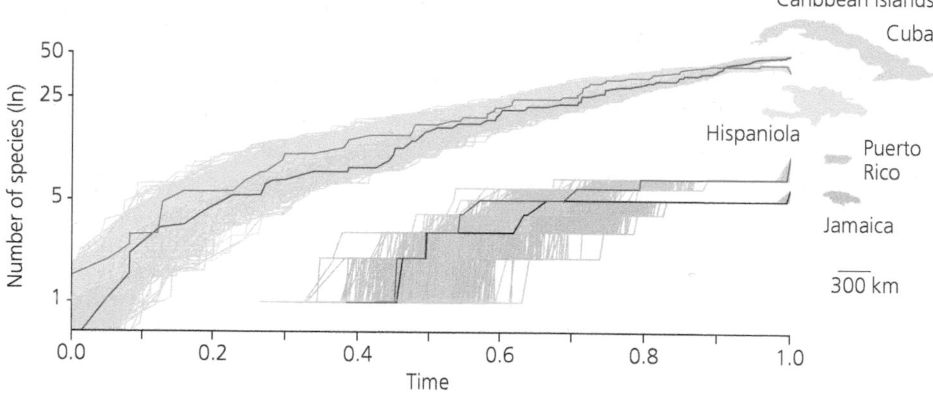

Figure 16.27 Diversification of *Anolis* species over time on four Caribbean islands For all islands, the rate at which new species are added (the slope of the curve) declines over time. Smaller islands have reached a plateau, while larger islands are approaching one. The formation of new species does not occur at a constant rate and must depend on the ecology of each island, particularly its size. From Rabosky and Glor (2010).

For each of four islands, the researchers used the *Anolis* phylogeny to construct lineage accumulation curves showing how the number of species increased over time for each island (Rabosky and Glor 2010). These curves appear in **Figure 16.27**. For the smaller islands of Jamaica and Puerto Rico, the curves have reached a plateau at a total number of species smaller than for the larger islands of Cuba and Hispaniola. Even for the larger islands, the overall shape of the curve suggests that a plateau has almost been reached. In the case of Cuban anoles, new species do not form at a constant rate but instead form less often as time progresses, suggesting that each island reaches a carrying capacity of species. The ecology of each island, especially its size, appears to control diversification.

Range Size and Diversification in Toads

The tendency of a lineage of organisms to diversify depends on several factors, including its dispersal ability and range size. Species with small range sizes and low dispersal are less likely to be exposed to new environments and selective pressures than those with larger ranges, but high dispersal can also act as a barrier to speciation due to continuing gene flow among populations. Ines Van Bocxlaer and collaborators investigated the importance of range size in the worldwide diversification of toads. They hypothesized that toad lineages with phenotypes that allowed them to expand their ranges were more likely to speciate because of introduction to novel environments (Van Bocxlaer et al. 2010).

To test their hypothesis, Van Bocxlaer and colleagues compared the phenotypes of extant toad species with large versus small ranges and defined a set of phenotypes shared only by toads with large ranges. This set included large body size, glands that enhance toxicity and rehydration ability, extra fat storage ability, the ability to lay eggs in diverse habitats, large clutch size, and larvae that consume food directly from the environment. Based on these phenotypes, the researchers created a single numerical value summarizing the range-expansion ability of each species. This value, called *p*, runs from zero, denoting no range-expansion characters to one, denoting the optimal range-expansion phenotype. The researchers then inferred the phylogeny of all toad species to see whether their range expansion values showed any patterns on the tree. An important element of analysis involved ancestral reconstruction, in which the researchers inferred the phenotypes of ancestral toad lineages based on the phenotypes of their descendants.

The results appear in **Figure 16.28**. Before dispersing beyond South America, toads lacked range-expanding phenotypes. However, before toads colonized new continents, the lineages that underwent rapid diversification acquired phenotypes associated with larger range sizes (compare nodes a, b, and c, as well as nodes d, e, and f). Furthermore, many lineages reverted to phenotypes less associated with range expansion near the end of the period of global colonization (nodes x, y, and z). This correlation between phenotype and diversification suggests that periods of

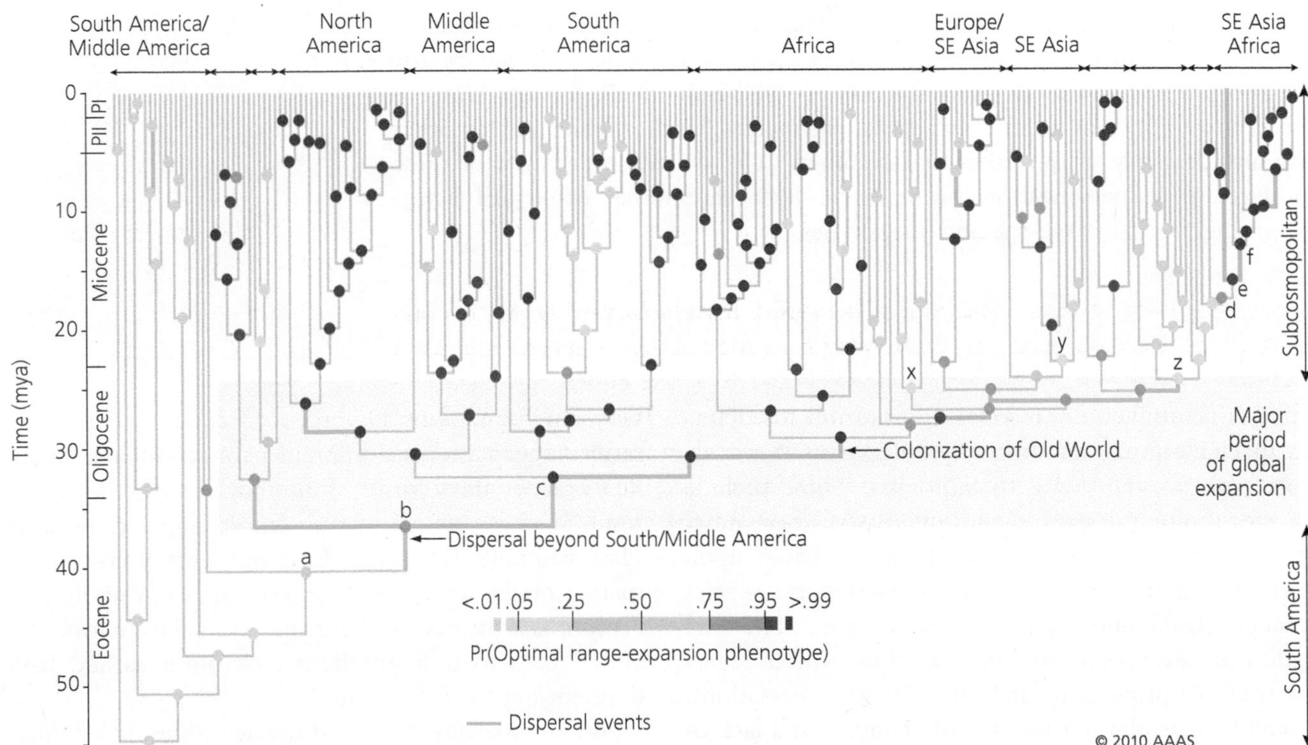

Figure 16.28 Diversification of toads over time correlates with their phenotype Toads with traits associated with large range sizes have higher values of *p*. At points on the phylogeny where rapid diversification occurred (gray shading), the toads had higher values of *p* than at other times. This correlation suggests that the rate of formation of new toad species has been influenced by the phenotypes toads possessed at that time. Redrawn from Van Bocxlaer et al. (2010). Van Bocxlaer, I., S. P. Loader, et al. 2010. "Gradual adaptation toward a range-expansion phenotype initiated the global radiation of toads." *Science* 327: 679–682. Reprinted with permission of the AAAS.

rapid speciation may not be driven primarily by environmental change. Instead, toad diversification may have increased 35 million years ago simply because certain toad lineages happened to become more likely to expand their ranges and colonize novel habitats, which eventually led to numerous speciation events.

Ecological versus Endogenous Controls of Diversification

Our lizard example shows that environmental parameters, such as island size, may determine speciation rates and how they change over time. This is an example of an external ecological control on diversification. Although diversification on each island was triggered by the arrival of the first anoles, the rate of diversification and the total number of species formed was determined by the habitat. The toad example, in contrast, shows how a property of an organism, rather than of its environment, can determine speciation rates. As toads attained phenotypic characters that enabled them to expand their ranges, a burst of diversification resulted. It was not the environment that changed and favored the formation of new toad species. It was the toads themselves that evolved phenotypes that allowed them to expand and subsequently diversify.

Rates of diversification may be driven by both ecological factors and by intrinsic properties of organisms.

Which of these two types of causes has been more common in driving diversification during the history of life on Earth? It is too soon to know, but recent research had demonstrated that both drivers of diversity are important, depending on the study system. In addition to the anole example, recent work by Paul Martin and colleagues (2010) shows that climatic changes during the past 3 million years have caused bird species at higher latitudes to overlap in range more frequently than in tropical latitudes, resulting in greater diversification in color pattern in breeding plumage. On the other end of the spectrum, the diversification of mormyrid electric fishes in Africa appears to have been caused by the evolution of specific brain anatomy that allowed individuals greater ability to discriminate between electric signals among species (Carlson et al. 2011), rather than any property of the fishes' environment. The overall implication of these findings is that diversification can be driven by both external, ecological factors and endogenous, lineage-specific phenotypes.

SUMMARY

There is no universally recognized criterion for defining species boundaries. Instead, multiple species concepts have proven useful to biologists. These include the morphological species concept (based upon morphology), the biological species concept (based upon reproductive isolation), and the phylogenetic species concept (based upon shared derived traits). The formation of new species requires isolation, which occurs as a result of physical separation (allopatric speciation), polyploidy or other chromosomal changes, or a lack of mating due to temporal, pollinator, or other forms of isolation. However, the establishment of isolation between populations does not necessarily mean that they will become different species. For speciation to occur, there must be a mechanism that promotes divergence between populations. Divergence frequently occurs as a result of adaptation to different habitats, such as in annual and perennial forms of monkeyflower, but can also result from assortative mating within populations, as in Hawaiian *Laupala* crickets. Environmental variation and mating preferences can act in concert to promote speciation, as is the case in the cichlids of Lake Victoria that are adapted to different light environments and have different nuptial coloration.

When two closely related species come into contact with each other, hybridization and gene flow between speciation often occurs. There are several possible results of interbreeding between species. One of these is reinforcement of prezygotic isolation, in which selection favors traits that reduce the occurrence of interbreeding because hybrids have low fitness. Hybrid speciation is also possible, in which a third entirely independent species is formed from the combination of two parental species. A third outcome is the formation of hybrid zones in particular geographic regions where the ranges of two species overlap.

The formation of new species over time does not occur at a constant rate, leading researchers to investigate the possible causes of diversification across evolutionary history. In some instances, adaptive radiations result from a lineage invading a new habitat such as an island, after which the rate of diversification is determined by ecological factors. However, rapid bursts of speciation can also be caused by the evolution of key phenotypes that permit a lineage to colonize new ecological niches or diversify in mating preferences.

QUESTIONS

1. What does it mean to say that species are "evolutionarily independent" or that "species form a boundary for gene flow"?
2. Compare and contrast the morphospecies concept, the biological species concept, and the phylogenetic species concept. What criterion does each use to identify species? What are the pros and cons of each?
3. The text discusses work on speciation in marine phytoplankton and European skates. In each case, did application of the phylogenetic species concept lead to the recognition of fewer, the same, or more species? Explain.
4. Explain the difference between dispersal versus vicariance. Why might dispersal or vicariance events initiate speciation?
5. When the Panama land bridge between North and South America was uncovered, some North American mammal lineages crossed to South America and underwent dramatic radiations. For terrestrial species, did the completion of the land bridge represent a vicariance or dispersal event? Did the recent construction of the Panama Canal represent a vicariance or dispersal event for terrestrial organisms? For marine organisms?
6. What does it mean to call a pair of closely related species "allopatric species" versus "sympatric species"? How do these definitions relate to the difference between the processes of allopatric versus sympatric speciation?
7. In the chapter, we described in detail two broad categories of isolation mechanisms between species: physical isolation and chromosomal isolation. But other categories exist, such as the temporal isolation described for the case of Japanese winter moths. Can you think of other mechanisms of isolation that may exist in nature that do not fit into the broad categories of physical separation or chromosomal incompatibilities?
8. Morphologically similar species of snapping shrimp that are each others' closest relatives live on either side of the Isthmus of Panama. Why does this observation support the hypothesis that speciation occurred via vicariance?
9. Phylogenetic analyses show that in many cases, closely related Hawaiian species occur in the following pattern: The more ancestral groups occupy older islands in the archipelago, while the more derived populations occupy younger islands. Why does this observation support the hypothesis that speciation has occurred via dispersal?
10. Would glaciation in northern Europe and North America have created vicariance events over the past 150,000 years? If so, how? (It may help to find a map showing the extent of the glaciers. Also think about the changes that took place in areas not covered with ice.) Which organisms might have been affected? For example, consider the different effects glaciation might have on small mammals, migratory birds, and trees.
11. It is often difficult to disentangle the roles of environmental factors and mating preferences in creating and maintaining species boundaries. This is why the cichlid example in Section 16.3 is so unique, and also why it can be difficult to understand. Study Figure 16.17 and answer the following questions by comparing the two most extreme island habitats of Marumbi (left column) and Makobe (right column):
    a. How do the two environments differ?
    b. What color are the male fish in these two habitats, and at what depth are the different colors of male fish found?
    c. What *LWS* opsin alleles are present in the two populations, and what does it mean for a female to have a "blue" or a "red" *LWS* opsin allele?

    Finally, ask yourself why the environmental context is so important for these cichlids. What colors of male fish would be present in Lake Victoria if all habitats were similar to Marumbi (low water clarity)?
12. What are the possible outcomes when species that have long been separated geographically come back into contact and begin hybridizing, and under what conditions does each outcome occur?
13. What is reinforcement? Is it an example of genetic drift, natural selection, or sexual selection?

14. Within the past 50 years, soapberry bug populations in the United States have diversified into populations distinguished by markedly different beak lengths. These bugs eat the seeds at the center of soapberry fruits. Native and recently introduced varieties of soapberries differ greatly in fruit size. Describe the experiments or observations you would make to launch an in-depth study of speciation in these bugs. What data would tell you whether they are separate populations evolving independently, or a single interbreeding population? Many museums contain insect specimens from decades ago. What would you examine in these old specimens? What information about the host plants would be useful?

15. Red crossbills (*Loxia curvirostra* species complex) are small finches specialized for eating seeds pried out of the cones of conifer trees. They fly thousands of kilometers each year in search of productive cone crops. Despite their mobility, crossbills have diverged into several "types" that differ in bill shape, body size, and vocalizations. Each type prefers to feed on a different species of conifer, and each species of conifer is found only in certain forests. Bill size and shape affect how efficiently a bird can open cones of a certain conifer species. Explain how a highly mobile animal such as the red crossbill could have diverged into different types in the absence of any geographic barrier. If crossbills could not fly, do you think speciation would occur more quickly or more slowly? If conifer species were not patchily distributed (i.e., in different forests), do you think crossbill speciation would occur more quickly or more slowly? Compare your answers to the analyses and data presented in Benkman (2003).

EXPLORING THE LITERATURE

16. Species boundaries within Bacteria and Archaea have been particularly difficult for biologists to define. Recent work with thermoacidophilic (specializing in hot, acidic environments) Archaea in a hot spring in the Russian Far East suggests that two closely related but genetically distinct Archaea lineages coexist in the same habitat. The authors argue that this finding is consistent with the biological species hypothesis. See:

    Cadillo-Quiroz, H., X. Didelot, et al. 2012. Patterns of gene flow define species of thermophilic Archaea. *PLoS Biology* 10: e1001265.

17. In Section 16.3, we explored how assortative mating can contribute to species boundaries in Hawaiian *Laupala* crickets. In particular, the data in Figure 16.15b indicate that the traits of male song pulse rate, and female preference for male song pulse rate, are genetically linked. Further experiments by Kerry Shaw's research group have explored the specific genetic basis of this genetic linkage:

    Wiley, C., C. K. Ellison, and K. L. Shaw. 2011. Widespread genetic linkage of mating signals and preferences in the Hawaiian cricket *Laupala*. *Proceedings of the Royal Society of London B* 279: 1203–1209.

18. When hybrids between closely related species have lower fitness than either parental type, it is known as extrinsic postzygotic isolation. Reinforcement is one form of extrinsic postzygotic isolation that involves assortative mating, as in the *Phlox* example described in this chapter. But assortative mating is not required for low hybrid fitness to contribute to isolation between groups in the early stages of speciation. For an example from butterflies, see:

    McBride, C. S., and M. C. Singer. 2010. Field studies reveal strong postmating isolation between ecologically divergent butterfly populations. *PLoS Biology* 8: e1000529.

19. Research by Peter Grant and Rosemary Grant and colleagues on Darwin's finches in the Galápagos Islands has contributed considerably to our understanding of natural selection and evolution. A recent paper by Grant and Grant tells a new story about what happens when species that diverged in allopatry (geographically separated) then come together in the same habitat:

    Grant, P. R., and B. R. Grant. 2009. The secondary contact phase of allopatric speciation in Darwin's finches. *Proceedings of the National Academy of Sciences, USA* 106: 20141–20148.

20. One mechanism of isolation that may lead to speciation is polyploidization, but the benefits to the individual of polyploidy are often unclear. Elsewhere in the book we discuss one inquiry (Figure 5.33, page 169). For another, see:

    Vamosi, J. C., and J. R. McEwen. 2012. Origin, elevation, and evolutionary success of hybrids and polyploids in British Columbia, Canada. *Botany* 91: 182–188.

21. Traits that influence mating behavior, such as those based upon sex pheromones, are fundamental to maintaining species boundaries in many groups of animals. The specific genetic mechanisms that underlie pheromone-based behaviors are largely unknown, but an intriguing example involving a fatty-acyl reductase gene comes from the European corn borer *Ostrinia nubilalis*:

    Lassance, J.-M., A. T. Groot, et al. 2010. Allelic variation in a fatty-acyl reductase gene causes divergence in moth sex pheromones. *Nature* 466: 486–489.

22. How do sex chromosomes (such as the X and Y chromosomes in humans) evolve, and are they involved in speciation? An example from a recently evolved sex chromosome system (or "neo-sex

chromosome") suggests that sex-chromosome evolution and speciation can be linked. See:

Kitano, J., J. A. Ross, et al. 2009. A role for a neo-sex chromosome in stickleback speciation. *Nature* 461: 1079–1083.

23. Can simple climatic variables influence overall rates of evolution? One possible example suggests the answer is yes, in the case of limited water availability in Australian deserts. See:

    Goldie, X., L. Gillman, et al. 2010. Evolutionary speed limited by water in arid Australia. *Proceedings of the Royal Society of London B* 277: 2645–2653.

24. Most snails have right-handed shells. When mutation leads to a reversal of coiling direction, the resulting left-handed snails are not only reproductively isolated from their right-handed relatives, they are also typically at a selective disadvantage. For one thing, they have trouble finding compatible mates. For another, they often suffer other deleterious developmental problems. See:

    Utsuno, H., T. Asami, et al. 2011. Internal selection against the evolution of left-right reversal. *Evolution* 65: 2399–2411.

    Given these inherent barriers, speciation by reversal of shell-coiling direction should be rare. Develop a hypothesis about conditions that might increase the probability of speciation by reversal of shell-coiling direction. Then see:

    Hoso, M., Y. Kameda, et al. 2010. A speciation gene for left-right reversal in snails results in anti-predator adaptation. *Nature Communications* 1: 133.

    Hoso, M. 2012. Non-adaptive speciation of snails by left-right reversal is facilitated on oceanic islands. *Contributions to Zoology* 81: 79–85.

CITATIONS

Abbott, R. J. 1992. Plant invasions, interspecific hybridization and the evolution of new plant taxa. *Trends in Ecology and Evolution* 7: 401–405.

Aldridge, G. 2005. Variation in frequency of hybrids and spatial structure among *Ipomopsis* Polemoniaceae contact sites. *New Phytologist* 167: 279–288.

Amato, A., W. Kooistra, et al. 2007. Reproductive isolation among sympatric cryptic species in marine diatoms. *Protist* 158: 193–207.

Beheregaray, L. B., J. P. Gibbs, et al. 2004. Giant tortoises are not so slow: Rapid diversification and biogeographic consensus in the Galápagos. *Proceedings of the National Academy of Sciences, USA* 101: 6514–6519.

Benkman, C. W. 2003. Divergent selection drives the adaptive radiation of crossbills. *Evolution* 57: 1176–1181.

Bonacum, J., P. M. O'Grady, et al. 2005. Phylogeny and age of diversification of the *planitibia* species group of the Hawaiian *Drosophila*. *Molecular Phylogenetics and Evolution* 37: 73–82.

Brander, K. 1981. Disappearance of common skate *Raia batis* from Irish Sea. *Nature* 290: 48–49.

Brelsford, A., and D. E. Irwin. 2009. Incipient speciation despite little assortative mating: The yellow-rumped warbler hybrid zone. *Evolution* 63: 3050–3060.

Brelsford, A., B. Milá, and D. E. Irwin. 2011. Hybrid origin of Audubon's warbler. *Molecular Ecology* 20: 2380–2389.

Brower, A. V. Z. 2010. Hybrid speciation in *Heliconius* butterflies? A review and critique of the evidence. *Genetica* 139: 589–609.

Campbell, D. R., N. M. Waser, et al. 2008. Lifetime fitness in two generations of *Ipomopsis* hybrids. *Evolution* 62: 2616–2627.

Carlson, B. A., S. M. Hasan, et al. 2011. Brain evolution triggers increased diversification of electric fishes. *Science* 332: 583–586.

Censky, E. J., K. Hodge, and J. Dudley. 1998. Over-water dispersal of lizards due to hurricanes. *Nature* 395: 556.

Chakrabarty, P., M. P. Davis, and J. S. Sparks. 2012. The first record of a trans-oceanic sister-group relationship between obligate vertebrate troglobites. *PLoS One* 7: e44083.

Chapman, A. D. 2009. *Numbers of Living Species in Australia and the World*. 2nd ed. Canberra: Australian Biological Resources Study.

Christiansen, D. G., and H. Reyer. 2009. From clonal to sexual hybrids: Genetic recombination via triploids in all-hybrid populations of water frogs. *Evolution* 63: 1754–1768.

Clark, R. W., W. S. Brown, et al. 2010. Roads, interrupted dispersal, and genetic diversity in timber rattlesnakes. *Conservation Biology* 24: 1059–1069.

Cohan, F. M. 1994. Genetic exchange and evolutionary divergence in prokaryotes. *Trends in Ecology and Evolution* 9: 175–180.

Cohan, F. M. 1995. Does recombination constrain neutral divergence among bacterial taxa? *Evolution* 49:164–175.

Coyne, J. A., and H. A. Orr. 2004. *Speciation*. Sunderland, MA: Sinauer.

Cunha, C., I. Doadrio, and M. M. Coelho. 2008. Speciation towards tetraploidization after intermediate processes of non-sexual reproduction. *Philosophical Transactions of the Royal Society of London B* 363: 2921–2929.

Davison, A., S. Chiba, et al. 2005. Speciation and gene flow between snails of opposite chirality. *PLoS Biology* 3: e282

Dawson, M. N., and W. M. Hamner. 2005. Rapid evolutionary radiation of marine zooplankton in peripheral environments. *Proceedings of the National Academy of Sciences, USA* 102: 9235–9240.

Dobzhansky, T. 1937. *Genetics and the Origin of Species*. New York: Columbia University Press.

Faria, R., and A. Navarro. 2010. Chromosomal speciation revisited: Rearranging theory with pieces of evidence. *Trends in Ecology and Evolution* 25: 660–669.

Giraud, T., G. Refrégier, et al. 2008. Speciation in fungi. *Fungal Genetics and Biology* 45: 791–802.

Gompert, Z., J. A. Fordyce, et al. 2006. Homoploid hybrid speciation in an extreme habitat. *Science* 314: 1923–1925.

Griffiths, A. M., D. W. Sims, et al. 2010. Molecular markers reveal spatially segregated cryptic species in a critically endangered fish, the common skate *Dipturus batis*. *Proceedings of the Royal Society of London B* 277: 1497–1503.

Hall, M. C., and J. H. Willis. 2006. Divergent selection on flowering time contributes to local adaptation in *Mimulus guttatus* populations. *Evolution* 60: 2466–2477.

Hedges, S. B., and S. Kumar. 2009. *The Timetree of Life*. New York: Oxford University Press.

Hegarty, M. J., G. L. Barker, et al. 2009. Extreme changes to gene expression associated with homoploid hybrid speciation. *Molecular Ecology* 18: 877–889.

Hopkins, R., and M. D. Rausher. 2011. Identification of two genes causing reinforcement in the Texas wildflower *Phlox drummondii*. *Nature* 469: 411–414.

Hoskin, C. J., M. Higgie, et al. 2005. Reinforcement drives rapid allopatric speciation. *Nature* 437: 1353–1356.

Hurt, C., A. Anker, and N. Knowlton. 2009. A multilocus test of simultaneous divergence across the Isthmus of Panama using snapping shrimp in the genus *Alpheus*. *Evolution* 63: 514–530.

Iglésias, S. P., L. Toulhoat, and D. Y. Sellos. 2010. Taxonomic confusion and market mislabelling of threatened skates: Important consequences for their conservation status. *Aquatic Conservation: Marine and Freshwater Ecosystems* 20: 319–333.

James, J. K., and R. J. Abbott. 2005. Recent, allopatric, homoploid hybrid speciation: The origin of *Senecio squalidus* Asteraceae in the British Isles from a hybrid zone on Mount Etna, Sicily. *Evolution* 59: 2533–2547.

Kandul, N. P., V. A. Lukhtanov, and N. E. Pierce. 2007. Karyotypic diversity and speciation in *Agrodiaetus* butterflies. *Evolution* 61: 546–559.

Knowlton, N., and L. A. Weigt. 1998. New dates and new rates for divergence across the Isthmus of Panama. *Proceedings of the Royal Society of London B* 265: 2257–2263.

Knowlton, N., L. A. Weigt, et al. 1993. Divergence in proteins, mitochondrial DNA, and reproductive compatibility across the Isthmus of Panama. *Science* 260: 1629–1632.

Konstantinidis, K. T., and J. M. Tiedje. 2005. Genomic insights that advance the species definition for prokaryotes. *Proceedings of the National Academy of Sciences, USA* 102: 2567–2572.

Larsen, P. A., M. R. Marchán-Rivadeneira, and R. J. Baker. 2010. Natural hybridization generates mammalian lineage with species characteristics. *Proceedings of the National Academy of Sciences, USA* 107: 11447–11452.

Lawrence, J. G., and H. Ochman. 1998. Molecular archaeology of the *Escherichia coli* genome. *Proceedings of the National Academy of Sciences, USA* 95: 9413–9417.

Levin, D. A. 1985. Reproductive character displacement in *Phlox*. *Evolution* 39: 1275–1281.

Lowry, D .B., and J. H. Willis. 2010. A widespread chromosomal inversion polymorphism contributes to a major life-history transition, local adaptation, and reproductive isolation. *PLoS Biology* 8: e1000500.

Lutes, A. A., D. P. Baumann, et al. 2011. Laboratory synthesis of an independently reproducing vertebrate species. *Proceedings of the National Academy of Sciences, USA* 108: 9910 –9915.

Machordom, A., and I. Doadrio. 2001. Evolutionary history and speciation modes in the cyprinid genus *Barbus*. *Proceedings of the Royal Society of London B* 268: 1297–1306.

Malenke, J. R., K. P. Johnson, and D. H. Clayton 2009. Host specialization differentiates cryptic species of feather-feeding lice. *Evolution* 63: 1427–1438.

Martin, P. R., R. Montgomerie, and S. C. Lougheed. 2010. Rapid sympatry explains greater color pattern divergence in high latitude birds. *Evolution* 64: 336–347.

Matute, D. R. 2010. Reinforcement of gametic isolation in *Drosophila*. *PLoS Biology* 8: e1000341.

Mavárez, J., C. A. Salazar, et al. 2006. Speciation by hybridization in *Heliconius* butterflies. *Nature* 441: 868–871.

Mayr, E. 1942. *Systematics and the Origin of Species, from the Viewpoint of a Zoologist*. Cambridge, MA: Harvard University Press.

Mayr, E. 1963. *Animal Species and Evolution*. Cambridge, MA: Belknap Press of Harvard University Press.

Melo, M. C., C. Salazar, et al. 2009. Assortative mating preferences among hybrids offers a route to hybrid speciation. *Evolution* 63: 1660–1665.

Mendelson, T. C., and K. L. Shaw. 2005. Sexual behaviour: Rapid speciation in an arthropod. *Nature* 433: 375–376.

Merrill, R. M., B. Van Schooten, et al. 2011. Pervasive genetic associations between traits causing reproductive isolation in *Heliconius* butterflies. *Proceedings of the Royal Society of London B* 278: 511–518.

Mora, C., D. P. Tittensor, et al. 2011. How many species are there on Earth and in the ocean? *PLoS Biology* 9(8): e1001127.

Rabosky, D. L., and R. E. Glor. 2010. Equilibrium speciation dynamics in a model adaptive radiation of island lizards. *Proceedings of the National Academy of Sciences, USA* 107: 22178–22183.

Rieseberg, L. H. 2003. Major ecological transitions in wild sunflowers facilitated by hybridization. *Science* 301: 1211–1216.

Rieseberg, L. H., B. Sinervo, et al. 1996. Role of gene interactions in hybrid speciation: Evidence from ancient and experimental hybrids. *Science* 272: 741–745.

Rieseberg, L. H., and J. H. Willis. 2007. Plant speciation. *Science* 317: 910–914.

Riginos, C. 2005. Cryptic vicariance in Gulf of California fishes parallels vicariant patterns found in Baja California mammals and reptiles. *Evolution* 59: 2678–2690.

Salazar, C., S. W. Baxter, et al. 2010. Genetic evidence for hybrid trait speciation in *Heliconius* butterflies. *PLoS Genetics* 6: e1000930.

Schemske, D. W. 2010. Adaptation and the origin of species. *American Naturalist* 176: S4–S25.

Schluter, D. 2009. Evidence for ecological speciation and its alternative. *Science* 323: 737–741.

Schwarz, D., B. M. Matta, et al. 2005. Host shift to an invasive plant triggers rapid animal hybrid speciation. *Nature* 436: 546–549.

Schwarz, D., K. D. Shoemaker, et al. 2007. A novel preference for an invasive plant as a mechanism for animal hybrid speciation. *Evolution* 61: 245–256.

Seehausen, O., Y. Terai, et al. 2008. Speciation through sensory drive in cichlid fish. *Nature* 455: 620–626.

Shaw, K. L. 2000. Interspecific genetics of mate recognition: Inheritance of female acoustic preference in Hawaiian crickets. *Evolution* 54: 1303–1312.

Shaw, K. L., Y. M. Parsons, and S. C. Lesnick. 2007. QTL analysis of a rapidly evolving speciation phenotype in the Hawaiian cricket *Laupala*. *Molecular Ecology* 16: 2879–2892.

Sobel, J. M., G. F. Chen, et al. 2010. The biology of speciation. *Evolution* 64: 295–315.

Stemshorn, K. C., F. A. Reed, et al. 2011. Rapid formation of distinct hybrid lineages after secondary contact of two fish species *Cottus* sp. *Molecular Ecology* 20: 1475–1491.

Suatoni, E., S. Vicario, et al. 2006. An analysis of species boundaries and biogeographic patterns in a cryptic species complex: The rotifer—*Brachionus plicatilis*. *Molecular Phylogenetics and Evolution* 41: 86–98.

Terai, Y., O. Seehausen, et al. 2006. Divergent selection on opsins drives incipient speciation in Lake Victoria cichlids. *PLoS Biology* 4: e433.

Thessen, A. E., H. A. Bowers, and D. K. Stoecker. 2009. Intra- and interspecies differences in growth and toxicity of *Pseudo-nitzschia* while using different nitrogen sources. *Harmful Algae* 8: 792–810.

Ueshima, R., and T. Asami. 2003. Evolution: Single-gene speciation by left–right reversal. *Nature* 425: 679.

Van Bocxlaer, I., S. P. Loader, et al. 2010. Gradual adaptation toward a range-expansion phenotype initiated the global radiation of toads. *Science* 327: 679–682.

van der Sluijs, I., O. Seehausen, et al. 2010. No evidence for a genetic association between female mating preference and male secondary sexual trait in a Lake Victoria cichlid fish. *Current Zoology* 56: 57–64.

Via, S. 2009. Natural selection in action during speciation. *Proceedings of the National Academy of Sciences, USA* 106: 9939–9946.

Weir, J. T., and D. Schluter. 2004. Ice sheets promote speciation in boreal birds. *Proceedings of the Royal Society of London B* 271: 1881–1887.

Wiens, J. J., R. A. Pyron, and D. S. Moen. 2011. Phylogenetic origins of local-scale diversity patterns and the causes of Amazonian megadiversity. *Ecology Letters* 14: 643–652.

Wiley, C., and K. L. Shaw. 2010. Multiple genetic linkages between female preference and male signal in rapidly speciating Hawaiian crickets. *Evolution* 64: 2238–2245.

Wu, C. A., and D. R. Campbell. 2005. Cytoplasmic and nuclear markers reveal contrasting patterns of spatial genetic structure in a natural *Ipomopsis* hybrid zone. *Molecular Ecology* 14: 781–792.

Yamamoto, S., and T. Sota. 2009. Incipient allochronic speciation by climatic disruption of the reproductive period. *Proceedings of the Royal Society of London B* 276: 2711–2719.

17

The Origins of Life and Precambrian Evolution

On January 14, 2005, the tiny *Huygens* probe, having hitched a ride aboard the *Cassini* spacecraft, parachuted onto the surface of Titan, Saturn's largest moon. As the lander descended, its gas chromatograph mass spectrometer sniffed the murky air. Data from the instrument, some of which is shown at lower right, confirmed that the main constituents of Titan's atmosphere are molecular nitrogen (N_2) and methane (CH_4). Minor constituents, detected by instruments aboard *Cassini,* include ethane (C_2H_6), acetylene (C_2H_2), propane (C_3H_8), benzene (C_6H_6), hydrogen cyanide (HCN), and a variety of other volatile compounds (Clark et al. 2010).

As this witches' brew of organic molecules attests, Titan is home to an active chemistry (Cable et al. 2012; Raulin et al. 2012). Photons from the Sun and electrons from Saturn's magnetosphere drive reactions converting molecular nitrogen and methane into more complex hydrocarbons and nitrogen-containing compounds. Solid products of these atmospheric reactions fall to the surface, where they accumulate as an organic veneer over a crust made of water ice. Ethane and methane form clouds, rain onto the dirty ice, and pool in liquid lakes. Experiments replicating Titan chemistry in labs on Earth have yielded products including amino acids and nucleotide bases (Hörst et al. 2012).

Titan, viewed against the rings and bulk of Saturn. Courtesy of NASA/JPL-Caltech/Space Science Institute. Titan's hazy shroud derives from reactions among the constituents of its atmosphere, below. Modified from Niemann et al. (2005).

From "The abundances of constituents of Titan's atmosphere from the GCMS instrument on the Huygens probe." H.B. Niemann, S.K. Atreya. *Nature* 438: 779–784. Copyright © 2005 Nature Publishing Group. Reprinted with permission.

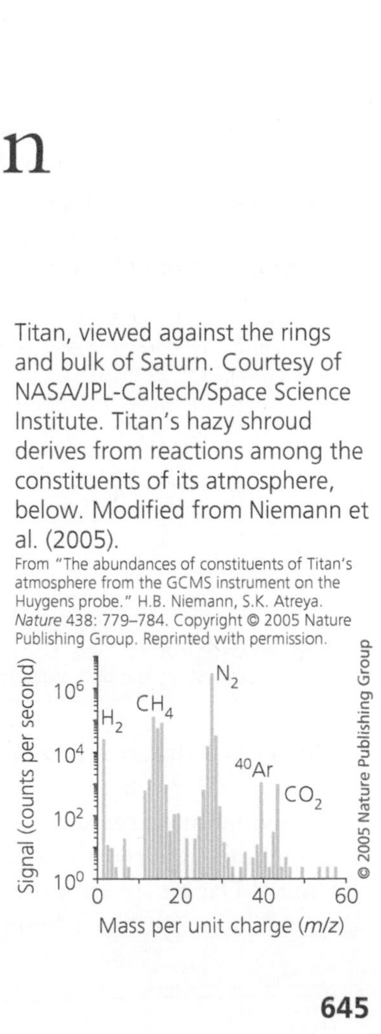

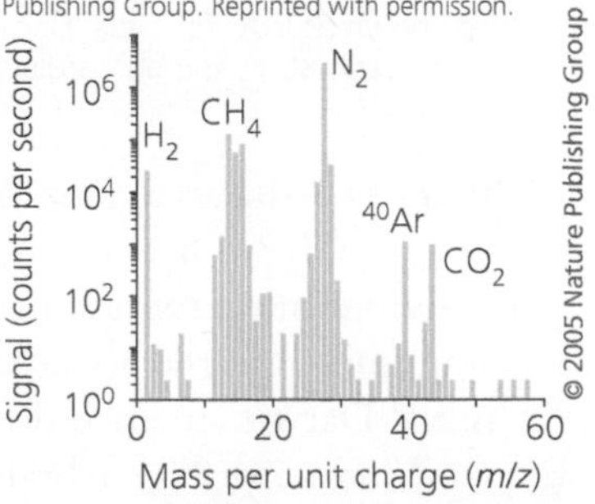

These observations and experimental results show that building blocks of molecular biology can form under natural conditions. And they raise the possibility that Titan might harbor life—albeit life rather different from that found on Earth (Norman and Fortes 2011). Some astrobiologists rank Titan as the top-priority world to search for extraterrestrial life in our solar system (Shapiro and Schulze-Makuch 2009). The discovery of an additional form of life would double our sample size, thereby dramatically improving our understanding of the conditions and probabilities under which life arises (Joyce 2012a).

In this chapter, we take up the origins of life and other mysteries concerning the big picture of life on Earth. We review work by scientists trying to answer some of the most intriguing, profound, and difficult questions in biology. The first two sections consider the nature of the first living thing and where it came from. Section 17.3 investigates the last common ancestor of all extant organisms and the shape of the tree of life. To conclude, Section 17.4 asks how the last common ancestor's descendants evolved into modern organismal forms.

These issues concern events of the far-distant past. Rocks dating from the time of Earth's formation do not exist on the planet's surface, but radiometric dating of meteorites yields an estimated age for the solar system, and hence Earth, of 4.5 to 4.6 billion years (see Badash 1989). The newborn Earth remained inhospitable for a few hundred million years. At first it was simply too hot. This is because the collisions of the planetesimals that coalesced to form Earth released enough heat to melt the entire planet all the way through (Wetherill 1990). Eventually, Earth's outer surface cooled and solidified to form a crust, and water vapor released from the planet's interior cooled and condensed to form the oceans. By the best estimates, life arose on the Earth a bit less than 4 billion years ago.

No physical record of the first biological events has survived. In contrast to the evolutionary processes we have investigated so far in this book, the origins of life must be reconstructed using indirect evidence alone. Consequently, biologists have turned to gathering disparate bits of information and fitting them together like pieces of a jigsaw puzzle. When more complete, this puzzle should present a clearer picture of life's origins.

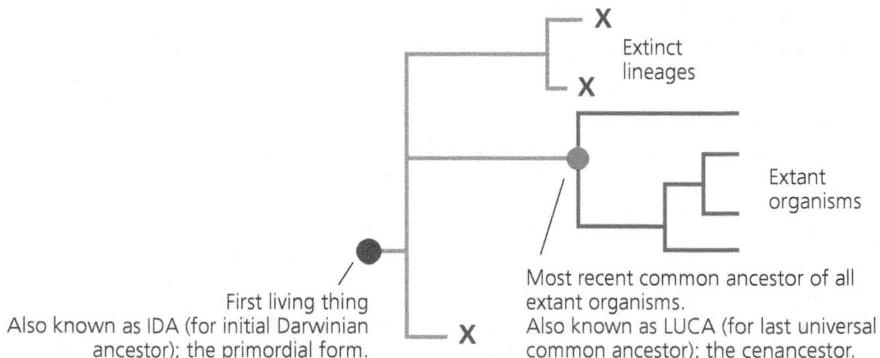

Figure 17.1 Cartoon of the tree of life The first living thing presumably had several descendant lineages, all but one of which died out. The most recent common ancestor of all living things is the organism whose immediate descendants diverged into the lineages that became all extant organisms.

Figure 17.1 shows a hypothetical history of life on Earth that will organize our discussion. Given that we are here to wonder what it was, there must have been a first living thing, represented by the purple dot at the left. The first living thing is sometimes referred to as the **primordial form** (Darwin 1859), or as **IDA,** for initial Darwinian ancestor (Yarus 2011). IDA presumably begat a diversity of descendant lineages (gray branches), most of which are long since extinct. Among

IDA's descendants was the last common ancestor of all extant organisms (orange dot), sometimes referred to as the **cenancestor** (Fitch and Upper 1987) or as **LUCA,** for last universal common ancestor (Forterre and Philippe 1999). The history of LUCA's descendants (blue branches) constitutes the tree of life.

It is important to keep in mind that the history in Figure 17.1 is hypothetical and subject to revision. For example, as we will see in Section 17.3, recent discoveries suggest that LUCA might not have been a single species, but instead a community of interbreeding forms.

17.1 What Was the First Living Thing?

In the early 1980s, two teams of scientists independently discovered small enzymes that could break and reform the chemical bonds that link nucleic acids together in chains. The enzymes did their job poorly. Compared to the hundreds of other such enzymes already known, they were slow at their catalytic task and showed little versatility. Yet the discovery has been recognized as among the most significant biological breakthroughs of the era. In 1989, the teams' leaders, Sidney Altman and Thomas Cech, shared the Nobel Prize.

Why were biologists so excited by the new enzymes? The answer is that the enzymes were made not of protein, but of nucleic acid—specifically RNA. Until 1982, all known enzymes were proteins. RNA was often considered to be DNA's poor cousin, relegated to the task of shuttling genetic information from DNA, where the information is stored, to proteins, which carry out the actual work of the cell. But Altman and Cech's discovery of RNA enzymes, or **ribozymes,** changed how biologists view the operations of the cell. Perhaps more important, the existence of ribozymes changed how biologists view the origin of IDA—how they think life originated and evolved on the early Earth.

The origin of life has been under investigation, via observation and experimentation, for many decades (see Fry 2006). Biologists have made artificial cells and cell membranes, and they have zeroed in on chemical reactions that could have built cellular materials from nonliving sources. Early on, however, a quandary became apparent. Which of its two most vital substances did life acquire first, proteins or DNA? Proteins can perform all sorts of complicated biological tasks, but there is no evidence that proteins can propagate themselves. They cannot store and transmit the information needed to replicate. DNA, on the other hand, is perfectly suited to store and transmit genetic information by complementary base pairing, but it was not known to be able to perform any biological work. Neither DNA nor proteins seems to be of any use without the other, but it is implausible that they appeared simultaneously.

This chicken-and-egg problem was essentially resolved with the discovery of catalytic RNA. Because RNA has both a capacity for information storage and transmission and the ability to perform biological work, researchers now think that it preceded both proteins and DNA in the origin of life. Was there once a time when life was based entirely on RNA—an RNA World (Gilbert 1986)? This question is the topic of Section 17.1.

An RNA World is appealing because it would possess many characteristics of modern life without the need for more than a few organic molecules in solution. The RNA World hypothesis is based on the realization, since the discovery of ribozymes, that RNA can possess both a genotype and a phenotype (Joyce 1989).

The RNA World hypothesis proposes that catalytic RNA molecules were a transitional form between nonliving matter and the earliest cells.

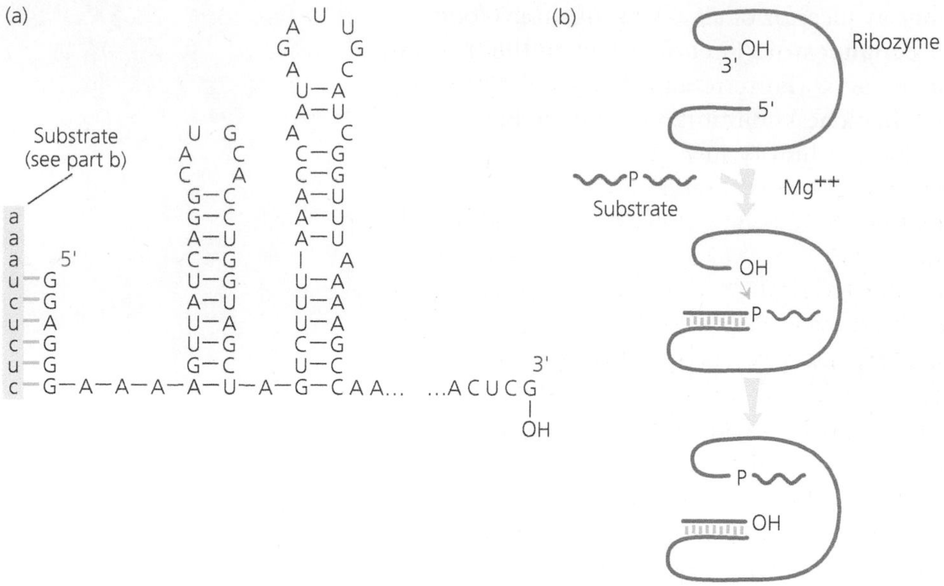

Figure 17.2 The ribozyme from *Tetrahymena thermophila* (a) The primary nucleotide sequence is the genotype. A secondary structure is formed when nucleotides pair as the molecule folds back on itself. (b) The catalysis performed by the ribozyme in vitro is the phenotype. A short oligonucleotide substrate (blue) binds to the 5′ end of the ribozyme (orange) through complementary base pairing (green ticks). The ribozyme catalyzes the breakage of a phosphoester bond in the substrate and the ligation of the 3′ fragment to its own 3′ end.

The genotype is the primary sequence of nucleotides along the RNA (**Figure 17.2a**), much like the genotype of a modern organism is the sequence of nucleotides along the DNA in the chromosome. Catalytic RNA, for example, contains between 30 and 1,000 ribonucleotides that form its primary sequence, and hence its genotype. The *Tetrahymena* ribozyme discovered by Cech and colleagues (Kruger et al. 1982; Zaug and Cech 1986) stretches some 400 nucleotides from head (the 5′ end) to tail (the 3′ end). This RNA is an intron (an intervening sequence between two genes) that separates two regions of the *Tetrahymena* genome that code for ribosomal RNA (rRNA) genes.

Unlike genomic DNA, which is usually double stranded, RNA typically exists as a single-stranded molecule that folds back on itself to form a three-dimensional structure. In ribozymes, this folded state can have an active site that enables the RNA to catalyze a chemical reaction on a substrate, like a protein enzyme. This reactivity gives RNA its phenotype. The ribozyme Cech and colleagues found has the catalytic capability to splice itself out from between the two adjacent rRNAs after they have been transcribed (Kruger et al. 1982). In test tubes, a shortened version of the ribozyme can catalyze a phosphoester transfer reaction on a short RNA substrate, called an oligonucleotide (a piece of single-stranded nucleic acid 5 to 30 nucleotides in length). In the reaction, the 3′ half of the substrate is broken off by the ribozyme and attached to the ribozyme's own 3′ end (Figure 17.2b).

Defining Life

All living organisms possess a genotype and a phenotype. In fact, when we consider what life really is, and how living systems can be distinguished from nonliving ones, the ability to store and transmit information (genotype) and the ability to express that information (phenotype) are perhaps the most important criteria that set life apart from nonlife. There is no neat list of characteristics that define life. Most biologists would include traits like growth and reproduction on such a list, but they cannot agree on what else should be used to exclude such lifelike systems as a growing salt crystal or a computer virus (if, indeed, these should be excluded). However, many now agree that the ability to evolve is a crucial component of any definition of life. Evolution—descent with modification—requires both the ability to record and make alterations in heritable information and a sorting process that distinguishes valuable changes from detrimental ones. The former is a property of genotype, while the latter occurs as a result of variation among individuals in phenotype.

Here is one way we might define life: If it forms populations capable of evolving by natural selection, then it is alive.

Dozens of naturally occurring ribozymes have been discovered (Gesteland et al. 1999). The phenotypes of most involve the formation and breaking of phosphoester bonds in RNA or DNA (Figure 17.2b). The chemistry of these reactions is precisely what is needed to replicate nucleic acids. This observation gives support to the idea of a primordial RNA World, where RNA would be responsible for replicating itself. If an RNA molecule could make a copy of itself while accommodating the possibility of mistakes—mutations—then it would exhibit many of the characteristics of modern life and could therefore be considered alive.

The Case for RNA as an Early Life-Form

The RNA World hypothesis posits that the primordial form was an RNA-based living system that later evolved into life-forms like those we see today, in which DNA stores biological information and proteins manifest this information. DNA is better suited as an information repository because it is chemically more stable than RNA. Especially when double stranded, DNA can better withstand high temperatures and spontaneous degradation by acids or bases.

What is the evidence that RNA is ancient? The existence of catalytic RNA is critical, but there are other indicators as well. One clue that RNA was involved in early life-forms is its role in the machinery cells use for replication and metabolism (Crick 1966; White 1976). The most conserved and universal component of the information processing machinery, for example, is the apparatus for translation of genetic information into protein: the ribosome (Harris et al. 2003; Koonin 2003). This apparatus, while it incorporates proteins, is built on a frame of RNA (rRNA). Ribosomes not only contain RNA themselves, they require RNA adaptors (tRNAs) to do their job. Furthermore, it is the RNA portion of ribosomes that actually carries out the catalytic steps in protein synthesis (Nissen et al. 2000; Steitz and Moore 2003). Another argument for the antiquity of RNA is that the basic currency for biological energy is ribonucleoside triphosphates, such as ATP and GTP (Joyce 1989). These molecules are involved in almost every energy-transfer operation of all cells and are even components of electron-transfer cofactors such as NAD (nicotinamide adenine dinucleotide), FAD (flavin adenine dinucleotide), and SAM (S-adenosyl methionine). With these ghosts of an RNA World in mind, we turn next to the question: Can RNA evolve?

The Experimental Evolution of RNA

RNA sequences can provide a blueprint for their own replication. For any RNA sequence, we can build a complementary sequence by base pairing. Thus RNA, like DNA, has the capacity to store heritable information that can be propagated. A good example is the life cycle of the HIV virus (see Chapter 1). HIV uses the protein enzyme reverse transcriptase to copy its RNA strand into a DNA complement, which can then be converted into double-stranded DNA. Given that RNA can store genetic information, populations of RNA molecules should be able to evolve. If an RNA molecule has a phenotype that involves catalyzing a specific chemical reaction, can we apply selection to improve or modify this phenotype and observe a heritable change?

One way researchers have tested the RNA World hypothesis is to check whether populations of RNA molecules can evolve by natural selection.

Following pioneering work by Donald Mills and colleagues (1967), Amber Beaudry and Gerald Joyce (1992) exploited the catalytic capacity of the *Tetrahymena* ribozyme to address this question. They used the ribozyme's ability to pick up a new 3′ tail, as shown in Figure 17.2b, to distinguish ribozymes that perform catalysis from those that do not.

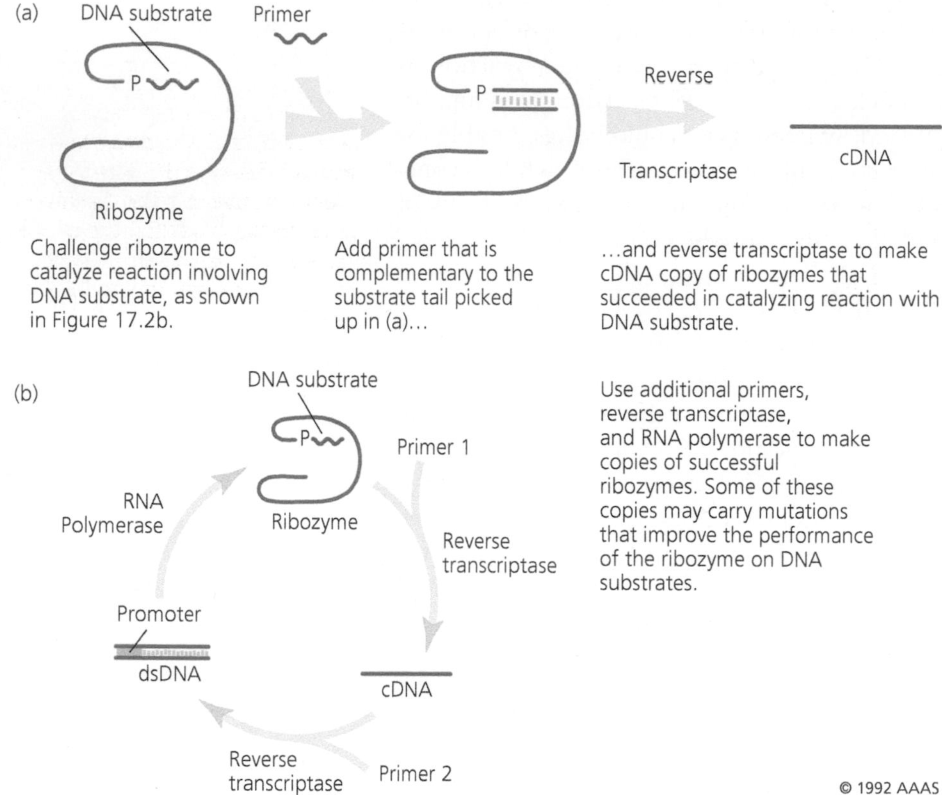
(a)
DNA substrate
Primer
P
Ribozyme
P
Reverse
Transcriptase
cDNA
Challenge ribozyme to catalyze reaction involving DNA substrate, as shown in Figure 17.2b.
Add primer that is complementary to the substrate tail picked up in (a)...
...and reverse transcriptase to make cDNA copy of ribozymes that succeeded in catalyzing reaction with DNA substrate.
(b)
DNA substrate
P
Primer 1
RNA Polymerase
Ribozyme
Reverse transcriptase
Promoter
dsDNA
cDNA
Reverse transcriptase
Primer 2
Use additional primers, reverse transcriptase, and RNA polymerase to make copies of successful ribozymes. Some of these copies may carry mutations that improve the performance of the ribozyme on DNA substrates.
© 1992 AAAS

Figure 17.3 Test-tube selection and reproduction of RNA (a) Selection. A pool of RNAs (orange), made by random mutagenesis of a ribozyme, is challenged to perform a desired chemical reaction. Only those that perform the reaction acquire a short "tail" of DNA nucleotides attached to their 3′ end (blue). (b) Reproduction. RNA sequences that have acquired a 3′ tail (top) bind primer 1 by complementary base pairing and are copied by reverse transcriptase into complementary DNA (cDNA). A second primer then binds to the cDNA so that the reverse transcriptase, which copies DNA in addition to RNA, can make the DNA double stranded. Primer 2 contains the promoter region for RNA polymerase, so that the double-stranded DNA can be copied back into RNA. From Beaudry and Joyce (1992).

From "Directed evolution of an RNA enzyme." *Science* 257: 635–641, Figure 1, page 636. Copyright © 1992. Reprinted with permission from AAAS.

Beaudry and Joyce (1992) first made a large population of RNA molecules by sprinkling random mutations throughout the *Tetrahymena* ribozyme at a rate of 5% per position. Then, this mutant population was challenged with a novel task to select certain genotypes **(Figure 17.3a)**. The task in this case was that the substrate oligonucleotide was provided in the form of DNA, not RNA. The naturally occurring sequence of the *Tetrahymena* ribozyme (the "wild type") used to start these experiments could cleave a DNA substrate only at a miserably slow rate. Beaudry and Joyce hoped that in the mutant pool there were variant sequences that, by chance, had an increased capacity for DNA cleavage.

The researchers incubated the mutant RNA population with a DNA substrate for an hour and then amplified the ribozyme RNA into many more copies by adding two protein enzymes—reverse transcriptase and RNA polymerase. Because ribozymes pick up a 3′ tail as a consequence of cleaving the substrate, a DNA primer for reverse transcriptase that is complementary to the 3′ tail discriminates sequences that catalyzed a reaction with the DNA substrate from sequences that did not. The 3′ tail is necessary to bind the primer, which in turn is necessary to initiate reverse transcription, which is in turn necessary to make more RNA (Figure 17.3b). The population of new RNAs that result from this cycle harbors novel variations resulting from copying errors and can be used to seed a completely new cycle.

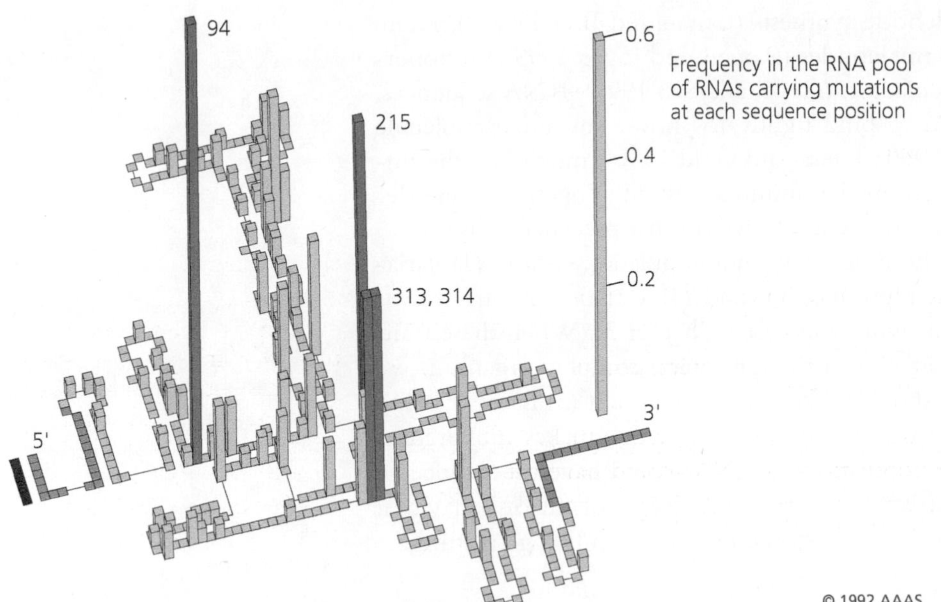

Figure 17.4 Allele frequency changes in an evolving RNA population This histogram, built on a crude representation of the folded secondary structure of the 413-nucleotide *Tetrahymena* ribozyme, shows the frequencies of mutations in the test-tube RNA population after nine rounds of selection and reproduction. Four mutations, at nucleotides 94, 215, 313, and 314, have risen to frequencies over 50% and are mainly responsible for the new phenotypic characteristics of the population. Reprinted with permission from Beaudry and Joyce (1992).

From "Directed evolution of an RNA enzyme." *Science* 257: 635–641, Figure 1, page 636. Copyright © 1992 AAAS. Reprinted with permission from AAAS.

After 10 such generations, the activity of the average RNA in the population in cleaving DNA substrates and attaching one of the resulting fragments to its own 3′ end had improved by a factor of 30. Importantly, Beaudry and Joyce could trace this phenotypic enhancement to changes in nucleotide sequence (**Figure 17.4**). Specific mutations at four nucleotide positions in the ribozyme's sequence were responsible for most of the catalytic improvement. Individual ribozymes carrying mutations at positions 94, 215, 313, and 314 proved to have a catalytic efficiency over 100 times greater than the ancestral sequence.

Populations of catalytic RNA molecules exhibit variation in nucleotide sequence. This variation is heritable when RNA is replicated. And researchers have devised experimental conditions under which sequence variation results in differences in survival.

This experiment demonstrated that RNA molecules in solution can possess features of living organisms that allow them to evolve. Each RNA can be ascribed a particular fitness, which is a function of both survival (substrate catalysis) and reproduction (ability to be reverse- and forward-transcribed). The fitness of the molecule is a reflection of its phenotype, which, in the case of ribozymes, is immediately specified by their primary sequence. Variation in an RNA population can be introduced at the outset by the randomization of a wild-type sequence, as was the case with the Beaudry and Joyce (1992) experiment. Alternatively, an investigator can rely on the intrinsic error rates of the protein enzymes used in RNA amplification and can even alter the chemical environment to make the error rates higher. With such online mutagenesis the system becomes truly evolutionary, and selection can operate on variants of variants over many generations. Thus, it is easy to see a parallel between an evolving population of RNA in a test tube and an evolving population of modern organisms in the natural environment (Lehman and Joyce 1993).

In test-tube experiments like these, researchers have evolved many ribozymes with either improved function or an entirely new function. The catalytic repertoire of RNA has greatly expanded (Joyce 1998), and we now know that RNA can catalyze such reactions as phosphorylation (Lorsch and Szostak 1994), aminoacyl transfer (Illangasekare et al. 1995), peptide-bond formation (Zhang and Cech 1997), and carbon–carbon bond formation (Tarasow et al. 1997; Fusz et al. 2005). Ribozymes have been designed that are allosteric, requiring a small-molecule cofactor to carry out catalysis (Tang and Breaker 1997). Ribozymes can be selected that can play a role in ribonucleotide synthesis (Unrau and Bartel 1998), retain activity with only three of four nucleotides (Rogers and Joyce 1999), and operate without divalent metal–ion cofactors (Geyer and Sen 1997). RNA sequences, called aptamers, can be selected to bind tightly to almost any other molecule desired (Ellington and Szostak 1990; Tuerk and Gold 1990), much like the immunoglobulin proteins of the mammalian immune system. Together, these developments implicate RNA as a possible living system that preceded cells.

Darwin deduced that when the individuals in a population exhibit (1) variation, (2) inheritance, (3) excess reproduction, and (4) variation in survival or reproductive success, populations will evolve (see Chapter 3). When these traits are stripped of the particular characteristics of an intact, complex organism, we realize that traits (1) and (2) are about having a genotype, trait (3) is about being self-replicating, and trait (4) is about having a phenotype that makes a difference. Consequently, a self-replicating population of RNA would have the essence of life, even without the cells or organelles or tissues or leaves or fur or behavioral characteristics, and so on, that we are accustomed to seeing in living creatures.

Self-Replication

From what we have discussed so far, there is a crucial piece conspicuously missing from the evidence that today's organisms could be descended from inhabitants of an RNA world. We know that RNA is a versatile molecule and that it can evolve under the right circumstances. In the experiments we have described, however, RNA was copied by protein enzymes. These proteins would not have existed in the RNA World. A main premise of the RNA World hypothesis is that RNA predates the use of proteins to do most biological work. The piece of evidence that we lack is thus the demonstration that RNA can copy itself. The "RNA-dependent RNA autoreplicase" remains an elusive quarry for origins-of-life research (Bartel and Unrau 1999; Müller 2006). Whether the RNA World used only one type of self-replicating RNA or a suite of interacting RNAs, an RNA with a replicase phenotype would be necessary (Bartel 1999). The acquisition of the ability to self-replicate by a collection of organic molecules, such as RNA, is arguably the point at which nonliving matter came to life.

Although populations of catalytic RNA molecules have most of the properties required for evolution by natural selection, they still cannot evolve on their own without considerable help from human researchers. This is because catalytic RNA molecules cannot yet copy themselves.

The hypothesis that an RNA molecule could replicate itself, serving as a simple proto-organism, is testable. If the hypothesis is correct, then we should be able to make a self-replicating RNA molecule in the lab. Although this has not been achieved to date, researchers have made significant advances. David Bartel and coworkers, for example, have used test-tube evolution to search for ribozymes capable of synthesizing RNA (Bartel and Szostak 1993; Ekland et al. 1995).

Figure 17.5 shows the selection scheme Bartel and Jack Szostak (1993) used to make ribozymes that catalyze the formation of a phosphoester bond to link a pair of adjacent RNA nucleotides. The researchers started with a large pool of RNA polynucleotides. This pool constituted the population to be subjected to selection. Every RNA in the pool had the same sequence on its 5′ and 3′ ends (represented by lines), plus a unique 220-nucleotide stretch of random sequence in the middle (represented by the box labeled Random 220). Figure 17.5 follows two RNAs from the pool: Random 220 A (left column) and Random 220 B (right).

Bartel and Szostak bound the pool RNAs to agarose beads by means of a base-pairing interaction on their 3′ ends. The scientists then bathed the pool RNAs in a solution containing many copies of a specific substrate polynucleotide (Figure 17.5a). This short RNA molecule had, on its 5′ end, a sequence of nucleotides

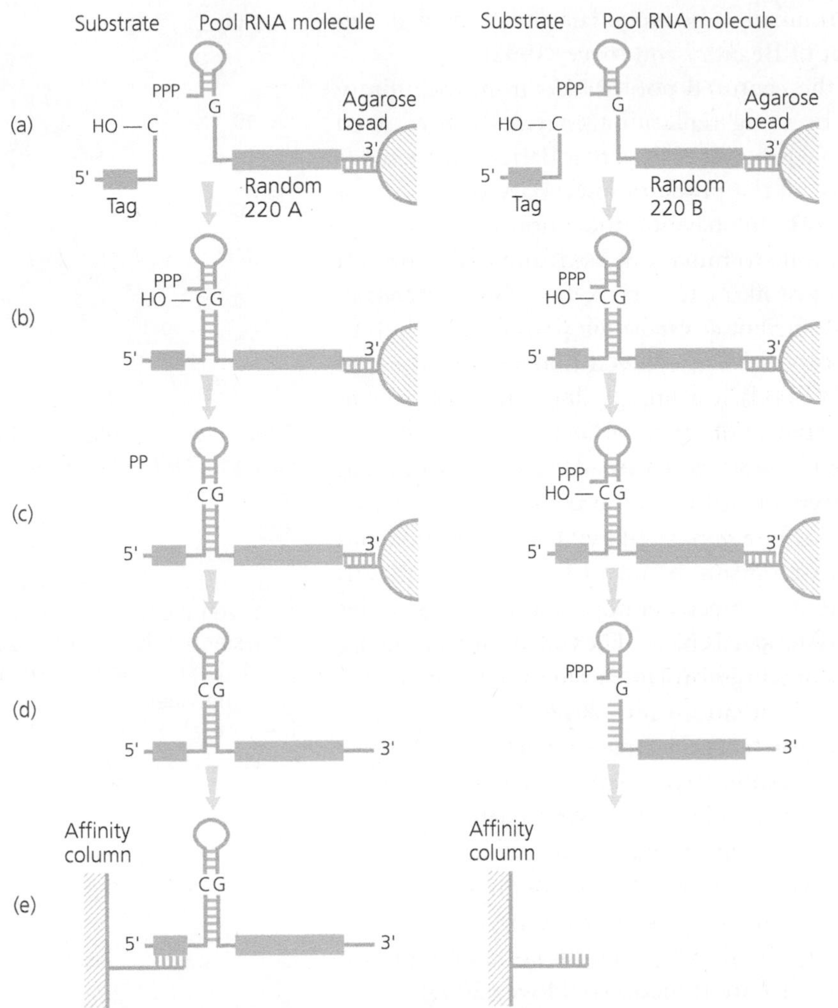

Figure 17.5 Test-tube selection scheme for identifying ribozymes that can synthesize RNA (a) A large pool of RNA molecules containing random sequences of 220 nucleotides are bound by base-pairing to agarose beads. Copies of a short substrate RNA are added. (b) The substrate binds by base-pairing to the free end of the pool RNAs. (c) If the random portion of a pool RNA's sequences confers appropriate ribozyme activity, the pool RNA permanently attaches the substrate to its own free end by forming a phosphoester bond. (d) Substrate RNAs not bound by phosophoester bonds are washed away, and the pool RNAs are released from their beads. (e) The pool RNAs are run through an affinity column that catches them by base-pairing with a tag on the free end of the substrate. Only pool RNAs with substrate bound are caught. After Bartel and Szostak (1993).

forming a tag, whose function will soon become clear. On its 3′ end, the substrate RNA had a sequence of nucleotides complementary to the free end of the pool RNA molecules. The substrate molecules quickly became bound, by base-pairing hydrogen bonds, to the pool RNAs (Figure 17.5b).

This annealing brought into adjacent position the triphosphate group (PPP) on the 5′ end of the pool RNA and the hydroxyl group (OH) on the 3′ end of the substrate RNA. If, by chance, the 220-nucleotide stretch of random sequence in a pool RNA molecule had some ability to catalyze the formation of phosphoester bonds, then it catalyzed the formation of such a bond between the substrate and pool RNA molecules. In Figure 17.5c, Random 220 A has catalyzed such a reaction, liberating a diphosphate molecule, whereas Random 220 B has not.

Researchers are performing selective breeding experiments with populations of catalytic RNAs in an effort to develop catalytic RNAs that can replicate themselves. If the researchers succeed, then by the definition we proposed earlier, they will have created life.

Bartel and Szostak then rinsed the pool RNAs under conditions that washed away any substrate RNAs not covalently bound (by phosphoester bonds) to pool RNAs, and liberated the pool RNAs from their agarose beads (Figure 17.5d). Random 220 A still has its substrate (with tag); Random 220 B does not.

Finally, the scientists ran the pool RNAs through an affinity column (Figure 17.5e). The affinity column caught hold of the tag sequence on the substrate RNA by base pairing. The column thus captured any pool RNA whose 220-nucleotide stretch of random sequence had catalytic activity (like Random 220 A) and let pass any pool RNA whose 220-nucleotide sequence did not. This selection step is analogous to the discrimination between tailed and untailed *Tetrahymena* ribozymes in the experiment of Beaudry and Joyce (1992).

Now Bartel and Szostak released the captured pool RNAs from the affinity column, made many copies of each by using replication enzymes that allowed some mutations, and repeated the whole process. Notice that Bartel and Szostak's protocol also gives, to the pool RNAs, all the properties necessary and sufficient for evolution by natural selection. The RNAs have reproduction with heritability (via the copying process), variation (due to mutation), and differential survival (in the affinity column). The RNAs most likely to survive from one generation to the next are the ones that are most efficient at catalyzing phosphoester bonds. After 10 rounds of selection, the RNA pool had evolved ribozymes that could catalyze the formation of phosphoester bonds at a rate 7 million times faster than such bonds form without a catalyst **(Figure 17.6)**.

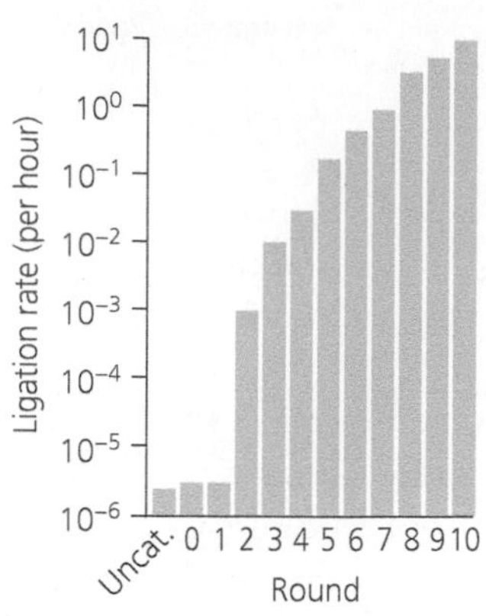

Figure 17.6 Evolution of catalytic ability in a laboratory population of ribozymes The graph shows the average rate at which the members of Bartel and Szostak's (1993) RNA pool catalyzed the formation of phosphoester bonds (ligation rate) as a function of round of selection. Note the logarithmic scale on the vertical axis. From Bartel and Szostak (1993).

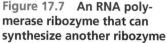

Aniela Wochner and colleagues (2011), using a strategy similar in spirit to that of Bartel and Szostak, produced a ribozyme that can synthesize another ribozyme. The researchers started with R18, a ribozyme generated by Wendy Johnston and colleagues (2001) that can use an RNA template to add 14 RNA nucleotides to an RNA primer. Wochner and colleagues subjected a population of R18 copies to selection for the ability to synthesize longer RNAs. They identified particular sequence changes responsible for improved performance, then used genetic engineering to incorporate these beneficial mutations into R18. The result was a new ribozyme, tC19, that can synthesize RNA molecules 95 nucleotides long—nearly half tC19's own size. The scientists subjected a second population of R18 copies to selection for the ability to copy a greater variety of template sequences, identified mutations conferring improved performance, and engineered them into tC19. The final result, tC19Z, proved able to synthesize, with 99.8% accuracy, functional copies of a different ribozyme **(Figure 17.7)**. Ribozyme tC19Z is not capable of self-replication. But it appears that biochemists are homing in on an RNA sequence, or set of sequences, that are (Lincoln and Joyce 2009).

As Gerald Joyce (1996) put it, "Once an RNA enzyme with RNA replicase activity is in hand, the dreaming stops and the fun begins." Given the right organic molecules to feed on, a population of self-replicating RNAs should evolve on its own by mutation and natural selection. Would a species of self-replicating RNA evolve a DNA genome with DNA replication and transcription? Would it invent proteins and translation? Would its machinery be anything like the machinery in naturally evolved organisms? Perhaps one day answers will come.

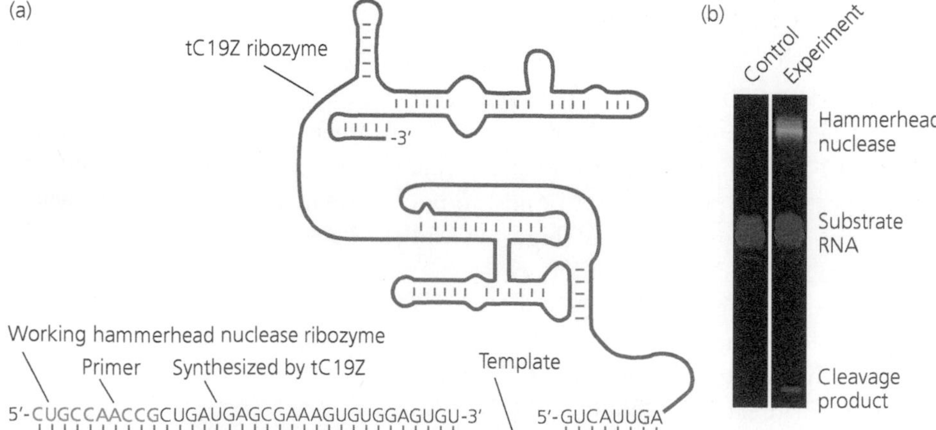

Figure 17.7 An RNA polymerase ribozyme that can synthesize another ribozyme (a) The secondary structure of tC19Z plus the template for another ribozyme—called a hammerhead nuclease—that, given a primer, tC19Z can copy. After Bentin (2011) and Wochner et al. (2011). (b) Electrophoresis gels demonstrating that the hammerhead nuclease synthesized by tC19Z can cleave an RNA substrate. From Wochner et al. (2011).

17.2 Where Did the First Living Thing Come From?

The RNA World has many attractive features, and it solves the problem of having to propose the advent of proteins before DNA existed to encode them. But an RNA World comes with troubles of its own. The fundamental problem is simple: How could RNA sequences of any kind arise in an abiotic environment?

The proposition that catalytic RNAs were a transitional form between nonliving matter and cellular life leaves many gaps. We must still explain where the first RNA molecules came from, and how a population of self-replicating RNA molecules evolved into DNA- and protein-based cells.

Chemists have studied the ways in which nucleic acids could be made without the aid of living systems. Certain aspects of the abiotic synthesis of nucleic acids turn out to be easy, and others turn out to be difficult. Many researchers have concluded that the RNA World was probably not the first self-replicating system. This is because the likelihood of making RNA abiotically is too minute. Later, we will talk about the challenges of RNA synthesis, but for now we note that RNA may be derived from a more primitive chemical system. In other words, IDA was perhaps not made of RNA, but of something else that preceded RNA.

Regardless of what the primordial form was made of, to reconstruct the advent of any information-containing organic molecule with the properties of self-replication, the following four issues need to be addressed:

1. Information-containing biomolecules need to be made from simple inorganic compounds. Where did these compounds come from?
2. The chemical reactions that construct larger molecules from simple inorganics must be favorable and have a source of energy. What were these reactions?
3. The building blocks must be able to self-assemble into polymers such as RNA and polypeptides. How did this happen?
4. Larger biomolecules must be protected from harsh environmental conditions. How was this accomplished?

A fifth issue concerns where IDA lived. The panspermia hypothesis suggests that life may have arisen elsewhere and traveled to Earth through space (Arrhenius 1908). Other worlds in our solar system that may be hospitable to life, or may have been hospitable in the past, include Venus, Mars, Jupiter's moon Europa, and Saturn's moons Enceladus, and—as we have already noted—Titan (Shapiro and Schulze-Makuch 2009). It is conceivable that microbes could survive an interplanetary trip (Abrevaya et al. 2011; Raggio et al. 2011). Earth's first life-forms might even have traveled from another star (Secker et al. 1994) or been sent here intentionally (Crick and Orgel 1973). However, following most researchers in the field, we will assume that the primordial form arose on Earth.

Where Did the Stuff of Life Come From?

On September 28, 1969, at about 11:00 a.m., a meteor entered Earth's atmosphere over Murchison, Australia, broke up, and scattered meteorites across 5 square miles **(Figure 17.8)**. Soon after, scientists collected some of the meteorites for chemical study (Kvenvolden et al. 1970). To their astonishment, the analyses revealed organic compounds in the interior of the rocks. In particular, the amino acids glycine, alanine, glutamic acid, valine, and proline were found in significant concentrations (1–6 micrograms of amino acid per gram of meteorite). These amino acids are in the kit used by modern organisms to make proteins. Amino acids had been found in meteorites before, but their presence was likely the result of contamination from human handling. The scientists who studied the Murchison

Figure 17.8 The Murchison meteorite A fragment of the 100 kg of material that fell near Murchison, Australia, in 1969.

meteorites fractured them in the lab and analyzed only the interior portions. In addition, the amino acids they found in the Murchison stones were racemic. They included roughly equal proportions of the D- and L-stereoisomers (mirror-image forms). By contrast, biological amino acids are almost purely of the L-form, and thus terrestrial life could not be the source of the Murchison compounds.

Why were the Murchison meteorites significant? The biomolecules of life, as well as their likely precursors, all require the elements carbon, hydrogen, oxygen, nitrogen, sulfur, and phosphorus in large amounts, plus trace quantities of other elements such as magnesium, calcium, and potassium. Moreover, these elements must be in a chemical form that allows them to be used in the construction of biological building blocks like amino acids, sugars, and carbohydrates. If these building blocks could have been synthesized on the primitive Earth, then presumably they would have been available for condensation into larger biomolecules. But if they could not have been made on Earth, we would have to look to extraterrestrial sources, such as meteors, to account for their presence.

The problem with terrestrial sources is that, 4 billion years ago, Earth's environment may or may not have been permissive for the synthesis of life's building blocks. In addition to temperature and pressure, a key feature of the environment is whether it was primarily oxidizing, with high abundances of molecular oxygen (O_2) and carbon dioxide (CO_2), or primarily reducing, with high concentrations of hydrogen (H_2), methane (CH_4), and ammonia (NH_3). Or it could have been intermediate in oxidizing activity. Which state it was in would have determined which chemical reactions were possible.

The composition of the early atmosphere remains uncertain (Lazcano and Miller 1996; Chyba 2005; Parker et al. 2011), and atmospheric chemists are looking for mechanisms by which organic molecules could have been synthesized, even in relatively unpermissive mixtures of gases (Kasting 1993). Some feel that geochemical evidence points to an atmosphere unfavorable for the generation of biologically important molecules, at least in the concentrations needed for the origins of life. Thus, many have explored an alternative hypothesis that certain critical biochemicals were made elsewhere in the solar system and delivered to Earth in vehicles such as the Murchison meteorite.

The young Earth experienced heavy bombardment by meteors and comets. **Figure 17.9** shows the history of very large impacts on both Earth and the Moon.

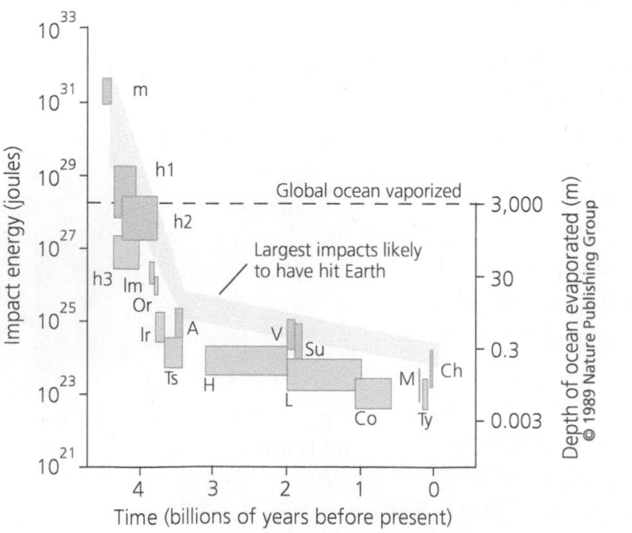

Terrestrial craters:
m = large impact thought to have formed Moon
h1 = hypothetical
h2 = hypothetical
A = Archaean spherule beds
V = Vredevort
Su = Sudbury
M = Manicougan
Ch= Chicxulub

Lunar craters:
h3 = hypothetical
Im = Imbrium
Or = Orientale
Ir = Iridium
Ts = Tsiolkovski
H = Hausen
L = Langrenus
Co = Copernicus
Ty = Tycho

Figure 17.9 The history of large impacts on Earth and the Moon Each box encloses the range of times during which a particular impact is estimated to have occurred, and the range of energies the impact is estimated to have fallen within. From Sleep et al. (1989).

"Annihilation of ecosystems by large asteroid impacts on the early Earth." N. H. Sleep, et al. *Nature* 342: 139–142. Copyright © 1989 The Nature Publishing Group. Reprinted with permission.

Like the Murchison meteorite, several carbonaceous chondrite meteorites, believed to be fragments of asteroids, have proven to contain an abundance of organic molecules (see Chyba et al. 1990; Lazcano and Miller 1996). Many comets also contain a variety of organic molecules (Chyba et al. 1990; Cruikshank 1997), including amino acids (Elsila et al. 2009). And organic compounds occur in interstellar space (Kwok and Zhang 2011). It is thus possible, at least in principle, that the stuff of life was delivered to Earth from elsewhere.

The simple organic molecules from which life was built may have formed in space and then fallen to Earth. Researchers have tested this idea by looking for amino acids and other organic molecules inside meteorites.

There is at least one difficulty with the hypothesis that life's building blocks came from space. When meteors and comets crash to Earth, friction with the atmosphere and collision with the ground generate tremendous heat (Anders 1989). This heat may destroy most or all of the organic molecules the meteors and comets carry (Chang 1999). Edward Anders (1989) notes that very small incoming particles are slowed gently enough by the atmosphere to avoid incinerating all of their organics; he suggests that dust may have been the primary source of the young Earth's organic molecules. Christopher Chyba (1990) and colleagues look instead to the possibility that the early atmosphere was dense with carbon dioxide. A dense CO_2 atmosphere may have provided a soft enough landing, even for large meteors and comets, for some of their organics to survive. The Murchison meteorites certainly provide direct evidence that at least some organic molecules can survive a descent to Earth.

The Oparin–Haldane Model

Originally, there was great hope that Earth itself could provide the "right stuff" for prebiotic synthesis. In 1953, Stanley Miller, then a graduate student in Harold Urey's laboratory at the University of Chicago, reported a simple and elegant experiment. He built an apparatus that boiled water and circulated the hot vapor through an atmosphere of methane, ammonia, and hydrogen, past an electric spark, and finally through a cooling jacket that condensed the vapor and directed it back into the boiling flask. Miller let the apparatus run for a week; the water inside turned deep red and cloudy. Using paper chromatography, Miller identified the cause of the red color as a mixture of organic molecules, most notably the amino acids glycine, α-alanine, and β-alanine. Since 1953, chemists working on the prebiotic synthesis of organic molecules have, in similar experiments, documented the formation of a tremendous diversity of organic molecules, including amino acids, nucleotides, and sugars (see Fox and Dose 1972; Miller 1992; Johnson et al. 2008; Powner et al. 2009; Parker et al. 2011).

The simple organic molecules from which life was built may also have formed on Earth. Researchers have tested this idea by trying to re-create the chemical conditions on the early Earth and replicate the chemical reactions that might have created amino acids and nucleotides.

Miller used methane, ammonia, and hydrogen as his atmosphere; in the 1950s, this highly reducing mixture was thought to model the atmosphere of the young Earth. The implication of Miller's result was that if lightning or UV radiation could have played the role that the spark did in his experiment, then the young Earth's oceans would have quickly become rich in biological building blocks.

Many atmospheric chemists believe that Earth's early atmosphere was not so reducing, being dominated by carbon dioxide rather than methane and by molecular nitrogen (N_2) rather than ammonia (Kasting 1993; Trail et al. 2011). This conclusion is based on the gases released by contemporary volcanoes, improved knowledge of chemical reactions in the upper atmosphere, and analysis of zircon crystals that are the oldest known terrestrial solids. Reaching a consensus on the prebiotic environment is important, because an atmosphere dominated by carbon dioxide and molecular nitrogen appears less conducive to the formation of certain organic molecules. However, the formation of aldehydes, especially formaldehyde (H_2CO), from carbon dioxide has been deemed plausible by several researchers—particularly in light of work by Feng Tian and colleagues (2005) suggesting that the early atmosphere might have contained as much as 30% molecular hydrogen (H_2). Aldehydes are necessary in the construction of the ribose sugars needed to make the nucleotides of an RNA World (Mojzsis et al. 1999).

The view that Earth possessed all the necessary ingredients for the origins of life is perhaps the most thoroughly investigated hypothesis and holds great appeal for many scientists. This opinion dates back to the efforts of A. Oparin and J. B. S. Haldane, in the first half of the 20th century, to reconstruct how life may have begun. These scientists and others (including Charles Darwin) created a lasting image of life arising in an aqueous environment brimming with biological building blocks. This was Darwin's "warm little pond" (Darwin 1887), the famous "prebiotic soup." There are many severe criticisms of this vision, not the least of which is whether liquid water existed on Earth at the time of life's origin. Nonetheless, this view remains as sort of a null model against which deviations can be tested, much like the Hardy–Weinberg equilibrium principle in population genetics. This scenario is often referred to as the Oparin–Haldane model.

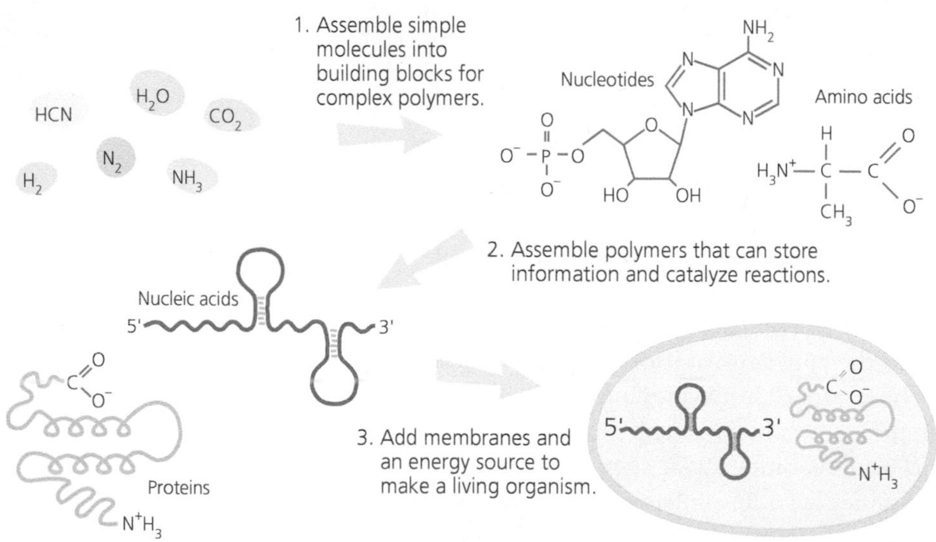

Figure 17.10 Synopsis of the Oparin–Haldane model The first stage would have been the formation of biological building blocks (nucleotides and amino acids) from existing inorganic material on the early Earth. The second stage would have been the polymerization of these building blocks to form biological macromolecules (nucleic acids and proteins). Finally, these macromolecules would have directed the formation of other biological structures, such as cell membranes.

We can break the Oparin–Haldane model into a series of steps that occurred sequentially in the waters or moist soil of the young Earth **(Figure 17.10)**. First, nonbiological processes synthesized organic molecules, such as amino acids and nucleotides, that would later serve as the building blocks of life. Then, the organic building blocks in the prebiotic soup were assembled into biological polymers, such as proteins and nucleic acids. Finally, some combination of biological polymers were assembled into a self-replicating organism that fed off of the existing organic molecules, much as we discussed earlier for the RNA World.

From Simple Inorganics to the Building Blocks of Life

Previously, we saw how easily amino acids can be made from simple inorganics like methane, ammonia, and hydrogen. What about nucleotides? A second monumental achievement in origins-of-life research was the demonstration by Juan Oró (1961) that the nitrogenous base adenine (a purine) could be made via a thermodynamically favorable reaction involving only ammonia and hydrogen cyanide (HCN). When these compounds are heated in water, adenine is produced in yields as high as 0.5%, which is significant if the early atmosphere was reducing and contained large amounts of ammonia and hydrogen cyanide. Miller referred to this reaction as the "rock of faith" for terrestrial prebiotic synthesis. Other chemists have had similar results for other purine bases. Pyrimidines (C, U, and T) are more difficult to construct abiotically, but chemists have had some successes (Voet and Schwartz 1982). Finally, the ribose sugars that form nucleotides can, at least under the right environmental conditions, be derived from a cascade of condensation reactions that begin only with formaldehyde.

The description of mostly independent, plausible chemical pathways that could have produced amino acids, nucleotides, and sugars leaves us far from fully formed building blocks on the verge of becoming a self-replicating system. One problem is exemplified by sugar formation. Not only does the sugar in nucleic acids (ribose) constitute a small percentage of all the sugars produced by formaldehyde condensation, there also exist multiple equally probable ways the purine bases can be attached to the sugar. Each of these produces a subtly, but importantly, different nucleotide isomer than that used by RNA. Pyrimidine bases cannot be made to attach to the sugar at all. To make matters worse, each building block needs to be activated, or chemically charged, before it can be incorporated into a polymer. Activation requires a preexisting source of chemical energy. Without cell membranes to concentrate this energy, it is challenging to understand how building blocks became activated in the RNA World (Orgel 1986).

Matthew Powner and colleagues (2009) discovered a clever solution **(Figure 17.11)**. The traditional method of synthesizing activated nucleotides (blue) has been to make the sugar and base in separate reactions, combine them to make a nucleoside, then phosphorylate the nucleoside to make an activated nucleotide (Szostak 2009). Powner and colleagues found that in the presence of inorganic phosphate, the same starting materials will form a different series of intermediates (green) and activated pyrimidine nucleotides will appear as the final product.

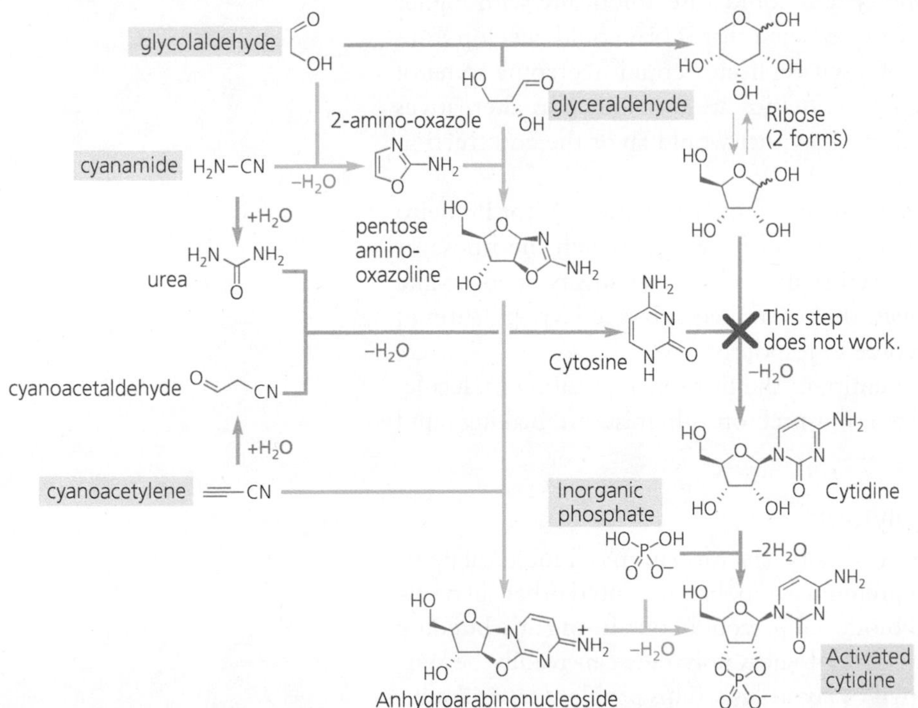

Figure 17.11 Prebiotic synthesis of activated pyrimidine ribonucleotides Starting materials are highlighted in gray. The final product is highlighted in lavender. The traditional method of making activated nucleotides, which does not work with pyrimidine bases, is shown in orange. The new method discovered by Powner and colleagues (2009) is shown in green. There is more than one way to skin a cat. Modified from Powner et al. (2009).

Based on this and other work, Powner and John Sutherland (2011) advocate a search for additonal one-pot recipes, using plausible reactants and under plausible prebiotic conditions, for the building blocks of life.

Another problem is the origin of chirality, or handedness. As noted earlier, living systems today use only one stereoisomer, or mirror-image form, of the amino acids in their proteins, and the same is true of nucleotides. In many of the chemical syntheses described by adherents to the Oparin–Haldane model, both mirror images of the building blocks are made in roughly equal quantities, and it is difficult to devise mechanisms that produce only one or the other. Exacerbating this problem is the fact that one mirror-image form would inhibit the polymerization of the other during any type of polymer self-replication (see Joyce et al. 1987). Prebiotic chemists are making progress on this front too. Jason Hein and colleagues (2011), for example, found that when they included amino acids in the mixture of reactants used to synthesize RNA precursors, the products were strongly biased toward the biologically natural stereoisomer.

An additional possibility under investigation for the origin of the RNA World is that the RNA World did not arise from scratch in a warm little pond. Instead, RNA may have been a later stage in an evolutionary lineage that derived from a simpler genetic system. Several non-RNA self-replicating systems have been proposed (see Orgel 2000; Joyce 2012b). Among them are the following: polymers made up not of ribonucleotides as we know them today, but of ribonucleotide analogs that have only one stereoisomer (Joyce et al. 1987); polymers made up of a hybrid between peptides and nucleic acids (Egholm et al. 1992); polymers made up of nucleotides composed of pyranose or tetrose sugar (Eschenmoser 1999; Schöning et al. 2000) or propylene glycol (Zhang et al. 2005) instead of ribose sugar; and even polymers made up of inorganic substances such as clay (Cairns-Smith et al. 1992). Christian deDuve (1991) has outlined a "Thioester World" in which information transfer is linked to the metabolic turnover of thioester linkages in a complex chemical milieu. All of these scenarios are based on the presumption that another self-replicating system could arise abiotically with higher probability than RNA. Some are envisioned such that RNA could develop from them; presumably, the preexistence of a self-replicator could overcome some of the challenges of RNA synthesis. Other scenarios are envisioned as alternatives to an RNA World, many formulated in a way that would favor the construction and use of catalysts other than RNA.

To demonstrate the plausibility of alternative genetic polymers, Vitor Pinheiro and colleagues (2012) synthesized half a dozen of them in which the ribose or deoxyribose of DNA or RNA have been replaced by other sugars or sugar-like compounds. For one of the alternatives, the researchers subjected a population of sequences to selection, and saw it evolve in response.

We do not know for certain the identity of the first self-replicating molecule, or where its building blocks came from, but prebiotic chemists are making rapid progress toward plausible answers.

The Assembly of Biological Polymers

The second step in the Oparin–Haldane theory, the formation of biological polymers from the building blocks in the prebiotic soup, has presented other theoretical and practical challenges. The prebiotic soup would contain organic building blocks dissolved in water, and although biological polymers can readily be synthesized in water, they also break down by hydrolysis. This problem raises doubts that polymers sufficiently long to serve as the basis of a self-replicating primordial organism would ever have formed in a simple organic soup (Ferris et al. 1996).

James Ferris and colleagues (1996), extending a tradition that dates from the 1940s and 1950s (see Ferris 1993), demonstrated a plausible mechanism to overcome the hydrolysis problem. Ferris et al. prepared a simple prebiotic soup in the lab and added the common clay mineral montmorillonite. Montmorillonite is a naturally occurring aluminum-silicate clay to which organic molecules readily adhere. When activated nucleotides (that is, nucleoside triphosphates) stick to montmorillonite, the clay acts as a catalyst and will join them together in a polynucleotide chain. While bound to the clay, the polynucleotides form more rapidly than they are hydrolyzed, and the researchers succeeded in encouraging the formation of polynucleotide chains containing 8–10 nucleotides in a row.

The building blocks of life may have been assembled into polymers on the surface of clay crystals. Adhering to clay helps a growing polymer avoid being broken apart by hydrolysis.

Ferris and colleagues then demonstrated that it was possible to prepare much longer polynucleotides by the daily addition of activated nucleotides to a premade oligonucleotide primer. They started with a polyadenylate primer 10 nucleotides long and let it bind to the montmorillonite. The scientists then added to the polyadenylate/clay solution a bath of activated adenosine nucleotides. The activated nucleotides reacted with the polyadenylate primers, adding themselves to the nucleotide chains. Ferris and colleagues then used a centrifuge to spin down the clay (and its attached nucleotide chains), poured off the spent solution, and added a fresh bath of activated nucleotides. By repeating this process, adding a fresh bath of activated nucleotides once each day, Ferris et al. synthesized polyadenylates over 40 nucleotides long (**Figure 17.12**). Ferris and his colleagues have since refined their recipe to the point that in one step, run over a single day, it can produce polynucleotides up to 50 nucleotides long (Huang and Ferris 2006).

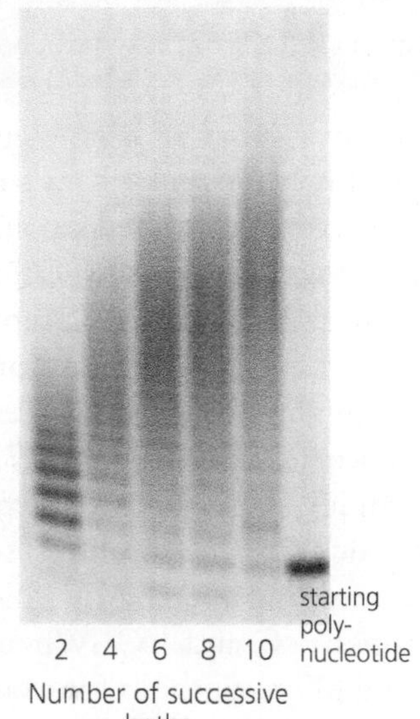

Figure 17.12 Synthesis of long nucleotide chains on clay This electrophoresis gel has separated mixtures of polyadenylates by size. The right lane contains a single band that corresponds to nucleotide chains 10 bases long; this was the starting point for Ferris et al.'s (1996) experiment. The left lane contains the mixture of polynucleotides produced when 10-nucleotide polyadenylates were allowed to bind to montmorillonite, then given two successive baths with activated adenosine nucleotides. Each successive band represents a one-nucleotide difference in length. The leftmost lane thus contains polyadenylates ranging from 11 to 20 nucleotides in length. The second lane from the left shows the results of four successive baths with activated nucleotides, and so on. Reactions run without montmorillonite failed to produce elongated nucleotide chains. From Ferris et al. (1996).

Ferris and Orgel have used repeated-bathing procedures to grow polypeptides up to 55 amino acids long on the minerals illite and hydroxylapatite (Ferris et al. 1996; Hill et al. 1998). The teams assert that their method models a mechanism by which biological polymers could have grown on the early Earth. Minerals in sediments that were repeatedly splashed with the prebiotic soup, or continuously bathed by it, could have nursed the formation of polymers that were long enough to become a self-replicating primordial form. Recombination among short polymers may have played a role as well (Lehman et al. 2011). This view has its critics (see Shapiro 2006), but the clay-catalysis research has given the second step of the Oparin–Haldane model at least some experimental support. We briefly discuss the third step in Section 17.3.

Protecting Life from the Environment

At this point, one can grapple with the possibility that we will discover a logical chain of events that led from simple inorganics, such as carbon dioxide, ammonia, and hydrogen cyanide, to nucleic acids. All these events could have taken place on Earth, or some could have taken place on extraterrestrial bodies. Some researchers have even suggested that certain chemical reactions might have occurred in the atmosphere itself, perhaps suspended in water droplets that rose and fell with the temperature. Regardless of the chemistry of the early atmosphere, early Earth probably offered many local opportunities for organic synthesis—hydrothermal environments, ocean water rich in ferrous iron, or the caldera of volcanoes, just to name a few. However, the final challenge for any model of the origins of life is not whether the early Earth would have provided for the needs of life, but whether it would have been hospitable enough to allow life to evolve.

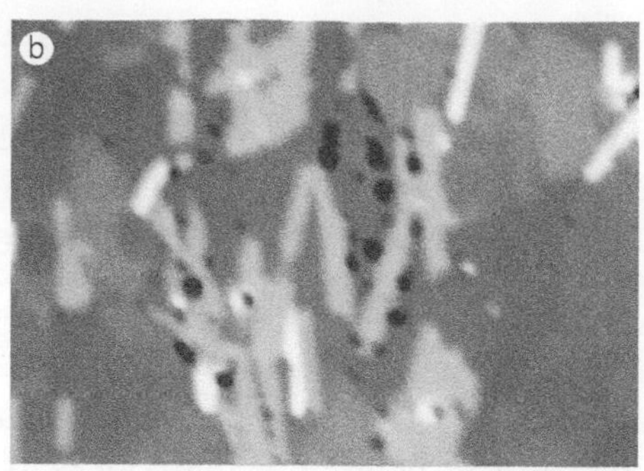

Figure 17.13 3.7-billion-year-old evidence suggesting life (a) This sedimentary rock is 3.7 billion years old. Note geologist's hammer for scale. (b) The rock contains microscopic graphite particles, which appear as black dots. The particles contain ratios of carbon isotopes that suggest they are derived from living cells. From Rosing (1999).

Sedimentary rocks from Isua, Greenland, contain evidence suggesting that life was already established on Earth by 3.7 billion years ago **(Figure 17.13a)**. The rocks, once part of the seafloor, have been exposed to high temperatures and pressures that have compacted the rocks and crystallized many of the minerals they contain. This transformation would have destroyed any microfossils the rocks might have originally harbored. The rocks do contain microscopic graphite globules, however (Figure 17.13b). Graphite is a mineral form of carbon.

Minick Rosing (1999) hypothesized that the graphite globules in the Greenland rocks are chemical fossils of ancient organisms. Rosing tested this hypothesis by measuring the isotopic composition of the graphite. Carbon has two stable isotopes: ^{12}C and ^{13}C. When organisms capture and fix environmental carbon, during photosynthesis for example, they harvest ^{12}C at a slightly higher rate than ^{13}C. As a result, carbonaceous material produced by biological processes has a slightly higher ratio of ^{12}C to ^{13}C than does carbonaceous material produced by nonbiological processes. When Rosing assayed the graphite globules, he found carbon isotope ratios characteristic of life. The results of subsequent analyses are consistent with this interpretation (Rosing and Frei 2004; Fedo et al. 2006; Nutman 2010), although Dominic Papineau and colleagues (2011) have called for confirmation that the graphite globules are as old as the rock around them, and Juske Horita (2005) has pointed out that there are nonbiological explanations for depleted ^{13}C.

Other researchers have examined rocks from Greenland that may be as much as 3.85 billion years old. Some have concluded that these rocks, too, contain chemical fossils of ancient life (Schidlowski 1988; Mojzsis et al. 1996). These conclusions have proven controversial (see Whitehouse et al. 2009; Lepland and Whitehouse 2011).

Are we likely to find evidence of life much earlier than the 3.7 (or 3.85) billion years ago demonstrated by the Greenland rocks? Probably not, for at least two reasons. First, erosion, plate tectonics, and volcanic eruptions have obliterated virtually all rocks from crust that might have existed earlier. Second, even if crust and oceans did exist earlier, continued bombardment of the planet by large meteors may have prevented life from being established much earlier than 3.7 to 3.85 billion years ago (see Figure 17.9). Large meteor impacts generate heat, create sun-blocking dust, and produce a blanket of debris. As time passed, and the largest planetesimals got swept up by Earth and other planets, the sizes of the largest impacts decreased. Norman Sleep and colleagues (1989) estimated that the last impact with sufficient energy to vaporize the entire global ocean, and thereby frustrate the emergence of any self-replicating system, probably happened between 4.44 and 3.8 billion years ago.

Although we still lack a complete scenario for how the first living things arose from nonliving matter, it appears that they did so quickly— almost as soon as early Earth was habitable.

Life's development may have been initiated many times over, if the intervals between sterilizing events were long enough for self-replication to re-evolve each time. Alternatively, life may have survived some of these impacts, sequestered in protective niches in the environment, such as deep-sea hydrothermal settings (Abramov and Mojzsis 2009). Regardless, we can estimate that the origins of life were threatened by an inhospitable environment until about 4 billion years ago. Whether the Oparin–Haldane model leading to an RNA World is correct, or some other scenario turns out to be more plausible, here is one last point before we consider cellular life: The origins of life occurred in a tumultuous abiotic environment. Paradoxically, Earth today is even more inhospitable to the origins of life. Life has become so successful at exploiting extreme niches that inorganic molecules have no sanctuaries left in which to reinvent self-replication before the first stages of these attempts are gobbled up by extant creatures.

17.3 What Was the Last Common Ancestor of All Extant Organisms, and What Is the Shape of the Tree of Life?

Once self-replicating systems evolved on Earth, at least one of them adapted to the use of DNA to store heritable information and to the use of proteins to express that information. This system eventually gave rise to all lineages of life on the planet today. We draw this conclusion because all life-forms (except some viruses) use DNA and proteins. In fact, all modern organisms use them in the same way. The same 20 amino acids and the same basic structure of the genetic code, as well as similar genetic and metabolic machinery encoded by correlated sequences, have been found in all creatures studied to date. On grounds of parsimony and explanatory power, we infer that all organisms share a common ancestry (Theobald 2010).

What Was the Most Recent Common Ancestor of All Living Things?

Because another shared feature of all extant life is the existence of cells, we also infer that the common ancestor was a cellular form. Technically speaking, we need to say that all life has descended from a population of interbreeding cells, because if portions of the primitive genome could be readily swapped, then life today cannot trace its ancestry to a single organism. The picture that emerges of the origins and early evolution of life on Earth can be diagrammed as in **Figure 17.14** (next page). The first cellular life whose descendants ultimately survived, the cenancestor (or cenancestors), appeared at least 2 billion years ago and probably much earlier. The advantages of cellular membranes as well as internal organellar membranes would have been enormous. Cells allow for compartmentalization. Certain chemicals can be concentrated inside the cell, and others can be pumped outside the cell. These capacities allowed life to accumulate its necessary constituents in much higher concentrations than they are found free in solution—activated nucleotides, for example. Cells also allowed genotypes and phenotypes to be linked, even after the latter had become the domain of proteins and not of the genetic material itself. It does a genotype little (evolutionary) good if the phenotype it encodes is free to diffuse to other genotypes.

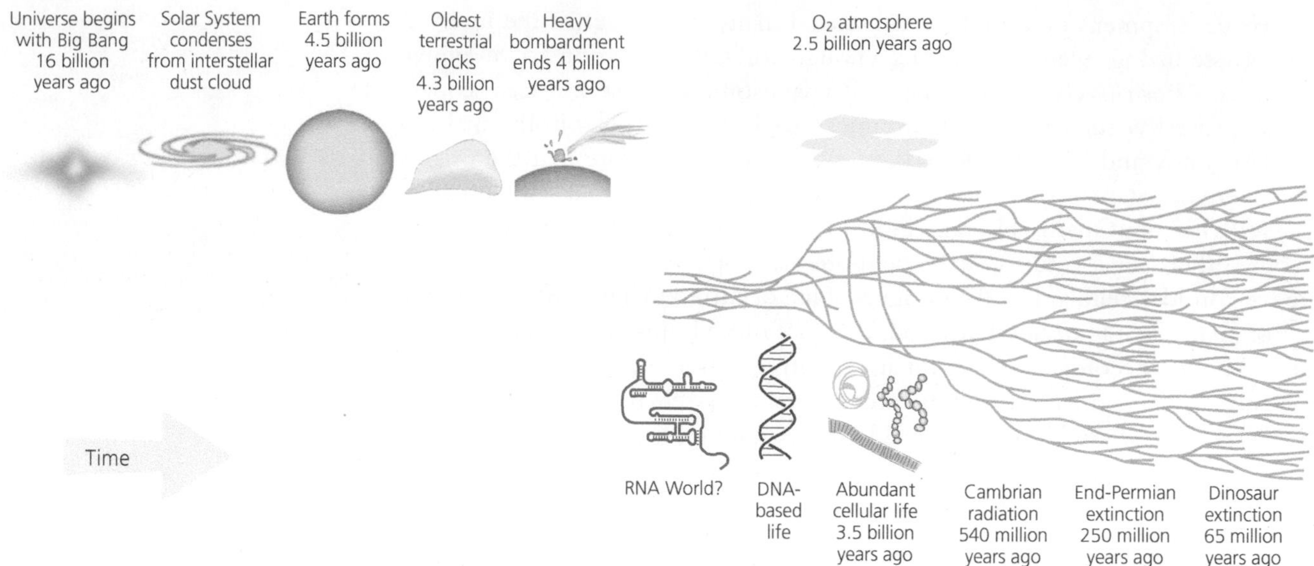

Figure 17.14 Overview of the evolution of life In the tree of life shown here (blue), the fusion of branches represents the acquisition of symbionts and other forms of horizontal gene transfer, phenomena we discuss later in the chapter. After Atkins and Gesteland (© 1998) in Gesteland et al. (1999); W. F. Doolittle (2000).

It is a long way from self-replicating RNA to the cenancestors, and many questions remain. For example, how did the earliest organisms acquire cellular form? One potential answer has come from the work of Sidney Fox and colleagues, who found that mixtures of polyamino acids in water or salt solution spontaneously organize themselves into microspheres with properties reminiscent of living cells (see Fox and Dose 1972; Fox 1988, 1991). Similarly, under the right conditions fatty acids and related molecules form vesicle-enclosing bilayers (Apel et al. 2002). As shown in **Figure 17.15**, such vesicles can serve as containers for nucleic acids. The vesicles can grow by absorbing more of their constituent molecules and will divide if squeezed through a small hole (Hanczyc et al. 2003). Sheref Mansy and colleagues (2008) demonstrated that fatty-acid-based vesicles are permeable to nucleic acids, which will diffuse into the interior and participate in template-based nucleic acid synthesis. Another potential answer holds that the precursors to cells were tiny compartments formed by inorganic mineral deposits (Martin and Russell 2003; Hansma 2010).

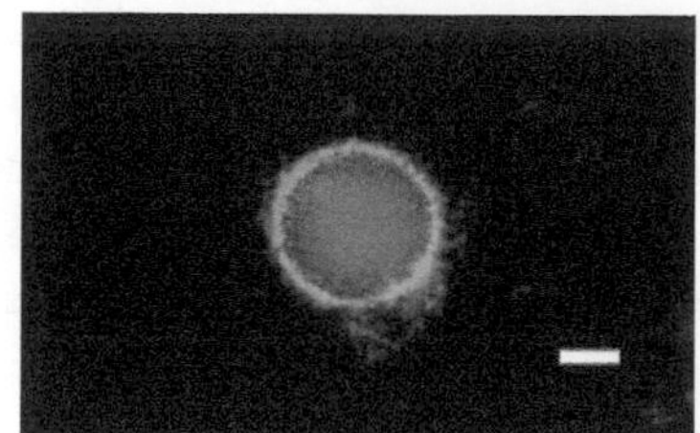

Figure 17.15 Model of an early cell This photo shows a vesicle made of the fatty-acid derviative myristoleate (green) containing fluorescently labeled RNA (orange). Scale bar = 1 micron. From Hanczyc et al. (2003).

Little is known about how the first self-replicating molecules evolved into cellular life-forms, although researchers have shown that structures reminiscent of cell membranes form spontaneously.

About the ancestral cellular lineage, like the first self-replicating system, we can ask what its general characteristics were, when it lived, and by what route its descendants evolved into today's orchids, ants, mushrooms, amoebae, and bacteria. Again, these events occurred early in Earth's history and much direct information has been lost. But if we know what questions to ask, the available data in the geological record can begin to remove the mystery of the first cellular life (Schopf 1994b).

The first place we might look in trying to identify the ancestral cells is the fossil record. In principle, a complete fossil record would allow us to trace lines of descent from living organisms all the way back to the cenancestors. However, it does not appear that the fossil record so far assembled can take us that deep into the past.

Several researchers have reported fossil cells preserved in rocks that are 3.2 to 3.5 billion years old (see Schopf 2006 for a review). For example, the fossils in

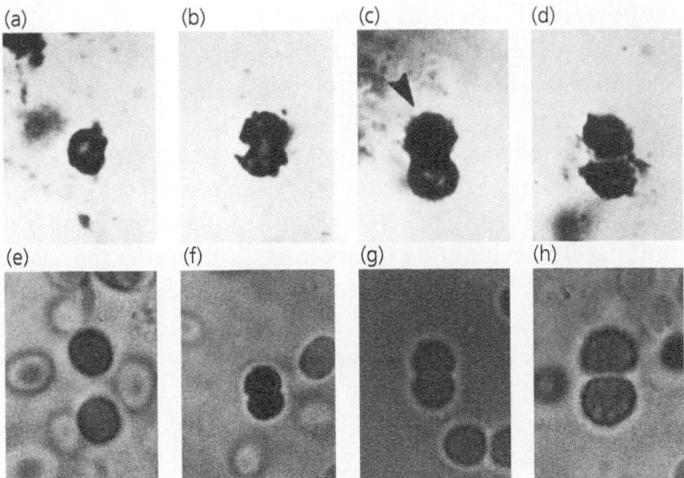

Figure 17.16 3.26-billion-year-old fossils of dividing cells The top row (a–d) shows microscopic fossils in 3.26-billion-year-old rocks from South Africa. The bottom row (e–h) shows living bacterial cells in various stages of division. Note the striking resemblance between the fossils and the living cells. From Knoll and Barghoorn (1977).

Figure 17.16a–d, discovered by Andrew Knoll and Elso Barghoorn (1977; see also Westall et al. 2001), are from a geological formation called the Swartkoppie chert in South Africa. Originally thought to be 3.4 billion years old, the cells are now estimated to be slightly younger. Among other reasons, Knoll and Barghoorn identified them as cells based on their carbon content, size distribution, location in sedimentary rocks, and resemblance to dividing bacteria (Figure 17.16e–h).

One way to learn about the characteristics of the earliest cells is to look for their fossils.

William Schopf (1993) reported fossils of what he believes are cyanobacteria from the slightly older Apex chert of Western Australia. Schopf's evidence has been the subject of considerable controversy (see Brasier et al. 2002; Dalton 2002; Kázmierczak and Kremer 2002; Kempe et al. 2002; Schopf et al. 2002a, 2002b; Pasteris and Wopenka 2002; De Gregorio et al. 2009; Marshall et al. 2011). Indeed, skeptics have questioned the biological origin of virtually all purported fossils more than 3 billion years old (Brasier et al. 2006). However, one of these skeptics coauthored a recent study reporting fossils of cells in rocks from the 3.4-billion-year-old Strelley Pool formation, also in Western Australia (Wacey et al. 2011; see also Javaux 2011).

Unfortunately, even if some or all of these purported fossils are genuine, they do not answer our present question. The fossil record for times earlier than 2.5 billion years ago is too spotty to allow paleontologists to trace lines of evolutionary descent from present-day organisms back to the fossils in the Swartkoppie or Apex cherts (Altermann and Schopf 1995). As a result, we have no direct way of knowing whether the organisms recorded in these rocks represent extinct or living branches of the tree of life, or whether they lived before or after the last common ancestor. If we want to discover the characteristics of ancestral cell lineages, we must use methods other than examination of the fossil record.

The Phylogeny of All Living Things

Another way to study the ancestral lineage is to reconstruct the phylogeny of all living things. A universal phylogeny should allow us to infer additional characteristics of the earliest life-forms beyond just their cellular nature (see Chapter 4). The first attempts to reconstruct the phylogeny of everything were based on the morphologies of organisms (see reviews in Woese 1991; Doolittle and Brown 1994). The morphological approach was productive for biologists interested in the branches of the tree of life that contain eukaryotes. Morphology was, historically, the basis of the phylogeny of many taxonomic groups. The morphological approach led only to frustration, however, for biologists interested in the branches of the universal phylogeny containing prokaryotes. Prokaryotes lack sufficient structural diversity to allow the reconstruction of morphology-based evolutionary trees.

When biologists developed methods for reading the sequences of amino acids in proteins, and the sequences of nucleotides in DNA and RNA, a new technique for estimating phylogenies quickly became established (Zuckerkandl and Pauling 1965). Some of the details of this technique are devilish (see Chapter 4), but the basic idea is straightforward. Imagine that we have a group of species, all carrying in their genomes a particular gene. We can read the sequence of nucleotides in this gene in each of the species, then compare the sequences. If species are closely related, their sequences ought to be fairly similar. If species occupy distant branches on the evolutionary tree, then their sequences ought to be less similar. As a result, we can use the relative similarity of the sequences of species to infer their evolutionary relationships. We place species with more similar sequences on neighboring branches of the evolutionary tree and species with less similar sequences on more distant branches.

Another way to learn about the earliest cells is to estimate the phylogeny of all living things, then infer the characteristics of the common ancestors.

The challenge in using sequence data to estimate the evolutionary tree for all living things is to find a gene that shows recognizable sequence similarities even between species as distantly related as *Escherichia coli* and *Homo sapiens* (Woese 1991). We need a gene that is present in all organisms and that encodes a product whose function is essential and thus subject to strong stabilizing selection. Without strong stabilizing selection, billions of years of genetic drift will have obliterated any recognizable similarities in the sequences of distantly related organisms. Also, the function of the gene must have remained the same in all organisms. This is because when a gene product's function shifts in some species but not in others, selection on the new function can cause a rapid divergence in nucleotide sequence that makes species look more distantly related than they actually are.

One gene that meets all the criteria for use in reconstructing the universal phylogeny is the gene that codes for the small-subunit ribosomal RNA (Woese and Fox 1977; Woese 1991). All organisms have ribosomes, and in all organisms the ribosomes have a similar composition, including both rRNA and protein. All ribosomes have a similar tertiary structure, including small and large subunits. In all organisms, the function of the ribosomes is the same: They are the machines responsible for translation. Translation is so vital, and organisms are under such strong natural selection to maintain it, that the ribosomal RNAs of humans and their intestinal bacteria show recognizable similarities in nucleotide sequence, even though humans and bacteria last shared a common ancestor billions of years ago. The small-subunit rRNA was the molecule chosen by Carl R. Woese, the chief pioneer of the use of molecular sequences in estimating the universal phylogeny (Fox et al. 1977; Woese and Fox 1977; see also Doolittle and Brown 1994). Though it is not a perfect solution, the small-subunit rRNA remains an informative resource for whole-life phylogenies.

Before presenting the tree of life as revealed by small-subunit rRNA sequences, it is worth recalling what biologists thought it looked like when Woese embarked on his project (**Figure 17.17**). According to the five-kingdom model (Whittaker 1969), the first split in the tree separates what will become the prokaryotes—the bacteria—on the left from what will become the eukaryotes on the right. The eukaryotes comprise three kingdoms containing the large multicellular organisms we are familiar with in daily life, plus a fourth kingdom of microorganisms.

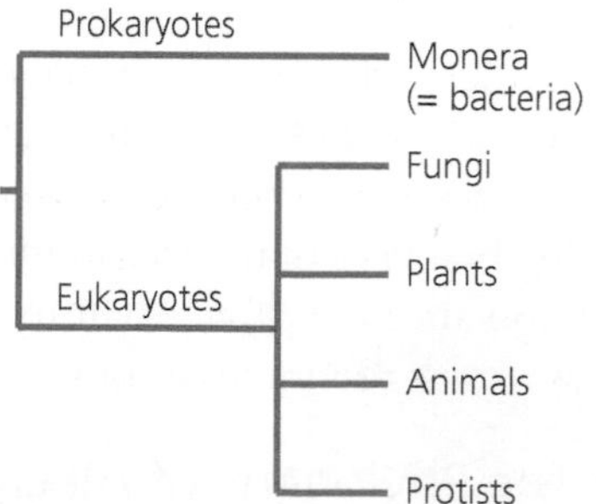

Figure 17.17 The tree of life, according to the five-kingdom scheme According to this scheme, the deepest node on the universal phylogeny is the split between the lineages that evolved into today's prokaryotes (bacteria) versus eukaryotes (fungi, plants, animals, and protists).

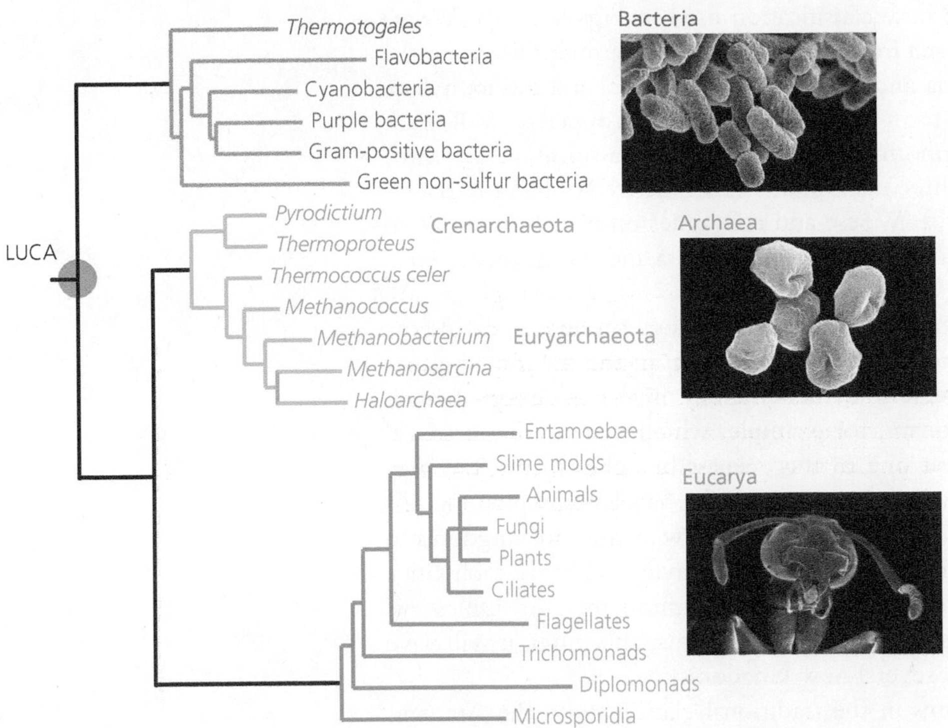

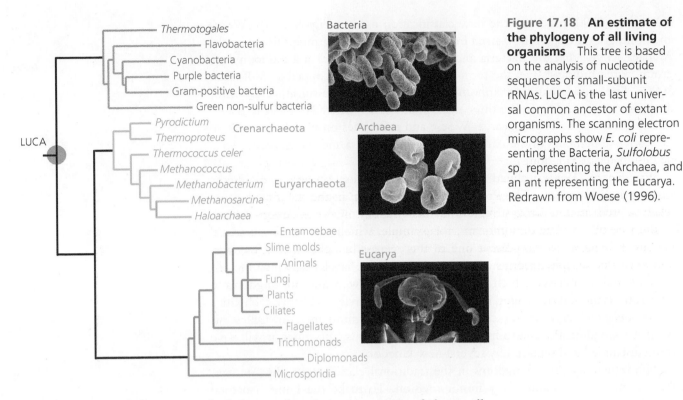

Figure 17.18 An estimate of the phylogeny of all living organisms This tree is based on the analysis of nucleotide sequences of small-subunit rRNAs. LUCA is the last universal common ancestor of extant organisms. The scanning electron micrographs show *E. coli* representing the Bacteria, *Sulfolobus* sp. representing the Archaea, and an ant representing the Eucarya. Redrawn from Woese (1996).

An estimate of the universal phylogeny based on sequences of the small-subunit rRNA appears in **Figure 17.18**. This whole-life rRNA phylogeny prompted a dramatic revision of our traditional view of the organization of life, because it reveals that the five-kingdom system of classification bears only a limited resemblance to actual evolutionary relationships (Woese et al. 1990; for a contrary view, see Margulis 1996).

The prokaryotes, for example, which are all grouped in kingdom Monera in the traditional classification, occupy two of the three main branches of the rRNA tree. One of these branches, the Bacteria, includes virtually all of the well-known prokaryotes. The Gram-positive bacteria, for instance, include *Mycobacterium tuberculosis,* the tuberculosis pathogen. The purple bacteria include *E. coli.* (The purple bacteria are so named because some of them are purple and photosynthetic, although *E. coli* is neither.) The cyanobacteria, all of which are photosynthetic, include *Nostoc,* an organism often seen in introductory biology labs.

The other prokaryote branch, the Archaea, is not as well known. Many of the Archaea live in physiologically harsh environments, are difficult to grow in culture, and were discovered only recently (see Madigan and Marrs 1997). Most of the Crenarchaeota, for example, are hyperthermophiles, living in hot springs at temperatures as high as 110°C. Many of the Euryarchaeota are anaerobic methane producers. Another group in the Euryarchaeota, the Haloarchaea, are highly salt dependent and are thus referred to as extreme halophiles.

The first whole-life phylogenies based on sequence data were estimated on the basis of small-subunit rRNA genes. These rRNA phylogenies revealed that the traditional five-kingdom system of classification offers a misleading view of evolutionary relationships.

Because of their prokaryotic cell structure, the Archaea were originally considered bacteria. When Woese and colleagues discovered that these organisms were only distantly related to the rest of the bacteria, they renamed them the archaebacteria (Fox et al. 1977; Woese and Fox 1977). Eventually biologists realized that, as the phylogeny in Figure 17.18 shows, the archaebacteria are in fact more closely related to the eukaryotes than they are to the true bacteria (see Bult et al. 1996; Olsen and Woese 1996). In recognition of this, Woese and colleagues (1990) proposed the new classification used in Figure 17.18. Woese and colleagues dropped the *bacteria* from archaebacteria, renaming this group the Archaea. Given that the Bacteria and the Archaea do not form a monophyletic group, some biologists feel the term *prokaryote* should be dropped as well (Pace 2006). The most inclusive taxonomic units in the new classification are three domains corresponding to the three main branches on the tree of life: the Bacteria, the Archaea, and the Eucarya. Woese and colleagues proposed that the two fundamental branches of the Archaea, the Crenarchaeota and the Euryarchaeota, be designated as kingdoms.

Woese et al. (1990) declined to offer a detailed proposal on how to divide the Eucarya into kingdoms. The Protista, a single kingdom in the traditional classification, are scattered across several fundamental limbs on the eukaryotic branch of the tree of life. The diplomonads, for example, which include the intestinal parasite *Giardia lamblia,* represent one of the deepest branches of the Eucarya. They are well separated from such other protists as the flagellates, which include *Euglena,* and the ciliates, which include *Paramecium.* If we want our kingdoms to be natural evolutionary groups, they should be monophyletic. That is, each kingdom should include all the descendants of a single common ancestor. Unless we want the kingdom Protista to include the animals, plants, and fungi, it will have to be disbanded and replaced by several new kingdoms.

The remaining three kingdoms in the traditional classification, the Animals, Plants, and Fungi, require only minor revision. To make the Fungi a natural group, for example, the cellular slime molds (such as *Dictyostelium,* a favorite of developmental biologists) has to be removed.

The universal rRNA phylogeny demonstrates, however, that the Animals, Plants, and Fungi, the kingdoms that have absorbed most of the attention of evolutionary biologists (and represent most of the examples in this book), are mere twigs on the tip of one branch of the tree of life. The multicellular, macroscopic organisms in these three kingdoms are newcomers on the evolutionary scene; they have a relatively recent last common ancestor. For genes shared among all organisms, such as the gene for the small-subunit rRNA, Animals, Plants, and Fungi appear to possess less than 10% of the nucleotide-level diversity observed on Earth (Olsen and Woese 1996).

An Examination of Early Cellular Life

Now that we have a universal phylogeny, what does it tell us about the earliest cellular life-forms? The orange dot in Figure 17.18 marks the last common ancestor of all extant organisms. According to this tree, LUCA's descendants diverged to become the Bacteria on one side and the Archaea–Eucarya on the other. Rooting the tree of life in this way was, and remains, a challenge because there is no outgroup to work with. The position of the root shown in Figure 17.18 is based on the work of several groups of researchers who used different analytical tricks **(Figure 17.19)**, but who all came up with approximately the same answer: The Archaea and Eucarya are more closely related to each other than either is to the Bacteria (Gogarten et al. 1989; Iwabe et al. 1989; Brown and Doolittle 1995; Baldauf et al. 1996). More recent data have yielded surprises, as we will discuss shortly. Estimating the location of the root remains an active area of research (see Zhaxybayeva et al. 2005; Dagan et al. 2010; Fournier and Gogarten 2010).

Assuming that we can accurately estimate a phylogeny that extends so far back in time, and that our placement of the root is reasonable, we can make

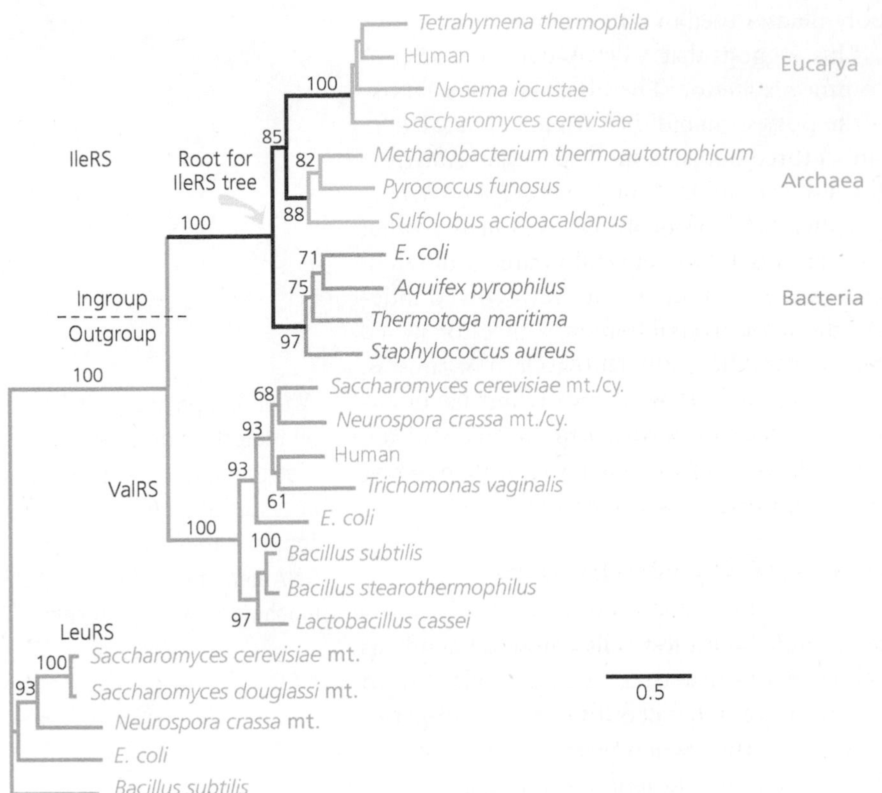

Figure 17.19 Rooting the universal phylogeny Rooting the tree of life, which has no organism to serve as an outgroup, requires an analytical trick. Here, gene families that arose in ancient duplications provide a molecular outgroup. The aminoacyl-tRNA synthetase gene family is our example. Keeping in mind that the phylogeny shown here is a gene tree, not an organism tree, look at the top portion, shown in black, blue, green, and gray. This portion is a phylogeny of the isoleucine aminoacyl-tRNA synthetase (IleRS) genes of organisms representing all three domains (the Bacteria, the Archaea, and the Eucarya). Genes for valine (ValRS) and leucine (LeuRS) aminoacyl-tRNA synthetase from a variety of bacteria and eukaryotes are the outgroups that root the IleRS tree. The outgroup phylogeny is shown in lavender lines. From Brown and Doolittle (1995).

inferences about when certain fundamental cellular traits evolved and in which lineages. Recall (from Chapter 4) that we can map character-state changes onto phylogenies using the principle of parsimony. For examples of how we might do this on the universal phylogeny, look at **Figure 17.20**. If a trait occurs in all three domains (Figure 17.20a), or if it occurs in the Bacteria and the Archaea but not in the Eucarya (Figure 17.20b), or if it occurs in the Bacteria and the Eucarya but not in the Archaea (Figure 17.20c), we can infer that the trait was present in the common ancestor and was lost on the lineage (if there is one) that lacks it. Alternative scenarios would require that the trait arose independently two or three times. If we assume that losses of fundamental cellular traits happen more readily than gains, then these alternative scenarios are less parsimonious.

We have already noted that the most conserved pieces of machinery inside cells function in the translation of genetic information from nucleic acids into proteins. Of the roughly 60 genes that occur in the genomes of all cellular organisms in all domains, 30 are ribosomal proteins and 15 are aminoacyl-tRNA synthetases—enzymes that attach amino acids to their tRNAs (Koonin 2003). We can infer that the last common universal ancestor had enzymes made of protein and a well-elaborated capacity for manufacturing them.

Can parsimony tell us whether the common ancestor of all extant organisms was already storing its genetic information in DNA? The fact that all extant organisms use DNA suggests that the common ancestor did the same. An alternative possibility is that the common ancestor stored its genetic information in some other molecule, such as RNA, but that storage in DNA was favored so strongly by natural selection that a conversion from RNA storage to DNA storage occurred independently in more than one domain. Use of DNA by the common ancestor appears more likely than this scenario of convergent evolution. One clue

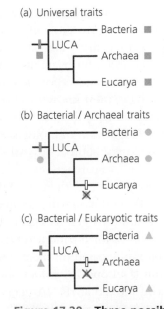

(a) Universal traits

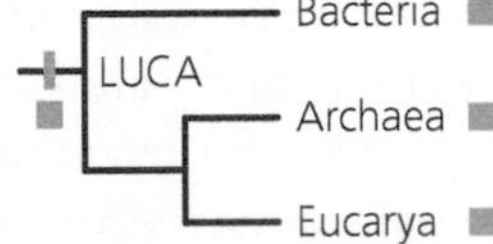

(b) Bacterial / Archaeal traits

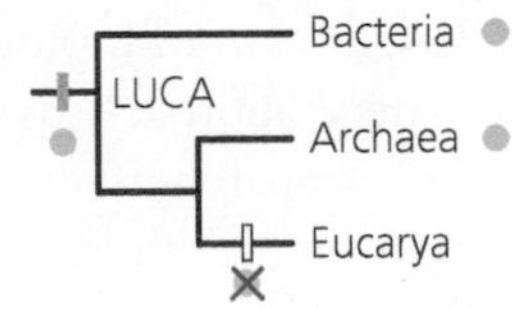

(c) Bacterial / Eukaryotic traits

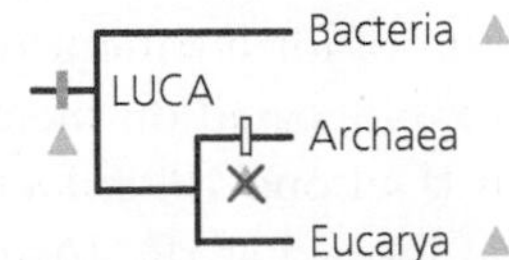

Figure 17.20 Three possible distributions of complex traits among the three domains of life Transitions marked by gray bars represent gains; transitions marked by open bars represent losses.

is that the DNA-dependent RNA polymerases used in transcription show strong similarities across all three domains. This suggests that a DNA-dependent RNA polymerase was present in the last common ancestor. The possession of a DNA-dependent RNA polymerase implies the possession of DNA (Benner et al. 1989). Likewise, DNA polymerases found in all three domains show enough similarities to suggest that the common ancestor also had a DNA polymerase. And where there was a DNA polymerase, again, there was probably DNA. On the other hand, some components of the machinery for DNA replication are so different in Bacteria versus Archaea and Eucarya that we can infer that they evolved independently (Leipe et al. 1999). Perhaps the last universal common ancestor stored its genetic information in DNA, but copied it differently than modern organisms.

Based on similar kinds of evidence and reasoning, many researchers have tentatively concluded that the most recent common ancestor was highly evolved and biologically sophisticated. Overall, the common ancestor appears in many ways to have been rather like a modern bacterium (Ouzounis et al. 2006).

Whole-life phylogenies based on molecular data suggest that the most recent common ancestor of all extant life was a sophisticated organism with a DNA genome and much of the machinery of modern cells....

Our Picture of the Tree of Life, and of the Earliest Cells, Continues to Evolve

Our understanding of the tree of life, and of the earliest cells whose descendants survive today, depends crucially on genetic sequence data. Such data allow us to estimate the universal phylogeny. Furthermore, sequence data provide much of the information about the traits of organisms that, when placed on the universal phylogeny, allow us to make inferences about the common ancestors. The amount of sequence data we have is growing explosively. Two trends in particular promise to yield many new insights.

First, our knowledge of the Archaea is increasing dramatically. As we mentioned earlier, many archaeans live in harsh and unusual environments. *Methanococcus jannaschii,* for example, lives anaerobically in deep-sea hydrothermal vents, at temperatures near 85°C and depths of at least 2,600 m (Jones et al. 1983). Not surprisingly, most known archaeans are difficult or impossible to grow in culture and thus are hard to study.

In 1984, a team of biologists working in the laboratory of Norman Pace pioneered a new approach to studying the environmental distribution of the Archaea. The researchers extracted DNA directly from mud and water samples collected in nature, then amplified and sequenced the DNA in the lab (Stahl et al. 1984). Following this approach, Edward DeLong and colleagues examined ribosomal RNA genes extracted from seawater collected in the Antarctic and off the coast of North America. DeLong and colleagues found many genes that were recognizable, based on their sequences, as belonging to previously unknown archaeans (DeLong 1992; DeLong et al. 1994). Susan Barns and colleagues (1994) likewise looked at rRNA genes extracted directly from mud in a hot spring in Yellowstone National Park. They also detected several rRNAs from previously unknown archaeans. Researchers in several laboratories are now pursuing similar studies (Service 1997; Schleper et al. 2005).

These environmental sequencing surveys have established that Archaea live not only in extreme envirnoments, but in moderate ones as well—including saltwater, freshwater, and soil. They are sufficiently abundant that they may turn out to play a substantial role in global energy and chemical cycles. And they include the only organisms capable of converting hydrogen and carbon dioxide into methane.

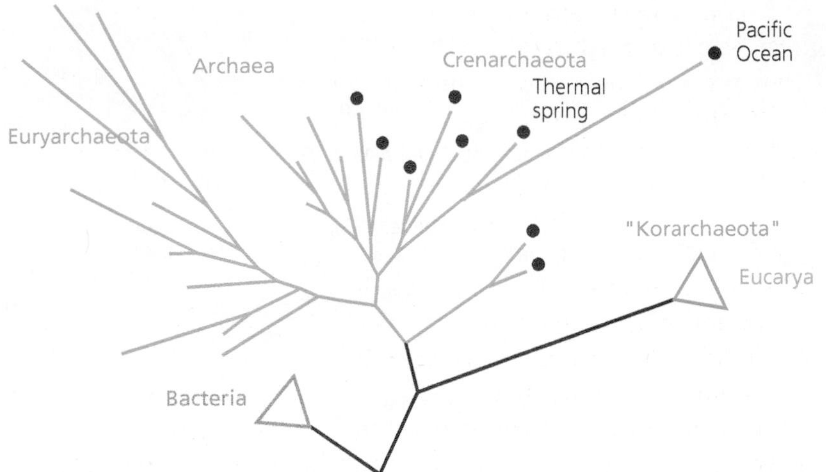

Figure 17.21 An estimate of the phylogeny of all extant organisms based on rRNA sequences Black dots indicate lineages known only from their rRNA genes. Branch lengths within the Archaea and among the domains are proportional to genetic distances; the genetic diversities among Bacteria and Eucarya are not to scale. Redrawn from Barns et al. (1996).

Barnes and colleagues (1996) used several new archaean rRNA sequences in the estimate of the whole-life phylogeny shown in **Figure 17.21**. This tree suggests the existence of a previously unknown kingdom of archaeans, the Korarchaeota. Given that the Archaea are one of the three fundamental groups of organisms, everything we learn about them improves our understanding of the universal phylogeny (Gribaldo and Brochier-Armanet 2006).

A second trend that will improve our understanding of both the universal phylogeny and the biology of the most recent common ancestor is the advent of whole-genome sequencing. When we wrote the first edition of this book, complete genomes had been sequenced for five organisms. When we wrote the second edition, 27 genomes were available; the third edition, 114. When we wrote the fourth edition, complete genomes had been sequenced for 389 organisms: 339 bacteria, 28 archaeans, and 22 eukaryotes—including a human, a mouse, a fruit fly, a roundworm, three plants, several fungi, and several protists. Rough drafts were complete for an additional 345 species, and genome projects were under way for 483 more (NCBI 2006). As we write the current edition, individuals with sufficiently deep pockets can have their own genomes sequenced—and the price is falling fast.

The availability of whole-genome sequences is allowing researchers to estimate evolutionary relationships with unprecedented accuracy, particularly for bacteria and archaea. Simon Harris and colleagues (2012), for example, used complete genomes to reconstruct the phylogeny of numerous strains of the bacterium *Chlamydia trachomatis,* which causes eye and urogenital infections in humans. Strains of *Chlamydia* have traditionally been classified using a standard panel of antibodies that react to a single protein. As the tree in **Figure 17.22** reveals, the traditional scheme often lumps together lineages that are only distantly related to each other. Note, for instance, the wide distribution on the tree of lineages traditionally placed in urogenital strain D.

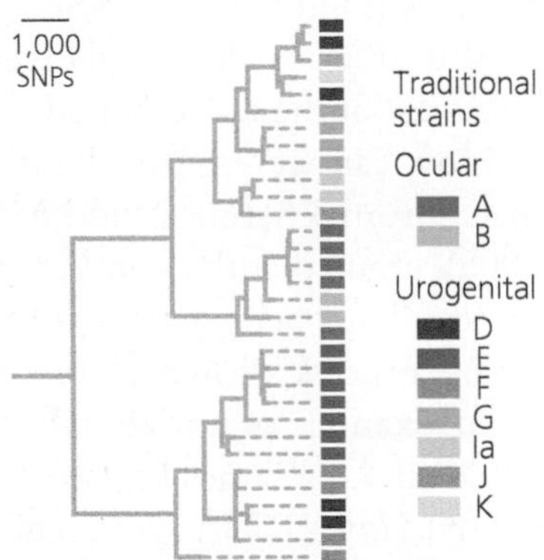

Figure 17.22 Relationships among lineages of *Chlamydia trachomatis* Modified from Harris et al. (2012).

The availability of whole-genome sequences also gives researchers the opportunity to estimate the universal phylogeny based on information from a great variety of genes. And this work, too, has produced surprises.

... However, whole-life phylogenies based on molecular data have also yielded surprises....

We would expect estimates of the universal phylogeny based on different genes to be broadly congruent. In fact, however, they are not. James R. Brown and W. Ford Doolittle (1997) estimated whole-life phylogenies based on some

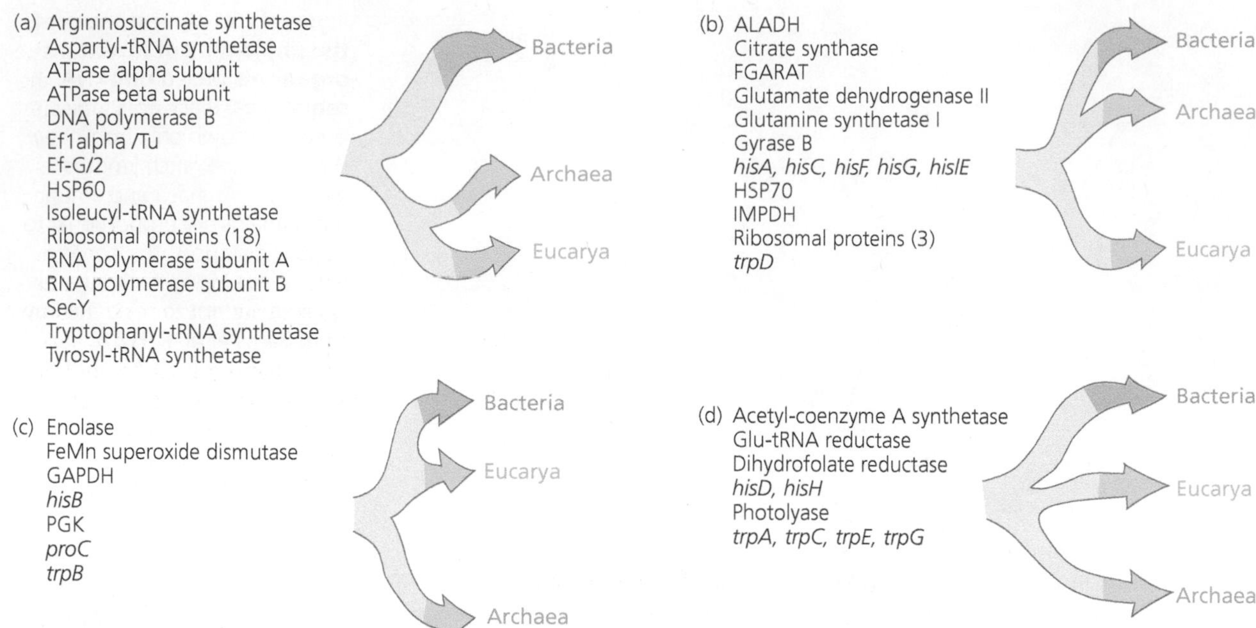

Figure 17.23 Different genes give different estimates of the universal phylogeny When James R. Brown and W. Ford Doolittle (1997) reconstructed the tree of life using a variety of genes, they found that different genes give different phylogenies. Some genes give trees in which Archaea and Eucarya are closest relatives (a); others give trees in which Bacteria and Archaea are closest relatives (b). Still other genes give trees in which Bacteria and Eucarya are closest relatives (c), or in which the relationships among the three domains are unresolved (d). From Brown and Doolittle (1997).

four dozen genes **(Figure 17.23)**. Genes for proteins involved in the storage and processing of genetic information often give a tree consistent with the small-subunit rRNA tree (Figure 17.23a). Genes for proteins involved in metabolism, however, often give a tree in which the Bacteria and the Archaea are closest relatives (Figure 17.23b). Still other genes give a tree in which the Bacteria and the Eucarya are closest relatives (17.23c), or in which there is an unresolved trichotomy of the three domains (Figure 17.23d).

How can we explain the discordance among the whole-life phylogenies estimated from different genes? Many researchers, Carl Woese included (1998, 2000, 2002), argue that the conflicts among data sets are too numerous and persistent to ignore. They believe the explanation is the horizontal movement of genes among taxa, a process known as **horizontal** (or **lateral) gene transfer.**

An example of horizontal gene transfer appears in **Figure 17.24**. This whole-life phylogeny, by James R. Brown (2001), is based on the gene for the β-subunit of phenylalanyl-tRNA synthetase, the enzyme that attaches the amino acid phenylalanine to its transfer RNA. Note the β-subunits from *Treponema pallidum* and *Borrelia burgdorferi*. These organisms are pathogenic spirochaetes. *Treponema pallidum* causes syphilis; *Borrelia burgdorferi* causes Lyme disease. They are unambiguously bacteria, and phylogenies based on most other components of the translation machinery put them where they belong. In **Figure 17.25**, for example, they appear deep within the bacteria (at about 5 o'clock). Yet on the tree in Figure 17.24, their genes for the β-subunit of phenylalanyl-tRNA synthetase appear to be archaeal. How could this be? The probable answer is that their β-subunit genes *are* archaeal. A common ancestor of the two spirochaetes lost its native bacterial β-subunit gene and replaced it with a gene from an archaean. We have discussed additional examples of lateral transfer elsewhere (Chapter 15).

... Chief among the surprises from whole-life phylogenies is that organisms appear to have swapped their genes more readily than anyone suspected. This means that the phylogenies of genes may be different from the phylogenies of the organisms that harbor them.

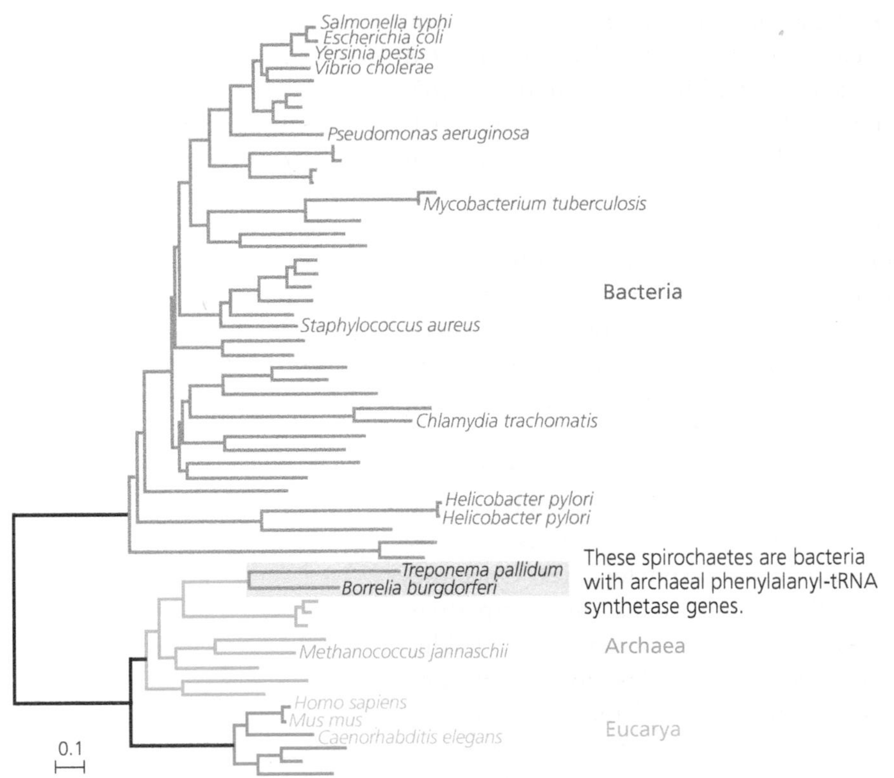

Figure 17.24 A universal phylogeny reveals lateral gene transfer This tree estimates the evolutionary relationships among genes for the phenylalanyl-tRNA synthetase β-subunit. The spirochaetes *Treponema pallidum* and *Borrelia burgdorferi* are bacteria, yet their genes for the β-subunit of the phenylalanyl-tRNA synthetase enzyme branch from within the archaea. The likely explanation is that the spirochaetes have lost their native bacterial version of the β-subunit gene and replaced it with a version of the gene picked up from an archaean. From Brown (2001).

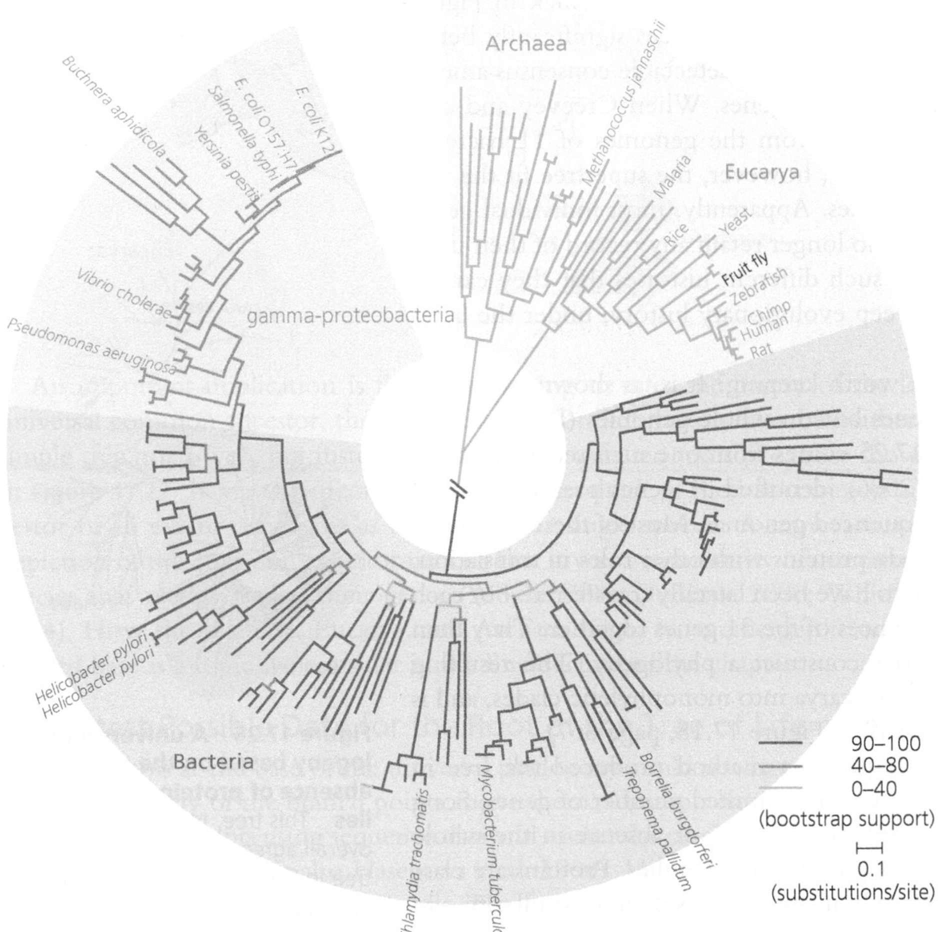

Figure 17.25 A universal phylogeny based on concatenated sequences of 31 universal genes, most for ribosomal proteins The tree resolves the Archaea, Bacteria, and Eucarya into monophyletic domains and is in good overall agreement with the rRNA tree in Figure 17.18. From Ciccarelli et al. (2006).

In summary, we can use the estimated universal phylogeny, along with geological and paleontological data, to bracket the time when the first branching in the universal phylogeny took place. The earliest possible date is when life on Earth began between 4.4 and 3.7 billion years ago; the most recent possible date, set by the oldest identifiable fossils, is at least 2 billion years ago.

17.4 How Did LUCA's Descendants Evolve into Today's Organisms?

In the preceding sections of this chapter, we have explored a hypothesized early stage of life, the RNA World. We have surveyed ideas on how the inhabitants of RNA World might have arisen from nonliving matter. And we have looked at reconstructions of the tree of life to see what they show us about the nature of its root. The researchers engaged in this last pursuit had set out to find LUCA, the last universal common ancestor, which they thought would be a single species of microbe. What they discovered instead was evidence of lateral gene transfer extensive enough to suggest that the last universal common ancestor was, in fact, a community.

In this last section of the chapter, we review a variety of ideas about how this ancestral community gave rise to the bacteria, archaeans, and eukaryotes that populate the Earth today. Each of the hypotheses we discuss offers a different scenario of lateral gene transfer to resolve apparent conflicts among the evolutionary histories of different genes (as illustrated in Figure 17.23). All of the hypotheses are speculative, and all are controversial.

The Universal Gene-Exchange Pool Hypothesis

The first biologist to reconstruct a universal phylogeny based on rRNA genes was, as we have discussed, Carl Woese. In building the first whole-life phylogeny, Woese discovered the Archaea. He was also among the first to recognize that the conflicts among universal phylogenies based on different genes were revealing something unexpected about the importance of lateral gene transfer in early evolution (Woese 1998).

Woese (2002, 2004) outlined a scenerio of early evolution in which lateral gene transfer was so rampant that it overshadowed vertical inheritance. Genomes, such as they existed, were modular in nature. That is, most ribozymes and proteins functioned independently of other ribozymes and proteins, and the genes encoding them could readily move from genome to genome. Organisms were assembled more by drawing genes from a universal gene-exchange pool (**Figure 17.31**), than by self-replication. Genealogical lineages, as we think of them today, did not exist. Nor did evolutionary trees.

As Woese himself asserts, the situation he describes is not conducive to evolution by natural selection. When genotypes and phenotypes are acquired rather than inherited, differential reproductive success is of limited consequence. Instead, Woese postulates a non-Darwinian mechanism of communal evolution. Gradually, as proteins became more interdependent, the modularity of genomes gave way to a more integrated and stable format. Individual genes could no longer move so easily among genomes. Self-replication now had the more prominent role in the generation of new organisms. At this point, which Woese calls the Darwinian threshold, populations began to evolve by natural selection.

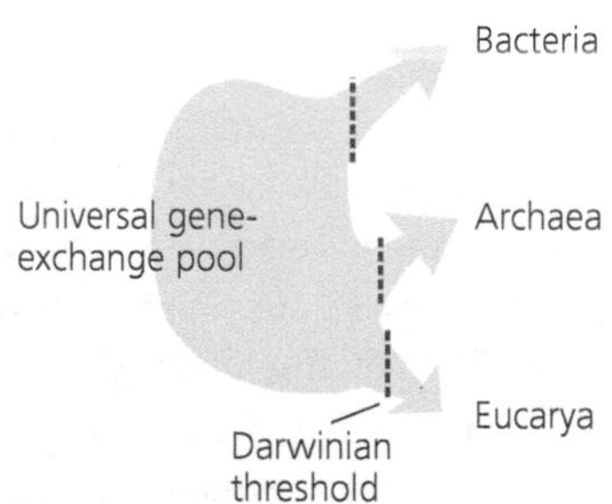

Figure 17.31 Carl Woese's conjecture on the origin of the three domains of life The last universal common ancestor was not a single species but rather a pool of readily exchanged and largely independent genes. Eventually, three cellular forms emerged with genomes stable enough to establish persistent lineages. At this point, which Woese calls the Darwinian threshold, populations began to evolve by natural selection. The three lineages became the three domains of life.

Woese believes that at least three stable lineages emerged independently from the universal gene-exchange pool. These were the ancestors of today's Bacteria, Archaea, and Eucarya. It is the order in which the three domains crossed their Darwinian thresholds that makes Archaea and Eucarya appear to be each other's closest kin in universal phylogenies based on rRNA (Figure 17.18), ribosomal proteins (Figure 17.25), and protein families (Figure 17.26). Bacteria crossed first, followed by Archaea, then Eucarya. Because they continued to draw from the universal gene exchange after the Bacteria had separated from it, the Archaea and Eucarya have more similar genes across most, but not all, of their genomes. But a deep phylogeny with a single unique root is a pattern we impose on the data, rather than a pattern that emerges from the data.

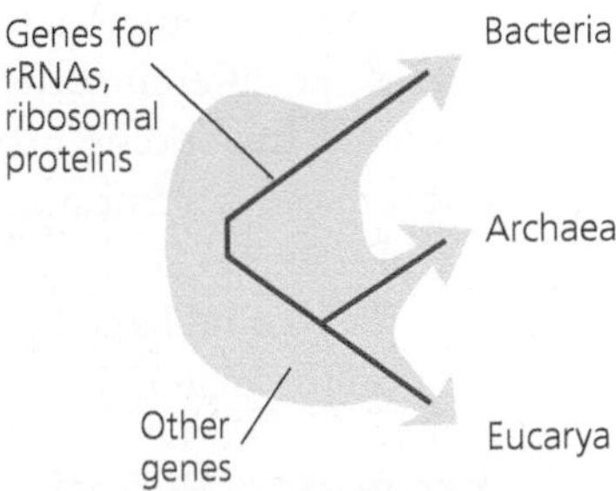

Figure 17.32 The deepest node on the universal phylogeny as a community We can reconstruct evolutionary trees from various kinds of data, as shown here for rRNA and ribosomal protein genes, but the most biologically accurate interpretation of the data may be to view the last common ancestor of the three domains as a community.

It is the non-Darwinian communal evolution at the heart of Woese's hypothesis that elicits skepticism from other biologists (see Whitfield 2004). What is the mechanism responsible? Peter Antonelli and colleagues (2003) developed a mathematical model of the universal gene-exchange pool and demonstrated that it was unstable. Woese asserts that other mathematical formulations of his verbal argument are possible. He and colleagues have developed a model showing that lateral gene transfer leads to convergence on a universal genetic code (Vestigian et al. 2006). But until a full quantitative model of the universal gene exchange is developed and shown to be workable, Woese's ancestral gene exchange will remain a rather abstract conjecture.

While they may not agree with the notion of non-Darwinian communal evolution, many other researchers do concur with Woese that lateral gene transfer was rampant enough during life's early history that we have to think of the last common ancestor of the three domains as a community rather than a single species (**Figure 17.32**; see Kurland et al. 2006).

The Ring-of-Life Hypothesis

Like researchers reconstructing the whole-life phylogeny from different genes, researchers comparing the genes of eukaryotes to those of bacteria and archaea have discovered a curious pattern. Christian Esser and colleagues (2004), for example, compared the amino acid sequences specified by more than 6,000 yeast genes to those encoded by more than 175,000 bacterial and archaean genes. For some 75% of the yeast genes, the most similar non-eukaryotic gene came from a bacterium; for the rest, it came from an archaean. In general, eukaryotic genes involved in the storage and use of genetic information, in processes such as transcription and translation, tend to be more similar to archaean genes. Eukaryotic genes involved in metabolic processes, such as the synthesis of amino acids, tend to be more similar to bacterial genes (Simonson et al. 2005).

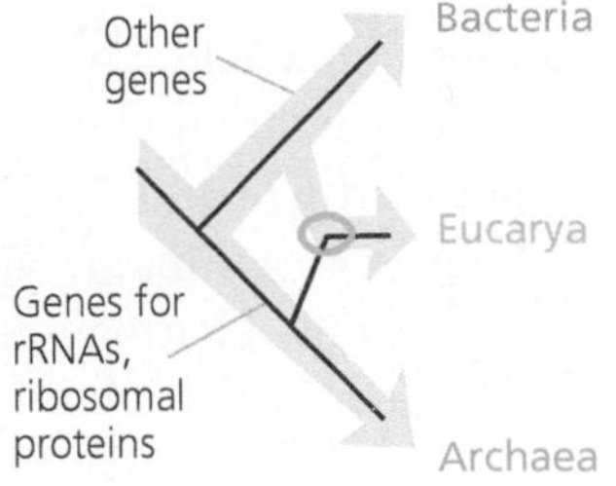

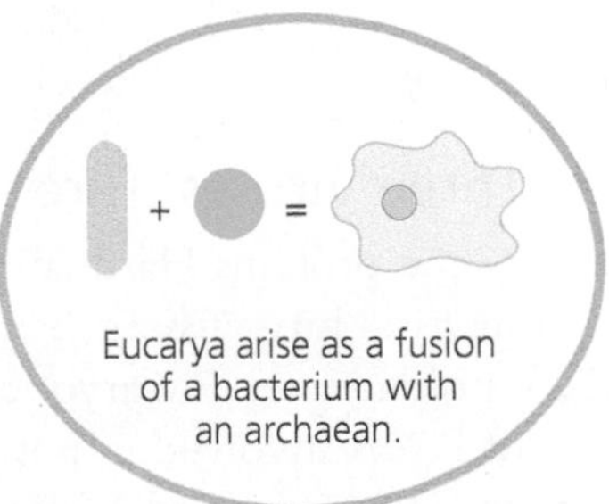

+
=
Eucarya arise as a fusion of a bacterium with an archaean.

Figure 17.33 The ring of life According to this idea, the Eucarya arose from the fusion of a bacterium with an archaean.

This pattern suggested to Maria Rivera and James Lake (2004) that the first eukaryote arose when a bacterium fused with an archaean (see also Horiike et al. 2002, 2004). The lineage that arose from this union retained the informational genes from the archaean and the metabolic genes from the bacterium. The whole-life phylogeny, as shown in **Figure 17.33**, has a ring at its center. In some versions of this hypothesis, the bacterial partner in the fusion that created the first eukaryote eventually became the mitochondrion (see Chapter 15); in other versions, the eukaryote lineage acquired the mitochondrion later.

If the bacterial partner in the fusion that created the first eukaryote was the ancestor of the mitochondrion, then the metabolic genes of eukaryotes should arise from within the α-proteobacteria, the bacterial clade known from rRNA phylogenies to be the source of the mitochondrion. Bjorn Canback and colleagues (2002) tested this prediction by reconstructing deep phylogenies for eight enzymes involved in glycolysis. In fact, none of the glycolytic enzymes of Eucarya are closely related to their α-proteobacterial homologs. If the Eucarya were born from the union of a bacterium and an archaean, that union long predated the acquisition of the mitochondrion.

An early fusion of a bacterium and an archaean is still a possibility. Critics argue, however, that the bacterial and archaeal genes present in the eukaryotic genome arose at different times from different lineages (Lester et al. 2005), that the ring-of-life hypothesis provides no explanation of where the hundreds of proteins found only in Eucarya came from (Kurland et al. 2006), and that because Archaea and Bacteria lack a cytoskeleton that enables phagocytosis, it is difficult to see how they could have fused in the first place (Kurland et al. 2006, including supporting online material).

The Chronocyte Hypothesis

Russell Doolittle (2000) is among the advocates of a scenario that offers a solution to the phagocytosis problem. In this scenario, outlined in **Figure 17.34**, the deepest fork in the tree of life separates a lineage that will become the Bacteria and the Archaea from the lineage that will become the Eucarya. Hyman Hartman calls this lineage the chronocytes (Hartman and Fedorov 2002). The chronocyte lineage evolved a cytoskeleton and the ability to eat other microbes by phagocytosis. A chronocyte then ate an archaean that resisted digestion and became an endosymbiont. This endosymbiont eventually evolved into an organelle: the nucleus. The nucleus preserved the information processing genes from its archaeal ancestor but incorporated cytoskeletal genes from its host.The chronocytes had spawned the Eucarya. The Eucarya later acquired the mitochondrion and the chloroplast in the same way.

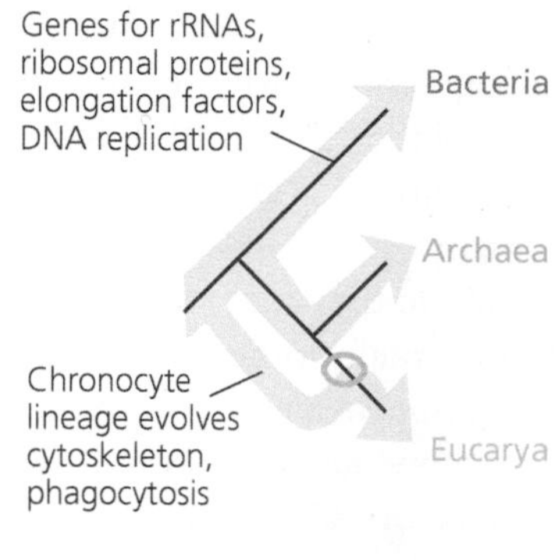

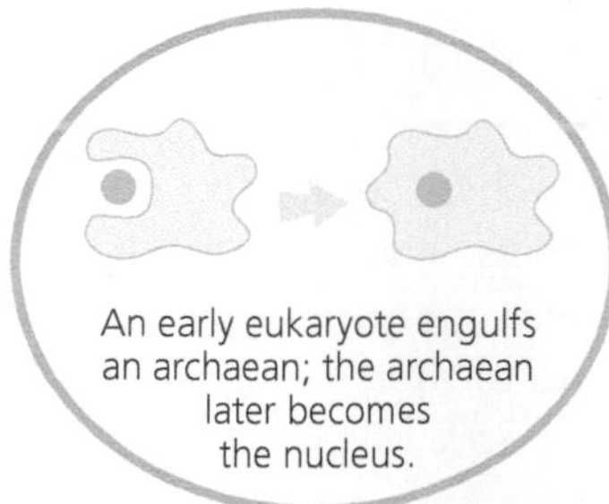

Figure 17.34 The chronocyte hypothesis According to this scheme, the deepest split in the tree of life separates the Bacteria and Archaea from the chronocytes, which eventually will become the Eucarya. After evolving a cytoskeleton and the ability to ingest other cells, a member of this lineage engulfed an archaean that resisted digestion and eventually became the eukaryotic nucleus. Later still, eukaryotes acquired the mitochondrion and chloroplast in a similar fashion.

One way to test the chronocyte hypothesis is to look for a living chronocyte. Such a creature would have a cytoskelton and feed on other cells, but it would lack a nucleus and mitochondria. To date, no such beast has been found.

Hyman Hartman and Alexei Fedorov (2002) assert, however, that they have found the next best thing. In an exhaustive search of whole genomes representing all three domains, Hartman and Fedorov identified 347 genes found in all eukaryotes but completely absent in the genomes of bacteria and archaeans. Among the 347, those with known function encode proteins that build and operate the cytoskeleton and inner membranes, modify RNA, and control various aspects of cellular physiology. Hartman and Fedorov believe that today's eukaryotes inherited these genes from their chronocyte ancestors.

The Three Viruses, Three Domains Hypothesis

Among the proteins Hartman and Fedorov found to be nearly universal among the Eucarya, but missing from Bacteria and Archaea, is an RNA-dependent RNA polymerase. Eukaryotic cells use this polymerase to replicate RNAi, a form of RNA involved in posttranscriptional gene regulation. On this and other evidence, Hartman and Fedorov suggest that chronocytes had RNA-based genomes. This notion contradicts the tentative inference, discussed in Section 17.3, that the last universal common ancestor stored its genetic information in DNA. And it raises the question of how the three domains made the transition to DNA from RNA.

Patrick Forterre has suggested an answer. This answer hinges on Forterre's view of where viruses came from. Viruses, the reader may have noticed, have been conspicuously absent from our discussion of the origin and history of life. This is a glaring omission, given that viruses vastly outnumber all other forms of life (Hamilton 2006). Virologists estimate, for example, that 1,200 different kinds of viruses inhabit the human gut, that a kilogram of marine sediment harbors a million distinguishable viral genotypes, that the Earth is home to 10^{31} individual virions, and that most of the genetic diversity among viruses remains undiscovered. To connect viruses to the tree of life, researchers have offered a full range of hypotheses on their origin (Forterre 2006a): Viruses are genes that have escaped from the genomes of cellular organisms; viruses are descended from cellular organisms that have evolved reduced genomes in association with a parasitic lifestyle; viruses are remnants of the earliest eras of life on Earth, including the RNA World and the early DNA world. Forterre believes the balance of recent evidence favors the last of these hypotheses. Among other reasons, viral genes are often not closely related to the homologous genes of their hosts—indeed, many viral genes have no known homologs—and structural similarities among viruses infecting all three domains suggest that they derive from a common ancestor that lived before the last universal common ancestor of cellular life.

If viruses evolved early, Forterre (2005, 2006a, 2006b) maintains, they provide a plausible explanation for how and why DNA-based life evolved from RNA-based life. The traditional view is that a switch to DNA was adaptive because DNA is more chemically stable and because mutations converting cytosine to uracil can be recognized and repaired in DNA but not in RNA. The trouble with this explanation is that the advantages it cites accrue over the long term, whereas natural selection happens in the short term. A more plausible scenario, according to Forterre, is that the switch from RNA to DNA first happened in viruses that made their living parasitizing cells with RNA-based genomes.

Cells that are parasitized by viruses evolve defenses. Among these defenses are enzymes that recognize viral genomes and chop them up. In turn, viruses that parasitize cells evolve counter-defenses. These include chemical modifications of the parasite's nucleic acids that prevent the host's defensive enzymes from recognizing and destroying the parasite's genome. Given that DNA is a chemically modified form of RNA, it seems plausible that DNA first appeared as an adaptation in a previously RNA-based virus engaged in an evolutionary arms race with an RNA-based host. Consistent with this scenario, extant viruses illustrate many of the required transitional forms. There are viruses with purely RNA-based genomes (and other means of defending themselves against their hosts). There are RNA viruses that replicate their genomes via DNA intermediates. There are DNA viruses that replicate their genomes through RNA-based intermediates. There are viruses with purely DNA-based genomes. There are even viruses with DNA-based genomes that use uracil instead of thymidine.

Finally, Forterre uses this scenario to explain how the cells that were the ancestors of the Bacteria, Archaea, and Eucarya themselves made the transition from RNA to DNA. Imagine that a DNA-based virus invades an RNA-based cell, loses the genes that encode its coat proteins, and thereby becomes an obligately intracellular extrachromosomal element. If the DNA virus carries a gene for reverse transcriptase, it may occasionally copy one of its host's genes into DNA and incorporate the gene into its own genome. Eventually the DNA genome will absorb all of the genes from the RNA genome, along the way ceasing to be a parasite and instead transmogrifying into a component of the host cell. Because the DNA genome replicates more efficiently, it will outcompete the RNA genome and ultimately cause its extinction. The RNA-based host cell has been converted to a DNA-based cell with an expanded genetic repertoire.

To explain the phylogenetic distributions of various cellular genes, Forterre postulates that the ancestors of the three domains of cellular life first diverged while still carrying their genetic information in RNA and that each was converted to DNA by a separate virus **(Figure 17.35)**. In particular, his hypothesis of three viruses for three domains accounts for the fact, mentioned in Section 17.3, that much of the machinery bacteria use to replicate their DNA appears unrelated to the machinery used by archaeans and eukaryotes. The viruses that carried DNA into the Archaea and Eucarya happened to be related to each other, but distantly related or unrelated to the virus that carried DNA into the Bacteria.

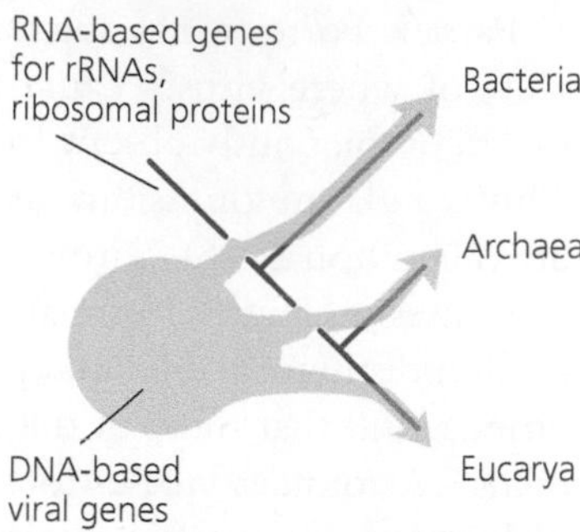

Figure 17.35 Three viruses, three domains According to this hypothesis, viruses infecting RNA-based cells first evolved DNA to counter their hosts' defenses. DNA was then transferred to cellular life when DNA-based viruses took up permanent residence inside their hosts.

Forterre asserts that his hypothesis also explains why there are only three domains of life. Once three lineages of DNA cells had evolved, they outcompeted and eliminated all other lineages of RNA-based cells. If, on the other hand, new domains of life can be generated by one of the endosymbiosis or fusion mechanisms discussed earlier, they should be appearing all the time.

Forterre notes that the best way to test his hypothesis would be to infect an RNA-based cell with a DNA-based virus and see if the host's descendants are ultimately transformed into DNA-based cells. Unfortunately, there are no known cellular organisms with RNA genomes. Forterre suggests, however, that it might be possible to use genetic engineering to make an RNA plasmid from the genome of an RNA virus, then insert it into a host cell whose genome encodes reverse transcriptase. If his hypothesis is correct, then RNA-based genes from the plasmid ought to turn up as DNA-based genes in the host cell's genome.

Researchers have proposed many hypotheses on how the three domains of life emerged. The most productive hypotheses make specific, testable predictions.

Another way to test the three viruses, three domains hypothesis is by reconstructing phylogenies of genes involved in managing DNA-based genomes. Forterre's hypothesis predicts that in such phylogenies, genes from the cellular domains of life will be derived from, and thus nested within, genes from viruses.

Figure 17.36 shows a phylogeny of DNA-dependent RNA polymerases from a sample of bacteria, archaeans, and eukaryotes, plus a variety of viruses. The tree was prepared by Didier Raoult and colleagues (2004) as a supplement to their

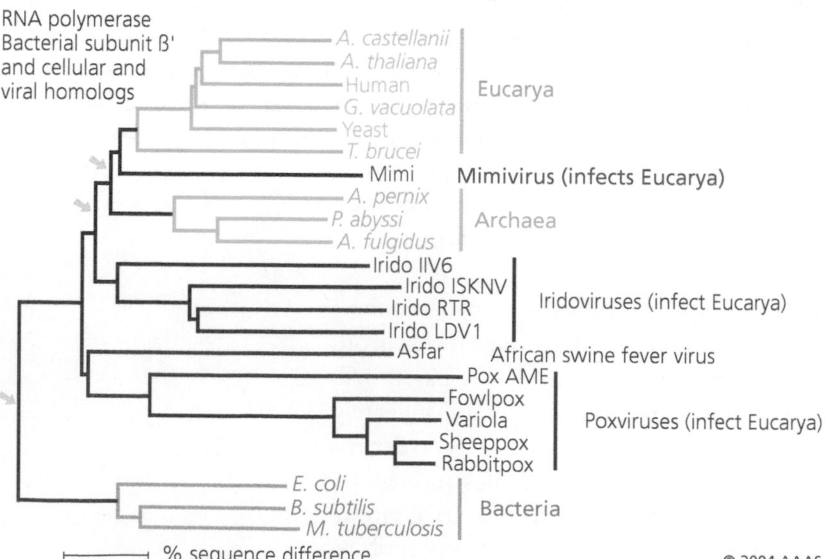

© 2004 AAAS

Figure 17.36 A phylogeny of cellular and viral DNA-dependent RNA polymerases The cellular RNA polymerases are interspersed with the viral RNA polymerases, consistent with the hypothesis that the cellular genes are derived from viral genes—but also with the hypothesis that the viral genes are derived from cellular genes. Gray arrows mark some plausible roots. Regardless of whether the root is cellular or viral, the tree requires at least three transitions. Redrawn from supplement to Raoult et al. (2004).

From "The 1.2-megabase genome sequence of Mimivirus." *Science* 306: 1344–1350, Figure S6, Supplementary Materials Page 17. Reprinted with permission from AAAS.

report on the complete genome of mimivirus. Mimivirus, which infects amoebae, is the largest virus yet discovered. Its genome is nearly 1.2 million base pairs long and appears to encode well over 1,000 genes. Mimivirus is, in fact, a virus. It lacks a ribozome and thus cannot reproduce without infecting a host cell. But it is more genetically complex than many bacteria.

The phylogeny is consistent with Forterre's hypothesis. The clades containing the three domains are interspersed with the viral clades. Mimivirus branches between the Eucarya and the Archaea. The iridoviruses, African swine fever virus, and the poxviruses branch between the bacteria and the other two cellular domains. One explanation for this pattern is that cellular RNA polymerases emerged, three times independently, from viral RNA polymerases.

Unfortunately, as Forterre (2005) points out, this is not the only possible explanation. The tree is also consistent with viral polymerases emerging three times independently from cellular polymerases. Note, for example, that among the places we could root the tree are the three gray arrows in the figure. Whether we assume the ancestor at the root was viral or cellular, the tree requires a minimum of three transitions.

Broader surveys of viral genomes might turn up genes that would allow us to reconstruct more extensive trees. These, in turn, might allow us to determine whether cellular genes evolved from viral ones, or vice versa. In the meantime, the three viruses, three domains hypothesis will remain controversial (Whitfield 2006; Zimmer 2006)—just like the other hypotheses we have reviewed in this section.

SUMMARY

Life arose from an abiotic environment a bit less than 4 billion years ago. Because of its extreme antiquity, the reconstruction of this event poses many challenges. Life may have begun only once and spread quickly over Earth. It may have arisen several times, each time only to be extinguished by the vaporization of Earth's water by the impact of meteorites. It may have evolved entirely on Earth or had its origins elsewhere in the Solar System.

Scientists have broken down the origins of life, regardless of its particulars, into three phases. The first phase would have been the synthesis of the building blocks of life, such as amino acids, nucleotides, and simple carbohydrates, from small inorganic molecules. Many plausible scenarios for these reactions exist, but significant uncertainties remain. The second phase would be the assembly of building blocks into a polymer, such as RNA, that contains and transmits information. Again, researchers have demonstrated that many of the details of such polymerization may be possible. And the third phase would be the advent of cellular compartmentalization, which would allow significant advances in phenotypic evolution and lead to the community of cells from which all current life is descended—the last universal common ancestors.

The study of life's origins is a highly collaborative venture, drawing on expertise from such diverse fields as astronomy, geology, chemistry, molecular biology, and evolutionary biology. It has forced us to consider exactly what *life* means. It is an excellent example of how science works, by formulating and testing hypotheses. It also reveals how great progress can be made in the absence, at this time, of a general consensus viewpoint. Notably, the editors of *Chemical and Engineering News* (December 6, 1999) asked prominent chemists what will be the major scientific questions for the next hundred years. Three responded that the origins of life would be one of the major topics of study. Rita R. Colwell, director of the National Science Foundation, remarked, "Chemists also will develop self-replicating molecular systems to provide insights into the molecular origins of life." This would be a milestone achievement, and yet it would be merely another piece in an elusive puzzle.

Evolutionary biologists attempt to assemble the big picture of evolution since the last common ancestors by reconstructing the universal phylogeny based on genomic sequence data. Sequence-based universal phylogenies have forced a dramatic revision of the fundamental organization of living things. Instead of five kingdoms of life, there are three domains: Bacteria, Archaea, and Eucarya. The first universal phylogeny indicated that among the three domains, Archaea and Eucarya are closest relatives. Comparison of phylogenies based on a variety of genes reveals, however, that there has been considerable horizontal gene transfer. Horizontal gene transfer may have been so rampant during life's history that it will force us to give up the idea of a single tree of life, and with it the idea of a single last common ancestor of all extant organisms.

How three domains of life emerged from the community of gene-exchanging organisms that now appears to have been the last universal common ancestor is the subject of much speculation and little consensus. As Russell Doolittle (2000) put it, "Vast amounts of sequence data notwithstanding, there are many things about early life on Earth that are not yet known."

QUESTIONS

1. The genesis of life is sometimes said to have required four things: energy, concentration, protection, and catalysis (for example, Cowen 1995). Explain why each of these four things was necessary for the generation of the primordial form.
2. What clues in the way RNA is used in modern cells hint that RNA may have an ancient role in cellular metabolism?
3. Briefly summarize two studies on evolution of RNA populations in the lab. In each experiment, what ability(ies) did the RNA population develop during evolution (i.e., what was the change in phenotype)? Do you think these RNA populations qualify as "life"? Do you think that a self-replicating RNA population will be developed in the lab in your lifetime?
4. Why was the gene for small-subunit RNA particularly well suited for studies of the phylogeny of all living things? Do you think this gene is also useful for studying relationships among living mammals, such as for elucidating the family tree of humans, chimpanzees, and gorillas? Why or why not?
5. Consider the classic five-kingdom model of life:

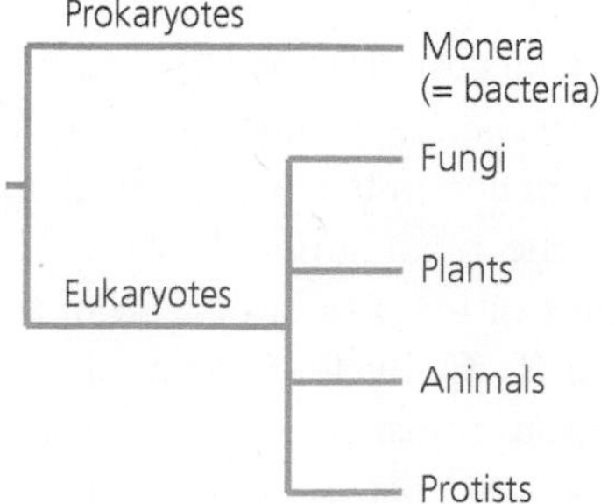

According to the "tree of life" first described by small-subunit rRNA analysis, which of the original "five kingdoms" need to be revised and which are still valid? Has this new tree of life stood the test of time, as other genes have been studied?
6. Briefly outline four possible hypotheses for the emergence of the three domains of life. Which is best supported (at present) from the evidence? Which is your favorite hypothesis (this need not be the one you think is most likely to be true!), and why?
7. It has been said that life could develop on Earth only because Earth is just the right distance from the Sun. Any closer, and Earth would have been too hot (like Mercury or Venus); any farther away, there would not have been sufficient solar energy for the evolution of living things. Recently, communities of organisms have been found in deep-sea vents on Earth. These communities seem to get all of their energy from the vents rather than the Sun. That is, the vent communities derive energy from the inner heat of Earth (which is provided ultimately by radioactivity). Even more recently, communities of bacteria have been found deep in subsurface rock. How does this discovery inform consideration of whether life might exist on other planets or moons that are not at "the right distance" from the Sun?
8. The SETI project (Search for Extraterrestrial Intelligence) is a research program that searches for intelligent life on other planets, using the assumptions that (a) intelligent life has probably evolved elsewhere in the universe, and (b) it should be detectable by scanning regions of the sky for anomalous radio signals. A fundamental uncertainty in this endeavor is the probability that any life at all will evolve on a planet, and if so, whether it will develop a civilization that uses radio waves. On Earth, how soon after Earth became habitable did life appear? How long did it take until eukaryotes appeared? How long until intelligent life appeared? The use of radio

waves? In your opinion, do the answers indicate that the evolution of life (of any kind) on other Earth-like planets is probable or improbable? How about the evolution of intelligent life? How about advanced civilization and radio communications?

9. The notion, suggested by Francis Crick and Leslie Orgel (1973), that Earth's first living things were sent here intentionally by an advanced civilization is known as the directed panspermia hypothesis. Orgel admitted to John Horgan (1991) that he and Crick intended the hypothesis as "sort of a joke." In their 1973 paper, however, Crick and Orgel treat the idea seriously enough to consider biological patterns that might serve as evidence. They point out, for example, that it is a little surprising that organisms with somewhat different [genetic] codes do not exist. The universality of the code follows naturally from an "infective" theory of the origins of life. Life on Earth would represent a clone derived from a single extraterrestrial organism.

   Since 1973, biologists have discovered that the genetic code is not universal and that organisms with "somewhat different codes" do, in fact, exist. Our mitochondria, for example, use a code slightly different from that used by our nuclei (see Knight et al. 1999). Many ciliates and other organisms also have slightly deviant codes (see Osawa et al. 1992). How strongly does the discovery that the genetic code is not universal refute the directed panspermia hypothesis? How strongly does it refute other versions of panspermia? Explain your reasoning. Can you think of other kinds of evidence that could (or do) either support or refute some version of panspermia?

10. Examine closely Figure 17.4. Recall that mutations were introduced at 140 randomized nucleotide positions. By the ninth round of selection, 4 nucleotides were responsible for most of the evolutionary change. Look at the other 136 nucleotides. Many of these have reverted to their original state. Why?

11. In the experiment diagrammed in Figure 17.5, why was it important for the researchers to include a tag on the end of the substrate RNAs?

12. In the chain of events leading from the abiotic synthesis of biological building blocks to the evolution of eukaryotes (Figure 17.10), which transition appears to be the least characterized? Why do you think this is the case?

13. Imagine an extremely primitive organism that has very primitive ribosomes with no proteins. Would it be possible to place this organism on the tree of life shown in Figure 17.18? Why or why not? How about an organism with no ribosomes? (Can you think of such an organism?) Is it conceivable that there are some as-yet-undiscovered primitive organisms that cannot be placed on these phylogenies? How would the discovery of such organisms affect our reconstruction of the cenancestor?

14. When biologists worked out the details of DNA replication in bacteria and eukaryotes, many researchers were surprised to discover that there are several different DNA polymerases, each with a different role. The machinery for replication seemed enormously complex, and every piece seemed essential if the whole system was to function at all. Many people found it hard to imagine how such a complex system of interdependent parts could have evolved by natural selection. Does the discovery of organisms with only one DNA polymerase (such as *Methanococcus jannaschii*) offer new insight into the evolution of replication? Why or why not?

15. Suppose you are trekking through remote Greenland on a day off from your summer job at a scientific camp, and you find an unusual layer of sedimentary rock that is not mapped on your geological charts. You suspect this rock might be even older than the 3.7-billion-year-old rocks from Isua (Figure 17.13). What would you do to determine whether these rocks have any evidence of ancient life? What results would show that life was indeed present before 3.7 billion years ago?

16. A recurring theme in literature of the past two centuries is that scientists should not try to "play God by trying to create life in the lab." Until recently, this phrase was just an unrealistic exaggeration used to make a point. Now, however, it appears that some scientists may be getting close to doing exactly that, by evolving self-replicating entities from abiotic molecules. Generally speaking, do you think these projects are worthwhile? What have they taught us about life and how it appeared on Earth?

17. a. A common objection to genetically modified food, and to genetic engineering in general, is that it is "not natural for genes to cross the species barrier." Comment on whether this argument is logically sound.

    b. Whether or not it is "natural" for genes to cross species barriers, many people have additional worries about genetically modified food. One such concern is the possibility that the genetically modified organisms might escape into the natural environment, where they could, conceivably, compete with other organisms and cause problems. Is this also a concern for research projects aimed at evolving self-replicating RNA populations? Which are more likely to survive if they escape into the natural environment: genetically engineered modern organisms or self-replicating RNA populations? Why?

    c. Do you think either of these research endeavors is unwise in any way? Why? If you were on a panel charged with developing guidelines for research, what safety measures would you support?

EXPLORING THE LITERATURE

18. For a review of the role of RNA in the origin of life, see this compilation of papers on the subject:

Gesteland, R. F., T. R. Cech, and J. F. Atkins (eds.). 2006. *The nature of modern RNA suggests a prebiotic RNA world. Cold Spring Harbor Monograph Series 43.* Cold Spring Harbor, NY: Cold Spring Harbor Laboratory Press.

And see this paper for an interesting computer model of what might happen once an RNA population manages to achieve self-replication:

Kuhn, C. 2005. A computer-glimpse of the origin of life. *Journal of Biological Physics* 31: 571–585.

19. For experimental evidence on a possible RNA precursor in a pre-RNA World, see:

Yu, H., S. Zhang, and J. C. Chaput. 2012. Darwinian evolution of an alternative genetic system provides support for TNA as an RNA progenitor. *Nature Chemistry* 4: 183–187.

20. Periodic meteor impacts were not the only hazard faced by Earth's early life. The Sun was much less luminous, which might have resulted in the Earth being frozen solid during the period that life apparently first arose (this is known as the Faint Young Sun Paradox). Yet at the same time, the Sun may also have produced more intense UV radiation, with UV doses at sea level on Earth more than 400 times as intense as today. See these papers for some ideas about these solar effects on the origin of life:

Karam, P. A. 2003. Inconstant sun: How solar evolution has affected cosmic and ultraviolet radiation exposure over the history of life on Earth. *Health Physics* 84: 322–333.

Pavlov, A. A., O. B. Toon, and T. Feng. 2006. Methane runaway in the early atmosphere—Two stable climate states of the Archean? *Astrobiology* 6: 161.

Bada, J. L., C. Bigham, and S. L. Miller. 1994. Impact melting of frozen oceans on the early Earth: Implications for the origin of life. *Proceedings of the National Academy of Sciences, USA* 91: 1248–2350.

21. For additional evidence that viruses were among the earliest forms of life, see:

Nasir, A., K. M. Kim, and G. Caetano-Anolles. 2012. Giant viruses coexisted with the cellular ancestors and represent a distinct supergroup along with superkingdoms Archaea, Bacteria and Eukarya. *BMC Evolutionary Biology* 12: 156.

22. The organisms on the deepest eukaryotic branches in Figure 17.18 were long thought to lack mitochondria. For a review of evidence suggesting that this belief is mistaken, see:

Palmer, J. D. 1997. Organelle genomes: Going, going, gone! *Science* 275: 790–791.

23. For a startling example in which a human parasite appears to have evolved from an ancestor that could photosynthesize, see:

Hannaert,V., E. Saavedra, et al. 2003. Plantlike traits associated with metabolism of *Trypanosoma* parasites. *Proceedings of the National Academy of Sciences, USA* 100: 1067–1071.

24. For evidence suggesting that life can exist in a cocktail of hydrocarbons like that on Titan, see:

Schulze-Makuch, D., S. Haque, et al. 2011. Microbial life in a liquid asphalt desert. *Astrobiology* 11: 241–258.

25. Among the intriguing new uses biologists are making of sequence data is to reconstruct the ancestral sequences of genes, then synthesize the sequences, translate them into proteins, and assess how the proteins function. For an example, see:

Gaucher E. A., S. Govindarajan, and O. K. Ganesh. 2008. Palaeotemperature trend for Precambrian life inferred from resurrected proteins. *Nature* 451: 704–707.

CITATIONS

Abramov, O., and S. J. Mojzsis. 2009. Microbial habitability of the Hadean Earth during the late heavy bombardment. *Nature* 459: 419–422.

Abrevaya, X. C., I. G. Paulino-Lima, et al. 2011. Comparative survival analysis of *Deinococcus radiodurans* and the haloarchaea *Natrialba magadii* and *Haloferax volcanii* exposed to vacuum ultraviolet irradiation. *Astrobiology* 11: 1034–1040.

Altermann, W., and J. W. Schopf. 1995. Microfossils from the Neoarchean Campbell Group, Griqualand West Sequence of the Transvaal Supergroup, and their paleoenvironmental and evolutionary implications. *Precambrian Research* 75: 65–90.

Anders, E. 1989. Prebiotic organic matter from comets and asteroids. *Nature* 342: 255–257.

Antonelli, P. L., L. Bevilacqua, and S. F. Rutz. 2003. Theories and models in symbiogenesis. *Nonlinear Analysis: Real World Applications* 4: 743–753.

Apel, C. L., D. W. Deamer, and M. N. Mautner. 2002. Self-assembled vesicles of monocarboxylic acids and alcohols: Conditions for stability and for the encapsulation of biopolymers. *Biochimica et Biophysica Acta* 1559: 1–9.

Arrhenius, S. 1908. *Worlds in the Making*. New York: Harper and Row.

Badash, L. 1989. The age-of-the-Earth debate. *Scientific American* 261: 90–96.

Baldauf, S. L., J. D. Palmer, and W. F. Doolittle. 1996. The root of the universal tree and the origin of eukaryotes based on elongation factor phylogeny. *Proceedings of the National Academy of Sciences, USA* 93: 7749–7754.

Bapteste, E., E. Susko, et al. 2005. Do orthologous gene phylogenies really support tree-thinking? *BMC Evolutionary Biology* 5: 33.

Barns, S. M., C. F. Delwiche, et al. 1996. Perspectives on archaeal diversity, thermophily and monophyly from environmental rRNA sequences. *Proceedings of the National Academy of Sciences, USA* 93: 9188–9193.

Barns, S. M., R. E. Fundyga, et al. 1994. Remarkable archaeal diversity detected in a Yellowstone National Park hot spring environment. *Proceedings of the National Academy of Sciences, USA* 91: 1609–1613.

Bartel, D. P. 1999. Recreating an RNA replicase. In *The RNA World* (2nd ed.), ed. R. F. Gesteland, T. R. Cech, and J. F. Atkins. Cold Spring Harbor, NY: Cold Spring Harbor Laboratory Press, 143–162.

Bartel, D. P., and J. W. Szostak. 1993. Isolation of new ribozymes from a large pool of random sequences. *Science* 261: 1411–1418.

Bartel, D. P., and P. J. Unrau. 1999. Constructing an RNA World. *Trends in Cell Biology* 9: M9–M13.

Beaudry, A. A., and G. F. Joyce. 1992. Directed evolution of an RNA enzyme. *Science* 257: 635–641.

Benner, S. A., A. D. Ellington, and A. Tauer. 1989. Modern metabolism as a palimpsest of the RNA world. *Proceedings of the National Academy of Sciences, USA* 86: 7054–7058.

Bentin, T. 2011. A ribozyme transcribed by a ribozyme. *Artificial DNA, PNA & XNA* 2: 40.

Brasier, M. D., O. R. Green, et al. 2002. Questioning the evidence for Earth's oldest fossils. *Nature* 416: 76–81.

Brasier, M., N. McLoughlin, et al. 2006. A fresh look at the fossil evidence for early Archaean cellular life. *Philosophical Transactions of the Royal Society B* 361: 887–902.

Brown, J. R. 2001. Genomic and phylogenetic perspectives on the evolution of prokaryotes. *Systematic Biology* 50: 497–512.

Brown, J. R. 2003. Ancient horizontal gene transfer. *Nature Reviews Genetics* 4: 121–132.

Brown, J. R., and W. F. Doolittle. 1995. Root of the universal tree of life based on ancient aminoacyl-tRNA synthetase gene duplications. *Proceedings of the National Academy of Sciences, USA* 92: 2441–2445.

Brown, J. R., and W. F. Doolittle. 1997. Archaea and the prokaryote-to-eukaryote transition. *Microbiology and Molecular Biology Reviews* 61: 456–502.

Bult, C. J., et al. 1996. Complete genome sequence of the methanogenic archaeon, *Methanococcus jannaschii*. *Science* 273: 1058–1073.

Cable, M. L., S. M. Hörst, et al. 2012. Titan tholins: Simulating Titan organic chemistry in the Cassini-Huygens era. *Chemical Reviews* 112: 1882–1909.

Cairns-Smith, A. G., A. J. Hall, and M. J. Russell. 1992. Mineral theories of the origin of life and an iron sulfide example. *Origins of Life and Evolution of the Biosphere* 22: 161–180.

Canback, B, S. G. E. Andersson, and C. G. Kurland. 2002. The global phylogeny of glycolytic enzymes. *Proceedings of the National Academy of Sciences USA* 99: 6097–6102.

Chang, S. 1999. Planetary environments and the origin of life. *Biological Bulletin* 196: 308–310.

Chyba, C. F. 2005. Rethinking Earth's early atmosphere. *Science* 308: 962–963.

Chyba, C. F., P. J. Thomas, et al. 1990. Cometary delivery of organic molecules to the early Earth. *Science* 249: 366–373.

Ciccarelli, F. D., T. Doerks, et al. 2006. Toward automatic reconstruction of a highly resolved tree of life. *Science* 311: 1283–1287.

Clark, R. N., J. M. Curchin, et al. 2010. Detection and mapping of hydrocarbon deposits on Titan. *Journal of Geophysical Research* 115: E10005.

Cowen, R. 1995. *History of Life*. 2nd ed. Cambridge: Blackwell Scientific Publications.

Creevey, C. J., D. A. Fitzpatrick, et al. 2004. Does a tree-like phylogeny only exist at the tips in the prokaryotes? *Proceedings of the Royal Society of London B* 271: 2551–2558.

Crick, F. H. C. 1966. The genetic code—yesterday, today, and tomorrow. *Cold Spring Harbor Symposia on Quantitative Biology* 31: 1–9.

Crick, F. H. C., and L. E. Orgel. 1973. Directed panspermia. *Icarus* 19: 341–346.

Cruikshank, D. P. 1997. Stardust memories. *Science* 275: 1895–1896.

Dagan, T., M. Roettger, et al. 2010. Genome networks root the tree of life between prokaryotic domains. *Genome Biology and Evolution* 2: 379–392.

Dalton, R. 2002. Squaring up over ancient life. *Nature* 417: 782–784.

Darwin, C. 1859. *On the Origin of Species by Means of Natural Selection*. London: John Murray.

Darwin, F. 1887. *The Life and Letters of Charles Darwin*, vol. 2. New York: Appleton.

DeDuve, C. 1991. *Blueprint for a Cell: The Nature and Origin of Life*. Burlington, NC: Patterson.

De Gregorio, B. T., T. G. Sharp, et al. 2009. Biogenic origin for Earth's oldest putative microfossils. *Geology* 37: 631.

DeLong, E. F. 1992. Archaea in coastal marine environments. *Proceedings of the National Academy of Sciences, USA* 89: 5685–5689.

DeLong, E. F., K. Y. Wu, et al. 1994. High abundance of Archaea in Antarctic marine picoplankton. *Nature* 371: 695–697.

Delsuc, F., H. Brinkman, and H. Philippe. 2005. Phylogenomics and the reconstruction of the tree of life. *Nature Reviews Genetics* 6: 361–375.

Doolittle, R. F. 2000. Searching for the common ancestor. *Research in Microbiology* 151: 85–89.

Doolittle, R. F. 2005. Evolutionary aspects of whole-genome biology. *Current Opinion in Structural Biology* 15: 248–253.

Doolittle, R. F., D.-F. Feng, et al. 1996. Determining divergence times of the major kingdoms of living organisms with a protein clock. *Science* 271: 470–477.

Doolittle, W. F. 1999. Phylogenetic classification and the universal tree. *Science* 284: 2124–2128.

Doolittle, W. F. 2000. Uprooting the tree of life. *Scientific American*, February, 90–95.

Doolittle, W. F., and J. R. Brown. 1994. Tempo, mode, the progenote, and the universal root. *Proceedings of the National Academy of Sciences, USA* 91: 6721–6728.

Egholm, M., O. Buchardt, et al. 1992. Peptide nucleic acids (PNA). Oligonucleotide analogues with an achiral peptide backbone. *Journal of the American Chemical Society* 114: 1895–1897.

Ekland, E. H., J. W. Szostak, and D. P. Bartel. 1995. Structurally complex and highly active RNA ligases derived from random RNA sequences. *Science* 269: 364–370.

Ellington, A. D., and J. W. Szostak. 1990. In vitro selection of RNA molecules that bind specific ligands. *Nature* 346: 818–822.

Elsila, J. E., D. P. Glavin, and J. P. Dworkin. 2009. Cometary glycine detected in samples returned by Stardust. *Meteoritics & Planetary Science* 44: 1323.

Eschenmoser A. 1999. Chemical etiology of nucleic acid structure. *Science* 284: 2118–2124.

Esser, C., N. Ahmadinejad, et al. 2004. A genome phylogeny for mitochondria among and a predominantly eubacterial ancestry of yeast nuclear genes. *Molecular Biology and Evolution* 21: 1643–1660.

Fedo, C. M., M. J. Whitehouse, and B. S. Kamber. 2006. Geological constraints on detecting the earliest life on Earth: A perspective from the Early Archaean (older than 3.7 Gyr) of southwest Greenland. *Philosophical Transactions of the Royal Society B* 361: 851–867.

Ferris, J. P. 1993. Catalysis and prebiotic RNA synthesis. *Origins of Life and Evolution of the Biosphere* 23: 307–315.

Ferris, J. P., A. R. Hill, Jr., et al. 1996. Synthesis of long prebiotic oligomers on mineral surfaces. *Nature* 381: 59–61.

Fitch, W. M., and K. Upper. 1987. The phylogeny of tRNA sequences provides evidence for ambiguity reduction in the origin of the genetic code. *Cold Spring Harbor Symposia on Quantitative Biology* 52: 759–767.

Fournier, G. P., and J. P. Gogarten. 2010. Rooting the ribosomal tree of life. *Molecular Biology and Evolution* 27: 1792–1801.

Forterre, P. 2005. The two ages of the RNA world, and the transition to the DNA world: A story of viruses and cells. *Biochimie* 87: 793–803.

Forterre, P. 2006a. The origin of viruses and their possible roles in major evolutionary transitions. *Virus Research* 117: 5–16.

Forterre, P. 2006b. Three RNA cells for ribosomal lineages and three DNA viruses to replicate their genomes: A hypothesis for the origin of cellular domain. *Proceedings of the National Academy of Sciences, USA* 103: 3669–3674.

Forterre, P., and H. Philippe. 1999. The last universal common ancestor (LUCA), simple or complex? *Biological Bulletin* 196: 373–377.

Fox, S. W. 1988. *The Emergence of Life: Darwinian Evolution from the Inside*. New York: Basic Books.

Fox, S. W. 1991. Synthesis of life in the lab? Defining a protoliving system. *Quarterly Review of Biology* 66: 181–185.

Fox, S. W., and K. Dose. 1972. *Molecular Evolution and the Origin of Life*. San Francisco: W. H. Freeman.

Fox, G. E., L. J. Magrum, et al. 1977. Classification of methanogenic bacteria by 16S ribosomal RNA characterization. *Proceedings of the National Academy of Sciences, USA* 74: 4537–4541.

Fry, I. 2006. The origins of research into the origins of life. *Endeavour* 30: 24–28.

Fusz, S., A. Eisenführ, et al. 2005. A ribozyme for the aldol reaction. *Chemistry & Biology* 12: 941–950.

Ge, F., L.-S. Wang, and J. Kim. 2005. The cobweb of life revealed by genome-scale estimates of horizontal gene transfer. *PLoS Biology* 3: e316.

Gesteland, R. F., T. R. Cech, and J. F. Atkins (eds.). 1999. *The RNA World*. 2nd ed. Cold Spring Harbor, NY: Cold Spring Harbor Laboratory Press.

Geyer, C. R., and D. Sen. 1997. Evidence for the metal-cofactor independence of an RNA phosphodiester-cleaving DNA enzyme. *Chemistry & Biology* 4: 579–593.

Gilbert, W. 1986. The RNA world. *Nature* 319: 618.

Gogarten, J. P., H. Kibak, et al. 1989. Evolution of the vacuolar H^+-ATPase: Implications for the origin of eukaryotes. *Proceedings of the National Academy of Sciences, USA* 86: 6661–6665.

Gogarten, J. P., W. F. Doolittle, and J. G. Lawrence. 2002. Prokaryotic evolution in light of gene transfer. *Molecular Biology and Evolution* 19: 2226–2238.

Gribaldo, S., and C. Brochier-Armanet. 2006. The origin and evolution of the Archaea: A state of the art. *Philosophical Transactions of the Royal Society B* 361: 1007–1022.

Hamilton, G. 2006. The gene weavers. *Nature* 441: 683–685.

Han, T.-M., and B. Runnegar. 1992. Megascopic eukaryotic algae from the 2.1-billion-year-old Negaunee Iron-Formation, Michigan. *Science* 257: 232–235.

Hanczyc, M. M., S. M. Fujikawa, and J. W. Szostak. 2003. Experimental models of primitive cellular compartments: Encapsulation, growth, and division. *Science* 302: 618–622.

Hansma, H. G. 2010. Possible origin of life between mica sheets. *Journal of Theoretical Biology* 266: 175–188.

Harris, J. K., S. T. Kelley, et al. 2003. The genetic core of the universal ancestor. *Genome Research* 13: 407–412.

Harris, S. R., I. N. Clarke, et al. 2012. Whole-genome analysis of diverse *Chlamydia trachomatis* strains identifies phylogenetic relationships masked by current clinical typing. *Nature Genetics* 44: 413–419, S1.

Hartman, H., and A. Fedorov. 2002. The origin of the eukaryotic cell: A genomic investigation. *Proceedings of the National Academy of Sciences, USA* 99: 1420–1425.

Hasegawa, M., W. M. Fitch, et al. 1996. Dating the cenancestor of organisms. *Science* 274: 1750–1753.

Hein, J. E., E. Tse, and D. G. Blackmond. 2011. A route to enantiopure RNA precursors from nearly racemic starting materials. *Nature Chemistry* 3: 704–706.

Hill, A. R., Jr., C. Böhler, and L. E. Orgel. 1998. Polymerization on the rocks: Negatively-charged α-amino acids. *Origins of Life and Evolution of the Biosphere* 28: 235–243.

Horgan, J. 1991. In the beginning. *Scientific American* 264(2): 116–125.

Horiike, T., K. Hamada, and T. Shinozawa. 2002. Origin of eukaryotic cell by symbiosis of archaea in bacteria supported by the newly clarified origin of functional genes. *Genes and Genetic Systems* 77: 369–376.

Horiike, T., K. Hamada, et al. 2004. The origin of eukaryotes is suggested as the symbiosis of *Pyrococcus* into γ-proteobacteria by phylogenetic tree based on gene content. *Journal of Molecular Evolution* 59: 606–619.

Horita, J. 2005. Some perspectives on isotope biosignatures for early life. *Chemical Geology* 218: 171–186.

Hörst, S. M., R. V. Yelle, et al. 2012. Formation of amino acids and nucleotide bases in a Titan atmosphere simulation experiment. *Astrobiology* 12: 809–817.

Huang, W., and J. P. Ferris. 2006. One-step, regioselective synthesis of up to 50-mers of RNA oligomers by montmorillonite catalysis. *Journal of the American Chemical Society* 128: 8914–8919.

Illangasekare, M., G. Sanchez, et al. 1995. Aminoacyl-RNA synthesis catalyzed by an RNA. *Science* 267: 643–647.

Iwabe, N., K.-i. Kuma, et al. 1989. Evolutionary relationship of archaebacteria, eubacteria, and eukaryotes inferred from phylogenetic trees of duplicated genes. *Proceedings of the National Academy of Sciences, USA* 86: 9355–9359.

Javaux, E. J., A. H. Knoll, and M. R. Walter. 2001. Morphological and ecological complexity in early eukaryotic ecosystems. *Nature* 412: 66–69.

Javaux, E. J. 2011. Palaeontology: Microfossils from early Earth. *Nature Geoscience* 4: 663–665.

Johnson, A. P., H. J. Cleaves, et al. 2008. The Miller volcanic spark discharge experiment. *Science* 322: 404.

Johnston, W. K., P. J. Unrau, et al. 2001. RNA-catalyzed RNA polymerization: Accurate and general RNA-templated primer extension. *Science* 292: 1319–1325.

Jones, W. J., J. A. Leigh, et al. 1983. *Methanococcus jannaschii* sp. nov., an extremely thermophilic methanogen from a submarine hydrothermal vent. *Archives of Microbiology* 136: 254–261.

Joyce, G. F. 1989. RNA evolution and the origins of life. *Nature* 338: 217–224.

Joyce, G. F. 1996. Ribozymes: Building the RNA world. *Current Biology* 6: 965–967.

Joyce, G. F. 1998. Nucleic acid enzymes: Playing with a fuller deck. *Proceedings of the National Academy of Sciences, USA* 95: 5845–5847.

Joyce, G. F. 2012a. Bit by bit: The Darwinian basis of life. *PLoS Biology* 10: e1001323.

Joyce, G. F. 2012b. Evolution: Toward an alternative biology. *Science* 336: 307–308.

Joyce, G. F., A. Schwartz, et al. 1987. The case for an ancestral genetic system involving simple analogues of the nucleotides. *Proceedings of the National Academy of Sciences, USA* 84: 4398–4402.

Kasting, J. F. 1993. Earth's early atmosphere. *Science* 259: 920–926.

Kázmierczak, J., and B. Kremer. 2002. Thermal alteration of the Earth's oldest fossils. *Nature* 420: 477–478.

Kempe, A., Schopf, J. W., et al. 2002. Atomic force microscopy of Precambrian microscopic fossils. *Proceedings of the National Academy of Sciences, USA* 99: 9117–9120.

Knight, R. D., S. J. Freeland, and L. F. Landweber. 1999. Selection, history, and chemistry: The three faces of the genetic code. *Trends in Biochemical Sciences* 24: 241–247.

Knoll, A. H. 1992. The early evolution of eukaryotes: A geological perspective. *Science* 256: 622–627.

Knoll, A. H. 1994. Proterozoic and Early Cambrian protists: Evidence for accelerating evolutionary tempo. *Proceedings of the National Academy of Sciences, USA* 91: 6743–6750.

Knoll, A. H., and E. S. Barghoorn. 1977. Archean microfossils showing cell division from the Swaziland system of South Africa. *Science* 198: 396–398.

Koonin, E. V. 2003. Comparative genomics, minimal gene-sets, and the last universal common ancestor. *Nature Reviews Microbiology* 1: 127–136.

Kruger, K., P. J. Grabowski, et al. 1982. Self-splicing RNA: Autoexcision and autocatalyzation of the ribosomal RNA intervening sequence of *Tetrahymena*. *Cell* 31: 147–157.

Kurland, C. G., C. Canback, and O. G. Berg. 2003. Horizontal gene transfer: A critical review. *Proceedings of the National Academy of Sciences, USA* 100: 9658–9662.

Kurland, C. G., L. J. Collins, and D. Penny. 2006. Genomes and the irreducible nature of Eukaryote cells. *Science* 312: 1011–1014.

Kvenvolden, K., J. Lawless, et al. 1970. Evidence for extraterrestrial amino acids and hydrocarbons in the Murchison meteorite. *Nature* 228: 923–926.

Kwok, S., and Y. Zhang. 2011. Mixed aromatic-aliphatic organic nanoparticles as carriers of unidentified infrared emission features. *Nature* 479: 80–83.

Lawrence, J. G., and H. Ochman. 1998. Molecular archaeology of the *Escherichia coli* genome. *Proceedings of the National Academy of Sciences, USA* 95: 9413–9417.

Lazcano, A., and S. L. Miller. 1996. The origin and early evolution of life: Prebiotic chemistry, the pre-RNA world, and time. *Cell* 85: 793–798.

Lehman, N., and G. F. Joyce. 1993. Evolution in vitro of an RNA enzyme with altered metal dependence. *Nature* 361: 182–185.

Lehman, N., C. D. Arenas, et al. 2011. Complexity through recombination: From chemistry to biology. *Entropy* 13: 17–37.

Leipe, D. D., L. Aravind, and E. V. Koonin. 1999. Did DNA replication evolve twice independently? *Nucleic Acids Research* 27: 3389–3401.

Lepland, A., and M. J. Whitehouse. 2011. Metamorphic alteration, mineral paragenesis and geochemical re-equilibration of early Archean quartz–amphibole–pyroxene gneiss from Akilia, Southwest Greenland. *International Journal of Earth Sciences* 100:1–22.

Lester, L., A. Meade, and M. Pagel. 2005. The slow road to the eukaryotic genome. *BioEssays* 28: 57–64.

Lincoln, T. A., and G. F. Joyce. 2009. Self-sustained replication of an RNA enzyme. *Science* 323: 1229–1232.

Lorsch, J. R., and J. W. Szostak. 1994. In vitro evolution of new ribozymes with polynucleotide kinase activity. *Nature* 371: 31–36.

Madigan, M. T., and B. L. Marrs. 1997. Extremophiles. *Scientific American* 276(4): 82–87.

Mansy, S. S., J. P. Schrum, et al. 2008. Template-directed synthesis of a genetic polymer in a model protocell. *Nature* 454: 122–125.

Margulis, L. 1996. Archaeal-eubacterial mergers in the origin of Eukarya: Phylogenetic classification of life. *Proceedings of the National Academy of Sciences, USA* 93: 1071–1076.

Marshall, C. P., J. R. Emry, and A. O. Marshall. 2011. Haematite pseudomicrofossils present in the 3.5-billion-year-old Apex chert. *Nature Geoscience* 4: 240–243.

Martin, W., and M. J. Russell. 2003. On the origins of cells: A hypothesis for the evolutionary transitions from abiotic geochemistry to chemoautotrophic prokaryotes, and from prokaryotes to nucleated cells. *Philosophical Transactions of the Royal Society of London B* 358: 59–85.

Miller, S. L. 1953. A production of amino acids under possible primitive Earth conditions. *Science* 117: 528–529.

Miller, S. L. 1992. The prebiotic synthesis of organic compounds as a step toward the origin of life. In *Major Events in the History of Life*, ed. J. W. Schopf. Boston: Jones and Bartlett, 1–28.

Mills, D. R., R. L. Peterson, and S. Spiegelman. 1967. An extracellular Darwinian experiment with a self-duplicating nucleic acid molecule. *Proceedings of the National Academy of Sciences, USA* 58: 217–220.

Mojzsis, S. J., G. Arrhenius, et al. 1996. Evidence for life on Earth before 3,800 million years ago. *Nature* 384: 55–59.

Mojzsis, S. J., R. Krishnamurthy, and G. Arrhenius. 1999. Before RNA and after: Geophysical and geochemical constraints on molecular evolution. In *The RNA World* (2nd ed.), ed. R. F. Gesteland, T. R. Cech, and J. F. Atkins. Cold Spring Harbor, NY: Cold Spring Harbor Laboratory Press. 1–47.

Müller, U. F. 2006. Re-creating an RNA world. *Cellular and Molecular Life Sciences* 63: 1278–1293.

Murzin, A. G., S. E. Brenner, et al. 1995. SCOP: A structural classification of proteins database for the investigation of sequences and structures. *Journal of Molecular Biology* 247: 536–540.

National Center for Biotechnology Information. 2006. Genome sequencing projects statistics. Available at http://www.ncbi.nlm.nih.gov/genomes/static/gpstat.html.

Nelson, K. E., R. A. Clayton, et al. 1999. Evidence for lateral gene transfer between Archaea and Bacteria from genome sequence of *Thermotoga maritima*. *Nature* 399: 323–329.

Niemann, H. B., S. K. Atreya, et al. 2005. The abundances of constituents of Titan's atmosphere from the GCMS instrument on the Huygens probe. *Nature* 438: 779–784.

Nissen, P., J. Hansen, et al. 2000. The structural basis of ribosome activity in peptide bond synthesis. *Science* 289: 920–930.

Norman, L. H., and A. D. Fortes. 2011. Is there life on ... Titan? *Astronomy & Geophysics* 52: 1-39 to 1-42.

Nutman, A. P., C. R. Friend, et al. 2010. ≥3700 Ma pre-metamorphic dolomite formed by microbial mediation in the Isua supracrustal belt (W. Greenland): Simple evidence for early life? *Precambrian Research* 183: 725–737.

Olsen, G. J., and C. R. Woese. 1996. Lessons from an Archaeal genome: What are we learning from *Methanococcus jannaschii*? *Trends in Genetics* 12: 377–379.

Orgel, L. E. 1986. RNA catalysis and the origins of life. *Journal of Theoretical Biology* 123: 127–149.

Orgel, L. 2000. A simpler nucleic acid. *Science* 290: 1306–1307.

Oró, J. 1961. Mechanism of synthesis of adenine from hydrogen cyanide under plausible primitive earth conditions. *Nature* 191: 1193–1194. .

Osawa, S., T. H. Jukes, et al. 1992. Recent evidence for evolution of the genetic code. *Microbiological Reviews* 56: 229–264.

Ouzounis, C. A., V. Kunin, et al. 2006. A minimal estimate for the gene content of the last universal common ancestor—exobiology from a terrestrial perspective. *Research in Microbiology* 157: 57–68.

Pace, N. R. 2006. Time for a change. *Nature* 441: 289.

Papineau, D., B. T. De Gregorio, et al. 2011. Young poorly crystalline graphite in the >3.8-Gyr-old Nuvvuagittuq banded iron formation. *Nature Geoscience* 4: 376–379.

Parker, E. T., H. J. Cleaves, et al. 2011. Primordial synthesis of amines and amino acids in a 1958 Miller H2S-rich spark discharge experiment. *Proceedings of the National Academy of Sciences USA* 108: 5526–5531.

Pasteris, J. D., and B. Wopenka. 2002. Images of the Earth's oldest fossils? *Nature* 420: 476–477.

Pinheiro, V. B., A. I. Taylor, et al. 2012. Synthetic genetic polymers capable of heredity and evolution. *Science* 336: 341–344.

Powner, M. W., B. Gerland, and J. D. Sutherland. 2009. Synthesis of activated pyrimidine ribonucleotides in prebiotically plausible conditions. *Nature* 459: 239–242.

Powner, M. W., and J. D. Sutherland. 2011. Prebiotic chemistry: A new modus operandi. *Philosophical Transactions of the Royal Society B* 366: 2870–2877.

Raoult, D., S. Audic, et al. 2004. The 1.2-megabase genome sequence of *Mimivirus*. *Science* 306: 1344–1350.

Raggio, J., A. Pintado, et al. 2011. Whole lichen thalli survive exposure to space conditions: Results of Lithopanspermia experiment with *Aspicilia fruticulosa*. *Astrobiology* 11: 281–292.

Raulin, F., C. Brassé, et al. 2012. Prebiotic-like chemistry on Titan. *Chemical Society Reviews* doi:10.1039/c2cs35014a.

Rivera, M. C., and J. A. Lake. 2004. The ring of life provides evidence for a genome fusion origin of eukaryotes. *Nature* 431: 152–155.

Rogers, J., and G. F. Joyce 1999. A ribozyme that lacks cytidine. *Nature* 402: 323–325.

Rosing, M. T. 1999. ^{13}C-depletedcarbon microparticles in >3700-Ma sea-floor sedimentary rocks from West Greenland. *Science* 283: 674–676.

Rosing, M. T., and R. Frei. 2004. U-rich Archaean sea-floor sediments form Greenland—indications of >3700 Ma oxygenic photosynthesis. *Earth and Planetary Science Letters* 217: 237–244.

Schidlowski, M. 1988. A 3,800-million-year isotopic record of life from carbon in sedimentary rocks. *Nature* 333: 313–318.

Schleper, C., G. Jurgens, and M. Jonusscheit. 2005. Genomic studies of uncultivated Archaea. *Nature Reviews Microbiology* 3: 479–488.

Schöning, K.-U., P. Scholz, et al. 2000. Chemical etiology of nucleic acid structure: The oligonucleotide system. *Science* 290: 1347–1351.

Schopf, J. W. 1993. Microfossils of the early Archean Apex chert: New evidence of the antiquity of life. *Science* 260: 640–646.

Schopf, J. W. 1994a. Disparate rates, differing fates: Tempo and mode of evolution changed from the Precambrian to the Phanerozoic. *Proceedings of the National Academy of Sciences, USA* 91: 6735–6742.

Schopf, J. W. 1994b. The early evolution of life: Solution to Darwin's dilemma. *Trends in Ecology and Evolution* 9: 375–378.

Schopf, J. W. 2006. Fossil evidence of Archaean life. *Philosophical Transactions of the Royal Society B* 361: 869–885.

Schopf, J. W., A. B. Kudryavtsev, et al. 2002a. Laser-Raman imagery of Earth's earliest fossils. *Nature* 416: 73–76.

Schopf, J. W., A. B. Kudryavtsev, et al. 2002b. Images of the Earth's oldest fossils? *Nature* 420: 477.

Secker, J., J. Lepock, and P. Wesson. 1994. Damage due to ultraviolet and ionizing radiation during the ejection of shielded micro-organisms from the vicinity of 1 M main sequence and red giant stars. *Astrophysics and Space Science* 219: 1–28.

Service, R. F. 1997. Microbiologists explore life's rich, hidden kingdoms. *Science* 275: 1740–1742.

Shapiro, R. 2006. Small molecule interactions were central to the origin of life. *Quarterly Review of Biology* 81: 105–125.

Shapiro, R., and D. Schulze-Makuch. 2009. The search for alien life in our solar system: Strategies and priorities. *Astrobiology* 9: 335–343.

Simonson, A. B., J. A. Servin, et al. 2005. Decoding the genomic tree of life. *Proceedings of the National Academy of Sciences, USA* 102: 6608–6613.

Sleep, N. H., K. J. Zahnle, et al. 1989. Annihilation of ecosystems by large asteroid impacts on the early Earth. *Nature* 342: 139–142.

Stahl, D. A., D. J. Lane, et al. 1984. Analysis of hydrothermal vent-associated symbionts by ribosomal RNA sequences. *Science* 224: 409–411.

Steitz, T. A., and P. B. Moore. 2003. RNA, the first macromolecular catalyst: The ribosome is a ribozyme. *Trends in Biochemical Sciences* 28: 411–418.

Szostak , J. W. 2009. Origins of life: Systems chemistry on early Earth. *Nature* 459: 171–172.

Tang, J., and R. R. Breaker. 1997. Rational design of allosteric ribozymes. *Chemistry and Biology* 4: 453–459.

Tarasow, T. M., S. L. Tarasow, and B. E. Eaton. 1997. RNA-catalyzed carbon–carbon bond formation. *Nature* 389: 54–57.

Theobald, D. L. 2010. A formal test of the theory of universal common ancestry. *Nature* 465: 219–222.

Tian, F., O. B. Toon, et al. 2005. A hydrogen-rich early atmosphere. *Science* 308: 1014–1017.

Trail, D., E. B. Watson, and N. D. Tailby. 2011. The oxidation state of Hadean magmas and implications for early Earth's atmosphere. *Nature* 480: 79–82.

Tuerk, C., and L. Gold. 1990. Systematic evolution of ligands by exponential enrichment: RNA ligands to bacteriophage T4 DNA polymerase. *Science* 249: 505–510.

Unrau, P. J., and D. P. Bartel. 1998. RNA-catalyzed nucleotide synthesis. *Nature* 395: 260–263.

Vestigian, K., C. Woese, and N. Goldenfeld. 2006. Collective evolution of the genetic code. *Proceedings of the National Academy of Sciences, USA* 103: 10696–10701.

Voet, A. B., and A. W. Schwartz. 1982. Uracil synthesis via HCN oligomerization. *Origins of Life* 12: 45–49.

Wacey, D., M. R. Kilburn, et al. 2011. Microfossils of sulphur-metabolizing cells in 3.4-billion-year-old rocks of Western Australia. *Nature Geoscience* 4: 698–702.

Westall, F., M. J. de Wit, et al. 2001. Early Archean fossil bacteria and biofilms in hydrothermally-influenced sediments from the Barberton greenstone belt, South Africa. *Precambrian Research* 106: 93–116.

Wetherill, G. W. 1990. Formation of the Earth. *Annual Review of Earth and Planetary Science* 18: 205–256.

White, H. B. 1976. Coenzymes as fossils of an earlier metabolic state. *Journal of Molecular Evolution* 7: 101–104.

Whitehouse, M. J., J. S. Myers, and C. M. Fedo. 2009. The Akilia controversy: Field, structural and geochronological evidence questions interpretations of >3.8 Ga life in SW Greenland. *Journal of the Geological Society* 166: 335–348.

Whitfield, J. 2004. Born in a watery commune. *Nature* 427: 674–676.

Whitfield, J. 2006. Base invaders. *Nature* 439: 130–131.

Whittaker, R. H. 1969. New concepts of kingdoms of organisms. *Science* 163: 150–160.

Wochner, A., J. Attwater, et al. 2011. Ribozyme-catalyzed transcription of an active ribozyme. *Science* 332: 209–212.

Woese, C. R. 1991. The use of ribosomal RNA in reconstructing evolutionary relationships among bacteria. In *Evolution at the Molecular Level*, ed. R. K. Selander, A. G. Clark, and T. S. Whittam. Sunderland, MA: Sinauer, 1–24.

Woese, C. R. 1996. Phylogenetic trees: Whither microbiology? *Current Biology* 6: 1060–1063.

Woese, C. R. 1998. The universal ancestor. *Proceedings of the National Academy of Sciences, USA* 95: 6854–6859.

Woese, C. R. 2000. Interpreting the universal phylogenetic tree. *Proceedings of the National Academy of Sciences, USA* 97: 8392–8396.

Woese, C. R. 2002. On the evolution of cells. *Proceedings of the National Academy of Sciences, USA* 99: 8742–8747.

Woese, C. R. 2004. A new biology for a new century. *Microbiology and Molecular Biology Reviews* 68: 173–186.

Woese, C. R., and G. E. Fox. 1977. Phylogenetic structure of the prokaryotic domain: The primary kingdoms. *Proceedings of the National Academy of Sciences, USA* 74: 5088–5090.

Woese, C. R., O. Kandler, and M. L. Wheelis. 1990. Towards a natural system of organisms: Proposal for the domains Archaea, Bacteria, and Eucarya. *Proceedings of the National Academy of Sciences, USA* 87: 4576–4579.

Yang, S., R. F. Doolittle, and P. E. Bourne. 2005. Phylogeny determined by protein domain content. *Proceedings of the National Academy of Sciences, USA* 102: 373–378.

Yarus, M. 2011. Getting past the RNA world: The initial Darwinian ancestor. *Cold Spring Harbor Perspectives in Biology* 3: a003590.

Zaug, A. J., and T. R. Cech. 1986. The intervening sequence RNA of *Tetrahymena* is an enzyme. *Science* 231: 470–475.

Zhang, B., and T. R. Cech. 1997. Peptide bond formation by in vitro selected ribozymes. *Nature* 390: 96–100.

Zhang, L., A. Peritz, and E. Meggers. 2005. A simple glycol nucleic acid. *Journal of the American Chemical Society* 127: 4174–4175.

Zhaxybayeva, O., P. Lapierre, and J. P. Gogarten. 2005. Ancient gene duplications and the root(s) of the tree of life. *Protoplasma* 227: 53–64.

Zimmer, C. 2006. Did DNA come from viruses? *Science* 312: 870–872.

Zuckerkandl, E., and L. Pauling. 1965. Molecules as documents of evolutionary history. *Journal of Theoretical Biology* 8: 357–366.

18

Evolution and the Fossil Record

Once the fundamental life processes of DNA replication, protein synthesis, respiration, and cell division had evolved, a spectacular diversification of life ensued. Innovations like photosynthesis and the nuclear envelope evolved. These events spanned some 3.2 billion years and created the deep branches on the tree of life. In rocks dated from 542 to about 488 Ma, most of the animal phyla living today appear: crustaceans and other arthropods, onychophorans, sipunculid worms, segmented worms, mollusks, and chordates. This time interval is called the Cambrian period, and the evolutionary profusion of body plans is known as the Cambrian explosion. The rapid appearance of so many large, complex animals ranks as one of the great events in the history of life.

The fossil record and phylogenetic analyses have confirmed other periods of rapid diversification over the past 542 million years, as well as five episodes of cataclysmic extinction. In addition, the fossil record documents profound morphological transitions, such as the emergence of mammals from reptile-like ancestors like *Thrinaxodon* (above and right). The time interval between the start of the Cambrian and the present is called the Phanerozoic ("visible life") eon. How did life diversify to reach its current level? How was diversification stunted by episodes of mass extinction? What can fossils tell us about the direction and rate

Above, twin *Thrinaxodon* babies. By Roger Smith; see Smith and Botha (2005). Below, a *Thrinaxodon* in its burrow. By Christian Sidor; see Sidor et al. (2008).

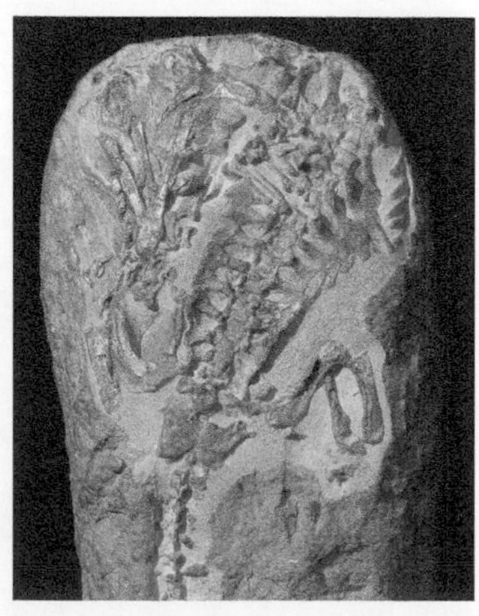

of evolutionary change? Questions like these are the focus of this chapter. Before addressing them, we need to look at the basics: how paleontologists read the fossil record and document the history of life. The chapter begins with sections on the nature of fossils and what they show us about how organisms have changed over time. Sections 18.3 and 18.4 look at diversification and extinction. The final two sections consider macroevolution and the integration of fossil and molecular data.

18.1 The Nature of the Fossil Record

Earlier we introduced the geological time scale established by paleontologists in the 19th century (Chapter 2). You may also recall that 21st-century geochronologists are using radioactive isotopes to estimate the absolute age of each eon, era, period, and epoch. For a deeper understanding of how life has changed over time, we review the process of fossilization, examine the strengths and weaknesses of the fossil record, and present a time line of major events in evolution.

(a) Termite in amber

(b) Woolly mammoth in permafrost

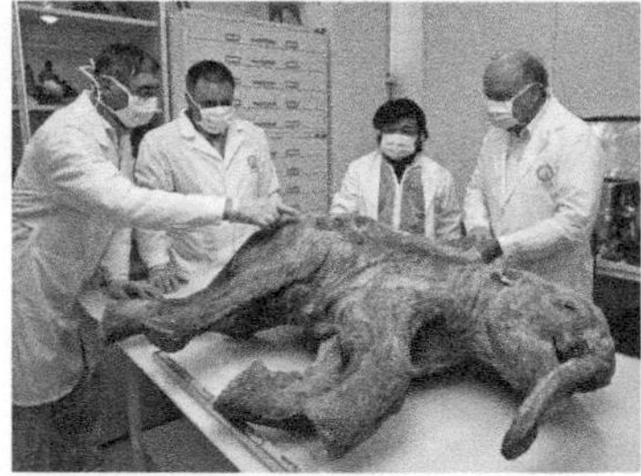

Figure 18.1 Minimally altered remains (a) A winged male termite preserved in amber. From the Upper Cretaceous of Canada, it is about 125 million years old. (b) A woolly mammoth calf frozen in permafrost. From the Yamal Peninsula in northern Russia, it is about 40,000 years old.

The Fossilization Process and Types of Fossils

A fossil is any trace left by an organism that lived in the past. Identified here are several types of fossils that differ in their method of formation, but keep in mind that this list is incomplete as the fossil record is enormously rich and diverse. There are two important issues to focus on: Which part of the organism is preserved and available for study? What kinds of habitats produce fossils?

- **Amber and freezing** (Figure 18.1) are among the least altered remains available to paleontologists, but they are rare. Viscous plant resins can harden into amber, preserving insects trapped inside so well that wing veins are visible. Woolly mammoths dug out of permafrost have fur and tissues preserved, and even incomplete sequences of DNA. How long can organic material remain unaltered? Two-thousand-year-old human cadavers from the Iron Age, buried in the highly acidic environment of peat bogs, have been recovered with flesh still intact (van der Plicht et al. 2004). Dried but otherwise unaltered 20,000-year-old dung from giant ground sloths can be found in protected, desiccating environments such as desert caves (Hansen 1978). These types of extraordinary preservation can give the most complete picture of ancient life.
- **Permineralization and replacement** (Figure 18.2) are common modes of fossilization and can form when structures are buried in sediments and dissolved minerals either replace the original mineral content or precipitate in and around it. The original shape of the fossil can be preserved (often to the level of cellular detail), but its composition is altered. In one extraordinary case, bones of a marine reptile from Australia were replaced with opal.

(a) *Tyrannosaurus rex*

(b) Petrified wood

Figure 18.2 Permineralized fossils Permineralized fossils are usually found in rock outcrops after they are exposed by weathering. (a) Skull of a predatory dinosaur. (b) Petrified wood at Petrified Forest National Park, Arizona.

- **Natural molds and casts** (Figure 18.3) originate when remains decay after being buried in sediment. Molds consist of unfilled spaces, whereas casts form when new material infiltrates the space, fills it, and hardens into rock. Molds and casts preserve information about surface shape, but not internal details.
- **Trace fossils** (Figure 18.4) differ from other types of fossils, which are collectively termed *body fossils,* in that trace fossils record behavior instead of form. Dinosaur trackways can tell us about an animal's stride length and thus yield an estimate of maximum speed. Coprolites, or fossilized feces, represent another type of trace fossil. A third type is a burrow or den, which can be subsequently in-filled by sediment during a flood, preserving a positive cast. Except in extraordinary cases when the trace and trace maker are preserved together, as with the *Thrinaxodon* burrow shown on the first page of the chapter, it is often difficult to determine with confidence which species made a given trace.

Figure 18.3 Casts and molds This horsetail stem, from the Carboniferous, is about 310 million years old.

Formation of most fossils depends on three key features of the specimen: durability, burial (usually in a water-saturated sediment), and lack of oxygen (which inhibits scavenging and bacterial breakdown). Each of these factors slows decomposition and makes fossilization more likely. As a result, the fossil record consists primarily of hard structures left in depositional environments such as river deltas, beaches, floodplains, marshes, lakeshores, and seafloors. By contrast, soft-bodied organisms, like slugs, have less chance of entering the fossil record. We discuss the study of factors that contribute to the formation of the fossil record next.

Taphonomy and Sampling Bias

Taphonomy (*taphos* = burial) is the study of the fossilization process. Taphonomic bias consists of the factors that contribute to the difference between what was once alive and its representation in the fossil record. One type of taphonomic bias is that organisms with hard parts preserve more easily than those that are entirely soft-bodied. This and other pre-fossilization filters affect what initially becomes fossilized. Another bias is caused by abundance. All else being equal, common species have better odds of being preserved at least once. Of course, some environments also offer better preservation potential. Bivalves that burrow into the seafloor live in the sediments that will eventually preserve them. At the opposite end of the spectrum, species that live in arid or rapidly eroding areas (such as mountains) are relatively unlikely to leave fossil representatives.

Beyond limiting what becomes fossilized, other factors affect what is eventually available for study. One important post-fossilization filter is the availability of fossiliferous rock. Sedimentary rocks from all time periods are not equally accessible. Rocks in some regions are covered by rain forest, whereas in others they form outcrops that can be mapped for kilometers. In addition to uneven geography, older rocks are more likely to have been destroyed by the ongoing subduction of Earth's crust. For this and other reasons, the fidelity of the fossil record improves as one approaches the modern. The improvement of the geological and fossil records through time is termed *the pull of the recent.* It is important to take this bias into account when interpreting paleontological data.

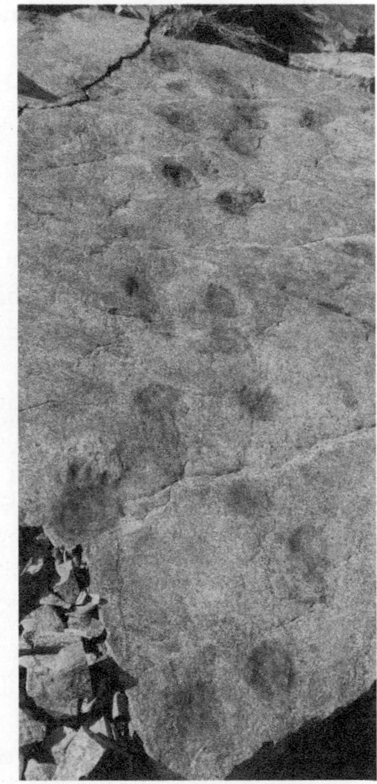

Figure 18.4 Trace fossils These 395-million-year-old footprints, found in a limestone quarry in Poland, may have been left by one of the earliest terrestrial tetrapods (Niedzwiedzki et al. 2010). Photo by Grzegorz Niedzwiedzki.

Other biases in the fossil record are the result of human activity. Collection bias comes in many forms: geographic (e.g., most marine invertebrate fossils have been collected in Europe and North America, where the most paleontologists live), taxonomic (certain groups of fossils are preferentially collected), and morphologic (certain parts of an organism are valued over others). For example, although modern paleontologists are much more likely to collect entire skeletons, the practice of head hunting (collecting only the skull) was common in the past. In addition, research effort is not evenly distributed. Fossil fish are vastly understudied when compared to carnivorous dinosaurs or to hominoids.

It is also important to realize that the fossil record typically documents change only at resolutions of tens to hundreds of thousands of years. This structural limitation makes direct comparisons of fossil versus living populations difficult.

The existence of biases and limitations in the sources of data is not unique to paleontology. Advances in developmental genetics depend on the generality of a few model systems, such as *Drosophila melanogaster,* the roundworm *Caenorhabditis elegans,* and the thale cress *Arabidopsis thaliana.* The important point is that the fossil record, like any source of data, has characteristics that restrict the information that can be retrieved and how broadly the data can be interpreted. The goal for paleontologists is to recognize the constraints and work creatively within them.

Like all sources of data, the fossil record has inherent strengths and limitations.

With these caveats in place, we can begin our intensive use of the fossil record with a broad look at the sequence of events during the Phanerozoic.

Life on an Evolving Earth

The geologic time scale is a hierarchy divided into eons, eras, periods, epochs, and stages. Each named interval is defined by a suite of diagnostic fossils and bounded by extinctions of varying severity. In general, the larger the subdivision of time, the larger the extinction. For example, mass extinctions mark the end of the Paleozoic and Mesozoic eras. When the scale was first formulated, in the early 1800s, the intervals were arranged by relative age only: Rocks were placed in younger-to-older sequence. Only much later, after the discovery of radioisotopes and the development of accurate dating techniques, were absolute times assigned to each interval. Consequently, named intervals span different amounts of time. The Paleozoic era lasted 292 million years and the Mesozoic 186 million. The geologic time scale is a work in progress. Estimates for the absolute ages improve as dating techniques become more sophisticated and more rocks are sampled.

Life evolved on a world that was itself changing. The primary driver of this change is plate tectonics, which constantly rearranges continents and oceans. During times of continental coalescence, climate patterns were extreme due to the large amount of continental interior relative to the moderating effects experienced by coastlines. In addition, the collision of plates can bring increased mountain building (orogeny), which through a pathway involving rock (especially silicate) weathering decreases carbon dioxide levels. Carbon dioxide is a greenhouse gas. Its atmospheric concentration correlates with global temperature (Royer et al. 2004) such that during times of high CO_2, the poles were devoid of glaciers. The physical changes set in motion by plate tectonics thus have important consequences for the environments in which the diversity of life evolved.

Figure 18.5 presents a time line for the eras that make up the Phanerozoic: the Paleozoic (ancient life), Mesozoic (middle life), and Cenozoic (recent life). In addition to giving a compact overview of the history of polar glaciation, carbon dioxide and oxygen levels, and continental positions, the figure should inspire questions about how life responded. For example, the diversification of flowering plants in the Cretaceous (Fiz-Palacios et al. 2011) coincided with a decrease in CO_2 and increase in O_2—why? (See Igamberdiev and Lea 2006.)

The history of Phanerozoic evolution includes both unique events and broad patterns. To introduce how research in contemporary paleontology is done and illustrate the most important concepts, we focus on a few of the broader patterns.

Figure 18.5 An overview of the Phanerozoic The geologic time scale appears at left, including eons, eras, periods, and epochs. The next column gives chronological age estimates based on radiometric dating. Redrawn from Gradstein and Ogg (2009); dates from ICS (2012). Following the chronological dates is a graph showing the extent of polar glaciation (blue), the concentration of carbon dioxide in the atmosphere (purple for estimates based on various data sources and gold for estimates from a comprehensive model) and the concentration of oxygen (green). From Berner (2009); Breecker et al. (2010). After the graph are various milestones in the history of life (First vertebrates: Zhang and Hou 2004; land plants: Rubinstein et al. 2010; jawed fish: Sansom et al. 2012; vascular plants: Wellman 2010; bony fish: Zhu et al. 2009; tetrapods: Niedzwiedzki et al. 2010; insects: Engel and Grimaldi 2004; winged insects: Knecht et al. 2011; synapsids: Angielczyk 2009; dinosaurs: Nesbitt et al. 2013; mammals: Luo 2007; birds: Lee and Worthy 2012; eutherian mammals: Luo et al. 2011; flowering plants: Sun et al. 1998; primates: Franzen et al. 2009; humans: Kimbel et al. 1996). Finally, globes show the changing locations of the continents. By Ron Blakey (http://jan.ucc.nau.edu/~rcb7/mollglobe.html).

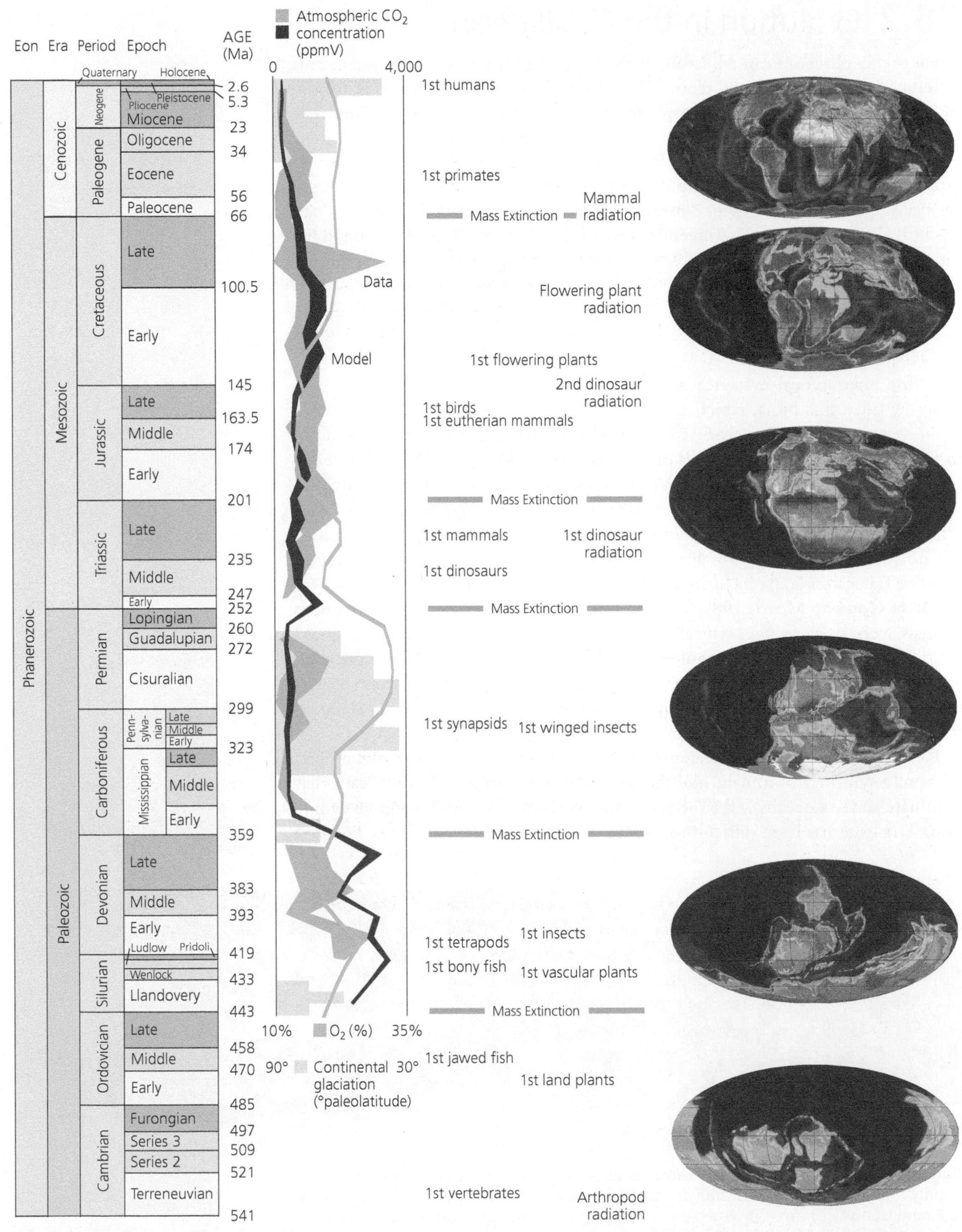

Eon Era Period Epoch
AGE (Ma)
Atmospheric CO_2 concentration (ppmV)
0
4,000
Phanerozoic
Cenozoic
Mesozoic
Paleozoic
Quaternary
Holocene
Pleistocene
Pliocene
Neogene
Miocene
Paleogene
Oligocene
Eocene
Paleocene
Cretaceous
Late
Early
Jurassic
Late
Middle
Early
Triassic
Late
Middle
Early
Permian
Lopingian
Guadalupian
Cisuralian
Carboniferous
Penn-sylva-nian
Late
Middle
Early
Mississippian
Late
Middle
Early
Devonian
Late
Middle
Early
Silurian
Ludlow
Pridoli
Wenlock
Llandovery
Ordovician
Late
Middle
Early
Cambrian
Furongian
Series 3
Series 2
Terreneuvian
2.6
5.3
23
34
56
66
100.5
145
163.5
174
201
235
247
252
260
272
299
323
359
383
393
419
433
443
458
470
485
497
509
521
541
Data
Model
10%
O_2 (%)
35%
90°
Continental glaciation (°paleolatitude)
30°
1st humans
1st primates
Mass Extinction
Mammal radiation
Flowering plant radiation
1st flowering plants
2nd dinosaur radiation
1st birds
1st eutherian mammals
Mass Extinction
1st mammals
1st dinosaur radiation
1st dinosaurs
Mass Extinction
1st synapsids
1st winged insects
Mass Extinction
1st tetrapods
1st insects
1st bony fish
1st vascular plants
Mass Extinction
1st jawed fish
1st land plants
1st vertebrates
Arthropod radiation

18.2 Evolution in the Fossil Record

What the fossil record can tell us about the evolutionary history of an organism is different from what we can discern within an extant system. DNA data is, for the most part, unavailable, and so morphology becomes a critical source of information. For example, fossil species are necessarily recognized on the basis of their anatomy (**morphospecies**) rather than the biological species concept, which emphasizes interbreeding. Paleontologists also use morphology to deduce the relationships among taxa, which means that convergent or parallel evolution can be difficult to recognize, and juveniles and adults can sometimes be confused for separate species if intervening growth stages are unavailable to link them. Despite these concerns, the fossil record provides data on aspects of evolution that can never be studied or replicated in the laboratory. Let us look at some examples.

The Ediacaran Biota

The first unequivocal evidence for macroscopic life in the fossil record comes from the Ediacaran biota, which are dated from about 565–544 Ma, placing them at the very end of the Proterozoic (early life) era. The first specimens were found in the 1940s in the Ediacara Hills of south Australia, but others have now been found at some 20 sites around the world. Most are preserved as impressions, and virtually none have shells or other hard parts. The Ediacaran fossils have been difficult to classify. Assignments have ranged from lichens and algae to forms of life no longer represented on Earth today. Most experts now agree, however, that the Ediacaran fauna includes sponges, jellyfish, and comb jelly relatives (**Figure 18.6**; Conway Morris 1989; Zhang et al. 2001; Xiao et al. 2002). Ediacaran animals were small in size, typically reaching only a few centimeters across, and relatively simple in their morphology. That said, paleontologists have a growing appreciation of the structural and physiological sophistication of Ediacaran organisms, including rangeomorphs like the one in Figure 18.6b (Laflamme et al. 2009; Vickers-Rich et al. 2013).

Recent analyses of the Ediacaran biota have sought to learn whether complex, bilaterally symmetric animals, like the mollusks and worms and crustaceans that dominate today's oceans and continents, were present this early in animal evolution. The issue has been difficult to resolve.

Figure18.6 Ediacaran fauna (a) *Dickinsonia*, a radially symmetric animal of uncertain identity that is common in Ediacaran deposits. Individuals were about 3.5 cm across. Photo by Simon Conway Morris. (b) "Frondlets" belonging to animals of uncertain identity that are generally referred to as *Rangea*. The scale bar is 0.25 cm. From Narbonne (2004).

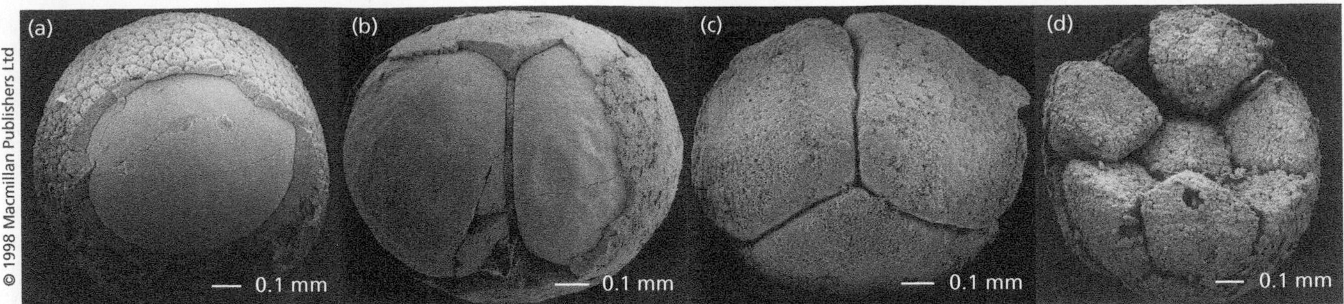

© 1998 Macmillan Publishers Ltd

Figure 18.7 Precambrian embryos? These microfossils from the Doushantuo formation may be (a) zygotes and (b–d) cleavage-stage embryos of a Proterozoic bilateral animal. A structure resembling an egg envelope surrounds each embryo, the cells occur in multiples of two, and the fossils show geometric patterns similar to the embryos of some living arthropods and flatworms. From Xiao et al. (1998).

(a–d) Reprinted by permission from Macmillan Publishers Ltd: *Nature* 391: 553–558, copyright 1998.

Fossilized embryos support the hypothesis that bilaterians evolved before the Cambrian. These come from the Doushantuo Formation of southern China, dated from 635 to 551 Ma (Condon et al. 2005). In Doushantuo rocks, phosphate minerals replaced soft tissue and created fossils that show fine anatomical details. Li et al. (1998) described sponges with larvae similar to those of living sponges. Xiao and coworkers (1998) found fossils that consist of two, four, eight, or more round structures that appear to be blastomeres in a cleaving embryo **(Figure 18.7)**. Because these structures resemble the first cells seen in a developing arthropod, the fossils have been interpreted as the embryos of a bilaterally symmetric species (Xiao et al. 1998). Other researchers disagree, suggesting that the fossils represent an encysting protist (Huldtgren et al. 2011; see also Conway Morris 1998a).

Controversy also exists over Ediacaran trace fossils. Some researchers have argued that linear burrows and tracks, like those in **Figure 18.8a**, were made by bilaterally symmetric organisms that had a head and tail region and moved in a line (Waggoner 1998). Other fossils offer stronger evidence for bilaterians in the late Precambrian: the mollusk-like *Kimberella* (Figure 18.8b; Fedonkin and Waggoner 1997) and the tiny *Vernanimalcula guizhouena* (Figure 18.8c; Chen et al. 2004). Taken together, the data argue that bilaterally symmetric animals were small in size but definitely present before the Cambrian period.

There is increasing evidence that small-bodied, bilaterally symmetric animals existed before the Cambrian explosion.

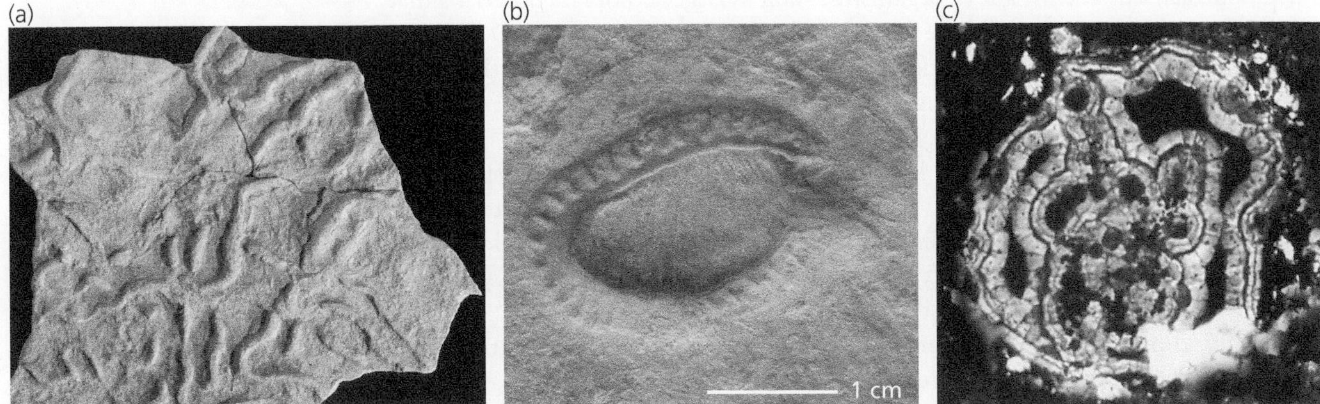

Figure 18.8 Evidence for bilaterally symmetric animals in the Precambrian (a) These trace fossils may be tracks of a worm-like burrowing animal that was bilaterally symmetrical. Each track is about 0.5 cm wide. By Guy Narbonne. A few whole-body fossils of bilaterally symmetrical animals are known from the Precambrian: (b) the mollusk-like *Kimberella* (by Eduard Solà Vázquez); and (c) the tiny *Vernanimalcula guizhouena*, shown here in cross section (Chen et al. 2004).

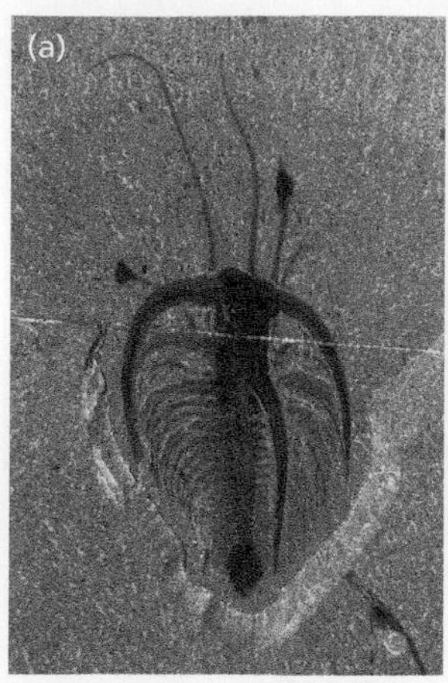

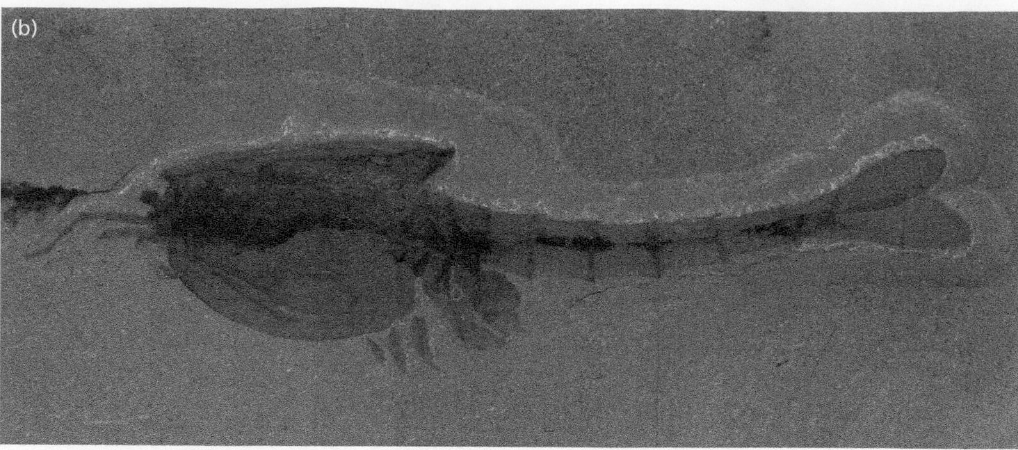

Figure 18.9 The Burgess Shale faunas The Burgess Shale and Chengjiang faunas are dominated by large, bilaterally symmetric animals with well-developed segmentation, heads, and appendages. (a) *Marrella splendens*. (b) *Waptia fieldensis*. Photos by Jean-Bernard Caron, Royal Ontario Museum.

The Burgess Shale Fauna

The contrast between the fossil records of the Precambrian and Cambrian could not be greater. Whereas a paucity of bilaterians appear among Ediacaran fossils in the Precambrian, the Cambrian period records an astonishing variety of large and complex bilaterally symmetric animals **(Figure 18.9)**. The most species-rich lineages of animals alive today—the arthropods, mollusks, vertebrates, and echinoderms—were all present in the Cambrian. How do we know, considering that many of these groups lack hard parts favoring preservation? In one of the luckiest paleontological discoveries ever, a **Lagerstätte** (an extraordinarily rich deposit) preserving soft-bodied Cambrian fossils was found in British Columbia, Canada. It came from a set of rocks called the Burgess Shale and has given us an unparalleled view of the early diversification of life in the Late Cambrian ocean.

The rocks of the Burgess Shale, which have been dated to about 505 Ma, and the Chengjiang biota from Yunnan Province in China (525–520 Ma) are arguably the most spectacular fossil deposits known (Conway Morris 1998b; Zhang et al. 2001). Both Lagerstätten preserve soft-bodied animals in extraordinary detail.

There is little overlap between species found in the Ediacaran and Burgess Shale deposits, but at least a few organisms, such as large, colonial cnidarians that are similar to modern sea pens, are present in both (Conway Morris 1998b). The Cambrian specimens include a wide array of complex and unusual arthropods, including trilobites, as well as segmented worms, wormlike priapulids and sipunculids, and a diversity of mollusks.

Remarkably, the deposits also harbor several chordates, including species of jawless vertebrates **(Figure 18.10)**. These early chordates had segmented trunk muscles and a supporting rod called a notochord. In overall morphology they resembled **extant** (living) jawless vertebrates—the hagfishes and lampreys (Chen et al. 1999; Shu et al. 1999; Shu et al. 2003).

Initially, though, many members of the Burgess Shale and Chengjiang faunas looked so odd that biologists were perplexed. Bizarre fossils were assigned to a jumbled group referred to as Problematica. Two examples appear in **Figure 18.11**. Some observers suggested that these species represented unique phyla, unlike any organisms living today. However, further study revealed that most or all are members or close relatives of living phyla. *Opabinia regalis* (Figure 18.11a), for

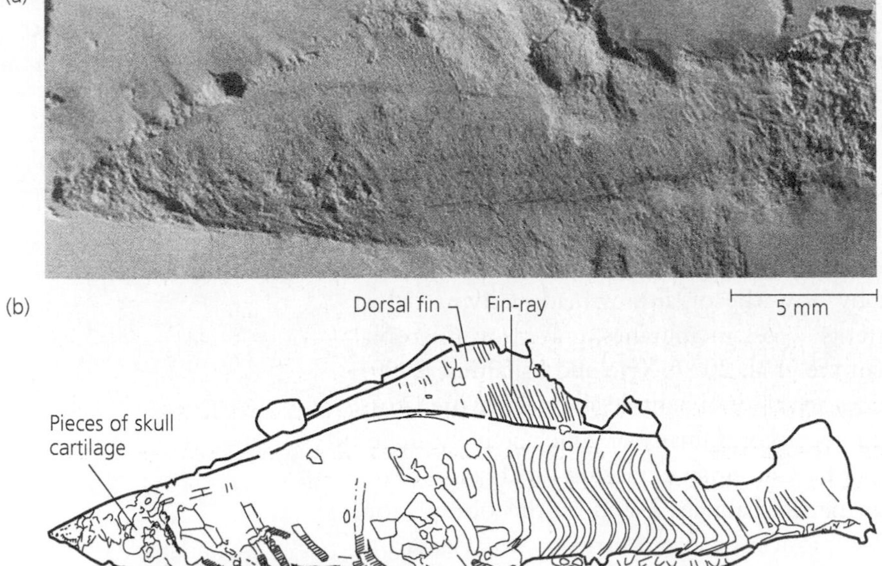

Figure 18.10 A vertebrate from the Cambrian: *Haikouichthys ercaicunensis* The photograph in part (a) shows the whole specimen. The interpretive drawing (b) shows some distinctive vertebrate characters, including a cartilaginous skull, gill skeleton, and fin-rays. From Shu et al. (1999).

example, was first described as an elongate bilaterian with serially repeated lateral plates, five eyes, and an elongate "nozzle" on the anterior end. This description left *Opabinia* out of the classification of living phyla. However, detailed examination of the best specimens suggests that the repeated lateral structures are homologous to arthropod limbs and that *Opabinia* is a close arthropod relative (Zhang and Briggs 2007). Similarly, *Wiwaxia corrugata* (Figure 18.11b) was formerly given phylum status as a problematic fossil consisting of spines and plates, but further study revealed that it is almost certainly a polychaete worm, a mollusk relative, or a member of the stem group of the clade including annelids and mollusks (Butterfield 1990; Conway Morris 1998b; Smith 2012). As a result of these analyses, the number of phyla that existed during the Cambrian is now recognized as roughly equivalent to the diversity observed today (Briggs et al. 1992; Wills et al. 1994).

Compared to animals present earlier in the fossil record, the hallmarks of the Cambrian fauna are a dramatic increase in body size, the origin of hard exoskeletons and complex body parts like limbs, and a diversification in basic body shapes and organization.

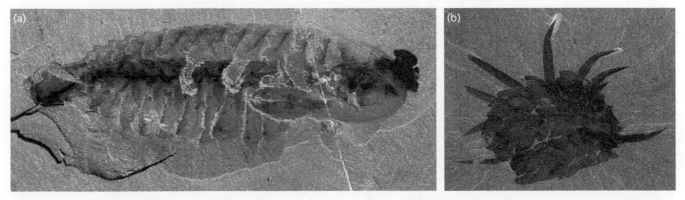

This is a remarkable conclusion. The earliest members of virtually all major animal lineages appeared relatively suddenly in the fossil record, at the same time, in geographically distant parts of the globe (Conway Morris 2000; Valentine 2002).

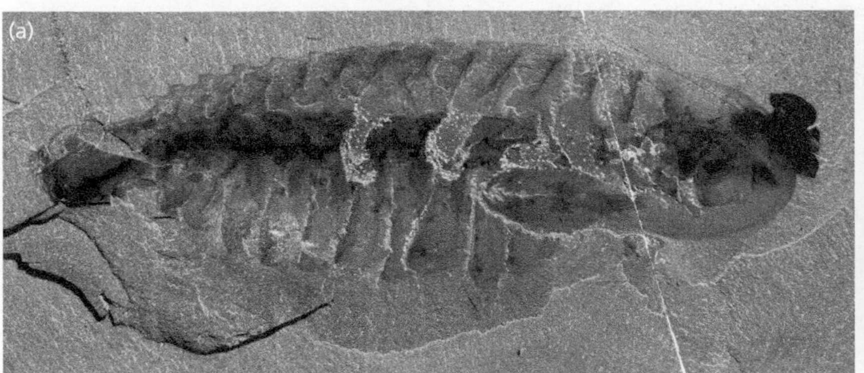

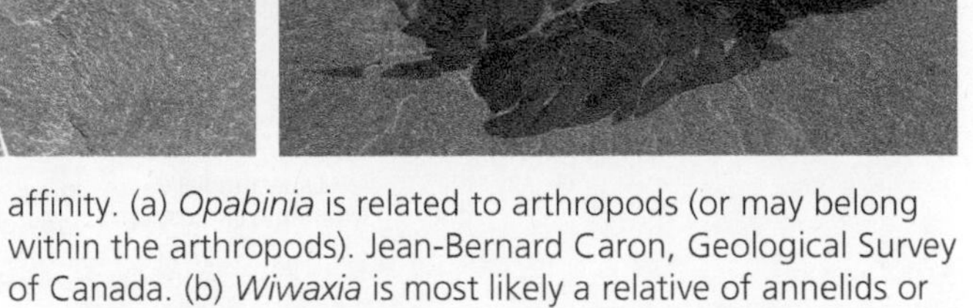

Figure 18.11 Problematica? Some species in the Burgess Shale fauna are so unusual they have been hard to classify. Many of these Problematica are now grouped with living phyla, or at least with early Phanerozoic fossils of known affinity. (a) *Opabinia* is related to arthropods (or may belong within the arthropods). Jean-Bernard Caron, Geological Survey of Canada. (b) *Wiwaxia* is most likely a relative of annelids or mollusks. Jean-Bernard Caron, Royal Ontario Museum.

The Burgess Shale fauna records an amazing variety and number of major morphological innovations, including large body size and the first segmented body plans, limbs, antennae, shells, external skeletons, and notochords. These animals sat, swam, burrowed, crawled, floated, and walked. They found food in almost every conceivable way, from filtering it out of the water to hunting it down.

Most animal phyla alive today make their first appearance in the fossil record during the Cambrian.

What Caused the Cambrian Explosion?

An astonishing variety of body plans, cell types, and developmental patterns evolved during the Cambrian explosion **(Figure 18.12)**. The radiation of bilaterally symmetric animals was driven by how the organisms made a living. Most Ediacaran organisms absorbed nutrients across membranes, grazed on microbial mats, or fed on organic debris (Laflamme et al. 2009; Xiao and Laflamme 2009). But the Burgess Shale fauna introduces a variety of benthic and pelagic predators, filter feeders, grazers, scavengers, and detritivores, many of which actively chased their prey (Bush and Bamach 2011). The Cambrian explosion filled many of the ecological niches present in shallow marine habitats. Based on this observation, the question of why the Cambrian explosion occurred turns on what environmental changes made all these novel ways of life possible.

Rising oxygen concentrations in seawater during the Proterozoic were key to the origin of multicellularity and large size (Johnston et al. 2012; Sahoo et al.

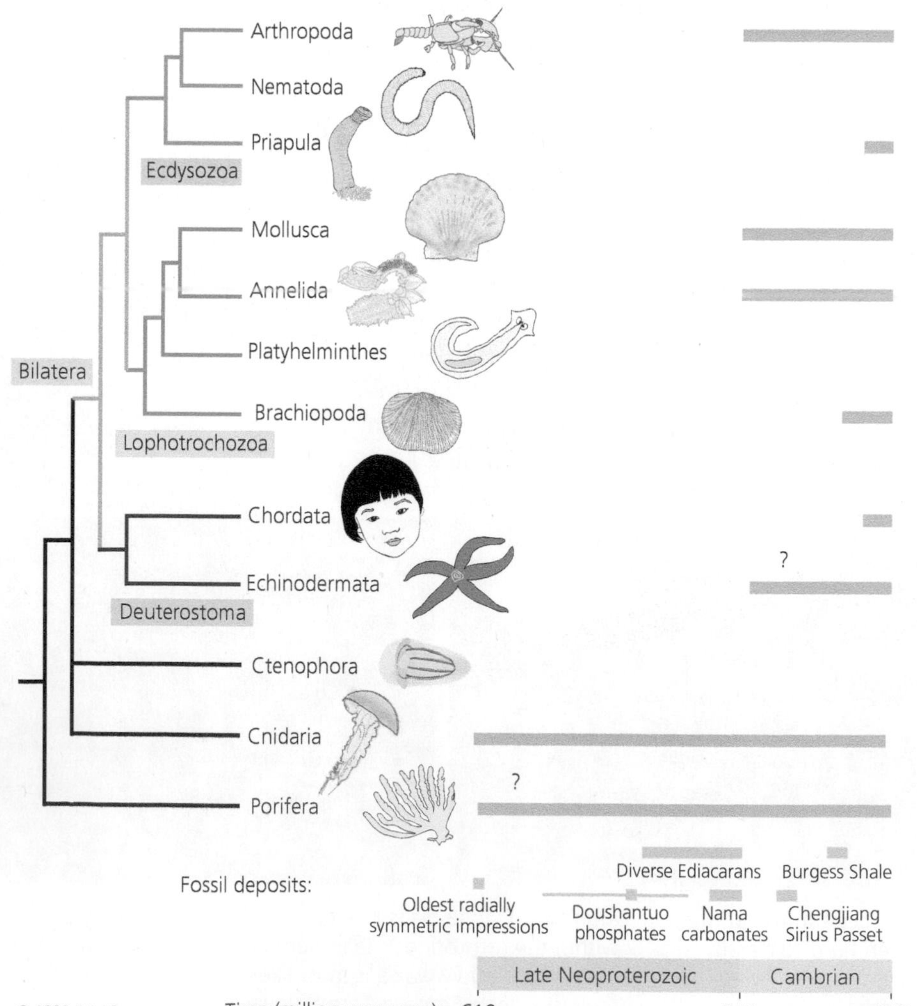

© 1999 AAAS

Figure 18.12 Phylogeny and fossil record for the earliest animals The phylogeny on the left shows the relationships among some of the major taxa represented in the earliest fossil faunas; it is based on molecular data from living members of these phyla (for example, Philippe et al. 2009; Telford and Copley 2011). The purple bars to the right of the tree indicate which phyla are represented in the fossil faunas identified in the time scale at the bottom; question marks indicate that the presence of a group at a certain time is controversial. After Knoll and Carroll (1999).

From A. Knoll and S. Carroll. 1999. "Early animal evolution: Emerging views from comparative biology and geology." *Science* 284: 2129–2137; figure 2, page 2131. Reprinted with permission from AAAS.

2012). More oxygen makes higher metabolic rates and bigger bodies possible. Larger size is a prerequisite for the evolution of tissues, and higher metabolic rates are required for active movement. Both traits first appear in Ediacaran faunas.

To explain the Cambrian explosion, Andrew Knoll and Sean Carroll (1999) suggested that a sudden increase in atmospheric oxygen occurred during the mid-Cambrian and made large size and rapid movement possible. They also posited that a mass extinction eliminated much of the Ediacaran fauna at the end of the Proterozoic, creating an opportunity for the tiny deuterostomes and protostomes present at the time to evolve in response to the changed conditions.

The leading hypothesis about the cause of the Cambrian explosion involves a mass extinction of Ediacaran fauna and an increase in atmospheric oxygen.

These hypotheses are currently being tested. If they are valid, then a series of predictions should be confirmed. Fossil evidence of small-bodied protostomes and deuterostomes should eventually be found in the Proterozoic, and analyses of molecular clocks (see Section 18.6) should be consistent with the claim that bilaterians arose hundreds of millions of years before the Cambrian explosion (Erwin et al. 2011). Fossil or geological evidence of a mass extinction event and a rise in available oxygen should come to light (Laflamme et al. 2013). Only time and further research will tell if the evolution of animals really was explosive and triggered by dramatic environmental change in the form of increased oxygen availability.

The Fish–Tetrapod Transition

The Devonian is often referred to as the "age of fishes" because of the astonishing diversity displayed by aquatic vertebrates during this interval. However, as is clear from Figure 18.5, terrestrial plants had evolved by the Ordovician, and insects began exploiting resources on land from the Silurian or earliest Devonian; the vertebrates would not be far behind. Over the past 20 years, a series of important papers have shed light on the sequence of character evolution associated with the first land-living, limbed vertebrates—the tetrapods **(Figure 18.13)**.

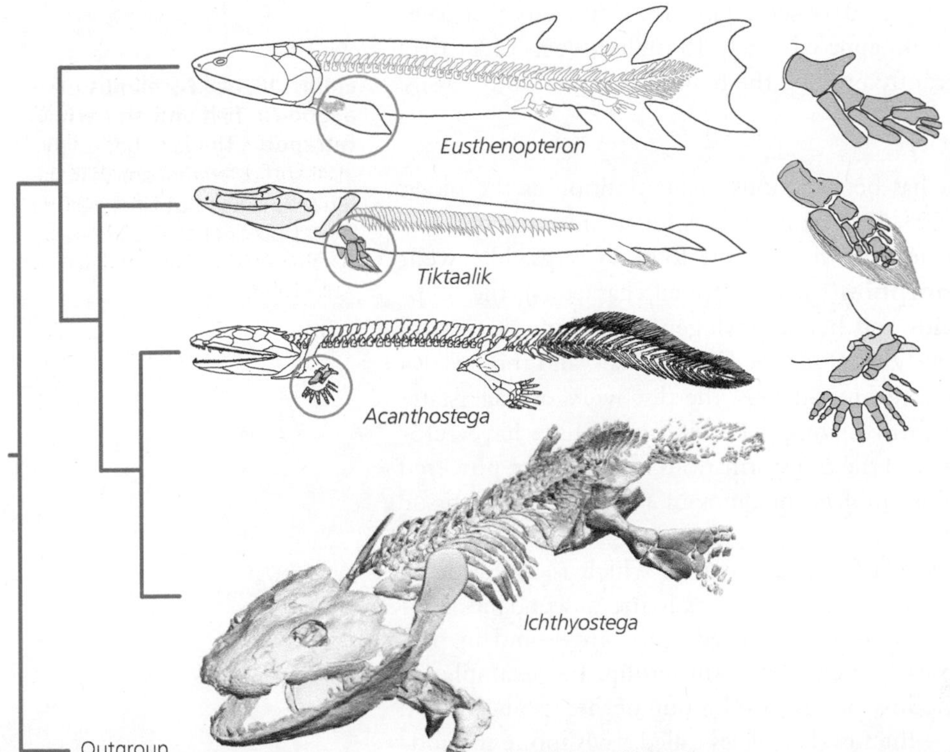

Figure 18.13 The fish-to-tetrapod transition All four species shown are sarcopterygians (lobe-fin bony fish). *Acanthostega* and *Ichthyostega*, more exclusively, are tetrapods. *Tiktaalik* is a transitional form—a "fishapod." *Tiktaalik* reconstruction after Daeschler et al. (2006) and Shubin et al. (2006). *Acanthostega* reconstruction by Michael Coates and Jennifer Clack (see Coates 1996). *Ichthyostega* reconstruction from Pierce et al. (2012).

Reprinted with permission from Macmillan Publishers Ltd. Pierce, S. E., J. A. Clack, and J. R. Hutchinson. "Three-dimensional limb joint mobility in the early tetrapod *Ichthyostega*." *Nature* 486: 523–526. Copyright © 2012 The Nature Publishing Group.

Figure 18.13 shows four Late Devonian vertebrates that span the fish-to-tetrapod transition. All are sarcopterygians (lobe-fin bony fish) characterized by a single proximal bone in the fin/limb recognized as the humerus (forelimb) or femur (hindlimb). In addition to modifying the fin into a limb, the transition from life in water to life on land necessitated structural changes to withstand the increased effects of gravity. Tetrapods such as *Acanthostega* show well-developed joints between consecutive vertebrae (zygapophyses) that are absent in aquatic taxa like *Eusthenopteron,* as well as enlarged rib attachments. Moreover, the pelvis was solidly connected to the vertebral column by means of novel sacral ribs. Many physiological changes undoubtedly also occurred (e.g., respiration, osmoregulation), but these are extremely difficult to gauge in the fossil record.

The loss of fin rays (lepidotrichia) and their replacement by digits **(Figure 18.14)** is a tetrapod hallmark, but there is increasing evidence that the earliest tetrapods spent much of their time in water (Coates et al. 2008; Pierce et al. 2012). For example, *Acanthostega* retained a caudal fin and had a fully formed gill skeleton. Equally surprising is the knowledge that many presumed tetrapod features were already present in derived aquatic sarcopterygians (such as *Tiktaalik*), like the loss of dorsal and anal fins and a flattened skull with dorsally facing orbits. Fossils of *Acanthostega, Ichthyostega,* and other early tetrapods were recovered from rocks interpreted as being freshwater, estuarine, or marginal marine in origin. The lack of tetrapod fossils from marine rocks helps to eliminate some prospective habitats for the origin of the group, but many aspects of tetrapod origins remain elusive.

Why did tetrapods invade dry land? The classic theory is largely based on the ecology of the modern Australian lungfish, which is known to move between ponds that shrink during the dry season. Climate models suggest that the Late Devonian was substantially warmer than today, but also important were untapped foods available on land, in the form of insects, which were in turn feeding on Devonian plants. The recent discovery of tetrapod trackways from the Early Devonian of Poland (Niedzwiedzki et al. 2010; see Figure 18.4) suggests that paleontologists may need to look in rocks approximately 18 million years older than the Late Devonian rocks that currently provide the bulk of tetrapod body fossils.

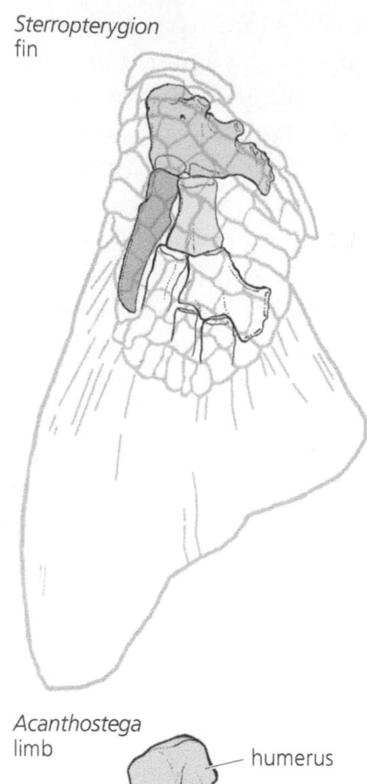

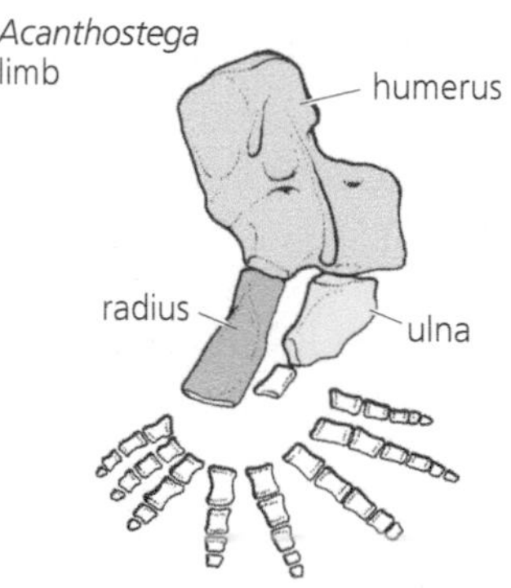

Figure 18.14 Forelimbs of a lobe-fin fish and an early tetrapod The joint geometry in *Acanthostega* meant that this tetrapod could not flex its elbow. From Coates et al. (2008).

The Dinosaur–Bird Transition

For over a century, *Archaeopteryx* has been famous for its position as the oldest definitive fossil bird (Late Jurassic, ~150 Ma) as well as for its mix of avian and "reptilian" features. Whereas the features linking the fossil to birds, like well-developed feathers, are **synapomorphies** (shared derived characters), the "reptilian" features are **plesiomorphies** and hence phylogenetically uninformative. Plesiomorphic features in *Archaeopteryx* include teeth, a long tail, and three claw-bearing fingers on the hand. Since the mid-1990s, the discovery of Lägerstatte deposits of Early Cretaceous age from Liaoning Province in China have refocused interest in the origin of birds and the transformations necessary for powered flight because of their incredible assemblage of dinosaur fossils with soft-tissue preservation (Norell and Xu 2005).

Figure 18.15 shows reconstructions of four theropods, which is the clade of bipedal, carnivorous dinosaurs that includes birds as its living descendants. Remarkably, many of the features commonly associated with birds—and in particular with flight—are present in early members of this group. For example, all theropods (including birds) have hollow bones. A subgroup of theropods, known as neotheropods, possess a furcula—the fused clavicles called a wishbone in modern birds. However, it was not until the discovery of the Liaoning fossils that

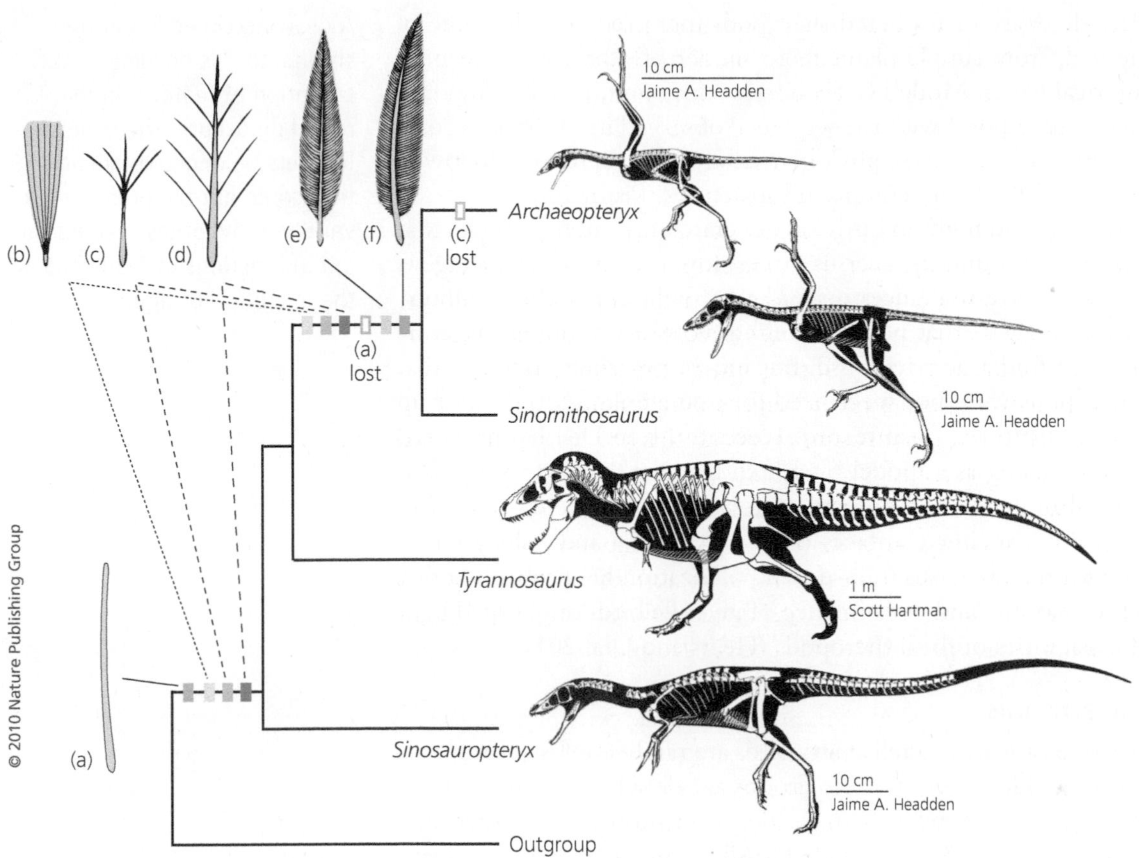

Figure 18.15 Evolution of theropod feathers (a–f) are derived feather types. (a) Thick monofilament. (b) Compound feather with multiple filaments joined at base. (c) Short barbs radiating from tip of a central filament. (d) Multiple filaments branching laterally along most of the length of a central filament. (e) Pennaceous feather with straight rachis and symmetrical vanes. (f) Pennaceous feather with curved rachis and asymmetrical vanes. Dashed lines indicate uncertainty over timing of evolution. After Xu et al. (2010).

Xu, X., X. Zheng, and H. You. 2010. "Exceptional dinosaur fossils show ontogenetic development of early feathers." *Nature* 464: 1338–1341. Copyright © 2010 Nature Publishing Group. Reprinted with permission.

paleontologists found something completely unexpected—feathers. And even more important, the feathers were from animals that clearly could not fly.

The discovery of *Sinosauropteryx,* which bears fuzzy, filamentous integumentary structures, provided the first evidence of a down-like covering in theropods. Subsequently, Liaoning fossils with feathers bearing all the hallmarks of modern flight feathers have come to light, which begs the question: Why did feathers first evolve, if not for flight? Several hypotheses have been proposed (Xu 2006), most of which center on display (feathers as sexual signaling devices; **Figure 18.16**) or thermoregulation (feathers for insulation or to aid in brooding eggs).

Figure 18.16 Plumage of the dinosaur *Anchiornis huxleyi* Quanguo Li and colleagues (2010) deduced the colors of the feathers from melanosomes preserved in fossils. Illustration by M. A. DiGiorgio. From Li et al. (2010).

Li, Q., K. Q. Gao, et al. 2010. "Plumage color patterns of an extinct dinosaur." *Science* 327: 1369–1372. Reprinted with permission of the AAAS.

Interestingly, as the diversity of feathered theropods increased, a wide range of feather forms emerged, from simple filaments to modern feathers with a central rachis and asymmetrical vanes. Models of feather development, based on modern experimental studies, correspond with the sequence observed in the fossil record: Single filamentous structures appear in phylogenetically basal theropods, followed by multiple filaments and finally by compound structures (Prum 2002).

A growing collection of transitional fossils documents the evolution of feathers in theropod dinosaurs. The earliest feathers belonged to animals that could not fly, prompting a variety of hypotheses about the initial functions of feathers and the evolution of flight.

The evolution of powered flight in birds has generated myriad hypotheses that can be dichotomized as "ground up" versus "trees down." The former suggests that birds evolved from terrestrial ancestors and that flight evolved to counteract gravity. The latter suggests that powered flight evolved via gliding. Paleontologists are unlikely to find a decisive fossil, but most researchers conclude that adult non-avian theropods were not specialized for tree climbing, so ground-up models appear a better fit to the fossil record. Recently, Ken Dial has proposed wing-assisted incline running as a model by which feathers that cannot be used in flight are still useful in generating downward force to help juvenile and adult birds scale trees and other inclined surfaces (Dial 2003). Dial and Ashley Heers also noted that the ground-up versus trees-down polarization has probably stifled research and that the anatomy and performance of juvenile birds might shed light on the functional capabilities of fossil theropods (Heers and Dial 2011).

The Origin of Mammals

Extant vertebrate groups are easy to tell apart: Birds are rarely confused for mammals, or turtles for frogs. However, as these groups are drawn back toward their common ancestors, we might expect them to look ever more alike as their distinguishing characteristics fade. With a detailed fossil record, a gradation of forms should document the steps in the divergence of groups. Such is the case with the **synapsid** lineage, which captures a nearly uninterrupted transition from reptile-like forms in the Carboniferous to the earliest mammals by the Early Jurassic (Rubidge and Sidor, 2001). For years, this group of fossils was called the *mammal-like reptiles* to emphasize their transitional nature, although the term is misleading because the fossils belong to the Synapsida (and not to the Reptilia). A better term might have been *reptile-like mammals*.

Dimetrodon **(Figure 18.17a)** was the top carnivore of the Early Permian. It had most features we associate with reptiles today (sprawling posture, simple teeth,

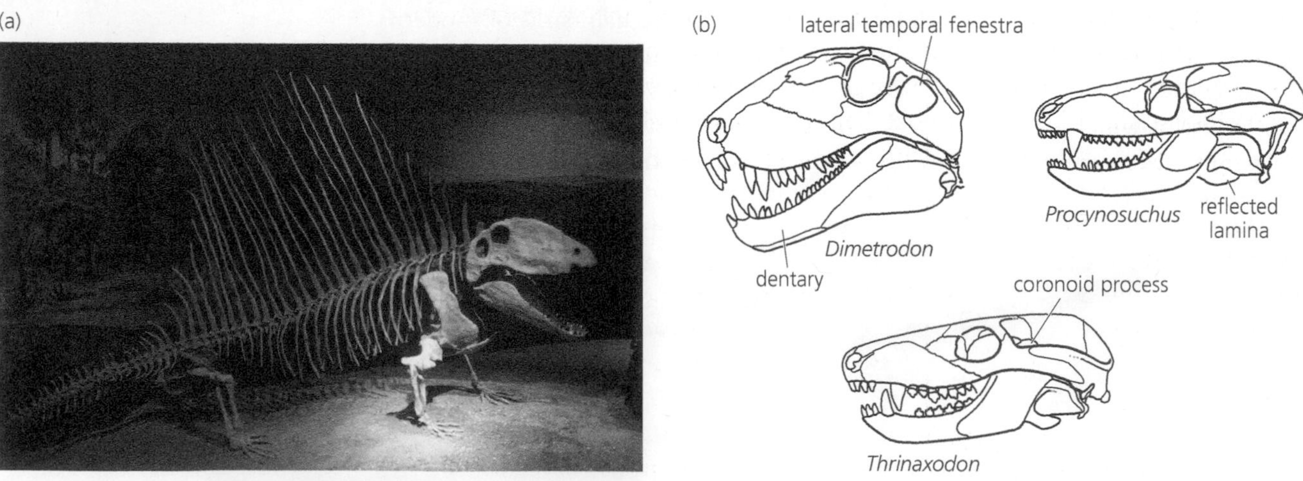

Figure 18.17 Derived traits of synapsids (a) *Dimetrodon* skeleton. (b) Skulls of *Dimetrodon*, *Procynosuchus*, and *Thrinaxodon*. From Sidor and Hopson (1998).

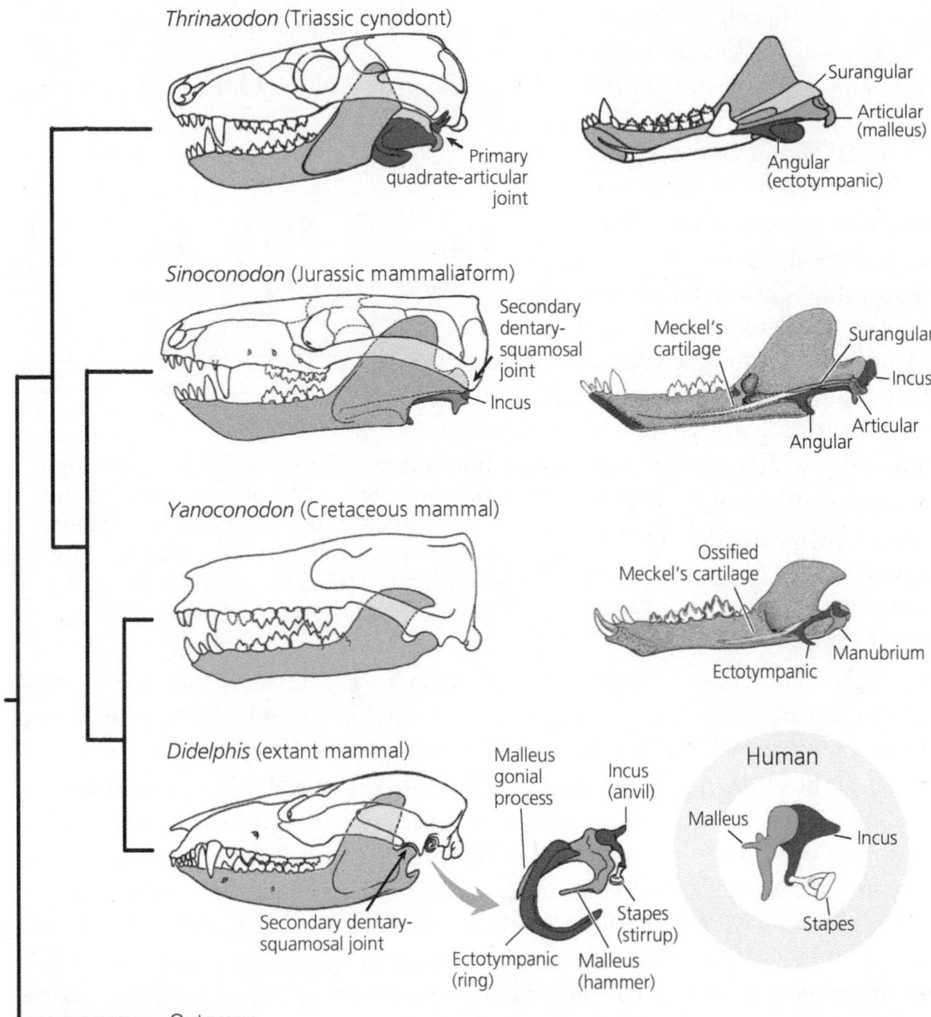

Figure 18.18 Transitional forms illustrating the evolution of the mammalian ear Modified from Luo (2011).

ectothermic physiology, multiple bones forming its lower jaw) but was an early member of the synapsid lineage by virtue of its lateral temporal fenestra, among other features. By the Late Permian and Early Triassic, more derived synapsids like *Procynosuchus* (Figure 18.17b) and *Thrinaxodon* had evolved mammalian characteristics like an enlarged dentary with a coronoid process, multicusped postcanine teeth, distinct thoracic and lumbar vertebrae, and a bony secondary palate that separated the breathing and chewing functions of the oral cavity.

Procynosuchus and *Thrinaxodon* both belong to a synapsid group called **cynodonts,** the clade that includes mammals as its living representatives. One of the most fascinating evolutionary transitions that took place within cynodonts concerns the middle ear. In mammals this structure includes an air-filled cavity with three bones (malleus, incus, and stapes), whereas in early synapsids (as well as in modern reptiles and birds) only a single element, the stapes, is present. Where did the two extra bones come from? An extraordinary series of fossils show that the two bones that form the jaw hinge in nonmammalian cynodonts were reduced in size and eventually incorporated into the middle ear (Luo 2011; **Figure 18.18**). At the same time, the tooth-bearing element of the lower jaw, the dentary, enlarged and took over the role of forming the lower jaw hinge. In modern mammals, the dentary forms the entire lower jaw, but this form represents the culmination of a long evolutionary trend (see **Computing Consequences 18.1**).

COMPUTING CONSEQUENCES 18.1

Evolutionary trends

Dinosaurs evolved larger and larger body sizes. Hominid brains got bigger and more complex. Trends seem to pervade paleontology—but are they more apparent than real? Dan McShea (1994) outlined methods to quantify evolutionary trends and to distinguish between the types of mechanisms that might underlie what is recorded in the fossil record.

An evolutionary trend can be defined as a persistent directional change in the average attribute of a clade over time. Therefore, depictions of evolutionary trends typically consist of two axes: geologic position (time) and the attribute, or state variable under consideration. For many state variables, McShea noted the possibility of lower or upper boundaries. For example, the maximum body size of insects is constrained by the architecture of their respiratory system, and the minimum size for any animal is theoretically a single cell. If the clade in question originated near one of these boundaries, then even if there were no bias in its direction of evolution, one would expect to see diffusion away from the bound as the clade diversified and evolved. This is what McShea termed a passive trend (Figure 18.19a).

In contrast, a driven trend requires a bias to be present (Figure 18.19b). This bias could be that some changes (i.e., toward greater complexity) are more probable than others, or that the size of the changes could be uneven (i.e., changes toward greater complexity are larger than changes toward reduced complexity).

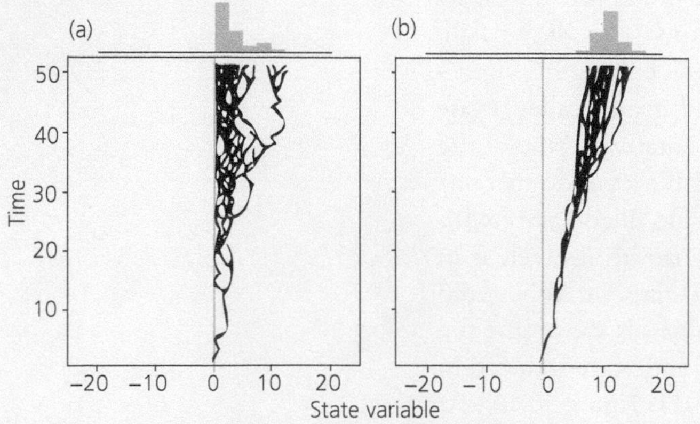

Figure 18.19 Simulated evolutionary trends (a) A passive trend. (b) A driven trend. From McShea (1994).

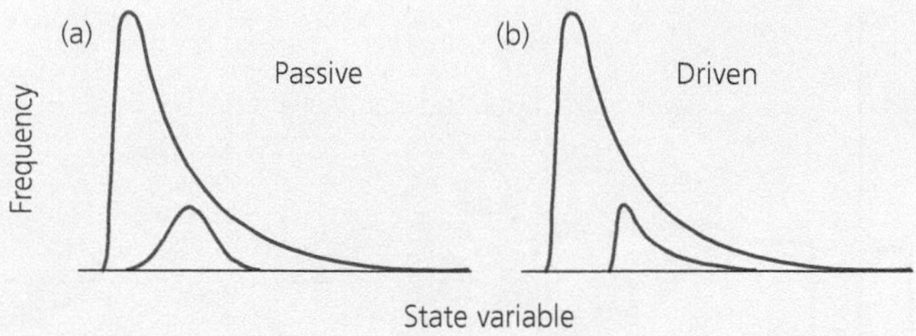

Figure 18.20 Subclade test for distinguishing between passive versus driven trends (a) If an evolutionary trend is passive, then a clade and its subclades will have frequency distributions with different shapes. (b) If an evolutionary trend is driven, then a clade and its subclades will have frequency distributions with similar shapes. From McShea (1994).

Instead of looking to the fossil record for empirical examples to test this dichotomy, McShea modeled branching evolution and was therefore able to modify the appropriate parameters directly. He proposed three tests to distinguish passive versus driven trends: (1) behavior of the minimum (in a driven trend, it tracks the maximum); (2) ancestor–descendant (in a driven trend, pairwise comparisons reveal a bias); and (3) subclade (Figure 18.20). Versions of these tests have been applied to fossil data sets and show evidence for both passive and driven systems (Figure 18.21). More important, determining the mechanism underlying an evolutionary trend can help focus attention on its possible causes: factors like developmental constraint, natural selection, or species selection.

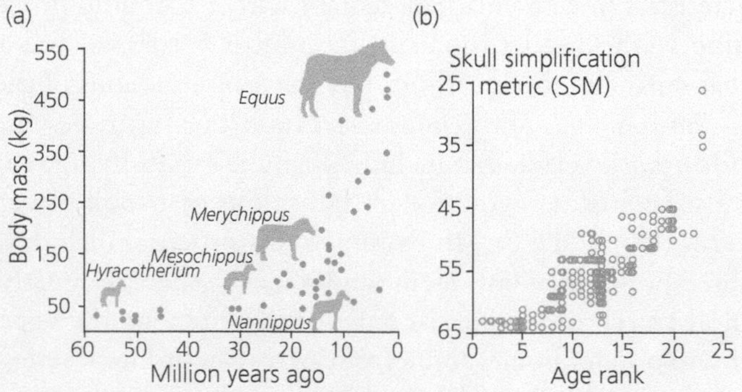

Figure 18.21 Evidence distinguishing driven versus passive evolutionary trends The behavior of the minimum test indicates driven trends in (a) body size in horses (from MacFadden 1986) and (b) synapsid skulls (from Sidor 2001).

18.3 Taxonomic and Morphological Diversity over Time

Documenting the Cambrian explosion and evolution's other "greatest hits" is only part of historical biology's portfolio. Searching for broad patterns in the fossil record is an equally important research program. Fossils document that the types of species on Earth have changed radically over vast reaches of time. As biologists analyze these changes, what patterns come to light? The literature on patterns of change across time is enormous; we can mention only a few major themes here.

Global Taxonomic Diversity

How has life diversified on Earth? The fossil record provides the most direct evidence to answer this fundamental question of historical biology and, as a result, paleontologists have studied it for over 100 years **(Figure 18.22a)**. However, most modern studies of Phanerozoic diversity can trace their origins to the work of one person. Jack Sepkoski spent most of his career gathering and analyzing diversity data for marine animals. Marine rocks yield more complete data than does the terrestrial fossil record, but they remain only a proxy for life on Earth as a whole. Using some of the earliest computer technology, Sepkoski first assembled a global database of marine orders and families, recording their first and last occurrence in the stratigraphic record. A subsequent version of the database focused on gathering genus-level data. In each case, he was able to quantify diversity dynamics, as well as extinction intensity, to a degree not seen before.

Sepkoski's database necessitated the development of new methods of analysis and new interpretations of their results. Paralleling ecological work on the carrying capacity of islands and other ecosystems, Sepkoski's data suggested that marine ecosystems showed logistic diversification, but that it was punctuated by mass extinctions. He suggested the presence of three evolutionary faunas that interacted with one another, creating a diversity curve that could be modeled with a coupled logistic equation (Sepkoski 1984; Figure 18.22b). Although more recent work has cast doubt on the cohesiveness of Sepkoski's evolutionary faunas, the ecological interpretation of paleontological data is now commonplace.

The diversity of life has generally increased over time, but there have been episodes of sharp decline.

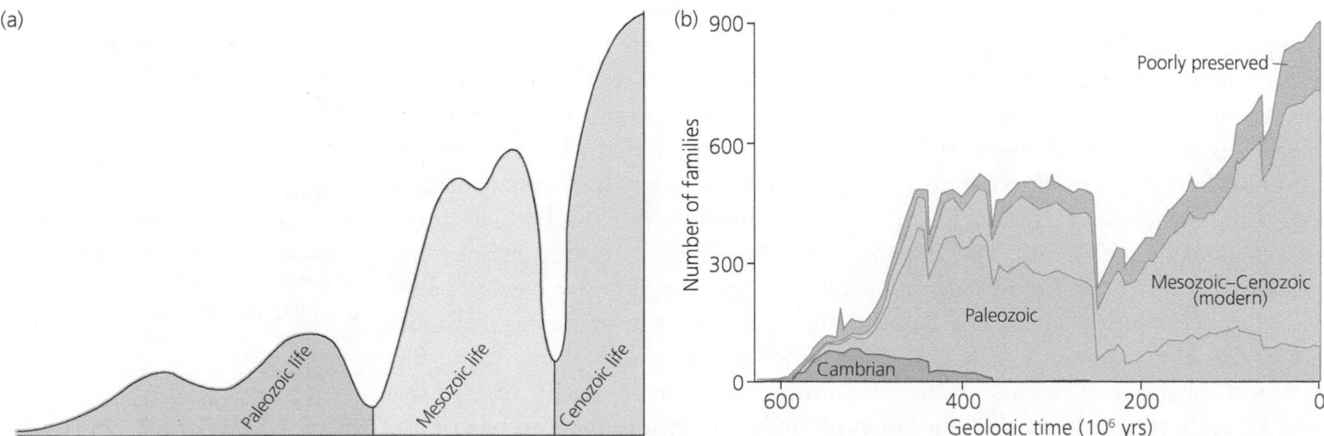

Figure 18.22 Diversity of life across the Phanerozoic Colors represent distinctive faunas. (a) Graph by John Phillips (1860). (b) Graph by Jack Sepkoski (1984).

One major criticism of Sepkoski's method was its reliance on taxonomic equivalency. For example, why should one family of ammonites count the same as one family of bivalves, when the number of species in each could be dramatically different? Moreover, many traditional vertebrate families consist of only one genus and are given family rank to emphasize morphological distinctiveness. Logically, species are the only taxonomic unit that can be considered equivalent between clades. Sepkoski understood the limitation of his data and considered families and genera only proxies for species-level dynamics. In addition, the higher-level taxonomic units were argued to be useful proxies for ecological "ways of making a living" and therefore a useful measure of ecospace usage. Through simulation studies, Sepkoski also showed that the inclusion of paraphyletic families would not necessarily undermine the overall patterns obtained.

New online databases such as the Paleobiology Database (paleodb.org) can address many of the deficiencies just noted and focus attention on research questions that once were intractable. For example, Sepkoski's database was global in scale, so it could not be used to investigate local or regional patterns or discover how each might contribute to the global signal. Regional data is very important when studying differential extinction or survivorship as well as for distinguishing between extinction mechanisms proposed for a given event. In addition, online databases draw on the expertise of many researchers from around the world and so have the potential for greater reliability and timely updates.

Measuring Morphological Diversity

The morphological species concept employed by paleontologists means that taxonomic diversity data has an intrinsic relationship with morphological diversity; increases in taxonomic diversity (richness) must necessarily accompany increases in morphological diversity. However, the nature of that relationship is not always uniform; what is considered a species-level distinction varies among fossil groups. Moreover, morphological diversity is not uniformly distributed among types of organisms. Instead of using taxonomic diversity as a proxy for morphological diversity (like Sepkoski), some paleontologists have focused on measuring morphological diversity directly. In particular, much work has compared the range of theoretically possible morphologies to what is observed in the fossil record. In this context morphological diversity is often termed **disparity,** and a **morphospace** is the actual or potential range of morphologies encompassed.

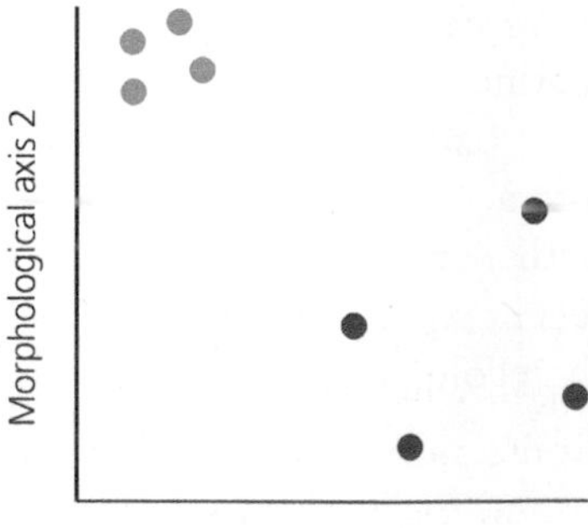

Figure 18.23 A simple morphospace defined by two traits The blue clade occupies a smaller area of morphospace than does the purple clade. This could be quantified with any of a number of measures of morphospace occupation: (1) area; (2) sum of ranges; (3) average pairwise distance; (4) variance.

A plot of lengths and widths for a set of eight species can define a simple morphospace **(Figure 18.23)**. In this case, each species is represented by one point in a two-dimensional space. However, to capture the shape of a species would probably require a greater number of measurements, with a corresponding increase in the dimensionality of the morphospace. It should be noted that many other attributes have been used to define a morphospace, including qualitative characters, three-dimensional landmarks, or any other attribute that can be quantified on a sample of specimens. The disparity of each species or clade can then be quantified by any number of statistics (such as range, variation, or average pairwise distance), either on the raw data or after dimensionality-reducing methods like principal components analysis are performed. How quickly does a clade fill its morphospace? Are certain regions of a morphospace preferentially affected by mass extinction? These are the kinds of evolutionary questions that can be posed.

One of the earliest morphospace studies was by David Raup (1967). He described the theoretical range of ammonoid shell shape by varying three coiling

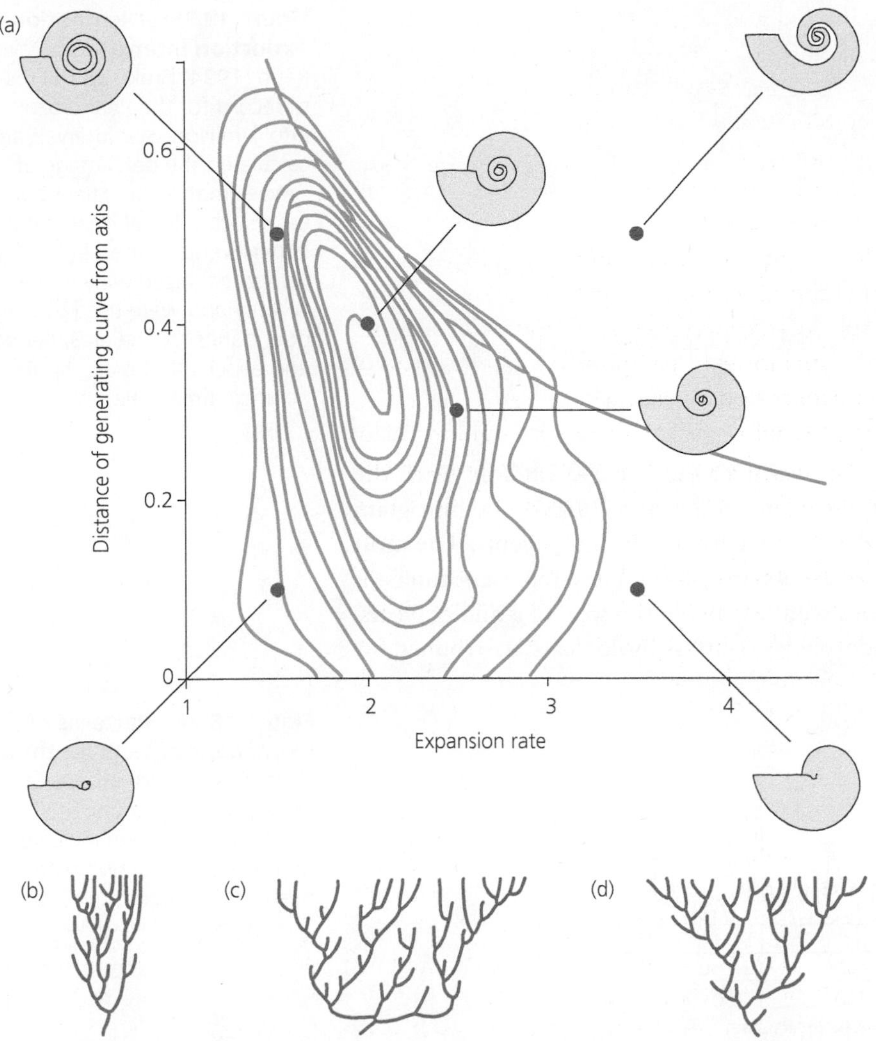

Figure 18.24 Explorations of morphological diversity (a) A morphospace for ammonoid shell structure defined by shell-coiling parameters. The topographic map shows the density of forms on the space among a sample of 405 species, along with six examples showing the overall range of variation. Modified from Raup (1967). (b–d) Patterns of change in morphological diversity and taxonomic diversity over time. Morphological diversity is represented by the horizontal spread of branches; taxonomic diversity by the number of branches at any given time. In (c) morphological diversity expands early while taxonomic diversity expands later. From Foote (1993).

parameters **(Figure 18.24a)**. The resulting plot produced shell designs corresponding to what is seen in nature or in the fossil record as well as others never observed. Some areas of the morphospace are empty because they represent structurally unsound shells. Others were once occupied, but later vacated by extinction.

Evolutionary changes in morphospace occupation in blastozoans, crinoids, and trilobites were explored in a series of papers by Michael Foote (see 1997 review). Despite the variety of taxonomic groups and data sets used, a common pattern that emerged was that disparity peaked before taxonomic diversity (Figure 18.24c, as compared to Figure 18.24b and d), suggesting that evolutionary transitions were larger in the early evolution of each clade, perhaps because there was more available ecospace to fill. Other disparity studies have shown that mass extinctions can affect taxonomic diversity more severely than disparity, suggesting that extinction was random with respect to the morphology analyzed (reviewed by Erwin 2007).

18.4 Mass and Background Extinctions

Extinction is the ultimate fate of all species. What patterns occur in the rate of extinction? Consider a plot that David Raup (1991, 1994) constructed by

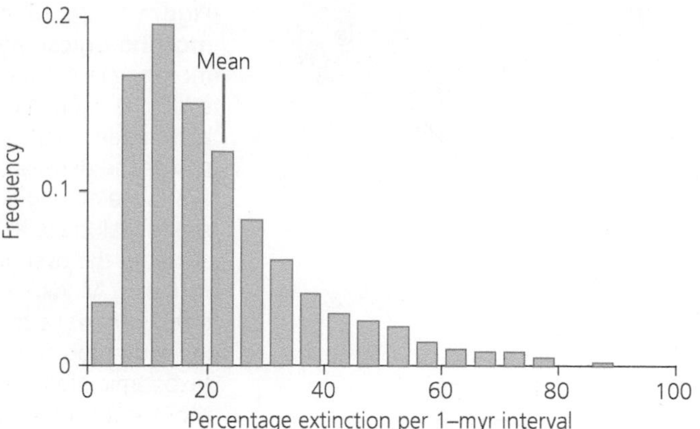

Figure 18.25 Distribution of extinction intensities David Raup (1994) broke up the fossil record for the Phanerozoic into 1-million-year intervals and calculated the percentage of species that went extinct during each such interval. Many of the larger events plotted in this figure were recognized early in the 19th century and were used to define boundaries for the eras, periods, and epochs that make up the geologic time scale.

calculating, for each 1-million-year interval over the last 543 million years, the percentage of taxa that went extinct in that interval **(Figure 18.25)**. The histogram has a pronounced right skew, created by a few particularly large events. The most extreme of events are referred to as **mass extinctions.** They represent intervals in which over 60% of the living species went extinct in the span of a million years. Because of their speed and magnitude, they qualify as biological catastrophes.

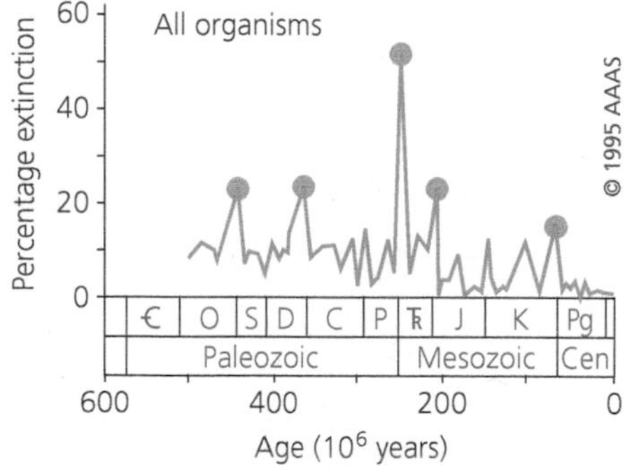

Figure 18.26 Patterns of extinctions of families through time Families are groups of closely related genera.The Big Five mass extinctions are indicated by orange dots. Redrawn with permission from Benton (1995).

"Diversification and extinction in the history of life." *Science* 268: 52–58. Copyright © 1995 AAAS. Reprinted with permission of the AAAS.

How many mass extinctions have occurred during the Phanerozoic? **Figure 18.26** plots the percentage of families that died out during each stage in the fossil record over the past 510 million years (Benton 1995). The five prominent spikes circled on the graph are traditionally recognized as mass extinctions and are referred to as the Big Five. On the geologic time scale, these events occurred at the terminal-Ordovician (ca. 444 Ma), Late Devonian (ca. 360 Ma), end-Permian (251 Ma), end-Triassic (ca. 200 Ma), and Cretaceous–Paleogene, or K–Pg (65.5 Ma). (Note that the Cretaceous is routinely symbolized with a K to distinguish it from the Carboniferous.)

It is important to recognize, however, that the Big Five are responsible for perhaps 4% of all extinctions during the Phanerozoic. The other 96% of extinctions recorded in Figures 18.25 and 18.26 are referred to as **background extinctions**—meaning that they occurred at normal rates. To distinguish mass extinctions from background extinctions, biologists point out that a mass extinction is global in extent, involves a broad range of organisms, and is rapid relative to the expected life span of taxa that are wiped out (Jablonski 1995). It is difficult to differentiate the two categories of extinction more precisely than this, however. As Raup's analysis makes clear, mass extinctions simply represent the tail of a continuous distribution of extinction events over time.

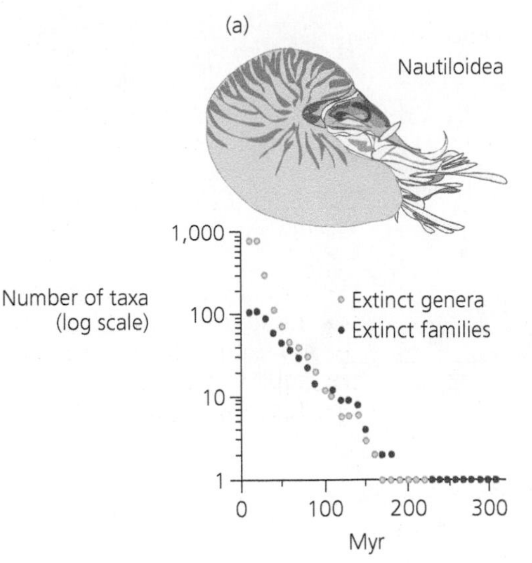

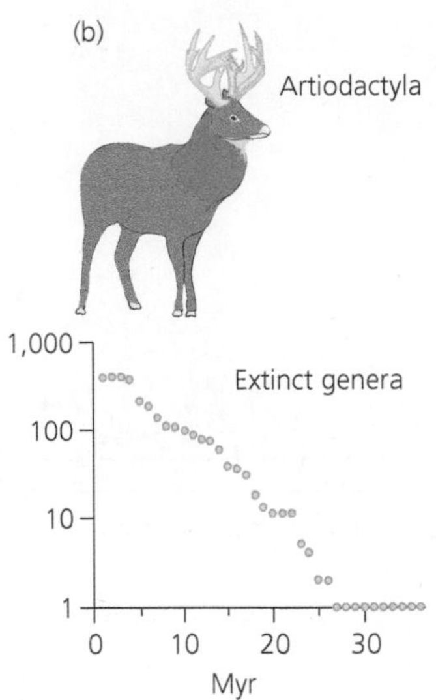

Figure 18.27 Lineage survivorship curves The first step in constructing these curves is to select a random sample of taxa from the fossil record of a particular clade—say, a family or class. The taxa included in the sample can come from any period. The logarithm of the number of genera or families in the clade that survive for different intervals is then plotted. The curves reproduced here are typical (Van Valen 1973): (a) is for genera and families of fossil marine invertebrates called nautiloids, and (b) is for genera in the deer family.

In this section we look at patterns that occur during times of background extinctions, delve into the causes of two mass extinctions, and finally ask whether a mass extinction event, caused by human beings, is now under way.

Background Extinction

Several interesting patterns have been resolved from data on background extinctions. First, within any particular group of organisms, the likelihood of particular lineages becoming extinct is constant and independent of how long the taxa have been in existence. Leigh Van Valen (1973) discovered this when he plotted simple survivorship curves for a wide variety of fossil groups. Survivorship curves show the proportion of an original sample that survives for a particular amount of time. For fossil taxa, Van Valen plotted the number of species, genera, or families from an order or phylum of fossil animals that survived for different intervals. He put the number surviving on a logarithmic scale, so the slope of the curve at any point equaled the probability of becoming extinct at that time. Virtually every plot he constructed, from many different fossil groups and eras, produced a straight line. This means that the probability of subgroups becoming extinct was constant over the life span of the larger clade.

The data in **Figure 18.27** are typical. Note that the slopes of the lines vary from taxon to taxon, meaning that rates of extinction vary dramatically between lineages. These data indicate that during background times, extinction rates are constant within clades, but highly variable across clades. Why background extinction rates vary among lineages is a question we return to later.

Extinctions have occurred throughout life's history on Earth. The rate of background extinction varies from clade to clade and appears to have slowly declined over geologic time.

Second, in marine organisms, several studies have documented a secular decline in global background origination and extinction rates through the Phanerozoic (Raup and Sepkoski 1982; Foote 2003). Although the general trend is punctuated by several spikes, which correspond to mass extinctions, marine biodiversity on the whole appears to be turning over more and more slowly **(Figure 18.28, next page)**. Several explanations for this pattern have been proposed. Biologically, Raup and Sepkoski (1982) suggested that optimization of fitness would lead to the observed pattern. If species in recent geological periods are better adapted, they will survive longer and evolve more slowly. Alternatively, the energy input to the Earth has been decreasing as solar luminosity has declined through the

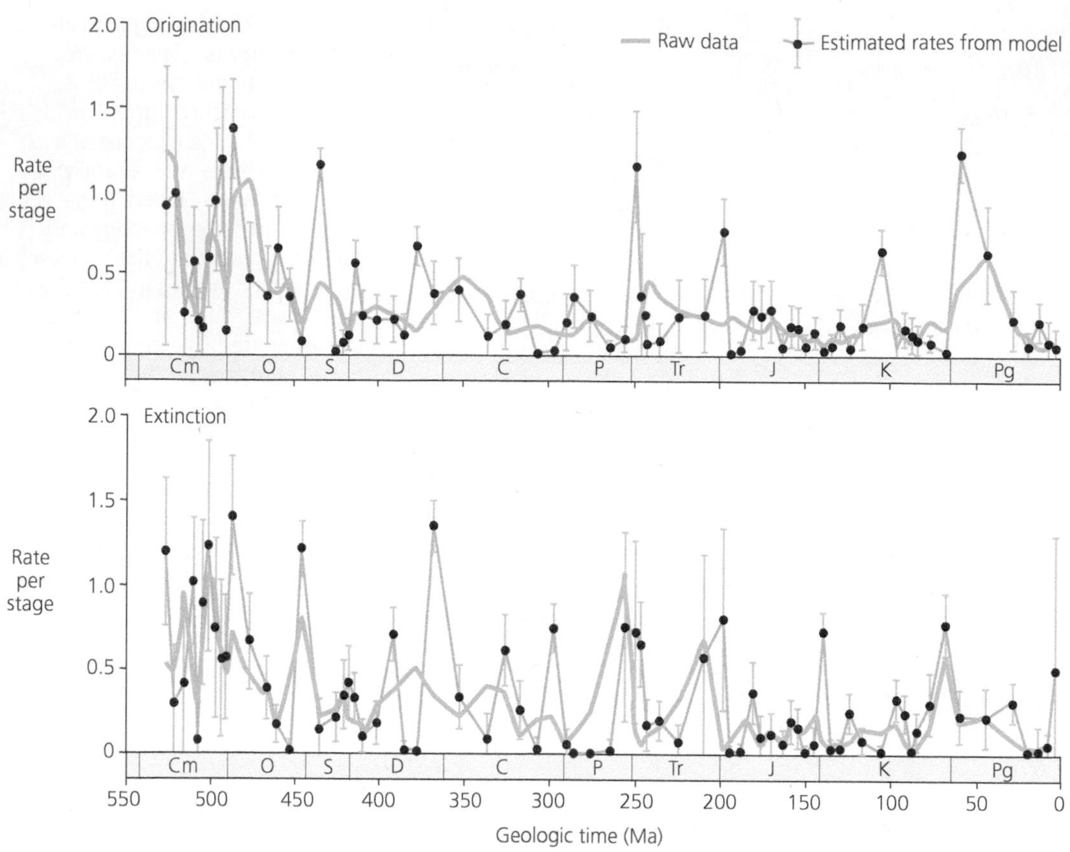

Figure 18.28 A decline in rate of both origination and extinction across the phanerozoic Estimated rates correct for variation in the quality of the fossil record by resampling from the data. From Foote (2003).

Phanerozoic. Thus decreases in energy flow may be responsible. Peters and Foote (2002) suggest that a decrease in the amount of exposed sedimentary rock might introduce a bias in estimates of extinction rates. Finally, David Jablonski (2001) suggested that the overall decline in origination and extinction rates might result from a tendancy of clades with intrinsically high rates of both processes to go extinct during crises, whereas less volatile clades persist. As groups with high turnover are replaced by groups with low turnover, the overall rate of turnover falls.

Cretaceous–Paleogene: High-Impact Extinction

Why do mass extinctions occur? The short answer is that they result from catastrophic episodes of environmental change. But current research suggests that the type of environmental change, and the underlying cause, is different for each of the Big Five. Here we examine the impact hypothesis for the extinction that marked the end of the Cretaceous period and the beginning of the Paleogene (Alvarez et al. 1980). The geological stratum that records the transition between these periods is known as the Cretaceous–Paleogene (K–Pg) boundary. The K–Pg extinction, which included the demise of the non-avian dinosaurs, is the most recent and best understood of the Big Five. The hypothesis that this mass extinction was caused by an enormous asteroid provoked intense debate and research.

Several times during life's history, environmental disasters have triggered mass extinctions.

Evidence for the Impact Event

The discovery of anomalous concentrations of the element iridium in sediments that were laid down at the K–Pg boundary **(Figure 18.29a)** was the first clue that

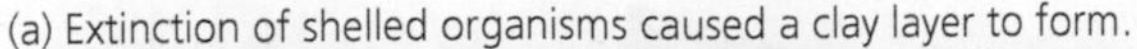
(a) Extinction of shelled organisms caused a clay layer to form.

(b) The clay layer contains high concentrations of iridium.

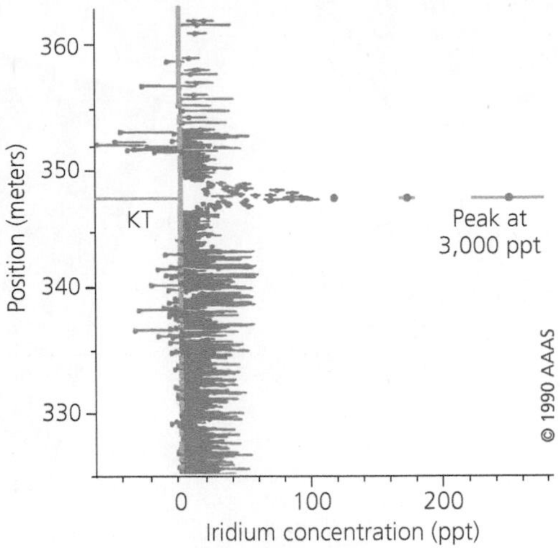

Figure 18.29 Evidence for an impact at the K–Pg boundary (a) The dark band is clay at the K–Pg boundary. The limestones on either side are made of shells of marine invertebrates. A mass extinction stopped limestone formation and allowed clay to build up. By Alessandro Montanari. (b) The concentration of iridium, in parts per trillion (ppt), in the strata in (a). The spike is in the clay. Modified from Alvarez et al. (1990).

(b) "Iridium profile for 10 million years across the Cretaceous-Tertiary boundary at Gubbio (Italy)." *Science* 250: 1700–1702. Copyright © 1990 AAAS. (Figure 1, page 1700). Reprinted with permission of the AAAS.

an asteroid hit Earth 65.5 Ma. Iridium is rare in Earth's crust but relatively abundant in meteorites and other extraterrestrial objects. Figure 18.29b shows a typical iridium spike found in strata that were laid down over the Cretaceous–Paleogene boundary. Over 350 K–Pg boundary sites are now recorded (Schulte et al. 2010). On the basis of estimates for the amount of iridium needed to produce the anomalies and the density of iridium in typical meteorites, Alvarez et al. (1980) suggested that the asteroid was on the order of 10 km wide; more recent estimates suggest that it may have been as large as 15 km wide. It was, quite literally, the size of a mountain.

The discovery of two unusual minerals in K–Pg boundary layers provides additional support for the hypothesis. Shocked quartz particles **(Figure 18.30a)**,

(a) Quartz grains shocked by intense pressure

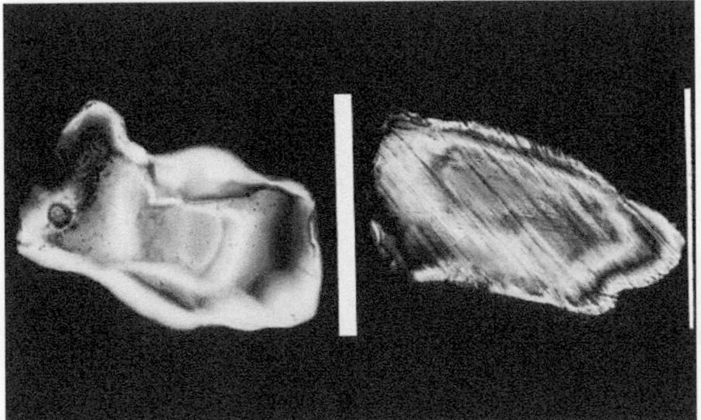

(b) Quartz grains melted into glassy microtektites

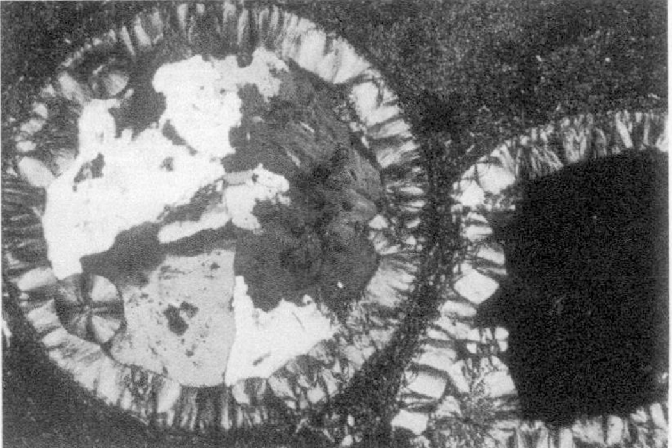

Figure 18.30 More evidence for an impact event at the K–Pg boundary (a) Small quartz grains (1–2 mm across) with parallel planes called lamellae are routinely found near meteorite strikes. The deformation is thought to be caused by the shock of impact. A shocked quartz grain is shown on the right, a normal grain on the left. (b) Microtektites are spherical or teardrop-shaped particles of glass associated with impact sites. These have been sectioned to show the interior. Photos by Glen A. Izett, U.S. Geological Survey.

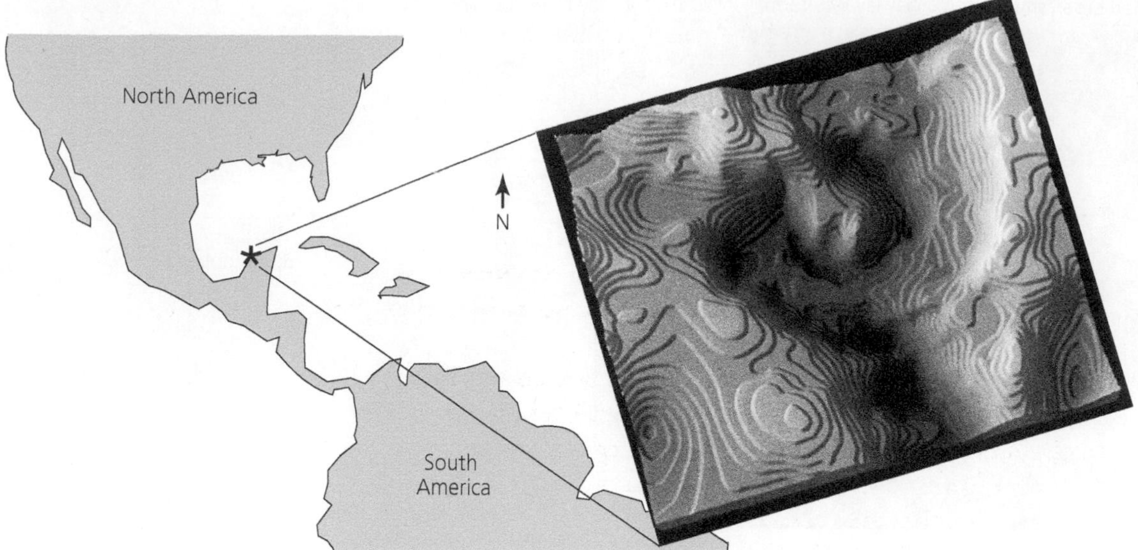

Figure 18.31 Location and shape of the Chicxulub crater Gravitational anomalies outline a 180-km diameter crater, buried beneath sediments near Chicxulub on the Yucatán Peninsula.The inset is a map of the impact site, showing the gravity field density (Schultz and D'Hondt 1996).

produced by intense, short-term pressure, had been found only on the margins of well-documented meteorite impact craters until they were discovered at K–Pg boundary sites. The other unusual structures are tiny glass particles called *microtektites* (Figure 18.30b). These have a variety of mineral compositions, depending on the source rock, but all originate as grains melted by the heat of an impact. If the melted particles are ejected from the crash site instead of being cooled in place, they are often teardrop or dumbbell shaped—a result of solidifying in flight.

The discovery of abundant shocked quartz and microtektites in K–Pg boundary layers from Haiti and other localities in the Caribbean helped investigators narrow the search for the crater. Then, in the early 1990s, a series of papers on magnetic and gravitational anomalies confirmed the existence of a crater 180 kilometers in diameter, centered near a town called Chicxulub (cheek-soo-LOOB) in the northwest part of Mexico's Yucatán Peninsula **(Figure 18.31)**. The shape of the crater suggested that it was created by an oblique impact—meaning that the asteroid hit at an angle and splashed material to the north and west.

Subsequent dating work confirmed that microtektites from the wall of the crater, recovered from cores drilled in the ocean floor, were 65 million years old (Swisher et al. 1992; but see Keller et al. 2004). This is a close match to dates for glasses ejected from the site and recovered in the Haitian K–Pg boundary layer.

The discovery of the crater was the long-sought smoking gun. It solidified a consensus among paleontologists, physicists, geologists, and astronomers that a large meteorite struck Earth at 65 Ma (Schulte et al. 2010). The existence of the impact is no longer controversial; the consequences of the impact are.

Killing Mechanisms

The mountain-sized asteroid that struck the ocean would have produced a series of events capable of affecting climate and atmospheric and oceanic chemistry all over the globe. The ocean floor near the impact site at Chicxulub consisted of carbonates, including large beds of anhydrite ($CaSO_4$). The distribution of shocked quartz and microtektites, far to the north and west of Chicxulub, confirms that a large quantity of this material was ejected from the site and that significant amounts were melted or vaporized by the heat generated at impact.

What consequences did the ejected material have? Vaporization of anhydrite and seawater would have contributed an enormous influx of sulfur dioxide (SO_2) and water vapor to the atmosphere. These two molecules would react to form sulfuric acid (H_2SO_4) and produce intense acid rain. Sulfur dioxide is also a strong scatterer of solar radiation in the visible spectrum, which would lead to global cooling (McKinnon 1992). The cooling effect would have been enhanced by dust-sized carbonate, granitic, and other particles. These were ejected in quantities large enough to block incoming solar radiation.

A variety of models suggest that the force of the impact was also sufficient to trigger massive earthquakes, perhaps as large as magnitude 11 on the Richter scale, and to set off volcanoes. The second-largest magma deposits of the Phanerozoic, the Deccan Traps of India, were contemporaneous with the K–Pg extinction, but it is now clear that they had already been erupting for hundreds of thousands of years before the impact. Nonetheless, researchers have suggested that the massive amount of climate-forcing gases injected into the atmosphere by the Deccan eruptions had likely destabilized ecosystems, such that the impact's effect was even more profound. Arens and West (2008) suggest that a press-pulse model, characterized by long-term ecological destabilization followed by an acute event like impact, might be a general phenomenon of mass extinctions.

Finally, the impact would have created an enormous tidal wave, or tsunami, in the Atlantic Ocean. If the asteroid was indeed 10 km wide, models suggest that the wave produced by the strike would have been as large as 4 km high. The mountain of rock made a mountain-sized splash. Joanne Bourgeois and colleagues (1988) provided evidence of the tsunami when they discovered a huge sandstone deposit along the Brazos River in Texas, which has been mapped throughout northeast Mexico. It is 300 km long and several meters thick and is now interpreted by most geologists as a product of the rapid and massive deposition typical of tsunamis. Additional evidence for a tsunami is found in the 65-Ma sediments of Haiti, where a thick jumble of coarse- and fine-grained particles are sandwiched between the iridium-enriched clay layer above and extensive tektite deposits below. Maurrasse and Sen (1990) interpret this middle stratum as the product of tsunami-induced mixing and deposition. This event occurred after the initial splash of microtektites, but before the fallout of iridium-enriched particulates from the atmosphere.

The physical consequences of asteroid impact are dramatic, and they undoubtedly led directly to the rapid elimination of marine and terrestrial biotas in the days or months immediately following the blow. However, a large proportion of the end-Cretaceous extinctions must have been caused by ecological interactions between organisms and their traumatized environment. The decline of many groups was not instantaneous, but drawn out over the 500,000 years following the impact. These extinctions were probably due to the disruption of ecological processes, biogeochemical cycles of nutrients, and interactions among species.

Permian–Triassic: The Biggest of the Big

The mass extinction at the end of the Permian is generally regarding as the biggest of the Big Five. Like the K–Pg, which ended the Mesozoic era, the Permian–Triassic (P–T) extinction marked the conclusion of the Paleozoic and had a substantial effect on reshaping future ecosystems (see Chen and Benton 2012). Gone were the trilobites, most brachiopods, and some archaic groups of tetrapods; echinoderms radiated, as did the precursors to dinosaurs. Unlike the K–Pg, however, identifying the cause of the P–T has been difficult, and only recently has a consensus emerged that large-scale volcanic activity, combined with lowered oxygen levels and runaway greenhouse conditions, was the likely culprit.

The biggest catastrophe in life's history was the mass extinction that marks the end of the Permian and the beginning of the Triassic.

Until about 2005, the timing of the P–T extinction was poorly understood. Was it a distinct catastrophic event like the K–Pg, or was the extinction a drawn-out process that lasted millions of years? The lack of reliable radiometric dates meant that paleontologists were not even sure if extinctions in the ocean and on land took place at the same time. In the past decade numerous high-precision dates have been recovered, and correlations between the marine and terrestrial fossil records have been solidified (Shen et al. 2011). The results of this work indicate that the extinction interval was less than 200,000 years, was synchronous on land and in the ocean, and was centered on a date of 252.3 Ma. The synchronous nature of the extinction suggests a common cause and greatly limits potential mechanisms.

For many years it was thought that one contributing factor to the decrease in biodiversity at the P–T was that the latest Permian was a time of pronounced regression (global shallowing of the ocean), which reduced the amount of continental shelf available for marine invertebrate communities to occupy. Subsequent transgression (oceanic deepening) restored available ecospace, but also featured the return of Lazarus taxa, or species that disappeared in the Late Permian and "came back from the dead" in the Triassic. Lazarus taxa show that a direct reading of the fossil record can be led astray by taphonomic influences. Late Permian regression has since been discounted, which suggests that the disappearance of coral reefs, as well as the terrestrial plants that led to coal deposits, until about 8–9 Ma after the extinction is better interpreted as the result of poor environmental conditions during Early Triassic times.

Recent work has shown that almost the entire Early Triassic (252–247 Ma) was characterized by massive fluctuations in the global carbon cycle, indicating long-term environmental instability. Payne and colleagues (2004) documented a marked reduction in gastropod body size during the same time interval and suggested an ecological relationship. The term *Lilliput effect* is commonly used to characterize body-size reductions in the wake of mass extinctions, which seem to be a common feature of the fossil record.

The fossil record of the terrestrial environment during the P–T crisis is best known from rocks in South Africa and Russia. Ward et al. (2000) showed that the extinction coincided with a large-scale restructuring in depositional style, from slow-flowing, meandering streams in the Permian to fast-flowing and relatively straight braided rivers in the Triassic. They hypothesized that these fluvial changes were caused by the extinction of ground vegetation, which generally stabilizes soils in addition to forming the primary trophic level in terrestrial ecosystems. The fossil record of South Africa documents that Early Triassic communities were remarkably uneven and numerically dominated by one genus, *Lystrosaurus*. The remains of this bizarre-looking animal **(Figure 18.32)** have also been found in Antarctica as well as India, China, and Russia, providing early evidence for the connection of these continents. Analyses of terrestrial community structure in Russia have shown similarly dramatic shifts across the P–T boundary (Benton et al. 2004).

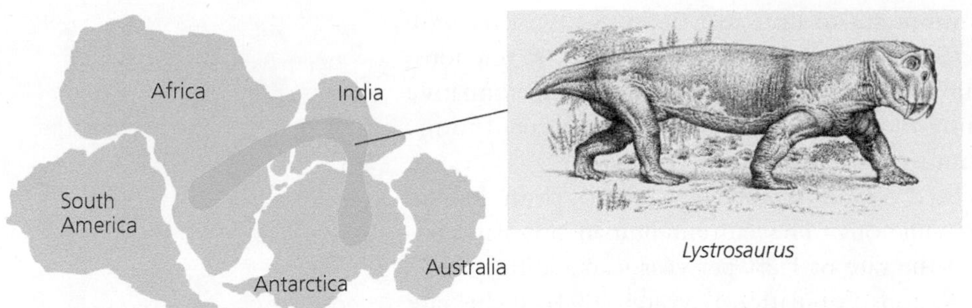

Figure 18.32 *Lystrosaurus* Fossils of *Lystrosaurus* are known from across Pangea, including modern-day Africa, India, and Antarctica as well as China and Russia (not shown). *Lystrosaurus* reconstruction from Benton (2003).

The magnitude of the P–T extinction has led many to draw parallels between it and the better-studied K–Pg extinction, especially in terms of potential mechanisms. There is no iridium spike at the P–T, but meteor fragments were reported from Antarctica in 2003 and a potential crater proposed north of Australia in 2004. Both of these claims have been greeted with skepticism from the scientific community, and no lab has been able to replicate some of the geochemical results associated with impact (Erwin 2003). Instead, the importance of massive basalt flows in Russia, the Siberian Traps, has increased substantially. Revised estimates of their timing (centered on 252 Ma) and size (2 million km^3, or the largest of the Phanerozoic) are now perfectly in line with triggering the extinction. The Siberian eruptions elevated CO_2 levels and produced short-term effects like acid rain, wildfires, global cooling (from dust and sulfates ejected in the atmosphere), and ocean acidification. At the same time, atmospheric models propose decreases in atmospheric oxygen levels (estimates as low as 13%, compared to 21% today). This information has led some researchers to hypothesize that terrestrial animals might have been restricted to low elevations, leaving much of the interior of continents off-limits and thereby reducing ecospace substantially (Huey and Ward 2005). Oceanic anoxia is also well documented in P–T strata.

The P–T was a profound extinction, but equally profound was the protracted duration of its recovery. Some rapidly evolving groups (e.g., ammonoids, conodonts) were able to diversify within 1–2 million years after the extinction, but for most, diversification to pre-extinction levels was delayed until the Middle Triassic. In other words, the first ~10 million years of the Early Triassic were a dead zone. Several studies have shown massive fluctuations in carbon isotope values, suggesting that marine ecosystems experienced continued perturbations that degraded the environment.

The Sixth Mass Extinction

Concern about widespread extinction is on the minds of people from all walks of life, from schoolchildren to heads of state. But despite celebrity examples like the dodo, passenger pigeon, and Carolina parakeet, is anything of special evolutionary significance going on now? That is, are we currently experiencing or contemplating an event anything like the Big Five in scale and speed? To find out, we examine data on extinctions that have occurred over the past 2,000 years.

Is a Mass Extinction Event Currently Under Way?

What are current rates of extinction, and how do they compare to what is seen in the fossil record? About 1.5 million species have been studied and named thus far, but only about 1,100 species have become extinct since 1600. Is concern about an impending mass extinction overblown?

To answer this question, it is important to note that the majority of recently extinct species inhabited islands (Smith et al. 1993). These island extinctions, in turn, usually resulted from human hunting or the introduction of nonnative predators or competitors. This introduction process has probably peaked in intensity and should be less important in the future.

Instead, current concern is focused on a different agent of extinction: habitat loss due to expanding human populations. The current human population is about 6.5 billion and is growing at the rate of 1.2% per year—an addition of 77 million people annually (United Nations Population Division 2004). If this rate continues, world population will exceed 12 billion by the year 2050. Unless human population growth declines rapidly, threats to natural habitats will grow in intensity over the next several decades.

Biologists have employed three types of approaches to predict how continued habitat destruction will affect extinction rates (May et al. 1995):

1. Multiply the number of species found per hectare in different environments by rates of habitat loss measured from satellite photos.
2. Quantify the rate that well-known species are moving from threatened to endangered to extinct status in the lists maintained by conservation groups.
3. Estimate the probability that all species currently listed as threatened or endangered will actually go extinct over the next 100 or 200 years.

These approaches suggest that extinctions are now occurring at 100 to 1,000 times the normal, or background, rate of extinction (May et al. 1995; Pimm et al. 1995). For example, the International Union for the Conservation of Nature estimates that the number of threatened and critically endangered species grew from slightly over 10,000 in 1996 to over 15,000 in 2004. If this rate continued, and if all of these rare species were to go extinct, it would take less than 100 years for 60% of living species to be wiped out. These data suffer from an ascertainment bias, however: The 2004 number is much higher than the 1996 number, simply because much more effort has been expended recently in studying endangered populations. Taken together, though, the analyses done to date suggest that if current rates of habitat destruction continue, the coming centuries or millennia will see a mass extinction on the same scale as the Big Five documented in the fossil record (Barnosky et al. 2011). A human meteor is hitting the Earth.

Human activities have increased extinction rates above background by orders of magnitude.

Where Is the Problem Most Acute?

If you asked a biologist where habitat destruction is causing the most severe threat to biodiversity, the answer is likely to be tropical rain forests. There are two reasons:

- Tropical rain forests are extraordinarily rich in species. E. O. Wilson (1988) recounts that he once collected 43 species of ants belonging to 26 genera from a single tree in a Peruvian rain forest. These numbers are roughly equivalent to the entire ant fauna of the British Isles. Similarly, Peter Ashton identified 700 different species of trees—the same number found in all of North America—in just ten 1-hectare sample plots from a rain forest in Borneo. With the exception of conifers, salamanders, and aphids, nearly every well-studied lineage on the tree of life shows a latitudinal gradient in diversity: The largest number of species reside in the tropics. Why this pattern occurs is not clear, but the results are striking. Tropical forests occupy less than 7% of Earth's land area but contain at least 50% of all plant and animal species.

- Tropical rain forests are presently under acute threat. Many nontropical habitats in the Northern Hemisphere, as well as most oceanic islands, have been continuously occupied by high densities of humans for several hundred years. As a result, the flora and fauna of these nontropical habitats have already sustained numerous extinctions. Andrew Balmford (1996) suggests that the long history of dense human occupation, combined with extinctions caused by radical climate change during the ice ages of the Pliocene and Pleistocene, have put nontropical biomes through an "extinction filter." The plant and animal communities now living in these regions are expected to be relatively resilient in the face of continued human impact. In contrast, many areas of the tropics have been relatively unaffected by humans in recent history and were less affected by glaciation and sea-level changes in the Pleistocene. The tropics are now experiencing the highest rates of growth in human populations and the highest rates of habitat loss.

The threat to these forests is grave. According to the United Nations Food and Agriculture Organization (FAO), total forest loss currently averages about 7 million hectares per year—an area about the size of Scotland. Most of these losses are occurring in the tropics. South and southeast Asia are of particular concern because the total forest area is relatively small. In this region, forest loss is averaging about 1.1% per year compared to about 0.7% per year in Africa and the Americas (Laurance 1999). The Brazilian Amazon is also a region of interest, because it is the largest continuous tropical forest in the world. Using satellite photographs, Brazil's National Institute of Space Research estimates that an average of about 25,000 square kilometers of forest was lost each year from 2002 to 2005—an annual loss equivalent in size to the state of Massachusetts. The recent pace of forest destruction is up from a rate of about 15,000 square kilometers lost annually in the Brazilian Amazon from 1978 to 1988 (Skole and Tucker 1993). Biologists also maintain that more than double the amount of cleared forest is adversely affected each year because of edge effects. Forested areas adjacent to clearings undergo dramatic changes: Light levels increase, soils dry, daily temperature fluctuations increase markedly, domestic livestock encroach, and hunting pressure by humans heightens. As Skole and Tucker (1993, p. 1909) note: the "Implications for biological diversity are not encouraging." Stopping the human meteor will take a combination of lower human population growth rates and sustainable development that preserves tropical forests.

18.5 Macroevolution

Natural selection happens when variation within a species ulitmately leads to differential survival and reproduction—this, along with its genetic underpinnings, is a reasonable summary of microevolution. **Macroevolution,** on the other hand, is a term used to cover two distinct phenomena. The first is large-scale evolutionary change, such as the examples of major morphological transitions discussed earlier (Sections 2.3 and 18.2). The second usage of *macroevolution*—evolutionary processes operating above the species level—was espoused by Steve Stanley (1975, 1982) and in its strictest form considers species to be the focal point of selection, akin to individuals in microevolution. This version of macroevolution was historically linked to the theory of punctuated equilibrium, and both draw critical data from the fossil record.

Punctuated Equilibrium and the Importance of Stasis

The fossil record contains many cases of new species that appear and then persist for millions of years without apparent change. Stated another way, evolution in some groups consists of long periods of stasis that are occasionally punctuated by speciation events that appear instantaneous in geological time. Darwin (1859) was well aware of these cases and considered them a problem for his theory. Because his ideas were an alternative to the theory of special creation, which predicts the instantaneous creation of new forms, Darwin repeatedly emphasized the gradual nature of evolution by natural selection. He attributed the sudden appearance of new taxa to the incompleteness of the fossil record and predicted that as specimen collections grew, the apparent gaps between species would be filled in by transitional forms. For a century thereafter, most paleontologists accepted Darwin's view.

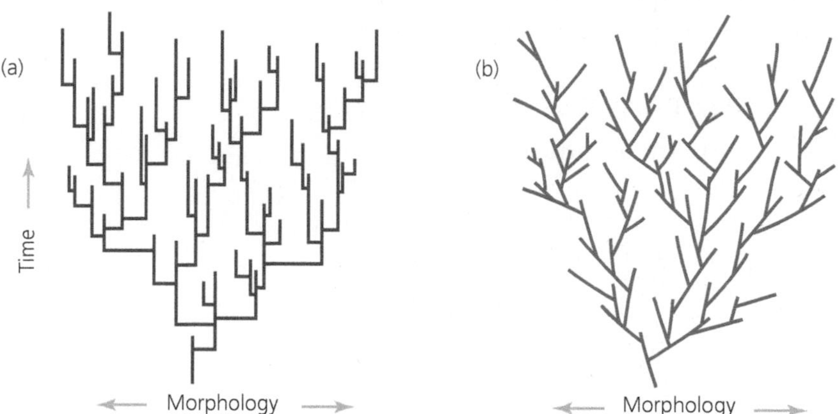

Figure 18.33 Stasis versus gradualism When time is plotted against morphology, two extreme patterns are possible, along with many intermediate or mixed patterns. (a) In punctuated equilibrium, all morphological variation occurs at the time of a speciation (branching) event; otherwise, there is stasis. (b) In phyletic gradualism, morphological change occurs gradually and is unrelated to speciation events. After Eldredge and Gould (1972).

In 1972, however, Niles Eldredge and Stephen Jay Gould broke from convention by claiming that stasis is a real pattern in the fossil record and that most morphological change occurs during speciation events. They called their proposal **punctuated equilibrium** **(Figure 18.33)**. The theory and its implications were hotly debated for 20 years.

Demonstrating Stasis

One benefit of the debate over punctuated equilibrium was that it spurred paleontologists to ask whether stasis is in fact real. Do the data support the claim that stasis, punctuated by morphological change at speciation events, is the predominant feature of species histories through time?

Gene Hunt (2007) analyzed more than 250 fossil sequences, looking for statistical evidence consistent with three evolutionary models: directional change, a random walk, and stasis. He found support for directional evolution in 5% of the fossil sequences. The remaining 95% of sequences showed evidence of random walks and stasis in roughly equal proportions. Stasis is clearly common.

It is not unusual for lineages to show little morphological change over long spans of time.

What about the claim that morphological change is associated with speciation? Before looking at data, it is important to clarify the requirements for testing the pattern. The goal is to follow changes in morphology in speciating clades through time and determine whether change occurs in conjunction with speciation events or independently, and whether rapid change is followed by stasis or continued change. As critics of the theory have emphasized, a rigorous test for stasis versus gradualism is difficult. This is because the theory of punctuated equilibrium can become tautological. Fossil species are defined on the basis of morphology, so it might be trivial to observe a strong correlation between speciation and morphological change. To avoid circularity, an acceptable test requires that

1. the phylogeny of the clade is known, so researchers can identify which species are ancestral and which descendant; and
2. ancestral species survive long enough to co-occur with the new species in the fossil record.

The second criterion is critical. If it is not fulfilled, it is impossible to know whether the new morphospecies is indeed a product of a splitting event or whether it is the result of rapid evolution in the ancestral form without speciation taking place. This second possibility is called **phyletic transformation,** or **anagenesis.**

These are demanding criteria, especially when compounded with other difficult practical issues: the problem of misidentifying cryptic species in the fossil record, the need for analyzing change at the level of species, the requirement of dense sampling in rocks with constant deposition, and the necessity of sampling multiple localities to distinguish normal, within-species geographic variation from authentically different morphospecies.

Stasis and Speciation in Bryozoans

Relatively few fossil series meet these stringent requirements (Jablonski 2000). One is a series of Late Cenozoic fossils of the marine invertebrate phylum Bryozoa. Experimental studies on cheilostome bryozoans alive today established that bryozoans that are identified as morphospecies also qualify as phylogenetic species (Jackson and Cheetham 1990, 1994). In addition to bryozoans being abundant in the fossil record of the past 100 million years, then, we can also be confident that species designations in this group actually reflect phylogeny.

Cheetham (1986) and Jackson and Cheetham (1994) performed a high-resolution analysis of speciation and morphologic change in cheilostomes from the Caribbean, starting about 15 Ma in the Miocene and ending with living taxa. They began by defining 19 morphospecies in the genus *Stylopoma,* based on an analysis of 15 skeletal characters. They estimated the phylogeny of the 19 morphospecies from differences in skeletal characters and scaled the tree so that the branch points and branch tips lined up with the dates of first and last appearance for fossil forms. They did a similar analysis for 19 living or extinct morphospecies in the genus *Metrarabdotos.* The resulting trees are pictured in **Figure 18.34**.

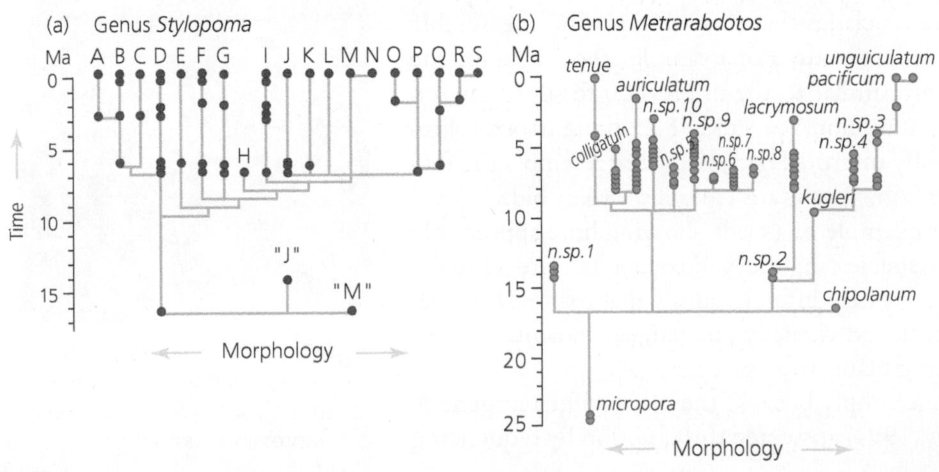

Figure 18.34 Punctuated change in cheilostome Bryozoa Phylogenies, for 19 living and fossil morphospecies in the genus *Stylopoma* (a) and 19 living and fossil morphospecies in the genus *Metrarabdotos* (b), were estimated from differences in skeletal characters (Jackson and Cheetham 1994). Each dot indicates a sampled population. None showed skeletal traits intermediate between species, and the characteristics of species were stable through time.

The phylogenies show an unequivocal pattern of stasis punctuated by rapid morphological change. The fact that ancestral and descendant species co-occur defends the idea that morphologic change was strongly associated with speciation events. This is an almost flawless example of stasis punctuated by evolutionary change at speciation.

What Is the Relative Frequency of Stasis and Gradualism?

How common is the pattern observed in cheilostome bryozoans? Doug Erwin and Robert Anstey (1995a, 1995b) reviewed a total of 58 studies conducted to test the theory of punctuated equilibrium. The analyses represent a wide variety of taxa and periods. Although the studies varied in their ability to meet the strict criteria required for a rigorous test of the theory, their sheer number may compensate somewhat. Erwin and Anstey's conclusion was that "Paleontological evidence overwhelmingly supports a view that speciation is sometimes gradual and sometimes punctuated, and that no one mode characterizes this very complicated process in the history of life." Furthermore, Erwin and Anstey noted that a quarter of the studies reported a third pattern: gradualism and stasis.

Morphological evolution and speciation are sometimes associated. Other times, however, they are not.

In the largest study to date, Melanie Hopkins and Scott Lidgard (2012) analyzed 635 traits in 153 lineages. They fit statistical models to characterize the mode of evolution for each trait in each lineage as either directional change, a random walk, or stasis. Consistent with earlier work, they found that within most lineages, different traits show different modes of evolution.

Once the controversy over punctuated equilibrium was resolved, biologists turned to other questions. Is it possible that different types of organisms exhibit distinct patterns of change through time? Researchers who have worked on the problem are beginning to argue that gradualist patterns tend to predominate in foraminifera, radiolarians, and other microscopic marine forms, while stasis occurs more often in macroscopic fossils such as marine arthropods, bivalves, corals, and bryozoans (Hunter et al. 1988; Benton and Pearson 2001). If so, why? Research continues.

Why Does Stasis Occur?

One of Eldredge and Gould's prominent claims about the fossil record is that "stasis is data." That is, lack of change is a pattern that needs to be explained. Jackson and Cheetham's study of bryozoans showed that virtually no change occurred in these sessile invertebrates over millions of years. Why would morphology remain unchanged for so long in these lineages? One approach to answering this question has focused on species or clades that show little or no measurable morphological change over extended periods. For example, the leaves of the living gingko tree in **Figure 18.35a** are similar to the impression fossils of gingko leaves in Figure 18.35b, which are 40 million years old. The living stromatolites pictured in **Figure 18.36a** were made by intertidal bacteria. They resemble the fossil stromatolites shown in Figure 18.36b, which are 1.8 billion years old.

Horseshoe crabs are a spectacular example of a clade showing little appreciable morphological change. The extant species, in the genus *Limulus,* are virtually identical in morphology to fossil species in a different family that existed 150 Ma. While some horseshoe crab lineages stayed virtually unchanged, the entire radiation of birds, mammals, and flowering plants took place.

Have these species failed to change simply because they lack sufficient genetic variation? John Avise and colleagues (1994) answered this question by sequencing

(a)

(b)

Figure 18.35 "Living fossils" Leaves from a living gingko tree (a) are similar to 40-million-year-old impression fossils (b).

(a) 

(b) 

Figure 18.36 More "living fossils" Contemporary stromatolite-forming bacteria from Australia (a) are similar to 1,800-million-year-old fossil forms from the Great Slave Lake area of Canada (b).

several genes in the mitochondrial DNA of horseshoe crabs and comparing the amount of genetic divergence they found to a previously published study of genetic distances in another arthropod clade: the king and hermit crabs (Cunningham et al. 1992). The result is striking: The horseshoe crabs show just as much genetic divergence as the king–hermit crab clade, even though far less morphological change has occurred **(Figure 18.37)**. This is strong evidence that stasis is not from a lack of genetic variability (Briggs et al. 2012).

Morphological stasis does not imply genetic stasis.

When stasis occurs despite ample genetic variation, the most obvious explanation is stabilizing selection. A striking example comes from an analysis by Gene Hunt and colleagues (2008) of a lineage of fossil sticklebacks (*Gasterosteus doryssus*) closely related to the threespine sticklebacks we have discussed elsewhere in the book. The data come from more than 5,000 individual fish from an extraordinarily high-resolution fossil record excavated from an open-pit diatomite mine in Nevada. The strata in the mine record change in the stickleback population across more than 7,000 generations. In the oldest strata, the population consisted of heavily armored fish typical of lineages that have recently invaded freshwater from the ocean. The population rapidly evolved reduced armor, following a

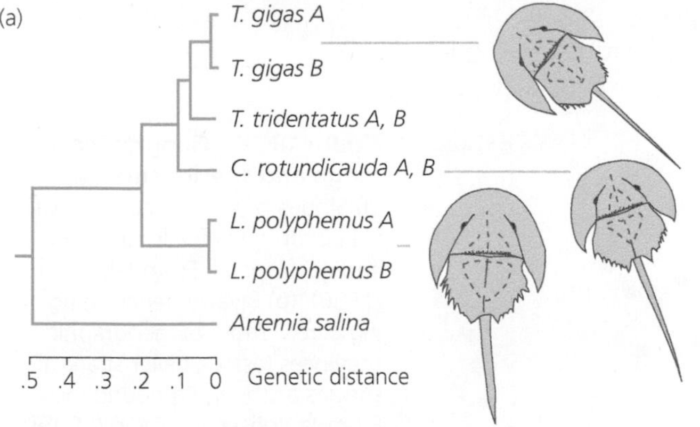

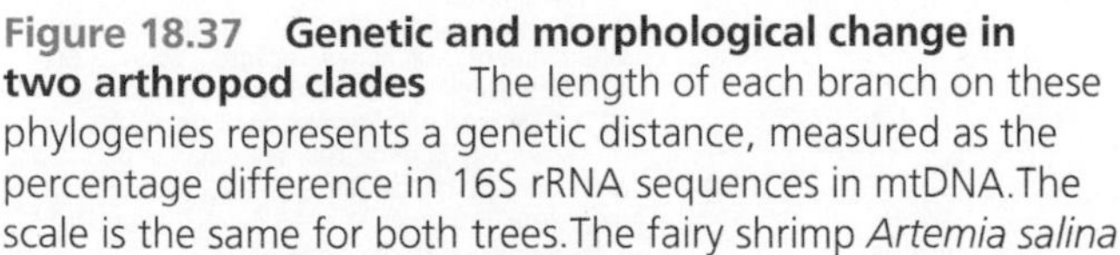

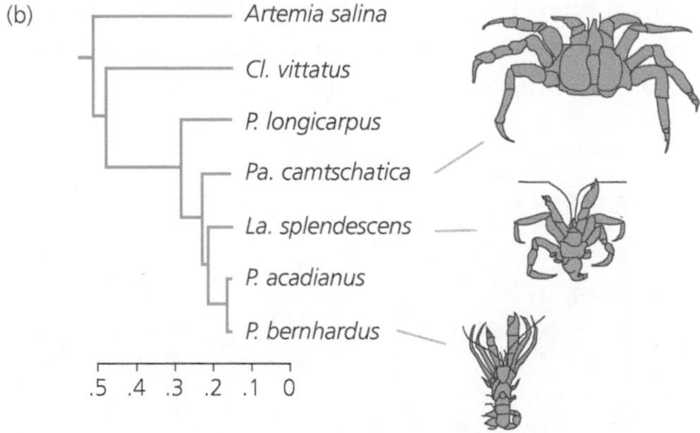

Figure 18.37 Genetic and morphological change in two arthropod clades The length of each branch on these phylogenies represents a genetic distance, measured as the percentage difference in 16S rRNA sequences in mtDNA. The scale is the same for both trees. The fairy shrimp *Artemia salina* was used as the outgroup to root each of these trees. Slightly more genetic divergence in 16S rRNA sequences has occurred in the horseshoe crab clade (a), even though much more morphological divergence has occurred in the clade that includes hermit crabs and allies (b). From Avise et al. (1994).

trajectory that matches what we would predict using a population genetics model of directional selection. Upon achieving a mean phenotype with low armor typical of freshwater sticklebacks free of predatory fish, the population thereafter remained unchanged. That is, it showed the stasis that would be expected under stabilizing selection around an optimal phenotype.

Evolution above the Species Level

Steven Stanley (1975, 1982) took the tenets of punctuated equilibrium and proposed that if speciation were effectively random with regard to natural selection, then evolution at the clade level would be decoupled from that within species. In other words, evolutionary patterns of clades are not predicted by the effects of natural selection acting on individuals in populations. In Stanley's view, if natural selection acts on variation within a species, then an analogous process should act on variation among species. Stanley termed the latter process *species selection* and suggested several parallels that could be drawn between it and natural selection (e.g., speciation would be akin to reproduction).

Paleontologists have long recognized clade-specific variation in the fossil record. For example, G. G. Simpson (1944) frequently compared bivalves and mammals in his classic book, *Tempo and Mode in Evolution,* to show their distinct intrinsic rates of net diversification: Bivalves always diversify much more slowly than mammals, regardless of background or mass extinction times. Average species durations vary in a similar way: On average, bivalve species persist for about 3–5 million years, whereas mammals last for about 1–2 million years. What are some other species or clade-level characteristics upon which species selection could act? Abundance, reproductive mode, generation time, geographic range, and aspects of life history (such as feeding ecology) have all been suggested. Research on macroevolution typically requires large data sets, which has limited the number of studies performed. Highlighted next are a few examples.

Some traits of species or clades, such as geographic range, are associated with extinction rate.

Geographic range is clearly an attribute of species or clades, and not of individuals, so much work has focused on how this feature might evolve over geologic time. David Jablonski (1986) studied how the geographic range of Cretaceous bivalves (clams and mussels) and gastropods (slugs and snails) related to their ability to survive extinction. He found that species with larger geographic ranges had a lower extinction probability during times of background extinction **(Figure 18.38a)**. During mass extinctions, there was a similar pattern for genera (Figure

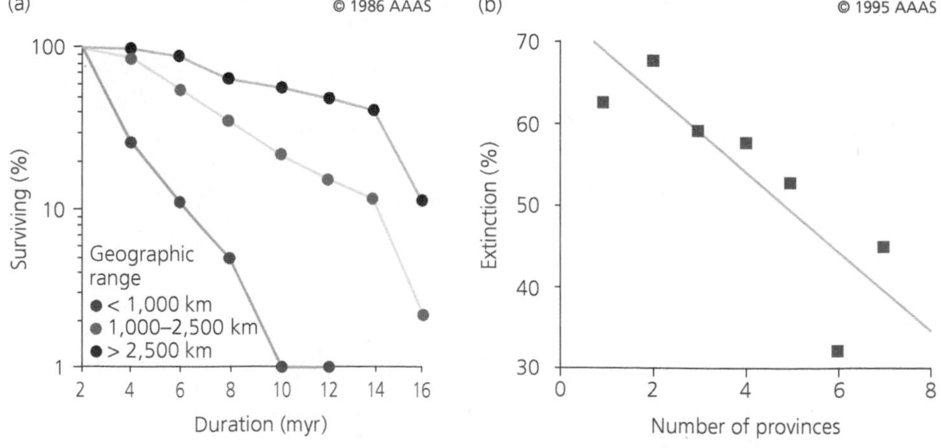

Figure 18.38 Geographic range and species survival (a) Slopes of survival curves give extinction rates for fossil bivalves and gastropods. From Jablonski (1986). (b) Bivalve genera ranging across more biogeographic provinces lost a smaller share of species at the K–Pg boundary. From Jablonski and Raup (1995).

(a) "Background and mass extinctions: The alternation of evolutionary regimes." *Science* 231: 129–329. Copyright © 1986 AAAS. (Figure 2a, 131). Reprinted with permission of the AAAS. (b) "Selectivity of end-Cretaceous marine bivalve extinctions." *Science* 268: 389–391. (Figure 3, page 390). Copyright © 1995 AAAS. Reprinted with permission of the AAAS.

18.38b; Jablonski and Raup 1995), but for species there was no significant relationship between survivorship and geographic range (Jablonski 2005). These results suggest that the accumulation of certain traits during background extinction times could be reset when patterns of selection change during mass extinctions.

A similar alternation between background and mass extinction times has been found for feeding type. Between the end-Triassic and end-Cretaceous mass extinctions, Jablonski (2005) showed that epifaunal suspension-feeding bivalves had significantly longer genus durations than infaunal suspension-feeding bivalves. But during times of mass extinction, there was no significant difference in extinction intensity between the two feeding types.

Extinction rates also vary with how far larvae disperse after eggs are fertilized and begin development. Jablonski (1986) came to this conclusion by studying extinction patterns in bivalve and gastropod species from the Gulf of Mexico and the Atlantic coastal plain region over the last 16 million years of the Cretaceous period. Jablonski found that marine invertebrate species with a planktonic larval stage survived longer, on average, than species whose young develop directly from the egg **(Figure 18.39)**. In living species, planktonic larvae are carried on currents and often disperse long distances. This gives them greater colonizing ability, which might reduce the frequency of extinction. Populations with this life history also tend to have larger ranges, which means that the ability to study dispersal ability and geographic range independently is difficult. The stark difference between the species-level traits that conferred extinction resistance during times of background extinction, and the failing of such traits to be beneficial during mass extinction, is what Jablonski (1986) termed the "alternation of macroevolutionary regimes."

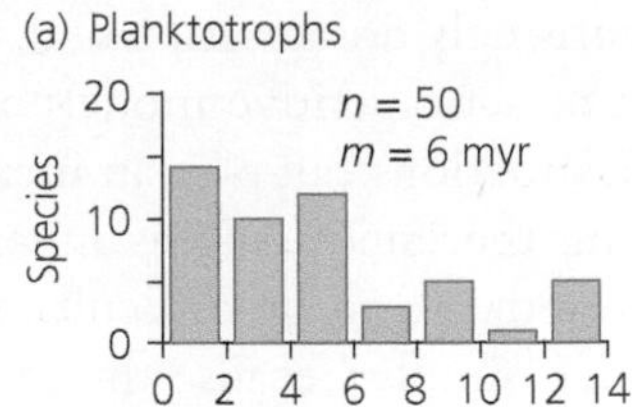

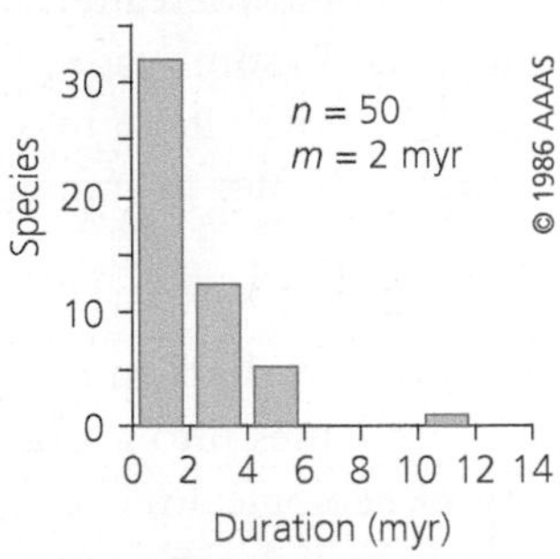

Figure 18.39 Persistence of marine bivalve species Planktotrophs are species with larvae that spend at least some time feeding in the plankton. Nonplanktotrophs have larvae that develop directly from the egg. n is the number of species and m the mean; myr stands for million years. Reprinted by permission from Jablonski (1986).

"Background and mass extinctions: The alternation of evolutionary regimes." *Science* 231: 129–329. Copyright © 1986 AAAS. Reprinted with permission of the AAAS.

Biotic Replacement

Individuals within a species can compete for resources, as can individuals within different species, but is it reasonable to infer that these processes scale up to clades competing over evolutionary timescales? This theory is implicit in Sepkoski's model of logistic growth over the Phanerozoic (see Section 18.3), but pointing to a pair of clades in the fossil record and demonstrating that they competed is often very difficult, because a variety of noncompetitive mechanisms could explain changes in their levels of diversity. Benton (1991) looked at competition and other forms of **biotic replacement** in the fossil record and noted that clade-level competition should ideally produce a "double wedge" pattern of diversity **(Figure 18.40)**. In addition, the pair of proposed competitors should be of similar ecologies and have lived at the same time in roughly the same geographic region.

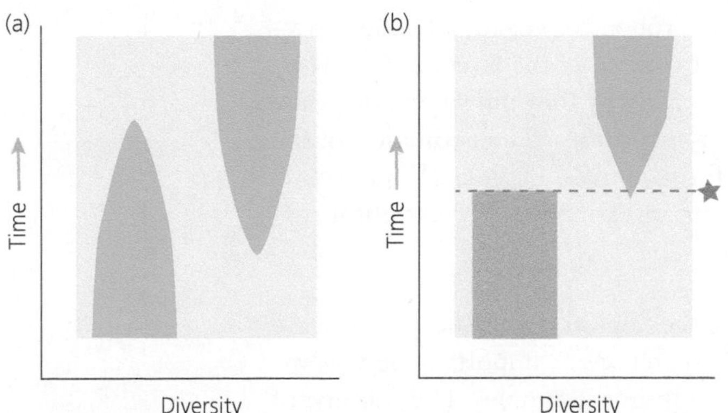

Figure 18.40 Patterns of biotic replacement Left, the double-wedge pattern expected under competitive replacement. Right, noncompetitive biotic replacement after a mass extinction. From Briggs (1998).

Surprisingly few examples satisfy all of these criteria, which led Benton to suggest that noncompetitive models of biotic replacement were more likely.

Extinctions can play an important role in fostering biotic replacement by mitigating the effects of incumbent taxa. Abundant and broadly distributed species are predicted to be difficult to dislodge, even by a competitively superior newcomer, simply because the incumbent species dominates the available resources. However, after a mass extinction, if one of the newcomers were lucky enough to survive it would thereafter compete with the incumbent on a much more level playing field. In some cases, the new species may experience an adaptive radiation (see next section) after a mass extinction has decreased levels of competition, but this outcome is not guaranteed.

Adaptive Radiation

An **adaptive radiation** occurs when a single or small group of ancestral species rapidly diversifies into a large number of descendant species that occupy a wide variety of ecological niches **(Figure 18.41)**. The Galápagos finches and Hawaiian *Drosophila,* which figure prominently in earlier chapters, are well-studied examples. But adaptive radiations have occurred in a wide array of groups at intervals throughout the history of life. They represent a prominent pattern. It is as if the tree of life suddenly sprouts a large number of highly diverse branches.

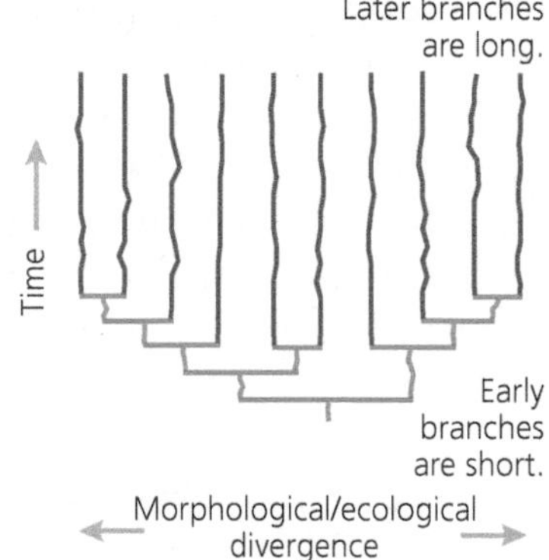

Figure 18.41 Adaptive radiation This diagram shows the branching pattern produced by a hypothetical adaptive radiation.

What factors trigger adaptive radiations? Why do only certain lineages diversify broadly and rapidly? The answers vary from time to time and clade to clade.

Ecological Opportunity as a Trigger

An ecological opportunity occurs when a small number of individuals or species are suddenly presented with a wide and abundant array of resources to exploit. The ancestors of the Hawaiian *Drosophila* and Galápagos finches, for example, colonized islands that had few competitors and a wide variety of resources and habitats to use. Such conditions favor rapid diversification and speciation.

Why certain populations colonize an area and undergo adaptive radiation is largely a matter of blind luck, however. For example, recent phylogenies of the Galápagos finches have shown that most of the closest living ancestors of the group live in the Caribbean (Sato et al. 2001; Burns et al. 2002). Thus, the leading hypothesis to explain the radiation is that a small population of birds happened to move from the Caribbean to the Galápagos and take up residence. Similarly, the ancestor of the Hawaiian *Drosophila* was a fruit-fly species that happened to be blown onto the islands millions of years ago; the ancestor of today's diverse array of Hawaiian silverswords was a tarweed native to California that probably arrived in Hawaii as a seed hitchhiking on a bird's foot or in its digestive tract.

Ecological opportunity is not created solely through colonization events, however. In the aftermath of the mass extinction at the end of the Cretaceous, mammals diversified rapidly. The leading hypothesis for why they did so was that they lacked competition, not that they had superior adaptations. The extinction of the dinosaurs created an ecological opportunity for mammals. Ecological opportunities can be created by dispersal and colonization or extinction of competitors.

Morphological Innovation as a Trigger

Not all adaptive radiations are associated with ecological opportunity; many are correlated with morphological innovations that represent important new adaptations. The diversification of arthropods is a prime example. The variety of ecological niches occupied by insects, crustaceans, and spiders and the number of species in these lineages are remarkable. Their success is closely associated with modifications and elaborations of their jointed limbs, which allowed species in these groups to move efficiently and find food. The genetic mechanisms responsible for the elaboration of arthropod limbs are a topic of ongoing research (see, for example, Briggs et al. 2012); here, the central point is that jointed limbs were a morphological innovation that is correlated with an adaptive radiation.

Other Examples: Adaptive Radiations in Land Plants

Adaptive radiations have occurred at different taxonomic levels during the evolution of land plants. Two of the most notable ones were unique events, similar to the Cambrian explosion of animal diversity. The first was the radiation of terrestrial plants from aquatic ancestors in the early Devonian, about 400 Ma. During this period, early terrestrial plants evolved key morphological features such as a waxy cuticle and the surface openings called stomata. They also evolved the life history, characterized by alternating gametophyte and sporophyte generations, observed in their living descendants (Bateman et al. 1998). These innovations are associated with the transition to terrestrial life.

Figure 18.42 Like the ancestral flower? The tropical shrub *Amborella* is the sister group to all other living flowering plants. If it has undergone less evolutionary change from the ancestral condition compared to other flowering plants, then *Amborella* may provide clues to the nature of the ancestral angiosperm. By Sandra Floyd, USDA ARS.

The second radiation in plant evolution was the Cretaceous explosion of flowering plants, or angiosperms, about 110 Ma. Work on the phylogeny of angiosperms and their close relatives showed that a little-known shrub from the island of New Caledonia called *Amborella* **(Figure 18.42)** is the sister taxon to all other flowering plants (Qiu et al. 1999; Soltis et al. 1999). Identifying *Amborella* as the sister to all other angiosperms offers a clue to what their common ancestor might have looked like and, by extension, what traits of that ancestor might have contributed to the early evolutionary success of flowering plants (Brown 1999). Over 250,000 species of angiosperms are alive today. They occupy a diversity of habitats and range from sprawling inhabitants of arctic tundra to the trees that dominate tropical rain forests. Because it made pollination so efficient, the flower is thought to be a morphological innovation that made this radiation possible.

Adaptive radiation has been a popular area of evolutionary analysis because it is spectacular: A great deal of evolution takes place in a relatively short amount of time. Another prominent pattern in the history of life is the opposite: lack of appreciable morphological change or speciation over long periods of time.

18.6 Fossil and Molecular Divergence Timing

Until relatively recently, estimates of clade origination and divergence were based entirely on the fossil record. Such estimates were known to be imperfect. One deficiency was that the occurrence of a certain fossil—a bird in Jurassic rocks, for example—is only a minimum age estimate. In other words, birds must have arisen by the Jurassic, but just how much older the group could be was impossible to know with certainty. A second deficiency concerned groups with little or no fossil record, where fossil-based age estimates were either very likely to be an underestimate (for example, when soft-bodied fossils are occasionally preserved) or paleontologists gave an opinion based on the fossil records of related groups.

Recent years have seen a complementary source of data on divergence times come from **molecular clock** studies. By comparing DNA sequences under a specified model of evolution, along with at least one fossil-based calibration point (see Parham et al. 2012), these studies produce time-calibrated trees (often called *time trees*) that yield minimum and maximum divergence time estimates. Time trees are not without controversy, however, because many early publications suggested divergences significantly older than the fossil data would suggest. For example, Kumar and Hedges (1998) suggested that many placental mammal orders arose deep within the Cretaceous, despite their fossil record being restricted to the Paleogene and younger. As discussed in more detail later, establishing the timing of mammalian divergences has significant implications for the evolutionary importance of the end-Cretaceous mass extinction.

Dating the tree of life is an exciting interdisciplinary research effort that links paleontology and molecular biology. However, when disparate fields become connected, the use of common terminology is important. In particular, two distinctions need to be recognized.

First, molecular data from extant organisms are used to estimate the divergence times of much larger clades. For example, the human–chicken divergence is the same as the mammal–bird split. In addition, because the mammal–bird split points to the origination time for the clade Amniota, it is equivalent to the mammal-turtle, mammal–snake, or mammal–crocodile split **(Figure 18.43)**.

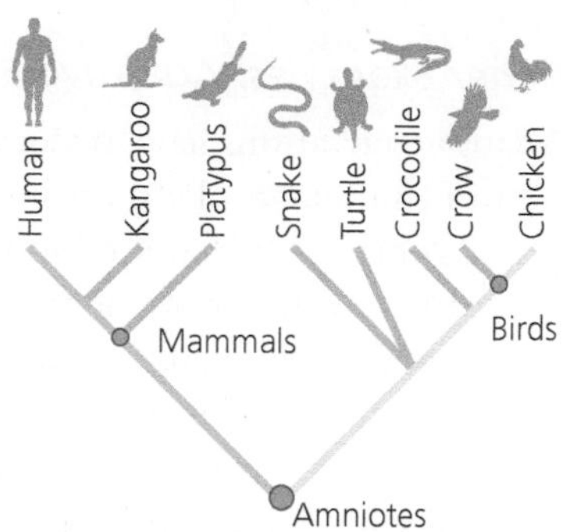

Figure 18.43 Equivalence of divergence times Divergence time for mammals versus birds is the same as for mammals versus snakes, mammals versus turtles, and mammals versus crocodiles.

Second, lineage divergence (cladogenesis) does not necessarily imply significant taxonomic or morphological diversification. It is possible for sister groups to have diverged and either left no fossils or be so morphologically similar as to be confused for a single lineage. Cladogenesis without subsequent diversification could yield a substantial mismatch between molecular and paleontological estimates of divergence times if morphological change was slow or the early fossil record of a clade was poor.

Figure 18.44 shows a time tree for vertebrates and insects assembled by Michael Benton and Philip Donoghue (2007).

How do fossil and molecular divergence times correspond in practice? Next we examine two examples from particularly important periods in the history of life.

Was the Cambrian Explosion Really Explosive?

Many phyla and morphological innovations first appear in the Cambrian. But it is important to recognize that these species and traits had to have existed for some time before being immortalized in the Burgess Shale and Chengjiang deposits. Just how long? To answer this question, evolutionary biologists have used molecular clocks to estimate when the earliest branches on the animal phylogeny occurred. Changes in DNA or protein sequences that are neutral with respect to selection should arise by mutation and then drift to fixation at a steady, clock-like rate (see Chapter 7). By observing the amount of selectively neutral genetic change that has occurred between taxa whose divergence is dated in the fossil or geological record, a molecular clock can be calibrated in terms of the amount of change expected per million years. This calibration, in turn, can then be used to date events that are not recorded in the fossil record.

To date the origin of the bilaterians, Bruce Runnegar (1982) analyzed differences among hemoglobin amino acid sequences in vertebrates and various invertebrate phyla. To translate these genetic distances into times of divergence, Runnegar used estimates of the rate of hemoglobin evolution among vertebrate groups with known fossil ages. He concluded that the earliest branches in Figure 18.12 occurred about 900 Ma—long before the Cambrian explosion. Greg Wray,

Jeff Levinton, and Leo Shapiro (1996) came to a similar conclusion using a different data set. Wray's group estimated that chordates and echinoderms diverged about 1,000 Ma, while protostomes and deuterostomes diverged about 1,200 Ma. This study included more genes and more taxa but used the same vertebrate fossil record for calibration as the Runnegar study. Both of these analyses admitted some uncertainty about the exact ages of the divergences among the bilaterally symmetric organisms but agreed that these divergences occurred hundreds of millions of years before their first appearance in the fossil record.

These papers generated considerable controversy, because such early divergence dates imply a long history of animal evolution before the Cambrian explosion. If the dates are correct, Proterozoic rocks should eventually yield fossils of deuterostomes, ecdysozoans, and lophotrochozoans. But with the exception of the Doushantuo embryos, *Kimberella,* and *Vernanimalcula,* they have not.

To resolve the discrepancy between the fossil record and the predictions of the molecular clock, Andrew Smith (1999) suggests that the lineages leading to the living Bilateria diverged from each other over a prolonged period in the Proterozoic, but that the vast majority of resulting species existed as small, larvalike organisms that left no trace in the fossil record (see also Erwin and Davidson 2002).

According to this point of view, the Cambrian explosion is an explosion of morphological diversity, but not necessarily an explosion of lineages, which occurred much earlier. The idea that the major animal lineages existed long before they diversified and produced large-bodied forms is captured in the quip that the Cambrian explosion had a "long fuse."

Did the End-Cretaceous Mass Extinction Release Mammals?

The replacement of dinosaur-dominated terrestrial ecosystems in the Cretaceous by those of mammals in the Paleogene is perhaps the best-known example of biotic replacement triggered by mass extinction. Because almost all of the traditional orders of placental mammals are found in the Paleogene or Eocene, but not beforehand, an evolutionary radiation in the early Cenozoic was long regarded as genuine. So it was a surprise when several molecular studies suggested that several interordinal splits occurred in the Cretaceous and that the mammalian radiation likely started much earlier (Springer 1997; Kumar and Hedges 1998).

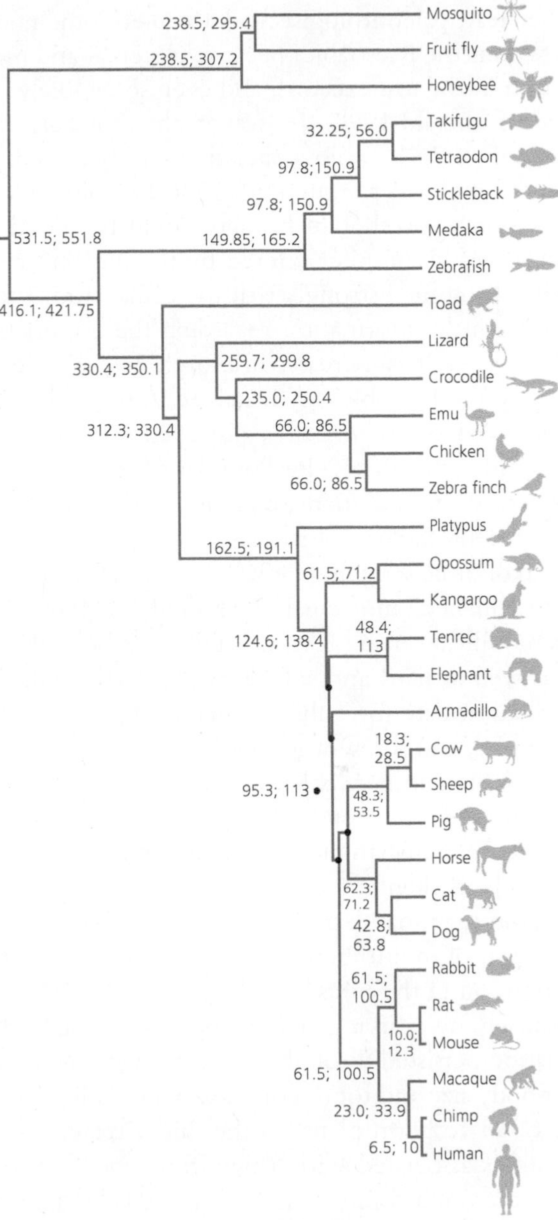

Figure 18.44 A time tree for vertebrates and insects The phylogeny is a consensus based on various kinds of evidence, including molecular sequence data; the dates at the nodes are estimated from the fossil record. The smaller number at each node is a minimum divergence time—in millions of years ago—based on the oldest fossil assigned to either of the lineages arising from the node. The larger number is a maximum divergence time based on the maximum ages of sister groups and on the absence of the taxa in question from underlying fossil-bearing rock formations. For example, the last common ancestor of mice and humans lived at least 61.5 million years ago—and possibly as long as 100.5 million years ago. Modified from Benton and Donoghue (2007).

To be clear, paleontologists had accepted some number of mammalian divergences within the Mesozoic, because placental and marsupial lineages were long recognized in the fossil record, and even some likely ungulate relatives had been reported (Archibald 1996). However, the molecular data seemed to suggest that a more profound Cretaceous radiation had occurred. John Alroy (1999) set out to parse the molecular arguments. First, he pointed to fossil data suggesting that at least seven living therian lineages could be traced to the Cretaceous, which was similar to the number offered by the molecular studies. Second, and more important, he argued strongly that molecular studies focused on lineage-splitting events offer little information regarding the important aspects of an evolutionary radiation, such as increases in morphological or ecological diversity. Using a precursor to the Paleobiology Database, Alroy was able to demonstrate that fossil mammals from North America displayed the characteristics of an evolutionary radiation only after the K–Pg boundary (e.g., increased numbers of species, increased speciation rate, increased range of body size).

More recent molecular studies of mammalian divergence dating have improved taxonomic sampling (>95% of extant families) and added more molecular data with enhanced molecular clock models (Meredith et al. 2011). Another important modification to previous studies has been the use of multiple calibration points as well as hard and soft dates. The Meredith et al. (2011) study showed the origination date for only one mammalian order, Eulipotyphlya (hedgehogs, shrews, moles, and some other traditional insectivorans) to be wholly within the Cretaceous, but another six had dates centered near the K–Pg boundary. Overall, new analyses seem to support Alroy's view that pre-Cenozoic splits were limited and that the mammalian radiation commenced only after the age of dinosaurs.

By combining data from the fossil record, molecular sequences, and other sources, scientists are addressing long-standing questions about life's history with new precision.

One study challenging this viewpoint came from an unlikely source, a team of Mesozoic mammal experts. Gregory Wilson and colleagues (2012) analyzed the evolution of multituberculates, an extinct group of mammals that were relatively common in the Jurassic and Cretaceous. They showed that this group was taxonomically diversifying and ecologically adapting during the Cretaceous, and this radiation persisted through the K–Pg event. Multituberculates increased their range of body size and tooth complexity (an indication of herbivory) in step with the rise of angiosperm plants in the mid-Cretaceous, suggesting that ecological opportunities associated with a new food source allowed at least one group of mammals the opportunity to radiate in the face of dinosaur competitors.

SUMMARY

The most efficient fossilization processes are compression, impression, casting, and permineralization. Because these events depend on the rapid burial of organic remains in water-saturated sediments, the fossil record is dominated by organisms with hard parts that lived in lowland or shallow-water marine environments. Thanks to new fossil finds and increasingly high-resolution dating techniques, the geological record of life on Earth is steadily improving.

Although some bilaterally symmetric animals were present in the Precambrian, most of the major animal lineages present today first appear in the fossil record during the Cambrian. The Cambrian explosion was characterized by the relatively sudden appearance of large and morphologically diverse animals that swam, crawled, or burrowed and that filled an array of ecological niches in shallow-water marine communities.

The Cambrian explosion is just the most spectacular of a series of adaptive radiations that characterize the rise of morphological complexity and diversity through the Phanerozoic. Adaptive radiations can be triggered by key morphological adaptations or chance events that create an ecological opportunity, such as colonizing a new habitat or surviving a mass extinction. Prolonged stasis is another pattern in evolutionary history. In some lineages, morphological stasis is punctuated by rapid change that occurs during speciation events.

The eventual fate of both new taxa and new morphological traits is extinction. The five most intense extinctions are designated as mass extinctions and are commonly distinguished from background extinctions. The K–Pg extinction is the best understood of the Big Five mass extinctions and was caused by an asteroid that slammed into Earth near Mexico's Yucatán peninsula. During both background and mass extinctions, geographically widespread species are less likely to go extinct. During recent times, a prominent extinction has been documented in the loss of bird species on Polynesian islands. Although dramatic, this event was too local to qualify as a mass extinction. But current projections of species loss due to rapid habitat destruction indicate that a mass extinction, caused by humans, may now be under way.

QUESTIONS

1. Explain why these are common in the fossil record:
   - marine-dwelling forms
   - burrowing species
   - recent organisms
   - pollen grains
2. Explain why these are rare in the fossil record:
   - desert-dwelling forms
   - species that were capable of flight
   - skeletal elements from sharks and rays
   - flowers
3. Define the Phanerozoic and Proterozoic eras.
4. What important events occurred during the following time intervals?
   - the boundary between the Silurian and Devonian
   - the Cambrian period
   - the boundary between the Permian and Triassic
   - the Cretaceous period
5. In what sense is the Cambrian period "explosive"? In what sense is the term *explosion* misleading?
6. Compare and contrast the Ediacaran and Burgess Shale faunas. What phyla are found in each? How did the species that were present differ in their morphology and ecology?
7. What is an adaptive radiation? State two hypotheses to explain why adaptive radiations occur.
8. Compare and contrast mass extinctions and background extinctions. How do their size and geographic and taxonomic extent differ?
9. Suppose that a species first appears in the fossil record 350 mya. Why is it logical to argue that this species actually existed before this date?
10. What data support the hypothesis that the origin of bilaterians, the deuterostome–protostome split, and the lophotrochozoan–ecdysozoan split all occurred long before the Cambrian explosion?
11. If data confirm that a mass extinction event occurred at the end of the Proterozoic era, what would be the consequences for our understanding of why the Cambrian explosion occurred?
12. Why would a rise in the availability of oxygen help explain why the Cambrian explosion occurred?
13. Give an example of an adaptive radiation. Provide evidence for the claim that the radiation originated with one or a few species, it was rapid, and the descendant groups occupy a wide array of ecological niches. Suggest a hypothesis to explain why the radiation occurred.
14. List the evidence in favor of the impact hypothesis for the K–Pg extinction. Which piece of evidence do you find most persuasive, and why?
15. Why would a meteor strike lead to global cooling, or what researchers call an "impact winter"?
16. Do you accept the hypothesis that a mass extinction event is currently under way? Why or why not?
17. Terrestrial fossils from a particular time (say, 230 mya) are patchily distributed around the world. Instead of being evenly distributed over the continents in a continuous thin layer, they often occur in narrow strips or pockets a few miles wide. Why is this?
18. Most fossils of Mesozoic birds are from marine diving birds. Relatively few terrestrial species are known. Does this mean that most Mesozoic birds were, in fact, marine diving birds? Explain your reasoning.
19. One of the (many) mysteries of the K–Pg extinction is the different fate of ammonites and nautiloids. These were mollusks with buoyant, chambered shells that lived in open-water habitats. Ammonites went extinct during the K–Pg extinction, but some nautiloids survived. The two groups had different reproductive strategies. Ammonites are thought to have produced many free-swimming young each year that fed near the ocean surface and grew rapidly. In contrast, a female nautilus produces just a few large eggs each year, each of which rests quietly in the depths for up to a year before hatching into a small, slow-growing nautilus. Based on these different reproductive strategies, suggest a possible hypothesis for why

the nautiloids, but not the ammonites, might have been able to survive an asteroid impact.

20. In 1996, Gregory Retallack announced that he had found shocked quartz crystals dating from the same time as the end-Permian extinction. What is the implication of this finding? Other geologists point out that no one has found any evidence of elevated iridium from these strata (despite much searching). What is the significance of the lack of iridium?

21. Suppose you are talking to a friend about extinctions, and you mention that humans are known to have caused thousands of extinctions in the last few millennia. Your friend responds, "So? Extinction is natural. Species have always gone extinct. So it's really not something we need to worry about." Is your friend correct that extinction is natural? Is the current rate of extinction typical? Is your friend correct that if extinctions are natural, then they are not a problem for the dominant life-forms on Earth?

EXPLORING THE LITERATURE

22. For a look at how ecological factors interacted with the asteroid impact at the K–Pg boundary, see:

Mitchell, J. S., P. D. Roopnarine, and K. D. Angielczyk. 2012. Late Cretaceous restructuring of terrestrial communities facilitated the end-Cretaceous mass extinction in North America. *Proceedings of the National Academy of Sciences, USA* 109: 18857–18861.

23. For an amibitious effort to combine data and estimate the date of origin of placental mammals, see:

O'Leary, M. A., J. I. Bloch, et al. 2013. The placental mammal ancestor and the Post-K–Pg radiation of placentals. *Science* 339: 662–667.

24. For a paper by our contributor that I will ask him to write a question for, see:

Sidor, C. A., D. A. Vilhena, et al. 2013. Provincialization of terrestrial faunas following the end-Permian mass extinction. *Proceedings of the National Academy of Sciences, USA* 110: 8129–8133.

25. For the surprising ubiquity of feathers on the hindlimbs of early birds, see:

Zheng, X., Z. Zhou, et al. 2013. Hind wings in basal birds and the evolution of leg feathers. *Science* 339: 1309–1312.

CITATIONS

Alroy, J. 1999. The fossil record on North American mammals: Evidence for a Paleocene evolutionary radiation. *Systematic Biology* 48: 107–118.

Alvarez, L. W., W. Alvarez, et al. 1980. Extraterrestrial cause for the Cretaceous–Tertiary extinction. *Science* 208: 1095–1108.

Alvarez, W., F. Asaro, and A. Montanari. 1990. Iridium profile for 10 million years across the Cretaceous–Tertiary boundary at Gubbio (Italy). *Science* 250: 1700–1702.

Angielczyk, K. 2009. *Dimetrodon* is not a dinosaur: Using tree thinking to understand the ancient relatives of mammals and their evolution. *Evolution: Education and Outreach* 2: 257–271.

Archibald, J. D. 1996. Fossil evidence for a Late Cretaceous origin of "hoofed" mammals. *Science* 272: 1150–1153.

Arens, N. C., and I. D. West. 2008. Press pulse: A general theory of mass extinction? *Paleobiology* 34: 456–471.

Avise, J. C., W. S. Nelson, and H. Sugita. 1994. A speciational history of "living fossils": Molecular evolutionary patterns in horseshoe crabs. *Evolution* 48: 1986–2001.

Balmford, A. 1996. Extinction filters and current resilience: The significance of past selection pressures for conservation biology. *Trends in Ecology and Evolution* 11: 193–196.

Barnosky, A. D., N. Matzke, et al. 2011. Has the Earth's sixth mass extinction already arrived? *Nature* 471: 51–57.

Bateman, R. M., P. R. Crane, et al. 1998. Early evolution of land plants: Phylogeny, physiology, and ecology of the primary terrestrial radiation. *Annual Review of Ecology and Systematics* 29: 263–292.

Benton, M. J. 1991. Extinction, biotic replacements, and clade interactions. In *The unity of evolutionary biology*, ed. E. C. Dudley. Portland, OR: Dioscorides Press, 89–102.

Benton, M. J. 1995. Diversification and extinction in the history of life. *Science* 268: 52–58.

Benton, M. J. 2003. *When life nearly died*. London: Thames & Hudson, p 336.

Benton, M. J., and P. C. J. Donoghue. 2007. Paleontological evidence to date the tree of life. *Molecular Biology and Evolution* 24: 26–53.

Benton, M. J., and P .N. Pearson. 2001. Speciation in the fossil record. *Trends in Ecology and Evolution* 16: 405–411.

Benton, M. J., V. P. Tverdokhlebov, and M. V. Surkov. 2004. Ecosystem remodelling among vertebrates at the Permian-Triassic boundary in Russia. *Nature* 432: 97–100.

Berner, R. A. 2009. Phanerozoic atmospheric oxygen: New results using the GEOCARBSULF model. *American Journal of Science* 309: 603–606.

Bourgeois, J., T. A. Hansen, et al. 1988. A tsunami deposit at the Cretaceous–Tertiary boundary in Texas. *Science* 241: 567–570.

Breecker, D. O., Z. D. Sharp, and L. D. McFadden. 2010. Atmospheric CO_2 concentrations during ancient greenhouse climates were similar to those predicted for A.D. 2100. *Proceedings of the National Academy of Sciences, USA* 107: 576–580.

Briggs, D. E. G., R. A. Fortey, and M. A. Wills. 1992. Morphological disparity in the Cambrian. *Science* 256: 1670–1673.

Briggs, D. E. G., D. J. Siveter, et al. 2012. Silurian horseshoe crab illuminates the evolution of arthropod limbs. *Proceedings of the National Academy of Sciences, USA* 109: 15702–15705.

Briggs, J. C. 1998. Biotic replacements: Extinction or clade interaction? *BioScience* 48: 389–395.

Brown, K. S. 1999. Deep Green rewrites evolutionary history of plants. *Science* 285: 990–991.

Burns, K. J., S. J. Hackett, and N. K. Klein. 2002. Phylogenetic relationships and morphological diversity in Darwin's finches and their relatives. *Evolution* 56: 1240–1252.

Bush, A. M., and R. K. Bambach. 2011. Paleoecologic megatrends in marine metazoa. *Annual Review of Earth and Planetary Sciences* 39: 241–269.

Butterfield, N. J. 1990. A reassessment of the enigmatic Burgess Shale [British Columbia, Canada] fossil *Wiwaxia corrugata* (Matthew) and its relationship to the polychaete *Canadia spinosa* Walcott. *Paleobiology* 16: 287–303.

Cheetham, A. H. 1986. Tempo of evolution in a Neogene bryozoan: Rates of morphologic change within and across species boundaries. *Paleobiology* 12: 190–202.

Chen, J.-Y., D. J. Bottjer, et al. 2004. Small bilaterian fossils from 40 to 55 million years before the Cambrian. *Science* 305: 218–222.

Chen, J.-Y., D.-Y. Huang, and C-W. Li. 1999. An early Cambrian craniate-like chordate. *Nature* 402: 518–522.

Chen, Z.-Q., and M. J. Benton. 2012. The timing and pattern of biotic recovery following the end-Permian mass extinction. *Nature Geoscience* 5: 375–383.

Coates, M. I. 1996. The Devonian tetrapod *Acanthostega gunnari* Jarvik: Postcranial anatomy, basal tetrapod interrelationships and patterns of skeletal evolution. *Transactions of the Royal Society of Edinburgh: Earth Sciences* 87: 363–421.

Coates, M. I., M. Ruta, and M. Friedman. 2008. Ever since Owen: Changing perspectives on the early evolution of tetrapods. *Annual Review of Ecology, Evolution, and Systematics* 39: 571–592.

Condon, D., M. Zhu, et al. 2005. U-Pb ages from the Neoproterozoic Doushantuo formation, China. *Science* 308: 95–98.

Conway Morris, S. 1989. Burgess Shale faunas and the Cambrian explosion. *Science* 246: 339–346.

Conway Morris, S. 1998a. Early metazoan evolution: Reconciling paleontology and molecular biology. *American Zoologist* 38: 867–877.

Conway Morris, S. 1998b. *The Crucible of Creation.* Oxford: Oxford University Press.

Conway Morris, S. 2000. The Cambrian "explosion": Slow-fuse or megatonnage? *Proceedings of the National Academy of Sciences, USA* 97: 4426–4429.

Cunningham, C. W., N. W. Blackstone, and L. W. Buss. 1992. Evolution of king crabs from hermit crab ancestors. *Nature* 355: 539–542.

Daeschler, E. B., N. H. Shubin, and F. A. Jenkins. 2006. A Devonian tetrapod-like fish and the evolution of the tetrapod body plan. *Nature* 440: 757–763.

Darwin, C. 1859. *On the Origin of Species By Means of Natural Selection.* 1st ed. London: John Murray.

Dial, K. P. 2003. Wing-assisted incline running and the evolution of flight. *Science* 299: 402–404.

Eldredge, N., and S. J. Gould. 1972. Punctuated equilibria: An alternative to phyletic gradualism. In *Models in Paleobiology*, ed. T. J. M. Schopf. San Francisco: Freeman, Cooper & Company, 82–115.

Engel, M. S., and D. A. Grimaldi. 2004. New light shed on the oldest insect. *Nature* 427: 627–630.

Erwin, D. H. 2003. Impact at the Permo-Triassic boundary: A critical evaluation. *Astrobiology* 3: 67–74.

Erwin, D. H. 2007. Disparity: Morphological pattern and developmental context. *Palaeontology* 50: 57–73.

Erwin, D. H., and R. L. Anstey. 1995a. Introduction. In *New Approaches to Speciation in the Fossil Record*, ed. D. H. Erwin and R. L. Anstey. New York: Columbia University Press, 1–8.

Erwin, D. H., and R. L. Anstey. 1995b. Speciation in the fossil record. In *New Approaches to Speciation in the Fossil Record*, ed. D. H. Erwin and R. L. Anstey. New York: Columbia University Press, 11–38.

Erwin, D. H., and E. H. Davidson. 2002. The last common bilaterian ancestor. *Development* 129: 3021–3032.

Erwin, D. H., M. Laflamme, et al. 2011. The Cambrian conundrum: Early divergence and later ecological success in the early history of animals. *Science* 334: 1091–1097.

Fedonkin, M. A., and B. M. Waggoner. 1997. The late Precambrian fossil *Kimberella* is a mollusc-like bilaterian organism. *Nature* 388: 868–871.

Fiz-Palacios, O., H. Schneider, et al. 2011. Diversification of land plants: Insights from a family-level phylogenetic analysis. *BMC Evolutionary Biology* 11: 341.

Foote, M. 1993. Discordance and concordance between morphological and taxonomic diversity. *Paleobiology* 19: 185–204.

Foote, M. 1997. The evolution of morphological diversity. *Annual Review of Ecology and Systematics* 28: 129–152.

Foote, M. 2003. Origination and extinction through the Phanerozoic: A new approach. *Journal of Geology* 111: 125–148.

Franzen, J. L., P. D. Gingerich, et al. 2009. Complete primate skeleton from the Middle Eocene of Messel in Germany: Morphology and paleobiology. *PLoS One* 4: e5723.

Gradstein, F. M., and J. G. Ogg. 2009. The geologic time scale. In *The TimeTree of Life*, ed. S. B. Hedges and S. Kumar. New York: Oxford University Press, 26–34.

Hansen, R. M. 1978. Shasta ground sloth food habits, Rampart Cave, Arizona. *Paleobiology* 4: 302–319.

Heers, A. M., and K. P. Dial. 2011. From extant to extinct: Locomotor ontogeny and the evolution of avian flight. *Trends in Ecology and Evolution* 27: 296–305.

Hopkins, M. J., and S. Lidgard. 2012. Evolutionary mode routinely varies among morphological traits within fossil species lineages. *Proceedings of the National Academy of Sciences, USA* 109: 20520–20525.

Huey, R. B., and P. D. Ward. 2005. Hypoxia, global warming, and terrestrial Late Permian extinctions. *Science* 308: 398–401.

Huldtgren, T., J. A. Cunningham, et al. 2011. Fossilized nuclei and germination structures identify Ediacaran "animal embryos" as encysting protists. *Science* 334: 1696–1699.

Hunt, G. 2007. The relative importance of directional change, random walks, and stasis in the evolution of fossil lineages. *Proceedings of the National Academy of Sciences, USA* 104: 18404–18408.

Hunt, G., M. A. Bell, and M. P. Travis. 2008. Evolution toward a new adaptive optimum: Phenotypic evolution in a fossil stickleback lineage. *Evolution* 62: 700–710.

Hunter, R. S. T., A. J. Arnold, and W. C. Parker. 1988. Evolution and homeomorphy in the development of the Paleocene *Planorotalites pseudomenardii* and the Miocene *Globorotalia* (*Globorotalia*) *maragritae* lineages. *Micropaleontology* 34: 181–192.

Igamberdiev, A. U., and P. J. Lea. 2006. Land plants equilibrate O_2 and CO_2 concentrations in the atmosphere. *Photosynthesis Research* 87: 177–194.

International Commission on Stratigraphy (ICS). 2012. *International Chronostratigraphic Chart.* Available at http://www.stratigraphy.org/column.php?id=Chart/Time%20Scale.

Jablonski, D. 1986. Background and mass extinctions: The alternation of macroevolutionary regimes. *Science* 231: 129–133.

Jablonski, D. 1995. Extinctions in the fossil record. In *Extinction Rates*, ed. J. H. Lawton and R. M. May. Oxford: Oxford University Press, 25–44.

Jablonski, D. 2000. Micro- and macroevolution: Scale and hierarchy in evolutionary biology and paleobiology. *Paleobiology* 26: S15–S52.

Jablonski, D. 2005. Mass extinctions and macroevolution. *Paleobiology* 31: 192–210.

Jablonski, D., and D. M. Raup. 1995. Selectivity of end-Cretaceous marine bivalve extinctions. *Science* 268: 389–391.

Jackson, J. B. C., and A. H. Cheetham. 1990. Evolutionary significance of morphospecies: A test with cheilostome bryozoa. *Science* 248: 579–583.

Jackson, J. B. C., and A. H. Cheetham. 1994. Phylogeny reconstruction and the tempo of speciation in the cheilostome Bryozoa. *Paleobiology* 20: 407–423.

Johnston, T., W. Poulton, et al. 2012. Late Ediacaran redox stability and metazoan evolution. *Earth and Planetary Science Letters* 335–336: 25–35.

Keller, G., T. Adatte, et al. 2004. Chicxulub impact predates the K-T boundary mass extinction. *Proceedings of the National Academy of Sciences, USA* 101: 3753–3758.

Kimbel, W. H., R. C. Walter, et al. 1996. Late Pliocene *Homo* and Oldowan tools from the Hadar Formation (Kada Hadar member), Ethiopia. *Journal of Human Evolution* 31: 549–561.

Knecht, R. J., M. S. Engel, and J. S. Benner. 2011. Late Carboniferous paleoichnology reveals the oldest full-body impression of a flying insect. *Proceedings of the National Academy of Sciences, USA* 108: 6515–6519.

Knoll, A. H., and S. B. Carroll. 1999. Early animal evolution: Emerging views from comparative biology and geology *Science* 284: 2129–2137.

Kumar, S., and S. B. Hedges. 1998. A molecular timescale for vertebrate evolution. *Nature* 392: 917–920.

Laflamme, M., S. A. Darroch, et al. 2013. The end of the Ediacara biota: Extinction, biotic replacement, or Cheshire Cat? *Gondwana Research* 23: 558–573.

Laflamme, M., S. Xiao, and M. Kowalewski. 2009. Osmotrophy in modular Ediacara organisms. *Proceedings of the National Academy of Sciences, USA* 106: 14438–14443.

Laurance, W. F. 1999. Reflections on the tropical deforestation crisis. *Biological Conservation* 91: 109–118.

Lee, M. S., and T. H. Worthy. 2012. Likelihood reinstates *Archaeopteryx* as a primitive bird. *Biology Letters* 8: 299–303.

Li, C.-W., J.-Y. Chen, and T.-E. Hua. 1998. Precambrian sponges with cellular structures. *Science* 279: 879–882.

Li, Q., K. Q. Gao, et al. 2010. Plumage color patterns of an extinct dinosaur. *Science* 327: 1369–1372.

Luo, Z.-X. 2007. Transformation and diversification in early mammal evolution. *Nature* 450: 1011–1019.

Luo, Z.-X. 2011. Developmental patterns in Mesozoic evolution of mammal ears. *Annual Review of Ecology, Evolution, and Systematics* 42: 355–380.

Luo, Z.-X., C. X. Yuan, et al. 2011. A Jurassic eutherian mammal and divergence of marsupials and placentals. *Nature* 476: 442–445.

MacFadden, B. J. 1986. Fossil horses from "*Eohippus*" (*Hyracotherium*) to *Equus*: Scaling, Cope's Law, and the evolution of body size. *Paleobiology* 12: 355–369.

Maurrasse, F. J.-M. R., and G. Sen. 1990. Impacts, tsunamis, and the Haitian Cretaceous–Tertiary boundary layer. *Science* 252: 1690–1693.

May, R. M., J. H. Lawton, and N. E. Stork. 1995. Assessing extinction rates. In *Extinction Rates*, ed. J. H. Lawton and R. M. May. Oxford: Oxford. University Press, 1–24.

McKinnon, W. B. 1992. Killer acid at the K–T boundary. *Nature* 357: 15–16.

McShea, D. W. 1994. Mechanisms of large-scale evolutionary trends. *Evolution* 48: 1747–1763.

Meredith, R. W., J. E. Janecka, et al. 2011. Impacts of the Cretaceous terrestrial revolution and KPg extinction on mammal diversification. *Science* 334: 521–524.

Narbonne, G. M. 2004. Modular construction of early Ediacaran complex life forms. *Science* 305: 1141–1144.

Nesbitt, S. J., P. M. Barrett, et al. 2013. The oldest dinosaur? A Middle Triassic dinosauriform from Tanzania. *Biology Letters* 9: 20120949.

Niedzwiedzki, G., P. Szrek, et al. 2010. Tetrapod trackways from the early Middle Devonian period of Poland. *Nature* 463: 43–48.

Norell, M. A., and X. Xu. 2005. Feathered dinosaurs. *Annual Review of Earth and Planetary Sciences* 33: 277–299.

Parham, J. F., P. C. Donoghue, et al. 2012. Best practices for justifying fossil calibrations. *Systematic Biology* 61: 337–345.

Payne, J. L., D. J. Lehrmann, et al. 2004. Large perturbations of the carbon cycle during recovery from the end-Permian extinction. *Science* 305: 506–509.

Peters, S. E., and M. Foote. 2002. Determinants of extinction in the fossil record. *Nature* 416: 420–424.

Philippe, H., R. Derelle, et al. 2009. Phylogenomics revives traditional views on deep animal relationships. *Current Biology* 19: 706–712.

Phillips, J. 1860. *Life on the Earth: Its Origin and Succession.* Cambridge, MA: Macmillan.

Pierce, S. E., J. A. Clack, and J. R. Hutchinson. 2012. Three-dimensional limb joint mobility in the early tetrapod *Ichthyostega*. *Nature* 486: 523–526.

Pimm, S. L., G. J. Russell, et al. 1995. The future of biodiversity. *Science* 269: 347–350.

Prum, R. O. 2002. Why ornithologists should care about the theropod origin of birds. *Auk* 119: 1–17.

Qiu, Y.-L., J. Lee, et al. 1999.The earliest angiosperms: Evidence from mitochondrial, plastid and nuclear genes. *Nature* 402: 404–407.

Raup, D. M. 1967. Geometric analysis of shell coiling: Coiling in ammonoids. *Journal of Paleontology* 41: 43–65.

Raup, D. M. 1991. A kill curve for Phanerozoic marine species. *Paleobiology* 17: 37–48.

Raup, D. M. 1994. The role of extinction in evolution. *Proceedings of the National Academy of Sciences, USA* 91: 6758–6763.

Raup, D. M., and J. J. Sepkoski. 1982. Mass extinctions in the marine fossil record. *Science* 215:1501–1503.

Royer, D. L., R. A. Berner, et al. 2004. CO_2 as a primary driver of Phanerozoic climate. *GSA Today* 14: 4–10.

Rubidge, B. S., and C. A. Sidor. 2001. Evolutionary patterns among Permo–Triassic therapsids. *Annual Review of Ecology and Systematics* 32: 449–480.

Rubinstein, C. V., P. Gerrienne, et al. 2010. Early Middle Ordovician evidence for land plants in Argentina (eastern Gondwana). *New Phytologist* 188: 365–369.

Runnegar, B. 1982. A molecular-clock date for the origin of the animal phyla. *Lethaia* 15: 199–205.

Sansom, I. J., N. S. Davies, et al. 2012. Chondrichthyan-like scales from the Middle Ordovician of Australia. *Palaeontology* 55: 243–247.

Sato, A., H. Tichy, et al. 2001. On the origin of Darwin's finches. *Molecular Biology and Evolution* 18: 299–311.

Shubin, N. H., E. B. Daeschler, and F. A. Jenkins. 2006. The pectoral fin of *Tiktaalik roseae* and the origin of the tetrapod limb. *Nature* 440: 764–771.

Schulte, P., L. Alegret, et al. 2010. The Chicxulub asteroid impact and mass extinction at the Cretaceous-Paleogene boundary. *Science* 327: 1214–1218.

Schultz, P. H., and S. D'Hondt. 1996. Cretaceous–Tertiary (Chicxulub) impact angle and its consequences. *Geology* 24: 963–967.

Sepkoski, J. J., Jr. 1984. A kinetic model of Phanerozoic taxonomic diversity. III: Post-Paleozoic families and multiple equilibria. *Paleobiology* 10: 246–267.

Sahoo, S. K., N. J. Planavsky, et al. 2012. Ocean oxygenation in the wake of the Marinoan glaciation. *Nature* 489: 546–549.

Shen, S.-Z., J. L. Crowley, et al. 2011. Calibrating the end-Permian mass extinction. *Science* 334: 1367–1372.

Shu, D.-G., H.-L. Luo, et al. 1999. Lower Cambrian vertebrates from south China. *Nature* 402: 42–46.

Shu, D.-G., S. Conway Morris, et al. 2003. Head and backbone of the early Cambrian vertebrate *Haikouichthys*. *Nature* 421: 526–529.

Sidor, C. A. 2001. Simplification as a trend in synapsid cranial evolution. *Evolution* 55: 1419–1442.

Sidor, C. A., and J. A. Hopson. 1998. Ghost lineages and mammalness: assessing the temporal pattern of character acquisition in the Synapsida. *Paleobiology* 24: 254–273.

Sidor, C. A., M. F. Miller, and J. L. Isbell. 2008. Tetrapod burrows from the Triassic of Antarctica. *Journal of Vertebrate Paleontology* 28: 277–284.

Simpson, G. G. 1944. *Tempo and mode in evolution.* New York: Columbia University Press.

Skole, D., and C. Tucker. 1993. Tropical deforestation and habitat fragmentation in the Amazon: Satellite data from 1978 to 1988. *Science* 260: 1905–1920.

Smith, A. B. 1999. Dating the origin of metazoan body plans. *Evolution and Development* 1: 138–142.

Smith, F. D. M., R. M. May, et al. 1993. How much do we know about the current extinction rate? *Trends in Ecology and Evolution* 8: 375–378.

Smith, M. R. 2012. Mouthparts of the Burgess Shale fossils Odontogriphus and *Wiwaxia*: Implications for the ancestral molluscan radula. *Proceedings of the Royal Society of London B* 279: 4287–4295.

Smith, R., and J. Botha. 2005. The recovery of terrestrial vertebrate diversity in the South African Karoo Basin after the end-Permian extinction. *Comptes Rendus Palevolution* 4: 623–636.

Soltis, P. S., D. E. Soltis, and M. W. Chase. 1999. Angiosperm phylogeny inferred from multiple genes as a tool for comparative biology. *Nature* 402: 402–404.

Springer, M. S. 1997. Molecular clocks and the timing of the placental and marsupial radiations in relation to the Cretaceous–Tertiary boundary. *Journal of Mammalian Evolution* 4: 285–302.

Stanley, S. M. 1975. A theory of evolution above the species level. *Proceedings of the National Academy of Sciences, USA* 72: 646–650.

Stanley, S. M. 1982. Macroevolution and the fossil record. *Evolution* 36: 460–473.

Sun, G., D. L. Dilcher, et al. 1998. In search of the first flower: A Jurassic angiosperm, *Archaefructus*, from northeast China. *Science* 282: 1692–1695.

Swisher, C. C. III, J. M. Grajales-Nishimura, et al. 1992. Coeval $^{40}Ar/^{39}Ar$ ages of 65.0 million years ago from Chicxulub crater melt rock and Cretaceous–Tertiary boundary tektites. *Science* 257: 954–958.

Telford, M. J., and R. R. Copley. 2011. Improving animal phylogenies with genomic data. *Trends in Genetics* 27: 186–195.

United Nations Population Division. 2004. *World Population Prospects, the 2000 Revision: Highlights.* Available at http://www.un.org/esa/population/publications/WPP2004/2004Highlights_finalrevised.pdf.

Valentine, J. W. 2002. Prelude to the Cambrian explosion. *Annual Review of Earth and Planetary Sciences* 30: 285–306.

Van der Plicht, J., W. A. B. van der Sanden, et al. 2004. Dating bog bodies by means of ^{14}C-AMS. *Journal of Archaeological Science* 3: 471–491.

Van Valen, L. 1973. A new evolutionary law. *Evolutionary Theory* 1: 1–30.

Vickers-Rich, P., A. Y. Ivantsov, et al. 2013. Reconstructing Rangea: New discoveries from the Ediacaran of southern Namibia. *Journal of Paleontology* 87: 1–15.

Waggoner, B. 1998. Interpreting the earliest metazoan fossils: What can we learn? *American Zoologist* 38: 975–982.

Ward, P. D., D. R. Montgomery, and R. M. H. Smith. 2000. Altered river morphology in South Africa related to the Permian-Triassic extinction. *Science* 289: 1740–1743.

Wellman, C. H. 2010. The invasion of the land by plants: When and where? *New Phytologist* 188: 306–309.

Wills, M. A., D. E. G. Briggs, and R. A. Fortey. 1994. Disparity as an evolutionary index: A comparison of Cambrian and Recent arthropods. *Paleobiology* 20: 93–130.

Wilson, E. O. 1988. The current state of biodiversity. In *Biodiversity*, ed. E. O. Wilson. Washington, DC: National Academy Press, 3–18.

Wilson, G. P., A. R. Evans, et al. 2012. Adaptive radiation of multituberculate mammals before the extinction of dinosaurs. *Nature* 483: 457–460.

Wray, G. A., J. S. Levinton, and L. H. Shapiro. 1996. Molecular evidence for deep Precambrian divergences among metazoan phyla. *Science* 274: 568–573.

Xiao, S., and M. Laflamme. 2009. On the eve of animal radiation: Phylogeny, ecology and evolution of the Ediacara biota. *Trends in Ecology and Evolution* 24: 31–40.

Xiao, S., X.Yuan, et al. 2002. Macroscopic carbonaceous compressions in a terminal Proterozoic shale: A systematic reassessment of the Miaohe biota, South China. *Journal of Paleontology* 76: 347–376.

Xiao, S., Y. Zhang, and A. H. Knoll. 1998. Three-dimensional preservation of algae and animal embryos in a Neoproterozoic phosphorite. *Nature* 391: 553–558.

Xu, X. 2006. Feathered dinosaurs from China and the evolution of major avian characters. *Integrative Zoology* 1: 4–11.

Xu, X., X. Zheng, and H. You. 2010. Exceptional dinosaur fossils show ontogenetic development of early feathers. *Nature* 464: 1338–1341.

Zhang, X., and D. E. G. Briggs. 2007. The nature and significance of the appendages of *Opabinia* from the Middle Cambrian Burgess Shale. *Lethaia* 40: 161–173.

Zhang, X. G., and X. G. Hou. 2004. Evidence for a single median fin-fold and tail in the Lower Cambrian vertebrate, *Haikouichthys ercaicunensis*. *Journal of Evolutionary Biology* 17: 1162–1166.

Zhang, X., D. Shu,Y. Li, and J. Han. 2001. New sites of Chengjiang fossils: Crucial windows on the Cambrian explosion. *Journal of the Geological Society of London* 158: 211–218.

Zhu, M., W. Zhao, et al. 2009. The oldest articulated osteichthyan reveals mosaic gnathostome characters. *Nature* 458: 469–474.

19

Development and Evolution

A key innovation in human history was the domestication of crops, particularly high-calorie cereals such as sorghum, rice, corn, and wheat. The wild relatives of these cereals share a trait called seed shattering. An abscission layer forms between the seed coat and the pedicel, or seed base, that lets mature seeds readily fall off the plant (right). Shattering facilitates seed dispersal, but it is problematic in domesticated crops because fallen seeds are hard to harvest. By gathering seeds that remained on the stalk, early farmers presumably selected for the non-shattering phenotypes now seen in all major cereal crops.

Zhongwei Lin and colleagues (2012) uncovered the genetic basis for the non-shattering phenotype in sorghum, the world's fifth most important cereal. Mutations in the *Shattering1* (*Sh1*) gene in domesticated sorghum are associated with failure to form an abcission layer during seed development. Mutations in *Sh1* are also associated with non-shattering phenotypes in domesticated corn and rice.

Charles Darwin would have been pleased, for Lin's study combines research in two areas that were of fundamental importance to his understanding of evolution and the evidence for it: artificial selection in domesticated plants and developmental biology. "Development," Darwin wrote (1872, p. 386), "is one of the most important subjects in the whole round of natural history."

Above, women harvesting sorghum in Burkina Faso. Below, due to differences in development, vigorous shaking scatters the seeds of wild sorghum (a and b), but leaves the seeds of domesticated sorghum firmly attached (c). From Lin et al. (2012).

Reprinted by permission of Macmillan Publishers Ltd. *Nature Genetics* 44: 720–724, copyright 2002.

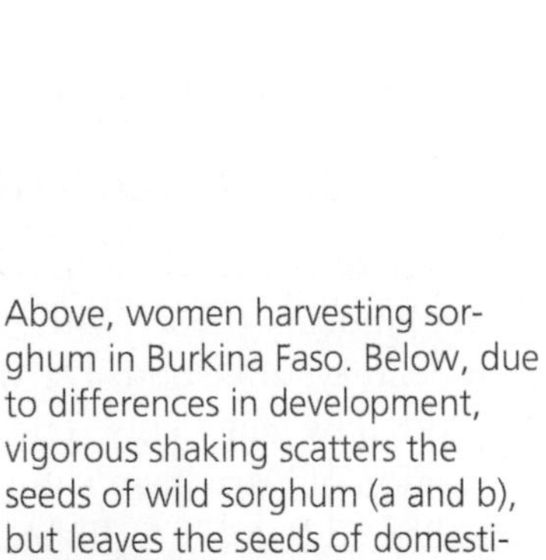

How are developmental biology and evolution related? Developmental biology is the study of the processes by which an organism grows from zygote to reproductive adult. Evolutionary biology is the study of changes in populations across generations. As with non-shattering cereals, evolutionary changes in form and function are rooted in corresponding changes in development. While evolutionary biologists are concerned with why such changes occur, developmental biology tells us how these changes happen. Darwin recognized that for a complete understanding of evolution, one needs to take account of both the "why" and the "how," and hence, of the "important subject" of developmental biology.

In Darwin's day, studies of development went hand in hand with evolution, as when Alexander Kowalevsky (1866) first described the larval stage of the sea squirt as having clear chordate affinities, something that is far less clear when examining their adults. Darwin himself (1851a,b; 1854a,b) undertook extensive studies of barnacles, inspired in part by Burmeister's description (1834) of their larval and metamorphic stages as allying them with the arthropods rather than the mollusks. If the intimate connection between development and evolution was so clear to Darwin and others 150 years ago, why is evolutionary developmental biology (or *evo-devo*) even considered a separate subject, and not completely integrated into the study of evolution? The answer seems to be historical. Although Darwin recognized the importance of development in understanding evolution, development was largely ignored by the architects of the 20th-century codification of evolutionary biology known as the modern evolutionary synthesis.

In Section 19.1 of this chapter, we consider why development was thus ignored. In Section 19.2 we look at how the two fields of study came back together. The third and fourth sections cover recent trends in evo-devo. Finally, in the last section, we look ahead to the future of evolutionary developmental biology.

19.1 The Divorce and Reconciliation of Development and Evolution

Unaware of the work of his contemporary, Gregor Mendel, Darwin developed his evolutionary theory without a clear understanding of inheritance. In the early 20th century, after both Darwin and Mendel had died, Mendel's work was rediscovered by geneticists. The modern evolutionary synthesis thus arose from the unification of Darwinian natural selection with Mendelian genetics. This unification allowed Ronald A. Fisher, J. B. S. Haldane, Sewell Wright, and others to lay the foundations of population genetics (see Mayr 1982). The term *modern synthesis* was coined by Julian Huxley to denote the compilation during the late 1930s and early 1940s, in the context of the new evolutionary genetic framework, of advances in the understanding of variation in natural populations, paleontology, and speciation. Notably missing from the synthesis was developmental biology.

The Divorce

According to Ernst Mayr (1982), another key figure in the modern synthesis, the reasons for development's omission from the modern synthesis were practical. Because the genetic and molecular mechanisms underlying development were so poorly understood at the time, a direct connection to evolutionary genetics could not be drawn. The belief was that development and genetics needed to be "properly separated" in order to make progress in both fields (Mayr 1982, p. 893).

A second explanation for the separation was the rejection of many Darwinian tenets by prominent developmental biologists and morphologists of the late 19th and early 20th centuries, sometimes characterized as *saltationists* and *structuralists*. For example, Darwin and the architects of the synthesis held that most evolutionary change was gradual, via almost imperceptible steps from one generation to the next. This conception followed one of Darwin's favorite adages, derived from Linnaeus (1751) and repeated often in *On the Origin of Species*: Nature does not make leaps. By contrast, "saltationists" believed that major evolutionary changes, such as the origin of new species, were the result of mutations of large effect.

Due to inadequate understanding of the mechanisms of development and to disagreements over theory, developmental biology was left out of the modern evolutionary synthesis.

Another fundamental Darwinian idea was that natural selection is the primary mechanism of change; mutation and variation merely provide the raw material for natural selection to act. Variations, in this view, are ever present, small in scale, and unbiased toward certain adaptations. "Structuralists," by contrast, proposed that physical and mathematical principles directed growth and form along defined pathways that could account for most of life's diversity (Thompson 1917).

Less well recognized was the comparative zoologist Libbie Hyman, whose comprehensive studies on the invertebrates during this period fluidly combined adult and developmental features into an encyclopedic understanding of adaptations and relationships (i.e., Hyman 1940). Nevertheless, most developmental and evolutionary biologists pursued their disciplines separately; it was only the advent of molecular biology in the late 20th century that began to heal the rift.

The Reconciliation

In the 1940s and 50s, the identification of DNA as the genetic material finally revealed the machinery of variation. The subsequent cracking of the genetic code tied the information content of DNA to the amino acid content of proteins.

Using the correspondence of DNA to proteins, Marie Claire King and Allan Wilson (1975) revisited a question controversial since Darwin's day: How closely related are humans and chimpanzees (**Figure 19.1**)? Comparing amino acid sequences, they estimated the genetic similarity at 99%. Charles Sibley and Jon Ahlquist (1987) used a different method to reach the same conclusion. More recently, our close kinship to chimps has been confirmed by genome sequencing.

Figure 19.1 Humans and chimpanzees are 99% similar genetically Debbie Cox has devoted her career to chimpanzee conservation. Despite being close relatives, humans and chimps differ in important ways, both in appearance and behavior. What is the nature of our 1% genetic difference, and how can it account for our many different physical and behavioral features? Evo-devo approaches will assist in our ability to one day answer these questions.

A second set of findings from molecular biology related to how genes are turned on and off. Francois Jacob and Jacques Monod (1961) elucidated the mechanism by which certain proteins regulate gene activity in bacteria. In 1963, Ed Lewis explicitly drew the connection from Jacob and Monod's findings to developmental biology, describing how **gene regulation** could orchestrate developmental processes in insects and thus explain for the first time how a single genome could produce a diversity of cells throughout a multicellular body.

The time was ripe for a reconciliation between evolution and development. In 1971, Roy Britten and Eric Davidson interpreted the new discoveries of gene regulation in an evolutionary and developmental context (p. 129): "It is clear... that alterations in the [regulatory] genes... could cause enormous changes in the developmental process and that this would be a potent source of evolutionary change." This proposal challenged the notion of gradualistic evolution at the heart of the modern synthesis by suggesting that relatively modest changes at the DNA level in regulatory genes could have profound impacts on development, and hence evolution. Given that human and chimp DNA is 99% similar, was it possible that relatively modest genetic changes in regulatory genes could lead to profound evolutionary changes after all?

Stephen Jay Gould (1977) drew on the discoveries of gene regulation to suggest that much evolutionary change could be attributed to alterations in the relative timing of developmental events (known as **heterochrony**), another example of modest changes in gene regulation leading to dramatic morphological change.

Previously, David Wake (1966) had discussed how evolutionary patterns in salamanders could be explained by their developmental trajectories. His student Pere Alberch postulated developmental rules that, along with selection, determine the direction of evolution. Alberch (1981) proposed that the evolution of large size in tree-dwelling salamanders—compare **Figure 19.2a versus b**—depended on a developmental modification in the relative length and shape of the distal finger and toe bones in smaller ancestral species (Figure 19.2c). This developmental change, and the associated origin of webbing between the digits, allowed for increased suction efficiency, which let larger salamanders climb trees without falling. A modest change in the rules of development—the relative growth of fingers and toes versus other body parts—could facilitate a macroevolutionary change. Alberch was offering a second-generation structuralist view of evolution in the context of a modern understanding of developmental mechanisms.

These works, while groundbreaking, initially did not have a strong impact either on developmental biology, which remained focused on a few well-studied organisms like fruit flies, roundworms, thale cress, chickens, and house mice, or on evolutionary biology, where the modern synthesis was still the dominant paradigm. Both fields continued to expand in scope throughout the 20th century, but they had distinct journals, different jargon, and limited integration.

And then came a discovery that suddenly had developmental biologists not only discussing evolution, but proposing bold hypotheses regarding major evolutionary issues such as the origin of animal phyla. Species from widely divergent branches on the animal phylogeny, fruit flies and house mice, appeared to build their segmented bodies in similar ways. The field of modern evo-devo was born.

(a) Giant palm salamander (*Bolitoglossa dofleini*)

(b) Atoyac salamander (*Bolitoglossa oaxacensis*)

(c) Hypothesized evolutionary scenario

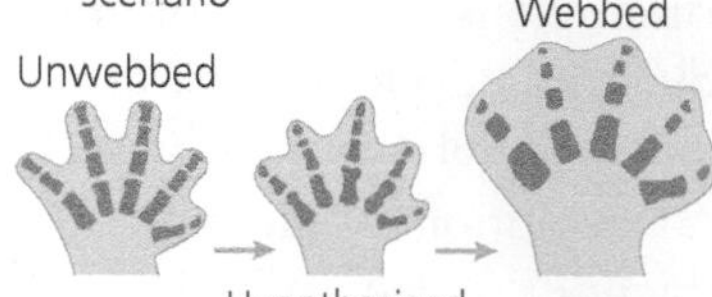

Figure 19.2 Insights on salamander evolution The arboreal giant palm salamander (a), the largest of the Bolitoglossine salamanders (~12 cm snout–vent length) has more extensive webbing between the fingers and toes than the Atoyac salamander (b), a small-bodied ground form (~5 cm snout–vent length) exemplifying the presumed ancestral condition. (c) A hypothesized transitional species had small body size and webbing. *B. dofleini* juveniles have hands and feet that resemble the hypothesized intermediate. (b) by David Wake; (c) after Alberch (1981).

19.2 Hox Genes and the Birth of Evo-Devo

As the tools of developmental biology became more sophisticated in the last decades of the 20th century, Britten and Davidson's hypothesis—that alterations in regulatory genes could cause enormous changes in development—found increasing support. It had long been known that mutations in embryonically active genes in the fruit fly *Drosophila melanogaster* could dramatically alter the formation of larval and adult structures. The most famous class of these mutations was discovered in the late 19th century by William Bateson (1894), often remembered for his saltationist views and resistance to the modern synthesis. In 1915, Calvin Bridges uncovered the genetic basis for one of Bateson's phenotypes. To Bateson, and later Ed Lewis (1978), "homeotic mutations" showed that evolution could make leaps. As we will see, they were partially correct but for the wrong reasons.

When **homeotic genes** are mutated, appendages appear in the wrong places. A mutant bithorax gene yields flies with four wings instead of the normal two, and a mutant *Antennapedia* gene yields flies with legs in place of antennae (**Figure 19.3, next page**). Lewis and others confirmed that homeotic genes were clustered in two locations in the genome: the bithorax complex (BX-C) and the Antennapedia complex (ANT-C). However, in the absence of DNA sequences and functional molecular studies, the nature of these clustered genes remained a mystery.

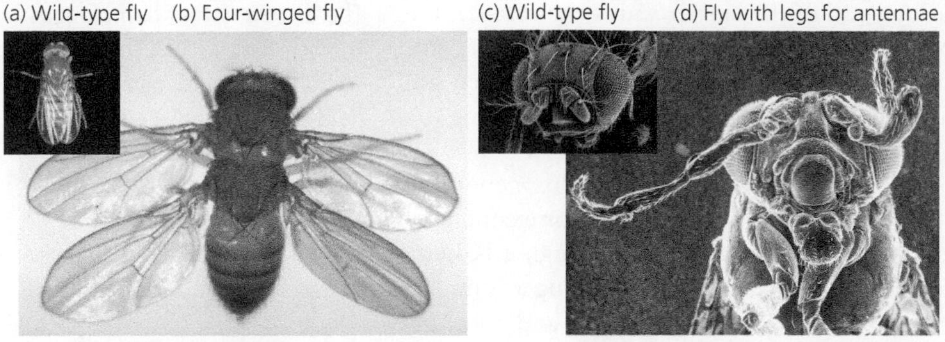

Figure 19.3 Homeotic mutants in *Drosophila* (a) Normal two-winged fly with wings on the second thoracic segment (T2); the third segment (T3) has balancer organs. (b) Four-winged *Ultrabithorax* (*Ubx*) mutant with identity of T3 transformed into T2. (c) A normal fly with small antennae. (d) *Antennapedia* (*Antp*) mutant with legs instead of antennae.

In 1983 that all changed. Scientists in Switzerland and the United States realized that the products of genes from both ANT-C and BX-C had a common stretch of amino acids, suggesting that they were evolutionarily related (McGinnis et al. 1984; Scott and Weiner 1984). Soon thereafter, the Switzerland group found multiple copies of this same amino acid sequence, the **homeodomain,** in a beetle, earthworm, frog, chicken, mouse, and human. Functional studies showed that the homeodomain directly interacted with DNA. The homeodomain proteins were identified as **transcription factors**—they regulate the transcription of other genes. Here were genuine developmental regulatory genes—the "Hox" genes—that were conserved in sequence across distantly related animals.

This finding surprised biologists. Most had assumed, based on the gradualism of the modern synthesis, that "the search for homologous genes is ... futile except in ... close relatives" (Mayr, 1963, p. 609). A bigger shock came when the mouse and fly Hox genes were found not only to be clustered on the chromosomes of both animals and expressed along the anterior–posterior body axis of both animals, but also expressed in spatial patterns that mirror the arrangement of the clustered genes on the chromosome (Gaunt 1988). As shown in **Figure 19.4a**, the

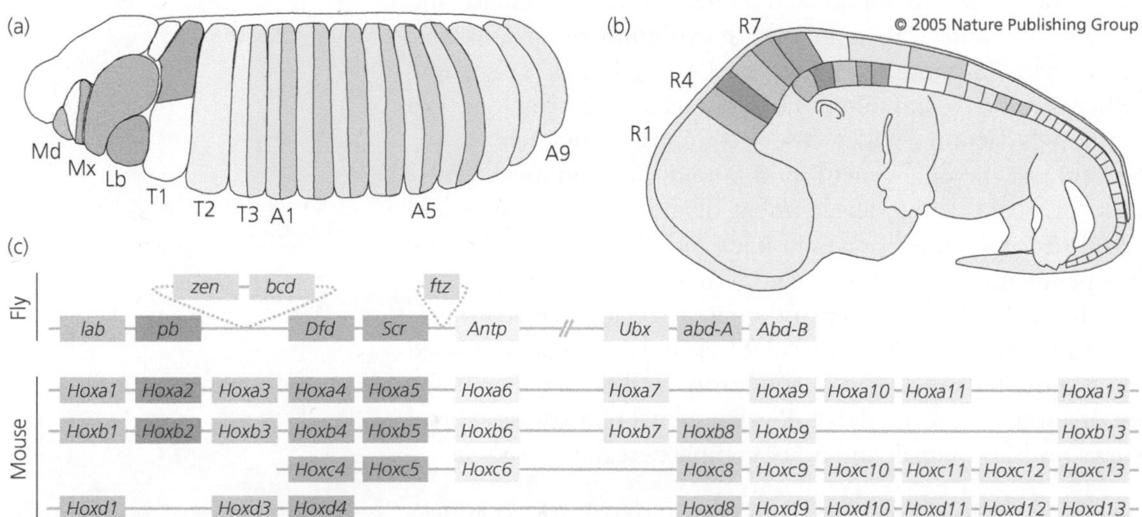

Figure 19.4 Hox genes in flies and mice (a) Fruit fly (*D. melanogaster*) and (b) mouse (*M. musculus*) embryos showing Hox gene expression on the anterior–posterior axis. (c) Hox gene arrangements. Flies have a split cluster—the Antennapedia complex (left of split) and the bithorax Complex (right) are separated by 9 million nucleotides. Mammals have four clusters, though to derive from two whole-cluster duplications followed by gene losses. The embryonic expression domains are less discrete than shown. Colors suggest orthology of fly and mouse genes for which strong evidence is, in some cases, lacking. Modified from Pearson et al. (2005).

Reprinted with permission from Macmillan Publishers Ltd. *Nature Review Genetics*. J. C. Pearson, D. Lemons, and W. McGinnis. "Modulating Hox gene functions during animal body patterning." *Nature Review Genetics* 6: 893 -904. Copyright © 2005 The Nature Publishing Group.

fruit fly *Deformed* (*Dfd*) gene is near one end of the cluster and is expressed in the developing head of the embryo, whereas the next gene, *Sex combs reduced* (*Scr*), is expressed just posterior to *Dfd,* and so on. This correspondence between genomic order of Hox gene loci and their spatial locations of expression along the body axis is known as **spatial colinearity.**

The discovery that homologous transcription factors influence fundamental aspects of development in insects and vertebrates sparked renewed interest among developmental biologists in evolution, and among evolutionary biologists in development.

Subsequent genetic studies showed that mutations in mouse Hox genes (see Figure 19.4b) also cause homeotic transformations of body parts, for example by adding extra neck vertebrae (Kessel et al. 1990). It looked as though a Rosetta stone of animal development and evolution had been found, showing a common developmental mechanism underlying the divergent bodies of flies and mice.

In the wake of these discoveries, students of development in the 1990s saw numerous false-color images of mouse and fly embryos like those in Figure 19.4. The similarities were striking, but was the evidence sufficient to suggest a specific evolutionary path? Are the anterior Hox genes of flies and mice really direct descendants of a single gene in their last common ancestor? What would that imply about the ancestor, which lived over 530 million years ago? Can we infer what it looked like, where its Hox genes were expressed, and what functions they had?

Flies and mosquitoes have essentially the same number of Hox genes. The *Deformed* (*Dfd*) gene occurs in the same position in the mosquito Hox cluster as in the fly. We can infer that the last common ancestor of flies and mosquitoes likely had one *Dfd* gene that was inherited faithfully by both species. Genes in different species derived from a common ancestor's gene are said to be **orthologous.**

In vertebrates there are four Hox clusters—Hoxa, Hoxb, Hoxc, and Hoxd—that contain most of the same genes in the same order. We can infer that *Hoxa4, Hoxb4, Hoxc4,* and *Hoxd4* arose from gene duplications in an ancestral vertebrate. Genes within a species that arise via duplication are said to be **paralogous.**

Can we also infer that the *Hox4* genes in vertebrates and the *Dfd* genes in insects are orthologous? Orthology determinations are difficult in Hox genes. They are based on just a few dozen shared amino acid sequences (e.g., Monteiro and Ferrier 2006), leaving us low statistical confidence in evolutionary relationships. Despite the color coding in illustrations like Figure 19.4, we do not really know in many cases whether specific fly and mouse Hox genes are true orthologs. We may never know, although data on a wider variety of animals could help.

The Hox genes were not the only genes found in both flies and mice. Other classes of genes were involved in the development of mouse and fly eyes and mouse and fly hearts. Previously, these organs in mice and flies had been thought to have evolved independently, due to their anatomical differences and differing embryonic origins. However, the discovery of similar developmental mechanisms forced reexamination of such ideas. In 1993, Slack and colleagues proposed that all animals would show Hox gene spatial colinearity, and that this would be the defining feature of animals. It seemed that developmental biology was poised to radically alter views on how animals and other complex creatures evolved.

Hox Paradox: The More Things Changed, the More They Stayed the Same?

There is a nagging problem with this view of conservation. If evolution is about change, does it not appear that the Hox gene story in Figure 19.4 implies sameness? From such conservation, how can we comprehend the diversity of life?

One resolution to the paradox comes when we look at the details more closely **(Figure 19.5)**. Since the initial Hox discoveries, the gene expression patterns,

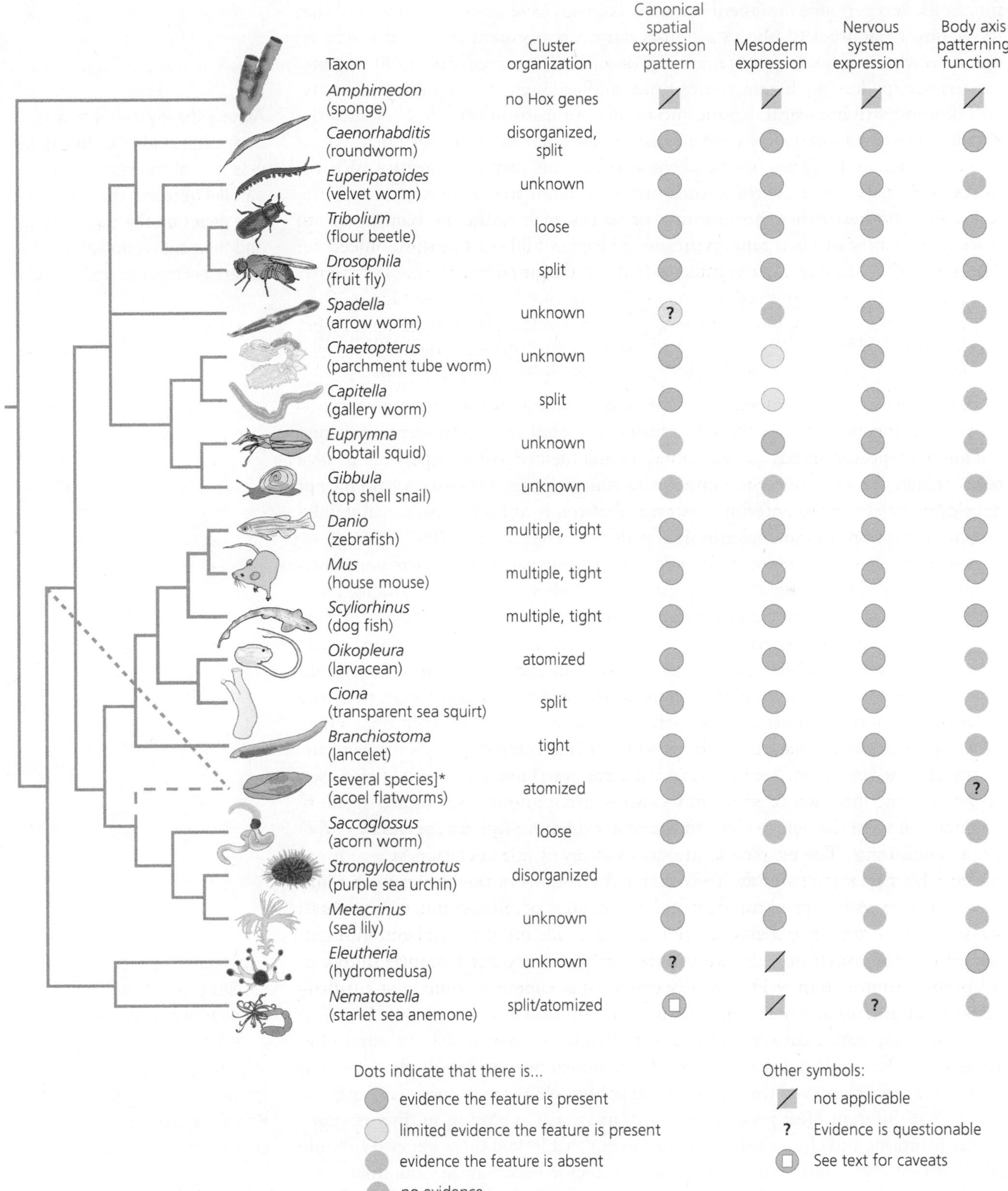

Taxon
Cluster organization
Canonical spatial expression pattern
Mesoderm expression
Nervous system expression
Body axis patterning function
Amphimedon (sponge)
no Hox genes
Caenorhabditis (roundworm)
disorganized, split
Euperipatoides (velvet worm)
unknown
Tribolium (flour beetle)
loose
Drosophila (fruit fly)
split
Spadella (arrow worm)
unknown
Chaetopterus (parchment tube worm)
unknown
Capitella (gallery worm)
split
Euprymna (bobtail squid)
unknown
Gibbula (top shell snail)
unknown
Danio (zebrafish)
multiple, tight
Mus (house mouse)
multiple, tight
Scyliorhinus (dog fish)
multiple, tight
Oikopleura (larvacean)
atomized
Ciona (transparent sea squirt)
split
Branchiostoma (lancelet)
tight
[several species]* (acoel flatworms)
atomized
Saccoglossus (acorn worm)
loose
Strongylocentrotus (purple sea urchin)
disorganized
Metacrinus (sea lily)
unknown
Eleutheria (hydromedusa)
unknown
Nematostella (starlet sea anemone)
split/atomized
Dots indicate that there is...
evidence the feature is present
limited evidence the feature is present
evidence the feature is absent
no evidence
Other symbols:
not applicable
? Evidence is questionable
See text for caveats

Figure 19.5 Diversity in Hox gene expression and function The phylogeny shown here is a consensus from the recent literature. The affinities of acoel flatworms are disputed; we show two alternative placements (dashed lines). *The information on acoel flatworm Hox genes is assembled from data on several species.

functions, and genomic arrangements of Hox genes have been elucidated in many more phyla. Figure 19.5 shows a conservative arrangement of the still-debated animal phylogeny, with representative groups that have well-studied Hox genes and/or occupy key phylogenetic positions, and is designed to highlight diversity.

The most striking feature of the mouse–fly comparison (Figure 19.4) was the parallel between Hox gene expression along the anterior–posterior body axis of and the order of the genes on the chromosome—the spatial colinearity. What is the connection between chromosomal position and spatial gene expression?

Among the mysteries to arise from early studies of Hox genes was spatial colinearity—a parallel between the order of Hox genes on the chromosome and their expression along the anterior–posterior body axis.

Let us hypothesize that chromosomal order is somehow the mechanistic key to spatial colinearity of Hox gene expression. How would we test this conjecture? The tools of modern evo-devo offer us two main approaches:

1. Manipulate the genomes of well-studied, amenable organisms;
2. Look for natural experiments by making comparisons between organisms that may show variation for the developmental phenomenon in question.

In the case of spatial colinearity, approach 1 can involve generating chromosomal rearrangements in Hox gene clusters to see if the body segments along the anterior–posterior axis are scrambled or maintained. Such experiments have been carried out extensively in vertebrates, roundworms, and insects, and proper development along the anterior–posterior axis tends to be faithfully maintained despite disruptions in the chromosomal order of genes (Ferrier 2007, 2011).

Approach 2 involves identifying species that show variation in Hox gene clustering compared to the tightly clustered, ordered Hox genes in vertebrates. How widespread is spatial colinearity among animals as a whole? Recall that the two fruit fly Hox complexes, ANT-C and BX-C, are in separate locations in the genome. In fact, the *Cluster organization* column in Figure 19.5 shows that tight clustering of Hox genes seems rare outside the vertebrates and lancelets. More common is a loose clustering, in which the Hox genes are gathered on a single part of one chromosome, but with non-Hox genes interspersed, as in many insects and an acorn worm. Split clusters, with different Hox genes widely dispersed in the genome into two or more sub-clusters, are found in a sea squirt, a gallery (segmented) worm, a nematode, and fruit flies; different fruit fly species have different split points. The purple sea urchin has a disorganized cluster, where more anterior Hox genes apparently have changed positions in the cluster with more posterior ones. And a planktonic larvacean and an acoel flatworm have atomized clusters, where the Hox genes are not linked at all. Nevertheless, in almost all cases, the spatial order of gene expression along the body axis is similar to that of the mouse. This pattern of Hox gene expression is sometimes called the **canonical spatial expression pattern** (see first column of symbols in Figure 19.5).

Experiments and comparative studies suggest that spatial colinearity may have been ancestral, but is not essential to proper Hox gene function. However, the spatial pattern of Hox gene expression is highly conserved.

These comparative data are consistent with a scenario in which ancestral Hox genes were clustered, and the ordered clustering has been lost multiple times. An alternative scenario, given the lack of a single Hox cluster in sponges and anemones, is that different Hox genes arose early in animal evolution in dispersed genomic locations and clustering was a later event that linked these genes (Duboule 2007). More comparative data are needed to decide between these alternatives.

Timothy Dubuc and colleagues (2012) published data on Hox genes for a relative of sea anemones, the coral *Acropora digitifera*. In *A. digitifera,* several Hox genes are tightly clustered together in one location of the genome, a finding that lends support to the ancestral cluster hypothesis. However, the coral and anemone Hox genes are particularly difficult to assign to specific orthology groups, so it is not yet possible to distinguish between a single ancestral cluster and tandem gene duplications in different lineages, yielding independent mini-clusters.

Despite all this diversity in the degree of clustering, almost all species examined show evidence of the canonical spatial expression pattern of their Hox genes. Thus, a paradoxical conclusion from Figure 19.5 is that a vertebrate-like spatial expression of Hox genes does not actually depend upon the order of the Hox genes along chromosomes. But what about Hox gene function? It is important to note that a gene expression pattern (i.e., the production of messenger RNA) is not always directly indicative of expression of the protein, which actually performs the cellular function of regulating transcription of other genes. As a result, we cannot confidently ascribe a developmental function to a Hox gene simply based upon when and where the gene is transcribed. In fact, even confirming the presence of the protein does not, in itself, confirm a predicted function there.

As such, additional studies are required to establish gene function. Functional studies could involve expressing a protein at inappropriate times or locations in the embryo or interfering with normal function through mutations or other manipulations. Such functional studies in insects, nematodes, and vertebrates typically confirm that Hox gene expression patterns correspond to regions of Hox protein function; nevertheless, there is one striking counterexample.

Sea squirts are invertebrate relatives of the vertebrates. They have a nonfeeding, tadpole stage with a clear anterior–posterior axis, a brain, and a notochord (**Figure 19.6a**). During a dramatic metamorphosis, the tail and notochord and many other tadpole organs are resorbed, and the adult emerges as a sessile filter feeder that superficially resembles a sponge more than a vertebrate (Figure 19.6b).

(a) Larva

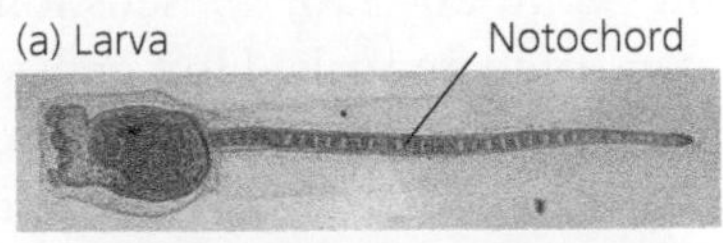

(b) Adult

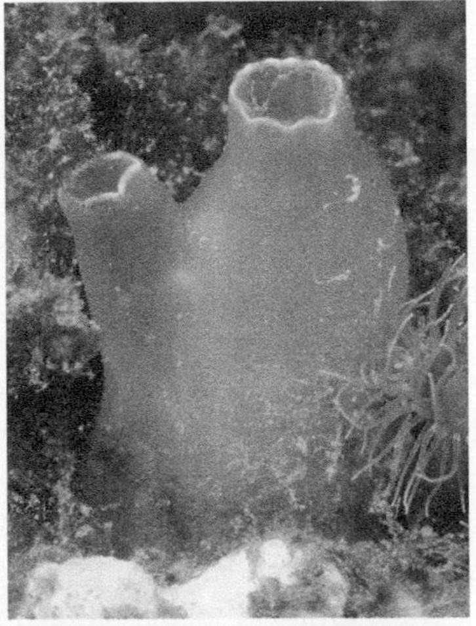

Figure 19.6 An invertebrate chordate (a) Nonfeeding tunicate larva, with readily identifiable chordate features such as a head, tail, and notochord. Photo by Daniel Clemens, Napa Valley College. (b) Filter-feeding adult tunicate (*Ciona intestinalis*), with yellow-fringed incurrent (top) and excurrent (left) siphons.

The sea squirt *Ciona intestinalis* has a split Hox cluster in which the canonical spatial expression pattern is more or less maintained. Thus the prediction was that sea squirt Hox genes would function as in other animals: Disruptions in Hox gene function would cause defects along the main body axis. Ikuta and colleagues (2010) tested this prediction in *C. intestinalis* using a technique called RNA interference (RNAi), where an injected double-stranded RNA molecule, targeted at a specific messenger RNA (mRNA), results in the failure of that mRNA to be translated into protein. Ikuta and colleagues blocked production of Hox proteins in *C. intestinalis* embryos, but saw only minor phenotypic changes and none of the major impacts expected of Hox genes. Controls showed that the RNAi technique reduced protein levels. Therefore, as far as the scientists could tell, Hox genes have no body-axis function in sea squirt embryonic development, despite being expressed in the canonical spatial pattern. Is it possible that the canonical spatial pattern of Hox gene expression is what has been conserved across animals, while the functions are—like the genomic positions—more evolutionarily labile?

Another clue comes from the echinoderms, a phylum of animals with odd, five-parted ("pentameral") body structures that includes sea stars, sea urchins, sea cucumbers, and sea lilies. Echinoderm embryos generally develop into a bilaterally symmetric larva. Then, after a period of growth and development, they undergo a dramatic metamorphosis to the pentameral adult. Surprisingly, some purple sea urchin (*Strongylocentrotus purpuratus*) Hox genes are not expressed at all during larval development, and others are expressed in the canonical spatial pattern in only a single body cavity (Arenas-Mena et al. 2000) that does not seem to have a key function in the initial morphogenesis of either the larval or adult body axis. *Metacrinus rotundus*, a sea lily, has almost the same pattern of expression in the homologous body cavity (Hara et al. 2006). In living echinoderms, as in the sea squirts, we see maintenance of the canonical spatial expression pattern despite the lack of apparent function in patterning along the primary body axes (see Mooi and David 2008 for a discussion of extinct echinoderms).

In some animals, Hox genes are also expressed in other locations in a pattern parallel to the canonical one, such as the urogenital system, gut, and limbs of some vertebrates, and the dorsal–ventral axis in an anemone (see Ryan et al. 2007).

Finally, the comparative data on pre-bilaterian Hox genes discussed earlier (Dubuc et al. 2012) indicates that after the Cnidaria (anemones, jellyfish, coral) split off from the line leading to the bilateral animals, and again after the bilateral animals diversified, the Hox gene sets must have expanded substantially. Thus, the evolutionary expansion of Hox clusters by gene duplication must have resulted in further subdivisions of the body into specific Hox gene expression domains. Again, this observation indicates a primacy of the canonical spatial expression patterns, perhaps independent of specific functions of the genes themselves.

With such diversity in Hox gene clustering, expression, and function, can we make any predictions about Hox function in early animals? Figure 19.5 reveals two things almost all animal Hox genes have in common: They show the canonical spatial expression pattern and are expressed in the nervous system. The most parsimonious hypothesis for the common ancestor of the bilateral animals (including beetles, snails, urchins, and humans) is that its Hox genes were expressed in the canonical spatial pattern in the nervous system (Samadi and Steiner 2010).

The canonical spatial expression pattern of Hox genes is most strongly conserved in the nervous system, suggesting that directing the nervous system's development was their ancestral function.

Is there a way to test this hypothesis of a nervous system origin of Hox gene expression in our deep animal ancestors? What did these ancestors look like? Did they have segments, a heart, limbs, eyes? If the ancestral function of Hox genes was in nervous system development, how can we account for the various functions that have evolved in different animal lineages since that time? And how can we explain the strikingly similar segmental patterns of Hox gene expression in vertebrates, segmented worms, and arthropods? For answers, we need to go beyond Hox genes and consider broader issues in evo-devo that have been discussed for years, but have found substantial experimental support only recently.

19.3 Post Hox: Evo-Devo 2.0

Homology and Homoplasy: The Eternal Recurrence

Organisms show curious similarities in structure, despite differences in function. The forelimbs of a mole and a bat have the same arrangement of bones, even though one serves as a shovel and the other as a wing (see Chapter 2). Darwin provided the first meaningful explanation for such similarities: common descent. Just as a child resembles her brother more than a randomly chosen classmate, a human resembles a chimp more than a lemur. Traits shared because they were present in, and inherited from, a common ancestor are called **homologous.**

However, similarity can also arise independently. Consider the similarities, shown in **Figure 19.7**, between New World cacti and African euphorbs, between tooth fungi from the genera *Hydnellum* and *Hydnum,* and between hedgehogs and Malagasi hedgehog tenrecs. These are plants, fungi, and mammals from distantly related families. Their similarities are due to independent evolution, not common descent. Another term for similar features in two organisms that were not present in, and inherited from, their most recent common ancestor is **homoplasy.**

Figure 19.7 Homoplasy in plants, fungi, and animals (a, b) Distantly related plants with similar growth forms. (c, d) Fungi with independently derived "tooth fungus" phenotypes, showing spinelike teeth instead of gills. (e, f) Mammals with similar morphologies. The lesser hedgehog tenrec shown in (f) is found only on Madagascar. Other tenrecs occur on Madagascar and in Africa. Different species of tenrecs resemble hedgehogs, shrews, and otters. Nevertheless, they belong to the order Afrosoricida, and are all more closely related to aardvarks and elephants than they are to true hedgehogs, shrews, or otters.

What does this have to do with developmental biology? Previously, we defined development as the processes by which an organism grows through its life cycle to produce the reproductive stage(s) and all of the stages in between. Let us here consider a related, yet slightly different and more technical definition. Every organism has a genotype, essentially the same in all cells in the body. Development is the process by which that genotype, in coordination with the environment, produces an organism's phenotypes through the life cycle **(Figure 19.8)**.

Figure 19.8 One conception of development Development is the process by which genes are expressed in an environmental context to yield phenotypes. Genotype is inherited, but phenotype is the target of selection.

In the examples of homoplasy shown in Figure 19.7, the phenotypes are strikingly similar. The standard explanation is that similar selection pressures in different taxa caused similar evolutionary changes to arise independently. But what if the developmental processes that build those phenotypes are also the same?

One such example can be seen in the independent evolution of mangroves in distantly related families of flowering plants **(Figure 19.9)**. Figure 19.9a and b show

Figure 19.9 Parallel evolution of developmental physiology in mangroves Mangroves, a shoreline-adapted growth form, occur in distantly related plant families, including (a) *Aegialitis* and (b) *Rhizophora*. The phylogeny in (c) shows a sample of mangrove genera in bold, and a few of their non-mangrove relatives. Background colors show taxonomic groups. Lineages indicated by blue lines show reproductive vivipary.

two such independently evolved mangroves from different plant families. Figure 19.9c shows a phylogenetic hypothesis for how these and other mangrove genera (blue branches on the phylogeny) are related to several non-mangrove genera (black branches). "Mangrove" describes a growth form (not a clade) of shoreline plants whose roots are often submerged in their brackish or saltwater habitats. Living in such an environment is challenging; mangroves need to rid their leaves of salt, and their seeds must either root quickly when dropped or remain buoyant until reaching a suitable spot to root. Both of these characteristics depend upon the plant hormone abscisic acid (ABA). High levels of ABA confer salt tolerance to leaves. Low levels of ABA in embryos cause them to start developing while the seed is still attached to the plant. This trait, known as vivipary, confers buoyancy and allows quick rooting.

Elizabeth Farnsworth and Jill Farrant (1998) examined ABA regulation in leaves and embryos of four independently evolved mangrove groups (including the two pictured in Figure 19.9) as well as in closely related non-mangroves. Every mangrove tested had high levels of ABA in leaves, and reduced levels of ABA in their embryos, compared to related non-mangroves. Independently evolved mangroves thus not only look superficially similar and exist in similar habitats, but their underlying developmental physiology is similar. This class of homoplasy, where similarity results from the same underlying developmental mechanism, is known as **parallel evolution.** This is contrasted with **convergence,** or similarity resulting from a different underlying developmental mechanism (Hodin 2000).

Similar traits that evolved independently may arise via the same or different developmental mechanisms.

It is important to recognize that the contrasting concepts of homology versus homoplasy and parallel versus convergent evolution can be applied at hierarchical levels ranging from genes to behavior. For example, one can identify parallel evolution at the level of amino acids, as in the independent origin of alanine-rich antifreeze proteins in arctic and antarctic fish, or at the level of colonial behavior, as in the independent origins of eusociality in different insect groups. It is thus crucial to specify the hierarchical level being discussed. For example, bat wings and pterodactyl wings are homologous as limbs, but homoplasious as flying limbs. Since this is a chapter on evo-devo, we are considering parallel and convergent evolution at the level of the developmental mechanism.

Parallel Evolution

As modern developmental biology is applied to comparative questions, many examples of parallel evolution are being uncovered. For example, independently evolved larval skeletons in two classes of echinoderms (sea urchins and brittle stars) involve a parallel embryonic activation of genes responsible for formation of the adult skeleton (Koga et al. 2010). Even more striking is the parallel evolution of juvenile attachment structures in three distant groups of chordates: sea squirts, frogs, and fish (Pottin et al. 2010). These attachment structures, shown in **Figure 19.10**, are clearly not homologous (Hall 2012). They have completely different embryonic origins, their morphologies are quite different, they reside in different locations on the respective larvae, and attachment organs are rare among chordates. Nevertheless, the formation of attachment organs in sea squirts, frogs, and fish involves the activation of related genes (*Bmp* and *Otx*). Moreover, the frog and fish adhesive organs receive neural inputs from the same part of the brain—the trigeminal ganglion—which processes a wide range of sensory information.

When evolution follows such similar trajectories using similar mechanisms, this is often seen as evidence for **developmental constraints,** defined as a bias

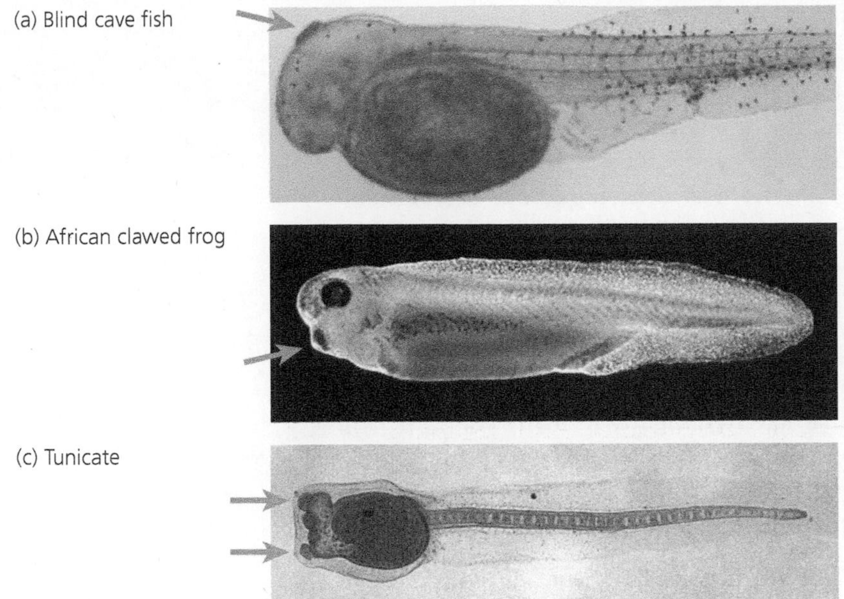

Figure 19.10 Parallel evolution of attachment organs in tadpole-like larvae of distantly related chordates (a) Blind cave fish larva, *Astyanax mexicanus* (subphylum Vertebrata, class Actinopterygii), and its dorsal attachment organ. From Protas and Jeffery (2012). (b) African clawed frog tadpole, *Xenopus laevis* (subphylum Vertebrata, class Amphibia), and its ventral attachment organ. Photo by Edgar Buhl, Bristol University. (c) Tunicate larva (subphylum Urochordata [=Tunicata], class Ascidiacea), and its anterior attachment organs. Photo by Daniel Clemens, Napa Valley College.

in the production of phenotypic variation due to developmental factors (Maynard Smith et al. 1985). Three things are important to keep in mind about constraints:

1. The term *bias* is crucial; *constraint* often conjures up notions of prohibition, whereas bias merely indicates likelihood and directionality of variation. For this reason, Wallace Arthur (2004) advocates the term **developmental bias.**
2. This notion of biased development challenges the concept espoused in the modern synthesis (following Darwin) that variation has no directionality and that all directional evolution is due to selection.
3. It is often difficult in practice to distinguish developmental constraint from strong directional selection.

We explore developmental constraints/biases, and the evidence for their existence, in the next section of this chapter.

Recent findings of unexpected levels of parallel evolution, such as the one illustrated in Figure 19.10, have spawned a robust discussion in the evo-devo community. If independently evolved phenotypes have similar underlying developmental mechanisms, does this imply that the developmental processes are in some sense homologous? If so, does this so-called deep homology (Shubin et al. 2009) suggest that parallel evolution is really a hybrid between homology (at the level of the developmental mechanism) and homoplasy (at the level of the phenotype)?

In a surprising number of cases, the homologies among developmental mechanisms appear to be much deeper than the homologies among the similar traits whose development they control.

One could, of course, argue that because all organisms use proteins to perform cellular functions, any homoplasy has a deep homology in the use of proteins. This extreme example illustrates the need to be clear about the hierarchical level at which we are defining homology or homoplasy. In the case of developmental mechanisms, genes are known to interact in networks of cross regulation, and some of these networks may be billions of years old. But we must be cautious in ascribing homologous functions to these networks. A screwdriver can be used to turn a screw, split glued boards along their seam, and open soda bottles. If individuals in Cameroon and Paraguay use screwdrivers to open bottles, that does not necessarily imply a meaningful homology in the mechanism of bottle opening. Screwdrivers might just have been seized independently as the best tool at hand.

Equal Variation under (Darwin's) Law?

In Darwin's time, natural selection was criticized as merely a negative mechanism. St. George Mivart (1872), for example, argued that because selection simply removes the unfit, it cannot explain the origin of more-fit individuals. Mivart noted the numerous examples of mimicry in insects. These include striking cases of masquerade in which insects look like fresh or decaying leaves or twigs and sometimes even act like vegetation—as when stick insects sway in a breeze. Mivart objected that small, imperceptible evolutionary changes, due to culling of less well-camouflaged individuals, could never add up to these striking resemblances.

In response, Darwin (1872), reasoned as follows:

> Assuming that an insect originally happened to resemble in some degree a dead twig or a decayed leaf, and that it varied slightly in many ways, then all the variations which rendered the insect at all more like any such object, and thus favoured its escape, would be preserved, whilst other variations... if they rendered the insect at all less like the imitated object...would be eliminated.

The key to Darwin's argument, as Darwin himself noted in an 1862 letter to Charles Lyell, is that variation is always present and is unbiased in direction. When the mean phenotype shifts toward closer mimicry, the population still shows variation in all directions—including even better mimicry.

But is variation really of equal probability in all directions and almost always present? In the case of insect mimicry, perhaps the variants are not of equal probability in all directions as Darwin supposed. What if certain types of variants are more likely to arise than others? What if insects are, for some reason, more likely to resemble a twig than a leaf? If variations are thus biased, then we must modify the Darwinian—and modern synthesis—view of natural selection as the predominant creative process of evolution, and accordingly elevate the prominence of internal processes such as developmental bias as an explanation for life's diversity. Can modern evo-devo help settle this age-old debate?

Many butterfly wings have striking patterns, known as eyespots, that distract bird predators by promoting sublethal attacks at the spots rather than at the body (Olofsson et al. 2010). They may also be sexually selected. One particularly well-studied species is the squinting bush brown butterfly from southern Africa, *Bicyclus anynana*. In 2008, Cerisse Allen, working with Paul Brakefield and colleagues, reported on experiments designed to test whether characters such as eyespot size and color could respond to selection in all directions, as hypothesized by Darwin.

Allen and colleagues selected on both eyespot size and eyespot color. As shown in **Figure 19.11a and b**, *B. anynana* has two forewing eyespots. Using the dry season morph of *B. anynana* (Figure 19.11a), the scientists imposed on lab populations 10 generations of artificial selection for four distinct spot size phenotypes:

1. Larger anterior and posterior eyespots (upper right in Figure 19.11c)
2. Smaller anterior and posterior eyespots (lower left in Figure 19.11c)
3. Larger anterior and smaller posterior eyespots (upper left in Figure 19.11c)
4. Smaller anterior and larger posterior eyespots (lower right in Figure 19.11c)

As is clear from the images, 10 generations were sufficient to independently alter both the anterior and posterior eyespot size, even though their sizes are normally correlated across *Bicyclus* species.

Allen and colleagues then tried selecting on hindwing eyespot color. This time they used the wet season form, which is more brightly colored (Figure 19.11b).

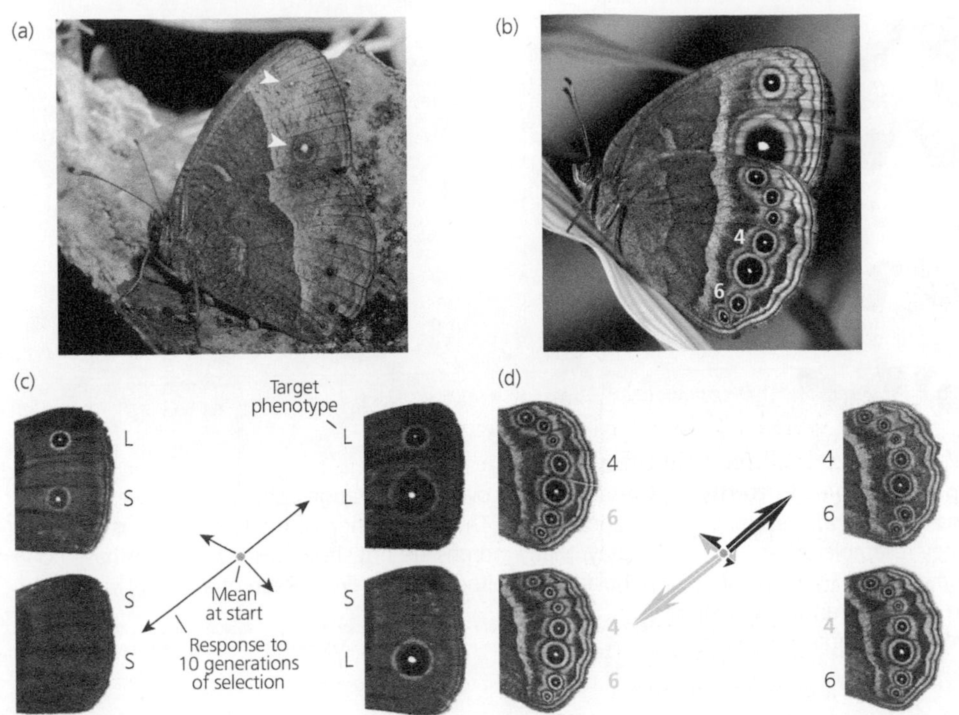

Figure 19.11 Constraint in butterfly wing eyespot color, but not size (a and b) Dry season and wet season morphs of the butterfly *Bicyclus anynana*. Photos by Antónia Monteiro, Yale University. (c) Allen and colleagues (2008) found that lab populations respond to selection for all combinations of larger and smaller forewing eyespot size. (d) In contrast, lab populations responded to selection for only two combinations of enhanced color in hindwing eyespots: enhanced black or enhanced yellow in both eyespot 4 and eyespot 6. But the populations did not respond to selection for enhancement of different colors in the two eyespots, an indication of a constraint on color variation. Photos in (c and d) from Allen et al. (2008).

B. anynana eyespots have three concentric colors: a gold ring surrounding a black ring with a white center. The scientists again imposed artificial selection for 10 generations, and separately on eyespots 4 and 6 (indicated on the hindwing image in Figure 19.11b), because these eyespots are approximately the same size. The researchers attempted to select for the following four eyespot color phenotypes:

1. More black color in eyespots 4 and 6 (upper right in Figure 19.11d)
2. More gold color in eyespots 4 and 6 (lower left in Figure 19.11d)
3. More black in eyespot 4 and more gold in 6 (upper left in Figure 19.11d)
4. More gold in eyespot 4 and more black in 6 (lower right in Figure 19.11d)

The butterfly populations responded to selection for enhanced gold or black color in both eyespots simultaneously, but not to selection for different color enhancements on the two eyespots. This result, in contrast to the eyespot size selection experiment, provides evidence for some constraint or bias, where certain types of variants are much more common than others.

Artificial selection experiments demonstrate that evolutionary change proceeds more readily in some directions than others. One explanation for this phenomenon is developmental bias—the notion that alterations in a developmental pathway can more easily produce some alterations in phenotype than others.

What is the mechanism underlying this biased pattern? The explanation may be related to developmental timing. Classic and modern experiments on the determinants of butterfly wing eyespot patterns indicate that the size of the eyespot is determined in forming wing tissue in the late larval (caterpillar) stage, whereas color is determined later, during the chrysalis stage (French and Brakefield 1995; Beldade et al. 2002, Monteiro et al. 2006). Antónia Monteiro (personal communication) speculates that by the time color is determined in the chrysalis, a convenient set of positional molecular signals that can differentiate the eyespots from each other may no longer be available.

Parallel Evolution and Biased Evolutionary Trajectories

Another dramatic example of wing-pattern variation in butterflies involves mimicry in the genus *Heliconius*. In Central and South America, *Heliconius erato* and *Heliconius melpomene* exist in "mimicry rings." Each species has more than a dozen

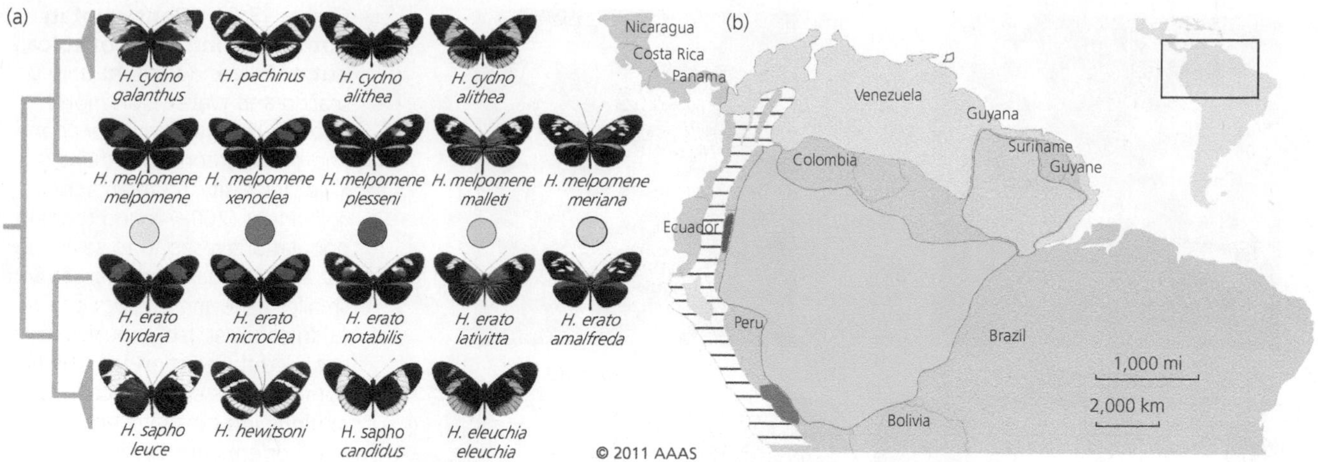

© 2011 AAAS

Figure 19.12 Co-mimicry in multiple *Heliconius* butterfly morphs (a) Morphs of *H. melpomene* and *H. erato* resemble each other across species more than they resemble other morphs in their own species. *H. melpomene* and *H. erato* are not sisters; the closest relatives of each (*H. cydno* and *H. sapho*, for example) look different from *H. melpomene* and *H. erato*. Co-mimics have overlapping ranges, shown by colored dots in (a) and map (b). The entire range of the two species is shown in gray. Many morphs are not shown. Hatched areas have no populations of either species. From Reed et al. (2011).

Reed RD, et al. 2011. Optix drives the repeated convergent evolution of butterfly wing pattern mimicry. 2011. *Science*, 333: 1137-41. Reprinted with permission of the AAAS.

genetically determined color morphs (**Figure 19.12**). Each morph is restricted to a particular locale, and each has an almost perfect co-mimic in the other species. The paired morphs are considered "co-mimics," rather than mimic–model pairs, because both species are unpalatable and both apparently benefit from their mutual resemblance. Note that *H. erato* and *H. melpomene* are not sister species (Figure 19.12a) and cannot hybridize. Thus *H. melpomene* morphs not only resemble their *H. erato* co-mimics more than they resemble closer relatives, they resemble their co-mimics more than they resemble other members of their own species.

How is the diversity of phenotypes maintained within each species? Different color morphs within each species hybridize, but offspring with intermediate phenotypes are quickly selected against by bird predators. Still, bird predation seems an insufficient explanation for the observed geographic diversity.

Recent evidence has provided clues to the mystery. The genetic determinants of wing-color pattern in *Heliconius* largely map to loci of single genes or tightly linked gene clusters: one each for black, yellow/cream, and red color patterns (Counterman et al. 2010; Joron et al. 2011). What are these genetic loci, and how does variation in each regulate so much diversity in wing patterns while at the same time promoting stable co-mimicry across the geographic ranges?

Robert Reed, Ricardo Papa, Owen McMillan, and colleagues (2011) made a remarkable discovery that provides the beginning of an answer. Variation in the expression pattern of a single homeobox transcription factor, called *optix*, accounts for variation in red color pattern in both *H. melpomene* and *H. erato* across their geographic ranges. Depending on the *optix* allele a butterfly carries, *optix* is expressed in different places on the wings during chrysalis development, the stage when color is determined. One can conceive of the expression of *optix* at this stage as similar to how an artist might make a pencil sketch on a canvas before executing a painting. In this way, the locations of *optix* transcription in the forming wing tissue of the chrysalis (as indicated by blue color in the right half of each of the image pairs in **Figure 19.13**) precisely matches the locations of red coloration in the adult wing (left half of each image pair). By contrast, *optix* expression zones

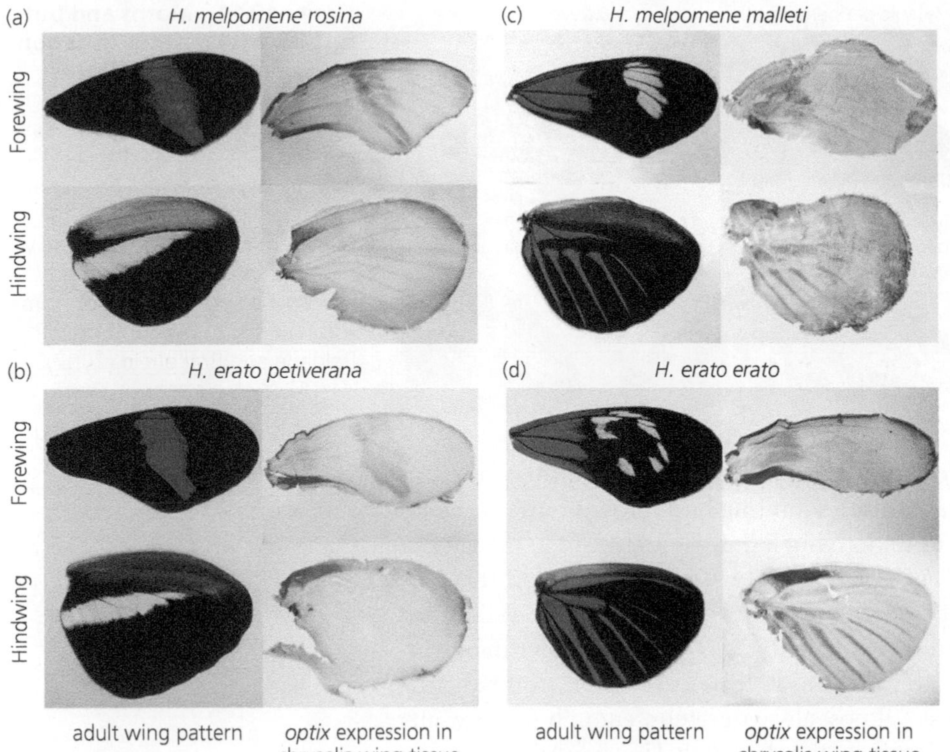

Figure 19.13 *Optix* expression patterns in chrysalis wing tissue predict where red color will form in the adult wing in both *H. melpomene* and *H. erato* (a) *H. m. rosina* and (b) its co-mimic *H. e. petiverana*. (c) *H. m. malleti* and (d) *H. e. erato*, a morph with a similar red color pattern. In each panel, the right side shows *optix* mRNA expression patterns revealed by in situ hybridization, a technique using a tagged RNA molecule synthesized to complement the nucleotide sequence of—and bind specifically to—*optix* mRNA. Wherever blue color is present in chrysalis wing tissue, the *optix* mRNA is also present. Note that the blue patterns in every right-side image match the red-colored regions in the corresponding adult wings (but not the black, yellow, or cream). From Reed et al. (2011).

in the chrysalis wing tissue do not predict the adult wing's black-, yellow-, or cream-colored regions; these map to different genetic loci (see below).

The correspondence of *optix* expression in pupal wing tissues and red color pattern in adult wings—in both species and across color morphs—seems an astonishing example of adaptive parallel evolution. But is it possible that all of the different color pattern alleles for optix were already present in the last common ancestor of *H. melpomene* and *H. erato*? If so, then the evolution of mimicry simply would have involved fixation of the same color pattern alleles in co-occurring populations of the two species, inherited in both lineages from their common ancestor. To distinguish between these "parallel evolution" and "ancient alleles" scenarios, Heather Hines and colleagues (2011) undertook a phylogenetic analysis of *optix* alleles from many populations in both species. The ancient alleles scenario would predict that co-mimics have similar or identical *optix* alleles. The parallel evolution scenario would predict unique *optix* alleles arising independently in the co-mimic pairs. Hines and colleagues' comparative analyses of *optix* sequences corresponding to the rayed wing-pattern phenotypes (for example, the rear wing patterns seen in Figure 19.13c and d) within and between species is inconsistent with the ancient alleles scenario. This is genuine parallel evolution.

Parallel evolution is the independent appearance of similar phenotypes via similar alterations of the same developmental mechanism.

The black color phenotype in *Heliconius* has been mapped to the *WntA* locus (Martin et al. 2012). Wnt proteins are secreted molecules with multiple functions in cell–cell signaling and developmental patterning, and they are active in many forms of cancer. Interestingly, the only other diffusible signaling molecule with a known function in animal color patterning comes from another *Wnt* gene called *wingless,* which is involved in butterfly wing patterning (Martin and Reed 2010) and also patterns of black spots on fruit fly wings (Werner et al. 2010).

The genetic identity of the third major *Heliconius* locus—the determinant of yellow- and cream-colored patterns—has not yet been identified, but it maps to

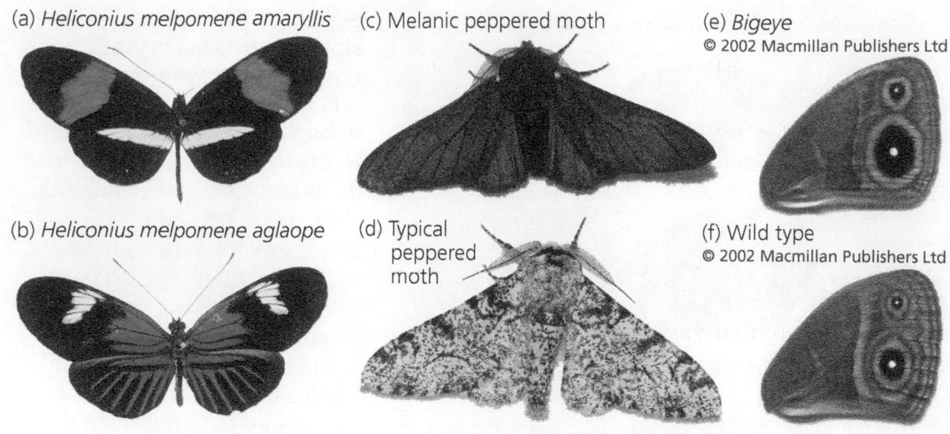

Figure 19.14 Moths and butterflies with pattern variation mapping to the same locus The gene or genes underlying cream and white color patterns in *Heliconius*—as in (a) and (b)—map to a single locus. The same locus may be responsible for melanism in *Biston betularia* (c vs. d), and the Bigeye mutant in *Bicyclus anynana* (e vs. f). (a, b) from Joron et al. (2006); (e, f) from Beldade and Brakefield (2002).

(e and f) Reprinted by permission of Macmillan Publishers Ltd: *Nature Reviews Genetics* 3: 442–452, copyright 2002.

the same genomic location as two wing-color pattern polymorphisms under active investigation: the *Bigeye* mutant in the squinting bush brown butterfly, *Bicyclus anynana;* and the *carbonaria* locus underlying industrial melanism in the famous British peppered moth, *Biston betularia* (**Figure 19.14**; Van't Hoff et al. 2011).

Taken together, what can we conclude from these findings on the evolutionary genetics of butterfly wing patterns? Allen's *Bicyclus* selection experiments indicate that some aspects of wing patterning show a bias in variation, contradicting Darwin's supposition of variation as "almost always present, enough to allow any amount of selected change." Furthermore, the data with *Heliconius* also points to some bias in evolutionary trajectories, because the same developmental-genetic mechanisms seem to underlie a startling array of butterfly wing-pattern variation, both within and among species.

Is it appropriate to think of these biases, or constraints, as evolutionary limitations? The spectacular examples of *Heliconius* mimicry certainly seem to suggest the opposite. Perhaps biases in developmental-genetic mechanisms actually favor the rapid and repeated evolution of pattern variations, making the course of adaptations in co-mimics more efficient and effective. In this conception, developmental bias or constraint is not a limitation at all, but instead an evolutionary opportunity for organisms to respond nimbly and rapidly to selection.

Pleiotropy and Developmental Trade-Offs

Another observation that follows from these and many other studies in evo-devo is that the same genes seem to function repeatedly at different times and places during development. This phenomenon is known as **pleiotropy.** Thus the *optix* gene has important functions in multiple tissues in insects including eye development and wing patterning. Hox genes pattern the main vertebrate body axis and also the limb proximal–distal axis. A plant homeobox gene called *KNOX* is involved in both primary and secondary leaf (or leaflet) patterning during multiple independent origins of compound leaf development in the orders Brassicales and Asterales. There are hundreds of similar examples in the literature.

An implication of such findings is that evolution involves reuse and repurposing of ancestral gene networks. A gene network can be thought of as a "food web" of genes with a complex, hierarchical series of interacting components, including feedbacks among levels of the hierarchy. In a simple food web, plants are eaten by grazers, which are eaten by predators, which eventually die and decay, feeding back as nutrients for the plants. In a gene network, environmental changes cause hormone release, which binds to transcription factors to activate a variety of genes, which carry out cellular functions that feed back to activate or repress release of the hormone. Different gene networks have evolved to carry out specific functions, such as setting up boundaries between regions of an embryo, or causing cells to move a certain way, or initiating an abcission layer in plant tissues. Such networks are modular and pleiotropic in the sense that they are used again and again within and among organisms across evolutionary time.

An assumption regarding pleiotropy is that selection for one function might limit or constrain selection on an alternative function. One of the big questions in evo-devo is, to what extent does pleiotropy in genes that belong to networks, or in entire gene networks themselves, limit potential variation? We do not yet know the answer, but it has significant implications for Darwin's concept of variation as "almost always present."

A concept related to pleiotropy is a **trade-off,** where one feature of an organism can be promoted only at the expense of another. This concept was discussed earlier in the book (see Chapter 10), but here we recast the issue as a question in evo-devo. Why are certain features of organisms traded off against others, while other features appear able to vary independently?

(a)

(b)

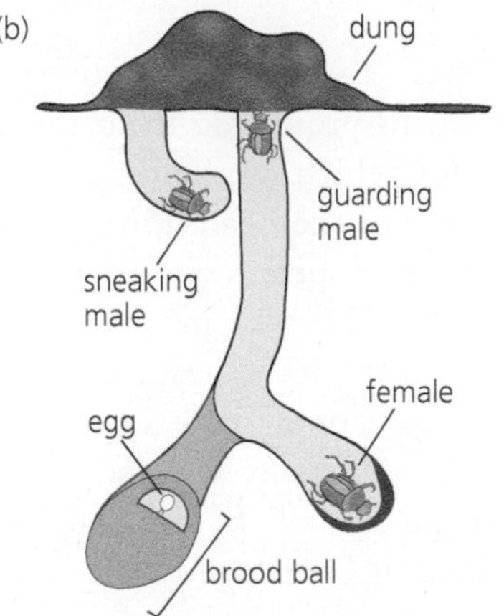

Figure 19.15 Alternate reproductive strategies in dung beetles (a) An *Onthophagus* dung beetle. (b) A female dung beetle digs a tunnel below a dung pile where she will lay her eggs, provisioning each with a ball of dung ("brood ball"). Her long-horned mate ("guarding male") guards the tunnel entrance, but small-horned males ("sneaking male") may dig side tunnels and surreptitiously mate with her. Redrawn from Emlen (2000).

Many groups of dung beetles are characterized by horns on the head or thorax of males, females, or both. The best-studied horned dung beetles are in the genus *Onthophagus* **(Figure 19.15a)**. In these beetles, the female buries dung balls to feed her larvae, and her large-horned mate uses his horns to guard the entrance to her tunnel, thus preventing other males from entering to mate with the female below. In many *Onthophagus* species, large-bodied males have relatively large horns, whereas small-bodied males have relatively small horns. Body size is determined by larval nutrition, so the size of horns in males is phenotypically plastic. A single genotype can produce either small or large horns, depending on its growth environment. This would seem to be disadvantageous to smaller males, but Doug Emlen (1997) discovered that hornless small males use an alternative mating strategy (Figure 19.15b): They dig their own tunnels to surreptitiously enter the burrow. The lack of horns allows the smaller males to dig tunnels that horned males cannot, because the horns would get in the way.

The consequences of large horns, and potential advantages of small horns, do not end there. Emlen (2001) identified trade-offs with not only horn size but also horn location in different *Onthophagus* species. For example, antenna size is negatively correlated with horn size, but only in species with horns projecting from the front of the head, not from the rear of the head or the thorax. A species

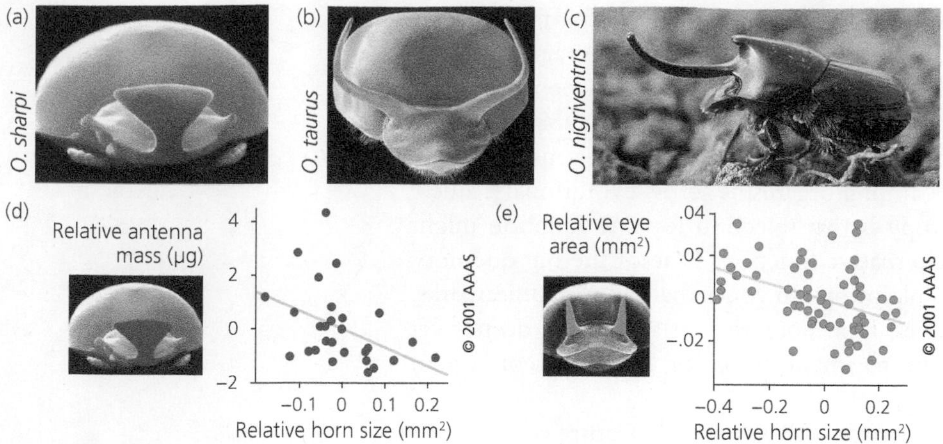

Figure 19.16 Developmental trade-offs in dung beetles (a, b, and c) Different species of *Onthophagus* have horns in different places. (d) In *O. sharpi*, the horn (colored red) is adjacent to the antennae, and the sizes of these structures are negatively correlated. (e) In this *Onthophagus* from Ecuador, eyes are adjacent to the paired horns (lavender), and horn size trades off with eye size. Photos by Doug Emlen; (d and e) from Emlen (2001).

(d, e) Emlen, D. J. 2001. "Costs and the Diversification of Exaggerated Animal Structures." *Science* 291: 1534–1536. Reprinted with permission of the AAAS.

with a horn on the front of its head, *O. sharpi,* appears in **Figure 19.16a**. One with horns on the back of its head, *O. taurus,* appears in Figure 19.16b. One with a horn projecting from the front of its thorax, *O. nigriventris,* appears in Figure 19.16c. Figure 19.16d documents the trade-off, in *O. sharpi,* between horn size and antenna size.

Emlen found other trade-offs as well. Eye size is negatively correlated with horn size, but only in species with horns projecting from the rear of the head (Figure 19.16e). Nocturnal species (for which large eyes aid vision under low light) tend not to have horns at the rear of the head. Finally, wing size is negatively correlated with horn size, but only in species with horns on the thorax.

In all of these cases, the horn size trade-off is with nearby structures: Eyes are at the rear of the head, antennae are at the front of the head, and wings emerge from the thorax. Are these merely correlations, or is there some mechanistic cause? To address this question, Emlen (1996) artificially selected *O. acuminatus* males, which have horns near their eyes at the rear of the head, for longer or shorter horns over seven generations. The resulting beetles showed a negative correlation with eye size. The increased horn size group had reduced eye size, while the decreased horn size group had increased eye size. Similar results are seen by giving unselected beetles juvenile hormone (JH) treatments during pupal development, the stage at which the size of these organs is determined. JH treatment results in increased horn size and decreased eye size at a given body size. JH levels are, likewise, known to differ between small and large males.

Taken together, these results suggest that some short-range signal or factor is involved in determining the size of organs in these beetles and that competition for a limited supply of this factor could explain the trade-offs. What could this factor be? One candidate is insulin-like growth factor, which may determine organ size in insects, is directly related to nutrition, and is regulated by insect juvenile hormone (Wu and Brown 2006; Emlen et al. 2012). In this conception, the growing horn tissue expresses high levels of insulin receptor, which acts as a sink, locally depleting circulating insulin and leaving less available for nearby tissues. The fact that these events occur during pupal development, when the organism is immobile, may explain how such processes of local depletion could occur.

Developmental trade-offs may arise because different body parts compete with each other.

However, not all beetle horn trade-offs are with nearby structures. Surgical removal of the developing genitalia results in increased horn size in adults, and gonad size also shows a negative correlation with horn size both within and among species (Moczek and Nijhout 2004). Therefore, we still await a definitive

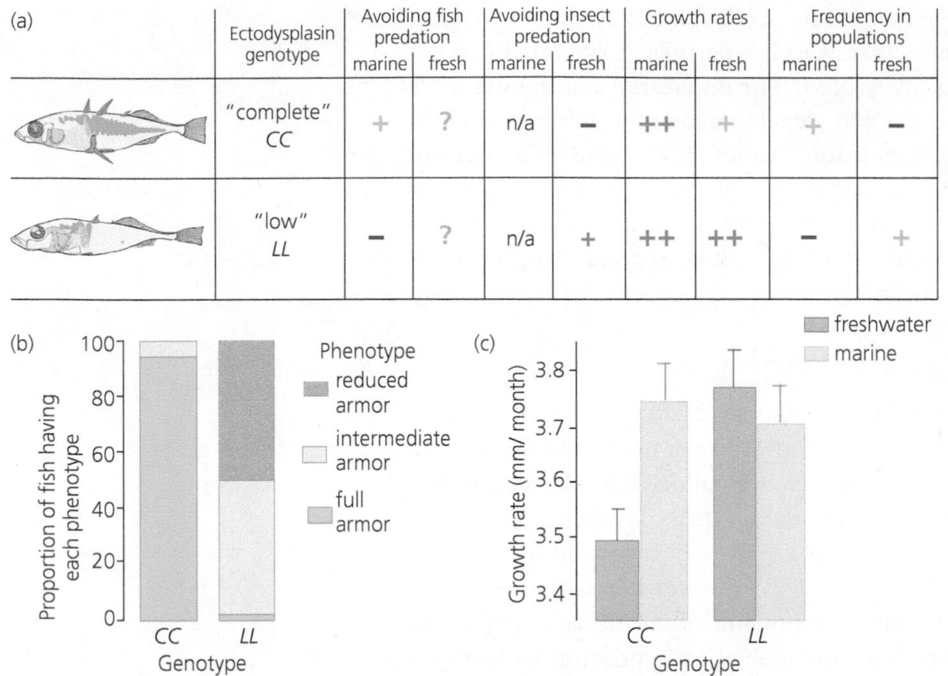

Figure 19.17 Pleiotropy in *Ectodysplasin* alleles in three-spine sticklebacks (a) Summary comparing the armored and reduced armor morphs, and their advantages and disadvantages in marine and freshwater habitats. The two *Ectodysplasin* alleles (*C*—complete; *L*—low) show partial dominance, but we present only data for homozygotes. Question marks indicate that the optimal phenotype depends on variable environmental factors, such as water clarity. (b) In the homozygous offspring of heterozygotes, *Ectodysplasin* genotype is closely related to phenotype (Barrett et al. 2009). (c) Reared in marine conditions, *CC* and *LL* fish grow at approximately the same rate. But in freshwater, *LL* fish grow faster. From Barrett et al. (2009).

mechanism to explain these trade-offs in beetles, if indeed there is a singular mechanism. In any case, the comparative results indicate that resource allocation trade-offs bias developmental as well as evolutionary trajectories.

Earlier, we introduced the concept of pleiotropy as related to trade-offs. Pleiotropy refers to multiple functions for the same gene within a single organism. The evolutionary implication is that there is a limitation to how specialized a gene can be for one function when it simultaneously has to perform another function. Such a situation appears in the threespine stickleback, *Gasterosteus aculeatus*.

Threespine stickleback populations are essentially one of two types. Fish in marine populations have life cycles similar to those of salmon: They live most of their lives in salt water, but swim into freshwater lakes and streams to reproduce. However, over the last several thousand years, various populations have come to spend their entire lives in freshwater. A notable feature of *G. aculeatus* from marine populations is their body armor: skeletal plates that offer protection from predation by other fish **(Figure 19.17a, upper row)**. In most freshwater populations, by contrast, the major predators are aquatic insects like dragonfly larvae, which rely on agility rather than crushing strength to capture juvenile sticklebacks (Marchinko 2009). Freshwater *G. aculeatus* populations have reduced body armor and instead can grow faster to their adult stage, where they are no longer subject to dragonfly predation (Figure 19.17a, lower row). These reduced-armor freshwater fish also show increased burst swimming speeds (Bergstrom 2002).

The connection to pleiotropy comes from evidence suggesting that changes in a single gene, *Ectodysplasin,* can account for both the reduction in body armor and the increase in growth rates in freshwater populations (Figure 19.17b, c; Barrett et al. 2008, 2009). This scenario suggests that body armor loss in freshwater populations is not merely a passive process akin to what Darwin and Wallace referred to as loss of features from "disuse." Instead, the multiple functions of *Ectodysplasin* seem to actively promote body armor loss in freshwater populations via an automatic and pleiotropic growth advantage.

Developmental trade-offs may also arise because of pleiotropy—the involvement of a gene in the development of traits.

In sum, we have seen evidence—ranging from selection experiments to comparative biology to genome analyses—that Darwin's postulate that variation is ever-present and omnidirectional was overstated. The implications of this evidence are significant, since it suggests that internal, developmental features of organisms guide evolution hand in hand with selection. Although the evidence does not topple the Darwinian pillar of the primacy of natural selection, it seems to validate a substantial modification of the concept. Next, we will evaluate whether findings in evo-devo may likewise call for restructuring a second Darwinian pillar: that evolution occurs only in small steps.

Nature Sometimes Makes Leaps

In *The Origin of Species* (1872, p. 156), Darwin wrote: "Natural selection acts only by taking advantage of slight successive variations; she can never take a great and sudden leap." This notion of gradual change was a hallmark of Darwinism and arguably the main organizing principle of the evolutionary synthesis in the early 20th century. Has this paradigm of gradual, continuous change held up?

The opposing concept of leaps in ("saltational") evolution has an uneven history in evolution and development. Most proponents of saltational evolution in the late 19th and early 20th centuries set themselves in opposition to Darwinian evolution. For example, Richard Goldschmidt's name evokes, in the minds of most evolutionary biologists, the concept of "hopeful monsters," where major mutational changes explain the origin of species, while microevolutionary changes below the species level are irrelevant to the evolution of life's diversity. Less well known is that Goldschmidt also believed that small changes in early development might propagate through ontogeny to yield large effects on the adult phenotype (Goldschmidt 1940), a notion not far outside the current orthodoxy.

William Bateson (1894, pp. 410–411), whom we met when discussing homeotic mutations, made a simple yet elegant point in the debate on continuous versus discontinuous evolution. Bateson noted that the antennae of long-horned beetles typically have 11 segments, and asked how 12-segmented antennae could arise via gradual acquisition of a new joint. "With evidence that transitions of this nature may be discontinuously effected," he noted, "the difficulty is removed."

Bateson thus demonstrated that evolution can proceed in leaps, in this case via the addition of segments, whether in the antennal segments of long-horned beetles or in the body-segment numbers of centipedes—which curiously are always an odd number, so must proceed in leaps of two. Another example is the direction of spiraling of shells (dextral versus sinistral): There are only two options, so any transition between the two is a leap. Therefore the question for us is not Does nature proceed in leaps? because it clearly does; the questions are, How big are the leaps? How often do they occur? and How do they occur?

Perhaps the most straightforward example of evolution by leaps is cross-species hybridization in plants, where pollen from one species fertilizes the ovum of a another, yielding a potential third species. Rapeseed (*Brassica napus*) offers an example. Rapeseed is the third most important oil crop in the world; canola is one variety. Rapeseed originated from a hybridization between wild cabbage (*B. oleracea*) and wild turnip (*B. rapa*) around the Middle Ages in Europe (Gupta and Pratap 2007). This event has been intentionally replicated many times by biologists. **Figure 19.18a** shows an individual from one such newly synthesized *B. napus* lineage, posed between plants from each of the parental species. Newly synthesized *B. napus* lineages show considerable genetic and phenotypic variation (Pires

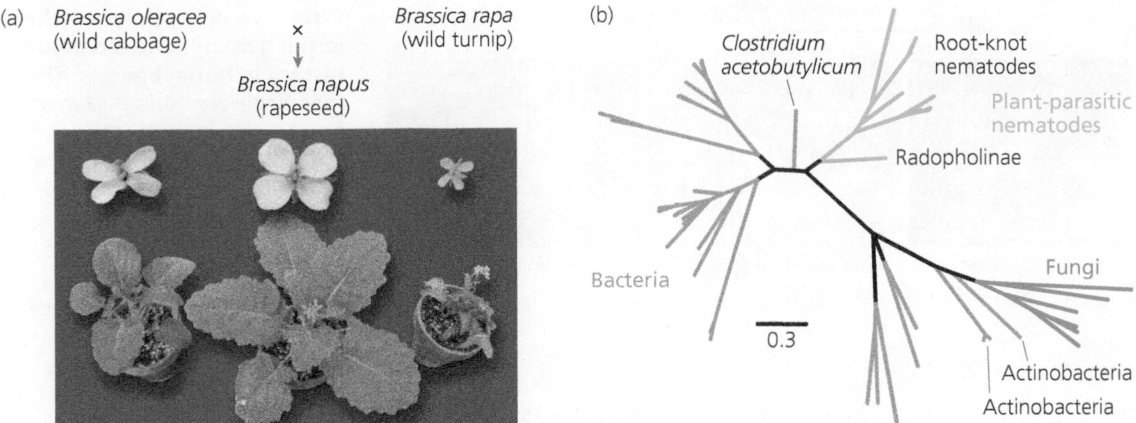

Figure 19.18 Evolutionary leaps (a) A newly synthesized rapeseed, the offspring of a wild cabbage and a wild turnip. Note the vigorous growth of the hybrid versus the parents. Photo by J. Chris Pires. (b) Unrooted phylogeny of hemicellulose-digesting xylanase genes in a variety of organisms. The xylanases of nematode worms that parasitize plants (green) branch from within the bacterial genes and are not closely related to the xylanases of the other eukaryotes known to make the enzyme, the fungi (blue). From Danchin et al. (2010).

et al. 2004; Gaeta et al. 2007). Warren Albertin and colleagues (2006, 2007) re-hybridized *B. oleracea* and *B. rapa* and examined the expression of over 1,600 stem and root proteins in the *B. napus* hybrid offspring. Many of the proteins showed non-additive effects, such as quantitative expression outside the range of both parents, and none of the proteins showed mis-expression or other obvious defects in gene regulation. These results suggest that hybrids with unique features can form stably in a single generation.

A second mechanism for evolutionary leaps is horizontal gene transfer, in which foreign DNA integrates stably into a new genome. Horizontal gene transfer is common in microbes, and the evidence for its importance in other kinds of organisms is growing (see Dunning Hotopp 2011). Striking examples are found in plant-parasitic roundworms, whose genomes encode cellulases and other cell-wall-degrading enzymes that aid the worms in exploiting their hosts. Phylogenetic analyses reveal that the genes for several of these enzymes, including the xylanases depicted in Figure 19.18b, came from bacteria (Danchin et al. 2010). Because horizontal gene transfers are all-or-none phenomena—the foreign gene is either integrated into the genome or not—they represent evolutionary jumps.

Via a variety of mechanisms, evolutionary change sometimes proceeds in discontinuous bursts.

The movement of genetic elements within genomes can also cause evolutionary leaps. Transposable elements, or transposons, are widespread across organisms and can increase rates of evolution by elevating mutation rates. Increased mutation rates can be harmful, and eukaryotic and prokaryotic genomes have mechanisms to suppress mobility of transposons. Nevertheless, genome sequencing has revealed that such immobilized transposons can subsequently perform important cellular functions. In one example, a transposon specific to tetrapods and our lobe-finned fish ancestors is found in multiple places in the genome and has acquired key functions, including regulation of a homeobox gene involved in neural development (Bejerano et al. 2006). In a follow-up study, Lindblad-Toh and colleagues (2011) analyzed noncoding DNA in 29 mammals with fully sequenced genomes, focusing on those sequences that showed a signature for positive selection. They determined that about 20% of such sequences are immobilized transposons. Lowe and colleagues (2007) found that such sequences showed preferential association with developmental regulatory genes.

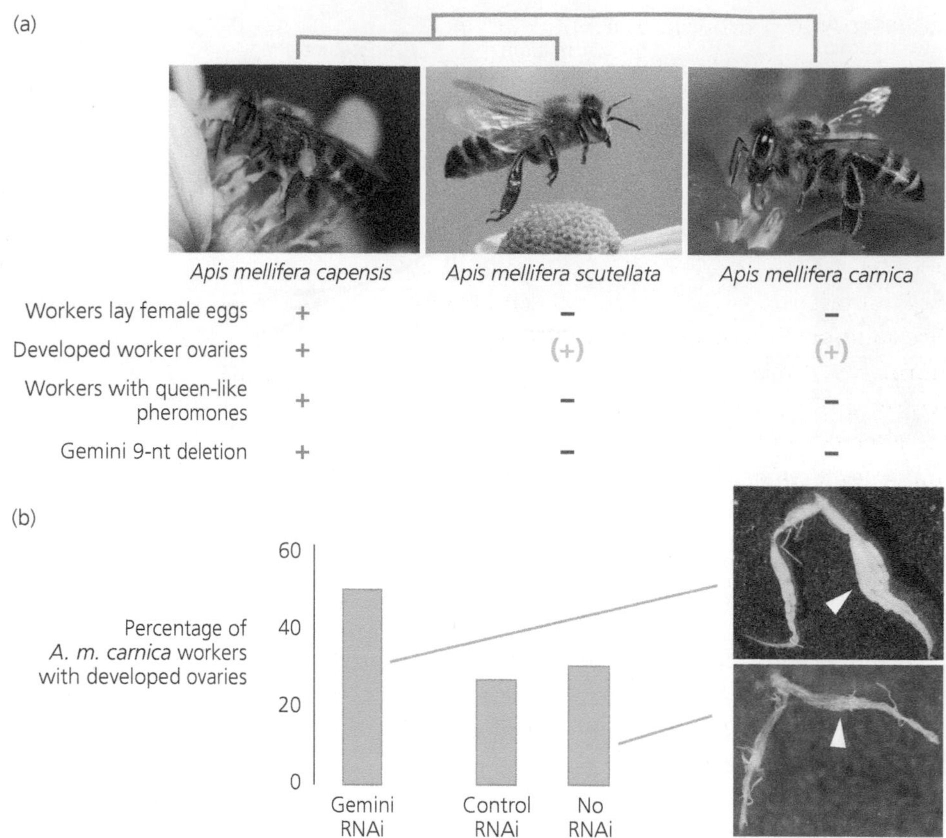

Figure 19.19 A polymorphism in the gemini gene alters life history in honeybee (a) The cape honeybee (*Apis mellifera capensis*) from South Africa's Cape of Good Hope, its neighbor and closest relative (*A. m. scutellata*), and the eastern European honeybee (*A. m. carnica*). The cape honeybee's unique traits are associated with a 9-nucleotide deletion in the gemini gene. (b) Results of feeding recently emerged *A. m. carnica* workers sugar water with: an RNA molecule designed to mimic the gemini deletion (gemini RNAi), an RNA with a scrambled sequence (control RNAi), or just sugar water (no RNAi). Images at right show a maturing ovary with active oogenesis (arrowhead pointing to bulge, top) compared to a non-maturing ovary with no active oogenesis (arrowhead, bottom). From Jarosch et al. (2011).

An additional challenge to Darwin's concept of evolution by slow steps is the finding that single alleles can have major life-history and population-level effects. *Optix* in *Heliconius* butterflies and *Ectodysplasin* in sticklebacks are two examples. A third involves the cape honeybee (*Apis mellifera capensis;* **Figure 19.19a**) from South Africa. In most honeybee colonies, a single queen reproduces. If the queen dies, the workers can lay eggs; but because the eggs are unfertilized, they yield only male (drone) offspring; the workers cannot re-queen the colony. This is due to bees' haplodiploid sex determination. Fertilized diploid eggs are female; unfertilized haploid eggs are male. The cape honeybee is an exception. Its female workers can asexually produce female eggs and hence re-queen a colony. When such a reproductively active *A. m. capensis* worker enters a colony of the ecologically dominant African honeybee (*A. m. scutellata;* Figure 19.19a), she starts laying female-determined eggs, and the local African honeybee workers treat her like a queen. The result is chaos: The cape honeybee's daughters themselves become pseudo-queens, and the colony collapses in disarray. Beekeepers have unwittingly spread this *A. m. capensis* social parasitism throughout southern Africa.

The genetic basis for these differences in reproductive life history is found in a gene orthologous to the *Drosophila* gene *gemini,* which encodes a CP2 family transcription factor involved in genital development and egg production (Jarosch et al. 2011). The *gemini* allele in *A. m. capensis* has a 9-nucleotide deletion, which results in a change in the protein products. All other African honeybee races and European honeybees so far examined have those nine nucleotides intact. The deletion seems to give honeybee workers more developed ovaries, the ability to give birth to queens, and a queen-like cuticle pheromone profile. Antje Jarosch and colleagues (2011) tested this hypothesis by feeding an RNA molecule designed to mimic the effects of the deletion to European honeybees (*A. m. carnica*). Compared to controls, treated bees showed increased ovary development (Figure 19.19b). We thus have a case in which a 9-nucleotide deletion apparently changed the life history of the subspecies in which it arose, spread through wild populations, and altered their dynamics, producing ecological disruption.

The reason a single gene—*gemini*—is thought to have such a wide range of effects on bee reproduction, physiology, and behavior is that it (like the Hox genes and *optix*) encodes a transcription factor, which itself interacts with and regulates numerous other genes. Several other classes of genes also can have effects on multiple other genes and are additional candidates for evolution in leaps. Among them are morphogenetic hormones, which are known to orchestrate animal and plant life histories. We discussed abscisic acid (ABA) in mangroves earlier in the chapter. Other cases include alterations in thyroid hormone metabolism or expression underlying the evolution of alternate life histories in salamanders, frogs, and sea urchins; changes in ecdysteroid cellular responses underlying the evolution of larval reproduction in flies; and changes in juvenile hormone metabolism underlying many aspects of insect evolution, such as seasonal morphs in butterflies, horn morphology and mating strategies in dung beetles, and worker and soldier caste difference in ants. Such hormones are likely targets for evolution because small changes in their timing or mode of action can have profound effects on the timing of life cycles (heterochrony) and their morphological and behavioral outcomes (reviewed in Heyland et al. 2005).

Among the mechanisms for evolutionary leaps is mutation in the genes encoding transcription factors.

In sum, while we cannot say for certain how frequent are jumps in evolution, we can be confident that they are neither impossible nor necessarily rare.

The "5 Percent of a Wing Problem" and the Evo-Devo Solution

Although we have shown how findings in evo-devo have led to revision of some tenets of the modern synthesis, most work in the field has confirmed the basic concepts of descent with modification and the mechanisms of evolution. Indeed, some of the most important discoveries in evo-devo have helped address a persistent mystery in evolution, one that Stephen J. Gould (2002, p. 1220) called the "5 percent of a wing" problem: "How can evolution ever make a wing in Darwin's gradualist and adaptationist mode if five percent of a wing can't possibly provide any benefit for flight?" As with so many other puzzles in evolution, Darwin himself (1872, p. 148) offered a key suggestion: "Bear in mind the probability of conversion from one function to another." Gould and Elizabeth Vrba (1982) coined the term **exaptation** to describe such conversions.

Novel structures sometimes evolve via change in function.

The proposed solution regarding wings is that the original small "proto-wings" were either adaptive for another function or even nonadaptive, but were not for flying. Their usefulness in flight came later as an exaptation. Joel Kingsolver and Mimi Koehl (1985) tested the hypothesis that the proto-wings of insects functioned in thermoregulation by building physical models on which they could vary the size of the wings at will. They found that even the slightest increase in wing size improves a model insect's ability to regulate its body temperature. Once these proto-wings get big enough, Kingsolver and Koehl's experimental models suggest that they begin to provide an aerodynamic function.

Although plausible, such conclusions remain tentative. We cannot go back in time and recreate the evolutionary history of wings. Nevertheless, the tools of evo-devo have allowed researchers to rigorously test some exaptation hypotheses, thus lending support to Darwin's solution to the 5 percent of a wing problem.

We can, for example, identify exaptations in protein function with some confidence. If 120 of 125 amino acids in two proteins are identical, likely the two are evolutionarily related—by either orthology or paralogy. With phylogenetic analysis and parsimony arguments, we can assess whether the function of a given protein has changed, and in what direction. An example appears in **Figure 19.20**.

Lenses are a common feature of complex eyes, including the independently evolved eyes of cephalopods and vertebrates. In both groups, the lenses are made of long-lived cells that lack a nucleus and most other organelles and whose contents are transparent and stable. The major structural proteins of lens cells are called crystallins. Many of these crystallins are well-known functional enzymes involved in basic metabolism; they are merely enriched in the lens (Wistow and Piatgorsky 1987). So, for example, the major crystallin in bird lenses is also a functioning urea cycle enzyme in the liver, and the major crystallin in elephant shrews is also an alcohol detoxifying enzyme—aldehyde dehydrogenase.

Neither of these enzyme functions, which are ancient and hence predate the evolution of eyes, are relevant when the enzymes are expressed at such high levels in the lens. These genes were presumably exapted due to their solubility at high concentrations, optical transparency, and longevity. In some taxa, the enzyme crystallin genes have duplicated. One daughter gene specializes in the lens function (and has lost enzyme activity) while the other gene continues to perform the original enzymatic function. In this sense exaptation, followed by gene duplication and functional divergence, may be a common mechanism by which new protein functions are acquired in evolution.

Although the lenses of vertebrates and invertebrates evolved independently, the major crystallin in both scallops and elephant shrews is an aldehyde dehydrogenase (Figure 19.20; Graham et al. 1996; Piatigorsky et al. 2000). Is this just happenstance? Probably not. The total set of possible proteins that fulfill all of the functional requirements for lens crystallin function (solubility, transparency, longevity) is a fraction of the total diversity of proteins available. It would therefore be expected for evolution to repeat itself now and again.

This finding raises a recurring issue in research on development and evolution. If we see a similarity between two organisms in some aspect of development, how can we decide if that similarity is due to homology or homoplasy? This choice was straightforward for aldehyde dehydrogenase crystallins. The original function was clear, and the vast phylogenetic separation of shrews and scallops makes their independent origins all but certain. But what about other cases? Comparisons of vertebrates and invertebrates (often mouse and fly) reveal that similar genes are involved in heart development, appendage development, anterior–posterior and dorsal–ventral axis development, and eye development. These are basic processes found in many groups of animals. How can we determine their evolutionary histories, and hence judge between homology and homoplasy?

Eric Davidson (2001, pp. 189–190) notes that although the heads, hearts, appendages, and eyes in vertebrates and invertebrates look superficially similar, their anatomy and underlying developmental processes are quite different. Nonetheless, "Over and over the same transcriptional regulators are found to be used for what appear at least externally to be similar purposes." We saw this in our discussion of the Hox paradox. It is perhaps the most unexpected finding of modern evo-devo. Anatomical structures that are classical examples of homoplasy—such as the octopus and the human eye—in fact use similar regulatory genes during their development.

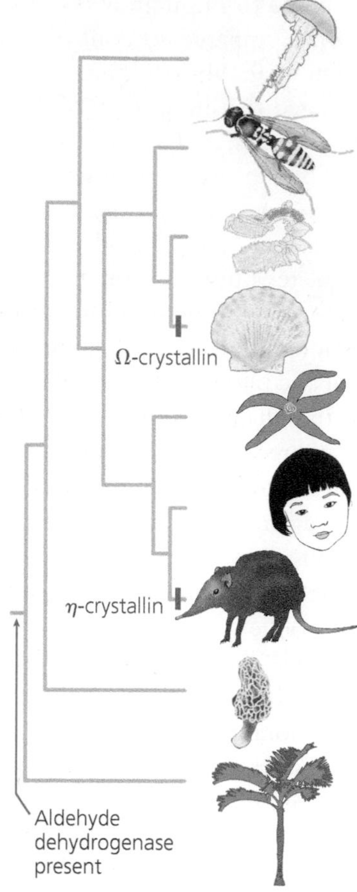

Figure 19.20 Ancestral and derived functions of proteins Aldehyde dehydrogenases are ancient enzymes that catalyze the oxidation of aldehydes in archaea, bacteria, and eukaryotes; their sequence similarities make it clear that they are evolutionarily related. Related (but not orthologous) aldehyde dehydrogenases have been independently recruited as major structural proteins—lens crystallins—in two divergent groups of animals: scallops and elephant shrews. In vertebrates, aldehyde dehydrogenases are enriched in the eyes, which has probably predisposed them for co-option as a lens crystallin in elephant shrews.

Davidson's resolution to this quandary inolves exaptation (emphasis added):

> However they are structured, brains must deploy neuronal differentiation programs, hearts need certain kinds of contractile cells; eyes need photo-receptor cells.... So a possible solution to our paradox is that the regulatory genes which we find [for example in insect and vertebrate hearts] originally ran the differentiation gene batteries required for [heart] functions, and since these genes were expressed in the right place *they could be coopted during evolution to produce successively more elaborate pattern formation functions, differently in each clade.*

When Davidson speculates about the original functions of the regulatory genes in question, he is imagining their functions in the last common ancestor of flies and vertebrates, an organism that must have lived more than half a billion years ago. One possibility is that this ancient ancestor had a rudimentary heart, and the development of that ancestral heart was regulated by the regulatory genes in question. This first scenario suggests that despite the anatomical differences and different embryonic origins, vertebrate and insect hearts are in fact homologous; we can call this the "ancient heart" scenario.

Davidson suggests an alternative possibility, one that does not require overturning the classical concept that insect and vertebrate hearts evolved independently. While we do not know if the insect-vertebrate ancestor had a heart, we can predict that it had structures that undergo rhythmic, pulsatile contractions such as gut peristalsis; such functionality is found not only in the diverse descendants of this ancestor, but also in the rhythmically contractile structures in more ancient animal lineages such as jellyfish, sea anemones, and possibly even sponges.

Therefore, Davidson's alternative scenario suggests that the original function of these regulatory genes in the insect-vertebrate ancestor was a basic function in rhythmic contractility. Then, independently in vertebrates and insects, hearts evolved and came under the control of the same generic regulator of contractility. In other words, just as in the scallop and elephant shrew lens, the similar gene regulation of insect and vertebrate hearts would have evolved in parallel.

To distinguish the ancient heart versus parallel evolution scenarios, we need to look more deeply at vertebrate and insect hearts and the genes known to control their development. A key homeobox-class transcriptional regulator underlying vertebrate heart cell specification is *Nkx-2.5,* a gene in the NK4 class. This gene is expressed in mesoderm that gives rise to the heart as well as associated gut cells. *Nkx-2.5* mutants in mice actually start forming a normal heart, but defects arise later in the heart tube. Remarkably, one of the NK4 family genes in *Drosophila* fruit flies, *tinman,* is also involved in fruit fly heart formation, though in *tinman* mutant flies, the heart does not form at all (see Olson 2006 for a review).

Involvement of NK4 genes in heart development is not the only such similarity. Like most transcriptional regulators, NK4 genes are part of regulatory gene networks, which in the case of *Nkx-2.5* in vertebrates include genes for proteins such as two transcription factors known as MEF2 and GATA and a signaling molecule from the BMP family. Orthologs of these genes function in insect heart development as well. Furthermore, hearts are often associated more broadly with branched vascular structures, and vascular endothelial growth factor receptor (VEGFR) is used in vascular development in both insects and vertebrates. Indeed, family members of some of these same genes (NK4, MEF2, VEGFR) are expressed during the development of the heart, or heart-like organs, and the associated vasculature, in other animal groups as well. These groups include squids,

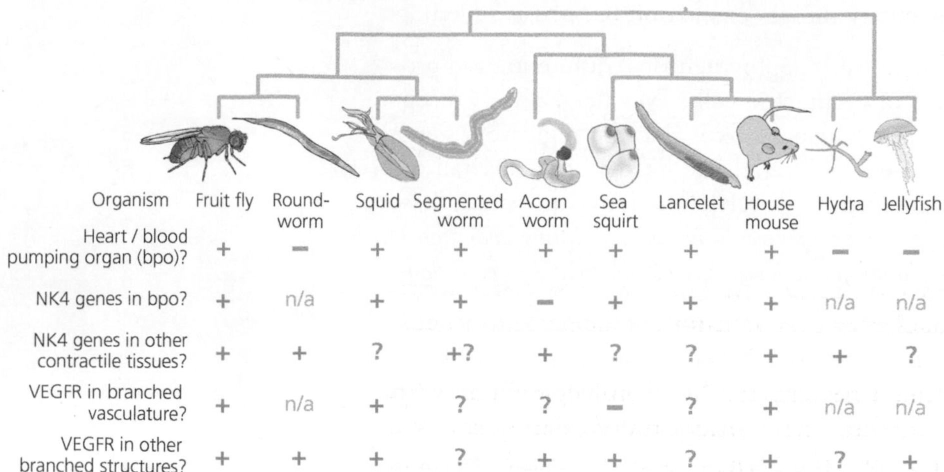

Organism	Fruit fly	Roundworm	Squid	Segmented worm	Acorn worm	Sea squirt	Lancelet	House mouse	Hydra	Jellyfish
Heart / blood pumping organ (bpo)?	+	−	+	+	+	+	+	+	−	−
NK4 genes in bpo?	+	n/a	+	+	−	+	+	+	n/a	n/a
NK4 genes in other contractile tissues?	+	+	?	+?	+	?	?	+	+	?
VEGFR in branched vasculature?	+	n/a	+	?	?	−	?	+	n/a	n/a
VEGFR in other branched structures?	+	+	+	?	+	+	?	+	?	+

Figure 19.21 NK4 and VEGFR: ancient heart or parallel evolution? Diagram shows relationships among animals with and without hearts or blood pumping organs (bpo), plus the correlation of NK4 and VEGFR genes with bpo and associated vasculature, and their underlying developmental processes. In all examined cases, NK4 expression is associated broadly with contractile structures, and VEGFR expression with diverse branching structures, from neuronal arborizations to cellular fillopodial extensions.

lancelets, and annelids (**Figure 19.21**; Yoshida et al. 2010)—all of which share the same ancestor that insects share with vertebrates.

All of the animals just mentioned have hearts or blood-pumping organs: rhythmically contractile structures that drive blood circulation in circulatory systems that are open (insect) or closed (lancelets, squid, annelid, vertebrate). Davidson's parallel evolution scenario suggests that the commonality is the ancient regulation of rhythmic contractility. Can we imagine evidence in modern organisms that would allow us to confirm or refute the parallel evolution scenario?

The roundworm *Caenorhabditis elegans* is another descendant of the same insect-vertebrate ancestor. It has no heart, but it does have an NK4 gene that functions in the development of the pharynx—a rhythmically contractile structure involved in digestion (Okkema et al. 1997). The hemichordate acorn worm has a heart-like contractile cardiac vesicle; its NK4 gene is not expressed there, but is again associated with pharynx development (Lowe et al. 2006). Still, if we want to infer the original function of NK4 in the insect-vertebrate ("bilaterian") ancestor, it would be best to have data on groups that diverged before the bilaterian ancestor appeared. Such data exist for the cnidarian *Hydra magnipapillata. Hydra* has no heart, but once again, its NK4-class gene is expressed in a contractile pharynx-like structure near the base of the stalk (Shimizu and Fujisawa 2003).

What about VEGFR? Its expression is also known from a jellyfish *Podocoryne carnea.* Jellyfish VEGFR is expressed in the branches of the digestive system (which also serves as a kind of vascular system) that extend into the tentacles. In roundworms, VEGFR is involved in chemosensory neuron branching. This unexpected finding has prompted a search for additional functions of VEGFR in other animals, and VEGFR orthologs have been found to be involved in neuronal path finding in vertebrates, in the branched circulatory system connecting individuals in a colonial sea squirt, and in tubular extensions in specialized "border cells" in fruit fly oocytes (reviewed in Ponnambalam and Alberghina 2011).

Data on more phyla with and without hearts would be useful, but based on available information (Figure 19.21) it seems likely that NK4 genes originally were involved in development of a rhythmically contractile structure, like the pharynx, and later exapted for the development of hearts and heart-like pumping organs. The lack of expression of NK4 in the simple hemichordate heart-like organ could be explained by loss of NK4 from heart-like development in hemichordates, or it could indicate the independent exaptation of NK4 genes for heart development in protostomes and deuterostomes. Data on NK4 and its network from a greater diversity of animals are needed. In particular, examining comb jellies (phylum Ctenophora) would be edifying. Like cnidarians, comb jellies evolved before the bilaterian ancestor. However, unlike cnidarians, comb jellies have more extensive mesoderm-like structures, and bilaterian hearts and blood pumps are mesodermal. As for VEGFR, the common denominator of all the invertebrate and vertebrate data to date is that this signaling system is specialized for formation of highly branched structures, whether neuronal, gastric, or vascular.

Thus the emerging details of the development and evolution of animal hearts is consistent with Davidson's parallel evolution scenario for similar gene regulation underlying dissimilar structures.

Can such findings help clarify the 5 percent of a wing problem? Remember that a transcription factor merely regulates the transcription of other genes. It is like the foreman of a construction crew. The foreman does not build anything, but coordinates the work of the carpenters, plumbers, and electricians. If we need a new building, we just contact the foreman—who brings along the whole team. If the parallel evolution scenario is correct, then in the course of evolution organisms did not have to completely reinvent the process of forming a fluid-pumping organ or a highly branched structure in every case. They simply may have recruited the NK gene, which brought along a gene network for constructing the pumping organ, and the VEGFR gene, which brought along a network for constructing branched structures. The exaptation of preexisting gene regulatory networks is an efficient way to evolve a complex structure.

Evolutionary changes in the function of regulatory gene networks is one explanation for the deeper homology of developmental mechanisms versus the structures they control.

These examples remind us that when discussing homology and homoplasy, we have to be clear about the hierarchical level. The comparative data on NK4 genes do not indicate that the roundworm pharynx is homologous to the vertebrate heart. They merely suggest that homologous transcription factors (and likely their associated gene regulatory networks) are used in the processes that underlie the formation of two similar, though nonhomologous fluid-pumping organs.

19.4 Hox Redux: Homology or Homoplasy?

Recall that we left our earlier discussion of Hox genes with the observation that the two principal commonalities across most bilaterally symmetric animals were spatial colinearity of Hox gene expression patterns along the main body axis and association of Hox genes with the nervous system. We also noted that in several different animal groups (sea anemones, lancelets, vertebrates, insects), expansions in the Hox gene cluster resulted in newly evolved genes that maintained spatial colinearity. This pattern suggests that spatial colinearity is a generic outcome of of Hox gene clustering, a finding that has gained support in observations of the crystal structures of Hox clusters in different parts of the mouse embryo undergoing active transcription (Noordermeer et al. 2011).

Nevertheless, we described many examples where inverted, split, or even atomized clusters retained the canonical spatial expression pattern. These latter findings suggest that additional buffering mechanisms, unrelated to the clustering per se, have evolved repeatedly across animals to ensure proper spatial expression of these genes. And finally, the surprising absence of any significant function in sea squirt Hox genes indicates that the highly buffered expression patterns of Hox genes may be more fundamental than the axial functions themselves.

Furthermore, the canonical spatial expression patterns of Hox genes are found in three cases that do not involve anterior–posterior axial patterning: in a coelomic compartment in sea urchins and sea lilies (Echinodermata); along the dorsal–ventral axis in sea anemones (Cnidaria); and in vertebrates along the limb axis, in the urogenital system, and in the gut. The vertebrate and echinoderm examples seem to be exaptations and suggest that Hox genes are ideal candidates to co-opt for regionalization along an alternate axis or an internal structure.

We are now equipped to address one of the hypotheses of Hox gene evolution: Does the strikingly similar expression and even function of Hox genes in conferring segmental identity in arthropods and chordates suggest that segmentation itself is homologous in these two groups, and thus that the common ancestor of the Bilateria was segmented? Instead of relying on broad, cross-phylum comparisons, we can address this question by looking at more closely related groups.

Although the definition of segmentation is disputed, a common one is "internal and external repetition of body structures and organs along the main body axis." Velvet worms are now accepted as the closest living relatives to the arthropods; but unlike arthropods, velvet worms are not segmented according to this definition. Their morphology suggests two alternative scenarios. Either the arthropod-velvet worm ancestor was not segmented, and segmentation arose during the early evolution of arthropods, or the arthropod-velvet worm ancestor was segmented, and segmentation was lost in the velvet worm lineage.

Often when a character is lost it leaves a remnant, some trace of its existence—like the hindlimb bones in a whale. We now have evidence of two cases where segmentation was lost in animals. Two groups of unsegmented worms—spoon worms (echiurans) and peanut worms (sipunculans)—appear to be derived from within the phylum of segmented annelid worms (Struck et al. 2007). If so, these two lineages must have lost segmentation sometime in their evolutionary history. Indeed, though they show no external signs of segmentation, the embryonic nervous systems of spoon worms and peanut worms are still segmented (Hessling 2002, 2003; Kristof et al. 2008): a clear vestige of their segmental past **(Figure 19.22a, b)**.

What about velvet worms? Unlike spoon and peanut worms, velvet worms show virtually no indication of segmentation in their developing nervous system, musculature, or other internal structures (Figure 19.22c; Mayer and Whitington 2009; Whitington and Mayer 2011). This evidence supports the notion that the arthropod-velvet worm ancestor was not segmented, and thus that segmentation arose independently in arthropods and vertebrates, as well as annelids.

What are the implications for Hox gene evolution? If segmentation in arthropods and chordates is an example of homoplasy, then the similar function of Hox genes in regulating segmental identity in vertebrates and insects is an example of parallel evolution, perhaps exapted from a regionalized expression of Hox genes in the central nervous system of some worm-like bilaterian ancestor.

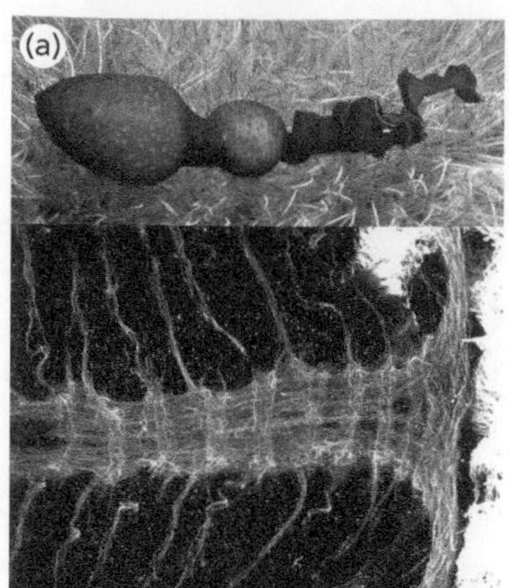

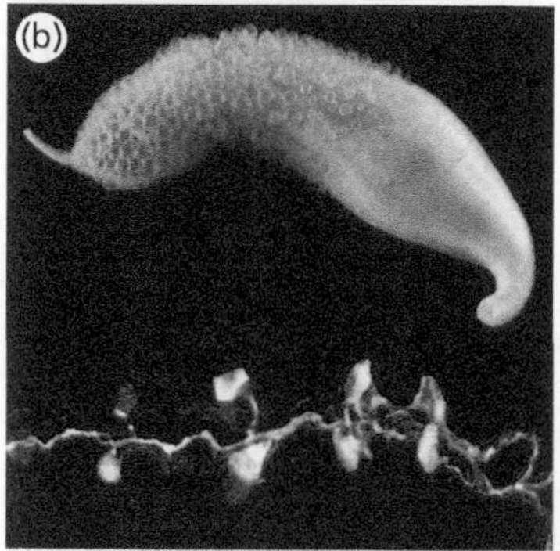

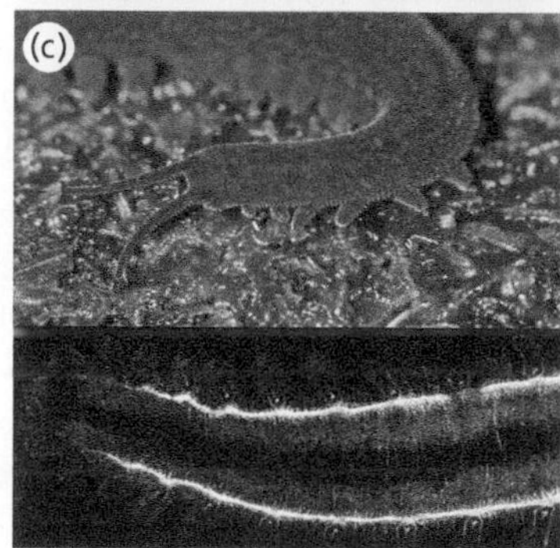

Figure 19.22 Vestigial nervous system segmentation In (a) spoon worms (echiurans) and (b) peanut worms (sipunculans), close examination of the nervous systems reveals vestiges of segmentation. Not so with (c) velvet worms (onycophora), which are also unsegmented but have repetitive appendages along their body. Examination of their nervous system reveals no clear vestiges of segmentation. Nervous system photo (a) by Rene Hessling; see Hessling (2002, 2003). Nervous system photos (b) and (c) from Kristof et al. (2008) and Mayer and Whitington (2009).

19.5 The Future of Evo-Devo

Recent decades have been exciting for evo-devo. We have gained profound insights into evolution through the application of developmental biology approaches and techniques. In particular, insights into evolution have come from studying the development of an ever-widening range of organisms. Less common has been the adoption of explicit comparative approaches, where evolutionary questions are framed and studied with carefully chosen taxa and independent contrasts, so that the generality of the conclusions can be assessed.

Thus far only two of the great multicellular taxa, animals and plants, have been studied to any appreciable degree. Although some single-celled organisms can be said to undergo a type of development during their life cycle, each appearance of multicellularity clearly involves the unique origin of a higher level of developmental complexity. But multicellularity, and hence complex developmental processes, arose only once each in plants and animals, so we are at risk of over-concentrating on provincial aspects of these two developmental systems. For full appreciation of how development shapes evolution, we need to explore the other great multicellular taxa—kelp, fungi, red algae, and green algae—some of which, themselves, show multiple origins of multicellularity and hence complex developmental processes. Because these poorly studied multicellular groups are important in global ecosystems, we have additional impetus to expand our horizons to include these taxa.

The integration of fields should extend beyond development and evolution. Although much of evo-devo focuses on gene function, few practitioners are well trained in biochemistry. To fully understand developmental biases in evolution, we need to better incorporate biochemistry. Furthermore, the rapid explosion in sequence information has spawned great advances in systems biology and network modeling. Ultimately, our understanding of organic evolution will have to synthesize all of these approaches. As such, evo-devo is really only a signpost on the road to a fully integrated biology. We thus welcome the day when we can drop the *devo* and fulfill Darwin's vision by calling it simply evolution.

SUMMARY

Darwin and his contemporaries recognized the intimate relationship between evolution and development, but Darwin's writings lacked a satisfying genetic mechanism. The rediscovery of Mendelian genetics in the 20th century led to the modern evolutionary synthesis, which excluded consideration of developmental biology. The explosion in molecular and genetic understanding since the late 20th century has heralded a reconciliation under the auspices of evo-devo.

The discovery of Hox genes across animals brought excitement, but Hox studies have often remained focused on commonalities among animals instead of explaining diversity. Indeed, the Hox gene story is more complex than originally thought, and it still yields interesting evolutionary insights.

Since the initial Hox discoveries, the field of evo-devo has matured and expanded, leading to reconsideration of these pillars of Darwinian thought: the predominance of gradual change in evolution, the near ubiquity of natural selection as the predominant explanation for life's diversity, and the notion that variation is ever present and unbiased. Evo-devo has not overturned these concepts but has elevated additional perspectives, such as mutations of large effect, and the surprising commonality of homoplasy as indicative of biases or constraints in evolution.

Nevertheless, most work in evo-devo has provided additional evidence and details about the functioning of evolution, quite in line with Darwinian thinking. For example, the multiple findings of co-option and exaptation in the origin of new and perhaps novel features of organisms were explicitly predicted by Darwin.

Evo-devo continues to yield surprising insights, such as the counterintuitive findings that expression patterns might be more stable evolutionarily than their canonical functions, as seen in the Hox genes and in insect segmentation.

That evo-devo remains a separate discipline speaks to the need to fully integrate developmental biology into evolutionary thinking.

QUESTIONS

1. Why did the evolutionary synthesis not include developmental biology? What discoveries initiated the reconciliation of development and evolution?
2. Can evolution proceed in jumps? Give examples to support your answer.
3. What is the canonical Hox gene expression pattern. Is is it maintained when the Hox genes are not found in a single cluster? What is the evidence?
4. How did Darwin explain the "5% of a wing problem"? Was his explanation correct? On what evidence?
5. In what sense are the lens crystallins of elephant shrews and scallops homologous? In what sense are they homoplasious? What about the red spots on the wings of *Heliconius melpomene xenoclea* and *H. erao microclea*?
6. Define exaptation and give an example. How do you know the trait you chose involved a change in function? Can you identify exaptations in your own body?
7. Do biases in developmental pathways limit evolutionary possibilities? How can this hypothesis be tested?
8. What is the Hox paradox? Can you suggest a solution?
9. Did the common ancestor of bilateral animals have a heart? Justify your answer by drawing an evolutionary tree and mapping hearts on it.
10. Do you think it would be possible, with artificial selection, to breed fully-armored freshwater sticklebacks that grow fast? Why or why not?
11. Why has it been useful to study Hox genes in many taxa? What has it suggested about their original function?

EXPLORING THE LITERATURE

12. Which came first, gene expression patterns or the complex structures they regulate? We discussed segmental differentiation along the anterior–posterior body axis, where the evidence suggests that expression patterns came first. But we must beware of hasty conclusions. Segmental gene expression patterns are seen in the unsegmented limbs of velvet worms as well as the segmented limbs of their sister group, the arthropods, suggesting that expression patterns came first. But recent fossil evidence shows that some ancient lobopods—presumed ancestors of velvet worms—had segmented limbs (though not segmented bodies). Thus the segmental gene expression pattern in extant velvet worm non-segmented limbs was probably inherited from a lobopod ancestor that had segmented limbs after all. See:

    Janssen, R., B. J. Eriksson, et al. 2010. Gene expression patterns in an onychophoran reveal that regionalization predates limb segmentation in pan-arthropods. *Evolution and Development* 12: 363–372.

    Liu, J., M. Steiner, et al. 2011. An armoured Cambrian lobopodian from China with arthropod-like appendages. *Nature* 470: 526–530.

    For analysis of another example, see:

    Pani, A. M., E. E. Mullarkey, et al. 2012. Ancient deuterostome origins of vertebrate brain signalling centres. *Nature* 483: 289–294.

13. One might suppose that the major transitions in evolution would have been accompanied by an increase in genetic complexity. For example, the appearance of practically all modern animal phyla in the Cambrian explosion over half a billion years ago was often assumed to have been accompanied by an explosion in genetic complexity. Was it? See:

    Domazet-Loso, T., J. Brajkovic, and D. Tautz. 2007. A phylostratigraphic approach to uncover the genomic history of major adaptations in metazoan lineages. *Trends in Genetics* 23: 533–539.

    Marshall, C. R., and J. W. Valentine. 2010. The importance of preadapted genomes in the origin of the animal bodyplans and the Cambrian explosion. *Evolution* 64: 1189–1201.

14. Although complex social behavior apparently only evolved once in an ancestor of modern ants, the division of labor into multiple worker castes has arisen independently in different ant lineages. An extreme example is the origin of specialized "supersoldier" ants with large bodies and powerful jaws. Not only have two independent origins of supersoldiers followed parallel modifications in hormonal regulation during soldier ant larval and pupal development, but related ants without supersoldiers can be induced to make them if these same hormone regulation pathways are artificially altered in their larvae. See:

    Rajakumar, R., D. S. Mauro, et al. 2012. Ancestral developmental potential facilitates parallel evolution in ants. *Science* 335: 79–82.

15. For experimental evidence on the involvement of a Hox gene in the evolution of tetrapod limbs from fish fins, see:

    Freitas, R., C. Gómez-Marín, et al. 2012. *Hoxd13* contribution to the evolution of vertebrate appendages. *Developmental Cell* 23: 1219–1229.

    Schneider, I., and N. Shubin. 2012. Making limbs from fins. *Developmental Cell* 23: 1121–1122.

16. Stomata (epidermal pores) in plants are a possible example of an evolutionary module—a quasi-separable entity within a multicellular organism that is a potential target for evolutionary change. For a review of the diversity, origin, and loss of stomata in land plants, the developmental-genetic control of stomata in thale cress (*Arabidopsis*), and comparative data on how variants in stomata are generated, see:

    Vatén, A., and D. C. Bergmann. 2012. Mechanisms of stomatal development: An evolutionary view. *EvoDevo* 3: 11.

CITATIONS

Alberch, P. 1981. Convergence and parallelism in foot morphology in the neotropical salamander genus *Bolitoglossa*. I. Function. *Evolution* 35: 84–100.

Albertin, W., K. Alix, et al. 2007. Differential regulation of gene products in newly synthesized *Brassica napus* allotetraploids is not related to protein function nor subcellular localization. *BMC Genomics* 8: 56.

Albertin, W., T. Balliau, et al. 2006. Numerous and rapid nonstochastic modifications of gene products in newly synthesized *Brassica napus* allotetraploids. *Genetics* 173: 1101–1113.

Allen, C. E., P. Beldade, et al. 2008. Differences in the selection response of serially repeated color pattern characters: Standing variation, development, and evolution. *BMC Evolutionary Biology* 8: 94.

Arenas-Mena, C, A. R. Cameron, and E. H. Davidson. 2000. Spatial expression of *Hox* cluster genes in the ontogeny of a sea urchin. *Development* 127: 4631–4643.

Arthur, W. 2004. *Biased Embryos and Evolution*. Cambridge: Cambridge University Press.

Barrett, R. D. H., S. M. Rogers, and D. Schluter. 2008. Natural selection on a major armor gene in threespine stickleback. *Science* 322: 255–257.

Barrett, R. D. H., S. M. Rogers, and D. Schluter. 2009. Environment specific pleiotropy facilitates divergence at the ectodysplasin locus in threespine stickleback. *Evolution* 63: 2831–2837.

Bateson, W. 1894. *Materials for the Study of Variation, Treated with Especial Regard to Discontinuity in the Origin of Species*. London: Macmillan.

Bejerano, G., C. B. Lowe, et al. 2006. A distal enhancer and an ultraconserved exon are derived from a novel retroposon. *Nature* 441: 87–90.

Beldade, P., K. Koops, and P. M. Brakefield. 2002. Modularity, individuality, and evo-devo in butterfly wings. *Proceedings of the National Academy of Sciences, USA* 99: 14262–14267.

Bergstrom, C. A. 2002. Fast-start swimming performance and reduction in lateral plate number in threespine stickleback. *Canadian Journal of Zoology* 80: 207–213.

Britten, R. J., and E. H. Davidson. 1971. Repetitive and non-repetitive DNA sequences and a speculation on the origins of evolutionary novelty. *Quarterly Review of Biology* 46: 111–138.

Burmeister, H. 1834. *Beiträge zur Naturgeschichte der Rankenfüsser (Cirripedia)*. Berlin: G. Reimer.

Counterman, B. A., F. Araujo-Perez, et al. 2010. Genomic hotspots for adaptation: The population genetics of Müllerian mimicry in *Heliconius erato*. *PLoS Genetics* 6: e1000796.

Danchin, E. G., M. N. Rosso, et al. 2010. Multiple lateral gene transfers and duplications have promoted plant parasitism ability in nematodes. *Proceedings of the National Academy of Sciences, USA* 107: 17651–17656.

Darwin, C. R. 1851a. *Living Cirripedia, A monograph on the sub-class Cirripedia, with figures of all the species. The Lepadidæ; or, pedunculated cirripedes*. Volume 1. London: The Ray Society.

Darwin, C. R. 1851b. *A monograph on the fossil Lepadidae, or, pedunculated cirripedes of Great Britain*. Volume 1. London: The Palaeontographical Society.

Darwin, C. R. 1854a. *Living Cirripedia, The Balanidæ, (or sessile cirripedes); the Verrucidæ*. Volume 2. London: The Ray Society.

Darwin, C. R. 1854b. *A monograph on the fossil Balanidae and Verrucidae of Great Britain*. Volume 2. London: The Palaeontographical Society.

Darwin, C. R. 1862. Letter to C. Lyell, 22 August 1862. In *The Life and Letters of Charles Darwin: Including an Autobiographical Chapter* (Volume 2, 1897), ed. Sir Francis Darwin. New York: Appleton, 180.

Darwin C. R. 1872. *The Origin of Species by Means of Natural Selection*. 6th edition. London: John Murray.

Davidson, E. H. 2001. *Genomic Regulatory Systems Development and Evolution*. San Diego, CA: Academic Press.

Duboule, D. 2007. The rise and fall of *Hox* gene clusters. *Development* 134: 2549–2560.

DuBuc, T., J. Ryan, et al. 2012. Coral comparative genomics reveal expanded *Hox* cluster in the cnidarian-bilaterian ancestor. *Integrative and Comparative Biology* 52: 835–841.

Dunning Hotopp, J. C. 2011. Horizontal gene transfer between bacteria and animals. *Trends in Genetics* 27: 157–163.

Emlen, D. J. 1996. Artificial selection on horn length-body size allometry in the horned beetle *Onthophagus acuminatus* (Coleoptera: Scarabaeidae). *Evolution* 50: 1219–1230.

Emlen, D. J. 1997. Alternative reproductive tactics and male-dimorphism in the horned beetle *Onthophagus acuminatus* (Coleoptera: Scarabaeidae). *Behavioral Ecology and Sociobiology* 41: 335–341.

Emlen, D. J. 2001. Costs and the diversification of exaggerated animal structures. *Science* 291: 1534–1536.

Emlen, D. J., I. A. Warren, et al. 2012. A mechanism of extreme growth and reliable signaling in sexually selected ornaments and weapons. *Science* 337: 860–864.

Farnsworth, E. J., and J. M. Farrant. 1998. Reductions in abscisic acid are linked with viviparous reproduction in mangroves. *American Journal of Botany* 85: 760–769.

Ferrier, D. E. K. 2007. Evolution of *Hox* gene clusters. In *HOX Gene Expression*, ed. S. Papageorgiou. Austin, TX: Landes Bioscience, 53–67.

Ferrier, D. E. K. 2011. Hox and ParaHox genes in evolution, development and genomics. *Genomics Proteomics and Bioinformatics* 9: 63–64.

French, V., and P. M. Brakefield. 1995. Eyespot development on butterfly wing: The focal signal. *Developmental Biology* 168: 112–123.

Gaeta, R. T., J. C. Pires, et al. 2007. Genomic changes in resynthesized *Brassica napus* and their effect on gene expression and phenotype. *Plant Cell* 19: 3403–3417.

Gaunt, S. J. 1988. Mouse homeobox gene transcripts occupy different but overlapping domains in embryonic germ layers and organs: A comparison of Hox-3.1 and Hox-1.5. *Development* 103: 135–144.

Goldschmidt R. 1940. *The Material Basis of Evolution*. New Haven: Yale University Press.

Gould, S. J. 1977. *Ontogeny and Phylogeny*. Cambridge: Harvard University Press.

Gould, S. J. 2002. *The Structure of Evolutionary Theory*. Cambridge: Harvard University Press.

Gould, S. J., and E. S. Vrba. 1982. Exaptation—A missing term in the science of form. *Paleobiology* 8: 4–15.

Graham, C, J. Hodin, and G. Wistow. 1996. A retinaldehyde dehydrogenase as a structural protein in a mammalian eye lens. Gene recruitment of eta-crystallin. *Journal of Biological Chemistry* 271: 15623–15628.

Gupta, S. K., and A. Pratap. 2007. History, origin and evolution. In *Advances in Botanical Research-Rapeseed Breeding* (Vol. 45), ed. S. K. Gupta. London: Academic Press.

Hall, B. K. 2012. Parallelism, deep homology, and evo-devo. *Evolution & Development* 14: 29–33.

Hara, Y., M. Yamaguchi, et al. 2006. Expression patterns of Hox genes in larvae of the sea lily *Metacrinus rotundus*. *Development Genes and Evolution* 216: 797–809.

Hessling, R. 2002. Metameric organisation of the nervous system in developmental stages of *Urechis caupo* (Echiura) and its phylogenetic implications. *Zoomorphology* 121: 221–234.

Hessling, R. 2003. Novel aspects of the nervous system of *Bonellia viridis* (Echiura) revealed by the combination of immunohistochemistry, confocal laser-scanning microscopy and three-dimensional reconstruction. *Hydrobiologia* 496: 225–239.

Heyland, A., J. Hodin, and A. M. Reitzel. 2005. Hormone signaling in evolution and development: A non-model system approach. *BioEssays* 27: 64–75.

Hines, H. M., B. A. Counterman, et al. 2011. Wing patterning gene redefines the mimetic history of *Heliconius* butterflies. *Proceedings of the National Academy of Sciences, USA* 108: 19666–19671.

Hodin, J. 2000. Plasticity and constraints in development and evolution. *Journal of Experimental Zoology B* 288: 1–20.

Hyman, L. H. 1940. The Invertebrates. Volume 1, Protozoa through *Ctenophora*. New York: McGraw-Hill.

Ikuta, T., N. Satoh, and H. Saiga. 2010. Limited functions of Hox genes in the larval development of the ascidian *Ciona intestinalis*. *Development* 137: 1505–1513.

Jacob, F., and J. Monod. 1961. Genetic regulatory mechanisms in the synthesis of proteins. *Journal of Molecular Biology* 3: 318–356.

Jarosch, A., E. Stolle, et al. 2011. Alternative splicing of a single transcription factor drives selfish reproductive behavior in honeybee workers (*Apis mellifera*). *Proceedings of the National Academy of Sciences, USA* 108: 15282–15287.

Joron, M., L. Frezal, et al. 2011. Chromosomal rearrangements maintain a polymorphic supergene controlling butterfly mimicry. *Nature* 477: 203–206.

Kessel, M., R. Balling, and P. Gruss. 1990. Variations of cervical vertebrae after expression of a Hox-1.1 transgene in mice. *Cell* 61: 301–306.

King, M. C., and A. C. Wilson. 1975. Evolution at two levels in humans and chimpanzees. *Science* 188: 107–116.

Kingsolver, J. G., and M. A. R. Koehl. 1985. Aerodynamics, thermoregulation, and the evolution of insect wings: Differential scaling and evolutionary change. *Evolution* 39: 488–504.

Koga, H., M. Matsubara, et al. 2010. Functional evolution of Ets in echinoderms with focus on the evolution of echinoderm larval skeletons. *Development Genes and Evolution* 220: 107–115.

Kowalevsky, A. O. 1866. *Entwicklungsgeschichte der einfachen Ascidien. Mémoires de l'Académie Impériale des Sciences de St. Pétersbourg* 10: 1–19.

Kristof, A., T. Wollesen, and A. Wanninger. 2008. Segmental mode of neural patterning in *Sipuncula. Current Biology* 18: 1129–1132.

Lemaire, P. 2011. Evolutionary crossroads in developmental biology: The tunicates. *Development* 138: 2143–2152.

Lewis, E. B. 1963. Genes and developmental pathways. *American Zoologist* 3: 33–56.

Lewis, E. B. 1978. A gene complex controlling segmentation in *Drosophila. Nature* 276: 565–570.

Lin, Z., X. Li, et al. 2012. Parallel domestication of the Shattering1 genes in cereals. *Nature Genetics* 44: 720–724.

Lindblad-Toh, K., M. Garber, et al. 2011. A high-resolution map of human evolutionary constraint using 29 mammals. *Nature* 478: 476–482.

Linnaeus, C. 1751. *Philosophia Botanica in qua explicantur Fundamenta Botanica etc.* Stockholm: Godofr Kiesewetter.

Lowe, C. B., G. Bejerano, and D. Haussler. 2007. Thousands of human mobile element fragments undergo strong purifying selection near developmental genes. *Proceedings of the National Academy of Sciences, USA* 104: 8005–8010.

Lowe, C. J., M. Terasaki, et al. 2006. Dorsoventral patterning in hemichordates: Insights into early chordate evolution. *PLoS Biology* 4: e291.

Marchinko, K. B. 2009. Predation's role in repeated phenotypic and genetic divergence of armor in threespine stickleback. *Evolution* 63: 127–138.

Martin, A., R. Papa, et al. 2012. Diversification of complex butterfly wing patterns by repeated regulatory evolution of a Wnt ligand. *Proceedings of the National Academy of Sciences, USA* 109: 12632–12637.

Martin, A., and R. D. Reed. 2010. *Wingless* and *aristaless2* define a developmental ground plan for moth and butterfly wing pattern evolution. *Molecular Biology and Evolution* 27: 2864–2878.

Mayer, G., and P. M. Whitington. 2009. Neural development in Onychophora (velvet worms) suggests a step-wise evolution of segmentation in the nervous system of Panarthropoda. *Developmental Biology* 335: 263–275.

Maynard Smith, J., R. Burian, et al. 1985. Developmental constraints and evolution. *Quarterly Review of Biology* 60: 265–287.

Mayr, E. 1963. *Animal Species and Evolution.* Cambridge, MA: Harvard University Press.

Mayr, E. 1982. *The Growth of Biological Thought: Diversity, Evolution, and Inheritance.* Cambridge, MA: Harvard University Press.

McGinnis, W., M. S. Levine, et al. 1984. A conserved DNA sequence in homoeotic genes of the *Drosophila* antennapedia and bithorax complexes. *Nature* 308: 428–433.

Mivart, St. George J. 1872. *On the Genesis of Species.* London: MacMillan.

Moczek, A. P., and H. F. Nijhout. 2004. Trade-offs during the development of primary and secondary sexual traits in a horn dimorphic beetle. *American Naturalist* 163: 184–191.

Monteiro, A. S., and D. E. K. Ferrier. 2006. Hox genes are not always colinear. *International Journal of Biological Sciences* 2: 95–103.

Monteiro, A., G. Glaser, et al. 2006. Comparative insights into questions of lepidopteran wing pattern homology. *BMC Developmental Biology* 6: 52.

Mooi, R., and B. David. 2008. Radial symmetry, the anterior/posterior axis, and echinoderm Hox genes. *Annual Review of Ecology, Evolution and Systematics* 39: 43–62.

Noordermeer, D., M. Leleu, et al. 2011. The dynamic architecture of Hox gene clusters. *Science* 334: 222–225.

Okkema, P. G., E. Ha, et al. 1997. The *Caenorhabditis elegans* NK-2 homeobox gene *ceh-22* activates pharyngeal muscle gene expression in combination with *pha-1* and is required for normal pharyngeal development. *Development* 124: 3965–3973.

Olofsson, M., A. Vallin, et al. 2010. Marginal eyespots on butterfly wings deflect bird attacks under low light intensities with UV wavelengths. *PLoS One* 5: e10798.

Olson, E. N. 2006. Gene regulatory networks in the evolution and development of the heart. *Science* 313: 1922–1927.

Pearson, J. C., D. Lemons, and W. McGinnis. 2005. Modulating Hox gene functions during animal body patterning. *Nature Reviews Genetics* 6: 893– 904.

Piatigorsky, J., Z. Kozmik, et al. 2000. Omega-crystallin of the scallop lens. A dimeric aldehyde dehydrogenase class 1/2 enzyme-crystallin. *Journal of Biological Chemistry* 275: 41064–41073.

Pires, J. C., J. Zhao, et al. 2004. Flowering time divergence and genomic rearrangements in resynthesized Brassica polyploids (Brassicaceae). *Biological Journal of the Linnean Society* 82: 675–688.

Ponnambalam, S., and M. Alberghina. 2011. Evolution of the VEGF-regulated vascular network from a neural guidance system. *Molecular Neurobiology* 43: 192–206.

Pottin, K., C. Hyacinthe, and S. Rétaux. 2010. Conservation, development, and function of a cement gland-like structure in the fish *Astyanax mexicanus. Proceedings of the National Academy of Sciences, USA* 107: 17256–17261.

Protas M and WR Jeffery. 2012. Evolution and development in cave animals: From fish to crustaceans. *WIREs: Developmental Biology* 1: 823–845.

Reed, R. D., R. Papa, et al. 2011. Optix drives the repeated convergent evolution of butterfly wing pattern mimicry. *Science* 333: 1137–1141.

Ryan, J. F., M. E. Mazza, et al. 2007. Pre-bilaterian origins of the Hox cluster and the Hox code: Evidence from the sea anemone, *Nematostella vectensis. PLoS One* 2: e153.

Samadi, L., and G. Steiner. 2010. Expression of Hox genes during the larval development of the snail, *Gibbula varia* (L.)—further evidence of non-colinearity in molluscs. *Development Genes and Evolution* 220: 161–172.

Scott, M. P., and A. J. Weiner. 1984. Structural relationships among genes that control development: Sequence homology between the Antennapedia, ultrabithorax, and fushi tarazu loci of *Drosophila. Proceedings of the National Academy of Sciences, USA* 81: 4115–4119.

Shimizu, H., and T. Fujisawa. 2003. Peduncle of *Hydra* and the heart of higher organisms share a common ancestral origin. *Genesis* 36: 182–186.

Shubin, N., C. Tabin, and S. Carroll. 2009. Deep homology and the origins of evolutionary novelty. *Nature* 457: 818–823.

Sibley, C. G., and J. E. Ahlquist. 1987. DNA hybridization evidence of hominoid phylogeny: Evidence from an expanded data set. *Journal of Molecular Evolution* 26: 99–121.

Slack, J. M. W., P. W. H. Holland, and C. F. Graham. 1993. The zootype and the phylotypic stage. *Nature* 361: 490–492.

Struck, T. H., N. Schult, et al. 2007. Annelid phylogeny and the status of *Sipuncula* and *Echiura. BMC Evolutionary Biology* 7: 57.

Thompson, D. W. 1917. *On Growth and Form.* Cambridge: Cambridge University Press.

van't Hof, A. E., N. Edmonds, et al. 2011. Industrial melanism in British peppered moths has a singular and recent mutational origin. *Science* 332: 958–960.

Wake, D. B. 1966. Comparative osteology and evolution of lungless salamanders, family Plethodontidae. *Memoirs of the Southern California Academy of Sciences* 4: 1–111.

Werner, T., S. Koshikawa, et al. 2010. Generation of a novel wing colour pattern by the Wingless morphogen. *Nature* 464: 1143–1148.

Whitington, P. M., and G. Mayer. 2011. The origins of the arthropod nervous system: Insights from the Onychophora. *Arthropod Structure & Development* 40: 193–209.

Wistow, G., and J. Piatigorsky. 1987. Recruitment of enzymes as lens structural proteins. *Science* 236: 1554–1556.

Wu, Q., and M. R. Brown. 2006. Signaling and function of insulin-like peptides in insects. *Annual Review of Entomology* 51: 1–24.

Yoshida, M. A., S. Shigeno, et al. 2010. Squid vascular endothelial growth factor receptor: A shared molecular signature in the convergent evolution of closed circulatory systems. *Evolution & Development* 12: 25–33.

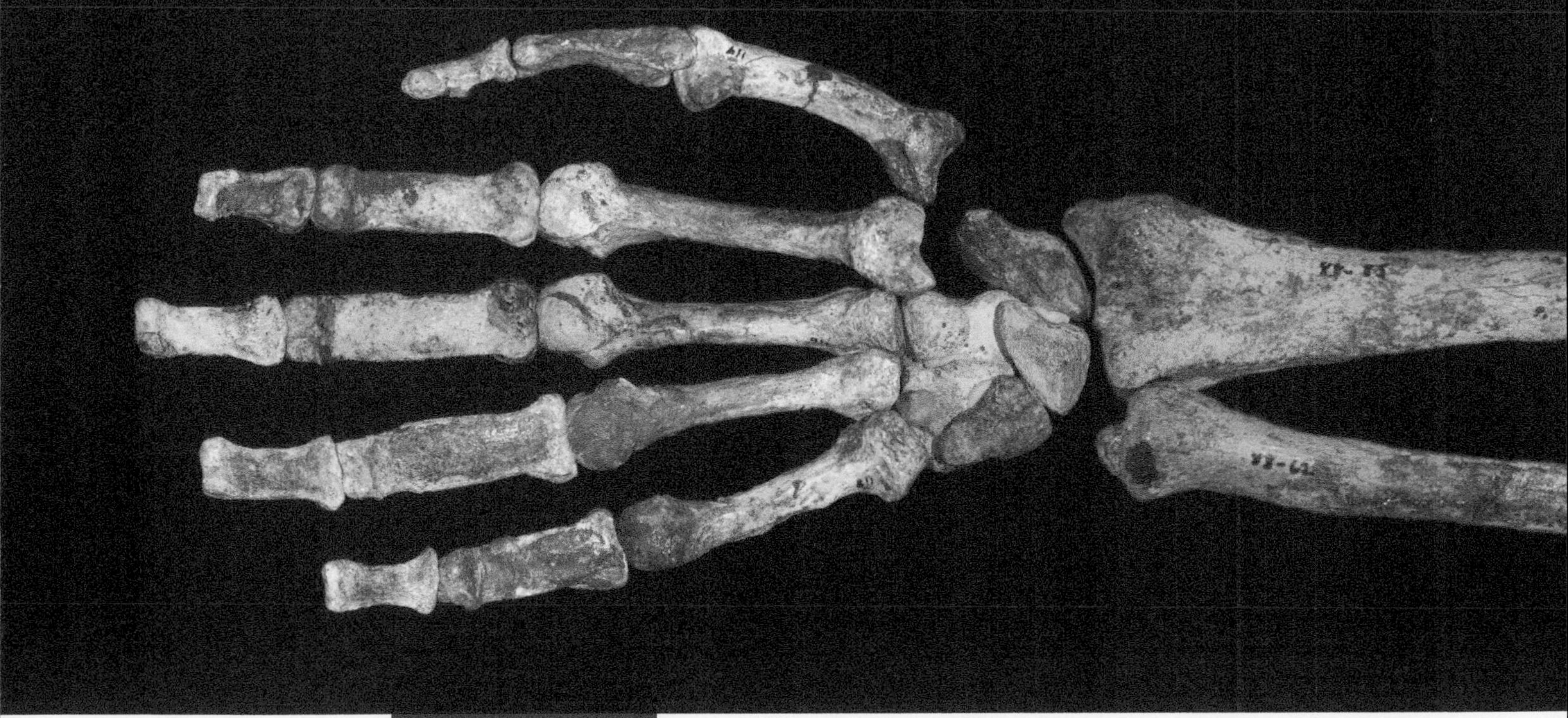

20

Human Evolution

The first printing of *On the Origin of Species* sold out on November 22, 1859, the first day Darwin's publisher, John Murray, offered it to booksellers. Among the profound implications that attracted such attention was what the book told its readers about themselves. Although Darwin saw this as clearly as anyone, his only explicit treatment of human evolution was a single paragraph in the last chapter, in which his strongest claim was that "Light will be thrown on the origin of man and his history". (Darwin 1859, page 488).

Not until 12 years later did Darwin reveal the depth and breadth of his thinking about humans. In 1871 he published a two-volume work, *The Descent of Man, and Selection in Relation to Sex*. In the introduction, Darwin explained his initial reticence on the subject of human evolution: "During many years I collected notes on the origin or descent of man, without any intention of publishing on the subject, but rather with the determination not to publish, as I thought that I should thus only add to the prejudices against my views" (Darwin 1871, page 1).

Darwin's apprehensions were well founded. The human implications of evolutionary biology have been, and remain, a cause of heated controversy. In 1925, Tennessee schoolteacher John T. Scopes was convicted of violating a new state law prohibiting the teaching of evolution (see Chapter 3). The Scopes case was

Australopithecus sediba, which lived some 2 million years ago, had thumbs more similar to those of a modern human than a chimp or gorilla. Photo by Peter Schmid, courtesy Lee R. Berger and the University of the Witwatersrand. Graph from Kivell et al. (2011).

From "*Australopithecus sediba* hand demonstrates mosaic evolution of locomotor and manipulative abilities." *Science* 333: 1411–1417. Reprinted with permission from AAAS.

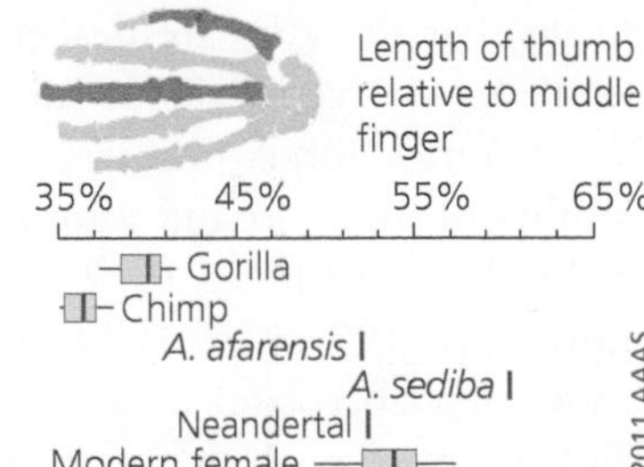

popularly known as the Monkey Trial, indicating that for many observers the central issue at stake was the origin of the human species. In 2004, the school board in Dover, Pennsylvania, adopted a policy requiring that a disclaimer be read to students in ninth-grade biology classes. The disclaimer admonished students to consider evolution as theory, not fact, and notified them that a book espousing intelligent design creationism was available as a reference. The Dover disclaimer became the subject of a widely publicized trial in which teaching intelligent design in public schools was ruled unconstitutional (see Chapter 3). Again, the origin of our species was a key issue for many involved. In a meeting leading up to the adoption of the disclaimer policy, board member William Buckingham declared, "It's inexcusable to have a book that says man descended from apes with nothing to counterbalance it" (Maldonado 2004; Jones 2005).

In this chapter, we explore research on the evolutionary history of our species. In Section 20.1 we review attempts to determine the evolutionary relationships among humans and the extant apes. In Section 20.2 we consider the fossil evidence, including Malapa Hominin 2, the female *Australopithecus sediba* whose hand appears on the previous page, bearing on the course of human evolution following the split between our lineage and the lineage of our closest living relatives. In Section 20.3 we look at fossil and molecular evidence on the emergence of *Homo sapiens*. Finally, in Section 20.4 we consider the evolutionary origins of some of our species' defining characteristics, including tool use and language. Our exploration illustrates that the subject of human evolution generates controversies within the scientific community that, while different in focus, are as heated as those it generates among the lay public.

20.1 Relationships among Humans and Extant Apes

Humans (*Homo sapiens*) belong to the primate taxon Catarrhini (Goodman et al. 1998), which includes the Old World monkeys, such as the baboons and macaques, and the apes (**Figure 20.1**). The apes include the gibbons (*Hylobates*) of southeast Asia and the great apes. The great apes include the orangutan (*Pongo pygmaeus*), also of southeast Asia, and three African species: the gorilla (*Gorilla gorilla*), the common chimpanzee (*Pan troglodytes*), and the bonobo, or pygmy chimpanzee (*Pan paniscus*).

Humans Belong to the Same Clade as the Apes

Scientists universally agree that humans evolved from within the apes. Humans share with the apes numerous derived characteristics (synapomorphies). These evolutionary innovations distinguish the apes from the rest of the Catarrhini and indicate that the apes are descended from a common ancestor (see Chapter 4). The shared derived traits of the apes include relatively large brains, the absence of a tail, a more erect posture, greater flexibility of the hips and ankles, increased flexibility of the wrist and thumb, and changes in the structure and use of the arm and shoulder (Andrews 1992; see also Groves 1986; Andrews and Martin 1987; Begun et al. 1997). In addition to this morphological evidence, the molecular analyses described later in this chapter also unequivocally demonstrate that in an evolutionary sense, humans are apes.

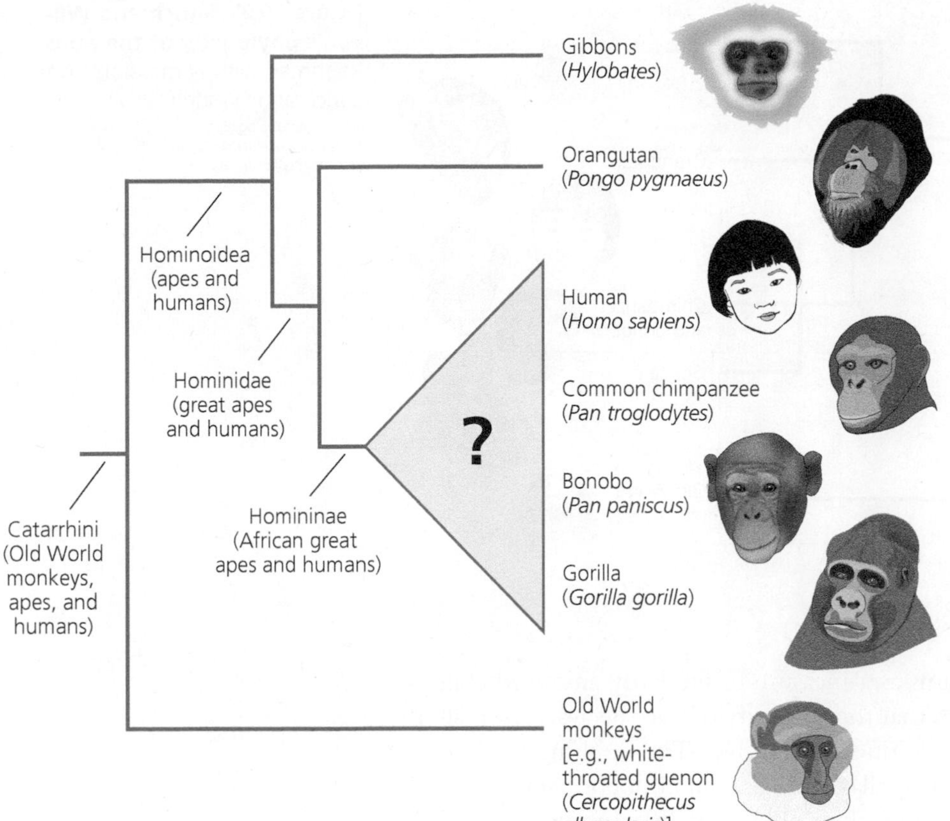

Figure 20.1 Phylogeny of the apes This evolutionary tree shows the relationships among the Old World monkeys, represented by a rhesus monkey, and the apes and humans. Among the apes, the gibbons branch off first, followed by the orangutan. The evolutionary relationships among the gorilla, the two chimpanzees, and humans (triangle with question mark) were long the subject of considerable dispute. For taxonomy, see Harrison (2010) and Wood and Harrison (2011).

Humans Belong to the Same Clade as the African Great Apes

Figure 20.1 includes a reconstruction of the phylogenetic relationships among the apes. This reconstruction places humans with the great apes, and more specifically with the African great apes. The reconstruction was first proposed by Thomas Henry Huxley (1863). Huxley's proposal raised dispute, but in recent years, as more data have been collected and analyzed, scientists in all fields have accepted the tree in Figure 20.1.

Cladistic analyses of morphology support the tree. Humans and the African great apes share a number of derived traits that distinguish them from the rest of the apes. These include elongated skulls, enlarged browridges, short but stout canine teeth, changes in the front of the upper jaw (premaxilla), fusion of certain bones in the wrist, enlarged ovaries and mammary glands, changes in muscular anatomy, and reduced hairiness (Ward and Kimbel 1983; Groves 1986; Andrews and Martin 1987; Andrews 1992; Begun et al. 1997; Lehtonen et al. 2011).

Morphological and molecular analyses demonstrate that humans are closely related to gorillas and chimpanzees.

Molecular analyses concur. They have, in fact, indicated a close relationship between humans and the African great apes since the beginnings of modern molecular systematics. Using a technique pioneered by George H. F. Nuttall (1904) and Morris Goodman (1962), Vincent Sarich and Allan Wilson (1967) took purified human serum albumin, a blood protein, and injected it into rabbits. After giving the rabbits time to make antibodies against the human albumin protein, Sarich and Wilson took blood serum from the rabbits. This serum contained rabbit antihuman antibodies. The researchers mixed the rabbit serum with purified serum albumin from a variety of apes and Old World monkeys. Sarich and Wilson used the strength of the immune reaction between the rabbit antihuman

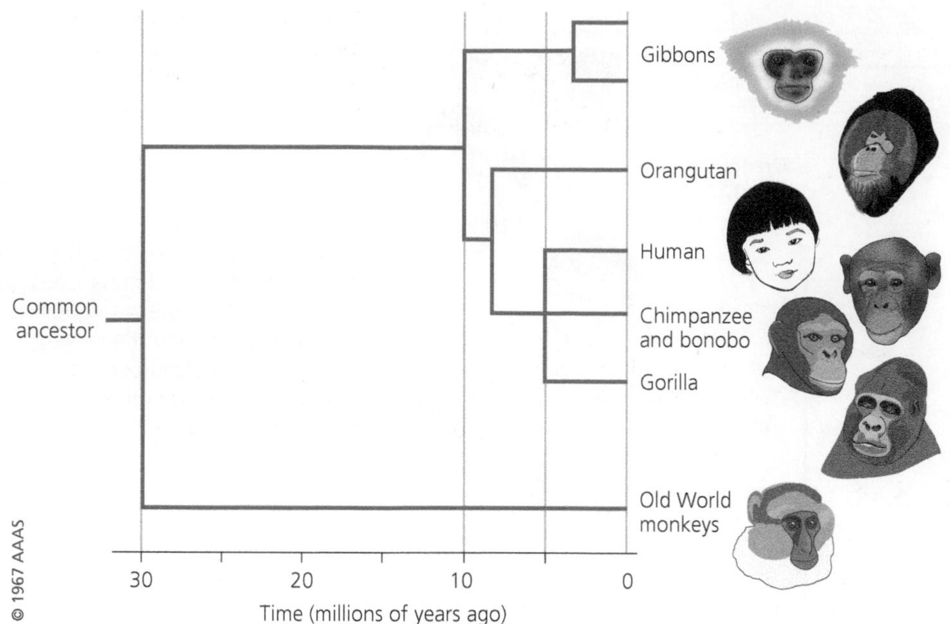

Figure 20.2 Sarich and Wilson's phylogeny of the apes Reprinted with permission from Sarich and Wilson (1967).

From "Immunological time scale for Hominid evolution." *Science* 158: 1200–1203. Reprinted with permission from AAAS.

antibodies and the primate albumins as a measure of similarity among the albumins they tested, and they assumed that the similarity of two species' serum albumin proteins reflects the species' evolutionary kinship. The resulting phylogeny shows that humans are close kin to gorillas, chimps, and bonobos **(Figure 20.2)**.

Sarich and Wilson put a time line on their phylogeny by assuming that serum albumin evolves at a constant rate and that, as indicated by the fossil record available at the time, the split between the apes and the Old World monkeys occurred 30 million years ago. The time line suggests that humans and the African great apes shared a common ancestor about 5 million years ago, which is more recently than previously suspected (see Lowenstein and Zihlman 1988). We review additional molecular phylogenies shortly; all are consistent with Sarich and Wilson's tree in showing close kinship between humans and the African great apes.

The phylogenies in Figures 20.1 and 20.2 show that humans, gorillas, and the two chimpanzees are close relatives, but they do not resolve the evolutionary relationships among these four species. The true phylogeny for humans, gorillas, and the two chimpanzees could be any one of the four trees shown in **Figure 20.3**. It is probably safe to say that more scientists have invested more effort in attempting to determine which of these trees is correct than has been invested in any other species-level problem in the history of systematics.

Humans, Gorillas, Chimpanzees, and Bonobos

After decades of debate, researchers have come to a consensus that the evolutionary relationships among humans and the African great apes are best characterized by the tree in Figure 20.3a. Humans and the chimpanzees are more closely related to each other than either is to gorillas. This consensus was slow in forming, for at least two reasons: There were conflicts among molecular data sets, and there were conflicts between molecular evidence and morphological evidence.

Molecular Evidence

Molecular biologists sought to resolve the human/African great ape phylogeny by analyzing DNA sequences. Maryellen Ruvolo and colleagues (1994), for

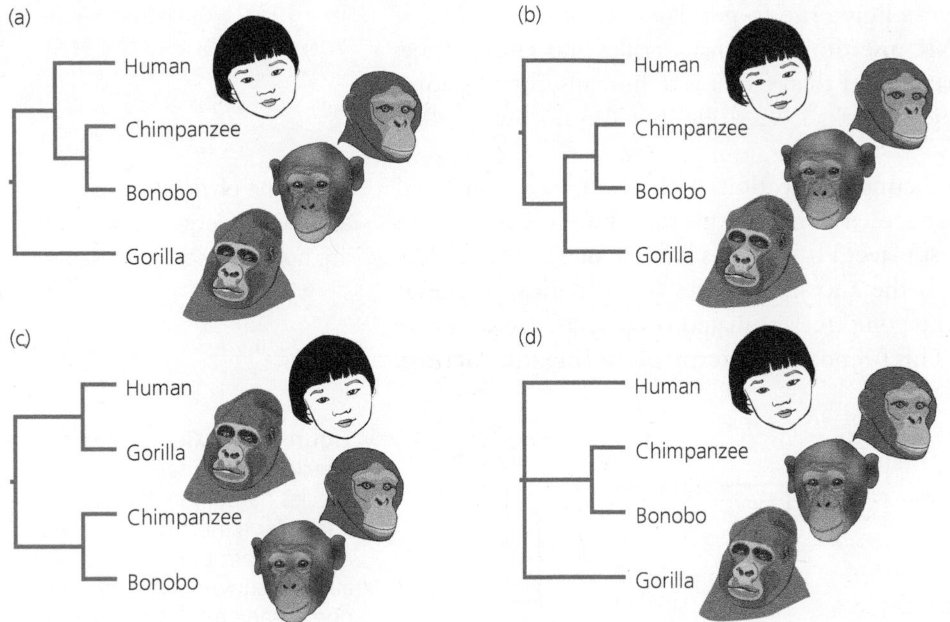

Figure 20.3 Possible phylogenies of humans and the African great apes Four possible resolutions of the evolutionary relationships among humans and the African great apes. All assume that the two species of chimpanzee are closest relatives. The true tree could have (a) humans and chimpanzees as closest relatives, (b) chimpanzees and gorillas as closest relatives, (c) humans and gorillas as closest relatives, or (d) a genuine simultaneous three-way split (polytomy).

example, reconstructed the evolutionary tree of the apes based on sequence data for a mitochondrial gene **(Figure 20.4)**. On the evidence of these data, humans and the chimpanzees diverged from each other only after their common ancestor diverged from the gorillas. Researchers have reconstructed the evolutionary history of the apes with data from a great variety of loci. Most analyses have produced trees like the one in Figure 20.4, in which humans and the chimpanzees are closest relatives (for examples, see Horai et al. 1992; Goodman et al. 1994; Kim and Takenaka 1996).

Molecular analyses indicate that humans and chimpanzees are closest relatives...

A persistent minority of analyses, however, have produced phylogenies in which gorillas and chimps, or even gorillas and humans, are closest relatives (Figure 20.3b and c). Madalina Barbulescu and colleagues (2001), for example, found a locus where, in the genome of gorillas and both chimpanzees, there exists an inserted nucleotide sequence. This insert is the genome of a retrovirus called human endogenous retrovirus K, or HERV-K. The implication is that this particular retroviral invasion happened long ago, in the common ancestor of gorillas and the chimpanzees. The same locus in humans lacks the HERV-K insert and appears never to have had it. This locus, taken on its own, suggests that humans

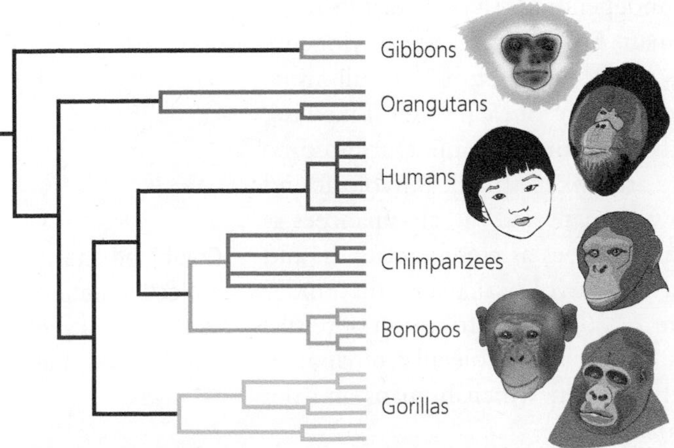

Figure 20.4 Phylogeny of mitochondrial cytochrome oxidase II alleles in humans and the African great apes Ruvolo and colleagues estimated this tree using the maximum parsimony method. From Ruvolo et al. (1994).

diverged from the lineage that would give rise to gorillas and the chimps before the latter acquired their HERV-K insertion, and that gorillas and chimpanzees are thus more closely related to each other than either is to humans (for additional examples, see Djian and Green 1989; Marks 1993, 1994 [but also Borowik 1995]; Deinard et al. 1998).

How can we reconcile the conflicting implications of these molecular analyses? The answer is that we should not necessarily have expected all molecular analyses to agree with each other in the first place. Phylogenies like the one in Figure 20.4 are gene trees, not species trees. If the ancestral species was genetically variable for the locus under study, then the gene tree estimated from sequence data may differ from the true species tree. This is known as **incomplete lineage sorting.**

...but the phylogenies of genes and the phylogenies of species are not necessarily the same.

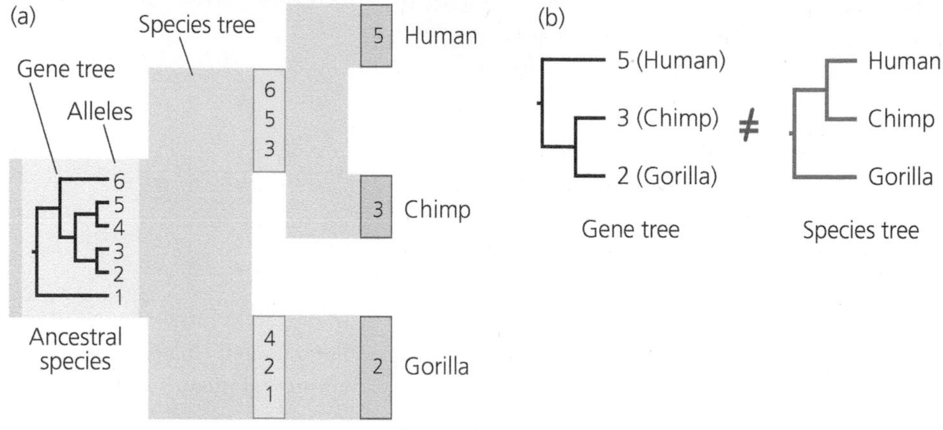

Figure 20.5 Gene trees versus species trees (a) An ancestral species harbors six alleles for a particular gene. The alleles are derived from a common ancestral allele as shown. After speciation, one lineage retains alleles 3, 5, and 6; the other retains 1, 2, and 4. After another speciation and further allele loss, humans retain allele 5, chimps allele 3, and gorillas allele 2. (b) The tree for the surviving alleles differs from the species tree. After Ruvolo (1994).

Figure 20.5 illustrates the reasoning. If different descendant species lose different ancestral alleles (Figure 20.5a), then we can end up reconstructing only a portion of the original gene tree. This portion may imply a different branching pattern than that of the true species tree (Figure 20.5b).

To determine the species tree from gene trees, we can reconstruct the phylogeny using many independent genes. Genes for which different alleles have persisted at random from a variable ancestral species will be equally likely to support a human–chimpanzee pairing, a human–gorilla pairing, and a chimpanzee–gorilla pairing. However, unless there has been a true three-way split (Figure 20.3d), genes whose tree matches the species tree should agree with each other and produce a clear signal against this random background.

Ruvolo (1995, 1997) reviewed and tallied independent data sets of DNA sequences that were informative about the human/African great ape phylogeny. She counted all mitochondrial DNA studies as a single data set, because all mitochondrial genes are linked and are thus not independent of each other. Likewise, any groups of nuclear genes that are near each other on the same chromosome counted as a single data set, because the genes are linked. In all, Ruvolo found 14 independent data sets. Eleven of these show humans and the chimpanzees as closest relatives, two show gorillas and the chimpanzees as closest relatives, and one shows humans and gorillas as closest relatives. Ruvolo calculated that under the null hypothesis of a true trichotomy (Figure 20.3d), this distribution of results has a probability of only 0.002. Ruvolo concluded that the molecular phylogeny data reject the trichotomous tree and favor the tree in which humans and the chimpanzees are closest relatives (Figure 20.3a).

Combined analyses of several molecular data sets strongly support the hypothesis that humans and chimps are closest relatives.

This resolution of conflicting molecular data sets is consistent with more recent analyses of even larger collections of loci (Satta et al. 2000; O'hUigin et al. 2002; Wildman et al. 2003; Rauum et al. 2005; Elango et al. 2006; Fabre et al. 2009). In a notable study, Abdel-Halim Salem and colleagues (2003) reconstructed the phylogeny of the apes using Alu elements. Alu elements are a type of short interspersed element, or SINE. SINEs are selfish DNA sequences that occasionally insert themselves into the chromosomes of their hosts (see Chapter 4). Because insertions are infrequent, and because deletion of a SINE is usually detectable from the sequence left behind, SINEs are nearly ideal derived traits for reconstructing evolutionary history. Salem and colleagues found one Alu element insertion that was shared by humans and gorillas, but absent in the chimpanzees, versus seven insertions shared by humans and the chimpanzees but absent in gorillas. Judging from the mutations it has accumulated, the insertion shared by humans and gorillas is older than the seven shared by humans and the chimps. Salem and colleagues infer that it reflects an ancestral polymorphism of the kind diagrammed in Figure 20.5. Overall, Salem's data strongly support the consensus that humans and the chimps form a monophyletic clade.

Svante Pääbo (2003) offered an elegant summary of the evidence provided by molecular analyses. Pääbo describes the human genome, and by implication the genomes of the African great apes, as mosaics. For each genomic segment, the homologous components from humans and the great apes have their own phylogeny. For some genomic segments, gorillas and the chimpanzees, or gorillas and humans, are closest relatives. For most genomic segments, however, the closest kin are humans and the chimpanzees.

Although humans and chimps are closest relatives, it is important to recognize that the human genome is a mosaic. In some portions of our genome, we are most closely related to gorillas.

The mosaic nature of the human, chimpanzee, and gorilla genomes was confirmed when Aylwyn Scally and colleagues (2012) finished sequencing the first complete genome of a gorilla. In comparing the gorilla sequence to sequences for a human and a common chimp, the researchers found that for roughly 30% of the genome, gorillas and chimps, or gorillas and humans, are closest relatives **(Figure 20.6a)**. For the remaining 70%, humans and chimps are closest relatives. As shown in Figure 20.6b, incomplete lineage sorting is reduced near protein-coding genes. This reduction is expected when genes are subject to selection,

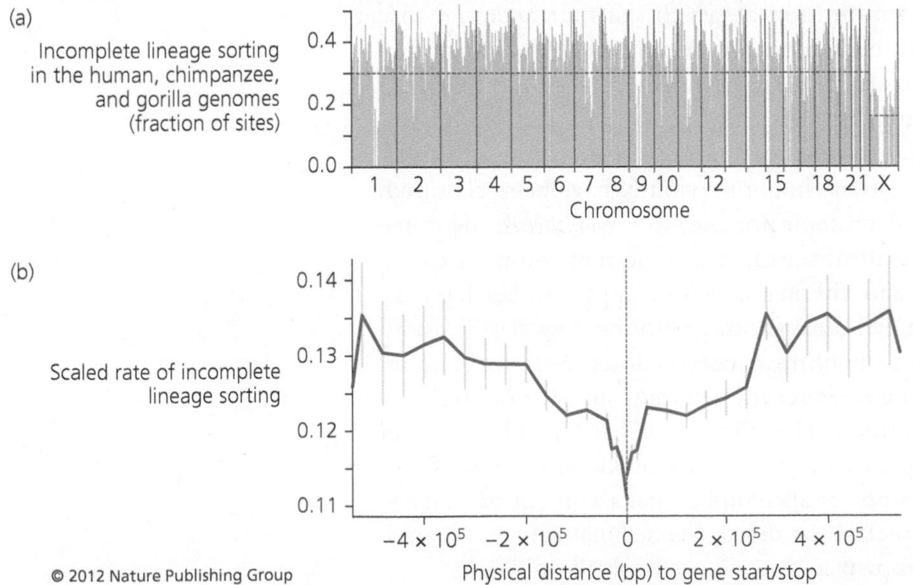

Figure 20.6 Incomplete lineage sorting in the human, chimpanzee, and gorilla genomes (a) Variation in incomplete lineage sorting across the genome. Each blue line shows the estimated incomplete lineage sorting for a 1-Mbp region. The dotted line shows the average. The roughly 30% incomplete lineage sorting is split between regions for which gorillas and chimps or gorillas and humans are closest relatives. (b) Incomplete lineage sorting (normalized by the mutation rate) near genes. From Scally et al. (2012).

"Insights into hominid evolution from the gorilla genome sequence." S. Scally, J.Y. Dutheil, *Nature* 483: 169–175. Copyright © 2012 Nature Publishing Group. Reprinted with permission.

because selection reduces the allelic diversity necessary for incomplete lineage sorting to occur. Scally and colleagues believe that much of the selection on the human, chimp, and gorilla genomes has been purifying. That is, it has involved the loss of deleterious alleles.

Morphological Evidence

Paleontologists sought to solve the human/chimp/gorilla puzzle with cladistic analyses of morphology. These researchers noted several features that are shared by gorillas and the two chimpanzees but absent in humans. These mainly include skeletal traits associated with knuckle walking (Andrews and Martin 1987). Knuckle walking is moving on four legs with the upper body supported on the backs of the middle phalanges—the middle of the three bones in each finger.

Knuckle walking is derived in the African great apes. The Asian great apes—the orangutans—do walk on their knuckles occasionally, but typically they walk on their fists. That is, they support their weight on the backs of the proximal phalanges—the bone in each finger that is directly connected to the hand. The African great apes—the gorillas and both species of chimpanzee—are dedicated knuckle walkers. They have specialized hand and wrist anatomy to go with the habit. Humans, of course, are not knuckle walkers.

Considering knuckle walking in isolation, the simplest explanation for its distribution is that humans diverged first from the lineage that would later produce the gorilla and the two chimpanzees (Figure 20.3b). This scenario requires only one appearance of knuckle walking, in a common ancestor of the gorilla and the chimps, and no losses. There is a catch, however. While the tree in Figure 20.3b gives a parsimonious explanation for knuckle walking, it requires that several other traits shared only by humans and the two chimps be interpreted either as ancestral traits that were lost in gorillas or as convergent derived traits that evolved independently in humans and the chimps. These traits include features of the teeth, skull, and limbs, delayed sexual maturity, and prominent labia minora in females and a pendulous scrotum in males (Groves 1986; Begun 1992). The tree in Figure 20.3b also conflicts with the consensus result from molecular evidence, which shows that humans and the chimpanzees are closest relatives.

Resolution of the human/chimp/gorilla evolutionary tree on morphological grounds thus depends on the identification of which traits are ancestral and which are derived. David R. Begun classified characteristics of the skulls of the great apes by including in his analysis an extinct European ape called *Dryopithecus,* known only from fossils about 10 million years old (Begun 1992; see also Begun 1995). *Dryopithecus* shares several cranial traits with gorillas that are absent in the two chimpanzees and humans. These traits might previously have been classified as uniquely derived in gorillas, but given their presence in *Dryopithecus,* the traits now appear to be ancestral. This, in turn, means that some traits thought to be ancestral or convergent in humans and chimpanzees now appear to be derived. When Begun reconstructed the ape evolutionary tree with the new classification of traits, he concluded that humans and chimpanzees are closest relatives (Figure 20.3a). This implies either (1) that the most recent common ancestor of humans, gorillas, and the chimpanzees was a knuckle walker, and that knuckle walking was subsequently lost in the human lineage, or (2) that knuckle walking evolved independently in gorillas and the chimps. It also implies that a number of characters of the teeth, skull, and limbs, as well as the delayed sexual maturity and shared genital anatomy of humans and chimpanzees, need have evolved only once.

Morphological analyses also suggest that humans and chimpanzees are closest relatives.

Some researchers were not convinced by Begun's reasoning, arguing that some of the skull features that Begun believes are shared derived traits in humans and chimpanzees may be ancestral or convergent and that knuckle walking may not be so readily evolved or lost as Begun's phylogeny requires (Andrews 1992; Dean and Delson 1992). However, more recent analyses of much-expanded data sets seem to confirm the close relationship among chimpanzees and humans (Shoshani et al. 1996; Begun et al. 1997; Gibbs et al. 2000; Gibbs et al. 2002; Lockwood et al. 2004; Strait and Grine 2004; Lehtonen et al. 2011).

Furthermore, Brian Richmond and David Strait (2000) compared the wrist bones of African fossils with those of living primates. They found evidence that at least two extinct species thought to be more closely related to humans than to chimpanzees or gorillas had anatomical features associated with knuckle walking (see also Collard and Aiello 2000; Corruccini and McHenry 2001; Dainton 2001; Lovejoy et al. 2001; Richmond and Strait 2001a, 2001b). This interpretation is consistent with the hypothesis that humans evolved from a knuckle-walking ancestor, and that humans and the chimpanzees are each other's closest living relatives (Figure 20.3a). Thus it appears that the morphological evidence is converging on the same conclusion as the molecular analyses.

Estimating the Divergence Times for Humans and the Apes

Working in the laboratory of S. Blair Hedges, a team led by undergraduate Rebecca L. Stauffer used molecular clocks to estimate the divergence times for humans and the apes (Stauffer et al. 2001). Based on the fossil record, the Old World monkeys diverged from the apes 23.3 million years ago. From comparisons of sequence differences in a variety of protein-coding genes from apes versus Old World monkeys, the researchers estimated the rate at which the genes have evolved since the two lineages diverged. Then, by counting the sequence differences in the same genes from, for example, humans versus chimpanzees, the team estimated the divergence times among the humans and the apes. To improve the accuracy of their estimates, Stauffer and colleagues combined data from dozens of genes. Their estimates are summarized in **Figure 20.7**. The lineage that would become today's gorillas diverged from the lineage that would become humans and the chimpanzees 6.4 $\pm$1.5 million years ago. The human and chimpanzee lineages split 5.4 $\pm$1.1 million years ago.

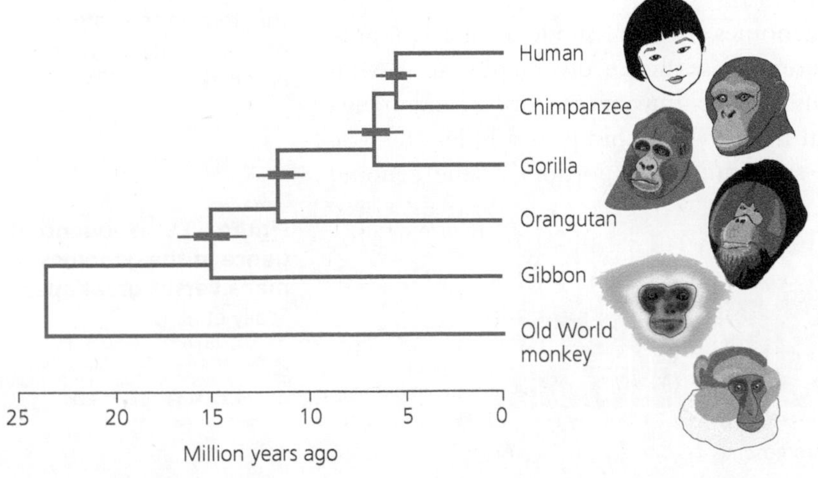

Figure 20.7 Divergence times for the apes Stauffer and colleagues estimated the dates of the common ancestors on this phylogeny by combining data from dozens of proteins used as molecular clocks. The heavy bars show $\pm$1 standard error around the time estimates; the lighter bars show 95% confidence intervals. From Stauffer et al. (2001).

Using similar methods, an even larger data set, and an ape–Old World monkey divergence time of 23.8 million years ago, Sudhir Kumar and colleagues (2005) estimated that the human and chimpanzee lineages diverged 4.98 million years ago, with a 95% confidence interval of 4.38 to 5.94 million years ago.

Richard Wilkinson and colleagues (2011) developed a method for combining molecular data with evidence from the fossil record, some of which we discuss in Section 20.2. They put the human/chimp divergence at 5.7 to 9.6 million years ago. Using a different method, data from whole genomes, and fossil evidence, Aylwyn Scally and colleagues (2012) put the human/chimp divergence at 5.5 to 7 million years ago and the human–chimp/gorilla divergence at 8.5 to 12 million years ago. Dating the divergence of humans and chimpanzees is still a work in progress (see Langergraber et al. 2012; Sun et al. 2012).

Before moving on to the next section, a natural question to ask is: How large are the genetic differences between humans and our closest relatives?

Genetic Differences between Humans, Chimpanzees, and Gorillas

The most obvious genetic difference between humans and the African great apes is in their karyotypes. Gorillas and chimpanzees have 24 pairs of chromosomes, whereas humans have only 23. The reason is that in the ancestors of humans, sometime after our lineage split from that of the chimpanzees, two chromosomes fused to become what we now know as chromosome 2 **(Figure 20.8)**. Genomic data are now allowing researchers to assess the genetic differences between ourselves and our closest kin on a much finer scale.

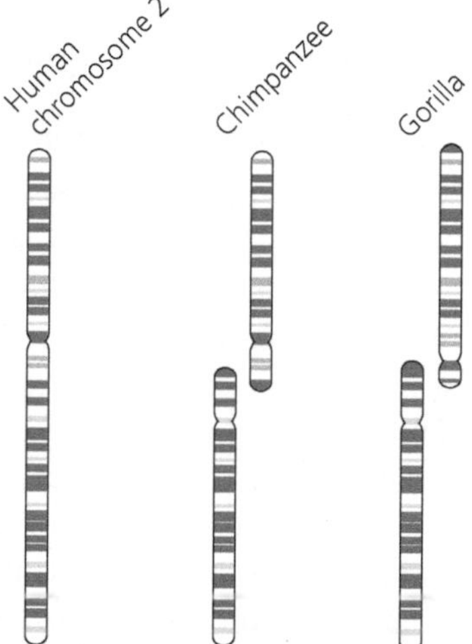

Figure 20.8 Human chromosome 2 and its homologs in chimpanzees and gorillas The banding patterns on stained chromosomes reveal that human chromosome 2 is derived from the fusion of two chromosomes that remain separate in the other great apes. Redrawn from Yunis and Prakash (1982).

Some of the differences are striking. Humans lack functional copies of some genes found in chimpanzees and gorillas. For example, a frameshift mutation in our gene for the enzyme CMP-sialic acid hydroxylase has resulted in the complete loss of a cell-surface sugar common in mammals (see Gagneux and Varki 2001). And humans lack regulatory sequences found in other mammals. One is an enhancer for the androgen receptor that, in chimpanzees and other mammals, triggers the development of sensory whiskers on the face and of spines on the penis (McLean et al. 2011)—although in the latter case, the resulting phenotypic difference between human and chimps is somewhat less dramatic than it might sound: The "spines" on a chimpanzee's penis are tiny keratinized bumps on the skin associated with sensory receptors (Hill 1946). Humans also have at least a few unique genes. One of these, called *JLJ33706,* arose as a result of mobile element insertions followed by nucleotide substitutions and encodes a protein expressed most strongly in the brain (Li et al. 2010).

Other differences among great ape genomes are more subtle. **Figure 20.9** provides an overview. The graph compares the sequence divergence, in 1-Mbp segments across the genome, for humans versus chimpanzees, gorillas, and orangutans. The overall divergence between humans and chimps is a little over 1%. This figure considers only nucleotide substitutions in portions of the genome

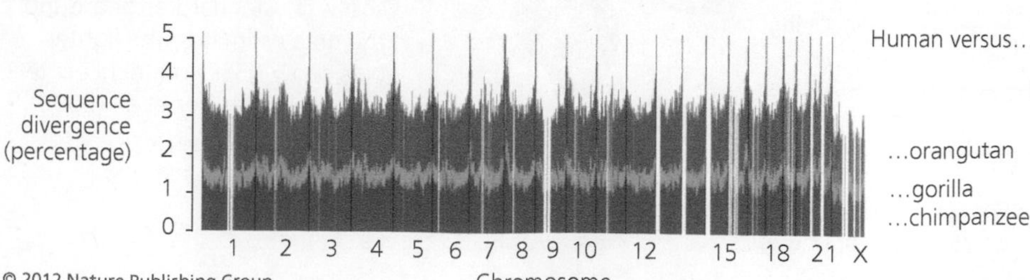

© 2012 Nature Publishing Group

Figure 20.9 Sequence divergence in the genomes of humans versus great apes From Scally et al. (2012).

Insights into hominid evolution from the gorilla genome sequence. A. Scally, J.Y. Dutheil, *Nature* 483: 169–175. Copyright © 2012 Nature Publishing Group. Used with permission.

where sequences can be aligned to one another (Britten 2002). Taking into account duplications, insertions, and deletions, about 4% of our genome differs from that of the chimpanzees (see Varki and Nelson 2007).

The phylogeny in **Figure 20.10** puts this number in perspective. The tree represents an estimate of evolutionary relationships based on shared features of whole genomes, rather than just alignable sequences (Sims et al. 2009a). Branch lengths are proportional to a measure of genomic distance. Humans and chimpanzees—and even humans and rhesus monkeys—are much more similar to each other than mice are to rats. The message is that the genomic divergence between humans and chimpanzees is rather small. Indeed, some biologists, including Derek Wildman and colleagues (2003), argue that humans and the chimpanzees are so closely related genetically that they belong together in the genus *Homo*.

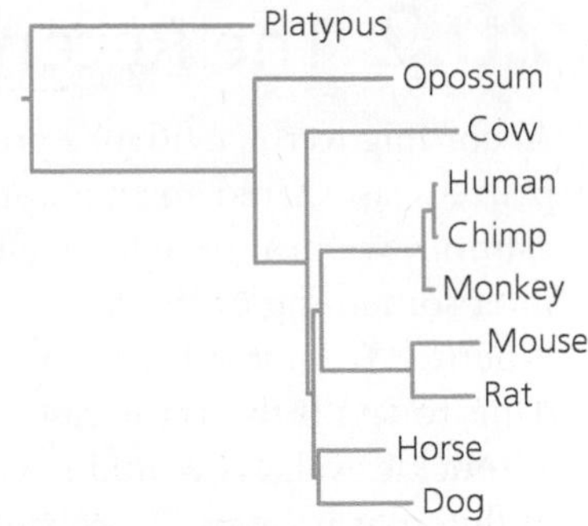

Figure 20.10 A whole-genome phylogeny of mammals Redrawn from Sims et al. (2009b).

As small as the differences between our genomes are, the genomes themselves are large. Upon completing a draft sequence of a chimpanzee genome, Tarjei Mikkelsen and colleagues (2005) reported that humans and chimpanzees are distinguished by about 35 million single-nucleotide substitutions, 5 million insertions and deletions, and an assortment of chromosomal rearrangements. About 29% of the proteins encoded in our genome are identical to the homologous protein in chimps. For the remaining proteins, the typical difference is two amino acid substitutions. Ever since, researchers have been working to determine which of these genetic differences between us and our closest relatives are the ones that make us human and them chimps (see Varki et al. 2008).

Some of these differences are among the relatively small number of genes that have been gained or lost in one lineage or the other. Xiaoxia Wang and colleagues (2006) identified 80 genes that are active in chimps but disabled in humans by loss-of-function mutations. These genes encode olfactory, taste and other chemoreceptors, and proteins involved in the immune response. Cécile Charrier and colleagues (2012) and Megan Dennis and colleagues (2012) investigated a gene unique to humans, *SRGAP2C,* that arose via partial duplication of another gene, *SRGAP2*. *SRGAP2* encodes a protein that influences the migration and maturation of nerve cells in the brain. *SRGAP2C* encodes a protein that binds to *SRGAP2*'s gene product and inhibits its activity. The novel human protein thus extends brain development and allows nerve cells to develop more dendrites (see Tyler-Smith and Xue 2012).

Other key differences between humans and chimps lurk among the amino acid substitutions in the two species' proteins. Rasmus Nielsen and colleagues (2005) combed the human and chimp genomes for genes with high ratios of nonsynonymous to synonymous substitutions. Many of the genes they turned up function in sensory perception, immune defense, tumor suppression, and spermatogenesis.

Still more of the genetic differences that make us human lie, as predicted decades ago by Marie-Claire King and Allan Wilson (1975), in the regulatory regions that control when, where, and in what amount each protein is made. A possible example, identified by Mehmet Somel and colleagues (2011), appears in **Figure 20.11**. The example involves a microRNA called miR-320b. MicroRNAs are short noncoding RNA molecules that bind to messenger RNAs and suppress their translation or mark them for degradation. Figure 20.11a shows that humans make more copies of miR-320b in the brain cells of the prefrontal cortex than do chimpanzees and rhesus macaques. Figure 20.11b shows the consequent reduction of expression of 18 known targets of miR-320b, eight of which are neuron-related genes. The phenotypic consequences remain to be discovered.

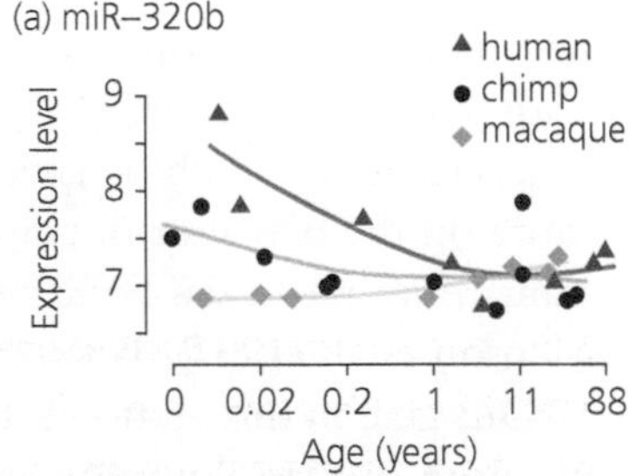

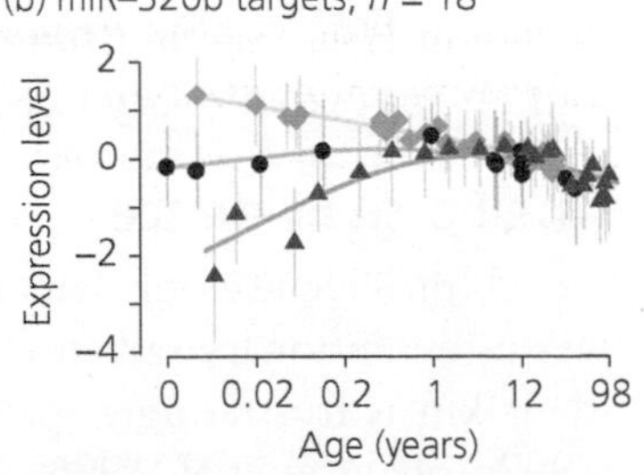

Figure 20.11 Expression of a microRNA and its targets in humans, chimps, and rhesus monkeys (a) Expression of miR-320b as a function of age. (b) Expression of miR-320b's target genes as a function of age. From Somel et al. (2011).

20.2 The Recent Ancestry of Humans

According to the evidence presented in Section 20.1, humans and the two chimpanzees last shared an ancestor roughly 5 to 7 million years ago. With appropriate caution, we can use what we know about humans, chimpanzees, and bonobos to infer something of the nature of that last common ancestor. It is probable that we inherited from it at least some of the behaviors that are shared by its three descendants today. If this is the case, then the last common ancestor, in addition to being a knuckle walker, would have had a broad, fruit-based diet and lived in a range of different habitats. It probably used tools to obtain and process food, and it may have hunted, as do living bonobos, chimpanzees, and humans (McGrew 2010b).

The last common ancestor also may have had culture—behavior that is taught and learned and varies among populations. Chimpanzees, like humans, exhibit cultural variation today (de Waal 1999; Whiten et al. 1999; Whiten 2005). Indeed, culture may have appeared well before our last common ancestor with chimps and bonobos, because orangutans have it too (van Schaik et al. 2003).

Other aspects of the last common ancestor's behavior are more elusive. Bonobos and common chimpanzees are equally close kin to humans, yet show striking differences in behavior (Parish and de Waal 2000). Elements of each's behavior resonate, for some observers, with the behavior shown by humans in at least some cultures. Chimpanzee societies, for example, are dominated by males that form strategic alliances, fight viciously, and sometimes stalk and kill their rivals. Bonobo societies, in contrast, are dominated by females that form strong bonds with each other, even when they are not kin, and that, while aggressive toward males, are less violent than chimpanzees. While chimpanzees sometimes engage in homosexual behaviors, they are primarily heterosexual. Bonobos, in contrast, have sex in all possible combinations, and for various reasons not directly connected with procreation. Perhaps the safest inference to draw is that we humans belong to a lineage in which behavior is flexible, both culturally and evolutionarily. For more on the behavior of chimpanzees and bonobos, and its possible relevance to human behavior, see Begun (1994), Parish (1994, 1996), de Waal (1997, 2005), Manson et al. (1997), Boesch and Tomasello (1998), and Wrangham (1999).

Our goal in this section is to review evidence on the pattern of evolution leading from our last common ancestor with the chimpanzees to ourselves. Has our history involved only the steady transformation of a single lineage, finally culminating in *Homo sapiens* **(Figure 20.12a)**, or have there been splits and extinctions in our recent evolutionary tree (Figure 20.12b)?

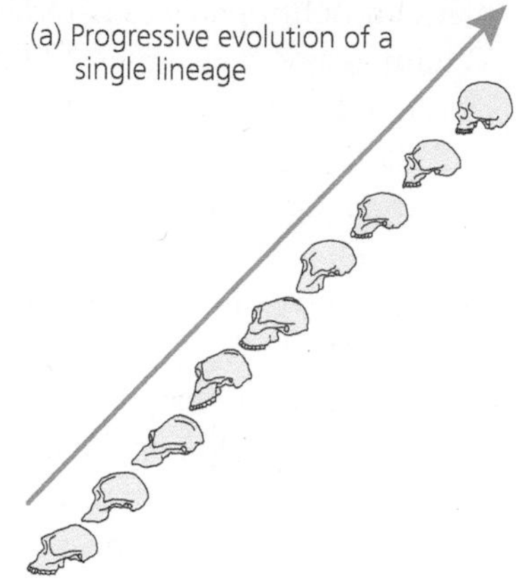

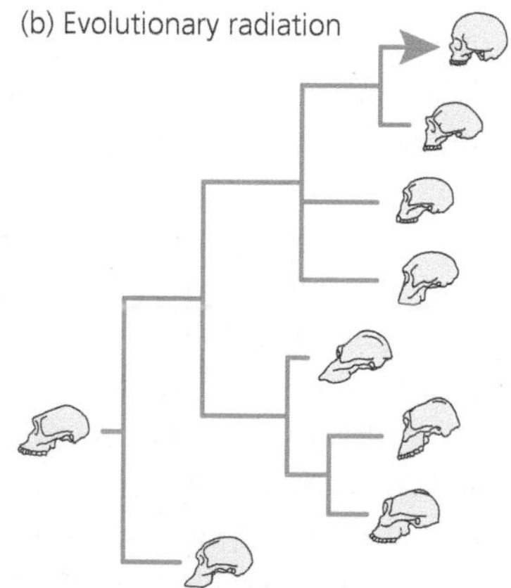

Figure 20.12 Two hypotheses about the pattern of human evolution since our lineage diverged from that of the chimpanzees

The Fossil Evidence

Fossils provide the only data available for distinguishing the steady transformation versus evolutionary radiation hypotheses. The fossil record for early humans and their kin is frustratingly sparse, but continually improving (see Tattersall 1995, 1997; Johanson et al. 1996; Wood and Leakey 2011). On the next several pages, we present illustrations and photos of some key specimens.

Paleoanthropologists disagree about the most appropriate names for some of these specimens. We use the names used by Wood (2010) in the belief that they will be the names most familiar to readers. In several cases, we note alternative names. We use the term *hominin* to describe any species more closely related to humans than to chimpanzees, but note that some paleoanthropologists still prefer the more traditional term *hominid* (see Gee 2001; Wood 2010).

Likewise, paleoanthropologists disagree about the number of species represented by the specimens in the figures (see Tattersall 1986, 1992; Wood 2010; Wood and Leakey 2011). For example, the specimens of *Homo habilis* and *Homo rudolfensis* we will see in Figure 20.19 are both from Koobi Fora, Kenya, and are both about 1.9 million years old. Some researchers consider them to be variants of the same species, whereas others consider them different species. As with the names, we have followed the classification used by Wood (2010).

The fossil record includes a diversity of hominins—species that lived after the human and chimpanzee lineages separated and are more closely related to humans than to chimpanzees.

Most of the time ranges noted in the figures are those given in Johanson et al. (1996). They differ somewhat from the estimates of other researchers, including those used in Figure 20.22 and those given by Strait et al. (1997) and used in Figure 20.23.

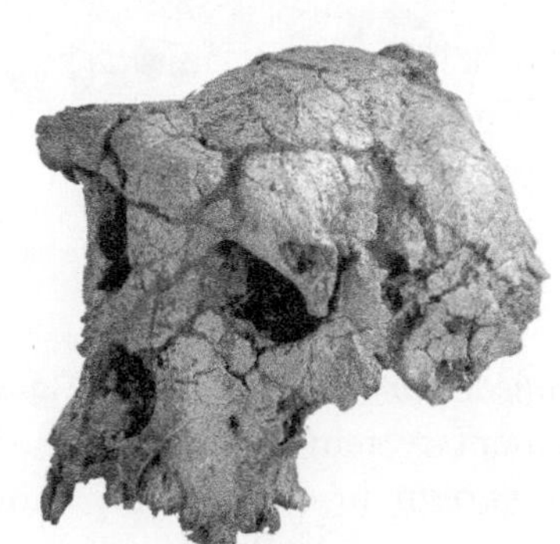

Name: *Sahelanthropus tchadensis*
Also known as: "*Toumaï*"
Specimen: TM 266-01-060-1
Age: 6–7 million years
Found by: Djimdoumalbaye Ahounta
Location: Djurab Desert, Chad
© 2002 Macmillan Publishers Ltd

Figure 20.13 A possible early hominin This 6- to 7-million-year-old skull, found by a member of a team led by Michel Brunet, may represent a close relative of our common ancestor with the chimpanzees. From Wood (2002).

Reprinted by permission from Macmillan Publishers Ltd: *Nature* 418: 133–135, copyright 2002.

We start with *Sahelanthropus tchadensis* **(Figure 20.13)**. Found in the Djurab Desert of Chad in July 2001 by Djimdoumalbaye Ahounta, a member of a team led by Michel Brunet, this nearly complete cranium stunned paleoanthropologists (Brunet et al. 2002; Gibbons 2002; see also Brunet et al. 2005). For one thing, it is 6 to 7 million years old. This places it toward the older end of the window during which molecular biologists estimate that humans diverged from chimpanzees. For another, it shows a curious mixture of traits. As Bernard Wood (2002) describes it, its small braincase (320–380 cm^3) makes it look, from the back, like a chimpanzee. From the front, however, its relatively flat face makes it look like an *Australopithecus, Kenyanthropus,* or *Homo* from as recently as 1.75 million years ago. In other words, it looks like a closer relative of humans than anyone expected in a fossil so old. *Sahelanthropus tchadensis* could be a close relative of the last common ancestor—or even, in principle, the last common ancestor itself.

Brunet and colleagues believe that *Sahelanthropus* is, indeed, an early hominin, a descendant of the last common ancestor on the human side of the evolutionary tree. Many other paleoanthropologists are inclined to agree (see Cela-Conde and Ayala 2003; Gibbons 2005; Wilford 2005). The view is not unanimous, however.

Among the dissenters are Brigitte Senut and Martin Pickford, discoverers of a rival candidate for the title of oldest known hominin. Their find, *Orrorin tugenensis,* lived about 6 million years ago in what is now Kenya. It is known primarily from three thighbones (Aiello and Collard 2001; Senut et al. 2001; Gibbons 2002). Shortly after Brunet and colleagues reported the discovery of *Sahelanthropus,* Senut and Pickford, along with Milford Wolpoff and John Hawks, suggested that *Sahelanthropus* belongs to the lineage that produced chimpanzees, or even the one that led to gorillas (Wolpoff et al. 2002). Brunet (2002) rejected this idea.

Part of the difficulty in interpreting *Sahelanthropus* is due to its crushed and distorted skull. To overcome this problem, Brunet collaborated with a team led by Chistoph Zollikofer and Marcia Ponce de León. The researchers X-rayed the skull with a CT scanner and used the resulting images to prepare a corrected three-dimensional reconstruction. The reconstructed *Sahelanthropus* resembles known hominins more strongly than it does chimpanzees or gorillas, bolstering Brunet's view that his fossil represents the human side of the family tree (Guy et al. 2005; Zollikofer et al. 2005). The dissenters have declined to concede (Wolpoff et al. 2006). The issue is likely to be resolved only by the discovery of additional fossils, including postcranial remains.

Name: ***Ardipithecus ramidus***
Originally named as: *Australopithecus ramidus*
Specimen: ARA-VP-1/128
Age: 4.4 million years
Found by: T. Assebework
Location: Aramis, Ethiopia
Color photo of same species:
Johanson et al. (1996), page 116

Species Time Range: ~4.4 mya

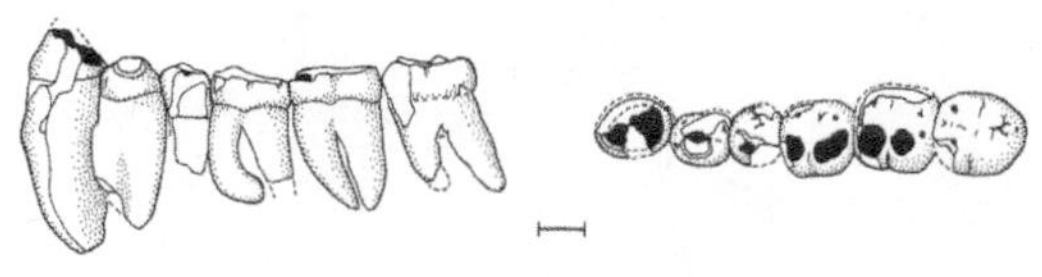

© 1994 Nature Publishing Group

Figure 20.14 A possible early hominin By Laszlo Meszoly, after Figure 3b in White et al. (1994); see also White et al. (1995). Scale bar = 1 cm.
"Australopithecus ramidus, a new species of early hominid from Aramis, Ethiopia." *Nature* 371: 306–312. Copyright © 1994 Nature Publishing Group. Reprinted with permission.

Another possible early hominin, *Ardipithecus ramidus,* appears in **Figure 20.14**. Originally described on the basis of fragmentary remains, including the teeth shown here, *Ardipithecus ramidus* is now known in considerably more detail (White et al. 1994; White et al. 2009). It stood about 120 cm tall; its face was small, and it had a brain the size of a chimpanzee's. Various features of its skeleton indicate that it was a good climber and was also capable of walking upright on the ground. The team that discovered and described *Ardipithecus ramidus* believes it is more closely related to humans than to the chimpanzees (White et al. 2006; White et al. 2009; White et al. 2010; see also Strait and Grine 2004). If *Ardipithecus ramidus* is a hominin, then its older congener *Ardipithecus kadabba* (Haile-Selassie et al. 2004) likely is too. Other researchers feel it is premature to draw conclusions about whether *Ardipithecus* is more closely related to humans or to the chimpanzees (Harrison 2010; Sarmiento 2010).

Figure 20.15 (opposite page) shows examples of undisputed early hominins: the gracile australopithecines and *Kenyanthropus.* The species depicted in Figure 20.15a and b, *Australopithecus africanus* and *Australopithecus afarensis,* have skulls with small braincases (400 to just over 500 cm^3) and relatively large, projecting faces (Johanson et al. 1996). Female *Australopithecus africanus* and *Australopithecus afarensis* grew to heights of about 1.1 meters (3′7″), whereas the males were some 1.4 to 1.5 meters (4′7″ to 4′11″) tall (but see Reno et al. 2003). Both species walked on two legs. Evidence for their erect posture comes from many bones of the skeleton, including the hips, knees, feet, limb proportions, and vertebral column, all of which are anatomically modified to permit upright posture and the support of the body mass on two rather than four feet. Other evidence for bipedal locomotion appears in the photo in **Figure 20.16**: fossilized footprints at Laetoli, Tanzania, of a pair of *A. afarensis* that walked side by side through fresh ash from the Sadiman volcano 3.6 million years ago (Stern and Susman 1983; White and Suwa 1987).

Kenyanthropus platyops (Figure 20.15c), 3.5 million years old, was discovered in August 1999 by J. Erus, an assistant to Meave Leakey and colleagues (2001). *Kenyanthropus platyops* has a brain the same size as that of *Australopithecus afarensis* (Figure 20.15b), which lived at the same time, and a variety of other ancestral

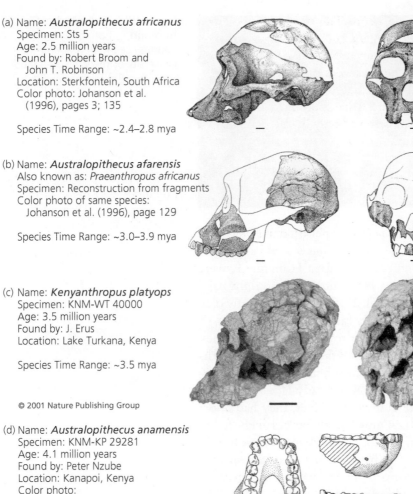

(a) Name: *Australopithecus africanus*
Specimen: Sts 5
Age: 2.5 million years
Found by: Robert Broom and John T. Robinson
Location: Sterkfontein, South Africa
Color photo: Johanson et al. (1996), pages 3; 135

Species Time Range: ~2.4–2.8 mya

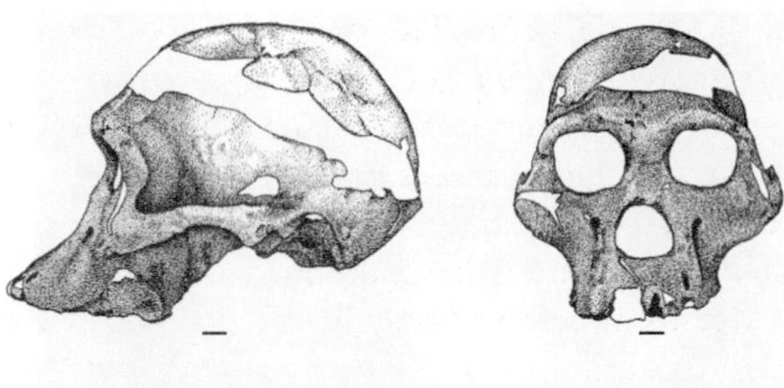

(b) Name: *Australopithecus afarensis*
Also known as: *Praeanthropus africanus*
Specimen: Reconstruction from fragments
Color photo of same species: Johanson et al. (1996), page 129

Species Time Range: ~3.0–3.9 mya

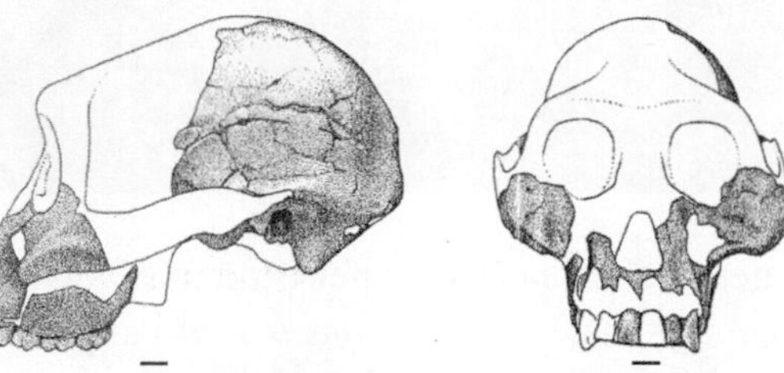

(c) Name: *Kenyanthropus platyops*
Specimen: KNM-WT 40000
Age: 3.5 million years
Found by: J. Erus
Location: Lake Turkana, Kenya

Species Time Range: ~3.5 mya

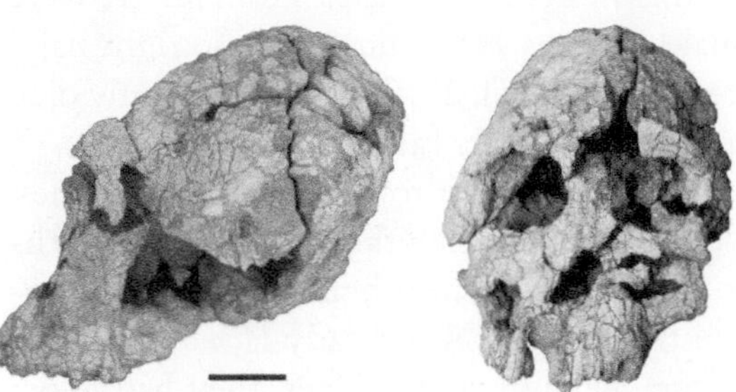

© 2001 Nature Publishing Group

(d) Name: *Australopithecus anamensis*
Specimen: KNM-KP 29281
Age: 4.1 million years
Found by: Peter Nzube
Location: Kanapoi, Kenya
Color photo: Johanson et al. (1996), page 123

Species Time Range: ~3.9–4.2 mya

© 1995 Nature Publishing Group

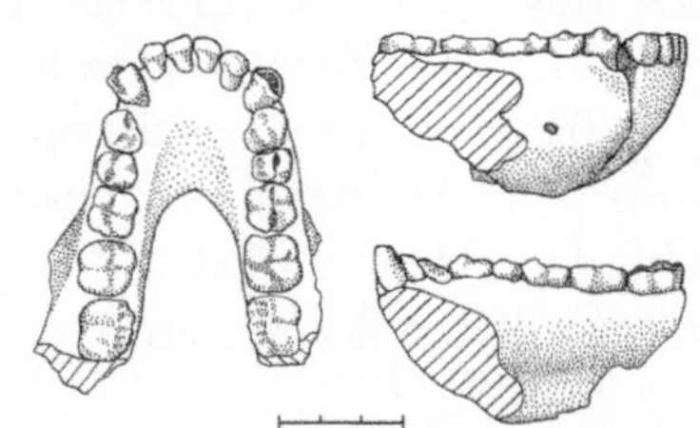

Figure 20.15 Archaic hominins (a) By Don McGranaghan, page 70 in Tattersall (1995). Scale bar = 1 cm. (b) By Don McGranaghan, page 146 in Tattersall (1995). Scale bar = 1 cm. (c) Figure 1 in Leakey et al. (2001). Scale bar = 3 cm. (d) By Laszlo Meszoly, after Figure 1a in Leakey et al. (1995). Scale unit = 1 cm.

(c) "Australopithecus ramidus, a new species of early hominid from Aramis, Ethiopia." M. G. Leakey, F. Spoor. *Nature* 371: 306–312. Copyright © 2001 Nature Publishing Group. Reprinted with permission. (d) "New 4-million-year-old hominid species from Kanapoi and Allia Bay, Kenya." M.G. Leakey, C.S. Feibel. *Nature* 376: 565–571. Copyright © 1995 Nature Publishing Group. Reprinted with permission.

skull characters. At the same time, *K. platyops* has smaller teeth and a flatter and more human-looking face than *A. afarensis*, or any other species traditionally classified as *Australopithecus*. Feeling that *K. platyops* does not fit into either *Australopithecus* or *Homo,* Leakey and colleagues assigned it to a new genus. Tim White (2003), in contrast, argues that the more human appearance of *K. platyops* is an illusion resulting because the skull has been fragmented and distorted by the rock it is preserved in. Were the skull not so distorted, White believes it would fall within the range of variation already known for other fossils of similar age that are assigned to genus *Australophithecus*. White maintains that all these fossils, *K. platyops* included, belong to a single lineage connecting *A. anamensis* (Figure 20.15d) to *A. afarensis* (Figure 20.15b). The fossils show considerable structural variation, but given the variation we can observe today among humans, among bonobos, and among chimpanzees, White asserts that we should expect the recent ancestors of these species to have been variable as well.

The species depicted in Figure 20.15d, *Australopithecus anamensis,* is less well known than the more recent species in the figure. The structure and size of a tibia from *A. anamensis* indicates that its owner was a biped somewhat larger than *A. afarensis* (Leakey et al. 1995).

Figure 20.16 Footprints of a pair of *Australopithecus afarensis* These 3.6-million-year-old footprints from Laetoli, Tanzania, were made by a pair of individuals who walked side by side through fresh ash from a volcanic eruption.

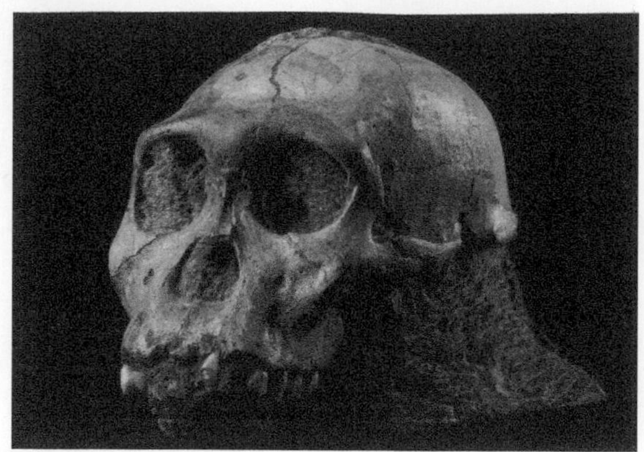

Name: *Australopithecus sediba*
Specimen: MH1
Age: 1.98 million years
Found by: Matthew Berger
Location: Malapa, South Africa

Figure 20.17 An archaic hominin Photo by Brett Eloff, courtesy Lee R. Berger and the University of the Witwatersrand. Age from Pickering et al. (2011).

Archaic hominins assigned to the genus *Australopithecus* persisted until less than 2 million years ago. The skull of a teenage male *Australopithecus sediba* appears in **Figure 20.17**. The skeletal remains of this individual, Malapa Hominin 1, were found next to those of the adult female, Malapa Hominin 2, whose right hand appears on the first page of this chapter (Berger et al. 2010). Both apparently died when they fell through a sinkhole onto the floor of a cave (Dirks et al. 2010).

Examples of three species formerly known as the robust australopithecines, now called *Paranthropus*, appear in **Figure 20.18a–c**. Like the gracile australopithecines in Figure 20.15, these species had relatively small braincases (in most instances between those of the gracile australopithecines and early *Homo* in relative size) and very large faces. Unlike the gracile australopithecines, they had enormous cheek teeth, robust jaws, and massive jaw muscles, sometimes anchored to a bony crest running along the centerline on the top of the skull (Johanson et al. 1996). These adaptations for powerful chewing have given one of the species, *Paranthropus boisei,* the nickname "nutcracker man." Its nickname notwithstanding, what *P. boisei* actually ate was grasses and sedges (Cerling et al. 2011). The robust australopithecines were about the same size as the gracile forms, and all were bipeds.

Berhane Asfaw and colleagues (1999) named the specimen shown in Figure 20.18d *Australopithecus garhi* because of the traits it shares with other members of the genus. Unlike those others, however, *A. garhi* has enormous teeth. Following Bernard Wood (2010), we have grouped it here with the other megadont, or large-toothed, archaic hominins.

Figure 20.19 shows examples of two forms whose taxonomic status is controversial. *Homo habilis* (Figure 20.19a) and *Homo rudolfensis* (Figure 20.19b) are transitional hominins that many researchers consider to be among the first humans. Both lived in the same place at about the same time, a factor that contributes to a dispute over whether they are different species or just large and small individuals of a single species. Working in Olduvai Gorge, Tanzania, Robert Blumenschine and colleagues (2003) discovered a 1.8-million-year-old set of jaws and teeth. The researchers assign the specimen to the species *H. habilis,* but note that it bears a decided resemblance to the *H. rudolfensis* shown in Figure 20.19b. Blumenschine and colleagues believe that their specimen argues against the designation of *H. rudolfensis* as a separate species, but it is unlikely to settle the issue (see Tobias 2003). Intriguingly, Blumenschine and colleagues found their *H. habilis* in association with stone tools, as well as animal bones bearing the marks of butchery.

Were *Homo habilis* and *Homo rudolfensis* human—as their traditional genus name, *Homo,* implies? The *H. habilis* specimen shown here has a braincase volume

(a) Name: ***Paranthropus robustus***
Also known as: *Australopithecus robustus*
Specimen: SK 48
Age: 1.5–2.0 million years
Found by: Fourie
Location: Swartkrans, South Africa

Species Time Range: ~1.0–2.0 mya

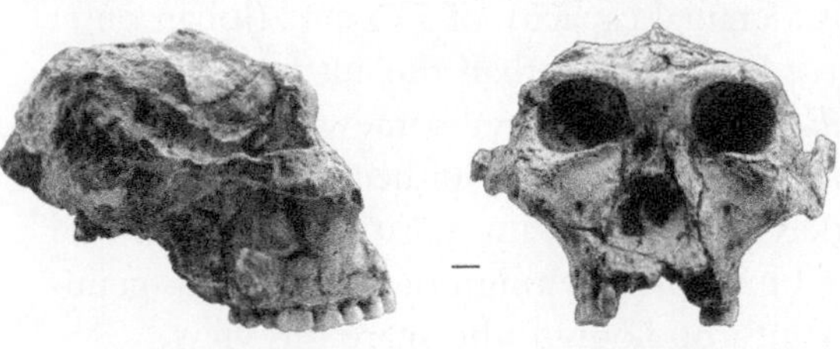

(b) Name: ***Paranthropus boisei***
Also known as: *Australopithecus boisei*
Specimen: KNM-ER 406
Age: 1.7 million years
Found by: Richard Leakey and H. Mutua
Location: Koobi Fora, Kenya
Color photo: Johanson et al. (1996), pages 54; 159; 160

Species Time Range: ~1.4–2.3 mya

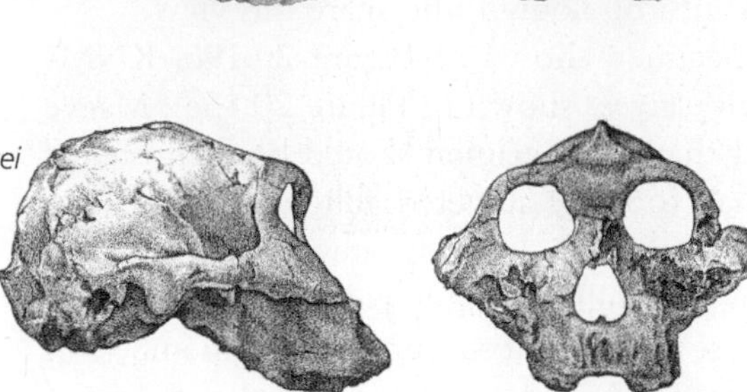

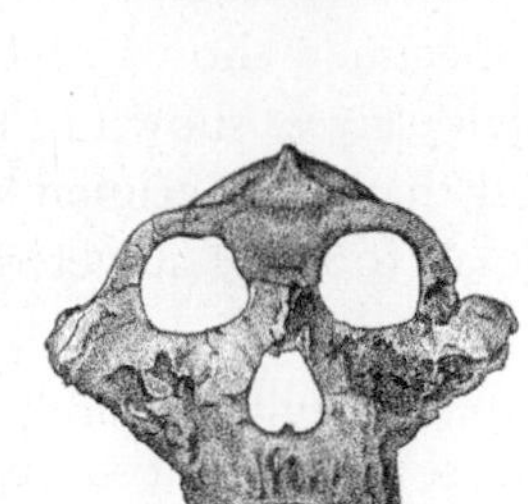

(c) Name: ***Paranthropus aethiopicus***
Also known as: *Australopithecus aethiopicus*
Specimen: KNM-WT 17000 (Black Skull)
Age: 2.5 million years
Found by: Alan C. Walker
Location: Lake Turkana, Kenya
Color photo: Johanson et al. (1996), pages 153; 154

Species Time Range: ~1.9–2.7 mya

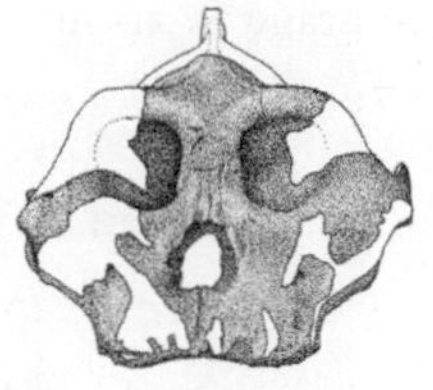

(d) Name: ***Australopithecus garhi***
Specimen: BOU-VP-12/130
Age: 2.5 million years
Found by: Yohannes Haile-Selassie
Location: Bouri Formation, Ethiopia

Species Time Range: ~2.5 mya

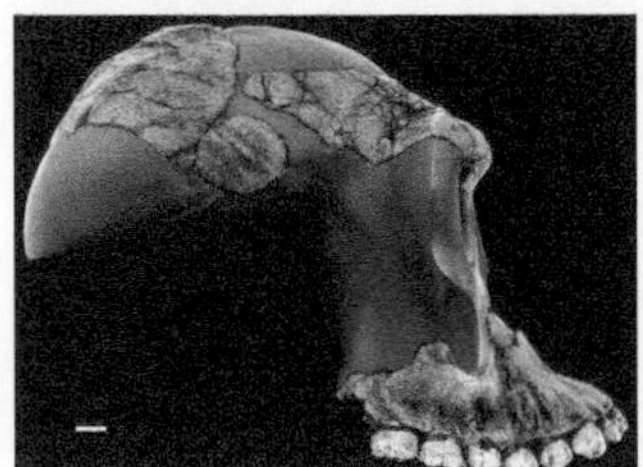

Figure 20.18 Megadont archaic hominins (a) Photos by David Brill, pages 108 and 150 in Johanson et al. (1996). Scale bar = 1 cm. (b) By Don McGranaghan, page 131 in Tattersall (1995). Scale bar = 1 cm. (c) By Don McGranaghan, page 195 in Tattersall (1995). Scale bar = 1 cm. (d) Photo by David Brill. Scale bar = 1 cm.

(a) Name: ***Homo habilis***
Also known as: *Australopithecus habilis*
Specimen: KNM-ER 1813
Age: 1.9 million years
Found by: Kamoya Kimeu
Location: Koobi Fora, Kenya
Color photo: Johanson et al. (1996), pages 6; 175

Species Time Range: ~1.6–1.9 mya

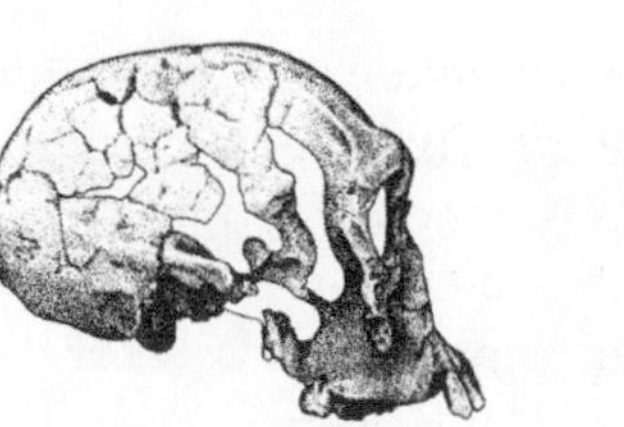

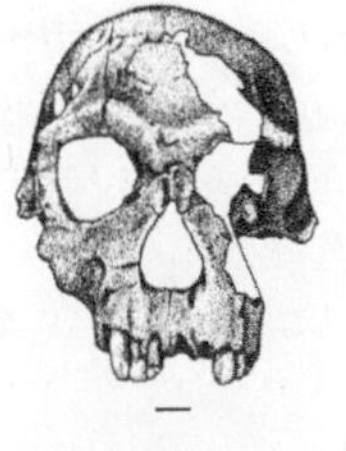

(b) Name: ***Homo rudolfensis***
Also known as: *H. habilis; A. rudolfensis; K. rudolfensis*
Specimen: KNM-ER 1470
Age: 1.8–1.9 million years
Found by: Bernard Ngeneo
Location: Koobi Fora, Kenya
Color photo: Johanson et al. (1996), pages 178; 179

Species Time Range: ~1.8–2.4 mya

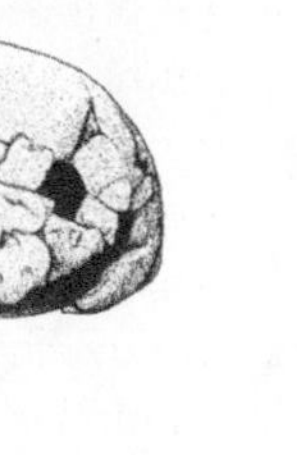

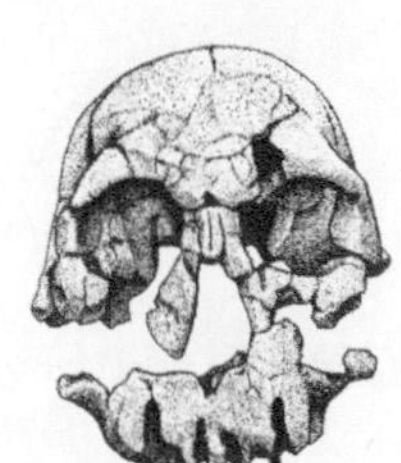

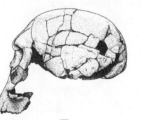

Figure 20.19 Transitional hominins (a) By Don McGranaghan, page 134 in Tattersall (1995). Scale bar = 1 cm. (b) By Don McGranaghan, page 133 in Tattersall (1995). Scale bar = 1 cm.

of just 510 cm^3; the *H. rudolfensis* has a cranial capacity of 775 cm^3 (Johanson et al. 1996). This means both forms have larger brains than the australopithecines, though just barely larger in *H. habilis*'s case. Both have somewhat flatter faces than the australopithecines, but they overlap the australopithecines in tooth and body size. Bernard Wood and Mark Collard (1999) argue that *H. habilis* and *H. rudolfensis* should not be considered human, and assign them instead to genus *Australopithecus*. There are other paleoanthropologists who share this view.

Given that the *Homo rudolfensis* specimen shown in Figure 20.19b, KNM-ER 1470, resembles the *Kenyanthropus platyops* shown in Figure 20.15c), Maeve Leakey and colleagues (2001) suggest that the specimen should be rechristened *Kenyanthropus rudolfensis*. This proposal, too, has attracted adherents (see Aiello and Collard 2001; Lieberman 2001). KNM-ER 1470, whatever it is, seems to bear the brunt of considerable pushing and pulling among paleoanthropologists.

Figure 20.20 shows examples of more recent, but still premodern members of the genus *Homo* from Africa, Europe, and Asia. It is undisputed that even the oldest of them, *Homo ergaster* (Figure 20.20d) should be considered **human** and thus assigned to our own genus. Its braincase volume is 850 cm^3 (Johanson et al. 1996). In present-day humans, the average is about 1,200 cm^3; some examples

(a) Name: *Homo neanderthalensis*
Specimen: Saccopastore 1
Age: ~120,000 years
Found by: Mario Grazioli
Location: Saccopastore quarry, Rome, Italy
Color photo: Johanson et al. (1996), pages 213; 214

Species Time Range: ~0.03–0.3 mya

(b) Name: *Homo heidelbergensis*
Specimen: Broken Hill 1
Age: ~300,000 years
Found by: Tom Zwigelaar
Location: Kabwe, Zambia
Color photo: Johanson et al. (1996), pages 209; 210

Species Time Range: ~0.2–0.6 mya

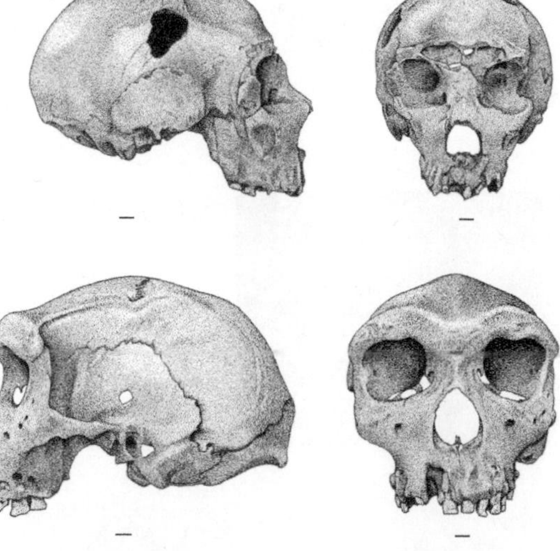

(c) Name: *Homo erectus*
Specimen: Sangiran 17
Age: ~800,000 years
Found by: Mr. Towikromo
Location: Sangiran, Java, Indonesia
Color photo: Johanson et al. (1996), pages 192; 193

Species Time Range: ~0.4–1.2 mya

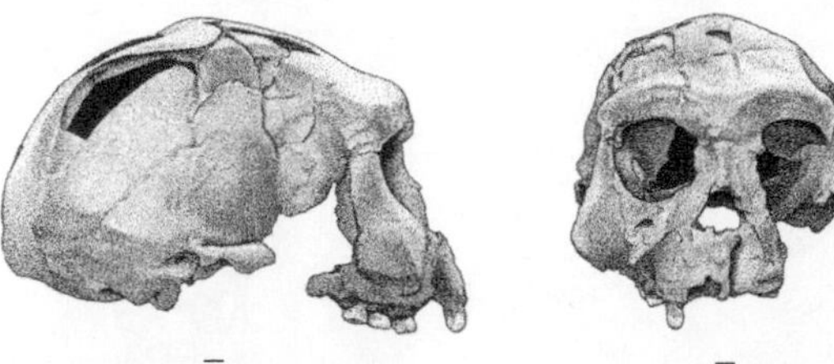

(d) Name: *Homo ergaster*
Also known as: (African) *Homo erectus*
Specimen: KNM-ER 3733
Age: 1.75 million years
Found by: Bernard Ngeneo
Location: Koobi Fora, Kenya
Color photo: Johanson et al. (1996), pages 180; 181

Species Time Range: ~1.5–1.8 mya

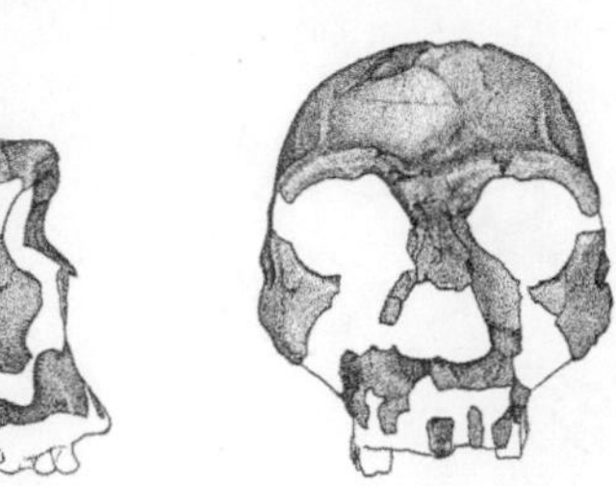

Figure 20.20 Premodern *Homo* (a) By Don Mc-Granaghan, page 83 in Tattersall (1995). Scale bar = 1 cm. (b) By Don McGranaghan, page 54 in Tattersall (1995). Scale bar = 1 cm. (c) By Don McGranaghan, page 172 in Tattersall (1995). Scale bar = 1 cm. (d) By Don McGranaghan, page 138 in Tattersall (1995). Scale bar = 1 cm.

of *H. sapiens* have braincase volumes as large as 2,000 cm^3. Even at two-thirds today's average, however, *H. ergaster* has a much larger brain than any of the other fossils we have discussed so far. Furthermore, compared to the fossils in Figures 20.13, 20.15, 20.17, 20.18, and 20.19, it has a number of other features characteristic of humans. These include a relatively smaller, flatter face; smaller teeth and jaws; greater height; longer legs; and reduced sexual size dimorphism.

Name: ***Homo sapiens***
Specimen: Cro-Magnon I
Age: 30,000 to 32,000 years
Found by: Louis Lartet and Henry Christy
Location: Abri Cro-Magnon,
Les Eyzies, France
Color photo: Johanson et al. (1996), pages 245; 246

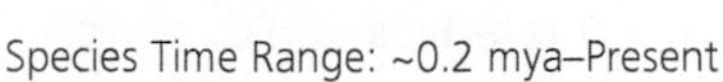
Species Time Range: ~0.2 mya–Present

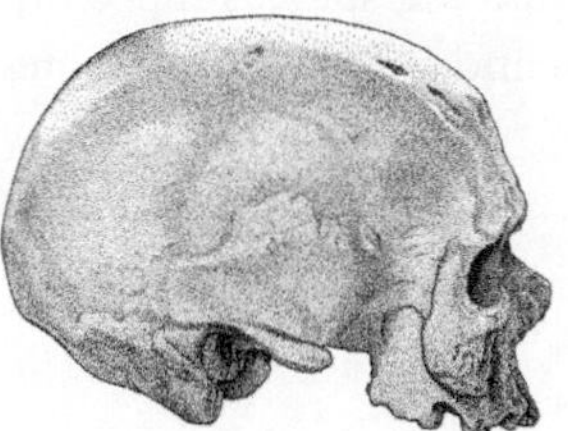

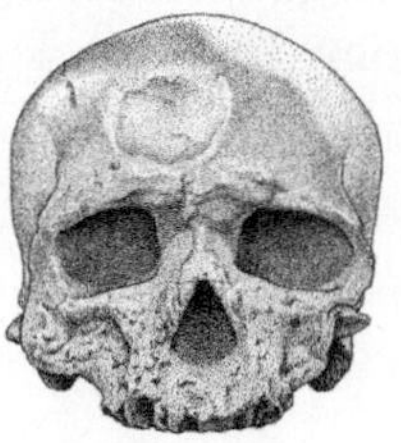

Figure 20.21 An anatomically modern *Homo sapiens* By Don McGranaghan, page 25 in Tattersall (1995). Scale bar = 1 cm.

Anatomically modern *Homo sapiens,* including Cro-Magnon I, whose skull appears in **Figure 20.21**, differ from earlier forms in a variety of traits (Johanson et al. 1996). Modern humans have large braincases. Cro-Magnon I's is over 1,600 cm^3, substantially over the present-day average. Associated with their large braincases, modern humans have high, steep foreheads. They also have relatively short, flat, vertical faces and prominent noses. Cro-Magnon I was a man who died in middle age about 30,000 years ago. His skeleton was found in a prepared grave along with those of two other adult men, an adult woman, and an infant. The group had been buried with an assortment of animal bones, jewelry, and stone tools.

Interpreting the Fossil Evidence

Figure 20.22 summarizes the fossil evidence we have discussed. On the upper left are the possible early hominins, apparently close relatives to our last common ancestor with the chimpanzee. In the center are the archaic hominins, most of them gracile australopithecines and all of them more closely related to us than to the chimpanzees. From the archaic hominins emerge two distinct groups.

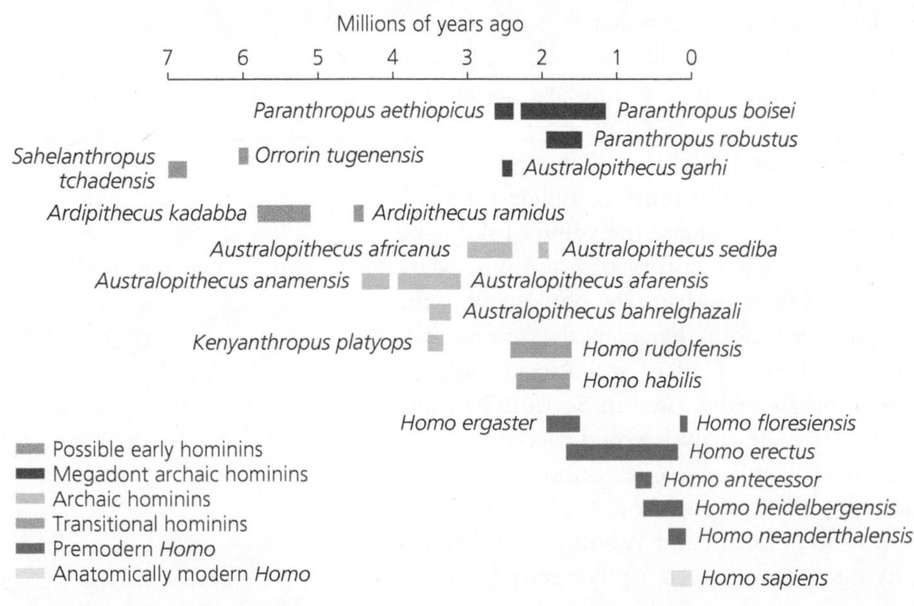

Figure 20.22 Summary of fossil evidence on the recent ancestry of humans The horizontal axis gives approximate time ranges for the species we have mentioned. Colors indicate rough groupings based on morphological and (inferred) behavioral and ecological similarity. After Wood (2010); Wood and Leakey (2011).

Above and to the right are the megadont hominins. Below and to the right are the transitional hominins, premodern members of genus *Homo* (that is, humans), and anatomically modern humans. Can we organize these fossils any more coherently, by arranging them on an evolutionary tree?

Paleoanthropologists have tried to reconstruct the evolutionary history of the hominins with a two-step process (Strait et al. 1997). First, the researchers use a cladistic analysis to estimate the evolutionary relationships among the various fossil species. Then they make educated guesses about which fossil species represent ancestors that lived at the branch points of the cladogram and which fossil species represent extinct side branches.

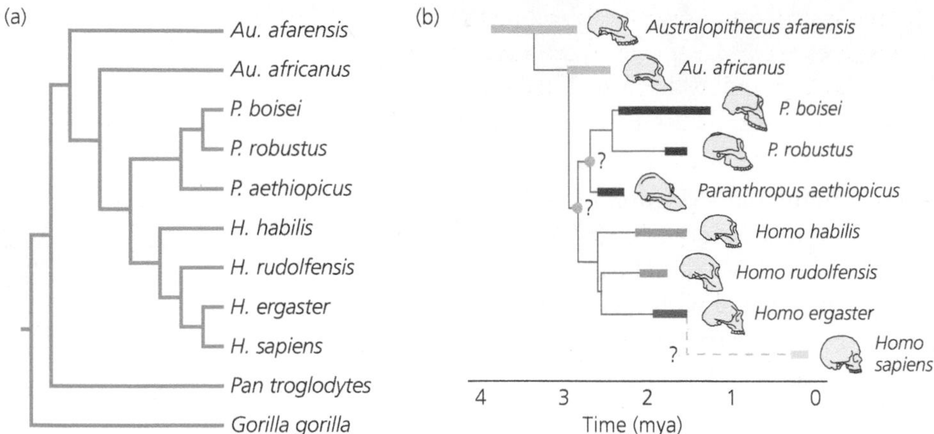

Figure 20.23 Cladogram and phylogeny of *Homo sapiens* and its recent ancestors and extinct relatives (a) Cladogram of three extant hominins (gorilla, common chimp, and human) and several extinct hominins. (b) A hypothesis about the ancestor–descendant relationships implied by the cladogram in (a). Heavy bars, colored as in Figure 20.22, show known time ranges; heavy dashes show suspected time ranges. After Strait et al. (1997).

Results of one such study, by David S. Strait and colleagues (1997), appear in **Figure 20.23**. Included are Strait et al.'s cladogram (Figure 20.23a) and a hypothesis about what the cladogram tells us concerning the phylogenetic relationships among the various species (Figure 20.23b). The cladogram is based on a variety of skull and tooth characters. Note that in the cladogram, the lengths of the branches are meaningless; the only information encoded in the cladogram is in the order of branching. The hypothesized phylogeny (Figure 20.23b) makes educated guesses about the actual lengths ascribed to the branches in the cladogram. For example, the branch leading to *Paranthropus aethiopicus* is nonzero, so that *P. aethiopicus* is a sister species to *P. boisei* and *P. robustus*. Another possibility, not shown here, is that the length of the branch leading to *P. aethiopicus* is zero, so that *P. aethiopicus* is the common ancestor of *P. boisei* and *P. robustus*.

The phylogenetic relationships among the species of fossil hominins have not been definitively established.

Nearly all such analyses performed to date rely heavily on characters of skulls and teeth. The reason is simply that the fossil record is more complete for skulls and teeth than for other parts of the skeleton. To evaluate the value of skull and tooth characters for reconstructing the evolutionary history of hominins, Mark Collard and Bernard Wood (2000) attempted to reconstruct the phylogeny of the living apes using a cladistic analysis of skull and tooth characters that are equivalent to the ones typically used for hominin fossils. Collard and Wood took the well-established molecular phylogeny of the apes (discussed in Section 20.1 and illustrated in Figure 20.4) as the "truth." If their analysis could reconstruct the truth as known from the molecules, then we can have some confidence in phylogenies of hominin fossils reconstructed by the same method.

The startling and depressing result was that Collard and Wood's cladistic analysis of skull and tooth characters in living apes produced a phylogeny in which gorillas and orangutans are closest relatives, chimpanzees are their next of kin, and humans branch first from the lineage that will become the other great apes. In other words, the analysis failed completely to recover the known phylogeny. This does not mean that cladistic analyses of morphology are a poor method in general for reconstructing the phylogeny of the apes or the hominins. Indeed, cladistic analyses of soft-tissue characters yield ape phylogenies that match the molecular phylogeny exactly (Gibbs et al. 2000, 2002). Nor does it mean that cladistic analyses of skull and tooth characters are unreliable for vertebrates in general. It simply suggests that cladistic analyses of skull and tooth characters may not produce a reliable answer to the question we are interested in here.

David Strait and Frederick Grine (2004) felt that this conclusion was too pessimistic. They suspected that a cladistic reconstruction using skulls and teeth could recover the true phylogeny of the living apes if extinct species were added to the analysis. They used methods similar to Collard and Wood's to infer the evolutionary relationships among gibbons, orangutans, the chimpanzees, humans, and most of the fossil taxa we have discussed in this section. The results were heartening. The relationships among the living species matched the molecular phylogeny exactly. This suggests we can have some confidence in what the tree suggests about the relationships among the fossil taxa, which was largely consistent with Figure 20.23. And it indicates that gathering more evidence will yield better answers. As David Begun (2004) observed, the mantra of every paleontologist is: "We need more fossils!"

Some Answers

Although the phylogeny of the fossil hominins is not yet known with certainty, the evidence we have reviewed gives a general answer to the question we posed at the beginning of this section. The pattern of evolution leading from our common ancestor with the chimpanzees to ourselves has not been simple. Instead, speciation has produced a diversity of lineages. Throughout most of the last 4 million years, multiple species, perhaps even as many as five at a time, have coexisted in Africa (see Tattersall 2000). For example, specimen KNM-ER 406 (Figure 20.18b) and specimen KNM-ER 3733 (Figure 20.20d) clearly represent different species **(Figure 20.24)**. Both were found at Koobi Fora, Kenya, in sediments of nearly the same age. *Paranthropus boisei* and *Homo ergaster* knew each other, but only one belonged to a lineage that persists today. We *Homo sapiens* are the lone survivors of an otherwise extinct radiation of bipedal African hominins.

The hominin fossil record is sufficiently detailed to allow us to conclude that *Homo sapiens* is the sole survivor among a diversity of species.

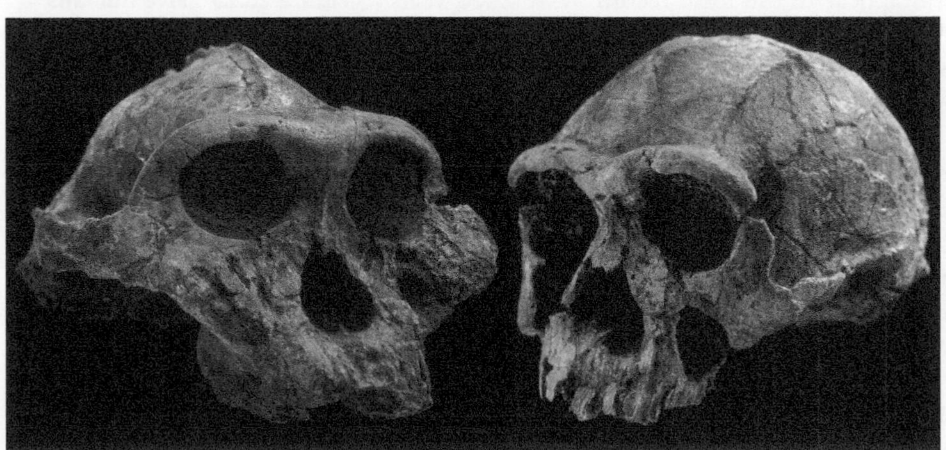

Figure 20.24 Evidence of a hominin radiation *Paranthropus boisei* (specimen KNM-ER 406, left) and *Homo ergaster* (specimen KNM-ER 3733) both lived in what is now Koobi Fora, Kenya, about 1.7 million years ago. Photos by David Brill, from Johanson, Edgar, and Brill (1996).

20.3 Origin of the Species *Homo sapiens*

In Section 20.2 we met five specimens that are uncontroversially human **(Figure 20.25)**. We have used for them the names *Homo ergaster, H. erectus, H. heidelbergensis, H. neanderthalensis,* and *H. sapiens.* There is uncertainty and debate, however, over how many species they actually represent and how modern humans, *Homo sapiens,* emerged from among the others.

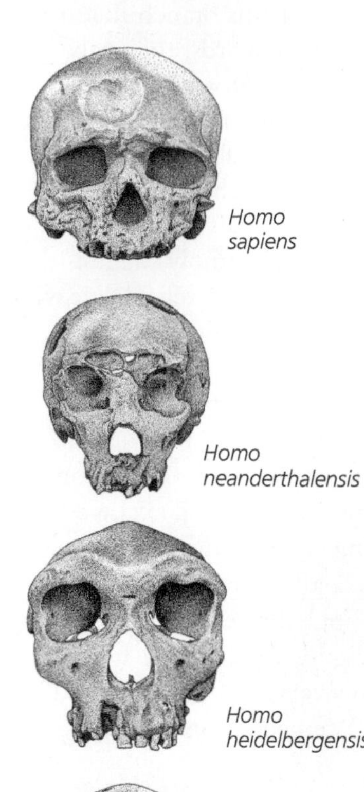

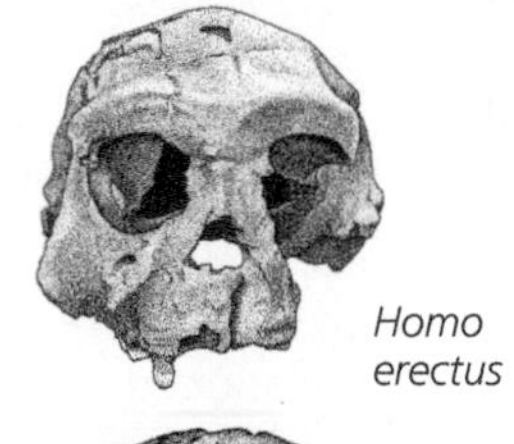

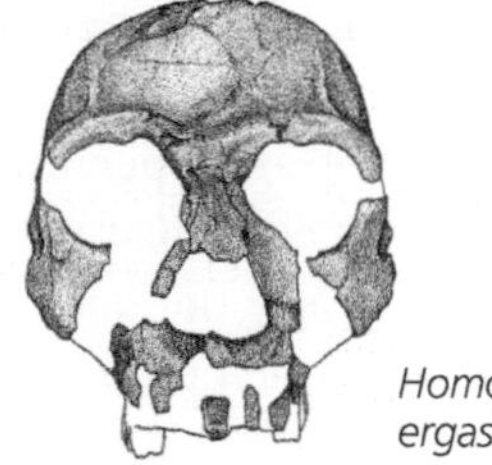

Figure 20.25 Five humans For details and credits, see Figures 20.20 and 20.21.

Controversies over the Origin of Modern Humans

Paleoanthropologists are divided on the taxonomic status of *H. ergaster* and *H. erectus.* Some researchers consider these two forms to be regional variants of a single species (*H. erectus*), whereas others consider *H. erectus* to be a distinct Asian species descended from the African species *H. ergaster.* Likewise, some researchers consider *H. neanderthalensis* and *H. heidelbergensis* to be regional variants of transitional forms between *H. erectus* and modern *H. sapiens.* Others consider them to be distinct species, with *H. heidelbergensis* descended from *H. ergaster,* and *H. neanderthalensis* descended from *H. heidelbergensis* (see Tattersall 1997). Yet another species, *Homo antecessor,* has been suggested to be the common ancestor of both Neandertals and modern humans (Bermúdez de Castro et al. 1997; Arsuaga et al. 1999). Paleoanthropologists generally agree that modern humans are the descendants of some or all of the populations in the *H. ergaster/erectus* group. However, how and where the transition from *H. ergaster/erectus* to *H. sapiens* took place is a matter of debate.

All hominins before *H. ergaster/erectus* were confined to Africa. The oldest examples of *H. ergaster/erectus,* however, appear in the fossil record nearly simultaneously at Koobi Fora in Africa, at Dmanisi in the Caucasus region of eastern Europe, at Longgupo Cave in China, and at Sangiran and Mojokerto in Java—all 1.6 to 1.9 million years ago (Gibbons 1994; Swisher et al. 1994; Gabunia and Vekua 1995; Huang Wanpo et al. 1995; Wood and Turner 1995; Gabunia et al. 2000). Because its immediate ancestors and closest relatives appeared to be restricted to Africa, most paleontologists had assumed that *H. erectus* evolved in Africa and then moved to Asia. The fossils at Longgupo Cave, China, however, are similar enough to African *H. habilis* and *H. ergaster* to suggest that *H. erectus* may have evolved in Asia from earlier migrants (Huang et al. 1995). Either way, before 2 million years ago the ancestors of our species within the genus *Homo* almost certainly lived in Africa.

Anatomically modern *H. sapiens* appear in the fossil record by 100,000 years ago in Africa and Israel and somewhat later throughout Europe and Asia (Stringer 1988; Valladas et al. 1988; Aiello 1993; White et al. 2003; but see McDougall et al. 2005). The range of hypotheses concerning the evolutionary transition from *H. ergaster/erectus* to *H. sapiens* is illustrated in **Figure 20.26**.

At one extreme is the African replacement, or out-of-Africa, model (Figure 20.26a). Under this model *H. sapiens* evolved in Africa and then migrated to Europe, Asia, and beyond, replacing *H. erectus* and *H. neanderthalensis* without interbreeding. No genes unique to these earlier forms persist in modern populations.

The origin of modern *Homo sapiens* has been controversial.

At the other extreme is the multiregional evolution model (Figure 20.26c). Under this model *H. sapiens* evolved concurrently in Europe, Africa, Asia, and beyond with sufficient gene flow among populations to maintain their continuity as a single species. Gene pools of all present-day human populations are derived from a mixture of local and distant archaic populations.

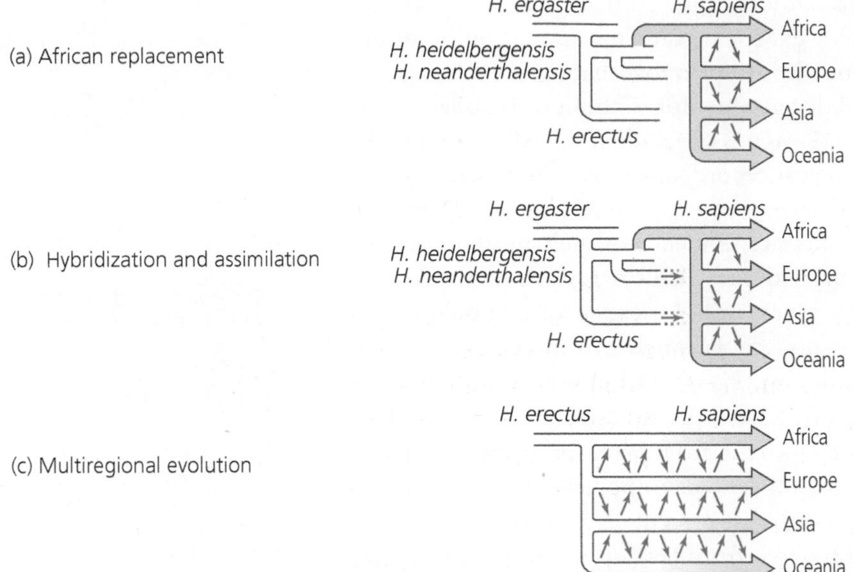

Figure 20.26 Hypotheses on the transition from archaic to anatomically modern humans White branches represent archaic humans; colored branches modern forms. Small orange arrows represent gene flow. Note that specimens identified as *H. heidelbergensis* have been found in Europe and Africa, and specimens identified as *H. neanderthalensis* have been found in Europe and the Middle East. After Aiello (1993); Ayala et al. (1994); Tattersall (1997).

Between the extremes is the hybridization and assimilation model (Figure 20.26b). It holds that modern humans evolved in Africa, then migrated to Europe, Asia, and beyond. *Homo sapiens* replaced archaic forms, but with hybridization between newcomers and established residents. As a result, some genes from archaic local populations were assimilated and persist in modern populations.

The model names and characterizations we use are based on those used by Aiello (1993), Ayala et al. (1994), and Tattersall (1997). Not all authors would endorse our definitions. Frayer et al. (1993), for example, apparently consider both models (b) and (c) to be variations of the multiregional model they favor.

At stake in the debate over the three models is the nature and antiquity of the present-day geographic races of humans. If the African replacement model is correct, then present-day racial variation is the result of recent geographic differentiation that occurred within the last 100,000 to 200,000 years, after anatomically modern *H. sapiens* emerged from Africa. If one of the other models is correct, then present-day racial variation represents a mixture of recent and ancient geographic differentiation. At least some of the differences among modern humans from different regions may derive from geographic differentiation among *H. ergaster/erectus* populations and could thus be as much as 1.5 to 2 million years old.

Argument over the models has been based on archaeological, paleontological, and genetic evidence. In much of the discussion that follows, we focus on the history of efforts to distinguish African replacement versus multiregional evolution. It is useful to keep in mind, however, that these models fall at the ends of a continuum. It also is useful to keep in mind a point we return to later: Whatever their origins, all living humans are extremely closely related.

African Replacement versus Multiregional Evolution: Archaeological and Paleontological Evidence

David Frayer and colleagues (1993) used archaeological and paleontological data to argue against the African replacement model (Figure 20.26a). This model holds that long-established populations of one or more tool-using, hunter–gatherer species (*H. erectus* and other archaic forms of *Homo*) were supplanted wholesale throughout Europe and Asia by populations of another tool-using, hunter–gatherer species (modern *H. sapiens* emerging from Africa). It is hard to imagine how this could have happened, except by direct competition between the invaders and the established residents. It is implausible that modern *H. sapiens* could have been such a relentlessly superior competitor without substantially better tools or weapons. Thus, Frayer and colleagues concluded, the African replacement model predicts that the archaeological record will show abrupt changes in technology in Europe and Asia as modern *H. sapiens* replaced archaic *Homo*. In fact, the researchers said, too many regions lack evidence of such abrupt changes.

Frayer and colleagues (1993) also argued that African replacement predicts that fossils of *Homo* populations in any given non-African region should show distinct changes in morphology when modern *H. sapiens* from Africa replaced local archaic *Homo*. To refute this prediction, Frayer and coworkers pointed to distinctive traits of regional populations that have persisted from the distant past to now. One-million-year-old fossils of *H. erectus* from Java, for example, have a straighter, more prominent browridge than their contemporaries elsewhere. This strong browridge remains a distinctive feature of present-day Australian aborigines, whose ancestors may have arrived by boat from Java up to 60,000 years ago. Likewise, many present-day Asians have shovel-shaped upper front teeth, a trait that characterizes virtually all fossil specimens of Asian *H. erectus* and *H. sapiens*. (For other examples of the continuity of distinctive regional traits, see Thorne and Wolpoff 1981; Li Tianyuan and Etler 1992; and Frayer et al. 1993.) On these and other grounds, Frayer and colleagues rejected the replacement model.

Diane Waddle (1994) and Daniel Lieberman (1995) used statistical and cladistic approaches, respectively, to evaluate predictions from the African replacement versus multiregional evolution models. The African replacement model predicts that all modern humans will be more closely related to each other than any is to any archaic species and that, among the archaic forms, those from Africa will be the most closely related to modern humans **(Figure 20.27a)**. In contrast, the multiregional evolution model predicts that the archaic and modern humans in each region will be each other's closest relatives (Figure 20.27b).

(a) Cladogram predicted by African replacement model

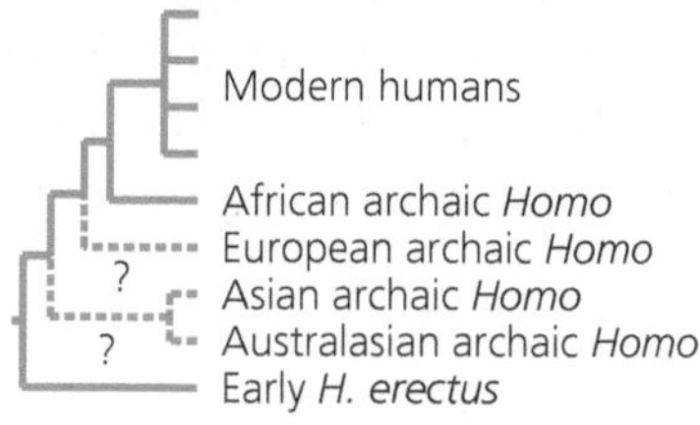

(b) Cladogram predicted by multiregional evolution model

Figure 20.27 Phylogenetic predictions of the African replacement model versus the multiregional evolution model After Lieberman (1995).

While Lieberman stressed that his data set of only 12 characters was too small to produce reliable inferences, both he and Waddle tentatively concluded that all modern humans are more closely related to archaic forms from Africa than regional groups are to local archaic forms. If Lieberman and Waddle were correct, then the examples described by Frayer and colleagues of apparent long-term continuity of regionally distinctive traits must be the result of convergent evolution in *H. erectus* and *H. sapiens*.

Morphological analyses, though not definitive, suggest that modern humans evolved in Africa, then replaced archaic humans elsewhere. However, the evidence also hints at complexities in the story.

Subsequent morphological analyses of Neandertal and pre-Neandertal fossils hinted that the story of modern human origins may be more complex than Lieberman's and Waddle's results suggested. Bermúdez de Castro and colleagues described human fossils and associated artifacts and animal remains from the Gran Dolina section of the Atapuerca locality in Spain (Carbonell et al. 1999 and papers cited therein). The researchers attributed the human fossils from this locality, which is dated to somewhere between 780,000 and 980,000 years ago, to a new species, *Homo antecessor*. Their analysis showed that the Gran Dolina specimens share features of both modern humans and Neandertals and suggest that it may be a common ancestor of both. At the opposite end of the time range for Neandertals, Trinkaus and colleagues found evidence of mixed Neandertal and modern human characteristics in two different specimens, one from France and the other from Portugal, both roughly 30,000 years old. Trinkaus and colleagues interpreted these fossils as possible evidence of hybridization between Neandertals and modern humans (Trinkaus et al. 1998; Duarte et al. 1999). This claim provoked acrimonious debate (see Tattersall and Schwartz 1999). Nonetheless, *Homo antecessor* and the possible Neandertal–modern human hybrids, all from western Europe, kept the multiregional evolution model alive.

African Replacement versus Multiregional Evolution: Molecular Evidence from Modern Humans

In principle, we could take Waddle's and Lieberman's approach and use it with DNA sequence data. If we had sequences of genes from both modern and archaic humans in all regions, we could estimate their phylogeny and see whether it most closely matches the tree predicted by the African replacement model or the tree predicted by the multiregional evolution model. The trouble is that genetic sequences from archaic humans that have been dead for tens of thousands of years are hard to come by. For a long time there were simply none available.

Working with DNA sequences of present-day humans only, researchers found it more difficult to design tests that distinguish the African replacement model from the multiregional evolution model (see, for example, the exchange between Sarah Tishkoff and colleagues [1996b] and Milford Wolpoff [1996]). The trouble was that from a genetic perspective, the two models are identical in most respects. Both describe a species originating in Africa, spreading to Europe, Asia, and elsewhere, and then differentiating into regionally distinct populations that nonetheless remain connected by gene flow (**Figure 20.28**). The only difference is that under the African replacement model this process began less than 200,000 years ago whereas under the multiregional evolution model it began over 1.8 million years ago. This means that any genetic patterns that might allow us to distinguish between the two models will involve quantitative differences rather than qualitative differences. **Table 20.1** lists four criteria that biologists have used in the effort to distinguish between African replacement versus multiregional evolution. We refer to the table in the paragraphs that follow.

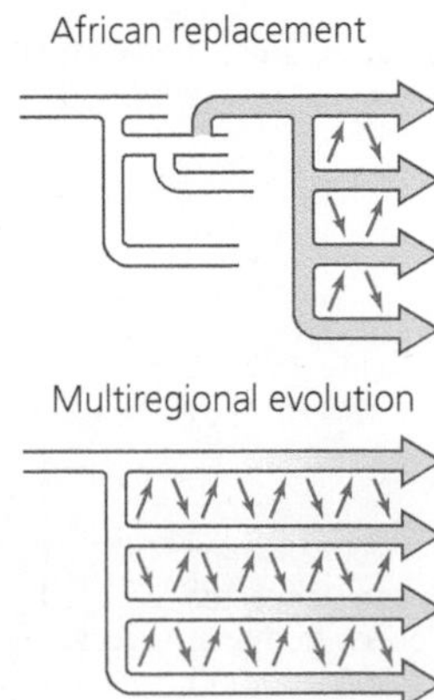

Figure 20.28 For modern humans, these two models differ only in the time of divergence

Table 20.1 Genetic predictions of African replacement versus multiregional evolution

Each criterion in the first column is a category of data we might use to distinguish the African replacement model from multiregional evolution. The next two columns predict the patterns in each type of data under each model. The last column explains why the distinctions implied by the predictions are not definitive. See text for more details.

Criteria	African replacement	Mulitregional evolution	Caveats
1. Location of ancestor of neutral alleles	Mostly Africa	Random	African origin of *H. ergaster/erectus* may bias location of alleles toward Africa even under multiregional evolution.
2. African vs. non-African divergence time	200,000 years or less	1 million years or more	Gene flow among regional populations can reduce the apparent age of population divergence under multiregional evolution.
3. Genetic diversity	Genetic diversity greater in Africa	Diversity roughly equal in all regions	African origin of *H. ergaster/erectus* and gene flow or selection may lead to greater diversity in Africa even under multiregional evolution.
4. Sets of neutral alleles	Alleles in non-African gene pools are subsets of those in Africa	No region's alleles are a subset of another's; all have some unique alleles	African origin of *H. ergaster/erectus* may mean alleles present in non-African gene pools are subsets of those in Africa even under multiregional evolution.

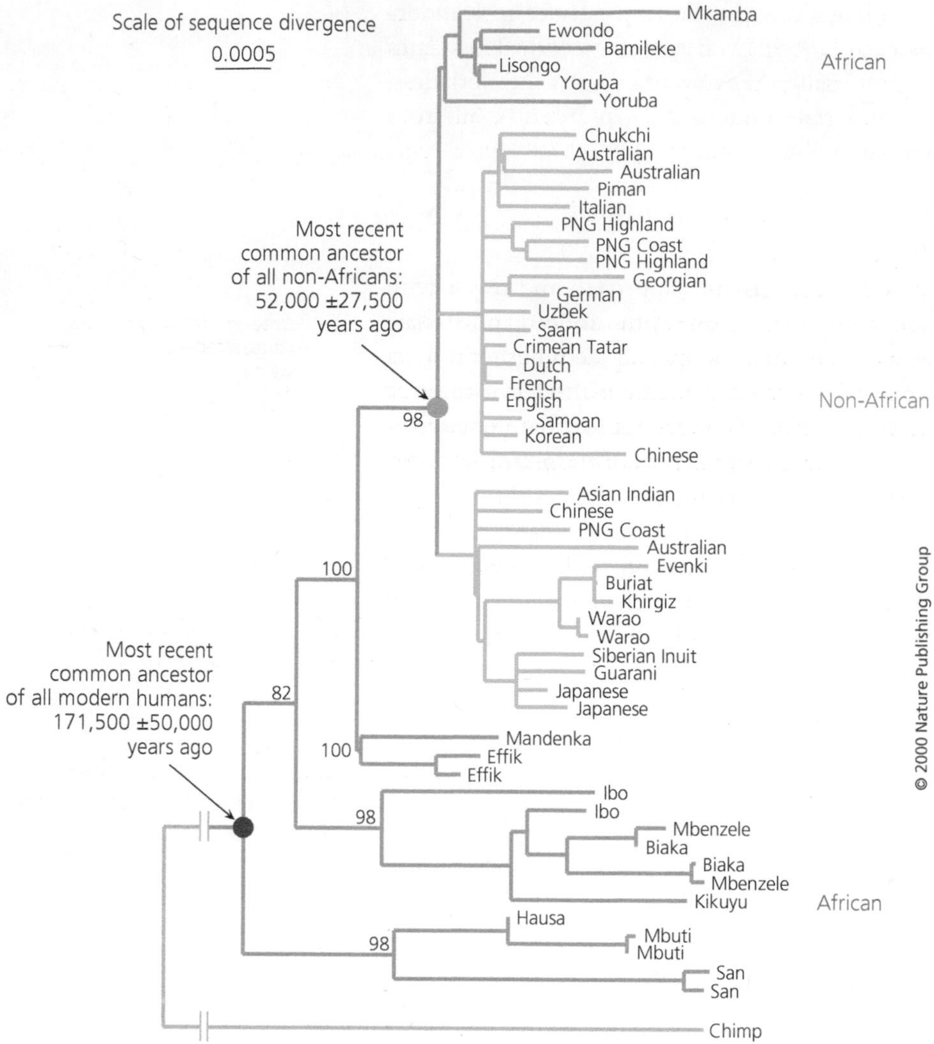

Figure 20.29 An evolutionary tree of complete mitochondrial DNAs of 53 humans Each branch tip represents a single individual, identified by his or her population. The tree was rooted by using a chimpanzee sequence as the outgroup. All non-African individuals branch from within the Africans. Numbers at nodes represent the strength of statistical support (percentage of 1,000 bootstrap replicates in which the node was present). From Ingman et al. (2000).

"Mitochondrial genome variation and the origin of modern humans." *Nature* 408: 708–713. Copyright © 2000 Nature Publishing Group. Reprinted with permission.

Max Ingman and colleagues (2000) attempted to distinguish between models by analyzing sequences of the entire mitochondrial genomes of 53 individuals from various places around the world. Their phylogeny has strong statistical support and shows all non-African sequences branching from within the African sequences **(Figure 20.29)**. It appears that the common ancestor of all present-day human mtDNAs lived in Africa. Does that help us decide, by criterion 1 of Table 20.1, between African replacement and multiregional evolution? Not by itself, because both models include a common ancestry for all present-day humans that traces back to Africa. We need to know when the common ancestor lived.

Ingman and colleagues estimated that the most recent common ancestor of all present-day mitochondrial DNAs, at the node highlighted in maroon in Figure 20.29, lived between 120,000 and 220,000 years ago. They arrived at this estimate by using a molecular clock. They verified for their data that mutations have accumulated in mtDNAs at a constant, clocklike rate. They assumed a human–chimpanzee divergence time of 5 million years ago and estimated the divergence time of all human sequences by comparing them to the sequence of a chimp. By the same method, Ingman and colleagues estimated that the most recent common ancestor of all non-Africans, at the node highlighted in blue in Figure 20.29, lived just 25,000 to 80,000 years ago (see Hedges 2000 for a commentary).

Analyses of modern human DNA suggest that modern humans evolved in Africa, then replaced archaic humans elsewhere.

At first glance these dates, which are consistent with the African replacement hypothesis, appear to refute the multiregional origin hypothesis by criterion 2 of Table 20.1. The dates suggest that non-African populations of humans diverged from African populations no more than a hundred thousand years ago and certainly nothing like a million years ago. Examination of **Figure 20.30**, however, shows that this is not necessarily the case. It is true that species cannot diverge any earlier than the divergence of any of their alleles, but populations connected by gene flow can.

Each diagram in the figure is a hypothetical gene phylogeny embedded within a species or population phylogeny. In scenario (a), a mutation creates a new allele to produce a split in the gene tree at the same time the species or population splits into two. One allele is then lost by drift or selection in each descendant species or population. The result is a gene tree that is exactly congruent with the species or population tree. If we use a molecular clock to estimate the divergence time for the species or populations, we will get the right answer.

In scenario (b), a mutation creates a new allele, producing a split in the gene tree. Some time later, the species or population splits into two. One allele is then lost by drift or selection in each descendant species or population. If we use a molecular clock for the gene tree to estimate the divergence time for the species or populations, the species or populations will appear to have diverged earlier than they actually did.

In scenario (c), a species first splits into two. Some time later, a mutation creates a new allele, producing a split in the gene tree. Finally, one of the alleles moves from one species to the other. This last step is impossible for species that do not exchange genes. A molecular clock will thus not make a split between species appear more recent than it was.

Scenario (d), however, shows that the last step in (c) is possible for populations within a species. First, a population splits into two. Some time later, a mutation creates a new allele, producing a split in the gene tree. Then a migrant carries the new allele to the other population (blue arrow). Finally, the new allele is lost in one population, and the ancestral allele is lost in the other population. If we use a molecular clock for the gene tree to estimate the divergence time for the population tree, the populations will appear to have diverged more recently than they actually did.

The lesson to keep in mind when we interpret Ingman et al.'s mitochondrial phylogeny is that the mitochondrial clock, which is effectively based on a single gene, might make the split between African and non-African populations look more recent than it actually was. A new mitochondrial genotype could have arisen after modern humans left Africa, been carried from one continent to another by a migrant, and then replaced older genotypes. We need to look at many loci at once and see whether, taken together, they tell the same story of a recent divergence between African and non-African populations.

A. M. Bowcock and colleagues (1994) looked at 30 nuclear microsatellite loci from people in each of 14 populations. Microsatellite loci are places in the genome where a short string of nucleotides, usually two to five bases long, is repeated in tandem. The number of repeats at a given locus is highly variable among individuals, meaning that each microsatellite locus has many alleles. Bowcock and colleagues calculated multilocus genetic distances among the 14 populations based on the allele frequencies at each of the 30 loci. They then used the genetic distances among populations to estimate the population phylogeny.

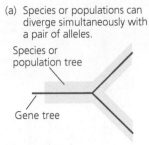

(a) Species or populations can diverge simultaneously with a pair of alleles.

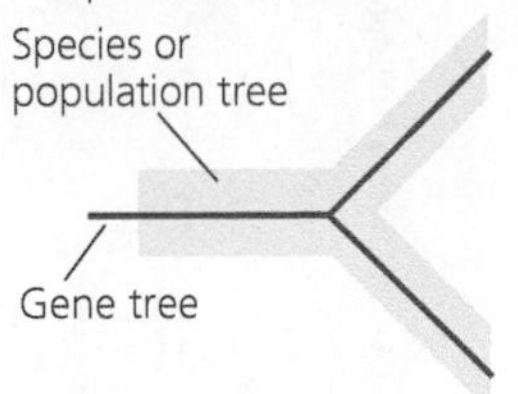

(b) Species or populations can diverge after a pair of alleles diverge.

(c) Species cannot diverge before a pair of alleles diverge...

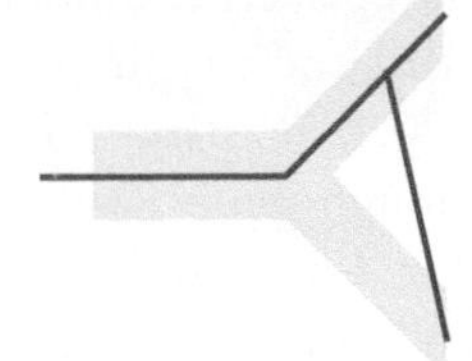

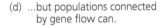

(d) ...but populations connected by gene flow can.

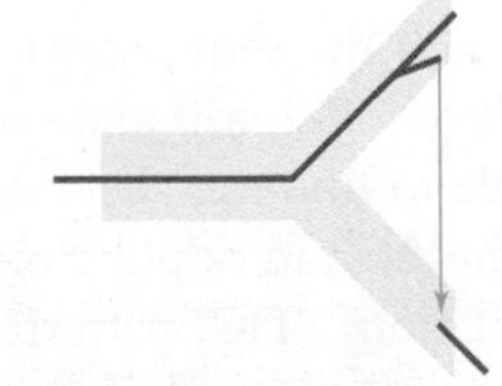

Figure 20.30 Divergence times of species trees, population trees, and gene trees The divergence times on gene trees may not coincide with the divergence times on species trees and population trees. Divergence times on gene trees can even be more recent than divergence times on population trees. See main text for details.

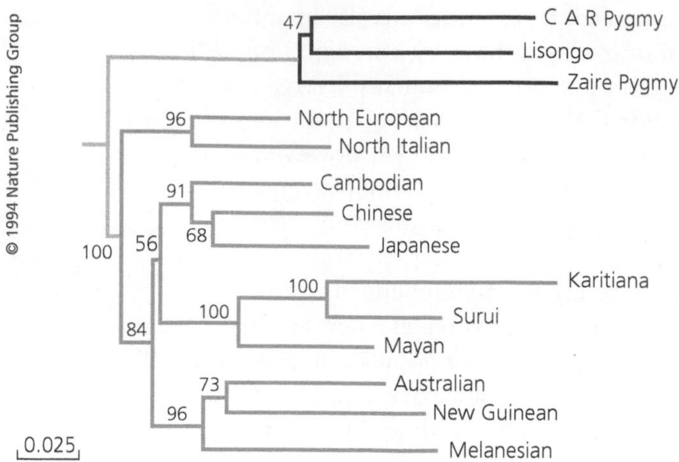

Figure 20.31 Phylogenetic tree for 14 human populations based on allele frequencies at 30 microsatellite loci Numbers at nodes indicate bootstrap support. The deepest split in the tree was present in 100% of the bootstrap replicates, indicating strong statistical support for the hypothesis that African (burgundy) versus non-African (orange) is the deepest division in this tree. From Bowcock et al. (1994).

"High resolution of human evolutionary trees with polymorphic microsatellites." *Nature* 368: 455–457. Copyright © 1994 The Nature Publishing Group. Reprinted with permission.

On Bowcock et al.'s phylogeny, shown in **Figure 20.31**, geographically neighboring populations cluster together. The deepest node separates African from non-African populations. From the same data, D. B. Goldstein and colleagues (1995) estimated that the split between African and non-African populations occurred 75,000 to 287,000 years ago. This range, which is consistent with the African replacement model, makes a more persuasive case than the mitochondrial clock that the multiregional evolution model can be rejected by criterion 2 of Table 20.1. It is still possible under multiregional evolution to argue that there was enough gene flow to make the population split look more recent than it was. But if there was that much gene flow, it is hard to explain how any regional differentiation of characters could persist for a million years or more (Nei 1995).

Finally, consider a study by Sarah Tishkoff and colleagues (1996a). These researchers examined allelic variation at a locus on chromosome 12 that is the site of a short tandem repeat (STR) polymorphism. This is a region of noncoding DNA in which the sequence TTTTC is repeated 4 to 15 times, producing a total of 12 alleles. Tishkoff and colleagues determined the genotypes of more than 1,600 people to find the allele frequencies in seven geographic regions **(Figure 20.32)**.

The African populations show much greater allelic diversity than non-African populations. This pattern is consistent with the African replacement model. If non-African populations were founded by small bands of people migrating out of Africa, then non-African populations should have reduced genetic diversity because of the founder effect (see Chapter 7).

Notice that the graphs in Figure 20.32 are arranged by travel distance from sub-Saharan Africa, with the closest regions at the bottom and the most distant regions at the top. Moving up from sub-Saharan Africa to northeast Africa, then to the Middle East and beyond, each region shows a set of alleles that is a subset of those present in the region below. Again, this pattern is consistent with African replacement. It is what we would expect if each more-distant region were settled by a small band of people picking up from where their ancestors lived and moving on. The pattern of allelic diversity is not only consistent with African replacement but also tends to refute multiregional evolution by criteria 3 and 4 of Table 20.1. This refutation is not definitive, however, because multiregional evolution postulates the same pattern of migration and settlement, just earlier.

Using additional genetic data from their study, Tishkoff and colleagues calculated the age of the non-African population (see **Computing Consequences 20.1**). They estimate that modern humans left Africa roughly 100,000 years ago.

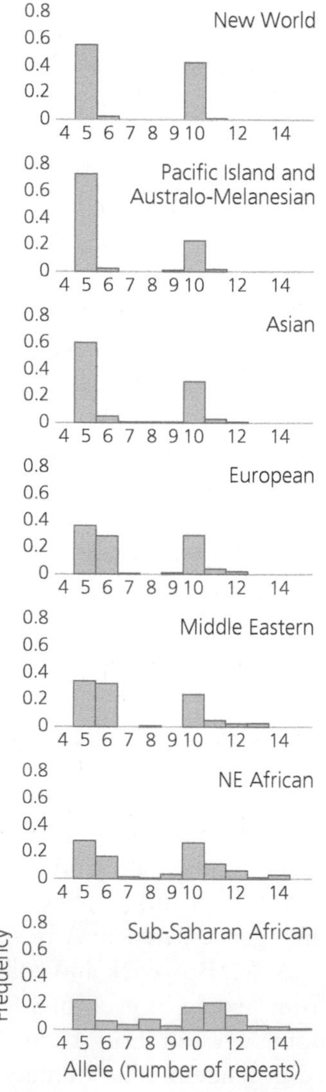

Figure 20.32 Genetic diversity at a single locus among the people of seven geographic regions Plotted from tables in Tishkoff et al. (1996a).

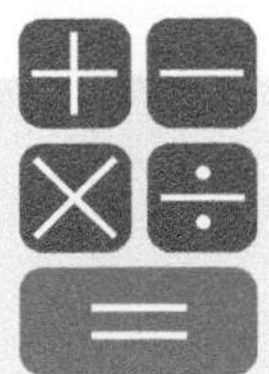

COMPUTING CONSEQUENCES 20.1

Using allele frequencies and linkage disequilibrium to date the modern human expansion from Africa

Near the short-tandem-repeat (STR) locus analyzed in Figure 20.32 is another locus with a nucleotide sequence known as an Alu element. This locus has two alleles: the ancestral, or *Alu*(+) allele; and a derived *Alu*(−) allele with a 256-base-pair deletion. Gorillas and chimps lack the *Alu*(−) allele, so it likely arose after the human lineage and the chimpanzee lineage diverged.

The deletion mutation that created the *Alu*(−) allele likely occurred only once, in Africa, in a chromosome that carried the six-repeat allele at the STR locus (Tishkoff et al. 1996a). At the moment the first *Alu*(−) chromosome appeared, the frequency of the six-repeat allele among *Alu*(−) chromosomes was 1. The Alu locus and the tandem repeat locus were in strong linkage disequilibrium (see Chapter 8).

Since then, as shown in Figure 20.33a, the linkage disequilibrium has largely broken down in sub-Saharan African populations. The distribution of frequencies of various repeat alleles is nearly the same among *Alu*(−) chromosomes as among *Alu*(+) chromosomes.

Two processes have generated *Alu*(−) chromosomes with alleles other than the six-repeat allele: mutation at the repeat locus and recombination between the repeat locus and the Alu locus. Sequence analysis implicates recombination in the origin of *Alu*(−) chromosomes with 10 or more repeats. Tiskoff and colleagues excluded them, and looked only at *Alu*(−) chromosomes with nine or fewer repeats. Among these, the frequency of the six-repeat allele has fallen from an initial value of 1 to a current value of 0.4 as a result of mutation.

For an allele declining in frequency due to mutation, the frequency in generation n, p_n, is given by

$$p_n = p_0 e^{-\mu n}$$

where p_0 is the initial frequency and μ is the mutation rate (see Computing Consequences 6.8). Substituting 1 for p_0, 0.4 for p_n, and rewriting in logarithmic form gives

$$-\mu n = \ln(0.4)$$

Figure 20.33b suggests that the first modern humans to leave Africa carried only *Alu*(−) chromosomes with the six-repeat allele. The linkage disequilibrium in non-African populations is still high. Among *Alu*(−) chromosomes with nine or fewer repeats the frequency of the six-repeat allele has fallen from 1 to 0.9814. For non-African populations we can write

$$-\mu n = \ln(0.9814)$$

Dividing the equation for sub-Saharan populations by the equation for non-African populations and canceling terms gives the relative age of *Alu*(−) chromosomes in sub-Saharan African versus non-African populations:

$$\frac{\text{sub-Saharan age}}{\text{non-African age}} = \frac{\ln(0.4)}{\ln(0.9814)} \approx 50$$

If the first *Alu*(−) chromosome arose in sub-Saharan Africa 5 million years ago, this equation suggests that the non-African population is 100,000 years old (see also Pritchard and Feldman 1996; Risch et al. 1996).

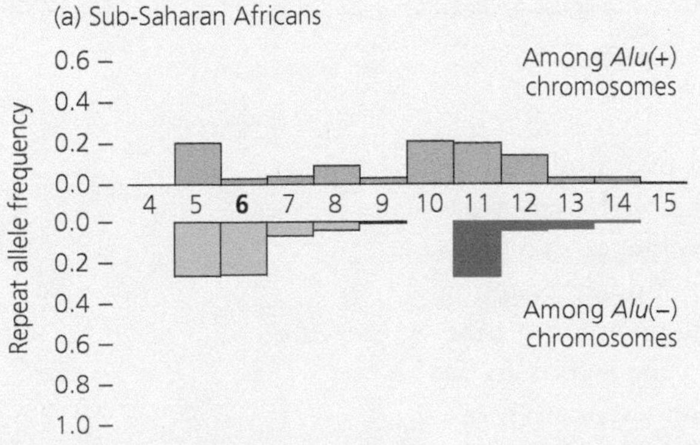

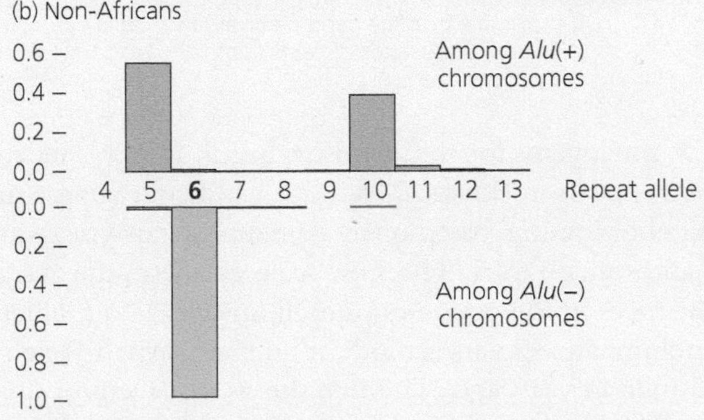

Figure 20.33 Repeat allele frequencies among *Alu*(+) and *Alu*(−) chromosomes Drawn from data in Table 3 in Tishkoff et al. (1996a).

Even allowing considerable leeway for various sources of error, this date is consistent with African replacement and tends, by criterion 2 of Table 20.1, to refute multiregional evolution.

The balance of evidence we have reviewed appears to favor the African replacement model for the origin of *H. sapiens*. None of the tests are definitive, so an intermediate model (Figure 20.26b) cannot be ruled out. But taken together, the genetic data and at least some of the morphological data suggest that (1) all present-day people are descended from African ancestors, and (2) all present-day non-Africans are descended from *H. sapiens* ancestors who left Africa within the last few hundred thousand years. Differences among races have arisen since then.

In recent years, however, new sources of data have yielded some surprises.

African Replacement versus Multiregional Evolution: Molecular Evidence from Premodern Humans

In 2010, teams led by Svante Pääbo published draft genome sequences representing two groups of premodern humans. The first group was Neandertals. Richard Green and colleagues (2010) assembled a Neandertal genome by sequencing DNA fragments painstakingly extracted from bones (found in Vindija Cave, Croatia) of three individuals who lived some 40,000 years ago.

The second group is known only from two molars and a fingertip bone found in Denisova Cave, Siberia (see Gibbons 2011). The fingertip belonged to a girl who lived 30,000 to 50,000 years ago. Upon sequencing her genome, David Reich and colleagues (2010) confirmed what they already suspected from the unusual shape of the molars. The girl and her population, though human, were neither Neandertal nor *Homo sapiens*. The researchers call them Denisovans.

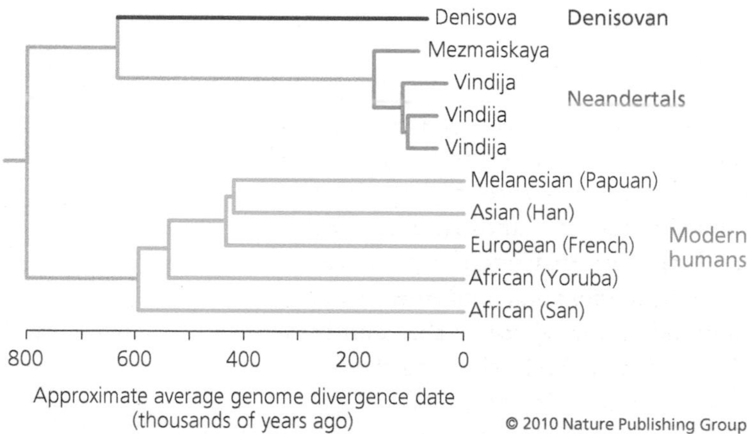

Figure 20.34 A molecular phylogeny for modern and premodern humans The time scale is calibrated to the genetic divergence between modern humans and chimpanzees, and assumes that this divergence occurred 6.5 million years ago. Redrawn from Reich et al. (2010).

"Genetic history of an archaic hominin group from Denisova Cave in Siberia." *Nature* 468: 1053–1060. Copyright © 2010 Nature Publishing Group. Reprinted with permission.

A phylogeny for the Denisovan girl, four Neandertals, and five modern humans appears in **Figure 20.34**. Reich and colleagues estimated the phylogeny with neighbor joining, based on assessments of pairwise sequence divergence in 56 Mb of autosomal DNA. The time scale estimates the average time of divergence for sequences in the genomes of each lineage. It is calibrated to the divergence time for chimpanzees versus modern humans, which Reich and colleagues took to be 6.5 million years ago. The tree shows that Denisovans and Neandertals are more closely related to each other than either is to modern humans.

The availability of complete genomes for premodern humans facilitates new tests distinguishing African replacement versus hybridization and assimilation.

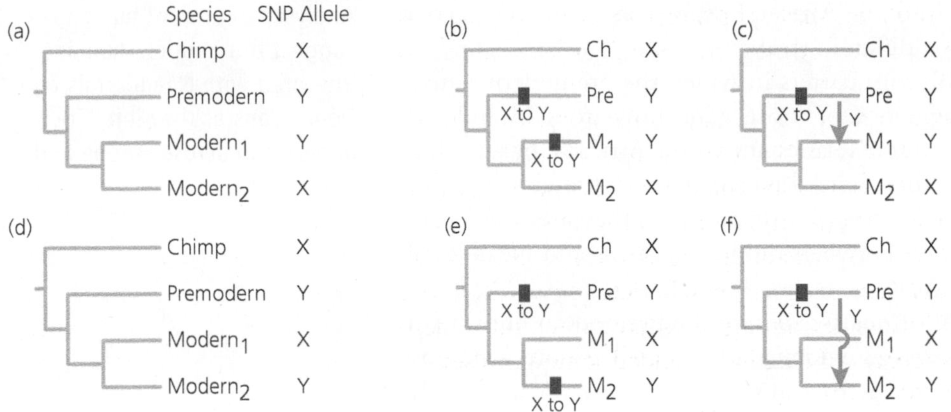

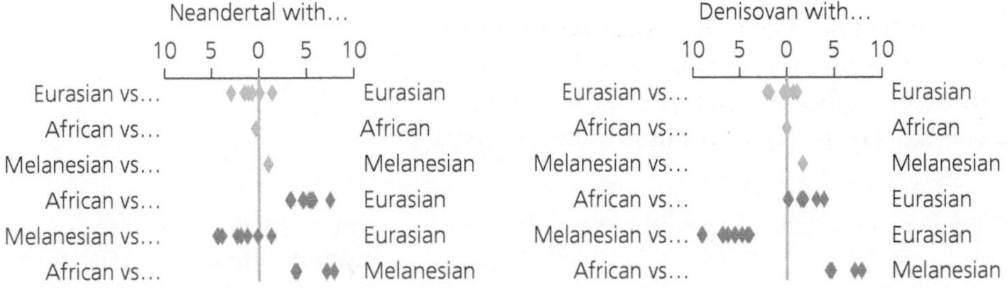

Figure 20.35 Tests for gene flow from premodern to modern humans (a) A two-allele SNP mapped onto the phylogeny for a quartet of individuals. The premodern human and one of the modern humans carry the same allele. (b) Without gene flow, two mutations are required to explain the distribution of alleles. (c) Gene flow from the premodern lineage to modern human lineage 1 provides an alternative explanation. (d–f) For this distribution of alleles, gene flow from the premodern lineage to modern human lineage 2 provides an alternative. (g) Data on the percentage excess of SNPs like that in (a) or (d) in comparisons of chimps, Neandertals or Denisovans, and the modern humans shown. Drawn from data in Table 1 of Reich et al. (2010).

Green and colleagues (2010) developed a method for detecting gene flow between premodern and modern humans. The method involves scanning aligned sequences from the genomes of a chimpanzee, a premodern human, and two modern humans for single-nucleotide polymorphisms (SNPs). We are looking for SNPs with just two alleles, where the premodern human and just one of the modern humans carry a different allele than the chimp.

Figure 20.35a shows one of the two patterns that meet the criteria. Without gene flow among lineages, the evolution of such a SNP requires at least two mutations. Figure 20.35b shows one scenario; others are possible. With gene flow, as shown in Figure 20.35c, there is an additional evolutionary path. Figure 20.35d shows the other pattern in which the premodern human shares an allele with only one modern human. Figure 20.35e and (f) show that without gene flow two mutations are required, whereas with gene flow there is an additional path.

If there has been no gene flow from the premodern human into either modern human lineage since the modern lineages diverged, then SNPs showing the pattern in (a) and the pattern in (d) will be equally common. If, on the other hand, there has been gene flow from the premodern lineage into modern human lineage 1 since the modern lineages diverged, SNPs showing the pattern in (a) will be more common. And if there has been gene flow from the premodern lineage into human lineage 2 since the modern lineages diverged, SNPs showing the pattern in (d) will be more common.

Figure 20.35g, drawn from results reported by Reich et al (2010) shows, for quartets in which the premodern human was a Neandertal (left) or the Denisovan (right), the excess of SNPs suggesting gene flow into one modern lineage versus the other as a percentage of the total number of SNPs meeting the criteria. The first three rows, with green symbols, function as controls. For example, they give no evidence of more gene flow from the Neandertal lineage into one European lineage versus another (top row). And there is no evidence of more gene flow from the Denisovan lineage into one African lineage versus another (second row). The bottom three rows tell a different story.

Look first at the orange symbols for quartets in which the premodern human was a Neandertal. They show evidence of more gene flow from Neandertals into Eurasians versus Africans and into Melanesians versus Africans, but not into Eurasians versus Melanesians. In other words, averaged across the genome, Neandertals are a bit more closely related to non-Africans than they are to Africans. This is consistent with hybridization between modern humans and Neandertals shortly after modern humans left Africa.

Now look at the pink symbols for quartets where the premodern human was the Denisovan girl. The first row shows evidence of somewhat more gene flow from Denisovans into Europeans versus Africans. This may be a legacy of the common ancestry of Denisovans and Neandertals. The last two rows show evidence of substantially more gene flow from Denisovans into Melanesians versus Eurasians and into Melanesians versus Africans. That is, averaged across the genome, Denisovans are a bit more closely related to Melanesians than to other modern humans. This pattern is consistent with hybridization between modern humans and Denisovans when the ancestors of today's Melanesians crossed Asia.

Analyses of ancient human DNA suggest that modern humans interbred with Neandertals and Denisovans as they left Africa and spread across Europe and Asia.

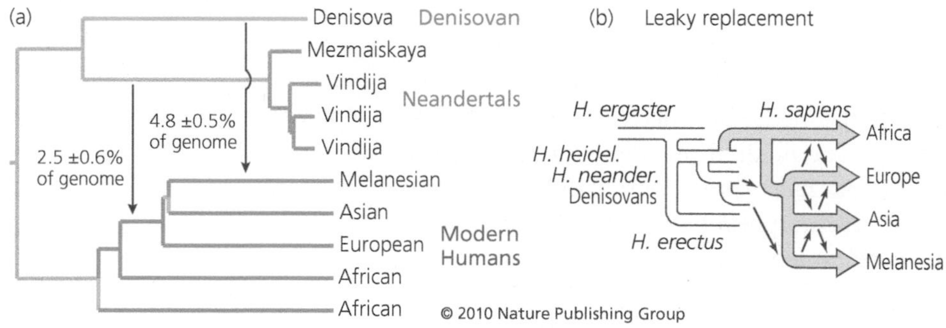

Figure 20.36 The origin of modern humans (a) Genomic analysis suggests gene flow from premodern to modern humans. After Reich et al. (2010). (b) This result is consistent with a "leaky replacement" scenario.

(a) "Genetic history of an archaic hominin group from Denisova Cave in Siberia." D. Reich, R.E. Green. *Nature* 468: 1053–1060. Copyright © 2010 Nature Publishing Group. Reprinted with permission.

Reich and colleagues estimate that 2.5% of the genome of non-African modern humans is derived from Neandertals, and that an additional 4.8% of the genome of Melanesians comes from Denisovans (**Figure 20.36a**). This result refutes the African replacement model. It is consistent with a scenario between replacement and hybridization and assimilation (Figure 20.36b). Pääbo describes the picture suggested by genomic analysis as "leaky replacement" (see Gibbons 2011).

Despite their mixed origins all modern humans are, as we noted earlier, extremely closely related. We now present data to support this claim.

Genetic Diversity among Living Humans

One way to put the genetic diversity among living humans into perspective is to compare the genetic diversity among humans to the genetic diversity among other great apes. Pascal Gagneux and colleagues (1999) examined mitochondrial DNA nucleotide sequences of 811 humans, 292 chimpanzees, 24 bonobos, and 26 gorillas. For each species, the researchers sorted all possible pairs of individuals by the percentage of nucleotides that were different between their sequences. The histograms in **Figure 20.37** show the distributions of pairwise differences. A randomly chosen pair of humans is likely to be about as different as a randomly chosen pair of eastern chimps and is likely to be substantially less different than a randomly chosen pair of central chimps, western chimps, bonobos, or gorillas.

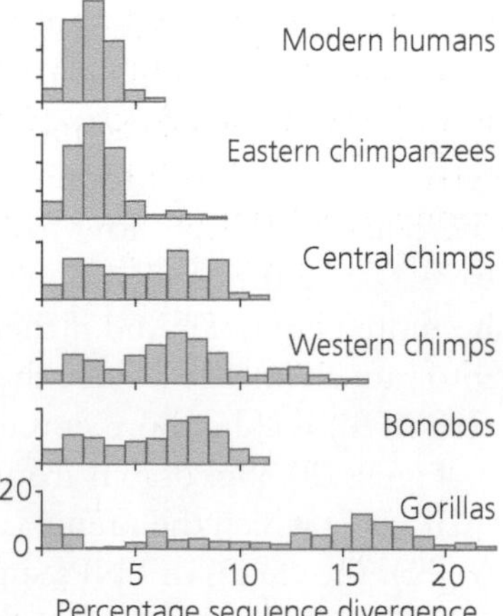

Figure 20.37 Genetic diversity in humans and great apes From Gagneux et al. (1999).

Indeed, the researchers found several cases in which bonobos or western chimpanzees living in the same social group were more genetically different than any two humans from anywhere in the world. By the standards of the other African great apes, all living humans are closely related. Furthermore, most of the genetic diversity among living humans occurs as differences among individuals within populations, rather than as differences between populations (Jorde et al. 2000).

Are there any genetic differences at all distinguishing present-day human populations? To find out, Noah Rosenberg and colleagues (2002) analyzed data on the genotypes, at 377 variable STR loci, of 1,056 individuals from 52 populations. They used a computer program to (1) define, using only the genotype data on the 1,056 individuals, five groups with distinct allele frequencies; and (2) assign each individual a partial membership in each group determined by how closely his or her genotype matches each group's membership criteria. If there are genetic differences that distinguish human populations, then individuals from the same geographic region should have similar group-membership profiles.

The results appear in **Figure 20.38**. Each individual is represented by a thin horizontal line. Each line is composed of five segments with different colors, representing the individual's partial membership in each of the five groups. It turns out that individuals from different geographic regions do, indeed, tend to have similar group-membership profiles.

Other teams of researchers conducting similar analyses have found similar patterns. Richard Redon and colleagues (2006) analyzed the genotypes of 210 individuals at 67 loci polymorphic for insertions, deletions, or duplications. Based on these genotypes, the researchers asked a computer to sort the individuals into groups derived from three ancestral populations. Nearly all the individuals in the sample ended up in a group with others from the same continent of origin. Michael Bamshad and colleagues (2003) analyzed the genotypes of 206 individuals at 60 polymorphic STR loci and 100 loci polymorphic for Alu transposable elements. Based solely on genotype, their computer assigned individuals to the correct continent of origin with 99% accuracy.

The lesson is that genetic differences among modern human populations do exist, but they are subtle enough that it takes extraordinarily large data sets and considerable computational effort to find them.

On the other hand, the volume of available data is increasing dramatically. So too is the power of computational resources. Researchers have begun to mine large data sets to identify particular alleles whose frequency varies among populations. For example, Angela Hancock and colleagues (2011) discovered, in a gene called *KRT77,* an amino acid substitution whose frequency is strongly associated, among populations in a variety of geographic regions around the world, with the intensity of summer solar radiation. *KRT77* encodes keratin 77, a protein made in the ducts of sweat glands.

Much human variation takes the form of rare alleles. Jacob Tennessen and colleagues (2012) sequenced more than 15,000 protein-coding genes in some 1,300 Europeans and 1,100 Africans. The team found over half a million single-nucleotide variants, the vast majority of which were previously unknown. Eighty-six percent of the variants had minor allele frequencies under 0.005, and 82% were found only in Europeans or only in Africans. The researchers estimate that a typical individual carries rare alleles likely to influence protein function in about 300 genes. A similar project by Matthew Nelson and colleagues (2012) yielded similar results. A great deal remains to be learned about human genetic diversity.

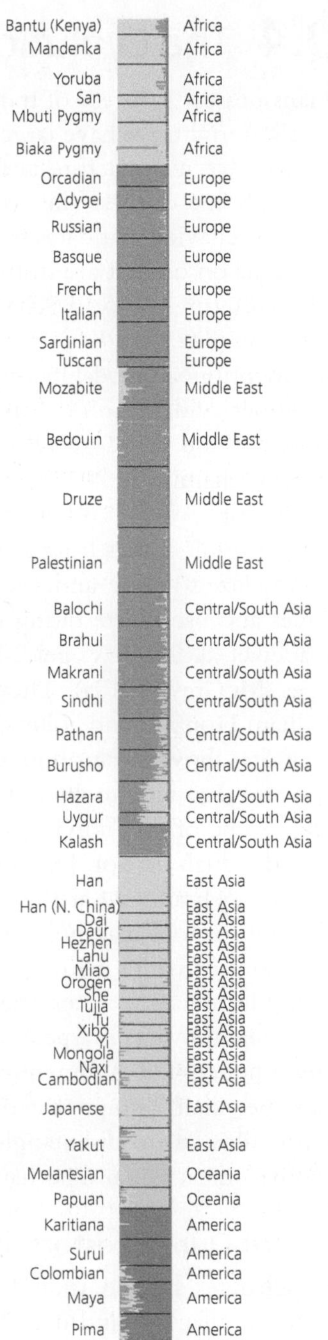

Figure 20.38 Evidence of geographic structure in living human populations From Rosenberg et al. (2002).

20.4 The Evolution of Distinctive Human Traits

Humans possess a variety of traits that distinguish us from other extant primates. We walk upright, we have large brains and culture, we use fire, we have reduced body hair, we manufacture and employ complex tools, we communicate with language. Some of these traits are unique to humans, such as our controlled use of fire—a behavior that dates from at least a million years ago (Berna et al. 2012). Other traits on our list are distinctive only in degree. Orangutans and chimpanzees have culture too (van Schaik et al. 2003; Lycett et al. 2007), but our culture is more complex.

We mentioned bipedal locomotion and brain size briefly in Section 20.2. Here we consider evidence on the use of complex tools and language.

Tool use and language rely on overlapping neural circuits in the brain (see Stout and Chaminade 2012). Evidence from comparative research hints that our capacities for tool use and language may be linked (see Steele et al. 2012). One intriguing result comes from a study of chimpanzees.

Chimpanzees make and use a variety of tools. They strip stems and twigs of leaves and use the resulting rods to fish termites out of termite mounds (see Mercader et al. 2002; Vogel 2002). They use leaves as umbrellas, napkins, and sponges (McGrew 2010a). They use rocks as anvils and hammers to open nuts.

William Hopkins and colleagues (2012) studied a captive population of chimpanzees that show variation in a form of social tool use also seen in wild chimps: aiming and throwing rocks and other missiles at other chimps or humans **(Figure 20.39a)**. Some of the individuals in the population reliably throw missiles, while other individuals do not. First Hopkins and colleagues scanned two areas of each chimpanzee's brain that are associated with hand movement in general and hand movement for tool manipulation in particular. The chimpanzees that throw had higher ratios of white matter to gray matter, indicating a higher number of myelinated interneurons. Then the researchers put the chimps through a battery of tests to measure various aspects of cognitive ability. On communication, but not on any other cognitive dimension, chimps that throw scored significantly higher, on average, than chimps that do not (Figure 20.39b). This observational study does not allow us to disentangle cause and effect. But it suggests that throwing and communication, and perhaps tool use and language, are deeply connected.

(a) A wild chimp throwing a rock at another chimp

(b) Cognitive abilities of chimps that reliably throw versus chimps that do not

Mean percentage correct (±s.e.) on tests of cognitive ability

0 0.2 0.4 0.6 0.8

Communication $p < 0.001$

Spatial memory and reasoning

Causal understanding

Quantitative reasoning

Theory of mind

Chimps that throw ($N = 39$)

Chimps that do not ($N = 52$)

Figure 20.39 Throwing and communication ability in chimpanzees (a) Bad chimpanzee. (b) Redrawn from Hopkins et al. (2012).

Which of Our Ancestors Made and Used Stone Tools?

What separates human tools from those of chimpanzees is the sophistication of their manufacture and use. The earliest tools to appear in the archaeological record that are beyond the capacities of chimpanzees are sharp-edged stone flakes and handheld chopping tools. An example appears in **Figure 20.40**.

A stone knapper making such tools begins by choosing an appropriate cobblestone from a riverbed, preferably one of fine-grained volcanic rock (Schick and Toth 1993). He or she then strikes the cobble with a second rock to chip off flakes. The flakes themselves are usable as cutting tools. Chipping many flakes off a cobble in an appropriate pattern produces a chopper. Tools of this style are said to belong to the Oldowan industrial complex because they were first discovered at Olduvai Gorge, Tanzania. Archaeologists have learned firsthand that making Oldowan-style stone tools requires skill and experience (Geribàs et al. 2010).

The earliest known stone tools predate the earliest known *Homo* specimens.

Beginning with an individual named Kanzi, bonobos have been taught to make and use Oldowan tools (see Wynn et al. 2011; Bril et al. 2012). However,

Figure 20.40 Oldowan stone tool from Melka Kunture, Ethiopia This 1.7-million-year-old hand-held chopper is an example of the earliest style of human tool making. Photo by Didier Descouens.

neither bonobos nor common chimpanzees are known to make them in the wild. And, despite years of practice, tools made by bonobos fall short in quality compared to the artifacts from Olduvai (Toth and Schick 2009).

The oldest known Oldowan tools are from Gona, Ethiopia. Found by Sileshi Semaw and colleagues (2003), the tools are 2.6 million years old (see also Semaw et al. 1997). Who were the stone knappers who made them?

An obvious candidate would be an early member of the genus *Homo*. The trouble is, we have no definitive evidence that any species of *Homo* had appeared by 2.6 million years ago. The oldest reliably dated *Homo* fossil is a 2.3-million-year-old upper jaw (maxilla) from Hadar, Ethiopia (Gibbons 1996; Kimbel et al. 1996). Which species this fossil represents is unclear. It could be *H. habilis, H. rudolfensis,* or some heretofore unknown species.

Circumstantial evidence suggests that the Hadar fossil represents the same species that made the 2.6-million-year-old Gona tools. Hadar is geographically close to Gona, 2.3 million years ago is geologically close to 2.6 million years ago, and the Hadar fossil was found near 34 Oldowan tools. It is possible that 2.6-million-year-old *Homo* fossils eventually will be found at Gona and that these early humans were the Gona stone knappers. However, as Bernard Wood (1997) points out, circumstantial evidence is not proof. If other hominins were present at the same time and place, then they are suspects, too. This is true even at Hadar, where the 2.3-million-year-old *Homo* jaw was found near Oldowan tools.

Wood notes that *Paranthropus* coexisted with early *Homo* in the same part of Africa over approximately the same time span as the Oldowan industrial complex. While there is no good circumstantial evidence to indicate that *Paranthropus* may have been responsible for Oldowan tools, some indirect evidence of their tool-using capabilities comes from their anatomy.

Randall L. Susman (1994) makes a case based on the anatomy of opposable thumbs. He starts by comparing the thumbs of humans versus chimpanzees (**Figure 20.41**). Humans have more elaborate musculature. They even have muscles, such as the flexor pollicis longus, that chimps lack (see also Diogo et al. 2012).

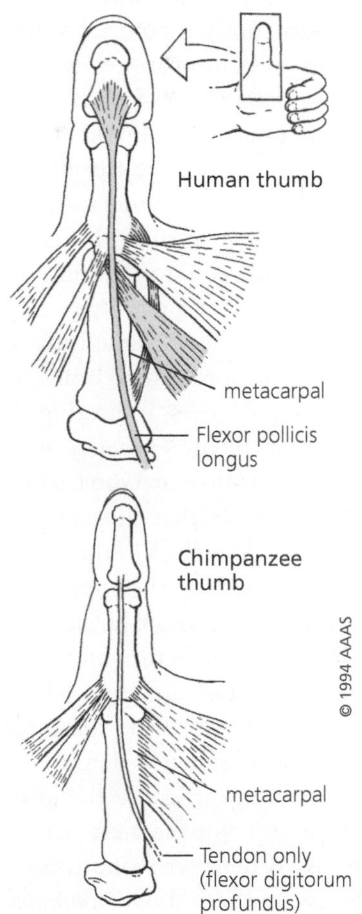

Figure 20.41 Human versus chimp thumbs Reprinted with permission from Susman (1994). From "Fossil evidence for early hominid tool use." *Science* 265: 1570–1573. AAAS. (Figure 3, page 1572.) Reprinted with permission from AAAS.

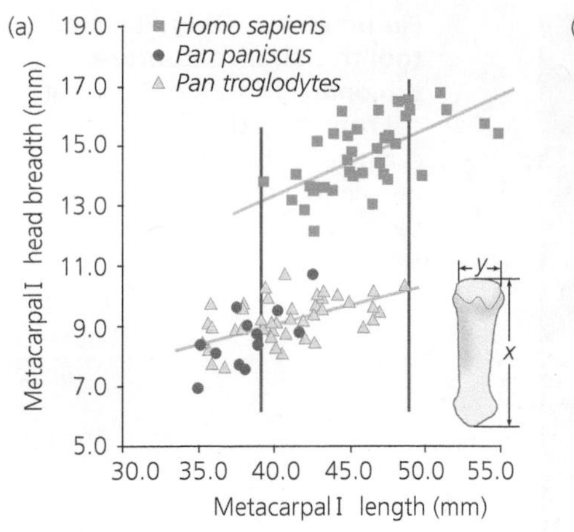

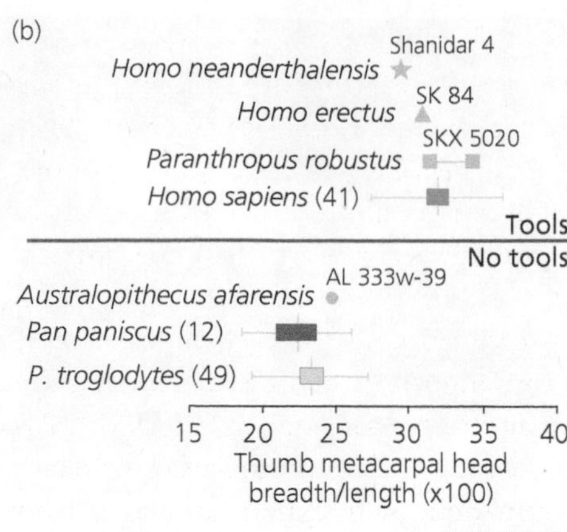

Figure 20.42 Thumb metacarpal bones in various hominins (a) Metacarpals of humans versus chimps. (b) Metacarpals of fossil and extant hominins. Labels on points are specimen numbers. For *Paranthropus*, breadth/width is given as an estimated range because the bone in question is not quite complete. Purple line separates species with versus without at least a temporal association with tools. From Susman (1994).

From "Fossil evidence for early hominid tool use." *Science* 265: 1570–1573. AAAS. (Figure 1, page 1571.) Reprinted with permission from AAAS.

Associated with their more elaborate musculature, humans have thicker metacarpal bones with broader heads **(Figure 20.42a)**. These differences in anatomy make the human hand more adept at precision grasping than the chimp hand. Susman argues that the modified anatomy of the human thumb evolved in response to selection pressures associated with the manufacture and use of complex tools.

Susman then compares the relative thickness of the thumb metacarpals in humans and chimpanzees with that in a variety of fossil hominins (Figure 20.42b). *Homo neanderthalensis, H. erectus,* and *P. robustus* resemble *H. sapiens* in having metacarpals with broad heads for their length. *Australopithecus afarensis,* which disappears from the fossil record before Oldowan tools appear, is like the chimpanzees in having metacarpals with narrow heads for their length. Susman asserts that we can use the thumb metacarpals to diagnose whether extinct hominins made and used stone tools. He concludes that both *H. erectus* and *P. robustus* were toolmakers. Susman's argument provoked debate (McGrew et al. 1995). Hamrick et al. (1998) argued that several australopithecine species—including *A. africanus* (the appropriate bones are not known for *A. afarensis*)—possessed powerful grasping thumbs, indicating that they may have used unmodified stones as tools.

Some morphological analyses suggest that both early *Homo* species and the robust australopithecines could have been toolmakers.

None of the evidence we have discussed establishes for certain whether the Oldowan tools at Gona were made by *Homo* or *Paranthropus*. Instead, the evidence argues that they could have been made by either or both. If we accept Susman's conclusion that the robust australopithecines were toolmakers, but the gracile australopithecines were not, and if we accept that the robust australopithecines were a sister lineage to *Homo*, then we must make one of two inferences: (1) The manufacture and use of complex stone tools originated in an undiscovered common ancestor of *Homo* and *Paranthropus;* or (2) the manufacture and use of complex stone tools originated independently in at least two hominin lineages.

In the absence of definitive proof—such as a fossil hand grasping a stone tool—we may never know the answer to our question. However, most paleoanthropologists favor the view that early *Homo* is responsible for most if not all Oldowan tools. Whenever stone tools are found in association with fossil humans, *Homo* is always there, but *Paranthropus* are often absent. (This includes Susman's fossils from Swartkrans, South Africa, where *Homo erectus* is also found.) And *Homo habilis,* if it was present at Hadar at 2.3 million years ago, comes very close to matching the time span of the Oldowan industrial complex. Three different species of robust australopithecine span the time from 2.5 to 1.0 million years ago, and there are no Oldowan tools after about 1.5 million years ago.

Most paleoanthropologists, however, still believe that most, if not all, Oldowan stone tools are the handiwork of *Homo*.

Which of Our Ancestors Had Language?

If the history of hominin tool use is murky, the history of hominin language is murkier. Language, like tool use, is a behavior. Because behaviors do not fossilize, we have no direct evidence of their history. We are left to examine circumstantial evidence in the archaeological and fossil record. Before the invention of writing, language left even less circumstantial evidence than tool use.

Language is a complex adaptation located in the neural circuitry of the brain. The vocabulary and particular grammatical rules of any given language are transmitted culturally, but the capacity for language and a fundamental grammar are, in present-day humans, both innate and universal (see Pinker 1994). Among the evidence for this assertion is the observation that communities of deaf children, if isolated from native signers, invent their own signed languages from scratch (Senghas et al. 2004; Sandler et al. 2005). After two generations of transmission to young children in the new deaf culture, these new sign languages develop all the hallmarks of genuine language. They have a standardized vocabulary and grammar, and their fluent users can efficiently communicate the full range of human ideas and emotions. Each of these new sign languages is unique, but all reflect the same universal grammar that linguists have identified in spoken languages.

Many of the brain's language circuits are concentrated in an area called the perisylvian cortex, usually in the left hemisphere (see Pinker 1994). These language circuits include Broca's area and Wernicke's area. Homologous structures exist in the brains of monkeys (Galaburda and Pandya 1982). The monkey homologues of Wernicke's area function in the recognition of sounds, including monkey calls. The monkey homologues of Broca's area function in controlling the muscles of the face, tongue, mouth, and larynx. However, neither of these structures plays a role in the production of monkey vocal calls. Instead, vocal calls are generated by circuits in the brain stem and limbic system. These same structures control nonlinguistic vocalizations in humans, such as laughing, sobbing, and shouting in pain. Thus, the human language organ appears to be a derived modification of neural circuits common to all primates. But among extant species, the nature of this modification—its specialization for linguistic communication—appears to be unique to humans. The implication is that the language organ, as such, evolved after our lineage split from the lineage of chimpanzees and bonobos.

How far back in our evolutionary lineage can we trace the existence of language, and on what evidence? Expert opinion is diverse. William Noble and Iain Davidson (1991), for example, assert that the only reliable evidence is located in the archaeological record. In their view, the hallmark of language is the use of arbitrary symbols, standardized within a culture, to represent objects and ideas. To find language, then, we must look for such symbols in the archaeological record. The first unequivocally arbitrary symbols occur in cave paintings, found in Germany and France, that are about 32,000 years old. Even Noble and Davidson cannot quite accept that language is as recent an innovation as that. They note that *Homo sapiens* had colonized Australia by 40,000 years ago—possibly as early as 60,000 years ago—and confess that they cannot imagine how people could build boats and cross the open ocean without language to facilitate planning and coordination. However, Noble and Davidson hold the line at about 40,000 years. This date would imply that *H. sapiens* is the only species ever to use language.

In contrast, after examining casts of the insides of the braincases of specimens of *Homo habilis,* Phillip Tobias (1987) suggested that language may be as much as 2 million years old. In addition to their sheer size, the endocasts revealed what

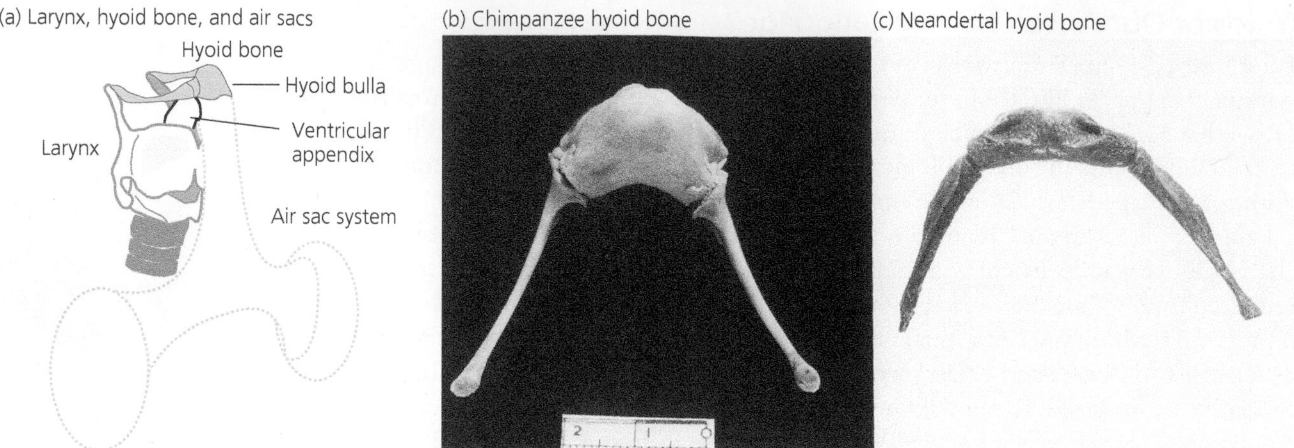

Figure 20.43 Hyoid bones and air sacs (a) The larynx, hyoid bone, and air sacs as they appear in a chimpanzee. From de Boer (2012). (b) and (c) Hyoid bones of a chimpanzee and a Neandertal. The scale bar, which applies to both hyoids, is in centimeters. The hyoid of a modern human is virtually identical to that of the Neandertal. Reproduced by permission from Arensburg, Schepartz, Tillier, Vandermeersch, and Rak (1990).

appear to be derived structural traits characteristic of our genus. Among them were enlargements of Broca's and Wernicke's areas. By his own account, this discovery converted Tobias from a skeptic to an advocate of the hypothesis that language first emerged, at least in rudimentary form, in the earliest *Homo* species.

Unfortunately, subsequent studies of the brains of great apes, and of variation among humans, has shown that external features of the brain visible in an endocast cannot be used to diagnose the capacity for language (Gannon et al. 1998). The same is true of other features of the skull that have been considered as indicators of language capacity, or at least the ability to speak (Fitch 2000). These include the angle of the base of the skull and the size of the canal that carries nerves from the brain to the tongue. Can the fossil record tell us anything?

Our vocal apparatus contains one hard part that fossilizes, the hyoid bone. It is located in the throat and provides an anchor for the muscles of the larynx and tongue **(Figure 20.43a)**. The hyoid bone of a chimpanzee is shaped differently than those of a Neandertal and a modern human (Figure 20.43b and c). In a chimp, the body of the hyoid is larger and shaped like an inverted cup (de Boer 2012). This cup-shaped structure is called the hyoid bulla. We are looking down on the chimpanzee hyoid bone in the photo and so cannot see the hollow space underneath. In a modern human, in contrast, the body of the hyoid is bar shaped.

This difference in hyoid bone shape appears to be related to the air sacs that branch from the vocal tract in chimpanzees but are absent in humans (de Boer 2012). In chimpanzees, the entrance to the air sacs passes under the hyoid bulla.

Because all other apes have air sacs too, their absence in humans appears to be derived. The adaptive significance of air sacs is unclear. Among their effects is to increase the low-frequency resonances of vocalizations, which makes the vocalizer sound bigger. The reason air sacs were lost in our ancestors is unclear as well. An obvious hypothesis is that their absence facilitates spoken communication.

Bart de Boer (2012) tested this hypothesis by building physical models of human vocal tracts with and without air sacs. He then played vowel sounds through them, contaminated with varying amounts of noise, and challenged volunteers to distinguish the vowels. As shown in **Figure 20.44**, the volunteers required cleaner signals to discriminate vowels played through a vocal tract with air sacs. This is

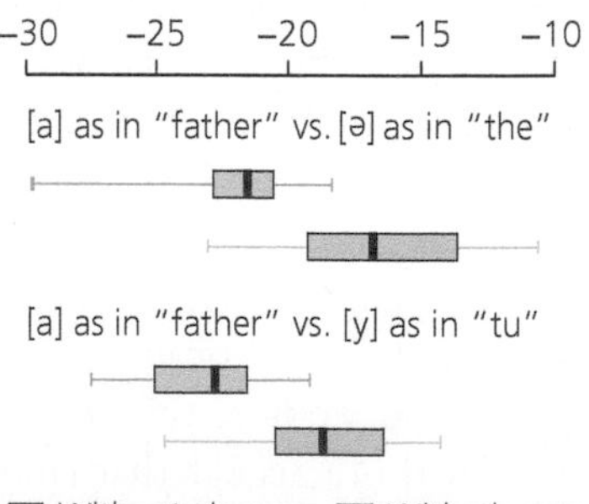

Figure 20.44 Absence of air sacs makes it easier to discriminate vowels Box plots indicate median, range of 2nd and 3rd quartile, and range of data. Both comparisons are highly significant. Redrawn from de Boer (2012).

consistent with the hypothesis that the loss of air sacs was adaptive for our ancestors because it made it easier for them to understand each other's utterances.

If this hypothesis is correct, and if the shape of the hyoid bone indeed reflects the presence or absence of air sacs, then we can bracket the advent of spoken language at least somewhat more tightly. The hyoid bone in Figure 20.43c is from a 60,000-year-old Neandertal skeleton found in Israel (Arensburg et al. 1989; Arensberg et al. 1990). It lacks a bulla. So do the hyoid bones from two 530,000-year-old specimens of *Homo heidelbergensis* found in Spain (Martínez et al. 2008). However, the hyoid of a 3.3-million-year old *Australopithecus afarensis* found in Ethiopia has a bulla (Alemseged et al. 2006). On this evidence, it appears that spoken language arose some time between 3.3 million and 530,000 years ago, probably in a member of genus *Homo*.

Analyses of fossil hyoid bones suggest that rudimentary language is at least half a million years old, and possibly considerably older.

SUMMARY

As Darwin predicted, an evolutionary perspective throws light on the origin and nature of humans. Humans are relatives of the great apes. Although consensus was slow to arrive, morphological and molecular studies indicate that our closest living relatives are the chimpanzees. Our most recent common ancestor with the chimpanzees lived roughly 5 to 7 million years ago.

Following its split from the chimpanzee lineage, our own lineage gave rise to multiple species of bipedal African hominins. Fossils provide strong evidence for the coexistence of at least two of these species and perhaps as many as five. We are the sole survivors of this evolutionary radiation.

The first members of genus *Homo* left Africa nearly 2 million years ago. Present-day non-African populations are descended from a more recent wave of emigrants that left Africa within the last 200,000 years. On their way out of Africa, probably in the Middle East, the modern human ancestors of non-Africans interbred with Neandertals. Neandertals thus contributed a small fraction of the genome of non-Africans. On their way across Asia, the modern human ancestors of today's Melanesians interbred with the Denisovans. Denisovans thus contributed a small fraction of the genome of Melanesians. The implication is that geographic variation among present-day humans largely, though perhaps not entirely, arose during the last 100,000 to 200,000 years. All modern humans are extremely closely related.

Among the derived traits unique to our species are the manufacture and use of complex tools and the capacity for language. Because behavior does not fossilize, researchers have to rely on circumstantial evidence to reconstruct the history of these traits. Stone knapping appeared at least 2.5 million years ago. It is most likely to have arisen in an early species of *Homo,* although *Paranthropus* may also have made stone tools. The evidence on language is more tenuous, but it suggests that spoken language also emerged in a species of *Homo,* at least half a million years ago. The evolution of tool use and language may have been linked.

QUESTIONS

1. What is the difference between a gene tree and a species tree? Explain in your own words how it is possible for gorillas and humans to share a genetic trait (such as a retroviral insertion) that chimpanzees do not share, if chimpanzees and humans are really each other's closest relatives. Given the lack of agreement between gene trees and species trees, how is it possible to reconstruct the true species tree?
2. Just a few decades ago, almost none of the fossils described in this chapter had been found. At that time, it was expected that when early hominin fossils were finally found, the first distinctly human feature—that is, the first derived trait of hominins that would distinguish them morphologically from the chimpanzee lineage—would prove to be an enlarged brain. It was thought that a large, human-sized brain must have evolved either before or simultaneously with bipedality. Now that we have the fossils to test this question, what do the fossils show? Which came first, large brains or bipedality?
3. How were Sarich and Wilson able to test the genetic

relationships of humans to great apes when at the time (1967) it was not possible to sequence DNA?

4. Explain why Ruvolo (1995, 1997) thought it was important to look at several nuclear genes, and not just the mtDNA genes, to study the relationships of humans and the great apes.
5. The data in this chapter show that humans and the chimpanzees are each other's closest relatives. Is it accurate to say that humans evolved from chimpanzees? How about that chimpanzees evolved from humans? Is it accurate to say that humans evolved from apes?
6. In a study of the phylogeny of Old World monkeys (Hayasaka et al. 1996), the three individual rhesus macaques that were studied did not form a monophyletic group. Instead, the mtDNA of one of the rhesus macaques was more similar to the mtDNA of Japanese and Taiwanese macaques (which are different species) than it was to the other rhesus macaques. How might this have happened? (There are at least two possibilities.) With this result as background, explain why it is useful that the phylogeny in Figure 20.4 includes several individuals from each species.
7. Look again at Figure 20.4, and this time focus on the diversity of sequences within each species. The length of the colored lines indicates the degree of genetic diversity within each species.
   a. Do humans have a large or small amount of genetic diversity, compared to the other primates studied? Which other species show similar patterns to humans? Why do you think some species have greater diversity than others?
   b. One of the human sequences is distinct from the other five. Can you guess the geographic location of this person?
8. Knuckle walking and fist walking are exceedingly rare methods of locomotion among animals. A similar type of locomotion occurs in a few other unusual cases such as the giant anteater and a few sloths. Why do you think it evolved in only these animals—what advantage does it bring? Why do you think humans lost this trait? What does this trait indicate about the different of animals have for their forelimbs?
9. Briefly describe (one sentence each) the three main models of the origin of *Homo sapiens* from *Homo erectus*. Which have been rejected, and why? What is the most likely scenario, given the evidence? What questions about the origins of present-day geographic variation remain?
10. What's in a name? Jared Diamond (1992) suggests that if we follow the naming traditions of cladistic taxonomy, then humans, chimpanzees, and bonobos should all be considered members of a single genus. Diamond proposes calling these species, respectively, *Homo sapiens, Homo troglodytes,* and *Homo paniscus*. Jonathan Marks (1994) objects to Diamond's taxonomic reasoning. Concerning the nature of humans and apes, Marks asserts that "Popular works tell us we are not merely genetically apes but that we are literally apes (e.g., Diamond 1992). Sometimes there is profundity in absurdity, but I don't think this is one of those times. It merely reflects the paraphyletic nature of the category 'apes'—humans are apes, but only in the same sense that pigeons are reptiles and horses are fish.... Focusing on the genetic relations obscures biologically significant patterns of phenotypic divergence." Do you think humans, chimpanzees, and bonobos should all be classified as members of the same genus? Is there more at stake in the disagreement between Diamond and Marks than just Latin names? If so, what?
11. Jared Diamond finds ethical dilemmas in the close kinship between humans and chimpanzees: "It's considered acceptable to exhibit caged apes in zoos, but it's not acceptable to do the same with humans. I wonder how the public will feel when the identifying label on the chimp cage in the zoo reads *Homo troglodytes*" (Diamond 1992, p. 29). Diamond finds the use of chimpanzees in medical research even more problematic. The scientific justification for the use of chimpanzees is that chimpanzee physiology is extremely similar to human physiology, so chimpanzees are the best substitute for human subjects. Diamond notes that jails are a very rough analogue to zoos, in the sense that they represent conditions under which we do consider it acceptable to keep people in cages without their consent (if not to display them). But there is no human analogue to research on chimpanzees: Under no conditions do we consider it acceptable to do medical experiments on humans without their consent. Is it ethically justified to keep animals in zoos? To use animals in medical research? Does the phylogenetic relationship between ourselves and the animals in question matter? If so, how and why?
12. Section 20.2 mentions that there is debate over the evolutionary affinities of *Sahelanthropus tchadensis*. Given the age and appearance of the skull in Figure 20.13, there are many possibilities: It could be our common ancestor with the chimpanzees; it could be a species more closely related to us than to the chimpanzees; it could be a species more closely related to the chimpanzees than to us; it could even be a species more closely related to gorillas than to either the chimpanzees or us. Suppose you are a paleoanthropologist who wants to figure out which of these possibilities is correct. What strata would you choose to search for more fossils? If you are lucky enough to find a skull, what features would tell you which hypothesis is closest to the truth?
13. One of the most heated aspects of human racial politics is the contention that human races are genetically distinct. How is this issue addressed by the African replacement

model versus the multiregional evolution model? That is, which model predicts that human races are more genetically different from each other? How different are people from different geographic regions?

14. We reviewed genetic studies showing that non-African human populations are descended from African populations. Some people might conclude from these data that modern African people are in some sense "primitive." What are the logical flaws in this thinking?

15. Different ethnic groups within Africa are more diverse than the ethnic groups on all other continents put together. What does this imply about the common U.S. practice of categorizing people into groups of so-called Africans, Caucasians, Asians, Hispanics, and Native Americans?

16. Work by C. Swisher and colleagues (1996) indicates that *Homo erectus* may have persisted in Java until 53,000 years ago at Sambungmachan and until 27,000 years ago at Ngandong. If correct, these dates imply that *H. erectus* and *H. sapiens* coexisted in Java. Is this finding relevant to the debate over the out-of-Africa model versus the multiregional origin model? Why or why not?

17. Describe the two lines of morphological evidence that have been used to infer whether extinct hominins like Neandertals or *Homo erectus* might have used spoken language. What did the data show? Do you find this to be convincing evidence?

18. For the sake of argument, adopt the proposition implied by Wood (1997) that early species of *Homo* did not participate in the production of Oldowan stone tools. What other puzzles must we now solve? Note that after the invention of Oldowan tools, the next advance in toolmaking is marked in the archaeological record by the appearance of Acheulean tools, which are substantially more sophisticated than Oldowan tools (Johanson et al. 1996). Acheulean tools appeared about 1.4 million years ago and persisted until less than 200,000 years ago.

19. Derek Bickerton (1995) and Charles Catania (1995) object to the suggestion that early humans had language. Bickerton writes, "If *H. habilis* already had all the necessary ingredients for language, what happened during the next million years?" And Catania writes, "I [deduce] that our hominin ancestors should have taken over the world 950,000 years ago if a million years ago they were really like us in their language competence. But they did not, so they were not; if they had been, those years would have been historic instead of prehistoric." How do you think Bart DeBoer would respond to Bickerton and Catania? Who do you think is right?

20. The ancestors of horses are each known from dozens, hundreds, or in some cases even thousands, of virtually complete specimens. Yet hominin species of the same age are often known only from one or a very few partial specimens, such as the crushed partial skull of *Sahelanthropus*. As a result, we understand equine evolution much better than we do hominid evolution. Speculate as to why this is so: Why are hominin fossils rare? Does the scarcity of hominin fossils invalidate the conclusions of paleontologists? Put another way, is it really possible to learn anything useful from a single bone fragment?

EXPLORING THE LITERATURE

21. In Sections 2.1 and 2.3, we discussed discoveries that have arisen from comparison of the gorilla, chimpanzee, Neandertal, Denisovan, and modern human genomes. Researchers have now sequenced a bonobo genome. See:

Prüfer, K., K. Munch, et al. 2012. The bonobo genome compared with the chimpanzee and human genomes. *Nature* 486: 527–531.

22. We discussed archaeological, paleontological, and experimental evidence on the age of human language. Another source of information is a language clock, analogous to the molecular clock. For an estimate of the age of the last common ancestor of extant spoken languages, based on a calibrated estimate of the rate at which languages diversify, see:

Perreault, C., and S. Mathew. 2012. Dating the origin of language using phonemic diversity. *PLoS One* 7: e35289.

23. Is tool use in animals hardwired or learned? See:

Matsusaka, T., H. Nishie, et al. 2006. Tool-use for drinking water by immature chimpanzees of Mahale: Prevalence of an unessential behavior. *Primates* 42 (2): 113–122.

Kruetzen, M., J. Mann, et al.. 2005. Cultural transmission of tool use in bottlenose dolphins. *Proceedings of the National Academy of Sciences, USA* 102 (25): 8939–8943.

Recent evidence suggests that certain birds not only make and use tools, but appear to understand what they are doing. See:

Weird, A. A. S., J. Chappell, and A. Kacelnik. 2002. Shaping of hooks in New Caledonian crows. *Science* 297: 981.

Videos of the New Caledonian crows making and using hook tools can be seen at http://users.ox.ac.uk/~kgroup/tools/movies.shtml.

24. New estimates of the human mutation rate, based on direct measurements from whole-genome sequences, are providing new data on which to estimate divergence times among human lineages. See:

Sankararaman, S., N. Patterson, et al. 2012. The date of interbreeding between Neandertals and modern humans. *PLoS Genetics* 8: e1002947.

Scally, A., and R. Durbin. 2012. Revising the human mutation rate: Implications for understanding human evolution. *Nature Reviews Genetics* 13: 745–753.

25. Might any nonhuman animals use a simple language? Most animals can produce vocalizations, but few show two fundamental aspects of human language: learned vocalizations, and the use of certain sounds to symbolize a concept. The ability to learn vocalizations is common in many birds, but rare in mammals. The use of sounds to symbolize concepts is rarer still. See these papers for two examples that might qualify:

Janik, V. M., L. S. Sayigh, and R. S. Wells. 2006. Signature whistle shape conveys identity information to bottlenose dolphins. *Proceedings of the National Academy of Sciences, USA* 103 (21): 8293–8297.

Seyfarth, R. M., D. L. Cheney, and P. Marler. 1980. Vervet monkey alarm calls—semantic communication in a free-ranging primate. *Animal Behaviour* 28: 1070–1094.

This paper reports evidence that dogs, though they obviously cannot produce language themselves, may have the capability to understand the symbolic meaning of up to 200 words, and can rapidly learn more—at least in the case of this border collie:

Kaminski, J., J. Call, and J. Fischer. 2004. Word learning in a domestic dog: Evidence for "fast mapping." *Science* 304: 1682–1683.

26. Why did hominins evolve bipedality? See this paper for evidence that endurance running may have played a major role in human evolution:

Bramble, D. M., and D. E. Lieberman. 2004. Endurance running and the evolution of *Homo*. *Nature* 432: 345–352.

For another (creative) idea on the adaptive significance of bipedal locomotion, see:

Sutou, S. 2012. Hairless mutation: A driving force of humanization from a human-ape common ancestor by enforcing upright walking while holding a baby with both hands. *Genes to Cells* 17: 264–272.

27. We noted that, genetically, humans and chimpanzees are startlingly similar. Are they similar enough that they could produce viable hybrids? How could we identify a hybrid if it existed? Claims that a sideshow performer named Oliver was a human–chimpanzee hybrid were credible enough to warrant scientific investigation. See:

Ely, J. J., M. Leland, et al. 1998. Technical note: Chromosomal and mtDNA analysis of Oliver. *American Journal of Physical Anthropology* 105: 395–403.

28. For a genetic mutation whose origin appears to correlate with morphological changes in the human lineage, see:

Stedman, H. H., B. W. Kozyak, et al. 2004. Myosin gene mutation correlates with anatomical changes in the human lineage. *Nature* 428: 415–418.

29. We discussed evidence that the human lineage underwent considerable diversification after it split from the chimpanzee lineage. For additional evidence of diversification among early members of genus *Homo*, see:

Leakey, M. G., F. Spoor, et al. 2012. New fossils from Koobi Fora in northern Kenya confirm taxonomic diversity in early *Homo*. *Nature* 488: 201–204.

For genetic evidence that the chimpanzee lineage also diversified following its split from the human lineage, and that this diversification may be ongoing, see:

Gonder, M. K., S. Locatelli, et al. 2011. Evidence from Cameroon reveals differences in the genetic structure and histories of chimpanzee populations. *Proceedings of the National Academy of Sciences, USA* 108: 4766–4771.

CITATIONS

Aiello, L. C. 1993. The fossil evidence for modern human origins in Africa: A revised review. *American Anthropologist* 95: 73–96.

Aiello, L. C., and M. Collard. 2001. Our newest oldest ancestor? *Nature* 410: 526–527.

Alemseged, Z., F. Spoor, et al. 2006. A juvenile early hominin skeleton from Dikika, Ethiopia. *Nature* 443: 296–301.

Andrews, P. 1992. Evolution and environment in the Hominoidea. *Nature* 360: 641–646.

Andrews, P., and L. Martin. 1987. Cladistic relationships of extant and fossil hominoids. *Journal of Human Evolution* 16: 101–118.

Arensburg, B., A. M. Tillier, et al. 1989. A middle paleolithic human hyoid bone. *Nature* 338: 758–760.

Arensburg, B., L. A. Schepartz, et al. 1990. A reappraisal of the anatomical basis for speech in Middle Paleolithic hominids. *American Journal of Physical Anthropology* 83: 137–146.

Arsuaga, J. L., I. Martínez, et al.. 1999. The human cranial remains from Gran Dolina lower-Pleistocene site (Sierra de Atapuerca, Spain). *Journal of Human Evolution* 37: 431–457.

Asfaw, B., T. White, et al. 1999. *Australopithecus gahri:* A new species of early hominid from Ethiopia. *Science* 284: 629–635.

Ayala, F. J., A. Escalante, et al. 1994. Molecular genetics of speciation and human origins. *Proceedings of the National Academy of Sciences, USA* 91: 6787–6794.

Bamshad, M. J., S. Wooding, et al. 2003. Human population genetic structure and inference of group membership. *American Journal of Human Genetics* 72: 578–589.

Barbulescu, M., G. Turner, et al. 2001. A HERV-K provirus in chimpanzees, bonobos, and gorillas, but not humans. *Current Biology* 11: 779–783.

Begun, D. R. 1992. Miocene fossil hominids and the chimp–human clade. *Science* 257: 1929–1933.

Begun, D. R. 1994. Relations among the great apes and humans: New interpretations based on the fossil great ape *Dryopithecus*. *Yearbook of Physical Anthropology* 37: 11–63.

Begun, D. R. 1995. Late-Miocene European orangutans, gorillas, humans, or none of the above? *Journal of Human Evolution* 29: 169–180.

Begun, D. R. 2004. The earliest hominins—Is less more? *Science* 303: 1478–1480.

Begun, D. R., C. V. Ward, and M. D. Rose. 1997. Events in hominoid evolution. In *Function, Phylogeny and Fossils: Miocene Hominoid Evolution and Adaptations*, ed. D. R. Begun, C. V. Ward, and M. D. Rose. New York: Plenum Publishing Company, 389–415.

Berger, L. R., D. J. de Ruiter, et al. 2010. *Australopithecus sediba*: A new species of *Homo*-like australopith from South Africa. *Science* 328: 195–204.

Bermúdez de Castro, J. M., J. L. Arsuaga, et al. 1997. A hominid from the lower Pleistocene of Atapuerca, Spain: Possible ancestor to Neandertals and modern humans. *Science* 276: 1392–1395.

Berna, F., P. Goldberg, et al. 2012. Microstratigraphic evidence of in situ fire in the Acheulean strata of Wonderwerk Cave, Northern Cape province, South Africa. *Proceedings of the National Academy of Sciences, USA* 109: E1215–E1220.

Bickerton, Derek. 1995. Finding the true place of *Homo habilis* in language evolution. Open peer commentary. In Wilkins, W. K., and J. Wakefield, Brain evolution and preconditions. *Behavioral and Brain Sciences* 18: 161–226.

Blumenschine, R. J., C. R. Peters, et al. 2003. Late Pliocene *Homo* and hominid land use from western Olduvai Gorge, Tanzania. *Science* 299: 1217–1221.

Boesch, C., and M. Tomasello. 1998. Chimpanzee and human cultures. *Current Anthropology* 39: 591–614.

Borowik, O. A. 1995. Coding chromosomal data for phylogenetic analysis: Phylogenetic resolution of the Pan-Homo-Gorilla trichotomy. *Systematic Biology* 44: 563–570.

Bowcock, A. M., A. Ruiz-Linares, et al. 1994. High resolution of human evolutionary trees with polymorphic microsatellites. *Nature* 368: 455–457.

Bril, B., J. Smaers, et al. 2012. Functional mastery of percussive technology in nut-cracking and stone-flaking actions: Experimental comparison and implications for the evolution of the human brain. *Philosophical Transactions of the Royal Society B* 367: 59–74.

Britten, R. J. 2002. Divergence between samples of chimpanzee and human DNA sequences is 5%, counting indels. *Proceedings of the National Academy of Sciences, USA* 99: 13633–13635.

Brunet, M. 2002. *Sahelanthropus* or "*Salhelpithecus*"? *Nature* 419: 582.

Brunet, M., F. Guy, et al. 2002. A new hominid from the upper Miocene of Chad, central Africa. *Nature* 418: 145–151.

Brunet., M., F. Guy, et al. 2005. New material from the earliest hominid from the Upper Miocene of Chad. *Nature* 434: 752–755.

Carbonell, E., J. M., Bermúdez de Castro, and J. L. Arsuaga. 1999. Preface: Special issue on Gran Dolina site: TD6 Aurora Stratum (Burgos, Spain). *Journal of Human Evolution* 37: 309–311.

Catania, A. C. 1995. Single words, multiple words, and the functions of language. Open peer commentary. In W. K. Wilkins and J. Wakefield, Brain evolution and neurolinguistic preconditions. *Behavioral and Brain Sciences* 18: 161–226.

Cela-Conde, C. J., and F. J. Ayala. 2003. Genera of the human lineage. *Proceedings of the National Academy of Sciences, USA* 100: 7684–7689.

Cerling, T. E., E. Mbua, et al. 2011. From the Cover: Diet of *Paranthropus boisei* in the early Pleistocene of East Africa. *Proceedings of the National Academy of Sciences, USA* 108: 9337–9341.

Charrier, C., K. Joshi, et al. 2012. Inhibition of SRGAP2 function by its human-specific paralogs induces neoteny during spine maturation. *Cell* 149: 923–935.

Collard, M., and L. C. Aiello. 2000. From forelimbs to two legs. *Nature* 404: 339–340.

Collard, M., and B. Wood. 2000. How reliable are human phylogenetic hypotheses? *Proceedings of the National Academy of Sciences, USA* 97: 5003–5006.

Corruccini, R. S., and H. M. McHenry. 2001. Knuckle-walking hominid ancestors. *Journal of Human Evolution* 40: 507–511.

Dainton, M. 2001. Did our ancestors knuckle-walk? *Nature* 410: 324–325.

Darwin, C. 1859. *On the Origin of Species by Means of Natural Selection, Or the Preservation of the Favoured Races in the Struggle for Life*. London: John Murray.

Darwin, C. 1871. *The Descent of Man, and Selection in Relation to Sex*. London: John Murray.

Dean, D., and E. Delson. 1992. Second gorilla or third chimp? *Nature* 359: 676–677.

de Boer, B. 2012. Loss of air sacs improved hominin speech abilities. *Journal of Human Evolution* 62: 1–6.

Deinard, A. S., G. S. Sirugo, and K. K. Kidd. 1998. Hominoid phylogeny: Inferences from a sub-terminal minisatellite analyzed by repeat expansion detection (RED). *Journal of Human Evolution* 35: 313–317.

Dennis, M. Y., X. Nuttle, et al. 2012. Evolution of human-specific neural SRGAP2 genes by incomplete segmental duplication. *Cell* 149: 912–922.

de Waal, F. B. M. 1997. *Bonobo: The Forgotten Ape*. Berkeley: University of California Press.

de Waal, F. B. M. 1999. Cultural primatology comes of age. *Nature* 399: 635–636.

de Waal, F. B. M. 2005. A century of getting to know the chimpanzee. *Nature* 437: 56–59.

Diamond, J. 1992. *The Third Chimpanzee*. New York: HarperCollins.

Dirks, P. H., J. M. Kibii, et al. 2010. Geological setting and age of *Australopithecus sediba* from southern Africa. *Science* 328: 205–208.

Diogo, R., B. G. Richmond, and B. Wood. 2012. Evolution and homologies of primate and modern human hand and forearm muscles, with notes on thumb movements and tool use. *Journal of Human Evolution* 63: 64–78.

Djian, P., and H. Green. 1989. Vectorial expansion of the involucrin gene and the relatedness of the hominoids. *Proceedings of the National Academy of Sciences, USA* 86: 8447–8451.

Duarte, C., J. Maurício, et al. 1999. The early upper Paleolithic human skeleton from the Abrigo do Laga Velho (Portugal) and modern human emergence in Iberia. *Proceedings of the National Academy of Sciences, USA* 96: 7604–7609.

Elango, N, J. W. Thomas, et al. 2006. Variable molecular clocks in hominoids. *Proceedings of the National Academy of Sciences, USA* 103: 1370–1375.

Fabre, P. H., A. Rodrigues, and E. J. Douzery. 2009. Patterns of macroevolution among Primates inferred from a supermatrix of mitochondrial and nuclear DNA. *Molecular Phylogenetics and Evolution* 53: 808–825.

Fitch, W. T. 2000. The evolution of speech: A comparative review. *Trends in Cognitive Science* 4: 258–267.

Frayer, D. W., M. H. Wolpoff, et al. 1993. Theories of modern human origins: The paleontological test. *American Anthropologist* 95: 14–50.

Gabunia, L., and A. Vekua. 1995. A Plio-Pleistocene hominid from Dmanisi, East Georgia, Caucasus. *Nature* 373: 509–512.

Gabunia, L., A. Vekua, et al. 2000. Earliest Pleistocene hominid cranial remains from Dmanisi, Republic of Georgia: Taxonomy, geological setting, and age. *Science* 288: 1019–1025.

Gagneux, P., and A. Varki. 2001. Genetic differences between humans and great apes. *Molecular Phylogenetics and Evolution* 18: 2–13.

Gagneux, P., C. Wills, et al. 1999. Mitochondrial sequences show diverse evolutionary histories of African hominoids. *Proceedings of the National Academy of Sciences, USA* 96: 5077–5082.

Galaburda, A. M., and D. N. Pandya. 1982. Role of architectonics and connections in the study of primate brain evolution. In *Primate Brain Evolution*, ed. E. Armstrong and D. Falk. New York: Plenum.

Gannon, P. J., R. L. Holloway, et al. 1998. Asymmetry of chimpanzee *planum temporale*: Humanlike pattern of Wernicke's brain language area homolog. *Science* 279: 220–222.

Gee, H. 2001. Return to the planet of the apes. *Nature* 412: 131–132.

Geribàs, N., M. Mosquera, and J. M. Vergès. 2010. What novice knappers have to learn to become expert stone toolmakers. *Journal of Archaeological Science* 37: 2857–2870.

Gibbons, A. 1994. Rewriting—and redating—prehistory. *Science* 263: 1087–1088.

Gibbons, A. 1996. A rare glimpse of an early human face. *Science* 274: 1298.

Gibbons, A. 2002. One scientist's quest for the origin of our species. *Science* 298: 1708–1711.

Gibbons, A. 2005. Facelift supports skull's status as oldest member of the human family. *Science* 308: 179–180.

Gibbons, A. 2011. Who were the Denisovans? *Science* 333: 1084–1087.

Gibbs, S., M. Collard, and B. Wood. 2000. Soft-tissue characters in higher primate phylogenetics. *Proceedings of the National Academy of Sciences, USA* 97: 11130–11132.

Gibbs, S., M. Collard, and B. Wood. 2002. Soft-tissue anatomy of the extant hominoids: A review and phylogenetic analysis. *Journal of Anatomy* 200: 2–40.

Goldstein, D. B., A. Ruiz Linares, et al. 1995. Genetic absolute dating based on microsatellites and the origin of modern humans. *Proceedings of the National Academy of Sciences, USA* 92: 6723–6727.

Goodman, M. 1962. Evolution of the immunologic species specificity of human serum proteins. *Human Biology* 34: 104–150.

Goodman, M., W. J. Bailey, et al. 1994. Molecular evidence on primate phylogeny from DNA sequences. *American Journal of Physical Anthropology* 94: 3–24.

Goodman, M., C. A. Porter, J. Czelusniak, et al. 1998. Toward a phylogenetic classification of primates based on DNA evidence complemented by fossil evidence. *Molecular Phylogenetics and Evolution* 9: 585–598.

Green, R. E., J. Krause, et al. 2010. A draft sequence of the Neandertal genome. *Science* 328: 710–722.

Groves, C. P. 1986. Systematics of the great apes. In D. R. Swindler and J. Erwin, eds. *Comparative Primate Biology, Volume 1: Systematics, Evolution, and Anatomy*. New York: Alan R. Liss, 187–217.

Guy, F., D. E. Lieberman, et al. 2005. Morphological affinities of the *Sahelanthropus tchadensis* (Late Miocene hominid from Chad) cranium. *Proceedings of the National Academy of Sciences, USA* 102: 18836–18841.

Haile-Selassie, Y., G. Suwa, and T. D. White. 2004. Late Miocene teeth from Middle Awash, Ethiopia, and early hominid dental evolution. *Science* 303: 1503–1505.

Hamrick M. W., S. E. Churchill, et al. 1998. EMG of the human *flexor pollicis longus* muscle: Implications for the evolution of Hominid tool use. *Journal of Human Evolution* 34: 123–136.

Hancock, A. M., D. B. Witonsky, et al. 2011. Adaptations to climate-mediated selective pressures in humans. *PLoS Genetics* 7: e1001375.

Harrison, T. 2010. Apes among the tangled branches of human origins. *Science* 327: 532–534.

Hayasaka, K., K. Fujii, and S. Horai. 1996. Molecular phylogeny of macaques: Implications of nucleotide sequences from an 896-base-pair region of mitochondrial DNA. *Molecular Biology and Evolution* 13: 1044–1053.

Hedges, S. B. 2000. A start for population genomics. *Nature* 408: 652–653.

Hill, W. C. 1946. Note on the male external genitalia of the chimpanzee. *Proceedings of the Zoological Society of London* 116: 129–132.

Hopkins, W. D., J. L. Russell, and J. A. Schaeffer. 2012. The neural and cognitive correlates of aimed throwing in chimpanzees: A magnetic resonance image and behavioural study on a unique form of social tool use. *Philosophical Transactions of the Royal Society B* 367: 37–47.

Horai, S., Y. Satta, et al. 1992. Man's place in the Hominoidea revealed by mitochondrial DNA genealogy. *Journal of Molecular Evolution* 35: 32–43.

Huang, W. P., R. G. Ciochon, et al. 1995. Early *Homo* and associated artifacts from Asia. *Nature* 378: 275–278.

Huxley, T. H. 1863. *Evidence as to Man's Place in Nature*. New York: D. Appelton.

Ingman, M., H. Kaessmann, et al. 2000. Mitochondrial genome variations and the origin of modern humans. *Nature* 408: 708–713.

Johanson, D. C., B. Edgar, and D. Brill. 1996. *From Lucy to Language*. New York: Simon & Schuster.

Jones, J. E., III. 2005. *Tammy Kitzmiller v. Dover Area School District, Memorandum Opinion*. U.S. District Court for the Middle District of Pennsylvania, Case No. 04cv2688.

Jorde, L. B., W. S. Watkins, et al. 2000. The distribution of human genetic diversity: A comparison of mitochondrial, autosomal, and Y-chromosome data. *American Journal of Human Genetics* 66: 979–988.

Kim, H.-S., and O. Takenaka. 1996. A comparison of TSPY genes from Y-chromosomal DNA of the great apes and humans: Sequence, evolution, and phylogeny. *American Journal of Physical Anthropology* 100: 301–309.

Kimbel, W. H., R. C. Walter, et al. 1996. Late Pliocene *Homo* and Oldowan tools from the Hadar Formation (Kada Hadar member), Ethiopia. *Journal of Human Evolution* 31: 549–561.

King, M.-C., and A. C. Wilson. 1975. Evolution at two levels in humans and chimpanzees. *Science* 188: 107–116.

Kivell, T. L., J. M. Kibii, et al. 2011. *Australopithecus sediba* hand demonstrates mosaic evolution of locomotor and manipulative abilities. *Science* 333: 1411–1417.

Kumar, S., A. Filipski, et al. 2005. Placing confidence limits on the molecular age of the human-chimpanzee divergence. *Proceedings of the National Academy of Sciences, USA* 102: 18842–18847.

Langergraber, K. E., K. Prüfer, et al. 2012. Generation times in wild chimpanzees and gorillas suggest earlier divergence times in great ape and human evolution. *Proceedings of the National Academy of Sciences, USA* 109: 15716–15721.

Leakey, M. G., C. S. Feibel, et al. 1995. New 4-million-year-old hominid species from Kanapoi and Allia Bay, Kenya. *Nature* 376: 565–571.

Leakey, M. G., F. Spoor, et al. 2001. New hominin genus from eastern Africa shows diverse middle Pliocene lineages. *Nature* 410: 433–440.

Lehtonen, S., I. E. Sääksjärvi, et al. 2011. Who is the closest extant cousin of humans? Total-evidence approach to hominid phylogenetics via simultaneous optimization. *Journal of Biogeography* 38: 805–808.

Li, T. Y., and D. A. Etler. 1992. New middle pleistocene hominid crania from Yunxian in China. *Nature* 357: 404–407.

Li, C. Y., Y. Zhang, et al. 2010. A human-specific de novo protein-coding gene associated with human brain functions. *PLoS Computational Biology* 6: e1000734.

Lieberman, D. E. 1995. Testing hypotheses about recent human evolution from skulls: Integrating morphology, function, development, and phylogeny. *Current Anthropology* 36: 159–197.

Lieberman, D. E. 2001. Another face in our family tree. *Nature* 410: 419–420.

Lockwood, C. A., W. H. Kimbel, and J. M. Lynch. 2004. Morphometrics and hominoid phylogeny: Support for a chimpanzee-human clade and differentiation among great ape subspecies. *Proceedings of the National Academy of Sciences, USA* 101: 4356–4360.

Lovejoy, C. O., K. G. Heiple, and R. S. Meindl. 2001. Did our ancestors knuckle-walk? *Nature* 410: 325–326.

Lowenstein, J., and A. Zihlman. 1988. The invisible ape. *New Scientist* 3 (December): 56–59.

Lycett, S. J., M. Collard, and W. C. McGrew. 2007. Phylogenetic analyses of behavior support existence of culture among wild chimpanzees. *Proceedings of the National Academy of Sciences, USA* 104: 17588–17592.

Maldonado, J. 2004. Dover schools still debating biology text. *York Daily Record*, June 9.

Manson, J. H., S. Perry, and A. R. Parish. 1997. Nonconceptive sexual behavior in bonobos and capuchins. *International Journal of Primatology* 8: 767–786.

Marks, J. 1993. Hominoid heterochromatin: Terminal C-bands as a complex genetic trait linking chimpanzee and gorilla. *American Journal of Physical Anthropology* 90: 237–246.

Marks, J. 1994. Blood will tell (won't it?): A century of molecular discourse in anthropological systematics. *American Journal of Physical Anthropology* 94: 59–79.

Martínez, I., J. L. Arsuaga, et al. 2008. Human hyoid bones from the middle Pleistocene site of the Sima de los Huesos (Sierra de Atapuerca, Spain). *Journal of Human Evolution* 54: 118–124.

McDougall, I., F. H. Brown, and J. G. Fleagle. 2005. Stratigraphic placement and age of modern humans from Kibish, Ethiopia. *Nature* 433: 733–736.

McGrew, W. C. 2010a. Chimpanzee technology. *Science* 328: 579–580.

McGrew, W. C. 2010b. In search of the last common ancestor: New findings on wild chimpanzees. *Philosophical Transactions of the Royal Society B* 365: 3267–3276.

McGrew, W. C., M. W. Hamrick, et al. 1995. Thumbs, tools, and early humans. *Science* 268: 586–589.

McLean, C. Y., P. L. Reno, et al. 2011. Human-specific loss of regulatory DNA and the evolution of human-specific traits. *Nature* 471: 216–219.

Mercader, J., M. Panger, and C. Boesch. 2002. Excavation of a chimpanzee stone tool site in the African rainforest. *Science* 296: 1452–1455.

Mikkelsen, T. S., L. W. Hillier, et al. 2005. Initial sequence of the chimpanzee genome and comparison with the human genome. *Nature* 437: 69–87.

Nei, M. 1995. Genetic support for the out-of-Africa theory of human evolution. *Proceedings of the National Academy of Sciences, USA* 92: 6720–6722.

Nelson, M. R., D. Wegmann, et al. 2012. An abundance of rare functional variants in 202 drug target genes sequenced in 14,002 people. *Science*: 337: 100–104.

Nielsen, R., C. Bustamante, et al. 2005. A scan for positively selected genes in the genomes of humans and chimpanzees. *PLoS Biology* 3: e170.

Noble, W., and I. Davidson. 1991. The evolutionary emergence of modern human behavior: Language and its archaeology. *Man* 26: 223–253.

Nuttall, G. H. F. 1904. *Blood, Immunity, and Blood Relationship: A Demonstration of Certain Blood-Relationships amongst Animals by Means of the Precipitin Test for Blood*. London: Cambridge University Press.

O'hUigin, C., Y. Satta, et al. 2002. Contribution of homoplasy and of ancestral polymorphism to the evolution of genes in anthropoid primates. *Molecular Biology and Evolution* 19: 1501–1513.

Pääbo, S. 2003. The mosaic that is our genome. *Nature* 421: 409–412.

Parish, A. R. 1994. Sex and food control in the "uncommon chimpanzee": How bonobo females overcame a phylogenetic legacy of male dominance. *Ethology and Sociobiology* 15: 157–179.

Parish, A. R. 1996. Female relationships in bonobos (*Pan paniscus*): Evidence for bonding, cooperation, and female dominance in a male-philopatric species. *Human Nature* 7: 61–96.

Parish, A. R., and F. B. M. de Waal. 2000. The other "closest living relative": How bonobos (*Pan paniscus*) challenge traditional assumptions about females, dominance, intra- and intersexual interactions, and Hominid evolution. *Annals of the New York Academy of Sciences* 907: 97–113.

Pickering, R., P. H. Dirks, et al. 2011. *Australopithecus sediba* at 1.977 Ma and implications for the origins of the genus *Homo*. *Science* 333: 1421–1423.

Pinker, S. 1994. *The Language Instinct*. New York: HarperCollins.

Pritchard, J. K., and M. W. Feldman. 1996. Genetic data and the African origin of humans. *Science* 274: 1548.

Raaum, R. L., K. N. Sterner, et al. 2005. Catarrhine primate divergence dates estimated from complete mitochondrial genomes: Concordance with fossil and nuclear DNA evidence. *Journal of Human Evolution* 48: 237–257.

Redon, R., S. Ishikawa, et al. 2006. Global variation in copy number in the human genome. *Nature* 444: 444–454.

Reich, D., R. E. Green, et al. 2010. Genetic history of an archaic hominin group from Denisova Cave in Siberia. *Nature* 468: 1053–1060.

Reno, P. L., R. S. Meindl, et al. 2003. Sexual dimorphism in *Australopithecus afarensis* was similar to that of modern humans. *Proceedings of the National Academy of Sciences, USA* 100: 9404–9409.

Richmond, B. G., and D. S. Strait. 2000. Evidence that humans evolved from a knuckle-walking ancestor. *Nature* 404: 382–385.

Richmond, B. G., and D. S. Strait. 2001a. Did our ancestors knuckle-walk? *Nature* 410: 326.

Richmond, B. G., and D. S. Strait. 2001b. Knuckle-walking hominid ancestor: A reply to Corruccini and McHenry. *Journal of Human Evolution* 40: 513–520.

Risch, N., K. K. Kidd, and S. A. Tishkoff. 1996. Genetic data and the African origin of humans—reply. *Science* 274: 1548–1549.

Rosenberg, N. A., J. K. Pritchard, et al. 2002. Genetic structure of human populations. *Science* 298: 2381–2385.

Ruvolo, M. 1994. Molecular evolutionary processes and conflicting gene trees: The Hominoid case. *American Journal of Physical Anthropology* 94: 89–113.

Ruvolo, M. 1995. Seeing the forest and the trees: Replies to Marks; Rogers and Commuzzie; Green and Djian. *American Journal of Physical Anthropology* 98: 218–232.

Ruvolo, M. 1997. Molecular phylogeny of the hominoids: Inferences from multiple independent DNA sequence data sets. *Molecular Biology and Evolution* 14: 248–265.

Ruvolo, M., D. Pan, et al. 1994. Gene trees and hominoid phylogeny. *Proceedings of the National Academy of Sciences, USA* 91: 8900–8904.

Salem, A.-H., D. A. Ray, et al. 2003. Alu elements and hominid phylogenetics. *Proceedings of the National Academy of Sciences, USA* 100: 12787–12791.

Sandler, W., I. Meir, et al. 2005. The emergence of grammar: Systematic structure in a new language. *Proceedings of the National Academy of Sciences, USA* 102: 2661–2665.

Sarich, V. M., and A. C. Wilson. 1967. Immunological time scale for Hominid evolution. *Science* 158: 1200–1203.

Sarmiento, E. E. 2010. Comment on the paleobiology and classification of *Ardipithecus ramidus*. *Science* 328: 1105; author reply 1105.

Satta, Y., J. Klein, and N. Takahata. 2000. DNA archives and our nearest relative: The trichotomy revisited. *Molecular Phylogenetics and Evolution* 14: 259–275.

Scally, A., J. Y. Dutheil, et al. 2012. Insights into hominid evolution from the gorilla genome sequence. *Nature* 483: 169–175.

Schick, K. D., and N. P. Toth. 1993. *Making Silent Stones Speak: Human Evolution and the Dawn of Technology*. New York: Simon & Schuster.

Semaw, S., P. Renne, et al. 1997. 2.5-Million-year-old stone tools from Gona, Ethiopia. *Nature* 385: 333–336.

Semaw, S., M. J. Rogers, et al. 2003. 2.6-Million-year-old stone tools and associated bones from OGS-6 and OGS-7, Gona, Afar, Ethiopia. *Journal of Human Evolution* 45: 169–177.

Senghas, A., S. Kita, and A. Ozyurek. 2004. Children creating core properties of language: Evidence from an emerging sign language in Nicaragua. *Science* 305: 1779–1782.

Senut, B. M. Pickford, et al. 2001. First hominid from the Miocene (Lukeino Formation, Kenya). *Comptes Rendus de l'Academie des Sciences Serie II Fascicule A-Sciences de la Terre et des Planetes* 332: 137–144.

Shoshani, J., C. P. Groves, et al. 1996. Primate phylogeny: Morphological vs. molecular results. *Molecular Phylogenetics and Evolution* 5: 101–153.

Sims, G. E., S. R. Jun, et al. 2009a. Alignment-free genome comparison with feature frequency profiles (FFP) and optimal resolutions. *Proceedings of the National Academy of Sciences, USA* 106: 2677–2682.

Sims, G. E., S. R. Jun, et al. 2009b. Whole-genome phylogeny of mammals: Evolutionary information in genic and nongenic regions. *Proceedings of the National Academy of Sciences, USA* 106: 17077–17082.

Somel, M., X. Liu, et al. 2011. MicroRNA-driven developmental remodeling in the brain distinguishes humans from other primates. *PLoS Biology* 9: e1001214.

Stauffer, R. L., A. Walker, et al. 2001. Human and ape molecular clocks and constraints on paleontological hypotheses. *Journal of Heredity* 92: 469–474.

Steele, J., P. F. Ferrari, and L. Fogassi. 2012. From action to language: Comparative perspectives on primate tool use, gesture and the evolution of human language. *Philosophical Transactions of the Royal Society B* 367: 4–9.

Stout, D., and T. Chaminade. 2012. Stone tools, language and the brain in human evolution. *Philosophical Transactions of the Royal Society B* 367: 75–87.

Stern, J. T., and R. L. Susman. 1983. The locomotor anatomy of Australopithecus afarensis. *American Journal of Physical Anthropology* 60: 279–317.

Strait, D. S., and F. E. Grine. 2004. Inferring hominoid and early hominid phylogeny using craniodental characters: The role of fossil taxa. *Journal of Human Evolution* 47: 399–452.

Strait, D. S., F. E. Grine, and M. A. Moniz. 1997. A reappraisal of early hominid phylogeny. *Journal of Human Evolution* 32: 17–82.

Stringer, C. 1988. The dates of Eden. *Nature* 331: 565–566.

Sun, J. X., A. Helgason, et al. 2012. A direct characterization of human mutation based on microsatellites. *Nature Genetics* 44: 1161–1165.

Susman, R. L. 1994. Fossil evidence for early hominid tool use. *Science* 265: 1570–1573.

Swisher, C. C., III, G. H. Curtis, et al. 1994. Age of the earliest known hominids in Java, Indonesia. *Science* 263: 1118–1121.

Swisher, C. C., III, W. J. Rink, et al. 1996. Latest *Homo erectus* of Java: Potential contemporaneity with *Homo sapiens* in Southeast Asia. *Science* 274: 1870–1874.

Tattersall, I. 1986. Species recognition in human paleontology. *Journal of Human Evolution* 15: 165–175.

Tattersall, I. 1992. Species concepts and species identification in human evolution. *Journal of Human Evolution* 22: 341–349.

Tattersall, I. 1995. *The Fossil Trail*. Oxford: Oxford University Press.

Tattersall, I. 1997. Out of Africa again . . . and again? *Scientific American* 276 (April): 60–67.

Tattersall, I. 2000. Once we were not alone. *Scientific American* 282 (January): 56–62.

Tattersall, I., and J. H. Schwartz. 1999. Hominids and hybrids: The place of Neanderthals in human evolution. *Proceedings of the National Academy of Sciences, USA* 96: 7117–7119.

Tennessen, J. A., A. W. Bigham, et al. 2012. Evolution and functional impact of rare coding variation from deep sequencing of human exomes. *Science*: 337: 64–69.

Thorne, A. G., and M. H. Wolpoff. 1981. Regional continuity in Australasian Pleistocene hominid evolution. *American Journal of Physical Anthropology* 55: 337–349.

Tishkoff, S. A., E. Dietzsch, et al. 1996a. Global patterns of linkage disequilibrium at the CD4 locus and modern human origins. *Science* 271: 1380–1387.

Tishkoff, S. A., K. K. Kidd, and N. Risch. 1996b. Interpretations of multiregional evolution—reply. *Science* 274: 706–707.

Tobias, P. V. 1987. The brain of *Homo habilis:* A new level of organization in cerebral evolution. *Journal of Human Evolution* 16: 741–761.

Tobias, P. V. 2003. Encore Olduvai. *Science* 299: 1193–1194.

Toth, N., and K. Schick. 2009. The Oldowan: The tool making of early hominins and chimpanzees compared. *Annual Review of Anthropology* 38: 289–305.

Trinkaus, E., C. B. Ruff, et al. 1998. Locomotion and body proportions of the Saint-Césaire 1 Châtelperronian Neandertal. *Proceedings of the National Academy of Sciences, USA*. 95: 5836–5840.

Tyler-Smith, C., and Y. Xue. 2012. Sibling rivalry among paralogs promotes evolution of the human brain. *Cell* 149: 737–739.

Valladas, H., J. L. Reyss, et al. 1988. Thermoluminescence dating of Mousterian "Proto-Cro-Magnon" remains from Israel and the origin of modern man. *Nature* 331: 614–616.

van Schaik, C. P., M. Ancrenaz, et al. 2003. Orangutan cultures and the evolution of material culture. *Science* 299: 102–105.

Varki, A., D. H. Geschwind, and E. E. Eichler. 2008. Explaining human uniqueness: genome interactions with environment, behaviour and culture. *Nature Reviews Genetics* 9: 749–763.

Varki, A., and D. L. Nelson. 2007. Genomic comparisons of humans and chimpanzees. *Annual Review of Anthropology* 36: 191–209.

Vogel, G. 2002. Can chimps ape ancient hominid toolmakers? *Science* 296: 1380.

Waddle, D. M. 1994. Matrix correlation tests support a single origin for modern humans. *Nature* 368: 452–454.

Wang, X., W. E. Grus, and J. Zhang. 2006. Gene losses during human origins. *PLoS Biology* 4: e52.

Ward, S. C., and W. H. Kimbel. 1983. Subnasal alveolar morphology and the systematic position of *Sivapithecus*. *American Journal of Physical Anthropology* 61: 157–171.

White, T. 2003. Early hominids—diversity or distortion? *Science* 299: 1994–1997.

White, T. D., B. Asfaw, et al. 2003. Pleistocene *Homo sapiens* from Middle Awash, Ethiopia. *Nature* 423: 742–747.

White, T. D., B. Asfaw, et al. 2009. *Ardipithecus ramidus* and the paleobiology of early Hominids. *Science* 326: 64–64, 75–86.

White, T. D., and G. Suwa. 1987. Hominid footprints at Laetoli: Facts and interpretations. *American Journal of Physical Anthropology* 72: 485–514.

White, T. D., G. Suwa, and B. Asfaw. 1994. *Australopithecus ramidus*, a new species of early hominid from Aramis, Ethiopia. *Nature* 371: 306–312.

White, T. D., G. Suwa, and B. Asfaw. 1995. Corrigendum: *Australopithecus ramidus*, a new species of early hominid from Aramis, Ethiopia. *Nature* 375: 88.

White, T. D., G. Suwa, and C. O. Lovejoy. 2010. Response to comment on the paleobiology and classification of *Ardipithecus ramidus*. *Science* 328: 1105.

White, T. D., G. Wolde Gabriel, et al. 2006. Asa Issie, Aramis and the origin of *Australopithecus*. *Nature* 440: 883–889.

Whiten, A. 2005. The second inheritance system of chimpanzees and humans. *Nature* 437: 52–55.

Whiten, A., J. Goodall, et al. 1999. Cultures in chimpanzees. *Nature* 399: 682–685.

Wildman, D. E., M. Uddin, et al. 2003. Implications of natural selection in shaping 99.4% nonsynonymous DNA identity between humans and chimpanzees: Enlarging genus *Homo*. *Proceedings of the National Academy of Sciences, USA* 100: 7181–7188.

Wilford, J. N. 2005. Fossils of apelike creature still stir lineage debate. *New York Times*, April 12.

Wilkinson, R. D., M. E. Steiper, et al. 2011. Dating primate divergences through an integrated analysis of palaeontological and molecular data. *Systematic Biology* 60: 16–31.

Wolpoff, M. H. 1996. Interpretations of multiregional evolution. *Science* 274: 704–706.

Wolpoff, M. H., J. Hawks, et al. 2006. An ape or the ape: Is the Toumaï cranium TM 266 a hominid? *PaleoAnthropology* 2006: 36–50.

Wolpoff, M., B. Senut, et al. 2002. *Sahelanthropus* or "*Sahelpithecus*"? *Nature* 419: 581–582.

Wood, B. 1997. The oldest whodunnit in the world. *Nature* 385: 292–293.

Wood, B. 2002. Hominid revelations from Chad. *Nature* 418: 133–135.

Wood, B. 2010. Reconstructing human evolution: Achievements, challenges, and opportunities. *Proceedings of the National Academy of Sciences, USA* 107 (Suppl 2): 8902–8909.

Wood, B., and M. Collard. 1999. The human genus. *Science* 284: 65–71.

Wood, B., and T. Harrison. 2011. The evolutionary context of the first hominins. *Nature* 470: 347–352.

Wood, B., and M. Leakey. 2011. The Omo-Turkana Basin fossil hominins and their contribution to our understanding of human evolution in Africa. *Evolutionary Anthropology* 20: 264–292.

Wood, B., and A. Turner. 1995. Out of Africa and into Asia. *Nature* 378: 239–240.

Wrangham, R. W. 1999. Evolution of coalitionary killing. *Yearbook of Physical Anthropology* 42: 1–30.

Wynn, T., R. A. Hernandez-Aguilar, et al. 2011. "An ape's view of the Oldowan" revisited. *Evolutionary Anthropology* 20: 181–197.

Yunis, J. J., and O. Prakash. 1982. The origin of man: A chromosomal pictorial legacy. *Science* 215: 1525–1530.

Zollikofer, C. P. E., M. S. Ponce de León, et al. 2005. Virtual cranial reconstruction of *Sahelanthropus tchadensis*. *Nature* 434: 755–759.

Glossary

adaptation A trait that increases the ability of an individual to survive or reproduce compared with individuals without the trait.

adaptation from new mutation Occurs when the genetic variants responsible for a new adaptive phenotype arise after a novel selective agent appears.

adaptation from standing genetic variation Occurs when the genetic variants responsible for an adaptive phenotype were already present when a novel selective agent appeared.

adaptive Describes a trait that increases the fitness of an individual relative to that of individuals lacking the trait.

adaptive constraint A trade-off in which a gene with two functions cannot evolve improved performance of one function without simultaneously evolving worse performance of the other function.

adaptive radiation The divergence of a clade into populations adapted to many different ecological niches.

additive effect The contribution an allele makes to the phenotype that is independent of the identity of the other alleles at the same or different loci.

additive genetic variation Differences among individuals in a population that are due to the additive effects of genes.

agent of selection Any factor that causes individuals with certain phenotypes to have, on average, higher fitness than individuals with other phenotypes.

alleles Variant forms of a gene, or variant nucleotide sequences at a particular locus.

allopatric model The hypothesis that speciation occurs when populations become geographically isolated and diverge because selection and drift act on them independently.

allopatry Living in different geographic areas.

allozymes Distinct forms of an enzyme, encoded by different alleles at the same locus.

altruism Behavior that decreases the fitness of the actor and increases the fitness of the recipient.

altruistic interaction An interaction between individuals resulting in a loss of fitness for the actor and an increase in fitness for the recipient.

Amniota A clade of vertebrate animals defined by, among other characters, an amniotic egg. Extant amniotes include the mammals, lizards, snakes, turtles, crocodiles, and birds.

anagenesis Descent with modification but no speciation.

ancestral character A character that was possessed by the common ancestor of the species in a clade; used in contrast with derived character.

antibiotic A chemical, typically extracted from a microorganism, that kills bacteria by disrupting a particular biochemical process.

antigenic site A portion of a foreign protein that is recognized by the immune system and initiates a response.

apomorphy A character present in one or more species in a clade that was not present in the clade's common ancestor; an evolutionary novelty. Also known as a derived character.

assortative mating Occurs when individuals tend to mate with other individuals with the same genotype or phenotype.

average excess The average excess for allele *a* is the difference between the mean fitness of individuals carrying allele *a* and the mean fitness of the entire population. If the average excess for an allele is positive, then the allele will rise in frequency.

back mutation A mutation that reverses the effect of a previous mutation; typically a mutation that restores function after a loss-of-function mutation.

background extinction Extinctions that are not part of mass extinction events; thought to be due to typical types and rates of environmental change or species interactions as opposed to the extraordinary environmental changes that occur during mass extinctions.

background selection The loss of neutral alleles due to selection against closely linked deleterious mutations.

Bateman gradient The slope of the best-fit line relating reproductive success to mating success. Measures the strength of sexual selection.

Bayesian phylogeny inference An approach to phylogeny inference based on computing the probability that a particular tree is correct, given a specific model of evolution for the characters being analyzed and the data observed.

best-fit line The line that most accurately represents the trend of the data in a scatterplot; typically best-fit lines are calculated by least-squares linear regression.

biogeography The study of where organisms live and how they got there.

biotic replacement Used in paleontology to describe events in which one clade disappears in the fossil record and another clade takes its place. Can be competitive, as when the new clade ecologically displaces the old, or non-competitive, as when a mass extinction is followed by an evolutionary radiation.

blending inheritance The hypothesis that heritable factors blend to produce a phenotype and are passed on to offspring in this blended form.

bootstrapping In phylogeny reconstruction, a technique for estimating the strength of the evidence in our data for the existence of a particular clade. Involves analyzing replicate data sets constructed by sampling, with replacement, from our actual data. Bootstrap support for a clade runs from 0 to 100, with higher values indicating stronger support.

bottleneck A large-scale but short-term reduction in population size followed by an increase in population size.

branch (of a phylogenetic tree) Lines that indicate a specific population or taxonomic group through time.

branch and bound In phylogeny inference, an algorithm for searching the space of possible evolutionary trees without exhaustively checking them all. Eliminates groups of trees upon discovery that all of their members are worse than the best tree found so far.

broad-sense heritability That fraction of the total phenotypic variation in a population that is caused by genetic differences among individuals.

C-value paradox The puzzling lack of association between an organism's apparent morpholgical complexity and the size of its genome.

canonical spatial expression pattern In evolutionary developmental biology, expression of Hox genes in an order along the body axis similar to that seen in the mouse.

catastrophism In geology, the view that most or all landforms are the product of catastrophic events. See *uniformitarianism*.

cenancestor The last common ancestor of all extant organisms.

chromosome inversion A region of DNA that has been flipped, so that the genes are in reverse order; results in lower rates of crossing-over and thus tighter linkage among loci within the inversion.

clade The set of species descended from a particular common ancestor; synonymous with monophyletic group. Loosely, a branch on an evolutionary tree and all of its twigs.

clade credibility In Bayesian phylogeny inference, the sum of the posterior probabilities of all the possible phylogenies in which a clade appears. Less precisely, the probability, given the data, that a hypothesized clade is real.

cladistics A classification scheme based on the historical sequence of divergence events (phylogeny); also used to identify a method of inferring phylogenies based on the presence of shared derived characters (synapomorphies).

cladogram An evolutionary tree reflecting the results of a cladistic analysis.

cline A systematic change along a geographic transect in the frequency of a genotype or phenotype.

clone An individual that is genetically identical to its parent, or a group of individuals that are genetically identical to each other.

coalescence In population genetics, the merging of allele lineages that we would see if we could trace alleles back in time to their common ancestors.

coding DNA The portion of a gene, or genome, that is both transcribed into mRNA and translated into protein.

codon A set of three bases in DNA that specifies a particular amino acid–carrying tRNA.

codon bias A nonrandom distribution of codon usage in a DNA sequence.

coefficient of inbreeding (*F*) The probability that the alleles at any particular locus in the same individual are identical by descent from a common ancestor.

coefficient of linkage disequilibrium (*D*) A calculated value that quantifies the degree to which genotypes at one locus are nonrandomly associated with genotypes at another locus.

coefficient of relatedness (*r*) The probability that the alleles at any particular locus in two different individuals are identical by descent from a common ancestor.

coevolution That which occurs when interactions between species over time lead to reciprocal adaptation.

common garden experiment An experiment in which individuals from different populations or treatments are reared together under identical conditions.

communal breeding Describes a situation in which several adults cooperatively rear a shared brood to which all have contributed offspring.

comparative method A research program that compares traits and environments across taxa and looks for correlations that test hypotheses about adaptation.

complementary base pairs Nucleotides that match up and form hydrogen bonds on opposite strands of a DNA molecule or DNA-RNA duplex. C complements G; A complements T or U.

confidence interval An indication of the statistical certainty of an estimate; if a study yielding an estimate is done repeatedly, and a 95% confidence interval is calculated for each estimate, the confidence interval will include the true value 95% of the time.

conjugation In bacterial genetics, the transfer of one or more genes from one cell to another via a plasmid that travels through a conjugation tube.

constraint Any factor that tends to slow the rate of adaptive evolution or prevent a population from evolving the optimal value of a trait.

control group A reference group that provides a basis for comparison; in an experiment, the control group is exposed to all conditions affecting the experimental group except one—the potential causative agent of interest.

convergence In general, the independent appearance in different lineages of similar evolutionary novelties. In evolutionary developmental biology, the indepedent appearance in different lineages of similar evolutionary novelties arising from different developmental mechanisms.

convergent evolution Similarity between species that is caused by a similar, but evolutionarily independent, response to a common environmental problem.

cooperation Describes behavior that is beneficial or costly to the actor, beneficial to the recipient, and selected for at least in part because the recipient benefits.

cooperative breeding Describes a situation in which individuals go without reproducing themselves, and instead assist in the reproduction of others.

cryptic species Species that are indistinguishable morphologically, but divergent in songs, calls, odor, or other traits.

cynodonts A clade of amniotes represented among extant organisms by the mammals. Nested within the larger therapsid clade, which is nested within the still-larger synapsid clade.

Darwinian fitness The extent to which an individual contributes genes to future generations, or an individual's score on a measure of performance expected to correlate with genetic contribution to future generations (such as lifetime reproductive success).

derived character A character present in one or more species in a clade that was not present in the clade's common ancestor; an evolutionary novelty; also known as an apomorphy; used in contrast with ancestral character.

deuterostome A lineage of animals that share a pattern of development, including radial cleavage and formation of the anus before the mouth. Includes echinoderms and chordates.

developmental bias An aspect of an organism's development that makes phenotypic variation more likely in some directions than in others.

developmental constraint Synonymous with developmental bias.

differential success A difference between the average survival, fecundity, or number of matings achieved by individuals with certain phenotypes versus individuals with other phenotypes.

dioecious Describes a species in which male and female reproductive function occurs in separate individuals; usually used with plants.

direct fitness Fitness that an individual attains from his or her own reproduction, without help from relatives.

directional selection That which occurs when individual fitness tends to increase or decrease with the value of phenotypic trait; can result in steady evolutionary change in the mean value of the trait in the population.

disparity In paleontology, that portion of the universe of theoretically possible morphologies actually occupied by a species or clade; the variation observed within a taxon as compared to the possible variation.

disruptive selection Occurs when individuals with more extreme values of a trait have higher fitness; can result in increased phenotypic variation in a population.

distance matrix methods In phylogeny inference, a group of methods in which the first step is the estimation from the data of pairwise evolutionary divergences. An evolutionary tree is then estimated from the table of divergences, or distance matrix.

dominance genetic variation Differences among individuals in a population that are due to the nonadditive effects of genes, such as dominance; typically means the genetic variation left over after the additive genetic variation has been taken into account.

drift Synonym for genetic drift.

ecdysozoan A lineage of protostome animals distinguished by the presence of molting.

effective population size The size of an ideal random mating population (with no selection, mutation, or migration) that would lose genetic variation via drift at the same rate as is observed in an actual population.

endosymbiosis A relationship where one organism lives inside the body or within the cells of another organism.

environmental genomics A research program aimed at understanding which genes are present in a particular environment, based on sequencing the entire genomes present. In most cases, the genes studied come from organisms that have never been identified or seen.

environmental sequencing A research program aimed at understanding which species are present in a particular environment, based on sequencing one or more genes directly from an environmental sample and using the data to place the organisms present on a phylogenetic tree. In most cases, the organisms that are identified have never been seen. Synonymous with direct sequencing.

environmental variation Differences among individuals in a population that are due to differences in the environments they have experienced.

epigenetic marks Chemical modifications of DNA, managed by enzymes encoded in the genome, that can influence phenotype by altering gene expression.

epitope The specific part of a protein that is recognized by the immune system and initiates a response. Synonymous with antigenic site and antigenic determinant.

escape from adaptive conflict A scenario in which a gene subject to an adaptive constraint is dupicated, after which different daughter genes become specialized for—and better at—different members of the original set of conflicting functions.

eugenics The study and practice of social control over the evolution of human populations; positive eugenics seeks to increase the frequency of desirable traits, whereas negative eugenics seeks to decrease the frequency of undesirable traits.

eusocial Describes a social system characterized by overlapping generations, cooperative brood care, and specialized reproductive and nonreproductive castes.

eusociality A social system characterized by overlapping generations, cooperative brood care, and specialized reproductive and nonreproductive castes.

evo-devo The study of how changes in genes that affect embryonic development could lead to important evolutionary changes; short for "evolutionary developmental biology."

evolution Originally defined as descent with modification, or change in the characteristics of populations over time. Currently defined as changes in allele frequencies over time.

evolutionarily conserved Describes a character that occurs as a homology in many distantly related species of a clade.

evolutionary arms race That which occurs when an adaptation in one species (a parasite, for example) reduces the fitness of individuals in a second species (such as a host), thereby selecting in favor of counter-adaptations in the second species. These counter-adaptations, in turn, select in favor of new adaptations in the first species, and so on.

evolutionary relationships Patterns of genealogical kinship among species resulting from descent with modification from common ancestors. Species *A* and *B* are more closely related to each other than either is to *C* if *A* and *B* share a more recent common ancestor than either shares with *C*.

evolutionary tree A diagram (typically an estimate) of the relationships of ancestry and descent among a group of species or populations; in paleontological studies the ancestors may be known from fossils, whereas in studies of extant species the ancestors may be hypothetical contructs. Also called a *phylogenetic tree* or *phylogeny*.

exaptation A character that has been coopted during evolution for a novel function.

exon A nucleotide sequence that occurs between introns and that remains in the messenger RNA after the introns have been spliced out.

expression In molecular biology, the production, from the information encoded in a gene, of a functional protein or RNA.

extant Living today.

extended haplotype homozygosity (EHH) A measure of the linkage disequilibrium between an allele at a locus of interest and alleles at other loci on the same chromosome. Allele *a*'s EHH to a particular distance *x* is the probability that two randomly chosen chromosomes carrying *a* will also have the same genotype at all marker loci between *a* and *x*.

fecundity The number of gametes produced by an individual; usually used in reference to the number of eggs produced by a female.

fitness The extent to which an individual contributes genes to future generations, or an individual's score on a measure of performance expected to correlate with genetic contribution to future generations (such as lifetime reproductive success).

fixation The elimination from a population of all the alleles at a locus but one; the one remaining allele, now at a frequency of 1.0, is said to have achieved fixation, or to be fixed.

fossil Any trace of an organism that lived in the past.

fossil record The complete collection of fossils, located in many institutions around the world.

founder effect A change in allele frequencies that occurs after a founder event, due to genetic drift in the form of sampling error in drawing founders from the source population.

founder event The establishment of a new population, usually by a small number of individuals.

frameshift mutation An insertion or deletion in a coding region of a gene in which the length of the inserted or deleted sequence is not a multiple of three; causes the codons downstream of the mutation to be translated in the wrong reading frame.

frequency The proportional representation of a phenotype, genotype, gamete, or allele in a population; if 6 out of 10 individuals have brown eyes, the frequency of brown eyes is 60%, or 0.6.

frequency-dependent selection Occurs when an individual's fitness depends on the frequency of its phenotype in the population; typically occurs when a phenotype has higher fitness when it is rare and lower fitness when it is common.

gamete pool The set of all copies of all gamete genotypes in a population that could potentially be contributed by the members of one generation to the members of the next generation.

gametic isolation Describes a situation in which, despite the mixture of gametes from different species, fertilizations occur preferentially among gametes of the same species.

gene duplication Generation of an extra copy of a locus, usually via unequal crossing-over or retroposition.

gene family A group of loci related by common descent and sharing identical or similar function.

gene flow The movement of alleles from one population to another population, typically via the movement of individuals or via the transport of gametes by wind, water, or pollinators.

gene genealogy An evolutionary tree for variants of a gene. The variants may be alleles within a population, paralogs within a genome, or homologs in different species. Synonymous with *gene tree*.

gene pool The set of all copies of all alleles in a population that could potentially be contributed by the members of one generation to the members of the next generation.

gene regulation The modulation, through any of a variety of molecular mechanisms, of gene expression.

gene tree An evolutionary tree for variants of a gene. The variants may be alleles within a population, paralogs within a genome, or homologs in different species. Synonymous with *gene genealogy*.

genetic distance A statistic that summarizes the number of genetic differences observed between populations or species.

genetic drift Change in the frequencies of alleles in a population resulting from sampling error in drawing gametes from the gene pool to make zygotes and from chance variation in the survival and/or reproductive success of individuals; results in nonadaptive evolution.

genetic hitchhiking Change in the frequency of an allele due to positive selection on a closely linked locus.

genetic load Reduction in the mean fitness of a population due to the presence of deleterious alleles.

genetic recombination The placement of allele copies into multilocus genotypes (on chromosomes or within gametes) that are different from the multilocus genotypes they belonged to in the previous generation; results from meiosis with crossing-over and sexual reproduction with outcrossing.

genetic variation Differences among individuals in a population that are due to differences in genotype.

genotype The combination of alleles an individual carries at one or more loci of interest.

genotype-by-environment interaction Differences in the effect of the environment on the phenotype displayed by different genotypes; for example, among people living in the same location some change their skin color with the seasons and others do not.

geologic column A composite, older-to-younger sequence of rock formations that describes geological events at a particular locality.

geologic time scale A sequence of eons, eras, periods, epochs, and stages that furnishes a chronology of Earth history.

greenbeard effect Occurs when an allele causes individuals carrying it to both recognize and be recognized by other carriers, and also to behave altruistically toward them.

h^2 Symbol for the narrow-sense heritability (see *heritability*).

half-life The time required for half of the atoms of a radioactive material, present at any time, to decay into a daughter isotope.

Hamilton's rule An inequality that predicts when alleles for altruism should increase in frequency.

haplodiploidy A reproductive system in which males are haploid and develop from unfertilized eggs, while females are diploid and develop from fertilized eggs.

haplotype Genotype for a suite of linked loci on a chromosome; typically used for mitochondrial genotypes, because mitochondria are haploid and all loci are linked.

Hardy–Weinberg equilibrium A situation in which allele and genotype frequencies in an ideal population do not change from one generation to the next, because the population experiences no selection, no mutation, no migration, no genetic drift, and random mating.

heritability In the broad sense, that fraction of the total phenotypic variation in a population that is caused by genetic differences among individuals; in the narrow sense, that fraction of the total variation that is due to the additive effects of genes.

hermaphroditic In general, describes a species in which male and female reproductive function occur in the same individual; with plants, describes a species with perfect flowers (that is, flowers with both male and female reproductive function).

heterochrony Change in the relative timing of events that occur during development.

heterozygosity That fraction of the individuals in a population that are heterozygotes.

heterozygote inferiority (underdominance) Describes a situation in which heterozygotes at a particular locus tend to have lower fitness than homozygotes.

heterozygote superiority (overdominance) Describes a situation in which heterozygotes at a particular locus tend to have higher fitness than homozygotes.

heuristic searches In phylogeny inference, a group of algorithms for searching the space of possible evolutionary trees without exhaustively checking them all. The algorithms look for trees better than the current leader by rearranging the leader in various ways.

histogram A bar chart that represents the variation among individuals in a sample; each bar represents the number of individuals, or the frequency of individuals, with a particular value (or within a particular range of values) for the measurement in question.

hitchhiking Change in the frequency of an allele due to positive selection on a closely linked locus. Also called a *selective sweep*.

homeodomain An amino acid sequence that forms a DNA-binding fold and defines a family of transcription factors that play a key role in early development.

homeotic genes Genes that induce the formation of particular parts or structures in animals and plants. More broadly, genes whose products provide positional information in a multicellular embryo.

homologous Describes characters derived from a common ancestor.

homology Classically defined as curious structural similarity between species despite differences in function. Today defined as similarity between species that results from inheritance of traits from a common ancestor.

homoplasy Similarity in the characters found in different species that is due to convergent evolution, parallelism, or reversal—not common descent.

homoploid hybrid speciation Formation of a new species by an interspecific mating in which the offspring have the same number of chromosome sets as the parents.

horizontal gene transfer The movement of genetic material across species barriers.

human Any member of the genus *Homo*—a clade of bipedal great apes characterized by large brain size and tool use.

hybrid zone A geographic region where differentiated populations interbreed.

IDA Initial Darwinian ancestor—the first living thing that is an ancestor of extant organisms; also known as the primordial form.

identical by descent Describes alleles, within a single individual or different individuals, that have been inherited from the same ancestral copy of the allele.

inbreeding Mating among kin.

inbreeding depression Reduced fitness in individuals or populations resulting from kin matings; often due to the decrease in heterozygosity associated with kin matings, either because heterozygotes are superior or because homozygotes for deleterious alleles become more common.

inclusive fitness An individual's total fitness; the sum of its indirect fitness, due to reproduction by relatives made possible by its actions, and direct fitness, due to its own unaided reproduction.

incomplete lineage sorting A situation in which a gene tree differs from the phylogeny of the species from which the genes were sampled. Occurs when the common ancestor is polymorphic, and different alleles are preserved in different descendants.

indel A type of mutation based on the insertion or deletion of one or more deoxyribonucleotides (bases).

independence (statistical) Lack of association among data points, such that the value of a data point does not affect the value of any other data point.

indirect fitness Fitness that is due to increased reproduction by relatives made possible by the focal individual's actions. See *direct fitness*.

inducible defense A character that is adaptive because of its role in evading predation and that develops only in the presence of a cue emanating from the predator.

inheritance of acquired characters The hypothesis that phenotypic changes in the parental generation can be passed on, intact, to the next generation.

interaction In genetics, occurs when the effect of an allele on the phenotype depends on the other alleles present at the same or different loci or on the environment; in statistics, occurs when the effect of a treatment depends on the value of other treatments.

intergenic regions The portions of a genome found between the protein-coding genes.

intersexual selection Differential mating success among individuals of one sex due to interactions with members of the other sex; for example, variation in mating success among males due to female choosiness.

intrasexual selection Differential mating success among individuals of one sex due to interactions with members of the same sex; for example, differences in mating success among males due to male–male competition over access to females.

intron (intervening sequence) A noncoding stretch of DNA nucleotides that occurs between the coding regions of a gene and that must be spliced out after transcription to produce a functional messenger RNA.

inversion A region of DNA that has been flipped, so that the genes are in reverse order; results in lower rates of crossing-over and thus tighter linkage among loci within the inversion.

iteroparous Describes a species or population in which individuals experience more than one bout of reproduction over the course of a typical lifetime; humans provide an example.

kin recognition The ability to discern the degree of genetic relatedness of other individuals.

kin selection Natural selection based on indirect fitness gains.

Lagerstätte A geological formation that yields copious well-preserved fossils.

lateral gene transfer Transfer of genetic material across species barriers.

law of succession The observation that fossil types are succeeded, in the same geographic area, by similar fossil or living species.

life history An individual's pattern of allocation, throughout life, of time and energy to various fundamental activities, such as growth, repair of cell and tissue damage, and reproduction.

likelihood The probability of a particular outcome given a model of the process that produced it. For example we might calculate the probability that a pair of parents will have an offspring with a particular multilocus genotype given a model specifying how closely the loci in question are linked. Or we might calculate the probability of getting a particular set of sequences given a phylogeny of the species from which we sampled them.

likelihood ratio Literally, the ratio of two likelihoods. Typically, the probability of a particular outcome given a model we are evaluating divided by the probability of the same outcome under a null model.

Lilliput effect A diminution in body size of the taxa that exist following a mass extinction.

LINEs Long interspersed elements—a group of reverse-transcriptase-encoding retrotransposons common in the genomes of eukaryotes.

lineage A group of ancestral and descendant populations or species that are descended from a common ancestor. Synonymous with *clade*.

linkage The tendency for alleles at different loci on a chromosome to be inherited together. Also called genetic linkage.

linkage (dis)equilibrium If, within a population, genotypes at one locus are randomly distributed with respect to genotypes at another locus, then the population is in linkage equilibrium for the two loci; otherwise, the population is in linkage disequilibrium.

LOD Literally, logarithm of the odds. The logarithm of a likelihood ratio.

lophotrochozoan A lineage of protostome animals, many of which have a feeding structure called a lophophore.

loss-of-function mutation A mutation that incapacitates a gene so that no functional product is produced; also called a forward, knockout, or null mutation.

LUCA Last universal common ancestor—the most recent shared ancestor of all extant organisms. Also known as the *cenancestor*.

macroevolution Large evolutionary change, usually in morphology; typically refers to the evolution of differences among populations that would warrant their placement in different genera or higher-level taxa.

mass extinction A large-scale, sudden extinction event that is geographically and taxonomically widespread.

maternal effect Variation among individuals due to variation in nongenetic influences exerted by their mothers; for example, chicks whose mothers feed them more may grow to larger sizes, and thus be able to feed their own chicks more, even when size is not heritable.

maximum likelihood In phylogeny inference, a method for choosing a preferred tree among many possible trees. The maximum likelihood tree is the phylogeny that, combined with a model of evolution, assigns the highest probability to our data.

mean heterozygosity In a population, either: (1) the average frequency across loci, of heterozygotes; or (2) the fraction of genes that are heterozygous in the genotype of the average individual.

Mendelian gene A locus whose alleles obey Mendel's laws of segregation and independent assortment.

methylation Chemical modification of a DNA nucleotide via the addition of a methyl group ($-CH_3$).

microevolution Changes in gene frequencies and trait distributions that occur within populations and species.

microtektites Tiny glass particles created when minerals are melted by the heat generated in a meteorite or asteroid impact.

midoffspring value The mean phenotype of the offspring within a family.

midparent value The mean phenotype of an individual's two parents.

migration In evolution, the movement of alleles from one population to another, typically via the movement of individuals or via the transport of gametes by wind, water, or pollinators.

mobile genetic elements DNA sequences that have the ability to move from location to location within a genome. Also known as *transposable elements* or *transposons*.

modern synthesis The broad-based effort, accomplished during the 1930s and 1940s, to unite Mendelian genetics with the theory of evolution by natural selection; also called the evolutionary synthesis.

molecular clock The hypothesis that base substitutions accumulate in populations in a clock-like fashion; that is, as a linear function of time.

monoecious Typically used for plants, to describe either: (1) a species in which male and female reproductive functions occur in the same individual; or (2) a species in which separate male and female flowers are present on the same individual (see also *hermaphroditic*).

monophyletic group The set of species (or populations) descended from a common ancestor.

morphology Structural form, or physical phenotype; also the study of structural form.

morphospace The universe of morphologies occupied by a taxon, or the universe of theoretically possible morphologies that could be occupied.

morphospecies Populations that are designated as separate species based on morphological differences.

multilevel selection A conceptulatization of adaptive evolution in which fitness is assigned to groups, and is a function of group composition. More broadly, an evolutionary process in which selection acts at multiple levels.

mutation accumulation Describes an experiment in which lineages of organisms are insulated as much as possible from natural selection, and are thus allowed to accumulate mutations by genetic drift. Also describes a hypothesis for the evolution of senescence in which late-acting deleterious mutations accumulate because they are unoppossed by selection.

mutation–selection balance Describes an equilibrium in the frequency of an allele that occurs because new copies of the allele are created by mutation at exactly the same rate that old copies of the allele are eliminated by natural selection.

mutualism An interaction between two individuals, typically of different species, in which both individuals benefit.

mutually beneficial interaction An interaction between individuals resulting in increased fitness for both the actor and the recipient.

narrow-sense heritability That fraction of the total phenotypic variation in a population that is due to the additive effects of genes.

natural selection A difference, on average, between the survival or fecundity of individuals with certain phenoypes compared with individuals with other phenotypes.

negative selection Selection against deleterious mutations. Also called *purifying selection*.

neighbor joining A distance matrix method for phylogeny inference in which a polytomy is resolved by sequentially pairing taxa so as to make, at each step, the greatest possible reduction in the total length of the tree.

neofunctionalization The acquistion, by one of the daughters of a gene duplication, of a novel adaptive function.

neutral (mutation) A mutation that has no effect on the fitness of the bearer.

neutral evolution (neutral theory) A theory that models the rate of fixation of alleles with no effect on fitness; also associated with the claim that the vast majority of observed base substitutions are neutral with respect to fitness.

node A point on an evolutionary tree at which a branch splits into two or more subbranches.

noncoding DNA The portion of a gene, or genome, that is not transcribed into mRNA and translated into protein.

nonsense mutation A mutation creating a new stop codon in the coding region of a gene.

nonsynonymous substitution A DNA substitution that changes the amino acid sequence specified by a gene.

null hypothesis The predicted outcome, under the simplest possible assumptions, of an experiment or observation; in a test of whether populations are different, the null hypothesis is typically that they are not different and that apparent differences are due to chance.

null model The set of simple and explicit assumptions that allows a researcher to state a null hypothesis.

outbreeding Mating among unrelated individuals.

orthologous Genes that diverged after a speciation event; describes the relationship among homologous genes found in different species.

outgroup A taxonomic group that diverged prior to the rest of the taxa in a phylogenetic analysis.

overdominance Describes a situation in which heterozygotes at a particular locus tend to have higher fitness than homozygotes.

***p* value** An estimate of the statistical support for a claim about a pattern in data, with smaller values indicating stronger support; an estimate of the probability that apparent violations of the null hypothesis are due to chance (see *statistically significant*).

paleontology The study of fossil organisms.

parallel evolution The indepedent appearance in different lineages of similar evolutionary novelties arising from the same developmental mechanism.

paralogous Duplicated genes found in the same genome; describes the relationship among members of the same gene family. A type of genetic homology.

paraphyletic group A set of species that includes a common ancestor and some, but not all, of its descendants.

parental investment Expenditure of time and energy on the provision, protection, and care of an offspring; more specifically, investment by a parent that increases the fitness of a particular offspring and reduces the fitness the parent can gain by investing in other offspring.

parsimony A criterion for selecting among alternative patterns or explanations based on minimizing the total amount of change or complexity.

parthenogenesis A reproductive mode in which offspring develop from unfertilized eggs.

percentage of polymorphic loci The fraction of genes in a population that have at least two alleles.

phenotype The set of traits an individual exhibits.

phenotypic plasticity Variation, under environmental influence, in the phenotype associated with a genotype.

phenotypic variation The total variation among the individuals in a population.

phyletic transformation The evolution of a new morpho-species by the gradual transformation of an ancestral species, without a speciation or splitting event taking place. Also called *anagenesis*.

phylogenetically independent contrasts The set of pairwise divergences that occurred within a clade as lineages arose from the nodes.

phylogenetic tree A diagram (typically an estimate) of the relationships of ancestry and descent among a group of species or populations; in paleontological studies the ancestors may be known from fossils, whereas in studies of extant species the ancestors may be hypothetical constructs. Also called an *evolutionary tree*.

phylogenomics The use of data from genome sequencing to answer questions about evolution.

phylogeny The evolutionary history of a group. Also used as a synonym for *evolutionary tree*.

phylogeography The use of evolutionary trees in answering questions about the geographic distribution of organisms.

plasmids Small loops of DNA that can replicate themselves; common in bacteria and observed in a small number of eukaryotes.

pleiotropic Describes a gene that influences more than one trait.

pleiotropy Occurs when a single gene influences more than one trait.

plesiomorphy An ancestral character; that is, a character present in the common ancestor of a clade.

point mutation Alteration of a single base in a DNA sequence.

polyandry A mating system in which at least some females mate with more than one male.

polygyny A mating system in which at least some males mate with more than one female.

polymorphic Describes a population, locus, or trait for which there is more than one phenotype or allele; variable.

polymorphism The existence within a population of more than one variant for a phenotypic trait, or of more than one allele.

polyphyletic group A set of species that are grouped by similarity, but not descended from a common ancestor.

polyploid Having more than two haploid sets of chromosomes.

polyploid hybrid speciation Formation of a new species by an interspecific mating in which the offspring have a different number of chromosome sets than the parents; results in genetic incompatibility between the daughter and parental species.

polytomy A node, or branch point, on a phylogeny with more than two descendant lineages emerging.

population For sexual species, a group of interbreeding individuals and their offspring; for asexual species, a group of individuals living in the same area.

population genetics The branch of evolutionary biology responsible for investigating processes that cause changes in allele and genotype frequencies in populations.

positive selection Selection in favor of advantageous mutations.

post-transcriptional silencing A cellular defense against mobile genetic elements in which the elements are destroyed after they have been transcribed into RNA.

posterior probability In Bayesian statistical inference, the probability of the hypothesis given the data. Calculated from the probability of the data given the hypothesis, the prior probability of the hypothesis, and the prior probability of the data.

postzygotic isolation Reproductive isolation between populations caused by dysfunctional development or sterility in hybrid forms.

pre-transcriptional silencing A cellular defense against mobile genetic elements in which the elements are prevented from being transcribed.

preadaptation A trait that changes due to natural selection and acquires a new function.

premutations Alterations in a DNA sequence, due to chemical degradation and replication errors, that may still potentially be detected and repaired.

prezygotic isolation Reproductive isolation between populations caused by differences in mate choice or timing of breeding, so that no hybrid zygotes are formed.

primordial form The first organism; the first entity capable of (1) replicating itself through the directed chemical transformation of its environment, and (2) evolving by natural selection.

prior probability In Bayesian statistical inference, the probability assigned to a hypothesis before considering the data.

processed pseudogene A pseudogene that originated when a messenger RNA from which the introns had already been removed was reverse-transcribed and inserted into the genome.

protostome A lineage of animals that share a pattern of development, including spiral cleavage and formation of the mouth before the anus. Includes arthropods, mollusks, and annelids.

proximate causation Explanations for how, in terms of physiological or molecular mechanisms, traits function.

pseudogene DNA sequences that are homologous to functioning genes, but are not transcribed.

punctuated equilibrium The hypothesis that evolution over geologic time has typically been characterized by rapid morphological change during speciation interspersed with long periods of stasis.

purifying selection Selection against deleterious mutations. Also called *negative selection.*

QTL Quantitative trait locus.

QTL mapping A collection of techniques that allow researchers to identify chromosomal regions containing loci that contribute to quantitative traits.

qualitative trait A trait for which phenotypes fall into discrete categories (such as affected versus unaffected with cystic fibrosis).

quantitative genetics The branch of evolutionary biology responsible for investigating the evolution of continuously variable traits that are influenced by the combined effects of genotype at many loci and the environment. That is, for investigating the evolution of traits not controlled by genotype at a single locus.

quantitative trait A trait for which phenotypes do not fall into discrete categories, but instead show continuous variation among individuals; a trait determined by the combined influence of the environment and many loci of small effect. See *qualitative trait.*

quantitative trait locus (QTL) A locus at which there is genetic variation that contributes to the phenotypic variation in a quantitative trait.

radiometric dating Techniques for assigning absolute ages to rock samples, based on the ratio of parent-to-daughter radioactive isotopes present.

random genetic drift Synonymous with genetic drift; change in allele frequencies due to sampling error.

reaction norm The pattern of phenotypic plasticity exhibited by a genotype.

reciprocity An exchange of fitness benefits, separated in time, between two individuals resulting in a net gain for both.

reciprocal altruism Reciprocity.

recombination rate (*r*) The frequency, during meiosis, of crossing-over between two linked loci; ranges from 0 to 0.5.

reinforcement Natural selection that results in assortative mating in recently diverged populations in secondary contact; also known as reproductive character displacement.

relatedness (*r*) The genetic similarity between individuals. May be caluculated as the probability that gene copies in the two indivdiuals are identical by descent, or as a function of the allele frequencies in the two individuals relative to that in the population as a whole.

relative dating Techniques for assigning relative ages to rock strata, based on assumptions about the relationships between newer and older rocks.

relative fitness The fitness of an individual, phenotype, or genotype compared with others in the population; can be calculated by dividing the individual's fitness by either (1) the mean fitness of the individuals in the population, or (2) the highest individual fitness found in the population; method 1 must be used when calculating the selection gradient.

replacement substitution A DNA substitution that changes the amino acid or RNA sequence specified by a gene. Also called a *nonsynonymous substitution.*

reproductive isolation Occurs when populations of organisms fail to hybridize regularly in nature, or fail to produce fertile offspring when they do hybridize.

reproductive success (RS) The number of viable, fertile offspring produced by an individual.

response to selection (R) In quantitative genetics, the difference between the mean phenotype of the offspring of the selected individuals in a population and the mean phenotype of the offspring of all the individuals.

retroposition (or retroduplication) The duplication of a genetic sequence resulting from the retrotranscription of a processed mRNA followed by integration into the genome.

retrotransposons Transposable elements that move via an RNA intermediate and contain the coding sequence for reverse transcriptase; closely related to retroviruses.

retrovirus An RNA virus whose genome is reverse transcribed to DNA by reverse transcriptase.

reversal An event that results in the reversion of a derived trait to the ancestral form.

ribozyme An RNA molecule that has the ability to catalyze a chemical reaction.

root The location on a phylogeny of the common ancestor of a clade.

sampling error A chance difference between the frequency of a trait in a subset of individuals from a population versus the frequency of the trait in the entire population. Sampling error is larger for small samples than for large ones.

secondary adaptation Additional adaptive evolution of a trait following co-option for a novel function.

secondary contact When two populations that have diverged in isolation from a common ancestor are reunited geographically.

segmental duplication Addition to the genome of a copy of a portion of a chromosome.

selection Synonym for *natural selection.*

selection coefficient A variable used in population genetics to represent the difference in fitness between one genotype and another.

selection differential (S) A measure of the strength of selection used in quantitative genetics; equal to the difference between the mean phenotype of the selected individuals (for example, those that survive to reproduce) and the mean phenotype of the entire population.

selection gradient A measure of the strength of selection used in quantitative genetics; for selection on a single trait, it is equal to the slope of the best-fit line in a scatterplot showing relative fitness as a function of phenotype.

selectionist theory The viewpoint that natural selection is responsible for a significant percentage of substitution events observed at the molecular level.

selective sweep Change in the frequency of an allele due to selection on a closely linked locus. Also called *hitchhiking*.

selfish interaction An interaction in which the actor gains fitness and the recipient loses fitness.

semelparous Describes a species or population in which individuals experience only one bout of reproduction over the course of a typical lifetime; salmon provide an example.

senescence A decline with age in reproductive performance, physiological function, or probability of survival.

sexual dimorphism A difference between the phenotypes of females versus males within a species.

sexual selection A difference, among members of the same sex, between the average mating success of individuals with a particular phenotype versus individuals with other phenotypes.

significant In scientific discussions, typically a synonym for statistically significant.

silent substitution (or silent-site substitution) A DNA substitution that does not change the amino acid or RNA sequence specified by the gene. Also called a *synonymous substitution*.

Simpson's paradox The appearance of a trend within subsets of a population that is reversed when the subsets are combined.

SINEs Short interspersed elements—a group of non-reverse-transcriptase-encoding retrotransposons.

sister species The species that diverged from the same ancestral node on a phylogenetic tree.

sister taxa Lineages that diverged from the same ancestral node on a phylogenetic tree. See *sister species*.

spatial colinearity Correspondence between the locations of Hox loci in the genome and the positions where they are expressed along the body axis during development.

speciation The process whereby an ancestral species gives rise to a pair of daughter species.

species Groups of interbreeding populations that are evolutionarily independent of other populations.

species tree A phylogeny showing the relationships among a suite of species; typically estimated from numerous independent genes.

spite Behavior that decreases the fitness of both the actor and the recipient.

spiteful interaction An interaction resulting in a loss of fitness for both actor and recipient.

stabilizing selection That which occurs when individuals with intermediate values of a trait have higher fitness; can result in reduced phenotypic variation in a population and can prevent evolution in the mean value of the trait.

standard deviation A measure of the variation among the numbers in a list; equal to the square root of the variance (see *variance*).

standard error The likely size of the error due to chance effects in an estimated value, such as the average phenotype for a population.

stasis Lack of change.

statistically significant Describes a claim for which there is a degree of evidence in the data; by convention, a result is considered statistically significant if the probability is less than or equal to 0.05 that the observed violation of the null hypothesis is due to chance effects.

subfunctionalization Occurs following duplication of a gene with two functions. One daughter experiences a mutation causing loss of one function. The other daughter experiences a mutation causing loss of the other function.

substitution Fixation of a new mutation in a population.

sympatric Living in the same geographic area.

synapomorphy A shared, derived character; in a phylogenetic analysis, synapomorphies are used to define clades and distinguish them from outgroups.

synapsids A clade of amniote vertebrates that ultimately gave rise to the mammals. Inaccurately described as *mammal-like reptiles*.

synonymous substitution A DNA substitution that does not change the amino acid or RNA sequence specified by the gene. Also called a *silent* (or *silent-site*) *substitution*.

systematics A scientific field devoted to the classification of organisms.

taphonomy Study of the fossilization process.

taxon Any named group of organisms (the plural form is taxa).

theory of evolution by natural selection The hypothesis that descent with modification is caused in large part by the action of natural selection.

time tree A phylogeny calibrated against the fossil record to estimate divergence times.

tips (of a phylogenetic tree) The ends of the branches on a phylogenetic tree, which represent extinct or living taxa.

trade-off An inescapable compromise between one trait and another.

transformation In genetics, the acquisition of DNA from the environment or another organism that becomes incorporated into an organism's genome.

transition In DNA, a mutation that substitutes a purine for a purine or a pyrimidine for a pyrimidine.

transitional form A species that exhibits traits common to ancestral and derived groups, especially when the groups are sharply differentiated.

transposable elements Any DNA sequence capable of transmitting itself or a copy of itself to a new location in the genome.

transpose Move from one location to another within a genome.

transposons Transposable elements that move via a DNA intermediate, and contain insertion sequences along with a transposase enzyme and possibly other coding sequences.

transversion In DNA, a mutation that substitutes a purine for a pyrimidine, or a pyrimidine for a purine.

ultimate causation Explanations for why, in terms of fitness benefits, traits evolved.

underdominance Describes a situation in which heterozygotes at a particular locus tend to have lower fitness than homozygotes.

unequal cross-over A crossing-over event between mispaired DNA strands that results in the duplication of sequences in some daughter strands and deletions in others.

uniformitarianism The assumption (sometimes called a "law") that processes identical to those at work today are responsible for events that occurred in the past; first articulated by James Hutton, the founder of modern geology.

uninformative character A character that fails to help distinguish among the possible trees in a parsimony analysis because its distribution requires the same number of evolutionary changes on all of them.

unrooted tree A phylogeny that encodes no information about the direction time flows along the branches.

variance A measure of the variation among the numbers in a list; to calculate the variance of a list of numbers, first square the difference between each number and the mean of the list, then take the sum of the squared differences and divide it by the number of items in the list. (For technical reasons, when researchers calculate the variance for a sample of individuals, they usually divide the sum of the squared differences by the sample size minus one).

vestigial traits (or structures) Rudimentary traits that are homologous to fully functional traits in closely related species.

vicariance Splitting of a population's former range into two or more isolated patches.

virulence The damage inflicted by a pathogen on its host; occurs because the pathogen extracts energy and nutrients from the host and because the pathogen produces toxic metabolic wastes.

wild type A phenotype or allele common in nature.

Credits

Text and Illustration Credits

CHAPTER 1 **1.1** Prepared with data from UNAIDS (2008). **1.2** Based on Figure 2.11, page 45, in UNAIDS. 2008. Report on the global HIV/AIDS epidemic 2008. **1.3a** Data from Pisani et al. (2003). **1.3b** Data from From Hall et al. (2008), Public Health Agency of Canada (2006), Health Protection Agency (2008). **1.4** Data from Hall et al. (2008). **1.6** Data from NIAID (2003) and Watkins (2008). **1.7** Data from UNAIDS (2008). **1.8** Data from Bartlett and Moore (1998), Brenchley et al. (2006), Pandrea et al. (2008). **1.9** Data from Appay and Sauce (2008); Pandrea et al. (2008); Douek et al. (2009); Silvestri (2009). **1.11** Data from Larder et al. (1989). **1.12** Data from Larder et al. (1989). **1.17** Data from Harrigan et al. (2005). **1.19** Redrawn from Figure 1, pg 1329, in Oyugi, J. O., F. C. Vouriot, et al. 2009. "A common CD4 gene variant is associated with an increased risk of HIV-1 infection in Kenyan female commercial sex workers." *Journal of Infectious Diseases* 199: 1327–1334. Reprinted with permission of Oxford University Press. **1.21** Data from Shankarappa et al. (1999). **1.24** Data from Metzker et al. (2002). **1.25a** Data from Hahn et al. (2000). **1.25b** Data from Plantier et al. (2009). **1.26** From "Timing the ancestor of the HIV-1 pandemic strains." *Science* 288: 1789–1796. Reprinted with permission from AAAS. **1.27** Data from Fraser et al. (2007). **1.28** Data from Leslie and colleagues (2004). **1.29** Data from Troyer et al. (2005). **1.30** Blaak, H., A. B. van't Wout, M. Brouwer, et al. 2000. "In vivo HIV-1 infection of CD45RA+CD4+ T cells is established primarily by syncytium-inducing variants and correlates with the rate of CD4+ T cell decline." *Proceedings of the National Academy of Sciences USA* 97: 1269–1274. **1.31** Lemey, P., S. L. Kosakovsky Pond et al. 2007. "Synonymous substitution rates predict HIV disease progression as a result of underlying replication dynamics." *PLoS Computational Biology* 3: e29. **1.32** From "Restriction of an extinct retrovirus by the human TRIM5a antiviral protein." *Science* 316: 1756–1758. Reprinted with permission from AAAS.

CHAPTER 2 **2.01** From "Science communication. Public acceptance of evolution." *Science* 313: 765–766. Reprinted with permission from AAAS. **2.3** Theodore Garland, Jr., Professor of Biology, University of California, Riverside. **2.6** Franks, S. J., S. Sim, and A. E. Weis. 2007. "Rapid evolution of flowering time by an annual plant in response to a climate fluctuation." *Proceedings of the National Academy of Sciences, USA* 104: 1278–1282. **2.10** Based on data in Table 1 in Bell et al. (2004). **2.12** Data from Duffy et al. (2007). **2.13** Data from Dodd, D. M. B. 1989. "Reproductive isolation as a consequence of adaptive divergence in *Drosophila pseudoobscura*." *Evolution* 43: 1308–1311. **2.14a** Data from Hendry et al (2009) and Nosil et al. (2009). **2.14b (diagram)** Data from Berner et al. (2008). **2.14b (graph)** Hendry, A. P., D. I. Bolnick et al. 2009. "Along the speciation continuum in sticklebacks." *Journal of Fish Biology* 75: 2000–2036. Reproduced with permission of John Wiley & Sons Ltd. **2.15b** Data from Irwin et al. (2005). **2.15c** From "Speciation by distance in a ring species." *Science* 307: 414–416. Reprinted with permission from AAAS. **2.15d** Irwin, D. E., M. P. Thimgan, and J. H. Irwin. 2008. "Call divergence is correlated with geographic and genetic distance in greenish warblers (*Phylloscopus trochiloides*): A strong role for stochasticity in signal evolution?" *Journal of Evolutionary Biology* 21: 435–448. Reproduced with permission of John Wiley & Sons Ltd. **2.19** Data from Hsieh (2010). **2.20b** Evolutionary tree based on Lloyd et al. (2008); Hu et al. (2009). Archaeopteryx reconstruction based on Longrich (2006). **2.22c** From "Evolution of the turtle body plan by the folding and creation of new muscle connections." *Science* 325: 193–196. Reprinted with permission from AAAS.

CHAPTER 3 **3.3** Data from Niklaus (1997). **3.4** Data from Hoekstra et al. (2006), Mullen and Hoekstra (2008), Mullen et al. (2009), Vignieri et al. (2010). **3.5b** Based on data from Table 3 of Genovart et al. (2010). **3.6** Based on data from Jones and Reithel (2001). **3.7** The evolutionary tree is based on data from Kenneth Petren and colleagues (2005). **3.9c** Data from Boag and Grant (1984a). **3.10** Data from Boag and Grant (1984b). **3.11** Based on Boag (1983). **3.12b** Based on Campàs et al. (2010). **3.13** From "Intense natural selection in a population of Darwin's finches (Geospizinae) in the Galápagos." *Science* 214: 82–85. Reprinted with permission from AAAS. **3.14** Based on Boag and Grant (1984b). **3.15** Based on Grant and Grant (2003). **3.16a** From "Evolution of character displacement in Darwin's finches." *Science* 313: 224–226. Reprinted with permission from AAAS. **3.16b–c** From "Unpredictable evolution in a 30-year study of Darwin's finches." *Science* 296: 707–711. Reprinted with permission from AAAS. **3.19** Data from Moose et al. (2004). **3.20b** Redrawn from Fig. 2, page 893, in Alcalá, R. E., N. A. Mariano, et al. 2010. "An experimental test of the defensive role of sticky traps in the carnivorous plant *Pinguicula moranensis* (Lentibulariaceae)." *Oikos* 119: 891–895. Reproduced with permission of John Wiley & Sons Ltd. **3.20c** Data from Heubl et al. (2006). **Quote (pg 94)** Dobzhansky, T. 1973. *American Biology Teacher* 35: 125–129. **3.22** Data from After Schaffer and Bolognia (2001). **3.24** Data from Modified Lamb et al. (2007). **3.25b** Mitchell, D. R. 2000. "*Chlamydomonas* flagella." *Journal of Phycology* 36: 261–273. Reprinted with permission from the Phycological Society of America. **3.26** Data from True and Carroll (2002). **Q7** T.G. Krontiris. 1995. "Minisatellites and Human Disease." *Science* 269. Reprinted with permission of the AAAS.

CHAPTER 4 **CO.4 (bottom)** Data from Baker et al. (1994). **4.1b** *On the Origin of Species* (1859). **4.2** After Werdelin and Olsson (1997). **4.3** After Werdelin and Olsson (1997). **4.4** After Werdelin and Olsson (1997). **4.5** Data from Gregory (2008). **4.6** The relationships among the lion, leopard, jaguar, tiger, and snow leopard (genus *Panthera*) are based on Davis et al. (2010). The relationships between *Panthera* and the other cats are based on Johnson et al. (2006). **4.10** Based on Gans and Clark (1976), Meylan (2001), Abilardi and Maderson (2003), Kearney (2003), Mindell and Brown (2005), Carroll (2007), Laurin and Reisz (2007), Claessens (2009), Hoffmann et al. (2010), Laurin (2011), and Laurin and Gauthier (2011). **4.16** Used with permission from the Paleontological Society. **4.17** Republished with permission of *Annual Review Of Earth and Planetary Sciences* 38: 189–219 from "The Origin(s) of Whales"; permission conveyed through Copyright Clearance Center. **4.19** Data from O'Leary and Geisler (1999). **4.20** Data from Schaeffer (1948) and Gingerich (2001). **4.21** Based on the data excerpted from a dataset made available in association with this paper: Gatesy, J., M. Milinkovitch, et al. 1999. "Stability of cladistic relationships between Cetacea and higher-level Artiodactyl taxa." *Systematic Biology* 48: 6–20. The dataset is here: http://hydrodictyon.eeb.uconn.edu/systbiol/issues/48_1/Gatesy1. **4.22** Based on the data excerpted from a dataset made available in association with this paper: Gatesy, J., M. Milinkovitch, et al. 1999. "Stability of cladistic relationships between Cetacea and higher-level Artiodactyl taxa." *Systematic Biology* 48: 6–20. The dataset is here: http://hydrodictyon.eeb.uconn.edu/systbiol/issues/48_1/Gatesy1. **4.26** Data from Swofford et al. (1996), Graur et al. (2000) and Huelsenbeck et al. (2001). **4.27a** Data from Felsenstein (2004). **4.28** Data from Baldauf (2003). **4.31** Data from Nikaido et al. (1999). **4.32 (*Ichthyolestes* and *Pakicetus*)** Reprinted by permission from Macmillan Publishers Ltd: *Nature*, J. G. M. Thewissen et al., 2001, "Skeletons of terrestrial cetaceans and the relationship of whales to artiodactyles," *Nature* 413: 277–281, Figure 2. Copyright © 2001 Macmillan Magazines Limited. **4.32 (*Rodhocetus*)** From "Origin of whales from early artiodactyls: Hands and feet of Eocene Protocetidae from Pakistan," *Science* 293: 2239–2242, Figure 3, illustration Douglas Boyer. © 2001 AAAS. Reprinted with permission from AAAS. **4.32 (*Rodhocetus* astragalus)** Republished with permission of the Annual Reviews, Inc., from Uhen, M. D. 2010. "The Origin(s) of Whales." *Annual Review Of Earth and Planetary Sciences* 38: 189–219; permission conveyed through Copyright Clearance Center, Inc. **4.33b** Data from Murgia et al. (2006). **4.34b** Data from Kittler et al. (2003, 2004). **4.36a** Drawn using online mapping facility provided by ODSN. Maps prepared from data assembled by Hay et al. (1999). **4.36c** The Royal Society, *Biology Letters* 7: 225–228., Author(s): T. M. Townsend, K. A. Tolley, © 2011. **4.37** Data from Murphy et al. (2001).

CHAPTER 5 **CO.5 (bottom)** Data from Ann McKellar and Andrew Hendry (2009). **5.4** Data from Watson (1976). **5.5** National Center for Biotechnology Information. **5.6b** Data from Newcomb et al. (2012). **5.8b** Data from Miyakawa et al. (2010). **5.11b** Viets, et al. 1993. "Temperature-dependent sex determination in the leopard gecko, *Eublepharis macularius*." *Journal of Experimental Zoology* 265: 679–683. Reproduced with permission of Wiley Inc. **5.11c** T. Rhen, A. Schroeder, J.T. Sakata, V. Huang, D. Crews, "Segregating variation for temperature-dependent sex determination in a lizard." *Heredity* 106: 649–660. Copyright © 2011 Nature Publishing Group. Reprinted with permission. **5.12** Data from Caspi et al. (2003). **5.13c** Data from Suzuki and Nijhout (2006). **5.14** From "Evolution of a polyphenism by genetic accommodation." *Science* 311: 650–652. Reprinted with permission from AAAS. **5.15a–c** Data from Watson (1977). **5.16** Madeleine Price Ball. **5.20** Data from Preston et al. (2010). **5.25** Avise, John C. 1994. *Molecular Markers, Natural History and Evolution.* New York: Chapman & Hall. Reprinted with kind permission from Springer Science+ Business Media B.V. **5.27a (bottom) & 5.27b** Reprinted by permission from Macmillan Publilshers Ltd.: Zhang,

J., Y. P. Zhang, and H. F. Rosenberg. 2002. "Adaptive evolution of a duplicated pancreatic ribonuclease gene in a leaf-eating monkey." *Nature Genetics* 30: 411–415. **5.28b** Data from Parker et al. (2009). **5.29** Data from Knowles and McLysaght (2009). **5.31** Data from Prevosti et al. (1988). **5.33b–c** Reprinted with permission of Ramsey, J. 2011. "Polyploidy and ecological adaptation in wild yarrow." *Proceedings of the National Academy of Sciences USA* 108: 7096–7101. **5.34** Prepared using estimates based on data from Lynch (2010a) and, for HIV, Manksy and Temin (1995). **5.35** Data from Conrad et al. (2011). **5.36a–b (graphs)** Permission courtesy of the Genetics Society of America. **5.37 (right)** Redrawn from Figure 2, pg 1829, in Rutter, M. T., F. H. Shaw, and C. B. Fenster. 2010. "Spontaneous mutation parameters for *Arabidopsis thaliana* measured in the wild." *Evolution* 64: 1825–1835. Reproduced with permission of John Wiley & Sons Ltd. **5.38 (illus)** Vassilieva, L. L., A. M. Hook, and M. Lynch. 2000. "The fitness effects of spontaneous mutations in *Caenorhabditis elegans*." *Evolution* 54: 1234–1246. Reproduced with permission of John Wiley & Sons Ltd.

CHAPTER 6 CO.6 (bottom) From "A localized negative genetic correlation constrains microevolution of coat color in wild sheep." *Science* 319: 318–320. Reprinted with permission from AAAS. **6.14** Cavener, D. R., and M. T. Clegg. 1981. "Multigenic response to ethanol in *Drosophila melanogaster*." *Evolution* 35: 1–10. Reproduced with permission of John Wiley & Sons Ltd. **6.19** Redrawn from Figure 1, page 154, in Dawson, P. S. 1970. "Linkage and the elimination of deleterious mutant genes from experimental populations." *Genetica* 41: 147–169. Reprinted with kind permission from Springer Science+Business Media B.V. **6.23** From "Chromosome rearrangements for the control of insect pests." *Science* 176: 875–880. Reprinted with permission from AAAS. **6.24b–c** Gigord, L. D. B., M. R. Macnair, and A. Smithson. 2001. "Negative frequency-dependent selection maintains a dramatic flower color polymorphism in the rewardless orchid *Dactylorhiza sambucina* (L.) Soò." *Proceedings of the National Academy of Sciences USA* 98: 6253–6255. **6.31a** Data from From Pier et al. (1998). **6.31b** Data from Lyczak et al. (2002). **6.33** Data from Chen, C. -H., H. Huang, et al. (2007). **6.35** Data from Chen, C. -H., H. Huang, et al. (2007).

CHAPTER 7 CO.7 (bottom) Data from Roelke et al. (1993). **7.5b** From R. B. King, et al. 1995. "Color-pattern variation in Lake Erie water snakes: The role of gene flow." *Evolution* 49: 885–896. Reproduced with permission of John Wiley & Sons Ltd. **7.6** Camin, J. H., and P. R. Ehrlich. 1958. "Natural selection in water snakes (*Natrix sipedon L.)* on islands in Lake Erie." *Evolution* 12: 504–511. Reprinted with permission of John Wiley & Sons Ltd. **7.12** Tinghitella, R. M., M. Zuk, et al. 2011. "Island hopping introduces Polynesian field crickets to novel environments, genetic bottlenecks and rapid evolution." *Journal of Evolutionary Biology* 24: 1199–1211. Reproduced with permission of John Wiley & Sons, Ltd. **7.16** Data from Buri (1956), Ayala and Kiger (1984). **7.17** Data from Buri (1956), Hartl (1981). **7.19** Reproduced by permission from Templeton et al., 1990: 77, pp. 13–27 (Missouri Botanical Garden Press, St. Louis, MO). **7.20 (graphs)** Data from Young et al. (1996). **7.22** Rich, S. S., A. E. Bell, and S. P. Wilson. 1979. "Genetic drift in small populations of Tribolium." *Evolution*: 579–584. Reproduced with permission of John Wiley & Sons Ltd. **7.23** Data from Charlesworth (2009). **7.24c** Data from Gojoburi et al. (1990). **7.26** Nei, M., Y. Suzuki, and M. Nozawa. 2010. "The Neutral Theory of Molecular Evolution in the Genomic Era." *Annual Review of Genomics and Human Genetics* 11: 265–289. Permission conveyed through Copyright Clearance Center, Inc. **7.27** Data from Charlesworth (2009). **7.28** Chao, L. and D. E. Carr. 1993. "The molecular clock and the relationship between population size and generation time." *Evolution* 47: 688–690. Reproduced with permission of John Wiley & Sons Ltd. **7.29** Reprinted by permission from Macmillan Publishers Ltd: Huttley, G. A., E. Easteal, M. C. Southey, et al. 2000. "Adaptive evolution of the tumor suppressor BRCA1 in humans and chimpanzees." *Nature Genetics* 25: 410–413. **7.31** Reprinted with permission of Sinauer Associates. **7.32** Data from Fay (2011). **7.38** Wolfe, L. M. 1993. "Inbreeding depression in *Hydrophyllum appendiculatum*: Role of maternal effects, crowding, and parental mating history." *Evolution* 47: 374–386. Reproduced with permission of John Wiley & Sons Ltd. **7.39** van Noordwijk, A. J., and W. Scharloo. 1981. "Inbreeding in an island population of the great tit." *Evolution* 35: 674–688. Reproduced with permission of John Wiley & Sons Ltd. **7.40** Data from Driscoll et al. (2002). **7.41a** From Johnson, et al. 2010. "Genetic restoration of the Florida panther." *Science* 329: 1641–1645. Reprinted with permission from AAAS. **7.41b** Benson, J. F., J. A. Hostetler, et al. 2011. "Intentional genetic introgression influences survival of adults and subadults in a small, inbred felid population." *Journal of Animal Ecology* 80: 958–967. Reproduced with permission of John Wiley & Sons Ltd. **7.42** Reprinted by permission from Macmillan Publishers Ltd: Seielstad, M. T., E. Minch, and L. L. Cavalli-Sforza. 1998. "Genetic evidence for a higher female migration rate in humans." *Nature Genetics* 20:278–280.

CHAPTER 8 CO.8 (bottom) From "Anciently asexual bdelloid rotifers escape lethal fungal parasites by drying up and blowing away." *Science* 327: 574–576. Reprinted with permission from AAAS. **8.7** Data from Hedrick (1983). **8.8** Permission courtesy of the Genetics Society of America. **8.10** Data from Wang et al. (2011). **8.11** Huff, C. D., D. Witherspoon, et al. 2012. "Crohn's disease and genetic hitch-hiking at IBD5." *Molecular Biology and Evolution* 29: 101–111. Reprinted with permission of Oxford University Press. **8.12** Data from Huff et al. (2012). **8.14** Reprinted by permission from Macmillan Publishers Ltd: Dawson, E., G. R. Abecasis, S. Bumpstead, et al. 2002. "A first-generation linkage disequilibrium map of human chromosome 22." *Nature* 418: 544–548. **8.16** Data from Diaz et al. (2000). **8.17** Data from Diaz et al. (2000). **8.18** From "Protecting against bad air." *Science* 293: 442–443. Reprinted with permission from AAAS. **8.19** Reprinted by permission from Macmillan Publishers Ltd: Sabeti, P. C., D. E. Reich, J. M. Higgins, et al. 2002. "Detecting recent positive selection in the human genome from haplotype structure." *Nature* 419: 832–837. **8.20** Based on Wu, D. D., W. Jin, et al. 2010. "Evidence for positive selection on the Osteogenin (BMP3) gene in human populations." *PLoS ONE* 5: e10959. **8.23** Data from John Maynard Smith (1978). **8.26** by permission from Macmillan Publishers Ltd: Morran, L. T., M. D. Parmenter, and P. C. Phillips. 2009. "Mutation load and rapid adaptation favour outcrossing over self-fertilization." *Nature* 462: 350–352. **8.27** Data from Maynard Smith (1988). **8.28b** Henry, L., T. Schwander, and B. J. Crespi. 2012. "Deleterious mutation accumulation in asexual timema stick insects." *Molecular Biology and Evolution* 29: 401–408. Reprinted with permission of Oxford University Press. **8.29** From "Running with the Red Queen: host-parasite coevolution selects for biparental sex." *Science* 333: 216–218. Reprinted with permission from AAAS. **8.31** Lively, C. M. 1992. "Parthenogenesis in a freshwater snail: Reproductive assurance versus parasitic release." *Evolution* 46: 907–913. Reproduced with permission of John Wiley & Sons Ltd. **8.32** Kohn, M. H., H.-J. Pelz, and R. K. Wayne. 2000. "Natural selection mapping of the warfarin-resistance gene." *Proceedings of the National Academy of Science USA* 97: 7911–7915.

CHAPTER 9 9.1c From "Substantial genetic influence on cognitive abilities in twins 80 or more years old." *Science* 276: 1560–1563, Figure 1A. Reprinted with permission from AAAS. **9.2** Data from East (1916) and Ayala (1982). **9.3** Data from East (1916) and Ayala (1982). **9.4** © Carnegie Institution. **9.5 (right)** From P. M. Beardsley, A. Yen, and R. G. Olmstead. 2003. "AFLP phylogeny of Mimulus section Erythranthe and the evolution of hummingbird pollination." Figure 5, *Evolution* 57: 1397–1410. Reprinted with permission of John Wiley & Sons Ltd. **9.8b** Reprinted by permission from Macmillan Publishers Ltd: A. H. Paterson, E. S. Lander, et al. 1988. "Resolution of quantitative traits into Mendelian factors by using a complete linkage map of restriction fragment length polymorphisms." *Nature* 335: 721–726, Figure 3. **9.9** Permission courtesy of the Genetics Society of America. **9.11** Data from Benjamin et al. (1996). **9.12** Reprinted by permission from Macmillan Publishers Ltd: Ogura et al., *Nature* 411: 603–606, Fig 2; "A frameshift mutation in NOD2 associated with susceptibility to Crohn's disease." Copyright © Macmillan Magazines Limited. **9.13** *Journal of Applied Physiology* by American Physiological Society (1887–) Reproduced with permission of American Physiological Society in the format reuse in a book/textbook via Copyright Clearance Center. **9.15** Data from Felsenstein, J. (2011). **9.16** From J. M. N. Smith and A. A. Dhondt. 1980. "Experimental confirmation of heritable morphological variation in a natural population of song sparrows." *Evolution* 34: 1155–1160. Copyright © 1980 Evolution. Reproduced with permission of John Wiley & Sons Ltd. **9.19** Data from Falconer (1989). **9.21** Galen, C. 1996. "Rates of floral evolution: Adaptation to bumblebee pollination in an alpine wildflower, *Polemonium viscosum*." *Evolution* 50: 120–125. Reproduced with permission of John Wiley & Sons Ltd. **9.23** Data from Galen (1996). **9.24** From Wilson, A. J., & Rambaut A. "Breeding racehorses: what price good genes?" *The Biology Letters* 4:173–175. © 2008. Reproduced with permission from The Royal Society. **9.26** From E. D. Brodie, III. 1992. "Correlational selection for color pattern and antipredator behavior in the garter snake *Thamnophis ordinoides*." *Evolution* 46: 1284–1298, Figure 1. Copyright © 1992 Evolution. Reprinted by permission John Wiley & Sons Ltd. **9.27** Data from Cavalli-Sforza and Bodmer (1971). **9.28** From A. E. Weis and W. G. Abrahamson. 1986. "Evolution of host plant manipulation by gall makers: Ecological and genetic factors in the solidago-eurosta system." *American Naturalist* 127: 681–695, Figures 2 and 3. Copyright © 1986 The University of Chicago Press. Reprinted by permission of the The University of Chicago Press. **9.29** Reprinted by permission from Macmillan Publishers Ltd: Bates Smith, T. 1993. "Disruptive selection and the genetic basis of bill size polymorphism in the African finch Pyrenestes." *Nature* 363: 618–620. **9.30** © Carnegie Institution. **9.33** From "Modern riding style improves horse racing times." *Science* 325: 289. Reprinted with permission from AAAS." **9.34** From "Role of genotype in the cycle of violence in maltreated children." *Science* 297: 851–854, Figure 1. Reprinted with permission from AAAS. **9.35** From "Association of anxiety-related traits with a polymorphism in the serotonin transporter gene regulatory regions." *Science* 274: 1527–1531, Figure 3. Reprinted with permission from AAAS.

CHAPTER 10 CO.10 (bottom) From "Floral signposts: testing the significance of visual 'nectar guides' for pollinator behaviour and plant fitness." *Proceedings of the Royal Society B* 279: 634–639. Reproduced with permission from The Royal Society. **10.2, 10.3, 10.4** Reprinted from P. Weeks. 2000. "Red-billed oxpeckers: vampires or tickbirds?" *Behavioral Ecology* 11(2): 154–160, Figure 1, by permission of the International Society for Behavioral Ecology. Copyright © 2000 International Society for Behavioral Ecology. Reprinted with permission of Oxford University Press. **10.7** From "A tephritid fly mimics the territorial displays of its jumping spider predators." *Science* 236: 310–312. Reprinted with permission from AAAS. **10.9** Nevo, E., Y. B. Fu, et al. 2012. "Evolution of wild cereals during 28 years of global warming in Israel." *Proceedings of the National Academy of Sciences, USA* 109: 3412–3415. **10.11** Data from Anderson et al. *BMC Evolutionary Biology* 2011, 11:157, http://www.biomedcentral.com/1471-2148/11/157. **10.12, 10.13, T10.1** Used by permission of the Ecological Society of America. **10.15** From D. J. Hosken. 1998. "Testes mass in megachiropteran bats varies in accordance with sperm competition theory." *Behavioral Ecology and Sociobiology* 44: 169–177. Copyright © 1998 Springer-Verlag GmbH & Co. KG. Reprinted with kind permission from Springer Science+Business Media B.V. **10.16** Data from Lauder et al. (1995). **10.18** From T. Garland, Jr. and S.C. Adolph. 1994. "Why not do two-species comparative studies: Limitations on inferring adaptation." *Physiological Zoology* 67: 797–828. Copyright © 1994 by The University of Chicago Press. Reprinted by permission of the University of Chicago Press. **10.19** From D. J. Hosken. 1998. "Testes mass in megachiropteran bats varies in accordance with sperm competition theory." *Behavioral Ecology and Sociobiology* 44: 169–177. Copyright © 1998 Springer-Verlag GmbH & Co. KG. Reprinted with kind permission from Springer Science+Business Media B.V. **10.20b** Bush, S. E., D. Kim, et al. 2010. "Evolution of Cryptic Coloration in Ectoparasites." *The American Naturalist* 176: 529–535. Reprinted with permission of the University of Chicago Press. **10.22** Data from De Meester (1996). **10.23** Data from Cousyn et al. (2001). **10.25** From D.W. Schemske and J. Agren. 1995. "Deceit pollination and selection on female flower size in Begonia involucrata: An experimental approach." *Evolution* 49: 209–214. Copyright © 1995 Evolution. Reprinted with permission of John Wiley & Sons Ltd. **10.26** From D.W. Schemske and J. Ågren. 1995. "Deceit pollination and selection on female flower size in Begonia involucrata: An experimental approach." *Evolution* 49: 209–214. Copyright © 1995 Evolution. Reproduced with permission of John Wiley & Sons Ltd. **10.28** From L. F. Delph and C. M. Lively. 1989. "The evolution of floral color change: Pollinator attraction versus physiological constraints in Fuchsia excorticate." *Evolution* 43: 1252–1262. Copyright © 1989 Evolution. Reproduced with permission of John Wiley & Sons Ltd. **10.29** Funk, D. J., D. J. Futuyma, G. Ortí, and A. Meyer. 1995. "A history of host associations and evolutionary diversification for *Ophraella* (Coleoptera: Chrysomelidae): New evidence from mitochondrial DNA." *Evolution* 49: 1008–1017. Reproduced with permission of John Wiley & Sons Ltd. **T10.3** From D. J. Futuyma, M. C. Keese, and D. J. Funk. 1995. "Genetic constraints on macroevolution:The evolution of host affiliation in the leaf beetle genus *Ophraella*." *Evolution* 49: 797–809, Table 7. Copyright © 1995 Evolution. Reproduced with permission of John Wiley & Sons Ltd. **10.30** From D. H. Clayton and K. P. Johnson. 2003. "Linking coevolutionary history to ecological process: doves and lice." *Evolution* 57: 2335–2341, Figure 4. Copyright © 2003. Reproduced with permission of John Wiley & Sons Ltd. **10.31** Data from Clayton et al. (2004). **10.32** From D. R. Taylor, C. Zeyl, and E. Cooke. 2002. "Conflicting levels of selection in the accumulation of mitochondrial defects in Saccharomyces cerevisiae." *Proceedings of the National Academy of Sciences USA* 99: 3690–3694, Figure 2. Copyright © 2002 National Academy of Sciences, U.S.A. **10.33** From "Evidence for selective advantage of pathogenic FGFR2 mutations in the male germ line." *Science* 301: 643–646, Figure 3. Reprinted with permission from AAAS.

CHAPTER 11 CO.11 (bottom) From "Signal perception in frogs and bats and the evolution of mating signals." *Science* 333: 751–752. Reprinted with permission from AAAS. **11.6** Hayward, A., and J. F. Gillooly. 2011. "The cost of sex: quantifying energetic investment in gamete production by males and females." *PLoS ONE* 6: e16557. **11.8** Pélissié, B., P. Jarne, and P. David. 2012. "Sexual selection without sexual dimorphism: Bateman gradients in a simultaneous hermaphrodite." *Evolution* 66: 66–81. Reproduced with permission of John Wiley & Sons Ltd. **11.10** Drawn from data from Adam G. Jones (Texas A&M University). **11.12** Plotted from data provided by Adam G. Jones (Texas A&M University). **11.14a** Used by permission of the Ecological Society of America. **11.14b** Wikelski, M., and F Trillmich. 1997. "Body size and sexual size dimorphism in marine iguanas fluctuate as result of opposing natural and sexual selection: An island comparison." *Evolution* 51: 922–936. Reproduced with permission of John Wiley & Sons Ltd. **11.16** Data from Trillmich (1983). **11.17** Data from Wikelski and Trillmich (1997). **11.18** From Wikelski, M., and S. Baurle. "Pre-copulatory ejaculation solves time constraints during copulations in marine iguanas." *Proceedings of the Royal Society of London B* 263: 439–444, © 1996. Reprinted with permission from The Royal Society. **11.19** Data from Gross (1991). **11.24** From S. R. Pryke and S. Andersson. 2005. "Experimental evidence for female choice and energetic costs of male tail elongation in red-collared widowbirds." *Biological Journal of the Linnean Society* 86: 35–43, Figure 3. Reproduced with permission of John Wiley & Sons Ltd. **11.25** Data from Pryke and Andersson (2005). **11.27a** Data from Gerhardt el al. (1996). **11.28b** Data from Arnold (1983). **11.28c** Data from Kirkpatrick (1987). **11.28d** Data from Falconer (1989). **11.29** Data from Kirkpatrick and Ryan (1991); Prum (2010). **11.30** Data from Arnold (1983), Kirkpatrick et al. (1991) and Prum (2010). **11.31** Data from Owen et al. (2012). **11.32b** From Wilkinson, Gerald, and P. R. Reillo. "Female choice response to artificial selection on an exaggerated male trait in a stalk-eyed fly." *Proceedings of the Royal Society of London B* 255: 1–6, © 1994. Reprinted with permission from The Royal Society. **11.34b** Grace, J. L., and K. L. Shaw. 2011. "Coevolution of male mating signal and female preference during early lineage divergence of the Hawaiian cricket, *Laupala cerasina*." *Evolution* 65: 2184–2196. Reproduced with permission of John Wiley & Sons Ltd. **11.35** Data from Kirkpatrick and Ryan (1991). **11.36** Data from Proctor (1991). **11.37** Data from Proctor (1992). **11.40** From R. Thornhill. 1976. "Sexual selection and nuptial feeding behavior in *Bittacus apicalis* (Insecta: Mecoptera)." *American Naturalist* 110: 529–548. Copyright © 1976 The University of Chicago Press. Reprinted by permission of The University of Chicago Press. **11.42** Data from Welch et al. (1998). **11.43** Data from Hoogland (1998). **11.44a,d–e** Data from Rosenqvist and Johansson (1995). **11.44b–c** Data from Rosenqvist (1990). **11.45c** Data from Vaughton and Ramsey (1998). **11.46** From "Floral evolution: Attractiveness to pollinators increases male fitness." *Science* 232: 1625–1727, Figure 1. Reprinted with permission from AAAS. **11.47a** From Delph et al. (1996). American Naturalist. Reprinted by permission of The University of Chicago Press. **11.47b** Data from Delph et al. (1996). **11.48** Rogers, A. R., and A. Mukherjee. 1992. "Quantitative genetics of sexual dimorphism in human body size." *Evolution* 46: 226–234. Reproduced with permission of John Wiley & Sons Ltd. **11.49a** Courtiol, A., J. E. Pettay, et al. 2012. "Natural and sexual selection in a monogamous historical human population." *PNAS*. Published online ahead of print. DOI: 10.1073/pnas.1118174109. **11.49b** Borgerhoff Mulder, M. 2009. "Serial Monogamy as Polygyny or Polyandry?" *Human Nature* 20: 130–150. Reprinted with kind permission from Springer Science+Business Media B.V. **11.50a** Data from Chagnon (1988). **11.50b** Beckerman, S., P. I. Erickson, et al. 2009. "Life histories, blood revenge, and reproductive success among the Waorani of Ecuador." *Proceedings of the National Academy of Sciences, USA* 106: 8134–8139. **11.51** Reprinted by permission from Macmillan Publishers Ltd: B. Pawlowski, R. I. M. Dunbar, and A. Lipowicz. 2000. "Tall men have more reproductive success." *Nature* 403: 156, Figure 1b. Copyright © 2000 Macmillan Magazines Limited. **11.52** From B. J. LeBoeuf and J. Reiter. 1988. "Lifetime reproductive success in northern elephant seals." In T. H. Cutton-Brock, ed., *Reproductive Success* (Chicago: University of Chicago Press), pp. 344–362. Copyright © 1988 by The University of Chicago Press. Reprinted by permission of The University of Chicago Press. **11.54** Redrawn from Figure 4, page 222, in Gangestad, S. W. 1993. "Sexual selection and physical attractiveness: Implications for mating dynamics." *Human Nature* 4: 205–235. Reprinted with kind permission from Springer Science+Business Media B.V. **11.55** Data from Gwynne et al. (1990).

CHAPTER 12 CO.12 (bottom) With kind permission from Springer Science+ Business Media. Gammie, S. C., T. J. Garland, and S. A. Stevenson. 2006. "Artificial selection for increased maternal defense behavior in mice." *Behavior Genetics* 36: 713–722. Copyright © 2006 Springer Science + Business Media. **12.3** Data from Riehl, C. (2011). **12.7** Based on data from Hawlena, H., F. Bashey, et al. (2010). **12.8** Based on Maynard Smith, J. (1998). **12.9** Based on Trivers, R. (1985). **12.11, 12.12** From J. L. Hoogland, *The Black-Tailed Prairie Dog: Social Life of a Burrowing Mammal* (Chicago: University of Chicago Press), pp. 173, 174. Copyright © 1995 by The University of Chicago Press. Reprinted by permission of The University of Chicago Press. **12.13b** Reprinted by permission from the Macmillan Publishers Ltd. *Nature Communications*. Gorrell, J. C., A. G. McAdam, et al. "Adopting kin enhances inclusive fitness in asocial red squirrels," 1:22. Copyright © 2010 The Nature Publishing Group. **12.14** Data from DeBruine, L. M. (2002). **12.15** DeBruine, L. M. 2002. "Facial Resemblance Enhances Trust." *Proceedings of the Royal Society of London B* 269: 1307–1312. Reprinted with permission. **12.17** Reprinted by permission from the Macmillan Publishers Ltd. *Nature*. Cornwallis, C. K., S. A. West, et al. "Promiscuity and the evolutionary transition to complex societies." *Nature* 466: 969–972. Copyright © 2010 The Nature Publishing Group. **12.19** Based on Grafen, A. (1985). **12.20** Data from Bashey, F., S. K. Young, et al. (2012). **12.22** Data from Kerr, B., and P. Godfrey-Smith (2002). **12.23** Data from J.S. Chuang et al. (2009). **12.24** Chuang, J. S., O. Rivoire, and S. Leibler. 2009. "Simpson's paradox in a synthetic microbial system." *Science* 323: 272–275. Copyright © 2008 AAAS. Reprinted with permission. **12.25** From Chuang, J. S., O. Rivoire, and S. Leibler. 2009. "Simpson's paradox in a synthetic microbial system." *Science* 323: 272–275. Copyright © 2009 AAAS. Reprinted with permission. **12.26** Data from C.J. Goodnight (1985). **12.27** Reprinted by permission from the Macmillan Publishers Ltd. *Nature*. From Apicella, C. L., F. W. Marlowe, et al. "Social networks and cooperation in hunter-gatherers." 481: 497–501. Copyright © 2012 The Nature Publishing Group. **12.29** Robert L. Trivers. "Parent-Offspring Conflict." *American Zoologist* 14:1. Copyright © 1974 Oxford University Press. Used with permission. **12.30 (right)** Briskie, J.V., C. T. Naugler, and S. M. Leech. 1994. "Begging intensity of nestling birds varies with sibling relatedness." *Proceedings of the Royal Society of London B* 258: 73–78. Reprinted with permission of the Royal Society Publishing. **12.31 (bottom)** Reprinted by permission from Macmillan Publishers Ltd. *Nature*, Emlen, S. T., and P. H. Wrege. 1992. "Parent–offspring conflict and the recruitment of helpers among bee-eaters." *Nature* 356: 331–333. Copyright © 1992. **12.32c** Data from L. W. Lougheed and D. J. Anderson (1999). **12.33 (bottom)** From Cheney, D. L., L. R. Moscovice, et al. 2010. "Contingent cooperation between wild female baboons." *Proceedings of the National Academy of Sciences USA* 107: 9562–9566. **12.34** Based on E. Fehr, S. Gachter (2002). **12.35** Henrich, J., R. McElreath, et al. 2006. "Costly punishment across human societies." *Science* 312: 1767–1770. Copyright © 2006 AAAS. Reprinted with permission. **12.37** Based on J. H. Hunt (1999).

CHAPTER 13 CO.13 (bottom) Skipped breeding vs. bred last year, in Velando, A., H. Drummond, and R. Torres. 2010. "Senescing sexual ornaments recover after a sabbatical." *Biology Letters* 6: 194–196. Reprinted with permission of the Royal Society Publishing. **13.2** E. L. Charnov and D. Berrigan (1993). **13.3** Based on data from E. G. King et al. (2011). **13.4a** Reprinted by permission of Macmillan Publishers Ltd. Gustafsson, L., and T. Pärt. 1990. "Acceleration of senescence in the collared flycatcher (*Ficedula albicollis*) by reproductive costs." *Nature* 347: 279–281. Copyright © 1990. **13.4b** Clutton-Brock, T. H., S. D. Albon, and F. E. Guinness. 1988. "Reproductive success in male and female red deer." In T. H. Clutton-Brock, ed. *Reproductive Success*. Chicago: University of Chicago Press. Copyright © 1988 The University of Chicago Press. Reprinted with permission. **13.4c** M. R. Rose. "Laboratory evolution of postponed senescence in *Drosophila melanogaster*." *Evolution*. Copyright © 1984 Society for the Study of Evolution. Reproduced with permission of John Wiley & Sons Ltd. **13.05** From N. Austad and K. E. Fischer. 1991. "Mammalian aging, metabolism and ecology: Evidence from the bats and marsupials." *Journal of Gerontology* 46. Copyright © 1991 Oxford University Press. Used with permission. **13.6** "Selection for delayed senescence in *Drosophila melanogaster*." L. S. Luckinbill, R. Arking, et al. *Evolution*. Copyright © 1984 Society for the Study of Evolution. Reproduced with permission of John Wiley & Sons Ltd. **13.7** From Heidinger, B. J., J. D. Blount, et al. 2012. "Telomere length in early life predicts lifespan." *Proceedings of the National Academy of Sciences USA* 109: 1743–1748. Used with permission. **13.8** "Comparative Biology of Mammalian Telomeres," N. M. Gomes, O. A. Ryder, et al. *Aging Cell*. Copyright © 2011 Blackwell Publishing Ltd/Anatomical Society. Reproduced with permission of John Wiley & Sons Ltd. **13.9a** Reprinted with permission of Macmillan Publishers Ltd. From Tyner, et al. 2002. "p53 mutant mice that display early ageing-associated phenotypes." *Nature* 415: 45–53, Figure 2a, page 47. Copyright © 2002. **13.11** Data from K. A. Hughes (2002). **13.12** Reprinted by permission of Macmillan Publishers Ltd. D. Reed and E. Bryant. 2000. "The evolution of senescence under curtailed life span in laboratory populations of *Musca domestica* (the housefly)." *Heredity* 85: 1115–1121. Copyright © 2000. **13.14** Reprinted with permissions of Macmillan Publishers, Ltd. D. W. Walker, et al. 2000. "Evolution of life span in *C. elegans*." *Nature* 405: 296–297. Copyright © 2000. **13.15** Based on R. J. Mockett and R. S. Sohal (2006). **13.16** Reprinted with permission of Macmillan Publishers, Ltd. From Gustafsson, L., and T. Pärt. 1990. "Acceleration of senescence in the collared flycatcher (*Ficedula albicollis*) by reproductive costs." *Nature* 347: 279–281. Copyright © 1990. **13.17** Data from T. P. Young (1990). **13.19** S. N. Austad. "Retarded senescence in an insular population of Virginia opossums (*Didelphus virginiana*)." *Journal of Zoology*. Copyright © 1993 The Zoological Society of London. Reproduced with permission of John Wiley & Sons Ltd. **13.20a** With kind permission from Springer Science+ Business Media. K. Hill and A. M. Hurtado. 1991. "The evolution of premature reproductive senescence and menopause in human females: An evaluation of the 'grandmother hypothesis.'" *Human Nature* 2(4): 313–351. Copyright © 1991 by Aldine Publishers. **13.21** Based on Austad (1994). **13.22** With kind permission from Springer Science+ Business Media. K. Hill and A. M. Hurtado. 1991. "The evolution of premature reproductive senescence and menopause in human females: An evaluation of the 'grandmother hypothesis." *Human Nature* 2(4): 313–351. Copyright © 1991 by Aldine Publishers. **13.23a** Reprinted with permission of British Ecological Society. **13.24** Reprinted with permission of Macmillan Publishers, Ltd. Lahdenperä, M., V. Lummaa, et al. 2004. "Fitness benefits of prolonged post-reproductive lifespan in women." *Nature* 428: 178–181. Copyright © 2004. **13.26 (left)** Based on D. A. Roff (1992). **13.26 (right)** Boyce, M. S., and C. M. Perrins. 1987. "Optimizing great tit clutch size in a fluctuating environment." *Ecology* 68: 142–153. Copyright © 1987 The Ecological Society of America. Reprinted with permission. **13.27** D. Schluter and L. Gustafsson. "Maternal Inheritance of Condition and Clutch Size in the Collared Flycatcher." *Evolution*. Copyright © 1993 Society for the Study of Evolution. Reproduced with permission of John Wiley & Sons Ltd. **13.28a** Reprinted with permission of the Entomological Society of America. **13.28b** Data from E. L. Charnov and S. W. Skinner (1985). **13.29** Charnov, E. L., and S. W. Skinner. 1985. "Complementary approaches to the understanding of parasitoid oviposition decisions." *Environmental Entomology* 14: 383–391. **13.30a** Elgar, M. A. 1990. "Evolutionary compromise between a few large and many small eggs: Comparative evidence in teleost fish." *OIKOS* 59: 283–287. Copyright © 1990 John Wiley & Sons. Reprinted with permission. **13.30b** Data from D. Berrigan (1991). **13.31** Based on C. C. Smith and S. D. Fretwell (1974). **13.32** From D. D. Heath, et al. 2003. "Rapid evaluation of egg size in captive salmon." *Science* 299: 1738–1740. Reprinted from permission from AAAS. **13.34** Based on C. C. Smith and S. D. Fretwell (1974). **13.35** From C. W. Fox, M. S. Thakar, and T. A. Mousseau. 1997. "Egg size plasticity in a seed beetle: An adaptive maternal effect." *American Naturalist* 149: 149–163. Copyright © 1997 by The University of Chicago. Reprinted with permission. **13.36** Reprinted by permissions of Macmillan Publishers Ltd. W. R. Rice. 1996. "Sexually antagonistic male adaptation triggered by experimental arrest of female evolution." *Nature* 381: 232–234. Copyright © 1996. **13.37** Reprinted with permission of Macmillan Publishers Ltd. From Mousseau, T. A., and D. A. Roff. 1987. "Natural selection and the heritability of fitness components." *Heredity* 1987: 181–197. Copyright © 1987. **13.39** R. K. Grosberg. "Life-history variation within a population of the colonial ascidian *Botryllus schlosseri I.* The genetic and environmental control seasonal variation." Copyright © 1988 Society for the Study of Evolution. Reproduced by permission of John Wiley & Sons Ltd. **13.40** L. M. Wolfe, J. A. Elzinga, and A. Biere. "Increased susceptibility to enemies following introduction the invasive plant *Silene latifolia*." *Ecology Letters*. Copyright © 2004 Blackwell Publishing Ltd/CNRS. Reproduced with permission of John Wiley & Sons Ltd. **13.42** From Lee, C. E., J. L. Remfert, and G. W. Gelembiuk. 2003. "Evolution of physiological tolerance and performance during freshwater invasions." *Integrative and Comparative Biology* 43: 439–449. Copyright © Oxford University Press. Reprinted with permission. **13.43** J. Werner and E. M. Griebeler (2011).

CHAPTER 14 CO.14 (bottom) W. Gilbert. "Pioneer maps of health and disease in England." *Geographical Journal*. Copyright © 1958 Royal Geographical Society. Reproduced with permission of John Wiley & Sons, Ltd. **14.1** U.S. Bureau of the Census, Washington DC. **14.2** Hendriksen, R. S., L. B. Price, et al. 2011. "Population genetics of *Vibrio cholerae* from Nepal in 2010: evidence on the origin of the Haitian outbreak." *MBio* 2: e00157–11. Copyright © *American Society for Microbiology*. Reprinted with permission. **14.3** Webster, R. G., W. J. Bean, et al. 1992. "Evolution and ecology of influenza A viruses." *Microbiological Review* 56: 152–179. Copyright © 1992 American Society for Microbiology. Reprinted with permission.. **14.4** "Codon bias and frequency-dependent selection on the hemagglutinin epitopes of influenza A virus." *Proceedings of the National Academy of Science USA* 100: 7152–7157. Copyright © 2003. **14.5** Data from W. M. Fitch et al. (1991). **14.6** Reprinted by permission of Macmillan Publishers Ltd. R. M. Bush. 2001. "Predicting adaptive evolution." *Nature Reviews Genetics* 2(5): 387–392. Copyright © 2001. **14.7** Gorman, O. T., W. J. Bean, et al. 1991. "Evolution of influenza A virus nucleoprotein genes: Implications for the origins of H1N1 human and classical swine viruses." *Journal of Virology* 65: 3704–3714. Copyright © 1991 American Society for Microbiology. Reprinted with permission. **14.8** Bean, W. J., M. Schell, et al. 1992. "Evolution of the H3 influenza virus hemagglutinin from human and nonhuman hosts." *Journal of Virology* 66: 1129–1138. Copyright © 1992 American Society for Microbiology. Reprinted with permission. **14.9** Redrawn from Gorman et al. (1991). **14.10** Reid, A. H., T. G. Fanning, et al. 2004. "Novel origin of the 1918 pandemic influenza virus nucleoprotein gene." *Journal of Virology* 78: 12462–12470. Copyright American Society for Microbiology. Reprinted with permission. **14.11** From dos Reis, M., A. J. Hay, and R. A. Goldstein. 2009. "Using non-homogeneous models of nucleotide substitution to identify host shift events: application to the origin of the 1918 'Spanish' influenza pandemic virus." *Journal of Molecular Evolution*. With kind permission of Springer Science + Business Media. **14.12** dos Reis, M., A. U. Tamuri, et al. "Charting the host adaptation of influenza viruses." *Molecular Biology and Evolution* 28: 1755–1767. Copyright © 2011 Oxford University Press. Reprinted with permission. **14.13a** From D. J. Austin, K. G. Kristinsson, and R. M. Anderson. 1999. "The relationship between the volume of antimicrobial consumption in human communities and the frequency of resistance." *Proceedings of the National Academy of Sciences USA* 96: 1152–1156. Copyright © 1999 National Academy of Sciences, U.S.A. **14.13b R**edrawn from Sun, L., E. Y. Klein, and R. Laxminarayan. "Seasonality and Temporal Correlation between Community Antibiotic Use and Resistance in the United States." *Clinical Infectious Diseases* 55:5. Copyright © 2012 Oxford University Press. Reprinted with permission. **14.14** S. J. Schrag, V. Perrot, and B. R. Levin. 1997. "Adaptation to the fitness costs of antibiotic resistance in *Escherichia coli*." *Proceedings of the Royal Society of London B* 264: 1287–1291. Reprinted with permission of Royal Society Publishing. **14.16** Data from S. L. Messenger et al. (1999). **14.17** From *Evolution of Infectious Disease* by Paul W. Ewald, Copyright © 1994 by Paul W. Ewald. Reprinted with permission. **14.18** P. W. Ewald, J. B. Sussman, et al. (1998). **14.19** Reprinted with permission of Macmillan Publishers Ltd. Hirschhorn, R., D. R. Yang, et al. 1996. "Spontaneous in vivo reversion to normal of an inherited mutation in a patient with adenosine deaminase deficiency." *Nature Genetics* 13: 290–295. Copyright © 1996. **14.20** From Y. Tao, J. Ruan, et al. 2011. "Rapid growth of a hepatocellular carcinoma and the driving mutations revealed by cell-population genetic analysis of whole-genome data." *Proceedings of the National Academy of Sciences USA* 108: 12042–12047. **14.22** Based on data from S. B. Eaton and L. Cordain (1997). **14.23** From Palsdottir, A., A. Helgason, et al. 2008. "A drastic reduction in the life span of cystatin C L68Q carriers due to life-style changes during the last two centuries." *PLoS Genetics* 4. **14.24** Reprinted by permission of Macmillan Publishers Ltd. H. M. Stewart, R. D. Sage, A. F. R. Stewart, and D. W. Cameron. 2000. "Breast cancer incidence highest in the range of one species of house mouse, *Mus domesticus*." *British Journal of Cancer* 82: 446–451. Copyright © 2000. **14.26** From B. I. Strassmann. 1999. "Menstrual cycling and breast cancer: An evolutionary perspective." *Journal of Women's Health* 8: 193–202. Copyright © 1991 Mary Ann Liebert Publishers, Inc. Reprinted with permission. **14.27a** Based on data from J. Komlos and M. Brabec (2011). **14.27b** National Center for Health Statistics. **14.28** Reprinted with permission of Macmillan Publishers Ltd. Turnbaugh, P. J., R. E. Ley, et al. 2006. "An obesity-associated gut microbiome with increased capacity for energy harvest." *Nature* 444: 1027–1131. **14.29b** Reprinted with permission of Macmillan Publishers Ltd. L. K. Vaughnn, H. A. Bernheim, and M. Kluger. 1974. "Fever in the lizard *Dipsosaurus dorsalis*." *Nature* 252. Copyright © 1974. **14.29c** From M. J. Kluger, D. H. Ringler, and M. R. Anver. 1975. "Fever and survival." *Science* 188: 166–168. Reprinted with permission of the AAAS. **14.30** From N. M. H. Graham, C. J. Burrell, R. M. Douglas, P. Debelle, and L. Davies. 1990. "Adverse effects of aspirin, acetaminophen, and ibuprofen on immune function, viral shedding, and clinical status in rhinovirus-infected volunteers." *Journal of Infectious Diseases* 162: 1277–1282. Copyright © 1990 Journal of Infectious Diseases. Reprinted with permission. **14.31** Reprinted by permission of Macmillan Publishers Ltd. Dixon, A., D. Ross, S. L. C. O'Malley, and T. Burke. 1994. "Paternal investment inversely related to degree of extra-pair paternity in the reed bunting." *Nature* 371: 698–700. Copyright © 1994. **14.32** Based on data from M.V. Flinn (1988). **14.33** Data from M.V. Flinn (1988). **14.34a–b** From M.V. Flinn and B. G. England. 1995. "Childhood stress and family environment." *Current Anthropology* 36: 854–866, Figures 3, page 859, and 7, page 862. Copyright © 1995 by The University of Chicago Press. Reprinted with permission. **14.34c** From M.V. Flinn and B. G. England. 1997. "Social economics of childhood glucocorticoid stress response." *American Journal of Physical Anthropology* 102: 33–53, Figure 5, page 43. Copyright © 1997 John Wiley & Sons, Inc. Reprinted with permission. **14.34d** Data from M.V. Flinn (1988). **14.36** Based on data from J. F. Anderson et al. (1999) and R. S. Lanciotti et al. (1999).

CHAPTER 15 CO.15 (bottom) From "Genome expansion and gene loss in powdery mildew fungi reveal tradeoffs in extreme parasitism." *Science* 330: 1543–1546. Reprinted with permission from AAAS. **15.1** Republished with permission of Annual Reviews Inc., from Lynch, M. 2006. "Streamlining and Simplification of Microbial Genome Architecture." *Annual Review of Microbiology* 60: 327–349; and Lynch, M., Bobay, L. M., Catania, F., Gout, J. F., and Rho, M. 2011. "The Repatterning of Eukaryotic Genomes by Random Genetic Drift." *Annual Review of Genomics and Human Genetics* 12: 347–366; permission conveyed through Copyright Clearance Center, Inc. **15.2** Republished with permission of Annual Reviews, Inc. from Lynch, M., Bobay, L. M., Catania, F., Gout, J. F., and Rho, M. 2011. "The Repatterning of Eukaryotic Genomes by Random Genetic Drift." *Annual Review of Genomics and Human Genetics* 12: 347–366.; permission conveyed through Copyright Clearance Center, Inc. **15.3** Republished with permission of Annual Reviews Inc., from Lynch, M. 2006. "Streamlining and Simplification of Microbial Genome Architecture." *Annual Review of Microbiology* 60: 327–349; and Lynch, M., Bobay, L. M., Catania, F., Gout, J. F., and Rho, M. 2011. "The Repatterning of Eukaryotic Genomes by Random Genetic Drift." *Annual Review of Genomics and Human Genetics* 12: 347–366; permission conveyed through Copyright Clearance Center, Inc. **15.4** After E.V. Koonin (2009). **15.6** Based on A. A. Aravin et al. (2007), T. Watanabe et al. (2008), J. Brennecke et al. (2007). **15.7a** Data from A. A. Aravin et al. (2007). **15.7b** Data from S. Kuramochi-Miyagawa et al. (2008). **15.8** Feng, S., et al. 2010. "Conservation and divergence of methylation patterning in plants and animals." *Proceedings of the National Academy of Sciences* 107: 8689–8694. **15.10** Based on H. Xiao, et al. (2008). **15.12 (top)** From "A Retrotransposon-Mediated Gene Duplication Underlies Morphological Variation of Tomato Fruit." *Science* 319: 1527–1530. Reprinted with permission from AAAS. **15.13a** Sung, W., M. S. Ackerman, et al. 2012. "Drift-barrier hypothesis and mutation-rate evolution." *Procedures of the National Academy of Science* 109: 18488–18492. Reprinted with permission. **15.13b** Saxer, G., P. Havlak, et al. 2012. "Whole genome sequencing of mutation accumulation lines reveals a low mutation rate in the social amoeba *Dictyostelium discoideum*." *PLoS One* 7: e46759. **15.14** Sung, W., M. S. Ackerman, et al. 2012. "Drift-barrier hypothesis and mutation-rate evolution." *Procedures of the National Academy of Science* 109: 18488–18492. Reprinted with permission. **15.15, 15.16** Loh, E., J. J. Salk, and L. A. Loeb. 2010. "Optimization of DNA polymerase mutation rates during bacterial evolution." *Proceedings of the National Academy of Sciences* 107: 1154–1159. **15.17** C. Pal et al. (2007). **15.18** Reprinted by permission from Macmillan Publishers Ltd: Innan, H., and F. Kondrashov. 2010. "The evolution of gene duplications: classifying and distinguishing between models." *Nature Reviews Genetics* 11: 97–108. **15.19** After M. Lynch (2007); Hahn, M.W. 2009. "Distinguishing Among Evolutionary Models for the Maintenance of Gene Duplicates." *Journal of Heredity* 100: 605–617. Reprinted with permission of Oxford University Press. **15.21** Deng, C., et al. 2010. "Evolution of an antifreeze protein by neofunctionalization under escape from adaptive conflict." *Proceedings of the National Academy of Sciences* 107: 21593–21598. **15.23** Data from C. Deng (2010). **15.24** Reprinted by permission from Macmillan Publishers Ltd: Perry, G. H., N. J. Dominy, et al. 2007. "Diet and the evolution of human amylase gene copy number variation." *Nature Genetics* 39: 1256–1260. **15.26** Redrawn/borrowed and rearranged from Figure 1, page 195, in Puillandre, N., M. Watkins, and B. M. Olivera. 2010. "Evolution of Conus Peptide Genes: Duplication and Positive Selection in the A-Superfamily." *Journal of Molecular Evolution* 70: 190–202. Reprinted with kind permission from Springer Science+Business Media B.V. **15.27** Redrawn from Figure 2, page 196, in Puillandre, N., M. Watkins, and B. M. Olivera. 2010. "Evolution of Conus Peptide Genes: Duplication and Positive Selection in the A-Superfamily." *Journal of Molecular Evolution* 70: 190–202. Reprinted with kind permission from Springer Science+Business Media B.V. **15.28 (map), 15.29, 15.30** Feldman, C. R., E. D. Brodie, E. D. Brodie, and M. E. Pfrender. 2009. "The evolutionary origins of beneficial alleles during the repeated adaptation of garter snakes to deadly prey." *Proceedings of the National Academy of Sciences* 106: 13415–13420. **15.31** From "Widespread Parallel Evolution in Sticklebacks by Repeated Fixation of Ectodysplasin Alleles." *Science* 307: 1928–1933. Reprinted with permission from AAAS. **15.32** Rogers, S. M., P. Tamkee, et al. 2012. "Genetic signature of adaptive peak shift in threespine stickleback." *Evolution* 66: 2439–2450. Reproduced with permission of John Wiley & Sons Ltd. **15.33a** Based on A. E. Novak et al. (2006). **15.33b** Based on M. E. Arnegard (2010).

CHAPTER 16 CO.16 (bottom) From Wiens, J. J., R. A. Pyron, and D. S. Moen. 2011. "Phylogenetic origins of local-scale diversity patterns and the causes of Amazonian megadiversity." *Ecology Letters* 14: 643–652. Copyright © 2011 Blackwell Publishing Ltd/CNRS. Reprinted with permission of John Wiley & Sons, Inc. **16.3** Redrawn from Amato et al. (2007). **16.4b** Iglésias, S. P., L. Toulhoat, and D.Y. Sellos. 2010. "Taxonomic confusion and market mislabelling of threatened skates: Important consequences for their conservation status." *Aquatic Conservation: Marine and Freshwater Ecosystems* 20: 319–333. Copyright © 2010 John Wiley & Sons, Ltd. Reprinted with permission. **16.5** From Griffiths, A. M., D.W. Sims, et al. 2010. "Molecular markers reveal spatially segregated cryptic species in a critically endangered fish, the common skate *Dipturus batis*." *Proceedings of the Royal Society of London B* 277: 1497–1503. Reprinted with permission. **16.7b–c** Based on J. Bonacum et. al (2005). **16.8b, 16.9** Hurt, C., A. Anker, and N. Knowlton. 2009. "A multilocus test of simultaneous divergence across the Isthmus of Panama using snapping shrimp in the genus Alpheus." *Evolution* 63: 514–530. Copyright © 2009 Society for the Study of Evolution. Reprinted with permission of John Wiley & Sons, Inc. **16.10** Yamamoto, S., and T. Sota. 2009. "Incipient allochronic speciation by climatic disruption of the reproductive period." *Proceedings of the Royal Society of London B* 276: 2711–2719. Reprinted with permission of the Royal Society Publishing. **16.11** Reprinted by permission of Macmillan Publishers Ltd. From Ueshima, R., and T. Asami. 2003. "Evolution: Single-gene speciation by left–right reversal." *Nature* 425: 679. Copyright © 2003. **16.13** From Lowry, D. B. and J. H. Willis. 2010. "A Widespread Chromosomal Inversion Polymorphism Contributes to a Major Life-History Transition, Local Adaptation, and Reproductive Isolation." *PLoS Biology* 8: e1000500. **16.14b** Shaw, K. L. "Interspecific Genetics of Mate Recognition." *Evolution*. Copyright © 2000 The Society for the Study of Evolution. Reproduced with permission of John Wiley & Sons Ltd. **16.15** From Wiley, C., and K. L. Shaw. 2010. "Multiple genetic linkages between female preference and male signal in rapidly speciating Hawaiian crickets." *Evolution* 64: 2238–2245. Copyright © 2010 Society for the Study of Evolution. Reprinted with permission of John Wiley & Sons, Inc. **16.17** Reprinted by permission of Macmillan Publishers Ltd. Seehausen, O., Y. Terai, et al. 2008. "Speciation through sensory drive in cichlid fish." *Nature* 455: 620–626. Copyright © 2008. **16.19** http://www.planiglobe.com/eula.html Source PDF: fig.16.19. Matute_2010_Plos Biol.pdf. **16.20** Matute, D. R. 2010. "Reinforcement of gametic isolation in Drosophila." *PLoS*. **16.21a** Based on J. R. Pannell (2012), R. Hopkins and M. D. Rauscher (2011). **16.21b** Data from D.A. Levin (1985). **16.22a** Based on R. Hopkins and M.D. Rauscher (2011). **16.22b–c, 16.23** Reprinted by permission of Macmillan Publishers Ltd. Hopkins, R., and M. D. Rausher. 2011. "Identification of two genes causing reinforcement in the Texas wildflower *Phlox drummondii*." *Nature* 469: 411–414. Copyright © 2011. **16.24** Brelsford, A., B. Milá, and D. E. Irwin. 2011. "Hybrid origin of Audubon's warbler." *Molecular Ecology* 20: 2380–2389. Copyright © 2011 Blackwell Publishing Ltd. Reprinted with permission of John Wiley & Sons, Inc. **16.25** A. Brelsford, B. Mila, and D. E. Irwin. "Hybrid Origin of Audobon's Warbler." *Molecular Ecology*. Copyright © 2011 Blackwell Publishing Ltd. Reproduced with permission of John Wiley & Sons, Ltd. **16.26** Campbell, D. R., N. M. Waser, et al. 2008. "Lifetime fitness in two generations of Ipomopsis hybrids." *Evolution* 62: 2616–2627. Copyright © 2008 John Wiley & Sons, Inc. Reprinted with permission. **16.27** Rabosky, D. L. and R. E. Glor. 2010. "Equilibrium speciation dynamics in a model adaptive radiation of island lizards." *Proceedings of the National Academy of Sciences USA* 107: 22178–22183. **16.28** Van Bocxlaer, I., S. P. Loader, et al. 2010. "Gradual adaptation toward a range-expansion phenotype initiated the global radiation of toads." *Science* 327: 679–682. Reprinted with permission of the AAAS.

CHAPTER 17 CO.17 (bottom) H. B. Niemann and S. K. Atreya. "The abundances of constituents of Titan's atmosphere from the GCMS instrument on the Huygens probe." *Nature* 438: 779–784. Copyright © 2005 Nature Publishing Group. Reprinted with permission. **17.3** From "Directed evolution of an RNA enzyme." *Science* 257: 635–641, Figure 1, page 636. Copyright © 1992. Reprinted with permission from AAAS. **17.4** From "Directed evolution of an RNA enzyme." *Science* 257: 635–641, Figure 1, page 636. Copyright © 1992 AAAS. Reprinted with permission from AAAS. **17.5** After Bartel and Szostak (1993). **17.7a** After Bentin (2011) and Wochner et al. (2011). **17.9** N. H. Sleep, et al. "Annihilation of ecosystems by large asteroid impacts on the early Earth." *Nature* 342: 139–142. Copyright © 1989 The Nature Publishing Group. Reprinted with permission. **17.14** After Atkins and Gesteland (© 1998) in Gesteland et al. (1999); W. F. Doolittle (2000). **17.18** Redrawn from Woese (1996). **17.19** Brown, J. R., and W. F. Doolittle. 1995. "Root of the universal tree of life based on ancient aminoacyl-tRNA synthetase gene duplications." *Proceedings of the National Academy of Sciences USA* 92: 2441–2445. **17.21** Redrawn from Barns et al. (1996). **17.22** Modified from Harris et al. (2012). **17.23** From J. R. Brown and W. F. Doolittle. 1997. "Archaea and the prokaryote-to-eukaryote transition." *Microbiology and Molecular Biology Reviews* 61: 456–502, Figure 4, page 463. Reprinted with permission from the American Society for Microbiology and Copyright Clearance Center. **17.24** J. R. Brown. 2001. "Genomic and phylogenetic perspectives on the evolution of prokaryotes." *Systematic Biology* 50: 497–512, Figure 2, page 504. Redrawn and reprinted with permission of Oxford University Press. **17.26** From S. Yang, R. F. Doolittle, and P. E. Bourne. 2005. "Phylogeny determined by protein domain content." *Proceedings of the National Academy of Sciences USA* 102: 373–378, Figure 4, page 376. Copyright © 2005 National Academy of Sciences,

U.S.A. **17.27** From W. F. Doolittle. 1999. "Phylogenetic classification and the universal tree." *Science* 284: 2124–2128, Figure 3, page 2127. Copyright © 1999 AAAS. Reprinted with permission from AAAS. **17.36** From "The 1.2-megabase genome sequence of Mimivirus." *Science* 306: 1344–1350, Figure S6, Supplementary Materials Page 17. Reprinted with permission from AAAS. **Quote (pg 683)** Chemical & Engineering News, Dec. 6, 1999.

CHAPTER 18 18.05 (geologic time scale) In *The Time Tree of Life*, eds. S. B. Hedges and S. Kumar. New York: Oxford University Press. Copyright © 2004. Reprinted with permission. **18.05 (oxygen concentration graph)** Berner, R. A. 2009. "Phanerozoic atmospheric oxygen: New results using the GEOCARBSULF model." *American Journal of Science* 309: 603–606, Fig. 2, pg 605. Phanerozoic atmospheric oxygen: New results using the GEOCARBSULF Model, R.A. Berner, 2009. Reprinted by permission of the *American Journal of Science*; Reprinted with permission of Robert Berner, PhD. **18.05 (CO_2 concentration)** "Atmospheric CO_2 concentrations during ancient greenhouse climates were similar to those predicted for A.D. 2100." *Proceedings of the National Academy of Sciences USA* 107: 576–580. Reprinted with permission. **18.05 (continental positions map)** Ron Blakey, Colorado Plateau Geosystems. Used with permission. **18.10b** Reprinted with permission from Macmillan Publishers Ltd. Nature, D. G. Shu, et al. 1999. "Lower Cambrian vertebrates from south China." *Nature* 402: 42–46, Figure 4c. Copyright © 1999 The Nature Publishing Group. **18.12** From A. Knoll and S. Carroll. 1999. "Early animal evolution: Emerging views from comparative biology and geology." *Science* 284: 2129–2137; figure 2, page 2131. Reprinted with permission from AAAS. **18.13 (Eusthenopteron)** Based on R. L. Caroll, (1988), E. Jarvik (1980). **18.13 (Tiktaalik)** Based on E. B. Daeschler, N. H. Shubin, F. A. Jenkins (2006). **18.13 (Acanthostega)** Reprinted with permission of Michael I. Coates. **18.13 (Acanthostega dorsal)** Reprinted with permission of Jennifer. A. Clack. **18.13 (Icthyostega)** Reprinted with permission from Macmillan Publishers Ltd. Pierce, S. E., J. A. Clack, and J. R. Hutchinson. "Three-dimensional limb joint mobility in the early tetrapod *Ichthyostega*." *Nature* 486: 523–526. Copyright © 2012 The Nature Publishing Group. **18.14** Coates, M. I., M. Ruta, and M. Friedman. 2008. "Ever Since Owen: Changing Perspectives on the Early Evolution of Tetrapods." *Annual Review of Ecology, Evolution, and Systematics* 39: 571–592. **18.15 (feathers)** Xu, X., X. Zheng, and H. You. 2010. "Exceptional dinosaur fossils show ontogenetic development of early feathers." *Nature* 464: 1338–1341. Copyright © 2010 Nature Publishing Group. Reprinted with permission. **18.15 (Archaeopteryx, Sinornithosaurus, Sinosauropteryx)** Reprinted with permission of Jaime Headden. **18.15 (Tyrannosaurus)** Copyright © Scott Hartman. **18.16** Li, Q., K. Q. Gao, et al. 2010. "Plumage color patterns of an extinct dinosaur." *Science* 327: 1369–1372. Reprinted with permission of the AAAS. **18.17b** Used with permission of the Paleontological Society. **18.18** "Developmental Patterns in Mesozoic Evolution of Mammal Ears." *Annual Review of Ecology, Evolution, and Systematics* 42: 355–380. **18.19, 18.20** McShea, "Mechanisms of large-scale evolutionary trends." *Evolution* 48: 1747–1763. Copyright © 1994 Society for the Study of Evolution. Reprinted with permission. **18.21a** Used with permission of the Paleontological Society. **18.21b** C. A. Sidor. Evidence of passive and driven evolutionary trends, in Figure 4a, page 1426, in "Simplification as a trend in synapsid cranial evolution." *Evolution* 55: 1419–1442. Copyright © 2001 The Society for the Study of Evolution. Reprinted with permission. **18.22a** J. Phillips (1860). **18.22b** Used with permission from the Paleontological Society. **18.24a** Raup, D. M. 1967. "Geometric analysis of shell coiling: coiling in ammonoids." *Journal of Paleontology* 41: 43–65. Copyright © 1967 Society for Sedimentary Geology. Reprinted with permission of the SEPM (Society for Sedimentary Geology). **18.24b** Foote (1993). **18.25** "The role of extinction in evolution." *Proceedings of the National Academy of Sciences USA* 91: 6758–6763. Copyright © 1994 National Academy of Sciences, U.S.A. Reprinted with permission. **18.26** "Diversification and extinction in the history of life." *Science* 268: 52–58. Copyright © 1995 AAAS. Reprinted with permission of the AAAS. Leigh Van Valen. 1973. "A new evolutionary law." *Evolutionary Theory* 1: 1–30. Copyright © 2011 Leigh Van Valen. Reprinted with permission. **18.28** M. Foote. "Origination and extinction through the Phanerozoic: A new approach." *Journal of Geology* 111: 125–148. Copyright © 2003 The University of Chicago Press. Reprinted with permission. **18.29b** "Iridium profile for 10 million years across the Cretaceous-Tertiary boundary at Gubbio (Italy)." *Science* 250: 1700–1702. Copyright © 1990 AAAS. (Figure 1, page 1700). Reprinted with permission of the AAAS. **18.31** Peter H. Schultz, Steven D'Hondt. **18.31 (crater map)** Schultz, P. H., and S. D'Hondt. 1996. "Cretaceous–Tertiary (Chicxulub) impact angle and its consequences." *Geology* 24: 963–967. **18.32 (map)** USGS. **18.32 (illus)** Reprinted with permission of J. A. Sibbick. **18.33** After N. Eldredge, S. J. Gould (1972). **18.34** Used with permission of the Paleontological Society. **18.37** J. C. Avise, W. S. Nelson, H. Sugita. "A speciational history of 'living fossils': Molecular evolutionary patterns in horseshoe crabs." *Evolution* 48: 1986–2001. Copyright © 1999 The Society for the Study of Evolution. Reprinted with permission. **18.38a** "Background and mass extinctions: The alternation of evolutionary regimes." *Science* 231: 129–329. Copyright © 1986 AAAS. (Figure 2a, page 131). Reprinted with permission of the AAAS. **18.38b** "Selectivity of end-Cretaceous marine bivalve extinctions." *Science* 268: 389–391. (Figure 3, page 390) Copyright © 1995 AAAS. Reprinted with permission of the AAAS. **18.39** "Background and mass extinctions: The alternation of evolutionary regimes." *Science* 231: 129–329. Copyright © 1986 AAAS. Reprinted with permission of the AAAS. **18.40** "Biotic replacements: Extinction or clade interaction?" *BioScience* 48: 389–395. Copyright © 1998 University of California Press and American Institute of Biological Sciences. Reprinted with permission. **18.44** M. J. Benton and P.C.J. Donoghue. "Paleontological Evidence to Date the Tree of Life." *Molecular Biology and Evolution* 24: 26–53. Copyright © 2007 Oxford University Press. Reprinted with permission.

CHAPTER 19 19.2c Pere Alberch. "Convergence and Parallelism in Foot Morphology in the Neotropical Salamander Genus Bolitoglossa I. Function." *Evolution*, Vol. 35, No. 1 (Jan. 1981), pp. 84–100. Copyright © 1981 Society for the Study of Evolution. **19.4** Reprinted with permission from Macmillan Publishers Ltd. *Nature Review Genetics*. J. C. Pearson, D. Lemons, and W. McGinnis. "Modulating Hox gene functions during animal body patterning." *Nature Review Genetics* 6: 893–904. Copyright © 2005 The Nature Publishing Group. **19.12** Reed, R. D., et al. 2011. "Optix drives the repeated convergent evolution of butterfly wing pattern mimicry." 2011. *Science* 333: 1137–1141. Reprinted with permission of the AAAS. **19.15b** Emlen, D. J. 2000. "Integrating development with evolution: a case study with beetle horns." *Bioscience* 50: 403–418. Copyright © 2000 University of California Press. **19.16d–e** Emlen, D. J. 2001. "Costs and the Diversification of Exaggerated Animal Structures." *Science* 291: 1534–1536. Reprinted with permission of the AAAS. **19.17b–c** Data from R.D.H. Barrett (2009). **19.18b** Danchin, E. G., M. N. Rosso, et al. 2010. "Multiple lateral gene transfers and duplications have promoted plant parasitism ability in nematodes." *Proceedings of the National Academy of Sciences USA* 107: 17651–17656. Reprinted with permission. **19.19b** Jarosch, A., et al. 2011. "Alternative splicing of a single transcription factor drives selfish reproductive behavior in honeybee workers (*Apis mellifera*)." *Proceedings of the National Academy of Science, USA*, 108: 15282–15287. Reprinted with permission.

CHAPTER 20 CO.20 (bottom) From "*Australopithecus sediba* hand demonstrates mosaic evolution of locomotor and manipulative abilities." *Science* 333: 1411–1417. Copyright © 2011 AAAS. Reprinted with permission from AAAS. **20.2** From "Immunological time scale for Hominid evolution." *Science* 158: 1200–1203. Reprinted with permission from AAAS. **20.4** From Ruvolo et al. (1994). **20.5** After Ruvolo (1994). **20.6** S. Scally, J. Y. Dutheil. "Insights into hominid evolution from the gorilla genome sequence." *Nature* 483: 169–175. Copyright © 2012 Nature Publishing Group. Reprinted with permission. **20.7** R. L. Stauffer, A. Walker, O. A. Ryder, et al. 2001. "Human and ape molecular clocks and constraints on paleontological hypotheses." *Journal of Heredity* 92: 469–474, by permission of Oxford University Press. (Figure 2, page 471.). **20.8** Data from Hacia, 2001. **20.9** A. Scally and J. Y. Dutheil. "Insights into hominid evolution from the gorilla genome sequence." *Nature* 483: 169–175. Copyright © 2012 Nature Publishing Group. Used with permission. **20.10** Sims, G. E., S. R. Jun, et al. 2009b. "Whole-genome phylogeny of mammals: evolutionary information in genic and nongenic regions." *Proceedings of the National Academy of Sciences, USA* 106: 17077–17082. **20.11** *PLoS Biology*. **20.14** "Australopithecus ramidus, a new species of early hominid from Aramis, Ethiopia." *Nature* 371: 306–312. Copyright © 1994 Nature Publishing Group. Reprinted with permission. **20.15a–b** Reprinted with permission of the Department of Anthropology, American Museum of Natural History. **20.15c** M. G. Leakey and F. Spoor. "Australopithecus ramidus, a new species of early hominid from Aramis, Ethiopia." *Nature* 371: 306–312. Copyright © 2001 Nature Publishing Group. Reprinted with permission. **20.15d** M. G. Leakey and C. S. Feibel. "New 4-million-year-old hominid species from Kanapoi and Allia Bay, Kenya." *Nature* 376: 565–571. Copyright © 1995 Nature Publishing Group. Reprinted with permission. **20.18a** Redrawn from Johanson, D. C., B. Edgar, and D. Brill. 1996. *From Lucy to Language*. New York: Simon & Schuster Editions. **20.18b–c, 20.19, 20.20, 20.21** Reprinted with permission of the Department of Anthropology, American Museum of Natural History. **20.22** After Wood (2010); Wood and Leakey (2011). **20.23** Data from Strait et al (1997). **20.25** Reprinted with permission of the Department of Anthropology, American Museum of Natural History. **20.26** After Aiello (1993), Ayala (1994), and Tattersall (1997). **20.27** Lieberman, D. E. 1995. "Testing hypotheses about recent human evolution from skulls: Integrating morphology, function, development, and phylogeny." *Current Anthropology* 36: 159–197. Reprinted with permission of the University of Chicago Press. **20.28** After Aiello (1993), Ayala (1994), and Tattersall (1997). **20.29** M. Ingman. "Mitochondrial genome variation and the origins of modern humans." *Nature* 408: 708–713. Copyright © 2000 Nature Publishing Group. Reprinted with permission. **20.31** "High resolution of human evolutionary trees with polymorphic microsatellites." *Nature* 368: 455–457. Copyright © 1994 The Nature Publishing Group. Reprinted with permission. **20.34, 20.36a** "Genetic history of an archaic hominin group from Denisova Cave in Siberia." *Nature* 468: 1053–1060. Copyright © 2010 Nature Publishing Group. Reprinted with permission. **20.37** From P. Gagneux, C. Wills, et al. 1999. "Mitochondrial sequences show diverse evolutionary histories of African hominoids." *Proceedings of the National Academy of Sciences USA* 96: 5077–5082, Figure 2, page 5080. Copyright © 1999 National Academy of Sciences, U.S.A. **20.39b** Based on Hopkins, W. D., J. L. Russell, and J. A. Schaeffer. 2012. "The neural and cognitive correlates of aimed throwing in chimpanzees: a magnetic resonance image and behavioural study on a unique form of social tool use." *Philosophical Transactions of the Royal Society of London B* 367: 37–47. **20.41** From "Fossil evidence for early hominid tool use." *Science* 265:1570–1573. AAAS. (Figure 3, page 1572.) Reprinted with permission from AAAS. **20.42** From "Fossil evidence for early hominid tool use." *Science* 265:1570–1573. (Figure 1, page 1571.). Reprinted with permission from AAAS. **20.43a, 20.44** Data from deBoer (2012).

Photo Credits

CHAPTER 1 CO.1 Horizons WWP/Alamy **1.14** Courtesy of Lori A. Kohlstaedt, University of California at Santa Barbara. Reproduced from L. Kohlstaedt and J. Cohen, "AIDS Research:The Mood Is Uncertain," *Science* 260:1254–1258, May 28, 1993.

CHAPTER 2 CO.2 Li, C., X.-C. Wu et al. 2008. "An ancestral turtle from the Late Triassic of southwestern China." *Nature* 456: 497–501. **CO.2 (bottom)** Li, C., X.-C. Wu et al. 2008. "An ancestral turtle from the Late Triassic of southwestern China." *Nature* 456: 497–501. **2.4 (left)** rafi/Fotolia. **2.4 (right)** Phil Banko/CORBIS. **2.7a (top)** Andrew Walmsley/Alamy. **2.7a (bottom)** Rogan Colbourne/Illustration & Graphics. **2.7b (top)** marioscy/Fotolia. **2.7b (bottom)** Simon D. Pollard/Getty Images. **2.8c** Mary Beth Angelo/Science Source. **2.9a–b** Cresko, W. A., K. L. McGuigan et al. 2007. "Studies of threespine stickleback developmental evolution: progress and promise." *Genetica* 129: 105–126. **2.11** Hans W. Ackermann. **2.14c** Daniel Berner. **2.14d** Schluter, D. 2010. "Resource competition and coevolution in sticklebacks." *Evolution: Education and Outreach* 3: 54–61. Photo by Todd Hatfield. **2.14e** Kitano, J., S. Mori, and C. L. Peichel. 2007. "Phenotypic divergence and reproductive isolation between sympatric forms of Japanese threespine sticklebacks." *Biological Journal of the Linnean Society* 91: 671–685. **2.15a** Wake, D. B. 2001. "Speciation in the round." *Nature* 409: 299–300. **2.16** Jon Herron. **2.17(top)** Tom McHugh/Science Source. **2.17(2nd from top)** Melba Photo Agency/Alamy. **2.17 (2nd from bottom)** Marco Tomasini/Fotolia. **2.17e (bottom)** National Geographic Image Collection/Alamy. **2.18a** S. Tonia Hsieh, Temple University. **2.18b** Hsieh S-TT. 2010. "A Locomotor Innovation Enables Water-Land Transition in a Marine Fish." *PLoS ONE* 5(6): e11197. doi:10.1371/journal.pone.0011197. **2.20a** Museum Fur Naturkunde. **2.21a** AP Photo. **2.21b (top)** Xu, X., X. Zheng, and H. You. 2010. "Exceptional dinosaur fossils show ontogenetic development of early feathers." *Nature* 464: 1338–1341. **2.21b (bottom)** Xu, X., X. Zheng, and H. You. 2010. "Exceptional dinosaur fossils show ontogenetic development of early feathers." *Nature* 464: 1338–1341. **2.22a** Sanchez-Villagra, M. R., H. Muller et al. 2009. "Skeletal development in the Chinese soft-shelled turtle Pelodiscus sinensis (Testudines: Trionychidae)." *Journal of Morphology* 270: 1381–1399. **2.22b** Riken Center for Developmental Biology. **2.33a** Ryan Somma. **2.33b** Daderot/Pterodactylus antiquus, Carnegie Museum of Natural History, Pittsburgh, Pennsylvania, USA.

CHAPTER 3 CO.3a imagebroker/Alamy. **3.1** Frary, A., T. C. Nesbitt, et al. 2000. "fw2.2: A quantitative trait locus key to the evolution of tomato fruit size." *Science* 289: 85–88. **3.2** Frary, A., T. C. Nesbitt, et al. 2000. "fw2.2: A quantitative trait locus key to the evolution of tomato fruit size." *Science* 289: 85–88. **3.5a** Georgios Alexandris/Shutterstock. **3.7** Petren, K., B. R. Grant, and P. R. Grant. 1999. "A phylogeny of Darwin's finches based on microsatellite DNA length variation." *Proceedings of the Royal Society of London B* 266: 321–329. **3.7 (mangrove finch)** Greg W. Lasley. **3.8 (top)** Peter R. Grant, Princeton University. **3.8 (bottom)** Peter R. Grant, Princeton University. **3.9b** Courtesy of Robert Podolsky. **3.12a (top)** Campàs, O., R. Mallarino et al. 2010. "Scaling and shear transformations capture beak shape variation in Darwin's finches." *Proceedings of the National Academy of Sciences USA* 107: 3356–3360. **3.12a (bottom)** Petren, K., B. R. Grant, and P. R. Grant. 1999. "A phylogeny of Darwin's finches based on microsatellite DNA length variation." *Proceedings of the Royal Society of London B* 266: 321–329. **3.20a** blickwinkel/Alamy. **3.20c (Drosophyllum)** Santos Fernández/agefotostock. **3.20c (Nepenthes)** Skynavin/Shutterstock. **3.20c (Dionaea)** Marco Uliana/Shutterstock. **3.21** From: Fig. 3, pg 7622: Langerhans, R.B., Layman, C.A. and DeWitt, T.J. "Male genital size reflects a tradeoff between attracting mates and avoiding predators in two livebearing fish species." *Proceedings of the National Academy of Sciences, USA* 102: 7618–7623. Copyright © 2005 National Academy of Sciences, U.S.A. **3.23** AP Images. **3.24** Lamb, T. D., S. P. Collin, and E. N. J. Pugh. 2007. "Evolution of the vertebrate eye: opsins, photoreceptors, retina and eye cup." *Nature Reviews Neuroscience* 8: 960–976. **3.24 (lancelet head)** David Ferrier. **3.24 (lancelet photoreceptor)** Fain, G. L., R. Hardie, and S. B. Laughlin. 2010. "Phototransduction and the evolution of photoreceptors." *Current Biology* 20: R114–24/with the permission of Cell Press, the authors, and Margery J. Fain, PhD (illustrator). **3.25a–b** Reprinted from D. R. Mitchell. 2000. "Cell and Developmental Biology, Upstate Medical University. Chlamydomonas flagella." *Journal of Phycology* 36: 261–273, Fig. 2. Reprinted with permission from the Journal of Phycology. **3.25c** From: *Journal of Cell. Science* 110: 85–91, Fig 3, pg 87. Woolley, D.M., 1997. Studies on the eel sperm flagellum. I. The structure of the inner dynein arm complex.

CHAPTER 4 CO.4 © Chris Huss/www.chrishuss.com. **4.1a** Charles Darwin. **4.15a** Mykhaylo Palinchak/Shutterstock. **4.15b** Jurgen Otto. **4.15c** Joe McDonald/Corbis. **4.15d** Peter Johnson/Corbis. **4.34a** Volker Steger/Science Source. **4.35** Hans Stieglitz.

CHAPTER 5 CO.5 NBAE/Getty Images. **5.1 (top)** Eric Knapp. **5.1 (middle)** John E Marriott/All Canada Photos/SuperStock. **5.1 (bottom)** Alison R. Davis Rabosky and Christian Cox. **5.2a** Cheryl Power/Science Source. **5.2b** Steve Gschmeissner/Science Source. **5.2c** Steve Gschmeissner/Science Photo Library. **5.3** Richard Wheeler (Zephyris) 2007. **5.8a** Christian Laforsch. **5.9** Elowitz, M. B., A. J. Levine, et al. 2002. "Stochastic gene expression in a single cell." *Science* 297: 1183–1186. **5.10a** Grant-Downton, R. T., and H. G. Dickinson. 2006. "Epigenetics and its implications for plant biology 2. The 'epigenetic epiphany': Epigenetics, evolution and beyond." *Annals of Botany* 97: 11–27. **5.10b** Grant-Downton, R. T., and H. G. Dickinson. 2006. "Epigenetics and its implications for plant biology 2. The 'epigenetic epiphany': Epigenetics, evolution and beyond." *Annals of Botany* 97: 11–27. **5.11a** Linn Currie/Shutterstock.com. **5.13a** Lori Schmidt/Alamy. **5.13b** Pennisi, E. 2006. "Evolution. Hidden genetic variation yields caterpillar of a different color." *Science* 311: 591. **5.13c** Suzuki, Y., and H. F. Nijhout. 2006. "Evolution of a polyphenism by genetic accommodation." *Science* 311: 650–652. **5.23** Nousbeck, J., B. Burger, et al. 2011. "A mutation in a skin-specific isoform of SMARCAD1 causes autosomal-dominant adermatoglyphia." *American Journal of Human Genetics* 89: 302–307. **5.24** Reprinted by permission from M. Samson, F. Libert, B.J. Doranz, et al. 1996. "Resistance to HIV-1 infection of Caucasian individuals bearing mutant alleles of the CCR5 chemokine receptor gene." *Nature* 382:722–725, Copyright 1996 Macmillan Magazines Limited. **5.27a** Arco Images GmbH/Alamy. **5.28a** Jon Herron. **5.33a** vaivirga/Shutterstock. **5.36a** L. Cora/Science Photo Library. **5.36b** Masur. **5.37 (left)** Janne Lempe and Suresh Balasubramanian/Max Planck Institute. **5.38** Bob Goldstein.

CHAPTER 6 CO.6a Arpat Ozgul. **6.1** NASA. **6.18** Courtesy of Susan J. Brown, Assoc. Professor/Kansas State University, Kansas. **6.24a** Science Photo Library/Alamy. **6.29** Bajmoczi, M., M. Gadjeva, et al. 2009. "Cystic fibrosis transmembrane conductance regulator and caveolin-1 regulate epithelial cell internalization of Pseudomonas aeruginosa." *American Journal of Physiology—Cell Physiology* 297: C263–C277. **6.30 (left & right)** James R. MacFall, Duke University Medical Center. **6.32b–c** Kambris, Z., H. Bilak, et al. 2003. "DmMyD88 controls dorsoventral patterning of the Drosophila embryo." *EMBO Reports* 4: 64–69.

CHAPTER 7 CO.7 Rodney Cammauf/National Park Service. **7.5a** Kristin Stanford. **7.11** Gerald McCormack, CINHT. **7.13a** NASA Headquarters. **7.13b** John Amato. **7.18** Alan Templeton. **7.20 (Eucalyptus)** Peter Woodward. **7.20 (Salvia)** Ervin Monn/Shutterstock. **7.20 (Scabiosa)** Martin Fowler/Shutterstock.com. **7.20 (Gentiana)** Science Photo Library/Alamy. **7.35** Richard R. Hansen/Science Source.

CHAPTER 8 CO.8a Damián H. Zanette. **CO.8b (inset)** Kent Loeffler, Kathie T. Hodge, and C. G. Wilson. **8.21** MedievalRich. **8.22a** Manfred Kage/Science Photo Library. **8.22b** M. I. Walker/Science Source. **8.22c** NHPA/SuperStock. **8.24** Bob Goldstein, UNC Chapel Hill. **8.28a** Joyce Gross.

CHAPTER 9 CO.9 Simon Bruty/Sports Illustrated/Getty Images. **9.1a** Peter Morenus, University of Connecticut. **9.5 (top left)** Alan Vernon/Getty Images. **9.5 (bottom left)** Joyce Gross. **9.6a–l** Courtesy Douglas W. Schemske, Michigan State University, Michigan/W. K. Kellogg Biological Station. **9.10a–b** Reprinted by permission from Nature: Monkeyflower. M. lewisii with M. cardinalis genotype at the UYP locus. About 70 times more attractive to hummingbirds than bees. Originally published as figure 1b on page 177 from: "Allele substitution at a flower colour locus produces a pollinator shift in monkeyflower," by H.D. Bradshaw Jr. & D.W. Schemske. *Nature* 426: 176–178. Copyright © 2007 Macmillan Magazines Limited. **9.12 (inset)** Reprinted by permission from Nature: Ogura et al., *Nature* 411: 603–606, Fig 2; A frameshift mutation in NOD2 associated with susceptibility to Crohn's disease. Copyright © Macmillan Magazines Limited. **9.20a** Richard Parker/Science Source. **9.20b** Stephen Dalton/Science Source. **9.33** Pfau, T., A. Spence, et al. 2009. "Modern riding style improves horse racing times." *Science* 325: 289.

CHAPTER 10 CO.10 Science Photo Library/Alamy. **10.1** Villiers Steyn/Shutterstock. **10.5** Erick Greene, University of Montana. **10.14** Carol Buchanan/age fotostock. **10.20a (top left)** Samu/Fotolia. **10.20a (top right)** Brooke Whatnall/Shutterstock.com. **10.20a (bottom left & right)** Bush, S. E., D. Kim, et al. 2010. "Evolution of Cryptic Coloration in Ectoparasites." *The American Naturalist* 176: 529–535. **10.21** Edward Kinsman/Science Source. **10.24a–b** Courtesy of Douglas W. Schemske, Michigan State University and Jon Agren, Uppsala University. **10.27** Lynda Delph, Indiana University.

CHAPTER 11 CO.11 Alexander T. Baugh. **11.1a** Anna Azimi/Shutterstock. **11.1b** Eric & David Hosking/Science Source. **11.1c** Dan L. Perlman/EcoLibrary.org. **11.2** Image Source/Alamy. **11.3a (top)** National Geographic Image Collection/Alamy. **11.3a (bottom)** SOBERKA Richard/hemis.fr/Getty Images. **11.3b** Gilles San Martin. **11.4** Tony Camacho/Science Source. **11.7** B. Borrell Casals/FLPA Image Broker/Newscom. **11.9** Stuart Wilson/Getty Images. **11.11a** Anders Berglund. **11.11b** Stefan von Bothmer/Anders Berglund. **11.13** Michael S Nolan/age fotostock. **11.15** Martin C. Wikelski, Princeton University. **11.22** Laura Romin & Larry Dalton/Alamy. **11.23** Gallo Images/Alamy. **11.26** Andrew McLachlan/All Canada Photos/SuperStock. **11.31** www.glofish.com. **11.32a** Gerald S. Wilkinson, University of Maryland. **11.34a** Photo by Jaime Grace. **11.38** John H. Christy, Smithsonian Tropical Research Institute. **11.39** Randy Thornhill, University of New Mexico. **11.45a** Dowery Orchard Nursery. **11.45b** Sharon Dahl.

CHAPTER 12 CO.12 mammalpix/Alamy. **12.2a** Glenn Bartley/Glow Images. **12.2b** Christina Riehl. **12.4** Mattias Hagman. **12.5** Richard R. Hansen/Science Source. **12.6** Majeed, H., O. Gillor, et al. 2011. Competitive interactions in Escherichia coli populations: the role of bacteriocins. *ISME Journal* 5: 71–81. **12.10** D. Robert & Lorri Franz/Corbis RF. **12.13a** Reprinted by permission from Macmillan Publishers Ltd: Gorrell, J. C., A. G. McAdam, et al. 2010. "Adopting kin enhances inclusive fitness in asocial red squirrels." *Nature Communications* 1: 22. Copyright © Macmillan Magazines Limited. **12.16** Auscape/UIG/Getty Images. **12.28** Sarah Blaffer Hrdy/Anthro-Photo File. **12.30 (left)** YK/Shutterstock. **12.31 (top)** Christian Heinrich/Newscom. **12.32a** Tier und Naturfotografie/Superstock. **12.32b** tamara bizjak/shutterstock. **12.33 (top)** Chris Kruger/Shutterstock. **12.36a** Dr. Morley Read/Shutterstock. **12.36b** angelshot/Shutterstock. **12.36c** J. E. Duffy. **12.36d** H Schmidbauer/age fotostock.

CHAPTER 13 CO.13 Design Pics Inc./Alamy. **13.1a** Reproduced by permission from E.A. Elbadry and M.S.F.Tawfik, 1966. "Life cycle of the mite Adactylidium sp. (Acarina: Pyemotidae), a predator of thrips eggs in the United Arab Republic," *Annals of the Entomological Society of America* 59(3):458–461, May 1966, p. 460, Fig. 6. **13.1b** Otorohanga Kiwi House, New Zealand. **13.9b–c** Reproduced by permission from Macmillan Publishers Ltd: Tyner, Donehower, et al. 2002. "Mutant mice that display early aging-associated phenotypes." *Nature* 415.5: 53. Copyright © Macmillan Magazines Limited. **13.23b** James F. O'Connell, University of Utah. **13.33** Timothy A. Mousseau, University of South Carolina. **13.38** Photo by Richard K. Grosberg, University of California-Davis. **13.41** Photo by Carol Eunmi Lee, University of Wisconsin.

CHAPTER 14 CO.14 Louisa Howard, Dartmouth College. **14.4** From: fig 3, pg 7156: Plotkin & Dushoff. "Codon bias and frequency-dependent selection on the hemagglutinin epitopes of influenza A virus." *PNAS* 100: 7152–7157. Copyright © 2003 National Academy of Sciences, U.S.A. **14.15** Dr. Morley Read/Science Source. **14.21** Ammit Jack/Shutterstock. **14.25** Charles & Josette Lenars/CORBIS.

CHAPTER 15 CO.15 Biophoto Associates/Science Source. **15.5** Sijen, T. & R.H.A. Plasterk. 2003. "Transposon silencing in the Caenorhabditis elegans germ line by natural RNAi." *Nature* 426: 310–314. **15.11** Xiao, H., et al. 2008. "A Retrotransposon-Mediated Gene Duplication Underlies Morphological Variation of Tomato Fruit." *Science* 319: 1527–1530. **15.12 (top & bottom)** Xiao, H., et al. 2008. "A Retrotransposon-Mediated Gene Duplication Underlies Morphological Variation of Tomato Fruit." *Science* 319: 1527–1530. **15.22a–c** Deng, C., et al. 2010. "Evolution of an antifreeze protein by neofunctionalization under escape from adaptive conflict." *Proceedings of the National Academy of Sciences* 107: 21593–21598. **15.25** UIG/Getty Images. **15.26** Puillandre, N., M. Watkins, and B. M. Olivera. 2010. "Evolution of Conus Peptide Genes: Duplication and Positive Selection in the A-Superfamily." *Journal of Molecular Evolution* 70: 190–202. **15.28** Feldman, C. R., E. D. Brodie, E. D. Brodie, and M. E. Pfrender. 2009. "The evolutionary origins of beneficial alleles during the repeated adaptation of garter snakes to deadly prey." *Proceedings of the National Academy of Sciences* 106: 13415–13420.

CHAPTER 16 CO.16 Stephan Roletto/Getty Images. **16.2a–b** Amato, A., W. Kooistra, et al. 2007. "Reproductive isolation among sympatric cryptic species in marine diatoms." *Protist* 158: 193–207. **16.4a** NHPA/SuperStock. **16.4c–d** Samuel Iglésias. **16.7a (left)** Karl Magnacca, DNA Barcoding Endemic Hawaiian Species Project, University of Hawaii at Hilo. **16.7a (middle)** Karl Magnacca, DNA Barcoding Endemic Hawaiian Species Project, University of Hawaii at Hilo. **16.7a (right)** Karl Magnacca, DNA Barcoding Endemic Hawaiian Species Project, University of Hawaii at Hilo. **16.8a** Carl C. Hansen/Nancy Knowlton/Smithsonian Institution Photo Services. **16.11 (top & middle)** Davison, A., S. Chiba, et al. 2005. "Speciation and gene flow between snails of opposite chirality." *PLoS Biology* 3: e282. **16.12** Lowry, D.B. & Willis, J.H. 2010. "A Widespread Chromosomal Inversion Polymorphism Contributes to a Major Life-History Transition, Local Adaptation, and Reproductive Isolation." *PLoS Biology* 8, e1000500. **16.14a (left)** Jaime Grace. **16.16** Dr. Ole Seehausen.

CHAPTER 17 CO.17 NASA/JPL-Caltech/Space Science Institute. **17.7a–b** Wochner, A., J. Attwater, et al. 2011. "Ribozyme-catalyzed transcription of an active ribozyme." *Science* 332: 209–212. **17.8** Russell Kempton, New England Meteoritical Services. **17.12** James P. Ferris, NY Center for Studies on the Origins of Life, Rensselaer Polytechnic Institute, Troy, NY 12180. **17.13a–b** Geologisk Museum. **17.15** Hanczyc, M. M., S. M. Fujikawa, and J. W. Szostak. 2003. "Experimental models of primitive cellular compartments: encapsulation, growth, and division." *Science* 302: 618–622. **17.16a–h** Andrew H. Knoll, Harvard University. **17.18 (top right)** Science Photo Library/SuperStock. **17.18 (middle right)** Eye of Science/Science Source. **17.18 (bottom right)** Robert Harding Picture Library/SuperStock. **17.28a–c** Andrew H. Knoll, Harvard University. **17.29** Han, T.-M., and B. Runnegar. 1992. "Megascopic eukaryotic algae from the 2.1-billion-year-old Negaunee Iron-Formation, Michigan." *Science* 257: 232–235. **17.30a–h** Schopf, J.W. 1994a. "Disparate rates, differing fates: Tempo and mode of evolution changed from the precambrian to the phanerozoic." *Proceedings of the National Academy of Sciences USA* 91: 6735–6742.

CHAPTER 18 CO.18 Roger Smith/Iziko Museums. **CO.18 (bottom)** Christian Sidor. **18.1a** John Koivula/Science Source. **18.1b** tkachev andrey/ITAR-TASS/Newscom. **18.2a** Francois Gohier/Science Source. **18.2b** Bill Florence/Shutterstock. **18.3** Ted Clutter/Science Source. **18.4** Grzegorz Niedzwiedzki. **18.6a** Simon Conway Morris. **18.6b** Narbonne, G. "Modular Construction of Early Ediacaran Complex Life Forms." *Science* 20, August 2004:Vol. 305. no. 5687, pp. 1141–1144. Copyright 2005 AAAS. **18.7a–d** Reproduced by permission S. Xiao, Y. Zhang, and A.H. Knoll, "Three-dimensional preservation of algae and animal embryos in a Neoproterozoic phosphorite." *Nature* 391:553–558 (February 5, 1998), fig. 5. Copyright © Macmillan Magazines Limited. **18.8a** Courtesy of Dr. Guy M. Narbonne/Queen's University, Department of Geological Sciences and Geological Engineering. **18.8b** Eduard Solà Vázquez/This work is licensed under a Creative Commons Attribution 3.0 Unported License. **18.8c** Jun-Yuan Chen, David J. Bottjer, et al. "Small bilaterian fossils from 40 to 55 million years before the Cambrian." *Science* 9:Vol. 305. no. 5681, pp. 218–222. Copyright 2000 AAAS. **18.9a–b** With permission of the Royal Ontario Museum and Parks Canada © ROM. J.B. Caron. **18.10a** Degan Shu, Northwest University, Xi'an, China. **18.11a** Opabinia regalis GSC 40251 © Geological Survey of Canada. Photo: Jean-Bernard Caron. **18.11b** With permission of the Royal Ontario Museum and Parks Canada © ROM. J.B. Caron. **18.17a** Superstock/Glow Images. **18.29a** Alessandro Montanari/Osservatorio Geologico di Coldigioco. **18.30a–b** Glen A. Izett/U.S. Geological Survey, Denver. **18.31 (inset)** Peter H. Schultz, Brown University; Steven D'Hondt, University of Rhode Island Graduate School of Oceanography. **18.35a** Ellie Nator/Fotolia. **18.35b** Sinclair Stammers/Science Source. **18.36a** LOOK Die Bildagentur der Fotografen GmbH / Alamy. **18.36b** Francois Gohier/Science Source. **18.42** USDA/ARS/Agricultural Research Service.

CHAPTER 19 CO.19 Joerg Boethling/Alamy. **CO.19 (bottom)** Reprinted with permission Lin, Z., X. Li, et al. 2012. "Parallel domestication of the Shattering1 genes in cereals." *Nature Genetics* 44: 720–724. Copyright © Macmillan Magazines Limited. **19.1** Penny Tweedie/Alamy. **19.2a** National Geographic Image Collection/Alamy. **19.2b** David B. Wake, Museum of Vertebrate Zoology. Photo: M. Garcia-Paris. **19.3a** Biology Pics/Science Source. **19.3b** Edward B. Lewis, California Institute of Technology. **19.3c** Dr. Jeremy Burgess. **19.3d** F. Rudolf Turner, Indiana University. **19.6a** Dan Clemens. **19.6b** Arco Images GmbH/Alamy. **19.7a** Frank Vincentz. **19.7b** Dr. David Midgley. **19.7c** James Lindsey. **19.7d** amadej trnkoczy. **19.7e** Igel. **19.7f** Mike Powles/Getty Images. **19.9a** M. Cheung, 2006. Used under a Creative Commons Attribution-ShareAlike license: http://creativecommons.org/licenses/by-sa/3.0/. **19.9b** tororo reaction/Fotolia. **19.10a** Protas, M., and W. R. Jeffery. 2012. "Evolution and development in cave animals: from fish to crustaceans." *Wiley Interdisciplinary Reviews: Developmental Biology* 1: 823–845. **19.10b** Dr. Edgar Buhl. **19.10c** Dan Clemens. **19.11a–b** William Piel & Antónia Monteiro. **19.11c–d** Allen, C. E., P. Beldade, B. J. Zwaan, and P. M. Brakefield. 2008. "Differences in the selection response of serially repeated color pattern characters: standing variation, development, and evolution." *BMC Evol Biol* 8: 94. **19.13a–d** Reed, R. D., et al. 2011. "Optix drives the repeated convergent evolution of butterfly wing pattern mimicry." *Science* 333: 1137–1141. **19.14a–b** Joron, M., C. D. Jiggins, et al. 2006. "Heliconius wing patterns: an evo-devo model for understanding phenotypic diversity." *Heredity (Edinb)* 97: 157–167. **19.14c–d** Olaf Leillinger. **19.14e–f** Reproduced by permission Beldade P. and P.M. Brakefield. 2002. "The genetics and evo–devo of butterfly wing patterns." *Nature Reviews Genetics* 3: 442–452. Copyright © Macmillan Magazines Limited. **19.15a** Photoshot/Newscom. **19.16a–c** Emlen, D. J. 2000. "Integrating development with evolution: a case study with beetle horns." *Bioscience* 50: 403–418. **19.16d–e** Emlen, D. J. 2001. "Costs and the Diversification of Exaggerated Animal Structures." *Science* 291: 1534–1536. **19.18a** Chris Pires. **19.19a (top left)** Hodin. **19.19a (top middle)** F1online digitale Bildagentur GmbH/Alamy. **19.19a (top right)** WILDLIFE GmbH/Alamy. **19.19b (top & bottom right)** Jarosch A, et al. 2011. "Alternative splicing of a single transcription factor drives selfish reproductive behavior in honeybee workers (Apis mellifera)." *Proc Natl Acad Sci USA* 108: 15282–15287. **19.22a (top)** Biosphoto/SuperStock. **19.22a (bottom)** Hessling R. 2002. "Metameric organisation of the nervous system in developmental stages of Urechis caupo (Echiura) and its phylogenetic implications." *Zoomorphology* 121: 221–234. **19.22b (top)** Kristof A, et al. 2008. "Segmental mode of neural patterning in Sipuncula." *Curr Biol* 18: 1129–1132. **19.22b (bottom)** David Shale/Nature Picture Library. **19.22c (top)** Pedro Bernardo / Shutterstock. **19.22c (bottom)** Mayer, G., P. M. Whitington. 2009. "Neural development in Onychophora (velvet worms) suggests a step-wise evolution of segmentation in the nervous system of Panarthropoda." *Developmental Biology* 335: 263–275.

CHAPTER 20 CO.20 Image created by Peter Schmid, courtesy Lee R. Berger and the University of the Witwatersrand. **20.13** Reprinted with permission from Nature, B. Wood, 2002, "Hominid revelations from Chad," *Nature* 418: 133–135, Figure 1, page 133. Copyright © 2002 Macmillan Magazines Limited. **20.16** John Reader/Science Source. **20.17** Photo by Brett Eloff, courtesy of Lee R. Berger and the University of the Witwatersrand. **20.18a 20.18b Don McGranaghan.** David L. Brill/Brill Atlanta. **20.18d** David L. Brill/Brill Atlanta. **20.24 (left & right)** David L. Brill/Brill Atlanta. **20.39a** Anup Shah/Nature Picture Library. **20.40** Didier Descouens. **20.43b–c** Arensburg, Schepartz, Tillier, Vandermeersch, and Rak. "A reappraisal of the anatomical basis for speech in middle paleolithic hominids." From: *American Journal of Physical Anthropology* 83:137–146.

COVER Auke Holwerda/Getty Images.

Index

Abecasis, Goncalo, 163
Abrahamson, Warren, 358
Abscisic acid (ABA), 746, 759
Abzhanov, Arhat, 85
Acacia greggii (catclaw acacia), 520, **520**
Acanthostega, **701**, 702, **702**
Acetaminophen, 566
Acetylene, 645
Ache Indians, **511**, 512, 572
Acheulean tools, 809
Achillea, **361**, 361–62
Achillea borealis (wild yarrow), 169, **169**
Achromatopsia, 245
Acid β-glucosidase (GBA), **307**, 307–8, **308**
A-conotoxin gene tree, 599–600, **600**
Acquired immune deficiency syndrome (AIDS), 1–2. *See also* Human immunodeficiency virus (HIV)
in Asia, **2**
CCR5-Δ32 allele frequency and, **200**, 200–1
combination therapies for, 13–14, **14**
infection mechanism, 6–8, **8**
long-term trends in, **3**
natural history of epidemic of, 2–8
neutral evolution and progression to, **28**, 28–29
in North America and Europe, **2**, 3
prevention successes and failures, 3–4
in sub-Saharan Africa, **2**, 3
viral load and progression of HIV to, **25**, 26
Acropora digitifera, 742–43
Actinobacteria, **757**
Acute bladder snail (*Physa acuta*), **412**, 412–13, **413**
Acute hemolytic anemia, 312
Acyrthosiphon pisum (pea aphid), 363, **363**
Adactylidium sp. (thrips egg mites), 492, **492**
Adaptation(s), 78, 369–406. *See also* Aging; Kin selection; Sexual selection
caveats in studying, 373
comparative analysis of, 382–87
constraints on, 389–97
defined, 370
experiments on, 373–78
design issues, 376, **376**
replication of, 376–78
genome duplication and, 169, **169**
to habitats, 623–25
hypothesis-formation strategies, 401–2
medical/health, 556–74
breast cancer, 559–62
environment and, 556–58
fever, 564–67
menstrual cycle, 560–62, **561**
myopia, 558–59
parenting, 567–74
molecular basis of, 601–6
adaptive mutations of small *vs.* large effect, 605–6
new mutation *vs.* standing genetic variation, 602–5
from new mutation, 602–603, **602–603**
observational studies of, 378–81
behavioral thermoregulation, **379**, 379–80
garter snake nighttime behavior, **380**, 380–81, 381, **381**
oxpeckers, 370–73
mutualist relationship with cattle and, hypothesis of, 370–72, **371**, **372**
parasitic relationship with cattle, hypothesis of, 372–73
phenotypic plasticity, 387–89
in water flea behavior, **387**, 387–89, **388**, **389**
secondary, 92
selection and, 370, 397–401
in Apert syndrome, 400–1, **401**
demonstration of, 398–400, **399**
transposable elements, 586–91
from standing genetic variation, 602, **604**, 604–5
trade-offs and constraints in, 389–97
begonia flower size, 389–91, **390**
fuchsia color change, 391–93, **392**
host shifts in feather lice, 396–97, **397**
host shifts in *Ophraella,* **394**, 394–96, *396*
Adaptive constraint, 596
Adaptive radiation, 638, **726**, 726–27, **727**, 730–31
Adaptive significance of sex, 314–24
environmentally-imposed selection and, **322**, 322–24, **323**, **324**
genetic drift and, 319–21
maintenance of males and, 316–18, **317**
mutation and, **319**, 319–21, **320**, **321**
population genetics consequences, 318–19
reproductive modes and, **315**, 315–16
Adaptive traits, 78, 370
Additive genetic variation (V_A), 344–45
Adenine, 149, **160**, 658–59
Adenosine deaminase (ADA), **553**, 553–54
Adenosine deaminase deficiency, 553–54
Adermatoglyphia, 161, **161**
Aegiceras corniculatum, **745**
Aeromonas hydrophila, 565
AFPIIIs, 596–98, **597**
Africa, HIV resistance in, 17
African Americans, heritability of IQ and, **362**, 362–63
African clawed frog (*Xenopus laevis*), 122, **747**
African great apes, 771–72
knuckle walking in, 776
African leaf-eating monkeys, 165
African replacement model, 790–98, **791**
archaeological and paleontological evidence of, 791–93, **792**
cladogram predicted by, **792**
genetic predictions distinguishing, 793, *793*
molecular evidence of, 793–800
Age, inbreeding depression and, 502–3, **503**
age-1 gene, 505
Aging, 495–512
in birds, **496**
cancer risk and, **499**, 499–500
evolutionary theory of, 500–10
deleterious mutations and, 500–3, **503**
natural experiments of, 507–10, **509**
trade-offs and, 500, 504–7, **505**, **506**, **507**
in insects, **496**
in mammals, **496**
rate-of-living theory of, 496–500
trade-offs and, **499**, 499–500
Agonistic parent–child interaction, 572–74
Agouti signaling protein (ASP), 95, **95**
Ågren, Jon, 390–91
Agrodiaetus, 621
Ahlquist, Jon, 737
Ahounta, Djimdoubalbaye, 781
AIDS. *See* Acquired immune deficiency syndrome
Air sacs, language and, **806**, 806–7
Aiuti, Alessandro, 554
Alarm calling, 458, **458**, **462**, 462–63
Alaska sea otters, 278, *278*
Alberch, Pere, 738
Albertin, Warren, 757
Albertson, Tina, 159
Alcohol dehydrogenase, 195
Aldehyde dehydrogenase, 760, **760**
Aldehydes, 658
Aldrin, Buzz, 180, **180**
Allele frequency(ies)
CCR5-Δ32, **200**, 200–1
chance events and, 240–42, **241**
changes in. *See also* Genetic drift; Migration; Mutation(s); Natural selection
evolution and, 90–91
different chromosome frequencies and, **293**, 293–94, **294**
in general case, **187**, 187–89, **188**
homogenization of, migration and, 235–37, **237**
human expansion from Africa and, 797
linkage disequilibrium and, 303, **303**, **305**, 305–6
migration and, 234–39, **235**
mutation and, **217**, 217–19, **218**
under mutation–selection balance, 220
numerical calculation, 183–87, **185**, **186**
selection and changes in, 192–95, **193**, **195**
selfing and, **275**, 275–76, *276*
in simulated life cycle, 181–83, **183**
statistical analysis of using chi-square test, 198–99
Alleles
calculating frequencies of, 162–63
codominant, 345
defined, 150
dominant, 345
with early benefit and late cost, 504, **504**
greenbeard, 468–69
Hardy–Weinberg equilibrium principle with more than two, 189
new, 157–61, **158**, **159**
alteration of protein function and, 159–61, **160**, **161**
random fixation of, **246**, 246–49, **247**
experiment on, 249–52
in natural populations, 252–55
probability of alleles to drift toward, 248
recessive, 345
selection on recessive and dominant, 202–6
Allelic richness, 254
Allelic segregation, 318
Allen, Cerisse, 748–49
Allopatric speciation, 617
Allosaurus, **53**
Allwood, Abigail, 66
α crystallins, 101
α-tubulin, 100
Alpha melanocyte-stimulating hormone (α-MSH), 95, **95**
Alpheus malleator (snapping shrimp), **619**, 619–20
Alpine skypilots (*Polemonium viscosum*), **351**, 351–53, **352**, **353**
Alroy, John, 730
Alter, Elizabeth, 274
Alternation of macroevolutionary regimes, 725
Alticus arnoldorum (Pacific leaping blenny), **50**, 50–51
Altman, Sidney, 647
Altruism, 94, 456, 458. *See also* Kin selection
in birds, 466–67, **467**
as central paradox of Darwinism, 459–60
Darwin on, 459–60
inclusive fitness and, 460
kin selection and evolution of, 459–69
greenbeard alleles, 468–69
inclusive fitness, 460–61
kin recognition, 465–66
under multilevel selection, 475, **475**
parent–offspring conflict and, 477–80, **478**
reciprocity, **481**, 481–83
Alu (+) chromosome, 797
Alu (−) chromosome, 797
Alu elements, 775
Amazon region, deforestation in, 719
Amber, 692, **692**
Amblyrhynchus cristatus (marine iguanas), **417**, 417–21, **418**, **419**, **420**
Amborella, 727, **727**
Ambulocetus, **136**
American robins (*Turdus migratorius*), 488
Amino acids, 149
in meteorites, 655–56
Amino acid substitutions, driven by positive selection, **271**, 271–72
Aminopenicillin, 546, **546**
Ammonoid shell shape, 708–9, **709**
Amniota, 54, **117**, 728
Ampicillin, 545
AMY1, 598, **598**
Anagenesis (phyletic transformation), 721
Analogy, 67
Anchiornis huxleyi, **703**
Anders, Edward, 657
Anderson, David, 480
Anderson, Jennifer, 318, 379
Anderson, Kermyt, 572
Andersson, Malte, 424–25
Andersson, Steffan, 424, 425

Note: Pages in **bold** locate figures; pages in *italic* locate tables.

Angiosperms
  Cretaceous explosion of, 727
  inbreeding depression in, 281
Anglerfish, 449
Anhydrobiosis, 291, **291**
Animals, 668
  diversification of, 698–700
  domestic, 74–76
  homoploid hybrid speciation in, 634
  phylogeny for earliest, **700**
Anolis sp., diversification of, 638, **638**
Anstey, Robert, 722
Antagonistic pleiotropy hypothesis, 504–7
Antarctic eelpout (*Lycodichthys dearborni*), **596**, 596–98, **597**
Antelopes, **118**, 118–20, **119**, **120**
Antennapedia complex (ANT-C), 738–39
Antennapedia gene, 738, **739**
Anthocyanin, **632**, 632–33, **633**
Antibiotic resistance, 545–48, **546**
  costs to bacteria of, 546–48, 547
  judicious use of antibiotics to inhibit, 548
  selection and, 545–46
Antibiotics, 536, 545
  effect on gut microbes, 563
  judicious use of, 548
Antifreeze proteins, 102
  evolution of, 596–98, **597**
Antigenic sites, 538–39
Antigen recognition site (ARS), 266
Antisocial behavior, 365, **365**
Antonelli, Peter, 679
Ants, 484, **484**
Anver, Miriam, 565
Apert syndrome, multilevel selection in, 400–1, **401**
Apes
  great, 770–72
    brain size in, **790**
    genetic diversity among, **800**
  phylogeny of, **771**, **772**, **773**
  relationship between humans and extant, 770–79
    divergence times, **777**, 777–78
    molecular evidence of, 772–76
    morphological evidence of, 776–77
Apex chert (Australia), 665
Aphids
  pea, 363, **363**
  sexual and asexual reproduction in, 314, **314**
Apicella, Coren, 476–77
Apis mellifera, **484**, **589**
Apis mellifera capensis (cape honeybee), 758, **758**
Apis mellifera carnica (European honeybees), **758**, 758–59
Apis mellifera (honeybees), **758**, 758–59
Apis mellifera scutellata, 758, **758**
Apollo 11, **180**
Apomorphy, 116, **116**
Apteryx australis mantelli (brown kiwi), 42, **42**, **492**, 492–93
Arabidopsis thaliana (thale cress), 169–70, **170**, 475–76, **589**, 694
  self-fertilizing in, 306, **306**
Archaea, 669, **669**, 672, **672**
  discovery of, 678
  in extreme environments, 670
  fusion with bacterium, **679**, 679–80
  hypotheses of origin of
    chronocyte hypothesis, 680, **680**
    ring-of-life hypothesis, **679**, 679–80
    three viruses, three domains hypothesis, 680–83, **682**
    universal gene-exchange pool hypothesis, **678**, 678–79, **679**
  phylogeny of all living things and, **667**, 667–68
  rRNA sequences of, 670–72, **671**
  speciation in, 612–13
Archaebacteria, 668
Archaeological record. *See also* Fossil record
  African replacement *vs.* multiregional evolution and, 791–93, **792**
  evidence of language in, 805–7
Archaeopteryx, **51**, 51–52, **53**, 702, **703**
Archosauria, 117, **117**
Ardipithecus kadabba, 782
Ardipithecus ramidus, 782, **782**
Arens, N. C., 715
Arg151Cys allele, 95
Argument from biochemical design, 99–103
Argument from design, 98
Armstrong, Neil, 180, **180**
Arrector pili, 42, **43**
Artemia salina (fairy shrimp), **723**
Arthur, Wallace, 747
Artificial selection, 39–40, **41**, 74–76, 91, **91**
Artiocetus clavis, 136
Artiodactyls, **711**
  whales and, **124**, 124–25, **125**, **127**, 130, 133, 134, 136, **136**
Asexual reproduction, 291, **314**, 314–15
  Muller's ratchet and, **320**, 320–21, **321**
  reproductive advantage of, **315**, 315–16
Asfaw, Berhane, 784
Ashkenazi population
  Δ*32* allele in, 162–63
  GBA-84GC mutation in, 307–11
  telomere length in, 498
Ashton, Peter, 718
Asia
  HIV/AIDS in, **2**
  tropical forest destruction in, 719
Asparagine, 26
Aspirin, 566
Assortative mating, **428**, 625–26
Asymmetric limits
  on fitness, behavioral consequences, 416–17
  on reproductive success, 411–16
Asymmetries in sexual reproduction, 410–11, **411**
Asterina miniata (bat stars), **148**
Asteroid, Cretaceous–Paleogene extinction and, 712–14, **713**, **714**
Astragalus, 125, **125**
Astrophytum asterias, **745**
Astyanax mexicanus (blind cave fish), **747**
Atlantic Ocean, age of, 62, 64
Atmosphere
  after asteroid impact at K–Pg boundary, 715
  composition of early, 657–58
Atoyac salamander (*Bolitoglossa oaxacensis*), **738**
Atta cephalotes (leafcutter ant), 483
Attention deficit/hyperactivity disorder (ADHD), 40
Atzmon, Gil, 498
Audubon's warbler (*Dendroica auduboni*), **635**, 635–36, **636**
Austad, Steven, 496, 507–10, 512
Austin, D. J., 546
Australian magpies, 466
Australopithecines, gracile, 782, 784
Australopithecus, 781
Australopithecus aethiopicus, **785**
Australopithecus afarensis, 782–83, **783**, 804, 807
Australopithecus africanus, 782, **783**, 804
Australopithecus anamensis, 783, **783**
Australopithecus boisei, **785**
Australopithecus garhi, 784, **785**
Australopithecus habilis, **785**
Australopithecus ramidus, **782**
Australopithecus robustus, **785**
Australopithecus rudolfensis, **785**
Australopithecus sediba, **769**, 770, 784, **784**
Average excess, 194
Aves, **117**
AVI, 150–51
Avise, John, 722–23
Azad, Priti, 219
Azidothymidine (AZT), 8
  blocking reverse transcription, 9, **9**
  evolution by natural selection illustrated by evolution of resistance to, **12**, 12–13
  patient physiology and, 9–10
  resistance to, **10**, 10–11, **11**
  virion population and, 10, **10**

B57 allele, 26
B5801 allele, 26
Baboons (*Papio hamadryas ursinus*), 481, **481**, 770
Background extinctions, 710, 711–12, **712**
Background mutations, evolution of cancer and, 555–56
Background selection, 270–71
Bacteria, 669, **669**, 672, **672**. *See also specific bacteria*
  antibiotics selecting for resistant, 545–46
  bacteriophage virus and mutation rates in, 594, **594**
  costs of antibiotic resistance to, 546–48, **547**
  fusion with archaean, **679**, 679–80
  hypotheses of origin of
    chronocyte hypothesis, 680, **680**
    ring-of-life hypothesis, **679**, 679–80
    three viruses, three domains hypothesis, 680–83, **682**
    universal gene-exchange pool hypothesis, **678**, 678–679, **679**
  multilevel selection and cooperation in, **474**, 474–75, **475**
  phylogeny of all living things and, 667, **667**
  random variation in protein production in, **153**
  speciation in, 612–13
Bacteriocins, 459, **459**, 470–71
Bacteriophage, 476
Bacteriophage f1, 550–51, **551**
Bacteriophage φ6, 45, **45**
Bacteriophage virus, bacterial mutation rates and, 594, **594**
Baja California, 620
Baker, Scott, 109
Bakker, Robert, 52
Balmford, Andrew, 718
Banet, Manuel, 565
Barbulescu, Madalina, 773
Barghoorn, Elso, 66, 665
Barley mildew (*Blumeria graminis*), 581, **581**
Barnacles, 736
Barns, Susan, 670
Barnshad, Michael, 801
Barn swallows (*Hirundo rustica*), 478
Bartel, David, 652–54
Bashey, Farrah, 470–71
Basilosaurus, 125, **136**
Bateman, Angus John, 410, 412, 413, 415, 416
Bateson, William, 738, 756
Batra, Susanne, 483
Bats
  fruit, **382**, 382–86, **385**
  testes size in, 382–86, **385**
Bat stars (*Asterina miniata*), **148**
Bäurle, Silke, 419, 420
Bayesian phylogeny inference, 133, **133**, 134
B cells, 6
  adenosine deaminase and, 553–54
Bdelloid rotifers, **291**, 291–92, 324
Beach mice, 605
Beak size in Galapagos finches, 346, **346**, 354–56, **355**
Bean, W. J., 542
Bears, 421
  black, 384–85, **385**, 492
Beaudry, Amber, 650–51
Beckerman, Stephen, 446–47
Bees, as pollinators, **334**, 335–36, 340, **340**, 390, 443–44
Beetles
  courtship in, 431–32
  dung, **753**, 753–55, **754**
  flour, **202**, 202–3, **203**
  host shifts in herbivorous, **394**, 394–96
  leaf, **394**, 394–96
  long-horned, 756
  phenotypic plasticity in egg size, 520–22, **521**
  red flour, 257–58, **258**
  seed, **520**, 520–22, **521**
Begonia (*Begonia involucrata*), 389–91, **390**, **391**
Begun, David R., 776–77, 789
Behavior
  asymmetry of sex and, 416–17
  cultural influences on, 568
  social. *See* Social behavior
Behavioral fever, **564**, 564–66
Behavioral thermoregulation, **379**, 379–80, 381
Behe, Michael, 100, 101, 103
Belding's ground squirrels (*Spermophilus beldingi*), 94, 458, **458**
Bell, Michael, 44
Bell-curve fallacy, 360–364
Bender, Catherine, 540

Benjamin, Jonathan, 341
Benson, John, 284
Benton, Michael J., 725, 728
Benzene, 645
Berglund, Anders, 440
Berner, David, 46
Bernheim, Harry, 564
Berrigan, David, 517
Berry, Andrew, 270
Bertram, B. C. R., 422
Best-fit line, **84**, **344**, **346**, **348**, **351**, **352**
β-casein, **126**, 126–27, **127**, 133, **133**, 134
βγ crystallins, 101
β-lactamase, 545
β-subunits, 672, **673**
β-tubulin, 100
Bias, **376**
  codon, 268–70, **269**
  developmental constraints and, 746–47
  in fossil record, 693–94
  preexisting sensory, 433–35
Biased evolutionary trajectories, 749–52
Bickerton, Derek, 809
Bicyclus anynana (squinting bush brown butterfly), **749**, 748–49, 752
Bigeye mutant, 752, **752**
Big Five extinctions, 709–19. *See also* Cretaceous–Paleogene (K–Pg) extinction
Bilaterians, 696–98, **697**, 728–29
Biochemical design, argument from, 99–103
Biogeography, 139
Biological invasions, 527–29
Biological polymers, assembly of, 660–61, **661**
Biological species concept (BSC), 45, 611–12
Biotic replacement, **725**, 725–26
Bipedal locomotion, 782, 783
Birds, **117**, 421. *See also specific species*
  aging in, **496**
  clutch size in, **513**, 513–15, **514**
  conflict over provisioning in, **477**, 477–79, **478**, **479**
  cooperative breeding in, **466**, 466–467, **467**
  dinosaurs and, 51–53, **53**, 702–4, **703**
  fossil of transitional, **51**, 51–52
  phylogeny, 114–15, **115**
  range overlap in, 640
  vulnerability to extinction, 529, **529**
Bird strains of influenza, 542–44, **543**, **544**
Bishai, William, 545
Bithorax complex (BX-C), 738–39
Bittacus apicalis (hangingflies), 435–36, **436**
Bivalves, 492, 693, **724**, 724–25, **725**
Blaak, Hetty, 27
Black bear (*Ursus americanus*), 384–85, **385**, 492
Black-bellied seedcracker (*Pyrenestes o. ostrinus*), 358–59, **359**
Black cherry (*Prunus serotina*), 492
Black cockatoo, 386
Black death, 310
Black rhinoceroses, 373
Black-tailed prairie dogs (*Cynomys ludovicianus*), **462**, 462–63
Blair, Amy, 527–28
Blending inheritance, 95
Blenniella gibbifrons, **50**, 50–51
Blind cave fish (*Astyanax mexicanus*), **747**
Bloch, Alan, 546
Blood-clotting cascade, 102
Blue-footed boobies, 480, **480**, 491, **491**
Blue jays (*Cyanocitta cristata*), 488
Blue skate (*Dipturus* cf. *flossada*), **615**
Blue snails, **57**, 57–58
Blumenschine, Robert, 784
Blumeria graminis (barley mildew), 581, **581**
BMP family, 761
Bmp gene, 746
BMP3 gene, **313**
Bmp4 gene, 85–86
Boag, Peter, 83–85, 86–87
Bobcat, 110, **111**, 111–12, **113**, **114**
Body fossils, 693
Body length, variation in, 147, **147**
Body lice (*Pediculus humanus corporis*), evolution of, **138**, 138–39
Body mass index, **562**, 562–63
Body plans, 93
Body size
  in humans
    brain size, 805–6
    sexual dimorphism in, 408, **444**, 444–45
  in marine iguanas, 417–21, **418**, **420**
Bogues, Muggsy, **147**
Bolitoglossa dofleini (giant palm salamander), **738**
Bolitoglossa oaxacensis (atoyac salamander), **738**
Bombus sp. (bumblebees), **351**, 351–53, **352**, **353**
Bonacum, James, 619
Boncoraglio, Giuseppe, 478
Bone morphogenic protein 4 (BMP4), 85–86, **86**
Bonobos (*Pan paniscus*), 59, 770, 780
  relationship between humans and, 772–79
  tool use and, 802–3
Boobies, 480, **480**, 491, **491**
Bootstrapping, **132**, 132–33, **133**
Borrelia burgdorferi, 672, **673**
Botryllus leachi (sea squirt), 527
Botryllus schlosseri (sea squirt), **526**, 526–27, **527**
Botswana, AIDS infection rate in, 3
Bourgeois, Joanne, 715
Boutin, Stan, 463
Bowcock, A. M., 795–96
Boyce, Mark, 513–14
Bradshaw, H. D., Jr., 334–41
Brain size, in humans, 805–6
Brakefield, Paul, 748–49
Branch and bound, 131
Branching events. *See* Speciation
Brassica napus (rapeseed), 756–57, **757**
Brassica oleracea acephala (kale), **76**
Brassica oleracea botrytis (cauliflower), **76**
Brassica oleracea (cabbage), 76, **76**
Brassica oleracea gemmifera (brussel sprouts), **76**
Brassica oleracea gongylodes (kohlrabi), **76**
Brassica oleracea italica (broccoli), **76**
Brassica oleracea oleracea (wild cabbage), **76**
Brassica oleracea (wild cabbage), 756, **757**
Brassica rapa (field mustard), **41**, 41–42, **42**
Brassica rapa (wild turnip), 756, **757**
Brazilian Amazon, deforestation in, 718
Brazos River, 715
BRCA1 gene, **266**, 266–67
Breast cancer, 559–62
  incidence of, 559, **560**
  menstrual cycle and, 560–62, **561**
  as viral disease, 559–60, **560**
Breeding, communal, 457
Brelsford, Alan, 635
Brewer's yeast (*Saccharomyces cerevisiae*), **170**, **171**, 171–72
Bridges, Calvin, 738
Briskie, James, 478–79
Brisson, Jennifer, 254
Britten, Ron, 737, 738
Broad-nosed pipefish (*Syngnathus typhle*), **415**, 415–16, 438, **440**, 440–41
Broad-sense heritability, 344
Broca's area, 805
Broccoli (*Brassica oleracea italica*), **76**
Brodie, Edmund, 356
Bronze Age frequency of *CCR5-Δ32* allele, 311
Brown, James R., 671–72, 672
Brown-headed cowbirds (*Molothrus ater*), 405
Brown kiwi (*Apteryx australis mantelli*), 42, **42**, **492**, 492–93
Brunet, Michel, 781–82
Brussel sprouts (*Brassica oleracea gemmifera*), **76**
Bryan, William Jennings, 97
Bryant, Edwin, 503
Bryozoans
  cheilostome, **721**, 722
  stasis and speciation in, **721**, 721–22
Bubonic plague, 310
Buchanan, Bryant, 427
Buckingham, William, 770
Buck v. Bell, 201–2
Bufo marinus. *See Rhinella marina* (cane toad)
Bull, James, 550
Bumblebees (*Bombus* sp.), **351**, 351–53, **352**, **353**
Burdick, Allan, 206–7
Burgess Shale fauna, **698**, 698–701, **699**, **700**
Buri, Peter, 249–51
Burmeister, H., 736
Bush, Robin, 539, 540
Bush, Sarah, 386–87
Bustamante, Carlos, 268
Butler Act, 97
Butterfly, 634
  wing-pattern variation in, 748–52, **749**, **750**, **751**, **752**
  wings, constraint in, 748–49, **749**
Butterfly (*Agrodiaetus*), 621
Butterwort (*Pinguicula moranensis*), 92, **92**
Byars, Sean, 512

C22orf45 gene, 166, **166**
C868T, 17
Cabbage (*Brassica oleracea*), 76, **76**, 756, **757**
Cactospiza heliobates (mangrove finch), 81, **81**
Cactospiza pallida (woodpecker finch), 81, **81**
Cactus ground finches (*Geospiza scandens*; *G. conirostris*), **81**, **86**, 87
Caecilians, 140
Caenorhabditis elegans, 694
  antagonistic pleiotropy in, **505**, 505–6
  base substitutions in, 170
  behavioral thermoregulation in, **379**, 379–380
  mutation and evolution in, **173**, 173–74
  NK4 gene in, 762
  outcrossing in, 322, **322**
  persistence of males in, **316**, 316–320, **319**
Caerulein, 122
Caimans, 121, **121**
California poppy (*Eschscholzia californica*), 492
California sea otters (*Enhydra lutris*), 276–78, **277**, *278*
Calumma tigris (Seychelles tiger chameleon), **139**, 139–40, **140**
Camarhynchus parvulus (tree finch), **81**
Camarhynchus pauper (tree finch), **81**
Camarhynchus psittacula (tree finch), **81**
Cambrian explosion, 691, 698–700, 730–31
  causes of, 700–1
  Ediacaran and Burgess Shale fauna, **696**, 696–700, **698**, **699**
  intensity of, 728–29
  morphology of, 698–700
  phylogeny of, 698–700, **700**
Camel, **124**, **125**, **132**
Campàs, Otger, 85
Campbell, Diane, 637
Canada lynx, 110, **111**, 111–12, **112**, **113**, **114**
Canback, Bjorn, 680
Cancer
  aging and risk of, **499**, 499–500
  breast, 559–62
  colon, 502
  endometrial, 562
  history of, 554–56
  liver, **555**, 555–56
  ovarian, 562
Candidate loci, 341–43
Cane toad (*Rhinella marina*), **457**, 457–58
Canine transmissible venereal tumor (CTVT), 137, **137**
Cannibalism, in cane toads, 457–58
Cannon, Carolyn, 224
Canonical spatial expression pattern, **741**, 742–44, 764
Cant, Michael, 512
Cape honeybee (*Apis mellifera capensis*), 758, **758**
Capuchin monkeys (*Cebus apella*), 481
Carbonaceous chondrite meteorites, 657
Carbon cycle, 716
Carey, Jane, 567
Caribbean lizards, 638, **638**
Carnivorous plants, 91–92, **92**
Carr, David, 264
Carrano, Matthew, 529
Carroll, Sean, 701
Caryophyllales, **92**
Caspi, Avshalon, 365
Cassini spacecraft, 645
Castle, William, 189
Castro, Bermúdez de, 792
Casts, 693, **693**

Cat
  domestic, **284**
  phylogeny, 110–14, **111**, **112**, **113**, **114**
Catania, Charles, 809
Catasetum, **441**, 441–42, 443–44
Catastrophism, 62
Catclaw acacia (*Acacia greggii*), 520, **520**
Caterpillars, 471
Cation-independent mannose-6-phosphate receptor (CI-MPR), 523
Cattle-oxpecker relationship, hypotheses on, 370–73, **371**, **372**
Cauliflower (*Brassica oleracea botrytis*), **76**
Cavener, Douglas, 195
CCR5, 6–7, **7**, 15–16, 18, 27, 180
CCR5-Δ32 allele, 180, 313
  age of, 311
  changes in frequency of, 191, **200**, 200–201
  frequency in future generations, 200
  frequency in Old World, **16**, 16–17, 18, 310
  history of, 310–11
  origin of, 311
CCR5 genotypes, 162, **162**
CD4, 5, **5**, 6–8, **7**, **8**, **17**, 17–18
Cebus apella (capuchin monkeys), 481
Cech, Thomas, 647, 648
Cells
  defenses against viruses, 681
  model of early, **664**
  selection at level of, 398–401, **399**, **401**
  senescence of, 498–99
  tissues as evolving populations of, 553–56
Cellular division, senescence of cells and telomere shortening with each, 498–99
Cellular life, 663–78
  advantages of, 663
  DNA and, 663
  examination of early, 668–70
  fossil record of, 664–65, **665**, 675–78, **676**, **677**
  See universal phylogeny
Cellular slime molds, 668
Cenancestor, 647, **675**
Cenozoic era, 694, **695**
Ceratitis capitata (Mediterranean fruit fly), 422, **422**
Cercidium floridum (palo verde), 520, **520**
Cercopithus, 29
Cereals, **557**, 735
Certhidea fusca (gray warbler finch), 81, **81**
Certhidea olivacea (olive warbler finch), 81, **81**
Cervus elaphus (red deer), **496**
Cetacea, 123–24. *See also* Whales
Chagnon, Napoleon, 446, 572
Chance events
  allele frequencies and, 240–42, **241**
  genotype frequencies and, 240–42, **241**
Chao, Lin, 264
Chaoborus flavicans, 152
Charcot-Marie-Tooth disease, 59
Charlesworth, Brian, 586
Charlesworth-Langley hypothesis, 586
Charnov, Eric, 515–16
Charrier, Cécile, 779
Chase-away sexual selection, 524
Chaung, John, 474–75
Cheetahs, **284**
Cheetham, A. H., 721, 722
Cheilostome bryozoans, **721**, 722
Chen, Chun-Hong, 224–27
Chen, Pei-ji, 52
Cheney, Dorothy, 481
Chengjiang biota, 698
Chestnut-crowned babbler (*Pomatostomus ruficeps*), 466
Chicxulub crater, **714**, 714–15
Children, parental discrimination and health of, 572–74
Chimpanzees (*Pan troglodytes*)
  arrectores pilorum, **43**
  genetic differences between humans, gorillas and, **778**, 778–79, **779**
  hyoid bone of, 806, **806**
  movement of HIV from, to humans, 21–23, **22**, 24
  PMP-22 gene and, 59, **59**
  positive selection on *BRCA1* gene in, **266**, 266–267
  relationship between humans and, 737, **737**, 772–79, 808
    divergence times, 777–78, **778**, **779**
    molecular evidence, 772–76
    morphological evidence, 776–77
  thumb in humans *vs.*, 804, **804**
  tool making by, 802–3
Chinook salmon (*Oncorhynchus tshawytscha*), 518–20, **519**
Chirality (handedness), 660
Chi-square (χ^2) test, 198–99
Chlamydia trachomatis, 671, **671**
Chlamydomonas, **100**, 101, **589**
Cholecystokinin gene, 122
Cholera (*Vibrio cholerae*), 535, **535**, 552
  evolutionary relationships among strains of, 536, **536**
Chondrodysplasia, 165, **165**
Chordates, 698, 729
  eyes, 98–99, **99**
  invertebrate, **743**
  parallel evolution of attachment organs, 746, **747**
Christy, John, 435
Chromosome, 149
  changes in, as barrier to gene flow, 621
  compound, 209–12
  polymorphic, 167
Chromosome frequencies, **293**, 293–94, **294**
Chromosome mutations, 166–69
  genome duplication, **168**, 168–69, **169**
  inversions, **167**, 167–68
  polyploidy, 168–69
Chromosome number, speciation and, 621
Chromosome position, spatial gene expression and, 742
Chromosome 7, **149**
Chronocyte hypothesis, 680, **680**
Chrysanthemum chlorotic mottle viroid, **170**
Chyba, Christopher, 657
Ciccarelli, Francesca, 674
Cichlid fish, **627**, 627–29, **628**
Cilia, eukaryotic, 100–1
Ciliary photoreceptor, 98–99, **99**
Ciliates, 668
Ciona intestinalis (sea squirt), **589**, 743, **743**
Clade credibility, 133
Clades, 116
  common skate, 615–16, **616**
  influenza, 541
Cladistic analysis
  of African replacement *vs.* multiregional evolution models, 792, **792**
  of hominoid fossil evidence, 787–89, **788**
  of human/African ape relationship, 790–92
Cladogenesis, 728
Cladogram. *See also* Phylogenetic trees
  of African replacement *vs.* multiregional evolution models, **792**
  of *Homo sapiens* and recent ancestors, **788**
Clam (*Lasaea subviridis*), 492
Clausen, Jens, 334, 361–62
Clayton, Dale, 396–97
Clegg, Michael, 195, 300–2
Clift, William, 49
Climate after asteroid impact at K–Pg boundary, 715
Cline, 167, **167**
Clock, molecular, 138, 140, 263, **263**, 727–28
Cloning, **314**, 318
Clostridium acetobutylicum, **757**
Clostridium difficile, 563
Clostridium tetani, 549
Clutch size
  in birds, **513**, 513–15, **514**
  in parasitoid wasps, **515**, 515–17, **516**
CMAH (CMP-N-acetylneuraminic acid hydroxylase), 43
CMT1A repeats, 58, **59**
CNGB3 gene, 245
Cnidaria, 698, 744
Coalescence, 272–74, **273**
  applied, 274
  defined, 272–74
Coccyx, 42, **43**
Cocos finch (*Pinaroloxias inornata*), **81**
Coding sequences. *See* Molecular evolution
Codominant alleles, 345
Codon, 149, 160
Codon bias, 268–270, **269**
Coefficient of inbreeding (*F*), 278–80
  computing, 280
Coefficient of linkage disequilibrium (*D*), 295
Coefficient of variation (CV), 147
Coevolution, 322, **322**
  sexual, 523–25, **524**
Cognitive ability, heritability of, 347
Coho salmon (*Oncorhynchus kisutch*), 421, **421**
Coincidental evolution hypothesis of virulence, 548–49
Coin toss probabilities, 378, **378**
Cold, fever and, **566**, 566–67
Cole, Alexander, 18
Collard, Mark, 786, 788–89
Collared flycatcher (*Ficedula albicollis*)
  aging in, **496**
  clutch size in, 514, **514**
  reproductive trade-off in, 507
Collared lizard (*Crotaphytus collaris*), **252**, 252–54, **253**
Colon cancer, hereditary nonpolyposis, 502
Colonization, dispersal and, **617**, 617–19, **618**
Color polymorphism, 212–14, **213**
Colosimo, Pamela, 604
Colwell, Rita R., 683
Combat, male-male
  among stalk-eyed flies, 432
  intrasexual selection through, 417–21
Combination therapies for AIDS, 9–15
Comb jelly, 696, 763
Comets, 656–67
Co-mimicry, 749–50, **750**
Common ancestry, 55–62. *See also* Homology(ies)
  predictive tests of, **58**
  significance of, 61–62
Common cold, fever and, **566**, 566–67
Common toadflax (*Linaria vulgaris*), 154, **154**
Communal breeding, 457
Communal evolution, 678–79
Comparative analysis, of adaptation, 382–87
Competition
  asexual *vs.* sexual reproduction and, 315–16
  male-male, 417–23
    alternative male mating strategies, 420–21
    among stalk-eyed flies, 432
    combat, 417–21
    homicides, 445–47, **446**
    infanticide, 422–23, **423**
    sperm competition, 421–22, **422**
Complementary base pairs, 158, **158**
Complexity in nature, 98–99, **99**
Compound chromosomes, 209–12
Compsognathus, 52
Compulsory sterilization, 201–2, 214–16
Cone snail (*Conus striatus*), **598**, 598–600
Conflict, cooperation and, 477–83
  among non-kin, **481**, 481–83, **482**, **483**
  parent–offspring conflict, **477**, 477–80, **478**, **479**
  siblicide, 480, **480**
Cong, Bin, 75–76
Conotoxin genes, 599–600, **600**
Conrad, Donald, 170–71
Conservation, of critically endangered fish, 614–16
Conspecific nest parasitism, estimating heritability and, 84, 85
Constraints on adaptation, 389–97
  fuchsia color change, **391**, 391–93, **392**, *393*
  host shifts in feather lice, 396–97, **397**
  host shifts in *Ophraella*, **394**, 394–96
Control groups, 376
Conus sp., relationships among, 598–99, **599**
Conus striatus (snail), **598**, 598–600
Convergence, 746
Convergent evolution, **121**, 121–23
Cooper, Vaughn, 551
Cooperation. *See also* Altruism; Eusociality
  conflict and, 477–83
    among non-kin, **481**, 481–83, **482**, **483**
    parent–offspring conflict, **477**, 477–80, **478**, **479**
    siblicide, 480, **480**
  defined, 471
  multilevel selection and, 471–77
    in bacteria, 474–75, **475**

human morality and, 471, 476–77, **477**
in plants, 475–76, **476**
Cooperative breeding, 486
in birds, **466**, 466–67, **467**
Copepod (*Eurytemora affinis*), **528**, 528–29
Coprolites, 693
Coral (*Acropora digitifera*), 742–43
Corcorax melanorhamphos (white-winged choughs), 466, **466**
Cordyceps fungus, **549**
Coreceptor, HIV infection and, 5, 6
Coreceptor inhibitors, 13
Corn, artificial selection, 91, **91**
Cornwallis, Charlie, 466–67
Correlated characters, selection on, 354–56
Cortisol, 573, **573**
Courtship
displays by males, evolution of, 424–25, 437. *See also* Female choice
in hangingflies, 435–36, **436**
in spotted cucumber beetle, 431–32
in water mite, 433–35, **434**
Cousyn, Christophe, 388–89
Cow, **124**, **133**
Cox, Debbie, **737**
COX10 protein, 58
Crassostrea gigas (oyster), 492
Creationism, 97–104
argument from biochemical design, 99–103
history of controversy, 97–98
other arguments used by, 103–104
perfection and complexity in nature, 98–99, **99**
Creevey, Christopher, 674
Crenarchaeota, 667, 668
Cretaceous, 694, **695**
Cretaceous–Paleogene (K–Pg) extinction, 710, 712–15, 731
impact event, evidence for, 712–14, **713**, **714**
killing mechanisms, 714–15
Crick, Francis, 158, 685
Crickets (*Gryllodes sigillatus*), 439
Crocodiles/Crocodylia, 117, **117**
Crohn's disease, **303**, 303–5, **304**, 341–42, **342**
Cro-Magnon I, 787
Crossland, Michael, 457–58
Crotaphytus collaris (collared lizard), **252**, 252–54, **253**
Crotophaga major (greater ani), **456**, 456–57, **457**
Crouzon syndrome, 400, **401**
Crundall, Reginald, 183, 187
Cryptic species, 610
in marine phytoplankton, **613**, 613–14, **614**
Crystallins, 101–2, **102**, 760
Cubitus interruptus Dominant (*ciD*), 270
Cultural evolution, 569
Culture, 780
influence on behavior, 568
Curie, Marie, 64
Cuvier, Georges, 49, 55
C-value paradox, 583
CXCR4, 27, **27**
Cyanobacteria, 667
fossil, 677, **677**
Cyanocitta cristata (blue jays), 488
Cynodonts, 705
Cynomys gunnisoni (Gunnison's prairie dogs), 438–39, **439**
Cynomys ludovicianus (black-tailed prairie dogs), **462**, 462–63
Cyrtodiopsis whitei (stalk-eyed flies), 408, 431, **431**, 432
Cystatin C gene, 557, **557**
Cystic fibrosis, 163, 216–17, 222–24, **223**, **224**
Cystic fibrosis transmembrane conductance regulator (CFTR), 163, 223–24
Cytochrome *c* oxidase proton pump, 102
Cytokines, 6
Cytosine, 149, 159
in influenza genomes, **543**, 543–44

D4 dopamine receptor, 341
Dactylorhiza sambucina (elderflower orchid), 212–14, **213**
Daly, Martin, 445–46, 573–74
Damselflies, 422, **422**
Danio rerio (zebrafish), 431, **431**, **589**
Daphne Major, 81–82, **82**, 354
Geospiza fortis on, **81**, 81–90, **82**, **83**, **84**, **86**
Geospiza magnirostris on, **81**, **86**, 89–90
Daphnia magna (water flea), **387**, 387–89, **388**, **389**
Daphnia pulex (water flea), 152–53, **153**
Darrow, Clarence, 97
Darwin, Charles, 1, 38–39, 49, 62, 536, 652
on age of Earth, 66
on altruism, 459–60
altruism of social insects and, 483
on conversion from one function to another, 759
on evolutionary relationships, 93
on evolution of human morality, 471
on female choice, 424
on gradual evolutionary change, 756
gradual evolutionary change and, 720, 737
human evolution and, 769
inheritance and, 96
loss of features due to disuse, 755
on natural selection, 97
natural selection and, 73, 77
on organs of perfection, 98
on peacock's tail, 410
on perfection and complexity in nature, 99–100
phylogenies and, 110
prebiotic soup, 658
relationship between evolution and development, 735, 736, 765
reproductive capacity of elephant and, 87
on sexual dimorphism, 410
sexual dimorphism and, 408
on structural homology, 55–56, 57
on variation, 748
Darwin, Erasmus, 38
Darwinian fitness, 78, **78**, 93–94
Darwinian medicine, 537
Darwinian threshold, 678, 679
Darwinism. *See also* Evidence for evolution; Natural selection
altruism as central paradox of, 459–60
evolution of, 94–97
age of Earth and, 96
inheritance and, 95–96
Modern Synthesis, 96–97
time and, 96
variation and, 94–95
view of life in, 97
Dating
potassium–argon, 65, 66
radiometric, 64–66, *65*, **65**
uranium–lead, 66
Davidson, Eric, 737, 738, 760–61, 762
Davidson, Iain, 805
Dawkins, Richard, 468, 569
Dawson, Elisabeth, 305
Dawson, Peter, 202
Deamination, spontaneous, 159, **159**
DeBoer, Bart, 806–7, 809
DeBruine, Lisa, 465–66
Decay, radioactive, **65**
Deccan Traps (India), 715
deDuve, Christian, 660
Deep homology, 747
Deer, **124**, **133**
astragalus of, **125**
extinct, **49**
linear survivorship curve, **711**
Deer mice (*Peromyscus maniculatus*), 492
DEFL1 gene, **589**, 589–91, **590**
Deforestation, 718–19
Deformed (*Dfd*) gene, 740
Deleterious alleles, mutation–selection balance and, 219–22
Deleterious mutations
evolutionary theory of aging and, 500–3, **503**
Muller's ratchet and, **320**, 320–21
Deleterious recessive alleles
extinction and, 283–84
inbreeding depression and, 280–81
mutation–selection balance for, 220
DeLong, Edward, 670
Delph, Lynda, 392–93, 443
Δ32 allele, 16–17
frequency of, 162–63
Dembski, William, 103, 104
De Meester, Luc, 387–89
Dendritic cells, 6
Dendroica auduboni (Audubon's warbler), **635**, 635–36, **636**
Dendroica coronata, 635, **635**
Dendroica goldmani, 635, **635**
Dendroica nigrifrons, 635, **635**
Deng, Cheng, 596–98
Denisova Cave (Siberia), 798
Denisovans, 798–800, **799**, 807
Dennis, Megan, 779
Denver, Dee, 170, 173
Depression, 156
Derived characters, 114–15, **115**
The Descent of Man, and Selection in Relation to Sex (Darwin), 424, 769
Descent with modification, 38, **39**, 58, 115
Desert iguana (*Dipsosaurus dorsalis*), **564**, 564–65
Design, argument from, 98
Determinism, genetic, 570
Deuterostomes, 701, 729
Development. *See also* Evo-devo
Darwin on, 735, 736
modern synthesis and, 736–37
relation between genotype and phenotype and, 745, **745**
Developmental bias, 747
Developmental constraints, 746–47
Developmental trade-offs, 752–56
Devonian, 701
Dhondt, André, 346
Diabrotica undecimpunctata howardii (spotted cucumber beetle), 431–32, **432**
Dial, Ken, 704
Diamond, Jared, 808
Diaz, George, 307, 309
Dickinsonia, **696**
Dictyostelium, 668
Dictyostelium discoideum (slime mold), 468–69, **469**
Didelphis, **705**
Didelphis virginiana (Virginia opossum), **493**, 493–94
Diet, hunter-gatherer *vs.* modern American, 556–57, **557**
Digital organisms, evolution of, 101
Di Masso, R. J., 348, 350
Dimetrodon, **704**, 704–5
Dimorphism, sexual. *See* Sexual dimorphism
Dinosaur-bird transition, 702–4, **703**
Dinosaurs
birds and, 51–53, **53**, 702–4, **703**
duck-billed, 58, **58**
skull, **692**
vulnerability to extinction, 529, **529**
Dionaea muscipula (Venus fly trap), 92
Diphtheria, 48, 575
Diplomonads, 668
Diprotodon, **49**
Dipsosaurus dorsalis (desert iguana), **564**, 564–65
Dipturus batis (European common skate), 615, **615**
Dipturus cf. flossada (blue skate), **615**
Dipturus cf. intermedia (flapper skate), **615**
Dipturus nidarosiensis (Norwegian skate), **615**
Dipturus oxyrinchus (long-nosed skate), **615**
Directed panspermia hypothesis, 685
Direct fitness, 461
Directional selection, 357, **357**, 358, 359
sex and, 322
on synonymous mutations, 268–70, **269**
Diseases. *See also* Cancer
of civilization, 557, 558
epidemic, 310
germ theory of, 535–36
origins of human, 48
vectorborne *vs.* transmitted, 552, **552**
Disparity, 708
Dispersal, **617**, 617–19, **618**
Dispersal hypothesis, 139–40
Disruptive selection, **357**, 358–59
Distance matrix methods, 130–31

Divergence
  between African and non-African populations, 790–91, 793, 797, 800
  of humans and apes, times for, **777**, 777–78, **778**
  mechanisms of, 623–29
    environmental factors and mating preferences, 626–29
    natural selection, 623–25
    sexual selection, 625–26
Diversity
  among genomes, 582–85
    variation in genome architecture, 584–85, **585**
    variation in genome size, **583**, 583–84, **584**
  in gene expression and function, 740–44, **741**
  genetic, in humans, **800**, 800–1
  morphological, **708**, 708–9, **709**
  taxonomic, **707**, 707–8
Dixon, Andrew, 570
Dizygotic twins, 347
Djurab Desert, 781
Dmanisi, 790
DNA, 149
  bases in, 149, **149**, **158**, 159
  cellular life and, 663
  coding, 583
  genome size and coding of, 583, **583**
  mitochondrial, 773, **773**, 774, 794, **794**
  mobile genetic element, 583–84, **584**, 586–91
  noncoding, 583
  repair systems, 159
  replication of, 159, **159**
  single-base substitutions in, 159, **159**, 160–61
  structure and function of, 149, **149**, 158, **158**
  three viruses, three domains hypothesis and transition to, 680–83, **682**
  transcription of, 160, **160**
  variation and, 737
DNA-dependent RNA polymerases, 670, **682**, 682–83
DNA methylation, 586–89, **589**
DNA mismatch errors, 502
DNA polymerase, 158–59, 162
DNA polymerase I, 593
DNA sequences
  African replacement *vs.* multiregional evolution and, 793–94, **794**
  detecting natural selection on, 265–71, **266**, **269**
  divergence patterns, **261**, 261–63, **262**
  phylogenetic analysis using, 125–27, **126**, **127**, **129**, 134–35
Dobzhansky, Theodosius, 1, 94, 154, 389, 629
Dodd, Dianne, 46
Dogon women, 561, **561**
Dogs, 117, **117**
  artificial selection and, 40, **41**
  astragalus of, **125**
  canine transmissible venereal tumor, 137, **137**
  fgf4 retrogene in, **165**, 165–66
Dolphin (*Lagenorhynchus obscurus*), **127**
Domains, origin of, **678**, 678–83, **679**, **680**, **682**
Domestic animals and plants, 74–76
Domestic cats, **284**
Dominance genetic variation, 344–45, **345**
Dominant alleles, 345
  selection on, 202–6
Donoghue, Philip, 728
Doolittle, Russell, 680
Doolittle, W. Ford, 671–72
dos Reis, Mario, 543–44
Double helix, 149, **149**, **158**
Douc langur (*Pygathrix nemaeaus*), **164**, 164–65
Doushantuo Formation (China), 697, **697**
Doves (*Columbicola*), 396–97, **397**
Drift. *See* Genetic drift
Driscoll, Carlos, 284
Driven trend, 706, **706**
Drosophila, **496**, 636
  adaptive radiation in, 726
  age and increasing inbreeding depression on, 502–3, **503**
  antagonistic sexual selection in, 524, **524**
  heart development, 761
  reduction of linkage disequilibrium in, 300–2, **302**
  reinforcement of postmating gametic isolation in, 631, **631**
  variation in egg size and number among, 517
Drosophila hemipeza, **618**
Drosophila heteroneura, **618**
Drosophila melanogaster, 170, **170**, 195, **195**, 267–68, 286, 412, **496**, 694
  age and increasing inbreeding depression in, 502–3, **503**
  antagonistic pleiotropy in, 506, **506**
  antagonistic sexual selection in, 524, **524**
  codon bias in, **269**
  DNA per cell, 583
  hitchhiking in, 270
  homeotic mutations in, 738–39, **739**
  Hox genes in, **739**, 740, 742
  life span increases in, 497, **497**
  random fixation and loss of heterozygosity in, 249–52, **250**, **251**
  rapid adaptive evolution in, 219, **219**
  selection on heterozygotes and homozygotes, 206–14, **207**, **210**
Drosophila planitibia, **618**
Drosophila pseudoobscura, 46, **46**
Drosophila santomea, 631, **631**
Drosophila simulans, 267–68, 270
Drosophila subobscura, **167**, 167–68
Drosophila yakuba, 267–68, 631, **631**
Dryopithecus, 776
Dubuc, Timothy, 742–43
Dudash, Michele, 281–82
Duffy, Patrick, 197, 198
Duffy, Siobain, 45, 48
Dung beetles, **753**, 753–55, **754**
Duplication, gene. *See* Gene duplication
Durodon atrox, 124, **124**
Dyson, Miranda, 426

Ear, evolution of mammalian, 705, **705**
Earth
  age of, 38, 62–66, **63**, 646
    Darwinism and, 96
    geologic time scale and, **63**, 64
    radiometric dating, 64–66, *65*, **65**
  early atmosphere, 657–58
  history of large impacts on, **656**, 656–57
Earthquakes, asteroid impact at K–Pg boundary and, 715
East, Edward M., 215, 331–33
Eastern Caroline Islands, 245, **245**
Eastern red cedars, 252–53
Eaton, S. Boyd, 556–57
Echinoderms, 729, 743–44, 746
Echinops telfairi, **745**
Echiurans, 764, **764**
Eciton sp., **484**
Ecological controls of diversification, 637–40
Ecological mortality, experiment on, 507–10, **509**
Ecology and life-history hypothesis, 484–86
Ectodysplasin, 44
Ectodysplasin gene, 604, 755, 758
Ectotherms, 380–81
Eda gene, 604, **604**
Ediacaran biota, **696**, 696–97, **697**
Edwards, Scott, 134
Edwards v. Aquillard, 98
Eel sperm flagella, **100**
Effective population size, 251
  of Florida panthers, 284
  mutation rates and, 591–92, **592**
  probability of fixation for mutation and, 263–65
Effector helper T cells, 6, 7
Effector T cells, 27
Egg size, phenotypic plasticity in, 520–22, **521**
EHH. *See* Extended haplotype homozygosity (EHH)
84GG, 307–8, **308**
Eisner, Thomas, 448–49
Ejaculation, prior, 420
Elderflower orchid (*Dactylorhiza sambucina*), 212–14, **213**
Eldredge, Niles, 720, 722
Electrophoresis, allozyme, 163
Elena, Santiago, 359–60
Elephant, reproductive capacity of, 87
Elephantiformes, **58**
Elephant seals, 449, **449**
Elephant shrews, 760, **760**
Elgar, Mark, 517
Elowitz, Michael, 153
Emberiza shoeniclus (reed buntings), 568, 570, **570**, 571
Embryos
  Cambrian and Precambian, 697, **697**
  fossil, 697, **697**
  homologies among, **54**
Emlen, Doug, 753–54
Emlen, Steve, 479
Empire Mine (Michigan), 676
Encephalitis, 576, **576**
Endangered species, 718
Endangered Species Act, U.S., 611, 614
End-Cretaceous extinction, 729–30
Endogenous controls of diversification, 638–40
Energy
  body size in marine iguanas and, 418
  mammal expenditure of, 497, **497**
  sources of, in hunter-gatherer diet, 556–57, **557**
  trade-offs between time and, 492–93
  trade-offs in energy allocation, 493–95, **494**
Engineering test of population genetics theory, 224–27, **225**, **226**
England, Barry, 572–73
Enhydra lutris (California sea otters), 276–78, **277**, *278*
Ensatina eschscholtzii (salamander), 48
Entophysalis, **677**
Environment
  changing-environment theories for sex, 322–24
  influence on quantitative traits, 331, 333, **334**
  mating preferences and, 626–29
  medical/health adaptation and, 556–68
  origins of life and, 661–63
  phenotypic plasticity and, 387–89
  prebiotic, 657
Environmental heterogeneity, diversification and, 637–40
Environmental stress, inbreeding effects and, 281
Environmental variation, 148, 150, 152–54, **153**, 344, 361
Enzyme reaction norms, **164**
Enzyme replacement therapy, 554
Eoentophysalis, **677**
Epidemic diseases, 310
Epigenetic inheritance, 154
Epigenetic marks, 154
Epitopes, 26
Eppeson v. Arkansas, 97
ε crystallins, 102
Equilibrium. *See also* Hardy–Weinberg equilibrium principle; Linkage disequilibrium
  heterozygote fitness and, 206–8
  linkage, 305, **305**, 318
  punctuated, **720**, 720–24
Ergothioneine transporter, **303**, 303–5, **304**
Erinaceus europaeus (European hedgehog), **745**
Erus, J., 782
Erwin, Doug, 722
Escape from adaptive conflict, 596
  evolution of antifreeze proteins, 596–98, **597**
Escherichia coli, 667
  bacteriocins and, 459, **459**
  bacteriophage f1 in, 550–51, **551**
  base substitution in, **170**
  codon bias in, **269**
  environmental variation in, 153
  evolution of mutation rates in, 593, **593**
  fitness effects of mutations in, 171
  genetic variation in fitness in, 359–60
  MG1655 strain, 674
  multilevel selection and cooperation in, **474**, 474–75, **475**
  streptomycin-resistant, 547, **547**
Eschrichtius robustus (gray whale), 274
Eschscholzia californica (California poppy), 492
Escobar, Juan, 503
Esser, Christian, 679
Estes, Suzanne, 173–74
Ethane, 645
Eublepharis macularius (leopard geckos), **155**, 155–56
Eucalyptus albens, **254**
Eucarya, 668, 669, **669**, 672, **672**
  fossils, **676**, 676–77

hypotheses of origins of
chronocyte hypothesis, 680, **680**
ring-of-life hypothesis, **679**, 679–80
three viruses, three domains hypothesis, 680–83, **682**
universal gene-exchange pool hypothesis, **678**, 678–79, **679**
phylogeny of all living things and, **667**, 668
Eugenic sterilization, 214–16
Euglena, 668
Euhadra 623, **623**
Eukaryotes
genome architectures for, **585**
lateral gene transfer in, 674, **674**, **675**
Eukaryotic cilium (flagellum), **100**, 100–1
Eulampis jugularis (purple-throated carib), 409, **409**
Eulipotyphlya, 730
Euphorbia obesa, **745**
Eurasian lynx, 112
Europe
CCR5-Δ32 allele in, 16, **16**
HIV/AIDS epidemic in, **2**, 3
European Americans, heritability of IQ and, **362**, 362–63
European common skate (*Dipturus batis*), 615, **615**
European hedgehog (*Erinaceus europaeus*), **745**
Eurosta solidaginis, 358, **358**
Euryarchaeota, 667, 668
Eurytemora affinis (copepod), **528**, 528–29
Eusocial, 483
Eusociality, 483–86
characteristics of, 483
ecology and life-history hypothesis, 484–86
haplodiploidy hypothesis, 484
monogamy hypothesis, 484
phylogenies to analyze, 484–85, **485**
in vespid wasps, 485–86
Eusthenopteron, 702
Evidence for evolution
age of Earth
geologic time scale, **63**, 64
radiometric dating, 64–66, *65*, **65**
change through time, 39–44
evidence from fossil record, 49–55
evidence from living species, 41–44
evidence from selective breeding, 39–40
common ancestry, 55–62
evolutionary trees. *See* Evolutionary trees
homology, 55–61
interspecies relationship, 61–62
ring species, 47–48
descent with modification, 38, **39**
Evo-devo, 735–68
beginnings of, 737–38
deep homology, 747
developmental trade-offs, **753**, 753–56, **754**
equal variation, 748–49, **749**
evolutionary leaps, 756–59, **757**
five percent of a wing problem, 759–63, **760**, **762**
future of, 764–65
homology, 744–46, 763–64
homoplasy, 744–46, **745**, 763–64, 765
Hox genes, 738–44, 765
canonical spatial expression pattern, **741**, 742–44
diversity in *Hox* gene expression and function, 740–42, **741**
Hox paradox, 740–44
parallel evolution, 746–47, **747**
biased evolutionary trajectories and, 749–52, **750**, **751**, **752**
pleiotropy, 752–56, **755**
Evolution, 37–72. *See also* Fossil record; Genome evolution; Molecular evolution
AZT resistance in HIV population and, **12**, 12–13
belief in theory of, 37, **38**
brief history of ideas on, 38
common ancestry, evidence of, 55–62
communal, 678–79
convergent, **121**, 121–23
cultural, 569
development and. *See* Evo-devo
drug design and, 13–14
duplicated genes and, 166
environmental variation and, 153–54
epigenetic inheritance and, 154
eugenic sterilization and, 214–16, **215**
evidence of speciation, 44–498
genetic drift and, 240
genetic variation and, 152
genotype-by-environment interaction and, **156**, 156–57, **157**
human. *See* Human evolution
of life, overview, **664**
macroevolution, 38, 49–55
mechanisms of, **190**. *See also* Mendelian genetics
microevolution, 38, 39–44
model of, by female choice, 427–29, **428**, **429**
in modern populations under selection on recessive and dominant alleles, **205**
mutations and, 173–74
by natural selection. *See* Natural selection
parallel, 746–47, **747**
parenthood and, 568–72
reasons to study, 1
Evolutionarily conserved, 588
Evolutionarily significant unit (ESU), 614
Evolutionary biology, evolution of, 94–97
Evolutionary engineering, 224–27, **225**, **226**
Evolutionary impact of mobile genetic elements, 586–91
Evolutionary leaps, 756–59, **757**, **758**
Evolutionary relationships, on evolutionary tree, 111–12, **112**
Evolutionary theory of aging, 500–10
deleterious mutations and, 500–3, **503**
natural experiments of, 507–10, **509**
trade-offs and, 504–7, **505**, **506**, **507**
Evolutionary trees, 109–46. *See also* Phylogeny(ies)
calculating likelihood of, 129, **129**
fish, **142**
HIV, **19**, 19–20, **20**, **21**
as hypotheses, 113–14, **114**
logic of inferring, 114–23, **115**, **116**, **117**, **118**, **119**, **120**, **121**, **123**
mammals, **141**
missing information on, 112
Old World primates, **142**
reading relationships on, **111**, 111–12, **112**
reading time on, **110**, 110–11
styles of, 112–13, **113**
Evolutionary trends, 706, **706**
Ewald, Paul, 551–52
Exaptation, 92, 759, 761
Exon, 59, 161, 584
Experimental design, 376
Expression, 152–53
Extant vertebrates, 698
Extended haplotype homozygosity (EHH), 312
Extinction, 49, 709–19
background, **711**, 711–12
Cretaceous–Paleogene (K–Pg), 710, 712–15, 731
impact event, evidence for, 712–14, **713**, **714**
killing mechanisms, 714–15
current, 717–18
rain forest, 718–19
genetic load and, 283
life-history traits and vulnerability to, 529, **529**
mass, 691–692, 694, **695**, 709–19, 731
Permian–Triassic (P–T), 715–17
during Phanerozoic, **712**
rates of, 710
Extinction vortex hypothesis, 283–84
Eye(s)
chordate, 98–99, **99**
cystallins of lenses of, 101–2, **102**
Eyespot color, 748–49, **749**
Eyre-Walker, Adam, 268

Facial resemblance, trust and, 465, **465**
Fairy shrimp (*Artemia salina*), **723**
Family lineages, inbreeding depression varying among, 280–81
Farnsworth, Elizabeth, 746
Farrant, Jill, 746
fas gene, 75–76
Fay, Justin, 271–72
Feather lice
color in, **386**, 386–87
host shifts in, 396–97, **397**
Feathers, evolution of, **703**, 703–4
Fecal microbiota transplantation, 563, **563**
Fedorov, Alexei, 680
Feeblemindedness, 215
Fehr, Ernst, 482
Feldman, Chris, 602–3
Felsenstein, Joseph, 103–4, 120, 132, 220, 383, 386
Female choice, 423–37
acquisition of resources and, 435–36
barn swallows and, 478
in gray tree frogs, 425–27, **426**, 437, **437**
offspring gene quality and, 436–37
preexisting sensory biases and, 433–35
in red-collared widowbirds, 424–25, **425**
in stalk-eyed flies, 431, **431**
Female promiscuity, evolutionary transitions in birds and, 467, **467**
Females, sexual selection and, 411–12, 439–41
Fenster, Charles, 172
Ferris, James, 661
Fever, 564–67
behavioral, **564**, 564–66
common cold and, **566**, 566–67
medical practice and, 567
fgf4 gene, **165**, 165–166
Ficedula albicollis (collared flycatchers), **496**, 507
clutch size in, 514, **514**
Fiddler crabs, 435, **435**
Field mustard (*Brassica rapa*), **41**, 41–42, **42**
Finches, Galápagos (Darwin's finches), **81**, 81–90, **82**, **83**, **84**, **86**, **87**, **88**, **89**
Fingerprints, lack of, 161, **161**
Fin/limb transition, 702, **702**
Fischer, Kathleen, 496
Fisher, Ronald A., 171, 215, 336, 378, 427, 429, 605, 736
Fish(es). *See also individual species*
conservation of critically endangered, 614–16
interaction between habitat and mating phenotypes in, **627**, 627–29, **628**
jawed, 99, **99**
offspring size in, 517, **517**
phylogeny, **142**
selection on offspring size in, 518–20, **519**
Fish–tetrapod transition, **701**, 701–2, **702**
Fitch, Walter, 537–39
Fitness. *See also* Reproductive success
aging and, 495
antagonistic sexual selection and, 524, **524**
asexual *vs.* sexual reproduction and, 315–16
asymmetric limits on, behavioral consequences of, 416–17
Darwinian, 78, **78**, 93–94
deleterious mutations and, 319–20
direct, 461
female choice and, 437, **437**
as function of group composition, **473**
of heterozygotes, 206–8
of homozygotes, 208–12
of hybrid populations, 636
hybrid zone and, 636–37
inclusive, 460–61
indirect, 461
maintenance of genetic variation for, 359–60
maternal
clutch size of parasitoid wasps and, **515**, 515–17, **516**
egg size and, 521, **521**
mean, of population, 205, 209
Mimulus ecotypes, **624**
multiple traits and, 354–56, **355**, **356**
mutation and, 171–73, **173**
of mutator and ancestor strains, **593**, 593–94
natural selection and, 78, 93–94
parent–offspring conflict over interests of, 478
phenotype and, 387–89
of RNA molecules, 651
selection modes and, 359–60
selfing and, 319
social behavior and, 456, **456**, 457

transposition and, 586
Five-kingdom model, 666, **666**
Fixation, random, 246–55
experimental study on, 249–52
loss of heterozygosity and, 248–49
in natural populations, 252–55
probability of allele to drift toward, 248
Fixation index, 249
Flagella (eukaryotic cilium), **100**, 100–1
Flagellates, 668
Flapper skate (*Dipturus* cf. *intermedia*), **615**
Flatfish, 93
Fleming, Alexander, 536
Flinn, Mark, 571–72, 572–73
Florida panther (*Puma concolor coryi*), 233, **233**, 283–84, **284**
conservation genetics of, 283–84, **284**
Flour beetle (*Tribolium castaneum*), **202**, 202–3, **203**, 257–58, **258**
Flower
color, 391–93, **392**, **393**, **632**, 632–33, **633**
orchid, **56**
size in alpine skypilots, **351**, 351–53, **352**, **353**
Flower, William H., 124
Flu virus. *See* Influenza A virus
Flying foxes (*Pteropus poliocephalus*), **382**, 382–86, **385**
fms-like tyrosine kinase 1, 197
Foote, M., 712
Foote, Michael, 709
Forelimbs
fin-limb transition, 702, **702**
vertebrate, 56, **56**
Formaldehyde, 658–59
Forterre, Patrick, 681–82, 683
Fortress nests, 485
Fossil and molecular divergence timing, 727–30
Fossilization processes, **692**, 692–93, **693**, 730
Fossil record, 49
ancestral cells in, 664–65, **665**
evolution above species level and, 724–27
evolution in, 696–706
Burgess Shale fauna, **698**, 698–701, **699**, **700**
dinosaur-bird transition, **703**, 702–4
Ediacaran biota, **696**, 696–97, **697**
evolutionary trends, 706
fish-tetrapod transition, **701**, 701–2, **702**
origin of mammals, **704**, 704–5, **705**
fossil and molecular divergence timing, 727–30, **728**, **729**
of life's origins, 664–65, **665**, 675–78, **676**, **677**
macroevolution, 719–24
mass and background extinctions, 709–19, 731
morphological diversity, **708**, 708–9, **709**
nature of, 692–95
rapid diversification and, 691–92
root for tree of life and, 675–78, **676**, **677**
taxonomic diversity, **707**, 707–8
Fossils
body, 693
defined, 49
hominin, 780–87
interpreting, 787–89
"living," **722**, 722–23, **723**
trace, 693, **693**, 697, **697**
transitional, 49–54, **50**, **51**, **53**, **54**
Foster, G. G., 208–12
Founder effect, 243–45, **244**, **245**, 796
genetic disease among Ashkenazi Jews and, 308
Fox, Charles, 520–22
Fox, Sidney, 664
Franks, Steven, 41–42
Frary, Anne, 75
Fraser, Christophe, 25
Frayer, David, 792–93
Free-recombination model, 338
Freezing, 692, **692**
Frequency, 162–63
Frequency-dependent selection, 212–14, 360
Freshwater snail (*Potamopyrgus antipodarum*), 323–24
Fretwell, Stephen, 517, 519, 520
Friedberg, Felix, 60
Frogs, 746
African clawed, 122, **747**
gray tree, 425–27, **426**, 437, **437**
Guenther's marsupial, 122
Litoria rothii, 458
magnificent tree, 122
tiger-striped leaf, **609**
túngara, 407, **407**
Frondlets, **696**
Fruit bats, **382**, 382–86, **385**
Fruit flies. *See Drosophila;* Tephritid fly (*Zonosemata*)
F_{ST}, 285–86
Functional constraints, 263
Fungi, 668
Fuchsia excorticata, **391**, 391–93, **392**, *393*
Fusion inhibitors, 13
Futuyma, Douglas, 394–96
fw2.2 gene, 75

Gabriel, Wilfried, 283
Gächter, Simon, 482
Gage, Matthew, 422
gag gene, 30
Gagneux, Pascal, 800
Galápagos Archipelago, **82**
Galápagos finches
adaptive radiation in, 726
evolution of beak shape in, **81**, 81–90, **82**, **83**, **84**, **86**, **87**, **88**, **89**, 354–56, **355**
Galápagos Islands, marine iguanas on, **417**, 417–21, **418**, **419**, **420**
Galen, Candace, 351–53
Galls, 358, **358**
Galton, Peter, 52
Galvani, Alison, 310
Gambusia affinis (mosquitofish), 92, **92**
Gametes, parental investment in, 411, **411**
Gametic isolation, reinforcement of postmating, 631
Gammie, Stephen, 455
Garland, Ted, 39
Garter snake (*Thamnophis elegans*), 356, **356**
adaptation from new mutation in, **602**, 602–3, **603**
nighttime behavior in, **380**, 380–81, *381*, **381**
Gasterosteus aculeatus (threespine stickleback), **43**, 43–44, **44**, 46–47, **47**, 122, **605**, 605–6, 755, **755**
Gasterosteus doryssus (sticklebacks), 723–24
Gastrin gene, 122
Gastropods, **724**, 724–25
Gastrotheca guentheri (Guenther's marsupial frog), 122
GATA, 761
Gatesy, John, 126–27
Gaucher disease, type 1, 307
GBA. *See* Acid β-glucosidase (GBA)
GBA-84GG mutation, 307–9, **308**
G+C content, of influenza genomes, **543**, 543–44
Ge, Fen, 674
Geckos, 403
Geffeney, Shana, 602
Geisler, Jonathan, 125
Gemini gene, 758–59
Gene, 149. *See also* Transposable elements (transposons); *specific genes*
co-option, 101–2, **102**
female choice and quality of, 436–37
homeotic. *See* Homeotic genes
new, 161–66
orthologous, 165, 740
paralogous, 165, 740
pseudogene, 164
reconstructing history of, **307**, 307–11, **308**
regulation of, 759–63
selfish, 569
Gene conservation, **595**
Gene co-option, 101–2, **102**
Gene duplication, 594–98
new genes from, 161–66
possible outcomes of, **595**, 595–96
preservation of duplicate genes, **595**, 595–98, **597**, **598**
retroposition, 164, **164**, **165**, 165–66
unequal crossing over, 161, **164**, 164–65
Gene families, 598–601, **599**, **600**
Gene flow, **234**, 234–35
in bacteria and Archaea, 612–13
physical isolation as barrier to, **617**, 617–20, **618**, **619**, **620**
polyploidy reducing, 621
from premodern to modern humans, 799, **799**
Gene genealogy, 272–74, **273**
Gene network, 752–53
Gene pool, 181–83, **182**, **185**
Generation time, population size *vs.*, 263–64, **264**
Gene regulation, 737–38
Gene therapy, 554
Genetic code, 61, **61**
Genetic determinism, 570
Genetic differences, among humans, chimpanzees and gorillas, **778**, 778–79, **779**
Genetic distance, 130
between human populations due to geographic distance, **285**, 285–86
Genetic diversity
among humans, **800**, 800–1
heterozygote inferiority and, 212
heterozygote superiority and, 207
population size and, **253**, 254
Genetic drift, 138, 183, **190**, 211, 237
adaptive significance of sex and, 319–21
CCR5-Δ32 gene and, 18, 310–11
cumulative effect of, 245–49, **246**
Florida panther population and, 283
Hardy–Weinberg equilibrium assumption of no, 190
linkage disequilibrium and, 299, **299**, 319–21
in Mendelian genetics of populations, 240–59
natural selection *vs.*, 257–59, **258**, **259**
population size and, 243, **243**
random fixation of alleles, 240–42, **241**
rate of evolution by, **255**, 255–57
sampling error and founder effect, 243–245, **244**, **245**
model of, 240–42, **241**
in molecular evolution, 260–74
DNA sequence divergence, **261**, 261–63, **262**
nearly neutral model of, **263**, 263–65, **264**
neutral theory of, 260–61, 265–71, **266**, **269**, **271**
mutation rates and, 591–92
in natural populations, **252**, 252–55, **253**, **254**
population size and, 243, **243**
probability of allele to drift, 258
rate of evolution and, **255**, 255–57
sex and, 319–21
Genetic hitchhiking, 303
Genetic isolation, 616–23
Genetic linkage. *See* Linkage disequilibrium; Linkage equilibrium
Genetic load, 283, 320–21
Genetic mapping, 338–39
Genetic polymorphism, **254**, 254–55
Genetic predictions, of African replacement *vs.* multiregional evolution, 793, *793*
Genetic recombination
adaptive significance of sex and, 318
artificial selection and, 91
linkage disequilibrium reduced by, 300, 301
randomness of, 93
sex in population and, 300
Genetics
modern synthesis and, 96–97
population, 179
Genetic variation, 148, 149, 150–51, 344, 361
adaptation from standing, 602, **604**, 604–5
additive *vs.* dominance, 344–45, **345**
evolution and, 152
female choice and, 428, **428**
gene duplications and, 164, 166, 169
genome duplication events and, 168–69
HIV and, 15–18, 152
maintenance of, 356–60, **357**, **359**, 525–27
mutation and, 219
in natural populations
extent of, 161–66
measuring, 162–63
from polyploidy, 168–69
reasons for, 166
selection modes, 356–60, **357**, **359**
tests for, *396*

Gene trees, 272–74, **273**, 599–600, **600**, **607**, 795, **795**
species trees *vs.*, **774**
Genome architecture, 582
variation in, 584–85, **585**
Genome duplication, **168**, 168–169, **169**
Genome evolution, 581–608
diversity among genomes, 582–85
variation in genome architecture, 584–85, **585**
variation in genome size, **583**, 583–84, **584**
evolution of mutation rates, **591**, 591–94, **592**, **593**, **594**
gene duplication, 594–98
preservation of duplicate genes, **595**, 595–98, **597**, **598**
gene families, 598–601, **599**, **600**
mobile genetic elements, 583–84, **584**, 586–91
molecular basis of adaptation, 601–6
adaptive mutations of small *vs.* large effect, 605–6
new mutation *vs.* standing genetic variation, 602–5
Genomes, 149
diversity among, 582–85
variation in genome architecture, 584–85, **585**
variation in genome size, **583**, 583–84, **584**
Genome size, **581**
mutation rates and, 591, **591**
Genomic imprinting, 522–23
Genotype, 150
determining, 162–63
development and, 745, **745**
human, 801
linkage disequilibrium and multilocus, 298, **298**
RNA, 648
Genotype-by-environment interaction, 148, 150, **155**, 155–56, **156**, **157**, 388
Genotype frequencies
chance events and, 240–42, **241**
general case, **187**, 187–89
inbreeding and, **275**, 275–76, *276*, 279, 283
linkage disequilibrium and, 293–95, **294**
migration and, 234–35, **235**
mutation and, **217**, 217–18
numerical calculation, **183**, 183–87, **184**
selection and changes in, 196–200, **197**, **200**
selfing and changes in, **275**, 275–76, *276*
in simulated life cycle, 181–83, **183**
statistical analysis of using chi-square test, 198–99
Genovart, Meritxell, 78
Genovesa, 419
Gentamycin, 563
Gentiana pneumonanthe, **254**
Geographic distance, genetic differentiation between populations and, **285**, 285–86
Geographic isolation, **618**, 618–20, **619**, **620**
through dispersal and colonization, 617, **617**, **618**, 618–19
through vicariance, 617, **617**, **619**, 619–20, **620**
Geographic range, species survival and, **724**, 724–25
Geographic structure, among humans, 801, **801**
Geologic time scale, **63**, 64, 694
Geometric relatedness, 470, **470**
Geospiza conirostris (cactus ground finch), **81**, **86**, 87
Geospiza difficilis (sharp-beaked ground finch), **81**, **86**
Geospiza fortis (ground finch), **81**, 81–90, **82**, **83**, **84**, **86**
Geospiza fuliginosa (ground finch), **81**, **86**
Geospiza magnirostris (ground finch), **81**, **86**, 89–90
Geospiza scandens (cactus ground finch), **81**, **86**
Gerhardt, H. Carl, 425–27
Germ theory of disease, 535–36
Giant palm salamander (*Bolitoglossa dofleini*), **738**
Giant scale insects, **408**
Giant waterbugs, 441
Giardia lamblia, 668
Gibbons (*Hylobates*), 770
Gibbs, Lisle, 438
Gigord, Luc, 212
Gilchrist, George, 168
Giles, Barbara, 235–37
Gillespie, John, 257
Gillooly, James, 411, 413
Ginkgo, 722, **722**
Gish, Duane, 103
Glades, 252–54
Gloeocapsa, **677**
Gloeodiniopsis, **677**
Glor, Richard, 638
Glucocerebrosidase, 307
Glucose-6-phosphate dehydrogenase deficiency, **311**, 311–12
Gluphisia septentrionis, 448–49
Glyptophonts, **49**
Goblet cells, 149
Goddard, Henry H., 215–16
Godfrey-Smith, Peter, 472, 473
Goldenrod (*Solidago altissima*), 358
Goldschmidt, Richard, 756
Goldstein, D. B., 796
Gomes, Nuno, 498–99
Gona tools, 803, 804
Gondwana, 139, **140**
Gonopodia, 92, **92**
Goodman, Morris, 771
Goodnight, Charles, 475–76
Goriely, Anne, 400–1
Gorilla (*Gorilla gorilla*), 770
genetic differences between humans, chimpanzees and, **778**, 778–79
relationship between humans and, 772–79
divergence times, **777**, 777–78
molecular evidence, 772–76
morphological evidence, 776–77
Gorman, Owen, 540
Gorrell, Jamieson, 463
Goudet, Jérôme, 235–37
Gould, Stephen Jay, 720, 722, 738, 759
G6PD-202A allele, 312–13, **313**
G6PD locus, **311**, 311–12
Grabowska-Zhang, Ada, 481
Grace, Jaime, 432
Gracile australopithecines, 782, 784
Gradualism, **720**, 720–22
Grafen's geometric relatedness, 470, **470**
Graham, Neil, 566–67
Grandmother hypothesis, menopause and, 511–12, **512**
Gran Dolina (Spain), 792
Grant, Peter, 81–82, 83, 86, 89–90, 103, 354–56
Grant, Rosemary, 81–82, 83, 86–87, 89–90, 103, 354–56
Graphite, 662
Gray, Asa, 410
Gray, Elizabeth, 438
Gray tree frogs (*Hyla versicolor*), 425–27, **426**, 437, **437**
Gray whale (*Eschrichtius robustus*), 274
Gray wolves, 76
Greater ani (*Crotophaga major*), **456**, 456–57, **457**
Greater kudu, **408**
Greater painted snipe, 438
Great frigate birds, **408**
Great Slave Lake (Canada), **723**
Great tit (*Parus major*), 282, **282**, 481, **513**, 513–15
Green, Richard, 798–99
Greenbeard alleles, 468–69
Greene, Erick, 374–78
Green fluorescent protein, 153
Greenwood, Paul, 282
Griebeler, Eva Maria, 529
Grine, Frederick, 789
Grizzly bears, 384–85, **385**
Grosberg, Richard, 526–27
Ground finches of Daphne Major, **81**, 81–90, **82**, **83**, **84**, **86**, **87**, **88**, **89**
Group selection, 473
Gryllodes sigillatus, 439
Gryllus firmus (sand crickets), **494**, 494–95
Grypania spiralis, 676
Guanine, 149
in influenza genomes, **543**, 543–44
Guenther's marsupial frog (*Gastrotheca guentheri*), 122
Gunnison's prairie dogs (*Cynomys gunnisoni*), 438–39, **439**
Gupta, Phalguni, 18
Gustafsson, Lars, 507, 514
Gut microbes, 563, **563**
Gypsy moth, 551

H3 (Hemagglutinin), 541–42
Haag-Liautard, Cathy, 170
HAART (Highly Active Anti-Retroviral Therapy), 14
Habitats
adaptation to different, 623–25
extinctions and loss of, 718
Habrotrocha elusa, 291
Hadar (Ethiopia) fossil, 803
Hadrosaurs, **58**, **529**
Hadza hunter-gatherers, 447, 572
cooperativity among, 476–77, **477**
grandmother hypothesis and, **511**, 511–12, **512**
Hagfish, 99, **99**, 698
Hahn, Beatrice, 21
Haig, David, 523
Haikouichthys ercaicunensis, **699**
Haldane, J. B. S., 488, 658, 736
Hall, David, 172
Haloarchaea, 667
Haloferax volcanii, **170**
Hamilton, William, 460, 469, 480
Hamilton's rule, **460**, 460–61, 464, 468, 469
Han, Tsu-Ming, 676
Hancock, Angela, 801
Hangingflies (*Bittacus apicalis*), 435–36, **436**
Hansen, Dennis, 369
Haplodiploidy, 484
Haplotype, 293
Hardy, Godfrey, H., 187
Hardy–Weinberg equilibrium, 191
Hardy–Weinberg equilibrium principle, 180–91
analysis using
adding inbreeding to, 275–79
adding migration to, **234**, 234–35, **235**
adding mutation to, **217**, 217–19, **218**
adding selection to, 192–93, **193**, 196–97, **197**
two-locus version of. *See* Two-locus version of Hardy–Weinbeg analysis
empirical research using, 197–200
general case, **187**, 187–189, **188**
with more than two alleles, 189
mutation as evolutionary mechanism and, **217**, 217–19, **218**, **219**
null model, 190–91
numerical calculation, **183**, 183–87, **184**, **185**, **187**
rate of evolution and, 205–6
selection
on heterozygotes and homozygotes, 206–14, **207**, **210**, **213**
on recessive and dominant alleles, **202**, 202–6, **203**
simulation, 181–83, **182**, **183**
use of, 189–91
Harrigan, Richard, 14
Harris, Simon, 671
Hartman, Hyman, 680
Hawaiian cricket, 432, **432**
Hawaiian cricket (*Laupala*), 619, **625**, 625–26, **626**
Hawaiian drosophilids, 618, **618**
Hawaiian islands, **618**, 618–19
Hawkes, Kristen, 511–12
Hawks, John, 781
Hawlena, Hadas, 459
Hay, Bruce, 224
Hayward, April, 411, 413
Hazard ratio, 14–15
Head lice (*Pediculus humanus capitis*), 138–39, 387
Health. *See* Medical applications
Hearnden, Mark, 458
Heart development, 761–63, **762**
Heath, Daniel, 518–20
Heckler, Margaret, 3
Hedgehogs, 744, **745**
Hedges, S. Blair, 728, 777
Hedrick, Philip, 284
Heers, Ashley, 704
Heidinger, Britt, 498
Height, variation in human, 147, **147**, 343, **344**, **444**, 447, **447**
Hein, Jason, 660
Helianthus, 634, 636
Heliconema, **677**
Heliconia bihai, 409
Heliconia caribaea, 409
Heliconius, 634
Heliconius cydno, **750**

Heliconius erato, 749–52, **750**, **751**
Heliconius melpomene, 749–52, **750**, **751**
Heliconius melpomene aglaope, **752**
Heliconius melpomene amaryllis, **752**
Heliconius sapho, **750**
Helper T cells, 29
Hemagglutinin, **538**, 538–39, 541–42, **542**
Hemoglobin, 148, 149, **149**
Hendriksen, Rene, 536
Hendry, Andrew, 41, 46, 147
Heng, Benjamin, 560
Hennig, Willi, 116
Henrich, Joseph, 482
Henry, Lee, 321
Hereditary nonpolyposis colon cancer, 502
Heritability, 83–84, **84**, 343–44
  broad-sense, 344
  defined, 343
  estimates of, 84–85
    from parents and offspring, **344**, 344–46, **346**
    from twins, 346–47, **347**
  of flower size in alpine skypilots, **351**, 351–53, **352**, **353**
  of IQ, **362**, 362–64, **363**
  misinterpretations of, 360–64, **361**
  narrow-sense, 344, 346
  predicting evolutionary response to selection and, 350–56, **351**
  of prize winnings in thoroughbred racehorses, 346, 353, **353**
  reasons to measure, 364, **364**
  of traits, **526**
  of variation, 79, 83–86, 91, **91**
Heritable traits, 12
Heritable variation, measuring, 343–47
Hermaphrodites, outcrossing, 321
Hermaphrodite, reproductive success in, 412–13
Hermit crabs, 723
Herpes simplex, **170**
Herrnstein, Richard J., 362
Heterocephalus glaber, **484**
Heterochrony, 738
Heterozygosity
  defined, 249
  inbreeding and, **275**, 275–80
  loss of, 246–55
    experimental study on, 249–52, **250**, **251**
    in natural populations, **252**, 252–55, **253**, **254**
    population size and, 251
  mean, 163
  in populations, 279
    random fixation and loss of, 246–55
Heterozygote inferiority (underdominance), 208–9, 211–12
Heterozygotes
  fitness of, 206–8
  selection on, 206–8, **207**
Heterozygote superiority, 207, 208–9
  decline in linkage disequilibrium and, 302
Heuristic searches, 131
Hexaploid, 168, 169
Hiccups, 55
Hiesey, William, 334, 361–62
Highly active antiretroviral therapy (HAART), 14
Hill, Emmeline, 342
Hill, Kim, 476, 512, 572
Hillis, David, 19, 20, 135
Hines, Heather, 751
Hippopotamuses
  eyes and ears of, 121, **121**
  whales and, **124**, **125**, 127, **127**, **128**, 130, 132, **132**, **133**, 134, 137
Hirschhorn, Rochelle, 553–54
Hirundo rustica (barn swallows), 478
Hitchhiking (selective sweep), 270–71, 272, 303
HIV. *See* Human immunodeficiency virus (HIV)
HLA. *See* Human leucocyte antigen (HLA)
H3N2, 541
H1N1 influenza virus, 542–43, 544
H5N1 influenza virus, 576
Hoggard, Patrick, 10
Holland, B., 525
Hollyhock weevil (*Rhopalapion longirostre*), 409, **409**
Homeodomain, 739
Homeotic genes, 738–44
Homicides
  parent–child, 573–74, **574**
  same-sex, 445–447, **446**
Hominid, 780
Hominin, 780–87, **784**, **785**
Homo, 781, 783
  premodern, **786**
Homo antecessor, 792–93
Homo erectus, **786**, 790, **790**, 792, 804, 809
Homo ergaster, **786**, 786–87, 789, **789**, 790, **790**
Homogenizing of allele frequencies, migration and, 235–237, **237**
Homo habilis, 781, 784–86, **785**, 803, 804, 805
Homo heidelbergensis, **786**, 790, **790**, 807
Homologous traits, 744
Homology(ies), 744–46, 763–64
  deep, 747
  distinguishing homoplasy from, 121–22
  as evidence for evolution, 55–61
    modern concepts, 61
    molecular homology, 58–61, **61**
    structural homology, 55–58, **56**
  model organisms and, 61–62
  structural, 55–58, **56**
  universal molecular, 61, **61**
Homo neanderthalensis, **786**, 790, **790**, 804
Homo paniscus, 808
Homoplasy, 121, 122, 744–46, **745**, 760, 763–64
Homoploid hybrid speciation, 634
Homo rudolfensis, 781, 784–86, **785**, 803
Homo sapiens, 787, **787**, **790**, 808
  cladogram, **788**
  language in, 805
  origin of, 790–801
    African replacement model *vs.* multiregional evolution, 790–800, **791**, **792**, *793*
    controversies over, 790–91
    genetic diversity among, **800**, 800–1, **801**
  phylogeny of, **798**
Homo troglodytes, 808
Homozygosity, extended haplotype, 312
Homozygotes
  fitness of, 208–12
  inbreeding and, **275**, 275–80
  selection on, 208–12, **210**
  selfing and frequency of, **275**, 275–80
Honeybees (*Apis mellifera*), 484, **484**, **758**, 758–59
Hoogland, John, 438–39, 462–63
Hopkins, Melanie, 722
Hopkins, Robin, 632–33
Hopkins, William, 802
Horgan, John, 685
Horita, Juske, 662
Horizontal (lateral) gene transfer, 672–75, **673**, 678–83
  in bacteria and archaea, 672–74, **673**
  bacterial speciation and, 612
  chronocyte hypothesis, 680, **680**
  evolutionary leaps and, 757
  phylogenetic evidence for, 672–75, **673**, **675**
  ring-of-life hypothesis, **679**, 679–80
  three viruses, three domains hypothesis, 680–83, **682**
  universal gene-exchange pool hypothesis, **678**, 678–79, **679**
Horse racing, 329
Horseshoe crabs, 722–23, **723**
Horsetail stem, **693**
Hosken, David, 384
Host-parasite arms race, sex and, 322–24, **323**
Host shifts
  in feather lice, 396–97, **397**
  in herbivorous beetles, **394**, 394–96
Housefly (*Musca domestica*), 374–75, **375**, 503, **503**
House mice (*Mus domesticus*), 455, **455**, 560, **560**
"How the Leopard Got Its Spots," 111
Hox genes, 738–44
  canonical spatial expression pattern, **741**, 742–44
  diversity in gene expression and function, 740–44, **741**
  in flies and mice, **739**, 740, 742
  homology *vs.* homoplasy, 763–64
  spatial colinearity and, 739–40, 742
Hrdy, Sarah, 438
Hsieh, S. Tonia, 50
Huaorani, **556**
Huey, Ray, 380–81
Huff, Chad, 304
Hughes, Austin, 262, 265–66
Hughes, Kimberly, 502
Hughes, William, 484
Human chromosome 22, **305**, 305–6
Human endogenous retrovirus K (HERV-K), 773–74
Human evolution, 769–814
  human–ape relationship, 770–79
    divergence times, **777**, 777–78
    genetic differences and, **778**, 778–79, **779**
    molecular evidence for, 772–76
    morphological evidence for, 776–77
  origin of *Homo sapiens,* 790–801
    African replacement *vs.* multiregional evolution models of, 790–800, **791**, **792**, *793*
    controversies over, 790–91
    genetic diversity among living humans, **800**, 800–1, **801**
  recent ancestry, 780–89
    fossil evidence, 780–87
    interpreting fossil evidence, 787–89
  as result of HIV pandemic, 15–18
  of uniquely human traits, 802–7
    language, 805–7
    stone tools, 802–4, **803**, 807
Human immunodeficiency virus (HIV), 1–31
  AIDS and, 1–2, 6–8, **8**
  AZT (azidothymidine) and temporary inhibition of, 9–10
  AZT resistance, 10–11, **11**
  base substitution rate in, **170**
  CCR5-Δ32 allele and, **16**, 16–17, 18, **200**, 200–1
  characteristics of, 4–6
  drug design and, 13–14
  escape mutation, 26, **26**
  evolutionary history of, **19**, 19–20, **20**
  evolution by natural selection and, 12–13
  evolution of human populations and, 15–18
  evolution of multiple drug resistance, **14**, 14–15
  geographic distribution of infections, **2**
  HIV resistance in African populations, 17
  immune system reaction to, 6, **6**
  infection mechanism, **5**, 5–8
  lethality of, 23–30
    short-sighted evolution and, 26–29
    transmission and, 23–26
  life cycle of, 5, **5**, 649
  long-term trends in, **3**
  movement from chimpanzees to humans, 21–23, **22**, 24
  multidrug therapies for, **1**, 9–15, **14**
  natural history of epidemic, 2–8
  origin of, 21–23
  phases of infection, 5–8
  phylogeny of, **19**, 19–20, **20**, 21, **21**
  prevention successes and failures, 3–4
  replication of, **5**, 5–6
  resistance to, 15–18
  susceptibility to, 180
  synonymous mutations and progressive to AIDS, **28**, 28–29
  transmission routes, 3–4, **4**
  unusual properties of, 29–30
  viral load
    progression to AIDS and, **25**, 26
    transmission of and, **25**, 25–26
  virions, 5, **5**, 7, 10–11, 12
Human immunodeficiency virus (HIV)-1, origin of, **22**, 22–23, 24, 25
Human immunodeficiency virus (HIV)-2, 22, 25
Human leucocyte antigen (HLA), 26, 306
Human papilloma virus, 137
Humans, 786
  chromosome 7, **149**
  compulsory sterilization, 201–2
  evolution of body lice and, **138**, 138–39
  extinctions caused by, 717–19
  falciparum malaria, 197–200
  founder effect among, 245, **245**

frequency of ergothioneine transporter allele in, 303–4, **304**
genetic differences between chimpanzees, gorillas and, **778**, 778–79, **779**
genetic diversity among living, **800**, 800–1, **801**
genotype-by-environment interaction in, 156, **156**
geographic structure in populations of, **801**
G6PD locus in, **311**, 311–12, **313**
inbreeding depression in, 281, **281**
kin selection and, 464–66
inherited wealth, 464–65
trust, **465**, 465–66, **466**
life cycle of, **151**
male *vs.* female differences in, **408**
menopause in, **510**, 510–12, **511**, **512**
movement of HIV from chimpanzees to, 21–23, **22**, 24
multilevel selection and morality in, 476–77, **477**
natural selection and, 174
novelty seeking in, 341
p53 as cancer suppressor in, 500
phylogeny of, **798**
PMP-22 gene and, 59, **59**
positive selection on *BRCA1* gene in, **266**, 266–67
quantitative traits in, **330**
reciprocity and punishment in, 481–83, **482**, **483**
relationship between extant apes and, 770–79
divergence times, **777**, 777–78
molecular evidence of, 772–76
morphological evidence of, 776–77
relation to chimpanzees, 737, **737**, 808
salivary amylase gene, 598, **598**
sexual dimorphism in, 408, **408**, **444**, 444–47, **445**, **446**, **447**
sperm competition in, 421
transposition events in, 586
variation in height among, 147, **147**, 343, **344**, **444**, 447, **447**
variation in skin color among, 148
variation in taste perception among, **150**, 150–51
vestigial structures in, 42, **43**
Human salivary amylase gene, 598, **598**
Hummel, Susanne, 311
Hummingbirds, as pollinators, **334**, 335–36, 340, **340**
Humpback whale, 109, **109**
Hunt, Gene, 720, 723
Hunt, James, 484–86
Hunter-gatherers, 476–77, **511**, 511–12, **556**, 556–57, **557**
Hurt, Carla, 620
Hurtado, Magdalene, 512
Huttley, Gavin, 266, 306
Hutton, James, 62, 64
Huxley, Julian, 736
Huxley, Thomas Henry, 52, 771
hx546 allele, 505–6
Hybridization, 629–37, **630**
creation of new species through, 634–36
in plants, 629
possible outcomes, **630**
reinforcement, 629–33
secondary contact and, 629
Hybridization and assimilation model, 791, **791**
Hybrid zones, 636–37
Hydnellum, 744, **745**
Hydnum, 744, **745**
Hydra, **314**, 315, 762
Hydrogen cyanide, 645
Hydrophyllum appendiculatum (waterleaf), 281–82, **282**
Hydroxylapatite, 661
Hyla versicolor (gray tree frogs), 425–27, **426**, 437, **437**
Hylobates (gibbons), 770
Hyman, Libbie, 737
Hymenoptera, 484–85
Hyoid bones, **806**, 806–7
Hypothesis/hypotheses
evolutionary trees as, 113–14, **114**
formation strategies, 401–2
null, 377
statistical testing of, 377–78

Ibuprofen, 566
Ichthyolestes pinfoldi, 136, **136**
Ichthyostega, **701**, 702
Idaho pumas, **284**
IDA (initial Darwinian ancestor), 646–47, 655
Identical by descent, 460
Identity by descent
calculating relatedness as probability of, 461, **461**
IGF-II. *See* Insulin-like growth factor II
Iguanas, 407
Ikuta, T., 743
Immune response
pathogenic evasion of, 537–45
to viral infection, 6, **6**
HIV infection, 6–8, **7**
virulence, 548–52
Impact hypothesis of K–Pg mass extinction, 712–14, **713**, **714**
Impala, **371**
Impatiens capensis (jewelweed), 281
Inbreeding, 275–83
coefficient of (F), 278–80
empirical research on, 276–278, **277**, *278*
general analysis of, 278–80
genotype frequencies, **275**
homozygotes and, 276–78, 279
inbreeding depression, 280–83, **281**, **282**
mechanisms to avoid, 282
selfing, **275**, 275–76, *276*, 281
sex and, 319–21
Inbreeding depression, 280–83, **281**, **282**
age and increase in, 502–3, **503**
Florida panther population and, 283–84
Inclusive fitness, 460–61, **463**
Incomplete lineage sorting, 774, **776**, 776–77
Indels, 161
Indirect effects, on synonymous mutations, 270–71
Indirect fitness, 461
Individuals
natural selection on, 90, **90**, 94
variation among, 83–87, 147–48, **148**
Individual selection, 473
Individual variation, 79
Inducible defense, 152, **153**
Infanticide, 94, 422–23, **423**
Influenza A virus, 537–40, **538**
molecular evolution in, **261**, 262
mortality from, 538
pandemic strains of, 538, 540–45
phylogeny of, 538–39, **539**, **540**, **541**, **542**, **543**
vaccines against, 540, 545
Ingman, Max, 794
Inheritance, 95–96
blending, 95
Inherited wealth, kin selection and, 464–65
Insects
aging in, **496**
mimicry in, 748–49
sociality in, 483–86, **484**
time tree for, **729**
Insulin-like growth factor II (IGF-II), 522–23
Integrase, 5, **5**
Integrase inhibitors, 13
Intelligent design creationism, 97–104
Interferon regulatory factor 1, 304–5
Intergenic, 584
Interleukin 5, 304–5
International Union for the Conservation of Nature, 614, 718
Intersexual selection, 417
Intrasexual selection, 417
Introns, 59, 161, 164, 584–85, **585**
Inuit, myopia among, 558
Inurois punctigera (Japanese winter moths), **622**, 622–23
Inversions, chromosome, **167**, 167–68
Ipomopsis **507**, 637, **637**
IQ, heritability of, **362**, 362–64
Iridium, 712–13, **713**
Iridoviruses, 683
Irish elk, 49
Irwin, Darren, 48
Isolation
genetic, mechanisms of, 616–23
chromosomal changes, 621
dispersal and colonization, 617, **617**, **618**, 618–19
physical isolation, 617–20
pollinator specialization, 621–22
shell coiling, 623, **623**
temporal isolation, **622**, 622–23
vicariance, 617, **617**, **619**, 619–20, **620**
postzygotic, 629, 631
prezygotic, 629, 631
reproductive, 611
Isoniazid resistance, 456, 545
Isotopes, used in radiometric dating, *65*
Isthmus of Panama, 619
Iteroparous, 526–27
Iva axillaris, 395
Ivy, Tracie, 439

Jablonski, David, 712, 724–25
Jackson, J. B. C., 721, 722
Jacob, Francois, 737
Jaguar, 110, **111**, 111–12, **112**, **113**, 114, **114**
Jaguarundi, 110, **111**, 111–12, **113**, **114**
Janis, Christine, 529
Japanese canopy plant (*Paris japonica*), 583
Japanese winter moths (*Inurois punctigera*), **622**, 622–23
Jarosch, Antje, 758–59
Javaux, Emmanuelle, 66
Jawed fish, 99, **99**
Jellyfish, 696
Jenkin, Fleeming, 95, 569
Jewelweed (*Impatiens capensis*), 281
Johannesson, Kerstin, 229
Johanson, D. C., 781
Johnson, Kevin, 396–97
Johnson, Warren, 113–14, 284
Johnston, Wendy, 654
Johnstone, Rufus, 512
Jones, Adam, 414, 415–16
Jones, John E., 106
Jones, Kristina Niovi, 79
Joseph, Sarah, 172
Joyce, Gerald, 650–51, 654
Jumping spiders, 373–78
Juvenile hormone acid methyltransferase, 153

Kairomone, 152–53, 154
Kaiser, Shari, 30
Kale (*Brassica oleracea acephala*), **76**
Kandul, Nikolai, 621
Kaplan, Hillard, 572
Karyotype
humans *vs.* African great apes, 778
speciation and differences in, 621
KatG gene, 545
Katydids, 441, 450, **450**
Keck, David, 334, 361–62
Keightley, Peter, 321
Keller, Lukas, 85
Keller, Marcel, 59
Kentucky Derby, 329
Kenyanthropus platyops, 781, 782–83, **783**, 786
Kenyanthropus rudolfensis, **785**, 786
Kerr, Benjamin, 472, 473, 476
Kia, John R., **97**
Killer T cells, 26
Kim, Un-kyung, 150
Kimberella, 697, **697**, 729
Kimura, Motoo, 257, 260, 263, 265
King, Elizabeth, 494
King, Marie-Claire, 737, 779
King, Richard B., 237
King crabs, **723**, 723
Kingdoms, 666–68
Kingman, John, 272
Kingsolver, Joel, 759
Kin selection, social behavior and, 459–71, 473
alarm calling, **462**, 462–63
in birds, **466**, 466–67, **467**
cooperative breeding in birds, **466**, 466–67, **467**
costs and benefits for adoptive mother squirrels, **463**, 463–64
defined, 461
greenbeard alleles, 468–69
in humans, 464–66, **465**, **466**
inclusive fitness, 460–61

parent–offspring conflict and, **477**, 477–80, **478**, **479**
siblicide, 480, **480**
spite and, 469–71, **471**
Kirkpatrick, Mark, 429
Kittler, Ralf, 138
Kitzmiller et al. v. Dover Area School District, 98, 106
Kiwi, brown, 42, **42**, 492–93, **493**
Kluger, Matthew, 564–65, 566
KNM-ER 406, 489
KNM-ER 3733, 789
Knoll, Andrew, 66, 665, 701
Knowles, David, 166
Knowlton, Nancy, 619
KNOX gene, 752
Knuckle walking, 776–77, 808
Koch, Robert, 535
Koehl, Mimi, 759
Kohlrabi (*Brassica oleracea gongylodes*), **76**
Kohn, Michael, 325
Koobi Fora, 789, 790
Korarchaeota, 671
Korber, Betty, 24
Kornberg, Arthur, 158
Kotukutuku, 391–93
Kowalevsky, Alexander, 736
Krebs citric acid cycle, 102
Kreitman, Martin, 267–68
KRT77 gene, 801
Kumar, Sudhir, 728, 778
!Kung people, **511**, 557

L68Q allele, 557, **557**
L503F, 303–5, **304**
Lack, David, 513
Lack's hypothesis, **513**, 513–17, **515**, **516**
Lactase, 557–58
Lactate dehydrogenase B, 102
Lagenorhynchus obscurus (dolphin), **127**
Lagerstätte, 698, 702
Lahdenperä, Mirkka, 512
Lake, James, 679
Lake Blankaart, 388
Lake Erie, 237, 617
Lake Gatún (Panama), 457
Lake Michigan, 528
Lake Victoria, 627
Lamarck, Jean-Baptiste, 38, 96
Lamarckism, 74
Lamb, Trevor, 99
Lamprey, 99, **99**, 698
Lancelet, 98–99, **99**
Lande, Russell, 429
Langerhans, Brian, 92
Langley, Charles, 586
Language
origin of, 805–7, 809
sign, 805
Langur monkey, **477**
Lapeirousia oreogena (South African iris), **369**, 369–70
Larder, Brendan, 10, 11
Lasaea subviridis (clam), 492
Lateral gene transfer. *See* Horizontal (lateral) gene transfer, 626
Laupala (cricket), 432, **432**, **625**, 625–626, **626**
Law of succession, 49
Lawrence, Jeffrey, 612, 674
Lawson, James, 560
Lcyc gene, **154**
Leaf beetles, **394**, 394–96
Leafcutter ant (*Atta cephalotes*), 483
Leakey, Meave, 782–83, 786
"Leaky replacement," 800
Leclerc, Georges-Louis (Comte de Buffon), 38
Lee, Carol E., 528–29
Leitner, Thomas, 20
Lemey, Philippe, 27, 28, 29
Lenses, 760
Lenski, Richard, 101, 359–60
Leopard, 110, 111, **111**, **112**, **113**, **114**, 287
Leopard geckos (*Eublepharis macularius*), **155**, 155–56
Lesch, Klaus-Peter, 366
Leslie, A. J., 26
Lesotho, AIDS infection rate in, 3
Lesser hedgehog tenrec (*Echinops telfairi*), **745**
Lethal dominant allele, mutation–selection balance for, 220, 221
Levin, Donald, 632
Levinton, Jeff, 729
Levy, Stuart, 548
Lewis, Ed, 737, 738
Li, C.-W., 697
Lice
body, **138**, 138–39
feather
color of, 386–87
host shifts in, 396–97, **397**
head, 138–39, 387
Lidgard, Scott, 722
Lidicker, William, 277
Lieberman, Daniel, 792
Life, defining, 648–49
Life, origins of, 645–90
biological polymers, 660–61, **661**
building blocks, 658–60
sources of, 658–59
cellular life, 663–78
advantages of, 663
DNA and, 663
examination of, 668–70
fossil record of, 675–78, **676**, 677
most recent common ancestor, 663–65
universal phylogeny, 665–68, **667**, **669**, **672**, **673**, **674**
environment and, 661–63
horizontal gene transfer, 672–75, **673**, 678–83
chronocyte hypothesis, 680, **680**
ring-of-life hypothesis, **679**, 679–80
three viruses, three domains hypothesis, 680–83, **682**
universal gene-exchange pool hypothesis, **678**, 678–78
nucleotides and, 658–60
Oparin–Haldane model of, 657–58, **658**
overview of, **664**
panspermia hypothesis of, 655
RNA enzymes (ribozymes) and, 647–49, **648**, 654
RNA World hypothesis, 647–54
case for, 649
tree of life, **646**, **666**, 670–75
Life cycle
HIV, 5, **5**, 649
human, **151**
Life histories, 491–534
aging, 594–512
deleterious mutations and, 500–3, **503**
evolutionary theory of, 500–10
natural experiments of evolution of, 507–10, **509**
trade-offs and, 500, 504–7, **505**, **506**, **507**
basic issues in, 493–95
in broader evolutionary context, 525–29
biological invasions, 527–29, **528**
maintenance of genetic variation, 525–27, **526**, **527**
vulnerability to extinction, 529, **529**
conflicts of interest between, 522–25
genomic imprinting, 522–23
sexual coevolution, 523–25, **524**
menopause, **510**, 510–12, **511**, **512**
number of offspring produced in year, 513–17
clutch size in birds, **513**, 513–15, **514**
clutch size in parasitoid wasps, **515**, 515–17, **516**
size of offspring, 517–22
phenotypic plasticity in egg size in beetle, 520–22, **521**
selection on, 517–20, **518**, **519**
Life-history analysis, 492, 493–95
Light-sensing organs, 98–99, **99**
Likelihood, 338
maximum, **128**, 128–29, **129**, 134
Likelihood ratio, 338
Lilliput effect, 716
Limbs
fin-limb transition, 702, **702**
forelimbs, 56, **56**, 702, **702**
Limulus, 722
Lin, Yi-Jyun, 506
Lin, Zhongwei, 735
Linanthus ciliatus (whiskerbrush), **148**
Linaria vulgaris (common toadflax), 154, **154**
LindbladToh, K. M., 757
Lindén, Mats, 514
Line breeding, 286–87
LINES (Long INterspersed Elements), 134–35, **135**
Linkage, 167
Linkage disequilibrium, 292–307
coefficient of (D), 295
creation of, 297–300
from genetic drift, 299, **299**, 319–21
from population admixture, **299**, 299–300
from selection on multilocus genotypes, 298, **298**
defined, 295–96
elimination of, **300**, 300–2, **302**
genetic drift and, 299, **299**, 319–21
genetic recombination and, 300–2
genotype frequencies and, 293–95, **294**
human expansion from Africa and, 797
in human leukocyte antigen (HLA) loci, 306
practical reasons to study, 307–13
gene and population history reconstruction, **307**, 307–11, **308**
positive selection detection, **311**, 311–13, **313**
selection and, 323
significance of, 302–7, **303**, **304**, **305**, **306**
single-locus studies and, 303, 305
Linkage equilibrium, 295, 305, **305**, 318
Linkage model with $r = 0.1$, 338
Linnaeus, 737
Lions, **284**, 287
infanticide among, 94, 422–23, **423**
phylogeny of, 110, **111**, 111–12, **112**, **113**, 114, **114**
Lipinski, Kendra, 170
Lissamphibia, **117**
Lister, Joseph, 536
Litoria (frog), 122
Liu, Kevin, 134
Liu, Rong, 16
Lively, Curtis, 323, 392–93
Living fossils, **722**, 722–23, **723**
Lizards, 117, 122, 421, 523
Locus(i), 149
variation among, 263
LOD (logarithm of the odds) scores, 338–39, **339**
Loeb, Lawrence, 593–94
Lohse, Nicolai, 14
Longflower tobacco (*Nicotiana longiflora*), 331–33
Longgupo Cave, 790
Long-horned beetles, 756
Long INterspersed Elements (LINES), 134–35, **135**
Long-nosed skate (*Dipturus oxyrinchus*), **615**
Lougheed, Lynn, 480
Lousy fly, 397, **397**
Lovastatin, 62
Lowe, C. J., 757
Lowry, David, 624–25
Loxia curvirostra species complex (red crossbills), 642
LUCA (last universal common ancestor), 647. *See also* Universal phylogeny
early cellular life and, 668–70
evolution of, 678–83
chronocyte hypothesis, 680, **680**
ring-of-life hypothesis, **679**, 679–80
three viruses, three domains hypothesis, 680–83, **682**
universal gene-exchange pool hypothesis, **678**, 678–79, **679**
fossil record and, 675–78, **676**, **677**
identifying, 663–65
Luckinbill Leo, 497
Lungfish, 702
LWS gene, 627–29
Lycaeides, 634
Lycodichthys dearborni (Antarctic eelpout), **596**, 596–98, **597**
Lyczak, Jeffrey, 224
Lyell, Charles, 62, 64, 748
Lymph nodes, AIDS and, 8
Lynch, Michael, 170, 173–74, 283, 591–92
Lyngbya, **677**
Lystrosaurus, 716, **717**

Macaques, 770
Macnair, Mark, 212, 213

Macroevolution, 38, 719–24
  adaptive radiations, **726**, 726–27
  biotic replacement, **725**, 725–26
  evidence of, 49–55
    extinction and succession, 49
    transitional forms, 49–54
  relative frequency of stasis and gradualism, **720**, 722
  significance of, 55
  species selection, 724–26
  stasis, 720–24
    in Bryozoans, **721**, 721–22
    demonstrating, 720–21
    reasons for, 722–24
Macrophages, 6, 7
Madrigal, Lorena, 512
Magnificent tree frog (*Litoria splendida*), 122
Major histocompatibility complex (MHC), 266
Majority-rule consensus phylogeny, 132–33
Malacoraja krefftii, **615**
Malapa Hominin 770, 784
Malaria, **311**, 312
  during pregnancy, 197–200
Malaria resistance, 224–27
Male–male competition, 417–23
  alternative male mating strategies, 420–21
  among stalk-eyed flies, 432
  body size and, 445
  combat, 417–21
  homicides, 445–47, **446**
  infanticide, 422–23, **423**
  sperm competition, 421–22, **422**
Males
  maintenance of, 316–18, **317**
  sexual selection and, 411–12, 439–41
Mammalia/mammals, **117**
  aging in, **496**
  end-Cretaceous extinction and, 729–30
  genomic imprinting in, 522–23
  lifetime energy expenditure in, 497, **497**
  origin of, **704**, 704–5, **705**
  parental investment in, 411
  phylogeny, **141**, **779**
  reproductive life span in, 492
  telomere shortening in, 499, **499**
  vulnerability to extinction, 529, **529**
Manakins, 433
Manduca sexta (tobacco hornworms), **156**, 156–57, **157**
Mangrove evolution, **745**, 745–46
Mangrove finch (*Cactospiza heliobates*), 81, **81**
Mansy, Sheref, 664
Maraviroc, 18
Marcotte, Edward, 61–62
Marine iguanas (*Amblyrhynchus cristatus*), **417**, 417–21, **418**, **419**, **420**
Marine snail, 229, **229**
Marks, Jonathan, 808
Marlowe, Frank, 447, 572
Martin, Paul, 640
Martinson, Jeremy, 162
Masked boobies, 480, **480**
Mass extinctions, 691–92, 694, **695**, 709–19
  Cretaceous–Paleogene (K–Pg), 710, 712–15, 731
    impact event, evidence for, 712–14, **713**, **714**
    killing mechanisms, 714–15
  current, 717–18
    rain forest, 718–19
  Permian–Triassic (P–T), 715–17
Maternal effects, 281
  estimating heritability and, 85
Maternal fitness
  clutch size of parasitoid wasps and, **515**, 515–17, **516**
  egg size and, 521, **521**
Mating. *See also* Sexual selection
  alternative male strategies for, 420–21
  assortative, 625–26
  asymmetry of sex and, 416–17
  in coho salmon, 421, **421**
  environmental factors and preferences in, 626–29
  in hangingfly, 435–36, **436**
  in marine iguanas, **419**, 419–21, **420**
  multiple, by females, 438–39
  nonrandom. *See* Inbreeding
  in water mites, 433–35, **434**
Mating call, in gray tree frog, **425**, 425–27
Mating success
  in hermaphrodite, 412–13
  in newt, 414–15
  in pipefish, **415**, 415–16
Matthew, Patrick, 73
Matute, Daniel, 631
Maurrasse, F. J.-M. R., 715
Maximum likelihood (ML), **128**, 128–29, **129**, 134
Mayr, Ernst, 45, 611, 736
Mazur, Allan, 447
McClearn, Gerald, 347
McCleery, R. H., 514–15
McCollum, F. C., 277
McDonald, John, 267–68
McDonald–Kreitman (MK) test, 267–68, 272
McElligott, Alan, 372–73
McGary, Kriston, 61–62
McGrady, Tracy, **147**
McKellar, Ann, 147
McLysaght, Aoife, 166
McMillan, Owen, 750
McShea, Dan, 706
Mean heterozygosity, 163
Measles, 48, 575
Medea, 224–26, **225**
Medical applications, 535–80
  adaptations
    breast cancer, 559–62, **560**, **561**
    environment and, 556–58
    fever, **564**, 564–57, **566**
    menstrual cycle
    myopia, 558–59
    obesity, **562**, 562–64, **563**
    parenting, 567–74, **570**, **571**, **572**, **573**, **574**
  pathogens, 537–52, 575
    antibiotic resistance, 545–48, **546**, **547**
    evasion of immune response, 537–45
    flu virus, **538**, 538–39, **539**
    pandemic flu strains, origin of, 540–45
    virulence, 548–52, **551**, **552**
  tissue as evolving populations of cells, 553–56
    history of cancer, 554–56, **555**
    patient's spontaneous recovery, 553–54
Medical practice, fever and, 567
Mediterranean fruit fly (*Ceratitis capitata*), 422, **422**
MEF2, 761
Megachiroptera, 382
Megadont hominins, **785**, 787–88
Melanesians, 800, 807
Melanic peppered moth, **752**
Melanocortin 1 receptors (MC1R), 95, **95**
Melanocytes, 95, **95**
Melanosomes, **95**
Meléndez-Obando, Mauricio, 512
Melospiza melodia (song sparrows), 346, **346**
Meme, 569
Memory helper T cells, 6, 7
Memory T cells, 27
Mendel, Gregor, 74, 95, 736
Mendelian genetics, 179–290
  genetic drift, 240–59
    molecular evolution and, 260–74
    natural selection *vs.*, 257–59, **258**, **259**
    population size and, 243, **243**
    random fixation of alleles, 240–42, **241**
    rate of evolution by, **255**, 255–57
    sampling error and founder effect, 243–45, **244**, **245**
  Hardy–Weinberg equilibrium principle, 180–91
    adding mutation to, **217**, 217–19, **218**
    adding selection to, 192–93, **193**, 196–97, **197**
    empirical research suing, 197–200
    general case, **187**, 187–89, **188**
    with more than two alleles, 189
    numerical calculation, **183**, 183–87, **184**, **185**, **186**
    simulation, 181–83, **182**, **183**
    use of, 189–91
  migration, 234–39
    defined, 234
    as evolutionary mechanism, **234**, 234–35, **235**
    homogenization across populations and, **235**, 235–37, **237**
    migration–selection balance, **237**, 237–39, **238**
  modern synthesis and, 736
  mutation, 216–24
    mutation–selection balance, 219–24, **223**, **224**
    selection and, 219, **219**
  nonrandom mating (inbreeding), 275–83
    coefficient of (F), 278–80
    empirical research on, 276–78, **277**, *278*
    general analysis of, 278–80
    homozygotes and, 275, 276–78, 279
    inbreeding depression, 280–83, **281**, **282**
    mechanisms to avoid, 282
  quantitative traits and, **331**, 331–33, **333**
  selection, 191–201
    added to Hardy–Weinberg analysis, 192–93, **193**, 196–97, **197**
    changes in allele frequencies and, 195, **195**, 198–99, **200**, 200–1
    frequency-dependent, 212–14, **213**
    general treatment of, 194
    genotype frequencies and, 196–200, **197**
    patterns of, 201–16
Menopause, **510**, 510–12, **511**, **512**
Menstrual cycle, breast cancer and, 560–62, **561**
Meredith, R. W., 730
Mesonychians, 124
Mesozoic era, 694, **695**
Messenger, Sharon, 550
Messenger RNA (mRNA), 160, **160**
Metabolism. *See also* Energy
  aging rate and rate of, 496–98
  body size in marine iguanas and, 418
Metacarpals, humans *vs.* chimpanzee thumb, 804, **804**
Metacrinus rotundus (sea lily), 743
Metastasis, 554
Meteorites
  amino acids in, 655–56
  carbonaceous chondrite, 657
  Murchison, **655**, 655–56
Meteors, Permian–Triassic extinction and, 717
Methanococcus jannaschii, 670, 685
Methuselah gene, 506, **506**
Methylation, 586–89, **589**
Metrarabdotos, 721, **721**
MHC. *See* Major histocompatibility complex (MHC)
M HIV-1, 23, 24, **24**, 29
Mice
  aging and cancer in, **499**, 499–500
  attacks on conspicuous, **73**, **77**
  Hox genes in, **739**, 740, 742
Microbes in gut, 563, **563**
Microevolution, 38
  evidence of, 39–44
  significance of, 44
MicroRNA, 779, **779**
Microsatellite, **244**, 244–45, 795–96, **796**
Microtektites, **713**, 714–15
Midoffspring value, 344
Midparent value, 344
Migration, **190**
  added to Hardy–Weinberg analysis, 234–35
  allele frequencies and, 234–39
  genotype frequencies and, 234–35
  Hardy–Weinberg equilibrium assumption of no, 190
  in Mendelian genetics of populations, 234–39
    defined, 234
    as evolutionary mechanism, **234**, 234–35, **235**
    homogenization across populations and, **235**, 235–37, **237**
    selection and, **237**, 237–39, **238**
  one-island model of, **234**, 234–35, 236
  two-island model of, 236
Mikkelsen, Tarjei, 779
Milk sugar, ability to digest in adulthood, 558
Miller, Jon D., 37
Miller, Stanley, 657, 659
Mills, Donald, 650
Mimicry, 748–49
Mimivirus, 683
Mimulus bicolor, **334**
Mimulus cardinalis, **334**, 334–40, **335**, *336*, **340**
Mimulus eastwoodiae, **334**
Mimulus guttatus (yellow monkeyflower), 282, 623–25, **624**

Mimulus lewisii, **334**, 334–40, **335**, *336*, **340**
Mimulus nelsonii, **334**
Mimulus parishii, **334**
Mimulus rupestris, **334**
Mimulus verbenaceus, **334**
miR-320b, 779, **779**
Mitochondria, selection at level of, 398–401, **399**, **401**
Mitochondrial DNA (mtDNA)
  common ancestor and, 794, **794**
  phylogeny of humans and African great apes based on, 773, **773**, 774
Mivart, St. George, 748
Miyakawa, Hitoshi, 152–53
Mobile genetic elements, 583–84, **584**, 586–91
  evolutionary impact of, 586–91
  as genomic parasites, 586
Mockett, Robin, 506
Modern Synthesis, 96–97, 736–37
Mojokerto, 790
Molds (fossils), 693, **693**
Molecular and fossil divergence timing, 727–30
Molecular basis of adaptation, 601–6
Molecular basis of reinforcement, 632–33
Molecular characters, parsimony with multiple, 127, **127**
Molecular clock, 138, 140, 727–28
  vertebrate, 263, **263**
Molecular evidence
  of evolutionary relationships, 125–34
  of relationship among humans, gorillas, chimpanzees and bonobos, 772–76
Molecular evolution
  genetic drift and, 260–74
    DNA sequence divergence, **261**, 261–63, **262**
    nearly neutral model of, **263**, 263–65, **264**
    neutral theory of, 260–61, 265–71
  mutation, selection and genetic drift in, 256
  neutral theory of, 256, 257, 260
Molecular homology, 58–59
  predictive test of common ancestry using, 59–61
  universal, 61, **61**
Molineux, Ian, 550
Møller, Anders, 514
Molothrus ater (brown-headed cowbirds), 405
Monkey cups (*Nepenthes*), 92
Monkeys
  African leaf-eating, 165
  brains of, 805
  capuchin, 481
  langur, **477**
  Old World, 772
Monoamine oxidase A (MAOA), 365
Monod, Jacques, 737
Monoecious, 389
Monogamy hypothesis, 484
Monophyletic group, **116**, 116–17, **117**, 122, 611
Monozygotic twins, 347
Monteiro, Antónia, 749
Montmorillonite, 661
Mooi, Richard, 122
Moorhens, 441
Morgan, Thomas Hunt, 95, 427
Mormyrid electric fishes, 640
Morphological change, adaptive radiation and, 726–27
Morphological diversity, **708**, 708–9, **709**
Morphological evidence, of relationship among humans, gorillas, chimpanzees, and bonobos, 776–77
Morphology, phylogeny and, 665–66
Morphospace, 708, **708**
Morphospecies, 696
  of *Pseudo-nitzschia,* **613**, 613–14
Morphospecies concept, 610
Morran, Levi, 316, 318, 319
Mortality
  ecological, 507–10
  physiological, 507
Mosquitoes, malaria resistant, 224–27, **225**, **226**
Mosquitofish (*Gambusia affinis*), 92, **92**
Mouse mammary tumor virus (MMTV), 560, **560**
Mousseau, T. A., 525
Mucigen, **148**, 149
Muehlenbachs, Atis, 197, 198, 200
Mueller, Ulrich, 447
Mukai, Terumi, 206–7
Muller, H. J., 320
Muller, Herman, 174
Muller's ratchet, **320**, 320–21
Mullins, James, 27
Multilevel selection
  cooperation and, 471–77
    in bacteria, 474–75, **475**
    human morality and, 471, 476–77, **477**
    in plants, 475–76, **476**
  Darwin and, 471
  numerical example of, **372**, 472–73
Multiregional evolution, 790–98, **791**
  archaeological and paleontological evidence of, 791–93, **792**
  cladogram predicted by, **792**
  genetic predictions distinguishing, 793, *793*
  molecular evidence of, 793–800
Murchison meteorite, **655**, 655–56
Murgia, Claudio, 137
Murphy, Robert, 13
Murray, Charles, 362
Murray, John, 769
Musca domestica (housefly), 374–75, **375**, 503, **503**
Mus domesticus (house mouse), 455, **455**, 560, **560**
Mus musculus, 560, **560**, **589**
  Hox genes in, **739**, 740, 742
Mutation accumulation, 169–70
Mutation accumulation hypothesis, 500–3
Mutation(s), 91, 156, 157–61, **190**, 216–24. *See also* Synonymous mutations
  adaptation from new, **602**, 602–3, **603**
  adaptive, of small *vs.* large effect, **605**, 605–6
  advantageous, 257
  alteration of protein function and, 159–61
  asexual reproduction and, 319–21, **320**, **321**
  beneficial, 260
  chromosome, 166–69
    genome duplication, **168**, 168–69, **169**
    inversions, **167**, 167–68
    polyploidy, 168–69
  by deamination, 159, **159**
  deleterious, 260
    evolutionary theory of aging and, 500–3, **503**
  evolution and, 173–74
  evolution of Darwinism and, 94–95
  fitness effects of, 171–73
  gene duplication, 161–66
  genome duplication, **168**, 168–169, **169**
  in HIV genome, 10–11, 13
  in *Hox* genes, 738, **739**
  isolation from, 621
  in Mendelian genetics of populations
    as evolutionary mechanism, **217**, 217–19, **218**, **219**
    mutation–selection balance, 219–24, **223**, **224**
    selection and, 219, **219**
  by misalignment, 159, **159**
  in molecular evolution, 256
  nature of, 157
  nearly neutral, **263**, 263–65, **264**
  neutral, 138, 260
  nonsense, 161
  pleiotropic, 505
  point, 160
  randomness of, 93
  rates of, 169–71
    for cystic fibrosis, 222, 223
    evolution of, 591–94
    of HIV, 13
    molecular clock and, 138
    for recessive alleles, 222
    for spinal muscular atrophy, 222
  replacement, 262
    selection on, 265–268, **266**
  selectionist theory and, 257
  sex and, **319**, 319–21
  silent-site (synonymous), 262
  speciation and, 621
  substitution *vs.,* 160–61, **255**, 255–57
Mutation–selection balance, 219–24, 360
  cystic fibrosis alleles and, 222–24
Mutually beneficial interaction, **456**, 456–57
Mycobacterium tuberculosis, 545, 667
Myopia, 557–58
Myostatin, 342

Nagashima, Hiroshi, 54
Naive helper T cells, 6
Naive T cells, 27
Naked mole rats, 484, **484**
Narrow-sense heritability, 344, 346
Natural selection, 12–13, 38, 73–108. *See also* Adaptation(s)
  action on individuals, 90, **90**, 94
  action on phenotypes, 90–91, **91**
  action on populations, 90, **90**
  backward-looking nature of, 91
  balance with mutation, **173**, 173–74
  beak shape in Galápagos finches, **81**, 81–90, **82**, **83**, **84**, **86**, **87**, **88**, **89**
  changes in allele frequencies and, 192–95, **193**, **195**
  cultural evolution and, 569
  detecting on DNA sequences, 265–71, **266**, **269**
  divergence and, 623–25
  evolution of irreducibly complex systems by, 101–3
  falsifiability of, 103
  fitness and, 78, 93–94
  genetic drift *vs.,* 257–59, **258**, **259**
  humans and, 174
  levels of, 397–401
  mutation rates, genetic drift, and, 591–92
  new traits from, **91**, 91–92, **92**
  non-perfect traits from, **92**, 92–93
  as nonrandom and nonprogressive, 93
  postulates of, **77**, 77–78
    heritability of variation, 79, 83–86
    individual variation in survival and reproduction success, 79, 86–87
    nonrandom survival and reproduction, 79, 87–88
    variability of population, 79, 83
  on replacement mutations, 265–68, **266**
  on replacement substitutions, **266**, 266–68
  on silent mutations, 268–71, **269**
  on silent substitutions, 267–68
  in snapdragons, 79–80
Nautiloids, **711**
Neandertals, 792–93, 798–800, **799**, 807
  hyoid bone, 806, **806**, 807
Nearly neutral mutations, **263**, 263–65, **264**
Nectar guides, 369–70
Nef protein, 29
Negative (purifying) selection, 262
Negative relatedness, 469–70, **470**
Nei, Masatoshi, 264–65, 265–66
Neighbor joining, 130–31, **131**, 134
Nelson, Craig, 444
Nelson, Karen E., 674
Nelson, Matthew, 801
Nematodes, RNA interference system, 587, **587**
Nematode worms. *See Caenorhabditis elegans*
Neofunctionalization, **595**, 595–96
Neopsittaconirums albus, 386
Neotherapods, 702–3
Nepenthes (monkey cups), 92
Nerodia sipedon (water snakes), **237**, 237–39, **238**, **239**
Nerophis ophidion, **440**, 440–41
Nervous system segmentation, vestigial, 764, **764**
Nesse, Randolph, 537
Nested pattern of traits, 58
Nettle, Daniel, 447
Neumania papillator (water mite), 433–35, **434**
Neuraminidase, 541
Neuroticism, 366, **366**
Neutral mutations, 138
Neutral theory of molecular evolution, 256, 257, 260–61, 265–71, 539
  current status of, **271**, 271–72
  nearly neutral model, **263**, 263–65, **264**
  as null hypotheses detecting natural selection on DNA sequences, 265–71, **266**, **269**
Nevo, Eviatar, 377
Newcomb, Richard, 150
Nicotiana longiflora (longflower tobacco), 331–33
Nielsen, Rasmus, 779

Nijhout, Frederik, 156–57
Nikaido, Masato, 135
NK4 genes, 761–63
Nkx-2.5 gene, 761
Noble, William, 805
Node, of evolutionary tree, 110
NOD2 gene, 342
No free lunch theorems, 103–4
Noncoding DNA, 583
Nonfunctionalization, **595**
Nonrandom mating, 275–83. *See also* Inbreeding
Nonrandom survival and reproduction, 79, 87–88
Nonsense mutation, 161
Nonsynonymous (replacement) mutations, 263
  selection on, 265–68, **266**
Nonsynonymous (replacement) substitution, 161, 265–68, **266**
Nordborg, Magnus, 306
Norell, Mark, 58
North America, HIV/AIDS in, **2**, 3
Norwegian skate (*Diputurus nidarosiensis*), **615**
Nostoc, 667
Notochord, 698, **699**, **743**
Nousbeck, Janna, 161
Novacek, Michael, 58
Novelty seeking, 341
Novembre, John, 310
Nuclear polyhedrosis virus, 551
Nucleic acids, abiotic synthesis of, 655
Nucleoproteins, in flu virus, 540–41, **541**, **542**, **543**, 543–44, **544**
Nucleotides, 149
  rates of substitution, 262, **262**
  synthesis of, 658–60, **659**
Null hypothesis, 180, 377
  Hardy–Weinberg principle and, 190–91
  neutral theory as, detecting natural selection on DNA sequences, 265–71, **266**, **269**
  of reproductive modes, 315–16
Nuttall, George H. F., 771

Obesity, **562**, 562–64, **563**
O'Brien, Stephen, 113–14
Observational studies
  of adaptation, 378–81
    behavioral thermoregulation, **379**, 379–80
    garter snake nighttime behavior, **380**, 380–81, *381*, **381**
Oceanic chemistry after steroid impact at K–Pg boundary, 715
Ocellus, 98
Ochman, Howard, 612, 674
Odontochelys semitestacea (turtle), 37, 54, **54**
Offspring. *See also* Parenting
  conflict over provisioning of, **477**, 477–79
  estimating heritability from, **344**, 344–46, **346**
  female choice and quality of, 436–37, **437**
  number produced in year, 513–17
    clutch size in birds, **513**, 513–15, **514**
    clutch size in parasitoid wasps, **515**, 515–17, **516**
  parent–offspring conflict, **477**, 477–80, **478**, **479**
  size of, 517–22
    phenotypic plasticity in egg size in beetle, 520–22, **521**
    selection on, 517–20, **518**, **519**
Ogura, Yasunori, 342
Ohta, Tomoko, 263
Oldowan stone tools, 802–4, **803**, 809
Olduvai Gorge, 784, 803
Old World monkeys, 772
Old World primates, phylogeny, **142**
O'Leary, Maureen, 124–25
Oligonucleotide, 648
Olsson, Lennart, 111, 112, 113
Oncorhynchus kisutch (coho salmon), 421, **421**
Oncorhynchus tshawytscha (chinook salmon), 518–20, **519**
One-island model of migration, **234**, 234–35, 236
On the Origin of Species by Means of Natural Selection (Darwin), 38, 51, 62, 73, 77, 97, 98, 110, 536, 737, 756, 769
Onthophagus, **753**, 753–55, **754**
Onycophora, 764, **764**
Opabinia regalis, 698–99, **699**
Oparin, A., 658
Oparin–Haldane model, 657–58, **658**
Ophraella, **394**, 394–96
Opossums, 522
Opposable thumb, anatomy of, **803**, 803–4, **804**
Opsin gene, 627–28
Optix gene, 750–51, **751**, 752, 758
Oral contraceptives, 562, 576
Orangutan (*Pongo pygmaeus*), 770, 802
  parental investment in, 411, **411**
Orca, **56**
Orchids (*Catasetum*), **56**, **441**, 441–42, 443–44
Ordovician, 701
Organic molecules, from comets, 657
Organisms, senescence of, 498–99
Orgel, Leslie, 685
Oró, Juan, 658
Orr, H. Allen, 336, 569
Orrorin tugenensis, 781
Orthologous genes, 165, 740
Oryza perennis, **507**, **589**
Ossowski, Stephan, 169, 172
Ostrom, John, 52
Otto, Sarah, 321
Otx gene, 746
Outgroup analysis, 118, **118**
Out of Africa model. *See* African replacement model
Overdominance, 207, **209**
Owen, Aaron, 431
Owen, Richard, 55
Oxford ragwort (*Senecio squalidus*), 634
Oxpeckers, 370–73, **371**
  mutualist relationship with cattle, hypothesis of, 370–72, **371**, **372**
  parasitic relationship with cattle, 372–73
Oxygen, Cambrian explosion and increases availability of, 700–1
Oyster (*Crassostrea gigas*), 492
Oyugi, Julius, 17
Ozark Mountains (Missouri), 252

P1 virus, 404
P3 protein, 45
p24 epitope, 26
p53 protein, 498, **499**, 499–500, 556
Pääbo, Svante, 775, 798
Pace, Norman, 670
Pacific leaping blenny (*Alticus arnoldorum*), **50**, 50–51
Packer, Craig, 422–23
Pakicetus attocki, 136, **136**
Pal, Csaba, 594
Paleobiology Database, 708
Paleolyngbya, **677**
Paleontological evidence, of African replacement *vs.* multiregional evolution, 791–93, **792**
Paleozoic era, 694, **695**
Paley, William, 98
Palo verde (*Cercidium floridum*), 520, **520**
Palsdottir, Astridur, 557
Palumbi, Stephen, 109
Pandemic flu viruses, 538, 540–45
Pangaea, 62, 64
Pan paniscus (bonobo), 59, 770, 780, 802–3
  relationship between humans and, 772–79
Panspermia hypothesis, 655
  directed, 685
Pan troglodytes. See Chimpanzees (*Pan troglodytes*)
Papa, Ricardo, 750
Papineau, Dominic, 662
Papio hamadryas ursinus (baboons), 481, **481**
PAP locus (1-phenyl-alanyl-1-proline peptidase), 277
Parallel evolution, 746–47, **747**, **762**
  biased evolutionary trajectories and, 749–52
Paralogous genes, 165, 740
Paramecium, 668
Paranthropus, 803, **804**
Paranthropus aethiopicus, **785**, 788
Paranthropus boisei, 784, **785**, 788, 789, **789**
Paranthropus robustus, **785**, 788, 804
Paraphyletic group, 116
Parasites
  attractiveness in mate choice and prevalence of, 449, **449**
  parasitic oxpecker-cattle relationship, hypothesis of, 372–73
  sex and parasite-host arms races, 322–24, **323**
Parasitism, conspecific, 84, 85
Parasitoid wasps, 358, **358**
  clutch size in, **515**, 515–17, **516**
Parental investment, 410–11, 439–40
Parenting, 567–74
  of biological *vs.* stepchildren, **571**, 571–74, **572**, **573**, **574**
  children's health and, 572–74, **573**, **574**
  evolution and, 568–72
  male *vs.* female investment in, 568–72
    sexual selection stronger for female than males and, 439–41
Parent–offspring conflict, **477**, 477–80, **478**, **479**
Parents, estimating heritability from, **344**, 344–46, **346**
Paris japonica (Japanese canopy plant), 583
Parker, Heidi, 165–66
Parsimony, 118–20, **119**, **120**, 134
  bootstrapping in, **132**, 132–33
  common ancestor and, 669–70
  evaluating alternative phylogenies with, 127, **127**
  phylogeny of fossil sand dollars by, **123**
Pärt, Tomas, 507
Parthenogenesis, 314
Parus major (great tit), 481
  clutch sizes of, **513**, 513–15
  inbreeding depression in, 282, **282**
Passive trend, 706, **706**
Pasteur, Louis, 535
Paternity, misidentified, 84, 85
Pathogens, 537–52, 575
  antibiotic resistance, 545–48, **546**, **547**
    costs to bacteria of, 546–48, **547**
    judicious use of antibiotics to inhibit, 548
    selection and, 545–46
  evasion of host immune response, 537–45
  flu virus, **538**, 538–39, **539**
    origin of pandemic strains, 540–45
  host-pathogen arms races, 292
  virulence, 548–52, **551**, **552**
    in humans, 551–52
Pauling, Linus, 138, 260
PAV, 150–51
Pawlowski, B., 447
Payne, J. L., 716
PB1, 542
Pea aphid (*Acyrthosiphon pisum*), 363, **363**
Peacocks, 121, **121**, 122
Peacock spiders, 121, **121**, 122
Peacock's train, **409**, 409–10
Peanut worms, 764, **764**
Pearson, Karl, 199
Peat bogs, 692
Peccary, **124**, 127, 128, **133**
Pediculus humanus capitis (head lice), 138–39, 387
Pediculus humanus corporis (body lice), **138**, 138–39
Pedigree, 280, **280**
  computing relatedness by, 461, **461**
Pélissié, Benjamin, 412–13
Penicillin-resistance, 546, **546**
Peppered moth (*Biston betularia*), 752, **752**
Percentage of polymorphic loci, 163
Perfection, in nature, 98–99, **99**
Peris, Joan, 171
Permafrost, fossils in, 692, **692**
Permian–Triassic (P–T) extinction, 715–17
Permineralization, 692, **692**
Peromyscus maniculatus (deer mice), 492
Perrins, C. M., 513–15
Personality trait, quantitative trait locus influencing, **341**
Peters, S. E., 712
Petrified wood, **692**
Pettifor, Richard, 514–15
pf14, 101
Pfau, Thilo, 364
Phanerozoic eon
  background extinctions during, 711–12, **712**
  diversity of life across, **707**
  mass extinctions during, 710, **710**
  overview of, 694, **695**
Phantom midges, 152–53

Phenotype, 150
blending inheritance and, 95
determination of, 192
development and, 745, **745**
fitness and, 387–89
mobile genetic elements and, **589**, 589–91, **590**
natural selection on, 90–91
Phenotypic plasticity, 156
in egg size in beetle, 520–22, **521**
environment and, 387–89
in humans, 570–71
in water flea behavior, **387**, 387–89, **388**, **389**
Phenotypic traits, genes and, 363–64
Phenotypic variation, 344
Phenylalanyl-tRNA synthetase, 672, **673**
Phenylthiocarbamide (PTC), **150**, 150–51
Pheomelanin, 95, **95**
Phillips, Patrick, 317
Phillomedusa tomopterna (tiger-striped leaf frog), **609**
Phlox, **632**, 632–33, **633**, 636
Photoreceptors, 98–99, **99**
Phototactic behavior, 387–89, **388**, **389**
Phyletic transformation, 721
Phylloscopus trochiloides (Siberian greenish warbler), 47–48, **48**
Phylogenetically independent contrasts, 383–86, **385**, **393**
Phylogenetic analysis, 575
Phylogenetics, in health-related research, 536
Phylogenetic species concept (PSC), 611, **611**
Phylogenetic trees, 109, **109**. *See also* Evolutionary trees
Phylogeny(ies), 109–46
to analyze eusociality, 484–85, **485**
answering questions using, 137–40
about canine transmissible venereal tumor, 137, **137**
about evolution of body lice, **138**, 138–39
about the distribution of Seychellean tiger chameleon, **139**, 139–40, **140**
of apes, **771**, **772**, **773**
convergence and reversal, **121**, 121–23, **123**
defined, 110
of earliest animals, **700**
of fish, **142**
Hox gene in animal, **741**
of influenza A virus, 538–39, **539**, **540**, **541**, **542**, **543**
majority-rule consensus, 132–33
of mammals, **141**, **779**
of Old World primates, **142**
phylogeny inference, 114–21, **115**, **116**, **117**, **118**, **119**, **120**
small-subunit rRNA and whole-life, 666–68
of snapping shrimp, **619**
universal, 665–68
based on concatenated sequences of 31 universal genes, 672, **673**
based on rRNA sequences, 671, **671**
discordance of different estimates of, 671–72, **672**
genetic sequencing approach to, 666–67
morphological approach to, 665–66
rooting, 668, **669**, 675, **675**
whole-genome sequencing in, **671**, 671–75
of whales, 109, **109**, 123–37
aligning sequences, **126**, 126–27, **127**
Bayesian phylogeny inference, 133, **133**
comparing phylogeny inference methods, 134
estimating uncertainty by bootstrapping, **132**, 132–33
evaluating alternative phylogenies with likelihood, **128**, 128–29
evaluating alternative phylogenies with parsimony, 127, **127**
molecular evidence of evolutionary relationships, 125–34
morphological evidence on origin of, **124**, 124–25, **125**
resolution on phylogeny of, 134–37, **135**, **136**
searching for best possible trees, 129–31
whole-life, **671**, 671–72, **672**
Phylogeny inference
Bayesian, 133, **133**, 134
comparing methods of, 134
key concepts, 116–17
in non-ideal cases, **118**, 118–21
number of possible trees, 121
outgroup analysis, 118, **118**
parsimony analysis, 118–20, **119**, **120**
Phylogeography, 139
Physa acuta (acute bladder snail), **412**, 412–13, **413**
Physalaemus pustulosus (túngara frogs), 407, **407**
Physconelloides, 396–97, **397**
Physiological mortality, 507
Phytoplankton, cryptic species in marine, **613**, 613–14, **614**
Pickford, Martin, 781
Pier, Gerald, 223, 224
Pigeons, 74
Pigs, **124**, **125**, 127, 128, **133**
Pig strains, of influenza, 542–44, **543**
Pike, M. C., 576
Pinaroloxias inornata (cocos finch), 81
Pingelapese people, 245, **245**
Pinguicola moranensis (butterwort), 92, **92**
Pinheiro, Vitor, 660
Pipefish, 316, 408, **415**, 415–16, **440**, 440–41
Pitnick, Scott, 384–86
Placental development, 523
Plantier, Jean-Christophe, 22
Plants, 668. *See also specific species*
adaptive radiations in, 727, **727**
carnivorous, 91–92, **92**
domestic, 74–76
genetic polymorphism among, **254**, 254–55
hybridization in, 629
multilevel selection in, 475–76, **476**
population size and genetic diversity in, **254**, 254–55
reproduction in, **314**, 315
reproductive life span, 492
reproductive trade-offs in, 507, **507**
sexual dimorphism in, **441**, 441–44, **443**
sexual selection in, 441–44, **442**
Plasmodium falciparum, **170**, 197–200
Plasterk, Ronald, 587
Plate tectonics, 694
Platyspiza crassirostris (vegetarian tree finch), 81, **81**
Pleiotropic mutation, 505
Pleiotropy, 752–56
antagonistic, 504–7, **505**
Plesiomorphy, 116, **116**, 702
PMP-22 (peripheral myelin protein-22) gene, 58–59
Pneumococcus, 546, **546**
Pogo, Beatriz G.-T., 560
Point mutations, 160
Polar bears, 384–85, **385**
Poliovirus, **170**, 549
Polistes metricus, 486
Pollen tubes, 393, *393*
Pollinator specialization, 621–22
Polyandry, 438–39
Polymers, assembly of biological, 660–61, **661**
Polymorphic chromosomes, 167
Polymorphism, 267
among plants, **254**, 254–55
evolutionary leaps and, **758**, 758–59
single-nucleotide, 262
Polynesian field crickets (*Teleogryllus oceanicus*), **244**, 244–45
Polynucleotides, 661
Polyphyletic group, 116
Polyploid hybrid speciation, 634
Polyploidy, 168–69
speciation and, 621
Polytomy, 117
Pomatostomus ruficeps (chestnut-crowned babbler), 466
Ponce de León, Marcia, 782
Pongo pygmaeus (orangutan), 770
parental investment in, 411, **411**
Population genetics, 179, 736. *See also* Mendelian genetics
consequences of sex, 318–19
engineering test of, 224–27, **225**, **226**
testing predictions of, 201–16
Population(s), 181
creation of linkage disequilibrium in, 297–300
elimination of linkage disequilibrium from, **300**, 300–2
growth of human, 718
heterozygosity in, 279
random fixation and loss of, 246–55
life cycle of idealized, **181**
linkage disequilibrium from admixture of, **299**, 299–300
reconstructing history of, **307**, 307–11, **308**
variability of, 79, 83, 90, **90**, **91**
Population size
effective. *See* Effective population size
generation time *vs.*, 263–64, **264**
genetic drift and, 243, **243**, 259
inbreeding and, 282
Muller's ratchet and, 321
Population trees, 795, **795**
Populus trichocarpa, **589**
Porphyria, 286
Positive selection, 262, **266**, 266–68
amino acid substitutions drive by, **271**, 271–72
detecting, **311**, 311–13, **313**
loci under, 268
Posterior probability, 133
Post-transcriptional silencing, 588
Postzygotic isolation, 629, 631
Potamopyrgus antipodarum (freshwater snail), 323–24
Potassium–argon dating, 65, 66
Powdery mildews, 581
Powner, Matthew, 659–60
Poxviruses, 683
Praealticus **50**, 50–51, 55
Prairie dogs
black-tailed, **462**, 462–63
Gunnison's, 438–39, **439**
Prebiotic soup, 658
Precision, **376**
Predictive tests, 180
Premutations, 159
Pre-transcriptional silencing, 588
Prezygotic isolation, 629, 631
Price, Trevor, 87
Príncipe, 631, **631**
Primordial form, 646–47
Principal component score, **330**
Prior ejaculation, 420
Prior probability, 133
Pritchard, Jonathan, 313
Probability(ies)
combining, 185
prior, 133
Problematica, 698–99, **699**
Processed pseudogenes, 59–60, **60**
Proctor, Heather, 433–35
Procynosuchus, **704**, 705
Prokaryotes, **585**, 666, 667–68
Propane, 645
Prosauropod, **529**
Prosoeca sp., 369
Protease inhibitors, 13
Protein co-option, 101–2, **102**
Proteins, 148–50
mutations and alteration of protein function, 159–61
Protein superfamilies, **674**, 674–75
α-Proteobacteria, 679–80
γ-Proteobacteria, 674
Protista, 668
Protostomes, 701, 729
Proto-wings, 759
Provisioning, conflict over, **477**, 477–79, **478**, **479**
Prum, Richard, 433
Prunus serotina (black cherry), 492
Pryke, Sarah, 424, 425
Przewalski's horse, 175
Pseudogenes, 164, 261
Pseudomonas aeruginosa, 216–17, 223, **223**
Pseudomonas fluorescens, 594
Pseudomonas pseudoalcaligenes, 45
Pseudomonas syringae, 45
Pseudo-nitzschia, **613**, 613–14
PTC. *See* Phenylthiocarbamide (PTC)
Pterodactylus, 69, **69**
Pteropus poliocephalus (flying foxes), **382**, 382–86, **385**
Pterosaurs, 68–69, **69**
PtERV1, 30

Public goods game, 477, 482, **482**
Puddling, 448–49
Puffins, 467
Puillandre, Nicolas, 599–600
Pull of the recent, 693
Puma concolor coryi (Florida panther), 233, **233**, 283–84, **284**
  conservation genetics of, 283–84, **284**
Punctuated equilibrium, **720**, 720–24
Pundamilia 627, **627**
Punishment, humans and, 481–83, **482**, **483**
Punnett, R. C., 215
Punnett squares, **183**, 183–84, **184**, **187**, **225**, **226**
Purifying selection, 262
Purine, 158, 160, 658
Purple sea urchin (*Strongylocentrotus purpuratus*), 743
Purple-throated carib (*Eulampis jugularis*), 409, **409**
Pusey, Anne, 422–23
Pygathrix nemaeaus (douc langur), **164**, 164–65
Pygmy armadillo, **49**
Pyrenestes o. ostrinus (black-bellied seedcracker), 358–59, **359**
Pyrimidine, 158, 160, 659, **659**
Python regius (royal python), **42**

QTL mapping, 334–41
  logic of, **337**
Qualitative traits, 330
  defined, 330
  loci contributing to, 334–43
    candidate loci, 341–43
    QTL mapping, **334**, 334–41, **335**, **337**, **339**, **340**
Qualitative traits loci (QTL), 334
  QTL mapping, **334**, 334–41, **335**, **337**, **339**, **340**
Quantitative genetics, 329–68
  measuring differential success, 348–50
  measuring heritable variation, 343–47
  modes of selection, 356–60, **357**
  predicting evolutionary response to selection, 350–56, **351**
  selection on multiple traits, 354–56, **355**
Quantitative trait loci (QTL) analysis, **605**, 605–6
Quantitative traits, 330–34
  environmental influence on, **334**
  in humans, **330**
  influencing personality trait, **341**
  Mendelian genetics and, **331**, 331–33, **333**
Quartz, shocked, **713**, 713–15
Queller, David, 468

R332Q, **30**
Rabosky, Daniel, 638
Radiation
  adaptive, 638, **726**, 726–27, **727**, 730–31
  plant, 727, **727**
Radioactive decay, **65**
Radiometric dating, 64–66, 65, **65**
Raja **615**
Rambaut, Andrew, 346, 353
Ramsey, Justin, 169
Ramsey, Mike, 443
Random fixation, loss of heterozygosity and, 246–55
Random genetic drift. *See* Genetic drift
Randomization, experiments and, 376
Random mating, **187**
  genotype frequencies produced by, 181–90, **182**, **183**, **184**, **185**
  Hardy–Weinberg equilibrium assumption of, 190
  linkage disequilibrium reduced by, 300–302, 305–306
Rangea, **696**
Rangeomorphs, 696
Range size, diversification and, 638–40, **639**
Raoult, Didier, 682–83
Rapeseed (*Brassica napus*), 756–57, **757**
Raphanus raphanistrum (wild radish), 442
Ratcliffe, Laurene, 86–87
Rate-of-living theory of aging, 496–500
  trade-offs and, **499**, 499–500
Rats
  Salmonella infection in, 565–66
  warfarin resistance in, 325–26, **326**
Raup, David, 708–10, 711
Rausher, Mark, 632–33
Ray, John, 38, 45
Reaction norm, 155–56, 388, 570–71
Recessive alleles, 345
  equilibrium frequency, 220
  estimating mutation rates for, 222
  selection on, 202–6
Reciprocity, 481, **481**
  in humans, 481–83, **482**, **483**
Recombination. *See* Genetic recombination
Red-billed oxpeckers. *See* Oxpeckers
Red bladder campion (*Silene dioica*), **235**, 235–37, **237**
Red blood cells, **148**, 148–49
Red-collared widowbirds, 407–8, 424–25, **425**
Red crossbills (*Loxia curvirostra* species complex), 642
Red deer (*Cervus elaphus*), **496**
Red flour beetle (*Tribolium castaneum*), 257–58, **258**
Redon, Richard, 801
Red Queen hypothesis, 323, 324
Red squirrels (*Tamiasciurus hudsonicus*), **463**, 463–64
Red-winged blackbirds, 438
Reed, David, 503
Reed, Robert, 750
Reed buntings (*Emberiza shoeniclus*), 568, 570, **570**, 571
Regulatory genes, function of, 759–63
Reich, David, 311, 312, 798, 799–800
Reid, Ann, 543
Reillo, Paul, 431
Reinforcement, 629–33, **630**
  molecular basis of, 632–33
  via gametic isolation, 631, **631**
Reithel, Jennifer, 79
Relatedness, 460
  calculating, as probability of identity by descent, 461, **461**
  geometric, 470, **470**
  negative, 469–70, **470**
Relative risk, 574
Replacement, fossils and, 692, **692**
Replacement (nonsynonymous) mutations, 262
  selection on, 265–68, **266**
Replacement substitutions, 161, 265–68, **266**
Reproduction. *See also* Sexual reproduction; Sexual selection
  asexual, **314**, 314–16, **315**
  comparison of modes of, **315**, 315–16
  nonrandom, 79, 87–88
  organisms with two modes of, **314**, 314–15, 316
  speciation and timing of, **622**, 622–23
  trade-off in current *vs.* future, 514
  trade-off in early *vs.* late in life, 507, **507**
Reproductive isolation, 611
Reproductive strategies, in dung beetles, **753**
Reproductive success, 491–92
  asymmetric limits on, 411–16
  in elephant seals, 449, **449**
  in hermaphrodite, 412–13
  individual variation in, 79, 86–87
  male height and, 447, **447**
  measuring differences in, 348–50
  in newt, **414**, 414–15
  in plants, 441–44, **442**
  for stepchildren *vs.* genetic children, 573, **573**
Reptilia, **58**, 117, **117**
Reserve transcriptase
  AZT-resistant, 10–11, **11**
Resistance
  Antibiotic. *See* Antibiotic resistance
  to AZT, **10**, 10–11, **11**
  to HIV, 15–18
  malaria, 224–27
  penicillin, 546, **546**
  warfarin, 325–26, **326**
Resources, female choice and acquisition of, 435–36
Retrocyclin, 17
Retrocyclin RC-101, 18
Retroposition/retroduplication, 164, **164**, **165**, 165–66
Retrotransposons, 59, 586
Retroviruses. *See* Human immunodeficiency virus (HIV)
Reversal, 121–23
  in SINE or LINE characters, 135
Reverse transcriptase, 5, **5**
  AZT and, 9, **9**
  retrotransposition and, 164
  transcription errors made by, 12
Reverse transcriptase inhibitors, 13
Reverse transcription, retrotransposons and, 164
Rhamphorhyncus, 68, **69**
Rhen, Turk, 155
Rhinella marina (cane toad), **457**, 457–58
Rhinoceros, **127**
Rhinovirus type 2, **566**, 566–67
Rhizophora mucronata, **745**
Rhoads, Allen, 60
Rhodopsin, **148**, 149
Rhopalapion longirostre (hollyhock weevil), 409, **409**
Ribosomal RNA (rRNA), 666–67, **667**, **670**, **671**, 671–72
  small-subunit, 666–68
Ribosome, 649
Ribozymes, 647–52, **648**, 654
  evolution of catalytic ability in, 654, **654**
  synthesizing ribozymes, 654, **654**
Rice, William, 524, 525
Rich, Stephen S., 257–58
Richmond, Brian, 777
Ricklefs, Robert, 480
Rider, **589**, 590
Riehl, Christina, 457
Rieseberg, Loren, 634
Rifampin-resistant tuberculosis, 545
Ringler, Daniel, 565
Ring-of-life hypothesis, **679**, 679–80
Ring species, 47–48
Risch, Neil, 308
Ritalin, 40
Rivera, Maria, 679
RNA, 149, **160**, 648
  antiquity of, 649
  catalytic, 647–48
  experimental evolution of, 649–52, **650**, **651**
    ribozymes and, **561**, 650–54, **653**, **654**
  genotype, 648
  genotype and phenotype possessed by, 647–48
  genotypic changes in population of, **651**
  precursor of, 658, 659–60
  ribosomal (rRNA), 666–67, **667**, **670**, **671**, 671–**72**
  self-replication of, 652–54, **653**
  small-subunit ribosomal, 666–68
  three viruses, three domains hypothesis and transition to DNA from, 680–83, **682**
  in vitro selection and reproduction of, **650**
RNA-dependent RNA polymerase, 680
RNA enzymes. *See* Ribozymes
RNA interference (RNAi), 587, 680, 743
RNA polymerase, 654, **654**
  DNA-dependent, 670, **682**, 682–83
  HIV and, **5**
  RNA-dependent, 680
RNASE1 gene, 164–65
  phylogeny of, **164**
RNA silencing, 587–88
RNA World hypothesis, 647–54
  case for, 649
Robust australopithecines, 784
Rod cells, **148**, 149
Rodhocetus kasrani, 136, **136**
Roff, D. A., 525
Rogers, Sean, 605–6
Rohwer, Sievert, 404
Roma tomato, **589**, 589–91, **590**
Romero, Gustav, 444
Root, of evolutionary tree, 110
Rosenberg, Noah, 801
Rosenqvist, Gunilla, 440
Rose pinks (*Sabatia angularis*), 281
Rosing, Minick, 662
Rostroraja alba (white skate), **615**
Rotiferophthora angustispora, 291
Rough-skinned newts (*Taricha granulosa*), **414**, 414–15
Roundworm. *See Caenorhabditis elegans*
Royal python (*Python regius*), **42**
rpsL gene, 547
Runaway sexual selection, 430

Runnegar, Bruce, 676, 728
Rutter, Mathew, 172, 173
Ruvolo, Maryellen, 772–73, 774

Sabatia angularis (rose pinks), 281
Sabbatini, Gloria, 481
Sabeti, Pardis, 310, 311, 312
Saccharomyces cerevisiae, **171**, 171–72, 398–400, **399**
  codon bias in, **269**
Sage grouse, 449
Sahelanthropus tchadensis, **781**, 781–82, 808, 809
St. Louis encephalitis virus, 576
Salamander (*Ensatina eschscholtzii*), 48
Salamander evolution, 738, **738**
Salem, Abdel-Halim, 775
Salmon
  chinook, 518–20, **519**
  coho, 421, **421**
  farm-raised, 76
Salmonella enteritidis, 565–66
Salmonella typhi, 223–24, **224**
Saltationists/saltational evolution, 737, 756
Salvia pratensis, **254**
Same-sex homicide, 445–46
Sample size, 376
Sampling bias, fossils and, 693–94
Sampling error, 241, 243
  founder effect and, 243–45, **244**, **245**. *See also* Genetic drift
Samson, Michel, 16, 18, 162
Sand crickets (*Gryllus firmus*), **494**, 494–95
Sand dollars, 122, **123**
Sangiran, 790
Sanitation, 536
São Tomé, 631, **631**
Sapelo Island, 508–9, **509**
Sarcopterygians, **701**
Sarich, Vincent, 771–72
SAS-B (sialic acid synthase B), 596–98, **597**
Saunders, Matthew, 313
Sauropod, **529**
Sawyer, Sara, 30
Scabiosa columbaria, **254**
Scallops, 760, **760**
Scally, Aylwyn, 775–76, 778
Scharloo, W., 282
Schemske, Douglas, 340, 390–91
Schilder, Rudolf, 495
Schistosomiasis, 449
Schluter, Dolph, 514
Schmidt, Richard, 18–19, 20–21
Schopf, William, 665
Schrag, Stephanie, 547–48
Schwartz, Joshua, 427
SCN4A gene, 607, **607**
Scopes, John T., 97, **97**, 769–70
Scopes Monkey Trial, 97, 769–70
Sculpin, 605–6, 634
Sea anemones, 742
Sea lily (*Metacrinus rotundus*), 743
Sear, Rebecca, 447
Sea squirt, 98–99, **99**, 736, 746
  Botryllus leachi, 527
  Botryllus schlosseri, **526**, 526–27, **527**
  Ciona intestinalis, **589**, 743, **743**
Secondary adaptations, 92
Secondary contact, 616, 629
Second Law of Thermodynamics, 103
Sedimentary rocks, 662, **662**
Seed beetle (*Stator limbatus*), **520**, 520–22, **521**
Seed shattering, 735
Seehausen, Ole, 627
Segmental duplication, 594
Selection, **190**, 191–201
  antibiotics as agents of, 545–46. *See also* Antibiotic resistance
  artificial, 39–40, **40**, 74–76, 91, **91**
  background, 270–71
  directional, 268–70, **269**, 322, 357, **357**, 358, 359
  disruptive, **357**, 358–59
  environmentally-imposed, 322–24
  evolutionary response to, 350–56, **351**
  frequency-dependent, 212–14, **213**, 360
  general treatment of, 194
  genotype frequencies and, 196–200, **197**
  group, 473
  Hardy–Weinberg equilibrium assumption of no, 189–90
  on heterozygotes and homozygotes, 206–14, **207**, **210**, **213**
  individual, 473
  intersexual, 417
  intrasexual, 417
    by combat, 417–21
    by infanticide, 422–23, **423**
    by sperm competition, 421–422, **422**
  linkage equilibrium and, **305**
  in Mendelian genetics
    added to Hardy–Weinberg analysis, 192–93, **193**, 196–97, **197**
    changes in allele frequencies and, 195, **195**, 198–99, **200**, 200–1
    frequency-dependent, 212–14, **213**
    genotype frequencies and, 196–200, **197**
  migration–selection balance, **237**, 237–39, **238**
  migration *vs.*, empirical research on, **237**, 237–39, **238**, **239**
  modes of, 356–60, **357**, **359**
  on molecular evolution, 256
  multilevel. *See* Multilevel selection
  on multilocus genotypes, 298, **298**
  on multiple traits and correlated characters, 354–56, **355**, **356**
  mutation and, 219
  mutation–selection balance, 219–24, **223**, **224**
  negative, 262
  on offspring size, **517**, 517–18, **518**
  operation on different levels, 397–401
    in Apert syndrome, 400–1, **401**
    demonstration of, 398–400, **400**
    selfish genes and, 569
    transposable elements, 586–91
  patterns of, 201–16
  phenotypes and, 90–91
  positive, 262
  predicting evolutionary response to, 350–56, **351**
  probability of fixation for mutation and, 263–65
  on recessive and dominant alleles, **202**, 202–6, **203**
  on replacement substitutions, **266**, 266–68
  runaway, 430
  stabilizing, 357, **357**, 358, **358**, 359
  strength of
    hybrid zone and, 636–37
    measuring, 348–50
Selection coefficient, 204
Selection differential (*S*), **348**, 348–49
  in alpine skypilots pollinated by bumblebees, **352**
  predicting evolutionary response to selection and, 350–51, **351**
Selection gradient, **348**, 348–50
Selectionist theory, 257
Selection thinking, 537, 556–64, 575
  breast cancer, 559–62
  environment and, 556–58
  myopia, 558–59
  obesity, **562**, 562–64
Selective breeding, 39–40, **40**
Selective sweep, 270–71
Selfing, **275**, 275–76, *276*, 281, 306, 318, 319, 320–21
The Selfish Gene (Dawkins), 569
Selfishness, 456
Self-replication, 652–54, **653**, **654**
Semaw, Sileshi, 803
Semelparous, 526–27
Sen, G., 715
Senecio (ragwort), 634
Senescence, 495. *See also* Aging
  of cells *vs.* organisms, 498–99
Sensory biases, female choice and, 433–35
Senut, Brigitte, 781
Sepkoski, Jack, 707–8, 711, 725
Serotonin transporter, 365–66, **366**
Serratia marcescens, 322
Service, Phillip, 497
SETI (Search for Extraterrestrial Intelligence) project, 684–85
Severe combined immunodeficiency, 553–54
Sex
  adaptive significance of, 314–24
    environmentally-imposed selection and, **322**, 322–24, **323**, **324**
    genetic drift and, 319–21
    maintenance of males and, 316–18, **317**
    mutation and, **319**, 319–21, **320**, **321**
    population genetics consequences, 318–19
    reproductive modes and, **315**, 315–16
    selection and, **322**, 322–24, **323**, **324**
  asymmetry in, 410–16
Sex combs reduced gene, 740
Sex determination, 326
Sex ratio, 228
  effective population size and, 251
Sex-role-reversed species, 440–41
Sexual coevolution, 523–25
Sexual dimorphism, 408–17
  Darwin on, 410
  in humans, 408, **408**, 444–47
    body size, **444**, 444–47, **447**
  in plants, **441**, 441–44, **442**, **443**
  in red-collared widowbirds, 407–8
Sexual reproduction
  adaptive significance of, 314–24
    comparison of asexual and sexual reproduction, **315**, 315–16
  asymmetries in, 410–11
  linkage disequilibrium reduced by, **300**, 300–2
  as paradox in evolution, 315
Sexual selection, 407–54
  antagonistic, 523–25
  asymmetries in sexual reproduction, 410–11, **411**
  asymmetry of sex
    behavioral consequences of, 416–17
    in fruit flies, 411–12
  chase-away, 524
  Darwin on, 408
  defined, 410
  female choice and, 423–37
    acquisition of resources and, 435–36
    in gray tree frogs, 425–27, **426**, 437, **437**
    models of female preference, 427–37, **428**, **429**
    offspring gene quality and, 436–37
    preexisting sensory biases and, 433–35
    in red-collared widowbirds, 424–25, **425**
    in stalk-eyed flies, 431, **431**
  intersexual selection, 417
  intrasexual selection, 417
  by male-male competition, 417–23
    alternative male mating strategies, 420–21
    among stalk-eyed flies, 432
    combat, 417–21
    homicides, 445–47, **446**
    infanticide, 422–23, **423**
    sperm competition, 421–22, **422**
  in plants, 441–44
  polyandry, 438–39
  runaway, 430
  stronger for females than males, 439–41
Seychelles tiger chameleon (*Calumma tigris*), **139**, 139–40, **140**
Shankarappa, Raj, 19, 27–28
Shapiro, Leo, 729
Shared environments, estimating heritability and, 84–85
Sharks, **56**, 523
Sharp, Paul, 267
Shaw, Frank, 172
Shaw, Kerry, 432, 626
Shea, Nicholas, 154
Sheep, **125**
Shell-coiling direction, 623, **623**
Sherman, Paul, 291–92, 458
Shine, Richard, 457–58
Shocked quartz, **713**, 713–15
Shore, Eileen, 206
Short interspersed elements (SINE), 134–35, **135**, 775
Shortsighted evolution hypothesis, 549, 550
Short tandem repeat (STR) polymorphism, 796, 797
Siberian greenish warbler (*Phylloscopus trochiloides*), 47–48, **48**
Siberian Traps, 717

Sibley, Charles, 737
Siblicide, 480, **480**
Sign language, 805
Sign test, 377
Sijen, Titia, 587
Silene dioica (red bladder campion), **235**, 235–37, **237**
Silene latifolia (snowy campion), 527–28, **528**
Silent mutations, natural selection on, 268–71, **269**
Silent sites, **261**, 261–63, **262**
Silent-site (synonymous) mutations, 262
"Silent" substitutions, 160, 267–68
Silurian, 701
Simian immunodeficiency viruses (SIVs), 21–23, **22**, 29
Similicaudipteryx, 52–53, **53**
Simpson, G. G., 724
Simpson's paradox, 472
SINES (Short INterspersed Elements), 134–35, **135**, 775
Single-nucleotide polymorphisms (SNPs), 262, 799
Sinoconodon, **705**
Sinornithosaurus, **703**
Sinosauropteryx, 52–53, **53**, 703, **703**
Sipunculans, 764, **764**
Sister species, 111
Site-directed mutagenesis, 171
SIVs. *See* Simian immunodeficiency viruses (SIVs)
Skate, 93, 615
Skin color, 95, **95**, 148
Skinner, Samuel, 515–16
Skole, D., 718
Slatkin, Montgomery, 310
Sleep, Norman, 662
Slime mold (*Dictyostelium discoideum*), 468–69, **469**
Smallpox, 2–3, 48, 310
Small RNAs, **587**, 587–89
Small-subunit ribosomal RNA, whole-life phylogenies and, 666–68
Smedley, Scott, 448–49
Smith, Andrew, 729
Smith, Christopher, 517, 519, 520
Smith, Gavin, 542
Smith, James, 346
Smith, John Maynard, 315–16, 321, 460
Smith, Martin, 464–65
Smith, Nick, 268
Smith, Thomas Bates, 358–59
Smithson, Ann, 212, 213
Snails
  Euhadra sp., 623, **623**
  evolutionary tree for, **57**, 57–58
Snakes, 117, 122
  garter, 356, **356**
  Sonora semiannulata, **148**
Snapdragons, evolution of color in, 79–80, **80**
Snapping shrimp (*Alpheus malleator*), **619**, 619–20
Snow, John, 535
Snow leopard, 110, **111**, 111–12, **112**, **113**, **114**
Snowy campion (*Silene latifolia*), 527–28, **528**
Soapberry bugs, 642
Soay sheep, 179, **179**
Social behavior, 455–90
  eusociality, 483–86
    characteristics of, 483
    ecology and life-history hypothesis, 484–86
    haplodiploidy and, 484
    monogamy hypothesis, 484
    in naked mole-rats, **484**
    phylogenies to analyze, 484–85, **485**
    in vespid wasps, 485–86
  kinds of, **456**, 456–59
    altruism, 458
    mutual benefit, 456–57
    selfishness, 457–58
    spite, 459
  kin selection and costly behavior, 459–71
    greenbeard alleles, 468–469
    inclusive fitness, 460–61
    kin recognition, 465–66
  multilevel selection and cooperation, 471–77, **472**, **475**, **476**, **477**
  parent–offspring conflict, **477**, 477–480, **478**, **479**
    siblicide and, 480, **480**
  reciprocity, 481, **481**
    in humans, 481–83, **482**, **483**
  in terms of fitness, 456, **456**, 457
Sodium channel gene
  gene tree, **607**
  phylogeny of *Thamnophis,* **603**
Sohal, Rajindar, 506
Solanum (tomato), 74–76, **75**
Solar system, age of, 646
Solidago altissima, 358
Somatic-cell gene therapy, 554
Somel, Mehmet, 779
Song sparrows (*Melospiza melodia*), heritability of beak size in, 346, **346**
Sonora semiannulata (ground snake), **148**
Sorghum, 735, **735**
Sota, Teiji, 622
South Africa, AIDS infection rate in, 3
South African iris (*Lapeirousia oreogena*), **369**, 369–70
South America, HIV/AIDS in, **2**
South American pumas, **284**
Spanu, Pietro, 581
Spatial colinearity, 739–40, 742, 763
Spaulding, Michelle, 137
Special creation, theory of, 38, **39**
Speciation, 38, 104, 609–44
  allopatric, 617
  in bryozoans, **721**, 721–22
  causes of, 637–40
    ecological controls on species richness in lizards, 638, **638**
    ecological *vs.* endogenous controls of diversification, 640
    range size and diversification in toads, 638–40, **639**
  divergence, 623–29
    environmental factors and mating preferences, 626–29
    natural selection, 623–25
    sexual selection, 625–26
  evidence of, 44–48
    from laboratory experiments, 45–46
    from natural populations, 46–48
  homoploid hybrid speciation, 634
  isolation mechanisms, 616–23
    chromosomal changes, 621
    dispersal and colonization of, **617**, **618**, 618–19
    physical, **617**, 617–20, **618**
    pollinator specialization, 621–22
    shell coiling, 623, **623**
    temporal isolation, **622**, 622–23
    vicariance, **617**, **619**, 619–20, **620**
  by natural selection, 623–25
  polyploid hybrid, 634
  polyploidy and, 168–69
  secondary contact, 616, 633
    hybridization, 629–37, **630**, **635**, **636**
    reinforcement, 629–33, **631**, **632**, **633**
  by sensory drive, 627–28, **628**
  by sexual selection, 625–26
  significance of, 48
  species concepts and, 610–16
Species
  cryptic, 610
  defined, 44–45
  definition of, 38
  morphospecies, 610
  number on Earth, 609
  phylogenetic, 611, **611**
  ring, 47–48
  silent and replacement changes within and between, 267–68
  sister, 111
  sympatric, 621
Species concepts, 610–16
  agreement between, **614**
  bacteria and Archaea, 612–13
  biological species concept, 611–12
  conservation of critically endangered fish, 614–16, **615**, **616**
  cryptic species in marine phytoplankton, **613**, 613–14, **614**
  morphospecies concept, 610
  phylogenetic species concept, 611, **611**
Species richness, in Caribbean lizards, 638, **638**
Species selection, 724–25
Species trees, 599, 795, **795**
  gene trees *vs.,* **774**
Spencer, Herbert, 93
Sperm, with acrosomal defects, 233, **233**
Sperm competition, 268, 382, 421–22, **422**
Spermophilus beldingi (Belding's ground squirrel), 94, 458, **458**
Spicer, Darcy V., 562, 576
Spiders, 421
  jumping, 373–78
  peacock, 121, **121**, 122
Spinal muscular atrophy, 221, 222
Spirulina, **677**
Spite, 456, 459, **459**
  kin selection and, 469–71
    enemy of my enemy is my friend, 470–71, **471**
    negative relatedness, 469–70, **470**
Sponge-dwelling shrimp, 484, **484**
Sponges, 696
Spoon worms, 764, **764**
Spotted cucumber beetle (*Diabrotica undecimpunctata howardii*), 431–32, **432**
Spotted hyenas, 487
Spotted sandpipers, 441
SPRY domain, 30
Squamata, 117, **117**
Squinting bush brown butterfly (*Bicyclus anynana*), 748–49, **749**, 752
Squirrels, 421
SRGAP2C gene, 779
SRGAP2 gene, 779
Stabilizing selection, 357, **357**, 358, **358**, 359
Stalk-eyed flies
  female choice in, 431, **431**
  male-male competition and, 432
  sexual dimorphism in, 408
Stanley, Steven, 719, 724
Stanton, Maureen, 442
Stasis, 720–24
  in bryozoans, **721**, 721–22
  demonstrating, 720–21
  reasons for, 722–24
State of Louisiana v. Richard J. Schmidt, 18–19, 20–21
Statistical hypothesis testing, 377–78
Stator limbatus (seed beetle), **520**, 520–22, **521**
Stauffer, Rebecca L., 777
Stepchildren, discriminatory parenting of, **571**, 571–74, **572**, **573**, **574**
Stephens, J. Claiborne, 310
Sterilization
  compulsory, 201–2, 214–16
  eugenic, 214–16
Sterropterygion, **702**
Stewart, Andrew, 317
Stewart, T. H. M., 560
Sticklebacks. *See also* Threespine stickleback (*Gasterosteus aculeatus*)
  adaptation from standing genetic variation in, 604, **604**
  egg stealing in, 404
  stasis in *Gasterosteus doryssus,* 723–24
Stomata, 727
Stomatolites, 722, **723**
Stone knapper, 802, 803
Stone tools, 802–4, 807, 809
Stotting, 487, **487**
Strait, David, 777, 788, 789
Strasburg, Jared, 254
Strassmann, Beverly, 561
Straus, Christian, 55
Strawberry, **314**
Strelley Pool formation (Australia), 665
Stremlau, Matt, 29–30
Streptomycin resistance, 547, **547**
Stress, illness and, 573, **573**
STR loci, 801
Strokes, 557
Strongylocentrotus purpuratus (purple sea urchin), 743
Structural homology, 55–57, **56**
  testing hypothesis of common ancestry using, 57–58
Structuralists, 737
Stylopoma, 721, **721**

Subclade test, 706
Subfunctionalization, **595**, 596
Sub-Saharan Africa
  HIV/AIDS in, **2**, 3
Substitution(s)
  base, **170**
  mutation *vs.*, **255**
  rates of, under genetic drift, 255–57
  replacement, 265–68, **266**
  silent, 267–68
  silent-site (synonymous), 160
  single-base, 159, **159**, 160–61
Succession, 49, **49**
Sugar formation, 659
Sulfur-crested cockatoo, **386**
Sulfur dioxide, 715
SUN locus, **589**, **590**, 590–91
Supergene, 167
Survival
  individual variation in, 79, 86–87
  nonrandom, 79, 87–88
Survival of the fittest, 93
Survival success, measuring differences in, **348**, 348–50
Survivorship curves, 711, **711**
Susman, Randall L., 803–4
Sutherland, John, 660
Suzuki, Yuichiro, 156–57
Swartkoppie chert (South Africa), 665
Swaziland, AIDS infection rate in, 3
Swisher, C., 809
Sympatric species, 621
Synalpheus regalis, **484**
Synapomorphy(ies), 116, 122, 702, 770
Synapsid, 704, **704**
Syngnathus typhle (broad-nosed pipefish), **415**, 415–16, 438, **440**, 440–41
Synonymous mutations, 262
  direct selection on, 268–70, **269**
  HIV progression to AIDS and, 28–29
  indirect effects on, 270–71
Synonymous (silent) substitution, 160
Szostak, Jack, 652–54

Taeniopygia guttata (zebra finches), 498, **498**
Tallamy, Douglas, 432
Tamiasciurus hudsonicus (red squirrels), **463**, 463–64
Tanksley, Steven, 75
Tanner, Steven, 426–27
Tao, Yong, 554–55
Taphonomy, 693–94
Tapir, astragalus of, **125**
Taricha granulosa (rough-skinned newts), **414**, 414–15
TAS2R38, 150–51
Taste, variation in perception of, **150**, 150–51
Taubenberger, Jeffery, 543
Taxonomic diversity, **707**, 707–8
Taxonomic equivalency, 707–8
Taylor, Douglas, 398
T cells, 6
  effector, 27
  effector helper, 6, 7
  helper, 29
  killer, 26
  memory, 27
  naive, 27
  naive helper, 6
Teikari, J. M., 557
Teleogryllus oceanicus (Polynesian field crickets), **244**, 244–45
Telomeres, aging and, **498**, 498–99, **499**
Temperature
  behavioral thermoregulation, **379**, 379–80
    nighttime behavior of garter snakes and, **380**, 380–81, **381**
  influence on coloration, **156**, 156–57, **157**
  influence on sex, **155**, 155–56
Templeton, Alan, 252–54
Tempo and Mode in Evolution (Simpson), 724
Temporal isolation, **622**, 622–23
Tennessen, Jacob, 801
Tephritid fly (*Zonosemata vittigera*), 373–78, **374**, **375**, 605, 634
Terai, Yohey, 627
Termites, 484, **692**
Testudinata, 117, **117**
Tetanus, 549
Tetherin, 29
Tetrahymena thermophila, ribozyme, 648, **648**, 650
Tetraploid, 168, **168**, 169
Tetrapods
  fish-to-tetrapod transition, **701**, 701–2, **702**
  major monophyletic groups of, 117, **117**
Tetrodotoxin (TTX), **602**, 602–3, **603**
Thale cress (*Arabidopsis thaliana*), 169–70, **170**, **172**, 172–73, 475–76, **589**, 694
  selfing in, 306, **306**
Thamnophis couchii, **602**, 602–3, **603**
Thamnophis elegans (garter snake)
  adaptation from new mutation in, **602**, 602–3, **603**
  antipredator defenses in, 356, **356**
  nighttime behavior in, **380**, 380–81, *381*, **381**
Theory of special creation, 38, **39**
Thermoregulation, behavioral, **379**, 379–81, **380**, **381**
Thermotoga maritima, 674
Theropods, 52–53, **529**, 702–3, **703**
Theta defensin, 17–18
Thioester World, 660
Thomson, William (Lord Kelvin), 66, 96
Thomson's gazelle, 487, **487**
Thoroughbred racehorses
  heritability of lifetime winnings, 346, 353, **353**
  speed and, 329, **329**, 342, **342**, 364, **364**
3020insC allele, 342, **342**
Threespine stickleback (*Gasterosteus aculeatus*)
  adaptive mutations in, **605**, 605–6
  pleiotropy in alleles in, 755, **755**
  reversals in, 122
  speciation among, 46–47, **47**
  vestigial traits in, **43**, 43–44, **44**
Three viruses, three domains hypothesis, 680–83, **682**
Threonine, 26
Thrinaxodon, 691, **691**, 693, **704**, 705, **705**
Thrips egg mites (*Adactylidium* sp.), 492, **492**
Thumbs, anatomy of opposable, **803**, 803–4, **804**
Thuny, Franck, 563
Thymine, 149, **160**
Tian, Feng, 658
Tigers, 110, 111, **111**, **112**, **113**, 114, **114**, 287
Tiger-striped leaf frog (*Phyllomedusa tomopterna*), **609**
Tiktaalik, **701**, 702
Time, on evolutionary tree, **110**, 110–11
Time-calibrated trees (time trees), 728, **728**, **729**
Timema walking stick, 321, **321**
Tinghitella, Robin, 244–45
Tinman gene, 761
Tishkoff, Sarah, 793, 796
Tissues, evolution of, 553–56
  cancer history, 554–56
  spontaneous recovery, 553–54
Titan, **645**, 645–46,
Toads, range size and diversification in, 638–40, **639**
Tobacco hornworms (*Manduca sexta*), **156**, 156–57, **157**
Tobias, Phillip, 805–6
Tomatoes, **589**, 589–90, **590**
  Solanum lycopersicum, 74–76, **75**
  Solanum pimpinellifolium, **75**
Tool use, 802–4
Toth, Amy, 486
Townsend, Ted, 139–40
Trace fossils, 693, **693**, 697, **697**
Trachops cirrhosus, 407
Trade-off hypothesis of virulence, 549, 550–52, **551**
Trade-offs
  in adaptation, 389–91
    begonia flower size, 389–91, **390**, **391**
  aging and, 504–7, **505**, **506**, **507**
  in allocation of energy, 493–95, **494**
  current *vs.* future reproduction, 514
  developmental, 752–56
  reproduction and survival in plants, 507, **507**
  reproduction early *vs.* late in life, 507, **507**
  size and number of offspring, **517**, 517–18, **518**
Trahan, Janet, 18–19, 20–21
Trait(s)
  Adaptive. *See* Adaptation(s)
  distinguishing homology from homoplasy, 121–22
  evolution of new, by natural selection, **91**, 91–92
  evolution of uniquely human, 802–7
    language, 805–7
    stone tools, 802–4
  heritability of. *See* Heritability
  homologous, 744
  molecular, 125–34
  morphological, 124–25
  non-perfect, **92**, 92–93
  normally distributed variation in, 343–44
  qualitative, 333
    loci contributing to, 334–43
  quantitative, **331**, 331–34, **333**
  selection on multiple, 354–56, **355**, **356**
Transcription factors, 739
Transfer RNA (tRNA), codon bias and, 269, **269**
Transition, 160, **160**
  on phylogenies, 110–11
Transitional forms, 49–54, **50**, **51**, **53**, **54**
Transmission chains, 20
Transmission disequilibrium test, 342
Transmitted diseases, virulence and, 552, **552**
Transposable elements (transposons), 586–91
  effect on phenotype, **589**, 589–91, **590**
  epigenetic marks and, 154
  evolutionary impact of, 586–91
    defending against spread of transposable elements, 586–88
    defenses against transposable elements, 588–89
  evolutionary leaps and, 757
  gene duplication and, 594
  as genomic parasites, 586
Transpose, 586
Transversion, 160, **160**
Tree finches, 81, **81**
Tree of life, **646**, **666**, 670–75
  five-kingdom scheme, **666**
Trematodes, 323–24, **324**
Treponema pallidum, 672, **673**
Tribolium castaneum (flour beetle), **202**, 202–3, **203**, 257–58, **258**
Tribulus cistoides, 87–88, 90
Trichogramma embryophagum, **515**, 515–16
Trichomes, 92, **92**
Trillmich, Fritz, 418, 420
Trillmich, Krisztina, 419–20
Trilobites, 698
TRIM5α, 29–30, **30**
Trinkaus, E., 792–93
Trivers, Robert, 410, 412, 413, 415, 477, 478, 481–82
Trochlea, 125
Tropical rain forests, threats to diversity and, 718–19
Troyer, Ryan, 27
Trust, kin selection and, **465**, 465–66, **466**
Tsunami, 715
Tuberculosis, antibiotic-resistant, 545–46
Tucker, C., 718
Tumors, 554–55
Tumpey, Terrence, 543
Túngara frogs (*Physalaemus pustulosus*), 407, **407**
Tunicate, **747**
Turdus migratorius (American robin), 488
Turnbaugh, Peter, 563
Turnip (*Brassica rapa*), 756, **757**
Turtles, 117
  Odontochelys semitestacea, 37
  transitional, 53–54
Twins, estimating heritability from, 346–347, **347**
Two-island model of migration, 236
Two-locus version of Hardy–Weinberg analysis
  adaptive significance of sex, 314–24
    environmentally-imposed selection and, **322**, 322–24, **323**, **324**
    genetic drift and, 319–21
    maintenance of males, **316**, 316–18
    mutation and, **319**, 319–21, **320**
    population genetics consequences of sex, 318–19
    reproductive modes and, **315**, 315–16
    selection and, **322**, 322–24
  chromosome frequencies and, 293–94, **294**
  linkage disequilibrium. *See* Linkage disequilibrium
  numerical example of, **293**, 293–95

Tyner, Stuart, 499
Type 1 Gaucher disease, 307
Typhoid fever, 224, **224**
Tyramine beta-monooxygenase, 153
Tyrannosaurus, 52, 403, **703**
Tyrosinemia, 286

Uganda, AIDS epidemic in, 4
Ultimatum game, 482–83, **483**
Ultrabithorax mutant, **739**
Underdominance, **209**, 211–12
Unequal crossing over, 161, **164**, 164–65
Uniformitarianism, 62
Uninformative characters, 119–20
United States, belief in evolution in, 37
U.S. Endangered Species Act (1973), 611, 614
U.S. Supreme Court, 97–98
Universal gene-exchange pool hypothesis, **678**, 678–79, **679**
Universal molecular homologies, 61, **61**
Universal phylogeny, 665–68
  based on concatenated sequences of 31 universal genes, 672, **673**
  based on rRNA sequences, 671, **671**
  discordance of different estimates of, 671–62, **672**
  genetic sequencing approach to, 666–67
  morphological approach to, 665–66
  rooting, 668, **669**, 675, **675**
  whole-genome sequencing in, **671**, 671–75
Unrooted trees, 130
Uracil, **160**
Uranium–lead dating, 66
Urey, Harold, 657
Ursus americanus (black bear), 384–85, **385**, 492

Vaccine, influenza, 540, 545
Van Bocxlaer, Ines, 638–40
Vancomycin, 563
van Heemst, Diana, 500
van Noordwijk, A. J., 282
Van Valen, Leigh, 124, 711
Variable number tandem repeat (VNTR), 105
Variation, 94–95. *See also* Genetic variation
  among individuals, 79, 83–85, 147–48, **148**
  Darwin on, 748
  DNA as mechanism of, 737
  environmental, 148, 150, 152–54, **153**
  equal, 748–49
  genotype-by-environment interaction, 148, 150, **155**, 155–57, **156**, **157**
  heritability of, 79, 83–86
  individual, 79
  in reproductive success, 79, 86–87
Vascular endothelial growth factor, 153
Vascular endothelial growth factor receptor, 761–63
Vascular endothelial growth factor receptor 1, 197, 199
Vassilieva, Larissa, 173–74
Vaughn, Linda, 564
Vaughton, Glenda, 443
Vectorborne diseases, virulence and, 552, **552**
Vegetarian tree finch (*Platyspiza crassirostris*), 81, **81**
Velazquez, John, 329
Velociraptor, 52, 53
Velvet worms, 764, **764**
Venus fly trap (*Dionaea muscipula*), 92
Vernanimalcula guizhouena, 697, **697**, 729
Vertebrates
  molecular clock, 263, **263**
  time tree for, **729**
Vespid wasps, 485–86
Vestigial structures, **42**, 42–44
Via, Sara, 363
Vibrio cholerae (cholera), 535, **535**, 536, **536**, 552
Vicariance, 617, **617**, **619**, 619–20, **620**
Vicariance hypothesis, 139–40
Vignieri, Sacha, 73
Virginia opossum (*Didelphis virginiana*), **493**, 493–94, **508**, 508–10, **509**
Virions, 5, **5**, 7, 10–11, 12
  AZT and, 10
Virulence, 548–52
  evolution of, 548–51
  in human pathogens, 551–52, **552**
Viruses. *See also* Human immunodeficiency virus (HIV)
  breast cancer and, 559–60, **560**
  hypotheses on origin of, 681
  influenza, **261**, 262
  three viruses, three domains hypothesis, 680–83, **682**
Vivipary, 746
VNTR. *See* Variable number tandem repeat (VNTR)
Voight, Benjamin, 313
Volvox, **314**, 315
vpu gene, 29
Vrba, Elizabeth, 759
Vrieze, Anna, 563

Waddle, Diane, 792
Wade, Michael, 475
Wake, David, 738
Wake, Marvalee, 140
Walker, David, 505
Wallace, Alfred Russel, 66, 74, 755
Wang, Xiaoxia, 779
Wang, Yue, 560
Waorani, **446**, 446–47
Warbler finches (*Certhidea olivacea, C. fusca*), 81, **81**
Ward, P. D., 716
Warfarin resistance, 325–26, **326**
Water fleas
  Daphnia magna, **387**, 387–389, **388**, **389**
  Daphnia pulex, 152–53, **153**
Waterleaf (*Hydrophyllum appendiculatum*), 281–82, **282**
Water mite (*Neumania papillator*), 433–35, **434**
Water snakes (*Nerodia sipedon*), **237**, 237–39, **238**, **239**
Watson, James, 158
Weaning conflict, **477**, 477–78
Weeks, Paul, 370–72
Weinberg, Wilhelm, 189
Weis, Arthur, 358
Welch, Allison, 437
Wells, W. C., 73
Welsh corgis, **165**
Werdelin, Lars, 111, 112, 113
Werner, Jan, 529
Wernicke's area, 805
West, I. D., 715
West, Stuart, 456, 471
West Nile virus, 576
Whales
  humpback, 109, **109**
  phylogeny of, 123–37
    evidence on origin of, **124**, 124–25, **125**
    identifying Cetacean, 123–24, **124**
    molecular evidence of evolutionary relationships, 125–34, **126**, **127**, **128**, **132**, **133**
    resolution of, 134–37, **135**, **136**
Whiskerbrush (*Linanthus ciliatus*), **148**
White, Tim, 783
White cockatoo, 386
White-fronted bee-eaters, **479**, 479–80
Whiteman, Noah, 397
White skate (*Rostroraja alba*), **615**
White-winged choughs (*Corcorax melanorhamphos*), 466, **466**
Whole-genome sequencing
  history of cancer and, 555–56
  universal phylogeny and, 671–72, **672**
Whole-life phylogenies, **671**, 671–72, **672**
Wiens, John, 609
Wikelski, Martin, 418–421
Wild barley, **377**, 377–78
Wild cabbage (*Brassica oleracea oleracea*), **76**
Wildman, Derek, 779
Wild radish (*Raphanus raphanistrum*), 442
Wild yarrow (*Achillea borealis*), 169, **169**
Wiley, Chris, 626
Wilkinson, Gerald, 431
Wilkinson, Richard, 778
Williams, George C., 500, 537
Willis, John, 624–25
Wilson, Alastair, 346, 353
Wilson, Allan, 737, 771–72, 779
Wilson, Christopher, 291–92
Wilson, David Sloan, 471–72
Wilson, E. O., 718
Wilson, Gregory, 730
Wilson, Margo, 445–46, 573–74
Wirth, Brunhilde, 221, 222
Wiwaxia corrugata, 699, **699**
Wnt locus, 751
Wochner, Aniela, 654
Woese, Carl R., 666–68, 672, 678–79
Wolfe, Lorne, 281, 527–28
Wolfe, Nathan, 48
Wolpoff, Milford, 781, 793
Wolves, 40, **41**
  gray, 76
Wombats, **49**
Wong, Karen, 126
Wood, Bernard, 781, 784, 786, 788–89, 803, 809
Wooding, Stephen, 151
Woodpecker finch (*Cactospiza pallida*), 81, **81**
Woodruff, Ron, 219
Woolly mammoth, **692**
Wray, Grey, 728–29
Wrege, Peter, 479
Wright, Sewall, 248, 249, 736
Wu, Carrie, 637
Wu, Dong-Dong, 313
Wurmbea dioica, **441**, 442, 443

X4 viruses, 27
Xenopus laevis (African clawed frog), 122, **747**
Xenorhabdus 459, 470–71
Xiao, Han, 589–90
Xiao, S., 697
Xylanases, 757

Yamamoto, Satoshi, 622
Yang, Song, 674–75
Yanoconodon, **705**
Yanomamö Indians, 446, **446**, **511**, 572
Yap, Melvyn, 30
Yarrow, 334
Yeast (*Saccaromyces cerevisiae*), 398–400, **399**
Yellow-legged gull, 78, **78**
Yellow monkeyflower (*Mimulus guttatus*), 282, 623–25, **624**
Yellow-rumped warblers, 634
Yellow-tailed cockatoo, **386**
Young, Andrew, 254–55
Young, Francis, 557–58
Young, Truman, 507
Yule, G. Udny, 186–87

Zebra finches (*Taeniopygia guttata*), 498, **498**
Zebrafish (*Danio rerio*), 431, **431**, **589**
Zera, Anthony, 495
ZFY locus, 270
Zhang, Jiajie, 543
Zhang, Jianzhi, 164–65
Zhang, Mingcai, 219
Zhang, Peng, 140
Zhang, Zhengdong, 43
Zhao, Zhangwu, 495
Zollikofer, Chistoph, 782
Zonosemata (tephritid fly), 373–78, **374**, **375**, 605, 634
Zuckerkandl, Emil, 138, 260